FIFTH EDITION

# TEXTBOOK OF
# MEDICAL
# PHYSIOLOGY

## ARTHUR C. GUYTON, M.D.

Professor and Chairman of the Department of Physiology
and Biophysics, University of Mississippi, School of Medicine

W. B. SAUNDERS COMPANY  Philadelphia  London  Toronto

W. B. Saunders Company:    West Washington Square
                           Philadelphia, Pa.   19105

                           1 St. Anne's Road
                           Eastbourne, East Sussex BN21 3UN, England

                           833 Oxford Street
                           Toronto, M8Z 5T9, Canada

Listed here are the latest translated editions of this book together with the language of the translation and the publisher.

Spanish—NEISA—Mexico City (4th edition)

Portuguese—Editora Guanabara Koogan—Rio de Janeiro (4th Edition)

Japanese—Kirokawa Publishing Company—Tokyo (4th Edition)

Serbo-Croatian—Medicinska Knjiga—Belgrade (4th Edition)

Textbook of Medical Physiology                                    ISBN   0-7216-4393-0

Last digit is the print number:    9    8    7    6    5    4    3    2

*Dedicated to*

MY FATHER
*for the uncompromising principles that guided his life*

MY MOTHER
*for leading her children into intellectual pursuits*

MY WIFE
*for her magnificent devotion to her family*

MY CHILDREN
*for making everything worthwhile*

# PREFACE

When I wrote the first edition of this textbook, I had the naive belief that once the book was completed, subsequent revisions would require only simple changes. However, with each new edition I find that progress in the field of physiology is so rapid that large portions of the text must be completely recast and rewritten, and even the emphasis of the subject matter must often be changed as our knowledge becomes more penetrating. Therefore, the reader will find hardly a chapter in this edition that has not been significantly altered. Also, most of the figures have been either changed in at least some feature or replaced, and new ones have been added. Indeed, the revisions have been very extensive.

If I should characterize the major change in this edition, it would be a striking trend toward more fundamental physiology. The reason for this is mainly the greater success of the research physiologist in probing more basic mechanisms of function than was true a few years age. Yet I have still attempted to present physiology as an integrated study of the body's functional systems, attempting to utilize the new fundamental knowledge to build a better understanding of the mechanisms upon which life depends.

I have also made a serious attempt to devise and achieve techniques of expression that will help bring the physical and chemical principles of the body's complexities into the medical student's realm of understanding. In pursuit of this goal, I have kept records of those types of material with which the student has difficulty; I have quizzed students in detail to determine their levels of comprehension; and I have attempted to note inconsistencies in logic that might appear in student discussions. All these data have been used to help choose the material and methods of presentation. Thus, it has been my desire to make this book a "teaching" text as well as one that covers essentially all the basic physiology required of a student of medicine.

A special emphasis of the book is a more detailed attention than is given in other textbooks of physiology to the body's many control systems. It is these that provide what physiologists call *homeostasis*. The reason for this special emphasis is that most disease conditions of the body result from abnormal function of one or more of the control systems. Therefore, the student's comprehension of "medical" physiology depends perhaps more on a knowledge of these systems than on any other facet of physiology.

Another goal in the preparation of this text has been to make the text as accurate as possible. To help attain this, suggestions and critiques from many physiologists, students, and clinicians throughout the United States and other parts of the world have been received and utilized in checking the factual accuracy of the text. Yet, even so, because of the likelihood of error in sorting through many thousands of bits of information, I wish to issue an invitation—in fact, much more than merely an invitation, actually a request—to all readers to send along criticisms of error or inaccuracy. Indeed, physiologists understand how important feedback is to proper function of the human body; so, too, is feedback equally important for progressive development of a textbook of medical physiology. I hope also that those many persons who have already helped will accept my sincerest appreciation for their efforts.

A word of explanation is needed about two features of the text—first, the references, and, second, the two print sizes. The references have been chosen primarily for their up-to-dateness and for the quality of their own bibliographies. Use of these references as well as of cross-references from them can give the student almost complete coverage of the entire field of physiology.

The print is set in two sizes. The material presented in small print is of several different kinds: first, anatomical, chemical, and other information that is needed for the immediate discussion but that most students will learn in more detail in other courses; second, information that is of special importance to certain clinical fields of medicine but that is not necessary to fundamental understanding of the body's basic physiologic mechanisms; and, third, information that will be of value to those students who wish to pursue a subject more deeply than does the average medical student. On the other hand, the material in large print, which represents about two-thirds of the text, constitutes the major bulk of physiological information that the student will require in his medical studies and that he will not obtain in other courses. For those teachers who would like to present a more limited course of physiology, the student's study can be restricted primarily to the large type.

Again, I wish to express my deepest appreciation to many others who have helped in the preparation of this book. I am particularly grateful to Mrs. Billie Howard for her secretarial services, to Miss Tomiko Mita and Mrs. Carolyn Hull for the new illustrations, and to the staff of the W. B. Saunders Company for its continued editorial and production excellence.

ARTHUR C. GUYTON

*Jackson, Mississippi*

# CONTENTS

## PART I
## INTRODUCTION TO PHYSIOLOGY:
## THE CELL AND GENERAL PHYSIOLOGY

# PART II
# BLOOD CELLS, IMMUNITY,
# AND BLOOD CLOTTING

# PART III
# NERVE AND MUSCLE

# PART IV
# THE HEART

CHAPTER 13

## HEART MUSCLE; THE HEART AS A PUMP

CHAPTER 14

## RHYTHMIC EXCITATION OF THE HEART

CHAPTER 15

## THE NORMAL ELECTROCARDIOGRAM

# PART V
# THE CIRCULATION

CHAPTER 18

## PHYSICS OF BLOOD, BLOOD FLOW, AND PRESSURE: HEMODYNAMICS .... 222

CHAPTER 19

## THE SYSTEMIC CIRCULATION ........... 237

CHAPTER 20

## LOCAL CONTROL OF BLOOD FLOW BY THE TISSUES; NERVOUS AND HUMORAL REGULATION .................... 250

CHAPTER 21

## REGULATION OF MEAN ARTERIAL PRESSURE: I. NERVOUS REFLEX AND HORMONAL MECHANISMS FOR RAPID PRESSURE CONTROL ............... 265

CHAPTER 22

## REGULATION OF ARTERIAL PRESSURE: II. THE RENAL-BODY FLUID SYSTEM FOR LONG-TERM PRESSURE CONTROL. MECHANISMS OF HYPERTENSION .... 279

# PART VI
# THE BODY FLUIDS AND KIDNEYS

# CONTENTS

**xix**

# PART VII
# RESPIRATION

# PART VIII
# AVIATION, SPACE, AND
# DEEP SEA DIVING PHYSIOLOGY

# PART IX
# THE NERVOUS SYSTEM

# PART X
# THE SPECIAL SENSES

# PART XI
# THE GASTROINTESTINAL TRACT

# PART XII
# METABOLISM AND TEMPERATURE REGULATION

# PART XIII
# ENDOCRINOLOGY AND REPRODUCTION

# PART I

# INTRODUCTION TO PHYSIOLOGY: THE CELL AND GENERAL PHYSIOLOGY

# 1

# Functional Organization of the Human Body and Control of the "Internal Environment"

Physiology is the study of function in living matter; it attempts to explain the physical and chemical factors that are responsible for the origin, development, and progression of life. Each type of life, from the monomolecular virus up to the largest tree or to the complicated human being, has its own functional characteristics. Therefore, the vast field of physiology can be divided into *viral physiology, bacterial physiology, cellular physiology, plant physiology, human physiology,* and many more subdivisions.

**Human Physiology.** In human physiology we attempt to explain the chemical reactions that occur in the cells, the transmission of nerve impulses from one part of the body to another, contraction of the muscles, reproduction, and even the minute details of transformation of light energy into chemical energy to excite the eyes, thus allowing us to see the world. The very fact that we are alive is almost beyond our own control, for hunger makes us seek food and fear makes us seek refuge. Sensations of cold make us provide warmth, and other forces cause us to seek fellowship and reproduce. Thus, the human being is actually an automaton, and the fact that we are sensing, feeling, and knowledgeable beings is part of this automatic sequence of life; these special attributes allow us to exist under widely varying condi-

tions, which otherwise would make life impossible.

## CELLS AS THE LIVING UNITS OF THE BODY

The basic living unit of the body is the cell, and each organ is actually an aggregate of many different cells held together by intercellular supporting structures. Each type of cell is specially adapted to perform one particular function. For instance, the red blood cells, twenty-five trillion in all, transport oxygen from the lungs to the tissues. Though this type of cell is perhaps the most abundant of any in the whole body, there are approximately another 50 trillion cells. The entire body, then, contains about 75 trillion cells.

However much the many cells of the body differ from each other, all of them have certain basic characteristics that are alike. For instance, each cell requires nutrition for maintenance of life, and all cells utilize almost identically the same types of nutrients. All cells use oxygen as one of the major substances from which energy is derived; the oxygen combines with carbohydrate, fat, or protein to release the energy required for cell function. The general mechanisms for changing nutrients into energy are basically the same in all cells, and all cells also deliver end-products of their chemical reactions into the surrounding fluids.

Almost all cells also have the ability to reproduce, and whenever cells of a particular type are destroyed for one cause or another, the remaining cells of this type usually divide again and again until the appropriate number is replenished.

## THE EXTRACELLULAR FLUID—THE INTERNAL ENVIRONMENT

About 56 per cent of the adult human body is fluid. Some of this fluid is inside the cells and is called, collectively, the *intracellular fluid*. The fluid in the spaces outside the cells is called the *extracellular fluid*. Among the dissolved constituents of the extracellular fluids are the ions and the nutrients needed by the cells for maintenance of life. The extracellular fluid is in constant motion throughout the body and is rapidly mixed by the blood circulation and by diffusion between the blood and the tissue fluids. Therefore, all cells live in essentially the same environment, the extracellular fluid, for which reason the extracellular fluid is often called the *internal environment* of the body, or the *milieu intérieur,* the French equivalent first used a hundred years ago by the great 19th century French physiologist, Claude Bernard.

Cells are capable of living, growing, and providing their special functions so long as the proper concentrations of oxygen, glucose, the different ions, amino acids, and fatty substances are available in the internal environment.

**Differences Between Extracellular and Intracellular Fluids.** The extracellular fluid contains large amounts of sodium, chloride, and bicarbonate ions, plus nutrients for the cells, such as oxygen, glucose, fatty acids, and amino acids. It also contains carbon dioxide, which is being transported from the cells to the lungs to be excreted, and other cellular products, which are being transported to the kidneys for excretion.

The intracellular fluid differs significantly from the extracellular fluid; particularly, it contains large amounts of potassium, magnesium, and phosphate ions instead of the sodium and chloride ions found in the extracellular fluid. Special mechanisms for transporting ions through the cell membranes maintain these dif-

ferences. These mechanisms will be discussed in detail in Chapter 4.

## "HOMEOSTATIC" MECHANISMS OF THE MAJOR FUNCTIONAL SYSTEMS

### HOMEOSTASIS

The term *homeostasis* is used by physiologists to mean *maintenance of static,* or *constant, conditions in the internal environment.* Essentially all the organs and tissues of the body perform functions that help to maintain these constant conditions. For instance, the lungs provide oxygen as it is required by the cells, the kidneys maintain constant ion concentrations, and the gut provides nutrients. A large segment of this text is concerned with the manner in which each organ or tissue contributes to homeostasis. To begin this discussion, the different functional systems of the body and their homeostatic mechanisms will be outlined briefly; then the basic theory of the control systems that cause the functional systems to operate in harmony with each other will be discussed.

### THE FLUID TRANSPORT SYSTEM

Extracellular fluid is transported to all parts of the body in two different stages. The first stage entails movement of blood around and around the circulatory system, and the second, movement of fluid between the blood capillaries and the cells. Figure 1–1 illustrates the overall circulation of blood, showing that the heart is actually two separate pumps, one of which propels blood through the lungs and the other through the systemic circulation. All the blood in the circulation traverses the entire circuit of the circulation an average of once each minute when a person is at rest and as many as six times each minute when he becomes extremely active.

As blood passes through the capillaries, continual exchange occurs between the plasma portion of the blood and the interstitial fluid in the spaces surrounding the capillaries. This process is illustrated in Figure 1–2. Note that the capillaries are porous so that large amounts of fluid can *diffuse* back and forth between the blood and the tissue spaces, as illustrated by the arrows. This process of diffusion is caused by kinetic motion of the molecules in both the

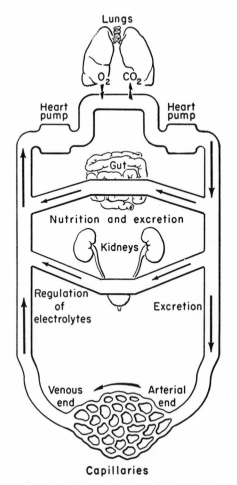

**Figure 1–1.** General organization of the circulatory system.

plasma and the extracellular fluid. That is, all fluid and dissolved molecules are continually moving and bouncing in all directions, through the pores, through the tissue spaces, and so forth. Almost no cell is located more than 25 to 50 microns from a capillary, which insures diffusion of almost any substance from the capillary to the cell within a few seconds. Thus, the extracellular fluid throughout the body is continually mixed and thereby maintains almost complete homogeneity.

## ORIGIN OF NUTRIENTS IN THE EXTRACELLULAR FLUID

**The Respiratory System.** Figure 1–1 shows that each time the blood passes through the body it also flows through the lungs. The blood picks up oxygen in the alveoli, thus acquiring the oxygen needed by the cells. The membrane

between the alveoli and the lumen of the pulmonary capillaries is only 0.4 to 2.0 microns in thickness, and oxygen diffuses through this membrane into the blood in exactly the same manner that water, nutrients, and excreta diffuse through the tissue capillaries.

**The Gastrointestinal Tract.** Figure 1–1 also shows that a large portion of the blood pumped by the heart passes through the walls of the gastrointestinal organs. Here, different dissolved nutrients, including carbohydrates, fatty acids, amino acids, and others, are absorbed into the extracellular fluid.

**The Liver and Other Organs that Perform Primarily Metabolic Functions.** Not all substances absorbed from the gastrointestinal tract can be used in their absorbed form by the cells. The liver changes the chemical compositions of many of these to more usable forms, and other tissues of the body—the fat cells, the gastrointestinal mucosa, the kidneys, and the endocrine glands—help to modify the absorbed substances or store them until they are needed at a later time.

**Musculoskeletal System.** Sometimes the question is asked: How does the musculoskeletal system fit into the homeostatic functions of the body? The answer to this is obvious and simple: Were it not for this system, the body could not move to the appropriate place at the appropriate time to obtain the foods required for nutrition. The musculoskeletal system also provides motility for protection against adverse surroundings, without which the entire body, and along with it all the homeostatic mechanisms, could be destroyed instantaneously.

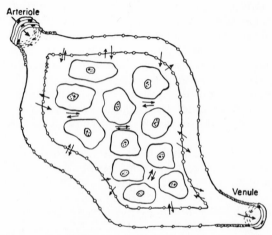

**Figure 1–2.** Diffusion of fluids through the capillary walls and through the interstitial spaces.

## REMOVAL OF METABOLIC END-PRODUCTS

**Removal of Carbon Dioxide by the Lungs.** At the same time that blood picks up oxygen in the lungs, carbon dioxide is released from the blood into the alveoli, and the respiratory movement of air into and out of the alveoli carries the carbon dioxide to the atmosphere. Carbon dioxide is the most abundant of all the end-products of metabolism.

**The Kidneys.** Passage of the blood through the kidneys removes most substances from the plasma that are not needed by the cells. These substances include especially different end-products of cellular metabolism and excesses of ions and water that might have accumulated in the extracellular fluids. The kidneys perform their function by, first, filtering large quantities of plasma through the glomeruli into the tubules and then reabsorbing into the blood those substances needed by the body, such as glucose, amino acids, large amounts of water, and many of the ions. However, substances not needed by the body, especially the metabolic end-products such as urea, generally are not reabsorbed but, instead, pass on through the renal tubules into the urine.

## REGULATION OF BODY FUNCTIONS

**The Nervous System.** The nervous system is composed of three major parts: the *sensory portion*, the *central nervous system* (or *integrative portion*), and the *motor portion*. Sensory receptors detect the state of the body or the state of the surroundings. For instance, receptors present everywhere in the skin apprise one every time an object touches him at any point. The eyes are sensory organs that give one a visual image of the surrounding area. The ears also are sensory organs. The central nervous system is comprised of the brain and spinal cord. The brain can store information, generate thoughts, create ambition, and determine reactions that the body should perform in response to the sensations. Appropriate signals are then transmitted through the motor portion of the nervous system to carry out the person's desires.

A large segment of the nervous system is called the *autonomic system*. It operates at a subconscious level and controls many functions of the internal organs, including the action of the heart, the movements of the gastrointestinal tract, and the secretion by different glands.

**The Hormonal System of Regulation.** Located in the body are eight major endocrine glands that secrete chemical substances, the *hormones*. Hormones are transported in the extracellular fluids to all parts of the body to help regulate function. For instance, thyroid hormone increases the rates of almost all chemical reactions in all cells. In this way thyroid hormone helps to set the tempo of bodily activity. Likewise, insulin controls glucose metabolism; adrenocortical hormones control ion and protein metabolism; and parathormone controls bone metabolism. Thus, the hormones are a system of regulation that complements the nervous system. The nervous system, in general, regulates rapid muscular and secretory activities of the body, whereas the hormonal system regulates mainly the slowly reacting metabolic functions.

## REPRODUCTION

Reproduction sometimes is not considered to be a homeostatic function. But it does help to maintain static conditions by generating new beings to take the place of ones that are dying. This perhaps sounds like a farfetched usage of the term homeostasis, but it does illustrate that, in the final analysis, essentially all structures of the body are so organized that they help to maintain continuity of life.

# THE CONTROL SYSTEMS OF THE BODY

The human body has literally thousands of control systems in it. Some of these operate within the cell to control intracellular function, a subject that will be discussed in detail in Chapter 3. Others operate within the organs to control functions of the individual parts of the organs, while others operate throughout the entire body to control the interrelationships between the different organs. For instance, the respiratory system, operating in association with the nervous system, regulates the concentration of carbon dioxide in the extracellular fluids. The liver and the pancreas regulate the concentration of glucose in the extracellular fluids. And the kidneys regulate the concentrations of hydrogen, sodium, potassium, phosphate, and other ions in the extracellular fluids.

## EXAMPLES OF CONTROL MECHANISMS

**Regulation of Oxygen and Carbon Dioxide Concentrations in the Extracellular Fluid.** Since oxygen is one of the major substances required for chemical reactions in the cells, the rates of the chemical reactions are, to a great extent, dependent on the concentration of oxygen in the extracellular fluid. For this reason a special control mechanism maintains an almost exact and constant oxygen concentration in the extracellular fluids. This mechanism depends principally on the chemical characteristics of *hemoglobin,* which is present in all the red blood cells. Hemoglobin combines with oxygen as the blood passes through the lungs. In the tissue capillaries the hemoglobin will not release oxygen into the tissue fluid if too much oxygen is already there, but if the oxygen concentration is too little, sufficient oxygen will be released to reestablish an adequate tissue oxygen concentration. Thus, the regulation of oxygen concentration in the tissues is vested principally in the chemical characteristics of hemoglobin itself. This regulation is called the *oxygen-buffering function of hemoglobin.*

Carbon dioxide is one of the major end-products of the oxidative reactions in cells. If all the carbon dioxide formed in the cells should continue to accumulate in the tissue fluids, the *mass action* of the carbon dioxide itself would soon halt all the energy-giving reactions of the cells. Fortunately, the following mechanism maintains a constant and reasonable concentration of carbon dioxide in the extracellular fluids: A very high carbon dioxide concentration *excites the respiratory center,* causing the person to breathe rapidly and deeply. This increases the rate of expiration of carbon dioxide and therefore increases its removal from the blood and extracellular fluid, and the process continues until the concentration returns to normal.

**Regulation of Arterial Pressure.** Several different systems contribute to the regulation of arterial pressure. One of these, the *baroreceptor system,* is very simple and an excellent example of a control mechanism. In the walls of most of the great arteries of the upper body, especially the bifurcation region of the carotids and the arch of the aorta, are many nerve receptors, called *baroreceptors,* which are stimulated by stretch of the arterial wall. When the arterial pressure becomes great, these baroreceptors are stimulated excessively, and impulses are transmitted to the medulla of the brain. Here the impulses inhibit the *vasomotor center,* which in turn decreases the number of impulses transmitted through the sympathetic nervous system to the heart and blood vessels. Lack of these impulses causes diminished pumping activity by the heart and increased ease of blood flow through the peripheral vessels, both of which lower the arterial pressure back toward normal. Conversely, a fall in arterial pressure relaxes the stretch receptors, allowing the vasomotor center to become more active than usual and thereby causing the arterial pressure to rise back toward normal.

## CHARACTERISTICS OF CONTROL SYSTEMS

The above examples of homeostatic control mechanisms are only a few of the many hundreds in the body, all of which have certain characteristics in common. These are explained in the following pages.

**Negative Feedback Nature of Control Systems.** The control systems of the body act by a process of *negative feedback,* which can be explained best by reviewing some of the homeostatic control systems mentioned above. In the regulation of carbon dioxide concentration, a high level of carbon dioxide in the extracellular fluid causes increased pulmonary ventilation, and this in turn causes decreased carbon dioxide concentration. In other words, the response is *negative* to the initiating stimulus. Conversely, if the carbon dioxide concentration falls too low, this causes feedback through the control system to raise the carbon dioxide concentration. This response also is negative to the initiating stimulus.

In the arterial pressure-regulating mechanism, a high pressure causes a series of reactions that promote a lowered pressure, or a low pressure causes a series of reactions that promote an elevated pressure. In both instances these effects are negative with respect to the initiating stimulus.

Therefore, in general, if some factor becomes excessive or too little, a control system initiates *negative feedback,* which consists of a series of changes that returns the factor toward a certain mean value, thus maintaining homeostasis.

*Amplification, or Gain, of a Control System.* The degree of effectiveness with which a control system maintains constant conditions is called the *amplification,* or *gain,* of the system. For instance, let us assume that a person has been in a room for 24 hours, the temperature of which has been 60° F. Then suddenly the room temperature is increased to 110° F., a total increase of 50°. However, the person's body temperature rises only from 98.0° to 99.0° F. because the body automatically controls its own temperature within very narrow limits rather than following the change in atmospheric temperature. In this case, the change in atmospheric temperature is 49° more than the change in body temperature, which changes only 1°. Therefore, we say that the *gain* of the control system is −49. In other words, for each degree change in body temperature that does occur, there would be 49 times that much additional change were it not for the control system.

The gain of the baroreceptor system for control of arterial pressure, as measured in dogs, is approximately −2. That is, an extraneous factor that tends to increase or decrease the arterial pressure does so only one-third as much as would occur if this control system were not present. Therefore, one can see that the temperature control system is much more effective than the baroreceptor system.

**Positive Feedback as a Cause of Death—Vicious Cycles.**   One might ask the question: why do essentially all control systems of the body operate by negative feedback rather than positive feedback? However, if he will consider the nature of positive feedback, he will immediately see that positive feedback does not lead to stability but to instability and often to death.

Figure 1–3 illustrates an instance in which death can ensue from positive feedback. This figure depicts the pumping effectiveness of the heart, showing that the heart of the normal human being pumps about 5 liters of blood per minute. However, if the person is suddenly bled 2 liters, the amount of blood in his body is decreased to such a low level that not enough is available for his heart to pump effectively. As a result, the arterial pressure falls, and the flow of blood to the heart muscle through the coronary vessels diminishes. This results in weakening of the heart, further diminished pumping, further decrease in coronary blood flow, and still more weakness of the heart; the cycle repeats itself again and again until death. Note that each cycle in the feedback results in further weakness of the heart. In other words, the initiating stimulus causes more of the same, which is *positive feedback*.

Positive feedback is better known as a "vicious cycle," but actually a mild degree of positive feedback can be overcome by the negative feedback control mechanisms of the body, and a vicious cycle will fail to develop. For instance, if the person in the above example were bled only 1 liter instead of 2 liters, the normal negative feedback mechanisms for controlling cardiac output and arterial pressure would

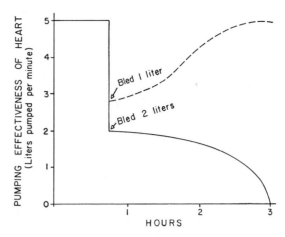

**Figure 1–3.**   Death caused by positive feedback when 2 liters of blood is removed from the circulation.

over-balance the positive feedback, and the person would recover, as shown by the dashed curve of Figure 1–3.

## AUTOMATICITY OF THE BODY

The purpose of this chapter has been to point out, first, the overall organization of the body and, second, the means by which the different parts of the body operate in harmony. To summarize, the body is actually an *aggregate of about 75 trillion cells* organized into different functional structures, some of which are called *organs*. Each functional structure provides its share in the maintenance of homeostatic conditions in the extracellular fluid, which is often called the *internal environment*. As long as normal conditions are maintained in the internal environment, the cells of the body will continue to live and function properly. Thus, each cell benefits from homeostasis, and in turn each cell contributes its share toward the maintenance of homeostasis. This reciprocal interplay provides continuous automaticity of the body until one or more functional systems lose their ability to contribute their share of function. When this happens, all the cells of the body suffer. Extreme dysfunction leads to death, while moderate dysfunction leads to sickness.

## APPENDIX

## BASIC PHYSICAL PRINCIPLES OF CONTROL SYSTEMS

Until recently the principles of control systems as applied to the human body have been taught only qualitatively rather than quantitatively. For instance, in the case of baroreceptor regulation of arterial pressure, physiologists have simply stated that an increase in pressure causes a reflex decrease in pressure back toward normal. However, this means very little unless we know *how much* effect occurs. But within the past 15 years, application of much more quantitative physical principles of control systems has begun to make physiology a much more exact science than it has been in the past.

**Basic Symbols Used in Control System Analysis.** Figure 1–4 illustrates common basic symbols used in control system analysis. These are the following:

*The Addition-Subtraction Symbol.* This symbol is shown in Figure 1–4. For instance, let x represent the rate of intake of salt in the solid food eaten each day, y the intake of salt in liquids drunk each day, and z the rate of loss of salt in the urine each day. Then the net rate of change of salt in the body will be x + y − z, as indicated by the arrow.

*The Multiplication Symbol.* Figure 1–4B illustrates multiplication of three quantities, x, y, and z, to give xyz. For instance, let us assume that three separate factors are affecting arterial pressure and that these factors multiply each other. Thus, the baroreceptor system might be causing x effect to elevate arterial pressure; a hormone secreted by one of the endocrine glands might be causing y effect to elevate arterial pressure; and hemorrhage might be causing z effect to decrease arterial pressure. The net effect would be xyz.

*Multiplication by a Constant Factor.* Figure 1–4C illustrates multiplication by a constant factor. Let us assume that x is the *concentration* of sodium in the extracellular fluids and K is the volume of extracellular fluid; the total *quantity* of sodium in extracellular fluid would then be Kx.

*The Division Symbol.* Figure 1–4D shows the value x divided by the value y to give x/y. For example, if x is the total *quantity* of sodium in the extracellular fluid and y is the extracellular fluid *volume,* then the output of this block, x/y, is the concentration of sodium in the extracellular fluid.

*The Integration or Accumulation Symbol.* Many functions of the body depend upon slow accumulation of some factor. Thus, in Figure 1–4E, if the *rate of change* of aldosterone in the body is represented by the differential term, dA/dt, the output of the block is the *quantity* of aldosterone that has accumulated in the body at any given time, t. The symbol in this block is called the *integration symbol,* or, in other words, the rate dA/dt is integrated to give A. If dA/dt is positive, the quantity of aldosterone in the body will be increasing, whereas if dA/dt is negative, the quantity of aldosterone in the body will be decreasing.

*Transfer Functions.* Figures 1–4F, G, and H illustrate transfer functions that show a quantity x entering a block and a quantity y leaving the block. Each block means that y is related to x in accordance with the function that is expressed inside the box. In Figure 1–4F this function is represented graphically. In Figure 1–4G it is represented by an algebraic equation. Figure 1–4H illustrates four inputs and one output, showing that three of the inputs add to each other and the other multiplies the first three.

As an example, Figure 1–4F might represent the relationship between glucose concentration x in the extracellular fluids and the rate of insulin secretion, y, by the pancreas. This transfer function shows that at low glucose concentrations essentially no insulin is secreted, but at high concentrations very large quantities of insulin are secreted.

**General Analysis of a Control System.** Figure 1–5 illustrates a general analysis that can be applied to essentially any control system in the body. The goal of all control systems is to keep some *controlled variable* at an almost constant value. In Figure 1–5 the controlled variable is represented by the arrow that projects to the right, and the essential components of the control system are represented by the other elements of this figure. The three components are represented by the three blocks as follows:

Block I illustrates the summation of the three different factors that determine the value of the controlled variable. The three inputs to this block are (1) the normal value for the controlled variable, that is, the mean level at which the control system attempts to maintain the variable; (2) any disturbance that is acting on the body to cause the variable to deviate from its normal value; and (3) any compensation that is caused by the control system to counteract the disturbance.

Block II calculates the difference between the ac-

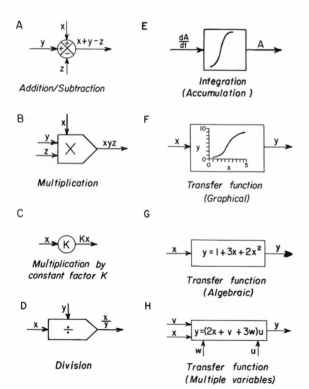

**Figure 1–4.** Standard symbols used in control system diagrams.

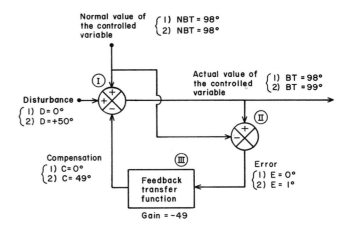

**Figure 1–5.** Static analysis of a control system. This figure, presenting an analysis of the control of body temperature, shows values for different variables in the system under two different conditions: (1) normal, and (2) when there is an increase in air temperature of 50° F, thereby creating a disturbance that attempts to raise the body temperature. (NBT = Normal body temperature; BT = Actual body temperature; E = Error; C = Compensation; D = Disturbance.)

tual value of the controlled variable and the normal value. The output of this block is called the *error*.

Block III is the feedback portion of the control system that responds to the error and determines the degree of compensation that will occur to overcome the disturbance.

***Application of the General Analysis to the Control of Body Temperature.*** Now let us apply this general analysis of a control system to an actual example, the control of body temperature. At each point in the system two different sets of values are shown. The first value in each instance represents the normal when there is no disturbance and no compensation by the control system. Thus, the normal values are as follows:

Normal value for the controlled variable 98°(F.)
Disturbance......................................... 0°
Actual value of the controlled variable.. 98°
Error............................................... 0°
Compensation.................................... 0°

Next, let us add a disturbance that attempts to change the body temperature. In this instance, we will assume that the air temperature increases by 50°. If the body temperature control system were not functioning, this would increase the body temperature by 50°. However, as the body temperature begins to rise, the actual temperature becomes different from the normal value, an error develops, and the control system provides sufficient compensation to overcome most of the disturbing effects of the abnormal air temperature. This compensation results from evaporation of the sweat from the surface of the skin, which in turn causes increased heat loss from the body. Therefore, the second set of values shown in Figure 1–5 is the following:

Normal value of the controlled variable 98°(F.)
Disturbance.................................. +50°
Actual value of the controlled variable.. 99°
Error............................................... 1°
Compensation............................... −49°

Thus, one sees that after the control system has become effective, the actual value of the controlled variable becomes equal to the sum of three different values: (1) the normal value for the controlled variable (98°), (2) the disturbance (+50°), and (3) the

compensation (−49°), giving an actual value of the controlled variable of 99°. That is, the abnormal disturbance of +50° causes a rise in body temperature of only 1°.

One can then calculate the gain of this system by dividing the compensation by the error (−49° divided by 1°), yielding a feedback gain of −49.

Figure 1–6 illustrates the effect of this 50° increase in air temperature under two conditions: (1) when the temperature control system is not functional, and (2) when it is functional. When the system is functioning, the compensation increases at the same time that the body temperature rises. Therefore, by the time the body temperature has risen from 98° to 99° the control system will have already initiated −49° of compensation. The body temperature, therefore, will rise only to 99° rather than to 148°, an increase that would occur if the control system did not exist.

**More Complex Analysis of a Control System— The Glucose Control System.** Figure 1–7 illustrates an analysis of the control system for regulating glucose concentration in the extracellular fluids. The basic system is the following: When a person eats increased quantities of glucose, the rising glucose concentration in the extracellular fluids causes the

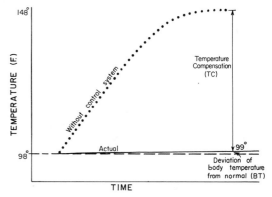

**Figure 1–6.** Effects on body temperature of suddenly increasing the air temperature 50° F, showing the hypothetical effect without a control system and the actual effect with a normal control system.

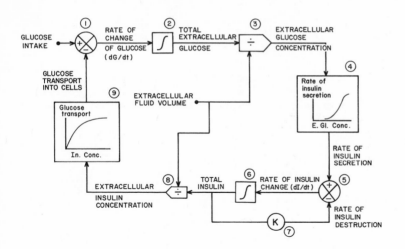

**Figure 1–7.** Analysis of the insulin control system for maintaining a constant glucose concentration in the extracellular fluid. By this analysis it is possible to predict transient as well as steady-state changes in variables of the system such as the readjustments of the system after sudden changes in the rate of glucose intake.

pancreas to secrete increased quantities of insulin. The insulin, in turn, causes increased transport of glucose through the cell membranes to the interior of the cells where the glucose is used for energy. This, obviously, returns the extracellular glucose back toward normal. A quantitative mathematical analysis is illustrated by the nine blocks of Figure 1–7; these blocks can be explained as follows:

Block 1 calculates the rate of change of glucose (dG/dt) in the extracellular fluids by subtracting rate of glucose transport into the cells from the rate of intake of glucose.

Block 2 integrates the rate of change of glucose with respect to time to give the total extracellular glucose.

Block 3 calculates the extracellular glucose concentration by dividing the total extracellular glucose by the extracellular fluid volume.

Block 4 illustrates the effect of extracellular glucose concentration on rate of insulin secretion.

Block 5 sums the rate of insulin secretion and the rate of insulin destruction to give the rate of insulin change (dI/dt).

Block 6 integrates the rate of insulin change to give the total insulin in the body at any one time.

Block 7 calculates rate of insulin destruction by multiplying the total insulin by the constant K.

Block 8 calculates the extracellular insulin concentration by dividing total insulin by extracellular fluid volume.

Block 9 illustrates the effect of extracellular insulin concentration on the rate of glucose transport into the body cells.

***Function of the Glucose Control System When Glucose is Infused into a Person.*** Figure 1–8 illustrates the effect on extracellular glucose concentration and also on insulin concentration when an infusion of glucose solution is suddenly started and maintained at a constant rate for many hours. The effect is shown for two different conditions: without the function of the control system (dashed lines) and with the function of the control system (solid line). Note that without the control system the glucose concentration rises to approximately 170 mg. per cent and stabilizes

at this level. Note, too, that there is also no increase in secretion of insulin.

On the other hand, when the control system is operative the glucose concentration begins to rise just as rapidly as before, but it soon stops rising and returns to a value not far from the normal glucose concentration level. This is caused by the effect of the glucose on the pancreas to cause insulin secretion, followed by buildup of insulin in the body fluids, until glucose transport through the cell membranes rises to equal the rate of glucose infusion into the blood. Note that the insulin concentration remains elevated as long as glucose continues to be infused, a condition that is necessary if the glucose concentration in the body fluids is to remain near normal. Note also that at first the glucose concentration slightly overshoots the final level at which it stabilizes. This is caused by the initial buildup of glucose in the extracellular fluid before the insulin secretion and insulin function have had time to become fully active.

To the right in Figure 1–8 the degree of correction caused by the glucose control system is illustrated by the downward-directed arrow. The final error is illustrated by the upward-directed arrow. The ratio of the lengths of these two arrows represents the gain of the system. Since the correction arrow is approximately

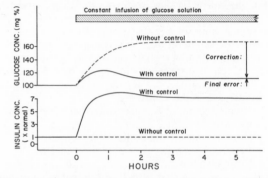

**Figure 1–8.** Transient changes in extracellular glucose and insulin concentrations following sudden changes in glucose intake, as predicted from the analysis of Figure 1–7 under two different conditions: (1) when the control system is not functioning, and (2) when it is functioning.

six times as long as the error arrow, the gain is approximately −6.

**Steady-state versus Transient Analyses of Control Systems.**    The analysis of the system for body temperature regulation shown in Figure 1–5 was a steady-state analysis. That is, it simply showed the initial conditions and the final conditions but did not characterize the transient events occurring in the control system during its activation. On the other hand, the analysis of glucose control in Figures 1–7 and 1–8 gives the transients through which all the different elements of the system pass in developing the final steady-state condition. In other words, the control system does not act instantly, but instead requires a certain amount of time to develop its compensation. Furthermore, the controlled variable often overshoots the final steady-state value before it stabilizes.

## OSCILLATION OF CONTROL SYSTEMS

Unfortunately, even feedback control systems sometimes become unstable and oscillate. Figure 1–9 illustrates blood pressure waves caused by oscillation of the baroreceptor system described earlier in the chapter. The cause of the oscillation is the following: Some extraneous factor causes the arterial pressure to become too high, and this activates the baroreceptor reflex. The pressure begins to fall and eventually falls even below normal. Then the very low arterial pressure activates the baroreceptor reflex in the opposite direction, causing the pressure to increase to a level above normal once again. This initiates a new cycle of oscillation, which can continue indefinitely.

**Damping of Control Systems.**    Fortunately, oscillation of the arterial pressure, as shown in Figure 1–9, does not occur very often but only under certain peculiar conditions such as decreased blood volume or compression of the brain. The reason for lack of oscillation is that this control system, as is also true of almost all other control systems of the body, is highly *damped*, which results from the basic organization of the system itself. For instance, the glucose control system of Figures 1–7 and 1–8 shows only slight overshoot in glucose concentration and insulin concentration, which represents a very highly damped oscillation that persists for only a single

overshoot cycle. The major reason for this very high degree of damping is the long periods of time required for both the glucose and the insulin to build up their concentrations in the body fluid. In the meantime, the other parts of the control system can adapt themselves almost completely to the changing glucose concentration, thereby preventing any overresponse of the control system.

Figure 1–9 shows *waxing oscillation, driving oscillation,* and *damped oscillation.* When the control system is very unstable—that is, has almost no damping—even the slightest disturbance can cause mild oscillation at first; this then grows to greater and greater oscillation. If the damping is moderate, the oscillations may continue as driving oscillation indefinitely, never increasing nor decreasing. On the other hand, if the damping is very great, the oscillation fades away.

A few control systems continue to oscillate indefinitely. One of these is the system that causes respiration, which results from oscillation, or "reverberation," of impulses within the respiratory center of the brain. Another continuous oscillation is responsible for the monthly sexual cycle of the female, in which hormones secreted by the anterior pituitary gland stimulate the ovaries to produce ovarian hormones, which in turn act on the pituitary to inhibit its secretion. Lack of the pituitary hormones then causes the ovaries to stop secreting their hormones. This in turn allows the pituitary to begin secreting once again. As a result, the cycle repeats itself again and again from puberty to the menopause.

## REFERENCES

Adolph, E. F.: Origins of Physiological Regulations. New York, Academic Press, Inc., 1968.

Adolph, E. F.: Physiological adaptations: Hypertrophies and superfunctions. *Amer. Sci.,* 60:608, 1972.

Bernard, C.: Lectures on the Phenomena of Life Common to Animals and Plants. Springfield, Ill., Charles C Thomas, Publisher, 1974.

Brown, J. H. U. (ed.): Engineering Principles in Physiology. New York, Academic Press, Inc., 1973, Vols. 1 and 2.

Cannon, W. B.: The Wisdom of the Body. New York, W. W. Norton & Company, Inc., 1932.

Guyton, A. C., and Coleman, T. G.: Quantitative analysis of the pathophysiology of hypertension. *Cir. Res.,* 14:I-1, 1969.

Hemker, H. C., and Hess, B.: Analysis and Simulation of Biochemical Systems. New York, American Elsevier Publishing Co., 1972.

Huffaker, C. B. (ed.): Biological Control. New York, Plenum Publishing Corp., 1974.

Iberall, A. S., and Guyton, A. C. (eds.): *Proc. Int. Symp. on Dynamics and Controls in Physiological Systems. Regulation and Control in Physiological Systems.* ISA, Pittsburgh, 1973.

Jones, R. W.: Principles of Biological Regulation: An Introduction to Feedback Systems. New York, Academic Press, Inc., 1973.

Milhorn, H. T.: The Application of Control Theory to Physiological Systems. Philadelphia, W. B. Saunders Company, 1966.

Miller, S. L., and Orgel, L. E.: The Origins of Life on the Earth. Englewood Cliffs, N.J., Prentice-Hall, Inc., 1974.

Parsegian, V. A.: Long-range physical forces in the biological milieu. *Ann. Rev. Biophys. Bioeng.,* 4:221, 1973.

Reeve, E. B., and Guyton, A. C.: Physical Bases of Circulatory Transport: Regulation and Exchange. Philadelphia, W. B. Saunders Company, 1967.

Söderberg, U.: Neurophysiological aspects of homeostasis. *Ann. Rev. Physiol.,* 26:271, 1964.

Toates, F. M.: Control Theory in Biology and Experimental Psychology. London, Hutchinson Education Ltd., 1975.

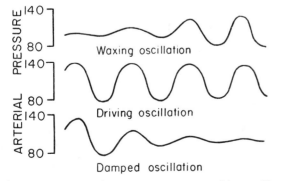

**Figure 1–9.**    Arterial pressure waves caused by oscillation of the baroreceptor system.

# | 2 |

# The Cell and Its Function

Each of the 75 trillion cells in the human being is a living structure that can survive indefinitely and in most instances can even reproduce itself, provided its surrounding fluids contain appropriate nutrients. To understand the function of organs and other structures of the body, it is essential that we first understand the basic organization of the cell and the functions of its component parts.

## ORGANIZATION OF THE CELL—PROTOPLASM

A typical cell, as seen by the light microscope, is illustrated in Figure 2–1. Its two major parts are the *nucleus* and the *cytoplasm*. The nucleus is separated from the cytoplasm by a *nuclear membrane,* and the cytoplasm is separated from the surrounding fluids by a *cell membrane*.

The different substances that make up the cell are collectively called *protoplasm*. Protoplasm is composed mainly of five basic substances: water, electrolytes, proteins, lipids, and carbohydrates.

**Water.**  The fluid medium of all protoplasm is water, which is present in a concentration between 70 and 85 per cent. Many cellular chemicals are dissolved in the water, while others are suspended in small particulate form. Chemical reactions take place among the dissolved chemicals or at the surface boundaries between the suspended particles and the water. The fluid nature of water allows both the dissolved and suspended substances to diffuse or flow to different parts of the cell, thereby providing transport of the substances from one part of the cell to another.

**Electrolytes.**  The most important electrolytes in the cell are *potassium, magnesium, phosphate, sulfate, bicarbonate,* and small quantities of *sodium, chloride,* and *calcium*. These will be discussed in much greater detail in Chapter 4, which will consider the interrelationships between the intracellular and extracellular fluids.

The electrolytes are dissolved in the water of the protoplasm, and they provide inorganic chemicals for cellular reactions. Also, they are necessary for operation of some of the cellular control mechanisms. For instance, electrolytes acting at the cell membrane allow transmission of electrochemical impulses in nerve and muscle fibers, and the intracellular electrolytes determine the activity of different enzymatically catalyzed reactions that are necessary for cellular metabolism.

**Proteins.**  Next to water, the most abundant substance in most cells is proteins, which normally constitute 10 to 20 per cent of the cell mass. These can be divided into two different types, *structural proteins* and *enzymes*.

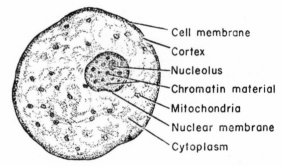

**Figure 2–1.**  Structure of the cell as seen with the light microscope.

12

To get an idea of what is meant by *structural proteins,* one needs only to note that leather is composed principally of structural proteins, and that hair is almost entirely a structural protein. Proteins of this type are present in the cell membrane, in the nuclear membrane, and in membranes surrounding special intracellular structures, such as the endoplasmic reticulum and the mitochondria. Thus, structural proteins hold the structures of the cell together. Most structural proteins are *fibrillar;* that is, the individual protein molecules are polymerized into long fibrous threads. The threads in turn provide tensile strength for the cellular structures.

*Enzymes,* on the other hand, are an entirely different type of protein, composed usually of individual protein molecules or at most of aggregates of a few molecules in a *globular* form rather than a fibrillar form. These proteins, in contrast to the fibrillar proteins, are often soluble in the fluid of the cell or are adsorbed or are adherent to the surfaces of membranous structures inside the cell. The enzymes come into direct contact with other substances inside the cell and catalyze chemical reactions. For instance, the chemical reactions that split glucose into its component parts and then combine these with oxygen to form carbon dioxide and water while at the same time providing energy for cellular function are catalyzed by a series of protein enzymes. Thus, enzyme proteins control the metabolic functions of the cell.

Special types of proteins are present in different parts of the cell. Of particular importance are the *nucleoproteins,* present both in the nucleus and in the cytoplasm. The nucleoproteins of the nucleus contain *deoxyribonucleic acid (DNA),* which constitutes the *genes,* and these control the overall function of the cell as well as the transmission of hereditary characteristics from cell to cell. These substances are so important that they will be considered in detail in Chapter 3. In addition, the chemical nature of proteins will be considered in Chapter 69, and the different structural and enzymatic functions of proteins will be subjects of discussion at numerous points throughout this text.

**Lipids.** Lipids are several different types of substances that are grouped together because of their common property of being soluble in fat solvents. The most abundant lipid of animal tissues is *triglyceride,* also called *neutral fat.* However, in addition to neutral fat two other types of lipids, *phospholipids* and *cholesterol,* are very common throughout cells.

The usual cell contains 2 to 3 per cent lipids which are dispersed throughout the cell but are present in especially high concentrations in the cell membrane, the nuclear membrane, and the membranes lining intracytoplasmic organelles, such as the endoplasmic reticulum and the mitochondria.

The special importance of lipids in the cell is that they are either insoluble or only partially soluble in water. They combine with structural proteins to form the different membranes that separate the different water compartments of the cell from each other. The lipids of each membrane form a boundary between the solutions on the two sides of the membrane, making it impervious to many dissolved substances.

The chemical natures of the different types of lipids and their functions in the body will be discussed in Chapter 68.

**Carbohydrates.** In general, carbohydrates have very little structural function in the cell, but they play a major role in nutrition of the cell. Most human cells do not maintain large stores of carbohydrates, usually averaging about 1 per cent of their total mass. However, carbohydrate, in the form of glucose, is always present in the surrounding extracellular fluid so that it is readily available to the cell. The small amount of carbohydrates stored in the cells is almost entirely in the form of *glycogen,* which is an insoluble polymer of glucose.

## PHYSICAL STRUCTURE OF THE CELL

The cell is not merely a bag of fluid, enzymes, and chemicals; it also contains highly organized physical structures called *organelles,* which are equally as important to the function of the cell as the cell's chemical constituents. For instance, without one of the organelles, the *mitochondria,* more than 95 per cent of the energy supply of the cell would cease immediately. Some principal organelles of the cell, are the *cell membrane, nuclear membrane, endoplasmic reticulum, mitochondria,* and *lysosomes.* These are illustrated in Figure 2–2. Others not shown in the figure are the *Golgi complex, centrioles, cilia,* and *microtubules.*

### MEMBRANES OF THE CELL

Essentially all physical structures of the cell are lined by a membrane composed primarily of lipids and proteins. All of the different membranes are similar in structure, and this common type of structure is called the "unit membrane," which will be described below in rela-

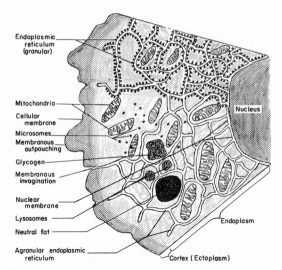

**Figure 2–2.** Organization of the cytoplasmic compartment of the cell.

tion to the cell membrane. The different membranes include the *cell membrane,* the *nuclear membrane,* the *membrane of the endoplasmic reticulum,* and the *membranes of the mitochondria, lysosomes, Golgi complexes,* and so forth.

**The Cell Membrane.**    The cell membrane is thin (approximately 75 to 100 Ångstroms) and elastic. It is composed almost entirely of proteins and lipids, with a percentage composition approximately as follows: proteins, 62 per cent; lipids, 35 per cent; polysaccharides, perhaps 3 per cent. The proteins in the cell are mainly proteins called *tektins,* insoluble structural proteins having elastic properties. The lipids are approximately 60 per cent phospholipids, 25 per cent cholesterol, and 15 per cent other lipids.

The precise molecular organization of the cell membrane is unknown, but the classical description of its structure is that illustrated in Figure 2–3A. This shows a central layer of lipids covered by protein layers and a thin mucopolysaccharide layer on the outside surface. The presence of protein and mucopolysaccharide on the surfaces makes the membrane *hydrophilic,* which means that water adheres easily to the membrane. The lipid center of the membrane makes the membrane mainly impervious to lipid-insoluble substances. The small knobbed structures lying at the bases of the protein molecules in Figure 2–3A are phospholipid molecules; the fat portion of the phospholipid molecule is attracted to the central lipid phase of the cell membrane, and the polar (ionized) portion of the molecule protrudes toward either surface, where it is

bound electrochemically with the inner and outer layers of proteins. The thin layer of mucopolysaccharide on the outside of the cell membrane helps to make the outside different from the inside, thus polarizing the membrane so that the chemical reactivities of the cell's inner surface are different from those of the outer surface.

Another interpretation of the cell membrane, and one that much recent evidence supports, is that the basic structure of the membrane is almost entirely lipid, with very little protein on the two surfaces. Instead, the protein seems to be dispersed throughout the lipid where it performs many important functions, including (1) contributing to the structural strength of the membrane; (2) acting as an enzyme to promote chemical reactions; (3) acting as a carrier protein for transport of substances through the membrane; and (4) providing molecular breaks in the lipid substances and, therefore, providing pores through the membrane.

***Enzymes in the Membrane.***    A number of protein enzymes are either dissolved in the cell membrane or are adherent to it. Many of these are specifically present in the inner protein layer of the membrane and, therefore, function at the boundary between the inner surface of the membrane and the cytoplasm to catalyze many chemical reactions. For instance, it will be pointed out in Chapter 4 that enzyme-catalyzed reactions of this type are responsible for transporting many important substances through the cell membrane. These enzymatic characteristics of the membrane also help to polarize the membrane, making the chemical reactivities of the inner surface different from those of the outer surface.

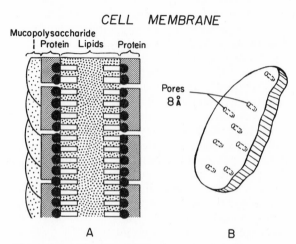

**Figure 2–3.**  (A) Postulated molecular organization of the cell membrane. (B) Pores in the cell membrane.

*The "Unit" Membrane.* The triple layer membrane described above, with lipid in the center and two layers of protein on the two sides of the lipid, is called the *unit membrane.* This same basic structure, though in slightly different forms, is found in all the different membranes of the cell, such as the nuclear membrane and the membranes of the mitochondria, endoplasmic reticulum, lysosomes, and so forth; these will be discussed in subsequent paragraphs.

*Pores in the Membrane.* The membrane is believed to have many minute pores that pass from one side to the other as shown in Figure 2–3B. These pores have never been demonstrated even with an electron microscope, but functional experiments to study the movement of molecules of different sizes between the extra- and intracellular fluids have demonstrated free diffusion of molecules up to a size of approximately 8 Ångstroms. It is through these pores that lipid-insoluble substances of very small sizes, such as water and urea molecules, are believed to pass with relative ease between the interior and exterior of the cell.

It is likely that the pores in the cell membrane are caused by the presence of large protein molecules that interrupt the lipid membrane structure and extend all the way through the membrane from one side to the other, thus providing direct watery passage through the interstices of the protein molecules. Furthermore, when we discuss the "pores" in nerve membranes in Chapter 10, we shall see that there are probably at least two varieties of pores—and perhaps still more—each allowing free movement of different types of substances.

**The Nuclear Membrane.** The nuclear membrane is actually two "unit" membranes, one surrounding the other and having a wide space in between. Each unit membrane is almost identical to the cell membrane, having lipids in its center and protein on its two surfaces, but having no mucopolysaccharide layer. Very large holes, or "pores," of several hundred Ångstroms diameter are present in the nuclear membrane as illustrated in Figure 2–9 so that almost all dissolved substances can move with ease between the fluids of the nucleus and those of the cytoplasm.

**The Endoplasmic Reticulum.** Figure 2–2 illustrates in the cytoplasm a network of tubular and vesicular structures, constructed of a system of unit membranes, called the *endoplasmic reticulum.* The detailed structure of this organelle is illustrated in Figure 2–4. The space

inside the tubules and vesicles is filled with *endoplasmic matrix,* a fluid medium that is different from the fluid outside the endoplasmic reticulum.

Electron micrographs show that the space inside the endoplasmic reticulum is connected with the space between the two membranes of the double nuclear membrane. Also, this space is continuous with the inner chambers of the Golgi complex, to be described below. In some instances the endoplasmic reticulum connects directly through small openings with the exterior of the cell. Substances formed in different parts of the cell enter the spaces of this vesicular system and are then conducted to other parts of the cell.

*Ribosomes and the Granular Endoplasmic Reticulum.* Attached to the outer surfaces of many parts of the endoplasmic reticulum are large numbers of small granular particles called ribosomes. Where these are present, the reticulum is frequently called the *granular endoplasmic reticulum.* The ribosomes are composed mainly of ribonucleic acid, which functions in the synthesis of protein in the cells, as discussed in the following chapter.

*The Agranular Endoplasmic Reticulum.* Part of the endoplasmic reticulum has no attached ribosomes. This part is called the *agranular,* or *smooth, endoplasmic reticulum.* The agranular reticulum helps in the synthesis of lipid substances and probably also plays an important role in glycogen resorption.

**Golgi Complex.** The Golgi complex, illustrated in Figure 2–5, is probably a specialized portion of the endoplasmic reticulum. It has membranes similar to those of the agranular endoplasmic reticulum and is usually composed of four or more layers of thin vesicles. Electron micrographs show direct connections between the endoplasmic reticulum and parts of the Golgi complex.

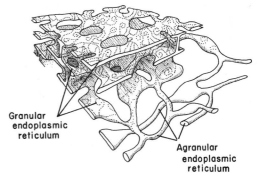

Granular endoplasmic reticulum

Agranular endoplasmic reticulum

**Figure 2–4.** Structure of the endoplasmic reticulum. (Modified from De Robertis, Saez, and De Robertis: Cell Biology, 6th Ed., 1975.)

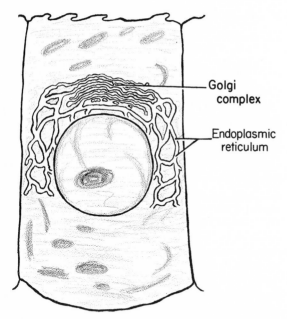

Golgi complex

Endoplasmic reticulum

**Figure 2–5.**   A typical Golgi complex.

The Golgi complex is very prominent in secretory cells; in these, it is located on the side of the cell from which substances will be secreted. Its function is believed to be temporary storage and condensation of secretory substances and preparation of these substances for final secretion. In addition, the Golgi complex also synthesizes carbohydrates and then combines these with protein in the Golgi cisternae to form glycoproteins. These are important secretory substances of many cells; especially important are the mucopolysaccharides that are major constituents of (1) mucus, (2) the ground substance of the interstitial spaces, and (3) the ground substance of both cartilage and bone.

Finally, the Golgi complex is involved in the formation of the lysosomes, cytoplasmic organelles that are important for digesting intracellular substances, as will be explained in a subsequent section of this chapter.

## THE CYTOPLASM AND ITS ORGANELLES

The cytoplasm is filled with both minute and large dispersed particles and organelles ranging in size from a few Ångstroms to 3 microns in size. The clear fluid portion of the cytoplasm in which the particles are dispersed is called *hyaloplasm;* this contains mainly dissolved proteins, electrolytes, glucose, and small quantities of phospholipids, cholesterol, and esterified fatty acids.

The portion of the cytoplasm immediately beneath the cell membrane is frequently gelled into a semi-solid called the *cortex,* or *ectoplasm.* The cytoplasm between the cortex and the nuclear membrane is liquefied and is called the *endoplasm.*

Among the large dispersed particles in the cytoplasm are neutral fat globules, glycogen granules, ribosomes, secretory granules, and two important organelles—the *mitochondria* and *lysosomes*—which are discussed below.

**Colloidal Nature of the Cytoplasm.**   All the particles dispersed in the cytoplasm, whether the large lysosomes and mitochondria or the small granules, are hydrophilic—that is, attracted to water—because of electrical charges on their surfaces. The lining membranes of the mitochondria and lysosomes have charges on their surfaces because of the protein molecules on the membrane surfaces. The particles remain dispersed in the cytoplasm mainly because of mutual repulsion of the charges on the different particles. Thus, the cytoplasm is actually a colloidal solution.

**Mitochondria.**   Mitochondria extract energy from the nutrients and oxygen and in turn provide most of the energy needed elsewhere in the cell for performing the cellular functions. These organelles are present in the cytoplasm of all cells, as illustrated in Figure 2–2, but the number per cell varies from a few hundred to many thousand, depending on the amount of energy required by each cell. Mitochondria are also very variable in size and shape; some are only a few hundred millimicrons in diameter and globular in shape while others are as large as 1 micron in diameter, as long as 7 microns, and filamentous in shape.

The basic structure of the mitochondrion is illustrated in Figure 2–6, which shows it to be composed mainly of two layers of unit membrane: an *outer membrane* and an *inner membrane.* Many infoldings of the inner membrane form *shelves* onto which the oxidative enzymes of the cell are attached. In addition, the inner cavity of the mitochondrion is filled with a gel *matrix* containing large quantities of dissolved enzymes that are necessary for extracting energy from nutrients. These enzymes operate in association with the oxidative enzymes on the shelves to cause oxidation of the nutrients, thereby forming carbon dioxide and water. The liberated energy is used to synthesize a high energy substance called *adenosine triphosphate (ATP).* ATP is then transported out of the mitochondrion, and it diffuses throughout the cell to release its energy wherever it is needed for performing cellular functions. The function

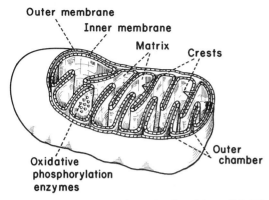

Outer membrane
Inner membrane
Matrix
Crests
Oxidative phosphorylation enzymes
Outer chamber

**Figure 2–6.** Structure of a mitochondrion. (Modified from De Robertis, Saez, and De Robertis: Cell Biology, 6th Ed., 1975.)

of ATP is so important to the cell that it is discussed in detail later in the chapter.

Mitochondria are probably self-replicative, which means that one mitochondrion can probably form a second one, a third one, and so on, whenever there is need in the cell for increased amounts of ATP. Indeed, the mitochondria contain a special type of deoxyribonucleic acid similar to, but different from, that found in the nucleus. In the following chapter we shall see that deoxyribonucleic acid is the basic substance that controls replication of the entire cell; it is reasonable, therefore, to suspect that this substance might play a similar role in the mitochondrion.

**Lysosomes.** The lysosomes provide an intracellular digestive system that allows the cell to digest and thereby remove unwanted substances and structures, especially damaged or foreign structures, such as bacteria. The lysosome, illustrated in Figure 2–2, is 250 to 750 millimicrons in diameter, and is surrounded by a unit membrane. It is filled with large numbers of small granules 55 to 80 Ångstroms in diameter which are protein aggregates of hydrolytic (digestive) enzymes. A hydrolytic enzyme is capable of splitting an organic compound into two or more parts by combining hydrogen from a water molecule with part of the compound and by combining the hydroxyl portion of the water molecule with the other part of the compound. For instance, protein is hydrolyzed to form amino acids, and glycogen is hydrolyzed to form glucose. More than a dozen different *acid hydrolases* have been found in lysosomes, and the principal substances that they digest are proteins, nucleic acids, mucopolysaccharides, and glycogen.

Ordinarily, the membrane surrounding the lysosome prevents the enclosed hydrolytic enzymes from coming in contact with other substances in the cell. However, many different conditions of the cell will break the membranes of some of the lysosomes, allowing release of the enzymes. These enzymes then split the organic substances with which they come in contact into small, highly diffusible substances, such as amino acids and glucose. Some of the more specific functions of lysosomes are discussed later in the chapter.

**Microtubules.** Located in many cells are fine tubular structures that are approximately 250 Å in diameter and that have lengths of 1 to many microns. These structures, called *microtubules*, are very thin in relation to their length, but they are usually arranged in bundles, which gives them, en masse, considerable structural strength. Furthermore, microtubules are usually stiff structures that break if bent too severely. Figure 2–7 illustrates some typical microtubules that have been teased from the flagellum of a sperm. Another example of microtubules is the tubular filaments that give cilia their structural strength, radiating upward from within the cell cytoplasm to the tip of the cilium.

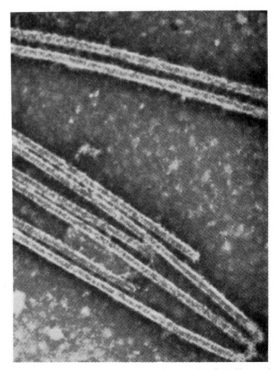

**Figure 2–7.** Microtubules teased from the flagellum of a sperm. (From Porter, K. R.: Ciba Foundation Symposium: Principles of Biomolecular Organization. Little, Brown and Co., 1966.)

The primary function of microtubules appears to be to act as a *cytoskeleton,* providing rigid physical structures for certain parts of cells such as the cilia just mentioned. However, the tubular nature of their structure also suggests that substances might be transported through the tubules. Indeed, cytoplasmic streaming has been observed in the vicinity of microtubules, indicating that these tubular structures might play a role in causing movement of cytoplasm.

**Secretory Granules.** One of the most important functions of many cells is secretion of special substances. The secretory substances are usually formed inside the cell and are held there until an appropriate time for release to the exterior. The storage depots within the cells are called secretory granules; these are illustrated by the dark spots in Figure 2–8, which shows pancreatic acinar cells that have formed and stored enzymes that will be secreted later into the intestinal tract.

Many secretory granules lie inside the tubules and vesicles of the endoplasmic reticulum and Golgi complex, while others are free in the cytoplasm and probably are formed in most if not all instances by the Golgi complex.

**Other Structures and Organelles of the Cytoplasm.** The cytoplasm of each cell contains two pairs of *centrioles,* which are small cylindrical structures that play a major role in cell division, as will be discussed in Chapter 3. Also, most cells contain small *lipid droplets* and *glycogen granules* that play important roles in

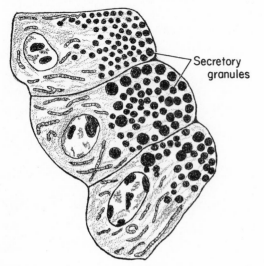

**Figure 2–8.** Secretory granules in acinar cells of the pancreas.

energy metabolism of the cell. And certain cells contain highly specialized structures such as the *cilia* of ciliated cells which are actually outgrowths from the cytoplasm, and the *myofibrils* of muscle cells. All of these are discussed in detail at different points in this text.

## THE NUCLEUS

The nucleus is the control center of the cell. It controls both the chemical reactions that occur in the cell and reproduction of the cell. Briefly, the nucleus contains large quantities of *deoxyribonucleic acid,* which we have called *genes* for many years. The genes determine the characteristics of the protein enzymes of the cytoplasm, and in this way control cytoplasmic activities. To control reproduction, the genes first reproduce themselves, and after this is accomplished the cell splits by a special process called *mitosis* to form two daughter cells, each of which receives one of the two sets of genes. These activities of the nucleus are considered in detail in the following chapter.

The appearance of the nucleus under the microscope does not give much of a clue to the mechanisms by which it performs its control activities. Figure 2–9 illustrates the light microscopic appearance of the interphase nucleus (period between mitoses), showing darkly staining *chromatin material* throughout the *nuclear sap.* During mitosis, the chromatin material becomes readily identifiable as part of the highly structured *chromosomes,* which can be seen easily with the light microscope. Even during the interphase of cellular activity the chromatin material is still organized into fibrillar chromosomal structures, but this is impossible to see except in a few types of cells.

**Nucleoli.** The nuclei of many cells contain one or more lightly staining structures called nucleoli. The nucleolus, unlike most of the organelles that we have discussed, does not have a limiting membrane. Instead, it is simply a protein structure that contains a moderate amount of *ribonucleic acid* of the type found in ribosomes. The nucleolus becomes considerably enlarged when a cell is actively synthesizing proteins. The genes of one of the chromosomes synthesize the ribonucleic acid and then store it in the nucleolus, beginning with a loose fibrillar RNA that later condenses to form the granular ribosomes. These in turn migrate into the cytoplasm where most of them become attached to the endoplasmic reticulum.

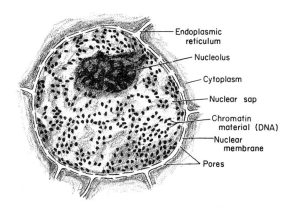

**Figure 2–9.** Structure of the nucleus.

# COMPARISON OF THE ANIMAL CELL WITH PRECELLULAR FORMS OF LIFE

Many of us think of the cell as the lowest level of animal life. However, the cell is a very complicated organism, which probably required several billion years to develop after the earliest form of life, some organism similar to the present-day *virus*, first appeared on Earth. Figure 2–10 illustrates the relative sizes of the smallest known virus, a large virus, a *rickettsia*, a *bacterium*, and a cell, showing that the cell has a diameter about 1000 times that of the smallest virus, and, therefore, a volume about 1 billion times that of the smallest virus. Correspondingly, the functions and anatomical organization of the cell are also far more complex than those of the virus.

The principal constituent of the very small virus is a *nucleic acid* embedded in a coat of protein. This nucleic acid is similar to that of the cell, and it is capable of reproducing itself if appropriate nutrients are available. Thus, the virus is capable of propagating its lineage from generation to generation, and, therefore, is a living structure in the same way that the cell and the human being are living structures.

As life evolved, other chemicals besides nucleic acid and simple proteins became integral parts of the organism, and specialized functions began to develop in different parts of the virus. A membrane formed around the virus, and inside the membrane a fluid matrix appeared. Specialized chemicals developed inside the matrix to perform special functions; many protein enzymes appeared which were capable of catalyzing chemical reactions, and, therefore, of controlling the organism's activities.

In still later stages, particularly in the rickettsial and bacterial stages, *organelles* developed inside the organism, these representing aggregates of chemical compounds that perform functions in a more efficient manner than can be achieved by dispersed chemicals throughout the fluid matrix. And, finally, in the cell, still more complex organelles developed, the most important of which is the *nucleus*. The nucleus distinguishes the cell from all lower forms of life; this structure provides a control center for all cellular activities, and it also provides for very exact reproduction of new cells generation after generation, each new cell having essentially the same structure as its progenitor.

# FUNCTIONAL SYSTEMS OF THE CELL

In the remainder of this chapter some of the more important functional systems of the cell are discussed. However, two of the most important functions, (1) control of protein synthesis and of other cellular functions by the genes of the nucleus, and (2) transport of substances through the cell membrane, are so important that the following two chapters are devoted to these subjects.

## *FUNCTIONS OF THE ENDOPLASMIC RETICULUM*

The endoplasmic reticulum exists in many different forms in different cells, sometimes highly granular with a large number of ribosomes on its surface, sometimes agranular, sometimes tubular, sometimes vesicular with large shelf-like surfaces, and so forth. Therefore, simply from anatomical considerations alone, it is certain that the endoplasmic reticulum performs a very large share of the cell's functions. Yet, understanding of its functions has been extremely slow to develop because of the difficulty of studying function with the electron microscope.

On the vast surfaces of the endoplasmic re-

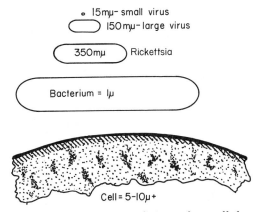

**Figure 2–10.** Comparison of sizes of precellular organisms with that of the average cell in the human body.

ticulum are adsorbed many of the protein enzymes of the cell, and perhaps still other enzymes are actually integral parts of the reticulum itself. Many of these enzymes synthesize substances in the cells, and others undoubtedly act to transport substances through the membrane of the endoplasmic reticulum from the hyaloplasm of the cytoplasm into the matrix of the reticulum or in the opposite direction. Some of the proved functions of the endoplasmic reticulum are the following:

**Secretion of Proteins by Secretory Cells.** Many cells, particularly cells in the various glands of the body, form special proteins that are secreted to the outside of the cells. The mechanism for this involves the endoplasmic reticulum and Golgi complex as illustrated in Figure 2–11. The ribosomes on the surface of the endoplasmic reticulum synthesize the protein that is to be secreted. This protein is either discharged directly into the tubules of the endoplasmic reticulum by the ribosomes or is immediately transported into the tubules to form small protein granules. These granules then move slowly through the tubules toward the Golgi complex, arriving there a few minutes to an hour or more later. In the Golgi complex the protein may be conjugated with carbohydrates to form glycoproteins which are a common secretory product. Also, the granules are condensed into coalesced granules that then are evaginated outward through the membrane of the Golgi complex into the cytoplasm of the cell to form *secretory granules*. Each of these granules carries with it part of the membrane of the Golgi apparatus which provides a membrane around the secretory granule and prevents it from dispersing in the cytoplasm. Gradually, the secretory granules move toward

the surface of the cell where their membranes become miscible with the membrane of the cell itself, and in some way not completely understood they expel their substances to the exterior. It is in this way that protein enzymes, for instance, are secreted by the exocrine glands of the gastrointestinal tract.

**Lipid Secretion.** Almost exactly the same mechanism applies to lipid secretion. The one major difference is that lipids are synthesized by the agranular portion of the endoplasmic reticulum. The Golgi complex also provides much the same function in lipid secretion as in protein secretion, for here lipids are stored for long periods of time before finally being extruded into the cytoplasm as lipid droplets and thence to the exterior of the cells.

**Release of Glucose from Glycogen Stores of Cells.** Electron micrographs show that glycogen is stored in cells as minute granules lying in close apposition to agranular tubules of the endoplasmic reticulum. The actual chemical reactions that cause polymerization of glucose to form glycogen granules probably occur in the hyaloplasm and not in the wall of the endoplasmic reticulum, but the reticulum seems to play a role in the breakdown of glycogen (glycogenolysis) because some of the enzymes for this process are found in the endoplasmic reticulum.

**Other Possible Functions of the Endoplasmic Reticulum.** The functions presented thus far for the endoplasmic reticulum have involved, first, synthesis of substances and, second, transport of these substances to the exterior of the cell. Many other substances are synthesized by the endoplasmic reticulum and are then secreted to the interior of the cell rather than to the exterior. The Golgi complex also plays a role in these internal secretory processes; indeed, large Golgi complexes are found in some cells, such as nerve cells, that do not perform any external secretory function.

Because of the multitude of different anatomical forms of the endoplasmic reticulum, it is certain that we will discover many more functional roles that it plays besides those few that have been discussed here.

### INGESTION BY THE CELL—PINOCYTOSIS

If a cell is to live and grow, it must obtain nutrients and other substances from the surrounding fluids. Substances can pass through a cell membrane in three separate ways: (1) by *diffusion* through the pores in the membrane or through the membrane matrix itself; (2) by *ac-*

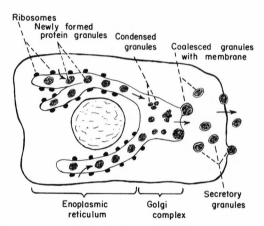

**Figure 2–11.** Function of the endoplasmic reticulum and Golgi complex in secreting proteins.

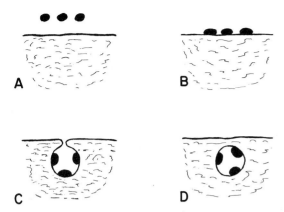

**Figure 2–12.**   Mechanism of pinocytosis.

*tive transport* through the membrane, a mechanism in which enzyme systems and special carrier substances "carry" the substances through the membrane; and (3) by *pinocytosis,* a mechanism by which the membrane actually engulfs some of the extracellular fluid and its contents. Transport of substances through the membrane is such an important subject that it will be considered in detail in Chapter 4, but one of these mechanisms of transport, pinocytosis, is a specialized cellular function that deserves mention here as well.

Figure 2–12 illustrates the mechanism of pinocytosis. Figure 2–12A shows three molecules of protein in the extracellular fluid approaching the surface of the cell. In Figure 2–12B these molecules have become attached to the surface, presumably by the simple process of adsorption. The presence of these proteins causes the surface tension properties of the cell surface to change in such a way that the membrane invaginates and closes over the proteins as shown in Figure 2–12C. Immediately thereafter, the invaginated portion of the membrane breaks away from the surface of the cell, forming a *pinocytic vesicle* which then penetrates deep into the cytoplasm away from the cell membrane.

Pinocytosis occurs only in response to certain types of substances that contact the cell membrane, the two most important of which are *proteins* and *strong solutions of electrolytes.* It is especially significant that proteins cause pinocytosis, because pinocytosis is the only means by which proteins can pass through the cell membrane.

**Phagocytosis.** Phagocytosis means the ingestion of large particulate matter by a cell, such as the ingestion of (a) a bacterium, (b) some other cell, or (c) particles of degenerating tissue. Pinocytosis was not discovered until the advent of the electron microscope because pinocytic vesicles are smaller than the resolution of the light microscope. However, phagocytosis has been known to occur from the earliest studies using the light microscope.

The mechanism of phagocytosis is almost identical with that of pinocytosis. The particle to be phagocytized must have a surface that can become adsorbed to the cell membrane. Many bacteria have membranes that, on contact with the cell membrane, actually become miscible with the cell membrane so that the cell membrane simply spreads around the bacterium and invaginates to form a *phagocytic vesicle* containing the bacterium, a vesicle that is essentially the same as the pinocytic vesicle shown in Figure 2–12 but much larger. Phagocytosis will be discussed at further length in Chapter 6 in relation to function of the white blood cells.

### THE DIGESTIVE ORGAN OF THE CELL— THE LYSOSOMES

Almost immediately after a pinocytic or phagocytic vesicle appears inside a cell, one or more lysosomes become attached to the vesicle and empty their hydrolases into the vesicle, as illustrated in Figure 2–13. Thus, a *digestive vesicle* is formed in which the hydrolases begin hydrolyzing the proteins, glycogen, nucleic acids, mucopolysaccharides, and other substances in the vesicle. The products of digestion are small molecules of amino acids, glucose, phosphates, and so forth that can then diffuse through the membrane of the vesicle into the cytoplasm. What is left of the digestive vesicle, called the *residual body,* is then either excreted or undergoes dissolution inside the cytoplasm.

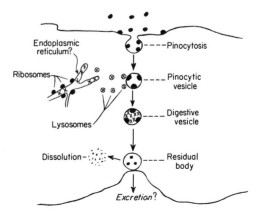

**Figure 2–13.**   Digestion of substances in pinocytic vesicles by enzymes derived from lysosomes. (Modified from C. De Duve, in Lysosomes, Ed. by Reuck and Cameron, Little, Brown and Co., 1963.)

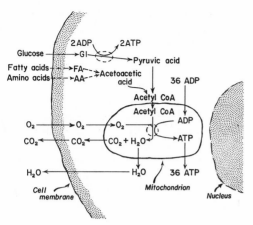

**Figure 2–14.** Formation of adenosine triphosphate in the cell, showing that most of the ATP is formed in the mitochondria.

Thus, the lysosomes may be called the *digestive organs* of the cells.

**Regression of Tissues and Autolysis of Cells.** Often, tissues of the body regress to a much smaller size than previously. For instance, this occurs in the uterus following pregnancy, in muscles during long periods of inactivity, and in mammary glands at the end of the period of lactation. Lysosomes are probably responsible for much if not most of this regression, for one can show that the lysosomes become very active at this time. However, the mechanism by which the lack of activity in a tissue causes the lysosomes to increase their activity is completely unknown.

Another very special role of the lysosomes is the removal of damaged cells or damaged portions of cells from tissues—cells damaged by heat, cold, trauma, chemicals, or any other factor. Damage to the cell causes lysosomes to rupture, and the released hydrolases begin immediately to digest the surrounding organic substances. If the damage is slight, only a portion of the cell will be removed, followed by repair of the cell. However, if the damage is severe the entire cell will be digested, a process called *autolysis*. In this way, the cell is completely removed and a new cell of the same type ordinarily is formed by mitotic reproduction of an adjacent cell to take the place of the old one.

## EXTRACTION OF ENERGY FROM NUTRIENTS—FUNCTION OF THE MITOCHONDRIA

The principal nutrients from which cells extract energy are oxygen and one or more of the foodstuffs—carbohydrates, fats, and proteins. In the human body essentially all carbohydrates are converted into glucose before they reach the cell, the proteins are converted into amino acids, and the fats are converted into fatty acids. Figure 2–14 shows oxygen and the foodstuffs—glucose, fatty acids, and amino acids—all entering the cell. Inside the cell, the foodstuffs react chemically with the oxygen under the influence of various enzymes that control their rates of reactions and channel the energy that is released in the proper direction.

**Formation of Adenosine Triphosphate (ATP).** The energy released from the nutrients is used to form adenosine triphosphate, generally called ATP, the formula for which is:

Note that ATP is a nucleotide composed of the nitrogenous base *adenine,* the pentose sugar *ribose,* and three *phosphate radicals.* The last two phosphate radicals are connected with the remainder of the molecule by so-called *high energy phosphate bonds,* which are represented by the symbol ~. Each of these bonds contains about 8000 calories of energy per mole of ATP under the physical conditions of the body (7000 calories under standard conditions), which is much greater than the energy stored in the average chemical bond of other organic compounds, thus giving rise to the term ''high energy'' bond. Furthermore, the high energy phosphate bond is very labile so that it can be split instantly on demand whenever energy is required to promote other cellular reactions.

When ATP releases its energy, a phosphoric acid radical is split away, and *adenosine diphosphate (ADP)* is formed. Then, energy derived from the cellular nutrients causes the ADP and phosphoric acid to recombine to form new ATP, the entire process continuing over and over again. For these reasons, ATP has been called the *energy currency* of the cell, for it can be spent and remade again and again.

***Chemical Processes in the Formation of ATP.*** On entry into the cells, glucose, fatty acids, and amino acids are subjected to enzymes in the cytoplasm or the nucleoplasm that convert glucose into *pyruvic acid* (a process called *glycolysis*), and convert fatty acids and most amino acids into *acetoacetic acid.* The chemical reactions of these conversions will be presented in Chapters 67 through 69. A small amount of ADP is changed into ATP by energy released during the conversion of glucose to pyruvic acid, but this amount plays only a minor role in the overall energy metabolism of the cell.

By far the major portion of the ATP formed in the cell is formed in the mitochondria. The pyruvic and acetoacetic acids are both converted into the compound *acetyl co-A* in the cytoplasm, and this is transported along with oxygen through the mitochondrial membrane into the matrix of the mitochondrion. Here this substance is acted upon by a series of enzymes and undergoes dissolution in a sequence of chemical reactions called the *tricarboxylic acid cycle,* or *Krebs cycle.* These chemical reactions will be explained in detail in Chapter 67.

In the tricarboxylic acid cycle, acetyl co-A is split into its component parts, hydrogen atoms and carbon dioxide. The carbon dioxide in turn diffuses out of the mitochondria and eventually

out of the cell. The hydrogen atoms combine with carrier substances and are carried to the surfaces of the shelves that protrude into the mitochondria, as shown in Figure 2–6. Attached to these shelves are the so-called *oxidative enzymes* which, by a series of sequential reactions, cause the hydrogen atoms to combine with oxygen. The enzymes are arranged on the surfaces of the shelves in such a way that the products of one chemical reaction are immediately relayed to the next enzyme, then to the next, and so on until the complete sequence of reactions has taken place. During the course of these reactions, the energy released from the combination of hydrogen with oxygen is used to manufacture tremendous quantities of ATP from ADP. The ATP is then transported out of the mitochondrion into all parts of the cytoplasm and nucleoplasm where its energy is used to energize the functions of the cell. (Some physiologists believe that the ATP formed in the mitochondrion does not leave the mitochondrion, but that it transfers its energy and a phosphate radical through the mitochondrial membrane to convert ADP on the outside to ATP, this ATP then performing the cytoplasmic functions.)

***Anaerobic Formation of ATP During Glycolysis.*** Note in Figure 2–14 that no oxygen is required for the splitting of glucose to form pyruvic acid, but, even so, a small amount of ATP is formed during this process. Therefore, this mechanism (glycolysis) for providing energy to the cell is called *anaerobic energy metabolism.*

***Oxidative Energy Metabolism in the Mitochondria.*** Note, on the other hand, that tremendous quantities of ATP are formed by the chemical reactions in the mitochondria. Because these reactions require oxygen, this mechanism for providing energy is called *oxidative energy metabolism.* Approximately 90 per cent of all ATP formed in the cell is formed in the mitochondria, in comparison with only 10 per cent outside the mitochondria: 5 per cent by anaerobic metabolism, as described previously, and 5 per cent by another method of oxidative metabolism that will be discussed in Chapter 67. Because of the tremendous quantity of ATP formed in the mitochondria and because of the need for ATP to supply energy for all cellular functions, the mitochondria are called the *power-houses* of the cell.

**Uses of ATP for Cellular Function.** ATP is used to promote three major categories of cellular functions: (1) *membrane transport,* (2) *synthesis of chemical compounds* throughout the cell, and (3) *mechanical work*. These three different uses of ATP are illustrated in Figure 2–15: (a) to supply energy for the transport of sodium through the membrane, (b) to promote protein synthesis by the ribosomes, and (c) to supply the energy needed during muscle contraction.

In addition to membrane transport of glucose, energy from ATP is required for transport of potassium ions and, in certain cells, calcium ions, phosphate ions, chloride ions, urate ions, hydrogen ions, and still many other special substances. Membrane transport is so important to cellular function that some cells utilize as much as 50 per cent of the ATP formed in the cells for this purpose alone.

In addition to synthesizing proteins, cells also synthesize phospholipids, cholesterol, purines, pyrimidines, and a great host of other substances. Synthesis of almost any chemical compound requires energy. For instance, a single protein molecule might be composed of as many as several thousand amino acids attached to each other by peptide linkages; the formation of each of these linkages requires the breakdown of three ATP molecules; thus many thousand ATP molecules must release their energy as each protein molecule is formed. Indeed, cells often utilize as much as 75 per cent of all the ATP formed in the cell simply to synthesize new chemical compounds; this is particularly true during the growth phase of cells.

The final major use of ATP is to supply energy for special cells to perform mechanical work. We shall see in Chapter 11 that each contraction of a muscle fibril requires expenditure of tremendous quantities of ATP. This is true whether the fibril is in skeletal muscle, smooth muscle, or cardiac muscle. Other cells perform mechanical work in two additional ways, by *ciliary* or *ameboid motion*, both of which will be described in the following section. The source of energy for all these types of mechanical work is ATP.

In summary, therefore, ATP is always available to release its energy rapidly and almost explosively wherever in the cell it is needed. To replace the ATP used by the cell, other much slower chemical reactions break down carbohydrates, fats, and proteins and use the energy derived from these to form new ATP.

## CELL MOVEMENT

By far the most important type of cell movement that occurs in the body is that of the specialized muscle cells in skeletal, cardiac, and smooth muscle, which comprise almost 50 per cent of the entire body mass. The specialized functions of these cells will be discussed in Chapter 11. However, two other types of movement occur in other cells, *ameboid movement and ciliary movement*.

**Ameboid Motion.** Ameboid motion means movement of an entire cell in relation to its surroundings, such as the movement of white blood cells through tissues. Typically ameboid motion begins with protrusion of a *pseudopodium* from one end of the cell. The pseudopodium projects far out away from the cell body, and then the remainder of the cell moves toward the pseudopodium. Formerly, it was believed that the protruding pseudopodium attached itself far away from the cell and then pulled the remainder of the cell toward it. However, in recent years, new studies have changed this idea to a "streaming" concept, as follows: Figure 2–16 illustrates diagrammatically an elongated cell, the right-hand end of which is a protruding pseudopodium. The membrane of this end of the cell is continually moving forward, while the membrane at the left-hand end of the cell is continually following along as the cell moves.

It is postulated that ameboid movement is caused in the following way: The outer portion of the cytoplasm is in a *gel* state and is called the *ectoplasm*, whereas the central portion of the cytoplasm is in a *sol* state and is called *endoplasm*. In the gel is a contractile protein

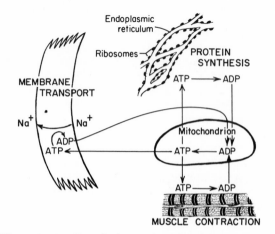

**Figure 2–15.** Use of adenosine triphosphate to provide energy for three major cellular functions: membrane transport, protein synthesis, and muscle contraction.

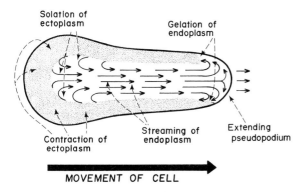

Solation of
ectoplasm

Gelation of
endoplasm

Contraction of
ectoplasm

Streaming of
endoplasm

Extending
pseudopodium

MOVEMENT OF CELL

**Figure 2–16.**   Ameboid motion by a cell.

called *myxomyosin,* which contracts in the presence of ATP and calcium ions. Therefore, normally there is a continual tendency for the ectoplasm to contract. However, in response to a "chemotaxic" stimulus the ectoplasm at one end of the cell becomes thin, causing a pseudopodium to bulge outward in the direction of the chemotaxic source. Thus, the pseudopodium moves progressively forward. However, this also causes contraction and thickening of the ectoplasm at the opposite end of the cell, which is accompanied by continuous solation of the inner portions of the ectoplasm; then the dissolved ectoplasm "streams" forward through the endoplasm toward the pseudopodium. Thus, the combination of contraction of the ectoplasm, as well as solation of its inner layers, forces a stream of endoplasm forward toward the pseudopodium, pushing the membrane of the pseudopodium progressively forward. On reaching the pseudopodial end of the cell, the endoplasm turns toward the sides of the cell and then becomes gelated to form new ectoplasm. Therefore, at the tail end of the cell, ectoplasm is continually being solated while new ectoplasm is being formed at the pseudopodial end. The continuous repetition of this process makes the cell move in the direction in which the pseudopodium projects. One can readily see that this streaming movement inside the cell is analogous to the revolving movement of the track of a Caterpillar tractor.

*Types of Cells That Exhibit Ameboid Motion.*   The most common cells to exhibit ameboid motion in the human body are the *white blood cells* moving out of the blood into the tissues in the form of *tissue macrophages* or *microphages.* However, many other types of cells can move by ameboid motion under certain circumstances. For instance, fibroblasts will move into any damaged area to help repair the damage, and even some of the germinal cells of the skin, though ordinarily completely sessile cells, will move by ameboid motion toward a cut area to repair the rent. Finally, ameboid motion is especially important in the development of the fetus, for embryonic cells often migrate long distances from the primordial sites of origin to new areas during the development of special structures.

*Control of Ameboid Motion—"Chemotaxis."*   The most important factor that usually initiates ameboid motion, presumably by causing potential changes in the cell membrane, is the appearance of certain chemical substances in the tissues. This phenomenon is called *chemotaxis,* and the chemical substance causing it to occur is called a *chemotaxic substance.* Most ameboid cells move toward the source of the chemotaxic substance—that is, from an area of lower concentration toward an area of higher concentration—which is called *positive chemotaxis.* However, some cells move away from the source, which is called *negative chemotaxis.*

**Movement of Cilia.**   A second type of cellular motion, *ciliary movement,* is the bending of cilia along the surface of cells in the respiratory tract and in the fallopian tubes of the reproductive tract. As illustrated in Figure 2–17, a cilium looks like a minute, sharp-pointed hair that projects 3 to 4 microns from the surface of the cell. Many cilia can project from a single cell.

The cilium is covered by an outcropping of the cell membrane, and it is supported by 11 microtubular filaments, 9 double tubular filaments located around the periphery of the cilium and 2 single tubular filaments down the center, as shown in the cross-section illustrated in Figure 2–17. Each cilium is an outgrowth of a structure that lies immediately beneath the cell membrane called the *basal body* of the cilium.

In the inset of Figure 2–17 movement of the cilium is illustrated. The cilium moves forward with a sudden rapid stroke, 10 to 17 times per second, bending sharply where it projects from the surface of the cell. Then it moves backward very slowly in a whiplike manner. The rapid forward movement pushes the fluid lying adjacent to the cell in the direction that the cilium moves, then the slow whiplike movement in the other direction has almost no effect on the fluid. As a result, fluid is continually propelled in the direction of the forward stroke. Since most ciliated cells have large numbers of cilia on their surfaces, and since ciliated cells on a surface are oriented in the same direction, this is a very

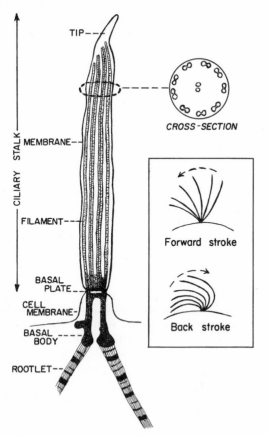

TIP---

CROSS-SECTION

CILIARY STALK

MEMBRANE---

FILAMENT---

Forward stroke

BASAL
PLATE---

CELL
MEMBRANE---

BASAL
BODY---

Back stroke

ROOTLET---

**Figure 2–17.** Structure and function of the cilium. (Modified from Satir: *Sci. Amer., 204*:108, 1961. Copyright © 1961 by Scientific American, Inc. All rights reserved.)

satisfactory means for moving fluids from one part of the surface to another, for instance, for moving mucus out of the lungs or for moving the ovum along the fallopian tube.

*Mechanism of Ciliary Movement.* Though the precise way in which ciliary movement occurs is unknown, we do know the following: First, the nine double tubules and the two single tubules are all linked to each other by a complex of protein cross-linkages; this total complex of tubules and cross-linkages is called the *axoneme.* Second, even after removal of the membrane and destruction of other elements of the cilium besides the *axoneme,* the cilium can still beat under appropriate conditions. Third, there are two necessary conditions for continued beating of the axomene after removal of the other structures of the cilium: (1) the presence of ATP, and (2) appropriate ionic conditions including especially appropriate concentrations of magnesium and calcium. Fourth, the cilium will continue to beat even after it has

been removed from the cell body. Fifth, the tubules on the front edge of the bending cilium slide outward toward the tip of the cilium while the tubules on the back edge of the bending cilium remain in place. Sixth, three protein arms with ATPase activity project from each set of peripheral tubules toward the next set.

Given the above basic information, it has been postulated that the release of energy from ATP in contact with the ATPase arms causes the arms to "crawl" along the surfaces of the adjacent pair of tubules. If this occurs simultaneously on the two sides of the axoneme in a synchronized manner, the front tubules can crawl outward while the back tubules remain stationary. Because of the elastic structure of the axoneme, this obviously will cause bending.

Since many cilia on a cell surface contract simultaneously in a wavelike manner, it is presumed that some synchronizing signal— perhaps an electrochemical signal over the cell surface—is transmitted from cilium to cilium. This is not necessary, however, to cause the actual beating of the cilia because this will occur even in the absence of the membrane. The ATP required for the ciliary movements is provided by mitochondria near the bases of the cilia from which the ATP diffuses into the cilia.

*Reproduction of Cilia.* Cilia have the peculiar ability to reproduce themselves. This is achieved by the *basal body,* which is almost identical to the centriole, an important structure in the reproduction of whole cells, as we shall see in the next chapter. The basal body, as does the centriole, has the ability to reproduce itself by means not yet understood. After it reproduces itself, the new basal body then grows an additional cilium from the surface of the cell.

## REFERENCES

Allison, A. C., and Davies, P.: Mechanisms of endocytosis and exocytosis. *Symp. Soc. Exp. Biol., (28)*:419, 1974.

Blake, J. R., and Sleigh, M. A.: Mechanics of ciliary locomotion. *Biol. Rev.,* 49:85, 1974.

Boardman, N. K. (ed.): The Autonomy and Biogenesis of Mitochondria and Chloroplasts. New York, American Elsevier Publishing Co., 1971.

Bourne, G. H.: Cytology and Cell Physiology. 3rd ed., New York, Academic Press, Inc., 1964.

Bulger, R. E., and Strum, J. M.: The Functioning Cytoplasm. New York, Plenum Publishing Corp., 1974.

Cohen, S. S.: Mitochondria and chloroplasts revisited. *Amer. Sci.,* 61:437, 1973.

Davson, H.: A Textbook of General Physiology, 4th Ed. New York, Churchill Livingstone, Div. of Longman Inc., 1970.

De Duve, C., and Wattiaux, R.: Functions of lysosomes. *Ann. Rev. Physiol.,* 28:435, 1966.

De Robertis, E. D. P., Saez, F. A., and De Robertis, E. M. F.: Cell Biology, 6th Ed. Philadelphia, W. B. Saunders Company, 1975.

Dingle, J. T.: Lysosomes in Biology and Pathology. New York, American Elsevier Publishing Co., 1973.

Dott, H. M.: Lysosomes and lysosomal enzymes in reproduction. *Adv. Reprod. Physiol.* 6:213, 1973.

Ebe, T., and Kobayashi, S.: Fine Structure of Human Cells and Tissues. New York, John Wiley & Sons, Inc., 1973.

Fawcett, D. W.: The Cell. Philadelphia, W. B. Saunders Company, 1966.

Fox, C. F.: The structure of cell membranes. *Sci. Amer., 226*:30, 1972.

Fridovich, I.: Evidence for the symbiotic origin of mitochondria. *Life Sci., 14*:819, 1974.

Gersh, I. (ed.): Submicroscopic Cytochemistry. New York, Academic Press, Inc., 1973.

Giese, A. C.: Cell Physiology, 4th Ed. Philadelphia, W. B. Saunders Company, 1973.

Goldman, R. D.: Fibrillar systems in cell motility. *Ciba Found. Symp., 14*:83, 1973.

Good, R. A., Day, S. B., and Yunis, J. J. (eds.): Molecular Pathology. Springfield, Ill., Charles C Thomas, Publisher, 1974.

Grundmann, E.: General Cytology. Baltimore, The Williams & Wilkins Co., 1966.

Harris, A. K.: Cell surface movements related to cell locomotion. *Ciba Found. Symp., 14*:3, 1973.

Harris, H.: Nucleus and Cytoplasm, 3rd Ed. New York, Oxford University Press, 1974.

Hayashi, T., and Szent-György, A.: Molecular Architecture in Cell Physiology. Englewood Cliffs, New Jersey, Prentice-Hall, Inc., 1966.

Inoue, S., and Stephens, R. E. (eds.): Molecules and Cell Movement. New York, Raven Press, 1975.

Jahn, T. L., and Bovee, E. C.: Protoplasmic movements within cells. *Physiol. Rev., 49*:793, 1969.

Kaplan, D. M., and Criddle, R. S.: Membrane structural proteins. *Physiol. Rev., 51*:249, 1971.

Locomotion of the Tissue Cells (CIBA Foundation Symposium 14): New York, American Elsevier Publishing Co., Inc., 1973.

Markham R., Bancroft, J. B., Davies, D. R., Hopwood, D. A., and Horne, R. W. (eds.): The Generation of Subcellular Structures. New York, American Elsevier Publishing Co., 1973.

Neutra, M., and Leblond, C. P.: The golgi apparatus. *Sci. Amer., 220*:100, 1969.

Porter, K. R., and Franzini-Armstrong, C.: The sarcoplasmic reticulum. *Sci. Amer., 212*:72, 1965.

Rabinowitz, M., and Swift, H.: Mitochondrial nucleic acids and their relation to the biogenesis of mitochondria. *Physiol. Rev., 50*:376, 1970.

Racker, E.: The membrane of the mitochondrion. *Sci. Amer., 218*:32, 1968.

Revel, J. P.: Contacts and junctions between cells. *Symp. Soc. Exp. Biol.,* (28):447, 1974.

Sharon, N.: The bacterial cell wall. *Sci. Amer., 220*:92, 1969.

Singer, S. J.: The molecular organization of membranes. *Ann. Rev. Physiol., 43*:805, 1974.

Threadgold, L. T.: The Ultrastructure of the Animal Cell. Oxford, England, Pergamon Press, Inc., 1967.

Toporek, M.: Basic Chemistry of Life, 2nd. Ed. New York, Appleton-Century-Crofts, 1975.

Wallach, D. F. H.: The Plasma Membrane: Dynamic Perspectives, Genetics, and Pathology. New York, Springer-Verlag New York, Inc., 1975.

Wessells, N. K., Spooner, B. S., and Luduena, M. A.: Surface movements, microfilaments, and cell locomotion. *Ciba Found. Symp., 14*:53, 1973.

Wilkinson, P. C.: Chemotaxis and Inflammation. New York, Churchill Livingstone, Div. of Longman Inc., 1973.

# 3

# Genetic Control of Cell Function— Protein Synthesis and Cell Reproduction

Almost everyone knows that the genes control heredity from parents to children, but most persons do not realize that the same genes control the reproduction of and the day-by-day function of all cells. The genes control function of the cell by determining what substances will be synthesized within the cell—what structures, what enzymes, what chemicals.

Figure 3–1 illustrates the general schema by which the genes control cellular function. The gene, which is a nucleic acid called *deoxyribonucleic acid (DNA),* automatically controls the formation of another nucleic acid, *ribonucleic acid (RNA),* which spreads throughout the cell and controls the formation of the different proteins. Some of these proteins are *structural proteins* which, in association with various lipids, form the structures of the various organelles that were discussed in the preceding chapter. But by far the majority of the proteins are *enzymes* that catalyze the different chemical reactions in the cells. For instance, enzymes promote all the oxidative reactions that supply energy to the cell, and they promote the synthesis of various chemicals such as lipids, glycogen, adenosine triphosphate, etc.

Though the number of genes in a nucleus of the human cell is unknown, it is known that the genes control the formation of at least several thousand separate types of protein essential to function of the different cells.

## THE GENES

The genes are contained in long, double-stranded, helical molecules of *deoxyribonucleic acid (DNA)*

having molecular weights usually measured in the millions. A very short segment of such a molecule is illustrated in Figure 3–2. This molecule is composed of several simple chemical compounds arranged in a regular pattern explained in the following few paragraphs.

**The Basic Building Blocks of DNA.** Figure 3–3 illustrates the basic chemical compounds involved in the formation of DNA. These include *phosphoric acid,* a sugar called *deoxyribose,* and four nitrogenous bases (two purines, *adenine* and *guanine,* and two pyrimidines, *thymine* and *cytosine*). The phosphoric acid and deoxyribose form the two helical strands of DNA, and the bases lie between the strands and connect them together.

**The Nucleotides.** The first stage in the formation of DNA is the combination of one molecule of phos-

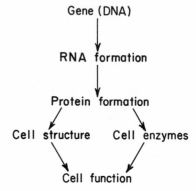

Figure 3–1. General schema by which the genes control cell function.

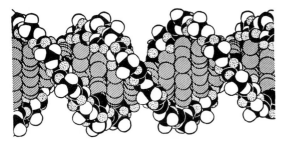

**Figure 3–2.** The helical, double-stranded structure of the gene. The outside strands are composed of phosphoric acid and the sugar deoxyribose. The internal molecules connecting the two strands of the helix are purine and pyrimidine bases; these determine the "code" of the gene.

phoric acid, one molecule of deoxyribose, and one of the four bases to form a nucleotide. Four separate nucleotides are thus formed, one for each of the four bases: *adenylic, thymidylic, guanylic,* and *cytidylic acids.* Figure 3–4A illustrates the chemical structure of adenylic acid, and Figure 3–4B illustrates simple symbols for all the four basic nucleotides that form DNA.

Note also in Figure 3–4B that the nucleotides are separated into two *complementary pairs.* Adenylic acid and thymidylic acid form one pair. Guanylic acid and cytidylic acid form the other pair. The bases of each pair can attach loosely (by hydrogen bonding) to each other, thus providing the means by which the two strands of the DNA helix are bound together: one nucleotide of a pair is on one strand of DNA, the other nucleotide of that pair is in a corresponding position on the other strand, and these are bound together by loose and reversible bonds between the nucleotide bases.

**Organization of the Nucleotides to Form DNA.** Figure 3–5 illustrates the manner in which multiple numbers of nucleotides are bound together to form DNA. Note that these are combined in such a way that phosphoric acid and deoxyribose alternate with each other in the two separate strands, and these strands are held together by the respective complementary pairs of bases. Thus, in Figure 3–5 the sequence of complementary pairs of bases is CG, CG, GC, TA, CG, TA, GC, AT, and AT. However, the bases are bound together by very loose hydrogen bonding, represented in the figure by dashed lines. Because of the looseness of these bonds, the two strands can pull apart with ease, and they do so many times during the course of their function in the cell.

Now, to put the DNA of Figure 3–5 into its proper physical perspective, one needs merely to pick up the two ends and twist them into a helix. Ten pairs of nucleotides are present in each full turn of the helix in the DNA molecule, as illustrated in Figure 3–2.

## *THE GENETIC CODE*

The importance of DNA lies in its ability to control the formation of other substances in the cell. It does

this by means of the so-called genetic code. When the two strands of a DNA molecule are split apart, this exposes a succession of purine and pyrimidine bases projecting to the side of each strand. It is these projecting bases that form the code.

Research studies in the past few years have demonstrated that the so-called *code words* consist of "triplets" of bases—that is, each three successive bases are a code word. And the successive code words control the sequence of amino acids in a protein molecule during its synthesis in the cell. Note in Figure 3–5 that each of the two strands of the DNA molecule carries its own genetic code. For instance, the top strand, reading from left to right, has the genetic code GGC, AGA, CTT, the code words being separated from each other by the arrows. As we follow this genetic code through Figures 3–6 and 3–7, we shall see that these three code words are responsible for placement of the three amino acids, *proline, serine,* and *glutamic acid,* in a molecule of protein. Furthermore, these three amino acids will be lined up in the protein molecule in exactly the same way that the genetic code is lined up in this strand of DNA.

It is also important that some code words do not cause amino acids to incorporate into proteins but, instead, perform other functions in the synthesis of protein molecules, such as initiating and stopping the

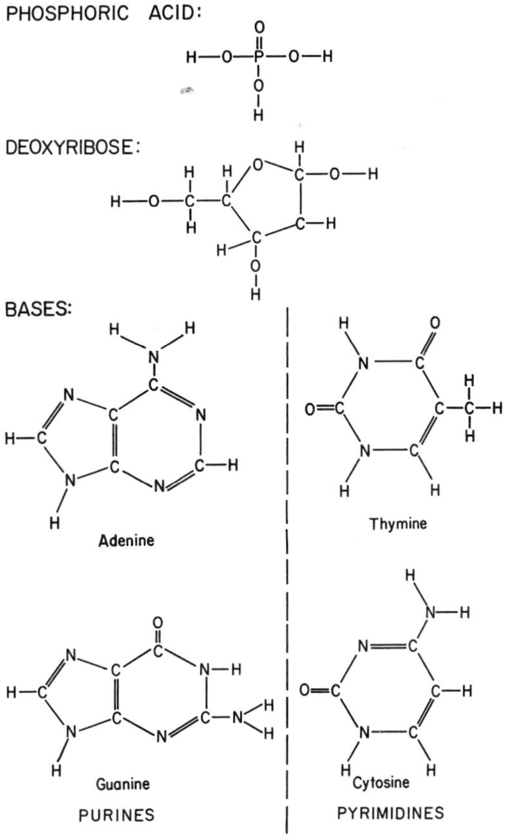

PHOSPHORIC ACID:

DEOXYRIBOSE:

BASES:

Adenine

Thymine

Guanine

Cytosine

PURINES

PYRIMIDINES

**Figure 3–3.** The basic building blocks of DNA.

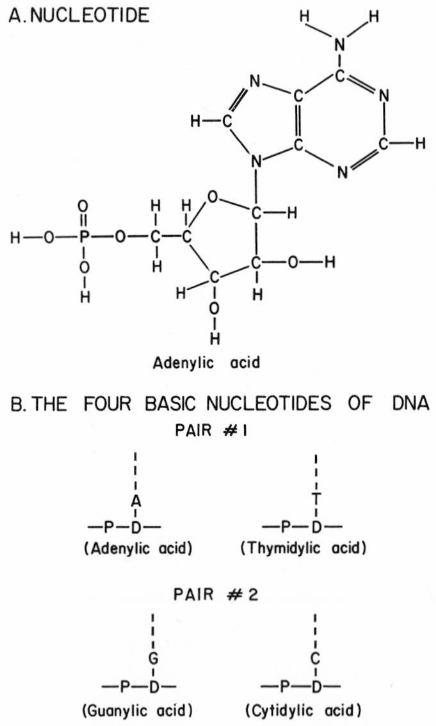

A. NUCLEOTIDE

Adenylic acid

B. THE FOUR BASIC NUCLEOTIDES OF DNA

PAIR #1

A
—P—D—
(Adenylic acid)

T
—P—D—
(Thymidylic acid)

PAIR #2

G
—P—D—
(Guanylic acid)

C
—P—D—
(Cytidylic acid)

**Figure 3–4.** Combinations of the basic building blocks of DNA to form nucleotides. (*A*, adenine; *C*, cytosine; *D*, deoxyribose; *G*, guanine; *P*, phosphoric acid; *T*, thymine.)

formation of a protein molecule. For instance, some DNA molecules of viruses have molecular weights of 40,000,000 or more and cause the formation of many more than a single protein molecule— 8, 10, 20, or more molecules. It is code words that do not cause incorporation of amino acids into a protein molecule that signal when to stop the formation of one protein and to start the formation of another.

# RIBONUCLEIC ACID (RNA)

Since almost all DNA is located in the nucleus of the cell and yet most of the functions of the cell are carried out in the cytoplasm, some means must be available for the genes of the nucleus to control the chemical reactions of the cytoplasm. This is done through the intermediary of another type of nucleic acid, ribonucleic acid (RNA), the formation of which is controlled by the DNA of the nucleus. The RNA is then transported into the cytoplasmic cavity where it controls protein synthesis.

Three separate types of RNA are important to protein synthesis: *messenger RNA, transfer RNA,* and *ribosomal RNA.* Before we describe the functions of these different RNAs in the synthesis of proteins, let us see how DNA controls the formation of RNA.

**Synthesis of RNA.** One strand of the DNA

molecule, which contains the genes, acts as a template for synthesis of RNA molecules. (The other strand of the DNA has no genetic function but does function for replication of the gene itself, which will be discussed later in the chapter.) The code words in DNA cause the formation of *complementary* code words (or *codons*) in RNA. The stages of RNA synthesis are as follows:

*The Basic Building Blocks of RNA.* The basic building blocks of RNA are almost the same as those of DNA except for two differences. First, the sugar deoxyribose is not used in the formation of RNA. In its place is another sugar of very slightly different composition, *ribose.* Referring back to Figure 3–3, the deoxyribose becomes ribose if the hydrogen radical to the right is replaced by a hydroxyl radical.

The second difference in the basic building blocks is that thymine is replaced by another pyrimidine, *uracil.* If in Figure 3–3 the methyl radical of the thymine is replaced by a single hydrogen atom, the structure then becomes uracil rather than thymine.

**Formation of RNA Nucleotides.** The basic building blocks of RNA first form nucleotides exactly as described above for the synthesis of DNA. Here again, four separate nucleotides are used in the formation of RNA. These nucleotides contain the bases *adenine, guanine, cytosine,* and *uracil,* respectively, the uracil replacing the thymine found in the four nucleotides that make up DNA. Also, uracil takes the place of thymine in pairing with the purine adenine, as we shall see in the following paragraphs.

**Activation of the Nucleotides.** The next step in the synthesis of RNA is activation of the nucleotides. This occurs by addition to each nucleotide of two phosphate radicals to form triphosphates. These last two phosphates are combined with the nucleotide by *high energy phosphate bonds* derived from the energy system of the cell.

The result of this activation process is that large quantities of energy are made available to each of the nucleotides, and it is this energy that is used in promoting the chemical reactions that eventuate in the formation of the RNA chain.

**Combination of the Activated Nucleotides with the DNA Strand.** The next stage in the formation of RNA is the splitting apart of the two strands of the DNA molecule. Then, activated nucleotides become attached to the bases on the DNA strand that contains the genes, as illustrated by the top panel in

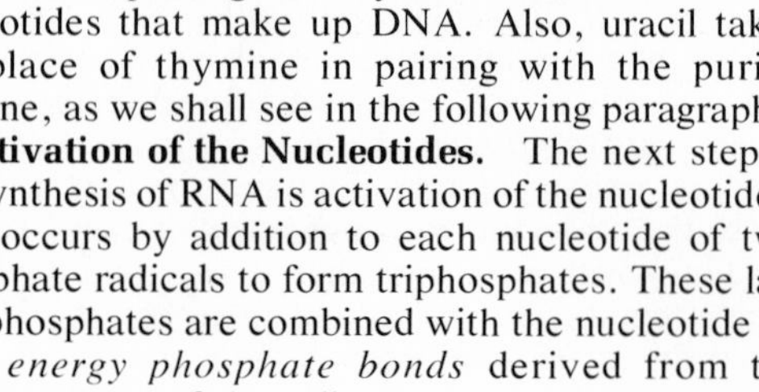

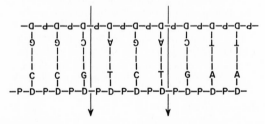

**Figure 3–5.** Combination of deoxyribose nucleotides to form DNA.

Figure 3–6. Note that ribose nucleotide bases always combine with the deoxyribose bases in the following combinations:

| *DNA base* | *RNA base* |
|---|---|
| guanine | cytosine |
| cytosine | guanine |
| adenine | uracil |
| thymine | adenine |

**Polymerization of the RNA chain.** Once the ribose nucleotides have combined with the DNA strand as shown in the top panel of Figure 3–6, they are then lined up in proper sequence to form code words (codons) that are complementary to those in the DNA molecule. At this time an enzyme, *RNA polymerase,* causes the two extra phosphates on each nucleotide to split away and, at the same time, to liberate enough energy to cause bonds to form between the successive ribose and phosphoric acid radicals. As this happens, the RNA strand automatically separates from the DNA strand and becomes a free molecule of RNA. This bonding of the ribose and phosphoric acid radicals and the simultaneous splitting of the RNA from the DNA is illustrated in the lower part of Figure 3–6.

Once the RNA molecules are formed, they diffuse into all parts of the cytoplasm where they perform further functions.

## *MESSENGER RNA—THE PROCESS OF TRANSCRIPTION*

The type of RNA that carries the genetic code to the cytoplasm for formation of proteins is called *messenger RNA.* Molecules of messenger RNA are usually composed of several hundred to several

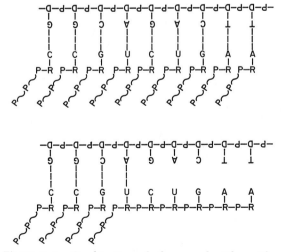

**Figure 3–6.** Combination of ribose nucleotides with a strand of DNA to form a molecule of ribonucleic acid (RNA) that carries the DNA code from the gene to the cytoplasm.

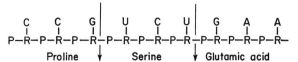

**Figure 3–7.** Portion of a ribonucleic acid molecule, showing three "code" words, CCG, UCU, and GAA, which represent the three amino acids *proline, serine,* and *glutamic acid.*

thousand nucleotides in a single strand containing codons that are exactly complementary to the code words of the genes. Figure 3–7 illustrates a small segment of a molecule of messenger RNA. Its codons are CCG, UCU, and GAA. These are the codons for proline, serine, and glutamic acid. If we now refer back to Figure 3–6, we see that the events recorded in this figure represent transfer of this particular genetic code from the DNA strand to the RNA strand. This process of transferring the genetic code from DNA to messenger RNA is called *transcription.*

Messenger RNA molecules are long straight strands that are suspended in the cytoplasm. These molecules migrate to the ribosomes where protein molecules are manufactured, which is explained below.

**RNA Codons.** Table 3–1 gives the RNA codons for the 20 common amino acids found in protein molecules. Note that several of the amino acids are represented by more than one codon; and, as was pointed out before, some codons represent such signals as "start manufacturing a protein molecule" or "stop manufacturing a protein molecule." In Table

**TABLE 3–1.   RNA Codons for the Different Amino Acids and for Start and Stop**

| *Amino Acid* | *RNA Codons* | | | | | |
|---|---|---|---|---|---|---|
| Alanine | GCU | GCC | GCA | GCG | | |
| Arginine | CGU | CGC | CGA | CGG | AGA | AGG |
| Asparagine | AAU | AAC | | | | |
| Aspartic acid | GAU | GAC | | | | |
| Cysteine | UGU | UGC | | | | |
| Glutamic acid | GAA | GAG | | | | |
| Glutamine | CAA | CAG | | | | |
| Glycine | GGU | GGC | GGA | GGG | | |
| Histidine | CAU | CAC | | | | |
| Isoleucine | AUU | AUC | AUA | | | |
| Leucine | CUU | CUC | CUA | CUG | UUA | UUG |
| Lysine | AAA | AAG | | | | |
| Methionine | AUG | | | | | |
| Phenylalanine | UUU | UUC | | | | |
| Proline | CCU | CCC | CCA | CCG | | |
| Serine | UCU | UCC | UCA | UCG | | |
| Threonine | ACU | ACC | ACA | ACG | | |
| Tryptophan | UGG | | | | | |
| Tyrosine | UAU | UAC | | | | |
| Valine | GUU | GUC | GUA | GUG | | |
| | | | | | | |
| Start (CI) | AUG | | | | | |
| Stop (CT) | UAA | UAG | UGA | | | |

3–1, these two codons are designated CI for "chain-initiating" and CT for "chain-terminating."

### TRANSFER RNA

Another type of RNA that plays a prominent role in protein systhesis is called *transfer RNA* because it transfers amino acid molecules to protein molecules as the protein is synthesized. There is a separate type of transfer RNA for each of the 20 amino acids that are incorporated into proteins. Furthermore, each type of transfer RNA usually combines with only one type of amino acid—one and no more. Transfer RNA then acts as a *carrier* to transport its specific type of amino acid to the ribosomes where protein molecules are formed. In the ribosomes, each specific type of transfer RNA recognizes a particular code word on the messenger RNA, as is described below, and thereby delivers the appropriate amino acid to the appropriate place in the chain of the newly forming protein molecule.

Transfer RNA, containing only about 80 nucleotides, is a relatively small molecule in comparison with messenger RNA. It is a folded chain of nucleotides with a cloverleaf appearance similar to that illustrated in Figure 3–8. At one end of the molecule is always an adenylic acid; it is to this that the transported amino acid attaches to a hydroxyl group of the ribose in the adenylic acid. A specific enzyme causes this attachment for each specific type of transfer RNA; this enzyme also determines the type of amino acid that will attach to the respective type of transfer RNA.

Since the function of transfer RNA is to cause attachment of a specific amino acid to a forming protein chain, it is essential that each type of transfer RNA also have specificity for a particular codon in the messenger RNA. The specific prosthetic group in the transfer RNA that allows it to recognize a specific codon is called an *anticodon,* and this is located approximately in the middle of the transfer RNA molecule (at the bottom of the cloverleaf configuration illustrated in Figure 3–8). During formation of a protein molecule, the anticodon bases combine loosely by hydrogen bonding with the codon bases of the messenger RNA. In this way the respective amino acids are lined up one after another along the messenger RNA chain, thus establishing the appropriate sequence of amino acids in the protein molecule.

### RIBOSOMAL RNA

The third type of RNA in the cell is that found in the ribosomes; it constitutes betwen 40 and 50 per cent of the ribosome. The remainder of the ribosome is protein.

The ribosomal RNA exists, along with the ribosomal protein, in particles of two different sizes. The transfer RNA, along with its attached amino acid, first binds with the smaller particle. Then as the messenger RNA passes through the ribosome, the

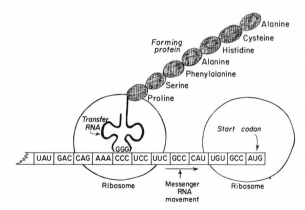

**Figure 3–8.** Postulated mechanism by which a protein molecule is formed in ribosomes in association with messenger RNA and soluble RNA.

amino acid is released to the forming protein while the transfer RNA is released back into the cytoplasm to combine again with another molecule of amino acid of the same type. The larger ribosomal particle provides the enzymes that promote peptide linkage between the successive amino acids, and various fractions of the ribosomal protein also enter into some of the intermediate chemical reactions that occur in the overall process. Thus the ribosome acts as a manufacturing plant in which the protein is formed.

**Formation of Ribosomes in the Nucleolus.** In the previous chapter it was pointed out that the *nucleolus* is a specialized structure of the nucleus that functions in association with one of the chromosomes. Also, the nucleolus contains a moderate concentration of RNA which is formed by the process of transcription by the DNA of the associated chromosome. Then, in the nucleolus the RNA is complexed with different types of proteins to form granular condensation products. These products are primordial forms of the *ribosomes*. The ribosomes are then released from the nucleolus and pass through the large pores of the nuclear membrane to migrate to almost all parts of the cytoplasm. In the cytoplasm most of the ribosomes attach to the outer surfaces of the endoplasmic reticulum, and it is here that they perform their function of manufacturing protein molecules. It is believed that, as many of the protein molecules are manufactured, they pass directly into the cisternae of the endoplasmic reticulum, then to be transported within the endoplasmic reticulum tubules to other parts of the cell, as was described in the previous chapter.

### FORMATION OF PROTEINS IN THE RIBOSOMES—THE PROCESS OF TRANSLATION

When a molecule of messenger RNA comes in contact with a ribosome, it travels through the ribosome, beginning at a predetermined end specified by the "start" (or "chain-initiating") codon. Then, as illustrated in Figure 3–8, while the messenger RNA travels through the ribosome, a protein molecule is formed—a process called *translation*. Thus, the ribosome reads the code of the messenger RNA in much the same way that a tape is "read" as it passes through the playback head of a tape recorder. Then, when a "stop" (or "chain-terminating") codon slips past the ribosome, the end of a protein molecule is signaled, and the entire molecule is freed into the cytoplasm.

**Polyribosomes.** A single messenger RNA molecule can form protein molecules in several different ribosomes at the same time, the molecule passing through a successive ribosome as it leaves the first, as shown in Figure 3–8. The protein molecules obviously are in different stages of development in each ribosome. As a result, clusters of ribosomes frequently occur, three to eight ribosomes being attached together by a single messenger RNA at the same time. These clusters are called polyribosomes.

It is especially important to note that a messenger RNA can cause the formation of a protein molecule in any ribosomes, and that there is no specificity of ribosomes for given types of protein. The ribosome seems to be simply the structure in which or on which the chemical reactions take place.

**Chemical Steps in Protein Synthesis.** Some of the chemical events that occur in synthesis of a protein molecule are illustrated in Figure 3–9. This figure shows representative reactions for three separate amino acids, $AA_1$, $AA_2$, and $AA_{20}$. The stages of the reactions are the following: (1) Each amino acid is *activated* by a chemical process in which ATP combines with the amino acid to form an *adenosine monophosphate complex with the amino acid*. (2) The activated amino acid, having an excess of energy, then *combines with its specific transfer RNA to form an amino acid–tRNA complex* and, at the same time, releases the adenosine monophosphate. (3) The transfer RNA carrying the amino acid complex then comes in contact with the messenger RNA molecule in the ribosome where the anticodon of the transfer RNA attaches temporarily to its specific codon of the messenger RNA thus lining up the amino acids in appropriate sequence to form a protein molecule. The energy from *guanosine triphosphate*, another high energy phosphate substance almost identical with ATP, is utilized to form the chemical bonds between the successive amino acids, thus creating the protein.

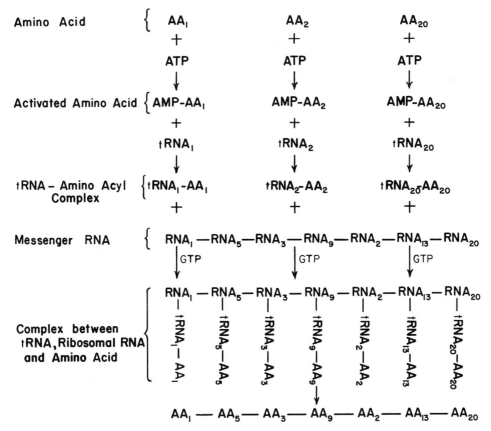

**Figure 3–9.** Chemical events in the formation of a protein molecule.

*Peptide Linkage.* The successive amino acids in the protein chain combine with each other according to the following typical reaction:

$$R-\overset{\overset{\displaystyle NH_2}{|}}{C}\overset{\overset{\displaystyle O}{\|}}{C}-OH + H-\overset{\overset{\displaystyle H}{|}}{N}-\overset{\overset{\displaystyle R}{|}}{C}-COOH \rightarrow$$

$$R-\overset{\overset{\displaystyle NH_2}{|}}{C}\overset{\overset{\displaystyle O}{\|}}{C}-\overset{\overset{\displaystyle H}{|}}{N}-\overset{\overset{\displaystyle R}{|}}{C}-COOH + H_2O$$

In this chemical reaction, a hydroxyl radical is removed from the COOH portion of one amino acid while a hydrogen of the $NH_2$ portion of the other amino acid is removed. These combine to form water, and the two reactive sites left on the two successive amino acids combine, resulting in a single molecule. This process is called *peptide linkage*.

## SYNTHESIS OF OTHER SUBSTANCES IN THE CELL

Many hundreds or perhaps a thousand or more protein enzymes formed in the manner just described control essentially all the other chemical reactions that take place in cells. These enzymes promote synthesis of lipids, glycogen, purines, pyrimidines, and hundreds of other substances. We will discuss many of these synthetic processes in relation to carbohydrate, lipid, and protein metabolism in Chapters 67 through 69. It is by means of all these different substances that the many functions of the cells are performed.

## CONTROL OF GENETIC FUNCTION AND BIOCHEMICAL ACTIVITY IN CELLS

There are basically two different methods by which the biochemical activities in the cell are controlled. One of these can be called *genetic regulation,* in which the activities of the genes themselves are controlled, and the other can be called *enzyme regulation,* in which the rates of activities of the enzymes within the cell are controlled.

### GENETIC REGULATION

Figure 3–10 illustrates the general plan by which function of the genes is regulated. At the top of the figure is a *regulatory gene* that has the ability to regulate the degree of activity of other genes. It does this by controlling the formation of a *repressor substance,* a compound of small molecular weight, which in turn represses the activities of other genes. The repressor substance does not act directly on

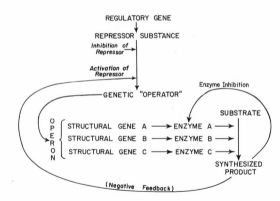

**Figure 3–10.** Control system that regulates function of genes. See text for detailed explanation.

these other genes but instead on a small part of the DNA strand called an *operator,* which lies adjacent to the genes that are to be controlled. The genetic operator excites the genes, but when the operator is repressed by the repressor substance, the genes become inactive.

The genes controlled by the operator are called *structural genes;* it is they that cause the eventual formation of enzymes in the cell. The group of structural genes under the influence of each genetic operator is known as an *operon.* In Figure 3–10, the structural genes initiate the formation of enzymes A, B, and C, which in turn control biochemical reactions in the cell leading to the formation of some specific chemical product.

As noted in Figure 3–10, the repressor substance from the regulatory gene can be either *inhibited* or *activated.* If inhibited, the genetic operator is no longer repressed and therefore becomes very active. As a result the biochemical synthesis proceeds unabated. On the other hand, if the repressor substance is activated, it immediately represses the genetic operator, and the events leading to the synthesized product cease within minutes or hours.

**Control of Cellular Constituents by Feedback Repression of the Genes.** The importance of the genetic regulatory system of Figure 3–10 is that it allows the concentrations of different cellular substances to be controlled. The mechanism of this is the following: The synthesized product often can *activate* the repressor substance, and the repressor substance then *inactivates* the genetic operator, leading to decreased or ceased production of the enzymes required for forming the synthesized product. Thus, negative feedback occurs in which the concentration of the synthesized product controls the product's rate of synthesis. When its concentration becomes too great, the rate of synthesis falls; when its concentration becomes too low, its rate of synthesis increases.

Genetic control systems of this type are especially important in controlling the intracellular concentrations of amino acids, amino acid derivatives, and some of the intermediary substances of carbohydrate, lipid, and protein metabolism.

**Induction of Genetic Activity and Enzyme Formation.** Note in Figure 3–10 that the repressor substance can be inhibited as well as activated. When it is inhibited, the repressor no longer holds the operon in abeyance, and the structural genes operate to their full capacity in forming enzymes. Therefore, a simple way to "induce" the formation of a particular group of enzymes is to inhibit the repressor. This is probably one of the means by which hormones act to increase or decrease specific cellular activities. For instance, the steroid hormone aldosterone increases the quantities of enzymes required for transport of sodium through membranes, and the steroid sex hormones induce the formation of enzymes required for protein synthesis in the sex organs. These effects will be discussed later at greater length in relation to the functions of these hormones.

Induction also occurs in response to substances inside the cell itself. For instance, cells grown in tissue culture in the absence of lactose do not have the ability to utilize lactose when first exposed to it. However, within a few hours after exposure, because of inhibition of the appropriate repressor substance, the enzyme required for splitting lactose into its monosaccharides appears in the cell. Thus, lactose induces a specific enzyme synthesis and, in so doing, makes it possible for the cell to utilize the lactose.

**Mechanism by Which the Repressor Substance Inactivates DNA.** Though it is possible that the repressor substance has a direct effect on the DNA molecule to repress its activity, a more likely proposed mechanism for this effect is the following: First, it is known that about half of the mass of the chromosome is composed of small molecular weight proteins called *histones*. These are loosely bound to the DNA strand. It is known that when the DNA becomes separated from its attachments to other substances, it automatically begins to assemble the building blocks of RNA and, therefore, to promote RNA function. Consequently, it has been suggested that one of the ways in which DNA can be made to begin forming RNA is simply to remove the histone molecules. Supposedly, the repressor substance promotes linkage between the histones and the DNA, whereas absence of the repressor substance allows separation and, therefore, permits rapid production of RNA.

**Inactivity of Most DNA.** It is rare that more than 1 to 5 per cent of the DNA is genetically active at any one time. This includes the DNA that codes the formation of all three types of RNA—messenger RNA, ribosomal RNA, and transfer RNA. It is possible that the remainder of the DNA has become repressed during the process differentiation, a process that will be discussed later in the chapter.

**Redundancy of Some Genetic Information in the Nuclear DNA.** Though most of the specific characteristic genes of the cell that code for the formation of messenger RNA are represented by only a single code word on only one DNA strand, this is not true for the genetic code for formation of both ribosomal and transfer RNA. In many cells literally thousands of such code words are available, all capable of forming the same types of RNA. It is also possible that duplicate DNA coding is available for promoting the formation of other intracellular substances that must be formed in large quantities.

## CONTROL OF ENZYME ACTIVITY

In the same way that inhibitors and activators can affect the genetic regulatory system, so also can the enzymes themselves be directly controlled by other inhibitors or activators. This, then, represents a second mechanism by which cellular biochemical functions can be controlled.

**Enzyme Inhibition.** A great many of the chemical substances formed in the cell have a direct feedback effect on the respective enzyme systems that synthesize them. Thus, in Figure 3–10, the synthesized product is shown to act directly back on enzyme A to inactivate it. If enzyme A is inactivated, then none of the substrate will begin to be converted into the synthesized product. Almost always in enzyme inhibition, the synthesized product acts on the first enzyme in a sequence, rather than on the subsequent enzymes. One can readily recognize the importance of inactivating this first enzyme: it prevents buildup of intermediary products that will not be utilized.

This process of enzyme inhibition is another example of negative feedback control; it is responsible for controlling the intracellular concentrations of some of the amino acids that are not controlled by the genetic mechanism as well as the concentrations of the purines, the pyrimidines, vitamins, and other substances.

**Enzyme Activation.** Enzymes that are either normally inactive or that have been inactivated by some inhibitor substance can often be activated. An example of this is the action of cyclic adenosine monophosphate (AMP) in causing glycogen to split so that the released glucose molecules can be used to form high energy ATP, as discussed in the previous chapter. When most of the adenosine triphosphate has been depleted in a cell, a considerable amount of cyclic AMP begins to be formed as a breakdown product of the ATP; the presence of this cyclic AMP in the cell indicates that the reserves of ATP have approached a low ebb. The cyclic AMP immediately activates the glycogen-splitting enzyme phosphorylase, liberating glucose molecules that are rapidly used for replenishment of the ATP stores. Thus, in this case the cyclic AMP acts as an enzyme activator and thereby helps to control intracellular ATP concentration.

Another interesting instance of both enzyme inhibition and enzyme activation occurs in the formation of the purines and the pyrimidines. These substances are needed by the cell in approximately equal quantities for formation of DNA and RNA. When purines are formed, they inhibit the enzymes that are required for formation of additional purines, but they activate the enzymes for formation of the pyrimidines. Conversely, the pyrimidines inhibit their own enzymes but activate the purine enzymes. In this way there is continual cross-feed between the synthesizing systems for these two substances, resulting in almost exactly equal amounts of the two substances in the cells at all times.

In summary, there are two different methods by which the cells control proper proportions and proper quantities of different cellular constituents: (1) the mechanism of genetic regulation, and (2) the mechanism of enzyme regulation. The genes can be either activated or inhibited, and, likewise, the enzymes can be either activated or inhibited. Furthermore, it is principally the substances synthesized by the enzymes that cause activation or inhibition; but on occasion, substances from without the cell (especially some of the hormones which will be discussed later in this text) also control the intracellular biochemical reactions.

# CELL REPRODUCTION

Most cells are continually growing and reproducing. The new cells take the place of the old ones that die, thus maintaining a complete complement of cells in the body.

Also, one can remove most types of cells from the human body and grow them in tissue culture where they will continue to grow and reproduce so long as appropriate nutrients are supplied and so long as the end-products of the cells' metabolism are not allowed to accumulate in the nutrient medium. Thus, the life lineage of most cells is indefinite.

As is true of almost all other events in the cell, reproduction also begins in the nucleus itself. The first step is *replication (duplication) of all DNA in the chromosomes.* The next step is division of the two sets of DNA between two separate nuclei. And the final step is splitting of the cell itself to form two new daughter cells, a process called *mitosis.*

The complete life cycle of a cell that is not inhibited in some way is about 10 to 30 hours from reproduction to reproduction, and the period of mitosis lasts for approximately one-half hour. The period between mitoses is called *interphase.* However, in the body there are almost always inhibitory controls that slow or stop the uninhibited life cycle of the cell and give cells life cycle periods that vary from as little as 10 hours for stimulated bone marrow cells to an entire lifetime of the human body for nerve cells.

## REPLICATION OF THE DNA

The DNA is reproduced several hours before mitosis takes place, and the duration of DNA replication is only about one hour. The DNA is duplicated only once. The net result is two exact duplicates of all DNA, which respectively become the DNA in the two new daughter cells that will be formed at mitosis. Following replication of the DNA, the nucleus continues to function normally for several hours before mitosis begins abruptly.

**Chemical and Physical Events.** The DNA is duplicated in almost exactly the same way that RNA is formed from DNA. First, the two strands of the DNA helix of the gene pull apart. Second, each of these strands combines with deoxyribose nucleotides of the four types described early in the chapter as the basic building blocks of DNA. Each of the bases on each strand of DNA in the chain attracts a nucleotide containing the appropriate *complementary* base. In this way, the appropriate nucleotides are lined up side by side. Third, enzyme mechanisms then provide energy and cause linkage of the nucleotides to form a new DNA strand. The only difference between this formation of the new strand of DNA and the formation of an RNA strand is that the new strand of DNA remains attached to the old strand that has formed it, thus forming a new double-stranded DNA helix. At the same time the other strand of the original helix forms its complementary DNA strand and thereby also forms a new double-stranded helix.

## THE CHROMOSOMES AND THEIR REPLICATION

The chromosomes consist of two major parts: the DNA and protein that consists of many small molecules of histones. The DNA is bound loosely with the protein and sometimes during the life cycle of the cell seems to become separated from the protein. The combination of the two is known as a *nucleoprotein.*

Recent experiments indicate that all the DNA of a particular chromosome is arranged in one long double helix and that the genes are attached end-on-end with each other. Such a molecule in the human being, if spread out linearly, would be approximately 7.5 cm. long or several thousand times as long as the diameter of the nucleus itself; but the experiments also indicate that this long double helix is folded or coiled like a spring and is held in this position by its linkages to protein molecules. The protein has nothing to do with the genetic potency of the chromosome, and the protein molecules can be replaced by new protein molecules without any alteration in the functions of the genes.

Replication of the chromosomes follows as a natural result of replication of the DNA strand. When the new double helix separates from the original double helix, it presumably carries some of the old protein with it or combines with new protein, the DNA acting as the backbone of the newly replicated chromosome and the protein acting only as an accessory to the chromosomal structure.

**Number of Chromosomes in the Human Cell.** Each human cell contains 46 chromosomes arranged in 23 pairs. In general, the genes in the two chromosomes of each pair are almost identical with each other, so that it is usually stated that the different genes exist in pairs, though occasionally this is not the case.

## MITOSIS

The actual process by which the cell splits into two new cells is called mitosis. Once the genes have been duplicated and each chromosome has split to form two new chromosomes, each of which is now called a *chromatid*, mitosis follows automatically, almost without fail, within a few hours.

**The Mitotic Apparatus.** The first event of mitosis takes place in the cytoplasm, occurring during the latter part of interphase in the small structures called *centrioles*. As illustrated in Figure 3–11, two pairs of centrioles lie close to each other near one pole of the nucleus. Each centriole is a small cylindrical body about 0.4 micron long and about 0.15 micron in diameter, consisting mainly of nine parallel, tubular-like structures arranged around the inner wall of the cylinder. The two centrioles of each pair lie at right angles to each other.

In so far as is known, the two pairs of centrioles remain dormant during interphase until shortly before mitosis is to take place. At that time, the two pairs begin to move apart from each other. This is caused by protein microtubules growing between the

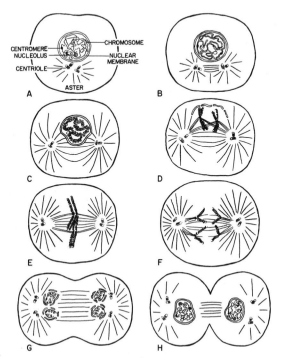

**Figure 3–11.** Stages in the reproduction of the cell. A, B, and C, prophase; D, prometaphase; E, metaphase; F, anaphase; G and H, telophase. (Redrawn from Mazia: *Sci. Amer.*, 205:102, 1961. Copyright © 1961 by Scientific American, Inc. All rights reserved.)

respective pairs and actually pushing them apart. At the same time, microtubules grow radially away from each of the pairs. Some of these penetrate the nucleus. The set of microtubules connecting the two centriole pairs is called the *spindle,* and the entire set of microtubules plus the two pairs of centrioles is called the *mitotic apparatus.*

**Prophase.** The first stage of mitosis, called *prophase,* is shown in Figure 3–11A, B, and C. While the spindle is forming, the *chromatin material* of the nucleus (the DNA), which in interphase consists of long loosely coiled strands, becomes shortened into well-defined chromosomes.

**Prometaphase.** During this stage (Figure 3–11D) the nuclear envelope dissolutes, and microtubules from the forming mitotic apparatus become attached to the chromosomes. This attachment always occurs at the same point on each chromosome, at a small condensed portion called the *centromere.*

**Metaphase.** During metaphase (Fig. 3–11E) the centriole pairs are pushed far apart by the growing spindle, and the chromosomes are thereby pulled tightly by the attached microtubules to the very center of the cell, lining up in the equatorial plane of the mitotic spindle.

**Anaphase.** With still further growth of the spindle, each pair of chromosomes is now broken apart, a stage of mitosis called anaphase (Fig. 3–11F). A microtubule connecting with one pair of centrioles pulls one chromatid and a microtubule connecting with the other centriole pair pulls the opposite chromatid. Thus, all 46 pairs of chromatids are separated, forming 46 daughter chromosomes that are pulled toward one mitotic spindle and another 46 duplicate chromosomes that are pulled toward the other mitotic spindle.

**Telophase.** In telophase (Figs. 3–11G and H) the mitotic spindle grows still longer, pulling the two sets of daughter chromosomes completely apart. Then the mitotic apparatus dissolutes and a new nuclear membrane develops around each set of chromosomes, this membrane perhaps being formed from portions of the endoplasmic reticulum that are already present in the cytoplasm. Shortly thereafter, the cell pinches in two midway between the two nuclei, for reasons totally unexplained at present.

Note, also, that each of the two pairs of centrioles are replicated during telophase, the mechanism of which is not understood. These new pairs of centrioles remain dormant through the next interphase until a mitotic apparatus is required for the next cell division.

## CONTROL OF CELL GROWTH AND REPRODUCTION

Cell growth and reproduction usually go together; growth normally leads to replication of the DNA of the nucleus, followed a few hours later by mitosis.

In the normal human body, regulation of cell growth and reproduction is mainly a mystery. We know that certain cells grow and reproduce all the time, such as the blood-forming cells of the bone marrow, the germinal layers of the skin, and the epithelium of the gut. However, many other cells, such as some muscle cells, do not reproduce for many years. And a few cells, such as the neurons, do not reproduce during the entire life of the person.

If there is an insufficiency of some types of cells in the body, these will grow and reproduce very rapidly until appropriate numbers of them are again available. For instance, seven-eighths of the liver can be removed surgically, and the cells of the remaining one-eighth will grow and divide until the liver mass returns almost to normal. The same effect occurs for almost all glandular cells, for cells of the bone marrow, the subcutaneous tissue, the intestinal epithelium, and almost any other tissue except highly differentiated cells, such as nerve and muscle cells.

We know very little about the mechanisms that maintain proper numbers of the different types of cells in the body. It is assumed that control substances are secreted by the different cells that cause feedback effects to stop or slow their growth when too many of them have been formed, though only a few such substances have been found. We know that cells of any type removed from the body and grown in tissue culture can grow and reproduce rapidly and indefinitely if the medium in which they grow is continually replenished. Yet they will stop growing when even small amounts of their own secretions are allowed to collect in the medium, which supports the idea that control substances limit cellular growth.

**Regulation of Cell Size.**    Cell size is determined almost entirely by the amount of DNA in the nucleus. If replication of the DNA does not occur, the cell grows to a certain size and thereafter remains at that size. On the other hand, it is possible, by use of the chemical *colchicine,* to prevent mitosis even though replication of the DNA continues. In this event, the nucleus then contains far greater quantities of DNA than normally, and the cell grows proportionately larger. It is assumed that this results simply from increased production of RNA and cell proteins, which in turn cause the cell to grow larger.

## *CANCER*

Cancer can occur in any tissue of the body. It results from a change in certain cells that allows them to disrespect normal growth limits, no longer obeying the feedback controls that normally stop cellular growth and reproduction after a given number of such cells have developed. As pointed out above, even normal cells when removed from the body and grown in tissue culture can grow and proliferate indefinitely if the growth medium is continually changed. Therefore, in tissue culture, normal tissue cells behave exactly as cancer cells, but in the body, normal tissue cells behave differently, for they are subject to limits, whereas cancer cells are not.

What is the difference between the cancer cell and the normal tissue cell that allows the cancer cell to grow and reproduce unabated? The answer to this question is not known, but researchers have found that the genetic make-up of most, if not all, cancer cells is different from that of normal cells. This has led to the idea that cancer almost invariably results from mutation of part of the genetic system in the nucleus (sometimes spontaneous but other times caused by irritants, irradiation, or the presence of a virus in the cell). The mutated "genome" eliminates the feedback mechanisms that normally limit growth and reproduction of the cell. Once even a single such cell is formed, it obviously will grow and proliferate indefinitely, its number increasing exponentially.

Cancerous tissue competes with normal tissues for nutrients, and because cancer cells continue to proliferate indefinitely, their number multiplying day-by-day, one can readily understand that the cancer cells will soon demand essentially all the nutrition available to the body. As a result, the normal tissues gradually suffer nutritive death.

## CELL DIFFERENTIATION

A special characteristic of cell growth and cell division is *cell differentiation,* which means changes in physical and functional properties of cells as they proliferate in the embryo to form the different bodily structures. However, our problem here is not to describe the stages of differentiation but simply to discuss the theories concerning the means by which the cells change their characteristics to form all the different organs and tissues of the body.

The earliest and simplest theory for explaining differentiation was that the genetic composition of the nucleus undergoes changes during successive generations of cells in such a way that one daughter cell develops one set of genetic characteristics while another daughter cell develops entirely different genetic characteristics.

However, this theory has now been almost completely disproved by the following simple experiment. The nucleus from an intestinal mucosal cell of a frog, when surgically implanted into a frog ovum from which the original nucleus has been removed, will often cause the formation of a completely normal frog. This demonstrates that even the intestinal mucosal cell, which is a reasonably well differentiated cell, still carries all the necessary genetic information for development of all structures required in the frog's body.

Therefore, the present idea is that instead of loss of genetic material during the process of differentiation, there occurs selective repression of different genetic operons. This presumably results from the buildup of different repressor substances in the cytoplasm, the repressor substances in one cell acting to repress one genetic characteristic and the repressor substances in another cell acting on a different set of genetic characteristics.

Embryological experiments show also that certain cells in an embryo control the differentiation of adjacent cells. For instance, the *primordial chor-*

*damesoderm* is called the *primary organizer* of the embryo because it forms a focus around which the rest of the embryo develops. It differentiates into a *mesodermal axis* containing segmentally arranged *somites* and, as a result of *inductions* in the surrounding tissues, causes formation of essentially all the organs of the body.

Another instance of induction occurs when the developing eye vesicles come in contact with the ectoderm of the head and cause it to thicken into a lens plate that folds inward to form the lens of the eye. It is possible that the entire embryo develops as a result of such inductions, one part of the body affecting another part, and this part affecting still other parts.

Thus, our understanding of cell differentiation is still hazy. We know many different control mechanisms by which differentiations *could* occur. Yet the overall basic controlling factors in cell differentiation are yet to be discovered.

# REFERENCES

Biswas, B. B., Mandal, R. K., Stevens, A., and Cohn, W. E. (eds.): Control of Transcription. New York, Plenum Publishing Corp., 1974.

Brawerman, G.: Eukaryotic messenger RNA. *Ann. Rev. Biochem.,* 43:621, 1974.

Brenner, S.: Theories of gene regulation. *Brit. Med. Bull.,* 21:244, 1965.

Britten, R. J., and Kohne, D. E.: Repeated segments of DNA. *Sci. Amer.,* 222:24, 1970.

Brock, D. J. H., and Mayo, O. (eds.): The Biochemical Genetics of Man. New York, Academic Press, Inc., 1972.

Brown, D. D.: The isolation of genes. *Sci. Amer.,* 229:20, 1973.

Brown, D. D., and Stern, R.: Methods of gene isolation. *Ann. Rev. Biochem.,* 43:667, 1974.

Busch, H.: The Cell Nucleus. New York, Academic Press, Inc., 1974.

Calvin, M.: Chemical evolution. *Amer. Sci.,* 63:169, 1975.

Cervenka, J., and Koulischer, L.: Chromosomes in Human Cancer. Springfield, Ill., Charles C Thomas, Publisher, 1973.

Chamberlin, M. J.: The selectivity of transcription. *Ann. Rev. Physiol.,* 43:721, 1974.

Clark, B. F. C., and Marcker, K. A.: How proteins start. *Sci. Amer.,* 218:36, 1968.

Dose, K., Fox, S. W., Deborin, G. A., and Pavlovskaya, T. E. (eds.): The Origin of Life and Evolutionary Biochemistry. New York, Plenum Publishing Corp., 1974.

Englesberg, E., and Wilcox, G.: Regulation: Positive control. *Ann. Rev. Genet.,* 8:219, 1974.

Fraser, G. R., and Mayo, O.: Textbook of Human Genetics. Philadelphia, J. B. Lippincott Company, 1975.

Frenster, J. H., and Herstein, P. R.: Gene de-repression. *N. Engl. J. Med.,* 288:1224, 1973.

Gendel, B. R., and Elsas, L. J., II: Medical genetics. In Sodeman, W. A., Jr., and Sodeman, W. A. (eds.): Pathologic Physiology: Mechanisms of Disease, 5th Ed. Philadelphia, W. B. Saunders Company, 1974, p. 40.

Goodenough, U. W., and Levine, R. P.: The genetic activity of mitochondria and chloroplasts. *Sci. Amer.,* 223:22, 1970.

Goss, R. J. (ed.): Regulation of Organ and Tissue Growth. New York, Academic Press, Inc., 1973.

Growth Control in Cell Cultures (Ciba Foundation Symposia): New York, Churchill Livingstone, Div. of Longman Inc., 1971.

Haggis, G. H.: Introduction to Molecular Biology, 2d Ed. New York, Halsted Press, Div. of John Wiley & Sons, Inc., 1974.

Hay, E. D., King, T. J., and Papaconstantinou, J. (eds.): Macromolecules Regulating Growth and Development. New York, Academic Press, Inc., 1974.

Kenney, F. T., Hamkalo, B. A., Favelukes, G., and August, J. T. (eds.): Gene Expression and Its Regulation. New York, Plenum Publishing Corp., 1974.

Knight, C. A.: Chemistry of Viruses. New York, Springer-Verlag New York, Inc., 1975.

Kohn, A., and Shatkay, A. (eds.): Control of Gene Expression. New York, Plenum Publishing Corp., 1974.

Kolber, A. R., and Kohiyama, M. (eds.): Mechanism and Regulation of DNA Replication. New York, Plenum Publishing Corp., 1974.

Lamerton, L. F.: The mitotic cycle and cell population control. *J. Clin. Pathol.* [Suppl.], 7:19, 1974.

Lewin, B. M. (ed.): Gene expression. London, New York, Wiley-Interscience, 1974.

Lewin, B. M.: Interaction of regulator proteins with recognition sequences of DNA. *Cell,* 2:1, 1974.

Lynch, H. T. (ed.): Cancer Genetics. Springfield, Ill., Charles C Thomas, Publisher, 1974.

McKusick, V. A.: The mapping of human chromosomes. *Sci. Amer.,* 224:104, 1971.

Marifield, R. B.: The automatic synthesis of proteins. *Sci. Amer.,* 218:56, 1968.

Mazia, D.: The cell cycle. *Sci. Amer.,* 230:54, 1974.

Meerson, F. Z.: Role of synthesis of nucleic acids and protein in adaptation to the external environment. *Physiol. Rev.,* 55:79, 1975.

Miller, O. L., Jr.: The visualization of genes in action. *Sci. Amer.,* 228:34, 1973.

Nicklas, R. B.: Mitosis. *Adv. Cell. Biol.,* 2:225, 1971.

Niu, M. C., and Segal, S. (eds.): The Role of RNA in Reproduction and Development. New York, American Elsevier Publishing Co., 1973.

Nomura, M.: Ribosomes. *Sci. Amer.,* 221:28, 1969.

Ptashne, M., and Gilbert, W.: Genetic repressors. *Sci. Amer.,* 222:36, 1970.

Reznikoff, W. S.: The operon revisited. *Ann. Rev. Genet.,* 6:133, 1972.

Salser, W. A.: DNA sequencing techniques. *Ann. Rev. Biochem.,* 43:923, 1974.

Schatz, G., and Mason, T. L.: The biosynthesis of mitochondrial proteins. *Ann. Rev. Biochem.,* 43:51, 1974.

Smith, J. D.: Genetics of transfer RNA. *Ann. Rev. Genet.,* 6:235, 1972.

Srb, A. M. (ed.): Genes, Enzymes, and Populations. New York, Plenum Publishing Corp., 1974.

Tate, W. P., and Caskey, C. T.: The mechanism of peptide chain termination. *Mol. Cell. Biochem.,* 5:115, 1974.

Temin, H. M.: RNA-directed DNA synthesis. *Sci. Amer.,* 226:24, 1972.

Thompson, J. A., and Thompson, M. W.: Genetics in Medicine, 2nd Ed. Philadelphia, W. B. Saunders Company, 1973.

Widdus, R., and Ault, C. R.: Progress in research related to genetic engineering and life synthesis. *Int. Rev. Cytol.,* 38:7, 1974.

Wolpert, I., and Lewis, J. H.: Towards a theory of development. *Fed. Proc.,* 34:14, 1975.

Wu, T. T., Fitch, W. M., and Margoliash, E.: The information content of protein amino acid sequences. *Ann. Rev. Biochem.,* 43:539, 1974.

Yielding, K. L.: Molecular biology. In Sodeman, W. A., Jr., and Sodeman, W. A. (eds.): Pathologic Physiology: Mechanisms of Disease, 5th Ed. Philadelphia, W. B. Saunders Company, 1974, p. 27.

# 4

# Transport Through the Cell Membrane

The fluid inside the cells of the body, called *intracellular fluid,* is very different from that outside the cells, called *extracellular fluid.* The extracellular fluid circulates in the spaces between the cells and also mixes freely through the capillary walls with the fluid of the blood. It is the extracellular fluid that supplies the cells with nutrients and other substances needed for cellular function. But before the cell can utilize these substances, they must be transported through the cell membrane.

Figure 4–1 gives the compositions of both the extracellular and intracellular fluids. Note that the extracellular fluid contains large quantities of *sodium* but only small quantities of *potassium.* Exactly the opposite is true of the intracellular fluid. Also, the extracellular fluid contains large quantities of chloride, while the intracellular fluid contains very little. But the concentration of phosphates in the intracellular fluid is considerably greater than that in the extracellular fluid. These differences between the components of the intracellular and extracellular fluids are extremely important to the life of the cell. It is the purpose of this chapter to explain how these differences are brought about by the transport mechanisms in the cell membrane.

Substances are transported through the cell membrane by two major processes, *diffusion* and *active transport.* Though there are many different variations of these two basic mechanisms, as we shall see later in this chapter, basically, diffusion means free movement of substances in a random fashion caused by the normal kinetic motion of matter, whereas active

transport means movement of substances in chemical combination with *carrier substances* in the membrane and also movement against an energy gradient such as from a low concentration state to a high concentration state, a process that requires chemical energy to cause the movement.

## DIFFUSION

All molecules and ions in the body fluids, including both water molecules and dissolved

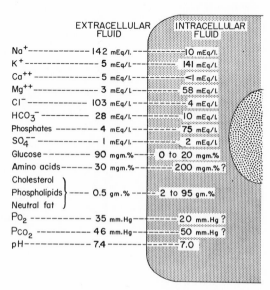

**Figure 4–1.** Chemical compositions of extracellular and intracellular fluids.

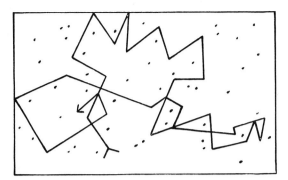

**Figure 4–2.** Diffusion of a fluid molecule during a fraction of a second.

substances, are in constant motion, each particle moving its own separate way. Motion of these particles is what physicists call heat—the greater the motion, the higher is the temperature—and motion never ceases under any conditions except absolute zero temperature. When a moving molecule, A, approaches a stationary molecule, B, the electrostatic forces of molecule A repel molecule B, momentarily adding some of the energy of motion to molecule B. Consequently, molecule B gains kinetic energy of motion while molecule A slows down, losing some of its kinetic energy. Thus, as shown in Figure 4–2, a single molecule in solution bounces among the other molecules first in one direction, then another, then another, and so forth, bouncing randomly hundreds to millions of times each second. At times it travels a far distance before striking the next molecule, but at other times only a short distance.

This continual movement of molecules among each other in liquids, or in gases, is called *diffusion*. Ions diffuse in exactly the same manner as whole molecules, and even suspended colloid particles diffuse in a similar manner, except that because of their very large sizes they diffuse far less rapidly than molecular substances.

## KINETICS OF DIFFUSION—THE CONCENTRATION DIFFERENCE

When a large amount of dissolved substance is placed in a solvent at one end of a chamber, it immediately begins to diffuse toward the opposite end of the chamber. If the same amount of substance is placed in the opposite end of the chamber, it begins to diffuse toward the first end, the same amount diffusing in each direction. As a result, the *net rate of diffusion* from one end to the other is zero. If, however, the concentration of the substance is greater at one end of the chamber than at the other end, the net rate of diffusion from the area of high concentration to low concentration is directly proportional to the larger concentration minus the lower concentration. The total concentration change along the axis of the chamber is called a *concentration difference,* and the concentration difference divided by the distance is called the *concentration* or *diffusion gradient*.

The rapidity with which a molecule diffuses from one point to another is less the greater the molecular size, because large particles are not impelled so intensely by collisions with other molecules. In gases, the rate of diffusion is approximately inversely proportional to the square root of the molecular weight, but in fluids this is determined primarily by the radius of the molecule.

If we consider all the different factors that affect the rate of diffusion of a substance from one area to another, they are the following: (1) The greater the concentration difference, the greater is the rate of diffusion. (2) The less the molecular radius, the greater is the rate of diffusion. (3) The shorter the distance, the greater is the rate. (4) The greater the cross-section of the chamber in which diffusion is taking place, the greater is the rate of diffusion. (5) The greater the temperature, the greater is the molecular motion and also the greater is the rate of diffusion. All these can be expressed in an approximate formula, found at the bottom of this page, for diffusion in solutions.

## DIFFUSION THROUGH THE CELL MEMBRANE

The cell membrane is essentially a sheet of lipid material, called the *lipid matrix,* perhaps partially covered on each surface by a layer of protein, the detailed structure of which was discussed in Chapter 2 and shown in the diagram of Figure 2–3. The fluids on each side of the membrane are believed to penetrate the protein portions of the membrane with ease, but the lipid portion of the membrane is an entirely dif-

$$\text{Diffusion rate} \propto \frac{\text{Concentration difference} \times \text{Cross-sectional area} \times \text{Temperature}}{\text{Molecular radius} \times \text{Distance}}$$

ferent type of fluid medium, acting as a limiting boundary between the extracellular and intracellular fluids.

Therefore, two different methods by which the substances can diffuse through the membrane are: (a) by becoming dissolved in the lipid and diffusing through it in the same way that diffusion occurs in water, or (b) by diffusing through minute pores that pass directly through the membrane at wide intervals over its surface; these "pores" are probably spaces created by protein molecules that penetrate all the way through the membrane, as was discussed in Chapter 2.

### Diffusion in the Dissolved State Through the Lipid Portion of the Membrane

A few substances are soluble in the lipid of the cell membrane as well as in water. These include oxygen, carbon dioxide, alcohol, fatty acids, and a few other less important substances. When one of these comes in contact with the membrane it immediately becomes dissolved in the lipid and continues to diffuse in exactly the same manner that it diffuses within the watery medium on either side of the membrane. In other words, the molecule continues its random motion within the substance of the membrane in exactly the same way that it undergoes random motion in the surrounding fluids.

**Effect of Lipid Solubility on Diffusion Through the Lipid Matrix.** The primary factor that determines how rapidly a substance can diffuse through the lipid matrix of a cell membrane is its solubility in lipids. If it is very soluble, it becomes dissolved in the membrane very easily and therefore passes on through. Indeed, if it is more soluble in lipid than in water, it actually diffuses even more rapidly through the membrane than in the water of the surrounding fluids. On the other hand, a substance that dissolves very poorly in lipids will be greatly retarded. For instance, oxygen, which is very soluble, passes through the lipid matrix many times as rapidly as through water, as shown in Figure 4–3, whereas water itself, which is almost completely insoluble in lipids, passes through the lipid matrix almost not at all.

**Carrier-Mediated or Facilitated Diffusion.** Some substances are very insoluble in lipids and yet can still pass through the lipid matrix by a process called carrier-mediated or facilitated diffusion. This is the means by which different sugars in particular cross the mem-

brane. The most important of these sugars is glucose, the membrane transport of which is illustrated in the lower part of Figure 4–3. This shows that glucose G1 combines with a *carrier* substance C at point 1 to form the compound CG1. This combination is soluble in the lipid so that it can diffuse (or simply move by rotation of the large carrier molecule) to the other side of the membrane, where the glucose breaks away from the carrier (point 2) and passes to the inside of the cell, while the carrier moves back to the outside surface of the membrane to pick up still more glucose and transport it also to the inside. Thus, the effect of the carrier is to make the glucose soluble in the membrane; without it, glucose cannot pass through the membrane.

The rate at which a substance passes through a membrane by facilitated diffusion depends on the difference in concentration of the substance on the two sides of the membrane, the amount of carrier available, and the rapidity with which the chemical reactions can take place. In the case of glucose transport, the overall rate is greatly increased by insulin, which is the primary hormone secreted by the pancreas. Large quantities of insulin can increase the rate of glucose transport about seven- to ten-fold, though it is not known whether this is caused by an effect of insulin to increase the quantity of carrier in the membrane or to increase the rate at which the chemical reactions take place between glucose and the carrier.

Very little is known about the carrier substance for transport of glucose through the membrane. It binds specifically with certain types of monosaccharides, but other monosaccharides are not transported at all or only slightly. A protein having a molecular weight of

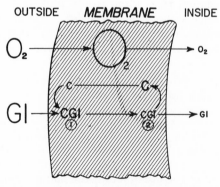

**Figure 4–3.** Diffusion of substances through the lipid matrix of the membrane. The upper part of the figure shows *free diffusion* of oxygen through the membrane and the lower part shows *carrier-mediated* or *facilitated diffusion* of glucose.

approximately 45,000 and also having the property of combining specifically and reversibly with the same types of monosaccharides that are transported has been discovered in the cell membrane. It is likely that this protein is the carrier substance for transport of glucose.

Many of the dynamics of facilitated diffusion are similar to those of active transport, which is discussed in much greater detail later in the chapter. The primary difference between the two is that active transport can move substances through a membrane from a low concentration on one side to a high concentration on the other side, whereas in facilitated diffusion, the substance can move only from a high concentration toward a low concentration.

An important difference between facilitated diffusion and simple diffusion of free particles through a membrane is that the net diffusion rate of free particles is almost exactly proportional to the difference between the concentrations of the particles on the two sides of the membrane. However, for a carrier-transported substance, only when the concentrations of the substance are very small will this relationship hold. When the concentrations become great, the system becomes *saturated*, which means that then the rate at which facilitated diffusion will occur is determined by the quantity of available carrier or by the rates at which the chemical reactions between the mediated substance and the carrier can take place.

### Diffusion Through the Membrane Pores

Some substances, such as water and many of the dissolved ions, seem to go through holes in the cell membrane called *membrane pores*. The nature of these pores is unknown. They might be simple holes lined by protein molecules, or they might be large protein molecules penetrating all the way through the cell membrane and providing pathways for movement of water-soluble substances along the axis of the molecule. At any rate, these so-called "pores" in the cell membrane behave as if they were minute round holes approximately 8 Å in diameter and as if the total area of the pores should equal approximately 1/1600 of the total surface area of the cell. Despite this very minute total area of the pores, molecules and ions diffuse so rapidly that the entire volume of fluid in some types of cells—the red blood cell for instance—can easily pass through the pores within a few hundredths of a second.

Figure 4–4 illustrates a postulated structure of a pore, indicating that its surface is probably

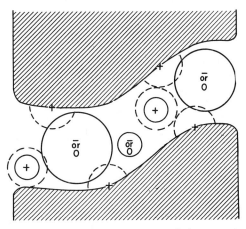

**Figure 4–4.** Postulated structure of the pore in the mammalian red cell membrane, showing the sphere of influence exerted by charges along the surface of the pore. (Modified from Solomon: *Sci. Amer.,* 203:146, 1960. Copyright © 1960 by Scientific American, Inc. All rights reserved.)

lined with positively charged prosthetic groups, probably positively charged proteins. This figure shows several small particles passing through the pore and also shows that the maximum size of the particle that can pass through is approximately equal to the size of the pore itself.

**Effect of Pore Size on Diffusion Through the Pore—Permeability.** Table 4–1 gives the effective diameters of various substances in comparison with the diameter of the pore, and it also gives the relative permeability of the pores for the different substances. *The permeability can be defined as the rate of transport through the membrane for a given concentration difference.* Note that some substances, such as the water molecule, urea molecule, and chloride ion, are considerably smaller in size than the pore. All these pass through the pore with great ease. For instance, the rate per second of diffusion of water in each direction through the pores of a cell is about one hundred times as great as the volume of the cell itself. It is fortunate that an identical amount of water normally diffuses in each direction, which keeps the cell from either swelling or shrinking despite the rapid rate of diffusion. The rates of diffusion of urea and chloride ions through the membrane are somewhat less than that of water, which is in keeping with the fact that their effective diameters are slightly greater than that of water.

Table 4–1 also shows that most of the sugars, including glucose, have effective diameters that are slightly greater than that of the pores. Obviously, not even a single molecule as large as these could go through a pore that is smaller than its size. For this reason essentially none of the sugars can pass through the pores; instead, those that do enter the cell pass through the lipid matrix by the process of facilitated diffusion.

**Effects of Electrical Charge on Transport of Ions Through the Membrane.** Positively charged ions, such as sodium and potassium ions, pass through the cell membrane with extreme difficulty, as indicated by their very low diffusion rates in Table 4–1. The reason for this is believed to be the presence of positive charges of proteins or of adsorbed positive ions, such as calcium ions, lining the pores. Figure 4–4 illustrates that each positive charge causes a sphere of electrostatic space charge to protrude into the lumen of the pore. A positive ion attempting to pass through a pore also exerts a sphere of positive electrostatic charge so that the two positive charges repel each other. This repulsion, therefore, blocks or greatly impedes movement of the positive ion through the pore.

In contrast to the positive ions, negatively charged ions pass through mammalian cell membrane pores much more easily. Thus, for the nerve membrane, chloride ions permeate

the membrane about two times as easily as potassium ions and 100 to 200 times as easily as sodium ions. Moreover, for the red cell membrane, chloride ions permeate the membrane still another half million times as easily. It is believed that these differences are caused by lack of negative charges lining the pores in contrast to strong positive charges.

In addition to electrical charge, the *effective diameter* of an ion also determines how easily it can pass through a membrane. For instance, the hydrated diameter of the sodium ion is approximately 30 per cent greater than that of the potassium ion, and because of this slight difference the sodium ion diffuses through the usual pore only $1/100$ as easily as the potassium ion.

**Effect of Different Factors on Pore Permeability.** Pore permeability does not always remain exactly the same under different conditions. For instance, excess *calcium* in the extracellular fluid causes the permeability to decrease, and diminished calcium causes considerably increased permeability. This is extremely important in the function of nerves, for the enlarged pores that occur in extracellular fluid calcium deficiency cause excessive diffusion of ions, which results in spurious nerve discharge throughout the body.

Another factor that has an important effect on pore permeability of many cells is *antidiuretic hormone,* which is secreted in the hypothalamus. This hormone has an especially important effect on the membranes of the cells lining the collecting ducts of the kidney. Increased quantities of the hormone increase the pore diameter, which allows water and other substances to diffuse out of the tubules and back into the blood with ease.

**TABLE 4–1.   Relationship of Effective Diameters of Different Substances to Pore Diameter and Relative Permeabilities***

| Substance | Diameter Å | Ratio to Pore Diameter | Approximate Relative Permeability |
|---|---|---|---|
| Water molecule | 3 | 0.38 | 50,000,000 |
| Urea molecule | 3.6 | 0.45 | 1,500,000 |
| Hydrated chloride ion | | | |
| (red cell) | 3.86 | 0.48 | 500,000 |
| (nerve membrane) | — | — | 0.06 |
| Hydrated potassium ion | | | |
| (red cell) | 3.96 | 0.49 | 1.1 |
| (nerve membrane) | — | — | 0.03 |
| Hydrated sodium ion | | | |
| (red cell) | 5.12 | 0.64 | 1 |
| (nerve membrane) | — | — | 0.0003 |
| Lactate ion | 5.2 | 0.65 | ? |
| Glycerol molecule | 6.2 | 0.77 | ? |
| Ribose molecule | 7.4 | 0.93 | ? |
| Pore size | 8 (Ave.) | 1.00 | — |
| Galactose | 8.4 | 1.03 | ? |
| Glucose | 8.6 | 1.04 | 0.4 |
| Mannitol | 8.6 | 1.04 | ? |
| Sucrose | 10.4 | 1.30 | ? |
| Lactose | 10.8 | 1.35 | ? |

* These data have been gathered from different sources but relate primarily to the red cell membrane. Other cell membranes have different characteristics.

## NET DIFFUSION THROUGH THE CELL MEMBRANE AND FACTORS THAT AFFECT IT

From the preceding discussion, it is evident that many different substances can diffuse either through the lipid matrix of the cell membrane or through the pores. It should be noted, however, that substances that diffuse in one direction can also diffuse in the opposite direction. Usually it is not the total quantity of substances diffusing in both directions through the cell membrane that is important to the cell but instead the *net quantity* diffusing either into or out of the cell.

In addition to the permeability of the membrane, which has already been discussed, three other factors determine the rate of net diffusion of a substance: the concentration difference of the substance across the membrane, the electrical potential difference across the membrane, and the pressure difference across the membrane.

**Effect of a Concentration Difference.** Figure 4–5A illustrates a membrane with a substance in high concentration on the outside and low concentration on the inside. The rate at which the substance diffuses *inward* is proportional to the concentration of molecules on the outside, for this concentration determines how many of the molecules strike the outside of the pore each second. On the other hand, the rate at which the molecules diffuse *outward* is proportional to their concentration inside the membrane. Obviously, therefore, the rate of net diffusion into the cell is proportional to the concentration on the outside *minus* the concentration on the inside, or

$$\text{Net diffusion} \propto P(C_1 - C_2)$$

## A. CONCENTRATION DIFFERENCE

OUTSIDE  *Membrane*  INSIDE

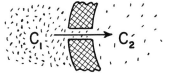

## B. ELECTRICAL POTENTIAL DIFFERENCE

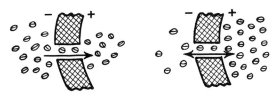

## C. PRESSURE DIFFERENCE

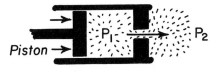

**Figure 4–5.** Effect of (A) concentration difference, (B) electrical difference, and (C) pressure difference on diffusion of molecules and ions through a cell membrane.

in which $C_1$, is the concentration on the outside, $C_2$ is the concentration on the inside, and P is the permeability of the membrane for the substance.

**Effect of an Electrical Potential Difference.** If an electrical potential is applied across the membrane as shown in Figure 4–5B, ions, because of their electrical charges, will move through the membrane even though no concentration difference exists to cause their movement. Thus, to the left in Figure 4–5B, the concentrations of negative ions are exactly the same on both sides of the membrane, but a positive charge has been applied to the right side and a negative charge to the left, creating an electrical gradient across the membrane. The positive charge attracts the negative ions while the negative charge repels them. Therefore, net diffusion occurs from left to right. After much time large quantities of negative ions will have moved to the right (if we neglect, for the time being, the disturbing effects of the positive ions of the solution) creating the condition illustrated on the right in Figure 4–5B, in which a concentration difference of the same ions has developed in the direction opposite to the electrical potential difference. Obviously, the concentration difference is tending to move the ions to the left, while the electrical difference is tending to move them to the right. When the concentration difference rises high enough, the two effects exactly balance each other. At normal body temperature (38° C.), the electrical difference that will exactly balance a given concentration difference of univalent ions can be determined from the following formula:

$$\text{EMF (in millivolts)} = \pm 61 \log \frac{C_1}{C_2}$$

in which EMF is the electromotive force (voltage) between side 1 and side 2 of the membrane, $C_1$ is the concentration on side 1, and $C_2$ is the concentration on side 2. The required polarity is $+$ for negative ions and $-$ for positive ions. This relationship is extremely important in understanding the transmission of nerve impulses, for which reason it is discussed in even greater detail in Chapter 10.

**Effect of a Pressure Difference.** At times considerable pressure difference develops between the two sides of a membrane. This occurs, for instance, at the capillary membrane, which has a pressure approximately 23 mm. Hg greater inside the capillary than outside. Pressure actually means the sum of all the forces of the different molecules striking a unit surface area at a given instant. Therefore, when the pressure is increased on one side of a membrane, this means that the sum of all the molecular forces striking the pores on that side of the membrane is greater than on the other side. This can result either from greater numbers of molecules striking the membrane per second or from greater kinetic energy of the average molecule striking the membrane. In either event, increased amounts of energy are available to cause net movement of molecules from the high pressure side toward the low pressure side. This effect is illustrated in Figure 4–5C, which shows a piston developing high pressure on one side of a cell membrane, thereby causing net diffusion through the membrane to the other side.

For the usual red cell membrane, 1 mm. Hg pressure difference causes approximately $10^{-4}$ cubic microns net diffusion of water through each square micron of membrane each second. This appears to be only a very minute rate of fluid movement through the membrane, but in relation to the normal cell size and the very large diffusion pressures that can develop at the cell membrane, as discussed later in the chapter, this rate can represent tremendous transport of fluid in only a few seconds.

### NET MOVEMENT OF WATER ACROSS CELL MEMBRANES—OSMOSIS ACROSS SEMIPERMEABLE MEMBRANES

By far the most abundant substance to diffuse through the cell membrane is water. It should be recalled again that enough water ordinarily diffuses in each direction through the red cell membrane per second to equal about *100 times the volume of the cell itself.* Yet, *normally,* the amount that diffuses in the two directions is so precisely balanced that not even the slightest *net* movement of water occurs. Therefore, the volume of the cell remains constant. However, under certain conditions, a *concentration difference for water* can develop across a membrane, just as concentration differences for other substances can also occur. When this happens, net movement of water does occur across the cell membrane, causing the cell either to swell or to shrink, depending on the direction of the net movement. This process of net movement of water caused by a concentration difference is called *osmosis.*

To give an example of osmosis, let us assume

that we have the conditions shown in Figure 4–6, with pure water on one side of the cell membrane and a solution of sodium chloride on the other side. Referring back to Table 4–1, we see that water molecules pass through the cell membrane with extreme ease while sodium ions pass through only with extreme difficulty. And chloride ions cannot pass through the membrane because the positive charge of the sodium ions holds the negatively charged chloride ions back to maintain a balance between the negative and positive charges in the solution, an effect called the *principle of electroneutrality.* Therefore, sodium chloride solution is actually a mixture of diffusible water molecules and non-diffusible sodium and chloride ions, and the membrane is said to be *semipermeable,* i.e., permeable to water but not to sodium and chloride ions. Yet, the presence of the sodium and chloride has reduced the concentration of water molecules to less than that of pure water. As a result, in the example of Figure 4–6, more water molecules strike the pores on the left side where there is pure water than on the right side where the water concentration has been reduced. Thus, net movement of water occurs from left to right—that is osmosis occurs from left to right.

### OSMOTIC PRESSURE

If in Figure 4–6 pressure were applied to the sodium chloride solution, osmosis of water into

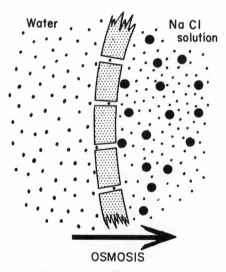

**Figure 4–6.** Osmosis at a cell membrane when a sodium chloride solution is placed on one side of the membrane and water on the other side.

this solution could be slowed or even stopped. The amount of pressure required to stop osmosis completely is called the osmotic pressure of the sodium chloride solution.

The principle of a pressure difference opposing osmosis is illustrated in Figure 4–7, which shows a semipermeable membrane separating two separate columns of fluid, one containing water and the other containing a solution of water and some solute that will not penetrate the membrane. Osmosis of water from chamber B into chamber A causes the levels of the fluid columns to become farther and farther apart, until eventually a pressure difference is developed that is great enough to oppose the osmotic effect. The pressure difference across the membrane at this time is the osmotic pressure of the solution containing the nondiffusible solute.

**Kinetics of Osmotic Pressure.** Figure 4–8 gives an analysis that can be very useful in understanding the basic cause of osmosis and osmotic pressure (even though the analysis is not strictly correct because of nonlinearities of the effects when one deals with solutions more concentrated than those of the body fluids). The actual pressure at the surface of a solution, such as a solution of sodium chloride, will be equal to atmospheric pressure. Thus, at sea level the pressure at the surface of a solution is 760 mm. Hg. Yet, strangely enough, pressure differences of many thousand mm. Hg frequently occur be-

tween the two sides of a semipermeable membrane when the solutions on the two sides of the membrane are different from each other. This ability of the solutions to cause such pressure can be expressed in terms of *potential pressure*. The potential pressure of each solution is usually expressed as *chemical potential* of the different molecules and ions present in the solution, and the amount of chemical potential of each type of molecule or ion is directly proportional to its fractional concentration in the solution.

Figure 4–8 illustrates some representative potential "pressures" as they occur at cell membranes. The first potential pressure, 5500 mm. Hg, represents the total potential pressure of the nondiffusible particles in extracellular fluid—such particles as sodium, potassium, magnesium, calcium, chloride, bicarbonate, and protein ions, as well as nonionic substances such as glucose.

In the middle of the figure are shown the potential pressures of water on both sides of the membrane. On the lower side there exists pure water, and 1,073,000 mm. Hg is the theoretical total potential pressure of pure water if its entire chemical potential were converted into actual pressure (and if the principles of ideal solutions applied at high concentrations). That is, if pure water were on one side of a semipermeable membrane and pure electrolyte were on the opposite side, water theoretically could move by osmosis through the semipermeable membrane with sufficient pressure to create an actual pressure across the membrane of 1,073,000 mm. Hg. On the top of the membrane, the potential pressure of water molecules is shown to be 1,067,500 mm. Hg, or 5500 mm. Hg less than that on the lower side. The reason for this is that the water on top has been diluted by the nondiffusible particles, making the molecular concentration of water above the membrane slightly less than that below.

Because of the greater potential pressure of water molecules below the membrane in Figure 4–8 than above, more water will diffuse from below upward than in the opposite direction, causing osmosis; and the amount of hydrostatic pressure that would be required on top of the membrane to prevent osmosis is 5500 mm. Hg. Note that this value is equal to the potential pressure of the nondiffusible particles in the upper solution.

Thus, the potential pressure of nondiffusible particles (caused by the chemical potential of these particles) is what is called osmotic pres-

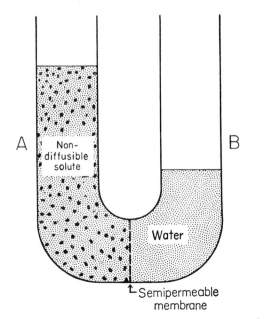

**Figure 4–7.** Demonstration of osmotic pressure on the two sides of a semipermeable membrane.

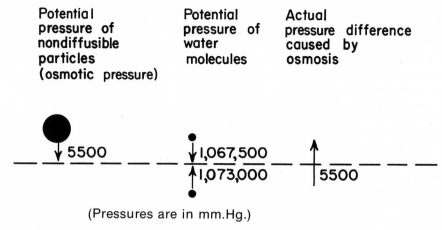

Potential pressure of nondiffusible particles (osmotic pressure)

Potential pressure of water molecules

Actual pressure difference caused by osmosis

↓5500          ↓1,067,500          ↑
               ↑1,073,000          5500

(Pressures are in mm.Hg.)

**Figure 4–8.** Kinetics of osmosis. Explained in the text.

sure of the solution. It causes osmosis of water in the direction opposite to the osmotic pressure. Therefore, the term "osmotic pressure" is actually a misnomer because it is not an actual pressure but only a potential pressure.

**Lack of Effect of Molecular and Ionic Mass on Osmotic Pressure —Importance of Numbers of Osmotic Particles.** The osmotic pressure exerted by nondiffusible particles in a solution, whether they be molecules or ions, is determined by the *numbers* of particles per unit volume of fluid and not the mass of the particles. The reason for this is that each particle in a solution, regardless of its mass, exerts, on the average, the same amount of pressure against the membrane. That is, all particles are bouncing among each other with, on the average, equal energy. If some particles have greater kinetic energy of movement than others, their impact with the low energy particles will impart part of their energy to these, thus decreasing the energy level of the high energy particles while increasing the energy level of the others. The large particles, which have greater mass (m) than the small particles, move at lower velocities (v), while the small particles move at higher velocities in such a way that their average energies, determined by the equation

$k = \dfrac{mv^2}{2}$, are equal to each other. Therefore,

on the average, the energy that a particle has when it strikes a membrane is approximately equal for all size particles, so that the factor that determines the osmotic pressure of a solution is the concentration of the solution in terms of numbers of particles and not in terms of mass of the solute.

**Effect of Interaction Between Dissolved Particles on Osmotic Pressure.** If particles interact with each other by either chemical or physical bonds, the osmotic effect becomes changed. Thus, sodium and chloride ions in solution have tremendous attractive power for each other so that neither of these ions can move totally unrestrained; therefore, their chemical activities are normally only about nine-tenths of what they would be were it not for this attraction. Because of interionic attraction (and other less powerful types of attraction) the amount of osmotic pressure in the body fluids is actually only 0.93 times the maximum amount that would be calculated on the basis of individual numbers of particles. However, since all the fluids of the body are affected approximately equally in this way, for practical purposes this interaction between particles is usually ignored in studying osmotic effects in the body.

**Osmolality.** Since the amount of osmotic pressure exerted by a solute is proportional to the concentration of the solute in numbers of molecules or ions, expressing the solute concentration in terms of mass is of no value in determining osmotic pressure. To express the concentration in terms of numbers of particles, the unit called the *osmol* is used in place of grams.

One osmol is the number of particles in 1 gram molecular weight of undissociated solute. Thus, 180 grams of glucose is equal to 1 osmol of glucose, because glucose does not dissociate. On the other hand, if the solute dissociates into two ions, 1 gram molecular weight of the solute equals 2 osmols, because the number of osmotically active particles is now twice as great as is the case in the undissociated solute. Therefore, 1 gram molecular weight of sodium chloride, 58.5 gm., is equal to 2 osmols.

A solution that has 1 osmol of solute dissolved in each kilogram of water is said to have an osmolality of 1 osmol per kilogram, and a solution that has $1/1000$ osmol dissolved per kilogram has an osmolality of 1 milliosmol per kilogram. The normal osmolality of the extracellular and intracellular fluids is about 300 milliosmols per kilogram.

**Osmolarity.** Because of the difficulty of measuring kilograms of water in a solution, when speaking of the osmotic characteristics of body fluids one usually uses another term, "osmolarity," which is the osmolar concentration expressed as *osmols per liter of solution* rather than osmols per kilogram of water. Though, strictly speaking, it is osmols per kilogram of water (osmolality) that determines the rate of osmosis, nevertheless, for dilute solutions such as occur in the body, the quantitative differences between osmolarity and osmolality are less than one per cent. Since it is far more practical to use the term "osmolarity" than the term "osmolality," this is the usual practice in almost all physiological studies.

**Relationship of Osmolarity to Osmotic Pressure.** At normal body temperature, 38° C, a concentration of 1 osmol per liter will cause 19,300 mm. Hg osmotic pressure in the solution. Likewise, 1 milliosmol per liter concentration is equivalent to 19.3 mm. Hg osmotic pressure.

*BULK FLOW OF WATER THROUGH PORES IN RESPONSE TO HYDROSTATIC AND OSMOTIC PRESSURE GRADIENTS*

When either a hydrostatic or an osmotic pressure gradient is created across a membrane, the rate of movement of water molecules down the gradient is often many times as great as the rate that can be accounted for by net diffusion down the gradient. This is caused by a phenomenon called "bulk flow," which means simply that since large numbers of molecules are moving in the same direction, they tend to "stream" through the pores in unison rather than to move randomly as occurs in pure diffusion. Because of the mutual drag of the adjacent molecules on each other, this results in flow of far more water down either the hydrostatic or the osmotic gradient than one would predict from simple diffusion kinetics. This is often important because it accounts for far more movement of water through some membranes than would be expected on the basis of movement by diffusion alone. For instance, the movement of fluid through the capillary membrane is

as much as ten times as great as can be accounted for by simple diffusion. This phenomenon will be discussed in greater detail in Chapter 30.

## ACTIVE TRANSPORT

Often only a minute concentration of a substance is present in the extracellular fluid, and yet a large concentration of the substance is required in the intracellular fluid. For instance, this is true of potassium ions. Conversely, other substances frequently enter cells and must be removed even though their concentrations inside are far less than outside. This is true of sodium ions.

From the discussion thus far it is evident that *no substances can diffuse against a concentration gradient,* or, as is often said, "uphill." To cause movement of substances uphill, energy must be imparted to the substance. This is analogous to the compression of air by a pump. After compression, the concentration of the air molecules is far greater than before compression, but to create this greater concentration, energy must be imparted to the air molecules by the piston of the pump as they are compressed. Likewise, as molecules are transported through a cell membrane from a dilute solution to a concentrated solution, energy must be imparted to the molecules. When a cell membrane moves molecules uphill against a concentration gradient (or uphill against an electrical or pressure gradient) the process is called *active transport.* Among the different substances that are actively transported through cell membranes are sodium ions, potassium ions, calcium ions, iron ions, hydrogen ions, chloride ions, iodide ions, ureate ions, several different sugars, and the amino acids.

*BASIC MECHANISM OF ACTIVE TRANSPORT*

The mechanism of active transport is believed to be similar for all substances and to be dependent on transport by *carriers.* Figure 4–9 illustrates the basic mechanism, showing a substance S entering the outside surface of the membrane where it combines with carrier C. At the inside surface of the membrane, S separates from the carrier and is released to the inside of the cell. C then moves back to the outside to pick up more S.

One will immediately recognize the similarity between this mechanism of active transport and

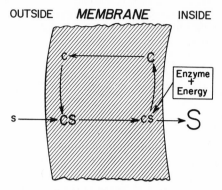

**Figure 4–9.** Basic mechanism of active transport.

that of facilitated diffusion discussed earlier in the chapter. The difference, however, is that *energy is imparted to the system* in the course of transport, so that transport can occur *against a concentration gradient* (or against an electrical or pressure gradient).

Though the mechanism by which energy is utilized to cause active transport is not entirely known, we do know some features of this process. First, the energy is delivered to the inside surface of the membrane from high energy substances, principally ATP, inside the cytoplasm of the cell. Second, active transport obeys the usual laws for chemical combination of one substance (the substance to be transported) with another substance (the carrier). Third, specific enzymes are required to promote active transport.

From this information, we can construct a theory for active transport, as follows: Referring again to Figure 4–9, we might suppose that the carrier has a natural affinity for the substance to be transported so that at the outer surface of the membrane the carrier and the substance readily combine. Then the combination of the two *diffuses* through the membrane to the inner surface. Here an enzyme-catalyzed reaction occurs, utilizing energy from ATP to split the substance away from the carrier. In other words, the enzyme-catalyzed reaction makes the affinity of the carrier for the transported substance very low and thereby displaces it from the combination. But the released substance, being insoluble in the membrane, cannot diffuse backward through the lipid matrix of the membrane. Therefore, it is released to the inside of the membrane while the carrier alone diffuses back to the outside surface to transport still another molecule of substance in the inward direction.

A special characteristic of active transport is

that the mechanism *saturates* when the concentration of the substance to be transported is very high. This saturation results from limitation either of quantity of carrier available to transport the substance or of enzymes to promote the chemical reactions that release the substance from the carrier. Thus, the principle of saturation of active transport is almost identical with that of saturation of carrier-mediated facilitated diffusion, which was discussed earlier in the chapter.

**Chemical Nature of Carrier Substances.** Carrier substances are believed to be either proteins or lipoproteins, the protein moiety providing a specific site for attachment of the substance to be transported and the lipid moiety providing solubility in the lipid phase of the cell membrane.

It has been suggested that the carrier might transport substances by a simple process of thermal motion, such as by rotating to expose its carrier site first to the outside surface of the cell membrane and then to the inside, or by sliding a substance from one reactive site of the carrier to another until the substance enters the cell. Whether or not a "shuttling" or a "fixed" type of carrier exists, nevertheless, essentially the same principles of transport through the membrane would still apply.

**Specificity of Carrier Systems.** Several different carrier systems exist in cell membranes, each of which transports only certain specific substances. One carrier system, for instance, transports sodium to the outside of the membrane and probably transports potassium to the inside at the same time. Another system actively transports sugars through the membranes of intestinal and renal tubular epithelial cells while still other specific carrier systems transport different ones of the amino acids.

The specificity of active transport systems for substances is determined either by the chemical nature of the carrier, which allows it to combine only with certain substances, or by the nature of the enzymes that catalyze the specific chemical reactions. Unfortunately, very little is known about the precise function of these two components of the carrier systems.

**Energetics of Active Transport.** The amount of energy required to transport a substance actively through a membrane (aside from energy lost as heat in the chemical reactions) is determined by the degree that the substance is concentrated during transport. Compared to the energy required to concentrate a substance 10-fold, to concentrate it 100-fold requires twice as much energy, and to concentrate it 1000-fold requires three times as much. In other words, the energy required is proportional to the logarithm of the degree that the substance is concentrated, as expressed by the following formula:

$$\text{Energy (in calories per osmol)} = 1400 \log \frac{C_1}{C_2}$$

Thus, in terms of calories, the amount of energy required to concentrate 1 osmol of substance 10-fold is about 1400 calories. One can see that the energy expenditure for concentrating substances in cells or for removing substances from cells against a concentration gradient can be tremendous. Some cells, such as those lining the renal tubules, expend as much as 50 per cent of their energy for this purpose alone.

### TRANSPORT THROUGH INTRACELLULAR MEMBRANES

Though much less is known about the transport processes of the membranes within the cell than those of the cell membrane, we do know that at least two of these membranes, that of the mitochondrion and that of the endoplasmic reticulum, obey the same laws of both diffusion and active transport as does the cell membrane. However, their permeabilities and specificities for transport are often different. Indeed, the mitochondrion has a double membrane, with an outer membrane that is much more permeable than the inner membrane. Undoubtedly, transport through these membranes plays important roles in function of the intracellular organelles, though most of these roles are still a mystery.

### ACTIVE TRANSPORT THROUGH CELLULAR SHEETS

In many places in the body substances must be transported through an entire *cellular layer* instead of simply through the cell membrane itself. Transport of this type occurs through the intestinal epithelium, the epithelium of the renal tubules, the epithelium of all exocrine glands, the membrane of the choroid plexus of the brain, and many other membranes.

One mechanism of active transport through a sheet of cells is illustrated in Figure 4–10. This figure shows two adjacent cells in a typical epithelial membrane as found in the intestine, in the gallbladder, and in certain other areas of the body. On the luminal surface of the cells is a brush border that is highly permeable to water and some solutes, allowing both of these to diffuse from the lumen of the intestine to the interior of the cell. Once inside the cell, some of the solutes are actively transported into the space between the cells. This space is closed at the brush border of the epithelium but is wide open at the base of the cells where they rest on the basement membrane. Furthermore, the basement membrane is extremely permeable. Therefore, substances transported into this channel between the cells flow toward the connective tissue.

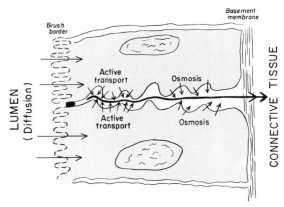

**Figure 4–10.** Basic mechanism of active transport through a layer of cells.

One of the most important substances actively transported in this manner is sodium ions; when sodium ions are transported into the space between the cells, their positive electrostatic charges pull negatively charged chloride ions through the membrane as well, illustrating once again the principle of electroneutrality. Then, when the concentration of sodium and chloride ions increases in the intercellular space, this in turn causes osmosis of water out of the cell and into the intercellular space. In consequence, the sodium and chloride flow along with the water into the connective tissue behind the basement membrane where the water and ions diffuse into the blood capillaries. These same principles of transport apply generally wherever such transport occurs through cellular sheets, whether in the intestine, gallbladder, kidneys, or elsewhere. Such transport will be described for several of these at later points in the text.

## ACTIVE TRANSPORT OF SPECIFIC SUBSTANCES

### ACTIVE TRANSPORT OF SODIUM, POTASSIUM, AND OTHER ELECTROLYTES

Referring back to Figure 4–1, one sees that the sodium concentration outside the cell is very high in comparison with its concentration inside, and the converse is true of potassium. Also, Table 4–1 shows that minute quantities of sodium and potassium can diffuse through the pores of the cell. If such diffusion should take place over a long period of time, the concentrations of the two ions would eventually become equal inside and outside the cell unless there were some means to remove the sodium from the inside and to transport potassium back in.

Fortunately, a system for active transport of

sodium and potassium is present probably in all cells of the body. The mechanism has been postulated to be that illustrated in Figure 4–11, which shows sodium (Na) inside the cell combining with carrier Y at the membrane surface to form large quantities of the combination NaY. This then moves to the outer surface where sodium is released and the carrier Y changes its chemical composition slightly to become carrier X. This carrier then combines with potassium K to form KX, which moves to the inner surface of the membrane where energy is provided to split K from X under the influence of the enzyme ATPase, the energy being derived from MgATP. At the same time carrier X is reconverted into Y, which transports a new sodium ion to the exterior, the cycle continuing indefinitely.

The transport mechanism in Figure 4–11 is believed to be more effective in transporting sodium than in transporting potassium, usually transporting one to three sodium ions for every one potassium ion. The carrier is probably a lipoprotein, and it is likely that this same lipoprotein molecule acts as the enzyme ATPase to release the energy required for transport.

The sodium transport mechanism is so important to many different functioning systems of the body—such as to nerve and muscle fibers for transmission of impulses, various glands for the secretion of different substances, and all cells of the body to prevent cellular swelling—

that it is frequently called the *sodium pump.* We will discuss the sodium pump at many places in this text.

**Importance of the Sodium Pump in Controlling Cell Size.** One of the most important functions of the sodium pump is to prevent continual swelling of the cells. This may be explained as follows: All cells form many intracellular substances that are nondiffusible through the cell membrane, such as protein molecules, phosphocreatine, and adenosine triphosphate. These substances tend to cause osmosis of water to the interior of the cell all the time. Also, electrolytes tend to leak along with the water to the inside of the cell. If some factor should not overcome the continual tendency for water and electrolytes to enter the cell, the cell would eventually swell until it should burst. However, the sodium transport mechanism opposes this tendency of the cell to swell by continually transporting sodium to the exterior, which initiates an opposite osmotic tendency to move water out of the cell. Whenever the metabolism of the cell ceases so that energy from adenosine triphosphate is not available to keep the sodium pump operating, the cell begins to swell immediately.

**Other Electrolytes.** Calcium and magnesium are probably transported by all cell membranes in much the same manner that sodium and potassium are transported, and certain cells of the body have the ability to transport still other ions. For instance, the glandular cell membranes of the thyroid gland can transport large quantities of iodide ion; the epithelial cells of the intestine can transport sodium, chloride, calcium, iron, hydrogen, and probably many other ions; and the epithelial cells of the renal tubules can transport hydrogen, calcium, sodium, potassium, and a number of other ions.

## ACTIVE TRANSPORT OF SUGARS

Facilitated diffusion of glucose and certain other sugars occurs in essentially all cells of the body, but active transport of sugars occurs in only a few places in the body. For instance, in the intestine and renal tubules, glucose and several other monosaccharides are continually transported into the blood even though their concentrations be minute in the lumens. Therefore, essentially none of these sugars is lost either in the intestinal excreta or the urine.

Though not all sugars are actively transported, almost all monosaccharides that are important to the body *are* actively transported, including *glucose, galactose, fructose, mannose, xylose, arabinose,* and *sorbose.* On the other hand, the disaccharides such as sucrose, lactose, and maltose are not actively transported at all.

As is true of essentially all other active trans-

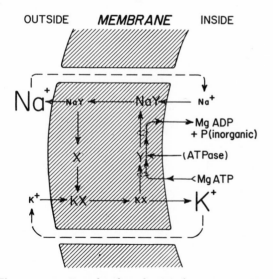

**Figure 4–11.** Postulated mechanism for active transport of sodium and potassium through the cell membrane, showing coupling of the two transport mechanisms and delivery of energy to the system at the inner surface of the membrane.

port mechanisms, the precise carrier system and chemical reactions responsible for transport of the monosaccharides are yet unknown. The one common denominator in transport of one large group of sugars, including especially glucose and galactose, is the necessity for an intact —OH group on the $C_2$ carbon of the monosaccharide molecule. It is presumed that the monosaccharide attaches to the carrier at this point. But in addition, fructose, another monosaccharide important to the body, is transported by a separate carrier system.

Transport of the monosaccharides will not occur when active transport of sodium is blocked. Because of this, it is believed that active transport of sodium ions, by means of coupled reactions with the glucose transport mechanism, provides the energy required to move the monosaccharides through the membrane. Therefore, it is stated that monosaccharide transport is *secondary active transport,* in contrast to *primary active transport* represented by the transport of sodium through the membrane. A theoretical mechanism for this interdependence between sodium and glucose transport is discussed in detail in Chapter 65 in relation to intestinal absorption.

### ACTIVE TRANSPORT OF AMINO ACIDS

Amino acids and the closely related substances, the amines, are actively transported through the membranes of all cells that have been studied. Active transport through the epithelium of the intestine, the renal tubules, and many glands is especially important. Probably at least four different carrier systems exist for transporting different groups of amino acids; this will be discussed in relation to intestinal absorption in Chapter 65. It is especially interesting that amino acid transport in the intestine, like sugar transport, is also dependent on energy from the sodium transport mechanism. One of the few known features of the amino acid carrier system is that transport of at least some amino acids depends on pyridoxine (vitamin $B_6$). Therefore, deficiency of this vitamin causes protein deficiency.

**Hormonal Regulation of Amino Acid Transport.** At least four different hormones are important in controlling amino acid transport: (1) *Growth hormone,* secreted by the adenohypophysis, increases amino acid transport into essentially all cells. (2) *Insulin* and (3) *glucocorticoids* increase amino acid transport at least into liver cells and possibly into other cells as well, though much less is known about

these. (4) *Estradiol,* the most important of the female sex hormones, causes rapid transport of amino acids into the musculature of the uterus, thereby promoting development of this organ.

Thus, several of the hormones exert much, if not most, of their effects in the body by controlling active transport of amino acids into all or certain cells.

### PINOCYTOSIS AND PHAGOCYTOSIS

It was pointed out in Chapter 2 that the cell membrane has the ability to imbibe small amounts of substances from the extracellular fluid by the process called *pinocytosis*, and *phagocytosis* occurs by essentially the same mechanism as pinocytosis, but phagocytosis means ingestion of large particulate matter, such as bacteria or cell fragments, that is free in the extracellular fluid.

The real importance of pinocytosis to the body is that this is the only known means by which very large molecules, such as those of protein, can be transported to the interior of cells. The importance of phagocytosis is that it is used by special cells, such as the white blood cells, to rid the body of bacteria and unwanted debris in the tissues.

# REFERENCES

Avery, J. (ed.): Membrane Structure and Mechanisms of Biological Energy Transduction. New York, Plenum Publishing Corp., 1974.

Azzone, G. F. (ed.): Membrane Proteins in Transport and Phosphorylation. New York, American Elsevier Publishing Co., 1974.

Bean, C. P.: The physics of porous membranes—neutral pores. *Membranes,* 1:1, 1972.

Bloch, K.: Comparative Biochemistry and Physiology of Transport. New York, American Elsevier Publishing Co., 1974.

Bolis, L. (ed.): Permeability and Function of Biological Membranes. New York, American Elsevier Publishing Co., Inc., 1972.

Braun, V., and Hantke, K.: Biochemistry of bacterial cell envelopes. *Ann. Rev. Biochem.,* 43:89, 1974.

Brinley, F. J., Jr.: Calcium and magnesium transport in single cells. *Fed. Proc.,* 32:1735, 1973.

Dahl, J. L., and Hokin, L. E.: The sodium-potassium adenosinetriphosphatase. *Ann. Rev. Biochem.,* 43:327, 1974.

Diamond, J. M., and Wright, E. M.: Biological membranes: The physical bases of ion and non-electrolyte selectivity. *Ann. Rev. Physiol.,* 31:581, 1969.

Gamble, J. L.: Chemical Anatomy, Physiology and Pathology of Extracellular Fluid: A Lecture Syllabus. 6th ed., Cambridge, Massachusetts, Harvard University Press, 1954.

Glynn, I. M., and Karlish, S. J. D.: The sodium pump. *Ann. Rev. Physiol.,* 37:13, 1975.

Hendler, R. W.: Biological membrane ultrastructure. *Physiol. Rev.,* 51:66, 1971.

Hope, A. B.: Ion Transport and Membranes. Baltimore, University Park Press, 1971.

Korn, E. D. (ed.): Methods in Membrane Biology. New York, Plenum Publishing Corp., 1974, Vols. 1 and 2.

Korn, E. D.: Transport. New York, Plenum Publishing Corp., 1975.

Kotyk, A.: Mechanisms of nonelectrolyte transport. *Biochem. Biophys. Acta,* 300:183, 1973.

Kotyk, A., and Janacek, K.: Cell Membrane Transport, 2nd Ed. New York, Plenum Publishing Corp., 1974.

Lehninger, A. L.: The molecular organization of mitochondrial membranes. *Adv. Cytopharmacol.,* 1:199, 1971.

Loewenstein, W. R.: Intercellular communication. *Sci. Amer.,* 222:78, 1970.

Loewenstein, W. R.: Membrane junctions in growth and differentiation. *Fed. Proc.,* 32:60, 1973.

Oschman, J. L., Wall, B. J., and Gupta, B. L.: Cellular basis of water transport. *Symp. Soc. Exp. Biol.,* (28): 305, 1974.

Oxender, D. L.: Membrane transport proteins. *Biomembranes,* 5:25, 1974.

Packer, L.: Biomembranes: Architecture, Biogenesis, Bioenergetics, and Differentiation. New York, Academic Press, Inc., 1975.

Parsons, D. S. (ed.): Biological Membranes. New York, Oxford University Press, 1975.

Poste, G., and Allison, A. C.: Membrane fusion. *Biochem. Biophys. Acta,* 300:421, 1973.

Rothfield, L. I. (ed.): Structure and Function of Biological Membranes. New York, Academic Press, Inc., 1971.

Schultz, S. G.: Principles of electrophysiology and their application to epithelial tissues. *In* International Review of Physiology. Baltimore, University Park Press, 1974, Vol. 4, p. 69.

Schultz, S. G., and Curran, P. F.: Role of sodium in non-electrolyte transport across animal cell membranes. *Physiologist, 12*:437, 1969.

Schultz, S. G., and Curran, P. F.: Coupled transport of sodium and organic solutes. *Physiol. Rev.,* 50:637, 1970.

Sen, A. K., and Post, R. L.: Stoichiometry and localization of adenosine triphosphate-dependent sodium and potassium transport in the erythrocyte. *J. Biol. Chem., 239*:345, 1964.

Siekevitz, P.: Biological membranes: The dynamics of their organization. *Ann. Rev. Physiol., 34*:117, 1972.

Singer, S. J.: The molecular organization of membranes. *Ann. Rev. Physiol., 43*:805, 1974.

Solomon, A. K.: Properties of water in red cell and synthetic membranes. *Biomembranes, 3*:299, 1972.

Structure and function of membranes (Symposium). *Brit. Med. Bull., 24*:99, 1968.

Tosteson, D. C.: Membrane transport of Na and K. *Physiologist, 9*:89, 1966.

Ussing, H. H., Erlij, D., and Lassen, U.: Transport in pathways in biological mechanisms. *Ann. Rev. Physiol., 36*:17, 1974.

Whittam, R., and Wheeler, K. P.: Transport across cell membranes. *Ann. Rev. Physiol., 32*:21, 1970.

# PART II

# BLOOD CELLS, IMMUNITY, AND BLOOD CLOTTING

# | 5 |

# Red Blood Cells, Anemia, and Polycythemia

With this chapter we begin a discussion of the blood cells and of other cells closely related to those of the blood: the cells of the reticuloendothelial system and of the lymphatic system. We will first present the functions of red blood cells, which are the most abundant of all the cells of the body and are necessary for delivery of oxygen to the tissues.

## THE RED BLOOD CELLS

The major function of red blood cells is to transport hemoglobin, which in turn carries oxygen from the lungs to the tissues. In some lower animals hemoglobin circulates as free protein in the plasma, not enclosed in red blood cells. However, when it is free in the plasma of the human being, approximately 3 per cent of it leaks through the capillary membrane into the tissue spaces or through the glomerular membrane of the kidney into Bowman's capsule each time the blood passes through the capillaries. Therefore, for hemoglobin to remain in the blood stream, it must exist inside red blood cells.

The red blood cells have other functions besides simply transport of hemoglobin. For instance, they contain a large quantity of carbonic anhydrase, which catalyzes the reaction between carbon dioxide and water, increasing the rate of this reaction about 250 times. The rapidity of this reaction makes it possible for blood to react with large quantities of carbon dioxide and thereby transport it from the tissues to the lungs. Also, the hemoglobin in the cells is an excellent acid-base buffer (as is true of most proteins), so that the red blood cells are responsible for approximately 70 per cent of all the buffering power of whole blood. These specific functions are discussed elsewhere in the text and, consequently, are merely mentioned at the present time.

**The Shape and Size of Red Blood Cells.** Normal red blood cells are biconcave disks having a mean diameter of approximately 8 microns and a thickness at the thickest point of 2 microns and in the center of 1 micron or less. The average volume of the red blood cell is 87 ($\pm$5) cubic microns.

The shapes of red blood cells can change remarkably as the cells pass through capillaries. Actually, the red blood cell is a "bag" that can be deformed into almost any shape. Furthermore, because the normal cell has a great excess of cell membrane for the quantity of material inside, deformation does not stretch the membrane, and consequently does not rupture the cell as would be the case with many other cells.

**Concentration of Red Blood Cells in the Blood.** In the normal man the average number of red blood cells per cubic millimeter is 5,200,000 ($\pm$300,000) and in the normal woman 4,700,000 ($\pm$300,000). The number of red blood cells varies in the two sexes and at different ages as shown in Figure 5-1. Also, the altitude at which the person lives affects the number of red blood cells; this is discussed later.

**Quantity of Hemoglobin in the Cells.** Red blood cells have the ability to concentrate hemoglobin in the cell fluid up to approximately 34 grams per 100 ml. of cells. The concentration never rises above this value, for this is a metabolic limit of the cell's hemoglobin-forming mechanism. Furthermore, in normal persons the percentage of hemoglobin is almost always near the maximum in each cell. How-

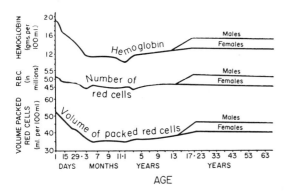

**Figure 5–1.** Relationship of age and sex to the hemoglobin content, red blood cell count, and hematocrit of the blood. (Redrawn from Wintrobe: Clinical Hematology. Lea & Febiger, 1967.)

ever, when hemoglobin formation is deficient in the bone marrow, the percentage of hemoglobin in the cells may fall to as low as 15 grams per cent or less.

When the hematocrit (the percentage of the blood that is cells—normally 40 to 45 per cent) and the quantity of hemoglobin in each respective cell are normal, the whole blood of men contains an average of 15 grams of hemoglobin per 100 ml. and of women an average of 14 grams per 100 ml. As will be discussed in connection with the transport of oxygen in Chapter 41, each gram of hemoglobin is capable of combining with approximately 1.33 ml. of oxygen. Therefore, in normal man, 20 ml. of oxygen can be carried in combination with hemoglobin in each 100 ml. of blood, and in normal woman 18 ml. of oxygen can be carried.

## PRODUCTION OF RED BLOOD CELLS

**Areas of the Body that Produce Red Blood Cells.** In the early few weeks of embryonic life, primitive red blood cells are produced in the yolk sac. During the middle trimester of gestation the liver is the main organ for production of red blood cells, and at the same time a reasonable quantity of red blood cells is also produced by the spleen and lymph nodes. Then, during the latter part of gestation and after birth, red blood cells are produced exclusively by the bone marrow.

As illustrated by Figure 5–2, the bone marrow of essentially all bones produces red blood cells until age 5, but the marrow of the long bones, except for the proximal portions of the humeri and tibiae, becomes quite fatty and produces no more red blood cells after approximately the age of 20. Beyond this age most red cells are produced in the marrow of the membranous bones, such as the vertebrae, the sternum, the ribs, and the pelvis. Even in these bones the marrow becomes less productive as age increases.

Sometimes when various factors stimulate the bone marrow to produce tremendous quantities of red blood cells, much of the marrow that has already stopped producing red blood cells can once again become productive, and marrow that is still producing red blood cells becomes greatly hyperplastic and produces far greater than normal quantities. Indeed, even the spleen and occasionally the liver as well may re-establish their hemopoietic functions long after birth when extreme stimuli persist for prolonged periods of time.

**Genesis of the Red Blood Cell.** The blood cells are derived from a cell known as the *hemocytoblast,* which is illustrated in Figure 5–3, and new hemocytoblasts are continually being formed from primordial *stem* cells located throughout the bone marrow. The hemocytoblast is also frequently called a *committed stem cell* or sometimes a *unipotential stem cell*. It, like the stem cell, can reproduce itself again and again. However, it is different from the stem cell in that it can lead only to the subsequent formation of red blood cells and not to the many other types of body cells.

As illustrated in Figure 5–3, the hemocytoblast first forms the *basophil erythroblast* which begins the synthesis of hemoglobin. The erythroblast then becomes a *polychromatophil erythroblast* so-called because of a mixture of basophilic material and the red hemoglobin. Following this, the nucleus of the cell shrinks while still greater quantities of hemoglobin are formed, and the cell becomes a *normoblast*. During the earlier stages the different cells continue to divide so that greater and greater numbers of cells are formed. Finally, after the cytoplasm of the normoblast has become filled with hemoglobin to a concentration of approximately 34 per cent, the nucleus becomes extremely small and is extruded. Therefore, the cell that is finally formed, the *erythrocyte,* almost never contains nuclear material when it passes by the process of diapedesis (squeezing through the pores of the membrane) into the blood capillaries.

Some, if not most, of the erythrocytes entering the blood contain small amounts of basophilic reticulum interspersed among the hemoglobin in the cytoplasm. This reticulum is chiefly the remains of the endoplasmic reticulum that produces the globin portion of the hemoglobin in the early cell, and hemoglobin continues to be produced as long as the reticulum persists, a period of up to two days. In this stage the cells are known as *reticulocytes*. Ordinarily, the total proportion of reticulocytes in the blood is less than 1.0 per cent.

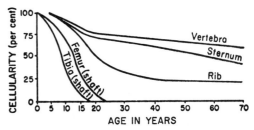

**Figure 5–2.** Relative rates of red blood cell production in the different bones at different ages. (Redrawn from Wintrobe: Clinical Hematology. Lea & Febiger, 1967.)

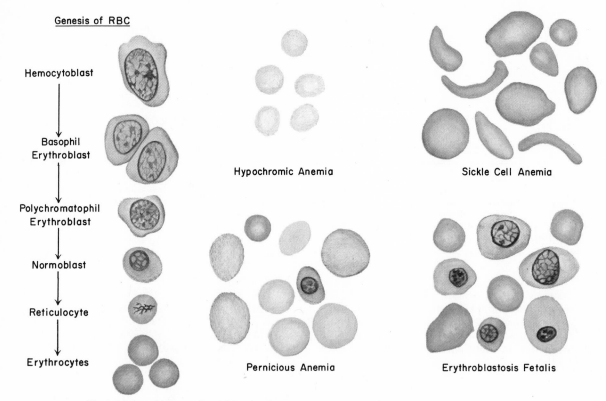

**Figure 5–3.** Genesis of red blood cells, and red blood cells in different types of anemias.

### Regulation of Red Blood Cell Production

The total mass of red blood cells in the circulatory system is regulated within very narrow limits so that an adequate number of red cells is always available to provide sufficient tissue oxygenation and, yet, so that the cells do not become so concentrated that they impede blood flow. The little we know about this control mechanism is the following:

**Tissue Oxygenation as the Basic Regulator of Red Blood Cell Production.** Any condition that causes the quantity of oxygen transported to the tissues to decrease ordinarily increases the rate of red blood cell production. Thus, when a person becomes extremely *anemic* as a result of hemorrhage or any other condition, the bone marrow immediately begins to produce large quantities of red blood cells. Also, destruction of major portions of the bone marrow by any means, especially x-ray therapy, causes hyperplasia of the remaining bone marrow, thereby supplying the demand for red blood cells in the body.

At very *high altitudes,* where the quantity of oxygen in the air is greatly decreased, insufficient oxygen is transported to the tissues, and red cells are produced so rapidly that their number in the blood is considerably increased. Therefore, it is obvious that it is not the concentration of red blood cells in the blood that controls the rate of red cell production, but instead it is the functional ability of the cells to transport oxygen to the tissues in relation to the tissue demand for oxygen.

Various diseases of the circulation that cause decreased blood flow through the peripheral vessels, and particularly those that cause failure of oxygen absorption by the blood as it passes through the lungs, also increase the rate of red cell production. This is especially apparent in prolonged *cardiac failure* and in many *lung diseases,* for the tissue hypoxia resulting from these conditions increases the rate of red cell production, with resultant increase in the hematocrit and usually some increase in the total blood volume.

**Erythropoietin, Its Response to Hypoxia, and Its Function in Regulating Red Blood Cell Production.** Erythropoietin, also called *erythropoietic stimulating factor* or *hemopoietin,* is a glycoprotein that has a molecular weight between 39,000 and 70,000 and that appears in the blood in response to hypoxia. Erythropoietin in turn acts on the bone marrow to increase the rate of red blood cell production.

There is no direct response of the bone marrow to hypoxia. Instead, hypoxia stimulates red blood cell production only through the mechanism of erythropoietin.

*Role of the Kidneys in the Formation of Erythropoietin—Renal Erythropoietic Factor.* The precise mechanism for formation of erythropoietin is not clearly understood, but it is known that only minute quantities of erythropoietin are formed in animals or persons whose kidneys have been removed. Yet, strangely enough, erythropoietin cannot be isolated from the kidneys. Therefore, at present it is believed that the relationship of the kidneys to erythropoietin formation is the following:

When the kidneys become hypoxic, it is believed that they release an enzyme called *renal erythropoietic factor.* This is secreted into the blood where it acts within a few minutes on one of the plasma proteins, a globulin, to split away the glycoprotein erythropoietin molecule. Erythropoietin, in turn, circulates in the blood for about one day and during this time acts on the bone marrow to cause erythropoiesis.

In the complete absence of the kidneys, minute amounts of erythropoietin are still formed, and the quantity that is formed is slightly increased in the presence of hypoxia. Therefore, it is clear that other tissues, probably the liver in particular, can form very slight amounts of erythropoietic factor that can lead to the formation of erythropoietin. Even so, in the absence of the kidneys a person usually becomes very anemic because of the extremely low levels of circulating erythropoietin.

*Effect of Erythropoietin on Erythrogenesis.* Though erythropoietin begins to be formed almost immediately upon placing an animal or person in an atmosphere of low oxygen, almost no new red blood cells appear in the circulating blood within the first two days; and it is only after five or more days that the maximum rate of new red cell production is reached. Thereafter, cells continue to be produced as long as the person remains in the low oxygen state or until he has produced enough red blood cells to carry adequate amounts of oxygen to his tissues despite the low oxygen.

Upon removal of a person from a state of low oxygen, his rate of oxygen transport to the tissues rises above normal, which causes his rate of erythropoietin formation to decrease to zero almost instantaneously and his rate of red blood cell production to fall essentially to zero within several days. Red cell production remains at this extremely low level until enough cells have lived out their life spans and degenerated, so that the tissues again receive only their normal complement of oxygen.

The site of action of the erythropoietin is mainly *at the stem cell and hemocytoblast level* to cause conversion of stem cells into hemocytoblasts (also called committed stem cells) and to cause proliferation of the hemocytoblasts themselves. The other stages of erythrogenesis are also accelerated, but this probably results mainly from the initial stimulatory effect of the erythropoietin at the stem cell and hemocytoblast level.

In the complete absence of erythropoietin, few red blood cells are formed by the bone marrow. At the other extreme, when extreme quantities of erythropoietin are formed, the rate of red blood cell production can rise to as high as 8 to 10 times normal. Therefore, the erythropoietin control mechanism for red blood cell production is an extremely powerful one.

*System Diagram for Control of Red Cell Numbers.* Figure 5–4 illustrates a control diagram for the feedback mechanism that regulates red blood cell concentration, drawn using the denotations described in Chapter 1. This shows that hypoxia causes production of erythropoietin. Erythropoietin then causes red blood cell production, and red blood cells finally cause a negative effect on the hypoxia—that is, they transport oxygen to the tissues, and this alleviates the hypoxia. Because of the extremely powerful control of erythropoietin over red cell production, this negative feedback mechanism can adjust the red blood cell concentration to almost precisely that amount required for adequate oxygenation of the tissues.

### Effect of Rate of Red Blood Cell Formation on the Type of Cell Released into the Blood

When the bone marrow produces red blood cells at a very rapid rate, many of the cells are released into the blood before they are mature erythrocytes. Thus,

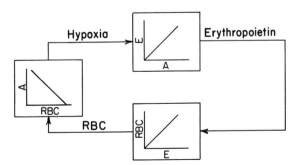

**Figure 5–4.** Control diagram, showing the negative feedback mechanism by which hypoxia regulates red blood cell concentration.

very rapid red cell production can cause the percentage of reticulocytes in the circulating blood to rise to as high as 30 to 50 per cent of the total number of red blood cells. If the rate of production is even greater, a large number of normoblasts may appear in the circulating blood. In some severe anemias, such as erythroblastosis fetalis, which is discussed later in the chapter and in greater detail in Chapter 8, the number of normoblasts may occasionally rise to as high as 5 to 20 per cent of all the circulating red blood cells, and a small number of erythroblasts and possibly even an occasional hemocytoblast may appear in the circulating blood.

### Vitamins Needed for Formation of Red Blood Cells

**The Maturation Factor—Vitamin $B_{12}$ (Cyanocobalamin).** Vitamin $B_{12}$ is an essential nutrient for all cells of the body, and growth of tissues in general is greatly depressed when this vitamin is lacking. This results from the fact that vitamin $B_{12}$ is required for conversion of ribose nucleotides into deoxyribose nucleotides, one of the essential steps in DNA formation. Therefore, lack of this vitamin causes failure of nuclear maturation and division. Since tissues that produce red blood cells are among the most rapidly growing and proliferating of all the body's tissues, lack of vitamin $B_{12}$ inhibits the rate of red blood cell production especially. Furthermore, the erythroblastic cells of the bone marrow, in addition to failing to proliferate rapidly, become larger than normal, developing into so-called *megaloblasts,* and the adult erythrocyte, called a *macrocyte,* has a flimsy membrane and is often irregular, large, and oval instead of the usual biconcave disk. These poorly formed macrocytes, after entering the circulating blood, are quite capable of carrying oxygen, but their fragility causes them to have a short life, measured in weeks instead of months as is true of normal cells. Therefore, it is said that vitamin $B_{12}$ deficiency causes *maturation failure* in the process of erythropoiesis.

The occurrence of maturation failure does not prevent normal formation of hemoglobin. Indeed, the hemoglobin *concentration* in the cell is approximately 34 grams per cent, the same as normal, and, moreover, the average *quantity* of hemoglobin in the macrocyte is considerably greater than normal because the average volume of each cell is greater than normal.

*Maturation Failure Caused by Poor Absorption of Vitamin $B_{12}$—Pernicious Anemia.* The most common cause of maturation failure is not a lack of vitamin $B_{12}$ in the diet but instead failure to absorb vitamin $B_{12}$ from the gastrointestinal tract. This often occurs in the disease called *pernicious anemia,* in which the basic abnormality is an *atrophic gastric mucosa* that fails to secrete normal gastric secretions. In the mucus secreted by the fundus and the body of the stomach is a mucopolysaccharide or mucopolypeptide (molecular weight about 50,000)

called *intrinsic factor* which combines with vitamin $B_{12}$ of the food and makes the $B_{12}$ available for absorption by the gut. It does this in the following way: First, the intrinsic factor binds tightly with the vitamin $B_{12}$. In this bound state the $B_{12}$ is protected from digestion by the gastrointestinal enzymes. Second, still in the bound state, the $B_{12}$ and intrinsic factor become adsorbed to the membranes of the mucosal cells in the ileum. Third, the combination is then transported into the cells in pinocytotic vesicles, and about four hours later free vitamin $B_{12}$ is released into the blood.

Lack of intrinsic factor, therefore, causes loss of much of the vitamin because of enzyme action in the gut, and also results in failure of absorption.

Once vitamin $B_{12}$ has been absorbed from the gastrointestinal tract it is stored in large quantities in the liver and then released slowly as needed to the bone marrow and other tissues of the body. The total amount of vitamin $B_{12}$ required each day to maintain normal red cell maturation is less than 1 microgram, and the normal store in the liver is about 1000 times this amount. Therefore, many months of defective $B_{12}$ absorption are required to cause maturation failure anemia.

*Relationship of Folic Acid (Pteroylglutamic Acid) to Vitamin $B_{12}$.* Occasionally a patient with maturation failure anemia responds equally as well to folic acid as to vitamin $B_{12}$, so that it has become apparent that this vitamin is also concerned with the maturation of red blood cells. Folic acid, like $B_{12}$, is required for formation of DNA but in a different way. It promotes the methylation of deoxyuridylate to form deoxythymidylate, one of the nucleotides required for DNA synthesis. In other ways, folic acid is also required for RNA synthesis.

## FORMATION OF HEMOGLOBIN

Synthesis of hemoglobin begins in the erythroblasts and continues through the normoblastic stage. Even when young red blood cells leave the bone marrow and pass into the blood stream, they continue to form minute quantities of hemoglobin for another day or so.

Figure 5–5 gives the basic chemical steps in the formation of hemoglobin. From tracer studies with isotopes it is known that the heme portion of hemoglobin is synthesized mainly from acetic acid and glycine and that most of this synthesis occurs in the mitochondria. It is believed that the acetic acid is changed in the Krebs cycle, which will be explained in Chapter 67, into $\alpha$-ketoglutaric acid, and then two molecules of this combine with one molecule of glycine to form a pyrrole compound. In turn, four pyrrole compounds combine to form a protoporphyrin compound. One of the protoporphyrin compounds, known as protoporphyrin III, then combines with iron to form the heme

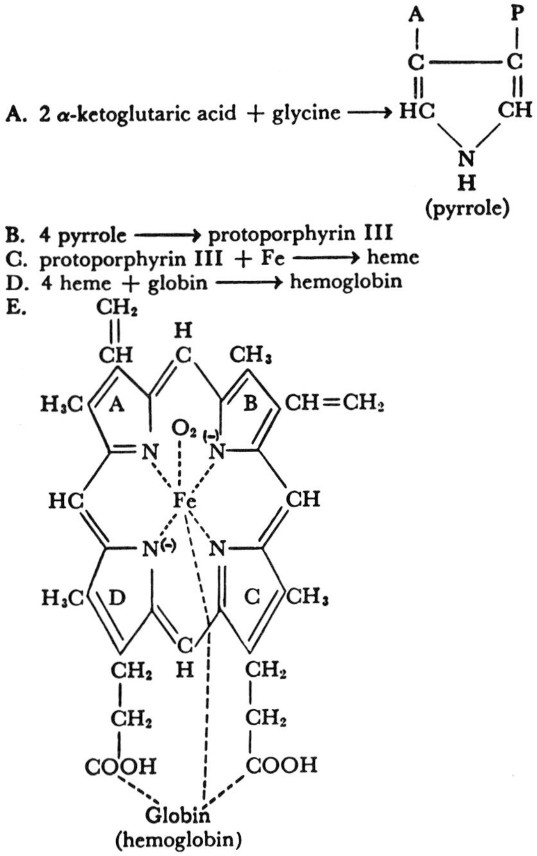

A. 2 α-ketoglutaric acid + glycine $\longrightarrow$

(pyrrole)

B. 4 pyrrole $\longrightarrow$ protoporphyrin III
C. protoporphyrin III + Fe $\longrightarrow$ heme
D. 4 heme + globin $\longrightarrow$ hemoglobin
E.

Globin
(hemoglobin)

**Figure 5–5.**  Formation of hemoglobin.

molecule. Finally, four heme molecules combine with one molecule of globin, a globulin that is synthesized in the ribosomes of the endoplasmic reticulum, to form hemoglobin, the formula for which is shown in Figure 5–5 E. Hemoglobin has a molecular weight of 68,000.

**Combination of Hemoglobin with Oxygen.** The most important feature of the hemoglobin molecule is its ability to combine loosely and reversibly with oxygen. This ability will be discussed in detail in Chapter 41 in relation to respiration, for the primary function of hemoglobin in the body depends upon its ability to combine with oxygen in the lungs and then to release this oxygen readily in the tissue capillaries where the gaseous tension of oxygen is much lower than in the lungs.

Oxygen *does not* combine with the two positive valences of the ferrous iron in the hemoglobin molecule. Instead, it binds loosely with two of the six "coordination" valences of the iron atom. This is an extremely loose bond so that the combination is easily reversible. Fur-

thermore, the oxygen does not become ionic oxygen but is carried as molecular oxygen to the tissues where, because of the loose, readily reversible combination, it is released into the tissue fluids in the form of dissolved molecular oxygen rather than ionic oxygen.

It should be recalled, also, that each molecule of hemoglobin contains four molecules of heme. Therefore, one molecule of hemoglobin contains four iron atoms and can carry four molecules of oxygen.

**Accessory Substances Needed for Formation of Hemoglobin.** In addition to amino acids and iron, which are needed directly for formation of the hemoglobin molecule, a number of other substances act as catalysts or enzymes during different stages of hemoglobin formation. For instance, an average human adult requires approximately 2 mg. of *copper* each day in his diet if normal hemoglobin formation is to take place, and addition of small quantities of copper to the diet of patients who have hypochromic anemia occasionally accelerates the rate of hemoglobin formation. This occurs even though copper is not one of the substances needed as a building stone in the formation of hemoglobin. Fortunately, the quantity of copper in the normal daily diet is sufficient so that copper deficiency anemia is almost unknown in the human being.

Lack of *pyridoxine* in the diet of some animals not only decreases the rate of red blood cell formation but also depresses the rate of hemoglobin formation to an even greater extent. Also, *cobalt* deficiency can greatly depress the formation of hemoglobin in some animals, whereas, strangely enough, a great excess of cobalt can cause formation of greater than normal numbers of red blood cells that contain normal quantities of hemoglobin. Finally, *nickel* has been found to take the place of cobalt to a moderate extent in aiding the synthesis of hemoglobin in the bone marrow.

Though the function of these different substances in the formation of hemoglobin is not known, the above listing of them serves mainly to emphasize the fact that hemoglobin formation results from a series of synthesis reactions, each of which depends upon appropriate building materials and also upon appropriate controlling catalysts and enzymes.

*IRON METABOLISM*

Because iron is important for formation of hemoglobin, myoglobin, and other substances such as the cytochromes, cytochrome oxidase, peroxidase, and catalase, it is essential to understand the means by which iron is utilized in the body.

The total quantity of iron in the body averages about 4 grams, approximately 65 per cent of which is present in the form of hemoglobin. About 4 per cent is present in the form of myoglobin, 1 per cent in the form of the various heme compounds that control intracellular oxidation, 0.1 per cent combined with the protein transferrin in the blood plasma, and 15 to 30 per cent stored mainly in the liver in the form of ferritin.

**Transport and Storage of Iron.** Transport, storage, and metabolism of iron in the body is illustrated in Figure 5–6, which may be explained as follows: When iron is absorbed from the small intestine, it immediately combines with a beta globulin, *transferrin*, with which it is transported in the blood plasma. The iron is very loosely combined with the globulin molecule and, consequently, can be released to any of the tissue cells at any point in the body. Excess iron in the blood is deposited in all cells of the body *but especially in the liver cells,* where about 60 per cent of the excess is stored. There it combines with protein, *apoferritin,* to form *ferritin.* Apoferritin has a molecular weight of approximately 460,000, and varying quantities of iron can combine in clusters of iron radicals with this large molecule; therefore, ferritin may contain only a small amount of iron or a relatively large amount. This iron stored in ferritin is called *storage iron.*

When the total quantity of iron in the body is more than the apoferritin storage pool can accommodate, some of it is stored in an extremely insoluble form called *hemosiderin.* Hemosiderin forms large clusters in the cells and consequently can be stained and observed as large particles in tissue slices by usual histologic techniques. Ferritin can also be stained, but the ferritin particles are so small and dispersed that they usually can be seen only with the electron microscope.

When the quantity of iron in the plasma falls very low, iron is removed from ferritin quite easily but less easily from hemosiderin. The iron is then transported to the portions of the body where it is needed.

When red blood cells have lived their life span and are destroyed, the hemoglobin released from the cells is ingested by the reticuloendothelial cells. There free iron is liberated, and it can then either be stored in the ferritin pool or be reused for formation of hemoglobin.

**Daily Loss of Iron.** About 0.6 mg. of iron is excreted each day by the male, mainly into the feces. Additional quantities of iron are lost whenever bleeding occurs. Thus, in the female, the menstrual loss of blood brings the average iron loss to a value of approximately 1.3 mg. per day.

Obviously, the average quantity of iron derived from the diet each day must at least equal that lost from the body.

**Absorption of Iron from the Gastrointestinal Tract.** Iron is absorbed almost entirely in the upper part of the small intestine, mainly in the duodenum. It is absorbed by an active absorptive process, though the precise mechanism of this active absorption is unknown.

Iron is absorbed mainly in the ferrous rather than ferric form, for, in general, the greater the degree of positive charge of any ion the more difficult it is to be absorbed from the intestinal tract. This has practical significance because it means that ferrous iron compounds are more effective in treating iron deficiency than are ferric compounds.

The rate of iron absorption is extremely slow, with a maximum rate of only a few milligrams per day. This means that when tremendous quantities of iron are present in the food, only

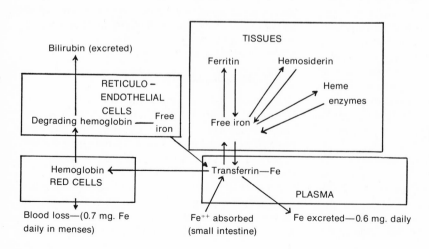

**Figure 5–6.** Iron transport and metabolism.

small proportions of this will be absorbed. On the other hand, if only minute quantities are present, far greater proportions will be absorbed.

**Regulation of Total Body Iron by Alteration of Rate of Absorption.**   When the body has become saturated with iron so that essentially all of the apoferritin in the iron storage areas is already combined with iron, the rate of absorption of iron from the intestinal tract becomes greatly decreased. On the other hand, when the iron stores have been depleted of iron, the rate of absorption becomes greatly accelerated, to as much as five or more times as great as when the iron stores are saturated. Thus, the total body iron is regulated to a great extent by alterations in rate of absorption.

*Feedback Mechanisms by which Iron Absorption is Regulated.*   When essentially all the apoferritin in the body has become saturated with iron, it becomes difficult for transferrin to release iron to the tissues. As a consequence, the transferrin, which is normally only one-third saturated with iron, now becomes almost fully bound with iron so that the transferrin accepts almost no new iron from the mucosal cells. Then, as a final stage of this process, the buildup of excess iron in the mucosal cells themselves depresses active absorption of iron from the intestinal lumen and at the same time enhances the rate of excretion of iron from the mucosa.

Though the details of the method by which excess iron depresses active absorption of iron by the mucosa are yet mostly unknown, some are beginning to emerge. When the epithelial cells of the mucosa are first formed in the crypts of the intestinal mucosa, the degree of their ability to absorb iron seems to be preordained. If there is excess iron already in the blood and tissues, small vesicles filled with ferritin, called *ferritin bodies*, are present in these newly formed epithelial cells. For reasons that are not understood, the presence of such bodies in the cells prevents or markedly reduces the active absorption of iron by the cells. The epithelial cells migrate in three to four days from the crypts outward over the villi of the intestines, at the end of which time they are desquamated into the intestinal lumen. During this entire period, their ability to absorb iron is greatly depressed.

On the other hand, if the body is deficient in iron, ferritin bodies do not appear in the newly formed epithelial cells, so that the rate of iron absorption through the intestinal mucosa becomes maximal. However, if the animal then suddenly comes into a great supply of iron, some of this combines slowly with apoferritin in the epithelial cells to form ferritin bodies, and these cells also become poor absorbers of iron. Thus, the presence of excess iron in the blood and tissues causes a *mucosal block* for the absorption of further iron.

The rate of iron excretion also increases when mucosal block occurs. The reason for this is that a considerable quantity of iron becomes stored in the epithelial cells, and these are later desquamated from the tips of the villi into the intestine and carry with them this stored iron. In addition, phagocytic cells phagocytize iron in the mucosal wall and are discharged along with the desquamating epithelial cells, thereby carrying still additional iron into the feces. Thus, the process of iron excretion also helps to regulate the concentration of iron in the body.

# DESTRUCTION OF RED BLOOD CELLS

When red blood cells are delivered from the bone marrow into the circulatory system they normally circulate an average of 120 days before being destroyed. Even though red cells do not have a nucleus, they do still have cytoplasmic enzymes for metabolizing glucose and other substances and for the utilization of oxygen, but many of these metabolic systems become progressively less active with time. As the cells become older they become progressively more fragile, presumably because their life processes simply wear out.

Once the red cell membrane becomes very fragile, it may rupture during passage through some tight spot of the circulation. Many of the red cells fragment in the spleen where the cells squeeze through the red pulp of the spleen. When the spleen is removed, the number of abnormal cells and old cells circulating in the blood increases considerably.

**Destruction of Hemoglobin.**   The hemoglobin released from the cells when they burst is phagocytized and digested almost immediately by reticuloendothelial cells, releasing iron back into the blood to be carried by transferrin either to the bone marrow for production of new red blood cells or to the liver and other tissues for storage in the form of ferritin. The heme portion of the hemoglobin molecule is converted by the reticuloendothelial cell, through a series of stages, into the bile pigment *bilirubin*, which is released into the blood and later secreted by the liver into the bile; this will be discussed in relation to liver function in Chapter 70.

# THE ANEMIAS

Anemia means a deficiency of red blood cells, which can be caused either by too rapid loss or too slow production of red blood cells. Some types of anemia and their physiological causes are the following:

**Blood Loss Anemia.**  After rapid hemorrhage the body replaces the plasma within one to three days, but this leaves a low concentration of red blood cells. If a second hemorrhage does not occur, the red blood cell concentration returns to normal within three to four weeks.

In chronic blood loss, the person frequently cannot absorb enough iron from the intestines to form hemoglobin as rapidly as it is lost. Therefore, red cells are frequently produced in too few numbers and with too little hemoglobin inside them, giving rise to *microcytic, hypochromic anemia.*

**Aplastic Anemia.**  *Bone marrow aplasia* means lack of a functioning bone marrow. For instance, the person exposed to gamma ray radiation from a nuclear bomb blast is likely to sustain complete destruction of his bone marrow, followed in a few weeks by lethal anemia. Likewise, excessive x-ray treatment, certain industrial chemicals, and even drugs to which the person might be sensitive can cause the same effect.

**Maturation Failure Anemia.**  From the earlier discussion in this chapter of vitamin $B_{12}$, folic acid, and intrinsic factor from the stomach mucosa, one can readily understand that loss of any one of these factors can lead to failure of maturation of the red blood cells. Thus, atrophy of the stomach mucosa, as occurs in *pernicious anemia,* or loss of the entire stomach as the result of total gastrectomy can lead to maturation failure. Also, patients who have intestinal sprue, in which folic acid, $B_{12}$, and other vitamin B compounds are poorly absorbed, often develop maturation failure. Because the bone marrow cannot proliferate rapidly enough to form normal numbers of red blood cells in maturation failure, the cells that are formed are oversized, of bizarre shapes, and have fragile membranes. Therefore, these cells rupture easily, leaving the person in dire need of an adequate number of red cells.

**Hemolytic Anemia.**  Many different abnormalities of the red blood cells, most of which are hereditarily acquired, make the cells very fragile, so that they rupture easily as they go through the capillaries, especially through the spleen. Therefore, even though the number of red blood cells formed is completely normal, the red cell life span is so short that serious anemia results. Some of these types of anemia are the following:

In *hereditary spherocytosis* the red cells are very small in size, and they are *spherical* in shape rather than being biconcave discs. These cells cannot be compressed because they do not have the normal loose bag-like cell membrane structure of the biconcave discs. Therefore, on passing through small capillaries they are easily ruptured by even slight compression.

In *sickle cell anemia,* which is present in about 0.3 per cent of West African and American Negroes, the cells contain an abnormal type of hemoglobin called *hemoglobin S,* caused by abnormal composition of the globin portion of the hemoglobin. When this hemoglobin is exposed to low concentrations of oxygen, it precipitates into long crystals inside the red blood cell. These crystals elongate the cell and give it the appearance of being a sickle rather than a biconcave disc. The precipitated hemoglobin also damages the cell membrane so that the cells become highly fragile, leading to serious anemia. Such patients frequently get into a vicious cycle in which low oxygen tension in the tissues causes sickling, which causes impediment of blood flow through the tissue, causing still further decrease in oxygen tension. Thus, once the process starts, it progresses rapidly, leading to a serious decrease in red blood cell mass within a few hours and, often, to death.

*Thalassemia,* which is also known as *Cooley's anemia* or *Mediterranean anemia,* is another hereditary type of hemolytic anemia in which the cells are small and have fragile membranes. Here again, the cells are easily ruptured upon passing through the tissues.

In *erythroblastosis fetalis,* Rh positive red blood cells in the fetus are attacked by antibodies from an Rh negative mother. These antibodies make the cells fragile and cause the child to be born with serious anemia. This will be discussed in detail in Chapter 8 in relation to the Rh factor of blood.

Hemolysis also occasionally results from transfusion reactions, from malaria, from reactions to certain drugs, and as an autoimmune process.

## EFFECTS OF ANEMIA ON THE CIRCULATORY SYSTEM

The viscosity of the blood, which will be discussed in detail in Chapter 18, is dependent almost entirely on the concentration of red blood cells. In severe anemia the blood viscosity may fall to as low as one and one-half times that of water rather than the normal value of approximately three times the viscosity of water. The greatly decreased viscosity decreases the resistance to blood flow in the peripheral vessels so that far greater than normal quantities of blood return to the heart. As a consequence, the cardiac output increases to as much as two or more times normal as a result of decreased viscosity.

Moreover, hypoxia due to diminished transport of oxygen by the blood causes the tissue vessels to dilate, allowing further increased return of blood to the heart, increasing the cardiac output to a higher level than ever. Thus, one of the major effects of anemia is greatly *increased work load on the heart.*

The increased cardiac output in anemia offsets many of the symptoms of anemia, for, even though each unit quantity of blood carries only small quantities of oxygen, the rate of blood flow may be increased to such an extent that almost normal quantities of oxygen are delivered to the tissues.

As long as an anemic person's rate of activity is low, he can live without fatal hypoxia of the tissues, even though his concentration of red blood cells may be reduced to one-fourth normal. However, when he begins to exercise, his heart is not capable of pumping much greater quantities of blood than it is already pumping. Consequently, during exercise, which greatly increases the tissue demands for oxygen, extreme tissue hypoxia results, and acute cardiac failure often ensues.

# POLYCYTHEMIA

**Secondary Polycythemia.** Whenever the tissues become hypoxic because of too little oxygen in the atmosphere or because of failure of delivery of oxygen to the tissues, as occurs in cardiac failure, the blood-forming organs automatically produce large quantities of red blood cells. This condition is called *secondary polycythemia,* and the red cell count commonly rises to as high as 6 to 8 million per cubic millimeter and occasionally to as high as 9 million.

A very common type of secondary polycythemia, called *physiologic polycythemia,* occurs in natives who live at altitudes of 14,000 to 17,000 feet. The blood count is generally 6 to 8 million per cubic millimeter; this is associated with an ability of these persons to perform high levels of continuous work even in the rarefied atmosphere.

**Polycythemia Vera (Erythremia).** In addition to those persons who have physiologic polycythemia, others have a condition known as *polycythemia vera* in which the red blood cell count may be as high as 8 to 9 million and the hematocrit as high as 70 to 80 per cent. Polycythemia vera is a tumorous condition of the organs that produce blood cells. It causes excess production of red blood cells in the same manner that a tumor of a breast causes excess production of a specific type of breast cell. It usually also causes excess production of white blood cells and platelets.

In polycythemia vera not only does the hematocrit increase but the total blood volume occasionally also increases to as much as twice normal. As a result, the entire vascular system becomes intensely engorged. In addition, many of the capillaries become plugged by the very viscous blood, for the viscosity of the blood in polycythemia vera sometimes increases from the normal of 3 times the viscosity of water to 15 times that of water.

## *EFFECT OF POLYCYTHEMIA ON THE CIRCULATORY SYSTEM*

Because of the greatly increased viscosity of the blood in polycythemia, the flow of blood through the vessels is extremely sluggish. In accordance with the factors that regulate the return of blood to the heart as discussed in Chapter 23, it is obvious that increasing the viscosity tends to decrease the rate of venous return to the heart. On the other hand, the blood volume is greatly increased in polycythemia, which tends to increase the venous return. Actually, the cardiac output in polycythemia is not far from normal because these two factors more or less neutralize each other.

Because the total blood volume in polycythemia is sometimes twice normal, the circulation time through the body may also be increased to twice normal. In other words, the mean circulation time occasionally is as great as 120 seconds instead of the normal of approximately 60 seconds. Thus, the velocity of blood flow in any given vessel is considerably decreased in polycythemia.

The arterial pressure is normal in most persons with polycythemia, though in approximately one third of them the pressure is elevated. This means that the blood pressure regulating mechanisms can usually offset the tendency for increased blood viscosity to increase peripheral resistance and thereby increase arterial pressure. Yet, beyond certain limits, these regulations fail.

The color of the skin is dependent to a great extent on the quantity of blood in the subpapillary venous plexus. In polycythemia vera the quantity of blood in this plexus is greatly increased. Furthermore, because the blood passes sluggishly through the skin capillaries before entering the venous plexus, a larger than normal proportion of the hemoglobin is reduced before the blood enters the plexus. The blue color of this reduced hemoglobin masks the red color of oxygenated hemoglobin. Therefore, a person with polycythemia vera ordinarily has a ruddy complexion but may at times have a bluish (cyanotic) tint to the skin. In secondary polycythemia, cyanosis is almost always evident because hypoxia is the usual cause of this type of polycythemia.

# REFERENCES

Adamson, J. W., and Finch, C. A.: Hemoglobin function, oxygen affinity, and erythropoietin. *Ann. Rev. Physiol.,* 37:351, 1975.

Atherton, R. W., and Timiras, P. S.: Erythropoietic and somatic development in chick embryos at high altitude. *Amer. J. Physiol.,* 218:75, 1970.

Boggs, D. R.: Homeostatic regulatory mechanisms of hematopoiesis. *Ann. Rev. Physiol.,* 28:39, 1966.

Callender, S. T.: The interstitial mucosa and iron absorption. *Brit. Med. Bull.,* 23:263, 1967.

Callender, S. T.: Iron absorption. *Biomembranes,* 5:761, 1974.

Castle, W. B.: Gastric intrinsic factor and vitamin $B_{12}$ absorption. *In* Handbook of Physiology. Baltimore, The Williams & Wilkins Co., 1968, Vol. III, Sec. 6, p. 1529.

Clarke, C. A.: The prevention of "rhesus" babies. *Sci. Amer.,* 219:46, 1968.

Crosby, W. H.: Iron absorption. *In* Handbook of Physiology. Baltimore, The Williams & Wilkins Co., 1968, Vol. III, Sec. 6, p. 1553.

Custer, R. P.: An Atlas of the Blood and Bone Marrow, 2nd Ed. Philadelphia, W. B. Saunders Company, 1975.

Erslev, A. J.: Renal biogenesis of erythropoietin. *Amer. J. Med.,* 58:25, 1975.

Erslev, A. J., and Gabuzda, T. G.: Pathophysiology of Blood. Philadelphia, W. B. Saunders Company, 1975.

Forth, W., and Rummel, W.: Iron absorption. *Physiol. Rev.,* 53:724, 1973.

Fried, W.: Erythropoietin. *Arch. Intern. Med., 131*:929, 1973.

Gordon, A. S., and Zanjani, E. D.: Humoral control of hemopoiesis. *Adv. Intern. Med., 18*:39, 1972.

Hardisty, R. M., and Westherall, D. J. (eds.): Blood and Its Disorders. Philadelphia, J. B. Lippincott Company, 1974.

Kass, L.: Bone Marrow Interpretation. Springfield, Ill., Charles C Thomas, Publisher, 1973.

Krantz, S. B.: Pure red-cell aplasia. *N. Engl. J. Med., 291*:345, 1974.

Linman, J. W.: Physiologic and pathophysiologic effects of anemia. *N. Engl. J. Med., 279*:819, 1968.

Matoth, Y. (ed.): Erythropoiesis: Regulatory Mechanisms and Developmental Aspects. New York, Academic Press, Inc., 1970.

Murayama, M., and Nalbandian, R. M.: Sickle Cell Hemoglobin. Boston, Little, Brown and Company, 1973.

Nalbandian, R. M. (ed.): Molecular Aspects of Sickle Cell Hemoglobin. Springfield, Ill., Charles C Thomas, Publisher, 1971.

Nathan, D. G., and Oski, F. A.: Hematology of Infancy and Childhood. Philadelphia, W. B. Saunders Company, 1975.

Platt, W. R.: Color Atlas and Textbook of Hematology. Philadelphia, J. B. Lippincott Company, 1975.

Rapaport, S. I.: Introduction to Hematology. New York, Harper & Row, Publishers, 1971.

Rodgers, G. M., Fisher, J. W., and George, W. J.: The role of renal adenosine 3′, 5′-monophosphate in the control of erythropoietin production. *Amer. J. Med., 58*:31, 1975.

Schade, S. G.: Iron metabolism. *Postgrad. Med., 55*:119, 1974.

Schleicher, E. M.: Bone Marrow Morphology and Mechanics of Biopsy. Springfield, Ill., Charles C Thomas, Publisher, 1974.

Stokstad, E. L. R., and Koch, J.: Folic acid metabolism. *Physiol. Rev., 47(1)*:83, 1967.

Surgenor, D. M.: The Red Blood Cell, 2nd Ed. New York, Academic Press, Inc., 1975, Vol. 21.

Tanaka, Y., and Goodman, J. R.: Electron Microscopy of Human Blood Cells. New York, Harper & Row, Publishers, 1972.

Wickramasinghe, S. N.: Kinetics and morphology of haemopoiesis in pernicious anaemia. *Br. J. Haematol., 22*:111, 1972.

York, J. L.: The Porphyrias. Springfield, Ill., Charles C Thomas, Publisher, 1972.

Yoshikawa, H., and Rapoport, S. M. (eds.): Cellular and Molecular Biology of Erythrocytes. Baltimore, University Park Press, 1974.

# 6

# Resistance of the Body to Infection—The Reticuloendothelial System, Leukocytes, and Inflammation

The body is constantly exposed to bacteria, these occurring especially in the mouth, the respiratory passageways, the colon, the mucous membranes of the eyes, and even the urinary tract. Many of these bacteria are capable of causing disease if they invade the deeper tissues. In addition, a person is intermittently exposed to highly virulent bacteria and viruses from outside the body which can cause specific diseases, such as pneumonia, streptococcal infections, and typhoid fever.

On the other hand, a group of tissues including the leukocytes and the *reticuloendothelial system* constantly combats any infectious agent that tries to invade the body. These tissues function in two different ways to prevent disease: (1) by actually destroying invading agents by the process of phagocytosis and (2) by forming antibodies and sensitized lymphocytes, one or both of which may destroy the invader. The present chapter is concerned with the first of these methods, while the following chapter is concerned with the second.

## THE LEUKOCYTES (WHITE BLOOD CELLS)

The leukocytes are the *mobile units* of the body's protective system. They are formed partially in the bone marrow (the *granulocytes* and

*monocytes,* and a few *lymphocytes*) and partially in the lymph tissue (*lymphocytes* and *plasma cells*), but after formation they are transported in the blood to the different parts of the body where they are to be used. The real value of the white blood cells is that most of them are specifically transported to areas of serious inflammation, thereby providing a rapid and potent defense against any infectious agent that might be present.

## GENERAL CHARACTERISTICS OF LEUKOCYTES

**The Types of White Blood Cells.** Six different types of white blood cells are normally found in the blood. These are *polymorphonuclear neutrophils, polymorphonuclear eosinophils, polymorphonuclear basophils, monocytes, lymphocytes* and *plasma cells.* In addition, there are large numbers of *platelets,* which are fragments of a seventh type of white cell found in the bone marrow, the *megakaryocyte.* The three types of polymorphonuclear cells have a granular appearance, as illustrated in Figure 6–1, for which reason they are called *granulocytes,* or in clinical terminology they are often called simply "polys."

The granulocytes and the monocytes protect the body against invading organisms by ingest-

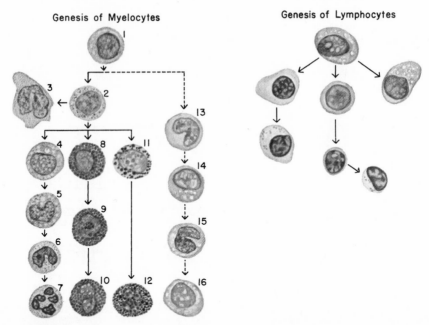

**Figure 6–1.** Genesis of the white blood cells. The different cells of the myelogenous series are: *1*, myeloblast; *2*, promyelocyte; *3*, megakaryocyte; *4*, neutrophil myelocyte; *5*, young neutrophil metamyelocyte; *6*, "band" neutrophil metamyelocyte; *7*, polymorphonuclear neutrophil; *8*, eosinophil myelocyte; *9*, eosinophil metamyelocyte; *10*, polymorphonuclear eosinophil; *11*, basophil myelocyte; *12*, polymorphonuclear basophil; *13–16*, stages of monocyte formation (Redrawn in part from Piney: A Clinical Atlas of Blood Diseases. The Blakiston Co.)

ing them—that is, by the process of *phagocytosis*. One of the functions of lymphocytes is to attach to specific invading organisms and to destroy them; this is part of the immunity system and will be discussed in the following chapter. Finally, the function of platelets is to activate the blood clotting mechanism, which will be discussed in Chapter 9. All these functions are protective mechanisms of one type or another.

**Concentrations of the Different White Blood Cells in the Blood.** The adult human being has approximately 7000 white blood cells per cubic millimeter of blood. The normal percentages of the different types of white blood cells are approximately the following:

| | |
|---|---|
| Polymorphonuclear neutrophils | 62.0% |
| Polymorphonuclear eosinophils | 2.3% |
| Polymorphonuclear basophils | 0.4% |
| Monocytes | 5.3% |
| Lymphocytes | 30.0% |

The number of platelets, which are only cell fragments, in each cubic millimeter of blood is normally about 300,000.

## GENESIS OF THE LEUKOCYTES

Figure 6–1 illustrates the stages in the development of white blood cells. The polymorphonuclear cells and monocytes are normally formed only in the bone marrow. On the other hand, lymphocytes and plasma cells are produced in the various lymphogenous organs, including the lymph glands, the spleen, the thymus, the tonsils, and various lymphoid rests in the gut and elsewhere.

Some of the white blood cells formed in the bone marrow, especially the granulocytes, are stored within the marrow until they are needed in the circulatory system. Then when the need arises, various factors that are discussed later cause them to be released.

As illustrated in Figure 6–1, megakaryocytes are also formed in the bone marrow and are part of the myelogenous group of bone marrow cells. These megakaryocytes fragment in the bone marrow, the small fragments known as *platelets* or *thrombocytes* passing then into the blood.

**Materials Needed for Formation of White Blood Cells.** In general, the white blood cells need essentially the same vitamins and amino acids as most of the other cells of the body for their formation. Especially does lack of folic acid, a compound of the vitamin B complex, block the formation of white blood cells as well as prevent maturation of red blood cells, which was pointed out in Chapter 5. Also, in extreme debilitation, production of white blood cells may be greatly reduced, despite the fact that these cells are needed more during such a state than usually.

## LIFE SPAN OF THE WHITE BLOOD CELLS

The main reason white blood cells are present in the blood is simply to be transported from the bone marrow or lymphoid tissue to the areas of the body where they are needed. Therefore, it is to be expected that the life of the white blood cells in the blood would be short.

The life span of the granulocytes in the blood averages about 12 hours, though this may be as short as only two to three hours during times of serious tissue infection.

The life span of the monocytes is difficult to assess, for monocytes wander back and forth between the tissues and the blood. The monocytes probably live for weeks or even months, expecially in the tissues, unless they are destroyed while combatting infectious or inflammatory processes.

Lymphocytes enter the circulatory system continually along with the drainage of fluid from the lymph nodes. The total number entering the blood from the thoracic duct in each 24 hours is usually several times the total number of lymphocytes present in the blood stream at any given time. Therefore, the span of time that the lymphocytes remain in the blood must be only a few hours. However, studies using radioactive lymphocytes have shown that almost all of these pass by diapedesis into the tissues, then re-enter the lymph and return to the blood again and again; thus, there is continual circulation of the lymphocytes through the tissues. And many of these cells have life spans of 100 to 300 days, or in some instances perhaps even years, but this also depends on the tissue's need for these cells.

The platelets in the blood are totally replaced approximately once every 10 days; in other words, about 30,000 platelets are formed each day for each cubic millimeter of blood.

## PROPERTIES OF WHITE BLOOD CELLS

**Diapedesis.** White blood cells can squeeze through the pores and even through the endothelial cells of the blood vessels by the process of diapedesis. That is, even though a pore is much smaller than the size of the cell, a small portion of the cell slides through the pore at a time, the portion sliding through being momentarily constricted to the size of the pore, as illustrated in Figure 6–2.

**Ameboid Motion.** Once the cells have entered the tissue spaces, the polymorphonuclear leukocytes especially, and the large lymphocytes and monocytes to a lesser degree, move through the tissues by ameboid motion, which was described in Chapter 2. Some of the cells can move through the tissues at rates as great as 40 microns per minute—that is, they can move at least three times their own length each minute.

**Chemotaxis.** A number of different chemical substances in the tissues cause the leukocytes to move either toward or away from the

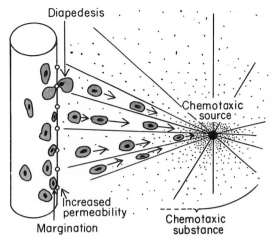

**Figure 6–2.** Movement of neutrophils by the process of *chemotaxis* toward an area of tissue damage.

source of the chemical. This phenomenon is known as chemotaxis. Degenerative products of inflamed tissues, especially tissue polysaccharides and also one of the reaction products of a complex of substances called "complement" (discussed in the following chapter) can *cause neutrophils and monocytes to move toward the area of inflammation*. In addition, a number of bacterial toxins can cause chemotaxis of leukocytes.

As illustrated in Figure 6–2, chemotaxis depends on the existence of a concentration gradient of the chemotaxic substance. The concentration is greatest near the source, and as the substance spreads by diffusion away from the source its concentration decreases approximately in proportion to the square of the distance. Therefore, on the side of the cell away from the source of the chemotaxic substance the concentration is less than on the side facing the source. The greater concentration on one side of the cell causes pseudopodia to project toward the source of the substance.

**Phagocytosis.** The most important function of the *neutrophils* and *monocytes* is phagocytosis.

Obviously, the phagocytes must be selective in the material that is phagocytized, or otherwise some of the structures of the body itself would be ingested. Whether or not phagocytosis will occur depends especially upon three selective procedures. First, if the surface of a particle is rough, the likelihood of phagocytosis is increased. Second, most natural substances of the body have electronegative surface charges and therefore repel the phagocytes, which also carry electronegative surface charges. On the other hand, dead tissues and foreign particles are frequently electropositive and are therefore subject to phagocytosis. Third, the body has a means for promoting phagocytosis of specific foreign materials by first combining them with globulin molecules called *opsonins.* After the opsonin has combined with the particle, it allows the phagocyte to adhere to the surface of the particle, which promotes phagocytosis. The special features of opsonization and its relationship to immunity is discussed in the following chapter.

*Phagocytosis by Neutrophils.* The neutrophils entering the tissues are already mature cells that can immediately begin phagocytosis. On approaching a particle to be phagocytized, the neutrophil projects pseudopodia in all directions around the particle, and the pseudopodia meet each other on the opposite side and fuse. This creates an enclosed chamber containing the phagocytized particle. Then the chamber invaginates to the inside of the cytoplasmic cavity, and the portion of the cell membrane that surrounds the phagocytized particle breaks away from the outer cell membrane to form a free-floating *phagocytic vesicle* inside the cytoplasm.

A neutrophil can usually phagocytize 5 to 20 bacteria before the neutrophil itself becomes inactivated and dies.

*Phagocytosis by Monocytes—Function of Macrophages.* When monocytes are discharged from the bone marrow into the blood they are still very immature cells. Within a few more hours the monocytes enter the tissues where marked changes begin to occur. The monocytes begin to swell, often increasing their diameters as much as five-fold. Also, extremely large numbers of lysosomes and mitochondria develop in the cytoplasm, giving the cytoplasm the appearance of a bag filled with granules. These cells are now called *macrophages,* and they represent the mature form of the monocytes. These macrophages are much more powerful phagocytes than the neutrophils, often capable of phagocytizing as many as 100 bacteria. They have the ability to engulf much larger particles and often five or more times as many particles as the neutrophils. And they can even phagocytize whole red blood cells or malarial parasites, whereas neutrophils are not capable of phagocytizing particles much larger than bacteria. Also, macrophages have much greater ability than neutrophils to phagocytize necrotic tissue, which is a very important function performed by these cells in chronic infection.

**Enzymatic Digestion of the Phagocytized Particles.** Once a foreign particle has been phagocytized, lysosomes immediately come in contact with the phagocytic vesicle, and their membranes fuse with those of the vesicle, thereby dumping the many acid hydrolase enzymes of the lysosomes into the vesicle. Thus, the phagocytic vesicle now becomes a *digestive vesicle,* and digestion of the phagocytized particle begins immediately.

Neutrophils and macrophages both have an abundance of lysosomes filled with *proteolytic enzymes* especially geared for digesting bacteria and other foreign protein matter. The lysosomes of macrophages also contain large amounts of *lipases,* which digest the thick lipid membranes possessed by tubercle bacteria, leprosy bacteria, and others.

In addition to the lysosomal enzymes that actually digest the ingested particles, the phagocytic cells also contain bactericidal agents that kill bacteria before they can multiply and destroy the phagocyte itself. The neutrophil, for instance, contains special vesicles of hydrogen peroxide which, upon entering the digestive vesicle, exerts an especially potent bactericidal effect based on its ability to oxidize the organic substances of the bacteria. Indeed, in a rare hereditary disease in which these peroxide vesicles are missing from the neutrophils, removal of some types of bacteria from the tissues is so incomplete that fulminating infection often leads to early death.

**Death of the Phagocytes as a Result of Phagocytosis.** Phagocytes continue to ingest and digest foreign particles until the toxic substances from the foreign particles and from the hydrolytic enzymes released by the lysosomes accumulate in the cytoplasm and kill the phagocytes themselves. Thus, a polymorphonuclear neutrophil is usually capable of phagocytizing about 5 to 25 bacteria before death occurs, but a macrophage sometimes engulfs as many as 100 bacteria before death.

# THE RETICULOENDOTHELIAL SYSTEM

In addition to the white blood cells, another group of cells distributed widely throughout the tissues and lining some of the blood and lymph channels also helps to protect the body against foreign invaders. This group of cells is mainly nonmotile and is collectively called the *reticuloendothelial system*. However, this term is used differently by different persons. Most frequently, the term includes two types of cells: (1) cells derived mainly from monocytes and that have enlarged to become *tissue macrophages*—these are present in the various tissues and are also adherent to the walls of blood and lymph channels; and (2) the *lymphocytic cells,* which are either wandering through the tissues or are entrapped in special lymphoid tissue, such as the lymph nodes.

Most of the functions of the lymphocytes, especially those functions related to immunity, will be discussed in the following chapter.

## THE RETICULOENDOTHELIAL CELLS DERIVED FROM MONOCYTES

Many of the monocytes, on entering the tissues, become fixed to the tissues and perform phagocytic activity in sessile positions. These have the general appearance of large macrophages except that instead of moving freely through the tissues, they are entrapped or adherent to the meshwork of the tissue. Some of these types of cell are the following:

**Tissue Macrophages (Histiocytes).** Many monocytes that wander into the tissues become fixed in the tissues and then swell to become fixed *tissue macrophages,* also called *histiocytes.* During the course of inflammation, these histiocytes can divide in situ and form more histiocytes. Frequently, they proliferate and form giant cell capsules around foreign particles that cannot be digested, such as particles of silica dust, carbon, and so forth, thus effectively isolating these particles from the remaining tissue. This "walling off" process also frequently occurs in response to certain chronic infections—tuberculosis, for instance—and therefore is an important mechanism for preventing spread of disease.

**The Macrophages of the Lymph Nodes.** Essentially no particulate matter that enters the tissues can be absorbed directly through the capillary membranes into the blood. Instead, if the particles are not destroyed locally in the tissues, they enter the lymph and flow through the lymphatic vessels to the lymph nodes located intermittently along the course of the lymphatics. The foreign particles are trapped there in a meshwork of sinuses that are lined by tissue macrophages.

Figure 6–3 illustrates the general organization of the lymph node, showing lymph entering by

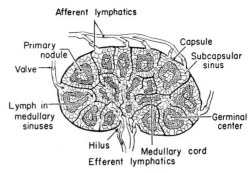

**Figure 6–3.** Functional diagram of a lymph node. (Redrawn from Ham: Histology. J. B. Lippincott Co., 1971.)

way of the *afferent lymphatics,* flowing through the *medullary sinuses,* and finally passing out of the *hilus* into the *efferent lymphatics.* Large numbers of tissue macrophages line the sinuses, and if any particles enter the sinuses, these cells phagocytize them and prevent general dissemination throughout the body.

**The Alveolar Macrophages.** Another route by which invading organisms frequently enter the body is through the respiratory system. Fortunately, large numbers of tissue macrophages are present as integral components of the alveolar walls. These can phagocytize particles that become entrapped in the alveoli. If the particles are digestible, the macrophages can also digest them and release the digestive products into the lymph. If the particle is not digestible the macrophages "wall off" the particles until such time, if ever, that they can be slowly dissoluted.

**The Tissue Macrophages (Kupffer Cells) in the Liver Sinuses.** Still another favorite route by which bacteria invade the body is through the gastrointestinal tract. Large numbers of bacteria constantly pass through the gastrointestinal mucosa into the portal blood. However, before this blood enters the general circulation, it must pass through the sinuses of the liver; these sinuses are lined with tissue macrophages called *Kupffer cells,* illustrated in Figure 6–4. These cells form such an effective particulate filtration system that almost none of the bacteria from the gastrointestinal tract succeeds in passing from the portal blood into the general systemic circulation. Indeed, motion pictures of phagocytosis by Kupffer cells have demonstrated phagocytosis of single bacteria in less than 1/100 second.

**The Macrophages of the Spleen and Bone Marrow.** If an invading organism does

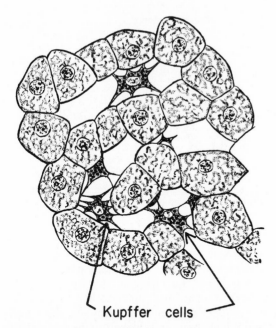

**Figure 6–4.** Kupffer cells lining the liver sinusoids, showing phagocytosis of India ink particles. (Redrawn from Copenhaver and Johnson: Bailey's Textbook of Histology. The Williams and Wilkins Co., 1969.)

succeed in entering the general circulation, there still remain other lines of defense by the reticuloendothelial system, especially by reticuloendothelial cells of the spleen and bone marrow. In both of these tissues, macrophages have become entrapped by the reticular meshworks of the two organs, and when foreign particles come in contact with them the particles are phagocytized.

The spleen is similar to the lymph nodes, except that blood, instead of lymph, flows through the substance of the spleen. Figure 6–5 illustrates the general structure of the spleen, showing a small peripheral segment of the spleen. Note that a small artery penetrates from the splenic capsule into the *splenic pulp,* and terminates in small capillaries. The capillaries are highly porous, allowing large numbers of whole blood cells to pass out of the capillaries into the *cords of the red pulp.* These cells then gradually *squeeze* through the tissue substance of the cords and eventually return to the circulation through the endothelial walls of the *venous sinuses.* The cords of the red pulp are loaded with macrophages, and in addition the venous sinuses are also lined with macrophages. This peculiar passage of blood through the cords of the red pulp provides an exceptional means for phagocytosis of unwanted debris in the blood,

especially old and abnormal red blood cells. The spleen is also an important organ for phagocytic removal of abnormal platelets, blood parasites, and any bacteria that might succeed in entering the general circulating blood.

In a similar way, macrophages of the bone marrow also help to remove unwanted debris and pathologic agents from the blood.

# INFLAMMATION AND THE FUNCTION OF LEUKOCYTES

## *THE PROCESS OF INFLAMMATION*

Inflammation is a complex of sequential changes in the tissues in response to injury. When tissue injury occurs, whether it be caused by bacteria, trauma, chemicals, heat, or any other phenomenon, the substance *histamine,* along with other humoral substances, is liberated by the damaged tissue into the surrounding fluids. This increases the local blood flow and also increases the permeability of the capillaries, allowing large quantities of fluid and protein, including fibrinogen, to leak into the tissues. Local extracellular edema results, and the extracellular fluid and lymphatic fluid both clot because of the coagulating effect of tissue exudates on the leaking fibrinogen. Thus, *brawny edema* develops in the spaces surrounding the injured cells.

**The "Walling Off" Effect of Inflammation.** It is clear that one of the first results of inflammation is to "wall off" the area of injury from the remaining tissues. The tissue spaces and the lymphatics in the inflamed area are blocked by fibrinogen clots so that fluid barely flows through the spaces. Therefore, walling off

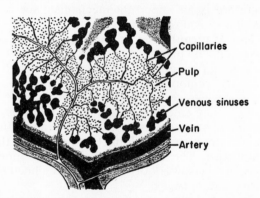

**Figure 6–5.** Functional structures of the spleen. (From Bloom and Fawcett: Textbook of Histology, 1968.)

the area of injury delays the spread of bacteria or toxic products.

The intensity of the inflammatory process is usually proportional to the degree of tissue injury. For instance, staphylococci invading the tissues liberate extremely lethal cellular toxins. As a result, the process of inflammation develops rapidly—indeed, much more rapidly than the staphylococci themselves can multiply and spread. Therefore, staphylococcal infection is characteristically ''walled off '' rapidly. On the other hand, streptococci do not cause such intense local tissue destruction. Therefore the walling off process develops slowly while the streptococci reproduce and migrate. As a result, streptococci have far greater tendency to spread through the body and cause death than do staphylococci, even though staphylococci are far more destructive to the tissues.

**Attraction of Neutrophils to the Area of Inflammation.** When tissues are damaged, several effects occur to cause movement of neutrophils into the damaged area. First, the neutrophils stick to the walls of the damaged capillary, causing the process known as ''margination,'' which is shown in Figure 6–2. Gradually, the cells pass by diapedesis into the tissue spaces.

The second effect is chemotaxis of the neutrophils toward the damaged area; this is caused by bacterial or cellular products that attract the neutrophils. Thus, within a few hours after tissue damage begins, the area becomes well supplied with neutrophils.

**Neutrophilia During Inflammation— Leukocytosis-Promoting Factor.** The term *neutrophilia* means an increase above normal in the number of neutrophils in the circulatory system, and the term *leukocytosis* means an excess total number of white blood cells.

A substance known as *leukocytosis-promoting factor* is believed to be liberated by inflamed tissues. This factor supposedly diffuses into the blood and finally to the bone marrow where it has two actions: First, it causes large numbers of granulocytes, especially neutrophils, to be released within a few minutes to a few hours into the blood from the storage areas of the bone marrow, thus increasing the total number of neutrophils per cubic millimeter of blood sometimes to as high as 20,000 to 30,000. Second, the rate of granulocyte production by the bone marrow increases either as a direct result of the factor or as an indirect result of the bone marrow release of the granulocytes. Within a day or two after onset of the inflamma-

tion, the bone marrow becomes hyperplastic and then continues to produce large numbers of granulocytes as long as leukocyte promoting factor is formed in the inflamed tissues. Yet it is only the stored white blood cells that provide the granulocytes for the first few days of inflammation until the bone marrow has had time to become hyperplastic. The quantity of stored neutrophils in the bone marrow fortunately is 30 to 40 times the quantity of these cells normally circulating in the blood; this obviously means that there is a large reserve available that can be called into service within hours.

## THE MACROPHAGE RESPONSE IN CHRONIC INFLAMMATION

The monocytic cells—including the tissue histiocytes and the blood monocytes—also play a major role in protecting the body against infection. First, the tissue histiocytes develop ameboid motion and migrate chemotaxically toward the area of inflammation. These cells provide the first line of defense against infection within the first hour or so, but their numbers are not very great. Within the next few hours, the neutrophils become the primary defense, reaching their maximum effectiveness in about 6 to 12 hours. By that time large numbers of monocytes have begun to enter the tissues from the blood. They change their characteristics drastically during the first few hours. They start to swell, to form greatly increased quantities of cytoplasmic lysosomes, to exhibit increased ameboid motion, and to move chemotaxically toward the damaged tissues.

The macrophages are several times as phagocytic as the neutrophils. Furthermore, they are large enough that they can engulf large quantities of necrotic tissue, including dead neutrophils themselves. Another reason the macrophagic response is of importance in chronic infection is that the affected area usually becomes acidic, and neutrophils cannot live in an acid environment, whereas macrophages live well under acid conditions, and their proteolytic enzymes actually become more active than ever. Consequently, after the initial stages of inflammation are over, the neutrophils are no longer nearly as effective phagocytes as the macrophages, and the shift from neutrophils to macrophages appears to be a purposeful reaction.

**Formation of Pus.** When the neutrophils and macrophages engulf large amounts of bacteria and necrotic tissue they themselves eventually die. After several days, a cavity is often excavated in the inflamed tissues containing varying portions of necrotic tissue, dead neutrophils, and dead macrophages. Such a mixture is commonly known as *pus*.

Ordinarily, pus formation continues until all infection is suppressed. Sometimes the pus cavity eats its way to the surface of the body or into an internal cavity and in this way empties itself. At other times the pus cavity remains enclosed even after tissue destruction has ceased. When this happens the dead cells and necrotic tissue in the pus gradually autolyze over a period of days, and the end-products of autolysis are usually absorbed into the surrounding tissues until most of the evidence of tissue damage is gone.

### NEUTROPHILIA CAUSED BY OTHER CONDITIONS BESIDES INFLAMMATION

Almost any factor that causes some degree of tissue destruction will cause neutrophilia. For instance, persons debilitated by cancer exhibit an increase in neutrophils from the normal of 4500 per cubic millimeter sometimes up to 15,000 or more. Even extreme fatigue can cause neutrophilia. Acute hemorrhage, poisoning, operative procedures, very slight hemorrhage into the peritoneal cavity, and injection of foreign protein into the body all cause considerable increase in the number of neutrophils in the circulatory system. In summary, neutrophilia results from almost any tissue-damaging process in the body, whether or not this process is associated with inflammation.

For instance, neutrophilia is one of the special diagnostic features of coronary thrombosis. Presumably, when the coronary vessel becomes blocked the ischemic musculature of the heart begins to necrose, and degenerative substances liberated into the blood promote the release of neutrophils from the bone marrow.

**Physiological Neutrophilia.** The number of neutrophils in the circulatory system can increase to as much as 2 times normal after a single minute of extremely hard exercise or after injection of norepinephrine. This can be explained as follows: When blood flow is sluggish through the tissues, large numbers of white blood cells, especially neutrophils, adhere to the walls of the capillaries—a process called *margination*—and, therefore, are sequestered from the usual circulation. Hard exercise or stimulation of the circulation by norepinephrine, with rapid flow of blood through essentially all capillaries, can mobilize the leukocytes.

Approximately one hour after physiological neutrophilia has resulted from exercise or any other stimulus, the number of leukocytes in the blood is usually back to normal because most of the leukocytes will again be sequestered in the capillaries.

### THE EOSINOPHILS

The eosinophils normally comprise 1 to 3 per cent of all the leukocytes. Eosinophils are weak phagocytes, and they exhibit chemotaxis, but in comparison with the neutrophils, it is doubtful that the eosinophils are of significant importance in protection against usual types of infection.

Eosinophils enter the blood in large numbers after foreign protein injection. Furthermore, many eosinophils are present in the mucosa of the intestinal tract and in the tissues of the lungs, where foreign proteins normally enter the body. It has been suggested that the function of these is to detoxify the proteins before they can cause damage to the body.

Eosinophils also migrate into blood clots where they probably release the substance *profibrinolysin*. This substance then becomes activated to form *fibrinolysin*, which is an enzyme that digests fibrin, a subject that will be discussed in Chapter 9. Therefore, eosinophils are possibly important for dissolution of old clots.

The total number of eosinophils increases greatly in the circulating blood during allergic reactions, and eosinophils collect at sites of antigen-antibody reactions in the tissues. It is possible that the allergic reactions release products from the tissues that in turn cause the number of eosinophils in the tissues and in the circulating blood to increase. It has also been suggested that the eosinophils remove and digest the antigen-antibody complex after the immune process has performed its functions.

Probably the most common cause of extremely large numbers of eosinophils in the blood is infection with parasites. In the condition known as *trichinosis,* which results from invasion of the muscles by the *Trichinella* parasite ("pork worm") after eating uncooked pork, the percentage of eosinophils in the circulating blood may rise to as high as 25 to 50 per cent of all the leukocytes. The means by which parasitic infections cause an increase in eosinophils also is not known, though if the function of eosinophils is to detoxify proteins, it is possible that parasites continually elaborate protein substances that must be detoxified.

### THE BASOPHILS

The basophils in the circulating blood are very similar to the large *mast* cells located immediately outside many of the capillaries in the body. These cells liberate *heparin* into the blood, a substance that can prevent blood coagulation and that can also speed the removal of fat particles from the blood after a fatty meal. Therefore, it is probable that the basophils in the circulating blood perform similar functions within the blood stream, or it is even possible that the blood simply transports basophils to tissues where they then become mast cells and perform the function of heparin liberation.

The total number of basophils in the blood is ordinarily very small, ranging about four such cells for every 1000 leukocytes. This number increases during the healing phase of inflammation, and it also increases a small amount during prolonged chronic inflammation. It is well-known that prolonged inflammation causes a tendency for red blood cells to adhere to each other. Therefore, it is possible that this increase in basophils in the blood and in the tissues might be a means, by releasing heparin that blocks the blood coagulative process, to combat this tendency for cells to adhere.

### THE LYMPHOCYTES

Until recently it was believed that the lymphocytes represent a homogeneous group of cells, with a possible distinction between the so-called small lymphocytes and large lymphocytes. However, it is now known that the cell known as the lymphocyte comprises a number of different types of cells, all of which have essentially the same staining characteristics. A *few* of these cells, probably produced primarily in the bone marrow, seem to be multipotential cells that are similar to the stem cell from which almost any other type of cell can be formed. These multipotential cells can possibly be changed under appropriate conditions into erythroblasts, myeloblasts, fibroblasts, and so forth.

However, the great bulk of the lymphocytes play special roles in the process of immunity. These will be discussed in detail in the following chapter.

# AGRANULOCYTOSIS

A clinical condition known as "agranulocytosis" occasionally occurs, in which the bone marrow stops producing neutrophils, leaving the body unprotected against bacteria and other agents that might invade the tissues.

Actually, the human body lives in symbiosis with many bacteria, for all the mucous membranes of the body are constantly exposed to large numbers of bacteria. The mouth almost always contains various spirochetal, fusiform, pneumococcal, and streptococcal bacteria, and these same bacteria are present to a lesser extent in the entire respiratory tract. The gastrointestinal tract is especially loaded with colon bacilli. Furthermore, one can almost always find bacteria in the eyes, the urethra, and the vagina. Therefore, any decrease in the number of neutrophils immediately allows invasion of the tissues by the bacteria that are already present in the body. Within two days after the bone marrow stops producing

white blood cells, ulcers may appear in the mouth and colon, or the person develops some form of severe respiratory infection. Bacteria from the ulcers then rapidly invade the surrounding tissues and the blood. Without treatment, death usually ensues three to six days after acute agranulocytosis begins.

Irradiation of the body by gamma rays caused by a nuclear explosion, or drugs and chemicals containing the benzene or anthracene nuclei is quite likely to cause aplasia of the bone marrow. Indeed, some of the common drugs, such as the sulfonamides, chloramphenicol, thiouracil (used to treat thyrotoxicosis), and even the various barbiturate hypnotics, on occasion cause agranulocytosis (or *bone marrow aplasia* in which no cells of any type— red cells included—are produced in the bone marrow), thus setting off the entire infective sequence of this malady.

After irradiation injury to the bone marrow, a large number of stem cells, myeloblasts, and hemocytoblasts usually remain undestroyed and are capable of regenerating the bone marrow, provided sufficient time is available. Therefore, the patient properly treated with antibiotics and other drugs to ward off infection will usually develop enough new bone marrow within several weeks to several months that his blood cell concentrations can return to normal.

# THE LEUKEMIAS

Uncontrolled production of white blood cells is caused by cancerous mutation of a myelogenous or a lymphogenous cell. It causes leukemia, which is usually characterized by greatly increased numbers of abnormal white blood cells in the circulating blood.

**Types of Leukemia.** Ordinarily, leukemias are divided into two general types: the *lymphogenous leukemias* and the *myelogenous leukemias*. The lymphogenous leukemias are caused by cancerous production of lymphoid cells, beginning first in a lymph node or other lymphogenous tissue and then spreading to other areas of the body. The second type of leukemia, myelogenous leukemia, begins by cancerous production of young myelogenous cells in the bone marrow and then spreads throughout the body so that white blood cells are produced in many extramedullary organs.

In myelogenous leukemia, the cancerous process occasionally produces reasonably differentiated cells, resulting in *neutrophilic leukemia, eosinophilic leukemia, basophilic leukemia,* or *monocytic leukemia.* More frequently, however, the leukemia cells are bizarre, and undifferentiated, and not identical with any of the normal white blood cells.

Leukemic cells are usually nonfunctional, so that they cannot provide the usual protection associated with white blood cells.

### EFFECTS OF LEUKEMIA ON THE BODY

The first effect of leukemia is metastatic growth of leukemic cells in abnormal areas of the body. The leukemic cells of the bone marrow may reproduce so greatly that they invade the surrounding bone, causing pain and eventually a tendency to easy fracture. Almost all leukemias spread to the spleen, the lymph nodes, the liver, and other especially vascular regions, regardless of whether the origin of the leukemia is in the bone marrow or in the lymph nodes. In each of these areas the rapidly growing cells invade the surrounding tissues, utilizing the metabolic elements of these tissues and consequently causing tissue destruction.

Very common effects in leukemia are the development of infections, severe anemia, and bleeding tendency caused by thrombocytopenia (lack of platelets). These effects result mainly from displacement of the normal bone marrow by the leukemic cells.

Finally, perhaps the most important effect of leukemia on the body is the excessive use of metabolic substrates by the growing cancerous cells. The leukemic tissues reproduce new cells so rapidly that tremendous demands are made on the body fluids for foodstuffs, especially the amino acids and vitamins. Consequently, the energy of the patient is greatly depleted, and the excessive utilization of amino acids causes rapid deterioration of the normal protein tissues of the body. Thus, while the leukemic tissues grow, the other tissues are debilitated. Obviously, after metabolic starvation has continued long enough, this alone is sufficient to cause death.

# REFERENCES

Archer, R. K.: The Eosinophil Leucocyte. Philadelphia, F. A. Davis Co., 1963.

Bellanti, J. A., and Dayton, D. H. (eds.): The Phagocytic Cell in Host Resistance. New York, Raven Press, 1975.

Carr, I.: The Macrophage. New York, Academic Press, Inc., 1973.

Durant, J. R., and Smalley, R. V.: The Chronic Leukemias. Springfield, Ill., Charles C Thomas, Publisher, 1972.

Elsbach, P.: On the interaction between phagocytes and microorganisms. N. Engl. J. Med., 289:846, 1973.

Golde, D. W., and Cline, M. J.: Regulation of granulopoiesis. N. Engl. J. Med., 291:1388, 1974.

Heller, J. H.: Host defence and the reticulo-endothelial system. In Bittar, E. Edward, and Bittar, Neville (eds.): The Biological Basis

of Medicine. New York, Academic Press, Inc., 1969, Vol. 4, p. 181.

Karnovsky, M. L.: Metabolic basis of phagocytic activity. Physiol. Rev., 42:143, 1962.

Kass, L., and Schnitzer, B.: Monocytes, Monocytosis and Monocytic Leukemia. Springfield, Ill., Charles C Thomas, Publisher, 1973.

Kleinschmidt, W. J.: Biochemistry of interferon and its inducers. Ann. Rev. Biochem., 41:517, 1972.

Lerner, R. A., and Dixon, F. J.: The human lymphocyte as an experimental animal. Sci. Amer., 228:82, 1973.

Metcalf, D.: Regulation of granulocyte and monocyte-macrophage proliferation by colony stimulating factor (CSF): A review. Exp. Hematol., 1:185, 1973.

Molander, D. W. (ed.): Lymphoproliferative Diseases. Springfield, Ill., Charles C Thomas, Publisher, 1974.

Nathan, D. G., and Baehner, R. L.: Disorders of phagocytic cell function. Prog. Hematol., 7:235, 1971.

Platt, W. R.: Color Atlas and Textbook of Hematology. Philadelphia, J. B. Lippincott Company, 1975.

Raab, S. O.: The spleen and reticuloendothelial system. In Sodeman, W. A., Jr., and Sodeman, W. A. (eds.): Pathologic Physiology: Mechanisms of Disease, 5th Ed. Philadelphia, W. B. Saunders Company, 1974, p. 665.

Ross, R.: Wound healing. Sci. Amer., 220:40, 1969.

Saba, T. M.: Physiology and pathophysiology of the reticuloendothelial system. Arch. Int. Med., 129:1031, 1970.

Speirs, R. S.: How cells attack antigens. Sci. Amer., 210:58, 1964.

Stossel, T. P.: Phagocytosis. N. Engl. J. Med., 290:833, 1974.

Stuart, A. E.: The Reticulo-Endothelial System. New York, Churchill Livingstone, Div. of Longman Inc., 1970.

Van Arman, C. G. (ed.): White Cells in Inflammation. Springfield, Ill., Charles C Thomas, Publisher, 1974.

Van Furth, R. (ed.): Mononuclear Phagocytes. Philadelphia, J. B. Lippincott Company, 1971.

Wagner, W. H., Hahn, H., and Evans, R. (eds.): Activation of Macrophages (Proceedings of the 2nd Workshop Conference, Hoechst, Schloss Reisenburg, 1973). New York, American Elsevier Publishing Co., 1974.

Ward, P. A.: Leukotaxis and leukotactic disorders. A review. Amer. J. Pathol., 77:520, 1974.

Weinstein, L., and Swartz, M. N.: Host responses to infection. In Sodeman, W. A., Jr., and Sodeman, W. A. (eds.): Pathologic Physiology: Mechanisms of Disease, 5th Ed. Philadelphia, W. B. Saunders Company, 1974, p. 473.

Weinstein, L., and Swartz, M. N.: Pathogenic properties of invading microorganisms. In Sodeman, W. A., Jr., and Sodeman, W. A. (eds.): Pathologic Physiology: Mechanisms of Disease, 5th Ed. Philadelphia, W. B. Saunders Company, 1974, p. 457.

Wilkinson, P. C.: Chemotaxis and Inflammation. New York, Churchill Livingstone, Div. of Longman Inc., 1973.

Winkelstein, J. A., and Drachman, R. H.: Phagocytosis: The normal process and its clinically significant abnormalities. Pediatr. Clin. North Amer., 21:551, 1974.

Zweifach, B. W., Grant, L., and McCluskey, R. T. (eds.): The Inflammatory Process, 2nd Ed. New York, Academic Press, Inc., 1974.

# | 7 |

# Immunity and Allergy

## INNATE IMMUNITY

The human body has the ability to resist almost all types of organisms or toxins that tend to damage the tissues and organs. This capacity is called *immunity*. Much of the immunity is caused by a special immunity system that forms antibodies and sensitized lymphocytes that attack and destroy the specific organisms or toxins. This type of immunity is *acquired immunity*. However, an additional portion of the immunity results from general processes rather than from processes directed at specific disease organisms. This is called *innate immunity*. It includes the following:

1. Phagocytosis of bacteria and other invaders by white blood cells and reticuloendothelial cells, as described in the previous chapter.

2. Destruction by the acid secretions of the stomach and by the digestive enzymes of organisms swallowed into the stomach.

3. Resistance of the skin to invasion by organisms.

4. Presence in the blood of certain chemical compounds that attach to foreign organisms or toxins and destroy them. Some of these are (a) *lysozyme,* a mucolytic polysaccharide that attacks bacteria and causes them to dissolute; (b) *basic polypeptides,* which react with and inactivate certain types of gram-positive bacteria; (c) *properdin,* a very large protein that can react directly with gram-negative bacteria and destroy them; and (d) naturally occurring antibodies in the blood that have the specific ability to destroy certain bacteria, viruses, or toxins (these antibodies are similar to those that will be described later in the chapter as part of the acquired immunity system, but they occur without the necessity of previous exposure to the invading agent).

This innate immunity makes the human body partially or completely resistant to such diseases as dysentery, some paralytic virus diseases of animals, hog cholera, cattle plague, and distemper, a viral disease that kills a large percentage of dogs that become afflicted with it. On the other hand, animals are resistant to many diseases, such as poliomyelitis, mumps, human cholera, measles, and syphilis, which are very destructive or even lethal to the human being.

## ACQUIRED IMMUNITY (OR ADAPTIVE IMMUNITY)

In addition to its innate immunity, the human body also has the ability to develop extremely powerful specific immunity against individual invading agents such as lethal bacteria, viruses, toxins, and even foreign tissues from other animals. This is called *acquired immunity* or *adaptive immunity*. It is with this immune mechanism and with a very closely allied sequela of the immune mechanism—allergy—that most of this chapter will be concerned.

This system of acquired immunity is important as a protection against invading organisms to which the body does not have an innate or natural immunity. The body does not block the invasion upon first exposure to the invader. However, within a few days to a few weeks after exposure the special immune system develops extremely powerful resistance to the invader. Furthermore, the resistance is highly specific for that particular invader and not for others. It is for this reason that this immunity is called "adaptive immunity" in addition to its more commonly used name "acquired immunity."

Acquired immunity can often bestow extreme protection. For instance, certain toxins such as the paralytic toxin of botulinum or the tetanizing toxin of tetanus can be protected against in doses as high as 100,000 times the amount that would be lethal without immunity. This is the reason the process known as "vaccination" is so extremely important in protecting human beings against disease and against toxins, as will be explained in the course of this chapter.

## TWO BASIC TYPES OF ACQUIRED IMMUNITY

Two basic, but closely allied, types of acquired immunity occur in the body. In one of these the body develops circulating *antibodies,* which are globulin molecules that are capable of attacking the invading agent. This type of immunity is called *humoral immunity*. The second type of acquired immunity is achieved through the formation of large numbers of highly specialized lymphocytes that are specifically sensitized against the foreign agent. These *sensitized lymphocytes* have the special capability to attach to the foreign agent and to destroy it. This type of immunity is called *cellular immunity* or, sometimes, *lymphocytic immunity*.

We shall see shortly that both the antibodies and the sensitized lymphocytes are formed in the lymphoid tissue of the body. First, let us discuss the initiation of the immune process by *antigens*.

## ANTIGENS

Since acquired immunity does not occur until after first invasion by a foreign organism or toxin, it is clear that the body must have some mechanism for recognizing the initial invasion. Each toxin or each type of organism contains one or more specific chemical compounds in its makeup that are different from all other compounds. In general, these are proteins, large polysaccharides, or large lipoprotein complexes, and it is they that cause the acquired immunity. These substances are called *antigens*.

Essentially all toxins secreted by bacteria are also proteins, large polysaccharides, or mucopolysaccharides, and they are highly antigenic. Also, the bodies of bacteria or viruses usually contain several antigenic chemical compounds. Likewise, tissues such as a transplanted heart  from other human beings or animals also contain numerous antigens that can elicit the immune process and cause subsequent destruction.

For a substance to be antigenic it usually must have a high molecular weight, 8,000 or greater. Furthermore, the process of antigenicity probably depends upon regularly recurring prosthetic radicals on the surface of the large molecule, which perhaps explains why proteins and polysaccharides are almost always antigenic, for they both have this type of stereochemical characteristic.

*Haptens.* Though substances with molecular weights less than 8,000 can only rarely act as antigens, immunity can nevertheless be developed against substances of low molecular weight in a very special way, as follows: If the low molecular weight compound, which is called a *hapten,* first combines with a substance that *is* antigenic, such as a protein, then the combination will elicit an immune response. The antibodies or sensitized lymphocytes that develop against the combination can then react either against the protein or against the hapten. Therefore, on second exposure to the hapten, the antibodies or lymphocytes react with it before it can spread through the body and cause damage.

The haptens that elicit immune responses of this type are usually drugs, chemical constituents in dust, breakdown products of dandruff from animals, degenerative products of scaling skin, various industrial chemicals, and so forth.

## ROLE OF LYMPHOID TISSUE IN ACQUIRED IMMUNITY

Acquired immunity is the product of the body's lymphoid tissue. In persons who have a genetic lack of lymphoid tissue or whose lymphoid tissue has been destroyed by radiation or by chemicals, no acquired immunity whatsoever can develop. And almost immediately after birth such a person dies of fulminating infection unless treated by heroic measures. Therefore, it is clear that the lymphoid tissue is essential to survival of the human being.

The lymphoid tissue is located most extensively in the lymph nodes, but it is also found in special lymphoid tissues such as that of the spleen, submucosal areas of the gastrointestinal tract, and, to a slight extent, in the bone marrow. The lymphoid tissue is distributed very advantageously in the body to intercept the invading organisms or toxins before they can spread too widely. For instance, the lymphoid tissue of the gastrointestinal tract is exposed immediately to antigens invading through the gut. The lymphoid tissue of the throat and pharynx (the tonsils and adenoids) is extremely well located to intercept antigens that enter by way of the upper respiratory tract. The lymphoid tissue in the lymph nodes is exposed to antigens that invade the peripheral tissues of the body. And, finally, the lymphoid tissue of the spleen and bone marrow plays the specific role of intercepting antigenic agents that have succeeded in reaching the circulating blood.

**Two Types of Lymphocytes that Promote, Respectively, Cellular Immunity and Humoral Immunity.** Though most of the lymphocytes in normal lymphoid tissue look alike when studied under the microscope, these cells are distinctly divided into two separate populations. One of the populations is responsible for forming the sensitized lymphocytes that provide cellular immunity and the other for forming the antibodies that provide humoral immunity.

Both of these types of lymphocytes are derived originally in the embryo from *lymphocytic stem cells in the bone marrow*. Then descendents of the stem cells eventually migrate to the lymphoid tissue. Before doing so, however, those lymphocytes that are eventually destined to form sensitized lymphocytes first migrate to and are preprocessed in the *thymus* gland, for which reason they are called ''T'' lymphocytes. These are responsible for cellular immunity.

The other population of lymphocytes—those that are destined to form antibodies—are processed in some unknown area of the body, possibly the liver and spleen. However, this population of cells was first discovered in birds in which the preprocessing occurs in the *bursa of Fabricius,* a structure not found in mammals. For this reason this population of lymphocytes is called the ''B'' lymphocytes, and they are responsible for humoral immunity.

To further emphasize the two separate populations of lymphocytes, they also tend to localize in separate parts of the lymphoid tissue. For instance, in the lymph nodes, the lymphocytes of the "B" system are mainly located in the cortical and germinal areas, whereas the "T" cells are located in the paracortical areas.

Figure 7–1 illustrates the two separate lymphocyte systems for the formation, respectively, of the sensitized lymphocytes and the antibodies.

## PREPROCESSING OF THE "T" AND "B" LYMPHOCYTES

Though all of the lymphocytes of the body originate from the lymphocytic stem cells of the bone marrow, these stem cells are themselves incapable of forming either sensitized lymphocytes or antibodies. Before they can do so, they must be further differentiated in appropriate processing areas in the thymus or in the "B" cell processing area.

**Role of the Thymus Gland for Preprocessing the "T" Lymphocytes.** Most of the preprocessing of the "T" lymphocytes of the thymus gland occurs shortly before birth of the baby and for a few months after birth. Therefore, beyond this period of time, removal of the thymus gland usually will not seriously impair the "T" lymphocytic immunity system, the system necessary for cellular immunity. However, removal of the thymus several months before birth can completely prevent the development of all cellular immunity. Since it is this cellular type of immunity that is mainly responsible for rejection of transplanted organs such as transplanted hearts, and kidneys, one can transplant organs with little likelihood of rejection if the thymus is removed from an animal a reasonable period of time before birth.

**Thymic Hormone.** In addition to preprocessing the "T" lymphocytes, the thymus probably also secretes a hormone that circulates through the body fluids and increases the activity of the "T" lymphocytes that have already left the thymus gland and have migrated to the lymphoid tissue. This hormone is believed to cause further proliferation and increased activity of these lymphocytes. Otherwise, little is known about either the nature or the function of this hormone.

**Role of the Bursa of Fabricius for Preprocessing "B" Lymphocytes in Birds.** It is during the latter part of fetal life that the bursa of Fabricus preprocesses the "B" lymphocytes and prepares them to manufacture antibodies. Here again, this process continues for a while after birth. In mammals, recent experiments indicate that it is lymphoid tissue in the fetal liver, and perhaps to a slight extent lymphoid tissue in the spleen, that performs this same function.

**Spread of Processed Lymphocytes to the Lymphoid Tissue.** After formation of processed lymphocytes in both the thymus and the bursa, these first circulate freely in the blood and then filter into the tissues. Then they enter the lymph and are carried to the lymphoid tissue. The lymphoid tissue contains reticulum cells that form a fine reticulum meshwork that filters the lymphocytes from the lymph, thereby entrapping them in the lymphoid tissue. Thus, the lymphocytes do not originate primordially in the lymphoid tissue, but, instead, are transported to this tissue by way of the preprocessing areas of the thymus and probably fetal liver.

## MECHANISMS FOR DETERMINING SPECIFICITY OF SENSITIZED LYMPHOCYTES AND ANTIBODIES— LYMPHOCYTE CLONES

Earlier in the chapter it was pointed out that the lymphocytes of the lymphoid tissue can form sen-

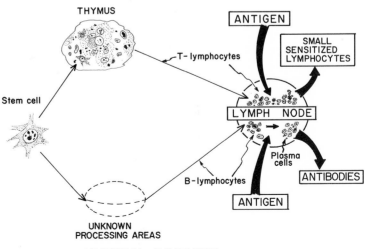

**Figure 7–1.** Formation of antibodies and sensitized lymphocytes by a lymph node in response to antigens. This figure also shows the origin of thymic ("T") and bursal ("B") lymphocytes that are responsible for the cellular and humoral immune processes of the lymph nodes.

CELLULAR IMMUNITY

THYMUS

ANTIGEN

SMALL SENSITIZED LYMPHOCYTES

T- lymphocytes

Stem cell

LYMPH NODE

Plasma cells

B- lymphocytes

ANTIBODIES

UNKNOWN PROCESSING AREAS

ANTIGEN

HUMORAL IMMUNITY

sitized lymphocytes and antibodies that are highly specific against particular types of invading agents. This effect is believed to occur in the following way:

**Specificity of the Sensitized Lymphocytes or Antibodies Formed by a Single Type of Lymphocyte—Lymphocyte Clones.** When a lymphocyte in the lymphoid tissue is stimulated to form either sensitized lymphocytes or antibodies, it always forms a sensitized lymphocyte or an antibody having specificity for a specific antigen. If more than one type of sensitized lymphocyte or antibody is to be formed, then a separate population of lymphocytes must be stimulated for each of these. Because it is known that the lymphocytes of the lymphoid tissue can form literally hundreds or thousands of different types of sensitized lymphocytes and antibodies all specific for different antigens, it is also almost certain that literally hundreds or thousands of different types of precursor lymphocytes pre-exist in the lymph nodes for formation of the many specific types of lymphocytes or antibodies.

All of the lymphocytes of one specific type in the lymphoid tissue—those that form one specific type of sensitized lymphocyte or one specific type of antibody—are called a *clone of lymphocytes*. That is, all of the lymphocytes in each clone are alike and are probably derived originally from one or a few early lymphocytes of the specific type.

**Origin of the Many Clones of Lymphocytes.** The way in which the many different clones of lymphocytes are originally formed is not known, but there are two main theories for the origin. The first theory suggests that each of the many clones is genetically determined, that is, that there is a separate gene for every clone. This theory proposes that in the thymus in which the "T" lymphocytes are processed and in the processing area for the "B" lymphocytes the respective genes for the different lymphocytic clones are brought to expression, causing differentiation of the stem cell lymphocytes into the multiple clones precommitted to forming single types of sensitized lymphocyte or single types of antibody.

The second theory assumes that the lymphocytes in the thymus or in the "B" cell processing area simply differentiate wildly into a whole host of random clones of lymphocytes.

**Excitation of a Clone of Lymphocytes.** Each clone of lymphocytes is responsive to only a single type of antigen (or to a group of antigens that have almost exactly the same stereochemical characteristics). When excited by the clone's specific antigen, all the cells of the clone proliferate madly, forming tremendous numbers of progeny, and these in turn lead to the formation of large quantities of antibodies if the clone is "B" lymphocytes, or to the formation of sensitized lymphocytes if the clone is "T" lymphocytes. During this process, the total number of lymphocytes in the lymphoid tissue increases markedly.

**Role of Macrophages in Stimulating the Clones of Lymphocytes.** Aside from the lymphocytes that are entrapped in the reticulum of the lymphoid tissue, many monocytes are also entrapped, and these swell to become tissue macrophages, as was discussed in the previous chapter. Furthermore, these lie in apposition to the lymph node lymphocytes. Most invading organisms that carry the antigens are first phagocytized by the macrophages, and the antigenic products are liberated from the invader. It is believed that these antigens then pass directly from the macrophages to the lymphocytes to stimulate the specific lymphocytic clones.

It is also believed that initial excitation of "T" lymphocytes can lead to secondary excitation of "B" lymphocytes so that both sensitized lymphocytes and antibodies can be formed against the same invading agent. In other words, there appears to be an element of cooperation between the two systems of acquired immunity.

## TOLERANCE OF THE ACQUIRED IMMUNITY SYSTEM TO ONE'S OWN TISSUES—ROLE OF THE THYMUS AND THE BURSA

Obviously, if a person should become immune to his own tissues, the process of acquired immunity would destroy his own body. Fortunately, the immune mechanism normally "recognizes" a person's own tissues as being completely distinctive from those of invaders, and his immunity system forms neither antibodies nor sensitized lymphocytes against his own antigens. This phenomenon is known as *tolerance* to the body's own tissues.

**Mechanism of Tolerance.** It is possible that tolerance to one's own tissues is determined genetically—that is, via absence of genes to form sensitized lymphocytes and antibodies against the person's own tissues. However, there is much reason to believe that tolerance develops during the processing of the lymphocytes in the thymus and in the "B" lymphocyte processing area. The reason for belief in this is that injecting a strong antigen into a fetus at the time that the lymphocytes are being processed in these two areas will prevent the development of clones of lymphocytes in the lymphoid tissue that are specific for the injected antigen. Also, experiments have shown that specific immature lymphocytes in the thymus, when exposed to a strong antigen, become lymphoblastic, proliferate considerably, and then combine with the stimulating antigen—an effect that causes the cells themselves to be destroyed before they can migrate to and colonize the lymphoid tissue.

Therefore, it is believed that during the processing of lymphocytes in the thymus and in the "B" lymphocyte processing area, all those clones of lymphocytes that are specific for the body's own tissues are self-destroyed because of their continual exposure to the body's antigens.

**Failure of the Tolerance Mechanism—Autoimmune Diseases.** Unfortunately, people frequently lose some of their immune tolerance to their own tissues. This occurs to a greater extent the older a person becomes. It usually results from destruction of some of the body's tissues, which releases considerable quantities of antigens that circulate in the body and cause acquired immunity either in the form of sensitized lymphocytes or antibodies. Some of these antigens perhaps combine with other proteins such as proteins from bacteria or viruses to form a new type of antigen that can then cause immunity. Then the resulting immune products, the sensitized lymphocytes and the antibodies, attack the body's own tissues. Also, it is believed that some of the proteins of the body are normally sequestered from the immune system during embryonic development of tolerance so that tolerance to these proteins never forms in the first place. For instance, the proteins of the cornea do not seem to circulate in the fluids of the fetus; this is also true of the thyroglobulin molecule of the thyroid; therefore, tolerance to these never develops. When damage occurs to either of these two tissues, these protein molecules can then elicit immunity, and the immunity in turn can attack the cornea in the first instance or the thyroid gland in the second instance to cause corneal opacity or thyroiditis.

Other diseases that result from autoimmunity include: *rheumatic fever,* in which the body becomes immunized against tissues in the heart and joints following exposure to a specific type of streptococcal toxin; *acute glomerulonephritis,* in which the person becomes immunized against the glomeruli, resulting from exposure to another specific type of streptococcal toxin; *myasthenia gravis,* in which immunity develops against muscles, causing paralysis; and *lupus erythematosis,* in which the person becomes immunized against many different body tissues at the same time, a disease that causes extensive damage, often causing rapid death.

## SPECIFIC ATTRIBUTES OF HUMORAL IMMUNITY—THE ANTIBODIES

**Formation of Antibodies by the Plasma Cells.** Prior to exposure to a specific antigen, the clones of "B" lymphocytes remain dormant in the lymphoid tissue. However, upon entry of a foreign antigen, the lymphocytes specific for that antigen immediately enlarge and take on the appearance of a *lymphoblast.* Some of these then further differentiate to form *plasmablasts,* which are the precursors of *plasma cells.* In these cells the cytoplasm expands, and the endoplasmic reticulum proliferates. They then begin to divide at a rate of approximately once every ten hours for about nine divisions, giving in four days a total population of about 500 cells for each original plasmablast. The mature plasma cell then produces gamma globulin antibodies at an extremely rapid rate—about 2000 molecules per second. The antibodies are secreted into the lymph and are carried to the circulating blood. This process continues for several days until death of the plasma cells.

**Formation of "Memory Cells—Difference Between the Primary Response and the Secondary Response.** Some of the lymphoblasts formed by activation of a clone of "B" lymphocytes do not go on to form plasma cells but, instead, form large numbers of new "B" lymphocytes similar to those of the original clone. In other words, the population of the specifically activated clone becomes greatly enhanced. And the new "B" lymphocytes are added to the original lymphocytes of the clone. These then remain dormant in the lymphoid tissue until activated once again by a new quantity of the same antigen. Obviously, subsequent exposure to the same antigen will then cause a much more rapid and much more potent antibody response. Thus, the differences between the so-called *primary response* and the *secondary response* are illustrated in Figure 7–2. Note the delay in the appearance of the primary response, its weak potency, and its short life. The secondary response, by contrast, begins rapidly after exposure to the antigen, is far more potent, and forms antibodies for many months rather than for only a few weeks.

The increased potency and duration of the secondary response explains why *vaccination* is usually accomplished by injecting antigen in multiple doses with periods of several weeks or several months between injections.

### The Nature of the Antibodies

The antibodies are gamma globulins called *immunoglobulins,* and they have molecular weights between approximately 150,000 and 900,000.

All of the immunoglobulins are composed of combinations of *light* and *heavy polypeptide chains,* most of which are a combination of two light and two heavy chains, as illustrated in Figure 7–3. Some of the immunoglobulins, though, have combinations of greater than two heavy and two light chains, which gives rise to the much larger molecular weight immunoglobulins. Yet, in all of these, each heavy chain is paralleled by a light chain at one of its ends, thus forming a heavy-light pair, and there are always at

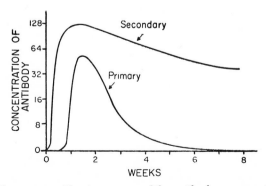

**Figure 7–2.** The time course of the antibody response to a *primary* injection of antigen and to a *secondary* injection several weeks later.

82 BLOOD CELLS, IMMUNITY, AND BLOOD CLOTTING

least two such pairs in each immunoglobin molecule.

Figure 7–3 shows a designated end of each of the light and each of the heavy chains called the "variable portion," and the remainder of each chain is called the "constant portion." The variable portion is different for each specificity of antibody, and it is this portion that allows the antibody to attach specifically to a particular type of antigen. The constant portion of the antibody determines the gross physical and chemical properties of the antibody, establishing such factors as mobility of the antibody in the tissues, adherence of the antibody to specific structures within the tissues, the ease with which the antibodies pass through membranes, and other biological properties of the antibody.

**Specificity of Antibodies.** Each antibody that is specific for a particular antigen has a different organization of amino acid residues in the variable portions of both the light and heavy chains. These have a specific steric shape for each antigen specificity so that when an antigen comes in contact with it, the prosthetic radicals of the antigen fit as a mirror image with those of the antibody, thus allowing a rapid and tight chemical bond between the antibody and the antigen.

The constant portions of the antibody, on the other hand, provide means for attachment of the antibody to cells or other tissues and also provide means by which the antibody can combine with other chemical substances, most particularly the *complement complex,* which will be discussed subsequently.

Note, especially, in Figure 7–3 that there are two variable sites on the antibody for attachment of antigens. Thus, most antibodies are *bivalent.* However, a small proportion of the antibodies, which have high molecular weight combinations of light and heavy chains, have more than two reactive sites.

**Classes of Antibodies.** There are five general classes of antibodies, respectively named *IgM, IgG, IgA, IgD,* and *IgE.* Ig stands for immunoglobulin, and the other five respective letters simply designate the respective classes of immunoglobulins.

For the purpose of our present discussion, two of

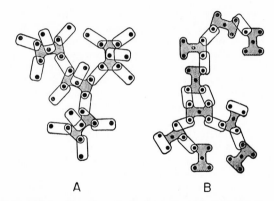

**Figure 7–4.** Reactions between antigens and antibodies. The antibodies are represented by bar structures, and the antigens are represented by stippled dumbbell structures. (A) Reaction when excess antibody is present. (B) Reaction when excess antigen is present. (Redrawn from Raffel: Immunity. Appleton-Century-Crofts, 1961.)

these classes of antibodies are of particular importance: IgG, which comprises about 75 per cent of the antibodies of the normal person; and IgE, which constitutes only a small per cent of the antibodies but which is especially involved in allergy.

### Mechanisms of Action of Antibodies

Antibodies can act in three different ways to protect the body against invading agents: (1) by direct attack on the invader, (2) by activation of the complement system that then destroys the invader, or (3) by activation of the anaphylactic system that changes the local environment around the invading antigen and in this way probably prevents its virulence.

**Direct Action of Antibodies on Invading Agents.** Figure 7–4 illustrates antibodies (designated by the bars) reacting with antigens (designated by the darkened dumbbells). Because of the bivalent nature of the antibodies and the multiple antigen sites on most invading agents, the antibodies can inactivate the invading agent in one of several ways, as follows:

1. *Agglutination,* in which multiple antigenic agents are bound together into a clump.

2. *Precipitation,* in which the complex of antigen and antibody becomes insoluble and precipitates.

3. *Neutralization,* in which the antibodies cover the toxic sites of the antigenic agent.

4. *Lysis,* in which some very potent antibodies are capable of directly attacking membranes of cellular agents and thereby causing rupture of the cell.

However, the direct actions of antibodies attacking the antigenic invaders probably, under normal conditions, are not strong enough to play a major role in protecting the body against the invader. Most of the protection probably comes through the *amplifying* effects of the complement and anaphylactic effector systems described below.

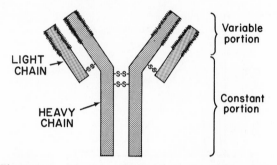

**Figure 7–3.** Structure of the typical IgG antibody, showing it to be composed of two heavy polypeptide chains and two light polypeptide chains. The antigen binds at two different sites on the variable portions of the chains.

**The Complement System for Antibody Action.** Complement is a system of nine different enzyme precursors (designated C–1 through C–9) which are found normally in the plasma and other body fluids, but the enzymes are also normally inactive. However, when an antibody combines with an antigen, the antigen-antibody complex then becomes an activator of the complement system, as illustrated symbolically in Figure 7–5. Only a few antigen-antibody combinations are required to activate large numbers of enzyme precursor molecules in the first stage of the complement system, and the enzymes thus formed then activate still far more of the enzymes in the later stages of the system. The activated enzymes then attack the invading agent in several different ways as well as initiate local tissue reactions that also provide protection against damage by the invader. Among the more important effects that occur are the following:

1. *Lysis.* The proteolytic enzymes of the complement system digest portions of the cell membrane, thus causing rupture of cellular agents such as bacteria or other types of invading cells.

2. *Opsonization and phagocytosis.* The complement enzymes attack the surfaces of bacteria and other antigens, making these highly susceptible to phagocytosis by neutrophils and tissue macrophages. This process is called *opsonization.* It often enhances the number of bacteria that can be destroyed many hundred-fold.

3. *Chemotaxis.* One or more of the complement products causes chemotaxis of neutrophils and macrophages, thus greatly enhancing the number of these phagocytes in the local region of the antigenic agent.

4. *Agglutination.* The complement enzymes also change the surfaces of some of the antigenic agents so that they adhere to each other, thus causing agglutination.

5. *Neutralization of viruses.* The complement enzymes frequently attack the molecular structures of viruses and thereby render them nonvirulent.

6. *Inflammatory effects.* The complement products elicit a local inflammatory reaction, leading to hyperemia, coagulation of proteins in the tissues, and other aspects of the inflammation process, thus preventing movement of the invading agent through the tissues.

**Activation of the Anaphylactic System by Antibodies.** Some of the antibodies, particularly the IgE antibodies, attach to the membranes of cells in the tissues and blood. Among the most important cells are the *mast cells* in tissues surrounding the blood vessels and the *basophils* circulating in the blood. When an antigen reacts with one of the antibody molecules attached to the cell, there is an immediate swelling and then rupture of the cell, with the release of a large number of factors that affect the local environment. These factors include:

1. *Histamine,* which causes local vasodilatation and increased permeability of the capillaries.

2. *Slow-reacting substance of anaphylaxis,* which causes prolonged contraction of certain types of smooth muscle such as the bronchi.

3. *Chemotaxic factor,* which causes chemotaxis of neutrophils and macrophages into the area of the antigen-antibody reaction. The chemotaxic factor, especially, causes chemotaxis of large numbers of eosinophils into the area. It has been suggested that eosinophils play a special role in phagocytizing the products of the antibody-antigen reactions.

4. *Lysosomal enzymes,* which elicit a local inflammatory reaction.

These anaphylactic reactions can frequently be very harmful to the body, often causing the harmful reactions of allergy, as will be discussed subsequently. However, it is also known that in persons who are genetically unable to respond with the anaphylactic reaction, many types of infection spread much more rapidly through the body than occurs when the reaction can take place. Therefore, this reaction presumably helps to immobilize the antigenic invader.

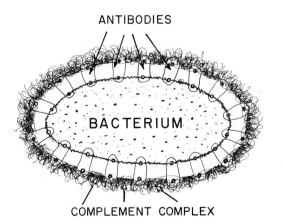

**Figure 7–5.** Attachment of antibodies to a bacterium and subsequent activation of the complement complex.

## SPECIAL ATTRIBUTES OF CELLULAR IMMUNITY

**Release of Sensitized Lymphocytes from Lymphoid Tissue and Formation of Memory Cells.** Upon exposure to the proper antigens, sensitized lymphocytes are released from lymphoid tissue in ways that parellel antibody release. The only real difference is that instead of releasing antibodies, whole sensitized lymphocytes are formed and released into the lymph. These then pass into the circulation where they remain a few minutes to a few hours, at most; instead, they filter out of the circulation into all the tissues of the body.

Also, lymphocyte *memory cells* are formed in the same way that memory cells are formed in the humoral antibody system. Thus, when T-lymphocytes are activated by an antigen, a large number of newly formed lymphocytes become additional T-lymphocytes of that specific clone and remain in the lymphoid tissue, thus greatly increasing the population of this type of T-lymphocyte. Therefore, upon subsequent exposure to the same antigen, the release of sensitized lymphocytes occurs much more rapidly and much more powerfully than in the first response.

**Persistence of Cellular Immunity.**  An important difference between cellular immunity and humoral immunity is its persistence. Humoral antibodies rarely persist more than a few months, or at most, a few years. On the other hand, sensitized lymphocytes probably have an indefinite life span and seem to persist until they eventually come in contact with their specific antigen. There is reason to believe that such sensitized lymphocytes might persist as long as ten years in some instances, a fact which makes cellular immunity far more persistent than humoral immunity.

**Types of Organisms Resisted by Sensitized Lymphocytes.**  Although the humoral antibody mechanism for immunity is especially efficacious against more acute bacterial diseases, the cellular immunity system is activated much more potently by the more slowly developing bacterial diseases such as tuberculosis, brucellosis, and so forth. Also, this system is active against cancer cells, cells of transplanted organs, and fungus organisms, all of which are far larger than bacteria. And, finally, the system is very active against some viruses.

Therefore, cellular immunity is especially important in protecting the body against some virus diseases, in destroying many early cancerous cells before they can cause cancer, and unfortunately in causing rejection of transplanted tissues from one person to another.

### Mechanism of Action of Sensitized Lymphocytes

The sensitized lymphocyte, on coming in contact with its specific antigen, combines with the antigen. This combination in turn leads to a sequence of reactions whereby the sensitized lymphocytes destroy the invader. As is also true of the humoral immunity system, the sensitized lymphocyte destroys the invader either directly or indirectly.

**Direct Destruction of the Invader.**  Figure 7–6 illustrates sensitized lymphocytes that have bound with antigens in the membrane of an invading cell such as a cancer cell, a heart transplant cell, or a parasitic cell of another type. The immediate effect of this attachment is swelling of the sensitized lymphocyte and release of cytotoxic substances from the lymphocyte to attack the invading cell. The cytotoxic substances are probably lysosomal enzymes manufactured in the lymphocytes. However, these direct effects of the sensitized lymphocyte in destroying the invading cell are probably relatively weak in comparison with the indirect effects, as follows:

**The Indirect "Amplifying" Mechanisms of Cellular Immunity.**  When the sensitized lymphocytes combine with their specific antigens, they release a number of different substances into the surrounding tissues that lead to a sequence of reactions. These reactions in turn are much more potent than the original attack on the invader. Some of these reactions are the following:

*Release of Transfer Factor.*  The sensitized lymphocytes release a polypeptide substance with a molecular weight of less than 10,000, called *transfer factor.* This then reacts with other small lymphocytes in the tissues that are of the nonsensitized vari-

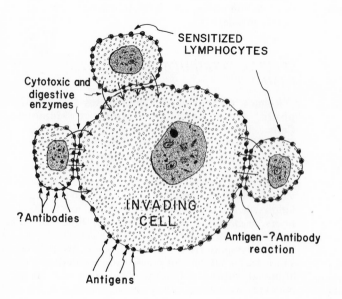

**Figure 7–6.**  Direct destruction of an invading cell by sensitized lymphocytes.

ety. On entering these lymphocytes, transfer factor causes them to take on the same characteristics as the original sensitized lymphocytes. Thus, transfer factor recruits additional lymphocytes having the same capability for causing the same cellular immunity reaction as the original sensitized lymphocytes. Furthermore, the newly sensitized lymphocytes are equally as specific for the original antigen as were the original sensitized lymphocytes. Thus, this mechanism multiplies the effect of the sensitized lymphocytes.

*Attraction and Activation of Macrophages.* A second product of the activated sensitized lymphocyte is a *macrophage chemotaxic factor* that causes as many as 1000 macrophages to enter the vicinity of the activated sensitized lymphocyte. A third factor, called *migration inhibition factor,* then stops the migration of the macrophages once they come into the vicinity of the activated lymphocyte. Thus, a single lymphocyte can collect as many as 1000 macrophages around it. Finally, a fourth substance increases the phagocytic activity of the macrophages. Therefore, the macrophages play a major role in removing the foreign antigenic invader.

Thus, it is by a combination of a weak direct effect of the sensitized lymphocytes on the antigen invader and much more powerful indirect reactions that the cellular immunity system destroys the invader.

## VACCINATION

The process of vaccination has been used for many years to cause acquired immunity against specific diseases. A person can be vaccinated by injecting dead organisms that are no longer capable of causing disease but which still have their chemical antigens. This type of vaccination is used to protect against typhoid fever, whooping cough, diphtheria, and many other types of bacterial diseases. Also, immunity can be achieved against toxins that have been treated with chemicals so that their toxic nature has been destroyed even though their antigens for causing immunity are still intact. This procedure is used in vaccinating against tetanus, botulism, and other similar toxic diseases. And, finally, a person can be vaccinated by infecting him with live organisms that have been "attenuated." That is, these organisms have been grown either in special culture mediums or have been passed through a series of animals until they have mutated enough that they will not cause disease but will still carry the specific antigens. This procedure is used to protect against poliomyelitis, yellow fever, measles, smallpox, and many other viral diseases.

## PASSIVE IMMUNITY

Thus far, all the acquired immunity that we have discussed has been *active immunity.* That is, the person's body develops either antibodies or sensitized lymphocytes in response to invasion of his body by a foreign antigen. However, it is possible to achieve temporary immunity in a person without injecting any antigen whatsoever. This is done by infusing antibodies, sensitized lymphocytes, or both into him from someone else or from some other animal that has been actively immunized against the antigen. The antibodies will last for two to three weeks, and during that time the person is protected against the invading disease. Sensitized lymphocytes will last for a few weeks if transfused from another person, and for a few hours to a few days if transfused from an animal. The transfusion of antibodies or lymphocytes to confer immunity is called *passive immunity.*

# INTERFERON—ANOTHER TYPE OF ACQUIRED IMMUNITY

Another type of acquired immunity has been discovered recently. Unfortunately, though, we still do not know how important it might be for protection against disease. This system is the following:

When cells in the body are attacked by viruses, many of them form a substance called *interferon* that specifically inactivates the virus that is attacking. This substance prevents the ribosomes from translating the messenger RNA of the virus and, therefore, inhibits its further damaging qualities. In addition, the interferon is released from the infected cells and is carried through the body fluids to cells elsewhere in the body where the interferon also prevents translation of the viral message. Therefore, infection of some cells by viral particles tends to protect other cells in the body from the same virus.

Unfortunately, it is not known yet whether this is an important mechanism of acquired immunity. However, purified interferon has been used in animals as an effective means for protecting against certain viral diseases.

# ALLERGY

One of the important side effects of immunity is the development, under some conditions, of allergy. There are at least three different types of allergy, two of which can occur in any person, and a third that occurs only in persons who have a specific allergic tendency.

## ALLERGIES THAT OCCUR IN NORMAL PEOPLE

**Delayed-Reaction Allergy.** This type of allergy frequently causes skin eruptions in response to certain drugs or chemicals, particularly some cosmetics and household chemicals, to which one's skin is often exposed. Another example of such an allergy is the skin eruption caused by exposure to poison ivy.

Delayed-reaction allergy is caused by sensitized lymphocytes and not by antibodies. In the case of poison ivy, the toxin of poison ivy in itself does not cause much harm to the tissues. However, upon repeated exposure it does cause the formation of sensitized lymphocytes. Then, following subsequent exposure to the poison ivy toxin, within a day or so the sensitized lymphocytes diffuse in sufficient numbers into the skin to combine with the poison ivy toxin and elicit a cellular immunity type of reaction. Remembering that cellular immunity can cause release of many toxic substances from the sensitized lymphocytes, as well as extensive invasion of the tissues by macrophages and their subsequent effects, one can well understand that the eventual result of some delayed-reaction allergies can be serious tissue damage.

**Allergies Caused by Reaction Between IgG Antibodies and Antigens.** When a person becomes strongly immunized against an antigen and has developed a very high titer of IgG antibodies (the most usual type of antibody), subsequent sudden exposure of that person to a high concentration of the same antigen can cause a serious tissue reaction. The antigen-antibody complex that is formed precipitates, and some of it deposits as granules in the walls of the small blood vessels. These granules also activate the complement system, setting off extensive release of proteolytic enzymes. The result of these two effects is severe inflammation and destruction of the small blood vessels.

This type of allergy is especially manifest in the reaction called the *Arthus reaction:* this occurs when a large amount of antigen is injected into the tissues of a person who is strongly immunized. The reaction between the IgG antibodies in the antigen elicits potent local vascular and other effects that cause almost total destruction of the local tissue—the damage beginning within a few minutes and leading to death and dissolution of the tissue within a few days.

Another manifestation of this type of reaction is *serum sickness.* Serum injected into a person can cause subsequent formation of IgG antibodies. When these begin to appear, they react with the protein of the injected serum and elicit a widespread antigen-antibody reaction throughout the body. Fortunately, this reaction occurs slowly over a period of days as the antibodies are formed, and usually it is not lethal. However, it can be lethal on occasion, and on other occasions it can cause widespread inflammation throughout the body with development of a circulatory shocklike syndrome.

## ALLERGIES IN THE "ALLERGIC" PERSON

Some persons have an "allergic" tendency. This phenomenon is genetically passed on from parent to child, and it is characterized by the presence of large quantities of *IgE antibodies.* These antibodies are called *reagins* or *sensitizing antibodies* to distinguish them from the more common IgG antibodies. When an *allergen* (defined as an antigen that reacts specifically with a specific type of IgE reagin antibody) enters the body, an allergen-reagin reaction takes place, and a subsequent allergic reaction takes place.

As was pointed out earlier in the discussion of immunity, the IgE antibodies (the reagins) attach to cells throughout the body; therefore, antigen-antibody reactions damage the cells. The result is *anaphylactoid types of immune reactions.* These result primarily from the rupture of eosinophils and basophils when the allergen (antigen) reacts with reagins attached to these cells. This reaction causes rupture of these cells, followed by the release of *histamine, slow-reacting substance of anaphylaxis, eosinophil chemotaxic substance, lysosomal enzymes,* and other less important substances.

Among the different types of allergic reaction of this type are:

**Anaphylaxis.** When a specific allergen is injected directly into the circulation it can react in widespread areas of the body with the basophils of the blood and the mast cells located immediately outside the small blood vessels. Therefore, the anaphylactic type of reaction occurs everywhere. The histamine released into the circulation causes widespread peripheral vasodilatation as well as increased permeability of the capillaries and marked loss of plasma from the circulation. Often, persons experiencing this reaction die of circulatory shock within a few minutes. But also released from the cells is the substance called slow-reacting substance of anaphylaxis, which sometimes causes spasm of the smooth muscle of the bronchioles, eliciting an asthmalike attack.

**Urticaria.** Urticaria results from antigen entering specific skin areas and causing localized anaphylactoid reactions. *Histamine* released locally causes (a) vasodilatation that induces an immediate *red flare* and (b) increased permeability of the capillaries that leads to swelling of the skin in another few minutes. The swellings are commonly called "hives." Administration of antihistamine drugs to a person prior to exposure will prevent the hives.

**Hay Fever.** In hay fever, the allergen-reagin reaction occurs in the nose. *Histamine* released in response to this causes local vascular dilatation with resultant increased capillary pressure, and it also causes increased capillary permeability. Both of these effects cause rapid fluid leakage into the tissues of the nose, and the nasal linings become swollen and secretory. Here again, use of antihistaminic drugs can prevent this swelling reaction. However, other products of the allergen-reagin reaction still cause irritation of the nose, still eliciting the typical sneezing syndrome despite drug therapy.

**Asthma.** In asthma, the allergen-reagin reaction occurs in the bronchioles of the lungs. Here, the most important product released from the mast cells seems to be the *slow-reacting substance of anaphylaxis,* which causes spasm of the bronchiolar smooth muscle. Consequently, the person has difficulty breathing until the reactive products of the allergic reaction have been removed. Unfortunately, administration of antihistaminics has little effect on the course of asthma, because histamine does not appear to be the major factor eliciting the asthmatic reaction.

# REFERENCES

Aas, K.: The Biochemical and Immunological Basis of Bronchial Asthma. Springfield, Ill., Charles C Thomas, Publisher, 1972.

Abramoff, P., and Duquesnoy, R. J.: Immunobiology. *In* Sodeman, W. A., Jr., and Sodeman, W. A. (eds.): Pathologic Physiology: Mechanisms of Disease, 5th Ed. Philadelphia, W. B. Saunders Company, 1974, p. 97.

Blanden, R. V.: T cell response to viral and bacterial infection. *Transplant. Rev.*, 19:56, 1974.

Boyse, E. A., and Abbott, J.: Surface reorganization as initial inductive event in the differentiation of prothymocytes to thymocytes. *Fed. Proc.*, 34:24, 1975.

Carpenter, P. L.: Immunology and Serology, 3rd Ed. Philadelphia, W. B. Saunders Company, 1975.

Cinader, B. (ed.): Regulation of the Antibody. Springfield, Ill., Charles C Thomas, Publisher, 1971.

Collins-Williams, C.: Pediatric Allergy and Clinical Immunology, 4th Ed. New York, Churchill Livingstone, Div. of Longman Inc., 1973.

Diener, E.: Antigen recognizing and processing cells. *Adv. Exp. Med. Biol.*, 29:197, 1973.

Duquesnoy, R. J., and Abramoff, P.: Immunodeficiency diseases and tumor immunobiology. *In* Sodeman, W. A., Jr., and Sodeman, W. A. (eds.): Pathologic Physiology: Mechanisms of Disease, 5th Ed. Philadelphia, W. B. Saunders Company, 1974, p. 124.

Edelman, G. M.: The structure and function of antibodies. *Sci. Amer.*, 223:34, 1970.

Edelman, G. M. (ed.): Cellular Selection and Regulation in the Immune Response. New York, Raven Press, 1974.

Eisen, H. N.: Immunology. New York, Harper & Row, Publishers, 1974.

Gasser, D. L., and Silvers, W. K.: Genetic determinants of immunological responsiveness. *Adv. Immunol.*, 18:1, 1974.

Goetzl, E. J., Wasserman, S. I., and Austen, F.: Eosinophil polymorphonuclear leukocyte function in immediate hypersensitivity. *Arch. Pathol.*, 99:1, 1975.

Hersh, E. M., Gutterman, J. U., and Mavligit, G.: Immunotherapy of Cancer in Man. Springfield, Ill., Charles C Thomas, Publisher, 1973.

Hilleman, M. R., and Tytell, A. A.: The induction of interferon. *Sci. Amer.*, 225:26, 1971.

Jerne, N. K.: The immune system. *Sci. Amer.*, 229:52, 1973.

Kabat, E. A., and Mayer, M. M.: Experimental Immunochemistry, 2nd Ed. Springfield, Ill., Charles C Thomas, Publisher, 1971.

Klinman, N. R., and Press, J. L.: Expression of specific clones during B cell development. *Fed. Proc.*, 34:47, 1975.

Lawrence, H. S.: Transfer factor in cellular immunity. *Harvey Lect.*, 68:239, 1974.

Lindahl-Kiessling, K., and Osoba, D. (eds.): Lymphocyte Recognition and Effector Mechanisms. New York, Academic Press, Inc., 1974.

McCluskey, R. T., and Cohen, S. (eds.): Mechanisms of Cell-Mediated Immunity. New York, John Wiley & Sons, Inc., 1974.

McDevitt, H. O., and Landy, M. (eds.): Genetic Control of Immune Responsiveness. New York, Academic Press, Inc., 1973.

Mansmann, H. C., Jr.: Allergy: Its nature and relationship to other immunologically induced disease states. *In* Sodeman, W. A., Jr., and Sodeman, W. A. (eds.): Pathologic Physiology: Mechanisms of Disease, 5th Ed. Philadelphia, W. B. Saunders Company, 1974, p. 445.

Mayer, M. M.: The complement system. *Sci. Amer.*, 229:54, 1973.

Mestecky, J., and Lawton, A. R., III (eds.): The Immunoglobulin A System. New York, Plenum Publishing Corp., 1974.

Nisonoff, A., Hopper, J. E., and Spring, S. B.: The Antibody Molecule. New York, Academic Press, Inc., 1975.

Notkins, A. L., and Koprowski, H.: How the immune response to a virus can cause disease. *Sci. Amer.*, 228:22, 1973.

Park, B. H., and Good, R. A.: Principles of Modern Immunobiology. Philadelphia, Lea & Febiger, 1974.

Patterson, R. (ed.): Allergic Diseases. Philadelphia, J. B. Lippincott Company, 1972.

Raff, M. C.: T and B lymphocytes and immune responses. *Nature*, 242:19, 1973.

Reite, O. B.: Comparative physiology of histamine. *Physiol. Rev.*, 52:778, 1972.

Robinson, W. D.: Rheumatic diseases. *In* Sodeman, W. A., Jr., and Sodeman, W. A. (eds.): Pathologic Physiology: Mechanisms of Disease, 5th Ed. Philadelphia, W. B. Saunders Company, 1974, p. 417.

Roitt, I. M.: Essential Immunology, 2nd Ed. Philadelphia, J. B. Lippincott Company, 1974.

Sela, M. (ed.): The Antigens. New York, Academic Press, Inc., 1973 and 1974, Vols. 1 and 2.

Sercarz, E. E., Williamson, A., and Fox, C. F. (eds.): The Immune System: Genes, Receptors, Signals. New York, Academic Press, Inc., 1974.

Sigel, M. M., and Good, R. A. (eds.): Tolerance, Autoimmunity and Aging. Springfield, Ill., Charles C Thomas, Publisher, 1972.

Siskind, G. W.: Manipulation of the immune response. *Pharmacol. Rev.*, 25:319, 1973.

Stanworth, D.: Immediate Hypersensitivity. New York, American Elsevier Publishing Co., 1973.

Swineford, O., Jr.: Asthma and Hay Fever. Springfield, Ill., Charles C Thomas, Publisher, 1973.

Tan, E. M.: Drug-induced autoimmune disease. *Fed. Proc.*, 33:1894, 1974.

Taylor, G.: Immunology in Medical Practice. Philadelphia, W. B. Saunders Co., 1975.

Turk, J. L.: Delayed Hypersensitivity, 2nd Ed. New York, American Elsevier Publishing Co., 1974.

Weinstein, L., and Swartz, M. N.: Host responses to infection. *In* Sodeman, W. A., Jr., and Sodeman, W. A. (eds.): Pathologic Physiology: Mechanisms of Disease, 5th Ed. Philadelphia, W. B. Saunders Company, 1974, p. 473.

# 8

# Blood Groups; Transfusion; Tissue And Organ Transplantation

## ANTIGENICITY AND IMMUNE REACTIONS OF BLOOD

When blood transfusions from one person to another were first attempted the transfusions were successful in some instances, but, in many more, immediate or delayed agglutination and hemolysis of the red blood cells occurred. Soon it was discovered that the bloods of different persons usually have different antigenic and immune properties so that antibodies in the plasma of one blood react with antigens in the cells of another. Furthermore, the antigens and the antibodies are almost never precisely the same in one person as in another. For this reason, it is easy for blood from a donor to be mismatched with that of a recipient. Fortunately, if proper precautions are taken, one can determine ahead of time whether or not appropriate antibodies and antigens are present in the donor and recipient bloods to cause a reaction, but, on the other hand, lack of proper precautions often results in varying degrees of red cell agglutination and hemolysis, resulting in a typical transfusion reaction that can lead to death.

**Multiplicity of Antigens in the Blood Cells.** As pointed out in Chapter 7, a person does not normally form antibodies against the antigens in his own cells, but if cells from one person are transfused into another person, antibodies will be developed against all the antigens not in the recipient's own blood. Fortunately, many of the antigens are common from one person to another, and bloods are grouped and typed on the basis of the major types of antigens appearing in the cells.

At least 30 commonly occurring antigens, each of which can at times cause antigen-antibody reactions, have been found in human blood cells, especially in the cell membranes. In addition to these, more than 100 others of less potency or that occur in individual families rather than having widespread occurrence are known to exist. Among the 30 or more common antigens, certain ones are highly antigenic and regularly cause transfusion reactions if proper precautions are not taken, whereas others are of importance principally for studying the inheritance of genes and therefore for establishing parentage, race, and so forth.

Two particular groups of antigens are more likely than the others to cause blood transfusion reactions. These are the so-called *O-A-B* system of antigens and the *Rh-Hr* system. Bloods are divided into different *groups* and *types* in accordance with the types of antigens present in the cells.

## O-A-B BLOOD GROUPS

### THE A AND B ANTIGENS— THE AGGLUTINOGENS

Two different but related antigens—type A and type B—occur in the cells of different persons. Because of the way these antigens are inherited, a person may have neither of them in his cells, or he may have one or both simultaneously.

As will be discussed below, some bloods also contain strong antibodies that react specifically with either the type A or type B antigens in the cells, causing agglutination and hemolysis. Because the type A and type B antigens in the cells make the cells susceptible to agglutination, these antigens are called *agglutinogens*. It is on the basis of the presence or absence of agglutinogens in the red blood cells that blood is grouped for the purpose of transfusion. The agglutinogens are also frequently called *group specific substances* because they are used to specify the O-A-B blood groups.

**The Four Major O-A-B Blood Groups.** In transfusing blood from one person to another, the bloods of donors and recipients are normally classified into four major O-A-B groups, as illustrated in Table 8–1, depending on the presence or absence of the two agglutinogens. When neither A nor B agglutinogen is present, the blood group is *group O*. When only type A agglutinogen is present, the blood is *group A*. When only type B agglutinogen is present, the blood is *group B*. And when both A and B agglutinogens are present, the blood is *group AB*.

*Relative Frequency of the Different Blood Types.* The prevalence of the different blood types among Caucasoids is approximately as follows:

| Type | Per cent |
|------|----------|
| O    | 47       |
| A    | 41       |
| B    | 9        |
| AB   | 3        |

It is obvious from these percentages that the O and A genes occur frequently but the B gene is infrequent.

**Genetic Determination of the Agglutinogens.** Genes on two adjacent chromosomes, one gene on each chromosome, determine the blood groups. These are allelomorphic genes that can be any one of three different types, but of only one type on each chromosome: type O, type A, or type B. There is no dominance among the three different allelomorphs. However, the type O gene is either functionless or almost functionless, so that it causes either no type O

agglutinogen in the cells or such a weak agglutinogen that it is normally insignificant. On the other hand, the type A and type B genes do cause strong agglutinogens in the cells. Therefore, if either of the genes on the two respective chromosomes is type A, the red blood cells will contain type A agglutinogen, and likewise, if either of the two genes is type B, the red blood cells will contain type B agglutinogen. Or, if the gene on one chromosome is type A and on the other is type B, the red blood cells will contain both A and B agglutinogens.

The six possible combinations of genes, as shown in Table 8–1, are OO, OA, OB, AA, BB, and AB. These different combinations of genes are known as the *genotypes,* and each person is one of the six different genotypes.

One can observe from the table that a person with genotype OO produces no agglutinogens at all, and, therefore, his blood group is O. A person with either genotype OA or AA produces type A agglutinogens and, therefore, has blood group A. Genotypes OB and BB give group B blood, and genotype AB gives group AB blood.

## THE AGGLUTININS

When type A agglutinogen *is not present* in a person's red blood cells, antibodies known as "anti-A" agglutinins develop in his plasma. Also, when type B agglutinogen *is not present* in the red blood cells, antibodies known as "anti-B" agglutinins develop in the plasma.

Thus referring once again to Table 8–1, it will be observed that group O blood, though containing no agglutinogens, does contain both *anti-A* and *anti-B agglutinins,* whereas group A blood contains type A agglutinogens and *anti-B agglutinins,* and group B blood contains type B agglutinogens and *anti-A agglutinins.* Finally, group AB blood contains both A and B agglutinogens but no agglutinins at all.

**Titer of the Agglutinins at Different Ages.** Immediately after birth the quantity of agglutinins in the plasma is almost zero. Two to eight months after birth, the infant begins to produce agglutinins—anti-A agglutinins when type A agglutinogens are not present in the cells and anti-B agglutinins when type B agglutinogens are not in the cells. Figure 8–1 illustrates the changing titer of alpha and beta agglutinins at different ages. A maximum titer is usually reached at 8 to 10 years of age, and this gradually declines throughout the remaining years of life.

**Origin of the Agglutinins in the Plasma.** The agglutinins are gamma globulins, as are other antibodies, and are produced by the same cells that produce antibodies to infectious

**TABLE 8–1. The Blood Groups, with Their Genotypes and Their Constituent Agglutinogens and Agglutinins**

| Genotypes | Blood Groups | Agglutinogens | Agglutinins |
|-----------|--------------|---------------|-------------|
| OO        | O            | —             | Anti-A and Anti-B |
| OA or AA  | A            | A             | Anti-B      |
| OB or BB  | B            | B             | Anti-A      |
| AB        | AB           | A and B       | —           |

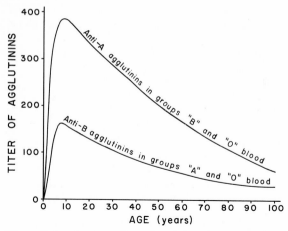

**Figure 8–1.** Average titers of anti-A and anti-B aggluti-nins in the blood of persons in group B and group A at different ages.

diseases. Most of them are IgM and IgG im-munoglobulin molecules.

It is difficult to understand how agglutinins are produced in individuals who do not have the respective antigenic substances in their red blood cells. However, small amounts of group A and B antigens are believed to enter the body in the food, in bacteria, or by other means, and these substances supposedly initiate the de-velopment of anti-A or anti-B agglutinins. One of the reasons for believing this is that injection of group A or group B antigen into a recipient of another blood type causes a typical immune re-sponse with formation of greater quantities of agglutinins than ever. Also, the newborn baby has few if any agglutinins, showing that the agglutinin formation occurs almost entirely after birth.

## THE AGGLUTINATION PROCESS IN TRANSFUSION REACTIONS

When bloods are mismatched so that anti-A or anti-B agglutinins are mixed with red blood cells containing A or B agglutinogens respec-tively, the red cells agglutinate by the following process: The agglutinins attach themselves to the red blood cells. Because the agglutinins are bivalent (IgG type) or polyvalent (IgM type), a single agglutinin can attach to two different red blood cells at the same time, thereby causing the cells to adhere to each other. This causes the cells to clump, and some of these clumps plug small blood vessels throughout the cir-culatory system. During the ensuing few hours

to few days, the phagocytic white blood cells and the reticuloendothelial system destroy the agglutinated cells, releasing hemoglobin into the plasma.

**Hemolysis in Transfusion Reactions.** Sometimes, usually when the titer of agglutinins is very high, immediate hemolysis of red cells occurs in the circulating blood. In this case the antibodies cause lysis of the red blood cells by activating the complement system. This in turn releases proteolytic enzymes that rupture the cell membranes, as was described in Chapter 7.

Direct intravascular hemolysis of this type is far less common than agglutination, because not only does there have to be a very high titer of antibodies for this to occur but also there seems to be a different type of antibody re-quired; these antibodies are called hemolysins. However, even agglutination eventually leads to hemolysis of the agglutinated red cells be-cause the phagocytic white blood cells and the reticuloendothelial cells rupture the aggluti-nated cells within a few hours after agglutina-tion has taken place.

## BLOOD TYPING

Prior to giving a transfusion, it is necessary to determine the blood group of the recipient and the group of the donor blood so that the bloods will be appropriately matched. This "typing" of blood is performed as follows:

Occasionally, a type A person has extremely strong anti-B agglutinins in his plasma, and, also occasionally, a type B person has ex-tremely strong anti-A agglutinins in his plasma. When such persons are found, their plasmas, after removal of the fibrinogen by clotting, are used as laboratory "typing sera." In other words, two sera are prepared, the first contain-ing a very strong titer of anti-A agglutinins and the second a very strong titer of anti-B aggluti-nins.

The usual method of blood typing is the slide technique. In using this technique a drop or more of blood is removed from the person to be typed. This is then diluted approximately 50 times with saline so that clotting will not occur. This leaves essentially a suspension of red blood cells in saline. Two separate drops of this suspension are placed on a microscope slide, and a drop of anti-A agglutinin serum is mixed with one of the drops of cell suspension while a drop of anti-B agglutinin serum is mixed with the second drop of cell suspension. After allow-ing several minutes for the agglutination process to take place, the slide is observed under a microscope to determine whether or not the cells have clumped. If they have clumped, one knows that an immune reaction has resulted between the serum and the cells.

Table 8–2 illustrates the reactions that occur with each of the four different types of blood. Group O red blood cells have no agglutinogens and, therefore, do not react with either the anti-A or the anti-B serum. Group A blood has A agglutinogens and therefore agglutinates with anti-A agglutinins. Group B blood has B agglutinogens and agglutinates with the anti-B serum. Group AB blood has both A and B agglutinogens and agglutinates with both types of serum.

### CROSS-MATCHING

If a person's life depends on immediate transfusion, it is not absolutely essential to type the bloods of the donor and the recipient, for one can simply test the bloods against each other to determine whether or not agglutination will occur. To do this, one prepares, first, a suspension of red cells from the donor, and, second, a small quantity of defibrinated serum from the recipient. Then he mixes the serum from the recipient with the cells from the donor to determine whether or not agglutination occurs. In a second test, the cells of the recipient are "cross-matched" against the serum of the donor. If no agglutination of either the donor's or the recipient's cells occurs, it can be assumed that the two bloods are probably compatible enough to proceed with a transfusion even though the actual blood types are unknown.

## THE Rh-Hr BLOOD TYPES

In addition to the O-A-B blood group system, several other systems are sometimes important

in the transfusion of blood, the most important of which is the Rh-Hr system. The one major difference between the O-A-B system and the Rh-Hr system is as follows: In the O-A-B system, the agglutinins responsible for causing transfusion reactions develop spontaneously, whereas in the Rh-Hr system spontaneous agglutinins almost never occur. Instead, the person must first be massively exposed to some antigen of the system, usually by transfusion of blood into him, before he will develop enough agglutinins to cause significant transfusion reaction.

**The Rh Agglutinogens.** There are at least eight different types of Rh agglutinogens, each of which is called an *Rh factor*. The symbols for these are given in Table 8–3. The table also shows the symbols for the genes responsible for inheritance of the agglutinogens and, too, the frequencies of the different Rh factors in Caucasoids. Note that three Rh factors, designated rh, $Rh_1$, and $Rh_2$, are common among the population, whereas the other five Rh factors are uncommon.

*Inheritance of the Rh Factors.* Though there exists some confusion about the exact means of inheritance of the Rh factors, it is now believed that a single pair of genes is responsible for this inheritance. The eight genes shown in Table 8–3 are "allelomorphic," which means that any one of these eight genes can occur at the appropriate gene site on either one of the respective chromosomes. The Rh factors are mendelian dominant, so that if either chromosome contains a particular gene, the corresponding Rh factor is present in the blood. Thus, if the gene *r* appears on one chromosome and the gene *r'* on the other chromosome, the red cells of that particular person would have both rh and rh' agglutinogens. This person would be *heterozygous* for rh factors. If he should have an *r* gene on each of the chromosomes, he would have only one agglutinogen, rh, in his red blood cells; and he would be *homozygous*.

**The Hr Factors.** A factor similar to the Rh factor, but having far less tendency to cause transfusion reactions, is the Hr factor. It has been suggested that for each type of Rh factor there is a corresponding Hr factor and that these are reciprocally related to the Rh factors. That is, when the Rh factor is missing, the Hr factor supposedly is present. However, only a few of the theoretically possible Hr factors have actually been demonstrated, the three most important of which are hr, hr', and hr".

**Typing Bloods for Rh Factors.** Table 8–3 illustrates the reactions of the eight different types of Rh agglutinogens with six different types of Rh and Hr antisera. Note from this chart that there is a great amount of cross-reactivity of the antisera for different ones of the Rh factors rather than reactivity for only specific factors. To determine the Rh type, the

**TABLE 8–2.  Blood Typing—Showing Agglutination of Cells of the Different Blood Groups with Anti-A and Anti-B Agglutinins**

| Red Blood Cells | Sera | |
|---|---|---|
| | Anti-A | Anti-B |
| O | − | − |
| A | + | − |
| B | − | + |
| AB | + | + |

**TABLE 8–3.  The Rh Series of Agglutinogens***

| | Agglutin-ogens | Frequencies among NYC Caucasoids (per cent) | Corre-sponding Genes | Gene Frequencies (per cent) | Reactions with Rh Antisera | | | Reactions with Hr Antisera | | |
| --- | --- | --- | --- | --- | --- | --- | --- | --- | --- | --- |
| | | | | | Anti-$Rh_0$ | Anti-rh' | Anti-rh'' | Anti-hr' | Anti-hr'' | Anti-hr |
| Rh Negative | rh | 61.6 | r | 38.0 | − | − | − | + | + | + |
| | rh' | 2.8 | r' | 1.4 | − | + | − | − | + | − |
| | rh'' | 1.0 | r'' | 0.5 | − | − | + | + | − | − |
| | $rh_y$ | 0.02 | $r^y$ | 0.01 | − | + | + | − | − | − |
| Rh Positive | $Rh_0$ | 6.3 | $R^0$ | 3.2 | + | − | − | + | + | + |
| | $Rh_1$ | 64.5 | $R^1$ | 40.4 | + | + | − | − | + | − |
| | $Rh_2$ | 29.3 | $R^2$ | 16.4 | + | − | + | + | − | − |
| | $Rh_z$ | 0.2 | $R^z$ | 0.1 | + | + | + | − | − | − |

* Modified from A. S. Wiener and I. B. Wexler: An Rh-Hr Syllabus. Grune & Stratton, 1963.

six different types of antisera listed in the table are each mixed with a saline suspension of red blood cells and a small amount of plasma protein. From the agglutination reactions that occur, one can use Table 8–3 to determine the different types of Rh factor present in the blood.

## Rh POSITIVE AND Rh NEGATIVE PERSONS

Certain Rh agglutinogens are very likely to cause transfusion reactions. These are capitalized in Table 8–3: $Rh_0$, $Rh_1$, $Rh_2$, and $Rh_z$. Note also in Table 8–3 that all these agglutinogens react with anti-$Rh_0$ antiserum. Therefore, any time a person's blood reacts with anti-$Rh_0$ antiserum, his blood contains an Rh factor that can cause a serious transfusion reaction. For this reason he is said to be Rh positive.

When a person's red blood cells will not react with anti-$Rh_0$ antiserum, he is not likely to have an Rh factor that can cause a significant transfusion reaction. Therefore, he is said to be Rh negative; however, his blood might contain some less antigenic Rh factor or Hr factor that on occasion can cause a mild reaction.

Approximately 85 per cent of all Caucasoids are Rh positive and approximately 15 per cent are Rh negative.

## THE Rh IMMUNE RESPONSE

**Formation of Anti-Rh Agglutinins.** When red blood cells containing Rh factor, or even protein breakdown products of such cells, are injected into an Rh negative person, anti-Rh agglutinins develop very slowly, the maximum concentration of agglutinins occuring approximately two to four months later. This immune response occurs to a much greater extent in some people than in others. On multiple exposure to the Rh factor, the Rh negative person eventually becomes strongly "sensitized" to the Rh factor—that is, he develops a very high titer of anti-Rh agglutinins.

**Characteristics of Rh Agglutination.** The anti-Rh agglutinins are similar to the anti-A and anti-B agglutinins discussed previously; they attach to the Rh positive red blood cells and cause them to agglutinate. However, less potent anti-Rh antibodies (IgG type) also frequently develop instead of or in addition to the more potent ones (IgM type). These will attach to the cells but are not potent enough to form bridges from cell to cell to cause agglutination. This attachment of these "incomplete" antibodies to the red cell antigens greatly reduces the available reactive sites for the agglutinins and especially interferes with serologic agglutination tests for detecting the presence of the Rh agglutinogens. Indeed, the Rh agglutination reaction can be detected by the usual methods of blood typing only when extremely high titers of agglutinins have formed. More often, agglutination will take place only when the red blood cells are suspended in a solution of plasma protein which causes the reaction to take place much more rapidly than it takes place in the absence of the protein.

Anti-Rh antibodies do not cause hemolysis

directly, but whenever agglutination of the cells occurs, the agglutinated cells are gradually destroyed by phagocytes during the next few hours to few days so that the final effect of the agglutination reaction is still hemolysis.

**Erythroblastosis Fetalis.** Erythroblastosis fetalis is a disease of the fetus and newborn infant characterized by progressive agglutination and subsequent phagocytosis of the red blood cells. In most instances of erythroblastosis fetalis the mother is Rh negative, and the father is Rh positive; the baby has inherited the Rh positive characteristic from the father, and the mother has developed anti-Rh agglutinins that have diffused into the fetus to cause red blood cell agglutination.

*Prevalence of the Disease.* An Rh negative mother having her first Rh positive child usually does not develop sufficient anti-Rh agglutinins to cause any harm. However, an Rh negative mother having her second Rh positive child often will have become "sensitized" by the first child and therefore will often develop anti-Rh agglutinins rapidly upon becoming pregnant with the second child. Approximately 3 per cent of these second babies exhibit some signs of erythroblastosis fetalis; approximately 10 per cent of the third babies exhibit the disease; and the incidence rises progressively with subsequent pregnancies.

The Rh negative mother develops anti-Rh agglutinins only when the fetus is Rh positive. Many of the Rh positive fathers are heterozygous (about 55 per cent), causing about one fourth of the offspring to be Rh negative. Therefore, after an erythroblastic child has been born, it is not certain that future children will also be erythroblastic, for some of them could be Rh negative and not susceptible. Also, it is probable that the anti-Rh immune response in the mother becomes less and less intense the longer the interval between pregnancies; as a result, long intervals between pregnancies might be considerably safer than short intervals when the mother is Rh negative and the father is positive.

*Prevention of Erythroblastosis in Babies with Rh Negative Mothers.* The Rh negative mother usually becomes sensitized to the Rh positive factor in her child during the period from a few days before birth of the baby to the first few days following birth. At this time large quantities of fetal tissues, especially degenerating products of the placenta, release their antigens into the blood of the mother, giving her a healthy dose of the baby's Rh positive antigen. If this Rh positive antigen can be destroyed at this time before it can initiate the antibody response, the mother will not become presensitized against the Rh factor for subsequent pregnancies. This result can be achieved by passively immunizing the mother against the Rh positive factor. This is done by injecting into the mother serum from another Rh negative person who has already formed anti-Rh agglutinins. These agglutinins circulate in the mother's blood for three to eight weeks and destroy all the Rh positive factor

from the baby, thus preventing development of the mother's own anti-Rh agglutinins. With this new treatment it should be possible almost to eliminate future occurrences of erythroblastosis fetalis.

*Effect of the Mother's Antibodies on the Fetus.* After anti-Rh antibodies have formed in the mother, they diffuse very slowly through the placental membrane into the fetus' blood. There they cause slow agglutination of the fetus' blood. The agglutinated red blood cells gradually hemolyze, releasing hemoglobin into the blood. The reticuloendothelial cells then convert the hemoglobin into bilirubin, which causes yellowness (jaundice) of the skin. The antibodies probably also attack and damage many of the other cells of the body.

*Clinical Picture of Erythroblastosis.* The newborn, jaundiced, erythroblastotic baby is usually anemic at birth, and the anti-Rh agglutinins from the mother usually circulate in the baby's blood for one to two months after birth, destroying more and more red blood cells. Therefore, the hemoglobin level of the erythroblastotic baby often falls for approximately the first 45 days after birth, and, if the level falls below 6 to 8 grams per cent, the baby usually dies.

The hemopoietic tissues of the baby attempt to replace the hemolyzing red blood cells. The liver and the spleen become greatly enlarged and produce red blood cells in the same manner that they normally produce red blood cells during the middle of gestation. Because of the very rapid production of cells, many early forms, including many nucleated blastic forms, are emptied into the circulatory system, and it is because of the presence of these in the blood that the disease has been called "erythroblastosis fetalis."

Though the severe anemia of erythroblastosis fetalis is usually the cause of death, it is not the only cause, for severely erythroblastotic children are very likely to have many other degenerative abnormalities throughout the body, including especially damage to the brain. Many children who barely survive from the anemia exhibit permanent mental impairment or damage to motor areas of the brain. Much of this damage is caused by precipitation of bilirubin in the neuronal cells, causing their destruction, a condition called *kernicterus.*

*Treatment of the Erythroblastotic Baby.* The usual treatment for erythroblastosis fetalis is to replace the newborn infant's blood with Rh negative blood. Approximately 400 ml. of Rh negative blood is infused over a period of one and a half or more hours while the baby's own Rh positive blood is being removed. Once the Rh negative blood is in the baby, little further destruction of red blood cells occurs, and second transfusions are usually unnecessary. By the time the Rh negative cells are replaced with the baby's own Rh positive cells, a process that requires six or more weeks, the anti-Rh agglutinins that had come from the mother will have been destroyed.

*Erythroblastosis Fetalis Occurring in Babies of Rh Positive Mothers.* Approximately 7 per cent of the babies who have erythroblastosis fetalis are born of

Rh positive mothers rather than Rh negative mothers. A very few cases are caused by anti-A or anti-B agglutinins that diffuse from the mother into the fetus and agglutinate the fetus' red blood cells. Some of the other instances result from the fact that the so-called "Rh positive" mother is sometimes Rh positive for only one or two Rh factors and yet Rh negative for the other Rh factors. Finally, similar reactions have also been noted on rare occasions as a result of the Hr factor or other less well-known antigenic factors in the fetus' blood, some of which are discussed in the following section.

## OTHER BLOOD FACTORS

Many antigenic proteins besides the O, A, B, Rh, and Hr factors are present in red blood cells of different persons, but these other factors very rarely cause transfusion reactions and, therefore, are mainly of academic and legal importance. Some of these different blood factors are the M, N, P, S, s, P, Kell, Lewis, Duffy, Kidd, Diego, and Lutheran factors. Occasionally, multiple transfusions of red cells containing one of these factors causes specific agglutinins to develop in the recipient, and transfusion reactions can then occur when the same type of blood is infused a second time. However, these different antigenic factors usually do not exhibit extreme degrees of antigenicity and, therefore, usually cause very weak transfusion reactions or no reactions at all.

Inheritance of most of the different blood factors follows almost the same principles as those observed for inheritance of the O-A-B agglutinogens and for the Rh and Hr factors. The genes for the different factors are almost all dominant, and whenever any one of the genes is inherited from a parent the factor is usually present in the child's blood.

**Method for Studying Obscure Blood Factors.** One of the means by which different blood factors, including the Rh factor, have been studied in the human being has been to immunize lower animals, such as rabbits, with human blood cells. Specific antibodies develop in the animal against the different antigens of the human cells, and specific immune sera prepared from the plasma of these animals can then be used for determining the presence or absence of the same antigens in red blood cells of other persons.

**Blood Typing in Forensic Medicine.** In the past three decades the use of blood typing has become an important legal procedure in cases of disputed parentage. There are three different genes in the O-A-B system of blood groups and eight major ones in the Rh-Hr system. Including all the other blood types as well, more than 30 different blood group genes can be determined for each person by blood-typing procedures. After the mother's and child's genes have been determined, many of the father's genes, and his corresponding blood factors, are then known im-

mediately, because any gene present in a child but not present in the mother must be present in the father. If a suspected man is missing any one of the necessary blood factors, he is not the father. A falsely charged man can be cleared in about 75 per cent of the cases using the usually available antisera for blood typing, and in a much higher percentage of cases using all possible types of antisera.

## TRANSFUSION

**Indications for Transfusion.** The most common reason for transfusion is decreased blood volume, which will be discussed in detail in Chapter 28 in relation to shock. Also, transfusions are often used for treating anemia or to supply the recipient with some other constituent of whole blood besides red blood cells, such as to supply a thrombocytopenic patient with new platelets. Also, hemophilic patients can be rendered temporarily nonhemophilic by plasma transfusion, and, occasionally, the quantity of "complement" in a recipient must be supplemented by fresh plasma infusions before certain antigen-antibody reactions can take place.

### TRANSFUSION REACTIONS RESULTING FROM MISMATCHED BLOOD GROUPS

If blood of one blood group is transfused to a recipient of another blood group, a transfusion reaction is likely to occur in which the red blood cells *of the donor blood* are agglutinated. However, even though the bloods might be mismatched, it is very rare that the transfused blood ever causes agglutination *of the recipient's cells* for the following reason: The plasma portion of the donor blood immediately becomes diluted by all the plasma of the recipient, thereby decreasing the titer of the infused agglutinins to a level too low to cause agglutination. On the other hand, the infused blood does not dilute the agglutinins in the recipient's plasma to a major extent. Therefore, the recipient's agglutinins can still agglutinate the donor cells.

**Universal Donor Blood.** Because the cells of group O blood contain neither of the two agglutinogens, small quantities of this blood can be transfused into almost any recipient without immediate agglutination occurring (unless the donor blood has an unusually high titer of agglutinins, which could agglutinate the recipient's cells). For this reason group O blood is sometimes called *universal donor blood.* However, transfusion of especially large amounts of group O blood into a mismatched recipient can cause either immediate or delayed agglutination of the recipient's own cells because the infused agglutinins then are not diluted sufficiently to prevent the reaction.

Nevertheless, for use in extreme emergencies, when sufficient time is not available for blood typing or cross-matching, some blood banks keep small quantities of group O blood specially available.

**The Universal Recipient.** Group AB persons are sometimes called universal recipients because their plasma contains neither alpha nor beta agglutinins. Small quantities of all other blood groups can usually be infused without causing a transfusion reaction. However, here again, if large quantities of these mismatched bloods should be administered, the agglutinins in the donor blood might accumulate in sufficient quantities to agglutinate the recipient's type AB cells.

**Hemolysis of Red Cells Following Transfusion Reactions.** All transfusion reactions resulting from mismatched blood groups eventually cause hemolysis of the red blood cells. Occasionally, the antibodies are potent enough and are composed of the appropriate class of immunoglobulins to cause immediate hemolysis, but more frequently the cells agglutinate first and then are mainly entrapped in the peripheral vessels. Over a period of hours to days the entrapped cells are phagocytized, thereby liberating hemoglobin into the circulatory system.

When the rate of hemolysis is rapid, the concentration of hemoglobin in the plasma can rise to extremely high values. A small quantity of hemoglobin, up to 100 mg./100 ml. of plasma, becomes attached to one of the plasma proteins, *haptoglobin,* and continues to circulate in the blood without causing any harm. However, above the threshold value of 100 mg./100 ml. of plasma, the excess hemoglobin remains in the free form and diffuses out of the circulation into the tissue spaces or through the renal glomeruli into the kidney tubules, as is discussed below. The hemoglobin remaining in the circulation or passing into the tissue spaces is gradually ingested by phagocytic cells and converted into bilirubin, which will be discussed in Chapter 70. The concentration of bilirubin in the body fluids sometimes rises high enough to cause *jaundice*—that is, the person's tissues become

tinged with yellow pigment. But, if liver function is normal, jaundice usually does not appear unless 300 to 500 ml. of blood is hemolyzed in less than a day.

**Acute Kidney Shutdown Following Transfusion Reactions.** One of the most lethal effects of transfusion reactions is acute kidney shutdown, which can begin within a few minutes to a few hours and continue until the person dies of renal failure.

The kidney shutdown seems to result from three different causes: First, the antigen-antibody reaction of the transfusion reaction releases toxic substances from the hemolyzing blood that cause powerful renal vasoconstriction. Second, the loss of circulating red cells along with production of toxic substances from the cells and from the immune reaction often causes circulatory shock; the arterial blood pressure falls very low and the renal blood flow and urinary output decrease. Third, if the total amount of free hemoglobin in the circulating blood is greater than that quantity which can bind with haptoglobin, much of the excess leaks through the glomerular membranes into the kidney tubules. If this amount is still slight, it can be reabsorbed through the tubular epithelium into the blood and will cause no harm, but, if it is great, then only a small percentage is reabsorbed. Yet water continues to be reabsorbed, causing the tubular hemoglobin concentration to rise so high that it precipitates and blocks many of the tubules; this is especially true if the urine is acidic. Thus, renal vasoconstriction, circulatory shock, and tubular blockage all add together to cause acute renal shutdown. If the shutdown is complete, the patient dies within a week to 12 days, as explained in Chapter 38.

**Physiological Principles of Treatment Following Transfusion Reactions.** A transfusion reaction can also cause considerable degrees of fever and certain amounts of general toxicity, but these effects, in the absence of renal shutdown, are almost never lethal. Therefore, treatment is usually directed toward preventing renal damage, especially toward preventing hemoglobin precipitation in the kidney tubules, as follows: (1) rapid infusion of dilute intravenous fluids to cause water diuresis, (2) administration of mannitol, an osmotic diuretic, to prevent reabsorption of water by the tubules and also to overcome the renal vasoconstriction, (3) alkalinization of the body fluids because alkaline tubular fluid can dissolve more hemoglobin than can acid tubular fluid, and (4) occasionally, transfusion of properly matched whole blood to overcome the circulatory shock.

### TRANSFUSION REACTIONS OCCURRING WHEN Rh FACTORS ARE MISMATCHED

If the recipient has previously received a blood transfusion containing a strongly antigenic Rh factor not present in his own blood, then he may have built up appropriate anti-Rh antibodies that can cause a transfusion reaction on subsequent exposure to the same type of Rh blood. However, since anti-Rh agglutination is usually weak, the intensity of the reaction is usually less severe than when there is an anti-A or anti-B reaction.

### OTHER TYPES OF TRANSFUSION REACTIONS

**Pyrogenic Reactions.** Pyrogenic reactions occur more frequently than either agglutination or hemolysis. These reactions cause fever in the recipient but do not destroy red blood cells. Most pyrogenic reactions probably result from the presence in the donor plasma of proteins to which the recipient is allergic, or they result from the presence of breakdown products in old, deteriorating donor blood. If the donor has eaten certain foods to which the recipient is allergic, allergens from these foods can sometimes be transmitted to the recipient, thereby causing allergic reactions. The usual pyrogenic reaction causes severe chills, and occasionally the blood pressure falls, but ordinarily no dire results follow the reaction.

**Transfusion Reactions Resulting from Anticoagulants.** The usual anticoagulant used for transfusion is a citrate salt. As discussed in the following chapter, citrate operates as an anticoagulant by combining with the calcium ions of the plasma so that these become nonionizable. Without the presence of ionizable calcium the coagulation process cannot take place. Normal nerve, muscle, and heart function also cannot occur in the absence of calcium ions. Therefore, if large quantities of blood containing citrate anticoagulant are administered rapidly, the recipient may experience typical tetany due to low calcium, which is discussed in detail in Chapter 79. Such tetany can kill the patient within a few minutes because of respiratory muscle spasm.

Ordinarily the liver can remove citrate from the blood within a few minutes and polymerize it to form glycogen or utilize it directly for energy. Therefore, when blood transfusions are given at slow rates (less than one liter per hour), the person is usually completely safe from the citrate type of reaction. On the other hand, if he has liver damage, the rate of transfusion must be decreased more than usually to prevent a low calcium ion reaction.

# TRANSPLANTATION OF TISSUES AND ORGANS

**Relation of Genotypes to Transplantation.** In this modern age of surgery, many attempts are being made to transplant tissues and organs from one person to another, or, occasionally, from lower animals to the human being. Many of the different antigenic proteins of red blood cells that cause transfusion reactions plus still many more are present in the other cells of the body. Consequently, any foreign cells transplanted into a recipient can cause immune responses and immune reactions. In other words, most recipients are just as able to resist invasion by foreign cells as to resist invasion by foreign bacteria.

**Autologous, Homologous, and Heterologous Transplants.** A transplant of tissue or of a whole organ from one part of a person's body to some other part of his body is called an *autologous transplant*. Such transplants can be skin grafts, bone grafts, cartilage grafts, and so forth. A transplant from one human being to another or from any other animal to another member of the same species is called a *homologous transplant*. Finally, a transplant from a lower animal to a human being or from an animal of one species to one of another species is called a *heterologous transplant*.

**Transplantation of Cellular Tissues.** In the case of autologous transplants, cells in the transplant will almost always live indefinitely if an adequate blood supply is provided, but, in the case of homologous and heterologous transplants, immune reactions almost always occur, causing death of all the cells in the transplant 3 to 10 weeks after transplantation unless some specific therapy is used to prevent the immune reaction. The cells of homologous transplants usually persist slightly longer than those of heterologous transplants because the antigenic structure of the homologous transplant is more nearly the same as that of the recipient's tissues than is true of the heterologous transplant. And when the tissues are properly "typed" and are very similar prior to transplant, completely successful homologous transplants occasionally result. The greater the difference in antigenic structure, the more rapid and the more severe are the immune reactions to the graft.

Transplants from one identical twin to another are one example in which homologous transplants of cellular tissues are usually successful. The reason for this is that the antigenic proteins of both twins are determined by identical genes derived originally from the single fertilized ovum.

Some of the different cellular tissues and organs that have been transplanted either experimentally or for temporary benefit from one person to another are skin, kidney, heart, liver, glandular tissue, bone marrow, and lung. A few kidney homologous transplants have been successful for as long as five years, a rare liver and heart transplant for one to two years, and lung transplants for one month.

**Transplantation of Noncellular Tissues.** Certain tissues that have no cells or in which the cells are unimportant to the purpose of the graft, such as the cornea, tendon, fascia, and bone, can usually be grafted from one person to another with considerable success. In these instances the grafts act merely as a supporting latticework into which or around which the surrounding living tissues of the recipient grow. Indeed, some such grafts—bone and fascia—are occasionally successful even when they come from a lower animal rather than from another human being.

A special problem arises in the case of a corneal transplant from one human eye to another, for, not only must it maintain structural continuity, but it must also remain transparent. About 50 to 90 per cent of the corneas remain transparent, the exact percentage depending principally on the preliminary condition of the recipient's eye. The remainder suddenly become cloudy at the end of two weeks to two years. This cloudiness is the result of an antigen-antibody reaction. It is believed that corneas that do not become cloudy fail to do so for any of three different reasons: First, their antigenicity might be low enough that they fail to cause the production of sufficient antibodies to elicit the response. Second, the vascularity of the cornea is so slight that the antigens of the cornea may never infuse the host in sufficient quantities to elicit an immune response. Or, third, it is possible that, if the reaction is delayed long enough, the corneal proteins adapt to the immunologic structure of the recipient so that a reaction can no longer occur against the corneal tissue.

## ATTEMPTS TO OVERCOME THE ANTIGEN-ANTIBODY REACTIONS IN TRANSPLANTED TISSUE

Because of the extreme potential importance of transplanting certain tissues and organs, such as skin, kidneys, and lungs, serious attempts have been made to prevent the antigen-antibody reactions associated with transplants. The following specific procedures have met with certain degrees of clinical or experimental success.

**Tissue Typing.**   In the same way that red blood cells can be typed to prevent reactions between recipient and donor, so also is it possible to "type" tissues to help prevent graft rejection, though thus far this procedure has met with far less success than has been achieved in red blood cell typing.

The most important antigens that cause graft rejection are a group of antigens called the HL-A antigens. These are a group of 25 or more different antigens in the tissue cell membranes that are determined by four separate genes at the so-called *HL-A genetic locus*. The genes are allelomorphic and, therefore, can code for only four of the HL-A antigens in any one person. That is, only four of the antigens can occur in any one individual, but they can be in a combination of any four of the more than 25 separate HL-A antigens.

The same HL-A antigens occur in the white blood cells. Therefore, the means for tissue typing is to "type" for these antigens in the membranes of lymphocytes separated from the person's blood. The lymphocytes are mixed with appropriate antisera and complement, and after appropriate incubation the cells are tested for membrane damage, usually by testing the rate of uptake by the lymphocytic cells of a supravital dye.

Fortunately, some of the HL-A antigens are not severely antigenic, for which reason precise match of some of the antigens between donor and recipient is not absolutely essential to allow homologous graft acceptance. This is very fortunate, because so many different combinations of these antigens occur in different persons that it is almost impossible ever to get a precise match. Nevertheless, by obtaining the best possible match between donor and recipient, the grafting procedure has become far less hazardous. The best success has been tissue-type matches between members of the same family. Of course, the match in identical twins is exact.

**Use of Anti-Lymphocyte Serum.**   It was pointed out in the previous chapter that grafted tissues are usually destroyed by lymphocytes that have become sensitized against the graft. These cells invade the graft and gradually destroy the grafted cells. Upon contact with the lymphocytes, the cells of the graft begin to swell, their membranes become very permeable, and finally they rupture. Simultaneously, macrophages move in to clean up the debris. Within a few days to a few weeks after this process begins, the tissue is completely destroyed, even though the graft had been completely viable and functioning normally for the few days to few weeks after its original transplantation.

Therefore, one of the most effective procedures in preventing rejection of grafted tissues has been to inoculate the recipient with *anti-lymphocyte serum*. This serum is made in horses by injecting human lymphocytes into them; the antibodies that develop in these animals will then attack and destroy the human lymphocytes. When serum from one of these animals is injected into the transplanted recipient, the number of circulating small lymphocytes can be decreased to as little as 5 to 10 per cent of the normal number, and there is corresponding decrease in the intensity of the rejection reaction. Unfortunately, the procedure does not continue to work very satisfactorily after the first few injections of the antiserum because the recipient soon begins to build up antibodies against the antiserum itself.

**Glucocorticoid Therapy (Cortisone, Hydrocortisone, and ACTH).**   The glucocorticoid hormones from the adrenal gland greatly suppress the formation of both antibodies and immunologically competent lymphocytes. Therefore, administration of large quantities of these, or of ACTH which causes the adrenal gland to produce glucocorticoids, often allows transplants to persist much longer than usual in the recipient. However, this is only a delaying process rather than a preventive one.

**Suppression of Antibody Formation.**   Occasionally, a human being has naturally suppressed antibody formation resulting from (a) congenital agammaglobulinemia, in which case gamma globulins are not produced, and (b) destructive diseases of the reticuloendothelial system. Transplants of homologous tissues into such individuals are occasionally successful, or at least their destruction is delayed. Also, irradiative destruction of most of the lymphoid tissue by either x-rays or gamma rays renders a person much more susceptible than usual to a homologous transplant. And treatment with certain drugs, such as azathioprine (Imuran) which suppresses antibody formation, also increases the likelihood of success with homologous transplants. Unfortunately, all of these procedures also leave the person unprotected from disease.

*Removal of the Thymus Gland to Prevent Antibody Formation.* It was pointed out in Chapter 7 that the thymus gland during fetal and early postnatal life forms thymic cells that migrate throughout the body and become the precursors of the sensitized lymphocytes. Also, for the first few weeks or months after birth, the thymus secretes a hormone that enhances the proliferation of lymphoid tissue throughout the body. Therefore, if the thymus gland is removed during fetal life or immediately after birth, the immune system is greatly impaired throughout the remainder of the animal's life. Homologous transplants are frequently successful in such animals, but the loss of immunity to disease makes it difficult for the animal to live.

To summarize, transplantation of living tissues in human beings up to the present has been hardly more than an experiment, though some degree of clinical success is now being recorded, especially for kidney transplants. But when someone succeeds in blocking the immune response of the recipient to a donor organ without at the same time destroying the recipient's specific immunity for disease, this story will change overnight.

# REFERENCES

Aminoff, D. (ed.): Blood and Tissue Antigens. New York, Academic Press, Inc., 1969.

Amos, D. B., and Ward, F. E.: Immunogenetics of the HL-A system. *Physiol. Rev.*, 55:206, 1975.

Anti-Lymphocytic Serum (CIBA Foundation Symposium): Boston, Little, Brown and Company, 1967.

Barrow, E. M., and Graham, J. B.: Blood coagulation factor VIII (antihemophilic factor): With comments on von Willebrand's disease and Christmas disease. *Physiol. Rev.*, 54:23, 1974.

Calne, R. Y.: Immunosuppression and clinical organ transplantation. *Transplant. Proc.*, 6:49, 1974.

Cerottini, J. C., and Brunner, K. T.: Cell-mediated cytotoxicity, allograft rejection, and tumor immunity. *Adv. Immunol.*, 18:67, 1974.

Clarke, C. A., and McConnell, R. B.: Prevention of Rh-Hemolytic Disease. Springfield, Ill., Charles C Thomas, Publisher, 1972.

David, J. R.: Lymphocyte mediators and cellular hypersensitivity. *N. Engl. J. Med.*, 288:143, 1973.

Erslev, A. J., and Gabuzda, T. G.: Pathophysiology of hematologic disorders. *In* Sodeman, W. A., Jr., and Sodeman, W. A. (eds.): Pathologic Physiology: Mechanisms of Disease, 5th Ed. Philadelphia, W. B. Saunders Company, 1974, p. 511.

Gamma globulins [Series of papers]. *N. Engl. J. Med.*, 275:480, 536, 591, 652, 709, 769, 826; 1966.

Gatti, R. A., Kersey, J. H., Yunis, E. J., and Good, R. A.: Graft-versus-host disease. *Prog. Clin. Pathol.*, 5:1, 1973.

Heppner, G. H., Griswold, D. E., DiLorenzo, J., Poplin, E. A., and Calabresi, P.: Selective immunosuppression by drugs in balanced immune responses. *Fed. Proc.*, 33:1882, 1974.

Kahan, B. D., and Reisfeld, R. A. (eds.): Transplantation Antigens. New York, Academic Press, Inc., 1973.

Krueger, G. R.: Morphology of chemical immunosuppression. *Adv. Pharmacol. Chemother.*, 10:1, 1972.

Lyle, L. R., and Parker, C. W.: New approaches to immunosuppression. *Fed. Proc.*, 33:1889, 1974.

Mollison, P. L.: Blood Transfusion in Clinical Medicine, 4th Ed. Philadelphia, F. A. Davis Co., 1967.

Petersen, V. P. et al.: Late failure of human renal transplants. An analysis of transplant disease and graft failure among 125 recipients surviving for one to eight years. *Medicine (Baltimore)*, 54:45, 1975.

Reisfeld, R. A., and Kahan, B. D.: Markers of biological individuality. *Sci. Amer.*, 226:28, 1972.

Rosenfield, R. E., Allen, F. H., Jr., and Rubinstein, P.: Genetic model for the Rh blood-group system. *Proc. Natl. Acad. Sci. U.S.A.*, 70:1303, 1973.

Rapaport, F. T.: Cross-reactive antigens, cancer, and transplantation. *Transplant. Proc.*, 6 (Suppl. 1):39, 1974.

Mollison, P. L.: Blood Transfusion in Clinical Medicine, 5th Ed. Philadelphia, J. B. Lippincott Company, 1972.

Mollison, P. L.: Clinical aspects of Rh immunization. Philip Levine Award Address. *Amer. J. Clin. Pathol.*, 60:287, 1973.

Najarian, J. S., and Simmons, R. L.: Transplantation. Philadelphia, Lea & Febiger, 1972.

Salmon, C. et al.: Current genetic problems in the ABO blood group system. *Biomedicine*, 18:375, 1973.

Terasaki, P. I., Opelz, G., and Mickey, M. R.: Histocompatibility and clinical kidney transplants. *Transplant. Proc.*, 6 (Suppl. 1): 33, 1974.

Unanue, E. R.: Cellular events following binding of antigen to lymphocytes. *Amer. J. Pathol.*, 77:2, 1974.

Wiener, A. S.: Elements of blood group nomenclature with special reference to the Rh-Hr blood types. *J.A.M.A.*, 199:985, 1967.

Wiener, A. S.: Modern blood group nomenclature. *J. Amer. Med. Tech.*, 30:174, 1968.

Wiener, A. S., and Socha, W. W.: Macro- and microdifferences in blood group antigens and antibodies. *Int. Arch. Allergy Appl. Immunol.*, 47:547, 1974.

Wiener, A. S., and Wexler, I. B.: An Rh-Hr Syllabus: The Types and Their Applications, 2nd Ed. New York, Grune & Stratton, Inc., 1963.

Yunis, E. J., Gatti, R. A., and Amos, D. B. (eds.): Tissue Typing and Organ Transplantation. New York, Academic Press, Inc., 1973.

Zipursky, A.: The universal prevention of Rh immunization. *Clin. Obstet. Gynecol.*, 14:869, 1971.

# 9

# Hemostasis and Blood Coagulation

## EVENTS IN HEMOSTASIS

The term hemostasis means prevention of blood loss. Whenever a vessel is severed or ruptured, hemostasis is achieved by several different mechanisms including (1) vascular spasm, (2) formation of a platelet plug, (3) blood coagulation, and (4) growth of fibrous tissue into the blood clot to close the hole in the vessel permanently.

### VASCULAR SPASM

Immediately after a blood vessel is cut or ruptured, the wall of the vessel contracts; this instantaneously reduces the flow of blood from the vessel rupture. The contraction results from both nervous reflexes and local myogenic spasm. The nervous reflexes presumably are initiated by pain impulses originating from the traumatized vessel or from nearby tissues. However, most of the spasm probably results from local myogenic contraction of the blood vessels. This is initiated by direct damage to the vascular wall, which presumably causes transmission of action potentials along the vessel wall for several centimeters and results in constriction of the vessel. The more of the vessel that is traumatized, the greater is the degree of spasm; this means that the sharply cut blood vessel usually bleeds much more than does the vessel ruptured by crushing. This local vascular spasm lasts for as long as 20 minutes to half an hour, during which time the ensuing processes of platelet plugging and blood coagulation can take place.

The value of vascular spasm as a means of hemostasis is illustrated by the fact that persons whose legs have been severed by crushing types of trauma sometimes have such intense spasm in vessels as large as the anterior tibial artery that there is no serious loss of blood.

### FORMATION OF THE PLATELET PLUG

The second event in hemostasis is an attempt by the platelets to plug the rent in the vessel. To understand this it is important that we first understand the nature of platelets themselves.

Platelets are minute round or oval discs about 2 microns in diameter. They are fragments of *megakaryocytes,* which are extremely large cells of the hemopoietic series formed in the bone marrow. The megakaryocytes disintegrate into platelets while they are still in the bone marrow and release the platelets into the blood. The normal concentration of platelets in the blood is between 200,000 and 400,000 per cubic millimeter.

**Mechanism of the Platelet Plug.** Platelet repair of vascular openings is based on several important functions of the platelet itself: When platelets come in contact with a *wettable* surface, such as the collagen fibers in the vascular wall, they immediately change their characteristics drastically. They begin to swell; they assume irregular forms with numerous irradiating processes protruding from their surfaces; they become sticky so that they stick to the collagen fibers; and they secrete large quantities of ADP. The ADP, in turn, acts on nearby platelets to activate them as well, and the stickiness of these additional platelets causes them to adhere to the originally activated platelets. Therefore, at the site of any rent in a vessel, the exposed collagen of the subendothelial tissues elicits a vicious cycle of activation of successively increasing numbers of platelets; these accumulate to form a *platelet plug*. This is a fairly loose plug, but it is usually successful in blocking the blood loss. During the process of blood coagulation, to be described in subsequent paragraphs, the substance *thrombin* is formed, and this further alters the platelets to make them bind together irreversibly, thus forming a tight and unyielding plug.

***Repair of Vascular Endothelium by Platelets.*** Minute vascular holes frequently occur even in the middle of the thin endothelial cells. In the absence of platelets, blood occasionally escapes through these holes rather than between the endothelial cells. On the other hand, when platelets are present in the blood, they immediately coalesce with the endothelial cell at the opening of the hole and repair it, presumably by providing structural materials to repair the membrane.

**Importance of the Platelet Method for Closing Vascular Holes.** If the rent in a vessel is small, the platelet plug by itself can stop blood loss completely, but if there is a large hole, a blood clot in addition to the platelet plug is required to stop the bleeding.

The platelet plugging mechanism is extremely important to close the minute ruptures in very small blood vessels that occur hundreds of times daily, including those through the endothelial cells themselves. A person who has very few platelets develops literally hundreds of small hemorrhagic areas under his skin and throughout his internal tissues, but this does not occur in the normal person. The platelet plugging mechanism does not occlude the vessel itself but merely plugs the hole, so that the vessel continues to function normally. This usually is not true when a blood clot occurs.

### CLOTTING IN THE RUPTURED VESSEL

The third mechanism for hemostasis is formation of the blood clot. The clot begins to develop in 15 to 20 seconds if the trauma of the vascular wall has been severe and in one to two minutes if the trauma has been minor. Activator substances both from the traumatized vascular wall and from platelets and blood proteins adhering to the collagen of the traumatized vascular wall initiate the clotting process. The physical events of this process are illustrated in

Figure 9–1, and the chemical events will be discussed in detail later in the chapter.

Within three to six minutes after rupture of a vessel, the entire cut or broken end of the vessel is filled with clot. After 30 minutes to an hour, the clot retracts; this closes the vessel still further. Platelets play an important role in this clot retraction, as will also be discussed later in the chapter.

### FIBROUS ORGANIZATION OF THE BLOOD CLOT

Once a blood clot has formed, it can follow two separate courses: it can become invaded by fibroblasts, which subsequently form connective tissue all through the clot; or it can dissolute. The usual course for a clot that forms in a small hole of a vessel wall is invasion by fibroblasts, beginning within a few hours after the clot is formed, and complete organization of the clot into fibrous tissue within approximately 7 to 10 days. On the other hand, when a large mass of blood clots, such as blood that has leaked into tissues, special substances within the tissue usually activate a mechanism for dissoluting most of the clot, which will be discussed later in the chapter.

## MECHANISM OF BLOOD COAGULATION

**Basic Theory.** Over 30 different substances that affect blood coagulation have been found in

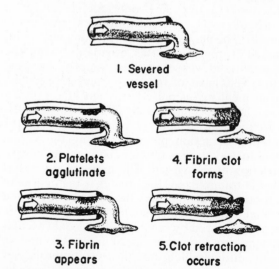

I. Severed vessel

2. Platelets agglutinate

3. Fibrin appears

4. Fibrin clot forms

5. Clot retraction occurs

**Figure 9–1.** The clotting process in the traumatized blood vessel. (Redrawn from Seegers: Hemostatic Agents. Charles C Thomas.)

the blood and tissues, some promoting coagulation, called *procoagulants,* and others inhibiting coagulation, called *anticoagulants*. Whether or not the blood will coagulate depends on the degree of balance between these two groups of substances. Normally the anticoagulants predominate and the blood does not coagulate, but when a vessel is ruptured the activity of the procoagulants in the area of damage becomes much greater than that of the anticoagulants, and then a clot does develop.

*General Mechanism.* Almost all research workers in the field of blood coagulation agree that clotting takes place in three essential steps:

**First, a substance called *prothrombin activator* is formed in response to rupture of the vessel or damage to the blood itself.**

**Second, the prothrombin activator catalyzes the conversion of prothrombin into *thrombin*.**

**Third, the thrombin acts as an enzyme to convert fibrinogen into *fibrin threads* that enmesh red blood cells and plasma to form the clot itself.**

Unfortunately, the detailed mechanisms by which prothrombin activator is formed are still doubtful. On the other hand, the mechanisms by which prothrombin is converted to thrombin and by which thrombin then acts to cause the formation of fibrin threads are much better understood. Therefore, let us first discuss the basic mechanism by which the blood clot is formed, beginning with the conversion of prothrombin to thrombin; then we will come back to the initiating stages in the clotting process by which prothrombin activator is formed.

## CONVERSION OF PROTHROMBIN TO THROMBIN

After prothrombin activator has been formed as a result of rupture of the blood vessel or as a result of damage to activator substances in the blood, it can cause conversion of prothrombin to thrombin, which in turn causes polymerization of fibrinogen molecules into fibrin threads within 10 to 15 seconds. Thus the rate limiting factor in causing blood coagulation is usually the formation of prothrombin activator and not the subsequent reactions beyond that point.

**Prothrombin and Thrombin.** Prothrombin is a plasma protein, an alpha$_2$ globulin, having a molecular weight of 68,700. It is present in normal plasma in a concentration of about 15 mg./100 ml. It is an unstable protein that can split easily into smaller compounds, one of which is *thrombin,* which has a molecular weight of 33,700, almost exactly half that of prothrombin.

Prothrombin is formed continually by the liver, and it is continually being used throughout the body for blood clotting. If the liver fails to produce prothrombin, its concentration in the plasma falls too low within 24 hours to provide normal blood coagulation. Vitamin K is required by the liver for normal formation of prothrombin; therefore, either lack of vitamin K or the presence of liver disease that prevents normal prothrombin formation can often decrease the prothrombin level so low that a bleeding tendency results.

**Effect of Prothrombin Activator to Form Thrombin from Prothrombin.** Figure 9–2 illustrates the conversion of prothrombin to thrombin under the influence of prothrombin activator and calcium ions. The rate of formation of thrombin from prothrombin is almost directly proportional to the quantity of prothrombin activator available, which in turn is approximately proportional to the degree of trauma to the vessel wall or to the blood. In turn, the rapidity of the clotting process is proportional to the quantity of thrombin formed.

## CONVERSION OF FIBRINOGEN TO FIBRIN—FORMATION OF THE CLOT

**Fibrinogen.** Fibrinogen is a high molecular weight protein (340,000) occurring in the plasma in quantities of 100 to 700 mg./100 ml. Most if not all of the fibrinogen in the circulating blood is formed in the liver, and liver disease occasionally decreases the concentration of circulating fibrinogen, as it does the concentration of prothrombin, which was pointed out previously.

Because of its large molecular size, fibrinogen normally leaks into the interstitial fluids to only a slight extent, and, since it is one of the

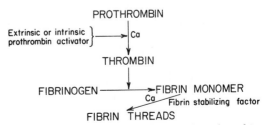

**Figure 9–2.** Schema for conversion of prothrombin to thrombin, and polymerization of fibrinogen to form fibrin threads.

essential factors in the coagulation process, interstitial fluids ordinarily coagulate poorly if at all. Yet, when the permeability of the capillaries becomes pathologically increased, fibrinogen does then appear in the tissue fluids and in lymph in sufficient quantities to allow clotting of these fluids in much the same way that plasma and whole blood clot.

**Action of Thrombin on Fibrinogen to Form Fibrin.** Thrombin is a protein *enzyme* with proteolytic capabilities. It acts on fibrinogen to remove two low molecular weight peptides from each molecule of fibrinogen, forming a molecule of *fibrin monomer* which has the automatic capability of polymerizing with other fibrin monomer molecules. Therefore, many fibrin monomer molecules polymerize within seconds into *long fibrin threads* that form the *reticulum* of the clot. In the early stages of this polymerization, the fibrin monomer molecules attach to each other by loose hydrogen and hydrophobic bonds; at this time the polymer chains are weak and can be broken apart with ease. However, immediately thereafter another plasma globulin factor, the *fibrin stabilizing factor,* acts as an enzyme to cause covalent bonding between the fibrin monomer molecules and also between the adjacent polymer chains. This adds tremendously both to the three dimensional meshwork of the fibrin threads and to the strength of the threads.

**The Blood Clot.** The clot is composed of a meshwork of fibrin threads running in all directions and entrapping blood cells, platelets, and plasma. The fibrin threads adhere to damaged surfaces of blood vessels; therefore, the blood clot becomes adherent to any vascular opening and thereby prevents blood loss.

**Clot Retraction—Serum.** Within a few minutes after a clot is formed, it begins to contract and usually expresses most of the fluid from the clot within 30 to 60 minutes. The fluid expressed is called *serum,* because all of its fibrinogen and most of the other clotting factors have been removed; in this way, serum differs from plasma. Serum obviously cannot clot because of lack of these factors.

For reasons not completely understood, large numbers of platelets are necessary for clot retraction to occur. Electron micrographs of platelets in blood clots show that they become attached to the fibrin threads in such a way that they actually bond different threads together. Furthermore, platelets contain tremendous quantities of the high energy compound adenosine triphosphate. It has been suggested that the energy in this compound plays a role in bonding the successive molecules in the fibrin threads or that it causes increased folding of the threads, thereby decreasing the lengths of the threads and thus expressing the serum from the clot.

As the clot retracts, the edges of the broken blood vessel are pulled together, thus possibly or probably contributing to the ultimate state of hemostasis.

## THE VICIOUS CYCLE OF CLOT FORMATION

Once a blood clot has started to develop, it normally extends within minutes into the surrounding blood. That is, the clot itself initiates a vicious cycle to promote more clotting. One of the most important causes of this is the fact that the proteolytic action of thrombin allows it to act on many of the other blood clotting factors in addition to fibrinogen. For instance, thrombin has a direct proteolytic effect on prothrombin itself, tending to split this into still more thrombin, and it acts on some of the blood clotting factors responsible for the formation of prothrombin activator. (These effects, to be discussed in subsequent paragraphs, include (a) acceleration of the actions of Factors VIII, IX, X, XI, and XII, and (b) aggregation of platelets.) Once a critical amount of thrombin is formed, a vicious cycle develops that causes still more blood clotting and more thrombin to be formed; thus, the blood clot continues to grow until something stops its growth.

## BLOCK OF CLOT GROWTH BY BLOOD FLOW

Fortunately, when a clot develops, the vicious cycle of continued clot formation occurs only where the blood is not moving because flowing blood carries the thrombin and the other procoagulants released during the clotting process away so rapidly that their concentrations cannot rise high enough to promote further clotting. Thus, extension of the clot almost always stops where it comes in contact with blood that is flowing faster than a certain velocity.

In addition, the clotting process fortunately is not self-propagating until the concentrations of the procoagulants rise above critical concentrations. At lower concentrations, many inhibitor substances in the blood, some of which will be discussed later in the chapter, continually block the actions of the procoagulants or destroy them. In addition, the reticuloendothelial system, particularly that of the liver and of the bone marrow, removes most of the circulating procoagulants within a few minutes.

## INITIATION OF COAGULATION: FORMATION OF PROTHROMBIN ACTIVATOR

Now that we have discussed the clotting process initiated by the conversion of prothrombin to thrombin, we must turn to the more complex mechanisms that initiate the activation of the prothrombin. These mechanisms can be set into play by trauma to the tissues, trauma to the blood, or contact of the blood with special substances, such as collagen outside the blood vessel endothelium. In each instance, they lead to the formation of *prothrombin activator,* which is the immediate initiator of prothrombin activation.

There are two basic ways in which the formation of prothrombin activator can be initiated and therefore also serve for the initiation of clotting: (1) by the *extrinsic pathway* that begins with trauma to the tissues outside the blood vessels, or (2) by the *intrinsic pathway* that begins in the blood itself.

In both the extrinsic and intrinsic pathways a series of different plasma proteins, especially beta globulins, play major roles. These, along with the other factors already discussed that enter into the clotting process, are called *blood clotting factors* and for the most part are designated by Roman numerals, as listed in Table 9–1. In the sections dealing with intrinsic and extrinsic pathways we will specifically discuss blood clotting factor V and factors VII through XII.

### The Extrinsic Mechanism for Initiating Clotting

The extrinsic mechanism for initiating the formation of prothrombin activator begins with blood coming in contact with traumatized tissues and occurs according to the following three basic steps, as illustrated in Figure 9–3.

(1) *Release of Tissue Factor and Tissue Phospholipids.* The traumatized tissue releases two factors that set the clotting process into motion. These are (a) *tissue factor,* which is a proteolytic enzyme, and (b) *tissue phospholipids,* which are mainly phospholipids of the tissue cell membranes.

**TABLE 9–1. Clotting Factors in the Blood and Their Synonyms**

| Clotting Factor | Synonym |
| --- | --- |
| Fibrinogen | Factor I |
| Platelets | Platelet factor 3 |
| Prothrombin | Factor II |
| Tissue thromboplastin | Factor III |
| Calcium | Factor IV |
| Factor V | Proaccelerin; labile factor; Ac-globulin; Ac-G |
| Factor VII | Serum prothrombin conversion accelerator; SPCA; proconvertin; autoprothrombin I |
| Factor VIII | Antihemophilic factor; AHF; antihemophilic globulin; AHG |
| Factor IX | Plasma thromboplastin component; PTC; Christmas factor; CF; autoprothrombin II |
| Factor X | Stuart factor; Stuart-Prower factor; Prower factor; autoprothrombin $I_c$ |
| Factor XI | Plasma thromboplastin antecedent; PTA |
| Factor XII | Hageman factor |
| Factor XIII | Fibrin stabilizing factor |
| Prothrombin activator | Thrombokinase; complete thromboplastin |

(2) *Activation of Factor X to Form Activated Factor X—Role of Factor VII and Tissue Factor.* The tissue factor proteolytic enzyme complexes with blood coagulation factor VII, and this complex, in the presence of tissue phospholipids, acts on factor X to form *activated factor X.*

(3) *Effect of Activated Factor X to Form Prothrombin Activator—Role of Factor V.* The activated factor X complexes immediately with the tissue phospholipids released from the traumatized tissue and also with factor V to form the complex called *prothrombin activator.* Within 10 to 15 seconds this splits prothrombin to form thrombin, and the clotting process proceeds as has already been explained.

### The Intrinsic Mechanism for Initiating Clotting

The second mechanism for initiating the formation of prothrombin activator, and therefore for initiating clotting, begins with trauma to the blood itself and continues through the following series of cascading reactions, as illustrated in Figure 9–4.

**EXTRINSIC PATHWAY**

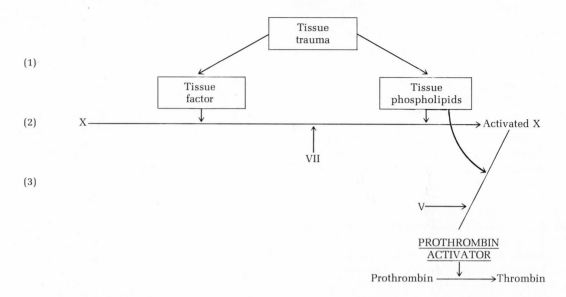

**Figure 9–3.**   The extrinsic pathway for initiating blood clotting.

**INTRINSIC PATHWAY**

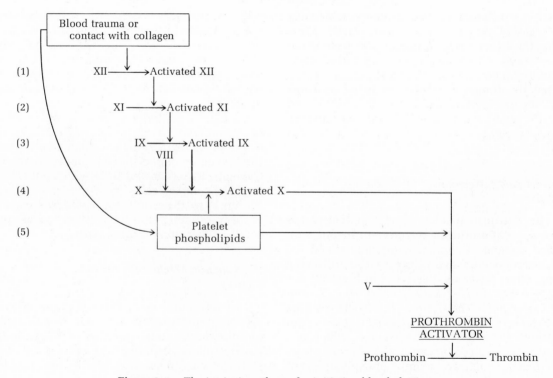

**Figure 9–4.**   The intrinsic pathway for initiating blood clotting.

(1) *Activation of Factor XII and Release of Platelet Phospholipids by Blood Trauma.* Trauma to the blood alters two important clotting factors in the blood—factor XII and the platelets. When factor XII is disturbed, such as by coming into contact with collagen or with a wettable surface such as glass, it takes on a new configuration that converts it into a proteolytic enzyme called "activated factor XII."

Simultaneously, the blood trauma also damages the platelets, either because of adherence to collagen or to a wettable surface (or by damage in other ways), and this releases platelet phospholipid, frequently called *platelet factor III*, which also plays a role in subsequent clotting reactions.

(2) *Activation of Factor XI.* The activated factor XII acts on factor XI to activate this as well, which is the second step in the intrinsic pathway.

(3) *Activation of Factor IX by Activated Factor XI.* The activated factor XI acts on factor IX to activate this factor, also.

(4) *Activation of Factor X—Role of Factor VIII.* The activated factor IX, acting in concert with factor VIII and with the platelet phospholipids from the traumatized platelets, activates factor X. It is clear that when either factor VIII or platelets are in short supply, this step is deficient. Factor VIII is the factor that is missing in the person who has classical *hemophilia,* for which reason it is called *antihemophilic factor.* Platelets are the clotting factor that is lacking in the bleeding disease called *thrombocytopenia.*

(5) *Action of Activated Factor X to Form Prothrombin Activator—Role of Factor V.* This step in the intrinsic pathway is essentially the same as the last step in the extrinsic pathway. That is, activated factor X combines with factor V and platelet phospholipids to form the complex called *prothrombin activator.* The only difference is that the phospholipids in this instance come from the traumatized platelets rather than from traumatized tissues. The prothrombin activator in turn initiates within seconds the cleavage of prothrombin to form thrombin, thereby setting into motion the final clotting process, as described earlier.

## Role of Calcium Ions in the Intrinsic and Extrinsic Pathways

Except for the first two steps in the intrinsic pathway, calcium ions are required for promotion of all of the reactions. Therefore, in the absence of calcium ions, blood clotting will not occur.

Fortunately, in the living body the calcium ion concentration never falls low enough to affect significantly the kinetics of blood clotting. The reason for this is that long before calcium ion concentration can fall this low, the diminished level of calcium ions will kill the person by causing muscle tetany throughout the body, especially of the respiratory muscles.

On the other hand, when blood is removed from a person, it can be prevented from clotting by reducing the calcium ion concentration below the threshold level for clotting, either by deionizing the calcium by reacting it with substances such as *citrate ion* or by precipitating the calcium with substances such as the *oxalate ion.*

## Summary of Blood Clotting Initiation

It is clear from the above schemas of the intrinsic and extrinsic systems for initiating blood clotting that clotting is initiated after rupture of blood vessels by both of the pathways. The tissue factor and tissue phospholipids initiate the extrinsic pathway, while contact of factor XII and the platelets with collagen in the vascular wall initiates the intrinsic pathway.

In contrast, when blood is removed from the body and maintained in a test tube, it is the intrinsic pathway alone that must elicit the clotting. This usually results from contact of factor XII and platelets with the wall of the vessel, which activates both of these and initiates the intrinsic mechanism. If the surface of the container is very "nonwettable", such as a siliconized surface, blood clotting can sometimes be prevented for an hour or more.

Finally, intravascular clotting sometimes results from diverse factors that activate the intrinsic pathway. For instance, antigen-antibody reactions sometimes initiate the clotting process, and the same is true of some drugs or circulating debris that might happen to enter the circulation.

An especially important difference between the extrinsic and intrinsic pathways is that the extrinsic pathway is explosive in nature; once initiated, its speed of occurrence is limited only by the amount of tissue factor and tissue phospholipids released from the traumatized tissues, and by the quantities of factors X, VII, and V in the blood. With severe tissue trauma, clotting can occur in as little as 15 seconds. On the other hand, the intrinsic pathway is much slower to proceed, usually requiring 1 to 3 minutes to cause clotting. Also, various inhibitors in the blood set up roadblocks along the way for the intrinsic pathway. At times, these inhibitors can block the intrinsic pathway entirely; some of them will be discussed in more detail later in the chapter.

### PREVENTION OF BLOOD CLOTTING IN THE NORMAL VASCULAR SYSTEM—THE INTRAVASCULAR ANTICOAGULANTS

**Endothelial Surface Factors.** Probably the two most important factors for preventing clotting in the normal vascular system are, first, the smoothness of the endothelium which prevents contact activation of the intrinsic clotting system and, second, a monomolecular layer of negatively charged protein adsorbed to the inner surface of the endothelium that repels the clotting factors and platelets, thereby preventing activation of clotting. When the endothelial wall is damaged, its smoothness and its negative electrical charge are both lost, which is believed to help activate factor XII and to set off the intrinsic pathway of clotting. And, if the factor XII comes in contact with the subendothelial collagen, the specific effect of this interaction is a powerful initiator of the clotting process.

**Antithrombin and the Antithrombin Action of Fibrin.** Among the most important anticoagulants in the blood itself are those that remove thrombin from the blood. The two most powerful of these are the *fibrin threads* that are formed during the process of clotting and an alpha globulin called *antithrombin III*.

While a clot is forming, approximately 85 to 90 per cent of the thrombin formed from the prothrombin becomes adsorbed to the fibrin threads as they develop. This obviously helps to prevent the spread of thrombin into the remaining blood and therefore prevents excessive spread of the clot.

The thrombin that does not adsorb to the fibrin threads combines with antithrombin, which by a binding process blocks the effect of the thrombin on the fibrinogen and then inactivates the bound thrombin during the next 12 to 20 minutes.

**Heparin.** Heparin, a powerful anticoagulant, is a conjugated polysaccharide found in the cytoplasm of many types of cells, including even the cytoplasm of unicellular animals. Therefore, heparin probably is produced by many different cells of the human body, though especially large quantities are formed by the basophilic *mast cells* located in the pericapillary connective tissue throughout the body. These mast cells continually secrete small quantities of heparin, and the heparin then diffuses into the circulatory system. The *basophil cells* of the blood, which seem to be functionally almost identical with the mast cells, also release minute quantities of heparin into the plasma.

Mast cells are extremely abundant in the tissue surrounding the capillaries of the lungs and to a less extent the capillaries of the liver. It is easy to understand why large quantities of heparin might be needed in these areas, for the capillaries of the lungs and liver receive many embolic clots formed in the slowly flowing venous blood; sufficient formation of heparin might prevent further growth of the clots.

The concentration of heparin in normal blood has been estimated to be as much as 0.01 mg./100 ml. of blood. Though this concentration is 10 to 100 times less than that often used clinically to prevent blood clotting, it is probably sufficient to aid in preventing blood coagulation in the normal circulatory system, for only minute quantities of procoagulants are normally formed, and only minute amounts of heparin are therefore needed to prevent clotting.

*Mechanism of Heparin Action.* We still do not know all the different mechanisms by which heparin prevents blood coagulation. However, it is known to have at least the following four effects:

First, heparin prevents the formation of prothrombin activator by the intrinsic pathway when the procoagulants of this pathway are present in low concentration. It acts at several of the stages of the pathway.

Second, heparin, in association with an albumin *cofactor,* inhibits the action of thrombin on fibrinogen and thereby prevents the conversion of fibrinogen into fibrin threads.

Third, heparin increases the rapidity with which thrombin interacts with antithrombin and therefore helps in the deactivation of thrombin.

And, fourth, heparin increases the amount of thrombin adsorbed by fibrin.

### LYSIS OF BLOOD CLOTS—PLASMIN

The plasma proteins contain a euglobulin called *plasminogen* or *profibrinolysin* which, when activated, becomes a substance called *plasmin* or *fibrinolysin*. Plasmin is a proteolytic enzyme that resembles trypsin, the most important digestive enzyme of pancreatic secretion. It digests the fibrin threads and also digests other substances in the surrounding blood, such as fibrinogen, factor V, factor VIII, prothrombin, and factor XII. Therefore, whenever plasmin is formed in the blood, it can cause lysis of a clot and also destruction of many of the clotting factors, thereby causing hypocoagulability of the blood.

**Activation of Plasmin and Lysis of Clots.** When a clot is formed a large amount of plasminogen is incorporated in the clot along with other plasma proteins. However, this will not become plasmin and will not cause lysis of the clot until it is activated. Fortunately, the tissues contain substances that can activate plasminogen to plasmin. In fact, even thrombin itself can activate plasminogen. Within a day or two after blood has leaked into a tissue and clotted, these activators cause the formation of enough plasmin that it in turn dissolves the clot.

Clots that occur inside blood vessels can also be dissolved, though this occurs less often than does dissolution of tissue clots. This illustrates that activator systems also occur within the blood itself.

An activator called *urokinase* is found in the urine and is believed to be important in the lysis of clots that develop in the renal tubules. It is possible that urokinase also plays a role as an intravascular activator before it is excreted into the urine by the kidneys.

Certain bacteria also release activator enzymes; for instance, streptococci release a substance called *streptokinase,* which acts on plasminogen to form plasmin. When streptococcal infection occurs in the tissues, the plasmin generated in response to the streptokinase dissolves clotted lymph and clotted tissue fluids and allows the streptococci to spread extensively through the tissues instead of being blocked by the body's "walling off" process described in Chapter 6.

**Significance of the Fibrinolysin System.** The lysis of blood clots allows slow clearing (over a period of several days) even of extraneous blood from the tissues and in a few instances even allows reopening of clotted vessels. Unfortunately, reopening of large vessels occurs only rarely. Perhaps an important function of the fibrinolysin system is to remove very minute clots from the millions of tiny peripheral vessels that eventually would all become occluded were there no way to cleanse them.

# CONDITIONS THAT CAUSE EXCESSIVE BLEEDING IN HUMAN BEINGS

Excessive bleeding can result from deficiency of any one of the many different blood clotting factors. Three particular types of bleeding tendencies that have been studied to the greatest extent will be discussed: (1) bleeding caused by vitamin K deficiency, (2) hemophilia, and (3) thrombocytopenia (platelet deficiency).

## DECREASED PROTHROMBIN, FACTOR VII, FACTOR IX, AND FACTOR X—VITAMIN K DEFICIENCY

*Hepatitis, cirrhosis, acute yellow atrophy,* and other diseases of the liver can all depress the formation of prothrombin and factors VII, IX, and X so greatly that the patient develops a severe tendency to bleed.

Another cause of depressed levels of these substances is vitamin K deficiency. Vitamin K is necessary for some of the intermediate stages in the formation of all of them. Fortunately, vitamin K is continually synthesized in the gastrointestinal tract by bacteria so that vitamin K deficiency rarely if ever occurs because of its absence from the diet. However, vitamin K deficiency does often occur as a result of poor absorption of fats from the gastrointestinal tract, because vitamin K is fat soluble and ordinarily is absorbed into the blood along with the fats.

One of the most prevalent causes of vitamin K deficiency is failure of the liver to secrete bile into the gastrointestinal tract (which occurs either as a result of obstruction of the bile ducts or as a result of liver disease), for lack of bile prevents adequate fat digestion and absorption. Therefore, liver disease often causes decreased production of prothrombin and the other factors both because of poor vitamin K absorption and because of dysfunctional liver cells. Because of this, vitamin K is injected into all patients with liver disease or obstructed bile ducts prior to performing any surgical procedure. Ordinarily, if vitamin K is given to a deficient patient four to eight hours prior to operation and the liver parenchymal cells are at least one-half normal in function, sufficient clotting factors will be produced to prevent excessive bleeding during the operation.

## HEMOPHILIA

The term hemophilia is loosely applied to several different hereditary deficiencies of coagulation, all of which cause bleeding tendencies hardly distinguishable from one another. The three most common causes of hemophilic syndrome are deficiency of (1) factor VIII (classical hemophilia)—about 83 per cent of the total, (2) factor IX—about 15 per cent, and (3) factor XI—about 2 per cent.

Many persons with hemophilia die in early life, though many others with less severe bleeding have a normal life span. Very commonly, the person's joints become severely damaged because of repeated joint hemorrhage following exercise or trauma.

Regardless of the precise type of hemophilia, transfusion of normal fresh plasma or of the appropriate purified protein clotting factor—factor VIII for classical hemophilia—into the hemophilic person usually relieves his bleeding tendency for a few days.

### THROMBOCYTOPENIA

Thrombocytopenia means the presence of a very low quantity of platelets in the circulatory system. Persons with thrombocytopenia have a tendency to bleed as do hemophiliacs, except that the bleeding is usually from many small capillaries rather than from larger vessels, as in hemophilia. As a result, small punctate hemorrhages occur throughout all the body tissues. The skin of such a person displays many small, purplish blotches, giving the disease the name *thrombocytopenic purpura.* It will be remembered that platelets are especially important for repair of minute breaks in capillaries and other small vessels. Indeed, platelets can agglutinate to fill such ruptures without actually causing clots.

Ordinarily, bleeding does not occur until the number of platelets in the blood falls below a value of approximately 50,000 per cubic millimeter rather than the normal of 200,000 to 400,000. Levels as low as 10,000 per cubic millimeter are frequently lethal.

Even without making specific platelet counts on the blood, one can sometimes suggest the existence of thrombocytopenia by simply noting whether or not a clot of the person's blood retracts, for, as pointed out earlier, clot retraction is normally dependent upon the presence of large numbers of platelets entrapped in the fibrin mesh of the clot.

Most persons with thrombocytopenia have the disease known as *idiopathic thrombocytopenia,* which means simply "thrombocytopenia of unknown cause." However, in the past few years it has been discovered that in most of these persons specific antibodies are destroying the platelets. Occasionally these have developed because of transfusions from other persons, but usually they result from development of autoimmunity to the person's own platelets, the cause of which, however, is not known.

In addition to idiopathic thrombocytopenia, the number of thrombocytes (platelets) in the blood may be greatly depressed by any abnormality that causes aplasia of the bone marrow. For instance, *irradiation injury* to the bone marrow, aplasia of the bone marrow resulting from *drug sensitivity,* and even *pernicious anemia* can cause sufficient decrease in the total number of platelets that thrombocytopenic bleeding results.

Relief from bleeding for one to four days can often be effected in the thrombocytopenic patient by giving him *fresh whole blood transfusions.* To do this, the blood is best removed from the donor into a siliconized chamber and then rapidly placed in the recipient so that the platelets are damaged as little as possible. *Cortisone,* which suppresses immune reactions, is often beneficial in the idiopathic type of thrombocytopenia. And, *splenectomy* often helps because the spleen removes large numbers of platelets, particularly damaged ones, from the blood.

# THROMBOEMBOLIC CONDITIONS IN THE HUMAN BEING

**Thrombi and Emboli.**   An abnormal clot that develops in a blood vessel is called a *thrombus.* Once a clot has developed, continued flow of blood past the clot is likely to break it away from its attachment, and such freely flowing clots are known as *emboli.* Emboli generally do not stop flowing until they come to a narrow point in the circulatory system. Thus, emboli originating in large arteries or in the left side of the heart eventually plug either smaller arteries or arterioles. On the other hand, emboli originating in the venous system of in the right side of the heart flow into the vessels of the lung to cause pulmonary embolism.

*Causes of Thromboembolic Conditions.*   The causes of thromboembolic conditions in the human being are usually two-fold: First, any *roughened endothelial surface of a vessel*—as may be caused by arteriosclerosis, infection, or trauma—is likely to initiate the clotting process. Second, blood often clots *when it flows very slowly* through blood vessels, for small quantities of thrombin and other procoagulants are always being formed. These are generally removed from the blood by the reticuloendothelial cells, mainly the Kupffer cells of the liver. If the blood is flowing too slowly, the concentrations of the procoagulants in local areas often rise high enough to initiate clotting, but when the blood flows rapidly these are rapidly mixed with large quantities of blood and are removed during passage through the liver.

### FEMORAL THROMBOSIS AND MASSIVE PULMONARY EMBOLISM

Because clotting almost always occurs when blood flow is blocked in any vessel of the body, the immobility of bed patients plus the practice of propping the knees up with underlying pillows often causes intravascular clotting because of blood stasis in one or more of the leg veins for as much as an hour at a time. Then the clot grows in all directions, especially in the direction of the slowly moving blood, sometimes growing the entire length of the leg veins and occasionally even into the common iliac vein and inferior vena cava. Then, about 1 time out of every 10, a large part of the clot disengages from its attachments to the vessel wall and flows freely with the venous blood into the right side of the heart and thence into the pulmonary arteries to cause *massive pulmonary embolism.* If the clot is large enough to occlude both the pulmonary arteries, immediate death ensues. If only one pulmonary artery or a smaller branch is blocked, death may not occur, or the embolism may lead to death a few hours to several days later because of further growth of the clot within the pulmonary vessels.

## DISSEMINATED INTRAVASCULAR CLOTTING

Occasionally, the clotting mechanism becomes activated in widespread areas of the circulation, giving rise to the condition called *disseminated intravascular clotting*. Frequently, the clots are small but numerous, and they plug a large share of the small peripheral blood vessels. This effect occurs especially in septicemic shock, in which either circulating bacteria or bacterial toxins—especially *endotoxins*—activate the clotting mechanisms. The plugging of the small peripheral vessels greatly diminishes the delivery of oxygen and other nutrients to the tissues—a situation which exacerbates the shock picture. It is partly for this reason that full-blown septicemic shock is lethal in 85 per cent or more of the patients.

A peculiar effect of disseminated intravascular clotting is that the patient frequently begins to bleed. The reason for this is that so many of the clotting factors are removed by the widespread clotting that too few procoagulants remain to allow normal hemostasis of the remaining blood.

# ANTICOAGULANTS FOR CLINICAL USE

In some thromboembolic conditions it is desirable to delay the coagulation process to a certain degree. Therefore, various anticoagulants have been developed for treatment of these conditions. The ones most useful clinically are heparin and Dicumarol.

## HEPARIN AS AN INTRAVENOUS ANTICOAGULANT

Commercial heparin is extracted from animal tissues from several organs and is prepared in almost pure form. Injection of relatively small quantities, approximately 0.5 to 1 mg. per kilogram of body weight, causes the blood clotting time to increase from a normal of approximately 6 minutes to 30 or more minutes. Furthermore, this change in clotting time occurs instantaneously, thereby immediately preventing further development of the thromboembolic condition.

The action of heparin lasts approximately three to four hours. It is believed that the injected heparin is destroyed by an enzyme in the blood known as *heparinase*. Also, much of the injected heparin is entrapped in the reticuloendothelial cells, diffuses into the interstitial fluids, and therefore becomes unavailable as a blood anticoagulant.

In the treatment of a patient with heparin, too much heparin is sometimes given, and *serious* bleeding crises occur. In these instances, *protamine* acts specifically as an antiheparin, and the clotting mechanism can be reverted to normal by administering this substance. This substance combines with heparin and inactivates it because it carries positive electrical charges, whereas heparin carries negative charges.

## DICUMAROL AS AN ANTICOAGULANT

When dicumarol is given to a patient, the plasma levels of prothrombin and factors VII, IX, and X, all formed by the liver, begin to fall, indicating that Dicumarol has a potent depressant effect on liver formation of all these compounds. Dicumarol possibly causes this effect by competing with vitamin K for reactive sites in the intermediate processes for formation of prothrombin and the other three clotting factors, thereby blocking the action of vitamin K.

After administration of an effective dose of Dicumarol, the coagulant activity of the blood decreases to approximately 50 per cent of normal by the end of 12 hours and to approximately 20 per cent of normal by the end of 24 hours. In other words, the coagulation process is not blocked immediately, but must await the consumption of the prothrombin and other factors already present in the plasma. Normal coagulation returns one to three days after discontinuing therapy.

## PREVENTION OF BLOOD COAGULATION OUTSIDE THE BODY

Though blood removed from the body normally clots in about six minutes, blood collected in *siliconized containers* often does not clot for as long as an hour or more. The reason for this delay in clotting is that preparing the surfaces of the containers with silicone prevents contact activation of factor XII, which initiates the intrinsic clotting mechanism. On the other hand, untreated glass containers allow contact activation and rapid development of clots.

*Heparin* can be used for preventing coagulation of blood outside the body as well as in the body, and heparin is occasionally used as an anticoagulant when blood is removed from a donor to be transfused later into a recipient. Also, heparin is used in all surgical procedures in which the blood is passed through a heart-lung machine and then back into the person. In this instance, the quantity of heparin that must be administered is 3 to 5 mg. per kilogram of body weight.

Various substances that *decrease the concentration of calcium ions* in the blood can be used for preventing blood coagulation outside the body. For instance, soluble *oxalate* compounds mixed in very small quantity with a sample of blood cause precipitation of calcium oxalate from the plasma and thereby decrease the ionic calcium level so much that blood coagulation is blocked.

A second calcium deionizing agent used for preventing coagulation is *sodium, ammonium,* or *potassium citrate*. The citrate ion combines with calcium in the blood to cause an unionized calcium compound, and the lack of ionic calcium prevents coagulation. Citrate anticoagulants have a very important advantage over the oxalate anticoagulants, for oxalate is toxic to the body, whereas moderate quantities of citrate can be injected intravenously. After injection, the citrate ion is removed from the blood within a few minutes by the liver and is polymerized into glucose, and then metabolized in the usual manner. Consequently, 500 ml. of blood that has been rendered incoagulable by sodium citrate can ordinarily be injected into a recipient within a few minutes without any dire consequences. If the liver is damaged or if large quantities of citrated blood or plasma are given too rapidly, the citrate ion may not be removed quickly enough, and the citrate can then greatly depress the level of calcium ion in the blood, which results in tetany and convulsive death.

# BLOOD COAGULATION TESTS

### BLEEDING TIME

When a sharp knife is used to pierce the tip of the finger or lobe of the ear, bleeding ordinarily lasts three to six minutes. However, the time depends largely on the depth of the wound and on the degree of hyperemia in the finger at the time of the test.

### CLOTTING TIME

Many methods have been devised for determining clotting times. The one most widely used is to collect blood in a chemically clean glass test tube and then to tip the tube back and forth approximately every 30 seconds until the blood has clotted. By this method, the normal clotting time ranges between five and eight minutes.

Procedures using multiple test tubes have been devised for determining clotting time more accurately. However, clotting times are also very dependent on the condition of the glass itself and even on the size of the tube, which makes a high degree of standardization necessary to obtain accurate results.

### PROTHROMBIN TIME

The prothrombin time gives an indication of the total quantity of prothrombin in the blood. Figure 9–5 shows the relationship of prothrombin concentration to prothrombin time. The means for determining prothrombin time is the following:

Blood removed from the patient is immediately oxalated so that none of the prothrombin can change into thrombin. At any time later, a large excess of calcium ion and tissue extract is suddenly mixed with the oxalated blood. The calcium nullifies the effect of the oxalate, and the tissue extract activates the prothrombin-to-thrombin reaction by means of the extrinsic clotting pathway. The time required for

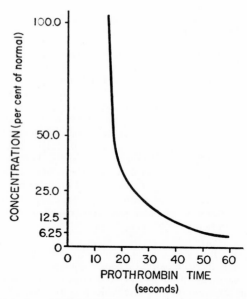

**Figure 9–5.** Relationship of prothrombin concentration in the blood to the prothrombin time. (Redrawn from Marple: Thromboembolic Conditions and Their Treatment with Anticoagulants. Charles C Thomas.)

coagulation to take place is known as the "prothrombin time." The normal prothrombin time is approximately 12 seconds, though this depends to a certain extent on the exact procedure employed. In each laboratory a curve relating prothrombin concentration to prothrombin time, such as that shown in Figure 9–5, is ordinarily drawn for the method used so that the significance of the prothrombin time can be evaluated.

### OTHER TESTS RELATING TO COAGULATION

The normal *platelet count* in the blood as measured by most platelet-counting techniques ranges between approximately 200,000 and 400,000 per cubic mm. Obviously, if a decreased platelet count is found, thrombocytopenia exists.

Tests similar to that for prothrombin time have been devised to determine the relative quantities of other clotting factors in the blood. In performing the tests excesses of all the factors besides the one being tested are added to oxalated blood all at once, and then the time of coagulation is determined in the same manner as the usual prothrombin time. If the factor is deficient, the time will be considerably prolonged. In determining the usual prothrombin time, a deficiency of some of these other factors can also prolong the measured time. Therefore, an increased prothrombin time as performed in the usual hospital laboratory does not always mean decreased quantity of prothrombin but may mean decreased quantity of some other factor, such as fibrinogen.

# REFERENCES

Ardlie, N. G.: Mechanism of platelet aggregation and release: their possible role in vascular injury. *Perspect. Nephrol. Hypertens.*, 2:891, 1973.

Biggs, R. (ed.): Human Blood Coagulation, Hemostasis and Thrombosis. Philadelphia, J. B. Lippincott Company, 1972.

Bowie, E. J., and Owen, C. A., Jr.: The bleeding time. *Prog. Hemostasis. Thromb.*, 2:249, 1974.

Copley, A. L.: Hemorheological aspects of the endothelium-plasma interface. *Microvasc. Res.*, 8:192, 1974.

Davidson, J. F., Samama, M. M., and Desnoyers, P. C. (eds): Progress in chemical Fibrinolysis and Thrombolysis. New York, Raven Press, 1975.

Davie, E. W., and Kirby, E. P.: Molecular mechanisms in blood coagulation. *Curr. Top. Cell. Regul.*, 7:51, 1973.

Douglas, A. S., and Ogston, D.: Clinical physiology of the fibrinolytic enzyme system. *J. Clin. Pathol.*, 25:615, 1972.

Erslev, A. J., and Gabuzda, T. G.: Pathophysiology of hematologic disorders. *In* Sodeman, W. A., Jr., and Sodeman, W. A. (eds): Pathologic Physiology: Mechanisms in Disease, 5th Ed. Philadelphia, W. B. Saunders Company, 1974, p. 511.

Goldsmith, H. L.: Blood flow and thrombosis. *Thromb. Diath. Haemorrh.*, 32:35, 1974.

Jaques, L. B.: Effects of hormones and drugs on hemostasis. *In* Drugs in Relation to Blood Coagulation, Haemostasis, and Thrombosis. New York, Pergamon Press, Inc., 1968.

Karpatkin, S., Garg, S. K., and Freedman, M. L.: Role of iron as a regulator of thrombopoiesis. *Amer. J. Med.*, 57:521, 1974.

Kazal, L. A.: Theories of blood coagulation. Properties and interactions of blood clotting factors. *Ann. Clin. Lab. Sci.*, 1:139, 1971.

Kline, D. L.: Blood coagulation: reactions leading to prothrombin. *Ann. Rev. Physiol.*, 27:285, 1965.

Lüscher, E. F., and Bettex-Galland, M.: Thrombosthenin, the contractile protein of blood platelets. New facts and problems. *Pathol. Biol. (Paris), 20 (Suppl.)*:89, 1972.

Mills, D. C. B.: Platelet aggregation. *In* Bittar, E., and Bittar, N. (eds): The Biological Basis of Medicine, New York, Academic Press, Inc., 1969, Vol. 3, p. 163.

Minna, J. D., Robboy, S. J., and Colman, R. W.: Disseminated Intravascular Coagulation in Man. Springfield, Ill., Charles C Thomas, Publisher, 1974.

Mobin-Uddin, K. (ed.): Pulmonary Thromboembolism. Springfield, Ill., Charles C Thomas, Publisher, 1974.

Nemerson, Y., and Pitlick, F. A.: The tissue factor pathway of blood coagulation. *Prog. Hemostasis Thromb.*, 1:1, 1972.

O'Brien, J. R.: Effect of lipids and hypolipaemic drugs on platelet aggregation and thrombosis. *Haemostasis*, 2:169, 1973–74.

Owren, P. A., and Stormorken, H.: The mechanism of blood coagulation. *Ergeb. Physiol.*, 68:1, 1973.

Quick, A. J.: The Hemorrhagic Diseases and the Pathology of Hemostasis. Springfield, Ill., Charles C Thomas, Publisher, 1974.

Ratnoff, O. D.: Some recent advances in the study of hemostasis. *Cir. Res.*, 35:1, 1974.

Schmer, G., and Strandjord, P. E. (eds.): Coagulation: Current Research and Clinical Applications. New York, Academic Press, Inc., 1972.

Seegers, W. H.: Prothrombin. Cambridge, Massachusetts, Harvard University Press, 1962.

Seegers, W. H. (ed): Blood Clotting Enzymology. New York, Academic Press, Inc., 1967.

Seegers, W. H.: Blood clotting mechanisms: three basic reactions. *Ann. Rev. Physiol.*, 31:269, 1969.

Stevens, D. J.: Vascular hemostasis: a review. *Amer. J. Med. Technol.*, 39:252, 1973.

Stormorken, H., Gjoennaess, H., and Laake, K.: Interrelations between the clotting and kinin systems. Activation of factor VII involving prekallikrein-kallikrein. A review. *Haemostasis*, 2:245, 1973–74.

Tullis, J. L.: Clot. Springfield, Ill., Charles C Thomas, Publisher, 1975.

Tyler, H. M.: Tissue transamidases, fibrin stabilization and clot lysis. *Ann. N.Y. Acad. Sci.*, 202:273, 1972.

Von Kaulla, K. N., and Davidson, J. F. (eds.): Synthetic Fibrinolytic Thrombolytic Agents. Springfield, Ill., Charles C Thomas, Publisher, 1974.

# PART III
# NERVE AND MUSCLE

# 10

# Membrane Potentials, Action Potentials, Excitation, and Rhythmicity

Electrical potentials exist across the membranes of essentially all cells of the body, and some cells, such as nerve and muscle cells, are "excitable"—that is, capable of transmitting electrochemical impulses along their membranes. And in still other types of cells, such as glandular cells, macrophages, and ciliated cells, changes in membrane potentials probably play significant roles in controlling many of the cell's functions. However, the present discussion is concerned with membrane potentials generated both at rest and during action by nerve and muscle cells.

## BASIC PHYSICS OF MEMBRANE POTENTIALS

Before beginning this discussion, let us first recall that the fluids both inside and outside the cells are electrolytic solutions containing approximately 155 mEq./liter of anions and the same concentration of cations. Generally, a very minute excess of negative ions (anions) accumulates immediately inside the cell membrane along its inner surface, and an equal number of positive ions (cations) accumulates immediately outside the membrane. The effect of this is the development of a *membrane potential*.

The two basic means by which membrane potentials can develop are: (1) active transport of ions through the membrane, thus creating an imbalance of negative and positive charges on the two sides of the membrane, and (2) diffusion of ions through the membrane as a result of a concentration difference between the two sides of the membrane, this also creating an imbalance of charges.

## MEMBRANE POTENTIALS CAUSED BY ACTIVE TRANSPORT—THE "ELECTROGENIC PUMP"

Figure 10–1A illustrates how the process of active transport can create a membrane potential. In this figure equal concentrations of anions, which are *negatively charged,* are present both inside and outside the nerve fiber. However, the sodium "pump," which was discussed in Chapter 4, has transported some of the *positively charged* sodium ions to the exterior of the fiber. Thus, more negatively charged anions than positively charged sodium ions remain inside the nerve fiber, causing negativity on the inside. On the other hand, outside the fiber there are more positively charged sodium ions than negatively charged anions, thus causing positivity outside the fiber. A pump such as this, which causes the development of a membrane potential, is called an *electrogenic pump.*

In most nerve and muscle cells potassium is pumped into the cell at the same time that sodium is pumped out; usually, however, two to five times as much sodium as potassium is pumped. Because of this imbalance of pumping, the pump is still considered electrogenic.

112

*The Cell Membrane as a Capacitor.* Note in Figure 10–1A that the negative and positive ionic charges are shown lined up against the membrane. This occurs because the negative charges inside the membrane and the positive charges on the outside pull each other together at the membrane barrier, thus creating an abrupt change in electrical potential across the cell membrane.

This alignment of electrical charges on the two sides of the membrane is exactly the same process that takes place when an electrical capacitor becomes charged with electricity. In the cell membrane, the lipid matrix of the cell is the *dielectric,* much as mica is frequently used as a dielectric in an electrical capacitor. It will be recalled that a capacitor's capacity for holding electrical charges is inversely proportional to the thickness of the membrane. Because of the extreme thinness of the cell membrane (70 to 100 Ångstroms), the capacitance of the cell membrane is tremendous for its area—about 1 microfarad per square centimeter.

Though the figures of this chapter will show large numbers of negative or positive ions (or charges) lined up against the nerve membrane, in actuality, very few positive or negative ions actually need to line up on the two sides of the membrane to create the usual membrane potential of nerves. In the case of the usual nerve fiber, only about 1/50,000 to 1/500,00 of the positive charges inside the nerve fiber need to be transferred to the outside of the fiber to create the normal nerve potential of −85 millivolts inside the fiber.

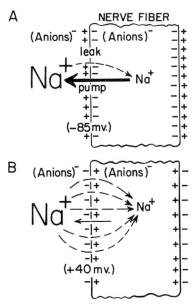

**Figure 10–1.** (A) Establishment of a membrane potential as a result of active transport of sodium ions. (B) Establishment of a membrane potential of opposite polarity caused by rapid diffusion of sodium ions from an area of high concentration to an area of low concentration.

## MEMBRANE POTENTIALS CAUSED BY DIFFUSION

Figure 10–1B illustrates the nerve fiber under another condition in which the permeability of the membrane to sodium has increased so much that diffusion of sodium through the membrane is now great in comparison with the transport of sodium by active transport. Since the sodium and anion concentrations are great outside the cell and slight inside, the positively charged sodium ions now move rapidly to the inside of the membrane, leaving an excess of negative charges outside, but creating an excess of positive charges inside.

**Relationship of the Diffusion Potential to the Concentration Difference—The Nernst Equation.** When a concentration difference of ions across a membrane causes diffusion of ions through the membrane, thus creating a membrane potential, the magnitude of the potential is determined by the ratio of tendency for the ions to diffuse in one direction versus the other direction, which is determined by the following formula (at body temperature, 38°C.):

$$\text{EMF (in millivolts)} = 61 \log \frac{\text{Conc. inside}}{\text{Conc. outside}} \quad (1)$$

Thus, when the concentration of ions on one side of a membrane is 10 times that on the other, the log of 10 is 1, and the potential difference calculates to be 61 millivolts. This equation is called the *Nernst Equation.*

However, two conditions are necessary for a membrane potential to develop as a result of diffusion: (1) The membrane must be semipermeable, allowing ions of one charge to diffuse through the pores more readily than ions of the opposite charge. (2) The concentration of the diffusible ions must be greater on one side of the membrane than on the other side.

**Calculation of the Membrane Potential when the Membrane is Permeable to Several Different Ions.** When a membrane is permeable to several different ions, the diffusion potential that will develop depends on three factors: (1) the polarity of the electrical charge of each ion, (2) the permeability of the membrane (*P*) to each ion, and (3) the concentration of the respective ions on the two sides of the membrane. Thus, the following formula, called the *Goldman equation,* gives the calculated membrane potential when two univalent cations (*C*) and two univalent anions (*A*) are involved.

EMF(in millivolts) =

$$-61 \log \frac{C_{1_i}P_1 + C_{2_i}P_2 + A_{3_o}P_3 + A_{4_o}P_4}{C_{1_o}P_1 + C_{2_o}P_2 + A_{3_i}P_3 + A_{4_i}P_4} \quad (2)$$

Note that a cation gradient from the inside ($i$) to the outside ($o$) of a membrane causes electronegativity inside the membrane, while an anion gradient in *exactly the opposite direction* also causes electronegativity on the inside.

## ORIGIN OF THE CELL MEMBRANE POTENTIAL

**Role of Sodium and Potassium Pumps and of Nondiffusible Anions in the Origin of the Cell Membrane Potential.**   Before attempting to explain the origin of the cell membrane potential, several basic facts need to be understood as follows:

1. The nerve membrane is endowed with a sodium and a potassium pump, sodium being pumped to the exterior and potassium to the interior. These pumps were discussed in Chapter 4.

2. The resting nerve membrane is normally 50 to 100 times as permeable to potassium as to sodium. Therefore, potassium diffuses with relative ease through the resting membrane, whereas sodium diffuses only with difficulty.

3. Inside the nerve fiber are large numbers of anions (negatively charged) that cannot diffuse through the nerve membrane at all or that diffuse very poorly. These anions include especially organic phosphate ions, sulfate ions, and protein ions.

Now, let us put the above facts together to see how the resting nerve membrane potential comes about. First, sodium is pumped to the outside of the fiber, while potassium is pumped to the inside. However, because two to five sodium ions are pumped out of the fiber for every potassium ion that is pumped in, more positive ions are continually being pumped out of the fiber than into it. Since most of the anions inside the nerve fiber are nondiffusible, the negative charges remain inside the nerve fiber so that the inside of the fiber becomes electronegative, while the outside becomes electropositive, as illustrated in Figure 10–2.

**Equilibration of Sodium Ion Transfer in the Two Directions Across the Membrane— Maximum Potential Caused by the Electrogenic Sodium Pump.**   As progressively more sodium ions are pumped out of the nerve

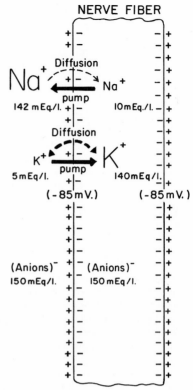

**Figure 10–2.** Establishment of a membrane potential of −85 millivolts in the normal resting nerve fiber, and development of concentration differences of sodium and potassium ions between the two sides of the membrane. The dashed arrows represent diffusion and the solid arrows represent active transport (pumps).

fiber, they begin to diffuse back into the nerve fiber for two reasons: (1) a sodium concentration gradient develops from the outside of the fiber toward the inside, and (2) a negative membrane potential develops inside the fiber and attracts the positively charged sodium ions inward. Eventually, there comes a point at which the inward diffusion equals the outward pumping by the sodium pump. When this occurs, the pump has reached its maximum capability to cause net transfer of sodium ions to the outside. This occurs when the sodium ion concentration inside the nerve fiber falls to about 10 mEg./liter (in contrast to 142 mEg./liter in the extracellular fluids) and when the membrane potential inside the fiber falls to approximately −85 millivolts. Therefore, this −85 millivolts is also the *resting potential* of the nerve membrane.

**Equilibration of Potassium Ion Transfer Across the Membrane.**   At the same time that the sodium pump pumps sodium ions to the exterior, a coupled potassium pump pumps about one third as many potassium ions to the interior. However, the resting membrane is 50 to 100 times as permeable to potassium ions as to sodium ions, which means that each time potassium ions are pumped in they tend to diffuse back out of the fiber almost immediately. Therefore, because the potassium pump is considerably weaker than the sodium pump and because the back leakage of potassium ions is very great, the potassium pump by itself can build up only a slight excess of potassium ions inside the nerve fiber.

Yet, it is known that potassium ions do build up to a high concentration inside the nerve fiber. How could this be, if it is not caused by the potassium pump? The answer is that there is a high degree of negativity created inside the fiber by the sodium pump. The −85 millivolts inside the fiber attracts the positive potassium ions from the exterior to the inside, and this attraction accounts for most of the buildup of potassium ions inside the fiber, while the potassium pump accounts for only a small proportion of the buildup.

Thus, from a quantitative point of view, once the membrane has reached a steady state condition, there are three factors that affect the equilibration of potassium ions on the two sides of the membrane: (1) the potassium pump pumping potassium ions *inward,* (2) the electrical gradient causing *inward* diffusion of potassium ions, and (3) the concentration gradient caused by the buildup of potassium ions inside the fiber causing diffusion of potassium ions *outward.* In the steady state condition, the above three factors all come to exact quantitative equilibrium; that is, the two factors causing inward movement of potassium become exactly equal to the single factor causing outward movement.

**Application of the Nernst Equation to the Dynamics of Potassium Transfer Equilibration Across the Nerve Fiber Membrane.** Since the resting membrane is quite permeable to potassium ions and since the pump is a very weak one, the factors that dominate the distribution of potassium ions across the membrane are the two factors that cause potassium ion diffusion, that is, (1) the concentration gradient that causes potassium ion diffusion out of the fiber and (2) the voltage gradient that causes potassium ion diffusion inward. If the pump were entirely inactive, the rates of diffusion of potassium ions in opposite directions caused by these two factors would be

exactly equal, and under these conditions the Nernst equation would give the exact relationship between the concentrations on the two sides of the membrane and the equilibrium membrane potential. Therefore, let us assume that the negative voltage of −85 millivolts has already been created by the sodium pump and that the extracellular fluid potassium ion concentration is 5 mEq./liter. Substituting these values in the Nernst equation, we obtain the following:

$$-85 = -61 \log \frac{K^+ \text{ Conc. (inside)}}{5}$$

Now, solving this equation for the concentration of potassium ions inside the nerve fiber, we find that the equilibrium level of potassium ions inside would be about 125 mEq./liter. Thus, of the 140 mEq./liter of potassium inside the nerve fiber, 125 of these are caused simply by diffusion into the nerve fiber resulting from the −85 millivolts that had previously been created by the sodium pump. The remaining 15 mEq./liter of potassium ions results from the action of the potassium pump.

We can also use the Nernst equation in still another way to test the degree of diffusional equilibration of potassium ions across the membrane. Given a concentration of potassium ions inside the fiber of 140 mEq./liter and a concentration outside of 5 mEq./liter what membrane potential would be required to establish diffusional equilibrium at the membrane, assuming that there were no potassium pump? Substituting these values in the Nernst equation and solving gives a membrane potential of −88.3 millivolts. This is only 3.3 millivolts more negative than the usual resting membrane potential of the nerve fiber. Therefore, under most conditions, the distribution of potassium ions between the inside of the fiber and the outside can be considered to be very nearly that required to satisfy the Nernst equation.

**The Nernst Potential for Sodium Distribution.** Now, let us calculate the Nernst potential for sodium as follows:

$$\text{Nernst potential} = -61 \log \frac{Na^+ \text{ (inside)}}{Na^+ \text{ (outside)}}$$
$$= -61 \log \frac{10}{142}$$
$$= +70 \text{ millivolts}$$

Thus, it is clear that the −85 millivolts of the

*resting nerve* is actually opposite to the +70 millivolts predicted by the Nernst equation. The reason for this is that the sodium pump completely dominates the picture in the resting state, and the diffusional factors (which cause development of the Nernst potential) are of very little significance. We shall see, however, that when a nerve impulse is being conducted, this is not true. During conduction of an impulse the membrane permeability to sodium increases several thousand-fold, sodium diffusion becomes very great, and the membrane potential inside the fiber then rises abruptly from −85 millivolts to about +45 millivolts, approaching the Nernst potential for sodium.

**Role of Chloride and Other Ions.** Chloride ions are not pumped through the nerve membrane in either direction. However, they readily diffuse through the membrane. Because there is no pump to force buildup of a concentration difference of chloride ions across the membrane, the distribution of chloride ions on the two sides of the membrane is determined entirely by electrical potential. That is, the negativity that develops inside the membrane repels chloride ions from inside the fiber and causes the chloride concentration to fall to a very low value, about 4 mEq./liter in comparison with an extracellular fluid concentration of about 103 mEq./liter. Because there is no chloride pump, the Nernst equation can be used to calculate the exact concentration ratio of chloride ions on the two sides of the membrane when the membrane potential is known and when the membrane is in an equilibrium state. Thus, chloride ions play only a passive role. Later in the chapter we will discuss instantaneous changes in membrane potential caused by the so-called action potential. Under these conditions, chloride ions move rapidly through the membrane and affect the duration and the magnitude of the action potential, but again their role is a passive one.

Essentially the same principles as those for sodium, potassium, and chloride distribution across the membrane also apply to other ions. Magnesium ions are affected in much the same way as potassium, and calcium ions in much the same manner as sodium ions. However, the concentrations and permeabilities of magnesium and calcium ions are small enough that the total number of electrical charges actually involved is also small. Yet both of these ions affect membrane potentials in another way, by altering the permeability of the membrane to other ions, a subject that is discussed at greater length later in the chapter.

**Membrane Potentials Measured in Nerve and Muscle Fibers.** Figure 10–3 illustrates a method that has been used for measuring the resting membrane potential. A micropipet is made from a minute capillary glass tube so that the tip of the pipet has a diameter of only 0.25 to

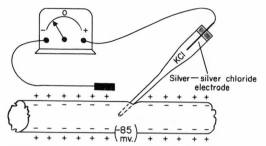

**Figure 10–3.** Measurement of the membrane potential of the nerve fiber using a microelectrode.

2 microns. Inside the pipet is a solution of potassium chloride, which acts as an electrical conductor. The fiber whose membrane is to be measured is pierced with the pipet and electrical connections are made from the pipet to an appropriate meter, as illustrated in the figure. The resting membrane potential measured in many different nerve and skeletal muscle fibers of mammals has usually ranged between −75 and −95 millivolts, with −85 millivolts as a reasonable average of the different measurements.

## THE ACTION POTENTIAL

So long as the membrane of the nerve fiber remains completely undisturbed, the membrane potential remains approximately −85 millivolts, which is called the *resting potential*. However, any factor that suddenly increases the permeability of the membrane to sodium is likely to elicit a *sequence of rapid changes* in membrane potential lasting a minute fraction of a second, followed immediately thereafter by return of the membrane potential to its resting value. This sequence of potential changes is called the *action potential*.

Some of the factors that can elicit an action potential are *electrical stimulation* of the membrane, *application of chemicals* to the membrane to cause increased permeability to sodium, *mechanical damage* to the membrane, *heat, cold,* or almost any other factor that momentarily disturbs the normal resting state of the membrane.

**Depolarization and Repolarization.** The action potential occurs in two separate stages called depolarization and repolarization, which may be explained by referring to Figure 10–4. Figure 10–4A illustrates the resting state of the membrane, with negativity inside and positivity outside. When the permeability of the mem-

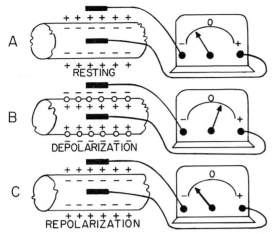

**Figure 10–4.** Sequential events during the action potential, showing: (A) the normal resting potential, (B) development of a reversal potential during depolarization, and (C) re-establishment of the normal resting potential during repolarization.

brane pores to sodium ions suddenly increases, many of the sodium ions rush to the inside of the fiber, carrying enough positive charges to the inside to cause complete disappearance of the normal resting potential and usually enough charges actually to develop a positive state inside the fiber instead of the normal negative state, as illustrated in Figure 10–4B. This positive state inside the fiber is called the *reversal potential*.

Almost immediately after depolarization takes place, the pores of the membrane again become almost totally impermeable to sodium ions. Because of this, the positive reversal potential inside the fiber disappears, and the normal resting membrane potential returns. This is called *repolarization* (Fig. 10–4C).

Now, let us explain in more detail *how* this sequence of events occurs.

## MEMBRANE PERMEABILITY CHANGES FOR SODIUM AND POTASSIUM DURING THE ACTION POTENTIAL

The action potential is caused by a sequence of changes in the membrane permeability for sodium and potassium.

The *first event* is a tremendous *increase in sodium permeability*—as much as a 5000-fold increase. Because of the very high concentration of sodium outside the fiber, sodium ions now diffuse rapidly to the inside, carrying positive charges and changing the inside electrical potential from negativity to positivity.

The *second event* is an *increase in potassium permeability* coupled with a simultaneous *decrease of the sodium permeability* back to normal. This allows large quantities of positively charged potassium ions to diffuse out of the fiber, a diffusion which returns the inside membrane potential back to its negative resting level.

The *third event* is a *decrease of the potassium permeability* back to normal.

The *fourth event* is *active transport of sodium back out of the nerve fiber* and, concurrently, both active transport and diffusion of potassium ions back into the nerve fiber, thus re-establishing the original state of the fiber.

**Sodium "Channels" and Potassium "Channels."** The sodium and potassium ions are believed to diffuse through pores in the membrane. However, the sodium ions are thought to move through one set of pores called *sodium channels* whereas the potassium ions move through another set of pores called *potassium channels*. The actual anatomical structures of these channels are not known, but some physiologists believe that they are actual physical pores. Other physiologists believe that they are openings within the latticeworks of complex protein molecules that penetrate all the way through the membrane; still others have differing concepts of the "pores." Yet, regardless of what these structures are, they have physiological properties *as if* the sodium channels were actual oval pores with dimensions of about 3 by 5 Ångstroms and the potassium channels were round pores with dimensions of about 3 by 3 Ångstroms.

Each channel is believed to be guarded by a "gate" that can open and close the channel. Under resting conditions both the sodium and potassium channels are almost completely closed; therefore, relatively few sodium and potassium ions can diffuse through the two respective channels (though the potassium channels are some 50 to 100 times as permeable as the sodium channels, which allows far more potassium diffusion than sodium, as already explained). Yet, when the gates open, the permeability of the sodium channels can increase as much as 5000-fold and the permeability of the potassium channels as much as 50-fold.

*Opening the Gates—the Gating Potential.* The gates themselves are believed to be positive charges that are fixed inside the channels, at their openings, or in the lipid matrix adjacent to the channels. These positive charges create a positive electric field that protrudes far into the channels and, therefore, blocks positive ion permeability. Alternatively, the field can be pulled back from the channel and thereby increase the channel permeability. It is believed that this closing and opening of the gates is caused by electrical potentials called *gating potentials* that occur in the lipid matrix of the cell membrane adjacent to the channels. That is, an appropriate electrical potential appearing in the lipid matrix near a pore can either close or open the gate, depending on the polarity of the potential.

With this background we can now discuss more intelligently the mechanisms for the changes in sodium and potassium permeabilities during the action potential.

**Sudden Increase in Sodium Permeability at the Onset of the Action Potential—"Activation" of the Membrane.** In the resting state the gate of the sodium channel is closed. However, any sudden change in electrical potential across the membrane that causes the potential inside the fiber to become suddenly less negative will also cause a simultaneous change in the gating potential of the sodium channel and increase the channel's permeability; that is, the gate will begin to open. Once the gate has begun this opening, the process becomes a *regenerative* vicious cycle: sodium ions diffuse to the interior; the internal negativity becomes still less; the gate opens even more widely; and the process continues until the gate is wide open. Therefore, the permeability of the sodium gate increases rapidly and drastically. This is illustrated by the lowermost curve of Figure 10–5, which shows that at the onset of the action potential the permeability of the membrane for sodium (measured as sodium "conductance") increases several thousand-fold in only a small fraction of a millisecond. This tremendous increase in sodium permeability is called *activation of the membrane.*

**Closure of the Sodium Gates and Decrease in Sodium Permeability—Membrane Inactivation.** Immediately after the sodium gates have opened widely, for reasons not now known, these gates immediately begin to close again. This is illustrated by the rapid decrease in sodium conductance in Figure 10–5. This closure of the sodium gates is called *membrane inactivation.*

Thus, at the onset of an action potential, the sodium channels open widely for a fraction of a millisecond and then they close almost equally as rapidly. Yet, during this short period of time, large quantities of sodium ions diffuse to the

interior of the fiber, thereby increasing the membrane potential from its normal negative value of −85 millivolts up to a positive value of approximately +45 millivolts.

Closure of the sodium channels—that is, inactivation of the membrane—does not in itself return the potential back to the negative value because once the sodium ions have moved to the interior, the inside of the fiber has become charged to a positive voltage, and will remain at this voltage until some other movement of ions through the membrane changes the voltage again. But, within a fraction of a millisecond potassium ions diffuse to the exterior, as described below, to return the membrane potential back to its resting value.

***Possible Role of Calcium Ions in Closing the Sodium Gates.*** When calcium ions are deficient

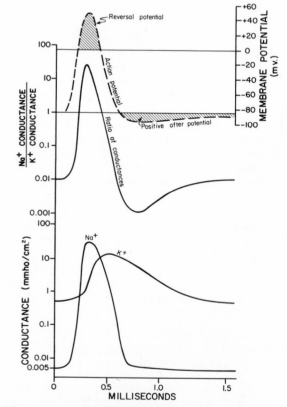

**Figure 10–5.** Changes in sodium and potassium conductances during the course of the action potential. Note that sodium conductance increases several thousand-fold during the early stages of the action potential, while potassium conductance increases only about 30-fold during the latter stages of the action potential and for a short period thereafter. (Curves constructed from data in Hodgkin and Huxley papers but transposed from squid axon to apply to the membrane potentials of large mammalian nerve fibers.)

in the extracellular fluids, the sodium gates will not close fully between action potentials, and the membrane remains very leaky to sodium ions—sometimes so much so that the membrane remains depolarized continuously or fires repetitively. Because of this effect, it is believed that calcium ions in some way play a significant role in the closure of the sodium gates. One theory is that the calcium ions bind with the outside of the cell membrane at or near the openings to the sodium channels. The presence of the electropositive charge created by the calcium ions then creates a positive field near the channels and blocks entry of sodium ions. At the onset of the next action potential, the first event is theoretically to remove these calcium ions, an occurrence which supposedly increases the permeability of the sodium channels to sodium ions.

**Delayed Increase, and Subsequent Decrease, in Potassium Permeability During the Action Potential.** Referring again to Figure 10–5, we see that the potassium permeability (as measured by conductance) does not change significantly during the first half of the action potential. Yet, toward the end of the action potential, the potassium permeability now increases about fifty-fold. This allows very rapid movement of potassium ions out of the fiber to the exterior. Because the sodium gates have already closed by this time, this loss of positive charges from the interior of the fiber causes rapid return of the membrane potential back to its normal resting value of −85 millivolts.

Figure 10–5 also shows that the potassium permeability remains elevated for a short period even after the major portion of the action potential is over, but within a millisecond or so it returns to normal. By this time the membrane potential has returned almost exactly to its original resting negative level.

Unfortunately, the factors that cause opening and closing of these potassium channels are even less well understood than those for the sodium channels. However, the potassium ion permeability does seem to be determined mainly by the electrical potential that exists at any given time across the membrane. When the potential becomes positive on the inside, the potassium ion permeability increases; when negative, it decreases. Therefore, when the action potential becomes positive, after a fraction of a millisecond lag time, potassium permeability automatically increases, and then when the action potential becomes negative again, after another millisecond or so lag, the potassium permeability decreases back to its original level.

Therefore, for the time being, without knowing the reasons why, we are forced simply to accept the major changes in sodium and potassium permeabilities during the course of the action potential. A simple way to remember these is that the sodium channels open widely for a fraction of a millisecond at the onset of the action potential, and at this time the inward flux of sodium ions increases the inside potential from negativity to positivity. Then, within another fraction of a millisecond the sodium channels close while the potassium channels open for about a millisecond; the resulting efflux of potassium ions returns the membrane potential back approximately to its resting negative level.

## SUMMARY OF THE EVENTS OF THE ACTION POTENTIAL

We can now summarize the action potential by referring specifically to the action potential shown at the top of Figure 10–5 and to the conductance changes for sodium and potassium at the bottom of this figure. When the sodium channels open and sodium ions pour to the inside of the membrane, the positive charges of the sodium ions not only neutralize the normal electronegativity inside the fiber but also create an excess of positive charges, thereby making the membrane potential inside the fiber positive. This is called the *reversal potential,* and for large myelinated nerve fibers it is normally about +45 millivolts.

Then, after a fraction of a millisecond, the sodium conductance decreases to its normal level. At the same time the potassium conductance *increases* and allows rapid flow of potassium ions outward through the membrane. This now transfers large numbers of positive charges to the outside, once again creating negativity inside the membrane and returning the membrane potential approximately to its original −85 millivolts.

**Quantity of Ions Lost from the Nerve Fiber During the Action Potential.** The actual quantity of ions that must pass through the nerve membrane to cause the action potential—that is, to cause a 135 millivolt increase in the membrane potential and then to return this potential back to its normal resting level—is extremely slight. For large myelinated nerve fibers only about 1/100,000 to 1/500,000 of the ions normally inside the fiber are exchanged during this process. Nevertheless, this does cause a very slight increase of sodium ions inside the fiber and a corresponding very slight decrease in potassium ions. As we shall see later in the chapter, the active transport processes restore these ions within a few milliseconds or seconds.

## SOME EXPERIMENTAL METHODS THAT HAVE BEEN USED TO STUDY THE ACTION POTENTIAL

The events of the action potential, as discussed in the preceding few pages, have been studied most widely in the *squid axon,* a nerve fiber that is sometimes as large as 1 mm. in diameter. This fiber is large enough that potentials are easily measured on its inside, and it is even possible to remove the axoplasm from inside the fiber and to replace this with artificial solution. Also, electrodes can be placed inside the fiber to excite it or to "clamp" the voltage across the membrane—that is, to fix the voltage across the membrane at a constant level by running current through the electrodes. Use of this clamping process has made it possible to study the effect of voltage changes on sodium and potassium fluxes and also on sodium and potassium permeabilities.

Two drugs have proved to be invaluable in the study of the sodium and potassium channels. One drug, *tetrodotoxin,* will block the sodium channels. However, it will block these channels only when applied to the outside of the nerve fiber. Nevertheless, when the sodium channels are blocked it is then possible to study the permeability of the potassium channels independently of the sodium channels. In a similar manner, application of *tetraethylammonium ion* to the inside of the fiber membrane blocks the potassium channels. In this state, the sodium channels can be studied independently of the potassium channels. Thus, by a painstaking process of elimination, the separate effects on the two types of channels are gradually being elucidated.

## PROPAGATION OF THE ACTION POTENTIAL

In the preceding paragraphs we have discussed the action potential as it occurs at one spot on the membrane. However, an action potential elicited at any one point on an excitable membrane usually excites adjacent portions of the membrane, resulting in propagation of the action potential. The mechanism of this is illustrated in Figure 10–6. Figure 10–6A shows a normal resting nerve fiber, and Figure 10–6B shows a nerve fiber that has been excited in its midportion—that is, the midportion has suddenly developed increased permeability to

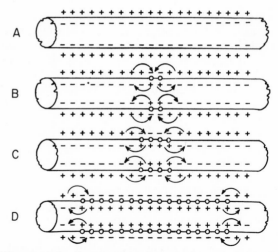

**Figure 10–6.**  Propagation of action potentials in both directions along a conductive fiber.

sodium. The arrows illustrate a *local circuit* of current flow between the depolarized and the resting membrane areas; current flows inward through the depolarized membrane and outward through the resting membrane, thus completing a circuit. In some way not understood, *the outward current flow through the resting membrane now increases the membrane's permeability to sodium,* which immediately allows sodium ions to diffuse inward through the membrane, thus setting up the vicious cycle of membrane activation discussed earlier in the chapter. As a result, depolarization occurs at this area of the membrane as well. Therefore, as illustrated in Figure 10–6C and D, successive portions of the membrane become depolarized. And these newly depolarized areas cause local circuits of current flow still farther along the membrane, causing progressively more and more depolarization. Thus, the depolarization process travels in both directions along the entire extent of the fiber. The transmission of the depolarization process along a nerve or muscle fiber is called a *nerve* or *muscle impulse.*

**Direction of Propagation.**  It is now obvious that an excitable membrane has no single direction of propagation, but that the impulse can travel in both directions away from the stimulus—and even along all branches of a nerve fiber—until the entire membrane has become depolarized.

**The All-or-Nothing Principle.**  It is equally obvious that, once an action potential has been elicited at any point on the membrane of a normal fiber, the depolarization process will travel over the entire membrane. This is called the all-or-nothing principle, and it applies to all normal excitable tissues. Occasionally, though, when the fiber is in an abnormal state the impulse will reach a point on the membrane at which the action potential does not generate sufficient voltage to stimulate the adjacent area of the membrane. When this occurs the spread of depolarization will stop. Therefore, for normal propagation of an impulse to occur, the ratio of action potential to threshold for excitation, called the *safety factor,* must at all times be greater than unity.

**Propagation of Repolarization.** The action potential normally lasts almost the same length of time at each point along a fiber. Therefore, repolarization normally occurs first at the point of original stimulus and then spreads progressively along the membrane, moving in the same direction that depolarization had previously spread. Figure 10–7 illustrates the same nerve fiber as that in Figure 10–6, showing that the polarization process is propagated in the same direction as the depolarization process but a few ten-thousandths of a second later.

## "RECHARGING" THE FIBER MEMBRANE—IMPORTANCE OF ENERGY METABOLISM

Since transmission of large numbers of impulses along the nerve fiber reduces the ionic concentration gradients, these gradients must be re-established. This is achieved by the action of the sodium and potassium pump in exactly the same way as that described in the first part of the chapter for establishment of the original resting potential. That is, the sodium ions that, during the action potentials, have diffused to the interior of the cell and the potassium ions that have diffused to the exterior are returned to their original state by the sodium and potassium pump. Since this pump requires energy for operation, this process of "recharging" the nerve fiber is an active metabolic one, utilizing energy derived from the adenosine triphosphate energy

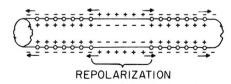

**Figure 10–7.** Propagation of repolarization in both directions along a conductive fiber.

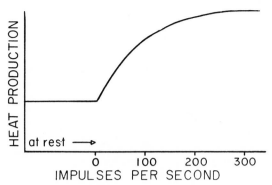

**Figure 10–8.** Heat production in a nerve fiber at rest and at progressively increasing rates of stimulation.

"currency" system of the cell.

**Heat Production by the Nerve Fiber.** Figure 10–8 illustrates the relationship of heat production in a nerve fiber to the number of impulses transmitted by the fiber each second. The rate of heat production is a measure of the rate of metabolism in the nerve, because heat is always liberated as a product of the chemical reactions of energy metabolism. Note that the heat production increases markedly as the number of impulses per second increases. It is this increased energy that causes the "recharging" process.

**Unimportance of Energy Metabolism During Actual Transmission of the Action Potential.** The role of the sodium and potassium pump to recharge the nerve fiber is to maintain adequate concentration differences of sodium and potassium across the nerve membrane. However, if metabolism in the nerve fiber suddenly ceased and the sodium and potassium ions that move through the membrane with each action potential were not returned to their original sides of the membrane, the nerve fiber would still be able to transmit perhaps fifty thousand to a million nerve impulses before it would be necessary to recharge the fiber again. The reason for this is that the quantities of sodium and potassium ions that diffuse through the nerve membrane with each action potential are extremely minute, allowing tremendous numbers of action potentials to occur before the recharging process is necessary.

Thus, the action potential itself is a passive process in contrast to the charging and recharging processes, which are active metabolic processes requiring the utilization of ATP as an energy source.

## THE SPIKE POTENTIAL AND THE AFTER-POTENTIALS

Figure 10–9 illustrates an action potential recorded with a much slower time scale than that illustrated in

Figure 10–5; many milliseconds of recording are shown in comparison with only the first 1.5 milliseconds of the action potential in Figure 10–5.

**The Spike Potential.** The initial very large change in membrane potential shown in Figure 10–9 is called the spike potential. In large type A myelinated nerve fibers it lasts for about 0.4 millisecond. The spike potential is analogous to the action potential that has been discussed in the preceding paragraphs, and it is the spike potential that is also called the *nerve impulse*.

**The Negative After-Potential.** At the termination of the spike potential, the membrane potential fails to return all the way to its resting level for another few milliseconds, as shown in Figure 10–9. This is called the *negative after-potential*. It is believed to result from a buildup of potassium ions immediately outside the membrane; this causes the concentration ratio of potassium across the membrane to be temporarily less than normal and therefore prevents full return of the normal resting membrane potential for a few additional milliseconds.

**The Positive After-Potential.** Once the membrane potential has returned to its resting value, it becomes a little more negative than its normal resting value; this excess negativity is called the *positive after-potential*. It is a fraction of a millivolt to, at most, a few millivolts more negative than the normal resting membrane potential, but it lasts from 50 milliseconds to as long as many seconds.

This positive after-potential is caused principally by the electrogenic pumping of sodium outward through the nerve fiber membrane, which is the recharging process that was discussed previously. If the active transport processes are poisoned, the positive after-potential is lost, though both the action potential and the negative after-potential continue to occur.

(The student might wonder why greater negativity in the resting membrane potential is called a positive rather than a negative after-potential, and likewise,

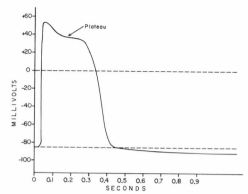

**Figure 10–10.** An action potential from a Purkinje fiber of the heart, showing a "plateau."

why the so-called negative after-potential is not named positive. The reason for this is that these potentials were first measured *outside* the nerve fibers rather than inside, and all potential changes on the outside are of exactly opposite polarity, whereas modern terminology expresses membrane potentials in terms of the inside potential rather than the outside potential.)

## PLATEAU IN THE ACTION POTENTIAL

In some instances the excitable membrane does not repolarize immediately after depolarization, but, instead, the potential remains on a plateau near the peak of the spike sometimes for many milliseconds before repolarization begins. Such a plateau is illustrated in Figure 10–10, from which one can readily see that the plateau greatly prolongs the period of depolarization. It is this type of action potential that occurs in the heart, where the plateau lasts for as long as two- to three-tenths second and causes contraction of the heart muscle during this entire period of time.

The cause of the action potential plateau is probably a combination of several different factors. First, there is delay in inactivation of the sodium channels, which allows sodium ions to continue flowing into the fiber for a long time after the onset of the action potential. Second, there is a small amount of calcium current flowing into the fiber at the same time, and these two currents together maintain the positive state inside the membrane that causes the plateau. However, probably equally as important is the fact that the permeability of the potassium channels decreases about five-fold at the onset of the action potential in excitable membranes that exhibit plateaus, and this prevents rapid outflow of potassium ions to the

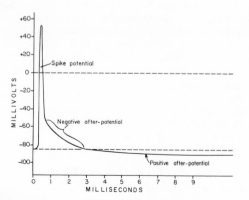

**Figure 10–9.** An idealized action potential, showing: the initial spike followed by a negative after-potential and a positive after-potential.

outside of the fiber and, therefore, delays the repolarization process.

Yet, when inactivation of the sodium channels does begin, the inactivation process proceeds unabated; simultaneously, the potassium permeability of the membrane increases a hundred-fold or more. Therefore, sodium and calcium ions stop diffusing to the interior of the fiber, while potassium ions diffuse outward extremely rapidly. Consequently, the membrane potential returns quickly to its normal negative level, as illustrated in Figure 10–10, by the rapid decline of the potential at the end of the plateau.

## RHYTHMICITY OF CERTAIN EXCITABLE TISSUES—REPETITIVE DISCHARGE

All excitable tissues can discharge repetitively if the threshold for stimulation is reduced low enough. For instance, even nerve fibers and skeletal muscle fibers, which normally are highly stable, discharge repetitively when they are veratrinized or when the calcium ion concentration falls below a critical value. Repetitive discharges, or rhythmicity, occur normally in the heart, in most smooth muscle, and probably also in some of the neurons of the central nervous system.

**The Re-excitation Process Necessary for Rhythmicity.**  For rhythmicity to occur, the membrane, even in its natural state, must be already permeable enough to sodium ions to allow automatic membrane activation and consequent depolarization. That is, (a) sodium ions flow inward, (b) this further increases the membrane permeability, (c) still more sodium ions flow inward, (d) the permeability increases more, and so forth, thus eliciting the regenerative process of membrane activation until an action potential is generated. Then, at the end of the action potential the membrane returns to its normal resting potential. Shortly thereafter, the activation process begins again and a new action potential occurs spontaneously—this cycle continuing again and again and causing self-induced rhythmic excitation of the excitable tissues.

Yet why does the membrane not remain depolarized all the time instead of re-establishing a membrane potential, only to beome depolarized again shortly thereafter? The answer to this can be found by referring back to Figure 10–5, which shows that toward the end of the action potential, and continuing for a short period thereafter, the membrane becomes excessively permeable to potassium. The excessive outflow of potassium carries tremendous numbers of positive charges to the outside of the membrane, creating inside the fiber considerably more negativity than would otherwise occur for a short period after the preceding action potential is over. This is a state called *hyperpolarization*. As long as this state exists, re-excitation will not occur; but gradually the excess potassium conductance (and the state of hyperpolarization) disappears, thereby allowing the onset of a new action potential.

Figure 10–11 illustrates this relationship between repetitive action potentials and potassium conductance. The state of hyperpolarization is established immediately after each preceding action potential; but it gradually recedes, and the membrane potential correspondingly increases until it reaches the *threshold* for excitation; then suddenly a new action potential results, the process occurring again and again.

## SPECIAL ASPECTS OF IMPULSE TRANSMISSION IN NERVES

**"Myelinated" and Unmyelinated Nerve Fibers.**  Figure 10–12 illustrates a cross-section of a typical small nerve trunk, showing a few very large nerve fibers that comprise most of the cross-sectional area and many more small fibers lying between the large ones. The large fibers are *myelinated* and the small ones are *unmyelinated*. The average nerve trunk contains about twice as many unmyelinated fibers as myelinated fibers.

Figure 10–13 illustrates a typical myelinated fiber. The central core of the fiber is the *axon*, and the membrane of the axon is the actual *conductive membrane*. The axon is filled in its center with *axoplasm*, which is a viscid intracellular fluid. Surrounding the axon is a *myelin*

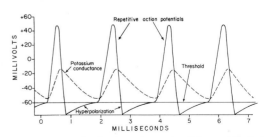

**Figure 10–11.**   Rhythmic action potentials, and their relationship to potassium conductance and to the state of hyperpolarization.

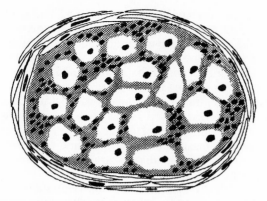

**Figure 10–12.** Cross-section of a small nerve trunk containing myelinated and unmyelinated fibers.

*sheath* that is approximately as thick as the axon itself, and about once every millimeter along the extent of the axon the myelin sheath is interrupted by a *node of Ranvier.*

The myelin sheath is deposited around the axon by Schwann cells in the following manner: The membrane of a Schwann cell first envelops the axon. Then the cell rotates around the axon several times, laying down multiple layers of cellular membrane containing the lipid substance *sphingomyelin.* This substance is an excellent insulator that prevents almost all flow of ions. However, at the juncture between each two successive Schwann cells along the axon, a small uninsulated area remains where ions can

flow with ease between the extracellular fluid and the axon. This area is the node of Ranvier.

**Saltatory Conduction in Myelinated Fibers.** Even though ions cannot flow to a significant extent through the thick myelin sheaths of myelinated nerves, they can flow with considerable ease through the nodes of Ranvier. Indeed, the membrane at this point is 500 times as permeable as the membranes of some unmyelinated fibers. Impulses are conducted from node to node by the myelinated nerve rather than continuously along the entire fiber as occurs in the unmyelinated fiber. This process, illustrated in Figure 10–14, is called *saltatory conduction.* That is, electrical current flows through the surrounding extracellular fluids and also through the axoplasm from node to node, exciting successive nodes one after another. Thus, the impulse jumps down the fiber, which is the origin of the term "saltatory."

Saltatory conduction is of value for two reasons: First, by causing the depolarization process to jump long intervals along the axis of the nerve fiber, this mechanism greatly increases the velocity of nerve transmission in myelinated fibers. Second, saltatory conduction conserves energy for the axon, for only the nodes depolarize, allowing several hundred times less loss of ions than would otherwise be necessary and therefore requiring little extra metabolism for retransporting the ions across the membrane.

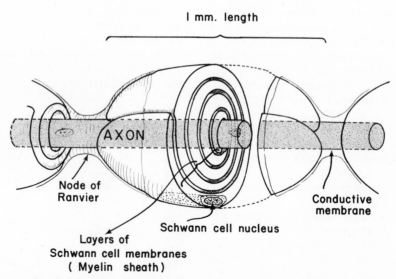

**Figure 10–13.** The myelin sheath and its formation by Schwann cells. (Modified from Elias and Pauley: Human Microanatomy, Davis Co., 1966.)

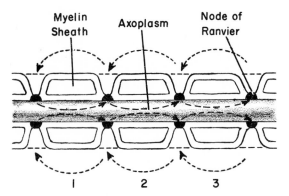

**Figure 10–14.**   Saltatory conduction along a myelinated axon.

## VELOCITY OF CONDUCTION IN NERVE FIBERS

The velocity of conduction in nerve fibers varies from as little as 0.5 meter per second in very small unmyelinated fibers up to as high as 130 meters per second (the length of a football field) in very large myelinated fibers. The velocity increases approximately with the fiber diameter in myelinated nerve fibers and approximately with the square root of fiber diameter in unmyelinated fibers.

## EXCITATION—THE PROCESS OF ELICITING THE ACTION POTENTIAL

**Chemical Stimulation.**   Basically, any factor that causes sodium ions to begin to diffuse inward through the membrane in sufficient numbers will set off the automatic, regenerative "activation" mechanism, noted earlier in the chapter, that eventuates in the action potential. Thus, certain chemicals can stimulate a nerve fiber by increasing the membrane permeability. Such chemicals include acids, bases, almost any salt solution of very strong concentration, and, most importantly, the substance *acetylcholine.* Many nerve fibers, when stimulated, secrete acetylcholine at their endings where they synapse with other neurons or where they end on muscle fibers. The acetylcholine in turn stimulates the successive neuron or muscle fiber. This is discussed in much greater detail in Chapter 12, and it is one of the most important means by which nerve and muscle fibers are stimulated. Likewise, *norepinephrine* secreted by sympathetic nerve endings can stimulate cardiac muscle fibers and some smooth muscle fibers, and still other hormonal substances can stimulate successive neurons in the central nervous system.

**Mechanical Stimulation.**   Crushing, pinching, or pricking a nerve fiber can cause a sudden surge of sodium influx and, for obvious reasons, can elicit an action potential. Even slight pressure on some specialized nerve endings can stimulate these; this will be discussed in Chapter 48 in relation to sensory perception.

**Electrical Stimulation.**   Electrical stimulation also can initiate an action potential. An electrical charge artificially induced across the membrane causes excess flow of ions through the membrane; this in turn can initiate an action potential. However, not all methods of applying electrical stimuli result in excitation, and, since this is the usual means by which nerve fibers are excited when they are studied in the laboratory, the process of electrical excitation deserves more detailed comment.

*Cathodal versus Anodal Currents.*   Figure 10–15 illustrates a battery connected to two electrodes on the surface of a nerve fiber. At the cathode, or negative electrode, the potential outside the membrane is negative with respect to that on the inside, and the current that flows outward through the membrane at this point is called *cathodal current.* At the anode, the electrode is positive with respect to the potential immediately inside the membrane, and the inward current flow at this point is called *anodal current.*

*A cathodal current excites the fiber whereas an anodal current actually makes the fiber more resistant to excitation than normal.* Though the cause of this difference between the two types of current cannot be explained completely, it is known that the normal impermeability of the membrane to sodium results partially from the high resting membrane potential across the membrane, and any condition that lessens this potential causes the membrane to become progressively more permeable to sodium. Obviously, at the cathode the applied voltage is opposite to the resting potential of the membrane, and this reduces the net potential. As a result, the membrane becomes far more permeable than usual to sodium followed by subsequent development of an action potential.

On the other hand, at the anode, the applied potential actually enhances the membrane potential. This makes the membrane less permeable to sodium than ever, resulting in increased resistance of the membrane to stimulation by other means.

*Threshold for Excitation and "Acute Subthreshold Potential."*   A very weak cathodal potential cannot excite the fiber. But, when this potential is progressively increased, there comes a point at which excitation takes place. Figure 10–16 illustrates the effects of successively applied cathodal stimuli of progressing strength. A very weak stimulus at point A causes the membrane potential to change from −85 to −80 millivolts, but this is not a sufficient change for the automatic regenerative processes of the action potential to develop. At point B the stimulus is greater, but, here again, the intensity still is not enough to set off the automatic action potential. Nevertheless, the membrane voltage is disturbed for

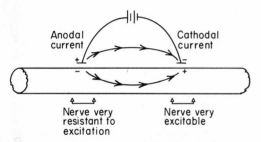

**Figure 10–15.** Effects of anodal and cathodal currents on excitability of the nerve membrane.

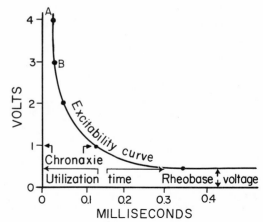

**Figure 10–17.** Excitability of a large myelinated nerve fiber.

as long as a millisecond or more after both of the weak stimuli; the potential changes during these small intervals of time are called *acute subthreshold potentials,* as illustrated in the figure.

At point C in Figure 10–16 the stimulus elicits an acute membrane potential that is not subthreshold but slightly more than the threshold value, and, after a short "latent period," it initiates an action potential. At point D the stimulus is still stronger, and the acute membrane potential initiates the action potential even sooner. Thus, this figure shows that even a very weak stimulus always causes a local potential change at the membrane, but that the intensity of the *local potential* must rise to a *threshold value* before the automatic action potential will be set off.

*"Accommodation" to Stimuli.* When a cathodal potential applied to a nerve fiber is made to increase very slowly, rather than rapidly, the threshold voltage required to cause firing is considerably increased. This phenomenon is called *accommodation;* in other words, the excitable membrane is said to "accommodate" itself to slowly increasing potentials rather than firing. It is probable that a slowly increasing stimulatory current allows time for ions to build up (or become depleted) in the areas immediately adjacent to the fibers, these changes in ion concentrations partially opposing the stimulus.

*Excitability Curve of Nerve Fibers.* A so-called "excitability curve" of a nerve fiber is shown in Figure 10–17. To obtain this curve a high voltage

stimulus (4 volts, in this instance) is first applied to the fiber, and the minimum duration of stimulus required to excite the fiber is found. The voltage and minimal time are plotted as point A. Then a stimulus voltage of 3 volts is applied, and the minimal time required is again determined; the results are plotted as point B. The same is repeated at 2 volts, 1 volt, 0.5 volt, and so forth, until the least voltage possible at which the membrane is stimulated has been reached. On connection of these points, the excitability curve is determined.

The excitability curve of Figure 10–17 is that of a large myelinated nerve fiber. The least possible voltage at which it will fire is called the *rheobase,* and the time required for this least voltage to stimulate the fiber is called the *utilization time.* Then, if the voltage is increased to twice the rheobase voltage, the time required to stimulate the fiber is called the *chronaxie;* this is often used as a means of expressing relative excitabilities of different excitable tissues. For instance, the chronaxie of a large type A fiber is about 0.0001 to 0.0002 second; of smaller myelinated nerve fibers, approximately 0.0003 second; of unmyelinated fibers, 0.0005 second; of skeletal muscle fibers, 0.00025 to 0.001 second; and of heart muscle, 0.001 to 0.003 second.

**The Refractory Period.** A second action potential cannot occur in an excitable fiber as long as the membrane is still depolarized from the preceding action potential. Therefore, even an electrical stimulus of maximum strength applied before the first spike potential is almost over will not elicit a second one. This interval of inexcitability is called the *absolute refractory period.* The absolute refractory period of large myelinated nerve fibers is about $1/2500$ second. Therefore, one can readily calculate that such a fiber can carry a maximum of about 2500 impulses per second. Following the absolute refractory period is a *relative refractory period* lasting about one quarter as long. During this period, stronger than normal stimuli are required to excite the fiber. In some types of

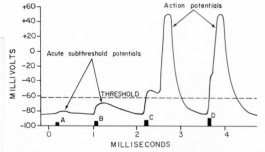

**Figure 10–16.** Effect of stimuli on the potential of the excitable membrane, showing the development of "acute subthreshold potentials" when the stimuli are below the threshold value required for eliciting an action potential.

fibers, a short period of supernormal excitability follows the relative refractory period.

## FACTORS THAT INCREASE MEMBRANE EXCITABILITY

Any condition that increases the natural permeability of the membrane usually causes it to become more excitable than usual. For instance, the drug *veratrine* has a direct action on the membrane to increase its permeability to sodium, and, as a consequence, the strength of stimulus needed to elicit an impulse is greatly reduced; or, on occasion, the fiber becomes so excitable that spontaneous impulses are generated without any extraneous excitation.

**Low Calcium Tetany.** An extremely important potentiator of excitability is low concentration of calcium ions in the extracellular fluids. Calcium ions normally decrease the permeability of the membrane to sodium. If sufficient calcium ions are not available, however, the permeability becomes increased, and, as a result, the membrane excitability greatly increases—sometimes so greatly that many spontaneous impulses result and cause muscular spasm. This condition is known as low calcium tetany. It often occurs in patients who have lost their parathyroid glands and who therefore cannot maintain normal calcium ion concentrations. This condition will be discussed in Chapter 79.

## INHIBITION OF EXCITABILITY— "STABILIZERS" AND LOCAL ANESTHETICS

Contrary to the factors that increase excitability, still others called *membrane stabilizing factors* can decrease excitability. For instance, a *high calcium ion concentration* decreases the membrane permeability and simultaneously reduces its excitability. Therefore, calcium ions are said to be a "stabilizer." Also, *low potassium ion concentration* in the extracellular fluids, because it increases the negativity of the resting membrane potential (a process called "hyperpolarization"), likewise acts as a stabilizer and reduces membrane excitability. Indeed, in a hereditary disease known as *familial periodic paralysis*, the extracellular potassium ion concentration is often so greatly reduced that the person actually becomes paralyzed, but he reverts to normalcy instantly after intravenous administration of potassium.

**Local Anesthetics and the "Safety Factor."** Among the most important stabilizers are the many substances used clinically as local anesthetics, including *cocaine, procaine, tetracaine,* and many other drugs. These act directly on the membrane, decreasing its permeability to sodium and, therefore, also reducing membrane excitability. When the excitability has been reduced so low that the ratio of *action potential strength to excitability threshold* (called the "safety factor") is reduced below unity, a

nerve impulse fails to pass through the anesthetized area.

# RECORDING MEMBRANE POTENTIALS AND ACTION POTENTIALS

**The Cathode Ray Oscilloscope.** Earlier in this chapter we have noted that the membrane potential changes very rapidly throughout the course of an action potential. Indeed, most of the action potential complex of large nerve fibers takes place in less than $1/1000$ second. In some figures of this chapter a meter has been shown recording these potential changes. However, it must be understood that any meter capable of recording them must be capable of responding extremely rapidly. For practical purposes the only type of meter that is capable of responding accurately to the very rapid membrane potential changes of most excitable fibers is the cathode ray oscilloscope.

Figure 10–18 illustrates the basic components of a cathode ray oscilloscope. The cathode ray tube itself is composed basically of an *electron gun* and a *fluorescent surface* against which electrons are fired. Where the electrons hit the surface, the fluorescent material glows. If the electron beam is moved across the surface, the spot of glowing light also moves and draws a fluorescent line on the screen.

In addition to the electron gun and fluorescent surface, the cathode ray tube is provided with two sets of plates: one set, called the *horizontal deflection plates,* positioned on either side of the electron beam, and the other set, called the *vertical deflection plates,* positioned above and below the beam. If a negative charge is applied to the left-hand plate and a positive charge to the right-hand plate, the electron beam will be repelled away from the left plate and attracted toward the right plate, thus bending the beam toward the right, and this will cause the spot of light on the fluorescent surface of the cathode ray

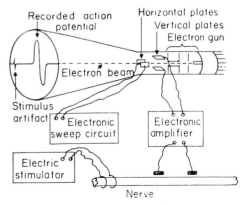

**Figure 10–18.** The cathode ray oscilloscope for recording transient action potentials.

screen to move to the right. Likewise, positive and negative charges can be applied to the vertical deflection plates to move the beam up or down.

Since electrons travel at extremely rapid velocity and since the plates of the cathode ray tube can be alternately charged positively or negatively within less than a millionth of a second, it is obvious that the spot of light on the face of the tube can also be moved to almost any position in less than a millionth of a second. For this reason, the cathode ray tube oscilloscope can be considered to be an inertialess meter capable of recording with extreme fidelity almost any change in membrane potential.

To use the cathode ray tube for recording action potentials, two electrical circuits must be employed. These are (1) an *electronic sweep circuit* that controls the voltages on the horizontal deflection plates and (2) an *electronic amplifier* that controls the voltages on the vertical deflection plates. The sweep circuit automatically causes the spot of light to begin at the left-hand side and move slowly toward the right. When the spot reaches the right side it jumps back immediately to the left-hand side and starts a new trace.

The electronic amplifier amplifies signals that come from the nerve. If a change in membrane potential occurs while the spot of light is moving across the screen, this change in potential will be amplified and will cause the spot to rise above or fall below the mean level of the trace, as illustrated in the figure. In other words, the sweep circuit provides the lateral movement of the electron beam while the amplifier provides the vertical movement in direct proportion to the changes in membrane potentials picked up by appropriate electrodes.

Figure 10–18 also shows an electric stimulator used to stimulate the nerve. When the nerve is stimulated, a small *stimulus artifact* usually appears on the oscilloscope screen prior to the action potential.

**Recording the Monophasic Action Potential.** Throughout this chapter monophasic action potentials have been shown in the different diagrams. To record these, an electrode such as that illustrated earlier in the chapter in Figure 10–3 must be inserted into the interior of the fiber. Then, as the action potential spreads down the fiber, the changes in the potential inside the fiber are recorded as illustrated earlier in the chapter in Figures 10–5, 10–9, and 10–10.

**Recording a Biphasic Action Potential.** When one wishes to record impulses from a whole nerve trunk, it is not feasible to place electrodes inside the nerve fibers. Therefore, the usual method of recording is to place two electrodes on the outside of fibers. However, the record that is obtained is then biphasic for the following reasons: When an action potential moving down the nerve fiber reaches the first electrode, it becomes charged negatively while the second electrode is still unaffected. This causes the oscilloscope to record in the negative direction. Then as the action potential passes beyond the first electrode and reaches the second electrode, the mem-

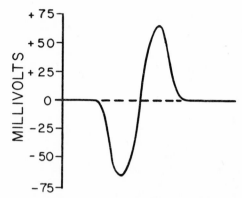

**Figure 10–19.**   Recording of a biphasic action potential.

brane beneath the first electrode becomes repolarized while the second electrode is still negative, and the oscilloscope records in the opposite direction. When these changes are recorded continuously by the oscilloscope, a graphic record such as that illustrated in Figure 10–19 is recorded, showing a potential change first in one direction and then in the opposite direction.

# REFERENCES

Abbott, B. C., and Howarth, J. V.: Heat studies in excitable tissues. *Physiol. Rev.*, 53:120, 1973.

Agin, D.: Negative conductance and electrodiffusion in excitable membrane systems. *Membranes*, 1:249, 1972.

Baker, P. F., Hodgkin, A. L., and Shaw, T. I.: Replacement of the axoplasm of giant nerve fibers with artificial solutions. *J. Physiol. (Lond.)*, 164:330, 1962.

Baker, P. F., Hodgkin, A. L., and Shaw, T. I.: The effects of changes in internal ionic concentrations on the electrical properties of perfused giant axons. *J. Physiol. (Lond.)*, 164:355, 1962.

Caldwell, P. C.: Factors governing movement and distribution of inorganic ions in nerve and muscle. *Physiol. Rev.*, 48:1, 1968.

Cohen, L. B.: Changes in neuron structure during action potential propagation and synaptic transmission. *Physiol. Rev.*, 53:373, 1973.

Cole, K. S.: Electrodiffusion models for the membrane of squid giant axon. *Physiol. Rev.*, 45:340, 1965.

De Weer, P.: Aspects of the recovery process in nerve. *In* Guyton, A. C. (ed.): MTP International Review of Science: Physiology. Baltimore, University Park Press, 1975, Vol. 3, p. 231.

Ehrenstein, G., and Lecar, H.: The mechanism of signal transmission in nerve axons. *Ann. Rev. Biophys. Bioeng.*, 1:347, 1972.

Evans, M. H.: Tetrodotoxin, saxitoxin, and related substances: their applications in neurobiology. *Int. Rev. Neurobiol.*, 15:83, 1972.

Glynn, I. M., and Karlish, S. J. D.: The sodium pump. *Ann. Rev. Physiol.*, 37:13, 1975.

Hodgkin, A. L.: The Conduction of the Nervous Impulse. Springfield, Ill., Charles C Thomas, Publisher, 1963.

Hodgkin, A. L., and Horowicz, P.: The effect of sudden changes in ionic concentrations on the membrane potential of single muscle fibres. *J. Physiol. (Lond.)*, 153:370, 1960.

Hodgkin, A. L., and Huxley, A. F.: Movement of sodium and potassium ions during nervous activity. *Cold Spr. Harb. Symp. Quant. Biol.*, 17:43, 1952.

Hodgkin, A. L., and Huxley, A. F.: Quantitative description of membrane current and its application to conduction and excitation in nerve. *J. Physiol. (Lond.)*, 117:500, 1952.

Hubbard, J. I. (ed.): The Peripheral Nervous System. New York, Plenum Publishing Corp., 1974.

Jack, J. J. B., Noble, D., and Tsien, R. W.: Electric Current Flow in Excitable Cells. New York, Oxford University Press, 1975.

Kater, S. B., Heyer, C. B., and Kaneko, C. R. S.: Invertebrate behavior and identifiable neurons. *In* Guyton, A. C. (ed.): MTP International Review of Science: Physiology. Baltimore, University Park Press, 1975, Vol. 3, p. 1.

Katz, B.: Nerve, Muscle, and Synapse. New York, McGraw-Hill Book Company, 1968.

Koketsu, K.: Calcium and the excitable cell membrane. *Neurosci. Res.*, 2:1, 1969.

Landowne, D., Potter, L. T., and Terrar, D. A.: Structure-function relationships in excitable membranes. *Ann. Rev. Physiol.*, 37:485, 1975.

Läuger, P., and Neumcke, B.: Theoretical analysis of ion conductance in lipid bilayer membranes. *Membranes*, 2:1, 1973.

Lerman, L., Watanabe, A., and Tasaki, I.: Intracellular perfusion of squid giant axons: recent findings and interpretations. *Neurosci. Res.*, 2:71, 1969.

Lieberstein, H. M.: Mathematical Physiology, Blood Flow, and Electrically Active Cells. New York, American Elsevier Publishing Co., Inc., 1973.

Martin, A. R., and Veale, J. L.: The nervous system at the cellular level. *Ann. Rev. Physiol.*, 29:401, 1967.

Mystrom, R. A.: Membrane Physiology. Englewood Cliffs, N.J., Prentice-Hall, Inc., 1973.

Nachmansohn, D.: Chemical and Molecular Basis of Nerve Activity. New York, Academic Press, Inc., 1975.

Narahashi, T.: Chemicals as tools in the study of excitable membranes. *Physiol. Rev.*, 54:813, 1974.

Nelson, P. G.: Nerve and muscle cells in culture. *Physiol. Rev.*, 55:1, 1975.

Noble, D.: Applications of Hodgkin-Huxley equations to excitable tissues. *Physiol. Rev.*, 46:1, 1966.

Noble, D.: Conductance mechanisms in excitable cells. *Biomembranes*, 3:427, 1972.

Peskoff, A., and Eisenberg, R. S.: Interpretation of some microelectrode measurements of electrical properties of cells. *Ann. Rev. Biophys. Bioeng.*, 4:65, 1973.

Reuter, H.: Divalent cations as charge carriers in excitable membranes. *Prog. Biophys. Mol. Biol.*, 26:1, 1973.

Ritchie, J. M.: Energetic aspects of nerve conduction: The relationships between heat production, electrical activity, and metabolism. *Prog. Biophys. Mol. Biol.*, 26:147, 1973.

Roche e Silva, M., and Suarez-Kurtz, G. (eds.): Concepts of Membranes in Regulation and Excitation. New York, Raven Press, 1975.

Schultz, S. G.: Principles of electrophysiology and their application to epithelial tissues. *In* Guyton, A. C. (ed.): MTP International Review of Science: Physiology. Baltimore, University Park Press, 1974, Vol. 4, p. 69.

Shanes, A. M.: Electrochemical aspects of physiological and pharmacological action in excitable cells. *Pharmacol. Rev.*, 10:59, 1958; 10:165, 1958.

Somjen, G. G.: Electrogenesis of sustained potentials. *Prog. Neurobiol.*, 1:201, 1973.

Spurway, N. C.: Mechanisms of anion permeation. A review of available data, principally on muscle cells, with the fixed charge concept in mind. *Biomembranes*, 3:363, 1972.

Tasaki, I., and Hallett, M.: Bioenergetics of nerve excitation. *J. Bioenerg.*, 3:65, 1972.

Thomas, R. C.: Electrogenic sodium pump in nerve and muscle cells. *Physiol. Rev.*, 52:563, 1972.

Thompson, R. F., and Patterson, M. M. (eds.): Bioelectric Recording Techniques, Part A, Cellular Processes and Brain Potentials. New York, Academic Press, Inc., 1973.

Trautwein, W.: Membrane currents in cardiac muscle fibers. *Physiol. Rev.*, 53:793, 1973.

Zachar, J.: Electrogenesis and Contractility in Skeletal Muscle Cells. Baltimore, University Park Press, 1973.

# 11

# Contraction of Skeletal Muscle

Approximately 40 per cent of the body is skeletal muscle and another 5 to 10 per cent is smooth or cardiac muscle. Many of the same principles of contraction apply to all these different types of muscle, but in the present chapter the function of skeletal muscle is considered mainly, while the specialized functions of smooth muscle will be discussed in the following chapter, and cardiac muscle in Chapter 13.

## PHYSIOLOGIC ANATOMY OF SKELETAL MUSCLE

### THE SKELETAL MUSCLE FIBER

All skeletal muscles of the body are made up of numerous muscle fibers ranging between 10 and 80 microns in diameter. In most muscles the fibers extend the entire length of the muscle, and except for about 2 per cent of the fibers, each of these is innervated by only one nerve ending located near the middle of the fiber.

**The Sarcolemma.** The sarcolemma is the cell membrane of the muscle fiber. However, the sarcolemma consists of a true cell membrane, called the *plasma membrane,* and a thin layer of polysaccharide material similar to that of the basement membrane surrounding blood capillaries; thin collagen fibrillae are also present in the outer layer of the sarcolemma. At the ends of the muscle fibers, these surface layers of the sarcolemma fuse with tendon fibers, which in turn collect into bundles to form the muscle tendons, and thence insert into the bones.

**Myofibrils; Actin and Myosin Filaments.** Each muscle fiber contains several hundred to several thousand *myofibrils,* which are illustrated by the small dots in the cross-sectional view of Figure 11–1. Each myofibril in turn has, lying side-by-side, about 1500 *myosin filaments* and two times this many *actin filaments,* which are large polymerized protein molecules that are responsible for muscle contraction. These can be seen in longitudinal view in the electron micrograph of Figure 11–2, and are represented diagrammatically in Figure 11–3. The thick filaments are *myosin* and the thin filaments are *actin.* Note that the myosin and actin filaments partially interdigitate and thus cause the myofibrils to have alternate light and dark bands. The light bands, which contain only actin filaments, are called *I bands* because they are mainly *isotropic* to polarized light. The dark bands, which contain the myosin filaments as well as the ends of the actin filaments where

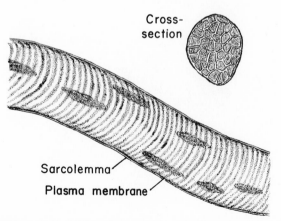

**Figure 11–1.** Side and cross-sectional views of a skeletal muscle fiber.

130

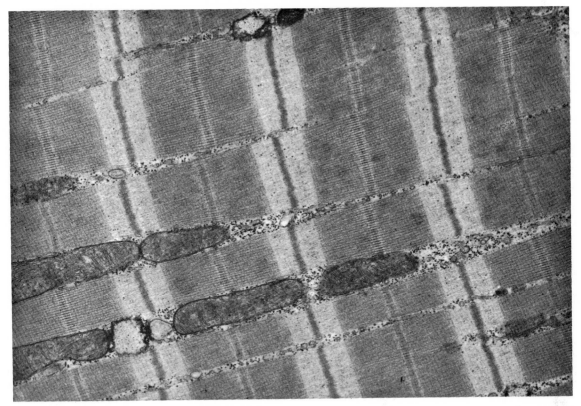

**Figure 11–2.** Electron micrograph of muscle myofibrils, showing the detailed organization of actin and myosin filaments in the fibril. Note the mitochondria lying between the myofibrils. (From Fawcett: The Cell, 1966.)

they overlap the myosin, are called *A bands* because they are *anisotropic* to polarized light. Note also the small projections from the sides of the myosin filaments. These are called *cross-bridges*. They protrude from the surfaces of the myosin filaments along the entire extent of the filament, except in the very center. It is interaction between these cross-bridges and the actin filaments that causes contraction.

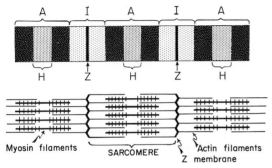

**Figure 11–3.** Arrangement of the myosin and actin filaments in the sarcomeres.

Figure 11–3 also shows that the actin filaments are attached to the so-called *Z membrane* or *Z line*, and the filaments extend on either side of the Z membrane to interdigitate with the myosin filaments. The Z membrane also passes from myofibril to myofibril, attaching the myofibrils to each other all the way across the muscle fiber. Therefore, the entire muscle fiber has light and dark bands, as is also true of the individual myofibrils. It is these bands that give skeletal and cardiac muscle their striated appearance.

The portion of a myofibril (or of the whole muscle fiber) that lies between two successive Z membranes is called a *sarcomere*. When the muscle fiber is at its normal fully stretched resting length, the length of the sarcomere is about 2.0 microns. At this length, the actin filaments completely overlap the myosin filaments and are just beginning to overlap each other. We shall see later that it is at this length that the sarcomere also is capable of generating its greatest force of contraction.

When a muscle fiber is stretched beyond its

resting length, the ends of the actin filaments pull apart, leaving a light area in the center of the A band. This light area, called the *H zone,* is illustrated in Figure 11–3. Such an H zone rarely occurs in the normally functioning muscle because normal sarcomere contraction occurs when the resting length of the sarcomere is from 2.0 microns to 1.6 microns. In this range the ends of the actin filaments not only overlap the myosin filaments but also overlap each other.

**The Sarcoplasm.**    The myofibrils are suspended inside the muscle fiber in a matrix called *sarcoplasm,* which is composed of usual intracellular constituents. The fluid of the sarcoplasm contains large quantities of potassium, magnesium, phosphate, and protein enzymes. Also present are tremendous numbers of *mitochondria* that lie between and parallel to the myofibrils, a condition which is indicative of the great need of the contracting myofibrils for large amounts of ATP formed by the mitochondria.

**The Sarcoplasmic Reticulum.**    Also in the sarcoplasm is an extensive endoplasmic reticulum, which in the muscle fiber is called the *sarcoplasmic reticulum.* This reticulum has a special organization that is extremely important in the control of muscle contraction, which will be discussed later in the chapter. The electron-micrograph of Figure 11–4 illustrates the arrangement of this sarcoplasmic reticulum and shows how extensive it can be. The more rapidly contracting types of muscle have especially extensive sarcoplasmic reticulums, indicating that this structure is important in causing rapid muscle contraction, as will also be discussed later.

One can see from the figure that the sarcoplasmic reticulum is composed of *longitudinal tubules* that lie parallel to the myofibrils. The reader can also note that the two ends of each longitudinal tubule terminate in *cisternae* that are bulbous structures quite different from the main body of the tubules. We shall see later that the two respective portions of the sarcoplasmic reticulum play specific roles in muscle contraction.

**The Transverse Tubule System—T Tubules.**    In addition to the sarcoplasmic reticulum, each muscle fiber also has another extensive tubular system called the *transverse tubules* or *T tubules.* These tubules run *perpendicular* to the myofibrils, in contrast to the sarcoplasmic reticulum that runs parallel. They are illustrated in Figure 11–4, which shows the T tubules cut in cross-sections (see arrows). Note that each T tubule is very small and lies between the ends of two successive longitudinal tubules with the cisternae of the longitudinal tubules abutting the T tubule. This area of contact between the sarcoplasmic reticulum and a T tubule is called a *triad* because it is composed of a central small tubule and, on its sides, two bulbous cisternae of the sarcoplasmic reticulum. In skeletal muscle, a triad occurs adjacent to each area where the actin and myosin filaments overlap, and since this occurs at each

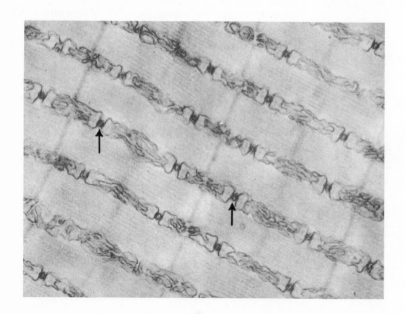

**Figure 11–4.** Sarcoplasmic reticulum surrounding the myofibril, showing the longitudinal system paralleling the myofibrils. Also shown in cross-section are the T tubules that lead to the exterior of the fiber membrane and that contain extracellular fluid (arrows). (From Fawcett: The Cell, 1966.)

end of each myosin filament, there are two triads per sarcomere. (However, in heart muscle there is only one triad per sarcomere, located adjacent to the Z membrane.)

Another important characteristic of the T tubules is that they pass all the way through the fiber from one side of the membrane to the other; they open to the exterior rather than to the interior of the cell and contain *extracellular fluid* that is continuous with the fluid outside the cell rather than intracellular fluid. When an action potential spreads over the muscle fiber membrane, electrical currents, probably in the form of an action potential, are transmitted to the interior of the muscle fiber by way of the T tubule system, and it is these currents that cause the muscle to contract, as we shall discuss shortly.

# MOLECULAR MECHANISM OF MUSCLE CONTRACTION

**Sliding Mechanism of Contraction.** Figure 11–5 illustrates the basic mechanism of muscle contraction. It shows the relaxed state of a sarcomere (above) and the contracted state (below). In the relaxed state, the ends of the actin filaments derived from two successive Z membranes barely overlap each other while at the same time completely overlapping the myosin filaments. On the other hand, in the contracted state these actin filaments have been pulled inward among the myosin filaments so that they now overlap each other to a major extent. Also, the Z membranes have been pulled by the actin filaments up to the ends of the myosin filaments. Indeed, the actin filaments can be pulled together so tightly that the ends of the myosin

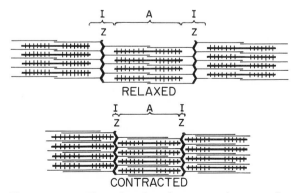

**Figure 11–5.** The relaxed and contracted states of a myofibril, showing sliding of the actin filaments into the channels between the myosin filaments.

filaments actually buckle during very intense contraction. Thus, muscle contraction occurs by a *sliding filament mechanism.*

But what causes the actin filaments to slide inward among the myosin filaments? Unfortunately, we do not completely know the answer to this question. Yet we do know that it is caused by attractive forces that develop between the actin and myosin filaments. Almost certainly, these attractive forces are the result of mechanical, chemical, or electrostatic forces generated by the interaction of the crossbridges of the myosin filaments with the actin filaments.

Under resting conditions, the attractive forces between the actin and myosin filaments are inhibited, but when an action potential travels over the muscle fiber membrane, this causes the release of large quantities of calcium ions into the sarcoplasm surrounding the myofibrils. These calcium ions activate the attractive forces between the filaments and contraction begins. But energy is also needed for the contractile process to proceed. This energy is derived from the high energy bonds of adenosine triphosphate (ATP), which is degraded to adenosine diphosphate (ADP) to give the energy required.

In the next few sections we will describe what is known about the details of the molecular processes of contraction. To begin this discussion, however, we must first characterize in detail the myosin and actin filaments.

## MOLECULAR CHARACTERISTICS OF THE CONTRACTILE FILAMENTS

**The Myosin Filament.** The myosin filament is composed of approximately 200 myosin molecules, each having a molecular weight of 450,000. Figure 11–6, section A illustrates an individual molecule; section B illustrates the organization of the molecules to form a myosin filament as well as its interaction with two actin filaments.

The myosin molecule is composed of two parts; one part called *light meromyosin* and the other part called *heavy meromyosin.* The light meromyosin consists of two peptide strands wound around each other in a helix. The heavy meromyosin in turn consists of two parts: first, a double helix similar to that of the light meromyosin; second, a head attached to the end of the double helix. The head itself is a composite of two globular protein masses.

It is believed that the myosin molecule is

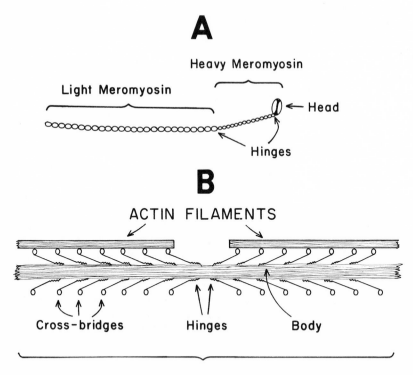

**Figure 11–6.** (A) The myosin molecule. (B) Combination of many myosin molecules to form a myosin filament. Also shown are the cross-bridges and the interaction between the heads of the cross-bridges with adjacent actin filaments.

especially flexible at two points—the juncture between the light meromyosin and the heavy meromysin and between the body of the heavy meromyosin and the head. These two areas are called hinges.

In section B of Figure 11–6 the central portion of a myosin filament is illustrated. The body of this filament is composed of parallel strands of light meromyosin from multiple myosin molecules. In fact, whenever myosin molecules are precipitated from solution, it is found that the light meromyosin portions of the myosin molecules have a natural tendency to aggregate together to form filaments almost precisely like those of the myosin filaments found in muscle. On the other hand, the heavy meromyosin portions of myosin molecules protrude from all sides of the myosin filament, as illustrated in the figure. These protrusions constitute the *cross-bridges*. The heads of the cross-bridges lie in apposition to the actin filaments, whereas the rod portions of the cross-bridges act as hinged arms that allow the heads to extend either far outward from the body of the myosin filament or to lie close to the body. The length of the head is 120 Ångstroms while the length of the hinged arm is 400 Ångstroms. Therefore, the combination of the

arm and the head allows the cross-bridges to extend far laterally. This is very important because during contraction of the muscle the actin and myosin filaments are spread progressively apart from each other by the engorgement of fluid in the shortened myofibrils.

Note also that the arms of the cross-bridges extend toward the two ends of the filament and away from the center of the filament. Therefore, in the very center of the myosin filament, for a length of about 0.2 micron, there are no cross-bridge heads.

The total length of the myosin filament is 1.6 microns, and the 200 myosin molecules allow the formation of 100 pairs of cross-bridges—50 pairs on each end of the myosin filament.

Now, to complete the picture, the myosin filament is twisted so that it makes one complete revolution for each three pairs of cross-bridges. Each twist has a length of 429 Ångstroms, and each pair of cross-bridges lies 143 Ångstroms from the next pair and is axially displaced from the previous pair by 120°.

**The Actin Filament.** The actin filament is also complex. It is composed of three different components; *actin, tropomyosin*, and *troponin*.

The backbone of the actin filament is a double-stranded F-actin protein molecule, illustrated in Figure 11–7. The two strands are wound in a helix, making a complete 360° revolution approximately every 700 Ångstroms.

Each strand of the double F-actin helix is composed of polymerized G-actin molecules, each having a molecular weight of 47,000 and a length of 54 Ångstroms. There are approximately 13 of these molecules in each revolution of each strand of helix. And attached to each one of the G-actin molecules is one molecule of ADP. It is believed that these ADP molecules are the active sites on the actin filaments with which the cross-bridges of the myosin filaments interact to cause muscle contraction. The active sites on the two F-actin strands of the double helix are staggered, giving one active site on the overall actin filament approximately every 27 Ångstroms.

*The Troponin-Tropomyosin Complex.* The actin filament also contains two additional protein strands that are polymers of tropomyosin molecules, each molecule having a molecular weight of 70,000. It is believed that these protein strands lie in the two grooves formed by the two F-actin strands of the actin helix. Each tropomyosin molecule in the tropomyosin strand extends for a length of 400 Ångstroms, and attached to each tropomyosin molecule is a molecule of still another protein, *troponin.* Troponin has a very high affinity for calcium ions. It is believed that the combination of calcium ions with troponin is the trigger that initiates muscle contraction, as will be discussed in the following section.

## INTERACTION OF MYOSIN AND ACTIN FILAMENTS TO CAUSE CONTRACTION

### Inhibition of the Actin Filament by the Troponin-Tropomyosin Complex; Activation

by Calcium Ions. A pure actin filament without the presence of the troponin-tropomyosin complex binds strongly with myosin molecules in the presence of magnesium ions and ATP, both of which are normally abundant in the myofibril. But, if the troponin-tropomyosin complex is added to the actin filament, this binding does not take place. Therefore, it is believed that the normal active sites on the normal actin filament of the relaxed muscle are inhibited (or perhaps physically covered) by the troponin-tropomyosin complex. Consequently, they cannot interact with the myosin filaments to cause contraction. Before contraction can take place the inhibitory effect of the troponin-tropomyosin complex must itself be inhibited.

Now, let us discuss the role of the calcium ions. In the presence of large amounts of calcium ions the inhibitory effect of the troponin-tropomyosin on the actin filaments is itself inhibited. The mechanism of this is not known, but one suggestion is the following: When calcium ions combine with troponin, which has an extremely high affinity for calcium ions even when they are present in minute quantities, the troponin molecule supposedly undergoes a conformational change that in some way tugs on the tropomyosin protein strand. This in turn supposedly moves the tropomyosin strand deeper into the groove between the two actin strands and thereby "uncovers" the active sites of the actin, thus allowing contraction to proceed. Though this is a hypothetical mechanism, nevertheless it does emphasize that the normal relationship between the tropomyosin-troponin complex and actin is altered by calcium ions—a condition which leads to contraction.

**Interaction Between the "Activated" Actin Filament and the Myosin Molecule—The Ratchet Theory of Contraction.** As soon as the actin filament becomes activated by the calcium ions, it is believed that the heads of the cross-bridges from the myosin filaments immediately become attracted to the active sites of the actin filament, and this in some way causes contraction to occur. Though the precise manner by which this interaction between the cross-bridges and the actin causes contraction is still unknown, a suggested hypothesis for which considerable circumstantial evidence exists is the so-called *ratchet theory of contraction.*

Figure 11–8 illustrates the postulated ratchet mechanism for contraction. This figure shows the heads of two cross-bridges attaching to and disengaging from the active sites of an actin

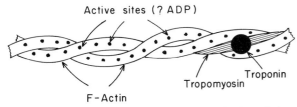

Active sites (? ADP)

Troponin

Tropomyosin

F–Actin

**Figure 11–7.** The actin filament, composed of two helical strands of F-actin. Also shown is a small portion of one of the two tropomyosin strands that lie in the grooves between the actin strands. On the surface of the tropomyosin one large molecule of troponin is shown schematically.

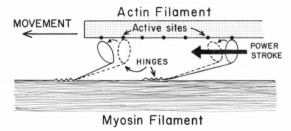

**Figure 11–8.** The ratchet mechanism for contraction of the muscle.

filament. It is postulated that when the head attaches to an active site this attachment alters the bonding forces between the head and its arm, thus causing the head to tilt toward the center of the myosin filament, and to drag the actin filament along with it. This tilt of the head of the cross-bridge is called the *power stroke*. Then, immediately after tilting, the head automatically splits away from the active site and returns to its normal perpendicular direction. In this position it combines with an active site further down along the actin filament; then, a similar tilt takes place again to cause a new power stroke, and the actin filament moves another step. Thus, the heads of the cross-bridges bend back and forth and step by step pull the actin filament toward the center of the myosin filament. Thus, the movements of the cross-bridges use the active sites of the actin filaments as cogs of a *ratchet*.

Each one of the cross-bridges is believed to operate independently of all others, each attaching and pulling in a continuous, alternating ratchet cycle. Therefore, the greater the number of cross-bridges in contact with the actin filament at any given time, the greater, theoretically, is the force of contraction.

**ATP as the Source of Energy for Contraction—Chemical Events in the Ratchet Cycle.** When a muscle contracts against a load, work is performed, and energy is required. It is found that large amounts of ATP are cleaved to form ADP during the contraction process. Furthermore, the greater the amount of work performed by the muscle, the greater is the amount of ATP that is cleaved. However, unfortunately, it is still not known exactly how ATP is used to provide the energy for contraction. Yet, the following is a sequence of events that has been suggested as the means by which this occurs:

1. When the inhibitory effect of the troponin-tropomyosin complex has itself been inhibited by calcium ions, the heads of the cross-bridges that are in the nontilted position bind with uncovered sites on the actin filament, as illustrated in Figure 11–8.

2. It is assumed that the bond between the head of the cross-bridge and the active site of the actin filament causes a conformational change in the head, thus causing the head to tilt and to provide the power stroke for pulling the actin filament. This power stroke is believed to result from energy that has already been stored in the heavy meromyosin molecule and not from energy derived from ATP.

3. Once the head of the cross-bridge is tilted, the conformational change in the head exposes a reactive site in the head where ATP can bind. Therefore, one molecule of ATP binds with the head, and this binding in turn causes detachment of the head from the active site.

4. Once the head bound with ATP splits away from the active site, the ATP is itself cleaved by a very potent ATPase activity of the heavy meromyosin. The energy released supposedly tilts the head back to its normal perpendicular condition and theoretically "cocks" the head in this position.

5. Then, when the "cocked" head, with its stored energy derived from the cleaved ATP, binds with a new active site on the actin filament, it becomes uncocked and once again provides the power stroke.

6. Thus, the process proceeds again and again until the actin filament pulls the Z membrane up against the ends of the myosin filaments or until the load on the muscle becomes too great for further pulling to occur.

### RELATIONSHIP BETWEEN ACTIN AND MYOSIN FILAMENT OVERLAP AND TENSION DEVELOPED BY THE CONTRACTING MUSCLE

Figure 11–9 illustrates the relationship between the length of sarcomere and the tension developed by the contracting muscle fiber. To the right are illustrated different degrees of overlap of the myosin and actin filaments at different sarcomere lengths. At point D on the diagram, the actin filament has pulled all the way out to the end of the myosin filament with no overlap at all. At this point, the tension developed by the activated muscle is zero. Then as the sarcomere shortens and the actin filament overlaps the myosin filament progressively more and more, the tension increases progressively until the sarcomere length decreases to

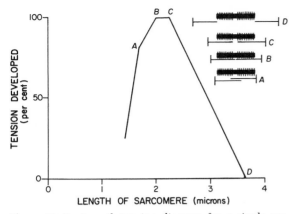

**Figure 11–9.** Length-tension diagram for a single sarcomere, illustrating maximum strength of contraction when the sarcomere is 2.0 to 2.2 microns in length. At the upper right are shown the relative positions of the actin and myosin filaments at different sarcomere lengths from point A to point D. (Modified from Gordon, Huxley, and Julian: *J. Physiol., 171*:28P, 1964.)

about 2.2 microns. At this point the actin filament has already overlapped all the cross-bridges of the myosin filament but has not yet reached the center of the myosin filament. Upon further shortening, the sarcomere maintains full tension until point B at a sarcomere length of approximately 2.0 microns. It is at this point that the ends of the two actin filaments begin to overlap. As the sarcomere length falls from 2 microns down to about 1.65 microns at point A, the strength of contraction decreases. It is at this point that the two Z membranes of the sarcomere abut the ends of the myosin filaments. Then, as contraction proceeds to still shorter sarcomere lengths, the ends of the myosin filaments are actually crumpled, but, as illustrated in Figure 11–9, the strength of contraction also decreases precipitously.

This diagram illustrates that maximum contraction occurs when there is maximum overlap between the actin filaments and the cross-bridges of the myosin filaments, and it supports the idea that the greater the number of cross-bridges pulling the actin filaments, the greater is the strength of contraction.

**Relation of Force of Contraction of the Intact Muscle to Muscle Length.** Figure 11–10 illustrates a diagram similar to that in Figure 11–9, but this time for the intact whole muscle rather than for the isolated muscle fiber. The whole muscle has a large amount of connective tissue in it; also the sarcomeres in different parts of the muscle do not necessarily contract exactly in unison. Therefore, the curve has

somewhat different dimensions from those illustrated for the individual muscle fiber, but it nevertheless exhibits the same form.

Note in Figure 11–10 that when the muscle is at its normal resting *stretched* length and is then activated, it contracts with maximum force of contraction. If the muscle is stretched to much greater than normal length prior to contraction, a large amount of *resting tension* develops in the muscle even before contraction takes place; that is, the two ends of the muscle are pulled toward each other by the elastic forces of the connective tissue, of the sarcolemma, the blood vessels, the nerves, and so forth. However, the *increase* in tension during contraction, called *active tension*, decreases as the muscle is stretched beyond its normal length.

Note also in Figure 11–10 that when the resting muscle is shortened to less than its normal fully stretched length, the maximum tension of contraction decreases progressively and reaches zero when the muscle has shortened to approximately 60 to 70 per cent of its maximum resting length.

### RELATION OF VELOCITY OF CONTRACTION TO LOAD

A muscle contracts extremely rapidly when it contracts against no load—to a state of full contraction in approximately $1/20$ second for the average muscle. However, when loads are applied, the velocity of contraction becomes progressively less as the load increases, as illustrated in Figure 11–11. When the load increases

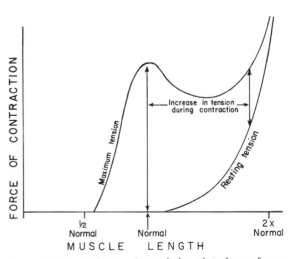

**Figure 11–10.** Relation of muscle length to force of contraction.

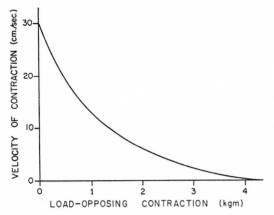

**Figure 11–11.** Relation of load to velocity of contraction in a skeletal muscle 8 cm. long.

to equal the maximum force that the muscle can exert, then the velocity of contraction becomes zero, and no contraction at all results, despite activation of the muscle fiber.

This decreasing velocity seems to be caused mainly by the fact that a load on a contracting muscle is a reverse force that opposes the contractile force caused by muscle contraction. Therefore, the net force that is available to cause velocity of shortening is correspondingly reduced.

# INITIATION OF MUSCLE CONTRACTION: EXCITATION-CONTRACTION COUPLING

## THE MUSCLE ACTION POTENTIAL

Initiation of contraction in skeletal muscle begins with action potentials in the muscle fibers. These elicit electrical currents that spread to the interior of the fiber where they cause release of calcium ions from the sarcoplasmic reticulum. It is the calcium ions that in turn initiate the chemical events of the contractile process.

Almost everything discussed in Chapter 10 regarding initiation and conduction of action potentials in nerve fibers applies equally well to skeletal muscle fibers, except for quantitative differences. Some of the quantitative aspects of muscle potentials are the following:

1. Resting membrane potential: Approximately −85 millivolts in skeletal muscle fibers—the same as in large myelinated nerve fibers of the type that excite skeletal muscle.

2. Duration of action potential: 1 to 5 milliseconds in skeletal muscle—about five times as long as in large myelinated nerves.

3. Velocity of conduction: 3 to 5 meters per second—about $1/18$ the velocity of conduction in the large myelinated nerve fibers that excite skeletal muscle.

**Excitation of Skeletal Muscle Fibers by Nerves.** In normal function of the body, skeletal muscle fibers are excited by large myelinated nerve fibers. These attach to the skeletal muscle fibers at the neuromuscular junction, which will be discussed in detail in the following chapter. Except for 2 per cent of the muscle fibers, there is only one neuromuscular junction to each muscle fiber; this junction is located near the middle of the fiber. Therefore, the action potential spreads from the middle of the fiber toward its two ends. This spreading is important because it allows nearly coincident contraction of all sarcomeres of the muscle so that they can all contract together rather than separately.

## SPREAD OF THE ACTION POTENTIAL THROUGH THE T-TUBULE SYSTEM

Not only does the action potential spread along the muscle fiber membrane but it also spreads through the T tubule system of the muscle fiber. As discussed earlier, the T tubules are a system of *transverse* tubules that interconnect with each other and spread all the way through the muscle fiber. A whole plane of such tubules spreads crosswise through the skeletal muscle fiber at two points in each sarcomere: the two points where the actin and myosin filaments overlap. It is at these points where the excitation activates the contractile process, for reasons that have been discussed in the past few pages.

Figure 11–12 illustrates the spread of the action potential in the form of local circuits of current, as illustrated by the dashed arrows along the sarcolemma and also into the T tubule system. Note that a local circuit of current also flows around each one of the T tubules. Though the T tubules are only 300 Ångstroms in diameter, the extensiveness of the T tubule system in each half of the sarcomere allows tremendous flow of electrical current between each T tubule and the fluid in the substance of the muscle fiber. Remember that the fluid in the T tubule system is *extracellular fluid* that is continuous with the outside of the muscle fiber.

There is only one significant difference be-

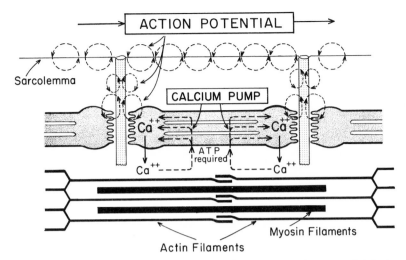

**Figure 11–12.**  Excitation-contraction coupling in the muscle, showing an action potential that causes release of calcium ions from the sarcoplasmic reticulum and then re-uptake of the calcium ions by a calcium pump.

tween the action potential that spreads through the T tubule system and that which spreads along the sarcolemma: the T tubule action potential spreads at a velocity of only 7 to 10 centimeters per second, which is about $^1/_{50}$ the velocity along the sarcolemma. Even so, the transverse distance across a muscle fiber is extremely slight, so that the action potential still reaches the deep recesses of the muscle fiber within less than 1 millisecond.

### RELEASE OF CALCIUM IONS BY THE CISTERNAE OF THE SARCOPLASMIC RETICULUM

Figure 11–12 shows that the action potential of the T tubule causes current flow through the cisternae of the sarcoplasmic reticulum. In some way not yet known, this causes the cisternae to release calcium ions into the surrounding sarcoplasmic fluid. Where the cisternae abut the T tubule, they project *junctional feet* toward the tubule; these structures surround the tubule, presumably facilitating the passage of electrical current from the T tubule into the cisternae. The electrical circuit is then completed by passage of current outward through the lateral sides of the cisternae and thence back into the T tubule system, as illustrated by the dashed arrows in the figure. The current flow through the walls of the cisternae is postulated to be the stimulus that causes release of calcium ions from the cisternae.

The calcium ions that are thus released from the cisternae diffuse to the adjacent myofibrils

where they bind strongly with troponin, as discussed in an earlier section, and this in turn supposedly elicits the muscle contraction, as has also been discussed. However, the calcium ions also bind less strongly with the myosin filaments, which could theoretically be another factor in initiating contraction.

**The Calcium Pump for Removing Calcium Ions from the Sarcoplasmic Fluid.**   Once the calcium ions have been released from the cisternae and have diffused to the myofibrils, muscle contraction will then continue as long as the calcium ions are still present in high concentration in the sarcoplasmic fluid. However, a continually active calcium pump located in the walls of the longitudinal tubules of the sarcoplasmic reticulum pumps calcium ions out of the sarcoplasmic fluid back into the vesicular cavities of the longitudinal tubules. This pump can concentrate the calcium ions about 2000-fold inside the tubules, a condition which allows massive buildup of calcium in the sarcoplasmic reticulum and also causes almost total depletion of calcium ions in the fluid of the myofibrils. Therefore, except immediately after an action potential, the calcium ion concentration in the myofibrils is kept at an extremely low level.

**The Excitatory "Pulse" of Calcium Ions.** The normal concentration (less than $10^{-7}$ molar) of calcium ions in the sarcoplasm that bathes the myofibrils is too little to elicit contraction. Therefore, in the resting state, the troponin-tropomyosin complex keeps the actin filaments inhibited and maintains a relaxed state of the muscle.

On the other hand, full excitation of the T tubule–sarcoplasmic reticulum system causes enough release of calcium ions to increase the concentration in the myofibrillar fluid to as high as $2 \times 10^{-4}$ molar concentration, which will cause maximum muscle contraction. Immediately thereafter, the calcium pump depletes the calcium ions again. The total duration of this calcium "pulse" in the usual skeletal muscle fiber lasts about $1/_{50}$ of a second, though it may last several times as long as this in some skeletal muscle fibers and be several times shorter in others (in heart muscle the pulse lasts for as long as 0.3 second). It is during this calcium pulse that muscle contraction occurs. If the contraction is to continue without interruption for longer intervals, a series of such pulses must be initiated by a continuous series of repetitive action potentials, as will be discussed in more detail later in the chapter.

## THE SOURCE OF ENERGY FOR MUSCLE CONTRACTION

We have already seen that muscle contraction is dependent upon energy supplied by ATP. Most of this energy is required to actuate the ratchet mechanism by which the cross-bridges pull the actin filaments, but small amounts are required for (1) pumping calcium from the sarcoplasm into the sarcoplasmic reticulum, and (2) pumping sodium and potassium ions through the muscle fiber membrane to maintain an appropriate ionic environment for the propagation of action potentials.

However, the amount of ATP that is present in the muscle fiber is sufficient to maintain full contraction for less than 1 second. Fortunately, after the ATP is broken into ADP, as was described in Chapter 2, the ATP is rephosphorylated to form new ATP within a fraction of a second. There are several sources of the energy required for this rephosphorylation:

The first source of energy that is used to reconstitute the ATP is the substance *creatine phosphate,* which carries a high energy phosphate bond similar to those of ATP. The high energy phosphate bond of the creatine phosphate is cleaved and the released energy causes bonding of a new phosphate ion to ADP to reconstitute the ATP. However, the total amount of creatine phosphate is also very little—only about five times as great as the ATP. Therefore, the combined energy of both the stored ATP and the creatine phosphate in the muscle is still capable of causing maximal muscle contraction for no longer than a few seconds.

The next source of energy used to reconstitute both the creatine phosphate and the ATP is energy released from the foodstuffs—from carbohydrates, fats, and proteins. Most of this energy is released in the course of oxidation of these foodstuffs. This oxidative release of energy takes place almost entirely in the mitochondria, which utilize the released energy to form new ATP. Thus, the ultimate source of energy for muscle contraction is the basic food substances and oxygen. The detailed mechanisms of these energetic processes are discussed in Chapters 67 through 71.

**Relation of Muscle Energy Expenditure to Work Performed—the "Fenn" Effect.** The shortening process of muscles can lift objects or move objects against force and thereby perform *work.* The amounts of oxygen and other nutrients consumed by the muscle increase greatly when the muscle performs work rather than simply contracting without causing work. This is called the "Fenn" effect. Though this seems to be an obvious effect that one would expect, nevertheless, the chemical basis for it has not been discovered. In some way the contraction of a muscle against a load causes the rate of breakdown of ATP to ADP to increase. This possibly results from the fact that increased numbers of reactive sites and cross-bridges must be activated to overcome the load.

**Efficiency of Muscle Contraction.** The "efficiency" of an engine or a motor is calculated as the percentage of energy input that is converted into work instead of heat. The percentage of the input energy to a muscle (the chemical energy in the nutrients) that can be converted into work is less than 20 to 25 per cent, the remainder becoming heat. Maximum efficiency can be realized only when the muscle contracts at a moderate velocity. If the muscle contracts very slowly, large amounts of *maintenance heat* are released during the process of contraction, thereby decreasing the efficiency. On the other hand, if contraction is too rapid, large proportions of the energy are used to overcome the viscous friction within the muscle itself, and this, too, reduces the efficiency of contraction. Ordinarily, maximum efficiency is developed when the velocity of contraction is about 30 per cent of maximum.

## CHARACTERISTICS OF A SINGLE MUSCLE TWITCH

Many features of muscle contraction can be especially well demonstrated by eliciting single *muscle twitches.* This can be accomplished by instantaneously exciting the nerve to a muscle or by passing a short electrical stimulus through the muscle itself, giving rise to a single, sudden contraction lasting for a fraction of a second.

**Isometric versus Isotonic Contraction.** Muscle contraction is said to be *isometric* when the muscle does not shorten during contraction and *isotonic* when it shortens but the tension on the muscle remains constant. Systems for recording the two types of muscle contraction are illustrated in Figure 11–13.

To the right is the isometric system in which the muscle is suspended between a solid rod and a lever of an electronic force transducer. This transducer records force with almost zero movement of the lever; therefore, in effect, the muscle is bound between fixed points so that it cannot contract significantly. To the left is shown an isotonic recording system: the muscle simply lifts a pan of weights so that the force against which the muscle contracts remains constant, though the length of the muscle changes considerably.

There are several basic differences between isometric and isotonic contractions. First, isometric contraction does not require sliding of myofibrils among each other. Second, in isotonic contraction a load is moved, which involves the phenomenon of inertia. That is, the weight or other type of object being moved must first be accelerated, and once a velocity has been attained the load has momentum that causes it to continue moving even after the contraction is over. Therefore an isotonic contraction is likely to last considerably longer than an isometric contraction of the same muscle. Third, isotonic contraction entails the performance of external work. Therefore, in accordance with the Fenn effect discussed previously, a greater amount of energy is used by the muscle.

In comparing the rapidity of contraction of different types of muscle, isometric recordings such as those illustrated in Figure 11-14 are usually used instead of isotonic recordings because the duration of an isotonic recording is almost as dependent on the inertia of the recording system as upon the contraction itself, and this makes it difficult to compare time relationships of contractions from one muscle to another.

Muscles can contract both isometrically and iso-

tonically in the body, but most contractions are actually a mixture of the two. When a person stands, he tenses his quadriceps muscles to tighten the knee joints and to keep the legs stiff. This is isometric contraction. On the other hand, when a person lifts a weight using his biceps, this is mainly an isotonic contraction. Finally, contractions of leg muscles during running are a mixture of isometric and isotonic contractions—isometric mainly to keep the limbs stiff when the legs hit the ground and isotonic mainly to move the limbs.

*The Series Elastic Component of Muscle Contraction.* When muscle fibers contract against a load, those portions of the muscle that do not contract—the tendons, the sarcolemmal ends of the muscle fibers where they attach to the tendons, and perhaps even the hinged arms of the cross-bridges—will stretch slightly as the tension increases. Consequently, the muscle must shorten an extra 3 to 5 per cent to make up for the stretch of these elements. The elements of the muscle that stretch during contraction are called the *series elastic component* of the muscle.

**Characteristics of Isometric Twitches Recorded from Different Muscles.** The body has many different sizes of skeletal muscles—from the very small stapedius muscle of only a few millimeters length and a millimeter or so in diameter up to the very large quadriceps muscle. Furthermore, the fibers may be as small as 10 microns in diameter or as large as 80 microns. And, finally, the energetics of muscle contraction vary considerably from one muscle to another. These different physical and chemical characteristics often manifest themselves in the form of different characteristics of contraction, some muscles contracting rapidly while others contract slowly.

**Fast versus Slow Muscle.** Figure 11–14 illustrates isometric contractions of three different types of skeletal muscles: an ocular muscle, which has a duration of contraction of less than $1/100$ second; the gastrocnemius muscle, which has a duration of contraction of about $1/30$ second; and the soleus muscle, which has a duration of contraction of about $1/10$ second. It is interesting that these durations of contraction are adapted to the function of each of the respective muscles, for ocular movements must be extremely rapid to maintain fixation of the eyes upon specific objects, the gastrocnemius muscle must contract moderately rapidly to provide sufficient velocity of limb movement for running and jumping, while the soleus muscle is concerned principally with slow reactions for continual support of the body against gravity.

Thus, there is a wide range of gradations from *fast* to *slow* muscles. In general, the slow muscles are used more for prolonged performance of work. Their muscle fibers generally are smaller, are surrounded by more blood capillaries, and have far more mitochondria than the fast muscles. They also have a large amount of *myoglobin* in the sarcoplasm (myoglobin is a substance similar to the hemoglobin in red

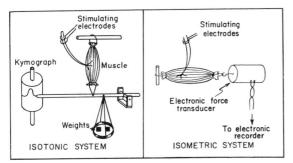

**Figure 11–13.** Isotonic and isometric recording systems.

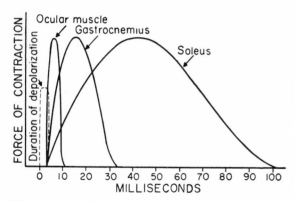

**Figure 11–14.** Duration of isometric contractions of different types of mammalian muscles, showing also a latent period between the action potential and muscle contraction.

blood cells and can combine with oxygen and store this oxygen inside the muscle cell until it is needed by the mitochondria). On the other hand, the fast muscles generally have a much more extensive sarcoplasmic reticulum, which allows very rapid release of calcium ions and then rapid re-uptake of the calcium ions so that the contraction can be rapid. The slow muscle is frequently called *red muscle* because of the reddish tint caused by the myoglobin as well as by the large amount of red blood cells in the capillaries; the fast muscle is called *white muscle* because of a whitish appearance caused by lack of these elements.

# CONTRACTION OF SKELETAL MUSCLE IN THE BODY

## *THE MOTOR UNIT*

Each motor neuron that leaves the spinal cord usually innervates many different muscle fibers, the number depending on the type of muscle. All the muscle fibers innervated by a single motor nerve fiber are called a *motor unit*. In general, small muscles that react rapidly and whose control is exact have few muscle fibers (as few as 2 to 3 in some of the laryngeal muscles) in each motor unit and have a large number of nerve fibers going to each muscle. On the other hand, the large muscles which do not require a very fine degree of control, such as the gastrocnemius muscle, may have as many as 1000 muscle fibers in each motor unit. An average figure for all the muscles of the body can be considered to be about 180 muscle fibers to the motor unit.

Usually muscle fibers of adjacent motor units

overlap, with small bundles of 10 to 15 fibers from one motor unit lying among similar bundles of the second motor unit. This interdigitation allows the separate motor units to contract in support of each other rather than entirely as individual segments.

**Macromotor Units.** Loss of some of the nerve fibers to a muscle belly causes the remaining nerve fibers to sprout forth and innervate many of the paralyzed muscle fibers. When this occurs, such as following poliomyelitis, one occasionally develops *macromotor units,* which can contain as many as 5 times the normal number of muscle fibers. This obviously decreases the degree of control that one has over his muscles, but, nevertheless, allows the muscle to regain function.

## *SUMMATION OF MUSCLE CONTRACTION*

Summation means the adding together of individual muscle twitches to make strong and concerted muscle movements. In general, summation occurs in two different ways: (1) by increasing the number of motor units contracting simultaneously and (2) by increasing the rapidity of contraction of individual motor units. These are called, respectively, *multiple motor unit summation* and *wave summation* (or spatial summation and temporal summation).

**Multiple Motor Unit Summation.** Figure 11–15 illustrates multiple motor unit summation, showing that the force of contraction increases progressively as the number of contracting motor units increases from 1 to 8.

Even within a single muscle, the numbers of muscle fibers and their sizes in the different motor units vary tremendously, so that one motor unit may be as much as 50 times as strong as another. The smaller motor units are far more easily excited than are the larger ones be-

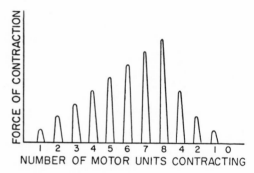

**Figure 11–15.** Multiple motor unit summation.

cause they are innervated by smaller nerve fibers whose cell bodies in the spinal cord have a naturally high level of excitability. This effect causes the gradations of muscle strength during weak muscle contraction to occur in very small steps, while the steps become progressively greater as the intensity of contraction increases.

**Wave Summation.** Figure 11–16 illustrates the principles of wave summation, showing in the lower left-hand corner a single muscle twitch followed by successive muscle twitches at various frequencies. When the frequency of twitches is 10 per second, the first muscle twitch is not completely over by the time the second one begins. Therefore, since the muscle is already in a partially contracted state when the second twitch begins, the degree of muscle shortening this time is slightly greater than that which occurs with the single muscle twitch. The third, fourth, and additional twitches add still more shortening.

At more rapid rates of contraction, the degree of summation of successive contractions becomes greater and greater, because the successive contractions appear at earlier times following the preceding contraction.

*Tetanization.* When a muscle is stimulated at progressively greater frequencies, a frequency is finally reached at which the successive contractions fuse together and cannot be distinguished one from the other. This state is called *tetanization,* and the lowest frequency at which it occurs is called the *critical frequency.*

Tetanization results partly from the viscous properties of the muscle and partly from the nature of the contractile process itself. The muscle fibers are filled with sarcoplasm, which is a viscous fluid, and the fibers are encased in fasciae and muscle sheaths that have a viscous resistance to change in length. Therefore, undoubtedly these viscous factors play a role in causing the successive contractions to fuse with each other.

But in addition to the viscous property of muscle, the activation process itself lasts for a definite period of time, and successive pulsatile states of activation of the muscle fiber can occur so rapidly that they fuse into a long continual state of activation; that is, free calcium ions persist continuously in the myofibrils and provide an uninterrupted stimulus for maintenance of contraction. Once the critical frequency for tetanization is reached, further increase in rate of stimulation increases the force of contraction only a few more per cent, as shown in Figure 11–16.

**Asynchronous Summation of Motor Units.** Actually it is rare for either multiple motor unit summation or wave summation to occur separately from each other in normal muscle function. Instead, special neurogenic mechanisms in the spinal cord normally increase both the impulse rate and the number of motor units firing at the same time. If a motor unit fires at all, it usually fires at least five times per second, but this can increase to as high as 50 per second for most muscles or much more than this for the very fast muscles—to frequencies sufficient to cause complete tetanization.

Yet, even when tetanization of individual motor units of a muscle is not occurring, the tension exerted by the whole muscle is still continuous and nonjerky because *the different motor units fire asynchronously;* that is, while one is contracting another is relaxing; then another fires, followed by still another, and so forth. Consequently, even when motor units fire as infrequently as five times per second, the muscle contraction, though weak, is nevertheless very smooth.

**Maximum Strength of Contraction.** The maximum strength of tetanic contraction of a muscle operating at a normal muscle length is about 3.5 kilograms per square centimeter of muscle, or 50 pounds per square inch. Since a quadriceps muscle can at times have as much as 16 square inches of muscle belly, as much as 800 pounds of tension may at times be applied to the patellar tendon. One can readily understand, therefore, how it is possible for muscles sometimes to pull their tendons out of the

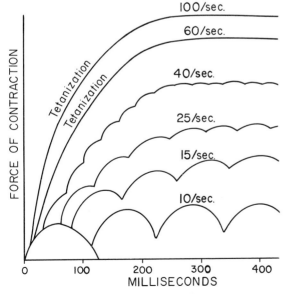

**Figure 11–16.** Wave summation and tetanization.

insertions in bones. This often occurs where the patellar tendon inserts in the tibia, and it occurs even more frequently where the Achilles tendon of the gastrocnemius muscle inserts at the heel.

**Changes in Muscle Strength at the Onset of Contraction—The Staircase Effect (Treppe).** When a muscle begins to contract after a long period of rest, its initial strength of contraction may be as little as one-half its strength 30 to 50 muscle twitches later. That is, the strength of contraction increases to a plateau, a phenomenon called the *staircase effect* or *treppe*. This phenomenon has interested physiologists greatly because it gives possible insights into the mechanism of muscle contraction.

Though all the possible causes of the staircase effect are not yet known, it is believed to be caused primarily by electrolyte changes that occur when a series of contractions begins. For instance, there is a net increase in calcium ions inside the muscle fiber because of movement of calcium ions inward through the membrane with each action potential. There is probably also further increase of calcium ions in the sarcoplasm because of release of these ions from the sarcoplasmic reticulum and failure to recapture the ions immediately. In addition, there is decreased potassium inside the cell as well as increased sodium; it has been suggested that the changes in these two ions increase the rate of liberation of calcium ions from the sarcoplasmic reticulum. Recalling the earlier discussion of the relationship of calcium ions to the contractile process, one can readily understand that progressive increase in calcium ion concentration in the sarcoplasm, caused either directly or as a consequence of sodium and potassium movement, could progressively increase the strength of muscle contraction, giving rise to the staircase effect.

## SKELETAL MUSCLE TONE

Even when muscles are at rest, a certain amount of tautness usually remains. This residual degree of contraction in skeletal muscle is called *muscle tone*. Since skeletal muscle fibers do not contract without an actual action potential to stimulate the fibers except in certain pathological conditions, it is believed that skeletal muscle tone results entirely from nerve impulses coming from the spinal cord. These in turn are controlled partly by impulses transmitted from the brain to the appropriate anterior motor neurons and partly by impulses that originate in *muscle spindles* located in the muscle itself.

Muscle spindles are sensory receptors that exist throughout essentially all skeletal muscles to detect the degree of muscle contraction. These will be discussed in detail in Chapter 51, but, briefly, they transmit impulses almost continually through the posterior roots into the spinal cord, where they excite the anterior motor neurons, which in turn provide the necessary nerve stimuli for muscle tone. Simply cutting the posterior roots, thereby blocking the muscle spindle impulses, usually reduces muscle tone to such a low level that the muscle becomes almost completely flaccid.

Many other neurogenic factors, originating especially in the brain, enter into the control of muscle tone. These will be discussed in detail in relation to muscle spindle and spinal cord function in Chapter 51.

## MUSCLE FATIGUE

Prolonged and strong contraction of a muscle leads to the well-known state of muscle fatigue. This results simply from inability of the contractile and metabolic processes of the muscle fibers to continue supplying the same work output. The nerve continues to function properly, the nerve impulses pass normally through the neuromuscular junction into the muscle fiber, and even normal action potentials spread over the muscle fibers, but the contraction becomes weaker and weaker because of depletion of ATP in the muscle fibers themselves.

Interruption of blood flow through a contracting muscle leads to almost complete muscle fatigue in a minute or more because of the obvious loss of nutrient supply.

## THE LEVER SYSTEMS OF THE BODY

Muscles obviously operate by applying tension to their points of insertion into bones, and the bones in turn form various types of lever systems. Figure 11–17 illustrates the lever system activated by the biceps muscle to lift the forearm. If we assume that a large biceps muscle has a cross-sectional area of 6 square inches, then the maximum force of contraction would be about 300 pounds. When the forearm is exactly at right angles with the upper arm, the tendon attachment of the biceps is about 2 inches anterior to the fulcrum at the elbow, and the total length of the forearm lever is about 14 inches. Therefore, the amount of lifting power that the biceps would have at the hand would be only one seventh of the 300 pounds force, or about 43 pounds. When the arm is in the fully extended position the attachment of the

**Figure 11–17.** The lever system activated by the biceps muscle.

biceps is much less than 2 inches anterior to the fulcrum, and the force with which the forearm can be brought forward is much less than 43 pounds.

In short, an analysis of the lever systems of the body depends on (a) a discrete knowledge of the point of muscle insertion and (b) its distance from the fulcrum of the lever, as well as (c) the length of the lever arm and (d) the position of the lever. Obviously, many different types of movement are required in the body, some of which need great strength and others large distances of movement. For this reason there are all varieties of muscles; some are long and contract a long distance and some are short but have large cross-sectional areas and therefore can provide extreme strengths of contraction over short distances. The study of different types of muscles, lever systems, and their movements is called *kinesiology* and is a very important phase of human physioanatomy.

**Accommodation of Muscle Length to the Length of the Lever System.** If a bone is broken and then heals in a shortened state, the force of contraction of the muscles lying along this broken bone would obviously become decreased because of the shortened lengths of muscles. However, muscles shortened in this manner undergo *physical shortening* during the next few weeks. That is, the muscle fibers actually shorten and re-establish new muscle lengths approximately equal to the maximum length of the lever system itself, thus re-establishing optimum force of contraction by the muscles.

The same shortening process also occurs in muscles of limbs immobilized for several weeks in casts if the muscles during this time are in a shortened position. When the cast is removed, the muscles must often be restretched over a period of weeks before full mobility is restored.

# SPECIAL FEATURES AND ABNORMALITIES OF SKELETAL MUSCLE FUNCTION

## MUSCULAR HYPERTROPHY

Forceful muscular activity causes the muscle size to increase, a phenomenon called hypertrophy. The diameters of the individual muscle fibers increase, and the fibers gain in total numbers of myofibrils as well as in various nutrient and intermediary metabolic substances, such as adenosine triphosphate, phosphocreatine, glycogen, and so forth. Briefly, muscular hypertrophy increases both the motive power of the muscle and the nutrient mechanisms for maintaining increased motive power.

Weak muscular activity, even when sustained over long periods of time, does not result in significant hypertrophy. Instead, hypertrophy results mainly from *very* forceful muscle activity, though the activity might occur for only a few minutes each day. For this reason, strength can be developed in muscles much more rapidly when "resistive" or "isometric" exercise is used rather than simply prolonged mild exercise. Indeed, essentially no new myofibrils develop unless the muscle contracts to at least 75 per cent of its maximum tension.

## MUSCULAR ATROPHY

Muscular atrophy is the reverse of muscular hypertrophy; it results any time a muscle is not used or even when a muscle is used only for very weak contractions. Atrophy is particularly likely to occur when limbs are placed in casts, thereby preventing muscular contraction. As little as one to two months of disuse can sometimes decrease the muscle size to one-half normal.

**Atrophy Caused by Muscle Denervation.** When a muscle is denervated it immediately begins to atrophy, and the muscle continues to decrease in size for several years. If the muscle becomes re-innervated during the first three to four months, full function of the muscle usually returns, but after four months of denervation some of the muscle fibers usually will have degenerated. Re-innervation after two years rarely results in return of any function at all. Pathological studies show that the muscle fibers have by that time been replaced by fat and fibrous tissue.

*Prevention of Muscle Atrophy by Electrical Stimulation.* Strong electrical stimulation of denervated muscles, particularly when the resulting contractions occur against loads, will delay and in some instances prevent muscle atrophy despite denervation. This procedure is used to keep muscles alive until re-innervation can take place.

*Physical Contracture of Muscle Following Denervation.* When a muscle is denervated, its fibers tend to shorten if the muscle is kept in a shortened position, and even the associated nerves and fasciae shorten. All this is a natural characteristic of protein fibers called "creep." That is, unless continual movement keeps stretching the muscle and other structures, they will creep toward a shortened length. This is one of the most difficult problems in the treatment of patients with denervated muscles, such as occur in poliomyelitis or nerve trauma. Unless passive stretching is applied daily to the muscles, they may become so shortened that even when re-innervated they will be of little value. But, more important, the shortening can often result in extremely contorted positions of different parts of the body.

## RIGOR MORTIS

Several hours after death all the muscles of the body go into a state of *contracture* called rigor mortis; that is, the muscle contracts and becomes rigid even without action potentials. It is believed that this rigidity is caused by loss of all the ATP, which is required to cause separation of the cross-bridges from the actin filaments during the relaxation process. The muscles remain in rigor until the muscle proteins are destroyed, which usually results from autolysis caused by enzymes released from the lysosomes some 15 to 25 hours later.

## FAMILIAL PERIODIC PARALYSIS

Occasionally, a hereditary disease called familial periodic paralysis occurs. In persons so afflicted, the extracellular fluid potassium concentration periodically falls to very low levels, causing various degrees of paralysis. The paralysis is caused in the following manner: A great decrease in extracellular fluid potassium increases the muscle fiber membrane potential to a very high value. This results in strong *hyperpolarization* of the membrane (a membrane potential more negative than the normal −85 millivolts) making the fiber almost totally inexcitable; that is, the membrane potential is so high that the normal stimulus at the neuromuscular junction is incapable of exciting the fiber.

## THE ELECTROMYOGRAM

Each time an action potential passes along a muscle fiber a small portion of the electrical current spreads away from the muscle as far as the skin. If many muscle fibers contract simultaneously, the summated electrical potentials at the skin may be very great. By placing two electrodes on the skin or inserting needle electrodes into the muscle, an electrical recording called the electromyogram can be made when the muscle is stimulated. Figure 11–18 illustrates a typical electromyographic recording from the gastrocnemius muscle during a moderate contraction. Electromyograms are frequently used clinically to discern abnormalities of muscle excitation. Two such abnormalities are muscle *fasciculation* and *fibrillation*.

**Muscle Fasciculation.** When an abnormal impulse occurs in a motor nerve fiber its whole motor unit contracts. This often causes sufficient contraction in the muscle that one can see a slight ripple in the skin over the muscle. This process is called fasciculation.

Fasciculation occurs especially following destruction of anterior motor neurons in poliomyelitis or fol-

lowing traumatic interruption of a nerve. As the peripheral nerve fibers die, spontaneous impulses are generated during the first few days, and fasciculatory muscle movements result in the muscle. Typical electromyographic records of weak periodic potentials can be obtained from the skin overlying the muscle.

**Muscle Fibrillation.** After all nerves to a muscle have been destroyed and the nerve fibers themselves have become nonfunctional, which requires three to five days, spontaneous impulses begin to appear in the denervated muscle fibers. At first, these occur at a rate of once every few seconds, but, after a few more days or a few weeks, the impulses become as rapid as 3 to 10 times per second. Thus, skeletal muscle fibers, when released from innervation, develop an intrinsic rhythmicity. After several more weeks the muscle fibers atrophy to such an extent that the fibrillatory impulses finally cease. To record an electromyogram of fibrillation, minute bipolar needle electrodes must be inserted into the muscle belly itself because adjacent muscle fibers do not fire simultaneously and, therefore, do not summate. As a result, the potentials are not strong enough to record from the surface of the skin.

## REFERENCES

Abbott, B. C., and Howarth, J. V.: Heat studies in excitable tissues. *Physiol. Rev.,* 53:120, 1973.
Astrand, P., and Rodahl, K.: Textbook of Work Physiology. New York, McGraw-Hill Book Company, 1968.
Basmajian, J. V.: Muscles Alive, 3rd Ed. Baltimore, The Williams & Wilkins Co., 1974.
Bourne, G. H. (ed.): The Structure and Function of Muscle, 2nd Ed. New York, Academic Press, Inc., 1973.
Buller, A. J.: The physiology of skeletal muscle. *In* Guyton, A. C. (ed.): MTP International Review of Science: Physiology. Baltimore, University Park Press, 1975, Vol. 3, p. 279.
Carlson, B. M.: The regeneration of skeletal muscle. A review. *Amer. J. Anat.,* 137:119, 1973.
Close, R. I.: Dynamic properties of mammalian skeletal muscles. *Physiol. Rev.,* 52:129, 1972.
Ebashi, S.: Calcium ions and muscle contraction. *Nature,* 240:217, 1972.
Ebashi, S.: Regulatory mechanism of muscle contraction with special reference to the Ca-troponin-tropomyosin system. *Essays Biochem.,* 10:1, 1974.
Falls, H. B.: Exercise Physiology. New York, Academic Press, Inc., 1968.
Fuchs, F.: Striated muscle. *Ann. Rev. Physiol.,* 36:461, 1974.
Grimby, G., and Saltin, B.: Physiological effects of physical training. *Scand. J. Rehabil. Med.,* 3:6, 1971.
Hess, A.: Vertebrate slow muscle fibers. *Physiol. Rev.,* 50:40, 1970.
Hill, A. L., et al.: Physiology of voluntary muscle. *Brit. Med. Bull.,* 12(Sept.): 1956.
Hoyle, G.: Comparative aspects of muscle. *Ann. Rev. Physiol.,* 31:43, 1969.
Hoyle, G.: How is muscle turned on and off? *Sci. Amer.,* 222:84, 1970.
Hughes, J. T.: Pathology of Muscle. Philadelphia, W. B. Saunders Company, 1975.
Huxley, H. E.: Muscular contraction and cell motility. *Nature,* 243:445, 1973.
Huxley, A. F., and Gordon, A. M.: Striation patterns in active and passive shortening of muscle. *Nature (Lond.),* 193:280, 1962.
James, S. L., and Brubaker, C. E.: Biomechanics of running. *Orthop. Clin. North. Amer.,* 4:605, 1973.

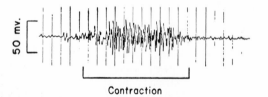

**Figure 11–18.** Electromyogram recorded during contraction of the gastrocnemius muscle.

Keul, J.: The relationship between circulation and metabolism during exercise. *Med. Sci. Sports,* 5:209, 1973.

Margaria, R.: The sources of muscular energy. *Sci. Amer., 226*:84, 1972.

Mommaerts, W. F. H. M.: Energetics of muscular contraction. *Physiol. Rev., 49*:427, 1969.

Murray, J. M., and Weber, A.: The cooperative action of muscle proteins. *Sci. Amer., 230*:58, 1974.

Nelson, P. G.: Nerve and muscle cells in culture. *Physiol. Rev., 55*:1, 1975.

Pearson, C. M. (ed.): The Striated Muscle. International Academy of Pathology, Monograph No. 2. Baltimore, The Williams & Wilkins Company, 1973.

Podolsky, R. J. (ed.): Mechanisms of Movement: A Symposium. Woods Hole, Mass., Sept. 1970. Englewood Cliffs, N.J., Prentice-Hall, Inc., 1972.

Sandow, A.: Skeletal muscle. *Ann. Rev. Physiol., 32*:87, 1970.

Spurway, N. C.: Mechanisms of anion permeation. A review of available data, principally on muscle cells, with the fixed charge concept in mind. *Biomembranes, 3*:363, 1972.

Taylor, E. W.: Chemistry of muscle contraction. *Ann. Rev. Biochem., 41*:577, 1972.

Tonomura, Y.: Muscle Proteins, Muscle Contraction, and Cation Transport. Baltimore, University Park Press, 1973.

Tonomura, Y., and Oosawa, F.: Molecular mechanism of contraction. *Ann. Rev. Biophys. Bioeng., 1*:159, 1972.

Walton, J. N. (ed.): Disorders of Voluntary Muscle. New York, Churchill Livingstone, Div. of Longman Inc., 1974.

Weber, A., and Murray, J. M.: Molecular control mechanisms in muscle contraction. *Physiol. Rev., 53*:612, 1973.

Zachar, J.: Electrogenesis and Contractility in Skeletal Muscle Cells. Baltimore, University Park Press, 1972.

# 12

# Neuromuscular Transmission; Function of Smooth Muscle

## TRANSMISSION OF IMPULSES FROM NERVES TO SKELETAL MUSCLE FIBERS: THE NEUROMUSCULAR JUNCTION

The skeletal muscles are innervated by large myelinated nerve fibers that originate in the large motoneurons of the anterior horns of the spinal cord. It was pointed out in the previous chapter that each nerve fiber normally branches many times and stimulates from 3 to 2000

skeletal muscle fibers. The nerve ending makes a junction, called the *neuromuscular junction* or the *myoneural junction,* with the muscle fiber approximately at the fiber's midpoint so that the action potential in the fiber travels in both directions. With the exception of about 2 per cent of the muscle fibers there is only one such junction per muscle fiber.

**Physiologic Anatomy of the Neuromuscular Junction.** Figure 12–1, Parts A and B, illustrates the neuromuscular junction between a large myelinated nerve fiber and a skeletal mus-

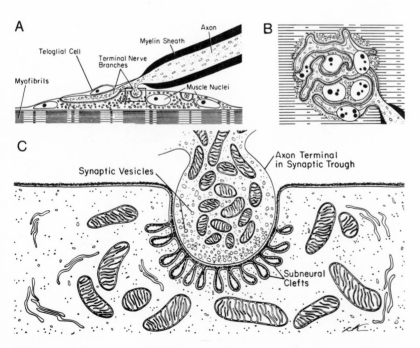

**Figure 12–1.** Different views of the motor end-plate. (A) Longitudinal section through the end-plate. (B) Surface view of the end-plate. (C) Electronmicrographic appearance of the contact point between one of the axon terminals and the muscle fiber membrane, representing the rectangular area shown in A. (From Bloom and Fawcett, as modified from R. Couteaux: A Textbook of Histology, 1975.)

cle fiber. The nerve fiber branches at its end to form a complex of branching nerve *terminals* called the *end-plate*, which invaginates into the muscle fiber but lies entirely outside the muscle fiber plasma membrane. The entire structure is covered by one or more Schwann cells that insulate the end-plate from the surrounding fluids.

Figure 12-1C shows an electronmicrographic sketch of the juncture between an axon terminal and the muscle fiber membrane. The invagination of the membrane is called the *synaptic gutter,* and the space between the terminal and the fiber membrane is called the *synaptic cleft*. The synaptic cleft is 200 to 300 Å wide and is filled with a gelatinous "ground" substance through which diffuses extracellular fluid. At the bottom of the gutter are numerous *folds* of the muscle membrane which form *secondary synaptic clefts* that greatly increase the surface area at which the synaptic transmitter can act. In the axon terminal are many mitochondria that supply energy mainly for synthesis of the excitatory transmitter *acetylcholine.* The acetylcholine is synthesized in the cytoplasm of the terminal, but is rapidly absorbed into many small synaptic vesicles, approximately 300,000 of which are normally in all the terminals of a single end-plate. On the surfaces of the folds in the synaptic gutter are aggregates of the enzyme *cholinesterase*, which is capable of destroying acetylcholine, as is explained in further detail below.

**Secretion of Acetylcholine by the Axon Terminals.** When a nerve impulse reaches the neuromuscular junction, some 200 to 300 vesicles of acetylcholine are released by the terminals into the synaptic clefts between the terminals and the muscle fiber membrane. It has been postulated that the nerve action potential causes calcium ions to move from the extracellular fluid into the membranes of the terminal, and that it is these ions that in turn cause the vesicles of acetylcholine to rupture through the membrane. In the absence of calcium or in the presence of excess magnesium, the release of acetylcholine is greatly depressed.

*Diffusion of the Released Acetylcholine Away from the Muscle Fiber Membrane, and Its Destruction by Cholinesterase.* Within approximately 2 to 3 milliseconds after acetylcholine is released by the axon terminal, some of it presumably diffuses out of the synaptic gutter and no longer acts on the muscle fiber membrane, and the remaining greater bulk of it is destroyed by the cholinesterase on the surfaces of the folds in the gutter. The very short period of time

that the acetylcholine remains in contact with the muscle fiber membrane—2 to 3 milliseconds—is almost always sufficient to excite the muscle fiber, and yet the rapid removal of the acetylcholine prevents re-excitation after the muscle fiber has recovered from the first action potential.

**The "End-Plate Potential" and Excitation of the Skeletal Muscle Fiber.** Even though the acetylcholine released into the space between the end-plate and the muscle membrane lasts for only a minute fraction of a second, nevertheless, even during this period of time it can affect the muscle membrane sufficiently to make it very permeable to sodium ions, allowing rapid influx of sodium into the muscle fiber. As a result, the membrane potential rises *in the local area of the end-plate* as much as 50 to 75 millivolts, creating a *local potential* called the *end-plate potential.*

The mechanism by which acetylcholine increases the permeability of the muscle membrane is probably the following: It is believed that the muscle membrane contains a special protein molecule called an *acetylcholine receptor substance* to which the acetylcholine binds. Under the influence of the acetylcholine this receptor supposedly undergoes a conformational change that increases the permeability of the membrane to ions, most importantly to sodium ions. Rapid influx of sodium ions ensues, and this elicits the end-plate potential that is responsible for initiation of the action potential at the muscle fiber membrane.

A typical end-plate potential is illustrated at point A in Figure 12–2. This potential was recorded after the muscle had been poisoned with

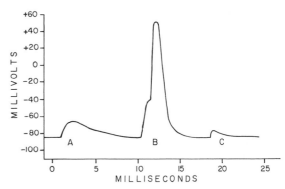

**Figure 12–2.** End-plate potentials: *A*, normal end-plate potential recorded in a curarized muscle so that the muscle fiber itself could not generate an action potential; *B*, end-plate potential eliciting a muscle action potential; and *C*, end-plate potential too weak to elicit a muscle action potential.

curare, which made the end-plate potential too low to excite an action potential in the muscle fiber. However, in the normal muscle fiber, the effect shown at point B occurs: The end-plate potential begins, but before it can complete its course the intense local current flow created by this potential initiates an action potential; this in turn spreads in both directions along the muscle fiber. The threshold voltage at which the muscle fiber is stimulated is approximately $-50$ millivolts. Thus, if the resting membrane potential of the muscle is at the normal level of about $-85$ millivolts, then an end-plate potential of $+35$ millivolts is required to elicit an action potential in the muscle fiber.

At point C, another situation is illustrated in which only a small quantity of acetylcholine is released at the end-plate. This also causes an end-plate potential that is less than the threshold value required to cause an action potential. Note that the potential dies in a few milliseconds. Thus, failure to transmit a signal to the muscle fiber can result from either diminished receptivity of the muscle membrane or decreased release of acetylcholine.

*''Safety Factor'' for Transmission at the Neuromuscular Junction; Fatigue of the Junction.* Ordinarily, each impulse that arrives at the neuromuscular junction creates an end-plate current flow about three to four times that required to stimulate the muscle fiber. Therefore, the normal neuromuscular junction is said to have a very high *safety factor*. However, artificial stimulation of the nerve fiber at rates greater than 150 times per second for many minutes often diminishes the number of vesicles of acetylcholine released with each impulse so that impulses then often fail to pass into the muscle fiber. This is called *fatigue* of the neuromuscular junction, and it is analogous to fatigue of the synapse in the central nervous system. However, under normal functioning conditions, fatigue of the neuromuscular junction almost never occurs, because almost never do the spinal nerves stimulate even the most active neuromuscular junctions more than 150 times per second.

**Drugs that Affect Transmission at the Neuromuscular Junction.** *Drugs that Stimulate the Muscle Fiber by Acetylcholine-Like Action.* Many different compounds, including *methacholine, carbachol,* and *nicotine,* have the same effect on the muscle fiber as does acetylcholine. The difference between these drugs and acetylcholine is that they are not destroyed by cholinesterase or are destroyed very slowly, so that when once applied to the muscle fiber the action persists for many minutes to several

hours. Moderate quantities of the above three drugs applied to a muscle fiber cause localized areas of depolarization, and every time the muscle fiber becomes repolarized elsewhere, these depolarized areas, by virtue of their leaking ions, cause new action potentials, thereby causing a state of spasm. On the other hand, when extreme dosages of these drugs are used, so much of the membranes becomes depolarized that the fibers can no longer pass impulses at all, and a state of flaccid paralysis exists instead of the spasm that occurs with moderate dosages.

*Drugs that Block Transmission at the Neuromuscular Junction.* One group of drugs, known as the *curariform drugs,* can prevent passage of impulses from the end-plate into the muscle. Thus, D-tubocurarine affects the membrane, probably by competing with acetylcholine for the receptor sites of the membrane, so that the acetylcholine cannot increase the permeability of the membrane sufficiently to initiate a depolarization wave.

A second group of drugs prevents transmission of impulses into muscle fibers by a different mechanism. These, of which *decamethonium* is a principal example, act in the same manner as *large doses* of nicotine, methacholine, and carbachol to depolarize the muscle fibers completely. As a result, no impulses can be transmitted over the muscle fiber membranes even though the end-plate of the motor nerve fiber functions entirely normally.

*Drugs that Stimulate the Neuromuscular Junction by Inactivating Cholinesterase.* Three particularly well-known drugs, *neostigmine, physostigmine,* and *diisopropyl fluorophosphate,* inactivate cholinesterase so that the cholinesterase normally in the muscle fibers will not hydrolyze acetylcholine released at the end-plate. As a result, acetylcholine increases in quantity with successive nerve impulses so that extreme amounts of acetylcholine can accumulate and repetitively stimulate the muscle fiber. This causes *muscular spasm* when even a few nerve impulses reach the muscle; this can cause death due to laryngeal spasm, which smothers the person.

Neostigmine and physostigmine combine with cholinesterase to inactivate it for several hours, after which they are displaced from the cholinesterase so that it once again becomes active. On the other hand, diisopropyl fluorophosphate, which has military potential as a very powerful ''nerve'' gas, actually inactivates cholinesterase for several weeks, which makes this a particularly lethal drug.

## MYASTHENIA GRAVIS

The disease *myasthenia gravis,* which occurs in rare instances in human beings, causes the person to become paralyzed because of inability of the neuromuscular junctions to transmit signals from the nerve fibers to the muscle fibers. Pathologically, the number of folds in the synaptic gutter is reduced and the synaptic cleft is widened. Also, antibodies that attack the muscle fibers have been demonstrated in

−60 millivolts, or about 25 millivolts less negative than in skeletal muscle.

**Action Potentials in Visceral Smooth Muscle.** Action potentials occur in visceral smooth muscle in the same way that they occur in skeletal muscle. However, action potentials probably do not normally occur in multiunit types of smooth muscle, as will be discussed in a subsequent section.

The action potentials of visceral smooth muscle occur in two different forms: (1) spike potentials and (2) action potentials with plateaus.

*Spike potentials.* Typical spike action potentials, such as those seen in skeletal muscle, occur in most types of visceral smooth muscle. The duration of this type of action potential is usually about 10 milliseconds, as illustrated in Figure 12–4A. Such action potentials can be elicited in many ways, such as by electrical stimulation, by the action of hormones on the smooth muscle, by the action of transmitter substances from nerve fibers, or as a result of spontaneous generation in the muscle fiber itself, as discussed below.

**Slow Wave Potentials in Visceral Smooth Muscle and Spontaneous Generation of Spike Potentials.** Some smooth muscle is self-excitatory. That is, action potentials arise within the smooth muscle itself without an extrinsic stimulus. This is usually associated with

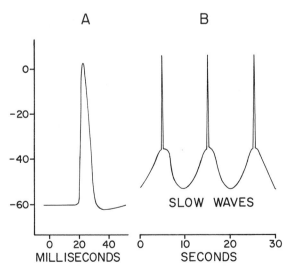

**Figure 12–4.** (A) A typical smooth muscle action potential (spike potential) elicited by an external stimulus. (B) A series of spike action potentials elicited by rhythmical slow electrical waves occurring spontaneously in the smooth muscle wall of the intestine.

a basic *slow wave rhythm* of the membrane potential. A typical slow wave of this type is illustrated in Figure 12–4B. The slow wave itself is not an action potential. It is not a self-regenerative process that spreads progressively over the membranes of the muscle fibers. Instead, it is a local property of the smooth muscle fibers that make up the muscle mass.

The cause of the slow wave rhythm is as yet unknown, though one suggestion for which there is considerable circumstantial evidence is that the slow waves are caused by waxing and waning of the pumping of sodium outward through the muscle fiber membrane; the membrane potential becomes more negative when sodium is pumped rapidly and less negative when the sodium pump becomes less active.

The importance of the slow waves lies in the fact that they can initiate action potentials. The slow waves themselves cannot cause muscle contraction, but when the potential of the slow wave rises above the level of approximately −35 millivolts (the approximate threshold for eliciting action potentials in visceral smooth muscle), an action potential develops and spreads over the visceral smooth muscle mass, and then contraction does occur. Figure 12–4B illustrates this effect, showing that at each peak of the slow wave, an action potential (a spike potential) occurs. Indeed, sometimes a series of action potentials occur on the peak while no action potentials at all occur during the trough. This effect can obviously promote a series of rhythmical contractions of the smooth muscle mass. Therefore, the slow waves are frequently called *pacemaker waves.* This type of activity is especially prominent in tubular types of smooth muscle masses, such as in the gut, the ureter, and so forth. In Chapter 63 we shall see that this type of activity controls the rhythmical contractions of the gut.

**Action Potentials with Plateaus.** Another type of action potential that occurs in some visceral smooth muscle is illustrated in Figure 12–5. The onset of this action potential is similar to that of the typical spike potential. However, instead of rapid repolarization of the muscle fiber membrane, the repolarization is delayed for several hundred milliseconds, mainly because of delay in the membrane inactivation process, as was explained in Chapter 10. This causes a plateau in the action potential before repolarization finally occurs and obviously provides a prolonged period of membrane depolarization that is also associated with prolonged contraction. This type of action potential

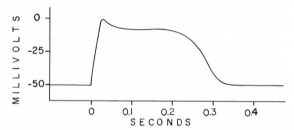

**Figure 12–5.** Monophasic action potential from a smooth muscle fiber of the rat uterus.

often occurs in the ureter and in the uterus under some conditions. It is also the type of action potential seen in cardiac muscle fibers, which also have a prolonged period of contraction.

**Spread of Action Potentials Through Visceral Smooth Muscle.** Because of the "tight junctions" that allow easy spread of electrical current between the smooth muscle fibers in visceral smooth muscle, once an action potential begins, it spreads slowly through the entire muscle mass. For instance, if it begins at the upper end of the gastrointestinal tract, it can spread downward along the intestinal wall, creating a constrictive ring that moves forward. The constrictive ring propels the intestinal contents forward. This process is called *peristalsis,* an important end result of smooth muscle function.

**Excitation of Visceral Smooth Muscle by Stretch.** When visceral smooth muscle is stretched sufficiently, spontaneous action potentials are usually generated. These result from a combination of the normal slow wave potentials plus a decrease in the membrane potential caused by the stretch itself. This response to stretch is an especially important function of visceral smooth muscle because it allows a hollow organ that is excessively stretched to contract automatically and therefore to resist the stretch. For instance, when the gut is overstretched by intestinal contents, a local automatic contraction sets up a peristaltic wave that moves the contents away from the super-stretched intestine.

**Depolarization of Multiunit Smooth Muscle Without Action Potentials.** The smooth muscle fibers of multiunit smooth muscle are normally contracted only in response to nerve stimuli. The nerve endings secrete acetylcholine in the case of some multiunit smooth muscles and norepinephrine in the case of others. In both instances, these transmitter substances cause depolarization of the smooth muscle membrane, and this response in turn elicits the contraction. However, action potentials most often do not develop. The reason for this is presumably that the fibers are too small to generate an action potential. In fact, when action potentials are elicited in visceral smooth muscle, as many as 30 to 40 smooth muscle fibers must depolarize simultaneously before a self-propagating action potential ensues. Yet, even without an action potential in the multiunit smooth muscle fibers, the local depolarization caused by the nerve transmitter substance itself spreads "electrotonically" over the entire fiber and is all that is needed to cause the muscle contraction.

**Role of Calcium Ions in Causing Smooth Muscle Action Potentials.** It will be recalled from the discussion in Chapter 10 that the depolarization process during the action potential of nerve fibers (and also of skeletal muscle fibers) is caused almost entirely by rapid influx of sodium ions to the interior of the cell membrane. In the action potential of smooth muscle fibers, the rapid influx of ions includes not only sodium ions but also a large quantity of calcium ions. Indeed, it is believed that for many types of smooth muscle, the onset of depolarization is caused mainly by influx of calcium ions rather than of sodium ions. This is particularly important in view of the fact that influx of calcium ions seems also to be the major means by which muscle contraction itself is elicited, as will be discussed in the following section.

## EXCITATION-CONTRACTION COUPLING—ROLE OF CALCIUM IONS

In the previous chapter it was pointed out that the actual contractile process in skeletal muscle is activated by calcium ions. This is also true in smooth muscle. However, the source of the calcium ions differs from smooth muscle to skeletal muscle because the sarcoplasmic reticulum of smooth muscle is poorly developed in contrast to the sarcoplasmic reticulum of skeletal muscle which is very extensive and is the source of the contraction-inducing calcium ions.

In smooth muscle, most of the calcium ions that cause contraction enter the muscle fiber from the extracellular fluid at the time of the action potential. There is a reasonably high concentration of calcium ions in the extracellular fluid, and as was pointed out in the previous section, the action potential is caused at least partly by influx of calcium ions into the muscle fiber. Because the smooth muscle fibers are extremely small (in contrast to the sizes of the skeletal muscle fibers), these calcium ions can diffuse to all parts of the smooth muscle and elicit the contractile process. The time required for this diffusion to occur is usually 200 to 300 milliseconds and is called the *latent period* before the contraction begins; this latent period is some 50 times as great as that for skeletal muscle contraction.

Yet, in some smooth muscle there are rudiments of a developing sarcoplasmic reticulum in which the cisternae of the reticulum abut the cell membrane. Therefore, it is believed that the membrane action potentials in these smooth muscle fibers cause additional release of calcium ions from these cisternae, thereby providing a greater degree of contraction than would occur on the basis of calcium ions entering through the cell membrane alone.

**The Calcium Pump.** To cause relaxation of the smooth muscle contractile elements, it is necessary to remove the calcium ions. This removal is achieved by a calcium pump that pumps the calcium ions out of the smooth muscle fiber and back into the extracellular fluid, or perhaps also pumps calcium ions into the rudimentary sarcoplasmic reticulum when it is present. However, this pump is very slow-acting in comparison with the fast-acting sarcoplasmic reticulum pump in skeletal muscle. Therefore, the duration of smooth muscle contraction is often in the order of seconds rather than in tens of milliseconds as occurs for skeletal muscle.

## GROSS CHARACTERISTICS OF SMOOTH MUSCLE CONTRACTION

From the foregoing discussion of the many different types of smooth muscle and the many different types of membrane depolarizations, one can readily understand why smooth muscle in different parts of the body has many different characteristics of contraction. For instance, the multiunit smooth muscle of the large blood vessels contracts mainly in response to nerve impulses, whereas, in many types of visceral smooth muscle—the smaller blood vessels, the ureter, the bile ducts, and other glandular ducts—a self-excitatory process causes continuous rhythmic contraction.

**Tone of Smooth Muscle—Summation of Individual Contractions.** Smooth muscle can maintain a state of long-term, steady contraction that has been called either *tonus* contraction of smooth muscle or simply *smooth muscle tone*. This is an important feature of smooth muscle contraction because it allows prolonged or even indefinite continuance of the smooth muscle function. For instance, the arterioles are maintained in a state of tonic contraction almost throughout the entire life of the person. Likewise, tonic contraction in the gut wall maintains steady pressure on the contents of the gut, and tonic contraction of the urinary bladder wall maintains a moderate amount of pressure on the urine in the bladder.

The tonic contraction of smooth muscle is caused by summation of individual contractile pulses. That is, each time an action potential occurs, a small quantity of calcium ions enters the smooth muscle cell. If another action potential occurs before the first calcium ions have been pumped out, then the concentration of calcium ions will rise still further. In addition, since a pulse of calcium ions in smooth muscle usually lasts for a second or more, any time action potentials occur at frequencies greater than 1 per second, summation usually ensues and the strength of contraction increases. This effect is illustrated in Figure 12–6. Furthermore, rhythmic contractions can be superimposed on the tonic contraction, as also illustrated in the figure. The rhythmic contractions are caused by waxing and waning of the frequency of stimulation and, therefore, waxing and waning of the summation of the contractile process.

**Degree of Shortening of Smooth Muscle During Contraction.** A special characteristic of smooth muscle—one that is also different from skeletal muscle—is its ability to shorten a far greater percentage of its length than can skeletal muscle. Skeletal muscle has a useful distance of contraction equal to only 25 to 35

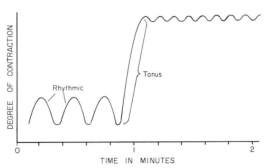

**Figure 12–6.** Record of rhythmic and tonic smooth muscle contraction

per cent of its length, while smooth muscle can contract quite effectively from a length two times its normal length to as short as one-half its normal length, giving as much as a four-fold distance of contraction. This allows smooth muscle to perform important functions in the hollow viscera—for instance, allowing the gut, the bladder, blood vessels, and other internal structures of the body to change their lumen diameters from almost zero up to very large values.

**Stress-Relaxation of Smooth Muscle.** A very important characteristic of smooth muscle is its ability to change length greatly without marked changes in tension. This results from a phenomenon called stress-relaxation, which may be explained as follows:

If a segment of smooth muscle 1 inch long is suddenly stretched to 2 inches, the tension between the two ends increases tremendously at first, but the extra tension also begins to disappear immediately, and within a few minutes it has returned almost to its level prior to the stretch, even though the muscle is now twice as long. This probably results from the fact that the actin and myosin filaments are arranged randomly in smooth muscle. Over a period of time, the filaments of the stretched muscle presumably rearrange their bonds and gradually allow the sliding process to take place, thus allowing the tension to return almost to its original amount.

Exactly the converse effect occurs when smooth muscle is shortened. Thus, if the 2 inch segment of smooth muscle is shortened back to 1 inch, essentially all tension will be lost from the muscle immediately. Gradually, over a period of 1 minute or more, the tension returns, this again presumably resulting from slow sliding of the filaments. This is called *reverse stress-relaxation*.

# NEUROMUSCULAR JUNCTIONS OF SMOOTH MUSCLE

**Physiologic Anatomy of Smooth Muscle Neuromuscular Junctions.** There are two major types of smooth muscle neuromuscular junctions: (1) the *contact* type, and (2) the *diffuse* type. The contact type of neuromuscular junctions is illustrated in Figure 12–7. This figure shows a nerve fiber arborizing into a reticulum of terminal fibrils that spread among smooth muscle fibers. These terminal fibrils are insulated by sheaths of Schwann cells, but

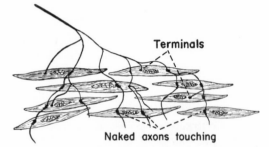

**Figure 12–7.** Junction of terminal nerve fibrils with smooth muscle fibers, showing a rare discrete terminal but many points where naked axons touch smooth muscle fibers, which points are also believed to act as transmission junctions.

where the fibers come in contact with muscle fibers the sheath is often interrupted, allowing the naked axons to touch the muscle fibers. Also, at these points the fibers form *varicosities* that contain acetylcholine or norepinephrine vesicles. In rare instances, a minute nerve terminal invaginates into the membrane of a smooth muscle fiber and is thus similar to the end-plate of the skeletal neuromuscular junction. However, it is believed that most transmission occurs at the points of contact between the varicosities and the smooth muscle fibers.

The diffuse type of smooth muscle neuromuscular junction is characterized by terminal nerve fibrils that never come into direct contact with the smooth muscle fibers. Instead, these terminal fibers either wend their way through the smooth muscle mass, partially protected by Schwann cells, or they even lie in the fibrous tissue adjacent to the smooth muscle mass, never penetrating at all. These fibers also have varicosities that release transmitter substances, and these substances then diffuse, often long distances, into the muscle mass. For instance, the innervation of portions of the gut and of some blood vessels is achieved in this way—simply secretion of the transmitter substances onto the surfaces of the smooth muscle masses or between layers of smooth muscle and the transmitter then diffusing to the other layers.

**Excitatory and Inhibitory Transmitter Substances at the Smooth Muscle Neuromuscular Junction.** Two different transmitter substances known to be secreted by the autonomic nerves innervating smooth muscle are *acetylcholine* and *norepinephrine*. Acetylcholine is an excitatory substance for smooth muscle fibers in some organs but an inhibitory substance for smooth muscle in other organs. And when acetylcholine excites a muscle fiber, norepinephrine ordinarily inhibits it, or when acetylcholine inhibits a fiber norepinephrine excites it.

It is believed that *receptor substances* in the membranes of the different smooth muscle fibers determine which will excite them, acetylcholine or norepinephrine. These receptor substances will be discussed in more detail in Chapter 57 in relation to function of the autonomic nervous system.

**Excitation of Action Potentials in Smooth Muscle Fibers—The Junctional Potential.** Transmission of impulses from terminal nerve fibers to smooth muscle fibers occurs in very much the same manner as transmission at the neuromuscular junction of skeletal muscle fibers except for temporal differences. When an action potential reaches the terminal of an excitatory nerve fibril, there is a latent period of about 50 milliseconds before any change in the membrane potential of the smooth muscle fiber can be detected. Then the potential rises to a maximal level in approximately 100 milliseconds. If an action potential does not occur, this potential gradually disappears at a rate of approximately one-half every 200 to 500 milliseconds. This complete sequence of potential changes is called the *junctional potential*; it is analogous to the end-plate potential of the skeletal muscle fibers except that its duration is 20 to 100 times as long.

If the junctional potential rises to the threshold level for discharge of the smooth muscle membrane, an action potential will occur in the smooth muscle fiber in exactly the same way that an action potential occurs in a skeletal muscle fiber. A typical smooth muscle fiber has a normal resting membrane potential of $-55$ to $-60$ millivolts, and the threshold potential at which the action potential occurs is about $-30$ to $-40$ millivolts.

*Inhibition at the Smooth Muscle Neuromuscular Junction.* When an inhibitory transmitter substance is released instead of an excitatory transmitter, the membrane potential of the muscle fiber becomes more negative than ever, for instance $-70$ millivolts; that is, it becomes *hyperpolarized* and therefore becomes much more difficult to excite than is usually the case.

**Excitation of Smooth Muscle by Hormones.** Many different hormones affect the degree of response of smooth muscle to nerve stimuli and other stimuli. Also, some hormones can cause direct stimulation of smooth muscle contraction. In addition to the usual transmitter substances acetylcholine and norepinephrine, the other hormonal factors that at times either excite or inhibit smooth muscle include epinephrine, serotonin, histamine, estrogens, progesterone, vasopressin, and oxytocin. The functions of these different substances will be discussed at different points in this text in relation to specific organs. To give an example, estrogens affect uterine muscle toward the end of pregnancy to decrease its membrane potential. This increases the excitability of the uterus, probably playing a role in the final onset of labor. On the other hand, progesterone has the opposite effect; it increases the membrane potential, causing a state of hyperpolarization and thus decreasing the excitability of the uterine musculature. Therefore, progesterone plays an important role in preventing expulsion of the fetus by the uterus during the early weeks and months of pregnancy.

# REFERENCES

Barbeau, A.: Drugs affecting movement disorders. *Ann. Rev. Pharmacol.*, 14:91, 1974.

Barr, L.: Smooth muscle as an electrical syncytium. *In* Fishman, A. P., and Hecht, H. H. (eds.): The Pulmonary Circulation and Interstitial Space. Chicago, University of Chicago Press, 1969, p. 161.

Bennett, M. R., and Burnstock, G.: Electrophysiology of the innervation of intestinal smooth muscle. *In* Handbook of Physiology. Baltimore, The Williams & Wilkins Co., 1968, Vol. 4, Sec. 6, p. 1709.

Bohr, D. F., and Uchida, F.: Activation of vascular smooth muscle. *In* Fishman, A. P., and Hecht, H. H. (eds): The Pulmonary Circulation and Interstitial Space. Chicago, University of Chicago Press, 1969, p. 133.

Bozler, E.: Role of calcium in initiation of activity of smooth muscle. *Amer. J. Physiol.*, 216:671, 1969.

Bradley, P. B.: Pharmacology of the central nervous system. *Brit. Med. Bull.*, 21:1, 1965.

Bulbring, E.: Smooth Muscle. Baltimore, The Williams & Wilkins Co., 1970.

Bulbring, E., Kostyuk, P. G., and Shuba, M. F. (eds.): Physiology of Smooth Muscles; Twenty-sixth International Congress of Physiological Sciences. New York, Raven Press, 1975.

Cohen, L. B.: Changes in neuron structure during action potential propagation and synaptic transmission. *Physiol. Rev.*, 53:373, 1973.

De Robertis, E., and Schacht, J. (eds.): Neurochemistry of Cholinergic Receptors. New York, American Elsevier Publishing Co., Inc., 1974.

Gerschenfeld, H. M.: Chemical transmission in invertebrate central nervous systems and neuromuscular junctions. *Physiol. Rev.*, 53:1, 1973.

Ginsborg, B. L.: Electrical changes in the membrane in junctional transmission. *Biochim. Biophys. Acta*, 300:289, 1973.

Greengard, P., and Kebabian, J. W.: Role of cyclic AMP in synaptic transmission in the mammalian peripheral nervous system. *Fed. Proc.*, 33:1059, 1974.

Guyton, A. C., and MacDonald, M. A.: Physiology of botulinus toxin. *Arch. Neurol. Psychiat.*, 57:578, 1947.

Guyton, A. C., and Reeder, R. C.: The dynamics of curarization. *J. Pharmacol. Exp. Ther.*, 97:322, 1949.

Hall, Z. W., Hildebrand, J. G., and Kravitz, E. A.: Chemistry of synaptic transmission. Newton, Mass., Chiron Press, 1974.

Hebb, C.: CNS at the cellular level: identity of transmitter agents. *Ann. Rev. Physiol., 32*:165, 1970.

Hubbard, J. I.: Microphysiology of vertebrate neuromuscular transmission. *Physiol. Rev., 53*:674, 1973.

Hubbard, J. I., and Quastel, D. M.: Micropharmacology of vertebrate neuromuscular transmission. *Ann. Rev. Pharmacol., 13*:199, 1973.

Kandel, E. R., and Kupfermann, I.: The functional organization of invertebrate ganglia. *Ann. Rev. Physiol., 32*:193, 1970.

McGeachie, J. K.: Smooth Muscle Regeneration; a Review and Experimental Study. Basel, S. Karger, A. G., 1975.

Minton, S. A., Jr.: Venom Diseases. Springfield, Ill., Charles C Thomas, Publisher, 1974.

Nachmansohn, D.: Chemical events in conducting and synaptic membranes during electrical activity. *Proc. Natl. Acad. Sci. U.S.A., 68*:3170, 1971.

Prosser, C. L.: Smooth muscle. *Ann. Rev. Physiol., 36*:503, 1974.

Radouco-Thomas, S.: Cellular and molecular aspects of transmitter release: calcium monoamine dynamics. *Adv. Cytopharmacol., 1*:457, 1971.

Reuter, H.: Divalent cations as charge carriers in excitable membranes. *Prog. Biophys. Mol. Biol., 26*:1, 1973.

Ruegg, J. C.: Smooth muscle tone. *Physiol. Rev., 51*:201, 1971.

Simpson, L. L., and Curtis, D. R. (eds.): Neuropoisons. New York, Plenum Publishing Co., 1974.

Somlyo, A. P.: Excitation-contraction coupling in vertebrate smooth muscle: correlation of ultrastructure with function. *Physiologist, 15*:338, 1972.

Tauc, L: Transmission in invertebrate and vertebrate ganglia. *Physiol. Rev., 47*:521, 1967.

Yang, C. C.: Chemistry and evolution of toxins in snake venoms. *Toxicon, 12*:1, 1974.

# PART IV
# THE HEART

# | 13 |

# Heart Muscle; the Heart as a Pump

With this chapter we begin discussion of the heart and circulatory system. The heart is a pulsatile, four-chamber pump composed of two atria and two ventricles. The atria function principally as entryways to the ventricles, but they also pump weakly to help move the blood on through the atria into the ventricles. The ventricles supply the main force that propels blood through the lungs and through the peripheral circulatory system.

In previous chapters the basic principles of membrane potentials, action potentials, and muscle contraction have been discussed. Special mechanisms in the heart maintain cardiac rhythmicity and transmit action potentials throughout the cardiac musculature to initiate its contraction. These mechanisms will be explained in detail in the following chapter. In the present chapter we wish to explain how the heart operates as a pump: that is, to explain the function of the muscle, of the valves, and of the various chambers of the heart. Therefore, we will discuss first the basic physiology of cardiac muscle itself, especially how it differs from skeletal muscle, which was discussed in Chapter 11.

## PHYSIOLOGY OF CARDIAC MUSCLE

The heart is composed of three major types of cardiac muscle: atrial muscle, ventricular muscle, and specialized excitatory and conductive muscle fibers. The atrial and ventricular types of muscle contract in much the same manner as skeletal muscle fibers. On the other hand, the specialized excitatory and conductive fibers contract only feebly because they contain few contractile fibrils; instead, they provide an excitatory system for the heart and a transmission system for rapid conduction of impulses throughout the heart.

### PHYSIOLOGIC ANATOMY OF CARDIAC MUSCLE

Figure 13–1 illustrates a typical histologic section of cardiac muscle, showing arrangement of the cardiac muscle fibers in a latticework, the fibers dividing, then recombining, and then spreading again. One notes immediately from this figure that cardiac muscle is *striated* in the same manner as typical skeletal muscle. Furthermore, cardiac muscle has typical myofibrils that contain *actin* and *myosin filaments* almost identical to those found in skeletal muscle, and these filaments interdigitate and slide along each other during the process of contraction in the same manner as occurs in skeletal muscle. (See Chap. 11.)

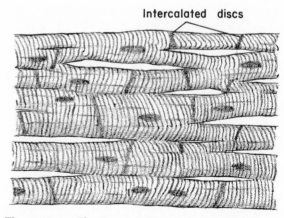

**Intercalated discs**

**Figure 13–1.** The "syncytial" nature of cardiac muscle.

160

**Cardiac Muscle as a "Functional Syncytium."** The angulated dark areas crossing the cardiac muscle fibers in Figure 13–1 are called *intercalated discs;* however, they are actually cell membranes that separate individual cardiac muscle cells from each other. In reality, the cardiac muscle fibers are a series of cardiac muscle cells connected in series with each other. Yet electrical resistance through the intercalated disc is only $1/400$ the resistance through the outside membrane of the cardiac muscle fiber. Therefore, from a functional point of view, ions flow with relative ease along the axes of the cardiac muscle fibers so that action potentials travel from one cardiac muscle cell to another, past the intercalated discs, without significant hindrance. Therefore, cardiac muscle is a *functional syncytium,* in which the cardiac muscle cells are so tightly bound that when one of these cells becomes excited, the action potential spreads to all of them, spreading from cell to cell and spreading throughout the latticework interconnections.

The heart is composed of two separate functional syncytiums, the *atrial syncytium* and the *ventricular syncytium.* These are separated from each other by the fibrous tissue surrounding the valvular rings, but an action potential can be conducted from the atrial syncytium into the ventricular syncytium by way of a specialized conductive system, the *A-V bundle,* which will be discussed in detail in the following chapter.

**All-or-Nothing Principle as Applied to the Heart.** Because of the syncytial nature of cardiac muscle, stimulation of any single atrial muscle fiber causes the action potential to travel over the entire atrial muscle mass, and, similarly, stimulation of any single ventricular fiber causes excitation of the entire ventricular muscle mass. If the A-V bundle is intact, the action potential passes also from the atria to the ventricles. This is called the all-or-nothing principle; and it is precisely the same as that discussed in Chapter 10 for nerve fibers, but because the cardiac muscle fibers interconnect with each other, the all-or-nothing principle applies to the entire functional syncytium of the heart rather than to single muscle fibers as in the case of skeletal muscle fibers.

## ACTION POTENTIALS IN CARDIAC MUSCLE

The *resting membrane potential* of normal cardiac muscle is approximately −80 to −85

millivolt (mv.) and approximately −90 to −100 mv. in the specialized conductive fibers, the Purkinje fibers, which are discussed in the following chapter.

The *action potential* recorded in cardiac muscle, shown in Figure 13–2, is 90 to 105 mv., which means that the membrane potential rises from its normally very negative value to a slightly positive value of about +20 mv. Because of this change of potential from negative to positive, the positive portion is called the *reversal potential.* The basic principles of the genesis of resting, action, and reversal potentials were discussed in Chapter 10.

Cardiac muscle has a peculiar type of action potential. After the initial *spike* the membrane remains depolarized for 0.15 to 0.3 second, exhibiting a *plateau* as illustrated in Figure 13–2, followed at the end of the plateau by abrupt repolarization. The presence of this plateau in the action potential causes the action potential to last 20 to 50 times as long in cardiac muscle as in skeletal muscle and causes a correspondingly increased period of contraction.

The cause of the plateau in the action potential is not fully known. However, the permeability of the membrane for potassium decreases about 5-fold after the onset of the spike and remains low during the period of the plateau; at the end of the plateau the potassium permeability increases greatly and the sodium permeability returns to normal (by a process called *membrane inactivation*). When this occurs, the normal repolarization process begins, the plateau disappears, and the membrane potential falls once more to the very low negative resting potential level.

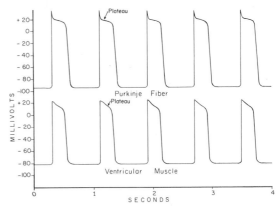

**Figure 13–2.** Rhythmic action potentials from a Purkinje fiber and from a ventricular muscle fiber recorded by means of microelectrodes.

Several theories have been offered to explain the low potassium permeability immediately following depolarization. One of these is that it is caused by excessive buildup of potassium immediately outside the membrane. Another is that calcium ions have diffused inward through the membrane during the depolarization process and that these in some way are responsible for the delay in the increase in potassium permeability that is necessary for the repolarization process to occur.

**Velocity of Conduction in Cardiac Muscle.** The velocity of conduction of the action potential in both atrial and ventricular muscle fibers is about 0.3 to 0.5 meter per second, or about $1/250$ the velocity in very large nerve fibers and about $1/10$ the velocity in skeletal muscle fibers. The velocity of conduction in the specialized conductive system varies from only a small fraction of a meter to several meters per second in different parts of the system, as is explained in the following chapter.

**Refractory Period of Cardiac Muscle.** Cardiac muscle, like all excitable tissue, is refractory to restimulation during the action potential. An extremely strong electrical stimulus can sometimes initiate a new spike at the very end of the action potential plateau, but the spike is not propagated along the muscle. Also, the normal cardiac impulse does not have the extreme stimulatory power of a high voltage electrical stimulus and cannot re-excite the muscle until after most of the repolarization process is complete. Therefore, the refractory period of the heart is usually stated in terms of the *functional refractory period,* which is the interval of time, as shown to the left in Figure 13–3, during which an action potential from

another part of the heart cannot re-excite an already excited area of cardiac muscle. The normal functional refractory period of the ventricle is 0.25 to 0.3 second, which is approximately the duration of the action potential. There is an additional *relative refractory period* of about 0.05 second during which the muscle is more difficult than normal to excite but nevertheless can be excited, as illustrated by the early premature contraction in Figure 13–3.

The functional refractory period of atrial muscle is much shorter than that for the ventricles (about 0.15 second), and the relative refractory period is another 0.03 second. Therefore, the rhythmical rate of contraction of the atria can be much faster than that of the ventricles.

## CONTRACTION OF CARDIAC MUSCLE

**Excitation-Contraction Coupling—Function of Calcium Ions.** It is not the action potential itself that causes the myofibrils of cardiac muscle to contract. Instead, it is movement of calcium ions into the myofibrils. This probably occurs in the following manner: When the action potential travels over the muscle, it also causes flow of action potentials to the interior of the fibers by way of the T tubules that were described in relation to skeletal muscle contraction in Chapter 11. These T tubule action potentials, in turn, are believed to cause release of calcium ions from the cisternae of the sarcoplasmic reticulum. These calcium ions diffuse rapidly into the myofibrils and there catalyze the chemical reactions that promote sliding of the actin and myosin filaments along each other; this in turn produces the muscle contraction. Immediately after the action potential is over, the calcium ions are transported back into the longitudinal tubules of the sarcoplasmic reticulum, so that within a few milliseconds the density of calcium ions around the myofibrils falls below that needed to maintain contraction; as a result, the muscle relaxes.

*Possible Transfer of Calcium Ions from the T Tubules During Excitation-Contraction Coupling.* In addition to the calcium released into the sarcoplasm from the cisternae of the sarcoplasmic reticulum, calcium ions probably also enter the sarcoplasm from the T tubules during the action potential. Several reasons for believing this are the following: (1) Moderate changes in calcium ion concentration in the extracellular fluid alter the strength of contraction of cardiac muscle fibers but have very little effect on the contraction of skeletal muscle fibers. (2) The size of the T tubules in cardiac muscle is about five times as great as that of skeletal muscle, whereas the degree

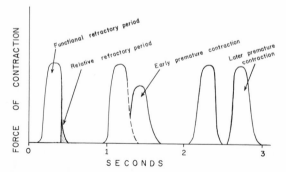

**Figure 13–3.** Contraction of the heart, showing: the durations of the functional refractory period and the relative refractory period, the effect of an early premature contraction, and the effect of a later premature contraction. Note that the premature contractions do not cause wave summation as occurs in skeletal muscle.

of development of the sarcoplasmic reticulum is considerably less in cardiac muscle fibers. (3) The T tubules of cardiac muscle are lined with a layer of mucopolysaccharide filaments that seem to be able to entrap calcium ions and to make these available to the sarcoplasm when an action potential travels through the T tubules.

Thus, it is believed that at least part of the calcium ions that excite the contractile process in cardiac muscle are derived from the T tubules rather than from the cisternae of the sarcoplasmic reticulum. And these calcium ions are supposedly pumped back through the wall of the T tubule into the extracellular fluid of this tubule in the same way that the calcium ions derived from the cisternae are pumped back into the sarcoplasmic reticulum.

Another difference between the tubules of cardiac muscle and those of skeletal muscle is that the T tubules in the skeletal muscle are located where the actin and myosin filaments overlap each other so that there are two T tubule systems to each sarcomere. On the other hand, in cardiac muscle there is only one T tubule system per sarcomere located at the Z line but composed of much larger T tubules. This difference in structure is compatible with the fact that cardiac muscle contracts much more slowly than skeletal muscle so that there is adequate time for calcium ions to diffuse from the Z line to the middle of the sarcomere where the contractile process occurs.

**Relation of Action Potential Strength to Strength of Contraction.** The greater the voltage of the action potential and the longer its duration, the greater also is the work output caused by the cardiac muscle contraction. The reason for this seems to be that either increased voltage or increased duration of action potential increases the quantity of calcium ions released into the myofibrils of the muscle fibers, which in turn enhances the chemical reactions of the contractile process itself.

**Duration of Contraction.** Cardiac muscle begins to contract a few milliseconds after the action potential begins but then continues to contract for a few milliseconds after the action potential ends. Therefore, the duration of contraction of cardiac muscle is mainly a function of the duration of the action potential—about 0.15 second in atrial muscle and 0.3 second in ventricular muscle.

**Effect of Heart Rate on Duration of Contraction.** When the heart rate increases, the duration of each total cycle of the heart, including both the contraction phase and the relaxation phase, obviously decreases. The duration of the action potential and the period of contraction also decrease but not as much percentagewise as does the relaxation phase. At a normal heart rate of 72 beats per minute, the period of contraction is about 0.4 of the entire cycle. At three times normal heart rate, this period is about 0.65 of the entire cycle, which means that the heart under some conditions does not remain relaxed long enough to allow complete filling of the cardiac chambers prior to the next contraction.

**Effect of Premature Contraction on Contraction Strength.** When the heart muscle is stimulated a second time soon after a previous contraction is over, the strength of the second contraction is usually depressed below that of the first, an effect that is illustrated in the second record of Figure 13–3. The presumptive reason for this is that when a second contraction occurs rapidly after the first some of the chemical substances necessary for contraction are slightly depleted, which results in diminished force of contraction. As the interval between two successive contractions increases, as illustrated in the third record of Figure 13–3, the strength of the second contraction rises to approach that of the first and sometimes actually exceeds it.

## THE CARDIAC CYCLE

The period from the end of one heart contraction to the end of the next is called the *cardiac cycle*. Each cycle is initiated by spontaneous generation of an action potential in the S-A node, as will be explained in detail in the following chapter. This node is located in the posterior wall of the right atrium near the opening of the superior vena cava, and the action potential travels rapidly through both atria and thence through the A-V bundle into the ventricles. However, because of a special arrangement of the conducting system from the atria into the ventricles, there is a delay of more than $1/_{10}$ second between passage of the cardiac impulse through the atria and then through the ventricles. This allows the atria to contract ahead of the ventricles, thereby pumping blood into the ventricles prior to the very strong ventricular contraction. Thus, the atria act as primer pumps for the ventricles, and the ventricles then provide the major source of power for moving blood through the vascular system.

### SYSTOLE AND DIASTOLE

The cardiac cycle consists of a period of relaxation called *diastole* followed by a period of contraction called *systole*.

Figure 13–4 illustrates the different events during the cardiac cycle. The top three curves show the pressure changes in the aorta, the left ventricle, and the left atrium, respectively. The fourth curve depicts the changes in ventricular volume, the fifth the electrocardiogram, and the sixth a phonocardiogram, which is a recording of the sounds produced by the heart as it pumps. It is especially important that the student study in detail the diagram of this figure and understand the causes of all the events illustrated. These are explained as follows:

### RELATIONSHIP OF THE ELECTROCARDIOGRAM TO THE CARDIAC CYCLE

The electrocardiogram in Figure 13–4 shows the *P, Q, R, S,* and *T* waves, which will be discussed in Chapters 15 through 17. These are electrical voltages generated by the heart and recorded by the electrocardiograph from the surface of the body. On looking also at the atrial pressure curve, it is evident that the P wave begins immediately prior to a rise in atrial pressure; that is, the *P wave* is caused by the *spread of depolarization* through the atria and this is followed by atrial contraction. Approximately 0.16 second after the onset of the P wave, the *QRS waves* appear as a result of depolarization of the ventricles, which initiates contraction of the ventricles and causes the ventricular pressure to begin rising, as illustrated in the figure. Therefore, the QRS complex begins slightly before the onset of ventricular systole.

Finally, one observes the *ventricular T wave*

in the electrocardiogram. This represents the stage of repolarization of the ventricles at which time the ventricular muscle fibers begin to relax. Therefore, the T wave occurs slightly prior to the end of ventricular contraction.

### FUNCTION OF THE ATRIA AS PUMPS

Blood normally flows continually from the great veins into the atria, and approximately 70 per cent of this flows directly from the atria into the ventricles even before the atria contract. Then, however, atrial contraction causes an additional 30 per cent filling. Therefore, the atria simply function as primer pumps that increase the effectiveness of the ventricles as pumps approximately 30 per cent. Yet, the heart can continue to operate quite satisfactorily under normal resting conditions even without this extra 30 per cent effectiveness because it normally has the capability of pumping 300 to 400 per cent more blood than is required by the body anyway. Therefore, the person is likely not to notice the difference unless he exercises, in which case he will develop acute signs of heart failure, especially shortness of breath.

**Pressure Changes in the Atria—The a, c,**

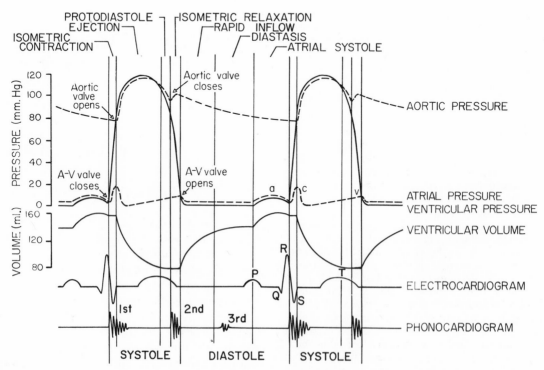

**Figure 13–4.** The events of the cardiac cycle, showing changes in left atrial pressure, left ventricular pressure, aortic pressure, ventricular volume, the electrocardiogram, and the phonocardiogram.

**and v Waves.**    If one observes the atrial pressure curve of Figure 13–4, he will note three major pressure elevations called the *a, c,* and *v atrial pressure waves.*

The *a wave* is caused by atrial contraction. Ordinarily, the *right* atrial pressure rises 4 to 6 mm. Hg during atrial contraction, while the *left* atrial pressure rises about 7 to 8 mm. Hg.

The *c wave* occurs when the ventricles begin to contract, and it is caused mainly by two factors: (1) bulging of the A-V valves toward the atria because of increasing pressure in the ventricles, and (2) pulling on the atrial muscle where it is attached to the ventricular muscle by the contracting ventricles.

The *v wave* occurs toward the end of ventricular contraction; it results from slow buildup of blood in the atria while the A-V valves are closed during ventricular contraction. Then, when ventricular contraction is over, the A-V valves open, allowing blood to flow rapidly into the ventricles and causing the v wave to disappear.

## FUNCTION OF THE VENTRICLES AS PUMPS

**Filling of the Ventricles.**    During ventricular systole large amounts of blood accumulate in the atria because of the closed A-V valves, and the atrial pressures become considerably elevated. Therefore, just as soon as systole is over and the ventricular pressures fall again to their low diastolic values, the high pressures in the atria immediately push the A-V valves open and allow blood to flow rapidly into the ventricles, as shown by the ventricular volume curve in Figure 13–4. This is called the *period of rapid filling of the ventricles.* The atrial pressures fall to within a fraction of a millimeter of the ventricular pressures because the normal A-V valve openings are so large that they offer almost no resistance to blood flow.

The period of rapid filling lasts approximately the first third of diastole. During the middle third of diastole only a small amount of blood normally flows into the ventricles; this is blood that continues to empty into the atria from the veins and passes on through the atria directly into the ventricles. This middle third of diastole, when the inflow of blood into the ventricles is almost at a standstill, is called *diastasis.*

During the latter third of diastole, the atria contract and give an additional thrust to the inflow of blood into the ventricles; this accounts for approximately 30 per cent of the filling of the ventricles during each heart cycle.

**Emptying of the Ventricles During Systole.**    *Period of Isometric Contraction.*    Immediately after ventricular contraction begins, the ventricular pressure abruptly rises, as shown in Figure 13–4, causing the A-V valves to close. Then an additional 0.02 to 0.03 second is required for the ventricle to build up sufficient pressure to push the semilunar (aortic and pulmonary) valves open against the pressures in the aorta and pulmonary artery. Therefore, during this period of time, contraction is occurring in the ventricles but there is no emptying. This period is called the period of isometric contraction, meaning by this term that tension is increasing in the muscle but no shortening of the muscle fibers is occurring.

*Period of Ejection.*    When the left ventricular pressure rises slightly above 80 mm. Hg and the right ventricular arterial pressure slightly above 8 mm. Hg, the ventricular pressures now push the semilunar valves open. Immediately, blood begins to pour out of the ventricles, about half of the emptying occurring during the first quarter of systole and most of the remaining half during the next two quarters. These three-quarters of systole are called the period of ejection.

*Protodiastole.*    During the last one-fourth of ventricular systole, almost no blood flows from the ventricles into the large arteries; yet the ventricular musculature remains contracted. This period is called protodiastole. The arterial pressure falls during this period, because almost no blood is entering the arteries even though large quantities of blood are flowing from the arteries through the peripheral vessels.

The ventricular pressure actually falls to a value *below* that in the aorta during protodiastole despite the fact that the ventricles are still contracting. The reason for this is that the blood flowing out of the ventricles has built up momentum. As this momentum is lost during the latter part of systole, the kinetic energy of the momentum is converted into pressure in the large arteries, which makes the arterial pressure slightly greater than the pressure inside the ventricles.

*Period of Isometric Relaxation.*    At the end of systole, ventricular relaxation begins suddenly, allowing the intraventricular pressures to fall rapidly. The elevated pressures in the large arteries immediately push blood back toward the ventricles, which snaps the aortic and pulmonary valves closed. For another 0.03 to 0.06

second, the ventricular muscle continues to relax, and the intraventricular pressures fall
rapidly back to their very low diastolic levels.
Then the A-V valves open to begin a new cycle
of ventricular pumping.

### THE VENTRICULAR VOLUME CURVE

Also shown in Figure 13–4 is a curve depicting the changes in ventricular volume during the
cardiac cycle. Such a curve can be recorded by
placing a cylinder, called an "oncometer,"
around the ventricle. The cylinder is closed at
one end and has a snug rubber gasket between
its other end and the atrioventricular ring of the
heart. Every time the ventricles increase in
size, they displace air from the cylinder. The
cylinder is connected by means of a rubber tube
to a tambour recording system, and the increase
and decrease in ventricular volume is recorded.

**End-Diastolic and End-Systolic Volumes of
the Ventricles.** During diastole, filling of the
ventricles normally increases the volume of
each ventricle to about 120 to 130 ml. This volume is known as the *end-diastolic volume*.
Then, as the ventricles empty during systole,
the volume decreases about 70 ml., which is
called the *stroke volume*. The remaining volume
in each ventricle, about 50 to 60 ml. is called the
*end-systolic volume*.

When the heart contracts strongly, the end-
systolic volume can fall to as little as 10 to 30
ml. On the other hand, when large amounts of
blood flow into the ventricles during diastole,
their end-diastolic volumes can become as great
as 200 to 250 ml. in the normal heart. And by
both increasing the end-diastolic volume and
decreasing the end-systolic volume, the stroke
volume output can be increased to more than
double normal.

### FUNCTION OF THE VALVES

**The Atrioventricular Valves.** The *A-V
valves* (the *tricuspid* and the *mitral* valves) prevent backflow of blood from the ventricles to
the atria during systole, and the *semilunar
valves* (the *aortic* and *pulmonary* valves) prevent backflow from the aorta and pulmonary
arteries into the ventricles during diastole. All
these valves, which are illustrated in Figure
13–5, close and open *passively*. That is, they
close when a backward pressure gradient
pushes blood backward, and they open when a
forward pressure gradient forces blood in the
forward direction. For obvious anatomical

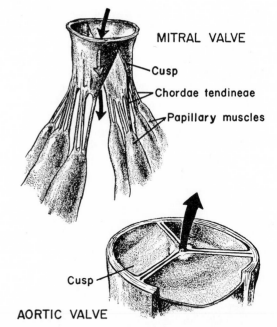

**Figure 13–5.** Mitral and aortic valves.

reasons, the thin, filmy A-V valves require almost no backflow to cause closure while the
much heavier semilunar valves require rather
strong backflow for a few milliseconds.

**Function of the Papillary Muscles.** Figure
13–5 also illustrates the papillary muscles that
attach to the vanes of the A-V valves by the
*chordae tendineae*. The papillary muscles contract when the ventricular walls contract, but,
contrary to what might be expected, they *do not*
help the valves to close. Instead, they pull the
vanes of the valves inward toward the ventricles to prevent their bulging too far backward
toward the atria during ventricular contraction.
If a chorda tendinea becomes ruptured or if one
of the papillary muscles becomes paralyzed, the
valve bulges far backward, sometimes so far
that it leaks severely and results in severe or
even lethal cardiac incapacity.

**The Aortic and Pulmonary Valves.** There
are differences between the operation of the
aortic and pulmonary valves and that of the
A-V valves. First, the high pressures in the arteries at the end of systole cause the semilunar
valves to snap to the closed position in comparison with a much softer closure of the A-V
valves. Second, the velocity of blood ejection
through the aortic and pulmonary valves is far
greater than that through the much larger A-V
valves. Because of the rapid closure and rapid
ejection, the edges of the semilunar valves are

subjected to much greater mechanical abrasion than are the A-V valves, which also are supported by the chordae tendineae. It is obvious from the anatomy of the aortic and pulmonary valves, as illustrated in Figure 13–5, that they are well adapted to withstand this extra physical trauma.

### THE AORTIC PRESSURE CURVE

The pressure that develops in the aorta depends on many other factors in addition to contraction of the heart. The aortic pressure curve will be discussed in detail in Chapter 19, but it is desirable now simply to point out those features of the curve that relate especially to the cardiac cycle.

When the left ventricle contracts, the ventricular pressure rises rapidly until the aortic valve opens. Then the pressure in the ventricle rises much less thereafter, as illustrated in Figure 13–4, because blood immediately flows out of the ventricle into the aorta.

The entry of blood into the arteries causes the walls of these arteries to stretch and the pressure to rise. Then, at the end of systole, after the left ventricle stops ejecting blood and the aortic valve closes, the elastic stretch of the arteries maintains high pressure in the arteries during diastole.

A so-called *incisura* occurs in the aortic pressure curve when the aortic valve closes. This is caused by a short period of backward flow of blood immediately prior to closure of the valve.

After the aortic valve has closed, pressure in the aorta falls slowly throughout diastole because blood stored in the distended elastic arteries flows continually through the peripheral vessels back to the veins. By the time the ventricle contracts again, the aortic pressure usually has fallen to approximately 80 mm. Hg (diastolic pressure), which is two-thirds the maximal pressure of 120 mm. Hg (systolic pressure) occurring in the aorta during ventricular contraction.

The pressure curve in the pulmonary artery is similar to that in the aorta, except that the pressures are much less, as will be discussed in Chapter 24.

### RELATIONSHIP OF THE HEART SOUNDS TO HEART PUMPING

When one listens to the heart with a stethoscope, he does not hear the opening of the valves, for this is a relatively slowly developing process that makes no

noise. However, when the valves close, the vanes of the valves and the surrounding fluids vibrate under the influence of the sudden pressure differentials which develop, giving off sound that travels in all directions through the chest. When the ventricles first contract, one hears a sound that is caused by closure of the A-V valves. The vibration is low in pitch and relatively long continued and is known as the *first heart sound.* When the aortic and pulmonary valves close, one hears a relatively rapid snap, for these valves close extremely rapidly, and the surroundings vibrate for only a short period of time. This sound is known as the *second heart sound.* The precise causes of these sounds will be discussed in Chapter 27 in relation to auscultation.

Occasionally, one can hear an *atrial sound* when the atria beat, presumably because of vibrations associated with the flow of blood into the ventricles. Also, a *third heart sound* sometimes occurs at the end of the first third of diastole or in the middle of diastole. This is said to be caused by blood flowing with a rumbling motion into the almost-filled ventricle. The atrial sound and the third heart sound can usually be recorded with special recording instruments but can be heard with the stethoscope only with great difficulty.

**The Terms "Systole" and "Diastole" in Clinical Usage.** Strictly speaking, systole means "contraction." Therefore, physiologically it is probably best that systole be considered to begin approximately with the closure of the A-V valves and to end approximately with the opening of the A-V valves.

Clinically, it is not possible to determine when the A-V valves open. Yet, because only a short interval elapses between the closure of the aortic and pulmonary valves and the opening of the A-V valves, the clinician measures systole as the time interval between the first heart sound and the second heart sound or, in other words, the time interval between closure of the A-V valves and closure of the semilunar valves. Diastole is considered to be the interval between closure of the aortic and pulmonary valves and closure of the A-V valves.

### STROKE VOLUME OUTPUT OF THE HEART

The stroke volume output of the heart is the quantity of blood pumped from each ventricle with each heartbeat. Normally this is about 70 ml., but under conditions compatible with life it can decrease to as little as a few ml. per beat and can increase to about 140 ml. per beat in the normal heart and to over 200 ml. per beat in persons with very large hearts, such as in some athletes.

### WORK OUTPUT OF THE HEART

The work output of the heart is the amount of energy that the heart converts to work while pumping blood into the arteries. This is in two forms: First, by far the major proportion is used to move the blood from the low pressure veins to the high pressure arteries. This is *potential energy of pressure*. Second, a minor proportion of the energy is used to accelerate the blood to its velocity of ejection through the aortic and pulmonary valves. This is *kinetic energy of blood flow*.

**Energy Expended to Create Potential Energy of Pressure.** The work performed by the left ventricle to raise the pressure of the blood during each heartbeat is equal to *stroke volume output × (left ventricular mean ejection pressure* minus *left atrial pressure)*. Likewise, the work performed by the right ventricle to raise the pressure of the blood is equal to *stroke volume output × (right ventricular mean ejection pressure* minus *right atrial pressure)*. When pressure is expressed in *dynes per square centimeter* and stroke volume output in *milliliters*, the work output is in *ergs*. Right ventricular work output is usually about one-seventh the work output of the left ventricle because of the difference in systolic pressure against which the two ventricles must pump.

**The Kinetic Energy of Blood Flow.** The work output of each ventricle required to create kinetic energy of blood flow is proportional to the mass of the blood ejected times the square of the velocity of ejection. That is,

$$\text{Kinetic energy} = \frac{mv^2}{2}$$

When the mass is expressed in *grams* of blood ejected and the velocity in *centimeters per second*, the work output is in *ergs*.

Ordinarily, the work output of the left ventricle required to create kinetic energy of blood flow is about 1 per cent of the total work output of the ventricle. Most of this energy is required to cause the rapid acceleration of blood during the first quarter of systole. In certain abnormal conditions, such as aortic stenosis, in which the blood flows with great velocity through the stenosed valve, as much as 50 per cent of the total work output may be required to create kinetic energy of blood flow.

### ENERGY FOR CARDIAC CONTRACTION

Heart muscle, like skeletal muscle, utilizes chemical energy to provide the work of contraction. This energy is derived mainly from metabolism of glucose and fatty acids with oxygen and, to a lesser extent, from metabolism of other nutrients with oxygen. The different reactions that liberate this energy will be discussed in detail in Chapters 67 and 68.

The amount of energy expended by the heart is related to its work load in the following manner: The energy expended is *approximately proportional to the degree of tension generated by the heart musculature during contraction × the amount of time that the tension is maintained*.

**Efficiency of Cardiac Contraction.** During muscular contraction most of the chemical energy is converted into heat and a small portion into work output. The ratio of work output to chemical energy expenditure is called the efficiency of cardiac contraction, or simply *efficiency of the heart*. The efficiency of the normal heart, beating under normal load, is usually low, some 10 per cent. However, during maximum work output it rises as high as 15 to 20 per cent in the normal heart and perhaps slightly higher in the heart of the well-trained athlete.

# REGULATION OF CARDIAC FUNCTION

When a person is at rest, the heart must pump only 4 to 6 liters of blood each minute. However, during severe exercise it may be required to pump as much as five times this amount. The present section discusses the means by which the heart can adapt itself to such extreme increases in cardiac output.

The two basic means by which the pumping action of the heart is regulated are (1) intrinsic autoregulation in response to changes in volume of blood flowing into the heart and (2) reflex control of the heart by the autonomic nervous system.

### INTRINSIC AUTOREGULATION OF CARDIAC PUMPING—THE FRANK-STARLING LAW OF THE HEART

In Chapter 23 we shall see that one of the major factors determining the amount of blood pumped by the heart each minute is the rate of blood flow into the heart from the veins, which is called *venous return*. That is, each peripheral tissue of the body controls its own blood flow, and whatever amount of blood flows through all the peripheral tissues returns by way of the veins to the right atrium. The heart in turn automatically pumps this incoming blood on into the systemic arteries so that it can flow around the circuit again. Thus, the heart must adapt itself from moment to moment or even second to second to widely varying inputs of blood, sometimes falling as low as 2 to 3 liters per minute and at other times rising to as high as 25 or more liters per minute.

This intrinsic ability of the heart to adapt itself to changing loads of inflowing blood is called the *Frank-Starling law of the heart*, in honor of Frank and Starling, two of the great physiologists of half a century ago. Basically, the Frank-Starling law states that the greater the heart is filled during diastole, the greater will be the quantity of blood pumped into the aorta. Or another way to express this law is: *within physiologic limits, the heart pumps all the blood that comes to it without allowing excessive damming of blood in the veins.* In other words, the heart can pump either a small amount of blood or a large amount, depending on the amount that flows into it from the veins; and it automatically adapts to whatever this load might be as long as the total quantity does not rise above the physiologic limit that the heart can pump.

**Mechanism of the Frank-Starling Law.** The primary mechanism by which the heart adapts to changing inflow of blood is the following: When the cardiac muscle becomes stretched an extra amount, as it does when extra amounts of blood enter the heart chambers, the unusually stretched muscle contracts with a greatly increased force, thereby automatically pumping the extra blood into the arteries. This ability of stretched muscle to contract with increased force is characteristic of all striated muscle and not simply of cardiac muscle. Referring back to Figure 11–10 in Chapter 11, one will see that stretching a skeletal muscle, within its physiological limit, also increases its force of contraction. As was also pointed out in Chapter 11, the increased force of contraction is probably caused by the fact that the actin and myosin filaments are brought to a more nearly optimal degree of interdigitation for achieving contraction. This ability of the heart to contract with increased force as its chambers are stretched is sometimes called *heterometric autoregulation of the heart.*

*Effect of Heart Rate and "Homeometric" Autoregulation.* In addition to the important effect of stretching the heart muscle, at least two other less important factors increase heart pumping effectiveness when its volume is increased. First, stretch of the right atrial walls increases the heart rate by as much as 10 to 15 per cent; this in itself can increase the amount of blood pumped each minute. Second, changes in heart metabolism that occur when the heart is stretched cause an additional increase in contractile strength. It takes approximately 30 seconds for this effect to develop fully. The effect is called *homeometric autoregulation* because the increased contractile strength returns the muscle fiber lengths nearly to their original lengths, thus giving a considerable increase in

output with very little change in heart muscle length.

**Failure of Arterial Pressure Load to Alter Cardiac Output.** One of the most important features of the Frank-Starling law of the heart is that, within reasonable limits, changes in arterial pressure load against which the heart is pumping have almost no effect on the amount of blood pumped by the heart each minute (the cardiac output). This effect is illustrated in Figure 13–6, which is a curve extrapolated to the human being from data in dogs in which the arterial pressure was progressively changed by constricting the arteries while the cardiac output was measured simultaneously. The significance of this effect is the following: Regardless of the arterial pressure, the most important determining factor in the amount of blood pumped by the heart is still the right atrial pressure generated by the entry of blood into the heart.

Figure 13–6 shows that when the arterial pressure rises above approximately 170 mm. Hg, the arterial pressure load then causes the heart to begin to fail. However, this figure also shows that the normal daily range of arterial pressures is between approximately 80 and 170 mm. Hg, again emphasizing that within the normal operating range, the output of the heart is independent of changes in arterial pressure.

The independence of the cardiac output of changes in arterial pressure load is partly caused by the fact that the heart is a two-stage pump, for the following reasons: Even if the left ventricle were to begin to fail moderately, this would raise the left atrial pressure a few millimeters Hg. This elevation of left atrial pressure, however, would raise the pulmonary arte-

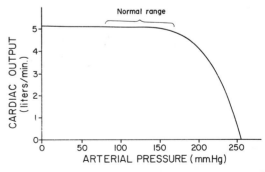

**Figure 13–6.** Constancy of cardiac output even in the face of wide changes in arterial pressure. Only when the arterial pressure rises above the normal operating pressure range does the pressure load cause the heart to begin to fail.

rial pressure only a fraction of a millimeter because left atrial pressure is not transmitted backwards through the lungs to a great extent until this pressure becomes abnormally elevated. Therefore, the right ventricle experiences almost no change in load and continues to pump normal quantities of blood into the lungs, forcing the left ventricle (even though it might be beginning to fail) to continue to pump a normal cardiac output.

**Ventricular Function Curves.** One of the best ways to express the functional ability of the ventricles to pump blood is by ventricular function curves, as shown in Figures 13–7 and 13–8. Figure 13–7 illustrates a type of ventricular function curve called the *stroke work output curve*. Note that as the atrial pressures increase, the stroke work output also increases until it reaches the limit of the heart's ability.

Figure 13–8 illustrates another type of ventricular function curve called the *minute ventricular output curve*. These two curves represent function of the two ventricles of the human heart based on data extrapolated from lower animals. As each atrial pressure rises, the respective ventricular volume output per minute also increases.

Thus, ventricular function curves are another way of expressing the Frank-Starling law of the heart. That is, as the ventricles fill to higher atrial pressures, the strength of cardiac contraction increases, causing the heart to pump increased quantities of blood into the arteries. In later chapters we shall see that ventricular function curves are exceedingly important in analyzing overall function of the circulation, for it is by such means that one can express the quantitative capabilities of the heart as a pump.

## CONTROL OF THE HEART BY NERVES

The heart is well supplied with both sympathetic and parasympathetic (vagal) nerves, as illustrated in Figure 13–9. These nerves affect cardiac pumping in two ways: (1) by changing the heart rate and (2) by changing the strength of contraction of the heart. The effect of nerve stimulation on heart rate and rhythm will be discussed in detail in the following chapter. For the present, suffice it to say that parasympathetic stimulation decreases heart rate and sympathetic stimulation increases heart rate. The range of control is from as little as 20 to 30 heart beats per minute with maximum vagal stimulation up to as high as 250 or, rarely, 300 heart beats per minute with maximum sympathetic stimulation.

**Effect of Heart Rate on Function of the Heart as a Pump.** In general, the more times the heart beats per minute, the more blood it can pump, but there are important limitations to this effect. For instance, once the heart rate rises above a critical level, the heart strength itself decreases, presumably because of overutilization of metabolic substrates in the cardiac muscle. In addition, the period of diastole between the contractions becomes so reduced that blood does not have time to flow adequately from the atria into the ventricles. For these reasons, when the heart rate is increased artificially by electrical stimulation, the heart has its peak ability to pump large quantities of blood at a heart rate between 100 and 150 beats per minute. On the other hand, when its rate is increased by sympathetic stimulation, it reaches its peak ability to pump blood at a heart rate between 170 and 250 beats per min-

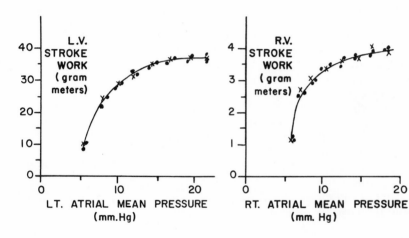

**Figure 13–7.** Left and right ventricular function curves in a dog, depicting ventricular stroke work output as a function of left and right mean atrial pressures. (Curves reconstructed from data in Sarnoff: *Physiol. Rev.*, 35:107, 1955.)

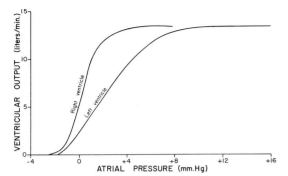

**Figure 13–8.** Approximate normal right and left ventricular output curves for the human heart as extrapolated from data obtained in dogs.

ute. The reason for this difference is that sympathetic stimulation not only increases the heart rate but also increases heart strength as well.

When the atrial pressures are high, the higher heart rates are especially effective in increasing the output of the heart.

**Nervous Regulation of Contractile Strength of the Heart.** The two atria are especially well supplied with large numbers of both sympathetic and parasympathetic nerves, but the ventricles are supplied mainly by sympathetic nerves and far fewer parasympathetic fibers. In general, sympathetic stimulation increases the strength of heart muscle contraction, whereas parasympathetic stimulation decreases the strength of contraction.

Under normal conditions the sympathetic nerve fibers to the heart continually discharge at a slow rate that maintains a strength of ventricular contraction about 20 per cent above its strength with no sympathetic stimulation at all.

Therefore, one method by which the nervous system can decrease the strength of ventricular contraction is simply to slow or stop the transmission of sympathetic impulses to the heart. On the other hand, maximal sympathetic stimulation can increase the strength of ventricular contraction to approximately 100 per cent greater than normal.

Maximal parasympathetic stimulation of the heart decreases ventricular contractile strength about 30 per cent. Thus, the parasympathetic effect is, by contrast with the sympathetic effect, relatively small.

**Effect of Autonomic Control of Cardiac Function as Depicted by Cardiac Function Curves.** Figure 13–10 illustrates four separate cardiac function curves. These are much the same as the ventricular function curves of Figures 13–7 and 13–8, except that they represent function of the entire heart rather than of a single ventricle; they show the relationship between the right atrial pressure at the input of the heart and cardiac output from the left ventricle.

The curves of Figure 13–10 demonstrate that at any given right atrial pressure, the cardiac output increases with increasing sympathetic stimulation and decreases with increasing parasympathetic stimulation. Note in particular that the changes in the function curves of Figure 13–10 are brought about by both *changes in heart rate* and *changes in contractile strength of the heart,* for both of these affect cardiac output.

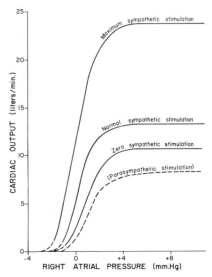

**Figure 13–10.** Effect on the cardiac output curve of different degrees of sympathetic and parasympathetic stimulation.

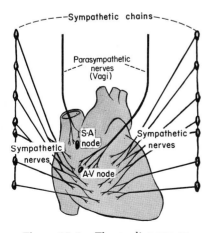

**Figure 13–9.** The cardiac nerves.

## EFFECT OF HEART DEBILITY ON CARDIAC FUNCTION–THE HYPOEFFECTIVE HEART

Any factor that damages the heart, whether it be damage to the myocardium, to the valves, to the conducting system, or otherwise, is likely to make the heart a poorer pump, and the heart under these conditions is called a hypoeffective heart. Figure 13–11 illustrates by the very dark curve the normal cardiac function curve and by the three curves below this the effect of different degrees of hypoeffectiveness on cardiac function. Obviously, the more serious the damage, the less will be the cardiac output at any given right atrial pressure.

Different factors that can cause a hypoeffective heart include:

> Myocardial infarction
> Valvular heart disease
> Vagal stimulation of the heart
> Inhibition of the sympathetics to the heart
> Congenital heart disease
> Myocarditis
> Cardiac anoxia
> Diphtheritic or other types of myocardial damage

## EFFECT OF EXERCISE ON THE HEART– THE HYPEREFFECTIVE HEART

Chronic heavy exercise over a period of many weeks or months leads to hypertrophy of the cardiac muscle and also to enlargement of the ventricular chambers. As a result, the overall strength of the heart becomes greatly enhanced, and the effectiveness of the heart as a pump increases. The upper three function curves of Figure 13–11 illustrate the effect of different degrees of cardiac hypereffectiveness on the cardiac function curve; maximal degrees of hypereffectiveness can increase pumping by the heart more than 100 per cent.

Other factors besides hypertrophy that can cause a hypereffective heart are:

> Sympathetic stimulation
> Parasympathetic inhibition

## EFFECT OF VARIOUS IONS ON HEART FUNCTION

In the discussion of membrane potentials in Chapter 10, it was pointed out that three particular cations—potassium, sodium, and calcium—all have marked effects on action potential transmission, and in Chapter 11 it was noted that calcium ions play an especially important role in initiating the muscle contractile process. Therefore, it is to be expected that the concentrations of these three ions in the extracellular fluids will also have marked effects on cardiac function.

The action of individual ions on the heart can be studied by perfusing the isolated heart, as illustrated in Figure 13–12, using the so-called *Langendorf preparation*. In making this preparation, the heart is removed rapidly from the animal's body, and the aorta is cannulated immediately. The cannula is connected to a perfusion bottle located approximately three feet higher than the heart itself so that the fluid in the aorta will be under pressure. The pressure causes the aortic valve to close and the fluid to flow through the coronary vessels. When the perfusion fluid contains appropriate nutrients (glucose and oxygen) and ions and the heart is kept appropriately warmed, the perfused mammalian heart will beat for many hours.

**Effect of Potassium Ions.** Excess potassium in the extracellular fluids causes the heart to become extremely dilated and flaccid and slows the heart rate. Very large quantities can also block conduction of the cardiac impulse from the atria to the ventricles through the A-V bundle. Elevation of potassium concentration to only 8 to 12 mEq./liter—two to three times

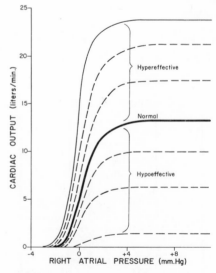

**Figure 13–11.** Cardiac output curves for various degrees of hypo- and hypereffective hearts. (From Guyton, Jones, and Coleman: Circulatory Physiology: Cardiac Output and Its Regulation, 1973.)

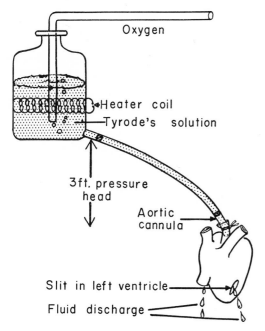

**Figure 13–12.** System for perfusing the heart.

the normal value—will usually cause such weakness of the heart that it will cause death.

All these effects of potassium excess are believed to be caused by decreased negativity of the resting membrane potentials which result from the high potassium concentration in the extracellular fluids. As the membrane potential decreases, the intensity of the action potential also decreases, which makes the contraction of the heart become progressively weaker and weaker, for, as discussed earlier in the chapter, the strength of the action potential determines to a great extent the strength of contraction.

**Effect of Calcium Ions.** An excess of calcium ions causes effects almost exactly opposite to those of potassium ions, causing the heart to go into spastic contraction. This is probably caused by the direct effect of calcium ions to excite the cardiac contractile process, as explained earlier in the chapter. Conversely, a deficiency of calcium ions causes cardiac flaccidity, similar to the effect of potassium.

However, the calcium ion concentration rarely changes sufficiently during life to alter cardiac function greatly, for greatly diminished calcium ion concentration will usually kill a person because of tetany before it will significantly affect the heart, and elevation of the calcium ion concentration to a level that will significantly affect the heart almost never oc-

curs because calcium ions are precipitated in bone or occasionally elsewhere in the body's tissues as insoluble calcium salts before such a level can be reached.

**Effect of Sodium Ions.** An excess of sodium ions depresses cardiac function, an effect similar to that of potassium ions but for an entirely different reason. Sodium ions compete with calcium ions at some yet unexplained point in the contractile process of muscle in such a way that the greater the sodium ion concentration in the extracellular fluids the less the effectiveness of the calcium ions in causing contraction when an action potential occurs.

However, from a practical point of view, the sodium ion concentration in the extracellular fluids probably never becomes abnormal enough even in serious pathological conditions to cause significant change in cardiac strength. However, very low sodium concentration, as occurs in water intoxication, often causes death because of cardiac fibrillation, a phenomenon explained in the following chapter.

### EFFECT OF TEMPERATURE ON THE HEART

The effect of temperature on the heart can also be studied using the perfusion system of Figure 13–12. Increased temperature causes greatly increased heart rate, and decreased temperature causes greatly decreased rate. These effects presumably result from increased permeability of the muscle membrane to the different ions, resulting in acceleration of the self-excitation process.

Contractile strength of the heart can be enhanced temporarily by moderate increase in temperature, but prolonged elevation of the temperature exhausts the heart and causes weakness.

### THE HEART-LUNG PREPARATION

A method frequently used to demonstrate the capabilities of the heart as a pump is the heart-lung preparation illustrated in Figure 13–13. In this preparation, blood leaving the heart by way of the aorta is diverted into an external system of tubes and then back again into the right atrium. The only parts of the animal's body that remain alive are the heart and lungs. The blood is oxygenated by artificial respiration of the lungs, and a glucose solution is added periodically to the blood to supply nutrition for cardiac contraction.

As blood flows through the external circuit, right atrial pressure, aortic pressure, and blood flow are measured. Also, an adjustable venous reservoir allows the pressure in the right atrium to be increased or decreased. A screw clamp is provided on the main outflow tube from the heart so that the arterial resistance can be increased or decreased. Finally, a special reservoir containing air in its upper part and

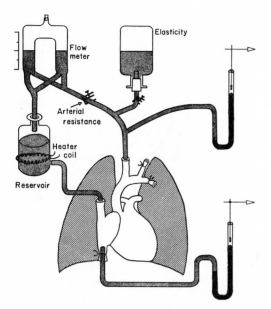

**Figure 13–13.** The heart-lung preparation.

blood in its lower part is used as an "elastic buffer" so that the blood pressure will not rise and fall excessively with each contraction of the heart. Some of the principles of cardiac function that can be demonstrated with the heart-lung preparation are the following:

First the *effect of end-diastolic volume or pressure on stroke volume output* can be demonstrated by raising and lowering the venous reservoir. When the reservoir is raised, blood flows into the heart rapidly, distending the cardiac chambers and causing increased stroke volume output, thus demonstrating the Frank-Starling law of the heart.

Second, one can show that *increasing the arterial resistance does not significantly decrease the stroke volume output as long as the right atrial pressure remains constant.* This illustrates that the aortic pressure, within physiological limits, has rather little effect on the output, though very high pressures can overload the heart so much that the stroke volume output does then become reduced.

Third, the *effect of different factors on heart rate* can be studied, such as a slight increase in heart rate that occurs when the load of inflowing blood increases, an increase in heart rate with increasing temperature, an increase in heart rate when epinephrine is injected into blood perfusing the heart, a decrease in heart rate when acetylcholine is injected, and changes in heart rate when the nerves to the heart are stimulated.

## ASSESSMENT OF CONTRACTILITY

Though it is very easy to determine the heart rate by simply timing the pulse, it has always been difficult to determine the strength of contraction of the heart, commonly called *cardiac contractility.* Very often the change in contractility is exactly opposite to the change in heart rate. Indeed, this effect occurs almost invariably in heart debilitating diseases.

One of the ways in which cardiac contractility can be determined with great precision is to record one or more of the cardiac function curves. However, this can be done only in experimental animals, and even then only with considerable difficulty. Therefore, many physiologists and clinicians have searched for a method to assess the cardiac contractility in a simple way. One of these methods is to determine the so-called dP/dt.

**dP/dt as a Measure of Cardiac Contractility.** dP/dt means the *rate of change of the ventricular pressure* with respect to time. The dP/dt record is generated by an electronic computer that differentiates the ventricular pressure wave, thus giving a record of the rate of change of the ventricular pressure. Figure 13–14 illustrates two separate recordings of the ventricular pressure wave as well as simultaneous recordings of the dP/dt. In the upper part of the figure the heart was beating normally, and in the lower part the heart had been stimulated by isoproterenol, a drug that has essentially the same effect on the heart as sympathetic stimulation, as will be discussed in more detail in Chapter 56.

Note in the upper record that at the same time that the ventricular pressure is increasing at its most rapid rate, the recording of the dP/dt record also reaches its greatest height. On the other hand, at the time that the ventricular pressure is falling most rapidly, the dP/dt record reaches its lowest level. When the ven-

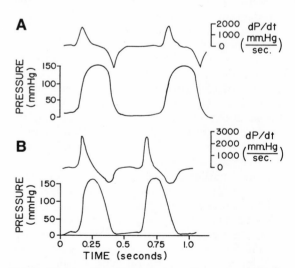

**Figure 13–14.** Simultaneous recordings of ventricular pressure and dP/dt. Panel A shows results from a normal heart and panel B from a heart stimulated by isoproterenol. (Modified from Mason et al., in Sodeman and Sodeman (Eds.): Pathologic Physiology, 5th Ed., 1974.)

tricular pressure is neither rising nor falling, the dP/dt record is at zero value.

Experimental studies have shown that the rate of rise of ventricular pressure, the dP/dt, in general correlates very well with the strength of contraction of the ventricle. This effect is illustrated by a comparison of the dP/dt record in the upper part of Figure 13–14 with that in the lower part. Note that the value for dP/dt in the upper record is only approximately 1800 mm. Hg per second, whereas in the lower record it rises to approximately 2600 mm. Hg per second, illustrating the stimulatory effect of isoproterenol on the contractility of the ventricle. Thus, the *peak* dP/dt is often used as a means for comparing the contractilities of hearts in different functional states.

Unfortunately, the quantitative value for peak dP/dt is also affected by some other factors that are not related to cardiac contractility. For instance, the value is increased by both increased input pressure to the left ventricle (the end-diastolic ventricular pressure) and the pressure in the aorta into which the heart is pumping the blood. Therefore, it is often difficult to use dP/dt as a measure of contractility in comparing hearts from one person to another because one of these factors may differ. For this reason other quantitative measures have also been used in attempts to assess cardiac contractility. One of these has been to use dP/dt divided by the instantaneous pressure in the ventricle, or (dP/dt)/P.

**Use of $V_{max}$ as a Measure of Contractility.**  Experimental studies have also shown that the rate of shortening of the cardiac muscle is often a good measure of the contractile strength. However, it is very difficult to measure this rate of shortening directly. On the other hand, several indirect methods for estimating the rate of shortening have been achieved. It will not be possible to explain these procedures here, but the peak estimated velocity of ventricular muscle shortening at zero intraventricular pressure (called $V_{max}$) is often used, especially in experimental situations, to assess cardiac contractility. However, here again, a number of factors besides cardiac contractility enter into determining the level of $V_{max}$.

Thus, both the dP/dt and the $V_{max}$ determinations are useful in comparative studies of ventricular contractility under many conditions. Since both of them also display serious inaccuracies in some cardiac states, however, they are at present considered to have considerable empirical experimental value but not necessarily to represent absolute measures of cardiac contractility.

# REFERENCES

Braunwald, E., Ross, J., and Sonnenblick, E.: Mechanism of Contractility of the Normal and Failing Heart. Boston, Little, Brown and Company, 1968.

Brecher, G. A., and Galletti, P. M.: Functional anatomy of cardiac pumping. In Hamilton, W. F. (ed.): Handbook of Physiology. Baltimore, The Williams & Wilkins Co., 1963, Vol 2., Sec. 2, p. 759.

Brooks, C. McC., and Lu, H.: The Sinoatrial Pacemaker of the Heart. Springfield, Ill., Charles C Thomas, Publisher, 1972.

Brutsaert, D. L., and Sonnenblick, E. H.: Cardiac muscle mechanics in the evaluation of myocardial contractility and pump function: problems, concepts, and directions. Prog. Cardiovasc. Dis., 16:337, 1973.

Cowley, A. W., Jr., and Guyton, A. C.: Heart rate as a determinant of cardiac output in dogs with arteriovenous fistula. Amer. J. Cardiol., 28:321, 1971.

Dodge, H. T.: Determination of left ventricular volume and mass. Radiol. Clin. North Amer., 9:459, 1971.

Fozzard, H. A., and Gibbons, W. R.: Action potential and contraction of heart muscle. Amer. J. Cardiol., 31:182, 1973.

Goody, R. S., and Mannherz, H. G.: The molecular basis of contractility. II. Basic Res. Cardiol., 69:204, 1974.

Guyton, A. C.: Determination of cardiac output by equating venous return curves with cardiac response curves. Physiol. Rev., 35:123, 1955.

Guyton, A. C., Jones, C. E., and Coleman, T. G.: Circulatory Physiology: Cardiac Output and Its Regulation, 2nd Ed. Philadelphia, W. B. Saunders Company, 1973.

Johnson, E. A., and Lieberman, M.: Heart: Excitation and contraction. Ann. Rev. Physiol., 33:479, 1971.

Katz, A. M.: Contractile proteins of the heart. Physiol. Rev., 50:63, 1970.

Kones, R. J.: The Molecular and Ionic Basis of Altered Myocardial Contractility. Westbury, N. Y., PJD Publications, 1973.

Korner, P. I.: Keynote address: present concepts about the myocardium. Adv. Cardiol., 12:1, 1974.

Langer, G. A.: Heart: Excitation-contraction coupling. Ann. Rev. Physiol., 35:55, 1973.

Langer, G. A., and Brady, A. J.: The Mammalian Myocardium. New York, John Wiley & Sons, Inc., 1974.

Legato, M. J.: The Myocardial Cell for the Clinical Cardiologist. New York, Futura Publishing Co., Inc., 1975.

Levy, M. N., and Berne, R. M.: Heart. Ann. Rev. Physiol., 32:373, 1970.

Levy, M. N., and Martin, P. J.: Cardiac excitation and contraction. In Guyton, A. C. (ed.): MTP International Review of Science: Physiology. Baltimore, University Park Press, 1974, Vol. 1, p. 49.

Lieberman, M., and Sano, T. (eds.): Developmental and Physiological Correlates of Cardiac Muscle. New York, Raven Press, 1975.

Linden, R. J.: Recent Advances in Physiology. New York, Churchill Livingstone, Div. of Longman Inc., 1974.

Mason, D. T., Zelis, R., Amsterdam, E. A., and Massumi, R. A.: Mechanisms of cardiac contraction: structural, biochemical, and functional relations in the normal and diseased heart. In Sodeman, W. A., Jr., and Sodeman, W. A. (eds.): Pathologic Physiology: Mechanisms of Disease, 5th Ed. Philadelphia, W. B. Saunders Company, 1974, p. 206.

Noble, M. I. M.: The contribution of blood momentum to left ventricular ejection in the dog. Circ. Res., 23:663, 1968.

Ross, J., Jr., and Sobel, B. E.: Regulation of cardiac contraction. Ann. Rev. Physiol., 34:47, 1972.

Sarnoff, S. J.: Myocardial contractility as described by ventricular function curves. Physiol. Rev., 35:107, 1955.

Spencer, M. P., and Greiss, F. C.: Dynamics of ventricular ejection. Circ. Res., 10:274, 1962.

Starling, E. H.: The Linacre Lecture on the Law of the Heart. London, Longmans Green & Co., 1918.

Stone, H. L., Bishop, V. S., and Guyton, A. C.: Cardiac function after embolization of coronaries with microspheres. Amer. J. Physiol., 204:16, 1963.

Sugimoto, T., Allison, J. L., and Guyton, A. C.: Effect of maximal work load on cardiac function. Jap. Heart J., 14:146, 1973.

Sugimoto, T., Sagawa, K., and Guyton, A. C.: Quantitative effect of low coronary pressure on left ventricular performance. Jap. Heart J., 9:46, 1968.

Trautwein, W.: Membrane currents in cardiac muscle fibers. Physiol. Rev., 53:793, 1973.

Van der Werf, T.: Proc. of the 2nd Workshop on Contractile Behavior of the Heart, Utrecht, 1973. New York, American Elsevier Publishing Co., 1974.

# | 14 |

# Rhythmic Excitation of the Heart

The heart is endowed with a special system (a) for generating rhythmical impulses to cause rhythmical contraction of the heart muscle and (b) for conducting these impulses throughout the heart. A major share of the ills of the heart is based on abnormalities of this special excitatory and conductive system.

## THE SPECIAL EXCITATORY AND CONDUCTIVE SYSTEM OF THE HEART

The adult human heart normally contracts at a rhythmic rate of about 72 beats per minute. Figure 14–1 illustrates the special excitatory and conductive system of the heart that controls these cardiac contractions. The figure shows: (A) the *S-A node* in which the normal rhythmic self-excitatory impulse is generated, (B) the *A-V node* in which the impulse from the atria is delayed before passing into the ventricles, (C) the *A-V bundle*, which conducts the impulse from the atria into the ventricles, and (D) the *left* and *right bundles of Purkinje fibers*, which conduct the cardiac impulse to all parts of the ventricles.

### THE SINO-ATRIAL NODE

The sino-atrial (S-A) node is a small crescent-shaped strip of specialized muscle approximately 3 mm. wide and 1 cm. long; it is located in the posterior wall of the right atrium immediately beneath and medial to the opening of the superior vena cava. The fibers of this node are each 3 to 5 microns in diameter in contrast to a diameter of 15 to 20 microns for the surrounding atrial muscle fibers. However, the S-A fibers are continuous with the atrial fibers so that any action potential that begins in the S-A node spreads immediately into the atria.

**Automatic Rhythmicity of the Sino-Atrial Fibers.** Most cardiac fibers have the capability of *self-excitation*, a process which can cause automatic rhythmical contraction. This is especially true of the fibers of the heart's specialized conducting system; the portion of this system that displays self-excitation to the greatest extent is the fibers of the S-A node. For this reason, the sino-atrial (S-A) node ordinarily

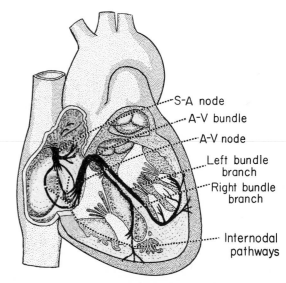

**Figure 14–1.** The S-A node and the Purkinje system of the heart. (Modified from Gray: Anatomy of the Human Body, Lea & Febiger, 1948.)

177

controls the rate of the beat of the entire heart, as will be discussed in detail later in this chapter. First, however, let us describe this automatic rhythmicity.

The sino-atrial fibers are somewhat different from most other cardiac muscle fibers, having a resting membrane potential of only −55 to −60 millivolts in comparison with −80 to −90 millivolts in most of the other fibers. This low resting potential is caused by a natural leakiness of the membranes to sodium ions. And it is also this leakage of sodium that causes the self-excitation of the S-A fibers. Figure 14–2 illustrates the rhythmical repetitiveness of the action potentials in the S-A node, which can be explained as follows:

Immediately after each action potential is over, the membrane is even more permeable to potassium ions than normally, as was discussed in Chapter 10, and this contrasts with a very low permeability for sodium ions. Recalling the Goldman equation from Chapter 10, one may note that when the ratio of potassium ion to sodium ion permeability is very high, the potential inside the membrane becomes very highly negative because potassium ions diffusing out of the cell carry positive charges away from the interior of the cell. Therefore, immediately after the action potential is over, the membrane potential reaches its greatest degree of negativity, a state called *hyperpolarization.* At this time the membrane potential is much too negative for the fiber to fire again. However, this does not last long because the high degree of negativity inside the membrane causes the membrane to become progressively less permeable to potassium during the next few tenths of a second, and the natural leakiness of the membrane to sodium causes the membrane potential to drift slowly back toward a less negative value. This is illustrated after each one of the action potentials in Figure 14–2. After a few tenths of a second, the so-called ''resting'' potential will have drifted enough for it finally to reach the threshold point for excitation of the fiber. When this occurs, the flux of sodium ions to the interior of the fiber accelerates, causing the fiber to become even more conductive to sodium. A self-regeneration process begins that leads to extremely high sodium permeability and rapid total depolarization of the membrane, followed by overshoot of the membrane potential to a positive potential of approximately +20 millivolts, called a *reversal potential,* as pointed out in Chapter 10. Then for slightly over 0.10 second, the membrane remains depolarized; then the permeability of the membrane for potassium increases while that for sodium decreases to a low value. After approximately 0.12 second the permeability for potassium is now great enough that a self-regenerative process begins in the opposite direction, with accelerating flux of potassium ions to the exterior; this recreates the negative potential inside the fiber. After the action potential is over, an especially high potassium permeability persists again for a few tenths of a second, thus maintaining a new state of hyperpolarization. Then the process begins again, with gradual decrease in permeability of the membrane to potassium, more and more leakage of sodium, and then a new regenerative cycle that causes depolarization. This process continues over and over throughout the life of the person, thereby providing rhythmical excitation of the S-A nodal fibers at a normal resting rate of about 72 times per minute.

## INTERNODAL PATHWAYS AND TRANSMISSION OF THE CARDIAC IMPULSE THROUGH THE ATRIA

The ends of the S-A nodal fibers fuse with the surrounding atrial muscle fibers, and action potentials originating in the S-A node travel outward into these fibers. In this way, the action potential spreads through the entire atrial muscle mass and eventually also to the A-V node. The velocity of conduction in the atrial muscle is approximately 0.3 meter per second. However, conduction is somewhat more rapid in several small bundles of atrial muscle fibers, some of which pass directly from the S-A node to the A-V node and conduct the cardiac impulse at a velocity of approximately 0.45 meter per second. Some physiologists claim that these bundles contain specialized conduction fibers similar to the Purkinje fibers of the ventricles, which will be discussed subsequently. How-

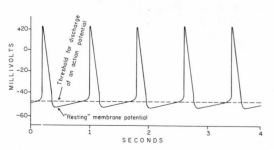

**Figure 14–2.**   Rhythmic discharge of an S-A nodal fiber.

ever, other physiologists believe that this more rapid conduction is caused simply by the greater mass of muscle and the more direct orientation of the muscle fibers in the special bundles. Nevertheless, these pathways do conduct the impulse from the S-A node to the A-V node more rapidly than the conduction in the general mass of atrial muscle; these pathways, called *internodal pathways,* are illustrated in Figure 14–1.

## THE ATRIOVENTRICULAR (A-V) NODE AND THE PURKINJE SYSTEM

**Delay in Transmission at the A-V Node.** Fortunately, the conductive system is organized so that the cardiac impulse will not travel from the atria into the ventricles too rapidly; this allows time for the atria to empty their contents into the ventricles before ventricular contraction begins. It is primarily the A-V node and its associated conductive fibers that delays this transmission of the cardiac impulse from the atria into the ventricles.

Figure 14–3 shows diagrammatically the different parts of the A-V node and its connections with the atrial internodal pathway fibers and the A-V bundle. The figure also shows the approximate intervals of time in fractions of a second between the genesis of the cardiac impulse in the S-A node and its appearance at different points in its conductive pathway. Note that the impulse, after traveling through the internodal pathway, reaches the A-V node approximately 0.04 second after its origin in the S-A node. However, between this time and the time that the impulse emerges in the A-V bundle, another 0.11 second elapses. About one half of this time

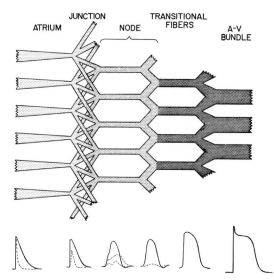

**Figure 14–4.** Functional diagram of the anatomical relationships in the region of the atrioventricular node. The action potentials at the bottom were recorded under normal conditions (solid lines) and under the influence of acetylcholine (dashed lines). (From Hoffman and Cranefield: Electrophysiology of the Heart. McGraw-Hill Book Co., 1960.)

lapse occurs in the *junctional fibers,* which are very small fibers that connect the normal atrial fibers with the fibers of the node itself (illustrated in Figures 14–3 and 14–4). The velocity of conduction in these fibers is about 0.01 meter per second (about one-fiftieth that in normal cardiac muscle), which greatly delays entrance of the impulse into the A-V node. After entering the node, the velocity of conduction in the *nodal fibers* is still quite low, only 0.1 meter per second, about one-fourth the conduction velocity in normal cardiac muscle. Therefore, a further delay in transmission occurs as the impulse travels through the A-V node into the *transitional fibers* and finally into the *A-V bundle,* also called the *bundle of His.*

Figure 14–4 illustrates a functional diagram of the A-V nodal region showing the atrial fibers leading into the minute junctional fibers, then progressive enlargement of the fibers again as they spread through the node, through the transitional region, and into the A-V bundle. The characteristics of the action potential in the different parts of the node are illustrated at the bottom.

The cause of the extremely slow conduction in the junctional and other A-V nodal fibers is probably 3-fold: (1) The fibers are very small, a fact which in itself makes the conduction rate very slow. (2) The number of tight junctions at

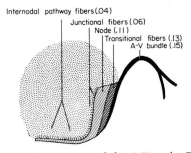

**Figure 14–3.** Organization of the A-V node. The numbers represent the interval of time from the origin of the impulse in the S-A node. The values have been extrapolated to the human being. (This figure is based on studies in lower animals discussed and illustrated in Hoffman and Cranefield: Electrophysiology of the Heart. McGraw-Hill Book Co., 1960.)

the intercalated discs from one cardiac cell to the next in these fibers is much less than that in the usual cardiac muscle fibers—that is, the number of *nexi* (points of tight fusion between the cell membranes) is greatly reduced between these cardiac muscle cells, thereby reducing the rapidity of ionic transport along the axis of the fiber. (3) These fibers are made up of a much more embryonic type of cell with much less differentiation than in the usual cardiac muscle cell, this also presumably reducing the capability of the cell to transmit the cardiac impulse.

## TRANSMISSION IN THE PURKINJE SYSTEM

The *Purkinje fibers* that lead from the A-V node through the A-V bundle and into the ventricles have functional characteristics quite the opposite of those of the A-V nodal fibers; they are very large fibers, even larger than the normal ventricular muscle fibers, and they transmit impulses at a velocity of 1.5 to 4.0 meters per second, a velocity about six times that in the usual cardiac muscle and 300 times that in the junctional fibers. This allows almost immediate transmission of the cardiac impulse throughout the entire ventricular system.

The very rapid transmission of action potentials by Purkinje fibers is probably caused by increased numbers of nexi between the successive cardiac cells that make up the Purkinje fibers. At these nexi, ions are transmitted easily from one cell to the next, thus enhancing the velocity of transmission. The Purkinje fibers also have very few myofibrils, which means that they barely contract during the course of impulse transmissions.

***Distribution of the Purkinje Fibers in the Ventricles.*** The Purkinje fibers, after originating in the A-V node, form the A-V bundle, which then threads between the valves of the heart and thence into the ventricular septum as shown in Figure 14–1. The A-V bundle divides almost immediately into the *left* and *right bundle branches* that lie beneath the endocardium of the respective sides of the septum. Each of these branches spreads downward toward the apex of the respective ventricle, but then divides into small branches and spreads around each ventricular chamber and finally back toward the base of the heart along the lateral wall. The terminal Purkinje fibers form swirls underneath the endocardium and penetrate about one-third of the way into the muscle mass to terminate on the muscle fibers.

From the time that the cardiac impulse first enters the A-V bundle until it reaches the terminations of the Purkinje fibers, the total time that lapses is only 0.03 second; therefore, once a cardiac impulse enters the Purkinje system, it spreads almost immediately to the entire endocardial surface of the ventricular muscle.

## TRANSMISSION OF THE CARDIAC IMPULSE IN THE VENTRICULAR MUSCLE

Once the cardiac impulse has reached the ends of the Purkinje fibers, it is then transmitted through the ventricular muscle mass by the ventricular muscle fibers themselves. The velocity of transmission is now only 0.4 to 0.5 meter per second, one-sixth that in the Purkinje fibers.

The cardiac muscle is arranged in whorls with fibrous septa between the whorls; therefore, the cardiac impulse does not necessarily travel directly outward toward the surface of the heart but instead angulates toward the surface along the directions of the whorls. Because of this, transmission from the endocardial surface to the epicardial surface of the ventricle requires as much as another 0.03 second, approximately equal to the time required for transmission through the entire Purkinje system. Thus, the total time for transmission of the cardiac impulse from the origin of the Purkinje system to the last of the ventricular muscle fibers in the normal heart is about 0.06 second.

One can see from this description of conduction in the ventricles that the first part of the ventricular muscle mass to be excited is the septum; this is followed rapidly by the endocardial surfaces of the apices and lateral walls of the ventricles and finally by the epicardial surfaces of the ventricles. The precise time intervals of spread of the impulse to the different anterior and posterior epicardial surfaces of the ventricles are illustrated in Figure 14–5, which shows actual measurements from human hearts.

## SUMMARY OF THE SPREAD OF THE CARDIAC IMPULSE THROUGH THE HEART

Figure 14–6 illustrates in summary form the transmission of the cardiac impulse through the human heart. The numbers on the figure represent the intervals of time in fractions of a second that lapse between the origin of the cardiac impulse in the S-A node and its appearance at

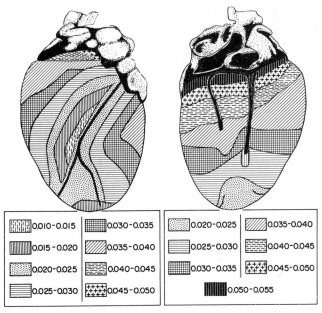

**Figure 14–5.** Sequence of activation of the anterior and posterior surfaces of the heart during the normal cardiac impulse. The numbers represent fractions of a second after first appearance of the impulse in the ventricular septum. (From Sodi-Pallares and Calder: New Bases of Electrocardiography. The C. V. Mosby Co.)

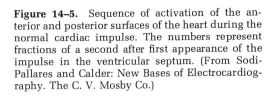

| | | |
|---|---|---|
| 0.010-0.015 | 0.030-0.035 | 0.020-0.025 | 0.035-0.040 |
| 0.015-0.020 | 0.035-0.040 | 0.025-0.030 | 0.040-0.045 |
| 0.020-0.025 | 0.040-0.045 | 0.030-0.035 | 0.045-0.050 |
| 0.025-0.030 | 0.045-0.050 | | 0.050-0.055 |

each respective point in the heart. Note that the impulse spreads at moderate velocity through the atria but is delayed more than 0.1 second in the A-V nodal region before appearing in the A-V bundle. Once it has entered the bundle, it spreads rapidly through the Purkinje fibers to the entire endocardial surfaces of the ventricles. Then the impulse spreads slowly through the ventricular muscle to the epicardial surfaces.

It is extremely important that the student learn in detail the course of the cardiac impulse through the heart and the times of its appearance in each separate part of the heart, for a quantitative knowledge of this process is essential to the understanding of electrocardiography, which is discussed in the following three chapters.

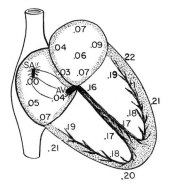

**Figure 14–6.** Transmission of the cardiac impulse through the heart, showing the time of appearance (in fractions of a second) of the impulse in different parts of the heart.

# CONTROL OF EXCITATION AND CONDUCTION IN THE HEART

### THE S-A NODE AS THE PACEMAKER OF THE HEART

In the above discussion of the genesis and transmission of the cardiac impulse through the heart, it was stated that the impulse normally arises in the S-A node. However, this need not be the case under abnormal conditions, for other parts of the heart can exhibit rhythmic contraction in the same way that the fibers of the S-A node can; this is particularly true of the A-V nodal and Purkinje fibers.

The A-V nodal fibers, when not stimulated from some outside source, discharge at an intrinsic rhythmic rate of 40 to 60 times per minute, and the Purkinje fibers discharge at a rate of somewhere between 15 to 40 times per minute. These rates are in contrast to the normal rate of the S-A node of 70 to 80 times per minute.

Therefore, the question that we must ask is: Why is it the S-A node that controls the heart's rhythmicity rather than the A-V node or the

Purkinje fibers? The answer to this is simply that the rate of the S-A node is considerably greater than that of either the A-V node or the Purkinje fibers. Each time the S-A node discharges, its impulse is conducted both into the A-V node and into the Purkinje fibers, discharging their excitable membranes. Then all these tissues recover from the action potential and become greatly hyperpolarized. But the S-A node loses this hyperpolarization much more rapidly than does either of the other two and emits a new impulse before either one of them can reach its own threshold for self-excitation. The new impulse again discharges both the A-V node and Purkinje fibers. This process continues on and on, the S-A node always exciting these other potentially self-excitatory tissues before self-excitation can actually occur.

Thus, the S-A node controls the beat of the heart because its rate of rhythmic discharge is greater than that of any other part of the heart. Therefore, it is said that the S-A node is the normal *pacemaker* of the heart.

**Abnormal Pacemakers—The Ectopic Pacemaker.** Occasionally some other part of the heart develops a rhythmic discharge rate that is more rapid than that of the S-A node. For instance, this frequently occurs in the A-V node or in the Purkinje fibers. In either of these cases, the pacemaker of the heart shifts from the S-A node to the A-V node or to the excitable Purkinje fibers. Under rare conditions a point in the atrial or ventricular muscle develops excessive excitability and becomes the pacemaker.

A pacemaker elsewhere than the S-A node is called an *ectopic pacemaker*. Obviously, an ectopic pacemaker causes an abnormal sequence of contraction of the different parts of the heart.

## ROLE OF THE PURKINJE SYSTEM IN CAUSING SYNCHRONOUS CONTRACTION OF THE VENTRICULAR MUSCLE

It is clear from the above description of the Purkinje system that the cardiac impulse arrives at almost all portions of the ventricles within a very narrow span of time, exciting the first ventricular muscle fiber only 0.06 second ahead of excitation of the last ventricular muscle fiber. Since the ventricular muscle fibers remain contracted for a total period of 0.30 second, one can see that this rapid spread of excitation throughout the entire ventricular muscle mass causes all portions of the ventricular muscle in both ventricles to contract at almost exactly the same time. Effective pumping by the two ventricular chambers requires this synchronous type of contraction. If the cardiac impulse traveled through the ventricular muscle very slowly, then much of the ventricular mass would contract prior to contraction of the remainder, in which case the overall pumping effect would be greatly depressed. Indeed, in some types of cardiac debilities, some of which will be discussed in Chapters 16 and 17, such slow transmission does indeed occur, and the pumping effectiveness of the ventricles is decreased perhaps as much as 20 to 30 per cent.

## FUNCTION OF THE PURKINJE SYSTEM IN PREVENTING ARRHYTHMIAS

Another extremely important function of the Purkinje system is to prevent the development of stray impulses in the heart that can cause ventricular fibrillation and other cardiac arrhythmias, as will be discussed later in this chapter and in Chapter 17 in relation to electrocardiography. This function may be explained as follows:

The refractory period of the cardiac muscle is approximately 0.3 second, which is also the duration of the muscle contraction, as was discussed in detail in the previous chapter. The relationship of this refractory period to the rapidity of transmission of the cardiac impulse throughout the ventricles plays an important role in preventing serious cardiac arrhythmias. The normal amount of time required for the cardiac impulse to travel from the A-V bundle all the way through the entire ventricular mass in the normal heart is only 0.06 second. Therefore, the first ventricular muscle that is stimulated is still very much in the refractory state even when the last ventricular muscle becomes stimulated. Consequently, the action potential in this last cardiac muscle now finds no remaining ventricular muscle to excite, and all the action potentials in all portions of the ventricular muscle simply die. We shall see later in this chapter that any serious delay in transmission of the impulse through the ventricle can make it possible for the impulse from the last excited ventricular muscle to re-enter the first muscle. This, in turn, sets up a re-entrance cycle in the ventricles that can continue on and on, creating the condition of ventricular fibrillation that causes the ventricles to remain in a semicontracted state indefinitely and, therefore, not to provide cardiac pumping, as will be discussed in detail later in the chapter.

**Refractory Period of the Purkinje Fibers and Its Functional Significance.** The action potential of the Purkinje fibers lasts about 25 per cent longer than the action potential of the usual ventricular cardiac muscle. Therefore, the refractory period of the Purkinje fibers is also somewhat longer than that of the muscle. This also has a very important functional importance for maintaining the normal heart rhythm, as follows: By the time the Purkinje fibers are no longer refractory, the ventricular muscle fibers have already long been out of the refractory state. As a result, all portions of the ventricular muscle are already available to accept a new action potential and, therefore, are ready to contract. We shall see later in the chapter that if some portions of the ventricular muscle are still in a state of refractoriness when a new depolarization wave travels over the ventricles, there is a great likelihood that the heart will develop ventricular fibrillation, a condition which means that there are incoordinate action potentials traveling in many diverse directions at the same time and never stopping. Therefore, this long refractory period of the Purkinje fibers in relation to the somewhat shorter refractory period of the ventricular muscle protects the ventricular muscle from being excited at too early a time in the cardiac cycle.

## CONTROL OF HEART RHYTHMICITY AND CONDUCTION BY THE AUTONOMIC NERVES

The heart is supplied with both sympathetic and parasympathetic nerves, as illustrated in Figure 13–5 of the previous chapter. The parasympathetic nerves are distributed mainly to the S-A and A-V nodes, to a lesser extent to the muscle of the two atria and even less to the ventricular muscle. The sympathetic nerves are distributed to these same areas but with a strong representation to the ventricular muscle as well as to the other parts of the heart.

**Effect of Parasympathetic (Vagal) Stimulation on Cardiac Function—Ventricular Escape.** Stimulation of the parasympathetic nerves to the heart (the vagi) causes the hormone acetylcholine to be released at the vagal endings. This hormone has two major effects on the heart. First, it decreases the rate of rhythm of the S-A node, and, second, it decreases the excitability of the A-V junctional fibers between the atrial musculature and the A-V node, thereby slowing transmission of the cardiac impulse into the ventricles. Very strong stimula-

tion of the vagi can completely stop the rhythmic contraction of the S-A node or completely block transmission of the cardiac impulse through the A-V junction. In either case, rhythmic impulses are no longer transmitted into the ventricles. The ventricles stop beating for 4 to 10 seconds, but then some point in the Purkinje fibers, usually in the A-V bundle, develops a rhythm of its own and causes ventricular contraction at a rate of 15 to 40 beats per minute. This phenomenon is called *ventricular escape.*

*Mechanism of the Vagal Effects.* The acetylcholine released at the vagal nerve endings greatly increases the permeability of the fiber membranes to potassium, which allows rapid leakage of potassium to the exterior. This causes increased negativity inside the fibers, an effect called *hyperpolarization,* which makes excitable tissue much less excitable, as was explained in Chapter 10.

In the A-V node, the state of hyperpolarization makes it difficult for the minute junctional fibers, which can generate only small quantities of current during the action potential, to excite the nodal fibers. Therefore, the *safety factor* for transmission of the cardiac impulse through the junctional fibers and into the nodal fibers decreases. A moderate decrease in this simply delays conduction of the impulse, but a decrease in safety factor below unity (which means so low that the action potential of one fiber cannot cause an action potential in the successive fiber) completely blocks conduction.

**Effect of Sympathetic Stimulation on Cardiac Function.** Sympathetic stimulation causes essentially the opposite effects on the heart to those caused by vagal stimulation as follows: First, it increases the rate of S-A nodal discharge. Second, it increases the excitability of all portions of the heart. Third, it increases greatly the force of contraction of all the cardiac musculature, both atrial and ventricular, as discussed in the previous chapter.

In short, sympathetic stimulation increases the overall activity of the heart. Maximal stimulation can almost triple the rate of heartbeat and can increase the strength of heart contraction as much as 2- to 3-fold.

*Mechanism of the Sympathetic Effect.* Stimulation of the sympathetic nerves releases the hormone norepinephrine at the sympathetic nerve endings. The precise mechanism by which this hormone acts on cardiac muscle fibers is still somewhat doubtful, but the present belief is that it increases the permeability of the fiber membrane to sodium and calcium. In the S-A node, an increase of sodium permeability would cause increased tendency for the resting membrane potential to decay to the threshold level for self-excitation, which obviously would accelerate the onset of self-excitation after each successive heartbeat and therefore increase the rate of the heart.

In the A-V node, increased sodium permeability would make it easier for each fiber to excite the succeeding fiber, thereby decreasing the conduction time from the atria to the ventricles.

The increase in permeability to calcium ions is probably at least partially responsible for the increase in contractile strength of the cardiac muscle under the influence of sympathetic stimulation because calcium ions play a powerful role in exciting the contractile process of the myofibrils.

# ABNORMAL RHYTHMS OF THE HEART

Abnormal cardiac rhythms can be caused by (1) abnormal rhythmicity of the pacemaker itself, (2) shift of the pacemaker from the S-A node to other parts of the heart, (3) blocks at different points in the transmission of the impulse through the heart, (4) abnormal pathways of impulse transmission through the heart, and (5) spontaneous generation of abnormal impulses in almost any part of the heart. Some of these will be discussed in Chapter 17 in relation to electrocardiographic analysis of cardiac arrhythmias, but the major disturbances and their causes are presented here to illustrate some aberrations that can occur in the function of the rhythmicity and conducting systems of the heart.

## PREMATURE CONTRACTIONS— ECTOPIC FOCI

Often, a small area of the heart becomes much more excitable than normal and causes an occasional abnormal impulse to be generated in between the normal impulses. A depolarization wave spreads outward from the irritable area and initiates a *premature contraction* of the heart. The focus at which the abnormal impulse is generated is called an *ectopic focus*.

The usual cause of an ectopic focus is an irritable area of cardiac muscle resulting from a local area of ischemia (too little coronary blood flow to the muscle), over-use of stimulants such as caffeine or nicotine, lack of sleep, anxiety, or other debilitating state.

**Shift of the Pacemaker to an Ectopic Focus.** Sometimes an ectopic focus becomes so irritable that it establishes a rhythmic contraction of its own at a more rapid rate than that of the S-A node. When this occurs the ectopic focus becomes the pacemaker of the heart. The most common point for development of an ectopic pacemaker is the A-V node itself or the A-V bundle, as will be discussed in more detail in Chapter 17.

## HEART BLOCK

Occasionally, transmission of the impulse through the heart is blocked at a critical point in the conductive system. One of the most common of these points is between the atria and the ventricles; this condition is called *atrioventricular block*. Another common point is in one of the *bundle branches* of the Purkinje system. Rarely a block also develops between the S-A node and the atrial musculature.

**Atrioventricular Block.** In the human being, a block between the atria and the ventricles can result from localized damage or depression of the *A-V junctional* fibers or of the *A-V bundle*. The causes include different types of infectious processes, excessive stimulation by the vagus nerves (which depresses conductivity of the junctional fibers), localized destruction of the A-V bundle as a result of a coronary infarct, pressure on the A-V bundle by arteriosclerotic plaques, or depression caused by various drugs.

Figure 14–7 illustrates a typical record of atrial and ventricular contraction while the A-V bundle was being progressively compressed to cause successive stages of block. During the first three contractions of the record, ventricular contraction followed in orderly sequence approximately 0.16 second after atrial contraction. Then, A-V bundle compression was begun, and the interval of time between the beginning of atrial contraction and the beginning of ventricular contraction increased steadily during the next five heartbeats from 0.16 to 0.32 second. Beyond this point further compression completely blocked impulse transmission. Thereafter, the atria continued to beat at their normal rate of rhythm, while the ventricles failed to contract at all for approximately 7 seconds. Then, "ventricular escape" occurred, and a rhythmic focus in the ventricles suddenly began to act as a ventricular pacemaker, causing ventricular contractions at a rate of approximately 30 per minute; these were completely dissociated from the atrial contraction.

## FLUTTER AND FIBRILLATION

Frequently, either the atria or the ventricles begin to contract extremely rapidly and often incoordinately. The low frequency, more coordinate contractions up to 200 to 300 beats per minute are generally called *flutter* and the very high frequency, incoordinate contractions *fibrillation*.

Two basic theories of flutter and fibrillation have been proposed; these are (1) a *single* or *multiple ectopic foci* emitting many impulses one after another

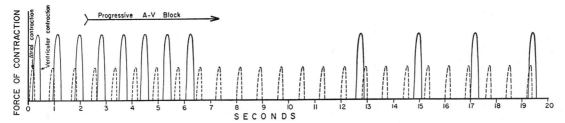

**Figure 14–7.** Contraction of the atria and ventricles of a heart, showing the effect of progressive A-V block. Note the progressive increase in the interval between the onset of atrial contraction and the onset of ventricular contraction. Note also the final complete asynchronism between atrial and ventricular contraction.

in rapid succession and (2) a *circus movement,* in which the impulse travels around and around through the heart muscle, never stopping.

**Ectopic Foci.** The ectopic focus theory is easy to understand. An area of the heart simply becomes so irritable that it keeps sending rapid impulses in all directions, resulting in rapid rates of contraction. There is reason to believe that at least some instances of atrial flutter may result from this cause.

**The Circus Movement (Re-entry of the Impulse).** There is also reason to believe that many instances of atrial flutter result from circus movements around the atria, and since only the circus movement theory has explained adequately the course of events in most instances of atrial and ventricular fibrillation, we will discuss it in detail.

Figure 14–8 illustrates several small cardiac muscle strips cut in the form of circles. If such a strip is stimulated at the 12 o'clock position *so that the impulse travels in only one direction,* the impulse spreads progressively around the circle until it returns to the 12 o'clock position. If the originally stimulated muscle fibers are still in a refractory state, the impulse then dies out, for refractory muscle cannot transmit a second impulse. However, there are three different conditions that could cause this im-

pulse to continue to travel around the circle, that is, to cause "re-entry" of the impulse into muscle that has already been excited.

First, if the *length of the pathway around the circle is long,* by the time the impulse returns to the 12 o'clock position the originally stimulated muscle will no longer be refractory, and the impulse will continue around the circle again and again.

Second, if the length of the pathway remains constant but the *velocity of conduction becomes decreased* enough, an increased interval of time will elapse before the impulse returns to the 12 o'clock position. By this time the originally stimulated muscle might be out of the refractory state, and the impulse can continue around the circle again and again.

Third, *the refractory period of the muscle might have become greatly shortened.* In this case, the impulse could also continue around and around the circle.

All three of these conditions occur in different pathological states of the human heart as follows: (1) A long pathway frequently occurs in dilated hearts. (2) Decreased rate of conduction frequently results from blockage of the Purkinje system, ischemia of the muscle, high blood potassium, and many other factors. (3) A shortened refractory period frequently occurs in response to various drugs, such as epinephrine, following intense vagal stimulation, or following repetitive electrical stimulation. Thus, in many different cardiac disturbances circus movements can cause abnormal cardiac rhythmicity that completely ignores the pacesetting effects of the S-A node.

**Atrial Flutter Resulting from a Circus Pathway.** Figure 14–9 illustrates a circus pathway around and around the atria from top to bottom, the pathway lying to the left of the superior and inferior venae cavae. Such circus pathways have been initiated experimentally in the atria of dogs' hearts, and electrocardiographic records, which will be discussed in Chapter 17, indicate that this type of circus pathway also develops in the human heart when the atria become greatly dilated as a result of valvular heart disease. The rate of flutter is usually 200 to 350 times per minute.

*Partial Block at the A-V Node During Atrial Flutter.* The functional refractory period of the Purkinje

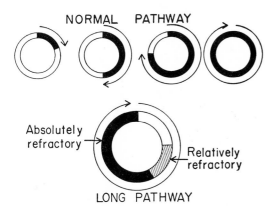

**Figure 14–8.** The circus movement, showing annihilation of the impulse in the short pathway and continued propagation of the impulse in the long pathway.

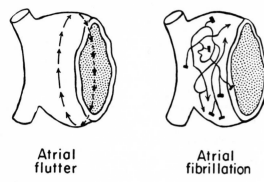

**Figure 14–9.** Pathways of impulses in atrial flutter and atrial fibrillation.

fibers and ventricular muscle is approximately 1/200 minute so that not over 200 impulses per minute can be transmitted into the ventricles. Therefore, when the atrium contracts as rapidly as 300 times per minute, only one out of every two impulses passes into the ventricles, thus causing the atria to beat at a rate two times that of the ventricles. The heart is then said to have a 2:1 rhythm. Occasionally, a 3:1 and, rarely, a 4:1 rhythm of the heart develops in the same manner.

**The "Chain Reaction" Mechanism of Fibrillation.** Fibrillation, whether it occurs in the atria or in the ventricles, is a very different condition from flutter. One can see many separate contractile waves spreading in different directions over the cardiac muscle at the same time in either atrial or ventricular fibrillation. Obviously, then, the circus movement in fibrillation is entirely different from that in flutter. One of the best ways to explain the mechanism of fibrillation is to describe the initiation of fibrillation by stimulation with 60 cycle alternating electrical current.

***Fibrillation Caused by 60 Cycle Alternating Current.*** At a central point in the ventricles of heart A in Figure 14–10 a 60 cycle electrical stimulus is applied through a stimulating electrode. The first cycle of the electrical stimulus causes a depolarization wave to spread in all directions, leaving all the muscle beneath the electrode in a refractory state. After about 0.25 second, this muscle begins to come out of the refractory state, some portions of the muscle coming out of refractoriness prior to other portions. This state of events is depicted in heart A by many light patches, which represent excitable cardiac muscle, and dark patches, which represent refractory muscle. New stimuli from the electrode can now cause impulses to travel in certain directions through the heart but not in all directions. It will be observed in heart A that certain impulses travel for short distances until they reach refractory areas of the heart and then are blocked. Other impulses, however, pass between the refractory areas and continue to travel in the excitable patches of muscle. Now, several events transpire in rapid succession, all

occurring simultaneously and eventuating in a state of fibrillation. These are:

First, block of the impulses in some directions but successful transmission in other directions creates one of the necessary conditions for a circus movement to develop—that is, *transmission of at least some of the depolarization waves around the heart in only one direction.* As a result, these waves do not annihilate themselves on the opposite side of the heart but can continue around and around the ventricles.

Second, the rapid stimulation of the heart causes two changes in the cardiac muscle itself, both of which predispose to circus movement: (1) The *velocity of conduction through the heart becomes decreased,* which allows a longer time interval for the impulses to travel around the heart. (2) The *refractory period of the muscle becomes shortened,* allowing re-entry of the impulse into previously excited heart muscle within a much shorter period of time than normally.

Third, one of the most important features of fibrillation is the *division of impulses,* as illustrated in heart A. When a depolarization wave reaches a refractory area in the heart, it travels to both sides around the area. Thus, a single impulse becomes two impulses. Then when each of these reaches another refractory area it, too, divides to form still two more impulses. In this way many different impulses are continually being formed in the heart by a progressive *chain reaction* until, finally, there are many impulses traveling in many different directions at the same time. Furthermore, this irregular pattern of impulse travel causes a *circuitous route for the impulses to travel, greatly lengthening the conductive pathway, which is one of the conditions leading to fibrillation.* It also results in a continual irregular pattern of patchy refractory areas in the heart. One can readily see that a vicious cycle has been initiated: more and more impulses are formed; these cause more and more patches of refractory muscle; and the refractory patches cause more and more division of the impulses. Therefore, any time a single area of

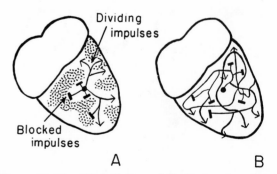

**Figure 14–10.** (A) Initiation of fibrillation in a heart when patches of refractory musculature are present. (B) Continued propagation of fibrillatory impulses in the fibrillating ventricle.

cardiac muscle comes out of refractoriness, an impulse is always close at hand to re-enter the area.

Heart B in Figure 14–10 illustrates the final state that develops in fibrillation. Here one can see many impulses traveling in all directions, some dividing and increasing the number of impulses while others are blocked entirely by refractory areas. In the final state of fibrillation, the number of new impulses being formed exactly equals the number of impulses that are being blocked by refractory areas. Thus, a steady state has developed with a certain average number of impulses traveling all the time in all directions through the cardiac syncytium.

As is the case with many other cardiac arrhythmias, fibrillation is usually confined to either the atria or the ventricles alone and not to both syncytial masses of muscle at the same time.

*Demonstration of the Chain Reaction Mechanism of Ventricular Fibrillation in the "Iron Heart."* In the human heart the chain reaction mechanism is difficult to demonstrate for two reasons: (1) It is impossible to follow exactly the wave fronts of the electrical impulses traveling through the heart muscle and (2) there are so many minute waves of contraction spreading at the same time that the eyes cannot follow these continually. An *iron heart model* has been developed in our laboratory that does show easily the chain reaction mechanism. This is a large iron bob suspended in nitric acid. Under appropriate conditions an oxide film develops on the surface of the iron. Then, a single electrical stimulus to the surface will cause a single excitation wave to travel over the entire surface comparable to normal stimulation of the heart. But multiple stimuli either all at once or in rapid succession will cause the typical chain reaction just described, with resultant fibrillation. One can easily see the chain reaction on the surface of the iron bob, and it can be recorded electrically or photographically. Also, typical fibrillatory bipolar and monopolar electrocardiographic patterns can be recorded.

**Atrial Fibrillation.** Atrial fibrillation is completely different from atrial flutter because the circus movement does not travel in a regular pathway. Instead, many different excitation waves can be seen to travel over the surface of the atria at the same time. Atrial fibrillation occurs very frequently when the atria become greatly overdilated—in fact, many times as frequently as flutter. When flutter does occur, it usually becomes fibrillation after a few days or weeks. To the right in Figure 14–9 are illustrated the pathways of fibrillatory impulses traveling through the atria.

Obviously, atrial fibrillation results in complete incoordination of atrial contraction so that atrial pumping ceases entirely.

*Effect of Atrial Fibrillation on the Overall Pumping Effectiveness of the Heart.* The normal function of the atria is to help fill the ventricles. However, the atria are probably responsible for not more than 25 to 30 per cent of the normal ventricular filling, which was explained in Chapter 13. Therefore, even when the atria fail to act as primer pumps because of atrial fibrillation, the ventricles can still fill enough that the effectiveness of the heart as a pump is reduced only 25 to 30 per cent, which is well within the "cardiac reserve" of all but severely weakened hearts. For this reason, atrial fibrillation can continue for many years without profound cardiac debility.

*Irregularity of Ventricular Rate During Atrial Fibrillation.* When the atria are fibrillating, impulses arrive at the A-V node rapidly but also irregularly. Since the A-V bundle will not pass a second impulse for approximately 0.3 second after a previous one, at least 0.3 second must elapse between one ventricular contraction and the next, and an additional interval of 0 to 0.6 second usually occurs before one of the irregular fibrillatory impulses happens to arrive at the A-V node. Thus, the interval between successive ventricular contractions varies from 0.3 second to about 0.9 second, causing a very irregular heart beat. In fact, this irregularity is one of the clinical findings used to diagnose the condition.

**Ventricular Fibrillation.** Ventricular fibrillation is extremely important because at least one quarter of all persons die in ventricular fibrillation. For instance, the hearts of most patients with coronary infarcts fibrillate shortly before death. In only a few instances on record have fibrillating human ventricles been known to return of their own accord to a rhythmic beat.

The likelihood of circus movements in the ventricles and, consequently, of ventricular fibrillation is greatly increased when the ventricles are dilated or when the rapidly conducting *Purkinje system is blocked* so that impulses cannot be transmitted rapidly. Also, *electric shock,* particularly with 60 cycle electric current, as discussed above, or *ectopic foci,* which are discussed below, are common initiating causes of ventricular fibrillation.

*Inability of the Heart to Pump Blood During Ventricular Fibrillation.* When the ventricles begin to fibrillate, the different parts of the ventricles no longer contract simultaneously. For the first few seconds the ventricular muscle undergoes rather coarse contractions which may pump a few milliliters of blood with each contraction. However, the impulses in the ventricles rapidly become divided into many much smaller impulses, and the contractions become so fine and asynchronous rather than coarse, that they pump no blood whatsoever. The ventricles dilate because of failure to pump the blood that is flowing into them, and within 60 to 90 seconds the ventricular muscle becomes too weak, because of lack of coronary blood supply, to contract strongly, even if coordinate contraction should return. Therefore, death is immediate when ventricular fibrillation begins.

*Irritable Foci as the Usual Cause of Ventricular Fibrillation.* The cause of ventricular fibrillation is most often an irritable focus in the ventricular muscle caused by coronary insufficiency or by compression from an arteriosclerotic plaque. Most impulses originating in an irritable focus travel around the heart in all directions, meet on the opposite side of the heart, and annihilate themselves. However, when impulses from a focus become frequent the refractory period of the stimulated muscle becomes shortened, impulse conduction becomes slowed, and patchy areas of refractoriness develop in the ventricles. These are the essential conditions for initiating the chain reaction mechanism, and fibrillation begins in the same manner as that described above for stimulation with a 60 cycle electrical current. Thus, any ectopic focus that emits rapid impulses is likely to initiate fibrillation.

*Electrical Defibrillation of the Ventricles.* Though a weak alternating current almost invariably throws the ventricles into fibrillation, a very strong electrical current passed through the ventricles for a short interval of time can stop fibrillation by throwing all the ventricular muscle into refractoriness simultaneously. This is accomplished by passing intense current through electrodes placed on two sides of the heart. The current penetrates most of the fibers of the ventricles, thus stimulating essentially all parts of the ventricles simultaneously and causing them to become refractory. All impulses stop, and the heart then remains quiescent for 3 to 5 seconds, after which it begins to beat again, with the S-A node or, often, some other part of the heart becoming the pacemaker. Occasionally, however, the same irritable focus that had originally thrown the ventricles into fibrillation is still present, and fibrillation begins again immediately.

When electrodes are applied directly to the two sides of the heart, fibrillation can usually be stopped with 70 to 100 volts of 60 cycle alternating current applied for $1/10$ second or 1000 volts direct current applied for a few thousandths of a second. When applied through the chest wall, as illustrated in Figure 14–11, approximately 440 volts of 60 cycle alternating current or several thousand volts direct current is required. In our laboratory the heart of a single anesthetized dog was defibrillated 130 times through the chest wall, and the animal remained in perfectly normal condition.

**Hand Pumping of the Heart ("Cardiac Massage") as an Aid to Defibrillation.** Unless defibrillated within one minute after fibrillation begins, the heart is usually too weak to be revived. However, it is still possible to revive the heart by preliminarily pumping it by hand. In this way small quantities of

blood are delivered into the aorta, and a renewed coronary blood supply develops. After a few minutes, electrical defibrillation often becomes possible. Indeed, fibrillating hearts have been pumped by hand as long as 90 minutes before defibrillation. In recent years, a technique of pumping the heart without opening the chest has been developed; this technique consists of intermittent, very powerful thrusts of pressure on the chest wall.

Lack of blood flow to the brain for more than 5 to 10 minutes usually results in permanent mental impairment or even total destruction of the brain. Even though the heart should be revived, the person might die from the effects of brain damage or live with permanent mental impairment.

## CARDIAC ARREST

When cardiac metabolism becomes greatly disturbed as a result of any one of many possible conditions, the rhythmic contractions of the heart occasionally stop. One of the most common causes of cardiac arrest is hypoxia of the heart, for severe hypoxia prevents the muscle fibers from maintaining normal ionic differentials across their membranes. Therefore, polarization of the membranes becomes reduced, and the excitability may be so affected that the automatic rhythmicity disappears.

Occasionally, patients with severe myocardial disease develop cardiac arrest, which obviously can lead to death. In some cases, however, rhythmic electrical impulses from an implanted electronic cardiac "pacemaker" have been successfully used in keeping patients alive for many years.

## REFERENCES

Bajusz, E., and Rona, G.: Myocardiology. Baltimore, University Park Press, 1971.
Cranefield, P. F.: The Conduction of the Cardiac Impulse. New York, Futura Publishing Co., Inc., 1975.
DeMaria, A. N., Vera, Z., Amsterdam, E. A., Mason, D. T., and Massumi, R. A.: Disturbances of cardiac rhythm and conduction induced by exercise. Diagnostic, prognostic, and therapeutic implications. *Amer. J. Cardiol.*, 33:732, 1974.
Dreifus, L. S., and Likoff, W.: Mechanisms and therapy of cardiac arrhythmias. In Dreifus, L. S., and Likoff, W. (eds.): Symposium on Cardiac Arrhythmias; the Twenty-fifth Hahnemann Symposium. New York, Grune & Stratton, 1973.
Fisch, C.: Relation of electrolyte disturbances to cardiac arrhythmias. *Circulation*, 47:408, 1973.
Gough, W., Dreifus, L. S., deAzevedo, I. M., and Katz, M. R.: Refractoriness of cardiac tissue. *Cardiovasc. Clin.*, 6:87, 1974.
Guyton, A. C., and Satterfield, J.: Factors concerned in electrical defibrillation of the heart, particularly through the unopened chest. *Amer. J. Physiol.*, 167:81, 1951.
Hayden, W. G., Hurley, E. J., and Rytand, D. A.: The mechanism of canine atrial flutter. *Circ. Res.*, 20:496, 1967.
Lau, S. H., and Damato, A. N.: Mechanisms of A-V block. *Cardiovasc. Clin.*, 2:49, 1970.
Levy, M. N., and Martin, P. J.: Cardiac excitation and contraction. *In* Guyton, A. C. (ed.): MTP International Review of Science: Physiology. Baltimore, University Park Press, 1974, Vol. 1, p. 49.

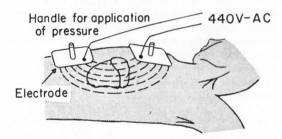

**Figure 14–11.** Application of alternating current to the chest to stop ventricular fibrillation.

Levy, M. N., Martin, P. J., Iano, T., and Zieske, H.: Paradoxical effect of vagus nerve stimulation on heart rate in dogs. *Circ. Res., 25*:303, 1969.

Levy, M. N., and Zieske, H.: Autonomic control of cardiac pacemaker activity and arterioventricular transmission. *J. Appl. Physiol., 27*:465, 1969.

Linden, R. J.: Recent Advances in Physiology. New York, Churchill Livingstone, Div. of Longman Inc., 1974.

Moe, G. K., and Mendez, C.: Physiologic basis of premature beats and sustained tachycardias. *N. Engl. J. Med., 288*:250, 1973.

Moss, A. J., and Patton, R. D.: Antiarrhythmic Agents. Springfield, Ill., Charles C Thomas, Publisher, 1973.

Pathak, C. L.: Autoregulation of chronotropic response of the heart through pacemaker stretch. *Cardiology, 58*:45, 1973.

Resnekov, L.: Circulatory effects of cardiac dysrhythmias. *Cardiovasc. Clin., 2*:23, 1970.

Rosen, M. R., Wit, A. L., and Hoffman, B. F.: Electrophysiology and pharmacology of cardiac arrhythmias. IV. Cardiac antiarrhythmic and toxic effects of digitalis. *Amer. Heart J. 89*:391, 1975.

Rusy, B. F.: Pharmacology of antiarrhythmic drugs. *Med. Clin. North Amer., 58*:987, 1974.

Samet, P.: Hemodynamic sequelae of cardiac arrhythmias. *Circulation, 47*:399, 1973.

Sawyer, D. C.: Effect of anesthetic agents on cardiovascular function and cardiac rhythm. *Vet. Clin. N. Amer., 3*:25, 1973.

Schler, A. M.: Excitation of the heart. *In* Hamilton, W. F. (ed.): Handbook of Physiology. Baltimore, The Williams & Wilkins Co., 1962, Vol. 1, Sec. 2, p. 287.

Scherf, D., and Cohen, J.: The Atrioventricular Node and Selected Cardiac Arrhythmias. New York, Grune & Stratton, Inc., 1964.

Tsagaris, T. J., Sutton, R. B., and Kuida, H.: Hemodynamic effects of varying pacemaker sites. *Amer. J. Physiol., 218*:246, 1970.

Tyberg, J. V., Parmley, W. W., and Sonnenblick, E. H.: In-vitro studies of myocardial asynchrony and regional hypoxia. *Circ. Res., 25*:569, 1969.

Vick, R. L.: Effects of increased transmural pressures upon atrial and ventricular rhythms in the dog heart-lung preparation. *Circ. Res., 13*:39, 1963.

Watanabe, Y., and Dreifus, L. S.: Arrhythmias—Mechanisms and pathogenesis. *In* Sodeman, W. A., Jr., and Sodeman, W. A. (eds.): Pathologic Physiology: Mechanisms of Disease, 5th Ed. Philadelphia, W. B. Saunders Company, 1974, p. 329.

# 15

# The Normal Electrocardiogram

Transmission of the depolarization wave, also commonly called the *cardiac impulse,* through the heart has been discussed in detail in Chapter 14. As the wave passes through the heart, electrical currents spread into the tissues surrounding the heart, and a small proportion of these spreads all the way to the surface of the body. If electrodes are placed on the body on opposite sides of the heart, the electrical potentials generated by the heart can be recorded; the recording is known as an *electrocardiogram.* A normal electrocardiogram for two beats of the heart is illustrated in Figure 15–1.

## CHARACTERISTICS OF THE NORMAL ELECTROCARDIOGRAM

The normal electrocardiogram is composed of a P wave, a "QRS complex," and a T wave. The QRS complex is actually three separate waves, the Q wave, the R wave, and the S wave.

The P wave is caused by electrical currents generated as the atria depolarize prior to contraction, and the QRS complex is caused by currents generated when the ventricles depolarize prior to contraction,

that is, as the depolarization wave spreads through the ventricles. Therefore, both the P wave and the components of the QRS complex are *depolarization waves.* The T wave is caused by currents generated as the ventricles recover from the state of depolarization. This process occurs in ventricular muscle 0.25 to 0.30 second after depolarization, and this wave is known as a *repolarization wave.* Thus, the electrocardiogram is composed of both depolarization and repolarization waves. The principles of depolarization and repolarization were discussed in Chapter 10. However, the distinction between depolarization waves and repolarization waves is so important in electrocardiography that further clarification is needed, as follows:

### DEPOLARIZATION WAVES VERSUS REPOLARIZATION WAVES

Figure 15–2 illustrates a muscle fiber in four different stages of depolarization and repolarization. During the process of "depolarization" the normal negative potential inside the fiber is lost and the membrane potential actually reverses; that is, it becomes slightly positive inside and negative outside.

In Figure 15–2A the process of depolarization, illustrated by positivity inside and negativity outside, is traveling from left to right, and the first half of the fiber has already depolarized while the remaining half is still polarized. Therefore, the left electrode on the fiber is in an area of negativity where it touches the outside of the fiber, while the right electrode is in an area of positivity; this causes the meter to record positively. To the right of the muscle fiber is illustrated a record of the potential between the electrodes as recorded by a high-speed recording meter at this particular stage of depolarization. Note that when depolarization has reached this halfway mark, the record has risen to a maximum positive value.

In Figure 15–2B depolarization has extended over the entire muscle fiber, and the recording to the right has returned to the zero base line because both elec-

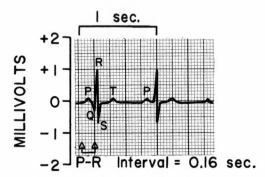

**Figure 15–1.**   The normal electrocardiogram.

190

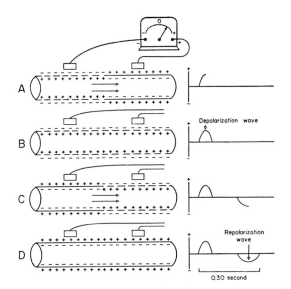

**Figure 15-2.** Recording the *depolarization wave* and the *repolarization* wave from a cardiac muscle fiber.

trodes are now in areas of equal negativity. The completed wave is a *depolarization wave* because it results from spread of depolarization along the extent of the muscle fiber.

Figure 15-2C illustrates the repolarization process in the muscle fiber, which has proceeded halfway along the extent of the fiber from left to right. At this point, the left electrode is in an area of positivity, while the right electrode is in an area of negativity. This is opposite to the polarity in Figure 15-2A. Consequently, the recording, as illustrated to the right, becomes negative.

Finally, in Figure 15-2D the muscle fiber has completely repolarized, and both electrodes are in areas of positivity so that no potential is recorded between them. Thus, in the recording to the right, the potential returns once more to the zero level. This completed negative wave is a *repolarization wave* because it results from spread of the repolarization process over the muscle fiber.

**Relationship of the Monophasic Action Potential of Cardiac Muscle to the QRS and T Waves.** The monophasic action potential of ventricular muscle, which was discussed in the preceding chapter, normally lasts between 0.25 and 0.30 second. The top part of Figure 15-3 illustrates a monophasic action potential recorded from a microelectrode inserted to the inside of a single ventricular muscle fiber. The upsweep of this action potential is caused by *depolarization,* and the return of the potential to the base line is caused by *repolarization.*

Note below the simultaneous recording of the electrocardiogram from this same ventricle, which shows the QRS wave appearing at the beginning of the monophasic action potential and the T wave appearing at the end. Note especially that *no potential at all*

*is recorded in the electrocardiogram when the ventricular muscle is either completely polarized or completely depolarized.* It is only when the muscle is partly polarized and partly depolarized that current flows from one part of the ventricles to another part and therefore also flows to the surface of the body to cause the electrocardiogram.

## RELATIONSHIP OF ATRIAL AND VENTRICULAR CONTRACTION TO THE WAVES OF THE ELECTROCARDIOGRAM

Before contraction of muscle can occur, a depolarization wave must spread through the muscle to initiate the chemical processes of contraction. The P wave results from spread of the depolarization wave through the atria, and the QRS wave from spread of the depolarization wave through the ventricles. Therefore, the P wave occurs immediately before the *beginning of contraction of the atria,* and the QRS wave occurs immediately before the *beginning of contraction of the ventricles.* Both the atria and the ventricles remain contracted until a few milliseconds after repolarization has occurred, that is, until after the end of the T wave.

The atria repolarize approximately 0.10 to 0.20 second after the depolarization wave. However, this is just at the moment that the QRS wave is being recorded in the electrocardiogram. Therefore, the atrial repolarization wave, known as the *atrial T wave,* is usually totally obscured by the much larger QRS wave. For this reason, an atrial T wave is rarely observed in the electrocardiogram.

On the other hand, the ventricular repolarization wave is the T wave of the normal electrocardiogram. Ordinarily, ventricular muscle begins to repolarize in

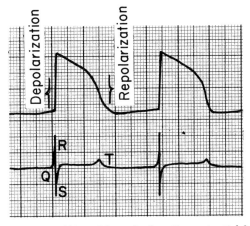

**Figure 15-3.** *Above:* Monophasic action potential from a ventricular muscle fiber during normal cardiac function, showing rapid depolarization and then repolarization occurring slowly during the plateau stage but very rapidly toward the end. *Below:* Electrocardiogram recorded simultaneously.

some fibers approximately 0.15 second after the beginning of the depolarization wave and completes its repolarization in all fibers approximately 0.30 second after onset of depolarization. Thus, the process of repolarization extends over a fairly long period of time, about 0.15 second. For this reason the T wave in the normal electrocardiogram is a fairly prolonged wave, but the voltage of the T wave is considerably less than the voltage of the QRS complex, partly because of its prolonged length.

## VOLTAGE AND TIME CALIBRATION OF THE ELECTROCARDIOGRAM

All recordings of electrocardiograms are made with appropriate calibration lines on the recording paper. Either these calibration lines are already ruled on the paper, as is the case when a pen recorder is used, or they are recorded on the paper at the same time that the electrocardiogram is recorded, which is the case with the photographic types of electrocardiographs.

As illustrated in Figure 15–1, the calibration lines are arranged so that 10 small divisions in the vertical direction in the standard electrocardiogram represent 1 millivolt, with positivity in the upward direction and negativity in the downward direction.

The vertical lines on the electrocardiogram are time calibration lines. Each inch in the horizontal direction is 1 second, and each inch in turn is usually broken into five segments by dark vertical lines; the intervals between these lines represent 0.20 second. These intervals are then broken into five smaller intervals by thin lines, and each of these represents 0.04 second.

*Normal Voltages in the Electrocardiogram.* The voltages of the waves in the normal electrocardiogram depend on the manner in which the electrodes are applied to the surface of the body. When one electrode is placed directly over the heart and the second electrode is placed elsewhere on the body, the voltage of the QRS complex may be as much as 3 to 4 millivolts (mv.). Even this voltage is very small in comparison with the monophasic action potential of 120 mv. recorded directly from heart muscle. When electrocardiograms are recorded from electrodes on the two arms or on one arm and one leg, the voltage of the QRS complex usually is approximately 1 mv. from the top of the R wave to the bottom of the S wave, the voltage of the P wave between 0.1 and 0.3 mv., and that of the T wave between 0.2 and 0.3 mv.

**The P-Q or P-R Interval.** The duration of time between the beginning of the P wave and the beginning of the QRS wave is the interval between the beginning of contraction of the atrium and the beginning of contraction of the ventricle. This period of time is called the P-Q interval. The normal P-Q interval is approximately 0.16 second. This interval is sometimes also called the P-R interval because the Q wave is frequently absent.

**The Q-T Interval.** Contraction of the ventricle lasts essentially between the beginning of the Q wave and the end of the T wave. This interval of time is called the Q-T interval and ordinarily is approximately 0.30 second.

**The Rate of the Heart as Determined from Electrocardiograms.** The rate of heartbeat can be determined easily from electrocardiograms, because the time interval between two successive beats is the reciprocal of the heart rate. If the interval between two beats as determined from the time calibration lines is 1 second, the heart rate is 60 beats per minute. The normal interval between two successive QRS complexes is approximately 0.83 second. This is a heart rate of 60/0.83 times per minute, or 72 beats per minute.

# METHODS FOR RECORDING ELECTROCARDIOGRAMS

The electrical currents generated by the cardiac muscle during each beat of the heart sometimes change potentials and polarity in less than 0.01 second. Therefore, it is essential that any apparatus for recording electrocardiograms be capable of responding rapidly to these changes in electrical potentials. In general, two different types of recording apparatuses are used for this purpose, as follows:

## THE PEN RECORDER

Most older types of electrocardiographic apparatus used optical and photographic methods for recording the electrocardiogram. In these a beam of light was focused on a mirror mounted on a galvanometer coil located in a powerful magnetic field. When current flowed through the coil, the coil rotated in the field, and the mirror swept the reflected beam of light across a moving photographic paper. However, because the records required photographic development before the recording could be viewed, direct pen writing recorders have become the vogue in recent years.

The pen writing recorders write the electrocardiogram with a pen directly on a moving sheet of paper. The pen is often a thin tube connected at one end to an inkwell, and its recording end is connected to a powerful electromagnet system that is capable of moving the pen back and forth at high speed. As the paper moves forward, the pen records the electrocardiogram. The movement of the pen in turn is controlled by means of appropriate amplifiers connected to electrocardiographic electrodes on the subject.

Other pen recording systems use special paper that does not require ink in the recording stylus. One such paper turns black when it is exposed to heat; the stylus itself is made very hot by electrical current flowing through its tip. Another type turns black when electrical current flows from the tip of the stylus through the paper to an electrode at its back.

This leaves a black line at every point on the paper that the stylus touches.

### RECORDING ELECTROCARDIOGRAMS WITH THE OSCILLOSCOPE

Electrocardiograms can also be viewed on the screen of an oscilloscope by the method discussed for nerve potentials in Chapter 10, or they can be photographed from the oscilloscopic screen. However, because of the cost of the oscilloscope, and because extremely high frequency electrical potentials do not need to be recorded, the less complicated and less expensive pen recorder just described is ordinarily used in clinical electrocardiography.

# FLOW OF CURRENT AROUND THE HEART DURING THE CARDIAC CYCLE

### RECORDING ELECTRICAL POTENTIALS FROM A PARTIALLY DEPOLARIZED MASS OF SYNCYTIAL CARDIAC MUSCLE

Figure 15–4 illustrates a syncytial mass of cardiac muscle that has been stimulated at its centralmost point. Prior to stimulation, all of the exteriors of the muscle cells had been positive and the interiors negative. However, for reasons presented in Chapter 10 in the discussion of membrane potentials, as soon as an area of the cardiac syncytium becomes depolarized, negative charges leak to the outside of the depolarized area, making this surface area electronegative, as represented by the negative signs, with respect to the remaining surface of the heart which is still polarized in the normal manner, as represented by the positive signs. Therefore, a meter connected with its negative terminal on the area of depolarization and its positive terminal on one of the still-polarized areas, as illustrated to the right in the figure, will record positively.

Two other possible electrode placements and meter readings are also illustrated in Figure 15–4. *These should be studied carefully.* Obviously, since the process of depolarization spreads in all directions through the heart, the potential differences shown in the figure last for only a few milliseconds, and the actual voltage measurements can be accomplished only with a high-speed recording apparatus.

### FLOW OF ELECTRICAL CURRENTS AROUND THE HEART IN THE CHEST

Figure 15–5 illustrates the ventricular muscle mass lying within the chest. Even the lungs, though filled with air, conduct electricity to a surprising extent, and fluids of the other tissues surrounding the heart conduct electricity even more easily. Therefore, the

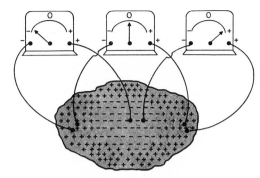

**Figure 15–4.** Instantaneous potentials developed on the surface of a cardiac muscle mass that has been depolarized in its center.

heart is actually suspended in a conductive medium. When one portion of the ventricles becomes electronegative with respect to the remainder, electrical current flows from the depolarized area to the polarized area in large circuitous routes, as noted in the figure.

It will be recalled from the discussion of the Purkinje system in Chapter 14 that the cardiac impulse first arrives in the ventricles in the septum and shortly thereafter on the endocardial surfaces of the remainder of the ventricles, as shown by the shaded areas and the negative signs in Figure 15–5. This provides electronegativity on the insides of the ventricles and electropositivity on the outer walls of the ventricles. If one algebraically averages all the lines of current flow (the elliptical lines in Figure 15–5), he finds that the average current flow is from the base of the heart toward the apex. During most of the remainder of the cycle of depolarization, the current continues to flow in this direction as the impulse

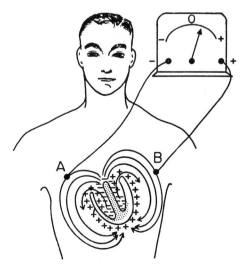

**Figure 15–5.** Flow of current in the chest around a partially depolarized heart.

spreads from the endocardial surface outward through the ventricular muscle. However, immediately before the depolarization wave has completed its course through the ventricles, the direction of current flow reverses for about 1/100 second, flowing then from the apex toward the base because the very last part of the heart to become depolarized is the outer walls of the ventricles near the base of the heart.

Thus, in the normal heart it may be considered that current flows primarily in the direction from the base toward the apex during almost the entire cycle of depolarization except at the very end. Therefore, if a meter is connected to the surface of the body as shown in Figure 15–5, the electrode nearer the base will be negative with respect to the electrode nearer the apex, so that the meter shows a slight potential between the two electrodes. In making electrocardiographic recordings, various standard positions for placement of electrodes are used, and whether the polarity of the recording during each cardiac cycle is positive or negative is determined by the orientation of electrodes with respect to the current flow in the heart. Some of the conventional electrode systems, commonly called *electrocardiographic leads,* are discussed below.

# ELECTROCARDIOGRAPHIC LEADS

### THE THREE STANDARD LIMB LEADS

Figure 15–6 illustrates electrical connections between the limbs and the electrocardiograph for recording electrocardiograms from the so-called "standard" limb leads. The electrocardiograph in each instance is illustrated by mechanical meters in the diagram, though the actual electrocardiograph is a high-speed recording meter.

**Lead I.** In recording limb lead I, the *negative terminal of the electrocardiograph is connected to the right arm* and the *positive terminal to the left arm.* Therefore, when the point on the chest where the right arm connects to the chest is electronegative with respect to the point where the left arm connects, the electrocardiograph records positively—that is, above the zero voltage line in the electrocardiogram. When the opposite is true, the electrocardiograph records below the line.

**Lead II.** In recording limb lead II, the *negative terminal of the electrocardiograph is connected to the right arm* and the *positive terminal to the left leg.* Therefore, when the right arm is negative with respect to the left leg, the electrocardiograph records positively.

**Lead III.** In recording limb lead III, the *negative terminal of the electrocardiograph is connected to the left arm* and the *positive terminal to the left leg.* This means that the electrocardiograph records positively when the left arm is negative with respect to the left leg.

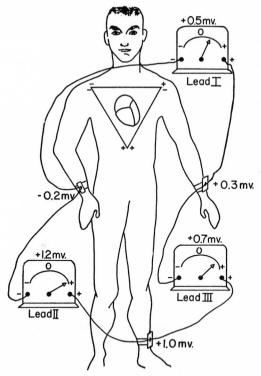

**Figure 15–6.** Conventional arrangement of electrodes for recording the standard electrocardiographic leads. Einthoven's triangle is superimposed on the chest.

**Einthoven's Triangle.** In Figure 15–6, a triangle, called *Einthoven's triangle,* is drawn around the area of the heart. This is a diagrammatic means of illustrating that the two arms and the left leg form apices of a triangle surrounding the heart. The two apices at the upper part of the triangle represent the points at which the two arms connect electrically with the fluids around the heart, and the lower apex is the point at which the left leg connects with the fluids.

*Einthoven's Law.* Einthoven's law states simply that if the electrical potentials of any two of the three standard electrocardiographic leads are known at any given instant, the third one can be determined mathematically from the first two by simply summing the first two. For instance, let us assume that momentarily, as noted in Figure 15–6, the right arm is 0.2 mv. negative with respect to the average potential in the body, the left arm is 0.3 mv. positive, and the left leg is 1.0 mv. positive. Observing the meters in the figure, it will be seen that lead I records a positive potential of 0.5 mv., lead III records a positive potential of 0.7 mv., and lead II records a positive potential of 1.2 mv., because these are the instantaneous potential differences between the respective pairs of limbs.

Note that *the sum of the voltages in leads I and III equals the voltage in lead II.* That is, 0.5 plus 0.7

equals 1.2. Mathematically, this principle, called Einthoven's law, holds true at any given instant while the electrocardiogram is being recorded.

**Normal Electrocardiograms Recorded by the Three Standard Leads.** Figure 15–7 illustrates simultaneous recordings of the electrocardiogram in leads I, II, and III. It is obvious from this figure that the electrocardiograms in these three standard leads are very similar to each other, for they all record positive *P* waves and positive *T* waves, and the major portion of the QRS complex is also positive in each electrocardiogram.

On analysis of the three electrocardiograms, it can be shown with careful measurements that at any given instant the sum of the potentials in leads I and III equals the potential in lead II, thus illustrating the validity of Einthoven's law.

Because all normal electrocardiograms are very similar to each other, it does not matter greatly which electrocardiographic lead is recorded when one wishes to diagnose the different cardiac arrhythmias, for diagnosis of arrhythmias depends mainly on the time relationships between the different waves of the cardiac cycle. On the other hand, when one wishes to determine the extent and type of damage in the ventricular or atrial muscle, it does matter greatly which leads are recorded, for abnormalities of the cardiac muscle change the patterns of the electrocardiograms markedly in some leads and yet may not affect other leads.

Electrocardiographic interpretation of these two types of conditions—cardiac myopathies and cardiac arrhythmias—are discussed separately in the following two chapters.

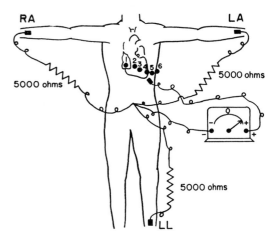

**Figure 15–8.** Connections of the body with the electrocardiograph for recording chest leads. (Modified from Burch and Winsor: A Primer of Electrocardiography. Lea & Febiger.)

## CHEST LEADS (PRECORDIAL LEADS)

Often electrocardiograms are recorded with one electrode placed on the anterior aspect of the chest over the heart, as illustrated by the six separate points in Figure 15–8. This electrode is connected to the positive terminal of the electrocardiograph, and the negative electrode, called the *indifferent electrode,* is normally connected simultaneously through electrical resistances to the right arm, left arm, and left leg, as also shown in the figure. Usually six different standard chest leads are recorded from the anterior chest wall, the chest electrode being placed respectively at the six points illustrated in the diagram. The different leads recorded by the method illustrated in Figure 15–8 are known as leads $V_1$, $V_2$, $V_3$, $V_4$, $V_5$, and $V_6$.

Figure 15–9 illustrates the electrocardiograms of the normal heart as recorded in the six standard chest leads. Because the heart surfaces are close to the chest wall, each chest lead records mainly the electrical potential of the cardiac musculature immediately beneath the electrode. Therefore, relatively minute abnormalities in the ventricles, particularly in the anterior ventricular wall, frequently cause

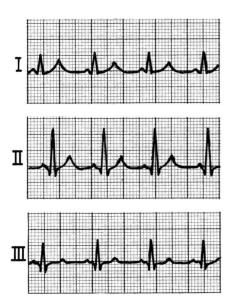

**Figure 15–7.** Normal electrocardiograms recorded from the three standard electrocardiographic leads.

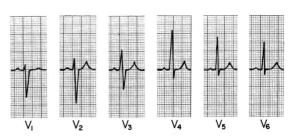

**Figure 15–9.** Normal electrocardiograms recorded from the six standard chest leads.

marked changes in the electrocardiograms recorded from chest leads.

In leads $V_1$ and $V_2$, the QRS recordings of the normal heart are mainly negative because, as illustrated in Figure 15–8, the chest electrode in these leads is nearer the base of the heart than the apex, which is the direction of electronegativity during most of the ventricular depolarization process. On the other hand, the QRS complexes in leads $V_4$, $V_5$, and $V_6$ are mainly positive because the chest electrode in these leads is nearer the apex, which is the direction of electropositivity during depolarization.

## *AUGMENTED UNIPOLAR LIMB LEADS*

Another system of leads in wide use is the "augmented unipolar limb lead." In this type of recording, two of the limbs are connected through electrical resistances to the negative terminal of the electrocardiograph while the third limb is connected to the positive terminal. When the positive terminal is on the right arm, the lead is known as the $aV_R$ lead; when on the left arm, the $aV_L$ lead; and when on the left leg, the $aV_F$ lead.

Normal recordings of the augmented unipolar limb leads are shown in Figure 15–10. These are all similar to the standard limb lead recordings except that the $aV_R$ lead is inverted. The reason for this inversion is that the polarity of the electrocardiograph in this instance is connected backward to the major direction of current flow in the heart during the cardiac cycle.

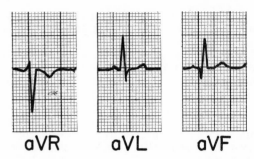

**Figure 15–10.** Normal electrocardiograms recorded from the three augmented unipolar limb leads.

Each augmented unipolar limb lead records the potential of the heart on the side nearest to the respective limb. Thus, when the recording in the $aV_R$ lead is negative, this means that the side of the heart nearest to the right arm is negative in relation to the remainder of the heart; when the recording in the $aV_F$ lead is positive, this means that the apex of the heart is positive with respect to the remainder of the heart.

## REFERENCES

See bibliography at end of Chapter 17.

# 16

# Electrocardiographic Interpretation in Cardiac Myopathies—Vectorial Analysis

From the discussion in Chapter 14 of impulse transmission through the heart, it is obvious that any change in the pattern of this transmission can cause abnormal electrical currents around the heart and, consequently, can alter the shapes of the waves in the electrocardiogram. For this reason, almost all serious abnormalities of the heart muscle can be detected by analyzing the contours of the different waves in the different electrocardiographic leads. The purpose of the present chapter is to discuss the various alterations in the electrocardiograms when either the muscle of the heart or the conduction system, especially of the ventricles, functions abnormally.

## PRINCIPLES OF VECTORIAL ANALYSIS OF ELECTROCARDIOGRAMS

### USE OF VECTORS TO REPRESENT ELECTRICAL POTENTIALS

Before it is possible for one to understand how cardiac abnormalities affect the contours of waves in the electrocardiogram, he must first become familiar with the concept of vectors and vectorial analysis as applied to electrical currents flowing in and around the heart.

Several times in the preceding chapter it was pointed out that heart currents flow in a particular direction at a given instant in the cardiac cycle. A vector is an arrow that points in the direction of current flow *with the arrowpoint in the positive direction*. Also, by convention, the length of the arrow is drawn *proportional to the voltage generated by the current flow*.

In Figure 16–1, several different vectors are shown between areas of negativity and positivity in a syncytial mass of cardiac muscle. Note, first, that each vector points from negative to positive and, second, that the length of each vector is proportional to the amount of charges causing the current flow.

**The Summated Vector in the Heart at Any Given Instant.** Figure 16–2 shows, via the shaded area and the negative signs, depolarization of the ventricular septum and parts of the lateral endocardial walls of the two ventricles. Electrical currents flow from the septum and from the lateral endocardial walls to the outside of the heart as indicated by the elliptical arrows. Currents also flow inside the heart chambers directly from the depolarized areas toward the polarized areas. Even though a small amount of current flows upward inside the heart, a considerably

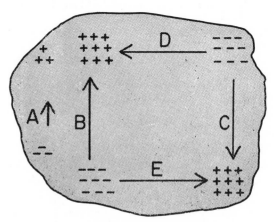

**Figure 16–1.** Vectors depicting current flow within a mass of cardiac muscle. Discussed in the text.

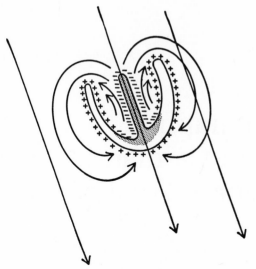

**Figure 16–2.** A summated vector through the partially depolarized heart. Two additional vectors are shown, one to each side of the heart. These have the same significance as the vector through the center of the heart (discussed in the text).

greater quantity flows downward on the outside of the ventricles toward the apex. Therefore, the summated vector of currents at this particular instant is drawn through the center of the ventricles in a direction from the base of the heart toward the apex. Furthermore, because these currents are considerable in quantity, the vector is relatively long.

**Transposition of Vectors in Space.** Two other vectors besides the one through the center of the heart are shown in Figure 16–2. These vectors extend in exactly the same direction as that through the center of the heart, and they are exactly the same length as the vector through the heart. They also have exactly the same meaning as the single vector drawn through the center of the heart, because a vector indicates only (1) the direction of current flow and (2) the potential caused by the current. Therefore, the vector can be drawn to pass directly through the center of the heart, or it can be drawn to the side of or at one end of the heart. As long as the direction and the length of the vector are appropriate, its meaning remains exactly the same regardless of its position with respect to the heart.

## DENOTING THE DIRECTION OF A VECTOR IN TERMS OF DEGREES

When a vector is horizontal and directed toward the subject's left side, it is said that the vector extends in the direction of 0 degrees, as is illustrated in Figure 16–3. From this zero reference point, the scale of vectors rotates clockwise: when the vector extends from above downward, it has a direction of

90 degrees; when it extends from the subject's left to his right, it has a direcron of 180 degrees; and, when it extends upward, it has a direction of −90 or +270 degrees.

In a normal heart the average direction of the vector of the heart during spread of the depolarization wave is approximately 59 degrees, which is illustrated by vector *A* drawn from the center of Figure 16–3 in the 59 degree direction. This means that during most of the depolarization wave, the apex of the heart remains positive with respect to the base of the heart, as is discussed later in the chapter.

## "AXIS" OF EACH OF THE STANDARD AND UNIPOLAR LEADS

In the preceding chapter the three standard leads and the three unipolar leads were described. Each lead is actually a pair of electrodes connected to the body on opposite sides of the heart, and the direction from the negative to the positive electrode is called the *axis of the lead*. Lead I is recorded from two electrodes placed respectively on the two arms. Since the electrodes lie in the horizontal direction with the positive electrode to the left, the axis of lead I is 0 degrees.

In recording lead II, electrodes are placed on the right arm and left leg. The right arm connects to the torso in the upper right-hand corner and the left leg to the lower left-hand corner. Therefore, the direction of this lead is approximately 60 degrees.

By a similar analysis it can be seen that lead III has an axis of approximately 120 degrees, lead $aV_R$ 210 degrees, $aV_F$ 90 degrees, and $aV_L$ −30 degrees. The directions of the axes of all these different leads are shown in Figure 16–4. Also, the polarities of the electrodes are illustrated by the plus and minus signs.

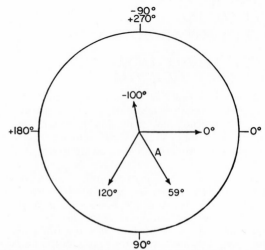

**Figure 16–3.** Vectors drawn to represent direction of current flow and potentials for several different hearts.

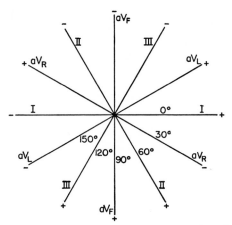

**Figure 16–4.** Axes of the three standard leads and of the three unipolar leads.

*The student must learn these axes and their polarities, particularly for the standard leads I, II, and III, if he wishes to understand the remainder of this chapter.*

## VECTORIAL ANALYSIS OF POTENTIALS RECORDED IN DIFFERENT LEADS

Now that we have discussed, first, the conventions for representing current flow and potentials across the heart by means of vectors and, second, the axes of the leads, it becomes possible to put these two together to determine the potential that will be recorded in each lead for a given vector in the heart.

Figure 16–5 illustrates a partially depolarized heart; vector *A* represents the direction of current flow in the heart and its potential. In this instance the direction of current flow is 55 degrees, and the potential will be assumed to be 2 mv. This vector *A* is now moved to a convenient point below the heart, but its direction and length remain exactly the same. Through its base is drawn the axis of lead I in the 0 degree direction. From the tip of vector *A* a perpen-

dicular is dropped to the lead I axis, and a so-called *resultant vector (B)* is drawn along the axis. The head of this vector is in the positive direction, which means that the record momentarily being recorded in the electrocardiogram of lead I will be positive. The voltage recorded will be equal to length of *B* divided by length of *A* times 2 mv., or approximately 1 mv.

In Figure 16–6, vector *A* represents the current flow at an instantaneous point during ventricular depolarization in another heart in which the left side of the heart becomes depolarized somewhat more rapidly than the right. In this instance the vector has a direction of 100 degrees, and the voltage is again 2 mv. To determine the potential in lead I, we drop a perpendicular to the lead I axis and find the resultant vector *B*. Vector *B* is very short and this time in the negative direction, indicating that at this particular instant the recording in lead I will be negative (below the zero line), and the voltage recorded will be slight. This figure illustrates that *when the vector in the heart is in a direction almost perpendicular to the axis of the lead, the voltage recorded in the electrocardiogram of this lead is very low.* On the other hand, *when the vector has almost exactly the same axis as the lead, essentially the entire voltage of the vector will be recorded.*

**Vectorial Analysis of Potentials in the Three Standard Leads.** In Figure 16–7, vector *A* depicts the instantaneous direction and potential of current in a partially depolarized ventricle. To determine the potential recorded at this instant in the electrocardiogram of each one of the three standard leads, perpendiculars are dropped to all the lines representing the different leads as illustrated in the figure. The resultant vector *B* depicts the potential recorded in lead I, vector *C* depicts the potential in lead II, and vector *D* in lead III. In each of these the record in the electrocardiogram is positive—that is, above the zero line—because the resultant vectors lie in the positive directions along the axes of the leads. The potential in lead I is approximately one-half that of the vector through the heart; in lead II it is almost exactly equal to that in the heart; and in lead III it is about one-third that in the heart.

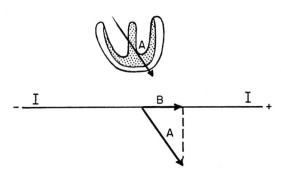

**Figure 16–5.** Determination of a resultant vector *B* along the axis of lead I when vector *A* represents the current flow in the ventricles.

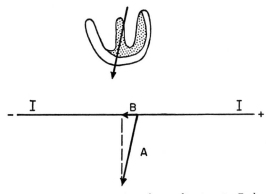

**Figure 16–6.** Determination of a resultant vector *B* along the axis of lead I when vector *A* represents the current flow in the ventricles.

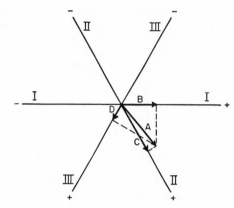

**Figure 16–7.** Determination of resultant vectors in leads I, II, and III when vector *A* represents the current flow in the ventricles.

An identical analysis can be used to determine potentials recorded in augmented limb leads, except that the respective axes of these leads (see Figure 16–4) are used in place of the standard lead axes used in Figure 16–7.

# VECTORIAL ANALYSIS OF THE NORMAL ELECTROCARDIOGRAM

## VECTORS OCCURRING DURING DEPOLARIZATION OF THE VENTRICLES — THE QRS COMPLEX

When the cardiac impulse enters the ventricles through the A-V bundle, the first part of the ventricles to become depolarized is the left endocardial surface of the septum. This depolarization spreads rapidly to involve both endocardial surfaces of the septum, as illustrated by the shaded portion of the ventricle in Figure 16–8A. Then the depolarization spreads along the endocardial surfaces of the two ventricles, as shown in Figures 16–8B and C. Finally, it spreads through the ventricular muscle to the outside of the heart, as shown progressively in Figures 16–8C, D, and E.

At each stage of depolarization of the ventricles in Figures 16–8A to E, the instantaneous, summated current flow is represented by a vector superimposed on the ventricle in each figure. Each of these vectors is analyzed by the method described in the preceding section to determine the voltages that will be recorded at each instant in each of the three standard electrocardiographic leads. To the right is shown the progressive development of the QRS complex. *Keep in mind that a positive vector in a lead will cause the recording in the electrocardiogram to be above the zero line, whereas a negative vector will cause the recording to be below the line.*

*Before proceeding with any further consideration of vectorial analysis, it is essential that this analysis of the successive normal vectors presented in Figure 16–8 be understood.* Each of these analyses should be studied in detail by the procedure given above. A short summary of this sequence is the following:

In Figure 16–8A the vector is short because only a small portion of the ventricles—the septum—is depolarized. Therefore, all electrocardiographic voltages are low at this instant. The voltage in lead II is greater than the voltages in leads I and III because the vector extends mainly in the same direction as the axis of lead II.

In Figure 16–8B the vector is long because much of the ventricles has now become depolarized. Therefore, the voltages in all electrocardiographic leads have increased.

In Figure 16–8C the vector is becoming shorter and the recorded EKG voltages less because the outside of the apex of the heart is now electronegative, neutralizing much of the negativity on the endocardial surfaces of the heart. Also, the axis of the vector is shifting toward the left side of the chest because the left ventricle is slightly slower to depolarize than the right. The ratio of the voltage in lead I to that in lead III is increasing.

In Figure 16–8D the vector points toward the base of the left ventricle, and it is short because only a minute portion of the ventricular muscle is still polarized. Because of the direction of the vector at this time, the voltages recorded in leads II and III are both negative—that is, below the line.

In Figure 16–8E the entire ventricular muscle mass is depolarized so that no current flows around the heart at all. The vector becomes zero, and the voltages in all leads become zero.

Thus, the QRS complexes are recorded in the three standard electrocardiographic leads. The QRS complex sometimes has a slight negative depression at its beginning, which is not shown in Figure 16–8; this is the Q wave. When it occurs, it is caused by initial depolarization of the septum nearer to the apex of the septum than to the base and/or depolarization of the left side of the septum before the right side, which occasionally creates a weak vector in the apex-to-base direction or from left to right for a minute fraction of a second before the usual base-to-apex vector occurs. The major positive deflection shown in Figure 16–8 is the R wave, and the final negative deflection is the S wave.

## THE ELECTROCARDIOGRAM DURING REPOLARIZATION— THE T WAVE

Once ventricular muscle becomes depolarized, approximately 0.15 second elapses before sufficient repolarization begins for it to be observed in the electrocardiogram, and repolarization is complete in about 0.30 second. It is this process of repolarization

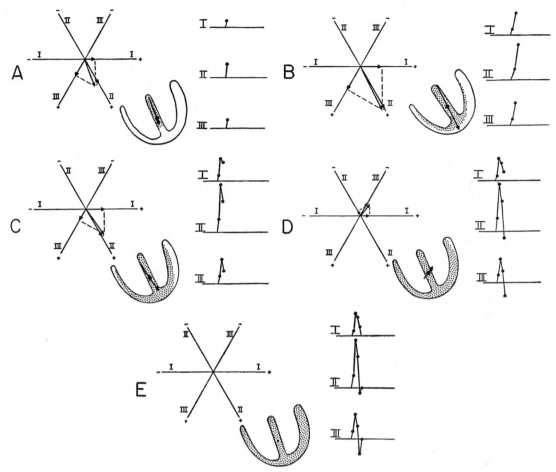

**Figure 16–8.** (A) The ventricular vectors and the QRS complexes 0.01 second after onset of ventricular depolarization, (B) 0.03 second after onset of depolarization, (C) 0.05 second after onset of depolarization, (D) 0.06 second after onset of depolarization, and (E) after depolarization of the ventricles is complete.

that causes the T wave in the electrocardiogram. Because the septum depolarizes first, it seems logical that it should repolarize first, but this is not the usual case because the septum has a longer period of contraction and depolarization than do other areas of the heart. Actually, many sections of the ventricles begin to repolarize almost simultaneously. Yet *the greatest portion of ventricular muscle to repolarize first is that located over the entire outer surface of the ventricles and especially near the apex of the heart.* And the endocardial surfaces, on the average, repolarize last. The reason for this abnormal sequence of repolarization is reputed to be that high pressure in the ventricles during contraction greatly reduces coronary blood flow to the endocardium, thereby slowing the repolarization process on the endocardial surfaces. The successive stages of repolarization are shown by the white areas in the ventricles of Figure 16–9.

Thus, *the predominant direction of the vector through the heart during* repolarization *of the ventri-*

*cles is from base to apex, which is also the predominant direction of the vector during* depolarization. *As a result, the T wave in the normal electrocardiogram is positive, which is also the polarity of most of the normal QRS complex.*

In Figure 16-9 five stages of repolarization of the ventricles are noted. In each of these, the vector extends from the base toward the apex. At first the vector is relatively small because the area of repolarization is small. Later the vector becomes stronger and stronger because of greater degrees of repolarization. Finally, the vector becomes weaker again because the areas of depolarization still persisting become so slight that the total quantity of current flow begins to decrease. These changes illustrate that the vector is greatest when approximately half the heart is in the polarized state and approximately half is depolarized.

The changes in the electrocardiograms of the three standard leads during the process of repolarization are noted under each of the ventricles, depicting the

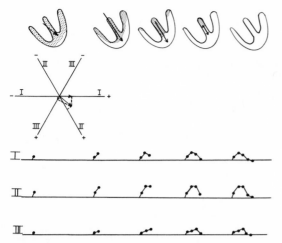

**Figure 16–9.** Generation of the T wave during repolarization of the ventricles, showing a vectorial analysis of the first stage.

progressive stages of repolarization. Over approximately 0.15 second, the period required for the whole process to take place, the T wave of the electrocardiogram is generated.

## DEPOLARIZATION OF THE ATRIA— THE P WAVE

Depolarization of the atria begins in the S-A node and spreads in all directions over the atria. Therefore, the point of original electronegativity in the atria is approximately at the base of the superior vena cava, and the direction of current flow in the atrium at the beginning of depolarization is in the direction noted in Figure 16–10. Furthermore, the vector remains generally in this direction throughout the process of atrial depolarization.

Thus, the vector of current flow during depolariza-

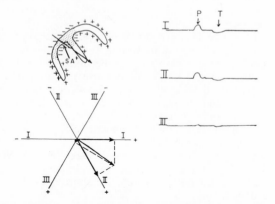

**Figure 16–10.** Depolarization of the atria and generation of the P wave, showing the vector through the atria and the resultant vectors in the three standard leads. To the right are shown the atrial P and T waves.

tion in the atria points in almost the same direction as that in the ventricles. And, because this direction is in the direction of the axes of the standard leads I, II, and III, the electrocardiogram recorded from the atria during the process of depolarization usually is positive in all three standard leads, as illustrated in Figure 16–10. The record of atrial depolarization is known as the P wave.

**Repolarization of the Atria—The Atrial T Wave.** Spread of the depolarization wave through the atrial muscle is *much slower than in the ventricles*. Therefore, the musculature around the S-A node becomes depolarized a long time before the musculature in distal parts of the atria. Because of this, *the area in the atria that becomes repolarized first is the area that had originally become depolarized first,* which is an entirely different situation from that in the ventricles. Thus, repolarization begins in the region surrounding the S-A node. At this time the region around the S-A node becomes positive with respect to the rest of the atria, and this same general direction of the vector is maintained throughout the repolarization process—that is, *backward to the vector of depolarization.* Note again that this is opposite to the effect that occurs in the ventricles. Therefore, as noted to the right in Figure 16–10, the so-called atrial T wave follows about 0.15 second after the atrial P wave, this T wave being on the opposite side of the zero reference line from the P wave—that is, it is normally negative rather than positive. In the normal electrocardiogram this T wave appears at approximately the same time that the QRS complex of the ventricle appears. Therefore, it is almost always totally obscured by the larger QRS complex, though in some abnormal states it does play a role in the recorded electrocardiogram.

## THE VECTORCARDIOGRAM

It was noted in the preceding discussions that the vector of current flow through the heart changes rapidly as the impulse spreads through the myocardium. It changes in two aspects: First, the vector increases and decreases in length because of the increasing and decreasing potential of the vector. Second, the vector changes its direction because of changes in the average direction of current flow around the heart. The so-called vectorcardiogram depicts these changes in the vectors at the different times during the cardiac cycle, as illustrated in Figure 16–11.

In the vectorcardiogram of Figure 16–11, point 5 is the zero reference point, and this point is the negative end of all the vectors. While the heart is quiescent, the positive end of the vector also remains at the zero point because there is no current flow. However, just as soon as current begins to flow through the heart, the positive end of the vector leaves the zero reference point.

When the septum first becomes depolarized, the vector extends downward toward the apex of the heart, but it is relatively weak, thus generating the

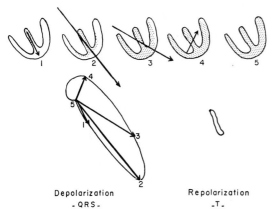

**Figure 16–11.** The QRS and T vectorcardiograms.

first portion of the vectorcardiogram, as illustrated by the positive end of vector 1. As more of the heart becomes depolarized, the vector becomes stronger and stronger, usually swinging slightly to one side. Thus vector 2 of Figure 16–11 represents the state of depolarization of the heart about 0.02 second after vector 1. After another 0.02 second, vector 3 represents the current flow in the heart, and vector 4 occurs in still another 0.01 second. Finally, the heart becomes totally depolarized, and the vector becomes zero once again, as shown at point 5.

The elliptical figure generated by the positive ends of the vectors is called the *QRS vectorcardiogram.*

Vectorcardiograms can be recorded instantaneously on an oscilloscope by connecting electrodes from above and below the heart to the vertical plates of the oscilloscope and connecting electrodes from each side of the heart to the horizontal plates. When the vector changes, the spot of light on the oscilloscope follows the course of the positive end of the changing vector, thus inscribing the vectorcardiogram on the oscilloscopic screen.

**The "T" Vectorcardiogram.** Changing vectors in the heart do not occur only during the depolarization process, for vectors depicting current flow around the ventricles reappear during repolarization. Therefore, a second, smaller vectorcardiogram—the *T vectorcardiogram*—is inscribed during repolarization of the muscular mass; this is depicted to the right in Figure 16–10. Also, a smaller *P vectorcardiogram* is inscribed during atrial depolarization.

# THE MEAN ELECTRICAL AXIS OF THE VENTRICLE

The vectorcardiogram of the ventricular depolarization wave (the QRS vectorcardiogram) shown in Figure 16–11 is that of a normal heart. Note from this vectorcardiogram that the preponderant direction of the vectors of the ventricles is normally toward the apex of the heart—that is, during most of the cycle of

ventricular depolarization, current flows from the base of the ventricles toward the apex. This preponderant direction of current flow during depolarization is called the *mean electrical axis of the ventricles.* The mean electrical axis of the normal ventricles is 59 degrees. However, in certain pathological conditions of the heart this direction of current flow is changed markedly—sometimes even to opposite poles of the heart.

## *DETERMINING THE ELECTRICAL AXIS FROM STANDARD LEAD ELECTROCARDIOGRAMS*

Clinically, the electrical axis of the heart is usually determined from the standard lead electrocardiograms. Figure 16–12 illustrates a method for doing this. After recording the various standard leads, one determines the maximum potential and polarity of the recording in two of the leads. In lead I of the figure the recording is positive and in lead III the recording is mainly positive, but negative during part of the cycle. If any part of the recording is negative, *this negative potential is subtracted from the positive potential.* After subtracting the negative portion of the QRS wave in lead III from the positive portion, each net vector is plotted on the axes of the respective leads, with the base of the vector at the point of intersection of the axes, as illustrated in Figure 16–12.

If the vector of lead I is positive, it is plotted in a positive direction along the line depicting lead I. On the other hand, if this vector is negative, it is plotted in a negative direction. Also, for lead III, the vector is placed with its base at the point of intersection, and, if positive, it is plotted in the positive direction along the line depicting lead III. If it is negative it is plotted in the negative direction.

In order to determine the actual vector of current flow in the heart, one draws perpendicular lines from the apices of the two vectors of leads I and III, respectively. The point of intersection of these two perpendicular lines represents, by vectorial analysis,

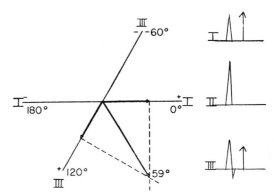

**Figure 16–12.** Plotting the mean electrical axis of the heart from two electrocardiographic leads.

the apex of the actual vector in the heart, and the point of intersection of the two lead axes represents the negative end of the actual vector. Therefore, the *actual vector* is drawn between these two points. The average potential generated in the direction of current flow is represented by the length of the actual vector, and the mean electrical axis is represented by the direction of the vector. Thus, the vector of the mean electrical axis of the normal heart, as determined in Figure 16–12, is 59 degrees.

The vector of current flow through the ventricles determined by this method does not determine the vector at a given instant in the depolarization process of the ventricles but, instead, determines approximately how much and in what direction the *average* current flows during the entire depolarization period.

## VENTRICULAR CONDITIONS THAT CAUSE AXIS DEVIATION

Though the mean electrical axis of the ventricles averages approximately 59 degrees, this average can swing to the left even in the normal heart to approximately 20 degrees or to the right to approximately 100 degrees. The causes of the normal variations are simply anatomical differences in the Purkinje fiber distribution or in the musculature itself of different hearts. Yet, a number of conditions can cause axis deviation even beyond these normal limits, as follows:

**Change in the Position of the Heart.** Obviously, if the heart itself is angulated to the left, the mean electrical axis of the heart will also shift to the left. Such shift occurs (a) during expiration, (b) when a person lies down, because the abdominal contents press upward against the diaphragm, and (c) quite frequently in stocky, fat persons whose diaphragms normally press upward against the heart.

Likewise, shift of the heart to the right causes the mean electrical axis of the ventricles to shift to the right. This condition occurs (a) during inspiration, (b) when a person stands up, and (c) normally in tall, lanky persons whose hearts hang downward.

**Hypertrophy of One Ventricle.** When one ventricle greatly hypertrophies, the axis of the heart shifts toward the hypertrophied ventricle for two reasons: First, far greater quantity of muscle exists on the hypertrophied side of the heart than on the other side, and this allows excess generation of electrical currents on that side. Second, more time is required for the depolarization wave to travel through the hypertrophied ventricle than through the normal ventricle. Consequently, the normal ventricle becomes depolarized considerably in advance of the hypertrophied ventricle, and this causes a strong vector from the normal side of the heart toward the hypertrophied side. Thus the axis deviates toward the hypertrophied ventricle.

*Left Axis Deviation Resulting from Hypertrophy of the Left Ventricle.* Figure 16–13 illustrates the three standard leads of an electrocardiogram in which an analysis of the axis direction shows left axis devia-

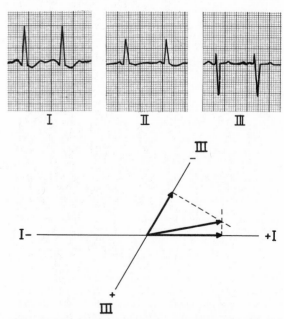

**Figure 16–13.** Left axis deviation in hypertensive heart disease. Note the slightly prolonged QRS complex.

tion with the mean electrical axis pointing in the −15 degree direction. This is a typical electrocardiogram resulting from increased muscular mass of the left ventricle. In this instance the axis deviation was caused by *hypertension* (high blood pressure), which caused the left ventricle to hypertrophy in order to pump blood against the elevated systemic arterial pressure. However, a similar picture of left axis deviation occurs when the left ventricle hypertrophies as a result of *aortic valvular stenosis, aortic valvular regurgitation,* or any of a number of *congenital heart conditions* in which the left ventricle enlarges while the right side of the heart remains relatively normal in size.

*Right Axis Deviation Resulting from Hypertrophy of the Right Ventricle.* The electrocardiogram of Figure 16–14 illustrates intense right axis deviation with an electrical axis of approximately 170 degrees, which is 111 degrees to the right of the normal mean electrical axis of the ventricles. The right axis deviation illustrated in this figure was caused by hypertrophy of the right ventricle as a result of *pulmonary stenosis.* However, right axis deviation may also occur in other congenital heart conditions, such as *tetralogy of Fallot* or *interventricular septal defect.* Also, hypertrophy of the right ventricle as a result of *increased pulmonary vascular resistance* can cause right axis deviation.

**Bundle Branch Block.** Ordinarily, the two lateral walls of the ventricles depolarize at almost the same time, because both the left and right bundle branches transmit the cardiac impulse to the endocardial surfaces of the two ventricular walls at almost the same instant. As a result, the currents flowing from the walls of the two ventricles almost neutralize each other. However, if one of the major bundle branches is blocked, depolarization of the two ventricles does not occur even nearly simultaneously, and the depolarization currents do not neutralize each other. As a result, axis deviation occurs as follows:

*Left Bundle Branch Block.*   When the left bundle branch is blocked, cardiac depolarization spreads through the right ventricle approximately three times as rapidly as through the left ventricle. Consequently, much of the left ventricle remains polarized for a long time after the right ventricle has become totally depolarized. Thus, the right ventricle is electronegative with respect to the left ventricle during most of the depolarization process, and a very strong vector projects from the right ventricle toward the left ventricle. In other words, there is intense left axis deviation because the positive end of the vector points toward the left ventricle. This is illustrated in Figure 16–15, which shows typical left axis deviation resulting from left bundle branch block. Note that the axis is approximately −50 degrees.

Because of slowness of impulse conduction when the Purkinje system is blocked, axis deviation resulting from bundle branch block greatly increases the duration of the QRS complex, as discussed in greater

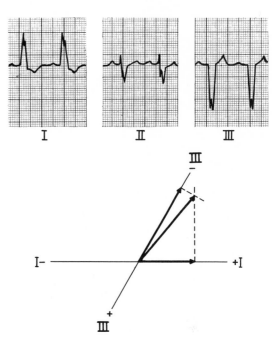

**Figure 16–15.**   Left axis deviation due to left bundle branch block. Note the greatly prolonged QRS complex.

detail later in the chapter. It is the prolonged QRS complex that differentiates this condition from axis deviation caused by hypertrophy.

*Right Bundle Branch Block.*   When the right bundle branch is blocked, the left ventricle depolarizes far more rapidly than the right ventricle so that the left becomes electronegative while the right remains electropositive. A very strong vector develops with its negative end toward the left ventricle and its positive end toward the right ventricle. In other words, intense right axis deviation occurs.

Right axis deviation caused by right bundle branch block is illustrated in Figure 16–16, which shows an axis of approximately 105 degrees and a prolonged QRS complex because of blocked conduction. Note also the great prolongation of the QRS complex.

**Axis Deviation Caused by Muscular Destruction.**   Very abnormal deviations of the axis of the ventricles can occur following heart attacks that result in destruction of part of the cardiac muscle and replacement of this muscle with fibrous tissue. The major causes of the axis deviation are two-fold: First, part of the muscular mass itself is destroyed and is replaced by scar tissue so that a smaller quantity of muscle may then be available on one side of the heart than on the other side to generate electrical currents. Second, and probably more important, is blockage of the conduction of depolarization at one or more local points in the smaller branches of the Purkinje system. When blockage occurs, the impulse must then be conducted by the muscle itself. This conduction occurs at a slow velocity, and it often has to detour around the scar tissue, allowing the muscular tissue

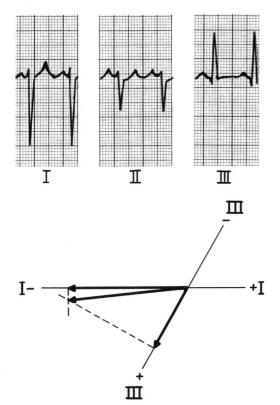

**Figure 16–14.**   High-voltage electrocardiogram due to pulmonary stenosis with right ventricular hypertrophy. Also, there is intense right axis deviation.

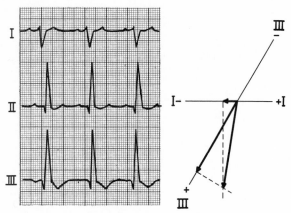

**Figure 16–16.** Right axis deviation due to right bundle branch block. Note the greatly prolonged QRS complex.

beyond the block to remain electropositive for long periods of time after other portions of the heart have already become totally depolarized. Thus, even local blocks in the ventricles can cause considerable shift of the axis of the heart.

Muscular destruction causes the axis to shift in the direction opposite to the side of destruction, but conduction block causes the axis to shift toward the side of destruction. Thus, the two effects may or may not neutralize each other. Consequently, old infarctions in the heart do not produce consistent changes in axis deviation. Therefore, when other signs of old infarction are present in the electrocardiogram, such as low voltage and a prominent Q wave (as shown in Figure 16–17), increased duration of the QRS complex, and bizarre spiking patterns of the QRS complex, the diagnosis is usually definite, and it is generally unimportant to determine the axis of the heart.

# CONDITIONS THAT CAUSE ABNORMAL VOLTAGES OF THE QRS COMPLEX

## INCREASED VOLTAGE IN THE STANDARD LEADS

Normally, the voltages in the three standard electrocardiographic leads, as measured from the peak of the R wave to the bottom of the S wave, vary between 0.5 and 2.0 mv., with lead III usually recording the lowest voltage and lead II the highest. However, these relationships are not invariably true even in the normal heart. In general, when the sum of the voltages of all the QRS complexes of the three standard leads is greater than 4 mv., one considers that the patient has a high-voltage electrocardiogram.

The cause of high-voltage QRS complexes is most often increased muscular mass of the heart, which ordinarily results from *hypertrophy of the muscle* in response to excessive load on one part of the heart or the other. For instance, the right ventricle hyper-

trophies when it must pump blood through a stenotic pulmonary valve, and the left ventricle hypertrophies in systemic hypertension. The increased quantity of muscle allows generation of increased quantities of current around the heart. As a result, the electrical potentials recorded in the electrocardiographic leads are considerably greater than normal, as shown in Figures 16–13 and 16–14.

## DECREASED VOLTAGE IN THE STANDARD LEADS

There are three major causes of decreased voltage in the electrocardiograms of the three standard electrocardiographic leads. These are, first, abnormalities of the cardiac muscle itself that prevent generation of large quantities of currents; second, abnormal conditions around the heart so that currents cannot be conducted from the heart to the surface of the body with ease; and, third, rotation of the apex of the heart to point toward the anterior chest wall so that the electrical currents of the heart flow mainly anteroposteriorly in the chest rather than in the frontal plane of the body.

**Decreased Voltage Caused by Cardiac Myopathies.** One of the most usual causes of decreased voltage of the QRS complex is a series of *old myocardial infarctions* associated with *diminished muscle mass*. This causes the depolarization wave to move through the ventricles slowly and prevents major portions of the heart from becoming massively depolarized all at once. Consequently, this condition causes moderate prolongation of the QRS complex along with decreased voltage. Figure 16–17 illustrates the typical low-voltage electrocardiogram with prolongation of the QRS complex, which one often finds after multiple small infarctions of the heart that have resulted in local blocks and loss of muscle mass throughout the ventricles.

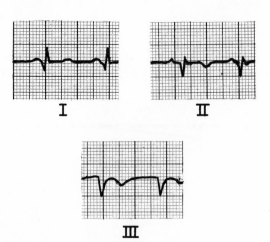

**Figure 16–17.** Low-voltage electrocardiogram, with evidences of local damage throughout the ventricles caused by old myocardial infarction.

**Decreased Voltage Caused by Conditions Surrounding the Heart.** One of the most important causes of decreased voltage in the electrocardiographic leads is *fluid in the pericardium*. Because extracellular fluid conducts electrical currents with great ease, a large portion of currents flowing out of the heart are conducted from one part of the heart to another through the pericardial effusion, so that the quantity of current reaching the surface of the body is greatly diminished. That is, the pericardial fluid ''short circuits'' the potentials generated by the heart. *Pleural effusion* to a lesser extent can also ''short'' the currents around the heart so that the voltages at the surface of the body and in the electrocardiograms are decreased.

*Pulmonary emphysema* can decrease the electrocardiographic potentials but by a different method from that of pericardial effusion. In pulmonary emphysema, conduction of electrical current through the lungs is considerably depressed because of the excessive quantity of air in the lungs. Also, the chest cavity enlarges, and the lungs tend to envelop the heart to a greater extent than normally. Therefore, the lungs act as an insulator to prevent spread of currents from the heart to the surface of the body, and this in general results in decreased electrocardiographic potentials in the various leads.

# PROLONGED AND BIZARRE PATTERNS OF THE QRS COMPLEX

## PROLONGED QRS COMPLEX AS A RESULT OF CARDIAC HYPERTROPHY OR DILATATION

The QRS complex lasts as long as the process of depolarization continues to spread through the ventricles—that is, as long as part of the ventricles is depolarized and part is still polarized. Therefore, the cause of a prolonged QRS complex is always *delayed conduction* of the impulse through the ventricles. Such delay often occurs when one or both ventricles are hypertrophied or dilated, owing to the longer pathway that the impulse must then travel. The normal QRS complex lasts about 0.06 second, whereas in hypertrophy or dilatation of the left or right ventricle the QRS complex may be prolonged to 0.09 second or occasionally 0.10 second.

## PROLONGED QRS COMPLEX RESULTING FROM PURKINJE SYSTEM BLOCKS

Block of the Purkinje fibers requires that the cardiac impulse be conducted by the ventricular muscle instead of through the specialized conduction system, thereby decreasing the velocity of impulse conduction to approximately one-third to one-fourth normal. Therefore, if complete block of one of the bundle branches occurs, the duration of the QRS complex is increased to 0.14 second or greater.

In general, a QRS complex is considered to be abnormally long when it lasts more than 0.08 second, and when it lasts more than 0.12 second the prolongation is almost certain to be caused by pathologic block of the conduction system somewhere in the ventricles, as illustrated by the electrocardiograms for bundle branch block in Figures 16–15 and 16–16.

## CONDITIONS CAUSING BIZARRE QRS COMPLEXES

Bizarre patterns of the QRS complex are most frequently caused by two conditions: first, destruction of cardiac muscle in various areas throughout the ventricular system with replacement of this muscle by scar tissue, and second, local blocks in the conduction of impulses by the Purkinje system.

Sometimes local blocks occur in both the right and left ventricles. If the blocks are such that the impulse reaches the blocked area in the right ventricle much later than it reaches the blocked area in the left ventricle, but the total quantity of blocked muscle in the right ventricle is greater than the total quantity in the left ventricle, then a situation might occur in which the axis of the heart first shifts to the left and then, when the left ventricle has become totally depolarized but a portion of the right ventricle still remains polarized, rapidly shifts diametrically oppositely toward the right ventricle. This effect causes double or even triple peaks in some of the standard electrocardiographic leads, such as those illustrated in Figure 16–15.

# CURRENT OF INJURY

Many different cardiac abnormalities, especially those that damage the heart muscle itself, often cause part of the heart to remain partially or totally *depolarized all the time*. When this occurs, current flows between the pathologically depolarized and the normally polarized areas. This is called the *current of injury*. Note especially that *the injured part of the heart is negative while the remainder is positive*.

Some of the abnormalities that can cause current of injury are: (1) *mechanical trauma,* which makes the membranes remain so permeable that full repolarization cannot take place, (2) *infectious processes* that damage the muscle membranes, and (3) *ischemia of local areas of muscle caused by coronary occlusion,* which is by far the most common cause of current of injury in the heart; during ischemia enough energy simply is not available to maintain normal function of the muscle.

## EFFECT OF CURRENT OF INJURY ON THE QRS COMPLEX

In Figure 16–18 a shaded area in the base of the left ventricle is newly infarcted. Therefore, during the

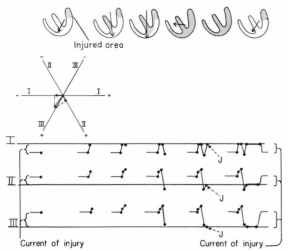

**Figure 16–18.** Effect of a current of injury on the electrocardiogram.

T-P interval—that is, when the normal ventricular muscle is polarized—current flows from the base of the left ventricle toward the rest of the ventricles. The vector of this "current of injury" is in a direction of approximately 125 degrees, with the base of the vector, the *negative end,* toward the injured muscle. As illustrated in the lower portions of the figure, even before the QRS complex begins, *this current flow causes an initial record in lead 1 below the zero potential line* because the resultant vector of the current of injury points toward the negative end of the lead I axis. In lead II the record is above the line because the resultant vector points toward the positive terminal of lead II. In lead III the vector of the current flow is also in the same direction as the polarity of lead III so that the record is positive. Furthermore, because the vector of the current of injury lies almost exactly along the axis of lead III, the potential of the current of injury in lead III is much higher than in either of the other two records.

As the heart then proceeds through its normal process of depolarization, part of the septum first becomes depolarized, and the depolarization spreads down to the apex and back toward the bases of the ventricles. The last portion of the ventricles to become totally depolarized is the base of the right ventricle, because the base of the left ventricle is already totally and permanently depolarized. By vectorial analysis, as illustrated in the figure, the electrocardiogram generated by the depolarization wave traveling through the ventricles may be constructed graphically, as shown in Figure 16–18.

When the heart becomes totally depolarized at the end of the depolarization process, as noted by the next to last stage in Figure 16–18, all of the ventricular muscle is in a negative state. Therefore, at this instant in the electrocardiogram, absolutely no current flows around the musculature of the ventricles,

because now the injured heart muscle and the contracting muscle are both completely depolarized.

As repolarization then takes place, all of the heart finally repolarizes except the area of permanent depolarization in the injured base of the left ventricle. Thus, the repolarization causes a return of the current of injury in each lead, as noted at the far right in Figure 16–18.

## THE J POINT—THE ZERO REFERENCE POTENTIAL OF THE ELECTROCARDIOGRAM

One would think that the electrocardiograph machines for recording electrocardiograms could determine when no current is flowing around the heart. However, many stray currents exist in the body, such as currents resulting from "skin potentials" and from differences in ionic concentrations in different parts of the body. Therefore, when two electrodes are connected between the arms or between an arm and a leg, these stray currents make it impossible for one to predetermine the exact zero reference level in the electrocardiogram. For these reasons, the following procedure must be used to determine the zero potential level: First, one notes *the exact point at which the wave of depolarization just completes its passage through the heart,* which occurs at the very end of the QRS complex. At exactly this point, all parts of the ventricles are depolarized so that no currents are flowing around the heart. Therefore, zero voltage is recorded in the electrocardiogram. This point is known as the "J point" in the electrocardiogram, as illustrated in Figures 16–18 and 16–19.

For further analysis of the electrical axis of the current of injury, a horizontal line is drawn through the electrocardiogram at the level of the J point, and this horizontal line is the zero potential line in the electrocardiogram from which all potentials caused by currents of injury must be measured.

**Use of the J Point in Plotting the Axis of a Current of Injury.** Figure 16–19 illustrates electrocardiograms recorded from leads I and III, both of which show currents of injury. In other words, the J point and the S-T segment of each of these two electrocardiograms are not on the same line as the T-P segment. A horizontal line has been drawn through the J point to represent the zero potential level in each of the two recordings. The potential of the current of injury in each lead is the difference between the level of the T-P segment of the electrocardiogram (which is recorded between heart beats) and the zero potential line, as illustrated by the arrows. In lead I the recorded current of injury is above the zero potential line and is, therefore, positive. On the other hand, in lead III the T-P segment is below the zero potential line; therefore, the current of injury in lead III is negative.

At the bottom in Figure 16–19, the potentials of the current of injury in leads I and III are plotted on the coordinates of these leads, and the vector of the cur-

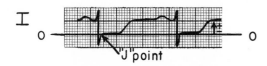

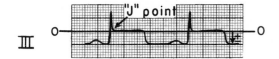

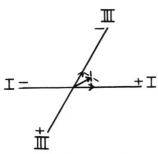

**Figure 16–19.** The "J" point as the zero reference voltage of the electrocardiogram. Also, method for plotting the axis of a current of injury.

rent of injury is determined by the method already described. In this instance, the vector of the current of injury extends from the right side of the body toward the left and slightly upward, with an axis of approximately −30 degrees.

If one places the vector of the current of injury directly over the ventricles, *the negative end of the vector points toward the permanently depolarized, "injured" area of the ventricles.* In the instance illustrated in Figure 16–19, the injured area would be in the lateral wall of the right ventricle.

## CORONARY ISCHEMIA AS A CAUSE OF CURRENT OF INJURY

Insufficient blood flow to the cardiac muscle depresses metabolism of the muscle for three different reasons: oxygen lack, excess carbon dioxide, and lack of sufficient nutrients. Consequently, complete, if any, polarization of the membranes cannot occur in areas of severe myocardial ischemia. Often, the heart muscle does not die, because the blood flow is sufficient to maintain life of the muscle even though it is not sufficient to cause repolarization of the membranes. As long as this state exists, a current of injury continues to flow during diastole.

Extreme ischemia of the cardiac muscle occurs following coronary occlusion, and strong currents of injury flow from the infarcted area of the ventricles

during the period between heartbeats, as is illustrated in Figures 16–20 and 16–21. Therefore, one of the most important diagnostic features of electrocardiograms recorded following acute coronary thrombosis is the current of injury.

**Acute Anterior Wall Infarction.** Figure 16–20 illustrates the electrocardiogram in the three standard leads and in one chest lead recorded from a patient with acute anterior wall cardiac infarction. The most important diagnostic feature of this electrocardiogram is the current of injury in the chest lead. If one draws a zero potential line through the J point of this lead, he finds a strong *negative* current of injury during diastole. In other words, the negative end of the current of injury vector is against the chest wall. This means that the current of injury is emanating from the anterior wall of the ventricles, which is the main reason for diagnosing this condition as anterior wall infarction.

If one analyzes the currents of injury in leads I and III, he finds a negative vector for the current of injury in lead I and a positive vector for the current of injury in lead III. This means that the resultant vector of the current of injury in the heart is approximately +150 degrees, with the negative end of the vector pointing toward the left ventricle and the positive end pointing toward the right ventricle. Thus, in this particular electrocardiogram the current of injury appears to be coming mainly from the left ventricle and from the anterior wall of the heart. There-

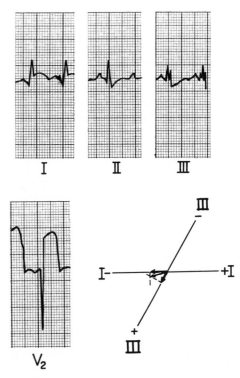

**Figure 16–20.** Current of injury in acute anterior wall infarction. Note the intense current of injury in lead $V_2$.

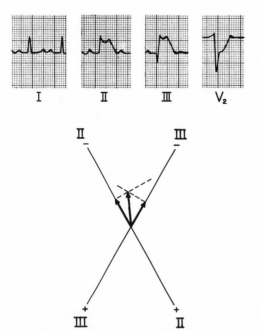

**Figure 16–21.**   Current of injury in acute posterior wall, apical infarction.

fore, one would suspect that this anterior wall infarction is probably caused by thrombosis of the anterior descending limb of the left coronary artery.

**Posterior Wall Infarction.**   Figure 16–21 illustrates the three standard leads and one chest lead from a patient with posterior wall infarction. The major diagnostic feature of this electrocardiogram is also in the chest lead. If a zero potential reference line is drawn through the J point of this lead, it is readily apparent that during the T-P interval the vector of the current of injury is positive. This means that the positive end of the vector is at the chest wall and the negative end (injured end) is away from the chest wall. In other words, the current of injury is coming from the opposite side of the heart from that portion adjacent to the chest wall, which is the reason why this type of electrocardiogram is the basis for diagnosing posterior wall infarction.

If one analyzes the currents of injury in leads II and III of Figure 16–21, it is readily apparent that the current of injury is negative in both leads. By vectorial analysis as shown in the figure, one finds that the vector of the current of injury is approximately −95 degrees with the negative end of the vector pointing downward and the positive end upward. Thus, because the infarct, as indicated by the chest lead, is on the posterior wall of the heart and, as indicated by the currents of injury in leads II and III, is in the apical portion of the heart, one would suspect that this infarct is close to the apex on the posterior wall of the left ventricle.

**Infarction in Other Parts of the Heart.**   By the same procedures as those illustrated in the preceding two discussions of anterior and posterior wall infarctions, it is possible to determine the locus of an infarcted area emitting a current of injury, regardless of which part of the heart is involved. In making such vectorial analyses, it must always be remembered that *the positive end of the vector points toward the normal cardiac muscle and the negative end points toward the abnormal portion of the heart that is emitting the current of injury.*

**Recovery from Coronary Thrombosis.**   Figure 16–22 illustrates a chest lead from a patient with posterior infarction, showing the change in this chest lead from the day of the attack to one week later, then three weeks later, and finally one year later. From this electrocardiogram it can be seen that the current of injury is strong immediately after the acute attack (T-P segment displaced positively from the J point and the S-T segment), but after approximately one week the current of injury has diminished considerably and after three weeks it is completely gone. After that, the electrocardiogram does not change greatly during the following year. This is the usual recovery pattern following cardiac infarction of moderate degree when the collateral coronary blood flow is sufficient to reestablish appropriate nutrition to most of the infarcted area.

On the other hand, when all the coronary vessels throughout the heart are fairly well sclerosed, it may not be possible for the adjacent coronary vessels to supply sufficient blood to the infarcted area for recovery. Therefore, in some patients with coronary infarction, the infarcted area never redevelops an adequate coronary blood supply, some of the heart muscle dies, and relative coronary insufficiency persists in this area of the heart indefinitely. If the muscle does not die and become replaced by scar tissue, it continually emits a current of injury as long as the relative ischemia exists, particularly during bouts of exercise when the heart is overloaded.

**Old Recovered Myocardial Infarction.**   Figure 16–23 illustrates leads I and III in anterior infarction and posterior infarction as these leads appear approximately a year after the acute episode. These are what might be called the "ideal" configurations of

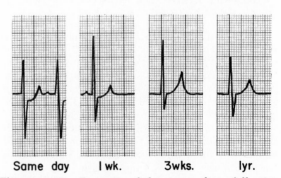

**Figure 16–22.**   Recovery of the myocardium following moderate posterior wall infarction, illustrating disappearance of the current of injury (lead $V_3$).

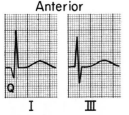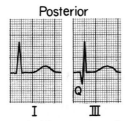

**Figure 16–23.** Electrocardiograms of old anterior and posterior wall infarctions, illustrating the Q wave in lead I in old anterior wall infarction and the Q wave in lead III in old posterior wall infarction.

the QRS complex in these types of recovered myocardial infarction. It is illustrated in these figures that usually a Q wave develops at the beginning of the QRS complex in lead I in anterior infarction because of loss of muscle mass in the anterior left wall of the left ventricle; whereas in posterior infarction a Q wave develops at the beginning of the QRS complex in lead III because of loss of muscle in the posterior apical part of the ventricle.

These configurations are certainly not those found in all cases of old anterior and posterior cardiac infarction. Local loss of muscle and local areas of conduction block can cause the following abnormalities of the QRS complex: bizarre patterns (the prominent Q waves, for instance), decreased voltage, and prolongation.

**Current of Injury in Angina Pectoris.** "Angina pectoris" means simply pain in the pectoral regions of the upper chest, this pain usually radiating down the left arm and into the neck. The pain is caused by *relative* ischemia of the heart. No pain may be felt as long as the person is perfectly quiet, but, just as soon as he overworks his heart, the pain appears.

A current of injury often occurs during an attack of severe angina pectoris, for the relative coronary insufficiency occasionally becomes great enough then to prevent adequate repolarization of the membranes in some areas of the heart during diastole.

# ABNORMALITIES IN THE T WAVE

Earlier in the chapter it was pointed out that the T wave is normally positive in all the standard leads and that this is caused by repolarization of the apex of the heart ahead of the endocardial surfaces of the ventricles. This is backward to the direction in which depolarization takes place. (If the basic principles of the upright T wave in the standard leads have not been understood by now, the student should become familiar with the earlier, more detailed discussion of this before proceeding to the following few sections.)

The T wave becomes abnormal when the normal sequence of repolarization does not occur. Several factors can change this sequence of repolarization, as follows:

## EFFECT OF SLOW CONDUCTION OF THE DEPOLARIZATION WAVE ON THE T WAVE

Referring back to Figure 16–15, note that the QRS complex is considerably prolonged. The reason for this prolongation is delayed conduction in the left ventricle as a result of left bundle branch block. The left ventricle becomes depolarized approximately 0.10 second after depolarization of the right ventricle. The refractory periods of the right and left ventricular muscle masses are not greatly different from each other. Therefore, the right ventricle begins to repolarize long before the left ventricle; this causes positivity in the right ventricle and negativity in the left ventricle. In other words, the mean axis of the T wave is from left to right, which is opposite to the mean electrical axis of the QRS complex in this same electrocardiogram. Thus, when conduction of the impulse through the ventricles is greatly delayed, the T wave is almost always of opposite polarity to that of the QRS complex.

In Figure 16–16 and in several figures in the following chapter, conduction also does not occur through the Purkinje system. As a result, the rate of conduction is greatly slowed, and in each instance the T wave is of opposite polarity to that of the QRS complex, whether the condition causing this delayed conduction happens to be left bundle branch block, right bundle branch block, ventricular extrasystole, or otherwise.

## PROLONGED ACTION POTENTIALS (PROLONGED DEPOLARIZATION) IN PORTIONS OF THE VENTRICULAR MUSCLE AS A CAUSE OF ABNORMALITIES IN THE T WAVE

If the apex of the ventricles should have an abnormally long period of depolarization, that is, a prolonged action potential, then repolarization of the ventricles would not begin at the apex as it normally does. Instead, the base of the ventricles would repolarize ahead of the apex, and the vector of repolarization would point from the apex toward the

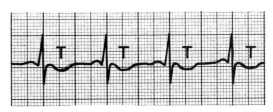

**Figure 16–24.** Inverted T wave resulting from mild ischemia of the apex of the ventricles.

base of the heart. This is opposite to the usual vector of repolarization, and, consequently, the T wave in all three standard leads would be negative rather than the usual positive. Thus, the simple fact that the apical muscle of the heart has a prolonged period of depolarization is sufficient to cause marked changes in the T wave, even to the extent of changing the entire polarity, as is illustrated in Figure 16–24.

*Ischemia* is by far the most common cause of increased period of depolarization of cardiac muscle, and, when the ischemia occurs in only one area of the heart, the depolarization period of this area increases out of proportion to that in other portions. As a result, definite changes in the T wave may take place. The ischemia may result from chronic, progressive coronary occlusion, acute coronary occlusion, or relative coronary insufficiency occurring during exercise.

One means for detecting mild coronary insufficiency is to have the patient exercise and then to record the electrocardiogram immediately thereafter, noting whether or not changes occur in the T waves. The changes in the T waves need not be specific, for any change in the T wave in any lead—inversion, for instance, or a biphasic wave—is often evidence enough that some portion of the ventricular muscle has increased its period of depolarization out of proportion to the rest of the heart, and this is probably caused by relative coronary insufficiency.

All the other conditions that can cause currents of injury, including pericarditis, myocarditis, and mechanical trauma of the heart, can also cause changes in the T wave. A current of injury occurs when the period of depolarization of some muscle is so long that the muscle fails to repolarize completely before the next cardiac cycle begins. Therefore, a current of injury is actually an exacerbated form of abnormal T wave, for both of these result from an increased depolarization period of one or more portions of cardiac muscle, and the difference is only a matter of degree.

**Effect of Digitalis on the T Wave.** As discussed in Chapter 26, digitalis is a drug that can be used

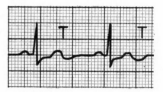

**Figure 16–25.** Biphasic T wave caused by digitalis toxicity.

during relative coronary insufficiency to increase the strength of cardiac muscular contraction. However, digitalis also increases the period of depolarization of cardiac muscle. It usually increases this period by approximately the same proportion in all or most of the ventricular muscle, but when overdosages of digitalis are given, the period of one part of the heart may be increased out of proportion to that of other parts. As a result, nonspecific changes, such as T wave inversion or biphasic T waves, may occur in one or more of the electrocardiographic leads. A biphasic T wave caused by excessive administration of digitalis is illustrated in Figure 16–25. There is a slight amount of current of injury, too. This probably results from continuous depolarization of part of the ventricular muscle.

Changes in the T wave during digitalis administration are the earliest signs of digitalis toxicity. If still more digitalis is given to the patient, strong currents of injury may develop, and, as noted in the following chapter, digitalis can also block conduction of the cardiac impulse to various portions of the heart, so that various arrhythmias result. It is desirable clinically to keep the effects of digitalis from going beyond the stage of mild T wave abnormalities. Therefore, the electrocardiograph is used routinely in following digitalized patients.

# REFERENCES

*See bibliography at end of Chapter 17*

# 17

# Electrocardiographic Interpretation of Cardiac Arrhythmias

The rhythmicity of the heart and some abnormalities of rhythmicity were discussed in Chapter 14. The major purpose of the present chapter is to discuss still other abnormalities of rhythm and to describe the electrocardiograms recorded in the conditions known clinically as "cardiac arrhythmias."

## ABNORMAL SINUS RHYTHMS

### TACHYCARDIA

The term "tachycardia" means fast heart rate, usually defined as faster than 100 beats per minute. An electrocardiogram recorded from a patient with tachycardia is illustrated in Figure 17–1. This electrocardiogram is normal except that the rate of heartbeat, as determined from the time intervals between QRS complexes, is approximately 150 per minute instead of the normal 72 per minute.

The three general causes of tachycardia are *increased body temperature, stimulation of the heart by the autonomic nerves,* and *toxic conditions of the heart.*

The rate of the heart increases approximately 10 beats per minute for each degree Fahrenheit increase in body temperature up to a body temperature of about 105°; beyond this the heart rate may actually decrease, owing to progressive weakening of the heart muscle as a result of the fever. Fever causes tachycardia because increased temperature increases the rate of metabolism of the S-A node, which in turn directly increases its excitability and rate of rhythm.

The various factors that can cause the autonomic nervous system to excite the heart will be discussed in relation to the reflexes that control the circulatory system in Chapter 22. For instance, when a patient loses blood and passes into a state of shock or semishock, reflex stimulation of the heart increases its rate to as great as 150 to 180 beats per minute. Also, simple weakening of the myocardium usually increases the heart rate because the weakened heart does not pump blood into the arterial tree to a normal extent, and this elicits reflexes to increase the rate of the heart.

### BRADYCARDIA

The term "bradycardia" means, simply, a slowed heart rate, usually defined as less than 60 beats per minute. Bradycardia is illustrated by the electrocardiogram in Figure 17–2.

**Bradycardia in Athletes.** The athlete's heart is considerably stronger than that of a normal person, a fact that allows his heart to pump a greater stroke volume output per beat. The excessive quantities of blood pumped into the arterial tree with each beat presumably initiate circulatory reflexes to cause the bradycardia.

**Vagal Stimulation as a Cause of Bradycardia.** Obviously, any circulatory reflex that stimulates the vagus nerve can cause the heart rate to decrease considerably. Perhaps the most striking example of this occurs in patients with the *carotid sinus syndrome.*

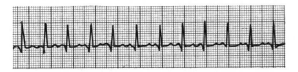

**Figure 17–1.**   Sinus tachycardia (lead I).

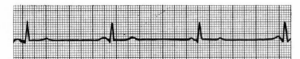

**Figure 17–2.**  Sinus bradycardia (lead III).

In these patients an arteriosclerotic process in the carotid sinus region of the carotid artery causes excessive sensitivity of the pressure receptors (baroreceptors) located in the arterial wall; as a result, mild pressure on the neck elicits a strong baroreceptor reflex, causing intense vagal stimulation of the heart and extreme bradycardia. Indeed, such pressure may actually stop the heart.

### SINUS ARRHYTHMIA

Figure 17–3 illustrates a *cardiotachometer* recording of the heart rate during normal and deep respiration. A cardiotachometer is an instrument that records by the height of successive spikes the duration of the interval between each two QRS complexes in the electrocardiogram. Note from this record that the heart rate increases and decreases approximately 5 per cent during the various phases of the normal respiratory cycle. However, during deep respiration, as is shown to the right in Figure 17–3, the heart rate increases and decreases by as much as 30 per cent.

Sinus arrhythmia is presumably the result of several circulatory reflexes that will be described in detail in Chapter 22. First, when the blood pressure rises and falls during each cycle of respiration, the *baroreceptors are alternately stimulated and depressed,* causing reflex slowing and speeding up of the heart. Second, during each respiratory cycle, the negative intrapleural pressure increases and decreases, which increases and decreases the effective pressure in the veins of the chest. This elicits a *waxing and waning Bainbridge reflex* that also increases and decreases the cardiac rate. Third, when the respiratory center of the medulla is excited during each respiratory cycle, some of the *impulses "spill over" from the respiratory center into the vasomotor center,* causing alternate increase and decrease in the number of impulses transmitted to the heart through the sympathetics and vagi.

# ABNORMAL RHYTHMS RESULTING FROM BLOCK IN IMPULSE CONDUCTION

### SINO-ATRIAL BLOCK

In rare instances the impulse from the S-A node is blocked before it enters the atrial muscle. This phenomenon is illustrated in Figure 17–4, which shows the sudden cessation of P waves with resultant standstill of the atrium. However, the ventricle picks up a new rhythm, the impulse usually originating in the A-V node or in the A-V bundle so that the ventricular QRS-T complex is not altered.

### ATRIOVENTRICULAR BLOCK

The only means by which impulses can ordinarily pass from the atria into the ventricles is through the *A-V bundle,* which is also known as the *bundle of His.* The different conditions that can either decrease the rate of conduction of the impulse through this bundle or can totally block the impulse are:

1. *Ischemia of the A-V junctional fibers* often delays or blocks conduction from the atria to the ventricles. Coronary insufficiency can cause ischemia of the A-V junction in the same manner that it can cause ischemia of the myocardium.

2. *Compression of the A-V bundle* by scar tissue or by calcified portions of the heart can depress or block conduction from the atria to the ventricles.

3. *Inflammation of the A-V bundle or fibers of the A-V junction* can depress conductivity between the atria and the ventricles. Inflammation results frequently from different types of myocarditis such as occur in diphtheria and rheumatic fever.

4. *Extreme stimulation of the heart by the vagi* in rare instances blocks impulse conduction through the A-V junctional fibers. Such vagal excitation occasionally results from strong stimulation of the baroreceptors in persons with the *carotid sinus syndrome,* which was just discussed in relation to bradycardia.

**Incomplete Heart Block.**  PROLONGED P-R (OR P-Q) INTERVAL—"FIRST DEGREE BLOCK." The normal lapse of time between the *beginning* of the P wave and the *beginning* of the QRS complex is approximately 0.16 second when the heart is beating at a normal rate. This P-R interval decreases in length with faster heartbeat and increases with slower

**Figure 17–3.**  Sinus arrhythmia as detected by a cardiotachometer. To the left is the recording taken when the subject was breathing normally; to the right, when breathing deeply.

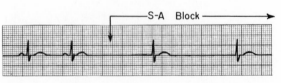

**Figure 17–4.**  S-A nodal block with A-V nodal rhythm (lead III).

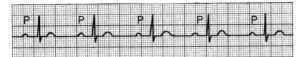

**Figure 17–5.** Prolonged P-R interval (lead II).

heartbeat. In general when the P-R interval increases above a value of approximately 0.20 second in the heart beating at normal rate, the P-R interval is said to be prolonged, and the patient is said to have *first degree incomplete heart block.* Figure 17-5 illustrates an electrocardiogram with a prolonged P-R interval, the interval in this instance being approximately 0.30 second.

The P-R interval rarely increases above 0.35 to 0.45 second, for by the time the rate of conduction through the A-V bundle is depressed to this extent, conduction stops entirely. Thus, when a patient's P-R interval is approaching these limits, additional slight increase in the severity of the condition will completely block impulse conduction rather than simply delay conduction further. Therefore, one of the means for determining the severity of some heart diseases—rheumatic fever, for instance—is to measure the P-R interval.

*Second Degree Block.* When conduction through the A-V junction is slowed until the P-R interval is 0.25 to 0.45 second, the action potentials traveling through the A-V junctional fibers are sometimes strong enough to pass on into the A-V node and at other times are not strong enough. Often the impulse passes into the ventricles on one heartbeat and fails to pass on the next one or two beats, thus alternating between conduction and nonconduction. In this instance, the atria beat at a considerably faster rate than the ventricles, and it is said that there are "dropped beats" of the ventricles. This condition is called *second degree incomplete heart block.*

Figure 17–6 illustrates P-R intervals of 0.30 second, and it also illustrates one dropped beat as a result of failure of conduction from the atria to the ventricles.

At times every other beat of the ventricles is dropped so that a "2:1 rhythm" develops in the heart, with the atria beating twice for every single beat of the ventricles. Sometimes other rhythms, such as 3:2 or 3:1, also develop.

**Complete Atrioventricular Block (Third Degree Block).** When the condition causing poor conduction in the A-V bundle becomes extremely severe, complete block of the impulse from the atria into the ventricles occurs. In this instance the P waves become completely dissociated from the QRS-T complexes, as illustrated in Figure 17–7. Note that the rate of rhythm of the atria in this electrocardiogram is approximately 100 beats per minute, while the rate of ventricular beat is less than 40 per minute. Furthermore, there is no relationship whatsoever between the rhythm of the atria and that of the ventricles, for the ventricles have "escaped" from control by the atria, and they are beating at their own natural rate.

*Stokes-Adams Syndrome—Ventricular Escape.* In some patients with atrioventricular block, the total block comes and goes—that is, all impulses are conducted from the atria into the ventricles for a period of time, and then suddenly no impulses at all are transmitted. The duration of total block may be a few seconds, a few minutes, a few hours, or even weeks or more before conduction returns. In particular, this condition occurs in hearts with borderline ischemia.

Immediately after A-V conduction is first blocked, the ventricles stop contracting entirely for about 5 to 10 seconds. Then some part of the Purkinje system beyond the block, usually in the A-V bundle itself, begins discharging rhythmically at a rate of 15 to 40 times per minute and acting as the pacemaker of the ventricles. This is called *ventricular escape.* Because the brain cannot remain active for more than 3 to 5 seconds without blood supply, patients usually faint between block of conduction and "escape" of the ventricles and then recover because even the slowly beating ventricles usually pump enough blood to sustain the person. These periodic fainting spells are known as the Stokes-Adams syndrome.

Occasionally the interval of ventricular standstill is so long that it either becomes detrimental to the health of the patient or causes death. Consequently, many of these patients would be better off with total heart block than constantly shifting back and forth between total heart block and atrial control of the ventricles. Such patients are frequently treated with drugs that depress conduction in the A-V bundle so that the total heart block persists all the time. Commonly, digitalis is the drug used for this purpose, but quinidine or barium ion can also convert a Stokes-Adams syndrome into total heart block.

Also, many of these patients are now provided with an *artificial pacemaker,* which is a small battery-operated electrical stimulator planted be-

Dropped beat

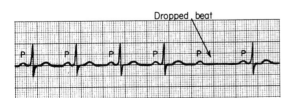

**Figure 17–6.** Partial atrioventricular block (lead V₃).

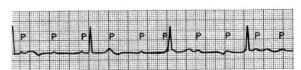

**Figure 17–7.** Complete atrioventricular block (lead II).

neath the skin, the electrodes from which are connected to the heart. This pacemaker provides continued rhythmic impulses that take over control of the heart. The batteries are replaced about once a year (once every 10 years for a new atomic powered battery).

## INCOMPLETE INTRAVENTRICULAR BLOCK—ELECTRICAL ALTERNANS

The same factors that can cause A-V block, except vagal stimulation, can also block impulse conduction in peripheral portions of the Purkinje system. At times, *incomplete* block occurs so that the impulse is sometimes transmitted and sometimes not transmitted, causing slowing or block of the impulse during some heart cycles and not during others. The QRS complex may be considerably abnormal during those cycles in which the impulse is slowed or blocked. Figure 17–8 illustrates the condition known as *electrical alternans,* which results from intraventricular block every other heartbeat. This electrocardiogram also illustrates tachycardia, which is probably the reason the block has occurred, for when the rate of the heart is very rapid, it may be impossible for portions of the Purkinje system to recover from the refractory period rapidly enough to respond during each succeeding heartbeat. Also, many conditions that depress the heart, such as ischemia, myocarditis, and digitalis toxicity, can cause incomplete intraventricular block and electrical alternans.

# PREMATURE BEATS

A premature beat is a contraction of the heart prior to the time that normal contraction would have been expected. This condition is also frequently called *extrasystole* or *ectopic beat.*

**Causes of Premature Beats.** Most premature beats result from *ectopic foci* in the heart, which emit abnormal impulses at odd times during the cardiac rhythm. The possible causes of ectopic foci are (1) local areas of ischemia, (2) small calcified plaques at different points in the heart, which press against the adjacent cardiac muscle so that some of the fibers are irritated, and (3) toxic irritation of the A-V node, Purkinje system, or myocardium caused by drugs, nicotine, caffeine, etc. Mechanical initiation of ec-

topic beats is also frequent during cardiac catheterization, large numbers of premature beats often occurring when the catheter enters the right ventricle and presses against the endocardium.

## ATRIAL PREMATURE BEATS

Figure 17–9 illustrates an electrocardiogram showing a single atrial premature beat. The P wave of this beat is relatively normal and the QRS complex is also normal, but the P-R interval and the interval between the preceding beat and premature beat is shortened. Also, the interval between the premature beat and the next succeeding beat is slightly prolonged. The reason for this is that the premature beat originated in the atrium some distance from the S-A node, and the impulse of the premature beat had to travel through a considerable amount of atrial muscle before it discharged the S-A node. Consequently, the S-A node discharged very late in the cycle and made the succeeding heartbeat also late in appearing.

Atrial premature beats occur frequently in healthy persons and, indeed, are often found in athletes or others whose hearts are certain to be in healthy condition. Yet mild toxic conditions resulting from such factors as excess smoking, lack of sleep, too much coffee, alcoholism, and use of various drugs can also initiate such beats.

**Pulse Deficit.** When the heart contracts ahead of schedule, the ventricles will not have filled normally, and the stroke volume output during that contraction is depressed or sometimes almost none at all. Therefore, the pulse wave passing to the periphery following a premature beat may be so weak that the pulse cannot be felt at all in the radial artery, whereas the succeeding pulse may be extra strong because of compensatory overfilling. Thus, a deficit in the number of pulses felt in the radial pulse occurs in relation to the number of beats of the heart.

**Bigeminal Pulse.** At times every other beat of the heart may be a premature beat. This causes the patient to have a bigeminal pulse—that is, two pulses close together, then a longer diastolic interval, then two again, etc.

## A-V NODAL OR A-V BUNDLE PREMATURE BEATS

Figure 17–10 illustrates a premature beat originating either in the A-V node or in the A-V bundle. The

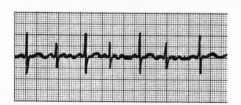

**Figure 17–8.** Partial intraventricular block—electrical alternans (lead III).

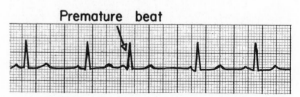

Premature beat

**Figure 17–9.** Atrial premature beat (lead I).

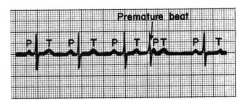

**Figure 17–10.**  A-V nodal premature beat (lead III).

P wave is missing from the record of the premature beat. Instead, the P wave is superimposed on the QRS-T complex of the premature beat; this distorts the complex, but the P wave itself cannot be discerned as such.

In general, A-V nodal premature beats have the same significance and causes as atrial premature beats.

## VENTRICULAR PREMATURE BEATS

The electrocardiogram of Figure 17–11 illustrates a series of ventricular premature beats alternating with normal beats. Ventricular premature beats result from an ectopic focus in the ventricles and cause several effects in the electrocardiogram, as follows: First, the QRS complex is usually considerably prolonged. The reason for this is that the impulse is conducted mainly through the slowly conducting muscle of the ventricle rather than through the Purkinje system.

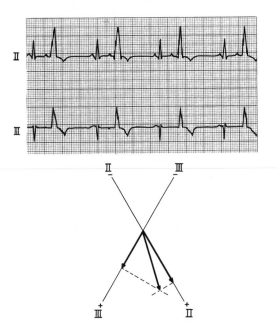

**Figure 17–11.**  Ventricular premature beats (leads II and III). Axis of the premature beats is plotted in accord with the principles of vectorial analysis explained in Chapter 16.

Second, the QRS complex has a very high voltage for the following reason: When the normal impulse passes through the heart, it passes through both ventricles approximately simultaneously; consequently, the depolarization waves of the two sides of the heart partially neutralize each other. However, when a ventricular premature beat occurs, the impulse travels in only one direction so that there is no such neutralization effect, and one entire side of the heart is depolarized while the other entire side is still polarized; this causes intense electrical potentials.

Third, following almost all ventricular premature beats the T wave has a potential opposite to that of the QRS complex, because the *slow conduction of the impulse* through the cardiac muscle causes the area first depolarized also to repolarize first. As a result, the direction of current flow in the heart during repolarization is opposite to that during depolarization, and the potential of the T wave is reversed to that of the QRS complex. This is not true of the normal T wave, as was explained in Chapter 16.

Some ventricular premature beats are relatively benign in their origin and result from simple factors such as cigarettes, coffee, lack of sleep, various mild toxic states, and even emotional irritability. On the other hand, a large share of ventricular premature beats results from actual pathology of the heart. For instance, many ventricular premature beats occur following coronary thrombosis because of stray impulses originating around the borders of the infarcted area of the heart. Therefore, presence of such beats is not to be taken lightly. Statistics show that persons with significant numbers of such beats have a much higher than normal chance of developing spontaneous lethal ventricular fibrillation, presumably initiated by a premature ventricular beat.

**Vector Analysis of the Origin of an Ectopic Ventricular Premature Beat.**  In Chapter 16 the principles of vectorial analysis were explained. Applying these principles one can determine from the electrocardiogram in Figure 17–11 the point of origin of the ventricular premature beat as follows: Note that the vectors of the premature beats in leads II and III are both strongly positive. Plotting these vectors on the axes of leads II and III and solving by vectorial analysis for the actual vector in the heart, one finds that the vector of the ectopic beat has its negative end (origin) at the base of the heart and its positive end toward the apex. Thus, the first portion of the heart to become depolarized during the premature beat lies near the base of the heart, which therefore is the locus of the ectopic focus.

## PAROXYSMAL TACHYCARDIA

Abnormalities in any portion of the heart, including the atria, the Purkinje system, and the ventricles, can sometimes cause rapid rhythmic discharge of impulses which spread in all directions throughout the heart. Because of the rapid rhythm in the irritable focus, this focus becomes the pacemaker of the heart.

The term "paroxysmal" means that the heart rate usually becomes very rapid in paroxysms, the paroxysms beginning suddenly and lasting for a few seconds, a few minutes, a few hours, or sometimes much longer. Then the paroxysms usually end as suddenly as they had begun, the pacemaker of the heart then shifting back to the S-A node.

Paroxysmal tachycardia can often be stopped by eliciting a vagal reflex. A strange type of vagal reflex sometimes elicited for this purpose is one that occurs when painful pressure is applied to the eyes. Also, pressure on the carotid sinuses can sometimes elicit enough vagal reflex to stop the tachycardia. Various drugs may also be used to stop the tachycardia. Two frequently used are quinidine and lidocaine, both of which inhibit the normal increase in sodium permeability of the cardiac muscle membrane during the generation of the action potential, thereby often blocking the rhythmic discharge of the irritable focus that is causing the paroxysmal attack.

### ATRIAL PAROXYSMAL TACHYCARDIA

Figure 17–12 illustrates a sudden increase in rate of heartbeat from approximately 95 beats per minute to approximately 150 beats per minute. On close observation of the electrocardiogram it will be seen that an inverted P wave occurs before each of the QRS-T complexes during the paroxysm of rapid heartbeat, though this P wave is partially superimposed on the normal T wave of the preceding beat. This indicates that the origin of this paroxysmal tachycardia is in the atrium, but, because the P wave is abnormal, the origin is not near the S-A node.

**A-V Nodal Paroxysmal Tachycardia.** Paroxysmal tachycardia very often results from an irritable focus in either the A-V node or the A-V bundle. This initiates normal impulses in the ventricles, thereby causing normal QRS-T complexes. However, P waves do not appear, because the atrial impulse travels backward from the A-V node through the atrium at the same time that the ventricular impulse travels through the ventricles. Therefore, the P wave is obscured by the QRS complex.

Atrial or A-V nodal paroxysmal tachycardia usually occurs in young, otherwise healthy persons. In general, atrial paroxysmal tachycardia frightens the individual tremendously and may cause him to be weak during the paroxysms, but no permanent harm comes from the attacks in most persons.

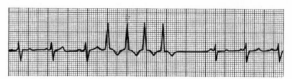

**Figure 17–13.**   Ventricular paroxysmal tachycardia (lead III).

### VENTRICULAR PAROXYSMAL TACHYCARDIA

Figure 17–13 illustrates a typical short paroxysm of ventricular tachycardia. The electrocardiogram of ventricular paroxysmal tachycardia has the appearance of a series of ventricular premature beats occurring one after another without any normal beats interspersed.

Ventricular paroxysmal tachycardia is usually a serious condition for two reasons. First, this type of tachycardia usually does not occur unless considerable damage is present in the ventricles. Second, ventricular tachycardia very frequently initiates ventricular fibrillation, which is almost invariably fatal. The reasons why ventricular tachycardia predisposes to ventricular fibrillation are: First, rapid stimulation decreases the refractory period of the heart and slows conduction, both of which predispose to circus movements. Second, one of the rapid ventricular impulses frequently occurs while part of the ventricular muscle is in a refractory state and part in a nonrefractory state; this causes the impulse to take a devious route through the heart, avoiding the refractory areas and initiating a long circus pathway, as was discussed in Chapter 14. The long circus pathway allows the impulse to continue around and around the ventricles, resulting in ventricular fibrillation.

Digitalis intoxication sometimes causes ventricular tachycardia. On the other hand, quinidine or lidocaine, which increase the refractory period of cardiac muscle and also increase its threshold for excitation, may be used to block irritable foci causing ventricular tachycardia.

## ABNORMAL RHYTHMS RESULTING FROM CIRCUS MOVEMENTS

The circus movement phenomenon was discussed in detail in Chapter 14, and it was pointed out that these movements can cause atrial flutter, atrial fibrillation, and ventricular fibrillation.

### ATRIAL FLUTTER

Figure 17–14 illustrates lead II of the electrocardiogram in atrial flutter. The rate of atrial contraction (P waves) is approximately 300 times per minute,

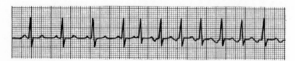

**Figure 17–12.**   Atrial paroxysmal tachycardia—onset in middle of record (lead I).

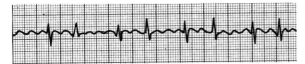

**Figure 17–14.**    Atrial flutter–2:1 and 3:1 rhythm (lead II).

while the rate of ventricular contraction (QRS-T waves) is only 125 times per minute. From the record it will be seen that sometimes a 2:1 rhythm occurs and sometimes a 3:1 rhythm. In other words, the atria beat two or three times for every one impulse that is conducted through the A-V bundle into the ventricles.

An important distinguishing feature of atrial flutter is the relatively high voltage, 0.2 to 0.3 mv., of the P waves in lead II and sometimes in lead III. Leads II and III record higher voltage than lead I because the circus movement pathway of the impulse causing atrial flutter passes around and around the atria from top to bottom. This means that the current flow in the atria extends up and down the body more or less parallel to the axes of leads II and III, which causes strong voltages in these leads but weak voltages in lead I.

## ATRIAL FIBRILLATION

Figure 17–15 illustrates the electrocardiogram during atrial fibrillation. As discussed in Chapter 14, numerous impulses spread in all directions through the atria during atrial fibrillation. The intervals between impulses arriving at the A-V node are extremely variable. Therefore, an impulse may arrive at the A-V node immediately after the node itself is out of its refractory period from its previous discharge, or it may not arrive there for several tenths of a second. Consequently, the rhythm of the ventricles is very irregular, many of the ventricular beats falling quite close together and many far apart, the overall ventricular rate being 125 to 150 beats per minute in most instances.

The fibrillatory process in the atria creates only minute voltages in the electrocardiogram because the impulses are traveling in all directions, and no major mass of muscle depolarizes at any one time. The potentials from the small fibrillatory impulses in the atria almost totally neutralize each other so that only a fine wavy record, or no waves at all, is recorded in

the electrocardiogram between ventricular beats, as shown in Figure 17–15. On the other hand, the QRS-T complexes are entirely normal unless there is some simultaneous pathology of the ventricles, but their timing is very irregular.

The pumping efficiency of the heart in atrial fibrillation is considerably depressed because the ventricles often do not have sufficient time to fill between beats. Therefore, one of the principles of treating patients with atrial fibrillation is to slow the rate of the ventricles. This is ordinarily done by administering digitalis, which causes a prolonged refractory period of the Purkinje system, probably by stimulating the vagal nerves to the A-V node and A-V bundle, thereby decreasing the number of impulses conducted from the atria into the ventricles and preventing any two ventricular beats from occurring close together. Heart function may be greatly improved when the rate is thus depressed to less than 100 beats per minute.

**Pulse Deficit in Atrial Fibrillation.**    In the same manner that premature beats can cause a pulse deficit in the radial pulse, so can ventricular beats occurring close together in atrial fibrillation also cause a pulse deficit. Therefore, before treatment of atrial fibrillation one frequently finds a pulse deficit, whereas, after slowing the ventricular rate, the pulse deficit will have disappeared.

## VENTRICULAR FIBRILLATION

In ventricular fibrillation the electrocardiogram is extremely bizarre, as shown in Figure 17–16, and ordinarily shows no tendency toward a rhythm of any type. In the early phases of ventricular fibrillation relatively large masses of muscle contract simultaneously, and this causes strong though irregular voltages in the electrocardiogram. However, after only a few seconds, the coarse contractions of the ventricles disappear, and the electrocardiogram changes into a new pattern of low-voltage, very irregular waves. Thus, no characteristic electrocardiographic pattern can be ascribed to ventricular fibrillation except that the electrical potentials constantly and spasmodically change, because the currents in the heart flow first in one direction, then in another, and rarely repeat any specific cycle.

The voltages of the waves in the electrocardiogram in ventricular fibrillation are usually about 0.5 mv. when ventricular fibrillation first begins, but these decay rapidly so that after 20 to 30 seconds they may be only 0.2 to 0.3 mv. Minute voltages of 0.1 mv. or

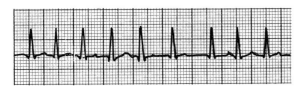

**Figure 17–15.**    Atrial fibrillation (lead I).

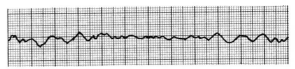

**Figure 17–16.**    Ventricular fibrillation (lead II).

less may be recorded for 10 minutes or longer after ventricular fibrillation begins. As already pointed out, ventricular fibrillation is lethal unless it is stopped by some heroic therapy such as immediate electroshock through the heart as explained in Chapter 14.

# REFERENCES

Abildskov, J. A.: The electrocardiogram. *In* Sodeman, W. A., Jr., and Sodeman; W. A. (eds.): Pathologic Physiology: Mechanisms of Disease, 5th Ed. Philadelphia, W. B. Saunders Company, 1974, p. 312.

Burch, G. E., and De Pasquale, N. P.: Electrocardiography in the Diagnosis of Congenital Heart Disease. Philadelphia, Lea & Febiger, 1967.

Burch, G. E., and Winsor, T.: A Primer of Electrocardiography, 6th Ed. Philadelphia, Lea & Febiger, 1972.

Caceres, C. A. (ed.): Clinical Electrocardiography and Computers: A Symposium. New York, Academic Press, Inc., 1970.

Chou, T., and Helm, R. A.: Clinical Vectorcardiography. New York, Grune & Stratton, Inc., 1967.

Chung, E. K.: Electrocardiography. Hagerstown, Maryland, Harper & Row, Publishers, 1974.

Dimond, E. G.: Electrocardiography and Vectorcardiography, 4th Ed. Boston, Little, Brown and Company, 1967.

Donoso, E. (ed.): Symposium on cardiac arrhythmias. *Circulation,* 47:1973.

Guyton, A. C., and Crowell, J. W.: A stereovectorcardiograph. *J. Lab. Clin. Med.,* 40:726, 1952.

Han, J. (ed.): Cardiac Arrhythmias. Springfield, Ill., Charles C Thomas, Publisher, 1972.

Hoffman, I.: Vectorcardiography. Philadelphia, J. B. Lippincott Co., 1966.

Hurst, J. W., and Myerburg, R. J.: Introduction to Electrocardiography. New York, McGraw-Hill Book Company, 1968.

Mangiola, S., and Ritota, M. C.: Cardiac Arrhythmias. Philadelphia, J. B. Lippincott Company, 1974.

Meyler, F. L., Robles de Medina, E. O., and Zimmerman, A. N. E.: Electrocardiography for Intensive Care Units. New York, American Elsevier Publishing Co., Inc., 1974.

Riseman, J. E. F.: P-Q-R-S-T: A Guide to Electrocardiogram Interpretation, 5th Ed. New York, The Macmillan Company, 1968.

Robles de Medina, E. O. (ed.): A New Coding System for Electrocardiography. New York, American Elsevier Publishing Co., Inc., 1972.

Scher, A. M., and Young, A. C.: Frequency analysis of the electrocardiogram. *Circ. Res.,* 8:344, 1960.

Simonson, E. (ed.): Measurement in Exercise Electrocardiography: The Ernst Simonson Conference. Springfield, Ill., Charles C Thomas, Publisher, 1969

Sodi-Pallares, D., and Calder, R. M.: New Bases of Electrocardiography. St. Louis, The C. V. Mosby Co., 1956.

Stephenson, H. E.: Cardiac Arrest and Resuscitation, 3rd Ed. St. Louis, The C. V. Mosby Co., 1969.

Watanabe, Y., and Dreifus, L. S.: Arrhythmias—mechanisms and pathogenesis. *In* Sodeman, W. A., Jr., and Sodeman, W. A. (eds.): Pathologic Physiology: Mechanisms of Disease, 5th Ed. Philadelphia, W. B. Saunders Company, 1974, p. 329.

Zywietz, Chr., and Schneider, B. (eds.): Computer Application on ECG and VCG Analysis. New York, American Elsevier Publishing Co., Inc., 1973.

# PART V
# THE CIRCULATION

# 18

# Physics of Blood, Blood Flow, and Pressure: Hemodynamics

## THE CIRCULATORY SYSTEM AS A "CIRCUIT"

The most important feature of the circulation that must always be kept in mind is that it is a continuous circuit. That is, if a given amount of blood is pumped by the heart, this same amount must also flow through each respective subdivision of the circulation. Furthermore, if blood is displaced from one segment of the circulation, some other segment of the circulation must expand unless fluid is lost from the circulation.

Figure 18–1 illustrates the general plan of the circulation, showing the two major subdivisions, the *systemic circulation* and the *pulmonary circulation*. In the figure the arteries of each subdivision are represented by a single distensible chamber and all the veins by another, even larger distensible chamber, and the arterioles and capillaries represent very small connections between the arteries and veins. Blood flows with almost no resistance in all the larger vessels of the circulation, but this is not the case in the arterioles and capillaries, where considerable resistance does occur. To cause blood to flow through these small "resistance" vessels, the heart pumps blood into the arteries under high pressure—normally at a systolic pressure of about 120 mm. Hg in the pulmonary system.

As a first step toward explaining the overall function of the circulation, this chapter will discuss the physical characteristics of blood itself and then the physical principles of blood flow through the vessels, including especially the interrelationships between pressure, flow, and resistance. The study of these interrelationships

and other basic physical principles of blood circulation is called *hemodynamics*.

## THE PHYSICAL CHARACTERISTICS OF BLOOD

Blood is a viscous fluid composed of *cells* and *plasma*. More than 99 per cent of the cells are red blood cells; this means that for practical purposes the white blood cells play almost no role in determining the physical characteristics of the blood. The fluid of the plasma is part of the extracellular fluid, and the fluid inside the cells is intracellular fluid.

### THE HEMATOCRIT

The hematocrit of blood is the per cent of the blood that is cells. Thus, if one says that a person has a hematocrit of 40, he means that 40 per cent of the blood volume is cells and the remainder is plasma. The hematocrit of normal man averages about 42, whereas that of normal woman averages about 38. These values vary tremendously, depending upon whether or not the person has anemia, upon his degree of bodily activity, and on the altitude at which he resides. These effects were discussed in relation to the red blood cells and their function in Chapter 5.

Blood hematocrit is determined by centrifuging blood in a calibrated tube such as that shown in Figure 18–2. The calibration allows direct reading of the per cent of cells.

PULMONARY  CIRCULATION

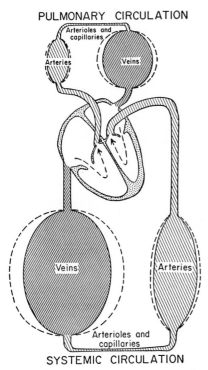

SYSTEMIC  CIRCULATION

**Figure 18–1.** Schematic representation of the circulation, showing both the distensible and the resistive portions of the systemic and pulmonary circulations.

**Effect of Hematocrit on Blood Viscosity.** The greater the percentage of cells in the blood—that is, the greater the hematocrit—the more friction there is between successive layers of blood, and it is this friction that determines viscosity. Therefore, the viscosity of blood increases drastically as the hematocrit increases, as illustrated in Figure 18–3. If we arbitrarily consider the viscosity of water to be 1, then the viscosity of whole blood at normal hematocrit is about 3 to 4; this means that 3 to 4 times as much pressure is required to force whole blood through a given tube than to force water through the same tube. Note that when the hematocrit rises to 60 or 70, which it often does in polycythemia as discussed in Chapter 5, the blood viscosity can become as great as 10 times that of water, and its flow through blood vessels is greatly retarded.

Another factor that affects blood viscosity is the concentration and types of proteins in the plasma, but these effects are so much less important than the effect of hematocrit that they are not significant considerations in most hemodynamic studies. The viscosity of blood plasma is 1.5 to 2.0 times that of water.

Since most resistance in the circulatory system occurs in the very small blood vessels, it is especially important to know how blood viscosity affects blood flow in these minute vessels. At least three additional factors besides hematocrit and plasma proteins affect blood viscosity in these vessels:

1. Blood flow in very minute vessels exhibits far less viscous effect than it does in large vessels. This effect, called the *Fahraeus-Lindqvist effect,* begins to appear when the vessel diameter falls below approximately 1.5 mm. In vessels as small as capillaries, this effect is so prominent that the viscosity of whole blood is as little as one-half that in large vessels. The Fahraeus-Lindqvist effect is probably caused by alignment of the red cells as they pass through the vessels. That is, the red cells, instead of moving randomly, line up and move through the vessels as a single plug, thus eliminating the viscous resistance that occurs internally in the blood itself. The Fahraeus-Lindqvist effect, however, is probably more than offset by the following two effects under most conditions.

2. The viscosity of blood increases drastically as its velocity of flow decreases. Since the velocity of blood flow in the small vessels is extremely minute, often less than 1 mm. per second, blood viscosity can increase as much as 10-fold from this factor alone. This effect is presumably caused by adherence of the

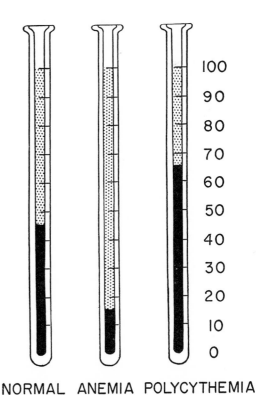

NORMAL  ANEMIA  POLYCYTHEMIA

**Figure 18–2.** Hematocrits in the normal person and in anemia and polycythemia.

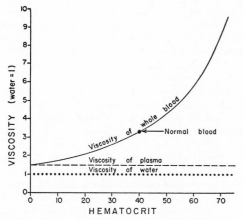

**Figure 18–3.** Effect of hematocrit on viscosity.

red cells to each other (formation of rouleaux and larger aggregates) and to the vessel walls.

3. Cells also often become stuck at constrictions in small blood vessels; this happens especially in capillaries where the nuclei of endothelial cells protrude into the capillary lumen. When this occurs, blood flow can become totally blocked for a fraction of a second, for several seconds, or for much longer periods of time, thus giving an apparent effect of greatly increased viscosity.

Because of these special effects that occur in the minute vessels of the circulatory system, it has been impossible to determine the exact manner in which hematocrit affects viscosity in the minute vessels, which is the place in the circulatory system where viscosity almost certainly plays its most important role. Nevertheless, because some of these effects tend to decrease viscosity and others tend to increase viscosity, it is perhaps best at present simply to assume that the overall viscous effects in the small vessels are approximately equivalent to those that occur in the larger vessels.

## PLASMA

Plasma is part of the extracellular fluid of the body. It is almost identical to the interstitial fluid found between the tissue cells except for one major difference: plasma contains about 7 per cent protein, whereas interstitial fluid contains an average of only 2 per cent protein. The reason for this difference is that plasma protein leaks only slightly through the capillary pores into the interstitial spaces. As a result, most of the plasma protein is held in the circulatory system, and that which does leak is eventually returned to the circulation by the lymph vessels. Therefore, the plasma protein concentration is about three and one-half times that of the fluid outside the capillaries.

**The Types of Protein in Plasma.** The plasma protein is divided into three major types, as follows:

|               | Grams Per Cent |
|---------------|------|
| Albumin       | 4.5  |
| Globulins     | 2.5  |
| Fibrinogen    | 0.3  |

The primary function of the *albumin* (and of the other types of protein to a lesser extent) is to cause osmotic pressure at the capillary membrane. This pressure, called *colloid osmotic pressure,* prevents the fluid of the plasma from leaking out of the capillaries into the interstitial spaces. This function is so important that it will be discussed in detail in Chapter 30.

The globulins are divided into three major types: alpha, beta, and gamma globulins. The *alpha* and *beta globulins* perform diverse functions in the circulation, such as transporting other substances by combining with them, acting as substrates for formation of other substances, and transporting protein itself from one part of the body to another. The *gamma globulins,* and to a lesser extent the *beta globulins,* play a special role in protecting the body against infection, for it is these globulins that are mainly the *antibodies* that resist infection and toxicity, thus providing the body with what we call *immunity*. The function of immunity was discussed in detail in Chapter 7.

The *fibrinogen* of plasma is of basic importance in blood clotting and was discussed in Chapter 9.

## INTERRELATIONSHIPS BETWEEN PRESSURE, FLOW, AND RESISTANCE

Flow through a blood vessel is determined entirely by two factors: (1) the *pressure difference* tending to push blood through the vessel and (2) the impediment to blood flow through the vessel, which is called vascular *resistance*. Figure 18–4 illustrates these relationships, showing a blood vessel segment located anywhere in the circulatory system.

$P_1$ represents the pressure at the origin of the vessel; at the other end the pressure is $P_2$. Since the vessel is relatively small, the blood experiences difficulty in flowing. This is the resistance. Stated another way, a *pressure difference* between the two ends of the vessel causes *blood* to *flow* from the high pressure area to the low pressure area while *resistance* impedes the flow. This can be expressed mathematically as follows:

$$Q = \frac{\Delta P}{R} \qquad (1)$$

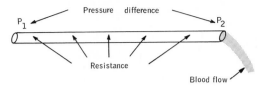

**Figure 18–4.** Relationship between pressure, resistance, and blood flow.

in which Q is blood flow, ΔP is the pressure difference ($P_1 - P_2$) between the two ends of the vessel, and R is the resistance.

It should be noted especially that it is the *difference* in pressure between the two ends of the vessel that determines the rate of flow and not the absolute pressure in the vessel. For instance, if the pressure at both ends of the segment were 100 mm. Hg and yet no difference existed between the two ends, there would be no flow despite the presence of the 100 mm. Hg pressure.

The above formula expresses the most important of all the relationships that the student needs to understand to comprehend the hemodynamics of the circulation. Because of the extreme importance of this formula the student should also become familiar with its other two algebraic forms:

$$\Delta P = Q \times R \qquad (2)$$

$$R = \frac{\Delta P}{Q} \qquad (3)$$

## BLOOD FLOW

Blood flow means simply the quantity of blood that passes a given point in the circulation in a given period of time. Ordinarily, blood flow is expressed in *milliliters* or *liters per minute,* but it can be expressed in milliliters per second or in any other unit of flow.

The overall blood flow in the circulation of an adult person at rest is about 5000 ml. per minute. This is called the *cardiac output* because it is the amount of blood pumped by each ventricle of the heart in a unit period of time. It is obvious that this same amount of blood must pass through both the systemic and pulmonary circulations.

**Methods for Measuring Blood Flow.** Many different mechanical or mechanoelectrical devices can be inserted in series with a blood vessel, or in some instances applied to the outside of the vessel, to measure flow. These are called simply *flowmeters.*

*The Rotameter.* Figure 18–5 illustrates a *rotameter.* In using this the blood vessel must be cut and the blood passed through the flowmeter. Also, an anticoagulant, such as heparin, must be used to prevent blood clotting.

Blood enters the rotameter from the bottom and flows upward through a cone-shaped cavity. In this cavity is a small float. When no blood at all is flowing, the float falls to the bottom of the cavity as shown in the figure, but as blood flows faster and faster, the float rises higher and higher. In using the rotameter, one determines the blood flow by measuring the height to which the float rises, or, if he wishes to record the blood flow continuously, an arrangement such as that shown in the figure can be used. A small steel shaft projects upward from the float into a coil of wire. The inductance of the coil increases as the steel shaft moves farther up into the coil, and by using appropriate electronic apparatus the blood flow can be recorded continuously.

*The Electromagnetic Flowmeter.* In recent years several new devices have been developed which can be used to measure blood flow in a vessel without opening it. One of the most important of these is the electromagnetic flowmeter, the principles of which are illustrated in Figure 18–6. Figure 18–6A shows generation of electromotive force in a wire that is moved rapidly through a magnetic field. This is the well-known principle for production of electricity by the electric generator. Figure 18–6B shows that exactly the same principle applies for generation of electromotive force in blood when it moves through a magnetic field. In this case, a blood vessel is placed between the poles of a strong magnet, and electrodes are placed on the two sides of the vessel perpendicu-

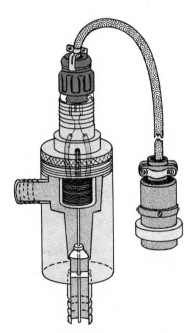

**Figure 18–5.** A flowmeter of the rotameter type.

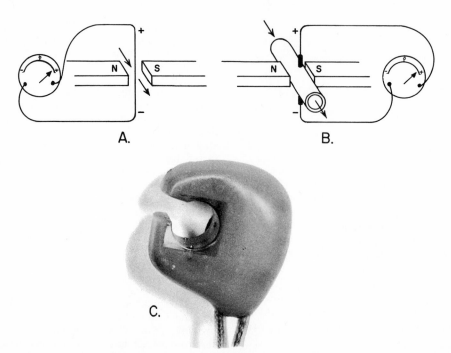

**Figure 18–6.** A flowmeter of the electromagnetic type, showing (A) generation of an electromotive force in a wire as it passes through an electromagnetic field, (B) generation of an electromotive force in electrodes on a blood vessel when the vessel is placed in a strong magnetic field and blood flows through the vessel, and (C) a modern electromagnetic flowmeter "probe" for chronic implantation around blood vessels.

lar to the magnetic lines of force. When blood flows through the vessel, electrical voltage proportional to the rate of flow is generated between the two electrodes, and this is recorded using an appropriate meter or electronic apparatus. Figure 18–6C illustrates an actual "probe" that is placed on a large blood vessel to record its blood flow. This probe contains both the strong magnet and the electrodes.

An additional advantage of the electromagnetic flowmeter is that it can record changes in flow that occur in less than 0.01 second, allowing accurate recording of pulsatile changes in flow as well as steady state changes.

*The Ultrasonic Flowmeter.* Another type of flowmeter that can be applied to the outside of the vessel and that has many of the same advantages as the electromagnetic flowmeter is the ultrasonic flowmeter illustrated in Figure 18–7. A minute piezoelectric crystal is mounted in the wall of each half of the device; one of these crystals transmits sound diagonally along the vessel and the other receives the sound. An electronic apparatus alternates the direction of sound transmission several hundred times per second, transmitting first downstream, then upstream. Sound waves travel downstream with greater velocity than upstream. An appropriate electronic apparatus measures the difference between these two velocities, which is a measure of blood flow. The sound frequency can be anywhere between 100,000 and 4,000,000 cycles per second.

*The Plethysmograph.* Figure 18–8 illustrates an apparatus known as a "plethysmograph," which is used for estimating the quantity of blood flow through the forearm. This apparatus operates as follows: The forearm is placed in a chamber, and an airtight seal is made between the proximal portion of the chamber and the forearm near the elbow. The chamber is connected to a tambour recorder so that an increase in volume of the arm inside the airtight chamber causes the membrane of the tambour to rise. To determine the quantity of blood flowing through the forearm, a blood pressure cuff on the arm above the plethysmograph is suddenly inflated to a pressure above venous blood pressure but below arterial blood pressure— 40 mm. Hg, for instance. The pressure in the cuff occludes the veins so that

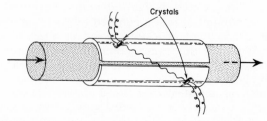

**Figure 18–7.** Basic construction of the ultrasonic flowmeter probe.

blood cannot return from the forearm into the general circulatory system, but blood does continue to flow through the unoccluded arteries into the forearm. Therefore, when the blood pressure cuff is first inflated, the forearm swells, and the rate of swelling is equal to the rate of blood flow into the forearm. This is registered on the recorder, as illustrated in Figure 18–8A and B. The plethysmograph is then calibrated by injecting into the chamber standard volumes of air or fluid from the syringe shown also in the diagram.

In making measurements of blood flow from the plethysmographic recording, a tangent is drawn to the recording at the very onset of the rise, as illustrated in Figure 18–8B. The reason for this is that, after a few seconds of blood flow into the arm, all the blood vessels of the forearm become distended and exert a considerable amount of back pressure which opposes further inflow of blood. Therefore, it is only immediately after the cuff pressure is elevated that the recording actually represents the normal rate of blood flow into the forearm. In the recording of the figure, measurements from the tangent show that the initial rate of blood flow is 30 ml. in 20 seconds, or 90 ml. per minute.

Plethysmographs can be used for recording blood flow in a leg, a toe, a finger, or even in a number of internal organs of the body, for plethysmographs known as "oncometers" can be placed around the spleen, the kidney, the liver, various glands, and even the heart. Unfortunately, technical difficulties in the use of plethysmographs are likely to make their recordings inaccurate—sometimes 20 per cent or more.

**Laminar Flow of Blood in Vessels.** When blood flows at a steady rate through a long, smooth vessel, it flows in *streamlines* and also in concentric layers within the vessel, with each layer remaining the same distance from the wall. Also, the central portion of the blood stays in the center of the vessel. This type of flow is opposite to *turbulent flow,* which is blood flowing in all directions in the vessels and continually mixing within the vessel, as will be discussed below.

**Parabolic Velocity Profile During Laminar Flow.** When laminar flow occurs, the velocity of flow in the center of the vessel is far greater than that toward the outer edges. This is illustrated by the experiment shown in Figure 18–9. In vessel A are two different fluids, the one to the left colored by a dye and the one to the right a clear fluid, but there is no flow in the vessel. Then the fluids are made to flow; and a parabolic interface develops between the two fluids, as shown 1 second later in vessel B, illustrating that the portion of fluid adjacent to the walls has hardly moved at all, the portion slightly away from the wall has moved a small

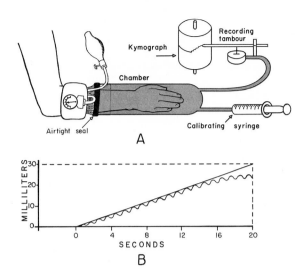

**Figure 18–8.** (A) The plethysmograph. (B) A plethysmographic record.

distance, and the portion in the center of the vessel has moved a long distance. This effect is call a parabolic profile for the velocity of blood flow.

The cause of the parabolic profile is the following: The fluid molecules touching the wall hardly move because of adherence to the vessel wall. The next layer of molecules slips over these, the third layer over the second, the fourth layer over the third, and so forth. Therefore, the fluid in the middle of the vessel can move rapidly because many layers of molecules exist between the middle of the vessel and the vessel wall, all of these capable of slipping over each other, while those portions of fluid near the wall do not have this advantage.

**Turbulent Flow of Blood Under Some Conditions.** When the rate of blood flow becomes too great, when it passes by an obstruction in a vessel, when it makes a sharp turn, or when it passes over a rough surface, the flow may then become *turbulent* rather than streamline. Turbulent flow means that the blood flows crosswise in the vessel as well as along the vessel, usually forming whorls in the blood called *eddy currents.* These are similar to the whirlpools

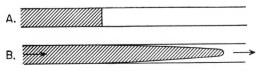

**Figure 18–9.** An experiment illustrating laminar blood flow, showing (A) two separate fluids before flow begins and (B) the same fluids 1 second after flow begins.

that one frequently sees in a rapidly flowing river at a point of obstruction.

When eddy currents are present, blood flows with much greater resistance than when the flow is streamline because the eddies add tremendously to the overall friction of flow in the vessel.

The tendency for turbulent flow increases in direct proportion to the velocity of blood flow, in direct proportion to the diameter of the blood vessel, and inversely proportional to the viscosity of the blood divided by its density in accordance with the following equation:

$$\text{Re} = \frac{v \cdot d}{\frac{\eta}{\rho}} \qquad (4)$$

in which Re is *Reynolds' number* and is the measure of the tendency for turbulence to occur, v is the velocity of blood flow (in centimeters per second), d is the diameter of the vessel (in centimeters), $\eta$ is the viscosity (in poises), and $\rho$ is density. The viscosity of blood is normally about $1/_{30}$ poise, and the density is only slightly greater than 1. When Reynolds' number rises above 200 (up to 400), turbulent flow will occur at the branches of vessels but will die out along the smooth portions of the vessels. However, when Reynolds' number rises above approximately 2000, turbulence can occur even in a straight, smooth vessel. Reynolds' number for flow in the vascular system rises above 200 (up to 400) in large arteries; as a result there is considerable turbulence of flow at the branches of these vessels. In the proximal portions of the aorta and pulmonary artery, Reynolds' number can rise to several thousand during the rapid phase of ejection by the ventricles; this causes marked turbulence in the proximal aorta and pulmonary artery where many conditions are appropriate for turbulence: (1) high velocity of blood flow, (2) pulsatile nature of the flow, (3) sudden change in vessel diameter, and (4) large vessel diameter.

However, in small vessels, Reynolds' number is almost never high enough to cause turbulence.

## BLOOD PRESSURE

**The Standard Units of Pressure.** Blood pressure is almost always measured in *millimeters of mercury (mm. Hg)* because the mercury manometer (shown in Figure 18–10) has been used as the standard reference for measuring blood pressure throughout the history of physiology. Actually, blood pressure means the *force exerted by the blood against any unit area of the vessel wall.* When one says that the pressure in a vessel is 50 mm. Hg, this means that the force exerted would be sufficient to push a column of mercury up to a level of 50 mm. If the pressure were 100 mm. Hg, it would push the column of mercury up to 100 mm.

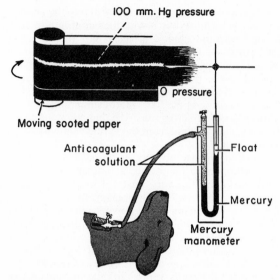

**Figure 18–10.** Recording arterial pressure with a mercury manometer.

Occasionally, pressure is measured in *centimeters of water.* A pressure of 10 cm. of water means a pressure sufficient to raise a column of water to a height of 10 cm. *One millimeter of mercury equals 1.36 cm. of water* because the specific gravity of mercury is 13.6 times that of water, and 1 cm. is 10 times as great as 1 mm. Dividing 13.6 by 10, we derive the factor 1.36.

**Measurement of Blood Pressure Using the Mercury Manometer.** Figure 18–10 illustrates a standard mercury manometer for measuring blood pressure. A cannula is inserted into an artery, a vein, or even into the heart, and the pressure from the cannula is transmitted to the left-hand side of the manometer where it pushes the mercury down while raising the right-hand mercury column. The difference between the two levels of mercury is approximately equal to the pressure in the circulation in terms of millimeters of mercury. (To be more exact, it is equal to 104 per cent of the true pressure because of the weight of the water on the left-hand column of mercury.)

**High-Fidelity Methods for Measuring Blood Pressure.** Unfortunately, the mercury in the mercury manometer has so much *inertia* that it cannot rise and fall rapidly. For this reason the mercury manometer, though excellent for recording steady pressures, cannot respond to pressure changes that occur more rapidly than approximately one cycle every 2 to 3 seconds. Whenever it is desired to record rapidly changing pressures, some other type of pressure recorder is needed. Figure 18–11 demonstrates the basic principles of three electronic pressure *transducers* commonly used for converting pressure into electrical signals and then recording the pressure on a high-speed electrical recorder. Each of these transducers employs a very thin and highly stretched metal membrane which forms one wall of the fluid chamber. The fluid chamber in turn is connected through a needle or catheter with the vessel in which the pressure is to be measured. Pressure variations in the vessel cause changes of pressure in the chamber beneath the membrane. When the pressure is high the membrane bulges outward slightly, and when low it returns toward its resting position. In Figure 18–11A a simple metal plate is placed a few thousandths of an inch above the membrane. When the membrane bulges outward, the *capacitance* between the plate and membrane increases, and this change in capacitance can be recorded by an appropriate electronic system. In Figure 18–11B a small iron slug rests on the membrane, and this can be displaced upward into a coil. Movement of the iron changes the *inductance* of the coil, and this, too, can be recorded electronically. Finally, in Figure 18–11C a very thin, stretched resistance wire is connected to the membrane. When this wire is greatly stretched its resistance increases, and when less stretched the resistance decreases. These changes also can be recorded by means of an electronic system.

It is possible also to connect a mirror to the edge of the membrane so that it angulates as the membrane bulges. A beam of light reflected from the mirror projects a spot of light onto a moving photographic paper, thus recording the rapidly changing pressure.

Using some of these high fidelity types of recording systems, pressure cycles up to 500 cycles per second have been recorded accurately, and in common use are recorders capable of registering pressure changes as rapidly as 20 to 100 cycles per second.

## RESISTANCE TO BLOOD FLOW

**Units of Resistance.**   Resistance is the impediment to blood flow in a vessel, but it cannot be measured by any direct means. Instead, resistance must be calculated from measurements of blood flow and pressure difference in the vessel. If the pressure difference between two points in a vessel is 1 mm. Hg and the flow is 1 ml./sec., then the resistance is said to be 1 *peripheral resistance unit,* usually abbreviated *PRU*.

***Expression of Resistance in CGS Units.***   Occasionally, a basic physical unit called the CGS unit is used to express resistance. This unit is *dyne seconds/centimeters⁵*. Resistance in these units can be calculated by the following formula:

$$R \left( in \frac{\text{dyne sec.}}{\text{cm.}^5} \right) = \frac{1333 \times \text{mm. Hg.}}{\text{ml./sec.}} \quad (5)$$

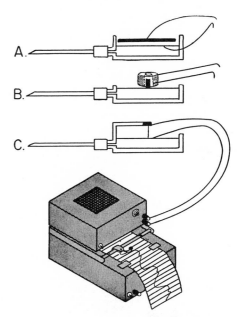

**Figure 18–11.**   Principles of three different types of electronic transducers for recording rapidly changing blood pressures.

**Total Peripheral Resistance and Total Pulmonary Resistance.**   The rate of blood flow through the circulatory system when a person is at rest is close to 100 ml./sec., and the pressure gradient from the systemic arteries to the systemic veins is about 100 mm. Hg. Therefore, in round figures the resistance of the entire systemic circulation, called the *total peripheral resistance,* is approximately 100/100 or 1 PRU. In some conditions in which the blood vessels

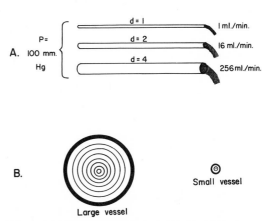

**Figure 18–12.**   (A) Demonstration of the effect of vessel diameter on blood flow. (B) Concentric rings of blood flowing at different velocities; the farther away from the vessel wall, the faster the flow.

throughout the body become strongly constricted, the total peripheral resistance rises to as high as 4 PRU, and when the vessels become greatly dilated it can fall to as little as 0.25 PRU.

In the pulmonary system the mean arterial pressure averages 13 mm. Hg and the mean left atrial pressure averages 4 mm. Hg, giving a net pressure difference of 9 mm. Therefore, in round figures the *total pulmonary resistance* at rest calculates to be about 0.09 PRU. This can increase in disease conditions to as high as 1 PRU and can fall in certain physiological states, such as exercise, to as low as 0.03 PRU.

**"Conductance" of Blood in a Vessel and Its Relationship to Resistance.** Conductance is a measure of the amount of blood flow that can pass through a vessel in a given time for a given pressure difference. This is generally expressed in terms of ml./sec./mm. Hg pressure, but it can also be expressed in terms of liters/sec./mm. Hg or in any other units of blood flow and pressure.

It is immediately evident that conductance is the reciprocal of resistance in accord with the following equation:

$$\text{Conductance} = \frac{1}{\text{Resistance}} \qquad (6)$$

**Effect of Vascular Diameter on Conductance.** Slight changes in diameter of a vessel cause tremendous changes in its ability to conduct blood when the blood flow is streamline. This is illustrated forcefully by the experiment in Figure 18–12A, which shows three separate vessels with relative diameters of 1, 2, and 4 but with the same pressure difference of 100 mm. Hg between the two ends of the vessels. Though the diameters of these vessels increase only four-fold, the respective flows are 1, 16, and 256 ml./mm., which is a 256-fold increase in flow. Thus, the conductance of the vessel increases in proportion to the *fourth power of the diameter,* in accord with the following formula:

$$\text{Conductance} \propto \text{Diameter}^4 \qquad (7)$$

The cause of this great increase in conductance with an increase in diameter can be explained by referring to Figure 18–12B. This illustrates cross-sections of a large and a small vessel. The concentric rings inside these vessels indicate that the velocity of flow in each ring is different from that in the other rings because of *laminar* flow, which was discussed earlier in the chapter. That is, the blood in the ring touching the wall of the vessel is flowing hardly at all because of its adherence to the vascular endothelium. The next ring of blood slips past the first

ring and, therefore, flows at a more rapid velocity. The third, fourth, fifth, and sixth rings likewise flow at progressively increasing velocities. Thus, the blood that is very near the wall of the vessel flows extremely slowly, while that in the middle of the vessel flows extremely rapidly.

In the small vessel essentially all of the blood is very near the wall so that the extremely rapidly flowing central stream of blood simply does not exist.

By integrating the velocities of all the concentric rings of flowing blood one can derive the following formula relating *mean velocity* of blood flow to vascular diameter:

$$v = \frac{\Delta P \cdot r^2}{8\eta l} \qquad (8)$$

in which v is velocity (in centimeters per second), $\Delta P$ is the pressure gradient (in dynes per square centimeter), r is the radius of the vessel (in centimeters), $\eta$ is the viscosity (in poises), and 1 is the length of the vessel (in centimeters).

**Poiseuille's Law.** The quantity of blood that will flow through a vessel in a given period of time is equal to the velocity of flow times the cross-sectional area according to the following equation:

$$Q = v\pi r^2 \qquad (9)$$

in which Q is the rate of blood flow (in milliliters per second) and $\pi r^2$ is the cross-sectional area (in square centimeters).

Now let us substitute the value for velocity of blood flow from Equation 8 into Equation 9. This gives the following equation, which is known as Poiseuille's law:

$$Q = \frac{\pi \Delta P r^4}{8\eta l} \qquad (10)$$

Note particularly in this equation that the rate of blood flow is directly proportional to the *fourth power of the radius* of the vessel, which illustrates once again that the diameter of a blood vessel plays by far the greatest role of all factors in determining the rate of blood flow through the vessel.

**Summary of the Different Factors that Affect Conductance and Resistance.** In the equation representing Poiseuille's law, Q represents flow, and $\Delta P$ represents the pressure difference. The remainder of the equation represents the conductance in accordance with the following equation:

$$C = \frac{\pi r^4}{8\eta l} \qquad (11)$$

And, since conductance is the reciprocal of resistance, the following equation shows the factors that affect resistance:

$$R = \frac{8\eta l}{\pi r^4} \qquad (12)$$

Thus, note that the resistance of a vessel is directly proportional to the *blood viscosity* ($\eta$) and *length* (l) of the vessel but inversely proportional to the *fourth power of the radius* (r).

**Resistance to Blood Flow Through Series Vessels.** In Figure 18–13A are two vessels connected in series, having resistances of $R_1$ and $R_2$. It is immediately evident that the total resistance is equal to the sum of the two, or

$$R_{(total)} = R_1 + R_2 \qquad (13)$$

Furthermore, it is equally evident that any number of resistances in series with each other must be added together. For instance, the total peripheral resistance is equal to the resistance of the arteries plus that of the arterioles plus that of the capillaries plus that of the veins.

**Resistance of Vessels in Parallel.** Shown in Figure 18–13B are four vessels, *connected in parallel*, with respective resistances of $R_1$, $R_2$, $R_3$, and $R_4$. However, the diameters of these vessels are not exactly the same. It is obvious that for a given pressure difference, far greater amounts of blood will flow through this system than through any one of the vessels alone. Therefore, the total resistance is far less than the resistance of any single vessel.

To calculate the total resistance in Figure 18–13B one first determines the *conductance* of each of the vessels, which is equal to the reciprocal of the resistance, or $\frac{1}{R_1}$, $\frac{1}{R_2}$, $\frac{1}{R_3}$, and $\frac{1}{R_4}$. The total conductance of all the vessels, $\frac{1}{R_{(total)}}$, is equal to the sum of the individual conductances:

$$\frac{1}{R_{(total)}} = \frac{1}{R_1} + \frac{1}{R_2} + \frac{1}{R_3} + \frac{1}{R_4} \qquad (14)$$

And resistance through the parallel circuit is the reciprocal of the total conductance, or

$$R_{(total)} = \cfrac{1}{\frac{1}{R_1} + \frac{1}{R_2} + \frac{1}{R_3} + \frac{1}{R_4}} \qquad (15)$$

### Effect of Pressure on Vascular Resistance– Critical Closing Pressure

Since all blood vessels are distensible, increasing the pressure inside the vessels causes the vascular diameters also to increase. This in turn reduces the resistance of the vessel. Conversely, reduction in vascular pressures increases the resistance.

The middle curve of Figure 18–14 illustrates the effect on blood flow through a small tissue vascular bed caused by changing the arterial

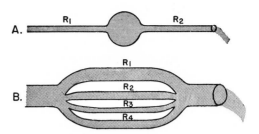

**Figure 18–13.** Vascular resistances: (A) in series and (B) in parallel.

pressure. As the arterial pressure falls from 130 mm. Hg, the flow decreases rapidly at first because of two factors: (1) the decreasing pressure difference between the artery and the vein of the tissue, and (2) the decreasing diameters of the vessels. At 20 mm. Hg blood flow ceases entirely. This point at which the blood stops flowing is called the *critical closing pressure,* because at this point the small vessels, the arterioles in particular, close so completely that all flow through the tissue ceases. The mechanism of this can be explained as follows:

**Mechanism of Critical Closing Pressure.** The vasomotor tone of the arterioles is always attempting to constrict these vessels to smaller diameters, while on the other hand the pressure inside the arterioles is tending to dilate them. As the pressure falls progressively lower and lower, it finally reaches a point at which the pressure inside the vessel is no longer capable of keeping the vessel open; this pressure is called the critical closing pressure.

A physical law, the *law of Laplace,* helps to ex-

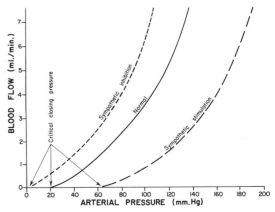

**Figure 18–14.** Effect of arterial pressure on blood flow through a blood vessel at different degrees of vascular tone, showing also the effect of vascular tone on critical closing pressure.

plain the closure of vessels at an exact critical pressure level. This law states that the circumferential force tending to stretch the muscle fibers in the vascular wall is proportional to the *diameter of the vessel* times the *pressure,* or F∝D·P. Therefore, as the pressure in the vessel falls, thus allowing the vascular diameter also to decrease, the force tending to keep the vascular wall stretched decreases extremely rapidly. There comes a pressure below which the elastic tension in the wall that is tending to close the vessel becomes greater than the stretching force caused by the pressure; then the vessel can no longer remain open.

Another factor that contributes to the cessation of blood flow when the arterial pressure falls low is the size of the blood cells themselves, for, when the arteriolar diameter falls below a certain critical diameter, the red cells cannot pass through, and they will actually block the flow of plasma as well. The average critical closing pressure is about 20 mm. Hg when whole blood is flowing through the vessels and 5 to 10 mm. Hg when only plasma is flowing through.

**Effect of Sympathetic Inhibition and Stimulation on Vascular Flow.**   Essentially all blood vessels of the body are normally stimulated by sympathetic impulses even under resting conditions. Different circulatory reflexes, which will be discussed in Chapter 21, can cause these impulses either to disappear, called *sympathetic inhibition,* or to increase to many times their normal rate, called *sympathetic stimulation*. Sympathetic impulses in most parts of the body cause an increase in tone of the vascular smooth muscle. Therefore, as illustrated by the dashed curves of Figure 18–14, sympathetic inhibition allows far more blood to flow through a tissue for a given pressure gradient than is normally true, whereas sympathetic stimulation vastly decreases the rate of blood flow.

Note also that sympathetic inhibition decreases the critical closing pressure, sometimes to as low as 5 mm. Hg. On the other hand, sympathetic stimulation can increase the critical closing pressure to as high as 100 or more mm. Hg, which means that strong sympathetic stimulation can actually stop blood flow through some tissues.

# VASCULAR DISTENSIBILITY— PRESSURE-VOLUME CURVES

The diameter of blood vessels, unlike that of metal pipes and glass tubes, increases as the internal pressure increases, because blood vessels are *distensible*. However, the vascular distensibilities differ greatly in different segments of the circulation, and, as we shall see, this affects significantly the operation of the circulatory system under many changing physiological conditions.

**Units of Vascular Distensibility.**   Vascular distensibility is normally expressed as the fractional increase in volume for each millimeter mercury rise in pressure in accordance with the following formula:

$$\text{Vascular distensibility} = \frac{\text{Increase in volume}}{\text{Increase in pressure} \times \text{Original volume}} \quad (16)$$

That is, if 1 mm. Hg causes a vessel originally containing 10 ml. of blood to increase its volume by 1 ml., then the distensibility would be 0.1 per mm. Hg. or 10 per cent per mm. Hg.

*Difference in Distensibility of the Arteries and the Veins.*   Anatomically, the walls of arteries are far stronger than those of veins. Consequently, the veins, on the average, are about 6 to 10 times as distensible as the arteries. That is, a given rise in pressure will cause about 6 to 10 times as much extra blood to fill a vein as an artery of comparable size.

In the pulmonary circulation the veins are very similar to those of the systemic veins. However, the pulmonary arteries, which normally operate under pressures about one-seventh those in the systemic arterial system, have distensibilities only about one-half those of veins, rather than one-eighth, as is true of the systemic arteries.

## VASCULAR COMPLIANCE (OR CAPACITANCE)

Usually in hemodynamic studies it is much more important to know the *total quantity of blood* that can be stored in a given portion of the circulation for each mm. Hg pressure rise than to know the distensibility of the individual vessels. This value is sometimes called the *overall distensibility* or *total distensibility,* or it can be expressed still more precisely by either of the terms *compliance* or *capacitance,* which are physical terms meaning the increase in volume caused by a given increase in pressure as follows:

$$\text{Vascular compliance} = \frac{\text{Increase in volume}}{\text{Increase in pressure}} \quad (17)$$

Compliance and distensibility are quite different. A highly distensible vessel which has a very slight volume may have far less compliance than a much less distensible vessel which has a very large volume, for *compliance is equal to distensibility×volume.*

The compliance of a vein is about 24 times that of its corresponding artery because it is about 8 times as distensible and it has a volume about 3 times as great ($8 \times 3 = 24$).

## PRESSURE-VOLUME CURVES OF THE ARTERIAL AND VENOUS CIRCULATIONS

A convenient method for expressing the relationship of pressure to volume in a vessel or in a large portion of the circulation is the so-called *pressure-volume curve*. The two solid curves of Figure 18–15 represent respectively the pressure-volume curves of the normal arterial and venous systems, showing that the arterial system, including the larger arteries, small arteries, and arterioles, contains approximately 750 ml. of blood when the mean arterial pressure is 100 mm. Hg but only 500 ml. when the pressure has fallen to zero.

The volume of blood normally in the entire venous tree is about 2500 ml., but only a slight change in venous pressure changes this volume tremendously.

**Difference in Compliance of the Arterial and Venous Systems.** Referring once again to Figure 18–15, one can see that a change of 1 mm. Hg increases the venous volume a very large amount but increases the arterial volume only slightly. That is, the *compliance of the venous system is far greater than the compliance of the arteries—about 24 times as great.*

This difference in compliance is particularly important because it means that tremendous amounts of blood can be stored in the veins with only slight changes in pressure. Therefore, the veins are frequently called the *storage areas* of the circulation.

**Effect of Sympathetic Stimulation or Sympathetic Inhibition on the Pressure-Volume Relationships of the Arterial and Venous Systems.** Also shown in Figure 18–15 are the pressure-volume curves of the arterial and venous systems during moderate sympathetic stimulation and during sympathetic inhibition. It is evident that sympathetic stimulation, with its concomitant increase in smooth muscle tone in the vascular walls, increases the pressure at each volume of the arteries or veins while, on the other hand, sympathetic inhibition decreases the pressure at each respective volume. Obviously, control of the vessels in this manner by the sympathetics can be valuable for diminishing or enhancing the dimensions of one segment of the circulation, thus transferring blood to or from other segments. For instance, an increase in vascular tone throughout the systemic circulation often causes large volumes of blood to shift out of the systemic vessels into the heart, which is a principal way in which pumping by the heart is controlled.

Sympathetic control of vascular capacity is also especially important during hemorrhage. Enhancement of the sympathetic tone of the vessels, especially of the veins, simply reduces the dimensions of the circulatory system, and the circulation continues to operate almost normally even when as much as 25 per cent of the total blood volume has been lost.

## "CIRCULATORY FILLING PRESSURE" AND PRESSURE-VOLUME CURVES OF THE ENTIRE CIRCULATORY SYSTEM

### THE CIRCULATORY FILLING PRESSURE

The circulatory filling pressure (also called the "mean circulatory pressure" or the "static pressure") is a measure of the degree of filling of the circulatory system. That is, it is the pressure that would be measured in the circulation if one could instantaneously stop all blood flow and bring all the pressures in the circulation immediately to equilibrium. The circulatory filling pressure has been measured reasonably accurately in dogs within 2 to 5 seconds after the heart has been stopped. To do this the heart is thrown into fibrillation by an electrical stimulus, and blood is pumped rapidly from the systemic arteries to the veins to cause equilib-

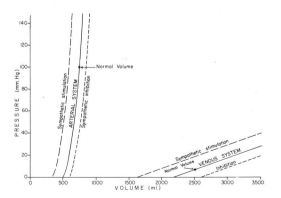

**Figure 18–15.** Pressure-volume curves of the systemic arterial and venous systems, showing also the effects of sympathetic stimulation and sympathetic inhibition.

rium between the two major chambers of the circulation.

The circulatory filling pressure measured in the above manner is almost exactly 7 mm. Hg and almost never varies more than 1 mm. from this value in the normal resting dog and is believed to be about this same value in the human being. However, many different factors can change it, including especially change in the blood volume and increased or decreased sympathetic stimulation.

The circulatory filling pressure is *one of the major factors that determines the rate at which blood flows from the vascular tree into the right atrium of the heart, which in turn determines the cardiac output.* This is so important that it will be explained in detail in Chapter 23; it will also be discussed in relation to blood volume regulation in Chapter 36.

### PRESSURE-VOLUME CURVES OF THE ENTIRE CIRCULATION

Figure 18–16 illustrates the changes in circulatory filling pressure as the total blood volume increases (1) under normal conditions, (2) during strong sympathetic stimulation, and (3) during complete sympathetic inhibition. The point marked by the arrow is the operating point of the normal circulation: a circulatory filling pressure of 7 mm. Hg and a blood volume of 5000 ml. However, *if blood is lost from the circulatory system,* the circulatory filling pressure falls to a lower value. If increased amounts of blood are added, the circulatory filling pressure rises accordingly.

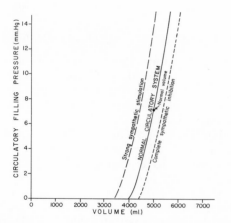

**Figure 18–16.** Pressure-volume curves of the entire circulation, illustrating the effect of strong sympathetic stimulation and complete sympathetic inhibition.

The *compliance* of the entire circulatory system in the human being, as estimated from experiments in dogs, is approximately 100 ml. for each 1 mm. rise in mean circulatory pressure.

*Sympathetic stimulation* and *inhibition* affect the pressure-volume curves of the entire circulatory system in the same way that they affect the pressure-volume curves of the individual parts of the circulation, as illustrated by the two dashed curves of Figure 18–16. That is, for any given blood volume, the circulatory filling pressure rises two- to fourfold with strong sympathetic stimulation and falls markedly when the sympathetics are inhibited. This is an extremely important factor in the regulation of blood flow into the heart and thereby for regulating the cardiac output. For instance, during exercise, sympathetic activity increases the circulatory filling pressure, in this way helping to increase the cardiac output.

### RELATIONSHIP BETWEEN CIRCULATORY FILLING PRESSURE, SYSTEMIC FILLING PRESSURE, AND PULMONARY FILLING PRESSURE

The term *circulatory filling pressure* refers to the pressure that would be measured in the entire circulatory system if all pressures were instantaneously brought to equilibrium. *Systemic filling pressure* is the pressure that would be measured in the systemic vessels if the root of the aorta and the great veins entering the heart were suddenly clamped and all pressures in the systemic system were brought instantaneously to equilibrium. Finally, *pulmonary filling pressure* is the pressure that would be measured in the pulmonary system if the pulmonary artery and large pulmonary veins were suddenly clamped and all pulmonary pressures brought instantaneously to equilibrium.

In measurements made in dogs, the normal systemic filling pressure has averaged about 7 mm. Hg, which is equal to the circulatory filling pressure. The pulmonary filling pressure has measured about 10 mm. Hg. However, these values change markedly in certain pathological conditions for the following reason:

If the right heart is much weaker than the left heart, blood becomes dammed in the systemic circulation, while it is pumped out of the pulmonary system. As a consequence, the systemic filling pressure rises while the pulmonary filling pressure falls. Because of the small compliance of the pulmonary system compared with that of the systemic system (about one seventh as great), the decrease in pulmonary filling pressure is about seven times as great as the rise in systemic filling pressure. Conversely, damage to the left heart causes exactly the opposite effect.

# DELAYED COMPLIANCE (STRESS-RELAXATION) OF VESSELS

The term "delayed compliance" means that a vessel whose pressure is increased by increased volume gradually loses much of this pressure over a period of many minutes because of progressive stretch of the vessel. Likewise, a vessel exposed to constantly increased pressure becomes progressively enlarged.

Figure 18–17 illustrates diagrammatically one of the effects of delayed compliance. In this figure, the pressure is being recorded in a small segment of a vein that is occluded at both ends. Then, an extra volume of blood is suddenly injected into the segment until the pressure rises from 5 to 12 mm. Hg. Even though none of the blood is removed after it is injected, the pressure nevertheless begins to fall immediately and approaches approximately 9 mm. Hg after several minutes. In other words, the volume of blood injected caused immediate *elastic* distention of the vein, but then the smooth muscle fibers of the vein began to adjust themselves to new tensions as a result of the stretch imposed on them. This is a characteristic of all smooth muscle tissue called *stress-relaxation*, which was explained in Chapter 12.

After the delayed increase in compliance had taken place in the experiment illustrated in Figure 18–17, the extra blood volume was suddenly removed, and the pressure immediately fell to a very low value. Subsequently, the smooth muscle fibers began to readjust their tensions back to their initial values, and after a number of minutes the normal vascular pressure of 5 mm. Hg returned.

Delayed compliance occurs only slightly in the arteries but to a much greater extent in the veins. As a result, prolonged elevation of venous pressure can often double the blood volume in the venous tree. This is a valuable mechanism by which the circula-tion can accommodate much extra blood when necessary, such as following too large a transfusion. Also, delayed compliance in the reverse direction is a valuable means by which the circulation automatically adjusts itself to diminished blood volume after serious hemorrhage.

# REFERENCES

Attinger, E. O.: Wall properties of veins. *I.E.E.E. Trans. Biomed. Eng.*, BME-16:253, 1969.

Bergel, D. H. (ed.): Cardiovascular Fluid Dynamics. New York, Academic Press, Inc., 1972.

Bergel, D. H., and Schultz, D. L.: Arterial elasticity and fluid dynamics. *Prog. Biophys. Mol. Biol.*, 22:1, 1971.

Braasch, D.: Red cell deformability and capillary blood flow. *Physiol. Rev.*, 51:679, 1971.

Caro, C. G., Pedley, T. J., and Seed, W. A.: Mechanics of the circulation. *In* Guyton, A. C. (ed.): MTP International Review of Science: Physiology. Baltimore, University Park Press, 1974, Vol 1, p. 1.

Fahraeus, R., and Lindqvist, T.: The viscosity of the blood in narrow capillary tubes. *Amer. J. Physiol.*, 96:562, 1931.

Fitz-Gerald, J. M.: Plasma motions in narrow capillary flow. *J. Fluid Mech.*, 51:463, 1972.

Fung, Y. C., Perrone, N., and Anliker, M. (eds.): Biomechanics: Its Foundations and Objectives. Englewood Cliffs, N. J., Prentice-Hall, Inc., 1972.

Fung, Y. C., and Sobin, S. S.: Theory of sheet flow in the lung alveoli. *J. Appl. Physiol.*, 26:472, 1969.

Gabelnick, H. L., and Litt, M. (eds.): Rheology of Biological Systems. Springfield, Ill., Charles C Thomas, Publisher, 1973.

Gaehtgens, P., Meiselman, H. J., and Wayland, H.: Erythrocyte flow velocities in mesenteric microvessels of the cat. *Microvasc. Res.*, 2:151, 1970.

Green, H. D.: Circulation: Physical principals. *In* Glasser, O. (ed.): Medical Physics. Chicago, Year Book Medical Publishers, 1944.

Guyton, A. C., Armstrong, G. G., and Chipley, P. L.: Pressure-volume curves of the entire arterial and venous systems in the living animal. *Amer. J. Physiol.*, 184:253, 1956.

Guyton, A. C., and Greganti, F. P.: A physiologic reference point for measuring circulatory pressures in the dog—particularly venous pressure. *Amer. J. Physiol.*, 185:137, 1956.

Harlan, J. C., Smith, E. E., and Richardson, T. Q.: Pressure volume curves of systemic and pulmonary circuit. *Amer. J. Physiol.*, 213:1499, 1967.

Holt, J. P.: Flow through collapsible tubes and through *in situ* veins. *I.E.E.E. Trans. Biomed. Eng.*, BME-16:274, 1969.

Intaglietta, M., Tompkins, W. R., and Richardson, D. R.: Velocity pressure, flow, and elastic properties in microvessels of cat omentum. *Amer. J. Physiol.*, 221:922, 1971.

Intaglietta, M., Tompkins, W. R., and Richardson, D. R.: Velocity measurements in the microvasculature of the cat omentum by on-line method. *Microvasc. Res.*, 2:462, 1970.

Johnson, P.C.: Hemodynamics. *Ann. Rev. Physiol.*, 31:331, 1969.

Klug, P. P., Lessin, L. S., and Radice, P.: Rheological aspects of sickle cell disease. *Arch. Intern. Med.*, 133:577, 1974.

Lighthill, M. J.: Physiological fluid dynamics: a survey. *J. Fluid Mech.*, 52:475, 1972.

McDonald, D. A.: Blood Flow in Arteries, 2nd Ed. Baltimore, The Williams & Wilkins Company, 1974.

Merrill, E. W.: Rheology of blood. *Physiol. Rev.*, 49:863, 1969.

Messmer, K., and Sunder-Plassmann, L.: Hemodilution. *Prog. Surg.*, 13:208, 1974.

Patel, D. J., Vaishnav, R. N., Gow, B. S., and Kot, P. A.: Hemodynamics. *Ann. Rev. Physiol.*, 36:125, 1974.

Porciuncula, C. I., Armstrong, G. G., Guyton, A. C., and Stone, H. L.: Delayed compliance in external jugular vein of the dog. *Amer. J. Physiol.*, 207:728, 1964.

Roberts, C.: Blood Flow Management. Baltimore, The Williams & Wilkins Company, 1973.

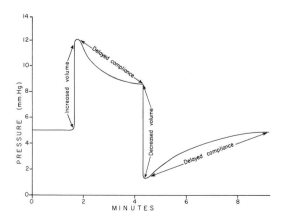

**Figure 18–17.** Effect on the intravascular pressure of injecting a small volume of blood into a venous segment, illustrating the principles of stress relaxation.

Schmid-Schonbein, H., Wells, R., and Goldstone, J.: Influence of deformability of human red cells upon blood viscosity. *Circ. Res.*, *25*:131, 1969.

Talbot, L., and Berger, S. A.: Fluid-mechanical aspects of the human circulation. *Amer. Sci.*, *62*:671, 1974.

Taylor, M. G.: An approach to an analysis of the arterial pulse wave. *Phys. Med. Biol.*, *1*:321, 1957.

Taylor, M. G.: Hemodynamics. *Ann. Rev. Physiol.*, *35*:87, 1973.

Wayland, H.: Rheology and the microcirculation. *Gastroenterology*, *52*:342, 1967.

Wiederhielm, C. A.: Distensibility characteristics of small blood-vessels. *Fed. Proc.*, *24*:1075, 1965.

Wolstenholme, G. E. W., and Knight, J. (eds.): Circulatory and Respiratory Mass Transport: A Ciba Foundation Symposium. Edinburgh, Churchill Livingstone, 1969.

# 19

# The Systemic Circulation

The circulation is divided into the *systemic circulation* and the *pulmonary circulation*. Since the systemic circulation supplies all the tissues of the body except the lungs with blood flow, it is also frequently called the *greater circulation* or *peripheral circulation*.

Though the local circulation in each part of the body has its own special characteristics, some general principles of vascular function nevertheless apply in all parts of the systemic circulation. It is the purpose of the present chapter to discuss these general principles.

**The Functional Parts of the Systemic Circulation.** Before attempting to discuss the details of function in the systemic circulation, it is important to understand the overall role of each of its parts, as follows:

The function of the *arteries* is to transport blood *under high pressure* to the tissues. For this reason the arteries have strong vascular walls, and blood flows rapidly in the arteries to the tissues.

The *arterioles* are the last small branches of the arterial system, and they act as *control valves* through which blood is released into the capillaries. The arteriole has a strong muscular wall that is capable of closing the arteriole completely or of allowing it to be dilated several fold, thereby vastly altering blood flow to the capillaries.

The function of the *capillaries* is to exchange fluid and nutrients between the blood and the interstitial spaces. For this role, the capillary walls are very thin and permeable to small molecular substances.

The *venules* collect blood from the capillaries; they gradually coalesce into progressively larger veins.

The *veins* function as conduits for transport of blood from the tissues back to the heart. Since the pressure in the venous system is very low, the venous walls are thin. Even so, they are muscular, and this allows them to contract or expand and thereby to store either a small or large amount of blood, depending upon the needs of the body.

## PHYSICAL CHARACTERISTICS OF THE SYSTEMIC CIRCULATION

**Quantities of Blood in the Different Parts of the Circulation.** By far the greater amount of the blood in the circulation is contained in the systemic veins. Thus, Figure 19–1 shows that approximately 84 per cent of the entire blood volume of the body is in the systemic circulation, with 59 per cent in the veins, 15 per cent in the arteries, and 5 per cent in the capillaries. The heart contains 7 per cent of the blood, and

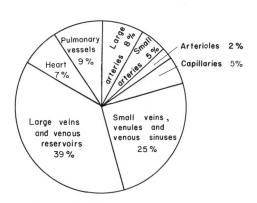

**Figure 19–1.** Percentage of the total blood volume in each portion of the circulatory system.

the pulmonary vessels contain 9 per cent. Most surprising is the very low blood volume in the capillaries of the systemic circulation, only about 5 per cent of the total, for it is here that the most important function of the systemic circulation occurs, namely, diffusion of substances back and forth between the blood and the interstitial fluid. This function is so important that it will be discussed in detail in Chapter 30.

**Cross-Sectional Areas and Velocities of Blood Flow.**   If all vessels of each type were put side by side, their total cross-sectional areas would be:

|  | $cm^2$ |
|---|---|
| Aorta | 2.5 |
| Small arteries | 20 |
| Arterioles | 40 |
| Capillaries | 2500 |
| Venules | 250 |
| Small veins | 80 |
| Venae cavae | 8 |

Note particularly the large cross-sectional areas of the veins, averaging about four times those of the corresponding arteries. This explains the very large storage of blood in the venous system in comparison with that in the arterial system.

The velocity of blood flow in each segment of the circulation is inversely proportional to its cross-sectional area. Thus, under resting conditions, the velocity averages 33 cm. per second in the aorta, but 1/1000 this in the capillaries, or about 0.3 mm per second. However, since the capillaries have a typical length of only 0.3 mm. to 1 mm., each segment of flowing blood remains in the capillaries for only 1 to 3 seconds, a very surprising fact, since all diffusion that is going to take place through the capillary walls must occur in this exceedingly short time.

**Pressures and Resistances in the Various Portions of the Systemic Circulation.**   Since the heart pumps blood continually into the aorta, the pressure in the aorta is obviously high, averaging approximately 100 mm. Hg. And, since pumping by the heart is intermittent, the arterial pressure fluctuates between a *systolic level* of 120 mm. Hg and a *diastolic level* of 80 mm. Hg, as illustrated in Figure 19–2. As the blood flows through the systemic circulation, its pressure falls progressively to approximately 0 mm. Hg by the time its reaches the right atrium.

The decrease in arterial pressure in each part of the systemic circulation is directly proportional to the vascular resistance. Thus, in the

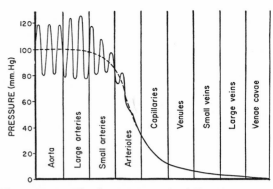

**Figure 19–2.**   Blood pressures in the different portions of the systemic circulatory system.

aorta the resistance is almost zero; therefore, the mean arterial pressure at the end of the aorta is still almost 100 mm. Hg. Likewise, the resistance in the large arteries is very slight so that the mean arterial pressure in arteries as small as 3 mm. in diameter is still 95 to 97 mm. Hg. Then the resistance begins to increase rapidly in the very small arteries, causing the pressure to drop to approximately 85 mm. Hg at the beginning of the arterioles.

The resistance of the *arterioles* is greatest of any part of the systemic circulation, accounting for about half the resistance in the entire systemic circulation. Thus, the pressure decreases about 55 mm. Hg in the arterioles so that the pressure of the blood as it leaves the arterioles to enter the capillaries is only about 30 mm. Hg. Arteriolar resistance is so important to the regulation of blood flow in different tissues of the body that it is discussed in detail later in the chapter and also in the following few chapters, which consider the regulation of the systemic circulation.

The pressure at the arterial ends of the *capillaries* is normally about 30 mm. Hg and at the venous ends about 10 mm. Hg. Therefore, the pressure decrease in the capillaries is only 20 mm. Hg, which illustrates that the capillary resistance is about two-fifths that of the arterioles.

The pressure at the beginning of the venous system, that is, at the *venules*, is about 10 mm. Hg, and this decreases to almost exactly 0 mm. Hg at the right atrium. This large decrease in pressure in the veins indicates that the veins have far more resistance than one would expect for vessels of their large sizes. Much of this resistance is caused by compression of the veins from the outside, which keeps many of them, especially the venae cavae, collapsed a large share of the time. This effect is discussed in detail later in the chapter.

# PRESSURE PULSES IN THE ARTERIES

Since the heart is a pulsatile pump, blood enters the arteries intermittently, causing *pressure pulses* in the arterial system. In the normal young adult the pressure at the height of a pulse, the *systolic pressure,* is about 120 mm. Hg and at its lowest point, the *diastolic pressure,* is about 80 mm. Hg. The difference between these two pressures, about 40 mm. Hg, is called the *pulse pressure.*

Figure 19–3 illustrates a typical, idealized *pressure pulse curve* recorded in the ascending aorta of a human being, showing a very rapid rise in arterial pressure during ventricular systole, followed by a maintained high level of pressure for 0.2 to 0.3 second. This is terminated by a sharp *incisura* or *notch* at the end of systole, followed by a slow decline of pressure back to the diastolic level. The incisura occurs immediately before the aortic valve closes, and is caused as follows: During systole the pressure in the arteries rises to a high value. When the ventricle relaxes the intraventricular pressure begins to fall rapidly, and backflow of blood from the aorta into the ventricle allows the aortic pressure also to begin falling. However, the backflow suddenly snaps the aortic valve closed. The momentum that has built up in the backflowing blood brings still more blood into the root of the aorta, raising the pressure again and thus giving the positive wave in the record immediately after the incisura.

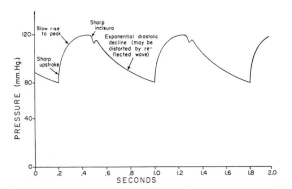

**Figure 19–3.** A normal pressure pulse contour recorded from the ascending aorta. (From Opdyke: *Fed. Proc.,* 11:734, 1952.)

After systole is over, the pressure in the central aorta decreases rapidly at first and then progressively more and more slowly as diastole proceeds. The reason for this is that blood flows out of the aorta through the peripheral vessels much more rapidly when the pressure is high than when it is low.

## *FACTORS THAT AFFECT THE PULSE PRESSURE*

**Effect of Stroke Volume and Arterial Compliance.** There are two major factors that affect the pulse pressure: (1) the *stroke volume output* of the heart and (2) the *compliance (total distensibility)* of the arterial tree; a third less important factor is the character of ejection from the heart during systole.

In general, the greater the stroke volume output, the greater is the amount of blood that must be accommodated in the arterial tree with each heartbeat and, therefore, the greater is the pressure rise during systole and the pressure fall during diastole, thus causing a greater pulse pressure.

On the other hand, the greater the compliance of the arterial system the less will be the rise in pressure for a given stroke volume of blood pumped into the arteries. In effect, then, the pulse pressure is determined approximately by the *ratio of stroke volume output to compliance of the arterial tree.* Therefore, any condition of the circulation that affects either of these two factors will also affect the pulse pressure.

**Factors that Affect the Pulse Pressure by Changing the Stroke Volume Output.** So many different circulatory conditions change the stroke volume output that only a few of these can be mentioned:

An *increase in heart rate* while the cardiac output remains constant causes the stroke volume output to decrease in inverse proportion to the increased rate, and the pulse pressure decreases accordingly.

A *decrease in total peripheral resistance* allows rapid flow of blood from the arteries to the veins. This increases the venous return to the heart and increases the stroke volume output. Therefore, the pulse pressure is also greatly increased.

An *increase in circulatory filling pressure,* if all other circulatory factors remain constant, increases the rate of venous return to the heart and consequently increases the stroke volume output. Here again the pulse pressure is accordingly increased.

The *character of ejection from the heart* affects the pulse pressure in two ways: First, if the duration of ejection is long, a large portion of the stroke volume output runs off through the systemic circulation while it is being ejected into the aorta; therefore, the magnitude of the stroke volume output effect on pulse pressure is decreased. Second, sudden ejection of blood from the heart causes the pressure in the initial portion of the aorta to rise very high before the blood can run off to the more distal portions of the aorta. Therefore, sudden ejection causes a greater pulse pressure than does more prolonged ejection.

**Factors that Affect the Pulse Pressure by Altering the Arterial Compliance.** In contrast to the many different conditions that can change the stroke volume output, there are only two major factors that often alter the arterial compliance. These are (1) *change in mean arterial pressure* and (2) *pathological changes that affect the distensibility of the arterial walls.*

In the preceding chapter it was pointed out that the compliance of a vascular segment is the volume of blood that can be accommodated for a given rise in pressure, which is expressed mathematically as $\Delta V / \Delta P$. As the arterial pressure rises from low to high values, the compliance decreases slightly. Therefore, in the low pressure range considerably more blood can be accommodated for a given rise in pressure than in the high pressure range. For this reason, *in a person who has high arterial pressure but a normal stroke volume output, the pulse pressure is considerably increased.*

In old age, the arterial walls lose much of their elastic and muscular tissues, and these are replaced by fibrous tissue and sometimes even calcified plaques that cannot stretch a significant amount. These changes greatly decrease the compliance of the arterial system, which in turn causes the arterial pressure to rise very high during systole and to fall greatly during diastole as blood runs off from the arteries to the veins.

## ABNORMAL PRESSURE PULSE CONTOURS

Some conditions of the circulation cause abnormal contours to the pressure pulse wave in addition to altering the pulse pressure. Especially distinctive among these are arteriosclerosis, patent ductus arteriosus, and aortic regurgitation.

**Arteriosclerosis.** In arteriosclerosis the arteries become fibrous and sometimes calcified, thereby resulting in greatly reduced arterial compliance and also resulting in markedly increased pulse pressure. Usually a mild degree of hypertension also accompanies arteriosclerosis so that not only does the pulse pressure rise but the mean pressure also. The middle curve of Figure 19–4 illustrates a characteristic aortic pressure pulse contour in arteriosclerosis, showing a markedly elevated systolic pressure, a slightly elevated diastolic pressure, and also a great increase in pulse pressure. The curve is more peaked than that of the normal curve, and the incisura lies at a lower point on the downslope of the curve than in the normal.

**Patent Ductus Arteriosus.** In patent ductus arteriosus, blood flows from the aorta through the open ductus into the pulmonary artery, allowing very rapid runoff of blood from the arterial tree after each heartbeat. However, this is compensated by a greater than normal stroke volume output, giving the pressure pulse contour shown by the lowest curve in Figure 19–4. Here one finds an elevated systolic

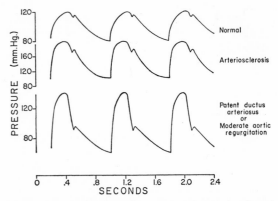

**Figure 19–4.** Pressure pulse contours in arteriosclerosis, patent ductus arteriosus, and moderate aortic regurgitation.

pressure, a greatly depressed diastolic pressure, a greatly increased pulse pressure, and the incisura occurring at a very low point on the downslope because an excessively large amount of blood runs off from the arteries even before systole is over.

**Aortic Regurgitation.** In aortic regurgitation much of the blood that is pumped into the aorta during systole flows back into the left ventricle during diastole. However, this backflow is compensated by a much greater than normal stroke volume output during systole. Thus, the condition is very similar to that of patent ductus arteriosus but not always identical, for in aortic regurgitation the valve sometimes fails entirely to close. When this is true the incisura illustrated in the lower curve of Figure 19–4 is entirely absent.

## TRANSMISSION OF THE PRESSURE PULSE TO THE PERIPHERY

When the heart ejects blood into the aorta during systole, only the proximal portion of the aorta becomes distended at first, and it is only in this portion of the arterial tree that the pressure rises immediately. The cause of this is the inertia of the blood in the aorta, which prevents its sudden movement away from the central arteries into the peripheral arteries.

However, the rising pressure in the central aorta rapidly overcomes the inertia of the blood, causing the pressure to rise progressively farther and farther out in the arterial tree. Figure 19–5 illustrates this *transmission of the pressure pulse* down the aorta, the more distal portions of the aorta becoming distended as the pressure wave moves forward.

The velocity of transmission of the pressure pulse along the normal aorta is 3 to 5 meters per second, along the large arterial branches 7 to 10 meters per second, and in the smaller arteries 15 to 35 meters per second. In general, the less the compliance of each vascular segment, the faster is the velocity of transmission, which explains the slowness of transmission in the aorta versus the rapidity of transmission in the far less compliant small, distal arteries.

The *velocity of transmission of the pressure pulse is much greater than the velocity of blood flow*. In transmission of the pressure pulse, only a small amount of blood entering the proximal aorta pushes the more distal blood forward sufficiently to elevate the pressure in the very distal arteries. Therefore, the actual blood ejected by the heart may have traveled only a few centimeters by the time the pressure wave has already reached the distal ends of the arteries. In the aorta, the velocity of the pressure pulse is approximately 15 times that of blood flow, while in the more distal arteries the velocity of the pressure pulse may be as great as 100 times the velocity of blood flow.

**Augmentation of the Pulse Pressure in the Peripheral Arteries.** An interesting phenomenon that often occurs in transmission of pressure pulses to the periphery is an increase in pulse pressure. This effect is illustrated in Figure 19–6, which shows considerable enhancement of the pulse pressure more and more peripherally. This effect, caused primarily by *reflection* of the pulse wave from peripheral arteries, may be explained as follows:

When a pressure pulse enters the peripheral arteries and distends them, the surge of pressure in these peripheral arteries causes the pulse wave to begin traveling backward along the same vessels from which it had come. This is analogous to a wave traveling in a bowl of water until it hits the edge. On striking the edge, the wave turns around and travels back onto the surface of the water. If the returning wave strikes an oncoming wave, the two "summate," causing a much higher wave than would otherwise occur. Such is the case in the arterial tree. The first portion of the pressure wave is reflected

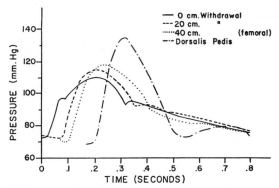

Figure 19–6. Pressure pulse contours in different segments of the arterial tree in man, showing (1) the delay in the pressure pulse as it spreads more and more peripherally and (2) augmentation of the pulse pressure especially in the dorsalis pedis artery. (Modified from Remington and Wood: *J. Appl. Physiol.,* 9:440, 1956.)

before the latter portion of the same wave ever reaches the peripheral arteries. Therefore, the first portion summates with the latter portion, causing higher pressures than would otherwise be recorded.

Though the significance of augmentation of the pulse pressure is disputed by different physiologists, this phenomenon must be remembered whenever pressure measurements are made in peripheral arteries, for systolic pressure is sometimes as much as 20 to 30 per cent above that in the central aorta, and diastolic pressure is often reduced as much as 10 to 15 per cent.

**Damping of the Pressure Pulse in the Small Arteries and Arterioles.** The pressure pulse becomes less and less intense as it passes through the small arteries and arterioles, until it becomes almost absent in the capillaries. This was illustrated in Figure 19–2 and is called *damping* of the pressure pulse.

Damping of the pressure pulse is caused mainly by a combined effect of vascular distensibility and vascular resistance. That is, for a pressure wave to travel from one area of an artery to another area, a small amount of blood must flow between the two areas. The resistance in the small arteries and arterioles is great enough that this flow of blood, and consequently the transmission of pressure, is greatly impeded. At the same time, the distensibility of the small vessels is very significant (particularly so when one considers that Distensibility = Compliance ÷ Vessel volume). Therefore, the small amount of blood that is caused to flow during each pressure pulse produces progressively less pressure rise the more distal the pulse wave proceeds in the vessels. Therefore, even though the pressure rises and falls quite significantly in the large arteries, the combined resistance-distensibility effects cause the pressure pulse to become progressively smaller and smaller as it proceeds farther and farther into the small vessels.

Figure 19–7 illustrates the extreme effect of resistance on damping. The curve to the left is a normal

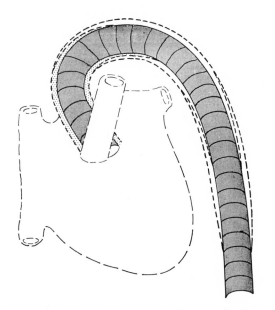

Figure 19–5. Progressive stages in the transmission of the pressure pulse along the aorta.

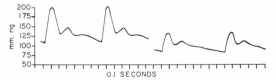

**Figure 19–7.** Pressure pulse contour in the dorsalis pedis artery of a dog recorded, first, under normal conditions and, second, during stimulation of the sympathetic nerves supplying the femoral artery. (Modified from Alexander and Kantowitz: *Surgery, 33*:42, 1953.)

pressure pulse contour recorded from the dorsalis pedis artery in a dog. To the right is the same pressure pulse contour recorded during contraction of the femoral artery. The damping effect of the increased resistance is readily apparent.

*Capillary Pulsation.* Pressure pulses are not completely damped by the time they reach the capillaries—the pulse pressure usually still averaging 1 mm. Hg or more. And under some conditions very severe capillary pulsation results. Two major factors can cause abnormal capillary pulsation: (1) It occurs when the central pressure pulse is greatly exacerbated, as when the *heart rate is very slow* or when the stroke volume output is greatly increased as a result of *aortic regurgitation, patent ductus arteriosus, or even extreme increase in venous* return. (2) Another cause of capillary pulsation is *extreme dilatation of the small arteries and arterioles,* which reduces the resistance of these vessels and thereby reduces the damping.

Abnormal capillary pulsation can be readily demonstrated by bending the anterior portion of the fingernail downward so that the blood is pressed out of the anterior capillaries of the nail bed. This causes blanching anteriorly while the posterior half of the nail remains red because of blood still in the capillaries. If significant pulsation is occurring in the capillaries, the border between the red and white areas will shift forward as more capillaries become filled during the high pressure phase and backward during the low pressure phase.

### THE RADIAL PULSE

Clinically, it has been the habit for many years for a physician to feel the radial pulse of each patient. This is done to determine the rate of the heartbeat or, frequently, because of the psychic contact that it gives the doctor with his patient. Under certain circumstances, however, the character of the pulse can also be of value in the diagnosis of circulatory diseases.

**Weak Pulse.** A weak pulse at the radial artery usually indicates either (1) greatly decreased central pulse pressure, such as occurs when the stroke volume output is low, or (2) increased damping of the pulse wave caused by vascular spasm; the latter occurs when the sympathetic nervous system becomes

overly active, for instance, following blood loss or when a person is having a chill.

**Pulsus Paradoxus.** Occasionally the strength of the pulse becomes strong, then weak, then strong, continuing in synchrony with the phases of respiration. This is caused by alternate increase and decrease in cardiac output with each respiration. During inspiration, all the blood vessels of the lungs increase in size because of increased negative pressure in the thorax. Therefore, blood collects in the lungs, and the stroke volume output and pulse strength decrease. During expiration, opposite effects occur. This is a normal phenomenon in all persons, but it becomes extremely distinct in some conditions, such as cardiac tamponade or very deep breathing.

**Pulse Deficit.** The rhythm of the heart is very irregular in atrial fibrillation or in the case of ectopic beats. In these arrhythmias, which were discussed in Chapter 17, two beats of the heart often come so close together that the second beat pumps no blood or very little blood because the left ventricle has too little time to fill between the two beats. In this circumstance, one can hear the second beat of the heart with a stethoscope applied directly over the heart but cannot feel a pulsation in the radial artery, an effect called *pulse deficit.* The greater the rate of pulse deficit each minute, the more serious, ordinarily, is the arrhythmia.

**Pulsus Alternans.** In a few conditions the heart beats strongly with one beat and then weakly with the next, causing similar alternation in the strength of the pulse, a condition called *pulsus alternans.* This alternation is frequently caused by enlarged ventricles, the stroke volume output alternating in amount from one heart beat to the next.

## THE ARTERIOLES AND CAPILLARIES

Blood flow in each tissue is controlled almost entirely by the degree of contraction or dilatation of the arterioles, and it is in the capillaries where the important process of exchange between blood and the interstitial fluid occurs. These two segments of the circulation are so important that they will receive special discussion in chapters to follow: arteriolar regulation of blood flow in Chapters 20 and 21, and capillary exchange phenomena in Chapters 30, 31, and 32.

Upon leaving the small arteries, blood courses through the arterioles, which are only a few millimeters in length and have diameters from 8 to 50 microns. Each arteriole branches many times to supply perhaps 10 to 100 capillaries.

There are approximately 10 billion capillaries in the peripheral tissues, having more than 100 square meters of surface area. The thickness of the capillary wall is usually less than 1 micron, and there are small pores in the wall (as will be explained in Chapter 30) through which substances can diffuse.

Though the capillary walls are very thin and, therefore, very weak, their diameters are also very small. It will be recalled from the discussion of vascular wall tension and vascular pressure in the previous chapter that the wall tension is directly proportional to the *pressure times the diameter* (law of Laplace). Therefore, since the diameter is extremely minute, the tension developed in the wall is also extremely minute, which explains why the very thin-walled capillaries can withstand the pressure therein.

## *EXCHANGE OF FLUID THROUGH THE CAPILLARY MEMBRANE*

Though the detailed dynamics of fluid exchange through the capillary membrane will be presented in Chapter 30, it is important to introduce this subject briefly here.

The capillary membrane is highly permeable to water as well as to all the substances dissolved in plasma and tissue fluids *except the plasma proteins*. This failure of the plasma proteins to go through the pores of the capillaries means that the proteins will cause osmotic pressure, called *colloid osmotic pressure*, at the membrane.

Therefore, two different types of pressure gradients can cause fluid movement through the capillary membrane: (1) the hydrostatic pressure gradient between the two sides of the membrane and (2) the colloid osmotic pressure gradient between the two sides. The greater the difference between the intracapillary hydrostatic pressure and the hydrostatic pressure in the tissue spaces surrounding the capillaries, the greater will be the tendency for fluid to move out of the capillaries into the interstitial spaces. On the other hand, the greater the difference between the plasma colloid osmotic pressure and the tissue fluid colloid osmotic pressure, the greater will be the osmotic tendency for fluid to move from the tissue spaces into the capillary. Under normal conditions, these two are approximately in equilibrium, so that net exchange of fluid volume across the capillary membrane is very slight—only a minute normal outward flow which provides barely enough fluid for formation of the lymph that returns by way of the lymphatic vessels once again to the circulation.

On the other hand, any significant increase in the capillary pressure from its normal value will cause loss of fluid out of the circulation into the tissue spaces, or a decrease in capillary pressure will cause movement of fluid into the circulation from the tissue spaces. Therefore, we will often refer to loss of fluid from the circulation when the capillary pressure rises too high or gain of fluid into the circulation when the capillary pressure falls below normal.

Another important feature of capillary function is two-way diffusion through the capillary pores of dissolved substances between the plasma and tissue fluids. Thus, sodium ions diffuse in both directions in approximately equal amounts so that the sodium concentration remains almost exactly the same in both the blood and the tissue fluids. Likewise, when the cells of the tissues deplete the oxygen in the tissue fluids, oxygen diffuses from the blood toward the cells. Conversely, when the cells form excess carbon dioxide it diffuses toward the blood. In this way, the capillaries provide nutrition to the cells and remove excreta from the cells. All these functions will be discussed in detail in Chapter 30, which is devoted specifically to capillary function.

# THE VEINS AND THEIR FUNCTIONS

For years the veins have been considered to be nothing more than passageways for flow of blood into the heart, but it is rapidly becoming apparent that they perform many functions that are necessary to the operation of the circulation. They are capable of constricting and enlarging, of storing large quantities of blood and making this blood available when it is required by the remainder of the circulation, of actually propelling blood forward by means of a so-called "venous pump," and even of helping to regulate cardiac output, a function so important that it will be described in detail in Chapter 23.

## *RIGHT ATRIAL PRESSURE AND ITS RELATION TO VENOUS PRESSURE*

To understand the various functions of the veins, it is first necessary to know something about the pressures in the veins and how they are regulated. Blood from all the systemic veins flows into the right atrium; therefore, the pressure in the right atrium is frequently called the *central venous pressure*. The pressures in the peripheral veins are to a great extent dependent on the level of this pressure, so that anything that affects this pressure usually affects venous pressure everywhere in the body.

*Right atrial pressure is regulated by a balance between,* first, *the ability of the heart to pump blood out of the right atrium* and, second, *the tendency for blood to flow from the peripheral vessels back into the right atrium.*

If the heart is pumping strongly, the right atrial pressure tends to decrease. On the other hand, weakness of the heart tends to elevate the right atrial pressure. Likewise, any effect that causes rapid inflow of blood into the right atrium from the veins tends to elevate the right atrial pressure. Some of the factors that increase this tendency for venous return are (1) increased blood volume, (2) increased large vessel tone throughout the body with resultant increased peripheral venous pressures, and (3) dilatation of the systemic small vessels, which decreases the peripheral resistance and allows rapid flow of blood from the arteries to the veins.

The same factors that regulate right atrial pressure also enter into the regulation of cardiac output, for the amount of blood pumped by the heart is dependent both on the ability of the heart to pump and the tendency for blood to flow into the heart from the peripheral vessels. Therefore, we will discuss the regulation of right atrial pressure in much more depth in Chapter 23 in connection with the regulation of cardiac output.

The *normal right atrial pressure* is approximately 0 mm. Hg, which is about equal to the atmospheric pressure around the body. However, it can rise to as high as 20 to 30 mm. Hg under very abnormal conditions, such as (a) serious heart failure or (b) following massive transfusion of blood, which will cause excessive quantities of blood to attempt to flow into the heart from the peripheral vessels.

The lower limit to the right atrial pressure is about −4 to −5 mm. Hg, which is the pressure in the pericardial and intrapleural spaces that surround the heart. The right atrial pressure approaches these very low values when the heart pumps with exceptional vigor or when the flow of blood into the heart from the peripheral vessels is greatly depressed, such as following severe hemorrhage.

## VENOUS RESISTANCE AND PERIPHERAL VENOUS PRESSURE

Large veins have almost no resistance *when they are distended.* However, as illustrated in

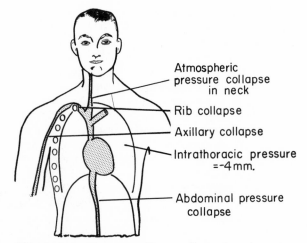

**Figure 19–8.** Factors tending to collapse the veins entering the thorax.

Figure 19–8, most of the large veins entering the thorax are compressed at many points so that blood flow is impeded. For instance, the veins from the arms are compressed by their sharp angulation over the first rib. Second, the pressure in the neck veins often falls so low that the atmospheric pressure on the outside of the neck causes them to collapse. Finally, veins coursing through the abdomen are often compressed by different organs and by the intra-abdominal pressure so that often they are almost totally collapsed. For these reasons the *large veins do usually offer considerable resistance to blood flow,* and because of this the pressure in the peripheral veins is usually 4 to 9 mm. Hg greater than the right atrial pressure.

Note, however, that the veins inside the thorax *are not collapsed* because the *negative pressure* inside the chest distends these veins.

**Effect of High Right Atrial Pressure on Peripheral Venous Pressure.** When the right atrial pressure rises above its normal value of 0 mm. Hg, blood begins to back up in the large veins and to open them up. The pressures in the peripheral veins do not rise until all the collapsed points between the peripheral veins and the large veins have opened up. This usually occurs when the right atrial pressure rises to about 4 to 6 mm. Hg. If the right atrial pressure then rises still further, the additional increase is reflected by a corresponding rise in peripheral venous pressure. Since the heart must be greatly weakened to cause a rise in right atrial pressure to as high as 4 to 6 mm. Hg, one often finds that the peripheral venous pressure is not elevated in the early stages of cardiac failure.

**Effect of Abdominal Pressure on Venous Pressures of the Leg.** The normal pressure in the peritoneal cavity averages about 2 mm. Hg, but at times it can rise to as high as 15 to 20 mm. Hg as a result of pregnancy, large tumors, or excessive ascites in the peritoneal cavity as explained in Chapter 32. When this happens, the pressure in the veins of the legs must rise *above* the abdominal pressure before the abdominal veins will open and allow the blood to flow from the legs to the heart. Thus, if the intra-abdominal pressure is 20 mm. Hg, the lowest possible pressure in the femoral veins is 20 mm. Hg.

## EFFECT OF HYDROSTATIC PRESSURE ON VENOUS PRESSURE

In any body of water, the pressure at the surface of the water is equal to atmospheric pressure, but the pressure rises 1 mm. Hg for each 13.6 mm. distance below the surface. This pressure results from the weight of the water and therefore is called *hydrostatic pressure.*

Hydrostatic pressure also occurs in the vascular system of the human being because of the weight of the blood in the vessels, as is illustrated in Figure 19–9. When a person is standing, the pressure in the right atrium remains ap-

proximately 0 mm. Hg because the heart pumps into the arteries any excess blood that attempts to accumulate at this point. However, in an adult *who is standing absolutely still* the pressure in the veins of the feet is approximately +90 mm. Hg simply because of the distance from the feet to the heart and the weight of the blood in the veins between the heart and the feet. The venous pressures at other levels of the body lie proportionately between 0 and 90 mm. Hg.

In the arm veins, the pressure at the level of the top rib is usually about +6 mm. Hg because of compression of the subclavian vein as it passes over this rib. The hydrostatic pressure down the length of the arm is then measured below the level of this rib. Thus, if the hydrostatic difference between the level of the rib and the hand is 29 mm. Hg, this hydrostatic pressure must be added to the 6 mm. Hg pressure caused in the vein by compression of the vein as it crosses the rib, making a total of 35 mm. Hg pressure in the veins of the hand.

The neck veins collapse almost completely all the way to the skull owing to atmospheric pressure on the outside of the neck. This collapse causes the pressure in these veins to remain zero along their entire extent. The reason for this is that any tendency for the pressure to rise above this level opens the veins and allows the pressure to fall back to zero, and any tendency for the pressure to fall below this level collapses the veins still more, which increases their resistance and again returns the pressure back to zero.

The veins inside the skull, however, are in a noncollapsible chamber, and they will not collapse. Consequently, *negative pressure can exist in the dural sinuses of the head;* in the standing position the venous pressure in the sagittal sinus is calculated to be approximately −10 mm. Hg because of the hydrostatic "suction" between the top of the skull and the base of the skull. Therefore, if the sagittal sinus is entered during surgery, air can be sucked immediately into this vein; and the air may even pass downward to cause air embolism in the heart so that the heart valves do not function satisfactorily, and death often ensues.

**Effect of the Hydrostatic Factor on Arterial and Other Pressures.** The hydrostatic factor also affects the peripheral pressures in the arteries and capillaries as well as in the veins. For instance, a standing person who has an arterial pressure of 100 mm. Hg at the level of his heart has an arterial pressure in his feet of about 190

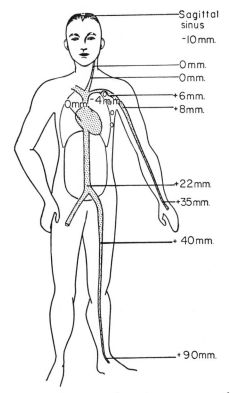

**Figure 19–9.** Effect of hydrostatic pressure on the venous pressures throughout the body.

mm. Hg. Therefore, any time one states that the arterial pressure is 100 mm. Hg, he generally means that this is the pressure at the hydrostatic level of the heart.

## VENOUS VALVES AND THE "VENOUS PUMP"

Because of hydrostatic pressure, the venous pressure in the feet would always be about +90 mm. Hg in a standing adult were it not for the valves in the veins. However, every time one moves his legs he tightens his muscles and compresses the veins either in the muscles or adjacent to them, and this squeezes the blood out of the veins. Yet, the valves in the veins, as illustrated in Figure 19–10, are arranged so that the direction of blood flow can be only toward the heart. Consequently, every time a person moves his legs or even tenses his muscles, a certain amount of blood is propelled toward the heart, and the pressure in the dependent veins of the body is lowered. This pumping system is known as the "venous pump" or "muscle pump," and it is efficient enough that under ordinary circumstances the venous pressure in the feet of a walking adult remains less than 25 mm. Hg.

If the human being stands perfectly still, the venous pump does not work, and the venous

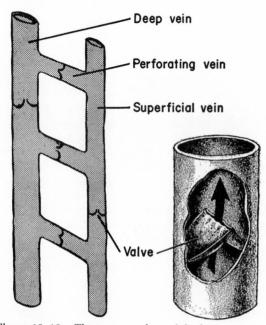

**Figure 19–10.**   The venous valves of the leg.

pressures in the lower part of the leg can rise to the full hydrostatic value of 90 mm. Hg in about 30 seconds. Under such circumstances the pressures within the capillaries also increase greatly, and fluid leaks from the circulatory system into the tissue spaces. As a result, the legs swell, and the blood volume diminishes. Indeed, as much as 15 to 20 per cent of the blood volume is frequently lost from the circulatory system within the first 15 minutes of standing absolutely still, as occurs when a soldier is made to stand at absolute attention.

**Varicose Veins.**   The valves of the venous system are frequently destroyed. This occurs particularly when the veins have been overstretched by an excess of venous pressure for a prolonged period of time, as occurs in pregnancy or when one stands on his feet most of the time. Stretching the veins, obviously, increases their cross-sectional areas, but the valves do not increase in size. Therefore, the valves will no longer block reverse blood flow in the enlarged veins. When such a situation develops, the pressure in the veins of the legs increases owing to failure of the venous pump; this further increases the size of the veins and finally destroys the function of the valves entirely. Thus, the person develops "varicose veins," which are characterized by large bulbous protrusions of the veins beneath the skin of the entire leg and particularly of the lower leg. The venous and capillary pressures become very high because of the incompetent venous pump, and leakage of fluid from the capillary blood into the tissues causes constant edema in the legs of these persons whenever they stand for more than a few minutes. The edema in turn prevents adequate diffusion of nutritional materials from the capillaries to the muscle and skin cells so that the muscles become painful and weak, and the skin frequently becomes gangrenous and ulcerates. Obviously, the best treatment for such a condition is continual elevation of the legs to a level at least as high as the heart, but tight binders on the legs are also of considerable aid in preventing the edema and its sequelae.

## REFERENCE POINT FOR MEASURING VENOUS AND OTHER CIRCULATORY PRESSURES

In previous discussions we have often spoken of right atrial pressure as being 0 mm. Hg and arterial pressure as being 100 mm. Hg, but we have not stated the hydrostatic level in the circulatory system to which this pressure is referred. There is one point in the circulatory system at which hydrostatic pressure factors do not affect the pressure measurement by more than 1 mm. Hg. This is at the level of the tricuspid valve, as shown by the crossed axes in Figure 19–11. Therefore, all pressure measurements discussed in this text are referred to the level of the

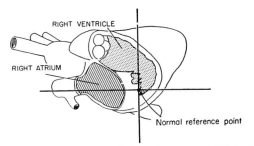

**Figure 19–11.** Location of the reference point for pressure measurement at the tricuspid valve.

tricuspid valve, which is called the *reference point for pressure measurement.*

The reason for lack of hydrostatic effects at the tricuspid valve is that the heart automatically prevents significant hydrostatic changes in pressure at this point in the following way:

If the pressure at the tricuspid valve rises slightly above normal, then the right ventricle fills to a greater extent than usual, causing the heart to pump blood more rapidly than usual and therefore to decrease the presssure at the tricuspid valve back toward the normal mean value. On the other hand, if the pressure at this point falls, the right ventricle fails to fill adequately, its pumping decreases, and blood dams up in the venous system until the tricuspid pressure again rises to a normal value. In other words, *the heart acts as a feedback regulator of pressure* at the tricuspid valve.

The three axes of the reference point, as extrapolated from animals to the human being, are (1) *back to the anterior chest wall approximately 61 per cent of the thickness of the chest,* (2) *almost exactly in the midline,* and (3) *approximately one quarter of the distance above the lower end of the sternum.* A person can be standing, lying on his back, lying on his stomach, lying on either side, or even in the head-down position, and his central venous pressure referred to this reference point will remain almost exactly constant regardless of the position of the body. This does not mean that the pressure at this point is always zero. It may be as low as −4 mm. Hg or as high as +20 mm. Hg, but, whatever its value, changing the position of the body does not alter it more than 1 mm. Hg.

In making arterial pressure measurements, the precise hydrostatic point in the chest to which pressures are referred matters little because percentagewise a hydrostatic error of as much as 10 or more centimeters (equivalent to 7.4 mm. Hg error) still does not affect the arterial pressure reading to a great extent. However, in venous pressure measurements the reference level must be very exact if the measurements are to be significant, for abnormalities of venous pressure as small as 1 mm. Hg can often result in changes in cardiac output as great as 50 to 100 per cent.

## PRESSURE PULSES IN THE VEINS

The pressure pulses in the arteries are almost completely damped out before they pass through the capillaries into the systemic veins. However, right atrial pulsations are sometimes transmitted backward to cause pressure pulses of an entirely different character in the large veins. This effect usually does not occur to a significant extent in the normal circulation because most of the veins, where they lead into the thoracic cavity, are compressed by surrounding tissues; this causes sufficient resistance to damp out the pulsations before they can be transmitted backward into the peripheral veins. However, whenever the mean right atrial pressure is high, especially in heart failure, the veins are then well filled with blood and can easily transmit the pulsations.

The atrial pulsations were discussed in Chapter 13. Briefly, they are (1) the *a wave,* caused by contraction of the atrium, (2) the *c wave,* caused by contraction of the ventricle, and (3) the *v wave,* caused by continued inflow of blood into the atrium when the A-V valves are closed during ventricular systole. It is these waves that are transmitted into the distended veins.

In severe cardiac failure the venous pressure waves usually become prominent enough that one can see the veins of the patient's neck pulsating. In earlier days these pulsations were recorded for diagnostic purposes. The *a-c interval* in the recording is approximately equal to the P-Q interval of the electrocardiogram, and before the days of electrocardiography this was often used as a measure of the delay between atrial excitation and ventricular excitation. These waves, therefore, were then of considerable value in diagnosing different degrees of heart block.

## MEASUREMENT OF VENOUS PRESSURE

**Clinical Estimation of Venous Pressure.** The venous pressure can be estimated by simply observing the distention of the peripheral veins—especially the neck veins. For instance, in the sitting position, the neck veins are never distended in the normal person. However, when the right atrial pressure becomes increased to as much as 10 mm. Hg, the lower veins of the neck begin to protrude even when one is sitting; when the right atrial pressure rises to as high as 15 mm. Hg, essentially all the veins in the neck become greatly distended, and the venous pulse becomes prominent in these veins.

Rough estimates of the venous pressure can also be made by raising or lowering an arm of a lying person while one observes the degree of distention of the antecubital or hand veins. As the arm is progressively raised, the veins suddenly collapse, and the level at which they collapse, when referred to the level of the heart, is a rough measure of the peripheral venous pressure.

***Direct Measurement of Venous Pressure and Right Atrial Pressure.*** Venous pressure can be measured with ease by inserting a syringe needle connected to a water manometer directly into a vein. The venous pressure is expressed in relation to the level of the tricuspid valve, i.e., the height in centimeters of water above the level of the tricuspid valve; this can be converted to mm. Hg by dividing by a factor of 1.36.

The only means by which *right atrial pressure* can be measured accurately is by inserting a catheter through the veins into the right atrium. This catheter can then be connected to a water manometer and the pressure measured as noted previously.

## BLOOD RESERVOIR FUNCTION OF THE VEINS

In discussing the general characteristics of the systemic circulation earlier in the chapter, it was pointed out that over 50 per cent of all the blood in the circulatory system is in the systemic veins. For this reason it is frequently said that the systemic veins act as a *blood reservoir* for the circulation. Also, relatively large quantities of blood are present in the veins of the lungs so that these, too, are considered to be blood reservoirs.

When blood is lost from the body to the extent that the arterial pressure begins to fall, pressure reflexes are elicited from the carotid sinuses and other pressure sensitive areas of the circulation, as will be detailed in Chapter 21; these in turn cause sympathetic nerve signals that constrict the veins, thus automatically taking up much of the slack in the circulatory system caused by the lost blood. Indeed, even after as much as 20 to 25 per cent of the total blood volume has been lost, the circulatory system often functions almost normally because of this variable reservoir system of the veins.

**Specific Blood Reservoirs.** Certain portions of the circulatory system are so extensive that they are specifically called "blood reservoirs." These include (1) the *spleen*, which can sometimes decrease in size sufficiently to release as much as 150 ml. of blood into other areas of the circulation, (2) the *liver*, the sinuses of which can release several hundred milliliters of blood into the remainder of the circulation, (3) the *large abdominal veins*, which can contribute as much as 300 ml., and (4) the *venous plexus beneath the skin*, which can probably contribute several hundred milliliters. The *heart* itself and the *lungs*, though not parts of the systemic venous reservoir system, must also be considered to be blood reservoirs. The heart, for instance, becomes reduced in size during sympathetic stimulation and in this way can contribute about 100 ml. of blood, and the lungs can contribute another 100 to 200 ml. when the pulmonary pressures fall to low values.

## ASSESSMENTS OF VENOUS FUNCTION BY MEASUREMENT OF CIRCULATORY FILLING PRESSURE

It was pointed out in the previous chapter that *circulatory filling pressure* can be measured by stopping the heart and pumping blood rapidly from the systemic arteries to the veins, bringing the pressures in the two major vascular reservoirs to equilibrium within a few seconds; this equilibrium pressure is the circulatory filling pressure. This pressure is actually a measure of how tightly the vascular system is filled with blood. Therefore, the greater the degree of filling (especially of the veins), the higher is the circulatory filling pressure. The normal value is approximately 7 mm. Hg, but following massive transfusion this value can rise to as high as 30 to 40 mm. Hg; and in congestive heart failure, in which the circulatory blood volume is greatly increased, it can rise to as high as 20 to 25 mm. Hg.

Measurements of circulatory filling pressure are also important in assessing venous contraction for the following reasons: The systemic venous system has many times as much compliance as all the remainder of the circulation put together. Therefore, the circulatory filling pressure is about three-fourths determined by venous tone and venous filling with blood in contrast with only about one-fourth by all the remainder of the circulation. Therefore, even slight changes in degree of venous contraction will cause marked changes in circulatory filling pressure, whereas even large changes in degree of contraction of other segments of the circulation, such as the arterioles, the pulmonary vessels, and so forth, have very little effect. Therefore, one of the most sensitive barometers of the degree of contraction of the venous system is the change that this causes in the circulatory filling pressure.

# REFERENCES

Alexander, R. S.: The peripheral venous system. In Hamilton, W. F. (ed.): Handbook of Physiology. Baltimore, The Williams & Wilkins Company, 1963, Vol. 2, Sec. 2, p. 1075.

Attinger, E. O.(ed.): Pulsatile Blood Flow. New York, McGraw-Hill Book Company, 1964.

Attinger, E. O.: Analysis of pulsatile blood flow. Adv. Biomed. Eng. Med. Phys., 1:1, 1968.

Baez, S.: Simultaneous measurements of radii and wall thickness of microvessels in the anesthetized rat. Circ. Res., 25:315, 1969.

Caro, C. G., Pedley, T. J., and Seed, W. A.: Mechanics of the circulation. In Guyton, A. C. (ed.): MTP International Review of Science: Physiology. Baltimore, University Park Press, 1974, Vol 1., p. 1.

D'Agrosa, L. S.: Patterns of venous vasomotion in the bat wing. Amer. J. Physiol., 218:530, 1970.

Davis, D. L.: Sympathetic stimulation and small artery constriction. Amer. J. Physiol., 206:262, 1964.

Folkow, B: Role of the nervous system in the control of vascular tone. *Circulation, 21*:706, 1960.

Friedman, S. M., and Friedman, C. L.: Effects of ions on vascular smooth muscle. *In* Hamilton, W. F. (ed.): Handbook of Physiology. Baltimore, The Williams & Wilkins Company, 1963, Vol. 2, Sec. 2, p. 1135.

Goodman, A. H., Guyton, A. C., Drake, R., and Loflin, J. H.: A television method for measuring capillary red cell velocities. *J. Appl. Physiol., 37*:126, 1974.

Guyton, A. C.: The venous system and its role in the circulation. *Mod. Conc. Cardiov. Dis., 27*:483, 1958.

Guyton, A. C.: Peripheral circulation. *Ann. Rev. Physiol., 21*:239, 1959.

Guyton, A. C.: Cardiac output and regional circulation. *In* Gordon, B. L. (ed.): Clinical Cardiopulmonary Physiology. New York, Grune & Stratton, Inc., 1969, pp. 28–38.

Guyton, A. C., and Jones, C. E.: Central venous pressure: physiological significance and clinical implications. *Amer. Heart J., 86*:431, 1973.

Guyton, A. C., Ross, J. M., Carrier, O., Jr., and Walker, J. R.: Evidence for tissue oxygen demand as the major factor causing autoregulation. *Circ. Res., 14*:60, 1964.

Haddy, F. J.: Local effects of sodium, calcium, and magnesium upon small and large blood vessels of the dog forelimb. *Circ. Res., 8*:57, 1960.

Herd, J. A.: Overall regulation of the circulation. *Ann. Rev. Physiol., 32*:289, 1970.

Korner, P. I.: Circulatory adaptations in hypoxia. *Physiol. Rev., 39*:687, 1959.

Lundgren, O., and Jodal, M.: Regional blood flow. *Ann. Rev. Physiol., 37*:395, 1975.

McDonald, D. A.: The relation of pulsatile pressure to flow in arteries. *J. Physiol., 127*:533, 1955.

Mellander, S.: Systematic circulation. *Ann. Rev. Physiol., 32*:313, 1970.

Nerem, R. M., and Seed, W. A.: An *in vivo* study of aortic flow disturbances. *Cardiovasc. Res., 6*:1, 1972.

Patel, D. J., de Frietas, F. M., Greenfield, J. C., and Fry, D. L.: Relationship of radius to pressure along the aorta in living dogs. *J. Appl. Physiol., 18*:1111, 1963.

Remington, J. W.: Contour changes of the aortic pulse during propagation. *Amer. J. Physiol., 199*:331, 1960.

Remington, J. W., and O'Brien, L. J.: Construction of aortic flow pulse from pressure pulse. *Amer. J. Physiol., 218*:437, 1970.

Rovick, A. A., and Randall, W. C.: Systemic circulation. *Ann. Rev. Physiol., 29*:225, 1967.

Schmidt-Nielsen, K., and Pennycuik, P.: Capillary density in mammals in relation to body size and oxygen consumption. *Amer. J. Physiol., 200*:746, 1961.

Seed, W. A., and Wood, N. B.: Velocity patterns in the aorta. *Cardiovasc. Res., 5*:319, 1971.

Sonnenschein, R. R., and White, F. N.: Systemic circulation. *Ann. Rev. Physiol., 30*:147, 1968.

Spencer, M. P.: Systematic circulation. *Ann. Rev. Physiol., 28*:311, 1966.

Spencer, M. P., Johnston, F. R., and Denison, A. B., Jr.: Dynamics of the normal aorta. *Circ. Res., 6*:491, 1958.

Stainsby, W. N.: Autoregulation of blood flow in skeletal muscle during increased metabolic activity. *Amer. J. Physiol., 202*:273, 1962.

Stainsby, W. N., and Renkin, E. M.: Autoregulation of blood flow in resting skeletal muscle. *Amer. J. Physiol., 207*:117, 1961.

Uchida, E., and Bohr, D. F.: Myogenic tone in isolated perfused resistance vessels from rats. *Amer. J. Physiol., 216*:1343, 1969.

Uchida, E., and Bohr, D. F.: Myogenic tone in isolated perfused vessels: occurrence among vascular beds and along vascular trees. *Circ. Res., 25*:549, 1969.

Wiedeman, M. P.: Dimensions of blood vessels from distributing artery to collecting vein. *Circ. Res., 12*:375, 1963.

Wood, E.: The Veins. Boston, Little, Brown and Company, 1965.

# 20

## Local Control of Blood Flow by the Tissues; Nervous and Humoral Regulation

The circulatory system is provided with a complex system for control of blood flow to the different parts of the body. In general, these are of three major types as follows:

1. Local control of blood flow in each individual tissue, the flow being controlled mainly in proportion to that tissue's need for blood perfusion.

2. Nervous control of blood flow, which often affects blood flow in large segments of the systemic circulation, such as shifting blood flow from the nonmuscular vascular beds to the muscles during exercise or changing the blood flow in the skin to control body temperature.

3. Humoral control, in which various substances dissolved in the blood such as hormones, ions, or other chemicals can cause either local increase or decrease in tissue flow or widespread generalized changes in flow.

## LOCAL CONTROL OF BLOOD FLOW BY THE TISSUES THEMSELVES

**Acute Local Regulation in Response to Tissue Need for Flow.** In most tissues the blood flow is controlled in proportion to the need for nutrition, such as the need for delivery of oxygen, especially, but also glucose, amino acids, fatty acids, and other nutrients. However, in some tissues the local flow performs other functions. In the skin its purpose is to transfer heat from the body to the surrounding air. In the

kidneys its purpose is to deliver substances to the kidneys for excretion. And in the brain it is to determine, to a great extent, the carbon dioxide and hydrogen ion concentrations of the brain fluids, which in turn play important roles in controlling the level of brain activity.

Fortunately, local blood flow can be increased in response to many different local factors in the tissues—at times to lack of oxygen, at times to excess of carbon dioxide or hydrogen ion concentration, and at other times to still other factors. It is these many different control factors that help to distribute the blood flow to the different parts of the body in proportion to their respective needs.

Table 20–1 gives the approximate *resting* blood flows through the different organs of the body. Note the tremendous flows through the brain, liver, and kidneys despite the fact that these organs represent only a small fraction of the total body mass. Yet, even under basal conditions, the need for flow in each one of these tissues is very great; in the liver to support its high level of metabolic activity, in the brain to provide nutrition and to prevent the carbon dioxide and hydrogen ion concentrations from becoming too great, and in the kidneys to maintain adequate excretion.

The skeletal muscle of the body represents 35 to 40 per cent of the total body mass, and yet, in the inactive state, the blood flow through all the skeletal muscle is only 15 to 20 per cent of the total cardiac output. This accords with the fact that inactive muscle has a very low metabolic

**TABLE 20–1.   Blood Flow to Different Organs and Tissues Under Basal Conditions***

|  | *Per cent* | *ml./min.* |
|---|---|---|
| Brain | 14 | 700 |
| Heart | 3 | 150 |
| Bronchial | 3 | 150 |
| Kidneys | 22 | 1100 |
| Liver | 27 | 1350 |
| Portal | (21) | (1050) |
| Arterial | (6) | ( 300) |
| Muscle (inactive state) | 15 | 750 |
| Bone | 5 | 250 |
| Skin (cool weather) | 6 | 300 |
| Thyroid gland | 1 | 50 |
| Adrenal glands | 0.5 | 25 |
| Other tissues | 3.5 | 175 |
| Total | 100.0 | 5000 |

* Based mainly on data compiled by Dr. L. A. Sapirstein.

rate. Yet, when the muscles become active, their metabolic rate sometimes increases as much as 50-fold, and the blood flow in individual muscles can increase as much as 20-fold, illustrating a marked increase in blood flow in response to the increased need of the muscle for nutrients.

**Functional Anatomy of the Systemic Microcirculation.** Though the student will recall that each tissue has its own characteristic vascular system, Figure 20–1 presents a typical capillary bed; this one is in the connective tissue of the mesentery, which is very easily studied and its components easily analyzed. This figure shows that blood enters the capillary bed through a small *arteriole* and leaves by way of a small *venule*. From the arteriole the blood usually divides and flows through several *metarterioles* before entering the *capillaries*. Some of the capillaries are very large, and they course almost directly to the venule. These are called *preferential channels*. However, most of the capillaries, called the *true capillaries,* branch mainly from the metarterioles and then finally terminate in a venule.

The arterioles have a strong muscular coat, and the metarterioles are surrounded by sparse but highly active smooth muscle fibers. In addition, at each point at which a capillary leaves a metarteriole, a small muscular *precapillary sphincter,* usually composed of a single spiral-

ing smooth muscle fiber, surrounds the origin of the capillary.

The venules also have a smooth muscle coat, but one that is much less extensive than that on the arteriole.

As will be discussed in more detail later in the chapter, the arterioles and venules are supplied by extensive innervation from the sympathetic nervous system, and the degree of contraction of these structures is strongly influenced by the intensity of sympathetic signals transmitted from the central nervous system to the blood vessels.

On the other hand, innervation of the metarterioles and the precapillary sphincters is usually very sparse, if there is any at all. Instead, the muscle fibers of these two structures are controlled almost entirely by the local humoral environment of the tissues, that is, by the concentrations of oxygen, carbon dioxide, hydrogen ions, electrolytes, and other factors in each individual tissue area. These local factors, therefore, are major controllers of blood flow in local tissue areas.

**Effect of Tissue Metabolism on Local Blood Flow.** Figure 20–2 illustrates the approximate quantitative effect on blood flow of increasing the rate of metabolism in a local tissue such as muscle. Note that an increase of metabolism up to eight times normal increases the blood flow about four-fold. The increase in flow at first is less than the increase in metabolism. However, once the metabolism rises high enough to remove most of the nutrients from the blood, further increase in metabolism can occur only with a concomitant increase in blood flow to supply the required nutrients.

**Local Blood Flow Regulation When Oxygen Availability Changes.** One of the most necessary of the nutrients is oxygen. Whenever the availability of oxygen to the tissues decreases, such as at high altitude, in pneumonia,

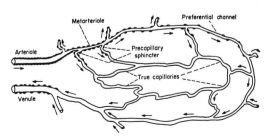

**Figure 20–1.**   Overall structure of a capillary bed. (From Zweifach: Factors Regulating Blood Pressure. Josiah Macy, Jr., Foundation, 1950.)

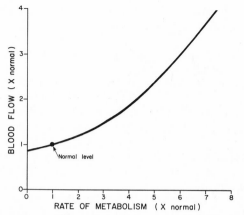

**Figure 20–2.** Effect of increasing rate of metabolism on tissue blood flow.

in carbon monoxide poisoning, or in cyanide poisoning, the blood flow through the tissues increases markedly. Figure 20–3 shows that as the arterial oxygen saturation falls to about 25 per cent of normal, the blood flow through an isolated leg increases about three-fold; that is, the blood flow increases almost enough, but not quite, to make up for the decreased amount of oxygen in the blood, thus automatically maintaining an almost constant supply of oxygen to the tissues. Cyanide poisoning of local tissue areas can cause a local blood flow increase as much as seven-fold, thus illustrating the extreme effect of oxygen deficiency in increasing blood flow.

**The Oxygen Demand Theory of Local Blood Flow Regulation.**  The discussion thus far has simply demonstrated the ability of each local tissue to regulate its own blood flow in proportion to its own individual needs but has not explained the mechanism by which this regulation occurs. Unfortunately, the precise

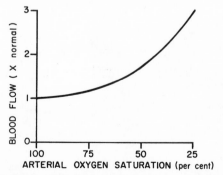

**Figure 20–3.** Effect of arterial oxygen saturation on blood flow through an isolated dog leg.

cause of local blood flow regulation is not known, but many experiments in recent years point mainly to oxygen concentration in the tissues as perhaps the most important regulator of local blood flow in most (but not all) organs—that is, the less the oxygen concentration, the greater the flow. Two of the reasons for believing this are:

1. Minute arteries isolated from a number of different tissues respond to changes in oxygen in exactly the same way that local tissues do. Increased oxygen in the blood causes constriction; decreased oxygen causes dilatation. Furthermore the smaller the artery, the greater the intensity of this effect, indicating that the minute arterioles, metarterioles, and precapillary sphincters could be extremely sensitive to slight alterations in local oxygen concentration.

2. Many different experiments have demonstrated that any factor that alters the availability of oxygen to the tissues, as illustrated in Figure 20–3, causes immediate and marked change in local blood flow.

*Mechanism by which Oxygen Concentration Could Regulate Blood Flow.*  Figure 20–4 illustrates what might be called a tissue unit, showing blood flowing into a tissue through a minute vessel and then entering the capillary; at the origin of the vessel is a precapillary sphincter. When one observes a local tissue under the microscope, he sees that the precapillary sphincters are normally either completely open or completely closed. And the number of precapillary sphincters that are open at any given time is approximately proportional to the requirements of the tissue for nutrition. In some tissues the precapillary sphincters open and close cyclically several times per minute, the duration of the open phases being approximately proportional to the metabolic needs of the tissues. This cyclic opening and closure of the sphincter is called *vasomotion.*

Now, let us explain how oxygen concentration in the local tissue *could* regulate blood flow through the area. It is well known that smooth muscle requires oxygen to remain contracted. Therefore, we might assume that the strength of contraction of the precapillary sphincter would increase with an increase in oxygen concentration. Consequently, when the oxygen concentration in the tissue should rise above a certain level, the precapillary sphincter presumably would close and remain closed until the tissue cells consumed the excess oxygen. Then, when the oxygen concentration fell low enough, the sphincter would open once more to begin the cycle again.

When many capillaries are present in the same local tissue area, this alternate opening and closure of precapillary sphincters does not occur. Instead, a certain proportion of the sphincters remains open while the others remain closed. Furthermore, the proportion that remains open is that number required to keep the oxygen concentration approximately midway between that necessary to cause closure of the sphincters and that required to cause opening.

Thus, the precapillary sphincter mechanism illustrated in Figure 20–4 could easily regulate local blood flow in such a way that the tissue oxygen concentration would remain almost exactly constant, whatever the requirements of the tissue for oxygen.

**The Vasodilator Theory for Local Blood Flow Regulation.** Another theory to explain regulation of local blood flow is the vasodilator theory. According to this theory, the greater the rate of metabolism or the less the availability of nutrients to a tissue, the greater becomes the rate of formation of a *vasodilator substance*. The vasodilator substance then supposedly diffuses back to the precapillary sphincters, metarterioles, and arterioles to cause dilatation. Some of the different vasodilator substances that have been suggested are carbon dioxide, lactic acid, adenosine, adenosine phosphate compounds, histamine, potassium ions, and hydrogen ions.

Some of the vasodilator theories assume that the vasodilator substance is released from the tissue in response to oxygen lack. For instance, it has been demonstrated that oxygen lack can cause both lactic acid and adenosine to be released from the tissues; these are substances

that can cause vasodilatation and therefore could be responsible, or partially responsible, for the local blood flow regulation.

*Possible Special Significance of Adenosine in Local Blood Flow Regulation.* Recently, many physiologists have suggested that the substance adenosine is an especially important local vasodilator that might play a major role in regulating local blood flow. For instance, vasodilator quantities of adenosine are released from heart muscle cells whenever coronary blood flow becomes too little, and it has been claimed that this causes local vasodilatation in the heart and thereby returns the blood flow back toward normal. Other physiologists have suggested that the same mechanism might occur though perhaps with slight alterations, in skeletal muscle and other tissues of the body.

On the other hand, recent studies have shown that the vasodilator effect of adenosine can be blocked with moderate concentrations of circulating theophylline. Yet, when theophylline is given to an animal to block the action of adenosine, local blood flow in the heart is still increased as much as or almost as much as ever when the heart becomes hypoxic. Therefore, this finding makes it dubious that adenosine is in truth an important regulator of local tissue blood flow.

Almost exactly the same story has been repeated for each of the different proposed local vasodilator substances. That is, each one of them has been shown to be a local dilator when present in high enough concentration. Yet, it has been difficult to prove that sufficient concentration normally occurs for the substance to be a physiological regulator of blood flow. Furthermore, when each one of them has been blocked from acting, local blood flow regulation, in general, continues normally or almost normally without its presence. It is this type of experience with each one of the vasodilator substances that has made some physiologists feel that control of local blood flow might be vested more in diminished availability of nutrients, especially oxygen, than in the release of vasodilator substances. However, regulation is likely achieved by a combination of both vasodilator factors and diminished availability of nutrients as well.

Thus far, oxygen seems to be the most important nutrient that enters into the regulation of blood flow. However, recent experiments have shown that the availability (or lack of availability) of glucose, and perhaps other nutrients, can also play an important role in regulating tissue flow.

**"Autoregulation" of Blood Flow When the Arterial Pressure Changes.** In dead organs or in tissues in which the local blood flow regulatory mechanisms are nonfunctional, an increase in arterial pressure always causes an increase in blood flow at least as great as the increase in pressure. However, in almost all normally functioning tissues of the body, the arterial pressure

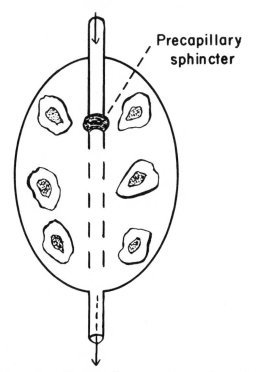

**Figure 20–4.** Diagram of a tissue unit area for explanation of local feedback control of blood flow.

can be changed over a very wide range, and the blood flow through the tissues will remain almost normal. This effect is illustrated by the dark curve in Figure 20–5, which shows the effect on blood flow through a muscle of increasing the arterial pressure progressively from a very low value to a very high value over a period of about 20 minutes. Note that between an arterial pressure of approximately 75 mm. Hg and 175 mm. Hg the blood flow remains within ± 10 to 15 per cent of the normal value. The dashed curve of this figure illustrates the long-term effects of arterial pressure on blood flow when the pressure changes slowly over a period of days or weeks, as will be discussed in more detail later in the chapter.

One can readily understand how autoregulation of blood flow can occur when the arterial pressure changes if he will simply apply the basic principles of local blood flow regulation, as already discussed in earlier sections. Thus, when the arterial pressure becomes too great, the excess flow will either provide too many nutrients to the tissues or will flush out all vasodilator substances, either of which will cause the blood vessels to constrict. Therefore, the increased pressure will not increase flow because the simultaneous constriction nullifies the effect of the pressure. On the other hand, whenever the arterial pressure falls too low, the diminished flow of nutrients into the tissues or the ischemic release of vasodilator substances causes vasodilatation, and the blood flow returns almost to normal despite the reduced arterial pressure.

**Reactive Hyperemia.** When the blood supply to a tissue is blocked for either a short or a long period of time and then is unblocked, the flow through the tissue usually increases to about 5 times normal for a few seconds if the block has lasted a few seconds and sometimes for as long as many hours if the blood flow has been stopped for an hour or more. This phenomenon is called *reactive hyperemia*. Reactive hyperemia is almost certainly another manifestation of the local blood flow regulation mechanism; that is, lack of flow sets into motion all of those factors that cause vasodilatation. Following short periods of vascular occlusion, the extra blood flow during the reactive hyperemia phase lasts long enough to repay almost exactly the tissue oxygen deficit that has accrued during the period of occlusion. This mechanism emphasizes the close connection between local blood flow regulation and delivery of nutrients to the tissues.

**Functional Hyperemia.** When any tissue becomes highly active, such as a muscle during exercise, one of the gastrointestinal glands during hypersecretory periods, or even the brain during rapid mental activity, the rate of blood flow through the tissue increases concomitantly. Here again, if one simply applies the basic principles of local blood flow control, he can easily understand this so-called *functional hyperemia*. The increase in local metabolism causes the cells to devour the tissue fluid nutrients extremely rapidly and possibly also to release large quantities of vasodilator substances. The result obviously would be to dilate the local blood vessels, and therefore, to increase greatly local blood flow. In this way, the highly active tissue will receive the nutrients that it requires to sustain its new level of function.

### SPECIAL TYPES OF LOCAL BLOOD FLOW REGULATION

In certain tissues, blood flow is regulated in proportion to other factors besides the requirement for nutrients. This is true particularly in the kidney and in the brain, as follows:

**"Autoregulation" in the Kidneys.** The rate of blood flow through the kidneys remains constant, even more constant than in most other tissues, despite marked changes in arterial pressure. Yet the availability of oxygen to the kidneys can be changed tremendously without changing renal blood flow significantly. Therefore, renal local blood flow regulation ("autoregulation" of flow) is undoubtedly different from the usual type of local blood flow regulation.

The two factors that can change renal blood flow to

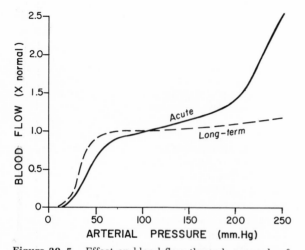

**Figure 20–5.** Effect on blood flow through a muscle of increasing arterial pressure. The solid curve shows the effect if the arterial pressure is raised over a period of a few minutes. The dashed curve shows the effect if the arterial pressure is raised extremely slowly over a period of many weeks.

the greatest extent are sodium concentration in the blood and the concentration of end-products of protein metabolism; an increase in either of these increases renal blood flow. Recent experiments suggest that it is the concentrations of some of these substances in the renal tubules that control renal blood flow. It has been postulated that feedback control of afferent arteriolar constriction occurs at the juxtaglomerular apparatus where the distal tubule comes in contact with the afferent arteriole. The degree of constriction of the afferent arteriole of each nephron supposedly changes in proportion to the concentrations of certain substances in the distal tubule. In Chapter 35 this mechanism will be discussed separately in relation to the physiology of the kidney.

**Local Regulation of Cerebral Blood Flow in Response to Tissue $CO_2$ and Hydrogen Ion Concentration.** The oxygen type of local blood flow regulation can occur in the brain in the same way that it occurs in most of the rest of the body. However, in addition to this type of regulation, a powerful regulation based on carbon dioxide and hydrogen ion concentrations also occurs in the brain. When the concentrations of these substances increase, the blood vessels dilate, allowing more rapid blood flow to remove much of the carbon dioxide from the tissues, and thereby reducing the $CO_2$ back toward a normal mean value. This also causes the hydrogen ion concentration to return toward normal because removal of $CO_2$ also removes carbonic acid, thereby removing hydrogen ions. Conversely, a decrease in $CO_2$ and hydrogen ions causes vasoconstriction, which allows these substances to accumulate in the tissues until they rise back toward the normal mean values. This is an important regulation because changes in $CO_2$ and hydrogen ion concentrations drastically change the degree of activity of all neurons.

# LONG-TERM LOCAL BLOOD FLOW REGULATION

The local blood flow regulation that has been discussed thus far occurs acutely, within a minute or more after local tissue conditions have changed. For instance, if the arterial pressure is suddenly increased from 100 mm. Hg to 150 mm. Hg, the blood flow increases almost instantaneously about 100 per cent, partly because the increase in pressure directly increases the flow and partly because the high pressure dilates the peripheral vessels. However, within one to two minutes the blood flow decreases back to only 15 per cent above the original control value. This illustrates the rapidity of the acute type of local regulation, but the regulation is still very incomplete because there remains a 15 per cent increase in blood flow.

However, over a period of hours, days, and weeks a long-term type of local blood flow regulation develops in addition to the acute regulation, and this long-term regulation gives far more complete regulation than does the acute mechanism. For instance, in the above example, if the arterial pressure should remain at 150 mm. Hg indefinitely, within a few weeks the blood flow through the tissues would gradually reapproach almost exactly the normal value. Figure 20–5 illustrates by the dashed curve the extreme effectiveness of this long-term local blood flow regulation.

Long-term regulation also occurs when the metabolic demands of a tissue change. Thus, if a tissue becomes chronically overactive and therefore requires chronically increased quantities of nutrients, the blood supply gradually increases to match the needs of the tissue.

**Long-Term Local Flow Regulation in Patients with Coarctation of the Aorta.** Coarctation of the aorta means occlusion of the aorta somewhere along its course. When such an occlusion occurs in the descending thoracic aorta, blood must flow to the lower part of the body through high-resistance collateral blood vessels in the chest wall. Therefore, the arterial pressure in the upper part of the body is very high—sometimes two times normal—while the arterial pressure in the lower body is slightly below normal. Yet, despite this tremendous difference in pressure between the upper and lower body, the blood flow per unit mass of tissue is essentially the same above and below the coarctation if the coarctation has existed for a long time. Studies of coarctation in animals have given us most of our quantitative information on the powerful nature of the long-term local flow regulatory mechanism. Indeed, this long-term mechanism seems to be several times as potent as is the mechanism of acute local flow regulation.

**Mechanism of Long-Term Regulation— Change in Tissue Vascularity.** The mechanism of long-term regulation is almost certainly a change in the degree of vascularity of the tissues. That is, if the arterial pressure falls to 60 mm. Hg and remains at this level for many weeks, the number and size of vessels in the tissue increase; if the pressure then rises to a very high level, the number and size of vessels decrease. Likewise, if the metabolism in a given tissue becomes elevated for a prolonged period of time, vascularity increases; or if the metabolism is decreased, vascularity decreases.

Thus, there is continual day-by-day reconstruction of the tissue vasculature to meet the needs of the tissues. This reconstruction occurs very rapidly (within days) in extremely young animals. It also occurs rapidly in new growth of tissue, such as in scar tissue or new growth of glandular tissue; but on the other hand, it occurs very slowly in old, well established tissues. Therefore, the time required for long-term regulation to take place may be only a few days in the newborn or as long as months or even years in the elderly person. Furthermore, the final degree of response is much greater in younger tissues than in older, so that in the newborn the vascularity will adjust to match almost exactly the needs of the tissue for blood flow; whereas in older tissues, vascularity frequently lags far behind the needs of the tissues.

*Role of Oxygen in Long-Term Regulation.* A probable stimulus for increased or decreased vascularity in many if not most instances is need of the tissue for oxygen. The reason for believing this is that hypoxia causes increased vascularity and hyperoxia causes decreased vascularity. This effect is demonstrated in newborn infants who are put into an oxygen tent for therapeutic purposes. The excess oxygen causes almost immediate cessation of new vascular growth in the retina of the eye and even degeneration of some of the capillaries that have already formed. Then when the baby is taken out of the oxygen, there is explosive overgrowth of new vessels to make up for the sudden decrease in available oxygen; indeed, there is so much overgrowth that the vessels grow into the vitreous humor and eventually cause blindness. (This condition is called *retrolental fibroplasia*.)

The mechanisms by which changes in oxygen availability also cause changes in vascularity are not yet clear. Lack of oxygen causes enlargement of the vessels already in the tissue and, under appropriate conditions, growth of new vessels as well. It is possible that the change in vascularity results secondarily to the acute local blood flow regulation mechanism in the following way: In the absence of oxygen, the blood vessels dilate. Because the tension in the vascular wall increases directly in proportion to the diameter of the vessel, the vascular wall elements then become subjected to far greater than normal stretch. This stretch theoretically could cause permanent enlargement of the vessel or outgrowth of new vessels. In support of this concept is the fact that venous occlusion causes more growth of new small

vessels than does arterial occlusion; both types of occlusion reduce tissue oxygenation, but only venous occlusion increases the intraluminal pressure.

## COLLATERAL CIRCULATION AS A PHENOMENON OF LONG-TERM LOCAL BLOOD FLOW REGULATION

The development of collateral circulation when blood flow to a tissue is blocked is a type of long-term local blood flow regulation. Thus, if the femoral artery becomes occluded, the small vessels that by-pass the femoral occlusion become greatly enlarged, and the leg, especially in younger persons, usually will develop an adequate blood supply. Within seconds after the occlusion, blood flow through these collateral vessels is only about one-eighth the normal blood flow to the leg. But acute dilatation of the collateral vessels occurs within the first 1 to 2 minutes, so that blood flow to the leg returns to approximately one-half normal. Then, over a period of a week or more the blood flow returns almost all the way to normal, indicating progressive opening of the collateral vessels.

Thus, the opening of collateral vessels follows almost identically the same pattern as that observed in other types of acute and long-term local blood flow regulation. Furthermore, the acute opening of collateral vessels seems to be metabolically initiated because reduction of blood flow to a tissue by reducing arterial pressure, without causing vascular occlusion, causes the collateral vessels to open completely; that is, insufficiency of blood flow to the tissues opens the collaterals, even though the main artery is not closed.

## SIGNIFICANCE OF LONG-TERM LOCAL REGULATION—THE METABOLIC MASS TO TISSUE VASCULARITY PROPORTIONALITY

From these discussions it should already be apparent to the student that there is a built-in mechanism in most tissues to keep the degree of vascularity of the tissue almost exactly that required to supply the metabolic needs of the tissue. Thus, one can state as a general rule that the vascularity of most tissues of the body is directly proportional to the local metabolism. If ever this proportionality constant becomes abnormal, the long-term local regulatory mechanism automatically readjusts the degree of vascularity over a period of weeks or months.

In young persons this degree of readjustment is usually very exact; in old persons it is only partial.

# NERVOUS REGULATION OF THE CIRCULATION

Superimposed onto the intrinsic local tissue regulation of blood flow are two additional types of regulation: (1) *nervous* and (2) *humoral*. These regulations are not necessary for most normal function of the circulation, but they do provide greatly increased effectiveness of control under special conditions such as exercise or hemorrhage.

There are two very important features of nervous regulation of the circulation: First, nervous regulation can function extremely rapidly, some of the nervous effects beginning to occur within one second and reaching full development within five to 30 seconds. Second, the nervous system provides a means for controlling large parts of the circulation simultaneously, often in spite of the effect that this has on the blood flow to individual tissues. For instance, when it is important to raise the arterial pressure temporarily, the nervous system can arbitrarily cut off, or at least greatly decrease, blood flow to major segments of the circulation despite the fact that the local blood flow regulatory mechanisms oppose this.

## THE AUTONOMIC NERVOUS SYSTEM

The autonomic nervous system will be discussed in detail in Chapter 56. However, it is so important to the regulation of the circulation that its specific anatomical and functional characteristics relating to the circulation deserve special attention.

By far the most important part of the autonomic nervous system for regulation of the circulation is the *sympathetic nervous system*. The *parasympathetic nervous system* is important only in its regulation of heart function, as we shall see later in the chapter.

**The Sympathetic Nervous System.** Figure 20–6 illustrates the anatomy of sympathetic nervous control of the circulation. Sympathetic vasomotor nerve fibers leave the spinal cord through all the thoracic and the first one to two lumbar spinal nerves. These pass into the sympathetic chain and thence by two routes to the blood vessels throughout the body: (1) through the *peripheral sympathetic nerves* and (2) through the *spinal nerves*. The precise path-

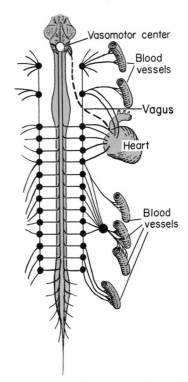

**Figure 20–6.** The vasomotor center and its control of the circulatory system through the sympathetic and vagus nerves.

ways of these fibers in the spinal cord and in the sympathetic chains will be discussed in Chapter 56. It suffices to say here that, with rare exceptions, all vessels of the body are supplied with sympathetic nerve fibers.

***Distribution of Sympathetic Nerve Fibers to the Peripheral Vasculature.*** Figure 20–7 illustrates the distribution of sympathetic nerve fibers to the peripheral blood vessels, and shows that all the vessels except the capillaries, sphincters, and most of the metarterioles are innervated.

The innervation of the arterioles, small arteries, venules, and small veins allows sympathetic stimulation to alter the *resistance* of the vessels and thereby to change the rate of blood flow through the tissues. The innervation of large vessels, particularly of the veins, makes it possible for sympathetic stimulation to change the volume of these vessels and thereby to alter the volume of the total circulatory system, which plays a major role in the regulation of cardiovascular function, as we shall see later in this chapter and subsequent chapters.

***Sympathetic Nerve Fibers to the Heart.*** In addition to sympathetic nerve fibers supplying the blood vessels, fibers also go to the heart. This

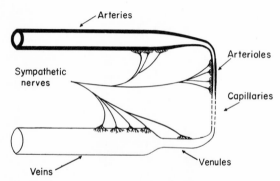

**Figure 20–7.**  Innervation of the systemic circulation.

innervation was discussed in Chapter 13. It will be recalled that sympathetic stimulation markedly increases the activity of the heart, increasing the heart rate and enhancing its strength of pumping.

**Parasympathetic Control of the Circulation—Control of the Heart.**  Though the parasympathetic nervous system is exceedingly important for many other autonomic functions of the body, it plays only a minor role in regulation of the circulation. Its only really important effect is its control of heart rate. It also has a slight influence on the control of cardiac contractility; however, this effect is far overshadowed by the sympathetic nervous system control of contractility. Parasympathetic nerves pass to the heart in the vagus nerve, as illustrated in Figure 20–6.

The effects of parasympathetic stimulation on heart function were discussed in detail in Chapter 13. Principally, parasympathetic stimulation causes a marked *decrease* in heart rate and slight decrease in contractility.

*The Sympathetic Vasoconstrictor System and Its Control by the Central Nervous System*

The sympathetic nerves carry both vasoconstrictor and vasodilator fibers, but by far the most important of these are the *sympathetic vasoconstrictor* fibers. Sympathetic vasoconstrictor fibers are distributed to essentially all segments of the circulation. However, this distribution is greater in some tissues than in others. It is rather poor in both skeletal and cardiac muscle and in the brain, while it is powerful in the kidneys, the gut, the spleen, and the skin.

**The Vasomotor Center and Its Control of the Vasoconstrictor System—Vasomotor Tone.**  Located bilaterally in the reticular sub-

stance of the lower third of the pons and upper two thirds of the medulla, as illustrated in Figure 20–8, is an area called the *vasomotor center*. This center transmits impulses downward through the cord and thence through the vasoconstrictor fibers to all the blood vessels of the body.

The upper and lateral portions of the vasomotor center are *tonically active*. That is, they have an inherent tendency to transmit nerve impulses all the time, thereby maintaining even normally a slow rate of firing in essentially all vasoconstrictor nerve fibers of the body at a rate of about one-half to two impulses per second. This continual firing is called *sympathetic vasoconstrictor tone*. These impulses maintain a partial state of contraction in the blood vessels, a state called *vasomotor tone*.

The brain stem can be severed above the lower third of the pons without significantly changing the normal activity of the vasomotor center. This center remains tonically active and continues to transmit approximately normal numbers of impulses to the sympathetic vasoconstrictor fibers throughout the body.

Figure 20–9 demonstrates the significance of vasoconstrictor tone. In the experiment of this figure, total spinal anesthesia was administered to an animal, which completely blocked all transmission of nerve impulses from the central

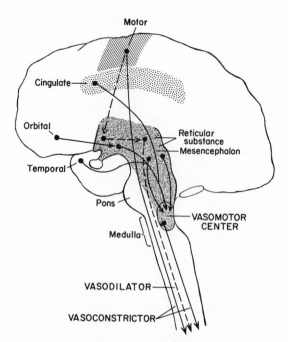

**Figure 20–8.**  Areas of the brain that play important roles in the nervous regulation of the circulation.

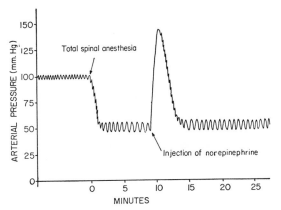

**Figure 20–9.** Effect of total spinal anesthesia on the arterial pressure, showing a marked fall in pressure resulting from loss of vasomotor tone.

nervous system to the periphery. As a result, the arterial pressure fell from 100 to 50 mm. Hg, illustrating the effect of loss of vasoconstrictor tone throughout the body. A few minutes later a small amount of the hormone norepinephrine was injected—norepinephrine is the substance secreted at the endings of sympathetic nerve fibers throughout the body. As this hormone was transported in the blood to all the blood vessels, the vessels once again became constricted, and the arterial pressure rose to a level even greater than normal for a minute or two until the norepinephrine was destroyed.

**The Inhibitory Area of the Vasomotor Center.** The medial and lower portion of the vasomotor center does not participate in excitation of the vasoconstrictor fibers. Instead, stimulation of this area transmits *inhibitory* impulses into the upper lateral parts of the vasomotor center, thereby *decreasing the degree of sympathetic vasoconstrictor tone* and consequently allowing dilatation of the blood vessels. Thus, the vasomotor center is composed of two parts, a bilateral *excitatory part* that can excite the vasoconstrictor fibers and cause vascular constriction and a medial *inhibitory part* that can inhibit vasoconstriction, thus allowing vasodilatation.

**Control of Heart Activity by the Vasomotor Center.** At the same time that the vasomotor center is controlling the degree of vascular constriction, it also controls heart activity. The lateral portions of the vasomotor center transmit excitatory impulses through the sympathetic nerve fibers to the heart to increase heart rate and contractility, while the medial portion of the vasomotor center, which lies in immediate

apposition to the *dorsal motor nucleus of the vagus nerve,* transmits impulses through the vagus nerve to the heart to decrease heart rate. Therefore, the vasomotor center can either increase or decrease heart activity, the activity ordinarily increasing at the same time that vasoconstriction occurs throughout the body and ordinarily decreasing at the same time that vasoconstriction is inhibited. However, these interrelationships are not invariable, because some nerve impulses that pass down the vagus nerves to the heart can bypass the vasomotor center.

**Control of the Vasomotor Center by Higher Nervous Centers.** Large numbers of areas throughout the *reticular substance* of the *pons, mesencephalon,* and *diencephalon* can either excite or inhibit the vasomotor center. This reticular substance is illustrated in Figure 20–8 by the diffuse shaded area. In general, the more lateral and superior portions of the reticular substance cause excitation, while the more medial and inferior portions cause inhibition.

The *hypothalamus* plays a special role in the control of the vasoconstrictor system, for it can exert powerful excitatory or inhibitory effects on the vasomotor center. The *posterolateral portions* of the hypothalamus cause mainly excitation, while the *anterior part* can cause mild excitation or inhibition, depending on the precise part of the anterior hypothalamus stimulated.

Many different parts of the *cerebral cortex* can also excite or inhibit the vasomotor center. Stimulation of the *motor cortex,* for instance, excites the vasomotor center because of impulses transmitted downward into the hypothalamus and thence to the vasomotor center. Also, stimulation of the *anterior temporal lobe,* the *orbital areas of the frontal cortex,* the *anterior part of the cingulate gyrus,* the *amygdala,* the *septum,* and the *hypocampus* can all either excite or inhibit the vasomotor center, depending on the precise portion of these areas that is stimulated and upon the intensity of the stimulus.

Thus, widespread areas of the brain can have profound effects on the vasomotor center and in turn on the sympathetic vasoconstrictor system of the body, either further enhancing the degree of vasoconstriction or inhibiting the vasoconstrictor tone and thereby causing vasodilatation.

**Norepinephrine—The Sympathetic Vasoconstrictor Transmitter Substance.** The substance secreted at the endings of the vasoconstrictor nerves is norepinephrine. Norepinephrine acts directly on the smooth muscle of the vessels to cause vasoconstriction, as will be discussed in Chapter 56.

**The Adrenal Medullae and Their Relationship to the Sympathetic Vasoconstrictor System.** Sympathetic vasoconstrictor impulses are transmitted to the adrenal medullae at the same time that they are transmitted to all the blood vessels. These impulses cause the medullae to secrete both norepinephrine and epinephrine into the circulating blood, as will be described in Chapter 56. These two hormones are carried in the bloodstream to all parts of the body, where they act directly on the blood vessels usually to cause vasoconstriction, but sometimes the epinephrine causes vasodilatation, as will be discussed later in the chapter.

### The Sympathetic Vasodilator System and Its Control by the Central Nervous System

The sympathetic nerves to skeletal muscles carry sympathetic vasodilator fibers as well as constrictor fibers. In lower animals, such as the cat, these fibers release *acetylcholine* at their endings, though in primates the vasodilator substance might be epinephrine. The vasodilator acts on the smooth muscle of the blood vessels to cause vasodilatation in contrast to the vasoconstrictor effect of norepinephrine.

The pathways for central nervous system control of the vasodilator system are illustrated by the dashed lines in Figure 20–8. The principal area of the brain controlling this system is the *anterior hypothalamus,* which transmits impulses to a relay station in the *subcollicular region* of the mesencephalon. From here impulses are transmitted down the cord to the *sympathetic preganglionic neurons* in the lateral horns of the cord. These then transmit vasodilator impulses to the blood vessels of the muscles.

Note also in Figure 20–8 that stimulation of the *motor cortex excites* the *sympathetic vasodilator system* by transmitting impulses downward into the hypothalamus.

**Importance of the Sympathetic Vasodilator System.** It is doubtful that the sympathetic vasodilator system plays a very important role in the control of the circulation, because complete block of the sympathetic nerves to the muscles hardly affects the ability of the muscles to control their own blood flow in response to their needs. Yet, it is possible, if not probable, that at the onset of exercise the sympathetic vasodilator system causes initial vasodilation in the skeletal muscles to allow an *anticipatory increase in blood flow* even before the muscles require increased nutrients.

## "PATTERNS" OF CIRCULATORY RESPONSES ELICITED BY DIFFERENT CENTRAL NERVOUS SYSTEM CENTERS

**Stimulation of the Vasomotor Center—The Mass Action Effect.** Stimulation of the lateral portions of the vasomotor center causes wide-spread activation of the vasoconstrictor fibers throughout the body, while stimulation of the medial portions of the vasomotor center causes widespread *inhibition of vasoconstriction.* In many conditions the entire vasomotor center acts as a unit, stimulating all vasoconstrictors throughout the body and also the heart, in addition to stimulating the adrenal medullae to secrete norepinephrine and epinephrine that circulate in the blood to excite the circulation still further. The results of this "mass action" are three-fold: First, the peripheral resistance increases in most parts of the circulation, thereby elevating the arterial pressure. Second, the capacity vessels, particularly the veins, are excited at the same time, greatly decreasing their capacity; this forces increased quantities of blood into the heart, thereby increasing the cardiac output. Third, the heart is simultaneously stimulated so that it can handle the increased cardiac output.

Thus, the overall effect of this "mass action" is to prepare the circulation for increased delivery of blood flow to the body.

**Stimulation of the Hypothalamus—The "Alarm" Pattern.** Under normal conditions, the hypothalamus probably does not transmit large numbers of impulses into the sympathetic vascular control system. However, on occasion the hypothalamus becomes strongly stimulated and can then activate both the vasoconstrictor and vasodilator systems or either of them separately.

*Diffuse stimulation* of the hypothalamus activates the vasodilator system to the muscles, thereby increasing the blood flow through the muscles; at the same time it causes intense vasoconstriction throughout the remainder of the body and an intense increase in heart activity. The arterial pressure rises, the cardiac output increases, the heart rate increases, and the circulation is ready to supply nutrients to the muscles if there be need. Also, impulses are transmitted simultaneously throughout the central nervous system to cause a state of generalized excitement and attentiveness, these often increasing to such a pitch that the overall pattern of the reaction is that of *alarm.*

Thus, the "alarm" pattern contains all the ingredients of the "mass action" pattern as well as the elements of muscle vasodilation and extreme psychic excitement. This pattern seems to have the purposeful effect of preparing the animal or person to perform on a second's notice whatever activity is required.

**The "Motor" Pattern of Circulatory Stimulation.** When the motor cortex transmits signals to the skeletal muscle to cause motor activities, it also sends signals to the circulatory system to cause the following effects: First, it excites the alarm system just described, and this generally activates the heart and increases arterial pressure, making the circulatory system ready to supply increased blood flow to the muscles. This effect also causes vasoconstriction in most nonmuscular parts of the body, such as the kidneys, the skin, and the gut, thereby forcing a larger share of the blood flow through the active muscles. Second, impulses pass directly from the motor cortex to the sympathetic neurons of the spinal cord. These enhance the vasoconstriction in the nonmuscular parts of the body, thereby raising the arterial pressure still more and helping to increase muscle blood flow. They possibly also inhibit the vasoconstrictor nerves to the excited muscles.

The motor pattern of circulatory responses can also be elicited by stimulation of several areas of the reticular substance in the brain stem. Stimulation of one of these areas, the *fields of Forel,* elicits a response almost precisely the same as that caused by direct activation of the motor cortex. Therefore, it is possible, if not likely, that this area of the reticular substance plays a major role in control of the circulatory system in exercise.

*Emotional Fainting—Vasovagal Syncopy.* A particularly interesting circulatory reaction occurs in persons who faint because of intense emotional experiences. In this condition, the muscle vasodilator system becomes powerfully activated so that blood flow through the muscles increases several-fold, and the heart slows markedly. This effect is called *vasovagal syncopy.* The arterial pressure falls instantly, which in turn reduces the blood flow to the brain and causes the person to lose consciousness. It is probable, therefore, that emotional fainting results from powerful stimulation of the anterior hypothalamic vasodilator center.

## REFLEX REGULATION OF THE CIRCULATION

In addition to the many ways in which signals transmitted from the brain to the blood vessels and heart can alter circulatory function, a system of specific circulatory reflexes can also regulate circulation. Most of these reflexes are concerned with the regulation of arterial pressure or with the regulation of blood flow to specific local areas of the body. Therefore, these will be discussed in detail in the following chapter in relation to arterial pressure regulation as well as at other points in the text. However, let us summarize here a few of the more important reflex regulations of the circulation.

**Arterial Pressure Reflexes.** The most important of the arterial pressure reflexes is the *baroreceptor reflex.* An increase in arterial pressure stretches the walls of the major arteries in the chest and neck and this in turn excites stretch receptors located in the vessel walls, especially stretch receptors located in the carotid sinuses and in the aortic arch. Signals are transmitted from these receptors to the vasomotor center of the brain stem, and reflex signals are transmitted back to the heart and to the blood vessels to slow the heart and to dilate the vessels, thereby reducing the arterial pressure toward normal. A decrease in arterial pressure causes exactly the opposite effects. Thus, the baroreceptor reflex helps to stabilize the arterial pressure.

**Reflexes for Control of Blood Volume.** When the blood volume increases, the volume of blood in the central veins and in the right and left atria usually also increases. This stretches the walls of the atria and large veins, again exciting stretch receptors. These stretch receptors also transmit signals into the vasomotor center, and reflex signals are then transmitted to the kidneys to increase urinary output of fluid. Also, signals are transmitted to the hypothalamus to decrease the rate of secretion of antidiuretic hormone, thereby also increasing the output of fluid by the kidneys. Thus, this reflex mechanism helps to control the total amount of fluid in the body, and in this way also helps to control blood volume.

**Reflexes for Control of Body Temperature.** When the body temperature rises too high, special neurons in the anterior hypothalamus become excited; these in turn send signals through the sympathetic nervous system to dilate the skin blood vessels, thereby allowing transfer of major amounts of internal body heat to the skin. This heat then passes through the skin to the surroundings. As a result, the body temperature falls back toward normal. This reflex is also associated with sweating reflexes and with other aspects of body temperature control, all of which will be discussed in Chapter 72.

## HUMORAL REGULATION OF THE CIRCULATION

The term humoral regulation means regulation by substances—such as hormones, ions, and so forth—in the body fluids. Among the most important of these factors are the following:

**Norepinephrine and Epinephrine.** Earlier in the chapter it was pointed out that the adrenal medullae secrete both norepinephrine and epinephrine. When the sympathetic nervous system throughout the body is stimulated to cause direct effects on the blood vessels, it also causes the adrenal medullae to secrete these two hormones, which then circulate everywhere in the body fluids and act on all vasculature. Norepinephrine has vasoconstrictor effects in almost all vascular beds of the body, and epinephrine has similar effects in some, but not all, beds. For instance, epinephrine causes vasodilatation in both skeletal and cardiac muscle. These two hormones will be discussed in more detail in Chapter 57 in the discussion of the autonomic nervous system.

**Angiotensin.** Angiotensin is the most powerful vasoconstrictor substance known. As little as *one ten-millionth* of a gram can increase the arterial pressure of a human being as much as 10 to 20 mm. Hg under some conditions. Since this substance is very important in relation to arterial pressure regulation, it will be discussed in detail in the following two chapters.

Briefly, either a decrease in arterial pressure or a decrease in quantity of sodium in the body fluids will cause the kidneys to secrete the substance *renin*. The renin in turn acts on one of the plasma proteins, *renin substrate,* to split away the vasoactive peptide angiotensin. The angiotensin in turn has a number of important effects on the circulation related to arterial pressure control: (1) it causes marked constriction of the peripheral arterioles; (2) it causes moderate constriction of the veins, thereby reducing the vascular volume and also probably decreasing vascular compliance; and (3) it causes constriction of the renal arterioles, thereby causing the kidneys to retain both water and salt, thus increasing the body fluid volume, which helps to raise the arterial pressure. Hence, an initial decrease in arterial pressure or decrease in sodium causes a compensatory buildup of body fluid and sodium as well as an increase in arterial pressure, thereby compensating for the original deficit.

**Bradykinin.** Several substances called *kinins* that can cause vasodilatation have been isolated from blood and tissue fluids. One of these substances is *bradykinin.*

The kinins are small polypeptides that are split away from alpha$_2$ globulins in the plasma or tissue fluids. Different types of proteolytic enzymes can split the kinins from the globulin. An enzyme of particular importance is *kallikrein,* which is present in the blood and tissue fluids in an inactive form. Kallikrein can be activated in several different ways, such as by maceration of the blood, dilution of the blood, contact of the blood with glass, and other similar chemical and physical effects on the blood. As kallikrein becomes activated, it acts immediately on the alpha$_2$ globulin to release bradykinin. Once formed, the bradykinin persists for only a few minutes, because it is digested by the enzyme *carboxypeptidase*. The activated kallikrein enzyme is destroyed by a kallikrein inhibitor also present in the body fluids.

Bradykinin causes very powerful *vasodilatation* and also *increased capillary permeability*. For instance, injection of 1 *microgram* of bradykinin into the brachial artery of a man increases the blood flow through the arm as much as six-fold, and even smaller amounts injected locally into tissues can cause marked edema because of the increase in capillary pore size.

Though unfortunately we know little about the function of the kinins in the control of the circulation, these extremely powerful effects, coupled with the fact that kinins can develop anywhere in the circulatory system or tissues with ease, indicate that they must play important roles in circulatory regulation. There is reason to believe that kinins play special roles in regulating blood flow, as well as capillary release of fluids in inflamed tissues. It has also been claimed—though it is doubtful—that bradykinin plays a role in regulating skin blood flow and blood flow in gastrointestinal glands.

**Vasopressin.** *Vasopressin* is a hormone formed in the hypothalamus but secreted through the posterior pituitary gland. It has a powerful effect on the arterioles that is similar to that of angiotensin, but it has almost no effect on the veins. The primary function of vasopressin is to control reabsorption of water from the renal tubules (see Chapter 36). However, recent experiments show that its concentration in the blood sometimes rises high enough to increase arterial pressure as much as 30 to 40 mm. Hg.

**Serotonin.** Serotonin (5-hydroxytryptamine) is present in large concentrations in the chromaffin tissue of the intestine and other abdominal structures. Also, it is present in high concentration in the platelets. Serotonin can have either a vasodilator or vasoconstrictor effect, depending on the condition or the area of the circulation. And, even though these effects can sometimes be powerful, the functions of serotonin in regulation of the circulation are almost entirely unknown. Occasionally, tumors composed of chromaffin tissue develop, called *carcinoid tumors*. These secrete tremendous quantities of serotonin and cause mottled areas of vasodilatation in the skin; but the very fact that these tremendous quantities of serotonin do not drastically disturb the circulation makes it doubtful that serotonin plays a widespread general role in regulation of circulatory function.

**Histamine.** Histamine is released by essentially every tissue of the body whenever it becomes damaged. Most of the histamine is probably derived from eosinophils and mast cells in the damaged tissues.

Histamine has a powerful vasodilator effect on the arterioles, and like bradykinin, also has a very potent effect on increasing capillary porosity, allowing leakage of both fluid and plasma protein into the tissues. Though the role of histamine in normal regulation of the circulation is unknown, in many pathological conditions the intense arteriolar dilatation and increased capillary porosity caused by histamine cause tremendous quantities of fluid to leak out of the circulation into the tissues, inducing edema. The effects of histamine were discussed in Chapter 7 in relation to allergic reactions.

**Prostaglandins.** Almost every tissue of the body contains small to moderate amounts of several chemically related substances called prostaglandins. These substances are released into the local tissue fluids and into the circulating bloods under both physiological and pathological conditions. Some of the prostaglandins cause vasoconstriction; still others cause vasodilation. Thus far, no specific pattern of function of the prostaglandins in circulatory control has been found. However, this widespread prevalence in the tissues and their myriad effects on the circulation make them ideal candidates for special roles in circulatory control, especially for control in local vascular areas. For this reason these substances are presently under intensive research investigation, though the unequivocal results thus far are not remarkable enough to justify further discussion here.

### EFFECTS OF CHEMICAL FACTORS ON VASCULAR CONSTRICTION

Many different chemical factors can either dilate or constrict local blood vessels. Though the roles of these substances in the overall *regulation* of the circulation generally are not known, their specific effects can be listed as follows:

An increase in *calcium ion* concentration causes vasoconstriction. This results from the general effect of calcium to stimulate smooth muscle contraction, as discussed in Chapter 12.

An increase in *potassium ion* concentration causes vasodilatation. This results from the general effect of potassium ions to inhibit smooth muscle contraction.

An increase in *magnesium ion* concentration causes powerful vasodilatation, for magnesium ions, like potassium ions, inhibit smooth muscle generally.

Increased *sodium ion* concentration causes arteriolar dilatation. This results from an increase in osmolality of the fluids rather than from a specific effect of sodium ion itself. *Increased osmolality* of the blood caused by increased quantities of *glucose* or other nonvasoactive substances also causes arteriolar dilatation. Decreased osmolality causes arteriolar constriction.

The only anions to have significant effects on blood vessels are *acetate* and *citrate*, both of which cause mild degrees of vasodilatation.

An *increase in hydrogen ion* concentration (decrease in pH) causes dilatation of the arterioles. A slight *decrease in hydrogen ion* concentration causes arteriolar constriction, but an intense decrease causes dilatation, which is the same effect as that which occurs with increased hydrogen ion concentration.

An increase in carbon dioxide concentration causes moderate vasodilatation in most tissues but marked vasodilatation in the brain. However, carbon dioxide, acting on the vasomotor center, has an extremely powerful indirect vasoconstrictor effect that is transmitted through the sympathetic vasoconstrictor system.

## REFERENCES

Altura, B. M.: Chemical and humoral regulation of blood flow through the precapillary sphincter. Microvasc. Res., 3:361, 1971.

Bloch, E. H.: The illumination, sensing and recording of the living microvascular system. Bibl. Anat., 11:185, 1973.

Charm, S. E., and Kurland, G. S.: Blood Flow and Microcirculation. New York, John Wiley & Sons, Inc., 1974.

Costa, E., Gessa, G. L., and Sandler, M. (eds.): Serotinin—New Vistas. New York, Raven Press, 1975.

Cowley, A. W., Jr., and Guyton, A. C.: Quantification of intermediate steps in the renin-angiotensin-vasoconstrictor feedback loop in the dog. Circ. Res., 30:557, 1972.

Donald, D. E., and Ferguson, D. A.: Study of the sympathetic vasoconstrictor nerves to the vessels of the dog hind limb. Circ. Res., 26:171, 1970.

Granger, H. J., and Guyton, A. C.: Autoregulation of the total systemic circulation following destruction of the central nervous system in the dog. Circ. Res., 25:379, 1969.

Guyton, A. C.: Integrative hemodynamics. In Sodeman, W. A., Jr., and Sodeman, W. A. (eds.): Pathologic Physiology: Mechanisms of Disease, 5th Ed. Philadelphia, W. B. Saunders Company, 1974, p. 149.

Guyton, A. C., Coleman, G., and Granger, J.: Circulation: overall regulation. Ann. Rev. Physiol., 34:13, 1972.

Haddy, F. J.: Local effects of sodium, calcium, and magnesium upon small and large blood vessels of the dog forelimb. Circ. Res., 8:57, 1960.

Hilton, S. M.: Hypothalamic regulation of the cardiovascular system. Brit. Med. Bull., 22:243, 1966.

Johnson, P. C.: The microcirculation and local and humoral control of the circulation. In Guyton, A. C. (ed.): MTP International Review of Science: Physiology. Baltimore, University Park Press, 1974, Vol.1, p. 163.

Kahlson, G., and Rosengren, E.: Biogenesis and Physiology of Histamine. Baltimore, The Williams & Wilkins Company, 1971.

Kahlson, G., and Rosengren, E.: Histamine: entering physiology. Experientia, 28:993, 1972.

Korner, P. I.: Circulatory adaptations in hypoxia. Physiol. Rev., 39:687, 1959.

Laragh, J. H., and Sealey, J.: The renin-angiotensin-aldosterone hormonal system and regulation of sodium, potassium, and blood pressure homeostasis. In Orloff, F., and Berliner, R. W. (eds.): Handbook of Physiology. Baltimore, The Williams & Wilkins Company, 1973, Sec. 8, p. 831.

Lee, J. B.: Prostaglandins. *Physiologist, 13*:379, 1970.

Lee, J. B.: Cardiovascular-renal effects of prostaglandins: the antihypertensive, natriuretic renal "endocrine" function. *Arch. Intern. Med., 133*:56, 1974.

Lefkowitz, R. J., and Haber, E.: Physiological receptors as physicochemical entities. *Circ. Res., 32 (Suppl. 1)*:46, 1973.

Mellander, S.: Systemic circulation. *Ann. Rev. Physiol., 32*:313, 1970.

Nakano, J.: Cardiovascular responses to neurohypophysial hormones. *In* Greep, R. O., and Astwood, E. B. (eds.): Handbook of Physiology. Baltimore, The Williams & Wilkins Company, 1974, Vol. 4, Sec. 7, Part 1, p. 395.

Nies, A. S., Shand, D. G., and Branch, R. A.: Hemodynamic drug interactions. *Cardiovasc. Clin., 6*:43, 1974.

Pike, J. E.: Prostaglandins. *Sci. Amer., 225*:84, 1971.

Rosenthal, S. L., and Guyton, A. C.: Hemodynamics of collateral vasodilatation following femoral artery occlusion in anesthetized dogs. *Circ. Res., 23*:239, 1968.

Schachter, M.: Kallikreins and kinins. *Physiol. Rev., 49*:509, 1969.

Scott, J. B., Rudko, M., Radawski, D., and Haddy, F. J.: Role of osmolarity, $K^+$, $H^+$, $Mg^{++}$, and $O_2$ in local blood flow regulation. *Amer. J. Physiol., 218*:338, 1970.

Share, L.: Blood pressure, blood volume, and the release of vasopressin. *In* Greep, R. O., and Astwood, E. B. (eds.): Handbook of Physiology. Baltimore, The Williams & Wilkins Company, 1974, Vol. 4, Sec. 7, Part 1, p. 243.

Shepherd, A. P., Granger, H. J., Smith, E. E., and Guyton, A. C.: Local control of tissue oxygen delivery and its contribution to the regulation of cardiac output. *Amer. J. Physiol., 225*:747, 1973.

Stainsby, W. N.: Local control of regional blood flow. *Ann. Rev. Physiol., 35*:151, 1973.

Tsakiris, A. G., Donald, D. E., Rutishauser, W. J., Banchero, N., and Wood, E. H.: Cardiovascular responses to hypertension and hypotension in dogs with denervated hearts. *J. Appl. Physiol., 27*:817, 1969.

Walker, J. R., and Guyton, A. C.: Influence of blood oxygen saturation on pressure-flow curve of dog hindleg. *Amer. J. Physiol., 212*:506, 1967.

Whalen, W. J.: Intracellular $PO_2$ in heart and skeletal muscle. *Physiologist, 14*:69, 1971.

Wolthuis, R. A., Bergman, S. A., and Nicogossian, A. E.: Physiological effects of locally applied reduced pressure in man. *Physiol. Rev., 54*:566, 1974.

# 21

# Regulation of Mean Arterial Pressure: I. Nervous Reflex and Hormonal Mechanisms for Rapid Pressure Control

In the previous chapter we pointed out that each tissue can control its own blood flow by simply dilating or constricting its local arterioles. For this mechanism to work, it is necessary that the arterial pressure remain constant or nearly constant because if the arterial pressure were very variable one would never know whether dilating the blood vessels would necessarily increase the local blood flow. Fortunately, the circulation has an intricate system for regulation of the arterial pressure. It maintains the normal mean arterial pressure within the rather narrow limits between 80 mm. Hg and 120 mm. Hg. Some of the pressure regulatory mechanisms (mainly nervous and hormonal mechanisms) act very rapidly, and some (mainly mechanisms related to kidney function and blood volume regulation) act very slowly. In the present chapter we will discuss the rapid nervous and hormonal pressure control mechanisms. In the following chapter we will discuss both the long-term regulation of arterial pressure, based primarily on renal and body-fluid mechanisms, and the clinical problem of hypertension or "high blood pressure," which is caused by abnormalities of the long-term pressure regulatory mechanisms.

But first, let us discuss some of the normal values of arterial pressure in a human being and the clinical method for measuring these pressures.

## NORMAL ARTERIAL PRESSURES

**Arterial Pressures at Different Ages.** Figure 21–1 illustrates the typical diastolic, systolic, and mean arterial pressures from birth to 80 years of age. From this figure it can be seen that the systolic pressure of a normal young adult averages about 120 mm. Hg and his diastolic pressure about 80 mm. Hg—that is, his arterial pressure is said to be 120/80. The shaded areas on either side of the curves depict the normal ranges of systolic and diastolic pres-

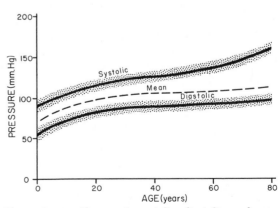

**Figure 21–1.** Changes in systolic, diastolic, and mean arterial pressures with age. The shaded areas show the normal range.

265

sures, showing considerable variation from person to person.

The increase in arterial pressures at older ages is usually associated with developing arteriosclerosis. The systolic pressure increases especially; in approximately one-tenth of all old people it eventually rises above 200 mm. Hg.

## THE MEAN ARTERIAL PRESSURE

The mean arterial pressure is the average pressure throughout the pressure pulse cycle. Offhand one might expect that it would be equal to the average of systolic and diastolic pressures, but this is not true; the arterial pressure usually remains nearer to diastolic level than to systolic level during a greater portion of the pulse cycle. Therefore, the mean arterial pressure is usually slightly less than the average of systolic and diastolic pressures, as is evident in Figure 21–1.

**Mean Arterial Pressure in the Human Being.** The mean arterial pressure of the normal young adult averages about 96 mm. Hg, which is slightly less than the average of his systolic and diastolic pressures, 120 and 80 mm. Hg, respectively. However, for purposes of discussion, the mean arterial pressure is usually considered to be 100 mm. Hg because this value is easy to remember.

The mean arterial pressure, like the systolic and diastolic pressures, is lowest immediately after birth, measuring about 70 mm. Hg at birth and reaching an average of about 110 mm. Hg in the normal old person or as high as 130 mm. Hg in the person with arteriosclerosis. From adolescence to middle age the mean arterial pressure does not vary greatly from the normal value of 100 mm. Hg.

**Significance of Mean Arterial Pressure.** Mean arterial pressure is the average pressure tending to push blood through the systemic circulatory system. Therefore, *from the point of view of tissue blood flow, it is generally the mean arterial pressure that is important.*

## CLINICAL METHODS FOR MEASURING SYSTOLIC AND DIASTOLIC PRESSURES

Obviously, it is impossible to use the various pressure recorders that require needle insertion into an artery as described in Chapter 18 for making routine pressure measurements in human patients, although they are used on occasion when special studies are necessary. Instead, the clinician determines systolic and diastolic pressures by indirect means, most usually by the auscultatory method.

**The Auscultatory Method.** Figure 21–2 illustrates the auscultatory method for determining systolic and diastolic arterial pressures. A stethoscope is placed over the antecubital artery while a blood pressure cuff is inflated around the upper arm. As long as the cuff presses against the arm with so little pressure that the artery remains distended with blood, no sounds whatsoever are heard by the stethoscope despite the fact that the blood pressure within the artery is pulsating. But when the cuff pressure is great enough to close the artery during part of the arterial pressure cycle, a sound is heard in the stethoscope with each pulsation. These sounds are called *Korotkow sounds.*

The exact cause of Korotkow sounds is still debated, but they are believed to be caused by blood jetting through the partly occluded vessel. The jet causes turbulence in the open vessel beyond the cuff, and this sets up the vibrations heard through the stethoscope.

In determining blood pressure by the auscultatory method, the pressure in the cuff is first elevated well above arterial systolic pressure. As long as this pressure is higher than systolic pressure, the brachial artery remains collapsed and no blood whatsoever flows into the lower artery during any part of the pressure cycle. Therefore, no Korotkow sounds are heard in the lower artery. But then the cuff pressure is gradually reduced. Just as soon as the pressure in the cuff falls below systolic pressure, blood slips through the artery beneath the cuff during

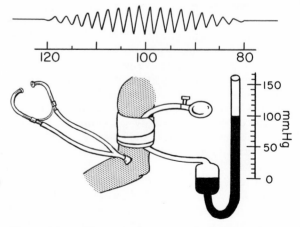

**Figure 21–2.** The auscultatory method for measuring systolic and diastolic pressures.

the peak of systolic pressure, and one begins to hear *tapping* sounds in the antecubital artery in synchrony with the heart beat. As soon as these sounds are heard, the pressure level indicated by the manometer connected to the cuff is approximately equal to the systolic pressure.

As the pressure in the cuff is lowered still more, the Korotkow sounds continue their tapping quality until the pressure in the cuff falls to equal diastolic pressure. Below this pressure level the artery no longer closes during diastole, which means that the basic factor causing the sound (the jetting of blood through a squeezed artery) is no longer present. Therefore, the tapping quality of the sound suddenly changes to a muffled quality, and the sounds usually disappear entirely after another 5 to 10 mm. drop in cuff pressure. One notes the manometer pressure when the Karotkow sound quality changes from tapping to muffled, and this pressure is approximately equal to the diastolic pressure.

The auscultatory method for determining systolic and diastolic pressures is not entirely accurate, but it usually gives values within 10 per cent of those determined by direct measurement from the arteries.

**Radial Pulse and Oscillometric Methods for Estimating Arterial Pressure.** Arterial blood pressure can also be estimated by feeling the radial pulse or by recording the pulsation in the lower arm with an oscillometer while a cuff is inflated over the upper arm. An oscillometer consists of a recording apparatus that can register pulsations in a lightly inflated blood pressure cuff around the forearm. A blood pressure recorder of the aneroid type makes an excellent oscillometer for estimating arterial pressure.

In applying these methods, the pressure in the cuff on the upper arm is raised well above the systolic level and then it is progressively reduced. No pulsation will be felt in the radial artery or registered by the oscillometer until the cuff pressure falls below the systolic pressure level. But, just as soon as this point is reached, a distinct radial pulse can be felt, or a distinct oscillation will be recorded by the oscillometer. Further decrease in the upper arm cuff pressure results, at first, in progressive increase in the intensity of radial or oscillometer pulsation. Then there is a slight decrease in these pulsations at approximately the diastolic level of pressure, but this change is so indistinct that estimation of diastolic pressure is likely to be in severe error.

The oscillometric method is valuable for measuring pressures when distinct sounds cannot be heard from the forearm arteries. This often results in (1) very young children or (2) adults when the arteries are in a state of spasm, such as when a patient is in *shock*. These instances are often among the most important in which it is essential to have at least some

estimation of arterial pressure, thus illustrating the value of knowing this accessory method for measuring systolic arterial pressure in addition to the auscultatory method.

## RELATIONSHIP OF ARTERIAL PRESSURE TO CARDIAC OUTPUT AND TOTAL PERIPHERAL RESISTANCE

Before discussing the overall regulation of arterial pressure, it is good to remember the basic relationship between arterial pressure, cardiac output, and total peripheral resistance, which was discussed in detail in Chapter 18, as follows:

$$\text{Arterial Pressure} = \text{Cardiac Output} \times \text{Total Peripheral Resistance}$$

It is obvious from this formula that any factor that increases either the cardiac output or total peripheral resistance will cause an increase in mean arterial pressure. Both of these factors are often manipulated in the control of arterial pressure, as we shall see in the remainder of this chapter.

## THE OVERALL SYSTEM FOR ARTERIAL PRESSURE REGULATION

Arterial pressure is not regulated by a single pressure controlling system but instead by several interrelated systems that perform specific functions. When a person bleeds severely and his pressure falls suddenly, two problems immediately confront the pressure control system. The first is to return the arterial pressure immediately to a high enough value that the person can continue to live. The second is to return the blood volume eventually to its normal level so that the circulatory system can re-establish full normality, including return of the arterial pressure back to its normal value. These two problems characterize two major types of arterial pressure control systems in the body: (1) a system of rapidly acting pressure control mechanisms, and (2) a system for long-term control of the basic arterial pressure level.

**The Rapidly Acting Pressure Control Mechanisms.** In general, the rapidly acting pressure control mechanisms operate through nervous or hormonal control of the circulation. Receptors in the arterial tree detect a change in arterial pressure and send an appropriate signal to the nervous system. The nervous system then sends signals back to the heart to increase its strength and rate of contraction and to the blood vessels to constrict the arterioles and veins; all these effects combine to raise the arterial pressure within seconds, so that even transient events, such as changing from the lying down to the standing position, will cause very minor changes in arterial pressure in comparison with the major changes that would occur without these nervous controls. The hormonal mechanisms, most importantly that of the hormone angiotensin, also respond rapidly to changes in pressure, and the responses help to return the pressure to normal. These mechanisms in general require from minutes to hours for a response and are therefore especially important for moderately long intervals of pressure control.

**The Long-term Mechanism for Arterial Pressure Regulation.** The nervous regulators of arterial pressure, though acting very rapidly and powerfully to correct acute abnormalities of arterial pressure, generally lose their power to control arterial pressure after a few hours to a few days because the nervous pressure receptors "adapt"; that is, they lose their responsiveness. Therefore, except under unusual circumstances, the nervous mechanisms for arterial pressure control do not play a major role in long-term regulation of arterial pressure. The long-term regulation, instead, is vested in a renal-fluid volume-pressure control mechanism. Basically, it works as follows: When the arterial pressure falls, the kidneys retain water and salt until the blood volume rises. This in turn raises the arterial pressure back to normal. Though many physiologists and clinicians have forgotten this very simple mechanism for arterial pressure regulation or have considered it to be unimportant in comparison with other mechanisms, recent experiments have demonstrated the extreme power of this mechanism and have shown that it is the basic long-term, day in and day out, month in and month out, mechanism for basic arterial pressure control. However, its operation can be modified greatly by various hormones, especially angiotensin and aldosterone. This overall system for long-term arterial pressure regulation will be presented in detail in the following chapter.

# RAPIDLY ACTING NERVOUS MECHANISMS FOR ARTERIAL PRESSURE CONTROL

## THE ARTERIAL BARORECEPTOR CONTROL SYSTEM—BARORECEPTOR REFLEXES

By far the best known of the mechanisms for arterial pressure control is the *baroreceptor reflex*. Basically, this reflex is initiated by pressure receptors, called either *baroreceptors* or *pressoreceptors,* located in the walls of the large systemic arteries. A rise in pressure causes the baroreceptors to transmit signals into the central nervous system, and other signals are in turn sent to the circulation to reduce arterial pressure back toward the normal level.

**Physiologic Anatomy of the Baroreceptors.** Baroreceptors are spray-type nerve endings lying in the walls of the arteries that are stimulated when stretched. A few baroreceptors are located in the wall of almost every large artery of the thoracic and neck regions; but, as illustrated in Figure 21–3, baroreceptors are extremely abundant, in (1) the walls of the internal carotid arteries slightly above the carotid bifurcations, areas known as the *carotid sinuses,* and (2) the walls of the aortic arch.

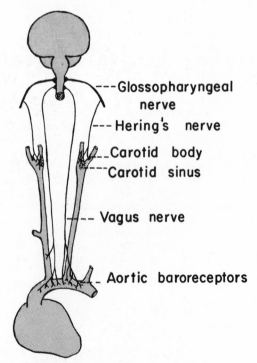

**Figure 21–3.** The baroreceptor system.

Figure 21–3 also shows that impulses are transmitted from each carotid sinus through the very small *Hering's nerve* to the glosso-pharyngeal nerve and thence to the medulla. Impulses from the arch of the aorta are transmitted through the vagus nerves to the medulla. Hering's nerve is especially important in physiologic experiments because baroreceptor impulses can be recorded from it with ease.

**Response of the Baroreceptors to Pressure.** Figure 21–4 illustrates the effect of different arterial pressures on the rate of impulse transmission in a Hering's nerve. Note that the baroreceptors are not stimulated at all by pressures between 0 and 60 mm. Hg, but above 60 mm. Hg they respond progressively more and more rapidly and reach a maximum at about 180 mm. Hg.

It is especially important that the increase in number of impulses for each unit change in arterial pressure, expressed as $\Delta I/\Delta P$ in the figure, is greatest at a pressure level near the normal mean arterial pressure. This means that the baroreceptors respond most markedly to changes in arterial pressure just at the level where the response needs to be most marked. That is, in the normal operating range of arterial pressure, even a slight change in pressure causes strong sympathetic reflexes to readjust the arterial pressure back toward normal.

The baroreceptors respond extremely rapidly to changes in arterial pressure; the number of impulses even increases during systole and decreases again during diastole. Furthermore, the baroreceptors *respond much more to a rising pressure* than to a stationary pressure and, also, respond less to a falling pressure. That is, if the mean arterial pressure is 150 mm. Hg but at that moment is rising rapidly, the rate of impulse transmission may be as much as twice that when the pressure is stationary at 150 mm. Hg. On the other hand, if the pressure is falling, the rate might be as little as one-quarter that for a stationary pressure.

**The Reflex Initiated by the Baroreceptors.** The baroreceptor impulses *inhibit the vasoconstrictor center* of the medulla and *excite the vagal center*. The net effects are (1) *vasodilatation* throughout the peripheral circulatory system and (2) *decreased cardiac rate* and *strength of contraction*. Therefore, excitation of the baroreceptors by pressure in the arteries reflexly *causes the arterial pressure to decrease*. Conversely, low pressure has opposite effects, reflexly causing the pressure to rise back toward normal.

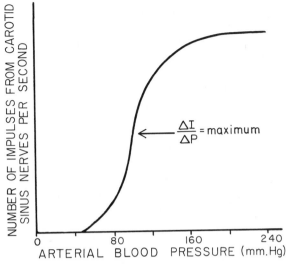

**Figure 21–4.** Response of the baroreceptors at different levels of arterial pressure.

Figure 21–5 illustrates a typical reflex change in arterial pressure caused by clamping the common carotids. This procedure reduces the carotid sinus pressure; as a result, the baroreceptors become inactive and lose their inhibitory effect on the vasomotor center. The vasomotor center then becomes much more active than usual, causing the arterial pressure to rise and to remain elevated as long as the carotids are clamped. Removal of the clamps allows the pressure to fall immediately to slightly below normal as a momentary overcompensation and then to return to normal in another minute or so.

*Function of the Baroreceptors During Changes in Body Posture.* The ability of the baroreceptors to maintain relatively constant arterial pressure is extremely important when a person sits or stands after having been lying down. Immediately upon standing, the arterial pressure in the head and upper part of the body obviously tends to fall, and marked reduction of this pressure can cause loss of consciousness. Fortunately, however, the falling pressure at the baroreceptors elicits an immediate reflex, resulting in strong sympathetic discharge throughout the body, and this minimizes the decrease in pressure in the head and upper body.

*The "Buffer" Function of the Baroreceptor Control System.* Because the baroreceptor system opposes increases and decreases in arterial pressure, it is often called a *buffer system*, and

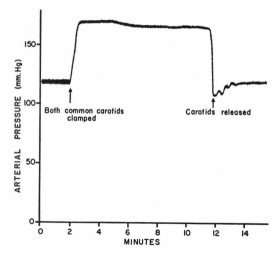

**Figure 21–5.** Typical carotid sinus reflex effect on arterial pressure caused by clamping both common carotids.

the nerves from the baroreceptors are called *buffer nerves*.

Figure 21–6 illustrates the importance of this buffer function of the baroreceptors. The lower panel of this figure shows an arterial pressure recording for 2 hours from a dog whose baroreceptor nerves from both the carotid sinuses and the aorta had previously been removed. Note the extreme variability of pressure caused by simple events of the day such as lying down, standing, excitement, eating, and so forth. The upper panel of this figure illustrates a recording from the same dog before the baroreceptor nerves had been removed, showing far less variability in the pressure. Figure 21–7 illustrates the frequency distributions of the arterial pressures recorded for a full 24-hour day. Note that when the baroreceptors were intact, the arterial pressure remained throughout the day within the narrow range from 80 mm. Hg to 120 mm. Hg—most of the day at almost exactly 100 mm. Hg. On the other hand, after removal of the baroreceptors the pressure range was increased 2½-fold, frequently falling to as low as 40 mm. Hg or rising to over 175 mm. Hg. Thus, one can see the extreme variability of pressure in the absence of the arterial baroreceptor system.

In summary, we can state that the primary purpose of the arterial baroreceptor system is to reduce the daily variation in arterial pressure to about one-half to one-third that which would occur were the baroreceptor system not present. But note, as shown in both Figures 21–6

and 21–7, the mean arterial pressure remains almost exactly the same whether the baroreceptors are present or not, illustrating the unimportance of the baroreceptor system for long-term regulation of arterial pressure even though it is extremely potent in preventing the rapid changes of arterial pressure that occur moment-by-moment or hour-by-hour.

**Unimportance of the Baroreceptor System for Long-Term Regulation of Arterial Pressure—Adaptation of the Baroreceptors.** The baroreceptor control system is probably of no importance whatsoever in long-term regulation of arterial pressure for a very simple reason: The baroreceptors themselves adapt in one to two days to whatever pressure level they are exposed. That is, if the pressure rises from the normal value of 100 mm. Hg up to 200 mm. Hg, extreme numbers of baroreceptor impulses are at first transmitted. During the next few seconds, the rate of firing diminishes considerably; then it diminishes very slowly during the

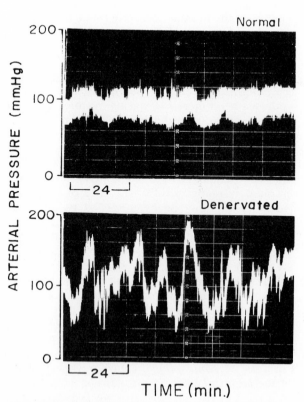

**Figure 21–6.** Two hour records of arterial pressure in a normal dog (above) and in the same dog (below) several weeks after the baroreceptors had been denervated. (Courtesy of Dr. Allen W. Cowley, Jr.)

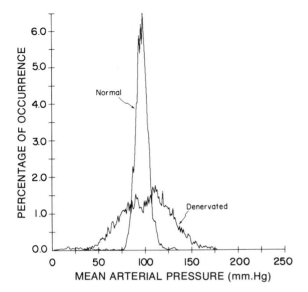

**Figure 21–7.** Frequency distribution curves of the arterial pressure for a 24-hour period in a normal dog and in the same dog several weeks after the baroreceptors had been denervated. (Courtesy of Dr. Allen W. Cowley, Jr.)

next one to two days, at the end of which time the rate will have returned essentially to the normal level despite the fact that the arterial pressure remains 200 mm. Hg. Conversely, when the arterial pressure falls to a very low value, the baroreceptors at first transmit no impulses at all, but gradually over a period of several days the rate of baroreceptor firing returns again to the original control level.

This adaptation of the baroreceptors obviously prevents the baroreceptor reflex from continuing to function as a control system for longer than a few days at a time. Therefore, prolonged regulation of arterial pressure requires other control systems, principally the renal-body fluid pressure control system to be discussed in the following chapter.

**The Carotid Sinus Syndrome.** Strong pressure on the neck over the bifurcations of the carotids in the human being can excite the baroreceptors of the carotid sinuses, causing the arterial pressure to fall as much as 20 mm. Hg in the normal person. In some older persons, particularly after calcified arteriosclerotic plaques have developed in the carotid arteries, pressure on the carotid sinuses often causes such strong responses that the heart stops completely, or at least the arterial pressure falls drastically. Even tight collars can cause the arterial pressure to fall low enough to cause fainting in these persons, an effect called *carotid sinus syncope*. Fortunately, when the reflex causes the heart to stop, the ventricles usually "escape" from the vagal inhibition

in approximately 7 to 10 seconds and begin to beat with their own intrinsic rhythm. However, occasionally, the ventricles fail to escape, and the patient dies of cardiac arrest. Treatment of the condition consists of surgically stripping the nerves from the carotid arteries above and below the bifurcations.

## ATRIAL AND PULMONARY ARTERY REFLEXES THAT HELP TO REGULATE ARTERIAL PRESSURE

Both the atria and the pulmonary arteries have stretch receptors, called *low pressure receptors,* in their walls similar to the baroreceptor stretch receptors of the large systemic arteries. When these low pressure receptors are intact, the arterial pressure changes far less in response to changes in blood volume than when they are not present. To give an example, if 300 ml. of blood is infused into a dog with no receptors at all—no pulmonary artery receptors, no atrial receptors, and no baroreceptors—the arterial pressure of the animal will increase approximately 120 mm. Hg. If the baroreceptors are intact, but the low pressure receptors are not, the arterial pressure will rise 50 mm. Hg. If all the receptors are intact, the pressure will rise only 15 mm. Hg.

Thus, one can see that even though the low pressure receptors in the pulmonary artery and in the atria cannot detect the systemic arterial pressure, these receptors, nevertheless, do detect the simultaneous increase in pressure in the low pressure areas of the circulation caused by the increase in volume, and they elicit reflexes parallel to the baroreceptor reflexes to make the total reflex system much more potent for control of arterial pressure.

The receptors in the pulmonary artery operate in almost identically the same way as the baroreceptors from the systemic arteries. On the other hand, the atrial receptors operate somewhat differently, as follows.

**Atrial Reflexes to Decrease Arterial Pressure.** Recent experiments have demonstrated that stretching the atria causes reflex vasodilatation of the peripheral arterioles. This in turn reduces the total peripheral resistance and, therefore, decreases the arterial pressure back toward the normal level. This effect also plays an important role in reducing the blood volume back toward normal in the following way: The decrease in arteriolar resistance causes rapid blood flow into the capillaries, thereby raising the capillary pressure. As a result, fluid filters out of the circulation into the tissue spaces so that some of the excess blood volume is temporarily dumped into the tissues. This mechanism will be discussed, especially in relation to blood volume regulation, in Chapter 36.

**Atrial Reflexes to the Kidneys—the Volume Reflex.** Stretch of the atria also causes reflex dilatation of the afferent arterioles in the kidneys, which is the same reflex effect that occurs in other peripheral arterioles. And signals are transmitted simultaneously to the hypothalamus to decrease the secretion of antidiuretic hormone, thereby affecting kidney function. The decreased afferent arteriolar resistance causes the glomerular capillary pressure to rise, with resultant increase in filtration of fluid into the kidney tubules. The diminution of antidiuretic hormone diminishes the reabsorption of water from the tubules. The combination of these two effects therefore causes rapid loss of fluid into the urine, which also serves as a powerful means to return the blood volume back toward normal.

Obviously, all these mechanisms that tend to return the blood volume back toward normal following a volume overload act indirectly as pressure controllers as well as volume controllers because excess volume drives the heart to greater cardiac output and leads therefore to greater arterial pressure.

**Atrial Reflexes for Control of Heart Rate (the Bainbridge Reflex).** An increase in atrial pressure also causes an increase in heart rate, sometimes increasing the heart rate as much as 100 per cent. Part of this increase in heart rate is caused by the direct effect of the increased atrial volume to stretch the S-A node; it was pointed out in Chapter 14 that such direct stretch can increase the heart rate as much as 20 to 70 per cent. An additional 30 to 40 per cent increase in rate is caused by a reflex called the *Bainbridge reflex*. The stretch receptors of the atria that elicit the Bainbridge reflex transmit their afferent signals through the vagus nerves to the medulla of the brain. Then, efferent signals are transmitted back through both the vagal and sympathetic nerves to increase the heart's rate and strength of contraction. Thus, this reflex helps to prevent damming of blood in the veins, the atria, and the pulmonary circulation. This reflex obviously has a different purpose from that of controlling arterial pressure and is actually detrimental to pressure control for short periods of time.

## CONTROL OF ARTERIAL PRESSURE BY THE VASOMOTOR CENTER IN RESPONSE TO DIMINISHED BRAIN BLOOD FLOW— THE CNS ISCHEMIC RESPONSE

When the arterial pressure falls *very low,* the brain becomes *ischemic,* meaning that there is insufficient blood flow to maintain normal metabolic function of the brain tissue. When this occurs, the vasomotor center becomes exceedingly active, and the systemic arterial pressure often rises to a level as high as the heart can possibly pump. This effect is believed to be caused by failure of the slowly flowing blood to carry carbon dioxide away from the vasomotor center; the local concentration of carbon dioxide increases greatly and has an extremely potent effect in stimulating the sympathetic nervous system. It is possible that other factors, such as the buildup of lactic acid and other acidic substances, also contribute to the marked stimulation of the vasomotor center and to the elevation in pressure. This arterial pressure elevation in response to cerebral ischemia is known as the *central nervous system ischemic response* or simply *CNS ischemic response.*

The magnitude of the ischemic effect on vasomotor activity is tremendous; it can elevate the mean arterial pressure sometimes to as high as 270 mm. Hg. *The degree of sympathetic vasoconstriction caused by intense cerebral ischemia is often so great that some of the peripheral vessels become totally or almost totally occluded.* The kidneys, for instance, will entirely cease their production of urine because of arteriolar constriction in response to the sympathetic discharge. Therefore, *the CNS ischemic response is one of the most powerful of all the activators of the sympathetic vasoconstrictor system.*

**Importance of the CNS Ischemic Response as a Regulator of Arterial Pressure.** Despite the extremely powerful nature of the CNS ischemic response, it does not become very active until the arterial pressure falls far below normal, down to *levels below 50 mm. Hg.* Therefore, it is not one of the mechanisms for regulating normal arterial pressure. Instead it operates principally as an *emergency arterial pressure control system that acts rapidly and extremely powerfully to prevent further decrease in arterial pressure whenever blood flow to the brain decreases dangerously close to the lethal level.*

**The Cushing Reaction.** The so-called Cushing reaction is a special type of CNS ischemic response that results from increased pressure in the cerebrospinal fluid system. When the cerebrospinal fluid pressure rises to equal the arterial pressure, it compresses the arteries in the cranial vault and cuts off the blood supply to the brain. Obviously, this initiates a CNS ischemic response, which causes the arterial pressure to rise. When the arterial pressure has risen to a level higher than the cerebrospinal fluid pressure, blood flows once again into the vessels of the brain to relieve the ischemia. Ordinarily, the blood pressure comes to a new equilibrium level slightly higher than the cerebrospinal fluid pressure, thus allowing blood to continue flowing to the brain at all times. A typical Cushing reaction is illustrated in Figure 21–8.

Obviously, the Cushing reaction helps to protect the vital centers of the brain from loss of nutrition if ever the cerebrospinal fluid pressure rises high enough to compress the cerebral arteries.

**Depressant Effect of Extreme Ischemia on the Vasomotor Center.** If cerebral ischemia becomes so severe that maximum rise in mean arterial pressure still cannot relieve the ischemia, the neuronal

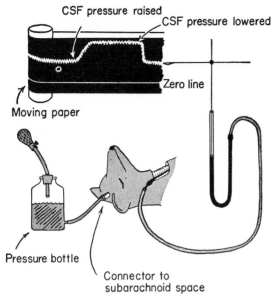

**Figure 21–8.** The Cushing reaction, showing a rise in arterial pressure resulting from increased cerebrospinal fluid pressure.

cells begin to suffer metabolically, and within 3 to 10 minutes they become totally inactive. The arterial pressure then falls to about 40 to 50 mm. Hg, which is the level to which the pressures falls when all tonic vasoconstrictor activity is lost. Therefore, it is fortunate that the ischemic response is extremely powerful so that arterial pressure can usually rise high enough to correct the ischemia before it causes nutritional depression and death of the neuronal cells.

### CHEMORECEPTOR REFLEXES– EFFECT OF OXYGEN LACK ON ARTERIAL PRESSURE

Located in the bifurcations of the carotid arteries and also along the arch of the aorta are several small bodies, 1 to 2 mm. in size, called respectively *carotid* and *aortic bodies*. These contain specialized sensory *receptors sensitive to oxygen lack* (and to carbon dioxide and hydrogen ion excess), called *chemoreceptors*. They stimulate nerve fibers that pass respectively along Hering's nerves and the vagus nerves into the vasomotor center. Each carotid or aortic body is supplied with an abundant blood flow through a small nutrient artery so that the chemoreceptors are always in close contact with arterial blood.

Whenever the oxygen concentration in the arterial blood or the blood flow to the chemoreceptors falls too low, the chemoreceptors become excited, and impulses are transmitted into the vasomotor center to *excite* the vasomotor center, thus reflexly elevating the arterial pressure. Obviously, this reflex helps to increase the quantity of oxygen carried to the tis-

sues whenever the arterial blood becomes deficient in oxygen or blood flow becomes too low because of low arterial pressure. The oxygen lack reflex is not powerful in the normal arterial pressure range, but it does exert reasonable feedback effect on arterial pressure when the pressure is in the range of 40 to 80 mm. Hg.

The chemoreceptors will be discussed in much more detail in Chapter 42 in relation to respiratory control.

### PARTICIPATION OF THE VEINS IN NERVOUS REGULATION OF THE CIRCULATION

Thus far, we have discussed primarily the ability of the nervous system to regulate arterial resistance and thereby to control arterial pressure. However, much of the regulatory effect of the nervous system is carried out through sympathetic vasoconstrictor fibers to the veins. Indeed, the veins constrict in response to even weaker sympathetic stimuli than do the arterioles and arteries.

**Sympathetic Alterations in Venous Capacity and Cardiac Output.** The veins offer relatively little resistance to blood flow in comparison with the arterioles and arteries. Therefore, sympathetic constriction of the veins does not significantly change the overall total peripheral resistance. Instead, the important effect of sympathetic stimulation of the veins is a *decrease in their capacity*. This means that the veins then hold less blood at any given venous pressure, which increases the translocation of blood out of the systemic veins into the heart, lungs, and systemic arteries. The distention of the heart in turn causes the heart to pump with increasing effectiveness in accordance with the Frank-Starling law of the heart, as discussed in Chapter 13. Therefore, the net effect of sympathetic stimulation of the veins is to increase the cardiac output, which in turn elevates the arterial pressure.

The veins participate in all of the reflexes and reactions that have been discussed thus far, including the baroreceptor reflex, the CNS ischemic response, the chemoreceptor reflex, and also the atrial reflexes. To comprehend the potency of the venous reaction in some of these nervous mechanisms, consider the following: The circulatory filling pressure (mainly determined by the degree of contraction of the veins) rises from the normal value of 7 mm. Hg to approximately 10 mm. Hg when the baroreceptor reflex is excited fully and rises to approximately 18 mm. Hg—enough to increase the pressure forcing blood into the heart by as much as 2½-fold—when a maximal CNS ischemic response is elicited.

## ROLE OF THE SKELETAL NERVES AND SKELETAL MUSCLES IN CIRCULATORY CONTROL

Though most nervous control of the circulation is effected through the autonomic nervous system, at least two conditions in which the skeletal nerves and muscles also play major roles in circulatory responses are the following:

**The Abdominal Compression Reflex.** When a baroreceptor or chemoreceptor reflex is elicited, or whenever almost any other factor stimulates the sympathetic vasoconstrictor system, the vasomotor center and other areas of the reticular substance of the brain stem transmit impulses simultaneously through skeletal nerves to skeletal muscles of the body, particularly to the abdominal muscles. This obviously increases the basal tone of these muscles, and their contraction compresses all the venous reservoirs of the abdomen, helping to translocate blood out of these reservoirs toward the heart. As a result, increased quantities of blood are made available to the heart to be pumped. This overall response is called the *abdominal compression reflex*. The resulting effect on the circulation is exactly the same as that caused by sympathetic vasoconstrictor impulses when they constrict the veins. The two effects operate together to increase the cardiac output.

The abdominal compression reflex is probably much more important than has been realized in the past, for it is well known that persons whose skeletal muscles have been paralyzed are considerably more prone to hypotensive episodes than are normal persons.

**Increased Systemic Filling Pressure Caused by Skeletal Muscle Contraction During Exercise.** When the skeletal muscles contract during exercise, they compress blood vessels throughout the body. And even anticipation of exercise tightens the muscles, thereby compressing the vessels. The resulting effect is to increase the *systemic filling pressure* from its normal value of about 7 mm. Hg to as high as 20 to 30 mm. Hg. We shall see in Chapter 23 that this high systemic filling pressure acts to translocate blood from the peripheral vessels into the heart and lungs and therefore to increase the cardiac output. This is probably an essential effect to help cause the 5- to 6-fold increase in cardiac output that sometimes occurs in severe exercise.

## VASOMOTOR WAVES—OSCILLATION OF THE PRESSURE REFLEX CONTROL SYSTEMS

Often when one is recording arterial pressure in an animal, he notes that the blood pressure rises 10 to 20 mm. Hg, then falls, then rises again, continuing cy-clically on and on. The usual duration of each cycle is from 15 to 40 seconds, averaging 26 seconds in a dog. These waves are called *vasomotor waves* and they are also frequently called ''Mayer waves'' or ''Traube-Hering waves.'' Such records are illustrated in Figure 21–9, showing the cyclical rise and fall in arterial pressure.

The cause of vasomotor waves is usually oscillation of one or more nervous pressure control mechanisms, some of which are the following:

**Oscillation of the Baroreceptor and Chemoreceptor Reflexes.** The vasomotor waves of Figure 21–9B are the common vasomotor waves that are seen almost daily in experimental pressure recordings. They are caused mainly by oscillation of the *baroreceptor reflex*. That is, a high pressure excites the baroreceptors; this then inhibits the sympathetic nervous system and lowers the pressure. The decreased pressure reduces the baroreceptor stimulation and allows the vasomotor center to become active once again, elevating the pressure to a high value. This high pressure then initiates another cycle, and the oscillation continues on and on.

The *chemoreceptor reflex* can also oscillate to give the same type of waves. Usually, this reflex oscillates simultaneously with the baroreceptor reflex. In fact, it probably plays the major role in causing vasomotor waves when the arterial pressure is in the range of 40 to 80 mm. Hg, because in this range chemoreceptor control of the circulation becomes powerful while baroreceptor control becomes weak.

**Oscillation of the CNS Ischemic Response.** The record in Figure 21–9A resulted from oscillation of the CNS ischemic pressure control mechanism. In this experiment the cerebrospinal fluid pressure was raised to 160 mm. Hg, which compressed the cerebral vessels and initiated a CNS ischemic response. When the arterial pressure rose above 160 mm. Hg, the ischemia was relieved, and the sympathetic nervous system became inactive. As a result, the arterial pressure fell rapidly back to a lower value, causing medullary ischemia once again. The ischemia then initiated another rise in pressure. Then, again the ischemia was relieved, and the pressure fell. This repeated itself cyclically as long as the cerebrospinal fluid pressure remained elevated.

Thus, any pressure control mechanism can oscillate if the intensity of "feedback" is strong enough. The vasomotor waves are of considerable theoretical

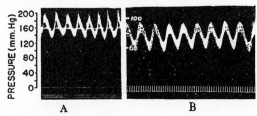

**Figure 21–9.** (A) Vasomotor waves caused by oscillation of the CNS ischemic response. (B) Vasomotor waves caused by pressoreceptor reflex oscillation.

importance because they show that the nervous reflexes that control arterial pressure obey identically the same principles as those applicable to mechanical and electrical control systems. For instance, if the feedback is too great in the guiding mechanism of an automatic pilot for an airplane, the plane oscillates from side to side instead of following a straight course.

## RESPIRATORY WAVES IN THE ARTERIAL PRESSURE

With each cycle of respiration, the arterial pressure rises and falls because of several different effects, some of which are reflex in nature, as follows:

First, many impulses arising in the respiratory center of the medulla "spill over" into the vasomotor center.

Second, every time a person inspires, the pressure in the thoracic cavity becomes more negative than usual, causing the blood vessels in the chest to expand. This reduces the quantity of blood returning to the left side of the heart and thereby momentarily decreases the cardiac output and arterial pressure.

Third, the pressure changes in thoracic vessels caused by respiration can excite baroreceptors.

Though it is difficult to analyze the exact relationship of all these factors in causing the so-called *respiratory pressure waves,* the net result during normal respiration is usually an increase in arterial pressure during the very late part of inspiration and early part of expiration and a fall in pressure during the remainder of the respiratory cycle. During deep respiration the blood pressure can rise and fall as much as 20 mm. Hg with each respiratory cycle.

# HORMONAL MECHANISMS FOR RAPID CONTROL OF ARTERIAL PRESSURE

In addition to the rapidly acting nervous mechanisms for control of arterial pressure, there are at least three hormonal mechanisms that also provide either rapid or moderately rapid control of arterial pressure. These mechanisms are:

1. The norepinephrine-epinephrine vasoconstrictor mechanism.
2. The renin-angiotensin vasoconstrictor mechanism.
3. The vasopressin vasoconstrictor mechanism.

## THE NOREPINEPHRINE-EPINEPHRINE VASOCONSTRICTOR MECHANISM

In the previous chapter it was pointed out that stimulation of the sympathetic nervous system not only causes direct nervous activation of the blood vessels and heart but also causes release by the adrenal medullae of norepinephrine and epinephrine into the circulating blood. These two hormones circulate to all parts of the body and cause essentially the same effects on the circulatory system as direct sympathetic stimulation. That is, they excite the heart, they constrict most of the blood vessels, and they constrict the veins.

Therefore, the different reflexes that regulate arterial pressure by exciting the sympathetic nervous system cause the pressure to rise in two ways: by direct circulatory stimulation and by indirect stimulation through the release of norepinephrine and epinephrine into the blood.

Norepinephrine and epinephrine circulate in the blood for 1 to 3 minutes before being destroyed, thus maintaining prolonged excitation of the circulation. Also, these hormones can reach some parts of the circulation that have no sympathetic nervous supply at all, including some of the very minute vessels such as the metarterioles. And these hormones have especially potent actions on some vascular beds, particularly the skin vasculature.

In general, the norepinephrine and epinephrine system can be considered to be a part of the total sympathetic mechanism for arterial pressure control.

## THE RENIN-ANGIOTENSIN VASOCONSTRICTOR MECHANISM FOR CONTROL OF ARTERIAL PRESSURE

The hormone *angiotensin II* is the most potent vasoconstrictor known. Whenever the arterial pressure falls very low, large quantities of angiotensin II appear in the circulation. This results from a special mechanism involving the kidneys and the release of the enzyme *renin* from the kidneys when the arterial pressure falls too low.

The overall schema for formation of angiotensin and the effect of angiotensin II to increase arterial pressure are illustrated in Figure 21–10. When blood flow through the kidneys is decreased, the *juxtaglomerular cells* (cells located in the walls of the afferent arterioles immediately proximal to the glomeruli) secrete renin into the blood. Renin itself is an enzyme that catalyzes the conversion of one of the plasma proteins, called *renin substrate,* into the peptide *angiotensin I.* The renin persists in the blood for as long as 1 hour and continues to form the angiotensin I during the entire time. Within a few seconds after formation of the angiotensin I, it is converted into another peptide, *angiotensin II*, by an enzyme called *converting enzyme* present mainly in the lungs. Angiotensin II persists in the blood for a minute or so but is rapidly inactivated by a number of different blood and tissue enzymes collectively called *angiotensinase*.

During its persistence in the blood, angiotensin II has several effects that can elevate arterial pressure. One of these effects occurs very rapidly—vasoconstriction especially of the arterioles and to a lesser extent of the veins at the same time. Constriction of the arterioles increases the peripheral resistance and thereby raises the arterial pressure back toward normal, as illustrated at the bottom of the schema in Figure 21–10. Also, constriction of the veins increases the circulatory filling pressure, sometimes as much as 50 per cent, and this promotes increased tendency for venous return of blood to the heart, thereby helping the heart to pump against the extra pressure load.

The other effects of angiotensin are mainly related to the body fluid volumes: (1) angiotensin has a direct effect on the kidneys to cause decreased excretion of both salt and water; and (2) angiotensin stimulates the secretion of aldosterone by the adrenal cortex, and this hormone in turn also acts on the kidneys to cause decreased excretion of both salt and water. Both these effects elevate the blood volume—

an important factor in long-term regulation of arterial pressure, as will be discussed in the following chapter.

**Rapidity of Action and Pressure Controlling Power of the Renin-Angiotensin-Vasoconstrictor System.** Figure 21–11 illustrates a typical experiment showing the effect of hemorrhage on the arterial pressure under two separate conditions: (1) with the renin-angiotensin system functioning, and (2) without the system functioning (the system was interrupted by a renin-blocking drug). Note that following hemorrhage, which caused an acute fall of the arterial pressure to 50 mm. Hg, the arterial pressure rose back to 83 mm. Hg when the renin-angiotensin system was functional. On the other hand, it rose to only 60 mm. Hg when the renin-angiotensin system was blocked. This illustrates that the renin-angiotensin system is powerful enough to return the arterial pressure at least halfway back to normal following severe hemorrhage. Therefore, it can sometimes be of life-saving service to the body, especially in circulatory shock.

Note that the renin-angiotensin-vasoconstrictor system requires approximately 20 minutes to become fully active. Therefore, the renin-angiotensin-vasoconstrictor system is far slower to act than are the nervous reflexes and the norepinephrine-epinephrine system; however, it also has a correspondingly longer duration of action.

There is reason to believe that the renin-angiotensin-vasoconstrictor system is much more powerful under some conditions than under others. For instance, some patients with diseased kidneys secrete tremendous quantities of renin, and the pressure control action of the system is then likely to be very powerful.

## ROLE OF VASOPRESSIN IN RAPID CONTROL OF ARTERIAL PRESSURE

When the arterial pressure falls low, the hypothalamus secretes large quantities of vasopressin by way of the posterior pituitary gland—a mechanism that will be discussed in detail in relation to the pituitary hormones in Chapter 75. The vasopressin in turn has a direct vasoconstrictor effect on the blood vessels, thereby increasing the total peripheral resistance and raising the arterial pressure back toward normal.

Until recently, most physiologists believed that the amount of vasopressin secreted in low blood pressure states was not sufficient to play a significant role in compensating for the low pressure. However, recent experiments in animals in which the baroreceptor pressure-controlling mechanism has been re-

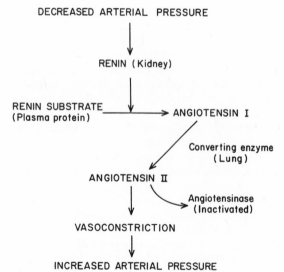

**Figure 21–10.** The renin-angiotensin-vasoconstrictor mechanism for arterial pressure control.

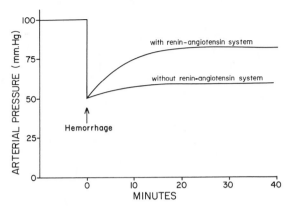

**Figure 21–11.** Pressure compensating effect of the renin-angiotensin-vasoconstrictor system following severe hemorrhage. (Data from experiments by Dr. Royce Brough.)

moved have shown that the circulating amounts of vasopressin found in the blood following hemorrhage can then increase the arterial pressure as much as 35 mm. Hg. This much effect cannot be seen in the normal animal, however, because the very potent control of pressure by the baroreceptors obscures the pressure rise that would otherwise occur. But the nervous reflexes usually fade out (adapt) after 24 to 48 hours' time, whereas the vasopressin mechanism does not do so. Therefore, it is very likely that the vasopressin system does indeed play an important role in arterial pressure regulation in more prolonged states of circulatory debility.

Vasopressin, also known as *antidiuretic hormone*, plays a second very important long-term role in arterial pressure regulation: this hormone acts on the kidneys to cause decreased excretion of water, an effect that helps to increase the blood volume any time that the arterial pressure falls too low. The increased blood volume then plays an especially important role in long-term regulation of arterial pressure, as will be discussed in the following chapter.

## INTERMEDIATE MECHANISMS FOR ARTERIAL PRESSURE REGULATION

In addition to the nervous and hormonal mechanisms for rapid control of arterial pressure, two intrinsic physical mechanisms of the circulation also help to control arterial pressure, usually beginning to act within a few minutes and reaching full function within a few hours. These are: (1) the capillary fluid shift mechanism, and (2) the vascular stress-relaxation mechanism.

**Capillary Fluid Shift.** When the arterial pressure changes, this is usually also associated with a similar change in capillary pressure. This causes fluid to begin moving across the capillary membrane between the blood and the interstitial fluid compartment. Within a few minutes to an hour a new state of equilibrium usually will be achieved, but in the meantime this shift of fluid has played a very beneficial role in the control of arterial pressure. For instance, if the arterial pressure rises too high, loss of fluid through the capillaries into the interstitial spaces causes the blood volume to fall and thereby causes return of arterial pressure back toward normal. The gain of this system is about 3, which means that it can return the arterial pressure about three-fourths of the way back toward its normal mean value; but it takes effect much more slowly than do the nervous reflex mechanisms.

**Stress-Relaxation.** When the arterial pressure falls, the pressure usually also falls in most of the blood storage areas such as in the veins, the liver, the spleen, the lungs, and so forth. Conversely, a rise in arterial pressure is often associated with a rise in pressure in the same storage areas. As pointed out in Chapter 18, a pressure change causes vessels gradually to adapt to a new size, thereby accommodating the amount of blood that is available. This phenomenon is called *stress-relaxation* or *reverse stress-relaxation*. Thus, following massive transfusion, the arterial pressure rises markedly at first, but because of relaxation of the circulation during the next 10 minutes to an hour, the arterial pressure returns essentially to normal even though the blood volume may be as great as 30 per cent above normal. Conversely, following severe bleeding, the reverse stress-relaxation mechanism can cause the blood vessels gradually to tighten around the amount of blood that is left, thereby again re-establishing almost normal circulatory dynamics.

Unfortunately, the stress-relaxation mechanism has very definite limits, so that acute changes in blood volume of more than approximately plus 30 per cent or minus 15 per cent cannot be corrected by this mechanism.

## REFERENCES

Bevan, J. A., and Su, C.: Sympathetic mechanisms in blood vessels: nerve and muscle relationships. *Ann. Rev. Pharmacol.,* 13:269, 1973.

Bhagat, B. D.: Role of Catecholamines in Cardiovascular Diseases. Springfield, Ill., Charles C Thomas, Publisher, 1974.

Braunwald, E.: Regulation of the circulation. I. *N. Engl. J. Med.,* 290:1124, 1974.

Braunwald, E.: Regulation of the circulation. II. *N. Engl. J. Med.,* 290:1420, 1974.

Cowley, A. W., Jr., and Guyton, A. C.: Baroreceptor reflex contribution in angiotensin-II-induced hypertension. *Circulation,* 50:61, 1974.

Cowley, A. W., Jr., Monos, E., and Guyton, A. C.: Interaction of vasopressin and the baroreceptor reflex system in the regulation of arterial pressure in the dog. *Circ. Res.,* 34:505, 1974.

Cushing, H.: Concerning a definite regulatory mechanism of the vasomotor center which controls blood pressure during cerebral compression. *Bull. Hopkins Hosp.,* 12:290, 1901.

De Burgh Daly, M.: Interaction of cardiovascular reflexes. *In* Gilliland, I. (ed.): Scientific Basis of Medicine Annual Reviews 1972. London, Athlone Press of the University of London, 1972, pp. 307–32.

Franz, G. N.: On blood pressure control. *Physiologist*, 17:73, 1974.

Gleser, M. A., and Grupp, G.: Mathematical model of response to carotid sinus nerve stimulation. *Amer. J. Physiol.*, 216:263, 1969.

Goetz, K. L., Bond, G. C., and Bloxham, D. D.: Atrial receptors and renal function. *Physiol. Rev.*, 55:157, 1975.

Guyton, A. C.: Acute hypertension in dogs with cerebral ischemia. *Amer. J. Physiol.*, 154:45, 1948.

Guyton, A. C., Batson, H. M., Smith, C. M., and Armstrong, G. G.: Method for studying competence of the body's blood pressure regulatory mechanisms and effect of pressoreceptor denervation. *Amer. J. Physiol.*, 164:360, 1951.

Guyton, A. C., and Satterfield, J. H.: Vasomotor waves possibly resulting from CNS ischemic reflex oscillation. *Amer. J. Physiol.*, 170:601, 1952.

Herd, J. A.: Behavior and cardiovascular function. *Physiologist*, 14:83, 1971.

Iriuchijima, J., and Koike, H.: Carotid flow, intrasinusal pressure, and collateral flow during carotid occlusion. *Amer. J. Physiol.*, 218:876, 1970.

Johnson, R. H., and Spaulding, J. M.: Disorders of the autonomic nervous system. Chap. 3. The nervous control of the circulation and its investigation. *Contemp. Neurol. Ser.*, 11:33, 1974.

Johnson, R. H., and Spaulding, J. M.: Disorders of the autonomic nervous system. Chap. 7. Some disorders of regional circulation. *Contemp. Neurol. Ser.* 11:114, 1974.

Korner, P. I.: Integrative neural cardiovascular control. *Physiol. Rev.*, 51:312, 1971.

Krieger, E. M.: Time course of baroreceptor resetting in acute hypertension. *Amer. J. Physiol.*, 218:486, 1970.

Laragh, J. H., Sealey, J. E., Buhler, F. R., Vaughan, E. D., Brunner, H. R., Gavras, H., and Baer, L.: The renin axis and vasoconstriction volume analysis for understanding and treating renovascular and renal hypertension. *Amer. J. Med.*, 58:4, 1975.

Miyakawa, K.: A Blood Pressure Oscillation (collected articles). Matsumoto, Nagaro-ken, Japan, Shinshu University Medical School, 1972.

Paintal, A. S.: Vagal sensory receptors and their reflex effects. *Physiol. Rev.*, 53:159, 1973.

Paintal, A. S.: Sensory mechanisms involved in the Bezold-Jarisch effect. *Aust. J. Exp. Biol. Med. Sci.*, 51:3, 1973.

Pelletier, C. L., and Shepherd, J. T.: Circulatory reflexes from mechanoreceptors in the cardio-aortic area. *Circ. Res.*, 33:131, 1973.

Rowell, L. B.: Human cardiovascular adjustments to exercise and thermal stress. *Physiol. Rev.*, 54:75, 1974.

Sagawa, K., Carrier, O., and Guyton, A. C.: Elicitation of theoretically predicted feedback oscillation in arterial pressure. *Amer. J. Physiol.*, 203:141, 1962.

Sagawa, K., Kumada, M., and Schramm, L. P.: Nervous control of the circulation. *In* Guyton, A. C. (ed.): MTP International Review of Science: Physiology. Baltimore, University Park Press, 1974, Vol. 1, p. 197.

Sagawa, K., Ross, J. M., and Guyton, A. C.: Quantitation of cerebral ischemic pressor response in dogs. *Amer. J. Physiol.*, 200:1164, 1961.

Sagawa, K., Taylor, A. E., and Guyton, A. C.: Dynamic performance and stability of cerebral ischemic pressor response. *Amer. J. Physiol.*, 201:1164, 1961.

Sagawa, K., and Watanabe, K.: Summation of bilateral carotid sinus signals in the barostatic reflex. *Amer. J. Physiol.*, 209:1278, 1965.

Scher, A. M., and Young, A. C.: Reflex control of heart rate in the unanesthetized dog. *Amer. J. Physiol.*, 218:780, 1970.

Smith, E. E., and Guyton, A. C.: Center of arterial pressure regulation during rotation of normal and abnormal dogs. *Amer. J. Physiol.*, 204:979, 1963.

Smith, O. A.: Reflex and central mechanisms involved in the control of the heart and circulation. *Ann. Rev. Physiol.*, 36:93, 1974.

Tarazi, R. C., and Gifford, R. W., Jr.: Systemic arterial pressure. *In* Sodeman, W. A., Jr., and Sodeman, W. A. (eds.): Pathologic Physiology: Mechanisms of Disease, 5th Ed. Philadelphia, W. B. Saunders Company, 1974, p. 177.

Thames, M. D., and Kontos, H. A.: Mechanisms of baroreceptor-induced changes in heart rate. *Amer. J. Physiol.*, 218:251, 1970.

Uther, J. B., and Guyton, A. C.: Cardiovascular regulation following changes in central nervous perfusion pressure in the unanesthetized rabbit. *Aust. J. Exp. Biol. Med. Sci.*, 51:295, 1973.

Warner, H. R., and Cox, A.: A mathematical model of heart rate control by sympathetic and vagus efferent information. *J. Appl. Physiol.*, 17:349, 1962.

Youmans, W. B., Murphy, Q. R., Turner, J. K., Davis, L. D., Briggs, D. I., and Hoye, A. S.: The Abdominal Compression Reaction. Baltimore, The Williams & Wilkins Company, 1963.

Youmans, W. B., Tjioe, D. T., and Tong, E. Y.: Control of involuntary activity of abdominal muscles. *Amer. J. Phys. Med.*, 53:57, 1974.

# 22

# Regulation of Arterial Pressure:
# II. The Renal-Body Fluid System
# for Long-term Pressure Control.
# Mechanisms of Hypertension

## SHORT-TERM VERSUS LONG-TERM PRESSURE CONTROL MECHANISMS

In the previous chapter, several short-term and intermediate-term arterial pressure control systems were described. In the present chapter we will discuss an extremely important long-term pressure control system based on the control of body fluid volume by the kidneys. First, however, let us compare these different control systems so that we can understand their relative importance under different conditions.

Figure 22–1 gives the time response for eight different arterial pressure control mechanisms. Note that the time base on the abscissa begins with seconds and then extends to minutes, hours, and days. The feedback gain for each of the pressure control mechanisms is given on the ordinate. It will be recalled from Chapter 1 that the feedback gain of a control system is a quantitative expression of its capability to maintain the variable—in this case the arterial pressure—near its normal mean value. Thus, the greater the feedback gain, the more important the mechanism as a pressure controller.

The pressure control mechanisms represented in Figure 22–1 can be divided into three different groups: (1) short-term mechanisms, (2) intermediate-term mechanisms, and (3) long-term mechanisms. The short-term mechanisms are the three nervous

feedback control mechanisms: (a) the baroreceptor mechanism, (b) the chemoreceptor mechanism, and (c) the CNS ischemic feedback mechanism, all of which were described in the previous chapter. Note that each of these mechanisms begins to act within seconds

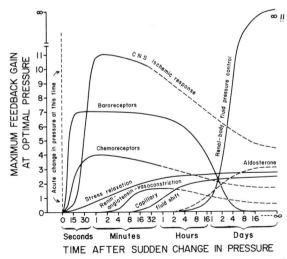

**Figure 22–1.** Potency of various arterial pressure control mechanisms at different time intervals after the onset of a disturbance to the arterial pressure. Note especially the infinite gain of the renal-body fluid pressure control mechanism that occurs after a few days' time.

whenever any disturbance attempts to change the arterial pressure either to a higher or a lower value than its normal level. Therefore, these nervous control mechanisms play an exceedingly important role in preventing the arterial pressure from bouncing to very high values or from falling to very low values from second to second or moment to moment. For instance, when a person stands up or sits down, when he exercises, when he squeezes his abdomen, or when he experiences mental disturbances that affect the function of his cardiovascular system, these nervous control mechanisms still keep the arterial pressure on an even keel.

The intermediate-term control mechanisms are represented by: (1) stress-relaxation of the vasculature, (2) the renin-angiotensin-vasoconstriction mechanism, and (3) the capillary fluid shift mechanism, all of which were also discussed briefly in the previous chapter. When a disturbance continues for minutes or hours tending to alter the arterial pressure, all of these mechanisms can help to keep the pressure close to its normal value, though they cannot respond in seconds. Thus, an excess of blood volume leads to slow stress-relaxation of the vasculature or to capillary fluid shift out of the circulation. A prolonged decrease in arterial pressure leads to the formation of angiotensin, and this in turn causes vasoconstriction and elevation of the arterial pressure back toward normal.

The long-term arterial pressure control mechanisms are represented by the renal-body fluid pressure control system and by the aldosterone control system. However, the aldosterone system works in conjunction with the renal-body fluid system so that in reality they are part of the same system, as we shall see later in the present chapter.

Note, especially, that the renal-body fluid pressure control system has two very distinctive characteristics. The first of these is that it is slow to act, usually requiring hours before it becomes very effective. Therefore, during the first few seconds, minutes, or several hours, the body is absolutely dependent upon the rapidly acting and intermediately acting control mechanism to maintain normal arterial pressures in the face of outside disturbances.

A second important characteristic of the renal-body fluid pressure control mechanism is that when it finally does become fully activated it has *infinite* feedback gain. Since the potency of a control system is directly proportional to its feedback gain, one can well understand that this renal-body fluid pressure control mechanism, with its infinite gain, in the long run completely dominates the long-term control of arterial pressure. For this reason, it is important that we spend most of this chapter discussing this all-important pressure control system.

## THE RENAL-BODY FLUID SYSTEM FOR ARTERIAL PRESSURE CONTROL

The basic mechanism of the renal-body fluid mechanism for arterial pressure control involves the following steps:

1. An increase in arterial pressure causes the kidneys to excrete increased quantities of fluid.

2. The increased loss of fluid through the kidneys reduces both the extracellular fluid volume and the blood volume.

3. The reduced blood volume reduces venous return of blood to the heart and therefore reduces the cardiac output.

4. The reduced cardiac output reduces the arterial pressure back to normal.

Conversely, when the arterial pressure falls too low, the kidneys retain fluid, the blood volume increases, the cardiac output increases, and the arterial pressure returns again to normal.

However, there are other complexities to this system. Therefore, let us now present the system in more detail.

**Details of the Renal-Body Fluid System.** Figure 22–2 illustrates in more detail the basic feedback mechanism of the renal-body fluid system for arterial pressure control. The steps in this mechanism are the following:

An increase in *arterial pressure* (Block 1) causes an increase in *renal output* of water and electrolytes (Block 2).

An increase in renal output causes the *extracellular fluid volume* to fall (the dashed arrow indicates a negative effect) (Block 3). On the other hand, an increase in the *net intake* of water and electrolytes (Block 7), which is defined as the fluid intake by mouth minus losses of fluid in all other ways except through the kidneys, causes the extracellular fluid volume to increase (Block 3). Thus, whether the extracellular fluid volume increases or decreases depends on the balance between fluid intake and renal output of fluid.

Since the extracellular fluid volume is distributed between the blood and the intersti-

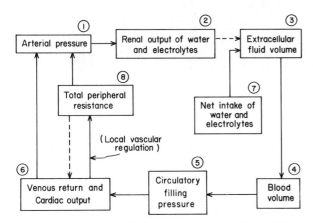

**Figure 22–2.** A block diagram of the renal-body fluid mechanism for long-term control of arterial pressure.

tial fluid spaces, an increase in extracellular fluid volume, up to a critical level, increases the blood volume (Block 4).

An increase in blood volume increases the *circulatory filling pressure* (Block 5), which is a measure of the fullness with which the blood fills the circulatory system.

The *venous return* (Block 6) is in general almost directly proportional to the circulatory filling pressure. Therefore, whenever excess blood volume fills the system, the venous return also increases. Since the heart must pump exactly the same amount of blood as that which enters it, the *cardiac output* is equal to the venous return.

An increase in cardiac output increases the arterial pressure (Block 1) in two different ways: (1) directly, because cardiac output times total peripheral resistance equals arterial pressure, and (2) indirectly, by increasing the *total peripheral resistance* (Block 8), which in turn further increases the arterial pressure. The cardiac output increases the total peripheral resistance by the process called *local vascular regulation* or *autoregulation*. That is, when the cardiac output increases, so also does the blood flow in all the local tissues throughout the body increase. As was discussed in Chapter 20, this increase in flow above the level required to deliver the needed flow to the tissues causes constriction of the local arterioles, thus increasing the total peripheral resistance. In long-term arterial pressure regulation, it is this second effect of cardiac output (the local autoregulatory increase in peripheral resistance) that is the more important in elevating the arterial pressure, because in the chronic state the total

peripheral resistance increases by this mechanism at least five times as much as the cardiac output itself increases.

**The Minuteness of the Fluid Volume Changes Required to Cause Marked Changes in Pressure.** One of the features of the renal-body fluid mechanism for arterial pressure control that is most often misunderstood is how small the changes in quantities of fluid in the body have to be to cause marked changes in pressure. To give an example, a 2 per cent chronic increase in blood volume can increase the circulatory filling pressure as much as 5 per cent, and this in turn can increase the venous return and cardiac output also as much as 5 per cent. Finally, an increase in cardiac output of 5 per cent can increase the total peripheral resistance 25 to 50 per cent; this figure multiplied by the 5 per cent increase in cardiac output shows an increase in the arterial pressure of as much as 30 to 57 per cent. Thus, one can understand very easily that a chronic increase of only a few hundred milliliters of extracellular fluid can lead to a serious hypertensive state. Indeed, in patients with hypertension, the arterial pressure can often be reduced back to normal by administering a diuretic that does nothing more than reduce the extracellular fluid volume by about 500 ml.

One of the reasons it has been difficult in the past to understand the minuteness of the fluid volumes involved is that large volumes of fluid can be infused into a person acutely without causing hypertension. However, it must be remembered that for the first few minutes to several hours after such an infusion the nervous reflex control mechanisms prevent a significant pressure rise. Before these mechanisms lose their potency for keeping the pressure low (which occurs 10 to 48 hours later), the fluid itself normally will have already been excreted by the kidneys.

**An Experiment Demonstrating the Renal-Body Fluid System for Arterial Pressure Control.** Figure 22–3 illustrates an experiment in dogs in which all the nervous reflex mechanisms for blood pressure control were blocked and the arterial pressure was then suddenly elevated by infusing 300 ml. of blood. Note the instantaneous increase in cardiac output to approximately double its normal level and the increase in arterial pressure to 135 mm. Hg above its resting level. Also shown, by the middle curve, is the effect of this increased arterial pressure on urinary output. The output increased 12-fold, and both the cardiac output and the arterial pressure

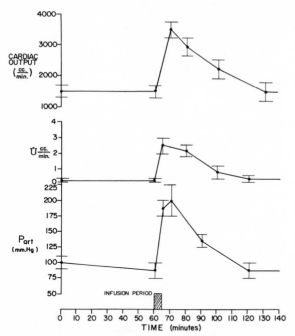

**Figure 22–3.** Increase in cardiac output, arterial pressure, and urinary output caused by increased blood volume in animals whose nervous pressure control mechanisms had been blocked. This figure shows the return of arterial pressure to normal after about an hour of fluid loss into the urine. (Courtesy of Dr. William Dobbs.)

error is one eighth. Therefore, the gain is seven. On the other hand, in the case of the renal-body fluid system, the compensation is *100 per cent of the way back to normal,* and the final error is zero. Thus, the gain of the system is 100 divided by 0, or *infinity,* which is the reason that this system is the all-important one for long-term control of arterial pressure.

## RELATIONSHIP OF ARTERIAL PRESSURE TO URINARY OUTPUT— THE URINARY OUTPUT CURVE

The key to the renal-body fluid system for pressure control is the extreme effect that a change in arterial pressure has on urinary output. The steep solid curve of Figure 22–4, called a *urinary output curve,* shows the relationship between urinary output and arterial pressure that is recorded when urinary output is made to increase day after day by infusing more and more water and salt into either an animal or human being. Note that a very minute increase in arterial pressure is associated with a tremendous increase in urinary output—an increase in arterial pressure of only 6 mm. Hg is associated with a seven-fold increase in urinary output. Thus, there is approximately a one-fold

returned back toward normal during the subsequent hour. Thus, one sees the extreme capability of the kidneys to readjust the blood volume and in so doing to return the arterial pressure back to normal.

**Infinite Gain of the Renal-Body Fluid Pressure Control System.** A special feature of this mechanism for pressure control is its ability to return the arterial pressure *all the way* back to normal—not merely a certain proportion of the way back. This is quite different from the nervous reflex mechanisms that were discussed in the previous chapter. For instance, the baroreceptor system is capable of returning the arterial pressure approximately seven eighths of the way back toward normal following a blood volume change, but never all the way. From the discussion of the effectiveness of a feedback control system in Chapter 1, remember that the "feedback gain" of a system is calculated by dividing the total amount of compensation caused by the feedback by the remaining error that fails to be compensated. In the case of the baroreceptor system, the compensation is seven eighths, and the remaining

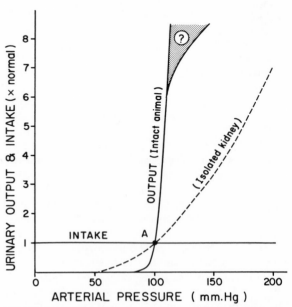

**Figure 22–4.** Graphical procedure for analyzing the long-term level of arterial pressure, based on the renal-body fluid mechanism for pressure control. The steep solid curve is the normal urinary output curve in an intact animal. (See explanation in the text.)

increase in urinary output for each 1 mm. Hg increase in pressure.

The dashed curve in Figure 22–4 shows the effect of arterial pressure on urinary output by the isolated kidneys soon after they have been removed from an animal and then perfused with blood at different arterial pressures. It is easy to see that in the intact animal a rise in arterial pressure is associated with about ten times as much increase in output as occurs for the isolated kidneys. The reason for this difference is not at present fully understood, but it probably results from the fact that in the intact animal there are a number of accessory control systems, such as the aldosterone mechanism for control of urinary output, that make the kidney far more sensitive to increases in pressure than it is in the isolated state.

The extreme steepness of the urinary output curve in the intact animal obviously provides the body with an extremely potent mechanism for control of arterial pressure. That is, whenever the arterial pressure in the normal person rises even slightly above 100 mm. Hg, the kidneys pour out fluid and the pressure falls back toward normal. Conversely, whenever the arterial pressure falls even slightly below 100 mm. Hg, the kidneys retain fluid until the pressure rises back up to the normal level.

## GRAPHICAL ANALYSIS OF THE FUNCTION OF THE KIDNEYS IN ARTERIAL PRESSURE CONTROL

Returning again to Figure 22–4, note the solid line labelled "intake," which illustrates the normal level for *net* intake of water and salt. The only point in which this normal intake can be in exact balance with urinary output is at Point A, or where the intake line crosses the urinary output curve. Let us assume, however, that the arterial pressure for some reason rises to 110 mm. Hg. This rise causes the urinary output to become far greater than normal. Therefore, the person will lose fluid at a rapid rate, the blood volume will decrease, and the arterial pressure will return to normal. When the arterial pressure has returned to exactly 100 mm. Hg, the urinary output will have returned to Point A on the output curve, and the output will again exactly balance the intake.

Thus, the system always automatically adjusts the output of urine to equal precisely the intake, and in so doing it also returns the arterial pressure to its exact original level. There-fore, this graphical analysis of pressure control again demonstrates the principle of infinite gain of this pressure control system.

**Effect of Increasing the Fluid Intake.** Referring again to the urinary output curve of a normal intact animal in Figure 22–4 one can readily see that, if the fluid intake of the animal or human being is elevated to some level above normal and maintained at this level, the intake line will cross the output curve at some level higher than normal. Therefore, the arterial pressure and the urinary output will both rise until the output rises after a few days of adjustment to equal exactly the intake. Thus, if we study the urinary output curve very carefully, we will see that an increase in the intake up to 3½ times normal will increase the arterial pressure from 100 mm. Hg to 106 mm. Hg while the urinary output increases to 3½ times normal, or exactly equal to the intake.

**Effect of Altering the Urinary Output Curve.** The kidneys of one person frequently function differently from those of other persons. This can result from pathological kidney conditions or from factors originating elsewhere in the body but acting on the kidneys to change its function, such as stimulation of the kidneys by excess aldosterone, stimulation by the sympathetic nerves, and so forth. We will discuss some of these in detail in relation to hypertension later in the chapter.

One can see readily from Figure 22–4 that any factor that shifts the urinary output curve to the right—that is, to a higher arterial pressure level—will also cause the point at which the intake line crosses the urinary output curve to rise to a higher arterial pressure as well. Therefore, any factor that alters the urinary output curve in this way will also, in the long run, alter the long-term level to which the arterial pressure of the person is controlled.

**The Two Determinants of the Long-term Arterial Pressure Level.** From the above discussion and from the principles illustrated by the graphical analysis of Figure 22–4, one can readily see that there are two primary factors that determine the long-term level of arterial pressure. These are called the *long-term determinants of arterial pressure;* they are: (1) the *urinary output curve,* and (2) the *net rate of fluid intake.* A change in either of these can change the long-term level of arterial pressure. Furthermore, the long-term level of arterial pressure cannot be changed in any way other than by changing one or both of these two determinants.

## ROLE OF TOTAL PERIPHERAL RESISTANCE AND CARDIAC OUTPUT IN THE LONG-TERM CONTROL OF ARTERIAL PRESSURE

It is frequently stated that arterial pressure is controlled by total peripheral resistance and cardiac output, because it is well known that arterial pressure is equal to the product of these two. However, in the long-term control of arterial pressure, these two factors are themselves increased or decreased by the renal-body fluid pressure control system to whatever values are required to adjust the arterial pressure. In other words, total peripheral resistance and cardiac output are variables that are utilized as pawns by the renal-body fluid system to control arterial pressure. This control is achieved in the following way:

1. The retention of fluid volume increases the venous return, and this increases the cardiac output.

2. The increase in cardiac output increases blood flow through all of the local tissues, which in turn increases the total peripheral resistance by the autoregulation mechanism. Thus, changes in both cardiac output and total peripheral resistance are an integral part of the renal-body fluid long-term mechanism for arterial pressure control.

**Failure of a Primary Change in Total Peripheral Resistance to Affect the Long-term Arterial Pressure Level—the Effect of Opening and Closing an Arteriovenous Fistula.** Even though the total peripheral resistance can be increased or decreased by the renal-body fluid pressure control system, and can thereby be utilized as a means for pressure control, a primary change in the total peripheral resistance will not affect the long-term arterial pressure level. This is contrary to what is frequently taught, but one can understand it very well by studying the effects of opening and closing an arteriovenous fistula, which can alter the total peripheral resistance as much as several hundred per cent. Figure 22–5 illustrates such an experiment. Note especially the arterial pressure in this figure. When the fistula was first opened, the total peripheral resistance fell instantaneously to one-half normal, and this caused an immediate fall in arterial pressure. However, within a day or so, the arterial pressure had returned all the way back to normal even though the total peripheral resistance still remained only one-half normal. A week later, when the fistula was closed, the sudden 100 per

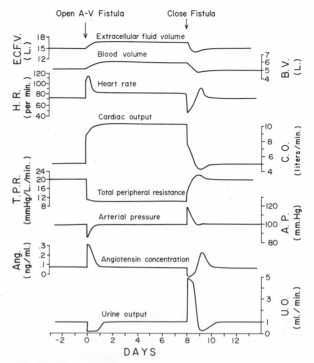

**Figure 22–5.** Effect on important circulatory variables caused by opening and closing an arteriovenous fistula. Note that after initial adjustment the arterial pressure was not at all affected by opening the fistula even though the total peripheral resistance was decreased to one-half normal. In addition, the pressure was not increased—except transiently—by closing the fistula even though this closure increased the total peripheral resistance by 100 per cent.

cent increase in total peripheral resistance caused an acute rise in arterial pressure of about 20 mm. Hg. Here again, however, during the next several days the arterial pressure returned exactly to its original level.

This experiment of opening and closing an arteriovenous fistula illustrates that a primary change in total peripheral resistance does not at all affect the long-term level of the arterial pressure. If the student will quickly reconsider the renal-body fluid mechanism for pressure control, he can readily understand why this is so. Opening a fistula does not in any way affect the function of the kidneys, nor does it alter the intake of fluid. Consequently, neither of the two primary determinants of the long-term level of the arterial pressure is changed by either opening or closing the fistula. Therefore, when the fistula is opened and the arterial pressure falls, output of the kidney decreases, extracellular fluid volume increases, blood volume increases, cardiac output increases, and arterial pressure returns to normal. All of these effects are illustrated in Figure 22–5.

**Importance of Renal Resistance in Determining the Long-term Level of Arterial Pressure.** Even though a primary change in total peripheral resistance—as illustrated by the above arteriovenous fistula experiment—cannot alter the long-term level of arterial pressure, a change in renal resistance usually will alter the pressure level. The reason for this is that a change in renal resistance alters the urinary output curve of the kidneys, and for reasons already discussed this can also alter the long-term arterial pressure level. Thus, an increase in renal resistance is one of the common causes of chronic high blood pressure.

Unfortunately, in both the experimental literature and the clinical literature increase in total peripheral resistance and increase in renal resistance are usually confused, since almost invariably it is stated that an increase in total peripheral resistance is the most common cause of high blood pressure. This is not a true statement. High blood pressure is usually *associated with* increased total peripheral resistance, though it is *not caused by* it, for two reasons: (1) usually, the same factors that increase peripheral resistance elsewhere in the body besides the kidneys simultaneously increase the renal resistance; (2) many of the mechanisms that cause hypertension cause a secondary increase in total peripheral resistance resulting from an initial increase in blood flow through the peripheral tissues, which in turn elicits an autoregulatory constriction of the tissue arterioles.

# HYPERTENSION (HIGH BLOOD PRESSURE)

Now that we have discussed the long-term control of arterial pressure, it is relatively simple to discuss the mechanisms by which a person can develop hypertension. *Any factor that increases the pressure level of the urinary output curve can also cause hypertension.* In the following pages, we will see how this occurs.

## SOME CHARACTERISTIC TYPES OF HYPERTENSION

Figure 22–6 illustrates the normal urinary output curve and four abnormal curves that frequently occur in hypertension. These abnormal curves are caused, respectively, by (1) reduced kidney mass, (2)

excess aldosterone secretion by the adrenal glands or excess circulating angiotensin, (3) increased renal arterial resistance (Goldblatt kidneys), and (4) reduced glomerular filtration coefficient. Now, let us see how each of these conditions causes hypertension.

## VOLUME-LOADING HYPERTENSION

**Hypertension Caused by Excess Water and Salt Intake in Patients with Low Kidney Mass.** The curve in Figure 22–6 labeled "reduced kidney mass" illustrates the relationship between arterial pressure and urinary output when approximately 70 per cent of the kidney mass has been removed or destroyed and the remaining kidney mass is still functioning normally. Point C illustrates the level to which the arterial pressure will be controlled when the fluid intake is normal, which is only 6 mm. Hg above normal. The reason for the smallness of this rise is that the remaining one-third of the kidney mass is still sufficient to excrete the normal amounts of water and salt ingested each day with only a slight increase in arterial pressure.

On the other hand, when the fluid intake level is progressively increased, finally a limit comes above which the kidneys are unable to excrete the excess ingested water and salt unless the arterial pressure rises markedly. Therefore, when a person with this condition ingests excess fluid, some of the fluid accumulates in the body, the blood volume increases, the cardiac output increases, and the arterial pressure finally rises high enough to cause the kidneys to excrete the increased water and salt load. Figure 22–6 shows that when the fluid intake is increased to three and one-half times normal, the pressure rises to point D where the elevated fluid intake line crosses the "reduced kidney mass" urinary output curve. Thus, the pressure must rise to 160 mm. Hg before balance between intake and output will be achieved. Once the pressure does rise to this level, it will stabilize there and continue at this level as long as the fluid intake remains high.

Figure 22–7 illustrates a typical experiment showing volume-loading hypertension in four dogs with 70 per cent of the renal mass removed. At the first circled point on the curve, the two poles of one of the kidneys were removed, and at the second circled point the entire opposite kidney was removed, leaving the animals with only 30 per cent of the renal mass. Note that removal of this amount of mass increased the arterial pressure an average of only 6 mm. Hg. However, at this point the dogs were required to drink salt solution instead of normal drinking water. Because salt solution fails to quench the thirst, the dogs also drank two to four times the normal amounts of fluid, and within a few days their average arterial pressure had adjusted to a much higher level. After two weeks the dogs were allowed to drink tap water again instead of salt solution; the pressure returned to normal within two days. Final-

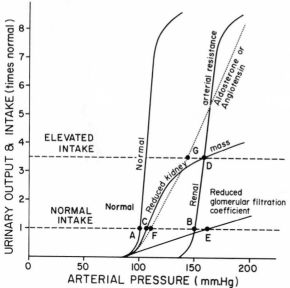

**Figure 22–6.** Abnormal urinary output curves and graphical analysis of the long-term level of arterial pressure in various hypertensive states caused by the abnormal renal function.

ly, at the end of the experiment, the dogs were required to drink salt solution again, and this time the pressure rose much more rapidly and to a higher level because the dogs had already learned to tolerate the salt solution and therefore drank much more. Thus, this experiment demonstrates the principles of volume-loading hypertension.

Returning to Figure 22–6, one can see that the

above experiment demonstrates the effects of changing the pressure control level successively from Point C to Point D, then back to Point C again, and finally once more to Point D, by simply changing the fluid intake. Note also in Figure 22–6, however, that a similar increase in fluid intake in an animal with normal kidneys would have increased the arterial pressure only 6 mm. Hg, which is exactly what is found when such an experiment is performed in normal animals.

**Volume-Loading Hypertension in Patients Who Have No Kidneys But Are Being Maintained on the Artificial Kidney.** When a patient is maintained on an artificial kidney, it is especially important to keep his body fluid volume at a normal level—that is, to remove the appropriate amount of water and salt each time that the patient is "dialyzed." Figure 22–8 illustrates a study in a patient whose body fluid volume increased only 3 liters. This is shown by the lowermost curve labeled "weight," which is the means by which one assesses sudden increases in fluid volume in these patients. Note that when the weight increased, the cardiac output also increased markedly, and this in turn caused the arterial pressure to rise from an original normal value of 95 mm. Hg to a hypertensive level of 140 mm. Hg.

**Role of "Autoregulation" in Volume-Loading Hypertension—Secondary Elevation of Total Peripheral Resistance.** It is important now to observe very carefully the changes that occurred in total peripheral resistance in the patient studied in Figure 22–8. When the body fluid volume first increased, the cardiac output and the arterial pressure increased, but at first the total peripheral resistance did not increase. This same effect has been demonstrated many times in dogs in which this same type of condition has been created. That is, the initial rise

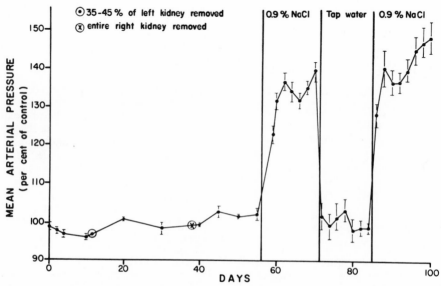

**Figure 22–7.** Effect on arterial pressure of requiring four dogs that had had 70 per cent of their renal tissue removed to drink 0.9 per cent saline solution instead of water. (Courtesy of Dr. Jimmy B. Langston.)

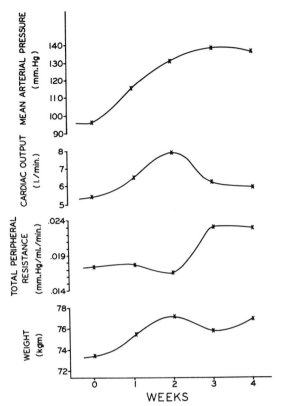

**Figure 22–8.** Hypertension in a patient whose two kidneys had been removed and whose body fluid volume had been increased by 3 liters. Note not only the increase in cardiac output at the onset of the hypertension, but also the rise in total peripheral resistance only *after* the hypertension had already developed. (Courtesy of Dr. Thomas Coleman.)

in arterial pressure is caused entirely by elevation of the cardiac output. Then, during the subsequent week or so, the total peripheral resistance rises while the cardiac output returns back toward normal. Thus, the total peripheral resistance increases *after,* rather than before, the arterial pressure has become elevated. To state this another way, the rise in total peripheral resistance is secondary, rather than primary, to the hypertension. This effect can be explained as follows:

During the initial onset of the hypertension, blood flow through all of the body's tissues is greatly increased. Because this flow is far greater than that required to supply the tissue's needs, the tissue vessels slowly constrict. Consequently, the total peripheral resistance increases while the cardiac output returns toward normal, a manifestation of the local tissue blood flow autoregulation mechanism discussed in Chapter 20. Thus, the effect of the increase in total peripheral resistance is not to cause the increased arterial pressure (because this has already risen before the resistance itself has risen) but to return the cardiac output back near to normal after the hypertension has occurred. Reduction of the cardiac output back to near-normal is very valuable to the body because it reduces the work load of the heart and therefore reduces the likelihood of heart failure.

**Relative Importance of Salt Retention and of Water Retention in the Causation of Volume-Loading Hypertension.** When one speaks of volume-loading hypertension, he usually means an increase in extracellular fluid volume; therefore, the quantity of both salt and water increase. But which of these is it that causes the hypertension? In animals with their kidneys removed and which are maintained by using an artificial kidney, one can increase the salt or the water in the body independently of one another. When the salt is allowed to increase so that its concentration rises to as much as 20 per cent above normal, no hypertension occurs unless the volume rises simultaneously. On the other hand, when the volume increases, but the total salt in the body does not, hypertension does ensue. Therefore, as would be predicted from an understanding of the fluid volume system for pressure control, it is the increase in volume and not the increase in salt that causes the hypertension.

However, so long as the kidneys are even partially functional it is usually almost impossible to increase the fluid volume of the body without a simultaneous increase in salt. The reason for this is simply that the kidneys will not retain water when there is not enough salt to go with it—a process that will be explained in Chapter 36. Furthermore, when a person with malexcreting kidneys ingests large amounts of salt, this ingestion automatically causes him to become thirsty and to drink a commensurate amount of water. Thus, salt is the usual factor that determines the volume of extracellular fluid in the body. For this reason, most clinicians and experimenters dealing with experimental hypertension are prone to think more in terms of salt retention rather than water retention as the cause of hypertension, even though experiments demonstrate that salt retention causes hypertension only when it also increases the volume as well.

## GOLDBLATT HYPERTENSION

When one kidney is removed and a constrictor is placed on the renal artery of the remaining kidney, as illustrated in Figure 22–9, the initial effect is greatly reduced renal arterial pressure, shown by the dashed curve in the lower portion of the figure. However, within a few minutes the systemic arterial pressure begins to rise and continues to rise for several days. The pressure usually rises rapidly for the first two hours, and during the next day returns to a slightly lower level, to be followed a day or so later by a second rise to a much higher pressure. When the

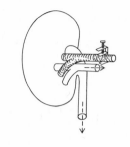

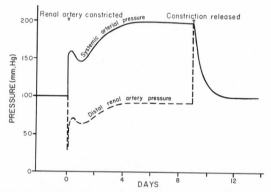

**Figure 22–9.** Changes in systemic arterial pressure and renal arterial pressure distal to the clamp following renal artery constriction.

systemic arterial pressure reaches its new stable pressure level, the renal arterial pressure will have returned almost all the way back to normal. The hypertension produced in this way is called *Goldblatt hypertension* in honor of Dr. Harry Goldblatt, who first studied the important quantitative features of hypertension caused by renal artery constriction.

The early rise in arterial pressure in Goldblatt hypertension is caused by the renin-angiotensin mechanism, one of the acute arterial pressure regulating mechanisms discussed in the previous chapter. Because of the poor blood flow through the kidney after acute application of the constrictor, large quantities of renin are secreted by the kidneys, and this causes angiotensin to be formed in the blood, as described in the previous chapter; the angiotensin in turn raises the arterial pressure acutely. However, the secretion of renin rises to a peak in a few hours but returns all the way back to normal within five to seven days because the renal arterial pressure by that time has also risen back to normal so that the kidney is no longer ischemic.

The second rise in arterial pressure is caused by fluid retention; within five to seven days the fluid volume has increased enough to raise the arterial pressure to its new sustained level. The quantitative value of this sustained pressure level is determined by the degree of constriction of the renal artery. Referring once again to Figure 22–6, one notes that the urinary output curve is shifted far to the right by

application of a constrictor to the renal artery (the curve labeled "renal arterial resistance"). That is, the increased renal arterial resistance requires that the aortic pressure rise to a much higher than normal level in order to make the renal arterial pressure distal to the constrictor rise high enough to cause normal urinary output.

**Hypertension Caused by Chronically Ischemic Kidneys and by Chronic Secretion of Renin.** Large numbers of patients who have diseased kidneys have varying levels of blood flow to different segments of the kidneys; one part of the kidney may be functioning perfectly normally while another part, because of scar tissue or some other abnormality, has depressed blood flow. In the areas of the kidney where the blood flow is reduced below normal, the kidney tissue secretes renin, as was explained in the previous chapter. The renin in turn causes the formation of angiotensin in the blood.

The circulating angiotensin then, in addition to constricting the peripheral arterioles, performs an essential function that causes chronic arterial pressure elevation. This essential function is to constrict the arterioles in the normal areas of both kidneys. Very recent experiments have shown that this effect greatly depresses the excretion of water and salt in the urine, thus elevating the pressure level of the urinary output curve, as illustrated in Figure 22–6 by the curve labeled "aldosterone or angiotensin." Consequently, the long-term level of arterial pressure rises.

## HYPERTENSION IN TOXEMIA OF PREGNANCY

During pregnancy many patients develop hypertension, which is one of the manifestations of the syndrome called *toxemia of pregnancy*. The pathological abnormality that causes this hypertension is thickening of the glomerular membranes, which reduces the rate of fluid filtration from the glomeruli into the renal tubules. For obvious reasons, the pressure level of the urinary output curve is elevated, and the long-term level of arterial pressure becomes correspondingly elevated. These patients are especially prone to hypertension when they eat large quantities of salt. This type of hypertension is illustrated in Figure 22–6 by the urinary output curve labeled "reduced glomerular filtration coefficient." Note especially that the pressure rises markedly as the net fluid intake increases.

## NEUROGENIC HYPERTENSION

Many nervous disorders can cause temporary hypertension and sometimes even permanent hypertension. Temporary hypertension can be caused a few hours at a time by sympathetic stimulation of the blood vessels, which increases the vascular resistance throughout the body. However, to cause long-term hypertension the renal arterioles must be constricted continuously by the sympathetic stimulation for days at a time. This will increase the pressure level of the urinary output curve; therefore, the renal mechanism for pressure control will then maintain the pressure at an elevated level as long as the sympathetic stimulation continues.

**Hypertension Caused by Artificial Stimulation of the Renal Sympathetic Nerves.** Direct stimulation of the sympathetic nerves to the kidneys through use of an electrical stimulator will increase the arterial pressure. When the stimulator is first turned on, the pressure begins to rise and reaches a new steady-state elevated level in approximately three days, and it remains at this level as long as the stimulus continues. After removal of the stimulus, the pressure returns to normal in about two days. This type of experiment demonstrates once again the importance of the kidneys even in hypertension that is of neurogenic origin.

Similar experiments have also been performed in animals in which the phrenic nerve has been anastomosed to the splanchnic nerve. Phrenic nerve fibers grow into the splanchnic nerves and deliver a continuous barrage of nerve impulses to the sympathetic nerves of the kidney because of the respiratory signals transmitted down the phrenic nerves. Animals prepared in this way also develop hypertension.

**Hypertension Caused by Frustration or Pain.** Prolonged sympathetic stimulation has been caused in animals by continual frustration or pain, such as by tantalizing the animal with food barely out of reach or by repeated electric shocks. Such animals develop mild to moderate hypertension as long as the abnormal conditions exist. However, within two to three weeks after the cause has been removed, the pressure usually returns to normal. Yet, it is believed by many clinicians that such abnormal sympathetic stimulation of the kidneys in patients over prolonged periods of time—perhaps for years—causes actual structural changes to occur gradually in the kidneys so that permanent pathological elevation of the urinary output curve develops. Then, even if the sympathetic stimulation is removed, the hypertension still persists. That is, neurogenic factors are believed finally to lead to hypertension that is sustained by a secondary renal abnormality.

## HYPERTENSION CAUSED BY PRIMARY ALDOSTERONISM

Small tumors that secrete large quantities of aldosterone occur occasionally in the adrenal glands. As will be discussed in detail in Chapter 35, the aldosterone increases the rate of reabsorption of salt and water in the distal tubules of the kidneys, thereby greatly reducing the rate of excretion of water and salt. Thus, the urinary output curve is shifted to a higher pressure level, as illustrated in Figure 22–6. Consequently, mild to moderate hypertension occurs even at normal fluid intake, and when the intake of water and salt is greatly increased, severe hypertension results.

## ESSENTIAL HYPERTENSION

Approximately 90 per cent of all persons who have hypertension are said to have "essential hypertension," meaning hypertension of unknown origin. A few years ago, many of the patients who are now known to have one of the types of hypertension described earlier in this chapter would have been said to have essential hypertension. Thus, the more we learn about hypertension the more we can diagnose the cause precisely.

One group of patients formerly diagnosed to have essential hypertension are now known to have hypertension caused by glomerulosclerosis. These patients in early life had mild acute glomerulonephritis. They all supposedly recovered completely from the glomerulonephritis; then, during the ensuing 20 years they gradually developed what was called essential hypertension. Recent biopsies on the kidneys of these patients have demonstrated a sclerotic process in their glomeruli, with obviously decreased filtering capacity by the glomeruli. The effect of this process on renal function is almost the same as that which occurs in toxemia of pregnancy, which was illustrated in Figure 22–6 by the curve labeled "reduced glomerular filtration coefficient." Thus, it is very clear that this type of patient, who would have been said to have essential hypertension a few years ago, in reality has a type of hypertension caused by a kidney abnormality.

**The Urinary Output Curve in Patients with Essential Hypertension.** Even though we do not know the cause of hypertension in most hypertensive patients and therefore must resort to the name "essential hypertension," we do know that the urinary output curve is shifted to a higher pressure level in these patients than in the normal person. This is illustrated in Figure 22–10 by the curve labeled "essential hypertension." The curve shows that the urinary output is exactly normal (Point B) in these patients as long as the arterial pressure is at the hypertensive level of 150 mm. Hg. However, if the patient bleeds or if any other factor causes his arterial pressure to fall to 110 mm. Hg (almost to the pressure level for a normal person), the urinary output falls to zero (Point C) even though in the normal person the urinary output would be excessive at this pressure.

Though the undisputed cause of the shifted pressure level of the urinary output curve in patients with essential hypertension is unknown, possible causes are: (1) thickening of the glomerular membranes or reduced surface areas of the glomeruli, either of which would cause a reduced glomerular filtration coefficient; or (2) increased afferent arteriolar resistance caused by vascular sclerosis. Either of these abnormalities would allow the patient with essential hypertension to have completely normal excretion of the urinary waste products, as will be explained in the following section.

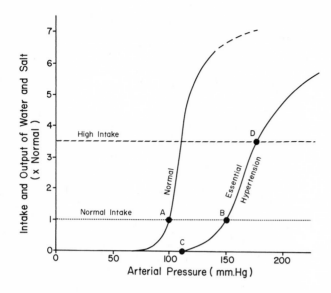

**Figure 22–10.**  Graphical analysis of elevated arterial pressure in essential hypertension. Note the shift of the urinary output curve to a much higher arterial pressure level in a patient with essential hypertension. (Reprinted from Guyton et al.: *Amer. J. Med.,* 52:584, 1972.)

**Normal Excretion of Waste Products by Patients with Essential Hypertension.**  Many clinicians have believed that essential hypertension does not have a renal origin because the kidneys of these patients excrete urinary waste products normally. Therefore, how could the kidneys be responsible for the hypertension, when the urinary function of the kidneys is so normal? The answer to this lies in the following observations: when the arterial pressure falls from the hypertensive level to normal, the kidneys of the patient with essential hypertension then retain the waste products in the blood, and the patient becomes uremic if the pressure is kept at the normal level. Thus, the elevated pressure is essential for the normal excretory function of the kidneys to occur, illustrating that the kidneys are not functioning normally.

Provided that the lesion affects all nephrons equally, the normal excretion of waste products by the kidneys is exactly what one would expect for any kidney lesion that occurs somewhere in the kidneys proximal to the tubules. The reason for this is that once the arterial pressure has risen to its hypertensive level, glomerular filtration in each glomerulus will then be exactly normal, and this glomerular filtrate will pass through the normal tubules and be processed normally. Therefore, one would expect normal urinary output of waste products—but only as long as the arterial pressure remains high enough to provide normal glomerular filtration.

**Treatment of Essential Hypertension.**  Essential hypertension can generally be treated by two different types of drugs: (1) a drug that will increase glomerular filtration, or (2) a drug that will decrease tubular reabsorption of salt and water. Those drugs that will increase glomerular filtration are the various vasodilator drugs. Some of these act by inhibiting sympathetic impulses to the kidney, and others act by direct paralysis of the smooth muscle in the walls of the blood vessels. The drugs that reduce reabsorption of water and salt by the tubules include, especially, drugs that block active transport of salt through the tubular wall; the blockage in turn prevents the osmotic reabsorption of water as well. Such substances that reduce reabsorption of salt (and consequently, of water as well) are called *natriuretics* or *diuretics;* they will be discussed in greater detail in Chapter 38.

## TYPES OF KIDNEY DISEASE THAT CAUSE HYPERTENSION; TYPES THAT CAUSE UREMIA

Not all types of kidney disease cause hypertension. Indeed, simple loss of kidney mass will cause hypertension only if the patient simultaneously ingests too much salt and water, which increases the body fluid volume and sets into play the typical sequence of volume-loading hypertension. However, loss of kidney mass always causes retention of the end-products of metabolism, such as urea, creatinine, uric acid, hydrogen ions, phosphates, and so forth.

From a physiological point of view, one can determine which type of kidney disease will cause uremia and which type will cause hypertension according to the following classification:

1. Any kidney disease that reduces kidney mass but allows all the remaining nephrons to remain normal always causes a uremic tendency but rarely causes hypertension.

2. Any kidney abnormality that tends to reduce glomerular filtration, such as diseases that decrease the glomerular filtration coefficient or that increase the renal vascular resistance, will cause hypertension but no uremic tendency if all the glomeruli are affected equally. That is, retention of water and salt

will occur until the arterial pressure rises high enough to form normal quantities of glomerular filtrate. Once glomerular filtration is normal, the tubules process the urine as usual, and no retention symptoms occur except for the retention of the water and salt that causes the hypertension.

3. Any factor that causes excess fluid reabsorption by the tubules, particularly excess reabsorption of salt and water, such as occurs in primary aldosteronism, will cause buildup of water and salt in the body, with resultant hypertension. If anything, these persons have lower than normal retention of the metabolic end-products. Thus, there is pure hypertension with no tendency toward uremia unless there is superimposed kidney damage as well.

4. Any kidney disease that causes pure destruction of tubular epithelium will cause decreased reabsorption of water and salt and, therefore, will cause a tendency to *hypotension* unless the person ingests large amounts of water and salt. This is a rare condition, but it does occur.

## EFFECTS OF HYPERTENSION ON THE BODY

Hypertension can be very damaging because of two primary effects: (1) increased work load on the heart and (2) damage to the arteries themselves by the excessive pressure.

**Effects of Increased Work Load on the Heart.** Cardiac muscle, like skeletal muscle, hypertrophies when its work load increases. In hypertension, the very high pressure against which the left ventricle must beat causes it to increase in weight to as great as 300 to 400 gm. instead of the usual weight of about 150 gm. This increase is not accompanied by quite as much increase in coronary blood supply as there is increase in muscle tissue itself. Therefore, *relative ischemia* of the left ventricle develops as the hypertension becomes more and more severe. In the late stages of hypertension, this can become serious enough that the person develops angina pectoris. Also, the very high pressure in the coronary arteries causes rapid development of coronary arteriosclerosis so that hypertensive patients tend to die of coronary occlusion at much earlier ages than do normal persons.

**Effects of the High Pressure in the Arteries.** High pressure in the arteries not only causes coronary sclerosis but also sclerosis of blood vessels throughout all the remainder of the body. The arteriosclerosis process causes blood clots to develop in the vessels and also causes the blood vessels to become weakened. Therefore these vessels frequently thrombose, or they rupture and bleed severely. In either case, marked damage can occur in organs throughout the body. The two most important types of damage which occur in hypertension are:

1. Cerebral hemorrhage, which means bleeding of a cerebral vessel with resultant destruction of local areas of brain tissue.

2. Hemorrhage of renal vessels inside the kidneys, which destroys large areas of the kidneys and therefore causes progressive deterioration of the kidneys and further exacerbation of the hypertension.

To illustrate the damaging effects of hypertension on the small vessels of the microcirculation, Figure 22–11 shows the retina of a person with malignant hypertension. This figure shows edema of the optic disc, spasm of the arterioles, dark areas in the retina that are the result of hemorrhages from the small arteries, and very light areas in the retina that represent protein that has exuded out of the blood vessels into the surrounding tissues.

## APPENDIX TO CHAPTER 22: SYSTEMS ANALYSIS OF THE RENAL AND FLUID VOLUME SYSTEM FOR ARTERIAL PRESSURE CONTROL

In the appendix to Chapter 1, some of the basic principles of systems analysis were presented. The renal-body fluid volume system for arterial pressure control is complex enough to cause it to be difficult to understand the detailed quantitative problems related to its function without the use of some such quantitative analytical procedure.

Figure 22–12 is a diagram of an abbreviated systems analysis of the renal-body fluid volume system for arterial pressure control. Each of its individual parts may be described as follows:

Block 1 shows the relationship between *arterial pressure* (AP) and *urinary output* (UO). This curve is precisely the same as the normal urinary output curve of Figure 22–4. Note especially that in the

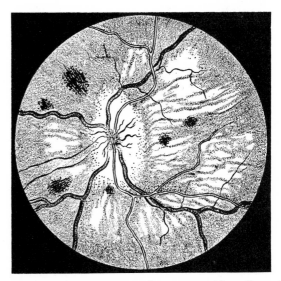

**Figure 22–11.** The retina of a person with malignant hypertension, illustrating edema of the optic disc, spasm of the arterioles, dark areas of retinal hemorrhages, and light areas of protein exudate in the retina.

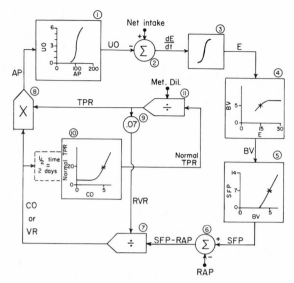

**Figure 22–12.** Composite systems analysis showing interaction between the local blood flow regulation mechanism and the fluid feedback control system.

range of arterial pressure from approximately 90 mm. Hg. to 120 mm. Hg extremely slight increases in arterial pressure will cause tremendous changes in urinary output. As was pointed out in the main text of this chapter, when the arterial pressure is elevated acutely, the urinary output does not increase nearly so much, but when the pressure is increased over a prolonged period of time, the curve in Block 1 is extremely steep. It is the steepness of this curve that makes the entire renal and fluid volume system for arterial pressure control so extremely effective in maintaining the long-term mean level of arterial pressure within a very narrow range—usually between 90 mm. Hg and 110 mm. Hg.

Block 2 subtracts urinary output from net intake of fluid, defined as the intake of fluid in all forms minus the non-renal losses of fluid. The output of this block is the rate of change of extracellular fluid volume (dE/dt).

Block 3 integrates the rate of change of extracellular fluid volume (that is, accumulates either a positive rate of change or a negative rate of change) to give the actual *extracellular fluid volume* (E).

Block 4 gives the relationship between extracellular fluid volume and *blood volume* (BV). Note that a change in extracellular fluid volume in all ranges below a volume of 22 liters causes a corresponding change in blood volume. However, once the extracellular fluid volume rises above approximately 22 liters in the average normal adult person, the blood volume does not continue to rise. The reason for this is that above a certain critical capillary pressure any excess fluid in the circulation simply spills through the capillary walls into the interstitial spaces. This effect will be discussed in detail in Chapter 31 in relation to the formation of edema.

Block 5 shows the relationship between blood vol-

ume and *systemic filling pressure* (SFP). At a normal blood volume of 5 liters, the systemic filling pressure is approximately 7 mm. Hg, as was discussed in Chapter 18. When the blood volume falls to about 3½ liters, the systemic filling pressure falls to zero, which means that the amount of blood is not sufficient to stretch the walls of the vessels. On the other hand, any increase in blood volume above this basal level increases the systemic filling pressure progressively toward higher values in almost linear proportion.

Block 6 subtracts *right atrial pressure* (RAP) from the systemic filling pressure. The output of this block gives the difference between systemic filling pressure and right atrial pressure (SFP − RAP), which is called the *pressure gradient for venous return*. As will be discussed in the following chapter, experiments have shown that the rate of blood flow into the heart, called the *venous return,* is directly proportional to this pressure gradient for venous return.

Block 7 divides the pressure gradient for venous return by the *resistance to venous return* (RVR) to give the venous return (VR). The resistance to venous return is an algebraic sum of all the resistances in the systemic circulation that impede return of blood to the heart. It is mainly the venous resistances. (This also will be discussed in the following chapter.) That is, the venous return is equal to the pressure gradient for venous return divided by the resistance to venous return. The output of Block 7 is also equal to the cardiac output (CO) because whatever amount of blood returns to the heart as venous return is also the amount that must be pumped by the heart (the cardiac output).

Block 8 calculates arterial pressure (AP) by multiplying cardiac output by total peripheral resistance (TPR).

**Function of the Primary Feedback Loop in the Systems Analysis.** The student will recognize that Blocks 1 through 8 represent a typical negative feedback mechanism. When the arterial pressure rises, the urinary output increases, and this initiates the following sequence of events: (1) negative rate of change of extracellular fluid volume, (2) decreased extracellular fluid volume, (3) decreased blood volume, (4) decreased systemic filling pressure, (5) decreased pressure gradient for venous return (SFP − RAP), (6) decreased venous return, (7) decreased cardiac output, and (8) therefore return of the arterial pressure back to normal. Conversely, if the arterial pressure falls, the fluid volumes then begin to increase, and the subsequent increase in cardiac output again returns the arterial pressure back to normal.

The *infinite gain* of this feedback loop is caused by the balance mechanism in Block 2. This system will not reach a new steady state until the urinary output becomes exactly equal to the net intake. If the net intake remains precisely constant, and if the shape of the urinary output curve in Block 1 also remains precisely the same, the arterial pressure will always return to precisely the same steady-state level after it has been displaced from this level—therefore, infinite correction of the abnormality, or "infinite gain" by the control system.

One can also see from the above discussion that the two mathematical determinants of the pressure control level to which the arterial pressure will always return are (1) the net intake of fluid, and (2) the shape of the urinary output curve. As explained in the main text of this chapter, the very steep slope of the curve in Block 1 means that in a normal person the quantitative significance of the urinary output curve far outweighs the quantitative significance of the net intake as a determinant of the pressure control level.

**Role of Total Peripheral Resistance and Autoregulation in the Pressure Feedback Control System.** Blocks 9, 10, and 11 represent the control of total peripheral resistance by the local tissue blood flow control system, commonly called the *autoregulation mechanism* by most persons working in the field of hypertension.

Block 10 shows the effect of cardiac output, operating through the autoregulation mechanism, on the total peripheral resistance. This process is not an instantaneous one, and it is shown to have an average half-time of approximately two days, which is the average half-time of the long-term autoregulation mechanism that plays a major role in readjusting the blood flow through tissues when it becomes too much or too little. Note that when the cardiac output rises above the normal value of 5 liters per minute, the total peripheral resistance increases markedly. The output of Block 10 is the normal total peripheral resistance.

Block 11 shows the effect of metabolic dilatation on total peripheral resistance; that is, when the tissues in any part of the body become overly active, the blood vessels dilate, and this decreases the total peripheral resistance to some value below the normal value. Thus, Block 11 divides normal total peripheral resistance by metabolic dilatation (Met. Dil.). And the output of this block is the actual momentary total peripheral resistance (TPR).

Block 9 calculates the resistance to venous return (RVR) which, under normal conditions, is approximately 7 per cent of the total peripheral resistance.

*Function of the Autoregulatory Loop in the Renal and Fluid Volume System for Pressure Control.* When the cardiac output rises, the total peripheral resistance slowly and gradually also rises to a much higher value. The increase in total peripheral resistance decreases the cardiac output back toward normal because the resistance to venous return rises along with the increase in total peripheral resistance. Therefore, in the early stages of operation of the renal and body fluid mechanism following volume loading, arterial pressure increases because of marked rise in cardiac output. Then, as the autoregulatory mechanism begins to take hold (with a half-time of approximately two days) the total peripheral resistance rises while the cardiac output returns back toward normal. The arterial pressure, however, remains elevated. Thus, the autoregulatory loop returns the blood flow

through the tissues back to that level required to supply the tissue needs. The importance of this mechanism is that it reduces the work load on the heart by adjusting the cardiac output to the minimum level required to supply the needs of the tissue for blood flow—no more and no less.

Thus, because of the autoregulatory mechanism, even hypertension caused by excess fluid loading is eventually accompanied by a very high total peripheral resistance and yet almost completely normal cardiac output. That is, the autoregulatory mechanism causes an increase in total peripheral resistance that is *secondary* to the increase in pressure.

**Further Extension of the Systems Analysis.** The systems analysis of Figure 22–12 has been extended to include all aspects of circulatory function, including the nervous reflex controls of the circulation, transfer of fluids into all segments of the tissues (such as into cells and into tissue gel), distribution of blood flow between the muscles and the non-muscle tissue, flow of blood through the lungs, and so forth. Systems diagrams having as many as 400 blocks have been explored in this way, and the function of the total circulatory system has been analyzed by solution of the system equations on digital computers. Utilizing such a systems analysis, one can predict the effects of almost any changes in the circulation on overall function of the system. In particular, it is possible to predict the way in which different abnormalities of the circulation can cause hypertension.

### Suggested Student Exercises Utilizing the Systems Analysis of Figure 22–12.

(The volumes in Figure 22–12 are expressed in liters, and the intake and output are expressed in milliliters per minute. The pressures are all expressed in mm. Hg, and the resistances are expressed in mm. Hg per liter of blood flow per minute.)

Suggested specific exercises:

1. Change the net intake from the normal value of 1 up to 6 ml. per minute and trace what effect this has on all of the different variables in the analysis.

2. Put into Block 1 the abnormal urinary output curves of Figure 22–6, one at a time. Set the intake at its normal value of 1 and trace what happens to all of the variables, especially the arterial pressure. Then, change the intake from 1 ml. per minute to 6 ml. per minute and then to 0.3 ml. per minute while considering each of the curves. Determine what will happen to each variable.

## REFERENCES

Brace, R. A., Jackson, T. E., Ferguson, J. D., Norman, R. A., Jr., and Guyton, A. C.: Pressure generated by scar tissue contraction: perinephritis hypertension. IRCS (Cardiovasc.; Conn. Tiss.; Kidney; Surg.), 2:1683, 1974.

Brunner, H. R., Gavras, H., and Laragh, J. H.: Specific inhibition of the renin-angiotensin system: A key to understanding blood pressure regulation. Prog. Cardiovasc. Dis., 17:87, 1974.

Coleman, T. G., Bower, J. D., Langford, H. G., and Guyton A. C.: Regulation of arterial pressure in the anephric state. *Circulation*, 42:509, 1970.

Coleman, T. G., Cowley, A. W., Jr., and Guyton, A. C.: Experimental hypertension and the long-term control of arterial pressure. In Guyton, A. C. (ed.): MTP International Review of Science: Physiology. Baltimore, University Park Press, 1974, Vol. 1, p. 259.

Coleman, T. G., Granger, H. J., and Guyton, A. C.: Whole-body circulatory autoregulation and hypertension. *Circ. Res., (Suppl. 2) 28, 29*:11, 1971.

Coleman, T. G., and Guyton, A. C.: Hypertension caused by salt loading in the dog. III. Onset transients of cardiac output and other circulatory variables. *Circ. Res.,* 25:153, 1969.

Coleman, T. G., Manning, R. D., Jr., Norman, R. A., Jr., Granger, H. J., and Guyton, A. C.: The role of salt in experimental and human hypertension. *Amer. J. Med. Sci.,* 264:103, 1972.

Cowley, A. W., Jr., Miller, J. P., and Guyton, A. C.: Open-loop analysis of the renin-angiotensin system in the dog. *Circ. Res.,* 28:568, 1971.

Crawford, M. P., Richardson, T. Q., and Guyton, A. C.: Renal servocontrol of arterial blood pressure. *J. Appl. Physiol.,* 22:139, 1967.

Douglas, B. H., Guyton, A. C., Langston, J. B., and Bishop, V. S.: Hypertension caused by salt loading. II: Fluid volume and tissue pressure changes. *Amer. J. Physiol.,* 207:669, 1964.

Folkow, B., Hallbäck, M., Lundgren, Y., Sivertsson, R., and Weiss, L.: Importance of adaptive changes in vascular design for establishment of primary hypertension, studied in man and in spontaneously hypertensive rats. *Circ. Res., (Suppl. 1)* 32:2, 1973.

Freis, E. D.: Hemodynamics of hypertension. *Physiol. Rev.,* 40:27, 1960.

Frohlich, E. D.: Clinical significance of hemodynamic findings in hypertension. *Chest,* 64:94, 1973.

Granger, H. J., and Guyton, A. C.: Autoregulation of the total systemic circulation following destruction of the central nervous system in the dog. *Circ. Res.,* 25:379, 1969.

Guyton, A. C.: Acute hypertension in dogs with cerebral ischemia. *Amer. J. Physiol.,* 154:45, 1948.

Guyton, A. C.: The ratio of arterial to venous resistance as a determinant of arterial pressure. *Amer. J. Physiol.,* 183:623, 1955.

Guyton, A. C.: Integrative hemodynamics. In Sodeman, W. A., Jr., and Sodeman, W. A. (eds.): Pathologic Physiology: Mechanisms of Disease, 5th Ed. Philadelphia, W. B. Saunders Company, 1974, p. 149.

Guyton, A. C., and Coleman T. G.: Long-term regulation of the circulation; interrelationships with body fluid volumes. In Physical Bases of Circulatory Transport Regulation and Exchange. Philadelphia, W. B. Saunders Company, 1967.

Guyton, A. C., and Coleman, T. G.: Quantitative analysis of the pathophysiology of hypertension. *Circ. Res.,* 24:I-1, 1969.

Guyton, A. C., Coleman, T. G., Bower, J. D., and Granger, H. J.: Circulatory control in hypertension. *Circ. Res., (Suppl. 2) 26* and *27*, II-135; II-147, 1970.

Guyton, A. C., Coleman, T. G., Cowley, A. W., Jr., Manning, R. D., Jr., Norman, R. A., Jr., and Ferguson, J. D.: A systems analysis approach to understanding long-range arterial blood pressure control and hypertension. *Circ. Res.,* 35:159, 1974.

Guyton, A. C., Coleman, T. G., Cowley, A. W., Jr., Norman, R. A., Jr., Manning, R. D., Jr., and Liard, J. F.: Relationship of fluid and electrolytes to arterial pressure control and hypertension: quantitative analysis of an infinite-gain feedback system. In Onesti, G., Kim, K. E., and Moyer, J. H. (eds.): Hypertension: Mechanisms and Management. Part II. Hemodynamics of Essential Hypertension. 1973, pp. 25–36.

Guyton, A. C., Coleman, T. G., Cowley, A. W., Jr., Scheel, K. W., Manning, R. D., Jr., and Norman R. A., Jr.: Arterial pressure regulation: overriding dominance of the kidneys in long-term regulation and in hypertension. *Amer. J. Med.,* 52:584, 1972.

Guyton, A. C., Coleman, T. G., Fourcade, J., and Navar, L. G.: Physiological control of arterial pressure. *Bull. N. Y. Acad. Med.,* 45:811, 1969.

Guyton, A. C., Cowley, A. W., Jr., and Coleman, T. G.: Interaction between the separate control systems in normal arterial pressure regulation and in hypertension. In Genest, J., and Koiw, E. (eds.): Hypertension '72. New York, Springer-Verlag New York, Inc., 1972, p. 384–393.

Guyton, A. C., Cowley, A. W., Jr., Coleman, T. G., DeClue, J. W., Norman, R. A., Jr., and Manning, R. D., Jr.: Hypertension: a disease of abnormal circulatory control. *Chest,* 65:328, 1974.

Guyton, A. C., Cowley, A. W., Jr., Coleman, T. G., Liard, J. F., McCaa, R. E., Manning, R. D., Jr., Norman, R. A., Jr., and Young, D. B.: Pretubular versus tubular mechanisms of renal hypertension. In Sambhi, M. P. (ed.): Mechanisms of Hypertension. New York, American Elsevier Publishing Co., Inc., 1974, p. 15–30.

Haas, E., and Goldblatt, H.: Kinetic constants of the human renin and human angiotensinogen reaction. *Circ. Res.,* 20:45, 1967.

Hollenberg, N. K., Adams, D. F., Solomon, H., Chenitz, W. R., Burger, B. M., Abrams, H. L., and Merrill, J. P.: Renal vascular tone in essential and secondary hypertension: hemodynamic and angiographic responses to vasodilators. *Medicine (Baltimore),* 54:29, 1975.

Horton, E. W.: Hypotheses on physiological roles of prostaglandins. *Physiol. Rev.,* 49:112, 1969.

Kezdi, P.: Baroreceptors and Hypertension. New York, Pergamon Press, Inc., 1968.

Langston, J. B., Guyton, A. C., Douglas, B. H., and Dorsett, P. E.: Effect of changes in salt intake on arterial pressure and renal function in partially nephrectomized dogs. *Circ. Res.,* 12:508, 1963.

Laragh, J. H., and Sealey, J.: The renin-angiotensin-aldosterone hormonal system and regulation of sodium, potassium, and blood pressure homeostasis. In Orloff, F., and Berliner, R. W. (eds.): Handbook of Physiology. Baltimore, The Williams & Wilkins Company, 1973, Sec. 8, p. 831.

Laragh, J. H., Sealey, J. E., Buhler, F. R., Vaughan, E. D., Brunner, H. R., Gavras, H., and Baer, L.: The renin axis and vasoconstriction volume analysis for understanding and treating renovascular and renal hypertension. *Amer. J. Med.,* 58:4, 1975.

Ledingham, J. M.: Blood pressure regulation in renal failure. *J. R. Coll. Physicians Lond.,* 5:103, 1971.

Lee, J. B.: Cardiovascular-renal effects of prostaglandins: the antihypertensive, natriuretic renal "endocrine" function. *Arch. Intern. Med.,* 133:56, 1974.

Liard, J. F., Cowley, A. W., Jr., McCaa, R. E., McCaa, C. S., and Guyton, A. C.: Renin-aldosterone, body fluid volumes, and baroreceptor reflex in the development and reversal of Goldblatt hypertension in conscious dogs. *Circ. Res.,* 34:549, 1974.

Luetscher, J. A., Boyers, D. G., Cuthbertson, J. G., and McMahon, D. F.: A model of the human circulation. Regulation by autonomic nervous system and renin-angiotensin system, and influence of blood volume on cardiac output and blood pressure. *Circ. Res.,* 32 (Suppl. 1):84, 1973.

McCubbin, J., and Page, I.: Renal Hypertension. Chicago, Year Book Medical Publishers, Inc., 1968.

Muirhead, E. E., Germain, G. S., Leach, B. E., Brooks, B., and Stephenson, P.: Renomedullary interstitial cells (RIC), prostaglandins (PG), and the antihypertensive function of the kidney. *Prostaglandins,* 3:581, 1973.

Pickering, G.: Hypertension, 2nd Ed. New York, Churchill Livingstone, Div. of Longman, Inc., 1974.

Sambhi, M. P. (ed.): Mechanisms of Hypertension (Proceedings of an International Workshop Conference held in Los Angeles, 1973). New York, American Elsevier Publishing Co., Inc., 1973.

Simpson, F. O.: Beta-adrenergic receptor blocking drugs in hypertension. *Drugs,* 7:85, 1974.

Tarazi, R. C., and Gifford, R. W., Jr.: Systemic arterial pressure. In Sodeman, W. A., Jr., and Sodeman, W. A. (eds.): Pathologic Physiology: Mechanisms of Disease, 5th Ed. Philadelphia, W. B. Saunders Company, 1974, p. 177.

Tobian, L.: How sodium and the kidney relate to the hypertensive arteriole. *Fed. Proc.,* 33:138, 1974.

Vander, A. J.: Control of renin release. *Physiol. Rev.,* 47:359, 1967.

van Zwieten, P. A.: The central action of antihypertensive drugs, mediated via central alpha-receptors. *J. Pharm. Pharmacol.,* 25:89, 1973.

# 23

# Cardiac Output, Venous Return, and Their Regulation

Cardiac output is perhaps the single most important factor that we have to consider in relation to the circulation, for it is cardiac output that is responsible for transport of substances to and from the tissues.

*Cardiac output* is the quantity of blood pumped by the left ventricle into the aorta each minute, and *venous return* is the quantity of blood flowing from the veins into the right atrium each minute. Obviously, over any prolonged period of time, venous return must equal cardiac output. However, for a few heart beats venous return and cardiac output need not be the same, since blood can temporarily increase or decrease in the central circulation.

## NORMAL VALUES FOR CARDIAC OUTPUT

The normal cardiac output for the young healthy male adult averages approximately 5.6 liters per minute. However, if we consider all adults, including older people and females, the average cardiac output is very close to 5 liters per minute. In general, the cardiac output of females is about 10 per cent less than that of males of the same body size.

**Cardiac Index.** The cardiac output changes markedly with body size. Therefore, it has been important to find some means by which the cardiac outputs of different sized persons can be compared with each other. Experiments have shown that the cardiac output increases approximately in proportion to the surface area of the body. Therefore, the cardiac output is frequently stated in terms of the cardiac index,

which is the *cardiac output per square meter of body surface area.* The normal human being weighing 70 kg. has a body surface area of approximately 1.7 square meters, which means that the normal average cardiac index for adults of all ages, both males and females, is approximately 3.0 liters per minute per square meter. The body surface area of persons of different heights and weights can be determined from the chart in Figure 71–7, Chapter 71.

**Effect of Age.** Figure 23–1 illustrates the change in cardiac index with age. Rising rapidly to a level greater than 4 liters per minute per square meter at 10

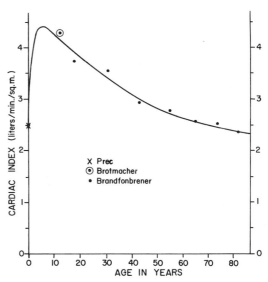

**Figure 23–1.** Cardiac index at different ages. (From Guyton, Jones, and Coleman: Circulatory Physiology: Cardiac Output and Its Regulation, 1973.)

years of age, the cardiac index declines to about 2.4 liters per minute at the age of 80.

**Effect of Posture.** When a person rises from the reclining to the standing position, the cardiac output falls about 20 per cent if he stands quietly. However, if his muscles become taut, as occurs when one prepares for exercise, the cardiac output rises 1 to 2 liters per minute. The causes of these changes are explained later in the chapter.

**Effect of Metabolism and Exercise.** The cardiac output usually remains almost proportional to the overall metabolism of the body. That is, the greater the degree of activity of the muscles and other organs, the greater also will be the cardiac output. This relationship is illustrated in Figure 23–2, which shows that as the work output during exercise increases, the cardiac output also increases in almost linear proportion. Note that in very intense exercise the cardiac output can rise to as high as 30 to 35 liters per minute in the young, well-trained athlete, which is about five to six times the normal control value.

Figure 23–2 also demonstrates that oxygen consumption increases in almost direct proportion to work output during exercise. We shall see later in the chapter that the increase in cardiac output probably results primarily from increased oxygen consumption.

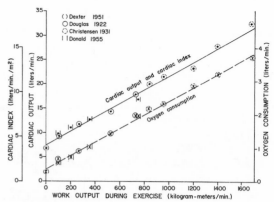

**Figure 23–2.** Relationship between cardiac output and work output (solid curve) and between oxygen consumption and work output (dashed curve) during different levels of exercise. (From Guyton, Jones, and Coleman: Circulatory Physiology: Cardiac Output and Its Regulation, 1973.)

# REGULATION OF CARDIAC OUTPUT

## PRIMARY ROLE OF THE PERIPHERAL CIRCULATION IN THE CONTROL OF CARDIAC OUTPUT; PERMISSIVE ROLE OF THE HEART

**Control of Cardiac Output by Venous Return—the Frank-Starling Law of the Heart.** It is worthwhile at this point to recall the Frank-Starling law of the heart, which was discussed in Chapter 13. This law states that the heart will pump whatever amount of blood enters the right atrium within the physiological limits of the heart's pumping capacity and without significant buildup of back pressure in the right atrium. In other words, the heart is an automatic pump that is capable of pumping far more than the 5 liters per minute that normally returns to it from the peripheral circulation. Consequently, the primary factor that determines how much blood will be pumped by the heart is the amount of blood that flows into the heart from the systemic circulation—not the pumping capacity of the heart. To state this another way, under normal circumstances the

major factor determining the cardiac output is the *rate of venous return*.

Therefore, a large share of our discussion in this chapter will concern the factors in the peripheral circulation that determine the rate at which blood returns to the heart. However, there are times when the amount of blood attempting to return to the heart is greater than the amount that the heart can pump. Under these conditions the heart then becomes the limiting factor in cardiac output control, and the heart is said to be "failing."

Thus, the heart normally plays a permissive role in cardiac output regulation. That is, it is capable of pumping a certain amount of blood each minute and, therefore, will *permit* the cardiac output to be regulated at any value below this given permitted level. Thus, the normal human heart, under *resting* conditions, permits a maximum heart pumping of about 13 to 15 liters per minute, but the actual cardiac output under resting conditions is only approximately 5 liters per minute because this is the normal level of venous return. It is the peripheral circulatory system, not the heart, that sets this level of 5 liters per minute.

The concept that the heart plays a permissive role in cardiac output regulation is so important that it needs to be explained in still another way, as follows: The normal heart, beating at a normal heart rate, and having a normal strength of contraction, neither excessively stimulated by the autonomic nervous system nor suppressed, will pump whatever amount of blood flows into the right atrium, up to about 13 to 15 liters per minute. If any more than this tries to flow into the right atrium, the heart will not pump it without cardiac stimulation. Under normal resting conditions the amount of blood that normally flows into the right atrium from the peripheral circulation is about 5 liters; and since this 5 liters is within the permissive range of heart pumping, it is pumped on into the aorta. Therefore, the venous return from the peripheral circulatory system controls the cardiac output whenever the permissive level of heart pumping is greater than venous return.

**Increase in Permissive Level for Heart Pumping in Hypertrophied Hearts.** Heavy athletic training causes the heart to enlarge sometimes as much as 50 per cent. Coincident with this enlargement is an increase in the permissive level to which the heart can pump. Thus, even under resting conditions, the permissive level for a well-trained athlete might be as great as 20 liters per minute, rather than the normal value of about 13 to 15 liters per minute.

**Increase in Permissive Level of Heart Pumping by Autonomic Stimulation of the Heart.** There are times when the cardiac output must rise temporarily to levels greater than the normal permissive level of the heart. For instance, in heavy exercise by well-trained athletes, cardiac outputs as high as 35 liters per minute have been measured. Obviously, the resting heart would not be able to pump this amount of blood. On the other hand, *stimulation of the heart by the sympathetic nervous system increases the permissive level of heart pumping to approximately double normal.* This effect comes about by autonomic enhancement of both heart rate and strength of heart contraction. Furthermore, the increase in permissive level of heart pumping occurs within a few seconds after exercise begins, even before most of the increase in venous return occurs.

**Reduction in Permissive Level for Heart Pumping in Heart Disease.** Though the normal permissive level for heart pumping is usually much higher than the venous return, this is not necessarily true when the heart is diseased. Such conditions as myocardial infarction, valvular heart disease, myocarditis, and congenital heart abnormalities can reduce the pumping effectiveness of the heart. In these instances, the permissive level for heart pumping may fall below 5 liters per minute, indeed for a few hours even as low as 2 to 3 liters per minute. When this happens, the heart becomes unable to cope with the amount of blood that is at-tempting to flow into the right atrium from the peripheral circulation. Therefore, the heart is said to *fail,* meaning simply that it fails to pump the amount of blood that is demanded of it. Under these conditions, the heart becomes the limiting factor in cardiac output control. However, this is not the normal state. (We will discuss cardiac failure in Chapter 26.)

### ROLE OF TOTAL PERIPHERAL RESISTANCE IN DETERMINING NORMAL VENOUS RETURN AND CARDIAC OUTPUT

Let us recall the formula for blood flow through blood vessels that was presented in Chapter 18:

$$\text{Blood Flow} = \frac{\text{Pressure (input)} - \text{Pressure (output)}}{\text{Resistance}}$$

Now if we apply this to venous return, the formula becomes:

$$\text{Venous Return} = \frac{\text{Arterial Pressure} - \text{Right Atrial Pressure}}{\text{Total Peripheral Resistance}}$$

Since the right atrial pressure remains very nearly zero, it is clear that the amount of blood that flows into the heart each minute (venous return) and that is pumped each minute (cardiac output) is determined by two prime factors: (1) the arterial pressure, and (2) the total resistance. When the arterial pressure remains normal, as it usually does, venous return and cardiac output are then inversely proportional to the total peripheral resistance. To state this still another way, every time a peripheral blood vessel dilates, the venous return and cardiac output increase. Furthermore, the more vessels that dilate in the peripheral circulation, the greater the cardiac output becomes.

**Cardiac Output Regulation as the Sum of the Local Blood Flow Regulations Throughout the Body.** As long as the arterial pressure remains normal, each local tissue in the body can control its own blood flow by simply dilating or constricting its local blood vessels. This mechanism, which was discussed in detail in Chapter 20, is the means by which the tissue protects its own nutrient supply, controlling the blood flow in response to its own needs.

Therefore, since the venous return to the heart is the sum of all the local blood flows through all the individual tissues of the body, all of the local blood flow regulatory mechanisms throughout the peripheral circulation are the true controllers of cardiac output under normal conditions. This is an automatic mechanism that allows the heart to respond instantaneously to the needs of each individual tissue. If some tissues need extra blood flow and their local blood vessels dilate, the venous return increases automatically, and the cardiac output increases by an equivalent amount. If most of the tissues throughout the body all require increased blood flow at the same time, the venous return becomes very great, and the cardiac output increases accordingly.

Thus, the whole theory of normal cardiac output regulation is that the tissues control the output in accordance with their needs. Again, it must be stated that it is not the heart that controls the cardiac output under normal conditions; instead, the heart plays a permissive role that allows the tissues to do the controlling. The heart does this by always maintaining a permissive pumping capacity that is somewhat above the actual venous return—that is, except when the heart fails.

**Effect of Local Tissue Metabolism on Cardiac Output Regulation.** The most important factor that controls the local blood flows in the individual tissues is the metabolic rates of the respective tissues. Therefore, venous return and cardiac output are normally controlled in relation to the level of metabolism of the body. This effect was illustrated in Figure 23–2 which showed that the cardiac output increases directly in relation to work output during exercise and that the cardiac output increases parallel to the increase in oxygen consumption, a measure of the rate of metabolism.

However, it is essential that the arterial pressure be maintained at a normal level if changes in metabolism are to regulate cardiac output. This is illustrated in Figure 23–3, which shows changes in cardiac output in response to approximately five-fold increase in tissue metabolism in a dog. The increase in metabolism was caused by the toxic substance dinitrophenol, which causes the metabolic systems to increase their metabolic use of oxygen tremendously and, simultaneously, to dilate the local blood vessels supplying the tissues. Note

in this figure that when the arterial pressure was normally controlled the cardiac output increased approximately three-fold. On the other hand, when the nervous mechanisms for control of arterial pressure were blocked, the cardiac output increased only one-fold.

Therefore, once again, we can state that, under normal conditions, venous return and cardiac output are determined almost entirely by the degree of dilatation of the local blood vessels in the tissues throughout the body. However, this mechanism operates properly only in case the mechanisms for maintaining a normal arterial pressure are also functioning properly.

## IMPORTANCE OF ARTERIAL PRESSURE REGULATION AS AN ADJUNCT TO CARDIAC OUTPUT CONTROL

If the arterial pressure falls when the local tissue blood vessels dilate, the increase in blood flow is far less than is achieved if the arterial pressure remains normal or rises. Therefore, the peripheral mechanism for control of cardiac output is seriously impaired when the arterial pressure fails to be regulated. This is illustrated by the dashed curves of Figure 23–3, which show that, in the absence of normal pressure regulation, acute peripheral vasodilatation causes far more of a fall in arterial pressure than a rise in cardiac output. Consequently, all the mechanisms for arterial pressure regulation that were discussed in the previous two chapters are about as important for cardiac output regulation as they are for arterial pressure regulation itself. Especially in heavy exercise the cardiac output will not rise high enough to supply the blood flow required by the vasodilated muscles; instead, the arterial pressure falls to about one-half normal, and the person becomes fatigued almost instantly.

## IMPORTANCE OF THE SYSTEMIC FILLING PRESSURE IN CARDIAC OUTPUT REGULATION

If the quantity of blood in the circulatory system is too little to fill the system adequately, blood will flow very poorly from the peripheral vessels into the heart. Therefore, the degree of filling of the circulation is one of the most important factors to determine the venous return

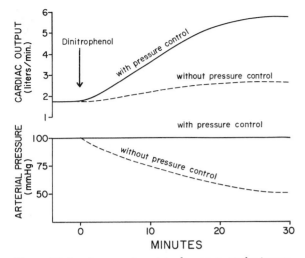

**Figure 23–3.** An experiment to demonstrate the importance of arterial pressure control as a prerequisite for cardiac output control. Note that with pressure control the metabolic stimulant dinitrophenol increases cardiac output; without pressure control, the arterial pressure falls and the cardiac output rises very little. (Data from experiments by Dr. M. Banet.)

to the heart and therefore also to determine the cardiac output.

The student may recall from Chapter 18, that the degree of filling of the circulatory system with blood can be expressed in terms of the *circulatory filling pressure,* and that the degree of filling of the systemic portion of the circulation can be expressed in terms of the *systemic filling pressure.* The systemic filling pressure is the pressure in all parts of the systemic circulation that is measured when blood flow in the circulatory system is suddenly stopped and the blood in the systemic vessels is redistributed so that the pressure in all the vessels is equal. Normally, the systemic filling pressure is 7 mm. Hg, but an increase in blood volume of 15 to 30 per cent doubles the systemic filling pressure, whereas a decrease in blood volume of this same amount reduces the systemic filling pressure to zero.

The systemic filling pressure is the *average* effective pressure of the blood in the peripheral circulation that tends to promote return of blood toward the heart. Experiments have demonstrated that the rate of blood flow from the systemic vessels through the veins to the heart is directly proportional to the *systemic filling pressure minus the right atrial pressure.* Therefore, whenever the systemic filling pressure falls to zero, the flow of blood to the heart

likewise approaches zero regardless of how much the peripheral circulatory system might become dilated.

Consequently, it is very important that the systemic filling pressure remain high enough at all times to supply the needed blood to the heart. For cardiac output to become greatly elevated, it is usually necessary for the systemic filling pressure to rise above normal. This is achieved by sympathetic stimulation of the peripheral vessels, particularly of the veins, which compresses the blood within the vessels and thereby elevates the pressure.

Thus, the systemic filling pressure is another important factor that must always be considered in cardiac output regulation. If the systemic filling pressure is inadequate, the arterial pressure cannot be regulated in a normal way, and, consequently, the normal mechanism of cardiac output control fails. The quantitative relationship of the systemic filling pressure to cardiac output regulation will be considered in much more detail later in the chapter in connection with graphical analysis of cardiac output regulation.

**Effect of Arteriovenous Fistulae on Cardiac Output.** The effect of an A-V fistula on cardiac output is discussed here because of the extremely important lessons that can be learned from this condition. An A-V fistula is a direct opening from an artery into a vein that allows rapid flow of blood directly from the artery to the vein. In exerimental animals, such a fistula can be opened and closed artificially, and one can study the changes in fistula flow, cardiac output, and arterial pressure. A typical experiment of this type in the dog is illustrated in Figure 23–4, showing an initial cardiac output of 1300 ml. per minute. After 15 seconds of recording normal conditions, a fistula is opened

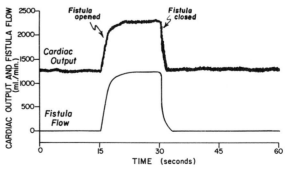

**Figure 23–4.** Effect of suddenly opening and suddenly closing an A-V fistula, showing changes in fistula flow and cardiac output.

through which 1200 ml. of blood flows per minute. Note that the cardiac output increases instantaneously about 1000 ml. per minute, or about 84 per cent as much increase in cardiac output as there is fistula flow. (If the fistula is left open for 24 hours or more, the cardiac output increases 100 per cent as much as the fistula flow because of renal retention of fluid that returns the initially decreased arterial pressure to normal.) Then, when the fistula is closed, the cardiac output returns back to its original level.

Now for the lesson to be learned from this study: When the fistula is opened, the pumping ability of the heart is not changed at all; there is only a change in the peripheral circulation—opening of the A-V fistula. Nevertheless, within a few heart beats after the fistula is opened, the cardiac output increases almost as much as there is increase in blood flow through the fistula. In other words, the fistula flow is added to the flow through the remainder of the body, and the two of these summate to determine the venous return and the cardiac output.

This experiment demonstrates the two cardinal features of normal cardiac output regulation: (1) almost all normal cardiac output regulation is determined by changes in total peripheral resistance, and (2) the normal heart has a permissive level of pumping that is considerably above the actual cardiac output. The heart, therefore, does not have to be stimulated for the increase in cardiac output to come about unless the venous return is excessive.

## REGULATION OF CARDIAC OUTPUT IN HEAVY EXERCISE, REQUIRING SIMULTANEOUS PERIPHERAL AND CARDIAC ADJUSTMENTS

Heavy exercise is one of the most stressful conditions to which the body is ever subjected. The tissues can require as much as 20 times normal amounts of oxygen and other nutrients, so that simply to transport enough oxygen from the lungs to the tissues sometimes demands a minimal cardiac output increase of five to six fold. This is greater than the amount of cardiac output the normal, unstimulated heart can pump. Therefore, to insure the massive increase in cardiac output that is required during heavy exercise, almost all factors that are known to increase cardiac output are called into play. These are:

1. Even before exercise begins, the thought of exercise stimulates the autonomic nervous system, which increases the heart rate and strength of heart contraction, thus immediately increasing the permissive level of heart pumping from the normal level of 13 to 15 liters per minute up to as high as 20 to 25 liters per minute (or as high as 35 liters in the trained athlete). Simultaneously, the sympathetic stimulation constricts the veins throughout the body, and this can increase the systemic filling pressure to as high as two and one-half times normal, thus pushing extra quantities of blood from the peripheral circulation toward the heart. Together, these effects increase the cardiac output instantaneously, even before exercise begins, as much as 50 per cent.

2. At the very onset of exercise, the motor cortex transmits signals through the sympathetic vasodilator nerve fibers (described in Chapter 20) to cause vasodilatation of the muscle blood vessels. This causes a further instantaneous increase in cardiac output, perhaps to as high as double normal.

3. Also, at the onset of exercise, the motor cortex transmits signals directly into the sympathetic nervous system to intensify further the degree of sympathetic activity, further enhancing heart activity, mean systemic pressure, and arterial pressure.

4. Still another effect at the very onset of exercise is tightening of the skeletal muscles and of the abdominal wall around the peripheral blood vessels. This additionally increases the systemic filling pressure, now to as high as four times normal. However, compression of blood vessels in the muscles during muscle contraction has a mechanical effect to increase their resistance, and this partially offsets the value of the additional increase in systemic filling pressure.

5. Finally, the most important effect of all is the direct effect of increased metabolism in the muscle itself. This increased metabolism causes a tremendously increased usage of oxygen and other nutrients, as well as formation of vasodilator substances, all of which cause marked local vasodilatation, thereby greatly increasing local blood flow. This local vasodilatation requires 3 to 15 seconds to reach full development after a person begins to exercise strongly; but, once it does reach full development, the marked decrease in total peripheral resistance caused by this vasodilatation is by all means the most important of all the factors that increase cardiac output during exercise.

In summary, there is tremendous setting of the background conditions of the circulatory system so that it can supply the required blood flow to the muscles during heavy exercise. These conditions include increased systemic filling pressure, increased arterial pressure, and greatly increased activity of the heart. But it is the local vasodilatation in the muscles themselves, occurring as a direct consequence of the muscle activity, that finally sets the level to which the cardiac output rises.

# CARDIAC OUTPUT IN ABNORMAL CONDITIONS

In many abnormal conditions the heart becomes unable to pump the amount of blood that attempts to flow into it. When this happens, the heart then becomes the limiting and determining factor in cardiac output regulation. Therefore, there are two entirely different sets of abnormal factors which can affect cardiac output: (1) abnormal factors affecting peripheral control of venous return and (2) abnormal factors affecting cardiac pumping ability.

## ABNORMALITIES AFFECTING PERIPHERAL CONTROL OF VENOUS RETURN

Figure 23–5 illustrates the cardiac output in different pathological conditions, showing excess cardiac output in some conditions and decreased cardiac output in others.

**Increased Cardiac Output Caused By Decreased Total Peripheral Resistance.** In almost all conditions in which there is excess cardiac output, one finds decreased total peripheral resistance. For instance, in *beriberi* the total peripheral resistance is sometimes reduced to one-half normal. The same is true in *A-V fistulae, anoxia, pulmonary disease, pregnancy,* and *Paget's disease.* In beriberi the decreased total peripheral resistance results from dilatation of the peripheral vessels that is caused by thiamine deficiency. In A-V fistula, decreased total peripheral resistance is caused by direct flow of blood from the arteries to the veins. In hyperthyroidism it is caused by increased peripheral metabolism, which results in local vasodilatation. In hypoxia (anoxia) it is caused by local vasodilatation resulting from lack of oxygen. In pregnancy it is caused by excess blood flow through the placenta. And in Paget's disease it is caused by large numbers of minute A-V anastomoses in the bones.

Obviously, all of these conditions can increase return of blood to the heart and thereby increase cardiac output to levels far above normal.

**Decrease in Cardiac Output Caused by Peripheral Factors—Decreased Blood Volume.** The only peripheral factor that usually decreases cardiac output is decreased systemic filling pressure, most often caused by decreased blood volume. This obviously reduces the peripheral pressures necessary to return blood to the heart. Ordinarily, rapid bleeding of a person of approximately 25 to 30 per cent of his blood volume will reduce the cardiac output to zero: and slow bleeding, over a period of an hour of 40 to 50 per cent, will do the same thing. This is such an important problem in the condition known as "circulatory shock" that it will be discussed in detail in Chapter 28.

## DECREASE IN CARDIAC OUTPUT CAUSED BY CARDIAC FACTORS

Whenever the heart becomes severely damaged, its permissive level of pumping falls below that needed for adequate blood flow to tissues. Therefore, the heart can then be the cause of *decreased* cardiac output, even though it ordinarily cannot by itself be the cause of *increased* cardiac output. Thus, note in Figure 23–5 that cardiac output is decreased in *myocardial infarction*, in severe *valvular heart disease*, and in *cardiac shock;* in all these conditions the heart simply is not capable of pumping an adequate cardiac output. These conditions will be discussed more fully in Chapter 26 in relation to cardiac failure.

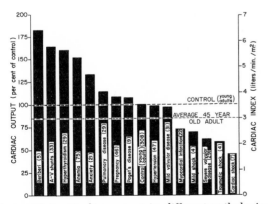

**Figure 23–5.** Cardiac output in different pathological conditions. (From Guyton, Jones, and Coleman: Circulatory Physiology: Cardiac Output and Its Regulation, 1973.)

# GRAPHICAL ANALYSIS OF CARDIAC OUTPUT REGULATION

The discussions of cardiac output regulation thus far are adequate for an understanding of the factors that control cardiac output in most simple conditions, and they also present a reasonably complete *qualitative* picture of cardiac output control. However, it is often important to obtain a much more quantitative understanding of cardiac output regulation. For instance, in severe abnormalities of the circulation, such as cardiac failure and circulatory shock, the qualitative type of analysis is only partially satisfactory. More quantitative analyses will allow one to understand many details of the clinical conditions that are impossible to comprehend through qualitative thinking alone.

To perform quantitative analyses of cardiac output regulation, it is necessary to distinguish separately the two primary factors that are concerned with cardiac output regulation: (1) the pumping ability of the heart, and (2) the peripheral factors that affect flow of blood into the heart. Then, these two factors, the cardiac and the peripheral, must be put together in a quantitative way to see how they interact with each other to determine cardiac output. This process has been frequently accomplished by using electronic computers, but it can also be done more simply by using a type of graphical analysis that one can carry in his mind and that can be used daily in solving some of the quantitative problems of cardiac output regulation. In the following sections we will discuss this type of analysis. To begin, we will first characterize the pumping ability of the heart by use of *cardiac output curves*, and then we characterize the peripheral factors that affect venous return by use of *venous return curves*.

## CARDIAC OUTPUT CURVES

Use the cardiac function curves to depict the ability of the heart to pump blood was discussed in Chapter 13, and the effects of several different factors on these function curves were demonstrated in Figures 13–10 and 13–11. For the purpose of understanding cardiac output regulation, the type of cardiac function curve required is that illustrated in Figures 23–6 and 23–7, which depicts the *cardiac output at different right atrial pressures* and is called the *cardiac output curve*. The normal cardiac output curve is labeled in each figure, while the other curves represent cardiac output in special conditions.

The basic factors that determine the precise characteristics of the cardiac output curve are (1) the *effectiveness of the heart as a pump* and (2) the *extracardiac pressure*.

**Effectiveness of the Heart as a Pump.** A heart that is capable of pumping greater quantities of blood than can the normal heart, such as a heart strongly stimulated by sympathetic impulses or a hyper-

trophied heart, is called a *hypereffective heart*. A weakened heart is called a *hypoeffective heart*. Figure 23–6 illustrates the effect of hypereffectivity or hypoeffectivity on the cardiac output curves, showing that the curve rises greatly in the case of the hypereffective heart and falls equally as much in the case of the hypoeffective heart. Obviously, there are all degrees of hypereffectivity and hypoeffectivity.

*Factors That Can Cause a Hypereffective Heart.* Factors that can make the heart a better pump than normal include:

1. *Sympathetic stimulation,* which can increase the cardiac function curve to approximately two times normal.

2. *Hypertrophy of the heart,* which can increase the cardiac function curve to perhaps as much as two times normal.

3. *Decreased systemic arterial pressure,* which allows the normal heart to pump perhaps as much as 10 per cent extra amounts of blood.

4. *Inhibition of the parasympathetics to the heart,* which removes the parasympathetic tone, allowing the heart rate to increase, and, therefore, increases the pumping effectiveness of the heart.

*Factors That Can Cause a Hypoeffective Heart.* Obviously, any factor that decreases the ability of the heart to pump blood can cause hypoeffectivity. Some of the numerous factors are:

*Myocardial infarction*
*Valvular heart disease*
*Parasympathetic stimulation of the heart*
*Inhibition of the sympathetics to the heart*
*Congenital heart disease*
*Myocarditis*
*Cardiac anoxia*
*Diphtheritic or other types of myocardial damage or toxicity*

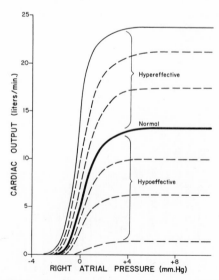

**Figure 23–6.** Cardiac output curves for the normal heart and for hypo- and hypereffective hearts. (From Guyton, Jones, and Coleman: Circulatory Physiology: Cardiac Output and Its Regulation, 1973.)

**Effect of Extracardiac Pressure on Cardiac Output Curves.** Figure 23–7 illustrates the effect of changes in extracardiac pressure on the cardiac output curve. The normal extracardiac pressure, which is also equal to the normal intrapleural pressure, is −4 mm. Hg. Note in the figure that a rise in intrapleural pressure to −2 mm. Hg shifts the entire curve to the right by this same amount. This shift occurs because an extra 2 mm. Hg right atrial pressure is now required to overcome the increased pressure on the outside of the heart. Likewise, an increase in intrapleural pressure to +2 mm. Hg requires a 6 mm. Hg increase in right atrial pressure, which obviously shifts the entire curve 6 mm. Hg to the right. And, in the same manner, a more negative intrapleural pressure shifts the curve to the left.

Some of the factors that alter the intrapleural pressure and thereby shift the cardiac output curve include:

1. *Cyclic changes during normal respiration.*
2. *Breathing against a negative pressure,* which shifts the curve to a more negative right atrial pressure (to the left).
3. *Positive pressure breathing,* which shifts the curve to the right.
4. *Opening the thoracic cage,* which increases the intrapleural pressure and shifts the curve to the right.
5. *Cardiac tamponade,* which means an accumulation of large quantities of fluid in the pericardial cavity around the heart with resultant increase in extracardiac pressure and shifting of the curve to the right. Note in Figure 23–7 that cardiac tamponade shifts the upper parts of the curves farther to the right than the lower parts because the extracardiac pressure rises to higher values as the chambers of the heart fill to increased volumes during high cardiac output.

**Combinations of Different Patterns of Cardiac Output Curves.** Figure 23–8 illustrates that the cardiac output curve can change as a result of simultaneous changes in extracardiac pressure and effectiveness of the heart as a pump. Thus, if one knows what is happening to the extracardiac pressure and

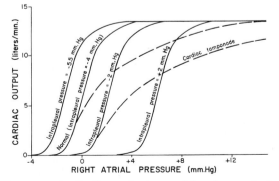

**Figure 23–7.** Cardiac output caused by changes in intrapleural pressure and by cardiac tamponade. (From Guyton, Jones, and Coleman: Circulatory Physiology: Cardiac Output and Its Regulation, 1973.)

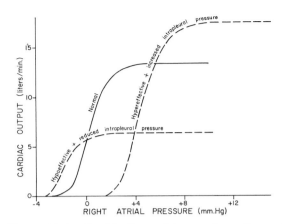

**Figure 23–8.** Combinations of two major patterns of cardiac output curves, showing the effect of alterations in both extracardiac pressure and effectiveness of the heart as a pump. (From Guyton, Jones, and Coleman: Circulatory Physiology: Cardiac Output and Its Regulation, 1973.)

also the approximate capability of the heart as a pump, he can express the momentary ability of the heart to pump blood by a single cardiac output curve.

## VENOUS RETURN CURVES

There still remains the entire systemic circulation that must be considered before a total analysis of circulatory function can be made. To analyze the function of the systemic circulation, we remove the heart and lungs from the circulation of an animal and replace these with a pump and artificial oxygenator system. Then, different factors, such as changes in blood volume and changes in right atrial pressure, are altered to determine how the systemic circulation operates in different circulatory states. In these studies, one finds two principal factors that affect the function of the systemic circulation in relation to cardiac output regulation. These are: (1) *the degree of filling of the systemic circulation,* which is measured by the *systemic filling pressure,* which was discussed in Chapter 18, and (2) the *resistance to blood flow* in the different segments of the systemic circulation.

**The Normal Venous Return Curve.** In the same way that the cardiac output curve relates cardiac output to right atrial pressure, the *venous return curve relates venous return to right atrial pressure.*

The very dark curve in Figure 23–9 is the normal venous return curve. This curve shows that as the right atrial pressure increases, it causes back pressure on the systemic circulation and thereby decreases venous return of blood to the heart. *If all circulatory reflexes are prevented from acting,* venous return decreases to zero when the right atrial pressure rises to equal the normal systemic filling pressure of about 7 mm. Hg. Such a slight rise in right atrial pressure causes a drastic decrease in ve-

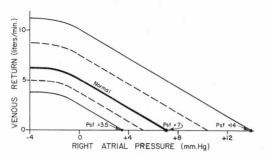

**Figure 23–9.** Venous return curves, showing the normal curve when the systemic filling pressure *(Psf)* is 7 mm. Hg, and showing the effect of altering the systemic filling pressure. (From Guyton, Jones, and Coleman: Circulatory Physiology: Cardiac Output and Its Regulation, 1973.)

nous return because the systemic circulation is a very distensible bag so that any increase in back pressure causes blood to dam up in this bag instead of returning to the heart. Lack of blood flow into the heart decreases the cardiac output, thereby decreasing the arterial pressure. Thus, at the same time that the right atrial pressure is rising, the arterial pressure is falling, and these two pressures come to equilibrium when all flow in the systemic circulation ceases at a pressure of 7 mm. Hg, which, by definition, is the systemic filling pressure.

*Plateau in the Venous Return Curve Caused by Collapse of the Veins.* When the right atrial pressure falls *below* zero—that is, below atmospheric pressure—venous return does not increase significantly. By the time the right atrial pressure has fallen to about −2 mm. Hg, the venous return will have reached a plateau; then it remains at this plateau level even though the right atrial pressure falls as low as −20 to −50 mm. Hg. This plateau is caused by collapse of the veins entering the chest. Low right atrial pressure sucks the walls of the veins together where they enter the chest, which prevents the negative pressure from sucking blood through the veins. Instead, the pressure in the veins immediately outside the chest remains almost exactly equal to atmospheric pressure (zero pressure). Therefore, for all practical purposes, the central venous pressure never falls below a value of 0 mm. Hg despite the fact that the right atrial pressure may fall to very negative values.

**Effect on the Venous Return Curve of Changes in Systemic Filling Pressure.** Figure 23–9 also illustrates the effect of increasing or decreasing the systemic filling pressure. By definition, the systemic filling pressure is the pressure in all parts of the systemic circulation when there is no flow in the circulation. Therefore, at the point where the venous return curve reaches the zero venous return level, the right atrial pressure becomes equal to the systemic filling pressure. Note in Figure 23–9 that the normal systemic filling pressure *(Psf)* is 7 mm. Hg. For the up-

permost curve in the figure, the systemic filling pressure is 14 mm. Hg, and for the lowermost curve it is 3.5 mm. Hg. These curves demonstrate that the greater the systemic filling pressure, which also means the greater the "tightness" with which the circulatory system is filled with blood, the more the venous return curve shifts upward and to the right. Conversely, the lower the systemic filling pressure, the more the curve shifts downward and to the left. To express this another way, the greater the degree of tightness with which the system is filled, the easier it is for blood to flow into the heart. And the less the tightness, the more difficult is it for blood to flow into the heart.

*Factors That Can Alter the Systemic Filling Pressure.* The factors that can alter the systemic filling pressure were discussed in Chapter 18. The three primary factors are: (1) The systemic filling pressure increases rapidly with an *increase in blood volume*— an acute increase in blood volume of 15 per cent or a chronic increase of about 30 per cent increasing the systemic filling pressure to approximately double normal. (2) Maximal *sympathetic stimulation* can increase the systemic filling pressure from 7 mm. Hg to about 17 mm. Hg, approximately a 2½-fold increase. (3) An *increase in contraction* of all the skeletal muscles throughout the body can increase the systemic filling pressure by compressing the blood vessels from the outside to about 25 mm. Hg.

Factors that decrease the systemic filling pressure include principally (1) *loss of sympathetic tone*, which decreases the systemic filling pressure from 7 mm. Hg to about 4 mm. Hg, or (2) *loss of blood volume*, which can decrease the systemic filling pressure to as little as zero.

*"Pressure Gradient for Venous Return."* When the right atrial pressure rises to equal the systemic filling pressure, all other pressures in the systemic circulation also approach this same pressure. Therefore, there becomes no pressure gradient for flow of blood toward the heart. However, when the right atrial pressure falls progressively lower than the systemic filling pressure, the flow to the heart increases proportionately, as one can see by studying any of the venous return curves in Figure 23–9. That is, *the greater the difference between the systemic filling pressure and the right atrial pressure the greater becomes the venous return.* Therefore, the difference between these two pressures is called the *pressure gradient for venous return.*

**Effect of Systemic Resistance on the Venous Return Curve.** Figure 23–10 illustrates the effect of different systemic resistances on the venous return curve, showing that a decrease in resistance to one-half normal *rotates the curve upward*, while an increase in resistance *rotates the curve downward*. The venous return still becomes zero when the right atrial pressure rises to equal the systemic filling pressure, because when there is no pressure gradient to cause flow of blood, it makes no difference what the resistance is in the circulation—the flow is still zero.

Obviously, any factor that dilates the peripheral

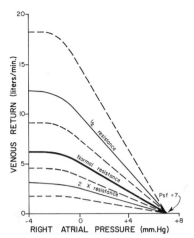

**Figure 23–10.** Venous return curves, depicting the effect of altering the "resistance to venous return." (From Guyton, Jones, and Coleman: Circulatory Physiology: Cardiac Output and Its Regulation, 1973.)

vessels decreases the resistance and rotates the function curve upward, while any factor that constricts the peripheral vessels rotates the curve downward. However, changes in venous resistance affect venous return about 8 times as much as similar changes in the arteries, because far more blood is stored in the distensible blood vessels proximal to the veins than in the vessels proximal to the arteries.

**Combinations of Venous Return Curve Patterns.** Figure 23–11 illustrates the effects on the systemic function curve caused by simultaneous changes in mean systemic pressure and resistance to venous return, illustrating that both of these factors can operate simultaneously.

## ANALYSIS OF CARDIAC OUTPUT AND RIGHT ATRIAL PRESSURE USING CARDIAC OUTPUT AND VENOUS RETURN CURVES

Up to this point we have discussed, first, the individual factors that affect the pumping ability of the heart as depicted by *cardiac output* curves and, second, the individual factors that affect the ability of blood to flow through the systemic circulation as depicted by *venous return* curves. In the complete circulation, however, the heart and the systemic circulation must operate together. This means: (1) The venous return from the systemic circulation must equal the cardiac output from the heart. (2) The right atrial pressure is the same for both the heart and for the systemic circulation.

Therefore, one can predict the cardiac output and right atrial pressure in the following way: (1) determine the momentary pumping ability of the heart and depict this in the form of a cardiac output curve, (2) determine the momentary state of the systemic circulation and depict this in the form of a venous return curve, and (3) "equate" these two curves against each other as shown in Figure 23–12.

The two solid curves of figure 23–12 depict both the *normal cardiac output curve* and the *normal venous return curve*. Obviously, there is only one point on the graph, point A, at which the venous return equals the cardiac output and at which the right atrial pressure is the same in relation to both the heart and the systemic circulation. Therefore, in the normal circulation the right atrial pressure, cardiac output, and venous return are all depicted by point A, called the *equilibrium point*.

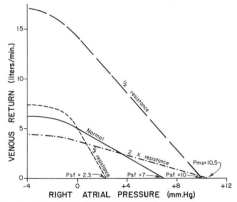

**Figure 23–11.** Combinations of the major patterns of venous return curves, illustrating the effects of simultaneous changes in systemic filling pressure *(Psf)* and in "resistance to venous return." (From Guyton, Jones, and Coleman: Circulatory Physiology: Cardiac Output and Its Regulation, 1973.)

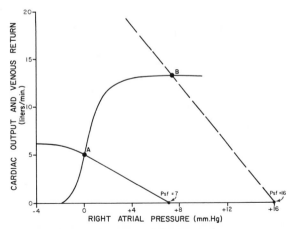

**Figure 23–12.** The two solid curves demonstrate an analysis of cardiac output and right atrial pressure when the cardiac output and venous return curves are known. Transfusion of blood equal to 20 per cent of the blood volume causes the venous return curve to become the dashed curve; as a result the cardiac output and right atrial pressure shift from point A to point B.

## Effect of Increased Blood Volume on Cardiac Output

A sudden increase in blood volume of about 20 per cent increases the cardiac output to about 2½ to 3 times normal. An analysis of this effect is illustrated by the dashed curve of Figure 23–12. Immediately upon infusing the large quantity of extra blood, the increased filling of the system causes the systemic filling pressure to rise to 16 mm. Hg, which shifts the venous return curve upward and to the right. At the same time, the increased blood volume distends the blood vessels, thus reducing their resistance and thereby rotating the curve upward. As a result of these two effects, the venous return curve of Figure 23–12 changes from the solid curve to the dashed curve. This new curve equates with the cardiac output curve at point B, showing that the cardiac output increases 2½ to 3 times and that the right atrial pressure rises to about +8 mm. Hg.

**Compensatory Effects Initiated in Response to the Increased Blood Volume.** The increased cardiac output caused by the increased blood volume lasts only a few minutes, because several different compensatory effects immediately begin to occur: (1) The increased cardiac output increases the capillary pressure so that fluid begins to transude out of the capillaries into the tissues, thereby returning the blood volume toward normal. (2) The increased pressure in the veins causes the veins to distend gradually by the mechanism called *stress-relaxation*, especially causing the venous blood reservoirs such as the liver and spleen to distend. These factors cause the systemic filling pressure to return back toward normal and also allow the resistance vessels of the systemic circulation to constrict once again. Therefore, gradually over a period of about 40 minutes, the cardiac output returns most of the way to normal.

## Effect of Sympathetic Stimulation on Cardiac Output

Sympathetic stimulation affects both the heart and the systemic circulation: (1) It makes the heart a stronger pump. (2) In the systemic circulation, it increases the systemic filling pressure because of contraction of the peripheral vessels, and it also increases the resistance to venous return. Figure 23–13 illustrates an analysis of the effects of moderate and maximal sympathetic stimulation on the cardiac output and right atrial pressure. The normal cardiac output and venous return curves are depicted by the very dark lines; moderate sympathetic stimulation is depicted by the long-dashed curves, and maximal sympathetic stimulation by the dot-dash curves.

Note that maximal sympathetic stimulation increases the systemic filling pressure (depicted by the point at which the venous return curve reaches the zero venous return level) to 17 mm. Hg, and it also increases the pumping effectiveness of the heart by about 70 per cent. As a result, the cardiac output rises from the normal value at equilibrium point A to

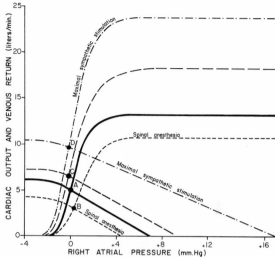

**Figure 23–13.** Analysis of the effect on cardiac output of (1) moderate sympathetic stimulation, (2) maximal sympathetic stimulation, and (3) sympathetic inhibition caused by total spinal anesthesia. (From Guyton, Jones, and Coleman: Circulatory Physiology: Cardiac Output and Its Regulation, 1973.)

approximately double normal at equilibrium point D. Thus, different degrees of sympathetic stimulation can increase the cardiac output progressively to about two times normal—at least for short periods of time.

## Effect of Sympathetic Inhibition

The sympathetic nervous system can be blocked completely by inducing total spinal anesthesia or by using some drug, such as hexamethonium, that blocks transmission of nerve impulses through the autonomic ganglia. The two lowermost curves in Figure 23–13, the short-dashed curves, show the effect of total sympathetic inhibition caused by spinal anesthesia, illustrating that (1) the systemic filling pressure falls to about 4 mm. Hg, and (2) the effectiveness of the heart as a pump decreases to about 80 per cent of normal. The cardiac output falls from point A to point B, which is a decrease to about 60 per cent of normal.

## Increase in Cardiac Output in Exercise

Several different factors operate in exercise to increase the cardiac output. These include: (1) intense sympathetic stimulation, which strengthens the heart and also increases the systemic filling pressure, (2) contraction of the muscles around the blood vessels, which further increases the systemic filling pressure, and (3) intense dilatation of the resistance vessels in the muscles, thereby decreasing the resistance to venous return. Figure 23–14 depicts the effects of these factors on the circulation, showing that (1)

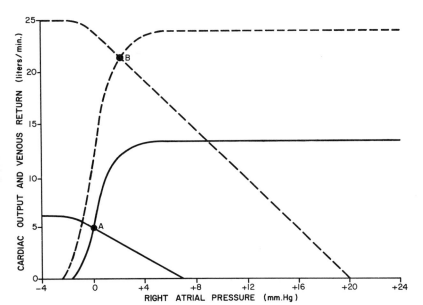

**Figure 23–14.** Graphical analysis of the changes in cardiac output and right atrial pressure with the onset of strenuous exercise.

compression of the blood vessels by the muscles plus sympathetic stimulation of the vessel walls increases the systemic filling pressure from 7 mm. Hg to 20 mm. Hg, (2) sympathetic stimulation increases the strength of the heart as a pump by about 70 per cent, and (3) the local vasodilatation in the muscles rotates the venous return curve upward to the right. As a result, the cardiac output and venous return curves in exercise (the dashed curves) equate at point B, which is a cardiac output of 22 liters per minute, or 4½ times normal. The right atrial pressure hardly changes.

# RIGHT VENTRICULAR OUTPUT VERSUS LEFT VENTRICULAR OUTPUT—BALANCE BETWEEN THE VENTRICLES

Obviously, the output of one ventricle must remain almost exactly the same as that of the other. Therefore, an intrinsic mechanism must be available for automatically adjusting the outputs of the two ventricles to each other. This operates as follows:

Let us assume that the strength of the left ventricle decreases suddenly so that left ventricular output falls below right ventricular output. This immediately allows more blood to be pumped into the lungs than is pumped into the systemic circulation. Consequently, the pulmonary filling pressure and the left atrial pressure rise while the systemic filling pressure and the right atrial pressure fall. If we apply the princi- ples of analysis presented previously to each of the ventricles separately, we will see that the increased *pulmonary filling pressure* increases left heart venous return and left ventricular output while the decrease in *systemic filling pressure* decreases right heart venous return and right ventricular output. This process continues until the output of the left ventricle rises to equal the falling output of the right ventricle. Thus, the outputs of the two ventricles become rebalanced with each other within a few beats of the heart. The same effects occur in the opposite direction when the strength of the right heart diminishes.

These problems of balance between the two ventricles are especially important when myocardial failure occurs in one ventricle or when valvular lesions cause poor pumping by one side of the heart. These conditions will be discussed in more detail in Chapter 26 in relation to cardiac failure.

# METHODS FOR MEASURING CARDIAC OUTPUT

In lower animals, the aorta, the pulmonary artery, or the great veins entering the heart can be cannulated and the flow measured by any type of flowmeter. Also, an electromagnetic or ultrasonic flowmeter can be placed on the aorta or pulmonary artery to measure cardiac output. However, in the human being, cardiac output is measured by indirect methods that do not require surgery. Two of these indirect methods commonly used are the *oxygen Fick method* and the *indicator dilution method.*

## PULSATILE OUTPUT OF THE HEART AS MEASURED BY AN ELECTROMAGNETIC OR ULTRASONIC FLOWMETER

Figure 23–15 illustrates a recording of blood flow in the root of the aorta made using an electromagnetic flowmeter. It demonstrates that the blood flow rises rapidly to a peak during systole and, then, at the end of systole, actually reverses for a small fraction of a second. It is this reverse flow that causes the aortic valve to close. And a minute amount of reverse flow continues throughout diastole to supply blood to the coronary vessels.

## MEASUREMENT OF CARDIAC OUTPUT BY THE OXYGEN FICK METHOD

The Fick procedure is best explained by Figure 23–16, which shows the absorption of 200 ml. of oxygen from the lungs into the pulmonary blood each minute and also illustrates that the blood entering the right side of the heart has an oxygen concentration of approximately 160 ml. per liter of blood, while that leaving the left side has an oxygen concentration of approximately 200 ml. per liter of blood. From these data we see that each liter of blood passing through the lungs picks up 40 ml. of oxygen. And, since the total quantity of oxygen absorbed into the blood from the lungs each minute is 200 ml., a total of 5 1-liter portions of blood must pass through the pulmonary circulation each minute to absorb this amount of oxygen. Therefore, the quantity of blood flowing through the lungs each minute is 5 liters, which is also a measure of the cardiac output. Thus, the cardiac output can be calculated by the following formula:

Cardiac output (liters/min.)=

$$\frac{\text{Oxygen absorbed per minute by the lungs (ml./min.)}}{\text{Arteriovenous oxygen difference (ml./liter of blood)}}$$

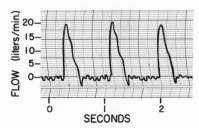

**Figure 23–15.** Pulsatile blood flow in the root of the aorta recorded by an electromagnetic flowmeter.

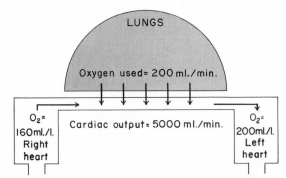

**Figure 23–16.** The Fick principle for determining cardiac output.

In applying the Fick procedure, it is the mean concentrations of oxygen in the venous and arterial bloods that must be used in the calculations. To obtain an accurate determination of venous oxygen concentration it is necessary to take the blood sample directly from the right ventricle or preferably the pulmonary artery, for blood in any one vein of the body usually has a concentration of oxygen different from that in other veins, and even blood in the right atrium usually has not mixed satisfactorily. To obtain such a sample of mixed venous blood a catheter is usually inserted up the brachial vein of the forearm, through the subclavian vein, down to the right atrium, and finally into the right ventricle or pulmonary artery.

Blood used for determining the oxygen saturation in arterial blood can be obtained from any artery in the body, because all arterial blood is thoroughly mixed before it leaves the heart and therefore has the same oxygen concentration.

The rate of oxygen absorption by the lungs is usually measured by a "respirometer," which will be described in Chapter 71. In essence, this device is a floating chamber containing oxygen that sinks in the water as oxygen is removed from it, thus measuring the use of oxygen by the rate at which it falls.

## THE INDICATOR DILUTION METHOD

In measuring the cardiac output by the indicator dilution method a small amount of indicator, such as a dye, is injected into a large vein or preferably into the right side of the heart itself. This then passes rapidly through the right heart, the lungs, the left heart, and finally into the arterial system. If one records the concentration of the dye as it passes through one of the peripheral arteries, a curve such as one of the solid curves illustrated in Figure 23–17 will be obtained. In each of these instances 5 mg. of "Cardio-Green" dye was injected at zero time. In the top recording none of the dye passed into the arterial tree until approximately 3 seconds after the injection, but then the arterial concentration of the dye rose rapidly to a maximum in approximately 6 to

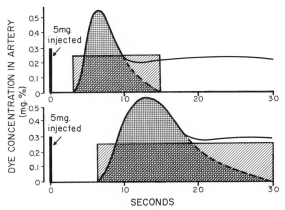

**Figure 23–17.** Dye concentration curves used to calculate the cardiac output by the dilution method. (The rectangular areas are the calculated average concentrations of dye in the arterial blood for the durations of the respective curves.)

7 seconds. After that, the concentration fell rapidly. However, before the concentration reached the zero point, some of the dye had already circulated all the way through some of the peripheral vessels and returned through the heart for a second time. Consequently, the dye concentration in the artery began to rise again. For the purpose of calculation, however, it is necessary to extrapolate the early downslope of the curve to the zero point, as shown by the dashed portion of the curve. In this way, the *time-concentration curve* of the dye in an artery can be measured in its first portion and estimated reasonably accurately in its latter portion.

Once the time-concentration curve has been determined, one can then calculate the mean concentration of dye in the arterial blood for the duration of the curve. In Figure 23–17, this was done by measuring the area under the entire curve, and then averaging the concentration of dye for the duration of the curve; one can see from the shaded rectangle straddling the upper curve of the figure that the average concentration of dye was approximately 0.25 mg./100 ml. blood and that the duration of the curve was 12 seconds. However, a total of 5 mg. of dye was injected at the beginning of the experiment. In order for blood carrying only 0.25 mg. of dye in each 100 ml. to carry the entire 5 mg. of dye through the heart and lungs in 12 seconds, it would be necessary for a total of 20 100-ml. portions of blood to pass through the heart during this time, which would be the same as a cardiac output of 2 liters per 12 seconds, or 10 liters per minute.

In the bottom curve of Figure 23–17, the blood flow through the heart was considerably slower, and the dye did not appear in the arterial system until approximately 6 seconds after it had been injected. It reached a maximum height in 12 to 13 seconds and was extrapolated to 0 at approximately 30 seconds.

Averaging the dye concentrations over the 24-second duration of the curve one finds again an average concentration of 0.25 mg. of dye in each 100 ml. of blood, but this time for a 24-second time interval instead of 12 seconds. To transport the total 5 mg. of dye, 20 100-ml. portions of blood would have had to pass through the heart during the 24-second time interval. Therefore, the cardiac output was 2 liters per 24 seconds, or 5 liters per minute.

To summarize, the cardiac output can be determined from the following formula:

Cardiac output (ml./min.) =

$$\frac{\text{Milligrams of dye injected} \times 60}{\left(\begin{array}{c}\text{Average concentration of dye}\\ \text{in each milliliter of blood}\\ \text{for the duration of the curve}\end{array}\right) \times \left(\begin{array}{c}\text{Duration of}\\ \text{the curve}\\ \text{in seconds}\end{array}\right)}$$

**Substances That Can Be Injected for Determining Cardiac Output by the Indicator Dilution Method.** Almost any substance that can be analyzed satisfactorily in the arterial blood can be injected when making use of the indicator dilution method for determining cardiac output. However, for optimum accuracy it is necessary that the injected substance not be lost into the tissues of the lungs during its passage to the sampling site. The most widely used substance is *Cardio-Green,* a dye that combines with the plasma proteins and, therefore, is not lost from the blood.

## ESTIMATION OF CARDIAC OUTPUT BY THE PULSE PRESSURE METHOD

A fairly reliable estimation of cardiac output can also be made from recordings of pressure pulse contours from the aorta. The basic theory of this method is the following: During diastole no blood flows into the aorta from the heart, but it does continue to flow from the central arteries into the peripheral blood vessels, thus allowing the pressure in the central arteries to decline. Obviously, the greater the rate of blood flow, the greater also is the rate of blood pressure decline during diastole. By using empirical formulas, the cardiac output can be calculated from this downward slope of the pressure during diastole, and other characteristics of the pressure pulse curve can be used to make the calculation even more valid.

The beauty of this method for estimating cardiac output is that a measurement of cardiac output can be made each beat of the heart, which allows one to observe rapid changes in cardiac output. Unfortunately, the characteristics of the pressure pulse curve depend on the distensibility of the arteries as well as on the rate of run-off of blood through the peripheral vessels, thus making the method sometimes seriously in error.

# REFERENCES

Banet, M., and Guyton, A.C.: Effect of body metabolism on cardiac output: role of the central nervous system. *Amer. J. Physiol., 220:*662, 1971.

Bishop, V. S., and Stone, H. L.: Quantitative description of ventricular output curves in conscious dogs. *Circ. Res., 20:*581, 1967.

Bishop, V. S., Stone, H. L., and Guyton, A. C.: Cardiac function curves in conscious dogs. *Amer. J. Physiol., 207:*677, 1964.

Brecher, G. A.: Venous Return. New York, Grune & Stratton, Inc., 1956.

Coleman, T. G., Manning, R. D., Jr., Norman, R. A., Jr., and Guyton, A. C.: Control of cardiac output by regional blood flow distribution. *Ann. Biomed. Eng., 2:*149, 1974.

Dobbs, W. A., Jr., Prather, J. W., and Guyton, A. C.: Relative importance of nervous control of cardiac output and arterial pressure. *Amer. J. Cardiol., 27:*507, 1971.

Dodge, H. T., and Kennedy, J. W.: Cardiac output, cardiac performance, hypertrophy, dilatation, valvular disease, ischemic heart disease, and pericardial disease. *In* Sodeman, W. A., Jr., and Sodeman, W. A. (eds.): Pathologic Physiology: Mechanisms of Disease, 5th Ed. Philadelphia, W. B. Saunders Company, 1974, p. 235.

Donald, D. E., and Shepherd, J. T.: Response to exercise in dogs with cardiac denervation. *Amer. J. Physiol., 205:*393, 1963.

Dow, P.: Estimations of cardiac output and central blood volume by dye dilution. *Physiol. Rev., 36:*77, 1956.

Fermoso, J. D., Richardson, T. Q., and Guyton, A. C.: Mechanism of decrease in cardiac output caused by opening the chest. *Amer. J. Physiol., 207:*1112, 1964.

Grodins, F. S.: Integrative cardiovascular physiology: a mathematical synthesis of cardiac and blood vessel hemodynamics. *Quart. Rev. Biol., 34:*93, 1959.

Guyton, A. C.: Determination of cardiac output by equating venous return curves with cardiac response curves. *Physiol. Rev., 35:*123, 1955.

Guyton, A. C.: Venous return. *In* Hamiton, W. F. (ed.): Handbook of Physiology. Sec. 2, Vol. 2. Baltimore, The Williams & Wilkins Company, 1963, p. 1099.

Guyton, A. C.: Regulation of cardiac output. *New Eng. J. Med., 277:*805, 1967.

Guyton, A. C., Abernathy, J. B., Langston, J. B., Kaufmann, B. N., and Fairchild, H. M.: Relative importance of venous and arterial resistance in controlling venous return and cardiac output. *Amer. J. Physiol., 196:*1008, 1959.

Guyton, A. C., Coleman, T. G., Cowley, A. W., Jr., Liard, J. F., Norman, R. A., Jr., and Manning, R. D., Jr.: Systems analysis of arterial pressure regulation and hypertension. *Ann. Biomed. Eng., 1:*254, 1972.

Guyton, A. C., Coleman, T. G., and Granger, H. J.: Circulation: overall regulation. *Ann. Rev. Physiol., 34:*13, 1972.

Guyton, A. C., Douglas, B. H., Langston, J. B., and Richardson, T. Q.: Instantaneous increase in mean circulatory pressure and cardiac output at onset of muscular activity. *Circ. Res., 11:*431, 1962.

Guyton, A. C., Granger, H. J., and Coleman, T. G.: Autoregulation of the total systemic circulation and its relation to control of cardiac output and arterial pressure. *Circ. Res., 28 (Suppl. 1):*93, 1971.

Guyton, A. C., Lindsey, A. W., Abernathy, J. B., and Richardson, T. Q.: Venous return at various right atrial pressures and the normal venous return curve. *Amer. J. Physiol., 189:*609, 1957.

Jones, C. E., Smith, E. E., and Crowell, J. W.: Cardiac output and physiological mechanisms in circulatory shock. *In* Guyton, A. C. (ed.): MTP International Review of Science: Physiology. Vol. 1. Baltimore, University Park Press, 1974, p. 233.

Keul, J.: The relationship between circulation and metabolism during exercise. *Med. Sci. Sports, 5:*209, 1973.

Knoop, A. A.: Physiological aspects of circulatory dynamics especially related to ageing as studied by displacement ballistocardiography and other cardiovascular methods. *Bibl. Cardiol., 30:*87, 1973.

Levy, M. N., and Zieske, H.: A closed circulatory system model. *Physiologist, 10:*419, 1967.

Mitchell, J. H., and Wildenthal, K.: Static (isometric) exercise and the heart: physiological and clinical considerations. *Ann. Rev. Med., 25:*369, 1974.

Nichols, W. W.: Continuous cardiac output derived from the aortic pressure waveform: a review of current methods. *Biomed. Eng., 8:*376, 1973.

Prather, J. W., Taylor, A. E., and Guyton, A. C.: Effect of blood volume, mean circulatory pressure, and stress relaxation on cardiac output. *Amer. J. Physiol., 216:*467, 1969.

Sarnoff, S., and Mitchell, J. H.: The regulation of the performance of the heart. *Amer. J. Med., 30:*747, 1961.

Stone, H. L., Bishop, V. S., and Dong, E., Jr.: Ventricular function in cardiac-denervated and cardiac-sympathectomized conscious dogs. *Circ. Res., 20:*587, 1967.

Sugimoto, T., Sagawa, K., and Guyton, A. C.: Effect of tachycardia on cardiac output during normal and increased venous return. *Amer. J. Physiol., 211:*288, 1966.

Varat, M. A., Adolph, R. J., and Fowler, N. O.: Cardiovascular effects of anemia. *Amer. Heart J., 83:*415, 1972.

Weisel, R. D., Berger, R. L., and Hechtman, H. B.: Current concepts measurement of cardiac output by thermodilution. *New Eng. J. Med., 292:*682, 1975.

# I 24 I

# The Pulmonary Circulation

The quantity of blood flowing through the lungs is essentially equal to that flowing through the systemic circulation. However, there are problems related to distribution of blood flow and other hemodynamics that are special to the pulmonary circulation. Therefore, the present discussion is concerned specifically with the peculiarities of blood flow in the pulmonary circuit and the function of the right side of the heart in maintaining this flow.

## PHYSIOLOGIC ANATOMY OF THE PULMONARY CIRCULATORY SYSTEM

**The Right Side of the Heart.** As illustrated in Figure 24–1, the right ventricle is wrapped halfway around the left ventricle. The cause of this is the difference in pressures developed by the two ventricles during systole. Because the left ventricle contracts with extreme force in comparison with the right ventricle, the left ventricle assumes a globular shape, and the septum protrudes into the right heart. Yet each side of the heart pumps essentially the same quantity of blood; therefore, the external wall of the right ventricle bulges far outward and extends around a large portion of the left ventricle in this way accommodating about the same quantity of blood as the left ventricle.

The muscle of the right ventricle is slightly more than one-third as thick as that of the left ventricle; this results from the difference in pressures between the two sides of the heart. Indeed, the wall of the right ventricle is only about three times as thick as the atrial walls, while the left ventricular muscle is about eight times as thick.

**The Pulmonary Vessels.** The pulmonary artery extends only 4 centimeters beyond the apex of the

right ventricle and then divides into the right and left main branches, which supply blood to the two respective lungs. The pulmonary artery is also thin, with a wall thickness approximately twice that of the venae cavae and one-third that of the aorta. The pulmonary arterial branches are all very short. However, all the pulmonary arteries, even the small arteries and arterioles, have much larger diameters than their counterpart systemic arteries. This, combined with the fact that the vessels are very thin and distensible, gives the pulmonary arterial tree a very large compliance, averaging 3 ml. per mm. Hg, which is almost equal to that of the entire systemic arterial

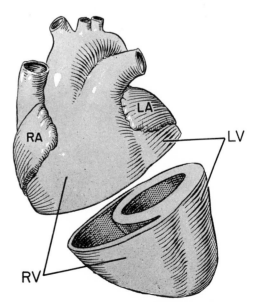

**Figure 24–1.** The anatomical relationship of the right ventricle to the left ventricle, showing the globular shape of the left ventricle and the half-moon shape of the right ventricle as it drapes around the left ventricle.

tree. This large compliance allows the pulmonary arteries to accommodate the stroke volume output of the right ventricle.

The pulmonary veins, like the pulmonary arteries, are also short, but their distensibility characteristics are similar to those of the veins in the systemic circulation.

**The Bronchial Vessels.** A minor accessory arterial blood supply to the lungs exists directly from the aorta through usually one bronchial artery to the right lung and two bronchial arteries to the left lung. The blood flowing in the bronchial arteries is *oxygenated* blood, in contrast to the partially deoxygenated blood in the pulmonary arteries. It supplies the supporting tissues of the lungs, including the connective tissue, the septa, and the large and small bronchi. After this bronchial arterial blood has passed through the supporting tissues, it empties into the pulmonary veins and *enters the left atrium* rather than passing back to the right atrium. An average of 1 to 2 per cent of the total cardiac output takes this route, thus making the left ventricular output slightly greater than the right ventricular output.

**The Lymphatics.** Lymphatics extend from all the supportive tissues of the lung, beginning in the perivascular and peribronchial spaces that lie in the angular junctions between the alveoli, and coursing to the hilum of the lung and thence mainly into the right lymphatic duct. Particulate matter entering the alveoli is usually removed rapidly via these channels, and protein is also removed from the lung tissues, thereby preventing edema.

# PRESSURES IN THE PULMONARY SYSTEM

## THE PRESSURE PULSE CURVE IN THE RIGHT VENTRICLE

The pressure pulse curves of the right ventricle and pulmonary artery are illustrated in the lower portion of Figure 24–2. These are contrasted with the much higher aortic pressure curve shown above. Approximately 0.16 second prior to ventricular systole, the atrium contracts, pumping a small quantity of blood into the right ventricle, and thereby causing about 4 mm. Hg initial rise in the right ventricular diastolic pressure even before the ventricle contracts. Immediately following this priming by the right atrium, the right ventricle contracts, and the right ventricular pressure rises rapidly until it equals the pressure in the pulmonary artery. The pulmonary valve opens, and for approximately 0.3 second blood flows from the right ventricle into the pulmonary artery. When the right ventricle relaxes, the pulmonary valve closes, and the right ventricular pressure falls back to its diastolic level of about zero.

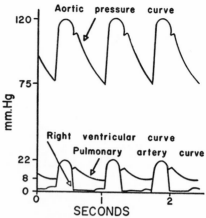

**Figure 24–2.** Pressure pulse contours in the right ventricle, pulmonary artery, and aorta. (Redrawn from Cournand: *Circulation,* 2:641, 1950.)

The systolic pressure in the right ventricle of the normal human being averages approximately 22 mm. Hg, and the diastolic pressure averages about 0 to 1 mm. Hg.

## PRESSURES IN THE PULMONARY ARTERY

During systole, the pressure in the pulmonary artery is essentially equal to the pressure in the right ventricle, as shown in Figure 24–2. However, after the pulmonary valve closes at the end of systole, the ventricular pressure falls, while the pulmonary arterial pressure remains elevated and then falls gradually as blood flows through the capillaries of the lungs.

As shown in Figure 24–3, the systolic pulmonary arterial pressure averages approximately 22 mm. Hg in the normal human being, the diastolic pulmonary arterial pressure approxi-

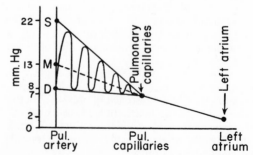

**Figure 24–3.** Pressures in the different vessels of the lungs. (Modified from Cournand: *Circulation,* 2:641, 1950.)

mately 8 mm. Hg, and the mean pulmonary arterial pressure 13 mm. Hg.

**Pulmonary Arterial Pulse Pressure.** The pulse pressure in the pulmonary arteries averages 14 mm. Hg, which is almost two thirds as much as the systolic pressure. In previous discussions of pulse pressure in the systemic circulation, it has been pointed out that the less the compliance of an elastic reservoir that receives pulsatile injections of blood, the greater will be the pulse pressure. If it were not for the large compliance of the thin pulmonary arteries, the pulmonary arterial pulse pressure would be even greater than it is. However, in addition, about one-half the blood ejected by the right ventricle runs off from the pulmonary arteries into the pulmonary veins and left atrium at the same time that it is being ejected, and this also keeps the pulse pressure low.

### PULMONARY CAPILLARY PRESSURE

The mean pulmonary capillary pressure, which is illustrated diagrammatically in Figure 24-3, is approximately 7 mm. Hg. This will be discussed in detail later in the chapter in relation to fluid exchange functions of the capillary. However, we should note here that the capillary pressure of 7 mm. Hg is almost exactly half-way between the mean pulmonary arterial pressure of 13 mm. Hg and the left atrial pressure of 2 mm. Hg, indicating that the arterial and venous resistances of the lungs are approximately equal. This relationship is in marked contrast to that in the systemic circulation in which the arterial resistance is four to seven times as great as the venous resistance.

### LEFT ATRIAL AND PULMONARY VENOUS PRESSURE

The mean pressure in the left atrium and in the major pulmonary veins averages approximately 2 mm. Hg in the human being. However, the pressure in the left atrium varies, even among normal individuals, from as low as 1 mm. Hg to as high as 4 mm. Hg.

The pulse waves in the left atrium are almost identical to those in the right ventricle, as described in Chapter 19, except that the pulse pressures in the left atrium are several times as great as those in the right atrium mainly because the left atrium and left ventricle contract with greater force than the right atrium and ventricle, but also because the combined compliance of the pulmonary veins and left atrium is considerably less than the combined compliance of the right atrium and the large systemic veins.

### THE BLOOD VOLUME OF THE LUNGS

The blood volume of the lungs is approximately 9 per cent of the total blood volume of the circulatory system. In other words, in the average human being the two lungs contain approximately 450 ml. of blood. About 70 ml. of this is in the capillaries, and the remainder is divided about equally between the arteries and veins.

**The Lungs as a Blood Reservoir.** Under different physiologic and pathologic conditions, the quantity of blood in the lungs can vary from as little as 50 per cent of normal up to as high as 200 per cent of normal. For instance, when a person blows air out so hard that he builds up high pressure in his lungs—such as when blowing a trumpet—as much as 250 ml. of blood can be expelled from the pulmonary circulatory system into the systemic circulation. Also, loss of blood from the systemic circulation by hemorrhage can be partly compensated by automatic shift of blood from the lungs into the systemic vessels.

***Shift of Blood Between the Pulmonary and Systemic Circulatory Systems as a Result of Cardiac Pathology.*** Failure of the left side of the heart or increased resistance to blood flow through the mitral valve as a result of mitral stenosis or mitral regurgitation causes blood to dam up in the pulmonary circulation, thus greatly increasing the pulmonary blood volume while decreasing the systemic volume. Concurrently, the pressures in the lung increase while the systemic pressures decrease.

On the other hand, exactly the opposite effects take place when the right side of the heart fails.

Because the volume of the systemic circulation is about seven times that of the pulmonary system, a shift of blood from one system to the other affects the pulmonary system greatly but usually has only mild systemic effects.

## BLOOD FLOW THROUGH THE LUNGS AND ITS DISTRIBUTION

The blood flow through the lungs is equal to the cardiac output (except for the 1 to 2 per cent that goes through the bronchial circulation). Therefore, the factors that control cardiac output—mainly peripheral factors, as discussed in Chapter 23— also control pulmonary blood flow. Under most conditions, the pulmonary vessels act as passive, distensible tubes that enlarge with increasing pressure and narrow with decreasing pressure. But, for the blood to be aerated adequately, it is important for it to be distributed as evenly as possible to all the different segments of the lung. Some of the problems related to this are the following:

**Maldistribution of Blood Flow Between the Top and Bottom of the Lung Because of Hydrostatic Pressure.** Hydrostatic pressure affects vascular pressures in the lungs in the same way as in the systemic circulation. Therefore, when the person is in the upright position, the pulmonary vascular pressures at the top of the lungs are about 10 mm. Hg less than they are at the level of the heart; and the pressures at the bottom of the lung are about 8 mm. Hg greater. This effect causes serious differences in blood flow between the upper and lower lung because the mean pulmonary arterial pressure is 13 mm. Hg at the level of the heart, about 3 mm. Hg at the apex of the lungs, and about 21 mm. Hg at the bottom. Consequently, blood flow in the bottom of the lung is moderately greater than at heart level, whereas blood flow in the apical lung tissue is very slight. Indeed, the pulmonary arterial pressure during diastole is too low to cause any flow whatsoever in the apices so that the flow that does occur happens only during systole. The gradation of flow at different levels of the lungs in the person standing at rest is illustrated by the lowermost curve of Figure 24–4.

Fortunately, there is also greater ventilation of the lowermost alveoli of the lungs in comparison with the

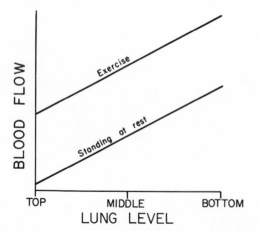

**Figure 24–4.** Blood flow at different levels in the lung of an upright person, both at rest and during exercise. Note that when the person is at rest, the blood flow is almost zero at the top of the lungs and most of the flow is through the lower lung.

ventilation in the lung apices, but not enough to keep step with the increased blood flow.

In the lying position, there is a different pattern of maldistribution of blood volume and blood flow, with extra blood volume and blood flow in the posterior parts of the lungs with respect to the anterior, but the effect is not nearly so severe because the differences in hydrostatic levels are much less.

Exercise partially overcomes this maldistribution, as illustrated by the upper curve of Figure 24–4, because the pulmonary arterial pressure then rises high enough to cause reasonable flow even in the lung apices. Indeed, a very good balance then develops between pulmonary blood flow and ventilation because there is at that time almost exactly the same amount of extra flow in the lower lungs as there is extra ventilation. After all, it is during exercise that the lungs are called upon to provide their greatest rates of oxygen uptake and carbon dioxide excretion.

*Effect of Hydrostatic Pressure on Pulmonary Capillary Filling.* Not only does an elevation in hydrostatic pressure increase blood flow through the lung tissue, but it also increases the volume of blood, especially the volume of blood in the perialveolar capillaries. For instance, in the resting person the perialveolar capillaries in the base of the lung are usually completely filled with blood while those in the apex are almost empty. During exercise, however, when the pulmonary vascular pressures increase at all levels of the lung, the capillaries of the apex then become well filled with blood. Because of this effect, during exercise the number of functional capillaries increases about two-fold, enhancing the rates of oxygen and carbon dioxide exchange between the pulmonary blood and the alveolar air by two to three-fold. This very important effect in exercise will be discussed in Chapter 40.

**Effect of Low Alveolar Oxygen Pressure on Pulmonary Vascular Resistance—Automatic Control of Local Pulmonary Blood Flow Distribution.** When alveolar oxygen concentration becomes very low, the adjacent blood vessels slowly constrict during the ensuing 5 to 10 minutes, the vascular resistance increasing to as much as double normal. It should be noted specifically that this effect of oxygen lack is *opposite to the effect* normally observed in systemic vessels, and its cause is yet unexplained. However, this constrictor effect of low oxygen concentration will not occur in a small pulmonary artery that has been isolated from all lung tissue. Therefore, it is believed that the low oxygen concentration causes some vasoconstrictive substance to be released from the lung tissue, with this substance in turn promoting small arterial and arteriolar constriction.

This effect of low oxygen on pulmonary vascular resistance has an important purpose: to distribute blood flow where it is most effective. That is, when some of the alveoli are poorly ventilated and the oxygen concentration in them is low, the local vessels constrict. This in turn causes most of the blood to flow through other areas of the lungs that are better aerated, thus providing an automatic control system for distributing blood flow through different pulmonary areas in proportion to their degrees of ventilation.

**Paucity of Autonomic Nervous Control of Blood Flow in the Lungs.** Though nerves innervate the lung tissues profusely, it is doubtful that these have a major function in normal control of pulmonary blood flow. Normally, stimulation of the vagal fibers to the lungs causes a slight decrease in pulmonary resistance, and stimulation of the sympathetics causes a slight increase in resistance. Likewise, injection of acetylcholine normally causes a slight decrease in resistance, while injection of norepinephrine or epinephrine causes a slight increase. But these effects are all so slight that they are difficult to demonstrate either in experimental animals or in the human being.

Often research workers describe reflexes in the pulmonary vascular system that might be of clinical importance. For instance, it has been claimed that small emboli occluding the small pulmonary arteries cause a reflex that elicits sympathetic vasoconstriction throughout the lungs, this vasoconstriction then leading to an intense increase in pulmonary arterial pressure. It has also been claimed that obstruction of a pulmonary vein causes a sympathetic reflex to constrict the arteries that supply the segment of the lungs feeding into the obstructed vein. Thus far, neither of these reflexes has been proved beyond doubt.

# CAPILLARY DYNAMICS IN THE LUNGS

## PULMONARY CAPILLARY PRESSURE

Unfortunately, no direct measurements of pulmonary capillary pressure have been made. However, "isogravimetric" measurement of pulmonary capillary pressure, using a technique described in Chapter 30, has given a value of 7 mm. Hg. This is probably very nearly correct because the mean left atrial pressure is about 2 mm. Hg and the mean pulmonary arterial pressure is only 13 mm. Hg, so that the mean pulmonary capillary pressure must be between these two values.

**Pulmonary Wedge Pressure.** By inserting a catheter through the pulmonary artery into one of the small branch arteries and then pushing the catheter until it wedges tightly in the artery, a pressure measurement called the "wedge pressure" can be made. The measured wedge pressure averages approximately 5 mm. Hg, and some physiologists have suggested that this represents a measurement of capillary pressure because the tip of the catheter is nearly at the capillaries. However, since the arterial blood flow is blocked by the catheter, but blood can still flow between the tip of the catheter and the pul-

monary veins through the capillaries, it is probable that wedge pressure is somewhere between left atrial pressure and capillary pressure.

## LENGTH OF TIME BLOOD STAYS IN THE CAPILLARIES

From histologic study of the total cross-sectional area of all the pulmonary capillaries, it can be calculated that when the cardiac output is normal, blood passes through the pulmonary capillaries in about 1 second. Increasing the cardiac output shortens this time sometimes to less than 0.4 second, but the shortening would be much greater were it not for the fact that additional capillaries, which normally remain collapsed, open up to accommodate the increased blood flow. Thus, in less than 1 second, blood passing through the capillaries becomes oxygenated and loses its excess carbon dioxide.

## CAPILLARY MEMBRANE DYNAMICS IN THE LUNGS

**Negative Interstitial Fluid Pressure in the Lungs and Its Significance.** Fluid exchange through the capillary membrane is discussed in detail in Chapter 30, where it is pointed out that the most significant difference between pulmonary capillary dynamics and capillary dynamics elsewhere in the body is the very low capillary pressure in the lungs, about 7 mm. Hg, in comparison with a considerably higher "functional" capillary pressure elsewhere in the body, probably about 17 mm. Hg. Because of the very low pulmonary capillary pressure, the hydrostatic force tending to push fluid out the capillary pores into the interstitial spaces is also very slight. Yet, the colloid osmotic pressure of the plasma, about 28 mm. Hg, is a large force tending to pull fluid into the capillaries. Therefore, there is continual osmotic tendency to dehydrate the interstitial spaces of the lungs. Referring to Chapter 30 again, it is calculated there that the normal interstitial fluid pressure of the human lungs is probably about −6 mm. Hg. That is, there is approximately 6 mm. Hg tending to pull the alveolar epithelial membrane toward the capillary membrane, thus squeezing the pulmonary interstitial space down to almost nothing. Electron micrographic studies have demonstrated exactly this fact, the interstitial space at times being so narrow that the basement membrane of the alveolar epithelium is fused with the basement membrane of the capillary endothelium. As a result, the distance between the air in the alveoli and the blood in the capillaries is minimal, averaging about 0.4 micron in distance; and this obviously allows very rapid diffusion of oxygen and carbon dioxide.

The details of fluid exchange through the pulmonary capillary membrane will be discussed in relation to overall capillary function in Chapter 30.

*Mechanism by which the Alveoli Remain Dry.* Another consequence of the negative pressure in the interstitial spaces is that it pulls fluid from the alveoli through the alveolar membrane and into the interstitial spaces, thereby normally keeping the aveoli dry.

**Pulmonary Edema.** Pulmonary edema means excessive quantities of fluid either in the pulmonary interstitial spaces or in the alveoli. The normal pulmonary interstitial fluid volume is approximately 20 per cent of the lung mass, but this can increase perhaps two-fold in pulmonary edema. In addition, many times this much fluid can enter the alveoli and cause *intra-alveolar edema;* intra-alveolar fluid sometimes reaches 500 to 1000 per cent of the normal interstitial fluid.

*Safety Factor Against Pulmonary Edema.* The most common cause of pulmonary edema is greatly elevated capillary pressure resulting from failure of the left heart and consequent damming of blood in the lungs. However, the pulmonary capillary pressure usually must rise to very high values before serious pulmonary edema will develop. The reason for this is the very high dehydrating force of the colloid osmotic pressure of the blood in the lungs. This effect is illustrated in Figure 24–5, which shows development of lung edema in dogs subjected to progressively increasing left atrial pressure. In this experiment no edema fluid developed in the lungs until left atrial pressure rose above 23 mm. Hg, which was approximately 3 mm. Hg greater than the colloid osmotic pressure of dog blood, about 20 mm. Hg. This experiment demonstrates that the hydrostatic pressure in the pulmonary capillary usually must rise a few mm. Hg above the colloid osmotic pressure before serious pulmonary edema can ensue. In the human being the colloid osmotic pressure is about 28 mm. Hg, so that pulmonary edema will rarely develop below 30 mm. Hg pulmonary capillary pressure. Thus, if the capillary pressure in the lungs is normally 7 mm. Hg and this pressure must usually rise above 30 mm. Hg before edema will occur, the lungs have a *safety factor against edema* of approximately 23 mm. Hg.

*Safety Factor Against Edema in Chronic Elevation of Capillary Pressure.* Patients with chronic elevation of pulmonary capillary pressure occasionally will not develop pulmonary edema even with pulmonary capillary pressures as high as 45 mm. Hg. The reason for this is probably extremely rapid run-off of fluid from the pulmonary interstitial spaces through the lymphatics, because when the pulmonary capillary pressure remains elevated for more than approximately two weeks, the pulmonary lymphatics enlarge as much as 6 to 10-fold, and lymph flow can increase

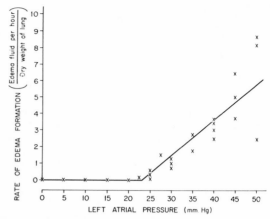

**Figure 24–5.** Rate of fluid loss into the lung tissues when the left atrial pressure (and also pulmonary capillary pressure) is increased.

perhaps 20-fold above the normal resting level. This extra lymph flow gives a safety factor perhaps 15 mm. Hg above that which one normally has against edema.

*Pulmonary Edema as a Result of Capillary Damage.* Pulmonary edema can also result from local capillary damage in the lungs. This effect is often caused by bacterial infection, such as occurs in pneumonia, or by irritant gases such as chlorine, phosgene, lewisite, or mustard gas. All of these directly damage the alveolar epithelium and the endothelium of the capillaries, allowing rapid transudation of both fluid and protein into the alveoli and interstitial spaces.

*Occurrence of Alveolar Fluid in Pulmonary Edema.* Though mild degrees of pulmonary edema can be limited to an increase only of the pulmonary interstitial fluid, serious edema almost always causes fluid transudation into the alveoli themselves. It seems that the alveolar epithelium does not have enough strength to resist any significant degree of positive pressure in the interstitial spaces. Therefore, minute or large ruptures occur in the epithelium, and fluid flows readily out of the interstitial spaces into the alveoli.

## EFFECT OF INCREASED CARDIAC OUTPUT ON THE PULMONARY CIRCULATION DURING HEAVY EXERCISE

During heavy exercise the lungs are frequently called upon to absorb up to 20 times as much oxygen into the blood as they do normally. This absorption is achieved in two ways: (1) by increasing the number of open capillaries so that oxygen can diffuse more readily between the alveolar gas and the blood (a mechanism discussed earlier in the chapter), and (2) by increasing cardiac output, with its concomitant increase in blood flow through the lungs—the blood picking up greater quantities of oxygen each minute.

Fortunately, the cardiac output can increase to four to six times normal before pulmonary arterial pressure becomes excessively elevated; this effect is illustrated in Figure 24–6. As the blood flow into the lungs becomes increased, the pulmonary arterioles and capillaries simply expand and allow the excess flow to pass on through the capillary system without excessive increase in pulmonary arterial pressure. It has already been pointed out that under resting conditions blood flow in large segments of the pulmonary capillary bed is completely dormant; indeed, many of the capillaries have almost no blood in them, particularly capillaries in the upper portions of the lungs. If the pulmonary arterial pressure increases even slightly as the cardiac output rises during exercise, most of these capillaries open so that the pulmonary vascular resistance decreases greatly. For this reason, the pulmonary arterial pressure increases relatively little even during extremely heavy exercise.

This ability of the lungs to accommodate greatly increased blood flow during exercise with relatively little increase in pulmonary vascular pressure is important for at least two reasons: (1) it obviously conserves the energy of the right heart, and (2) it prevents a significant rise in pulmonary capillary pressure and therefore also prevents development of pulmonary edema during the increased cardiac output.

### FUNCTION OF THE PULMONARY CIRCULATION WHEN THE LEFT ATRIAL PRESSURE RISES AS A RESULT OF LEFT HEART FAILURE

When the left atrial pressure is increased, blood becomes dammed up in the pulmonary system. The initial effect is to increase the volume of blood in the pulmonary veins. Then, with further increase in left

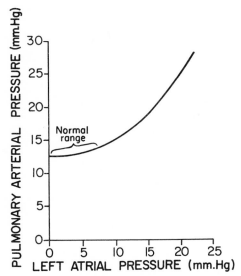

**Figure 24–7.**  Effect of left atrial pressure on pulmonary arterial pressure.

atrial pressure, the capillaries and the pulmonary arteries also become excessively filled with blood, and this damming of blood in the pulmonary arteries eventually raises the pulmonary arterial pressure.

**Quantitative Relationship Between Left Atrial Pressure and Pulmonary Arterial Pressure.**  Figure 24–7 shows the approximate relationship between the level of left atrial pressure and its effect on the pulmonary arterial pressure, provided that the pulmonary blood flow is normal. This figure shows that an increase in left atrial pressure from zero up to about 7 mm. Hg has relatively little effect on the pulmonary arterial pressure. The reason for this is that this moderate increase in left atrial pressure causes more and more opening up of veins, capillaries, and small arteries. This opening of blood vessels progressively reduces the pulmonary vascular resistance, which mainly compensates for the back pressure effect of the rising left atrial pressure. However, once all of these vessels have been opened, they do not expand more with ease. Therefore, any further rise in left atrial pressure causes a marked rise in pulmonary arterial pressure, as shown in the figure, as the atrial pressure rises above 7 mm. Hg.

**Failure of Increasing Load on the Normal Left Ventricle to Affect Significantly the Pulmonary Circulation.**  When the work load of the left ventricle is greatly increased as a result of increased systemic arterial pressure or as a result of increased work output of the heart, the left atrial pressure rises a few mm. Hg—up to perhaps 5 to 6 mm. Hg under most maximal loads. However, as long as the left ventricle does not actually fail in its job of pumping blood, the left atrial pressure will not rise high enough to alter pulmonary circulatory function significantly. Also, Figure 24–7 illustrates that these small increases in left atrial pressure have little back pressure effect on pulmonary arterial pressure and

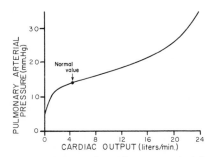

**Figure 24–6.**  Effect on the pulmonary arterial pressure of increasing the cardiac output.

therefore do not measurably increase the load on the right ventricle.

**Right Ventricular Compensation and Failure as the Pulmonary Arterial Pressure Rises.** When the left heart does fail, the left atrial pressure can then rise to values as high as 15 to 40 mm. Hg. The mean pulmonary arterial pressure also rises to high levels, placing a considerably greater than normal pressure load on the right ventricle. Up to 30 to 40 mm. Hg mean pulmonary arterial pressure, the right ventricle continues to pump essentially normal quantities of blood with only a slight rise in right atrial pressure. However, above this pressure the right ventricle also begins to fail, so that further increases in pulmonary arterial pressure cause inordinate increases in right atrial pressure inducing the development of peripheral edema and also decreasing the cardiac output.

## PATHOLOGIC CONDITIONS THAT OBSTRUCT BLOOD FLOW THROUGH THE LUNGS

**Removal of Lung Tissue.** When one *entire lung* is removed, the flow of blood through the remaining lung usually is well within the limits of compensation as long as the patient remains inactive. However, the patient thereafter has far less *pulmonary circulatory reserve* than does the normal person, for if the cardiac output increases more than 100 per cent above normal, his pulmonary arterial pressure begins to rise rapidly. This compares with the normal individual whose pulmonary circulatory reserve is about 300 per cent above normal before the pressure rises seriously.

**Massive Pulmonary Embolism.** One of the most severe postoperative calamities in surgical practice is massive pulmonary embolism. Patients lying immobile in bed tend to develop extensive clots especially in the veins of the legs because of sluggish blood flow. Also, women, after delivery of babies, frequently develop massive clots in the hypogastric veins. Such clots often break away from the initial sites of formation, particularly when the patient first walks after a long period of immobilization. The clots then flow to the right side of the heart and into the pulmonary artery. Such a free-moving clot is called an "embolus."

Total blockage of only one of the major branches of the pulmonary artery usually is not immediately fatal because the opposite lung can accommodate all the blood flow. However, blood clots, as was discussed in Chapter 9, have a tendency to grow. Consequently, the embolus becomes larger and larger, and, as it extends into the other major branch of the pulmonary artery, the few remaining vessels that do not become plugged are taxed beyond their limit, and death ensues because of an inordinate rise in pulmonary arterial pressure and right-sided heart failure.

If the blood is rendered less coagulable by administration of anticoagulants so that the clot cannot grow, the life of the patient who has suffered massive pulmonary embolism can often be saved; also, surgical removal of the clot is sometimes successful.

**Diffuse Pulmonary Embolism.** Sometimes many small blood clot emboli or small emboli of other substances besides blood clots enter the venous blood and eventually clog the vessels of the lungs. One such instance is *fat embolism*. This occurs when a large volume of fatty tissue, such as that in the breast, becomes traumatized or infected so that the fatty material liquefies and enters the venous blood. It passes to the lungs, and because of the high surface tension of fat, the pulmonary arterial pressure often cannot force the fat globules through the small capillary vessels. Consequently, many of the small vessels of the lungs become blocked.

Another instance is *air embolism*. The air, like fat, has a large surface tension at the interface between the blood and air so that the globules of air cannot be deformed enough to be pushed through the capillaries.

The physical effects of diffuse pulmonary embolism on the pulmonary circulatory system are similar to those of massive pulmonary embolism—that is, increased pulmonary circulatory resistance with resulting increase in pulmonary arterial pressure and failure of the right side of the heart. However, with diffuse embolism, the pulmonary vessels appear to develop considerable vasospasm which adds additional resistance to flow besides that caused by the emboli themselves. It is believed that this vasospasm is caused by a sympathetic reflex, and, clinically, the stellate ganglion is often anesthetized to block the reflex.

Patients with this type of embolism, exhibit a rapid respiratory rate because of local irritation by the emboli in the lungs and also because of resultant ischemia throughout the body.

**Emphysema.** Pulmonary emphysema literally means too much air in the lungs, and it is usually characterized by destruction of many of the alveolar walls. This causes the adjacent alveoli to become confluent, thereby forming large *emphysematous cavities* rather than the usual small alveoli. Obviously, loss of the alveolar septa greatly decreases the total alveolar surface area of the lungs and hinders gas exchange between the alveoli and the blood. This effect will be discussed in connection with the gas exchange functions of the lungs in Chapter 43.

Emphysema also has an important effect on the pulmonary vasculature, for each time an alveolar wall is destroyed, some of the small blood vessels of the pulmonary system are also destroyed, thus progressively increasing the pulmonary resistance and elevating the pulmonary arterial pressure.

In addition, a *physiologic increase in vascular resistance* also occurs in emphysema in the following way: Some of the emphysematous alveoli have poor exchange of air with the atmosphere. This poor ventilation results in alveolar hypoxia so that the adjacent blood vessels come continually under the influence of the hypoxic stimulus discussed earlier in the chapter, resulting in chronic vasoconstriction, which compounds the difficulties already caused by the pathologic destruction of pulmonary vessels. Furthermore, the generalized hypoxia throughout the body causes increased cardiac output, and this, combined with the increased pulmonary resistance, sometimes results in serious pulmonary hypertension and right heart failure.

The physiologic increase in resistance can often be treated effectively by simple oxygen therapy. Therefore, emphysematous patients in right heart failure often experience rapid reversal of the failure along with decrease of the dyspneic symptoms after only a few hours of oxygen breathing.

Unfortunately, the prevalence of emphysema is increasing rapidly because of smoking. Therefore, the physiologic principles of its treatment are becoming progressively more important.

**Diffuse Sclerosis of the Lungs.** A number of pathologic conditions cause excessive fibrosis in the supportive tissues in the lungs, and the fibrous tissue in turn contracts around the vessels. Some of these conditions are silicosis, tuberculosis, syphilis, and, to a lesser extent, anthracosis.

In early stages of diffuse sclerosis, the pulmonary arterial pressure is normal as long as the person is not exercising, but just as soon as he performs even mild exercise the pulmonary arterial pressure often rises inordinately because the vessels do not have the ability to expand as much as normal pulmonary vessels do. In late stages of diffuse sclerosis the pulmonary arterial pressure remains elevated constantly; as a result, the right ventricle hypertrophies, and it may fail. The alveolar membranes also thicken, or their surface area decreases so much that the rate of gaseous exchange between the alveoli and the blood is diminished.

**Atelectasis.** Atelectasis is the clinical term for collapse of a lung or part of a lung. This occurs often when the bronchi become plugged because the blood in the capillaries rapidly absorbs the air in the entrapped alveoli, which causes the alveoli to collapse.

Atelectasis also occurs when the chest cavity is opened to atmospheric pressure, for, when air is allowed to enter the pleural space, the elastic nature of the lungs causes them to collapse immediately.

When the elastic tissues of the lungs contract during atelectasis, they constrict not only the alveoli but also the blood vessels. This automatically decreases blood flow in the atelectatic portions of the lungs to about one-fourth normal, shifting the remaining three-fourths to the aerated portions. This is an important safety mechanism, for it prevents flow of major quantities of blood through collapsed, nonaerated pulmonary areas.

# REFERENCES

Aviado, D. M.: The Lung Circulation. Vols. 1 and 2. New York, Pergamon Press, Inc., 1965.

Bakhle, Y. S., and Vane, J. R.: Pharmacokinetic function of the pulmonary circulation. Physiol. Rev., 54:1007, 1974.

Bergofsky, E. H.: Mechanisms underlying vasomotor regulation of regional pulmonary blood flow in normal and disease states. Amer. J. Med., 57:378, 1974.

Cournand, A.: Some aspects of the pulmonary circulation in normal man and in chronic cardiopulmonary diseases. Circulation, 2:641, 1950.

Cumming, G.: The pulmonary circulation. In Guyton, A. C. (ed.): MTP International Review of Science: Physiology. Baltimore, University Park Press, Vol. 1, 1974, p. 93.

Fishman, A. P.: Dietary pulmonary hypertension. Circ. Res., 35:657, 1974.

Gaar, K. A., Taylor, A. E., Owens, J., and Guyton, A. C.: Pulmonary capillary pressure and filtration coefficient in the isolated perfused lung. Amer. J. Physiol., 213:910, 1967.

Guyton, A. C.: Introduction to Part I: Pulmonary alveolar-capillary interface and interstitium. In Fishman, A. P., and Hecht, H. H. (eds.): The Pulmonary Circulation and Interstitial Space. Chicago, University of Chicago Press, 1969, p. 3.

Harlan, J. C., Smith, E. E., and Richardson, T. Q.: Pressure-volume curves of systemic and pulmonary circuits. Amer. J. Physiol, 213:1499, 1967.

Hebb, C.: Motor innervation of the pulmonary blood vessels of mammals. In Fishman, A. P., and Hecht, H. H. (eds.): The Pulmonary Circulation and Interstitial Space. Chicago, University of Chicago Press, 1969, p. 195.

Levine, O. R., Mellins, R. B., and Senior, R. M.: Extravascular lung water and distribution of pulmonary blood flow in the dog. J. Appl. Physiol., 28:166, 1970.

McIntyre, K. M., and Sasahara, A. A.: Hemodynamic and ventricular responses to pulmonary embolism. Prog. Cardiovasc. Dis. 17:175, 1974.

Meyer, B. J., Meyer, A., and Guyton, A. C.: Interstitial fluid pressure V: Negative pressure in the lungs. Circ. Res., 22:263, 1968.

Mobin-Uddin, K. (ed.): Pulmonary Thromboembolism. Springfield, Ill., Charles C Thomas, Publisher, 1974.

Permutt, S., Howell, J. B., Proctor, D. F., and Riley, R. L.: Effect of lung inflation on static pressure-volume characteristics of pulmonary vessels. J. Appl. Physiol., 16:64, 1961.

Racz, G. B.: Pulmonary blood flow in normal and abnormal states. Surg. Clin. North Amer., 54:967, 1974.

Reed, J. H., Jr., and Wood, E. H.: Effect of body position on vertical distribution of pulmonary blood flow. J. Appl. Physiol., 28:303, 1970.

Reeves, J. T.: Pulmonary vascular response to high altitude residence. Cardiovasc. Clin., 5:81, 1973.

Staub, N. C.: Pulmonary edema. Physiol. Rev., 54:678, 1974.

Stein, M., and Levy, S. E.: Reflex and humoral responses to pulmonary embolism. Prog. Cardiovasc. Dis., 17:167, 1974.

Taylor, A. E., Gaar, K., and Guyton, A. C.: Na$^{24}$ space, D$_2$O space, and blood volume in isolated dog lung. Amer. J. Physiol., 211:66, 1966.

Taylor, A. E., Guyton, A. C., and Bishop, V. S.: Permeability of the alveolar membrane to solutes. Circ. Res., 16:353, 1965.

Webb, W. R., Wax, S. D. Kusajima, K. Kamiyama, T. M.: Microscopic studies of the pulmonary circulation in situ. Surg. Clin. North Amer., 54:1067, 1974.

West, J. B.: Ventilation/Blood Flow and Gas Exchange. Oxford, Blackwell Scientific Publications Ltd., 1965.

West, J. B.: Blood flow to the lung and gas exchange. Anesthesiology, 41:124, 1974.

# | 25 |

# The Coronary Circulation and Ischemic Heart Disease

Approximately one third of all deaths result from coronary artery disease, and almost all elderly persons have at least some impairment of coronary artery circulation. For this reason, the normal and pathologic physiology of the coronary circulation is one of the most important subjects in the entire field of medicine. The purpose of this chapter is to present the subject of coronary circulation, emphasizing also the physiology of coronary occlusion and myocardial infarction.

## NORMAL CORONARY BLOOD FLOW AND ITS VARIATIONS

### PHYSIOLOGIC ANATOMY OF THE CORONARY BLOOD SUPPLY

Figure 25–1 illustrates the heart with its coronary blood supply. Note that the main coronary arteries lie on the surface of the heart, and small arteries penetrate into the cardiac muscle mass. It is almost entirely through these arteries that the heart receives its nutritive blood supply. Only the inner 0.5 mm. or less of the muscle mass can obtain nutrition directly from the blood in the cardiac chambers.

The *left coronary artery* supplies mainly the left ventricle, and the *right coronary artery* supplies mainly the right ventricle but usually also a major part of the left ventricle as well. In about one-half of all human beings, more blood flows through the right coronary artery than through the left, whereas the left artery predominates in only 20 per cent.

Most of the venous blood flow from the left ventricle leaves by way of the *coronary sinus*—about 75 per cent of the total coronary blood flow—and most of the venous blood from the right ventricle flows through the small *anterior cardiac veins,* which empty directly into the right atrium and are not connected with the coronary sinus. A small amount of coronary blood flows back into the heart through the *thebesian veins,* which empty directly into all chambers of the heart.

### NORMAL CORONARY BLOOD FLOW

The resting coronary blood flow in the human being averages approximately 225 ml. per minute, which is about 0.8 ml. per gram of heart muscle, or 4 to 5 per cent of the total cardiac output.

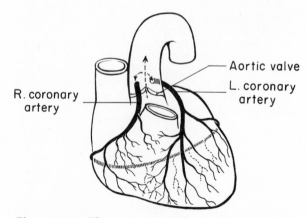

**Figure 25–1.** The coronary vessels.

R. coronary artery

Aortic valve

L. coronary artery

In strenuous exercise the heart increases its cardiac output as much as four- to six-fold, and it pumps this blood against a higher than normal arterial pressure. Consequently, the work output of the heart under severe conditions may increase as much as six- to eight-fold. The coronary blood flow increases four- to five-fold to supply the extra nutrients needed by the heart. Obviously, this increase is not quite as much as the increase in work load, which means that the ratio of coronary blood flow to energy expenditure by the heart decreases. However, the "efficiency" of cardiac contraction increases to make up for this relative deficiency of blood supply.

**Phasic Changes in Coronary Blood Flow—Effect of Cardiac Muscle Compression.** Figure 25–2 illustrates the average blood flow *through the small nutrient vessels* of the coronary system in milliliters per minute in the human heart during systole and diastole as *calculated* from experiments in lower animals. Note from this diagram that the blood flow through the capillaries of the left ventricle falls to a low value during *systole,* which is opposite to the flow in all other vascular beds of the body. The reason for this is the strong compression of the cardiac muscle around the intramuscular vessels during systole.

During *diastole,* the cardiac muscle relaxes completely and no longer obstructs the blood flow through the left ventricular capillaries, so that blood now flows rapidly during all of diastole.

Blood flow through the coronary capillaries of the right ventricle undergoes phasic changes similar to those in the coronary capillaries of

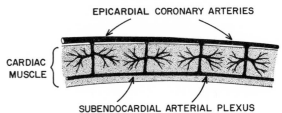

**Figure 25–3.** Diagram showing the epicardial, intramuscular, and subendocardial coronary vasculature.

the left ventricle during the cardiac cycle, but, because the force of contraction of the right ventricle is far less than that of the left ventricle, these phasic changes are relatively mild compared with those in the left ventricle, as is shown in Figure 25–2.

**Epicardial versus Subendocardial Blood Flow—Effect of Intramyocardial Pressure.** During cardiac contraction all the cardiac muscle squeezes toward the centers of the ventricles, creating high pressure in the blood of the ventricles. However, the contraction also increases the pressure in the muscle tissue itself, especially in the subendocardial muscle. That is, the ventricular muscle adjacent to the ventricular chambers (the subendocardial muscle) squeezes the blood in the ventricle; the muscle in the middle layer of the ventricle squeezes the blood in the ventricle and also the subendocardial muscle; and the outermost muscle squeezes both the middle and the subendocardial muscle as well as the blood in the ventricle. Therefore, during systole, a gradient of *intramyocardial pressure* develops, with the pressure in the subendocardial muscle having a pressure almost as great as the pressure inside the ventricle, while the pressure in the outer layer of the heart is only slightly above atmospheric pressure. The importance of this pressure gradient is that the intramyocardial pressure compresses the subendocardial blood vessels far more than it does the outer vessels.

Figure 25–3 illustrates the special arrangement of the coronary vessels at different depths in the heart, showing on the surface of the cardiac muscle the large epicardial coronary arteries that supply the heart. Smaller intramuscular arteries penetrate the muscle, supplying the needed nutriends en route to the endocardium. Then immediately outside the endocardium lies a plexus of subendocardial arteries. During systole, blood flow through the subendocardial plexus of the left ventricle, where the contractile force of the muscle is very great, falls almost to zero. To compensate for this almost lack of flow during systole, the subendocardial arteries are much larger than the nutrient arteries in the middle and outer layers of the heart. Therefore, during diastole, blood flow in the subendocardial arteries is considerably greater than is blood flow in the outermost arteries. Later in the chapter we shall see that this peculiar

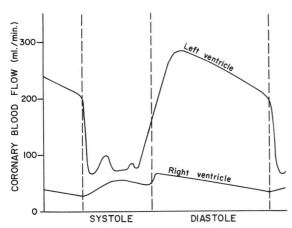

**Figure 25–2.** Phasic flow of blood through the coronary capillaries of the left and right ventricles.

difference between blood flow in the epicardial and subendocardial arteries plays an important role in certain types of coronary ischemia.

# CONTROL OF CORONARY BLOOD FLOW

## LOCAL METABOLISM AS THE PRIMARY CONTROLLER OF CORONARY FLOW

Blood flow through the coronary system is regulated almost entirely by intrinsic vascular response to local needs of the cardiac musculature for nutrition. This mechanism works equally well when the nerves to the heart are intact or are removed. Whenever the vigor of contraction is increased, regardless of cause, the rate of coronary blood flow simultaneously increases, and, conversely, decreased activity is accompanied by decreased coronary flow. It is immediately obvious that this local regulation of blood flow is almost identical with that which occurs in many other tissues, especially in the skeletal muscles of all the body.

**Oxygen Demand as a Major Factor in Local Blood Flow Regulation.** Blood flow in the coronaries is regulated almost exactly in proportion to the need of the cardiac musculature for oxygen. Even in the normal resting state, about 65 per cent of the oxygen in the arterial blood is removed as the blood passes through the heart; and, because not much oxygen is left in the blood, little additional oxygen can be removed from the blood unless the flow increases. Fortunately, the blood flow does increase, and almost directly in proportion to the need of the heart for oxygen. Therefore, it is believed that *oxygen lack dilates the coronary arterioles.*

Yet the means by which oxygen lack causes coronary dilatation has not been determined. The two principal possibilities that have been suggested are: (1) Decreased oxygen tension in the cardiac tissues reduces the amount of oxygen available to the coronary vessels themselves and this causes the coronaries to become weakened and, therefore, to dilate automatically. (2) Oxygen lack causes vasodilator substances, such as adenosine compounds, to be released by the tissues. (Small amounts of adenosine are released during oxygen lack, but it has not been shown that it is released in sufficient quantity to cause the extreme degree of vasodilatation that results. Also, theophylline, which blocks the vasodilator effect of

adenosine, fails to alter significantly the effectiveness of the coronary blood flow regulatory process.) Therefore, for the present we can simply say that oxygen lack is followed by coronary vasodilatation and in this way regulates blood flow to the cardiac musculature in proportion to the metabolic need for oxygen by the muscle fibers.

**The Determinants of Oxygen Consumption.** Since the rate of oxygen consumption is the major factor that determines coronary blood flow, it is important to know the different factors that can alter myocardial oxygen consumption.

In general, the rate of cardiac oxygen consumption is closely related to work performed by the heart—the greater the work, the greater the oxygen consumption, and consequently the greater the coronary flow. However, this is not exactly true for the following reasons: Work output of the heart is determined by the arterial pressure against which the heart is pumping × the cardiac output pumped per minute. When the pressure increases, the oxygen consumption increases almost directly in proportion to the increase in pressure. On the other hand, when the cardiac output increases without an increase in pressure, the oxygen consumption increases only a slight to moderate amount rather than in direct proportion to the increase in cardiac output.

*Tension × Time as the Primary Determinant of Oxygen Consumption.* Perhaps the best relationship that has yet been found between cardiac function and oxygen consumption is: *oxygen consumption is proportional to myocardial muscle tension × time of contraction.* Thus, when the arterial pressure rises, the muscle tension increases and oxygen consumption also increases. Likewise, when the heart dilates, which makes it necessary (because of the law of Laplace) for the muscle to generate increased tension to pump against even a normal arterial pressure, oxygen consumption also increases even though the work output of the heart does not increase.

*Other Causes of Increased Oxygen Consumption.* Other factors that increase cardiac oxygen consumption are stimulation of the heart by epinephrine, norepinephrine, thyroxine, digitalis, calcium ions, or increased temperature of the heart. All these factors increase the metabolic activity of the cardiac muscle fibers themselves, which in turn increases the rate of oxygen usage even though they might not increase the work output of the heart. They also increase coronary blood flow approximately in proportion to the increase in rate of oxygen usage.

## Importance of the Increase in Coronary Blood Flow in Response to Myocardial Oxygen Usage.

The resting heart extracts most of the oxygen from the coronary blood as it flows through the heart muscle, and very little of the heart's oxygen need can be met by additional extraction of oxygen from the coronary blood. Therefore, the only significant way in which the heart can be supplied with additional amounts of oxygen is through an increase in blood flow. Consequently, it is essential that the coronary blood flow increase whenever the heart muscle demands additional oxygen. When the coronary blood flow fails to increase appropriately, the strength of the muscle diminishes rapidly and drastically, often causing acute heart failure. Also, this *relative* ischemia of the muscle can cause severe pain, called *anginal pain,* which will be discussed later in the chapter.

## Other Factors That Affect Local Regulation of Coronary Blood Flow.

Though the preponderance of evidence indicates that coronary blood flow is determined mainly by the heart's need for oxygen, other factors that affect coronary flow are: (1) carbon dioxide released from the metabolizing muscle, (2) lactate and pyruvate ions released from the muscle, (3) potassium ions released during muscle activity, and (4) adenosine (discussed above in relation to cardiac muscle hypoxia). The role of each of these is presumably almost the same as in other tissues. These factors were discussed in much more detail in Chapter 20.

## Reactive Hyperemia in the Coronary System.

If the coronary flow to the heart is completely occluded for a few seconds to a few minutes and then suddenly unoccluded, the blood flow increases to as high as six times normal, as shown in Figure 25–4. It remains high for a few seconds to several minutes, depending on the period of occlusion. This extra flow of blood is called reactive hyperemia, which was explained in Chapter 20. During the period of excess flow, the heart removes a large amount of extra oxygen from the blood, as shown by the bottom curve of the figure, to make up for the deficiency of oxygen during the period of occlusion. Reactive hyperemia is simply another manifestation of the ability of the coronary system to adjust its flow to the metabolic needs of the heart.

## NERVOUS CONTROL OF CORONARY BLOOD FLOW

Stimulation of the autonomic nerves to the heart can affect coronary blood flow in two ways—directly and indirectly. The direct effects result from direct action of the nervous transmitter substances, acetylcholine and norepinephrine, on the coronary vessels themselves. The indirect effects result from secondary changes in coronary blood flow caused by increased or decreased activity of the heart.

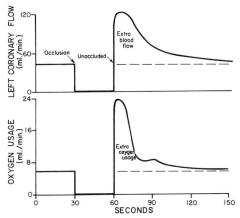

**Figure 25–4.** Reactive hyperemia in the coronary system caused by a 30 second period of coronary occlusion. Note the extra blood flow called "reactive hyperemia" and the extra oxygen usage after the period of occlusion was over. (Reconstructed from data in Gregg and Fisher: Handbook of Physiology, Section II, Vol. I. The Williams & Wilkins Co.)

The indirect effects play the more important role in control of coronary blood flow. Thus, sympathetic stimulation increases both heart rate and heart contractility as well as its rate of metabolism. In turn, the increased activity of the heart sets off local blood flow regulatory mechanisms for increasing coronary blood flow, the blood flow increasing approximately in proportion to the metabolic needs of the heart muscle. In contrast, parasympathetic stimulation slows the heart and also has a slight depressive effect on cardiac contractility. Both of these effects decrease cardiac oxygen consumption and therefore also reduce the coronary blood flow.

## Direct Effects of Nervous Stimuli on the Coronary Vasculature.

The distribution of parasympathetic (vagal) nerve fibers to the ventricular coronary system is so slight that parasympathetic stimulation has almost negligible direct effect on coronary blood flow. On the other hand, there is extensive sympathetic innervation of the coronary vessels. In Chapter 56 we shall see that the sympathetic transmitter substances, norepinephrine and epinephrine, can have either dilator or constrictor effects depending on the presence or absence of specific receptors in the blood vessel walls. The constrictor receptors are called *alpha receptors,* and the dilator receptors are called *beta receptors.* Both alpha and beta receptors are known to exist in the coronary vessels. In general, the epicardial coronary vessels have a preponderance of alpha receptors, whereas the intramuscular arteries have a preponderance of beta receptors. Therefore, sympathetic stimulation can cause moderate degrees of constriction of the epicardial arteries but simultaneous moderate dilatation of the intramuscular arteries. Yet, since most of the resistance to blood flow occurs in the intramuscular arteries, the overall net effect of sympathetic stimulation is moderate dilatation of the coronary vasculature and therefore moderate increase in blood flow. On the other hand, it is believed that alpha constriction of the epicardial arteries can predominate under some conditions, and that this can cause temporary myocardial ischemia, with resultant anginal pain.

It must be pointed out again, however, that metabolic factors—especially the myocardial oxygen need—are the major controllers of myocardial blood flow. Therefore, whenever nervous stimulation alters the coronary blood flow, the metabolic factors usually within seconds return the coronary blood flow most of the way back toward normal.

## THE SUBSTRATES OF CARDIAC METABOLISM

The general principles of cellular metabolism, which will be discussed in Chapters 67 through 70, apply to cardiac muscle, though there are some quantitative differences. Most importantly, cardiac muscle normally utilizes fats mainly for its energy, with approximately 75 per cent of the normal metabolism being derived from fatty acids. However, as is true of other tissues, under anaerobic or ischemic conditions cardiac metabolism shifts mainly to the anaerobic glycolysis mechanism for its energy. This, unfortunately, utilizes tremendous quantities of the blood glucose and at the same time forms large amounts of lactic acid in the cardiac tissue, which is probably one of the causes of cardiac pain in cardiac ischemic conditions, as will be discussed later in the chapter.

As is true in other tissues, more than 95 per cent of the metabolic energy liberated from the foods is used to form ATP in the mitochondria. This ATP in turn transfers its energy through the mitochondrial membrane to ATP and creatinine phosphate in the hyaloplasm. Finally, these latter two substances supply the energy for cellular function. In coronary ischemia, the ATP degrades to ADP, AMP, and adenosine. Because the cell membrane is permeable to adenosine, much of this is rapidly lost from the hyaloplasm into the circulating blood. This released adenosine is believed by many physiologists to be one of the substances that causes dilatation of the coronary arterioles during coronary hypoxia. However, the loss of adenosine also has a very serious cellular consequence as well. Within as little as half an hour of severe coronary ischemia, such as can occur after a myocardial infarct or during cardiac arrest, essentially all of the adenine base can be lost from the cardiac cellular hyaloplasm. Furthermore, this can be replaced by a new synthesis of adenosine at a rate of only 2 per cent per hour. Therefore, once a serious bout of ischemia has persisted for half an hour or so, relief of the coronary ischemia may be too late to save the lives of the cardiac cells. This is almost certainly one of the major causes of cardiac cellular death following myocardial ischemia and also one of the most important causes of cardiac debility in the late stages of circulatory shock, as will be discussed in Chapter 28.

# ISCHEMIC HEART DISEASE

The single most common cause of death is ischemic heart disease, which results from insufficient coronary blood flow. Approximately 35 per cent of all human beings in the United States die of this cause. Some deaths occur suddenly as a result of an acute coronary occlusion or of fibrillation of the heart, whereas others occur slowly over a period of weeks to years as a result of progressive weakening of the heart pumping process. In the present chapter we will discuss the coronary ischemia problem itself, as well as acute coronary occlusion and myocardial infarction. In the following chapter we will discuss congestive heart failure, the most frequent cause of which is progressive coronary ischemia.

**Atherosclerosis as the Cause of Ischemic Heart Disease.** The most frequent cause of diminished coronary blood flow is atherosclerosis. Although the atherosclerotic process will be discussed in connection with lipid metabolism in Chapter 68, briefly, this process is the following: In certain persons who have a genetic predisposition to atherosclerosis or in persons who eat excessive quantities of cholesterol and other fats, the cholesterol and fats gradually become deposited beneath the intima at many points in the arteries. Later, these areas of deposit become invaded by fibrous tissue, and they also frequently become calcified. The net result is the development of both "atherosclerotic plaques" and very stiff arterial walls that can be neither constricted nor dilated. The internal diameter of the artery almost always becomes greatly diminished, thereby decreasing the rate of blood flow.

A very common site for development of atherosclerotic plaques is the first few centimeters of the coronary arteries. Therefore, a usual sequela of atherosclerosis is ischemic heart disease.

**Acute Coronary Occlusion.** Acute occlusion of the coronary artery frequently occurs in a person who already has serious underlying atherosclerotic coronary heart disease, but almost never in a person with a normal coronary circulation. This condition can result from any one of several different effects, as follows:

1. The atherosclerotic plaque can cause a local blood clot called a *thrombus,* which in turn occludes the artery. The thrombus usually begins where the plaque has grown so much that it has broken through the intima, thus coming in contact with the flowing blood. Because the plaque presents an unsmooth surface to the blood, platelets begin to adhere to it, fibrin

begins to be deposited, and blood cells become entrapped and form a clot that grows until it occludes the vessel. Often the clot breaks away from its attachment on the atherosclerotic plaque and flows to a more peripheral branch of the coronary arterial tree where it blocks the artery at that point. A thrombus that flows along the artery in this way and occludes the vessel more distally is called an *embolus*.

2. Often a small nutrient artery supplying blood to the arterial wall adjacent to the atherosclerotic plaque hemorrhages into the plaque and causes the plaque to protrude further into the lumen of the artery. This protrusion can decrease the arterial blood flow to a level considerably below normal and perhaps occasionally can totally occlude the artery.

3. It is believed by many clinicians that local spasm of a coronary artery can also cause sudden occlusion. The spasm might result from irritation of the smooth muscle of the arterial wall by the edges of an arteriosclerotic plaque, or it might result from nervous reflexes that cause the involuntary contraction.

**Collateral Circulation in the Heart.** The degree of damage to the heart caused either by slowly developing atherosclerotic constriction of the coronary arteries or by sudden occlusion is determined to a great extent by the degree of collateral circulation that is already developed or that can develop within a short period of time after the occlusion.

Unfortunately, in a normal heart, relatively few communications exist among the larger coronary arteries. But there do exist many anastomoses among the smaller arteries of sizes 20 to 250 microns in diameter, as shown in Figure 25–5.

When a sudden occlusion occurs in one of the larger coronary arteries, the sizes of the minute anastomoses increase to their maximum physical diameters within a few seconds. The blood flow through these minute collaterals is less than one-half that needed to keep alive the cardiac muscle that they supply; and unfortunately the diameters of the collateral vessels do not enlarge further for the next 8 to 24 hours. But then collateral flow begins to increase, doubling in the next one to two days, often reaching normal or almost normal coronary supply within about one month; and it is capable even of increasing further with increased metabolic loads. It is because of these developing collateral channels that a patient recovers from the various types of coronary occlusion.

The coronary arteries also anastomose through minute channels with the arteries of the parietal pericardium, these anastomoses traveling along the pericardial reflections around the major vessels of the heart. It will be recalled that the heart floats within the pericardial cavity, and the only possible communication between the coronary vessels and other vessels outside the pericardium would be by way of these reflections. A few patients at autopsy have been found to have both major coronary arteries totally occluded, and yet the small anastomotic vessels in the pericardial reflections have become large enough to supply the entire heart with blood.

When the coronary arteries are slowly constricted over a period of many years rather than suddenly, collateral vessels can develop at the same time that the atherosclerosis does. Therefore, the person may never experience an acute episode of cardiac dysfunction. Eventually, however, the sclerotic process develops beyond the limits of even the collateral blood supply to provide the needed blood flow. When this occurs, the heart muscle becomes severely limited in its work output, sometimes so much so that the heart cannot pump even the normally required amounts of blood flow. This is the most common cause of cardiac failure.

## MYOCARDIAL INFARCTION

Immediately after an acute coronary occlusion, blood flow ceases in the coronary vessels beyond the occlusion except for small amounts of collateral flow from surrounding vessels. The area of muscle that has either zero flow or so little flow that it cannot sustain cardiac muscle function is said to be *infarcted*. The overall process is called a *myocardial infarction*.

Soon after the onset of the infarction small amounts of collateral blood flow seep into the infarcted area, and this, combined with progressive dilatation of the local blood vessels, causes the area to become overfilled with stagnant blood. Simultaneously, the muscle fibers utilize the last vestiges of the oxygen in the blood, causing the hemoglobin to become totally reduced and very dark blue in color. Therefore, the infarcted area takes on a dark bluish hue, and the blood vessels of the area appear to be

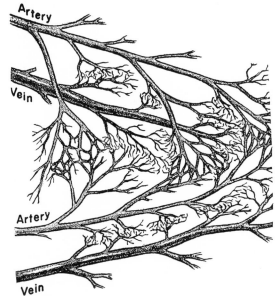

**Figure 25–5.** Minute anastomoses of the coronary arterial system.

engorged despite the lack of blood flow. In later stages, the vessels become highly permeable and leak fluid, the tissue becomes edematous, and the cardiac muscle cells begin to swell because of diminished cellular metabolism. Finally, many of the cells die.

Cardiac muscle requires approximately 1.3 ml. of oxygen per 100 grams of muscle tissue per minute simply to remain alive. This is in comparison with approximately 8 ml. of oxygen per 100 grams delivered to the normal resting heart each minute. Therefore, if there is even as much as 10 to 15 per cent of normal resting coronary blood flow, the muscle will not die. In the central portion of a large infarct, however, the blood flow is usually even less than this so that this muscle does die.

**Myocardial Infarction Caused by Myocardial Ischemia but WITHOUT Coronary Occlusion.** In studying the hearts of patients who have died from myocardial infarctions, pathologists have found many patients with no evident signs of occlusion. Therefore, it is believed that such patients develop myocardial infarctions simply because of reduced coronary perfusion but not because of occlusion. It is presumed that a vicious cycle develops, as follows:

1. The first event is that coronary perfusion of an isolated area in the heart becomes so low that some of the heart muscle becomes nonfunctional.

2. The nonfunctioning muscle causes diminished ventricular pumping, which in turn causes increased stretch of the surrounding muscle and resultant increase in work load and oxygen consumption. Also, the blood vessels in the nonfunctional muscle dilate and "steal" blood flow from the surrounding muscle. Consequently, because of greater oxygen need but lesser oxygen supply this surrounding muscle also becomes nonfunctional if it too has some limitation of coronary blood flow.

3. Once the vicious cycle has been initiated, the process continues until all the heart muscle in the area where the blood supply is poor has become nonfunctional and infarcted.

It is also possible that sympathetic nerve reflexes enter into this vicious cycle in the following way: When some of the muscle becomes infarcted and the output of the heart diminishes, sympathetic reflexes tend to drive the heart to greater pumping capacity. In so doing, the muscle requires still far more oxygen, which makes the muscle in the infarcting area progressively more ischemic.

Thus, it is possible—if not highly probable—that many myocardial infarctions are caused in some manner such as that described rather than by frank acute occlusion of the coronary artery. However, infarction will almost never occur unless the internal diameter of one or more of the major coronary arteries has been reduced by at least 75 per cent. This emphasizes the fact that pre-existing coronary artery stenosis is generally a prerequisite to the occurrence of myocardial infarction.

**Subendocardial Myocardial Infarction.** Myocardial infarction frequently occurs in the sub-endocardial muscle even when the epicardial portions of the heart muscle remain uninfarcted and also in the absence of demonstrable occlusion. This form of infarction especially occurs when the diastolic arterial pressure is very low or when the diastolic intraventricular pressure is very high. From the discussion earlier in the chapter of the subendocardial arterial plexus, it was pointed out that most blood flow into this area of the heart occurs during diastole. Therefore, when the diastolic arterial pressure is very low—as occurs in patients who have aortic regurgitation, patent ductus arteriosus, or to a lesser extent arteriosclerosis—one can expect a high incidence of subendocardial myocardial infarction.

Likewise, a high intraventricular pressure during diastole compresses the subendocardial arteries and also reduces blood flow during the diastolic interval. In patients who already have severe ischemic coronary arterial lesions, this can precipitate the vicious cycle of myocardial infarction.

## CAUSES OF DEATH FOLLOWING ACUTE CORONARY OCCLUSION

The four major causes of death following acute myocardial infarction are decreased cardiac output; damming of blood in the pulmonary or systemic veins with death resulting from edema, especially pulmonary edema; fibrillation of the heart; and, occasionally, rupture of the heart.

**Decreased Cardiac Output—Systolic Stretch.** When some of the cardiac muscular fibers are not functioning at all and others are too weak to contract with great force, the overall pumping ability of the affected ventricle is proportionally depressed. Indeed, the overall pumping strength of the heart is often decreased more than one might expect because of the phenomenon of *systolic stretch,* which is illustrated in Figure 25–6. When the normal portions of the ventricular muscle contract, the ischemic muscle, whether this be dead or simply nonfunctional, instead of contracting is actually forced outward by the pressure that develops inside the ventricle. If this area did not stretch, the pumping force of the ventricle would be directly proportional to the strength of contraction of the remainder of the ventricle. However, owing to stretch of the ischemic area, much of the pumping force of the ventricle is dissipated by bulging of the area of nonfunctional cardiac muscle.

When the heart becomes incapable of contracting with sufficient force to pump enough blood into the arterial tree, cardiac failure and death of the peripheral tissues ensue as a result of peripheral ischemia. This condition is called *coronary shock, cardiac shock,* or *low cardiac output failure.* It will be discussed in the following chapter.

**Damming of Blood in the Venous System.** When the heart is not pumping blood forward, it must be damming blood in the venous system of the lungs or the systemic circulation. In the early stages of myocardial infarction this is not usually as serious as it is later, for, as will be discussed in the following

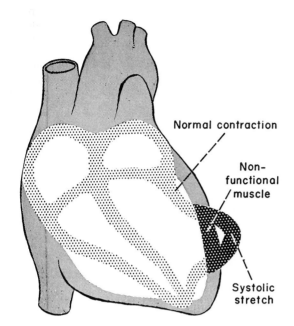

**Figure 25–6.**  Systolic stretch in an area of ischemic cardiac muscle.

chapter in relation to cardiac failure, during the first few days after a myocardial infarction more and more fluid collects in the body because of renal shutdown and adds progressively to the venous congestive symptoms. When the congestion becomes intense, death frequently results from pulmonary edema or, rarely, from systemic congestive symptoms.

**Rupture of the Infarcted Area.**  During the first day of an acute infarct there is little danger of rupture of the ischemic portion of the heart, but a few days after a large infarct occurs, the dead muscle fibers begin to degenerate, and the dead heart musculature is likely to become very thin. If this happens, the degree of systolic stretch becomes greater and greater until finally the heart ruptures. In fact, one of the means used in assessing the progress of a severe myocardial infarction is to record by x-ray whether the degree of systolic stretch is getting worse or better.

When a ventricle does rupture, the loss of blood into the pericardial space causes rapid development of *cardiac tamponade*—that is, compression of the heart from the outside by blood collecting in the pericardial cavity. Because the heart is compressed, blood cannot flow into the right atrium with ease, and the patient dies of suddenly decreased cardiac output.

**Fibrillation of the Ventricles Following Myocardial Infarction.**  Many persons who die of coronary occlusion die because of ventricular fibrillation. The tendency to develop fibrillation is especially great following a large infarction, but fibrillation occasionally occurs even following a small occlusion.

There are two especially dangerous periods during which fibrillation is most likely to occur. The first of these is during the first 10 minutes after the infarction occurs. Then there is a period of relative safety, followed by a secondary period of cardiac irritability beginning three to five hours after the infarction and lasting for hours to days thereafter.

At least four different factors enter into the tendency for the heart to fibrillate:

First, acute loss of blood supply to the cardiac muscle causes rapid depletion of potassium from the ischemic musculature. This increases the potassium concentration in the extracellular fluids surrounding the cardiac muscle fibers. Experiments in which potassium has been injected into the coronary system have demonstrated that an elevated extracellular potassium concentration increases the irritability of the cardiac musculature.

Second, ischemia of the muscle causes an "injury current," which was described in Chapter 16 in relation to electrocardiograms in patients with acute myocardial infarction. The ischemic musculature cannot repolarize its membranes so that this muscle remains negative with respect to the normal polarized cardiac muscle membrane. Therefore, electrical current flows from this ischemic area of the heart to the normal area and can elicit abnormal impulses which can cause fibrillation.

Third, powerful sympathetic reflexes develop following massive infarction, principally because the heart does not pump an adequate volume of blood into the arterial tree. The sympathetic stimulation also increases the irritability of the cardiac muscle and thereby predisposes to fibrillation.

Fourth, the myocardial infarction itself often causes the ventricle to dilate excessively. This increases the pathway length for impulse conduction in the heart and also frequently causes abnormal conduction pathways around the infarcted area of the cardiac muscle. Both of these effects predispose to development of circus movements. As was discussed in Chapter 14, any prolongation of the conduction pathway in the ventricles will allow an impulse to re-enter muscle that is already recovering from refractoriness, thereby initiating a subsequent cycle of excitation and causing the process to continue on and on.

## THE STAGES OF RECOVERY FROM ACUTE MYOCARDIAL INFARCTION

The upper part of Figure 25–7 illustrates the effects of acute coronary occlusion, on the left, in a patient with a small area of muscle ischemia and, on the right, in a patient with a large area of ischemia. When the area of ischemia is small, little or no death of the muscle cells may occur, but part of the muscle often does become temporarily nonfunctional because of inadequate nutrition to support muscle contraction.

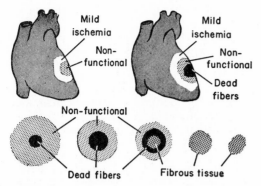

**Figure 25–7.** *Top:* small and large areas of coronary ischemia. *Bottom:* stages of recovery from myocardial infarction.

When the area of ischemia is large, some of the muscle fibers in the very center of the area die rapidly, within about one hour in an area of total cessation of coronary blood supply. Immediately around the dead area is a nonfunctional area because of failure of impulse conduction or of contraction. Then, extending circumferentially around the nonfunctional area, is an area that is still contracting, but weakly so because of mild ischemia. In this mildly ischemic area the small arterial collaterals are sufficient to supply the cardiac musculature, provided that the patient is kept at rest, but even this area may become nonfunctional if the coronary blood flow is diverted to normal musculature during exercise. This illustrates the necessity for rest in patients following coronary heart attacks.

**Replacement of Dead Muscle by Scar Tissue.** In the lower part of Figure 25–7, the various stages of recovery following a myocardial infarction are illustrated. Shortly after the occlusion, the muscle fibers die in the very center of the ischemic area. Then during the ensuing days, this area of dead fibers grows because many of the marginal fibers finally succumb to the prolonged ischemia. At the same time, owing to the enlargement of the collateral arterial channels growing into the outer rim of the infarcted area, the nonfunctional area of muscle becomes smaller and smaller. After approximately two to three weeks most of the nonfunctional area of muscle becomes functional again or dead—one or the other. In the meantime, fibrous tissue begins developing among the dead fibers, for ischemia stimulates growth of fibroblasts and promotes development of greater than normal quantities of fibrous tissue. Therefore, the dead muscular tissue is gradually replaced by fibrous tissue. Then, because it is a general property of fibrous tissue to undergo progressive elastomeric contraction and dissolution, the size of the fibrous scar becomes smaller and smaller over a period of several months to a year.

During progressive recovery of the infarcted area of the heart, the development of a strong, fibrous scar which becomes smaller and smaller finally stops

the original systolic stretch, and the functional musculature once again becomes capable of exerting its entire force for pumping blood rather than for stretching the dead area of the heart. Furthermore, the normal areas of the heart gradually hypertrophy to compensate at least partially for the lost cardiac musculature. By these means the heart recovers.

**Value of Rest in Treating Myocardial Infarction.** The degree of cellular death is determined by the *degree of ischemia × the degree of metabolism* of the heart muscle. When the metabolism of the heart muscle is greatly increased, such as during exercise, in severe emotional strain, or as a result of fatigue, the heart musculature needs increased oxygen and other nutrients for sustaining its life. Furthermore, anastomotic blood vessels that supply blood to ischemic areas of the heart must still supply the areas of the heart that they normally supply. When the heart becomes excessively active, the vessels of the normal musculature become greatly dilated, and this allows most of the blood flowing into the coronary vessels to flow through the normal muscle tissue, thus leaving little blood to flow through the small anastomotic channels into the ischemic areas. This condition is called the "coronary steal" syndrome. Consequently, one of the most important factors in the treatment of a patient with myocardial infarction is observance of absolute rest. The greater the degree of rest during the first few weeks following sudden infarction, the less the degree of cellular death.

### FUNCTION OF THE HEART FOLLOWING RECOVERY FROM MYOCARDIAL INFARCTION

Occasionally, a heart that has recovered from a large myocardial infarction returns to full functional capability, but more frequently its pumping capability is decreased below that of a normal heart. This does not mean that the person is necessarily a cardiac invalid or that his cardiac output is depressed below normal, because the normal person's heart is capable of pumping 400 per cent more blood per minute than his body requires—that is, he has a "cardiac reserve" of 400 per cent. Even when the cardiac reserve is reduced to as little as 100 per cent, the person can still perform normal activity of a quiet, restful type, but he cannot perform strenuous exercise that would overload his heart.

The effect of myocardial infarction and other myocardial debilities on cardiac output is discussed in the following chapter in relation to the regulation of cardiac output in cardiac failure.

## PAIN IN CORONARY DISEASE

Normally, a person cannot "feel" his heart, but ischemic cardiac muscle does exhibit pain sensation. Exactly what causes this pain is not known, but it is believed that ischemia causes the muscle to release acidic substances such as lactic acid or other pain-promoting products such as histamine or kinins that are not removed rapidly enough by the slowly moving blood. The high concentrations of these abnormal products then stimulate the pain endings in the cardiac muscles, and pain impulses are conducted through the sympathetic afferent nerve fibers into the central nervous system.

## ANGINA PECTORIS

In most persons who develop progressive constriction of their coronary arteries, cardiac pain, called *angina pectoris,* begins to appear whenever the load on the heart becomes too great in relation to the coronary blood flow. This pain is usually felt beneath the upper sternum and is often also transferred to surface areas of the body, most often to the left arm and left shoulder but also frequently to the neck and even to the side of the face or to the opposite arm and shoulder. The reason for this distribution of pain is that the heart originates during embryonic life in the neck, as do the arms. Therefore, both of these structures receive pain nerve fibers from the same spinal cord segments.

In general, most persons who have chronic angina pectoris feel the pain only when they exercise or when they experience emotions that increase metabolism of the heart. Usually the pain lasts for only a few minutes. However, some patients have such severe and lasting ischemia that the pain is present all the time. The pain is frequently described as dull, pressing, and constricting; it is of such quality that it usually makes the patient stop whatever he is doing and come to a complete state of rest.

When a person has frequent attacks of angina, his likelihood of developing an acute coronary occlusion is generally very great.

**Treatment with Vasodilator Drugs.** Several vasodilator drugs, when administered during an acute anginal attack, will often give immediate relief from the pain. Two commonly used drugs are nitroglycerin and amyl nitrite. The original theory for use of these drugs was that they caused coronary vasodilatation and that this relieved the pain by increasing coronary blood flow. However, in the normal person the drugs have almost no effect on dilatation of the cornary arterioles and do not cause a measurable increase in the coronary blood flow. But they do have two other effects that probably explain their efficacy in relieving anginal pain. First, they cause marked venous dilatation throughout the body, which reduces venous return to the heart, cardiac output, and arterial pressure. Therefore, the work load of the heart is immediately diminished, thereby decreasing the formation of the ischemic products that are presumed to be the cause of the pain. Second, even though these drugs do not dilate the coronary arterioles, they do dilate the large epicardial coronary arteries. In the person who has severely constricted epicardial arteries, this slight dilatation

can probably be of additional benefit in relieving the anginal pain even though in the normal heart such dilatation causes no measurable increase in coronary blood flow.

**Treatment with Sympathetic Blocking Drugs.** Another medical procedure for relieving anginal pain is to block the sympathetic nervous stimulation of the heart. This is most frequently accomplished by administering *propanolol,* which blocks the beta adrenergic receptors. It is stimulation of these receptors that enhances cardiac activity. Therefore, when they are blocked, the level of cardiac muscle contractility decreases, and the cardiac work output also decreases; at the same time, the heart's usage of oxygen diminishes as much as 20 per cent. In this way, the ratio of oxygen need to oxygen supply is reduced enough to eliminate the pain.

## SURGICAL TREATMENT OF CORONARY DISEASE

**Treatment of the Pain of Angina Pectoris.** For treating intractable angina pectoris the sympathetic chain is occasionally removed from T-2 through T-5 to block the pathway of cardiac pain fibers into the spinal cord. Such an operation occasionally is successful when performed only on the left side, but often it is necessary to remove the sympathetic chain on both sides of the vertebral column.

**Treatment of Ischemia—Aortic-Coronary By-Pass Surgery.** Various attempts have been made to create a new blood supply for the heart. In many patients with coronary ischemia, the constricted areas of the coronary vessels are located at only a few discrete points, and the coronary vessels beyond these points are of normal or almost normal size. A surgical procedure has been developed in the past few years, called *aortic-coronary by-pass,* for anastomosing small vein grafts to the aorta and to the sides of the more peripheral coronary vessels. Usually, one to three such grafts are performed during the operation, each of which supplies a peripheral coronary artery beyond a block. However, on occasion, as many as five grafts have been inserted in a single person. The vein that is used for the graft is usually the long superficial saphenous vein removed from the leg of the recipient.

The acute results from this type of surgery have been especially good, causing this to be the most common cardiac operation performed. Anginal pain is relieved in most patients. Unfortunately, it is still too early to determine whether the operative procedure will prolong the lives of the patients because other complications, such as secondary closure of the grafts, may in the long run be more detrimental than the original disease.

## MEASUREMENT OF CORONARY BLOOD FLOW IN MAN

Obviously, it would be almost impossible to insert a flowmeter into the coronary circulatory system of man. However, a satisfactory indirect method for measuring coronary flow has been achieved, the principles of which are the following:

The blood flow through any organ of the body can be determined by means of the so-called Fick principle, which was discussed in Chapter 23 in relation to the measurement of cardiac output. That is, if a constituent of the blood is removed as it flows through an organ, then the blood flow can be determined from two measurements: first, the amount removed from each milliliter of blood as it goes through the organ and, second, the total quantity of the substance that is removed by the organ in a given period of time. The total blood flow during the period of measurement can be calculated by dividing the amount of substance removed from each milliliter of blood into the total quantity removed.

In measuring blood flow through the heart the subject suddenly begins to breathe nitrous oxide of a given concentration. Samples of arterial blood are removed from any artery in the body and samples of venous blood are obtained from the coronary sinus through a catheter that has been passed into the sinus by way of the venous system. The concentrations of the nitrous oxide in both of these bloods are then plotted for 10 minutes, as shown in Figure 25–8. The quantity of nitrous oxide lost to the heart by each milliliter of blood as it passes through the heart at each instant of the experiment can be determined from the difference in height of the two curves. This is called the *A-V difference*. The A-V difference is determined each minute for a period of 10 minutes and then averaged.

Now, to determine the amount of nitrous oxide that is removed from the blood by the cardiac muscle, one proceeds as follows: It will be noted from the curves of Figure 25–8 that the concentrations of nitrous oxide begin to reach an equilibrium level toward the end of the 10 minute experiment. It is also known that the total amount of nitrous oxide absorbed by the cardiac muscle is directly proportional to its concentration in the arterial blood at the end of the 10 minute run. A *proportionality factor* has been determined in experiments on isolated cardiac muscle so that at the end of 10 minutes the total amount of nitrous oxide absorbed by each 100 grams of cardiac muscle can be determined by multiplying the arterial concentration of nitrous oxide times the proportionality factor.

Once the average *A-V difference* and the *amount of nitrous oxide absorbed* by each 100 grams of heart muscle have been determined, the coronary blood flow per 100 grams of muscle is then calculated by dividing the average A-V difference into the quantity absorbed per minute. As an example, if 8 ml. of nitrous oxide is absorbed by 100 grams of heart in 10 minutes and the average A-V difference during this period of time is 0.01 ml. for each ml. of blood, a total of 800 ml. of blood would have passed through the 100 grams of heart during the 10 minute interval. In other words, the blood flow would be 80 ml. per 100 grams of heart per minute. Assuming a heart mass of 280 grams, the blood flow would be 224 ml. per minute.

# REFERENCES

Amsterdam, E. A., Hughes, J. L., DeMaria, A. N., Zelis, R., and Mason, D. T.: Indirect assessment of myocardial oxygen consumption in the evaluation of mechanisms and therapy of angina pectoris. *Amer. J. Cardiol.*, 33:737, 1974.

Baltaxe, H. A., Amplatz, K., and Levin, D. C.: Coronary Angiography. Springfield, Ill., Charles C Thomas, Publisher, 1974.

Becker, L., and Pitt, B.: Regional myocardial blood flow, ischemia and antianginal drugs. *Ann. Clin. Res.*, 3:353, 1971.

Bell, J. R., and Fox, A. C.: Pathogenesis of subendocardial ischemia. *Amer. J. Med. Sci.* 268:3, 1974.

Beneken, J. E. W., Guyton, A. C., and Sagawa, K.: Coronary perfusion pressure and left ventricular function. *Pflugers Arch.*, 305:76, 1969.

Berne, R. M., and Rubio, R.: Regulation of coronary blood flow. *Adv. Cardiol.*, 12:303, 1974.

Berne, R. M., and Rubio, R.: Adenine nucleotide metabolism in the heart. *Circ. Res.*, 35:109, 1974.

Bloor, C. M.: Functional significance of the coronary collateral circulation. A review. *Amer. J. Pathol.*, 76:561, 1974.

Chandler, A. B.: Mechanisms and frequency of thrombosis in the coronary circulation. *Thromb. Res.*, 4 (Suppl. 1): 3, 1974.

Cosby, R. S., Giddings, J. A., and See, J. R.: Coronary collateral circulation. *Chest*, 66:27, 1974.

Fox, S. M., 3rd., Naughton, J. P., and Gorman, P. A.: Physical activity and cardiovascular health. I. Potential for prevention of coronary heart disease and possible mechanisms. *Mod. Concepts Cardiovasc. Dis.*, 41:17, 1972.

Gensini, G. G. (ed.): The Study of the Systemic, Coronary, and Myocardial Effects of Nitrates. Springfield, Ill., Charles C Thomas, Publisher, 1972.

Gregg, D. E.: Coronary Circulation in Health and Disease. Philadelphia, Lea & Febiger, 1950.

Gregg, D. E.: The natural history of coronary collateral development. *Circ. Res.*, 35:335, 1974.

Grover, R. F.: Mechanisms augmenting coronary arterial oxygen extraction. *Adv. Cardiol.*, 9:89, 1973.

Hatch, F. T.: Interactions between nutrition and heredity in coronary heart disease. *Amer. J. Clin. Nutr.*, 27:80, 1974.

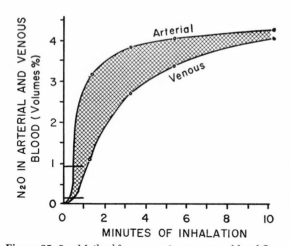

**Figure 25–8.** Method for measuring coronary blood flow in man as explained in the text. (Modified from Kety: Methods in Medical Research, 1948.)

Kirk, E. S., Urschel, C. W., and Sonnenblick, E. H.: Problems in cardiac performance: regulation of coronary blood flow and the physiology of heart failure. In Guyton, A. C. (ed.): MTP International Review of Science: Physiology, Vol. 1. Baltimore, University Park Press, 1974, p. 299.

Maseri, A.: Pathophysiological, diagnostic and methodological problems in the study of myocardial blood flow in ischaemic heart disease. J. Nucl. Biol. Med., 16:259, 1972.

Morrison, L. M., and Schjeide, O. A.: Coronary Heart Disease and the Mucopolysaccharides (Glycosaminoglycans). Springfield, Ill., Charles C Thomas, Publisher, 1974.

Naughton, J., and Hellerstein, H. K. (eds.): Exercise Testing and Exercise Training in Coronary Heart Disease. New York, Academic Press, Inc., 1973.

Neely, J. R., and Morgan, H. E.: Relationship between carbohydrate and lipid metabolism and the energy balance of the heart. Ann. Rev. Physiol., 36:413, 1974.

Pothuizen, L. M. (ed.): Coronary Artery Surgery (Proceedings of Symposium organized by the Netherlands, Society of Cardiology, 1974). New York, American Elsevier Publishing Company, 1974.

Rakusan, K.: Quantitative morphology of capillaries of the heart. Number of capillaries in animal and human hearts under normal and pathological conditions. Methods Achiev. Exp. Pathol., 5:272, 1971.

Rowe, G. G.: Responses of the coronary circulation to physiologic changes and pharmacologic agents. Anesthesiology, 41:182, 1974.

Rubio, R., and Berne, R. M.: Release of adenosine by the normal myocardium in dogs and its relationship to the regulation of coronary resistance. Circ. Res., 25:407, 1969.

Sabiston, D. C., Jr.: The William F. Rienhoff, Jr. lecture: The coronary circulation. Johns Hopkins Med. J., 134:314, 1974.

Sarnoff, S. J., Braunwald, E., Welch, G. H., Jr., Case, R. B., Stainsby, W. N., and Macruz, R.: Hemodynamic determinants of oxygen consumption of the heart with special reference to the tension-time index. Amer. J. Physiol., 192:148, 1958.

Schaper, W.: The Collateral Circulation of the Heart. New York, American Elsevier Publishing Company, 1971.

Shaper, A. G.: Diet in the epidemiology of coronary heart disease. Proc. Nutr. Soc., 31:297, 1972.

Stone, H. L., Bishop, V. S., and Guyton, A. C.: Cardiac function after embolization of coronaries with microspheres. Amer. J. Physiol., 204:16, 1963.

Stone, H. L., Bishop, V. S., and Guyton, A. C.: Progressive changes in cardiovascular function after unilateral heart irradiation. Amer. J. Physiol., 206:289, 1964.

Sugimoto, T., Sagawa, K., and Guyton, A. C.: Quantitative effect of low coronary pressure on left ventricular performance. Jap. Heart J., 9:46, 1968.

# 26

# Cardiac Failure

Perhaps the most important ailment that must be treated by the physician is cardiac failure, which can result from any heart condition that reduces the ability of the heart to pump blood. Usually the cause is decreased contractility of the myocardium caused by diminished coronary blood flow, but failure to pump adequate quantities of blood can also be caused by damage to the heart valves, external pressure around the heart, vitamin deficiency, primary cardiac muscle disease, or any other abnormality that makes the heart a hypoeffective pump. In the present chapter, we will discuss primarily cardiac failure caused by ischemic heart disease. In the following chapter we will discuss valvular and congenital heart disease.

**Definition of Cardiac Failure.** The term cardiac failure means simply *failure of the heart to pump blood adequately*. This does not mean that the cardiac output in all instances of failure is less than normal, for the output can be normal or sometimes even above normal provided the tendency for venous return is high enough to offset the diminished strength of the heart. Therefore, cardiac failure may be manifest in either of two ways: (1) by a decrease in cardiac output or (2) by an increase in either left or right atrial pressure even though the cardiac output is normal or above normal.

**Unilateral versus Bilateral Cardiac Failure.** Since the left and right sides of the heart are two separate pumping systems, it is possible for one of these to fail independently of the other. For instance, unilateral failure can result from coronary thrombosis in one or the other of the ventricles. Because debilitating thrombosis occurs approximately 30 times as often in the left ventricle as in the right ventricle, there is a tendency among clinicians to view failure following myocardial infarction as almost always primarily left-sided. Occasionally, however, right-sided failure does occur with no left-sided failure at all, but this happens most frequently in persons with pulmonary stenosis or some other congenital disease affecting primarily the right heart.

Usually, though, when one side of the heart be-

comes weakened, this causes a sequence of events that makes the opposite side of the heart also fail. For instance, in left-sided failure the left atrial pressure increases greatly, resulting in considerable back pressure in the pulmonary system and a rise in pulmonary arterial pressure sometimes to two to three times normal. This loads the right ventricle, causing combined failure of both ventricles, even though the initiating cause began in the left side of the heart only. On the other hand, primary damage to the right ventricle sometimes reduces the overall cardiac output, resulting in diminished coronary blood flow and at least a mild degree of left heart weakening.

In the first part of this chapter we will consider the whole heart failing as a unit and then return to the specific features of unilateral left- and right-sided failure.

## DYNAMICS OF THE CIRCULATION IN CARDIAC FAILURE

In the following discussion we will consider the progressive changes in the circulation following acute cardiac failure, first, when the failure is only moderately severe and, second, when it is almost lethal.

### ACUTE EFFECTS OF MODERATE CARDIAC FAILURE

If a heart suddenly becomes severely damaged in any way, such as by myocardial infarction, the pumping ability of the heart is immediately depressed. As a result, two essential effects occur: (a) reduced cardiac output and (b) increased systemic venous pressure. These effects are shown in Figure 26–1, which illustrates, first, a normal cardiac output curve, depicting the state of the circulation prior to the cardiac damage. Point A represents the normal state of the circulation, showing that the normal cardiac output

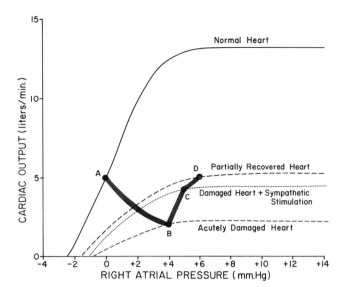

**Figure 26–1.** Progressive changes in the cardiac output curve following acute myocardial infarction. The cardiac output and right atrial pressure change progressively from point A to point D, as explained in the text.

under resting conditions is 5 liters per minute and the right atrial pressure is 0 mm. Hg.

Immediately after the heart becomes damaged, the cardiac output curve becomes greatly reduced, falling to the lower, long-dashed curve. Within a few seconds after the acute heart attack, a new circulatory state is established at point B rather than point A, showing that the cardiac output has fallen to 2 liters per minute, or about two-fifths normal, while the right atrial pressure has risen to 4 mm. Hg because blood returning to the heart is damming up in the right atrium. This low cardiac output is still sufficient to sustain life, but it is likely to be associated with fainting. Fortunately, this acute stage lasts for only a few seconds because sympathetic reflexes occur immediately that can compensate to a great extent for the damaged heart as follows:

**Compensation for Acute Cardiac Failure by Sympathetic Reflexes.** When the cardiac output falls precariously low, many of the different circulatory reflexes discussed in Chapter 21 immediately become active. The best known of these reflexes is the baroreceptor reflex, which is activated by diminished arterial pressure. It is probable that the chemoreceptor reflex, the central nervous system ischemic response, and possibly even reflexes originating in the damaged heart itself also become activated. Whatever all the reflexes might be, the sympathetics become strongly stimulated within a few seconds, and the parasympathetics become reciprocally inhibited at the same time.

Strong sympathetic stimulation has two major effects on the circulation, first, on the heart itself, and, second, on the peripheral vasculature. Even a damaged myocardium usually responds with increased force of contraction following sympathetic stimulation. If all the musculature is diffusely damaged, sympathetic stimulation usually strengthens this damaged musculature. Likewise, if part of the muscle is totally nonfunctional while part of it is still normal, the normal muscle is strongly stimulated by sympathetic stimulation, in this way compensating for the nonfunctional muscle. Thus, *the heart one way or another becomes a stronger pump, often as much as 100 per cent stronger, under the influence of the sympathetic impulses.* This effect is also illustrated in Figure 26–1, which shows elevation of the cardiac output curve after sympathetic compensation (the short-dashed curve).

Sympathetic stimulation also increases the tendency for venous return, for it increases the tone of most of the blood vessels of the circulation, especially of the veins, *raising the systemic filling pressure* to 12 to 14 mm. Hg, almost 100 per cent above normal. As will be recalled from the discussion in Chapter 23, this greatly increases the tendency for blood to flow back to the heart. Therefore, the damaged heart becomes primed with more inflowing blood than usual, and the right atrial pressure rises still further, which helps the heart to pump larger quantities of blood. Thus, in Figure 26–1 the new circulatory state is depicted by Point C, showing a cardiac output of 4.2 liters per minute and a right atrial pressure of 5 mm. Hg.

The sympathetic reflexes become maximally developed in about 30 seconds, and the resulting increased ability of the heart and increased tendency for venous return add together to cause the cardiac output to become once again almost normal. As a result, the person who has a sudden moderate heart attack might experience nothing more than cardiac pain and a few seconds of fainting. Shortly thereafter, with the aid of the sympathetic reflex compensation, his circulation may be completely adequate as long as he remains quiet, though the pain might persist.

## THE CHRONIC STAGE OF FAILURE

After the first few minutes of an acute heart attack, a prolonged secondary stage begins. This is characterized mainly by two events: (1) retention of fluid by the kidneys, and (2) progressive recovery of the heart itself, over a period of several weeks to months, as was discussed in the previous chapter.

**Compensation by Renal Retention of Fluid.** A low cardiac output has a profound effect on renal function, sometimes causing anuria when the cardiac output falls to as low as one-half to two-thirds normal. In general, the urinary output is less than normal as long as the cardiac output is significantly less than normal, and it usually does not return to normal after an acute heart attack until the cardiac output rises either all the way back to normal or almost to normal. This relationship of renal function to cardiac output is one of the most important of all the factors affecting the dynamics of the circulation in chronic cardiac failure.

**Causes of Renal Retention of Fluid in Cardiac Failure.** There are three known causes of the reduced renal output during cardiac failure, all of which are perhaps equally important.

*1. Decreased Glomerular Filtration.* A decrease in cardiac output has a tendency to reduce the glomerular pressure in the kidneys because of (a) *reduced arterial pressure* and (b) *intense sympathetic constriction of the afferent arterioles of the kidney.* As a consequence, the glomerular filtration rate becomes less than normal, and it will become evident from the discussion of kidney function in Chapter 35 that even a slight decrease in glomerular filtration often decreases urine output markedly. When the cardiac output falls to about one-half normal, this factor alone can result in almost complete anuria.

*2. Activation of the Renin-Angiotensin System and Increased Reabsorption of Water and Salt by the Kidneys.* The reduced blood flow to the kidneys causes marked increase in renin output, and this in turn causes the formation of angiotensin by the mechanism described in Chapter 21. The angiotensin has a direct effect on the arterioles of the kidneys to decrease further the blood flow through the kidneys. Also, increased efferent arterial resistance decreases peritubular capillary pressure, thus promoting greatly increased reabsorption of both water and salt from the renal tubules. Therefore, the net loss of water and salt into the urine is greatly decreased, and the quantities of salt and water in the body fluids increase.

*3. Increased Aldosterone Secretion.* In the chronic stage of heart failure, large quantities of aldosterone are secreted by the adrenal cortex. Most physiologists have taught that this results from stimulation of the adrenal glands by angiotensin. However, recent research indicates that it might result more from a slight tendency for potassium concentration to rise in the body fluids as a result of the reduced renal function in cardiac failure. Regardless of the cause of the increased aldosterone, it increases the reabsorption of sodium from the renal tubules, and this in turn leads to a secondary increase in water reabsorption for two reasons: First, as the sodium is reabsorbed, it reduces the osmotic pressure in the tubules while increasing the osmotic pressure in the renal interstitial fluids; these changes promote osmosis of water into the blood. Second, the absorbed sodium increases the osmotic concentration of the extracellular fluid and elicits *antidiuretic hormone* secretion by the supraoptico-hypophyseal system, which is discussed in Chapter 36. The antidiuretic hormone then promotes increased tubular reabsorption of water.

**Effect of Fluid Retention on Cardiovascular Dynamics.** Fluid retention does not have any significant effect on the pumping ability of the heart; but it does have an extreme effect on the tendency for venous return. The fluid retained by the kidneys causes a progressive increase in extracellular fluid volume, and part of this increased fluid usually remains in the blood, causing an increase in blood volume as well. For two different reasons this increases the tendency for venous return to the heart: First, both the increased extracellular fluid and blood volume increase the systemic filling pressure, which *increases the pressure gradient for flow of blood toward the heart.* Second, *reduced venous resistance* caused by distention of the veins allows increased flow of blood toward the heart.

*Detrimental Effects of Fluid Retention in the Late Stages of Cardiac Failure.* In moderate cardiac failure, retention of fluid helps to compensate for the failure by allowing more blood to flow into the heart and thereby priming the heart with increased quantities of blood. In severe cardiac failure, however, the amount of blood priming the heart often becomes so great that the cardiac muscle fibers are stretched beyond their normal physiological lengths, and this sometimes decreases the effectiveness of the heart as a pump. Thus, in the late stages, as will be discussed in subsequent sections of the chapter, further fluid retention can be very detrimental.

**Recovery of the Myocardium Following Myocardial Infarction.** After a heart becomes suddenly damaged as a result of myocardial infarction, the natural reparative processes of the body begin immediately to help restore normal cardiac function. Thus, a new collateral blood supply begins to penetrate the peripheral portions of the infarcted area, often completely restoring the muscle function. Also, the undamaged musculature hypertrophies, in this way offsetting much of the cardiac damage.

Obviously, the degree of recovery depends on the type of cardiac damage, and it varies from no recovery at all to almost complete recovery. Ordinarily, after myocardial infarction the heart will have achieved most of its final state of recovery within four to six months.

**Cardiac Output Curve After Partial Recovery.** The notched curve of Figure 26–1 illustrates function of the partially recovered heart a week or so after the acute myocardial infarction. By this time, considerable fluid has been retained in the body, and the tendency for venous return has increased markedly; therefore, the right atrial pressure has also risen. As a result, the state of the circulation is now changed from Point C to Point D, which represents a *normal* cardiac output of 5 liters per minute but with a right atrial pressure elevated to 6 mm. Hg.

Since the cardiac output has returned to normal, renal output also will have returned to normal and no further fluid retention will occur. Therefore, except for the high right atrial pressure represented by point D in the figure, the person now has essentially normal cardiovascular dynamics as long as he remains at rest.

If the heart itself recovers to a significant extent and if adequate fluid retention occurs, the sympathetic stimulation gradually abates toward normal for the following reasons: The partial recovery of the heart can do the same thing for the cardiac curve as sympathetic stimulation, and fluid retention in the circulatory system can do the same thing for venous return as sympathetic stimulation. Thus, as these two factors develop, the fast pulse rate, cold skin, sweating, and pallor resulting from sympathetic stimulation in the acute stage of cardiac failure gradually disappear.

## SUMMARY OF THE CHANGES THAT OCCUR FOLLOWING ACUTE CARDIAC FAILURE— "COMPENSATED HEART FAILURE"

To summarize the events discussed in the past few sections describing the dynamics of circulatory changes following an acute, moderate heart attack, we may divide the stages into (1) the instantaneous effect of the cardiac damage, (2) compensation by the sympathetic nervous system, and (3) chronic compensations resulting from partial cardiac recovery and renal retention of fluid. All these changes are shown graphically by the very heavy line in Figure 26–1. The progression of this line shows the normal state of the circulation (point A), the state a few seconds after the heart attack but before sympathetic reflexes have occurred (point B), the rise in cardiac output toward normal caused by sympathetic stimulation (point C), and final return of the cardiac output to normal following several days to several weeks of cardiac recovery and fluid retention (point D). This final state is called compensated heart failure.

**Compensated Heart Failure.** Note especially in Figure 26–1 that the pumping ability of the heart, as depicted by the cardiac function curve, is still depressed to less than one-half normal. This illustrates that factors that increase the right atrial pressure (principally retention of fluid) can return the cardiac output to normal despite continued weakness of the heart itself. However, one of the results of chronic cardiac weakness is this chronic increase in right atrial pressure itself; in Figure 26–1 it is shown to be 6 mm. Hg. There are many persons, especially in old

age, who have completely normal resting cardiac outputs but mildly to moderately elevated right atrial pressures because of compensated heart failure. These persons may not know that they have cardiac damage because the damage more often than not has occurred a little at a time, and the compensation has occurred concurrently with the progressive stages of damage.

The slight elevation of right atrial pressure has little harmful effect on the circulatory system (perhaps a slightly enlarged liver and tendency toward ankle edema). However, the person with compensated heart failure certainly does not have a normal circulatory system, for should he try to exercise strongly or should any other stress be placed on his circulatory system, such as might occur in some disease condition, his heart would be unable to respond normally. The reason for this is that the compensatory mechanisms normally used to increase the cardiac output are already partially in use simply to provide a normal output. Therefore, the remaining compensation that can be invoked is much below normal. To state this another way, the normal person has far greater *cardiac reserve* than the person with compensated heart disease. Cardiac reserve is discussed at greater length later in the chapter.

## DYNAMICS OF SEVERE CARDIAC FAILURE—DECOMPENSATED HEART FAILURE

If the heart becomes severely damaged, then no amount of compensation, either by sympathetic nervous reflexes or by fluid retention, can make this weakened heart pump a normal cardiac output. As a consequence, the cardiac output cannot rise to a high enough value to bring about return of normal renal function. Fluid continues to be retained, the person develops progressively more and more edema, and this state of events eventually leads to his death. This is called decompensated heart failure. The basis of decompensated heart failure is *failure of the heart to pump sufficient blood to make the kidneys function adequately.*

**Analysis of Decompensated Heart Failure.** Figure 26–2 illustrates a greatly depressed cardiac output curve, depicting the function of a heart that has become extremely weakened and cannot be strengthened. Point A on this curve represents the approximate state of the circulation before any compensation has occurred in the venous return, that is, before any sympathetic stimulation of the peripheral vessels and before any increase in fluid volume. In this state, the cardiac output is approximately 3 liters per minute—much too low to provide normal circulatory function. However, this state lasts for only a few seconds, because sympathetic stimulation of the peripheral veins and other vessels immediately increases the systemic filling pressure; this increase forces additional amounts of blood into the heart and increases the right atrial pressure.

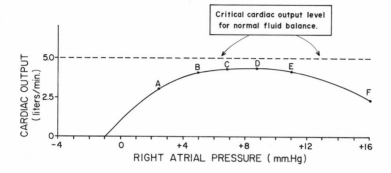

**Figure 26–2.** Greatly depressed cardiac output curve that indicates decompensated heart disease. Progressive fluid retention raises the right atrial pressure, and the cardiac output progresses from point A to point F, as explained in the text.

Thus, at Point B the right atrial pressure has risen to 5 mm. Hg and the cardiac output to 4 liters per minute. The person appears to be in reasonably good condition, but this state will not remain static for the following reason: the cardiac output has not risen quite high enough to cause adequate kidney excretion of fluid; therefore, fluid retention continues unabated and can eventually be the cause of death. These events can be explained quantitatively in the following way:

Note the dashed line at a cardiac output level of 5 liters in Figure 26–2. This is the critical cardiac output level that is required for re-establishment of normal fluid balance. At any cardiac output below this level, all the fluid-retaining mechanisms discussed in the earlier section remain in play, and the body fluid volumes will increase progressively. Because of this progressive increase in fluid volume, the systemic filling pressure continues to rise, and this forces progressively increasing quantities of blood into the right atrium, thus increasing the right atrial pressure. After a day or so, the state of the circulation changes from Point B in Figure 26–2 to Point C—the right atrial pressure rising to about 7 mm. Hg and the cardiac output to 4.2 liters per minute. However, note again that the cardiac output is still not high enough to cause normal renal output of fluid; therefore, fluid continues to be retained, and after another day or so the right atrial pressure rises to 9 mm. Hg, and the circulatory state becomes that depicted by Point D on the curve. Still, the cardiac output is not enough to establish normal fluid balance.

After another few days the right atrial pressure has risen still further, but by now the cardiac function curve is beginning to decline back toward a lower level. This decline is caused by overstretch of the heart, edema of the heart muscle, and other factors that diminish the pumping performance of the heart. It is especially evident, however, that further retention of fluid from then on will be only detrimental to the condition. Still, the cardiac output is not high enough to bring about normal fluid balance, and fluid retention not only continues but actually accelerates because of the falling cardiac output. Consequently, within a few days the state of the circulation has reached Point F on the curve, with the cardiac output

less than 2.5 liters per minute and the right atrial pressure 16 mm. Hg. This is a state that has now reached incompatibility with life, and the patient dies in *decompensation.*

Thus, one can see from this analysis that failure of the cardiac output ever to rise to the critical level required for normal renal function results in (a) progressive retention of fluid, which causes (b) progressive elevation of the systemic filling pressure, and (c) progressive elevation of the right atrial pressure until finally the heart is so overstretched or so edematous that it becomes unable to pump even moderate quantities of blood, and, therefore, fails completely. Clinically, one detects this serious condition of decompensation principally by the progressive edema, especially edema of the lungs, which leads to bubbling rales and dyspnea (air hunger). All clinicians know that failure to institute appropriate therapy when this state of events occurs leads to rapid death.

**Treatment of Decompensation.** The two ways in which the decompensation process can often be stopped are: (1) by strengthening the heart in any one of several ways, especially by administration of a cardiotonic drug, such as digitalis, so that it can pump adequate quantities of blood to make the kidneys function normally again, or (2) by administering diuretics and reducing water and salt intake, which brings about a balance between fluid intake and output despite the low cardiac output.

Both methods stop the decompensation process by re-establishing normal fluid balance so that at least as much fluid leaves the body as enters it.

## UNILATERAL CARDIAC FAILURE

In the discussions thus far in this chapter, we have considered failure of the heart as a whole. Yet in a large number of patients, especially those with early acute failure, left-sided failure predominates over right-sided failure, and in rare instances, especially in congenital heart disease, the right side may fail without significant failure of the left side. Therefore, we now need to discuss the special features of unilateral cardiac failure.

## UNILATERAL LEFT HEART FAILURE

When the left side of the heart fails without concomitant failure of the right side, blood continues to be pumped into the lungs with usual right heart vigor while it is not pumped adequately out of the lungs into the systemic circulation. As a result, the *pulmonary filling pressure* rises while the *systemic filling pressure* falls because of shift of large volumes of blood from the systemic circulation into the pulmonary circulation.

As the volume of blood in the lungs increases, the pulmonary vessels enlarge, and, if the pulmonary capillary pressure rises above 28 mm. Hg, that is, above the colloid osmotic pressure of the plasma, fluid begins to filter out of the capillaries into the interstitial spaces and alveoli, resulting in pulmonary edema.

Thus, among the most important problems of left heart failure are *pulmonary vascular congestion* and *pulmonary edema,* which are discussed in detail in Chapters 24 and 31 in relation to the pulmonary circulation and capillary dynamics. As long as the pulmonary capillary pressure remains less than the normal colloid osmotic pressure of the blood, about 23 to 33 mm. Hg, the lungs remain ''dry.'' But even a few millimeters' rise in capillary pressure above this critical level causes progressive transudation of fluid into the interstitial spaces and alveoli, leading rapidly to death. Pulmonary edema can occur so rapidly that it can cause death after only 20 to 30 minutes of severe acute left heart failure.

### Course of Events in Chronic Left Heart Failure

In chronic left heart failure, one additional feature must be added to the acute picture. This is retention of fluid resulting from reduced renal function. In moderate acute left heart failure the pulmonary capillary pressure sometimes rises only to 15 to 20 mm. Hg, not enough to cause pulmonary edema. Yet following retention of fluid for the next few days, the blood volume increases, and more blood is pumped into the lungs by the right ventricle. Then, the pulmonary capillary pressure often rises above the colloid osmotic pressure, resulting in severe pulmonary edema, as shown in Figure 26–3. Indeed, this is a common occurrence, the patient suddenly developing severe pulmonary edema a week or so after the acute attack and dying a respiratory death—not a death resulting from diminished cardiac output.

## UNILATERAL RIGHT HEART FAILURE

In unilateral right heart failure, blood is pumped normally by the left ventricle from the lungs into the systemic circulation, but it is not pumped adequately from the systemic circulation into the lungs. Therefore, blood shifts from the lungs into the systemic circulation, causing systemic congestion. However,

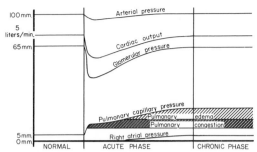

**Figure 26–3.** The overall effects, acute and chronic, of left-sided heart failure.

in acute right heart failure this congestion is hardly noticeable for the following reason: The total amount of blood in the lungs is only about one-eighth that in the systemic circulation. Therefore, even in severe right heart failure, the systemic blood volume increases only a few per cent because of blood shift from the lungs, and this is not sufficient to cause significant systemic congestion.

**Low Cardiac Output in Acute Right Heart Failure.** On the other hand, acute right heart failure can cause greater depression of the cardiac output than acute left heart failure of the same degree. This again stems from the far greater compliance of the systemic circulation than of the pulmonary circulation. Not enough blood can transfer from the lungs into the systemic vessels to raise the systemic pressures to a very high level, and these pressures are not enough to make the weakened right ventricle pump adequate quantities of blood. Therefore, in those rare conditions in which the right heart does fail acutely, the cardiac output falls greatly, often leading to such low cardiac output that death ensues rapidly.

**The Chronic Stage of Unilateral Right Heart Failure.** The chronic stage of unilateral right heart failure is much the same as that discussed earlier for the entire heart. The depressed cardiac output results in progressively more and more retention of fluid by the kidneys until the cardiac output either rises back nearly to normal or until the person goes into decompensation and dies. Figure 26–4 illustrates the progressive changes that occur in chronic right-sided heart failure, showing a gradual return of cardiac output to normal or near normal, a return of arterial and glomerular pressures along with increased urinary output, and, finally, progressive development of peripheral congestion and edema.

# ''HIGH CARDIAC OUTPUT FAILURE''—OVERLOADING OF THE HEART

A condition called ''high cardiac output failure'' frequently occurs in persons who have cardiac out-

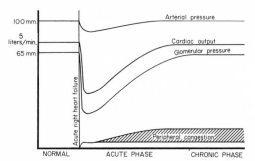

**Figure 26–4.** The overall effects, acute and chronic, of right-sided heart failure.

puts much higher than normal but who appear to be in cardiac failure because the left or right or both atrial pressures are very high. Sometimes there is also accumulation of edema. The true problem is often not failure of the pumping ability of the heart at all but *overloading of the heart with too much venous return.* The basic causes of increased cardiac output are (1) decreased systemic resistance or (2) increased systemic filling pressure. These *may* or *may not* be associated with a weak heart, depending on the condition causing the high output failure.

**Decreased Systemic Resistance.** Diseases that cause high cardiac output failure by decreasing the systemic resistance include (a) *arteriovenous fistulae,* in which the blood is shunted directly from the arteries to the veins, (b) *beriberi heart disease,* in which avitaminosis of the B vitamins results in profound systemic vasodilatation, and (c) *thyrotoxicosis,* which, because of its stimulatory effect on metabolism, causes generalized systemic vasodilatation. Obviously, the decreased resistance results in rapid venous return, often causing more blood to flow into the heart than can be properly handled and resulting in greatly increased right and left atrial pressures but at the same time resulting in a cardiac output sometimes as much as two to three times normal.

**Increased Systemic Filling Pressure.** The conditions that cause high cardiac output failure by increasing the systemic filling pressure include (a) too rapid and too much *transfusion* and (b) fluid retention resulting from excess secretion of different *steroid hormones*—especially aldosterone. These *increase the pressure gradient for venous return,* thereby forcing large excesses of blood into the heart and overloading it.

# CARDIOGENIC SHOCK

In the description of acute heart failure, it was pointed out that the cardiac output can fall very low immediately after heart damage occurs. This obviously leads to greatly diminished blood flow throughout the body and can lead to a typical picture of circulatory shock as described in Chapter 28. This means, simply, that the cardiac output falls too low to supply the body with adequate blood flow. As a result, the tissues deteriorate rapidly and death ensues. Sometimes death comes in less than an hour; at other times, it comes over a period of several days. The circulatory shock that is caused by inadequate cardiac pumping is called *cardiogenic shock* or *cardiac shock,* and it is sometimes also called the *power failure syndrome.*

Cardiogenic shock is extremely important to the clinician because approximately one-tenth of all patients who have acute myocardial infarction will have enough power failure to die of circulatory shock before the physiologic compensatory measures can come into play to save life. Once cardiac shock has become well established after myocardial infarction, all the typical events occur that also occur in the late stages of other types of circulatory shock, as described in Chapter 28, especially rapid deterioration of almost all bodily functions.

**Vicious Cycle of Cardiac Deterioration in Cardiogenic Shock.** The discussion of circulatory shock in Chapter 28 will emphasize the tendency for the heart itself to become progressively damaged when its coronary blood supply is reduced during the course of shock. That is, the low arterial pressure that occurs during shock reduces the coronary supply, which makes the heart still weaker, which makes the shock still worse, the process eventually becoming a vicious cycle of cardiac deterioration. In cardiogenic shock caused by myocardial infarction, this problem is greatly compounded by the already existing coronary thrombosis. For instance, in a normal heart, the arterial pressure usually must be reduced below about 45 mm. Hg before cardiac deterioration sets in. However, in a heart that already has a major coronary vessel blocked, deterioration will set in when the arterial pressure falls as low as 80 to 90 mm. Hg. In other words, even the minutest amount of fall in arterial pressure can set off a vicious cycle of cardiac deterioration following myocardial infarction, though this cycle will occur in a normal heart only in the very late stages of shock. For this reason, in treating myocardial infarction it is extremely important to prevent even short periods of hypotension.

**Physiology of Treatment.** Often a patient dies of cardiogenic shock before the various compensatory processes can return his cardiac output to a life-sustaining level. Therefore, treatment of this condition is one of the most important problems in the management of acute heart attacks. Immediate digitalization of the heart is often employed for strengthening the heart, though this is of value only in case the non-infarcted muscle is also chronically ischemic. But more frequently, intravenous infusion of whole blood, plasma, or a blood pressure raising drug is used to sustain the arterial pressure. If this is done, the coronary blood flow can often be elevated to a high enough value to prevent the vicious cycle of deterioration until appropriate compensatory mechanisms in the body have corrected the shock.

Even with the best therapy, however, once the shock syndrome has begun, with the arterial pressure remaining as much as 20 mm. Hg below normal for as long as an hour, 85 per cent of the patients die.

# EDEMA IN PATIENTS WITH CARDIAC FAILURE

Much of the preceding discussion has emphasized the edema that develops in cardiac failure. On first thought one would think that it should be easy to understand the cause of edema when the heart fails. One's natural logic directs that when the heart fails, blood becomes dammed up behind the heart, thereby increasing the venous and capillary pressures until fluid transudes out of the capillaries into the tissues. However, peripheral edema will not occur until large quantities of water and salt are retained by the kidneys. Consequently, there are special quirks of logic that must be explained before one can understand some of the problems related to edema in cardiac failure.

**Role of the Kidneys in Cardiac Edema—Inability of Acute Cardiac Failure to Cause Peripheral Edema.** Though acute left heart failure can cause terrific congestion of the lungs, with rapid development of pulmonary edema, acute heart failure of any type never causes immediate development of peripheral edema. This can be explained best by reference to Figure 26–5. When a previously normal heart acutely fails as a pump, the cardiac output falls, the arterial pressure falls, and the right atrial pressure rises. These approach each other at an equilibrium value of about 13 mm. Hg. It is obvious that capillary pressure must also fall from its normal value of 17 mm. Hg to the equilibrium pressure of 13 mm. Hg. Thus, severe *acute cardiac failure actually causes a fall in capillary pressure rather than a rise.* And animal experiments as well as experience in human beings show that acute cardiac failure does not cause immediate development of peripheral edema. In fact, during the first hours fluid is actually absorbed into the capillaries from the interstitial spaces.

The edema, on the other hand, develops during the succeeding days *because of fluid retention by the kidneys.* The retention of fluid increases the systemic filling pressure, resulting in increased tendency for blood to return to the heart. This now elevates the right atrial pressure to a still higher value, and it also returns the arterial pressure to normal. Therefore, the rising right atrial pressure *now causes the capillary pressure to rise markedly,* thus causing loss of fluid into the tissues and development of severe edema.

The mechanisms of fluid retention by the kidneys were discussed earlier in the chapter, but these are so important that they need to be relisted here for our present discussion: (1) decreased glomerular pressure because of slightly decreased arterial pressure and especially because of sympathetic constriction of the afferent arterioles; (2) formation of angiotensin, which causes efferent arteriolar constriction with resultant excessive reabsorption of water and salt by the renal tubules; and (3) marked increase in aldosterone secretion, which also causes greatly increased salt and water reabsorption by the renal tubules.

## *ACUTE PULMONARY EDEMA IN CARDIAC FAILURE*

**Acute Pulmonary Edema Following Myocardial Infarction.** Immediately following myocardial infarction, severe acute pulmonary edema occasionally occurs within 15 to 20 minutes. This response can so decrease oxygen uptake and $CO_2$ excretion by the lungs that immediate death results. The reason it is possible to develop acute pulmonary edema but not peripheral edema following acute heart failure derives from the seven times as great compliance of the systemic circulation as that of the pulmonary circulation. Following left heart failure, tremendous quantities of blood transfer from the very compliant systemic circulation into the relatively noncompliant pulmonary circulation and therefore increase the pulmonary pressures tremendously, even under acute conditions. As explained above, not enough blood is available from the lungs to cause a similar situation in the systemic circulation following right heart failure.

It is especially interesting that simultaneous acute failure of both the right heart and the left heart is often less likely to be lethal than is pure left heart failure, because the right heart failure reduces the degree of pulmonary edema.

**Acute Pulmonary Edema Occurring in Chronic Heart Failure—A Lethal Vicious Cycle.** A very frequent cause of death is acute pulmonary edema

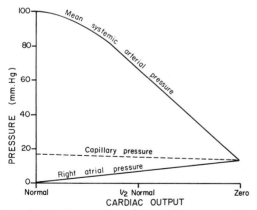

**Figure 26–5.** Progressive changes in mean systemic arterial pressure and right atrial pressure as the cardiac output falls from normal to zero. This figure also shows the effect of decreased cardiac output on capillary pressure.

occurring in patients who have had chronic heart failure for a long period of time. Sometimes this edema is caused by a new myocardial infarction, but in many instances no new pathology is found in the heart.

When acute pulmonary edema occurs in a person without new cardiac damage, it usually follows some temporary overload of the heart, such as might result from a bout of heavy exercise, some emotional experience, or even a severe cold. This acute pulmonary edema is believed to result from the following vicious cycle:

1. A temporarily increased load on the already weak left ventricle results from increased venous return from the peripheral circulation. Because of the limited pumping capacity of the left heart, blood begins to dam up in the lungs.

2. The increased blood in the lungs increases the pulmonary capillary pressure, and a small amount of fluid begins to transude into the lung tissues and alveoli.

3. The increased fluid in the lungs diminishes the degree of oxygenation of the blood.

4. The decreased oxygen in the blood carried to the peripheral tissues causes peripheral vasodilatation.

5. The peripheral vasodilatation increases venous return from the peripheral circulation still more.

6. The increased venous return further increases the damming of the blood in the lungs, leading to still more transudation of fluid, more arterial oxygen desaturation, more venous return, and so forth. Thus, a vicious cycle has been established.

7. Once this vicious cycle has proceeded beyond a certain critical point, it will continue until death of the patient unless heroic therapeutic measures are employed. The types of heroic therapeutic measures that can reverse the process and save the life of the patient include:

a. Putting tourniquets on all four limbs to sequester much of the blood in the veins and therefore to decrease the work load on the left heart.

b. Bleeding the patient.

c. Giving a rapidly acting diuretic such as furosemide to cause rapid loss of fluid from the body.

d. Giving the patient pure oxygen to breathe to reverse the blood desaturation and the peripheral vasodilatation.

e. Giving the patient a rapidly acting cardiotonic drug such as ouabain to strengthen the heart.

Unfortunately, this vicious cycle of acute pulmonary edema can proceed so rapidly that death can occur within 30 minutes to an hour. Therefore, any procedure that is to be successful must be instituted immediately.

# DECREASED MYOCARDIAL CONTRACTILITY IN CHRONIC HEART FAILURE

Though the initiating cause of heart failure is most frequently some extrinsic factor such as ischemia of the muscle, long-term failure eventually leads to moderate depression of the contractile state of the muscle itself. This results from a number of different factors:

1. Excessive stretch of the cardiac muscle, with evidence of actual physical damage to the muscle fibers and sometimes growth of excessive fibrous tissue in the ventricular walls.

2. Diminished number of capillaries in relation to the myocardial mass. This occurs especially when some parts of the heart muscle have been destroyed and other parts have become hypertrophied to compensate for the lost muscle; unfortunately, the capillaries fail to grow in proportion to the new muscle.

3. Diminished stores of norepinephrine vesicles in the sympathetic nerve endings. This results from reduction in the enzyme tyrosine hydroxylase, perhaps resulting from prolonged excessive stimulation of the sympathetic nervous system. Lack of the norepinephrine vesicles decreases the sympathetic drive to the myocardium, which is especially useful in times of stress.

4. Slightly defective excitation-contraction coupling, probably caused by diminished activity of the ATP-energized pump for calcium absorption into the sarcoplasmic reticulum.

Thus, a number of abnormalities of the myocardium eventually occur in prolonged cardiac failure, and these further exacerbate the failure itself.

**Use of Cardiotonic Drugs in Cardiac Failure.** Cardiotonic drugs such as digitalis have very little effect on increasing the contractile strength of normal cardiac muscle. On the other hand, these same drugs, when administered to a person with chronic cardiac failure, can frequently increase the strength of the failing myocardium as much as twofold. Therefore, they are the mainstay of therapy in chronic heart failure.

Unfortunately, the mechanism by which digitalis and the other cardiotonic drugs strengthen heart contraction is still debated. The most likely possibility is that it in some way increases the quantity of calcium ions released from the sarcoplasmic reticulum during the process of excitation-contraction coupling. The increased quantities of calcium ions then act directly on the actin and myosin filament contractile mechanism to cause increased force of contraction.

# PHYSIOLOGICAL CLASSIFICATION OF CARDIAC FAILURE

From the above discussions, it is apparent that the symptoms of cardiac failure fall into the following three physiological classifications:

Low cardiac output
Pulmonary congestion
Systemic congestion.

*Low cardiac output* usually occurs immediately after a heart attack. If the attack is mainly right-sided, this may be the only symptom. If the acute heart attack is mainly left-sided, concurrent pulmonary congestion essentially always occurs along with the low cardiac output, but the pulmonary congestion may be mild (without pulmonary edema) until after considerable fluid has been retained by the kidneys. Thus, low cardiac output may be the only significant clinical effect observed in many persons who have sudden heart attacks. This results in the following symptoms:

1. Generalized weakness
2. Fainting
3. Symptoms of increased sympathetic activity such as high heart rate, thready pulse, cold skin, sweating, and so forth.

*Pulmonary congestion* may be the only effect in patients with pure left-sided chronic heart failure, for, after the fluid that shifts from the systemic circulation into the lungs during acute left-sided heart failure has been replenished by renal retention of fluid, the heart often pumps normal quantities of blood, and the right side of the heart may not fail at all. Therefore, pulmonary congestive symptoms alone can occur with essentially no systemic congestion nor low cardiac output.

*Systemic congestion* alone can occur in pure right-sided chronic heart failure. In this condition there is no pulmonary congestion, and, if sufficient fluids have been retained in the blood to prime the heart sufficiently, the heart may pump a normal cardiac output.

Obviously, all the above classes of heart failure can occur together or in any combination.

# CARDIAC RESERVE

Fortunately, the normal heart can increase its output to four times normal under conditions of stress in most younger persons and to six times normal in endurance athletes. The maximum percentage that the cardiac output can increase above normal is called the *cardiac reserve*. Thus, in the normal young adult the cardiac reserve is slightly more than 300 per cent. In the athletically trained person it is occasionally as high as 500 per cent, while in the asthenic person it may be as low as 200 per cent. As an example, during severe exercise, the cardiac output of the normal adult can rise to about four times normal; this is an increase above normal of 300 per cent—that is, the normal young adult has a cardiac reserve of about 300 per cent.

Any factor that prevents the heart from pumping blood satisfactorily decreases the cardiac reserve. Such decrease can result from ischemic heart disease, primary myocardial disease, diphtheritic damage to the myocardium, valvular heart disease, and many other factors, some of which are illustrated in Figure 26–6.

**Diagnosis of Low Cardiac Reserve—The Exercise Test.** So long as a person with low cardiac reserve remains in a state of rest, he probably will not

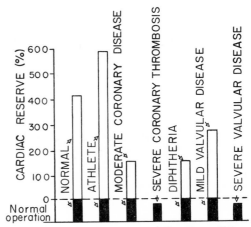

**Figure 26–6.** Cardiac reserve in different conditions.

know that he has heart disease. However, a diagnosis of low cardiac reserve can usually be made easily by requiring the person to exercise either on a treadmill or by walking up and down steps. The increased load on the heart rapidly uses up the small amount of reserve that is available, and the cardiac output fails to rise high enough to sustain the body's new level of activity. The acute effects are:

1. Immediate and sometimes extreme shortness of breath (dyspnea) resulting from the heart's not pumping sufficient blood to the tissues, thereby causing tissue ischemia and creating a sensation of air hunger.

2. Extreme muscle fatigue resulting from muscle ischemia, thus limiting the person's ability to continue with the exercise.

Therefore, exercise tests are a regular part of the armamentarium of the cardiologist. These tests take the place of cardiac output measurements that unfortunately cannot be made with ease in most clinical settings.

# A QUANTITATIVE GRAPHICAL METHOD FOR ANALYSIS OF CARDIAC FAILURE

Though it is possible to understand most of the general principles of cardiac failure using mainly qualitative logic, one can grasp with far greater depth the importance of the different factors in cardiac failure by using more quantitative approaches. One such approach is the graphical method for analysis of cardiac output regulation, which was introduced in Chapter 23. In the following few paragraphs, we will analyze several aspects of cardiac failure utilizing this graphical technique.

**Graphical Analysis of Acute Heart Failure and of Chronic Compensation.** Figure 26–7 illustrates cardiac output and venous return curves for different states of the heart and of the peripheral circulation.

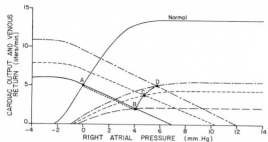

**Figure 26–7.** Diagram showing the progressive changes in cardiac output and right atrial pressure during different stages of cardiac failure. This figure is a composite of the three preceding figures.

The two solid curves are (1) the normal cardiac output curve, and (2) the normal venous return curve. As pointed out in Chapter 23, there is only one point on each of the curves at which the circulatory system can operate so long as these two curves remain normal. This point is where the two curves cross at Point A. Therefore, the normal state of the circulation is a cardiac output and venous return of 5 liters per minute and a right atrial pressure of 0 mm. Hg.

During the first few seconds after a moderately severe heart attack, the cardiac output curve falls to the lowermost long-dashed curve. During this few seconds, the venous return curve still has not changed because the peripheral circulatory system is still operating normally. Therefore, the new state of the circulation is depicted by Point B, where the new cardiac output curve crosses the normal venous return curve. Thus, the right atrial pressure rises to 4 mm. Hg while the cardiac output falls to 2 liters per minute.

Within the next 30 seconds, the sympathetic reflexes become very active. These affect both of the curves, raising them to the short-dashed curves. Sympathetic stimulation can increase the plateau level of the cardiac output curve as much as 100 per cent. It can also increase the systemic filling pressure (depicted by the point where the venous return curve crosses the zero venous return axis) by several mm. Hg—in this figure from a normal value of 7 mm. Hg up to 10.5 mm. Hg. This increase in systemic filling pressure shifts the entire venous return curve to the right and upward. The new cardiac output and venous return curves now equilibrate at Point C, that is, at a right atrial pressure of 5 mm. Hg and at a cardiac output of 4 liters per minute.

During the ensuing week, the cardiac output and venous return curves rise to the dot-dashed curves because of (1) some recovery of the heart and (2) renal retention of salt and water, which raises the systemic filling pressure still further—this time up to 12 mm. Hg. The two new curves, the dot-dashed curves, now equilibrate at Point D. Thus, the cardiac output has now returned entirely to normal. The right atrial pressure, however, has risen still further to 6

mm. Hg. Because the cardiac output is now normal, renal output is also normal so that a new state of equilibrated fluid balance has been achieved. Therefore, the circulatory system will continue to function at Point D and will remain stable, with a normal cardiac output and an elevated right atrial pressure, until some additional extrinsic factor changes either the cardiac output curve or the venous return curve.

Utilizing this technique for analysis, one can see especially the importance of fluid retention and how it eventually leads to a new stable state of the circulation; one can also see the interrelationship between the systemic filling pressure and cardiac pumping at various degrees of cardiac failure.

Note especially that the events described in Figure 26–7 are the same as those presented in Figure 26–1, but presented in a more quantitative manner.

**Graphical Analysis of "Decompensated" Cardiac Failure.** The cardiac output curve in Figure 26–8 is the same as that in Figure 26–2, a very low curve that in this case has already reached a degree of recovery as great as this heart can achieve. In this figure we have added venous return curves that occur during successive days following the acute fall of the cardiac output curve to this very low level. At Point A the curve equates with the normal venous return curve to give a cardiac output of approximately 3 liters per minute. However, stimulation of the sympathetic nervous system, caused by this low cardiac output, increases the systemic filling pressure within 30 seconds from 7 mm. Hg to 10.5 mm. Hg and shifts the venous return curve upward and to the right to give the curve labeled "autonomic compensation." Thus, the new venous return curve equates with the cardiac output curve at Point B. The cardiac output has been improved up to a level of 4 liters per minute but at the expense of an additional rise in right atrial pressure to 5 mm. Hg.

The cardiac output of 4 liters per minute is still too low to cause the kidneys to function normally. Therefore, fluid continues to be retained, the systemic filling pressure rises from 10.5 almost to 13 mm. Hg, and the venous return curve becomes that labeled "second day." This equilibrates with the cardiac output curve at Point C, and the cardiac output rises to 4.2 liters per minute while the right atrial pressure rises to 7 mm. Hg.

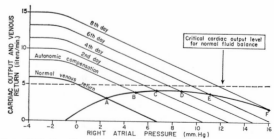

**Figure 26–8.** Graphical analysis of decompensated heart disease, showing progressive shift of the venous return curve to the right as a result of fluid retention.

During the succeeding days the cardiac output never rises high enough to re-establish normal fluid balance. Therefore, fluid continues to be retained, the systemic filling pressure continues to rise, the venous return curve continues to shift to the right, and the equilibrium point between the venous return curve and the cardiac output curve shifts progressively to Point D, to Point E, and finally to Point F. The equilibration process is now on the downslope of the cardiac output curve so that further retention of fluid causes only a detrimental effect on cardiac output. Therefore, the condition accelerates in a downhill direction until death occurs.

Thus, the process of the "decompensation" results from the fact that the cardiac output curve never rises up to the critical level of 5 liters per minute required to re-establish normal balance between fluid input and output.

**Treatment of Decompensated Heart Disease with Digitalis.** Let us assume that the stage of decompensation has already reached Point E in Figure 26–8, and let us proceed to the same point E in Figure 26–9. At this point, digitalis is given to strengthen the heart and to raise the cardiac output curve up to the heavy curve in Figure 26–9. This does not change the venous return curve. Therefore, the new cardiac output curve equates immediately with the venous return curve at Point G. The cardiac output is now 5.7 liters per minute, a value greater than the critical level of 5 liters required for normal fluid balance. Therefore, the kidneys eliminate far more fluid than normally, causing diuresis, a well known effect of digitalis.

The progressive loss of fluid over a period of several days reduces the systemic filling pressure back down to 11.5 mm. Hg, and the new venous return curve becomes the heavy curve labeled "several days later." This curve equates with the cardiac output curve of the digitalized heart at Point H, at an output of 5 liters per minute and a right atrial pressure of 4.6 mm. Hg. This cardiac output is precisely that required for normal fluid balance. Therefore, no additional fluid will be lost, and none will be gained. Consequently, the circulatory system has now stabilized, or, in other words, the decompensation of

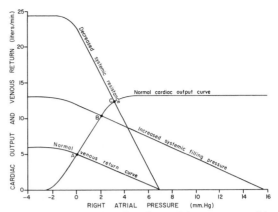

**Figure 26–10.** Graphical analysis of the two types of factors that can cause high cardiac output failure.

the heart failure has been "compensated." To state this another way, the final steady-state condition of the circulation is defined by the crossing point of three different curves: the cardiac output curve, the venous return curve, and the critical level for normal fluid balance. The compensatory mechanisms automatically stabilize the circulation when all of these three curves cross at the same point.

**Graphical Analysis of High Cardiac Output Failure.** Figure 26–10 shows an analysis of two types of high cardiac output failure. In this analysis the cardiac output curve remains normal at all times. As long as the venous return curve is also normal, the cardiac output and venous return curves equate with each other at point A, at a normal cardiac output of 5 liters per minute and a normal right atrial pressure of 0 mm. Hg.

Now, let us assume that the systemic resistance (the total peripheral resistance) becomes greatly decreased (such as might occur with the opening of a very large arteriovenous fistula). The venous return curve rotates upward to give the curve labeled "decreased systemic resistance." This venous return curve equates with the cardiac output curve at point C, with a cardiac output of 12.5 liters per minute and a right atrial pressure of 3 mm. Hg. Thus, the cardiac output has become greatly elevated, the right atrial pressure is slightly elevated, and there are mild signs of peripheral congestion. If the person attempts to exercise, he will have very little cardiac reserve because the heart is already being used almost to maximum capacity simply to pump the extra blood flow through the arteriovenous fistula. Therefore, this condition resembles a failure condition and is called "high output failure," but in reality the heart is simply overloaded by excess venous return.

The second cause of high output failure is excessive blood volume, which increases the systemic filling pressure to a level far too great. The venous return curve, labeled "increased systemic filling pressure," is caused by overtransfusion of a person,

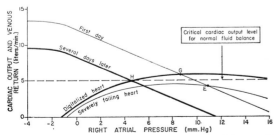

**Figure 26–9.** Treatment of decompensated heart disease, showing the effect of digitalis in elevating the cardiac output curve, this in turn causing progressive shift of the venous return curve to the left.

causing a systemic filling pressure of 16 mm. Hg and a venous return curve shifted far to the right and elevated. This curve equilibrates with the cardiac output curve at point B, with a cardiac output of 11 liters per minute and a right atrial pressure of 2 mm. Hg. Here again, mild symptoms of congestion occur, which in some ways resemble cardiac failure, but in reality this is also a condition of overloading the heart.

In other instances of high cardiac output failure, the cardiac output curve may actually be reduced below normal, but because of enhanced venous return the cardiac output is still greater than normal. This frequently occurs in beriberi heart disease and in thyrotoxicosis. The student should draw into Figure 26–10 such a cardiac function curve to show how high cardiac output can actually occur in a person even in the event of a weakened heart.

# REFERENCES

Barger, A. C.: The kidney in congestive heart failure. *Circulation*, 21:124, 1960.

Braunwald, E.: The determinants of myocardial oxygen consumption. *Physiologist*, 12:65, 1969.

Braunwald, E.: Mechanics and energetics of the normal and failing heart. *Trans. Assoc. Amer. Physicians*, 84:63, 1971.

Braunwald, E., Ross, J., and Sonnenblick, E.: Mechanism of Contractility of the Normal and Failing Heart. Boston, Little, Brown and Company, 1968.

Brender, D., Vanhoutte, P. M., and Shepherd, J. T.: Potentiation of adrenergic venomotor responses in dogs by cardiac glycosides. *Circ. Res.*, 25:597, 1969.

Burch, G. E.: Interesting aspects of geriatric cardiology. *Amer. Heart J.*, 89:99, 1975.

Conradsson, T. B., and Werkö, L.: Management of heart disease in pregnancy. *Prog. Cardiovasc. Dis.*, 16:407, 1974.

Dock, W.: Cardiomyopathies of the senescent and senile. *Cardiovasc. Clin.* 4:362, 1972.

Friedberg, C. K.: Diseases of the Heart, 3rd Ed. Philadelphia, W. B. Saunders Company, 1966.

Guyton, A. C.: The systemic venous system in cardiac failure. *J. Chronic Dis.*, 9:465, 1959.

Haber, E.: The role of antibodies and physiological receptors in cardiovascular diagnosis, therapy, and research. *Pharmacol. Rev.*, 25:215, 1973.

Harrison, T. R., and Reeves, T. J.: Principles and Problems of Ischemic Heart Disease. Chicago, Year Book Medical Publishers, Inc., 1968.

Hurst, W.: The Heart. New York, McGraw-Hill Book Company, 1966.

Kirk, E. S., Urschel, C. W., and Sonnenblick, E. H.: Problems in the cardiac performance regulation of coronary blood flow and the physiology of heart failure. In Guyton, A. C. (ed.): MTP International Review of Science: Physiology. Vol. 1. Baltimore, University Park Press, 1974, p. 299.

Kones, R. J.: The catecholamines: reappraisal of their use for acute myocardial infarction and the low cardiac output syndromes. *Crit. Care Med.*, 1:203, 1973.

Kones, R. J.: Cardiogenic shock: therapeutic implications of altered myocardial energy balance. *Angiology*, 25:317, 1974.

Kones, R. J.: Cardiogenic Shock: Mechanisms and Management. New York, Futura Publishing Co., Inc., 1975.

Marks, B. H., and Weissler, A. M. (eds.): Basic and Clinical Pharmacology of Digitalis. Springfield, Ill., Charles C Thomas, Publisher, 1972.

Resnekov, L.: Hemodynamic effects of acute myocardial infarction. *Med. Clin. North Amer.* 57:243, 1973.

Stone, H. L., Bishop, V. S., and Guyton A. C.: Progressive changes in cardiovascular function after unilateral heart irradiation. *Amer. J. Physiol.*, 206:289, 1964.

Stone, H. L., Bishop, V. S., and Guyton, A. C.: Ventricular function following radiation damage of the right ventricle. *Amer. J. of Physiol.*, 211:1209, 1966.

Swan, H. J. C., and Parmley, W. W.: Congestive heart failure. In Sodeman, W. A., Jr., and Sodeman, W. A. (eds.): Pathologic Physiology: Mechanisms of Disease, 5th Ed. Philadelphia, W. B. Saunders Company, 1974, p. 273.

Zelis, R., Longhurst, J., Capone, R. J., and Lee, G.: Peripheral circulatory control mechanisms in congestive heart failure. *Amer. J. Cardiol.*, 32:481, 1973.

# 27

# Heart Sounds; Dynamics of Valvular and Congenital Heart Defects

## THE HEART SOUNDS

The function of the heart valves was discussed in Chapter 13, and it was pointed out that closure of the valves is associated with audible sounds, though no sounds whatsoever occur when the valves open. The purpose of the present section is to discuss the factors that cause the sounds in the heart, under both normal and abnormal conditions.

### NORMAL HEART SOUNDS

When one listens with a stethoscope to a normal heart, he hears a sound usually described as "lub, dub, lub, dub ---." The "lub" is associated with closure of the A-V valves at the beginning of systole and the "dub" with closure of the semilunar valves at the end of systole. The "lub" sound is called the *first heart sound* and the "dub" the *second heart sound* because the normal cycle of the heart is considered to start with the beginning of systole.

**Causes of the First and Second Heart Sounds.** Closure of the valves in any pump system usually causes a certain amount of noise because the valves close solidly and suddenly over some opening, setting up vibrations in the fluid or walls of the pump. In the heart, the valves are cushioned by blood, so that it is difficult to understand why these valves create as much sound as they do.

The earliest suggestion for the cause of the heart sounds was that the slapping together of the vanes of the valves sets up vibrations, but this has now been shown to cause little if any of the sound because of the cushioning effect of the blood. Instead, the cause is *vibration of the taut valves immediately after closure, and vibration of the walls of the heart and major vessels around the heart.* For instance, contraction of the ventricle causes sudden backflow of blood against the A-V valves, causing the valves to close and to bulge toward the atria. The elastic tautness of the valves then causes the backsurging blood to bounce forward again into each respective ventricle. This effect sets the ventricles as well as the valves into vibration. The vibrations then travel to the chest wall where they can be heard as sound by the stethoscope. This is the cause of the major part of the first heart sound.

A possible cause of a small portion of the first heart sound is contraction of the heart muscle itself, for intrinsic rearrangement of muscular elements during contraction in any muscle causes sounds that can be heard through an overlying stethoscope. However, this contribution to heart sounds is extremely slight because the first heart sound disappears almost entirely when the valves are removed.

The second heart sound results from vibration of the taut, closed semilunar valves and from vibration of the walls of the pulmonary artery, the aorta, and, to much less extent, the ventricles. When the semilunar valves close, they bulge backward toward the ventricles, and their elastic stretch recoils the blood in the bulge back into the arteries, which causes a short period of reverberation of blood back and forth between the walls of the arteries and the valves. The vibrations set up in the arterial walls are then transmitted along the arteries at the velocity of the pulse wave. These vibrations are not transmitted as compression waves characteristic of sound but instead as lateral vibrations of the arterial walls. However, when the vibrations of the walls come into contact with a "sounding board," such as the chest wall, they create sound that can be heard.

**Durations and Frequencies of the First and Second Heart Sounds.** The duration of each of the heart sounds is slightly more than 0.1 second; the first sound lasts about 0.14 second and the second about 0.11 second. And both of them are described as very low-pitch sounds, the first lower than the second.

Sound consists of vibrations of different frequencies. Figure 27–1 illustrates by the shaded area the amplitudes of the different frequencies in the heart sounds and murmurs, illustrating that these are composed of frequencies ranging all the way from a few cycles per second to more than 1000 cycles per second, with the maximum amplitude of vibration occurring at a frequency of about 24 cycles per second.

Also shown in Figure 27–1 is another curve called the "threshold of audibility," which depicts the capability of the ear to hear sounds of different amplitudes. Note that in the very low frequency range the heart vibrations have a high degree of amplitude, but the threshold of audibility is so high that ordinarily the heart vibrations below approximately 30 to 50 cycles per

second are not heard by the ears. Then above about 500 cycles per second, the heart sounds are so weak that, despite a low threshold of audibility, no frequencies in this range are heard. For practical purposes, then, we can consider that all the *audible* heart sounds lie in the range of approximately 40 to 500 cycles per second despite the fact that the maximum amplitude of vibration occurs at the very low frequency of 24 cycles per second.

Both the first and second heart sounds have a mixture of frequencies in the entire audible range of the heart sounds, though the first sound has slightly more low frequency sound than does the second heart sound.

The reason the frequency of the first heart sound is lower than that of the second sound is probably two-fold: First, the *elastic modulus* of the A-V valves and of the walls of the ventricles is far less than the elastic modulus of the semilunar valves and the arterial walls. It is well known that any mechanical vibrating system having a low elastic modulus oscillates at a lower frequency than a system having a greater modulus. Second, the *mass of blood* in the great vessels is less than that in the ventricles, which means that the inertia of the vibrating mass is less. This also would cause the second heart sound to have a higher frequency than the first.

The reason the second heart sound is shorter in duration than the first is probably that the second sound is "damped" out by the vascular walls much more rapidly than is the first heart sound by the ventricular walls.

**Loudness of the First and Second Heart Sounds.** The loudness of the first and second heart sounds is almost directly proportional to the *rate of change* of the respective pressure differences across the A-V and semilunar valves. For instance, when the onset of systole is very rapid, the intraventricular pressure rises very rapidly during the isometric period of ventricular contraction; yet the pressure on the opposite side of the A-V valves remains only a few mm. Hg. Thus, the rate of change of the pressure difference between the two sides of the valves is very great, and as a result the first heart sound is loud. Also, when the heart is very active, such as during and immediately following exercise, the force of contraction of the ventricle is greatly enhanced, so that the first heart sound is in this instance also greatly accentuated. Conversely, in a weakened heart in which the onset of contraction is sluggish, the loudness of the first sound is greatly diminished.

In the case of the second heart sound, it is the rate of decrease in ventricular pressure at the

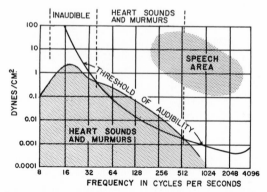

**Figure 27–1.** Amplitude of different frequency vibrations in the heart sounds and heart murmurs in relation to the threshold of audibility, showing that the range of sounds that can be heard is between about 40 and 500 cycles per second. (Modified from Butterworth, Chassin, and McGrath: Cardiac Auscultation. Grune & Stratton, 1960.)

end of systole that determines the loudness. The magnitude of this rate of decrease is determined mainly by the level of ventricular systolic pressure at the time the valve closes. In a person who has hypertension, the pressure at the time the aortic valve closes may be as great as 200 mm. Hg, so that the intraventricular pressure falls twice as rapidly as normally, all the way to zero in a few hundredths of a second. Therefore, the aortic sound is markedly accentuated. Likewise, in pulmonary hypertension the pulmonic sound is greatly accentuated. On the other hand, when the arterial pressure is low, such as in shock or in the terminal stages of cardiac failure, the second heart sound is diminished to a very low intensity.

**The Third Heart Sound.** Occasionally a third heart sound is heard at the beginning of the middle third of diastole. The most logical explanation of this sound is oscillation of blood back and forth between the walls of the ventricles initiated by inrushing blood from the atria. This is analogous to running water from a faucet into a sack, the inrushing water reverberating back and forth between the walls of the sack to cause vibrations in the walls.

The third heart sound is an extremely weak rumble of such low frequency that it usually cannot be heard with a stethoscope, but it can be recorded frequently in the phonocardiogram. The very low frequency of this sound presumably results from the flaccid, inelastic condition of the heart during diastole. Also, the reason the third heart sound does not occur until after the first third of diastole is over is presumably that in the early part of diastole the heart is not filled sufficiently to create even the small amount of elastic tension in the ventricles necessary for reverberation. The reason the third heart sound does not continue into the latter part of diastole is presumably that little blood flows into the ventricles during the latter part of diastole, so that no initiating stimulus then exists for causing reverberation.

**The Atrial Heart Sound (Fourth Heart Sound).** An atrial heart sound can be recorded in many persons in the phonocardiogram, but it can almost never be heard with a stethoscope because of its low frequency—usually 20 cycles per second or less. This sound occurs when the atria contract, and presumably it is caused by an inrush of blood into the ventricles, which initiates vibrations similar to those of the third heart sound.

## AREAS FOR AUSCULTATION OF NORMAL HEART SOUNDS

Listening to the sounds of the body, usually with the aid of a stethoscope, is called *auscultation*. Figure 27–2 illustrates the areas of the chest wall from which the heart sounds from each valve are usually best heard. With the

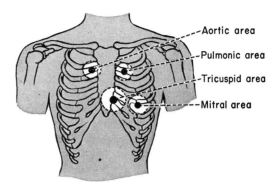

**Figure 27–2.** Chest areas from which each valve sound is best heard. (Redrawn from Adams: Physical Diagnosis. The Williams & Wilkins Co., 1958.)

stethoscope placed in any one of the special valvular areas, the sounds from all the other valves can still be heard, though the sound from the special valve is as loud, *relative to the other sounds,* as it ever will be. The cardiologist distinguishes the sounds from the different valves by a process of elimination; that is, he moves the stethoscope from one area to another, noting the loudness of the sounds in different areas and gradually picking out the sound components from each valve.

The areas for listening to the different heart sounds are not directly over the valves themselves. The aortic area is upward along the aorta, the pulmonic area is upward along the pulmonary artery, the tricuspid area is over the right ventricle, and the mitral area is over the apex of the heart, which is the only portion of the left ventricle near the surface of the chest because the heart is rotated so that most of the left ventricle lies behind the right ventricle. In other words, the sounds caused by the A-V valves are transmitted to the chest wall through each respective ventricle, and the sounds from the semilunar valves are transmitted especially along the great vessels leading from the heart. This transmission of sounds is in keeping with the concept that vibrations in the ventricles or large arteries are the cause of the heart sounds.

## THE PHONOCARDIOGRAM

If a microphone specially designed to detect low-frequency sound waves is placed on the chest, the heart sounds can be amplified and recorded by a high-speed recording apparatus, such as an oscilloscope or a high-speed pen recorder, which were described in Chapters 10 and 15. The recording is called a *phonocardiogram,* and the heart sounds appear as waves in the record as illustrated in Figure 27–3. Record A illustrates a recording of normal heart sounds, showing the vibrations of the first, second, and third heart sounds and even the atrial sound. Note specifically that the third and atrial heart sounds are each a very low rumble. The third heart sound can be recorded in only one third to one half of all persons, and the atrial heart sound can be recorded in perhaps one fourth of all persons.

## VALVULAR LESIONS

**Rheumatic Valvular Lesions.** By far the greatest number of valvular lesions results from rheumatic fever. Rheumatic fever is an autoimmune or allergic disease in which the heart valves are likely to be damaged or destroyed. It is initiated by streptococcal toxin in the following manner:

The entire sequence of events almost always begins with a preliminary streptococcal infection (caused by group A streptococci) such as a sore throat, scarlet fever, or middle ear infection. The streptococci release several different proteins against which antibodies are formed, the most important of which is the streptolysin O antigen. The antibodies then react with many different tissues of the body, causing either immunologic or allergic damage. These reactions continue to take place as long as the antibodies persist in the blood—six months or more. As a result, rheumatic fever causes damage in many parts of the body but especially in certain very susceptible areas such as the heart valves. The degree of heart valve damage is directly correlated with the titer of antistreptolysin O antibodies and with the persistence of these antibodies. Principles of immunity relating to this type of reaction were discussed in Chapter 7, and it is also noted in Chapter 38 that acute glomerular nephritis has a similar basis.

In rheumatic fever, large hemorrhagic, fibrinous, denuded bulbous lesions grow along the inflamed edges of the heart valves. Because the mitral valve receives more trauma during valvular action than do any of the other valves, this valve is the one most often affected, and the aortic valve is second most frequently involved. The tricuspid and pulmonary valves are also frequently involved, but much less severely, probably because the stresses acting on these valves are slight compared with those in the left ventricle.

**Scarring of the Valves.** The lesions of acute rheumatic fever frequently occur on adjacent valve leaflets simultaneously so that the edges of the leaflets become stuck together. Then, in the late stages of rheumatic fever, the lesions become scar tissue, permanently fusing portions of the leaflets. Also, the free edges of the leaflets, which are normally filmy and free-flapping, become solid, scarred masses.

A valve in which the leaflets adhere to each other so extensively that blood cannot flow through satisfactorily is said to be *stenosed*. On the other hand, when the valve edges are so destroyed by scar tissue that they cannot close together, *regurgitation*, or backflow, of blood occurs when the valve should be closed. Stenosis usually does not occur without coexistence of at least some degree of regurgitation, and vice versa. Therefore, when a person is said to have stenosis or regurgitation, it is usually meant that one predominates over the other.

**Other Causes of Valvular Lesions.** Until the last few years, a prominent cause of aortic valvular destruction was vascular syphilis. However, with the advent of modern treatment, syphilis now rarely proceeds to such a severe state. In syphilitic lesions, aortic regurgitation almost always occurs without any stenosis of the valve, differing in this respect from regurgitation caused by rheumatic fever.

Stenosis or lack of one or more leaflets of a valve frequently occurs as a congenital defect. Complete lack of leaflets is rare, though stenosis is common, as is discussed later in this chapter.

## ABNORMAL HEART SOUNDS CAUSED BY VALVULAR LESIONS

As illustrated by the phonocardiograms of Figure 27–3, many abnormal heart sounds, known as "murmurs," occur when there are abnormalities of the valves, as follows:

**The Murmur of Aortic Stenosis.** In aortic stenosis, blood is ejected from the left ventricle through only a small opening through the aortic valve. Because of the resistance to ejection, the pressure in the left ventricle rises sometimes to as high as 350 mm. Hg while the pressure in the aorta is still normal. Thus, a nozzle effect is created *during systole*, with blood jetting at tremendous velocity through the small opening of the valve. This causes severe *turbulence* in the root of the aorta. The turbulent blood impinging against the aortic walls causes intense vibration, and a loud murmur is transmitted throughout the upper aorta and even into the larger

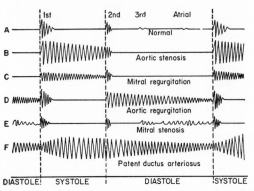

**Figure 27–3.** Phonocardiograms from normal and abnormal hearts.

vessels of the neck. This sound is harsh and occasionally so loud that it can be heard several feet away from the patient. Also, the sound vibrations can often be felt with the hand on the upper chest and lower neck, a phenomenon known as a "thrill."

**The Murmur of Aortic Regurgitation.** In aortic regurgitation no sound is heard during systole, but *during diastole* blood flows backward from the aorta into the left ventricle, causing a "blowing" murmur of relatively high pitch and with a swishing quality heard maximally over the left ventricle. This murmur results from the *turbulence* of blood jetting backward into the blood already in the left ventricle.

The sound of aortic regurgitation is not nearly so loud as that of aortic stenosis, mainly because the pressure differential between the aorta and left ventricle in regurgitation is not so great as it can sometimes be in stenosis.

If the aortic valve is so badly destroyed that essentially all the return of blood from the arterial tree into the heart takes place during the first portion of diastole, the sound of aortic regurgitation may not be heard at all during the latter portion of diastole. Therefore, loud aortic regurgitation murmurs lasting throughout diastole sometimes mean less severely damaged valves than weaker murmurs heard only during the early part.

**The Murmur of Mitral Regurgitation.** In mitral regurgitation blood flows backward through the mitral valve *during systole*. This also causes a high frequency "blowing," swishing sound, which is transmitted most strongly into the left atrium, but the left atrium is so deep within the chest that it is difficult to hear this sound directly over the atrium. As a result, the sound of mitral regurgitation is transmitted to the chest wall mainly through the left ventricle, and it is usually heard best at the apex of the heart.

The blowing, swishing quality of the mitral regurgitation murmur, like that of aortic regurgitation, is presumably caused by the turbulence of blood ejected backward through the mitral valve into the blood of the left atrium or against the atrial wall. The quality of the mitral regurgitation murmur is almost exactly the same as that of aortic regurgitation, but it occurs during systole rather than diastole.

**The Murmur of Mitral Stenosis.** In mitral stenosis, blood passes with difficulty from the left atrium into the left ventricle and, because the pressure in the left atrium rarely rises above 35 mm. Hg except for short periods of time, a great pressure differential forcing blood from the left atrium into the left ventricle never develops. Consequently, the abnormal sounds heard in mitral stenosis are usually weak.

During the early part of diastole, the ventricle has so little blood in it and its walls are so flabby that blood does not reverberate back and forth between the walls of the ventricle. For this reason, even in severe mitral stenosis, no murmur at all might be heard during the first third of diastole. However, after the first third the ventricle is stretched enough

for blood to reverberate, and a low rumbling murmur then often begins. This murmur is of such low pitch that it is difficult to hear, but, with the aid of a proper stethoscope (the "bell" type), one can usually discern very low frequency sounds of 30 to 50 cycles per second. In mild stenosis the murmur lasts only during the first half of the middle third of diastole, but in severe stenosis it can begin early in diastole and persist for the whole of diastole. On the other hand, one can often feel low frequency vibrations, called a "thrill," over the apex of the heart despite the fact that the sound itself may be weak. The reason for this is that the frequency of vibration in the ventricle is often so low that it cannot be heard but yet can be felt.

Often in early stages of mitral stenosis a *presystolic murmur* may be heard. This presystolic murmur is caused by the momentarily increased left atrial pressure resulting from atrial contraction.

**Phonocardiograms of Valvular Murmurs.** Phonocardiograms B, C, D, and E of Figure 27–3 illustrate, respectively, idealized records obtained from patients with aortic stenosis, mitral regurgitation, aortic regurgitation, and mitral stenosis. It is obvious from these phonocardiograms that the aortic stenotic lesion causes the loudest of all these murmurs, and the mitral stenotic lesion causes the weakest, a murmur of very low frequency and rumbling quality. The phonocardiograms show how the intensity of the murmurs varies during different portions of systole and diastole, and the relative timing of each murmur is also evident. Note especially that the murmurs of aortic stenosis and mitral regurgitation occur only during systole, while the murmurs of aortic regurgitation and mitral stenosis occur only during diastole—if the student does not understand this timing, he should pause a moment until he does understand it.

# ABNORMAL CIRCULATORY DYNAMICS IN VALVULAR HEART DISEASE

## DYNAMICS OF THE CIRCULATION IN AORTIC STENOSIS AND AORTIC REGURGITATION

*Aortic stenosis* means a constricted aortic valve with a valvular opening too small for easy ejection of blood from the left ventricle. *Aortic regurgitation* means failure of the aortic valve to close completely and, therefore, failure to prevent backflow of blood from the aorta into the left ventricle during diastole. In aortic stenosis the left ventricle fails to empty adequately, while in aortic regurgitation blood returns to the ventricle after it has been emptied. Therefore, in either case, the *net* stroke volume output of the heart is reduced, and this in turn tends to reduce the cardiac output, resulting eventually in typical circulatory failure.

However, several important compensations take place that can ameliorate the severity of the circulatory defects. Some of these are the following:

**Hypertrophy of the Left Ventricle.** In both aortic stenosis and regurgitation, the left ventricular musculature hypertrophies, and in regurgitation the ventricle also enlarges to hold all the regurgitant blood. Sometimes the left ventricle muscle mass increases as much as four- to five-fold, creating a tremendously large left heart. When the aortic valve is seriously stenosed, this hypertrophied muscle allows the left ventricle to develop as much as 450 mm. Hg intraventricular pressure during occasional periods of peak activity; even at rest the pressure differential across the stenotic valve is occasionally 150 mm Hg. In severe aortic regurgitation, the hypertrophied muscle allows the left ventricle sometimes to pump a stroke volume output as high as 300 ml., though as much as three-fourths of this blood on occasion returns to the ventricle during diastole.

**Increase in Blood Volume.** Another effect that helps to compensate for the diminished net pumping by the left ventricle is increased blood volume. This results from a slight decrease in arterial pressure plus peripheral circulatory reflexes that this decrease induces, both of which diminish renal output of urine. Also, red cell mass increases because of a slight degree of tissue hypoxia.

The increase in blood volume tends to increase venous return to the heart. This in turn increases the ventricular end-diastolic volume, causing the left ventricle to pump its very high pressure in aortic stenosis or its very high stroke volume output in aortic regurgitation.

### Eventual Failure of the Left Ventricle, and Development of Pulmonary Edema

In the early stages of aortic stenosis or aortic regurgitation, the intrinsic ability of the left ventricle to adapt to increasing loads prevents significant abnormalities in circulatory function other than increased work output required of the left ventricle. Therefore, marked degrees of aortic stenosis or aortic regurgitation often occur before the person knows that he has serious heart disease, such as resting left ventricular systolic pressures as high as 200 mm. Hg in aortic stenosis, or left ventricular stroke volume outputs as high as double normal in aortic regurgitation.

However, beyond critical stages of development of these two aortic lesions, the left ventricle finally cannot keep up with the work demand, and as a consequence the left ventricle dilates and cardiac output begins to fall while blood simultaneously dams up in the left atrium and lungs behind the failing left ventricle. The left atrial pressure rises progressively, and at pressures above 30 to 40 mm. Hg mean pressure edema appears in the lungs, as is discussed in detail in Chapter 31.

The progressive changes in circulatory dynamics are shown diagrammatically in Figures 27–4 and 27–5 for aortic stenosis and regurgitation, respectively.

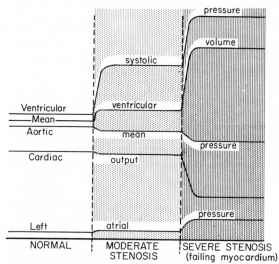

**Figure 27–4.**  Circulatory dynamics in aortic stenosis.

**Relative Myocardial Ischemia in Aortic Valvular Disease.** Because of the very high intraventricular pressure during systole in aortic stenosis, very little blood flows through the myocardial vessels during systole. Therefore, extra blood flow is required during diastole to make up the difference. Unfortunately, the hypertrophied muscle of the left ventricle also frequently has a relatively deficient coronary vasculature. Finally, the intraventricular pressure sometimes remains high during diastole, thereby compressing the inner layers of the heart and also diminishing coronary blood flow. For all these reasons, the patient frequently experiences severe degrees of coronary ischemia and angina.

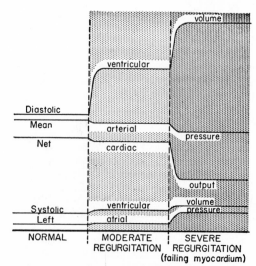

**Figure 27–5.**  Circulatory dynamics in aortic regurgitation.

In aortic regurgitation the problem is especially compounded because the diastolic aortic pressure also frequently falls very low. Since most of the left ventricular coronary blood flow occurs during diastole, this low pressure can be particularly detrimental to the flow. The effect is especially serious for the subendocardial myocardium in which almost zero flow occurs during systole. Therefore, again, relative coronary ischemia occurs with concomitant anginal pain.

## DYNAMICS OF MITRAL STENOSIS AND MITRAL REGURGITATION

In mitral stenosis blood flow from the left atrium into the left ventricle is impeded, and in mitral regurgitation much of the blood that has flowed into the left ventricle leaks back into the left atrium during systole rather than being pumped into the aorta. Therefore, the effect is reduced *net* movement of blood from the left atrium into the left ventricle.

**Pulmonary Edema in Mitral Valvular Disease.** Obviously, the buildup of blood in the left atrium causes progressive increase in left atrial pressure, and this can result eventually in the development of serious pulmonary edema. Ordinarily, lethal edema will not occur until the mean left atrial pressure rises at least above 30 mm. Hg; more often it must rise to as high as 40 mm. Hg because the lymphatic vasculature enlarges many-fold and can carry fluid away from the lung tissues extremely rapidly.

**Enlarged Left Atrium and Atrial Fibrillation.** The high left atrial pressure also causes progressive enlargement of the left atrium, which increases the distance that the cardiac impulse must travel in the atrial walls. Eventually, this pathway becomes so long that it predisposes to the development of circus movements. Therefore, in late stages of mitral valvular disease, especially stenosis, atrial fibrillation usually occurs. This state further reduces the pumping effectiveness of the heart and, therefore, causes still further cardiac debility.

**Compensations in Mitral Valvular Disease.** As also occurs in aortic valvular disease and in many types of congenital heart disease, the blood volume increases in mitral valvular disease. This increases venous return to the heart, thereby helping to overcome the effect of the cardiac debility to reduce cardiac output. Therefore, cardiac output does not fall more than minimally until the late stages of mitral valvular disease.

As the left atrial and pulmonary capillary pressures rise, blood also begins to dam up in the pulmonary artery, and the incipient edema of the lungs causes intense pulmonary arteriolar constriction, these two effects together then increasing pulmonary arterial pressure sometimes to as high as 60 mm. Hg. This, in turn, causes hypertrophy of the right heart, which partially compensates for its increased work load.

Figure 27–6 illustrates the progressive changes in circulatory dynamics in mitral stenosis. Essentially

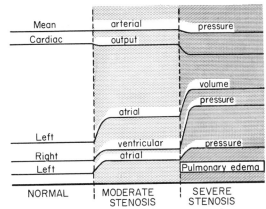

**Figure 27–6.** Circulatory dynamics in mitral stenosis.

the same effects occur in mitral regurgitation. Indeed, these two conditions more often than not occur together in varying degrees.

## CIRCULATORY DYNAMICS DURING EXERCISE IN PATIENTS WITH VALVULAR LESIONS

During exercise very large quantities of venous blood are returned to the heart from the peripheral circulation. Therefore, all of the dynamic abnormalities that occur in the different types of valvular heart disease become tremendously exacerbated. Even in mild valvular heart disease, in which the symptoms may be completely unrecognizable at rest, severe symptoms often develop during heavy exercise. For instance, in patients with aortic valvular lesions, exercise can cause acute left ventricular failure followed by acute pulmonary edema. Also, in patients with mitral disease, exercise can cause so much damming of blood in the lungs that serious pulmonary edema ensues within minutes.

Even in the mildest cases of valvular disease, the patient finds that his cardiac reserve is diminished in proportion to the severity of the valvular dysfunction. That is, the cardiac output does not increase as it should during exercise. Therefore, the muscles of the body fatigue rapidly.

# ABNORMAL CIRCULATORY DYNAMICS IN CONGENITAL HEART DEFECTS

Occasionally, the heart or its associated blood vessels are malformed during fetal life; the defect is called a *congenital anomaly*. Basically, there are three major types of congenital anomalies of the heart and its associated vessels: (1) *stenosis* of the channel of blood flow at some point in the heart or in a closely allied major vessel, (2) an abnormality that allows blood to flow directly from the left heart or aorta to the right heart or pulmonary artery, thus by-passing the systemic circulation—this is called a *left-to-right shunt*—and (3) an abnormality that allows blood to flow from the right heart or pulmonary artery directly into the left heart or aorta, thus by-passing the lungs—this is called a *right-to-left shunt*.

The three most common types of stenotic lesions are *coarctation of the aorta, congenital pulmonary stenosis,* and *congenital aortic stenosis.* The effect of coarctation of the aorta on the circulation in relation to the regulation of arterial pressure was discussed in Chapter 22, and congenital aortic stenosis causes almost exactly the same alterations in circulatory dynamics as those noted above for acquired aortic stenosis. Therefore, in this present section only the dynamics of pulmonary stenosis are discussed.

In all left-to-right shunts, blood is shunted directly from the left heart back to the right heart without going through the systemic circulation, and compensatory effects occur in the circulatory system to increase the total cardiac output, thereby making up for most of the extra blood flow through the shunt. For instance, if 50 per cent of all the blood pumped by the left heart is shunted directly back to the right heart, then the compensatory mechanisms normally increase the cardiac output to almost double normal so that the amount of blood flowing through the systemic vessels is still almost normal despite the shunt. However, this causes an extra load on the heart and therefore usually causes the heart to fail at an early age. Also, it reduces the cardiac reserve of the person throughout his life. Some of the types of left-to-right shunts are *patent ductus arteriosus, interatrial septal defect, interventricular septal defect,* and a *direct communication between the aorta and pulmonary artery at their bases.*

In all right-to-left shunts, much of the blood bypasses the lungs and therefore fails to become oxygenated. Consequently, large quantities of venous blood enter the arterial system directly, and the patient is usually cyanotic (blue) all the time. The most common type of right-to-left shunt is the *tetralogy of Fallot,* which is described later; several other right-to-left shunts are some variation of this abnormality.

**Causes of Congenital Anomalies.**   One of the most common causes of congenital heart defects is a virus infection of the mother during the first trimester of pregnancy when the fetal heart is being formed. Defects are particularly prone to develop when the mother contracts German measles at this time—so often indeed that obstetricians advise termination of pregnancy if German measles occurs in the first trimester. However, some congenital defects of the heart are believed to be hereditary because the same defect has been known to occur in identical twins and also in succeeding generations. Children of patients surgically treated for congenital heart disease have ten times as much chance of having congenital heart disease as do other children. Congenital defects of the heart are frequently also associated with other congenital defects of the body.

## PATENT DUCTUS ARTERIOSUS— A LEFT-TO-RIGHT SHUNT

During fetal life the lungs are collapsed, and the elastic factors that keep the alveoli collapsed also keep the blood vessels collapsed. Therefore, in the collapsed lung the resistance to blood flow is about 5 times as great as it is in the inflated lung. For this reason, the pulmonary arterial pressure is high in the fetus. Because of very low resistance through the large vessels of the placenta, the pressure in the aorta is lower than in the pulmonary artery, causing almost all the pulmonary arterial blood to flow through the ductus arteriosus into the aorta rather than through the lungs. This allows immediate recirculation of the blood through the systemic arteries of the fetus. Obviously, this lack of blood flow through the lungs is of no detriment to the fetus because the blood is oxygenated by the placenta of the mother.

**Closure of the Ductus.**   As soon as the baby is born, his lungs inflate; and not only do the alveoli fill, but the resistance to blood flow through the pulmonary vascular tree decreases tremendously, allowing pulmonary arterial pressure to fall. Simultaneously, the aortic pressure rises because of sudden cessation of blood flow through the placenta. Thus, the pressure in the pulmonary artery falls, while that in the aorta rises. As a result, forward blood flow through the ductus ceases suddenly at birth, and blood even flows backward from the aorta to the pulmonary artery. This new state of blood flow causes the ductus arteriosus to become occluded within a few hours to a few days in most babies so that blood flow through the ductus does not persist. The possible causes of ductus closure will be discussed in Chapter 83. In many instances it takes several months for the ductus to close completely, and in about 1 out of every 5500 babies the ductus never closes, causing the condition known as *patent ductus arteriosus,* which is illustrated in Figure 27–7.

**Dynamics of Persistent Patent Ductus.**   During the early months of an infant's life a patent ductus usually does not cause severely abnormal dynamics because the blood pressure of the aorta then is not

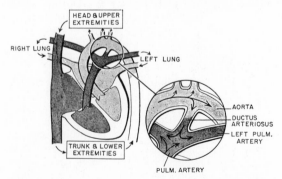

**Figure 27–7.**  Patent ductus arteriosus, illustrating the degree of blood oxygenation in the different parts of the circulation.

much higher than the pressure in the pulmonary artery, and only a small amount of blood flows backward into the pulmonary system. However, as the child grows older, the differential between the pressure in the aorta and that in the pulmonary artery progressively increases, with corresponding increase in the backward flow of blood from the aorta to the pulmonary artery. Also, the diameter of the partially closed ductus often increases with time, making the condition worse.

*Recirculation Through the Lungs.*    In the older child with a patent ductus, as much as half to three fourths of the aortic blood flows into the pulmonary artery, then through the lungs, into the left atrium, and finally back into the left ventricle, passing through this lung circuit two or more times for every one time that it passes through the systemic circulation.

These persons do not show cyanosis until the heart fails or until the lungs become congested. Indeed, the arterial blood is often better oxygenated than normally because of the extra times of passage of the blood through the lungs. Furthermore, in the early stages of patent ductus arteriosus the quantity of blood flowing into the systemic aorta remains essentially normal because the quantity of blood returning to the heart from the peripheral circulatory system is normal. Yet, because of the tremendous accessory flow of blood around and around through the lungs and left side of the heart, the output of the left ventricle in patent ductus arteriosus is usually two to four times normal.

*Diminished Cardiac and Respiratory Reserve.*    The major effects of patent ductus arteriosus on the patient are low cardiac and respiratory reserve. The left ventricle is already pumping approximately two or more times the normal cardiac output, and the maximum that it can possibly pump is about four to six times normal. Therefore, during exercise the cardiac output can be increased much less than usual. Under basal conditions, the patient usually appears normal except for possible heaving of his chest with each beat of the heart, but just as soon as he tries to perform even moderately strenuous exercise he is likely to become weak and occasionally even faint from momentary heart failure. Also, the high pressures in the pulmonary vessels soon lead to pulmonary congestion.

The entire heart usually hypertrophies greatly in patent ductus arteriosus. The left ventricle hypertrophies because of the excessive work load that it must perform in pumping a far greater than normal cardiac output, while the right ventricle hypertrophies because of increased pulmonary arterial pressure resulting from, first, increased flow of blood through the lungs caused by the extra blood from the patent ductus and, second, increased resistance to blood flow through the lungs caused by progressive sclerosing of the vessels as they are exposed year in and year out to excessive pulmonary blood flow.

As a result of the increased load on the heart and because of the pulmonary congestion, most patients with patent ductus die between the ages of 20 and 40 unless the defect is corrected by surgery.

**The Machinery Murmur.**    In the infant with patent ductus arteriosus, occasionally no abnormal heart sounds are heard because the quantity of reversed blood flow may be insignificant. As the baby grows older, reaching the age of one to three years, a harsh, blowing murmur begins to be heard in the pulmonic area of the chest. This sound is much more intense during systole when the aortic pressure is high and much less intense during diastole, so that it waxes and wanes with each beat of the heart, creating the so-called "machinery murmur." The idealized phonocardiogram of this murmur is shown in Figure 27–3F.

**Surgical Treatment.**    Surgical treatment of patent ductus arteriosus is extremely simple, for all one needs to do is to ligate the patent ductus or to divide it and sew the two ends.

## INTERVENTRICULAR SEPTAL DEFECT— A LEFT-TO-RIGHT SHUNT

Because the systolic pressure in the left ventricle is normally about six times that in the right ventricle, a large amount of blood flows from the left to the right ventricle whenever an interventricular defect is present. As a result of elevated pressure in the right ventricle, the right ventricle hypertrophies, sometimes to such an extent that its muscular wall approximately equals that of the left ventricle.

Diagnosis of an interventricular septal defect is based on (1) the presence of a systolic blowing murmur heard over the anterior projection of the heart, (2) high right ventricular systolic pressure recorded from a catheter, and (3) the presence of oxygenated blood in a blood sample removed through the catheter from the right ventricle, this blood having leaked backward from the left ventricle.

Blood flowing from the left ventricle into the right ventricle passes one or more times through the lungs and then back to the left ventricle again before finally entering the peripheral circulatory system. This condition, therefore, is analogous to patent ductus arteriosus except that the flow of blood from the systemic circulation to the pulmonary circulation occurs only during systole rather than during both systole and diastole as in patent ductus.

The septal defect can be treated surgically by placing a patch over the defect.

## INTERATRIAL SEPTAL DEFECTS— A LEFT-TO-RIGHT SHUNT

**Closure of the Foramen Ovale.**    During fetal life much of the blood entering the right atrium fails to pass into the right ventricle but instead courses directly through the foramen ovale into the left atrium and thence out into the systemic circulation. This mechanism aids the ductus arteriosus in shunting blood around the lungs, thereby relieving the fetal heart from the unnecessary load. Immediately after birth of the child, the pressure in the pulmonary artery decreases while that in the aorta increases, as discussed above. These changes decrease and increase respectively the loads on the right and left ventricles so that the right atrial pressure decreases and the left atrial pressure increases. As a result, blood then attempts to flow from the left atrium back into the right atrium. However, the foramen ovale is covered by a small valvelike vane, which closes over the foramen. In two thirds of all persons the foramen later becomes totally occluded by fibrous tissue, but in one third the foramen never becomes totally occluded. Yet this nonoccluded opening does not cause any physiological abnormality in the heart because the left atrial pressure remains 1 to 3 mm. Hg higher than the right atrial pressure, keeping the valve permanently closed.

**Dynamics of Interatrial Defects.** Occasionally, the valve does not cover the foramen ovale, and a hole persists permanently between the left atrium and the right atrium. If this hole is small, only a small amount of blood passes from the left atrium back to the right atrium, but occasionally almost the entire wall between the two atria is missing. In this case a tremendous quantity of blood flows from the left atrium into the right atrium and recirculates, sometimes as much as three times, through the right ventricle and lungs before finally passing into the systemic circulation. Therefore, large atrial septal defects greatly reduce the cardiac reserve, and early death results from right ventricle failure and pulmonary congestion.

The diagnosis of an interatrial septal defect is difficult to make, for no murmurs can be heard, the patient's arterial blood is well oxygenated, and the usual x-ray studies are relatively nonspecific. The diagnosis can be made accurately by (1) angiocardiograms (x-rays made after injecting a radiopaque dye into the heart) or (2) finding oxygenated blood in a catheter specimen from the right atrium in the absence of a detectable murmur.

As in interventricular septal defect, in interatrial septal defect a surgically applied patch over the opening can cause dramatic reversal of the course of the disease.

## TETRALOGY OF FALLOT— A RIGHT-TO-LEFT SHUNT

Tetralogy of Fallot is illustrated in Figure 27–8, from which it will be noted that four different abnormalities of the heart occur simultaneously.

First, the aorta originates from the right ventricle rather than the left, or it overrides the septum as shown in the figure.

Second, the pulmonary artery is stenosed so that a much less than normal amount of blood passes from the right side of the heart into the lungs; instead the blood passes into the aorta.

Third, blood from the left ventricle flows through a

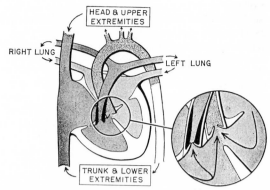

**Figure 27–8.** Tetralogy of Fallot, illustrating the degree of blood oxygenation in the different parts of the circulation.

ventricular septal defect into the right ventricle and then into the aorta or directly into the overriding aorta.

Fourth, because the right side of the heart must pump large quantities of blood against the high pressure in the aorta, its musculature is highly developed, causing an enlarged right ventricle.

**Abnormal Dynamics.** It is readily apparent that the major physiological difficulty caused by tetralogy of Fallot is the shunting of blood past the lungs without its becoming oxygenated. As much as 75 percent of the venous blood returning to the heart may pass directly from the right ventricle into the aorta without becoming oxygenated. Tetralogy of Fallot is the major cause of cyanosis in babies ("blue babies").

A diagnosis of tetralogy of Fallot is usually based on (1) the fact that the baby is blue, (2) records of high systolic pressure in the right ventricle recorded through a catheter, (3) characteristic changes in the x-ray silhouette of the heart showing an enlarged right ventricle, and (4) angiograms showing abnormal blood flow through the interventricular septal defect, the overriding aorta, and the pulmonary stenotic area.

**Surgical Treatment.** In recent years tetralogy of Fallot has been treated relatively successfully by surgery. One type of treatment is to create an artificial ductus arteriosus by making a small opening between the aorta and pulmonary artery, thereby correcting the cyanosis and increasing the life expectancy of the patient with tetralogy of Fallot from approximately 1 to 10 years to an age of perhaps 50 or more years. Another operation is to open the pulmonary stenosis, to close the septal defect, and to reconstruct the flow pathway into the aorta in those cases in which this is possible.

## PULMONARY STENOSIS

Often a child is born with pulmonary stenosis but without other congenital abnormalities. Such severe pulmonary stenosis occasionally occurs that the right side of the heart is likely to fail at an early age because blood flow from the right ventricle into the lungs is greatly impeded, the right ventricular systolic pressure rising to 75 to 100 mm. Hg instead of the normal 22 mm. Hg. The right side of the heart dilates, and the muscle becomes greatly hypertrophied in order to withstand the load. Also, a loud stenotic murmur is heard over the pulmonary valve area. In many cases of pulmonary stenosis, the stenotic area can be enlarged by surgical maneuvers so that the heart resumes normal function.

## USE OF EXTRACORPOREAL CIRCULATION IN CARDIAC SURGERY

It is almost impossible to repair intracardiac defects while the heart is still pumping. Therefore, many different types of artificial *heart-lung machines* have been developed to take the place of the heart and lungs during the course of operation. Such a system is called an *extracorporeal circulation*. The system consists principally of (1) a pump and (2) an oxygenating device. Almost any type of pump that does not cause hemolysis of the blood seems to be suitable.

The different principles that have been used for oxygenating blood are (1) bubbling oxygen through the blood and then removing the bubbles from the blood before passing it back into the patient, (2) dripping the blood downward over the surfaces of large areas of screen wire, (3) passing the blood over the surfaces of rotating discs, and (4) passing the blood between thin membranes or through thin tubes that are porous to oxygen and carbon dioxide.

The different oxygenators have been fraught with many difficulties, including hemolysis of the blood, development of small clots in the blood, likelihood of small bubbles of oxygen or small emboli of antifoaming agent passing into the arteries of the patient, necessity for large quantities of blood to prime the entire system, failure to exchange adequate quantities of oxygen, and the necessity to use heparin in the system to prevent blood coagulation, the heparin also preventing adequate hemostasis during the surgical procedure. Yet, despite these difficulties, in the hands of experts, patients can be kept on artificial heart-lung machines for many hours while operations are performed on the inside of the heart.

## HYPERTROPHY OF THE HEART IN VALVULAR AND CONGENITAL HEART DISEASE

Hypertrophy of cardiac muscle is one of the most important mechanisms by which the heart adapts to increased work loads, whether these loads be caused by increased pressure against which the heart muscle must contract or by increased volume that must be pumped. Some investigators believe that it is the increased work load itself that causes the hypertrophy; others believe the increased metabolic rate of the muscle to be the primary stimulus and the work load simply to be the cause of the increase in the metabolic rate. Regardless of which of these is correct, one can calculate approximately how much hypertrophy will occur in each chamber of the heart by multiplying ventricular output times the pressure against which the ventricle must work, with special emphasis on the pressure. Thus, hypertrophy occurs in most types of valvular and congenital disease, as follows:

In *aortic stenosis* and *aortic regurgitation,* the left ventricular musculature hypertrophies tremendously, sometimes to as much as four to five times normal, so that the weight of the heart on occasion may be as great as 1000 grams instead of the normal 300 grams.

In *mitral stenosis,* the left atrium hypertrophies and dilates, but no left ventricular hypertrophy occurs. In *mitral regurgitation* moderate to severe hypertrophy of the left ventricle occurs, and hypertrophy of the right ventricle develops owing to back pressure effects through the lungs causing elevation of the pulmonary arterial pressure.

In *patent ductus arteriosus,* the work load of both ventricles is increased. The left ventricle pumps an average of twice the quantity of blood that it normally pumps; therefore, it would be expected to hypertrophy. On the other hand, the right ventricle must pump its blood against a much higher than normal pulmonary arterial pressure because of the large quantity of blood refluxing from the aorta into the pulmonary artery. Consequently, the right ventricle also hypertrophies.

In *tetralogy of Fallot,* one of the cardinal signs is the marked right ventricular hypertrophy, for the right ventricle must pump against the pressure in the aorta, and it also must pump an increased volume of blood. On the other hand, the work load of the left ventricle is actually less than normal because of the reduced blood flow from the lungs into the left heart. Therefore, the right ventricle in tetralogy of Fallot is often larger than the normal left ventricle, while the left ventricle may be relatively small.

**Detrimental Effects of the Late Stages of Hypertrophy.** Though physiological hypertrophy of heart muscle is usually very beneficial to cardiac function, extreme degrees of hypertrophy sometimes lead to failure. One of the reasons for this is that coronary blood flow does not always increase to the same extent as the mass of muscle. A second reason is that fibrosis often develops in the muscle, the fibrous tissue replacing degenerating muscle fibers. Because of the sometimes disproportionate increase in muscle mass relative to coronary flow, relative ischemia sometimes develops as the muscle hypertrophies, and coronary insufficiency easily ensues. Therefore, anginal pain is a frequent accompaniment of many valvular and congenital heart diseases.

# REFERENCES

Alpert, N. (ed.): Cardiac Hypertrophy. New York, Academic Press, Inc., 1971.

Aravanis, C.: Exercise and the heart. *Cardiovasc. Clin.*, 2:181, 1971.

Badeer, H. S.: Development of cardiomegaly. *Cardiology*, 57:247, 1972.

Bayer, L. M., and Honzik, M. P.: Children with Intracardiac Defects. Springfield, Ill., Charles C Thomas, Publisher, 1974.

Bellhouse, B. J.: Fluid mechanics of a model mitral valve and left ventricle. *Cardiovasc. Res.*, 6:199, 1972.

Butterworth, J. S., Chassin, M. R., McGrath, R., and Reppert, E. H.: Cardiac Auscultation–Including Audio-Visual Principles, 2nd Ed. New York, Grune & Stratton, Inc., 1960.

Cooley, D. A., and Hallman, G. L.: Surgical Treatment of Congenital Heart Disease. Philadelphia, Lea & Febiger, 1966.

Galletti, P. M., and Brecher, G. A.: Heart-Lung Bypass–Principles and Techniques of Extracorporeal Circulation. New York, Grune & Stratton, Inc., 1962.

Grossman, W. (ed.): Cardiac Catheterization and Angiography. Philadelphia, Lea & Febiger, 1974.

Gyepes, M. T., and Vincent, W. R.: Cardiac Catheterization and Angiocardiography in Severe Neonatal Heart Disease. Springfield, Ill., Charles C Thomas, Publisher, 1974.

Heymann, M. A., and Rudolph, A. M.: Control of the ductus arteriosus. *Physiol. Rev.*, 55:62, 1975.

James, P. M., Jr.: Clinical uses of central venous cannulation. *Postgrad. Med.*, 55:155, 1974.

Mendel, D.: A Practice of Cardiac Catheterization, 2nd Ed. Philadelphia, J. B. Lippincott Company, 1973.

Mitchell, J. H., and Wildenthal, K.: Static (isometric) exercise and the heart: physiological and clinical considerations. *Ann. Rev. Med.*, 25:369, 1974.

Pomerance, A., and Davies, M. J. (eds.): Pathology of the Heart. Philadelphia, J. B. Lippincott Company, 1975.

Ravin, A.: Auscultation of the Heart. Baltimore, The Williams & Wilkins Company, 1967.

Selzer, A.: Principles of Clinical Cardiology. Philadelphia, W. B. Saunders Company, 1973.

Spaeth, E. E.: Blood oxygenation in extracorporeal devices: theoretical considerations. *CRC Crit. Rev. Bioeng.*, 1:383, 1973.

Stapleton, J. F., and Harvey, W. P.: Heart sounds, murmurs, and precordial movements. *In* Sodeman, W. A., Jr., and Sodeman, W. A. (eds.): Pathologic Physiology: Mechanisms of Disease, 5th Ed. Philadelphia, W. B. Saunders Company, 1974, p. 295.

Taussig, H.: Congenital Malformations of the Heart. Vol. 1: General Considerations, 2nd Ed. Cambridge, Mass., Harvard University Press, 1960.

Taussig, H.: Congenital Malformations of the Heart. Vol. 2: Specific Malformations, 2nd Ed. Cambridge, Mass., Harvard University Press, 1961.

Taylor, D. E. M., and Wade, J. D.: The pattern of flow around the atrioventricular valves during diastolic ventricular filling. *J. Physiol.*, 207, 71P, 1970.

Wood, E. H.: Diagnostic applications of indicator-dilution technics in congenital heart disease. *Circ. Res.*, 10:531, 1962.

# | 28 |

# Circulatory Shock and
# Physiology of Its Treatment

*Circulatory shock* means generalized inadequacy of blood flow throughout the body, to the extent that the tissues are damaged because of lack of adequate tissue blood flow. Even the cardiovascular system itself—the heart musculature, the walls of the blood vessels, the vasomotor system, and other circulatory parts—begins to deteriorate so that the shock becomes progressively worse.

In this discussion, shock will be divided into three major stages: (1) a *nonprogressive stage* (sometimes called the *compensated stage*), (2) a *progressive stage,* and (3) an *irreversible stage.* In the nonprogressive stage, tissue perfusion is deficient but not deficient enough to cause a vicious cycle of cardiovascular deterioration. In the progressive stage, the shock has progressed to the point that the circulatory system begins to deteriorate, thus leading to a vicious cycle that eventuates in death unless treatment is instituted. In the irreversible stage, the shock has progressed to the point that all forms of therapy will be inadequate to save the life of the person even though the person is at the moment still alive.

**Cardiac Output in Shock.** Usually, the cause of inadequate tissue perfusion in circulatory shock is inadequate cardiac output, meaning simply that not enough blood is pumped by the heart to supply adequate blood flow to the tissues. However, under special circumstances the cardiac output itself is normal or even greater than normal but still is inadequate to supply the tissue needs. This condition may result from too high a rate of metabolism in the tissues or from abnormal flow patterns in the peripheral vasculature that prevent adequate diffusion of nutrients and other substances between the circulatory system and the tissue cells.

Yet, most times, one can consider shock to be caused by too low a cardiac output in relation to the tissue metabolism.

**Arterial Pressure in Shock.** There are times when a person is in severe shock and still has a normal arterial pressure because of nervous reflexes that keep the pressure from falling. At other times the arterial pressure can fall to as low as one-half normal but the person can still have normal tissue perfusion and not be in shock.

However, in many types of shock, especially that caused by severe blood loss, the arterial pressure does usually fall at the same time that the cardiac output decreases. Therefore, measurements of arterial pressure are of major value in assessing the degree of shock.

## PHYSIOLOGICAL CAUSES OF SHOCK

Since shock usually results from inadequate cardiac output, any factor that can reduce cardiac output can also cause shock. The different factors that can do this were discussed in Chapter 23 and were grouped into two categories:

1. Those that decrease the ability of the heart to pump blood.

2. Those that tend to decrease venous return.

Thus, serious myocardial infarction or any

other factor that damages the heart so severely that it cannot pump adequate quantities of blood can cause a type of shock called *cardiogenic shock,* which was discussed in Chapter 26. Also, all the factors that reduce venous return, including (a) diminished blood volume, (b) decreased vasomotor tone, or (c) greatly increased resistance to blood flow, can result in shock. The present chapter will deal primarily with these factors.

# SHOCK CAUSED BY HYPOVOLEMIA— HEMORRHAGIC SHOCK

Hypovolemia means diminished blood volume, and hemorrhage is perhaps the most common cause of hypovolemic shock. Therefore, a discussion of hemorrhagic shock will serve to explain many of the basic principles of the shock problem.

Hemorrhage *decreases the systemic filling pressure* and as a consequence decreases venous return. As a result, the cardiac output falls below normal, and shock ensues. Obviously, all degrees of shock can result from hemorrhage, from the mildest diminishment of cardiac output to almost complete cessation of output.

## *RELATIONSHIP OF BLEEDING VOLUME TO CARDIAC OUTPUT AND ARTERIAL PRESSURE*

Figure 28–1 illustrates the effect on both cardiac output and arterial pressure of removing blood from the circulatory system over a period of about half an hour. Approximately 10 per cent of the total blood volume can be removed with no significant effect on arterial pressure or cardiac output, but greater blood loss usually diminishes the cardiac output first and later the pressure, both of these falling to zero when about 35 to 45 per cent of the total blood volume has been removed.

**Sympathetic Reflex Compensation in Shock.** Fortunately, the decrease in arterial pressure caused by blood loss initiates powerful sympathetic reflexes that stimulate the sympathetic vasoconstrictor system throughout the body, resulting in three important effects: (1) The arterioles constrict in most parts of the body, thereby greatly increasing the total peripheral resistance. (2) The veins and venous reservoirs constrict, thereby helping to maintain adequate venous return despite diminished blood volume. And (3) heart activity increases markedly, sometimes increasing the heart rate from the normal value of 72 beats per minute to as much as 200 beats per minute.

*Value of the Reflexes.* In the absence of the sympathetic reflexes, only 15 to 20 per cent of the blood volume can be removed over a period of half an hour before a person will die; this is in contrast to 30 to 40 per cent when the reflexes are intact. Therefore, the reflexes extend the amount of blood loss that can occur without causing death to about two times that which would be possible in their absence.

*Greater Effect of the Reflexes in Maintaining Arterial Pressure Than in Maintaining Output.* Referring again to Figure 28–1, one sees that the arterial pressure is maintained at or near normal levels in the hemorrhaging person longer than is the cardiac output. The reason for this is that the sympathetic reflexes are geared more for maintenance of arterial pressure than for maintenance of output. They increase the arterial pressure to a great extent by increasing the total peripheral resistance, which has no beneficial effect on cardiac output; instead, cardiac output is regulated more by local factors in the circulation than by the nervous system, as was discussed in Chapter 23.

Especially interesting are the two plateaus in the arterial pressure curve of Figure 28–1. The failure of the arterial pressure to decrease when one first begins to remove blood from the circulation is caused mainly by strong stimulation of the *baroreceptor reflexes.* Then, as the arterial pressure falls to about 30 mm. Hg, a second plateau results from activation of the *CNS ischemic response,* which was discussed in Chapter 21. This effect of the CNS ischemic response

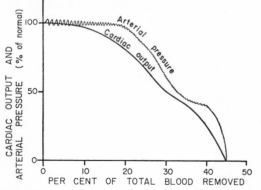

**Figure 28–1.** Effect of hemorrhage on cardiac output and arterial pressure.

can be called the "last-ditch stand" of the sympathetic reflexes in their attempt to keep the arterial pressure from falling too low.

***Protection of Coronary and Cerebral Blood Flow by the Reflexes.*** A special value of the maintenance of normal arterial pressure even in the face of decreasing cardiac output is protection of blood flow through the coronary and cerebral circulatory systems. Sympathetic stimulation does not cause significant constriction of either the cerebral or cardiac vessels—if anything, it causes slight vasodilatation of the coronary vessels. In addition, in both these vascular beds local autoregulation is excellent, which prevents moderate changes in arterial pressure from significantly affecting their blood flows. Therefore, blood flow through the heart and brain is maintained essentially at normal levels as long as the arterial pressure does not fall below about 70 mm. Hg, despite the fact that blood flow in many other areas of the body might be decreased almost to zero because of vasospasm.

## NONPROGRESSIVE AND PROGRESSIVE HEMORRHAGIC SHOCK

Figure 28–2 illustrates an experiment in dogs that demonstrates the effects of different degrees of hemorrhage on the subsequent course of arterial pressure. The dogs were bled rapidly until their arterial pressures fell to different levels. Those dogs whose pressures fell immediately no lower than 45 mm. Hg (Groups I, II, and III) all eventually recovered; the recovery occurred rapidly if the pressure fell only slightly (Group I) and occurred slowly if it fell almost to the 45 mm. Hg level (Group III). On the other hand, when the arterial pressure fell below 45 mm. Hg (Groups IV, V, and VI), all the dogs died, though many of them hovered between life and death for many hours before the circulatory system began to deteriorate.

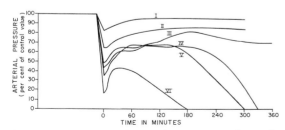

**Figure 28–2.** Course of arterial pressure in dogs after different degrees of acute hemorrhage. Each curve represents the average results from six dogs.

This experiment demonstrates that the circulatory system can recover as long as the degree of hemorrhage is no greater than a certain critical amount. However, crossing this critical amount by even a few milliliters of blood loss makes the eventual difference between life and death. Thus, hemorrhage beyond a certain critical level causes shock to become *progressive.* That is, *the shock itself causes still more shock,* the condition becoming a vicious cycle that leads eventually to complete deterioration of the circulation and to death.

### Nonprogressive Shock–Compensated Shock

If shock is not severe enough to cause its own progression, the person eventually recovers. Therefore, shock of this lesser degree can be called *nonprogressive shock.* It is also frequently called *compensated shock,* meaning that the sympathetic reflexes and other factors have compensated enough to prevent deterioration of the circulation.

The factors that cause a person to recover from moderate degrees of shock are the negative feedback control mechanisms of the circulation that attempt to return cardiac output and arterial pressure to normal levels. These include:

1. The *baroreceptor reflexes,* which elicit powerful sympathetic stimulation of the circulation

2. The *central nervous system ischemic response,* which elicits even more powerful sympathetic stimulation throughout the body but is not activated until the arterial pressure falls below 50 mm. Hg

3. *Reverse stress-relaxation of the circulatory system,* which causes the blood vessels to contract down around the diminished blood volume so that the blood volume that is available will more adequately fill the circulation

4. *Formation of angiotensin,* which constricts the peripheral arteries, constricts the veins, and causes increased conservation of water and salt by the kidneys, all of which help to prevent progression of the shock

5. *Compensatory mechanisms that return the blood volume back toward normal,* including absorption of large quantities of fluid from the intestinal tract, absorption of fluid from the interstitial spaces of the body, and increased thirst and increased appetite for salt which make the person drink water and eat salty foods if he is able

The sympathetic reflexes provide immediate

help toward bringing about recovery, for they become maximally activated within 30 seconds after hemorrhage. The reverse stress relaxation that causes contraction of the blood vessels and venous reservoirs around the blood requires some 10 minutes to an hour to occur completely, but, nevertheless, this aids greatly in increasing the systemic filling pressure and thereby increasing the return of blood to the heart. Finally, the readjustment of blood volume by absorption of fluid from the interstitial spaces and from the intestinal tract, as well as the ingestion and absorption of additional quantities of fluid and salt, may require from 1 to 48 hours, but eventually recovery takes place provided the shock does not become severe enough to enter the progressive stage.

### Progressive Shock–The Vicious Cycle of Cardiovascular Deterioration

Once shock has become severe enough, the structures of the circulatory system themselves begin to deteriorate, and various types of positive feedback develop that can cause a vicious cycle of progressively decreasing cardiac output. Figure 28–3 illustrates some of these different types of positive feedback that further de-

press the cardiac output in shock. These are the following:

**Cardiac Depression.** When the arterial pressure falls low enough, coronary blood flow decreases below that required for adequate nutrition of the myocardium itself. This obviously weakens the heart and thereby decreases the cardiac output still more. As a consequence, the arterial pressure falls still further, and the coronary blood flow decreases more, making the heart still weaker. Thus, a positive feedback cycle has developed whereby the shock becomes more and more severe.

Figure 28–4 illustrates cardiac output curves from experiments in dogs, showing progressive deterioration of the heart at different times following the onset of shock. The dog was bled until the arterial pressure fell to 30 mm. Hg, and the pressure was held at this level by further bleeding or retransfusion of blood as required. Note that there was little deterioration of the heart during the first two hours, but by four hours the heart had deteriorated about 40 per cent; then, rapidly, during the last hour of the experiment the heart deteriorated almost completely.

Thus, one of the important features of progressive shock, whether it be hemorrhagic

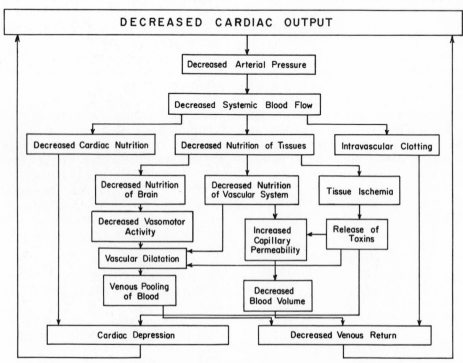

**Figure 28–3.** Different types of feedback that can lead to progression of shock.

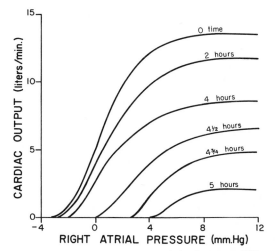

**Figure 28–4.** Function curves of the heart at different times after hemorrhagic shock begins. (These curves are extrapolated to the human heart from data obtained in dog experiments by Dr. J. W. Crowell.)

shock or any other type of shock, is eventual progressive deterioration of the heart. In the early stages of shock, this plays very little role in the condition of the person, partly because deterioration of the heart itself is not very severe during the first hour or so of shock but mainly because the heart has tremendous reserve that makes it normally capable of pumping 300 to 400 per cent more blood than is required by the body for adequate nutrition. However, in the late stages of shock, deterioration of the heart is probably by far the most important factor in the further progression of the shock.

Aside from the myocardial depression caused by poor coronary blood flow in shock, the myocardium can also be depressed by toxic factors transported to the heart from other parts of the body, especially by the factor called *myocardial toxic factor* (MTF) but also by such substances as excess lactic acid, bacterial toxins from the gut, degeneration products from dying tissues, and so forth. Some of these factors will be discussed further later in this chapter.

**Vasomotor Failure.** In the early stages of shock various circulatory reflexes cause intense activity of the sympathetic nervous system. This, as discussed above, helps to delay depression of the cardiac output and especially helps to prevent decreased arterial pressure. However, there comes a point at which diminished blood flow to the vasomotor center itself so de-

presses the center that it becomes progressively less active and finally totally inactive. For instance, complete circulatory arrest to the brain causes, during the first four to eight minutes, the most intense of all sympathetic discharges, resulting in a typical CNS ischemic response, but by the end of 10 to 15 minutes the vasomotor center becomes so depressed that no evidence of sympathetic discharge at all can be demonstrated. Fortunately, though, the vasomotor center does not usually fail in the early stages of shock—only in the late stages.

**Vascular Failure.** In addition to vascular dilatation caused by vasomotor failure, dilatation can also result from direct diminishment of nutrients to the vessels during shock. *Arteriolar dilatation* can reduce the arterial pressure and thereby reduce the blood flow to the heart and brain, in this way causing further progression of the shock. And *venous dilatation* can cause blood to "pool" in the veins, thereby depressing cardiac output.

**Thrombosis of the Minute Vessels— Sludged Blood.** Recently, several research workers have shown that thrombosis of many of the minute vessels in the circulatory system can be one of the causes of progression of shock. That is, blood flow through many tissues becomes extremely sluggish, but tissue metabolism continues so that large amounts of acid, either carbonic acid or lactic acid, continue to empty into the blood. This acid, plus other deterioration products from the ischemic tissues, causes blood agglutination or actual blood clots, thus leading to minute plugs in the small vessels. Even if the vessels do not become plugged, the agglutination of cells makes it more difficult for blood to flow through the microvasculature, giving rise to the term *sludged blood*.

**Increased Capillary Permeability.** After many hours of capillary hypoxia the permeability of the capillaries gradually increases and large quantities of fluid begin to transude into the tissues. This decreases the blood volume, with resultant further decrease in cardiac output, thus making the shock still more severe.

Fortunately, capillary hypoxia does not cause increased capillary permeability until the very late stages of extremely prolonged shock. Therefore, this factor plays a significant role in only few instances of shock.

**Release of Toxins by Ischemic Tissues.** Throughout the history of research in the field of shock, it has been suggested time and again that shock causes tissues to release toxic substances that then cause further deterioration of the circulatory system. For instance, during the First World War, it was suggested that a so-called H substance (a substance that supposedly resembled histamine in its actions) was released from essentially all the ischemic tissues of the body and that this substance caused arteriolar paralysis and increased capillary permeability with resultant loss of fluid from the circulation. The release of this substance theoretically made the shock progress very rapidly, ending soon in death. And true enough, it is now known that tissues suffering from shock do indeed release small amounts of histamine, as well as serotonin and many tissue enzymes. Critical quantitative studies, however, have failed to prove the significance of histamine and serotonin in shock. On the other hand, two toxic factors are now known to play important roles in at least some types of shock. These are *myocardial toxic factor* and *endotoxin.*

*Myocardial Toxic Factor (MTF).* During the course of shock, the splanchnic arterioles become strongly constricted, and the arterioles of the pancreas appear to become constricted even more than those in other areas of the abdomen. The extreme ischemia that occurs in the pancreas allows activation of some of the pancreatic enzymes, including trypsin itself. This sets into play degenerative processes within the pancreatic tissues, and several toxic factors are released into the circulating blood. One of these is a peptide with a molecular weight of 500 to 1000 that is called *myocardial toxic factor,* more commonly known as *MTF.* The prominent effect of MTF is that of a direct depressant on the heart itself, frequently depressing cardiac contractility as much as 50 per cent. The toxin seems to interfere with the function of calcium ions in the excitation-contraction coupling process because increase in calcium ions in the blood or administration of ouabain, one of the cardiac glycosides that causes increased concentrations of calcium in the cardiac muscle cell, nullifies the depressant effect of MTF.

Regardless of the precise mechanism by which MTF reduces cardiac contractility, its generation during the course of shock further exacerbates the shock syndrome and, therefore, is part of the positive feedback deteriorative process that causes progression of the shock.

*Endotoxin.* Another toxic factor that probably plays a role in the progression of shock is *endotoxin,* which is a toxin released from the bodies of dead gram-negative bacteria in the intestines. Diminished blood flow to the intestines causes enhanced absorption of this toxic substance from the intestines, and it then causes extensive vascular dilatation and cardiac depression. Though this toxin can play a major role in some types of shock, especially septic shock as is discussed later in the chapter, it is still not clear how much endotoxin is released during hemorrhagic shock and whether or not it is an important factor in the progression of this type of shock.

**Generalized Cellular Deterioration.** As shock becomes very severe, many signs of generalized cellular deterioration occur throughout the body. One organ especially affected is the *liver,* primarily because of the normally high rate of metabolism in liver cells, but partly because of the extreme vascular exposure of the liver cells to any toxic or other abnormal metabolic factors in shock. Among the different effects that are known to occur are:

1. Active transport of sodium and potassium through the cell membrane is greatly diminished. As a result, sodium and chloride accumulate in the cells and potassium is lost from the cells. In addition the cells begin to swell.

2. Mitochondrial activity in the liver cells, as well as in many other tissues of the body, becomes severely depressed.

3. Lysosomes begin to split in widespread tissue areas, with intracellular release of hydrolases that cause further intracellular deterioration.

4. Cellular metabolism of nutrients such as glucose becomes greatly depressed. Some of the activities of the hormones are depressed as well, including as much as 200-fold depression in the action of insulin.

Obviously, all of these effects contribute to further deterioration of many different organs of the body, particularly of the liver, and also of the heart, thereby further depressing the contractility of the heart.

*Tissue Necrosis in Severe Shock—Effect of Pattern of Vascular Blood Flow in Different Organs.* Not all cells of the body are equally damaged by shock because some tissues have better blood supplies than others. For instance, the cells adjacent to the arterial ends of capillaries receive better nutrition than the cells adjacent to the venous ends of the same capillaries. Therefore, one would expect more nutritive deficiency around the venous ends of capillaries than elsewhere. This is precisely the effect that

Crowell has found in studying tissue areas in many parts of the body. Figure 28–5 illustrates necrosis in the center of a liver lobule, the portion of the lobule that is last to be bathed by the blood as it passes through the liver sinusoids. The first sign of damage is swelling of the cells, and this swelling then compresses the sinusoids, often causing total blockage of blood flow in their central ends; the result is rapidly accelerating deterioration of the central portions of the liver lobules.

Similar punctate lesions occur in heart muscle, though here a definite repetitive pattern such as occurs in the liver cannot be demonstrated. Nevertheless, the cardiac lesions probably play an important role in leading to the final irreversible stage of shock. Similar deteriorative lesions occur in the kidneys, especially in the kidney tubules, leading to kidney failure and subsequent uremic death several days later.

**Acidosis in Shock.** Most of the metabolic derangements that occur in shocked tissue can lead to acidosis. Especially important is the poor delivery of oxygen to the tissues, which greatly diminishes oxidative metabolism of the foodstuffs. When this occurs, the cells obtain their energy by the anaerobic process of glycolysis that leads to tremendous quantities of *excess lactic acid* in the blood. In addition, the poor blood flow through the tissues prevents normal removal of carbon dioxide; the carbon

dioxide reacts locally in the cells with water to form very high concentrations of intracellular carbonic acid; this in turn reacts with the various tissue buffers to form still other intracellular acidic substances.

Thus, another deteriorative effect of shock is both generalized and local tissue acidosis, leading to still further progression of the shock itself.

**Relationship of Positive Feedback to Vicious Cycles in Shock.** All the factors just discussed that can lead to further progression of shock are types of *positive feedback*. That is, each increase in the degree of shock causes a further increase in the shock.

However, positive feedback does not always necessarily lead to a vicious cycle. Whether or not a vicious cycle develops depends on the intensity of the positive feedback. In mild degrees of shock, the negative feedback mechanisms— the sympathetic reflexes, the plasticity of the circulation, the recovery of blood volume, and others—can easily overcome the positive feedback influences and can therefore cause recovery. In severe degrees of shock, however, the positive feedback mechanisms become more and more powerful, thus leading to such rapid deterioration of the circulation that the negative feedback systems cannot return the cardiac output to normal.

Thus, if one will consider once again the principles of positive feedback as discussed in Chapter 1, he will see that a vicious cycle of deterioration occurs in shock only when the positive feedback effect to decrease cardiac output further is greater than the initial decrease in output. This explains why there is a critical cardiac output level above which a person in shock recovers and below which he goes into a vicious cycle of circulatory deterioration.

### IRREVERSIBLE SHOCK

After shock has progressed to a certain stage, transfusion or any other type of therapy becomes incapable of saving the life of the person. Therefore, the person is then said to be in the *irreversible stage of shock*. Paradoxically, therapy can on occasion still return the arterial pressure and even the cardiac output to normal for short periods of time, but the circulatory system nevertheless continues to deteriorate and death ensues in another few minutes to hours.

Figure 28–6 illustrates this effect, showing that transfusion during the irreversible stage

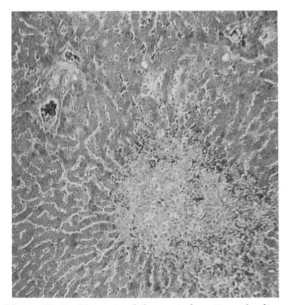

**Figure 28–5.** Necrosis of the central portion of a liver lobule in severe shock. (Courtesy of Dr. Jack Crowell.)

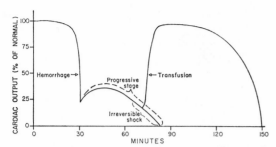

**Figure 28–6.** Failure of transfusion to prevent death in irreversible shock.

can sometimes cause the cardiac output (as well as the arterial pressure) to return to normal. However, the cardiac output soon begins to fall again, and subsequent transfusions cause less and less effect. Thus, something changes in the overall function of the circulatory system during shock that does not affect the *immediate* ability of the heart to pump blood but over a long period of time does depress this ability and eventuates in death. Now the question remains: What factor or factors lead to the eventual total deterioration of circulatory function?

The answer to this question seems to be, simply, that beyond a certain point so much tissue damage has occurred, so many destructive enzymes have been released into the body fluids, so much acidosis has developed, and so many other destructive factors are now in progress that even a normal cardiac output cannot reverse the continuing deterioration. Therefore, frequently a person in severe shock reaches a stage beyond which he is destined to die even though vigorous therapy can still return the cardiac output to normal for short periods of time.

**Depletion of Cellular High Energy Phosphate Reserves in Shock.** The high energy phosphate reserves in the tissues of the body, especially in the liver and in the heart, are greatly diminished in severe degrees of shock. Essentially all of the creatinine phosphate is degraded, and almost all of the sarcoplasmic ATP has been degraded to ADP, AMP, or adenosine (though the intramitochondrial ATP still seems to persist even in rather severe stages of shock). Much of the adenosine that is derived from degradation of the ATP diffuses out of the cells into the circulating blood and is converted into uric acid, a substance that cannot re-enter the cells to reconstitute the adenosine phosphate system. Unfortunately, new ATP can be synthesized at the rate of only about 2 per cent an hour, meaning that, once depleted, the high energy phosphate stores of the cells are difficult to replenish. Therefore, one of the most impor-

tant of the end-results of deterioration in shock, and one that is perhaps the most significant of all in the development of the final state of irreversibility, is this cellular depletion of the high energy compounds.

**Deterioration of the Heart as a Primary Cause of Irreversible Shock.** Though it is clear that deterioration can occur in many different organ systems in shock and that the degeneration in any of these systems could become so severe that it would eventually be incompatible with continued life, there is reason to believe that in most instances it is deterioration of the heart itself that makes the shock irreversible. The reason for believing this is the following: Modern therapy is very effective in producing adequate venous return. Administration of blood and other substitution fluids can almost always provide adequate inflow pressure to the heart even in the most severe degrees of shock. But, still, in the late stages of shock, the heart fails to pump this attempted inflow of blood. Therefore, the heart is the weak link in the system.

On the other hand, if one does not use all forms of available therapy, a person can die of shock because of peripheral abnormalities such as continued loss of fluid into the tissues, pooling of blood in greatly distended blood vessels, and so forth. Unfortunately, in such instances, the patient dies of shock that is still reversible if adequate therapy is provided. Therefore, we can state again that, with modern therapy, the heart is usually the final weak link that leads to the irreversibility.

**Special Role of Oxygen Deficiency in Irreversibility.** Though poor blood flow leads to tissue deterioration because of deficiency of many different nutrients, deficiency of oxygen almost certainly is the most important. For instance, in a large series of dogs, Crowell measured the accumulated deficit of oxygen usage of animals (a) in mild shock, (b) in moderate shock, and (c) in very severe shock. In some animals, this deficit accumulated slowly, whereas in others it accumulated very rapidly. But in each group of animals, when the average accumulated oxygen deficit reached 120 milliliters of oxygen per kilogram of body mass, 50 per cent of the animals died regardless of how long it took to accumulate this amount of oxygen deficit.

Therefore, it seems to be clear that the one most important nutrient necessary to prevent cellular deterioration and death during shock is oxygen.

### OTHER CAUSES OF HYPOVOLEMIC SHOCK

**Plasma Loss Shock.** Loss of plasma from the circulatory system can sometimes be severe enough to reduce the total blood volume markedly, in this way causing typical hypovolemic shock similar in almost all details to that caused by hemorrhage. Severe plasma loss occurs in the following conditions:

1. *Intestinal obstruction* is often a cause of reduced plasma volume. The resulting distention of the intestine causes fluid to leak from the intestinal capillaries into the intestinal walls and intestinal lumen. This increased loss of fluid might result from elevated capillary pressure caused by increased resistance in the stretched veins over the surface of the intestine, or it might be caused by direct damage to the capillaries themselves. Nevertheless, the lost fluid has a very high content of protein, thereby reducing the total plasma protein as well as the plasma volume.

2. Often, in patients who have *severe burns* or other denuding conditions of the skin, so much plasma is lost through the exposed areas that the plasma volume becomes markedly reduced.

The hypovolemic shock that results from plasma loss has almost the same characteristics as the shock caused by hemorrhage, except for one additional complicating factor—the blood viscosity increases greatly as a result of plasma loss, and this further exacerbates the sluggishness of blood flow.

3. Loss of fluid from all fluid compartments of the body is called *dehydration;* this, too, can reduce the blood volume and cause hypovolemic shock very similar to that resulting from hemorrhage. Some of the causes of this type of shock are: (a) excessive sweating; (b) fluid loss in severe diarrhea or vomiting; (c) excess loss of fluid by nephrotic kidneys; (d) inadequate intake of fluid and electrolytes; (e) destruction of the adrenal cortices, with consequent failure of the kidneys to reabsorb sodium, chloride, and water; and (f) loss of the secretion of antidiuretic hormone by the supraoptico-hypophyseal system.

# NEUROGENIC SHOCK— INCREASED VASCULAR CAPACITY

Occasionally shock results without any loss of blood volume whatsoever. Instead, the *vascular capacity* increases so much that even the normal amount of blood becomes incapable of adequately filling the circulatory system. One of the major causes of this is *loss of vasomotor tone* throughout the body, and the resulting condition is then known as *neurogenic shock.*

The relationship of vascular capacity to blood volume was discussed in Chapter 18, where it was pointed out that either an increase in vascular capacity or a decrease in blood volume *reduces the systemic filling pressure,* which in turn reduces the ve-

nous return to the heart. This effect is often called "venous pooling" of blood.

**Importance of Body Position in Neurogenic Shock.** Ordinarily, sudden loss of vasomotor tone throughout the body will not cause shock if the person is lying in a slightly head-down position (Trendelenburg position). If he is lying in a completely horizontal position, only mild to moderate degrees of reduced cardiac output result. But, if the person is in the upright position, the vessels of the lower part of the body become so distended that the blood "pools" and fails to flow uphill in sufficient quantities to maintain cardiac output. As a result, extremely severe shock results.

**Causes of Neurogenic Shock.** Some of the different factors that can cause sudden loss of vasomotor tone include:

1. *Deep general anesthesia* often depresses the vasomotor center enough to cause vasomotor collapse, with resulting neurogenic shock.

2. *Spinal anesthesia,* especially when this extends all the way up the spinal cord, blocks the sympathetic outflow from the nervous system and is a common cause of neurogenic shock.

3. *Brain damage* is often a cause of vasomotor collapse. Many patients who have had brain concussion or contusion of the basal regions of the brain develop profound neurogenic shock. Also, even though short periods of medullary ischemia cause extreme vasomotor activity, prolonged ischemia can result in inactivation of the vasomotor neurons and can cause development of severe neurogenic shock.

4. In *fainting,* the peripheral blood vessels become greatly dilated. As a result, blood pools, and the cardiac output falls drastically. If such a person is held in an upright position, he will go into the progressive stage of shock and can die as a result. Fortunately, on fainting, a person usually falls to a horizontal position so that almost normal cardiac output ordinarily re-ensues almost immediately.

**Vasovagal Syncope—Emotional Fainting.** The circulatory collapse that results from "emotional" fainting usually is not caused by vasomotor failure but instead by strong emotional excitation of the parasympathetic nerves to the heart and of the vasodilator nerves to the skeletal muscles, thereby slowing the heart and reducing the arterial pressure. Therefore, the fainting that results from an emotional disturbance is called *vasovagal syncope* to differentiate it from the other types of fainting which result from diminished sympathetic activity throughout the body or from other causes of reduced cardiac output.

# ANAPHYLACTIC SHOCK

"Anaphylaxis" is an allergic condition in which the cardiac output and arterial pressure often fall drastically. This was discussed in Chapter 7. It results primarily from an antigen-antibody reaction that takes place all through the body immediately after an antigen to which the person is sensitive has entered the circulatory system. Such a reaction is detrimental to the circulatory system in several important ways, as illustrated in Figure 28–7. First, if the antigen-antibody reaction takes place in direct contact with the vascular walls or cardiac musculature, damage to these tissues presumably can result directly. Second, cells damaged anywhere in the body by the antigen-antibody reaction release several highly toxic substances into the blood. Among these is *histamine* or a *histamine-like substance* that has a strong vasodilator effect. The histamine in turn causes (1) an increase in vascular capacity because of venous dilatation, (2) dilatation of the arterioles with resultant greatly reduced arterial pressure, and (3) greatly increased capillary permeability with rapid loss of fluid into the tissue spaces. Unfortunately, all the precise relationships of the above factors in anaphylaxis have not been determined, but the sum total is a great reduction in venous return and often such serious shock that the person dies within minutes.

Intravenous injection of large amounts of histamine causes "histamine shock," which has characteristics almost identical with those of anaphylactic shock, though usually less severe.

## SEPTIC SHOCK

The condition that was formerly known by the popular name of "blood poisoning" is now called *septic shock* by most clinicians. This means simply widely disseminated infection in many areas of the body, the infection often being borne through the blood from one tissue to another and causing extensive damage. Actually, there are many different varieties of septic shock because of the many different types of bacterial infection that can cause it and also because infection in one part of the body will produce different effects from those caused by infection elsewhere in the body.

Septic shock is extremely important to the clinician because it is this type of shock that, more frequently than any other kind of shock besides cardiogenic shock, causes death in the modern hospital. Some of the typical causes of septic shock include:

1. Peritonitis caused by spread of infection from the uterus and fallopian tubes, frequently resulting from instrumental abortion

2. Peritonitis resulting from rupture of the gut, sometimes caused by intestinal disease and sometimes by wounds

3. Generalized infection resulting from spread of a simple skin infection such as streptococcal or staphylococcal infection

4. Generalized gangrenous infection resulting specifically from gas gangrene bacilli, spreading first through the tissues themselves and finally by way of the blood to the internal organs, especially to the liver

**Special Features of Septic Shock.** Because of the multiple types of septic shock, it is difficult to give categorically an exact picture of this condition. However, some features often seen in septic shock are the following:

1. High fever

2. Marked vasodilatation throughout the body, especially in the infected tissues

3. High cardiac output in perhaps half of the patients, probably caused by the vasodilatation in the infected tissues and also by high metabolic rate and vasodilatation elsewhere in the body resulting from the high body temperature

4. Sludging of the blood, presumably caused by red cell agglutination in response to bacterial toxins or in response to degenerating tissues

In the early stages of septic shock the patient usually does not have signs of circulatory collapse but, instead, simply signs of the bacterial infection itself. However, as the infection becomes more severe, the circulatory system usually becomes involved either directly or as a secondary result of toxins from the bacteria, and *there finally comes a point at which deterioration of the circulation becomes progressive in the same way that progression occurs in all other types of shock. Therefore, the end stages of septic shock are not greatly different from the end stages of hemorrhagic shock,* even though the initiating factors are markedly different in the two conditions.

**Endotoxin Shock.** A special type of septic shock is known as endotoxin shock. It frequently occurs

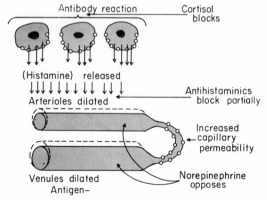

**Figure 28–7.** Basic mechanisms of anaphylactic shock and effects of therapy with cortisol, antihistaminics, and norepinephrine.

when a large segment of the gut becomes strangulated and loses most of its blood supply. The gut rapidly becomes gangrenous, and the bacteria in the gut multiply rapidly. Most of these bacteria are so-called "gram-negative" bacteria, mainly colon bacilli, that contain a toxin called *endotoxin*.

On entering the circulation, endotoxin causes an effect very similar to that of anaphylaxis, sometimes resulting in severe shock. Indeed, the cause of this shock is possibly an anaphylactic reaction to the endotoxin itself. However, endotoxin also has direct effects on the circulation as well, including (1) toxic depression of the heart and (2) vascular dilatation.

Severe shock also occurs during the course of diseases caused by other gram-negative bacterial infections, such as dysentery, tularemia, brucellosis, and typhoid fever. It is likely that this shock results at least partially from endotoxin released from the gram negative bacteria causing these conditions.

# TRAUMATIC SHOCK

One of the most common causes of circulatory shock is trauma to the body. Often the shock results simply from hemorrhage caused by the trauma, but it can also occur even without hemorrhage, for contusion of the body can often damage the capillaries sufficiently to allow excessive loss of plasma into the tissues. This results in greatly reduced plasma volume with resultant hypovolemic shock. Thus, whether or not hemorrhage occurs when a person is severely traumatized, the blood volume can still be markedly reduced.

The *pain* associated with serious trauma can be an additional aggravating factor in traumatic shock, for pain sometimes strongly inhibits the vasomotor center, thereby increasing the vascular capacitance and reducing the venous return. Various attempts have also been made to implicate toxic factors released by the traumatized tissues as one of the causes of shock following trauma. However, cross-transfusion experiments have failed to show any such toxic element.

In summary, traumatic shock seems to result mainly from hypovolemia, though there may also be a moderate degree of concomitant neurogenic shock caused by the pain.

# EFFECTS OF SHOCK ON
# THE BODY

**Decreased Metabolism.** The decreased cardiac output in shock reduces the amount of oxygen and other nutrients available to the different tissues, and this in turn reduces the level of metabolism that can be maintained by the different cells of the body. Usually a person can continue to live for only a few hours if his cardiac output falls to as low as 40 per cent of normal.

**Muscular Weakness.** One of the earliest symptoms of shock is severe muscular weakness which is also associated with profound and rapid fatigue whenever the patient attempts to use his muscles. This obviously results from the diminished supply of nutrients—especially oxygen—to the muscles.

**Body Temperature.** Because of the depressed metabolism in shock, the amount of heat liberated in the body is reduced in shock. As a result, the body temperature tends to decrease if the body is exposed to even the slightest cold.

**Mental Function.** In the early stages of shock the person is usually conscious, though he may exhibit signs of mental haziness. As the shock progresses, he falls into a state of stupor, and in the last stages of shock even his subconscious mental functions, including vasomotor control and respiration, fail.

A person who recovers from shock usually exhibits no permanent impairment of mental functions. However, following complete circulatory arrest, in which the blood flow is completely cut off from the brain for many minutes, the brain does often suffer severe permanent impairment, the cause of which is discussed in a separate section on circulatory arrest later in the chapter.

**Reduced Renal Function.** The very low blood flow during shock greatly diminishes urine output because glomerular pressure falls below the critical value required for filtration of fluid into Bowman's capsule, as explained in Chapters 33 and 34. Also, the kidney has such a high rate of metabolism and requires such large amounts of nutrients that the reduced blood flow often causes *tubular necrosis,* which means death of the tubular epithelial cells, with subsequent sloughing and blockage of the tubules, causing total loss of function of the respective nephrons. This is often a serious aftereffect of shock that occurs during major surgical operations; the patient sometimes survives the shock associated with the surgical procedure and then dies a week or so later of uremia.

# PHYSIOLOGY OF TREATMENT
# IN SHOCK

## REPLACEMENT THERAPY

**Blood and Plasma Transfusion.** If a person is in shock caused by hemorrhage, the best possible therapy is usually transfusion of whole blood. If the shock is caused by plasma loss, the best therapy is administration of plasma; when dehydration is the cause, administration of the appropriate electrolytic solution can correct the shock.

Unfortunately, whole blood is not always available, such as under battlefield conditions. However, plasma can usually substitute adequately for whole blood because it increases the blood volume and restores normal hemodynamics. Plasma cannot restore a normal hematocrit, but the human being can usually stand a decrease in hematocrit to about one-third normal before serious consequences result. Therefore, in acute conditions it is reasonable to use plasma in place of whole blood for treatment of hemorrhagic and most other types of hypovolemic shock.

Sometimes plasma also is unavailable. For these instances, various *plasma substitutes* have been developed that perform almost exactly the same hemodynamic functions as plasma. One of these is the following:

**Dextran Solution as a Plasma Substitute.** The principal requirement of a truly effective plasma substitute is that it remain in the circulatory system—that is, not filter through the capillary pores into the tissue spaces. But, in addition, the solution must be nontoxic and must contain appropriate electrolytes to prevent derangement of the extracellular fluid electrolytes on administration. To remain in the circulation the plasma substitute must contain some substance that has a large enough molecular size to exert colloid osmotic pressure.

One of the most satisfactory substances that has been developed thus far for this purpose is dextran, a large polysaccharide polymer of glucose. Certain bacteria secrete dextran as a by-product of their growth, and commercial dextran is manufactured by a bacterial culture procedure. By varying the growth conditions of the bacteria, the molecular weight of dextran can be controlled to any desired value. Dextrans of appropriate molecular size do not pass through the capillary pores and, therefore, can replace plasma proteins as colloid osmotic agents.

Fortunately, few toxic reactions have been observed when using dextran to provide colloid osmotic pressure; therefore, solutions of this substance have proved to be a satisfactory substitute for plasma in fluid replacement therapy.

## TREATMENT OF SHOCK WITH SYMPATHOMIMETIC AND SYMPATHOLYTIC DRUGS OR OTHER THERAPY

A *sympathomimetic drug* mimics sympathetic stimulation. These drugs include norepinephrine, epinephrine, and a large number of long-acting drugs that have the same effects as epinephrine and norepinephrine. A *sympatholytic drug,* on the other hand, blocks the action of normal sympathetic stimuli on the circulatory system.

Both sympathomimetic and sympatholytic drugs have been used in treating different types of shock, including hemorrhagic shock. However, there is no proof that either plays any very beneficial role in this type of shock. The sympathomimetic drugs usually have little benefit because the circulatory reflexes are usually already maximally activated in severe hypovolemic shock and are already causing marked endogenous secretion of norepinephrine and epinephrine. On the other hand, the proponents of the use of sympatholytic drugs have claimed that these

drugs increase blood flow through the tissues and thereby prevent nutritional damage. Unfortunately, these drugs also usually cause paralysis of the venous vasculature, resulting in increased pooling of blood in the venous system, which reduces the cardiac output and negates much if not all of the direct effect of the drugs in increasing local blood flow. Yet a combination of *transfusion plus sympatholytic drugs* probably is beneficial in the treatment of many patients with shock; the transfusion fills up the expanded vascular bed and prevents the pooling, while the sympatholytic drug allows rapid blood flow to the tissues.

## OTHER THERAPY

**Treatment with Glucocorticoids.** Glucocorticoids are frequently given to patients in severe shock for several reasons: (1) experiments have shown empirically that glucocorticoids frequently increase the strength of the heart in the late stages of shock; (2) glucocorticoids stabilize the lysosomal membranes and prevent release of lysosomal enzymes into the cytoplasm of the cells, thus preventing deterioration from this source; (3) glucocorticoids have a specific action to oppose release of MTF from the pancreas; and (4) glucocorticoids might also aid in the metabolism of glucose by the severely damaged cells.

**Treatment of Neurogenic Shock.** In neurogenic shock, sympathomimetic drugs actually reverse the basic cause of the shock itself by increasing the vasomotor tone throughout the body, but even in this condition these drugs may not be required, for, as long as the patient is in the horizontal or head-down position, adequate venous return to the heart usually still occurs and maintains a sufficient cardiac output to prevent progressive shock.

**Treatment of Anaphylactic Shock.** Unfortunately, anaphylactic shock occurs so rapidly that one often cannot institute any therapy before death ensues, but if therapy can be instituted the condition can often be ameliorated or almost completely reverted by rapid administration of norepinephrine or some other sympathomimetic drug. This does not correct the basic cause of the anaphylaxis, but it does cause vasoconstriction, which opposes the vasodilatation caused by histamine in anaphylaxis. Also, if one suspects that anaphylaxis might occur in a patient, its severity can usually be reduced by preliminary administration of *cortisol,* which attenuates the allergic reaction responsible for the anaphylaxis, or preliminary administration of *antihistaminics,* which reduce the effects of the histamine released during anaphylaxis. However, in both these instances the therapy must be given *before* anaphylaxis takes place, for which reason they are usually of little value in therapy of anaphylaxis.

**Pain-Relieving Drugs in Traumatic Shock.** The use of pain-relieving drugs, such as morphine, has been found clinically to reduce the severity of most traumatic shock, supposedly by removing the neurogenic element of the shock.

# CIRCULATORY ARREST

A condition closely allied to circulatory shock is circulatory arrest, in which all blood flow completely stops. This occurs frequently on the surgical operating table as a result of *cardiac arrest* or of *ventricular fibrillation*.

Ventricular fibrillation can usually be stopped by the following procedure: The heart is pumped by forceful, thrusting compression of the sternum over the heart for a few minutes to restore adequate coronary flow to the ventricular muscle. Then a strong electrical current is passed through the chest and heart from two electrode plates on the chest wall. This throws essentially all the heart into a state of refractoriness, and usually a normal cardiac rhythm is restored thereafter. The basic principles of this procedure were described in Chapter 14.

Cardiac arrest usually results from too little oxygen in the anesthetic gaseous mixture or from a depressant effect of the anesthesia itself. A normal cardiac rhythm can usually be restored by removing the anesthetic and then pumping the heart for a few minutes while supplying the patient's lungs with adequate quantities of ventilatory oxygen.

## EFFECT OF CIRCULATORY ARREST ON THE BRAIN

The real problem in circulatory arrest is usually not to restore cardiac function but instead to prevent detrimental effects in the brain as a result of the circulatory arrest. In general, four to five minutes of circulatory arrest causes permanent brain damage in over half the patients, and circulatory arrest for as long as 10 minutes almost universally destroys most, if not all, of the mental powers.

For many years it has been taught that these detrimental effects on the brain are caused by the cerebral hypoxia that occurs during circulatory arrest. However, recent studies by Crowell and others have shown that dogs can almost universally stand 30 minutes of circulatory arrest without permanent brain damage *if the blood is removed from the brain circulation prior to the arrest*. On the basis of these studies, it is postulated that the circulatory arrest causes vascular *clots* to develop throughout the brain and that these cause permanent or semipermanent ischemia of brain areas. This accords well with the results in human beings who have undergone long periods of circulatory arrest, for complete destruction of large areas in one side of the brain often occurs while corresponding areas in the opposite side of the brain, which should also be affected if hypoxia were the cause of the damage, are not affected even in the slightest.

# REFERENCES

Bergentz, S. E., and Leandoer, L.: Disseminated intravascular coagulation in shock. *Ann. Chir. Gynaecol. Fenn.*, 60:175, 1971.

Chien, S.: Role of the sympathetic nervous system in hemorrhage. *Physiol. Rev.*, 47:214, 1967.

Crowell, J. W.: Cardiac deterioration as the cause of irreversibility in shock. *In* Mills, L. J., and Moyer, J. H. (eds.): Shock and Hypotension: Pathogenesis and Treatment. Grune & Stratton, Inc., 1965, p. 605.

Crowell, J. W., Bounds, S., and Johnson, W. W.: The effect of varying the hematocrit ratio on the susceptibility to hemorrhagic shock. *Amer. J. Physiol.*, 192:171, 1958.

Crowell, J. W., and Guyton, A. C.: Evidence favoring a cardiac mechanism in irreversible hemorrhagic shock. *Amer. J. Physiol.*, 201:893, 1961.

Crowell, J. W., Jones, C. E., and Smith, E. E.: Effect of allopurinol on hemorrhagic shock. *Amer. J. Physiol.*, 216:744, 1969.

Crowell, J. W., and Read, W. L.: *In vivo* coagulation—a probable cause of irreversible shock. *Amer. J. Physiol.*, 183:565, 1955.

Crowell, J. W., Sharpe, G. P., Lambright, R. L., and Read, W. L.: The mechanism of death after resuscitation following acute circulatory failure. *Surgery*, 38:696, 1955.

Crowell, J. W., and Smith, E. E.: Oxygen deficit and irreversible hemorrhagic shock. *Amer. J. Physiol.*, 206:313, 1964.

Gruber, U. F., and Rittmann, W. W.: Hypovolaemic shock—therapy of hypovolaemia and respiratory insufficiency. *Triangle*, 13:91, 1974.

Guyton, A. C., and Crowell, J. W.: Dynamics of the heart in shock. *Fed. Proc.*, 20:51, 1961.

Halmagyi, D. F. J., Goodman, A. H., and Neering, I. R.: Hindlimb blood flow and oxygen usage in hemorrhagic shock. *J. Appl. Physiol.*, 27:508, 1969.

Hershey, S. G., and Altura, B. M.: Vasopressors and low-flow states. *Clin. Anesth.*, 10:31, 1974.

Irving, M.: Current studies in surgical shock. *Proc. R. Soc. Med.* 65:1116, 1972.

Jones, C. E., Crowell, J. W., and Smith, E. E.: Significance of increased blood uric acid following extensive hemorrhage. *Amer. J. Physiol.*, 214:1374, 1968.

Jones, C. E., Crowell, J. W., and Smith, E. E.: A cause-effect relationship between oxygen deficit and irreversible hemorrhagic shock. *Surgery*, 127:93, 1968.

Jones, C. E., Smith, E. E. and Crowell, J. W.: Cardiac output and physiological mechanisms in circulatory shock. *In* Guyton, A. C. (ed.): MTP International Review of Science: Physiology. Vol. 1. Baltimore, University Park Press, 1974, p. 233.

Lefer, A. M., Cowgill, R., Marshall, F. F., Hall, L. M., and Brand, E. D.: Characterization of a myocardial depressant factor present in hemorrhagic shock. *Amer. J. Physiol.*, 213:492, 1967.

Lefer, A. M., and Martin, J.: Mechanism of the protective effect of corticosteroids in hemorrhagic shock. *Amer. J. Physiol.*, 216:314, 1969.

Lefer, A. M., and Martin, J.: Relationship of plasma peptides to the myocardial depressant factor in hemorrhagic shock in cats. *Circ. Res.*, 26:59, 1970.

Massion, W. H., and Dietzel, W.: Effect of anesthetics on surgical blood loss. *Inter. Anesthesiol. Clin.*, 12:203, 1974.

Moyer, C. A., and Butcher, H. R.: Burns, Shock, and Plasma Volume Regulation. St. Louis, The C. V. Mosby Co., 1967.

Rothe, C. F., Love, J. R., and Selkurt, E. E.: Control of total vascular resistance in hemorrhagic shock in the dog. *Circ. Res.*, 12:667, 1963.

Schildt, B. E.: The role of the RES in shock states. *Ann. Chir. Gynaecol. Fenn.*, 60:165, 1971.

Schumer, W., and Nyhus, L. M. (eds.): Treatment of Shock. Philadelphia, Lea & Febiger, 1974.

Shoemaker, W. C.: Pattern of pulmonary hemodynamic and functional changes in shock. *Crit. Care Med.*, 2:200, 1974.

Smith, E. E., and Crowell, J. W.: Effect of hemorrhagic hypotension on oxygen consumption of dogs. *Amer. J. Physiol.*, 207:647, 1964.

Weil, M. H., and Shubin, H.: Shock. Baltimore, The Williams & Wilkins Company, 1967.

Wilkinson, A. W.: Body Fluids in Surgery, 4th Ed. New York, Churchill Livingstone, Div. of Longman, Inc., 1973.

Wolthuis, R. A., Bergman, S. A., and Nicogossian, A. E.: Physiological effects of locally applied reduced pressure in man. *Physiol. Rev.*, 54:566, 1974.

Zweifach, B. W.: Mechanisms of blood flow and fluid exhange in microvessels: hemorrhagic hypotension model. *Anesthesiology*, 41:157, 1974.

# 29

# Muscle Blood Flow During Exercise; Cerebral, Splanchnic, and Skin Blood Flows

The blood flow in many special areas of the body, such as the lungs and the heart, has already been discussed in previous chapters, and the circulation in the kidney, in the uterus, and in the fetus will be discussed later in the text. In the present chapter the characteristics of blood flow in some of the other important tissues, such as the muscles, the brain, the splanchnic system, and the skin, are presented.

## BLOOD FLOW THROUGH SKELETAL MUSCLES AND ITS REGULATION IN EXERCISE

Very strenuous exercise is the most stressful condition that the normal circulatory system faces. This is true because the blood flow in muscles can increase as much as 20-fold (perhaps more than in any other tissue of the body) and also because there is such a very large mass of skeletal muscle in the body. The product of these two factors indicates that the total muscle blood flow can become so great that it increases the cardiac output in the normal young adult to as much as five times normal and in the rare well trained athlete to as much as seven times normal.

### RATE OF BLOOD FLOW THROUGH THE MUSCLES

During rest, blood flow through skeletal muscle averages 4 to 7 ml. per minute per 100 grams

of muscle. However, during extreme exercise in the well trained athlete this rate can increase as much as 12- to 18-fold, rising to 50 to 75 ml. per 100 grams of muscle.

**Intermittent Flow During Muscle Contraction.** Figure 29–1 illustrates a study of blood flow changes in the calf muscles of the human leg during strong rhythmic contraction. Note that the flow increases and decreases with each muscle contraction, decreasing during the contraction phase and increasing between contractions. At the end of the rhythmic contractions, the blood flow remains very high for a minute or so and then gradually fades toward normal.

The cause of the decreased flow during sustained muscle contraction is compression of the blood vessels by the contracted muscle. During

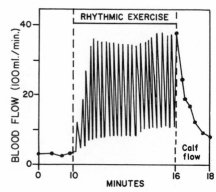

**Figure 29–1.** Effect of muscle exercise on blood flow in the calf of a leg during strong rhythmic contraction. The blood flow was much less during contraction than between contractions. (From Barcroft and Dornhorst: *J. Physiol.,* 109:402, 1949.)

370

strong *tetanic* contraction, blood flow can be almost totally stopped.

**Opening of Muscle Capillaries During Exercise.**    During rest, only 12 to 20 per cent of the muscle capillaries are open. But during strenuous exercise all the capillaries open up, which can be demonstrated by studying histologic specimens removed from muscles appropriately stained during exercise. It is this opening up of dormant capillaries that allows most of the increased blood flow. It also diminishes the distance that oxygen and other nutrients must diffuse from the capillaries to the muscle fibers and contributes a much increased surface area through which nutrients can diffuse from the blood.

## CONTROL OF BLOOD FLOW THROUGH THE SKELETAL MUSCLES

**Local Regulation.**    The tremendous increase in muscle blood flow that occurs during skeletal muscle activity is caused primarily by local effects in the muscles acting directly on the arterioles to cause vasodilatation.

This local increase in blood flow during muscle contraction is probably caused by several different factors all operating at the same time. One of the most important of these is reduction of dissolved oxygen in the muscle tissues. That is, during muscle activity the muscle utilizes oxygen very rapidly, thereby decreasing its concentration in the tissue fluids. This in turn causes vasodilatation either because the vessel walls cannot maintain contraction in the absence of oxygen or because oxygen deficiency causes release of some vasodilator material. The vasodilator material that has been suggested most widely in recent years has been adenosine.

Other vasodilator substances released during muscle contraction include potassium ions, acetylcholine, adenosine triphosphate, lactic acid, and carbon dioxide. Unfortunately, we still do not know quantitatively how much role each of these plays in increasing muscle blood flow during muscle activity, but this subject was discussed in more detail in Chapter 20.

**Nervous Control of Muscle Blood Flow.**    In addition to the local tissue regulatory mechanism, the skeletal muscles are also provided with both sympathetic vasconstrictor nerves and sympathetic vasodilator nerves.

*Sympathetic Vasoconstrictor Nerves.*    The sympathetic vasoconstrictor nerve fibers secrete norepinephrine and when maximally stimulated can decrease blood flow through the muscles to about one-fourth normal. This represents rather poor vasoconstriction in comparison with that caused by sympathetic nerves in some other areas of the body in which blood flow can be almost completely blocked. Yet under some conditions even this degree of vasoconstriction may be of physiological importance, such as during shock and other periods of stress when it is desirable to reduce blood flow through the many muscles of the body.

In addition to the norepinephrine secreted at the sympathetic vasoconstrictor nerve endings, the adrenal medullae secrete large amounts of additional norepinephrine and epinephrine into the circulating blood during strenuous exercise. The circulating norepinephrine acts on the muscle vessels to cause a vasoconstrictor effect similar to that caused by direct sympathetic nerve stimulation. The epinephrine, on the other hand, often has a vasodilator effect. It is believed that this results from the fact that epinephrine excites the *beta* receptors of the vessels, which are vasodilator receptors, in contrast to the *alpha* receptors excited by the norepinephrine. These receptors are discussed in Chapter 57.

*Sympathetic Vasodilator Fibers.*    Maximal stimulation of the sympathetic vasodilator fibers to the skeletal muscles can increase blood flow by 400 per cent. These fibers are activated by a special nervous pathway beginning in the cerebral cortex in close association with the motor areas for control of muscular activity and passing downward through the hypothalamus and brain stem into the spinal cord, as was explained in Chapter 20. When the motor cortex initiates muscle activity it simultaneously excites the vasodilator fibers to the active muscles, and vasodilatation occurs immediately, several seconds before the local regulatory vasodilatation can take place. Thus, it seems that this vasodilator system has the important *function of initiating extra blood flow through the muscles at the onset of muscular activity.* However, the vasodilator fibers are probably of little or no importance in maintaining increased blood flow during prolonged muscular activity, for both in lower animals and in human beings the final degree of vasodilatation which occurs during exercise is not discernibly different when the muscle is normally innervated with sympathetics or denervated.

In some lower animals (the cat, for instance) the vasodilator fibers secrete acetylcholine, and this hormone causes the vasodilatation. However, secretion of acetylcholine has not been proved to occur in the human being; some studies suggest that norepinephrine is secreted instead and that this causes the vasodilatation by acting on *beta* receptors (vasodilator receptors) in the muscle blood vessels.

## CIRCULATORY READJUSTMENTS DURING EXERCISE

Three major effects that are essential for the circulatory system to supply the tremendous blood flow required by the muscles occur during exercise. These effects are: (1) mass discharge of the sympathetic nervous system throughout the body with consequent stimulatory effects on the circulation, (2) increase in cardiac output, and (3) increase in arterial pressure.

**Mass Sympathetic Discharge.**    At the onset of exercise, signals are transmitted not only from the brain to the muscles to cause muscle contraction but also from the higher levels of the brain into the vasomotor center to initiate mass sympathetic discharge. Simultaneously, the parasympathetic signals to the heart are greatly attenuated. Therefore, two major circulatory effects result. First, the heart is stimulated to greatly increased activity as a result of the sympathetic drive to the heart as well as release of the heart from the normal parasympathetic inhibition. Second, all the blood vessels of the peripheral circulation are strongly contracted except the vessels in the active muscles, which are strongly vasodilated by the local vasodilator effects in the muscles themselves. Thus, the heart is stimulated to supply the increased blood flow required by the muscles, and blood flow through most nonmuscular areas of the body is temporarily reduced, thereby temporarily "lending" their blood supply to the muscles. This effect is exceedingly important when one thinks of the wild animal running for its life, because even a fractional increase in running speed may make the difference between life and death. However, two of the organ circulatory systems, the coronary and cerebral systems, are spared this vasoconstrictor effect. This sparing effect occurs because both of these circulatory areas have very poor vasoconstrictor innervation, fortunately so because both the heart and the brain are just as essential to exercise as are the skeletal muscles themselves.

**Increase in Cardiac Output.**    The increase in cardiac output that occurs during exercise results mainly from the intense local vasodilatation in active muscles. As was explained in Chapter 23 in relation to the basic theory of cardiac output regulation, the local vasodilatation increases the venous return of blood back to the heart. The heart in turn pumps this extra returning blood and sends it immediately back to the muscles through the arteries. Thus, it is mainly the muscles themselves that determine the amount of increase in cardiac output—up to the limit of the heart's ability to respond.

An additional factor that is necessary to cause the large increase in cardiac output is the strong sympathetic stimulation of the veins. This stimulation greatly increases the systemic filling pressure, sometimes to as high as 30 mm. Hg (4 times normal), and is therefore important in increasing venous return.

*Mechanisms by Which the Heart Increases Its Output.*    One of the principal mechanisms by which the heart increases its output during exercise is the Frank-Starling mechanism, which was discussed in Chapter 13. Via this mechanism, when increased quantities of blood flow from the veins into the heart and dilate its chambers, the heart muscle contracts with increased force, thus also pumping an increased volume of blood with each heart beat. However, in addition to this basic intrinsic cardiac mechanism, the heart is also strongly stimulated by the sympathetic nervous system, and the normal parasympathetic inhibition is reduced or eliminated. The net effects are greatly increased heart rate (occasionally to as high as 180 beats per minute) and almost doubling of the cardiac muscle strength of contraction. These two effects combine to make the heart capable of pumping at least 100 per cent more blood than would be true based on the Frank-Starling mechanism alone.

**Increase in Arterial Pressure.**    The mass sympathetic discharge throughout the body during exercise and the resultant vasoconstriction of most of the blood vessels besides those in the active muscles almost always increase the arterial pressure during exercise. This increase can be as little as 20 mm. Hg or as great as 80 mm. Hg, depending on the conditions under which the exercise is performed. For instance, when a person performs exercise under very tense conditions but uses only a few muscles, the sympathetic response still occurs throughout the body but vasodilatation occurs in only a few muscles. Therefore, the net effect is mainly one of vasoconstriction, often increasing the mean arterial pressure to as high as 180 mm. Hg. Such a condition occurs in a person standing on a ladder and nailing with a hammer on the ceiling above. The tenseness of the situation is obvious, and yet the amount of muscle vasodilatation is relatively slight.

On the other hand, when a person performs whole-body exercise, such as running or swimming, the increase in arterial pressure is usually only 20 to 40 mm. Hg. The lack of a tremendous rise in pressure results from the extreme vasodilatation occurring in large masses of muscle.

In rare instances, persons are found in whom the sympathetic nervous system is absent, either because of congenital absence or because of surgical removal. When such a person exercises, instead of arterial pressure rising, the pressure actually falls—sometimes to as low as one-half normal, and the cardiac output rises only about one-third as much as it does normally. Therefore, one can readily understand the major importance of increased sympathetic activity during exercise.

*Importance of the Arterial Pressure Rise During Exercise.* In the well trained athlete it has been calculated that muscle blood flow can increase as much as 20-fold. Though most of this increase results from vasodilatation in the active muscles, the increase in arterial pressure also plays an important role. If one remembers that an increase in pressure not only forces extra blood through the muscle because of the increased pressure itself but also dilates the blood vessels, he can see that as little as a 20 to 40 mm. Hg rise in pressure can at times actually double peripheral blood flow.

It is especially important to note that in animals or human beings who do not have a sympathetic nervous system, the fall in arterial pressure that occurs during exercise has a strong effect on negating the rise in cardiac output that normally occurs. In such instances, the cardiac output can almost never be increased more than two-fold, instead of the four- to seven-fold that can occur when the arterial pressure rises above normal.

**The Cardiovascular System as the Limiting Factor in Heavy Exercise.** The capability of an athlete to enhance his cardiac output and thereby to deliver increased quantities of oxygen and other nutrients to his tissues is the major factor that determines the degree of prolonged heavy exercise that the athlete can sustain. For instance, the speed of a marathon runner is almost directly proportional to his ability to enhance his cardiac output. Therefore, the ability of the circulatory system to adapt to exercise is equally as important as the muscles themselves in setting the limit for the performance of muscle work.

# THE CEREBRAL CIRCULATION

## NORMAL RATE OF CEREBRAL BLOOD FLOW

The normal blood flow through brain tissue averages 50 to 55 ml. per 100 grams of brain per minute. For the entire brain of the average adult, this is approximately 750 ml. per minute, or 15 per cent of the total resting cardiac output. This blood flow, even under extreme conditions, usually does not vary greatly from the normal value because the control systems are especially geared to maintain constant cerebral blood flow. Some exceptions to this occur when the brain is subjected to excess carbon dioxide, hydrogen ions, or to lack of oxygen, as is discussed subsequently.

## REGULATION OF CEREBRAL CIRCULATION

As in most other tissues of the body, cerebral blood flow is highly related to the metabolism of the cerebral tissue. At least three different metabolic factors have been shown to have very potent effects on cerebral blood flow. These are carbon dioxide concentration, hydrogen ion concentration, and oxygen concentration. An increase in either the carbon dioxide or the hydrogen ion concentration increases cerebral blood flow, whereas a decrease in oxygen concentration increases the flow.

**Regulation of Cerebral Blood Flow in Response to Excess Carbon Dioxide or Hydrogen Ion Concentration.** An increase in carbon dioxide concentration in the arterial blood perfusing the brain greatly increases cerebral blood flow. This is illustrated in Figure 29–2, which shows that doubling the arterial $P_{CO_2}$ by breathing carbon dioxide also approximately doubles the blood flow.

Carbon dioxide increases cerebral blood flow by combining with water in the body fluids to form carbonic acid, with subsequent dissociation to form hydrogen ions. The hydrogen ions then cause vasodilatation of the cerebral

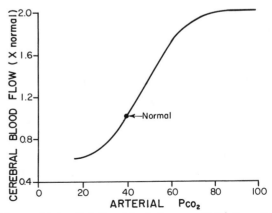

**Figure 29–2.** Relationship between arterial $P_{CO_2}$ and cerebral blood flow.

vessels—the dilatation being almost directly proportional to the increase in hydrogen ion concentration.

Any other substance that increases the acidity of the brain tissue, and therefore also increases the hydrogen ion concentration, increases blood flow as well. Such substances include lactic acid, pyruvic acid, or any other acidic material formed during the course of metabolism.

*Value of the Carbon Dioxide and Hydrogen Control of the Cerebral Blood Flow.* Increased hydrogen ion concentration greatly depresses neuronal activity; conversely, diminished hydrogen ion concentration greatly increases neuronal activity. Therefore, it is fortunate that an increase in hydrogen ion concentration causes an increase in the blood flow, which in turn carries both carbon dioxide and dissolved acids away from the brain tissues at an increased rate. Loss of the carbon dioxide removes carbonic acid from the tissues, and this, along with the removal of other acids, reduces the hydrogen ion concentration back toward normal. Thus, this mechanism helps to maintain a very constant hydrogen ion concentration in the cerebral fluids and therefore also to maintain a normal level of neuronal activity.

**Oxygen Deficiency as a Regulator of Cerebral Blood Flow.** Except during short periods of intense brain activity, the utilization of oxygen by the brain tissue remains within very narrow limits—within a few per cent of 3.5 ml of oxygen per 100 grams of brain tissue per minute. If the blood flow to the brain ever becomes insufficient to supply this needed amount of oxygen, the oxygen deficiency mechanism for vasodilatation, which was discussed in Chapter 20 and which functions in essentially all tissues of the body, immediately causes vasodilatation, bringing the blood flow and transport of oxygen to the cerebral tissues back near to normal. Thus, this local blood flow regulatory mechanism is much the same as that which exists in the coronary and skeletal muscle circulations and in many other circulatory areas of the body.

Experiments have shown that a decrease in cerebral *venous* blood $Po_2$ below approximately 30 mm. Hg (normal value is about 35 mm. Hg) will begin to increase cerebral blood flow. It is very fortunate that flow does respond at this critical level because brain function begins to become deranged at not much lower values of $Po_2$, such as at approximately 20 mm. Hg. Thus, the oxygen mechanism for local regulation of cerebral blood flow is a very important

protective response against diminished cerebral neuronal activity and, therefore, against derangement of mental capability.

**Effect of Cerebral Activity on Blood Flow.** Only rarely does neuronal activity increase sufficiently to increase the overall rate of metabolism in the brain. And, in those few instances in which this does occur, the cerebral blood flow increases only a moderate amount. For instance, a convulsive attack in an animal, which causes extreme activity throughout the entire brain, can result in an overall increase in cerebral blood flow of as much as 50 per cent. On the other hand, administration of anesthetics sometimes reduces brain metabolism and also cerebral blood flow as much as 30 to 40 per cent.

Blood flow in localized portions of the brain can increase as much as 40 to 50 per cent as a result of intense localized neuronal activity even though the total cerebral flow is hardly affected. This effect is illustrated in Figure 29–3, which shows the effect of shining an intense light into the eyes of a cat; this results in increased blood flow in the occipital cortices that become excited.

**Effect of Arterial Pressure on Cerebral Blood Flow.** Figure 29–4 shows cerebral blood flow measured in human beings having different blood pressures. This figure shows especially the extreme constancy of cerebral blood flow between the limits of 60 and 180 mm. Hg mean arterial pressure. Cerebral blood flow suffers only when the arterial pressure falls below approximately the 60 mm. Hg mark.

**Minor Importance of Autonomic Nerves in Regulating Cerebral Blood Flow.** Sympathetic nerves from the cervical sympathetic chain pass upward along the cerebral arteries to supply at least some of the cerebral vasculature. Also, parasympathetic fibers from the great superficial petrosal and facial nerves supply some of the vessels. However, transection of either the sympathetic or parasympathetic

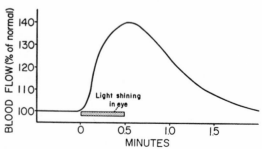

**Figure 29–3.** Increase in blood flow to the occipital regions of the brain when a light is flashed in the eyes of an animal.

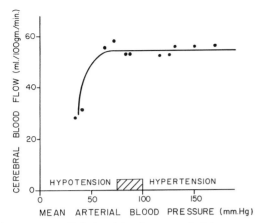

**Figure 29-4.** Relationship of mean arterial pressure to cerebral blood flow in normotensive, hypotensive, and hypertensive persons. (Modified from Lassen: *Physiol. Rev.*, 39:183, 1959.)

nerves causes no measurable effect on cerebral blood flow. On the other hand, stimulation does affect blood flow slightly, the sympathetics causing mild vasoconstriction of the large vessels but not of the small vessels, and the parasympathetics causing mild vasodilatation. It remains to be proved that either of these has any functional importance in regulation of cerebral blood flow.

### MEASUREMENT OF CEREBRAL BLOOD FLOW IN THE HUMAN BEING

Cerebral blood flow in the human being can be measured in the same way that coronary flow is measured, as explained in Chapter 25. That is, the person breathes nitrous oxide for a period of ten minutes, and samples are removed from the arterial blood supplying the brain and from the venous blood flowing out of the brain. By using these values and by employing an appropriate integrative procedure, as explained for measurement of coronary blood flow, one can then calculate the cerebral blood flow.

Recently, cerebral blood flow has most often been measured by use of a radioactive gas technique, usually employing radioactive krypton or xenon. With this technique, the gas is breathed for a few minutes until its concentration in the brain substance has reached an equilibrium state. Then the person stops breathing the gas, and the radioactivity emanating from the brain is recorded by a scintillation counter. The more rapid the blood flow through the brain, the more rapid the removal of the radioactive gas from the brain tissue. Using an appropriate mathematical formula, one can calculate the blood flow in the brain tissue from the rate of decrease in radioactivity. As many as 20 separate scintillation counters have been used simultaneously to detect blood flow in different areas of the brain at the same time. By using these counters one may either summate the blood flow for

the entire brain or determine regional distribution of blood flow, which in itself can be of value in spotting disturbed flows caused by tumors or other abnormalities.

## THE SPLANCHNIC CIRCULATION

A large share of the cardiac output flows through the vessels of the intestines and through the spleen, finally coursing into the portal venous system and then through the liver, as illustrated in Figure 29-5. This is called the portal circulatory system, and it, plus the arterial blood flow into the liver, is called the splanchnic circulation.

### BLOOD FLOW THROUGH THE LIVER

About 1100 ml. of portal blood enter the liver each minute. This flows through the *hepatic sinuses* in close contact with the cords of liver parenchymal cells. Then it enters the *central veins* of the liver and from there flows into the vena cava.

In addition to the portal blood flow, approximately 350 ml. of blood flows into the liver each minute through the hepatic artery, making a total hepatic flow of almost 1500 ml. per minute, or an average of 29 per cent of the total cardiac output. The hepatic arterial blood flow maintains nutrition of the connective tissue and

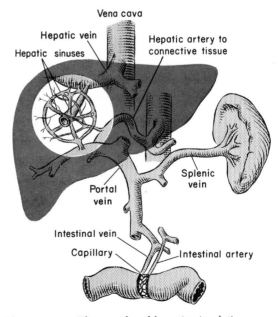

**Figure 29-5.** The portal and hepatic circulations.

especially of the walls of the bile ducts. Therefore, loss of hepatic arterial flow can be lethal because this often causes necrosis of the basic liver structures. The blood from the hepatic artery, after it supplies the structural elements of the liver, empties into the hepatic sinuses to mix with the portal blood.

**Control of Liver Blood Flow.** Three quarters of the blood flow through the liver is derived from the portal blood flow into the liver; this flow is controlled by the various factors that determine flow through the gastrointestinal tract and spleen—factors that will be discussed in subsequent sections of this chapter.

The additional one quarter of the blood flow through the liver is derived from the hepatic arteries; its flow rate is determined primarily by local metabolic factors in the liver itself. For instance, a decrease in oxygen in the hepatic blood causes an increase in hepatic arterial blood flow, indicating that the need to deliver nutrients to the liver tissues has a direct vasodilating effect.

**Reservoir Function of the Liver.** Because the liver is an expandable and compressible organ, large quantities of blood can be stored in its blood vessels. Its normal blood volume, including both that in the hepatic veins and hepatic sinuses, is about 500 ml., or 10 per cent of the total blood volume. However, when high pressure in the right atrium causes back pressure in the liver, the liver expands, and as much as 1 liter of extra blood occasionally is thereby stored in the hepatic veins and sinuses. This occurs especially in cardiac failure with peripheral congestion, which was discussed in Chapter 26.

Thus, in effect, the liver is a large expandable venous organ capable of acting as a valuable blood reservoir in time of excess blood volume and capable of supplying extra blood in times of diminished blood volume.

*Control of Blood Storage in the Liver by the Sympathetic Nervous System.* The storage portions of the liver, particularly the large veins and much less so the sinusoids, are vasoconstricted by sympathetic stimulation. Therefore, in times of circulatory stress when the sympathetic nervous system is discharging strongly, a large portion of blood stored in the liver is expelled into the general circulation within one to four minutes. In the normal human being this can amount to as much as 350 ml. of blood. Thus, the liver is probably the single most important source of blood to fill the other portions of the circulation in times of

need, especially during heavy exercise or after severe hemorrhage.

**Permeability of the Hepatic Sinuses.** The hepatic sinuses are lined with an endothelium similar to that of the capillaries, but its permeability is extreme in comparison with that of usual capillaries—so much so that even the proteins of the blood diffuse into the extravascular spaces of the liver almost as easily as fluids. Fortunately, the pressure in the sinuses is only 6 to 8 mm. Hg so that most of the proteins that diffuse out of the sinuses also diffuse back in. Yet, the remainder passes into the liver lymphatics, so that liver lymph normally contains almost as great a protein concentration as that of the plasma itself. Also, outflow of lymph from the liver accounts for one-half to two-thirds of the total body lymph flow.

This extreme permeability of the liver sinuses brings the fluids of the hepatic blood into extremely close contact with the liver parenchymal cells, thus facilitating rapid exchange of nutrient materials between the blood and the liver cells.

**The Blood Cleansing Function of the Liver.** Blood flowing through the intestinal capillaries picks up many bacteria from the intestines. Indeed, a sample of blood from the portal system almost always grows colon bacilli when cultured, whereas growth of colon bacilli from blood in the systemic circulation is extremely rare. Special high-speed motion pictures of the action of Kupffer cells, the large phagocytic cells that line the hepatic sinuses, have demonstrated that these cells can cleanse blood extremely efficiently as it passes through the sinuses; when a bacterium comes into momentary contact with a Kupffer cell, in less than 0.01 second the bacterium passes inward through the wall of the Kupffer cell to become permanently lodged therein until it is digested. Probably not over 1 per cent of the bacteria entering the portal blood from the intestines succeeds in passing through the liver into the systemic circulation.

**Measurement of Blood Flow Through the Liver in Man.** Several indirect methods for measuring blood flow through the liver similar to those used for measuring blood flow through the heart and brain have been developed. The most important of these is the following: The substance bromosulfophthalein, when injected into the blood, is excreted into the gut by the liver, and almost none of it is lost from the blood in any other way. This material is injected continuously into the circulation and blood samples are removed from two points in the circulation: (1) from a catheter inserted into a peripheral vein and (2) from another catheter inserted through a peripheral vein all the way into the hepatic vein. The difference between the concentrations in a usual peripheral vein and in the hepatic vein represents the quantity of bromosulfophthalein removed from each unit of blood passing through the liver. To determine the quantity of bromosulfophthalein removed by the liver each minute, the substance is injected into the circulation continuously and its rate of injection is adjusted so that its concentration in the blood is neither increasing nor decreasing. When this is achieved the amount being injected is equal to the rate of liver excretion. On dividing the A-V difference into the rate being secreted, one determines the total liver blood flow per minute.

## BLOOD FLOW THROUGH THE INTESTINAL VESSELS

About four-fifths of the portal blood flow originates in the intestines and stomach (about 850 ml. per minute), and the remaining one-fifth originates in the spleen and pancreas.

**Control of Gastrointestinal Blood Flow.** Blood flow in the gastrointestinal tract seems to be controlled in almost exactly the same way as in most other areas of the body: that is, mainly by local regulatory mechanisms. Furthermore, blood flow to the mucosa and submucosa, where the glands are located and where absorption occurs, is controlled separately from blood flow to the musculature. When glandular secretion increases, so does mucosal and submucosal blood flow. Likewise, when motor activity of the gut increases, blood flow in the muscle layers increases.

However, the precise mechanisms by which alterations in gastrointestinal activity alter the blood flow are not completely understood. It is known that decreased availability of oxygen to the gut increases local blood flow in the same way that this occurs elsewhere in the body, so that local regulation of blood flow in the gut might occur entirely secondarily to changes in metabolic rate. On the other hand, it has been claimed that stimulation of most gastrointestinal glands causes the glands to release an enzyme that acts on a plasma protein to form the substance bradykinin. The bradykinin then supposedly causes local vasodilatation, thereby increasing the glandular blood flow. Though this mechanism has received much publicity, especially in relation to salivary glands, recent experiments in which the bradykinin mechanism has been blocked have shown that, in spite of the block, salivary blood flow increases just as much as normally during glandular secretion. Therefore, it is doubtful that this mechanism is a correct one.

*Nervous Control of Gastrointestinal Blood Flow.* Stimulation of the parasympathetic nerves (the vagi) to the *stomach* and *lower colon* increases local blood flow at the same time that it increases glandular secretion. However, this increased flow probably results from the increased glandular activity. On the other hand, parasympathetic stimulation has little effect on blood flow in the small intestine where glandular secretion is controlled primarily by local mechanisms rather than by nerve stimulation.

Sympathetic stimulation, in contrast, has a direct effect on essentially all blood vessels of the gastrointestinal tract to cause intense vasoconstriction. However, after a few minutes of this vasoconstriction, the flow returns to or almost to normal via a mechanism called "autoregulatory escape." That is, the local metabolic vasodilator mechanisms that are elicited by ischemia become prepotent over the sympathetic vasoconstriction and, therefore, cause return of the necessary nutrient blood flow to the gastrointestinal glands and muscle.

A major value of sympathetic vasoconstriction in the gut is that it allows shutting off of splanchnic blood flow for short periods of time during heavy exercise when increased flow is needed by the skeletal muscle and heart.

## PORTAL VENOUS PRESSURE

The liver offers a moderate amount of resistance to blood flow from the portal system to the vena cava. As a result, the pressure in the portal vein averages 10 mm. Hg, which is considerably higher than the almost zero pressure in the vena cava. Because of this high portal venous pressure, the pressures in the portal venules and capillaries have a much greater tendency to become abnormally high than is true elsewhere in the body.

**Blockage of the Portal System.** Frequently, extreme amounts of fibrous tissue develop within the liver structure, destroying many of the parenchymal cells and eventually contracting around the blood vessels, thereby greatly impeding the flow of portal blood through the liver. This disease process is known as *cirrhosis of the liver.* It results most frequently from alcoholism, but it can also follow ingestion of poisons such as carbon tetrachloride, virus diseases such as infectious hepatitis, or infectious processes in the bile ducts.

The portal system is also occasionally blocked by a large clot developing in the portal vein or in its major branches.

When the portal system is suddenly blocked, the return of blood from the intestines and spleen to the systemic circulation is tremendously impeded, the capillary pressure rising as much as 15 to 20 mm. Hg, and the patient often dies within a few hours because of excessive loss of fluid from the capillaries into the lumina and walls of the intestines.

***Collateral Circulation from the Portal Veins to the Systemic Veins.*** When portal blood flow is blocked slowly rather than suddenly, such as occurs in slowly developing cirrhosis of the liver, large collateral vessels develop between the portal veins and the systemic veins. The most important of these are collaterals from the splenic to the esophageal veins and these collaterals frequently become so large that they protrude deeply into the lumen of the esophagus and are then called *esophageal varicosities*. The esophageal mucosa overlying these varicosities eventually becomes eroded; and in many if not most of these patients, the erosion finally penetrates all the way into the varicosity and causes the person to bleed severely, often to bleed to death.

***Ascites as a Result of Portal Obstruction.*** Ascites is free fluid in the peritoneal cavity. It results from exudation of fluid either from the surface of the liver or from the surfaces of the gut and its mesentery. Ascites usually will develop only in case outflow of blood from the liver into the inferior vena cava is blocked. This causes extremely high pressure in the liver sinusoids, which in turn causes fluid to weep from the surfaces of the liver. The weeping fluid is almost pure plasma, containing tremendous quantities of protein. The protein, because it causes a high colloid osmotic pressure in the abdominal fluid, then pulls by osmosis additional fluid from the surfaces of the gut and mesentery.

On the other hand, obstruction of the portal vein, without directly involving the liver, rarely causes ascites. If obstruction occurs acutely, the person is likely to die of shock within hours because of fluid loss into the gut; if it occurs slowly, collateral vessels can usually develop enough to prevent ascites.

## THE SPLENIC CIRCULATION

**The Spleen as a Reservoir.** The capsule of the spleen in many lower animals contains large amounts of smooth muscle; and sympathetic stimulation causes intense contraction of the spleen. On the other hand, sympathetic inhibition results in considerable splenic expansion with consequent storage of blood.

In man, the splenic capsule is nonmuscular, but even so, dilatation of vessels within the spleen can still cause the spleen to store several hundred milliliters of blood at times, and then, under the influence of sympathetic stimulation, constriction of the vessels will express most of this blood into the general circulation. Unfortunately, these effects in man are poorly understood.

As illustrated in Figure 29–6, two areas exist in the spleen for the storage of blood; these are the venous sinuses and the pulp. Small vessels flow directly into the venous sinuses, and when the spleen distends, the venous sinuses swell, storing large quantities of blood.

In the splenic pulp, the capillaries are very permeable, so that much of the blood passes first into the pulp and then oozes through this before entering the venous sinuses. As the spleen enlarges, many cells become stored in the pulp. Therefore, the net quantity of red blood cells in the general circulation decreases slightly when the spleen enlarges. The spleen can store enough cells that splenic contraction can cause the hematocrit of the systemic blood to increase as much as 3 to 4 per cent.

This ability of the spleen to store and release blood is important, at least in lower animals, in times of stress such as during strenuous exercise, for the circulatory system then needs an increased volume of blood so that rapid return of blood to the heart can be maintained. At the same time, the blood that flows through the muscles needs to carry larger quantities of oxygen than normally. Therefore, the increased hematocrit of the circulating blood resulting from splenic contraction is also an aid to the body during these periods of stress.

**The Blood Cleansing Function of the Spleen—Removal of Old Cells.** Blood passing through the splenic pulp before it enters the sinuses undergoes thorough squeezing. Indeed, it is supposed that the red blood cells re-enter the venous sinuses from the pulp by "diapedesis"; that is, by squeezing through pores much smaller than the size of the cells themselves. Under these circumstances it is to be expected that fragile red blood cells would not withstand the trauma. For this reason, many of the red blood cells destroyed in the body have their final demise in the spleen. After the cells rupture, the released hemoglobin and

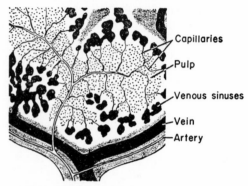

**Figure 29–6.** The functional structures of the spleen. (Modified from Bloom and Fawcett: Textbook of Histology, 1975.)

the cell stroma are ingested by the reticuloen-dothelial cells of the spleen.

*Reticuloendothelial Cells of the Spleen.*   The pulp of the spleen contains many large phagocytic re-ticuloendothelial cells, and the venous sinuses are lined with similar cells. These cells act as a cleansing system for the blood, similar to that in the venous sinuses of the liver. When the blood is invaded by infectious agents, the reticuloendothelial cells of the spleen rapidly remove debris, bacteria, parasites, etc. Also, in many infectious processes the spleen enlarges in the same manner that lymph glands en-large and performs its cleansing function even more adequately.

Much of the spleen is filled with *white pulp,* which is in reality a large quantity of lymphocytes and plasma cells. These function in exactly the same way in the spleen as in the lymph glands to cause either humoral or lymphocytic immunity against toxins, bacteria, and so forth, as described in Chapter 7.

**The Spleen as a Hemopoietic Organ.**   During fetal life, the splenic pulp produces blood cells in exactly the same manner that the red bone marrow in the adult produces cells. As the normal fetus ap-proaches birth, the spleen normally loses this ability to produce cells, but, in some diseases, the spleen continues to produce cells even after birth. For in-stance, in the disease *erythroblastosis fetalis,* which results from excessive destruction of red blood cells by abnormal antibodies in the plasma, as discussed in Chapter 5, the fetus must produce 10 or more times as many red blood cells as normally. As a result, the hemopoietic function of the spleen persists for sev-eral weeks after birth.

# CIRCULATION IN THE SKIN

## *PHYSIOLOGIC ANATOMY OF THE CUTANEOUS CIRCULATION*

Circulation through the skin subserves two major functions: first, *nutrition of the skin tis-sues* and, second, *conduction of heat* from the internal structures of the body to the skin so that the heat can be removed from the body. To perform these two functions the circulatory ap-paratus of the skin is characterized by two major types of vessels, illustrated diagrammati-cally in Figure 29–7: (1) the usual nutritive ar-teries, capillaries, and veins and (2) vascular structures concerned with heating the skin, consisting principally of (a) an extensive *sub-cutaneous venous plexus,* which holds large quantities of blood that can heat the surface of the skin, and (b) in some skin areas, *arteriove-nous anastomoses,* which are large vascular communications directly between the arteries and the venous plexuses. The walls of these anastomoses have strong muscular coats inner-

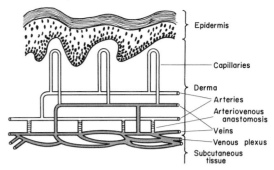

**Figure 29–7.**  Diagrammatic representation of the skin circulation.

vated by sympathetic vasoconstrictor nerve fibers that secrete norepinephrine. When con-stricted, they reduce the flow of blood into the venous plexuses to almost nothing; or when maximally dilated, they allow extremely rapid flow of warm blood into the plexuses. The ar-teriovenous anastomoses are found principally in the volar surfaces of the hands and feet, the lips, the nose, and the ears, which are areas of the body most often exposed to maximal cool-ing.

**Rate of Blood Flow Through the Skin.**   The rate of blood flow through the skin is among the most variable of any part of the body, because the flow required to regulate body temperature changes markedly in response to, first, the rate of metabolic activity of the body, and, second, the temperature of the surroundings. This will be discussed in detail in Chapter 72. The blood flow required for nutrition is slight, so that this plays almost no role in controlling normal skin blood flow. At ordinary skin temperatures, the amount of blood flowing through the skin ves-sels to subserve heat regulation is about 10 times as much as that needed to supply the nu-tritive needs of the tissues. But, when the skin is exposed to extreme cold, the blood flow may become so slight that nutrition begins to suffer—even to the extent, for instance, that the fingernails grow considerably more slowly in arctic climates than in temperate climates.

Under ordinary cool conditions the blood flow to the skin is about 0.25 liter/sq. meter of body surface area, or a total of about 400 ml. per minute, in the average adult. On the other hand, when the skin is heated until maximal vasodilatation has resulted, the blood flow can be as much as 7 times this value, or a total of about 2.8 liters per minute, thus illustrating both the extreme variability of skin blood flow and the great drain on cardiac output that can occur under hot conditions. Indeed, many persons with borderline cardiac failure develop severe failure in hot weather because of the extra load on the heart and then revert from failure in cool weather.

## REGULATION OF BLOOD FLOW IN THE SKIN

**Nervous Control of Cutaneous Blood Flow.** Since most of the blood flow through the skin is to control body temperature, and since this function in turn is regulated by the nervous system, the blood flow through the skin is principally regulated by nervous mechanisms rather than by local regulation, which is opposite to the regulation in most parts of the body.

*Temperature Control Center of the Hypothalamus.* Located in the preoptic region of the anterior hypothalamus is a small center that is capable of controlling body temperature. Heating this area causes vasodilation of essentially all the skin vessels of the body and also causes sweating. Cooling of the center causes vasoconstriction and cessation of sweating. The detailed function of this center will be discussed in Chapter 72 in relation to body temperature. The important point in the present discussion is that the hypothalamus controls blood flow through the skin in response to changes in body temperature by two mechanisms: (1) a sympathetic vasoconstrictor mechanism and (2) a sympathetic vasodilator mechanism.

*The Sympathetic Vasoconstrictor Mechanism.* The skin throughout the body is supplied with sympathetic vasoconstrictor fibers that secrete norepinephrine at their endings. This constrictor system is extremely powerful in the feet, hands, lips, nose, and ears, which are the areas most frequently exposed to severe cold and which are also the areas where large numbers of arteriovenous anastomoses are found. At normal body temperature the sympathetic vasoconstrictor nerves keep these anastomoses almost totally closed, but when the body becomes overly heated the number of sympathetic impulses is greatly reduced so that the anastomoses dilate and allow large quantities of warm blood to flow into the venous plexuses, thereby promoting loss of heat from the body.

In the remainder of the body—that is, over the surfaces of the arms, legs, and body trunk—almost no arteriovenous anastomoses are present, but nevertheless vasoconstrictor control of the nutritive vessels can still effect major changes in blood flow. When the body becomes overheated, the sympathetic vasoconstrictor impulses cease, and the blood flow to the skin vessels increases about two-fold.

*Extreme Sensitivity of the Skin Blood Vessels to Circulating Norepinephrine and Epinephrine.* In addition to the direct sympathetic vasoconstrictor effect in the skin, the skin blood vessels are also extremely sensitive to circulating norepinephrine and epinephrine. Therefore, even in areas of the skin that might have lost their sympathetic innervation, any mass discharge of the sympathetic nervous system will still cause intense skin vasoconstriction. Sometimes the sensitivity of the skin to circulating norepinephrine and epinephrine is so great that the vasoconstriction can actually damage the skin. This response will be discussed in more detail later in the chapter in relation to *Raynaud's disease.*

*The Sympathetic Vasodilator Mechanism.* When the body temperature becomes excessive and sweating begins to occur, the blood flow through the skin of the forearms and trunk increases an additional three-fold, which occurs as a result of so-called "active" vasodilatation of the vessels. However, the basic mechanism by which this active dilation occurs is not completely known. Some physiologists believe that sympathetic vasodilator fibers that secrete acetylcholine at their endings exist in the skin of human beings. Yet this additional increase in blood flow does not occur except in the presence of sweating, and it does not occur in lower animals that do not have sweat glands. Therefore, another theory is that the sympathetic fibers that secrete acetylcholine in the sweat glands cause a secondary vasodilatation as follows: The increased activity of the sweat glands is postulated to cause these glands to release an enzyme called *kallikrein,* which in turn splits the polypeptide *bradykinin* from globulin in the interstitial fluids. Bradykinin in turn is a powerful vasodilator substance that could account for the greatly increased blood flow when sweating begins to occur. In opposition to this theory, however, is the fact that inhibition of the bradykinin mechanism does not block the increased blood flow associated with sweating.

**Effect of Cold on the Skin Circulation.** When cold is applied directly to the skin, the skin vessels constrict more and more down to a temperature of about 15° C., at which point they reach their maximum degree of constriction. This constriction results primarily from increased sensitivity of the vessels to nerve stimulation, but it probably also results at least partly from a reflex that passes to the cord and then back to the vessels. At temperatures below 15° C., the vessels begin to dilate. This dilation is caused by a direct local effect of the cold on the vessels themselves—probably paralysis of the contractile mechanism of the vessel wall or block of the nerve impulses coming to the vessels. In any event, at temperatures approaching 0° C. the vessels frequently reach maximum vasodilation. This intense vasodilatation in severe degrees of cold plays a purposeful role in preventing freezing of the exposed portions of the body, particularly the hands and ears.

**Local Regulation of Blood Flow in the Skin.** Though local regulation of blood flow usually plays a small role in skin blood flow control, it does have an effect in those few instances in which skin blood flow becomes greatly decreased. For instance, if one sits on his buttocks for 30 minutes or more and then stands, he will note intense reddening of the affected skin. That is, the blood flow to the skin area has now increased to a marked extent, which is typical *reactive hyperemia* resulting from diminished availability of nutrients to the tissues during the period of compression. Thus, the local regulatory mechanism present in essentially all other tissues of the body is also present in the skin and can be called into play when needed to prevent nutritional damage to the tissues.

**Shift of Blood from the Skin to the Remainder of the Circulation in Times of Circulatory Stress.** The skin venous plexuses of the entire body, including those of the hands, feet, arms, legs, and trunk, are all strongly supplied with sympathetic vasoconstrictor innervation. In times of circulatory stress, such as during exercise, following severe hemorrhage, or even in states of anxiety, sympathetic stimulation of these venous plexuses can force large quantities of blood, estimated to be as much as 5 to 10 per cent of the blood volume, into the internal vessels. Thus, the subcutaneous veins of the skin act as an important blood reservoir, often providing blood to subserve other circulatory functions when needed.

## COLOR OF THE SKIN IN RELATION TO SKIN TEMPERATURE

Much of the skin color is due to the color of the blood in the skin capillaries and veins. Therefore, when the skin is hot and arterial blood is flowing rapidly into these vessels, the skin is usually red. On the other hand, when the skin is cold and the blood is flowing extremely slowly, most of the oxygen is removed for nutritive purposes before the blood can leave the capillaries. Consequently, the capillaries and veins then contain large amounts of dark, deoxygenated blood that gives the skin a bluish hue.

Another cutaneous vascular effect that often affects skin color is severe constriction of the cutaneous vessels, which expresses most of the blood out of the skin into other parts of the circulation. When this occurs the skin takes on the color of the subcutaneous connective tissue, which is composed principally of collagen fibers and has a whitish hue. Thus, in conditions in which the cutaneous vessels become greatly constricted, the skin has an ashen white pallor.

# PHYSIOLOGY OF VASCULAR DISEASES OF THE LIMBS

## RAYNAUD'S DISEASE

The local vasoconstrictor reflex that occurs normally during exposure to cold becomes especially marked in the limbs of some persons, particularly in many women who live in northern climates. This effect, called *Raynaud's disease,* occurs most frequently in the hands, though occasionally also in the feet. When the hands become even slightly cooled, they are likely to exhibit extreme vascular spasm—so much so that the hands do not receive sufficient flow of blood even for maintaining adequate metabolism. The tissues then develop *ischemic pain* in exactly the same manner that the heart develops pain in angina pectoris and as other tissues develop pain during ischemia. After the ischemia persists for several minutes, *hyperemia* supervenes, causing dilatation of all the vessels of the hands so that the final stage is sometimes *intense reddening* of the hands. This effect is probably caused by the same factors that cause reactive hyperemia following temporary vascular occlusion, as discussed in Chapter 20. Strangely enough, the pain becomes even more severe at this moment, owing either to mobilization of the abnormal metabolic products that cause pain or, perhaps, in part to the greatly increased blood pressure within blood vessels upon dilatation of all the arterioles.

Though Raynaud's disease in its early stages is merely a physiologic entity that passes through the phases of blanching of the skin, pain, and reddening of the skin, after it progresses for years the vessels may become continuously constricted. In severe cases the blood flow can become decreased so much that the fingers become gangrenous.

**Sympathectomy for Raynaud's Disease.** To treat Raynaud's disease the sympathetic reflex is often blocked. This is done best by cutting the preganglionic fibers in the sympathetic chain at T-2 and T-3, which interrupts the sympathetic nerve impulses from the spinal cord to the hand. The postganglionic fibers to the hand originate mainly in the stellate ganglion, but the stellate ganglion is not removed, for removal of postganglionic sympathetic fibers allows the blood vessels to become excessively sensitized to circulating norepinephrine and epinephrine. If this happens the hands will still exhibit Raynaud's syndrome every time the adrenal glands are excited. This phenomenon of sensitization will be discussed in Chapter 57 in connection with the various reactions of the sympathetic nervous system.

## BUERGER'S DISEASE

Buerger's disease, also known as *thromboangiitis obliterans*, is caused by an inflammatory reaction of the small blood vessels in the triangular sheaths containing the nerve, the artery, and the vein. This condition occurs most frequently in the legs but occasionally in the arms. Even though the disease is caused by an inflammatory process that constricts the vessels, it is made much worse by the effects of nicotine incident to smoking.

The patient with Buerger's disease suffers a symptom known as *intermittent claudication*—that is, when he walks, his calves become extremely painful because of ischemic products collecting in his muscles and not removed by adequate blood flow. The ischemic pain is similar to that of Raynaud's disease, and also to the ischemic cardiac pain of angina pectoris.

**Sympathectomy for Buerger's Disease.** Buerger's disease can be benefited greatly by sectioning the sympathetic nerves to the area afflicted. This is done for a leg by removing the L-1, L-2, and L-3 ganglia of the sympathetic chain. Even though the pathologic changes of Buerger's disease are not caused by excessive sympathetic activity, removal of the sympathetic nerves blocks normal vasomotor tone, which allows vasodilatation in the affected areas. Approximately 90 per cent of the patients who otherwise might lose toes or even legs from Buerger's disease can save these by sympathectomy.

## PERIPHERAL ARTERIOSCLEROSIS

Arteriosclerosis, which causes intra-arterial plaques to develop, has already been discussed in connection with the coronary arteries. This same process can occur throughout the arterial tree and not simply in the coronaries. Therefore, in old age many of the peripheral arteries will often have their intimal surfaces almost completely covered with lipid deposits or calcified plaques.

Progressive obliteration of the peripheral arterial tree by arteriosclerosis results in all the symptoms that one would expect. It causes pain in the tissues because of ischemia, and it is likely to cause gangrene and ulceration of the skin because of poor blood supply.

A special difference between arteriosclerotic peripheral vascular disease and either Buerger's disease or Raynaud's disease is that sympathectomy is not nearly so effective for treatment as it is in the other two diseases. This is true probably because this vascular disease is caused by obstruction of the large arteries rather than the small arteries, and the small arteries, which are affected most by the sympathetics, probably remain maximally vasodilated in the ischemic tissues whether the sympathetics are present or not.

# REFERENCES

Abramson, D. I.: Vascular Disorders of the Extremities, 2nd Ed. Hagerstown, Maryland, Harper & Row, Publishers, 1974.

Alpert, J. S., and Coffman, J. D.: Effect of intravenous epinephrine on skeletal muscle, skin, and subcutaneous blood flow. *Amer. J. Physiol.*, 216:156, 1969.

Altura, B. M., and Altura, B. T.: Effects of local anesthetics, antihistamines, and glucocorticoids on peripheral blood flow and vascular smooth muscle. *Anesthesiology*, 41:197, 1974.

Barker, W. F.: Peripheral Arterial Disease, 2nd Ed. Philadelphia, W. B. Saunders Company, 1975.

Betz, E.: Cerebral blood flow: its measurement and regulation. *Physiol. Rev.*, 52:595, 1972.

Bevegard, B. S., and Shepherd, J. T.: Regulation of the circulation during exercise in man. *Physiol. Rev.*, 47:178, 1967.

Chapman, C. B.: The physiology of exercise. *Sci. Amer.*, 212:88, 1965.

Clarke, D. H.: Exercise Physiology. Englewood Cliffs, N.J., Prentice-Hall, Inc., 1975.

Glass, H. I.: Theory of current techniques of measuring cerebral blood flow. *J. Nucl. Biol. Med.*, 16:267, 1972.

Grayson, J.: The gastrointestinal circulation. In Guyton, A. C. (ed.): MTP International Review of Science: Physiology. Vol. 4. Baltimore, University Park Press, 1974, p. 105.

Green, H. D., and Kepchar, J. H.: Control of peripheral resistance in major systemic vascular beds. *Physiol. Rev.*, 39:617, 1959.

Greenway, C. V., and Stark, R. D.: Hepatic vascular bed. *Physiol. Rev.*, 51:23, 1971.

Haddy, F. J., Molnar, J. I., Borden, C. W., and Texter, E. C., Jr.: Comparison of direct effects of angiotensin and other vasoactive agents on small and large blood vessels in several vascular beds. *Circulation*, 25:239, 1962.

Heistad, D. D., and Abboud, F. M.: Factors that influence blood flow in skeletal muscle and skin. *Anesthesiology*, 41:139, 1974.

Jacobson, E. D.: Recent advances in the gastrointestinal circulation and related areas: comments on a symposium. *Gastroenterology*, 52:332, 1967.

Jacobson, E. D.: The gastrointestinal circulation. *Ann. Rev. Physiol.*, 30:133, 1968.

Keul, J.: The relationship between circulation and metabolism during exercise. *Med. Sci. Sports*, 5:209, 1973.

Kontos, H. A.: Mechanisms of regulation of the cerebral microcirculation. *Current Concepts Cerebrovasc. Dis.*, 10:7, 1975.

Korner, P. I.: Control of blood flow to special vascular areas: brain, kidney, muscle, skin, liver, and intestine. In Guyton, A. C. (ed.): MTP International Review of Science: Physiology. Vol. 1. Baltimore, University Park Press, 1974, p. 123.

Larson, C. P. Jr., Mazze, R. I., Cooperman, L. H., and Wollman, H.: Effects of anesthetics on cerebral, renal, and splanchnic circulations: recent developments. *Anesthesiology*, 41:169, 1974.

Lassen, N. A.: Control of cerebral circulation in health and disease. *Circ. Res.*, 34:749, 1974.

Lassen, N. A., and Ingvar, D. H.: Radioisotopic assessment of regional cerebral blood flow. *Prog. Nucl. Med.*, 1:376, 1972.

Lundgren, O., and Jodal, M.: Regional blood flow. *Ann. Rev. Physiol.*, 37:395, 1975.

Macpherson, A. I. S., Richmond, J., and Stuart, A. E.: The Spleen. Springfield, Ill., Charles C Thomas, Publisher, 1973.

Meyer, J. S., Reivich, M., Lechner, H., and Eichhorn, O. (eds.): Research on the Cerebral Circulation. Springfield, Ill., Charles C Thomas, Publisher, 1973.

Morse, R. L. (ed.): Exercise and the Heart. Springfield, Ill., Charles C Thomas, Publisher, 1974.

Moskalenko, Y. E., Kislyakov, Y. Y., Vainshtein, G. B., and Zelikson, B. B.: Biophysical aspects of the intracranial circulation. *Amer. Heart J.* 83:401, 1972.

Nicoll, P. A., and Cortese, T. A., Jr.: The physiology of skin. *Ann. Rev. Physiol.*, 34:177, 1972.

Nutter, D. O., Schlant, R. C., and Hurst, J. W.: Isometric exercise and the cardiovascular system. *Mod. Concepts Cardiovasc. Dis.*, 41:11, 1972.

Purves, M. J.: The Physiology of the Cerebral Circulation. (Physiological Society Monographs). New York, Cambridge University Press, 1972.

Reynolds, S. R. M.: Blood and lymph vascular systems of the ovary. *In* Greep, R. O., and Astwood, E. B. (eds.): Handbook of Physiology. Sec. 7, Vol. 2, Part 1. Baltimore, The Williams & Wilkins Company, 1973, p. 261.

Ross, G.: The regional circulation. *Ann. Rev. Physiol., 33:*445, 1971.

Rowell, L. B.: Regulation of splanchnic blood flow in man. *Physiologist, 16:*127, 1973.

Rowell, L. B.: Human cardiovascular adjustments to exercise and thermal stress. *Physiol. Rev., 54:*75, 1974.

Sedgewick, C. E., and Poulantzas, J. K.: Portal Hypertension. Boston, Little, Brown & Company, 1967.

Sejrsen, P.: Cutaneous blood flow and tissue blood exchange. *Adv. Biol. Skin, 12:*191, 1972.

Shephard, R. J.: Men at Work. Springfield, Ill., Charles C Thomas, Publisher, 1974.

Shepherd, J. T.: Physiology of Circulation in the Limbs. Philadelphia, W. B. Saunders Company, 1963.

Simonson, E. (ed.): Physiology of Work Capacity and Fatigue. Springfield, Ill., Charles C Thomas, Publisher, 1971.

Stainsby, W. N.: Local control of regional blood flow. *Ann. Rev. Physiol., 35:*151, 1973.

Taylor, A. W.: Application of Science and Medicine to Sport. Springfield, Ill., Charles C Thomas, Publisher, 1974.

Taylor, C. R.: Exercise and thermoregulation. *In* Guyton, A. C. (ed.): MTP International Review of Science: Physiology. Vol. 7. Baltimore, University Park Press, 1974, p. 163.

# PART VI

# THE BODY FLUIDS AND KIDNEYS

# | 30 |

# Capillary Dynamics, and Exchange of Fluid Between the Blood and Interstitial Fluid

It is in the capillaries that the most purposeful function of the circulation occurs, namely, interchange of nutrients and cellular excreta between the tissues and the circulating blood. About 10 billion capillaries, having a total surface area probably about 500 square meters, provide this function. Indeed, it is rare that any single functional cell of the body is more than 20 to 30 microns away from a capillary.

It is the purpose of this chapter to discuss the transfer of substances between the blood and interstitial fluid and especially to discuss the factors that affect the transfer of fluid volume itself between the circulating blood and the interstitial fluids.

## STRUCTURE OF THE CAPILLARY SYSTEM

Figure 30–1 illustrates the structure of a "unit" capillary bed, illustrating that blood enters the capillaries through an *arteriole* and leaves by way of a *venule*. Blood from the arteriole passes into a series of *metarterioles,* which have a structure midway between that of arterioles and capillaries. After leaving the metarteriole, the blood enters the *capillaries,* some of which are large and are called *preferential channels* and others of which are small and are *true capillaries.* After passing through the capillaries the blood enters the venule and returns to the general circulation.

The arterioles are highly muscular, and their diameters can change many-fold, as was discussed in Chapter 18. The metarterioles do not have a continuous muscular coat, but smooth muscle fibers encircle the vessel at intermediate points, as illustrated in Figure 30–1 by the large black dots to the sides of the metarteriole.

At the point where the true capillaries originate from the metarterioles a smooth muscle fiber usually encircles the capillary. This is called the *precapillary sphincter.* This sphincter can open and close the entrance to the capillary.

The venules are considerably larger than the arterioles and have a much weaker muscular coat. Yet, it must be remembered that the pressure in the venules is much less than that in the arterioles so that the venules can perhaps contract as much as can the arterioles.

This typical arrangement of the capillary bed is not found in all parts of the body; however, some similar arrangement is found, beginning with arterioles, passing through metarterioles and capillaries, and returning through venules.

**Structure of the Capillary Wall.** Figure 30–2 illustrates an electron micrograph of the capillary wall. Note that the wall is primarily composed of a thin, unicellular layer of endothelial cells and is surrounded by a thin basement membrane on the outside. The total thickness of the membrane is about 0.5 micron. Within the endothelial cells are found the usual intracellular structures, such as mitochondria, pinocytic vesicles, and so forth.

The diameter of the capillary is 7 to 9 microns, barely large enough for red blood cells and other blood cells to squeeze through.

*Pores in the Capillary Membrane.* On studying Figure 30–2 again, one can see a minute passageway connecting the interior of the capillary

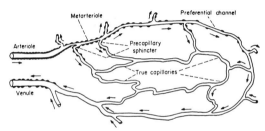

**Figure 30–1.** Overall structure of a capillary bed. (From Zweifach: Factors Regulating Blood Pressure. Josiah Macy, Jr., Foundation, 1950.)

with the exterior. These passageways are actually slitlike spaces between the adjacent endothelial cells where they contact each other. However, these slits are located far apart from one another, and they represent no more than 0.001 of the total surface area of the capillary. These passageways are called *pores* of the capillary membrane, and it is through these that water and many dissolved substances pass both from the lumen of the capillary into the interstitial spaces and in the opposite direction, as is discussed later in the chapter. These slit-pores have a width between 80 and 90 Ångstroms.

## FLOW OF BLOOD IN THE CAPILLARIES—VASOMOTION

Blood usually does not flow at a continuous rate through the capillaries. Instead, it flows intermittently. The cause of this intermittency is the phenomenon called *vasomotion,* which means intermittent contraction of the metarterioles and precapillary sphincters. These con-

strict and relax in an alternating cycle from 10 times per minute to less than 1 time per minute.

**Regulation of Vasomotion.** The most important factor found thus far to affect the degree of opening and closing of the metarterioles and precapillary sphincters is the concentration of *oxygen* in the tissues. When the oxygen concentration is very low, the intermittent periods of blood flow occur more often, and the duration of each period of flow lasts for a longer time, thereby allowing the blood to carry increased quantities of oxygen (as well as other nutrients) to the tissues. It follows also that the less the amount of oxygen in the blood, the greater will be the amount of blood allowed to flow through the tissue. And, too, the greater the use of oxygen by the tissue, the greater will be the amount of blood that flows. Thus, by this intermittent opening and closing of the precapillary sphincters and metarterioles, the blood flow to the tissue is *autoregulated,* as was discussed in Chapter 20.

### AVERAGE FUNCTION OF THE CAPILLARY SYSTEM

Despite the fact that blood flow through each capillary is intermittent, so many capillaries are present in the tissues that their overall function becomes averaged. That is, there is an *average rate of blood flow* through each tissue capillary bed, an *average capillary pressure* within the capillaries, an *average rate of transfer of substances* between the blood of the capillaries and the surrounding interstitial fluid. In the remainder of this chapter, we will be concerned with these averages, though one must remember that the average functions are in reality the functions of literally billions of individual capillaries, each operating intermittently.

## EXCHANGE OF NUTRIENTS AND OTHER SUBSTANCES BETWEEN THE BLOOD AND INTERSTITIAL FLUID

### DIFFUSION THROUGH THE CAPILLARY MEMBRANE

By far the most important means by which substances are transferred between the plasma and interstitial fluids is by diffusion. Figure 30–3 illustrates this process, showing that as the blood traverses the capillary, tremendous numbers of water molecules and dissolved particles

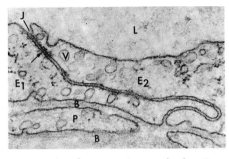

**Figure 30–2.** An electron micrograph showing a long slitlike pore that exists at the junction (*J*) between adjacent capillary endothelial cells (*E₁* and *E₂*). (Approximate magnification 30,000×.) *L,* capillary lumen; *B,* basement membrane; *P,* pericyte; *V,* pinocytic vesicle. (Courtesy of Drs. Guido Majno and Ramzi Cotran.)

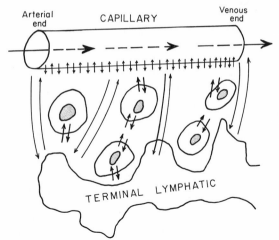

**Figure 30–3.** Diffusion of fluid and dissolved substances between the capillary and interstitial fluid spaces.

diffuse back and forth through the capillary wall, providing continual mixing between the interstitial fluids and the plasma, as was explained in Chapter 4. Diffusion results from thermal motion of the water molecules and dissolved substances in the fluid, the different particles moving first in one direction, then another, moving randomly in every direction.

**Diffusion of Lipid-Soluble Substances Through the Capillary Membrane.** If a substance is lipid-soluble, it can diffuse directly through the cell membranes of the capillary without having to go through the pores. Lipid-soluble substances normally transported in the blood include especially oxygen and carbon dioxide. Since these substances can permeate all areas of the capillary membrane, their rates of transport through the membrane are several hundred times the rates for most lipid-insoluble substances such as sodium ions, glucose, and so forth.

Other lipid-soluble substances that are rapidly transported through the capillary membrane include various anesthetic gases and alcohol.

**Diffusion of Water Molecules Through the Capillary Membrane.** After the lipid-soluble molecules, water molecules are the next most rapid to diffuse through the capillary membrane. Part of the water diffusion occurs through the pores of the endothelial *cellular* membranes, first into the intracellular fluid of the endothelial cells and then out of the membrane on the other side of the cell. Another portion of the water diffusion can occur almost

without hindrance through the slit-pores between the endothelial cells.

Even though water molecules can diffuse through the endothelial cells, very little *net* water movement occurs by this route for two reasons. First, the surface area of the endothelial cellular membrane pores is probably no greater than about $1/2000$ of the total surface area, meaning that water diffusion by this route is considerably restricted. Second, diffusion of water without simultaneous diffusion of its dissolved solutes causes the fluid on the side from which the water diffuses to become concentrated. This creates a strong osmotic gradient that immediately pulls the water molecules back through the membrane. Therefore, for practical purposes, essentially all of the *net* movement of water molecules takes place through the slit-pores that are large enough to allow simultaneous movement of most of the dissolved substances as well.

**Diffusion of Water-Soluble, Lipid-Insoluble Substances through the Capillary Membrane.** Many substances needed by the tissues are soluble in water but are completely insoluble in the lipid membranes of the capillaries; such substances include sodium ions, chloride ions, glucose, and so forth. These substances diffuse between the plasma and interstitial fluids only through the slit-pores, which themselves are filled with water.

Despite the fact that not over $1/1000$ of the surface area of the capillaries is represented by slit-pores, the velocity of thermal motion is so great that even this small area is sufficient to allow tremendous diffusion of water and water-soluble substances through the pores. To give one an idea of the extreme rapidity with which substances diffuse, *the rate at which water molecules diffuse through the capillary membrane is approximately 80 times as great as the rate at which plasma itself flows linearly along the capillary.* That is, the water of the plasma is exchanged with the water of the interstitial fluids 80 times before the plasma can go the entire distance through the capillary. Yet, despite this rapid rate of diffusion, the rates of diffusion out of the capillary and into the capillary are so nearly equal that the rate of net movement of fluid volume through the capillaries is thousands of times less than the rate of two-way diffusional exchange.

*Effect of Molecular Size on Pore Permeability.* The width of the capillary slit-pores, 80 to 90 Ångstroms, is about 25 times the diameter of the water molecule, which is the smallest molecule that normally passes through the capillary pores. However, the diameters of plasma protein molecules are slightly greater than that of the pores, and other substances, such as sodium ions, chloride ions, glucose, and urea, have intermediate diameters. Therefore, it is obvious that the permeability of the capillary pores for different substances will vary according to their molecular diameters.

Table 30–1 gives the relative capillary permeabilities for substances commonly encountered by the capillary membrane, illustrating for instance that the permeability for glucose molecules is 0.6 times as great as that for water molecules, whereas the permeability for albumin molecules is less than $1/10,000$ that for water molecules. Thus, the membrane is almost impermeable to albumin, which causes a significant concentration difference to develop between the albumin of the plasma and that of the interstitial fluid, as will become evident later in the chapter.

**Effect of Concentration Difference on Net Rate of Diffusion Through the Capillary Membrane.** As explained in Chapter 4, the "net" rate of diffusion of a substance through any membrane is proportional to the *concentration difference* between the two sides of the membrane. That is, the greater the difference between the concentrations of any given substance on the two sides of the capillary membrane, the greater will be the net movement of the substance through the membrane. Thus, the concentration of oxygen in the blood is normally greater than that in the interstitial fluids. Therefore, large quantities of oxygen normally move from the blood toward the tissues. Conversely, the concentration of carbon dioxide is greater in the tissues than in the blood, which obviously causes carbon dioxide to move into the blood and to be carried away from the tissues.

Fortunately, the rates of diffusion through the capillary membranes of most nutritionally important substances are so great that only slight concentration differences suffice to cause more than adequate transport between the plasma and interstitial fluid. For instance, the concentration of oxygen in the interstitial fluid immediately outside the capillary is normally no more than 1 per cent less than the concentration in the blood, and yet this 1 per cent difference causes enough oxygen to move from the blood into the interstitial spaces to provide all the oxygen required for tissue metabolism.

**Effect of a Hydrostatic Pressure Difference on the Net Rate of Diffusion Through the Capillary Membrane.** Not only can a concentration difference cause net diffusion through a capillary membrane, but so also can a hydrostatic pressure difference. Pressure increases the rate of diffusion by increasing kinetic energy of the molecules, causing the side with the higher pressure also to have a higher tendency for diffusion.

**Bulk Flow through Capillary Pores.** When one measures the rate of net movement of water molecules and low molecular weight solutes through the capillary membrane caused either by a colloid osmotic pressure or by a hydrostatic pressure, he finds this movement to be about ten times as great as can be predicted on the basis of net diffusion alone. The explanation for this excess movement of the water and its dissolved substances is simply the following: When these molecules move through pores that are far larger than their own diameters, as is the case for the slit-pores of the capillary membrane, a large amount of mutual drag occurs between the molecules. Therefore, diffusion of some of the molecules drags others along at the same time so that the molecules actually "flow" together. This phenomenon is called *bulk flow*, and it can increase the movement of both water and low molecular weight solutes many-fold.

### TRANSPORT THROUGH THE CAPILLARY MEMBRANE BY PINOCYTOSIS

Another means by which minute quantities of substances can be transported through the capillary membrane is by pinocytosis, a phenomenon that was explained in Chapter 4. Basically, this process means that the endothelial cells of the capillary wall ingest

**TABLE 30–1.  Relative Permeability of the Capillary Pores to Different Size Molecules**

| Substance | Molecular Weight | Permeability |
|---|---|---|
| Water | 18 | 1.00 |
| NaCl | 58.5 | 0.96 |
| Urea | 60 | 0.8 |
| Glucose | 180 | 0.6 |
| Sucrose | 342 | 0.4 |
| Inulin | 5,000 | 0.2 |
| Myoglobin | 17,600 | 0.03 |
| Hemoglobin | 68,000 | 0.01 |
| Albumin | 69,000 | <0.0001 |

Modified from Pappenheimer.

small amounts of plasma or interstitial fluid and form small intracellular vesicles of the ingested fluid; these then migrate from one surface of the endothelial cell to the other where the fluid is released.

Dynamic studies have shown that low molecular weight substances, such as water, sodium ions, chloride ions, glucose and urea, are transported through the capillary membrane at rates more than *10 million times* as great as can be accounted for by pinocytosis. On the other hand, transport of high molecular weight substances—for instance, plasma proteins, glycoproteins and large polysaccharides such as dextran—might well occur to a significant extent by pinocytosis.

# DISTRIBUTION OF FLUID VOLUME BETWEEN THE PLASMA AND INTERSTITIAL FLUIDS

Despite the tremendous rates of diffusion of substances both out of the capillary into the interstitial spaces and in the opposite direction, these rates in both directions so nearly equal each other that the rate of net volume movement across the capillary membrane is normally very low. Consequently, the volumes of both the blood and the interstitial fluids normally change very little from hour to hour or even from day to day. Yet, under abnormal conditions, fluid can leak rapidly out of the circulation into the interstitial spaces, sometimes causing circulatory shock because of decreased blood volume or tissue edema because of excess fluid in the interstitial spaces.

The pressure in the capillaries continuously tends to force fluid and its dissolved substances through the capillary pores into the interstitial spaces. But, in contrast, osmotic pressure caused by the plasma proteins (the *colloid* osmotic pressure) tends to cause fluid movement by osmosis from the interstitial spaces into the blood; it is mainly this osmotic pressure that prevents continual loss of fluid volume from the blood into the interstitial spaces. Yet, the process is much more complicated than this and includes the role of the lymphatic system to return back to the circulation the small amounts of fluid that do leak continuously into the interstitial spaces. In the following few paragraphs we will discuss all of the factors that play a significant role in the movement of fluid volume through the capillary membrane, and in the following chapter we will discuss the role of the lymphatic system in this overall mechanism.

**The Four Primary Factors That Determine Fluid Movement Through the Capillary Membrane.** Figure 30–4 illustrates the four

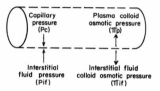

**Figure 30–4.** Forces operative at the capillary membrane tending to move fluid either outward or inward through the membrane.

primary factors that determine whether fluid will move out of the blood into the interstitial fluid or in the opposite direction; these are:

1. The *capillary pressure* (Pc), which tends to move fluid outward through the capillary membrane.

2. The *interstitial fluid pressure* (Pif), which tends to move fluid inward through the capillary membrane.

3. The *plasma colloid osmotic pressure* ($\Pi$p), which tends to cause osmosis of fluid inward through the membrane.

4. The *interstitial fluid colloid osmotic pressure* ($\Pi$if), which tends to cause osmosis of fluid outward through the membrane.

The regulation of fluid volumes in the blood and interstitial fluid is so important that each of these factors is discussed in turn in the following sections.

## CAPILLARY PRESSURE

Unfortunately, the exact capillary pressure is not known because it has been impossible to measure capillary pressure under absolutely normal conditions. Yet two different methods have been used to estimate the capillary pressure: (1) *direct cannulation of the capillaries,* which has given an average mean capillary pressure of about 25 mm. Hg, and (2) *indirect functional measurement of the capillary pressure,* which has given a capillary pressure averaging about 17 mm. Hg. These methods are the following:

**Cannulation Method for Measuring Capillary Pressure.** To measure pressure in a capillary by cannulation, a microscopic glass pipet is thrust directly into the capillary, and the pressure is measured by an appropriate micromanometer system. Using this method, capillary pressures have been measured in capillaries of exposed tissues of lower animals and in large capillary loops of the eponychium at the base of the fingernail in human beings. These measurements have given pressures of 30 to 40 mm. Hg in the arterial ends of the capillaries, 10 to 15 mm. Hg in the venous ends, and about 25 mm. Hg in the middle.

**Isogravimetric and Isovolumetric Methods for Indirectly Measuring Mean Capillary Pressure.** Figure 30–5 illustrates an *isogravimetric* method for estimating capillary pressure. This figure shows a section of gut held by one arm of a gravimetric balance. Blood is perfused through the gut. When the arterial pressure is decreased, the resulting decrease in capillary pressure allows the osmotic pressure of the plasma proteins to cause absorption of fluid out of the gut wall and makes the weight of the gut decrease. This immediately causes displacement of the balance arm. However, to prevent this weight decrease, the venous pressure is raised an amount sufficient to overcome the effect of decreasing the arterial pressure.

In the lower part of the figure, the changes in arterial and venous pressures that exactly nullify their respective effects on the weight of the gut are illustrated. The arterial and venous curves meet each other at a value of 17 mm. Hg. Therefore, the capillary pressure must have remained at this same level of 17 mm. Hg throughout these maneuvers. Thus, in a roundabout way, the "functional" capillary pressure is measured to be about 17 mm. Hg.

The *isovolumetric* method for measuring capillary pressure is essentially the same as the isogravimetric method, except that the *volume* of the tissue is recorded rather than the weight. The arterial and venous pressures are gradually

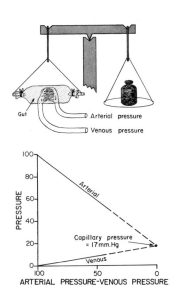

**Figure 30–5.** Isogravimetric method for measuring capillary pressure (explained in the text).

brought toward each other and are continually adjusted so that the tissue neither gains nor loses volume. On extrapolating the arterial and venous pressures, one again finds that the capillary pressure estimated by this means is considerably lower than that measured by direct cannulation, again averaging about 17 mm. Hg.

**Functional Capillary Pressure.** Since the cannulation method does not give the same pressure measurement as the isogravimetric and isovolumetric methods, one must decide which of these measurements is probably the true functional capillary pressure of the tissues. The isovolumetric and isogravimetric measurements are probably much nearer to the normal values for capillary pressure than are the micropipet measurements, for several reasons:

First, the metarterioles and precapillary sphincters of the capillary system are normally closed during a greater part of the vasomotion cycle than they are open. When they are closed the pressure in the entire capillary system beyond the closures should be almost exactly equal to the pressure at the venous ends of the capillaries, about 10 mm. Hg. Therefore, if one considers a *weighted* average of the pressures in all capillaries, one would expect the *functional* mean capillary pressure to be much nearer to the pressure in the venous ends of the capillaries than to the pressure in the arterial ends.

Second, the surface area of the venous capillaries is several times as great as the surface area of the arterial capillaries. Therefore, the mean pressure in the venous capillaries, 10 mm. Hg, plays far more role in determining the *functional* capillary pressure than does the mean arterial capillary pressure, 35 mm. Hg.

Third, the venous capillaries are several times as plentiful as the arterial capillaries. Therefore, for determining fluid movement through the capillary membrane, the venous capillary pressures are much more important than the arterial capillary pressures.

Thus, there are many reasons for believing *that the normal functional mean capillary pressure is about 17 mm. Hg.*

What is the *average* functional pressure in the arterial ends of the capillaries? This we do not know, but it must be considerably below the 30 to 40 mm. Hg measured with the micropipet in *open* arterial capillaries, because even the arterial ends of the capillaries are closed off most of the time because of metarteriole and precapillary sphincter closure. We might estimate this pressure to be about 25 mm. Hg, which is somewhat below the pipet measurements and somewhat above the functional mean capillary pressure.

## INTERSTITIAL FLUID PRESSURE

The interstitial fluid pressure, like capillary pressure, has been difficult to measure, primarily because the maximum width of interstitial spaces is about 1 micron. Thus far, it has been impossible to cannulate these spaces directly to make measurements of the pressure. Therefore, several indirect methods have been used to estimate the interstitial fluid pressure. These are the following:

**Measurement of Interstitial Fluid Pressure in Implanted Perforated Spheres.** Figure 30–6 illustrates an indirect method for measuring interstitial fluid pressure which may be explained as follows: A small hollow plastic sphere perforated by several hundred small holes is implanted in the tissues, and the surgical wound is allowed to heal for approximately one month. At the end of that time, tissue will have grown inward through the holes to line the inner surface of the sphere. Furthermore, the cavity is filled with fluid that flows freely through the perforations back and forth between the fluid in the interstitial spaces and the fluid in the cavity. Therefore, the pressure in the cavity should equal the pressure in the interstitial fluid spaces. A needle is inserted through the skin and through one of the perforations to the interior of the cavity, and the pressure is measured by use of an appropriate manometer.

Interstitial fluid pressure measured by this method in normal tissues averages about −6.3 mm. Hg. That is, the pressure is actually *less than atmospheric pressure* or, in other words, is a semi-vacuum or a suction. The significance of

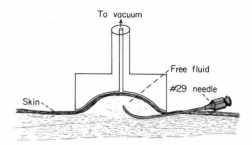

**Figure 30–7.** The equilibrium method for measuring interstitial fluid pressure.

the negativity of interstitial fluid pressure is that it causes *suction of fluid out of the capillaries,* as we shall see in subsequent sections of this chapter. The mechanism of its development is discussed in the following chapter in relation to interstitial fluid dynamics.

**Measurement of Interstitial Fluid Pressure in Fluid Spaces Developed Underneath the Skin.** Figure 30–7 illustrates another method that has been used to measure the interstitial fluid pressure. A small vacuum cup is placed over the skin, and the skin is sucked up into the cup. Gradually, over about 24 hours, free fluid collects beneath the mound of skin, and its pressure comes to equilibrium. A needle is then inserted into the tissue so that its tip is within this free fluid, and the pressure is recorded. Since the fluid is in free communication with the fluid in the surrounding tissues, the pressure measured here should equal the pressure in the other interstitial spaces. The pressure measured by this method is also negative, averaging again about −6 to −7 mm. Hg.

**Measurement of Interstitial Fluid Pressure by Means of a Cotton Wick.** A new method used recently is to insert into a tissue a small Teflon tube with about eight cotton fibers protruding from its end. Each cotton fiber is itself a very minute tube containing large numbers of small side openings. The fluid inside the cotton fibers, therefore, makes excellent contact with the tissue fluids and transmits interstitial fluid pressure into the Teflon tube; the pressure can then be measured from the tube by usual manometric means. Pressures measured by this technique have also been negative, similar to those measured by the two techniques previously discussed.

## PLASMA COLLOID OSMOTIC PRESSURE

**Colloid Osmotic Pressure Caused by Proteins.** The proteins are the only dissolved substances of the plasma that do not diffuse readily into the interstitial fluid. Furthermore, when small quantities of protein do diffuse into the interstitial fluid, these are soon removed from the interstitial spaces by way of the lymph ves-

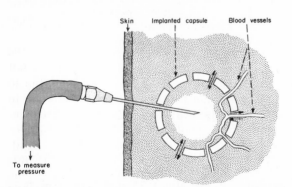

**Figure 30–6.** The perforated capsule method for interstitial fluid pressure.

sels. Therefore, the concentration of protein in the plasma averages almost four times as much as that in the interstitial fluid, 7.3 gm./100 ml. in the plasma versus 2 gm./100 ml. in the interstitial fluid.

In the discussion of osmotic pressure in Chapter 4, it was pointed out that only those substances that fail to pass through the pores of a semipermeable membrane exert osmotic pressure. Since the proteins are the only dissolved constituents that do not readily penetrate the pores of the capillary membrane, it is the dissolved proteins of the plasma and interstitial fluids that are responsible for the osmotic pressure at the capillary membrane. To distinguish this osmotic pressure from that which occurs at the cell membrane, it is called *colloid osmotic pressure* or *oncotic pressure*. The term "colloid" osmotic pressure is derived from the fact that a protein solution resembles a colloidal solution despite the fact that it is actually a true solution. The osmotic pressure that results at the cell membrane is often called *total osmotic pressure* because essentially all dissolved substances of the body fluids exert osmotic pressure at the cell membrane.

***Effect of the Donnan Equilibrium on the Colloid Osmotic Pressure.*** A special effect, called the *Donnan equilibrium*, causes the colloid osmotic pressure to be about 50 per cent greater than that caused by the proteins alone. This results from the fact that the proteins are negative ions, and to balance these negative ions a large number of positively charged ions (cations), mainly sodium ions, are held by the electronegative charges of the proteins on the same side of the membrane as the proteins. These extra cations increase the number of osmotically active substances and increase the total colloid osmotic pressure. Even more important, however, (for mathematical reasons that cannot be explained here), the Donnan equilibrium effect becomes progressively more significant the higher the concentration of proteins. This means, as illustrated in Figure 30–8, that the initial few grams of protein in each 100 ml. of plasma or interstitial fluid has much less colloid osmotic effect than do the next few grams.

**Normal Values for Plasma Colloid Osmotic Pressure.** The colloid osmotic pressure of normal human plasma averages approximately 28 mm. Hg; 19 mm. of this is caused by the dissolved protein, and 9 mm. by the cations held in the plasma by the Donnan effect of the proteins, as was just discussed.

Note particularly that the 28 mm. Hg colloid

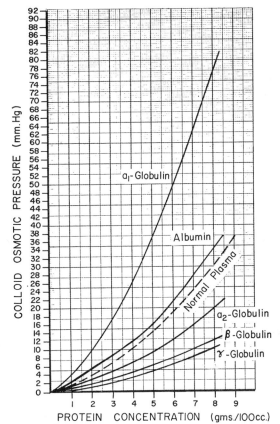

**Figure 30–8.** The osmotic pressure of five fractions of the plasma proteins at different concentrations. Also, the dashed line shows the osmotic pressure of normal plasma proteins, which are a mixture of the others. (Modified from Ott: *Klin. Wschr.*, 34:1079, 1956.)

osmotic pressure that can develop at the capillary membrane is only about $1/200$ the total osmotic pressure that would develop at a *cell* membrane if normal interstitial fluid were on one side of the membrane and pure water on the other side. Thus, the colloid osmotic pressure of the plasma is actually a weak osmotic force; but, even so, it plays an exceedingly important role in the maintenance of normal blood and interstitial fluid volumes, as will be evident in the remainder of this chapter and in the following chapters.

**Effect of the Different Plasma Proteins on Colloid Osmotic Pressure.** The plasma proteins are a mixture of proteins that contains albumin, with an average molecular weight of 69,000; globulins, 140,000; and fibrinogen, 400,000. Thus, 1 gram of globulin contains only half as many molecules as 1 gram of albumin, and 1 gram of fibrinogen contains only one-sixth as many molecules as 1 gram of albumin. (It will be recalled from the discussion of osmotic pressure in Chapter 4 that the osmotic pressure is determined by the *number of molecules* dissolved in a fluid rather than by the weight of these molecules.) Furthermore, the average relative concentrations of these different types of proteins in the plasma are:

|  | gm. % |
|---|---|
| Albumin | 4.5 |
| Globulins | 2.5 |
| Fibrinogen | 0.3 |
| TOTAL | 7.3 |

Because each gram of albumin exerts twice the osmotic pressure of a gram of globulins and because there is almost twice as much albumin in the plasma as globulins, about 70 per cent of the total colloid osmotic pressure of the plasma results from the albumin fraction and only about 30 per cent from the globulins and fibrinogens. Therefore, from the point of view of capillary dynamics, it is mainly albumin that is important.

Figure 30–8 illustrates graphically the colloid osmotic pressure exerted by different concentrations of albumin and four different fractions of globulins.

## INTERSTITIAL FLUID COLLOID OSMOTIC PRESSURE

Though the size of the usual capillary pore is smaller than the molecular sizes of the plasma proteins, this is not true of all the pores. Therefore, small amounts of plasma proteins do leak into the interstitial spaces. Albumin molecules, because they are smaller than most globulin molecules, normally leak 1.6 times as readily as the globulins, causing the proteins in the interstitial fluids to have a disproportionately high albumin to globulin ratio.

The total quantity of protein in the entire 12 liters of interstitial fluid of the body is slightly greater than the total quantity of protein in the plasma itself, but since this volume is four times the volume of plasma the average protein *concentration* of the interstitial fluid is only a little more than one-fourth that in plasma, or approximately 2 grams per cent. Referring to the diagram of Figure 30–8, one finds that the average colloid osmotic pressure for this concentration of proteins in the interstitial fluids is approximately 5 mm. Hg.

## EXCHANGE OF FLUID VOLUME THROUGH THE CAPILLARY MEMBRANE

Now that the different factors affecting capillary membrane dynamics have been discussed, it is possible to put all these together to see how normal capillaries function.

The capillary pressure at the arterial ends of the capillaries is 15 to 20 mm. Hg greater than at the venous ends. Because of this difference, fluid "filters" out of the capillaries at their arterial ends and then is reabsorbed into the capillaries at their venous ends. Thus, a small amount of fluid actually "flows" through the tissues from the arterial ends of the capillaries to the venous ends. The dynamics of this flow are the following:

**Analysis of the Forces Causing Filtration at the Arterial End of the Capillary.** The forces operative at the arterial end of the capillary to cause movement through the capillary membrane are:

Forces tending to move fluid outward:

|  | mm. Hg |
|---|---|
| Capillary pressure | 25.0 |
| *Negative* interstitial fluid pressure | 6.3 |
| Interstitial fluid colloid osmotic pressure | 5.0 |
| TOTAL OUTWARD FORCE | 36.3 |

Force tending to move fluid inward:

|  | mm. Hg. |
|---|---|
| Plasma colloid osmotic pressure | 28.0 |
| TOTAL INWARD FORCE | 28.0 |

Summation of forces:

|  | mm. Hg. |
|---|---|
| Outward | 36.3 |
| Inward | 28.0 |
| NET OUTWARD FORCE | 8.3 |

Thus, the summation of forces at the arterial end of the capillary shows a net *filtration pressure* of 8.3 mm. Hg tending to move fluid out of the arterial ends of the capillaries into the interstitial spaces.

This 8.3 mm. Hg filtration pressure causes about 0.3 per cent of the plasma entering the capillaries to filter out of the arterial ends of the capillaries into the interstitial spaces. (This percentage varies tremendously from very, very low in the brain to very high in the liver.)

**Analysis of Reabsorption at the Venous End of the Capillary.** The low pressure at the venous end of the capillary changes the balance of forces in favor of absorption as follows:

Force tending to move fluid inward:

|  | mm. Hg |
|---|---|
| Plasma colloid osmotic pressure | 28.0 |
| TOTAL INWARD FORCE | 28.0 |

Forces tending to move fluid outward:

|  | mm. Hg |
|---|---|
| Capillary pressure | 10.0 |
| *Negative* interstitial fluid pressure | 6.3 |
| Interstitial fluid colloid osmotic pressure | 5.0 |
| TOTAL OUTWARD FORCE | 21.3 |

Summation of forces:

|  | mm. Hg |
|---|---|
| Inward | 28.0 |
| Outward | 21.3 |
| NET INWARD FORCE | 6.7 |

Thus, the force that causes fluid to move into the capillary, 28 mm. Hg, is greater than that opposing reabsorption, 21.3 mm. Hg. The difference, 6.7 mm. Hg, is the *reabsorption pressure*.

The reabsorption pressure causes about nine-tenths of the fluid that has filtered out the arterial ends of the capillaries to be reabsorbed at the venous ends. The other one-tenth flows into the lymph vessels, as is discussed in the following chapter.

**Flow of Fluid Through the Interstitial Spaces.** The 0.3 per cent of the plasma that filters out of the arterial ends of the capillaries *flows* through the tissue spaces to the venous ends of the capillaries where all but about $1/_{10}$ of it is reabsorbed. (A much higher proportion than this is reabsorbed in the muscles where very little protein leaks through the capillary membranes, and much less is reabsorbed in the liver where tremendous amounts of protein leak.)

**Distinction Between Filtration and Diffusion.** It is especially important to distinguish the difference between *filtration* and *diffusion* through the capillary membrane. Diffusion occurs in both directions, while filtration is the *net* movement of fluid out of the capillaries at the arterial ends. The rate of diffusion of water through all the capillary membranes of the entire body is about 240,000 ml./minute, while the normal rate of filtration at the arterial ends of all the capillaries is only 16 ml./minute, a difference of about 15,000-fold. Thus, the quantitative rate of diffusion of water and nutrients back and forth between the capillaries and the interstitial spaces is tremendous in comparison with the minute rate of "flow" of fluid through the tissues.

## THE STARLING EQUILIBRIUM OF CAPILLARY EXCHANGE

E. H. Starling pointed out three-quarters of a century ago that under normal conditions a state of near-equilibrium exists at the capillary membrane whereby the amount of fluid filtering outward through the arterial capillaries exactly equals that quantity of fluid that is returned to the circulation by reabsorption at the venous ends of the capillaries. This near-equilibrium is caused by near-equilibration of the *mean* forces tending to move fluid through the capillary membranes. If we assume that the mean capillary pressure is 17 mm. Hg, the normal mean dynamics of the capillary are the following:

Mean forces tending to move fluid outward:

|  | mm. Hg |
|---|---|
| Mean capillary pressure | 17.0 |
| *Negative* interstitial fluid pressure | 6.3 |
| Interstitial fluid colloid osmotic pressure | 5.0 |
| TOTAL OUTWARD FORCE | 28.3 |

Mean force tending to move fluid inward:

|  | mm. Hg |
|---|---|
| Plasma colloid osmotic pressure | 28.0 |
| TOTAL INWARD FORCE | 28.0 |

Summation of mean forces:

|  | mm. Hg |
|---|---|
| Outward | 28.3 |
| Inward | 28.0 |
| NET OUTWARD FORCE | 0.3 |

Thus, we find a near-equilibrium but nevertheless a slight imbalance of forces at the capillary membranes that causes slightly more filtration of fluid into the interstitial spaces than reabsorption. This slight excess of filtration is called the *net filtration*, and it is balanced by fluid return to the circulation through the lymphatics. The normal rate of net filtration in the entire body is about 1.7 ml. per minute. This figure also represents the rate of fluid flow into the lymphatics each minute.

**The Filtration Coefficient.** In the above example there is an average net imbalance of forces at the capillary membranes of approximately 0.3 mm. Hg tending to cause filtration of fluid out of the circulation into the tissue spaces; this causes a net rate of fluid filtration in the entire body of 1.7 ml./min. Or, expressing this per mm. Hg, one finds a net filtration rate of 5.67 ml. fluid/min./mm. Hg for the entire body. This expression is the *filtration coefficient*.

The filtration coefficient can also be expressed for different areas of the body in terms of the rate of filtration/min./mm. Hg/100 gm. of tissue. On this basis the filtration coefficient of the average tissue is 0.008 ml./min./mm. Hg/100 gm. of tissue. However, because of extreme differences in permeabilities of the capillary systems of different tissues, this coefficient varies greatly from one tissue to another. It is very small in both brain and muscle, moderately great in subcutaneous tissue, large in the intestine, and extreme in the liver where the pores of the liver sinusoids are almost wide open. By the same token, the permeation of proteins through the capillary membranes varies in approximately the same way. The concentration of protein in the interstitial fluid of muscles is about 1.5 gm. per cent, in subcutaneous tissue 2 gm. per cent, in intestine 4 gm. per cent, and in liver 6 gm. per cent.

**Effect of Imbalance of Forces at the Capillary Membrane.** If the mean capillary pressure rises above 17 mm. Hg the net force tending to cause filtration of fluid into the tissue spaces obviously rises. Thus a 20 mm. Hg rise in mean capillary pressure causes an increase in the net filtration pressure from 0.3 mm. Hg to 20.3 mm. Hg, which results in 68 times as much net filtration of fluid into the interstitial spaces as normally occurs, and this would require also 68 times the normal flow of fluid into the lymphatic system, an amount that is usually too much for the lymphatics to carry away. As a result, fluid begins to accumulate in the interstitial spaces, and edema results.

Conversely, if the capillary pressure falls very low, net reabsorption of fluid into the capillaries occurs instead of net filtration, and the blood volume increases at the expense of the interstitial fluid volume. The effects of these imbalances at the capillary membrane are discussed in the following chapter in relation to the formation of edema.

# REFERENCES

Charm, S. E., and Kurland, G. S.: Blood Flow and Microcirculation. New York, John Wiley & Sons, Inc., 1974.

Chinard, F. P.: Starling's hypothesis in the formation of edema. *Bull. N.Y. Acad. Med.*, 38:375, 1962.

Duling, B. R., and Berne, R. M.: Propagated vasodilation in the microcirculation of the hamster cheek pouch. *Circ. Res.*, 26:163, 1970.

Granger, H. J., Taylor, A. E., and Guyton, A. C.: Quantitative analysis of the permeability characteristics of membranes isolated from chronically implanted subcutaneous capsules. *Microvasc. Res.*, 2:240, 1970.

Gross, J. F., Aroesty, J.: Mathematical models of capillary flow: a critical review. *Biorheology*, 9:225, 1972.

Guyton, A. C.: Concept of negative interstitial pressure based on pressures in implanted perforated capsules. *Circ. Res.*, 12:399, 1963.

Guyton, A. C.: Interstitial fluid pressure: II. Pressure-volume curves of interstitial space. *Circ. Res.*, 16:452, 1965.

Guyton, A. C.: Interstitial fluid pressure-volume relationships and their regulation. In Wolstenholme, G. E. W., and Knight, J. (eds.): Ciba Foundation Symposium on Circulatory and Respiratory Mass Transport. London, J. & A. Churchill Ltd., 1969, p. 4.

Guyton, A. C., Prather, J., Scheel, K., and McGehee, J.: Interstitial fluid pressure: IV. Its effect on fluid movement through the capillary wall. *Circ. Res.*, 19:1022, 1966.

Guyton, A. C., Scheel, K., and Murphree, D.: Interstitial fluid pressure: III. Its effect on resistance to tissue fluid mobility. *Circ. Res.*, 19:412, 1966.

Guyton, A. C., Taylor, A. E., Granger, H. J., and Coleman, T. G.: Interstitial Fluid Pressure. *Physiol. Rev.*, 51:527, 1971.

Guyton, A. C., Taylor, A. E., and Granger, H. J.: Circulatory Physiology II. Dynamics and Control of the Body Fluids. Philadelphia, W. B. Saunders Company, 1975.

Intaglietta, M.: The measurement of pressure and flow in the microcirculation. *Microvasc. Res.*, 5:357, 1973.

Intaglietta, M., and Zweifach, B. W.: Microcirculatory basis of fluid exchange. *Adv. Biol. Med. Phys.*, 15:111, 1974.

Johnson, P. C.: Renaissance in the microcirculation. *Circ. Res.*, 31:817, 1972.

Johnson, P. C.: The microcirculation and local and humoral control of the circulation. In Guyton, A. C. (ed.): MTP International Review of Science: Physiology. Vol. 1. Baltimore, University Park Press, 1974, p. 163.

Landis, E. M.: Capillary pressure and capillary permeability. *Physiol. Rev.*, 14:404, 1934.

Landis, E. M., and Papenheimer, J. R.: Exchange of substances through the capillary walls. In Hamilton, W. F. (ed.): Handbook of Physiology. Sec. 2, Vol. 2. Baltimore, The Williams & Wilkins Company, 1963, p. 961.

Leonard, E. F., and Jorgensen, S. B.: The analysis of convection and diffusion in capillary beds. *Ann. Rev. Biophys. Bioeng.*, 3:293, 1974.

Mayerson, H. S.: The physiologic importance of lymph. In Hamilton, W. F. (ed.): Handbook of Physiology. Sec. 2, Vol. 2. Baltimore, The Williams & Wilkins Company, 1963, p. 1035.

Pappenheimer, J. R.: Passage of molecules through capillary walls. *Physiol. Rev.*, 33:387, 1953.

Prather, J. W., Gaar, K. A., Jr., and Guyton, A. C.: Direct continuous recording of plasma colloid osmotic pressure of whole blood. *J. Appl. Physiol.*, 24:602, 1968.

Stromme, S. B., Maggert, J. E., and Scholander, P. F.: Interstitial fluid pressure in terrestrial and semiterrestrial animals. *J. Appl. Physiol.*, 27:123, 1969.

Taylor, A. E., Gibson, W. H., Granger, H. J., and Guyton, A. C.: The interaction between intercapillary forces in the overall regulation of interstitial fluid volume. *Lymphology*, 6:192, 1973.

Tombs, M., and Peacocke, A. R.: The Osmotic Pressure of Biological Macromolecules. New York, Oxford University Press, 1974.

Wayland, H.: Photosensor methods of flow measurement in the microcirculation. *Microvasc. Res.*, 5:336, 1973.

Zweifach, B. W.: Microcirculation. *Ann. Rev. Physiol.*, 35:117, 1973.

Zweifach, B. W.: Mechanisms of blood flow and fluid exchange in microvessels: hemorrhagic hypotension model. *Anesthesiology*, 41:157, 1974.

# 31

# The Lymphatic System, Interstitial Fluid Dynamics, Edema, and Pulmonary Fluid

## THE LYMPHATIC SYSTEM

The lymphatic system represents an accessory route by which fluids can flow from the interstitial spaces into the blood. And, most important of all, the lymphatics can carry proteins and large particulate matter away from the tissue spaces, neither of which can be removed by absorption directly into the blood capillary. We shall see that this removal of proteins from the interstitial spaces is an absolutely essential function without which we would die within about 24 hours.

### THE LYMPH CHANNELS OF THE BODY

All tissues of the body with the exception of a very few have lymphatic channels that drain excess fluid directly from the interstitial spaces. The exceptions include the superficial portions of the skin, the central nervous system, deeper portions of peripheral nerves, the endomysium of muscles, and the bones. However, even these tissues have minute interstitial channels through which interstitial fluid can flow; eventually this fluid flows into lymphatic vessels along the periphery of the tissue or, in the case of the brain, flows into the cerebrospinal fluid and thence directly back into the blood.

Essentially all the lymph from the lower part of the body—even some of that from the legs—flows up the *thoracic duct* and empties into the venous system at the juncture of the *left* internal jugular vein and subclavian vein, as illustrated in Figure 31–1. However, small amounts of lymph from the lower part of the body can enter the veins in the inguinal region and perhaps also at various points in the abdomen.

Lymph from the left side of the head, the left arm, and left chest region also enters the thoracic duct before it empties into the veins. Lymph from the right side of the neck and head, from the right arm, and from parts of the right thorax enters the *right lymph duct,* which then empties into the venous system at the juncture of the *right* subclavian vein and internal jugular vein.

**The Lymphatic Capillaries and Their Permeability.** Most of the fluid filtering from the arterial capillaries flows among the cells and is finally reabsorbed back into the *venous* capillaries; but, on the average, about *one-tenth* of the fluid enters the *lymphatic* capillaries and returns to the blood through the lymphatic system rather than through the venous capillaries.

The minute quantity of fluid that returns to the circulation by way of the lymphatics is extremely important because substances of high molecular weight such as proteins cannot pass with ease through the pores of the venous capillaries, but they can enter the lymphatic capillaries almost completely unimpeded. The reason for this is a special structure of the lymphatic capillaries, illustrated in Figure 31–2. This figure shows the endothelial cells of the capillary attached by *anchoring filaments* to the connective tissue between the surrounding tissue cells. However, at the junctions of adjacent endothelial cells there are no connections between the cells. Instead, the edge of one endothelial cell simply overlaps the edge of the adjacent one in such a way that the overlapping edge is free to flap inward, thus forming a minute valve that opens to the interior of the capil-

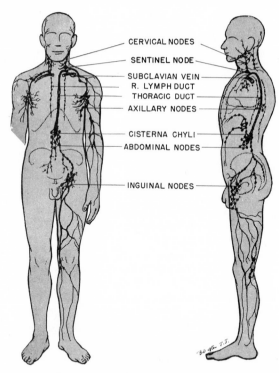

CERVICAL NODES
SENTINEL NODE
SUBCLAVIAN VEIN
R. LYMPH DUCT
THORACIC DUCT
AXILLARY NODES
CISTERNA CHYLI
ABDOMINAL NODES
INGUINAL NODES

**Figure 31–1.** The lymphatic system.

lary. Interstitial fluid, along with its suspended particles, can push the valve open and then flow directly into the capillary. But this fluid cannot leave the capillary once it has entered because any backflow will close the flap valve. Thus, the lymphatics have valves at the very tips of the terminal lymphatic capillaries as well as valves along their larger vessels up to the point where they empty into the blood circulation.

### FORMATION OF LYMPH

Lymph is interstitial fluid that flows into the lymphatics. Therefore, lymph has almost the identical composition as the tissue fluid in the part of the body from which the lymph flows.

The protein concentration in the interstitial fluid averages about 2 gm. per cent, and this is also the protein concentration of lymph flowing from most of the peripheral tissues. On the other hand, lymph formed in the liver has a protein concentration as high as 6 gm. per cent, and lymph formed in the intestines has a protein concentration as high as 3 to 5 gm. per cent. Since more than half of the lymph is derived from the liver and intestines, the thoracic lymph, which is a mixture of lymph from all

areas of the body, usually has a protein concentration of 3 to 5 gm. per cent.

The lymphatic system is also one of the major routes for absorption of nutrients from the gastrointestinal tract, being responsible principally for the absorption of fats, as will be discussed in Chapter 65. Indeed, after a fatty meal, thoracic duct lymph sometimes contains as much as 1 to 2 per cent fat.

Finally, even large particles, such as bacteria, can push their way between the endothelial cells of the lymphatic capillaries and in this way enter the lymph. As the lymph passes through the lymph nodes, these particles are removed and destroyed, as was discussed in Chapter 6.

**The Mechanism by which Proteins Become Concentrated in Interstitial Fluids.** Fluid filtering from the arterial ends of the capillaries in peripheral tissues such as the subcutaneous tissue usually has a protein concentration of less than 0.2 per cent, though the average concentration of protein in the interstitial fluids is some 10 times this value. The reason for this difference is that almost none of the protein that leaks into the tissue spaces is reabsorbed at the venous ends of the capillaries, while most of the water is reabsorbed. In this way, the proteins become concentrated in the interstitial fluid before it flows into the lymphatics.

### TOTAL RATE OF LYMPH FLOW

Approximately 100 ml. of lymph flow through the thoracic duct of a resting man per hour, and perhaps another 20 ml. of lymph flow into the circulation each hour through other channels,

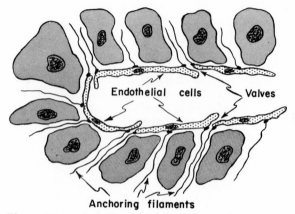

Endothelial cells        Valves

Anchoring filaments

**Figure 31–2.** Special structure of the lymphatic capillaries that permits passage of substances of high molecular weight back into the circulation.

making a total estimated lymph flow of perhaps 120 ml. per hour. This is about 1/120,000 of the calculated rate of fluid *diffusion* back and forth through the capillary membranes, and it is only one-tenth the rate of fluid *filtration* from the arterial ends of the capillaries into the tissue spaces in the entire body. These facts illustrate that the flow of lymph is relatively small in comparison with the total exchange of fluid between the plasma and the interstitial fluid.

### Factors That Determine the Rate of Lymph Flow

**Interstitial Fluid Pressure.**    Elevation of interstitial fluid pressure above its normal level of −6.3 mm. Hg increases the flow of interstitial fluid into the lymphatic capillaries and consequently also increases the rate of lymph flow. The increase in flow is very great until the interstitial fluid pressure rises slightly above zero mm. Hg, at which point the flow rate has risen to 10 to 50 times normal. Therefore, any factor (besides obstruction of the lymphatic system itself) that tends to increase interstitial pressure increases the rate of lymph flow. Such factors include:

> Elevated capillary pressure
> Decreased plasma colloid osmotic pressure
> Increased interstitial fluid protein
> Increased permeability of the capillaries

**The Lymphatic Pump.**    Valves exist in all lymph channels, even down to the tips of the lymphatic capillaries, as shown in Figure 31–2. In the large lymphatics, valves exist every few millimeters, and in the smaller lymphatics the valves are much closer together than this, which illustrates the widespread existence of the valves. Every time the lymph vessel is compressed by pressure from any source, lymph tends to be squeezed in both directions, but because the valves open only in the central direction, the fluid moves unidirectionally. The lymph vessels can be compressed either by contraction of the walls of the lymphatics themselves or by pressure from surrounding structures.

Motion pictures taken of exposed lymph vessels, both in animals and man, have shown that any time a lymph vessel becomes stretched with fluid the smooth muscle in the wall of the vessel automatically contracts. Furthermore, each segment of the lymph vessel between successive valves functions as a separate automatic pump. That is, filling of a segment causes it to contract, and the fluid is then pumped through the next valve into the following lymphatic segment. This fills the subsequent segment so that within a few seconds it too contracts, the process continuing all along the lymphatic until the fluid is finally emptied from the lymph vessel into the veins. In a large lymph vessel this lymphatic pump can generate pressure as high as 25 mm. Hg.

In addition to the pumping caused by intrinsic contraction of the lymph vessel walls, any external factor that compresses the lymph vessel can also cause pumping. In order of their importance, such factors are:

> Contraction of muscles
> Passive movements of the parts of the body
> Arterial pulsations
> Compression of the tissues by objects outside the body

Obviously, the lymphatic pump becomes very active during exercise, often increasing lymph flow as much as 5- to 15-fold. On the other hand, during periods of rest lymph flow is very sluggish.

**The Lymphatic Capillary Pump.**    In addition to the lymphatic pump of the larger lymph vessels, even the lymphatic capillary has its own intricate pumping mechanism. Though the lymphatic capillary does not have any smooth muscle cells in its wall, the lymphatic capillary endothelial cells do contain contractile fibers called *myoendothelial fibers.* Under some conditions these cause the lymphatic capillaries to contract several times a minute. With each contraction, fluid is pumped from the capillary through the first lymphatic valve into a *collecting lymph vessel.* During this pumping process the flap valves between the endothelial cells close so that fluid will not move backwards out of the capillary into the tissue spaces. Following each cycle of pumping the anchoring filaments pull the lymphatic capillary back to an expanded state. This creates a mild degree of suction inside the lymphatic capillary and causes fluid to flow from the surrounding tissue spaces into the capillary through the flap valves, the endothelial edges of which flap to the interior of the capillary and thereby provide wide-open communication for the fluid inflow.

In addition to the pumping caused by intrinsic contraction of the lymphatic capillary wall, all of the same factors that cause compression of the larger lymph vessels from the outside can also compress the lymphatic capillaries and provide additional pumping in this way as well.

**Flow of Lymph into the Lymphatic Capillary Despite Negative Pressure in the Interstitial Spaces.**    It has been difficult for many students of physiology to understand how fluid can flow out of the interstitial spaces into the lymphatic capillary in the face of the negative pressure in the interstitial spaces averaging $-6$ mm. Hg, as was discussed in the previous chapter. The resolution to this difficulty lies in the fact that the lymphatic capillaries, during their periodic expansion process, can create small amounts of suction. Indeed, this has even been demonstrated to occur in some of the large lymphatic vessels as well. Another way in which fluid can move from the tissues into the lymphatic vessels despite the negative interstitial fluid pressure is the following: Every time a tissue is compressed, interstitial fluid pressure in the local area of compression rises temporarily to a positive value. This causes minute amounts of fluid to move into the lymphatics and subsequently to be pumped away from the tissues. Then upon removal of the compression, the recoiling of the elastic structures in the tissues, particularly of the filaments of the tissue gel, creates suction in the tissue spaces. Thus, except during the momentary periods of compression, a negative pressure is maintained in the tissue spaces.

**Summary of Factors that Determine Lymph Flow.**    From the above discussion one can see that the two primary factors that determine lymph flow are the interstitial fluid pressure and the activity of the lymphatic pump. Therefore, one can state that, roughly, *the rate of lymph flow is determined by the product of interstitial fluid pressure and the activity of the lymphatic pump.*

**Maximum Limit to the Rate of Lymph Flow.**    Figure 31–3 illustrates the relationship between interstitial fluid pressure ($P_T$) and the rate of lymph flow. Note that at the normal interstitial fluid pressure of $-6$ to $-7$ mm. Hg the lymph flow is very low. However, as the interstitial fluid pressure rises to slightly greater than 0 mm. Hg, the flow increases approximately 20 times, but at this point it reaches a plateau beyond which it will not rise any more even though the interstitial fluid pressure continues to rise.

There are two major reasons that lymph flow reaches a maximum limit: (1) Once the tissues become edematous, the lymphatic capillaries also become tremendously dilated. This causes the flap valves between the endothelial cells of the capillaries to become separated from each other so that they are no longer reliable; therefore, the lymphatic capillary pump begins to fail. (2) The interstitial fluid pressure presses on the outside of the lymph channels to cause them to collapse; therefore, the input pressure at the tips of the lymphatic capillaries—also the interstitial fluid pressure—is opposed by the compression of the lymphatic walls of equal magnitude.

This maximum limit of lymph flow is of major

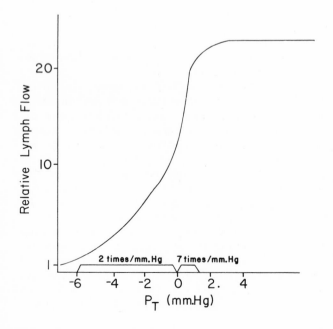

**Figure 31–3.**   Relationship between interstitial fluid pressure and lymph flow. Note that lymph flow reaches a maximum as the interstitial pressure rises slightly above atmospheric pressure (0 mm. Hg). (Courtesy of Drs. Harry Gibson and Aubrey Taylor.)

significance because it demonstrates that most of the compensation to prevent edema caused by increased lymph flow occurs before the edema actually appears. That is, this mechanism most often keeps us from having edema without our ever knowing that it is in effect. It is only those persons whose abnormalities have gone beyond the limits of this lymphatic compensating mechanism who actually develop edema.

# CONTROL OF INTERSTITIAL FLUID PROTEIN CONCENTRATION AND INTERSTITIAL FLUID PRESSURE

The fact that interstitial fluid pressure is negative (that is, subatmospheric) was discovered only a few years ago, though it has now been confirmed by a number of different independent methods described in the previous chapter. Even so, it has been difficult for many students and even professional physiologists to understand the significance of the negative pressure. To explain this, it is first necessary to discuss the regulation of interstitial fluid protein concentration because the problem of interstitial fluid pressure is inextricably bound with that of interstitial fluid protein concentration, as we shall see in the following paragraphs.

## REGULATION OF INTERSTITIAL FLUID PROTEIN BY LYMPHATIC PUMPING

Since protein continually leaks from the capillaries into the interstitial fluid spaces, it must also be removed continually, or otherwise the tissue colloid osmotic pressure will become so high that normal capillary dynamics can no longer continue. Unfortunately, only a small proportion of the protein that leaks into the tissue spaces can diffuse back into the capillaries because there is almost four times as much protein in the plasma as in the interstitial fluid. Therefore, by far the most important of all the lymphatic functions is the maintenance of low protein concentration in the interstitial fluid. The mechanism of this is the following:

As fluid leaks from the arterial ends of the capillaries into the interstitial spaces, only small quantities of protein accompany it, but then, as fluid is reabsorbed at the venous ends of the capillaries, most of the protein is left behind. Therefore, *protein progressively accumulates in the interstitial fluid* and this in turn *increases the tissue colloid osmotic pressure*. The osmotic pressure decreases reabsorption of fluid by the capillaries, thereby *promoting increased tissue fluid volume* and *increased interstitial fluid pressure*. The increased pressure then forces interstitial fluid into the lymphatic capillaries, and the fluid carries with it the excess protein that has accumulated. This continual washout of the protein keeps its concentration at a low level.

To summarize, an increase in tissue fluid protein increases the rate of lymph flow, and this washes the proteins out of the tissue spaces, automatically returning the protein concentration to its normal low level.

The importance of this function of the lymphatics cannot be stressed too strongly, for *there is no other route besides the lymphatics through which excess proteins can return to the circulatory system*. If it were not for this continual removal of proteins, the dynamics of the capillaries would become so abnormal within only a few hours that life could no longer continue. There is certainly no other function of the lymphatics that can even approach this in importance.

## MECHANISM OF NEGATIVE INTERSTITIAL FLUID PRESSURE

Until recent measurements of the interstitial fluid pressure demonstrated that the interstitial fluid pressure is negative rather than positive, as explained in the preceding chapter, it had been taught that the normal interstitial fluid pressure ranges between +1 and +4 mm. Hg, and it has still been difficult to understand how negative pressure can develop in the interstitial fluid spaces. However, we can explain this negative interstitial fluid pressure by the following considerations:

First, it was pointed out above that fluid can flow into lymphatic vessels from interstitial spaces even when the interstitial fluid pressure is negative, mainly because the lymphatic pump can create slight degrees of suction. The continual movement of interstitial fluid into the lymphatics keeps the protein concentration of the interstitial fluid at a low value and thereby keeps the colloid osmotic pressure also at a low value, usually at about 5 mm. Hg.

Second, the negativity of the interstitial fluid pressure can then be explained mainly on the basis of the balance of forces at the capillary membrane. If we add all the other forces besides the interstitial fluid pressure that cause movement of fluid across the capillary membrane, we find the following:

Outward force:

|                                        | mm. Hg |
|----------------------------------------|--------|
| Capillary pressure                     | 17     |
| Interstitial fluid colloid osmotic pressure | 5  |
| TOTAL                                  | 22     |

Inward force:

|                                        | mm. Hg |
|----------------------------------------|--------|
| Colloid osmotic pressure               | 28     |
| DIFFERENCE (Interstitial fluid pressure) | −6   |

Thus, we see that the interstitial fluid pressure required to balance the other forces across the capillary membrane is −6 mm. Hg. Thus, −6 mm. of the negative interstitial fluid pressure is caused by this imbalance of forces at the capillary membrane. Another −0.3 mm. Hg is caused by the continual pumping of fluid into the lymphatic vessels, giving a total negativity of −6.3 mm. Hg.

**Role of the Lymphatic System in the Regulation of Interstitial Fluid Pressure.** One might at first think that the lymphatic pump simply creates all of the negative pressure by pumping fluid volume itself away from the tissue spaces at a rate rapid enough to maintain the negative pressure. However, one can calculate that the normal rate of lymph flow is sufficient to create no more than −0.3 mm. Hg negative pressure. On the other hand, removal of protein from the interstitial spaces decreases the tissue fluid colloid osmotic pressure, which allows the venous ends of the capillaries to reabsorb tremendous quantities of fluid from the tissue spaces by osmosis. It is this effect that creates the major share of the negative interstitial fluid pressure.

Therefore, though the lymphatic pump is the prime mover in creating the negative interstitial fluid pressure, only a minor part of this creation of negative pressure occurs as a result of the pumping of fluid volume from the tissues. Instead, the major share results indirectly from the removal of protein from the interstitial spaces.

**Significance of Negative Interstitial Fluid Pressure as a Means for Holding the Body Tissues Together.** In the past it has been assumed that the different tissues of the body are held together entirely by connective tissue fibers. However, at many places in the body, connective tissue fibers are absent. This occurs particularly at points where tissues slide over each other. Yet, even at these places, the tissues are held together by the negative interstitial fluid pressure, which is actually a partial vacuum. When the tissues lose their negative pressure, fluid accumulates in the spaces, and the condition known as *edema* occurs, which is discussed later in the chapter.

**Significance of the Normally "Dry" State of the Interstitial Spaces.** The normal tendency for the capillaries to absorb fluid from the interstitial spaces and thereby to create a partial vacuum causes all the minute structures of the interstitial spaces to be *compacted*. Figure 31–4 illustrates a physical model of the tissues constructed to illustrate this effect. To the left, positive pressure is present and excessive quantities of fluid are present in the "interstitial spaces." To the right, negative pressure has been applied through the perforated tube, which simulates a capillary, and the tissue elements are pulled tightly together. This represents a "dry" state; that is, no *excess* fluid is present besides that required simply to fill the crevices between the tissue elements.

The "dry" state of the tissues is particularly important for optimal nutrition of the tissues, because nutrients pass from the blood to the cells by diffusion; and the rate of diffusion between two points is inversely proportional to the distance between the cells and the capillaries. Therefore, it is essential that the distances be maintained at a minimum, or otherwise nutritive damage to the cells can result.

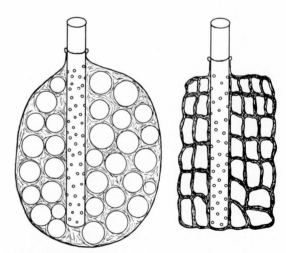

**Figure 31–4.** Physical model of the tissues constructed of a rubber bag, a perforated tube to simulate a capillary, balloons filled with water to simulate cells, and cotton between the balloons to simulate intercellular elements. *Left:* edematous state. *Right:* nonedematous state.

# EDEMA

Edema means the presence of excess interstitial fluid in the tissues. Thus, referring again to Figure 31–4, one sees that the left-hand portion of the figure represents the edematous state, while the right-hand portion represents the nonedematous state.

Obviously, any factor that increases the interstitial fluid pressure high enough can cause excess interstitial fluid volume and thereby cause edema. However, to explain the conditions under which edema develops, we must first characterize the *pressure-volume curve* of the interstitial fluid spaces.

## *PRESSURE-VOLUME CURVE OF THE INTERSTITIAL FLUID SPACES*

Figure 31–5 illustrates the average relationship between pressure and volume in the interstitial fluid spaces in the human body as extrapolated from measurements in the dog. The shape of the curve was determined in the following manner: A dog's leg was removed from its body and then perfused with concentrated dextran solution having a colloid osmotic pressure about twice that of normal plasma. This high colloid osmotic force inside the capillaries

caused absorption of fluid from the interstitial spaces and caused the weight of the leg to decrease. Measurement of this change in weight provided a means for measuring the decrease in interstitial fluid volume. Simultaneously, the pressure in the interstitial fluid spaces was measured using the implanted capsule method described in the preceding chapter. Later in the experiment, fluid having no colloid osmotic pressure was perfused through the limb, and this caused tremendous quantities of fluid to leak out of the capillaries into the interstitial spaces, thereby increasing the interstitial fluid volume. The curve of Figure 31–5 represents the average results obtained in experiments of this type.

**The Slight Interstitial Fluid Volume Change in the Negative Pressure Range.** One of the most significant features of the curve in Figure 31–5 is that so long as the interstitial fluid pressure remains in the negative range there is little change in interstitial fluid volume despite marked change in pressure. Therefore, edema will not occur so long as the interstitial fluid pressure remains negative. Indeed, in several hundred measurements of interstitial fluid pressure made in experimental animals, no edema has ever been recorded in the presence of negative interstitial pressure.

**Tremendous Increase in Interstitial Fluid Volume When the Interstitial Fluid Pressure Becomes Positive.** Note in Figure 31–5 that just as soon as the interstitial fluid pressure rises to equal atmospheric pressure (zero pressure), the slope of the pressure-volume curve suddenly changes and the volume increases precipitously. An additional increase in interstitial fluid pressure of only 1 to 3 mm. Hg now causes the interstitial fluid volume to increase several hundred per cent. Finally, at the very top of the figure, the skin begins to be stretched so that the volume now increases much less rapidly.

**Similarity of the Tissue Spaces to a Bag.** If one will think for a moment, he will realize that a pressure-volume curve similar to that illustrated in Figure 31–5 can also be recorded from almost any collapsible bag, such as a rubber balloon. When negative pressure is applied to the balloon, its volume remains almost nothing. But when the pressure is increased above atmospheric pressure, the balloon suddenly begins to expand. Almost no additional pressure is then required to fill the balloon until its walls begin to stretch. Then, further filling is impeded by the stretching wall. These are precisely the characteristics illustrated in Figure 31–5 for the

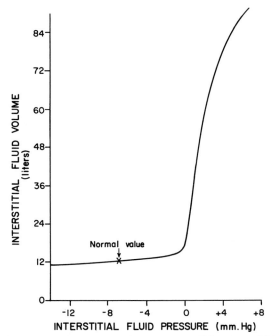

**Figure 31–5.** Pressure-volume curve of the interstitial spaces. (Extrapolated to the human being from data obtained in dogs.)

interstitial fluid spaces of the body. These spaces are in effect a collapsed bag which expands greatly when the interstitial fluid pressure rises above the surrounding atmospheric pressure.

**Tissue Space Compliance in Different Pressure Ranges.** Another way to express the pressure-volume characteristics of the interstitial fluid spaces is in terms of compliance, which is defined as the change in volume for a given change in pressure. In the negative pressure range, the compliance of the interstitial spaces is slight, about 400 ml./mm. Hg for the entire human body (as extrapolated from measurements in dogs). But, just as soon as interstitial fluid pressure rises into the positive range, this compliance increases tremendously, rising to about 10,000 ml./mm. Hg. Thus, the compliance increases approximately 25-fold between the negative pressure range and the positive pressure range.

## POSITIVE INTERSTITIAL FLUID PRESSURE AS THE PHYSICAL BASIS FOR EDEMA

After studying the pressure-volume curve of Figure 31–5, one can readily see that whenever the interstitial fluid pressure rises above the surrounding atmospheric pressure, the tissue spaces begin to swell. Therefore, *the physical cause of edema is positive pressure in the interstitial fluid spaces.*

**Degree of Edema in Relation to the Degree of Positive Pressure.** The solid curve of Figure 31–6 is the same pressure-volume curve as that shown in Figure 31–5, but an edema scale has been added to the figure. 1+ edema means edema that is barely detectable, and 4+ edema means edema in which the limbs are swollen to diameters 1½ to 2 times normal.

Note in Figure 31–6 that edema usually is not detectable in tissues until the interstitial fluid volume has risen to about 30 per cent above normal. And note also that the interstitial fluid volume increases to several hundred per cent above normal in seriously edematous tissues.

**Stretch of the Tissue Spaces in Chronic Edema.** If edema persists even for a few hours, and especially if it persists for weeks, months, or years, the tissue spaces gradually become stretched. As a result, the pressure-volume curve changes from the solid curve in Figure 31–6 to the progressively higher dashed curves. In other words, in chronic edema the tissue "bag" expands, which increases the ease with which the tissues can develop severe edema. Even a 1 to 2 mm. Hg pressure rise above atmospheric pressure can cause 4+ edema once the tissue spaces have been stretched for many days. This phenomenon of

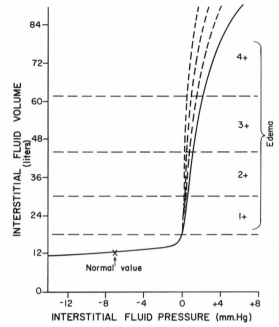

**Figure 31–6.** Relationship of edema to the pressure-volume curve of the interstitial spaces. The dashed curves show the effect of prolonged edema on the pressure-volume curve.

stretch of the tissues is called *delayed compliance* or *stress-relaxation of the tissue spaces.*

**The Phenomenon of "Pitting" Edema.** If one presses his finger on the skin over an edematous area and then suddenly removes the finger, a small depression called a "pit" remains. Gradually, within 5 to 30 seconds, the pit disappears. The cause of "pitting" is that edema fluid has been translocated away from the area beneath the pressure point. The fluid simply flows through the tissue spaces to other tissue areas. Then, when the finger is removed, 5 to 30 seconds is required for the fluid to flow back into the area from which it had been displaced.

**Nonpitting Edema.** Occasionally, the fluid in seriously edematous tissues cannot be mobilized to other areas by pressure. The usual cause of this is coagulation of the fluid in the tissues. For instance, in an infected or traumatized area, large quantities of fluid can collect, but coagulation of the fluid prevents the fluid from being expelled by pressure. This type of edema is frequently called *brawny* edema.

## THE CONCEPT OF A "SAFETY FACTOR" BEFORE EDEMA DEVELOPS

**Safety Factor Caused by the Normal Negative Interstitial Fluid Pressure.** One can readily see from Figures 31–5 and 31–6 that the interstitial fluid pressure must rise from the normal value of −6.3 to above zero mm. Hg before edema begins to appear. Thus, there is a safety factor of 6.3 mm. Hg caused by the normal negative interstitial fluid pressure before edema appears.

**Safety Factor Caused by Flow of Lymph from the Tissues.** Another safety factor that helps to prevent edema is increased lymph flow. When the interstitial fluid pressure rises above the normal value of −6.3 mm. Hg, lymph flow increases greatly, which removes a portion of the extra fluid entering the interstitial spaces. And this obviously helps to prevent development of edema.

One can estimate that maximally increased lymph flow gives approximately a 7 mm. Hg safety factor, for the extra lymph flow can carry away from the tissues approximately the extra amount of fluid that is formed by a 7 mm. Hg excess in capillary pressure.

**Safety Factor Caused by Washout of Protein from the Interstitial Spaces.** In addition to removal of fluid volume from the interstitial fluid spaces, increased lymph flow also washes out most of the proteins from the interstitial fluid spaces, decreasing the colloid osmotic pressure of the interstitial fluid from the normal value of 5 mm. Hg down to about 1 mm. Hg. This provides another 4 mm. Hg safety factor.

**Total Safety Factor and Its Significance.** Now, let us add all the above safety factors:

|  | *mm. Hg* |
|---|---|
| Negative interstitial fluid pressure | 6.3 |
| Lymphatic flow | 7.0 |
| Lymphatic washout of proteins | 4.0 |
| TOTAL | 17.3 |

Thus we find that a total safety factor of about 17 mm. Hg is present to prevent edema. This means that the capillary pressure must rise about 17 mm. Hg above its normal value—that is, above 34 mm. Hg—before edema begins to appear. Or the plasma colloid osmotic pressure must fall from the normal level of 28 mm. Hg to below 11 mm. Hg before edema begins to appear. This explains why the normal human being does not become edematous until severe abnormalities occur in his circulatory system.

## EDEMA RESULTING FROM ABNORMAL CAPILLARY DYNAMICS

From the discussions of capillary and interstitial fluid dynamics in the preceding and present chapters, it is already evident that several different abnormalities in these dynamics can increase the tissue pressure and in turn cause extracellular fluid edema. The different causes of extracellular fluid edema are:

**Increased Capillary Pressure as a Cause of Edema.** Figure 31–7A shows the effect of increased mean capillary pressure on the dynamics of fluid exchange at the capillary membrane. When the mean capillary pressure first becomes abnormally high, more fluid flows out of the capillary than returns into the capillary and, therefore, it collects in the tissue

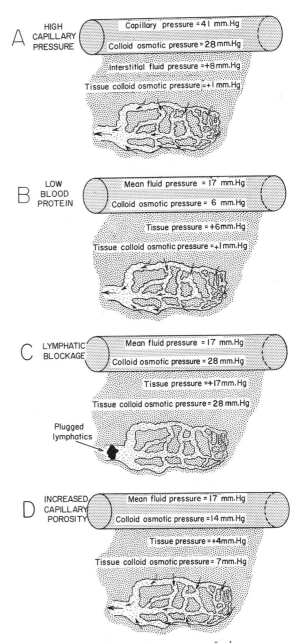

**Figure 31–7.** Various causes of edema.

spaces until the interstitial fluid pressure rises high enough to balance the excessive level of pressure in the capillaries. In Figure 31–7A the mean capillary pressure is 41 mm. Hg instead of the usual normal 17 mm. Hg. Consequently, in this instance enough fluid flows into the tissue spaces to raise the interstitial fluid pressure to +8 mm. Hg. This is far above atmospheric pressure, and therefore causes progressive enlargement of the tissue spaces with tremendous expansion of the extracellular fluid volume.

*Causes of Increased Capillary Pressure.* Increased capillary pressure can result from any clinical condition that causes either venous obstruction or arteriolar dilatation. Large venous blood clots frequently cause local areas of venous obstruction, which block the return of blood to the heart and promote edema in the tissues normally drained by the obstructed veins.

More frequently, capillary pressure is increased by obstruction of venous return due to cardiac failure, for, when the heart no longer pumps blood out of the veins with ease, blood dams up in the venous system. The capillary pressure rises, and serious "cardiac edema" occurs. The dynamics of this type of edema are complicated, however, and were discussed in detail in Chapter 26.

When arteriolar dilatation occurs in localized areas of the body, blood flows rapidly through the locally dilated arterioles and the capillary pressure increases tremendously. Therefore, local edema results. Such local edema occurs commonly in allergic conditions and in the condition known as "angioneurotic edema." Allergic reactions (discussed in Chapter 7) cause the release of histamine into the tissues; histamine relaxes the smooth muscle of arterioles and, when present in large amounts, constricts the venules. The localized edematous areas that result are called "hives" or *urticaria*.

*Angioneurotic edema* apparently is caused by localized decrease in arteriolar tone due to abnormal vascular control by the autonomic nervous system. When a person is emotionally upset, such edema frequently occurs in the larynx and causes hoarseness.

**Decreased Plasma Proteins as a Cause of Edema.** Figure 31–7B illustrates the abnormal dynamics that occur at the capillary membrane when the quantity of plasma protein falls to abnormally low values. The major effect is a markedly lowered colloid osmotic pressure of the plasma. Consequently, the capillary pressure far overbalances the colloid osmotic pressure, increasing the tendency for fluid to leave the capillaries and enter the tissue spaces. As a result, fluid collects in the tissue spaces, and the interstitial fluid pressure rises. As long as the pressure remains excessively elevated, the tissue spaces continually enlarge, with the edema becoming progressively worse.

As is also the case with changes in capillary pressure, the decrease in plasma colloid osmotic pressure must be extreme before edema begins to develop. Since the normal safety factor is about 17 mm. Hg, one can calculate that edema begins to appear when

the plasma colloid osmotic pressure falls below approximately 11 mm. Hg.

*Conditions that Decrease the Plasma Protein Concentration.* Albumin is often lost from the plasma in large quantities when the skin is extensively burned. Therefore, one of the complications of severe burns is not only severe edema in the tissues surrounding the burned area but also edema throughout the body because of lowered colloid osmotic pressure.

Often large quantities of protein, especially albumin, are lost through the kidneys into the urine in the disease known as *nephrosis*. Sometimes as much as 20 to 30 grams of albumin are lost each day, and the colloid osmotic pressure of the plasma may fall to one-half normal or even less. This results in severe edema, and the edema itself is likely to contribute to death by means that are discussed later in the chapter.

Finally, persons who do not have sufficient protein in their diets are unable to form adequate quantities of plasma protein and, therefore, are likely to develop protein deficiency edema, which is called *nutritional edema*. This occurs frequently in war-torn areas.

**Lymphatic Obstruction as a Cause of Edema.** A small amount of protein leaks continually from the capillaries into the tissue spaces, but this protein cannot be reabsorbed into the circulatory system through the capillary membrane. The only route by which the protein can be returned to the circulatory system is through the lymphatics. If the lymphatic drainage from any area of the body becomes blocked, more and more protein collects in the local tissue spaces until finally the concentration of this protein may approach the concentration of protein in the plasma. As shown in Figure 31–7C, the colloid osmotic pressure of the tissue fluids may theoretically rise to as high as 28 mm. Hg, and, to balance this, fluid collects in the tissues until the interstitial fluid pressure rises to a value equal to the capillary pressure, about 17 mm. Hg. Such elevated tissue pressure rapidly expands the tissue spaces, with resultant edema of the severest kind.

*Causes of Lymphatic Obstruction.* One of the most common causes of lymphatic obstruction is *filariasis*—that is, infection by nematodes of the superfamily Filarioidea. The disease is widespread in the tropics, where larvae (microfilariae) are transmitted to human hosts by mosquitoes. The larvae pass out of the capillaries into the interstitial fluid and then by way of the lymph into the lymph nodes. Subsequent inflammatory reactions progressively obstruct the lymphatic channels of these nodes with scar tissue. After several years, the lymphatic drainage from one of the peripheral parts of the body may become almost totally occluded. Thus, a leg can swell to such a size that it might weigh as much as all the remainder of the body. Because of this extreme degree of edema, the swollen condition is frequently called *elephantiasis*. A very interesting type of elephantiasis is that which occasionally occurs in the scrotum, which has been known to enlarge so much that the person must carry it in a wheelbarrow in order to move about.

Lymphatic obstruction also occurs following operations for removal of cancerous tissue. Because the lymph nodes draining a cancerous area of the body must be removed in order to prevent possible spread of the cancer, the return of lymph to the circulatory system from that area will be blocked. Occasionally a radical mastectomy for removal of a cancerous breast causes the corresponding arm to swell to as much as twice its normal size, but usually the swelling regresses during the following two to three months as new lymph channels develop.

**Increased Permeability of the Capillaries as a Cause of Edema.** Figure 31–7D illustrates a capillary whose membrane has become so permeable that even protein molecules pass from the plasma into the interstitial spaces with ease. The protein content of the plasma decreases while that of the interstitial spaces increases. In the example of the figure the tissue pressure rises to +7 mm. Hg in order to balance the changes in plasma and tissue colloid osmotic pressure occasioned by the leakage of protein. The elevated interstitial fluid pressure in turn causes progressive edema.

*Cause of Increased Capillary Permeability.* Capillaries become excessively permeable when any factor destroys the integrity of the capillary endothelium. Burns are a frequent cause of increased permeability of the capillaries because overheated capillaries become friable, and their pores enlarge. Allergic reactions also frequently cause the release of histamine or various polypeptides that damage the capillary membranes and cause increased permeability.

A bacterial toxin produced by *Clostridium oedematiens* can often cause such extreme increase in capillary permeability that plasma loss into the tissues kills the patient within a few hours.

Recently it has been shown that overincreasing the pressure in the capillaries can stretch the capillary walls so greatly that the pores themselves enlarge. This is called the *stretched pore phenomenon.* Even a few minutes of overstretching can cause the condition, and at least 4 to 10 hours are required for the pores to return to normal condition.

### EDEMA CAUSED BY KIDNEY RETENTION OF FLUID

When the kidney fails to excrete adequate quantities of urine, and the person continues to drink normal amounts of water and ingest normal amounts of electrolytes, the total amount of extracellular fluid in the body increases progressively. This fluid is absorbed from the gut into the blood and elevates the capillary pressure. This in turn causes most of the fluid to pass into the interstitial fluid spaces, elevating the interstitial fluid pressure as well. Therefore, simple retention of fluid by the kidneys can result in extensive edema. Furthermore, if the retained fluid is mainly water, intracellular edema will also result, as will be discussed in Chapter 33.

## THE PRESENCE AND IMPORTANCE OF GEL IN THE INTERSTITIAL SPACES

Up to this point we have talked about interstitial fluid as if it were entirely in a mobile state. But, both biochemical and physical studies indicate that under normal conditions almost all of the interstitial fluid is held in a gel that fills the spaces between the cells. This gel contains large quantities of mucopolysaccharides, the most abundant of which is *hyaluronic acid.*

Most of the principles of interstitial fluid hemodynamics still hold true even though most of the fluid is in a gel state. For instance, diffusion of small molecules can occur through the gel more than 95 per cent as effectively as in free fluid. Therefore, nutrients can diffuse from the capillaries to the cells almost equally as well through the gel as through free fluid.

On the other hand, the fluid in the gel cannot be readily moved from one part of the tissue to another part. That is, the gel structure holds the interstitial fluid in place, except for extremely slow flow through the gel itself (measured in millimeters per hour) and minute rivulets of free interstitial fluid that move along the surfaces of the gel between the gel and the cells.

**Relation of Edema Fluid to the Gel.** When edema occurs, a small portion of the edema fluid is imbibed by the gel, but once the extracellular volume has increased beyond approximately 30 to 50 per cent, large quantities of freely moving fluid appear in the tissues. When this occurs, pressure on one tissue area will cause fluid to move freely to another area, which is the phenomenon called *pitting.* Also, the fluid flows downward in the tissues because of the pull of gravity, helping to cause the phenomenon called *dependent edema;* that is, edema in the lower parts of the body.

Figure 31–8 illustrates the relationship between free interstitial fluid, gel fluid (non-mobile fluid), and total interstitial fluid. Note that under normal conditions, when the interstitial fluid pressure is in its normal negative pressure range, there is almost imperceptible free fluid in the tissues. Instead, almost all of the fluid is in the gel phase, and it is also highly immobile. On the other hand, as the interstitial fluid pressure rises and as the state of edema approaches, the gel swells some 30 to 50 per cent, after which it swells no more. With still more increase in fluid volume, all the additional fluid that accumulates is free fluid that is highly mobile through the tissue spaces.

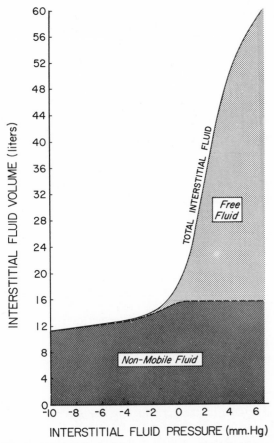

**Figure 31–8.** Effect of increasing interstitial fluid pressure on the volumes of total interstitial fluid, nonmobile fluid (gel fluid), and free fluid. Note that significant amounts of free fluid occur only when the interstitial fluid pressure becomes positive. (Reprinted from Guyton, Granger, and Taylor: *Physiol. Rev.*, 51:527, 1971.)

**Theoretical Importance of Having a Gel in the Interstitial Spaces.** One can readily understand that if all cells abutted directly onto each other without any interstitial spaces, those cells not directly in contact with capillaries could receive nutrition only through other cells; many substances, such as glucose, amino acids, and so forth, would not be able to reach the outlying cells in adequate quantities. At the opposite extreme, the presence of tremendous quantities of extracellular fluid, with consequent very large tissue spaces, would increase the distances from the capillaries to the outlying cells so much that again the diffusion of nutrients to outlying cells would be diminished. Obviously, therefore, there is a certain interstitial fluid space size for optimum transport of substances from the capillaries to the cells. It has been postulated that the function of the gel in the interstitial spaces is to create interstitial fluid spaces of appropriate size for optimum cell nutrition to occur.

Two other important functions of gel in the interstitial spaces are: (1) prevention of fluid flow through the tissue spaces into the dependent areas of the body as a result of gravity, a mechanism mentioned above, and (2) prevention of spread of infection in the tissues, that is, prevention of spread of bacteria or other agents from one tissue area to another.

**Relationship of Interstitial Fluid Gel to the Regulation of Interstitial Fluid Volume.** Since approximately 16 per cent of the average tissue is composed of interstitial fluid but almost none of this is in the mobile state, one can derive the following theory for the regulation of interstitial fluid volume. The mechanism previously described for creating a negative pressure in the tissue spaces is actually a "drying" mechanism that attempts all the time to remove any excess fluid that appears in the tissues. Thus, essentially all *mobile* fluid would be removed. The question remains then, why does this drying mechanism not also remove fluid from the gel? The answer to this has two components: First, the fine reticular filaments of the gel are composed of hyaluronic acid molecules that are coiled like springs and are compressed against each other. Therefore, the elastic forces of these molecules prevent further compression in the same way that cotton fibers in absorbent cotton prevent compression beyond a certain point. Second, the gel has a slight amount of osmotic pressure caused by the Donnan equilibrium effect, that is, the gel reticulum has negative electrostatic charges that hold small mobile positive ions—mainly sodium ions—within the gel. These ions in turn cause osmosis of water into the gel. The quantity of mucopolysaccharides in the tissue gel is sufficient to give an osmotic absorptive pressure in the gel that is calculated to be about 1 mm. Hg. The elastic recoil of the hyaluronic acid "springs" gives approximately another 7 mm. Hg of recoil force, making a total of about 8 mm. Hg to resist dehydration caused by the −6.3 mm. Hg pressure in the free fluid of the tissue spaces.

# RELATIONSHIP OF INTERSTITIAL FLUID PRESSURE TO *SOLID TISSUE PRESSURE* AND *TOTAL TISSUE PRESSURE*

Throughout the present chapter and preceding chapter, the term interstitial fluid pressure has been used to designate the pressure of the *free fluid* in the minute tissue spaces. This pressure has been called simply *tissue pressure* by many physiologists. However, there are at least two other types of tissue pressure, called *solid tissue pressure* and *total tissue pressure*.

Figure 31–9 illustrates these three different types of pressure. In the top of Section A, pressure is applied by a surface against a fluid medium, and this

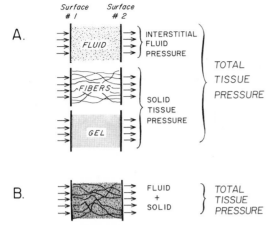

**Figure 31–9.** (A) Transmission of pressure separately through fluid, fibers, and gel. (B) Transmission of pressure through all media simultaneously. (Reprinted from Guyton, Granger, and Taylor: *Physiol. Rev.,* 51:527, 1971.)

causes pressure transmission to a second surface simply as a result of the hydraulic forces developed in the fluid; the pressure transmitted in this way is the *interstitial fluid pressure.* Next, pressure is applied against solid elements such as collagen fibers in the tissues, reticular fibers in the interstitial gel, and semi-solid cells. In these cases the elastic recoil of the solid elements provides transfer of pressure from the first surface to the second surface. Pressure transmitted in this way is the *solid tissue pressure.* In Section B, pressure is applied to both the fluid and the solid elements. Therefore, in this instance, the pressure is transmitted through both media, and pressure transmitted is then the *total tissue pressure.*

Thus, it is clear that the total tissue pressure is an algebraic sum of the interstitial fluid pressure and the solid pressure.

*Importance of interstitial fluid pressure.* The interstitial fluid pressure, which results from the kinetic motion of the fluid molecules themselves, is the pressure that causes molecules to diffuse through pores and also that causes flow of fluid through the tissue spaces.

*Importance of solid tissue pressure.* Solid tissue pressure represents the force exerted by the solid elements of the tissues upon each other. It is these forces that cause the cells and other solid structures to resist compression when negative pressure in the interstitial fluid sucks the solid structures against each other. It also causes much of the transmission of pressure from the skin to blood vessels or from the skin to cells and other structures deep within the tissues. For instance, when a blood pressure cuff is inflated around the arm, pressure is transmitted into the tissues both by the fluid and by the solid elements.

*Total tissue pressure.* Total tissue pressure is the sum of the fluid and solid tissue pressures. At every

membrane surface both the fluid and the solid elements can cause pressure. Therefore, the type of tissue pressure that causes membranes to move or to be compressed is the total tissue pressure—that is, the sum of both the fluid and the solid pressures. For instance, the pressure that causes compression of the blood vessels or that causes compression of cells is the total tissue pressure and not the interstitial fluid pressure or the solid pressure alone.

**Effect of the Interstitial Fluid Volume on the Quantitative Levels of the Different Types of Tissue Pressure.** Figure 31–10 illustrates the relationship between the interstitial fluid volume of a tissue and the quantitative values of the three different types of tissue pressure. Note that when the volume is normal the interstitial fluid pressure is −6.3 mm. Hg, the solid tissue pressure is +7.3 mm. Hg and the sum of these gives a total tissue pressure of +1 mm. Hg. However, as progressively more fluid enters the interstitial spaces, the solid elements are pushed farther and farther apart until finally the solid tissue pressure falls almost to zero. Thereafter, the total tissue pressure and the interstitial fluid pressure become almost equal. That is, in the edematous state, the solid tissue pressure is relatively unimportant, even though in normal tissues the solid elements are compressed to the point that they have a compressive elastic recoil of approximately +7.3 mm. Hg. This can be demonstrated by removing the skin over any subcutaneous tissue area and simply superfusing the tissue with saline. The tissue immediately imbibes fluid, principally because of the solid elements in the tissues, and this causes obvious swelling of the subcutaneous tissue that can be readily observed.

**Measurement of Total Tissue Pressure—The Balloon and Needle Methods.** Total tissue pressure has been measured in several different ways and is +1 to +3 mm. Hg. This is in contrast to the −6.3 mm. Hg for interstitial fluid pressure.

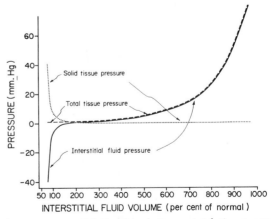

**Figure 31–10.** Relationship between total tissue pressure, solid tissue pressure, and interstitial fluid pressure in normal subcutaneous tissues. (Reprinted from Guyton, Taylor, and Granger: Dynamics of the Body Fluids. 1975.)

The most easily understandable method for measuring total tissue pressure is to implant a minute flaccid balloon into a tissue and then measure the pressure that is required barely to begin filling the balloon. Since both fluid and solid elements compress the outside of this balloon, the pressure required to push the walls of the balloon apart is equal to the total tissue pressure, and the value measured in this way is +1 to +3 mm. Hg.

However, the method that has been used most frequently in the past to measure total tissue pressure has been to insert a small needle into the tissue and then apply increasing pressure to the fluid in the needle until it *barely* begins to move into the tissue. Here again, this pressure measures to be about +1 to +3 mm. Hg. If flow of fluid through the needle is stopped for more than 10 to 15 seconds, the pressure measured through the needle is likely to become progressively negative or to become very erratic because the tip of the needle becomes blocked. Therefore, it is essential to make this pressure measurement within 5 to 30 seconds after a small amount of fluid has moved into the tissue.

**Calculation of Solid Tissue Pressure.** It has not been possible to measure solid tissue pressure, but one can calculate this by subtracting interstitial fluid pressure from total tissue pressure. Assuming that the average total pressure is +2 mm. Hg and that the average interstitial fluid pressure is −6.3 mm. Hg, one then calculates the average solid tissue pressure to be +8.3 mm. Hg. That is, the solid elements of the tissues are compacted against each other with a force of +8.3 mm. Hg. Obviously, this solid tissue pressure is the summation of large numbers of minute forces exerted at the contact points between the solid elements of the tissues. (If these solid contact points cover a significant proportion of the surface area of the tissue spaces, the equation for the interrelationship between interstitial fluid pressure, solid tissue pressure, and total tissue pressure has to be modified slightly because the interstitial fluid pressure then acts only on part of interstitial space surface rather than on all of it.)

# PULMONARY INTERSTITIAL FLUID DYNAMICS

The dynamics of the pulmonary interstitial fluid are essentially the same as those for fluid in the peripheral tissues except for the following important quantitative differences:

1. The pulmonary capillary pressure is very low in comparison with the systemic capillary pressure, being approximately 7 mm. Hg in comparison with 17 mm. Hg.

2. The pulmonary capillaries are relatively leaky to protein molecules so that the protein concentration of lymph leaving the lungs is rela-

tively high, averaging about 4 gm. per cent instead of 2 gm. per cent in the peripheral tissues.

3. The rate of lymph flow from the lungs is also very high, mainly because of the continuous pumping motion of the lungs.

4. The interstitial spaces of the alveolar portions of the lungs are very narrow, represented by the minute spaces between the capillary endothelium and the alveolar epithelium.

5. The alveolar epithelia are not strong enough to resist very much positive pressure. They are probably ruptured by any positive pressure in the interstitial spaces greater than +2 mm. Hg, which allows dumping of fluid from the interstitial spaces into the alveoli.

Now let us see how these quantitative differences affect pulmonary fluid dynamics.

**Interrelationship Between Interstitial Fluid Pressure and Other Pressures in the Lung.** Figure 31–11 illustrates a pulmonary capillary, a pulmonary alveolus, and a lymphatic capillary draining the interstitial space between the capillary and the alveolus. Note the balance of forces at the capillary membrane as follows:

Forces tending to cause movement of fluid out of the capillaries:

| | mm. Hg |
|---|---|
| Capillary pressure | 7 |
| Interstitial fluid colloid osmotic pressure | 16 |
| TOTAL OUTWARD FORCE | 23 |

Forces tending to cause absorption of fluid into the capillaries:

| | mm. Hg |
|---|---|
| Plasma colloid osmotic pressure | 28 |
| Interstitial fluid pressure | −6 |
| TOTAL INWARD FORCE | 22 |

Note that the normal outward forces are slightly greater than the inward forces. The *net mean filtration pressure* at the pulmonary capillary membrane can be calculated as follows:

| | mm. Hg |
|---|---|
| Total outward force | +23 |
| Total inward force | −22 |
| NET MEAN FILTRATION PRESSURE | +1 |

This net filtration pressure causes a slight continual flow of fluid from the pulmonary capillaries into the interstitial spaces, and this fluid is

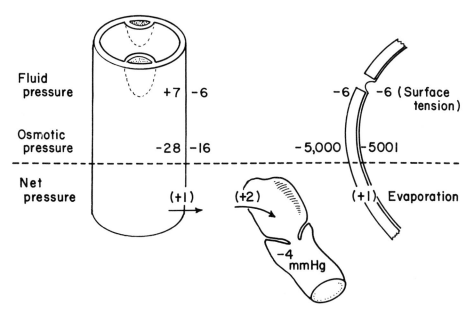

**Figure 31–11.**  Hydrostatic and osmotic forces at the capillary (left) and alveolar membrane (right) of the lungs. Also shown is a lymphatic (center) that pumps fluid from the pulmonary interstitial spaces. (Reprinted from Guyton, Taylor, and Granger: Dynamics of the Body Fluids. 1975.)

pumped back to the circulation through the pulmonary lymphatic system.

**Fluid Exchange at the Pulmonary Alveolar Membrane; the Mechanism for Keeping the Alveoli "Dry."**    The alveolar epithelial membrane is quite different from the pulmonary capillary membrane in the following way: The pulmonary capillaries, like other capillaries of the body, have very large slit-pores between the adjacent endothelial cells. Ions such as sodium, chloride, and potassium as well as crystalloid molecules such as glucose, urea, and so forth can pass through these large capillary pores with ease. On the other hand, the alveolar epithelial membrane contains no such large openings. Therefore, all of the above molecules can cause osmotic pressure effects at the alveolar membrane even though they have no such effect at the pulmonary capillary membrane. For instance, when water enters the alveoli, the high concentration of the different dissolved substances in the pulmonary interstitial fluid causes almost instantaneous osmosis of the water from the alveoli into the interstitial fluid, and the fluid is then absorbed into the pulmonary capillaries because of the *colloid* osmotic pressure of the plasma. Indeed, in a person who drowns in fresh water, enough fluid can be absorbed from the alveoli into the blood within

two to three minutes to cause fibrillation of the heart because of dilution of the blood electrolytes.

In addition to osmosis of fluid from the alveoli, small amounts of fluid can also be moved from the alveoli into the interstitial spaces as a result of suction by the negative pressure in these spaces. Even saline solution, the ions of which prevent its osmosis into the interstitial fluid, moves slowly from the alveoli into the interstitial spaces in this way and is then transported either into the blood or into the pulmonary lymph vessels.

But much more important than the question of fluid absorption from the alveoli is the following question: How is it that the fluid normally present in the interstitial spaces is prevented from flooding the alveoli? The answer to this is related to the negative interstitial fluid pressure of approximately −6 mm. Hg, which continually tends to pull fluid inward through the alveolar membrane and therefore also prevents fluid loss in the outward direction.

The only fluid that does go outward through the alveolar membrane is that small amount that moves by the mechanism of *capillarity* through the cellular pores of the epithelial cells and then creeps along the lining surfaces of the alveoli to keep them moist.

## PULMONARY EDEMA

Pulmonary edema occurs in the same way that it occurs elsewhere in the body. Any factor that causes the pulmonary interstitial fluid pressure to rise from the negative range into the positive range will cause sudden filling of the fluid spaces with large amounts of free fluid.

The usual causes of pulmonary edema are:

1. Left heart failure or mitral valvular disease with consequent great increase in pulmonary capillary pressure and flooding of the interstitial spaces

2. Damage to the pulmonary capillary membrane caused by breathing noxious substances such as chloride gas and sulfur dioxide gas

3. Decrease in plasma colloid osmotic pressure to a low enough level that fluids transude from the blood into the pulmonary interstitial spaces (occurs only rarely)

**Pulmonary "Interstitial Fluid" Edema versus Pulmonary "Alveolar" Edema.** The interstitial fluid volume of the lungs usually cannot increase more than about 50 per cent (representing less than 100 milliliters of fluid) before the alveolar epithelial membranes rupture and fluid begins to pour from the interstitial spaces into the alveoli. The cause of this is simply the almost infinitesimal tensional strength of the pulmonary alveolar epithelium, that is, any positive pressure more than 1 to 3 mm. Hg (as estimated from experiments in dogs) in the interstitial fluid spaces will cause rupture of the alveolar epithelium.

Therefore, except in the mildest cases of pulmonary edema, edema fluid always enters the alveoli; if this edema becomes severe enough, it can cause death by suffocation, as has already been discussed in Chapters 26 and 27.

**Safety Factor Against Pulmonary Edema.** All of the same factors that tend to prevent edema in the peripheral tissues also tend to prevent edema in the lungs. That is, before positive interstitial fluid pressure can occur and cause edema the following must occur: (1) the normal negativity of the interstitial fluid pressure of the lungs must be overcome, (2) the lymphatic pumping of fluid out of the interstitial spaces must be exceeded, and (3) the tendency for the protein concentration in the interstitial fluid to decrease when the lymph flow increases must also be overcome. In experiments in animals it has been found that the pulmonary capillary pressure normally must rise to a value at least equal to the plasma colloid osmotic pressure before significant pulmonary edema will occur. Thus, in the human being, who normally has a plasma colloid osmotic pressure of 28 mm. Hg, one can predict that the pulmonary capillary pressure must rise from the normal level of 7 mm. Hg to over 28 mm. Hg to cause pulmonary edema, giving a safety factor against edema of about 21 mm. Hg.

*Safety factor in chronic conditions.* When the pulmonary capillary pressure is elevated chronically (for at least more than two weeks), the lung becomes even more resistant to pulmonary edema because the lymph vessels expand greatly, increasing their capability for carrying fluid away from the interstitial spaces as much as 10-fold. Therefore, a patient with chronic mitral stenosis frequently has a pulmonary capillary pressure as great as 40 to 45 mm. Hg without having significant pulmonary edema.

Thus, in chronic pulmonary edema, the safety factor against edema can rise to as high as 35 to 40 mm. Hg in comparison with a normal value of 21 mm. Hg under acute conditions.

**Rapidity of Death in Acute Pulmonary Edema.** When the pulmonary capillary pressure does rise above the safety factor level, lethal pulmonary edema can occur within hours if it is only slightly above the safety factor, and within 20 to 30 minutes if it is as much as 25 to 30 mm. Hg above the safety factor level. Thus, in acute left heart failure, in which the pulmonary capillary pressure occasionally rises to as high as 50 mm. Hg, death frequently ensues within 30 minutes from acute pulmonary edema.

# REFERENCES

Allen, L.: Lymphatics and lymphoid tissues. *Ann. Rev. Physiol.,* 29:197, 1967.

Burleigh, P. M. C., and Poole, A. R.: Dynamics of Connective Tissue Macromolecules. New York, American Elsevier Publishing Company, 1974.

Burleigh, P. M. C., and Poole, A. R. (eds.): Dynamics of the Connective Tissue Macromolecules (Proceedings of the International Symposium, Cambridge, Eng., July 1–3, 1974). New York, American Elsevier Publishing Company, 1974.

Drinker, C. K.: The Lymphatic System. Stanford, Stanford University Press, 1942.

Gaar, K. A., Jr., Taylor, A. E., and Guyton A. C.: Effect of lung edema on pulmonary capillary pressure. *Amer. J. Physiol.,* 216:1370, 1969.

Gaar, K. A., Jr., Taylor, A. E., Owens, L. J., and Guyton, A. C.: Effect of capillary pressure and plasma protein on development of pulmonary edema. *Amer. J. Physiol.,* 213:79, 1967.

Gauer, O. H., Henry, J. P., and Behn, C.: The regulation of extracellular fluid volume. *Ann. Rev. Physiol.,* 32:547, 1970.

Guyton, A. C.: Interstitial fluid pressure: II. Pressure-volume curves of interstitial space. *Circ. Res.,* 16:452, 1965.

Guyton, A. C.: Interstitial fluid pressure-volume relationships and their regulation. *In* Wolstenholme, G. E. W., and Knight, J.

(eds.): CIBA Foundation Symposium on circulatory and Respiratory Mass Transport. London, J. & A. Churchill, Ltd., 1969, p. 4.

Guyton, A. C.: Introduction to Part I: Pulmonary alveolar-capillary interface and interstitium. *In* Fishman, A. P., and Hecht, H. H. (eds.): The Pulmonary Circulation and Interstitial Space, Chicago, University of Chicago Press, 1969, p. 3.

Guyton, A. C., and Coleman, T. G.: Regulation of interstitial fluid volume and pressure. *Ann. N.Y. Acad. Sci., 150*:537, 1968.

Guyton, A. C., Granger, H. J., and Taylor, A. E.: Interstitial fluid pressure. *Physiol. Rev., 51*:527, 1971.

Guyton, A. C., and Lindsey, A. W.: Effect of elevated left atrial pressure and decreased plasma protein concentration on the development of pulmonary edema. *Circ. Res., 7*:649, 1959.

Guyton, A. C., Taylor, A. E., and Granger, H. J.: Circulatory Physiology II. Dynamics and Control of the Body Fluids. Philadelphia, W. B. Saunders Company, 1975.

Kalima, T. V.: Ultrastructure of the intestinal lymphatics in regard to absorption. *Scand. J. Gastroenterol., 8*:193, 1973.

Leak, L. V., and Burke, J. F.: Ultrastructural studies on the lymphatic anchoring of filaments. *J. Cell Biol., 36*:129, 1968.

Marks, J., and Shuster, S.: Disorders of capillary permeability. *Brit. J. Dermatol., 88*:619, 1973.

Mayerson, H. S.: The physiologic importance of lymph. *In* Hamilton, W. F. (ed.): Handbook of Physiology. Sec. 2, Vol. 2. Baltimore, The Williams & Wilkins Co., 1963, p. 1035.

Meyer, B. J., Meyer, A., and Guyton, A. C.: Interstitial fluid pressure. V. Negative pressure in the lungs. *Circ. Res., 22*:263, 1968.

Sterns, E. E.: Current concepts of lymphatic transport. *Surg. Gynecol. Obstet., 138*:773, 1974.

Visscher, M. B., Haddy, F. J., and Stephens, G.: The physiology and pharmacology of lung edema. *Pharm. Rev., 8*:389, 1956.

Yoffey, J. M., and Courtice, F. C. (eds.): Lymphatics, Lymph, and Lymphomyeloid Complex. New York, Academic Press, Inc., 1970.

# | 32 |

# The Special Fluid Systems of the Body—Cerebrospinal, Ocular, Pleural, Pericardial, Peritoneal, and Synovial

Several special fluid systems exist in the body, each performing functions peculiar to itself. For instance, the cerebrospinal fluid supports the brain in the cranial vault, the intraocular fluid maintains distention of the eyeballs so that the optical dimensions of the eye remain constant, and the potential spaces, such as the pleural and pericardial spaces, provide lubricated chambers in which the internal organs can move. All these fluid systems have characteristics that are similar to each other and that are also similar to those of the interstitial fluid system. However, they are also sufficiently different that they require special consideration.

## THE CEREBROSPINAL FLUID SYSTEM

The entire cavity enclosing the brain and spinal cord has a volume of approximately 1650 ml., and about 150 ml. of this volume is occupied by cerebrospinal fluid. This fluid, as shown in Figure 32–1, is found in the ventricles of the brain, in the cisterns around the brain, and in the subarachnoid space around both the brain and the spinal cord. All these chambers are connected with each other, and the pressure of the fluid is regulated at a constant level.

### CUSHIONING FUNCTION OF THE CEREBROSPINAL FLUID

A major function of the cerebrospinal fluid is to cushion the brain within its solid vault. Were it not for this fluid, any blow to the head would cause the brain to be juggled around and severely damaged. However, the brain and the cerebrospinal fluid have approximately the same specific gravity, so that the brain simply floats in the fluid. Therefore, blows on the head move the entire brain simultaneously, causing no one portion of the brain to be momentarily contorted by the blow.

**Contrecoup.** When a blow to the head is extremely severe, it usually does not damage the brain on the side of the head where the blow is struck, but, instead, on the opposite side. This phenomenon is known as "contrecoup," and the reason for this effect is the following: When the blow is struck, the fluid on the struck side is so incompressible that, as the skull moves, the fluid pushes the brain at the same time. However, on the opposite side, the sudden movement of the skull causes it to pull away from the brain momentarily, creating for a short instant a vacuum in the cranial vault at this point. Then, when the skull is no longer being accelerated by the blow, the vacuum suddenly collapses and the brain strikes the inner surface of the skull. For instance, the damage to the brain of a boxer usually does not occur in the frontal regions but, instead, in the occipital regions.

**Cerebral Damage from Rotational Blows.** Even though the cerebrospinal fluid can cushion the brain against linear blows, it is not capable of preventing damage from *rotational acceleration* of the skull. The brain, like all objects that have mass, has inertia. Therefore, when the skull is suddenly rotated, the brain tends to remain stationary inside the cranial

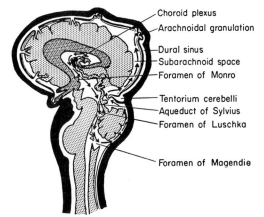

**Figure 32–1.** Pathway of cerebrospinal fluid flow. (Modified from Ranson and Clark: Anatomy of the Nervous System.)

vault. This causes a shearing effect where the brain is attached to the skull, sometimes resulting in tearing of the arachnoidal tissues or even rupture of the diploic blood vessels passing from the substance of the brain through the meninges into the skull. This ability of rotational movement to damage the surface structures of the brain probably explains why sudden rotational movements of the head when a person has a headache are likely to aggravate the headache, and it possibly also explains why a rotational blow to the head during a boxing match, such as an uppercut or a sidewise blow to the chin, is likely to knock out the opponent even though a linear blow fails.

## FORMATION OF CEREBROSPINAL FLUID

Cerebrospinal fluid is formed in several different ways; the major portion is formed by the choroid plexuses in the ventricles, smaller portions by the blood vessels of the meningeal and ependymal linings of the cerebrospinal fluid chambers, and still smaller portions by the blood vessels of the brain and spinal cord.

**Formation of Cerebrospinal Fluid by the Choroid Plexus.** The choroid plexus, which is illustrated in Figure 32–2, is a cauliflower growth of blood vessels covered by a thin coat of epithelial cells. This plexus projects into (a) the temporal horns of the lateral ventricles, (b) the posterior portions of the third ventricle, and (c) the roof of the fourth ventricle.

Cerebrospinal fluid continually exudes from the surface of the choroid plexus. This fluid is not exactly like the other extracellular fluids. Instead, its concentration of sodium is 7 per cent greater than the concentration in extracellular fluid; its glucose concentration is 30 per cent less and potassium 40 per cent less. Obviously, then, the fluid arising from the

choroid plexus is not a simple filtrate from the capillaries, but a *choroid secretion.*

The probable mechanism by which the choroid plexus secretes fluid is the following: The cuboidal epithelial cells of the choroid plexus actively secrete sodium ions, which develop a positive charge in the cerebrospinal fluid. This in turn pulls negatively charged ions, particularly chloride ions, also into the cerebrospinal fluid. Thus an excess of ions develops in the fluids of the ventricles. As a result, the osmotic pressure of the ventricular fluid is elevated to approximately 160 mm. Hg greater than that of the plasma, and this osmotic force causes large quantities of water and other dissolved substances to move through the choroidal membrane into the cerebrospinal fluid. Since glucose is not as diffusible as water, its concentration remains somewhat low. The low concentration of potassium is probably caused by potassium transport in the opposite direction through the epithelial cells. The major basis for this osmotic theory is that the concentration of osmotically active substances in the cerebrospinal fluid is 9 milliosmols greater than that of plasma; this difference would exert the 160 mm. Hg osmotic force noted above.

The rate of choroidal secretion is estimated to be about 840 ml. each day, which is five to six times as much as the total volume of fluid in the entire cerebrospinal cavity.

**Diffusion into the Cerebrospinal Fluid Through the Meningeal and Ependymal Surfaces.** The surfaces of the ventricles are lined with a thin cuboidal epithelium called the *ependyma,* and the cerebrospi-

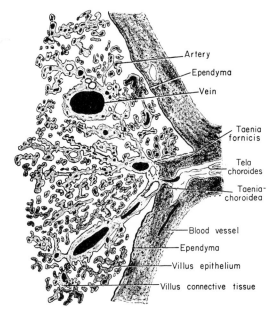

**Figure 32–2.** The choroid plexus. (Modified from Clara: Das Nervensystem des Menschen. Barth.)

nal fluid is in contact with this surface at all points. In addition, cerebrospinal fluid fills the *subarachnoid space* between the *pia mater* that covers the brain and the *arachnoidal membrane*. The cerebrospinal fluid, therefore, is in contact with large surface areas of ependyma and meninges, and diffusion occurs continually between the cerebrospinal fluid and the brain substance beneath the ependyma and also between the cerebrospinal fluid and the blood vessels of the meninges, especially the arachnoidal vessels. Some substances diffuse through these membranes extremely rapidly, but others diffuse very poorly. For instance, lipid-soluble substances, such as alcohol, diffuse so rapidly that equilibrium occurs between alcohol in the cerebrospinal fluid and alcohol in the substance of the brain within less than an hour; other substances such as proteins, sucrose, and so forth, may never reach equilibrium between their concentrations in the cerebrospinal fluid and in the interstitial fluid of the brain substance.

**The Perivascular Spaces and Cerebrospinal Fluid.** The blood vessels entering the substance of the brain pass first along the surface of the brain and then penetrate inward, carrying a layer of *pia mater* with them, as shown in Figure 32–3. The pia is only loosely adherent to the vessels, so that a space, the *perivascular space,* exists between it and each vessel. Perivascular spaces follow both the arteries and the veins into the brain as far as the arterioles and venules but not to the capillaries.

*The Lymphatic Function of the Perivascular Spaces.* As is true elsewhere in the body, a small amount of protein leaks out of the parenchymal capillaries into the interstitial spaces of the brain, and since no true lymphatics are present in brain tissue, this protein leaves the tissues mainly through the perivascular spaces but partly also by direct diffusion through the pia mater into the subarachnoid spaces. On reaching the subarachnoid spaces, the protein flows along with the cerebrospinal fluid to be absorbed through the *arachnoidal villi* into the cerebral veins, as is dis-

cussed subsequently. Therefore, the perivascular spaces, in effect, are a modified lymphatic system for the brain.

In addition to transporting fluid and proteins, the perivascular spaces also transport extraneous particulate matter from the brain into the subarachnoid space. For instance, whenever infection occurs in the brain, dead white blood cells are carried away through the perivascular spaces.

## ABSORPTION OF CEREBROSPINAL FLUID—THE ARACHNOIDAL VILLI

Almost all the cerebrospinal fluid formed each day is reabsorbed into the blood through special structures called *arachnoidal villi* or *granulations,* which project from the subarachnoid spaces into the venous sinuses of the brain and, rarely, also into the veins of the spinal canal. The arachnoidal villi are actually arachnoidal trabeculae that protrude through the venous walls, resulting in extremely permeable areas that allow relatively free flow of cerebrospinal fluid as well as protein molecules and even small particles (less than 1 micron in size) into the blood.

## FLOW OF FLUID IN THE CEREBROSPINAL FLUID SYSTEM

The main channel of fluid flow from the choroid plexuses of the ventricles to the arachnoidal villi is illustrated in Figure 32–1. Fluid formed in the lateral ventricles passes into the third ventricle through the *foramina of Monro,* combines with that secreted in the third ventricle, and passes along the *aqueduct of Sylvius* into the fourth ventricle, where still more fluid is formed. It then passes into the *cisterna magna* through two lateral *foramina of Luschka* and a midline *foramen of Magendie.* From here it flows through the *subarachnoid spaces* upward toward the cerebrum where almost all of the arachnoidal villi are located; but first the fluid must pass through the subarachnoid spaces of the small *tentorial opening* around the mesencephalon. Finally, the fluid reaches the arachnoidal villi and empties into the venous sinuses.

## CEREBROSPINAL FLUID PRESSURE

The normal pressure in the cerebrospinal fluid system when one is lying in a horizontal position averages 130 mm. water (10 mm. Hg), though this may be as low as 70 mm. water or as high as 180 mm. water even in the normal person. These values are considerably greater than the −6.3 mm. Hg pressure in the interstitial spaces elsewhere in the body.

The cerebrospinal fluid pressure is regulated by the product of, first, the *rate of fluid formation* and, second, the *resistance to absorption through the arachnoidal villi.* When either of these is increased, the pressure rises; and when either is decreased, the pressure falls.

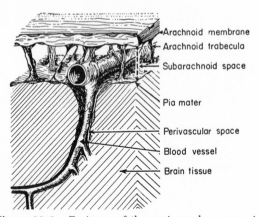

Arachnoid membrane
Arachnoid trabecula
Subarachnoid space
Pia mater
Perivascular space
Blood vessel
Brain tissue

**Figure 32–3.** Drainage of the perivascular spaces into the subarachnoid space. (From Ranson and Clark: Anatomy of the Nervous System.)

**Cerebrospinal Fluid Pressure in Pathologic Conditions of the Brain.** Often a large *brain tumor* elevates the cerebrospinal fluid pressure by decreasing the rate of absorption of fluid. For instance, if the tumor is above the tentorium and becomes so large that it compresses the brain downward, the upward flow of fluid through the tentorial opening may become blocked and the absorption of fluid greatly curtailed. As a result, the cerebrospinal fluid pressure can rise to as high as 500 mm. water (37 mm. Hg) or more. Also, if the tumor is near the surface of the brain, it can cause inflamed meninges, which then exude large quantities of fluid and protein into the cerebrospinal fluid, thereby raising the pressure in this way as well.

The pressure also rises considerably when *hemorrhage* or *infection* occurs in the cranial vault. In both of these conditions, large numbers of cells suddenly appear in the cerebrospinal fluid, and these can almost totally block the small channels for absorption through the arachnoidal villi. This sometimes elevates the cerebrospinal fluid pressure to as high as 400 to 600 mm. water.

Many babies are born with very high cerebrospinal fluid pressure. Most frequently it is caused by excess formation of fluid by the choroid plexuses, but abnormal absorption is also an occasional cause. This is discussed later in connection with *hydrocephalus.*

**Measurement of Cerebrospinal Fluid Pressure.** The usual procedure for measuring cerebrospinal fluid pressure is the following: First, the subject lies exactly horizontally on his side so that the spinal fluid pressure is equal to the pressure in the cranial vault. A spinal needle is then inserted into the lumbar spinal canal below the lower end of the cord and is connected to a glass tube. The spinal fluid is allowed to rise in the tube as high as it will. If it rises to a level 100 mm. above the level of the needle, the pressure is said to be 100 mm. water pressure or, dividing this by 13.6, which is the specific gravity of mercury, about 7.5 mm. Hg pressure.

**Effect of High Cerebrospinal Fluid Pressure on the Optic Disc—Papilledema.** Anatomically the dura of the brain extends as a sheath around the optic nerve and then connects with the sclera of the eye. When the pressure rises in the cerebrospinal fluid system, it also rises in the optic nerve sheath. The retinal artery and vein pierce this sheath a few millimeters behind the eye and then pass with the optic nerve into the eye itself. The high pressure in the optic sheath impedes the flow of blood in the retinal veins, thereby also increasing the retinal capillary pressure throughout the eye, which results in retinal edema. The tissues of the optic disc are much more distensible than those of the remainder of the retina, so that the disc becomes far more edematous than the remainder of the retina and swells into the cavity of the eye. The swelling of the disc, which can be observed with an ophthalmoscope, is called *papilledema,* and neurologists can estimate the cerebrospinal fluid pressure by assessing the extent to which the optic disc protrudes into the eyeball.

## OBSTRUCTION TO THE FLOW OF CEREBROSPINAL FLUID

**Hydrocephalus.** "Hydrocephalus" means excess water in the cranial vault. This condition is frequently divided into *communicating hydrocephalus* and *noncommunicating hydrocephalus.* In communicating hydrocephalus fluid flows readily from the ventricular system into the subarachnoid space, whereas in noncommunicating hydrocephalus fluid flow out of one or more of the ventricles is blocked.

Usually the noncommunicating type of hydrocephalus is caused by a block in the aqueduct of Sylvius so that fluid is formed by the choroid plexuses of the two lateral and the third ventricles, the volumes of these three ventricles increase greatly. This flattens the brain into a thin shell against the skull. In newborn babies the increased pressure also causes the whole head to swell. This type of hydrocephalus can occasionally be treated satisfactorily by making an artificial opening between the ventricular system and the subarachnoid space, such as between the third ventricle and the interpeduncular cistern, or through a plastic tube from one of the ventricles into the cisterna magna.

The communicating type of hydrocephalus is usually caused by overdevelopment of the choroid plexuses in the newborn infant. As a result, far more fluid is formed than can re-enter the venous system through the arachnoidal villi. Fluid therefore collects both inside the ventricles and on the outside of the brain, causing the head to swell tremendously and often damaging the brain severely. To correct this condition the choroid plexuses in both lateral ventricles and sometimes in the third ventricle are destroyed. Thereafter, far less fluid is formed, the cerebrospinal fluid pressure falls, and the hydrocephalus does not become any worse.

**Blockage of Fluid Flow in the Spinal Canal.** Only small quantities of cerebrospinal fluid can be absorbed into the lymphatics and capillaries of the spinal canal. Instead, most of the fluid formed in the spinal regions flows all the way to the arachnoidal granulations in the cranial vault to be absorbed. Consequently, when the spinal canal is blocked by a tumor, a blood clot, or cord transection, the spinal fluid pressure below the block usually rises from the normal value of 130 mm. water to as high as 300 to 500 mm. water, even though the pressure above the block, measured by passing a spinal needle beneath the skull into the cisterna magna, remains normal.

Another important aid for diagnosing spinal canal block is *Queckenstedt's test.* To perform this, the observer presses tightly on both sides of the neck to compress the internal jugular veins. This temporarily impedes the return of blood from the venous system of the brain and causes the venous sinuses in the cranial vault to swell. The swelling sinuses in turn press against the cerebrospinal fluid to increase the cerebrospinal fluid pressure. If the spinal canal is not blocked, the pressure in the lower canal, as measured by a spinal fluid manometer, immediately rises to double or triple normal. On the other hand, if the canal is blocked, the pressure does not change.

### THE BLOOD–CEREBROSPINAL FLUID AND BLOOD-BRAIN BARRIERS

It has already been pointed out that the constituents of the cerebrospinal fluid are not exactly the same as those of the extracellular fluid elsewhere in the body. Furthermore, many large molecular substances hardly pass at all from the blood into the cerebrospinal fluid or into the interstitial fluids of the brain even though these same substances pass readily into the usual interstitial fluids of the body. Therefore, it is said that barriers, called the *blood-cerebrospinal barrier* and *blood-brain barrier,* exist between the blood and the cerebrospinal fluid and brain fluid, respectively. These barriers exist in the choroid plexus and in essentially all areas of the brain parenchyma *except the hypothalamus,* where substances diffuse with ease into the tissue spaces. This ease of diffusion is very important because the hypothalamus responds to many different changes in the body fluids, such as changes in osmolality, glucose concentration, and so forth; these responses provide the signals for feedback regulation of each of the factors.

In general, the blood-cerebrospinal fluid and blood-brain barriers are highly permeable to water, carbon dioxide, and oxygen, slightly permeable to the electrolytes, such as sodium, chloride, and potassium, and almost totally impermeable to substances such as arsenic, sulfur, and gold. Though these latter substances are not of physiological importance, they are important occasionally for certain types of drug therapy; the blood-cerebrospinal fluid and blood-brain barriers often make it impossible to achieve effective concentrations of some such drugs in the cerebrospinal fluid or parenchyma of the brain.

The cause of the low permeability of the blood-cerebrospinal fluid and blood-brain barriers is the manner in which the endothelial cells of the capillaries are joined to each other: They are joined by so-called tight junctions. That is, the membranes of the adjacent endothelial cells are almost fused with each other rather than having slit-pores between them, as is the case in most other capillaries of the body. However, in addition to the tight junctions between the endothelial cells, the capillaries are also surrounded by glial "feet" that abut the outsides of the capillaries. Some physiologists have suggested that these, too, decrease the permeability of the capillaries, but others have demonstrated that there are sufficient openings between the glial feet to allow almost normal diffusion of fluid between the capillary blood and the interstitial fluid in the brain tissue.

**Interstitial Fluid of the Brain.** Research studies have shown that the interstitial fluid in brain tissue is about 12 per cent of the tissue weight, in contrast to 17 per cent elsewhere in the body. The cause of this is lack of collagen fibers between the cells.

# THE INTRAOCULAR FLUID

The eye is filled with intraocular fluid which maintains sufficient pressure in the eyeball to keep it distended. Figure 32–4 illustrates that this fluid can be divided into two portions, the *aqueous humor,* which lies in front and to the sides of the lens, and the *vitreous humor,* which lies between the lens and the retina. The aqueous humor is a freely flowing clear fluid, while the vitreous humor, sometimes called the *vitreous body,* is a gelatinous mass held together by a fine fibrillar network. Substances can *diffuse* slowly in the vitreous humor, but there is little *flow* of fluid.

Aqueous humor is continually being formed and reabsorbed. The balance between formation and reabsorption of aqueous humor regulates the total volume and pressure of the intraocular fluid.

### FORMATION OF AQUEOUS HUMOR BY THE CILIARY BODY

Aqueous humor is formed in the human eye *at an average rate of 2 to 5 cubic millimeters each minute.* Essentially all of this is secreted by the *ciliary processes,* which are linear folds projecting from the *ciliary body* into the space behind the iris where the lens ligaments also attach to the eyeball. A cross-section of these ciliary processes is illustrated in Figure 32–5, and their relationship to the fluid chambers of the eye can be seen in Figure 32–4. Because of their folding architecture, the total surface area of the ciliary processes is approximately 6 sq. cm. in each eye, a large area, considering the small size of the ciliary body. The surfaces of these processes are covered by epithelial cells, and immediately beneath these is a highly vascular area.

Aqueous humor is formed by the ciliary processes in much the same manner that cerebrospinal fluid is formed by the choroid plexus. The theoretical mechanism is the following: It is believed that the ciliary epithelium actively secretes sodium ions into the aqueous humor, and negatively charged chloride

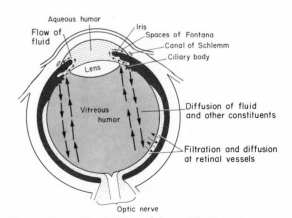

**Figure 32–4.** Formation and flow of fluid in the eye.

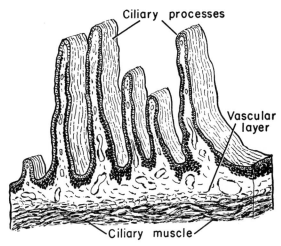

**Figure 32–5.** Anatomy of the ciliary processes.

and bicarbonate ions are probably pulled along with the sodium by the positive charge of the sodium. These substances increase the osmolar activity of the aqueous humor, which in turn causes osmosis of large amounts of water through the ciliary epithelium to the inside of the eye. The reasons for postulating this mechanism are: First, the osmolar activity of the aqueous humor has been measured to be 3 to 6 milliosmols greater than that of plasma, which could exert an osmotic pressure at the ciliary membrane of 50 to 100 mm. Hg, far more than required to promote adequate transfer of water through the membrane. Second, direct studies of the epithelial membrane of the ciliary processes have demonstrated active transport of sodium from the plasma into the aqueous humor, thus providing the necessary osmotic activity to cause osmosis of water. A few other substances are also actively transported, including especially the amino acids.

## DIFFUSION INTO INTRAOCULAR FLUID FROM OTHER STRUCTURES OF THE EYE

In addition to the capillaries of the ciliary body, there are three other capillary systems in the eye: that of the choroid, that of the retina, and that of the iris. Insofar as is known, the capillaries of the retina function purely for nutrition of the retina, similar to capillary function elsewhere in the body. The choroidal capillaries, which have a very high blood flow, also supply nutrition to the outer portions of the rods and cones, but they probably also serve to maintain a constant temperature of the retina in the event of radiant heat or cold entering the pupil of the eye. However, in addition, small amounts of the fluid from both the retinal and the choroidal capillaries diffuse into the aqueous humor.

The capillary system of the iris also is involved principally with nutrition of the iridic structures, but

because of the location of the iris in the very middle of the aqueous humor, the rates of diffusion of substances both into and out of the aqueous humor from the iris are rapid, accounting for as much as 20 to 70 per cent of some of the ions that enter the aqueous humor. In addition, the iris is covered with an epithelium that is capable of phagocytizing proteins and small particulate matter from the aqueous humor, thereby helping to maintain a perfectly clear fluid.

## OUTFLOW OF AQUEOUS HUMOR FROM THE EYE

After aqueous humor is formed by the ciliary processes, it flows, as shown in Figure 32–4, *between the ligaments of the lens,* then *through the pupil,* and finally *into the anterior chamber of the eye.* Here, the fluid flows into the *angle between the cornea and the iris* and thence through a latticework of *trabeculae,* finally into the *canal of Schlemm.* Figure 32–6 illustrates the anatomical structures at the irido-corneal angle, showing that the spaces between the trabeculae extend all the way from the anterior chamber to the canal of Schlemm. The canal of Schlemm in turn is a thin-walled vein that extends circumferentially all the way around the eye. Its endothelial membrane is so porous that even large protein molecules, as well as small particulate matter, can pass from the anterior chamber into the canal of Schlemm. Even though the canal of Schlemm is actually a venous blood vessel, so much aqueous humor normally flows into it that it is filled only with aqueous humor rather than with blood. Also, the small veins that lead from the canal of Schlemm to the larger veins of the eye usually contain only aqueous humor, and these are called *aqueous veins.*

**Removal of Protein from the Aqueous Humor.** It has previously been pointed out that all extravascular fluid chambers of the body must have a

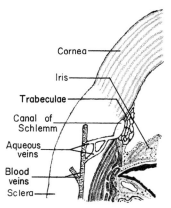

**Figure 32–6.** Anatomy of the irido-corneal angle, showing the system for outflow of aqueous humor into the conjunctival veins.

means for removing the protein that leaks into the fluid from the blood; otherwise, the colloid osmotic pressure will eventually rise high enough to cause abnormal dynamics. In most areas of the body this function is subserved by the lymphatics, but in the eye the extreme permeability of the canal of Schlemm allows the proteins to flow directly into the venous system. This means of protein removal, obviously, is different from the method of protein return in most other structures of the body, but is very much like the return of protein from the cerebrospinal fluid through the arachnoidal villi.

## INTRAOCULAR PRESSURE

The average normal intraocular pressure is approximately 15 mm. Hg, with a range from 10 to 30 mm. Hg.

**Tonometry.** Because it is impractical to pass a needle into a patient's eye for measurement of intraocular pressure, this pressure is measured clinically by means of a tonometer, the principle of which is illustrated in Figure 32–7. The cornea of the eye is anesthetized with a local anesthetic, and the footplate of the tonometer is placed on the cornea. A small force is then applied to a central plunger, causing the central part of the cornea to be displaced inward. The amount of displacement is recorded on the scale of the tonometer, and this in turn is calibrated in terms of intraocular pressure.

**Regulation of Intraocular Pressure.** The intraocular pressure of the normal eye remains almost exactly constant throughout life, illustrating that the pressure regulating mechanism is very effective. But the precise operation of this mechanism is not clear. The pressure is regulated mainly by the outflow resistance from the anterior chamber into the canal of Schlemm, presumably in the following way: The "trabeculae" guarding the entrance of the fluid into the canal of Schlemm are shown in cross-section in Figure 32–6. When studied in three dimensions, these

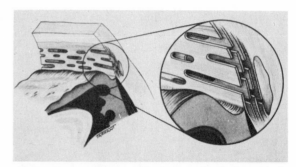

**Figure 32–8.** Perforated laminar plates that overlie the canal of Schlemm. (From Ashton, Brini, and Smith: *Br. J. Ophthalmol.*, 40:257, 1956.)

are actually laminar plates that lie one on top of the other, as illustrated in Figure 32–8. Each of the plates is penetrated by numerous small holes. When the plates are compressed against each other, each successive plate blocks the holes in the next plate. An increase in pressure above normal distends the spaces between the plates and therefore opens the holes, thus causing rapid flow into the canal of Schlemm and decrease in pressure back to normal. On the other hand, a decrease in pressure below normal allows the plates to impinge upon each other, thus preventing fluid loss until the pressure rises again back to normal. Thus, this mechanism acts as an automatic feedback regulatory system for keeping the intraocular pressure at a very constant level day in and day out.

**Glaucoma.** Glaucoma is a disease of the eye in which the intraocular pressure becomes pathologically high, sometimes rising to as high as 70 mm. Hg. Pressures rising above as little as 30 to 40 mm. Hg can cause severe pain and, when maintained for long periods of time, even blindness. As the pressure rises, the retinal artery, which enters the eyeball at the optic disc, is compressed, thus reducing the nutrition to the retina. This often results in permanent atrophy of the retina and optic nerve with consequent blindness.

Glaucoma is one of the most common causes of blindness. Very high pressures lasting only a few days can at times cause total and permanent blindness, but, in cases with only mildly elevated pressures, the blindness may develop progressively over a period of many years.

In essentially all cases of glaucoma the abnormally high pressure results from increased resistance to fluid outflow at the irido-corneal junction. In most patients, the cause of this is unknown, but in others it results from infection or trauma to the eye. In these persons large quantities of red blood cells, white blood cells, and tissue debris collect in the aqueous humor and then flow into the trabecular spaces of the irido-corneal angle where they block the outflow of fluid, thereby greatly increasing the intraocular pressure.

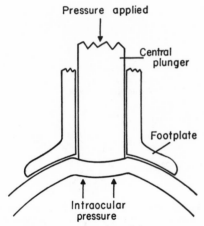

**Figure 32–7.** Principles of the tonometer.

Administration of *Diamox* is usually effective in reducing the pressure in glaucoma. This drug acts by reducing the rate of formation of aqueous humor, possibly because of an effect that reduces sodium secretion.

Sometimes, when all other therapeutic procedures fail to reinstitute adequate outflow of aqueous humor, a *trephine* operation is performed—that is, a small hole is made through the corneoscleral junction into the anterior chamber, a portion of the iris is removed, and the conjunctiva is sewed over the hole. Fluid thereafter leaks from the anterior chamber into the areolar tissue beneath the conjunctiva and is absorbed into the lymphatic channels.

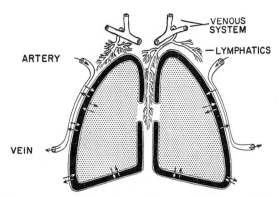

**Figure 32–9.** Dynamics of fluid exchange in the intrapleural spaces.

# FLUID CIRCULATION IN THE POTENTIAL SPACES OF THE BODY

Among the potential spaces of the body are the *pleural cavity,* the *pericardial cavity,* the *peritoneal cavity,* the cavity of the *tunica vaginalis* surrounding the testis, and the *synovial cavities* including both the joint cavities and the bursae. Normally, these are all empty except for a few milliliters of viscous fluid that lubricates movement of the surfaces against each other. But in certain abnormal conditions the spaces can swell to contain tremendous quantities of fluid, which is the reason they are called "potential" spaces.

## FLUID EXCHANGE BETWEEN THE CAPILLARIES AND THE POTENTIAL SPACES

The membrane of a potential space usually does not offer significant resistance to the passage of fluid and electrolytes back and forth between the space and the interstitial fluid of the surrounding tissues. Consequently, fluid leaving a capillary adjacent to the potential space membrane, as shown in Figure 32–9, not only diffuses into the interstitial fluid but also into the potential space. Likewise, fluid can diffuse back out of the space into the interstitial fluid and thence into the capillary.

**Lymphatic Drainage of the Potential Spaces.** In the same manner that protein collects in the interstitial spaces because of leakage out of the capillaries, so does it tend to collect in the potential spaces, and it must be removed through lymphatics or some other channel. Fortunately, each potential space is either directly or indirectly connected with lymphatic drainage systems, which will be discussed in connection with each different type of potential space.

**Function of the Lymphatic Pump in the Drainage of the Potential Spaces.** Almost all the potential spaces are located where there is continual movement: the joints are moved, the bursae are compressed, the heart moves in the pericardial cavity, the lungs move in the pleural cavity with respira-

tion, and the peritoneal spaces are also squeezed with each respiratory cycle or with other body movement. Each movement compresses fluid into the lymphatics surrounding the potential spaces, and, because of the valves in the lymphatics and the pumping mechanisms of the lymphatics, the fluid flows continually away from the space. Therefore, anytime that excess protein collects in the potential space and osmotically pulls any excess fluid whatever into the space, this fluid automatically washes the protein into the lymphatics and from there back into the blood stream. Loss of the protein reduces the osmotic pressure, and the space returns to the collapsed state.

**Altered Capillary Dynamics as a Cause of Increased Pressure and Fluid in the Potential Spaces.** Any abnormal changes that can occur in capillary dynamics to cause extracellular edema of the tissue spaces, as described in the preceding chapter, can also cause increased pressure and fluid in the potential spaces. Thus, *increased capillary permeability, increased capillary pressure, decreased plasma colloid osmotic pressure,* and *blockage of the lymphatics* from a potential space can all cause swelling of the space. The fluid that collects is called a *transudate.*

One of the most common causes of swelling in a potential space is *infection.* White blood cells and other debris caused by the infection block the lymphatics, resulting in (1) buildup of protein in the space, (2) increased colloid osmotic pressure, and (3) consequent failure of fluid reabsorption.

## THE PLEURAL CAVITY

Figure 32–9 shows specifically the diffusion of fluid into and out of the pleural cavity at the parietal and the visceral pleural surfaces. This occurs in precisely the same manner as in the usual tissue spaces, except that a very porous mesenchymal *serous membrane,* the *pleura,* is interspersed between the capillaries and the pleural cavity.

Large numbers of lymphatics drain from the

mediastinal and lateral surfaces of the parietal pleura, and with each expiration the intrapleural pressure rises, forcing small amounts of fluid into the lymphatics; also the respiratory movements alternately compress the lymphatic vessels, promoting continuous flow along the lymphatic channels.

**Maintenance of a Negative Pressure in the Pleural Cavity.** The visceral pleura of the lungs continually absorbs fluid with considerable "absorptive force." This is caused by the low capillary pressure—about 5 to 10 mm. Hg—in the pulmonary system. In contrast to this low pressure, the plasma proteins exert about 28 mm. Hg colloid osmotic pressure, causing an absorption pressure at the visceral pleura of up to 20 mm. Hg at all times. As a result, the pressure of the *fluid* in the intrapleural space remains negative at all times, averaging −8 to −10 mm. Hg. This negative pressure is much greater than the elastic force of the lungs (−4 mm. Hg) that tends to collapse the lungs and to pull the lungs away from the chest wall. Therefore, it keeps the lungs expanded.

### THE PERICARDIAL CAVITY

The space around the heart operates with essentially the same dynamics as those of the pleural cavity. The pressure of the fluid in the pericardial cavity, like that in the pleural cavity, is negative. Here again, during expiration as well as during excessive filling of the heart, the pericardial pressure rises intermittently, forcing excess fluid into lymphatic channels of the mediastinum.

### THE PERITONEAL CAVITY

The peritoneal cavity of the abdomen is subject to the same fluid dynamics as are the other potential spaces; fluid is filtered into the peritoneal space through the serous membrane called the *visceral peritoneum* covering the viscera and the *parietal peritoneum* lining the outer walls of the abdominal cavity. Fluid is also absorbed through the peritoneum.

The peritoneal cavity is more susceptible to the development of excess fluids than are most of the other cavities, for two reasons: (1) Any time the pressure in the liver sinusoids rises more than 5 to 10 mm. Hg, fluid containing large amounts of protein begins to transude through the liver surface into the abdominal cavity, and (2) the capillary pressure in the visceral peritoneum is probably higher than elsewhere in the body; this higher pressure is caused by the resistance to portal blood flow through the liver. High venous pressure caused by heart failure or extra resistance in the liver in pathologic states, such as *cirrhosis, carcinoma,* or *portal vein obstruction,* frequently results in marked transudation of fluid into the abdomen; the transudate in this case is called *ascites.*

Numerous large lymphatic channels lead from the peritoneal cavity, especially from the lower surface of the diaphragm. With each diaphragmatic movement relatively large quantities of lymph flow out of the peritoneal cavity into the thoracic duct. This can be shown effectively by injecting radioactive red blood cells into the abdomen. A large proportion of these cells, still in the whole form, is found in the blood within 10 to 20 minutes. Occasionally, however, cancer spreads so widely throughout the abdomen that it blocks the lymphatics, thereby preventing return of protein to the blood. As a result, the colloid osmotic pressure also rises and severe chronic ascites ensues.

### THE SYNOVIAL CAVITIES

The joint cavities and the bursae are known as *synovial cavities.* The synovial membrane is not a true membrane at all but only a collection of dense fibrous tissue cells that line the surface between the interstitial spaces and the cavities. For this reason these cavities might be considered to be nothing more than enlarged tissue spaces. However, the synovial cavities do contain large amounts of mucopolysaccharides, much more than normally present in the interstitial fluids. The origin of this is not known, though presumably it is secreted by the surrounding connective tissue cells.

In the synovial cavities, as in the other potential spaces, excess proteins are likely to collect, and these must be returned to the circulatory system through the lymphatics; otherwise the space swells. Since the synovial membrane offers little or no barrier to the transfer of fluid into the surrounding tissues, the protein can flow into the lymphatics of the area.

The pressure in joint cavities often measures as low as −8 to −10 mm. Hg. This negative pressure presumably results from the same factors that cause negative pressures throughout most of the interstitial spaces of the body; these factors were discussed in detail in the preceding chapter.

## REFERENCES

Agostoni, E.: Mechanics of the pleural space. *Physiol. Rev.,* 52:57, 1972.
Allen, L., and Weatherford, T.: Role of fenestrated basement membrane in lymphatic absorption from peritoneal cavity. *Amer. J. Physiol.,* 197:551, 1959.
Bell, W. E., and McCormick, W. F.: Increased intracranial pressure in children. *Major Probl. Clin. Pediatr.,* 8:3, 1972.
Bowsher, D.: Cerebrospinal Fluid Dynamics in Health and Disease. Springfield, Illinois, Charles C Thomas, Publisher, 1960.
Cserr, H. F.: Physiology of the choroid plexus. *Physiol. Rev.,* 51:273, 1971.
Davson, H.: The Physiology of the Cerebrospinal Fluid. Boston, Little, Brown and Company, 1967.
Davson, H.: Physiology of the Eye, 3rd Ed. New York, Churchill Livingstone, Div. of Longman, Inc., 1972.
Dhopeshwarkar, G. A., and Mead, J. F.: Uptake and transport of fatty acids into the brain and the role of the blood-brain barrier system. *Adv. Lipid Res.,* 11:109, 1973.

Guyton, A. C., Taylor, A. E., and Granger, H. J.: Circulatory Physiology. II. Dynamics and Control of the Body Fluids. Philadelphia, W. B. Saunders Company, 1975.

Hamerman, D., Barland, P., and Janis, R.: The structure and chemistry of the synovial membrane in health and disease. *In* Bittar, E. E., and Bittar, N. (eds.): The Biological Basis of Medicine. Vol. 3. New York, Academic Press, Inc., 1969, p. 269.

Holman, B. L.: The blood brain barrier: anatomy and physiology. *Prog. Nucl. Med.*, 1:236, 1972.

Katzman, R., and Pappius, H.: Brain Electrolytes and Fluid Metabolism. Baltimore, The Williams & Wilkins Company, 1973.

Milhorat, T. H.: Hydrocephalus and the Cerebrospinal Fluid. Baltimore, The Williams & Wilkins Company, 1972.

Millen, J. W., and Woollam, D. H. M.: The Anatomy of the Cerebrospinal Fluid. New York, Oxford University Press, 1962.

Nicholls, J. G., and Kuffler, S. W.: Extracellular space as a pathway for exchange between blood and neurons in the central nervous system of the leech: ionic composition of glial cells and neurons. *J. Neurophysiol.*, 27:645, 1964.

Oldendorf, W. H.: Blood-brain barrier permeability to drugs. *Ann. Rev. Pharmacol.*, 14:239, 1974.

Pappenheimer, J. R., Heisey, S. R., Jordan, E. F., and Downer, J. D.: Perfusion of the cerebral ventricular system in unanesthetized goats. *Amer. J. Physiol.*, 203:763, 1962.

Podos, S. M.: Glaucoma. *Invest. Ophthalmol.*, 12:3, 1973.

# | 33 |

# Partition of the Body Fluids: Osmotic Equilibria Between Extracellular and Intracellular Fluids

The body fluids are so important to the basic physiology of bodily function that we discuss them at several points in this text but each time in different contexts. In Chapter 4 it was pointed out that the body fluids can be divided mainly into extracellular and intracellular fluids, and the basic differences between these were discussed, as well as the manner in which these differences come about as the result of cell membrane transport. In Chapter 22 the relationship of blood volume to the overall regulation of the circulation was presented, and in Chapters 30 through 32 the interrelationships between capillary dynamics and interstitial fluids were discussed.

In the present chapter and in the following chapters on the kidneys, we will be concerned with the body fluids from the total point of view, including regulation of body fluid volume, regulation of the constituents in the extracellular fluid, regulation of acid-base balance, and factors that govern gross interchange of fluid between the extracellular and the intracellular compartments—especially the osmotic relationships between these compartments. The present chapter will discuss the general distribution of body fluids and their osmotic interrelationships.

## TOTAL BODY WATER

The total amount of water in a man of average weight (70 kg.) is approximately 40 liters (see Fig. 33–1), averaging 57 per cent of his total body weight. In a newborn infant this may be as high as 75 per cent of the body weight, but it progressively decreases from birth to old age, most of the decrease occurring in the first 10 years of life. Also, obesity decreases the percentage of water in the body, sometimes down to as low as 45 per cent.

## INTAKE VERSUS OUTPUT OF WATER

**Daily Intake of Water.** Most of our daily intake of water enters by the oral route. Approximately two-thirds is in the form of pure water or some other beverage, and the remainder is in the food that is eaten. A small amount is also synthesized in the body as the result of oxidation of hydrogen in the food; this quantity ranges between 150 and 250 ml. per day, depending on the rate of metabolism. The normal intake of fluid, including that synthesized in the body, averages about 2400 ml. per day.

**Daily Loss of Body Water.** Table 33–1 shows the routes by which water is lost from the body under different conditions. Normally, at an atmospheric temperature of about 68 degrees, approximately 1400 ml. of the 2400 ml. of water intake is lost in the *urine*, 100 ml. is lost in the *sweat*, and 200 ml. in the *feces*. The remaining 700 ml. is lost by *evaporation through the lungs* or by *diffusion through the skin*.

*Insensible Water Loss.* Loss of water by diffusion through the skin and by evaporation from the lungs is known as *insensible water loss* because the person does not know that he is actually losing water at the time that it is leaving the body.

424

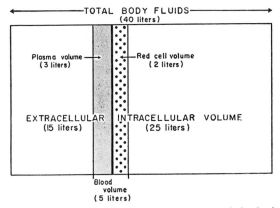

**Figure 33–1.** Diagrammatic representation of the body fluids, showing the extracellular fluid volume, intracellular fluid volume, blood volume, and total body fluids.

The average loss of water by diffusion through the skin is approximately 300 to 400 ml. per day; this amount is lost even in a person who is born with congenital lack of sweat glands. In other words, the water molecules actually diffuse through the cells of the skin themselves. Fortunately, the cholesterol-filled, cornified layer of the skin acts as a protector against still much greater loss of water by diffusion. However, when the cornified layer becomes denuded, such as after extensive burns, the rate of evaporation can increase to as much as 3 to 5 liters each day.

All air that enters the lungs becomes totally saturated with moisture, to a vapor pressure of approximately 47 mm. Hg, before it is expelled. Since the vapor pressure of the inspired atmospheric air is usually far less than 47 mm. Hg, the average water loss through the lungs is about 300 to 400 ml. per day. The atmospheric vapor pressure normally decreases with decreasing temperature so that the loss is greatest in very cold and least in very warm weather. This explains the dry feeling in the respiratory passages in cold weather.

**TABLE 33–1.  Daily Loss of Water**
**(in Milliliters)**

|  | Normal Temperature | Hot Weather | Prolonged Heavy Exercise |
|---|---|---|---|
| Insensible Loss: |  |  |  |
| Skin | 350 | 350 | 350 |
| Lungs | 350 | 250 | 650 |
| Urine | 1400 | 1200 | 500 |
| Sweat | 100 | 1400 | 5000 |
| Feces | 200 | 200 | 200 |
| Total | 2400 | 3400 | 6700 |

*Loss of Water in Hot Weather and During Exercise.* In very hot weather, water loss in the sweat is occasionally increased to as much as 3.5 liters an hour, which obviously can rapidly deplete the body fluids. Sweating will be discussed in Chapter 72.

Exercise increases the loss of water in two ways: First, it increases the rate of respiration, which promotes increased water loss through the lungs in proportion to the increased ventilatory rate. Second, and much more important, exercise increases the body heat and consequently is likely to result in excessive sweating.

# BODY FLUID COMPARTMENTS

## THE INTRACELLULAR COMPARTMENT

About 25 of the 40 liters of fluid in the body are inside the approximately 75 trillion cells of the body and are collectively called the *intracellular fluid.* The fluid of each cell contains its own individual mixture of different constituents, but the concentrations of these constituents are reasonably similar from one cell to another. For this reason the intracellular fluid of all the different cells is considered to be one large fluid compartment, though in reality it is an aggregate of trillions of minute compartments.

## THE EXTRACELLULAR FLUID COMPARTMENT

All the fluids outside the cells are called *extracellular fluid,* and these fluids are constantly mixing, as was explained in Chapter 1. The total amount of fluid in the extracellular compartment averages 15 liters in a 70 kg. adult.

The extracellular fluid can be divided into *interstitial fluid, plasma, cerebrospinal fluid, intraocular fluid, fluids of the gastrointestinal tract,* and *fluids of the potential spaces.*

**Interstitial Fluid.** The interstitial fluid lies in the spaces between the cells. A minute portion of it is free in the form of actual flowing fluid, while probably more than 99 per cent of it is held in the gel of the interstitial spaces, as discussed in Chapter 31. The gel prevents all but the most minute *flow* of fluid through the tissue spaces. Yet, dissolved substances can still move through the spaces in great quantity by the process of *diffusion,* which was explained in detail in Chapters 4 and 30.

**Plasma.** The plasma is the noncellular portion of blood. It is part of the extracellular fluid and communicates continually with the interstitial fluid through pores in the capillaries. Loss of plasma from the circulatory system through the capillary pores is minimized by the colloid osmotic pressure exerted by the plasma proteins, which was explained in Chapter 30. Yet the capillaries are porous enough for most dissolved substances and water molecules to *diffuse* through them freely, allowing constant mixing between the plasma and interstitial fluid of almost all substances except the protein.

The plasma volume averages 3 liters in the normal adult.

**Fluid in Other Extracellular Spaces.** The *cerebrospinal fluid* comprises all the fluid in the ventricles of the brain and in the subarachnoid spaces surrounding both the brain and spinal cord. This fluid has slightly different constituents from the interstitial fluid and plasma because of somewhat restricted diffusion back and forth between it and plasma and because of active secretion of a few substances by the choroid plexus. Nevertheless, from the point of view of the present discussion, cerebrospinal fluid is so nearly like interstitial fluid that it is considered to be actually a part of it. The special dynamics of the cerebrospinal fluid were presented in detail in Chapter 32.

The *intraocular fluid,* the fluid in the eyes, has properties similar to those of the cerebrospinal fluid, and here again the fluid is a product of both diffusion and secretion. These fluids, too, were discussed in detail in Chapter 32, and, for the purpose of the present discussion, are considered to be part of the interstitial fluid.

Many spaces exist in the body that normally contain little fluid but under special conditions can become filled with large amounts. These are called *potential spaces*. An example of a potential space is the space between the visceral and parietal pleurae of the lungs. Normally only 10 to 15 ml. of very viscid fluid is present in this space, but under abnormal circumstances the amount can become as great as several liters. Other potential spaces are the *peritoneal cavity,* the *pericardial cavity,* all the *joint spaces,* and the *bursae.* The fluids of these potential spaces communicate freely with the surrounding interstitial fluids and therefore are considered to be part of the interstitial fluid.

Finally, moderate amounts of extracellular fluid are normally inside the *gastrointestinal tract.* The quantity of these fluids varies greatly at different times of the day in relation to the intake and digestion of food. As much as a liter of digestive juices is sometimes in the gastrointestinal tract at once, and under certain pathologic conditions, such as gastrointestinal obstruction, this can become as much as 10 liters. The gastrointestinal fluids, except for a few of the glandular secretions, have electrolyte compositions similar to those of the interstitial fluid, and these fluids are usually considered to be part of the extracellular fluid.

## BLOOD VOLUME

Blood contains both extracellular fluid (the fluid of the plasma) and intracellular fluid (the fluid in the red blood cells). However, since blood is contained in a closed chamber all its own—the circulatory system—its volume and its special dynamics are exceedingly important.

The average blood volume of a normal adult is almost exactly 5000 ml. On the average, approximately 3000 ml. of this is plasma, and the remainder, 2000 ml., is red blood cells. However, these values vary greatly in different individuals; also, sex, weight, and many other factors affect the blood volume.

**Effect of Weight and Sex on Blood Volume.** In persons who have a minimum of adipose tissue the blood volume varies almost directly in proportion to the body weight, normally averaging about 79 ml./kg. ±10% for both *lean* males and *lean* females. However, the greater the obesity, the less the blood volume per unit weight, because fat tissue has little vascular volume. Figure 33–2 illustrates this effect of obesity, showing that the heavier the person becomes, the less on the average becomes the blood volume in relation to weight. Furthermore, in very heavy females, who are especially prone to have high ratios of fat tissue to lean tissue, the blood volume per kilogram decreases much more than for males. The average female, because of a far greater fat-to-lean tissue ratio than that of the average male, has a blood volume per unit weight about 20 per cent below that of the average male.

**The Hematocrit.** The hematocrit is the percentage of red blood cells in the blood as determined by centrifuging blood in "hematocrit tubes" until the cells become packed tightly in the bottoms of the tubes. The percentage of red blood cells in the blood can be determined roughly from the levels of the packed cells. Such centrifuged blood in hematocrit tubes was illustrated in Figure 18–2 of Chapter 18. Unfortunately, it is impossible for the red blood cells to be packed completely together; about 3 to 8

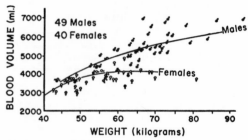

**Figure 33–2.** Relationship of blood volume to sex and weight (Redrawn from Gibson and Evans: *J. Clin. Invest.,* 16:317, 1937.)

per cent plasma remains entrapped among the cells. Therefore, the true cell percentage (H) averages about 96 per cent of the measured hematocrit (Hct); that is:

$$H = 0.96 \text{ Hct}$$

The normal hematocrit (H) is approximately 40 for a man and 36 for a woman.

In severe anemia, the hematocrit may fall to as low as 10, but this small quantity of red blood cells is barely sufficient to sustain life. On the other hand, a few conditions cause excessive production of red blood cells, resulting in polycythemia. In these instances, the hematocrit often rises to 65, and occasionally to 80. Obviously, there is an upper limit to the level of the hematocrit in polycythemic blood because excessive hematocrit causes the blood to become so viscous that death ensues as a result of multiple plugging of the peripheral vascular tree.

*The Body Hematocrit.* The hematocrit of the blood in the capillaries, arterioles, and other very small vessels of the body is considerably less than that in the large veins and arteries. The cause of this is *axial streaming* of blood cells in blood vessels, which was explained in Chapter 18. In general, red cells cannot flow near the walls of the vessels nearly so easily as can plasma. Therefore, the cells tend to migrate to the center of the vessels, while a large portion of the plasma remains near the walls. In the very large vessels the ratio of wall surface to total volume is slight, so that the accumulation of plasma near the walls does not affect the hematocrit significantly. However, in the small vessels this ratio of wall surface to volume is great, causing the ratio of plasma to cells to be far greater than in the large vessels.

If one averages the hematocrit in both the large and small vessels, he determines a value called *body hematocrit*. For normal man the body hematocrit ($H_0$) averages 91 per cent of the large vessel hematocrit; that is:

$$H_0 = 0.91 \text{ H} = 0.87 \text{ Hct}$$

# MEASUREMENT OF BODY FLUID VOLUMES

## THE DILUTION PRINCIPLE FOR MEASURING FLUID VOLUMES

The volume of any fluid compartment of the body can be measured by placing a substance in the compartment, allowing it to disperse evenly throughout the fluid, and then measuring the extent to which the substance becomes diluted.

Figure 33–3 illustrates this "dilution" principle for measuring the volume of any fluid compartment of the body. A small quantity of dye or other foreign substance is placed in fluid chamber A, and the substance is allowed to disperse throughout the chamber until it becomes mixed in equal concentrations in all areas, as shown in Chamber B. Then a sample of the dispersed fluid is removed and the concentration of the substance is analyzed chemically, photoelectrically, or by any other means. The volume of the chamber can then be determined from the following formula:

Volume in ml.

$$= \frac{\text{Quantity of test substance instilled}}{\text{Concentration per ml. of dispersed fluid}}$$

Note that all one needs to know is (1) the *total quantity of the test substance* put into the chamber and (2) the *concentration in the fluid after dispersement.*

## DETERMINATION OF BLOOD VOLUME

**Substances Used in Determining Blood Volume.** A substance used for measuring blood volume must be capable of dispersing throughout the blood with ease, and it must remain in the circulatory system long enough for measurements to be made. The two major groups of substances that satisfy these conditions for measurement of blood volume are substances that combine with the red blood cells or substances that combine with the plasma proteins, for both the red blood cells and the plasma proteins remain reasonably well in the circulatory system, and any foreign substance

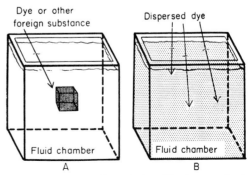

**Figure 33–3.** Principles of the dilution method for measuring fluid volumes (explained in the text).

that combines with either of them likewise remains in the blood stream.

Substances that combine with red blood cells and that are used for determining blood volume are *radioactive iron, radioactive chromium,* and *radioactive phosphate.* Substances that combine with plasma proteins are the *vital dyes* and *radioactive iodine.*

**Radioactive Red Blood Cells.** The method most often used to make red blood cells radioactive is to tag the red blood cells with radioactive chromium ($Cr^{51}$). A small quantity of $Cr^{51}$ is mixed with a few milliliters of blood removed from the person, and this is incubated at 36° C. for half an hour or more. After this time, most of the $Cr^{51}$ will have entered the red blood cells, but to remove the extra chromium from the mixture, the red blood cells are washed in saline. Their total content of $Cr^{51}$ is then determined with a Geiger or scintillation counter (apparatuses for measuring the total number of radioactive disintegrations occurring in the sample per minute). Then the radioactive cells are reinjected into the person. After mixing in the circulatory system has continued for approximately 10 minutes, blood is removed from the circulatory system, and the radioactivity in this blood is determined. Using the above dilution formula, the total blood volume is calculated.

To be accurate, this calculated blood volume must now be corrected to determine the true value, because the blood sample is removed from the veins where the hematocrit is not equal to the body hematocrit, as discussed earlier in the chapter. This correction is made as follows:

Actual blood volume
$$= 1.1 \times \text{Measured blood volume}$$

**Vital Dyes for Measurement of Plasma Volume.** A number of dyes, generally known as "vital dyes," have the ability to combine with proteins. When such a dye is injected into the blood, it immediately forms a tight union with the plasma proteins. Thereafter, the dye travels where the proteins travel.

The dye almost universally used for measuring plasma volume is *T-1824,* also called *Evans blue.* In making determinations of plasma volume, a known quantity of the dye is injected, and it immediately combines with the proteins and disperses throughout the circulatory system within approximately 10 minutes. A sample of the blood is then taken, and the red blood cells are removed from the plasma by centrifugation. Then, by spectrophotometric analysis of the plasma, one can determine the exact quantity of dye in the sample of plasma. From the determined quantity of dye in each milliliter of plasma and the known quantity of dye injected, the *plasma volume* is calculated using the dilution formula noted above.

To be even more exact in measuring the plasma volume, the rate of loss of dye from the circulatory system during the interval of mixing must also be considered. On the average, T-1824 is lost from the circulatory system at a rate of about 5 per cent per hour, part of this being excreted into the urine and part being carried into the interstitial spaces by the leakage of plasma proteins through the capillary walls. Under abnormal conditions the rate of loss can be as great as 20 to 50 per cent per hour, which can cause considerable error in the measurement of plasma volume. To offset this error, three different samples of blood are usually removed from the circulation at 10, 20, and 30 minutes. Then the plasma volumes are calculated from these samples and plotted on semilog graph paper, as illustrated in Figure 33–4. A straight line is drawn through the measured points and extrapolated back to zero time. Since the percentage loss is approximately the same during the first 10 minutes as during each of the two succeeding 10-minute intervals, the extrapolated point indicates approximately the true plasma volume. In Figure 33–4 this value is 2500 ml.

Note that neither T-1824 nor any other vital dye enters the red blood cells. Therefore, this method *does not measure the total blood volume.* However, the blood volume can be calculated from the plasma volume, provided that the hematocrit is determined by using the following formula:

Blood volume
$$= \text{Plasma volume} \times \frac{100}{100 - 0.87 \text{ Hematocrit}}$$

**Radioactive Protein.** If a sample of plasma is allowed to incubate with radioactive iodine ($I^{131}$) for 30 minutes or more, some of the protein combines with the iodine, and the iodinated protein can be separated from the remaining iodine by dialysis. The radioactive protein is then injected into the subject, and plasma and blood volumes are determined in the same manner as that discussed for the vital dyes.

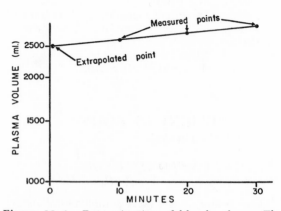

**Figure 33–4.** Determination of blood volume. The measured points represent plasma volumes calculated from blood samples removed 10, 20, and 30 minutes after injection of T-1824. The extrapolated point represents the calculated blood volume at the instant of dye injection.

## MEASUREMENT OF THE EXTRACELLULAR FLUID VOLUME

To use the dilution principle for measuring the volume of the extracellular fluid one injects into the blood stream a substance that can diffuse readily throughout the entire extracellular fluid chamber, passing easily through the capillary membranes but as little as possible through the cell membranes into the cells. After half an hour or more of mixing, a sample of extracellular fluid is obtained by removing blood and separating the plasma from the cells by centrifugation. The plasma, which is actually a part of the extracellular fluid, is then analyzed for the injected substance.

**Substances Used in Measuring Extracellular Fluid Volume—The Concept of "Fluid Space."** Substances that have been used for measuring extracellular fluid volume are *radioactive sodium, radioactive chloride, radioactive bromide, thiosulfate ion, thiocyanate ion, inulin,* and *sucrose.* Some of these, sucrose and inulin especially, do not diffuse readily into all the out-of-the-way places of the extracellular fluid compartment. Therefore, the volume of extracellular fluid measured with these is likely to be lower than the actual volume of the compartment. On the other hand, others of these substances—radioactive chloride, radioactive bromide, radioactive sodium, and thiocyanate ion, for instance—are likely to penetrate into the cells to a slight extent and, therefore, are likely to measure a space somewhat in excess of the extracellular fluid volume.

Because there is no single substance that measures the exact extracellular fluid volume, one usually speaks of the *sodium space,* the *thiocyanate space,* the *inulin space,* and so forth, rather than the extracellular volume. At present it is impossible to say exactly which "space" most nearly approximates the true extracellular volume. Measurements for the normal 70 kg. adult, when different ones of the above substances have been used, have ranged from 8 liters to 22 liters; but the average measurement has been about 15 liters.

**Correction for Loss of the Test Substance.** In measuring the extracellular fluid volume of a normal person, at least 30 minutes is required for almost complete dispersion of the test substance throughout the extracellular compartment, and, if major amounts of the substance are lost from the compartment during this time, an error will be made in the measurement. For this reason, measurements employing each of the different test substances require different systems of correction but they generally employ an extrapolation back to zero time as explained above for plasma volume determination.

## MEASUREMENT OF TOTAL BODY WATER

The total body water can be measured in exactly the same way as the extracellular fluid volume except that a substance must be used that will diffuse into the cells as well as throughout the extracellular compartment. The substances that give best results are either *tritiated water* (which can be analyzed with radiation-measuring instruments) or *heavy water* (which can be analyzed quantitatively either by accurate specific gravity measurements of water samples or by infrared spectrophotometry). After administration of the "tagged" water, several hours are required for complete mixing with all the water of the body, and appropriate corrections must be made for any fluid that is lost either into the urine or otherwise during this period of mixing. The concentration of the tagged water in the total body water can, at the end of the period of mixing, be determined by simply measuring the tagged water concentration in the plasma.

Another substance that has proved satisfactory for measuring total body water is *antipyrine,* which diffuses almost uniformly into all cells of the body and which can be analyzed readily by chemical means.

Measurements of total body water in the 70 kg. adult have ranged from as low as 30 liters up to as high as 50 liters, with a reasonable average of approximately 40 liters, or 57 per cent of the total body mass.

## CALCULATION OF INTERSTITIAL FLUID VOLUME

Since any substance that passes into the interstitial fluid also passes into almost all other portions of the extracellular fluid, there is no direct method for measuring interstitial fluid volume separately from the entire extracellular fluid volume. However, if the extracellular fluid volume and plasma volume have both been measured, the interstitial fluid volume can be approximated by *subtracting the plasma volume from the total extracellular fluid volume.* This calculation gives a normal interstitial fluid volume of 12 liters in the 70 kg. adult.

# CONSTITUENTS OF EXTRACELLULAR AND INTRACELLULAR FLUIDS

Figure 33–5 illustrates diagrammatically the major constituents of the extracellular and intracellular fluids. (The actual values for these constituents were given in tabular form in Fig. 4–1 of Chap. 4.) The quantities of the different substances are represented in Figure 33–5 in *milliequivalents* or *millimols per liter.* However, the protein molecules and some of the nonelectrolyte molecules are extremely large in comparison with the more numerous small ions. Therefore, *in terms of mass,* the proteins and nonelectrolytes actually comprise about 90 per cent of the dissolved constituents in the plasma, about 60 per cent of those in the interstitial fluid, and about 97 per cent of those in the intracellular fluid.

Figure 33–6 illustrates the distribution of the nonelectrolytes in the plasma, and most of these same substances are also present in almost equal concentrations in the interstitial fluid, except that some of the fatty compounds exist in

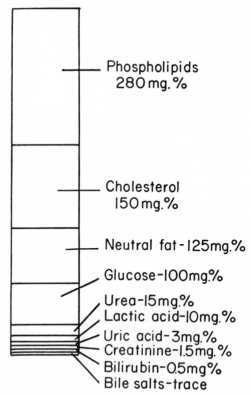

**Figure 33–6.** The nonelectrolytes of the extracellular fluid.

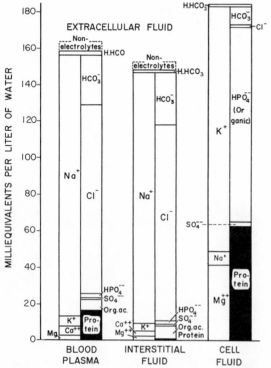

**Figure 33–5.** The compositions of plasma, interstitial fluid, and intracellular fluid. (Modified from Gamble: Chemical Anatomy, Physiology, and Pathology of Extracellular Fluid: A Lecture Syllabus. Harvard University Press, 1954.)

the blood in large suspended particles, the *lipoproteins,* and, therefore, do not pass to any significant extent into the interstitial spaces.

From the point of view of most chemical and physical reactions of the fluids, it is usually the *concentration of molecules or ions* of a particular substance that is important. Therefore, in this text all concentrations are generally expressed as presented in Figure 33–5; that is, either in millimols of nonelectrolytes or milliequivalents of electrolytes.

**Constituents of the Extracellular Fluid.** Referring again to Figure 33–5, one sees that extracellular fluid, both that of the blood plasma and of the interstitial fluid, contains large quantities of *sodium* and *chloride ions,* reasonably large quantities of *bicarbonate ion,* but only small quantities of potassium, calcium, magnesium, phosphate, sulfate, and organic acid ions. In addition, plasma contains a large amount of protein while interstitial fluid contains much less. (The proteins in these fluids and their significance were discussed in detail in Chap. 30.)

In Chapter 1 it was pointed out that the extracellular fluid is called the *internal environment* of the body and that its constituents are accurately regulated so that the cells remain bathed continually in a fluid containing the proper electrolytes and nutrients for continued cellular function. The regulation of most of these constituents will be presented in Chapter 36.

**Constituents of the Intracellular Fluid.** From Figure 33–5 it is also readily apparent that the intracellular fluid contains only small quantities of sodium and chloride ions and almost no calcium ions; but it does contain large quantities of *potassium* and *phosphate* and moderate quantities of *magnesium* and *sulfate ions,* all of which are present in only small concentrations in the extracellular fluid. In addition, cells contain large amounts of protein, approximately four times as much as the plasma.

# OSMOTIC EQUILIBRIA AND FLUID SHIFTS BETWEEN THE EXTRACELLULAR AND INTRACELLULAR FLUIDS

One of the most troublesome of all problems in clinical medicine is maintenance of adequate body fluids and proper balance between the extracellular and intracellular fluid volumes in seriously ill patients. The purpose of the following discussion, therefore, is to explain the interrelationships between extracellular and intracellular fluid volumes and the osmotic factors that cause shifts of fluid between the extracellular and intracellular compartments.

## BASIC PRINCIPLES OF OSMOSIS AND OSMOTIC PRESSURE

The basic principles of osmosis and osmotic pressure were presented in Chapter 4. However, these principles are so important to the following discussion that they are reviewed briefly here.

Whenever a membrane between two fluid compartments is permeable to water but not to some of the dissolved solutes (this is called a *semipermeable membrane*) and the concentration of nondiffusible substances is greater on one side of the membrane than on the other, water passes through the membrane toward the side with the greater concentration of nondiffusible substances. This phenomenon is called *osmosis*.

Osmosis results from the kinetic motion of the molecules in the solutions on the two sides of the membrane and can be explained in the following way: The individual molecules on both sides of the membrane are equally active because the temperature, which is a measure of the kinetic activity of the molecules, is the same on both sides. However, the nondiffusible solute on one side of the membrane displaces some of the water molecules, thereby reducing the concentration of water molecules. As a result, the so-called *total chemical activity* of water molecules on this side is less than on the other side, so that fewer water molecules strike each pore of the membrane each second on the solute side of the pore than on the pure water side, resulting in net diffusion of water molecules from the water side to the solute side. This net rate of diffusion is the *rate of osmosis.*

Osmosis of water molecules can be opposed by applying a pressure across the semipermeable membrane in the direction opposite to that of the osmosis. The amount of pressure required exactly to oppose the osmosis is called the *osmotic pressure.*

**Relationship of the Molecular Concentration of a Solution to Its Osmotic Pressure.** Each nondiffusible molecule dissolved in water dilutes the "activity" of the water molecules by a given amount. Consequently, the tendency for the water in the solution to diffuse through a membrane is reduced in direct proportion to the concentration of nondiffusible molecules. And, as a corollary, the osmotic pressure of the solution is also proportional to the concentration of nondiffusible molecules in the solution. This relationship holds true for all nondiffusible molecules almost regardless of their molecular weight. For instance, one molecule of albumin with a molecular weight of 70,000 has the same osmotic effect as a molecule of glucose with a molecular weight of 180.

**Osmotic Effect of Ions.** Nondiffusible ions cause osmosis and osmotic pressure in exactly the same manner as do nondiffusible molecules. Furthermore, when a molecule dissociates into two or more ions, each of the ions then exerts osmotic pressure individually. Therefore, to determine the osmotic effect, all the nondiffusible ions must be added to all the nondiffusible molecules; but note that a bivalent ion, such as calcium, exerts no more osmotic pressure than does a univalent ion, such as sodium.

**Osmols.** The ability of solutes to cause osmosis and osmotic pressure is measured in terms of "osmols;" the osmol is a measure of the total number of particles. *One gram mol of nondiffusible and nonionizable substance is equal to 1 osmol.* On the other hand, if a substance ionizes into two ions (sodium chloride into sodium and chloride ions, for instance), then 0.5 gram mol of the substance would equal 1 osmol. The obvious reason for using the osmol is that osmotic pressure is determined by the number of particles instead of the mass of the solute.

In general, the osmol is too large a unit for satisfactory use in expressing osmotic activity of solutes in the body. Therefore, the term *milliosmol,* which equals $^1/_{1000}$ osmol, is commonly used.

**Osmolality and Osmolarity.** The osmolal concentration of a solution is called its *osmolality* when the concentration is expressed in osmols per kilogram of water; it is called *osmolarity* when it is expressed as osmols per liter of solution. The term "osmolality" is generally preferred because the osmotic pressure of a solution is considerably more closely related to its osmolality than to its osmolarity in very concentrated solutions. However, in the very dilute solutions of the normal human body, the differences are so slight that the terms are frequently used interchangeably. Furthermore, it is so much easier to express the body fluid quantities in liters than in kilograms of water that almost all calculations are based on osmolarities rather than on osmolalities. It is common practice for many physiologists to speak in terms of osmolality even though they make their calculations in terms of osmolarities.

**Relationship of Osmotic Pressure to Osmolarity.** The osmotic pressure of a solution *at body temperature* can be determined approximately from the following formula:

Osmotic pressure (mm. Hg)
$$= 19.3 \times \text{Osmolarity (milliosmol/liter)}$$

## OSMOLALITY OF THE BODY FLUIDS

Table 33–2 lists the osmotically active substances in plasma, interstitial fluid, and intracellular fluid. The milliosmols of each of these per liter of water is given. Note especially that approximately four-fifths of the total osmolality of the interstitial fluid and plasma is caused by sodium and chloride ions, while approximately half of the intracellular osmolality is caused by potassium ions, the remainder being divided among the many other intracellular substances.

**TABLE 33–2. Osmolar Substances in Extracellular and Intracellular Fluids**

|  | Plasma (mOsmol./L. of $H_2O$) | Interstitial (mOsmol./L. of $H_2O$) | Intracellular (mOsmol./L. of $H_2O$) |
|---|---|---|---|
| $Na^+$ | 144 | 137 | 10 |
| $K^+$ | 5 | 4.7 | 141 |
| $Ca^{++}$ | 2.5 | 2.4 | 0 |
| $Mg^{++}$ | 1.5 | 1.4 | 31 |
| $Cl^-$ | 107 | 112.7 | 4 |
| $HCO_3^-$ | 27 | 28.3 | 10 |
| $HPO_4^{--}$ $H_2PO_4^-$ | 2 | 2 | 11 |
| $SO_4$ | 0.5 | 0.5 | 1 |
| Phosphocreatine |  |  | 45 |
| Carnosine |  |  | 14 |
| Amino acids | 2 | 2 | 8 |
| Creatine | 0.2 | 0.2 | 9 |
| Lactate | 1.2 | 1.2 | 1.5 |
| Adenosine triphosphate |  |  | 5 |
| Hexose monophosphate |  |  | 3.7 |
| Glucose | 5.6 | 5.6 |  |
| Protein | 1.2 | 0.2 | 4 |
| Urea | 4 | 4 | 4 |
| TOTAL mOsmol. | 303.7 | 302.2 | 302.2 |
| Corrected osmolar activity (mOsmol.) | 282.6 | 281.3 | 281.3 |
| Total osmotic pressure at 37° C. (mm. Hg) | 5453 | 5430 | 5430 |

As noted at the bottom of Table 33–2, the total osmolality of each of the three compartments is approximately 300 milliosmols per liter, with that of the plasma 1.3 milliosmols greater than that of the interstitial and intracellular fluids. This slight difference between plasma and interstitial fluid is caused by the osmotic effect of the plasma proteins, which maintains about 23 mm. Hg greater pressure in the capillaries than in the surrounding interstitial fluid spaces, as was explained in Chapter 30.

**Corrected Osmolar Activity of the Body Fluids.**    At the bottom of Table 33–2 is shown a corrected osmolar activity of plasma, interstitial fluid, and intracellular fluid. The reason for this correction is the following: All molecules and ions in solution exert either *intermolecular attraction* or *intermolecular repulsion,* and these two effects can cause, respectively, a decrease or an increase in the osmotic "activity" of the dissolved substance. In general, there is more intermolecular attraction than repulsion, so that the overall osmotic activity of the substances is only about 93 per cent of that which one would calculate from the number of milliosmols present. For this reason, the actual osmotic pressure of the body fluids is proportional to the corrected osmolar activity, which amounts to approximately 280 milliosomols/liter.

*However, since osmotic effects in the body are usually determined by the relative, rather than the absolute, osmolar concentrations, the correction factor is often ignored.*

*Total Osmotic Pressure Exerted by the Body Fluids.*    At the bottom of Table 33–2 is shown the total osmotic pressure in mm. Hg that would be exerted by each of the different fluids if it were placed on one side of a cell membrane with pure water on the other side. Note that this total pressure averages about 5450 mm. Hg and also that the osmotic pressure of plasma is 23 mm. Hg greater than that of the interstitial fluids, this difference equaling the approximate hydrostatic pressure difference between the pressure of the blood inside the capillaries and the negative pressure in the interstitial fluid outside the capillaries.

## MAINTENANCE OF OSMOTIC EQUILIBRIUM BETWEEN EXTRACELLULAR AND INTRACELLULAR FLUIDS

The tremendous osmotic pressure that can develop across the cell membrane when one

side is exposed to pure water—about 5400 mm. Hg—illustrates how much force can become available to push water molecules through the membrane when the solutions on the two sides of the membrane are not in osmotic equilibrium. For instance, in Figure 33–7A, a cell is placed in a solution that has an osmolality far less than that of the intracellular fluid. As a result, osmosis of water begins immediately from the extracellular fluid to the intracellular fluid, causing the cell to swell and diluting the intracellular fluid while concentrating the extracellular fluid. When the fluid inside the cell becomes diluted sufficiently to equal the osmolal concentration of the fluid on the outside, further osmosis then ceases. This condition is shown in Figure 33–7B. In Figure 33–7C, a cell is placed in a solution having a much higher concentration outside the cell than inside. This time, water passes by osmosis to the exterior, diluting the extracellular fluid while concentrating the intracellular fluid. In this process the cell shrinks until the two concentrations become equal, as shown in Figure 33–7D.

**Rapidity of Attaining Extracellular and Intracellular Osmotic Equilibrium.**    The transfer of water through the cell membrane by osmosis occurs so rapidly that any lack of osmotic equilibrium between the two fluid compartments in any given tissue is usually corrected within a few seconds and at most within a min-

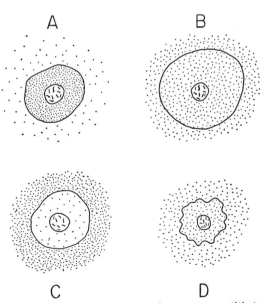

**Figure 33–7.**  Establishment of osmotic equilibrium when cells are placed in a hypo- or hypertonic solution.

ute or so. However, this rapid transfer of water does not mean that complete equilibration occurs between the extracellular and intracellular compartments throughout the whole body within this same short period of time. The reason for this is that fluid usually enters the body through the gut and must then be transported by the blood to all tissues before complete equilibration can occur. In the normal person it may take as long as 30 minutes to an hour to achieve reasonably good equilibration everywhere in the body after drinking water.

**Isotonicity, Hypotonicity, and Hypertonicity.** A fluid into which normal body cells can be placed without causing either swelling or shrinkage of the cells is said to be *isotonic* with the cells. A 0.9 per cent solution of sodium chloride or a 5 per cent glucose solution is approximately isotonic.

A solution that will cause the cells to swell is said to be *hypotonic;* any solution of sodium chloride with less than 0.9 per cent concentration is hypotonic.

A solution that will cause the cells to shrink is said to be *hypertonic;* sodium chloride solutions of greater than 0.9 per cent concentration are all hypertonic.

# CHANGES IN THE VOLUMES AND OSMOLALITIES OF THE EXTRACELLULAR AND INTRACELLULAR FLUID COMPARTMENTS IN ABNORMAL STATES

## CALCULATION OF FLUID SHIFTS BETWEEN THE EXTRACELLULAR AND INTRACELLULAR FLUID COMPARTMENTS

Among the different factors that can cause extracellular or intracellular volumes to change markedly are ingestion of water, dehydration, intravenous infusion of different types of solutions, loss of large quantities of fluids from the gastrointestinal tract, or loss of abnormal quantities of fluid by sweating or through the kidneys.

The changes in both extracellular and intracellular fluid volumes can be calculated easily if one will keep the following two basic principles in mind:

1. *The osmolalities of the extracellular and intracellular fluids remain exactly equal to each other except for a few minutes after a change in one of the fluids occurs.*

2. *The number of osmols of osmotically active substance in each compartment, in the extracellular*

*fluid or in the intracellular fluid, remains constant unless one of the osmotically active substances moves through the cell membranes to the other compartment or is lost from or added to one of the two compartments in some other way.*

Using these two basic principles, we can now analyze the effects of different abnormal fluid conditions on extracellular and intracellular fluid volumes and osmolalities.

## EFFECT OF ADDING WATER TO THE EXTRACELLULAR FLUID

Water can be added to the extracellular fluid by injection into the bloodstream, injection beneath the skin, or by ingesting water and this followed by absorption from the gastrointestinal tract into the blood. The water dilutes the extracellular fluid, causing it to become hypotonic with respect to the intracellular fluids. Osmosis begins immediately at the cell membranes, with large amounts of water passing to the interiors of the cells. Within a few minutes the water becomes distributed almost evenly among all the extracellular and intracellular fluid compartments.

**Calculation of the Changes in Extracellular and Intracellular Volumes and Osmolalities.** Table 33–3 illustrates the progressive changes that would result from injecting 10 liters of water into the extracellular fluid. This table gives the volumes and milliosmols per liter of the extracellular fluid, the intracellular fluid, and the total body water.

Initially the extracellular fluid volume is 15 liters, the intracellular fluid volume 25 liters, and the total body water 40 liters; the milliosomols per liter is 300 in each of these. The solution that is added, 10 liters of water, has no osmolality, and when first injected it is added both to the extracellular fluid and to the total body water, but not to the intracellular fluid.

The third line of the table shows the instantaneous effect on the volumes and osmolalities caused by addition of this fluid (before any osmosis occurs). The extracellular fluid volume rises to 25 liters, and its concentration becomes diluted immediately to 180 milliosmols/liter. During this instantaneous interval of time nothing has happened to the intracellular fluids. But within a few seconds to a few minutes, the dilute extracellular fluid diffuses throughout the interstitial spaces of the body and comes in contact with the cells, allowing large portions of the water to pass by osmosis into the intracellular fluid. Osmosis will not stop until the milliosmolalities of the two fluids become equal. As a result, the milliosmolality of the extracellular fluid rises upward from 180 while that of the intracellular fluid falls downward from 300 until the two equal each other. After a few moments the conditions in the last line of the table obtain, with the extracellular fluid volume 18.75 liters, the intracellular volume 31.25 liters, and a milliosmolality in both compartments of 240.

To calculate the changes occurring in the above

**TABLE 33–3. Effect of Administering 10 Liters of Water Intravenously**

| | Extracellular | | | Intracellular | | | Total Body Water | | |
|---|---|---|---|---|---|---|---|---|---|
| | Volume (Liters) | Concentration (mOsmol./L.) | Total mOsmol. | Volume (Liters) | Concentration (mOsmol./L.) | Total mOsmol. | Volume (Liters) | Concentration (mOsmol./L.) | Total mOsmol. |
| Initial | 15 | 300 | 4500 | 25 | 300 | 7500 | 40 | 300 | 12000 |
| Solution added | 10 | 0 | 0 | 0 | 0 | 0 | 10 | 0 | 0 |
| Instantaneous effect | 25 | 180 | 4500 | 25 | 300 | 7500 | 50 | No equilibrium | 12000 |
| After osmotic equilibrium | 18.75 | 240 | 4500 | 31.25 | 240 | 7500 | 50 | 240 | 12000 |

example, one needs only to keep accurate accounting of the total number of milliosmols in each fluid compartment and also in the total body water in the following manner: After the 10 liters of water are added, the total body water becomes 50 liters instead of 40, but the total milliosmols in the whole body remain the same, 12,000. Dividing 50 into 12,000, the average milliosmolar concentration in each liter of the body water under the new conditions is found to be 240. One can readily understand that, after osmotic equilibrium has occurred throughout the body, this calculated value of 240 is the milliosmolality in both the extracellular and intracellular fluid compartments. Then, dividing 240 milliosmols into the total milliosmols in the extracellular fluid compartment, 4500, one finds that the new volume of extracellular fluid is 18.75 liters. Dividing 240 into the total intracellular milliosmols, 7500, gives the new intracellular fluid volume of 31.25 liters.

## EFFECT OF DEHYDRATION

Water can be removed from the body by evaporation from the skin, evaporation from the lungs, or excretion of a very dilute urine. In all these conditions the water leaves the extracellular fluid compartment, but on doing so some of the intracellular water passes immediately into the extracellular compartment by osmosis, thus keeping the osmolalities of the extracellular and intracellular fluids equal to each other. The overall effect is called *dehydration*.

## EFFECT OF ADDING SALINE SOLUTION TO THE EXTRACELLULAR FLUID

If an *isotonic* saline solution is added to the extracellular fluid compartment, the osmolality of the extracellular fluid does not change, and no osmosis

results. The only effect is an increase in extracellular fluid volume.

However, if a *hypertonic* solution is added to the extracellular fluid, the osmolality increases and causes osmosis of water out of the cells into the extracellular compartment.

Finally, if a *hypotonic* solution is added, the osmolality of the extracellular fluid decreases, and some of the extracellular fluids pass into the cells.

Table 33–4 gives the calculations of the effects which would occur if 2 liters of 4.4 per cent (five times isotonic concentration) sodium chloride solution were added to the extracellular fluid compartment. The added solution contains a large total number of milliosmols, 3000, which are added to the extracellular fluid compartment and also to the total body fluids, and the volume of the total body water rises from 40 to 42. Dividing 42 into the new total milliosmols, 15,000, one finds that the new milliosmolar concentration is 357. Now, dividing 357 into the total milliosmols in the extracellular fluid, 7500, one finds that the new extracellular fluid volume is 21 liters. And, dividing 357 into the total intracellular milliosmols, 7500, the intracellular fluid is found to be 21 liters.

The student should become completely familiar with this method of calculation, for an understanding of the mathematical aspects of osmotic equilibria between the two compartments is essential to the understanding of any fluid problem of the body.

## EFFECT OF INFUSING HYPERTONIC GLUCOSE, MANNITOL, OR SUCROSE SOLUTIONS

Very concentrated glucose, mannitol, or sucrose solutions are often injected into patients to cause

**TABLE 33–4.   Effect of Adding 2 Liters of 4.4 Per Cent Sodium Chloride Solution**

| | Extracellular | | | Intracellular | | | Total Body Water | | |
|---|---|---|---|---|---|---|---|---|---|
| | Volume (Liters) | Concentration (mOsmol./L.) | Total mOsmol. | Volume (Liters) | Concentration (mOsmol./L.) | Total mOsmol. | Volume (Liters) | Concentration (mOsmol./L.) | Total mOsmol. |
| Initial | 15 | 300 | 4500 | 25 | 300 | 7500 | 40 | 300 | 12000 |
| Solution added | 2 | 1500 | 3000 | 0 | 0 | 0 | 2 | 1500 | 3000 |
| Instantaneous effect | 17 | 441 | 7500 | 25 | 300 | 7500 | 42 | No equilibrium | 15000 |
| After osmotic equilibrium | 21 | 357 | 7500 | 21 | 357 | 7500 | 42 | 357 | 15000 |

immediate decrease in intracellular fluid volume. For instance, often in severe cerebral edema the patient dies because of too much pressure in the cranial vault, which obstructs the flow of blood to the brain. The condition can be relieved, however, in a few minutes by injecting a hypertonic solution of a substance that will not enter the intracellular compartment. The dynamics of the resulting changes are shown in Figure 33–8, illustrating that the intracellular fluid volume can be decreased by several liters in a few minutes.

But glucose, mannitol, and sucrose are all excreted rapidly by the kidneys, and glucose is also metabolized by the cells for energy. Therefore, within two to four hours the osmotic effects of these substances are lost so that large quantities of water can then rediffuse into the intracellular compartment. Thus, this procedure is temporarily beneficial, lasting only a few hours. Nevertheless, from the standpoint of saving a life of a patient it is often very valuable.

## GLUCOSE AND OTHER SOLUTIONS ADMINISTERED FOR NUTRITIVE PURPOSES

Many different types of solutions are often administered intravenously to provide nutrition to patients who cannot otherwise take adequate amounts of food. Especially used are glucose solutions, to a lesser extent amino acid solutions, and, rarely, homogenized fat solutions. In administering all these, their concentrations are adjusted nearly to isotonicity, or they are given slowly enough that they do not upset the osmotic equilibria of the body fluids. However, after the glucose or other nutrient is metabolized, an excess of water often remains. Ordinarily, the kidneys excrete this in the form of a very dilute urine. Thus, the net result is only addition of the nutrient to the body.

But occasionally the kidneys are functioning poorly, such as often occurs after a surgical operation, and the body becomes greatly overhydrated, resulting sometimes in "water intoxication," which is characterized by mental irritability and even convulsions.

## SOLUTIONS USED FOR PHYSIOLOGIC AND CLINICAL REPLACEMENT PURPOSES

Below are listed the compositions of several solutions commonly used for replacing fluids lost from the body or for nutrition:

Isotonic saline (normal saline) solution for extracellular fluid replacement:

$Na^+$     155 mEq./l.          $Cl^-$          155 mEq./l.

Glucose solutions for nutrition:

I. Glucose   2.5, 5, or 10%

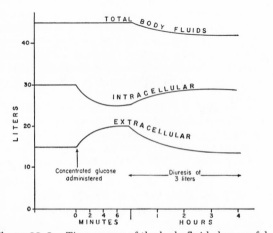

**Figure 33–8.**   Time course of the body fluid changes following infusion of very concentrated glucose into the extracellular fluids.

II. Glucose   2.5, 5, or 10%
Na⁺   155 mEq./l.   Cl⁻   155 mEq./l.

Tyrode's solution for extracellular fluid replacement:

| Na⁺ | 149.4 mEq./l. | Cl⁻ | 145.1 mEq./l. |
|---|---|---|---|
| K⁺ | 2.7 | HCO₃⁻ | 12.0 |
| Ca⁺⁺ | 3.6 | HPO₄⁻⁻ | 0.7 |
| Mg⁺⁺ | 2.1 | | |

Glucose 0.1%

Ringer's solution for frog and turtle experiments:

| Na⁺ | 115 mEq./l. | Cl⁻ | 106 mEq./l. |
|---|---|---|---|
| K⁺ | 1 | HCO₃⁻ | 12 |
| Ca⁺⁺ | 2 | | |

Ringer's solution for mammalian extracellular fluid replacement:

| Na⁺ | 146 mEq./l. | Cl⁻ | 155.4 mEq./l. |
|---|---|---|---|
| K⁺ | 4 | | |
| Ca⁺⁺ | 5.4 | | |

Modified Hartmann's solution for extracellular fluid replacement:

| Na⁺ | 128 mEq./l. | Cl⁻ | 110.6 mEq./l. |
|---|---|---|---|
| K⁺ | 4 | Lactate⁻ | 25 |
| Ca⁺⁺ | 3.6 | | |

Darrow's solution for potassium and extracellular fluid replacement:

| Na⁺ | 122 mEq./l. | Cl⁻ | 104 mEq./l. |
|---|---|---|---|
| K⁺ | 3.5 | Lactate⁻ | 52 |

Sodium lactate solution (¹/₆ molar) for correcting acidosis:

Na⁺   167 mEq./l.   Lactate⁻   167 mEq./l.

Ammonium chloride solution (¹/₆ molar) for correcting alkalosis:

NH₄⁺   167 mEq./l.   Cl⁻   167 mEq./l.

Amino acid solution for nutrition:

Amino acid   5 or 10%
Glucose   5%

Plasma for nutrition or replacement:

Protein 7%
(for other constituents, see Figures 33–5 and 33–6.)

Fat emulsion for nutrition:

Fat 15%

# REFERENCES

Adolph, E. F.: Physiology of Man in the Desert. New York, Interscience Publishers, 1947.
Albert, S. N.: Blood volume studies, Clin. Anesth., 9:171, 1973.
Ben-Naim, A.: Water and Aqueous Solutions. New York, Plenum Publishing Corporation, 1974.
Borut, A., and Shkolnik, A.: Physiological adaptations to the desert environment. Int. Rev. Physiol., 7:185, 1974.
Bradbury, M. W.: Physiology of body fluids and electrolytes. Brit. J. Anaesth., 45:937, 1973.
Brown, G. H.: Liquid crystals and their roles in inanimate and animate systems. Amer. Sci., 60:64, 1973.
Burton, R. F.: The significance of ionic concentrations in the internal media of animals. Biol. Rev., 48:195, 1973.
Coleman, T. G., Manning, R. D., Jr., Norman, R. A., Jr., and Guyton, A. C.: Dynamics of water-isotope distribution. Amer. J. Physiol., 223:1371, 1972.
Cooke, R., and Kuntz, I. D.: The properties of water in biological systems. Ann. Rev. Biophys. Bioeng., 3:95, 1974.
Derrick, J. R., and Guest, M. M. (eds.): Dextrans: Current Concepts of Basic Actions and Clinical Applications. Springfield, Ill., Charles C Thomas, Publisher, 1971.
Epstein, A. N., Kissileff, H. R., and Stellar, E. (eds.): The Neuropsychology of Thirst. Washington, V. H. Winston & Sons, Inc., 1973.
Guyton, A. C., Taylor, A. E., and Granger, H. J.: Circulatory Physiology II: Dynamics and Control of Body Fluids. Philadelphia, W. B. Saunders Company, 1975.
House, C. R.: Water transport in Cells and Tissues. London, Edward Arnold (Publishers) Ltd., 1974.
Kay, R. L. (ed.): The Physical Chemistry of Aqueous Systems. New York, Plenum Publishing Corporation, 1974.
Lee, H. A. (ed.): Parenteral Nutrition in Acute Metabolic Illness. New York, Academic Press, Inc., 1974.
Lightfoot, E. N.: Transport Phenomena and Living Systems; Biomedical Aspects of Momentum and Mass Transport. New York, John Wiley & Sons, Inc., 1974.
Ling, G. N., Miller, C., and Ochsenfeld, M. M.: The physical state of solutes and water in living cells according to the association-induction hypothesis. Ann. N.Y. Acad. Sci., 204:6, 1973.
Marty, A. T.: Hyperoncotic albumin therapy. Surg. Gynecol. Obstet., 139:105, 1974.
Mason, E. E.: Fluid, Electrolyte, and Nutrient Therapy in Surgery. Philadelphia, Lea & Febiger, 1974.
Moore, F. D., Olesen, K. H., McMurrey, J. D., Parker, H. V., Ball, M. R., and Boyden, C. M.: The Body Cell Mass and Its Supporting Environment. Philadelphia, W. B. Saunders Company, 1963.
Parsa, M. H., Ferrer, J. M., and Habif, D. V.: Safe Central Venous Nutrition. Springfield, Ill., Charles C Thomas, Publisher, 1974.
Potts, W. T. W.: Osmotic and ionic regulation. Ann. Rev. Physiol., 30:73, 1968.
Scheuplein, R. J.: Properties of the skin as a membrane. Adv. Biol. Skin, 12:125, 1972.
Solomon, A. K.: The state of water in red cells. Sci. Amer., 2:88, 1971.
Vinnars, E.: Recent advances in parenteral nutrition. Crit. Care Med., 2:143, 1974.
Wilkinson, A. W.: Body Fluids in Surgery, 4th Ed. New York, Churchill Livingstone, Div. of Longman, Inc., 1973.
Wills, J. H.: Percutaneous absorption. Adv. Biol. Skin, 12:169, 1972.
Wolf, A. V., and Crowder, N. A.: Introduction to Body Fluid Metabolism. Baltimore, The Williams & Wilkins Company, 1964.

# 34

## Formation of Urine by the Kidney: Glomerular Filtration, Tubular Function, and Plasma Clearance

The kidneys perform two major functions: first, they excrete most of the end-products of bodily metabolism, and, second, they control the concentrations of most of the constituents of the body fluids. The purpose of the present chapter is to discuss the principles of urine formation and especially the mechanisms by which the kidneys excrete the end-products of metabolism.

### PHYSIOLOGIC ANATOMY OF THE KIDNEY

The two kidneys together contain about 2,400,000 nephrons, and each nephron is capable of forming urine by itself. Therefore, in most instances, it is not necessary to discuss the entire kidney but merely the activities in the single nephron to explain the function of the kidney.

The nephron is composed basically of (1) a *glomerulus* from which fluid is filtered, and (2) a long *tubule* in which the filtered fluid is converted into urine on its way to the *pelvis* of the kidney. Figure 34–1 shows the general organizational plan of the kidney, illustrating especially the distinction between the *cortex* of the kidney and the *medulla*. Figure 34–2 illustrates the basic anatomy of the nephron, which may be described as follows: Blood enters the glomerulus through the *afferent arteriole* and

then leaves through the *efferent arteriole*. The glomerulus, illustrated in Figure 34–3, is a network of up to 50 parallel capillaries covered by epithelial cells and encased in *Bowman's capsule*. Pressure of the blood in the glomerulus

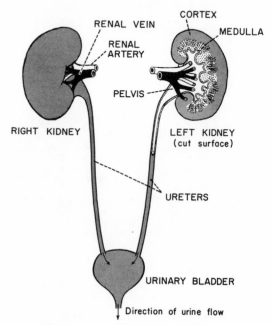

**Figure 34–1.** The general organizational plan of the urinary system.

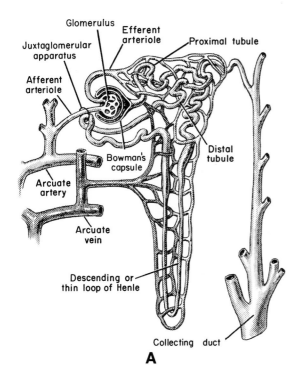

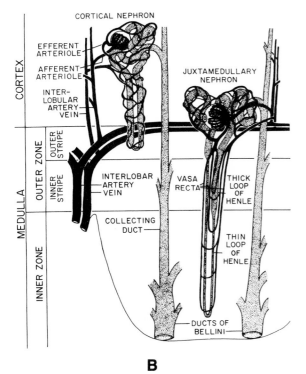

**Figure 34–2.** (A) The nephron. (From Smith: The Kidney: Structure and Functions in Health and Disease. Oxford University Press, 1951.) (B) Differences between a cortical and a juxtamedullary nephron. (From Pitts: Physiology of the Kidney and Body Fluids. Year Book Medical Pubs., Inc., 1974.)

causes fluid to filter into Bowman's capsule, from which it flows first into the *proximal tubule* that lies in the *cortex* of the kidney along with the glomerulus. From there the fluid passes into the *loop of Henle.* Those nephrons that have glomeruli lying very close to the renal *medulla* are called the *juxtamedullary nephrons,* and they have long extended loops of Henle that dip deep into the medulla; the lower portion of the loop has a very thin wall and therefore is called the *thin segment* of the loop of Henle. From the loop of Henle the fluid flows next into the *distal tubule,* which again is in the renal cortex. Finally, the fluid flows into the *collecting tubule,* also called the *collecting duct,* which collects fluid from several nephrons. The collecting tubule passes from the cortex back downward through the medulla, paralleling the loops of Henle. Then it empties into the pelvis of the kidney.

As the glomerular filtrate flows through the tubules, most of its water and varying amounts of its solutes are reabsorbed into the *peritubular capillaries.* The water and solutes that are not reabsorbed become urine.

Nephrons that have glomeruli lying close to the surface of the kidney are called *cortical nephrons;* these have very short, thin segments of their loops of Henle, and their loops of Henle fail to penetrate all the way into the medulla. Other than this difference, these cortical nephrons are much the same as the juxtamedullary nephrons.

After blood passes into the efferent arteriole from the glomerulus, most of it flows through the *peritubular capillary network* that surrounds the cortical portions of the tubules. The remainder of the blood from the efferent arterioles, most of it coming from the juxtamedullary glomeruli, flows into straight capillary loops called *vasa recta* that extend downward into the medulla to envelop the lower parts of the thin segments before looping back upward to empty into the cortical veins.

**Functional Diagram of the Nephron.** Figure 34–4 illustrates a simplified diagram of the "physiologic nephron." This diagram contains most of the nephron's functional structures, and it is used in the present discussion to explain many aspects of renal function.

## BASIC THEORY OF NEPHRON FUNCTION

The basic function of the nephron is to clean, or "clear," the blood plasma of unwanted substances as it passes through the kidney. The

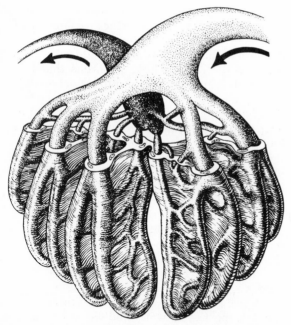

**Figure 34–3.** The human glomerulus showing the epithelial membranes enveloping the glomerular capillaries. (Reproduced from Barger and Herd, as modified from Elias *et al.*: Handbook of Physiology, Sec. 8, p. 255, Williams & Wilkins Co., 1973.)

substances that must be cleared include particularly the end-products of metabolism such as urea, creatinine, uric acid, and urates. In addition, many other substances, such as sodium ions, potassium ions, chloride ions, and hydrogen ions tend to accumulate in the body in excess quantities; it is the function of the nephron also to clear the plasma of the excesses.

The principal mechanism by which the nephron clears the plasma of unwanted substances is: (1) It filters a large proportion of the plasma, usually about one-fifth of it, through the glomerular membrane into the tubules of the nephron. (2) Then, as this filtered fluid flows through the tubules, the unwanted substances fail to be reabsorbed while the wanted substances, especially the water and many of the electrolytes, are reabsorbed back into the plasma of the peritubular capillaries. In other words, the wanted portions of the tubular fluid are returned to the blood, while the unwanted portions pass into the urine.

A second mechanism by which the nephron clears the plasma of unwanted substances is by *secretion.* That is, substances are secreted from the plasma directly through the epithelial cells lining the tubules and into the tubular fluid. Thus, the urine that is eventually formed is composed of *filtered* substances and *secreted* substances.

# RENAL BLOOD FLOW AND PRESSURES

## *BLOOD FLOW THROUGH THE KIDNEYS*

The rate of blood flow through both kidneys of a 70 kg. man is about 1200 ml./minute.

The portion of the total cardiac output that passes through the kidneys is called the *renal fraction.* Since the normal cardiac output of a 70 kg. adult male is about 5600 ml./minute, and the blood flow through both kidneys is about 1200 ml./minute, one can calculate that the normal renal fraction is about 21 per cent. This can vary from as little as 12 per cent to as high as 30 per cent even in the normal resting person.

**Special Aspects of Blood Flow Through the Nephron.**　Note in Figure 34–4 that there are two capillary beds supplying the nephron: (1) the *glomerulus* and (2) the *peritubular capillaries.* The glomerular capillary bed receives its blood from the *afferent arteriole,* and this bed is separated from the peritubular capillary bed by the *efferent arteriole,* which offers considerable resistance to blood flow. As a result, the

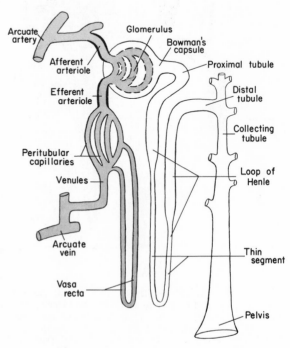

**Figure 34–4.**　The functional nephron.

glomerular capillary bed is a *high pressure bed* while the peritubular capillary bed is a *low pressure bed*. Because of the high pressure in the glomerulus, it functions in much the same way as the usual arterial ends of the tissue capillaries, with fluid filtering continually out of the glomerulus into Bowman's capsule. On the other hand, the low pressure in the peritubular capillary system causes it to function in much the same way as the venous ends of the tissue capillaries, with fluid being absorbed continually into the capillaries.

**The Vasa Recta.** A special portion of the peritubular capillary system is the vasa recta, which are a network of capillaries that descend around the lower portions of the loops of Henle. These capillaries form loops in the medulla of the kidney and then return to the cortex before emptying into the veins. The vasa recta play a special role in the formation of concentrated urine, as is discussed in the following chapter.

Only a small proportion of the total renal blood flow, about 1 to 2 per cent, flows through the vasa recta. In other words, blood flow through the medulla of the kidney is sluggish in contrast to the rapid blood flow in the cortex.

At the outer rim of the medulla where the medulla comes in contact with the cortex there is an extensive peritubular capillary plexus through which approximately 10 per cent of the renal blood flows. Unfortunately, the significance of this plexus is yet unknown, but the large blood flow in it is in sharp contrast to the poor blood flow (only 1 to 2 per cent of the total) to the inner medulla through the vasa recta.

## PRESSURES IN THE RENAL CIRCULATION

Figure 34–5 gives the approximate pressures in the different parts of the renal circulation, showing an initial pressure of approximately 100 mm. Hg in the large arcuate arteries and about 8 mm. Hg in the veins into which the blood finally drains. The two major areas of resistance to blood flow through the nephron are (1) the *afferent arteriole* and (2) the *efferent arteriole*. In the afferent arteriole the pressure falls from 100 mm. Hg at its arterial end to an estimated mean pressure of about 60 mm. Hg in the glomerulus. (This pressure is still in serious doubt, having been calculated to be as high as 70 mm. Hg in the dog and measured to be as low as 45 mm. Hg in the rat. Therefore, 60 mm. Hg is a reasonable average estimate.) As the blood flows through the efferent arterioles from the glomerulus to the peritubular capillary system,

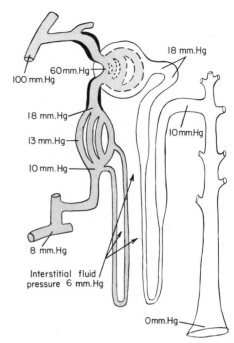

**Figure 34–5.** Pressures at different points in the vessels and tubules of the functional nephron and in the interstitial fluid.

the pressure falls another 47 mm. Hg to a mean peritubular capillary pressure of 13 mm. Hg. Thus, the high pressure capillary bed in the glomerulus operates at a mean pressure of about 60 mm. Hg while the low pressure capillary bed in the peritubular capillary system operates at a mean capillary pressure of about 13 mm. Hg.

## INTRARENAL PRESSURE AND RENAL INTERSTITIAL FLUID PRESSURE

The kidney is encased in a tight *fibrous capsule*. When a needle is inserted into the kidney and the pressure in the needle is gradually raised until fluid flows into the kidney tissue, the pressure at which flow begins is between 10 and 18 mm. Hg, averaging perhaps 12 mm. Hg. This "needle" pressure is called the *intrarenal pressure*. It was pointed out in Chapter 30 that needle pressures do not measure the interstitial fluid pressure but instead measure the *total tissue pressure* (the pressure that tends to collapse blood vessels and tubules).

Recent attempts to measure the *interstitial fluid pressure* of the kidney utilizing implanted perforated capsules as described in Chapter 30 have given an average mean value of 6 mm. Hg, which, at present, is probably the best estimate of interstitial fluid pressure of the kidney.

## FUNCTION OF THE PERITUBULAR CAPILLARIES

Tremendous quantities of fluid, about 180 liters each day, are filtered through all the glomeruli; all but slightly over 1 liter of this is reabsorbed from the tubules into the renal interstitial spaces and thence into the peritubular capillaries. This represents about four times as much fluid as that reabsorbed at the venous ends of all the other capillaries of the entire body. Therefore, one can readily see that reabsorption of fluid into the peritubular capillaries presents a special problem. However, the peritubular capillaries are extremely porous in comparison with those in other body tissues, so that extremely rapid osmosis of fluid resulting from the colloid osmotic pressure of the plasma proteins probably can account for the rapid absorption that is required.

## GLOMERULAR FILTRATION AND THE GLOMERULAR FILTRATE

**The Glomerular Membrane and Glomerular Filtrate.** The fluid that filters through the glomerulus into Bowman's capsule is called *glomerular filtrate,* and the membrane of the glomerular capillaries is called the *glomerular membrane.* Though, in general, this membrane is similar to that of other capillaries throughout the body, it has several differences. First, it has three major layers: (1) the endothelial layer of the capillary itself, (2) a basement membrane, and (3) a layer of epithelial cells that line the surfaces of Bowman's capsule. Yet, despite the number of layers, the permeability of the glomerular membrane is from 100 to 1000 times as great as that of the usual capillary.

The tremendous permeability of the glomerular membrane is caused by its special structure, which is illustrated in Figure 34–6. The capillary *endothelial cells* lining the glomerulus are perforated by literally thousands of small holes called *fenestrae.* Then, outside the endothelial cells is a basement membrane composed mainly of a meshwork of mucopolysaccharide fibrillae. A final layer of the glomerular membrane is a layer of epithelial cells that line the outside of the glomerulus. However, these cells are not continuous but instead consist mainly of finger-like projections that cover the outer surface of the basement membrane. These "fingers" form slits called *slit-pores* through which the glomerular filtrate filters. Thus, the glomerular

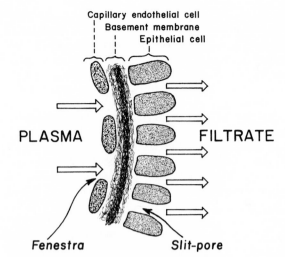

**Figure 34–6.** Functional structure of the glomerular membrane.

filtrate must pass through three different layers before entering Bowman's capsule.

Experiments have shown that the fenestrae of the capillary endothelial cells are small enough in diameter that they prevent the filtration of all particles with an average size greater than 160 Å. The meshwork of the basement membrane prevents filtration of all particles greater in size than 110 Å. And the slit-pores prevent filtration of all particles with diameters greater than 70 Å. Since plasma proteins are slightly larger than the 70 Å diameter, it is possible for the glomerular membrane to prevent the filtration of all substances with molecular weights equal to or greater than those of the plasma proteins. Yet, the great numbers of fenestrae and slit-pores allow tremendously rapid filtration of fluid and small molecular weight substances from the plasma into Bowman's capsule.

Overall, the permeability of the glomerular membrane to substances of different molecular weights (expressed as the ratio of concentration of the dissolved substance on the filtrate side of the membrane to its concentration on the plasma side) is approximately as follows:

| Molecular Weight | Permeability | Example Substance |
|---|---|---|
| 5,000 | 1.00 | Inulin |
| 30,000 | 0.5 | Very small protein |
| 69,000 | 0.005 | Albumin |

This means that at a molecular weight of 5000, the dissolved substance filters just as easily as water, but at a molecular weight of 69,000 only 0.005 per cent of the dissolved substance filters. Note that the molecular weight of the smallest plasma protein, albumin, is 69,000. Therefore, for practical purposes the glomerular membrane is almost completely impermeable to all plasma proteins but is highly permeable to essentially all other dissolved substances in normal plasma.

**Composition of the Glomerular Filtrate.** The glomerular filtrate has almost exactly the same composition as the fluid that filters from the arterial ends of the capillaries into the interstitial fluids. It contains no red blood cells and about 0.03 per cent protein, or less than $1/200$ the protein in the plasma.

The electrolyte and other solute composition of glomerular filtrate is also similar to that of the interstitial fluid. Because of the paucity of the negatively charged protein ions in the filtrate, a Donnan equilibrium effect occurs that causes the concentration of the other negative ions, including chloride and bicarbonate ions, to be about 5 per cent higher in both interstitial fluid and glomerular filtrate than in plasma; and the concentration of positive ions is about 5 per cent lower. Also, the concentrations of the nonionized substances, such as urea, creatinine, and glucose, are increased about 4 per cent because of the almost total lack of proteins.

To summarize: for all practical purposes, glomerular filtrate is the same as plasma except that it has no significant amount of proteins.

### THE GLOMERULAR FILTRATION RATE

The quantity of glomerular filtrate formed each minute in all nephrons of both kidneys is called the *glomerular filtration rate*. In the normal person this averages approximately 125 ml./min.; however, in different normal functional states of the kidneys, it can vary from a few milliliters to 200 ml./min. To express this differently, the total quantity of glomerular filtrate formed each day averages about 180 liters, or more than two times the total weight of the body. Over 99 per cent of the filtrate is usually reabsorbed in the tubules, the remainder passing into the urine, as explained later in the chapter.

**The Filtration Fraction.** The filtration fraction is the fraction of the renal plasma flow that becomes glomerular filtrate. Since the normal plasma flow through both kidneys is 650 ml./min. and the normal glomerular filtration rate in both kidneys is 125 ml., *the average filtration fraction is approximately* $125/650$, *or 19 per cent.* Here again, this value can vary tremendously, both physiologically and pathologically.

### DYNAMICS OF GLOMERULAR FILTRATION

Glomerular filtration occurs in almost exactly the same manner that fluid filters out of any high pressure capillary in the body. That is, *pressure inside the glomerular capillaries* causes filtration of fluid through the capillary membrane into Bowman's capsule. On the other hand, *colloid osmotic pressure in the blood and pressure in Bowman's capsule* oppose the filtration. Ordinarily, the amount of protein in Bowman's capsule is too slight to be of any significance, but if this ever becomes increased to a significant amount, its colloid osmotic pressure will obviously also be active at the membrane, promoting increased filtration of fluid through the membrane.

**Glomerular Pressure.** The glomerular pressure is the average pressure in the glomerular capillaries. This unfortunately has been measured directly only in one mammal—the rat. The average value is about 45 mm. Hg. From various indirect measurements it has also been calculated to be about 65 mm. Hg in the dog. Because man is a large mammal, *a reasonable average value can be considered to be 60 mm. Hg,* though, as noted below, this can increase or decrease considerably under varying conditions.

**Pressure in Bowman's Capsule.** In lower animals, pressure measurements have actually been made in Bowman's capsule and at different points along the renal tubules by inserting micropipets into the lumen. On the basis of these studies, *capsular pressure in the human being is estimated to be 18 mm. Hg.*

**Colloid Osmotic Pressure in the Glomerular Capillaries.** Because approximately one fifth of the plasma in the capillaries filters into the capsule, the protein concentration increases about 20 per cent as the blood passes from the arterial to the venous ends of the glomerular capillaries. If the normal colloid osmotic pressure of blood entering the capillaries is 28 mm. Hg, it rises to approximately 36 mm. Hg by the time the blood reaches the venous ends of the capillaries, and the average colloid osmotic pressure is about 32 mm. Hg. (See Fig. 30–8 in Chap. 30, which shows the relationship between protein concentration and colloid osmotic pressure.)

**Filtration Pressure, Filtration Coefficient, and Glomerular Filtration Rate.** The filtration pressure is the net pressure forcing fluid through the glomerular membrane, and this is *equal to the glomerular pressure minus the sum of glomerular colloid osmotic pressure and capsular pressure.* In Figure 34–7A, *the normal filtration pressure is shown to be about 10 mm. Hg.*

The filtration coefficient, called $K_f$, is a constant; it is the glomerular filtration rate for both kidneys per mm. Hg of filtration pressure. That is, the glomerular filtration rate is equal to the filtration pressure times the filtration coefficient, or

$$GFR = \text{Filtration pressure} \cdot K_f$$

The normal filtration coefficient is 12.5 ml. per min. per mm. Hg of filtration pressure. Thus, at a normal mean filtration pressure of 10 mm. Hg, the total filtration rate of both kidneys is 125 ml. per min.

## FACTORS THAT AFFECT THE GLOMERULAR FILTRATION RATE

It is clear from the above equation that the filtration pressure and the filtration coefficient determine the glomerular filtration rate. The filtration coefficient probably does not change greatly from normal except when the kidneys become diseased.

On the other hand, the three factors that determine filtration pressure—(1) glomerular pressure, (2) plasma colloid osmotic pressure, and (3) Bowman's capsule pressure—do play

very significant roles in determining glomerular filtration rate. In general, the greater the glomerular pressure, the greater will be the filtration rate; conversely, the greater the plasma colloid osmotic pressure or Bowman's capsule pressure, the less will be the glomerular filtration rate.

**Effect of Renal Blood Flow on Glomerular Filtration Rate.** The glomerular filtration rate is affected to a great extent by the rate of blood flow through the nephrons. This effect can be explained as follows: It is not the plasma colloid osmotic pressure in the arterial blood that determines filtration through the glomerulus but instead is the colloid osmotic pressure of the plasma in the glomerulus itself. Since a very large proportion of the plasma is filtered through the glomerular membrane, the colloid osmotic pressure in the glomerulus rises very high, and this increase opposes further filtration. Therefore, once a certain proportion of the plasma has filtered, no more will filter until new plasma flows into the glomerulus. Consequently, the greater the rate of flow of plasma into the glomerulus, the greater the glomerular filtration rate.

For mathematical reasons that cannot be explained here, the greater the filtration coefficient of the glomeruli, the greater the effect of blood flow on glomerular filtration rate. On the other hand, the lower the filtration coefficient, the greater the effect glomerular pressure has on filtration rate.

**Effect of Afferent Arteriolar Constriction on Glomerular Filtration Rate.** Afferent arteriolar constriction decreases the rate of blood flow into the glomerulus and also decreases the glomerular pressure, both of these effects de-

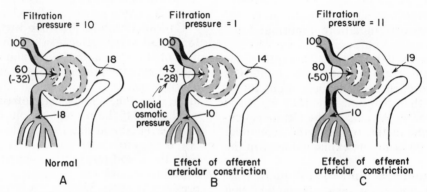

**Figure 34–7.** (A) Normal pressures at different points in the nephron, and the normal filtration pressure. (B) Effect of afferent arteriolar constriction on pressures in the nephron and on filtration pressure. (C) Effect of efferent arteriolar constriction on pressures in the nephron and on filtration pressure.

creasing the filtration rate. This effect is illustrated in Figure 34–7B. Conversely, dilatation of the afferent arteriole increases the glomerular pressure, with a corresponding increase in glomerular filtration rate.

**Effect of Efferent Arteriolar Constriction.** Constriction of the efferent arteriole increases the resistance to outflow from the glomeruli. This obviously increases the glomerular pressure and usually increases the glomerular filtration rate as well, which is illustrated in Figure 34–7C. However, the blood flow decreases at the same time, and if the degree of efferent arteriolar constriction is severe, the plasma will remain for a long period of time in the glomerulus, and extra large portions of plasma will filter out. This will increase the plasma colloid osmotic pressure to excessive levels and will cause glomerular filtration to fall paradoxically to a low value despite the elevated glomerular pressure.

**Effect of Sympathetic Stimulation.** During mild to moderate sympathetic stimulation of the kidneys, the afferent arterioles constrict preferentially, thereby decreasing the glomerular filtration rate (unless the arterial pressure rises simultaneously, as is common during sympathetic stimulation).

With strong sympathetic stimulation, glomerular blood flow and the glomerular pressure are reduced so greatly that the glomerular filtration rate falls almost to zero.

**Effect of Arterial Pressure.** One would expect an increase in arterial pressure to cause a proportionate increase in all pressures in the nephron and therefore to increase the glomerular filtration rate to a great extent. In actual fact, this effect is greatly blunted because of the phenomenon called *autoregulation,* which is explained in the following chapter. Briefly, when the arterial pressure rises, afferent arteriolar constriction occurs automatically; this prevents a major rise in glomerular pressure despite the rise in arterial pressure. Therefore, the glomerular filtration rate increases only 15 to 20 per cent even when the arterial pressure rises from its normal value of 100 mm. Hg to as high as 200 mm. Hg.

Nevertheless, we shall also see in the following chapter that even a small percentage increase in glomerular filtration rate can cause a many-fold increase in urinary output. Therefore, an increase in arterial pressure can greatly increase urinary output even though it affects glomerular filtration rate only slightly.

# REABSORPTION AND SECRETION IN THE TUBULES

The glomerular filtrate entering the tubules of the nephron flows (1) through the *proximal tubule,* (2) through the *loop of Henle,* (3) through the *distal tubule,* and (4) through the *collecting tubule* into the pelvis of the kidney. Along this course, substances are selectively reabsorbed or secreted by the tubular epithelium, and the resultant fluid entering the pelvis is *urine.* Reabsorption plays a much greater role than does secretion in this formation of urine, but secretion is especially important in determining the amounts of potassium ions, hydrogen ions, and a few other substances in the urine, as is discussed later.

Ordinarily, more than 99 per cent of the water in the glomerular filtrate is reabsorbed as it passes through the tubules. Therefore, if some dissolved constituent of the glomerular filtrate is not reabsorbed at all along the entire course of the tubules, this reabsorption of water obviously concentrates the substance more than 99-fold. On the other hand, some constituents, such as glucose and amino acids, are reabsorbed almost entirely so that their concentrations decrease almost to zero before the fluid becomes urine. In this way the tubules separate substances that are to be conserved in the body from those that are to be eliminated in the urine.

## *BASIC MECHANISMS OF ABSORPTION AND SECRETION IN THE TUBULES*

The basic mechanisms for transport through the tubular membrane are essentially the same as those for transport through other membranes of the body. These can be divided into *active transport* and *passive transport* (or *diffusion*). The basic essentials of these mechanisms are described here, but for additional details refer to Chapter 4.

**Active Transport Through the Tubular Wall.** Figure 34–8 illustrates, by way of example, the mechanism for active transport of sodium from the lumen of the proximal tubule into the peritubular capillary. Note, first, the character of the epithelial cells that line the tubule. Each epithelial cell has a "brush" border on its luminal surface. This brush is composed of literally thousands of very minute microvilli that multiply the surface area of luminal exposure of the cell about 20-fold. The base of

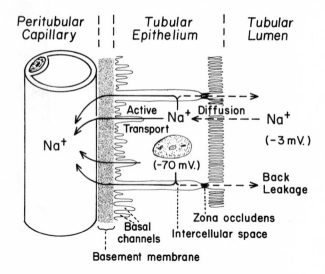

**Figure 34–8.** Mechanism for active transport of sodium from the tubular lumen into the peritubular capillary, illustrating active transport at the base and sides of the epithelial cell and diffusion through the luminal border of the cell.

the cell sits on the basement membrane, but it is pocked by an extensive system of *basal channels* that multiply the basal surface area many fold as well. The epithelial cells are attached to each other only where they touch each other adjacent to the brush border, an attachment area called the *zona occludens*.

Active transport of sodium occurs from inside the epithelial cell through its basal and side membranes into the basal channels and from there into the spaces between the cells. This transport outward from the cell diminishes the sodium concentration inside the cell. Then, because this low concentration inside the cell establishes a sodium ion diffusion gradient from the tubule to the inside of the cell, sodium ions diffuse from the tubule into the cell. Once inside the cell, the sodium is carried by the active transport process the rest of the way into the peritubular fluid of the basal channels and the spaces between the cells.

Note in the figure that the electrical potential inside the epithelial cell, caused by the continual active transport of the sodium out of the cells, is approximately −70 millivolts. This very negative intracellular voltage is important because it, as well as the low concentration of sodium inside the epithelial cell, is one of the important factors that causes sodium diffusion from the tubular lumen into the cell. These two factors together are called the "electrochemical" gradient. The rapid diffusion of sodium into the cell is also facilitated by a very high permeability of the brush border membrane to sodium, primarily because of the extensive surface area of the thousands of microvilli.

Once the sodium has been transported into the basal channels of the epithelial cells and also into the intercellular spaces, the sodium can then move on into the peritubular capillary, or it can leak backwards through the zona occludens into the tubular lumen. The way it will go is determined by several different factors, including especially the hydrostatic pressures in both the tubule and the peritubular capillary and the colloid osmotic pressure in the peritubular capillary. The effects of these factors will be discussed in the following chapter.

Other substances besides sodium that are actively absorbed through the tubular epithelial cells include *glucose, amino acids, calcium ions, potassium ions, phosphate ions, urate ions,* and others.

In addition, some substances are actively *secreted* into all or some portions of the tubules; these include especially *hydrogen ions, potassium ions,* and *urate ions.* Active secretion occurs in the same way as active absorption except that the cell membrane transports the secreted substance in the opposite direction; for some of the actively secreted substances, it is the brush border rather than the base and sides of the cell that provides the active transport mechanism.

**Passive Absorption of Urea and Other Nonactively Transported Solutes by the Process of Diffusion.** When water is reabsorbed by osmosis, the concentration of urea in the tubular fluid rises, which obviously establishes a concentration difference for urea between the tubular and peritubular fluids. This in turn causes urea also to diffuse from the tubular fluid into the peritubular fluid. This same effect also occurs for other tubular solutes that are not actively reabsorbed but that are diffusible through the tubular membrane.

The rate of resorption of a nonactively reabsorbed solute is determined by (1) the amount of water that is reabsorbed, because this determines the tubular concentration of the solute, and (2) the permeability of the tubular membrane for the solute. The permeability of the membrane for urea is far less than that for water, which means that less urea is reabsorbed than water. Therefore, a large proportion of the urea remains in the tubules and is lost in the urine—usually about 50 per cent of all that enters the glomerular filtrate. The permeability of the tubular membrane for reabsorption of creatinine, inulin (a large polysaccharide), mannitol (a disaccharide), and sucrose is zero, which means that once these substances have filtered into the glomerular filtrate, 100 per cent of that which enters the glomerular filtrate passes on into the urine.

**Diffusion Caused by Electrical Differences Across the Tubular Membrane.** The active absorption of sodium from the tubule creates negativity inside the tubule with respect to the peritubular fluid, as was illustrated in Figure 34–8. In the proximal tubule, this electrical potential is approximately −3 millivolts. In the distal tubule, it ranges from −10 to −70 millivolts.

Note particularly that the peritubular fluid is *positive* with respect to the negative voltage in the tubular fluid. This causes *negative ions,* such as *chloride, phosphate,* and *bicarbonate ions,* all to be attracted from the tubular fluid toward the peritubular fluid. The factor that determines how rapidly these ions will diffuse along this electrical gradient is the degree of permeability of the epithelial cells to the different ions. In the proximal tubules, the permeability to chloride ions is especially great, so that chloride ions diffuse in this manner to a great extent.

**Diffusion and Osmosis of Water Through the Tubular Membrane.** Water diffuses through the tubular membrane in all parts of the tubular system, but it diffuses much more readily in some portions than in others, as will become apparent in subsequent discussions. In those portions of the tubules that are highly permeable to water, such as the proximal tubules, rapid osmosis of water occurs any time the osmolar concentration of solutes on one side of the membrane becomes different from

that on the other side. For instance, referring again to Figure 34–8, one finds that when sodium ions are transported from the tubules into the intercellular spaces and basal channels, this transport causes increased osmotic pressure in the peritubular fluids but decreased osmotic pressure in the tubules. Since the proximal tubular membrane is also highly permeable to water, water moves by the process of osmosis almost instantaneously into the intercellular spaces and into the basal channels to reequilibrate the osmolar concentrations on the two sides of the membrane. Thus, as sodium is transported through each cellular membrane of the tubular epithelium, so also is water transported by the process of osmosis.

**Passive Secretion by the Process of Diffusion.** A few substances are secreted by diffusion in a manner similar to the reabsorption of chloride ions, water, and other substances by diffusion. For instance, ammonium ions are synthesized inside the epithelial cells and then diffuse into the tubular lumen; they help to control the degree of acidity of the tubular fluid, as will be discussed in Chapter 37. Also, potassium ions diffuse through the luminal border of the epithelial cells in the distal tubules and collecting ducts, and in this way are "secreted." Active transport, however, plays a major role in the movement of potassium ions from the peritubular blood into the interior of the epithelial cells through their basal borders. This will be discussed further in Chapter 36.

## ABSORPTIVE CAPABILITIES OF DIFFERENT TUBULE SEGMENTS

In subsequent sections of this chapter the absorption and secretion of specific substances in different segments of the tubular system will be discussed. However, it is first important to point out basic differences between the absorptive and secretory capabilities of the different types of tubules.

**Proximal Tubular Epithelium.** Figure 34–9 illustrates the cellular characteristics of the tubular membrane in (1) the proximal tubule, (2) the thin segment of the loop of Henle, (3) the distal tubule, and (4) the collecting tubule. The proximal tubular cells have the appearance of being highly metabolic cells, having large numbers of mitochondria to support extremely rapid active transport processes; and, true enough, one finds that about 65 per cent of all reabsorptive and secretory processes that occur in the

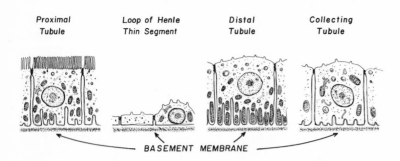

| Proximal Tubule | Loop of Henle Thin Segment | Distal Tubule | Collecting Tubule |

BASEMENT MEMBRANE

**Figure 34–9.** Characteristics of the epithelial cells in different tubular segments.

tubular system take place in the proximal tubules. That is, only 35 per cent of the glomerular filtrate normally passes the whole distance through the proximal tubules, the remaining 65 per cent being reabsorbed before reaching the loops of Henle. As already described in relation to sodium transport, the proximal tubular epithelial cells have an extensive brush border. They also have an extensive labyrinth of intercellular and basal channels through which large quantities of fluid can flow.

**Thin Segment of the Loop of Henle.** The epithelium of the thin segment of the loop of Henle, as the name implies, is very thin. There is also an extensive pore system between the epithelial cells so that the permeability is great. These cells also have no brush border and a very scanty complement of mitochondria, indicating a minimal level of metabolic activity. Thus, the thin segment of the loop of Henle appears to be adapted primarily for simple diffusion of substances through its walls (though some physiologists do ascribe an important sodium transport process to this membrane, as will be discussed in the following chapter).

**Thick Segment of the Loop of Henle and the Distal Tubule Epithelium.** The distal tubule is considered to begin at the point where the tubular system has returned from the loop of Henle to the level of the glomerulus. However, the last portion of the ascending limb of the loop of Henle, known as the *thick segment,* has thickened epithelial cells with physical characteristics almost identical to those of the distal tubular cells. These cells are similar to the cells in the proximal tubules, except that they have only a rudimentary brush border and have much tighter zona occludens where the cells attach to each other. The cells are specifically adapted for active transport of sodium against high concentration and electrical gradients. Also, the epithelium is poorly permeable to water and almost completely impermeable to urea. To provide energy for pumping sodium against a high electrochemical gradient, large

numbers of mitochondria lie in close proximity to the basal membrane of the epithelial cell.

**Collecting Tubule Epithelium.** The collecting tubule is composed of two important segments that are similar in physical characteristics but distinctly different in functional characteristics: the *cortical portion* and the *medullary portion.* Both of these have epithelial cells that are nearly cuboidal in shape, with smooth surfaces and a paucity of mitochondria and other organelles. It is in the collecting tubules that the urine becomes either highly concentrated or highly dilute, highly acidic or highly basic. The collecting tubule epithelium appears to be well designed to resist these extremes of tubular fluid characteristics.

The cortical portion of the collecting tubules is almost totally impermeable to urea. On the other hand, the medullary portion is moderately permeable to urea so that large quantities of urea are normally reabsorbed from the medullary collecting tubules into the medullary interstitium, an effect that is very important for concentrating the urine, as will be discussed in the following chapter.

The permeability of the epithelium to water in both portions of the collecting tubule is determined mainly by the concentration of antidiuretic hormone in the blood. When large quantities of antidiuretic hormone are present, the collecting tubule becomes very permeable to water, and most of the water will be reabsorbed from the tubules and returned to the blood; in the absence of antidiuretic hormone, very little water is reabsorbed, most of it instead passing on into the urine.

## REABSORPTION AND SECRETION OF INDIVIDUAL SUBSTANCES IN DIFFERENT SEGMENTS OF THE TUBULES

**Transport of Water and Flow of Fluid at Different Points in the Tubular System.** Water transport occurs entirely by osmotic diffusion. That is, whenever some solute in the glomerular filtrate is absorbed either by active absorption or by diffusion caused by an electrochemical gradient, the resulting decreased concentration of solute in the tubular fluid and increased concentration in the peritubular fluid causes osmosis of water out of the tubules.

Figure 34–10 depicts the rates of water flow axially along the tubular system at different points in the tubules. In both kidneys of man, the total fluid volumes flowing into each segment of the tubular system each minute (under normal resting conditions) are the following:

|  | ml./min. |
|---|---|
| Glomerular filtrate | 125 |
| Flowing into the loops of Henle | 45 |
| Flowing into the distal tubules | 25 |
| Flowing into the collecting ducts | 12 |
| Flowing into the urine | 1 |

From this chart one can also deduce the per cent of the glomerular filtrate water that is reabsorbed in each segment of the tubules, as follows:

|  | Per Cent |
|---|---|
| Proximal tubules | 65 |
| Loop of Henle | 15 |
| Distal tubules | 10 |
| Collecting ducts | 9.3 |
| Passing into the urine | 0.7 |

We shall see later in the chapter that these

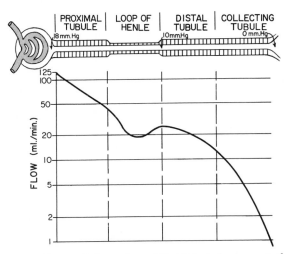

**Figure 34–10.** Volume flow of fluid in each segment of the tubular system per minute. Note that the flow is plotted on a semi-logarithmic scale, illustrating the tremendous difference in flow between the earlier and later segments of the tubules.

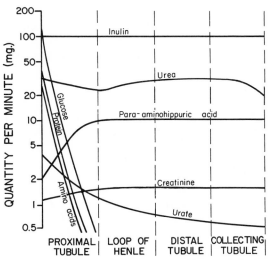

**Figure 34–11.** Rates of flow of important organic substances through the tubules at each point along the tubular course. This figure demonstrates reabsorption of all nutritionally important substances in the proximal tubules, and poor reabsorption of the metabolic end-products in all segments of the tubules. Note the total absence of reabsorption of inulin and the secretion of para-aminohippuric acid into the proximal tubules.

values vary greatly under different operational conditions of the kidney, particularly when the kidney is forming very dilute or very concentrated urine.

**Reabsorption of Substances of Nutritional Value to the Body—Glucose, Proteins, Amino Acids, Acetoacetate Ions, and Vitamins.** Five different substances in the glomerular filtrate of particular importance to bodily nutrition are glucose, proteins, amino acids, acetoacetate ions, and the vitamins. Normally all of these are completely or almost completely reabsorbed by active processes in the *proximal tubules* of the kidney, as shown in Figure 34–11 for glucose, protein, and amino acids. Therefore, almost none of these substances remain in the tubular fluid entering the loop of Henle.

*Special Mechanism for Absorption of Protein.* As much as 30 grams of protein filter into the glomerular filtrate each day. This would be a great metabolic drain on the body if the protein were not returned to the body fluids. Because the protein molecule is much too large to be transported by the usual transport processes, protein is absorbed through the brush border of the proximal tubular epithelium by pinocytosis, which means simply that the protein attaches itself to the membrane and this portion of the membrane then invaginates to the interior of the cell. Once inside the cell the protein is probably digested into its constituent amino acids which are then actively absorbed through the base and sides of the cell into the peritubular fluids. Details of the pinocytosis mechanism were discussed in Chapter 4.

**Poor Reabsorption of the Metabolic End Products: Urea, Creatinine, and Others.** Figure 34–11 also illustrates the rates of flow of three major metabolic end-products in the different segments of the tubular system—urea, creatinine, and urate ions. Note, especially, that only moderate quantities of *urea*—about 50 per cent of the total—are reabsorbed during the entire course through the tubular system.

Creatinine is not reabsorbed in the tubules at all; indeed, small quantities of creatinine are actually secreted into the tubules by the proximal tubules so that the total quantity of creatinine increases about 20 per cent.

The *urate ion* is absorbed much more than urea—about 86 per cent reabsorption. But even so, large quantities of urate remain in the fluid that finally issues into the urine. Several other end-products, such as *sulfates, phosphates,* and *nitrates,* are transported in much the same way as urate ions. All of these are normallly reabsorbed to a far less extent than is water so that their concentrations become greatly increased as they flow along the tubules. Yet, *each is actively reabsorbed to some extent,* which keeps their concentrations in the extracellular fluid from ever falling too low.

**Transport of Inulin and Para-aminohippuric Acid by the Tubules.** Note also in Figure 34–11 that the rate of flow of *inulin,* a large polysaccharide, remains exactly the same throughout the entire tubular system. The cause of this is simply that inulin is neither reabsorbed nor secreted in any segment of the tubules.

Finally, Figure 34–11 shows that the rate of flow of *para-aminohippuric acid* (PAH) increases about five-fold as the tubular fluid passes through the proximal tubules; then its rate of flow remains constant in the other tubules. This is because large quantities of PAH are *secreted* into the tubular fluid by the proximal tubular epithelial cells, and it is not reabsorbed in any segment of the tubular system.

These two substances play an important role in experimental studies of tubular function, as will be discussed later in the chapter.

**Transport of Different Ions by the Tubular Epithelium—Sodium, Chloride, Bicarbonate, and Potassium.** Figure 34–12 illustrates the rates of flow of different important ions—sodium, chloride, bicarbonate, and potassium—in different segments of the tubular system. Note that all these decrease markedly because of reabsorption as the tubular fluid progresses from glomerular filtrate to urine.

Positive ions are generally transported through the tubular epithelium by active transport processes, whereas negative ions are usually transported passively as a result of electrical differences developed across the membrane when the positive ions are transported. For instance, when sodium ions are transported out of the tubular fluid, the resulting electronegativity that develops in the tubular fluid causes chloride ions to follow in the wake of the sodium ions. But, despite the general rule that negative ions are usually passively reabsorbed, under some conditions, such as in extreme metabolic alkalosis, chloride ions probably can be actively reabsorbed in the distal tubules. Also, chloride ions are probably actively reabsorbed, even normally, through the thick segment of the ascending limb of the loop of Henle.

The increased flow of all ions in the first half of the loop of Henle is caused by *diffusion* of ions into the tubules from the peritubular fluids of the medulla, which is explained in the following chapter in relation to the mechanism for concentrating urine.

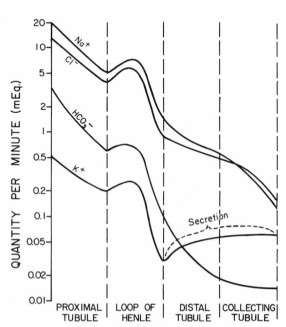

**Figure 34–12.** Rates of flow of the different ions at different points in the tubules each minute. Note not only the reabsorption of the ions but also the secretion of potassium into the distal and collecting tubules.

***Secretion of Potassium and Hydrogen Ions in the Distal Tubules and Collecting Tubules.***    Note also in Figure 34–12 that the flow of potassium ions normally increases as the tubular fluid passes through the distal tubules and collecting tubules, as a result of *secretion of potassium.* This will be discussed at greater length in Chapter 36 in relation to the regulation of potassium concentration in the extracellular fluids.

*Hydrogen ions* are actively secreted in the proximal tubules, distal tubules, and collecting tubules. This secretion is controlled by the hydrogen ion concentration of the extracellular fluid, which will be discussed in Chapter 37.

***Special Aspects of Bicarbonate Ion Transport.*** Bicarbonate ion is primarily reabsorbed in the form of carbon dioxide rather than in the form of bicarbonate ion itself. This occurs as follows: The bicarbonate ions first combine with hydrogen ions that are secreted into the tubular fluid by the epithelial cells. The reaction forms carbonic acid which then dissociates into water and carbon dioxide. The carbon dioxide, being highly lipid-soluble, diffuses rapidly through the tubular membrane into the peritubular capillary blood. When more bicarbonate ions are present than there are available hydrogen ions, almost all of the excess bicarbonate ions flow on into the urine because the tubules are only slightly permeable to them.

***Transport of Other Ions.***    Though we know much less about transport of other ions besides the four illustrated in Figure 34–12, in general essentially all of them can be reabsorbed either by active transport or as a result of electrical differences across the membrane. Thus, calcium, magnesium, and other positive ions are actively reabsorbed, and many of the negative ions are reabsorbed as a result of electrical differences that develop when the positive ions are reabsorbed. In addition, certain negative ions—urate, phosphates, sulfates, and nitrates—can be reabsorbed by active transport, this occurring primarily in the proximal tubules.

## CONCENTRATION OF DIFFERENT SUBSTANCES AT DIFFERENT POINTS IN THE TUBULES

Whether or not a substance becomes concentrated in the tubular fluid as it moves along the tubules is determined by the *relative reabsorption of the substance versus the reabsorption of water.* If a greater percentage of water is reabsorbed, the substance becomes progressively more and more concentrated. Conversely, if a greater percentage of the substance is reabsorbed, it becomes more and more dilute. Therefore, we can combine the curves of Figures 34–10, 34–11, and 34–12 to determine which substances become concentrated in the tubular system and which become diluted. Figure 34–13 illustrates the results of this combination, showing three different classes of substances as follows:

First, the nutritionally important substances—glucose, protein, and amino acids—are reabsorbed so much more rapidly than water that their concentrations fall extremely rapidly in the proximal tubules and remain essentially zero throughout the remainder of the tubular system as well as in the urine.

Second, the concentrations of the metabolic end-products as well as of the artificially injected substances, inulin and para-aminohippuric acid (PAH), become progressively greater throughout the tubular system, because all of these substances are reabsorbed to a far less extent than is water. Note that potassium ions also fall into this category because a far greater proportion of potassium ions is normally

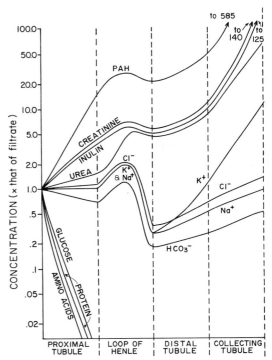

**Figure 34–13.** Composite figure, showing average concentrations of different substances at different points in the tubular system.

removed each day from the extracellular fluid than is water.

Third, many other ions are normally excreted into the urine in concentrations not greatly different from those in the glomerular filtrate and extracellular fluid. That is, sodium, chloride, and bicarbonate ions, on the average, are normally reabsorbed from the tubules in not too dissimilar proportions from water.

Table 34–1 summarizes the concentrating ability of the tubular system for different substances. It also gives the actual quantities of the different substances normally handled by the tubules each minute.

**Variability of Concentrating Power.** A word of caution must be given at this point: the quantitative values for reabsorption of different substances as given in Figures 34–10, 34–11, and 34–12 and for concentrating power of the tubules as given in Figure 34–13 and Table 34–1 are all *average* values. These values change greatly in different physiological states of the kidney. For instance, when essentially no antidiuretic hormone is secreted by the neurohypophyseal mechanism, reabsorption of water by the collecting tubules becomes greatly reduced; the presence of the extra water in the urine dilutes essentially all the substances listed in Table 34–1 by as much as four-fold. Conversely, secretion of large quantities of antidiuretic hormone causes such excessive reabsorption of water that all the substances can be concentrated to as much as four to five times the values given in Table 34–1.

**TABLE 34–1.** **Relative Concentrations of Substances in the Glomerular Filtrate and in the Urine**

| | Glomerular Filtrate (125 ml./min.) | | Urine (1 ml./min.) | | Conc. Urine/ Conc. Plasma (Plasma Clearance per Minute) |
|---|---|---|---|---|---|
| | Quantity/min. | Concentration | Quantity/min. | Concentration | |
| Na+ | 17.7 mEq. | 142 mEq./l. | 0.128 mEq. | 128 mEq./l. | 0.9 |
| K+ | 0.63 | 5 | 0.06 | 60 | 12 |
| Ca++ | 0.5 | 4 | 0.0048 | 4.8 | 1.2 |
| Mg++ | 0.38 | 3 | 0.015 | 15 | 5.0 |
| Cl⁻ | 12.9 | 103 | 0.134 | 134 | 1.3 |
| $HCO_3^-$ | 3.5 | 28 | 0.014 | 14 | 0.5 |
| $H_2PO_4^-$ $HPO_4^{--}$ | 0.25 | 2 | 0.05 | 50 | 25 |
| $SO_4^{--}$ | 0.09 | 0.7 | 0.033 | 33 | 47 |
| Glucose | 125 mg. | 100 mg.% | 0 mg. | 0 mg.% | 0.0 |
| Urea | 33 | 26 | 18.2 | 1820 | 70 |
| Uric acid | 3.8 | 3 | 0.42 | 42 | 14 |
| Creatinine | 1.4 | 1.1 | 1.96 | 196 | 140 |
| Inulin | .... | .... | .... | .... | 125 |
| Diodrast | .... | .... | .... | .... | 560 |
| PAH | .... | .... | .... | .... | 585 |

# THE CONCEPT OF "PLASMA CLEARANCE"

The term "plasma clearance" is used to express the ability of the kidneys to clean, or "clear," the plasma of various substances. Thus, if the plasma passing through the kidneys contains 0.1 gram of a substance in each 100 ml., and 0.1 gram of this substance also passes into the urine each minute, then 100 ml. of the plasma are cleaned or "cleared" of the substance per minute.

Referring again to Table 34–1, note that the normal concentration of urea in each milliliter of plasma and glomerular filtrate is 0.26 mg., and the quantity of urea that passes into the urine each minute is approximately 18.2 mg. Therefore, the equivalent quantity of plasma that completely loses its entire content of urea each minute can be calculated by dividing the quantity of urea entering the urine each minute by the quantity of urea in each milliliter of plasma. Thus, $18.2 \div 0.26 = 70$; that is, 70 ml. of plasma, by filtering through the glomerulus and then being processed by the tubules, *is cleaned* or *cleared* of urea each minute. This is known as the *plasma clearance* of urea.

Plasma clearance for any substance can be calculated by the formula at the bottom of the page.

The concept of plasma clearance is important because it is an excellent measure of kidney function, and the clearance of different substances can be determined by simply analyzing the concentrations of the substances simultaneously in the plasma and in the urine while also measuring the rate of urine formation.

The plasma clearances of the usual constituents of urine are shown in the last column of Table 34–1.

## INULIN CLEARANCE AS A MEASURE OF GLOMERULAR FILTRATION RATE

Inulin is a polysaccharide that has the specific attributes of not being reabsorbed to any extent by the tubules of the nephron and yet being of small enough molecular weight (about 5400) that it passes through the glomerular membrane as freely as the crystalloids and water of the plasma. Also, inulin is not actively secreted even in the minutest amount by the tubules. Consequently, glomerular filtrate contains the same concentration of inulin as does plasma, and as the filtrate flows down the tubules all the inulin continues on into the urine. Thus, *all the glomerular filtrate formed is cleared of inulin, and this is equal to the amount of plasma that is simultaneously "cleared."* Therefore, the plasma clearance per minute of inulin is also equal to the glomerular filtration rate.

As an example, let us assume that it is found by chemical analysis that the inulin concentration in the plasma is 0.1 gram in each 100 ml., and that 0.125 gram of inulin passes into the urine per minute. Therefore, by dividing 0.1 into 0.125, one finds that 1.25 *100-milliliter portions* of glomerular filtrate must be formed each minute in order to deliver to the urine the analyzed quantity of inulin. In other words, 125 ml. of glomerular filtrate is formed per minute, and this is also the plasma clearance of inulin.

Inulin is not the only compound that can be used for determining the quantity of glomerular filtrate formed each minute, for the plasma clearance of any other substance that is totally diffusible through the glomerular membrane but is neither absorbed nor secreted by the tubular walls is equal to the glomerular filtration rate. *Mannitol* is another polysaccharide that is frequently used instead of inulin for such measurements.

## PARA-AMINOHIPPURIC ACID (PAH) CLEARANCE AS A MEASURE OF PLASMA FLOW AND BLOOD FLOW THROUGH THE KIDNEYS

PAH, like inulin, passes through the glomerular membrane with perfect ease along with the remainder of the glomerular filtrate. However, it is different from inulin in that almost all the PAH remaining in the plasma after the glomerular filtrate is formed is secreted into the tubules by the tubular epithelium if the plasma concentration of PAH is very low. Indeed, only about one tenth of the original PAH remains in the plasma when the blood leaves the kidneys.

One can use the clearance of PAH for estimating the *flow of plasma* through the kidneys. As an example, let us assume that 1 mg. of PAH is present in each 100 ml. of plasma, and that 5.85 mg. of PAH passes into the urine per minute. Consequently, 585 ml. of plasma is cleared of PAH each minute. Obviously, if this much plasma is cleared of PAH, *at least this much plasma must have passed through the kidneys* in this same period of time. And, since it is known that almost all the PAH is cleared from the blood as it passes through the kidneys, 585 ml. would be a reasonable first approximation of the actual plasma flow per minute.

Yet, to be still more accurate, one can correct for the average amount of PAH that is still in the blood when it leaves the kidney. In many different experiments the PAH clearance has averaged 91 per cent of the plasma load of PAH entering the kidneys. Therefore, the 585 ml. of plasma calculated above would be only 91 per cent of the total amount of plasma flowing through the kidneys. Dividing 585 by 0.91 gives a total plasma flow per minute of approximately 650 ml.

$$\text{Plasma clearance (ml./min.)} = \frac{\text{Quantity of urine (ml./min.)} \times \text{Concentration in urine}}{\text{Concentration in plasma}}$$

The 91 per cent removal of PAH as it passes through the kidneys is called the *extraction ratio*.

One can calculate the *total blood flow* through the kidneys each minute from the plasma flow and the hematocrit (the percentage of red blood cells in the blood). If the hematocrit is 45 per cent and the plasma flow 650 ml./min., the total blood flow through both kidneys is 650 × (100/55), or 1182 ml./min.

Clearance of *Diodrast* can also be used for determining plasma flow or blood flow through the two kidneys. However, the extraction ratio of Diodrast averages about 0.85, slightly less than that of PAH.

### CALCULATING THE FILTRATION FRACTION FROM PLASMA CLEARANCES

To calculate the filtration fraction—that is, the fraction of the plasma that filters through the glomerular membrane—one must determine (1) the plasma flow through the two kidneys (PAH clearance) and (2) the glomerular filtration rate per minute (inulin clearance). Using 650 ml. plasma flow and 125 ml. glomerular filtration rate as values, we find that the calculated filtration fraction is 125/650, or, to express this as a percentage, 19 per cent.

## EFFECT OF "TUBULAR LOAD" AND "TRANSFER MAXIMUM" ON URINE CONSTITUENTS

**Plasma Load and Tubular Load.** The *plasma load* of a substance means the total amount of the substance in the plasma that passes through the kidney each minute. For instance, if the concentration of glucose in the plasma is 100 mg./100 ml., and 600 ml. of plasma passes through both kidneys each minute, the plasma *load of glucose* is 600 mg./min.

Normally a fraction of the plasma load filters into the glomeruli, and this portion is called the *tubular load*. In the example just given, if 125 ml. of glomerular filtrate is formed each minute with a glucose concentration of 100 mg. per cent, the tubular load of glucose is 100 ml. × 1.25, or 125 mg. of glucose per minute. Similarly, the load of sodium that enters the tubules each minute is approximately 18 mEq./min., the load of chloride ion is about 13 mEq./min., and the load of urea is approximately 33 mg./min., etc.

**Maximum Rate of Transport of Actively Reabsorbed or Secreted Substances—The "Transfer Maximum" (Tm).** Since each substance that is actively reabsorbed requires a specific transport system in the tubular epithelial cells, the maximum amount that can be reabsorbed often depends on the maximum rate at which the transport system itself can operate, and this in turn depends on the total amounts of carrier and specific enzymes available. Consequently, for almost every actively reabsorbed

substance, there is a maximum rate at which it can be reabsorbed; this is called the *transfer maximum* or *transport maximum* for the substance, and the abbreviation is *Tm*. For instance, the Tm for glucose averages 320 mg./min. for the adult human being, and if the tubular load becomes greater than 320 mg./min., the excess above this amount is not reabsorbed but instead passes on into the urine. Ordinarily, though, the tubular load of glucose is only 125 mg./min., so that for practical purposes, all of it is reabsorbed.

Figure 34–14 demonstrates the relationship between tubular load of glucose, transfer maximum (transport maximum) for glucose, and rate of glucose loss into the urine. Note that when the tubular load is at its normal level of 125 mg./min. there is no detectable loss of glucose into the urine. However, when the tubular load rises above about 220 mg./min., significant quantities of glucose begin to appear in the urine. And, once the load has risen above approximately 400 mg./min., the loss into the urine thereafter remains *320 mg./min. less than the tubular load.* Thus, at 400 mg./min. tubular load, the loss is 80 mg./min., and at 800 mg./min. tubular load it is 480 mg./min. In other words, 320 mg./min. of the tubular load, which represents the transfer maximum for glucose, is reabsorbed, and all the remainder is lost into the urine.

***Threshold for Substances That Have a Transfer Maximum.*** Every substance that has a transfer maximum also has a "threshold" concentration in the plasma below which none of it appears in the urine and above which progressively larger quantities appear.

Thus, Figure 34–14 shows that glucose begins to spill into the urine when its tubular load exceeds 220 mg./min. The threshold concentration of glucose in plasma that gives this tubular load is 180 mg. per cent

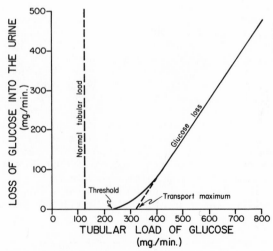

**Figure 34–14.** Relationship of tubular load of glucose to loss of glucose into the urine.

when the kidneys are operating at their normal glomerular filtration rate of 125 ml./min.

***Transfer Maximums of Important Substances Absorbed from the Tubules.*** Some of the important transfer maximums for substances absorbed from the tubules are the following:

| | | |
|---|---|---|
| Glucose | 320 | mg./min. |
| Phosphate | 0.1 | mM./min. |
| Sulfate | 0.06 | mM./min. |
| Amino acids | 1.5 | mM./min. |
| Vitamin C | 1.77 | mg./min. |
| Urate | 15 | mg./min. |
| Plasma protein | 30 | mg./min. |
| Hemoglobin | 1 | mg./min. |
| Lactate | 75 | mg./min. |
| Acetoacetate | variable | (about 30 mg./min) |

It is particularly significant that sodium ions *do not* exhibit a transfer maximum.

***Transfer Maximums for Secretion.*** Many substances secreted by the tubules also exhibit transfer maximums as follows:

| | *mg./min.* |
|---|---|
| Creatinine | 16 |
| PAH | 80 |
| Diodrast | 57 (of iodine) |
| Phenol red | 56 |

## DETERMINATION OF THE TRANSFER MAXIMUM

The transfer maximum for any substance can be determined by loading the tubular cells with far more of the substance than can be transported and then analyzing the amount that actually is transported.

The secretory transfer maximum of either Diodrast ($Tm_D$) or para-aminohippuric acid ($Tm_{PAH}$) is often measured to estimate the amount of active tubular tissue in both kidneys. The normal $Tm_D$ is 57 mg./ min. (measured as iodine content), while the normal $Tm_{PAH}$ is 80 mg./min. in the human adult. Whenever the transfer maximum for either of these substances is less than normal, this is a sign of diminished total amount of tubular cells or tubular enzyme carrier systems for transport of the substance being measured. Therefore, transfer maximum measurements can often give one an excellent assessment of tubular function in the kidneys.

A typical calculation of transfer maximum for Diodrast is the following: When the concentration of Diodrast in the plasma rises above approximately 5 mg. per cent (measured in terms of its iodine content), the tubular walls become unable to excrete all of it from the plasma because the plasma load will have exceeded the transfer maximum. To determine the transfer maximum, inulin clearance must first be determined to measure the quantity of glomerular filtrate formed per minute. Let us assume that 125 ml. of glomerular filtrate is formed each minute. Also, analysis of the plasma Diodrast iodine shows a concentration of 40 mg./100 ml., and analysis of the urine shows 107 mg. of Diodrast iodine entering the urine each minute. From these values it can be calculated that the amount of Diodrast iodine entering the urine by way of the glomerular filtrate is 40 mg. $\times$ 1.25, or 50 mg./min. Subtracting this quantity from the total 107 mg./min., one finds that the Diodrast transfer maximum ($Tm_D$) is 57 mg./minute.

The transfer maximum for glucose or for other substances that are actively reabsorbed can be determined by loading the plasma so greatly that the tubular load becomes greater than that which can be reabsorbed. For instance, if the tubular load of glucose is 400 mg./minute and the amount that passes into the urine is 80 mg./minute, then the transfer maximum for glucose would be 320 mg./min.

## REFERENCES

*See bibliography at end of Chapter 35.*

# | 35 |

# Renal Mechanisms For Concentrating And Diluting The Urine; Urea, Sodium, Potassium, And Fluid Volume Excretion

In the previous chapter we discussed the mechanisms by which glomerular filtrate is formed and the way in which it is then processed in the tubules to become urine. The present chapter will be devoted to the kidney's capability to change the composition of urine from moment to moment to reflect the needs of the body for excreting different substances. For instance, the body at times has great excesses of water and, therefore, must excrete a dilute urine; at other times, it has a deficit of water and, therefore, must excrete a minimum amount of water but still excrete the end-products of metabolism as well as other substances that must be removed from the body fluids.

In this chapter we will discuss principally the ways in which the kidneys can change (1) the osmolar concentration of the urine, (2) the volume of both water and dissolved substances excreted each day, and (3) the rates of excretion of urea, sodium, and potassium. In Chapter 36 we will discuss the feedback systems that control these various excretory functions and in Chapter 37 the ability of the kidneys to alter the urine concentrations of hydrogen ions, chloride ions, and bicarbonate ions and thereby to control the acid-base balance of the extracellular fluid.

## CONCENTRATING AND DILUTING MECHANISM OF THE KIDNEY— THE COUNTER-CURRENT MECHANISM

Since water is reabsorbed from the tubules entirely passively while many solutes in the tubular fluid are absorbed by active absorption, it is easy to understand how a dilute urine could be formed: This could be achieved by simply absorbing the solutes from the glomerular filtrate and leaving the water behind to be excreted in the urine. This is basically the means by which a dilute urine is formed, as we shall see in the following paragraphs.

However, the process for concentrating urine is not nearly so simple, and there are times when it is exceedingly important to concentrate the urine as much as possible to eliminate as many waste products as possible while at the same time conserving water in the body—for instance, when one is exposed to desert conditions with an inadequate supply of water. Fortunately, the kidneys have developed a special mechanism for concentrating the urine, called the *counter-current mechanism*.

The counter-current mechanism depends on a special anatomical arrangement of the loops of Henle and vasa recta. The *loops of Henle* of approximately one fifth of the nephrons, called the *juxtamedullary nephrons,* dip deep into the medulla and then return to the cortex, some dipping all the way to the tips of the papillae that project into the renal pelvis. Also, straight peritubular capillaries called the *vasa recta* loop down into the medulla and then back out of the medulla to the cortex. These arrangements of the different parts of the juxtamedullary nephron and of the vasa recta are illustrated diagrammatically in Figure 35–1.

**Hyperosmolality of the Medullary Fluid.** The normal osmolality of the glomerular filtrate as it enters the proximal tubules is about 300 milliosmols/liter. However, as shown by the numbers in Figure 35–1, the osmolality of the interstitial fluid in the medulla of the kidney becomes progressively greater the more deeply one goes into the medulla, increasing from 300 milliosmols/liter in the cortex to 1200 milliosmols/liter (occasionally as high as 1400 milliosmols/liter) at the pelvic tip of the medulla.

The cause of this greatly increased osmolality of the medullary interstitial fluid is active transport of both sodium and chloride ions out of the thick portion of the loop of Henle's ascending limb and into the medullary interstitial fluid. The very dark arrows through the wall of the ascending limb illustrate this active absorption of sodium and chloride ions. In addition, a slight amount of sodium is actively absorbed through the walls of the collecting tubule into the interstitial fluid—this too, helps to concentrate sodium and chloride ions in the medullary interstitium.

The question that must now be asked is: Why does the osmolar concentration of sodium chloride rise so high in the medulla but not in the fluid of the cortex where equally large quantities of sodium and chloride ions are reabsorbed from the proximal and distal tubules? The answer to this is two-fold: First, the blood flow through the vasa recta is extremely sluggish. Only about 1 to 2 per cent of the total renal blood flow passes through these vessels in contrast to 90 per cent through the cortical peritubular capillaries. Obviously, the sluggish flow in the medulla would help to prevent removal of the sodium chloride that is being continually absorbed into the medullary interstitial fluid.

The second cause of the high concentration of sodium chloride in the medulla is the presence of counter-current mechanisms operating in both the loop of Henle and the vasa recta. These may be explained as follows:

**Counter-Current "Multiplier" of the Loop of Henle.** A counter-current fluid mechanism is one in which fluid flows through a long U

**Figure 35–1.** The "counter-current" mechanism for concentrating the urine. (Numerical values are in milliosmols.)

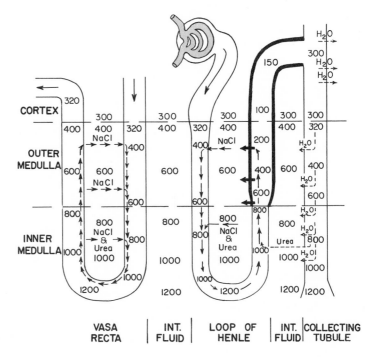

| VASA RECTA | INT. FLUID | LOOP OF HENLE | INT. FLUID | COLLECTING TUBULE |

tube, with the two arms of the U lying in close proximity to each other so that exchange of constituents can take place between the two arms. When the fluids in the two parallel streams of flow interact appropriately with each other, tremendous concentrations of solute can be built up at the tip of the loop. As an example, let us first observe the loop of Henle in Figure 35–1. The descending limb of the loop is highly permeable to sodium chloride, while the thick portion of the ascending limb (some physiologists believe the thin portion, as well) has a strong transport mechanism to remove sodium chloride from the tubule into the interstitial fluid. Therefore, each time sodium chloride is transported out of the ascending limb, it almost immediately diffuses into the descending limb, thereby increasing the concentration of sodium chloride in the tubular fluid that flows downward toward the tip of the loop. As illustrated by the arrows in the diagram, this sodium chloride flows on around the loop and is then actively transported again out of the ascending limb. In addition, new sodium chloride is entering the tubular system in the glomerular filtrate and is also transported out of the ascending limb. Thus, by this retransport again and again of sodium chloride, plus constant addition of more and more sodium chloride, one can readily see how the concentration of sodium chloride in the medullary fluid rises very high. This mechanism of the loop of Henle is called a counter-current multiplier because each time the sodium chloride makes the circuit around the loop of Henle, this "multiplies" the concentration of the sodium chloride in the medulla.

**Counter-Current Mechanism in the Vasa Recta.**    Another factor that helps to explain the very high concentration of sodium chloride and urea in the medulla is the *counter-current flow of blood in the vasa recta*. This is illustrated to the left in Figure 35–1. As blood flows down the descending limbs of the vasa recta, sodium chloride and urea diffuse into the blood from the interstitial fluid, causing the osmolar concentration to rise progressively higher, to a maximum concentration of 1200 milliosmols/liter at the tips of the vasa recta. Then as the blood flows back up the loops, the extreme diffusibility of sodium chloride through the capillary membrane allows essentially all of the extra sodium chloride and urea to diffuse once again back into the interstitial fluid. Therefore, by the time the blood finally leaves the medulla, its osmolar concentration is only slightly greater than that of the blood that had initially entered the vasa

recta. As a result, blood flowing through the vasa recta carries only a minute amount of sodium chloride away from the medulla.

## EXCRETION OF A DILUTE URINE

**Dilution of the Tubular Fluid in the Ascending Limb of the Loop of Henle.**    The thick portion of the ascending limb of the loop of Henle (and perhaps also the thin portion) is highly impermeable to water, and yet tremendous quantities of sodium chloride are actively transported out of the ascending limb into the interstitial fluid. Therefore, as illustrated in Figure 35–1, the fluid in the ascending limb of the loop of Henle becomes progressively more dilute as it ascends toward the cortex, decreasing to an osmolality of about 100 milliosmols/liter before leaving the loop of Henle.

**Mechanism for Excreting A Dilute Urine.**    Now we can easily explain how the kidney can secrete a dilute urine. Since the tubular fluid has a concentration of only 100 milliosmols/liter as it leaves the ascending limb of the loop of Henle, the kidney can excrete a dilute urine by simply allowing this fluid to empty directly into the pelvis of the kidney. And the kidney has a mechanism for doing this. The distal tubules are very poorly permeable to water. Also, in the absence of *antidiuretic hormone,* which is secreted by a hypothalamic-neurohypophyseal mechanism, the epithelial cells of the collecting tubules are almost completely impermeable to water as well. Therefore, in the absence of this hormone almost no water is absorbed from the distal tubules and collecting tubules, and the dilute tubular fluid from the ascending limb of the loop of Henle flushes rapidly on into the urine. In addition, a slight amount of sodium and a few other substances may be actively absorbed in both the distal tubules and the collecting tubules, which further reduces the osmolality—sometimes to as low as 70 milliosmols/liter—by the time the fluid enters the urine; this is about one-fourth the osmolality of the plasma and glomerular filtrate.

## MECHANISM FOR EXCRETING A CONCENTRATED URINE

The mechanism for excreting a concentrated urine is exactly opposite to that for excreting a dilute urine. In this case, large quantities of *antidiuretic hormone* are secreted into the body fluids, and this in turn greatly increases the permeability of the collecting tubules for water.

The very dilute fluid leaving the ascending limb of the loop of Henle rapidly loses water by osmosis in the cortical portion of the collecting tubule, and the osmolality of the tubular fluid rises to equilibrium with that in the cortical interstitial fluids, rising to 300 milliosmols/liter, as illustrated in Figure 35–1. Now, the tubular fluid passes through the medulla a second time, this time downward through the medullary portion of the collecting tubule. Here it is exposed to the hyperosmolality of the medullary interstitial fluid. Therefore, as illustrated in Figure 35–1, large quantities of water are absorbed by osmosis into the medullary interstitial fluid from the collecting tubule, and the concentration of the fluid in the tubule rises to approach the osmolality of the interstitial fluid, about 1200 milliosmols per liter. Thus, the fluid leaving the collecting tubule to enter the pelvis of the kidney is very concentrated urine. The maximum concentration of the urine that can usually be achieved is about four times the osmolality of the plasma and glomerular filtrate.

**Summary of the Osmolar Concentration Changes in the Different Segments of the Tubules.** Figure 35–2 illustrates the changes in osmolality of the fluid as it passes through the different segments of the tubules. In the proximal tubules, the tubular membrane is so highly permeable to water that any time a solute is transported across the membrane almost exactly a proportionate amount of water crosses the membrane by osmosis at the same time; therefore, the osmolality of the fluid remains almost exactly equal to that of the glomerular filtrate, 300 milliosmols/liter, throughout the entire extent of the proximal tubule.

In the descending limb of the loop of Henle the osmolality rises rapidly because of (a) diffusion of sodium and chloride ions into the tubular fluid as it descends in the loop, (b) buildup of urea in the interstitium and diffusion of some of this into the loop, and (c) osmosis of water out of the loop into the interstitial fluid, all of these effects occurring at the same time.

In the ascending limb of the loop of Henle the transport of large quantities of sodium and chloride out of the fluid, and yet impermeability of the membrane to water, causes the osmolality to fall drastically, to almost exactly 100 milliosmols/liter at the point where the fluid leaves the ascending limb of the loop of Henle.

Beyond the loop of Henle the osmolality of the tubular fluid depends on the absorptive processes of the distal and collecting tubules and on the presence or absence of antidiuretic hormone. In the absence of antidiuretic hormone the fluid actually becomes a little more dilute, as shown by the lower border of the darkened area in Figure 35–3, because of active transport of sodium and other substances through the membranes of the distal tubule and collecting duct, falling to a minimum of about 70 milliosmols/liter. In the presence of excess antidiuretic hormone, water diffuses through the cortical portion of the collecting tubule so rapidly that the tubular fluid comes almost to complete equilibrium with the cortical interstitial fluid; that is, to an osmolality of 300 milliosmols/liter. Then, as the fluid passes down the collecting tubule its osmolality comes almost to complete equilibrium with the osmolality of the interstitial fluid of the renal medulla, rising to an osmolality of 1200 milliosmols/liter as it issues into the urine.

All degrees of concentration of the urine can occur between the two extremes of 70 and 1200 milliosmols/liter. This, of course, depends mainly on the quantity of antidiuretic hormone present in the body fluid at any given time. The importance of antidiuretic hormone and its relationship to control of osmolality of the body fluids is discussed in the following chapter.

**Recirculation of Urea Between the Collecting Tubule and the Loop of Henle—A Mechanism for Concentrating Urea in the Medulla.** Sodium

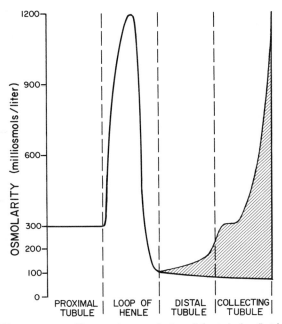

**Figure 35–2.** Changes in osmolarity of the tubular fluid as it passes through the tubular system.

chloride is not the only substance that becomes concentrated in the medulla. Urea is also highly concentrated, but by a different mechanism. This results from the fact that urea is highly concentrated in the collecting tubules and diffuses through the walls of the collecting tubules into the medullary interstitial fluid. From the interstitium the urea is then reabsorbed into the loop of Henle and flows with the tubular fluid up the ascending limb of the loop of Henle, through the distal tubule, and back once again into the collecting tubule. Ordinarily, urea makes this circuit several times before it finally flows out the distal end of the collecting tubule into the urine; in the process, it causes urea to accumulate in high concentration in the medullary interstitium. Figure 35–3 illustrates this recirculation process, showing that the urea osmolality in the glomerular filtrate is only 4.5 milliosmols, but the urea osmolality, even under normal conditions, rises to approximately 300 milliosmols in the lower part of the inner medulla and to approximately 350 milliosmols in the urine. When a person becomes severely dehydrated and the volume of urine flow is very low, these values can rise to more than twice as high.

The osmolality of the urea in the medullary interstitial fluid adds to the osmolality of the sodium chloride to cause the very high osmolality of this fluid. The osmolar values shown in Figure 35–1, rising to as high as 1200 milliosmols in the lower portion of the inner medulla, represent the combined molalities of the sodium chloride, the urea, and also other less important substances when a person is in a slightly dehydrated state.

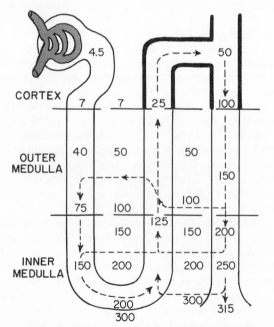

**Figure 35–3.** Recirculation of urea from collecting tubule, then into the loop of Henle, and finally back to the collecting tubule. (Numerical values represent milliosmolalities of urea.)

**Importance of the Urea-Concentrating Mechanism.** The recirculation of urea, which actually constitutes another type of counter-current multiplier system, plays at least one very important and well established role in the process of concentrating the urine, and it probably plays a second role, as well. The well established role is that of causing urea to become far more concentrated in the urine than would otherwise be possible. This concentration results from the continual, simultaneous buildup of urea both in the collecting tubule and in the medullary interstitium. Therefore, before urine issues from the collecting tubule into the pelvis, the urea concentration becomes very high, thus requiring loss of extremely small amounts of water, while nevertheless excreting large quantities of urea.

The second probable importance of this mechanism is that it can (at least theoretically) cause increased sodium chloride concentration in the inner medulla to an even greater extent than can be achieved by the loop of Henle's sodium chloride multiplier system alone. This increased concentration results from the continual dynamic diffusion of urea from the collecting tubule into the medullary interstitium and finally into the loop of Henle, thereby creating a steady state urea concentration gradient across the loop of Henle wall. This, in turn, causes osmosis of water out of the loop of Henle into the interstitium, which concentrates the sodium chloride in the loop of Henle. Consequently, sodium chloride then diffuses from the loop of Henle into the interstitium. In the final stage of this process, as urea diffuses into the loop of Henle it causes osmosis of water back into the loop of Henle, leaving a lagging hyperconcentration of sodium chloride in the interstitium. Though all the steps of this postulated mechanism have not been proved, the mechanism could explain why the sodium chloride becomes even more concentrated in the inner medulla than in the outer medulla, even though the epithelium of the thin segment of the loop of Henle probably cannot actively transport sodium chloride. (There is not unanimity in these beliefs, since some physiologists still suggest that the thin segment epithelium *can* transport sodium chloride despite its appearance of having no significant intracellular metabolic energy system.)

## OSMOLAR CLEARANCE; FREE WATER CLEARANCE

One can calculate the clearance of osmolar substances in terms of the volume of plasma cleared per minute in the same way that one can calculate the clearance of any single substance. That is, osmolar clearance ($C_{osm}$) can be calculated by means of the following formula:

$$C_{osm} = \frac{\text{Osmols entering urine per minute}}{\text{Plasma osmolar concentration}}$$

For instance, if the plasma osmolarity is 300

mosmol./liter, and the total milliosmols entering the urine per minute is 1.5, then the osmolar clearance will be 1.5/300 liters/min., or 5 ml./min.

**Free Water Clearance.**    When the kidney forms urine that is osmotically more dilute than plasma, it is obvious that a higher proportion of the water in the glomerular filtrate is excreted than of the osmolar substances. The excess water that is excreted is called *free water,* and the total plasma volume that is cleared of this excess water each minute is called the *free water clearance*. The free water clearance can be calculated by first determining the osmolar clearance and then subtracting this rate from the rate of urine flow per minute. Thus, the formula for free water clearance ($C_{H_2O}$) is:

$$C_{H_2O} = \text{Urine volume per minute} - C_{osm}$$

The free water clearance is important because it determines how rapidly the kidneys are changing the balance between water and osmotic substances in the body fluid—that is, the effect of renal excretion on the rate of change of body fluid osmolarity. The free water clearance can be either positive, in which case excess water is being removed; or it can be negative, in which case excess solutes are being removed.

# UREA EXCRETION

The body forms an average of 25 to 30 grams of urea each day—more than this in persons who eat a very high protein diet and less in persons who are on a low protein diet. All this urea must be excreted in the urine; otherwise, it will accumulate in the body fluids. Its normal concentration in plasma is approximately 26 mg./100 ml., but it has been recorded in rare abnormal states to be as high as 800 mg./100 ml.; patients with renal insufficiency frequently have levels as high as 200 mg./100 ml.

The two major factors that determine the rate of urea excretion are (1) the concentration of urea in the plasma, and (2) the glomerular filtration rate. These factors increase urea excretion mainly because the load of urea entering the proximal tubules is equal to the product of the plasma urea concentration and the glomerular filtration rate. In general, the quantity of urea that passes on through the tubules into the urine is approximately proportional to the urea load that enters the proximal tubules, averaging 50 to 60 per cent. However, this holds true only when the glomerular filtration rate does not vary too greatly from normal.

**Effect of Extremes of Glomerular Filtration Rate on Urea Excretion.**    When the glomerular filtration rate is very low, the filtrate remains in the tubules for a prolonged period of time before it finally becomes urine. Because all the tubules are at least slightly permeable to urea, the longer the tubular fluid remains in the tubules, the greater is the percentage of reabsorption of the urea into the blood; the proportion of the filtered urea that reaches the urine decreases greatly.

On the other hand, when the glomerular filtration is very high, the fluid passes through the tubular system so rapidly that very little urea is reabsorbed. Thus, at extremely high glomerular filtration rates, almost 100 per cent of the tubular load of urea finally passes into the urine.

An important lesson to be learned from these relationships is that in patients with renal insufficiency it is important for the glomerular filtration rate to be maintained at a high level. When the glomerular filtration rate falls too low, the blood urea concentration rises to a proportionately higher level.

# SODIUM EXCRETION

**Sodium Reabsorption from the Proximal Tubules and Loops of Henle.**    From the previous chapter it will be recalled that approximately 65 per cent of glomerular filtrate is reabsorbed in the proximal tubules. This reabsorption is caused primarily by the active transport of sodium through the proximal tubular epithelium. When the sodium is reabsorbed it causes diffusion of negative ions through the membrane as well, and the cumulative reabsorption of ions creates an osmotic pressure that then moves water through the membranes, too. The epithelium is so permeable to water that almost identically the same proportion of water and sodium ions is reabsorbed.

In the thin segments of the loops of Henle, both sodium and water are again absorbed in approximately equal proportions. However, in the thick segments of the ascending limbs of the loops of Henle both sodium and chloride ions are actively reabsorbed but with very little water reabsorption, and the concentrations of sodium and chloride in the tubular fluid often fall to as low as 1/3 to 1/5 their concentrations in the original glomerular filtrate. Therefore, on the average, less than 10 per cent of the sodium chloride in the original glomerular filtrate still remains in the tubules by the time the fluid reaches the distal tubules.

**Sodium Reabsorption in the Distal Tubules and Collecting Tubules.** Sodium reabsorption in the distal tubules and collecting tubules is highly variable, depending mainly on the concentration of aldosterone, a hormone secreted by the adrenal cortex. In the presence of large amounts of aldosterone, almost the last vestiges of the tubular sodium are reabsorbed from the distal tubules and collecting tubules so that essentially none of the sodium issues into the urine. On the other hand, in the absence of aldosterone, a very large proportion of the sodium is not reabsorbed and passes on into the urine. Thus, the sodium excretion may be as little as a few tenths of a gram per day or as great as 30 to 40 grams. This ability of the tubular system to reabsorb almost all the sodium that filters through the glomeruli is a remarkable feat when one recognizes that an average of 600 grams of sodium filter each day. Indeed, the importance of this conservation of sodium is even more apparent when one realizes that almost 10 times as much sodium enters the glomerular filtrate each day as is present in the entire body.

**Mechanism of Sodium Transport Through the Distal and Collecting Tubular Epithelium.** The mechanism for sodium transport through the distal tubular epithelium is illustrated in Figure 35–4. The movement is caused primarily by active transport of sodium from the interior of the tubular epithelial cell into the lateral intercellular spaces and into the extensive channels at the base of the cell. The transport out of the cell creates a very low concentration of sodium inside the cell and also creates a negative electrical potential of approximately −70 millivolts within the cell. The low sodium

concentration and the very negative potential create a high electrochemical gradient that causes sodium ions to diffuse from the tubular lumen into the cell, thus replacing the sodium as it is transported into the peritubular fluid.

The active transport of sodium out of the epithelial cell is coupled at least partially with active transport of potassium into the cell—potassium being exchanged for sodium. However, the exchange process is generally in favor of more sodium transport than potassium transport, which is the cause of the very negative potential inside the cell. The rates of transport of both sodium and potassium in this exchange process are determined almost entirely by the concentration of aldosterone in the body fluids. Thus, the rates of sodium and potassium transport through both the distal and collecting tubular epithelium are almost entirely under the control of this hormone, a very important factor in the control of both sodium and potassium ion concentration in the extracellular fluids, as will be discussed in the following chapter.

**Mechanism by Which Aldosterone Enhances Sodium and Potassium Transport.** Upon entering a tubular epithelial cell, aldosterone combines with a *receptor protein;* this combination diffuses within minutes into the nucleus where it activates the DNA molecules to form one or more types of messenger RNA. The RNA is then believed to cause formation of carrier proteins or protein enzymes that are necessary for the sodium and potassium transport process.

Ordinarily, aldosterone has no effect on sodium and potassium transport for the first 45 minutes after it is administered; after this time the specific proteins important for transport begin to appear in the epithelial cells, followed by progressive increase in transport during the ensuing few hours.

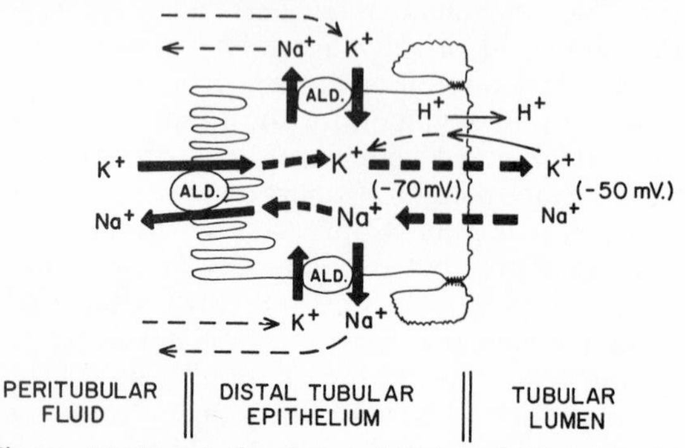

## POTASSIUM EXCRETION

**Potassium Transport in the Proximal Tubules and Loops of Henle.** Potassium is transported through the epithelium of the proximal tubules and of all sections of the loops of Henle in almost exactly parallel fashion to the transport of sodium. The potassium and sodium are transported from the tubule into the blood. They are not, however, transported in an exchange transport process, as occurs in the distal and collecting tubules. Approximately 65 per cent of the potassium in the glomerular filtrate is

**Figure 35–4.** Mechanisms of Na$^+$ and K$^+$ transport through the distal tubular epithelium.

absorbed in the proximal tubules (as is also true for sodium), and another 25 per cent is absorbed in the loops of Henle so that by the time the tubular fluid reaches the distal tubules, the total quantity of potassium delivery to the distal tubules each minute is less than 10 per cent of that in the original glomerular filtrate (as is also true for sodium).

**Active Secretion of Potassium in the Distal Tubules and Collecting Tubules.** Figure 35–4 shows that as sodium is transported from the cytoplasm of the distal tubular epithelial cell into the peritubular fluid, potassium is simultaneously transported into the epithelial cell. Then the potassium diffuses from the cell into the tubular lumen because of its high concentration in the epithelial cell.

This secretory transport of potassium into the distal tubules is extremely important for the control of plasma potassium concentration because of the following simple reason: The total quantity of potassium delivered from the loops of Henle into the distal tubules each day averages only about 70 mEq. Yet, the human being regularly eats this much potassium each day and on occasion eats as much as several hundred mEq. per day. Even if all of the 70 mEq. that enters the distal tubule should pass on into the urine, this still would not be enough potassium elimination. Therefore, it is essential that the excess potassium be removed by the process of secretion; otherwise, death might ensue from potassium toxicity. Indeed, cardiac arrhythmias usually appear when the potassium concentration rises from the normal value of 4 to 5 mEq./liter up to a level of 8 mEq./liter. A slightly higher potassium concentration than this ends in cardiac fibrillation.

**The Occasional Reabsorption of Potassium by the Distal and Collecting Tubules.** In addition to the secretion of potassium by the distal and collecting tubules, there is also continual active reabsorption of potassium through the luminal membrane of the epithelial cell to the inside of the cell, as illustrated by the light solid arrow in Figure 35–4. When sodium transport is suppressed so that the electrical negativity in the tubule is gone, this condition can cause movement of potassium from the tubular lumen into the peritubular fluid, this causing potassium reabsorption rather than secretion.

This reabsorption of potassium from the tubule is usually far overshadowed by potassium secretion into the tubule. However, in the absence of aldosterone, in which case potassium secretion falls essentially to zero, the reabsorptive process then becomes dominant. Thus, in the presence of aldosterone potassium secretion occurs, and in its absence potassium reabsorption occurs.

**Importance of Aldosterone Control of Potassium Secretion.** It is exceedingly important to emphasize that aldosterone plays equally as important a role in determining potassium secretion as in determining sodium reabsorption from the distal and collecting tubules even though the potassium secretion is mainly a secondary result of the sodium reabsorption. Indeed, in the following chapter when we discuss the control of the concentrations of these two ions in the extracellular fluids, we shall see that aldosterone actually plays far more of a role in controlling extracellular fluid potassium concentration than in controlling sodium concentration.

**Role of Tubular Sodium in Potassium Secretion.** Though under normal conditions potassium secretion is determined almost entirely by the aldosterone concentration in the body fluids, under abnormal conditions another factor also plays a significant role in potassium secretion: the amount of sodium delivered in the tubular fluid to the distal and collecting tubules. This determines how much sodium ions can be exchanged for potassium ions in the transport process. If the glomerular filtration rate is very low so that only small quantities of sodium are delivered, or if for any other reason too little sodium is available for exchange, then potassium secretion becomes very deficient.

# FLUID VOLUME EXCRETION

Up to this point we have considered the intrarenal mechanisms that determine the *concentrations* of various substances in the urine—water, urea, sodium, and potassium. Now it is important to consider the different factors that determine the rate of fluid *volume* excretion.

**Glomerulotubular Balance and Its Relationship to Fluid Volume Excretion.** Unfortunately, the term *glomerulotubular balance* means different things to different physiologists. The precise meaning of the words themselves is: exact balance between glomerular filtration rate and tubular reabsorption rate—in which case there would be zero urine output. However, by the term glomerulotubular balance most physiologists mean that whenever the glomerular filtration rate increases, all of the *additional* filtrate is reabsorbed and does not pass into the urine. This is *almost* true for the normal kidney, so that it is usually said that the kidney normally obeys the principle of glomerulotubular balance.

**Glomerulotubular "Imbalance" and Its Importance to Fluid Volume Regulation.** Though the concept of glomerulotubular balance is important to explain the way in which glomerular filtration rate and rate of tubular reabsorption automatically adjust to one another, very precise measurements show that 100 per cent glomerulotubular balance, even in the restricted sense employed by most physiologists, very rarely occurs. For instance, the following table gives approximate values for glomerular filtration rates, rates of fluid reabsorption, and rates of urine output for the average human adult:

| Glomerular Filtration Rate | Rate of Tubular Reabsorption | Rate of Urine Output |
|---|---|---|
| ml. | ml. | ml. |
| 50 | 49.8 | 0.2 |
| 75 | 74.7 | 0.3 |
| 100 | 99.5 | 0.5 |
| 125 | 124.0 | 1.0 |
| 150 | 145.0 | 5.0 |
| 175 | 163.0 | 12.0 |

If we examine these figures critically, we see that glomerular filtration rate and rate of tubular reabsorption actually do appear to parallel each other very closely. On the other hand, the degree of imbalance that occurs causes far greater change, proportionately, in urine output than in either glomerular filtration rate or tubular reabsorption rate. For instance, let us study the increase in glomerular filtration rate from 100 ml./min. up to 150 ml./min. The rate of reabsorption increases from 99.5 ml./min. up to 145 ml./min, representing only slight glomerulotubular imbalance. Nevertheless, this 50 per cent increase in glomerular filtration rate causes a 1000 per cent increase in rate of urine output! Thus, even a very slight degree of glomerulotubular imbalance leads to a tremendous increase in urine output when the glomerular filtration rate is increased. Also, very slight changes in rate of reabsorption of tubular fluid can cause equally as great alterations in urine output.

Therefore, the various factors that can alter either glomerular filtration rate or rate of tubular reabsorption are also the factors that play significant roles in determining the rate of fluid volume excretion. The five most important of these are: (1) the tubular osmolar clearance, (2) the plasma colloid osmotic pressure, (3) the degree of sympathetic stimulation of the kidneys, (4) the arterial pressure, and (5) the effect of antidiuretic hormone on tubular reabsorption.

## 1. EFFECT OF TUBULAR OSMOLAR CLEARANCE ON RATE OF FLUID VOLUME EXCRETION

Under most normal conditions, approximately proportional quantities of solutes and water are reabsorbed from the tubules. That is, whenever an osmolar substance is reabsorbed, so also will a proportionate amount of water be reabsorbed because the change in osmotic gradient across the tubular wall causes osmosis of water from the tubules into the peritubular spaces. Conversely, the greater the quantity of osmolar substances that fails to be reabsorbed by the tubules, the greater the quantity of water that fails to be reabsorbed. To state this another way, when the osmolar clearance is great, the volume of urine usually increases by approximately the same percentage.

**Osmotic Diuresis.** The effect of increased osmolar clearance to cause a proportionate amount of water to be lost in the urine at the same time is called *osmotic diuresis*. It will occur when any nonreabsorbed osmotic substance passes through the tubular system into the urine because the osmotic substance acts osmotically inside the tubules to prevent water reabsorption.

In diabetes mellitus a particularly interesting type of osmotic diuresis occurs in which the proximal tubules fail to reabsorb all the glucose, as normally occurs. Instead, the nonreabsorbed glucose passes the entire distance through the tubules and carries with it a large portion of the tubular water. Therefore, in diabetes mellitus (the word "diabetes" means diuresis) the urine output occasionally increases to as high as 4 to 5 liters per day.

Osmotic diuresis also occurs when substances that cannot be reabsorbed by the tubules are filtered in excessive quantities from the plasma into the glomerular filtrate. For instance, sucrose, mannitol, and urea, when in the circulating plasma in large quantities, are all filtered into the glomerular filtrate and cause large tubular loads of osmotic substances that either are not reabsorbed at all or are reabsorbed very poorly. Therefore, they too cause extreme osmosis diuresis.

## 2. EFFECT OF PLASMA COLLOID OSMOTIC PRESSURE ON RATE OF FLUID VOLUME EXCRETION

Another factor that greatly affects the rate of volume excretion is the plasma colloid osmotic pressure. A sudden increase in plasma colloid osmotic pressure instantaneously decreases the rate of fluid volume excretion. The cause of this effect is two-fold: (1) an increase in plasma colloid osmotic pressure *decreases* glomerular filtration rate, and (2) an increase in plasma colloid osmotic pressure *increases* tubular reabsorption. Both of these effects add together to decrease greatly the urine volume excretion.

**Decreased Glomerular Filtration Rate.** An increase in plasma colloid osmotic pressure of 1 mm. Hg decreases glomerular filtration rate as much as 5 to 10 per cent. The cause of this is simply the decrease in glomerular filtration pressure.

**Increased Tubular Reabsorption.** The effect of increased plasma colloid osmotic pressure to increase tubular reabsorption is less well understood, but the mechanism is believed to be that which is illustrated in Figure 35–5. This figure shows transport of sodium and water from the tubule into the spaces surrounding the epithelial cells. Once the sodium and water (and the other substances that move along with these two) have entered the intercellular spaces and the labyrinth of channels at the bases of the epithelial cells, the fluid can then go one of two ways: either it can leak backward into the tubule through the junctions between the epithelial cells, or it can move forward through the basement membrane into the peritubular capillary.

One of the major factors that determines whether or not the sodium and water will leak backward into the tubular lumen or will move forward into the peritubular capillary is the balance of hydrostatic and colloid osmotic pressures at the capillary membrane. At the bottom of Figure 35–5 the normal balance of pressures is depicted, showing a peritubular capillary pressure of 13 mm. Hg and an interstitial fluid pressure of 6 mm. Hg. This gives a hydrostatic pressure difference of 7 mm. Hg in the outward direction. On the other hand, the plasma colloid osmotic pressure is −32 mm. Hg and the tissue colloid osmotic pressure −15 mm. Hg. This gives a colloid osmotic pressure difference of −17 mm. Hg, which causes fluid movement into the peritubular capillary. Summing the hydrostatic and colloid osmotic forces, one finds that the colloid osmotic forces dominate, giving a net absorption pressure of −10 mm. Hg.

If any one of the four factors affecting the balance at the capillary membrane changes, this obviously will also change the net absorption pressure. Thus, an increase in the plasma osmotic pressure will increase the rate at which the capillary pulls fluid through the basement membrane, which also decreases the rate at which fluid leaks backward into the tubular lumen. To a very great extent, therefore, the rate of fluid absorption from the tubular lumen into the peritubular capillaries is determined by the plasma colloid osmotic pressure.

## 3. EFFECT OF SYMPATHETIC STIMULATION ON RATE OF FLUID VOLUME EXCRETION

Sympathetic stimulation has an especially powerful effect on constriction of the afferent arterioles. It greatly decreases the glomerular pressure and simultaneously decreases glomerular filtration rate. As has already been pointed out, a decrease in glomerular filtration rate often causes 10 times as much decrease (in terms of percentage) in urine output because of the slight degree of glomerulotubular imbalance

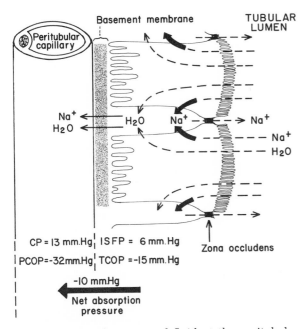

**Figure 35–5.** Absorption of fluid at the peritubular capillary membrane; the effect of this on absorption from the tubule.

that occurs even normally, as was discussed earlier.

Conversely, a decrease of sympathetic stimulation to below normal causes a mild degree of afferent arteriolar dilatation, which increases the glomerular filtration rate a slight amount. Consequently decreased sympathetic stimulation leads to greatly increased urine volume excretion.

## 4. EFFECT OF ARTERIAL PRESSURE ON RATE OF FLUID VOLUME EXCRETION

If all other factors remain constant but the renal arterial pressure is changed, the rate of urine output changes markedly. This effect is illustrated in Figure 35–6 by the dashed curve, which shows that when the arterial pressure rises from 100 mm. Hg to 200 mm. Hg the increase in urine output is approximately sevenfold. Conversely, when the arterial pressure falls from 100 mm. Hg to 60 mm. Hg, the urine output falls either to zero or near to zero. We have already pointed out in Chapter 22 that this pressure effect on urine output plays an extremely important role in the feedback regulation of arterial pressure. It also plays an extremely important role in the feedback regulation of body fluid volume, as we shall discuss in the following chapter. Now, however, let us discuss the mechanism of the increased urine output caused by the increase in arterial pressure. This results from two separate effects: (1) an increase in arterial pressure increases glomerular pressure, which in turn increases glomerular filtration rate, thus leading to increased urine output; (2) an increase in arterial pressure which also leads to increased peritubular capillary pressure, thereby leading to decreased tubular reabsorption. The combination of these two effects causes considerable glomerulotubular imbalance and therefore also causes marked increase in urine output, as illustrated in Figure 35–6.

Which of the above two effects is more important in increasing the urine output—the increase in glomerular filtration or the decrease in reabsorption? Data that are presently available show that an increase in arterial pressure from 100 mm. Hg to 200 mm. Hg increases glomerular filtration rate from approximately 125 ml./min. to 145 ml./min., a total increase of approximately 20 ml./min. At the same time, the urine increases from 1 ml./min. to 7 ml./min., a total increase of 6 ml. Therefore, the tubular reabsorption obviously increases from 124 ml. to 138 ml., an increase of 14 ml./min.

Thus, it is clear from the above mathematical relationships that the increase in urine output usually comes entirely from the increased glomerular filtration. Yet, the simultaneous increase in peritubular capillary pressure undoubtedly prevents still greater increase in tubular reabsorption than the 14 ml./min. Therefore, without the effect of increased peritubular capillary pressure to reduce tubular reabsorption, the increase in urine output probably would be much less.

Regardless of which of the two factors is quantitatively more important, it is nevertheless clear that an increase in arterial pressure causes an extreme increase in rate of fluid volume excretion.

**Long-Term Relationship between Arterial Pressure and Urinary Output.** Figure 35–6 also illustrates that the relationship between arterial pressure and urinary output under short-term experimental conditions is very different from this same relationship under long-term conditions. As pointed out above, the *dashed curve* illustrates the effect of increasing arterial pressure on urinary output in an acute experiment. The very steep *solid curve*, on the other hand, shows the relationship between arterial pressure and urinary output when increasing volumes of fluid are infused into an animal day after day, causing the arterial pressure to rise slowly over a period of days and at the same time causing increased urinary output.

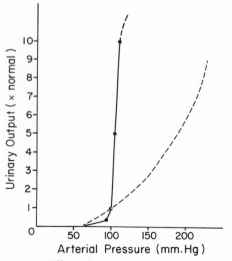

**Figure 35–6.** Effect of arterial pressure on urinary output when the pressure is changed acutely (dashed curve) or chronically (solid curve).

Note that the arterial pressure rises only a very slight amount even when there is extreme increase in urinary output. In experiments of this type in our laboratory, a rise in arterial pressure of only 6 mm. Hg has been found to be associated with about a seven-fold increase in urinary output. Thus, there appears to be an extremely marked increase in urinary output for very small increase in pressure.

There is still a question, however, as to why the long-term relationship is different from that found in the acute experiment. This difference may be caused by extrinsic feedback mechanisms such as sympathetic stimulation, ADH secretion, aldosterone secretion, and so forth that operate in the chronic stituation but do not operate in the acute experiment. However, another possibility is a change in intrinsic formation and function of angiotensin in the kidneys as the arterial pressure changes, which can theoretically make the curve very steep under chronic conditions. This effect will be discussed later in the chapter.

However, for the time being, it should be made very clear that, over a long period of time, the relationship between arterial pressure and urinary output is depicted by an extremely steep curve, indicating that extremely slight increases in arterial pressure might be all that is needed over the long-term to promote extreme increases in urine output.

## 5. EFFECT OF ANTIDIURETIC HORMONE ON RATE OF FLUID VOLUME EXCRETION

When excess antidiuretic hormone is secreted by the hypothalamic-posterior pituitary system, this secretion causes an acute effect to decrease the urinary volume output. The reason for this is that antidiuretic hormone causes increased water reabsorption from the collecting tubules and perhaps to a slight extent from the distal tubules as well. Therefore, less urinary volume is excreted; on the other hand, the urine that is excreted is highly concentrated.

When excess antidiuretic hormone is secreted for long periods of time, the acute effect to decrease urinary output is not sustained. The reason for this is that other factors, such as the arterial pressure, colloid osmotic pressure, and concentrations of the osmolar substances in the glomerular filtrate change—all of which lead eventually to a urinary volume output equal to the daily intake of fluid minus the losses of fluid in other ways from the body. In other words,

balance between body fluid intake and body fluid output is re-established. Therefore, contrary to what is often taught, long-term secretion of excess antidiuretic hormone plays only a small role in the regulation of body fluid volume. As we shall see in discussion of the condition called "inappropriate ADH syndrome" in the following chapter, continued long-term secretion of very large quantities of antidiuretic hormone causes severe abnormalities in body fluid ionic composition even though the body fluid volume is only slightly altered.

**Mechanism by Which Antidiuretic Hormone Increases Water Reabsorption.** The precise mechanism by which antidiuretic hormone increases water reabsorption by the collecting tubules is not known. However, several established facts about the mechanism are the following: Stimulation of the epithelial cells of the collecting tubules by antidiuretic hormone activates *adenyl cyclase* either in the epithelial cell membrane or in the cytoplasm of the cell, and this causes rapid formation of cyclic adenosine monophosphate (cyclic AMP). The increase in cyclic AMP is then associated—for reasons that are not yet known—with marked increase in permeability of the lumenal border of the epithelial cells to water; this in turn is responsible for the increase in water reabsorption by the collecting tubules.

## SUMMARY OF THE CONTROL OF FLUID VOLUME EXCRETION

From the preceding few sections, it is clear that many different factors have an effect on the regulation of urine volume excretion. Some of the factors cause extreme acute changes in urinary output—for instance, the acute decrease in urine volume output caused by antidiuretic hormone or the acute increase in urine volume output caused by increased tubular osmotic loads. However, over a longer period of time, other longer acting factors, such as changes in arterial pressure and colloid osmotic pressure, tend to nullify the extreme acute effects of these first factors.

Especially important in the long-term regulation of urinary output is the increase in output caused by elevated arterial pressure, an effect that seems to be sustained indefinitely. Indeed, present data indicate that this effect, instead of fading, actually increases with time. Therefore, the level of arterial pressure is probably the single most important effect that determines the long-term level of urinary volume excretion.

# AUTOREGULATION OF RENAL BLOOD FLOW AND OF GLOMERULAR FILTRATION RATE

Even though a change in arterial pressure causes marked change in urinary output, the change in arterial pressure does not change either the glomerular filtration rate or the renal blood flow very greatly. This effect is illustrated in Figure 35–7, which shows that an acute increase in arterial pressure from 100 mm. Hg to 200 mm. Hg increases renal blood flow only about 7 per cent and glomerular filtration rate only about 15 per cent, though at the same time it increases urinary output by 600 per cent. This ability of the renal blood flow and glomerular filtration rate to resist severe change is called *autoregulation of renal blood flow and autoregulation of glomerular filtration rate.*

**Mechanism of Autoregulation of Renal Blood Flow and Glomerular Filtration Rate—Possible Function of the Juxtaglomerular Apparatus.** The exact mechanism by which both renal blood flow and glomerular filtration rate are autoregulated is still a mystery. However, there is much reason to believe from recent experiments that this effect results from a feedback effect that occurs at the juxtaglomerular apparatus. The juxtaglomerular apparatus, illustrated in Figure 35–8, shows that the early part of the distal tubule makes contact with the pole of the glomerulus. Furthermore, those epithelial cells of the distal tubule that come in contact with the afferent arteriole are more dense than the other tubular epithelial cells and are collectively called the *macula den-*

*sa.* These cells appear to be secreting some substance toward the afferent arterioles; the Golgi apparatus, a cellular secretory organ, is directed toward the afferent arteriole and not toward the lumen of the tubule as in all the other tubular epithelial cells. On the other hand, those smooth muscle cells of the afferent arteriole that come in close contact with the macula densa are different from the other smooth muscle cells of the afferent arteriole. They are swollen and contain heavy granules composed mainly of inactive *renin*. These cells are called *juxtaglomerular cells;* the whole complex of macula densa and juxtaglomerular cells is called the juxtaglomerular apparatus.

The anatomical structure of the juxtaglomerular apparatus suggests very strongly that the fluid in the distal tubules in some way plays an important role in helping to control blood flow through the afferent arteriole. Also, physiological experiments have shown that an increase in rate of tubular fluid flow through the tubular system causes afferent arteriolar constriction. A few physiologists believe that this results from release of renin, which in turn causes formation of angiotensin and subsequent constriction of the afferent arteriole in response to the angiotensin. However, other physiologists believe that the constriction is caused by some other yet undetermined feedback effect from the macula densa; one suggestion is that it is the sodium concentration in the tubular fluid.

Nevertheless, the very fact that increased fluid flow through the tubular system causes afferent arteriolar constriction indicates that in each nephron there is a feedback mechanism for control of glomerular filtration rate, an effect that allows delivery into the tubules of an amount of glomerular filtrate that the tubules can process—not too much and not little.

Obviously, at the same time that the glomerular filtrate is being autoregulated, blood flow is also autoregulated because the feedback mechanism operates by controlling blood flow, which in turn determines glomerular filtration.

**Importance of Autoregulation of Glomerular Filtration Rate to Maintain Glomerulotubular Balance.** Autoregulation of glomerular filtration rate, which means that the glomerular filtration rate neither falls nor rises drastically in response to arterial pressure changes, plays an extremely important role in the maintenance of glomerulotubular balance. That is, the rate of formation of glomerular filtrate is maintained in an optimum range for proper processing of the filtrate by the tubular system.

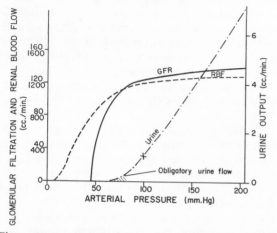

**Figure 35–7.** Autoregulation of renal blood flow *(RBF)* and glomerular filtration rate *(GFR)*, and lack of autoregulation of urine flow.

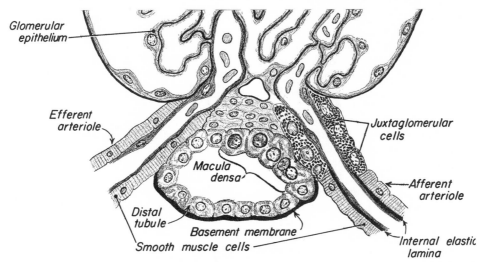

**Figure 35–8.** Structure of the juxtaglomerular apparatus, illustrating its possible feedback role in the control of nephron function. (Modified from Ham: Histology. J. B. Lippincott Co.)

Failure to maintain this optimum glomerular filtration rate causes serious consequences: First, when the glomerular filtration rate is too high, both the urine volume output and the urine sodium output become extreme and causes excessive loss of fluid and salt from the body. Second, even a small decrease in glomerular filtration rate below normal can decrease urea output so much that urea retention begins to occur in the body fluids.

Consequently, maintenance of the glomerular filtration rate at a very constant level seems to be necessary if the kidney is to perform its desired function of retaining needed substances such as water and salt but excreting unwanted elements such as urea, creatinine, and so forth.

## INTRARENAL FUNCTION OF THE RENIN-ANGIOTENSIN SYSTEM

In the above discussion of the juxtaglomerular apparatus it was pointed out that the juxtaglomerular cells in the wall of the afferent arterioles contain granules filled with inactive renin. This renin probably plays an important role in conserving sodium in hypotensive states and in controlling fluid volume excretion.

**Release of Renin from the Juxtaglomerular Granules.** The three physiological effects most likely to cause release of renin from the juxtaglomerular cells are: (1) reduced glomerular filtration rate, (2) reduced glomerular pressure, and (3) increased sympathetic stimulation of the kidneys. It is possible that all three of these actually are one and the same stimulus,

(resulting simply from reduced glomerular filtration rate) because the last two always cause reduced glomerular filtration rate in addition to other effects. One theory is that reduced glomerular filtration rate causes reduced sodium concentration in the tubular fluid as it flows past the macula densa, and that this low sodium is the stimulus that activates release of renin from the granules of the juxtaglomerular cells.

**Formation of Angiotensin.** Once renin is released from the juxtaglomerular cells (also called JG cells) it diffuses into the blood of the afferent arterioles and then circulates throughout the body. In the blood it splits angiotensin I from *renin substrate,* an alpha-2 globulin. When the *angiotensin I,* which is a decapeptide, passes through the lungs, it is further split to the octapeptide *angiotensin II* under the influence of *converting enzyme* located in the lung tissues. Angiotensin II is a very powerful vasoconstrictor agent and can cause extensive vasoconstriction throughout the body.

**Intrarenal Formation of Angiotensin II.** In addition to the formation of angiotensin in the circulating blood, experiments have shown that a reasonable amount of angiotensin II is also formed in the kidney (probably in the JG cells themselves) under the influence of renin and converting enzyme. Therefore, it is reasonable to believe that this angiotensin II has an important intrarenal function.

**Constriction of the Efferent Arterioles by Angiotensin II.** Infusion of extremely minute amounts of angiotensin II into the renal artery causes marked constriction of the efferent arterioles and less constriction of the afferent arterioles. In the normal kidney, in which angiotensin II is formed at the lower end of the afferent arterioles, one would expect the angiotensin to have its effect almost entirely on the efferent arterioles and probably to have an insignificant effect on the afferent arterioles.

**Consequences of Efferent Arteriolar Constriction.** Efferent arteriolar constriction tends to increase the glomerular pressure, but it also decreases renal blood flow. The increase in pressure tends to increase glomerular filtration rate. On the other hand, the decrease in blood flow tends to decrease glomerular filtration rate. Thus, the two opposing effects tend to nullify each other so that, in actuality, efferent arteriolar constriction usually causes little change in glomerular filtration rate.

On the other hand, the decreased blood flow into the peritubular capillaries decreases peritubular capillary pressure. This effect and the increased colloid osmotic pressure of the plasma entering the peritubular capillaries combine to cause *marked increase in tubular reabsorption.* Therefore, the net effect of efferent arteriolar constriction is greatly reduced excretion of fluid volume.

**Effect of the Renin-Angiotensin System in Causing Water and Salt Retention and Yet Normal Excretion of Metabolic Waste Products.** Excretion of metabolic waste products, such as urea and creatinine, occurs primarily by glomerular filtration. When the efferent arterioles are constricted, the glomerular filtration rate remains almost normal. Therefore, excretion of waste products continues almost the same as before. Yet, at the same time, the efferent arteriolar constriction causes marked retention of both water and salt because of enhanced tubular reabsorption. Thus, the specific effect of intrarenal angiotensin is probably to conserve salt and water in the body while still allowing normal excretion of metabolic waste products.

There is much reason to believe that this ability of the renin-angiotensin system to cause specific reabsorption of water and salt plays an essential role in the control of arterial blood pressure. That is, any time the arterial pressure falls, an intrarenal angiotensin-mediated effect occurs to promote water and salt retention while at the same time allowing normal excretion of waste products. The retention of water and salt then helps to return the arterial pressure back to normal.

**Possible Role of the Renin-Angiotensin System to Maintain Balance of Nephron Function in the Two Kidneys.** With 2.4 million nephrons in the human kidneys, it is important to have some mechanism that will maintain approximately equal function among all of these nephrons. Angiotensin can possibly play at least part of this role. It supposedly does so by helping to provide each nephron with an individual feedback control mechanism for maintaining waste product excretion, as described above. However, in addition, the release of renin into the circulating blood by ischemic nephrons, and the subsequent formation of angiotensin II, distribute angiotensin to all the other nephrons that themselves may not be forming large amounts. This angiotensin arrives first at the afferent arterioles and causes a moderate degree of afferent arteriolar constriction and then constricts the efferent arterioles as well; both of these processes cause water and salt retention. Therefore, if one set of nephrons becomes ischemic and their function is reduced as a result of this ischemia, the circulating angiotensin can cause the other nephrons also to retain water and salt. Thus, ischemia of one kidney can cause water and salt retention not only by the ischemic kidney but also by the opposite kidney as well, an effect that has been demonstrated to occur when only one renal artery is constricted. The water and salt retention then causes increased arterial pressure, which helps to relieve the ischemia of the malfunctioning nephrons. This is sometimes of life-saving importance because the increased elimination of waste products by these marginal nephrons is often required to prevent uremia.

# REFERENCES

Barger, A. C., and Herd, J. A.: Renal vascular anatomy and distribution of blood flow *In* Orloff, F., and Berliner, R. W. (eds.): Handbook of Physiology. Sec. 8. Baltimore, The Williams & Wilkins Company, 1973. p. 249.

Bevan, D. R.: The sodium story; effects of anaesthesia and surgery on intrarenal mechanisms concerned with sodium homeostasis. *Proc. R. Soc. Med.,* 66:1215, 1973.

Bianchi, C.: Measurement of the glomerular filtration rate. *Prog. Nucl. Med.,* 2:21, 1972.

Brenner, B. M., and Berliner, R. W.: Transport of potassium. *In* Orloff, F., and Berliner, R. W. (eds.): Handbook of Physiology. Sec. 8. Baltimore, The Williams & Wilkins Company, 1973. p. 497.

Brenner, B. M., and Deen, W. M.: The physiological basis of glomerular ultrafiltration. *In* Guyton, A. C. (ed.): MTP International Review of Science: Physiology. Vol. 6. Baltimore, University Park Press, 1974, p. 335.

Cohen, J. J., and Barac-Nieto, M.: Renal metabolism of substrates in relation to renal function. *In* Orloff, F., and Berliner, R. W. (eds.): Handbook of Physiology. Sec. 8. Baltimore, The Williams & Wilkins Company, 1973, p. 909.

Deetjen, P., Boylan, J. W., and Kramer, K.: Physiology of the Kidney and of Water Balance. New York, Springer-Verlag New York, Inc., 1975.

de Rouffignac, C.: Editorial: Physiological role of the loop of Henle in urinary concentration. *Kidney Int.,* 2:297, 1972.

de Rouffignac, C., and Bonvalet, J. P.: Heterogeneity of nephron population. *In* Guyton, A. C. (ed.): MTP International Review of

Science: Physiology. Vol. 6. Baltimore, University Park Press, 1974, p. 411.

De Wardener, H. E.: The Kidney, 3rd Ed. Boston, Little, Brown and Company, 1968.

Edmonds, C. J.: Salts and water. *Biomembranes,* 4B:711, 1974.

Fourman, J., and Moffat, D. B.: The Blood Vessels of the Kidney. Philadelphia, J. B. Lippincott Company, 1971.

Fromter, E.: Electrophysiology and isotonic fluid absorption of proximal tubules of mammalian kidney. *In* Guyton, A. C. (ed.): MTP Internation Review of Science: Physiology. Vol. 6. Baltimore, University Park Press, 1974, p. 1.

Giebisch, G., and Windhager, E.: Electrolyte transport across renal tubular membranes. *In* Orloff, F., and Berliner, R. W. (eds.): Handbook of Physiology. Sec. 8. Baltimore, The Williams & Wilkins Company, 1973, p. 315.

Gilmore, J. P.: Renal Physiology. Baltimore, The Williams & Wilkins Company, 1972.

Glynn, I. M., and Karlish, S. J. D.: The sodium pump. *Ann. Rev. Physiol.,* 37:13, 1975.

Gottschalk, C. W., Lassiter, W. E., and Mylle, M.: Localization of urine acidification in the mammalian kidney. *Amer. J. Physiol.,* 198:581, 1960.

Gottschalk, C. W., and Lassiter, W. E.: Micropuncture methodology. *In* Orloff, F., and Berliner, R. W. (eds.): Handbook of Physiology. Sec. 8. Baltimore, The Williams & Wilkins Company, 1973, p. 129.

Gottschalk, C. W., Lassiter, W. E., Mylle, M., Ulrich, K. J., Schmidt-Nielsen, B., O'Dell, R., and Pehling, G.: Micropuncture study of composition of loop of Henle fluid in desert rodents. *Amer. J. Physiol.,* 204:532, 1963.

Hierholzer, K., and Lange, S.: The effects of adrenal steroids on renal function. *In* Guyton, A. C. (ed.): MTP International Review of Science: Physiology. Vol. 6. Baltimore, University Park Press, 1974, p. 273.

Knox, F. G.: Effect of increased proximal delivery on furosemide natriuresis. *Amer. J. Physiol.,* 218:819, 1970.

Knox, F. G., and Davis, B. B.: Role of physical and neuroendocrine factors in proximal electrolyte reabsorption. *Metabolism,* 23:793, 1974.

Lassiter, W. E.: Kidney. *Ann. Rev. Physiol.,* 37:371, 1975.

Latta, H.: Ultrastructure of the glomerulus and juxtaglomerular apparatus. *In* Orloff, F., and Berliner, R. W. (eds.): Handbook of Physiology. Sec. 8. Baltimore, The Williams & Wilkins Company, 1973, p. 1.

Malvin, R. L., Wilde, W. S., and Sullivan, L. P.: Localization of nephron transport by stop flow analysis. *Amer. J. Physiol.,* 194:135, 1958.

Maude, D. L.: Mechanism of tubular transport of salt and water. *In* Guyton, A. C. (ed.): MTP International Review of Science: Physiology. Vol. 6. Baltimore, University Park Press, 1974, p. 39.

Maunsbach, A. B.: Ultrastructure of the proximal tubule. *In* Orloff, F., and Berliner, R. W. (eds.): Handbook of Physiology. Sec. 8. Baltimore, The Williams & Wilkins Company, 1973, p. 31.

Mercer, P. F., Maddox, D. A., and Brenner, B. M.: Current concepts of sodium chloride and water transport by the mammalian nephron. *West J. Med.,* 120:33, 1974.

Morel, F., and de Rouffignac, C.: Kidney. *Ann. Rev. Physiol.,* 35:17, 1973.

Mudge, G. H., Berndt, W. O., and Valtin, H.: Tubular transport of urea, glucose, phosphate, uric acid, sulfate, and thiosulfate. *In* Orloff, F., and Berliner, R. W. (eds.): Handbook of Physiology. Sec. 8. Baltimore, The Williams & Wilkins Company, 1973, p. 587.

Navar, L. G., Guyton, A. C., and Langston, J. B.: Effect of alterations in plasma osmolality on renal blood flow autoregulation. *Amer. J. Physiol.,* 211:1387, 1966.

Nechay, B. R.: Relationship between inhibition of renal $Na^+$ plus $K^+$-ATPase and natriuresis. *Ann. N.Y. Acad. Sci.,* 242:501, 1974.

Orloff, J. and Burg, M.: Kidney. *Ann. Rev. Physiol.,* 33:83, 1971.

Osvaldo-Decima, L.: Ultrastructure of the lower nephron. *In* Orloff, F., and Berliner, R. W. (eds.): Handbook of Physiology. Sec. 8. Baltimore, The Williams & Wilkins Company, 1973, p. 81.

Pitts, R.: Physiology of the Kidney and Body Fluids, 3rd Ed. Chicago, Year Book Medical Publishers, Inc., 1974.

Renkin, E. M., and Gilmore, J. P.: Glomerular filtration. *In* Orloff, F., and Berliner, R. W. (eds.): Handbook of Physiology. Sec. 8. Baltimore, The Williams & Wilkins Company, 1973, p. 185.

Renkin, E. M., and Robinson, R. R.: Glomerular filtration. *New Engl. J. Med.,* 290:785, 1974.

Rouiller, C., and Muller, A. F., (ed.): The Kidney. New York, Academic Press, Inc., 1969.

Schrier, R. W.: Effects of adrenergic nervous system and catecholamines on systemic and renal hemodynamics, sodium and water excretion, and renin secretion. *Kidney Int.,* 6:291, 1974.

Smith, H. W.: The Kidney: Structure and Function in Health and Disease. New York, Oxford University Press, Inc., 1951.

Sullivan, L. P.: Physiology of the Kidney. Philadelphia, Lea & Febiger, 1974.

Thurau, K., Valtin, H., and Schnermann, J.: Kidney. *Ann. Rev. Physiol.,* 30:441, 1968.

Walser, M.: Divalent cations: physiocochemical state in glomerular filtrate and renal excretion. *In* Orloff, F., and Berliner, R. W. (eds.): Handbook of Physiology. Sec. 8. Baltimore, The Williams & Wilkins Company, 1973, p. 555.

Weinstein, S. W.: Functional renal metabolism. *In* Guyton, A. C. (ed.): MTP Internation Review of Science: Physiology. Vol. 6. Baltimore, University Park Press, 1974, p. 129.

Windhager, E. E.: Kidney, water, and electrolytes. *Ann. Rev. Physiol.,* 31:117, 1969.

Wolf, A. V.: Urinary concentrative properties. *Amer. J. Med.,* 32:329, 1962.

Wright, F. S.: Intrarenal regulation of glomerular filtration rate. *N. Eng. J. Med.,* 291:135, 1974.

Wright, F. S.: Potassium transport by the renal tubule. *In* Guyton, A. C. (ed.): MTP International Review of Science: Physiology. Vol. 6. Baltimore, University Park Press, 1974, p. 79.

# | 36 |

# Regulation of Blood Volume, Extracellular Fluid Volume, and Extracellular Fluid Composition by the Kidneys and by the Thirst Mechanism

The kidneys, more than any other organ, play determining roles in regulating important characteristics of the body fluids, including: (1) blood volume, (2) extracellular fluid volume, (3) osmolality of the body fluids—that is, the ratio of water to dissolved substances, (4) specific concentrations of the different ions, and (5) degree of acidity of the body fluids. The first four of these will be discussed in the present chapter, and the last—regulation of hydrogen ion concentration—will be discussed in Chapter 37. In controlling some of these characteristics the thirst mechanism also plays a key role; this mechanism will also be discussed.

## CONTROL OF BLOOD VOLUME

**Constancy of the Blood Volume.** The extreme degree of precision with which the blood volume is controlled is illustrated in Figure 36–1. This shows the effect of changing the daily fluid intake, including both water and dissolved electrolytes, from very low values to

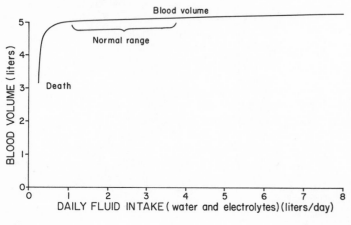

**Figure 36–1.** Effect on blood volume of marked changes in daily fluid intake. Note the precision of blood volume control in the normal range.

very high values. It indicates that almost no change in blood volume occurs despite tremendous changes in intake (except when the intake becomes so low that it is not sufficient to make up for fluid losses caused by evaporation from the surface of the body, evaporation from the lungs, and a small amount of "obligatory" loss of fluid by the kidneys needed for continued excretion of minimal amounts of waste products). To state this another way, even when the intake of water and salt is increased many-fold, the blood volume is hardly altered. Conversely, a decrease in fluid intake to as little as one-third normal causes hardly a change.

## BASIC MECHANISM FOR BLOOD VOLUME CONTROL

The basic mechanism for blood volume control is illustrated in Figure 36–2. This is essentially the same as the basic mechanism for arterial pressure control that was presented in Chapter 22. It was pointed out in the discussion of this earlier chapter that extracellular fluid volume, blood volume, cardiac output, arterial pressure, and urine output are all controlled by a single common basic feedback mechanism. The basic features of this common mechanism are illustrated again in Figure 36–2, but this time to emphasize those factors that are important mainly in blood volume and extracellular fluid volume regulation. Therefore, let us explain the six major steps in this mechanism, illustrated by each of the six blocks in the figure:

Block 1 shows that an increase in blood volume increases cardiac output.

Block 2 illustrates the relationship between cardiac output and arterial pressure, showing that an increase in cardiac output increases arterial pressure.

Block 3 shows that an increase in arterial pressure increases urinary output. The dashed curve is the effect that occurs in acute experiments when the arterial pressure is raised, and the solid curve is the more chronic effect, illustrating the extreme increase in urinary output when the arterial pressure rises only slightly above normal, as was explained in the previous chapter.

Block 4 gives a summation of the fluid intake minus the fluid losses from the body—that is, urinary output and other fluid loss. The output of this block is the rate of change of extracellular fluid volume. If the intake is greater than the output, the rate of change will be positive; if the output is greater than the intake, the rate of change will be negative.

Block 5 integrates the rate of change of extracellular fluid volume—that is, it gives the accumulation of greater or lesser fluid volume with time, depending on whether the rate of change is positive or negative. The output of Block 5 is the actual extracellular fluid volume.

Block 6 gives the relationship between extracellular fluid volume and blood volume, showing that, in general, as the extracellular fluid volume increases, the blood volume also increases. We will discuss this basic relationship in greater detail later in connection with extracellular fluid volume control.

**Summary of the Basic Blood Volume Control Mechanisms.** To summarize the principles illustrated in Figure 36–2, we can trace what happens when the blood volume becomes abnormal. When the blood volume becomes too great, the cardiac output and arterial pressure increase. This, in turn, has a profound effect on the kidneys, causing loss of fluid from the body and returning the blood volume back to normal. Conversely, if the blood volume falls below normal, the cardiac output and arterial pressure decrease, the kidneys retain fluid, and progressive accumulation of the fluid intake builds the blood volume eventually back to normal. Obviously, parallel processes will also occur to reconstitute red cell mass, plasma proteins, and so forth if these have become abnormal at the same time. (If the red cell volume remains abnormal, however, the plasma volume will simply make up the difference, the volume becoming essentially normal despite the low red cell mass.)

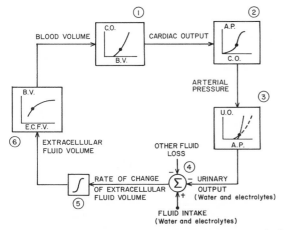

**Figure 36–2.** Basic feedback mechanism for control of blood volume and extracellular fluid volume (the points on each of the curves represent normal values).

**Reason for the Precision of the Blood Volume Regulating Mechanism.** If one studies the diagram of Figure 36–2 very carefully, he can see why the blood volume remains almost exactly constant, as illustrated in Figure 36–1, despite extreme changes in daily fluid intake. The reason for this is that the slopes of the curves in Blocks 1, 2, and 3 are all very steep, meaning that a slight change in blood volume causes a marked change in cardiac output, a slight change in cardiac output causes a marked change in arterial pressure, and a slight change in arterial pressure causes a marked change in urinary output. These factors all multiply together to give an extremely high gain for the feedback control of blood volume.

## ROLE OF THE VOLUME RECEPTORS IN BLOOD VOLUME CONTROL

It was pointed out in Chapter 21 that "volume receptor" reflexes help to control blood volume. These modify the function of the above basic mechanism for blood volume control in a very specific way, as follows:

The basic mechanism of Figure 36–2 is slow to come to equilibrium. That is, when the blood volume increases, the effects on cardiac output, arterial pressure, and urinary output are slow to build up, often requiring many hours to develop full effect; even after they do develop full effect, additional hours are still required for the excess fluid to be eliminated from the body. However, the volume receptor reflexes can greatly accelerate this process.

The volume receptors are mainly stretch receptors located in the walls of the right and left atria. When the blood volume becomes excessive, a large share of this volume accumulates in the central veins of the thorax and causes increased pressure in the two atria. The resultant stretch of the atrial walls transmits nerve signals into the brain, and these in turn elicit responses that accelerate the return of blood volume to normal. The various responses that occur include the following:

1. The sympathetic nervous signals to the kidneys are inhibited, greatly increasing the rate of urinary output.

2. The secretion of antidiuretic hormone by the supraopticohypophyseal system is reduced, allowing increased free water clearance by the kidneys.

3. The peripheral arterioles throughout the body are dilated, increasing capillary pressure and allowing much of the excess blood volume to filter temporarily into the tissue spaces for a few hours until the excess fluid can be excreted through the kidneys.

In most instances, these volume receptor reflex effects can cause the blood volume to return almost all the way to normal within an hour or so, but the final determination of the precise level to which the blood volume will be adjusted is still a function of the basic volume control mechanism illustrated in Figure 36–2. The reason for this condition is that over a period of one to three days the volume receptors finally adapt completely so that they no longer transmit any corrective signals. Therefore, they are of value only to help readjust the volume during the first few hours or few days after an abnormality occurs, but not for long-term monitoring of volume or for precise adjustments of the long-term level of blood volume.

**Role of the Baroreceptors and Other Stretch Receptors in the Volume Receptor Response.** Ordinarily, when the blood volume increases, the systemic arterial pressure and the pulmonary arterial pressure also increase at least to some extent. Therefore, the baroreceptors of the carotid and aortic regions and also of the pulmonary arteries are excited. These, too, cause essentially the same effects as the atrial volume receptors. Therefore, the baroreceptors add still more to the volume receptor reflex effect.

## OTHER FACTORS THAT HELP TO CONTROL BLOOD VOLUME

**Capacity of the Circulation.** Any change in the basic capacity of the circulation will automatically change the level to which the blood volume is regulated. For instance, when a person develops very severe varicose veins, the total capacity of the circulatory system often increases as much as 1 liter. Thereafter, the volume controlling mechanisms will automatically control the blood volume to this higher level.

Other factors that sometimes change the capacity of the system include vasoconstrictors, vasodilators, aneurysms, and so forth. For instance, when a person is under the chronic influence of strong sympathetic vasoconstrictor stimulation or of vasoconstrictor agents such as norepinephrine, the blood volume becomes regulated to a lower level. That is, the volume control system is geared to adjust to the volume of blood to fill the capacity of the system itself.

**Effect of Antidiuretic Hormone and Aldosterone on Blood Volume.** The two hormones antidiuretic hormone and aldosterone both play important roles in some aspects of fluid and electrolyte economy of the body. Therefore, both of them have been extolled as very important blood volume regulators. However, actual measurements show that these normally affect blood volume relatively little. Therefore, let us discuss only briefly their effects on blood volume regulation.

*Antidiuretic hormone.* Complete absence of antidiuretic hormone in the condition called *diabetes insipidus* usually will cause no measurable decrease in blood volume even though it may increase the output of urine by as much as 3- to 10-fold. The thirst mechanism simply causes the person to drink enough water to make up the difference. On the other hand, if the person is prevented from obtaining water to drink, then the blood volume does decrease drastically, and this causes circulatory shock; this combination of effects almost never occurs.

When the secretion of antidiuretic hormone is tremendous, as occurs in the conditions called *inappropriate ADH syndrome,* the blood volume again increases almost imperceptibly—perhaps as much as 3 to 8 per cent. The reason for this very minute increase in blood volume is that the slight volume increase that does occur simply increases the arterial pressure enough (it requires only a minute rise in pressure) to overcome the effect of the antidiuretic hormone to cause water retention.

*Aldosterone.* Aldosterone causes excessive salt reabsorption from the distal and collecting tubules of the kidneys, and this in turn causes osmotic reabsorption of water. Therefore, the immediate effect is to decrease urine output greatly. The extracellular fluid volume and the blood volume both increase. However, before these can increase more than a few per cent, the basic feedback mechanism for blood volume control (Figure 36–2) comes into play and overbalances the retention of fluid by the kidneys. It does this by raising the arterial pressure a few mm. Hg, increasing glomerular filtration rate a few per cent, and therefore increasing the urinary output back to equal the intake of fluid.

Consequently, even in persons who have tremendous secretion of aldosterone (patients with *primary aldosteronism)* the extracellular fluid volume and blood volume rarely rise more than 5 to 10 per cent, at most. On the other hand, in persons who have no secretion of aldosterone (patients with *Addison's disease)* the distal and collecting tubules fail to reabsorb salt and water, and the kidneys lose tremendous quantities of fluids into the urine. If the person eats enough salt and drinks a concomitant amount of water, his blood volume will still regulate in the normal range. But if his salt and water intake are not sufficient, he can develop severe dehydration and hypovolemia with resultant circulatory shock.

# CONTROL OF EXTRACELLULAR FLUID VOLUME

It is already clear from the above discussion of the basic mechanism for blood volume control that extracellular fluid volume is controlled at the same time. That is, when fluid is reabsorbed by the kidneys or is ingested by mouth, the fluid first goes into the blood, but it rapidly becomes distributed between the interstitial spaces and the plasma. Though it is the blood volume that initiates the events to cause increased urinary output and not interstitial fluid volume, fluid will not remain in the circulatory system until the interstitial spaces are also appropriately filled with fluid. Therefore, it is impossible to control blood volume to any given

level without controlling the extracellular fluid volume at the same time. Yet, the relative distribution of volume between the interstitial spaces and the blood can vary greatly, depending on the physical characteristics of the circulatory system and of the interstitial spaces. Under normal conditions, the interstitial spaces are in a relatively "dry" state. That is, the fluid in the interstitial spaces is bound in a gel-like matrix of hyaluronic acid molecules, and there is essentially no free fluid. At other times, however, abnormal conditions can cause edema to occur. These abnormal conditions were discussed in detail in Chapter 31. The principal factors that can cause edema are: (1) increased capillary pressure, (2) decreased plasma colloid osmotic pressure, (3) increased tissue colloid osmotic pressure, and (4) increased permeability of the capillaries. Whenever any one of these conditions occurs, a very high proportion of the extracellular fluid volume becomes distributed to the interstitial spaces, and if the blood volume is still controlled at its normal level, then the interstitial fluid volume and extracellular fluid volume will be abnormally high.

**Normal Distribution of Fluid Volume Between the Interstitial Spaces and the Vascular System.** Figure 36–3 illustrates the approximate normal relationship between extracellular fluid volume and blood volume. In the usual normal operating range for both the circulatory system and the interstitial fluid system, an increase in extracellular fluid volume is associated with an increase in blood volume of about one-fourth to one-third as much. The re-

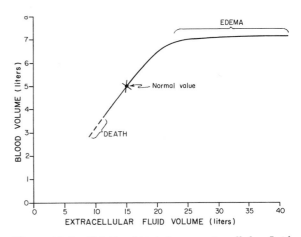

**Figure 36–3.** Relationship between extracellular fluid volume and blood volume, showing a nearly linear relationship in the normal range but indicating failure of the blood volume to continue rising when the extracellular fluid volume becomes excessive.

mainder of the fluid is distributed to the interstitial spaces. However, when the extracellular fluid volume rises considerably above normal, there comes a point at which very little of the additional fluid will remain in the blood—almost all of it instead going into the interstitial spaces. This effect occurs when the interstitial fluid pressure rises from its normal negative value (subatmospheric) to a positive value, because when the interstitial fluid pressure becomes positive, the compliance of the tissue spaces becomes tremendous, so that they can then hold as much as 20 to 40 liters of fluid with very little rise in the interstitial pressure. Therefore, the interstitial fluid spaces, under these conditions, literally become an "overflow" reservoir for excess fluid. This obviously causes edema, but it also acts as an important overflow release valve for the circulatory system, a well known phenomenon that is utilized daily by the clinician to allow him to administer almost unlimited quantities of intravenous fluid and yet not force the heart into cardiac failure.

To summarize, extracellular fluid volume is controlled simultaneously with the control of blood volume, but the relative ratio of the extracellular fluid volume to blood volume depends upon the physical properties of the circulation and of the interstitial spaces, including their compliances and their dynamics.

# CONTROL OF EXTRACELLULAR FLUID SODIUM CONCENTRATION AND EXTRACELLULAR FLUID OSMOLALITY

**Relationship of Sodium Concentration to Extracellular Fluid Osmolality.** The osmolality of the extracellular fluids (and also of the intracellular fluids, since they remain in osmotic equilibrium with the extracellular fluids) is determined almost entirely by the extracellular fluid sodium concentration. The reason for this is that sodium is by far the most abundant cation of the extracellular fluid. Furthermore, the acid-base control mechanisms of the kidneys, which will be discussed in the following chapter, adjust the anion concentrations of the body fluids to equal those of the cations. Finally, sodium is the only significant extracellular cation that exerts much osmotic pressure at the cellular membrane because the other cations—potassium, magnesium, and calcium ions—rep-

resent less than 8 milliosmols of ions anyway, in comparison with sodium at 142 milliosmols. Also, the glucose and urea in the extracellular fluids normally represent only 3 per cent of the osmotic substances, and the urea exerts very little effective osmotic pressure because it penetrates cells far more easily than does sodium. Therefore, in effect, the sodium ion of the extracellular fluid controls 90 to 95 per cent of the *effective* osmotic pressure of the extracellular fluid. Consequently, we can generally talk in terms of control of sodium concentration and control of osmolality at the same time.

Two separate control systems operate in close association to regulate extracellular sodium concentration and osmolality. These are: (1) the osmo-sodium receptor-antidiuretic hormone system, and (2) the thirst mechanism.

## THE OSMO-SODIUM RECEPTOR–ANTIDIURETIC HORMONE FEEDBACK CONTROL SYSTEM

Figure 36–4 illustrates the osmo-sodium receptor–antidiuretic hormone system for control of extracellular fluid sodium concentration and osmolality. It is a typical feedback control system that operates by the following steps:

1. An increase in osmolality (excess sodium) excites *osmoreceptors* located in the supraoptic nuclei of the hypothalamus.

2. Excitation of the supraoptic nuclei causes release of antidiuretic hormone.

3. The antidiuretic hormone increases the permeability of the collecting tubules, as explained in the previous chapter, and therefore causes increased reabsorption of water by the kidneys.

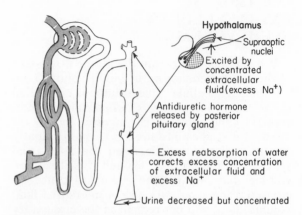

**Figure 36–4.** Control of extracellular fluid osmolality and sodium ion concentration by the osmo-sodium receptor–antidiuretic hormone feedback control system.

4. The increased reabsorption of water causes retention of water in the body while still allowing sodium and other osmotically active substances to be excreted in the urine.

5. The retention of water but loss of sodium causes dilution of the sodium and other osmotically active substances remaining in the extracellular fluid, thus correcting the initial excessively concentrated extracellular fluid.

Conversely, when the extracellular fluid becomes too dilute (hypo-osmotic), less antidiuretic hormone is formed, and excess water is lost in comparison with the extracellular fluid solutes, thus concentrating the body fluids back toward normal.

**The Osmoreceptors (or Osmo-sodium Receptors).** Located in the supraoptic nuclei of the anterior hypothalamus, as shown in Figure 36–5, are specialized neuronal cells called *osmoreceptors*. These respond to changes in osmolality (sodium concentration) of the extracellular fluid. When the osmolality becomes very low, osmosis of water into the osmoreceptors causes them to swell. This decreases their rate of impulse discharge. Conversely, increased osmolality in the extracellular fluid pulls water out of the osmoreceptors, causing them to shrink and thereby to increase their rate of discharge.

The osmoreceptors respond to changes in sodium concentration but not to changes in potassium concentration and only slightly to changes in urea and glucose concentrations. Therefore, it must be emphasized again that, for all practical purposes, the osmoreceptors are actually sodium concentration receptors.

The impulses from the osmoreceptors are transmitted from the supraoptic nuclei through the pituitary stalk into the posterior pituitary gland where they promote the release of antidiuretic hormone (ADH), the details of which will be discussed in Chapter 75 in relation to the endocrinology of the pituitary gland.

Thus, ADH secretion is controlled by the osmolality (or sodium concentration) of the extracellular fluid—the greater the osmolality (excess sodium concentration), the less the rate of ADH secretion. Figure 36–6 illustrates the effect of different levels of extracellular fluid osmolality on antidiuretic hormone concentration in the body fluids, showing a very marked and very important effect on ADH concentration of relatively slight changes in osmolality. (The second curve of this figure illustrates the effect of hemorrhage on ADH concentration. Note especially that the loss of blood volume must be

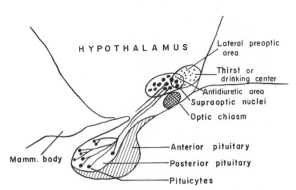

**Figure 36–5.** The supraoptico-pituitary antidiuretic system and its relationship to the thirst center in the hypothalamus.

very marked before this causes a significant increase in the ADH concentration; usually a 10 per cent loss in blood volume is required before the effect can be observed.)

**Water Diuresis.** When a person drinks a large amount of water, a phenomenon called *water diuresis* ensues, a typical record of which is shown in Figure 36–7. In this example, a person drank 1 liter of water, and approximately 30 minutes later his urine output had increased to eight times normal. It remained at this level for two hours—that is, until the osmolality of the extracellular fluid had returned essentially to normal. The delay in onset of water diuresis is

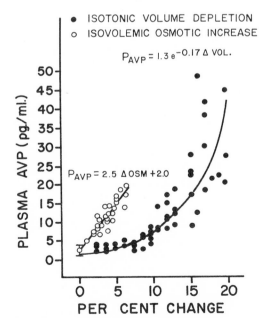

**Figure 36–6.** Effect of changes in plasma osmolality or blood volume on the level of plasma ADH (arginine vasopressin—AVP). (From Dunn et al.: *J. Clin. Invest.*, 52:3212, 1973.)

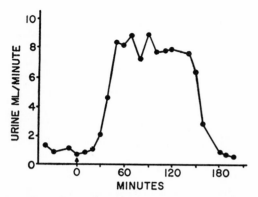

**Figure 36–7.** Water diuresis in a human being following ingestion of 1000 ml. of water. (Redrawn from Smith: The Kidney: Structure and Functions in Health and Disease. Oxford University Press, 1951.)

caused partly by delay in absorption of the water from the gastrointestinal tract but mainly by the time required for destruction of the antidiuretic hormone that has already been released by the pituitary gland prior to drinking the water.

**Diabetes Insipidus.** Destruction of the supraoptic nuclei or high level destruction of the nerve tract (above the median eminence) from the supraoptic nuclei to the posterior pituitary gland causes antidiuretic hormone secretion to cease or at least to become greatly reduced. When this happens, the person thereafter secretes a dilute urine, and his daily urine volume is increased to 5 to 15 liters per day—a condition called *diabetes insipidus*. In diabetes insipidus, the body fluid volumes remain almost exactly normal so long as the thirst mechanism is still functional because this ordinarily makes the person drink water often enough to make up for the increased clearance of water in the urine. On the other hand, any factor that prevents adequate intake of fluid, such as unconsciousness, results in a state of dehydration, tremendous hyperosmolality, and excessive concentration of sodium in the extracellular fluid.

The different lesions in the hypothalamus or pituitary stalk that can cause diabetes inspidus are discussed in Chapter 75.

**Inappropriate ADH Syndrome.** Certain types of tumors, especially of the bronchus or of the basal regions of the brain, can cause excessive secretion of either antidiuretic hormone or a similar hormone. This condition is called *inappropriate ADH syndrome*. The excess ADH ordinarily causes only a slight increase in extracellular fluid volume. Instead, its principal effect is *to decrease greatly the sodium concentration of the extracellular fluid*. The explanation of this effect is the following: The slight increase in extracellular fluid volume causes a decrease in urine output at first and a simultaneous slight increase in blood volume. This in turn activates the basic mechanism for blood volume control, illustrated in Figure 36–2, that is, a slight increase in arterial pressure that increases the urine output back to normal. However, the urine that is excreted is tremendously concentrated because of the tendency of the kidneys to retain water. Consequently, the kidneys excrete extreme amounts of sodium into the urine but keep the water in the extracellular fluids. Therefore, the sodium concentration becomes seriously reduced, sometimes falling from a normal value of 142 mEq./liter down to as low as 110 to 120 mEq./liter. At values this low, patients frequently die sudden deaths because of cardiac fibrillation.

This disease is especially instructive because it illustrates the extreme importance of the antidiuretic hormone mechanism for control of sodium concentration, and yet its relatively mild effect on control of body fluid volume.

## THIRST, AND ITS ROLE IN CONTROLLING SODIUM CONCENTRATION AND OSMOLALITY

The phenomenon of thirst is equally as important for regulating body water and sodium concentration as is the osmoreceptor-renal mechanism discussed above, because the amount of water in the body at any one time is determined by the balance between both *intake* and *output* of water each day. Thirst, the primary regulator of the intake of water, is defined as the *conscious desire for water*.

### Neural Integration of Thirst— the "Thirst" Center

Referring again to Figure 36-5, one sees a small area located slightly anterior to the supraoptic nuclei in the lateral preoptic area of the hypothalamus called the thirst center. Electrical stimulation of this center by implanted electrodes causes an animal to begin drinking within seconds and to continue drinking until the electrical stimulus is stopped. Also, injection of hypertonic salt solutions into the area, which causes osmosis of water out of the neuronal cells and the cells to shrink, also causes drinking. Thus, the neuronal cells of the thirst center function in almost identically the same way as the osmoreceptors of the supraoptic nuclei.

**Basic Stimulus for Exciting the Thirst Center—Intracellular Dehydration.** Any factor that will cause *intracellular dehydration* will in general cause the sensation of thirst. The most common cause of this is increased osmolar concentration of the extracellular fluid, which causes osmosis of fluid from the neuronal cells of the thirst center. However, another important cause is excessive loss of potassium from the body, which reduces the intracellular potassium of the thirst cells and therefore decreases their volume.

### Other Stimuli That Lead to Thirst

**Hemorrhage and Low Cardiac Output.** A small amount of hemorrhage ordinarily does not cause thirst, but loss of as much as 10 per cent of the blood volume, which also reduces the cardiac output significantly, usually does lead to a sensation of thirst. This is almost identically the same effect that hemorrhage has on the secretion of ADH, as was discussed earlier in the chapter.

Also, cardiac failure leads to thirst, especially during the early stages when the person is accumulating large quantities of edema fluid. A possible cause of this effect is loss of potassium from the cells, because low delivery of nutrients to cells is invariably accompanied by potassium loss and intracellular dehydration.

**Dryness of the Mouth.** Almost everyone is aware that a dry mouth is often associated with thirst. The probable explanation for this is that the same factors that cause intracellular dehydration and therefore stimulate the thirst center also cause a dry mouth. Therefore, we have come to associate a dry mouth with the thirst sensation. However, in opposition to this concept is the fact that in animal experiments in which a dry mouth has been achieved by blocking secretion by the salivary glands, the animals do not drink excessively except under one condition: when they are eating food. This seems to result from the need for lubricating the food and not from thirst. The same effect occurs in human beings whose salivary glands do not secrete saliva.

### Temporary Relief of Thirst Caused by the Act of Drinking

A thirsty person receives relief from his thirst immediately after drinking water even before the water has been absorbed from the gastrointestinal tract. In fact, in persons who have esophageal fistulae (a condition in which the water never goes into the gastrointestinal tract), partial relief of thirst still occurs following the act of drinking, but this relief is only temporary, and the thirst returns after 15 minutes or more. If the water does enter the stomach, distension of the stomach and other portions of the upper gastrointestinal tract provides still further temporary relief from thirst. For instance, simple inflation of a balloon in the stomach can relieve thirst for 5 to 30 minutes.

One might wonder what the value of this temporary relief from thirst could be, but there is good reason for its occurrence as follows: After a person has drunk water, as long as one-half to one hour may be required for all of the water to be absorbed into the tissue fluids. Were his thirst sensation not temporarily relieved after the drinking of water, he would continue to drink more and more. When all this water should finally become absorbed, his body fluids would be far more diluted than normal, and he would have created an abnormal condition opposite to that which he was attempting to correct. It is well known that a thirsty animal almost never drinks more than the amount of water needed to relieve his state of dehydration. Indeed, it is uncanny that he usually drinks almost exactly the right amount.

### Role of Thirst in Controlling Osmolality and Sodium Concentration of the Extracellular Fluid

**Threshold for Drinking—the Tripping Mechanism.** The kidneys are continually excreting fluid, and water is also lost by evaporation from the skin and lungs. Therefore, a person is continually being dehydrated, causing the volume of extracellular fluid to decrease and its concentration of sodium and other osmolar elements to rise. When the sodium concentration rises approximately 2 mEq./liter above normal (or the osmolality rises approximately 4 mOsm./liter above normal) the drinking mechanism becomes "tripped" because the person has then reached a level of thirst that is strong enough to activate the necessary motor effort to cause drinking. He ordinarily drinks precisely the proper amount of fluid to bring him back to normal—that is, to a state of *satiety.* Then the process of hyperconcentration of sodium begins again, dehydration eventually occurs once more, and the drinking act is tripped again, the process continuing on and on indefinitely.

In this way, both the sodium concentration and the osmolality of the extracellular fluid are very precisely controlled.

## COMBINED ROLES OF THE ANTIDIURETIC AND THIRST MECHANISMS FOR CONTROL OF EXTRACELLULAR FLUID SODIUM CONCENTRATION AND OSMOLALITY

When either the antidiuretic hormone mechanism or the thirst mechanism fails, the other ordinarily can still control both sodium concentration and extracellular fluid osmolality with reasonable effectiveness. On the other hand, if both of them fail simultaneously, neither sodium nor osmolality is then adequately controlled.

Figure 36–8 gives a dramatic demonstration of the overall capability of the ADH-thirst system to control extracellular fluid sodium concentration. This figure demonstrates the relative ability of the same animal to control its extracellular fluid sodium concentration in two different conditions: (1) in the normal state, and (2) after both the antidiuretic hormone and thirst

mechanisms had been blocked. Note that in the normal animal a six-fold increase in sodium intake caused the sodium concentration to change only two-thirds of 1 per cent (from 142 mEq./liter up to 143 mEq./liter)—an excellent degree of sodium concentration control. Now note the dashed curve of the figure, which shows the change in sodium concentration when the ADH-thirst system was blocked. In this case, sodium concentration increased 10 per cent with only a five-fold increase in sodium intake (a change in sodium concentration from 137 mEq./liter up to 151 mEq./liter), which is an extreme change in sodium concentration when one realizes that the normal sodium concentration rarely rises or falls more than 1 per cent from day to day. Mathematically, the control by the animal in the normal state was 17 times as effective as was true when the ADH-thirst system was blocked.

Therefore, the major feedback mechanism for control of sodium concentration (and also for extracellular osmolality) is the ADH-thirst mechanism. That is, in the absence of this mechanism there is no feedback mechanism that will cause the body to increase water ingestion or water conservation when excess sodium enters the body. Therefore, the sodium concentration simply increases.

## EFFECT OF ALDOSTERONE ON SODIUM CONCENTRATION

A second hormonal system that plays a *small* role in controlling extracellular fluid sodium concentration is the aldosterone feedback system. Figure 36–9 illustrates how slight the effect of the aldosterone system is in controlling plasma sodium concentration. This figure shows the effect on sodium concentration of more than a six-fold increase in sodium intake in the same dog (a) under normal conditions, and (b) after the aldosterone system had been blocked—that is, the adrenal glands had been removed and the animals were infused at a constant rate of aldosterone that could neither change upward nor downward. Note that in the normal state the sodium concentration changed exactly 1 per cent, while when the aldosterone system was blocked it changed exactly 2 per cent. In other words, even without a functional aldosterone feedback system (because the aldosterone could neither change upward nor downward) sodium concentration was still very well regulated.

Because of the great effect that aldosterone has on tubular sodium reabsorption, this lack of importance of aldosterone for regulation of sodium concentration seems to be a paradox, but it results from the following simple effect: When the aldosterone causes sodium reabsorption from the tubules, as was dis-

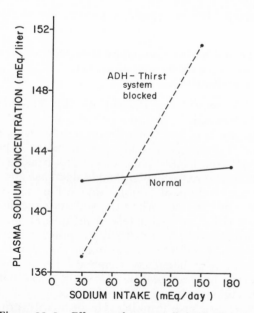

**Figure 36–8.** Effect on the extracellular fluid sodium concentration in a series of dogs of tremendous changes in sodium intake (1) under normal conditions, and (2) after the antidiuretic hormone and thirst feedback systems had been blocked. This figure shows lack of sodium ion control in the absence of these systems. (Courtesy of Dr. David B. Young.)

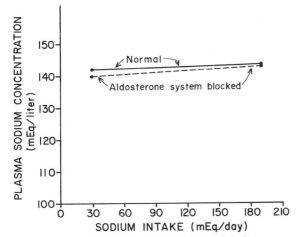

**Figure 36–9.** Effect on extracellular fluid sodium concentration in a series of dogs of tremendous changes in sodium intake (1) under normal conditions, and (2) after the aldosterone feedback system had been blocked. Note that sodium is exceedingly well controlled with or without aldosterone feedback control. (Courtesy of Dr. David B. Young.)

cussed in the previous chapter, this causes a simultaneous reabsorption of water and an increase in extracellular fluid volume. An increase of only a few per cent in the extracellular fluid volume eventually leads to an increase in arterial pressure, and the increase in arterial pressure to increased glomerular filtration rate, a well known effect in the presence of excess aldosterone. The rapid flow of filtrate down the tubular system then compensates for the excessive reabsorptive effect of the aldosterone and thereby almost nullifies the effect of aldosterone on extracellular fluid sodium concentration.

Furthermore, as was explained above, the ADH-thirst system is an extremely powerful controller of sodium concentration—much more powerful than the aldosterone feedback system—so that the ADH-thirst system greatly overshadows the aldosterone system for sodium control under normal conditions. Indeed, even in patients who have primary aldosteronism (these patients secrete tremendous quantities of aldosterone) the sodium concentration still rises only 2 to 3 mEq. above normal.

## CONTROL OF SODIUM INTAKE— APPETITE AND CRAVING FOR SALT

Maintenance of normal extracellular sodium requires not only the control of sodium excretion but also the control of sodium intake. Unfortunately, we know very little about this except that salt-depleted persons (or persons who have lost blood) develop a desire for salt; this occurs especially in persons who have *Addison's disease,* a condition in which the adrenal cortices no longer secrete aldosterone, so that the salt stores of the body become very depleted. The

salt-depleted person craves and eats naturally salty foods. Likewise, it is well known that animals living in areas far removed from the seashore actively search out "salt licks." This craving for salt is analogous to thirst, and it is also analogous to appetite for other types of foods, which is still another homeostatic mechanism that will be discussed in Chapter 73.

## CONTROL OF EXTRACELLULAR POTASSIUM CONCENTRATION —ROLE OF ALDOSTERONE

**Importance of the Aldosterone Feedback System for Control of Potassium Concentration.** In contrast to the very small effect that the aldosterone mechanism plays in the control of sodium concentration, this system plays a very powerful role in the control of extracellular potassium concentration. Indeed, without a functioning aldosterone feedback system, an animal can easily die from either hypopotassemia or hyperpotassemia.

Figure 36–10 illustrates this potent effect of the aldosterone feedback system in control of potassium concentration. In the experiment illustrated in this figure, a series of dogs was subjected to an almost seven-fold increase in potassium intake in two different states: (1) the normal state, and (2) after the aldosterone feedback system had been blocked by removing the adrenal glands and the animals had been given a fixed rate of aldosterone infusion.

Note that in the normal animal the seven-fold increase in potassium intake caused an increase in plasma potassium concentration of only 2.4 per cent—from a concentration of 4.2 mEq./liter to 4.3 mEq./liter. Thus, when the aldosterone feedback system was functioning normally, the potassium concentration remained very precisely controlled despite the tremendous change in potassium intake.

On the other hand, the dashed curve of Figure 36–10 shows the effect after the aldosterone system had been blocked. Note that the same increase in potassium intake now caused a 26 per cent increase in potassium concentration! Thus, the control of potassium concentration in the normal animals was more than ten times as effective as in the animals without an aldosterone feedback mechanism.

**Effect of Primary Aldosteronism and Addison's Disease on Extracellular Fluid Potassium Concentration.** Primary aldosteronism is caused by a tumor of the zona glomerulosa of one of the adrenal glands, the tumor secreting

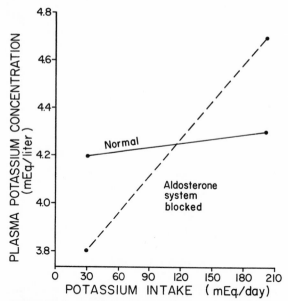

**Figure 36–10.** Effect on extracellular fluid potassium concentration of tremendous changes in potassium intake (1) under normal conditions, and (2) after the aldosterone feedback system had been blocked. This figure demonstrates that potassium concentration is very poorly controlled after block of the aldosterone system. (Courtesy of Dr. David B. Young.)

tremendous quantities of aldosterone. One of the most important effects of this disease is a severe decrease in extracellular fluid potassium concentration, so much so that many of these patients experience paralysis caused by failure of nerve transmission resulting from hyperpolarization of the nerve membranes.

Conversely, in the patient with Addison's disease, whose adrenal glands have been destroyed, the extracellular fluid potassium concentration frequently rises to as high as double normal. This is often the cause of death in these patients, resulting in cardiac arrhythmias that can lead to cardiac fibrillation.

### EFFECT OF POTASSIUM ION CONCENTRATION ON RATE OF ALDOSTERONE SECRETION

In a properly functioning feedback system, the factor that is controlled almost invariably has a feedback effect to control the controller; this is precisely true for the aldosterone potassium control system because the rate of aldosterone secretion is controlled very strongly by the extracellular fluid potassium concentration. Figure 36–11 illustrates the results from a series of dogs in which different rates of potassium

infusion were maintained for several weeks while the aldosterone secretion rate was measured. Note the almost linear relationship between the change in potassium concentration and the aldosterone secretion rate; note also the tremendous increase in aldosterone secretion rate caused by a very minute increase in potassium ion concentration. This extreme feedback effect is the hallmark of a very potent feedback control system.

### BASIC MECHANISM FOR ALDOSTERONE CONTROL OF POTASSIUM CONCENTRATION

In the previous chapter it was pointed out that aldosterone is the primary controller of potassium secretion by the kidneys. Putting this together with the effects illustrated in Figure 36–11, one can construct a very simple system for negative feedback control of potassium concentration, as illustrated in Figure 36–12. That is, an increase in potassium concentration causes an increase in aldosterone concentration (Block 1). The increase in aldosterone concentration then causes a marked increase in potassium excretion by the kidneys (Block 2). The increased potassium excretion then decreases the extracellular fluid potassium concentration back toward normal (Blocks 3 and 4).

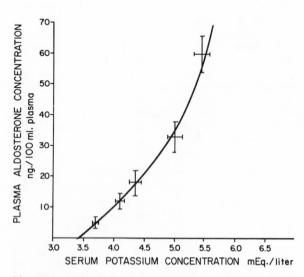

**Figure 36–11.** Effect on extracellular fluid aldosterone concentration of potassium ion concentration changes. Note the extreme change in aldosterone concentration for very minute changes in potassium concentration. (Courtesy of Dr. R. E. McCaa.)

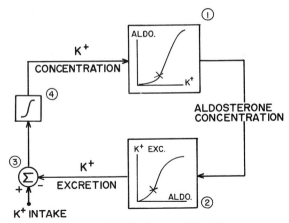

**Figure 36–12.** Simplified schema of the aldosterone system for control of extracellular fluid potassium concentration.

## OTHER FACTORS THAT AFFECT POTASSIUM ION CONCENTRATION

Other factors that affect the extracellular fluid potassium ion concentration include (1) changes in the hydrogen ion concentration (because hydrogen competes with potassium for secretion by the kidney tubules), and (2) the level of sodium intake (because sodium is reciprocally transported through the distal and collecting tubule epithelium in exchange for potassium). However, the simple experiment illustrated in Figure 36–10, in which potassium ion concentration became almost totally uncontrolled in the absence of a functional aldosterone feedback system, illustrates that the aldosterone mechanism is the prepotent one. In experiments in which the sodium intake has been altered markedly, it has been shown that the potassium ion concentration remains almost exactly normal as long as the aldosterone feedback control system functions normally. However, if the aldosterone feedback control system does not function properly, then a marked increase in tubular sodium load will cause excessive excretion of potassium and a serious decrease in extracellular fluid potassium concentration. This again illustrates the prepotency of the aldosterone system for control of potassium concentration.

## OTHER FACTORS THAT CONTROL THE RATE OF ALDOSTERONE SECRETION

Three other factors that are known to affect aldosterone secretion at least to some extent are (1) angiotensin II, (2) adrenocorticotropic hormone (ACTH), and (3) sodium ion concentration. The effects of these on aldosterone secretion will be discussed in detail in Chapter 77 in relation to function of the adrenal glands. It is sufficient to say now that all of these effects appear to be much less potent than the effect of potassium ion. However, one of them, angiotensin II, needs to be mentioned here. Based on

older methods of measurement of aldosterone secretion, unfortunately it is widely taught that angiotensin is the primary controller of aldosterone secretion rate. However, with modern methods of experimentation, it has now been found that a continuous infusion of angiotensin causes a marked transient increase in aldosterone secretion lasting for 24 to 48 hours. However, it causes only slightly increased aldosterone secretion for longer periods of time and therefore probably is not very important as a long-term controller of aldosterone secretion. In contrast, when excess potassium ions stimulate aldosterone secretion, the effect becomes more pronounced with time rather than less pronounced.

# CONTROL OF THE EXTRACELLULAR CONCENTRATIONS OF OTHER IONS

The hydrogen ion concentration of the extracellular fluids, which determines the degree of acidity of the fluids, is so important as a controlling factor in many metabolic systems of the body that it deserves special consideration. Therefore, the following chapter will be devoted entirely to this. Also, in the process of controlling hydrogen ion concentration the chloride and bicarbonate ion concentrations are controlled at the same time; these will be discussed, too, in the following chapter.

Other important ions that require discussion include calcium, magnesium, and phosphate.

**Regulation of Calcium Ion Concentration.** A low concentration of calcium ions in the extracellular fluids causes spontaneous impulses to originate in the peripheral nerves, resulting in tetanic contraction of muscles throughout the body and in death due to spastic respiratory paralysis.

The calcium ion concentration is controlled principally by the parathyroid glands in the following manner: A low level of calcium in the extracellular fluids has a direct effect on the parathyroid glands to promote increased secretion of parathyroid hormone, which in turn acts directly on the bones to increase the rate of reabsorption of bone salts. This releases large amounts of calcium and phosphate ions into the extracellular fluids and thereby elevates the calcium level to normal.

However, calcium is also regulated to some extent by the kidney tubules and gastrointestinal mucosa as well, for large amounts of calcium are known to be lost into the urine and in the feces when its extracellular concentration is high, while very little is lost when the concentration is low. Unfortunately, the total mechanism for this regulation is unknown, though at least part of it results from a direct effect of parathyroid hormone on the renal tubules and on the gastrointestinal mucosa to increase calcium reabsorption (see Chap. 79).

**Regulation of Magnesium Ion Concentration.** Very little is understood about the regulation of magnesium ion concentration. However, magnesium is known to have a transfer maximum. Therefore, whenever its concentration in the glomerular filtrate is very great, the tubular load of magnesium exceeds the transfer maximum and large quantities of magnesium are lost into the urine.

It is possible that there are also active feedback mechanisms for control of magnesium concentration. The similarity of handling of magnesium by many cells of the body to the handling of potassium suggests that some aspects of magnesium control might be similar to those for control of potassium.

**Regulation of Phosphate Concentration.** Phosphate concentration is regulated primarily by an *overflow* mechanism, which can be explained as follows: The renal tubules have a normal transfer maximum of 0.1 millimol of phosphate per minute. When less than this "load" of phosphate is present in the glomerular filtrate, all of it is reabsorbed. When more than this amount is present, the excess is excreted. Therefore, normally, phosphate ion spills into the urine when its concentration in the plasma is above the threshold value of approximately 0.8 millimol/liter. Any time the concentration falls below this value, all the phosphate is conserved in the plasma, and the daily ingested phosphate accumulates in the extracellular fluid until its concentration rises above the threshold. On the other hand, whenever the phosphate concentration rises above this level, the excess is excreted into the urine. Since most people ingest large quantities of phosphate day in and day out, either in milk or in meat, the concentration of phosphate is usually maintained at a level of about 1.0 millimol/liter, a level at which there is continual overflow of excess phosphate into the urine.

*Role of Parathyroid Hormone in Phosphate Ion Regulation.* Parathyroid hormone, which plays a major role in regulation of calcium ion concentration as explained above, also affects phosphate ion concentration in two different ways. First, parathyroid hormone promotes bone reabsorption, thereby dumping large quantities of phosphate ions into the extracellular fluid from the bone salts. Second, this hormone decreases the transfer maximum for phosphate by the renal tubules so that a greater proportion of the tubular phosphate is lost in the urine; also phosphate increases glomerular filtration. The combination of these factors causes marked loss of phosphates in the urine. The interrelationships between phosphate, calcium, and parathyroid hormone control will be discussed in Chapter 79.

**Regulation of Other Negative Ions.** Other less important negative ions in the body fluid include sulfates, nitrates, urates, lactates, and the amino acids. Essentially all of these, like phosphate, have definite transfer maximums. When the concentration of each is below its respective threshold, it is conserved in the extracellular fluid, but when above this threshold the excess begins to spill into the urine. Thus, the concentrations of most of these negative ions are regulated by the overflow mechanism in the same way that phosphate ion concentration is regulated.

# REFERENCES

Adolph, E. F.: Regulation of water intake in relation to body water content. In Code, C. F., and Heidel, W. (eds.): Handbook of Physiology. Sec. 6, Vol. 1. Baltimore, The Williams & Wilkins Company, 1967, p. 163.

Crawford, M. P., Richardson, T. Q., and Guyton, A. C.: Renal servocontrol of arterial blood pressure. J. App. Physiol., 22:139, 1967.

Davis, J. O.: The control of renin release. Amer. J. Med., 55:333, 1973.

Denton, D. A.: Salt appetite. Proc. Int. Union of Physiol. Sci., 6:61, 1968.

DeRubertis, F. R., Michelis, M. F., and Davis, B. B.: "Essential" hypernatremia. Report of three cases and review of the literature. Arch. Intern. Med., 134:889, 1974.

de Wardener, H. E.: The control of sodium excretion. In Orloff, F., and Berliner, R. W. (eds.): Handbook of Physiology. Sec. 8. Baltimore, The Williams & Wilkins Company, 1973, p. 677.

Earley, L. E., and Schrier, R. W.: Intrarenal control of sodium excretion by hemodynamic and physical factors. In Orloff, F., and Berliner, R. W. (eds.): Handbook of Physiology. Sec. 8. Baltimore, The Williams & Wilkins Company, 1973, p. 721.

Fitzsimons, J. T.: Thirst. Physiol. Rev., 52:468, 1972.

Fourcade, J. C., Navar, L. G., and Guyton, A. C.: Possibility that angiotensin resulting from unilateral kidney disease affects contralateral renal function. Nephron, 8:1, 1971.

Gertz, K. H., and Boylan, J. W.: Glomerular-tubular balance. In Orloff, F., and Berliner, R. W. (eds.): Handbook of Physiology, Sec. 8. Baltimore, The Williams & Wilkins Company, 1973, p. 763.

Goetz, K. L., Bond, G. C., and Bloxham, D. D.: Atrial receptors and renal function. Physiol. Rev., 55:157, 1975.

Gottschalk, C. W., and Mylle, M.: Micropuncture study of the mammalian urinary concentrating mechanism: evidence for the countercurrent hypothesis. Amer. J. Physiol., 196:927, 1959.

Grantham, J. J.: Action of antidiuretic hormone in the mammalian kidney. In Guyton, A. C. (ed.): MTP International Review of Science: Physiology. Vol. 6. Baltimore, University Park Press, 1974, p. 247.

Guyton, A. C., Langston, J. B., and Navar, G.: Theory for renal autoregulation by feedback at the juxtaglomerular apparatus. Circ. Res., 14:187, 1964.

Handler, J. S., and Orloff, J.: The mechanism of action of antidiuretic hormone. In Orloff, F., and Berliner, R. W. (eds.): Handbook of Physiology. Sec. 8. Baltimore, The Williams & Wilkins Company, 1973, p. 791.

Hayward, J. N.: Neural control of the posterior pituitary. Ann. Rev. Physiol., 37:191, 1975.

Hierholzer, K., and Lange, S.: The effects of adrenal steroids on renal function. In Guyton, A. C. (ed.): MTP International Review of Science: Physiology. Vo.l. 6. Baltimore, University Park Press, 1974, p. 273.

Hope, D. B., and Pickup, J. C.: Neurophysins. In Greep. R. O., and Astwood, E. B. (eds.): Handbook of Physiology. Sec. 7, Vol., 4, Part 1. Baltimore, The Williams & Wilkins Company, 1974, p. 173.

Jamison, R. L.: Countercurrent system. In Guyton, A. C. (ed.): MTP International Review of Science: Physiology. Vol. 6. Baltimore, University Park Press, 1974, p. 199.

Klahr, S., and Slatopolsky, E.: Renal regulation of sodium excretion. Function in health and in edema-forming states. Arch. Intern. Med., 131:780, 1973.

Lee, J. B.: Prostaglandins and the renal antihypertensive and natriuretic endocrine function. Recent Prog. Horm. Res., 30:481, 1974.

Lee, J., and deWardener, H. E.: Neurosecretion and sodium excretion. Kidney Int., 6:323, 1974.

Levinsky, N. G.: Natriuretic hormones. Adv. Metab. Disord., 7:37, 1974.

McCaa, R. E., McCaa, C. S., Read, D. G., Bower, J. D., and Guyton, A. C.: Increased plasma aldosterone concentration in response to hemodialysis in nephrectomized man. *Circ. Res.*, 31:473, 1972.

Moses, A. M., and Miller, M.: Osmotic influences on the release of vasopressin. *In* Greep, R. O., and Astwood, E. B. (eds.): Handbook of Physiology. Sec. 7, Vol. 4, Part 1. Baltimore, The Williams & Wilkins Company, 1974, p. 225.

Navar, L. G., Guyton, A. C., and Langston, J. B.: Effect of alterations in plasma osmolality on renal blood flow autoregulation. *Amer. J. Physiol.*, 211:1387, 1966.

Ott, C. E., Navar, L. G., and Guyton, A. C.: Pressures in static and dynamic states from capsules implanted in the kidney. *Amer. J. Physiol.*, 221:392, 1971.

Robertson, G. L.: Vasopressin in osmotic regulation in man. *Ann. Rev. Med.*, 25:315, 1974.

Samueloff, S.: Metabolic aspects of desert adaptation (man). *Adv. Metab. Disord.*, 7:95, 1974.

Sawyer, W. H.: The mammalian antidiuretic response. *In* Greep, R. O., and Astwood, E. B. (eds.): Handbook of Physiology. Sec. 7, Vol. 4, Part 1. Baltimore, The Williams & Wilkins Company, 1974, p. 443.

Schmid, H. E., Jr., and Spencer, M. P.: Characteristics of pressure-flow regulation by the kidney. *J. Appl. Physiol.*, 17:201, 1962.

Schmidt-Nielsen, B., and Laws, D. F.: Invertebrate mechanisms for diluting and concentrating the urine. *Ann. Rev. Physiol.*, 25:631, 1963.

Schnermann, J.: Physical forces and transtubular movement of solutes and water. *In* Guyton, A. C. (ed.): MTP International Review of Science: Physiology. Vol. 6. Baltimore, University Park Press, 1974, p. 157.

Share, L., and Claybaugh, J. R.: Regulation of body fluids. *Ann. Rev. Physiol.*, 34:235, 1972.

Sharp, G. W. G., and Leaf, A.: Effects of aldosterone and its mechanism of action on sodium transport. *In* Orloff, F., and Berliner, R. W. (eds.): Handbook of Physiology. Sec. 8. Baltimore, The Williams & Wilkins Company, 1973, p. 815.

Stein, J. H., and Reineck, H. J.: Effect of alterations in extracellular fluid volume on segmental sodium transport. *Physiol. Rev.*, 55:127, 1975.

Thurau, K.: The intrarenal function of the juxtaglomerular apparatus. *In* Guyton, A. C. (ed.): MTP International Review of Science: Physiology. Vol. 6. Baltimore, University Park Press, 1974, p. 357.

Verney, E. B.: Absorption and excretion of water; antidiuretic hormone. *Lancet*, 2:739, 1946.

Wayner, M. J.: The lateral hypothalamus and adjunctive drinking. *Prog. Brain Res.*, 41: 371, 1974.

Windhager, E. E.: Glomerulo-tubular balance of salt and water. *Physiologist*, 11:103, 1968.

Wirz, H., and Dirix, R.: Urinary concentration and dilution. *In* Orloff, F., and Berliner, R. W. (eds.): Handbook of Physiology. Sec. 8. Baltimore, The Williams & Wilkins Company, 1973, p. 415.

Wolf, A. V.: Thirst: Physiology of the Urge to Drink and Problems of Water Lack. Springfield, Ill., Charles C Thomas, Publishers, 1958.

Wolf, G., McGovern, J. F., and Dicara, L. V.: Sodium appetite: some conceptual and methodologic aspects of a model drive system. *Behav. Biol.*, 10:27, 1974.

Wright, F. S.: Potassium transport by the renal tubule. *In* Guyton, A. C. (ed.): MTP International Review of Physiology. Vol. 6. Baltimore, University Park Press, 1974, p. 79.

# | 37 |

# Regulation of Acid-Base Balance

When one speaks of the regulation of acid-base balance he actually means regulation of hydrogen ion concentration in the body fluids. The hydrogen ion concentration in different solutions can vary from less than $10^{-14}$ equivalents per liter to higher than $10^1$, which means a total variation of more than a quadrillion-fold. On a logarithmic basis, the hydrogen ion concentration in the human body is approximately midway between these two extremes.

Only slight changes in hydrogen ion concentration from the normal value can cause marked alterations in the rates of chemical reactions in the cells, some being depressed and others accelerated. For this reason the regulation of hydrogen ion concentration is one of the most important aspects of homeostasis. Later in the chapter the overall effects of high hydrogen ion concentration (acidosis) and low hydrogen ion concentration (alkalosis) are discussed. In general, when a person becomes acidotic he is likely to die in coma, and when he becomes alkalotic he may die of tetany or convulsions.

**Normal Hydrogen Ion Concentration and Normal pH of the Body Fluids—Acidosis and Alkalosis.** The hydrogen ion concentration in the extracellular fluid is normally regulated at a constant value of approximately $4 \times 10^{-8}$ Eq./liter; this value can vary from as low as $1.6 \times 10^{-8}$ to as high as $1.2 \times 10^{-7}$.

From these values, it is already apparent that expressing hydrogen ion concentration in terms of actual concentrations is a cumbersome procedure. Therefore, the symbol *pH* has come into usage for expressing the concentration, and pH is related to the actual hydrogen ion concentration by the following formula (when $H^+$ conc. is expressed in equivalents per liter):

$$pH = \log \frac{1}{H^+ \text{ conc.}} = -\log H^+ \text{ conc.} \quad (1)$$

Note from this formula that a low pH corresponds to a high hydrogen ion concentration, which is called *acidosis;* and, conversely, a high pH corresponds to a low hydrogen ion concentration, which is called *alkalosis*.

The normal pH of arterial blood is 7.4, while the pH of venous blood and of interstitial fluids is about 7.35 because of extra quantities of carbon dioxide which form carbonic acid in these fluids.

Since the normal pH of the arterial blood is 7.4, a person is considered to have acidosis whenever the pH is below this value and to have alkalosis when it rises above 7.4. The lower limit at which a person can live more than a few hours is about 7.0, and the upper limit approximately 7.7.

**Intracellular pH.** On the basis of indirect measurements, it has been found that the intracellular pH usually ranges between 4.5 and 7.4 in different cells, perhaps averaging about 7.0. A *rapid rate of metabolism* in cells increases the rate of carbon dioxide formation and consequently decreases the pH. Also, *poor blood flow* to any tissue causes carbon dioxide accumulation and a decrease in pH.

## DEFENSE AGAINST CHANGES IN HYDROGEN ION CONCENTRATION

To prevent acidosis or alkalosis, several special control systems are available: (1) All the body fluids are supplied with acid-base *buffer systems* that immediately combine with any acid or alkali and thereby prevent excessive changes in hydrogen ion concentration. (2) If the hydrogen ion concentration does change measurably, the *respiratory center is immediately stimulated* to alter the rate of pulmonary ventilation. As a result, the rate of carbon dioxide removal from the body fluids is automatically changed, and, for reasons that will be presented later, this causes the hydrogen ion concentration to return toward normal. (3) When the hydrogen ion concentration changes from normal, *the kidneys excrete either an acid or alkaline urine,* thereby also helping to readjust the hydrogen ion concentration of the body fluids back toward normal.

The buffer systems can act within a fraction of a second to prevent excessive changes in hydrogen ion concentration. On the other hand, it takes one to three minutes for the respiratory system to readjust the hydrogen ion concentration after a sudden change has occurred. Finally, the kidneys, though providing the most powerful of all the acid-base regulatory systems, require several hours to a day or more to readjust the hydrogen ion concentration.

## FUNCTION OF ACID-BASE BUFFERS

An acid-base buffer is a solution of two or more chemical compounds that prevents marked changes in hydrogen ion concentration when either an acid or a base is added to the solution. As an example, if only a few drops of concentrated hydrochloric acid are added to a beaker of pure water, the pH of the water might immediately fall to as low as 1.0. However, if a satisfactory buffer system is present, the hydrochloric acid combines instantaneously with the buffer, and the pH falls only slightly. Perhaps the best way to explain the action of an acid-base buffer is to consider an actual simple buffer system, such as the bicarbonate buffer, which is extremely important in regulation of acid-base balance in the body.

### THE BICARBONATE BUFFER SYSTEM

A typical bicarbonate buffer system consists of a mixture of carbonic acid ($H_2CO_3$) and sodium bicarbonate ($NaHCO_3$) in the same solution. It must first be noted that carbonic acid is a very weak acid for two reasons: First, its degree of dissociation into hydrogen ions and bicarbonate ions is poor in comparison with that of many other acids. Second, about 999 parts out of 1000 of any carbonic acid in a solution almost immediately dissociate into carbon dioxide and water, the net result being a high concentration of dissolved carbon dioxide but only a weak concentration of acid.

When a strong acid, such as hydrochloric acid, is added to a buffer solution containing bicarbonate salt, the following reaction takes place:

$$HCl + NaHCO_3 \rightarrow H_2CO_3 + NaCl \qquad (2)$$

From this equation it can be seen that the strong hydrochloric acid is converted into the very weak carbonic acid. Therefore, the HCl lowers the pH of the solution only slightly.

On the other hand, if a strong base, such as sodium hydroxide, is added to a buffer solution containing carbonic acid, the following reaction takes place:

$$NaOH + H_2CO_3 \rightarrow NaHCO_3 + H_2O \qquad (3)$$

This equation shows that the hydroxyl ion of the sodium hydroxide combines with the hydrogen ion from the carbonic acid to form water and that the other product formed is sodium bicarbonate. The net result is exchange of the strong base NaOH for the weak base $NaHCO_3$.

Though this bicarbonate buffer system has been illustrated in the above reactions as a mixture of carbonic acid and *sodium* bicarbonate, any other salt of bicarbonate besides sodium bicarbonate has identically the same function. Therefore, the small quantities of potassium bicarbonate, calcium bicarbonate, and magnesium bicarbonate in the extracellular fluids are also effective in the bicarbonate buffer system. In the intracellular fluid, where little sodium bicarbonate is present, the bicarbonate ion is provided mainly as potassium and magnesium bicarbonate.

### Quantitative Dynamics of Buffer Systems

**Dissociation of Carbonic Acid.** All acids are ionized to a certain extent, and the percentage of ionization is called the *degree of dissociation.* Equation 4 illustrates the reversible relationship between undissociated carbonic acid and the two ions that it forms, $H^+$ and $HCO_3^-$.

$$H_2CO_3 \rightleftarrows H^+ + HCO_3^- \qquad (4)$$

A physicochemical law that has been found to apply to the dissociation of all acids is expressed by the following formula:

$$\frac{H^+ \times HCO_3^-}{H_2CO_3} = K' \qquad (5)$$

This formula states that in any given solution the concentration of hydrogen ions times the concentration of bicarbonate ions divided by the concentration of undissociated carbonic acid is equal to a constant, K .

However, it is almost impossible to measure the concentration of undissociated carbonic acid in a solution because it is also continually in reversible equilibrium with dissolved carbon dioxide in the solution. Ordinarily, the amount of dissolved carbon dioxide is approximately 1000 times the concentration of the undissociated acid. On the other hand, it is possible to measure the total amount of dissolved carbon dioxide, and, since the amount of undissociated carbonic acid is proportional to the amount of dissolved carbon dioxide, Formula 5 above can also be expressed as follows:

$$\frac{H^+ \times HCO_3^-}{CO_2} = K \tag{6}$$

The only real difference between the above two formulas is that the constant $K'$ is approximately 1000 times the constant K.

Formula 6 can be changed into the following form:

$$H^+ = K \cdot \frac{CO_2}{HCO_3^-} \tag{7}$$

If we take the logarithm of each of the two sides of Formula 7, it becomes the following:

$$\log H^+ = \log K + \log \frac{CO_2}{HCO_3^-} \tag{8}$$

Now, the signs of the log H$^+$ and of log K are changed from positive to negative and the carbon dioxide and bicarbonate are inverted in the last term, which is the same as changing its sign also, giving the following formula:

$$-\log H^+ = -\log K + \log \frac{HCO_3^-}{CO_2} \tag{9}$$

It will be recalled from earlier in the chapter that $-\log H^+$ is equal to the pH of the solution. Likewise, *-log K is called the pK of a buffer.* Therefore, this formula can be changed still further to the following:

$$pH = pK + \log \frac{HCO_3^-}{CO_2} \tag{10}$$

**The Henderson-Hasselbalch Equation.** For the bicarbonate buffer system the pK is 6.1, and Formula 10 may be expressed as follows:

$$pH = 6.1 + \log \frac{HCO_3^-}{CO_2} \tag{11}$$

This is called the Henderson-Hasselbalch equation, and by using it one can calculate the pH of a solution

with reasonable accuracy if he knows the molar concentrations of bicarbonate ion and dissolved carbon dioxide. If the bicarbonate concentration is equal to the dissolved carbon dioxide concentration, the second member of the right-hand portion of the equation becomes log of 1, which is equal to zero. Therefore, under these conditions the pH of the solution is equal to the pK.

From the Henderson-Hasselbalch equation one can readily see that an increase in bicarbonate ion concentration causes the pH to rise, or, in other words, shifts the acid-base balance toward the alkaline side. On the other hand, an increase in the concentration of dissolved carbon dioxide decreases the pH, or shifts the acid-base balance toward the acid side. It will be apparent later in this chapter that one can change the concentration of dissolved $CO_2$ in the body fluids by increasing or decreasing the rate of respiration. In this way the respiratory system can change to a certain extent the pH of the body fluids. On the other hand, the kidneys can increase or decrease the concentration of bicarbonate ion in the body fluids, in this way increasing or decreasing the pH. Thus, these two major hydrogen ion regulatory systems operate principally by altering one or the other of the two elements of the bicarbonate buffer system.

**The Reaction Curve of the Bicarbonate Buffer System.** Figure 37–1 shows the changes in pH of the body fluids when the ratio of bicarbonate ion to carbon dioxide changes. Note that when the concentrations of the two elements of the buffer are equal, the pH of the solution is 6.1, which is equal to the pK of the bicarbonate buffer system. When base is added to the buffer a large proportion of the dissolved carbon dioxide and carbonic acid is converted into bicarbonate ions, and the ratio is altered. As a result, the pH rises as indicated by the upward slope of the curve. On the other hand, when acid is added, a large proportion of the bicarbonate ion is converted first into carbonic acid and then into dissolved carbon dioxide so that the ratio changes in favor of the acidic

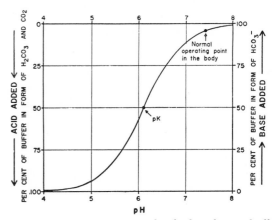

**Figure 37–1.** Reaction curve for the bicarbonate buffer system.

side, and the pH falls, as illustrated by the downslope of the curve

***Buffering Power of the Bicarbonate Buffer System.*** Referring once again to Figure 37–1, note that at the central point of the curve addition of a slight amount of acid or base causes minimal change in pH. However, toward each end of the curve addition of a slight amount of acid or base causes the pH to change greatly. Thus, the so-called *buffering power* of the buffer system *is greatest when the pH is equal to the pK,* which is in the exact center of the curve. The buffering power is still reasonably effective until the ratio of one element of the buffer system to the other reaches as much as 8:1 or 1:8, but beyond these limits the buffering power diminishes rapidly. And, when all the carbon dioxide has been converted into bicarbonate ion or when all the bicarbonate ion has been converted into carbon dioxide, the system has no more buffering power at all.

A second factor that determines the buffering power is the concentrations of the two elements in the buffer solution. Obviously, if the concentrations are slight, only a small amount of acid or base need be added to the solution to change the pH considerably. Thus, *the buffering power of a buffer is directly proportional to the concentrations of the buffer substances.*

## THE BUFFER SYSTEMS OF THE BODY FLUIDS

The three major buffer systems of the body fluids are the *bicarbonate buffer,* which has been described above, the *phosphate buffer,* and the *protein buffer.* Each of these performs major buffering functions under different conditions.

**The Bicarbonate Buffer System.** The bicarbonate system is not an exceedingly powerful buffer for two reasons. First, the pH in the extracellular fluids is about 7.4, while the pK of the bicarbonate buffer system is 6.1. This means that approximately 20 times as much of the bicarbonate buffer is in the form of bicarbonate ion as in the form of dissolved carbon dioxide. For this reason, the system is operating on a portion of its buffering curve where the buffering power is poor. Second, the concentrations of the two elements of the bicarbonate buffer system $CO_2$ and $HCO_3^-$, are not great.

*Yet, despite the fact that the bicarbonate buffer system is not especially powerful, it is probably equally as important as all the others in the body because the concentrations of each of the two elements of the bicarbonate system can be regulated,* carbon dioxide by the respiratory system and bicarbonate ion by the kidneys. As a result, the pH of the blood can be shifted up or down by the respiratory and renal regulatory systems.

**The Phosphate Buffer System.** The phosphate buffer system acts in almost identically the same manner as the bicarbonate buffer system, but it is composed of the following two elements: $H_2PO_4^-$ and $HPO_4^{--}$. When a strong acid, such as hydrochloric acid, is added to a mixture of these two substances, the following reaction occurs:

$$HCl + Na_2HPO_4 \rightarrow NaH_2PO_4 + NaCl \quad (12)$$

The net result of this reaction is that the hydrochloric acid is removed, and in its place an additional quantity of $NaH_2PO_4$ is formed. Since $NaH_2PO_4$ is only weakly acidic, a strong acid has been traded for a very weak acid, and the pH changes relatively slightly.

Conversely, if a strong base, such as sodium hydroxide, is added to the buffer system, the following reaction takes place:

$$NaOH + NaH_2PO_4 \rightarrow Na_2HPO_4 + H_2O \quad (13)$$

Here sodium hydroxide is decomposed to form water and $Na_2HPO_4$. That is, a strong base is traded for the very weak base $Na_2HPO_4$, allowing only a slight shift in pH toward the alkaline side.

The phosphate buffer system has a pK of 6.8, which is not far from the normal pH of 7.4 in the body fluids; this allows the phosphate system to operate near its maximum buffering power. However, despite the fact that this buffer system operates in a reasonably good portion of the buffer curve, its concentration in the extracellular fluid is only one-sixth that of the bicarbonate buffer. Therefore, its total buffering power *in the extracellular fluid* is even less than that of the bicarbonate system.

On the other hand, the phosphate buffer is especially important in the tubular fluids of the kidneys for two reasons: First, phosphate usually becomes greatly concentrated in the tubules, thereby also greatly increasing the buffering power of the phosphate system. Second, the tubular fluid usually becomes more acidic than the extracellular fluid, bringing the operating range of the buffer closer to the pK of the system.

The phosphate buffer is also extremely important in the intracellular fluids because the concentration of phosphate in these fluids is many times that in the extracellular fluids, and also because the pH of the intracellular fluids is usually closer to the pK of the phosphate buffer system than is the pH of the extracellular fluid.

**The Protein Buffer System.** By far the most plentiful buffer of the body is the proteins of the cells and plasma. There is a slight amount of diffusion of hydrogen ions through the cell membrane, and even more important, carbon dioxide can diffuse readily through cell membranes and bicarbonate ions can diffuse to some extent (they require several hours to come to equilibrium in most cells other than the red blood cells). The diffusion of the two elements of the bicarbonate buffer system causes the pH in the intracellular fluids to change approximately in proportion to the changes in pH in the extracellular fluids. Thus, all the buffer systems inside the cells help to buffer the extracellular fluids as well. These include the extremely large amounts of proteins inside the cells. Indeed, experimental studies have shown that at least three quarters of all the *chemical* buffering power of the body fluids is inside the cells, and most of this results from the intracellular proteins.

The method by which the protein buffer system operates is precisely the same as that of the bicarbonate buffer system. It will be recalled that a protein is composed of amino acids bound together by peptide linkages, but some of the different amino acids have free acidic radicals in the form of $-COOH$, and these can dissociate into $-COO^-$ and $H^+$. Also, some have free basic radicals in the form of $-NH_3OH$, which can dissociate into $-NH_3^+$ and $OH^-$. The $OH^-$ in turn can react with hydrogen ions to form water, in this way reducing the hydrogen ion concentration. Thus, proteins can operate in both *acidic* and *basic* buffering systems, the basic systems operating oppositely to the acidic systems. Furthermore, the pK's of the different weak acidic and basic buffering systems are not far from 7.4. This, too, helps to make the protein buffering systems by far the most powerful of the body.

## THE ISOHYDRIC PRINCIPLE

Each of the above buffer systems has been discussed as if it could operate individually in the body fluids. However, they all actually work together, for the hydrogen ion is common to the chemical reactions of all the systems. Therefore, whenever any condition causes the hydrogen ion concentration to change, it causes the balance of all the buffer systems to change at the same time. This phenomenon, called the *isohydric principle,* is represented by the following formula:

$$H^+ = \frac{K_1 \times HA_1}{A_1^-} = \frac{K_2 \times HA_2}{A_2^-} = \frac{K_3 \times HA_3}{A_3^-} \quad (14)$$

in which $K_1$, $K_2$, and $K_3$ are the dissociation constants of three respective acids, $HA_1$, $HA_2$, and $HA_3$, and $A_1^-$, $A_2^-$, and $A_3^-$ are the concentrations of the free negative ions of the acids.

The important feature of this principle is that any condition that changes the balance of any one of the buffer systems also changes the balance of all the others, for *the buffer systems actually buffer each other.*

## RESPIRATORY REGULATION OF ACID-BASE BALANCE

In the discussion of the Henderson-Hasselbalch equation, it was noted that an increase in carbon dioxide concentration in the body fluids lowers the pH toward the acidic side, whereas a decrease in carbon dioxide raises the pH toward the alkaline side. It is on the basis of this effect that the respiratory system is capable of altering the pH either up or down.

**Balance Between Metabolic Formation of Carbon Dioxide and Pulmonary Expiration of Carbon Dioxide.** Carbon dioxide is continually being formed in the body by the different intracellular metabolic processes, the carbon in the foods being oxidized by oxygen to form carbon dioxide. This in turn diffuses into the interstitial fluids and blood, and is transported to the lungs where it diffuses into the alveoli and is transferred to the atmosphere by pulmonary ventilation. However, several minutes are required for this passage of carbon dioxide from the cells to the atmosphere. Since the carbon dioxide is not removed instantaneously, an average of 1.2 millimols/liter of dissolved carbon dioxide is normally in the extracellular fluids.

If the rate of metabolic formation of carbon dioxide becomes increased, its concentration in the extracellular fluids is likewise increased. Conversely, decreased metabolism decreases the carbon dioxide concentration.

On the other hand, if the rate of pulmonary ventilation is increased, the rate of expiration of carbon dioxide becomes increased, and this de-

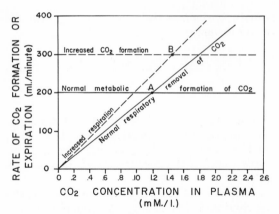

**Figure 37–2.** Relationship between metabolic formation of carbon dioxide and rate of carbon dioxide expiration by the lungs.

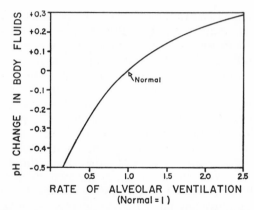

**Figure 37–3.** Approximate change in body fluid pH caused by increased or decreased rate of alveolar ventilation.

creases the amount of accumulated carbon dioxide in the extracellular fluids.

Figure 37–2 illustrates the relationship between metabolic formation of carbon dioxide and its expiration by the lungs. The horizontal solid line shows that under resting conditions, the normal metabolic formation of carbon dioxide is approximately constant at about 200 ml./minute. The diagonal solid curve illustrates that, with normal respiration, the rate of carbon dioxide expiration is directly proportional to the carbon dioxide concentration in the plasma. The point at which these two curves cross, point A, is the actual operating conditions in the body, because at this point the rate of expiration of carbon dioxide equals the rate of metabolic formation. Thus, as shown by the figure, the normal carbon dioxide level in the plasma is 1.2 millimols/liter.

The dashed lines of Figure 37–2 illustrate what happens when the rate of carbon dioxide formation and the rate of expiration change. In this instance the rate of formation has risen to 300 ml./minute, and the pulmonary ventilation has increased approximately 25 per cent. At point B where these two curves equilibrate with each other, the $CO_2$ concentration is approximately 1.4 millimols/liter. Thus, it is evident that the carbon dioxide concentration in the extracellular fluids is controlled by a balance between the rate of metabolism and the rate of pulmonary excretion.

**Effect of Increasing or Decreasing the Alveoloar Ventilation on pH of the Extracellular Fluids.** If we assume that the rate of metabolic formation of carbon dioxide remains constant, then the only factor that affects the carbon dioxide concentration in the body fluids

is the rate of alveolar ventilation as expressed by the following formula:

$$CO_2 \propto \frac{1}{\text{Alveolar ventilation}} \qquad (15)$$

And, since an increase in carbon dioxide decreases the pH, changes in alveolar ventilation also change the hydrogen ion concentration.

Figure 37–3 illustrates the approximate change in pH in the blood that can be effected by increasing or decreasing the rate of alveolar ventilation. Note that an increase in alveolar ventilation to two times normal raises the pH of the extracellular fluids by about 0.23 pH unit. This means that if the pH of the body fluids has been 7.4 with normal alveolar ventilation, doubling the ventilation raises the pH to 7.63. Conversely, a decrease in alveolar ventilation to one-quarter normal reduces the pH 0.4 pH unit. That is, if at normal alveolar ventilation the pH had been 7.4, reducing the ventilation to one-quarter normal reduces the pH 0.4 unit to 7.0. Since alveolar ventilation can be reduced to zero ventilation or increased to about 15 times normal, one can readily understand how much the pH of the body fluids can be changed by alterations in the activity of the respiratory system.

### EFFECT OF HYDROGEN ION CONCENTRATION ON ALVEOLAR VENTILATION

Not only does the rate of alveolar ventilation affect the hydrogen ion concentration of the body fluids, but, in turn, the hydrogen ion concentration can affect the rate of alveolar ventilation. This results from a *direct action of hydrogen ions on the respiratory center in the medulla oblongata* that controls breathing.

Figure 37–4 illustrates the changes in alveolar ventilation caused by changing the pH of arterial blood from 7.0 to 7.6. From this graph it is evident that a decrease in pH from the normal value of 7.4 to the strongly acidic range can increase the rate of alveolar ventilation to as much as four to five times normal, while an increase in pH into the alkaline range can decrease the rate of alveolar ventilation to as little as 50 to 75 per cent of normal.

**Feedback Regulation of Hydrogen Ion Concentration by the Respiratory System.** Because of the ability of the respiratory center to respond to hydrogen ion concentration, and because changes in alveolar ventilation in turn alter the hydrogen ion concentration in the body fluids, the respiratory system acts as a typical feedback regulatory system for controlling hydrogen ion concentration. That is, any time the hydrogen ion concentration becomes high, the respiratory system becomes more active, and alveolar ventilation increases. As a result, the carbon dioxide concentration in the extracellular fluids decreases, thus reducing the hydrogen ion concentration back toward a normal value. Conversely, if the hydrogen ion concentration falls too low, the respiratory center becomes depressed, alveolar ventilation also decreases, and the hydrogen ion concentration rises back toward normal.

*Efficiency of Respiratory Regulation of Hydrogen Ion Concentration.* Unfortunately, respiratory control cannot return the hydrogen ion concentration all the way to the normal value of 7.4 when some abnormality outside the respiratory system has altered the pH from the normal. The reason for this is that, as the pH returns toward normal, the stimulus that has been causing either increased or decreased respiration will itself begin to be lost. Ordinarily, the respiratory mechanism for regulation of hydrogen ion concentration has a control effectiveness of between 50 and 75 per cent (a feedback gain of 1 to 3). That is, if the hydrogen ion concentration should suddenly be decreased from 7.4 to 7.0 by some extraneous factor, the respiratory system, in a minute or more, returns the pH to a value of about 7.2 to 7.3.

*Buffering Power of the Respiratory System.* In effect, respiratory regulation of acid-base balance is a *physiological type of buffer system* having almost identically the same importance as the chemical buffering systems of the body discussed earlier in the chapter. The overall "buffering power" of the respiratory system is one to two times as great as that of all the chemical buffers combined. That is, one to two times as much acid or base can normally be buffered by this mechanism as by the chemical buffers.

# RENAL REGULATION OF HYDROGEN ION CONCENTRATION

In the earlier discussion of the Henderson-Hasselbalch equation it was pointed out that the kidneys regulate hydrogen ion concentration principally by increasing or decreasing the bicarbonate ion concentration in the body fluid. To do this, a complex series of reactions occurs in the renal tubules, including reactions for hydrogen ion secretion, sodium ion reabsorption, bicarbonate ion excretion into the urine, and ammonia secretion into the tubules. The following sections describe the interplay of these different tubular mechanisms and the manner in which they help to regulate the hydrogen ion concentration of the body fluids.

## *TUBULAR SECRETION OF HYDROGEN IONS*

The epithelial cells of the proximal tubules, distal tubules, collecting tubules, and even the thick portion of the loop of Henle all secrete hydrogen ions into the tubular fluid. The mechanism by which this occurs is illustrated in Figure 37–5. The secretory process *begins with carbon dioxide* in the tubular epithelial cells.

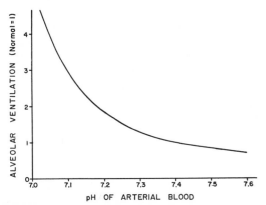

**Figure 37–4.** Effect of blood pH on the rate of alveolar ventilation. (Constructed from data obtained by Gray: Pulmonary Ventilation and Its Regulation. Charles C Thomas.)

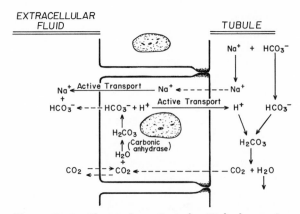

**Figure 37–5.** Chemical reactions for (1) hydrogen ion secretion, (2) sodium ion absorption in exchange for a hydrogen ion, and (3) combination of hydrogen ions with bicarbonate ions in the tubules.

The carbon dioxide, under the influence of an enzyme, *carbonic anhydrase,* combines with water to form *carbonic acid.* This then dissociates into *bicarbonate ion* and *hydrogen ion,* and the hydrogen ion is secreted by active transport through the lumenal border of the cell membrane into the tubule.

In the collecting tubules hydrogen ion secretion can continue until the concentration of hydrogen ions in the tubules becomes as much as 900 times that in the extracellular fluid or, in other words, until the pH of the tubular fluids falls to about 4.5. This represents a limit to the ability of the tubular epithelium to secrete hydrogen ions.

About 84 per cent of all the hydrogen ions secreted by the tubules are secreted in the proximal tubules, but the maximum concentration gradient that can be achieved here is only about three- to four-fold instead of the 900-fold that can be achieved in the collecting tubules. That is, the pH can be decreased only to about 6.9, 0.5 pH unit below 7.4, the pH of the glomerular filtrate. The distal tubules can decrease the pH to about 6.0 to 6.5; such a decrease is between that achieved by the proximal tubules and collecting tubules.

**Reabsorption of Sodium Ions.** Note in Figure 37–5 that sodium ions are reabsorbed at the same time that hydrogen ions are secreted. However, the driving force for this sodium reabsorption is active transport at the basal and lateral sides of the epithelial cells rather than at the lumenal borders. Though this active transport of sodium is not directly coupled with the active secretion of hydrogen ions, in general, one sodium ion is reabsorbed for each hydrogen ion that is secreted. This maintains appropriate electrical balance between the anions and cations in both the tubular fluid and the extracellular fluid.

**Regulation of Hydrogen Ion Secretion by the Carbon Dioxide Concentration in the Extracellular Fluid.** Since the chemical reactions for secretion of hydrogen ions begin with carbon dioxide, the greater the carbon dioxide concentration in the extracellular fluid, the more rapidly the reactions proceed, and the greater becomes the rate of hydrogen ion secretion. Therefore, any factor that increases the carbon dioxide concentration in the extracellular fluids, such as decreased respiration or increased metabolic rate, also increases the rate of hydrogen ion secretion. Conversely, any factor that decreases the carbon dioxide, such as excess pulmonary ventilation or decreased metabolic rate, decreases the rate of hydrogen ion secretion.

At normal carbon dioxide concentrations, the rate of hydrogen ion secretion is about 3.5 millimols per minute, but this rises or falls directly in proportion to changes in extracellular carbon dioxide.

**Interaction of Bicarbonate Ions with Hydrogen Ions in the Tubules—"Reabsorption" of Bicarbonate Ions.** It is already clear from previous discussions that the bicarbonate ion concentration in the extracellular fluid plays an extremely important role in the acid-base buffer system and, therefore, in the control of extracellular fluid hydrogen ion concentration. Therefore, it is important that the kidney tubules help to regulate the extracellular fluid bicarbonate ion concentration. Yet, the tubules are almost entirely impermeable to bicarbonate ion because it is a large ion and also is electrically charged. However, the bicarbonate ion can, in effect, be "reabsorbed" by the special process, which is also illustrated in Figure 37–5.

The "reabsorption" of bicarbonate ions is initiated by a reaction in the tubules between the bicarbonate ions and the hydrogen ions secreted by the tubular cells, as illustrated in the figure. The carbonic acid then dissociates into carbon dioxide and water. The water becomes part of the tubular fluid, while the carbon dioxide, having the capability to diffuse extremely readily through all cellular membranes, instantaneously diffuses into the blood where it combines with water to form new bicarbonate ions. *If an excess of hydrogen ions is secreted by the tubules, the bicarbonate ions will be almost completely removed from the tubules* so that, for practical purposes, *none* remain to pass into the urine—that is, if sufficient hydrogen ions are available.

If we now note in Figure 37–5 the chemical reactions that are responsible for formation of hydrogen ions in the epithelial cells, we will see that each time a hydrogen ion is formed a bicarbonate ion is formed inside these cells by the dissociation of $H_2CO_3$. This bicarbonate ion then diffuses into the peritubular fluid in combination with the sodium ion that has been absorbed from the tubule.

The net effect of all these reactions is a mechanism for "reabsorption" of bicarbonate ions from the tubules, though the bicarbonate ions that enter the peritubular fluid are not the same bicarbonate ions that are removed from the tubular fluid.

**Normal Rates of Bicarbonate Ion Filtration and Hydrogen Ion Secretion into the Tubules—Titration of Bicarbonate Ions Against Hydrogen Ions.**   Under normal conditions, the rate of hydrogen ion secretion is about 3.50 millimols/minute, and the rate of filtration of bicarbonate ions is about 3.49 millimols/minute. Therefore, essentially all the bicarbonate ions are normally "reabsorbed" from the tubules by the process just described, while a slight excess of hydrogen ions remains in the tubules to react with other substances and to be excreted into the urine.

Note that under these normal conditions there is almost precise balance between the rate of hydrogen ion secretion and the rate of bicarbonate ion filtration into the glomerular filtrate. These two ions combine with each other and actually annihilate each other, the end products being carbon dioxide and water. Therefore, it is said that the bicarbonate ions and hydrogen ions normally "titrate" each other in the tubules.

However, note also that this titration process is not quite complete, for usually a slight excess of hydrogen ions (the acidic component) remains in the tubules to be excreted in the urine. The reason for this is that under normal conditions a person's metabolic processes continually form a small amount of excess acid that gives rise to the slight excess of hydrogen ions over bicarbonate ions in the tubules.

On rare occasions the bicarbonate ions are in excess, as we shall see in subsequent discussions. When this occurs, the titration process again is not quite complete; this time, excess bicarbonate ions (the basic component) are left in the tubules to pass into the urine.

Thus, the basic mechanism by which the kidney corrects either acidosis or alkalosis is by incomplete titration of hydrogen ions against bicarbonate ions, leaving one or the other of these to pass into the urine and therefore to be removed from the extracellular fluid.

From 80 to 99 per cent of this titration process occurs in the proximal tubules, and the carbonic acid formed by the titration reaction is then split very rapidly into its end products, carbon dioxide and water. The water passes down the tubules, and the carbon dioxide diffuses into the extracellular fluid. To promote this rapid dissociation of carbonic acid into carbon dioxide and water, the luminal brush border surface of the proximal tubules (but not of the other tubules) has a large amount of attached carbonic anhydrase that accelerates the reaction.

### RENAL CORRECTION OF ALKALOSIS– DECREASE IN BICARBONATE IONS IN THE EXTRACELLULAR FLUID

Now that we have described the mechanisms by which the renal tubules secrete hydrogen ions and reabsorb bicarbonate ions, we can explain the manner in which the kidneys readjust the pH of the extracellular fluids when it becomes abnormal.

The initial step in this explanation is to understand what happens to the concentrations of carbon dioxide and bicarbonate ions in the extracellular fluids in alkalosis and acidosis. First, let us consider *alkalosis*. Referring again to Equation 11, the Henderson-Hasselbalch equation, we see that the *ratio* of bicarbonate ions to dissolved carbon dioxide molecules increases when the pH rises into the alkalosis range above 7.4. The effect of this on the titration process in the tubules is to increase the *ratio* of bicarbonate ions filtered into the tubules to hydrogen ions secreted. This increase occurs because the high extracellular bicarbonate ion concentration also increases its concentration in the glomerular filtrate, and the low carbon dioxide concentration decreases the secretion of hydrogen ions. Therefore, the fine balance that normally exists in the tubules between the hydrogen and bicarbonate ions no longer occurs. Instead, far greater quantities of bicarbonate ions than hydrogen ions now enter the tubules. Since no bicarbonate ions can be reabsorbed without first reacting with hydrogen ions, all the excess bicarbonate ions pass into the urine and carry with them sodium ions or other positive ions.

Thus, in effect, sodium bicarbonate is removed from the extracellular fluid.

Loss of sodium bicarbonate from the extracellular fluid decreases the bicarbonate ion portion of the bicarbonate buffer system, and, in accordance with the Henderson-Hasselbalch equation, this shifts the pH of the body fluids back in the acid direction. Furthermore, because of the isohydric principle, all of the body buffers shift back in the acid direction. Thus, the alkalosis is corrected.

## RENAL CORRECTION OF ACIDOSIS– INCREASE IN BICARBONATE IONS IN THE EXTRACELLULAR FLUID

In acidosis, the *ratio* of carbon dioxide to bicarbonate ions in the extracellular fluid increases, which is exactly opposite to the effect in alkalosis. Therefore, in acidosis, the *rate of hydrogen ion secretion* rises to a level far greater than the *rate of bicarbonate ion filtration* into the tubules. As a result, excess hydrogen ions are secreted into the tubules and have no bicarbonate ions to react with. These combine with the buffers in the tubular fluid, as explained in the following paragraphs, and are excreted into the urine.

Figure 37–5 shows that each time a hydrogen ion is secreted into the tubules two other effects occur simultaneously: first, a bicarbonate ion is formed in the tubular epithelial cell and, second, a sodium ion is absorbed from the tubule into the epithelial cell. The sodium ion and bicarbonate ion then diffuse together from the epithelial cell into the peritubular fluid. Thus, *the net effect of secreting excess hydrogen ions into the tubules is to increase the quantity of sodium bicarbonate in the extracellular fluid.* This increases the bicarbonate salt portion of the bicarbonate buffer system, which, in accordance with the Henderson-Hasselbalch equation and the isohydric principle, shifts all of the buffers in the alkaline direction, increasing the pH in the process, and thereby correcting the acidosis.

## COMBINATION OF THE EXCESS HYDROGEN IONS WITH TUBULAR BUFFERS AND THEIR TRANSPORT INTO THE URINE

When excess hydrogen ions are secreted into the tubules, only a small portion of these can be carried in the free form by the tubular fluid into the urine. The reason for this is that the maximum hydrogen ion concentration that the tubular system can achieve is $10^{-4.5}$ molar, which corresponds to a pH of 4.5. At normal daily urine flows, this concentration represents only 1 per cent of the daily excretion of excess hydrogen ions.

Therefore, to carry the excess hydrogen ions into the urine the hydrogen ions must combine with buffers in the tubular fluid to keep the hydrogen ion concentration itself from rising too high. Otherwise, the high concentration of hydrogen ions would limit further secretion of hydrogen ions by the tubules because the hydrogen ion secreting process is "gradient limited." That is, as the hydrogen ion concentration of the tubular fluid approaches the maximum limit that can be achieved (a concentration of $10^{-4.5}$ molar), the rate of secretion of hydrogen ions falls to near-zero.

To state this another way, in the absence of buffer systems in the tubular fluid to carry the excess hydrogen ion into the urine, the hydrogen ion secreting process simply cannot occur to a significant extent.

The tubular fluids have two very important buffer systems for transport of the excess hydrogen ions into the urine: (1) the phosphate buffer and (2) the ammonia buffer. In addition, there are a number of weak buffer systems such as urate, citrate, and other similar systems, as well as the bicarbonate buffer system that plays the very highly specialized titration role discussed above.

**Transport of Excess Hydrogen Ions Into the Urine by the Phosphate Buffer.** The phosphate buffer is composed of a mixture of $HPO_4^{--}$ and $H_2PO_4^-$. Both of these become considerably concentrated in the tubular fluid because of their relatively poor reabsorption and because of removal of water from the tubular fluid. Therefore, even though the phosphate buffer is very weak in the blood, it is a much more powerful buffer in the tubular fluid.

The quantity of $HPO_4^{--}$ in the glomerular filtrate is normally about four times as great as that of $H_2PO_4^-$. Excess hydrogen ions entering the tubules combine with the $HPO_4^{--}$, as illustrated in Figure 37–6, forming $H_2PO_4^-$, which passes on into the urine. Sodium ion is absorbed into the extracellular fluid in place of the hydrogen ion involved in the reaction, and at the same time a *bicarbonate ion,* formed in the process of secreting the hydrogen ion, is also released into the extracellular fluid. Thus, the net effect of

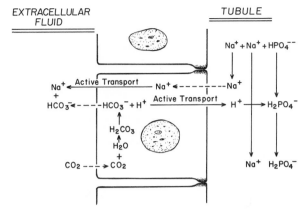

**Figure 37–6.** Chemical reactions in the tubules involving hydrogen ions, sodium ions, and the phosphate buffer system.

this reaction is to increase the amount of sodium bicarbonate in the extracellular fluids, which is the kidney's way of reducing the degree of acidosis in the body fluids.

**Transport of Excess Hydrogen Ions Into the Urine by the Ammonia Buffer System.** Another very potent buffer system of the tubular fluid is composed of ammonia ($NH_3$) and the ammonium ion ($NH_4^+$). The epithelial cells of all the tubules besides those of the thin segment of the loop of Henle continually synthesize ammonia, and this diffuses into the tubules. The ammonia then reacts with hydrogen ions, as illustrated in Figure 37–7, to form ammonium ions. These are then excreted into the urine in combination with chloride ions and other tubular anions. Note in the figure that the net effect of these reactions is, again, *to increase the sodium bicarbonate concentration* in the extracellular fluid.

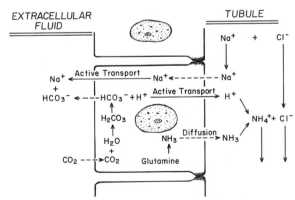

**Figure 37–7.** Secretion of ammonia by the tubular epithelial cells, and reaction of the ammonia with hydrogen ions in the tubules.

This ammonium ion mechanism for transport of excess hydrogen ions in the tubules is especially important for two different reasons: (1) Each time an ammonia molecule combines with a hydrogen ion to form an ammonium ion the concentration of ammonia in the tubular fluid becomes decreased, which causes still more ammonia to diffuse from the epithelial cells into the tubular fluid. Thus, the rate of ammonia secretion into the tubular fluid is actually controlled by the amount of excess hydrogen ions to be transported. (2) Most of the negative ions of the tubular fluid are chloride ions. Only a few hydrogen ions could be transported into the urine in direct combination with chloride, because hydrochloric acid is a very strong acid and the tubular pH would fall rapidly to the critical value of 4.5, below which further hydrogen ion secretion would cease. However, when hydrogen ions combine with ammonia and the resulting ammonium ions then combine with chloride, the pH does not fall significantly because ammonium chloride is only very weakly acidic.

Sixty per cent of the ammonia secreted by the tubular epithelium is derived from *glutamine,* and the remaining 40 per cent from different ones of the amino acids, particularly glycine and alanine.

*Enhancement of the Ammonia Buffer System in Chronic Acidosis.* If the tubular fluids remain highly acidic for long periods of time, the formation of ammonia steadily increases during the first two to three days, rising as much as 10-fold. For instance, immediately after acidosis begins, as little as 30 millimols of ammonia might be secreted each day, but after two days as much as 200 to 300 millimols can be secreted, illustrating that the ammonia-secreting mechanism can adapt readily to handle greatly increased loads of acid elimination.

## RAPIDITY OF ACID-BASE REGULATION BY THE KIDNEYS

Figure 37–8 illustrates the effect of extracellular fluid pH on the rate at which bicarbonate ions are lost from or gained by the body fluids each minute. For instance, at a pH of 7.0, approximately 2.3 millimols of bicarbonate ions is gained each minute, but as the pH returns toward the normal value of 7.4, the rate of gain falls off markedly. Then, when the pH becomes significantly greater than normal, bicarbonate ions are lost by the extracellular fluids. For in-

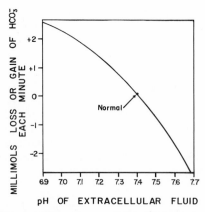

**Figure 37–8.** Effect of extracellular fluid pH on the rate of loss or gain of bicarbonate ions in the body fluids each minute.

stance, at a pH of 7.6 about 1.5 millimols of bicarbonate ions is lost each minute.

The total amount of buffers in the entire body (within the range of pH 7.0 to 7.7, at which life is possible) is approximately 1000 millimols. If all these should be suddenly shifted to the alkaline or acidic side by injecting an alkali or an acid, the kidneys would be able to return the pH of the body fluids back almost to normal in one to three days. Though this mechanism is slow to act, it continues acting until the pH returns almost exactly to normal rather than a certain percentage of the way. Therefore, the real value of the renal mechanism for regulating hydrogen ion concentration is not rapidity of action but instead its ability in the end to neutralize completely any excess acid or alkali that enters the body fluids.

Ordinarily, the kidneys can remove up to about 500 millimols of acid or alkali each day. If greater quantities than this enter the body fluids, the kidneys are unable to cope with the extra load, and severe acidosis or alkalosis ensues.

**Range of Urinary pH.** In the process of adjusting the hydrogen ion concentration of the extracellular fluid, the kidneys often excrete urine at pH's as low as 4.5 or as high as 8.0. When acid is being excreted the pH falls, and when alkali is being excreted the pH rises. Even when the pH of the extracellular fluids is at the normal value of 7.4, a fraction of a millimol of acid is still lost each minute. The reason for this is that about 50 to 100 millimols more acid than alkali are formed in the body each day, and this acid must be removed continually. Because of the presence of this excess acid in the urine, the normal urine pH is about 6.0 instead of 7.4, the pH of the blood.

## RENAL REGULATION OF PLASMA CHLORIDE CONCENTRATION—THE CHLORIDE TO BICARBONATE RATIO

In the above discussions we have emphasized the ability of the kidneys to conserve bicarbonate ion in the extracellular fluids whenever a state of acidosis develops, or to remove bicarbonate ions in a state of alkalosis. Thus, the bicarbonate ion is shuttled back and forth between high and low values as one of the principal means of adjusting the acid-base balance of the extracellular buffer systems and therefore also for adjusting the extracellular fluid pH.

However, in the process of juggling the extracellular fluid concentration of bicarbonate ion, it is essential to remove some other anion from the extracellular fluids each time the bicarbonate is increased, or to increase some other anion when the bicarbonate concentration is decreased. In general, the anion that is reciprocally juggled up or down with the bicarbonate ion is the chloride ion because this is the anion in greatest concentration in the extracellular fluid.

**Function of the Ammonia Buffer System in Controlling the Bicarbonate Ion to Chloride Ion Ratio.** It was pointed out above that the ammonia buffer system plays an extremely important role in removal of excess hydrogen ions from the tubules. Now, let us study Figure 37–7 once again. We see that in the process of transporting excess hydrogen ions into the urine in combination with ammonia, for each hydrogen ion transported a chloride ion also passes into the urine and a bicarbonate ion simultaneously enters the extracellular fluid. Thus, this ammonia system substitutes a bicarbonate ion in the extracellular fluid for a chloride ion that is lost from the extracellular fluids. Conversely, when a person is alkalotic, the ammonia system becomes inoperative; bicarbonate ions instead of chloride ions then pass into the urine, and a concomitant excess of chloride is reabsorbed.

Thus, in the process of controlling the pH of the body fluids, the renal acid-base regulating system also regulates the ratio of chloride ions to bicarbonate ions in the extracellular fluid.

# CLINICAL ABNORMALITIES OF ACID-BASE BALANCE

## RESPIRATORY ACIDOSIS AND ALKALOSIS

From the discussions earlier in the chapter it is obvious that any factor that decreases the rate of pulmonary ventilation increases the concentration of dissolved carbon dioxide in the extracellular fluid, which in turn leads to increased carbonic acid and hydrogen ions, thus resulting in acidosis. Because this type of acidosis is caused by an abnormality of respiration, it is called *respiratory acidosis.*

On the other hand, excessive pulmonary ventilation reverses the process and decreases the hydrogen ion concentration, thus resulting in alkalosis; this condition is called *respiratory alkalosis.*

A person can cause respiratory acidosis in himself by simply holding his breath, which he can do until the pH of the body fluids falls to as low as perhaps 7.0. On the other hand, he can voluntarily overbreathe and cause alkalosis to a pH of about 7.8.

*Respiratory acidosis* frequently results from pathologic conditions. For instance, damage to the respiratory center in the medulla oblongata, causing reduced breathing, obstruction of the passageways in the respiratory tract, pneumonia, decreased pulmonary surface area, and any other factor that interferes with the exchange of gases between the blood and alveolar air, results in respiratory acidosis.

On the other hand, only rarely do pathologic conditions cause *respiratory alkalosis.* However, an occasional psychoneurosis causes a person to overbreathe to such an extent that he becomes alkalotic. A physiological type of respiratory alkalosis occurs when a person ascends to a *high altitude.* The low oxygen content of the air stimulates respiration, which causes excess loss of carbon dioxide and development of mild respiratory alkalosis.

## METABOLIC ACIDOSIS AND ALKALOSIS

The terms "metabolic acidosis" and "metabolic alkalosis" refer to all other abnormalities of acid-base balance besides those caused by excess or insufficient carbon dioxide in the body fluids. Use of the word "metabolic" in this instance is unfortunate, because carbon dioxide is also a metabolic product. Yet, by convention, carbonic acid resulting from dissolved carbon dioxide is called a *respiratory acid* while any other acid in the body, whether it be formed by metabolism or simply ingested by the person, is called a *metabolic acid.*

**Causes of Metabolic Acidosis.** Metabolic acidosis can result from (1) failure of the kidneys to excrete the metabolic acids normally formed in the body, (2) formation of excessive quantities of metabolic acids in the body, (3) intravenous administration of metabolic acids, or (4) addition of metabolic acids by way of the gastrointestinal tract. Metabolic acidosis can result also from (5) loss of alkali from the body fluids. Some of the specific conditions that cause metabolic acidosis are the following:

*Diarrhea.* Severe diarrhea is one of the most frequent causes of metabolic acidosis for the following reasons: The gastrointestinal secretions normally contain large amounts of sodium bicarbonate. Therefore, excessive loss of these secretions during a bout of diarrhea is exactly the same as excretion of large amounts of sodium bicarbonate into the urine. In accordance with the Henderson-Hasselbalch equation, this results in a shift of the bicarbonate buffer system toward the acid side and results in metabolic acidosis. In fact, acidosis resulting from severe diarrhea can be so serious that it is one of the most common causes of death in young children.

*Vomiting.* A second cause of metabolic acidosis is vomiting. Vomiting of gastric contents alone, which occurs rarely, causes a loss of acid and leads to alkalosis, but vomiting of contents from deeper in the gastrointestinal tract, which often occurs, causes loss of alkali and results in metabolic acidosis.

*Uremia.* A third common type of acidosis is uremic acidosis, which occurs in severe renal disease. The cause of this is failure of the kidneys to rid the body of even the normal amounts of acids formed each day by the metabolic processes of the body.

*Diabetes Mellitus.* A fourth and extremely important cause of metabolic acidosis is diabetes mellitus. In this condition, lack of insulin secretion by the pancreas prevents normal use of glucose for metabolism. Instead, the stored fats are split into acetoacetic acid, and this in turn is metabolized by the tissues for energy in place of glucose. Simultaneously, the concentration of acetoacetic acid in the extracellular fluids often rises very high, and large quantities of it are excreted in the urine, sometimes as much as 500 to 1000 millimols per day.

*Acidosis Caused by Carbonic Anhydrase Inhibitors.* Administration of the common carbonic anhydrase inhibitor, acetazolamide (Diamox), a drug that is frequently used to cause diuresis, also causes a mild degree of acidosis. This occurs because inhibition of the carbonic anhydrase on the lumenal surface of the proximal tubular epithelium prevents adequate reabsorption of bicarbonate ions. Loss of these into the urine causes a decrease in bicarbonate ions in the extracellular fluid, thus leading to acidosis.

*Acidosis Caused by High Extracellular Fluid Potassium Concentration.* Potassium ions, like hydrogen ions, are also secreted by the distal tubules and collecting tubules, as was explained in Chapter 34. When the plasma concentration of potassium ions is high, the tubular epithelium secretes extra large quantities of these into the tubules, and these compete with hydrogen ions to combine with the available buffer anions of the tubular fluid. Therefore, an excess of potassium decreases the quantity of excess hydrogen ions that can be carried into the urine, and this in turn *decreases* the secretion of hydrogen ions. Consequently, normal removal of hydrogen ions cannot occur, thereby leading to acidosis.

**Causes of Metabolic Alkalosis.** Metabolic alkalosis does not occur nearly as often as metabolic acidosis. However, there are several common causes of metabolic alkalosis, as follows:

*Alkalosis Caused by Administering Diuretics (Except the Carbonic Anhydrase Inhibitors).* All diuretics cause increased flow of fluid along the tubules, and this increase usually causes a great excess of sodium ions to flow into the distal and collecting tubules, leading also to rapid reabsorption of sodium ions from these tubules. This rapid reabsorption also leaves behind an excess of buffer anions in the tubules. These buffers combine readily with hydrogen ions and remove excessive amounts of these from the body, thus resulting in extracellular fluid alkalosis.

**Excessive Ingestion of Alkaline Drugs.**   Perhaps the second most common cause of alkalosis is excessive ingestion of alkaline drugs, such as sodium bicarbonate, for the treatment of gastritis or peptic ulcer.

**Alkalosis Caused by the Loss of Chloride Ions.**   Excessive vomiting of gastric contents without vomiting of lower gastrointestinal contents causes excessive loss of hydrochloric acid secreted by the stomach mucosa. This loss leads to reduction of chloride ions and enhancement of bicarbonate ions in the extracellular fluid. The bicarbonate ions are derived from the stomach glandular cells that secrete the hydrogen ions; these form bicarbonate ion in the same way that the tubular cells do when they secrete hydrogen ions. The net result is loss of acid from the extracellular fluids and development of metabolic alkalosis.

**Alkalosis Caused by Excess Aldosterone.**   When excess quantities of aldosterone are secreted by the adrenal glands, the extracellular fluid becomes slightly alkalotic. This is caused in the following way: The aldosterone promotes extensive reabsorption of sodium ions from the tubules, leaving behind in the tubules a higher than normal quantity of buffer ions. Hydrogen ions then combine with these buffers and are transported into the urine, causing increased secretion of hydrogen ions and loss from the extracellular fluids and thus promoting alkalosis.

## EFFECTS OF ACIDOSIS AND ALKALOSIS ON THE BODY

**Acidosis.**   The major effect of acidosis is depression of the *central nervous system*. When the pH of the blood falls below 7.0, the nervous system becomes so depressed that the person first becomes disoriented and, later, comatose. Therefore, patients dying of diabetic acidosis, uremic acidosis, and other types of acidosis usually die in a state of coma.

*In metabolic acidosis the high hydrogen ion concentration causes increased rate and depth of respiration.* Therefore, one of the diagnostic signs of *metabolic* acidosis is increased pulmonary ventilation. On the other hand, *in respiratory acidosis, respiration is usually depressed* because this is the cause of the acidosis, which is opposite to the effect in metabolic acidosis.

**Alkalosis.**   The major effect of alkalosis on the body is *overexcitability of the nervous system*. This effect occurs both in the central nervous system and in the peripheral nerves, but usually the peripheral nerves are affected before the central nervous system. The nerves become so excitable that they automatically and repetitively fire even when they are

not stimulated by normal stimuli. As a result, the muscles go into a state of *tetany,* which means a state of tonic spasm. This tetany usually appears first in the muscles of the forearm, then spreads rapidly to the muscles of the face, and finally all over the body. Alkalotic patients may die from tetany of the respiratory muscles.

Occasionally an alkalotic person develops severe symptoms of central nervous system overexcitability. The symptoms may manifest themselves as extreme nervousness or, in susceptible persons, as convulsions. For instance, in persons who are predisposed to epileptic fits, simply overbreathing often results in an attack. Indeed, this is one of the clinical methods for assessing one's degree of epileptic predisposition.

## RESPIRATORY COMPENSATION OF METABOLIC ACIDOSIS OR ALKALOSIS

It was pointed out above that the high hydrogen ion concentration of metabolic acidosis causes increased pulmonary ventilation, which in turn results in rapid removal of carbon dioxide from the body fluids and reduces the hydrogen ion concentration. Thus, this respiratory effect helps to compensate for the metabolic acidosis. However, this compensation occurs only partway. Ordinarily, the respiratory system is capable of compensating between 50 and 75 per cent. That is, if the metabolic factor makes the pH of the blood fall to 7.0 at normal pulmonary ventilation, the rate of pulmonary ventilation normally increases sufficiently to return the pH of the blood to 7.2 to 7.3, as was pointed out earlier in the chapter.

The opposite effect occurs in metabolic alkalosis. That is, alkalosis diminishes the pulmonary ventilation, which in turn causes increased hydrogen ion concentration. Here again the compensation can take place to about 50 to 75 per cent.

## RENAL COMPENSATION OF RESPIRATORY ACIDOSIS OR ALKALOSIS

If a person develops respiratory acidosis that continues for a prolonged period of time, the kidneys will secrete an excess of hydrogen ions, resulting in an increase in sodium bicarbonate in the extracellular fluids. After half a day or more, the pH of the body fluids will have returned most of the way to normal even though the person continues to breathe poorly.

Exactly the converse effect occurs in respiratory alkalosis. Large amounts of sodium bicarbonate are lost into the urine, decreasing the extracellular bicarbonate ion and thereby decreasing the pH also almost back to normal.

## PHYSIOLOGY OF TREATMENT IN ACIDOSIS OR ALKALOSIS

Obviously, the best treatment for acidosis or alkalosis is to remove the condition causing the abnormality, but, if this cannot be effected, different drugs can be used to neutralize the excess acid or alkali.

To neutralize excess acid, large amounts of sodium bicarbonate can be ingested by mouth. This is absorbed into the blood stream and increases the bicarbonate ion portion of the bicarbonate buffer, thereby shifting the pH to the alkaline side. Sodium bicarbonate is occasionally used also for intravenous therapy, but this has such strong and often dangerous physiological effects that other substances are often used instead, such as sodium lactate, sodium gluconate, or other organic compounds of sodium. The lactate and gluconate portions of the molecules are metabolized in the body, leaving the sodium in the extracellular fluids in the form of sodium bicarbonate, and thereby shifting the pH of the fluids in the alkaline direction.

For treatment of alkalosis, ammonium chloride is often administered by mouth. When this is absorbed into the blood, the ammonia portion of the ammonium chloride is converted by the liver into urea; this reaction liberates hydrochloric acid, which immediately reacts with the buffers of the body fluids to shift the hydrogen ion concentration in the acid direction. Occasionally, ammonium chloride is infused intravenously, but the ammonium ion is highly toxic and this procedure can be dangerous. Another substance occasionally used is *lysine monohydrochloride*.

## CLINICAL MEASUREMENTS FOR STUDY OF ACID-BASE ABNORMALITIES

**pH Measurements.** In studying a patient with acidosis or alkalosis one needs preferably to know the pH of his body fluids. This can be determined easily by measuring the pH of the plasma with a glass electrode pH meter. However, extreme care must be exercised in removing the plasma and in making the measurement, because even the slightest diffusion of carbon dioxide out of the plasma into the air shifts the bicarbonate buffer system in the alkaline direction, resulting in a much elevated pH measurement.

**The Concept of "Buffer Base."** The term "buffer base" means the anion components of the buffer system. Thus, in plasma the principal buffer base is bicarbonate ion. In whole blood it is bicarbonate ion plus the hemoglobin ion in the red blood cell. In the entire body it is mainly the bicarbonate ion plus all of the proteins in the body. In addition, however, there are many buffer bases of lesser quantity such as the phosphate ion, the urate ion, and so forth.

The normal quantity of buffer base in whole blood is approximately 45 mEq./liter. About half of this is usually bicarbonate ion and the other half the hemoglobin buffer base. Whenever the bicarbonate portion of the buffer base is greater than normal, one suspects metabolic alkalosis, but when less than normal, metabolic acidosis. However, the total quantity of buffer base is determined to a great extent by the hemoglobin concentration of the blood as well as

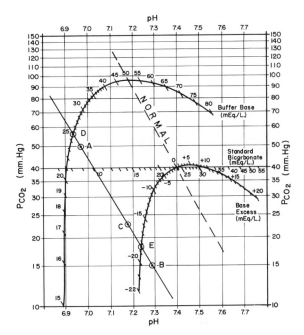

**Figure 37–9.** The Siggaard-Andersen nomogram, a method widely used clinically for diagnosis of acid-base abnormalities. (See text for explanation.)

by the bicarbonate concentration. Therefore, the total buffer base must be corrected for the hemoglobin concentration before it is very meaningful in a determination of the degree of metabolic acidosis or alkalosis.

**The Astrup and Siggaard-Andersen Procedure for Calculating Abnormalities of Buffer Base—the Concept of Base Excess.** Astrup and Siggaard-Andersen developed a method that is very simple to use in clinical patients for determing both the buffer base content of blood and a quantity called "base excess." The base excess is the quantity of extra base above normal (principally bicarbonate ions) after exclusion of the hemoglobin content. In other words, the deviation of the buffer base content from normal is calculated, and then this is corrected for the hemoglobin content to give the value called base excess. If the base excess is a positive number, then the person has an excess of metabolic base in his body fluids and therefore has metabolic alkalosis. If the base excess is a negative number, then he has a deficit of metabolic base, or in other words, he has an excess of metabolic acid in his blood and thus has metabolic acidosis.

The Astrup and Siggaard-Andersen procedure for determining base excess is the following: The nomogram illustrated in Figure 37–9 has been calculated and constructed to determine both the buffer base and the base excess in the blood. To use this nomogram several steps must first be followed: A small amount of blood is removed from the finger of a patient. Then the blood is equilibrated with carbon dioxide in a small chamber at a known carbon dioxide tension, and the pH of the blood is measured while equilibrated at this tension. Next, the carbon dioxide tension is changed to another value and the pH measured again. Points A and B in Figure 37–9 illustrate a pair of such measurements for the blood from a given patient. Now, a line is drawn through these two points—the line that is shown to intersect with three different curves on the nomogram. One of these curves is labeled "buffer base"; the point at which the experimental line crosses this curve gives the buffer base of the blood. The point where the experimental line crosses the "base excess" curve gives the base excess for the blood. Then, the point at which the experimental line crosses the line labeled "standard bicarbonate" gives the bicarbonate ion concentration in the blood that would be measured were the carbon dioxide tension at a normal level of 40 mm. Hg. Finally, the pH of the blood as it first issues from the finger also is measured and this point can be plotted on the experimental line to tell what the $Pco_2$ of the blood of that particular patient was at the time of measurement. Point C illustrates this.

Now let us summarize the information that one can derive from the nomogram: Points A and B are the experimental determinations when the blood is equilibrated with different known gaseous tensions of $Pco_2$. Point C gives the blood pH and also the $Pco_2$ of the fresh blood. If this $Pco_2$ is above 40 mm. Hg, then the person has a degree of respiratory acidosis, that is, too little respiration with accumulation of carbonic acid in the blood. If the $Pco_2$ is below 40 mm. Hg, then the patient has respiratory alkalosis. In this case the $Pco_2$ is 23 mm. Hg; therefore, there is definite respiratory alkalosis. Point D illustrates the buffer base in the blood, a value of 25 mEq./liter. Point E gives the base excess in the blood. Note that this is $-18.5$ mEq./liter, which means that the person has a marked degree of metabolic acidosis.

**Therapy Based on the Astrup and Siggaard-Andersen Nomogram.** One of the principal values of this type of nomogram is that it can be used readily and quickly as a guide for treating metabolic acidosis or alkalosis in a patient because the degree of abnormality is given directly in terms of base excess or base deficiency. To calculate the *total body* base excess or base deficiency one multiplies the base excess or deficiency per liter of blood by 0.3 and by body weight in kilograms. This value can then be used as an estimate of the total amount of acid or base that will have to be given therapeutically to correct the whole body acid-base abnormality.

# REFERENCES

Chan, J. C.: The influence of dietary intake on endogenous acid production. Theoretical and experimental background. *Nutr. Metab.,* 16:1, 1974.

Christensen, H. N.: Body Fluids and the Acid-Base Balance, 2nd Ed. Philadelphia; W. B. Saunders Company, 1964.

Gottschalk, C. W., Lassiter, W. E., and Mylle, M.: Localization of urine acidification in the mammalian kidney. *Amer. J. Physiol.,* 198:581, 1960.

Hills, A. G.: Acid-Base Balance. Baltimore, The Williams & Wilkins Company, 1973.

Irvine, R. D.: Diet and drugs in renal acidosis and acid-base regulation. *Prog. Biochem. Pharmacol.,* 7:146, 1972.

Kintner, E. P.: Acid base, blood gas, and electrolyte balances. *Prog. Clin. Pathol.,* 4:143, 1972.

Lemann, J., Jr., and Lennon, E. J.: Role of diet, gastrointestinal tract, and bone in acid-base homeostasis. *Kidney Int.,* 1:275, 1972.

Malnic, G.: Tubular handling of H. *In* Guyton, A. C. (ed.): MTP International Review of Science: Physiology. Vol. 6. Baltimore, University Park Press, 1974, p. 107.

Malnic, G., and Giebisch, G.: Symposium on acid-base homeostasis. Mechanism of renal hydrogen ion secretion. *Kidney Int.,* 1:280, 1972.

Pitts, R. F.: The renal metabolism of ammonia. *Physiologist,* 9:97, 1966.

Pitts, R. F.: Production and excretion of ammonia in relation to acid-base regulation. *In* Orloff, F., and Berliner, R. W. (eds.): Handbook of Physiology. Sec. 8. Baltimore, The Williams & Wilkins Company, 1973, p. 455.

Rahn, H.: Body temperature and acid-base regulation. (Review article). *Pneumonologie,* 151:87, 1974.

Rattenborg, C. C.: Acid-base monitoring. *Clin. Anesth.,* 9:139, 1973.

Rector, F. C., Jr.: Acidification of the urine. *In* Orloff, F., and Berliner, R. W. (eds.): Handbook of Physiology. Sec. 8. Baltimore, The Williams & Wilkins Company, 1973, p. 431.

Relman, A. S.: Metabolic consequences of acid-base disorders. *Kidney Int.,* 1:347, 1972.

Robinson, J. R.: Fundamentals of Acid-Base Regulation, 4th Ed. Philadelphia, J. B. Lippincott Company, 1972.

Siesjö, B. K.: Symposium on acid-base homeostasis. The regulation of cerebrospinal fluid pH. *Kidney Int.,* 1:360, 1972.

Steinmetz, P. R.: Cellular mechanisms of urinary acidification. *Physiol. Rev.,* 54: 890, 1974.

Waddell, W. J., and Bates, R. G.: Intracellular pH. *Physiol. Rev.,* 49:285, 1969.

Weiner, I. M.: Transport of weak acids and bases. *In* Orloff, F., and Berliner, R. W. (eds.): Handbook of Physiology. Sec. 8. Baltimore, The Williams & Wilkins Company, 1973, p. 521.

# | 38 |

# Micturition, Renal Disease, and Diuresis

## MICTURITION

Micturition is the process by which the urinary bladder empties itself when it becomes filled. Basically the bladder (1) progressively fills until the tension in its walls rises above a threshold value, at which time (2) a nervous reflex called the "micturition reflex" occurs that either causes micturition or, if it fails in this, at least causes a conscious desire to urinate.

### PHYSIOLOGIC ANATOMY OF THE BLADDER AND ITS NERVOUS CONNECTIONS

The urinary bladder, which is illustrated in Figure 38–1, is mainly a smooth muscle vesicle composed of two principal parts: (a) the *body,* which is comprised mainly of the *detrusor muscle,* and (b) the *trigone,* a small triangular area near the mouth of the bladder through which both the *ureters* and the *urethra* pass. During bladder expansion the body of the bladder stretches, and during the micturition reflex the detrusor muscle contracts to empty the bladder.

The trigonal muscle is interlaced around the opening of the urethra and maintains tonic closure of the urethral opening until the pressure in the bladder rises high enough to overcome the tone of the trigonal muscle. The trigonal muscle, therefore, is called the *internal sphincter of the bladder.* A few centimeters beyond the bladder, the urethra passes through the *urogenital diaphragm,* the muscle of which constitutes the *external sphincter* of the bladder. This muscle is a voluntary skeletal muscle in contrast to the muscle of the bladder, which is entirely smooth muscle. Normally, this external sphinc-

ter remains tonically contracted, which prevents constant dribbling of urine, but it can be reflexly or voluntarily relaxed at the time of micturition.

Figure 38–1 also illustrates the basic nervous connections with the spinal cord for bladder control. Parasympathetic excitation causes contraction of the detrusor muscle and perhaps some dilatation of the internal sphincter. Sensory nerve fibers leave the bladder mainly in association with the parasympathetic nerves and enter the spinal cord through the pelvic nerves and sacral plexus. The external sphincter of the bladder, which is a voluntary skeletal muscle, is controlled by the *pudic nerve,* which has its origin in the first two sacral segments of the spinal cord. Figure 38–1 also shows sympathetic nerve fibers passing from the lumbar regions of the spinal cord through the hypogastric plexus to the bladder. Stimulation of these nerves causes the bladder to relax, which is opposite to the effect of parasympathetic stimulation. However, the sympathetic nerves do not normally enter into micturition control.

### TRANSPORT OF URINE THROUGH THE URETERS

The ureters are small smooth muscle tubes that originate in the pelves of the two kidneys and pass downward to enter the bladder. Each ureter is innervated by both sympathetic and parasympathetic nerves, and each also has an intramural plexus of neurons and nerve fibers that extends along its entire length.

As urine collects in the pelvis, the pressure in the pelvis increases and initiates a peristaltic contraction beginning in the pelvis and spreading down along the ureter to force urine toward the bladder. A peristaltic wave, traveling at a

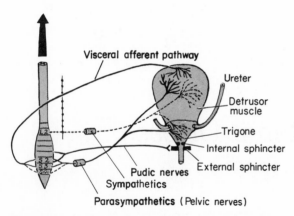

**Figure 38–1.** The urinary bladder and its innervation.

velocity of about 3 cm./sec., occurs from once every 10 seconds to once every two to three minutes. The peristaltic wave can move urine against an obstruction with a pressure as high as 25 mm. Hg. Parasympathetic stimulation increases and sympathetic stimulation decreases the frequency. Transmission of the peristaltic wave is probably caused mainly by nerve impulses passing along the intramural plexus in the same manner that the intramural plexus functions in the gut.

At the lower end, the ureter penetrates the bladder obliquely through the *trigone,* as illustrated in Figure 38–1. The ureter courses for several centimeters under the bladder epithelium so that pressure in the bladder compresses the ureter, thereby helping to prevent back-flow of urine when pressure builds up in the bladder during micturition.

**Pain Sensations from the Ureters, and the Ureterorenal Reflex.** The ureters are well supplied with pain nerve fibers. Any time the ureter becomes blocked, such as by a ureteral stone, intense reflex constriction, associated with very severe pain, occurs. In addition, the pain impulses probably cause a sympathetic reflex back to the kidney to constrict the renal arterioles, thereby decreasing urinary output from that kidney. This effect is called the *ureterorenal reflex;* it obviously is important to prevent excessive flow of fluid into the pelvis of a kidney whose ureter is blocked.

## TONUS OF THE BLADDER, AND THE CYSTOMETROGRAM DURING BLADDER FILLING

The solid curve of Figure 38–2 is called the *cystometrogram* of the bladder. It shows the changes in intravesical pressure as the bladder fills with urine. When no urine at all is in the bladder, the intravesical pressure is approximately zero, but, by the time 100 ml. of urine has collected, the pressure will have risen to about 10 cm. water. Additional urine up to 300 to 400 ml. can collect without significant further rise in pressure; the pressure remains still at 10 cm. water, this caused by *intrinsic tone* of the bladder wall itself. Beyond this point, collection of more urine causes the pressure to rise very rapidly.

Superimposed on the tonic pressure changes during filling of the bladder are periodic acute increases in pressure, which last from a few seconds to more than a minute. The pressure can rise only a few centimeters of water or it can rise to over 100 cm. water. These are *micturition waves* in the cystometrogram caused by the micturition reflex, which is discussed below.

## THE MICTURITION REFLEX

Referring again to Figure 38–2, one sees that as the bladder approaches the full state, many superimposed *micturition contractions* begin to appear. These are the result of a stretch reflex initiated by stretch receptors in the bladder wall and proximal urethra. Sensory signals are conducted to the sacral segments of the cord through the pelvic nerves and then back again to the bladder through the parasympathetic fibers in these same nerves.

Once a micturition reflex begins, it is "self-regenerative." That is, initial contraction of the

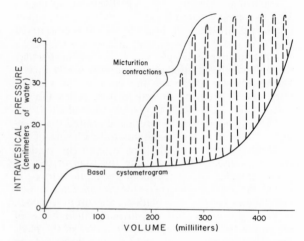

**Figure 38–2.** A normal cystometrogram showing also acute pressure waves (the dashed curves) caused by micturition reflexes.

bladder further activates the receptors to cause still further increase in afferent impulses from the bladder, which causes further increase in reflex contraction of the bladder, the cycle thus repeating itself again and again until the bladder has reached a strong degree of contraction. Then, after a few seconds to more than a minute, the reflex begins to fatigue, and the regenerative cycle of the micturition reflex ceases, allowing rapid reduction in bladder contraction. In other words, the micturition reflex is a single complete cycle of (a) progressive and rapid increase in pressure, (b) a period of sustained pressure, and (c) return of the pressure back to the basal tonic pressure of the bladder. Once a micturition reflex has occurred and has not succeeded in emptying the bladder, the nervous elements of this reflex usually remain in an inhibited state for at least a few minutes to sometimes as long as an hour or more before another micturition reflex occurs. However, as the bladder becomes more and more filled, micturition reflexes occur more and more often and more and more powerfully.

**Control of Micturition by the Brain.**  The micturition reflex is a completely automatic cord reflex, but it can be inhibited or facilitated by centers in the brain. These include (a) strong *facilitatory and inhibitory centers in the brain stem,* and (b) several *centers located in the cerebral cortex* that are mainly inhibitory but can at times become excitatory.

The fiber pathways of the cord through which micturition signals are transmitted to and from the brain are not known. However, some studies have indicated that the afferent fibers travel in or in close association with the lateral spinothalamic tracts and that the efferent fibers lie immediately deep to the lateral corticospinal tract.

The micturition reflex is the basic cause of micturition, but the higher centers normally exert final control of micturition by the following means:

1. The higher centers keep the micturition reflex partially inhibited all the time except when it is desired to micturate.

2. The higher centers prevent micturition, even if a micturition reflex occurs, by continual tonic contraction of the external urinary sphincter until a convenient time presents itself.

3. When the time to urinate arrives, the cortical centers can (a) facilitate the sacral micturition centers to initiate a micturition reflex and (b) inhibit the external urinary sphincter so that urination can occur.

## ABNORMALITIES OF MICTURITION

**The Atonic Bladder.**  Destruction of the sensory nerve fibers from the bladder to the spinal cord prevents transmission of stretch signals from the bladder and, therefore, also prevents micturition reflex contractions. Therefore, the person loses all control of his bladder despite intact efferent fibers from the cord to the bladder and despite intact neurogenic connections with the brain. Instead of emptying periodically, the bladder fills to capacity and overflows a few drops at a time through the urethra. This is called *overflow dribbling.*

The atonic bladder was a common occurrence when syphilis was widespread, because syphilis frequently causes constrictive fibrosis around the dorsal nerve root fibers where they enter the spinal cord and, subsequently, destroys these fibers. This condition is called *tabes dorsalis,* and the resulting bladder condition is called a *tabetic bladder.* Another common cause of this condition is crushing injuries to the sacral region of the cord.

**The Automatic Bladder.** If the spinal cord is damaged above the sacral region but the sacral segments are still intact, typical micturition reflexes still occur. However, they are no longer controllable by the brain. During the first few days to several weeks after the damage to the cord has occurred, the micturition reflexes are completely suppressed because of sudden loss of facilitatory impulses from the brain stem and cerebrum. However, the excitability of the micturition centers in the sacral cord gradually increases until typical micturition reflexes return. Sometimes these reflexes are elicited by approximately normal amounts of urine in the bladder, whereas at other times the threshold for eliciting the reflex may be low or high. The factor that causes this difference is the treatment the patient receives during the early stages of the convalescent period. If the bladder is kept completely empty by continuous drainage, contracture of the bladder causes a low threshold for development of micturition reflexes. But if the bladder is allowed to remain overly filled for long periods of time, it becomes progressively stretched, and a large quantity of urine is then necessary to elicit micturition reflexes.

It is especially interesting that scratching or otherwise stimulating the skin in the genital region can sometimes elicit a micturition reflex in this condition, thus providing a means by which some patients can still control urination.

**The Uninhibited Neurogenic Bladder.**  Another common abnormality of micturation is the so-called "uninhibited neurogenic bladder," which results in frequent and relatively uncontrollable micturition. This condition derives from damage in the spinal cord or brain stem that interrupts most of the inhibitory signals. Therefore, facilitatory impulses passing continually down the cord keep the sacral centers so excitable that even a very small quantity of urine will elicit an uncontrollable micturition reflex and thereby promote urination.

# RENAL DISEASE

The physiology of renal disease can be classified into five different physiological categories: (1) *acute renal failure,* in which the kidneys stop working entirely or almost entirely, (2) *chronic renal failure,* in which progressively more nephrons are destroyed until the kidneys simply cannot perform all the necessary functions, (3) *hypertensive kidney disease,* in which vascular or glomerular lesions cause hypertension but not renal failure, (4) *nephrotic syndrome,* in which the glomeruli have become far more permeable than normal so that large amounts of protein are lost into the urine, and (5) *specific tubular abnormalities* that cause abnormal reabsorption or lack of reabsorption of certain substances by the tubules.

## ACUTE RENAL FAILURE

Almost any condition that seriously interferes with kidney function can cause acute renal failure. Two of the most common causes are (1) acute glomerulonephritis and (2) acute damage and obstruction of the tubules.

**Renal Failure Caused by Acute Glomerular Nephritis.** Acute glomerular nephritis is a disease that results from an antigen-antibody reaction and in which the glomeruli become markedly inflamed. Large numbers of white blood cells collect in the inflamed glomeruli, and the endothelial cells on the vascular side of the glomerular membrane, as well as the epithelial cells on the Bowman's capsule side of the membrane, proliferate, sometimes completely filling the glomeruli and the capsule. These inflammatory reactions can cause total or partial blockage of large numbers of glomeruli, and many of those glomeruli that are not blocked develop greatly increased permeability of the glomerular membrane, allowing large amounts of protein to leak into the glomerular filtrate. Also, rupture of the membrane in severe cases often allows large numbers of red blood cells to pass into the glomerular filtrate. In the severest cases total renal shutdown occurs.

The inflammation of acute glomerular nephritis almost invariably occurs one to three weeks following, elsewhere in the body, an infection caused by certain types of group A beta streptococci, such as a streptococcal sore throat, streptococcal tonsillitis, scarlet fever, or even streptococcal infection of the skin. The infection itself does not cause damage to the kidneys, but when antibodies develop against the streptococcal antigen, the antibodies and antigen react with each other to form a precipitate that becomes entrapped in the middle of the glomerular membrane. The reactivity of this precipitated complex leads to the inflamed glomeruli.

The acute inflammation of the glomeruli most often subsides in 10 days to two weeks, and the nephrons may return to normal function. Sometimes, however, the inflammatory reactions are so severe that many of the glomeruli will have been destroyed permanently.

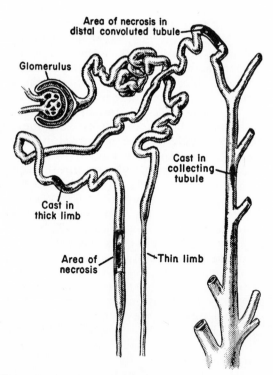

**Figure 38–3.** Damage to the distal tubules as a result of shock. (Modified from MacLean: Acute Renal Failure. Charles C Thomas.)

**Tubular Necrosis as a Cause of Acute Renal Failure.** Another common cause of acute renal shutdown is *tubular necrosis,* which means destruction of epithelial cells in the tubules, as illustrated in Figure 38–3. Common causes of tubular necrosis are (1) *various poisons* that destroy the tubular epithelial cells, and (2) *severe acute ischemia* of the kidneys.

*Renal Poisons.* Among the different renal poisons are *carbon tetrachloride* and the heavy metals such as the *mercuric ion.* These substances have specific nephrotoxic action on the tubular epithelial cells, causing death of many of them. As a result, the epithelial cells slough away from the basement membrane and plug the tubules. In some instances the basement membrane also is destroyed, but, if not, new tubular epithelial cells can usually grow along the surface of the membrane so that the tubule becomes repaired within 10 to 20 days.

*Severe Acute Renal Ischemia.* Severe ischemia of the kidney is likely to result from *severe circulatory shock.* In shock, the heart simply fails to pump sufficient amounts of blood to supply adequate nutrition to the different parts of the body, and renal blood flow is particularly likely to suffer because of strong sympathetic constriction of the renal vessels. Therefore, lack of adequate nutrition often destroys many tubular epithelial cells and thereby plugs many of the nephrons.

**Transfusion Reaction as a Cause of Acute Renal Failure.** A transfusion reaction normally results in hemolysis of large amounts of red blood cells with release of hemoglobin into the plasma. The size of the hemoglobin molecule is slightly less than that of the pores in the glomerular membrane so that much of the hemoglobin passes through the membrane into the glomerular filtrate. Therefore, after a transfusion reaction the tubular load of hemoglobin becomes far greater than can be reabsorbed by the proximal tubules. The excess hemoglobin is likely to precipitate in the nephron and in this way cause blockage. In addition, hemolysis of red blood cells also seems to release vasoconstrictor agents into the blood stream, and it is believed that this can promote poor blood supply to the tubules, this acting as a further cause of tubular damage.

## CHRONIC RENAL FAILURE–DECREASE IN NUMBER OF FUNCTIONAL NEPHRONS

In many types of kidney disease large numbers of nephrons are destroyed or damaged so severely that the remaining nephrons simply cannot perform the normal functions of the kidneys. Some of the different causes of this include *chronic glomerulonephritis, traumatic loss of kidney tissue, congenital absence of kidney tissue, congenital polycystic disease* (in which large cysts develop in the kidneys and destroy surrounding nephrons by compression), *urinary tract obstruction* resulting from renal stones, *pyelonephritis,* and diseases of the renal vasculature.

**Chronic Glomerulonephritis.** Chronic glomerulonephritis is caused by any one of several different diseases that damage principally the glomeruli. The basic glomerular lesion is usually very similar to that which occurs in acute glomerulonephritis. It seems to begin with accumulation of precipitated antigen-antibody complex in the glomerular membrane followed by inflammation of the glomeruli. The glomerular membrane becomes progressively thickened and is eventually invaded by fibrous tissue. In the later stages of the disease, the glomerular filtration coefficient becomes greatly reduced because of decreased numbers of filtering capillaries in the glomerular tufts and because of thickened glomerular membranes. In the final stages of the disease many of the glomeruli are completely replaced by fibrous tissue, and the function of these nephrons is thereafter lost forever.

**Pyelonephritis.** Pyelonephritis is an infectious and inflammatory process that usually begins in the renal pelvis but extends progressively into the renal parenchyma. The infection can result from many different types of bacteria, but especially from the colon bacillus that originates from fecal contamination of the urinary tract. Invasion of the kidneys by these bacteria results in progressive destruction of renal tubules, glomeruli, and any other structures in the path of the invading organisms. Consequently, large portions of the functional renal tissue are lost.

A particularly interesting feature of pyelonephritis is that the invading infection usually affects the medulla of the kidney before it affects the cortex. Since one of the primary functions of the medulla is to provide the counter-current mechanism for concentrating the urine, patients with pyelonephritis frequently have reasonably normal renal function except for inability to concentrate their urine.

**Destruction of Nephrons by Renal Vascular Disease—Benign Nephrosclerosis.** Though many different types of vascular lesions can lead to renal ischemia and death of the renal tissue, the most common of these are: (1) *atherosclerosis* of the larger renal arteries with progressive sclerotic constriction of the vessels, (2) *fibromuscular hyperplasia* of one or more of the larger arteries, which also causes occlusion of these large vessels, and (3) *benign nephrosclerosis*, a very common condition caused by sclerotic lesions of the smaller arteries and arterioles.

Arteriosclerotic or hyperplastic lesions of the larger arteries frequently affect one kidney more than the other and therefore cause unilaterally diminished renal function.

*Benign nephrosclerosis* is believed to begin with leakage of plasma through the intimal membrane of the small arteries and arterioles. This causes collection of a fibrinoid deposit in the medial layers of these vessels followed by progressive fibrous tissue invasion that eventually constricts the vessel—in many instances totally occluding it. Since there is essentially no collateral circulation among the smaller renal arteries, destruction of each one of these also causes destruction of a comparable amount of renal mass. Therefore, much of the kidney tissue becomes replaced by minute areas of fibrous tissue; the kidneys become greatly reduced in size and progressively develop a nodular surface. This process occurs at least to some extent in almost all persons in old age and causes progressive decrease in both renal blood flow and renal plasma clearance. Figure 38–4 illustrates the diodrast clearance of the kidneys (as a measure of renal blood flow) of otherwise normal persons at different ages, showing that even in the "normal" person, kidney blood flow decreases on the average to about 45 per cent of normal by the age

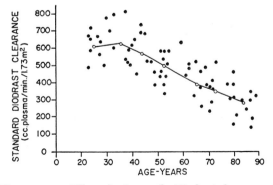

**Figure 38–4.** Effect of aging on the Diodrast clearance of the kidneys. (Modified from Wolstenholme et al.: Ciba Foundation Colloquia on Ageing. Little, Brown and Co.)

of 80, and there is accompanying reduction of excretory function.

## Abnormal Nephron Function in Chronic Renal Failure

**Inability of Failing Kidneys to Excrete Excess "Loads" of Excretory Products.** The most important effect of renal failure is the inability of the kidneys to cope with large "loads" of electrolytes or other substances that must be excreted. Normally, one-third of the nephrons can eliminate essentially all the normal load of waste products from the body without serious accumulation of any of these in the body fluids. However, further reduction of numbers of nephrons leads to urinary retention, and death ensues when the number of nephrons falls to 10 to 20 per cent of normal.

**Function of the Remaining Nephrons in Renal Failure.** In renal failure, the still functioning nephrons usually become greatly overloaded in several different ways. First, for reasons not understood, the blood flow through the glomerulus and the amount of glomerular filtrate formed each minute by each nephron often increase to as much as double normal. Second, extra large amounts of excretory substances, such as urea, phosphates, sulfates, uric acid, and creatinine, accumulate in the extracellular fluid. On entering the glomerular filtrate, these constitute markedly increased tubular loads of substances that are poorly reabsorbed. These increasing loads are partially compensated for by as much as 50 per cent increase in reabsorptive power of each tubule, but even this increase is often far too little to keep pace with the loads, which may be increased as much as 1000 per cent for each nephron. Therefore, only a small fraction of the solutes are reabsorbed, and the remaining solutes act like an osmotic diuretic, resulting in rapid flushing of tubular fluid through the tubules. Consequently, the volume of urine formation by each nephron can rise to as much as 20 times normal, and a person will occasionally have *as much as three times normal total urine output* despite the fact that he has significant renal insufficiency. This paradoxical situation is caused by a greater increase in urine volume output per nephron than reduction in numbers of nephrons.

*Isosthenuria.* Another effect of rapid flushing of fluid through the tubules is that the normal concentrating and diluting mechanisms of the kidney do not have time to function properly, especially the concentrating mechanism. Therefore, as progressively more nephrons are destroyed, the specific gravity of the urine approaches that of the glomerular filtrate, approximately 1.008. These effects are illustrated in Figure 38–5, which gives the approximate upper and lower limits of urine specific gravity as the number of nephrons decreases. Since the concentrating mechanism is impaired more than the diluting mechanism, an important renal function test is to determine how well the kidneys can concentrate urine when the person is dehydrated for 12 or more hours.

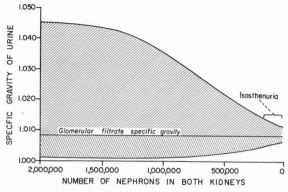

**Figure 38–5.** Development of *isothenuria* in patients with decreased numbers of active nephrons.

## Effects of Renal Failure on the Body Fluids—Uremia

The effect of acute or chronic renal failure on the body fluids depends to a great extent on the water and food intake of the person. Assuming that the person continues to ingest moderate amounts of water and food, the concentration changes of different substances in the extracellular fluid are approximately those shown in Figure 38–6. The most important effects are (1) *generalized edema* resulting from water and salt retention, (2) *acidosis* resulting from failure of the kidneys to rid the body of normal acidic products, (3) *high concentrations of the nonprotein nitrogens*, especially *urea*, resulting from failure of the body to excrete the metabolic endproducts, and (4) *high concentration of other urinary retention products* including *creatinine, uric acid, phenols, guanidine bases, sulfates, phosphates,* and *potassium*. This condition is called *uremia* because of the high concentrations of normal urinary excretory products that collect in the body fluids.

**Water Retention and Edema.** If treatment is begun immediately after acute renal failure, the total

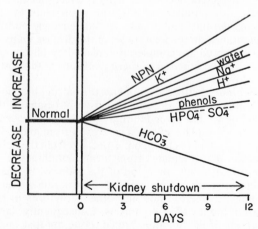

**Figure 38–6.** Effect of kidney shutdown on extracellular fluid constituents.

body fluid content may not change at all. However, when the patient drinks water in response to his normal desire, the body fluids begin increasing immediately and rapidly. If he takes in no electrolytes at the same time, approximately five-eighths of the water enters the cells and three-eighths remains in the extracellular fluids. Thus, a patient with edema in renal failure will usually have both *intracellular* and *extracellular edema*. On the other hand, if the patient should ingest large amounts of salt along with the water, the edema might be only extracellular because of greatly increased amounts of osmotically active substances in the extracellular compartment.

**Acidosis in Renal Failure.** Each day the metabolic processes of the body normally produce 50 to 100 millimols more metabolic acid than metabolic alkali. Therefore, any time the kidneys fail to function, acid begins to accumulate in the body fluids. Normally, the buffers of the fluids can buffer up to a total of 500 to 1000 millimols of acid without severe depression of the extracellular fluid pH, and the phosphate compounds in the bones can buffer an additional few thousand millimols, but gradually this buffering power is used up so that the pH falls drastically. The patient becomes *comatose* at about this same time, and it is believed that this is partly caused by the acidosis, as is discussed below.

**Elevated Potassium Concentration in Uremia.** The amount that the potassium concentration increases depends principally on the rate of protein catabolism and on potassium intake after failure has occurred. Breakdown of the cellular proteins releases potassium from combination with the proteins, and the extra potassium then passes into the extracellular fluid. Obviously, failure of the kidneys to excrete the potassium causes an elevated potassium concentration in the extracellular fluid. When the potassium concentration rises to 8 mEq./liter, it begins to have a cardiotoxic effect, resulting in arrhythmias and dilatation of the heart, and at higher levels it leads to acute cardiac failure and death.

**Increase in Urea and Other Nonprotein Nitrogens (Azotemia) in Uremia.** The nonprotein nitrogens include urea, uric acid, creatinine, and a few less important compounds. These, in general, are the end-products of protein metabolism and must be removed from the body continually to insure continued protein metabolism in the cells. The concentrations of these, particularly of urea, can rise to as high as 10 times normal during one to two weeks of renal failure. However, even these high levels do not seem to affect physiological function nearly so much as the high concentrations of hydrogen and potassium ions and some of the other less obvious substances such as very toxic guanidine bases, ammonium ions, and others. Yet one of the most important means for assessing the degree of renal failure is to measure the concentrations of the nonprotein nitrogens.

**Uremic Coma.** After a week or more of renal failure the sensorium of the patient becomes clouded, and he soon progresses into a state of coma. The acidosis is believed to be one of the principal factors

responsible for the coma because acidosis caused by other conditions, such as severe diabetes mellitus, also causes coma. However, many other abnormalities could also be contributory—the generalized edema, the high potassium concentration, and possibly even the high nonprotein nitrogen concentration.

The respiration usually is deep and rapid in coma, which is a respiratory attempt to compensate for the metabolic acidosis. In addition to this, during the last day or so before death, the arterial pressure falls progressively, then rapidly in the last few hours. Death occurs usually when the pH of the blood falls to about 6.9.

**Anemia in Renal Failure.** A patient with chronic renal failure of severe degree almost always also develops severe *anemia*. The probable cause of this is the following: The kidneys normally secrete an enzyme that splits the substance *erythropoietin* from a plasma *protein*. The erythropoietin in turn stimulates the bone marrow to produce red blood cells. Obviously, if the kidneys are seriously damaged they are unable to initiate adequate production of this substance, which leads to diminished red blood cell production and consequent anemia. However, other factors such as the high concentration of urea, hydrogen ions, and potassium, might also play important roles in causing the anemia.

**Osteomalacia in Renal Failure.** Prolonged renal failure also causes osteomalacia, a condition in which the bones are partially absorbed and therefore greatly weakened, as will be explained in relation to the physiology of bone in Chapter 79. The most important cause of this condition is the following: Vitamin D must be converted by a two-stage process, first in the liver and then in the kidney, into 1, 25-dihydroxycholecalciferol before it is able to promote calcium absorption from the intestine. Therefore, serious damage to the kidneys greatly reduces the availability of calcium to the bones.

### Dialysis of Uremic Patients with the Artificial Kidney

Artificial kidneys have now been used for about 25 years to treat patients with severe renal failure. In certain types of acute renal failure, such as that following mercury poisoning or following circulatory shock, the artificial kidney is used simply to tide the patient over for a few weeks until the renal damage heals so that the kidneys can resume function. Also, the artificial kidney has now been developed to the point that several thousand persons with permanent renal failure or even total kidney removal are being maintained in health for years at a time, their lives depending entirely on the artificial kidney.

The basic principle of the artificial kidney is to pass blood through very minute blood channels bounded by thin membranes. On the other sides of the membranes is a *dialyzing fluid* into which unwanted substances in the blood pass by diffusion.

Figure 38–7 illustrates an artificial kidney in which

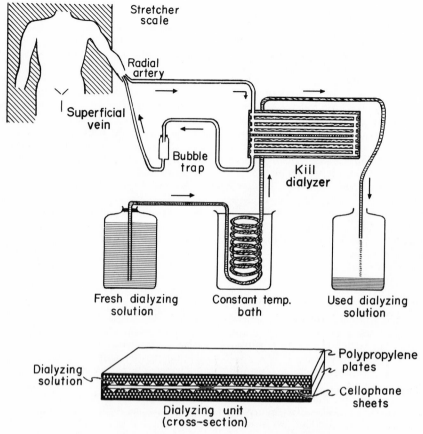

**Figure 38–7.** Schematic diagram of the artificial kidney.

blood flows continually between two thin sheets of cellophane; on the outside of the sheets is the dialyzing fluid. The cellophane is porous enough to allow all constituents of the plasma except the plasma proteins to diffuse freely in both directions—from plasma into the dialyzing fluid and from the dialyzing fluid back into the plasma. If the concentration of a substance is greater in the plasma than in the dialyzing fluid, there will be net transfer of the substance from the plasma into the dialyzing fluid. The amount of the substance that transfers depends on: (1) the difference between the concentrations on the two sides of the membrane, (2) molecular size, the smaller molecules diffusing more rapidly than larger ones, and (3) the length of time that the blood and the fluid remain in contact with the membrane.

In normal operation of the artificial kidney, blood continually flows from an artery, through the kidney, and back into a vein. The total amount of blood in the artificial kidney at any one time is usually less than 500 ml., the rate of flow may be several hundred ml. per minute, and the total diffusing surface is usually between 10,000 and 20,000 square centimeters. To prevent coagulation of blood in the artificial kidney, heparin is infused into the blood as it enters the "kidney." Then, to prevent bleeding in the patient as a result of the heparin, an anti-heparin substance, protamine, is infused into the blood as it is returned to the patient.

**The Dialyzing Fluid.** Table 38–1 compares the constituents in a typical dialyzing fluid with those in normal plasma and uremic plasma. Note that sodium, potassium, and chloride concentrations in the dialyzing fluid and in normal plasma are identical, but in uremic plasma the potassium concentration is considerably greater. This ion diffuses through the dialyzing membrane so rapidly that its concentration falls to equal that in the dialyzing fluid within only three to four hours' exposure to the dialyzing fluid.

On the other hand, there is no phosphate, urea, urate, sulfate, or creatinine in the dialyzing fluid. Therefore, when the uremic patient is dialyzed, these substances are lost in large quantities into the dialyzing fluid, thereby removing major proportions of them from the plasma.

Thus, the constituents of the dialyzing fluid are chosen so that those substances in excess in the extracellular fluid in uremia can be removed at rapid rates, while the normal electrolytes remain essentially normal.

**Effectiveness of the Artificial Kidney.** The effectiveness of an artificial kidney is expressed in

**TABLE 38–1.   Comparison of Dialyzing Fluid with Normal and Uremic Plasma**

| Constituent | Normal Plasma | Dialyzing Fluid | Uremic Plasma |
|---|---|---|---|
| *Electrolytes* | | | |
| *(mEq./liter)* | | | |
| Na$^+$ | 142 | 142 | 142 |
| K$^+$ | 5 | 5 | 7 |
| Ca$^{++}$ | 3 | 3 | 2 |
| Mg$^{++}$ | 1.5 | 1.5 | 1.5 |
| Cl$^-$ | 107 | 107 | 107 |
| HCO$_3^-$ | 27 | 27 | 14 |
| Lactate$^-$ | 1.2 | 1.2 | 1.2 |
| HPO$_4^{--}$ | 3 | 0 | 9 |
| Urate$^-$ | 0.3 | 0 | 2 |
| Sulfate$^{--}$ | 0.5 | 0 | 3 |
| *Nonelectrolytes* | | | |
| *(mg. %)* | | | |
| Glucose | 100 | 125 | 100 |
| Urea | 26 | 0 | 200 |
| Creatinine | 1 | 0 | 6 |

terms of the amount of plasma that can be cleared of different substances each minute, which, as will be recalled from Chapter 35, is also the principal means for expressing the functional effectiveness of the kidneys themselves. Most artificial kidneys can clear 100 to 200 ml. of plasma per minute of urea, which shows that, at least in the excretion of this substance, the artificial kidney can function about two times as rapidly as the two normal kidneys together, whose urea clearance is only 70 ml. per minute. Yet the artificial kidney can be used for no more than 12 hours every three to four days because of danger from excess heparin, hemolysis of blood, and infection. Therefore, the overall plasma clearance is still somewhat limited when the artificial kidney replaces the normal kidneys.

## HYPERTENSIVE KIDNEY DISEASE

Most of the same types of kidney disease that lead to chronic renal failure can also cause hypertension. However, this is not invariably true because damage to certain portions of the kidney is very prone to cause hypertension, whereas damage to other portions will cause uremia without hypertension. A classification of renal disease relative to its hypertensive or nonhypertensive effects is the following:

**Hypertensive Renal Lesions.** Essentially all renal lesions that cause *diminished vascular supply* or *diminished glomerular filtration* will cause hypertension. Both of these conditions tend to decrease the glomerular filtration rate, which in turn leads to a sequence of events (explained in Chapter 22) that begins with retention of salt and water and eventually leads to hypertension. Once the hypertension has developed, the glomerular filtration rate may return entirely to normal. If all the tubules are normal, the

filtrate is then processed normally in these so that urinary excretion may be completely normal, and there may be absolutely no sign of renal failure.

**Renal Diseases that Lead to Uremia but Might Not Lead to Hypertension.** Loss of large numbers of whole nephrons, such as occurs from loss of one kidney and part of another kidney, will always lead to uremia if the amount of kidney tissue lost is great enough. However, if the remaining nephrons are completely normal, this condition frequently will not cause hypertension because even a slight rise in arterial pressure will increase the glomerular filtration rate sufficiently to promote rapid water and salt loss in the urine. On the other hand, if a patient with this type of kidney abnormality eats large quantities of salt, he will then develop very severe hypertension because the kidneys simply cannot clear adequate quantities of salt under these conditions.

**Kidney Lesions that Tend to Cause Hypotension.** An occasional person has a type of kidney disease that leads to tremendous salt and water loss in the urine even though there may be simultaneous retention of waste products and a state of uremia. This most frequently results from interstitial kidney disease, such as *medullary pyelonephritis*. As long as the person ingests sufficient quantities of salt, his pressure will remain normal. Placing a patient of this type on a normal salt regime, or on treatment with a diuretic, however, can lead to arterial pressure decreased to hypotensive levels, decreased urine output, and increased symptoms of uremia, sometimes causing death.

**Hypertension Caused by Renal Secretion of Renin.** When one part of the kidney mass is ischemic while the remainder is nonischemic, (such as occurs when one renal artery is severely constricted) the ischemic renal tissue secretes large quantities of renin. This secretion leads to the formation of angiotensin II, which in turn leads to hypertension. The most probable causes of the chronic hypertension, as was discussed in Chapter 22, are (1) the ischemic kidney tissue itself excretes less than normal amounts of water and salt, and (2) the angiotensin affects the nonischemic kidney tissue, causing it also to retain water and salt. (Many physiologists have believed in the past that peripheral vasoconstriction caused by the angiotensin is the cause of this type of hypertension. However, recent experiments have demonstrated almost conclusively that this is not the case, as was discussed in Chapter 22.)

## THE NEPHROTIC SYNDROME— INCREASED GLOMERULAR PERMEABILITY

Large numbers of patients with renal disease develop a so-called *nephrotic syndrome,* which is characterized especially by *loss of large quantities of plasma proteins into the urine.* In some instances this occurs without evidence of any other abnormality of renal function, but more often it is associated with some degree of renal failure.

The cause of the protein loss in the urine is increased permeability of the glomerular membrane. Therefore, any disease condition that can increase the permeability of this membrane can cause the nephrotic syndrome. Such diseases include some types of *chronic glomerulonephritis* (in the previous discussion, it was noted that this disease primarily affects the glomeruli and causes a greatly increased permeability of the glomerular membrane), *amyloidosis,* which results from deposition of an abnormal proteinoid substance in the walls of blood vessels and seriously damages the basement membrane of the glomerulus, and *lipoid nephrosis,* a disease found mainly in young children.

**Lipoid Nephrosis.** Lipoid nephrosis is very common in children, occurring most often before the age of four but occurring occasionally in adults as well. Its basic cause is unknown, but the resulting renal lesion increases permeability of the glomerular membrane and causes loss of proteins into the urine. This lesion develops in the following way: The epithelial cells that line the outer surface of the glomerulus are defective. They are greatly swollen, and they fail to form the usual foot processes that cover the Bowman's capsule surface of the glomerulus. (It will be recalled from the discussion of the glomerular membrane in Chapter 34 that the openings between these foot processes are very small, and it is normally the smallness of these openings that prevents the passage of protein through the glomerular membrane.) Therefore, in the absence of the foot processes, tremendous quantities of protein leak into the tubules even though larger elements of the blood, such as red cells, are still completely prevented from leaking.

The name ''lipoid nephrosis'' is derived from the fact that large quantities of lipid droplets are found in the epithelial cells and also from the fact that the concentration of lipid substances in the blood is usually increased (this is also true of some other types of nephrosis to a lesser extent). The lipid deposits in the tubules apparently play no role in the disease.

Administration of glucocorticoids such as hydrocortisone will usually cause complete remission of the disease, although they will not cause remission of most other types of nephrotic syndrome.

**Protein Loss.** In the nephrotic syndrome, as much as 30 to 40 grams of plasma proteins can be lost into the urine each day. Though the resulting low plasma protein concentration stimulates the liver to produce far more plasma proteins than usual, nevertheless, the liver often cannot keep up with the loss. Therefore, in severe nephrosis the colloid osmotic pressure sometimes falls extremely low, often from the normal level of 28 mm. Hg to as low as 6 to 8 mm. Hg.

**Edema.** The low colloid osmotic pressure in turn allows large amounts of fluid to filter into the interstitial spaces and also into the potential spaces of the body, thus causing serious *edema.* The nephrotic person has been known on occasion to develop as much as 40 liters of excess extracellular fluid, and as much as 15 liters of this has been *ascites* in the abdomen. Also, the joints swell, and the pleural cavity and the pericardium become partially filled with fluid.

In severe nephrosis, the loss of fluid into the interstitial spaces causes the blood volume to diminish, and this in turn elicits a sequence of events described in Chapter 76 that leads to increased aldosterone secretion by the adrenal glands. The aldosterone then causes the kidneys to reabsorb excessive quantities of sodium chloride and water, which further compounds the seriousness of the edema.

A nephrotic person can be greatly benefited by intravenous infusion of large quantities of concentrated plasma proteins. Yet this is of only temporary benefit because enough protein can be lost into the urine in only a day to return the person to his original predicament.

## SPECIFIC TUBULAR DISORDERS

In the discussion of active reabsorption and secretion by the tubules in Chapter 34, it was pointed out that the active transport processes are carried out by various carriers and enzymes in the tubular epithelial cells. Furthermore, there are a number of different carrier mechanisms for the different individual substances. In Chapter 3 it was also pointed out that each cellular enzyme and probably also each carrier substance are formed in response to a respective gene in the nucleus. If any of the respective genes happens to be absent or abnormal, the tubules might be deficient in one of the appropriate enzymes or carriers. For this reason many different specific tubular disorders are known to occur for the transport of individual or special groups of substances through the tubular membrane. Essentially all these are hereditary disorders. Some of the more important ones are:

**Renal Glycosuria.** In this condition the transport mechanism for reabsorbing glucose is either deficient or lacking. The blood glucose concentration may be completely normal, but the transport maximum for reabsorption of glucose each minute is greatly limited. Consequently, despite the normal blood glucose level, large quantities of glucose pass into the urine each day. Because one of the tests for diabetes mellitus (which results from lack of insulin secretion by the pancreas) is the presence of glucose in the urine, renal glycosuria (a benign condition which causes essentially no dysfunction of the body) must always be ruled out before making a diagnosis of diabetes mellitus.

**Nephrogenic Diabetes Insipidus.** Occasionally the renal tubules do not respond completely to antidiuretic hormone secreted by the supraopticohypophyseal system, and as a consequence large quantities of dilute urine are continually excreted. As long as the person is supplied with plenty of water, this condition rarely causes any severe difficulty. However, when he fails to receive adequate quantities of water, he rapidly becomes dehydrated.

**Renal Tubular Acidosis.** In this condition the person is unable to secrete adequate quantities of hydrogen ions, and, as a result, large amounts of sodium bicarbonate are continually lost into the urine for reasons discussed in the preceding chapter. This causes a continual state of metabolic acidosis. However, adequate replacement therapy by continual administration of alkali can maintain normal bodily function.

**Renal Hypophosphatemia.** In renal hypophosphatemia the renal tubules fail to reabsorb adequate quantities of phosphate ions even when the phosphate concentration of the body fluids falls very low. This condition does not cause any serious immediate abnormalities, because the phosphate level of the extracellular fluids can vary widely without significant cellular dysfunction. However, over a long period of time the low phosphate level results in diminished calcification of the bones and causes the person to develop rickets. Furthermore, this type of rickets is refractory to vitamin D therapy in contrast to the rapid response of the usual type of rickets, which is discussed in Chapter 79.

**Amino-Acidurias.** Some amino acids share mutual carrier systems for reabsorption, while other amino acids have distinct carrier systems of their own. Rarely, a condition called *generalized aminoaciduria* results from deficient reabsorption of all amino acids, but, more frequently, deficiencies of specific carrier systems may result in (1) *essential cystinuria*, in which large amounts of cystine fail to be reabsorbed and often crystallize in the urine to form renal stones, (2) simple *glycinuria*, in which glycine fails to be reabsorbed, or (3) *beta-aminoisobutyric aciduria*, which occurs in about 5 per cent of all people but apparently has no clinical significance.

# RENAL FUNCTION TESTS

The renal function tests can be divided into three categories: (1) determination of renal clearances, (2) measurement of substances in the blood that are normally excreted by the kidneys, and (3) chemical and physical analyses of the urine.

**Renal Clearance Tests.** Any of the renal clearance tests, including clearance of paraaminohippuric acid, Diodrast, inulin, mannitol, or other substances, as described in Chapter 35, can be used as a renal function test. Indeed, if all these are run, one can determine the glomerular filtration rate, the effective blood flow through the kidney per minute, the filtration fraction, and many other characteristics of renal function. However, most of these clearance tests are not satisfactory for routine clinical use because of their complexities. Instead, three other types of renal clearance tests are normally used. These are urea clearance, phenolsulfonphthalein clearance, and x-ray or radioactive measurements of Diodrast clearance into the renal pelves.

*Urea Clearance.* Urea clearance is especially easily measured because all one needs to measure is the blood urea concentration and the total amount of urea entering the urine each minute. Employing the plasma clearance formula of Chapter 35, *urea clearance* equals *quantity of urea entering the urine each minute* divided by *plasma urea concentration*. Normal urea clearance is 70 ml. of plasma per minute for the average young adult. Values less than this indicate decreased renal function. Clinical symptoms of renal failure usually begin to show when the urea clearance falls below approximately 20 ml. per minute.

*Phenolsulfonphthalein Clearance.* Phenolsulfonphthalein, an alkaline dye, is injected either intravenously or subcutaneously, and successive samples of urine are collected. This dye is cleared into the urine both by the glomeruli and by tubular secretion. When injected intravenously, at least 60 per cent, and usually much more, of the dye is returned to the urine during the first hour if the kidneys are normal. The quantity of dye in the urine is determined by alkalinizing the urine with sodium hydroxide to bring out the color of the dye and then checking this against a color standard. In severe cases of renal insufficiency the amount cleared into the urine during the first hour is often depressed to as low as 5 to 10 per cent of the quantity originally injected.

*Intravenous Pyelography.* Several substances containing large quantities of iodine in their molecules—Diodrast, Hippuran, and Iopax—are excreted into the urine both by glomerular filtration and by active tubular secretion. Consequently, their concentration in the urine becomes very high within a few minutes after intravenous injection. Also, the iodine in the compounds makes them relatively opaque to x-rays. Therefore, x-ray pictures can be made showing shadows of the renal pelves, of the ureters, and even of the urinary bladder. Ordinarily, a sufficient quantity is excreted within five minutes after injection to give good shadows of the kidney pelves. Failure to show a distinct shadow within this time indicates depressed renal function.

*Radioactive Pyelography.* If any of the above substances is prepared with *radioactive* iodine, one can measure the radioactivity from both renal pelves by placing appropriate radioactivity counters over the kidneys. Only a trace amount of the substance need be injected intravenously and the degree of radioactivity recorded during the following few minutes. Simultaneous records from the two kidneys demonstrate the functional status of the two kidneys.

The value of both x-ray and radioactive pyelography is that they measure function of each kidney independently of the other rather than total function of both kidneys together, as is measured by the other renal function tests.

**Blood Analyses as Tests of Renal Function.** One can also estimate how well the kidneys are performing their functions by measuring the concentrations of various substances in the blood. For instance, the normal concentration of *urea* in the blood is 26 mg./100 ml., but in severe cases of renal insufficiency this can rise to as much as 300 mg. per cent. The normal concentration of *creatinine* in the blood is 1.1 mg. per cent, but this, too, can increase as much as 10-fold. To determine the degree of metabolic acidosis resulting from renal dysfunction, one can measure the *base excess* in the blood, as discussed in the preceding chapter. Though these different tests are not so satisfactory as clearance tests for determining the functional capabilities of the kidneys, they are easy to perform, and they do tell the physician how seriously the internal environment has been disturbed.

**Physical Measurements of the Urine as Renal Function Tests.** Obviously, one of the most important urinary measurements is the *volume of urine* formed each day. In acute renal failure this can fall to zero, and in chronic renal failure it usually is diminished. On the other hand, moderate renal failure may actually increase the urinary output, as was described above, because of vast overfunction of the remaining nephrons when the majority have been destroyed.

A second factor frequently measured is the *specific gravity* of the urine. Depending on the types of substances being cleared, this can vary tremendously; its upper extreme can go as high as 1.045 and it can fall to as little as 1.002. To test the ability of the kidneys to dilute the urine, the patient drinks large quantities of water, and measurements are made of the minimal specific gravity that can be attained. Then, at another time, he goes without water for 12 to 24 hours, and the maximum concentration of the urine is measured. Referring back to Figure 38–5, note that the concentrating ability of the kidneys is especially compromised as the number of nephrons decreases.

# DIURETICS AND MECHANISMS OF THEIR ACTIONS

A diuretic is a substance that increases the rate of urine output. Diuretics can act either by increasing the glomerular filtration rate or by decreasing the rate of reabsorption of fluid from the tubules.

The principal use of diuretics is to reduce the total amount of fluid in the body. They are especially important in treating edema and hypertension.

When using a diuretic it is usually important that the rate of sodium loss in the urine also be increased as well as the rate of water loss. The reason for this is the following: If water alone were removed from the body fluids, the fluids would become hypertonic and elicit an osmoreceptor response, followed by marked secretion of antidiuretic hormone. Consequently, large amounts of water would be reabsorbed immediately from the tubules, which would nullify the effect of the diuretic. However, if sodium is lost along with the water, this nullification will not result. Therefore, all valuable diuretics cause marked *natriuresis* (sodium loss) as well as diuresis.

Diuretics can be classified into three different physiologic types: (1) those that increase glomerular filtration, (2) those that increase the tubular osmotic load, and (3) those that inhibit the secretion of antidiuretic hormone.

## DIURETICS THAT INCREASE GLOMERULAR FILTRATION

The glomerular filtration rate can be increased by *increasing the arterial pressure,* by *dilating the afferent arterioles,* by *constricting the efferent arterioles,* or by *decreasing the colloid osmotic pressure.* Some of the different agents that can increase the glomerular filtration rate in one of these ways are the following:

Agents that increase the arterial pressure include *norepinephrine,* other sympathomimetic drugs, and sometimes large infusions of fluid. However, the pressor drugs, such as norepinephrine, usually have a dual effect on fluid output. *The increase in arterial pressure caused by norepinephrine tends to increase the output,* while *direct constriction of the afferent arterioles tends to reduce urine output.* In low and moderate pressure ranges the pressure effect is more important, so that norepinephrine normally causes diuresis, whereas in high pressure ranges, above about 160 mm. Hg, the urine output begins to fall.

Administration of *digitalis* to a patient with congestive heart failure frequently benefits the circulation so greatly that the glomerular pressure increases. And this increases the glomerular filtration rate, resulting in diuresis.

Certain chemicals directly dilate the afferent arterioles, including *theophylline* and *caffeine,* which increase the glomerular filtration rate and result in increased urinary output.

In general, the agents that increase glomerular filtration rate can increase the rate of urinary output only about two- to four-fold at most.

## DIURETICS THAT INCREASE THE TUBULAR OSMOTIC LOAD

**The Osmotic Diuretics.** Injection into the blood stream of *urea, sucrose, mannitol,* or any other substance not easily reabsorbed by the tubules causes a great increase in the osmotically active substances in the tubules. The osmotic pressure of these prevents water reabsorption, so that large amounts of tubular fluid flush on into the urine.

The same effect occurs when the glucose concentration of the blood rises to very high levels in diabetes mellitus. Above a glucose concentration of about 250 mg. per cent, very little glucose is reabsorbed by the tubules; instead it acts as an osmotic diuretic and causes rapid loss of fluid into the urine. The name "diabetes" refers to the prolific urine flow.

**Diuretics that Diminish Active Reabsorption.** Any substance that inhibits carrier systems in the tubular epithelial cells and thereby diminishes active reabsorption of tubular solutes increases the tubular osmotic pressure and causes osmotic diuresis. The most common drugs of this type are the following:

*The "Loop" Diuretics, Furosemide and Ethacrynic Acid.* Furosemide and ethacrynic acid are the most powerful of all the clinically used diuretics. They are called "loop" diuretics because they block the absorption of sodium and chloride ions from the ascending limb of the loop of Henle. This blockage causes diuresis for two reasons: (1) It allows greatly increased quantities of sodium and other ions to be delivered from the loop of Henle into the distal tubules, and these act as osmotic agents to prevent water reabsorption in both the distal and collecting tubules. (2) Failure to absorb sodium and chloride ions from the loop of Henle into the medullary interstitium decreases the concentration of the medullary interstitial fluid. Consequently, the concentrating ability of the kidney is greatly decreased so that the reabsorption of fluid is further decreased. Because of these two effects as much as 20 per cent of the glomerular filtrate may be delivered into the urine, causing under acute conditions urine outputs as great as 25 times normal for a period of a few minutes.

*Chlorothiazide.* Chlorothiazide and other thiazide derivatives act primarily on the distal tubules and cortical portions of the collecting tubules to prevent active sodium reabsorption; under favorable conditions they cause as much as 8 per cent of of the glomerular filtrate to pass into the urine.

*Carbonic Anhydrase Inhibitors—Acetazolamide.* Acetazolamide (Diamox) and other carbonic anhydrase inhibitors primarily block reabsorption of bicarbonate ions from the proximal tubules. They do this by inhibiting the carbonic anhydrase that lines the lumenal border of the tubular epithelial cells and that normally catalyzes the dissociation of carbonic acid into water and carbon dioxide. This prevents removal of bicarbonate ions from the tubular fluid, and these remain in the tubules to act as an osmotic diuretic. However, use of this drug also causes some degree of acidosis because of excessive loss of bicarbonate ions from the body fluids.

*Competitive Aldosterone Inhibitors—Spironolactone.* Spironolactone and several other similar substances compete with aldosterone for receptor sites in the tubular epithelial cells and thereby block the sodium reabsorption promoting effect of aldosterone. As a consequence, the sodium remains in the tubules and acts as an osmotic diuretic. However, these drugs also block the effect of aldosterone to promote potassium secretion into the tubules. Therefore, in some instances, the extracellular fluid potassium concentration becomes dangerously elevated.

## DIURETICS THAT INHIBIT THE SECRETION OF ANTIDIURETIC HORMONE

The single diuretic that is most important in inhibiting the secretion of antidiuretic hormone (ADH) is *water*. This was discussed in Chapter 36. When large amounts of water are ingested, the body fluids become diluted and ADH is no longer secreted by the supraoptico-posterior pituitary system. As a result, water fails to be reabsorbed by the distal tubules and collecting ducts, and large amounts of dilute urine flow. In addition to water, various psychic factors as well as certain drugs that affect the central nervous system, such as *narcotics, hypnotics,* and *anesthetics,* can inhibit ADH secretion. Also, *alcohol* inhibits ADH secretion. For this reason, these can all cause increases in urinary output.

# REFERENCES

Bailey, G. L. (ed.): Hemodialysis: Principles and Practice. New York, Academic Press, Inc., 1973.

Berlyne, G. M.: A Course in Renal Diseases, 4th Ed. Philadelphia, J. B. Lippincott Company, 1973.

Black, D. (ed.): Renal Disease, 3rd Ed. Philadelphia, J. B. Lippincott Company, 1973.

Boyarsky, S. (ed.): Urodynamics. New York, Academic Press, Inc., 1971.

Boyarsky, S.: Ureteral Dynamics. Baltimore, The Williams & Wilkins Company, 1972.

Burkholder, P. M.: Atlas of Human Glomerular Pathology. Hagerstown, Maryland, Harper & Row, Publishers, 1974.

Caine, M., and Raz, S.: Some clinical implications of adrenergic receptors in the urinary tract. Arch. Surg., 110: 247, 1975.

Coleman, T. G., Bower, J. D., and Guyton, A. C.: Chronic hemodialysis and circulatory function. Simulation, Nov.:222, 1970.

DeLuca, H. F.: The kidney as an endocrine organ involved in calcium homeostasis. Kidney Int., 4:80, 1973.

Gennari, F. J., and Kassirer, J. P.: Osmotic diuresis. New Engl. J. Med., 291:714, 1974.

Goldberg, M.: The renal physiology of diuretics. In Orloff, F., and Berliner, R. W. (eds.): Handbook of Physiology. Sec. 8. Baltimore, The Williams & Wilkins Company, 1973, p. 1003.

Harrington, J. T., and Cohen, J. J.: Clinical disorders of urine concentration and dilution. Arch. Intern. Med., 131:810, 1973.

Hepinstall, R. H.: Pathology of the Kidney, 2nd Ed. Boston, Little, Brown and Company, 1975.

Kiil; F., and Setekleiv; J.: Physiology of the ureter and renal pelvis. In Orloff, F., and Berliner, R. W. (eds.): Handbook of Physiology. Sec. 8. Baltimore, The Williams & Wilkins Company, 1973, p. 1033.

Kincaid-Smith, P.: Coagulation and renal disease. Kidney Int., 2:183, 1972.

Kincaid-Smith, P.: The role of coagulation in the obliteration of glomerular capillaries. Perspect. Nephrol. Hypertens., 2:871, 1973.

Kincaid-Smith, P., Mathew, T. H., and Becker, E. L. (eds.): Glomerulonephritis: Morphology, Natural History, and Treatment. New York, John Wiley & Sons, Inc., 1973.

Kountz, S. L.: Clinical renal transplantation. Transplant Proc., 6 (4 Suppl. 1):99, 1974.

Kuru, M.: Nervous control of micturition. Physiol. Rev., 45:425, 1965.

Lant, A. F., and Wilson, G. M. (eds.): Modern Diuretic Therapy in the Treatment of Cardiovascular and Renal Disease. New York, American Elsevier Publishing Company, 1973.

Lindheimer, M. D., and Katz, A. I.: Sodium and diuretics in pregnancy. New Engl. J. Med., 288:891, 1973.

Natochin, Y. V.: Renal pharmacology: comparative, developmental, and cellular aspects. Ann. Rev. Pharmacol., 14:75, 1974.

Nergårdh, A.: Autonomic receptor functions in the lower urinary tract: a survey of recent experimental results. J. Urol., 113:180, 1975.

Nowinski, W., and Goss, R. J.: Compensatory Renal Hypertrophy. New York, Academic Press, Inc., 1970.

Rubin, M. I., and Barratt, T. M. (eds.): Pediatric Nephrology. Baltimore, The Williams & Wilkins Company, 1975.

Segal, S., and Thier, S. O.: Renal handling of amino acids. *In* Orloff, F., and Berliner, R. W. (eds.): Handbook of Physiology. Sec. 8. Baltimore, The Williams & Wilkins Company, 1973, p. 653.

Stenzel, K. H., Cheigh, J. S., Sullivan, J. F., Tapia, L., Riggio, R. R., and Rubin, A. L.: Clinical effects of bilateral nephrectomy. *Amer. J. Med., 58*:69, 1975.

Strauss, M. B., and Walt, L. G.: Diseases of the Kidney. Boston, Little, Brown and Company, 1963.

Vander, A. J., and Carlson, J.: Mechanism of the effects of furosemide on renin secretion in anesthetized dogs. *Circ. Res., 25*:145, 1969.

Whelpton, D. (ed.): Renal Dialysis. Philadelphia, J. B. Lippincott Company, 1974.

# PART VII
# RESPIRATION

# 39

# Pulmonary Ventilation

The process of respiration can be divided into four major categories: (1) pulmonary ventilation, which means the inflow and outflow of air between the atmosphere and the alveoli, (2) diffusion of oxygen and carbon dioxide between the alveoli and the blood, (3) transport of oxygen and carbon dioxide in the blood and body fluids to and from the cells, and (4) regulation of ventilation and other facets of respiration. The present chapter and the three following discuss, respectively, these four major aspects of respiration. In subsequent chapters pulmonary disorders and special respiratory problems related to aviation medicine and deep sea diving physiology are discussed to illustrate some of the basic principles of respiratory physiology.

## MECHANICS OF RESPIRATION

### BASIC MECHANISMS OF LUNG EXPANSION AND CONTRACTION

The lungs can be expanded and contracted by (1) downward and upward movement of the diaphragm to lengthen or shorten the chest cavity and (2) elevation and depression of the ribs to increase and decrease the anteroposterior diameter of the chest cavity. Figure 39–1 illustrates these two methods.

It is readily evident that contraction of the diaphragm pulls the lower boundary of the chest cavity downward and therefore increases its longitudinal length. Upward movement of the diaphragm during normal respiration is caused by simple relaxation of the diaphragm, thus allowing the elastic recoil of the lungs to pull the diaphragm upward; during heavy breathing it is

also caused by active contraction of the abdominal muscles, which forces the abdominal contents upward against the bottom of the diaphragm.

Elevation of the anterior portion of the chest cage causes the anteroposterior dimension of the chest cavity to increase by the following mechanism: During expiration the ribs are pulled in a downward direction from the spinal column, but when the sternum is lifted upward, the ribs then extend more directly forward rather than downward. This makes the anteroposterior diameter of the chest about 20 per cent greater during inspiration than during expiration. Therefore, muscles that elevate the chest cage can be classified as muscles of inspiration, and muscles that depress the chest cage as muscles of expiration.

Normal quiet breathing is accomplished almost entirely by movement of the diaphragm;

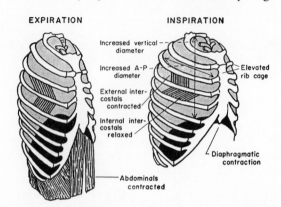

**Figure 39–1.** Expansion and contraction of the thoracic cage during expiration and inspiration, illustrating especially diaphragmatic contraction, elevation of the rib cage, and function of the intercostals.

but during maximal breathing, increase in chest thickness can account for as much as half of the lung enlargement.

## THE MUSCLES OF INSPIRATION AND EXPIRATION

The different muscles of inspiration and expiration are the following:

MUSCLES OF INSPIRATION
  Diaphragm
  External intercostals
  Sternocleidomastoids
  Scapular elevators plus anterior serrati
  Scaleni
  Erectus muscles of the spine

MUSCLES OF EXPIRATION
  Abdominals (major muscles of expiration)
  Internal intercostals
  Posterior inferior serrati

**The Diaphragm and the Abdominal Muscles.** Normal *inspiration* is caused principally by contraction of the diaphragm. This muscle is bell shaped so that contraction of any of its muscle fibers pulls it downward to cause inspiration.

Ordinarily, expiration is an entirely passive process; that is, when the diaphragm relaxes, the elastic structures of the lung, chest cage, and abdomen, as well as the tone of the abdominal muscles, force the diaphragm upward. However, if forceful expiration is required, the diaphragm can also be pushed upward powerfully by active contraction of the abdominal muscles against the abdominal contents. Thus, all the abdominal muscles combined together represent the major muscles of expiration.

**Muscles that Raise and Lower the Chest Cage.** Three different groups of muscles cause inspiration by elevating the entire chest cage. The *sternocleidomastoid muscles* lift upward on the sternum; the *anterior serrati* lift many of the ribs; and the *scaleni* lift the first two ribs. Because the anterior margins of all of the upper ribs are connected to the sternum, lifting one portion of the anterior chest cage lifts it all.

To cause expiration, the abdominal recti, in addition to helping to compress the abdominal contents upward against the diaphragm, also pull downward on the lower ribs, thereby decreasing the anteroposterior diameter of the chest. Thus, these muscles act as muscles of expiration both by depressing the rib cage and by compressing the abdominal contents upward.

The external and internal intercostals are also important muscles of respiration despite their seemingly small size. Figure 42–1 illustrates the mechanism by which these act to cause inspiration or expiration. To the left, the ribs during expiration are angled downward and the external intercostals are stretched in a forward and downward direction. As they contract, they pull the upper ribs forward in relation to the lower ribs, and this causes leverage on the ribs to raise them upward. Conversely, in the inspiratory position, the internal intercostals are stretched, and their contraction pulls the upper ribs backward in relation to the lower ribs. This causes leverage in the opposite direction and lowers the chest cage.

## RESPIRATORY PRESSURES

**Intra-alveolar Pressure.** The respiratory muscles cause pulmonary ventilation by alternatively compressing and distending the lungs, which in turn causes the pressure in the alveoli to rise and fall. During inspiration the intra-alveolar pressure becomes slightly negative *with respect to atmospheric pressure,* normally less than −1 mm. Hg, and this causes air to flow inward through the respiratory passageways. During normal expiration, on the other hand, the intra-alveolar pressure rises to almost +1 mm. Hg, which causes air to flow outward through the respiratory passageways. Note especially, how little pressure is required to move air into and out of the lungs.

During maximum expiratory effort with the glottis closed the intra-alveolar pressure can be increased to over 100 mm. Hg in the strong healthy male, and during maximum inspiratory effort it can be reduced to as low as −80 mm. Hg.

**Fluid Pressure in the Intrapleural Cavity.** The lungs are physically attached to the body only at their hila, while their outer surfaces have no attachment whatsoever to the chest wall. However, the membranes of the intrapleural space constantly absorb any gas or fluid that enters this space; this absorption creates a partial vacuum that makes the lungs adhere closely to the inner walls of the chest cavity.

The normal pressure *of the fluid* in the intrapleural space is about −10 mm. Hg. This negative pressure acts as a suction force to hold the visceral pleura of the lungs tightly against the parietal pleura of the chest wall. When the chest cavity enlarges, this suction causes the lungs also to enlarge. And when the chest cavity becomes smaller, the lungs likewise become smaller. The lungs slide up and down in the chest cavity during respiration, the visceral pleura sliding over the parietal pleura, lubricated by a few milliliters of mucopolysaccharide-containing fluid in the intrapleural space.

The cause of the very negative intrapleural

*fluid* pressure is the continual tendency of the pleural capillaries to absorb fluid from the intrapleural spaces. This is particularly true of the visceral pleural capillaries because these are part of the pulmonary circulatory system and have a very low capillary pressure of about 7 mm. Hg that causes rapid absorption of fluid.

**Recoil Tendency of the Lungs and the Intrapleural Pressure.** The lungs have a continual elastic tendency to collapse and therefore to recoil away from the chest wall. This elastic tendency is caused by two different factors. First, throughout the lungs are many *elastic fibers* that are stretched by lung inflation and therefore attempt to shorten. Second, and even more important, the *surface tension* of the fluid lining the alveoli also causes a continual elastic tendency for the alveoli to collapse. This effect is caused by intermolecular attraction between the surface molecules of the fluid that tends continually to reduce the surface areas of the individual alveoli; all these minute forces added together tend to collapse the whole lung and therefore to cause its recoil away from the chest wall.

Ordinarily, the elastic fibers in the lungs account for about one-third of the recoil tendency, and the surface tension phenomenon accounts for about two-thirds.

The total recoil tendency of the lungs can be measured by the amount of negative pressure in the intrapleural spaces required to prevent collapse of the lungs, and this pressure is called the *intrapleural pressure* or, occasionally, the *recoil pressure*. It is normally about −4 mm. Hg, a value quite different from the intrapleural *fluid* pressure discussed above. That is, when the alveolar spaces are open to the atmosphere through the trachea so that their pressure is at atmospheric pressure, a pressure of −4 mm. Hg in the intrapleural space is required to keep the lungs expanded to normal size. When the lungs are stretched to very large size, such as at the end of deep inspiration, the intrapleural pressure required then to expand the lungs may be as great as −9 to −12 mm. Hg.

**Explanation of the Difference Between Intrapleural Fluid Pressure and Intrapleural Pressure.** The recent discovery that the fluid pressure in the intrapleural space is considerably more negative than the so-called intrapleural pressure has been confusing to many students as well as to some physiologists. However, if the student will return to Chapter 31 and study the discussion of interstitial fluid pressure, solid tissue pressure, and total tissue pressure, he will see that the pressure dynamics of the intrapleural space are essentially the same as those in the usual interstitial fluid spaces. The intrapleural fluid pressure is analogous to the interstitial fluid pressure, and the intrapleural pressure is analogous to the total tissue pressure. The contact pressure of the visceral pleura against the parietal pleura, which is called by respiratory physiologists *pleural surface pressure,* is analogous to solid tissue pressure. This pressure calculates to be about +6 mm. Hg under normal resting conditions. That is, *intrapleural pressure* (−4 mm. Hg) = *pleural fluid pressure* (−10 mm. Hg) + *pleural surface pressure* (+6 mm. Hg).

**"Surfactant" in the Alveoli, and Its Effect on the Collapse Tendency.** A lipoprotein mixture called "surfactant" is secreted by special *surfactant-secreting cells* that are component parts of the alveolar epithelium. This mixture, containing especially the phospholipid *dipalmitoyl lecithin,* acts in much the same way that a detergent acts. That is, it decreases the surface tension of fluids lining the alveoli and respiratory passages. In the absence of surfactant, lung expansion is extremely difficult, which illustrates that the presence of surfactant is exceedingly important to minimize the effect of surface tension in causing collapse of the lungs.

A few newborn babies do not secrete adequate quantities of surfactant, which makes lung expansion difficult. Some of these die soon after birth because of inadequate respiration. Others, with less severe lack of surfactant, often live only with difficulty and frequently die in early infancy. This condition is called *hyaline membrane disease* or *respiratory distress syndrome.*

Surfactant acts by forming a monomolecular layer at the interface between the fluids lining the alveoli and the air in the alveoli. This prevents the development of a water-air interface, which has 7 to 14 times as much surface tension as does the surfactant-air interface.

**Role of Surfactant in Preventing Collapse of the Alveoli.** In the absence of surfactant, as much as 20 to 30 mm. Hg of expansion pressure is required to keep the alveoli open. The reason for this is that the surface tension of water is high enough to create large amounts of contractile force along the surfaces of the alveoli, thus tending to cause the alveoli to collapse. Furthermore, the collapse tendency caused by the surface tension increases as the alveoli become smaller because the collapse pressure caused by the surface tension is proportional to the inverse of the radius of the alveoli.

However, surfactant reduces the surface tension and therefore also reduces the required expansion pressure to only 3 to 5 mm. Hg to maintain normal-sized alveoli. Furthermore, surfactant has a special property of decreasing the surface tension still more as the alveoli become smaller, which to a great extent nullifies the increasing collapse tendency of the alveoli as they become smaller. Consequently, surfactant is very important in maintaining the equality of size of the alveoli—the large alveoli developing greater surface tension and therefore contracting, while the smaller alveoli develop less surface tension and therefore tend to enlarge.

## EXPANSIBILITY OF THE LUNGS AND THORAX: "COMPLIANCE"

The lungs are viscoelastic structures. Therefore, a small amount of intra-alveolar pressure causes them to expand to a certain volume, more pressure causes them to expand to a still greater volume, and so forth. Likewise, the thorax has viscoelastic properties so that the greater the pressure in the lungs, the greater becomes the expansion of the chest. The elastic properties of the lungs, as pointed out previously, are caused, first, by the surface tension of the fluids lining the alveoli and, second, by elastic fibers throughout the lung tissue itself. The elastic properties of the thorax are caused by the natural elasticity of the muscles, tendons, and connective tissue of the chest. Therefore, part of the effort expended by the inspiratory muscles during breathing is simply to stretch the elastic structures of the lungs and thorax.

The expansibility of the lungs and thorax is called compliance. This is expressed as the *volume increase in the lungs for each unit increase in intra-alveolar pressure*. The compliance of the normal lungs and thorax combined is 0.13 liter per centimeter of water pressure. That is, every time the alveolar pressure is increased by 1 cm. water, the lungs expand 130 ml.

**Compliance of the Lungs Alone.** The lungs alone, when removed from the chest, are almost twice as distensible as the lungs and thorax together because the thoracic cage itself must also be stretched when the lungs are expanded in situ. Thus, the compliance of the normal lungs when removed from the thorax is about 0.22 liter per cm. water. This illustrates that the

muscles of inspiration must expend energy not only to expand the lungs but also to expand the thoracic cage around the lungs.

**Measurement of Lung Compliance.** Compliance of the lungs is measured in the following way: The person's glottis must first be completely open and remain so. Then air is inspired in steps of approximately 50 to 100 ml. at a time, and pressure measurements are made from an intra-esophageal balloon (which measures almost exactly the intrapleural pressure) at the end of each step until the total volume of air in the lungs is equal to the normal tidal volume of the person. Then the air is expired also in steps, until the lung volume returns to the expiratory resting level. The relationship of lung volume to pressure is then plotted as illustrated in Figure 39–2. This graph shows that the plot during inspiration is a different curve from that during expiration. The compliance of the lungs is usually determined from the slope of the lower straight portion of the expiration curve. Thus, in the figure the lung volume decreases 400 ml. for a change in trans-lung pressure (atmospheric pressure in the lung minus intra-esophageal pressure) of 1 cm. water. Therefore, the compliance in this instance is 0.4 liter per cm. water.

Slight modifications of this procedure can be used to measure the compliance of the lungs and thoracic cage together.

**Factors that Cause Abnormal Compliance.** Any condition that destroys lung tissue, causes it to become fibrotic or edematous, blocks the alveoli, or in any other way impedes lung expansion and contraction causes decreased lung compliance. When considering the

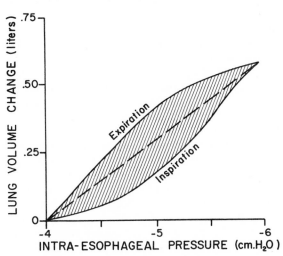

**Figure 39–2.** Compliance diagram in a large subject. This diagram shows the compliance of the lungs alone.

compliance of both the lung and thorax together, one must also include any abnormality that reduces the expansibility of the thoracic cage. Thus, deformities of the chest cage, such as kyphosis, severe scoliosis, and other restraining conditions, such as fibrotic pleurisy or paralyzed and fibrotic muscles, can all reduce the expansibility of the lungs and thereby reduce the total pulmonary compliance.

## THE "WORK" OF BREATHING

When the lungs are expanded, energy is expended by the respiratory muscles to cause the expansion. This is part of the "work" of breathing. But, in addition to the work required simply to expand the lungs, work is needed to overcome two other factors that impede the expansion and contraction of the lungs. These are (1) *viscosity* of the pulmonary tissues, which is also called *nonelastic tissue resistance,* and (2) *airway resistance*.

**Nonelastic Tissue Resistance (Viscosity).** Nonelastic tissue resistance simply means that a certain amount of energy (or work) is required to rearrange the molecules in the tissues of the lungs and thoracic cage in order to change them to new dimensions. This requires the slipping of molecules over each other, which constitutes a type of viscous resistance.

The effect of nonelastic tissue resistance of the lungs on lung function can be measured by using the compliance diagram of Figure 39–2. As was pointed out above, the inspiratory compliance curve follows a different path from the expiratory curve. For mathematical reasons that cannot be discussed in detail here, but which will be obvious to a student of mathematics, the *area* between these curves is proportional to the work required to overcome the nonelastic tissue resistance. When the lungs are edematous or when they have lost most of their elastic properties because of pulmonary fibrotic changes, this area becomes greatly widened, indicating that the nonelastic tissue resistance has also greatly increased.

**Airway Resistance.** A certain amount of energy is also required to move the air along the respiratory passageways. In normal quiet breathing, this amount of energy is slight, but, when the airway becomes obstructed as a result of asthma, obstructive emphysema, diphtheria, or so forth, the airway resistance may become so greatly increased that a large amount of extra energy must be expended by the respiratory muscles simply to force the air back and forth through the narrow passageways.

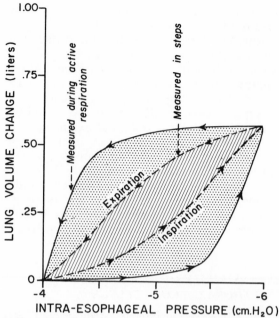

**Figure 39–3.** *Outer loop:* Pressure-volume diagram recorded continuously during a cycle of respiration. *Inner loop:* Static pressure-volume curve of Figure 39–2. The stippled area represents the increased work over that shown in Figure 39–2; it is caused by airway resistance.

Airway resistance can be measured from a modified compliance diagram recorded dynamically during the course of respiration. In Figure 39–3, the outer loop shows a curve made while air was actually flowing in and out of the lungs during the recording in a person who has airway obstruction, and it will be noted that the area between the curves is greatly increased. The extra area over and above that recorded in Figure 39–2—this extra area denoted by the dotted shading—is a measure of the work expended to overcome the airway resistance. In Chapter 43 several pulmonary diseases will be discussed, especially asthma, in which airway resistance is the principal pulmonary problem.

**Energy Required for Respiration.** During normal quiet respiration, only 2 to 3 per cent of the total energy expended by the body is required to energize the pulmonary ventilatory process. During very heavy exercise, the absolute amount of energy required for pulmonary ventilation can increase as much as twenty-five–fold. However, this still does not represent a significant increase in *percentage* of total energy expenditure because the total energy release in the body increases at the same time as much as fifteen- to twenty-fold. Thus, even in heavy exercise only 3 to 4 per cent of the total energy expended is used for ventilation.

On the other hand, pulmonary diseases that decrease the pulmonary compliance, that increase airway resistance, or that increase the viscosity of the lung or chest wall can at times increase the work of breathing so much that one-third or more of the total energy expended by the body is for respiration alone. Such respiratory diseases can proceed to the point that this excess work load alone is the cause of death.

# THE PULMONARY VOLUMES AND CAPACITIES

Figure 39–4 gives a graphical representation of changes in lung volume under different conditions of breathing. For ease in describing the events of pulmonary ventilation, the air in the lungs has been subdivided at different points on this diagram into four different *volumes* and four different *capacities,* which are as follows:

## THE PULMONARY "VOLUMES"

To the left in Figure 39–4 are listed four different pulmonary lung "volumes" which, when added together, equal the maximum volume to which the lungs can be expanded. The significance of each of these volumes is the following:

1. The *tidal volume* is the volume of air inspired and expired with each normal breath, and it amounts to about 500 ml. in the normal young male adult.

2. The *inspiratory reserve volume* is the extra volume of air that can be inspired over and beyond the normal tidal volume, and it is usually equal to approximately 3000 ml. in the young male adult.

3. The *expiratory reserve volume* is the amount of air that can still be expired by forceful expiration after the end of a normal tidal expiration; this normally amounts to about 1100 ml. in the young male adult.

4. The *residual volume* is the volume of air still remaining in the lungs after the most forceful expiration. This volume averages about 1200 ml. in the young male adult.

## THE PULMONARY "CAPACITIES"

In describing events in the pulmonary cycle, it is sometimes desirable to consider two or more of the above volumes together. Such combinations are called pulmonary capacities. To the right in Figure 39–4 are listed the different pulmonary capacities, which can be described as follows:

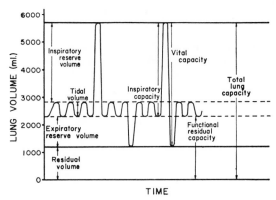

**Figure 39–4.** Diagram showing respiratory excursions during normal breathing and during maximal inspiration and maximal expiration.

1. The *inspiratory capacity* equals the *tidal volume* plus the *inspiratory reserve volume.* This is the amount of air (about 3500 ml.) that a person can breathe beginning at the normal expiratory level and distending his lungs to the maximum amount.

2. The *functional residual capacity* equals the *expiratory reserve volume* plus the *residual volume.* This is the amount of air remaining in the lungs at the end of normal expiration (about 2300 ml.).

3. The *vital capacity* equals the *inspiratory reserve volume* plus the *tidal volume* plus the *expiratory reserve volume.* This is the maximum amount of air that a person can expel from his lungs after first filling his lungs to their maximum extent and then expiring to the maximum extent (about 4600 ml.).

4. The *total lung capacity* is the maximum volume to which the lungs can be expanded with the greatest possible inspiratory effort (about 5800 ml.).

*All pulmonary volumes and capacities are about 20 to 25 per cent less in the female than in the male,* and they obviously are greater in large and athletic persons than in small and asthenic persons.

**Resting Expiratory Level.** Normal pulmonary ventilation is accomplished almost entirely by the muscles of inspiration. On relaxation of the inspiratory muscles the elastic properties of the lungs and thorax cause the lungs to contract passively. Therefore, when all inspiratory muscles are completely relaxed the lungs return to a relaxed state called the resting expiratory level. The volume of air in the lungs at this level is equal to the functional residual capacity, or about 2300 ml. in the young male adult.

## SIGNIFICANCE OF THE PULMONARY VOLUMES AND CAPACITIES

In normal persons the volume of air in the lungs depends primarily on body size and build. Furthermore, the different "volumes" and "capacities" change with the position of the body, most of them decreasing when the person lies down and increasing when he stands. This change with position is caused by two major factors: first, a tendency for the abdominal contents to press upward against the diaphragm in the lying position, and, second, increase in the pulmonary blood volume in the lying position, which correspondingly decreases the space available for pulmonary air.

**Significance of the Residual Volume.** The residual volume represents the air that cannot be removed from the lungs even by forceful expiration. This is important because it provides air in the alveoli to aerate the blood even between breaths. Were it not for the residual air, the concentrations of oxygen and carbon dioxide in the blood would rise and fall markedly with each respiration, which would certainly be disadvantageous to the respiratory process.

**Significance of the Vital Capacity.** Other than the anatomical build of a person, the major factors that affect vital capacity are (1) the position of the person during the vital capacity measurement, (2) the strength of the respiratory muscles, and (3) the distensibility of the lungs and chest cage, which is called "pulmonary compliance."

The average vital capacity in the young adult male is about 4.6 liters, and in the young adult female about 3.1 liters, though these values are much greater in some persons of the same weight than in others. A tall, thin person usually has a higher vital capacity than an obese person, and a well-developed athlete may have a vital capacity as great as 30 to 40 per cent above normal—that is, 6 to 7 liters.

*Vital Capacity Following Paralysis of the Respiratory Muscles.* Paralysis of the respiratory muscles, which occurs often following spinal cord injuries or poliomyelitis, can cause a great decrease in vital capacity, to as low as 500 to 1000 ml.—barely enough to maintain life. This decrease may be even lower in the case of respirator patients.

*Decreased Vital Capacity Caused by Diminished Pulmonary Compliance.* Obviously, any factor that reduces the ability of the lungs to expand also reduces the vital capacity. Thus, tuberculosis, emphysema, chronic asthma, lung cancer, chronic bronchitis, and fibrotic pleurisy can all reduce the pulmonary compliance and thereby decrease the vital capacity. For this reason vital capacity measurements are among the most important of all clinical respiratory measurements for assessing the progress of different types of pulmonary fibrotic diseases.

*Changes in Vital Capacity Resulting from Pulmonary Congestion.* In left heart disease or any other disease that causes pulmonary vascular congestion and edema, the vital capacity becomes reduced because excess fluid in the lungs decreases lung compliance.

Vital capacity measurements made periodically in left-sided heart disease are a good means for determining whether the person's condition is progressing or getting better, for these measurements can indicate the degree of pulmonary edema.

## MEASUREMENT OF PULMONARY VOLUMES AND CAPACITIES

**Spirometry.** A simple method by which most of the pulmonary volumes and capacities can be measured is spirometry. A typical spirometer is illustrated in Figure 39–5. This consists of a drum inverted over a chamber of water, the drum counterbalanced by a weight. In the drum is a breathing mixture of gases, usually air or oxygen, and a tube connects the mouth with this gas chamber. On breathing in and out of the chamber the drum rises and falls, and an appropriate recording is made on another drum. When using this spirometer only a few breaths can be recorded because of the buildup of carbon dioxide and loss of oxygen from the gas chamber. However, a *respirometer* such as that illustrated in Figure 71–4 of Chapter 71 can be used for continual recording of the spirogram. This more elaborate apparatus chemically removes the carbon dioxide as it is formed.

Figure 39–4, discussed earlier in this chapter, shows a typical *spirogram* recorded by use of the spirometer. From this figure it is obvious that most of the lung volumes and capacities can be measured by simple spirometry. The volumes that *cannot* be measured this way are: residual volume, functional residual capacity, and total lung capacity. Failure of the spirometer to measure these results from its inability to measure the residual volume still remaining in the lungs after a forceful expiration.

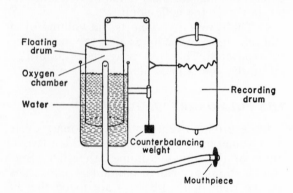

**Figure 39–5.** A spirometer.

**Determination of Residual Volume, Functional Residual Capacity, and Total Lung Capacity by the Nitrogen Washout Method.** The functional residual capacity can be measured by the following indirect procedure:

At the end of a normal expiration, the subject suddenly switches from breathing normal air to breathing pure oxygen, and he continues to breathe the pure oxygen for several minutes. During this period of time all the nitrogen in his lungs is "washed out" into the expired air. By collecting the expired air and analyzing the total amount of nitrogen that has been washed out of the lungs, one can calculate the amount of air in the lungs at the beginning of the test, which equals the functional residual capacity, by the following formula:

Functional residual capacity =

$$\text{Volume of nitrogen washed out} \times \frac{100}{78}$$

This formula is based on the fact that 78 per cent of the gases in the lungs is nitrogen and 22 per cent is oxygen, carbon dioxide, and water vapor. (To be completely accurate in the measurement, additional minor corrections must be made for nitrogen diffusion out of the blood into the lungs.)

Once the functional residual capacity has been determined, the residual volume can be determined by subtracting the expiratory reserve volume from the functional residual capacity. Also, the total lung capacity can be determined by adding the inspiratory capacity to the functional residual capacity.

# THE MINUTE RESPIRATORY VOLUME—RESPIRATORY RATE AND TIDAL VOLUME

The *minute respiratory volume* is the total amount of new air moved into the respiratory passages each minute, and this is equal to the *tidal volume* × the *respiratory rate*. The normal tidal volume of a young adult male, as pointed out above, is 500 ml., and the normal respiratory rate is approximately 12 breaths per minute. Therefore, the *minute respiratory volume averages about 6 liters per minute*. A person can occasionally live for short periods of time with a minute respiratory volume as low as 1.5 liters per minute and with a respiratory rate as low as two to four breaths per minute. Obviously, a greatly increased tidal volume can compensate for a markedly reduced respiratory rate.

The respiratory rate occasionally rises to as high as 40 to 50 per minute, and the tidal volume can become as great as the vital capacity, about 4600 ml. in the young adult male. However, at rapid breathing rates, a person usually cannot sustain a tidal volume greater than about one-half the vital capacity.

**Maximum Breathing Capacity.** A young male adult forcing himself to breathe as much volume of air as possible can usually breathe 150 to 170 liters per minute for about 15 seconds. This is called the maximum breathing capacity. On the average, this same person can maintain for periods of strenuous exercise a minute respiratory volume as high as 100 to 120 liters per minute. It is evident, then, that the respiratory system has a marked reserve, being capable of increasing its minute respiratory volume as much as 28-fold for short periods of time and as much as 20-fold for prolonged periods of time. Yet, despite this tremendous reserve, pathologic conditions of the lungs can often so decrease the breathing capacity that a person becomes a respiratory invalid. Some of these pathologic conditions will be discussed in Chapter 43.

**Maximum Expiratory Flow.** When a person expires with progressively increasing force, the expiratory air flow reaches a maximum despite still further increase in expiratory force. This effect can be explained by reference to Section A of Figure 39–6. When pressure is applied to the lungs by chest cage compression, the same amount of pressure is applied to the outsides of both the alveoli and the respiratory passageways, as indicated by the arrows.

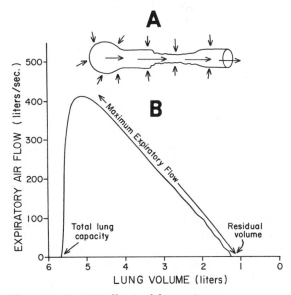

**Figure 39–6.** (A) Collapse of the respiratory passageway during maximum expiratory effort, an effect that limits the expiratory flow rate. (B) Effect of lung volume on the maximum expiratory air flow, showing decreasing maximum expiratory air flow as the lung volume becomes smaller.

Therefore, not only is the pressure increased in the alveoli to force air to the exterior, but the bronchioles and small bronchi are compressed at the same time to increase airway resistance. Beyond a certain compressive force, these two effects have equal but opposing results on air flow, thus preventing further increase in flow.

Section B of Figure 39–6 illustrates the effect of this limiting factor on the maximum expiratory flow. The curve recorded in Section B is the expiratory flow achieved by a normal person who first inhales as much air as possible and then expires with maximum expiratory effort until he can expire no more. Note that he quickly reaches an expiratory air flow of over 400 liters/sec. Furthermore, it does not matter how much additional expiratory effort he exerts, this is still the *maximum expiratory flow* that he can achieve.

Note also that as the lung volume gets smaller and smaller this maximum expiratory flow becomes less and less. The main reason for this response is that in the enlarged lung the bronchi are held open partially via elastic pull on their outsides by lung structural elements; however, as the lung becomes smaller, these structures are relaxed so that the bronchi become compressed more easily.

In many disease conditions the maximum expiratory air flow is greatly decreased. This occurs especially in emphysema and asthma—in emphysema because of the natural loss of elasticity of the lung tissue and in asthma because of an extra compressive force on the bronchi caused by bronchial muscle spasm.

# VENTILATION OF THE ALVEOLI

The truly important factor of the entire pulmonary ventilatory process is the rate at which the alveolar air is renewed each minute by atmospheric air; this is called *alveolar ventilation*. One can readily understand that alveolar ventilation per minute is not equal to the minute respiratory volume because a large portion of the inspired air goes to fill the respiratory passageways, the membranes of which are not capable of significant gaseous exchange with the blood.

## THE DEAD SPACE

**Effect of Dead Space on Alveolar Ventilation.** The air that goes to fill the respiratory passages with each breath is called *dead space air*. On inspiration, much of the new air must first fill the different dead space areas—the nasal passageways, the pharynx, the trachea, and the bronchi—before any reaches the alveoli. Then, on expiration, all the air in the dead space is expired first before any of the air from the alveoli reaches the atmosphere. *The volume*

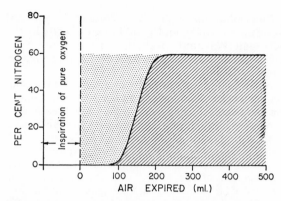

**Figure 39–7.** Continuous record of the changes in nitrogen concentration in the expired air following a previous inspiration of pure oxygen. This record can be used to calculate dead space as discussed in the text.

*of air that enters the alveoli with each breath, therefore, is equal to the tidal volume minus the dead space volume.*

**Measurement of the Dead Space Volume.** A simple method for measuring dead space volume is illustrated by Figure 39–7. In making this measurement the subject first breathes normal air and then suddenly takes an inspiration of oxygen. This, obviously, fills his entire dead space with pure oxygen, and some of the oxygen also mixes with the alveolar air. Then he expires through a rapidly recording nitrogen meter, which makes the record shown in the figure. The first portion of the expired air contains only pure oxygen, and the per cent nitrogen is zero, but about one-quarter of the way into expiration, as the alveolar air reaches the nitrogen meter, the nitrogen concentration rises, in this instance up to 60 per cent, and then levels off. The total volume of air expired in this instance was 500 ml., and it can readily be seen that the area covered by the dots represents the air that has no nitrogen in it. Therefore, this area also represents the dead space portion of the expired air. The area covered by the hatching represents the air containing nitrogen, and therefore is the alveolar portion of the expired air. Thus, one can determine the amount of dead space air from the following equation:

Dead space air =
$$\frac{\text{Area of the dots} \times \text{Total air expired}}{\text{Area of the hatching} + \text{Area of the dots}}$$

Let us assume, for instance, that the area of the dots on the graph is 30 cm.², and the area of the hatching is 70 cm.², and the total volume expired is 500 ml. The dead space then would be

$$\frac{30}{30 + 70} \times 500, \text{ or } 150 \text{ ml.}$$

**Normal Dead Space Volume.** The normal dead space air in the young male adult is about 150 ml. This increases slightly with age.

**Anatomical versus Physiological Dead Space.** The method just described for measuring the dead space measures the volume of all the spaces of the respiratory system besides the gas exchange areas, the alveoli and terminal ducts; this is called the *anatomical dead space*. On occasion, however, some of the alveoli themselves are not functional and, therefore, must be considered also to be dead space because there is no blood flow through the adjacent pulmonary vessels. Also, at other times the *ratio of pulmonary blood flow to ventilation* in certain alveoli is so low that these alveoli are only partially functional, so that they too, can be considered to be partially dead space. When the alveolar dead space is included in the total measurement of dead space this is then called *physiological dead space* in contradistinction to the anatomical dead space. In the normal person, the anatomical and the physiological dead spaces are nearly equal because all alveoli are functional in the normal lung, but in persons with partially functional or nonfunctional alveoli in some parts of the lungs, the physiological dead space is sometimes as much as 10 times the anatomical dead space, or as much as 1 to 2 liters. These problems will be discussed further in Chapter 43 in relation to certain pulmonary diseases.

### RATE OF ALVEOLAR VENTILATION

Alveolar ventilation per minute is the total volume of new air entering the alveoli each minute. It is equal to the respiratory rate times the amount of new air that enters the alveoli with each breath:

Alveolar ventilation per minute =
    Respiratory rate × (Tidal volume − Physiological dead space volume)

Thus, with a normal tidal volume of 500 ml., a normal physiological dead space of 150 ml., and a respiratory rate of 12 times per minute, alveolar ventilation equals 12 × (500 − 150), or 4200 ml. per minute.

Theoretically, when the tidal volume falls to equal the dead space volume, no new air at all enters the alveoli with each breath, and the al-

veolar ventilation per minute becomes zero however rapidly the person breathes. (This relationship is not entirely true because all the dead space air is never completely expired before some of the alveolar air begins to be expired, and the same is true for inspiration. Therefore, there can be a slight amount of alveolar ventilation even with tidal volumes as little as 60 to 75 ml.)

On the other hand, when the tidal volume is several liters, the effect of dead space volume on alveolar ventilation is almost insignificant, and under these conditions the alveolar ventilation per minute almost equals the minute respiratory volume. For instance, if the tidal volume is 4 liters and the dead space volume is 200 ml., the alveolar ventilation per minute is 95 per cent of the minute respiratory volume.

Alveolar ventilation is one of the major factors determining the concentrations of oxygen and carbon dioxide in the alveoli. Therefore, almost all discussions of gaseous exchange problems in the following chapters emphasize alveolar ventilation. *The respiratory rate, the tidal volume, and the minute respiratory volume are of importance only insofar as they affect alveolar ventilation.*

## FUNCTIONS OF THE RESPIRATORY PASSAGEWAYS

### FUNCTIONS OF THE NOSE

As air passes through the nose, three distinct functions are performed by the nasal cavities: First, the *air is warmed* by the extensive surfaces of the turbinates and septum, which are illustrated in Figure 39–8. Second, the *air is moistened* to a considerable extent even before it passes beyond the nose. Third, the *air is filtered*. All these functions together are called the *air conditioning function* of the upper respiratory passageways. Ordinarily, the air rises to within 2 to 3 per cent of body temperature and within 2 to 3 per cent of full saturation with water vapor before it reaches the lower trachea. When a person breathes air through a tube directly into his trachea (as through a tracheostomy), the cooling and, especially, the drying effect in the lower lung can lead to lung infection.

**Filtration Function of the Nose.** The hairs at the entrance to the nostrils are important for the removal of large particles. Much more important, though, is the removal of particles by *turbulent precipitation*. That is, the air passing through the nasal passageways hits many obstructing vanes: the turbinates, the septum, and the pharyngeal wall. Each time air hits one of these obstructions it must change its direction of movement; the particles suspended in the air, hav-

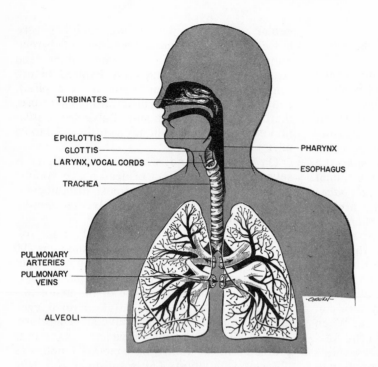

TURBINATES

EPIGLOTTIS
GLOTTIS
LARYNX, VOCAL CORDS

TRACHEA

PHARYNX

ESOPHAGUS

PULMONARY
ARTERIES
PULMONARY
VEINS

ALVEOLI

**Figure 39–8.** The respiratory passages.

ing far more mass and momentum than air, cannot change their direction of travel as rapidly as can the air. Therefore, they continue forward, striking the surfaces of the obstructions.

All the surfaces of the nose are coated with a thin layer of *mucus,* which is secreted by the *mucous membrane* covering the surfaces. Furthermore, the epithelium of the nasal passageways is ciliated, and these *cilia* constantly beat toward the pharynx. Therefore, after particles are entrapped in the mucus, this mucus is moved like a sliding sheet toward the pharynx and finally is either expectorated or swallowed.

*Size of Particles Entrapped in the Respiratory Passages.* The nasal turbulence mechanism for removing particles from air is so effective that almost no particles larger than 4 to 6 microns in diameter enter the lungs through the nose. This size is smaller than the size of red blood cells. Consequently, by far the greater proportion of dust and other large particles in air is removed before the air finally reaches the lungs. Many of the remaining particles *diffuse* against the walls of the alveoli where flow of the respiratory air is negligible, and these particles adhere to the alveolar fluid. But most of the particles smaller than 0.5 micron in diameter usually remain suspended even in the alveolar air and are expelled from the lungs during expiration. The particles of cigarette smoke, for example, having a particle size of approximately 0.3 micron, pass to the alveoli with almost no loss of particles, and two-thirds or more of them then return from the lungs to be expelled along with the expired air.

Particles that do become entrapped in the alveoli are slowly removed mainly by macrophages. An ex-

cess of particles causes growth of fibrous tissue in the alveolar septa, leading to permanent debility.

## THE COUGH REFLEX

The cough reflex is almost essential to life, for the cough is the means by which the passageways of the lungs are maintained free of foreign matter.

The bronchi and the trachea are so sensitive that any foreign matter or other cause of irritation initiates the cough reflex. The larynx and carina (the point where the trachea divides into the bronchi) are especially sensitive, and the terminal bronchioles and alveoli are especially sensitive to corrosive chemical stimuli such as sulfur dioxide gas and chlorine. Afferent impulses pass from the respiratory passages mainly through the vagus nerve to the medulla. There, an automatic sequence of events is triggered by the neuronal circuits of the medulla, causing the following effects:

First, about 2.5 liters of air is inspired. Second, the epiglottis closes, and the vocal cords shut tightly to entrap the air within the lungs. Third, the abdominal muscles contract forcefully, pushing against the diaphragm while other expiratory muscles, such as the internal intercostals, also contract forcefully. Consequently, the pressure in the lungs rises to as high as 100 or more mm. Hg. Fourth, the vocal cords and the epiglottis suddenly open widely so that air under pressure in the lungs *explodes* outward. Indeed, this air is sometimes expelled at velocities as high as 75 to 100 miles an hour. Furthermore, and very important, the strong compression of the lungs also collapses the bronchi and trachea (the noncartilaginous part of the trachea invaginating inward) so that the exploding air actually passes through *bronchial* and *tracheal slits*. The rapidly moving air usually carries with it any foreign matter that is present in the bronchi or trachea.

## THE SNEEZE REFLEX

The sneeze reflex is very much like the cough reflex except that it applies to the nasal passageways instead of to the lower respiratory passages. The initiating stimulus of the sneeze reflex is irritation in the nasal passageways, the afferent impulses passing in the fifth nerve to the medulla where the reflex is triggered. A series of reactions similar to those for the cough reflex takes place; however, the uvula is depressed so that large amounts of air pass rapidly through the nose, as well as through the mouth, thus helping to clear the nasal passages of foreign matter.

## ACTION OF THE CILIA TO CLEAR RESPIRATORY PASSAGEWAYS

In addition to the cough mechanism, the respiratory passageways of the trachea and lungs are lined with a ciliated, mucus-coated epithelium that aids in clearing the passages, for the cilia beat toward the pharynx and move the mucus as a continually flowing sheet. Thus, small foreign particles and mucus are mobilized at a velocity of as much as a centimeter per minute along the surface of the trachea toward the pharynx, in the same manner that foreign matter in the nasal passageways is also mobilized toward the pharynx. Once this material enters the pharynx, either as the result of a cough or as the result of action by the cilia, it is either expectorated or swallowed. Indeed, in persons with tuberculosis, tubercle bacilli often cannot be found in the sputum from the lungs but can be found in stomach washings, for the constant movement of foreign matter from the lungs into the throat and then into the stomach provides a ready source of tubercle bacilli.

## VOCALIZATION

Speech involves the respiratory system particularly, but it also involves (1) specific speech control centers in the cerebral cortex, which will be discussed in Chapter 55, (2) respiratory centers of the brain stem, and (3) the articulation and resonance structures of the mouth and nasal cavities. Basically, speech is composed of two separate mechanical functions: (1) *phonation,* which is achieved by the larynx, and (2) *articulation,* which is achieved by the structures of the mouth.

**Phonation.**    The larynx is specially adapted to act as a vibrator. The vibrating element is the *vocal cords,* which are folds along the lateral walls of the larynx that are stretched and positioned by several specific muscles within the confines of the larynx itself.

Figure 39–9A illustrates the basic structure of the larynx, showing that each vocal cord is stretched between the *thyroid cartilage* and an *arytenoid cartilage.* The specific muscles within the larynx that position and control the degree of stretch of the vocal cords are also shown. Thus, one can see in the figure that contraction of the *posterior cricoarytenoid* muscles pulls the arytenoid cartilages away from the thyroid cartilage and thereby stretches the vocal cords. The *transverse arytenoid muscle* pulls the arytenoid cartilages together and, therefore, approximates the two vocal cords so that they vibrate in a stream of expired air. Conversely, contraction of the *lateral cricoarytenoid muscles* pulls the arytenoid cartilages forward and apart to allow normal respiration.

The *thyroarytenoid muscles* are made up of many small slips of muscle controlled separately by different nerve fibers. The slips of muscle adjacent to the edges of the vocal cord can contract separately from those adjacent to the wall of the larynx, and other individual portions of these muscles can also contract independently of each other. The contractions control the *shapes* of the vocal cords during different types of phonation.

*Vibration of the Vocal Cords.*    One might suspect that the vocal cords would vibrate in the direction of the flowing air. However, this is not the case. Instead, they vibrate laterally. The cause of the vibration is the following: When the vocal cords are approximated and air is expired, pressure of the air from below first pushes the vocal cords apart, which allows rapid flow of air between their margins. The rapid flow of air then immediately creates a partial vacuum between the vocal cords, which pulls them once again toward each other. This stops the flow of air, pressure builds up behind the cords, and the

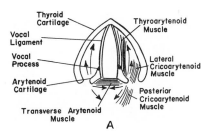

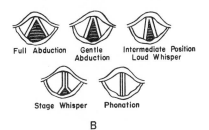

**Figure 39–9.**   Laryngeal function in phonation. (Modified from Greene: The Voice and Its Disorders, 3rd Ed., J. B. Lippincott Co., 1972.)

cords open once more, thus continuing in a vibratory pattern.

*Frequency of Vibration.* The pitch of the sound emitted by the larynx can be changed in two different ways. First, a change can be achieved by *stretching or relaxing the vocal cords*. The mechanisms involved were partly explained above in the discussion of the intrinsic laryngeal muscles; but, in addition to the effects of the intrinsic muscles, the muscles attached to the external surfaces of the larynx can also pull against the cartilages and thereby help to stretch or relax the vocal cords. For instance, the entire larynx is moved upward by the external laryngeal muscles, which helps to stretch the vocal cords when one wishes to emit a very high frequency sound, and the larynx is moved downward, with corresponding loosening of the vocal cords, when one wishes to emit a very bass sound.

The second means for changing the sound frequency is *to change the shape and mass of the vocal cord edges*. When very high frequency sounds are emitted, different slips of the thyroarytenoid muscles contract in such a way that the edges of the vocal cords are sharpened and thinned, whereas when bass frequencies are emitted, the thyroarytenoid muscles contract in a different pattern so that broad edges with a large mass are approximated. Figure 39–9B shows some of the positions and shapes of the vocal cords during different types of phonation.

**Articulation and Resonance.** The three major organs of articulation are the *lips*, the *tongue*, and the *soft palate*. These need not be discussed in detail because all of us are familiar with their movements during speech and other vocalizations.

The resonators include the *mouth*, the *nose and associated nasal sinuses*, the *pharynx*, and even the *chest cavity* itself. Here again we are all familiar with the resonating qualities of these different structures. For instance, the function of the nasal resonators is illustrated by the change in quality of the voice when a person has a severe cold.

# ARTIFICIAL RESPIRATION

## MOUTH-TO-MOUTH BREATHING

A very successful method of artificial respiration is mouth-to-mouth breathing, in which the operator rapidly inspires a deep breath and then breathes into the mouth of the subject. This method has often been shunned in the past because of the belief that the expired air of the operator would not be beneficial to the subject. This is not true, because normal expired air usually still has an adequate amount of oxygen to sustain life in almost anyone. Furthermore, the carbon dioxide in the expired air is sometimes actually desirable because it helps to stimulate the respiratory center of the subject.

## MECHANICAL METHODS OF ARTIFICIAL RESPIRATION

**The Resuscitator.** Many types of resuscitators are available, and each has its own characteristic principles of operation. Basically, the resuscitator, illustrated in Figure 39–10A, consists of a supply of oxygen or air, a mechanism for applying intermittent positive pressure and with some machines negative pressure as well, and a mask which fits over the face of the patient. This apparatus forces air through the mask into the lungs of the patient during the positive pressure cycle and then either allows the air to flow out of the lungs during the remainder of the cycle or pulls the air out by negative pressure.

Earlier resuscitators often caused such severe damage to the lungs because of excessive positive pressure that their usage was at one time greatly deprecated. However, most resuscitators now have safety valves which prevent the positive pressure from rising usually above +14 mm. Hg and the negative pressure from falling below −9 mm. Hg. These pressure limits are adequate to cause excellent artificial respiration of *normal lungs* and yet are slight enough to prevent damage.

**The Tank Respirator.** Figure 39–10B illustrates the usual tank respirator with a patient's body inside the tank and his head protruding through a flexible but airtight collar. At the end of the tank opposite to the patient's head is a motor driven leather diaphragm that moves back and forth with sufficient excursion to raise and lower the pressure inside the tank. As the leather diaphragm moves inward, positive pressure develops around the body and causes expiration; and as the diaphragm moves outward, negative pressure causes inspiration. Check valves

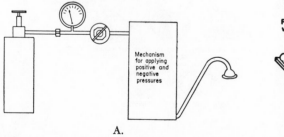

A.

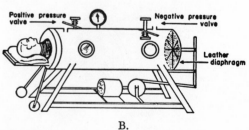

B.

**Figure 39–10.**   (A) The resuscitator. (B) Tank respirator.

on the respirator control the positive and negative pressures. Ordinarily these pressures are adjusted so that the negative pressure that causes inspiration falls to $-10$ to $-20$ cm. water and the positive pressure rises to 0 to $+5$ cm. water.

**Effect of the Resuscitator and the Tank Respirator on Venous Return.**    When air is forced into the lungs under positive pressure, or when the pressure around the patient's body is greatly reduced, as in the case of the tank respirator, the pressure inside the chest cavity becomes greater than the pressure everywhere else in the body. Therefore, the flow of blood into the chest from the peripheral veins becomes impeded. As a result, use of excessive pressures with either the resuscitator or the tank respirator can reduce the cardiac output—sometimes to lethal levels. A person can usually survive as much as 20 mm. Hg positive pressure in the lungs, but exposure for more than a few minutes to greater than 30 mm. Hg usually causes death.

# REFERENCES

Agostoni, E.: Thickness and pressure of the pleural liquid. In Fishman, A. P., and Hecht, H. H. (eds.): The Pulmonary Circulation and Interstitial Space. Chicago, University of Chicago Press, 1969, p. 65.

Clements, J. A., and Tierney, D. F.: Alveolar instability associated with altered surface tension. In Fenn, W. O., and Rahn, H. (eds.): Handbook of Physiology. Sec. 3, Vol. 2. Baltimore, The Williams & Wilkins Company, 1965, p. 1565.

Comroe, J. H., Jr., et al.: The Lung: Clinical Physiology and Pulmonary Function Tests, 2nd Ed. Chicago, Year Book Medical Publishers, Inc., 1963.

Cumming, G.: Alveolar ventilation: recent model analysis. In Guyton, A. C. (ed.): MTP International Review of Science: Physiology. Vol. 2. Baltimore, University Park Press, 1974, p. 139.

Davis, J. N.: Control of the muscles of breathing. In Guyton, A. C. (ed.): MTP International Review of Science: Physiology. Vol. 2. Baltimore, University Park Press, 1974, p. 221.

Fink, B. R.: The Human Larynx: A Functional Study. New York, Raven Press, 1975.

Forster, R. E.: Pulmonary ventilation and blood gas exchange. In Sodeman, W. A., Jr., and Sodeman, W. A. (eds.): Pathologic Physiology: Mechanisms of Disease, 5th Ed. Philadelphia, W. B. Saunders Company, 1974, p. 371.

Greene, M.: The Voice and Its Disorders, 2nd Ed. Philadelphia, J. B. Lippincott Company, 1965.

Guyton, A. C.: Electronic counting and size determination of particles in aerosols. J. Indust. Hyg. Toxicol., 28:133, 1946.

Guyton, A. C.: Analysis of respiratory patterns in laboratory animals. Amer. J. Physiol., 150: 78, 1947.

Guyton, A. C.: Measurement of the respiratory volumes of laboratory animals. Amer. J. Physiol., 150:70, 1947.

Heinemann, H. O., and Fishman, A. P.: Nonrespiratory functions of mammalian lung. Physiol. Rev., 49:1, 1969.

Hong, S. K., Cerretelli, P., Cruz, J. C., and Rahn, H.: Mechanics of respiration during submersion in water. J. Appl. Physiol., 27:535, 1969.

Hoppin, G. G., Jr., Green, I. D., and Mead, J.: Distribution of pleural surface pressures in dogs. J. Appl. Physiol., 27:863, 1969.

Horsfield, K.: The regulation between structure and function in the airways of the lung. Brit. J. Dis. Chest, 68:145, 1974.

Hyatt, R. E., and Black, L. F.: The flow-volume curve. A current perspective. Amer. Rev. Respir. Dis., 107:191, 1973.

Kao, F. F.: An Introduction to Respiratory Physiology. New York, American Elsevier Publishing Company, 1972.

Lapp, N. L.: Physiological approaches to detection of small airway disease. Environ. Res. 6:253, 1973.

Luchsinger, R., and Arnold, G.: Voice-Speech-Language: Clinical Communicology–Its Physiology and Pathology. Wadsworth Publishing Co., Inc., 1965.

Macklem, P. T.: Airway obstruction and collateral ventilation. Physiol. Rev., 51:368, 1971.

Mead, J.: Respiration: pulmonary mechanics. Ann. Rev. Physiol., 35:169, 1973.

Milic-Emili, J.: Pulmonary statics. In Guyton, A. C. (ed.): MTP International Review of Science: Physiology. Vol. 2. Baltimore, University Park Press, 1974, p. 105.

Murray, J. F.: The Normal Lung. Philadelphia, W. B. Saunders Company, 1975.

Nagaishi, C.: Functional Anatomy and Histology of the Lung. Baltimore, University Park Press, 1973.

Rahn, H., Otis, A. B., Chadwick, L. F., and Fenn, W. O.: The pressure-volume diagram of the thorax and lung. Amer. J. Physiol., 146:161, 1946.

Scarpelli, E., and Auld, P. A. M. (eds.): Pediatric Pulmonary Physiology. Philadelphia, Lea & Febiger, 1975.

Staub, N. C.: Respiration. Ann. Rev. Physiol., 31:173, 1969.

Strang, L. B.: Fetal and newborn lung. In Guyton, A. C. (ed.): MTP International Review of Science: Physiology. Vol. 2. Baltimore, University Park Press, 1974, p. 31.

Stuart, B. O.: Deposition of inhaled aerosols. Arch. Intern. Med., 131:60, 1973.

Thurlbeck, W. M., and Wang, N. S.: The structure of the lungs. In Guyton, A. C. (ed.): MTP International Review of Science: Physiology. Vol. 2. Baltimore, University Park Press, 1974, p. 1.

Tierney, D. F.: Lung metabolism and biochemistry. Ann. Rev. Physiol., 36:209, 1974.

West, J. B.: Respiration. Ann. Rev. Physiol., 34:91, 1972.

West, J. B.: Respiratory Physiology. Baltimore, The Williams & Wilkins Company, 1974.

Wyke, B. D. (ed.): Ventilatory and Phonatory Control Systems. New York, Oxford University Press, 1974.

# ▌ 40 ▌

# Physical Principles of Gaseous Exchange; Diffusion of Oxygen and Carbon Dioxide Through the Respiratory Membrane

After the alveoli are ventilated with fresh air, the next step in the respiratory process is *diffusion* of oxygen from the alveoli into the pulmonary blood and diffusion of carbon dioxide in the opposite direction—from the pulmonary blood into the alveoli. The process of diffusion is simple, involving merely random molecular motion of molecules, these intertwining their ways in both directions through the respiratory membrane. However, in respiratory physiology we are concerned not only with the basic mechanism by which diffusion occurs but also with the *rate* at which it occurs, and this is a much more complicated problem, requiring a rather deep understanding of the physics of gases.

The major problems discussed in this chapter, therefore, are, first, the physical factors that determine the alveolar concentrations of gases, particularly of oxygen and carbon dioxide, and, second, the factors that affect the rate at which these gases can diffuse through the respiratory membrane. Since the following discussions entail many basic physical principles of gases, a brief review of this subject is presented as a prelude to the main text of the chapter. The student who is already familiar with these basic principles should go directly to the discussion of the composition of alveolar air.

## THE PHYSICS OF GASES

**Relationship of Pressure to Volume—Boyle's Law.** If the mass and temperature of a gas in a chamber remain constant but the pressure is increased or decreased, the volume of the gas varies inversely with the pressure. That is,

$$\text{Volume} = \frac{\text{Constant}}{\text{Pressure}}$$

Thus, as the pressure increases from 1 to 2 atmospheres (760 to 1520 mm. Hg), the volume of 1 gram-mol of gas at 0° C. is initially 22.4 liters but then decreases to 11.2 liters. On the other hand, if the pressure of this same gram-mol of gas is decreased from 760 to 380 mm. Hg, the total volume increases from 22.4 to 44.8 liters.

**Relationship of Temperature to Volume—Gay-Lussac's Law (Charles' Law).** If the pressure of a given quantity of gas remains constant but the temperature is varied, the volume of the gas increases directly in proportion to the increase in temperature. When the temperature is expressed in absolute (Kelvin) degrees, the following relationship holds:

$$\text{Volume} = \text{Constant} \times \text{Temperature (K)}$$

For example, at one atmosphere pressure, 1 gram-mol of gas has a volume of 22.4 liters at 273° K, which is 0° C. At body temperature of 37° C. (310° K) 1 gram-mol of gas has a volume of 25.4 liters.

**The Gas Law.** Combining Boyle's law and Gay-Lussac's law, the following gas law applies:

$$PV = nRT$$

in which $P$ is pressure, $V$ is volume, $n$ is the quantity of gas, $R$ is a constant depending on the units of measure used for the other factors in the formula, and $T$ is temperature. When $P$ is expressed in millimeters Hg, $V$ in liters, $n$ in gram-mols, and $T$ in absolute degrees, the value of the constant $R$ is 62.36.

### THE VAPOR PRESSURE OF WATER

All gases in the body are in direct contact with water. Therefore, all gaseous mixtures in the body are saturated with water vapor, and this must always be considered when the dynamics of gaseous exchange are discussed.

Figure 40–1 illustrates to the left a chamber containing normal dry air that has in it only nitrogen and oxygen. The pressure in this chamber is the normal atmospheric pressure of 760 mm. Hg, which is illustrated on the scale of the manometer. In the second chamber a few drops of water are added. Immediately, some of the water vaporizes, this vaporization increasing the mass of gas in the chamber; if the temperature in the chamber is body temperature (37° C.), the pressure increases by 47 mm. Hg (the vapor pressure of water at 37° C.) to 807 mm. Hg, as shown to the right in Figure 40–1.

The vapor pressure of water depends entirely on the temperature. The greater the temperature, the greater is the activity of the molecules in the water, and the greater is the likelihood these molecules will escape from the surface of the water into the gaseous phase. On the other hand, the lower the temperature, the greater is the condensation of water from the gaseous phase. When dry air is suddenly mixed with water, the water vapor pressure is zero at first, but water molecules immediately begin escaping from the surface of the water into the air. As the air becomes progressively more humidified, an equilibrium vapor pressure is approached at which the rate of condensation of water becomes equal to the rate of

**TABLE 40–1.  Vapor Pressure of Water**

| Temp. (°C.) | Vapor pressure (mm.) | Temp. (°C.) | Vapor pressure (mm.) |
|---|---|---|---|
| 0 | 4.6 | 39 | 52.0 |
| 10 | 9.1 | 40 | 54.9 |
| 20 | 17.4 | 50 | 92.0 |
| 30 | 31.5 | 60 | 148.9 |
| 35 | 41.8 | 70 | 233.3 |
| 36 | 44.2 | 80 | 354.9 |
| 37 | 46.6 | 90 | 525.5 |
| 38 | 49.3 | 100 | 760.0 |

water vaporization. Because the kinetic activity of molecules in water increases with increased temperature, the equilibrium pressure (the *water vapor pressure*) also increases with increased temperatures. The water vapor pressure at various temperatures from 0° to 100° C. is given in Table 40–1. The most important value to remember is the vapor pressure at body temperature, 47 mm. Hg; this value will appear in most of our subsequent discussions.

### THE SOLUTION OF GASES IN WATER (HENRY'S LAW)

Figure 40–2 illustrates three chambers, with chamber A having oxygen in the top of the chamber at a pressure of 100 mm. Hg and pure water in the bottom of the chamber. Large numbers of oxygen molecules continually strike the surface of the water and some of them enter the water to become dissolved. However, the quantity of dissolved oxygen in chamber A is zero because the process is just beginning. In chamber B a fairly large number of molecules of oxygen has already been dissolved in the water so that the number of molecules striking from below upward against the surface of the water and leaving the water to enter the gaseous phase is equal to approximately half the number of molecules striking the surface from above and entering the dissolved state. Therefore, greater numbers of oxygen molecules are still entering the dissolved state than

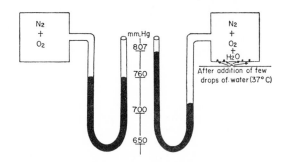

**Figure 40–1.**  Demonstration of 47 mm. Hg vapor pressure in the chamber to the right.

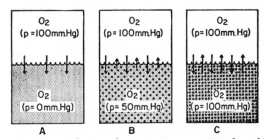

**Figure 40–2.**  Solution of oxygen in water: (A) when the oxygen first comes in contact with pure water, (B) after the dissolved oxygen is halfway to equilibrium with the gaseous oxygen, and (C) after equilibrium has been established.

are leaving the dissolved state. In chamber C the amount of oxygen in the dissolved state has become great enough that the number of molecules leaving the surface of the water to enter the gaseous phase is equal to the number of molecules entering the water to become dissolved. Thus, at this point a state of equilibrium exists.

Two factors determine the quantity of a gas that will be dissolved in water or other fluid when the equilibrium state has been reached. These factors are (1) the pressure of the gas surrounding the water and (2) the solubility coefficient of the gas in the fluid at the temperature of the fluid. When the gas has reached the equilibrium state, as in Figure 40–2C, the total quantity of dissolved gas at any particular temperature may be expressed by the following formula, which is Henry's law:

$$\text{Volume} = \text{Pressure} \times \text{Solubility coefficient}$$

When volume is expressed in volumes of gas dissolved in each volume of water at 0° C. and pressure is expressed in atmospheres, the solubility coefficients for important respiratory gases at body temperatures are the following:

| | |
|---|---|
| Oxygen | 0.024 |
| Carbon dioxide | 0.57 |
| Carbon monoxide | 0.018 |
| Nitrogen | 0.012 |
| Helium | 0.008 |

### PRESSURE OF DISSOLVED GASES

If the oxygen should be removed suddenly from the chamber above the water in Figure 40–2A, no oxygen molecules would be striking the surface of the water from beneath and passing into the gaseous phase. Therefore, there would be no upward pressure against the surface of the water exerted by oxygen. In chamber B of Figure 40–2, half as many molecules of oxygen are pushing upward from the surface of the water as are striking the surface of the water from above. Because pressure is the instantaneous sum of the forces of impact of molecules against a unit area of surface, if half as many molecules are striking upward against the surface of the fluid as are striking downward against the surface of the fluid, the pressure exerted upward is equal to one-half the pressure downward. In other words, in Figure 40–2B the pressure of the oxygen in the gaseous phase is 100 mm. Hg and in the dissolved phase 50 mm. Hg. In chamber C the total force of oxygen molecules striking upward is equal to the total force of molecules striking downward, which means that the pressure exerted by the dissolved oxygen is equal to the pressure exerted by the gaseous oxygen. At this point the pressure of the oxygen in the gaseous phase is in equilibrium with the pressure of the oxygen in the dissolved phase, and both are equal to 100 mm. Hg.

The pressure exerted by a gas as it attempts to pass outward from the surface of a liquid is generally written as $P_{O_2}$, $P_{CO_2}$, $P_{N_2}$, $P_{He}$, etc., for each of the gases dissolved in the liquid.

The pressure of each individual dissolved gas is proportional to the quantity of that gas dissolved in the fluid divided by the solubility coefficient of the gas in that particular fluid. Molecules of gases that are highly soluble are attracted by the fluid molecules so that, as they strike the surface of the fluid, this attraction prevents many of these highly soluble molecules from leaving the surface. Consequently, the total number of molecules that must be dissolved to exert a given pressure is considerably greater for very soluble gases than that for gases not so soluble.

### PARTIAL PRESSURES

An understanding of partial pressures is an important step toward understanding gaseous diffusion from the alveoli to the pulmonary blood, for it is the partial pressure of a gas that determines the force it exerts in attempting to diffuse through the pulmonary membrane.

Figure 40–3 illustrates four separate chambers, each of which has a capacity of 100 volumes of gas. The second chamber is divided by a partition, and 79 volumes of nitrogen at 760 mm. Hg pressure (1 atmosphere of pressure) are placed above the partition and 21 volumes of oxygen at 760 mm. Hg below. Then all the nitrogen in the upper part of chamber 2 is moved to chamber 1, which has a total capacity of 100 volumes. In other words, 79 volumes of nitrogen are expanded to 100 volumes. As this nitrogen is expanded, it can be shown from Boyle's law that the pressure falls to 79/100 of 760 mm. Hg—that is, to 600 mm. Hg.

The 21 volumes of oxygen at 760 mm. Hg are moved from the second chamber to the third chamber. Here again it can be shown from Boyle's law that this expansion causes the pressure of the oxygen to fall to 21/100 of 760 mm. Hg—that is, to 160 mm. Hg.

Finally, the nitrogen in the first chamber and the oxygen in the third chamber are mixed together in the fourth chamber. The original pressure of the nitrogen is 600 mm. Hg, and the pressure of the oxygen is 160 mm. Hg, but the total pressure in chamber 4 after mixing is 760 mm. Hg—the normal atmospheric pressure level. It is obvious that 600 mm. of this 760 mm. Hg results from the presence of nitrogen in the chamber, and 160 mm. results from the presence of oxygen in the chamber. The 600 mm. Hg pressure exerted by nitrogen and the 160 mm. Hg pressure exerted by the oxygen are known as the *partial pressures* of the respective gases in the mixture.

The partial pressures of the gases in a mixture are designated by the same terms used to designate pressures in liquids, i.e., $P_{O_2}$, $P_{CO_2}$, $P_{N_2}$, $P_{He}$, etc.

From the kinetic theory of gases we know that pressure against any membrane or against any other

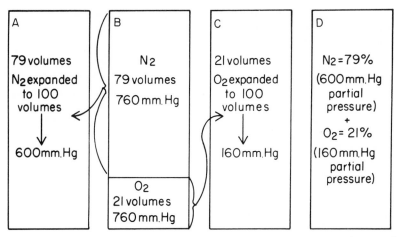

**Figure 40–3.** Relationship of partial pressures to gaseous percentages in mixtures of gases.

surface is determined by the number of molecules striking a unit area of the membrane at any given instant times the average kinetic energy of the molecules. Therefore, the partial pressure of a gas in a mixture is in reality the sum of the instantaneous forces of impact of the molecules of that particular gas against the surface. In other words, in chamber 4 the total force of nitrogen molecules striking against the chamber walls is sufficient to elevate the mercury in a manometer to a level of 600 mm. The total force exerted by the oxygen molecules striking against the walls of the chamber is sufficient to elevate the mercury in a manometer to a level of 160 mm. With both these gases exerting force against the walls of the chamber at the same time, the mercury is elevated to a total level of 760 mm. Hg.

**Pressures of Dissolved Gases in Equilibrium with a Mixture of Gases.** Because each gas in a mixture exerts its own partial pressure in proportion to the concentration of its molecules, when the gases of a mixture dissolve in a liquid and come to equilibrium with the gaseous phase of the mixture, the pressure of each dissolved gas is equal to the partial pressure of the same gas in the gaseous mixture. In other words, *each gas is independent of the others in its ability to dissolve in a liquid;* obviously, this principle also applies to the solution of gases in blood. Thus, an increase in quantity of carbon dioxide dissolved in the body fluids does not significantly affect the quantity of oxygen that can be dissolved in the same fluids. (However, the various dissolved gases often do interfere with the chemical reactions of each other, as is discussed in the following chapter in connection with the reactions of oxygen and carbon dioxide with hemoglobin.)

## DIFFUSION OF GASES THROUGH LIQUIDS

Figure 40–4 illustrates a chamber filled with water. The water at one end of this chamber contains a large amount of dissolved oxygen. Because of their kinetic energy, the dissolved gaseous molecules are constantly undergoing molecular motion, bouncing among the molecules of the solvent. Thus, in Figure 40–4 the oxygen molecules are bouncing in all directions, and this process is called *diffusion*. More of them bounce from the area of high concentration toward the area of low concentration for the following reasons:

**The Pressure Gradient for Diffusion.** The molecules in the area of high pressure in Figure 40–4, because of their greater number, have more *statistical chance* of bouncing into the area of low pressure than do molecules attempting to go in the other direction. However, some molecules do bounce from the area of low pressure toward the area of high pressure. Therefore, the *net diffusion* of gas from the area of high pressure to the area of low pressure is equal to the number of molecules bouncing in this direction minus the number bouncing in the opposite direction, and this in turn is proportional to the gas pressure difference between the two areas. The pressure in area A of Figure 40–4 minus the pressure in area B divided by the distance of diffusion is known as the *pressure gradient for diffusion* or simply the *diffusion gradient*. The rate of net gas diffusion from area A to area B is directly proportional to this gradient.

This principle of diffusion from an area of high pressure to an area of low pressure holds true for diffusion of gases in a gaseous mixture, diffusion of dissolved gases in a solution, and even diffusion of

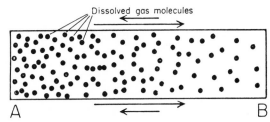

**Figure 40–4.** Net diffusion of oxygen from one end of a chamber to the other.

gases from the gaseous phase into the dissolved state in liquids. That is, *there is always net diffusion from areas of high pressure to areas of low pressure.*

As more and more gas diffuses from area A to area B in Figure 40–4, the pressure in area A falls while that in area B rises so that the two pressures approach each other; as a result, the net rate of diffusion becomes less and less. After a reasonable length of time, the gaseous pressures in both ends of the chamber become essentially equal, and, thereafter, no net diffusion of gas occurs from one end to the other end. This does not mean that no molecules of gas diffuse, but merely that as many molecules then diffuse in one direction as the other.

**Rate of Diffusion.**   In addition to the pressure difference, several other factors affect the rate of gas diffusion in a fluid. These are (1) the solubility of the gas in the fluid, (2) cross-sectional area of the fluid, (3) the distance through which the gas must diffuse, (4) the molecular weight of the gas, and (5) the temperature of the fluid. In the body, the temperature remains reasonably constant and usually need not be considered.

Obviously, the greater the solubility of the gas, the greater will be the number of molecules available to diffuse for any given pressure difference. Also, the greater the cross-sectional area of the chamber, the greater will be the total number of molecules to diffuse. On the other hand, the greater the distance that the molecules must diffuse, the longer it will take the molecules to diffuse the entire distance. Finally, the greater the velocity of kinetic movement of the molecules, which at any given temperature is inversely proportional to the square root of the molecular weight, the greater is the rate of diffusion of the gas. All of these factors can be expressed in a single formula, as follows:

$$DR \propto \frac{PD \times A \times S}{D \times \sqrt{MW}}$$

in which $DR$ is the diffusion rate, $PD$ is the pressure difference between the two ends of the chamber, $A$ is the cross-sectional area of the chamber, $S$ is the solubility of the gas, $D$ is the distance of diffusion, and $MW$ is the molecular weight of the gas.

It is obvious from this formula that the characteristics of the gas itself determine two factors of the formula: solubility and molecular weight. Therefore, the *diffusion coefficient*–that is, the rate of diffusion through a given area for a given distance and pressure difference—for any given gas is proportional to $S/\sqrt{MW}$. Considering the diffusion coefficient for oxygen to be 1, the diffusion coefficients for different gases of respiratory importance in the body fluids are:

| | |
|---|---|
| Oxygen | 1.0 |
| Carbon dioxide | 20.3 |
| Carbon monoxide | 0.81 |
| Nitrogen | 0.53 |
| Helium | 0.95 |

## DIFFUSION OF GASES THROUGH TISSUES

The gases that are of respiratory importance are highly soluble in lipids and, consequently, are also highly soluble in cell membranes. Because of this, these gases diffuse through the cell membranes with very little impediment. Instead, the major limitation to the movement of gases in tissues is the rate at which the gases can diffuse through the tissue fluids instead of through the cell membranes. Therefore, diffusion of gases through the tissues, including through the pulmonary membrane, is almost equal to the diffusion of gases through water, as given in the above list of diffusion rates for the important respiratory gases.

# COMPOSITION OF ALVEOLAR AIR

Alveolar air does not have the same concentrations of gases as atmospheric air by any means, which can readily be seen by comparing the alveolar air composition in column 3 of Table 40–2 with the composition of atmospheric air in column 1. There are several reasons for the differences. First, the alveolar air is only partially replaced by atmospheric air with each breath. Second, oxygen is constantly being absorbed from the alveolar air. Third, carbon dioxide is constantly diffusing from the pulmonary blood into the alveoli. And, fourth, dry atmospheric air that enters the respiratory passages is humidified even before it reaches the alveoli.

**Humidification of the Air as It Enters the Respiratory Passages.**   Column 1 of Table 40–2 shows that atmospheric air is composed almost entirely of nitrogen and oxygen; it normally contains almost no carbon dioxide and little water vapor. However, as soon as the atmospheric air enters the respiratory passages, it is exposed to the fluids covering the respiratory surfaces. Even before the air enters the alveoli, it becomes totally humidified.

The partial pressure of water vapor at normal body temperature of 37° C. is 47 mm. Hg, which, therefore, is the partial pressure of water in the alveolar air. Since the total pressure in the alveoli cannot rise to more than the atmospheric pressure, this water vapor simply expands the volume of the air and thereby *dilutes* all the other gases in the inspired air. In column 2 of Table 40–2 it can be seen that humidification of the air has diluted the oxygen partial pressure at sea level from an average of 159 mm. Hg in atmospheric air to 149 mm. Hg in the

**TABLE 40–2.  Partial Pressures of Respiratory Gases as They Enter and Leave the Lungs (at Sea Level)—Per Cent Concentrations Are Given in Parentheses**

|  | Atmospheric Air* (mm. Hg) | | Humidified Air (mm. Hg) | | Alveolar Air (mm. Hg) | | Expired Air (mm. Hg) | |
|---|---|---|---|---|---|---|---|---|
| $N_2$ | 597.0 | (78.62%) | 563.4 | (74.09%) | 569.0 | (74.9%) | 566.0 | (74.5%) |
| $O_2$ | 159.0 | (20.84%) | 149.3 | (19.67%) | 104.0 | (13.6%) | 120.0 | (15.7%) |
| $CO_2$ | 0.3 | (0.04%) | 0.3 | (0.04%) | 40.0 | (5.3%) | 27.0 | (3.6%) |
| $H_2O$ | 3.7 | (0.50%) | 47.0 | (6.20%) | 47.0 | (6.2%) | 47.0 | (6.2%) |
| TOTAL | 760.0 | (100.0%) | 760.0 | (100.0%) | 760.0 | (100.0%) | 760.0 | (100.0%) |

*On an average cool, clear day.

humidified air, and it has diluted the nitrogen partial pressure from 597 to 563 mm. Hg.

## RATE AT WHICH ALVEOLAR AIR IS RENEWED BY ATMOSPHERIC AIR

In the preceding chapter it was pointed out that the *functional residual capacity* of the lungs, which is the amount of air remaining in the lungs at the end of normal expiration, measures approximately 2300 ml. Furthermore, only 350 ml. of new air is brought into the alveoli with each normal respiration, and the same amount of old alveolar air is expired. Therefore, the amount of alveolar air replaced by new atmospheric air with each breath is only one seventh of the total, so that many breaths are required to exchange most of the alveolar air. Figure 40–5 illustrates this slow rate of renewal of the alveolar air. In the first alveolus of the figure is a highly concentrated foreign gas that has been placed momentarily in all the alveoli and that is not dissolved in the blood. The second alveolus shows slight dilution of this foreign gas with the first breath; the next alveolus shows still further dilution with the second breath, and so forth for the third, fourth, eighth, twelfth, and sixteenth breaths. Note that even at the end of 16

breaths the foreign gas still has not been completely removed from the alveoli.

Figure 40–6 illustrates graphically the rate at which a foreign gas in the alveoli is normally removed, showing that with normal alveolar ventilation approximately half the foreign gas is removed in 17 seconds. When a person's rate of alveolar ventilation is only half normal, half the gas is removed in 34 seconds, and, when his rate of ventilation is two times normal, half is removed in about 8 seconds.

**Value of the Slow Replacement of Alveolar Air.**  This slow replacement of alveolar air is of particular importance in preventing sudden changes in gaseous concentrations in the blood.

This makes the respiratory control mechanism much more stable than it would otherwise be and helps to prevent excessive increases and decreases in tissue oxygenation, tissue carbon dioxide concentration, and tissue pH when the respiration is temporarily interrupted.

## OXYGEN CONCENTRATION IN THE ALVEOLI

Oxygen is continually being absorbed into the blood of the lungs, and new oxygen is continu-

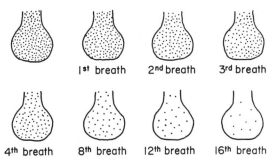

**Figure 40–5.**  Expiration of a foreign gas from the alveoli with successive breaths.

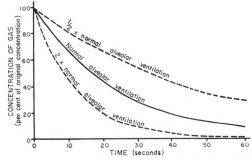

**Figure 40–6.**  Rate of removal of a foreign gas from the alveoli.

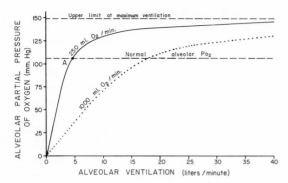

**Figure 40–7.** Effect of alveolar ventilation and of rate of oxygen absorption from the alveoli on the alveolar $P_{O_2}$.

ally entering the alveoli from the atmosphere. The more rapidly oxygen is absorbed, the lower becomes its concentration in the alveoli; on the other hand, the more rapidly new oxygen is brought into the alveoli from the atmosphere, the higher becomes its concentration. Therefore, oxygen concentration in the alveoli is controlled by, first, the rate of absorption of oxygen into the blood and, second, the rate of entry of new oxygen into the lungs by the ventilatory process.

Figure 40–7 illustrates the effect both of alveolar ventilation and of rate of oxygen absorption into the blood on the alveolar partial pressure of oxygen ($P_{O_2}$). The dark curve represents oxygen absorption at a rate of 250 ml. per minute, and the dotted curve represents 1000 ml. per minute. At a normal ventilatory rate of 4.2 liters per minute and an oxygen consumption of 250 ml. per minute, the normal operating point in Figure 40–7 is point A. However, during exercise, the rate of oxygen utilization is increased in proportion to the intensity of the exercise. From Figure 40–7 it can be seen that when 1000 ml. of oxygen is being absorbed each minute, the rate of alveolar ventilation must increase four-fold to maintain the alveolar $P_{O_2}$ at the normal value of 104 mm. Hg, and at still higher rates of oxygen absorption, the rate of ventilation must rise proportionately to maintain normal alveolar $P_{O_2}$.

Another effect illustrated in Figure 40–7 is that an extremely marked increase in alveolar ventilation can never increase the alveolar $P_{O_2}$ above 149 mm. Hg. as long as the person is breathing normal atmospheric air, for this is the maximum content of oxygen in humidified air. However, if the person breathes gases containing concentrations of oxygen higher than 149 mm. Hg, the alveolar $P_{O_2}$ can approach these

higher concentrations when alveolar ventilation approaches maximum.

## $CO_2$ CONCENTRATION IN THE ALVEOLI

Carbon dioxide is continually being formed in the body, then discharged into the alveoli; and it is continually being removed from the alveoli by the process of ventilation. Therefore, the two factors that determine carbon dioxide partial pressure ($P_{CO_2}$) in the lungs are (1) the rate of excretion of carbon dioxide from the blood into the alveoli and (2) the rate at which carbon dioxide is removed from the alveoli by alveolar ventilation.

Figure 40–8 illustrates the effects on the alveolar $P_{CO_2}$ of both alveolar ventilation and the rate of carbon dioxide excretion. The dark curve represents normal rate of carbon dioxide excretion of 200 ml. per minute. At the normal rate of alveolar ventilation of 4.2 liters per minute, the operating point for alveolar $P_{CO_2}$ is at point A in Figure 40–8—that is, 40 mm. Hg.

Two other facts are also evident from Figure 40–8: First, *the alveolar $P_{CO_2}$ increases directly in proportion to the rate of carbon dioxide excretion,* as represented by the dotted curve for 800 ml. $CO_2$ excretion per minute. Second, *the alveolar $P_{CO_2}$ decreases in inverse proportion to alveolar ventilation.* Therefore, the concentration of both oxygen and carbon dioxide in the alveoli is determined by the rate of absorption or excretion, respectively, of these two gases and also by the alveolar ventilation.

## EXPIRED AIR

Expired air is a combination of dead space air and alveolar air, and its overall composition is, therefore, determined by, first, the proportion of the expired air that is dead space air and the proportion that is alveo-

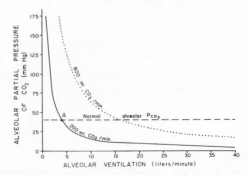

**Figure 40–8.** Effect on alveolar $P_{CO_2}$ of alveolar ventilation and rate of carbon dioxide excretion from the blood.

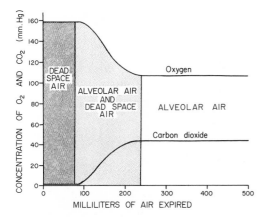

**Figure 40-9.** Oxygen and carbon dioxide concentrations in the various portions of expired air.

lar air. Figure 40–9 shows the progressive changes in oxygen and carbon dioxide concentrations in the expired air during the course of expiration. The very first portion of this air, the dead space air, is typical humidified air as shown in column 2 of Table 40–2. Then, progressively more and more alveolar air becomes mixed with the dead space air until all the dead space air has finally been washed out and nothing but alveolar air remains. Thus, one of the means for collecting alveolar air for study is simply to collect a sample of the last portion of expired air.

Normal expired air, containing both dead space air and alveolar air, has gaseous concentrations approximately as shown in column 4 of Table 40–2—that is, concentrations somewhere between those of humidified atmospheric air and alveolar air, under normal resting conditions about two-thirds alveolar air and one-third dead space air.

# DIFFUSION OF GASES THROUGH THE RESPIRATORY MEMBRANE

**The Respiratory Unit.** Figure 40–10 illustrates the respiratory unit, which is comprised of a *respiratory bronchiole, alveolar ducts, atria,* and *alveolar sacs* or *alveoli* (of which there are about 300 million in the two lungs, each alveolus having an average diameter of about 0.25 mm.). The alveolar walls are extremely thin, and within them is an almost solid network of interconnecting capillaries, as illustrated in Figure 40–11. Indeed, the flow of blood in the alveolar wall has been described as a "sheet" of flowing blood. Thus, it is obvious that the alveolar gases are in close proximity to the blood of the capillaries. Consequently, gaseous exchange between the alveolar air and the pulmonary blood occurs through the mem-

branes of all the terminal portions of the lungs. These membranes are collectively known as the *respiratory membrane,* also called the *pulmonary membrane.*

**The Respiratory Membrane.** Figure 40–12 illustrates the ultrastructure of the respiratory membrane. It also shows the diffusion of oxygen from the alveolus into the red blood cell and diffusion of the carbon dioxide in the opposite direction. Note the following different layers of the respiratory membrane:

1. A layer of fluid lining the alveolus and containing a mixture of phospholipids and perhaps other substances that reduces the surface tension of the alveolar fluid.

2. The alveolar epithelium comprised of very thin epithelial cells.

3. An epithelial basement membrane.

4. A very thin interstitial space between the alveolar epithelium and capillary membrane.

5. A capillary basement membrane that in many places fuses with the epithelial basement membrane.

6. The capillary endothelial membrane.

Despite the large numbers of layers, the overall thickness of the respiratory membrane in some areas is as little as 0.2 micron and averages perhaps 0.5 micron.

From histologic studies it has been estimated that the total surface area of the respiratory membrane is approximately 70 square meters in

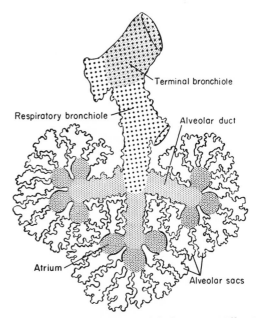

**Figure 40–10.** The respiratory lobule. (From Miller: The Lung. Charles C Thomas.)

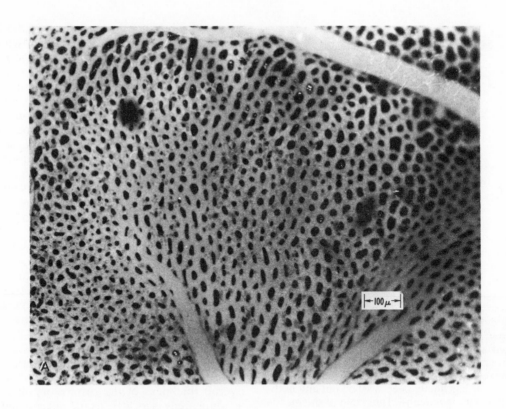

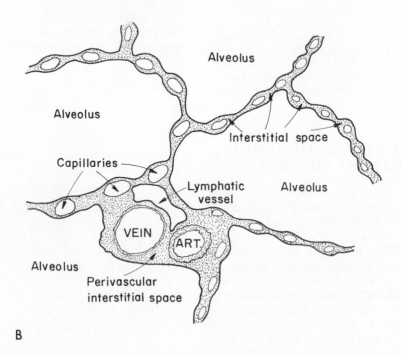

**Figure 40–11.** (A) Surface view of capillaries in an alveolar wall. (From Maloney and Castle: *Resp. Physiol.*, 7:150, 1969. Reproduced by permission of ASP Biological and Medical Press, North-Holland Division.) (B) Schematic cross-sectional view of alveolar walls and their vascular supply.

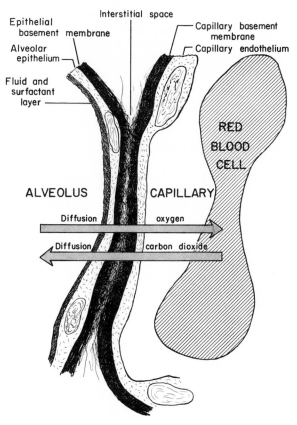

Interstitial space

Epithelial
basement membrane

Alveolar
epithelium

Fluid and
surfactant
layer

Capillary basement
membrane

Capillary endothelium

RED
BLOOD
CELL

ALVEOLUS                                   CAPILLARY

Diffusion          oxygen

Diffusion          carbon dioxide

**Figure 40–12.** Ultrastructure of the respiratory membrane.

the normal adult. This is equivalent to the floor area of a room approximately 30 feet long by 25 feet wide. The total quantity of blood in the capillaries of the lung at any given instant is 60 to 140 ml. If this small amount of blood were spread over the entire surface of a 25 by 30 foot floor, one could readily understand how respiratory exchange of gases occurs as rapidly as it does.

The average diameter of the pulmonary capillaries is only 8 microns, which means that red blood cells must actually squeeze through them. Therefore, the red blood cell membrane usually touches the capillary wall so that oxygen and carbon dioxide need not pass through significant amounts of plasma as they diffuse between the alveolus and the red cell. Obviously, this increases the rapidity of diffusion

**Permeability of the Respiratory Membrane.** The limiting factor in the permeability of the respiratory membrane to gases is the rate at which the gases can diffuse through the water in the membrane. All gases of respiratory importance are highly soluble in the lipid sub-

stances of the cell membranes and for this reason can diffuse through these with great ease. Earlier in the chapter it was pointed out that the rate of carbon dioxide diffusion in water is about 20 times as rapid as the rate of oxygen diffusion. Therefore, it follows that the rate of diffusion of carbon dioxide through the respiratory membrane is also about 20 times as rapid as the diffusion of oxygen, as would be predicted.

*Permeability of the Respiratory Membrane to Solid Solutes.* In contrast to the extreme permeability of the respiratory membrane for respiratory gases, this membrane is much less permeable to most dissolved crystalloid substances in the interstitial fluid. The pulmonary *capillary membrane* has the same high permeability as capillary membranes elsewhere in the body, but the *alveolar epithelial membrane* is only about 1/100 as permeable, except for lipid-soluble substances, such as alcohol and dinitrophenol, that can dissolve in the epithelial cell membranes and therefore pass through with ease. Likewise, the membrane is very permeable to water and reasonably permeable to urea and potassium but very poorly permeable to sodium and glucose. By referring to the discussion of cell membrane permeability in Chapter 4, one can see that these are characteristics similar to those exhibited by all cell membranes. In other words, the alveolar epithelial layer of the respiratory membrane seems to be a continuous layer of epithelial cells with only sparse pores between their borders. This is in contrast to the pulmonary capillary membrane, which has large pores between the individual endothelial cells through which sodium, glucose, urea, and other water-soluble substances diffuse at rates a hundred or more times as rapidly as they diffuse through the alveolar epithelium.

## FACTORS AFFECTING GASEOUS DIFFUSION THROUGH THE RESPIRATORY MEMBRANE

Referring to the above discussion of diffusion through water, one can apply the same principles and same formula to diffusion of gases through the respiratory membrane. Thus, (1) the *thickness of the membrane,* (2) the *surface area of the membrane,* (3) the *diffusion coefficient* of the gas in the substance of the membrane—that is, in water, and (4) the *pressure difference* between the two sides of the membrane, all combined, determine how rapidly a gas will pass through the membrane.

The *thickness of the respiratory membrane* occasionally increases—often as a result of edema fluid in the interstitial space of the membrane. Also, fluid may collect in the alveoli, so that the respiratory gases must diffuse not only through the membrane but also through this fluid. Finally, some pulmonary diseases cause fibrosis of the lungs, which can increase the thickness of some portions of the respiratory membrane. Because the rate of diffusion through the membrane is inversely proportional to the thickness of the membrane, any factor that increases the thickness more than two to three times above normal can interfere markedly with normal respiratory exchange of gases.

The *surface area of the respiratory membrane* may be greatly decreased by many different conditions. For instance, removal of an entire lung decreases the surface area to half normal. Also, in *emphysema* many of the alveoli coalesce, with dissolution of many alveolar walls. Therefore, the new chambers are much larger than the original alveoli, but the total surface area of the respiratory membrane is considerably decreased because of loss of the alveolar walls. When the total surface area is decreased to approximately a third to one-fourth normal, exchange of gases through the membrane is impeded to a significant degree *even under resting conditions*. And, during heavy exercise, even the slightest decrease in surface area of the lungs can be a detriment to respiratory exchange of gases.

The *diffusion coefficient* for the transfer of each gas through the respiratory membrane depends on its *solubility* in the membrane and inversely on the *square root* of its *molecular weight*. The rate of diffusion in the respiratory membrane is almost exactly the same as that in water, for reasons explained above. Therefore, carbon dioxide diffuses through the membrane about 20 times as rapidly as oxygen. Oxygen in turn diffuses about two times as rapidly as nitrogen and at almost the same rate as helium.

The *pressure difference* across the respiratory membrane is the difference between the partial pressure of the gas in the alveoli and the pressure of the gas in the blood. The partial pressure represents a measure of the total number of molecules of a particular gas striking a unit area of the alveolar surface of the membrane, and the pressure of the gas in the blood represents the number of molecules striking the same area of the membrane from the opposite side. Therefore, the difference between these two pressures is a measure of the *net tendency* for the gas to move through the membrane. Obviously, when the partial pressure of a gas in the alveoli is greater than the pressure of the gas in the blood, as is true for oxygen, net diffusion from the alveoli into the blood occurs, but, when the pressure of the gas in the blood is greater than the partial pressure in the alveoli, as is true for carbon dioxide, net diffusion from the blood into the alveoli occurs.

## DIFFUSING CAPACITY OF THE RESPIRATORY MEMBRANE

The overall ability of the respiratory membrane to exchange a gas between the alveoli and the pulmonary blood can be expressed in terms of its *diffusing capacity*, which is defined as the *volume of a gas that diffuses through the membrane each minute for a pressure difference of 1 mm. Hg.*

Obviously, all the factors discussed above that affect diffusion through the respiratory membrane can affect the diffusing capacity.

**The Diffusing Capacity for Oxygen.** In the average young male adult the diffusing capacity for oxygen under resting conditions averages 21 ml. per minute. The mean oxygen pressure difference across the respiratory membrane during quiet breathing is approximately 11 mm. Hg. Multiplication of this pressure by the diffusing capacity ($11 \times 21$) gives a total of about 230 ml. of oxygen diffusing through the respiratory membrane each minute.

*Change in Oxygen-Diffusing Capacity During Exercise.* During strenuous exercise, or during other conditions that greatly increase pulmonary activity, the diffusing capacity for oxygen increases in young male adults to a maximum of about 65 ml. per minute, which is three times the diffusing capacity under resting conditions. This increase is caused by three different factors: (1) opening up of a number of previously dormant pulmonary capillaries, thereby increasing the surface area of the blood into which the oxygen can diffuse, (2) dilatation of all the pulmonary capillaries that were already open, thereby further increasing the surface area, and (3) stretching of the alveolar membranes, which increases their surface area and decreases their thickness. Therefore, during exercise, the oxygenation of the blood is increased not only by increased alveolar ventilation but also by a greater capacity of the respiratory membrane for transmitting oxygen into the blood.

**Diffusing Capacity for Carbon Dioxide.** The diffusing capacity for carbon dioxide has never been measured because of the following technical difficulty: Carbon dioxide diffuses through the respiratory membrane so rapidly that the average $P_{CO_2}$ in the pulmonary blood is not far different from the $P_{CO_2}$ in the alveoli—the difference is less than 1 mm. Hg—and with the available techniques, this difference is too small to be measured.

Nevertheless, measurements of diffusion of other gases have shown that the diffusing capacity varies directly with the diffusion coefficient of the particular gas. Since the diffusion coefficient of carbon dioxide is 20 times that of oxygen, one would expect a diffusing capacity for carbon dioxide under resting conditions of about 400 to 450 ml. and during exercise of about 1200 to 1300 ml. per minute.

The importance of these high diffusing capacities for carbon dioxide is this: When the respiratory membrane becomes progressively damaged, its capacity for transmitting oxygen into the blood is always impaired enough to cause death of the person long before significant impairment of carbon dioxide diffusion occurs. The only time that a low diffusing capacity for carbon dioxide causes any significant difficulty is when lung damage is far beyond that which ordinarily causes death but when the person's life is being maintained by intensive oxygen therapy that overcomes the reduction in oxygen-diffusing capacity.

Figure 40–13 compares the diffusing capacities of oxygen, carbon dioxide, and carbon monoxide at rest and during exercise, showing the extreme diffusing capacity of carbon dioxide and also the effect of exercise on the diffusing capacities of all the gases.

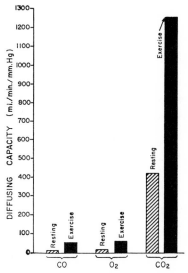

**Figure 40–13.** Diffusing capacities of carbon monoxide, oxygen, and carbon dioxide in the normal lungs.

**Measurement of Diffusing Capacity.** The oxygen-diffusing capacity can be estimated from measurements of (1) alveolar $P_{O_2}$, (2) $P_{O_2}$ in the pulmonary capillary blood, and (3) the rate of oxygen utilization. To measure the $P_{O_2}$ in the pulmonary blood is a difficult procedure and is usually accomplished by measuring the $P_{O_2}$ in the pulmonary arterial blood and the $P_{O_2}$ in the aortic arterial blood and then mathematically obtaining a reasonable average for pulmonary capillary blood. This average must be a time-integrated average rather than a simple average, because pulmonary arterial blood on entering the capillaries becomes rapidly oxygenated during the first fraction of its transit through the capillaries and then progressively more slowly during the latter stages of transit. Obviously, this mathematical procedure is complicated, and one never knows exactly how accurate the results are.

To obviate the difficulties encountered in measuring oxygen-diffusing capacity, physiologists usually measure carbon monoxide-diffusing capacity instead, and calculate the oxygen-diffusing capacity from this. The principle of the carbon monoxide method is the following: A small amount of carbon monoxide is breathed into the alveoli, and the partial pressure of the carbon monoxide in the alveoli is measured from appropriate alveolar air samples. Under most conditions it can be assumed that the carbon monoxide pressure in the blood is essentially zero because the hemoglobin combines with this so rapidly that the pressure never has time to build up. Therefore, the pressure difference of carbon monoxide across the respiratory membrane is equal to its partial pressure in the alveoli. Then, by measuring the volume of carbon monoxide absorbed in a short period of time and dividing this by the carbon monoxide partial pressure, one can determine accurately the carbon monoxide-diffusing capacity.

To convert carbon monoxide-diffusing capacity to oxygen-diffusing capacity, the value is multiplied by a factor of 1.23 because the diffusion coefficient for oxygen is 1.23 times that for carbon monoxide. Thus, the average diffusing capacity for carbon monoxide in young male adults is 17 ml. per minute, and the diffusing capacity for oxygen is 1.23 times this, or 21 ml. per minute.

## DIFFUSION OF GASES WITHIN THE RESPIRATORY UNIT

Not only must the respiratory gases diffuse between the alveolar air and the blood but they must also diffuse within the passageways of the respiratory unit for the following reason: When a person inspires, the new air entering the lungs travels into the respiratory passageways only as far as the terminal bronchioles. Therefore, were there no diffusion occurring within the gases of the alveolar ducts and respiratory bronchioles, no new air would reach the alveoli. However, diffusion within the respiratory unit is so great that the gases in the terminal bronchioles normally come to equilibrium with those in the alveoli within a fraction of a second. Therefore, in most discussions of alveolar ventilation, one simply assumes that the new air does indeed flow into the alveoli.

Under some abnormal conditions air flow to some sections of the lungs may be greatly impeded, in which case the new air does not even reach the terminal bronchioles. If the distance for diffusion is very great, alveolar air exchange obviously becomes seriously curtailed. This will be discussed in relation to diseases of the lungs in Chapter 43.

# REFERENCES

Burrows, B.: Arterial oxygenation and pulmonary hemodynamics in patients with chronic airways obstruction. *Amer. Rev. Respir. Dis.* 110:64, 1974.

Cumming, G.: Alveolar ventilation: recent model analysis. *In* Guyton A. C. (ed.): MTP International Review of Science: Physiology. Vol. 2. Baltimore, University Park Press, 1974, p. 139.

Cunningham, D. J.: Time patterns of alveolar carbon dioxide and oxygen: the effects of various patterns of oscillations on breathing in man. *Sci. Basis Med.*, 333, 1972.

Forster, R. E.: Diffusion of gases. *In* Fenn, W. O., and Rahn, H. (eds.): Handbook of Physiology. Sec. 3, Vol. 1. Baltimore, The Williams & Wilkins Company, 1964, p. 839.

Forster, R. E.: Interpretation of measurements of pulmonary diffusing capacity. *In* Fenn, W. O., and Rahn, H. (eds.): Handbook of Physiology. Sec. 3, Vol. 2. Baltimore, The Williams & Wilkins Company, 1965, p. 1435.

Forster, R. W.: Pulmonary ventilation and blood gas exchange. *In* Sodeman, W. A., Jr., and Sodeman, W. A. (eds.): Pathologic Physiology: Mechanisms of Disease, 5th Ed. Philadelphia, W. B. Saunders Company, 1974, p. 371.

Guyton, A. C., Nichols, R. J., and Farish, C. A.: An arteriovenous oxygen difference recorder. *J. Appl. Physiol.*, 10:158, 1957.

Hughes, J. M.: Proceedings: regional differences in gas exchange. *Proc. R. Soc. Med.*, 66:974, 1973.

Johansen, K.: Comparative physiology: gas exchange and circulation in fishes. *Ann. Rev. Physiol.*, 33:569, 1971.

Milhorn, H. T., Jr., and Pulley, P. E., Jr.: A theoretical study of pulmonary capillary gas exchange and venous admixture. *Biophys. J.*, 8:337, 1968.

Moran, F., and Pack, A. I.: Proceedings: measurement of ventilation-perfusion distribution. *Proc. R. Soc. Med.*, 66:975, 1973.

Morrow, P. E.: Alveolar clearance of aerosols. *Arch. Intern. Med.*, 131: 101, 1973.

Otis, A. B.: Quantitative relationships in steady-state gas exchange. *In* Fenn, W. O., and Rahn, H. (eds.): Handbook of Physiology. Sec. 3, Vol. 1. Baltimore, The Williams & Wilkins Company, 1964, p. 681.

Piiper, J., and Scheid, P.: Respiration: alveolar gas exchange. *Ann. Rev. Physiol.*, 33:131, 1971.

Radford, E. P., Jr.: The physics of gases. *In* Fenn, W. O., and Rahn, H. (eds): Handbook of Physiology. Sec. 3, Vol. 1. Baltimore, The Williams & Wilkins Company, 1964, p. 125.

Rahn, H., and Farhi, L. E.: Ventilation, perfusion, and gas exchange—the Va/Q concept. *In* Fenn, W. O., and Rahn, H. (eds.): Handbook of Physiology. Sec. 3, Vol. 1. Baltimore, The Williams & Wilkins Company, 1964, p. 735.

Rahn, H., and Fenn, W. O.: A Graphical Analysis of Respiratory Gas Exchange. Washington, American Physiological Society, 1955.

Riley, R. L., and Permutt, S.: The four-quandrant diagram for analyzing the distribution of gas and blood in the lung. *In* Fenn, W. O., and Rahn, H. (eds.): Handbook of Physiology. Sec. 3, Vol. 2. Baltimore, The Williams & Wilkins Company, 1965, p. 1413.

Thurlbeck, W. M., and Wang, N. S.: The structure of the lungs. *In* Guyton, A. C. (ed.): MTP International Review of Science: Physiology. Vol. 2. Baltimore, University Park Press, 1974, p. 1.

Weibel, E. R.: Morphological basis of alveolar capillary gas exchange. *Physiol. Rev.*, 53:419, 1973.

West, J. B.: Ventilation/Blood Flow and Gas Exchange, 2nd Ed. Philadelphia, J. B. Lippincott Company, 1970.

# 41

# Transport of Oxygen and Carbon Dioxide in the Blood and Body Fluids

Once oxygen has diffused from the alveoli into the pulmonary blood, it is transported principally in combination with hemoglobin to the tissue capillaries where it is released for use by the cells. The presence of hemoglobin in the red cells of the blood allows the blood to transport 30 to 100 times as much oxygen as could be transported simply in the form of dissolved oxygen in the water of the blood.

In the tissue cells oxygen reacts with various foodstuffs to form large quantities of carbon dioxide. This in turn enters the tissue capillaries and is transported by the blood back to the lungs. Carbon dioxide, like oxygen, also combines with chemical substances in the blood that increase carbon dioxide transport 15- to 20-fold.

The purpose of the present chapter, therefore, is to present both qualitatively and quantitatively the physical and chemical principles of oxygen and carbon dioxide transport in the blood and body fluids.

## PRESSURE DIFFERENCES OF OXYGEN AND CARBON DIOXIDE FROM THE LUNGS TO THE TISSUES

In the discussions of the preceding chapter it was pointed out that gases can move from one point to another by diffusion and that the cause of this movement is always a pressure difference from the first point to the other. Thus, oxygen diffuses from the alveoli into the pulmonary capillary blood because of a pressure difference—that is, the oxygen pressure ($Po_2$) in the alveoli is greater than the $Po_2$ in the pulmonary blood. Then the pulmonary blood is transported by way of the circulation to the peripheral tissues. There the $Po_2$ is lower in the cells than in the arterial blood entering the capillaries. Here again, the much higher $Po_2$ in the capillary blood causes oxygen to diffuse out of the capillaries and through the interstitial spaces to the cells.

Conversely, when oxygen is metabolized with the foods in the cells to form carbon dioxide, the carbon dioxide pressure ($Pco_2$) in the cells rises to a high value, which causes carbon dioxide to diffuse into the tissue capillaries. Once in the blood the carbon dioxide is transported to the pulmonary capillaries where it diffuses out of the blood into the alveoli because the $Pco_2$ in the alveoli is lower than that in the pulmonary capillary blood.

Basically, then, the transport of oxygen and carbon dioxide by the blood depends on both diffusion and the movement of blood. We now need to consider quantitatively the factors responsible for these effects as well as their significance in the overall physiology of respiration.

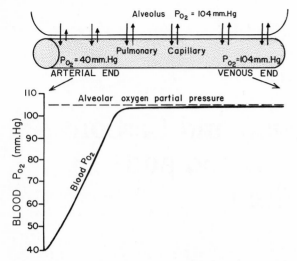

**Figure 41–1.** Uptake of oxygen by the pulmonary capillary blood. (The curve in this figure was constructed from data in Milhorn and Pulley: *Biophys. J.*, 8:337, 1968.)

## *UPTAKE OF OXYGEN BY THE PULMONARY BLOOD*

The top part of Figure 41–1 illustrates a pulmonary alveolus adjacent to a pulmonary capillary, showing diffusion of oxygen molecules between the alveolar air and the pulmonary blood. However, the $Po_2$ of the venous blood entering the capillary is only 40 mm. Hg because a large amount of oxygen has been removed from this blood as it has passed through the tissue capillaries. The $Po_2$ in the alveolus is 104 mm. Hg, giving an initial pressure difference for diffusion of oxygen into the pulmonary capillary of 104 − 40 or 64 mm. Hg. Therefore, far more oxygen diffuses into the pulmonary capillary than in the opposite direction. The curve below the capillary shows the progressive rise in blood $Po_2$ as the blood passes through the capillary. This curve illustrates that the $Po_2$ rises essentially to equal that of the alveolar air before reaching the midpoint of the capillary, becoming approximately 104 mm. Hg.

The average pressure difference for oxygen diffusion through the pulmonary capillary during normal respiration is about 11 mm. Hg. This is a "time-integrated" average and not simply an average of the 64 mm. Hg pressure difference at the beginning of the capillary and the final zero pressure difference at the end of the capillary, because the initial pressure difference lasts for only a fraction of the transit time in the pulmonary capillary, while the low pressure difference lasts for a long time.

**Uptake of Oxygen by the Pulmonary Blood During Exercise.** During strenuous exercise, a person's body may require as much as 20 times the normal amount of oxygen. However, because of the increased cardiac output, the time that the blood remains in the capillary is greatly reduced. Therefore, oxygenation of the blood might suffer for two reasons: First, the blood remains in contact with the alveoli for short periods of time and, second, far larger quantities of oxygen are needed to oxygenate the blood. Yet, because of a great *safety factor* for diffusion of oxygen through the pulmonary membrane, the blood is still *almost completely saturated* with oxygen when it leaves the pulmonary capillaries. The reasons for this are the following:

First, it was pointed out in the previous chapter that the diffusing capacity for oxygen increases about three-fold during exercise; this results mainly from increased numbers of capillaries participating in the diffusion and also from dilatation of both the alveoli and the capillaries.

Second, even if the diffusing capacity should not increase, quantities of oxygen far greater than normal can still enter the blood because of the special nature of diffusion in the pulmonary capillaries. This can be explained by referring once again to Figure 41–1. Note that during normal resting pulmonary blood flow the blood becomes almost saturated with oxygen by the time it has passed through one-third of the pulmonary capillary, and little additional oxygen enters the capillary blood during the latter two-thirds of its transit. That is, the blood normally stays in the lungs about three times as long as necessary to cause adequate oxygenation anyway. Therefore, even with the shortened time of exposure in exercise, the blood still can become fully oxygenated.

**Effects of Pulmonary Shunts on the $Po_2$ of Arterial Blood.** A small amount of blood, usually 1 to 2 per cent of the total cardiac output, fails to pass through alveolar capillaries but instead is shunted through poorly aerated vessels either in the lungs themselves or in adjacent tissues. This blood mixes with the aerated blood in the left heart and slightly reduces the $Po_2$ of the blood before it enters the arterial tree. This is called *venous admixture* of blood, and its effect is illustrated in Figure 41–2. Note that the blood leaving the pulmonary capillaries has a $Po_2$ of approximately 104 mm. Hg, but the venous admixture reduces this to approximately 95 mm. Hg in the systemic arterial blood. It seems strange that only 1 to 2 per cent venous admixture could cause this much decrease in $Po_2$, but later in the chapter we will see that pressures above 90 mm. Hg are also mainly above the level at which oxygen combines with hemoglobin. In this range even slight changes in

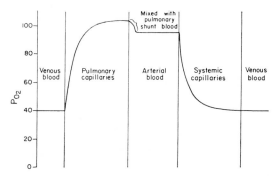

**Figure 41–2.** Changes in $P_{O_2}$ in the pulmonary capillary blood, the arterial blood, and the systemic capillary blood, illustrating the effect of "venous admixture."

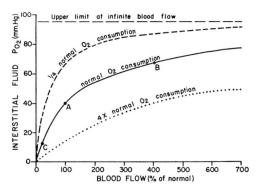

**Figure 41–4.** Effect of blood flow and rate of oxygen consumption on tissue $P_{O_2}$.

oxygenation can change the $P_{O_2}$ greatly even though it does not change the *total quantity* of oxygen in the blood a significant amount.

In some diseases of the pulmonary circulation the shunted blood amounts to more than 50 per cent of the total cardiac output. In these instances the arterial $P_{O_2}$ can be decreased to as low as 40 to 50 mm. Hg.

## DIFFUSION OF OXYGEN FROM THE CAPILLARIES TO THE INTERSTITIAL FLUID

At the tissue capillaries, oxygen diffuses into the tissues by a process essentially the reverse of that which takes place in the lungs, as illustrated in Figure 41–3. That is, the $P_{O_2}$ in the interstitial fluid immediately outside a capillary is low and, though very variable, averages about 40 mm. Hg, while that in the arterial blood is high, about 95 mm. Hg. Therefore, at the arterial end of the capillary, a pressure difference of 55 mm. Hg causes diffusion of oxygen. As illustrated in the figure, by the time the blood has passed through the capillary a large portion of the oxygen has diffused into the tissues, and the capillary $P_{O_2}$ has approached the 40 mm. Hg oxygen pressure in the tissue fluids. Consequently, the venous blood leaving the tissue capillaries contains oxygen at essentially the same pressure as that immediately outside the tissue capillaries, 40 mm. Hg.

**Effect of Rate of Blood Flow on Interstitial Fluid $P_{O_2}$.** If the blood flow through a particular tissue becomes increased, greater quantities of oxygen are transported into the tissue in a given period of time, and the tissue $P_{O_2}$ becomes correspondingly increased—an effect illustrated in Figure 41–4. Note that an increase in flow to 400 per cent of normal increases the $P_{O_2}$ from Point A to Point B, that is, from 40 mm. Hg to 66 mm. Hg. However, the upper limit to which the $P_{O_2}$ can rise, even with maximal blood flow, is 95 mm. Hg, which is the oxygen pressure in the arterial blood.

**Effect of Rate of Tissue Metabolism on Interstitial Fluid $P_{O_2}$.** If the cells utilize far more oxygen for metabolism than normally, this tends to reduce the interstitial fluid $P_{O_2}$; or, conversely, the $P_{O_2}$ rises to approach the $P_{O_2}$ of arterial blood when tissue metabolism ceases. Figure 41–4 also illustrates these effects, showing that increased tissue oxygen consumption greatly reduces the interstitial fluid $P_{O_2}$, whereas reduced consumption greatly increases the $P_{O_2}$.

**Effect of Hemoglobin Concentration on Interstitial Fluid $P_{O_2}$.** Because approximately 97 per cent of the oxygen transported in the blood is carried by hemoglobin, a decrease in hemoglobin concentration has the same effect on interstitial fluid $P_{O_2}$ as does a decrease in blood flow. Thus, reducing the hemoglobin concentration to one quarter normal while maintaining normal blood flow reduces the interstitial fluid $P_{O_2}$ to point C in Figure 41–4—that is, to about 13 mm. Hg.

In summary, tissue $P_{O_2}$ is determined by a balance between (a) the rate of oxygen transport to the tissues in the blood and (b) the rate at which the oxygen is utilized by the tissues.

## DIFFUSION OF OXYGEN FROM THE INTERSTITIAL FLUIDS INTO THE CELLS

Since oxygen is always being used by the cells, the intracellular $P_{O_2}$ remains lower than the interstitial fluid $P_{O_2}$. When the interstitial

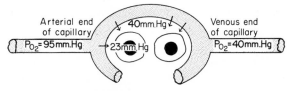

**Figure 41–3.** Diffusion of oxygen from a tissue capillary to the cells.

fluid $Po_2$ falls, the intracellular $Po_2$ also falls, and, conversely, an increase in interstitial $Po_2$ causes a similar increase in intracellular $Po_2$.

As pointed out in Chapter 4, oxygen diffuses through cell membranes extremely rapidly. Therefore, the intracellular $Po_2$ is almost equal to that of the interstitial fluids. Yet, in many instances, there is considerable distance between the capillaries and the cells. Therefore, the normal intracellular $Po_2$ ranges from as low as 5 mm. Hg to as high as 60 mm. Hg, averaging (by direct measurement in lower animals) 23 mm. Hg, which is the value given for the cell in Figure 41–3. Since only 1 to 5 mm. Hg oxygen pressure is required for full support of the metabolic processes of the cell, one can see that even this low cellular $Po_2$ is adequate and actually provides a considerable safety factor.

## DIFFUSION OF CARBON DIOXIDE FROM THE CELLS TO THE TISSUE CAPILLARIES

Because of the continual large quantities of carbon dioxide formed in the cells, the intracellular $Pco_2$ tends to rise. However, carbon dioxide diffuses about 20 times as easily as oxygen, diffusing from the cells extremely rapidly into the interstitial fluids and thence into the capillary blood. Thus, in Figure 41–5 the intracellular $Pco_2$ is shown to be 46 mm. Hg, while that in the interstitial fluid immediately adjacent to the capillaries is about 45 mm. Hg, a pressure differential of only 1 mm. Hg.

Arterial blood entering the tissue capillaries contains carbon dioxide at a pressure of approximately 40 mm. Hg. As the blood passes through the capillaries, the blood $Pco_2$ rises to approach the 45 mm. Hg $Pco_2$ of the interstitial fluid. Because of the very large diffusion coefficient for carbon dioxide, the $Pco_2$ of venous blood is also about 45 mm. Hg, within a fraction of a millimeter of reaching complete equilibrium with the $Pco_2$ of the interstitial fluid.

**Effect of Tissue Metabolism and Blood Flow on the Interstitial Fluid $Pco_2$.** Blood

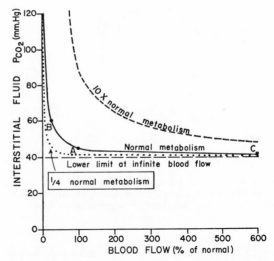

**Figure 41–6.** Effect of blood flow and metabolic rate on tissue $Pco_2$.

flow and tissue metabolism affect tissue $Pco_2$ in a way exactly opposite to the way that they affect tissue $Po_2$. Figure 41–6 shows the effects of normal metabolism, one-quarter normal metabolism, and 10 times normal metabolism on interstitial fluid $Pco_2$ at different rates of blood flow. Note that the lower limit to which the interstitial fluid $Pco_2$ can possibly fall is the $Pco_2$ of the arterial blood entering the tissue capillaries—normally about 40 mm. Hg. The elevation of tissue $Pco_2$ above this value depends on both blood flow and metabolic rate. Point A illustrates the $Pco_2$ when both the blood flow and metabolism are normal. Point B shows that a decrease in blood flow to one-quarter normal increases the tissue $Pco_2$ to 60 mm. Hg, while point C shows that an increase in blood flow to six times normal reduces the tissue $Pco_2$ to 41 mm. Hg, almost to its lower limit of 40 mm. Hg. It can also be seen that increasing the rate of metabolism, especially at low blood flow rates, can increase the tissue $Pco_2$ to extremely high levels, while reducing the rate of metabolism correspondingly reduces the tissue $Pco_2$.

## REMOVAL OF CARBON DIOXIDE FROM THE PULMONARY BLOOD

On arriving at the lungs, the $Pco_2$ of the venous blood is about 45 mm. Hg while that in the alveoli is 40 mm. Hg. Therefore, as illustrated in Figure 41–7, the initial pressure difference for diffusion is only 5 mm. Hg, which is far less than that for diffusion of oxygen across the

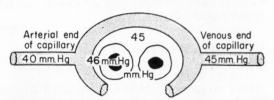

**Figure 41–5.** Uptake of carbon dioxide by the blood in the capillaries.

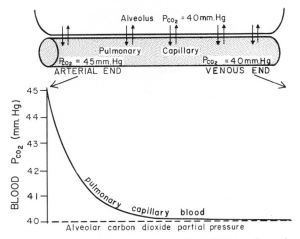

**Figure 41–7.** Diffusion of carbon dioxide from the pulmonary blood into the alveolus. (This curve was constructed from data in Milhorn and Pulley: *Biophys. J.*, 8:337, 1968.)

membrane. Yet, even so, because of the 20 times as great diffusion coefficient for carbon dioxide as for oxygen, the excess carbon dioxide in the blood is rapidly transferred into the alveoli. Indeed, the figure shows that the $P_{CO_2}$ of the pulmonary capillary blood becomes almost equal to that of the alveoli within the first four-tenths of the blood's transit through the pulmonary capillary.

# TRANSPORT OF OXYGEN IN THE BLOOD

Normally, about 97 per cent of the oxygen transported from the lungs to the tissues is carried in chemical combination with hemoglobin in the red blood cells, and the remaining 3 per cent is carried in the dissolved state in the water of the plasma and cells. Thus, *under normal conditions* the transport of oxygen in the dissolved state is negligible. However, when a person breathes oxygen at very high pressures, as much oxygen can sometimes be transported in the dissolved state as in chemical combination with hemoglobin. Therefore, the present discussion will consider the transport of oxygen first in combination with hemoglobin and then in the dissolved state under special conditions.

## THE REVERSIBLE COMBINATION OF OXYGEN WITH HEMOGLOBIN

The chemistry of hemoglobin was presented in Chapter 5, where it was pointed out that the

oxygen molecule combines loosely and reversibly with the heme portion of the hemoglobin. When the $P_{O_2}$ is high, as in the pulmonary capillaries, oxygen binds with the hemoglobin, but, when the $P_{O_2}$ is low, as in the tissue capillaries, oxygen is released from the hemoglobin. This is the basis for oxygen transport from the lungs to the tissues.

**The Oxygen-Hemoglobin Dissociation Curve.** Figure 41–8 illustrates the oxygen-hemoglobin dissociation curve, which shows the progressive increase in the per cent of the hemoglobin that is bound with oxygen as the $P_{O_2}$ increases. This is called the *per cent saturation of the hemoglobin.* Since the blood leaving the lungs usually has a $P_{O_2}$ of about 100 mm. Hg, one can see from the dissociation curve that the *usual oxygen saturation of arterial blood is about 97 per cent.* On the other hand, the $P_{O_2}$ is about 40 mm. Hg and *the saturation of the hemoglobin is about 70 per cent in normal venous blood.*

**Maximum Amount of Oxygen that Can Combine with the Hemoglobin of the Blood.** The blood of a normal person contains approximately 15 grams of hemoglobin in each 100 ml. of blood, and each gram of hemoglobin can bind with a maximum of about 1.34 ml. of oxygen. Therefore, on the average, the hemoglobin in 100 ml. of blood can combine with a total of almost exactly 20 ml. of oxygen when the extent of saturation is 100 per cent. This is usually expressed as 20 *volumes per cent.* The oxygen-hemoglobin dissociation curve for the normal person, therefore, can be expressed in terms of volume per cent of oxygen, as shown in Figure 41–9, rather than per cent saturation of hemoglobin.

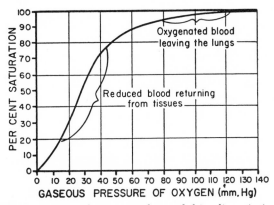

**Figure 41–8.** The oxygen-hemoglobin dissociation curve.

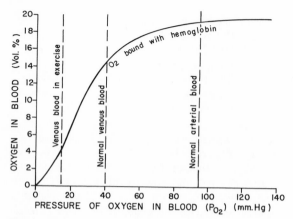

**Figure 41–9.** Effect of blood $P_{O_2}$ on the quantity of oxygen bound with hemoglobin in each 100 ml. of blood.

**Amount of Oxygen Released from the Hemoglobin in the Tissues.** The total quantity of oxygen *bound with hemoglobin* in normal arterial blood, which is normally 97 per cent saturated, is approximately 19.4 ml. This is illustrated in Figure 41–9. However, on passing through the tissue capillaries, this amount is reduced to 14.4 ml. ($P_{O_2}$ of 40 mm. Hg, 75 per cent saturated), or a total loss of 5 ml. of oxygen from each 100 ml. of blood. Then, when the blood returns to the lungs, the same quantity of oxygen diffuses from the alveoli to the hemoglobin, and this too is carried to the tissues. Thus, *under normal conditions about 5 ml. of oxygen is transported by each 100 ml. of blood during each cycle through the tissues.*

**Transport of Oxygen during Strenuous Exercise.** In heavy exercise the muscle cells utilize oxygen at a rapid rate, which causes the interstitial fluid $P_{O_2}$ to fall to as low as 15 mm. Hg. At this pressure only 4.4 ml. of oxygen remains bound with the hemoglobin in each 100 ml. of blood, as shown in Figure 41–9. Thus 19.4−4.4, or 15 ml., is the total quantity of oxygen transported by each 100 ml. of blood during each cycle through the tissues. This, obviously, is three times as much as that normally transported by the same amount of blood, illustrating that simply an increase in rate of oxygen utilization by the tissues causes an automatic increase in the rate of oxygen release from the hemoglobin.

**The Utilization Coefficient.** The fraction of the blood that gives up its oxygen as it passes through the tissue capillaries is called the *utilization coefficient.* Normally, this is approximately 0.25, or 25 per cent, of the blood. That is, *the normal utilization coefficient is approxi-*

*mately one-fourth.* During strenuous exercise, as much as 75 to 85 per cent of the blood can give up its oxygen; the utilization coefficient is then approximately 0.75 to 0.85. These values are about the highest utilization coefficients that can be attained in the overall body even when the tissues are in extreme need of oxygen. However, in local tissue areas where the blood flow is very slow or the metabolic rate very high, utilization coefficients approaching 100 per cent have been recorded—that is, essentially all the oxygen is removed.

## THE OXYGEN BUFFER FUNCTION OF HEMOGLOBIN

Though hemoglobin is necessary for transport of oxygen to the tissues, it performs still another major function essential to life. This is the function of hemoglobin as an "oxygen buffer" system. That is, the hemoglobin in the blood is mainly responsible for controlling the oxygen pressure in the tissues. This can be explained as follows:

**Value of Hemoglobin for Maintaining Constant $P_{O_2}$ in the Tissue Fluids.** Even under basal conditions the tissues require about 5 ml. of oxygen from every 100 ml. of blood passing through the tissue capillaries. Referring back to the oxygen-hemoglobin dissociation curve in Figure 41–9, one can see that for 5 ml. of oxygen to be released from each 100 ml. of blood, the $P_{O_2}$ must fall to about 40 mm. Hg. Therefore, tissue capillary $P_{O_2}$ ordinarily does not rise above approximately 40 mm. Hg, for if such should occur, the oxygen needed by the tissues could not be released from the hemoglobin. In this way, the hemoglobin normally sets an upper limit on the gaseous pressure in the tissues at approximately 40 mm. Hg.

On the other hand, in heavy exercise the $P_{O_2}$ in the tissue capillaries falls, causing the hemoglobin to release extra quantities of oxygen to the tissues. This in turn helps to prevent the capillary $P_{O_2}$ from falling too low; therefore, it rarely falls below 20 mm. Hg.

It can be seen, then, that hemoglobin automatically delivers oxygen to the tissues at a pressure between approximately 20 and 40 mm. Hg. This seems to be a wide range of $P_{O_2}$'s in the interstitial fluid, but, when one considers how much the interstitial fluid $P_{O_2}$ might possibly change during exercise and other types of stress, this range of 20 to 40 mm. Hg is relatively narrow.

**Value of Hemoglobin for Maintaining Constant Tissue $P_{O_2}$ When Atmospheric Oxygen Concentration Changes Markedly.** The normal $P_{O_2}$ in the alveoli is approximately 104 mm. Hg, but, as one ascends a mountain or goes high in an airplane, the $P_{O_2}$ falls considerably. Or, when one enters areas of compressed air, such as deep below the sea or in tunnels, the $P_{O_2}$ may rise to very high values. It will be seen from the oxygen-hemoglobin dissociation curve of Figure 41–8 that, when the $P_{O_2}$ is decreased to as low as 60 mm. Hg, the hemoglobin is still 89 per cent saturated, only 8 per cent below the normal saturation of 97 per cent. Furthermore, the tissues still remove approximately 5 ml. of oxygen from every 100 ml. of blood passing through the tissues; to remove this oxygen, the $P_{O_2}$ of the venous blood falls to only slightly less than 40 mm. Hg. Thus, the tissue $P_{O_2}$ hardly changes despite the marked fall in alveolar $P_{O_2}$ from 104 to 60 mm. Hg.

On the other hand, when the alveolar $P_{O_2}$ rises far above the normal value of 104 mm. Hg, the maximum oxygen saturation of hemoglobin can never rise above 100 per cent. Therefore, even though the oxygen in the alveoli should rise to a partial pressure of 500 mm. Hg, or even more, the increase in the saturation of hemoglobin would be only 3 per cent because, even at 104 mm. Hg $P_{O_2}$, 97 per cent of the hemoglobin is already combined with oxygen; and only a small amount of additional oxygen dissolves in the fluid of the blood, as will be discussed subsequently. When the hemoglobin that has been subjected to $P_{O_2}$'s of several hundred millimeters in the alveoli passes to the tissue capillaries, it still loses several milliliters of oxygen to the tissues, and this loss automatically reduces the $P_{O_2}$ of the capillary blood to a value only a few millimeters greater than the normal 40 mm. Hg.

Consequently, the atmospheric oxygen content may vary greatly—from 60 to more than 500 mm. Hg $P_{O_2}$—and still the $P_{O_2}$ of the tissue capillaries does not vary more than a few millimeters from normal.

**Buffering of Tissue $P_{O_2}$ by the Hemoglobin Mechanism when Pulmonary Ventilation Changes.** When pulmonary ventilation decreases to approximately one-half normal, arterial $P_{O_2}$ falls to 50 to 60 mm. Hg, thereby decreasing the arterial oxygen saturation about 15 per cent. Then as the blood passes through the peripheral capillaries, the saturation falls another 25 per cent—down to a final level of about 60 per cent of normal. At this saturation, the $P_{O_2}$ of the oxygen delivered from the hemoglobin to the tissues is approximately 35 mm. Hg, only 5 mm. Hg less than the normal level.

Conversely, when a person breathes even to the maximal extent, he still cannot saturate his hemoglobin beyond the 100 per cent mark. Therefore, oxygen is delivered to the tissues still at a $P_{O_2}$ of 40 mm. Hg. Thus, it is clear that between the limits of one-half normal pulmonary ventilation and maximal ventilation the $P_{O_2}$ in the tissues remains very nearly constant.

## FACTORS THAT CAUSE THE HEMOGLOBIN DISSOCIATION CURVE TO SHIFT

The hemoglobin dissociation curves of Figures 41–8 and 41–9 are those for normal average blood. However, a number of different factors can displace the hemoglobin dissociation curve in one direction or the other in the manner illustrated in Figure 41–10. This figure shows that when the blood becomes slightly acidic, with the pH decreased from the normal value of 7.4 to 7.2, the hemoglobin dissociation curve shifts on the average about 15 per cent to the right. On the other hand, an increase in the pH to 7.6 shifts the curve a similar amount to the left.

In addition to pH changes, several other factors are also known to shift the curve. Three of these, all of which shift the curve to the right, are (1) increased carbon dioxide concentration, (2) increased blood temperature, and (3) increased 2,3-diphosphoglycerate, a phosphate compound normally present in the blood but in differing concentrations under different conditions.

A factor that shifts the dissociation curve to the left is the presence of large quantities of *fetal hemoglobin*, a type of hemoglobin present in the fetus before birth and which is different from the normal hemoglobin called *adult hemoglobin*. Shift of the curve to the left when fetal hemoglobin is present is important for oxygen delivery to the fetal tissues under the hypoxic conditions in which the fetus exists. This will be discussed further in Chapter 83.

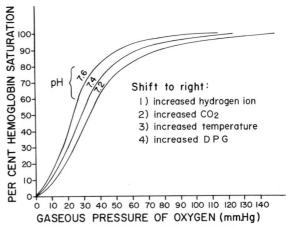

**Figure 41–10.** Shift of the oxygen-hemoglobin dissociation curve to the right by increases in (1) hydrogen ions, (2) $CO_2$, (3) temperature, or (4) DPG.

**Effect of Carbon Dioxide to Shift the Hemoglobin Dissociation Curve—the Bohr Effect.** Shift of the hemoglobin dissociation curve by changes in the blood $CO_2$ is important to enhance oxygenation of the blood in the lungs and also to enhance release of oxygen from the blood in the tissues. This is called the *Bohr effect,* and it can be explained as follows: As the blood passes through the lungs, carbon dioxide diffuses from the blood into the alveoli. This reduces the $P_{CO_2}$ and the pH of the blood; both of these effects shift the hemoglobin dissociation curve to the left and upward. Therefore, at any given oxygen pressure the quantity of oxygen that binds with the hemoglobin becomes considerably increased, thus allowing greater oxygen transport to the tissues. Then when the blood reaches the tissue capillaries, exactly opposite effects occur. Carbon dioxide entering the blood from the tissues displaces oxygen from the hemoglobin and therefore delivers oxygen to the tissues at a higher $P_{O_2}$ than would otherwise occur. That is, the dissociation curve shifts to the right in the tissues—exactly opposite to the leftward shift in the lungs.

(Most of the Bohr effect results from the increase in hydrogen ion concentration in the blood that occurs when the $P_{CO_2}$ increases. That is, the carbon dioxide combines with water to form carbonic acid, and the hydrogen ions from this acid are responsible for almost all of the dissociation curve shift.)

**Effect of 2,3-Diphosphoglycerate (DPG).** The normal DPG in the blood keeps the hemoglobin dissociation curve shifted slightly to the right all the time. However, in hypoxic conditions that last longer than a few hours, the quantity of DPG in the blood increases considerably, thus shifting the hemoglobin dissociation curve even further to the right. This causes oxygen to be released to the tissues at as much as 10 mm. Hg higher oxygen pressure than would be the case without this increased DPG. Therefore, students have been taught for the past few years that this might be an important mechanism for adaptation to hypoxia. However, more critical studies have shown that the presence of the excess DPG also makes it difficult for the blood to absorb oxygen in the lungs, thereby often creating as much harm as good.

**Shift of the Dissociation Curve During Exercise.** In exercise, several different factors shift the dissociation curve considerably to the right. The exercising muscles release large quantities of carbon dioxide and this plus acids released by the exercising muscles increases the hydrogen ion concentration in the muscle capillary blood. In addition, the temperature of the muscle often rises as much as 3 to 4 degrees, and, finally, phosphate compounds are also released. All of these factors acting together shift the hemoglobin dissociation curve *of the muscle capillary blood* considerably to the right. Therefore, oxygen can sometimes be released to the muscle at $P_{O_2}$'s as great as 40 mm. Hg (the normal resting value) even though as much as 75 to 85 per cent of the oxygen is removed from the hemoglobin. Then, in the lungs,

the shift proceeds most of the way back in the opposite direction, thus allowing pickup of normal amounts of oxygen from the alveoli.

## TOTAL RATE OF OXYGEN TRANSPORT FROM THE LUNGS TO THE TISSUES

If under resting conditions about 5 ml. of oxygen is transported by each 100 ml. of blood, and if the normal cardiac output is approximately 5000 ml. per minute, the calculated total quantity of oxygen delivered to the tissues each minute is about 250 ml. This is also the amount measured by a respirometer.

This rate of oxygen transport to the tissues can be increased to about 15 times normal in heavy exercise and in other instances of excessive need for oxygen (and very rarely to as high as 20 times normal in the best trained athletes). Oxygen transport can be increased to three times normal simply by an increase in utilization coefficient, and it can be increased another five-fold as a result of increased cardiac output, thus accounting for the total 15-fold increase. Therefore, the maximum rate of oxygen transport to the tissues is about $15 \times 250$ ml., or 3750 ml., per minute in the normal young adult. Special adaptations in athletic training, such as an increase in blood hemoglobin concentration and an increase in maximum cardiac output, can sometimes increase this value to as high as 4.5 to 5 liters per minute.

**Effect of Hematocrit on Oxygen Transport to the Tissues.** An increase in the blood hematocrit much above the normal level of 40 reduces the cardiac output because of increased blood viscosity, and the percentage reduction in the cardiac output is often more than the percentage increase in oxygen-carrying capacity of the blood. Since the total amount of oxygen that can be carried to the tissues each minute is the product of these two, the rate of oxygen transport is actually *reduced* by an excessive rise in hematocrit.

On the other hand, in anemia the oxygen-carrying capacity of the blood is reduced in proportion to the decrease in hematocrit; there is some compensatory increase in cardiac output, but this is not so great as the decrease in oxygen-carrying capacity. Here again, there is an overall reduction in the rate of oxygen transport to the tissues. The solid curve in Figure 41–11 illustrates this effect, showing that *maximum oxygen transport occurs at a hematocrit of about 40* and that *oxygen transport falls off at either high or low hematocrits.* Therefore, it is fortunate that the normal hematocrit of 40 is approximately optimal for oxygen transport.

The ''optimum hematocrit'' for oxygen transport rises when an animal develops a high hematocrit during acclimatization to high altitudes, as shown by the

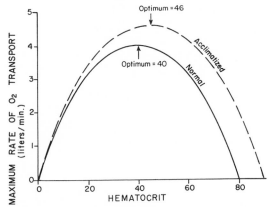

Figure 41–11. Depressant effect on oxygen transport of either an elevated or a decreased hematocrit. (Courtesy of Drs. J. W. Crowell and E. E. Smith.)

dashed curve in Figure 41–11. In this figure, the optimum hematocrit has risen from 40 to 46. The probable cause of this increase in optimal hematocrit is increase in the numbers and sizes of the peripheral small vessels, which prevents a decrease in cardiac output when the hematocrit rises. Note also in the figure that, in addition to the shift in optimum hematocrit, the maximum rate of oxygen transport at each hematocrit also increases because of the increased vascular dimensions.

## METABOLIC USE OF OXYGEN BY THE CELLS

**Relationship Between Intracellular $Po_2$ and Rate of Oxygen Usage.** Only a minute level of oxygen pressure is required in the cells for normal intracellular chemical reactions to take place. The reason for this is that the respiratory enzyme systems of the cell, which will be discussed in Chapter 67, are geared so that when the cellular $Po_2$ is more than 3 to 5 mm. Hg, oxygen availability is no longer a limiting factor in the rates of the chemical reactions. Instead, the main limiting factor then is the *concentration of adenosine diphosphate* (ADP) in the cells, as was explained in Chapter 3. This effect is illustrated in Figure 41–12, which shows the relationship between intracellular $Po_2$ and rate of oxygen usage. Note that whenever the intracellular $Po_2$ is above 3 to 5 mm. Hg the rate of oxygen usage becomes constant for any given concentration of ADP in the cell. On the other hand, when the ADP concentration is altered, the rate of oxygen usage changes in proportion to the change in ADP concentration.

It will be recalled from the discussion in Chapter 3 that when adenosine triphosphate

(ATP) is utilized in the cells to provide energy, it is converted into ADP. The increasing concentration of ADP, in turn, increases the metabolic usage of oxygen and the various nutrients that combine with oxygen to release energy. This energy is used to reform the ATP. Therefore, *under normal operating conditions the rate of oxygen utilization by the cells is controlled by the rate of energy expenditure within the cells—that is, by the rate at which ADP is formed from ATP—and not by the availability of oxygen to the cells.*

**Effect on Metabolic Use of Oxygen of the Diffusion Distance from the Capillary to the Cell.** Cells are rarely more than 50 microns away from a capillary, in which case oxygen can diffuse readily enough from the capillary to the cell to supply all the required amounts of oxygen for metabolism. However, occasionally, cells are located farther than this from the capillaries, and the rate of oxygen diffusion to these cells is so low that the intracellular $Po_2$ falls below the critical level of 3 to 5 mm. Hg required to maintain maximum intracellular metabolism. Thus, under these conditions, oxygen utilization by the cells is *diffusion limited* and is no longer determined by the amount of ADP formed in the cells.

**Effect of Blood Flow on Metabolic Use of Oxygen.** The total amount of oxygen available each minute for use in any given tissue is determined by (1) the quantity of oxygen transported in each 100 ml. of blood and (2) the rate

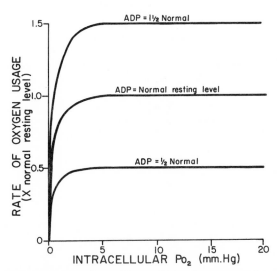

Figure 41–12. Effect of intracellular $Po_2$ on rate of oxygen usage by the cells. Note that increasing the intracellular concentration of adenosine diphosphate *(ADP)* increases the rate of oxygen usage.

of blood flow. If the rate of blood flow falls to zero, the amount of oxygen available for metabolism obviously also falls to zero. Thus, there are times when the rate of blood flow through a tissue can be so low that the tissue $Po_2$ falls below the critical 3 to 5 mm. Hg required for maximal intracellular metabolism. Under these conditions, the rate of tissue utilization of oxygen is *blood flow limited*.

## TRANSPORT OF OXYGEN IN THE DISSOLVED STATE

At the normal arterial $Po_2$ of 95 mm. Hg, approximately 0.29 ml. of oxygen is dissolved in every 100 ml. of water in the blood. When the $Po_2$ of the blood falls to 40 mm. Hg in the tissue capillaries, 0.12 ml. of oxygen remains dissolved. In other words, 0.17 ml. of oxygen is transported in the dissolved state to the tissues by each 100 ml. of blood water. This compares with about 5.0 ml. transported by the hemoglobin. Therefore, the amount of oxygen transported to the tissues in the dissolved state is normally slight, only about 3 per cent of the total as compared with the 97 per cent transported by the hemoglobin. During heavy exercise, when the utilization coefficient rises, the quantity transported in the dissolved state falls to as little as 1.5 per cent. In Figure 41–12 the quantitative relationships of the amount of oxygen that can be transported in the dissolved state versus that bound with hemoglobin can be seen. This figure shows that the total oxygen in the blood under normal conditions is accounted for almost entirely by that bound with hemoglobin, while only a minute portion is dissolved in the fluids. Yet under abnormal conditions the amount transported in the dissolved state can become tremendous as follows:

**Effect of Extremely High $Po_2$'s on Oxygen Transport.** When the $Po_2$ in the blood rises above 100 mm. Hg, the amount of oxygen dissolved in the water of the blood increases markedly. This effect is illustrated in Figure 41–13, which depicts the same oxygen-hemoglobin dissociation curve as in Figure 41–9 except that the $Po_2$'s in the lungs are extended to over 3000 mm. Hg. Also shown is volume of oxygen dissolved in the blood at each $Po_2$ level. Note that in the normal range of alveolar $Po_2$'s almost none of the total oxygen in the blood is accounted for by dissolved oxygen, but as the pressure rises progressively into the thousands of mm. Hg, a significant portion of the total oxygen is dissolved rather than bound with hemoglobin.

**Effect of High Pulmonary $Po_2$ on Tissue $Po_2$.** Let us assume that the $Po_2$ in the lungs is about 3000 mm. Hg (4 atmospheres pressure). Referring to Figure 41–13, one finds that this represents a total oxygen content in each 100 ml. of blood of about 29 volumes per cent, as illustrated by point A in the figure. As this blood passes through the tissue capillaries and the tissues utilize their normal amount of oxygen, about 5 ml. from each 100 ml. of blood, the

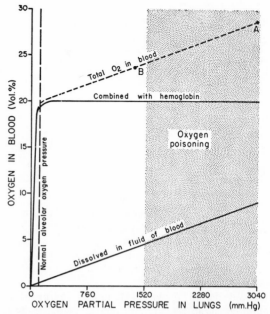

**Figure 41–13.** Quantity of oxygen dissolved in the water of the blood and in combination with hemoglobin at very high $Po_2$'s.

oxygen content on leaving the tissue capillaries is 24 volumes per cent (point B in the figure). At this point, the $Po_2$ in the blood leaving the tissue capillaries is still about 1200 mm. Hg, which means that oxygen is delivered to the tissues at this extremely high pressure instead of at the normal value of 40 mm. Hg. Thus, very high $Po_2$'s breathed by the lungs can also cause very high tissue $Po_2$'s, but only if these pressures are high enough for transport of excessive quantities of oxygen in the dissolved state.

**Oxygen Poisoning.** When the $Po_2$ in the tissues rises too high, it has a tendency to change many of the chemical reactions within the cells, sometimes to such an extent that it actually "poisons" the tissues. This will be discussed in Chapter 45 in relation to the effects of high gas pressures on the body.

**Experimental Treatment of Anemia with Hyperbaric Oxygen.** An obvious implication of this discussion is that a person with severe anemia whose blood could not carry sufficient quantities of oxygen to the tissues in combination with hemoglobin would be aided greatly by breathing pure oxygen or by being placed in a pressure chamber containing a high $Po_2$, for then a large quantity of oxygen could be carried to the tissues in the dissolved state rather than combined with hemoglobin. Indeed, in human beings treated in such pressure chambers (hyperbaric oxygen chambers), it has been possible to transport enough dissolved oxygen to supply all that is needed by the tissues. Yet, whenever more oxygen than that used by the tissues is supplied in the dissolved state, no oxygen is removed from the hemoglobin, and the oxygen buffer function of the hemoglobin is lost. In order for hemoglobin to control the level of oxygen in the tissues, at least several per cent of the hemoglobin must lose its oxygen. Then the oxygen-hemoglobin dissociation curve begins to operate along the steep portion of the curve, thus buffering the $P_{O_2}$ of the tissue fluids as explained above. Therefore, administration of oxygen under extreme pressure may lead to death, not because of failure to transport enough oxygen but because of failure of the oxygen buffer function of hemoglobin.

### COMBINATION OF HEMOGLOBIN WITH CARBON MONOXIDE

Carbon monoxide combines with hemoglobin at the same point on the hemoglobin molecule as does oxygen. Furthermore, it binds with 210 to 250 times as much tenacity as oxygen, which is illustrated by the carbon monoxide-hemoglobin dissociation curve in Figure 41–14. This curve is almost identical with the oxygen-hemoglobin dissociation curve, except that the pressures of the carbon monoxide shown on the abscissa are at a level 1/210 of those in the oxygen-hemoglobin dissociation curve of Figure 41–8. Therefore, a carbon monoxide pressure of only 0.5 mm. Hg in the alveoli, 1/210 that of the alveolar oxygen, allows the carbon monoxide to compete equally with the oxygen for combination with the hemoglobin and causes half the hemoglobin in the blood to become bound with carbon monoxide instead of with oxygen. A carbon monoxide pressure of 0.7 mm. Hg (a concentration of about 0.1 per cent) can be lethal.

A patient severely poisoned with carbon monoxide can be advantageously treated by administering pure oxygen, for oxygen at high alveolar pressures displaces carbon monoxide from its combination with hemoglobin far more rapidly than can oxygen at the low pressure of atmospheric oxygen.

The patient can also be benefited by simultaneous administration of a few per cent carbon dioxide because this strongly stimulates the respiratory center, as is discussed in the following chapter. This increases alveolar ventilation and reduces the alveolar

carbon monoxide concentration, which allows increased carbon monoxide release from the blood.

With intensive oxygen and carbon dioxide therapy, carbon monoxide can be removed from the blood 10 to 20 times as rapidly as without therapy.

## TRANSPORT OF CARBON DIOXIDE IN THE BLOOD

Transport of carbon dioxide is not nearly so great a problem as transport of oxygen, because even in the most abnormal conditions carbon dioxide can usually be transported by the blood in far greater quantities than can oxygen. However, the amount of carbon dioxide in the blood does have much to do with acid-base balance of the body fluids, which was discussed in detail in Chapter 37. Under normal resting conditions *an average of 4 ml. of carbon dioxide is transported from the tissues to the lungs in each 100 ml. of blood.*

### CHEMICAL FORMS IN WHICH CARBON DIOXIDE IS TRANSPORTED

To begin the process of carbon dioxide transport, carbon dioxide diffuses out of the tissue cells in the gaseous form (but not to a significant extent in the bicarbonate form because the cell membrane is far less permeable to bicarbonate than to the dissolved gas). On entering the capillary, the chemical reactions illustrated in Figure 41–15 occur immediately; the quantitative aspects of these can be described as follows:

**Transport of Carbon Dioxide in the Dissolved State.** A small portion of the carbon dioxide is transported in the dissolved state to the lungs. It will be recalled that the $P_{CO_2}$ of venous blood is 45 mm. Hg and that of arterial blood is 40 mm. Hg. The amount of carbon dioxide dissolved in the fluid of the blood at 45 mm. Hg is about 2.7 ml. per 100 ml. (2.7 volumes per cent). The amount dissolved at 40 mm. Hg is about 2.4 ml., or a difference of 0.3 ml. Therefore, only about 0.3 ml. of carbon dioxide is transported in the form of dissolved carbon dioxide by each 100 ml. of blood. This is about 7 per cent of all the carbon dioxide transported.

**Transport of Carbon Dioxide in the Form of Bicarbonate Ion.** *Reaction of Carbon Dioxide with Water in the Red Blood Cells—Effect of Carbonic Anhydrase.* The dissolved carbon dioxide in the blood reacts with water to form carbonic acid. However, this reaction would occur too slowly to be of importance were it not for the fact that inside the red blood cells the enzyme called *carbonic anhydrase* catalyzes the reaction between carbon dioxide and water,

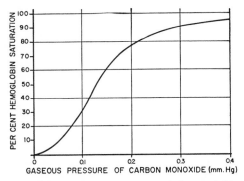

**Figure 41–14.** The carbon monoxide-hemoglobin dissociation curve.

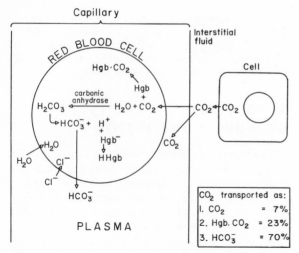

**Figure 41–15.** Transport of carbon dioxide in the blood.

accelerating its rate about 5000-fold. Therefore, instead of requiring many seconds to occur, as is true in the plasma, the reaction occurs so rapidly in the red blood cells that it reaches almost complete equilibrium within a fraction of a second. This allows tremendous amounts of carbon dioxide to react with the red cell water even before the blood leaves the tissue capillaries.

*Dissociation of Carbonic Acid into Bicarbonate Ions.* In another small fraction of a second the carbonic acid formed in the red cells dissociates into hydrogen and bicarbonate ions. This dissociation is about 99.9 per cent complete, so that only an infinitesimal fraction of the carbonic acid remains in the undissociated form.

As illustrated in Figure 41–15, most of the hydrogen ions formed inside the red blood cells react rapidly with hemoglobin, which is a powerful acid-base buffer. In reacting with hemoglobin the hydrogen ions are removed from the fluid and leave, dissolved in the fluid, a large quantity of *bicarbonate ions.* Many of these ions then diffuse into the plasma while chloride ions diffuse into the red cells to take their place. Thus, the chloride content of venous red blood cells is greater than that of arterial cells, a phenomenon called the *chloride shift.*

The reversible combination of carbon dioxide with water in the red blood cells under the influence of carbonic anhydrase accounts for an average of about 70 per cent of all the carbon dioxide transported from the tissues to the lungs. Thus, this means of transporting carbon dioxide is by far the most important of all the methods for transport. Indeed, when a carbonic

anhydrase inhibitor (acetazolamide) is administered to an animal to block the action of carbonic anhydrase in the red blood cells, carbon dioxide transport from the tissues becomes very poor—so poor that the tissue $P_{CO_2}$ rises to as high as 80 mm. Hg instead of the normal 45 mm. Hg.

**Transport of Carbon Dioxide in Combination with Hemoglobin and Plasma Proteins—Carbaminohemoglobin.** In addition to reacting with water, carbon dioxide also reacts directly with hemoglobin. The combination of carbon dioxide with hemoglobin is a reversible reaction that occurs with a very loose bond. The compound formed by this reaction is known as "carbaminohemoglobin." A small amount of carbon dioxide also reacts in this same way with the plasma proteins, but this is much less significant because of the one-fourth as great quantity of these proteins in the blood.

The *theoretical* quantity of carbon dioxide that can be carried from the tissues to the lungs in combination with hemoglobin and plasma proteins is approximately 30 per cent of the total quantity transported—that is, about 1.5 ml. of carbon dioxide in each 100 ml. of blood. However, this reaction is much slower than the reaction of carbon dioxide with water inside the red blood cells. Therefore, it is doubtful that this mechanism provides transport of more than 15 to 25 per cent of the total quantity of carbon dioxide.

## THE CARBON DIOXIDE DISSOCIATION CURVE

It is apparent that carbon dioxide can exist in blood in many different forms, (1) as free carbon dioxide and (2) in chemical combinations with water, hemoglobin, and plasma protein. The total quantity of carbon dioxide combined with the blood in all these forms depends on the $P_{CO_2}$. The curve shown in Figure 41–16 depicts this dependence of total blood $CO_2$ in all its

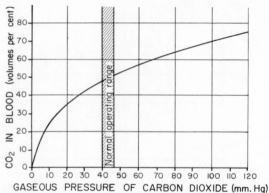

**Figure 41–16.** The carbon dioxide dissociation curve.

forms on $P_{CO_2}$; this curve is called the *carbon dioxide dissociation curve*.

Note that the normal blood $P_{CO_2}$ ranges between the limits of 40 mm. Hg in arterial blood and 45 mm. Hg in venous blood, which is a very narrow range. Note also that the normal concentration of carbon dioxide in the blood is about 50 volumes per cent but that only 4 volumes per cent of this is actually exchanged in the process of transporting carbon dioxide from the tissues to the lungs. That is, the concentration rises to about 52 volumes per cent as the blood passes through the tissues, and falls to about 48 volumes per cent as it passes through the lungs.

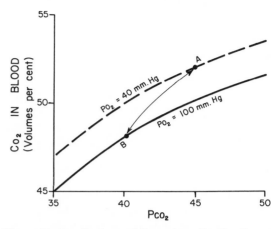

**Figure 41–17.** Portions of the carbon dioxide dissociation curves when the $P_{O_2}$ is 100 mm. Hg and 40 mm. Hg, respectively. The arrow represents the Haldane effect on the transport of carbon dioxide, as explained in the text.

## EFFECT OF THE OXYGEN-HEMOGLOBIN REACTION ON CARBON DIOXIDE TRANSPORT—THE HALDANE EFFECT

Earlier in the chapter it was pointed out that an increase in carbon dioxide in the blood will cause oxygen to be displaced from the hemoglobin and that this is an important factor in promoting oxygen transport. The reverse is also true: binding of oxygen with hemoglobin tends to displace carbon dioxide from the blood. Indeed, this effect, called the *Haldane effect,* is quantitatively far more important in promoting carbon dioxide transport than is the Bohr effect in promoting oxygen transport.

The Haldane effect results from the simple fact that combination of oxygen with hemoglobin causes the hemoglobin to become a stronger acid. This in turn displaces carbon dioxide from the blood in two ways: (1) The more highly acidic hemoglobin has less tendency to combine with carbon dioxide to form carbaminohemoglobin, thus displacing much of the carbon dioxide that is present in this form in the blood. (2) The increased acidity of the hemoglobin causes a general increase in the acidity of all the fluids both in the red blood cells and in the plasma. The increased hydrogen ions combine with the bicarbonate ions of the blood to form carbonic acid, which then dissociates and releases carbon dioxide from the blood.

Therefore, in the tissue the Haldane effect causes increased pickup of carbon dioxide because of oxygen removal from the hemoglobin, and in the lungs it causes increased release of carbon dioxide because of oxygen pickup by the hemoglobin.

Figure 41–17 illustrates the significance of the Haldane effect on the transport of carbon dioxide from the tissues to the lungs. This figure shows small portions of two separate carbon dioxide dissociation curves, the solid curve when the $P_{O_2}$ is 100 mm. Hg, which is the case in the lungs, and the dashed curve when the $P_{O_2}$ is 40 mm. Hg, which is the case in the tissues. Point A on the dashed curve shows that the normal $P_{CO_2}$ of 45 mm. Hg in the tissues causes 52

volumes per cent of carbon dioxide to combine with the blood. On entering the lungs the $P_{CO_2}$ falls to 40 mm. Hg while the $P_{O_2}$ rises to 100 mg. Hg. If the carbon dioxide dissociation curve did not change, the carbon dioxide content of the blood would fall only to 50 volumes per cent, which would be a loss of only 2 volumes per cent of carbon dioxide. However, the increase in $P_{O_2}$ in the lungs decreases the carbon dioxide dissociation curve from the dashed curve to the solid curve of the figure, so that the carbon dioxide content falls to 48 volumes per cent (point B). This represents an additional 2 volumes per cent loss of carbon dioxide. Thus, the Haldane effect approximately doubles the amount of carbon dioxide released from the blood in the lungs and approximately doubles the pickup of $P_{CO_2}$ in the tissues.

## CHANGE IN BLOOD ACIDITY DURING CARBON DIOXIDE TRANSPORT

The carbonic acid formed when carbon dioxide enters the blood in the tissues decreases the blood pH. Fortunately, though, the reaction of this acid with the buffers of the blood prevents the hydrogen ion concentration from rising greatly. Ordinarily, arterial blood has a pH of approximately 7.40, and, as the blood acquires carbon dioxide in the tissues, the pH falls to approximately 7.36. In other words, a pH change of 0.04 unit takes place. The reverse occurs when carbon dioxide is released from the blood in the lungs, the pH rising to the arterial value once again. In exercise, in other conditions of high metabolic activity, or when the blood flow through the tissues is sluggish, the decrease in pH in the tissue blood (and in the tissues themselves) can be as much as 0.5, or on occasion even more than this, thus causing tissue acidosis.

## RELEASE OF CARBON DIOXIDE IN THE LUNGS

In the lungs the $P_{CO_2}$ in the alveoli is slightly less than that of the blood, as explained earlier in the chapter, which causes carbon dioxide to diffuse from the blood into the alveoli. This decreases the $P_{CO_2}$ of the red blood cells so that the carbonic acid of the cells, under the influence of carbonic anhydrase, changes back into water and carbon dioxide, and carbaminohemoglobin also releases its carbon dioxide. In other words, exactly the reverse reactions occur in a split second in the lung capillaries as in the systemic capillaries, because all the reactions for transport of carbon dioxide are reversible.

## THE RESPIRATORY EXCHANGE RATIO

The discerning student will have noted that normal transport of oxygen from the lungs to the tissues by each 100 ml. of blood is about 5 ml., while normal transport of carbon dioxide from the tissues to the lungs is about 4 ml. Thus, under normal resting conditions only about 80 per cent as much carbon dioxide is expired from the lungs as there is oxygen uptake by the lungs. The ratio of carbon dioxide output to oxygen uptake is called the respiratory exchange ratio (R). That is,

$$R = \frac{\text{Rate of carbon dioxide output}}{\text{Rate of oxygen uptake}}$$

The value for R changes under different metabolic conditions. When a person is utilizing entirely carbohydrates for body metabolism, R rises to 1.00. On the other hand, when the person is utilizing fats almost entirely for his metabolic energy, the level falls to as low as 0.7. The reason for this difference is that when oxygen is metabolized with carbohydrates one molecule of carbon dioxide is formed for each molecule of oxygen consumed, while when oxygen reacts with fats a large share of the oxygen combines with hydrogen atoms to form water instead of carbon dioxide. Therefore, the quantity of carbon dioxide released from these reactions is far less than the quantity of oxygen consumed. In other words, the *respiratory quotient of the chemical reactions* in the tissues is now about 0.70 instead of 1.00 as is the case when carbohydrates are being utilized. The tis-

sue respiratory quotient will be discussed in Chapter 71.

For a person on a normal diet consuming average amounts of carbohydrates, fats, and proteins, the average value for R is considered to be 0.825.

## REFERENCES

Adamson, J. W., and Finch, C. A.: Hemoglobin function, oxygen affinity, and erythropoietin. *Ann. Rev. Physiol.*, 37:351, 1975.

Brighton, C. T., and Friedenberg, Z. B.: Electrical stimulation and oxygen tension. *Ann. N.Y. Acad. Sci.*, 238:314, 1974.

Bruley, D. F.: Mathematical considerations for oxygen transport to tissue. *Adv. Exp. Med. Biol.*, 37:749, 1973.

Chance, B.: Regulation of intracellular oxygen. *Proc. Int. Union of Physiol. Sci.*, 6:13, 1968.

Cherniack, N. S., and Longobardo, G. S.: Oxygen and carbon dioxide gas stores of the body. *Physiol. Rev.*, 50:196, 1970.

Crowell, J. W., and Smith, E. E.: Determinants of the optimal hematocrit. *J. Appl. Physiol.*, 22:501, 1967.

Dutton, P. L., and Wilson, D. F.: Redox potentiometry in mitochondrial and photosynthetic bioenergetics. *Biochim. Biophys. Acta*, 346:165, 1974.

Forster, R. E.: $CO_2$: chemical, biochemical, and physiological aspects. *Physiologist*, 13:398, 1970.

Forster, R. E.: Pulmonary ventilation and blood gas exchange. In Sodeman, W. A., Jr., and Sodeman, W. A. (eds.): Pathologic Physiology: Mechanisms of Disease, 5th Ed. Philadelphia, W. B. Saunders Company, 1974, p. 371.

Grodins, F. S., and Yamashiro, S. M.: Optimization of the mammalian respiratory gas transport system. *Ann. Rev. Biophys. Bioeng.*, 4:115, 1973.

Haldane, J. S., and Priestley, J. G.: Respiration. New Haven, Yale University Press, 1935.

Hayaishi, O.: Molecular Oxygen in Biology. New York, American Elsevier Publishing Company, 1974.

Jöbsis, F. F.: Intracellular metabolism of oxygen. *Amer. Rev. Respir. Dis.*, 110:58, 1974.

Jones, C. E., Crowell, J. W., and Smith, E. E.: Determination of mean tissue oxygen tensions by implanted perforated capsules. *J. Appl. Physiol.*, 26:630, 1969.

Kessler, M., Bruley, D. F., Clark, L. C., Lubbers, D. W., Silver, I. A., and Strauss, J. (eds.): Oxygen Supply. Baltimore, University Park Press, 1973.

Kilmartin, J. V., and Rossi-Bernardi, L.: Interaction of hemoglobin with hydrogen ions, carbon dioxide, and organic phosphates. *Physiol. Rev.*, 53:836, 1973.

Lehman, H., and Huntsman, R. G.: Man's Hemoglobins, 2nd Ed. Philadelphia, J. B. Lippincott Company, 1974.

Maren, T. H.: Carbonic anhydrase: chemistry, physiology, and inhibition. *Physiol. Rev.*, 47:595, 1967.

Michel, C. C.: The transport of oxygen and carbon dioxide by the blood. In Guyton, A. C. (ed.): MTP International Review of Science: Physiology. Vol. 2. Baltimore, University Park Press, 1974, p. 67.

Oski, F. A., and Gottlieb, A. J.: The interrelationships between red blood cell metabolites, hemoglobin, and the oxygen-equilibrium curve. *Prog. Hematol.*, 7:33, 1971.

Root, W. S.: Carbon monoxide. In Fenn, W. O., and Rahn, H. (eds.): Handbook of Physiology. Sec. 2, Vol. 2. Baltimore, The Williams & Wilkins Company, 1965, p. 1087.

Stainsby, W. N., and Barclay, J. K.: Oxygen uptake by striated muscle. *Muscle Biol.*, 1:273, 1972.

Wagner, P. D.: The oxyhemoglobin dissociation curve and pulmonary gas exchange. *Semin. Hematol.*, 11:405, 1974.

Whalen, W. J.: Intracellular $P_{O_2}$ in heart and skeletal muscle. *Physiologist*, 14:69, 1971.

Wittenberg, J. B.: Myoglobin-facilitated oxygen diffusion: role of myoglobin in oxygen entry into muscle. *Physiol. Rev.*, 50:559, 1970.

# 42

# Regulation of Respiration

The nervous system adjusts the rate of alveolar ventilation almost exactly to the demands of the body so that the blood oxygen pressure ($P_{O_2}$) and carbon dioxide pressure ($P_{CO_2}$) are hardly altered even during strenuous exercise or other types of respiratory stress.

The present chapter describes the operation of this neurogenic system for regulation of respiration.

## THE RESPIRATORY CENTER

The so-called "respiratory center" is a widely dispersed group of neurons located bilaterally in the reticular substance of the medulla oblongata and pons, as illustrated in Figure 42–1A. It is divided into three major areas: (1) the medullary rhythmicity area, (2) the apneustic area, and (3) the pneumotaxic area. The medullary rhythmicity area seems to be by far the most important. Therefore, it deserves particular attention.

**The Medullary Rhythmicity Area.** The medullary rhythmicity area, which is also frequently referred to simply as the *medullary respiratory center,* is located diffusely in the reticular substance of the medulla. Microelectrodes inserted into this center detect some neurons throughout the center that discharge during inspiration and others that discharge during expiration. There have been attempts to show that part of this center is primarily related to inspiration and part primarily related to expiration; however, most efforts in this direction have not been really successful. Thus, in general, *inspiratory neurons* and *expiratory neurons* intermingle in the medullary rhythmicity center.

When respiration increases, both inspiratory and expiratory neurons become excited far above normal, transmitting greatly increased numbers of inspiratory signals to the respiratory muscles during inspiration and greatly increased numbers of expiratory signals during expiration. Most of this chapter is concerned

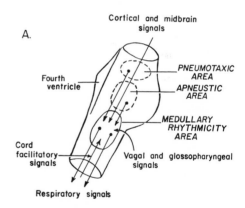

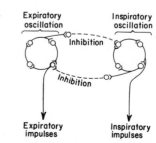

**Figure 42–1.** (A) The respiratory center located bilaterally in the lateral reticular substance of the medulla and lower pons. (B) Theoretical mechanism for the rhythmicity of the respiratory center.

with the factors that increase and decrease the degree of activity of these neurons.

*Basic Rhythmicity in the Medullary Rhythmicity Area.* It is in the medullary rhythmicity area that the basic rhythm of respiration is established. In the normal resting person, inspiration usually lasts for about two seconds and expiration for about three seconds. Both of these are correspondingly shortened during increased respiration and lengthened during decreased respiration.

When the medulla is transected immediately above the medullary rhythmicity area and also immediately below this area, one still finds a basic rhythmicity occurring in at least some of the inspiratory neurons. Therefore, it is certain that there is a basic oscillatory circuit located in this medullary rhythmicity area and that it is capable of causing repetitive inspiration and expiration. However, the medullary rhythmicity area is not capable by itself of giving a normal smooth pattern of respiration. If the medulla is transected immediately above the rhythmicity area but with this area still connected to the spinal cord, respiration occurs in gasps rather than in normal smooth inspiration and expiration. Furthermore, the rhythmical activity of the medullary rhythmicity center is very weak when afferent signals do not reach it from other sources. Thus, in Figure 42–1A, signals are shown entering the medullary rhythmicity area from the spinal cord, from the cerebral cortex and midbrain, from the pneumotaxic area in the upper pons, and from the apneustic area in the lower pons. All of these signals modify the rhythm of respiration and contribute to the normal smooth pattern of respiration.

**Function of the Apneustic and Pneumotaxic Areas.** The apneustic and pneumotaxic areas are located in the reticular substance of the pons. They are not necessary for maintenance of the basic rhythm of respiration. However, when the apneustic area is still connected to the medullary rhythmicity area but the pons has been transected between the apneustic and pneumotaxic areas, the animal breathes with the pattern of prolonged inspiration and very short expiration, which is exactly opposite to the pattern that occurs when breathing is accomplished entirely by the medullary rhythmicity area alone. This *apneustic pattern* becomes especially marked when the afferent nerve fibers of the vagus and of the glossopharyngeal nerves have been transected.

Finally, if the pneumotaxic area is also connected to the rhythmicity center, the pattern of respiration becomes essentially that normally observed, having a reasonable balance between inspiration and expiration. Also, stimulation of the pneumotaxic area can change the rate of respiration, which is the reason for its name, pneumotaxic area.

**Basic Mechanism of the Rhythmicity Observed in the Medullary Rhythmicity Area.** For years, physiologists have been attempting to explain the mechanism that causes the inspiratory neurons to become active during inspiration and the expiratory neurons to become active during expiration. Though the final answer to this is unknown, the following is a recent theory that is at least partially substantiated by experimental studies:

Figure 42–1B shows four expiratory neurons to the left and four inspiratory neurons to the right. Each of these neuronal groups is arranged in an oscillating network that functions in the following manner: Let us first consider the inspiratory neurons. If one of the neurons becomes excited, it excites the next one, which in turn excites the third one, and so forth. Thus, this excitation continues around and around the "inspiratory" circuit, and as it does so, collateral inspiratory impulses are transmitted to the inspiratory muscles. This process continues until the oscillation stops after about 2 seconds. The factor that stops the oscillation is perhaps fatigue of the neurons themselves, for we know that neurons cannot transmit large numbers of impulses in rapid succession without decreasing their excitability.

In addition to transmitting impulses to the inspiratory muscles, the inspiratory neurons also transmit *inhibitory* impulses into the expiratory network. This keeps the expiratory network from oscillating so long as the inspiratory network is oscillating. But just as soon as inspiratory oscillation ceases, there is no longer inhibition of the expiratory network. In the absence of this inhibition, the natural excitability of the expiratory neurons causes the expiratory network to begin oscillating. It then oscillates for about 3 seconds until its neurons likewise fatigue. During this oscillation, expiratory impulses are transmitted to the expiratory muscles, and *inhibitory* impulses are transmitted into the inspiratory network to keep it from oscillating. However, when the expiratory network has fatigued and stopped oscillating, the inhibitory impulses to the inspiratory network cease, and the inspiratory network begins to oscillate once again.

In summary, two oscillating circuits are postulated: one for inspiration and one for expiration. However, the two circuits cannot oscillate simultaneously because they mutually inhibit each other. Thus, when the inspiratory neurons are active the expiratory neurons are inactive, and the reverse occurs when the expiratory neurons become active. Therefore, alternation occurs back and forth between inspiratory signals and expiratory signals, this process continuing indefinitely and causing the act of respiration.

### Effect on the "Respiratory Center" of Nervous Signals from Other Parts of the Nervous System

All of us are familiar with the fact that respiration is altered (a) when a person speaks, (b) when he suddenly enters a cold shower, (c) when he is excited by a pin prick, or (d) when almost any other stressful condition suddenly occurs. The respective changes in respiration result from nerve impulses arriving in the respiratory center from other parts of the nervous system. Several special types of nervous activity that play major roles in control of respiration are described here, and still another is described later in the chapter in relation to the regulation of respiration during muscular exercise.

**Facilitatory Effect of Impulses from the Spinal Cord.** When the spinal cord is cut shortly below the medulla, the respiratory center becomes only weakly active. Yet stimulation of the cut fiber tracts leading from the transected cord to the medulla will re-establish normal respiratory rhythmicity. Therefore, impulses from the spinal cord play a facilitatory role in keeping the respiratory center active. For this reason, it is believed that the overall degree of stimulation of different peripheral sensory receptors throughout the body play an important role in maintaining normal respiration. It is interesting that when a person has a depressed respiratory center, depressed to the point that respiration has ceased, almost any peripheral stimulus, such as slapping the skin or applying cold water to the skin, can often start respiration once again.

**The Hering-Breuer Reflexes—The Inflation Reflex and the Deflation Reflex.** In the lungs are many receptors that detect stretch. These are located mainly in the bronchi and bronchioles. When the lungs become stretched, the stretch receptors transmit impulses through the vagus nerves into the *tractus solitarius* of the brain stem and thence into the respiratory center where they inhibit inspiration and thereby prevent further inflation. The effect is called the *Hering-Breuer inflation reflex*. This reflex prevents overdistention of the lungs.

A *Hering-Breuer deflation reflex* has also been described: that is, as the stretch receptors become unstretched during expiration the impulses from these receptors cease, which allows inspiration to begin again. Also, it is postulated that compression receptors (not well defined as yet but possibly in the alveolar septa) transmit impulses that inhibit expiration in the same manner that the stretch receptors inhibit inspiration. Regardless of the precise mechanism, the degree of lung deflation is reduced by this reflex. However, this deflation reflex is normally much less active than the inflation reflex if it exists at all.

The major effects of the Hering-Breuer reflexes are (1) a decrease in tidal volume and (2) a compensatory increase in respiratory rate. These effects result as follows: The inspiratory phase of respiration is cut off at an earlier point than would occur were it not for the inflation reflex. Likewise the expiratory phase is perhaps shortened by the deflation reflex. Thus, the depth of respiration is greatly decreased, but there is a compensatory increase in respiratory rate.

**Influence of the Vasomotor Center on Respiration.** Most factors that increase the activity of the vasomotor center, thereby increasing blood pressure, also increase the pulmonary ventilatory rate. Perhaps the vasomotor center itself excites the respiratory center, or it is possible—if not probable—that the same factors excite both the vasomotor center and the respiratory center at the same time—particularly since these centers are actually intermingled among each other in the reticular substance of the brain stem. For instance, inhibition of the baroreceptors increases the rate of pulmonary ventilation as well as increases vasomotor activity. Therefore, in many if not most instances, parallel degrees of activity occur in the vasomotor and respiratory centers at the same time.

**Effect of Body Heat on Respiratory Regulation—Panting.** Lower animals have very little ability to lose heat from the surfaces of their bodies for two reasons: First, the surfaces of their bodies are usually covered with fur, which prevents rapid loss of heat, and, second, most of the skin of lower animals is not supplied with sweat glands, which prevents evaporative loss of heat from the body surfaces. Therefore, a substitute mechanism, that of *panting,* is used by many lower animals as a means for dissipating heat.

The phenomenon of panting is initiated by the hypothalamic thermoregulator centers, which will be discussed in detail in Chapter 72. In essence, when the blood becomes overheated, the hypothalamus initiates a series of neurogenic reactions to decrease the body temperature. One of these in lower animals is to initiate panting. The actual panting process is then controlled by centers closely related to the pneumotaxic center of the pons.

When an animal pants, he breathes in and out rapidly so that large quantities of new air from the exterior come in contact with all the respiratory passages, thus cooling these passages by water evaporation from their surfaces. Yet, panting does not greatly increase the alveolar ventilation because each breath is extremely shallow, so that most of the air that enters the alveoli is dead space air. This prevents excess loss of carbon dioxide from the blood and thereby prevents serious respiratory alkalosis.

# HUMORAL REGULATION OF RESPIRATION

The ultimate goal of respiration is to maintain proper concentrations of oxygen, carbon dioxide, and hydrogen ions in the body fluids. It is fortunate, therefore, that respiratory activity is highly responsive to even slight changes in any one of these in the fluids. Therefore, when one speaks of *humoral* regulation of respiration, he is referring primarily to the regulation of respiratory activity by changes in concentrations of oxygen, carbon dioxide, or hydrogen ions in the body fluids.

Carbon dioxide and hydrogen ions control respiration primarily by acting directly on the respiratory center in the brain; both of these excite respiratory activity. On the other hand, a *decrease* in oxygen concentration excites respiratory activity by acting on peripheral chemoreceptors located in the carotid and aortic bodies rather than on the respiratory center itself. The chemoreceptors in turn transmit signals to the brain to excite the respiratory activity.

## QUANTITATIVE EFFECTS OF CARBON DIOXIDE, HYDROGEN ION, AND OXYGEN CONCENTRATIONS ON ALVEOLAR VENTILATION

For the moment, without discussing the precise mechanisms of control, let us first see the quantitative effect different degrees of change in blood carbon dioxide, hydrogen ion, or oxygen have on alveolar ventilation. Figure 42–2 illustrates the effect of each of these factors separately. In this figure the $P_{CO_2}$, $P_{O_2}$, or pH is

changed through the maximum range that occurs in the body fluids; each of the curves of this figure was derived in experiments in which *only one factor at a time* was changed while the other two factors were kept absolutely constant at normal values.

Note that an *increase* in carbon dioxide concentration or an *increase* in hydrogen ion concentration (decrease in pH) increases alveolar ventilation, while on the other hand a *decrease* in oxygen concentration increases alveolar ventilation. Furthermore, each of these factors can increase alveolar ventilation almost as much as any one of the others. But, *in the usual range of arterial oxygen concentrations*—between $P_{O_2}$'s of 80 mm. Hg and 140 mm. Hg—there is little effect of oxygen on alveolar ventilation, and a major oxygen effect occurs only at very low oxygen pressures. On the other hand, even in the normal range of both $P_{CO_2}$ and pH, there is a strong effect of each on alveolar ventilation. For this reason, as will be explained later in the chapter, normal respiration is controlled almost entirely by $P_{CO_2}$ and pH, though in many abnormal states of the respiratory system alveolar ventilation is controlled to a major extent, and sometimes almost entirely, by $P_{O_2}$ rather than by the other two factors.

However, remember that the curves of Figure 42–2 were all derived when *only a single factor was changed* by itself and the other two factors were artificially controlled so that they could not change from their normal values. When the other two factors are *not* artificially controlled, the results are markedly different,

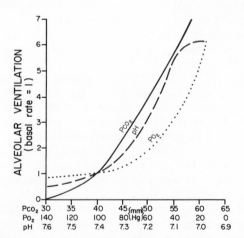

**Figure 42–2.** Approximate effects on alveolar ventilation of changing the concentrations of carbon dioxide, hydrogen ions, and oxygen in the arterial blood when only one of the humoral factors is changed at a time and the other two are maintained at absolutely normal levels.

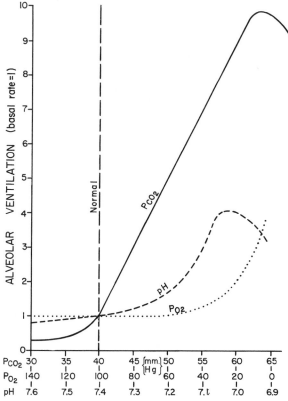

**Figure 42–3.** Effects of increased arterial $P_{CO_2}$, decreased arterial $P_{O_2}$, and decreased arterial pH on the rate of alveolar ventilation.

as illustrated in Figure 42–3. One sees that the effect of $P_{CO_2}$ on alveolar ventilation is now tremendous, the effect of pH is intermediate, and the effect of $P_{O_2}$ is slight. The purpose of the following few sections, therefore, will be to explain these differences.

## CARBON DIOXIDE, THE MAJOR CHEMICAL FACTOR REGULATING ALVEOLAR VENTILATION

Note in Figure 42–3 that an increase in alveolar $P_{CO_2}$ from the normal level of 40 mm. Hg to 63 mm. Hg causes a 10-fold increase in alveolar ventilation; this is about one and a half times as great as that which occurs when both pH and $P_{O_2}$ are artificially kept at normal values. The cause of this difference is that an increase in $P_{CO_2}$ causes an almost proportional increase in hydrogen ion concentration in all the body fluids, as was explained in Chapter 37. Therefore, an increase in $P_{CO_2}$ stimulates alveolar ventilation not only directly but also indirectly

through its effect on hydrogen ion concentration.

In contrast to the enhanced effect of changes in $P_{CO_2}$ on ventilation when the $P_{CO_2}$ rises above normal, the effect of $P_{CO_2}$ changes on ventilation at low $P_{CO_2}$'s is *decreased*. The reason for this is that the decreased carbon dioxide and hydrogen ion concentrations reduce ventilation and cause the alveolar and arterial $P_{O_2}$ to fall to critically low values. These low $P_{O_2}$'s then have a direct stimulatory effect on the peripheral chemoreceptors to enhance alveolar ventilation, thus acting as a brake on the depressing effect of low $P_{CO_2}$. Consequently, instead of ventilation falling to zero, this hypoxic drive usually maintains a minimum level of respiration.

**Value of Carbon Dioxide as a Regulator of Alveolar Ventilation.** Since carbon dioxide is one of the end-products of metabolism, its concentration in the body fluids greatly affects the chemical reactions of the cells and also affects the tissue pH. For these reasons, the tissue fluid $P_{CO_2}$ must be regulated exactly. In the preceding chapter, it was pointed out that blood and interstitial fluid $P_{CO_2}$ are determined to a great extent by the rate of alveolar ventilation. Therefore, stimulation of the respiratory center by carbon dioxide provides an important feedback mechanism for regulation of the concentration of carbon dioxide throughout the body. That is, (1) an increase in $P_{CO_2}$ stimulates the respiratory center; (2) this increases alveolar ventilation and reduces the alveolar carbon dioxide; (3) as a result, the tissue $P_{CO_2}$ returns most of the way back toward normal. In this way, the respiratory center maintains the $P_{CO_2}$ of the tissue fluids at a relatively constant level and, therefore, might well be called a "carbon dioxide pressostat."

**Effect on Respiration Caused by Breathing Progressively Increasing Concentrations of Carbon Dioxide.** When one breathes air containing carbon dioxide, the alveolar and tissue $P_{CO_2}$'s rise above normal, and as a consequence alveolar ventilation increases. The increased ventilation, in turn, keeps the alveolar $P_{CO_2}$ from rising as high as it otherwise would. Therefore, 1 to 2 per cent carbon dioxide in the inspired air hardly changes the carbon dioxide concentration of the tissue fluids, but as the carbon dioxide concentration rises above 5 per cent, even tremendous alveolar ventilation will not keep the tissue fluid $P_{CO_2}$ from rising excessively. Maximum ventilation is reached at approximately 9 per cent carbon dioxide. Beyond this level, a further increase in $P_{CO_2}$ begins to depress the respiratory center, causing progressive reduction in respiratory activity rather than a further increase. The person begins to pass into coma at 15 to 20 per cent carbon dioxide, becomes completely anesthetized at 30 to 40 per cent, and dies at 40 to 50 per cent—indeed, at much lower values if the exposure is longer than a few minutes.

## CONTROL OF ALVEOLAR VENTILATION BY THE HYDROGEN ION CONCENTRATION OF THE EXTRACELLULAR FLUIDS

Figure 42–3 illustrates that changes in hydrogen ion concentration (pH) have considerably *less* effect on alveolar ventilation when the other humoral factors are not controlled than when they are controlled. This is opposite to the uncontrolled effect of $P_{CO_2}$. The reason for this is that an increase in hydrogen ion concentration causes the person to breathe more rapidly than usual, which decreases the carbon dioxide concentration in the body fluids and at the same time increases the oxygen concentration. The decreased carbon dioxide *inhibits* the respiratory center and the increased oxygen *inhibits* the chemoreceptors. Therefore, opposing effects are caused by these other two factors, thus causing a "braking" effect on the hydrogen ion stimulation of respiration.

Nevertheless, despite the "braking" effect of the $P_{CO_2}$ and $P_{O_2}$ changes, a decrease in pH to approximately 7.1 can increase alveolar ventilation about four-fold, and an increase in pH to 7.6 can decrease ventilation to about 80 per cent of normal.

## CONTROL OF ALVEOLAR VENTILATION BY ARTERIAL OXYGEN SATURATION

**Unimportance of Oxygen Regulation of Respiration Under Normal Conditions—The "Braking" Effect of the $P_{CO_2}$ and pH Regulatory Mechanisms.** Figure 42–3 illustrates that changes in alveolar $P_{O_2}$ have extremely little effect on alveolar ventilation when $P_{CO_2}$ and pH are not controlled. This is in marked contrast to the effect when these other two factors are controlled, as was illustrated in Figure 42–2. This lack of importance of oxygen in normal regulation of respiration is caused by two effects, as follows:

First, even when both $P_{CO_2}$ and pH are controlled, the arterial $P_{O_2}$ must fall below 70 mm. Hg before a marked effect on alveolar ventilation occurs, because arterial blood does not develop any significant degree of desaturation

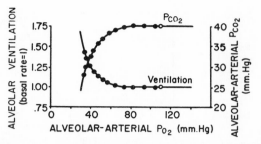

**Figure 42–4.** Effect of arterial $P_{O_2}$ on alveolar ventilation and on the subsequent decrease in arterial $P_{CO_2}$. (From Gray: Pulmonary Ventilation and Its Physiological Regulation. Charles C Thomas.)

until the $P_{O_2}$ falls below 70. Since it is the failure of blood to *deliver* oxygen to the chemoreceptors that stimulates them, one can readily see that decreasing the alveolar $P_{O_2}$ from 100 to 70, in which range the saturation of the arterial blood hardly changes, has little effect in stimulating the chemoreceptors, as was illustrated in Figure 42–2.

The second cause of the poor effect of $P_{O_2}$ changes on respiratory control is a "braking" effect caused by *both* the carbon dioxide and the hydrogen ion control mechanisms. The cause of this is illustrated in Figure 42–4, which shows a slight increase in ventilation as the alveolar $P_{O_2}$ is decreased. The increase in ventilation blows off carbon dioxide from the blood and therefore decreases the $P_{CO_2}$, which is also illustrated in the figure; at the same time it also decreases the hydrogen ion concentration. Therefore, two powerful respiratory inhibitory effects are caused by (a) diminished carbon dioxide and (b) diminished hydrogen ions. These two exert an inhibitory braking effect that opposes the excitatory effect of the diminished oxygen. As a result, they keep the decreased oxygen from causing a marked increase in ventilation until the $P_{O_2}$ falls to 20 to 40 mm. Hg, a range that is incompatible with life for more than a few minutes. Therefore, the maximum effect of decreased alveolar oxygen on alveolar ventilation, in the range compatible with life, is normally only a 66 per cent increase. This is in contrast to about a 400 per cent increase caused by decreased pH and about a 1000 per cent increase caused by increased carbon dioxide.

Thus, one can see that for the control of the usual normal respiration the $P_{CO_2}$ and pH feedback control mechanisms are extremely powerful in relation to the $P_{O_2}$ feedback control of respiration. Indeed, under normal conditions the $P_{O_2}$ mechanism is of almost no significance in the control of respiration.

Yet, under some abnormal conditions the $P_{CO_2}$ and hydrogen ion concentrations *increase at the same time that the arterial $P_{O_2}$ decreases.* Under these conditions, all three of the feedback mechanisms support each other, and the $P_{O_2}$ mechanism then exerts its full share of respiratory stimulation, sometimes becoming even more potent as a controller of respiration than are the $P_{CO_2}$ and hydrogen ion mechanisms.

***Reason Oxygen Regulation of Respiration Is Not Normally Needed.*** On first thought, it seems strange that oxygen should play so little role in the normal regulation of respiration, particularly since one of the primary functions of the respiratory center is to provide adequate intake of oxygen. However, oxygen control of respiration is not needed under most normal circumstances for the following reason:

The respiratory system ordinarily maintains an alveolar $P_{O_2}$ high enough to saturate either fully or almost fully the hemoglobin of the arterial blood. It does not matter whether alveolar ventilation is normal or 10 times normal, the blood will still be essentially fully saturated. Also, alveolar ventilation can decrease to as low as one-half normal, and the blood still remains within 10 per cent of complete saturation. Therefore, one can see that alveolar ventilation can change tremendously without significantly affecting oxygen transport to the tissues. Instead, the factors that play the major roles in controlling $P_{O_2}$ in the tissues under normal conditions are the hemoglobin-oxygen buffer system and the blood flow to the tissues, which were discussed in the preceding chapter.

On the other hand, changes in alveolar ventilation do have a tremendous effect on tissue carbon dioxide concentration, as was explained above and illustrated in Figure 42-4. Therefore, it is exceedingly important that carbon dioxide—not oxygen—be the major controller of respiration under normal conditions.

**Conditions Under Which Diminished Oxygen Plays a Major Role in the Regulation of Respiration.** *Pneumonia, Emphysema, Etc.* In peumonia, emphysema, and other lung ailments in which gases are not readily exchanged between the atmosphere and the pulmonary blood, the oxygen regulatory system *does* then play a major role in the regulation of respiration. Contrary to the normal effect, the increased ventilation caused by oxygen lack is not followed by reduced arterial $P_{CO_2}$ and hydrogen ion concentration because the pulmonary disease also diminishes carbon dioxide exchange as well as oxygen exchange. Instead, the $CO_2$ builds up in the blood, and this build-up in turn enhances the hydrogen ion concentration. Therefore, the "braking" effect of these other two control systems on the oxygen lack system is not present. As a result, the oxygen lack system develops its full power and can increase alveolar ventilation as much as five- to seven-fold.

***Effect of the Oxygen Lack Mechanism at High Altitudes.*** When a person first ascends to high altitudes or in any other way is exposed to a rarefied atmosphere, the diminished oxygen in the air stimulates his oxygen lack control system of respiration. The respiration at first increases to a maximum of about two-thirds above normal, which is a comparatively slight increase. Once again, the cause of this slight increase is the tremendous "braking" effect of the carbon dioxide and hydrogen ion control mechanisms on the oxygen lack mechanism.

However, over about a week, the respiratory center gradually becomes "adapted" to the diminished carbon dioxide and hydrogen ion concentrations in the blood so that these then depress the respiratory center very little. Thus, the "braking" effect on the oxygen control is gradually lost, and alveolar ventilation then rises to as high as five to seven times normal. This is part of the acclimatization that occurs as a person slowly ascends a mountain, thus allowing the person to adjust his respiration gradually to a level fitted for the higher altitude. The cause of the adaptation is not known, but a possible mechanism will be discussed later in the chapter.

## COMPOSITE EFFECTS OF $P_{CO_2}$, pH, AND $P_{O_2}$ ON RESPIRATORY ACTIVITY

Now that the effects of each of the individual humoral factors on respiratory activity has been discussed, we now need to see how all of them function together in controlling respiration. We have already seen that the different factors oppose each other in some instances but at other times support each other. One of the best attempts to show how these factors interact is illustrated in Figure 42-5, which shows two separate "families" of curves relating alveolar $P_{CO_2}$ to pulmonary ventilation. The "family" of curves represented by the solid lines is that determined in a person at different alveolar $P_{O_2}$'s from 40 to 100 mm. Hg and with the pH con-

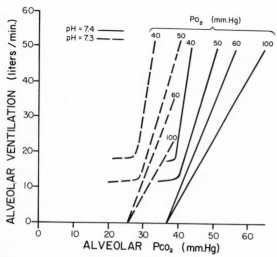

**Figure 42–5.** A composite diagram showing the interrelated effects of $P_{CO_2}$, $P_{O_2}$, and pH on alveolar ventilation. (Drawn from data presented in Cunningham and Lloyd: The Regulation of Human Respiration. F. A. Davis Co.)

stant at the normal level of 7.4. The dashed "family" of curves was determined for the same person after he had developed metabolic acidosis, with a shift of his blood pH from 7.4 to 7.3. Both these families of curves show that the *slope* of the curve relating $P_{CO_2}$ to pulmonary ventilation increases markedly as the $P_{O_2}$ decreases. To express this another way, a decrease in $P_{O_2}$ in the alveoli *multiplies* the effects on alveolar ventilation of carbon dioxide concentration *changes*.

Figure 42–5 shows also that acidosis shifts the entire family of curves to the left. This means that increased hydrogen ion concentration decreases the carbon dioxide concentration required to stimulate the respiratory center.

The curves in Figure 42–5 are of importance when one wishes to analyze respiratory control from a quantitative point of view because, in most instances in which respiratory control becomes altered, more than one of the individual humoral factors is changed at the same time.

# BASIC MECHANISMS OF HUMORAL STIMULATION OF THE RESPIRATORY CENTER

**Hydrogen Ion Concentration as the Basic Stimulus of the Respiratory Center.** It is believed that the basic stimulus for exciting the respiratory neurons is the hydrogen ions of the

fluids that bathe the neurons. It should also be recalled that any time the carbon dioxide concentration increases, this also increases the hydrogen ion concentration because carbon dioxide combines with water to form carbonic acid, the carbonic acid then dissociating into hydrogen ions and bicarbonate ions. Indeed, in the acute situation (but not in the chronic condition, as we shall see later) an increase in carbon dioxide concentration is accompanied by an almost exactly proportional increase in hydrogen ion concentration.

Therefore, either an increase in hydrogen ions in the respiratory center or an acute increase in carbon dioxide, which indirectly increases the hydrogen ions, will stimulate the respiratory neurons. Furthermore, this stimulation is known to occur in two different ways: (1) mainly by direct diffusion of carbon dioxide and hydrogen ions from the blood into the respiratory center, and (2) to much less extent by changes in the hydrogen ion concentration of the cerebrospinal fluid surrounding the brain stem. The first of these methods for stimulation of the respiratory neurons is easily understood, but stimulation by way of the cerebrospinal fluid exhibits some peculiar characteristics that deserve special explanation, as follows.

**Stimulation of Respiration by an Increase in Hydrogen Ion Concentration in the Cerebrospinal Fluid.** Increasing the hydrogen ion concentration of the cerebrospinal fluid at the level of the medulla will stimulate respiration; this effect has been further localized to two small bilateral areas of the medulla located near the entry of cranial nerves IX and X into the medulla. Hydrogen ions diffuse a fraction of a millimeter into the tissue of the medulla and directly affect the hydrogen ion concentration of the local medullary interstitial fluids. On the average, between 1/4 and 1/6 of the stimulation of the chemosensitive respiratory neurons is effected through this cerebrospinal fluid pathway, while 3/4 to 5/6 is effected by the direct pathway from the blood to the medullary interstitial fluid and thence to the respiratory neurons.

**Stimulation of Respiration by Blood Carbon Dioxide Acting Through the Cerebrospinal Fluid Pathway.** Even though it is hydrogen ions in the cerebrospinal fluid that stimulate the respiratory center, strangely enough, changes in carbon dioxide concentration in the blood have a much more rapid and also a greater effect on stimulation of the respiratory neurons by way of the CSF pathway than do changes in blood hydrogen ion concentration. This can be explained as follows: Hydrogen ions, because of their strong electrical charge, diffuse very poorly from the blood through the blood-CSF barrier into the cerebrospinal fluid. On the other hand, carbon dioxide diffuses with great ease, the diffusion occurring especially rapidly through the meningeal and arachnoidal vessel walls. On entering the CSF, the carbon dioxide combines with water to form carbonic acid, and, as elsewhere in the body, the carbonic acid dissociates to form hydrogen ions and bicarbonate ions. This dissociation obeys the following equation, which was discussed in detail in Chapter 37 in relation to acid-base equilibrium in the body fluids.

$$H^+ = K' \cdot \frac{CO_2}{HCO_3^-}$$

in which $K'$ is a constant. Thus, one can see that as long as the bicarbonate ion concentration of the cerebrospinal fluid remains constant, the hydrogen ion concentration is directly related to the blood carbon dioxide concentration. Therefore, every time the carbon dioxide concentration increases in the circulating blood, the hydrogen ion concentration of the cerebrospinal fluid increases almost immediately thereafter, and this in turn plays its usual role in stimulating respiration.

*Delayed Stimulation of Respiration by Blood Hydrogen Ion Concentration Acting Through the CSF Pathway.* Though a sudden increase in hydrogen ions in the blood caused by some blood acid other than carbonic acid will not immediately stimulate respiration through the CSF pathway, over a period of several hours the increased blood hydrogen ion concentration will finally increase CSF hydrogen ion concentration and eventually stimulate respiration. Even so, the increase in hydrogen ions in the CSF still is only about half that in the blood, so that this CSF mechanism never becomes a highly potent means for hydrogen ion control of respiration. This effect is in contrast with the direct stimulation of the respiratory center by hydrogen ions, which is a potent controlling influence.

One of the consequences of this delayed stimulation is that the effects of either acute metabolic acidosis or acute metabolic alkalosis on respiration become progressively more intense over a period of hours. In fact, the CSF pathway for respiratory control actually acts as a *brake* on respiratory changes at first, in the following way: When the blood first becomes acidic and the hydrogen ions reach the respiratory neurons by the direct interstitial pathway, the person begins to breathe more heavily; this reduces the blood carbon dioxide concentration. The reduction in carbon dioxide concentration has an instantaneous effect through the CSF pathway, reducing CSF hydrogen ion concentration rather than increasing it. Consequently, the CSF pathway acts as a brake on respiration rather than helping to increase it. Yet, after a number of hours, when the delayed stimulatory response of the CSF mechanism to the increase in blood hydrogen ions occurs, this brake disappears and the respiration increases markedly—

an effect that is slow but that finally is achieved. A similar but opposite slow response occurs in acute alkalosis.

**Progressive Diminishment of the Carbon Dioxide Drive on Respiration.** In contrast to the increasing stimulation of respiration by metabolic acid over a period of hours, the respiratory drive caused by increased carbon dioxide is greatest within the first few minutes after exposure and becomes progressively less during the next few hours to a day or so. This effect can be ascribed to several different factors as follows:

The immediate effect of increased carbon dioxide is to increase hydrogen ion concentration in the blood and also in both the interstitial fluids and the cerebrospinal fluid so that respiration is strongly stimulated. However, the kidneys begin immediately to compensate for this respiratory acidotic condition. They do this by excreting hydrogen ions while retaining bicarbonate ions, as explained in Chapter 37. The progressive buildup of bicarbonate ions decreases the hydrogen ion concentration of the body fluids in accordance with the equation given above. Indeed, this occurs even though the carbon dioxide concentration remains elevated. Thus, over a period of time the hydrogen ion concentration in the interstitial fluids bathing the respiratory neurons falls back toward normal and respiratory activity also decreases back toward normal. Then over a period of several more hours the bicarbonate ion concentration of the CSF also becomes elevated, and this too reduces the carbon dioxide drive still further.

Thus, one can see that an acute increase in carbon dioxide causes a marked increase in respiration at first; there is then progressively less respiratory drive during the ensuing hours to a day or more.

**Increasing Respiratory Drive of Hypoxia Caused by Adaptation of the Carbon Dioxide Stimulatory Mechanism.** In earlier discussions of this chapter it has already been pointed out that acute hypoxia does not increase respiration to a great extent. The reason for this is that even a slight increase in respiration resulting from hypoxia causes carbon dioxide to be blown off from the blood, and the decreased blood carbon dioxide inhibits respiration. However, during the next few hours to a day or so, most of this inhibitory effect caused by the carbon dioxide mechanism wears off, as described above. Consequently, since it is much less opposed by the carbon dioxide mechanism, the hypoxia now increases alveolar ventilation markedly. For example, hypoxia, as experienced by mountain climbers, at first increases respiration no more than 50 to 60 per cent; within a day or so, however, it gradually increases respiration as much as 600 to 700 per cent—a tremendously important effect in the acclimatization process. This change in responsiveness does not at all seem to be attributable to enhanced excitability to oxygen but instead entirely to the decreased effect of the carbon dioxide braking mechanism.

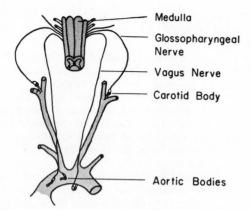

Figure 42–6.   Respiratory control by the carotid and aortic bodies.

# THE CHEMORECEPTOR SYSTEM FOR CONTROL OF RESPIRATORY ACTIVITY— ROLE OF OXYGEN IN RESPIRATORY CONTROL

Aside from the direct sensitivity of the respiratory center itself to humoral factors, special chemical receptors called *chemoreceptors*, located outside the central nervous system are responsive to changes in oxygen, carbon dioxide, and hydrogen ion concentrations. These transmit signals to the respiratory center to help regulate respiratory activity. These chemoreceptors are located primarily in association with the large arteries of the thorax and neck; most of them are in the *carotid* and *aortic bodies,* which are illustrated in Figure 42–6 along with their afferent nerve connections to the respiratory center. The *carotid bodies* are located bilaterally in the bifurcations of the common carotid arteries, and their afferent nerve fibers pass through Hering's nerves to the glossopharyngeal nerves and thence to the medulla. The *aortic bodies* are located along the arch of the aorta; their afferent nerve fibers pass to the medulla through the vagi. Each of these bodies receives a special blood supply through a minute artery directly from the adjacent arterial trunk

**Stimulation of the Chemoreceptors by Decreased Arterial Oxygen.**   Changes in arterial oxygen concentration have *no* direct stimulatory effect on the respiratory center itself, but when the oxygen concentration in the arterial blood falls below normal, the chemoreceptors become strongly stimulated. This effect is illustrated in Figure 42–7, which shows the relation-ship between arterial $P_{O_2}$ and rate of nerve impulse transmission from a carotid body. Note that the impulse rate is particularly sensitive to changes in arterial $P_{O_2}$ in the range between 60 and 30 mm. Hg, which is the range in which the arterial hemoglobin saturation with oxygen decreases rapidly.

**Effect of Carbon Dioxide and Hydrogen Ion Concentration on Chemoreceptor Activity.**   An increase in either carbon dioxide concentration or hydrogen ion concentration also excites the chemoreceptors and in this way indirectly increases respiratory activity. However, the direct effects of both of these factors in the respiratory center itself are so much more powerful than their effects mediated through the chemoreceptors that for most practical purposes one can disregard the indirect effects through the chemoreceptors. In the case of oxygen, on the other hand, this is not true, because diminished oxygen in the arterial blood can affect the respiration significantly *only* by acting through the chemoreceptors.

**Basic Mechanism of Stimulation of the Chemoreceptors by Oxygen Deficiency.**   The blood flow through the carotid and aortic bodies is extremely high, the highest that has been found for any tissue in the body. Because of this, the A-V oxygen difference is only about one volume per cent, which means that the venous blood leaving the carotid bodies still has a $P_{O_2}$ nearly equal to that of the arterial blood. It also means that the $P_{O_2}$ of the tissues in the carotid and aortic bodies at all times remains also almost equal to that of the arterial blood. Therefore, it is the arterial $P_{O_2}$ that normally determines the degree of stimulation of the chemoreceptors. The curve illustrated in Figure 42–7 shows the normal relationship between the arterial $P_{O_2}$ and the degree of chemoreceptor stimulation.

Yet, in hypotension, particularly when the mean arterial pressure falls below 80 mm. Hg (and even more so when it falls below 60 mm. Hg), the blood flow through the carotid and aortic bodies then be-

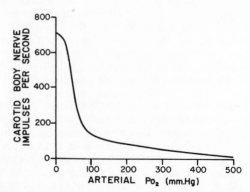

Figure 42–7.   Effect of arterial $P_{O_2}$ on impulse rate from the carotid body of a cat. (Curve drawn from data from several sources, but primarily from Von Euler.)

comes sluggish, and the tissue $P_{O_2}$ falls considerably below the arterial $P_{O_2}$. At these lower pressures, therefore, the chemoreceptors now do become stimulated in response to the hypotension even when the arterial $P_{O_2}$ is normal. This gives rise to reflexes that enhance respiration and that cause also peripheral vasoconstriction that increases the arterial pressure back toward normal.

# REGULATION OF RESPIRATION DURING EXERCISE

In strenuous exercise, oxygen utilization and carbon dioxide formation can increase as much as 20-fold (Fig. 42–8). Yet, except in very heavy exercise, alveolar ventilation ordinarily increases almost the same amount so that the blood $P_{O_2}$ and $P_{CO_2}$ remain almost exactly normal.

In trying to analyze the factors that cause increased ventilation during exercise, one is tempted immediately to ascribe this to the chemical alterations in the body fluids during exercise, including increase of carbon dioxide and hydrogen ions and decrease of oxygen. However, this is not valid, for measurements of arterial $P_{CO_2}$, pH, and $P_{O_2}$ show that none of these usually changes significantly and certainly not enough to account for the increase in ventilation. Indeed, even if a very high $P_{CO_2}$ should develop during exercise, this still would be sufficient to account for only two-thirds of the increased ventilation of heavy muscular exercise, for, as shown in Figure 42–9, the minute respiratory volume in exercise is about 50 per cent greater than that which can be effected by maximal carbon dioxide stimulation.

Therefore, the question must be asked: What

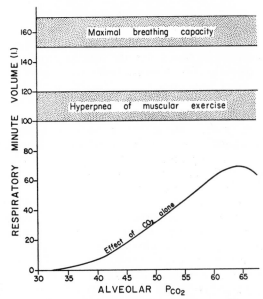

**Figure 42–9.** Relationship of hyperpnea caused by muscular exercise to that caused by increased alveolar $P_{CO_2}$. (Modified from Comroe: The Lung: Clinical Physiology and Pulmonary Function Tests. Year Book Publishers, 1962.)

is it during exercise that causes the intense respiration? This question has not been answered, but at least two different effects seem to be predominantly concerned as follows:

1. The cerebral cortex, on transmitting impulses to the contracting muscles, is believed to transmit collateral impulses into the reticular substance of the brain stem to excite the respiratory center. This is analogous to the stimulatory effect that causes the arterial pressure to rise during exercise when similar collateral impulses pass to the vasomotor center.

2. During exercise, the body movements, especially of the limbs, are believed to increase pulmonary ventilation by exciting joint proprioceptors that then transmit excitatory impulses to the respiratory center. The reason for believing this is that even passive movements of the limbs often increase pulmonary ventilation several-fold.

It is possible that still other factors are also important in regulating respiration during exercise. For instance, some experiments even suggest that hypoxia developing in the muscles during exercise elicits afferent nerve signals to the respiratory center to excite respiration. However, since the increase in ventilation begins immediately upon the initiation of exercise, most of the increase in respiration probably results from the two neurogenic factors noted

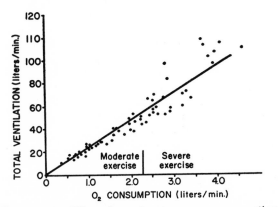

**Figure 42–8.** Effect of exercise on oxygen consumption and ventilatory rate. (From Gray: Pulmonary Ventilation and Its Physiological Regulation. Charles C Thomas.)

above, namely *stimulatory impulses from the brain* and *proprioceptive stimulatory reflexes*.

**Interrelationship Between Humoral Factors and Nervous Factors in the Control of Respiration During Exercise.** Figure 42–10 illustrates diagrammatically the different factors that operate in the control of respiration during exercise, showing two neurogenic factors, (1) direct stimulation of the respiratory center by the cerebral cortex and (2) indirect stimulation by proprioceptors, and showing also the three humoral factors, (1) carbon dioxide, (2) hydrogen ions, and (3) oxygen.

Most times when a person exercises, the nervous factors stimulate the respiratory center almost exactly the proper amount to supply the extra oxygen requirements for the exercise and to blow off the extra carbon dioxide. But, occasionally, the nervous signals are either too strong or too weak in their stimulation of the respiratory center. Then, the humoral factors play a very significant role in bringing about the final adjustment in respiration required to keep the carbon dioxide and hydrogen ion concentrations of the body fluids as nearly normal as possible. This effect is illustrated in Figure 42–11, which shows changes in alveolar ventilation and in arterial $P_{CO_2}$ during a one-minute period of exercise and then for another minute after the exercise is over. Note that the alveolar ventilation increases without an initial increase in arterial $P_{CO_2}$; this increase is caused by a stimulus that originates in the brain at the same time that the brain excites the muscles. Often, the increase in alveolar ventilation is so great that it actually *decreases* arterial $P_{CO_2}$ below normal, as shown in the figure, even though the exercising muscles are beginning to form large amounts of carbon dioxide. The reason for this is that the ventilation forges ahead of the increase in carbon dioxide formation. Thus, the brain provides an "anticipatory" effect at the onset of exercise, causing excessive alveolar ventilation even before this is needed.

Note also in the figure that after a short delay the decrease in arterial $P_{CO_2}$ finally acts on the respiratory center to reduce the alveolar ventilation back toward normal for a fraction of a minute; soon thereafter, the large amount of carbon dioxide formed by the muscles gradually increases the $P_{CO_2}$ up toward the normal level and the alveolar ventilation increases once again.

The anticipatory signal from the brain thus initiates the increase in alveolar ventilation during exercise; then humoral factors (mainly arterial $P_{CO_2}$) make additional adjustments, either upward or downward, to balance the rate of alveolar ventilation with the rate of metabolism in the body. Because of the direct brain stimulus, the final level of $P_{CO_2}$, hydrogen ion concentration, and $P_{O_2}$ in the arterial blood is almost exactly normal during exercise—sometimes slightly above normal and sometimes slightly below normal.

Finally, note in Figure 42–11 that exactly the opposite effects occur when exercise is over. The shutting-off of the brain stimulus returns the alveolar ventilation toward normal instantaneously, but this gets ahead of the decrease in carbon dioxide formation by the muscles so that a sequence of humoral-directed feedback readjustments is necessary before the alveolar ventilation finally becomes stabilized.

Now for a final comment regarding the brain factor for stimulation of respiration during exercise: Many experiments indicate that this brain factor is partly a learned response. That is, with repeated periods of exercise of the same degree of strenuousness, the brain factor seems to become progressively more able to establish the proper amount of brain signal to maintain the humoral factors at their normal levels during exercise. Therefore, there is much reason to believe that some of the higher centers of learning in the brain are important in this brain factor—probably even the cerebral cortex.

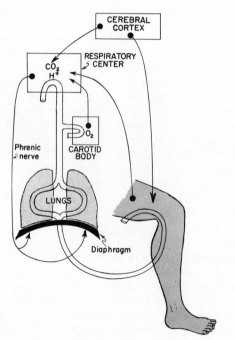

**Figure 42–10.** The different factors that enter into regulation of respiration during exercise.

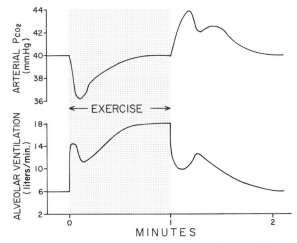

**Figure 42–11.** Changes in alveolar ventilation and arterial $P_{CO_2}$ during a 1-minute period of exercise and also following termination of the exercise. (Extrapolated to the human being from data in dogs; from Bainton: *J. Appl. Physiol.*, 33:778, 1972.)

# ABNORMALITIES OF RESPIRATORY CONTROL

## *RESPIRATORY DEPRESSION*

**Brain Edema.** The activity of the respiratory center may be depressed or totally inactivated by a number of different factors, one of the most common of which is brain edema resulting from brain concussion. For instance, the head might be struck against some solid object, following which the damaged brain tissues swell, compressing the cerebral arteries against the cranial vault and thus totally or partially blocking the cerebral blood supply. As a result, the neurons of the respiratory center first become inactive and later die. In this manner brain edema may either depress or totally inactivate the respiratory center.

Occasionally, respiratory depression resulting from brain edema can be relieved by intravenous injection of hypertonic solutions. These solutions osmotically remove some of the intracellular fluids of the brain, thus relieving intracranial pressure and sometimes re-establishing respiration within a few minutes.

**Pressure Conus.** A special condition that occurs frequently in brain injuries and in brain tumors is known as "pressure conus." That is, pressure in the cranial vault forces the brain downward through the foramen magnum, compressing the large upper portion of the medulla into the small opening of the foramen. As a result, blood flow into the medulla is partially or totally occluded so that the respiratory center becomes partially or totally inactive. Many patients with brain tumors die precipitously because of the development of such a conus. Such a situation occasionally occurs following removal of fluid from the spinal canal in patients with high cerebrospinal fluid pressure, for as soon as fluid is removed from the canal, the fluid in the cranial vault pushes the medulla into the foramen magnum, and the patient dies a few minutes later.

**Anesthesia.** Perhaps the most prevalent cause of respiratory depression and respiratory arrest is overdosage of anesthetics or narcotics. The best agent to be used for anesthesia is one that depresses the respiratory center the least while depressing the cerebral cortex the most. Such an anesthetic allows the anesthetist to block completely the conscious elements of the brain and yet not disturb the automatic function of respiration. Ether is among the best of the anesthetics by these criteria, though halothane, cyclopropane, ethylene, nitrous oxide, and a few others have almost the same value. On the other hand, sodium pentobarbital is a poor anesthetic because it depresses the respiratory center considerably more than the above agents. In early days morphine was used as an anesthetic, but this drug is now used only as an adjunct to anesthetics because it greatly depresses the respiratory center.

**Respiratory Stimulants to Offset Depression.** A few drugs, like picrotoxin and Metrazol, directly stimulate the respiratory center and can to a certain extent overcome the depressant effects of anesthetics, narcotics, and neuronal damage. Coramine is still another drug that stimulates respiration, but it does so indirectly. This drug excites the carotid and aortic chemoreceptors, which in turn stimulate the respiratory center. Likewise, caffeine, Benzedrine, theophylline, and many other general central nervous system stimulants can aid in stimulating the depressed respiratory center.

## *PERIODIC BREATHING*

An abnormality of respiration called periodic breathing occurs in a number of different disease conditions. The person breathes deeply for a short interval of time and then breathes slightly or not at all for an additional interval, the cycle repeating itself over and over again.

The most common type of periodic breathing, *Cheyne-Stokes breathing,* is characterized by slowly waxing and waning respiration, occurring over and over again every 45 seconds to 3 minutes. A second important type of periodic breathing occasionally seen is *Biot's breathing,* which is characterized by "runs" of several normal respirations—1, 2, 3, 4, or more at a time—followed suddenly by a period of complete cessation of respiration, then another series of normal respirations, and so forth, the cycle repeating itself again and again. The duration of the cycle is extremely variable, sometimes as short as 10 seconds and sometimes as long as a minute.

**Basic Mechanism of Cheyne-Stokes Breathing.** Let us assume that the respiration becomes much more rapid and deeper than usual. This causes the $P_{CO_2}$ in the pulmonary blood to *decrease.* A few seconds later the pulmonary blood reaches the brain, and the decreased $P_{CO_2}$ inhibits respiration. As a re-

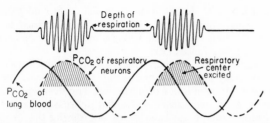

**Figure 42–12.** Cheyne-Stokes breathing, showing the changing $P_{CO_2}$ in the pulmonary blood (solid line) and the delayed changes in $P_{CO_2}$ of the fluids of the respiratory center (dashed line).

sult, the pulmonary blood $P_{CO_2}$ gradually *increases*. After another few seconds the blood carrying the increased $CO_2$ arrives at the respiratory center and stimulates respiration again, thus making the person overbreathe once again and initiating a new cycle, the cycles thus continuing on and on, causing Cheyne-Stokes periodic breathing.

The successive changes in pulmonary and respiratory center $P_{CO_2}$ during Cheyne-Stokes breathing are illustrated in Figure 42–12.

Obviously, an increase and decrease in the blood concentration of oxygen could also account partially for the periodic feedback required to perpetuate Cheyne-Stokes oscillations. However, Cheyne-Stokes breathing can occur even after denervation of the chemoreceptors, illustrating that stimulation caused by oxygen lack is not a necessary feature of the condition.

Since the basic feedback mechanism for causing Cheyne-Stokes respiration is present in every person, the real question that must be answered is not why Cheyne-Stokes breathing develops under certain pathological conditions but, instead, why is it not present all the time in everyone? The reason for this is that the feedback is highly "damped," which may be explained as follows:

**Damping of the Normal Respiratory Control Mechanism.** In order for Cheyne-Stokes oscillation to occur, sufficient time must occur during the hyperpneic phase for the body fluid $P_{CO_2}$ to fall considerably below the mean value. Then this decrease in $P_{CO_2}$ initiates the apneic phase, which also must last long enough for the tissue $P_{CO_2}$ to rise well above the mean value. To express this another way, the total body fluid contains tremendous quantities of stored gases, and it takes a long period of hyperventilation or hypoventilation to change the concentrations of these gases significantly. In the normal person, before significant changes can occur the respiratory center ordinarily readjusts the breathing back toward normal, thereby preventing the extreme excursions of $P_{CO_2}$ in the tissue fluids that are necessary to cause Cheyne-Stokes breathing. Thus, the tremendous storage capacity of the tissues for carbon dioxide functions as an important "damping" factor to prevent Cheyne-Stokes respiration in the normal person.

Yet even in the normal person the feedback is not completely damped. Therefore, when a person purposely overbreathes for a minute or more and then allows his involuntary respiratory control mechanisms to take over, he will first go into apnea and then go through one to two highly damped cycles of Cheyne-Stokes breathing.

A number of different conditions can overcome the damping of the feedback mechanism and cause it to oscillate spontaneously. Two of these are: (1) increased delay time in the flow of blood from the lungs to the brain and (2) increased feedback gain of the respiratory center mechanisms for control of respiration.

***Cheyne-Stokes Breathing Caused by Increased Circulatory Delay from the Lungs to the Brain.*** If the blood is long delayed in its transmission from the lungs to the brain, the damping factor discussed above cannot operate satisfactorily to block the feedback, for the body fluid gases can now change drastically before the respiratory center detects the altered blood gas concentration. As a result, the effective amplification or "gain" of the feedback circuit is greatly enhanced, and the system oscillates spontaneously. Experiments have shown that a one- to two-minute delay in the passage of blood from the lungs to the brain causes most experimental animals to go into spontaneous Cheyne-Stokes respiration.

***Cheyne-Stokes Breathing Caused by Increased Feedback Gain in the Respiratory Center.*** Occasionally, damage to the brain stem increases the feedback gain of the respiratory center in response to changes in blood $P_{CO_2}$ and blood pH. This does not mean that the sensitivity of these centers to these two humoral factors necessarily increases but instead that a minute concentration *change* in one of the two humoral factors causes a drastic *change* in ventilation. Indeed, the respiratory center is often severely depressed even though the gain is very high. An example of this would be complete cessation of ventilation when the $P_{CO_2}$ is 30 mm. Hg, but one-half normal ventilation when the $P_{CO_2}$ is 40 mm. Hg. In this case the percentage *change* in ventilation is infinite—from zero to a finite value. Therefore, the gain is infinite even though the ventilation is only one-half normal. Consequently, the respiratory control system can oscillate back and forth between apnea and breathing, which is the typical pattern of Cheyne-Stokes breathing. This effect probably explains why many patients with brain damage develop a type of Cheyne-Stokes breathing.

**Cheyne-Stokes Respiration in Cardiac Failure.** The most common cause of Cheyne-Stokes breathing is cardiac failure, which sometimes causes this type of breathing continually for many months at a time. There seem to be several factors operative in causing the Cheyne-Stokes breathing in this condition:

First, the slowed circulation in cardiac failure increases the delay time for transmission of blood from the lungs to the brain, thus reducing the effectiveness of the damping system as discussed above.

Second, the left heart volume is sometimes greatly increased. This, too, increases the circulation time from the lungs to the brain. Combining this factor with the sluggish rate of blood flow, the delay time from the lungs to the brain can increase as much as six-fold, the total interval increasing to 30 to 60 seconds instead of the normal 5 to 10 seconds. This approaches the range that causes spontaneous oscillations in experimental dogs.

Third, another factor that perhaps enters into the cardiac failure type of Cheyne-Stokes breathing is impaired oxygen uptake by the edematous lungs. As a result, the arterial oxygen saturation sometimes operates in a range low enough to allow significant feedback by the oxygen-lack-chemoreceptor mechanism, thus greatly increasing the net gain of the feedback in the Cheyne-Stokes oscillatory system. One of the special reasons for believing this to be true is that cardiac Cheyne-Stokes breathing can often be stopped by oxygen therapy.

**Periodic Breathing Following Brain Damage—Biot's Breathing.** Many patients with very high cerebrospinal fluid pressure, with contusion of the brain tissues, with direct compression of the brain tissues, or with destructive diseases of the brain often develop periodic breathing which usually lasts only a short time before death supervenes. This type of breathing is usually of the Biot's type, though occasionally it is of the Cheyne-Stokes type. That is, the respirations usually come in couples, triples, quadruples, and so forth rather than waxing and waning.

The cause of Biot's breathing is unknown, but one would suspect that it results from direct abnormality of the basic rhythmic mechanism of the respiratory center itself. The short periodicity of Biot's breathing makes it almost certain that it is not caused by the same feedback system that causes Cheyne-Stokes breathing.

# REFERENCES

Bainton, C. R.: Effect of speed versus grade and shivering on ventilation in dogs during active exercise. *J. Appl. Physiol.*, 33:778, 1972.

Biscoe, T. J.: Carotid body: structure and function. *Physiol. Rev.*, 51:427, 1971.

Burns, B. D.: The central control of respiratory movements. *Brit. Med. Bull.*, 19:7, 1963.

Carroll, D.: Sleep, periodic breathing, and snoring in the aged: control of ventilation in the aging and diseased respiratory system. *J. Amer. Geriatr. Soc.*, 22:307, 1974.

Cherniack, N. S., and Longobardo, G. S.: Cheyne-Stokes breathing. An instability in physiologic control. *N. Engl. J. Med.*, 288:952, 1973.

Cohen, M. I.: Discharge patterns of brain-stem respiratory neurons in relation to carbon dioxide tension. *J. Neurophysiol.*, 31:142, 1968.

Coleridge, H. M., Coleridge, J. C. G., and Howe, A.: Thoracic chemoreceptors in the dog: a histological and electrophysiological study of the location, innervation and blood supply of the aortic bodies. *Circ. Res.*, 26:235, 1970.

Cunningham, D. J. C.: Integrative aspects of the regulation of breathing: a personal view. *In* Guyton, A. C. (ed.): MTP International Review of Science: Physiology. Vol. 2. Baltimore, University Park Press, 1974, p. 303.

Cunningham, D. J. C., and Lloyd, B. B. (eds.): The Regulation of Human Respiration. Philadelphia, F. A. Davis Company, 1963.

Fencl, V.: Notes on the centrogenic drive in respiration. *Physiologist*, 16:589, 1973.

Fenn, W. O., and Rahn, H. (eds.): Handbook of Physiology. Sec. 3, Vol. 1, Sec. 3, Vol. 2. Baltimore, The Williams & Wilkins Company, 1964; 1965.

Guyton, A. C., Crowell, J. W., and Moore, J. W.: Basic oscillating mechanism of Cheyne-Stokes breathing. *Amer. J. Physiol.*, 187:395, 1956.

Guz, A.: Regulation of respiration in man. *Ann. Rev. Physiol.*, 37:303, 1975.

Heinemann, H. O., and Goldring, R. M.: Bicarbonate and the regulation of ventilation. *Amer. J. Med.*, 57:361, 1974.

Karczewski, W. A.: Organization of the brainstem respiratory complex. *In* Guyton, A. C. (ed.): MTP International Review of Science: Physiology. Vol. 2. Baltimore, University Park Press, 1974, p. 197.

Koepchen, H. P., and Umbach, W. (eds.): Central Rhythmic and Regulation; Circulation, Respiration, Extrapyramidal Motor System. Stuttgart, Hippokrates-Verlag GMBH, 1974.

Leusen, I.: Regulation of cerebrospinal fluid composition with reference to breathing. *Physiol. Rev.*, 52:1, 1972.

Loeschcke, H. H.: Respiratory chemosensitivity in the medulla oblongata. *Acta Neurobiol. Exp.*, 33:97, 1973.

Loeschcke, H. H.: Central nervous chemoreceptors. *In*: Guyton, A. C. (ed.): MTP International Review of Science: Physiology. Vol. 2. Baltimore, University Park Press, 1974, p. 167.

Milhorn, H. T., Jr., Benton, R., Ross, R., and Guyton, A. C.: A mathematical model of the human respiratory control system. *Biophys. J.*, 5:27, 1965.

Milhorn, H. T., Jr., and Guyton, A. C.: An analog computer analysis of Cheyne-Stokes breathing. *J. Appl. Physiol.*, 20:328, 1965.

Mitchell, R. A.: Respiration. *Ann. Rev. Physiol.*, 32:415, 1970.

Mitchell, R. A., and Berger, A. J.: Neural regulation of respiration. *Amer. Rev. Respir. Dis.*, 111:206, 1975.

Paintal, A. S.: Vagal sensory receptors and their reflex effects. *Physiol. Rev.*, 53:159, 1973.

Pearson, S. B., and Cunningham, D. J.: Some observations on the relation between ventilation, tidal volume, and frequency in man in various steady and transient states. *Acta Neurobiol. Exp.*, 33:177, 1973.

Sykes, M. K., McNicol, M. W., and Campbell, E. J. M.: Respiratory Failure. Philadelphia, J. B. Lippincott Company, 1974.

Torrance, R. W.: Arterial chemoreceptors. *In* Guyton, A. C. (ed.): MTP International Review of Science: Physiology. Vol. 2. Baltimore, University Park Press, 1974, p. 247.

Widdicombe, J. G.: Reflex control of breathing. *In* Guyton, A. C. (ed.): MTP International Review of Science: Physiology. Vol. 2, Baltimore, University Park Press, 1974, p. 273.

# | 43 |

# Respiratory Insufficiency

The diagnosis and treatment of most respiratory disorders has come to depend highly on an understanding of the basic physiological principles of respiration and gas exchange. Some diseases of respiration result from inadequate ventilation, while others result from abnormalities of diffusion through the pulmonary membrane or of transport from the lungs to the tissues. In each of these instances, the therapy is often entirely different so that it is no longer satisfactory simply to make a diagnosis of "respiratory insufficiency."

**Definitions.** In describing different respiratory disorders the following descriptive terms are used:

*Eupnea* means normal breathing; *tachypnea* means rapid breathing; and *bradypnea* means slow breathing.

*Hyperpnea* means a rate of alveolar ventilation great enough to cause over-respiration. However, this term is commonly used also to indicate simply a very high level of alveolar ventilation without necessarily implying over-respiration. *Hypopnea* is the opposite of hyperpnea, indicating under-respiration.

*Anoxia* means total lack of oxygen, but this term is more frequently used to mean simply decreased oxygen. A more correct term for decreased oxygen is *hypoxia*.

*Anoxemia* means lack of oxygen in the blood, but the term is generally used to mean simply reduced oxygen in the blood. A better term for this is *hypoxemia*.

*Hypercapnia* means excess carbon dioxide in the blood, and *hypocapnia* means depressed carbon dioxide. The term *acapnia* is also frequently used to imply hypocapnia, but, as is true with anoxemia, during life the absolute state of acapnia with no carbon dioxide at all in the body fluids never exists.

## METHODS FOR STUDYING RESPIRATORY ABNORMALITIES

### Study of Respiratory Ventilation.

The measurement of vital capacity, tidal air, functional residual capacity, and other lung volumes and capacities was discussed in Chapter 39. From these, two calculations are frequently made, (1) the *breathing reserve* and (2) the *ventilatory reserve*.

**The Breathing Reserve.** The breathing reserve is the difference between the normal minute respiratory volume and the maximum breathing capacity. Thus, if a person has a respiratory minute volume of 6 liters/minute and a maximum breathing capacity of 150 liters/minute, his breathing reserve is 144 liters/minute. In respiratory disease, the normal respiratory minute volume might be 20 liters/minute and his maximum breathing capacity 40 liters/minute. In this instance the breathing reserve is 20 liters/minute.

**The Ventilatory Reserve.** The ventilatory reserve is the maximum *percentage* increase in alveolar ventilation that a person can achieve above his normal alveolar ventilation. Thus, the normal person with a normal alveolar ventilation of 4.2 liters/minute and a maximum alveolar ventilation of about 140 liters/minute has a ventilatory reserve of more than 3000 per cent. If, in the course of a disease, the normal alveolar ventilation becomes 20 liters/minute and the maximum alveolar ventilation 40 liters/minute, the ventilatory reserve is then only 100 per cent. Ordinarily, when the ventilatory reserve falls below 100 per cent the person experiences dyspnea or "air hunger," which is described later in the chapter.

**Use of the Body Plethysmograph to Study Pulmonary Function.** A body plethysmograph, which

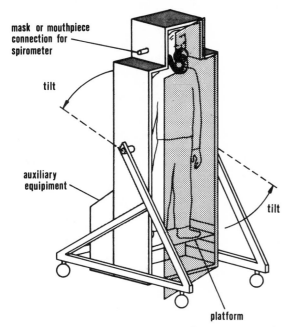

mask or mouthpiece
connection for
spirometer

tilt

auxiliary
equipiment

tilt

platform

**Figure 43–1.** A body plethysmograph. (From Ray (ed.): "Instrumentation for Pulmonary Function," in Medical Engineering. Copyright © 1974 by Year Book Medical Publishers, Inc., Chicago. Used by permission. Adapted from "Pulmo-Box," Med-Science Electronics, Inc., St. Louis, Missouri.)

is illustrated in Figure 43–1, is simply an airtight tank into which a person is placed. It can be used in different ways to measure many different aspects of pulmonary function, as described below:

1. The body plethysmograph can be used *as a spirometer*. The person breathes through a tube to the exterior of the tank. The excursions of his chest cause displacement of air volume within the tank, and an appropriate recording device connected to the tank can record this displacement in the form of volume changes, thus giving the usual spirometric recordings.

2. The body plethysmograph can be used *to measure lung volumes*. The person does not breathe to the exterior of the tank but instead breathes the air within the tank while the tank is completely closed. He then expires with a forceful expiration against a manometer, and the expiratory pressure is recorded. The increased pressure within his chest decreases the volume of gases in his chest, also allowing his chest dimensions to decrease. Therefore, the person's body requires less volume within the body plethysmograph. An appropriate volume recorder connected to the tank records the amount of compression of the gases in the lungs. This measure, combined with the recording of pressure change within the chest, can be used to calculate the instantaneous volume of gas in the lung.

3. The body plethysmograph can be used *to determine residual volume* of the lung. The volume of air in the lung is measured by the above method and then

as much volume as possible is expired. Subtracting the volume of the expired air from the previously measured lung volume gives the residual volume.

4. The body plethysmograph can be used *to measure airway resistance*. The person is allowed to breathe the air within the tank while volume changes in the tank are recorded. During each expiration the gases in the lung become compressed so that the person's body occupies less space within the chamber. This decreases the tank volume, and this is recorded. Knowing the lung volume, as measured above, as well as the *change* in volume during expiration, one can then calculate the pressure of the gas within the lung. This pressure minus the pressure of the gas in the tank gives the pressure difference that forces air through the airways. If the expiratory air flow is measured at the same time, one can then use these two bits of information to calculate the airway resistance.

5. The body plethysmograph can be used *to measure pulmonary blood flow*. The tank is completely closed from the surrounding atmosphere and the person inhales a gas mixture containing nitrous oxide from a flaccid balloon. As blood flows through the lung the nitrous oxide is absorbed into the pulmonary blood. This reduces the total quantity of gases in the tank and reduces its volume. One can then calculate the rate of pulmonary blood flow from (a) the rate of reduction of the volume in the tank, (b) the concentration of the nitrous oxide in the alveoli, and (c) the solubility coefficient of nitrous oxide in the blood.

Thus, the body plethysmograph is an extremely versatile instrument that has become an important tool for the clinical respiratory physiologist.

## STUDY OF BLOOD GAS CONCENTRATIONS

The physician often assesses the quantity of oxygen and carbon dioxide in the blood by observing various signs in the patient. A decrease in blood oxygen to a sufficiently low level causes the skin to become bluish because deoxygenated hemoglobin is blue. This condition, called *cyanosis*, is discussed in greater detail later in the chapter. When the blood concentration of carbon dioxide or hydrogen ions increases, the rate and depth of respiration usually also increases, which is a useful indicator in respiratory diseases. In addition to these clinical signs, blood gas analyses can be made as follows:

**Determination of Blood Oxygen Saturation.** The concentration of oxygen in the blood is determined in a Van Slyke gas analysis apparatus or other suitable device in which a special reagent causes the hemoglobin to release its oxygen. Then the amount of released oxygen is measured.

To determine the *per cent oxygen* saturation of the blood, a similar blood sample is exposed to air until it becomes 100 per cent saturated with oxygen. The oxygen concentration of this blood is then measured as described above. Dividing this value into the initial concentration of oxygen and multiplying by 100 gives the percentage of the hemoglobin in the original blood sample that is bound with oxygen.

The normal concentration of oxygen in arterial blood is approximately 20 volumes per cent, and the normal arterial oxygen saturation is 97 per cent.

**Measurement of Blood and Tissue Oxygen Pressures by Polarography.** As explained in Chapter 41, the per cent saturation of hemoglobin is different from the actual pressure of the oxygen in the blood. A particularly valuable method has been developed for measuring the oxygen pressure ($Po_2$) of solutions, based on an electrochemical procedure called "polarography." Basically, this apparatus applies a negative electrical voltage to a minute platinum electrode, and the electrical current flowing from the electrode to the solution reduces the gaseous oxygen in contact with the electrode to an ionic form. Since the rate of current flow depends on the $Po_2$ of the solution, the $Po_2$ can be read directly from an electrical current flow meter (microammeter) in the circuit.

When this apparatus is used for measuring blood or tissue $Po_2$, a very thin membrane that allows oxygen diffusion but no diffusion of the plasma proteins is usually placed over the electrode to prevent difficulties with the proteins, red cells, or tissue elements. Such electrodes have been made as small as 1 micron in diameter, small enough to measure $Po_2$ inside individual cells.

**Measurement of Carbon Dioxide Concentration and $Pco_2$ of the Blood.** By using different reagents in the Van Slyke gas analysis apparatus, one can liberate carbon dioxide instead of oxygen from the blood, and its concentration in the blood can thereby be measured. Also, a modified glass electrode pH meter can be used to measure the carbon dioxide pressure ($Pco_2$) in the following way: A very thin membrane permeable to carbon dioxide but impermeable to bicarbonate ion is stretched over the glass electrode, and a solution of bicarbonate ions is placed in the thin space between the membrane and the electrode. The pH of this bicarbonate solution will vary as a function of the concentration of the carbon dioxide that diffuses through the membrane, and this in turn is determined by the $Pco_2$.

# PHYSIOLOGIC TYPES OF RESPIRATORY INSUFFICIENCY

In general, the different types of abnormalities that cause respiratory insufficiency can be divided into three major categories: (1) those that cause inadequate ventilation of the alveoli, (2) those that reduce gaseous diffusion through the respiratory membrane, and (3) those that decrease oxygen transport from the lungs to the tissues.

## ABNORMALITIES THAT CAUSE ALVEOLAR HYPOVENTILATION

**Paralysis of the Respiratory Muscles.** It is obvious how paralysis of the respiratory muscles can decrease pulmonary ventilation. Fortunately, paralysis of any single muscle of respiration is compensated by overactivity of those that are still normal. Indeed, patients with polio or cervical transection of the spinal cord resulting from a broken spinal column can sometimes maintain respiration despite complete paralysis of the diaphragm and abdominal muscles, utilizing only the neck muscles, or even helping respiration by forcing air into the lungs with the cheek muscles. This illustrates the great reserve of the respiratory muscular system.

**Diseases That Increase the "Work" of Ventilation.** Much more common than paralysis of the respiratory muscles are the many pulmonary diseases that increase the work required to move air in and out of the lungs. These can be divided into three different types of abnormalities: (1) those that increase the airway resistance, (2) those that increase the tissue resistance, and (3) those that decrease the compliance of the lungs and thorax.

*Increased Airway Resistance.* One can readily understand how obstruction of the respiratory passageway can reduce a person's maximum breathing capacity. But, in addition, airway resistance is especially increased in two diseases that affect the bronchioles, *asthma* and *emphysema,* which are discussed in further detail later in the chapter. In both of these, air flow is impeded in the smaller bronchioles, which causes far greater breathing effort during expiration than inspiration for the following reasons:

When the muscles of respiration expand the lungs, the expansion process not only opens the alveoli but also expands the small bronchioles at the same time. Consequently, air flows easily into the alveoli. On the other hand, during expiration the pressure of the thoracic cage against the lungs not only compresses the alveoli but also compresses the small bronchi, in this way reducing their diameter and thereby increasing the airway resistance. As a result, the airway resistance sometimes becomes many times as great during expiration as during inspiration, and expiration is correspondingly prolonged. Also, because of this differential between expiratory and inspiratory resistance, the lungs become greatly distended, with a marked increase in functional residual capacity and residual volume.

*Increased Tissue Resistance.* Another factor that increases the work of respiration is often loss of normal pulmonary elasticity. In normal respiration, the inspiratory muscles stretch the lungs and thorax, and then the elastic recoil of the lungs and thorax causes the lungs to deflate during expiration. In this way, no effort is usually required for expiration. Many diseases of the lungs, particularly such destructive diseases as *emphysema, pulmonary fibrosis, tuberculosis,* and *various infections,* often destroy much of the elasticity of the lungs so that little recoil occurs. As a result, effort must then be expended by the respiratory muscles to deflate the lungs as well as to inflate them.

*Decreased Compliance of the Lungs and Chest Wall.* The term "compliance" means the change in lung volume caused by a unit change in distending pressure, as explained in Chapter 39. Normally, a rise in intra-alveolar pressure of 1 cm. water increases the lung volume 0.13 liter. But any disease that makes the tissues of the lungs or of the chest wall less distensible than usual—diseases such as *silicosis, asbestosis, sarcoidosis, tuberculosis, cancer, pneumonia,* or even *skeletal abnormalities of the chest*—reduces the compliance and, therefore, increases the work that must be expended by the respiratory muscles to expand the lungs. Also, any disease that decreases the total amount of functional lung tissue reduces the compliance and, therefore, increases the effort of respiration.

## DISEASES THAT DECREASE PULMONARY DIFFUSING CAPACITY

Three different types of abnormalities can decrease the diffusing capacity of the lungs. These are: (1) decreased area of the respiratory membrane, (2) increased thickness of the respiratory membrane, called *alveolocapillary block,* and (3) abnormal ventilation-perfusion ratio in some parts of the lungs.

**Decreased Area of the Respiratory Membrane.** Diseases or abnormalities that decrease the area of the respiratory membrane include *removal of part or all of one lung, tuberculous destruction of the lung, cancerous destruction,* and *emphysema* which causes gradual destruction of alveolar septa.

Also, any acute condition that fills the alveoli with fluid or otherwise prevents air from coming in contact with the alveolar membrane, such as *pneumonia, pulmonary edema,* and *atelectasis,* can temporarily reduce the surface area of the respiratory membrane.

**Increased Thickness of the Respiratory Membrane—Alveolocapillary Block.** The most common acute cause of increased thickness of the respiratory membrane is *pulmonary edema* resulting from left heart failure or pneumonia. However, *silicosis, tuberculosis,* and *many other fibrotic conditions* can cause progressive deposition of fibrous tissue in the interstitial spaces between the alveolar membrane and the pulmonary capillary membrane, thereby increasing the thickness of the respiratory membrane. This is usually called *alveolocapillary block,* or, occasionally, *interstitial fibrosis.* Since the rate of gaseous diffusion through the respiratory membrane is inversely proportional to the distance that the gas must diffuse, it is readily understood how alveolocapillary block can reduce the diffusing capacity of the lungs.

**Abnormal Ventilation-Perfusion Ratio.** Abnormal ventilation-perfusion ratio of the lungs can cause decreased pulmonary diffusing capacity. This is one of the abnormalities that is more difficult to understand, but it can be explained as follows: If the blood flow to many of the alveoli is blocked or partially blocked, the ventilation of these alveoli is of little or no benefit in the aeration of the blood, and yet, these alveoli take ventilation away from the normal alveoli.

Likewise, when the blood flow is normal but ventilation of many alveoli is blocked, the blood flowing past the blocked alveoli fails to be aerated.

Therefore, it is obvious that optimum aeration of blood requires approximately proportional distribution of ventilation and blood flow to the respective alveoli.

A small amount of ventilation-perfusion ratio abnormality occurs even normally in the upright person because the lower portion of the lung is far better perfused with the blood than the upper portion, as was discussed in Chapter 24. However, during exercise the blood flow in the upper portion of the lung increases greatly; therefore, when maximum amounts of gas exchange through the pulmonary membrane are required during heavy exercise, the ventilation-perfusion ratio becomes almost optimum.

In lung diseases the ventilation-perfusion ratio can become so severely abnormal in different parts of the lung that the diffusing capacity of the lung for this reason alone may be reduced to as little as one-fifth normal, this reduction often occurring even though both total ventilation and total perfusion of the lungs are entirely normal.

When the ventilation-perfusion ratio is abnormal because of poor ventilation of many alveoli, the blood perfusing these alveoli does not become aerated. Therefore, the blood is said to be *shunted* through the lungs—that is, it passes through the lungs without being exposed to functional alveoli.

When the abnormality is caused by poor blood flow to alveoli that are normally ventilated, the ventilation of these alveoli is wasted. Therefore, the *physiological dead space* of the lungs increases.

Special physiological tests have been devised to distinguish between pulmonary shunting and increased physiological dead space. These tests help to determine whether or not the abnormality is in the airways or in the vascular distribution system.

Diseases that cause abnormal ventilation-perfusion ratios include *thrombosis of a pulmonary artery, excessive airway resistance to some alveoli (emphysema), reduced compliance of one lung without concomitant abnormality of the other lung,* and many other conditions that cause diffuse damage throughout the lungs.

## ABNORMALITIES OF OXYGEN TRANSPORT FROM THE LUNGS TO THE TISSUES

Different conditions that reduce oxygen transport from the lungs to the tissues include *anemia,* in which the total amount of hemoglobin available to transport the oxygen is reduced, *carbon monoxide poisoning,* in which a large proportion of the hemoglobin becomes unable to transport oxygen, and *decreased blood flow to the tissues* caused by either low cardiac output or localized tissue ischemia. These have been discussed in Chapters 5, 41, and 23, respectively.

# PHYSIOLOGIC PECULIARITIES OF SPECIFIC PULMONARY ABNORMALITIES

## *CHRONIC EMPHYSEMA*

A disease called chronic emphysema is becoming highly prevalent, partly because it is a disease of old age and the average age of our population is increasing each year, but mainly because of the effects of smoking. It results from two major pathophysiological changes in the lungs. First, air flow through many of the terminal bronchioles is obstructed. Second, much of the lung parenchyma is destroyed.

Many clinicians believe that chronic emphysema begins with chronic infection in the lung that causes *bronchiolitis,* which means inflammation of the small air passages of the lungs. It is likely that this inflammation also involves alveolar septa and destroys many of these or that obstruction to expiration causes excess expiratory pressures in the alveoli and

that these pressures rupture the alveolar septa. At any rate, many bronchioles become irreparably obstructed, and the total surface of the respiratory membrane also becomes greatly decreased, as illustrated in Figures 43–2 and 43–3, sometimes to less than one-quarter normal.

The physiological effects of chronic emphysema are extremely varied, depending on the severity of the disease and on the relative degree of bronchiolar obstruction versus lung parenchymal destruction. However, among the different abnormalities are the following:

First, the bronchiolar obstruction greatly *increases airway resistance* and results in greatly increased work of breathing. It is especially difficult for the person to move air through the bronchioles during expiration because the compressive force on the outside of the lung not only compresses the alveoli but also compresses the bronchioles, which further increases their resistance.

Second, the marked loss of lung parenchyma greatly *decreases the diffusing capacity* of the lung, which reduces the ability of the lungs to oxygenate the blood.

Third, the obstructive process is frequently much worse in some parts of the lungs than in others so that some portions of the lungs are well ventilated while

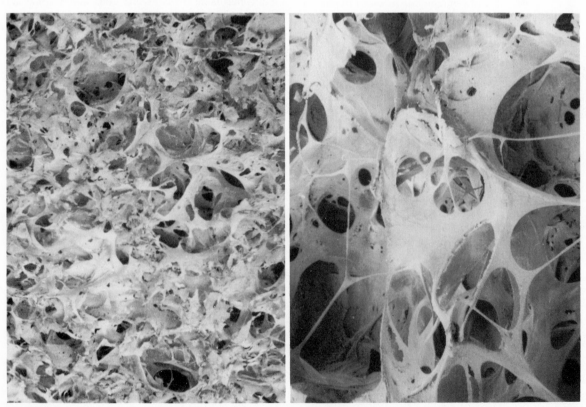

**Figure 43–2.** Contrast of the emphysematous lung (right) with the normal lung (left), showing extensive alveolar destruction. (Reproduced with permission of Patricia Delaney and the Department of Anatomy, The Medical College of Wisconsin.)

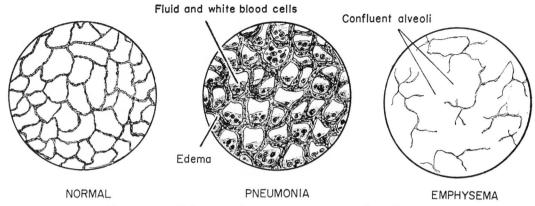

**Figure 43–3.**  Pulmonary changes in pneumonia and emphysema.

other portions are poorly ventilated. This causes an *abnormal ventilation-perfusion ratio,* which also causes a pulmonary *shunt,* resulting in poor aeration of the blood as was explained earlier in the chapter.

Fourth, loss of large portions of the lung parenchyma also decreases the number of pulmonary capillaries through which blood can pass. As a result, the pulmonary vascular resistance increases markedly, causing pulmonary hypertension. This in turn overloads the right heart and frequently causes right-heart failure.

Chronic emphysema usually progresses slowly over many years. The person develops hypoxia and hypercapnia because of hypoventilation of many alveoli and because of loss of lung parenchyma. The net result of all of these effects is severe and prolonged air hunger that can last for years until the hypoxia and hypercapnia cause death—a very high penalty to pay for smoking.

## PNEUMONIA

The term pneumonia describes any inflammatory condition of the lung in which the alveoli are usually filled with fluid and blood cells, as shown in Figure 43–3. A common type of pneumonia is *bacterial pneumonia,* caused most frequently by pneumococci. This disease begins with infection in the alveoli; the pulmonary membrane becomes inflamed and highly porous so that fluid and often even red and white blood cells pass out of the blood into the alveoli. Thus, the infected alveoli become progressively filled with fluid and cells, and the infection spreads by extension of bacteria from alveolus to alveolus. Eventually, large areas of the lungs, sometimes whole lobes or even a whole lung, become "consolidated," which means that they are filled with fluid and cellular debris.

The pulmonary function of the lungs during pneumonia changes in different stages of the disease. In the early stages, the pneumonia process might well be localized to only one lung, and alveolar ventilation may be reduced even though blood flow through the lung continues normally. This results in two major

pulmonary abnormalities: (1) reduction in the total available surface area of the respiratory membrane and (2) abnormal ventilation-perfusion ratio. Both these effects cause reduced diffusing capacity, which results in hypoxemia.

Figure 43–4 illustrates the effect of an abnormal ventilation-perfusion ratio in pneumonia, showing that the blood passing through the aerated lung becomes 97 per cent saturated while that passing through the unaerated lung remains only 60 per cent saturated, causing the mean saturation of the aortic blood to be about 78 per cent, which is far below normal. Fortunately, in some stages of peneumonia the blood flow through the diseased areas decreases concurrently with the decrease in ventilation. This gives much less debility than that resulting when the blood flow through the unventilated lung is normal.

## ATELECTASIS

Atelectasis means collapse of the alveoli. It can occur in a localized area of a lung, in an entire lobe, or in an entire lung. Its causes are three-fold: (1) obstruction of the airway, (2) external compression of the lung, or (3) lack of surfactant in the fluids lining the alveoli.

**Airway Obstruction.**  The airway obstruction type of atelectasis usually results from blockage of

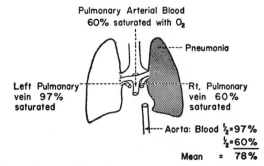

**Figure 43–4.**  Effect of pneumonia on arterial blood oxygen saturation.

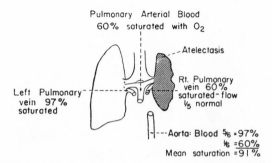

Pulmonary Arterial Blood
60% saturated with O₂

-- Atelectasis

Left Pulmonary
vein 97%
saturated

Rt. Pulmonary
vein 60%
saturated-flow
⅕ normal

---Aorta: Blood ⅚ =97%
⅙ =60%
Mean saturation =91%

**Figure 43–5.** Effect of atelectasis on arterial blood oxygen saturation.

many small bronchi with mucus or obstruction of a major bronchus by either a large mucous plug or some solid object. The air entrapped beyond the block is absorbed within minutes to hours by the blood flowing in the pulmonary capillaries, which pulls the walls of the alveoli together.

The effects on overall pulmonary function caused by *massive collapse* (atelectasis) of an entire lung are illustrated in Figure 43–5. Collapse of the lung tissue not only occludes the alveoli but also increases the resistance to blood flow through the pulmonary vessels as much as five-fold. This resistance increase occurs partially because of the collapse itself, which compresses and folds the vessels as the volume of the lung decreases. Anoxia in the collapsed alveoli then causes additional vasoconstriction, as was explained in Chapter 24.

Because of the vascular constriction, blood flow through the atelectatic lung becomes slight. Essentially all the blood is routed through the ventilated lung and therefore becomes well aerated. In the situation shown in Figure 43–5, five-sixths of the blood passes through the aerated lung and only one-sixth through the unaerated lung. As a result of the low flow through the collapsed lung, the aortic blood becomes almost fully saturated with oxygen despite the massive collapse of an entire lung.

**Compressed Lung Tissue.** If one portion of a lung is compressed more than other portions, the alveoli of this region will be smaller than the other alveoli. One can show from the physics of fluid lined bubbles, which are analogous to the alveoli, that the smaller a bubble becomes the more likely will the surface tension of the fluid cause it to collapse. Therefore, in an area partially compressed by outside pressure, this surface tension effect sometimes becomes great enough to cause the alveoli to collapse.

**Lack of Surfactant.** In Chapter 39 it was pointed out that the substance surfactant is secreted by the alveolar membrane into the fluids that line the alveoli. This substance decreases the surface tension in the alveoli 3- to 8-fold. In the normal lung this plays a major role in preventing alveolar collapse. However, in a number of different conditions the quantity of surfactant secreted by the alveoli is greatly decreased. Sometimes this is severe enough to cause

atelectasis. For instance, in the condition called *hyaline membrane disease* or *respiratory distress syndrome*, which occurs in newborn babies, the quantity of surfactant secreted by the alveoli is greatly depressed. This effect (along with other lung abnormalities) causes a serious tendency for the lungs of these babies to collapse; many of the infants die of suffocation as increasing portions of the lungs become atelectatic.

Atelectasis also frequently results from decreased surfactant secretion following extracorporeal perfusion of the lungs during open-heart surgery. The reason for the reduced surfactant secretion in this instance is not understood.

## BRONCHIAL ASTHMA

Bronchial asthma is usually caused by allergic hypersensitivity of the person to foreign substances in the air—especially to plant pollens. The allergic reaction causes localized edema in the walls of the small bronchioles, secretion of thick mucus into their lumens, and spasm of their smooth muscle walls. These effects greatly increase the airway resistance. For reasons discussed earlier in the chapter, the bronchiolar diameter becomes especially reduced during expiration and then opens up to a certain extent during inspiration. Therefore, the asthmatic person usually can inspire quite adequately but has great difficulty expiring. This results in dyspnea, or "air hunger," which is discussed later in the chapter.

The functional residual capacity and the residual volume of the lung become greatly increased during the asthmatic attack because of the difficulty in expiring air from the lungs. Over a long period of time the chest cage becomes permanently enlarged, causing a "barrel chest," and the functional residual capacity and residual volume also become permanently increased.

## TUBERCULOSIS

In tuberculosis the tubercle bacilli cause a peculiar tissue reaction in the lungs including, first, invasion of the infected region by macrophages and, second, walling off of the lesion by fibrous tissue to form the so-called "tubercle." This walling-off process helps to limit further transmission of the tubercle bacilli in the lungs and, therefore, is part of the protective process against the infection. However, in approximately 3 per cent of all persons who contract tuberculosis, the walling-off process fails and tubercle bacilli spread throughout the lungs, causing fibrotic tubercles in many areas. In the late stages of tuberculosis, secondary infection by other bacteria causes extensive destruction of the lungs. Thus, tuberculosis causes many areas of fibrosis throughout the lungs, and, secondly, it reduces the total amount of functional lung tissue. These effects cause (1) increased effort on the part of the respiratory muscles to cause pulmonary ventilation and therefore *re-*

*duced vital capacity and maximum breathing capacity,* (2) *reduced total respiratory membrane surface* and *increased thickness of the respiratory membrane,* these causing progressively diminished pulmonary diffusing capacity, and (3) *abnormal ventilation-perfusion ratio* in the lungs, further reducing the pulmonary diffusing capacity.

## PULMONARY EDEMA

The circulatory aspects of pulmonary edema were discussed in Chapter 24. Briefly, any abnormality that reduces the pumping capability of the left side of the heart causes blood to dam up in the pulmonary circulation. And, if the pulmonary capillary pressure rises above approximately 28 mm. Hg, which is the amount of the colloid osmotic pressure of the plasma proteins, fluid immediately begins to transude out of the capillaries into the pulmonary interstitial spaces and even into the alveoli. This results in acute thickening of the respiratory membrane, thereby reducing the diffusing capacity of the lungs, but only by a small amount. However, it also causes actual blockage of many of the alveoli as a result of total filling with fluid. This greatly reduces the diffusing capacity by reducing the total area of the respiratory membrane. Finally, the presence of large quantities of fluid in the lungs increases the *tissue resistance* of the lungs so much that they cannot be expanded and contracted with ease; this factor increases the work load of the respiratory muscles, though it is not so detrimental an effect as the greatly reduced pulmonary diffusing capacity.

# HYPOXIA

Obviously, almost any of the conditions discussed in the past few sections of this chapter can cause serious degrees of cellular hypoxia. In some of these, oxygen therapy is of great value; in others it is of moderate value, while in still others it is of almost no value. Therefore, it is important to classify the different types of hypoxia; then from this classification we can readily discuss the physiological principles of therapy. The following is a descriptive classification of the different causes of hypoxia:

1. Inadequate oxygenation of the lungs because of extrinsic reasons
   a. Deficiency of oxygen in atmosphere
   b. Hypoventilation (neuromuscular disorders)
2. Pulmonary disease
   a. Hypoventilation due to increased airway resistance or decreased pulmonary compliance
   b Uneven alveolar ventilation-perfusion ratio (including increased physiological dead space and pulmonary shunt)
   c. Diminished diffusing capacity
3. Venous-to-arterial shunts

4. Inadequate transport and delivery of oxygen
   a. Anemia, abnormal hemoglobin
   b. General circulatory deficiency
   c. Localized circulatory deficiency (peripheral, cerebral, coronary vessels)
5. Inadequate tissue oxygenation or oxygen use
   a. Tissue edema
   b. Abnormal tissue demand
   c. Poisoning of cellular enzymes

This classification of the different types of hypoxia is mainly self-evident from the discussions earlier in the chapter. Only two of the types of hypoxia in the above classification need further elaboration; these are the hypoxia caused by venous-to-arterial shunts and inadequate tissue oxygenation.

**Hypoxia Caused by Venous-to-Arterial Shunts.** A venous-to-arterial shunt means a direct pathway for blood to flow from the systemic veins to the systemic arteries without going through the lungs. This occurs most frequently in the congenital heart disease called tetralogy of Fallot, which was discussed in Chapter 27. In this condition, the right heart pumps as much as 80 per cent of the venous blood directly into the aorta, without its having gone through the lungs at all.

**Inadequate Tissue Oxygenation.** The most classic type of inadequate tissue oxygenation is caused by cyanide poisoning in which the action of cytochrome oxidase is completely blocked—to such an extent that the tissues simply cannot utilize the oxygen even though plenty is available. This type of hypoxia is frequently also called *histotoxic hypoxia.*

A common type of tissue hypoxia occurs in tissue edema, which causes increased distances through which the oxygen must diffuse before it can reach the cells. This type of hypoxia can become so severe that the tissues in edematous areas actually die, as is often illustrated by serious ulcers in edematous skin.

Finally, tissues can become hypoxic when the cells themselves demand more oxygen than can be supplied to them by the normal respiratory and oxygen transport systems. For instance, in strenuous exercise, the major limiting factor to the degree of exercise that can be performed is the tissue hypoxia that develops.

**Effects of Hypoxia on the Body.** Hypoxia, if severe enough, can actually cause death of the cells, but in less severe degrees it results principally in (1) depressed mental activity, sometimes culminating in coma and (2) reduced work capacity of the muscles. These effects are discussed in the following chapter in relation to high altitude physiology and, therefore, are only mentioned here.

## CYANOSIS

The term "cyanosis" means blueness of the skin, and its cause is excessive amounts of deoxygenated hemoglobin in the skin blood vessels, especially in the capillaries. This deoxygenated hemoglobin has an intense dark blue color that is transmitted through the skin.

The presence of cyanosis is one of the most common clinical signs of different degrees of respiratory insufficiency, and for this reason it is important to understand the factors that determine the degree of cyanosis. These are the following:

1. One of the most important factors determining the degree of cyanosis is the *quantity of deoxygenated hemoglobin in the arterial blood*. It is not the percentage deoxygenation of the hemoglobin that causes the bluish hue of the skin, but principally the *concentration of deoxygenated hemoglobin without regard to the concentration of oxygenated hemoglobin*. The reason for this is that the red color of oxygenated blood is weak in comparison with the dark blue color of deoxygenated blood. Therefore, when the two are mixed together, the oxygenated blood has relatively little coloring effect in comparison with that of the deoxygenated blood.

In general, definite cyanosis appears whenever the arterial blood contains more than 5 grams per cent of deoxygenated hemoglobin, and mild cyanosis can frequently be discerned when as little as 3 to 4 grams per cent deoxygenated hemoglobin is present. In polycythemia, cyanosis is very common because of the large amount of hemoglobin in the blood whereas in anemia, cyanosis is rare because it is difficult for there to be enough deoxygenated hemoglobin to produce the blue color.

2. Another important factor that affects the degree of cyanosis is the *rate of blood flow through the skin,* for the following reasons: Principally, the blood in the capillaries determines skin color; the blueness of this capillary blood is determined by two factors: (1) the concentration of deoxygenated hemoglobin in the arterial blood entering the capillaries and (2) the amount of deoxygenation that occurs as the blood passes through the capillaries. Ordinarily, the metabolism of the skin is relatively low so that little deoxygenation occurs as the blood passes through the skin capillaries. However, if the blood flow becomes extremely sluggish, even a low metabolism can cause marked desaturation of the blood and therefore can cause cyanosis. This explains the cyanosis that appears in very cold weather, particularly in children who have thin skins.

3. A final factor that affects the blueness of the skin is *skin thickness* because this determines the intensity of the color that is transmitted from the deeper vascular tissues. For instance, in newborn babies, who have very thin skin, cyanosis occurs readily, particularly in highly vascular portions of the body such as the heels. Also, in adults (as well as babies) the lips and fingernails often appear cyanotic before the remainder of the body shows any blueness.

## *DYSPNEA*

Dyspnea means primarily a desire for air or mental anguish associated with the act of ventilating enough to satisfy the air demand. A common synonym is "air hunger."

At least three different factors often enter into the development of the sensation of dyspnea. These are: (1) abnormality of the respiratory gases in the body fluids, especially hypercapnia and to much less extent hypoxia, (2) the amount of work that must be performed by the respiratory muscles to provide adequate ventilation, and (3) the state of the mind itself. A person becomes very dyspneic especially from excess buildup of carbon dioxide and reduction of oxygen in his body fluids.

At times, however, the levels of both carbon dioxide and oxygen in the body fluids are completely normal, but to attain this normality of the respiratory gases, the person has to breathe forcefully. In these instances the forceful activity of the respiratory muscles gives the person a sensation of air hunger. Indeed, the dyspnea can be so intense, despite normal gaseous concentrations in the body fluids, that clinicians often overemphasize this cause of dyspnea while forgetting that dyspnea can also result from abnormal gaseous concentrations.

Finally, the person's respiratory functions may be completely normal, and still he experiences dyspnea because of an abnormal state of mind. This is called *neurogenic dyspnea* or, sometimes, *emotional dyspnea*. For instance, almost anyone momentarily thinking about his act of breathing will suddenly find himself taking breaths a little more deeply than ordinarily because of a feeling of mild dyspnea. This feeling is greatly enhanced in persons who have a psychic fear of not being able to receive a sufficient quantity of air. For example, many persons on entering small or crowded rooms immediately experience emotional dyspnea, and patients with "cardiac neurosis" who have heard that dyspnea is associated with heart failure frequently experience severe psychic dyspnea even though the blood gases are completely normal. Neurogenic dyspnea has been known to be so intense that the person over-respires and causes alkalotic tetany.

## HYPERCAPNIA

Hypercapnia means excess carbon dioxide in the body fluids and especially refers to excess carbon dioxide at the cellular level.

One might suspect on first thought that any respiratory condition that causes hypoxia also causes hypercapnia. However, hypercapnia usually occurs in association with hypoxia only when the hypoxia is caused by *hypoventilation* or by *circulatory deficiency*. The reasons for this are the following:

Obviously, hypoxia caused by *too little oxygen in the air,* by *too little hemoglobin,* or by *poisoning of the oxidative enzymes* has to do only with the availability of oxygen or use of oxygen by the tissues. Therefore, it is readily understandable that hypercapnia is *not* a concomitant of these types of hypoxia.

Also, in hypoxia resulting from poor diffusion through the pulmonary membrane or through the tissues, serious hypercapnia usually does not occur because carbon dioxide diffuses 20 times as rapidly as oxygen. Therefore, even when oxygen diffusion is greatly depressed, carbon dioxide diffusion usually is still sufficient for adequate carbon dioxide transfer.

However, in hypoxia caused by hypoventilation, including hypoventilation of only some of the alveoli (abnormal ventilation-perfusion ratio), carbon dioxide transfer between the alveoli and the atmosphere is affected as much as is oxygen transfer. Therefore, hypercapnia always results along with hypoxia. And in circulatory deficiency, diminished flow of blood decreases the removal of carbon dioxide from the tissues, resulting in tissue hypercapnia. However, the transport capacity of the blood for carbon dioxide is about three times that for oxygen, so that even here the hypercapnia is much less than the hypoxia.

### EFFECTS OF HYPERCAPNIA ON THE BODY

When the alveolar $P_{CO_2}$ rises above approximately 60 to 75 mm. Hg, dyspnea usually becomes intolerable, and as the $P_{CO_2}$ rises to 70 to 80 mm. Hg, the person becomes lethargic and sometimes even semicomatose. Total anesthesia and death result when the $P_{CO_2}$ rises to 100 to 150 mm. Hg.

## OXYGEN THERAPY IN THE DIFFERENT TYPES OF HYPOXIA

Oxygen can be administered by (1) placing the patient's head in a "tent" which contains air fortified with oxygen, (2) allowing the patient to breathe either pure oxygen or high concentrations of oxygen from a mask, or (3) administering oxygen through an intranasal tube.

Oxygen therapy is of great value in certain types of hypoxia but of almost no value at all in other types. However, if one will simply recall the basic physiological principles of the different types of hypoxia, he can readily decide when oxygen therapy is of value and, if so, how valuable. For instance:

In *atmospheric hypoxia,* oxygen therapy can obviously completely correct the depressed oxygen level in the inspired gases and, therefore, provide 100 per cent effective therapy.

In *hypoventilation hypoxia,* a person breathing 100 per cent oxygen can move five times as much oxygen into the alveoli with each breath as he can when breathing normal air. Therefore, here again oxygen therapy can be extremely beneficial, increasing the available oxygen to as much as 400 per cent above normal. (But this does nothing for the hypercapnia also caused by the hypoventilation.)

In *hypoxia caused by impaired diffusion,* essentially the same result occurs as in hypoventilation

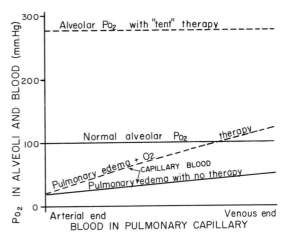

**Figure 43–6.** Absorption of oxygen into the pulmonary capillary blood in pulmonary edema with and without oxygen therapy.

hypoxia, for oxygen therapy can increase the $P_{O_2}$ in the lungs from a normal value of about 100 mm. Hg to as high as 600 mm. Hg. This causes a greatly increased diffusion gradient between the alveoli and the blood, the gradient rising from a normal value of 60 mm. Hg to as high as 560 mm. Hg, or an increase of more than 800 per cent. This highly beneficial effect of oxygen therapy in diffusion hypoxia is illustrated in Figure 43–6, which shows that the pulmonary blood in a patient with pulmonary edema picks up oxygen up to eight times as rapidly as it would with no therapy.

In *hypoxia caused by anemia, carbon monoxide poisoning,* or *any other abnormality of hemoglobin transport,* oxygen therapy is of slight to moderate value because the amount of oxygen transported in the dissolved form in the fluids of the blood can still be increased above normal even though that transported by the hemoglobin is hardly altered. This is illustrated in Figure 43–7, which shows that an in-

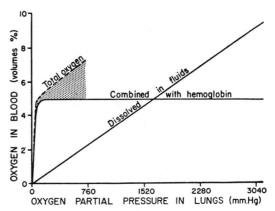

**Figure 43–7.** Effects of oxygen therapy in hypohemoglobinemic hypoxia.

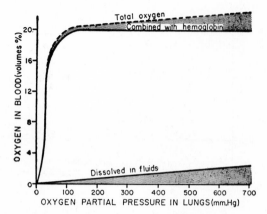

**Figure 43–8.** Effects of oxygen therapy in ischemic hypoxia.

crease in alveolar $P_{O_2}$ from a normal value of 100 mm. Hg to 600 mm. Hg increases the total oxygen in the blood (both combined with hemoglobin and dissolved) from 5 volumes per cent to 6.5 volumes per cent. This represents a 30 per cent increase in the amount of oxygen transported to the tissues, and 30 per cent is often the difference between life and death. However, Figure 43–6 also shows that oxygen therapy in a *hyperbaric pressure chamber* at pressures above atmospheric pressure (760 mm. Hg) can be of still much greater value.

In *hypoxia caused by circulatory deficiency,* also called *ischemic hypoxia,* the value of oxygen therapy is usually very slight, because the problem here is sluggish flow of blood and not insufficient oxygen. Figure 43–8 illustrates, however, that even normal blood can carry a small amount of extra oxygen to the tissues (about 10 per cent extra) when the alveolar oxygen concentration is increased to 600 mm. Hg. Here again, this 10 per cent difference may save the life of the patient, as, for example, following an acute heart attack which causes the cardiac output to fall very low.

In the different types of *hypoxia caused by inadequate tissue use of oxygen,* there is no abnormality of oxygen pickup by the lungs or of transport to the tissues. Instead, the tissues simply need more oxygen than can reach them or than the enzymes can utilize. Therefore, it is doubtful that oxygen therapy is of any benefit at all.

### DANGER OF HYPERCAPNIA DURING OXYGEN THERAPY

Much of the stimulus that helps to maintain ventilation in hypoxia results from hypoxic stimulation of the aortic and carotid chemoreceptors. In the preceding chapter it was noted that in chronic hypoxia, oxygen lack becomes a far more powerful stimulus to respiration than usual, sometimes increasing the ventilation as much as 5- to 7-fold. Therefore, during oxygen therapy, relief of the hypoxia occasionally causes pulmonary ventilation to decrease so low that lethal levels of hypercapnia develop. For this reason, oxygen therapy in hypoxia is sometimes contraindicated, particularly in conditions that otherwise tend to cause hypercapnia, such as uneven ventilation-perfusion ratio or depressed respiratory center activity.

## ABSORPTION OF ENTRAPPED AIR

Whenever air becomes entrapped anywhere in the body, whether in the alveoli as a result of bronchial obstruction, in the gastrointestinal tract, in one of the nasal or auditory cavities, or simply as air injected beneath the skin, it is usually absorbed in a few hours or days.

**Physical Principles of Air Absorption.** Figure 43–9 shows the progressive stages of air absorption from a cavity. The cavity to the left is filled with recently injected air which has become humidified almost instantaneously by the fluids surrounding the cavity. Therefore, this cavity contains the normal concentrations of oxygen, nitrogen, and water vapor found in humidified air at sea level. Note that the total pressure of these three gases added together equals 760 mm. Hg. Furthermore, the pressure in the cavity remains exactly the same during the entire process of absorption because the atmospheric pressure continually presses against the body, and the body's tissues are flexible enough that the atmospheric pressure is transmitted also to almost any gas cavity beneath the skin.

At first, the $P_{O_2}$ in the injected air is much greater than the $P_{O_2}$ in the interstitial fluid, so that oxygen begins to be absorbed rapidly. On the other hand, the nitrogen pressure ($P_{N_2}$) in the cavity is actually less than that in the interstitial fluid because nitrogen, unlike oxygen, is not metabolized by the tissues of the body. As a result, minute amounts of nitrogen actually diffuse into the cavity at first. After large amounts of oxygen have been absorbed, the $P_{O_2}$ falls

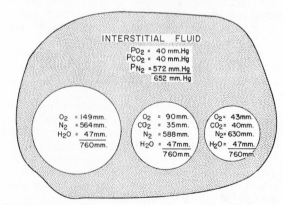

**Figure 43–9.** Absorption of air from an occluded cavity of the body.

to as little as 90 mm. Hg, as shown in the second cavity of Figure 43–9. This absorption of oxygen makes the cavity become smaller, now increasing the $P_{N_2}$ to a value greater than that in the interstitial fluid. Also, carbon dioxide diffuses into the cavity.

The final equilibrium state between the gases in the cavity and in the interstitial fluid is shown in the third cavity; the $P_{O_2}$ has decreased to about 43 mm. Hg, the $P_{CO_2}$ has increased to 40 mm. Hg, and the $P_{N_2}$ has increased to 630 mm. Hg. From then on, both oxygen and nitrogen are absorbed continually. As this reduces the size of the cavity, the carbon dioxide and water pressures increase to slightly higher than the pressures of the surrounding fluids. As a consequence, the carbon dioxide and water vapor are also absorbed. This process continues until all the gases leave the cavity and the cavity collapses totally.

# REFERENCES

Avery, M. E., Wang, N., and Taeusch, H. W., Jr.: The lung of the newborn infant. *Sci. Amer., 228:*74, 1973.

Bass, H.: Assessment of regional pulmonary function with radioactive gases. *Prog. Nucl. Med., 3:*67, 1973.

Bendixen, H., et al.: Respiratory Care, 2nd Ed. St. Louis, The C. V. Mosby Company, 1969.

Cohen, A. B., and Gold, W. M.: Defense mechanisms of the lungs. *Ann. Rev. Physiol., 37:*325, 1975.

Comroe, J. H., Jr., Forster, R. E., II, Dubois, A. B., Briscoe, W. A., and Carlsen, E.: The Lung: Clinical Physiology and Pulmonary Function Tests, 2nd Ed. Chicago, Year Book Medical Publishers, Inc., 1962.

Crofton, J., and Douglas, A.: Respiratory Diseases. Philadelphia, J. B. Lippincott Company, 1975.

Cumming, G., and Semple, S. J.: Disorders of the Respiratory System. Philadelphia, J. B. Lippincott Company, 1973.

Domm, B. M., and Vassallo, C. L.: Pulmonary function testing. *Clin. Anesth., 9:*191, 1973.

Filley, G.: Pulmonary Insufficiency and Respiratory Failure. Philadelphia, Lea & Febiger, 1967.

Gelb, A. F., and MacAnally, B. J.: Early detection of obstructive lung disease by analysis of maximal expiratory flow-volume curves. *Chest, 64:*749, 1973.

Gross, N. J.: Bronchial Asthma. Hagerstown, Maryland, Harper & Row, Publishers, 1974.

Guyton, A. C., and Farish, C. A.: A rapidly responding continuous oxygen consumption recorder. *J. Appl. Physiol., 14:*143, 1959.

Hemingway, A.: Measurement of Airway Resistance with the Body Plethysmograph. Springfield, Ill., Charles C Thomas, Publisher, 1973.

Killough, J. H.: Protective mechanisms of the lungs; pulmonary disease; pleural disease. *In* Sodeman, W. A., Jr., and Sodeman, W. A. (eds.): Pathologic Physiology: Mechanisms of Disease, 5th Ed. Philadelphia, W. B. Saunders Company, 1974, p. 393.

Middleton, E., Jr.: Autonomic imbalance in asthma with special reference to beta adrenergic blockade. *Adv. Intern. Med., 18:*177, 1972.

Moore, F. D.: Postoperative pulmonary insufficiency: anoxia, the shunted lung, and mechanical assistance. *Cardiovasc. Clin., 3:*121, 1971.

Nilsson, N. J.: Oximetry. *Physiol. Rev., 40:*1, 1960.

Rochester, D. F., and Enson, Y.: Current concepts in the pathogenesis of the obesity-hypoventilation syndrome. Mechanical and circulatory factors. *Amer. J. Med., 57:*402, 1974.

Schonell, M.: Respiratory Medicine. New York, Churchill Livingstone, Div. of Longman, Inc., 1974.

Secker-Walker, R. H., and Evens, R. G.: The clinical application of computers in ventilation-perfusion studies. *Prog. Nucl. Med., 3:*166, 1973.

Siegel, B. A., and Potchen, E. J.: Radio nuclide studies of pulmonary function. Anatomic and physiologic considerations. *Prog. Nucl. Med., 3:*49, 1973.

Slonim, N. B., Bell, B. P., and Christensen, S. E.: Cardiopulmonary Laboratory Basic Methods and Calculations. Springfield, Ill., Charles C Thomas, Publisher, 1974.

Staub, N. C.: Pulmonary edema. *Physiol. Rev., 54:*678, 1974.

Tysinger, D. S., Jr.: The Clinical Physics and Physiology of Chronic Lung Disease, Inhalation Therapy, and Pulmonary Function Testing. Springfield, Ill., Charles C Thomas, Publisher, 1973.

# PART VIII

# AVIATION, SPACE, AND DEEP SEA DIVING PHYSIOLOGY

# 44

# Aviation, High Altitude, and Space Physiology

As man has ascended to higher and higher altitudes in aviation, in mountain climbing, and in space vehicles, it has become progressively more important to understand the effects of altitude and low gas pressures on the human body. In the early days of aviation only two factors were of concern: (1) the effects of hypoxia on the body and (2) the effects of physical factors of high altitude, such as temperature and ultraviolet radiation. When airplanes were further developed, it was soon learned that they could be built to withstand acceleratory forces far greater than the human body can stand. Now, with the space age at hand, all these problems have become multiplied to the point that the physical conditions in the spacecraft must be created artificially.

The present chapter deals with all these problems: first, the hypoxia at high altitudes, second, the other physical factors affecting the body at high altitudes, and, third, the tremendous acceleratory forces that occur in both aviation and space physiology.

## EFFECTS OF LOW OXYGEN PRESSURE ON THE BODY

**Barometric Pressures at Different Altitudes.** As a prelude to discussing the effects of low oxygen pressure on the body, we must recall that the total pressure of all the gases in the air, the *barometric pressure,* decreases as one rises to progressively higher and higher altitudes. Table 44–1 gives the pressures at different altitudes, showing that at sea level the pressure is 760 mm. Hg, while at 10,000 feet it is only 523 mm. Hg, and at 50,000 feet, 87 mm. Hg. This decrease in barometric pressure is the basic cause of all the hypoxia problems in high altitude physiology, for as the barometric pressure decreases,

the oxygen pressure decreases proportionately, remaining at all times slightly less than 21 per cent of the total barometric pressure.

**Oxygen Partial Pressures at Different Elevations.** Table 44–1 also shows that the partial pressure of oxygen ($Po_2$) in dry air at sea level is approximately 159 mm. Hg, though this can be decreased as much as 10 mm. when large amounts of water vapor exist in the air. The $Po_2$ at 10,000 feet is approximately 110 mm. Hg, at 20,000 feet 73 mm. Hg, and at 50,000 feet 18 mm. Hg.

## *ALVEOLAR $Po_2$ AT DIFFERENT ELEVATIONS*

Obviously, when the $Po_2$ in the atmosphere decreases at higher elevations, a decrease in alveolar $Po_2$ is also to be expected. At low altitudes the alveolar $Po_2$ does not decrease quite so much as the atmospheric $Po_2$ because increased pulmonary ventilation helps to compensate for the diminished atmospheric oxygen. But at higher altitudes the alveolar $Po_2$ decreases even more than atmospheric $Po_2$ for peculiar reasons that are explained as follows:

**Effect of Carbon Dioxide and Water Vapor on Alveolar Oxygen.** Even at high altitudes carbon dioxide is continually excreted from the pulmonary blood into the alveoli. Also, water vaporizes into the alveolar space from the respiratory surfaces. Therefore, these two gases dilute the oxygen and nitrogen already in the alveoli, thus reducing the oxygen concentration.

The presence of carbon dioxide and water vapor in the alveoli becomes exceedingly important at high altitudes because the total barometric pressure falls to low levels while the pressures of carbon dioxide and water vapor do not fall comparably. Water vapor pressure remains 47 mm. Hg as long as the body

586

**TABLE 44–1.**  Effects of Low Atmospheric Pressures on Alveolar Gas Concentrations and on Arterial Oxygen Saturation

| Altitude (ft.) | Barometric Pressure (mm. Hg) | $P_{O_2}$ in Air (mm. Hg) | Breathing Air | | | Breathing Pure Oxygen | | |
|---|---|---|---|---|---|---|---|---|
| | | | $P_{CO_2}$ in Alveoli (mm. Hg) | $P_{O_2}$ in Alveoli (mm. Hg) | Arterial Oxygen Saturation (%) | $P_{CO_2}$ in Alveoli (mm. Hg) | $P_{O_2}$ in Alveoli (mm. Hg) | Arterial Oxygen Saturation (%) |
| 0 | 760 | 159 | 40 | 104 | 97 | 40 | 673 | 100 |
| 10,000 | 523 | 110 | 36 | 67 | 90 | 40 | 436 | 100 |
| 20,000 | 349 | 73 | 24 | 40 | 70 | 40 | 262 | 100 |
| 30,000 | 226 | 47 | 24 | 21 | 20 | 40 | 139 | 99 |
| 40,000 | 141 | 29 | 24 | 8 | 5 | 36 | 58 | 87 |
| 50,000 | 87 | 18 | 24 | 1 | 1 | 24 | 16 | 15 |

temperature is normal, regardless of altitude; and the pressure of carbon dioxide falls from about 40 mm. Hg at sea level to about 24 mm. Hg at extremely high altitudes because of increased respiration.

Now let us see how the pressures of these two gases affect the available space for oxygen. Let us assume that the total barometric pressure falls to 100 mm. Hg; 47 mm. Hg of this must be water vapor, leaving only 53 mm. Hg for all the other gases. Under acute exposure to high altitude, 24 mm. Hg of the 53 mm. Hg must be carbon dioxide, leaving a remaining space of only 29 mm. Hg. If there were no use of oxygen by the body, one-fifth of this 29 mm. Hg would be oxygen, and four-fifths would be nitrogen; or, the $P_{O_2}$ in the alveoli would be 6 mm. Hg. However, by this time the person's tissues would be almost totally anoxic so that even most of this last remaining amount of alveolar oxygen would be absorbed into the blood, leaving not more than 1 mm. Hg oxygen pressure in the alveoli. Therefore, at a barometric pressure of 100 mm. Hg, the person could not possibly survive when breathing air. But the effect is very much different if the person is breathing pure oxygen, as we shall see in the following discussions.

A simple formula for approximating alveolar $P_{O_2}$ is the following:

$$\text{Alveolar } P_{O_2} = \frac{P_B - P_{CO_2} - 47}{5} - P_{O_2} \text{ LOSS}$$

In this formula $P_B$ is barometric pressure, the value 47 is the vapor pressure of water, and $P_{O_2}$ LOSS is the oxygen pressure decrease caused by oxygen uptake into the blood.

**Aveolar $P_{O_2}$ at Different Altitudes.**  Table 44–1 also shows the $P_{O_2}$'s in the alveoli at different altitudes when one is breathing air and when breathing pure oxygen. When breathing air, the alveolar $P_{O_2}$ is 104 mm. Hg. at sea level; it falls to approximately 67 mm. Hg at 10,000 feet and to only 1 mm. Hg at 50,000 feet.

**Saturation of Hemoglobin With Oxygen at Different Altitudes.**  Figure 44–1 illustrates arterial

oxygen saturation at different altitudes when breathing air and when breathing oxygen, and the actual per cent saturation at each 10,000 foot level is given in Table 44–1. Up to an altitude of approximately 10,000 feet, even when air is breathed, the arterial oxygen saturation remains at least as high as 90 per cent. However, above 10,000 feet the arterial oxygen saturation falls progressively, as illustrated by the left-hand curve of the figure, until it is only 70 per cent at 20,000 feet altitude and still less at higher altitudes.

## EFFECT OF BREATHING PURE OXYGEN ON THE ALVEOLAR $P_{O_2}$ AT DIFFERENT ALTITUDES

Referring once again to Table 44–1, note that when a person breathes air at 30,000 feet, his alveolar $P_{O_2}$ is only 21 mm. Hg even though the barometric pressure is 226 mm. Hg. This difference is caused primarily by the fact that a considerable proportion of his alveolar air is nitrogen. But if he breathes pure oxygen instead of air, most of the space in the alveoli formerly occupied by nitrogen now becomes oc-

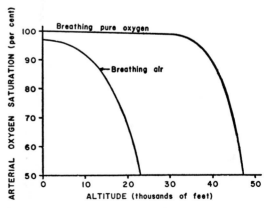

**Figure 44–1.**  Effect of low atmospheric pressure on arterial oxygen saturation when breathing air and when breathing pure oxygen.

cupied by oxygen instead. Theoretically, at 30,000 feet the aviator could have an alveolar Po$_2$ of 139 mm. Hg instead of the 21 mm. Hg that he has when he breathes air.

The second curve of Figure 44–1 illustrates the arterial oxygen saturation at different altitudes when one is breathing pure oxygen. Note that the saturation remains above 90 per cent until the aviator ascends to approximately 39,000 feet; then it falls rapidly to approximately 50 per cent at about 47,000 feet.

## THE "CEILING" WHEN BREATHING AIR AND WHEN BREATHING OXYGEN IN AN UNPRESSURIZED AIRPLANE

Comparing the two arterial oxygen saturation curves in Figure 44–1, one notes that an aviator breathing oxygen can ascend to far higher altitudes than one not breathing oxygen. For instance, the arterial saturation at 47,000 feet when breathing oxygen is about 50 per cent and is equivalent to the arterial oxygen saturation at 23,000 feet when breathing air. And, because a person ordinarily can remain conscious until the arterial oxygen saturation falls to 40 to 50 per cent, the ceiling for an unacclimatized aviator in an unpressurized airplane when breathing air is approximately 23,000 feet and when breathing pure oxygen about 47,000 feet, provided the oxygen-supplying equipment operates perfectly.

## EFFECTS OF HYPOXIA

Probably the earliest effect of hypoxia on bodily function is decreased proficiency of night vision. The amount of light needed for an aviator to see his surroundings must be increased approximately 23 per cent above normal at 5000 feet, 59 per cent at 10,000 feet, and 140 per cent at 16,000 feet. Thus, even the slightest decrease in oxygen saturation of the arterial blood depresses the function of the rods in the retina.

The rate of pulmonary ventilation ordinarily does not increase significantly until one has ascended to about 8000 feet. At this height the arterial oxygen saturation has fallen to approximately 93 per cent, at which level the chemoreceptors respond significantly. Above 8000 feet the chemoreceptor stimulatory mechanism progressively increases the ventilation until one reaches approximately 16,000 to 20,000 feet, at which altitude the ventilation has reached a maximum of approximately 65 per cent above normal. Further increase in altitude does not further activate the chemoreceptors.

Other effects of hypoxia, beginning at an altitude of approximately 12,000 feet, are drowsiness, lassitude, mental fatigue, sometimes headache, occasionally nausea, and sometimes euphoria. Most of these symptoms increase in intensity at still higher altitudes, the headache often becoming especially prominent and the cerebral symptoms sometimes progressing to the stage of twitchings or convulsions

and ending, above 23,000 feet in the unacclimatized person, in coma.

One of the most important effects of hypoxia is decreased mental proficiency, which decreases judgment, memory, and the performance of discrete motor movements. Ordinarily these abilities remain absolutely normal up to approximately 9000 feet, and may be completely normal for a short time up to elevations of 15,000 feet. But, if the aviator is exposed to hypoxia for a long time, his mental proficiency, as measured by reaction times, handwriting, and other psychological tests, may decrease to 80 per cent of normal even at altitudes as low as 11,000 feet. If an aviator stays at 15,000 feet for one hour, his mental proficiency ordinarily will have fallen to approximately 50 per cent of normal, and after 18 hours at this level to approximately 20 per cent of normal.

**Respiratory Center Depression as the Cause of Death in Hypoxia.** When a person develops severe hypoxia—that is, to the stage of coma—the respiratory center itself often becomes depressed within another few minutes because of metabolic deficit in its neuronal cells. This counteracts the stimulatory effect of the chemoreceptor mechanism, and the respiration, instead of increasing further, decreases precipitously until it actually ceases.

**Effects of Sudden Exposure to Low Po$_2$.** At times the aviator who has been flying at a very high altitude with special oxygen equipment or in a pressurized cabin suddenly becomes detached from his oxygen supply, or the cabin decompresses. The alveolar Po$_2$ falls within a few seconds to a low value, but because oxygen is stored in his body fluids (combined with hemoglobin and with various oxygen carriers of the tissues), a short time elapses before the body suffers dire results from oxygen lack.

Figure 44–2 illustrates graphically the time that ordinarily elapses, first, before the aviator shows signs

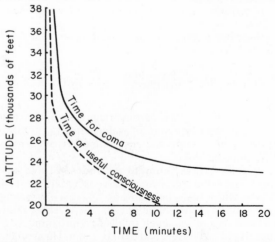

**Figure 44–2.** Time of exposure to low oxygen concentrations required to cause diminished consciousness or coma. (From Armstrong: Principles and Practice of Aviation Medicine. The Williams & Wilkins Co., 1943.)

of diminished consciousness, and, second, before actual coma results. Note from this figure that at 38,000 feet diminished consciousness begins in approximately 30 seconds, and coma results in approximately one minute, whereas at 28,000 feet diminished consciousness begins in approximately one minute, and coma occurs in approximately three minutes.

## ACCLIMATIZATION TO LOW $Po_2$

If a person remains at high altitudes for days, weeks, or years, he gradually becomes acclimatized to the low $Po_2$ so that it causes fewer and fewer deleterious effects to his body and also so that it becomes possible for him to work harder or to ascend to still higher altitudes. The five principal means by which acclimatization comes about are: (1) increased pulmonary ventilation, (2) increased hemoglobin in the blood, (3) increased diffusing capacity of the lungs, (4) increased vascularity of the tissues, and (5) increased ability of the cells to utilize oxygen despite the low $Po_2$.

**Increased Pulmonary Ventilation.** On immediate exposure to low $Po_2$, the hypoxic stimulation of the chemoreceptors increases alveolar ventilation to a maximum of about 65 per cent. This is an immediate compensation for the high altitude, and it alone allows the person to rise several thousand feet higher than would be possible without the increased ventilation. Then, if he remains at a very high altitude for several days, his ventilation gradually increases to as much as five to seven times normal. The basic cause of this gradual increase is the following:

The immediate 65 per cent increase in pulmonary ventilation on rising to a high altitude blows off large quantities of carbon dioxide, reducing the $Pco_2$ and increasing the pH of the body fluids. Both of these changes *inhibit* the respiratory center and thereby *oppose the stimulation by the hypoxia*. However, during the ensuing three to five days, this inhibition fades away, allowing the respiratory center now to respond with full force to the chemoreceptor stimuli resulting from hypoxia, and the ventilation increases to about five to seven times normal. The cause of this fading inhibition is unknown, but there is some evidence that it results from reduced CSF bicarbonate ion and perhaps also from reduced brain tissue bicarbonate ion, and this in turn decreases the pH. (See Chap. 42.)

**Rise in Hemoglobin During Acclimatization.** It will be recalled from Chapter 5 that hypoxia is the principal stimulus for causing an increase in red blood cell production. Ordinarily, in full acclimatization to low oxygen the hematocrit rises from a normal value of 40 to 45 to an average of 60 to 65, with an average increase in hemoglobin concentration from the normal of 15 gm. per cent to about 22 gm. per cent.

In addition, the blood volume also increases, often by as much as 20 to 30 per cent, resulting in a total increase in circulating hemoglobin of as much as 50 to 90 per cent.

Unfortunately, this increase in hemoglobin and blood volume is a slow one, having almost no effect until after two to three weeks, reaching half development in a month or so, and becoming fully developed only after many months.

*Decreased Affinity of the Hemoglobin for Oxygen Under Hypoxic Conditions.* Within a few hours after the blood is first exposed to hypoxia at high altitudes, increased quantities of phosphate compounds are formed inside the red blood cells, and some of these combine with the hemoglobin to decrease its affinity for oxygen. One of these, 2,3-diphosphoglycerate, commonly called 2,3-DPG, is especially significant. Because of the reduced affinity for oxygen, the hemoglobin delivers the oxygen to the tissue cells at a higher $Po_2$. Therefore, when this phenomenon was first discovered a few years ago, it was believed that the increase in 2,3-DPG played an important role in promoting tissue oxygenation. However, two factors have subsequently been discovered that have partially changed this view. First, the decreased affinity for oxygen also decreases the pickup of oxygen by the hemoglobin in the lungs, and therefore it decreases the overall availability of oxygen to the tissues, which is more significant than the increased tendency for oxygen release by the hemoglobin at the tissue level. Second, the increase in 2,3-DPG usually does not last for a long period of time because of other acclimatization effects. Therefore, though a number of investigators have suggested that this phenomenon is an important mechanism of acclimatization, its importance has not been proved and is in serious doubt.

**Increased Diffusing Capacity During Acclimatization.** It will be recalled that the normal diffusing capacity for oxygen through the pulmonary membrane is approximately 21 ml. per mm. Hg pressure gradient per minute, and this diffusing capacity can increase as much as three-fold during exercise. A similar increase in diffusing capacity occurs at high altitude. Part of the increase probably results from greatly increased pulmonary capillary blood volume, which expands the capillaries and increases the surface through which oxygen can diffuse into the blood. Another part results from an increase in lung volume, which perhaps expands the surface area of the alveolar membrane. A final part results from an increase in pulmonary arterial pressure; this forces blood into greater numbers of alveolar capillaries than normally—especially in the upper parts of the lungs, which are poorly perfused under usual conditions.

**The Circulatory System in Acclimatization— Increased Vascularity.** The cardiac output often increases as much as 20 to 30 per cent immediately after a person ascends to a high altitude, but it usually falls back to normal within a few days. In the meantime, though, the blood flow through certain organs, such as the skin and kidneys, decreases, while the flow through the muscles, heart, brain, and other organs that normally require large quantities of oxygen increases. Furthermore, histologic studies of animals that have been exposed to low oxygen levels for months or years show *greatly increased vascularity* (increased numbers and sizes of capillaries) of the hypoxic tissues. This helps to explain what happens to the 20 to 30 per cent increase in blood volume, and it means that the blood comes into much closer contact with the tissue cells than normally.

**Cellular Acclimatization.** In animals that are native to altitudes of 13,000 to 17,000 feet, mitochondria and certain cellular oxidative enzyme systems are more plentiful than in sea level inhabitants. Therefore, it is presumed that acclimatized human beings as well as these animals can utilize oxygen more effectively than can their sea level counterparts.

## NATURAL ACCLIMATIZATION OF NATIVES LIVING AT HIGH ALTITUDES

Many natives in the Andes and in the Himalayas live at altitudes above 13,000 feet—one group in the Peruvian Andes actually living at an altitude of 17,500 feet and working a mine at an altitude of 19,000 feet. Many of these natives are born at these altitudes and live there all their lives. In all of the aspects of acclimatization listed above, the natives are superior to even the best acclimatized lowlanders, even though the lowlanders might have also lived at high altitudes for ten or more years. This process of acclimatization of the natives begins in infancy. The chest size, especially, is greatly increased whereas the body size is somewhat decreased, giving a high ratio of ventilatory capacity to body mass. In addition, their hearts, particularly the right heart which provides a high pulmonary arterial pressure to pump blood through a greatly expanded pulmonary capillary system, are considerably larger than the hearts of lowlanders.

The delivery of oxygen by the blood to the tissues is also highly facilitated in these natives. For instance, Figure 44–3 shows the hemoglobin-oxygen dissociation curves for natives who live at sea level and for their counterparts who live at 15,000 feet. Note that the arterial oxygen $Po_2$ in the natives at higher altitude is only 40 mm. Hg, but because of the greater quantity of hemoglobin the quantity of oxygen in the arterial blood is actually greater than in the blood of the natives at the lower altitude. Note also that the venous $Po_2$ in the high altitude natives is only 10 mm. Hg less than the venous $Po_2$ for the lowlanders, despite the low arterial $Po_2$ and the fact that the oxygen consumption in the highlanders is actually greater than that in the lowlanders, indicating that the tissues of the naturally acclimatized high altitude natives can utilize oxygen with greatly increased efficiency.

## WORK CAPACITY AT HIGH ALTITUDES; THE EFFECT OF ACCLIMATIZATION

Aside from the mental depression caused by hypoxia, as discussed earlier, the work capacity of all the muscles is also greatly decreased in hypoxia. This includes not only the skeletal muscles but also the cardiac muscle so that even the maximum level of cardiac output is reduced. In general, the work capacity is reduced in direct proportion to the decease in maximum rate of oxygen uptake that the body can achieve. Therefore, it is obvious that an acclimatized person has far greater work capacity at high altitudes than an unacclimatized person.

To give an idea of the importance of acclimatization for work capacity, consider this: The work capacities in per cent of sea level maximum for a normal person at an altitude of 17,000 feet are the following:

|  | *per cent* |
| --- | --- |
| Unacclimatized | 50 |
| Acclimatized for two months | 68 |
| Native living at 13,200 feet but working at 17,000 feet | 87 |

Thus, naturally acclimatized natives can achieve a daily work output even at these high altitudes almost equal to that of a normal person at sea level, but even well acclimatized lowlanders can almost never achieve this result.

To emphasize further the importance of natural acclimatization of natives, the Sherpas of the Himalayas can probably survive without oxygen for hours at altitudes as high as Mt. Everest—over 29,000 feet!

## CHRONIC MOUNTAIN SICKNESS

An occasional person who remains at high altitude too long develops chronic mountain sickness, in which the following effects occur: (1) the red cell

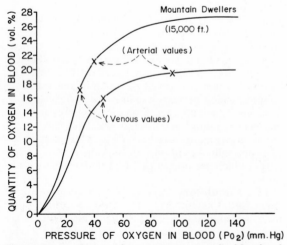

**Figure 44–3.** Oxygen-dissociation curves for bloods of high-altitude and sea-level residents, showing the respective arterial and venous $Po_2$'s and oxygen contents as recorded in their native surroundings. (From "Oxygen-dissociation curves for bloods of high-altitude and sea-level residents." PAHO Scientific Publication No. 140, Life at High Altitudes, 1966.)

mass and hematocrit become exceptionally high, (2) the pulmonary arterial pressure becomes elevated even more than the normal elevation that occurs during acclimatization, (3) the right heart becomes greatly enlarged, (4) the peripheral arterial pressure begins to fall, (5) congestive failure ensues, and (6) death often follows unless the person is removed to a lower altitude. The cause of this sequence of events is probably the excess red cell mass and the vasoconstrictor effect that occurs in the lungs during hypoxia, a mechanism that was discussed in Chapter 24 in which the blood vessels supplying hypoxic alveoli constrict in an effort to divert the blood toward aerated alveoli. Fortunately, most of these persons recover promptly when they are removed to a lower altitude.

# DECOMPRESSION EFFECTS AT HIGH ALTITUDES

**Decompression Sickness.** In addition to the detrimental effects of hypoxia at high altitudes, a rapid decrease in barometric pressure can directly harm the body by causing gas bubbles in the body fluids. All degrees of damage can occur, ranging from mild pain to death. This condition is most frequently called *decompression sickness,* though it is sometimes also known as *dysbarism.*

Decompression sickness occurs much more frequently and much more severely in deep sea divers than in aviators. Therefore, its basic causes and effects are discussed in detail in the following chapter. Briefly, it is caused by a decrease in the barometric pressure below the total pressure of all the gases dissolved in the body fluids, and this allows the gases to escape from the dissolved state to form gas bubbles in the tissue fluids, as will be explained in the next chapter.

The usual cause of decompression sickness in aviation is rapid ascent in the unpressurized airplane—ascending to over 25,000 to 30,000 feet in a few minutes. Slower ascents rarely cause the sickness.

**Explosive Decompression.** Experimentally, human beings have been decompressed explosively from sea level pressure to barometic pressures equal to those above 50,000 feet, and animals have been explosively decompressed to an almost complete vacuum. Essentially no harm results from the suddenness of the change in pressure. Indeed, even the ears ordinarily do not suffer any damage because the increasing volume of air in the middle ear automatically opens the eustachian tube so that the air flows outward without damage to the tympanum. Perhaps the most significant effect is the sudden increase in volume of gastrointestinal gases, for any gas in the gastrointestinal tract is expanded several-fold when one is explosively decompressed at a level of 50,000 ft. Ordinarily, most of these gases can be expelled as flatus, for most gastrointestinal gases are in the colon, which is in direct communication with the anus, and the remainder which is ordinarily in the stomach, can be released by belching.

**Boiling of the Body Fluids.** The barometric pressure falls to 47 mm. Hg at an altitude of approximately 63,000 feet. Therefore, if an aviator should be subjected to the atmosphere above 63,000 feet, his total alveolar pressure would be less than the vapor pressure of water in the alveoli. Consequently, the water in the alveoli would immediately boil. Also, water vapor bubbles would appear all through the body fluids. In fact, in an animal suddenly subjected to a pressure equal to that at 70,000 feet, approximately 2.5 per cent of his body weight boils away in the three minutes that he remains alive. This means that in a human being explosively decompressed to an elevation of 70,000 feet, approximately 4 pounds of water would probably boil away before death ensued in approximately three minutes.

# EFFECTS OF ACCELERATORY FORCES ON THE BODY IN AVIATION AND SPACE PHYSIOLOGY

Because of rapid changes in velocity and direction of motion in airplanes and space ships, several types of acceleratory forces often affect the body during flight. At the beginning of flight, simple linear acceleration occurs; at the end of flight, deceleration; and, every time the airplane turns, angular and centrifugal acceleration occur. In aviation physiology it is usually centrifugal acceleration that demands greatest consideration because the structure of the airplane is capable of withstanding much greater centrifugal acceleration than is the human body.

## *CENTRIFUGAL ACCELERATORY FORCES*

When an airplane makes a turn, the force of centrifugal acceleration is determined by the following relationship;

$$f = \frac{mv^2}{r}$$

in which $f$ is the centrifugal acceleratory force, $m$ is the mass of the object, $v$ is the velocity of travel, and $r$ is the radius of curvature of the turn. From this formula it is obvious that as the velocity increases, the force of centrifugal acceleration increases in proportion to the square of the velocity. It is also obvious that the force of acceleration is directly proportional to the sharpness of the turn.

**Measurement of Acceleratory Force—"G."** When a subject is simply sitting in his seat, the force with which he is pressing against the seat results from the pull of gravity, and it is equal to his weight. The intensity of this force is 1 "G" because it is equal to the pull of gravity. If the force with which he presses against his seat becomes five times his normal weight during a pull-out from a dive, the force acting upon the seat is 5 G.

If the airplane goes through an outside loop so that the pilot is held down by his seat belt, *negative G* is applied to his body, and, if the force with which he is thrown against his belt is equal to the weight of his body, the negative force is —1 G.

**Effects of Centrifugal Acceleratory Force on the Body.** *Effects on the Circulatory System.* The most important effect of centrifugal acceleration is on the circulatory system because blood is mobile and can be translocated by centrifugal forces. Centrifugal forces also tend to displace the tissues, but, because of their more solid structure, they only sag—ordinarily not enough to cause abnormal function.

When the aviator is subjected to *positive G* his blood is centrifuged toward the lower part of his body. Thus, if the centrifugal acceleratory force is 5 G and the subject is in a standing position, the hydrostatic pressure in the veins of the feet is five times normal, or approximately 450 mm. Hg, and even in the sitting position this pressure is nearly 300 mm. Hg. As the pressure in the vessels of the lower part of the body increases, the vessels passively dilate, and a major proportion of the blood from the upper part of the body is translocated into these lower vessels. Because the heart cannot pump unless blood returns to it, the greater the quantity of blood "pooled" in the lower body the less becomes the cardiac output.

Figure 44–4 illustrates the effect of different degrees of acceleration on systemic arterial pressure in an aviator in the sitting position, showing that when

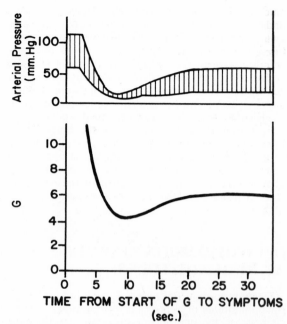

**Figure 44–5.** *Lower curve:* Time required at different G to cause blackout. *Top curve:* Changes in systolic and diastolic arterial pressures following abrupt and continuing exposure to an angular acceleratory force of 3.3 G. (Modified from Martin and Henry: *J. Aviat Med.,* 22:382, 1951.)

the acceleration rises to 4 G the systemic arterial pressure at the level of the heart falls to approximately 40 mm. Hg. As a result, positive G diminishes blood flow to the brain. Acceleration greater than 4 to 6 G ordinarily causes "blackout" of vision within a few seconds and then unconsciousness shortly thereafter. The time required for these symptoms to appear is shown in Figure 44–5. The upper part of this figure shows that the arterial pressure recovers considerably after about 15 seconds of exposure to 3.3 G of acceleration; this recovery is caused by activation of the baroreceptor reflexes

*Effects on the Vertebrae.* Extremely high acceleratory forces for even a fraction of a second can fracture the vertebrae. The degree of positive acceleration that the average person can withstand in the sitting position before vertebral fracture occurs is approximately 20 G. This force also approaches the limit of safety for the structural elements of most airplanes. Therefore, any attempt to utilize such intense positive G as this in maneuvers would be extremely dangerous for reasons other than simply the effects on the body.

**Transverse G.** The human body can withstand tremendous transverse acceleratory forces (forces applied along the antero-posterior axis of the body —lying down in the airplane, for instance). If the transverse acceleratory forces are applied rather uniformly over large areas of the body, as much as 100 G can be withstood for a fraction of a second, and as

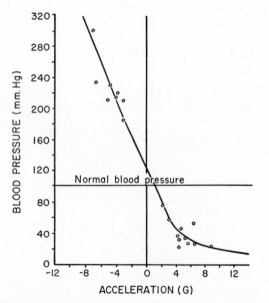

**Figure 44–4.** Effect of angular acceleratory forces on arterial pressure measured at the heart. (From Armstrong: Principles and Practice of Aviation Medicine. The Williams & Wilkins Co., 1943.)

much as 15 to 25 G can be withstood for many seconds without serious effects other than occasional collapse of a lung, which is not a lethal effect. Therefore, when very large acceleratory forces are to be involved, the aviator or astronaut flies in the semi-reclining or lying position.

**Negative G.** The effects of negative G on the body are less dramatic acutely but possibly more damaging permanently than the effects of positive G. An aviator can usually go through outside loops up to negative acceleratory forces of −4 to −5 G without harm except for intense momentary hyperemia of the head, though occasionally psychotic disturbances lasting for 15 to 20 minutes occur thereafter as a result of brain edema.

Figure 44–4 shows that the arterial pressure at the level of the heart increases greatly, as would be expected, and the pressure in the head obviously increases far more, to as high as 400 mm. Hg. An interesting but paradoxical effect is often caused by this very high pressure: it elicits such an extreme baroreceptor reflex that severe vagal slowing of the heart occurs, sometimes actually stopping the heart for 5 to 10 seconds.

If the arterial pressure becomes great enough, it can cause some of the small vessels on the surface of the head and possibly in the brain to rupture. However, the vessels inside the cranium show less tendency for rupture than would be expected, for the cerebrospinal fluid is centrifuged toward the head at the same time that blood is centrifuged toward the cranial vessels, and the greatly increased pressure of this fluid acts as a cushioning buffer on the outside of the brain to prevent vascular rupture. Even so, animals exposed to negative acceleratory forces of 20 to 40 G have developed subarachnoid hemorrhages, and undoubtedly such effects could occur in the human being.

Because the eyes are not protected by the cranium, intense hyperemia occurs in them during negative G. As a result, the eyes often become temporarily blinded with "redout."

**Protection of the Body Against Centrifugal Acceleratory Forces.** Specific procedures and apparatus have been developed to protect aviators against the circulatory collapse that occurs during positive G. First, if the aviator tightens his abdominal muscles to an extreme degree and leans forward to compress his abdomen, he can prevent some of the pooling of blood in the large vessels of the abdomen, thereby delaying the onset of blackout. Also, special "anti-G" suits have been devised to prevent pooling of blood in the lower abdomen and legs. The simplest of these applies positive pressure to the legs and abdomen by inflating compression bags as the G increases. Theoretically, a pilot submerged in a tank or suit of water could withstand very high degrees of acceleration, both positive and negative, for the pressures developed in the water during centrifugal acceleration almost exactly balance the forces acting on the body. Unfortunately, however, the presence of air in the lungs still allows displacement of the heart, the lung tissues, and the diaphragm into seriously abnormal positions despite submersion in water. Therefore, even when this procedure is used, the limits of safety are about 15 to 20 G.

## EFFECTS OF LINEAR ACCELERATORY FORCES ON THE BODY

**Acceleratory Forces in Space Travel.** In contrast to aviation physiology, a spacecraft cannot make rapid turns; therefore, centrifugal acceleration is of little importance except when the spacecraft goes into abnormal gyrations. On the other hand, blast-off acceleration and landing deceleration might be tremendous; both of these are types of linear acceleration.

Figure 44–6 illustrates a typical profile of the acceleration during blast-off in a three-stage spacecraft, showing that the first stage booster causes acceleration as high as 9 G and the second stage booster as high as 8 G. In the standing position the human body could not withstand this much acceleration, but in a semi-reclining position *transverse to the axis of acceleration,* this amount of acceleration can be withstood with ease despite the fact that the acceleratory forces continue for as long as five minutes at a time. Therefore, we see the reason for the reclining seats used by the astronauts.

Problems also occur during deceleration when the spacecraft re-enters the atmosphere. A person traveling at Mach 1 (the speed of sound and of fast airplanes) can be safely decelerated in a distance of approximately 0.12 mile, whereas a person traveling at a speed of Mach 100 (a speed possible in interplanetary space travel) requires a distance of about 10,000 miles for safe deceleration. The principal reason for this difference is that the total amount of energy that must be dispelled during deceleration is proportional to the *square* of the velocity, which alone increases the distance 10,000-fold. But, in addition to this, a human being can withstand far less

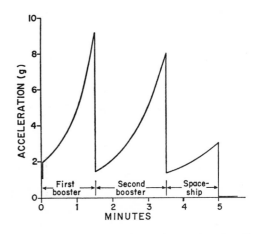

**Figure 44–6.** Acceleratory forces during the take-off of a spacecraft.

deceleration if it lasts for a long time than for a short time. Therefore, deceleration must be accomplished much more slowly from the very high velocities than is necessary at the slower velocities.

**Deceleratory Forces Associated with Parachute Jumps.** When the parachuting aviator leaves the airplane, his velocity of fall is exactly 0 feet per second at first. However, because of the acceleratory force of gravity, within 1 second his velocity of fall is 32 feet per second (if there is no air resistance); in two seconds it is 64 feet per second; etc. However, as his rate of fall increases, the air resistance tending to slow his fall also increases. Finally, the deceleratory force of the air resistance is equal to the acceleratory force of gravity so that by the time he has fallen for approximately 12 seconds and a distance of 1400 feet, he will be falling at a "terminal velocity" of 109 to 119 miles per hour (175 feet per second). This is illustrated by the curve in Figure 44-7. However, if the pilot jumps from a very high altitude, where the atmosphere offers little resistance, the terminal velocity is much greater than 175 feet per second, but it slows to 175 feet per second as the body reaches the higher density atmosphere close to earth.

If the parachutist has already reached the terminal velocity of fall before he opens his parachute, an "opening shock load" of approximately 1200 pounds occurs on the parachute shrouds.

The usual size parachute slows the fall of the parachutist to approximately one-ninth the terminal velocity. In other words, the speed of landing is approximately 20 feet per second, and the force of impact against the earth is approximately 1/81 the impact force without a parachute. Even so, the force of impact is still great enough to cause considerable damage to the body unless the parachutist is properly trained in landing, because it is not evident as one descends how rapidly the earth is approaching. Actually, the force of impact with the earth is approximately the same as that which would be experienced from jumping from a height of about 7 feet. If the parachutist is not careful, his senses will allow him to strike the earth with his legs still extended, and this will result in tremendous deceleratory forces along the skeletal axis of his body, resulting in fracture of the pelvis, of a vertebra, or a leg. Consequently, the trained parachutist strikes the earth with knees bent to cushion the shock of landing.

## PERCEPTIONS OF EQUILIBRIUM AND TURNING IN BLIND FLYING

The various sensations of equilibrium will be discussed in Chapter 52, but, in essence, any time the position of the head is not along the vertical axis of forces applied to the body, the otoliths in the utricles of the labyrinths apprise the psyche of this lack of equilibrium. However, the degree of proficiency of equilibrium perceptions has certain limits. For instance, the body must lean forward as much as 5 to 10 degrees before the utricles sense this forward leaning. When the aviator is not flying blind, he can perceive forward leaning of the airplane by observing the ground, but, when he is flying blind, it is necessary that appropriate instruments be available to indicate the rate of descent. Also, the utricle may fail to apprise the pilot of ascent up to an error of as much as 24 degrees. Obviously, then, blind flying often results in stall of the airplane unless appropriate instruments are available.

Perhaps the least effective of the perceptive organs in blind flying are those for turning, because the organs that perceive turning—the semicircular canals—are excited only for the first few seconds after a turn begins and only for the first few seconds after the turn ends. Consequently, an airplane may gradually go into a turn and the aviator might perceive that he is going into the turn, but once in it he usually loses all perception of continuing to turn. Furthermore, if he goes into the turn slowly—at an angular acceleratory rate less than 2 degrees per second²—he may not even perceive the fact that he enters the turn. Consequently, of especial importance among the instruments for blind flying are those that apprise the aviator of the direction of travel of the airplane and the rate of turn.

# PROBLEMS OF TEMPERATURE IN AVIATION AND SPACE PHYSIOLOGY

The cold temperatures of the upper atmosphere involve essentially the same physiological problems as cold temperatures on the surface of the earth; these problems will be discussed in Chapter 72. Therefore, for the present it shall only be pointed out how the temperatures change as one ascends to higher and higher elevations. If the temperature at the earth's surface is approximately 20°C., temperatures at different altitudes are approximately the following:

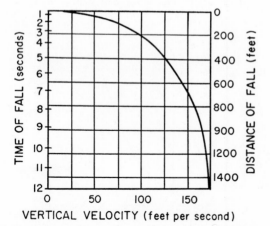

**Figure 44–7.** Velocity of fall of a human body from a high altitude, showing the attainment of a "terminal velocity." (From Armstrong: Principles and Practice of Aviation Medicine. The Williams & Wilkins Co., 1943.)

| feet | °C. |
|---|---|
| 0 | 20 |
| 10,000 | 0 |
| 20,000 | −22 |
| 30,000 | −44 |
| 40,000 | −55 |

It is obvious from these values that special clothing or special heating apparatus must be designed for flying at high altitudes.

**Temperature in Space.** Though the temperature of the air falls to about −55° C. several miles above the earth, the temperature rises to a very high value in space. In the ionosphere several hundred miles above the earth and in space, the kinetic energy of the few molecules, atoms, and ions is extreme. The reason for this is that any particles that can escape this far away from the gravitational pull of the celestial bodies must have a great velocity. As a result, by the time the spacecraft has reached an altitude above 350 miles the temperature of the surrounding particles is about 3000°C. Yet, strangely enough, this has almost no effect on the temperature of the spacecraft because of the *sparsity* of these particles—that is, they are far too few to impart any significant amount of heat to the spacecraft. Instead, the temperature of the spacecraft is determined by the relative *absorption of radiant energy* from the sun versus the *re-radiation of energy* away from the spacecraft. Different coatings for the spacecraft have different absorptive and radiation characteristics. Therefore, with appropriate coating and by orientation of the spacecraft appropriately with respect to the sun, the temperature inside can be made almost any desired value.

# RADIATION AT HIGH ALTITUDES AND IN SPACE

**Radiation in the Atmosphere.** The electromagnetic radiations change considerably between the surface of the earth and at high altitudes because the atmosphere filters some of the radiations from the sun's rays before they reach the earth's surface. Ordinarily most of the ultraviolet light is absorbed before it reaches the earth's surface. Consequently, a person exposed to the sun in the upper atmosphere is many times as likely to develop sunburn as on the earth's surface.

Approximately 18 per cent of the visible light is filtered from the sun's rays before they strike the surface of the earth even on a perfectly clear day. Therefore, the brightness of the sun is 1.2 times as great in the upper atmosphere as on earth. This makes the earth less distinct to the aviator than otherwise for two reasons: first, he is looking from an area of greater brightness toward an area of lesser brightness, and, second, light is reflected from the atmosphere into his eyes. This reflected light especially blocks his vision of the horizon.

In aviation there is no significant hazard from gamma and x-rays, but this does become a problem in space physiology.

**Radiation Hazards in Space Physiology.** Large quantities of cosmic particles are continually bombarding the earth's upper atmosphere, some originating from the sun and some from outer space. The magnetic field of the earth traps many of these cosmic particles in two major belts around the earth called *Van Allen radiation belts*, illustrated in Figure

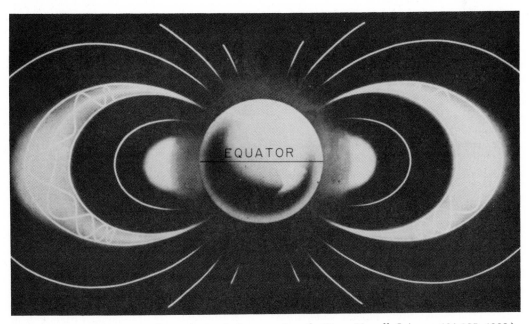

**Figure 44–8.** The hazardous Van Allen radiation belts around earth. (From Newell: *Science,* 131:385, 1960.)

44-8. The inner belt begins at an altitude of about 300 miles and extends to about 3000 miles. The outer belt begins at about 6000 miles and extends to 20,000 miles. As also illustrated in Figure 44–8, the inner belt extends only 30 degrees on each side of the equatorial plane, and the outer belt 70 to 75 degrees.

The types of radiation in the two Van Allen belts are almost entirely high energy electrons and protons, the outer belt being comprised almost entirely of electrons and the inner belt of both. The energy level in the inner belt is extremely high, many of the particles having energies as high as 40 Mev., which makes it almost impossible to shield a spacecraft adequately against this radiation. Even with best possible shielding, a person traversing these two belts in an interplanetary space trip would be expected to receive as much as 10 roentgens of radiation, which is about one-fortieth the lethal dose; and a person in a spacecraft orbiting the earth within one of these two belts could receive enough radiation in only a few hours to cause death. During solar flares (large amounts of "sun spot" activity), the intensity of the outer Van Allen belt increases tremendously, which could make space travel through this belt at this time extremely dangerous.

Thus, it is important to orbit spacecraft below an altitude of 200 to 300 miles, an altitude at which the radiation hazard is slight. Also, it is possible to minimize the radiation hazard during interplanetary space travel by leaving the earth or returning to earth near one of the poles rather than near the equator.

## "ARTIFICIAL CLIMATE" IN THE SEALED SPACECRAFT

Since there is no atmosphere in outer space, an artificial atmosphere and other artificial conditions of climate must be provided. The ability of a person to survive in this artificial climate depends entirely on appropriate engineering design.

Most important of all, the oxygen concentration must remain high enough and the carbon dioxide concentration low enough. The gas mixture most suitable for spacecraft travel is one of high oxygen concentration but one having just enough nitrogen to prevent fire hazards. The total pressure is maintained as low as 300 to 400 mm. Hg. This allows more than adequate oxygenation of the astronauts, and yet the low pressure minimizes the hazard of explosive decompression of the spacecraft.

For space travel lasting more than several weeks, it will be impractical to carry along an adequate oxygen supply and enough carbon dioxide absorbent. For this reason, "recycling techniques" have been developed for use over and over again of the same oxygen. These techniques also frequently include re-use of the same food and water. Basically, they involve (1) a method for removing oxygen from carbon dioxide, (2) a method for removing water from the human excreta, and (3) use of the human excreta

for resynthesizing or regrowing an adequate food supply.

Large amounts of energy are required for these processes, and the real problem at present is to derive enough energy from the sun's radiation to energize the necessary chemical reactions. Some recycling processes depend on purely physical procedures, such as distillation, electrolysis of water, and capture of the sun's energy by solar batteries, whereas others depend on biologic methods, such as use of algae, with its large store of chlorophyll, to generate foodstuffs by photosynthesis. Unfortunately, a completely practical system for recycling is yet to be achieved. The problem is the weight of the equipment that must be carried.

## WEIGHTLESSNESS IN SPACE

A person in an orbiting satellite or in any non-propelled spacecraft experiences weightlessness. That is, he is not drawn toward the bottom, sides, or top of the spacecraft but simply floats inside its chambers. The cause of this is not failure of gravity to pull on the body, because gravity from any nearby heavenly body is still active. However, the gravity acts on both the spacecraft and the person at the same time, and since there is no resistance to movement in space, both are pulled with exactly the same acceleratory forces and in the same direction. For this reason, the person simply is not attracted toward any wall of the spacecraft.

Weightlessness is mainly an engineering problem in regard to providing special techniques for eating and drinking (since food and water will not stay in open plates or glasses), special waste disposal systems, and adequate hand holds or other means for stabilizing the person in the spacecraft so that he can adequately control the operation of the ship.

**Physiological Problems of Weightlessness.** Fortunately, the physiological problems of weightlessness have not proved to be severe. Most of the problems that do occur appear to be related to two effects of the weightlessness: (1) translocation of fluids within the body because of failure of gravity to cause hydrostatic pressures, and (2) diminishment of physical activity because no strength of muscle contraction is required to oppose the force of gravity.

The observed effects of prolonged stay in space are the following: (1) decrease in blood volume, (2) decrease in red cell mass, (3) decreased work capacity, (4) decrease in maximum cardiac output, and (5) loss of calcium from the bones. Essentially these same effects also occur in persons lying in bed for an extended period of time. For this reason an extensive exercise program was carried out during the most recent sojourn by three astronauts in the Space Laboratory, and all of the above effects were greatly reduced. In previous Space Lab expeditions, in which the exercise program had been less vigorous, the astronauts had considerably lower work capacities for the first few days after returning to earth.

The effects of weightlessness that do occur usually reach their maximum within the first few weeks after the astronaut enters the space environment and fortunately are not progressive thereafter. Therefore, it appears that with an appropriate exercise program the physiological effects of weightlessness will not be a serious problem even during prolonged space voyages.

# REFERENCES

Adey, W. R.: The physiology of weightlessness. *Physiologist, 16:* 178, 1974.

Andrews, H. L.: Radiation Biophysics, 2nd Ed. Englewood Cliffs, N.J., Prentice-Hall, Inc., 1974.

Armstrong, H. G.: Aerospace Medicine, Baltimore, The Williams & Wilkins Company, 1961.

Brown, J. H. U.: Physiology of Man in Space. New York, Academic Press, Inc., 1963.

Bullard, R. W.: Physiological problems of space travel. *Ann. Rev. Physiol., 34:*205, 1972.

Burton, R. R., Leverett, S. D., Jr., and Michaelson, E. D.: Man at high sustained +Gz acceleration: a review. *Aerosp. Med., 45:*1115, 1974.

Ciba Foundation Symposia: High Altitude Physiology. New York, Churchill Livingstone, Div. of Longman, Inc., 1971.

Dill, D. B., Horvath, S. M., Dahns, T. E., Parker, R. E., and Lynch, J. R.: Hemoconcentration at altitude. *J. Appl. Physiol., 27:*514, 1969.

Frisancho, A. R.: Functional adaptation to high altitude hypoxia. *Science, 187:*313, 1975.

Hock, R. J.: The physiology of high altitude. *Sci. Amer., 222:*52, 1970.

Kellogg, R. H.: Altitude acclimatization, a historical introduction emphasizing the regulation of breathing. *Physiologist, 11:*37, 1968.

Korner, P. I.: Circulatory adaptations in hypoxia. *Physiol. Rev., 39:*687, 1959.

Lahiri, S.: Physiological responses and adaptations to high altitude. *In* Guyton, A. C. (ed.): MTP International Review of Science: Physiology. Vol. 7. Baltimore, University Park Press, 1974, p. 271.

Lee, D. H. K. (ed.): Physiology, Environment, and Man. New York, Academic Press, Inc., 1970.

McCally, M.: Hypodynamics and Hypogravics, The Physiology of Inactivity and Weightlessness. New York, Academic Press, Inc., 1969.

Pace, N.: Respiration at high altitude. *Fed. Proc., 33:* 2126, 1974.

Painter, R. B.: The action of ultraviolet light on mammalian cells. *Photophysiology, 5:*169, 1970.

Stickney, J. C.: Some problems of homeostasis in high-altitude exposure. *Physiologist, 15:* 349, 1972.

Smith, E. E., and Crowell, J. W.: Influence of hematocrit ratio on survival of unacclimatized dogs at simulated high altitude. *Amer. J. Physiol., 205:*1172, 1963.

Smith, E. E., and Crowell, J. W.: Role of the hematocrit in altitude acclimatization. *Aerospace Med., 38:*39, 1966.

# | 45 |

# Physiology of Deep Sea Diving and Other High Pressure Operations

When a person descends beneath the sea, the pressure around him increases tremendously. To keep his lungs from collapsing, air must be supplied also under high pressure, which exposes the blood in his lungs to extremely high alveolar gas pressures. Beyond certain limits these high pressures can cause tremendous alterations in the physiology of the body, which explains the necessity for the present discussion.

Also exposed to high atmospheric pressures are caisson workers who, in digging tunnels beneath rivers or elsewhere, often must work in a pressurized area to keep the tunnel from caving in. Here again, the same problems of excessively high gas pressures in the alveoli occur.

Before explaining the effects of high alveolar gas pressures on the body, it is necessary to review some physical principles of pressure and volume changes at different depths beneath the sea.

**Relationship of Sea Depth to Pressure.** A column of fresh water 34 feet high (sea water, 33 feet) exerts the same pressure at its bottom as all the atmosphere above the earth. Therefore, a person 33 feet beneath the ocean surface is exposed to a pressure of 2 atmospheres, 1 atmosphere of pressure caused by the air above the water and the second atmosphere by the weight of the water itself. At 66 feet the pressure is 3 atmospheres, and so forth, in accord with the table in Figure 45–1.

**Effect of Depth on the Volume of Gases.** Another important effect of depth is the compression of gases to smaller and smaller volumes. Figure 45–1 also illustrates a bell jar at sea level containing 1 liter of air. At 33 feet beneath the sea where the pressure is 2 atmospheres, the volume has been compressed to only one-half liter. At 100 feet, where the pressure is

4 atmospheres, the volume has been compressed to one-fourth liter, and at 8 atmospheres (233 feet) to one-eighth liter. This is an extremely important effect in diving because it can cause the air chambers of the diver's body, including the lungs, to become so small in some instances that serious damage results, as is discussed later in the chapter.

Many times in this chapter it is necessary to refer to *actual volume* versus *sea level volume*. For instance, we might speak of an actual volume of 1 liter at a depth of 300 feet; this is the same quantity of air as a sea level volume of 10 liters. Therefore, these two terms must be kept in mind while studying this chapter.

## EFFECT OF HIGH PARTIAL PRESSURES OF GASES ON THE BODY

The three gases to which a diver breathing air is normally exposed are nitrogen, oxygen, and carbon dioxide. However, helium is often substituted for nitrogen in the diving mixture; therefore, the effects of this gas under high pressure must also be considered.

**Nitrogen Narcosis at High Nitrogen Pressures.** Approximately four-fifths of the air is nitrogen. At sea level pressure this has no known effect on bodily function, but at high pressures it can cause varying degrees of narcosis. When the diver remains beneath the sea for many hours and is breathing compressed air, the depth at which the first symptoms of mild narcosis appear is approximately 130 to 150 feet, at which level he begins to exhibit joviality and to lose many of his cares. At 150 to 200

| Depth (feet) | Atmosphere(s) |
|---|---|
| Sea level | 1 |
| 33 | 2 |
| 66 | 3 |
| 100 | 4 |
| 133 | 5 |
| 166 | 6 |
| 200 | 7 |
| 300 | 10 |
| 400 | 13 |
| 500 | 16 |

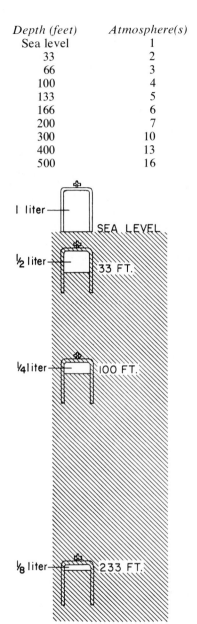

**Figure 45–1.** Effect of depth on gas volumes.

feet, he becomes drowsy. At 200 to 250 feet, his strength wanes considerably, and he often becomes too clumsy to perform the work required of him. Beyond 300 feet (10 atmospheres pressure), the diver usually becomes almost useless as a result of nitrogen narcosis. It should be noted, however, that *an hour or more of the high pressure is usually required* before enough nitrogen dissolves in the body to cause these effects.

Nitrogen narcosis has characteristics very similar to those of alcohol intoxication, and for this reason it has frequently been called "raptures of the depths."

The mechanism of the narcotic effect is believed to

be the same as that of essentially all the gas anesthetics. That is, nitrogen dissolves freely in the fats of the body, and it is presumed that it, like most other anesthetic gases, dissolves in the membranes or other lipid structures of the neurons and because of its *physical* effect on altering electrical charge transfer reduces their excitability.

**Oxygen Toxicity at High Pressures.** *Acute Oxygen Poisoning.* Breathing oxygen under very high partial pressures can be detrimental to the central nervous system, sometimes causing epileptic convulsions followed by coma. Indeed, exposure to 3 atmospheres pressure of oxygen ($P_{O_2} = 2280$ mm. Hg) will cause convulsions and coma in most persons after about one hour. These convulsions often occur without any warning, and they obviously are likely to be lethal to a diver submerged beneath the sea. In an extensive study of persons who have developed oxygen toxicity at still lower than 3 atmospheres pressure, the following frequencies of different symptoms were encountered:

| | per cent |
|---|---|
| Nausea | 40 |
| Muscular twitchings | 21 |
| Dizziness | 17 |
| Disturbances of vision | 6 |
| Restlessness and irritability | 6 |
| Numbness and pins-and-needles sensations | 6 |
| Convulsive seizures and coma | 4 |

Exercise greatly increases the diver's susceptibility to oxygen toxicity at high pressures, causing symptoms to appear much earlier and with far greater severity than in the resting person.

Figure 45-2 gives the so-called *oxygen safe tolerance curve* for persons performing moderate amounts of work at different depths under the sea while breathing 100 per cent oxygen. This shows that

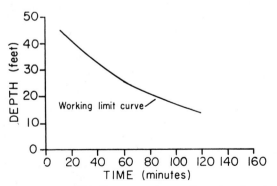

**Figure 45–2.** The "oxygen tolerance" curve, showing the length of time that a person can remain without danger at different depths when breathing pure oxygen. (Modified from Submarine Medicine Practice, Department of the Navy. U.S. Gov't. Printing Office, 1956.)

a person performing work at a depth of only 40 feet (slightly over 2 atmospheres) is reasonably safe for a maximum period of only 23 minutes. At a depth of 30 feet, he is safe for about 45 minutes and at 20 feet for about 1½ hours. However, there is tremendous variability in the tolerance of different persons and in the tolerance of the same person on different days. Therefore, since a convulsion underneath the sea might be lethal, even this tolerance curve is possibly too liberal for absolute safety.

The cause or causes of oxygen toxicity are yet unknown, but some of the experimental data and suggested causes are: (1) Following severe oxygen toxicity the concentrations of some of the oxidative enzymes in the tissues are considerably reduced. Therefore, it has been postulated that excess oxygen inactivates some oxidative enzymes and causes toxicity by decreasing the ability of the tissues to form high energy phosphate bonds. (2) Blood flow through the brain decreases 25 to 50 per cent when a person breathes high concentrations of oxygen. This presumably results from the local tissue blood flow control mechanism explained in Chapter 20. That is, the greater the amount of oxygen available in the tissues, the greater is the degree of constriction of the blood vessels, which normally helps to regulate the amount of oxygen delivered to the tissues. Yet, the decrease in blood flow can decrease the availability of other nutrients required by the cerebral tissues or can decrease the removal of excreta, such as carbon dioxide and nitrogenous end-products. Therefore, it has been postulated that either lack of certain nutrients or buildup of metabolic end-products can cause the convulsions of oxygen toxicity. (3) Still other experiments have shown that excess oxygen in the tissues can cause the development of large concentrations of oxidizing free radicals that could cause abnormal oxidative destruction of many essential elements of the cells, thereby damaging the cells' metabolic systems.

*Chronic Oxygen Poisoning as a Cause of Pulmonary Disability.*   A person can be exposed to 100 per cent oxygen at normal atmospheric pressure almost indefinitely without developing the *acute* oxygen toxicity described above. However, after 12 hours or so of this exposure, he begins to develop *lung passageway congestion and edema* caused by damage to the linings of the bronchi and alveoli. Here again, it appears that these local tissues of the lungs are damaged as a result of oxidative destruction of some of the essential elements of the tissues. The reason this effect occurs in the lungs and not in the other tissues is that the lungs are directly exposed to the high oxygen pressure ($Po_2$), whereas oxygen is delivered to the other tissues at essentially normal $Po_2$ because of the hemoglobin oxygen buffer system described in Chapter 41. When the $Po_2$ in the air rises above about 1500 mm. Hg, this buffer system fails, which then allows the $Po_2$ of all tissues to rise and to cause the acute oxygen poisoning described in the previous section.

**Carbon Dioxide Toxicity at Great Depths.**   If the diving gear is properly designed and also functions properly, the diver has no problem from carbon dioxide toxicity, for depth alone does not increase the carbon dioxide partial pressure in the alveoli. This is true because carbon dioxide is manufactured in the body, and as long as the diver continues to breathe a normal tidal volume, he continues to expire the carbon dioxide as it is formed, maintaining his alveolar carbon dioxide partial pressure at a normal value.

Unfortunately, though, in certain types of diving gear, such as the diving helmet and the different types of rebreathing apparatuses, carbon dioxide can frequently build up in the dead space air of the apparatus and be rebreathed by the diver. Up to a carbon dioxide pressure ($Pco_2$) of about 80 mm. Hg, two times that of normal alveoli, the diver tolerates this buildup, his minute respiratory volume increasing up to a maximum of 6- to 10-fold to compensate for the increased carbon dioxide. However, beyond the 80 mm. Hg level the situation becomes intolerable, and eventually the respiratory center begins to be depressed rather than excited; the diver's respiration then actually begins to fail rather than to compensate. In addition, the diver develops severe respiratory acidosis, and varying degrees of lethargy, narcosis, and finally anesthesia ensue, as was discussed in Chapter 42.

**Effects of Helium at High Pressures.**   In deep dives, helium is used to replace the nitrogen because it has only one-fourth to one-fifth the narcotic effect of nitrogen, exhibiting essentially no narcotic effect to a depth of over 650 feet. Furthermore, three other properties make it desirable in the diving gas mixture under some conditions: (1) Because of its low atomic weight, its density is slight, which reduces the airway resistance of the diver. (2) Also because of its low atomic weight, helium diffuses through the tissues much more rapidly than nitrogen, which allows more rapid removal of helium than nitrogen from the body fluids under some conditions. (3) Helium is less soluble in the body fluids than nitrogen, which reduces the quantity of bubbles that can form in his tissues when the diver is decompressed after a prolonged dive. Yet there are problems in the use of helium that make it undesirable under some conditions, as is discussed below in relation to decompression.

## DECOMPRESSION OF THE DIVER AFTER EXPOSURE TO HIGH PRESSURES

When a person breathes air under high pressure for a long time, the amount of nitrogen dissolved in his body fluids becomes great. The reason for this is the following: The blood flowing through the pulmonary capillaries becomes saturated with nitrogen to the same pressure as that in the breathing mixture. Over several hours, enough nitrogen is carried to all the tissues of the body to saturate them also with dissolved nitrogen. And, since nitrogen is not metabolized by the body, it remains dissolved until the nitrogen pressure in the lungs decreases, at which time the nitrogen is then removed by the reverse respiratory process.

**Volume of Nitrogen Dissolved in the Body Fluids at Different Depths.** At sea level almost 1 liter of nitrogen is dissolved in the entire body. A little less than half of this is dissolved in the water of the body and a little more than half in the fat of the body. This is true despite the fact that fat constitutes only 15 per cent of the normal body, and it is explained by the fact that nitrogen is five times as soluble in fat as in water.

After the diver has become totally saturated with nitrogen the *sea level volume of nitrogen* dissolved in his body fluids at the different depths is:

| feet | liters |
|------|--------|
| 33 | 2 |
| 100 | 4 |
| 200 | 7 |
| 300 | 10 |

However, several hours are required for the gaseous pressures of nitrogen in all the body tissues to come to equilibrium with the gas pressure of nitrogen in the alveoli simply because the blood does not flow rapidly enough and the nitrogen does not diffuse rapidly enough to cause instantaneous equilibrium. The nitrogen dissolved in the water of the body comes to almost complete saturation in about one hour, but the fat, requiring much more nitrogen for saturation and also having a relatively poor blood supply, reaches saturation only after several hours. For this reason, if a person remains at deep levels for only a few minutes not much nitrogen dissolves in his fluids and tissues, whereas if he remains at a deep level for several hours his fluids and tissues become almost completely saturated with nitrogen.

**Decompression Sickness (Synonyms: Compressed Air Sickness, Bends, Caisson Disease, Diver's Paralysis, Dysbarism).** If a diver has been beneath the sea long enough so that large amounts of nitrogen have dissolved in his body, and then he suddenly comes back to the surface of the sea, significant quantities of nitrogen bubbles can develop in his body fluids either intracellularly or extracellularly, and these can cause minor or serious damage in almost any area of the body, depending on the amount of bubbles formed.

The principles underlying this effect are shown in Figure 45–3. To the left, the diver's tissues have become equilibrated to a very high nitrogen pressure. However, as long as the diver remains deep beneath the sea, the pressure against the outside of his body (5000 mm. Hg) compresses all the body tissues sufficiently to keep the dissolved gases in solution. Then, when the diver suddenly rises to sea level, the pressure on the outside of his body becomes only 1 atmosphere (760 mm. Hg), while the pressure inside the body fluids is the sum of the pressures of water

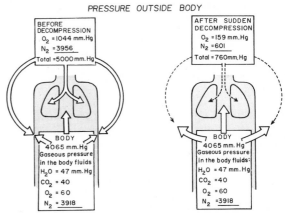

**Figure 45–3.** Gaseous pressures responsible for bubble formation in the body tissues.

vapor, carbon dioxide, oxygen, and nitrogen, or a total of 4065 mm. Hg, which is far greater than the pressure on the outside of the body. Therefore, the gases can escape from the dissolved state and form actual bubbles inside the tissues.

Exercise hastens the formation of bubbles during decompression because of increased agitation of the tissues and fluids. This is an effect analogous to that of shaking an opened bottle of soda pop to release the bubbles.

Fortunately, the phenomenon of "supersaturation" normally allows nitrogen to remain dissolved and not to form significant quantities of bubbles if the nitrogen pressure in the body fluids does not rise to more than 3.0 times the pressure on the outside of the body. Therefore, a diver can theoretically be brought immediately from a depth of 66 feet beneath the sea (3 atmospheres pressure) to sea level (1 atmosphere pressure) without significant bubble formation and without developing decompression sickness, even though on arrival at sea level the gas pressure in his body fluids is almost three times the pressure on the outside of his body. Yet for safety's sake the diver is rarely allowed to push this theoretical limit in his ascent from beneath the sea.

*Symptoms of Decompression Sickness.* In persons who have developed decompression sickness, symptoms have occurred with the following frequencies:

| | per cent |
|---|---|
| Local pain in the legs or arms | 89 |
| Dizziness | 5.3 |
| Paralysis | 2.3 |
| Shortness of breath ("the chokes") | 1.6 |
| Extreme fatigue and pain | 1.3 |
| Collapse with unconsciousness | 0.5 |

From the above list of symptoms of decompression sickness it can be seen that the most serious problems are usually related to bubble formation in the nervous system. Bubbles sometimes actually disrupt important pathways in the brain or spinal cord; the bubbles in the peripheral nerves can cause severe pain. Unfortunately, formation of large bubbles in the central nervous system occasionally even leads to permanent paralysis or permanent mental disturbances.

But the nervous system is not the only locus of damage in decompression sickness, for bubbles can also form in the blood and become caught in the capillaries of the lungs; these bubbles block pulmonary blood flow and cause "the chokes," characterized by serious shortness of breath. This is often followed by severe pulmonary edema, which further aggravates the condition and can cause death.

The symptoms of decompression sickness usually appear within a few minutes to an hour after sudden decompression. However, occasional symptoms of decompression sickness develop as long as six or more hours after decompression.

**Rate of Nitrogen Elimination from the Body. Decompression Tables.** Fortunately, if a diver is brought to the surface slowly, the dissolved nitrogen is eliminated through his lungs rapidly enough to prevent decompression sickness. Figure 45–4 illustrates the rate at which nitrogen is liberated from the water of the body, from the fat of the body, and from both these sources when a person is saturated to a depth of 33 feet and then brought to sea level suddenly. Approximately two-thirds of the total nitrogen is liberated in one hour and about 90 per cent in six hours. However, some excess nitrogen is still present in the body fluids for many more hours, and the diver is not completely safe for as long as 9 to 12 hours. Therefore, a diver must be decompressed sometimes for many hours if he has been deep beneath the sea for a long time.

The rate at which a diver can be brought to the surface depends on, first, the *depth* to which he has descended and, second, the *amount of time* he has

been there. If he remains at deep levels for only a short time, the body fluids do not become saturated, and, therefore, the decompression time can be reduced accordingly.

Table 45–1 gives a typical decompression table used by the U.S. Navy when the diver breathes compressed air. Note that only 20 minutes at a depth of 300 feet requires over two and a half hours decompression time (45 minutes at 300 feet requires over five hours). On the other hand, a person can remain at 50 feet for as long as three hours and yet be decompressed in only 12 minutes.

The "optimal time on the bottom," as given in Table 45–1, represents the optimum exposure time at each depth for the best balance between length of work period and amount of useful work the average diver can perform. Note how short these times are at great depths; this is caused principally by (1) the nitrogen narcosis effect, (2) the labored breathing that results from increased density of gases in the lungs, and (3) the time required for decompression.

*Oxygen Administration for More Rapid Decompression.* If oxygen is pumped to the diver in higher than normal concentrations as he ascends closer to the surface of the sea, the nitrogen partial pressure in his alveoli is considerably reduced, and, as a consequence, the rate of nitrogen removed from his body fluids is correspondingly increased. Therefore, a diver can be brought to the surface far more rapidly when oxygen is pumped to him once he has come close enough to the surface to tolerate the necessary oxygen partial pressures. Different decompression tables are used when oxygen is supplied.

**Decompression in a Tank and Treatment of Decompression Sickness.** Another procedure for decompression, used especially in heavily polluted waters and when climatic situations require it, involves bringing the diver to the surface immediately and then placing him in a decompression tank within five minutes after arriving at the surface. Pressure is reapplied, and an appropriate decompression table that prevents bubble formation is used.

A person who begins to develop symptoms of decompression sickness can also be treated by placing him in such a decompression tank for a long time, several times as long as the usual decompression times, and allowing the nitrogen to be released from his body slowly.

**Use of Helium-Oxygen Mixtures in Deep Dives.** In deep dives helium has advantages over nitrogen, including (1) decreased decompression time, (2) lack of narcotic effect, and (3) decreased airway resistance in the lungs. The decreased decompression time results from two of its properties: (a) Only 40 per cent as much helium dissolves in the body as does nitrogen. (b) Because of its small atomic size it diffuses through the tissues at a velocity about 2½ times that of nitrogen.

However, helium has not proved to be as advantageous as was once thought because of another property that is different from nitrogen: bubbles begin to form when the pressure of helium in the

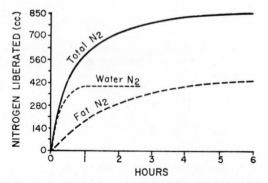

**Figure 45–4.** Rate of nitrogen liberation from the body when a person has come to sea level from prolonged exposure to compressed air at 33 feet depth, showing separately the rate of nitrogen release from the whole body, from the water of the body, and from the fat. (From Armstrong: Principles and Practice of Aviation Medicine. The Williams & Wilkins Co., 1943.)

**TABLE 45–1.**

| 1 | 2 | 3 | | | | | | | | | | 4 |
|---|---|---|---|---|---|---|---|---|---|---|---|---|
| Depth of dive (feet) | Optimal time on bottom (minutes) | Stops (feet and minutes) | | | | | | | | | | Approximate total decompression time (minutes) |
| | | Feet 90 | Feet 80 | Feet 70 | Feet 60 | Feet 50 | Feet 40 | Feet 30 | Feet 20 | Feet 10 | |
| 40 | 240 | | | | | | | | | 4 | 6 |
| 50 | 190 | | | | | | | | | 9 | 12 |
| 60 | 150 | | | | | | | | 5 | 15 | 24 |
| 70 | 120 | | | | | | | | 13 | 16 | 33 |
| 80 | 115 | | | | | | | | 22 | 26 | 53 |
| 90 | 95 | | | | | | | 2 | 27 | 21 | 56 |
| 100 | 85 | | | | | | | 6 | 28 | 21 | 61 |
| 110 | 75 | | | | | | | 14 | 27 | 37 | 84 |
| 120 | 65 | | | | | | | 13 | 28 | 32 | 80 |
| 130 | 60 | | | | | | | 13 | 28 | 28 | 76 |
| 140 | 55 | | | | | | | 15 | 28 | 32 | 82 |
| 150 | 50 | | | | | | | 16 | 28 | 32 | 84 |
| 160 | 45 | | | | | | | 17 | 28 | 43 | 96 |
| 170 | 40 | | | | | | | 19 | 28 | 46 | 102 |
| 185 | 35 | | | | | | | 19 | 28 | 46 | 102 |
| 200 | 35 | | | | | | | 22 | 28 | 46 | 106 |
| 210 | 30 | | | | | | 5 | 16 | 28 | 40 | 100 |
| 225 | 27 | | | | | | 22 | 26 | 35 | 48 | 143 |
| 250 | 25 | | | | | 2 | 23 | 26 | 35 | 51 | 150 |
| 300 | 20 | | | | | 9 | 23 | 26 | 35 | 51 | 159 |

body fluids is only 1.7 times the pressure on the outside of the body. This compares with 3.0 for nitrogen. Therefore, a diver cannot be brought up as far at a time with helium as with nitrogen. And still another factor makes nitrogen better than helium for shallow dives: the rapid diffusion of helium allows far more helium than nitrogen to become dissolved in the body fluids in a short time. Therefore, for short dives at moderate depths, nitrogen is still preferable.

If one calculates the relative advantages of helium versus nitrogen, he finds that long, deep dives favor the use of helium while short, shallow dives favor the use of nitrogen. Figure 45–5 illustrates the dividing line for most effective use of the two types of gas mixtures. For instance, point A illustrates a dive of 150 feet for 120 minutes. This is a long deep dive, and in this instance helium is far more satisfactory than nitrogen. On the other hand, point B shows a dive of 200 feet for 10 minutes, and here the person can be decompressed for a shorter time when using nitrogen than helium. Beyond 300 feet, nitrogen cannot be used at all because of nitrogen narcosis, which can develop rapidly.

# SOME PHYSICAL PROBLEMS OF DIVING

Aside from the effects of high gas pressures on the body, still other physical factors place limitations on diving. These are based principally on changes in gas volumes from sea level to greater depths and include the following:

**Volume of Air that Must Be Pumped to the Diver—Relationship to Rate of Carbon Dioxide Elimination.** To blow off carbon dioxide from the lungs, the tidal volume of air flowing in and out of the lungs with each breath must remain the same regardless of the depth of the dive. A tidal volume of 0.5 liter at 300 feet depth (10 atmospheres pressure) would be a sea level volume of 5 liters. Therefore, a compressor operating at sea level must pump 5 liters of air to the diver at 300 feet depth for each breath

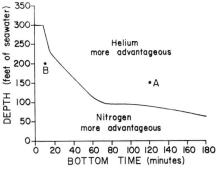

**Figure 45–5.** Depths and times beneath the sea at which it is more advantageous to use helium versus the depths and times at which it is more advantageous to use nitrogen. (Modified from Submarine Medicine Practice, Department of the Navy. U.S. Gov't. Printing Office, 1956.)

that the diver takes in order to wash the carbon dioxide out of his lungs, and an additional amount must be pumped to wash the carbon dioxide out of the diving gear. Stating this another way, the amount of air that must be pumped to the diver to keep his alveolar carbon dioxide normal is directly proportional to the pressure under which he is operating. At sea level the working diver requires about 1.5 cubic feet of air per minute for adequate carbon dioxide washout from his diving helmet. Therefore, the sea level volumes of air that must be pumped each minute for different depths of operation are the following:

| feet | cubic feet |
|------|------------|
| Sea level | 1.5 |
| 33 | 3 |
| 66 | 4.5 |
| 100 | 6 |
| 200 | 10.5 |
| 300 | 15 |

**Change in Density of the Air—Effect on Maximum Breathing Capacity.** The density of the air increases in proportion to the pressure, which means that the density is four times as great at 100 feet depth as at sea level and seven times as great at 200 feet.

The resistance to air flow through the respiratory passageways increases directly in proportion to the density of the breathing mixture. Therefore, one can readily see that the increased density of the air will increase the work of breathing and, as a corollary, will decrease the maximum breathing capacity (the amount of air that one can breathe each minute). The following table gives the *maximum breathing capacity* in per cent of normal at different depths when breathing air and when breathing a mixture in which helium replaces the nitrogen of the air.

| Depth (ft.) | Air (% of normal) | Oxygen-helium (% of normal) |
|-------------|-------------------|------------------------------|
| 25 | 75 | 100+ |
| 50 | 60 | 100+ |
| 100 | 50 | 86 |
| 200 | 35 | 63 |
| 400 | 24 | 48 |

**Effect of Rapid Descent—"The Squeeze."** On rapid descent, the volumes of all gases in the body become greatly reduced because of increasing pressure applied to the outside of the body. If additional quantities of air are supplied to the gas cavities during descent—especially to the lungs—no harm is done, but, if the person continues to descend without addition of gas to his cavities, the volume becomes greatly reduced, and serious physical damage results; this is called "the squeeze."

The most damaging effects of the squeeze occur in the lungs, for the smallest volume that the lungs can normally achieve is approximately 1.5 liters. Even if the diver inspires a maximal breath prior to descending, he can go down no farther than 100 feet before his chest begins to cave in. Therefore, to prevent lung squeeze, the diver must inspire additional air as he descends.

When air becomes entrapped in the middle ear during descent, the squeeze can cause a ruptured tympanum, and when air is entrapped in one of the nasal sinuses intense pain results. Occasionally, also, when a diver loses air pressure to his helmet his body is literally squeezed upward into the helmet, and, when air volume is lost from the mask of a free diving apparatus, the eyes can actually pop out into the mask and the face can become greatly distorted to fill the mask.

**Overexpansion of the Lungs on Rapid Ascent—Air Embolism.** Exactly the opposite pulmonary effects occur on rapid ascent if the person fails to expel air from his lungs on the way up. Unfortunately, panic can frequently cause a person to close his glottis spastically, which can result in serious damage to the lungs. When the lungs become expanded to their limit the pressure continues to rise, and above an *excess* alveolar pressure of 80 to 100 mm. Hg, air is forced into the pulmonary capillaries, causing air embolism in the circulation and often resulting in death. Also, increased pressure in the lungs frequently blows out large blebs on the surfaces of the lungs or ruptures the lungs to cause pneumothorax.

In rare instances, in a diver who has been deep below the sea for a long time and in whom large amounts of gas have accumulated in the abdomen, rapid ascent can also cause serious trauma in the gastrointestinal tract.

Rapid ascent can be especially serious when a person is attempting to escape from a submarine and must rise to the surface rapidly without appropriate diving gear. It also occurs frequently when the diver loses control of his diving gear and his suit balloons up so greatly that he "blows up" to the surface.

# SCUBA DIVING (SELF-CONTAINED UNDERWATER BREATHING APPARATUS)

In recent years a diving apparatus that does not require connections with the surface has been perfected and is probably best known under the trade name "Aqualung." The two basic types of self-contained underwater breathing apparatuses from which various modifications have been made are (1) the open circuit demand system and (2) the closed circuit system.

**The Open Circuit Demand System.** Figure 45–6 illustrates an open circuit demand type of underwater breathing apparatus showing the following compo-

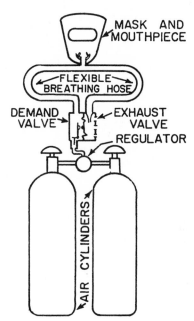

**Figure 45–6.** An open circuit demand system type of self-contained underwater breathing apparatus. (Modified from Submarine Medicine Practice, Department of the Navy. U.S. Gov't. Printing Office, 1956.)

nents: (1) tanks of compressed air or other breathing mixture, (2) a regulator valve for reducing the pressure from the tanks, (3) a "demand" valve which allows air to be pulled into the mask with slight negative pressure in the system, (4) a mask and tube system with small "dead space," and (5) an exhalation valve located in close contiguity with the demand valve.

Basically, the demand system operates as follows: With each inspiration, slight negative pressure in the mask pulls the diaphragm of the demand valve inward, and this automatically releases air from the compressed air containers into the mask. In this way only the amount of air needed for inhalation enters the system. Then, on expiration, the air cannot go back into the tank but instead is expired through the expiration valve.

The most important problem in use of the self-contained underwater breathing apparatus is the time limit that one can remain beneath the surface; only a few minutes are possible at great depths because tremendous airflow from the tanks is required to wash carbon dioxide out of the lungs—the greater the depth, the greater the airflow required, as discussed earlier.

**The Closed Circuit System.** In the simplest type of closed circuit system a person breathes pure oxygen. This system contains the following elements: (1) a tank of pure oxygen, (2) a rubber bellows into which the diver can breathe back and forth, (3) a valve system for allowing oxygen to flow from the oxygen tank into the bellows as needed to keep it

moderately filled, (4) a canister containing soda lime through which the rebreathed air passes to absorb the carbon dioxide, and (5) an appropriate mask system with valves to keep the gaseous mixture flowing through the canister for carbon dioxide removal. Thus, the closed circuit system is similar to a standard anesthetic machine in which oxygen is continually rebreathed—except that no anesthetic is used.

The most important problem in use of this closed circuit system is limitation in depth that a person can remain beneath the sea because of poor oxygen tolerance. Figure 45–2 gives these tolerances for depth and time. A safe working rule is no more than 30 minutes at 30 feet depth.

Several new closed circuit systems are available on an experimental basis. These utilize mixtures of oxygen and helium, and they have separate supply tanks for each gas. As oxygen is utilized from the breathing mixture, an electrical oxygen sensor and control mechanism replenish the oxygen from the oxygen tank but keep the oxygen concentration from rising too high; at the same time carbon dioxide is removed by a carbon dioxide absorbent. Since this gaseous mixture can be rebreathed again and again without ever blowing off any of it into the sea, none of the gases are wasted. Therefore, very small tanks of oxygen and helium will allow the diver to remain under the sea for many hours, thus allowing very deep dives as well. Also, the presence of helium in the mixture prevents both oxygen and nitrogen toxicity at great depths and also allows more rapid ascent after long, deep dives.

## SPECIAL PHYSIOLOGICAL PROBLEMS OF SUBMARINES

**Escape from Submarines.** Essentially the same problems as those of deep sea diving are often met in relation to submarines, especially when it is necessary to escape from a submerged submarine. Escape is possible from as deep as 300 feet even without using any special type of apparatus. Proper use of rebreathing devices using helium or hydrogen could theoretically allow escape from as deep as 600 feet.

One of the major problems of escape is prevention of air embolism. As the person ascends, the gases in his lungs expand and sometimes rupture a major pulmonary vessel, allowing the gases to enter into the pulmonary vascular system to cause embolism of the circulation. Therefore, as the person ascends, he must exhale continually.

Expansion and exhalation of gases from the lungs during ascent, even without breathing, is often rapid enough to blow off the accumulating carbon dioxide in the lungs. This keeps the concentration of carbon dioxide from building up in the blood and keeps the person from having the desire to breathe. Therefore, he can hold his breath for an extremely long time during ascent.

**Health Problems in the Submarine Internal Environment.** Except for escape, submarine medicine generally centers around several engineering problems to keep hazards out of the internal environment of the submarine. In atomic submarines there exists the problem of radiation hazards, but, with appropriate shielding, the amount of radiation received by the crew submerged beneath the sea has actually been less than the normal radiation received above the surface of the sea from cosmic rays. Therefore, no essential hazard results from this unless some failure in the apparatus causes unexpected release of radioactive materials.

Second, poisonous gases on occasion escape into the atmosphere of the submarine and must be controlled exactly. For instance, during several weeks' submergence, cigarette smoking by the crew can liberate sufficient amounts of carbon monoxide, if it is not removed from the air, to cause carbon monoxide poisoning, and on occasion even Freon gas has been found to diffuse through the tubes in refrigeration systems in sufficient quantity to cause dangerous toxicity. Finally, it is well known that chlorine and other poisonous gases are released when salt water comes in contact with batteries in the old type submarines.

A highly publicized factor of submarine medicine has been the possibility of psychological problems caused by prolonged submergence. Fortunately, this has turned out to be more a figment of the public's imagination than truth, for the problems here are the same as those relating (1) to any other confinement or (2) to any other type of danger. Psychological screening has been used to great advantage to keep such problems almost to zero even in month-long submergence.

# DROWNING

**Asphyxia in Drowning.** Approximately 30 per cent of all persons who drown do not inhale water, and even after recovery of the body from the water the lungs are still dry. These persons simply die from asphyxiation. The reason they do not inhale water is that water, in attempting to enter the trachea, elicits a powerful laryngeal reflex that causes spastic closure of the vocal cords.

**Inhalation of Fresh Water—Cardiac Fibrillation.** When a drowning person does inhale fresh water, water is absorbed through the alveolar membrane into the blood extremely rapidly by osmosis because of the total osmotic pressure of the pulmonary capillary blood (which gives an absorption gradient of about 5400 mm. Hg). Several liters of water

are absorbed into the blood within one to three minutes. This greatly dilutes the electrolytes of the blood and even causes hemolysis of many red cells, with spillage of large quantities of potassium into the plasma. The resulting changes in electrolyte composition of the plasma, plus anoxia, cause the heart to fibrillate within one to three minutes after the person first inhales the water, causing death much earlier than would have occurred had the water not been inhaled.

**Effect of Inhaling Salt Water.** If a drowning person inhales salt water instead of fresh water, osmosis of water occurs in the opposite direction through the pulmonary membrane because the total osmotic pressure of salt water is several times that of blood. Loss of water out of the blood causes marked hemoconcentration but not hemolysis or cardiac fibrillation. Instead, the person dies of asphyxia in five to eight minutes.

# REFERENCES

Anderson, H. T.: Physiological adaptations in diving vertebrates. *Physiol. Rev.*, 46:212, 1966.

Behnke, A. R., Jr., and Lanphier, E. H.: Underwater physiology. In Fenn, W. O., and Rahn, H. (eds.): Handbook of Physiology. Sec. 3, Vol. 2. Baltimore, The Williams & Wilkins Company, 1965, p. 1159.

Bennett, P. B.: The Aetiology of Compressed Air Intoxication and Inert Gas Narcosis. (International Series of Monographs in Pure and Applied Biology. Zoology Div., Vol. 31.) New York, Pergamon Press, Inc., 1966.

Bennett, P. B., and Elliott, D. H.: The Physiology and Medicine of Diving and Compressed Air Work, 2nd Ed. Baltimore, The Williams & Wilkins Company, 1975.

Gamarra, J. A.: Decompression Sickness. Hagerstown, Maryland, Harper & Row, Publishers, 1974.

Gooden, B. A.: Drowning and the diving reflex in man. *Med. J. Aust.*, 2:583, 1972.

Greene, D. G.: Drowning. In Fenn, W. O., and Rahn, H. (eds.): Handbook of Physiology. Sec. 3, Vol. 2. Baltimore, The Williams & Wilkins Company, 1965, p. 1195.

Haugaard, N.: Cellular mechanisms of oxygen toxicity. *Physiol. Rev.*, 48:311, 1968.

Hochachka, B. W., and Storey, K. B.: Metabolic consequences of diving in animals and man. *Science*, 187:613, 1975.

Lambertsen, C. J.: Proceedings of the Third Symposium on Underwater Physiology. Baltimore, The Williams & Wilkins Company, 1967.

Lanphier, E. H.: Human respiration under increased pressures. *Symp. Soc. Exp. Biol.*, 26:379, 1972.

Miles, S.: Underwater Medicine, 2nd Ed. Philadelphia, J. B. Lippincott Company, 1966.

Submarine Medicine Practice, Department of the Navy. Washington, D.C., U.S. Government Printing Office, 1956.

Vail, E. G.: Hyperbaric respiratory mechanics. *Aerosp. Med.*, 42:536, 1971.

Wyndam, C. H.: Physiological problems of deep level mining. *Proc. Inter. Union Physiol. Sci.*, 6:45, 1968.

Zimmerman, A. M. (ed.): High pressure effects on cellular processes. New York, Academic Press, Inc., 1970.

# PART IX
# THE NERVOUS SYSTEM

# | 46 |

# Organization of the Nervous System; Basic Functions of Synapses

The nervous system, along with the endocrine system, provides most of the control functions for the body. In general, the nervous system controls the rapid activities of the body, such as muscular contractions, rapidly changing visceral events, and even the rate of secretion of some endocrine glands. The endocrine system, by contrast, regulates principally the metabolic functions of the body.

The nervous system is unique in the vast complexity of the control actions that it can perform. It receives literally thousands of bits of information from the different sensory organs and then integrates all these to determine the response to be made by the body. The purpose of this chapter is to present, first, a general outline of the overall mechanisms by which the nervous system performs such functions. Then we will discuss the function of central nervous system synapses, the basic structures that control the passage of signals into, through, and then out of the nervous system. In succeeding chapters we will analyze in detail the functions of the individual parts of the nervous system. Before beginning this discussion, however, the reader should refer to Chapters 10 and 12, which present, respectively, the principles of membrane potentials and transmission of impulses through neuromuscular junctions.

## GENERAL DESIGN OF THE NERVOUS SYSTEM

### THE SENSORY DIVISION— SENSORY RECEPTORS

Most activities of the nervous system are originated by sensory experience emanating from *sensory receptors,* whether these be visual receptors, auditory receptors, tactile receptors on the surface of the body, or other kinds of receptors. This sensory experience can cause an immediate reaction, or its memory can be stored in the brain for minutes, weeks, or years and then can help to determine the bodily reactions at some future date.

Figure 46–1 illustrates a portion of the sensory system, the *somatic* portion that transmits sensory information from the receptors of the entire surface of the body and deep structures. This information enters the nervous system through the spinal nerves and is conducted into (a) the spinal cord at all levels, (b) the reticular substance of the medulla, pons, and mesencephalon, (c) the cerebellum, (d) the thalamus, and (e) the somesthetic areas of the cerebral cortex. But in addition to these "primary sensory" areas, signals are then relayed to essentially all other segments of the nervous system.

### THE MOTOR DIVISION— THE EFFECTORS

The most important ultimate role of the nervous system is control of bodily activities. This is achieved by controlling (a) contraction of skeletal muscles throughout the body, (b) contraction of smooth muscle in the internal organs, and (c) secretion of both exocrine and endocrine glands in many parts of the body. These activities are collectively called *motor functions* of the nervous system, and the muscles and glands are called *effectors* because they perform the functions dictated by the nerve signals. That portion of the nervous system directly

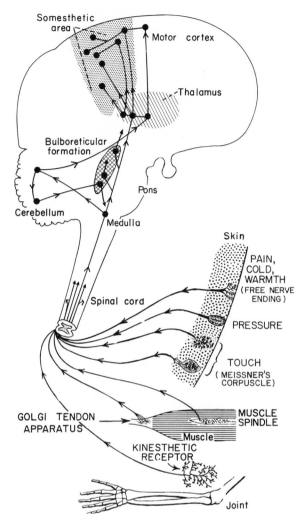

**Figure 46–1.** The somatic sensory axis of the nervous system.

plays its own specific role in the control of body movements, the lower regions being concerned primarily with automatic, instantaneous responses of the body to sensory stimuli and the higher regions with deliberate movements controlled by the thought processes of the cerebrum.

## PROCESSING OF INFORMATION

The nervous system would not be at all effective in controlling bodily functions if each bit of sensory information caused some motor reaction. Therefore, one of the major functions of the nervous system is to process incoming information in such a way that *appropriate* motor responses occur. Indeed, more than 99 per cent of all sensory information is continually discarded by the brain as unimportant. For instance, one is ordinarily totally unaware of the parts of his body that are in contact with his clothes and is also unaware of the pressure on his seat when he is sitting. Likewise, his attention is drawn only to an occasional object in his field of vision, and even the perpetual noise of

concerned with transmitting signals to the muscles and glands is called the motor division of the nervous system.

Figure 46–2 illustrates the *motor axis* of the nervous system for controlling skeletal muscle contraction. Operating parallel with this axis is another similar system for control of the smooth muscles and glands; it is the *autonomic nervous system,* which will be presented in detail in Chapter 57. Note in Figure 46–2 that the skeletal muscles can be controlled from many different levels of the central nervous system, including (a) the spinal cord, (b) the reticular substance of the medulla, pons, and mesencephalon, (c) the basal ganglia, (d) the cerebellum, and (e) the motor cortex. Each of these different areas

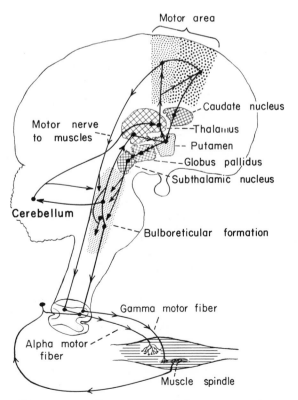

**Figure 46–2.** The motor axis of the nervous system.

his surroundings is usually relegated to the background.

After the important sensory information has been selected, it must be channeled into proper motor regions of the brain to cause the desired responses. Thus, if a person places his hand on a hot stove, the desired response is to lift the hand, plus other associated responses, such as moving the entire body away from the stove and perhaps even shouting with pain. Yet even these responses represent activity by only a small fraction of the total motor system of the body.

**Role of Synapses in Processing Information.** The synapse is the junction point from one neuron to the next and, therefore, is an advantageous site for control of signal transmission. Later in this chapter we will discuss the details of synaptic function. However, it is important to point out here that the synapses determine the directions that the nervous signals spread in the nervous system. Some synapses transmit signals from one neuron to the next with ease, whereas others transmit signals only with difficulty. Also, facilitatory and inhibitory signals from other areas in the nervous system can control synaptic activity, sometimes opening the synapses for transmission and other times closing them. In addition, some post-synaptic neurons respond with large numbers of impulses, whereas others respond with only a few. Thus, the synapses perform a selective action, often blocking the weak signals while allowing the strong signals to pass, often selecting and amplifying certain weak signals, and often channeling the signal in many different directions rather than simply in one direction. The basic principles of this processing of information by the synapses are so important that they are discussed in detail in the latter part of this chapter and in the entire following chapter.

## STORAGE OF INFORMATION—MEMORY

Only a small fraction of the important sensory information causes an immediate motor response. Much of the remainder is stored for future control of motor activities and for use in the thinking processes. Most of this storage occurs in the *cerebral cortex*, but not all, for even the basal regions of the brain and perhaps even the spinal cord can store small amounts of information.

The storage of information is the process we call *memory*, and this too is a function of the synapses. That is, each time a particular sensory signal passes through a sequence of synapses, the respective synapses become more capable of transmitting the same signal the next time, which process is called *facilitation*. After the sensory signal has passed through the synapses a large number of times, the synapses become so facilitated that signals from the "control center" of the brain can also cause transmission of impulses through the same sequence of synapses even though the sensory input has not been excited. This gives the person a perception of experiencing the original sensation, though in effect it is only a memory of the sensation.

Unfortunately, we do not know the precise mechanism by which facilitation of synapses occurs in the memory process, but what is known about this and other details of the memory process will be discussed in Chapter 55.

Once memories have been stored in the nervous system, they become part of the processing mechanism. The thought processes of the brain compare new sensory experiences with the stored memories; the memories help to select the important new sensory information and to channel this into appropriate storage areas for future use or into motor areas to cause bodily responses.

## THE THREE MAJOR LEVELS OF NERVOUS SYSTEM FUNCTION

The human nervous system has inherited specific characteristics from each stage of evolutionary development. From this heritage, there remain three major levels of the nervous system that have special functional significance: (1) the spinal cord level, (2) the lower brain level, and (3) the higher brain or cortical level.

### THE SPINAL CORD LEVEL

The spinal cord of the human being still retains many functions of the multisegmental animal. Sensory signals are transmitted through the spinal nerves into each *segment* of the spinal cord, and these signals can cause localized motor responses either in the segment of the body from which the sensory information is received or in adjacent segments. Essentially all the spinal cord motor responses are automatic and occur almost instantaneously in response to

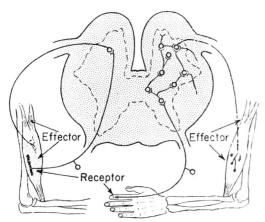

**Figure 46–3.** *Left:* The simple stretch reflex. *Right:* A withdrawal reflex.

the sensory signal. In addition, they occur in specific patterns of response called *reflexes*.

Figure 46-3 illustrates two of the simpler cord reflexes. To the left is the neural control of the *muscle stretch reflex*. If a muscle suddenly becomes stretched, a sensory nerve receptor in the muscle called the *muscle spindle* becomes stimulated and transmits nerve impulses through a sensory nerve fiber into the spinal cord. This fiber synapses directly with a motoneuron in the anterior horn of the cord gray matter, and the motoneuron in turn transmits impulses back to the muscle to cause the muscle, the effector, to contract. The muscle contraction opposes the original muscle stretch. Thus, this reflex acts as a *feedback* mechanism, operating from a receptor to an effector, to prevent sudden changes in length of the muscle. This allows a person to maintain his limbs and other parts of his body in desired positions despite sudden outside forces that tend to move the parts out of position.

To the right in Figure 46–3 is illustrated the neural control of another reflex called the *withdrawal reflex*. This is a protective reflex that causes withdrawal of any part of the body from an object that is causing pain. For instance, let us assume that the hand is placed on a sharp object. Pain signals are transmitted into the gray matter of the spinal cord, and, after appropriate selection of information by the synapses, signals are channeled to the appropriate motoneurons to cause flexion of the biceps muscle. This obviously lifts the hand away from the sharp object.

We see, then, that the withdrawal reflex is

much more complex than the stretch reflex, for it involves many neurons in the gray matter of the cord, and signals are transmitted to many adjacent segments of the cord to cause contraction of the appropriate muscles.

**Cord Functions After the Brain Is Removed.**   The many reflexes of the spinal cord will be discussed in Chapter 51; however, the following list of important cord reflex functions that occur even after the brain is removed illustrates the many capabilities of the spinal cord:

1. The animal can under certain conditions be made to stand up. This is caused primarily by reflexes initiated from the pads of the feet. Sensory signals from the pads cause the extensor muscles of the limbs to tighten, which in turn allows the limbs to support the animal's body.

2. A spinal animal held in a sling so that its feet hang downward often begins walking or galloping movements involving one, two, or all its legs. This illustrates that the basic patterns for causing the limb movements of locomotion are present in the spinal cord.

3. A flea crawling on the skin of a spinal animal causes reflex to-and-fro scratching by the paw, and the paw can actually localize the flea on the surface of the body.

4. Cord reflexes exist to cause emptying of the urinary bladder and of the rectum.

5. Segmental temperature reflexes are present throughout the body. Local cooling of the skin causes vasoconstriction, which helps to conserve heat in the body. Conversely, local heating in the skin causes vasodilatation, resulting in loss of heat from the body.

This list of some of the segmental and multisegmental reflexes of the spinal cord demonstrates that many of our day-by-day and moment-by-moment activities are controlled locally by the respective segmental levels of the spinal cord, and the brain plays only a modifying role in these local controls.

### THE LOWER BRAIN LEVEL

Many if not most of what we call subconscious activities of the body are controlled in the lower areas of the brain—in the medulla, pons, mesencephalon, hypothalamus, thalamus, cerebellum, and basal ganglia. Subconscious control of arterial blood pressure and respiration is achieved primarily in the reticular substance of the medulla and pons. Control of equilibrium is a combined function of the older portions of the cerebellum and the reticular substance of the medulla, pons, and mesencephalon. The coordinated turning movements of the head, of the entire body, and of the eyes are controlled by specific centers located in the mesencephalon, paleocerebellum, and lower basal ganglia. Feeding reflexes, such as salivation in response to taste of food and licking of the lips, are controlled by areas in the medulla, pons, mesencephalon, amygdala, and hypothalamus. And many emotional patterns, such as anger, excitement, sexual activities, reactions to pain, or reactions of pleasure, can occur in animals without a cerebral cortex.

In short, the subconscious but coordinate functions of the body, as well as many of the life processes themselves—arterial pressure and respiration, for instance—are controlled by the lower regions of the brain, regions that usually, but not always, operate below the conscious level.

## THE HIGHER BRAIN OR CORTICAL LEVEL

We have seen from the above discussion that many of the intrinsic life processes of the body are controlled by subcortical regions of the brain or by the spinal cord. What then is the function of the cerebral cortex? The cerebral cortex is primarily a vast information storage area. Approximately three quarters of all the neuronal cell bodies of the entire nervous system are located in the cerebral cortex. It is here that most of the memories of past experiences are stored, and it is here that many of the patterns of motor responses are stored, which information can be called forth at will to control motor functions of the body.

**Relation of the Cortex to the Thalamus and Other Lower Centers.** The cerebral cortex is actually an outgrowth of the lower regions of the brain, particularly of the thalamus. For each area of the cerebral cortex there is a corresponding and connecting area of the thalamus, and activation of a minute portion of the thalamus activates the corresponding and much larger portion of the cerebral cortex. It is presumed that in this way the thalamus can call forth cortical activities at will. Also, activation of regions in the mesencephalon transmits diffuse signals to the cerebral cortex, partially through the thalamus and partially directly, to activate the entire cortex. This is the process that we call *wakefulness*. On the other hand, when these areas of the mesencephalon become inactive, the thalamic and cortical regions also become inactive, which is the process we call *sleep*.

**Function of the Cerebral Cortex in Thought Processes.** Some areas of the cerebral cortex are not directly concerned with either sensory or motor functions of the nervous system—for example, the prefrontal lobe and large portions of the temporal and parietal lobes. These areas are set aside for the more abstract processes of thought, but even they also have direct nerve connections with the lower regions of the brain.

Large areas of the cerebral cortex can be destroyed without blocking the subconscious, and even some of the involuntary conscious, activities of the body. For instance, destruction of the somesthetic cortex does not destroy one's ability to feel objects touching his skin, but it does destroy his ability to distinguish the shapes of objects, their character, and the precise points on the skin where the objects are touching. Thus, the cortex is not required for perception of sensation, but it does add immeasurably to its depth of meaning. Likewise, destruction of the prefrontal lobe does not destroy one's ability to think, but it does destroy his ability to think in abstract terms. In other words, each time a portion of the cerebral cortex is destroyed, a vast amount of information is lost to the thinking process and some of the mechanisms for processing this information are also lost. Therefore, total loss of the cerebral cortex causes a vegetative type of existence rather than a "living" existence.

**Telencephalization.** In the process of evolution, the higher regions of the human nervous system have taken over many sensory and motor functions performed by lower regions of the brain in lower animals. For instance, if the spinal cord of an opossum is cut in the midthoracic region, the opossum can still walk perfectly well on both his forelimbs and hindlimbs, except that the hindlimbs are then unsynchronized with the forelimbs. A similar transection of the spinal cord in the human being causes complete loss of ability to use the legs for locomotion.

This process by which the progressively higher centers of the brain have taken over more and more of the function of the lower centers is called *telencephalization*. Yet, despite telencephalization, some of the very basic functions of the lower centers still remain active or at least partially active in the human being, such as many of the cord reflexes and the stereotyped control systems of the basal regions of the brain that were just discussed.

# COMPARISON OF THE NERVOUS SYSTEM WITH AN ELECTRONIC COMPUTER

When electronic computers were first developed in many different laboratories of the world by as many different engineers, it soon became apparent that all these machines have many features in common with the nervous system. First, they all have input circuits which are comparable to the sensory portion of the nervous system and output circuits which are comparable to the motor portion of the nervous system. In the conducting pathway between the inputs and the outputs are the mechanisms for performing the different types of computations.

In simple computers, the output signals are controlled directly by the input signals, operating in a manner similar to that of the simple reflexes of the spinal cord. But, in the more complex computers, the output is determined both by the input signals and by information that has already been stored in the computer, which is analogous to the more complex reflex and processing mechanisms of our nervous system. Furthermore, as the computers become even more complex it is necessary to add still another unit, called the *programming unit,* which determines the sequence of computational operations. This unit is analogous to the mechanism in our brain that allows us to direct our attention first to one thought, then to another, and so forth, until complex sequences of thought take place.

Figure 46–4 illustrates a simple block diagram of a modern computer. Even a rapid study of this diagram will demonstrate its similarity to the nervous system. Furthermore, general purpose computers designed and built by many different scientific groups have ended up with almost identically the same basic components. The fact that these components are analogous to those of the human nervous system demonstrates that the brain is basically a computer that continuously collects sensory information and uses this along with stored information to compute the daily course of bodily activity.

# FUNCTION OF NEURONAL SYNAPSES

Every medical student is aware that information is transmitted in the central nervous system through a succession of neurons, one after another. However, it is not immediately apparent that the impulse may be (a) blocked in its transmission from one neuron to the next, (b) changed from a single impulse into repetitive impulses, or (c) integrated with impulses from other neurons to cause highly intricate patterns of impulses in successive neurons. All these

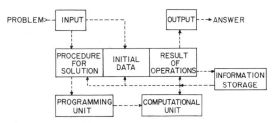

**Figure 46–4.**   Block diagram of a general-purpose electronic computer, showing the basic components and their interrelationships.

functions can be classified as *synaptic functions of neurons.*

## *PHYSIOLOGIC ANATOMY OF THE SYNAPSE*

The juncture between one neuron and the next is called a *synapse.* Figure 46–5 illustrates a typical *motoneuron* in the anterior horn of the spinal cord. It is comprised of three major parts, the *soma,* which is the main body of the neuron; a single *axon,* which extends from the soma into the peripheral nerve; and the *dendrites,* which

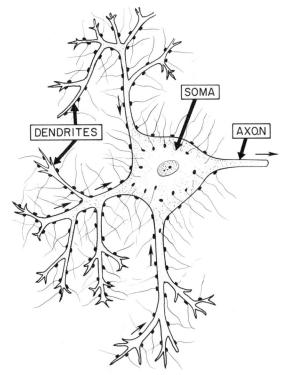

**Figure 46–5.**   A typical motoneuron, showing synaptic knobs on the neuronal soma and dendrites. Note also the single axon.

are thin projections of the soma that extend up to 1 mm. into the surrounding areas of the cord.

It should be noted, also, that literally hundreds to many thousands of small knobs called *synaptic knobs* lie on the surfaces of the dendrites and soma, approximately 80 to 90 per cent of them on the dendrites. These knobs are the terminal ends of nerve fibrils that originate in many other neurons, and usually not more than a few of the knobs are derived from any single previous neuron. Later it will become evident that many of these synaptic knobs are *excitatory* and secrete a substance that excites the neuron, whereas still others are *inhibitory* and secrete a substance that inhibits the neuron.

Neurons in other parts of the cord and brain differ markedly from the motoneuron in (1) the size of the cell body, (2) the length, size, and number of dendrites, ranging in length from almost none at all up to as long as one meter (the peripheral sensory nerve fiber), (3) the length and size of the axon, and (4) the number of synaptic knobs, which may range from only a few up to many thousand. It is these differences that make neurons in different parts of the nervous system react differently to incoming signals and therefore to perform different functions, as will be explained in subsequent chapters.

**The Synaptic Knobs.** Electron microscope studies of the synaptic knobs show that they have varied anatomical forms, but most resemble small round or oval knobs and therefore are frequently called *terminal knobs, boutons, end-feet,* or simply *presynaptic terminals*.

Figure 46–6 illustrates the basic structure of the synaptic knob (presynaptic terminal). It is separated from the neuronal soma by a *synaptic cleft* having a width usually of 200 to 300 Ångstroms. The knob has two internal struc-

tures important to the excitatory or inhibitory functions of the synapse: the *synaptic vesicles* and the *mitochondria*. The synaptic vesicles of the excitatory knobs contain an *excitatory transmitter,* which, when released into the synaptic cleft, excites the neuron; the vesicles of the inhibitory knobs contain an *inhibitory transmitter* that inhibits the neuron. The mitochondria provide ATP, which is required to synthesize new transmitter substance. This transmitter must be synthesized extremely rapidly because the amount stored in the vesicles is sufficient to last for only a few seconds to a few minutes of maximum activity.

When an action potential spreads over a presynaptic terminal, the membrane depolarization causes emptying of a small number of vesicles into the cleft; and the released transmitter in turn causes an immediate change in the permeability characteristics of the subsynaptic neuronal membrane, which leads to excitation or inhibition of the neuron, depending on the type of transmitter substance.

**Mechanism by Which the Synaptic Knob Action Potential Causes Release of Transmitter Vesicles.** Unfortunately, we can only guess at the mechanism by which an action potential on reaching the synaptic knob causes the vesicles to release transmitter substance into the synaptic cleft. However, the number of vesicles released with each action potential is greatly reduced (a) when the quantity of calcium ions in the extracellular fluid is diminished, (b) when the quantity of sodium ions in the extracellular fluid is diminished, (c) when the quantity of magnesium ions in the extracellular fluids is increased, or (d) when the membrane of the synaptic knob has already been partially depolarized prior to transmission of the action potential so that the action potential is weaker than usual. On the basis of these characteristics, it has been suggested that the spread of the action potential over the membrane of the knob causes small amounts of calcium ions to leak into the knob. The calcium ions then supposedly attract the transmitter vesicles to the membrane and simultaneously cause one or more of them to rupture, thus allowing spillage of their contents into the synaptic cleft.

At any rate, it is known that each time an action potential travels over the surface of the synaptic knob, one or more vesicles of transmitter substance are emptied into the synaptic cleft. It may be only one vesicle, which is the average, or several vesicles, depending upon the factors mentioned above.

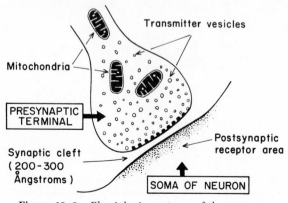

**Figure 46–6.** Physiologic anatomy of the synapse.

One transmitter substance that occurs in certain parts of the nervous system is acetylcholine, as is discussed later. It has been calculated that about 3000 molecules of acetylcholine are present in each vesicle, and enough vesicles are present in the synaptic knob on a neuron to transmit about 10,000 impulses.

**Synthesis of New Transmitter Substance.** Fortunately, the synaptic knobs have the capability of continually synthesizing new transmitter substance. Were it not for this ability, synaptic transmission would become completely ineffective within a few minutes. The synthesis is believed to occur mainly in the cytoplasm of the synaptic knobs, and then the newly synthesized transmitter is immediately absorbed into the vesicles and stored until needed. Thus, each time a vesicle empties its contents into the synaptic cleft, soon thereafter it becomes filled again with new transmitter. It is also possible that the vesicle wall itself plays a role in the synthesis of some of the transmitters.

In the case of the synthesis of acetylcholine, this substance is synthesized from acetyl-CoA and choline in the presence of the enzyme *choline acetyltransferase,* an enzyme that is present in abundance in the cytoplasm of the cholinergic type of synaptic knob. When acetylcholine is released from the knob into the synaptic cleft, it is rapidly split again to acetate and choline by the enzyme cholinesterase that is adherent to the outer surface of the knob. Then the choline is actively transported back into the knob to be used once more for synthesis of new acetylcholine. Thus, the vesicles are used again and again. But even so, both the vesicles and the mitochondria that supply the energy for transmitter synthesis eventually disintegrate. Fortunately, new vesicles and mitochondria are continually transported from the cell soma down the axon to the synaptic knob, moving along the axon at a velocity of about 10 cm. per day, and replenishing the supply in the knobs.

The formation, release, and re-uptake of norepinephrine by the presynaptic terminals of the sympathetic nervous system will be discussed in detail in Chapter 56 in relation to function of the autonomic nervous system.

**Action of the Transmitter Substance on the Postsynaptic Neuron.** The membrane of the postsynaptic neuron where a synaptic knob abuts it is believed to contain specific receptor molecules that bind the transmitter substance. These receptors are probably proteins that respond to the transmitter by changing their shapes or activities in such a way that they increase the membrane permeability especially to sodium ions when the transmitter is excitatory and increase the permeability to potassium and chloride ions when the transmitter is inhibitory.

The ribosomes and the endoplasmic reticulum are both increased in the area immediately beneath the synapse, as illustrated by the density of the postsynaptic area in Figure 46–6. It is possible that the degree of development of this postsynaptic receptor area increases with the intensity of activity of the synapse; therefore, this is possibly a means by which the synapse can subserve the memory function, as will be discussed in detail in Chapter 55.

## CHEMICAL NATURES OF THE TRANSMITTER SUBSTANCES

**The Excitatory Transmitters.** From the discussion in Chapter 12 on the neuromuscular junction, a peripheral type of synapse, it will be recalled that the excitatory transmitter at that synapse is acetylcholine. It is also almost certain that the excitatory transmitter at the synapses in the autonomic ganglia is *acetylcholine.* The major reasons for believing this are: (a) acetylcholine-like substances will stimulate the postganglionic neurons in the autonomic ganglia, (b) substances that prevent the destruction of acetylcholine by cholinesterase at the synapses will potentiate the transmission through the autonomic ganglia, and (c) measurable quantities of acetylcholine can be recovered from fluids perfusing the autonomic ganglia after strong stimulation of the preganglionic neurons.

In the case of the central nervous system, at least several, and perhaps many, different excitatory transmitter substances are secreted by different types of excitatory synaptic knobs. At least one of these excitatory transmitters is acetylcholine, but how extensively this transmitter is used throughout the nervous system is not certain. It is known that synaptic knobs in widespread areas of the nervous system do contain vesicles of acetylcholine as well as choline acetyltransferase, which is required for synthesis of acetylcholine. They also contain cholinesterase for splitting the acetylcholine after it is secreted into the synaptic clefts.

Other excitatory substances for which specific synthesizing enzymes have been proved in different central nervous system neurons include:

1. Norepinephrine
2. Dopamine
3. Serotonin

**The Inhibitory Transmitters.** At least two substances appear to be important inhibitory transmitters:

1. Gamma aminobutyric acid (GABA), which is present especially in many brain nerve terminals but to less extent also in the spinal cord. This substance has been proved to be an inhibitory transmitter in some lower animals, thereby making it almost certain that it also acts as an inhibitory transmitter in mammals.

2. Glycine, one of the simple amino acids. This is highly concentrated in some synaptic knobs in the spinal cord and probably functions as an inhibitory transmitter at many synapses in that area.

**Other Possible Transmitter Substances.** To find an explanation for the many varied functions of synapses in different parts of the central nervous system, a vast search is under way for still other transmitter substances. Many substances that will excite or inhibit neurons have been found, but none of these has been proved with real certainty to be a functional transmitter. Indeed, the exact function of even several of the substances already mentioned is in doubt with the probable exception of acetylcholine, norepinephrine, and dopamine. Among the still newer substances that have been suggested as possible transmitters are:

1. Excitatory transmitters: L-*glutamate* and L-*aspartate*

2. Inhibitory transmitters: *taurine* and *alanine*

3. Possible excitatory or inhibitory transmitters under different circumstances: *histamine, prostaglandins*, and *P-substances*, which are a series of polypeptides that are found in the nervous system and that may have long-term excitatory or inhibitory effects on neurons.

It is especially to be noted that some of the transmitter substances might have prolonged effects of neuronal excitation or inhibition, perhaps lasting for minutes or hours. The presence of these effects gives an entirely new dimension to synaptic control over postsynaptic neuronal function. For instance, some of the small polypeptides when injected into the brain can at times cause prolonged sleep.

**Excitatory versus Inhibitory Neurons.** A single neuron can secrete only one type of transmitter substance at its nerve terminals. Therefore, even though the axon from a neuron divides a thousand times it will still secrete only the one type of transmitter substance. For this reason a neuron that secretes an excitatory transmitter substance will cause excitation wherever the terminal fibrils synapse with the next neuron. These neurons are therefore called *excitatory neurons*. Likewise, the terminals of neurons that secrete inhibitory substance can cause only inhibition, and these neurons are called *inhibitory neurons*.

## ELECTRICAL EVENTS DURING NEURONAL EXCITATION

The electrical events in neuronal excitation have been studied mainly in the large motoneuron of the anterior horn of the spinal cord. Therefore, the events to be described in the following few sections pertain essentially to these neurons. However, except for some quantitative differences, they also apply to most other neurons of the nervous system as well.

**Resting Membrane Potential of the Neuronal Soma.** Figure 46–7 illustrates the soma of a motoneuron, showing the resting membrane potential to be about −70 millivolts. This is somewhat less than the −85 millivolts found in large peripheral nerve fibers and in skeletal muscle fibers; the lower voltage is important, however, because it allows both positive and negative control of the degree of excitability of the neuron. That is, decreasing the voltage to a less negative value makes the membrane of the neuron more excitable, whereas increasing this voltage to a more negative value makes the neuron less excitable. This is the basis of the two modes of function of the neuron—either excitation or inhibition—as we will explain in detail in the following sections.

*Concentration Differences of Ions Across the Neuronal Somal Membrane.* Figure 46–7 also illustrates the concentration differences across the neuronal somal membrane of the three ions that are most important for neuronal function: sodium ions, potassium ions, and chloride ions.

At the top, the sodium ion concentration is shown to be very great in the extracellular fluid but very low inside the neuron. This sodium concentration gradient is caused by a very strong sodium pump that continually pumps sodium out of the neuron. The capability of the pump to move sodium ions through the membrane is so great that it is the dominant factor in determining the distribution of sodium ions across the membrane; that is, the sodium pump maintains a very low sodium concentration inside the neuron despite slight back leakage of sodium through the pores, as illustrated in the figure.

The figure also shows that the potassium ion concentration is very great inside the neuronal soma but very low in the extracellular fluid. It illustrates that there is a weak potassium pump that tends to pump potassium to the interior while there is a very high degree of permeability to potassium. The pump is relatively unimportant because potassium ions leak through the neuronal somal pores so readily that this nullifies most of the effectiveness of the pump. Therefore, for most purposes one can consider the membrane to be relatively highly permeable to potassium and the potassium pump to be of only slight importance.

Figure 46–7 shows the chloride ion to be of high concentration in the extracellular fluid but of low concentration inside the neuron. It also shows that the membrane is highly permeable to chloride ions and that there is no chloride pump. Therefore, the chloride ions become distributed across the membrane passively. The reason for the low concentration of chloride ions inside the neuron is the −70 millivolts in the neuron. That is, this negative voltage repels the negatively charged chloride ions, forcing them outward through the pores until the concentration difference is so great that its tendency to move chloride ions inward exactly balances the tendency of the electrical difference to move them outward. That is, the chloride ions become distributed across the membrane in accordance with the Nernst equation for equilibrium conditions. This equation was discussed in detail in Chapter 10.

*Origin of the Resting Membrane Potential of the Neuronal Soma.* The basic cause of the −70 millivolt resting membrane potential of the neuronal soma is the sodium pump. This pump causes the extrusion of positively charged sodium ions to the exterior. Since there are large numbers of negatively charged ions inside the soma that cannot diffuse through the membrane—protein ions, phosphate ions, and many others—extrusion of the positively charged sodium ions to the exterior leaves all these nondiffusible negative ions unbalanced by positive ions on the inside. Therefore, the interior of the neuron becomes negatively charged as the result of the sodium pump. This principle was discussed in more detail in Chapter 10 in relation to the resting membrane potential of nerves.

*Uniform Distribution of the Potential Inside the Soma.* The interior of the neuronal soma contains a very highly conductive electrolytic solution, the intracellular fluid of the neuron. Furthermore, the diameter of the neuronal soma is very large (from 10 to 80 microns in diameter) causing there to be almost no resistance to conduction of electrical current from one part of the somal interior to another part. Therefore, any change in potential in any part of the soma causes an almost exactly equal change in potential at all other points in the soma. This is an important principle because it plays a major role in the summation of signals entering the neuron from multiple sources, as we shall see in subsequent sections of this chapter.

**Effect of Excitatory Transmitter on the Postsynaptic Membrane—the Excitatory Postsynaptic Potential.** Figure 46–8A illustrates the resting neuron with an unexcited synaptic knob resting upon its surface. The resting membrane potential everywhere in the soma is −70 millivolts.

Figure 46–8B illustrates an excitatory knob

**Figure 46–7.** Distribution of sodium, potassium, and chloride ions across the neuronal somal membrane; origin of the intrasomal membrane potential.

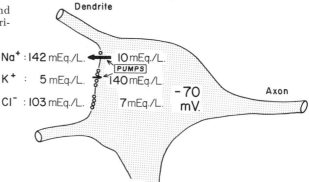

**A**

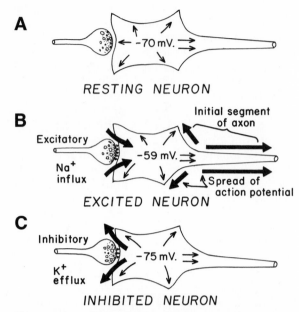

RESTING NEURON

**B**

EXCITED NEURON

**C**

INHIBITED NEURON

**Figure 46–8.** Three states of a neuron: (A) a resting neuron, (B) a neuron in an excited state, with increased intraneuronal potential caused by sodium influx, and (C) a neuron in an inhibited state, with decreased intraneuronal membrane potential caused by potassium ion efflux.

that has secreted an excitatory transmitter into the cleft between the knob and the neuronal somal membrane. This excitatory transmitter acts on the membrane receptor to increase the membrane's permeability to all ions. This causes sodium ions in particular to flow to the interior of the neuron because of the large electrochemical gradient that tends to move sodium inward.

The rapid influx of the positively charged sodium ions to the interior of the neuron neutralizes part of the negativity of the resting membrane potential. Thus, in Figure 46–8B the resting membrane potential has been increased from −70 millivolts to −59 millivolts. This increase in voltage above the normal resting neuronal potential—that is, to a less negative value—is called the *excitatory postsynaptic potential* because when this potential rises high enough it will elicit an action potential in the neuron, thus exciting it.

However, we need to issue a word of warning at this point. Discharge of a single excitatory synaptic knob can never increase the neuronal potential from −70 millivolts up to −59 millivolts. Instead, an increase of this magnitude requires the simultaneous discharge of many excitatory knobs—tens to hundreds, usually—

at the same time. This occurs by a process called *summation*, which will be discussed in detail in the following sections.

**Excitation at the Initial Segment of the Axon—Threshold for Excitation.** When the membrane potential inside the neuron rises high enough, there comes a point that this increase initiates an action potential in the neuron. However, the action potential does not begin on the somal membrane adjacent to the excitatory knobs. Instead, it begins in the initial segment of the axon. This may be explained as follows: Any factor that increases the potential inside the soma at any single point also increases this potential everywhere in the soma at the same time. Yet, because of physical differences in the membrane and differences in geometrical arrangement of the membrane in different parts of the neuron, the intrasomal voltage that will elicit an action potential is also different at different points on the neuronal membrane. The most excitable part of the neuron by far is the initial segment of the axon—that is, the first 50 to 100 microns of the axon beyond the point where it leaves the neuronal soma. The excitatory postsynaptic potential that will elicit an action potential at this point on the neuron is approximately 11 millivolts. This is in contrast to approximately 30 millivolts required to elicit an action potential on the soma itself. Therefore, the new action potential originates in the initial segment of the axon and not on the soma. Once the action potential begins, it travels peripherally along the axon and also travels backwards over the soma of the neuron (it usually does not travel backwards very far into the dendrites, however).

Thus, in Figure 46–8B, it is shown that under normal conditions the *threshold* for excitation of the neuron is −59 millivolts, which represents an excitatory postsynaptic potential of +11 millivolts—that is, 11 millivolts more positive than the normal resting neuronal potential of −70 millivolts.

## ELECTRICAL EVENTS IN NEURONAL INHIBITION

**Effect of Inhibitory Transmitter on the Postsynaptic Membrane—the Inhibitory Postsynaptic Potential.** It was pointed out above that the excitatory transmitter increases the permeability of the somal membrane to all ions—including sodium, potassium, and chloride. The inhibitory transmitter, in contrast, increases the permeability of the postsynaptic membrane only to potassium and chloride ions. Therefore, influx of sodium ions does not occur. However, potassium efflux does occur, as illustrated in Figure 46–8C. The reason potassium flows outward through the membrane is the following: In the resting state, a weak potassium pump has been pumping a slight excess of potassium ions to the interior of the neuron. Therefore, there are a few too many potassium ions on the inside to be in an exact equilibrium state as described by the Nernst equation for equilibrium conditions. Consequently, opening of the pores causes outward diffusion of some of the excess potassium ions, thereby decreasing the positive ions inside the neuron and leaving still more of the nondiffusible negative ions of the neuron (protein ions, phosphate ions, and others). This concentration of negative ions makes the internal potential of the neuron more negative than ever, as illustrated by the −75 millivolt potential inside the neuron in Figure 46–8C. This is called a *hyperpolarized state*. And the 5 millivolt decrease in intraneuronal voltage below the normal resting potential of −70 millivolts caused by the inhibitory transmitter is called the *inhibitory postsynaptic potential*.

Obviously, the increased negativity of the membrane potential (−75 millivolts) makes the neuron less excitable than it is normally. Since the potential must rise to −59 millivolts to excite the neuron, an excitatory postsynaptic potential must now be 16 millivolts instead of the normal 11 millivolts to cause excitation. Thus, in this way the inhibitory transmitter inhibits the neuron.

**"Clamping" of the Resting Membrane Potential as a Means to Inhibit Neurons.** Sometimes, excitation of the inhibitory synaptic knobs does not cause an inhibitory postsynaptic potential, but it still inhibits the neuron. The reason that the potential does not change is that in some neurons the concentration differences across the membrane for both the potassium and the chloride ions are already exactly in equilibrium states in accord with the Nernst equation. That is, the normal resting potential is exactly the voltage that each of these concentration differences will create when the membrane becomes permeable to the ions. Therefore, when the inhibitory pores open, there is no net outward flow of potassium ions to cause an inhibitory postsynaptic potential. Yet, both the potassium and the chloride ions do diffuse bidirectionally through the wide open pores many times as rapidly as normally, and this high

flux of these two ions inhibits the neuron in the following way: When excitatory knobs fire and sodium ions flow into the neuron, this raises the intraneuronal voltage far less than usual because any tendency for the membrane potential to change to a value that is not equal to the Nernst potential for the potassium and chloride ions is immediately opposed by rapid flux of these ions in the appropriate directions through the inhibitory pores to bring the potential back to the Nernst value. Therefore, the amount of influx of sodium ions required to cause excitation may be increased to as much as 5 to 20 times normal.

This tendency for the potassium and chloride ions to maintain a membrane potential near the resting value when the inhibitory pores are wide open is called "clamping" of the potential.

To express the phenomenon for clamping more mathematically, one needs to recall the Goldman equation from Chapter 10. This equation indicates that the membrane potential is determined by summation of the tendencies for the different ions to carry electrical charges through the membrane in the two directions. The membrane potential will approach the Nernst equilibrium potential for those ions that permeate the membrane to the greatest extent. When the inhibitory pores are wide open, the chloride and potassium ions permeate the membrane very greatly. Therefore, when the excitatory knobs fire, it is difficult to raise the neuronal potential up to the threshold value for excitation.

### Presynaptic Inhibition

In addition to the inhibition caused by inhibitory knobs operating at the synapses, called *postsynaptic inhibition,* another type of inhibition often occurs before the signal reaches the synapse. This type of inhibition, called *presynaptic inhibition,* is believed to occur in the following way:

In presynaptic inhibition, instead of the postsynaptic neuron being inhibited, the presynaptic terminal fibrils and synaptic knobs are inhibited. This inhibition is caused by the presence of inhibitory knobs lying directly on the terminal fibrils and excitatory knobs themselves. These presynaptic inhibitory knobs are believed to come from interneurons that are excited by the incoming afferent signals; they secrete a transmitter substance that partially depolarizes the terminal fibrils and excitatory synaptic knobs. Consequently, the voltage of the action potential that occurs at the membrane of the excitatory knob is depressed, and, as has already been pointed out, this greatly decreases the amount of excitatory transmitter released by the knob. Therefore, the degree of excitation of the neuron is greatly suppressed.

The nature of the transmitter substance released by the interneurons to cause presynaptic inhibition is not known. However, presynaptic inhibition is blocked by *picrotoxin* and also by *bicuculline*. Since these two substances also block the action of the inhibitory transmitter substance, GABA, it has been suggested that the transmitter substance of the knobs that cause the presynaptic inhibition is GABA. On the other hand, since GABA normally causes hyperpolarization rather than depolarization, some neurophysiologists believe that the transmitter substance from these interneurons is one of the excitatory transmitters.

Presynaptic inhibition occurs especially at the more peripheral synapses of the sensory pathways. For instance, when an incoming sensory fiber to the spinal cord excites the second neuron in the sensory pathway, it simultaneously causes presynaptic inhibition of the surrounding neurons. This "sharpens" the spatial boundaries of the signal transmitted to the brain; that is, it decreases or eliminates other signals on each side of the major signal so that the major signal stands out more clearly at its terminus in the brain. We shall see the importance of this sharpening process, also called "contrast enhancement," in discussions of subsequent chapters.

## SUMMATION OF POSTSYNAPTIC POTENTIALS

**Time Course of Postsynaptic Potentials.** When the excitatory transmitter is released from an excitatory synaptic knob, the neuronal membrane becomes highly permeable for only about 1 millisecond. During this time sodium ions diffuse rapidly to the interior of the cell to increase the intraneuronal potential, thus creating the *excitatory postsynaptic potential*. This potential then persists for about 15 milliseconds because this is the time required for the sodium ions to be pumped out or for potassium ions to leak out to reestablish the normal resting membrane potential. In the large anterior motoneuron, the potential disappears with a half-time of approximately 4 milliseconds. That is, each 4 milliseconds it decreases by one-half.

Precisely the same effect occurs for the inhibitory postsynaptic potential. That is, the inhibitory transmitter increases the permeability of the membrane to potassium and chloride ions

for approximately one millisecond, and this effect usually decreases the intraneuronal potential to a more negative value than normal, thereby creating the *inhibitory postsynaptic potential*. This potential also persists for about 15 milliseconds, disappearing with a half-time of approximately 4 milliseconds.

**Spatial Summation of the Postsynaptic Potentials.** It has already been pointed out that excitation of a single synaptic knob on the surface of a neuron will almost never excite the neuron. The reason for this is that sufficient excitatory transmitter substance is released by a single knob to cause an excitatory postsynaptic potential usually no more than a fraction of a millivolt instead of the required 10 millivolts or more that is the usual threshold for excitation. However, during an excitatory state in a neuronal pool of the nervous system, many excitatory knobs are usually stimulated at the same time, and even though these knobs are spread over wide areas of the neuron, their effects can still summate. The reason for this summation is the following: It has already been pointed out that a change in the potential at any single point within the soma will cause the potential to change everywhere in the soma almost exactly equally. Therefore, for each excitatory knob that discharges simultaneously, the intrasomal potential rises another fraction of a millivolt. When the excitatory postsynaptic potential becomes great enough, the threshold for firing will be reached, and an action potential will generate at the axon hillock. This effect is illustrated in Figure 46–9, which shows several excitatory postsynaptic potentials. The bottom postsynaptic potential in the figure was *theoretically* caused by stimulation of only four ex-

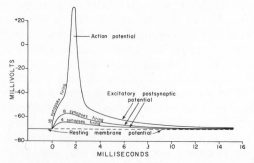

**Figure 46–9.** Excitatory postsynaptic potentials, showing that simultaneous firing of only a few synapses will not cause sufficient summated potential to elicit an action potential, but that simultaneous firing of many synapses will raise the summated potential to the threshold for excitation and cause a superimposed action potential. (The exact number of synapses firing is hypothetical.)

citatory knobs; then the next higher potential was caused by stimulation of two times as many knobs; finally, a still higher excitatory postsynaptic potential was caused by stimulation of four times as many knobs. This time an action potential was generated at the axon hillock.

This effect of summing simultaneous postsynaptic potentials created by excitation of multiple knobs on widely spaced areas of the membrane is called *spatial summation*.

**Temporal Summation.** Most synaptic knobs can fire in rapid succession only a few milliseconds apart. Each time a knob fires, the released transmitter substance opens the membrane pores for a millisecond or so. Since the postsynaptic potential lasts for up to 15 milliseconds, a second opening of the same pore can increase the postsynaptic potential to a still greater level so that the more rapid the rate of knob stimulation, the greater the effective postsynaptic potential. Thus, successive postsynaptic potentials of individual synaptic knobs, if they occur rapidly enough, can summate in the same way that postsynaptic potentials can summate from widely distributed knobs over the surface of the neuron. This summation of rapidly repetitive postsynaptic potentials is called *temporal summation*.

**Simultaneous Summation of Inhibitory and Excitatory Postsynaptic Potentials.** Obviously, if an inhibitory postsynaptic potential is tending to decrease the membrane potential to a more negative value while an excitatory postsynaptic potential is tending to increase the potential to a more positive value at the same time, these two effects can either completely nullify each other or partially nullify each other. Also, inhibitory "clamping" of the membrane potential can nullify much of an excitatory potential. Thus, if a neuron is currently being excited by an excitatory postsynaptic potential, then an inhibitory signal from another source can easily reduce the postsynaptic potential below the threshold value for excitation, thus turning off the activity of the neuron.

**Facilitation of Neurons.** Often the summated postsynaptic potential is excitatory in nature but has not risen high enough to reach the threshold for excitation. When this happens the neuron is said to be *facilitated*. That is, its membrane potential is nearer the threshold for firing than normally, but it is not yet to the level of firing. Nevertheless, a signal entering the neuron from some other source can then excite the neuron very easily. Diffuse signals in the nervous system often facilitate large groups of neurons so that they can respond quickly and easily to signals arriving from secondary sources.

## SPECIAL FUNCTIONS OF DENDRITES IN EXCITING NEURONS

**The Large Spatial Field of Excitation of the Dendrites.** The dendrites of the anterior motoneurons extend for one-half to one millimeter in all directions from the neuronal soma. Therefore, these dendrites can receive signals from a fairly large spatial area around the motoneuron. This provides vast opportunity for summation of signals from many separate presynaptic neurons.

It is also important to note that between 80 and 90 per cent of all the synaptic knobs terminate on the dendrites of the anterior motoneuron in contrast to only 10 to 20 per cent terminating on the neuronal soma. Therefore, the preponderant share of the excitation of the neuron is provided by signals transmitted over the dendrites.

**Failure of Many Dendrites to Transmit Action Potentials.** Many dendrites fail to transmit action potentials because their thresholds for excitation are very high. Yet they do transmit *electrotonic current* down the dendrites to the soma. (Transmission of electrotonic current means the direct spread of current by electrical conduction in the fluids of the dendrites with no generation of new current by action potentials.) Stimulation of the neuron by this current has special characteristics, as follows:

*Decrement of Electrotonic Conduction in the Dendrites—Greater Excitation Near the Soma.* In Figure 46–10 a number of excitatory and inhibitory knobs are shown stimulating the dendrites of a neuron. On the two dendrites to the left in the figure are shown excitatory effects near the ends of the dendrites; note the high levels of the excitatory postsynaptic potentials at these ends. However, a large share of the excitatory postsynaptic potential is lost before it reaches the soma. The reason for this is that the dendrites are long and thin, and their membranes are also very leaky to electrical current. Therefore, before the excitatory potentials can reach the soma, a large share of the potential is lost by leakage through the membrane. This decrease in membrane potential as it spreads along dendrites toward the soma is called *decremental conduction* in the dendrites.

It is also obvious that the nearer the excitatory knob is to the soma of the neuron, the less will be the decrement of conduction from that point to the soma. Therefore, those knobs that lie near the soma have far more excitatory effect than those that lie far away from the soma. This is a very important principle necessary to an understanding of the function of neurons in neuronal pools, because knobs from presynaptic fibrils that terminate near the soma of the postsynaptic neuron always have a much higher chance of causing the postsynaptic neuron to fire than do knobs that terminate on the dendrites at a distance from the soma.

*Rapid Re-excitation of the Neuron by the Dendrites After the Neuron Fires.* When an action potential is generated in a neuron, this action potential spreads back over the soma but not always over the dendrites. Furthermore, the postsynaptic potentials in the dendrites often remain relatively undisturbed by the action potential. Therefore, just as soon as the action potential is over, the electrical charges still existing in the dendrites are ready and waiting to flow into the soma of the neuron and, therefore, ready to re-excite the neuron. Thus, the dendrites have a "holding capacity" for the excitatory signal from presynaptic sources.

**Summation of Excitation and Inhibition in Dendrites.** The uppermost dendrite of Figure 46–10 is shown to be stimulated by both excitatory and inhibitory knobs. At the tip end of the dendrite there is a strong excitatory postsynaptic potential, but much of this is lost by decremental conduction along the dendrite. Then near the soma two inhibitory knobs are stimulating the same dendrite. These inhibitory knobs provide a hyperpolarizing voltage in the dendrite that completely nullifies the excitatory effect and indeed transmits a small amount of inhibition, by decremental conduction, toward the soma. Thus, dendrites can summate excitatory and inhibitory postsynaptic potentials in the same way that the soma can. In those instances when action potentials do develop in dendrites, inhibitory knobs located near the soma can completely block the action potentials and can prevent their ever entering the soma of the neuron. It is especially interesting that the inhibitory knobs tend to terminate on or near the soma.

## RELATION OF STATE OF EXCITATION OF THE NEURON TO THE RATE OF FIRING

**The "Central Excitatory State."** The "central excitatory state" of a neuron is defined as the degree of excitatory drive to the neuron. If there is a higher degree of excitation than inhibition of the neuron at any given instant, then it is said that there is a *central excitatory state*. On the other hand, if there is more inhibition than excitation, then it is said that there is a *central inhibitory state*.

A quantitative definition of the central excitatory state is the increase in postsynaptic potential above the resting membrane potential of the soma that *would* develop in the neuron if action potentials did not occur to interfere with this postsynaptic potential. Thus, a postsynaptic potential of 5 millivolts above the resting membrane potential would be a central excitatory state of 5 millivolts. A postsynaptic potential of 50 millivolts above the resting membrane potential would be a central excitatory state of 50 millivolts. Unfortunately, however, when a postsynaptic potential this high occurs, an action potential supervenes immediately so that the central excitatory state is then in reality a theoretical potential rather than an actual potential.

When the central excitatory state of a neuron rises above the threshold for excitation, then the neuron will fire repetitively as long as the excitatory state remains at this level. However, the rate at which it will fire is determined by how much above threshold value the central excitatory state is. To explain this, we must first consider what happens to the neuronal somal potential during and following the action potential.

**Changes in Neuronal Somal Potential During and Following the Action Potential.**

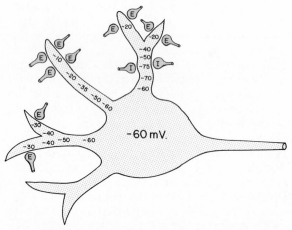

**Figure 46–10.** Stimulation of a neuron by synaptic knobs located on dendrites, showing, especially, decremental conduction of electrotonic potentials in the two dendrites to the left and inhibition of dendritic excitation in the dendrite that is uppermost.

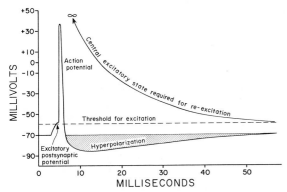

**Figure 46–11.** A neuronal action potential followed by a prolonged period of neuronal hyperpolarization. Also shown is the central excitatory state required for re-excitation of the neuron at given intervals after the action potential is over.

Figure 46–11 illustrates an action potential which is initiated by an excitatory postsynaptic potential that rises above the threshold level of −59 millivolts. Following the action potential there is a very long *positive after-potential* lasting for many milliseconds. This is a state of hyperpolarization during which the somal membrane potential falls below the normal resting membrane potential of −70 millivolts. It is caused by a very high degree of permeability of the neuronal membrane to potassium ions that persists for many milliseconds after the action potential is over. Potassium ions from inside the neuron diffuse to the exterior carrying positive charges out of the neuron and, therefore, creating an excessively high degree of negativity inside the neuron.

The importance of this state of hyperpolarization after the action potential is that the neuron remains in an *inhibited state* during this period of time. Therefore, a far greater central excitatory state is required during this period of time than normally to cause re-excitation of the neuron.

**Relationship of Central Excitatory State to Rate of Firing.** The curve shown at the top of Figure 46–11, labeled "central excitatory state required for re-excitation," depicts the level of the central excitatory state required at each instant after an action potential is over to re-excite the neuron. Note that very soon after an action potential is over a very high central excitatory state is required to cause re-excitation. That is, a very large number of excitatory synaptic knobs must be firing simultaneously. Then, after many milliseconds have passed and the state of hyperpolarization of the neuron has begun to disappear, the central excitatory state

required for re-excitation becomes greatly reduced.

Therefore, it is immediately evident that when the central excitatory state is high a second action potential will appear very soon after a previous one. Still a third action potential will appear soon after the second, and this process will continue indefinitely. Thus, at a very high central excitatory state the rate of firing of the neuron is great.

On the other hand, when the central excitatory state is only barely above threshold, the neuron must recover for many milliseconds before it can fire again. Therefore, the rate of neuronal firing is very slow.

(Note that the increase in the central excitatory state that is required for re-excitation is far greater than the decrease in the neuronal potential during the state of hyperpolarization. The reason for this is that at the same time that the neuron is hyperpolarized its voltage is also "clamped" to a great extent by the high permeability of the membrane to potassium and chloride ions. This clamping mechanism of neuronal inhibition was discussed earlier in the chapter.)

**Response Characteristics of Different Neurons to Increasing Levels of Central Excitatory State.** Histological study of the nervous system immediately convinces one of the widely varying types of neurons in different parts of the nervous system. And, physiologically, the different types of neurons perform different functions. Therefore, as would be expected, the ability to respond to stimulation by the presynaptic knobs varies from one type of neuron to another.

Figure 46–12 illustrates theoretical responses

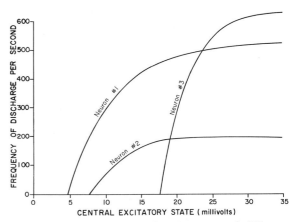

**Figure 46–12.** Response characteristics of different types of neurons to progressively increasing central excitatory states.

of three different types of neurons to varying levels of central excitatory state. Note that neuron number 1 will not discharge at all until the excitatory state rises to 5 millivolts. Then, as the excitatory state rises progressively to 35 millivolts the frequency of discharge rises to slightly over 500 per second.

Neuron number 2 is quite a different type, having a threshold for excitation of 8 millivolts and a maximum rate of discharge of only 190 per second. Finally, neuron number 3 has a high threshold for excitation, about 18 millivolts, but its discharge rate rises rapidly to over 600 per second as the central excitatory state rises only slightly above threshold.

Thus, different neurons respond differently, have different thresholds for excitation, and have widely differing maximal frequencies of discharge. With a little imagination one can readily understand the importance of having neurons with many different types of response characteristics to perform the widely varying functions of the nervous system.

# SOME SPECIAL CHARACTERISTICS OF SYNAPTIC TRANSMISSION

**Forward Conduction through Synapses.** From the above discussion, it should be evident by now that impulses are conducted through synapses only from the synaptic knobs to the successive neurons and never in the reverse direction. This is called the *principle of forward conduction*.

**Synaptic Delay.** In transmission of an impulse from a synaptic knob to a postsynaptic neuron, a certain amount of time is consumed in the process of (a) discharge of the transmitter substance by the knob, (b) diffusion of the transmitter to the subsynaptic neuronal membrane, (c) action of the transmitter on the membrane, and (d) inward diffusion of sodium to raise the excitatory postsynaptic potential to a high enough value to elicit an action potential. The *minimal* period of time required for all these events to take place, even when large numbers of presynaptic terminals are stimulated simultaneously, is approximately 0.5 millisecond. This is called the *synaptic delay*. It is important for the following reason: Neurophysiologists can measure the *minimal* delay time between an input volley of impulses and an output volley and from this can estimate the number of series neurons in the circuit.

**Fatigue of Synaptic Transmission.** When the presynaptic knobs are repetitively stimu-

lated at a rapid rate, the number of discharges by the postsynaptic neuron is at first very great but becomes progressively less in succeeding milliseconds or seconds. This is called *fatigue* of synaptic transmission. Fatigue is an exceedingly important characteristic of synaptic function, for when areas of the nervous system become overexcited, fatigue causes them to lose this excess excitability after a while. For example, fatigue is probably the most important means by which the excess excitability of the brain during an epileptic fit is finally subdued so that the fit ceases. Thus, the development of fatigue is a protective mechanism against excess neuronal activity. This will be discussed further in the description of reverberating neuronal circuits in the following chapter.

The mechanism of fatigue is mainly exhaustion of the stores of transmitter substance in the synaptic knobs, particularly since it has been calculated that the excitatory knobs can store enough excitatory transmitter for only 10,000 normal synaptic transmissions, an amount that can be exhausted in only a few seconds to a few minutes.

**Post-Tetanic Facilitation.** When a rapidly repetitive (tetanizing) stimulus is applied to the synaptic knobs for a period of time and then a rest period is allowed, the neuron will usually be even more responsive to subsequent stimulation than normally. This is called *post-tetanic facilitation*.

One of the most likely explanations of post-tetanic facilitation is that the repetitive stimulation alters the membranes of the knobs to cause more rapid emptying of transmitter vesicles.

The physiological significance of post-tetanic facilitation is still very doubtful, and it may have no real significance at all. However, since post-tetanic facilitation can last from a few seconds in some neurons to many hours in others, it is immediately apparent that neurons could possibly store information by this mechanism. Therefore post-tetanic facilitation might well be a mechanism of "short-term" memory in the central nervous system. This possibility will be discussed at further length in Chapter 55 in relation to the memory function of the cerebral cortex.

**Effect of Acidosis and Alkalosis on Synaptic Transmission.** The neurons are highly responsive to changes in pH of the surrounding interstitial fluids. *Alkalosis greatly increases neuronal excitability.* For instance, a rise in arterial pH from the normal of 7.4 to about 7.8 often causes cerebral convulsions because of increased excitability of the neurons. This can be demonstrated especially well by having a person who is normally predisposed to epileptic fits overbreathe. The overbreathing elevates the pH of the blood only momentarily, but even this short interval can often precipitate an epileptic convulsive attack.

On the other hand, *acidosis greatly depresses neuronal activity;* a fall in pH from 7.4 to below 7.0 usually causes a comatose state. For instance, in very severe diabetic or uremic acidosis, coma always develops.

**Effect of Hypoxia on Synaptic Transmission.** Neuronal excitability is also highly dependent on an adequate supply of oxygen. Cessation of oxygen supply for only a few seconds can cause complete inexcitability of the neurons. This is often seen when the cerebral circulation is temporarily interrupted, for within 3 to 5 seconds the person becomes unconscious.

**Effect of Drugs on Synaptic Transmission.** Many different drugs are known to increase the excitability of neurons, and others are known to decrease the excitability. For instance, caffeine, theophylline, and theobromine, which are found in coffee, tea, and cocoa, respectively, all increase neuronal excitability, presumably by reducing the threshold for excitation of the neurons. However, strychnine, which is one of the best known of all the agents that increase the excitability of neurons, does not reduce the threshold for excitation of the neurons at all but, instead, *inhibits the action of at least some of the inhibitory transmitters* on the neurons, probably especially the inhibitory effect of glycine in the spinal cord. In consequence, the effects of the excitatory transmitter from the excitatory knobs become overwhelming, and the neurons become so excited that they go into rapidly repetitive discharge, resulting in severe convulsions.

The hypnotics and anesthetics have long been believed to increase the threshold for excitation of the neurons and thereby to decrease neuronal activity throughout the body. This is based principally on the fact that most of the volatile anesthetics are chemically inert compounds but are lipid soluble. Therefore, it has been reasoned that these substances might change the physical characteristics of the neuronal membranes, making them less excitable to excitatory agents. It still remains a possibility, however, that at least some hypnotics and anesthetics cause their anesthetic actions by (a) reducing the quantity of excitatory transmitter released or (b) enhancing the inhibitory effects of the inhibitory transmitter.

# REFERENCES

Baldessarini, R. J., and Karobath, M.: Biochemical physiology of central synapses. *Ann. Rev. Physiol.*, 35:273, 1973.

Barondes, S. H.: Synaptic macromolecules: identification and metabolism. *Ann. Rev. Biochem.*, 43:147, 1974.

Bennett, M. V. L. (ed.): Synaptic Transmission and Neuronal Interaction (Society of General Physiologists Series, Vol. 28). New York, Raven Press, 1974.

Calvin, W. H.: Generation of spike trains in CNS neurons. *Brain Res.*, 84:1, 1975.

Costa, E., Gessa, G. L., and Sandler, M. (eds.): Serotonin, New Vistas: Biochemistry and Behavioral and Clinical Studies. New York, Raven Press, 1974.

Costa, E., Iversen, L. L., and Paoletti, R. (eds.): Studies of Neurotransmitters at the Synaptic Level. (Advances in Biochemical Psychopharmacology Series, Vol. 6.) New York, Raven Press, 1972.

Costa, E., and Meek, J. L.: Regulation of biosynthesis of catecholamines and serotonin in the CNS. *Ann. Rev. Pharmacol.*, 14:491, 1974.

Curtis, D. R.: Bicuculline, GABA, and central inhibition. *Proc. Aust. Assoc. Neurol.*, 9:145, 1973.

De Robertis, E., and Schacht, J. (eds.): Neurochemistry of Cholinergic Receptors. New York, Raven Press, 1974.

Fink, B. R. (ed.): Molecular Mechanisms of Anesthesia. Vol. 1: Progress in Anesthesiology. New York, Raven Press, 1975.

Fonnum, F.: Recent developments in biochemical investigations of cholinergic transmission. *Brain Res.*, 62:497, 1973.

Gerschenfeld, H. M.: Chemical transmissions in invertebrate central nervous systems and neuromuscular junctions. *Physiol. Rev.*, 53:1, 1973.

Hebb, C.: Biosynthesis of acetylcholine in nervous tissue. *Physiol. Rev.*, 52:918, 1972.

Hubbard, J. I., Linas, R., and Quastel, D. M.: Electrophysiological Analysis of Synaptic Transmission. Baltimore, The Williams & Wilkins Company, 1969.

Johnson, J. L.: Glutamic acid as a synaptic transmitter in the nervous system. A review. *Brain Res.*, 37:1, 1972.

Kandel, E. R.: Nerve cells and behavior. *Sci. Amer.*, 223:57, 1970.

Karlin, A.: The acetylcholine receptor: progress report. *Life Sci.*, 14:1385, 1974.

Kuno, M.: Quantum aspects of central and ganglionic transmission in vertebrates. *Physiol. Rev.*, 51:647, 1971.

Leibovic, K. N.: Nervous System Theory: An Introductory Study. New York, Academic Press, Inc., 1973.

Noback, C. R.: The Human Nervous System; Basic Principles of Neurobiology, 2nd Ed. New York, McGraw-Hill Book Company, 1975.

Obata, K.: The inhibitory action of –aminobutyric acid, a probable synaptic transmitter. *Inter. Rev. Neurobiol.*, 15:167, 1972.

Pappas, G. D., and Purpura, D. P. (eds.): Structure and Function of Synapses. New York, Raven Press, 1972.

Roberts, E., Chase, T., and Tower, D. B. (eds.): GABA in Nervous System Function. New York, Raven Press, 1975.

Sidman, R. L., and Rakic, P.: Neuronal migration, with special reference to developing human brain: a review. *Brain Res.*, 62:1, 1973.

Somjen, G. G.: Electrophysiology of neuroglia. *Ann. Rev. Physiol.*, 37:163, 1975.

Storm-Mathisen, J.: Proceedings: GABA as a transmitter in the central nervous system of vertebrates. *J. Neural Transm.*, 11:227, 1974.

Torda, C.: Model of molecular mechanism able to generate a depolarization-hyperpolarization cycle. *Inter. Rev. Neurobiol.*, 16:1, 1974.

Werman, R.: CNS cellular level. *Ann. Rev. Physiol.*, 34:337, 1972.

Willows, A. O. D.: Giant brain cells in mollusks. *Sci. Amer.*, 224:68, 1971.

# 47

# Transmission and Processing of Information in the Nervous System

## INFORMATION, SIGNALS, AND IMPULSES

The term *information,* as it applies to the nervous system, means a variety of different things, such as knowledge, facts, quantitative values, intensity of pain, intensity of light, temperature, and any other aspect of the body or its immediate surroundings that has meaning. Thus, pain from a pin prick is information, pressure on the bottom of the feet is information, degree of angulation of the joints is information, and a stored memory in the brain is information.

The primary function of the nervous system is (a) to transmit information from one point to another and (b) to process this information so that it can be used advantageously or so that its meaning can become clear to the mind. However, information cannot be transmitted in its original form but only in the form of nerve impulses. Thus, a part of the body that is subjected to pain must first convert this information into *nerve impulses;* and specific areas of the brain convert abstract thoughts also into nerve impulses that are then transmitted either elsewhere in the brain or into peripheral nerves to motor effectors throughout the body. The retina of the eye converts vision into nerve impulses, the nerve endings of the joints convert degree of angulation of the joints into nerve impulses, and so forth. The nerve impulses in turn are transmitted to the opposite ends of the neurons where they cause subsequent effects either at the synapses of other neurons or at the ends of motor nerves. This is what is meant by the transmission of information.

**Signals.** In the transmission of information, it is frequently not desirable to speak in terms of the individual impulses but instead in terms of the overall pattern of impulses; this pattern is called a *signal.* As an example, when pressure is applied to a large area of skin, impulses are transmitted by large numbers of parallel nerve fibers, and the total pattern of impulses transmitted by all these fibers is called a signal. Thus, we can speak of visual signals, auditory signals, somesthetic sensory signals, motor signals, and so forth.

## TRANSMISSION OF SIGNALS IN NERVE TRACTS

A signal is almost never transmitted only in a single nerve fiber but instead it is transmitted in a group of parallel fibers bound together in peripheral nerves or in *nerve tracts* in the central nervous system.

### SIGNAL STRENGTH

**Spatial Summation (Multiple Fiber Summation).** Figure 47–1 illustrates a sensory nerve leading from a segment of skin, showing extensive arborization of pain nerve fibers in the skin. A pin or other object stimulates nerve

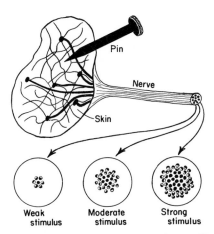

**Figure 47–1.** Pattern of stimulation of pain fibers in a nerve trunk leading from an area of skin pricked by a pin.

endings from a large number of different fibers, because each separate fiber divides and spreads over an area of skin sometimes as large as 5 cm. in diameter called the *receptive field*. The number of nerve endings from each fiber is large in the very center of its "field" but becomes less and less toward the periphery. Therefore, in Figure 47–1, a pin pricking the skin stimulates the nerve fiber strongly if the point of stimulation is in the center of its field of reception, but it stimulates the fiber only weakly if the point of stimulus is in the periphery of the field.

In the lower part of Figure 47–1 is shown a cross-section of the nerve leading from the skin, showing also the fibers stimulated by the pinprick. Note in the left-hand circle that if the pinprick is slight, only a single fiber is stimulated strongly (the black fiber in the center), and several immediately adjacent fibers are stimulated moderately (half-black fibers), while all other fibers are not stimulated at all. When the stimulus becomes more intense, the number of fibers strongly stimulated becomes progressively greater. Thus, the signal spreads to more and more fibers. This phenomenon is called *spatial summation*, which means simply that one of the means by which signals of increasing strength are transmitted in the nervous system is by utilization of progressively greater numbers of fibers. The increase in number of fibers as the strength of signal increases is called *recruitment* of the additional fibers.

**Temporal Summation (Frequency Modulation).** A second means by which signals of increasing strength are transmitted through nerve tracts is by increasing the *frequency* of nerve impulses in each fiber. This occurs in ad-

dition to the increase in number of excited fibers. This process is called *temporal summation,* which means simply that during any given interval of time, the number of impulses transmitted is directly related to the strength of the signal.

Figure 47–2 illustrates temporal summation, showing in the upper part a changing strength of signal and in the lower part the actual impulses transmitted by the nerve fiber. This figure demonstrates that the frequency increases and decreases with the strength of the signal. Thus, this is actually a process of frequency modulation, the frequency of impulses being "modulated" in proportion to the strength of the signal.

## DETECTION OF A SIGNAL AT THE TERMINUS OF THE NERVE PATHWAY

The nerve signals obviously are nothing more than a series of pulses rather than a smooth signal. However, all terminal points in the nervous system have means for averaging impulses and thereby smoothing out the signal. Let us consider as an example skeletal muscles. Impulses controlling a skeletal muscle are transmitted through large numbers of parallel nerve fibers. If 50 nerve fibers innervate a muscle and one impulse is transmitted per second, but nonsynchronously, the 50 separate impulses cause a weak but smooth contraction because the pulses in the different nerve fibers are spread out over a period of time. If each nerve fiber transmits 200 impulses per second, making a total of 10,000 pulses per second, the 10,000 pulses give a strong and still smooth contraction.

Similar averaging processes occur in the

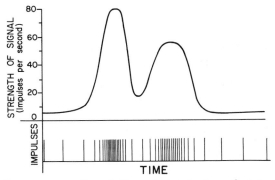

**Figure 47–2.** Translation of signal strength into a frequency-modulated series of nerve impulses, showing *above* the strength of signal and *below* the separate nerve impulses.

central nervous system; most postsynaptic neurons in the nervous system receive literally hundreds to thousands of impulses each second from the presynaptic terminals, and the synaptic membrane is capable of summating and averaging these impulses to give smooth changes in the excitatory state of the neuron, as was explained in the previous chapter.

**"Granularity" of the Detected Signal.** Despite the ability of the muscles and synapses in the central nervous system to average impulse rates, under some conditions the detected signal has unwanted ripples, which is called *granularity* of the signal. Two causes of this are, first, slow impulse rates and, second, synchrony of impulses in the parallel fibers through which the signal is transmitted.

A slow impulse rate causes granularity of the signal because of the nature of the averaging system itself. The synapses in the nervous system, the skeletal muscle, and the other signal detectors react to an impulse for a given period of time. Thus, the contractile twitch of a typical muscle lasts about one-twentieth second. If a second impulse comes along before this twitch is over, it causes the process of contraction to continue. Therefore, rapidly successive impulses cause smooth contraction of the muscle called *tetanization.* On the other hand, if the impulse rate is less than 30 per second, the muscle contracts in a series of twitches, and the signal is said to be granular. The same principles apply to the averaging mechanism of the central nervous system synapses.

The second cause of granularity is *synchrony* of impulses in the parallel fibers transmitting the signal. For instance, if 100 fibers are transmitting one impulse per second and the impulses are spread out in time in the different fibers, the signal will be smooth despite the low impulse frequency. But if all these impulses are transmitted in synchrony, a very granular signal will reach the detector at a frequency of one per second.

**Fidelity of the Temporal Pattern of a Signal.** The "temporal pattern" of a signal means the changes in signal strength with time. Unfortunately, extremely rapid changes in signal strengths cannot be transmitted through nerve pathways. The reason for this is that the averaging mechanisms of the detectors require a considerable number of impulses before the average can be determined. Therefore, the change in signal strength in the detector always *lags* a few milliseconds behind the change in signal strength in the nerve tract itself.

The fidelity of the temporal pattern is different for different types of nerve fibers. The large type A myelinated nerve fibers that can transmit as many as 2000 impulses per second are usually associated with rapidly responding sensory receptors and rapidly responding detectors. Therefore, they can transmit changes in signal strengths that occur as rapidly as 700 cycles per second (particularly those nerve fibers associated with the pacinian corpuscles). On the

other hand, the minute unmyelinated type C fibers that transmit only a few impulses per second are usually associated with slowly responding receptors and detectors. The frequency of varying signals transmitted through these pathways is often limited to as little as one or less than one cycle per second (as is the case for autonomic signals controlling the internal organs).

## SPATIAL ORIENTATION OF SIGNALS IN FIBER TRACTS

How does the brain detect the precise position on the body that is receiving a sensory stimulus, and how does the brain transmit impulses precisely to individual skeletal muscle bundles? The answer to this is by transmitting their signals in a precise spatial pattern through the nerve tracts. All the different nerve tracts, both in the peripheral nerves and in the fiber tracts of the central nervous system, are spatially organized. For instance, in the dorsal columns of the spinal cord the sensory fibers from the feet lie toward the midline, while those fibers entering the dorsal columns at higher levels of the body lie progressively more toward the lateral sides of the dorsal columns. This spatial organization is maintained with precision throughout the sensory pathway all the way to the somesthetic cortex. Likewise, the fiber tracts within the brain and those extending into motor nerves are spatially oriented in the same way.

As an example, Figure 47–3 illustrates to the left three separate pins stimulating the skin; to the right the spatial orientation of the stimulated fibers in the nerve are shown. Each pin in this example stimulates a single fiber strongly and adjacent fibers less strongly. This spatial orientation of the three separate bundles of fibers is maintained all the way to the cerebral cortex and obviously allows the brain to localize the three different pinpricks to their points of stimulation.

## TRANSMISSION AND PROCESSING OF SIGNALS IN NEURONAL POOLS

The central nervous system is made up of literally hundreds of separate neuronal pools, some of which are extremely small and some very large. For instance, the entire cerebral cortex could be considered to be a single large neuronal pool. If all the surface area of the cerebral cortex were flattened out, including the

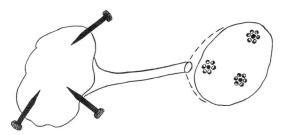

**Figure 47–3.**  Spatial pattern of nerve fiber stimulation in a nerve trunk following stimulation of the skin by three separate but simultaneous pinpricks.

surfaces of the penetrating folds, the total area of this large flat pool would be several square feet. It has many separate fiber tracts coming to it (afferent fibers) and others leaving it (efferent fibers). Furthermore it maintains the same quality of spatial orientation as that found in the nerve bundles, individual points of the cortex connecting with specific points elsewhere in the nervous system or connecting through the peripheral nerves with specific points in the body. However, within this pool of neurons are large numbers of short nerve fibers whereby signals spread horizontally from neuron to neuron within the pool itself.

Other neuronal pools include the different basal ganglia and the specific nuclei in the thalamus, cerebellum, mesencephalon, pons, and medulla. Also, the entire dorsal gray matter of the spinal cord could be considered to be one long pool of neurons, and the entire anterior gray matter another long neuronal pool. Each pool has its own special characteristics of organization which cause it to process signals in its own special way. It is these special characteristics of the different pools that allow the multitude of functions of the nervous system. Yet, despite their differences in function, the pools also have many similarities of function which are described in the following pages.

**Organization of Neurons in the Neuronal Pools.**  Figure 47–4 is a schematic diagram of the organization of neurons in a neuronal pool, showing "input" fibers to the left and "output" fibers to the right. Each input fiber divides hundreds to thousands of times and provides an average of a thousand or more terminal fibrils that spread over a large area in the pool to synapse with the dendrites or cell bodies of the neurons in the pool. The area into which the endings of each incoming nerve fiber spread is called its *stimulatory field*. Note that each input fiber arborizes so that large numbers of its

synaptic knobs lie on the centermost neurons in its "field," but progressively fewer knobs lie on the neurons farther from the center of the field.

**Threshold and Subthreshold Stimuli—Facilitation.**  Going back to the discussion of synaptic function in the previous chapter, it will be recalled that stimulation of a single excitatory synaptic knob almost never stimulates the postsynaptic neuron. Instead, large numbers of knobs must discharge on the same neuron either simultaneously or in rapid succession to cause excitation. For instance, let us assume that 6 separate knobs must discharge simultaneously or in rapid succession to excite any one of the neurons in Figure 47–4. Note that input fiber 1 contributes a total of 10 synaptic knobs to neuron *a*. Therefore, an incoming impulse in fiber 1 will cause neuron *a* to discharge. This same incoming fiber has two knobs on neuron *b* and three on neuron *c*. Neither of these two neurons will fire.

Incoming nerve fiber 2 has 12 synaptic knobs on neuron *d*, three on neuron *c*, and two on neuron *b*. In this case, an impulse in fiber 2 excites only neuron *d*.

Though fiber 1 fails to stimulate neurons *b* and *c*, discharge of the synaptic knobs changes the membrane potentials of these neurons so that they can be more easily excited by other incoming signals. Thus, there are two types of stimuli that enter a neuronal pool, *excitatory stimuli* and *subthreshold stimuli*. A sub-

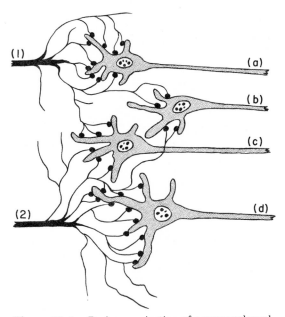

**Figure 47–4.**  Basic organization of a neuronal pool.

threshold stimulus fails to excite the neuron but does make the neuron more excitable to impulses from other sources. The neuron that is made more excitable but does not discharge is said to be *facilitated*.

**"Convergence" of Subthreshold Stimuli to Cause Excitation.**    Let us now assume that both fibers 1 and 2 in Figure 47–4 transmit concurrent impulses into the neuronal pool. In this case, neuron *c* is stimulated by six synaptic knobs simultaneously, which is the required number to cause excitation. Thus, subthreshold stimuli can *converge* from several sources and *summate* at a neuron to cause an excitatory stimulus.

**Summation of Facilitation.**    Now, let us consider neuron *b* of Figure 47–4 when both input nerve fibers are stimulated. In this case each fiber excites two synaptic knobs on neuron *b,* but the total of four excited knobs still is not up to the threshold value required to excite the neuron. However, the degree of facilitation does become greatly increased so that now only two additional knobs excited from another source need fire to cause discharge.

**The "Field" of Terminals.**    It must be recognized that Figure 47–4 represents a highly condensed version of the neuronal pool, for each input nerve fiber gives off terminals that spread to perhaps 100 or more separate neurons, and the surface of each neuron is covered by many hundred or many thousand synapses. In the centralmost portion of this *field* of terminals, almost all the neurons are stimulated by the incoming fiber; whereas farther toward the periphery of the field, the neurons are facilitated but do not discharge. Figure 47–5 illustrates this effect, showing the field of a single input nerve fiber. The area in the neuronal pool in which all the neurons discharge is called the *discharge* or *excited* or *liminal zone,* and the area to either side in which the neurons are facilitated but do not discharge is called the *facilitated* or *subthreshold* or *subliminal* zone.

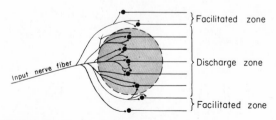

**Figure 47–5.**    "Discharge" and "facilitated" zones of a neuronal pool.

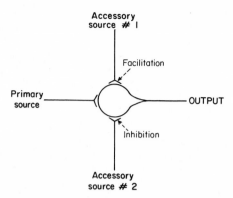

**Figure 47–6.**    Basic neuronal circuit for facilitation or inhibition.

**Facilitation of the Neuronal Pool by Signals from Accessory Sources.**    A neuronal pool frequently receives input nerve fibers from many different sources. Thus, in Figure 47–6 a neuron in a given pool receives impulses from a *primary source* and from two *accessory sources.* The nerve fiber from accessory source #1 is an excitatory fiber which secretes excitatory transmitter at its synaptic knobs. If enough knobs from this source are stimulated, the postsynaptic neuron discharges. However, in many parts of the nervous system the accessory sources supply only a few nerve fibers to the pool, not enough usually to cause excitation but yet enough to *facilitate* the neurons. Thus, in Figure 47–6, if a facilitatory signal is entering the pool from accessory source #1, a much weaker than usual signal from the primary source is able to excite the postsynaptic neuron. In this way the signal from accessory source #1 controls the ease with which signals pass from the primary source to the output, which explains one of the means by which different parts of the nervous system control the degree of activity of other parts.

**Inhibition of a Neuronal Pool by Signals from an Accessory Source.**    It will be recalled from the previous chapter that some neurons of the central nervous system secrete an inhibitory transmitter substance instead of an excitatory transmitter substance. Thus, in Figure 47–6, stimulation of the inhibitory fibers from accessory source #2 strongly inhibits the neuronal pool so that a strong signal from the primary source is required to cause normal output.

**Mechanism By Which a Single Input Signal Can Cause both Excitation and Inhibition— The Inhibitory Circuit.**    Figure 47–7 illustrates the so-called inhibitory circuit which can

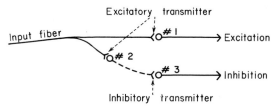

**Figure 47–7.** Inhibitory circuit. Neuron #2 is an inhibitory neuron.

change an excitatory signal into an inhibitory signal. In this figure the input fiber divides and secretes excitatory transmitter at both its endings. This causes excitation of both neurons 1 and 2. However, neuron 2 is an inhibitory neuron that secretes inhibitory transmitter at its terminal nerve endings. Excitation of this neuron therefore inhibits neuron 3.

In short, the usual means, if not the only means, to cause inhibition is for a signal *to be transmitted through an inhibitory neuron,* which then secretes the inhibitory transmitter substance.

**Convergence.** The term "convergence" means control of a single neuron by the converging signals from two or more separate input nerve fibers. One type of convergence was illustrated in Figure 47–4 in which two excitatory input nerve fibers from the same source converged upon several separate neurons to stimulate them. This type of convergence from a single source is illustrated again in Figure 47–8A.

However, convergence can also result from input signals (excitatory or inhibitory) from several different sources, which is illustrated in Figure 47–8B. For instance, the interneurons of the spinal cord receive converging signals from

(a) peripheral nerve fibers entering the cord, (b) propriospinal fibers passing from one segment of the cord to another, (c) corticospinal fibers from the cerebral cortex, and (d) probably several other long pathways descending from the brain into the spinal cord. Then the signals from the interneurons converge on the motoneurons to control muscle function.

Such convergence allows summation of information from different sources, and the resulting response is a summated effect of all these different types of information. Obviously, therefore, convergence is one of the important means by which the central nervous system correlates, summates, and sorts different types of information.

**Divergence.** Divergence means that excitation of a single input nerve fiber stimulates multiple output fibers from the neuronal pool. The two major types of divergence are illustrated in Figure 47–9 and may be described as follows:

An *amplifying* type of divergence often occurs, illustrated in Figure 47–9A. This means simply that an input signal spreads to an increasing number of neurons as it passes through successive pools of a nervous pathway. This type of divergence is characteristic of the corticospinal pathway in its control of skeletal muscles as follows: Stimulation of a single large pyramidal cell in the motor cortex transmits a single impulse into the spinal cord. Yet, under appropriate conditions, this impulse can stimulate perhaps several hundred interneurons, and these in turn stimulate perhaps an equal number of anterior motoneurons. Each of these then stimulates as many as 100 to 300 muscle fibers. Thus, there is a total divergence, or amplification, of as much as 10,000-fold.

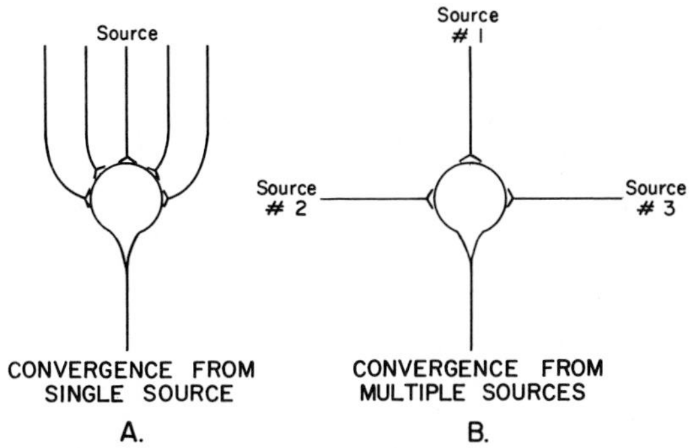

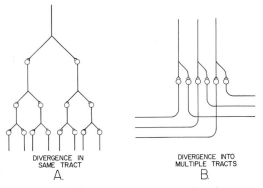

**Figure 47–9.** "Divergence" in neuronal pathways: (A) Divergence within a pathway to cause "amplification" of the signal. (B) Divergence into multiple tracts to transmit the signal to separate areas.

**Figure 47–8.** "Convergence" of multiple input fibers on a single neuron: (A) Input fibers from a single source. (B) Input fibers from multiple sources.

The second type of divergence, illustrated in Figure 47–9B, is *divergence into multiple tracts*. In this case, the signal is transmitted in two separate directions from the pool. This allows the same information to be transmitted to several different parts of the nervous system where it is needed. For instance, information transmitted in the dorsal columns of the spinal cord takes two courses in the lower part of the brain, (1) into the cerebellum and (2) on through the lower regions of the brain to the thalamus and cerebral cortex. Likewise, in the thalamus almost all sensory information is relayed both into deep structures of the thalamus and to discrete regions of the cerebral cortex.

## TRANSMISSION OF SPATIAL PATTERNS THROUGH SUCCESSIVE NEURONAL POOLS

Most information is transmitted from one part of the nervous system to another through several successive neuronal pools. For instance, sensory information from the skin passes first through the peripheral nerve fibers, then through second order neurons that originate either in the spinal cord or in the cuneate and gracile nuclei of the medulla, and finally through third order neurons originating in the thalamus to the cerebral cortex. Such a pathway is illustrated at the top of Figure 47–10. Note that the sensory nerve endings in the skin overlap each other tremendously; and the terminal fibrils of each nerve fiber, on entering each neuronal pool, spread to many adjacent neurons, innervating perhaps 100 or more separate neurons. On first thought, one would expect signals from the skin to become completely mixed up by this haphazard arrangement of terminal fibrils in each neuronal pool. For statistical reasons, however, this does not occur, which can be explained as follows:

First, if a single point is stimulated in the skin, the nerve fiber with the most nerve endings in that particular spot becomes stimulated to the strongest extent, while the immediately adjacent nerve fibers become stimulated less strongly, and the nerve fibers still farther away become stimulated only weakly. When this signal arrives at the first neuronal pool, the stimulus spreads in many directions in the terminal fibrils of the neuronal pool. Yet the *greatest number* of *excited presynaptic terminals* lies very near the point of entry of the most excited input nerve fiber. Therefore, the neuron closest to this central point is the one that be-

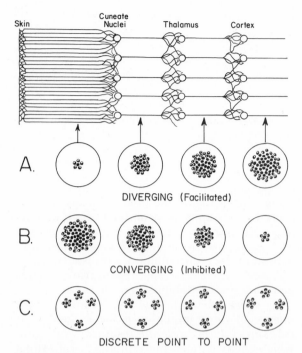

**Figure 47–10.** Typical organization of a sensory pathway from the skin to the cerebral cortex. *Below:* The patterns of fiber stimulation at different points in the pathway following stimulation by a pinprick when the pathway is (A) facilitated, (B) inhibited, and (C) normally excitable.

comes stimulated to the greatest extent. Exactly the same effect occurs in the second neuronal pool in the thalamus and again when the signal reaches the cerebral cortex.

Yet, it is true that a signal passing through *highly facilitated neuronal pools* could diverge so much that the spatial pattern at the terminus of the pathway would be completely obscured. This effect is illustrated in Figure 47–10A, which shows successively expanding spatial patterns of neuron stimulation in such a facilitated, diverging pathway.

However, the degree of facilitation of the different neuronal pools varies from time to time. Under some conditions the degree of facilitation is so low that the pathway becomes converging, as illustrated in Figure 47–10B. In this case, a broad area of the skin is stimulated, but the signal loses part of its fringe stimuli as it passes through each successive pool until the breadth of the stimulus becomes contracted at the opposite end. One can achieve this type of stimulation by pressing ever so lightly with a flat object on the skin. The signal converges to give the person a sensation of almost a point contact.

In Figure 47–10C, four separate points are simultaneously stimulated on the skin, and the degree of excitability in each neuronal pool is exactly that amount required to prevent either divergence or convergence. Therefore, a reasonably true spatial pattern of each of the four points of stimulation is transmitted through the entire pathway.

**Centrifugal Control of Neuronal Facilitation in the Sensory Pathways.** It is obvious from the above discussion that the degree of facilitation of each neuronal pool must be maintained at exactly the proper level if faithful transmission of the spatial pattern is to occur. Recent discoveries have demonstrated that the degrees of facilitation of most—indeed, probably all—neuronal pools in the different pathways are controlled by *centrifugal nerve fibers* that pass from the respective sensory areas of the cortex downward to the separate neuronal pools. Thus, these nerve fibers undoubtedly help to control the faithfulness of signal transmission.

**Inhibitory Circuits to Provide Contrast in the Spatial Pattern.** When a single point of the skin or other sensory area is stimulated, not only is a single fiber excited but a number of "fringe" fibers are excited less strongly at the same time, as already explained above. Therefore, the spatial pattern is blurred even before the signal begins to be transmitted through the pathway. However, in many pathways—if not all—such as in the visual pathway and in the somesthetic pathway, lateral *inhibitory circuits* inhibit the fringe neurons and re-establish a truer spatial pattern.

Figure 47–11A illustrates this circuit, showing that the nerve fibers of a pathway give off collateral fibers that excite inhibitory neurons. These inhibitory neurons in turn inhibit the less

excited fringe neurons in the signal pathway. The effect of this on transmission of the spatial pattern is illustrated in Figure 47–11B, which shows the same point-to-point transmission pattern that was illustrated in Figure 47–10C. The left-hand pattern illustrates four strongly excited fibers; penumbras of fringe excitation surround each of these. To the right is illustrated removal of the penumbras by the lateral inhibitory circuits; and obviously, this increases the contrast in the signal and helps the faithfulness of transmission of the spatial pattern. Unfortunately, we still know too little about this lateral inhibitory mechanism to be sure how effective it actually is. Furthermore, much of the lateral inhibition results from presynaptic inhibition; this was discussed in the previous chapter. The presynaptic type of lateral inhibition is probably the more prevalent mechanism in more peripheral areas of the nervous system, whereas postsynaptic inhibition is the mechanism in the cortical area.

**Relationship of Numbers of Fibers in a Pathway to the Total Amount of Spatial Information that Can Be Transmitted.** The upper part of Figure 47–10 illustrates that far more sensory nerve receptors transmit information into the spinal cord than there are nerve fibers to transmit the same information upward from the spinal cord to the cortex. Furthermore, the total number of points that can be recognized as individual and separate areas on the surface of the body is probably many times the number of nerve fibers in the sensory pathways ascending the spinal cord. The question must be asked, can a transmission system carry more discrete bits of spatial information than there are nerve fibers themselves?

In considering once again the organization of the transmission pathway shown in the top of Figure 47–10, we see that each sensory nerve fiber leads to a particular *place* in the first neuronal pool. Some fibers end almost directly on a second order neuron while others end *between* second order neurons. If a fiber lies directly on the neuron, that neuron is stimulated far more than the adjacent one. If the fiber lies between two neurons, both neurons are stimulated approximately equally. If the fiber lies three-quarters the way toward one neuron but one-quarter toward the second, the first neuron is stimulated perhaps three times as much as the second neuron. Thus, the spatial position of the incoming neuron can be transmitted by the second order neurons by the *ratios* of signal strengths in the output fibers. One ratio means one position, another ratio another position, and so forth. In this way, a large amount of information from many different sensory receptors probably can be condensed into much fewer nerve fibers. Here again, we do not know how highly this principle of transmission is developed in the nervous system. But, to give an example, each eye has about 125,000,000 re-

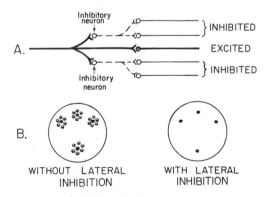

**Figure 47–11.** (A) Lateral inhibitory circuit by which an excited fiber of a neuronal pool can cause inhibition of adjacent fibers. (B) Increase in contrast of the stimulus pattern caused by the inhibitory circuit.

ceptors, and it can detect about 5,000,000 individual point areas in the visual field. Yet the total number of nerve fibers transmitting information in the optic nerve is believed to be only one million, many of which are probably concerned with other functions besides transmission of the spatial pattern. Therefore, we can guess that the fibers in this tract can carry 10 or more times as many bits of spatial information as there are fibers.

## PROLONGATION OF A SIGNAL BY A NEURONAL POOL— "AFTER-DISCHARGE"

Thus far, we have considered signals that are transmitted instantaneously through a neuronal pool. However, in many instances, a signal entering a pool causes a prolonged output discharge, called *after-discharge,* even after the incoming signal is over. The three basic mechanisms by which after-discharge occurs are as follows:

**Synaptic After-Discharge.** When presynaptic terminals discharge on the surfaces of dendrites or the soma of a neuron, a postsynaptic potential develops in the neuron and lasts for many milliseconds—in the anterior motoneuron for about 15 milliseconds, though perhaps much longer in other neurons. As long as this potential lasts it can excite the neuron, causing it to transmit output impulses as was explained in the previous chapter. Thus, as a result of this synaptic after-discharge mechanism alone, it is possible for a single instantaneous input to cause a sustained signal output (a series of repetitive discharges) lasting as long as 15 milliseconds.

**The Parallel Circuit Type of After-Discharge.** Figure 47–12 illustrates a second type of neuronal circuit that can cause short periods of after-discharge. In this case, the input signal spreads through a series of neurons in the neuronal pool, and from many of these neurons impulses keep converging on an output neuron. It will be recalled that a signal is delayed at each synapse for at least 0.5 millisecond, which is called the *synaptic delay.* Therefore, signals that pass through a succession of intermediate neurons reach the output neuron one by one after varying periods of delay. Therefore, the output neuron continues to be stimulated for many milliseconds.

It is doubtful that more than a few dozen successive neurons ordinarily enter into a parallel after-discharge circuit. Therefore, one would suspect that this type of after-discharge circuit could cause after-discharges that last for no more than perhaps 25 to 50 milliseconds. Yet, this circuit does represent a means by which a single input signal, lasting less than 1 millisecond, can be converted into a sustained output signal lasting many milliseconds.

**The Reverberating (Oscillatory) Circuit as a Cause of After-Discharge.** Probably one of the most important of all circuits in the entire nervous system is the reverberating, or oscillatory, circuit, several different varieties of which are illustrated in Figure 47–13. The simplest

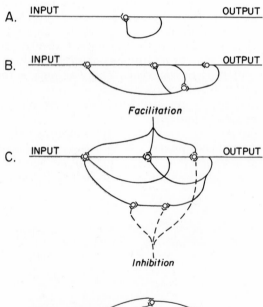

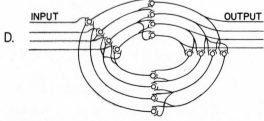

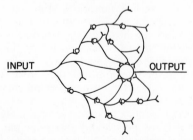

Figure 47–12. The parallel after-discharge circuit.

**Figure 47–13.** Reverberatory circuits of increasing complexity.

theoretical reverberating circuit, even though such may not actually exist in the nervous system, is that illustrated in Figure 47–13A, which involves a single neuron. In this case, the output neuron simply sends a collateral nerve fiber back to its own dendrites or soma to restimulate itself; therefore, once the neuron should discharge, the feedback stimuli could theoretically keep the neuron discharging for a long time thereafter.

Figure 47–13B illustrates a few additional neurons in the feedback circuit, which would give a longer period of time between the initial discharge and the feedback signal. Figure 47–13C illustrates a still more complex system in which both facilitatory and inhibitory fibers impinge on the reverberating pool. A facilitatory signal increases the ease with which reverberation takes place, while an inhibitory signal decreases the ease with which reverberation takes place.

Figure 47–13D illustrates that most reverberating pathways are constituted of many parallel fibers, and at each cell station the terminal fibrils diffuse widely. In such a system the total reverberating signal can be either weak or strong, depending on how many parallel nerve fibers are momentarily involved in the reverberation.

Finally, reverberation need not occur only in a single neuronal pool, for it can occur through a circuit of successive pools.

*Characteristics of After-Discharge from a Reverberating Circuit.* Figure 47–14 illustrates postulated output signals from a reverberating after-discharge circuit. The input stimulus need last only 1 millisecond or so, and yet the output can last for many milliseconds or even minutes. The figure demonstrates that the intensity of the output signal increases to a reasonably high value early in the reverberation, then decreases to a critical point, and suddenly ceases entirely. Furthermore, the duration of the after-discharge is determined by the degree of inhibition or facilitation of the neuronal pool. In this way, signals from other parts of the brain can control the reaction of the pool to the input stimulus.

Almost these exact patterns of output signals can be recorded from the motor nerves exciting a muscle involved in the flexor reflex (discussed in Chap. 51), which is believed to be caused by a reverberating type of after-discharge following stimulation of a pain fiber.

*Importance of Synaptic Fatigue in Determining the Duration of Reverberation.* It was pointed

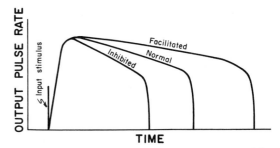

**Figure 47–14.** Typical pattern of the output signal from a reverberatory circuit following a single input stimulus, showing the effects of facilitation and inhibition.

out in the previous chapter that synapses fatigue if stimulated for prolonged periods of time. Therefore, one of the most important factors that determines the duration of the reverberatory type of after-discharge is probably the rapidity with which the involved synapses fatigue. Rapid fatigue would obviously tend to shorten the period of after-discharge and slow fatigue to lengthen it.

Furthermore, the greater the number of neurons in the reverberatory pathway and the greater the number of collateral feedback fibrils, the easier it would be to keep the reverberation going. Therefore, it is to be expected that longer reverberating pathways would in general sustain after-discharges for longer periods of time.

*Duration of Reverberation.* Typical after-discharge patterns of different reverberatory circuits have durations from as short as 10 milliseconds to as long as several minutes, or perhaps even hours. Indeed, as will be explained in Chapter 54, wakefulness may be an example of reverberation of neuronal circuits in the basal region of the brain. In this theory of wakefulness, "arousal impulses" are postulated to set off the wakefulness reverberation at the beginning of each day and thereby to cause sustained excitability of the brain, this excitability lasting 14 or more hours.

## CONTINUOUS SIGNAL OUTPUT FROM NEURONAL POOLS

Some neuronal pools emit output signals continuously even without excitatory input signals. At least two different mechanisms can cause this effect, (1) intrinsic neuronal discharge and (2) reverberatory signals.

**Continuous Discharge Caused by Intrinsic Neuronal Excitability.** Neurons, like other excitable tissues, discharge repetitively if their membrane potentials rise above certain threshold levels. The membrane potentials of some neurons are even normally high enough to cause these neurons to emit impulses continually. This occurs especially in some of the cells of the cerebellum and in some of the interneurons of the spinal cord. The rates at which these cells emit impulses can be increased by facilitatory signals or decreased by inhibitory signals; the latter can sometimes decrease the rate to extinction.

**Continuous Signals Emitted from Reverberation Circuits.** Obviously, a reverberating circuit that never fatigues to extinction could also be a source of continual impulses. Facilitatory impulses entering the reverberating pool, as illustrated in Figure 47–13C, could increase the output signal, and inhibition could decrease or even extinguish the output signal.

Figure 47–15 illustrates a continual output signal from a pool of neurons, whether it be a pool emitting impulses because of intrinsic neuronal excitability or as a result of reverberation. Note that an excitatory (or facilitatory) input signal greatly increases the output signal, whereas an inhibitory input signal greatly decreases the output. Those students who are familiar with radio transmitters will recognize this to be a *carrier wave* type of information transmission. That is, the excitatory and inhibitory control signals are not the *cause* of the output signal, but they do *control* it. Note that this carrier wave system allows decrease in signal intensity as well as increase, whereas, up to this point, the types of information transmission that we have discussed have been only positive information rather than negative information. This type of information transmission is used by the autonomic nervous system to control such functions as vascular tone, gut tone, degree of constriction of the iris, heart rate, and others.

### RHYTHMIC SIGNAL OUTPUT

Many neuronal circuits emit rhythmic output signals—for instance, the rhythmic respiratory signal originating in the reticular substance of the medulla and pons. This repetitive rhythmic signal continues throughout life, while other rhythmic signals, such as those that cause scratching movements by the hind leg of a dog or the walking movements in an animal, require input stimuli into the respective circuits to initiate the signals.

Rhythmic signals probably result from reverberating pathways. One can readily understand that each time a signal passes around a reverberatory loop, collateral impulses could be transmitted into the output pathway. Thus, the rhythmic respiratory signals could result from such a reverberatory mechanism, as was suggested in the description of respiratory center function in Chapter 42.

Obviously, facilitatory or inhibitory signals can affect rhythmic signal output in the same way that they can affect continual signal outputs. Figure 47–16, for instance, illustrates the rhythmic respiratory signal in the phrenic nerve. However, when the carotid body is stimulated by arterial oxygen deficiency, the frequency and amplitude of the rhythmic signal pattern increase progressively.

## INSTABILITY AND STABILITY OF NEURONAL CIRCUITS

Almost every part of the brain connects either directly or indirectly with every other part, and this creates a serious problem. If the first part excites the second, the second the third, the third the fourth, and so on until finally the signal re-excites the first part, it is clear that an excitatory signal entering any part of the brain would set off a continuous cycle of re-excitation of all parts. If this should occur, the brain would be inundated by a mass of uncontrolled reverberating signals—signals that would be transmitting no information but, nevertheless, would be consuming the circuits of the brain so that none of the informational signals could be transmitted. Such an effect actually occurs in widespread areas of the brain during *epileptic fits.*

How does the central nervous system prevent this from happening all the time? The answer

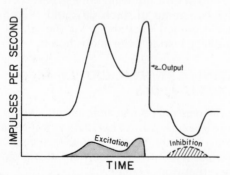

**Figure 47–15.** Continuous output from either a reverberating circuit or from a pool of intrinsically discharging neurons. This figure also shows the effect of excitatory or inhibitory input signals.

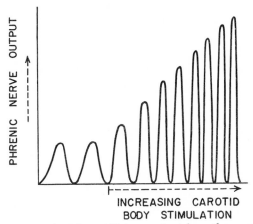

**Figure 47–16.** The oscillatory output from the respiratory center, showing that progressively increasing stimulation of the carotid body increases both the intensity and frequency of oscillation.

seems to lie in two basic mechanisms that function throughout the central nervous system: (1) inhibitory circuits, and (2) decremental transmission through synapses.

## INHIBITORY CIRCUITS AS A MECHANISM FOR STABILIZING NERVOUS SYSTEM FUNCTION

The phenomenon of lateral inhibition that was discussed earlier in the chapter prevents signals in an informational pathway from spreading diffusely everywhere. In addition, two other types of inhibitory circuits in widespread areas of the brain help to prevent excessive spread of unwanted signals: (a) inhibitory feedback circuits that return from the termini of pathways back to the initial excitatory neurons of the same pathways—these inhibit the input neurons when the termini become overly excited, and (b) some neuronal pools that exert gross inhibitory control over widespread areas of the brain—for instance, many of the basal ganglia exert inhibitory influences throughout the motor control system.

It is easy to understand how the inhibitory feedbacks can provide negative feedback signals for limiting the degree of excitability when an area of the brain tends to become too excited. Unfortunately, the inhibitory circuits do not seem to be widespread enough to stabilize nervous system function by themselves. Instead, the equally as important (if not even much more important) phenomenon of decremental conduction through synapses stabilizes nervous system function.

## DECREMENTAL CONDUCTION THROUGH SYNAPSES AS A MEANS FOR STABILIZING THE NERVOUS SYSTEM (HABITUATION, FATIGUE)

Decremental conduction through synapses means simply that the signal becomes progressively weaker the more prolonged the period of excitation. This phenomenon has often been called *fatigue,* which is similar to the fatigue of synapses that was discussed in the previous chapter, and it has also been called *habituation.* Figure 47–17 illustrates typical decremental conduction, showing a muscle response during the so-called flexor reflex. That is, stimulation of the pain sensory endings in a limb causes reflex contraction of the flexor muscles of the limb, leading to withdrawal of the limb from the painful stimulus. However, if this stimulus is repeated at close intervals, as was done in Figure 47–17, the response progressively decreases. But how does this decremental conduction, as illustrated by the decreasing response of the flexor reflex, play a role in stabilizing nervous system function? The answer to this is the following: When any given synapse is excited too much, the phenomenon of decremental conduction progressively depresses the signal pathway. Therefore, any tendency for reverberatory feedbacks to occur becomes automatically blocked before reverberation can become widespread.

**Synaptic Pathways That Exhibit Decremental Conduction; Those That Do Not.** Fortunately, some synaptic pathways exhibit very little decremental conduction, whereas others exhibit extreme degrees. One type of pathway that exhibits almost no decremental conduction

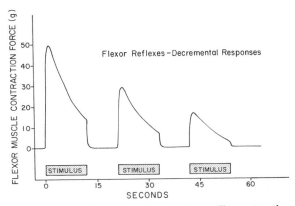

**Figure 47–17.** Successive flexor reflexes illustrating decrement in conduction of the signals through the reflex pathway.

is the fast-conducting sensory pathways from the peripheral parts of the body to the brain, such as the pathways through the dorsal columns to the thalamus and thence to the somesthetic cortex. This seems to be of purposeful benefit to nervous function because it allows faithful transmission of sensory information into the brain.

On the other hand, the interconnecting links through the short interneurons between the sensory system and the motor system frequently exhibit extreme degrees of decremental conduction. This was illustrated in Figure 47-17 for the flexor reflex pathway, because it is the interneurons between the sensory input and the motor output of the spinal cord that fatigue. At least in certain parts of the nervous system, this decremental conduction has been shown to result from exhaustion of transmitter substance in the synaptic knobs.

Obviously, if the connecting links between major portions of the nervous system exhibit decremental conduction, reverberatory instability in the nervous system is generally prevented.

**Recovery From Decremental Conduction.** After a major degree of decrement has occurred in the conductive properties of a synapse, the conduction recovers if an adequate period of rest between stimuli is allowed. Thus, when a pathway is not used often, it becomes progressively more excitable until full conduction returns.

**Automatic Adjustment of Sensitivity of Pathways by the Decremental Conduction Mechanism.** It is clear by now that if the excitability of a pathway in the nervous system is too great, then too many impulses will be transmitted and the mechanism of decremental conduction will automatically adjust the sensitivity of the pathway to a lower level. Conversely, if the sensitivity is too low, then too few impulses will be transmitted, and the sensitivity of the system will automatically adjust to a higher level. Therefore, this decremental conduction mechanism, and its recovery during rest, is probably an extremely important mechanism for maintaining a proper balance among the sensitivities of the respective pathways throughout the brain.

## ALERTING SIGNALS ("DIFFERENTIAL" SIGNALS)

A special type of output signal that occurs commonly in neuronal pools of the nervous system is the *alerting signal*. The alerting signal is a very strong output signal from a neuronal pool that occurs immediately after the input signal enters the pool; the alerting signal then dies away, however, even though the input signal continues indefinitely. When the input signal is suddenly removed, a second alerting response is likely to occur.

Figure 47-18 illustrates typical alerting signals that occur in the output from a retinal ganglion cell in response to changes in degree of retinal receptor stimulation. Thus, when light first impinges on the eye and the rods and cones become excited, the ganglion cells, which fire normally at a slow rate of about 5 per second, suddenly begin to fire at a much higher rate. However, once the signal from the rods and cones reaches a steady state level, the output signal from the ganglion cell falls back essentially to its original level. Nevertheless, the brain has been "alerted" to the onset of the light stimulus. Then, when the light is turned off, the signal from the ganglion cell falls below normal (actually all the way to zero, at times), but this too lasts for only a short time. Therefore, once again the brain is alerted to the change in the light stimulus.

Similar alerting responses occur in response to changes in pressure on the surface of the body: to noise, to movement of images across the retinal field, to changes in pitch in the auditory signal, to onset of a new taste sensation, to onset of smell stimuli, and so forth.

*Decremental Conduction as the Cause of the Alerting Response.* The alerting response almost certainly occurs from *very fast* decremental conduction through synapses. That is, at the

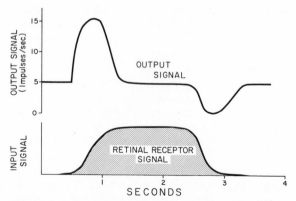

**Figure 47-18.** "Differentiation" of the signal from the retinal receptors (rods and cones) as it passes through the retinal neuronal circuit.

onset of the input signal conduction is very great, but then it fades very quickly.

### Differentiation of Signals

The two curves in Figure 47–18 illustrate that the output curve is very nearly the *differential* of the input curve that depicts the retinal receptor signal, though there is always a very slight lag in the differential signal. Nevertheless, it is clear that the nervous system does have the capability of "differentiating" input signals, which is a simple mathematical way of depicting *changes* in signals without being bothered by the continuous signal that lasts thereafter. In this way the brain can become alerted, but still, it can discard a tremendous amount of chaff or noise in the incoming signals to the nervous system after extracting most of their meaningful information.

The nervous system can also differentiate spatial signal patterns as well as time signal patterns. For instance, everywhere that a sharp boundary in light intensity occurs on the retina of the eye, the retinal neuronal circuit transmits a very powerful signal to the brain, whereas it transmits only weak signals from areas of the retina that are illuminated with noncontrasting borders. Here again, this alerts the person to the important features of the signal. Indeed, it explains how it is possible for one to recognize a person's image simply from a line drawing even though the line drawing is certainly not a true picture of the person. It just so happens that the "mind's eye" actually sees images almost as if they were line drawings even when the object is not a series of lines.

## REFERENCES

Bennett, M. V. L. (ed.): Synaptic Transmission and Neuronal Interaction. New York, Raven Press, 1975.

Blumenthal, R., Homsy, Y. M., Katchalsky, A. K., and Rowland, V. (eds.): Dynamic Patterns of Brain Cell Assemblies. Neurosciences Research Program Bulletin. Vol. 12, No. 1. Cambridge, Mass., Massachusetts Institute of Technology, 1974.

Chung, S. H.: In search of the rules for nerve connections. *Cell*, 3:201, 1974.

Freides, D.: Human information processing and sensory modality: cross-modal functions, information complexity, memory, and deficit. *Psychol. Bull.*, 81:284, 1974.

Heimer, L.: Pathways in the brain. *Sci. Amer.*, 225:48, 1971.

Kater, S. B., Heyer, C. B., and Kaneko, C. R. S.: Identifiable neurons and invertebrate behavior. *In* Guyton, A. C. (ed.): MTP International Review of Science: Physiology. Vol. 3. Baltimore, University Park Press, 1974, p. 1.

Kennedy, D.: Crayfish interneurons. *Physiologist*, 14:5, 1971.

Kreutzberg, G. W. (ed.): Physiology and Pathology of Dendrites. (Advances in Neurology, Vol. 12.) New York, Raven Press, 1975.

Martin, A. R.: Synaptic transmissions. *In* Guyton, A. C. (ed.): MTP International Review of Science: Physiology. Vol. 3. Baltimore, University Park Press, 1974, p. 53.

Nicholls, J. C., and Van Essen, D.: The nervous system of the leech. *Sci. Amer.*, 230:38, 1974.

Shepherd, G. M.: The Synaptic Organization of the Brain: an Introduction. New York, Oxford University Press, 1974.

Simpson, L. L.: The use of neuropoisons in the study of cholinergic transmission. *Ann. Rev. Pharmacol.*, 14:305, 1974.

Stenevi, U., Björklund, A., and Moore, R. Y.: Morphological plasticity of central adrenergic neurons. *Brain Behav. Evol.*, 8:110, 1973.

Szentagothai, J., Arbib, M. A., and Homsy, Y. M. (eds.): Conceptual Models of Neural Organization. Neurosciences Research Program. Cambridge, Mass., Massachusetts Institute of Technology, 1974.

Ungar, G., and Irwin, L. N.: Chemical correlates of neural function. *Neurosci. Res.*, 1:73, 1968.

Zippel, H. P. (ed.): Symposium on Memory and Transfer of Information, Göttingen, Mercksche Gesellschaft für Kunst und Wissenschaft, 1972.

Zippel, H. P. (ed.): Memory and Transfer of Information. New York, Plenum Publishing Corporation, 1973.

*See also bibliographies at the ends of Chapters 46 and 51.*

# 48

# Sensory Receptors and Their Basic Mechanisms of Action

Input to the nervous system is provided by the sensory receptors that detect such sensory stimuli as touch, sound, light, cold, warmth, and so forth. The purpose of the present chapter is to discuss the basic mechanisms by which these receptors change sensory stimuli into nerve signals and, also, how both the type of sensory stimulus and its strength are detected by the brain.

## TYPES OF SENSORY RECEPTORS AND THE SENSORY STIMULI THEY DETECT

The student of medical physiology will have already studied many different anatomical types of nerve endings, and Table 48–1 gives a list and classification of most of the body's sensory receptors. This table shows that there are basically five different types of sensory receptors: (1) *mechanoreceptors*, which detect mechanical deformation of the receptor or of cells adjacent to the receptors; (2) *thermoreceptors*, which detect changes in temperature, some receptors detecting cold and others warmth; (3) *nociceptors*, which detect damage in the tissues, whether it be physical damage or chemical damage; (4) *electromagnetic receptors*, which detect light on the retina of the eye; and (5) *chemoreceptors*, which detect taste in the mouth, smell in the nose, oxygen level in the arterial blood, osmolality of the body fluids, carbon dioxide concentration, and perhaps other factors that make up the chemistry of the body.

This chapter will discuss especially the function of specific types of receptors, primarily the peripheral mechanoreceptors, to illustrate some of the basic principles by which receptors in general operate. Other receptors will be discussed in relation to the sensory systems that they subserve, which will be presented mainly in the next few chapters. Figure 48–1 illustrates some of the different types of mechanoreceptors found in the skin or in the deep structures of the body, and Table 48–1 gives their respective sensory functions. All of these receptors will be discussed in the following chapters in relation to the respective sensory systems. However, the functions of some of these are described briefly, as follows:

*Free nerve endings* are found in all parts of the body. A very large proportion of these detect pain. However, other free nerve endings detect warmth, cold, and crude touch sensations.

Several of the more complex receptors listed in Figure 48–1 detect tissue deformation. These include the *Merkel's discs*, the *tactile hairs*, *pacinian corpuscles*, *Meissner's corpuscles*, *Krause's corpuscles*, and *Ruffini's end-organs*. In the skin, it is these receptors that detect the tactile sensations of touch and pressure. In the deep tissues, they detect stretch, deep pressure, or any other type of tissue deformation—even the stretch of joint capsules and ligaments to determine the angulation of a joint.

The *Golgi tendon apparatus* detects degree of tension in tendons, and the *muscle spindle* detects the length of muscle. These receptors will be discussed in Chapter 51 in relation to the muscle and tendon reflexes.

640

**TABLE 48–1.  Classification of Sensory Receptors**

---

*Mechanoreceptors*

---

Skin tactile sensibilities (epidermis and dermis)
  Free nerve endings
  Expanded tip endings
    Ruffini's endings
    Merkel's discs
    Plus several other variants
  Encapsulated endings
    Meissner's corpuscles
    Krause's corpuscles
    Pacinian corpuscles
  Hair end-organs
Deep tissue sensibilities
  Free nerve endings
  Expanded tip endings
    Ruffini's endings
    Plus a few other variants
  Encapsulated endings
    Pacinian corpuscles
    Plus a few other variants
  Specialized endings
    Muscle spindles
    Golgi tendon receptors
Hearing
  Sound receptors of cochlea
Equilibrium
  Vestibular receptors
Arterial pressure
  Baroreceptors of carotid sinuses and aorta

---

*Thermoreceptors*

---

Cold
  Probably free nerve endings
Warmth
  Probably free nerve endings

---

*Nociceptors*

---

Pain
  Free nerve endings

---

*Electromagnetic Receptors*

---

Vision
  Rods
  Cones

---

*Chemoreceptors*

---

Taste
  Receptors of taste buds
Smell
  Receptors of olfactory epithelium
Arterial oxygen
  Receptors of aortic and carotid bodies
Osmolality
  Probably neurons of supraoptic nuclei
Blood $CO_2$
  Receptors in or on surface of medulla and in aortic and
    carotid bodies
Blood glucose, amino acids, fatty acids
  Receptors in hypothalamus

---

## DIFFERENTIAL SENSITIVITY OF RECEPTORS

The first question that must be answered is how do two types of sensory receptors detect different types of sensory stimuli? The answer to this: By virtue of differential sensitivities. That is, each type of receptor is very highly sensitive to one type of stimulus for which it is designed and yet is almost nonresponsive to normal intensities of the other types of sensory stimuli. Thus, the rods and cones are highly responsive to light but are almost completely nonresponsive to heat, cold, pressure on the eyeballs, or chemical changes in the blood. The osmoreceptors of the supraoptic nuclei in the hypothalamus detect minute changes in the osmolality of the body fluids but yet have never been known to respond to sound. Finally, pain receptors in the skin are almost never stimulated by usual touch or pressure stimuli but do become highly active the moment tactile stimuli become severe enough to damage the tissues.

**Modality of Sensation—Law of Specific Nerve Energies.**  Each of the principal types of sensation that we can experience—pain, touch, sight, sound, and so forth—is called a *modality* of sensation. Yet, despite the fact that we experience these different modalities of sensation, nerve fibers transmit only impulses. Therefore, how is it that different nerve fibers transmit different modalities of sensation?

The answer to this is that each nerve tract terminates at a specific point in the central nervous system, and the type of sensation felt when a nerve fiber is stimulated is determined by this specific area in the nervous system to which the fiber leads. For instance, if a pain fiber is stimulated, the person perceives pain regardless of what type of stimulus excites the fiber. This stimulus can be electricity, heat, crushing, or stimulation of the pain nerve ending by damage to the tissue cells. Yet, whatever the means of stimulation, the person still perceives pain. Likewise, if a touch fiber is stimulated by exciting a touch receptor electrically or in any other way, the person perceives touch because touch fibers lead to specific touch areas in the brain. Similarly, fibers from the retina of the eye terminate in the vision areas of the brain, fibers from the ear terminate in the auditory areas of the brain, and temperature fibers terminate in the temperature areas.

This specificity of nerve fibers for transmitting only one modality of sensation is called the *law of specific nerve energies*.

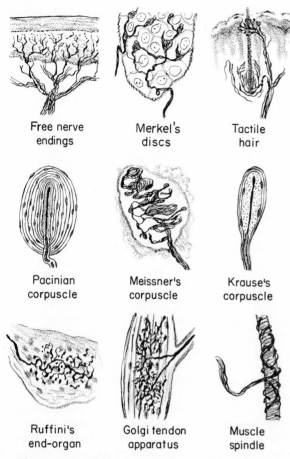

**Figure 48–1.** Several types of somatic sensory nerve endings. (Modified from Ramon y Cajal: Histology. William Wood and Co.)

Free nerve endings

Merkel's discs

Tactile hair

Pacinian corpuscle

Meissner's corpuscle

Krause's corpuscle

Ruffini's end-organ

Golgi tendon apparatus

Muscle spindle

# TRANSDUCTION OF SENSORY STIMULI INTO NERVE IMPULSES

## *LOCAL CURRENTS AT NERVE ENDINGS—RECEPTOR POTENTIALS*

All sensory receptors studied thus far have one feature in common. Whatever the type of stimulus that excites the ending, it first causes a local potential called a *receptor potential* in the neighborhood of the nerve endings, and it is *local flow of current* caused by the receptor potential that in turn excites action potentials in the nerve fiber.

There are two different ways in which receptor potentials can be elicited. One of these is to deform or chemically alter the terminal nerve ending itself. This causes ions to diffuse through the nerve membrane, thereby setting up the receptor potential.

The second method for causing receptor potentials involves specialized receptor cells that lie adjacent to the nerve endings. For instance, when sound enters the cochlea of the ear, specialized *receptor cells* called *hair cells* that lie on the basilar membrane develop local potentials which are receptor potentials that stimulate the terminal nerve fibrils entwining the hair cells.

(Some physiologists prefer to use the term *generator potentials* to designate the receptor potentials elicited in terminal nerve endings because the nerve fibers themselves actually "generate" the potentials, and they reserve the term "receptor potentials" only for those potentials that arise in specialized receptor cells of non-nervous tissue origin such as taste cells, hair cells of the ear, and so forth. However, because both of these potentials subserve the same function and because of the confusion that has developed by use of two separate terms, it is probably best to use the single term "receptor potentials" as we shall do here.)

**The Receptor Potential of the Pacinian Corpuscle.** The pacinian corpuscle is a very large and easily dissected sensory receptor. For this reason, one can study in detail the mechanism by which tactile stimuli excite it and by which it causes action potentials in the sensory fiber leading from it. Note in Figure 48–1 that the pacinian corpuscle has a central non-myelinated tip of a nerve fiber extending through its core. Surrounding this fiber are many concentric capsule layers so that compression on the outside of the corpuscle tends to elongate, shorten, indent, or otherwise deform the central core of the fiber, depending on how the compression is applied. The deformation causes a sudden change in membrane potential, as illustrated in Figure 48–2. This perhaps results from stretching the nerve fiber membrane, thus increasing its permeability and allowing

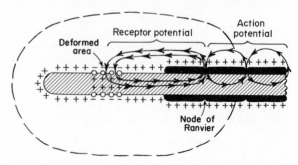

**Figure 48–2.** Excitation of a sensory nerve fiber by a generator potential produced in a pacinian corpuscle. (Modified from Loewenstein: *Ann. N.Y. Acad. Sci.*, 94:510, 1961.)

positively charged sodium ions to leak to the interior of the fiber. This change in local potential causes a local circuit of current flow that spreads along the nerve fiber to its myelinated portion. At the first node of Ranvier, which itself lies inside the capsule of the pacinian corpuscle, the local current flow initiates action potentials in the nerve fiber. That is, the current flow through the node depolarizes it, and this then sets off a typical saltatory transmission of an action potential along the nerve fiber toward the central nervous system, as was explained in Chapter 10.

**Electrotonic Nature of the Receptor Potential.** Note especially that the receptor potential has a different electrical character from the action potential. It will be recalled from Chapter 10 and also from Chapter 46 that the action potential is a self-regenerative cyclic event that begins with a resting negative potential, then changes to a positive potential and finally returns back to a negative potential. On the other hand, the receptor potential is an "electrotonic" potential that causes "tonic" flow of current without proceeding through the regenerative events of an action potential. It is a local potential just as are the end-plate potential of muscle fibers and the postsynaptic potential of neurons. If the receptor potential is great enough, it will elicit one or more action potentials at the first node of Ranvier. On the other hand, if the potential does not reach *threshold* level for excitation of an action potential, it will simply exist locally and will spread only a short distance along the fiber; the spreading will be by the process of electrotonic conduction, not by means of a self-regenerative action potential.

**Relationship Between Receptor Potential and Stimulus Strength.** Figure 48–3 illustrates the effect on the amplitude of the receptor potential caused by progressively stronger stimuli applied to the central core of the pacinian corpuscle. Note that the amplitude increases rapidly at first but then progressively less rapidly at high stimulus strengths. The maximum amplitude that can be achieved by receptor potentials is around 100 millivolts. That is, a receptor potential can have almost as high a voltage as an action potential.

**Receptor Potentials Recorded from Other Sensory Receptors.** Receptor potentials have been recorded from many other sensory receptors, including most notably the muscle spindles, the hair cells of the ear, and the rods and cones of the eyes and many others. In all of these, the amplitude of the potential increases as the strength of stimulus increases, but the additional response usually becomes progressively less as the strength of stimulus becomes great.

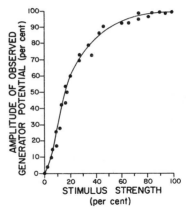

**Figure 48–3.** Relationship of amplitude of receptor (generator) potential to strength of a stimulus applied to a pacinian corpuscle. (From Loewenstein: *Ann. N.Y. Acad. Sci.,* 94:510, 1961.)

Yet, the mechanism for causing the receptor potential is not the same in different receptors. For instance, in the rods and cones of the eye changes in certain intracellular chemicals caused by exposure to light alter the membrane potential, resulting in the receptor potential. In this case, the basic mechanism eliciting the receptor potential is a chemical one in contrast to mechanical deformation that causes the receptor potential in the pacinian corpuscle. In the case of thermal receptors, it is believed that changes in rates of chemical reaction at or near the membrane alter the membrane potential and thereby create a receptor potential. In the case of the hair cells of the ear, bending of cilia protruding from the hairs probably causes the receptor potentials. Thus, the mechanisms for eliciting receptor potentials are individualized for each type of receptor.

**Relationship of Amplitude of Receptor Potential to Nerve Impulse Rate.** Referring once again to Figure 48–2, we see that the receptor potential generated in the core of the pacinian corpuscle causes a local circuit of current flow through the first node of Ranvier. When an action potential occurs at the node, this does not affect the receptor potential being emitted by the core of the pacinian corpuscle. Instead, the core continues to emit its current as long as an effective mechanical stimulus is applied. As a result, when the node of Ranvier repolarizes after its first action potential is over, it discharges once again, and action potentials continue as long as the receptor potential persists, which, in the case of the pacinian corpuscle, is only a few thousandths or hundredths of a second.

The frequency of action potentials in the nerve fiber (impulse rate) is almost directly proportional to the amplitude of the receptor potential. This relationship is illustrated in Figure 48–4, which shows the impulse rate corresponding to different voltages of receptor ("generator") potential recorded from a muscle spindle; there is an almost exact proportional relationship. This same relationship between receptor potential and impulse rate is approximately true for most sensory receptors.

### ADAPTATION OF RECEPTORS

A special characteristic of all sensory receptors is that they *adapt* either partially or completely to their stimuli after a period of time. That is, when a continuous sensory stimulus is first applied, the receptors respond at a very high impulse rate at first, then progressively less rapidly until finally many of them no longer respond at all.

Figure 48–5 illustrates typical adaptation of certain types of receptors. Note that the pacinian corpuscle and the receptor at the hair bases adapt extremely rapidly, while joint capsule and muscle spindle receptors adapt very slowly.

Furthermore some sensory receptors adapt to a far greater extent than others. For example,

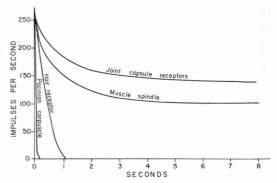

**Figure 48–5.** Adaptation of different types of receptors, showing rapid adaptation of some receptors and slow adaptation of others.

the pacinian corpuscles adapt to "extinction" within a few thousandths to a few hundredths of a second, and the hair base receptors adapt to extinction within a second or more.

**Mechanisms by which Receptors Adapt.** Adaptation of receptors is an individual property of each type of receptor in much the same way that development of a receptor potential is an individual property. For instance, in the eye, the rods and cones adapt by changing their chemical compositions (which will be discussed in Chapter 59). In the case of the mechanoreceptors, the receptor that has been studied in greatest detail is the pacinian corpuscle. Adaptation occurs in this receptor in two ways. First, the corpuscular structure itself very rapidly adapts to the deformation of the tissue. This can be explained as follows: The pacinian corpuscle is a viscoelastic structure so that when a distorting force is suddenly applied to one side of the corpuscle it is transmitted by the viscous component of the corpuscle directly to the same side of the central core, thus eliciting a receptor potential. However, within a few thousandths to a few hundredths of a second the fluid within the corpuscle redistributes so that the pressure becomes essentially equal all through the corpuscle; this applies an even pressure on all sides of the central core fiber, so that the receptor potential is no longer elicited. Thus, a receptor potential appears at the onset of compression but then disappears within a small fraction of a second. Then, when the distorting force is removed from the corpuscle, essentially the reverse events occur. The sudden removal of the distortion from one side of the corpuscle allows rapid expansion on that side and a corresponding distortion of the central core occurs once more. Again, within milliseconds, the pressure becomes equalized all

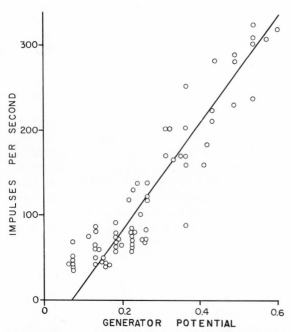

**Figure 48–4.** Relationship between the receptor (generator) potential of a muscle spindle and the frequency of sensory impulses transmitted from the spindle. [From Katz: *J. Physiol. (Lond.)*, 111:161, 1950.]

through the corpuscle and the stimulus is lost. Thus, the pacinian corpuscle signals the onset of compression and again signals the offset of compression.

The second mechanism of adaptation of the pacinian corpuscle results from a process of accommodation that occurs in the nerve fiber itself. That is, even if by chance the central core fiber should continue to be excited, as can be achieved after the capsule has been removed and the core is compressed with a stylus, the tip of the nerve fiber itself gradually becomes "accommodated" to the stimulus. This perhaps results from redistribution of ions across the nerve fiber membrane.

Presumably, these same two general mechanisms of adaptation apply to other types of receptors. That is, part of the adaptation results from readjustments in the structure of the receptor itself, and part results from accommodation in the terminal nerve fibril.

**Function of the Poorly Adapting Receptors—The "Tonic" Receptors.** The poorly adapting receptors (receptors that adapt very slowly and do not adapt to extinction) continue to transmit impulses to the brain as long as the stimulus is present (or at least for many minutes or hours). Therefore, they keep the brain constantly apprised of the status of the body and its relation to its surroundings. For instance, impulses from the slowly adapting joint capsule receptors allow the person to "know" at all times the degree of bending of the joints and therefore the positions of the different parts of his body. And impulses from the muscle spindles and Golgi tendon apparatuses allow the central nervous system to know respectively the status of muscle contraction and the load on the muscle tendon at each instant.

Other types of poorly adapting receptors include the receptors of the macula in the vestibular apparatus, the sound receptors of the ear, the pain receptors, the pressoreceptors of the arterial tree, the chemoreceptors of the carotid and aortic bodies, and some of the tactile receptors, such as the Ruffini endings and the Merkel's discs.

Because the poorly adapting receptors can continue to transmit information for many hours, they are also called *tonic* receptors. It is probable that many of these poorly adapting receptors would adapt to extinction if the intensity of the stimulus should remain absolutely constant over many days. Fortunately, because of our continually changing bodily state, the tonic receptors could almost never reach a state of complete adaptation.

**Function of the Rapidly Adapting Receptors—The Rate Receptors or Movement Receptors or Phasic Receptors.** Obviously, receptors that adapt rapidly cannot be used to transmit a continuous signal because these receptors are stimulated only when the stimulus strength changes. Yet they react strongly *while a change is actually taking place*. Furthermore, the number of impulses transmitted is directly related to the *rate at which the change takes place*. Therefore, these receptors are called *rate* receptors, *movement* receptors, or *phasic* receptors. Thus, in the case of the pacinian corpuscle, sudden pressure applied to the skin excites this receptor for a few milliseconds, and then its excitation is over even though the pressure continues. But then it transmits a signal again when the pressure is released. In other words, the pacinian corpuscle is exceedingly important in transmitting information about rapid changes in pressure against the body, but it is useless in transmitting information about constant pressure applied to the body.

*Importance of the Rate Receptors–Their Predictive Function.* If one knows the rate at which some change in his bodily status is taking place, he can predict ahead to the state of the body a few seconds or even a few minutes later. For instance, the receptors of the semicircular canals in the vestibular apparatus of the ear detect the rate at which the head begins to turn when one runs around a curve. Using this information, a person can predict that he will turn 10, 30, or some other number of degrees within the next 10 seconds, and he can adjust the motion of his limbs *ahead of time* to keep from losing his balance. Likewise, pacinian corpuscles located in or near the joint capsules help to detect the rates of movement of the different parts of the body. Therefore, when one is running, information from these receptors allows the nervous system to predict ahead of time where the feet will be during any precise fraction of a second, and appropriate motor signals can be transmitted to the muscles of the legs to make any necessary anticipatory corrections in limb position so that the person will not fall. Loss of this predictive function makes it impossible for the person to run.

## PSYCHIC INTERPRETATION OF STIMULUS STRENGTH

The ultimate goal of most sensory stimulation is to appraise the psyche of the state of the body and its surroundings. Therefore, it is important that we discuss briefly some of the principles related to the transmission of sensory stimulus strength to the higher levels of the nervous system.

The first question that comes to mind is, "How is it possible for the sensory system to transmit sensory experiences of tremendously varying intensities?" For instance, the auditory system can detect the weakest possible whisper but can also discern the meanings of an explosive sound only a few feet away, even though the sound intensities of these two experiences can vary as much as a trillion-fold; the eyes can see visual images with light intensities that vary as much as a million-fold; or the skin can detect pressure differences of ten thousand- to one hundred thousand-fold.

As a partial explanation of these effects, note in Figure 48–3 the relationship of the receptor potential ("generator" potential) produced by the pacinian corpuscle to the strength of stimulus. At low stimulus strength, very slight changes in stimulus strength increase the potential markedly; whereas at high levels of stimulus strength, further increases in receptor potential are very slight. Thus, the pacinian corpuscle is capable of accurately measuring extremely minute changes in stimulus strength at low intensity levels and is also capable of detecting much larger changes in stimulus strength at high intensity levels.

The transduction mechanism for detecting sound by the cochlea of the ear illustrates still another method for separating gradations of stimulus intensity. When sound causes vibration at a specific point on the basilar membrane, weak vibration stimulates only those hair cells in the very center of the vibratory point. But, as the vibration intensity increases, not only do the centralmost hair cells become more intensely stimulated, but hair cells in each direction farther away from the central point also become stimulated. Thus, signals transmitted over progressively increasing numbers of cochlear nerve fibers as well as increasing intensity of signal strength in each nerve fiber are two mechanisms by which stimulus strength is transmitted into the central nervous system. These two mechanisms, which multiply each other, make it possible for the ear to operate reasonably faithfully at stimulus intensity levels changing as much as a trillion-fold.

**Importance of the Tremendous Intensity Range of Sensory Reception.** Were it not for the tremendous intensity range of sensory reception that we can experience, the various sensory systems would more often than not operate in the wrong range. This is illustrated by the attempt of most persons to adjust the light exposure on a camera without using a light meter. Left to intuitive judgment of light intensity, a person almost always overexposes the film on very bright days and greatly underexposes the film at twilight. Yet, his eyes are perfectly capable of discriminating with great detail the objects around him in both very bright sunlight and at twilight; the camera cannot do this because of the narrow critical range of light intensity required for proper exposure of film.

## JUDGMENT OF STIMULUS STRENGTH

Physiopsychologists have evolved numerous methods for testing one's judgment of sensory stimulus strength, but only rarely do the results from the different methods agree with each other. For instance, one testing method requires a person to select a weight that is exactly 100 per cent heavier than another. But he usually selects a weight that is about 50 per cent heavier instead of 100 per cent heavier. Thus, the weak stimulus is underestimated and the strong stimulus is overestimated. In still another test procedure, a person is given a weight to hold and is then required to select the minimum amount of additional weight that must be added for him to detect a difference. In this case he might be holding a 50 gram weight and find out that an additional 5 grams are necessary to detect a difference. Then he holds a 500 gram weight and finds that 50 grams of additional weight are required. In this instance the discriminatory ability is far greater at the low intensity level than at the high intensity level. Thus, the results of these two different types of tests are exactly opposite to each other, which means that the real argument lies in the meaning of the tests themselves. Therefore, at present no real agreement exists as to a proper method for measuring one judgment of stimulus strength. Yet, two principles are widely discussed in the physiopsychology field of sensory interpretation: the *Weber-Fechner principle* and the *power principle*.

**The Weber-Fechner Principle—Detection of "Ratio" of Stimulus Strength.** In the mid–eighteen-hundreds, Weber first and Fechner later proposed the principle that *gradations of stimulus strength are discriminated approximately in proportion to the logarithm of stimulus strength*. This law is based primarily on one's ability to judge minimal changes in stimulus strength that can be detected, utilizing the second test described in the previous section; that is, in this example just given, a person can barely detect a 5 gram increase in weight when holding 50 grams, or a 50 gram increase when holding 500 grams. Thus, the *ratio* of the change in stimulus strength required for perception of a change remains essentially constant, which is what the logarithmic principle means.

Because the Weber-Fechner principle offers a ready explanation for the tremendous range of stimulus strength that our nervous system can discern, it unfortunately became widely accepted for all types of sensory experience and for all levels of background sensory intensity. More recently it has become evident that this principle applies mainly to higher levels of visual, auditory, and cutaneous sensory experience and that it applies only poorly to most other types of sensory experience.

Yet, the Weber-Fechner principle is still a good one to remember because it emphasizes that the greater the background sensory stimulus, the greater also must be the additional change in stimulus strength in order for the psyche to detect the change.

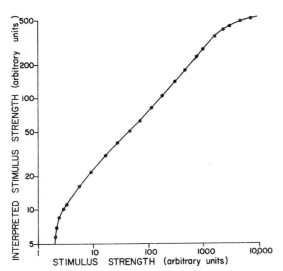

**Figure 48–6.** Graphical demonstration of the "power law" relationship between actual stimulus strength and strength that the psyche interprets it to be. Note that the power law does not hold at either very weak or very strong stimulus strengths.

**The Power Law.** Another attempt by physio-psychologists to find a good mathematical relationship between actual stimulus strength and interpretation of stimulus strength is the following formula, known as the power law:

Interpreted strength = K · (Stimulus strength)$^{y}$

In this formula K is a constant, and y is the power to which the stimulus strength is raised. The exponent y and the constant K are different for each type of sensation.

When this power law relationship is plotted on a graph using double logarithmic coordinates, as illustrated in Figure 48–6, a linear curve can be attained between interpreted stimulus strength and actual stimulus strength over a large range. However, as illustrated in the figure, even this power law relationship fails to hold satisfactorily at both low and high stimulus strengths.

Unfortunately, the power law is not of great philosophical importance for a very simple mathematical reason. Almost any curve (even the curve of a hill) that has a progressive change in slope in the same direction can be fitted over most of its range to the power law equation, provided one simply finds proper values for the exponent y and constant K. Therefore, use of the power law is more an exercise in mathematical curve fitting than it is a valuable tool in understanding sensory experience.

# PHYSIOLOGICAL CLASSIFICATION OF NERVE FIBERS

Unfortunately, two separate classifications of nerve fibers are in general use. The first of these is a general classification given in Table 48–2. The fibers are divided into types A, B, and C; and the type A fibers are further subdivided into $\alpha$, $\beta$, $\gamma$, and $\delta$ fibers.

Type A fibers are the typical myelinated fibers of spinal nerves. The type B fibers differ from very small type A fibers only in the fact that they do not display a negative after-potential following stimulation. However, they also are myelinated like type A fibers. They are the preganglionic autonomic nerve fibers.

Type C fibers are the very small unmyelinated nerve fibers that conduct impulses at low velocities. These constitute more than half the sensory fibers in

**TABLE 48–2.   Properties of Different Mammalian Nerve Fibers**

| Type of Fiber | Diameter of Fiber ($\mu$) | Velocity of Conduction (meters/sec.) | Duration of Spike (msec.) | Duration of Negative After-potential (msec.) | Duration of Positive After-potential (msec.) | Function |
|---|---|---|---|---|---|---|
| A ($\alpha$) | 13–22 | 70–120 | 0.4–0.5 | 12–20 | 40–60 | Motor, muscle proprioceptors |
| A ($\beta$) | 8–13 | 40–70 | 0.4–0.6 | (?) | (?) | Touch, pressure, kinesthesia |
| A ($\gamma$) | 4–8 | 15–40 | 0.5–0.7 | (?) | (?) | Touch, motor excitation of muscle spindles |
| A ($\delta$) | 1–4 | 5–15 | 0.6–1.0 | (?) | (?) | Pain. heat, cold, pressure |
| B | 1–3 | 3–14 | 1.2 | None | 100–300 | Preganglionic autonomic |
| C | 0.2–1.0 | 0.2–2 | 2.0 | 50–80 | 300–1000 | Pain, itch, heat(?), cold(?), pressure(?), postganglionic autonomic, smell |

Compiled from various sources but mainly from Grundfest.

most peripheral nerves and also all of the post-ganglionic autonomic fibers.

The sizes, velocities of conduction, and functions of the different nerve fibers are given in Table 48–2. Note that the very large fibers can transmit impulses at velocities as great as 120 meters per second, a distance in one second that is longer than a football field. On the other hand, the smallest fibers transmit impulses as slowly as 0.2 meter per second, requiring several seconds to go from the big toe to the spinal cord.

Over two-thirds of all the nerve fibers in peripheral nerves are type C fibers. Because of their great number, these can transmit tremendous amounts of information from the surface of the body, even though their velocities of transmission are very slow. Utilization of type C fibers for transmitting this great mass of information represents an important economy of space in the nerves, for use of type A fibers would require peripheral nerves the sizes of large ropes and a spinal cord almost as large as the body itself.

**Alternate Classification Used by Sensory Physiologists.** Certain recording techniques have made it possible to separate the type Aα fibers into two subgroups; and, yet, these same recording techniques cannot distinguish easily between Aβ and Aγ fibers. Therefore, the following classification is frequently used by sensory physiologists:

*GROUP I A.* Fibers from the annulospiral endings of muscle spindles. (About 17 microns in diameter. These are alpha type A fibers in the classification of Table 48–2.)

*GROUP I B.* Fibers from the Golgi tendon apparatuses. (About 16 microns in diameter; these also are alpha type A fibers.)

*GROUP II.* Fibers from the discrete cutaneous tactile receptors and also from the flower-spray endings of the muscle spindles. (Average about 8 microns in diameter; these are beta and gamma type A fibers in the other classification.)

*GROUP III.* Fibers carrying temperature, crude touch, and pain sensations. (Average about 3 microns in diameter; these are delta type A fibers in the other classification.)

*GROUP IV.* Unmyelinated fibers carrying pain, itch, crude temperature, and crude touch sensations. (0.2 to 1 micron in diameter; called type C fibers in the other classification.)

# REFERENCES

Anderson, D. J., Hannam, A. G., and Matthews, B.: Sensory mechanisms in mammalian teeth and their supporting structures. Physiol. Rev., 50:171, 1970.
Appelle, S.: Perception and discrimination as a function of stimulus orientation: the "oblique effect" in man and animals. Psychol. Bull., 78:266, 1972.
Babel, J., Bischoff, A., and Spoendlin, H.: Ultrastructure of the Peripheral Nervous System and Sense Organs. New York, Churchill Livingstone, Div. of Longman, Inc., 1971.
Bullock, T. H.: Seeing the world through a new sense: electroreception in fish. Amer. Sci., 61:316, 1973.
Catton, W. T.: Mechanoreceptor function. Physiol. Rev., 50:297, 1970.
Galun, R. (ed.): Sensory Physiology and Behavior. New York, Plenum Publishing Corporation, 1975.
Goldberg, J. M., and Lavine, R. A.: Nervous system: afferent mechanisms. Ann. Rev. Physiol., 30:319, 1968.
Granit, R.: Receptors and Sensory Perception. New Haven, Yale University Press, 1955.
Gray, J. A. B.: Initiation of impulses at receptors. In Magoun, H. W. (ed.): Handbook of Physiology. Sec. 1, Vol. 1. Baltimore, The Williams & Wilkins Company, 1959.
Halata, Z.: The Mechanoreceptors of the Mammalian Skin. New York, Springer-Verlag New York, Inc., 1975.
Hunt, C. C., and Takeuchi, A.: Responses of the nerve terminal of the pacinian corpuscle. J. Physiol. (Lond.), 160:1, 1962.
Jacobson, M., and Hunt, R. K.: The origins of nerve-cell specificity. Sci. Amer., 228:26, 1973.
Kruger, L., and Kenton, B.: Quantitative neural and psychophysical data for cutaneous mechanoreceptor function. Brain Res., 49:1, 1973.
Leibowitz, H. W., and Harvey, L. O., Jr.: Perception. Ann. Rev. Psychol., 24:207, 1973.
Loewenstein, W. R.: The generation of electric activity in a nerve ending. Ann. N.Y. Acad. Sci., 81:367, 1959.
Loewenstein, W. R.: Biological transducers. Sci. Amer., 203:98, 1960.
Loewenstein, W. R.: Excitation and inactivation in a receptor membrane. Ann. N.Y. Acad. Sci., 94:510, 1961.
Lynn, B.: Somatosensory receptors and their CNS connections. Ann. Rev. Physiol., 37:105, 1975.
Perl, E. R., and Boivie, J. G.: Neural substrates of somatic sensation. In Guyton, A. C. (ed.): MTP International Review of Science: Physiology. Vol. 3. Baltimore, University Park Press, 1974, p. 303.
Somjen, G.: Sensory Coding in the Mammalian Nervous System. New York, Appleton-Century-Crofts, 1972.
Tamar, H.: Principles of Sensory Physiology. Springfield, Ill., Charles C Thomas, Publisher, 1972.
Thompson, R. F., and Patterson, M. (eds.): Bioelectric Recording Techniques, Part C. Receptor and Effector Processes. New York, Academic Press, Inc., 1974.
Thorson, J., and Biederman-Thorson, M.: Distributed relaxation processes in sensory adaptation. Science, 183:161, 1974.
Winkelmann, R. K.: Nerve Endings in Normal and Pathologic Skin. Springfield, Illinois, Charles C Thomas, Publisher, 1960.
Zotterman, Y.: Sensory Mechanisms. Progress in Brain Research. Vol. 23. New York, American Elsevier Publishing Co., Inc., 1967.

# | 49 |

# Somatic Sensations: I. The Mechanoreceptive Sensations

The *somatic senses* are the nervous mechanisms that collect sensory information from the body. These senses are in contradistinction to the *special senses,* which mean specifically sight, hearing, smell, taste, and equilibrium.

## CLASSIFICATION OF SOMATIC SENSATIONS

The somatic senses can be classified into three different physiological types: (1) the *mechanoreceptive somatic senses,* stimulated by mechanical displacement of some tissue of the body, (2) the *thermoreceptive senses,* which detect heat and cold, and (3) the *pain sense,* which is activated by any factor that damages the tissues. The present chapter deals with the mechanoreceptive somatic senses, and the following chapter deals with the thermoreceptive and pain senses.

The mechanoreceptive senses include *touch, pressure,* and *vibration* senses (which are frequently called the *tactile senses*) and the *kinesthetic sense,* which determines the relative positions and rates of movement of the different parts of the body.

**Other Classifications of Somatic Sensations.** Different physiological types of somatic sensations are grouped together in special classes that are not necessarily mutually exclusive, as follows:

*Exteroceptive sensations* are those from the surface of the body.

*Proprioceptive sensations* are those having to do with the physical state of the body, including kinesthetic sensations, possible tendon and muscle sensations, pressure sensations from the bottom of the feet, and even the sensation of equilibrium, which is generally considered to be a "special" sensation rather than a somatic sensation.

*Visceral sensations* are those from the viscera of the body; in using this term one usually refers specifically to sensations from the internal organs.

The *deep sensations* are those that come from the deep tissues, such as from the bone, fasciae, and so forth. These include mainly "deep" pressure, pain, and vibration.

## DETECTION AND TRANSMISSION OF TACTILE SENSATIONS

**Interrelationship Between the Tactile Sensations of Touch, Pressure, and Vibration.** Though touch, pressure, and vibration are frequently classified as separate sensations, they are all detected by the same types of receptors. The only differences among these three types of sensations are (1) touch sensation generally results from stimulation of tactile receptors in the skin or in tissues immediately beneath the skin, (2) pressure sensation generally results from deformation of deeper tissues, and (3) vibration sensation results from rapidly repetitive sensory signals, but some of the same types of receptors as those for both touch and pressure are utilized—specifically the rapidly adapting types of receptors.

**The Tactile Receptors.** At least six entirely different types of tactile receptors are known, but many more similar to these probably also exist. Some of these receptors were illustrated in Figure 48–1 of Chapter 48, and their special characteristics are the following:

First, some *free nerve endings,* which are found everywhere in the skin and in many other tissues, can detect touch and pressure. For instance, even light contact with the cornea of the eye, which contains no other type of nerve ending besides free nerve endings, can nevertheless elicit touch and pressure sensations.

Second, a touch receptor of special sensitivity is *Meissner's corpuscle,* an encapsulated nerve ending that excites a large myelinated sensory nerve fiber. Inside the capsulation are many whorls of terminal nerve filaments. These receptors are particularly abundant in the fingertips, lips, and other areas of the skin where one's ability to discern spatial characteristics of touch sensations is highly developed. These receptors, along with the expanded tip receptors described subsequently, are probably responsible for the ability to recognize exactly what point of the body is touched and to recognize the texture of objects touched. Meissner's corpuscles probably adapt within a second or perhaps even less after they are stimulated, which means that they are particularly sensitive to movement of very light objects over the surface of the skin and also to low-frequency vibration.

Third, the fingertips and other areas that contain large numbers of Meissner's corpuscles also contain *expanded tip tactile receptors,* one type of which is *Merkel's discs.* These receptors differ from Meissner's corpuscles in that they transmit an initial strong but partially adapting signal and then a continuing weaker signal that adapts only slowly. Therefore, they are probably responsible for giving steady state signals that allow one to determine continuous touch of objects against the skin. The hairy parts of the body contain almost no Meissner's corpuscles but do contain a few expanded tip receptors.

Fourth, slight movement of any hair on the body stimulates the nerve fiber entwining its base. Thus, each hair and its basal nerve fiber, called the *hair end-organ,* is also a type of touch receptor. This receptor adapts readily and, therefore, like Meissner's corpuscles, detects mainly movement of objects on the surface of the body.

Fifth, located in the deeper layers of the skin and also in deeper tissues of the body are many *Ruffini's end-organs,* which are multibranched endings, as described and illustrated in the previous chapter. These endings adapt very little and, therefore, are important for signaling continuous states of deformation of the skin and deeper tissues, such as heavy and continuous touch signals and pressure signals. They are also found in joint capsules and signal the degree of joint rotation.

Sixth, many *pacinian corpuscles,* which were discussed in detail in Chapter 48, lie both beneath the skin and also deep in the tissues of the body. These are stimulated only by very rapid movement of the tissues because these receptors adapt in a small fraction of a second. Therefore, they are particularly important for detecting tissue vibration or other extremely rapid changes in the mechanical state of the tissues.

**Transmission of Tactile Sensations in Peripheral Nerve Fibers.** The specialized sensory receptors, such as Meissner's corpuscles, expanded tip endings, pacinian corpuscles, and Ruffini's endings, all transmit their signals in beta type A nerve fibers that have transmission velocities of 30 to 60 meters per second. On the other hand, free nerve ending tactile receptors and probably some of the hair end-organs transmit signals via the small delta type A nerve fibers that conduct at velocities of 6 to 15 meters per second. Some of the tactile free nerve endings transmit via type C fibers at velocities of about 1 meter per second; it has not yet been proved that these send signals to the brain that can be detected consciously. Thus, the more critical types of sensory signals—those that help to determine precise localization on the skin, minute gradations of intensity, or rapid changes in sensory signal intensity—are all transmitted in the rapidly conducting types of sensory nerve fibers. On the other hand, the cruder types of signals, such as crude pressure and poorly localized touch, are transmitted via much slower nerve fibers, fibers that also require much less space in the nerves.

## DETECTION OF VIBRATION

All the different tactile receptors are involved in detection of vibration, though different receptors detect different frequencies of vibration. Pacinian corpuscles can discern vibrations up to as high as 400 to 500 cycles per second, because they respond extremely rapidly to minute and rapid deformations of the tissues, and they also transmit their signals over beta type A nerve fibers, which can transmit more than 1000 impulses per second.

Low frequency vibrations up to 80 cycles per

second, on the other hand, stimulate other tactile receptors—especially Meissner's corpuscles, which are less rapidly adapting in preference to pacinian corpuscles.

## POSSIBILITY OF A "MUSCLE SENSE"

Special sensory receptors are found in both skeletal muscles and tendons, *muscle spindles* in muscles and *Golgi tendon apparatuses* in tendons. These receptors transmit their signals into the spinal cord and cerebellum to control reflex contraction of the muscles. But, from what is known at present, these signals operate entirely at a subconscious level. These specialized receptors and the reflexes that they subserve will be considered in detail in relation to spinal cord reflexes and control of muscle contraction in Chapter 51.

However, some psychological effects have pointed to the existence of a conscious "muscle sense." For instance, one is capable of determining how heavy an object is, and this has long been believed to be detected by the specialized muscle spindles or tendon organs; but critical experiments show this not to be true. Instead, this type of information is probably signaled in two ways: (1) by usual tactile signals in the deep tissues and (2) by signals from the motor portions of the brain indicating intensity of motor signals to the muscles required to lift an object.

## KINESTHETIC SENSATIONS

The term "kinesthesia" means conscious recognition of the orientation of the different parts of the body with respect to each other as well as of the rates of movement of the different parts of the body. These functions are subserved principally by extensive sensory endings in the joint capsules and ligaments.

**The Kinesthetic Receptors.** Three major types of nerve endings have been described in the joint capsules and ligaments about the joints. (1) By far the most abundant of these are spray type *Ruffini endings,* one of which was illustrated in Figure 48–1 of Chapter 48. These endings are stimulated strongly when the joint is suddenly moved; they adapt slightly at first but then transmit a steady signal thereafter. (2) A second type of ending resembling the stretch receptors found in muscle tendons (called *Golgi tendon receptors*) is found particularly in the ligaments about the joints. Though far less numerous than the Ruffini endings they have essentially the same response properties. (3) A few *pacinian corpuscles* are also found in the tissues around the joints. These adapt extremely rapidly and presumably help to detect *rate of rotation* at the joint.

**Detection of the Degree of Joint Rotation by the Joint Receptors.** Figure 49–1 illustrates the excitation of seven different nerve fibers leading from separate joint receptors in the capsule of a cat's knee joint. Note that at 180 degrees of joint rotation one of the receptors is stimulated; then at 150 degrees still another is stimulated; at 140 degrees two are stimulated, and so forth. The information from these joint receptors continually apprises the central nervous system of the momentary rotation of the joint. That is, the rotation determines *which* receptor is stimulated and how much it is stimulated, and from this the brain knows how far the joint is bent.

**Detection of Rate of Movement at the Joint.** Because pacinian corpuscles are especially adapted for detecting movement of tissues, it is tempting to suggest that rate of movement at the joints is detected by the pacinian corpuscles. However, the number of pacinian corpuscles in the joint tissues is small, for which reason rate of movement at the joint is probably detected mainly in the following way: The Ruffini and Golgi endings in the joint tissues are stimulated very strongly at first by the process of joint movement, but within a fraction of a second this strong level of stimulation fades to a lower, steady state rate of firing. Nevertheless, this early overshoot in receptor stimulation is directly proportional to the rate of joint movement and is believed to be the signal used by the brain to discern the rate of movement. However, it is likely that the few pacinian corpuscles also play at least some role in this process.

**Transmission of Kinesthetic Signals in the Peripheral Nerves.** Kinesthetic signals, like those from the critical tactile sensory receptors, are transmitted almost entirely in the beta type

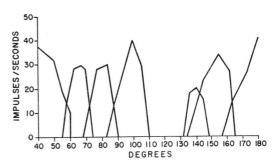

**Figure 49–1.** Responses of seven different nerve fibers from knee joint receptors in a cat at different degrees of rotation. (Modified from Skogland: *Acta Physiol. Scand.*, Suppl. 124, 36:1, 1956.)

A sensory nerve fibers, which carry signals very rapidly to the cord and, thence, to the brain. This rapid transmission of kinesthetic signals is particularly important when parts of the body are moving rapidly, because it is essential for the central nervous system to "know" at each small fraction of a second the exact locations of the different parts of the body; otherwise one would not be capable of controlling further movements.

## THE DUAL SYSTEM FOR TRANSMISSION OF MECHANORECEPTIVE SOMATIC SENSATIONS IN THE CENTRAL NERVOUS SYSTEM

All sensory information from the somatic segments of the body enters the spinal cord through the posterior roots. On entering the cord, most large sensory nerve fibers (mainly the beta type A fibers) immediately enter the *dorsal columns* of the cord and ascend the entire length of the cord. The smaller sensory fibers (type C and delta type A) as well as lateral collaterals from the larger fibers travel upward for one to six segments and downward for one to four segments, and then synapse with dorsal horn cells that give rise to the *ventral* and *lateral spinothalamic tracts*. These tracts ascend to the brain in the *anterior* and *lateral columns* of the spinal cord. This separation of the fibers at the dorsal roots represents a separation of the pathways for transmission of sensory impulses: the dorsal column pathway gives rise to the *dorsal column system,* and the spinothalamic tracts give rise to the *spinothalamic system.*

**Comparison of the Dorsal Column System With the Spinothalamic System.** The dorsal column system is composed of large, heavily myelinated nerve fibers that transmit signals to the brain at velocities of 35 to 70 meters per second. Also, there is a high degree of spatial orientation of the nerve fibers with respect to their origin on the surface of the body. On the other hand, the spinothalamic system is composed mainly of small fibers, some of which are not myelinated at all, or are poorly myelinated, and which transmit impulses at velocities perhaps as low as one meter per second but not over 15 meters per second. Some of these fibers are spatially oriented but only poorly so.

These differences in the two systems immediately characterize the types of sensory information that can be transmitted by the two pathways. First, sensory information that must be transmitted rapidly and with temporal fidelity is transmitted in the dorsal column system, while that which does not need to be transmitted rapidly is transmitted mainly in the spinothalamic system. Second, those sensations that detect fine gradations of intensity are transmitted in the dorsal column system, while those that lack the fine gradations are transmitted in the spinothalamic system. And, third, sensations that are discretely localized to exact points in the body are transmitted in the dorsal column system, while those transmitted in the spinothalamic system can be localized much less exactly. On the other hand, the spinothalamic system has a special capability that the dorsal column system does not have: the ability to transmit a broad spectrum of sensory modalities—pain, warmth, cold, and crude tactile sensations; the dorsal column system is limited to mechanoreceptive sensations alone. With this differentiation in mind we can now list the types of sensations transmitted in the two systems.

### The Dorsal Column System

1. Touch sensations requiring a high degree of localization of the stimulus.
2. Touch sensations requiring transmission of fine gradations of intensity.
3. Phasic sensations, such as vibratory sensations.
4. Sensations that signal movement against the skin.
5. Kinesthetic sensations.
6. Pressure sensations having to do with fine degrees of judgment of pressure intensity.

### The Spinothalamic System

1. Pain.
2. Thermal sensations, including both warm and cold sensations.
3. Crude touch and pressure sensations capable of very little localizing ability on the surface of the body and having very little capability for intensity discrimination.
4. Tickle and itch sensations.
5. Sexual sensations.

## TRANSMISSION IN THE DORSAL COLUMN SYSTEM

Because the dorsal column system is large and its fibers numerous, we know much about transmission of sensory signals in this system in comparison with very little about transmission of signals in the spinothalamic system. Therefore, the dorsal column system will be discussed at length and then the known differences between this and the spinothalamic system will be pointed out.

## ANATOMY OF THE DORSAL COLUMN PATHWAY

The dorsal column system is illustrated in Figure 49–2. The nerve fibers entering the *dorsal columns* pass all the way up these columns to the medulla, where they synapse in the *dorsal column nuclei* (the *cuneate* and *gracile nuclei*). From here, *second order neurons* decussate immediately to the opposite side and then pass upward to the thalamus through bilateral pathways called the *medial lemnisci*. Each medial lemniscus terminates in a *ventrobasal complex of nuclei* located in the ventral posterolateral nucleus of the *thalamus*. In its pathway through the hindbrain, the medial lemniscus is joined by additional fibers from the *main sensory nucleus of the trigeminal*

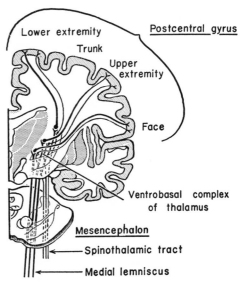

**Figure 49–3.** Projection of the dorsal column system from the thalamus to the somesthetic cortex. (Modified from Brodal: Neurological Anatomy in Relation to Clinical Medicine. Oxford University Press, 1969.)

*nerve and from the upper portion of its descending nuclei;* these fibers subserve the same sensory functions for the head that the dorsal column fibers subserve for the body.

From the ventrobasal complex, *third order neurons* project, as shown in Figure 49–3, mainly to the *postcentral gyrus* of the *cerebral cortex* called *somatic sensory area I*. But, in addition, neurons also project to closely associated regions of the cortex behind and in front of the postcentral gyrus. Finally, a few fibers project to the lowermost lateral portion of each parietal lobe, an area called *somatic sensory area II*.

**Spatial Orientation of the Nerve Fibers in the Dorsal Column Pathway.** All the way from the origin of the dorsal columns to the cerebral cortex, a distinct spatial orientation of the fibers from individual parts of the body is maintained. The fibers from the lower parts of the body lie toward the center of the dorsal columns, while those that enter the dorsal columns at progressively higher and higher levels form successive layers on the lateral sides of the dorsal columns.

In the thalamus, the distinct spatial orientation is still maintained, with the tail end of the body represented by the most lateral portions of the ventrobasal complex and the head and face represented in the medial component of the complex. However, because of the crossing of the medial lemnisci in the medulla, the left side of the body is represented in the right side of the

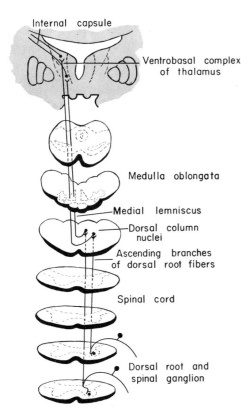

**Figure 49–2.** The dorsal column pathway for transmitting critical types of mechanoreceptive signals. (Modified from Ransom and Clark: Anatomy of the Nervous System, 1959.)

thalamus and the right side of the body is represented in the left side of the thalamus. In a similar manner, the fibers passing to the cerebral cortex also are spatially oriented so that a single part of the cortex receives signals from a discrete area of the body, as is described below.

## THE SOMESTHETIC CORTEX

The area of the cerebral cortex to which the primary sensory impulses are projected is called the *somesthetic cortex*. In the human being, this area lies mainly in the anterior portions of the parietal lobes. Two distinct and separate areas are known to receive direct afferent nerve fibers from the relay nuclei of the thalamus; these, called *somatic sensory area I* and *somatic sensory area II,* are illustrated in Figure 49–4. However, somatic sensory area I is so much more important to the sensory functions of the body than is somatic sensory area II that in popular usage the term somesthetic cortex is almost always used to designate area I exclusive of area II. Yet, to keep these two areas separated, we will henceforth refer to them separately as somatic sensory area I and somatic sensory area II.

**Projection of the Body in Somatic Sensory Area I.** Somatic sensory area I lies in the postcentral gyrus of the human cerebral cortex. A distinct spatial orientation exists in this area for reception of nerve signals from the different areas of the body. Figure 49–5 illustrates a cross-section through the brain at the level of the postcentral gyrus, showing the representations of the different parts of the body in separate regions of somatic sensory area I. Note, however, that each side of the cortex receives sensory information exclusively from the oppo-

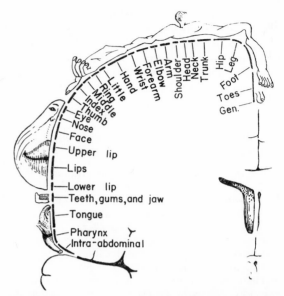

**Figure 49–5.** Representation of the different areas of the body in the somatic sensory area I of the cortex. (From Penfield and Rasmussen: Cerebral Cortex of Man: A Clinical Study of Localization of Function. The Macmillan Co., 1968.)

site side of the body (with the possible exception of a small amount of sensory information from the same side of the face).

Some areas of the body are represented by large areas in the somatic cortex—the lips by far the greatest area of all, followed by the face and thumb—while the entire trunk and lower part of the body are represented by relatively small areas. The sizes of these areas are directly proportional to the number of specialized sensory receptors in each respective peripheral area of the body. For instance, a great number of specialized nerve endings are found in the lips and thumb, while only a few are present in the skin of the trunk.

Note also that the head is represented in the lower or lateral portion of the postcentral gyrus, while the lower part of the body is represented in the medial or upper portion of the postcentral gyrus.

*Modality Differentiation in Somatic Sensory Area I.* Not only is there spatial projection of the body in somatic sensory area I, but there is also a moderate degree of modality separation between tactile signals and kinesthetic signals. Tactile signals stimulate neurons mainly in the anteriormost portion of the postcentral gyrus where it dips into the central sulcus. On the other hand, kinesthetic signals stimulate neurons mainly in the posteriormost portion of

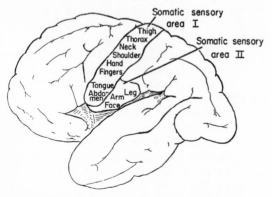

**Figure 49–4.** The two somesthetic cortical areas, somatic sensory areas I and II.

the postcentral gyrus. This separation of modalities is important because it helps us to understand how the brain dissects different types of information from incoming sensory signals, a subject that will be discussed in more detail later in the chapter.

**Somatic Sensory Area II.** The second cortical area to which somatic afferent fibers project, somatic sensory area II, lies posterior and inferior to the lower end of the postcentral gyrus and on the upper wall of the lateral fissure, as shown in Figure 49–4. The degree of localization of the different parts of the body is far less acute in this area than in somatic sensory area I. The face is represented anteriorly, the arms centrally, and the legs posteriorly.

So little is known about the function of somatic sensory area II that it cannot be discussed intelligently. It is known that signals enter this area from both the dorsal column system and the spinothalamic system, though much more from the latter. In fact, it has been suggested that somatic sensory area II might be the cortical terminus of pain information. Also, stimulation of somatic sensory area II in some instances causes complex body movements, for which reason it possibly plays a role in sensory control of motor functions.

**Excitation of Vertical Columns of Neurons in the Somesthetic Cortex.** The cerebral cortex contains *six* separate layers of neurons, and, as would be expected, the neurons in each layer perform functions different from those in other layers. Also, the neurons are arranged in vertical columns extending all the way through the six layers of the cortex, each column having a diameter of about one-half to one millimeter and containing hundreds of neuronal cell bodies in each layer. Unfortunately, we still know relatively little about the functions of these columns of cells but we are certain about the following facts:

1. The incoming sensory signal excites mainly neuronal layer IV first; then the signal spreads toward the surface of the cortex and also toward the deeper layers.

2. Layers I and II receive a diffuse, nonspecific input from the reticular activating system. This input perhaps controls the overall level of excitability of the cortex.

3. The neurons in layers V and VI send axons to other parts of the nervous system— some to other areas of the cortex, some to deeper structures of the brain, such as the thalamus or brain stem, and some even to the spinal cord.

Similar vertical columns of neurons exist in all other areas of the cortex as well as in the somesthetic cortex. In particular, they have been shown to be very important for function of the visual cortex and the motor cortex.

Each vertical column of neurons seems to be able to decipher a specific quality of information from the sensory signal. For instance, in the visual cortex one specific column will detect a line oriented in a particular direction, whereas an adjacent column will detect a similar line oriented in a slightly different direction. Presumably, in the somesthetic cortex each column detects separate qualities of signals (angles of orientation of rough spots, lengths of rough spots, perhaps roundness of objects, perhaps sharpness of objects, and so forth) from specific surface areas of the body.

**Functions of Somatic Sensory Area I.** The functional capabilities of different areas of the somesthetic cortex have been determined by selective excision of the different portions. Widespread excision of somatic sensory area I causes loss of the following types of sensory judgment:

1. The person is unable to localize discretely the different sensations in the different parts of the body. However, he can localize these sensations very crudely, such as to a particular hand, which indicates that the thalamus or parts of the cerebral cortex not normally considered to be concerned with somatic sensations can perform some degree of localization.

2. He is unable to judge critical degrees of pressure against his body.

3. He is unable to judge exactly the weights of objects.

4. He is unable to judge shapes or forms of objects. This is called *astereognosis.*

5. He is unable to judge texture of materials, for this type of judgment depends on highly critical sensations caused by movement of the skin over the surface to be judged.

6. He is unable to judge fine gradations in temperature.

7. He is unable to recognize the relative orientation of the different parts of his body with respect to each other.

Note in the above list that nothing has been said about loss of pain. However, in the absence of somatic sensory area I, the appreciation of pain is often altered, sometimes even the intensification of the pain. But more important, the pain that does occur is poorly localized, indicating that pain localization is probably dependent mainly upon simultaneous stimulation of tactile stimuli that use the topographical map of the body in somatic sensory area I to localize the source of the pain.

## SOMATIC ASSOCIATION AREAS

Brodmann areas V and VII of the cerebral cortex, which are located in the parietal cortex immediately behind somatic sensory area I and above somatic sensory area II, play especially important roles in deciphering the sensory information that enters the somatic sensory areas. Therefore, these areas are called the *somatic association areas*.

Electrical stimulation in the somatic association area can occasionally cause a person to experience a complex somatic sensation, sometimes even the "feeling" of an object such as a knife or a ball. Therefore, it seems clear that the somatic association area combines information from multiple points in the somatic sensory area to decipher its meaning. This also fits with the anatomical arrangement of the neuronal tracts that enter the somatic association area, for it receives signals from (a) the primary somatic areas, (b) the ventrobasal complex of the thalamus, and (c) adjacent areas of the thalamus which themselves receive input from the ventrobasal complex.

**Effect of Removing the Somatic Association Area—Amorphosynthesis.** When the somatic association area is removed, the person especially loses his ability to recognize complex objects and complex forms that he feels. In addition, he loses a great deal of the sense of form of his own body. An especially interesting fact is that loss of the somatic association area on one side of the brain causes the person sometimes to be oblivious of the opposite side of his body—that is, to forget that it is there. Likewise, when he is feeling objects he will tend to feel only one side of the object and to forget that the other side even exists. This complex sensory deficit is called *amorphosynthesis*.

## CHARACTERISTICS OF TRANSMISSION IN THE DORSAL COLUMN PATHWAY

**Faithfulness of Transmission.** The most important functional characteristic of the dorsal column pathway is its *faithfulness* of transmission. That is, each time a point in the periphery is stimulated, a signal ordinarily is transmitted all the way to the somesthetic cortex. Also, if this peripheral stimulus increases in intensity, the intensity of the signal at the cerebral cortex

increases proportionately. And, finally, when a discrete area of the body is stimulated, the signal from this area is transmitted to a discrete area of the cerebral cortex. Thus, the dorsal column pathway is adequately organized for transmission of accurate information from the periphery to the sensorium. Furthermore, the responsiveness of this system can be altered only moderately by stimuli from other areas of the nervous system, and it cannot be depressed to a significant extent by extraneous agents, such as moderate degrees of anesthesia.

**Basic Neuronal Circuit and Discharge Pattern in the Dorsal Column System.** The lower part of Figure 49–6 illustrates the basic organization of the neuronal circuit of the dorsal column system, showing that at each synaptic stage divergence occurs; indeed, there is as much as a 10-fold divergence at the dorsal column nuclei alone. However, the upper part of Figure 49–6 shows that a single receptor stimulus on the skin does not cause all the cortical neurons with which that receptor connects

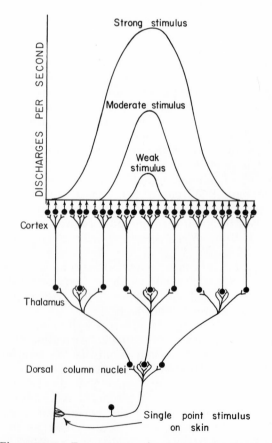

**Figure 49–6.** Transmission of a pinpoint stimulus signal to the cortex.

to discharge at the same rate. Instead, the cortical neurons that discharge to the greatest extent are those in a central part of the cortical "field" for each respective receptor. Thus, a weak stimulus causes only the centralmost neurons to fire. A moderate stimulus causes still more neurons to fire, but those in the center still discharge at a considerably more rapid rate than do those farther away from the center. Finally, a strong stimulus causes widespread discharge in the cortex but again a much more rapid discharge of the centralmost neurons in comparison with the peripheral neurons.

**Localization of Signals From the Body.** Referring once again to Figure 49–6, it is evident that, when a single receptor is stimulated in the skin, a particular point in the somesthetic cortex is excited more intensively than are all the surrounding areas. This does not mean that adjacent areas of the somesthetic cortex are not also stimulated. However, these adjacent areas are stimulated *less* intensively than a central point in the cortical field of the stimulated receptor. Thus, by this means, the sensory cortex is capable of detecting the precise localization of signals from the body, despite the spread of excitation in the cortex.

**Two-Point Discrimination.** A method frequently used to test a person's tactile capabilities is to determine his so-called "two-point discriminatory ability." In this test, two needles are pressed against the skin, and the subject determines whether he feels two points of stimulus or one point. On the tips of the fingers a person can distinguish two separate points even when the needles are as close together as 2 mm. However, on the back, the needles must usually be as far apart as 20 to 50 mm. before one can detect two separate points. The reason for this is that there are many specialized tactile receptors in the tips of the fingers in comparison with a small number in the skin of the back. Referring back to Figure 49–5, we can see also that the portions of the body that have a high degree of two-point discrimination have a correspondingly large cortical representation in somatic sensory area I.

Figure 49–7 illustrates the probable mechanism by which the dorsal column pathway transmits two-point discriminatory information. This shows two adjacent points on the skin that are strongly stimulated, and it shows the small area of the somesthetic cortex (greated enlarged) that is excited by signals from the two stimulated points in the skin. The two dashed curves show the individually ex-

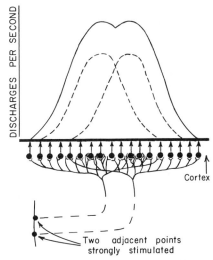

**Figure 49–7.** Transmission of signals to the cortex from two adjacent pinpoint stimuli.

cited cortical fields, and the solid curve shows the resultant cortical excitation when both the skin points are stimulated simultaneously. Note that the resultant zone of excitation has two separate peaks. It is believed to be these two peaks, separated by a valley, that allow the sensory cortex to detect the presence of two stimulatory points rather than a single stimulatory point.

**Increase in Contrast in the Spatial Pattern Caused by Lateral Inhibition.** In Chapter 47 it was pointed out that contrast in sensory patterns is increased by inhibitory signals transmitted laterally in the sensory pathway. This effect was illustrated in Figure 47–11 of that chapter.

In the case of the dorsal column system, an excited sensory receptor in the skin transmits not only excitatory signals to the somesthetic cortex but also inhibitory signals laterally to adjacent fiber pathways. These inhibitory signals help to block lateral spread of the excitatory signal, a process called *lateral inhibition* or *afferent inhibition*. As a result, the peak of excitation stands out, and much of the surrounding diffuse stimulation is blocked. Obviously, this mechanism accentuates the contrast between the areas of peak stimulation and the surrounding areas, thus greatly increasing the contrast or sharpness of the perceived spatial pattern.

**Transmission of Rapidly Changing and Repetitive Sensations.** The dorsal column system is of particular value for apprising the sensorium of rapidly changing peripheral conditions. This system can "follow" changing stimuli up to at least 100 cycles per second and can "detect" changes as high as 400 to 500 cycles. This rapid response allows one to direct his attention immediately to any point of contact, which, in turn, allows him to make necessary corrections before damage can be done.

*Vibratory Sensation.* Because vibratory signals are rapidly repetitive (can be detected up to 400 to 500 cycles per second), they can be transmitted only in the dorsal column system and not at all in the slowly transmitting spinothalamic system. For this reason, application of vibration with a tuning fork to different peripheral parts of the body is an important tool used by the neurologist for testing functional integrity of the dorsal column system.

## PROCESSING OF INFORMATION IN THE DORSAL COLUMN SYSTEM

Despite the faithfulness of transmission of signals from the periphery to the sensory cortex through the dorsal column system, there is also some processing of sensory information at lower synaptic levels before it reaches the cerebral cortex. For instance, the signal pattern from kinesthetic receptors changes as it passes progressively up the dorsal column system. Figure 49–1 showed that individual joint receptors are stimulated maximally at specific degrees of rotation of the joint, with the intensity of stimulation decreasing on either side of the maximal point for each receptor. However, the kinesthetic signal is quite different at the level of the ventrobasal complex of the thalamus, as can be seen by referring to Figure 49–8. This

figure shows that the ventrobasal neurons are of two types: (1) those that are maximally stimulated when the joint is at full rotation and (2) those that are maximally stimulated when the joint is at minimal rotation. In each case, as the degree of rotation changes, the rate of stimulation of the neuron either decreases or increases depending on the direction in which the joint is being rotated. Furthermore, the intensity of neuronal excitation changes over angles of 40 to 60 degrees of angulation in contrast to 20 to 30 degrees for the individual receptors, as was illustrated in Figure 49–1. Thus, the signals from the individual joint receptors are integrated in the space domain by the thalamic neurons, giving a progressively stronger signal as the joint moves in only one direction rather than giving a peaked signal as occurs in stimulation of individual receptors.

Another instance of information processing in the dorsal column system is physical separation of signals of different modalities, with cutaneous sensibilities terminating anteriorly in somatic area I and kinesthetic sensibilities terminating posteriorly. It is presumably these types of changes in signal patterns and separation of modalities that allow the sensorium to separate out all the ingredients of sensory information.

# TRANSMISSION IN THE SPINOTHALAMIC SYSTEM

It was pointed out earlier in the chapter that the spinothalamic system transmits sensory signals that do not require rapid transmission nor highly discrete localization in the body. These include pain, heat, cold, crude tactile sensations, and sexual sensations. In the following chapter, pain and temperature sensations are discussed specifically, but the present chapter is concerned principally with transmission of the crude tactile sensations in the spinothalamic system.

## THE SPINOTHALAMIC PATHWAY

Most anatomists divide the spinothalamic pathway into three different tracts: (1) the ventral spinothalamic tract, (2) the lateral spinothalamic tract, and (3) the spinoreticular tract. In the following discussion we will follow this convention, though these tracts overlap each other greatly, and they course diffusely through the anterolateral quadrants of the white columns of the cord. For this reason, many physiologists simply speak of the anterolateral spinothalamic pathway.

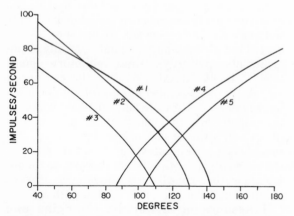

**Figure 49–8.** Typical responses of five different neurons in the knee joint receptor field of the ventrobasal complex when the knee joint is moved through its range of motion. (The curves were constructed from data in Mountcastle: *J. Neurophysiol.*, 26:807, 1963.)

**Ventral Spinothalamic Tract.**  Most of the spinothalamic fibers subserving the tactile sense (about 2500 of them) enter the dorsal columns along with the fibers of the dorsal column system, as shown in Figure 49–9, and then travel upward as much as six segments or downward as much as two segments before ending on second order neurons in the dorsal horns. Then fibers from these neurons cross through the anterior commissure to the opposite anterior column and form the *ventral spinothalamic tract* that passes all the way to the thalamus.

The ventral spinothalamic tract terminates, along with the terminations of the dorsal column system, in the ventrobasal nuclear complex of the thalamus, though many of the fibers also give off collaterals that terminate in the bulbar and mesencephalic reticular areas.

The fibers of the ventral spinothalamic tract are mainly larger size delta type A nerve fibers in contrast to smaller fibers found in the lateral spinothalamic tract. Furthermore, there is crude localization of touch and pressure sensations transmitted through this tract, indicating a moderate degree of *spatial orientation* of the fibers in the tract.

**Lateral Spinothalamic Tract.**  On entering the cord through the posterior roots, the *pain* and *temperature* fibers travel upward in the *tract of Lissauer*

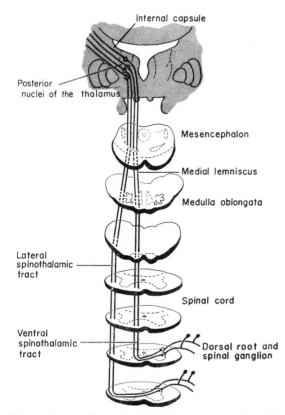

**Figure 49–9.**  The spinothalamic pathways. (Modified from Ranson and Clark: Anatomy of the Nervous System, 1959.)

for one or two segments and then terminate in the gray matter of the dorsal horns. At this point, the signals probably pass through one or more local neurons, the last of which sends long fibers that cross through the anterior commissure of the cord gray matter, pass to the opposite lateral column of the spinal cord where they form the *lateral spinothalamic tract,* and eventually terminate mainly in the intralaminar nuclei of the thalamus, a different terminus from that of the ventral spinothalamic tract.

However, a few of the fibers (especially the delta pain fibers that will be discussed in the following chapter) terminate in the most caudal portion of the ventrobasal complex, an area that is the human analogue of the posterior nuclei of the thalamus of lower animals. Many of the fibers also give rise to collaterals that terminate in the reticular substance of the bulbar and mesencephalic areas.

The fibers of the lateral spinothalamic tract are mainly delta type A or type C fibers.

**Spinoreticular Tract.**  In addition to the fibers of the ventral and lateral spinothalamic tracts, still other fibers originate in the dorsal horns of the spinal cord along with the fibers of these two tracts but never extend as far as the thalamus. Instead, these fibers terminate in the reticular substance of the bulbar and mesencephalic areas and therefore constitute the spinoreticular tract. Here they synapse with one or more successive neurons that eventually send fibers into the intralaminar nuclei of the thalamus, terminating in the same areas as many of the fibers from the lateral spinothalamic tract; from here signals are transmitted to other parts of the thalamus, the hypothalamus, and the cortex. The fibers of this tract are probably all of the type C variety, and they are concerned either entirely or almost entirely with the transmission of pain signals.

Fibers also enter the lateral and the ventral spinothalamic tracts and the spinoreticular tract from the spinal nucleus of the fifth nerve. These fibers transmit respective types of sensory information from the head.

**Projection of Spinothalamic Signals from the Thalamus to the Cortex.**  Third order neurons from the ventrobasal complex of the thalamus, which carry tactile signals from the ventral spinothalamic tract, project primarily to somatic area I and, to a lesser extent, to somatic area II. However, on the basis of inadequate information, it has been suggested that the lateral spinothalamic system, which transmits pain and temperature signals, and which synapses in the intralaminar thalamic nuclei, probably projects to somatic area II, though even these projections are weak ones.

**Separation of Modalities in the Spinothalamic Tracts.**  It has already been pointed out in the above description of the spinothalamic pathway that the nerve fibers that subserve the senses of touch and pressure are transmitted in the ventral spinothalamic tract, and those that subserve pain and temperature are transmitted in the lateral spinothalamic tract. In addition, the lateral spinothalamic tract is at least partially divided into *superficial* and *deep divisions,* the superficial division transmitting pain signals and the deep division transmitting temperature signals. Thus, the process of separating the different modalities of sensation actually begins in the spinal cord.

**Characteristics of Transmission in the Spinothalamic Tracts.** The precise nature of transmission in the spinothalamic tracts is not nearly so well known as in the dorsal column system. In general, the same principles apply to transmission in this tract as in the dorsal column system, except for the following differences: (a) the velocities of transmission in the spinothalamic tracts are only one-fiftieth to one-quarter those in the dorsal column system; (b) the spatial localization of signals transmitted in the spinothalamic tracts is poor, especially in the lateral spinothalamic tract that transmits pain and temperature signals; (c) the gradations of intensities are also far less acute, most of the sensations being recognized in 10 to 20 gradations of strength rather than the 100 or more gradations for the dorsal column system; and (d) the ability to transmit rapidly repetitive sensations is almost nil in the spinothalamic system.

Thus, it is evident that the spinothalamic tracts are by far a cruder type of transmission system than is the dorsal column system. However, it must be realized that certain types of sensations are, nevertheless, transmitted only by the spinothalamic tracts, including pain, thermal sensations, and sexual sensations. Only the mechanoreceptive sensations are transmitted by both systems, the dorsal column system having to do with the critical types of mechanoreceptive sensations, and the ventral spinothalamic tract having to do with the cruder types of mechanoreceptive sensations.

## SOME SPECIAL ASPECTS OF SENSORY FUNCTION

**Function of the Thalamus in Somatic Sensations.** The major function of the thalamus in tactile sensation is probably to relay information to the cortex. However, since a person with his somesthetic cortex removed can redevelop a mild degree of tactile sensibility, it must be assumed that the thalamus can do much more than simply relay tactile information to the somesthetic cortex. The thalamus seems to be even more important in pain sensation, as pointed out in the following chapter.

**Cortical Control of Sensory Sensitivity.** Almost all sensory information that enters the cerebrum, except that from the olfactory system, but including sensory information from the eyes, the ears, the taste receptors, and all the somatic receptors, is relayed through one or another of the thalamic nuclei. Furthermore, the conscious brain is capable of directing its attention to different segments of the sensory system. This function is believed to be mainly achieved through facilitation or inhibition of the cortical receptive areas.

But, in addition, ''corticofugal'' signals are transmitted from the cortex to the lower relay stations in the sensory pathways to *inhibit* transmission. For instance, corticofugal pathways control the sensitivity of the synapses in the thalamus as well as in the dorsal column nuclei and in the posterior horn relay station of the spinothalamic system. Also, similar systems are known for the visual, auditory, and olfactory systems, which are discussed in later chapters. Each corticofugal pathway begins in the cortex where the sensory pathway that it controls terminates. Thus, a feedback loop exists for each sensory pathway.

Obviously, corticofugal control of sensory input could allow the cerebral cortex to alter the threshold

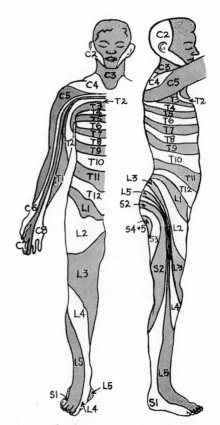

**Figure 49–10.** The dermatomes. (Modified from Grinker and Sahs: Neurology. Charles C Thomas, 1966.)

for different sensory signals. Also, it might help the brain focus its attention on specific types of information, which is an important and necessary quality of nervous system function.

**Segmental Fields of Sensation—The Dermatomes.** Each spinal nerve innervates a "segmental field" of the skin called a *dermatome*. The different dermatomes are illustrated in Figure 49–10. However, these are shown as if there were distinct borders between the adjacent dermatomes, which is far from true because much overlap exists from segment to segment. Indeed, because of the great overlap, the posterior roots from an entire segment of the spinal cord can be destroyed without causing significant loss of sensation in the skin.

Figure 49–10 shows that the anal region of the body lies in the dermatome of the most distal cord segment. In the embryo, this is the tail region and is the most distal portion of the body. The legs develop from the lumbar and upper sacral segments rather than from the distal sacral segments, which is evident from the dermatomal map. Obviously, one can use a dermatomal map such as that illustrated in Figure 49–10 to determine the level in the spinal cord at which various cord injuries may have occurred when the peripheral sensations are disturbed.

# REFERENCES

Adrian, E. D.: The Physical Background of Perception. Oxford, Clarendon Press, 1947.

Albe-Fessard, D., and Iggo, A. (eds.): Somatosensory System. Berlin, New York, Springer-Verlag, 1973.

Asanuma, H.: Recent developments in the study of the columnar arrangement of neurons within the motor cortex. *Physiol. Rev.,* 55:143, 1975.

Bishop, P. O.: Central nervous system: afferent mechanisms and perception. *Ann. Rev. Physiol.,* 29:427, 1967.

Boudreau, J. C., and Tsuchitani, C.: Sensory Neurophysiology: with Special Reference to the Cat. New York, Van Nostrand Reinhold Company, 1973.

Darian-Smith, I.: Somatic sensation. *Ann. Rev. Physiol.,* 31:417, 1969.

Emmers, R., and Tasker, R. R.: The Human Somesthetic Thalamus. New York, Raven Press, 1975.

Frigyesi, T. L., Rinvik, E., and Yahr, M. D. (eds.): Corticothalamic Projections and Sensorimotor Activities. New York, Raven Press, 1972.

Lapresle, J., and Haguenau, M.: Anatomico-chemical correlation in focal thalamic lesions. *Z. Neurol.,* 205:29, 1973.

Lloyd, D. P.: Action in primary afferent fibers in the spinal cord. *Inter. J. Neurosci.,* 1:1, 1970.

Lynn, B.: Somatosensory receptors and their CNS connections. *Ann. Rev. Physiol.,* 37:105, 1975.

Mountcastle, V. B., Poggio, G. F., and Werner, G.: The relation of thalamic cell response to peripheral stimuli varied over an intensive continuum. *J. Neurophysiol.,* 26:807, 1963.

Omer, G. E., Jr.: Sensation and sensibility in the upper extremity. *Clin. Orthop.,* 104:30, 1974.

Perl, E. R., and Boivie, J. G.: Neural substrates of somatic sensation. *In* Guyton, A. C. (ed.): MTP International Review of Science: Physiology. Vol. 3. Baltimore, University Park Press, 1974, p. 303.

Persson, H. E.: Development of Somatosensory Cortical Functions; an Electrophysiological Study in Prenatal Sheep. Stockholm, Department of Physiology, Karolinska Institute, 1973.

Rodieck, R. W.: Central nervous system: afferent mechanisms. *Ann. Rev. Physiol.,* 33:203, 1971.

Sinclair, D. C.: Cutaneous Sensation. New York, Oxford University Press, 1967.

Wall, P. D., and Dubner, R.: Somatosensory pathways. *Ann. Rev. Physiol.,* 34:315, 1972.

Werner, G., and Mountcastle, V. B.: Neural activity in mechanoreceptive cutaneous afferents: stimulus-response relations, Weber functions, and information transmission. *J. Neurophysiol,* 28:359, 1965.

Whitsel, B. L., Petrucelli, L. M., and Werner, G.: Symmetry and connectivity in the map of the body surface in somatosensory area II of primates. *J. Neurophysiol.,* 32:170, 1969.

*See also bibliographies at the ends of Chapters 48 and 50.*

# | 50 |

# Somatic Sensations:
# II. Pain, Visceral Pain, Headache,
# and Thermal Sensations

Many, if not most, ailments of the body cause pain. Furthermore, one's ability to diagnose different diseases depends to a great extent on a knowledge of the different qualities of pain, a knowledge of how pain can be referred from one part of the body to another, how pain can spread from the painful site, and, finally, what the different causes of pain are. For these reasons, the present chapter is devoted mainly to pain and to the physiologic basis of some of the associated clinical phenomena.

**The Purpose of Pain.** Pain is a protective mechanism for the body; it occurs whenever any tissues are being damaged, and it causes the individual to react to remove the pain stimulus. Even such simple activities as sitting for a long time on the ischia can cause tissue destruction because of lack of blood flow to the skin where the skin is compressed by the weight of the body. When the skin becomes painful as a result of the ischemia, the person shifts his weight unconsciously. A person who has lost his pain sense, such as after spinal cord injury, fails to feel the pain and therefore fails to shift his weight. This eventually results in ulceration at the areas of pressure unless special measures are taken to move the person from time to time.

## QUALITIES OF PAIN

Pain has been classified into three different major types: pricking, burning, and aching pain. Other terms used to describe different types of pain include throbbing pain, nauseous pain, cramping pain, sharp pain, electric pain, and others, most of which are well known to almost everyone.

*Pricking pain* is felt when a needle is stuck into the skin or when the skin is cut with a knife. It is also often felt when a widespread area of the skin is diffusely but strongly irritated.

*Burning pain* is, as its name implies, the type of pain felt when the skin is burned. It can be excruciating and is the most likely of the pain types to cause suffering.

*Aching pain* is not felt on the surface of the body, but, instead, is a deep pain with varying degrees of annoyance. Aching pain of low intensity in widespread areas of the body can summate into a very disagreeable sensation.

It is not necessary to describe these different qualities of pain in great detail because they are well known to all persons. The real problem, and one that is only partially solved, is what causes the differences in quality. Pricking pain results from stimulation of delta type A pain fibers, whereas burning and aching pain results from stimulation of the more primitive type C fibers, which will be discussed later in the chapter.

## METHODS FOR MEASURING THE PERCEPTION OF PAIN

The intensity of a stimulus necessary to cause pain can be measured in many different ways, but the most used methods have been pricking the skin with a pin at measured pressures, pressing a solid object against a protruding bone with measured force, or heating the skin with measured amounts of heat. The latter methods has proved to be especially accurate from a quantitative point of view.

Figure 50–1 illustrates the basic principles of a heat apparatus used for measuring pain threshold. An intense light is focused by a large condenser lens onto a black spot painted on the forehead of the subject, and the heat intensity delivered by the light is controlled by a rheostat. In determining the subject's threshold for pain, the intensity of the heat is increased in progressive steps, and the length of time required for the forehead to heat sufficiently to elicit pain is recorded for each heat intensity. These data are then plotted in the form of a "strength-duration curve" to express pain threshold, as follows.

**Strength-Duration Curve for Expressing Pain Threshold.** Figure 50–2 illustrates a typical strength-duration curve obtained by using the above procedure for measuring pain threshold. This curve is almost identical with the strength-duration curves discussed in Chapter 10 for stimulation of nerve fibers by electrical currents of increasing intensities. Note that a very intense stimulus applied for only a second elicits a sensation of pain while a stimulus of much less intensity may require many seconds. The lowest intensity of stimulus that will excite the sensation of pain when the stimulus is applied for a prolonged period of time is called the *pain threshold.*

**The Intensity of Pain—"JND's."** The increase in stimulus intensity that will barely cause a detectable difference in degree of pain is called a *just noticeable difference* (JND). By applying all different stimulus intensities between the level of no pain at all and the most intense pain that a person can distinguish, it has been found that approximately 22 JND's can be discerned by the average person.

Figure 50–3 illustrates the relationship of stimulus intensity (applied by the apparatus of Fig. 50–1) to the intensity of pain expressed in JND's.

**Uniformity of Pain Threshold in Different People.** Figure 50–4 shows graphically the

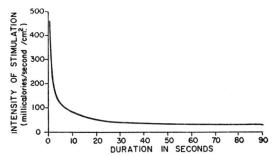

**Figure 50–2.** Strength-duration curve for depicting pain threshold. (From Hardy: *J. Chronic Dis.,* 4:20, 1956.)

skin temperature at which pain is first perceived by different persons. By far the greatest number of subjects perceive pain when the skin temperature reaches almost exactly 45° C., and almost everyone perceives pain before the temperature reaches 47° C. In other words, it is almost never true that some persons are unusually sensitive or insensitive to pain. Indeed, measurements in people as widely different as Eskimos, Indians, and whites have shown no significant differences in their *thresholds for pain.* However, different people do *react* very differently to pain, as is discussed below.

# THE PAIN RECEPTORS AND THEIR STIMULATION

**Free Nerve Endings as Pain Receptors.** The pain receptors in the skin and other tissues are all free nerve endings. They are widespread in the superficial layers of the *skin* and also in certain internal tissues, such as the *periosteum,* the *arterial walls,* the *joint surfaces,* and the *falx* and *tentorium* of the cranial vault. Most of the other deep tissues are not extensively supplied with pain endings but are weakly supplied; nevertheless, any widespread tissue

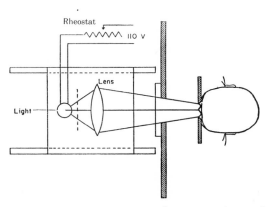

**Figure 50–1.** Heat apparatus for measuring pain threshold. (From Hardy, Wolff, and Goodell: *J. Clin. Invest,* 19:649, 1940.)

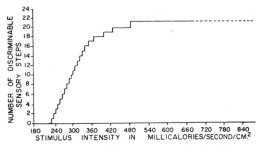

**Figure 50– 3.** Pain intensity, expressed in JND's, at different intensities of stimulus. (From Hardy: *J. Chronic Dis.,* 4:22, 1956.)

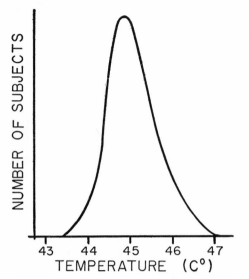

**Figure 50–4.** Distribution curve obtained from a large number of subjects of the minimal skin temperature that causes pain. (Modified from Hardy: *J. Chronic Dis.*, 4:22, 1956.)

damage can still summate to cause the aching type of pain in these areas.

**Nonadapting Nature of Pain Receptors.** In contrast to most other sensory receptors of the body the pain receptors adapt either not at all or almost not at all. In fact, under some conditions, the threshold for excitation of the pain fibers becomes progressively lower and lower as the pain stimulus continues, thus allowing these receptors to become progressively more activated with time. This increase in sensitivity of the pain receptors is called *hyperalgesia*.

One can readily understand the importance of this failure of pain receptors to adapt, for it allows them to keep the person apprised of a damaging stimulus that causes the pain as long as it persists.

### RATE OF TISSUE DAMAGE AS THE CAUSE OF PAIN

The average critical temperature of 45° C. at which a person first begins to perceive pain is also the temperature at which the tissues begin to be damaged by heat; indeed, the tissues are eventually completely destroyed if the temperature remains at this level indefinitely. Therefore, it is immediately apparent that pain resulting from heat is closely correlated with the ability of heat to damage the tissues.

Furthermore, in studying soldiers who had been severely wounded in World War II, it was found that the majority of them felt little or no pain except for a short time after the severe wound had been sustained. This, too, indicates that *pain generally is not felt after damage has been done* but only *while damage is being done*.

**Bradykinin and Similar Polypeptides as Possible Stimulators of Pain Endings.** The precise mechanism by which tissue damage stimulates pain endings is not known. However, many research workers have shown that extracts from damaged tissues cause intense pain when injected beneath the normal skin. Therefore, it is almost certain that some chemical substance or substances released from damaged tissues excite the pain nerve endings.

There are many reasons to believe that the substance *bradykinin* or some similar polypeptide might be the principal substance that stimulates pain endings. For instance, when this substance is injected in extremely minute quantities underneath the skin, severe pain is felt. Furthermore, cell damage releases proteolytic enzymes that almost immediately split bradykinin and other similar substances from the globulins in the interstitial fluid. And, finally, bradykinin and similar substances can be found in the skin when painful stimuli are applied.

Thus, the postulated mechanism for eliciting pain: damage to cells releases proteolytic enzymes that then split bradykinin and associated substances from globulin, and these in turn stimulate the nerve endings.

**Tissue Ischemia and Muscle Spasm as Causes of Pain.** When blood flow to a tissue is blocked, the tissue becomes very painful within a few minutes. And the greater the rate of metabolism of the tissue, the more rapidly the pain appears. For instance, if a blood pressure cuff is placed around the upper arm and inflated until the arterial blood flow ceases, exercise of the forearm muscles can cause severe muscle pain within 15 to 20 seconds. In the absence of muscle exercise, the pain will not appear for three to four minutes. Cessation of blood flow to the skin, in which the metabolic rate is very low, usually does not cause pain for about 20 to 30 minutes.

*Muscle spasm* is also a frequent cause of pain. The reason for this is probably two-fold. First, the contracting muscle compresses the intramuscular blood vessels and either reduces or cuts off the blood flow. Second, muscle contraction increases the rate of metabolism of the muscle. Therefore, muscle spasm probably causes relative muscle ischemia so that typical ischemic pain results.

The cause of pain in ischemia is yet unknown; however, it is relieved by supplying oxygen to the ischemic tissue. Flow of unoxygenated blood to the tissue will not relieve the pain.

One of the suggested causes of pain in ischemia is accumulation of large amounts of lactic acid in the tissues, formed as a consequence of the anaerobic metabolism (metabolism without oxygen) that occurs during ischemia. However, it is also possible that other chemical agents, such as bradykinin and the polypeptides, are formed in the tissues because of muscle cell damage and that these, rather than lactic acid, stimulate the pain nerve endings.

### TICKLING AND ITCH

The phenomenon of tickling and itch has often been stated to be caused by very mild stimulation of pain nerve endings, since whenever pain is blocked by anesthesia of a nerve or by compressing the nerve, the phenomenon of tickling and itch also disappears. However, recent neurophysiologic studies have demonstrated the existence of very sensitive free nerve endings that elicit only the itch sensation. Furthermore, these endings are found almost exclusively in the superficial layers of the skin, which is also the only tissue from which the itch sensation can usually be elicited. Furthermore, exciting itch receptors in animals initiates scratch reflexes, which contrasts with the effect of exciting pain nerve endings that always causes withdrawal reflexes instead.

Therefore, it seems clear that the itch and tickle sensations are transmitted by very small type C fibers similar to those that transmit the burning type of pain; these fibers, however, are distinctly separate from the pain fibers. Furthermore, the endings of these fibers are readily excited by light mechanoreceptive stimuli, indicating that they are mechanoreceptors in contradistinction to the probable chemoreceptor nature of pain fibers.

The purpose of the itch sensation is presumably to call attention to mild surface stimuli such as a flea crawling on the skin or a fly about to bite, and the elicited signals then lead to scratching or other maneuvers that rid the host of the irritant.

The relief of itch by the process of scratching occurs only when the irritant is removed or when the scratch is strong enough to elicit pain. The pain signals are believed to suppress the itch signals in the cord by a process of inhibition that will be described later in the chapter.

## TRANSMISSION OF PAIN SIGNALS INTO THE CENTRAL NERVOUS SYSTEM

**"Fast" Pain Fibers and "Slow" Pain Fibers.** Pain signals are transmitted by small delta type A fibers at velocities of between 3 and 20 meters per second and also by type C fibers at velocities between 0.5 and 2 meters per second. When the delta type A fibers are blocked without blocking the C fibers by moderate compression of the nerve trunk, the pricking type of pain disappears. On the other hand, when the type C fibers are blocked without blocking the delta fibers by low concentrations of local anesthetic, the burning and aching types of pain disappear.

Therefore, a sudden onset of painful stimulus gives a "double" pain sensation: a fast pricking pain sensation followed a second or so later by a slow burning pain sensation. The pricking pain presumably apprises the person very rapidly of a damaging influence and, therefore, plays an important role in making the person react immediately to remove himself from the stimulus. On the other hand, the slow burning sensation tends to become more and more painful over a period of time. It is this sensation that gives one the intolerable suffering of pain.

**Transmission in the Spinothalamic and Spinoreticular Tracts.** Pain fibers enter the cord through the dorsal roots, ascend or descend one to two segments in the *tract of Lissauer,* and then terminate on neurons in the dorsal horns of the cord gray matter. Here the signals probably pass through one or more additional short-fibered neurons, the last of which gives rise to long fibers that cross immediately to the opposite side of the cord in the *anterior commissure* and pass upward to the brain via the spinothalamic and spinoreticular tracts. As discussed in the previous chapter, most of the pain fibers lie in the lateral portions of the spinothalamic and spinoreticular tracts; some pain fibers, however, are found in all areas of the entire anterolateral quadrant of the cord white columns.

We shall see later that the intensity of pain signals can be modified markedly as they pass through the neuronal synapses of the gray matter of the dorsal horns, especially in response to simultaneous signals transmitted by mechanoreceptor sensory nerve fibers and in response to signals entering the dorsal horns from the brain via corticofugal fibers.

As the pain pathways pass into the brain they separate into two separate pathways: *the pricking pain pathway* composed almost entirely of small type A delta fibers, and *the burning pain pathway* composed almost entirely of the slow type C fibers.

**The Pricking Pain Pathway.** Figure 50–5 illustrates that the pricking pain pathway terminates in the caudalmost portion of the ventrobasal complex, an area that is the human analogue of the posterior nuclear group in lower animals. From here signals are transmitted into other areas of the thalamus and to the somatic sensory cortex. Most of the fibers from this area probably go to somatic sensory area II, as was explained in the previous chapter. However, it is possible that a few pass also to somatic sensory area I.

**The Burning Pain Pathway—Stimulation of the Reticular Activating System.** Figure 50–5 shows that the burning and aching pain fibers terminate in the reticular area of the brain stem and in the intralaminar nuclei of the thalamus, the nonspecific nuclei located among the thalamic specific nuclei. Both the reticular area of the brain stem and the intralaminar nuclei are parts of the reticular activating system. In Chapter 54 we will discuss in detail the func-

tions of this system; briefly, however, it transmits activating signals into essentially all parts of the brain, especially upward through the thalamus to all areas of the cerebral cortex and also into the basal regions of the brain around the thalamus including very importantly the hypothalamus.

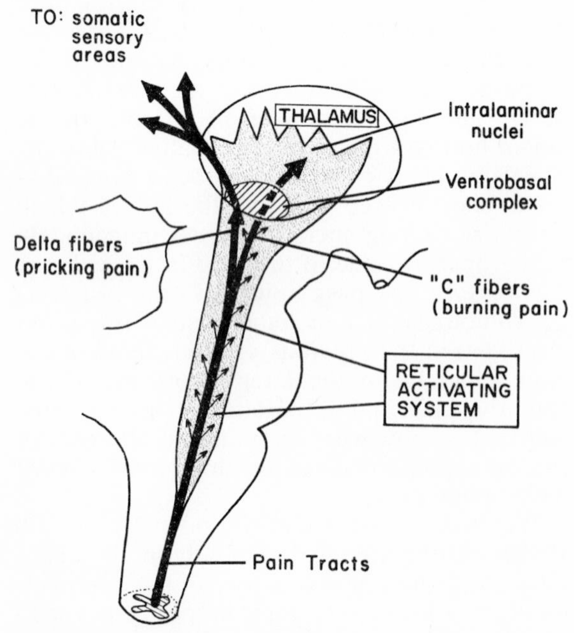

Thus, the burning and aching pain fibers, because they do excite the reticular activating system, have a very potent effect on activating essentially the entire nervous system, that is, to arouse one from sleep, to create a state of excitement, to create a sense of urgency, and to promote defense and aversion reactions designed to rid the person or animal of the painful stimulus.

The signals that are transmitted through the burning pain pathway can be localized only to very gross areas of the body, if they can be localized at all. Therefore, these signals are designed almost entirely for the single purpose of calling one's attention to injurious processes in the body. They create suffering that is sometimes intolerable. Their gradation of intensity is poor; instead, weak pain signals can summate over a period of time by a process of temporal summation to create an unbearable feeling even though the same pain for short periods of time may be relatively mild.

**Function of the Thalamus and Cerebral Cortex in the Appreciation of Pain.** Complete removal of the somatic sensory areas of the cerebral cortex does not destroy one's ability to perceive pain. Therefore, it is believed that pain impulses entering only the thalamus and lower centers cause at least some conscious perception of pain. However, this does not mean that the cerebral cortex has nothing to do with normal pain appreciation; indeed, electrical stimulation of the somesthetic cortical areas causes a person to perceive mild pain in approximately 3 per cent of the stimulations. Furthermore, lesions in these areas, particularly in somatic sensory area II, at times give rise to severe pain. Thus, there is reason to believe that somatic sensory area II may be more concerned with pain sensation than is somatic sensory area I.

**Localization of Pain in the Body.** Most localization of pain probably results from simultaneous stimulation of tactile receptors along with the pain stimulation. However, the pricking type of pain, transmitted through delta type A fibers, can be localized perhaps within 10 to 20 cm. of the stimulated area. On the other hand, the burning and aching types of pain,

**Figure 50–5.** Transmission of pain signals into the hindbrain, thalamus, and cortex.

transmitted through type C fibers, is localizable only very grossly, perhaps to a major part of the body such as a limb but certainly not to small areas. This is in keeping with the fact that these fibers terminate extremely diffusely in the hindbrain and thalamus.

*Surgical Interruption of Pain Pathways.* Often a person has such severe and intractable pain (this often results from rapidly spreading cancer), that it is necessary to relieve the pain. To do this the pain pathway can be destroyed at any one of several different points. If the pain is in the lower part of the body, a *cordotomy* in the upper thoracic region usually relieves the pain. To do this, the spinal cord on the side opposite to the pain is sectioned almost entirely through its anterolateral quadrant, which interrupts the spinothalamic tracts.

A cordotomy does not relieve pain in the upper part of the body because many pain fibers from the upper body do not cross to the opposite side of the spinal cord before reaching the medulla. In these patients, the lateral spinothalamic tract is sectioned in the brain stem where it passes over the inferior olive. This procedure is called a *bulbar tractotomy.*

Finally, if the pain originates in the neck or face, even a bulbar tractotomy is not satisfactory. Pain sometimes originates within the thalamus or elsewhere in the central pain pathways beyond the tractotomy site, as will be discussed later in the chapter. To relieve these types of pain, extirpation of either somatic sensory area I or somatic sensory area II has been attempted, but almost always without success—indeed, sometimes with exacerbation of the pain. Recently, however, it has been claimed that destruction of specific portions of the intralaminar nuclei in the thalamus can relieve the suffering elicited by pain while still leaving intact one's appreciation of the pricking quality of pain, which remains an important protective mechanism. Unfortunately, such operations are so difficult to perform without damaging other vital areas of the thalamus that they have been successful in only a few instances.

# THE REACTION TO PAIN

Even though the threshold for recognition of pain remains approximately equal from one person to another, the degree to which each one reacts to pain varies tremendously. Stoical persons, such as members of the American Indian race, react to pain far less intensely than do more emotional persons. In the preceding chapter, it was pointed out that conditioning impulses entering the sensory areas of the central nervous system from various portions of the central and peripheral nervous systems can determine whether incoming sensory impulses will be transmitted extensively or weakly to other areas of the brain. It is probably some

such mechanism as this that determines how much one reacts to pain.

Pain causes both reflex motor reactions and psychic reactions. Some of the reflex actions occur directly from the spinal cord, for pain impulses entering the gray matter of the cord can directly initiate "withdrawal reflexes" that remove the body or a portion of the body from the noxious stimulus, as will be discussed in Chapter 51. These primitive spinal cord reflexes, though important in lower animals, are mainly suppressed in the human being by the higher centers of the central nervous system. Yet, in their place, much more complicated and more effective reflexes from the motor cortex are initiated by the pain stimuli to eliminate the painful stimulus.

The psychic reactions to pain are likely to be far more subtle; they include all the well-known aspects of pain such as anguish, anxiety, crying, depression, nausea, and excess muscular excitability throughout the body. These reactions vary tremendously from one person to another following comparable degrees of pain stimuli.

## THE "GATING" THEORY FOR CONTROL OF THE REACTION TO PAIN

It is common experience that two types of non-pain signals from other parts of the nervous system can greatly alter the degree of transmission of pain through the spinal cord.

1. Stimulation of the large sensory fibers from peripheral mechanoreceptors greatly depresses the transmission of pain signals either from the same area of the body or from areas sometimes located many segments away.

2. Corticofugal signals from the brain can decrease pain sensitivity.

A special quality of the suppression caused both by the cutaneous mechanoreceptors and by the brain is that even after the signals are over the pain sometimes remains suppressed for an hour or more thereafter.

**Mechanism of Pain Gating.** Unfortunately, the detailed mechanism of pain gating is not known. However, the general mechanism is probably the following: After the pain fibers enter the spinal cord and pass into the tract of Lissauer, within one or two segments they terminate in the *substantia gelatinosa,* a group of small neurons located near the tip of the dorsal horn. The pain signals are then transmitted through one or more additional neurons before passing upward through the long pathways to the brain. However, there are other inputs to the substantia gelatinosa in addition to the pain fibers. For instance, collaterals from the mechanoreceptors terminate in the mid-portion of the dorsal horns, and from here second order neurons send short axons dorsally in the dorsal horns to terminate in the substantia gelatinosa. Most of these terminate on the pain fiber axons before they synapse with the neurons of the substantia gelatinosa, and they cause *presynaptic inhibition,* a mechanism explained in Chapter 46. In a similar manner, corticofugal pathways also terminate in the substantia gelatinosa where they, too, cause presynaptic inhibition.

Thus, signals either from the mechanoreceptors or from the corticofugal fibers can greatly depress the transmission of pain signals from the pain fibers into the successive levels of the pain pathway.

**Treatment of Pain by Exciting the Dorsal Column Sensory Pathway.** Experiments have shown that mild electrical excitation of the large sensory fibers of peripheral nerves or even electrical excitation of the large fibers of the dorsal columns in the spinal cord can often suppress pain. When the large peripheral fibers are stimulated, the suppression probably results from presynaptic inhibition in the substantia gelatinosa. When the dorsal columns are excited, a similar type of inhibition probably occurs at the brain stem level. It is possible that future advances in this area of "counter-stimulation" will allow relief from many types of presently unbearable pain.

**A Possible Physiologic Basis for Acupuncture.** One will recognize that the gating theory of pain control provides a partial physiologic basis for the claims of acupunture. That is, stimulation of non-painful fibers, the mechanoreceptor sensory afferents, can suppress pain signals. In addition, strong conditioning signals from the brain can also suppress pain. A combination of these two could help to provide an explanation for the success of acupuncture in some people. Also, it can explain the efficacy of skin irritants or heat to relieve pain—for instance, the use of liniments on the skin to relieve pain of overly exercised muscles.

# REFERRED PAIN

Often a person feels pain in a part of his body that is considerably removed from the tissues causing the pain. This pain is called *referred pain.* On occasion, pain can even be referred from one surface area of the body to another, but more frequently it is initiated in one of the visceral organs and referred to an area of the body surface. Also, pain may originate in a viscus and be referred to another deep area of the body not exactly coincident with the location of the viscus producing the pain. A knowledge of these different types of referred pain is extremely important because many visceral ailments cause no other signs except referred pain.

**Mechanism of Referred Pain.** Figure 50–6 illustrates the most generally accepted mech-

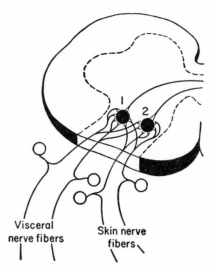

**Figure 50–6.** Mechanism of referred pain and referred hyperalgesia.

anism by which most pain is referred. In the figure, branches of visceral pain fibers are shown to synapse in the spinal cord with some of the same second order neurons that receive pain fibers from the skin. When the visceral pain fibers are stimulated intensely, pain sensations from the viscera spread into some of the neurons that normally conduct pain sensations only from the skin, and the person has the feeling that the sensations actually originate in the skin itself. It is also possible that some referred pain results from convergence of visceral and skin impulses at the level of the thalamus rather than in the spinal cord.

## *REFERRED PAIN CAUSED BY REFLEX MUSCULAR SPASM*

Some types of referred pain are caused secondarily by reflex muscular spasm. For instance, pain in a ureter can cause reflex spasm of the lumbar muscles. Often the pain from the ureter itself is hardly felt at all, but instead almost all the pain results from spasm of the lumbar muscles.

Many back pains and some types of headache also appear to be caused by muscular spasm, the spasm originating reflexly from much weaker pain impulses originating elsewhere in the body.

## VISCERAL PAIN

In clinical diagnosis, pain from the different viscera of the abdomen and chest is one of the few criteria that can be used for diagnosing visceral inflammation, disease, and other ailments.

In general, the viscera have sensory receptors for no other modalities of sensation besides pain, and visceral pain differs from surface pain in many important aspects.

One of the most important differences between surface pain and visceral pain is that highly localized types of damage to the viscera rarely cause severe pain. For instance, a surgeon can cut the gut entirely in two in a patient who is awake without causing significant pain. On the other hand, any stimulus that causes *diffuse stimulation of pain nerve endings* throughout a viscus causes pain that can be extremely severe. For instance, occluding the blood supply to a large area of gut stimulates many diffuse pain fibers at the same time and can result in extreme pain.

### CAUSES OF TRUE VISCERAL PAIN

Any stimulus that excites pain nerve endings in diffuse areas of the viscera causes visceral pain. Such stimuli include ischemia of visceral tissue, chemical damage to the surfaces of the viscera, spasm of the smooth muscle in a hollow viscus, distention of a hollow viscus, or stretching of the ligaments. Essentially all of the true visceral pain originating in the thoracic and abdominal cavities is transmitted through sensory nerve fibers that run in the sympathetic nerves. These fibers are small type C fibers and, therefore, can transmit only burning and aching types of pain. The pathways for transmitting true visceral pain will be discussed in detail later in the chapter.

**Ischemia.**   Ischemia causes visceral pain in exactly the same way that it does in other tissues, presumably because of the formation of acidic metabolic end-products or tissue degenerative products, such as bradykinin, that stimulate the pain nerve endings.

**Chemical Stimuli.**   On occasion, damaging substances leak from the gastrointestinal tract into the peritoneal cavity. For instance, proteolytic acidic gastric juice often leaks through a ruptured gastric or duodenal ulcer. This juice causes widespread digestion of the visceral peritoneum, thus stimulating extremely broad areas of pain fibers. The pain is usually extremely severe.

**Spasm of a Hollow Viscus.**   Spasm of the gut, the gallbladder, a bile duct, the ureter, or any other hollow viscus can cause pain in exactly the same way that spasm of the skeletal muscle causes pain. This presumably results from diminished blood flow to the muscle, combined with increased metabolic need of the muscle for nutrients. Thus, *relative* ischemia develops, which causes severe pain.

Often, pain from a spastic viscus occurs in the form of *cramps,* the pain increasing to a high degree of severity and then subsiding, this process continuing rhythmically once every few minutes. The rhythmic cycles result from rhythmic contraction of smooth muscle. For instance, each time a peristaltic wave travels along an overly excitable spastic gut, a cramp occurs. The cramping type pain frequently occurs in gastroenteritis, constipation, menstruation, parturition, gallbladder disease, or ureteral obstruction.

**Overdistention of a Hollow Viscus.**   Extreme overfilling of a hollow viscus also results in pain, presumably because of overstretch of the tissues themselves. However, overdistention can also collapse the blood vessels that encircle the viscus, or that pass into its wall, thereby perhaps promoting ischemic pain.

### "PARIETAL" PAIN CAUSED BY VISCERAL DAMAGE

In addition to true visceral pain, some pain sensations are also transmitted from the viscera through nerve fibers that innervate the parietal peritoneum, pleura, or pericardium. The parietal surfaces of the visceral cavities are supplied mainly by spinal nerve fibers that penetrate from the surface of the body inward.

**Characteristics of Parietal Visceral Pain.**   When a disease affects a viscus, it often spreads to the parietal wall of the visceral cavity. This wall, like the skin, is supplied with extensive innervation including the "fast" delta fibers, which are different from the fibers in the true visceral pain pathways of the sympathetic nerves. Therefore, pain from the parietal wall of a visceral cavity is frequently very sharp and pricking in quality, though it can also have burning and aching qualities if the pain stimulus is diffuse. Thus, a knife incision through the *parietal* peritoneum is very painful, even though a similar cut through the visceral peritoneum or through a gut is not painful.

### INSENSITIVE VISCERA

A few visceral areas are almost entirely insensitive to pain of any type. These include the parenchyma of the liver and the alveoli of the lungs. Yet the liver *capsule* is extremely sensitive to both direct trauma and stretch, and the *bile ducts* are also sensitive to pain. In the lungs, even though the aveoli are insensitive, the bronchi and the parietal pleura are both very sensitive to pain.

## LOCALIZATION OF VISCERAL PAIN– REFERRED VISCERAL PAIN

Pain from the different viscera is frequently difficult to localize for a number of reasons. First, the brain does not know from firsthand experience that the different organs exist, and, therefore, any pain that is localized internally can be localized only generally. Second, sensations from the abdomen and thorax are transmitted by two separate pathways to the central nervous system—the *true visceral pathway* and the *parietal pathway*. The true visceral pain is transmitted via sensory fibers of the autonomic nervous system, and the sensations are *referred* to surface areas of the body often far from the painful organ. On the other hand, parietal sensations are conducted *directly* from the parietal peritoneum, pleura, or pericardium, and the sensations are usually *localized directly over the painful area.*

**The Visceral Pathway for Transmission of Pain.**   Most of the internal organs of the body are supplied by type C pain fibers that pass along the visceral sympathetic nerves into the spinal cord and thence up the lateral spinothalamic tract along with the pain fibers from the body's surface. A few visceral pain fibers— those from the distal portion of the colon, from the rectum, and from the bladder—enter the spinal cord through the sacral parasympathetic nerves, and some enter the central nervous system through various cranial nerves. These include fibers in the glossopharyngeal and vagus nerves, which transmit pain from the pharynx, trachea, and upper esophagus. And fibers from the surfaces of the diaphragm as well as from the lower esophagus are carried in the phrenic nerves.

*Localization of Referred Pain Transmitted by the Visceral Pathways.*   The position in the cord to which visceral afferent fibers pass from each organ depends on the segment of the body from which the organ developed embryologically. For instance, the heart originated in the neck and upper thorax. Consequently, the heart's visceral pain fibers enter the cord all the way from C-3 down to T-5. The stomach had its origin approximately from the seventh to the ninth thoracic segments of the embryo, and consequently the visceral afferents from the stomach

enter the spinal cord between these levels. The gallbladder had its origin almost entirely in the ninth thoracic segment, so that the visceral afferents from the gallbladder enter the spinal cord at T-9.

Because the visceral afferent pain fibers are responsible for transmitting referred pain from the viscera, the location of the referred pain on the surface of the body is in the dermatome of the segment from which the visceral organ was originally derived in the embryo. Some of the areas of referred pain on the surface of the body are shown in Figure 50–7.

**The Parietal Pathway for Transmission of Abdominal and Thoracic Pain.**   Where spinal nerves overlie the abdomen or thorax, pain fibers penetrate inward to innervate the parietal peritoneum, parietal pleura, and parietal pericardium, Also, retroperitoneal visceral organs and perhaps portions of the mesentery are innervated to some extent by parietal pain fibers. The kidney, for instance, is supplied by both visceral and parietal fibers.

Pain from the viscera is frequently localized in two surface areas of the body because of the dual pathways for transmission of pain. Figure 50–8 illustrates dual transmission of pain from an inflamed appendix. Impulses pass from the appendix through the sympathetic visceral pain fibers into the sympathetic chain and then into the spinal cord at approximately T-10 or T-11; this pain is referred to an area around the um-

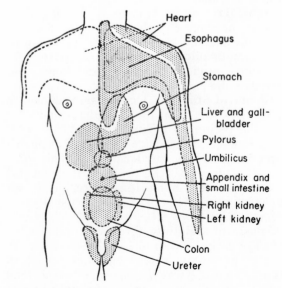

**Figure 50–7.**   Surface areas of referred pain from different visceral organs.

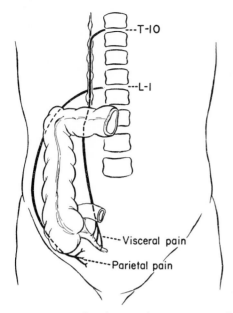

**Figure 50–8.** Visceral and parietal transmission of pain from the appendix.

bilicus and is of the aching, cramping type. On the other hand, pain impulses also often originate in the parietal peritoneum where the inflamed appendix touches the abdominal wall, and these impulses pass directly through the spinal nerves into the spinal cord at a level of approximately L-1 or L-2. This pain is localized directly over the irritated peritoneum in the right lower quadrant of the abdomen and is of the sharp type.

## VISCERAL PAIN FROM VARIOUS ORGANS

**Cardiac Pain.**   Almost all pain that originates in the heart results from ischemia secondary to coronary sclerosis, which was discussed in Chapter 25. This pain is referred mainly to the base of the neck, over the shoulders, over the pectoral muscles, and down the arms. Most frequently, the referred pain is on the left side rather than on the right—probably because the left side of the heart is much more frequently involved in coronary disease than is the right side—but occasionally mild referred pain occurs on the right side of the body as well as on the left.

The pain impulses are conducted through nerves passing to the middle cervical ganglia, to the stellate ganglia, and to the first four or five thoracic ganglia of the sympathetic chains. Then the impulses pass into the spinal cord through the second, third, fourth and fifth thoracic spinal nerves.

*Direct Parietal Pain from the Heart.*   When coro-nary ischemia is extremely severe, such as immediately after a coronary thrombosis, intense cardiac pain sometimes occurs directly underneath the sternum simultaneously with pain referred to other areas. This direct pain from underneath the sternum is difficult to explain on the basis of the visceral nerve connections. Therefore, it is highly probable that sensory nerve endings passing from the heart through the pericardial reflections around the great vessels conduct this direct pain.

In addition to pain from the heart, other sensations may accompany coronary thrombosis. One of these is a tight, oppressive sensation beneath the sternum. The exact cause of this is unknown, but a possible cause is reflex spasm of blood vessels, bronchioles, or muscles in the chest region.

*Relief of Referred Cardiac Pain by Sympathectomy.*   To interrupt pain impulses from the heart one can either cut the sympathetic nerves that pass from the heart to the sympathetic chains or, as is usually performed, cut nerve fibers as they pass through the sympathetic chains into the spinal cord. Relief of cardiac pain can frequently be accomplished simply by removing the sympathetic chain from T-2 through T-5 on only the left side, but sometimes it is necessary to remove the fibers on both sides in order to obtain satisfactory results.

**Esophageal Pain.**   Pain from the esophagus is usually referred to the pharynx, to the lower neck, to the arms, or to midline chest regions beginning at the upper portion of the sternum and ending approximately at the lower level of the heart. Irritation of the gastric end of the esophagus may cause pain directly over the heart, though the pain has nothing whatsoever to do with the heart. Such pain may be caused by spasm of the cardia, the area where the esophagus empties into the stomach, which causes excessive dilatation of the lower esophagus, or it may result from chemical, bacterial, or other types of inflammatory irritations.

**Gastric Pain.**   Pain arising in the fundus of the stomach—usually caused by gastritis—is referred to the anterior surface of the chest or upper abdomen from slightly below the heart to an inch or so below the xyphoid process. These pains are frequently characterized as burning pains; and they, or pains from the lower esophagus, cause the condition known as "heartburn."

Most peptic ulcers occur within 1 to 2 inches on either side of the pylorus in the stomach or in the duodenum, and pain from such ulcers is usually referred to a surface point approximately midway between the umbilicus and the xyphoid process. The origin of ulcer pain is almost undoubtedly chemical, because when the acid juices of the stomach are not allowed to reach the pain fibers in the ulcer crater the pain does not exist. This pain is characteristically intensely burning.

**Biliary and Gallbladder Pain.**   Pain from the bile ducts and gallbladder is localized in the midepigastrium almost coincident with pains caused by peptic ulcers. Also, biliary and gallbladder pain is often burning, like that from ulcers, though cramps often occur too.

Biliary disease, in addition to causing pain on the abdominal surface, frequently refers pain to a small area at the tip of the right scapula. This pain is transmitted through sympathetic afferent fibers that enter the ninth thoracic segment of the spinal cord.

**Pancreatic Pain.** Lesions of the pancreas, such as acute or chronic pancreatitis in which the pancreatic enzymes eat away the pancreas and surrounding structures, promote intense pain in areas both anterior to and behind the pancreas. It should be remembered that the pancreas is located beneath the parietal peritoneum and that it receives many sensory fibers from the posterior abdominal wall. Therefore, the pain is frequently localized directly behind the pancreas in the back and is severe and burning in character.

**Renal Pain.** The kidney, kidney pelvis, and ureters are all retroperitoneal structures and receive most of their pain fibers directly from skeletal nerves. Therefore, pain is usually felt directly behind the ailing structure. However, pain occasionally is referred via visceral afferents to the anterior abdominal wall below and about 2 inches to the side of the umbilicus.

Pain from the bladder is felt directly over the bladder, presumably because the bladder is well innervated by parietal pain fibers. However, pain is also frequently referred into the groin and testicles because some nerve fibers from the bladder apparently synapse in the cord in association with fibers from the genital areas.

**Uterine Pain.** Both parietal and visceral afferent pain may be transmitted from the uterus. The low abdominal cramping pains of dysmenorrhea are mediated through the sympathetic afferents, and an operation to cut the hypogastric nerves between the hypogastric plexus and the uterus will in many instances relieve this pain. On the other hand, lesions of the uterus that spread into the adnexa around the uterus, or lesions of the fallopian tubes and broad ligaments, usually cause pain in the lower back or side. This pain is conducted over parietal nerve fibers and is usually sharper in nature rather than resembling the diffuse cramping pain of true dysmenorrhea.

# SOME CLINICAL ABNORMALITIES OF PAIN AND OTHER SENSATIONS

## HYPERALGESIA

A pain pathway may become excessively excitable; this gives rise to hyperalgesia, which means hypersensitivity to pain. The basic causes of hyperalgesia are: (1) excessive sensitivity of the pain receptors themselves, which is called *primary hyperalgesia,* or facilitation of sensory transmission, which is called *secondary hyperalgesia.*

An example of primary hyperalgesia is the extreme sensitivity of sunburned skin. Secondary hyperalgesia frequently results from lesions in the spinal cord or in the thalamus. Several of these will be discussed in subsequent sections.

## THE THALAMIC SYNDROME

Occasionally the posterolateral branch of the posterior cerebral artery, a small artery supplying the posteroventral portion of the thalamus, becomes blocked by thrombosis, so that the nuclei of this area of the thalamus degenerate, while the medial and anterior nuclei of the thalamus remain intact. The patient suffers a series of abnormalities, as follows: First, loss of almost all sensations from the opposite side of the body occurs because of destruction of the relay nuclei. Second, ataxia (inability to control movements precisely) may be evident because of loss of kinesthetic impulses normally relayed through the thalamus to the cortex. Third, after a few weeks to a few months some sensory perception in the opposite side of the body returns, but strong stimuli are usually necessary to elicit this. When the sensations do occur, they are poorly—if at all—localized, almost always very painful, sometimes lancinating, regardless of the type of stimulus applied to the body. Fourth, the person is likely to perceive many affective sensations of extreme unpleasantness or, rarely, extreme pleasantness; the unpleasant ones are often associated with emotional tirades.

The medial nuclei of the thalamus are not destroyed by thrombosis of the artery. Therefore, it is believed that these nuclei become facilitated and give rise to the enhanced sensitivity to pain transmitted through the reticular system as well as to the affective sensations.

## HERPES ZOSTER

Occasionally a virus probably identical with that of chickenpox infects a dorsal root ganglion. This causes severe pain in the dermatomal segment normally subserved by the ganglion, thus eliciting a segmental type of pain that circles around the body. This pain is commonly called "shingles."

There are two possible causes of the pain of herpes zoster. One of these is that the disease destroys mainly the large mechanoreceptor afferent sensory fibers. This theoretically could reduce the normal inhibitory effect of these fibers on the pain pathway in the substantia gelatinosa of the dorsal horn. Therefore, pain signals could become exacerbated in the absence of this inhibition.

A second possibility is that the irritated neuronal cell bodies of the root ganglion are stimulated to excessive activity and thereby cause the pain.

## TIC DOULOUREUX

Lancinating pains occur in some persons over one side of the face in part of the sensory distribution area of the fifth or ninth nerves; this phenomenon is called tic douloureux. The pains feel like sudden electric shocks, and they may appear for only a few seconds at a time or they may be almost continuous. Often, they are set off by exceedingly sensitive "trigger areas" on the surface of the head, in the mouth, or in the nose—almost always by a mechanoreceptive stimulus instead of a pain stimulus. For instance, when the patient swallows a bolus of food, as the food touches a tonsil it might set off a severe lancinating pain in the mandibular portion of the fifth nerve.

The pain of tic douloureux can usually be blocked by cutting the peripheral nerve from the hypersensitive area. The fifth nerve is often sectioned immediately inside the cranium, where the motor and sensory roots of the fifth nerve can be separated so that the motor portions, which are needed for many of the jaw movements, are spared while the sensory elements are destroyed. Obviously, this operation leaves the side of the face anesthetic, which in itself may be annoying. Furthermore, it is sometimes unsuccessful, indicating that the lesion might be more central than the nerves themselves. Therefore, still another operation is often performed in the medulla for relief of this pain, as follows:

The sensory nuclei of the fifth nerve divide into the main sensory nucleus, which is located in the pons, and the spinal sensory nucleus, which descends into the upper part of the spinal cord. It is the spinal nucleus that subserves the function of pain. Therefore, the spinal tract of the fifth nerve can be cut in the medulla as it passes to the spinal nucleus. This blocks pain sensations from the side of the face but does not block the sensations of touch and pressure.

## THE BROWN-SÉQUARD SYNDROME

Obviously, if the spinal cord is transected entirely, all sensations and motor functions distal to the segments of transection are blocked, but if only one side of the spinal cord is transected, the socalled Brown-Séquard syndrome occurs. The following effects of such a transection occur, and these can be predicted from a knowledge of the cord fiber tracts illustrated in Figure 50–9: All motor functions are blocked on the side of the transection in all segments below the level of the transection; while only some of the modalities of sensation are lost on the transected side and others are lost on the opposite side. The sensations of pain, heat, and cold are lost on the opposite side of the body in all dermatomes two to six segments below the level of the transection. The sensations that are transmitted only in the dorsal columns—kinesthetic sensation, vibration sensation, discrete localization, and two-point discrimination—are lost entirely on the side of the transection in all dermatomes below the level of the transection. Touch is impaired on the side of the transection because the principal pathway for transmission of light touch, the dorsal columns, is transected. Yet, "crude touch," which is poorly localized, still persists be-

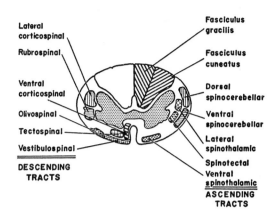

**Figure 50–9.** Cross-section of the spinal cord, showing principal ascending tracts on the right and principal descending tracts on the left.

cause of transmission in the opposite ventral spinothalamic tract.

# HEADACHE

Headaches are actually referred pain to the surface of the head from the deep structures. Many headaches result from pain stimuli arising inside the cranium, but equally as many probably result from pain arising outside the cranium.

## HEADACHE OF INTRACRANIAL ORIGIN

**Pain-Sensitive Areas in the Cranial Vault.** The brain itself is almost totally insensitive to pain. Even cutting or electrically stimulating the somesthetic centers of the cortex only occasionally causes pain; instead, it causes tactile paresthesias on the area of the body represented by the portion of the somesthetic cortex stimulated. Therefore, it is obvious from the outset that much or most of the pain of headache probably is not caused by damage within the brain itself.

On the other hand, *tugging on the venous sinuses, damaging the tentorium,* or *stretching the dura at the base of the brain* can all cause intense pain that is recognized as headache. Also, almost any type of traumatizing, crushing, or stretching stimulus to the *blood vessels of the dura* can cause headache. One of the most sensitive structures of the entire cranial vault is the middle meningeal artery, and neurosurgeons are careful to remain clear of this artery as much as possible when performing brain operations under local anesthesia.

**Areas of the Head to which Intracranial Headache is Referred.** Stimulation of pain receptors in the intracranial vault above the tentorium, including the upper surface of the tentorium itself, initiates impulses in the fifth nerve and, therefore, causes re-

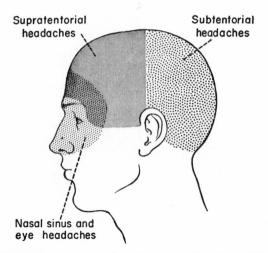

**Figure 50–10.** Areas of headache resulting from different causes.

ferred headache on the outside of the head in the area supplied by the fifth cranial nerve. This area includes the upper part of the head anterior to the ear, as outlined by the dark shaded area of Figure 50–10. Thus, pain arising above the tentorium causes what is called "frontal headache."

On the other hand, pain impulses from beneath the tentorium enter the central nervous system mainly through the second cervical nerve, which also supplies the scalp behind the ear. Therefore, subtentorial pain stimuli cause "occipital headache" referred to the posterior part of the head shown in Figure 50–10.

**Types of Intracranial Headache.** *Headache of Meningitis.* One of the most severe headaches of all is that resulting from meningitis, which causes inflammation of all the meninges, including the sensitive areas of the dura and the sensitive areas around the venous sinuses. Such intense damage as this can cause extreme headache pain referred over the entire head.

*Headache Resulting from Direct Meningeal Trauma.* Following a brain operation one ordinarily has intense headache for several days to several weeks. Though part of this headache may result from the trauma of the brain itself, experiments indicate that most of it results from meningeal irritation.

Another type of meningeal trauma that almost invariably causes headache is the meningeal irritation resulting from brain tumor. Usually, tumor headache is referred to a localized area of the head, the exact area depending on the portion of the meninges affected by the tumor. Since any tumor above the tentorium refers its pain to the frontal areas and any tumor below the tentorium refers its pain to the occipital region of the skull, the general location of an intracranial tumor can often be predicted from the area of the headache.

*Headache Caused by Low Cerebrospinal Fluid Pressure.* Removing as little as 20 ml. of fluid from the spinal canal, particularly if the person remains in the upright position, often causes intense intracranial headache. Removing this quantity of fluid removes the flotation for the brain that is normally provided by the cerebrospinal fluid. Therefore, the weight of the brain stretches the various dural surfaces and thereby elicits the pain which causes the headache.

*Migraine Headache.* Migraine headache is a special type of headache that is thought to result from abnormal vascular phenomena, though the exact mechanism is unknown.

Migraine headaches often begin with various prodromal sensations, such as nausea, loss of vision in part of the fields of vision, visual aura, or other types of sensory hallucinations. Ordinarily, the prodromal symptoms begin half an hour to an hour prior to the beginning of the headache itself. Therefore, any theory that explains migraine headache must also explain these prodromal symptoms.

One of the theories of the cause of migraine headaches is that prolonged emotion or tension causes reflex vasospasm of some of the arteries of the head, including arteries that supply the brain itself. The vasospasm theoretically produces ischemia of portions of the brain, and this is responsible for the prodromal symptoms. Then, as a result of the intense ischemia, something happens to the vascular wall to allow it to become flaccid and incapable of maintaining vascular tone for 24 to 48 hours. The blood pressure in the vessels causes them to dilate and pulsate intensely, and it is supposedly the excessive stretching of the walls of the arteries—including the extracranial arteries such as the temporal artery—that causes the actual pain of migraine headaches. However, it is possible that diffuse after-effects of ischemia in the brain itself are responsible for this type of headache.

*Alcoholic Headache.* As many people have experienced, a headache usually follows an alcoholic binge. It is most likely that alcohol, because it is toxic to tissues, directly irritates the meninges and causes the cerebral pain.

*Headache Caused by Constipation.* Constipation causes headache in many persons. Because it has been shown that constipation headache can occur in persons whose spinal cords have been cut, we know that this headache is not caused by nervous impulses from the colon. Therefore, it possibly results from absorbed toxic products or from changes in the circulatory system. Indeed, constipation sometimes does cause temporary loss of plasma into the wall of the gut, and a resulting poor flow of blood to the head could be the cause of the headache.

## EXTRACRANIAL TYPES OF HEADACHE

**Headache Resulting from Muscular Spasm.** Emotional tension often causes many of the muscles of the head, including especially those muscles attached to the scalp and the neck muscles attached to the occiput, to become moderately spastic, and it is postulated that this causes headache. The pain of the spastic head muscles supposedly is referred to the overlying areas of the head and gives one the same type of headache as do intracranial lesions.

**Headache Caused by Irritation of the Nasal and Accessory Nasal Structures.** The mucous membranes of the nose and also of all the nasal sinuses are sensitive to pain, but not intensely so. Nevertheless, infection or other irritative processes in widespread areas of the nasal structures usually cause headache that is referred behind the eyes or, in the case of frontal sinus infection, to the frontal surfaces of the forehead and scalp, as illustrated in Figure 50–10. Also, pain from the lower sinuses—such as the maxillary sinuses—can be felt in the face.

**Headache Caused by Eye Disorders.** Difficulty in focusing one's eyes clearly may cause excessive contraction of the ciliary muscles in an attempt to gain clear vision. Even though these muscles are extremely small, tonic contraction of them can be the cause of retroorbital headache. Also, excessive attempts to focus the eyes can result in reflex spasm in the various facial and extraocular muscles. Tonic spasm in these muscles is also a possible cause of headache.

A second type of headache originating in the eyes occurs when the eyes are exposed to excessive irradiation by ultraviolet light rays. Watching the sun or the arc of an arc-welder for even a few seconds may result in headache that lasts from 24 to 48 hours. The headache sometimes results from ''actinic'' irritation of the conjunctivae, and the pain is referred to the surface of the head or retroorbitally. However, focusing intense light from an arc or the sun on the retina can actually burn the retina, and this could result in headache.

# THERMAL SENSATIONS

## *THERMAL RECEPTORS AND THEIR EXCITATION*

The human being can perceive different gradations of cold and heat, progressing from *cold* to *cool* to *indifferent* to *warm* to *hot;* some persons perceive even freezing cold and burning hot.

The different thermal gradations in the body are discriminated by at least three different types of sensory end-organs, the *cold receptors,* the *warm receptors,* and the *pain receptors.* The pain receptors are stimulated only by extreme degrees of heat or cold and therefore will not be considered as thermal receptors but merely as pain receptors that become stimulated when either heat or cold becomes sufficiently severe to cause damage.

Both heat and cold can be perceived from areas of the body that contain only free nerve endings, such as the cornea. Therefore, it is known that free nerve endings can subserve these sensory functions. Thus far, no complex specialized receptors for either cold or warmth have been discovered.

**Stimulation of Thermal Receptors— Sensations of Cold, Warmth, and Hot.** Figure 50–11 illustrates the responses to different temperatures of three different types of nerve fibers: (1) a cold fiber, (2) a warm fiber, and (3) a pain fiber. Note especially that these fibers respond differently at different levels of temperature. For instance, in the *very* cold region only the pain fibers are stimulated (if the skin becomes even colder so that it nearly freezes or actually does freeze, even these fibers cannot be stimulated). As the temperature rises to 10 to 15° C., pain impulses cease, but the cold receptors begin to be stimulated. Then, above about 25° C. the warm end-organs become stimulated while the cold end-organs fade out at about 35° C. Finally, at around 45° C. the warm endings become nonresponsive once more; but the cold endings begin to respond again, and the pain endings also begin to be stimulated.

One can understand from Figure 50–11, therefore, that a person determines the different gradations of thermal sensations by the relative degrees of stimulation of the different types of endings. For instance, at 20° C. only cold endings are stimulated, whereas at 40° C. only warm endings are stimulated; at 33° C. both cold and warm endings are stimulated, and at 50° C. both cold and pain endings are stimulated. One can understand also from this figure why extreme degrees of cold or heat can be painful and why both these sensations, when intense enough, may give almost exactly the same quality of sensations—freezing cold and burning hot sensations feel almost alike to the person.

It is particularly interesting that a few areas of

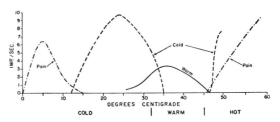

**Figure 50–11.** Frequencies of discharge of a cold receptor, a warm receptor, and a pain nerve fiber at different temperatures. (The responses of the cold and warm receptors are modified from Zotterman: *Ann. Rev. Physiol.,* 15:357, 1953.)

the body, such as the tip of the penis, do not contain any warmth receptors; but these areas can experience the sensations of cold with ease and of "hot," which depends on stimulating cold and pain endings, but never the sensation of warmth.

**Stimulatory Effects of Rising and Falling Temperature—Adaptation of Thermal Receptors.** When a cold receptor is suddenly subjected to an abrupt fall in temperature, it becomes strongly stimulated at first, but this stimulation fades rapidly during the first minute and progressively more slowly during the next half hour or more. In other words, the receptor adapts to a great extent; this is illustrated in Figure 50–12, which shows that the frequency of discharge of a cold receptor rose approximately four-fold when the temperature fell suddenly from 32° to 30° C., but in less than a minute the frequency fell about five sixths of the way back to the original control value. Later, the temperature was suddenly raised from 30° to 32° C. At this point the cold receptor stopped firing entirely for a short time, but after adaptation returned to its original control level. The warm receptors, on the other hand, have not yet been shown to adapt, though psychological experience such as the entering of a hot tub of water indicates that the warmth signals do in some way adapt.

Thus, it is evident that the thermal senses respond markedly to *changes in temperature* in addition to being able to respond to steady states of temperature, as was depicted in Figure 50–11. This means, therefore, that when the temperature of the skin is actively falling, a person feels much colder than when the temperature remains at the same level. Conversely, if the temperature is actively rising the person feels much warmer than he would at the same temperature if it were constant.

The response to changes in temperature explains the extreme degree of heat that one feels when he first enters a hot tub of water and the extreme degree of cold he feels when he goes from a heated room to the out-of-doors on a cold day. The adaptation of the thermal senses explains the ability of the person to become gradually accustomed to his new temperature environmant.

**Mechanism of Stimulation of the Thermal Receptors.** It is believed that the thermal receptors are stimulated by changes in their metabolic rates, these changes resulting from the fact that temperature alters the rates of intracellular chemical reactions about 2.3 times for each 10° C. change. In other words, thermal detection probably results not from direct physical stimulation but instead from chemical stimulation of the endings as modified by the temperature.

**Spatial Summation of Thermal Sensations.** The number of cold or warmth endings in any small surface area of the body is very slight, so that it is difficult to judge gradations of temperature when small areas are stimulated. However, when a large area of the body is stimulated, the thermal signals from the entire area summate. Indeed, one reaches his maximum ability to discern minute temperature variations when his entire body is subjected to a temperature change all at once. For instance, rapid changes in temperature as little as 0.01° C. can be detected if this change affects the entire surface of the body simultaneously. On the other hand, temperature changes 100 times this great might not be detected when the skin surface affected is only a square centimeter or so in size.

## TRANSMISSION OF THERMAL SIGNALS IN THE NERVOUS SYSTEM

Thermal signals are transmitted by small myelinated fibers of the delta group with diameters of only 2 to 5 microns. It is also possible that some very crude thermal sensations, such as "burning hot" or "freezing cold," but with very poor differentiation of temperature gradations, are transmitted by type C fibers.

On entering the central nervous system, the thermal fibers travel in the lateral spinothalamic tract. The exact point in the thalamus at which the impulses synapse is not known, but since other delta spinothalamic fibers terminate

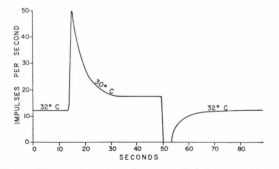

**Figure 50–12.** Response of a nerve fiber from a cold receptor following, first, instantaneous change in skin temperature from 32° to 30° C. and, second, instantaneous change back to 32° C. Note the adaptation of the receptor and also the higher steady state level of discharge at 30° than at 32°.

mainly in the caudalmost portion of the ventrobasal complex of the thalamus, it is believed that the thermal tracts do too. Then, third order neurons travel to the somesthetic cortex, but the portion of the cerebral cortex to which the major share of thermal fibers radiate is also unknown.

Occasionally, a neuron in somatic sensory area I has been found by microelectrode studies to be directly responsive to either cold or warm stimuli in specific areas of the skin. Furthermore, it is known that removal of the postcentral gyrus in the human being reduces his ability to distinguish different gradations of temperature. Therefore, it is possible that somatic sensory area I, rather than area II, is the major somesthetic area concerned with thermal sensations.

# REFERENCES

Bloedel, J. R.: The substrate for integration in the central pain pathways. *Clin. Neurosurg., 21*:194, 1974.

Bonica, J. J. (ed.): International Symposium on Pain. New York, Raven Press, 1974.

Bonica, J. J., Procacci, P., and Pagni, C. A.: Recent Advances in Pain. Springfield, Ill., Charles C Thomas, Publisher, 1974.

Casey, K. L.: Pain: a current view of neural mechanisms. *Amer. Sci., 61*:194, 1973.

Ciba Foundation Symposium: Touch, Heat and Pain. Boston, Little, Brown and Company, 1965.

Crue, B. L.: Pain: Research and Treatment. Further observations from City of Hope National Medical Center. New York, Academic Press, Inc., 1975.

Dalessio, D. J.: Wolff's Headache and Other Head Pain, 3rd Ed. New York, Oxford University Press, 1972.

Guyton, A. C., and Reeder, R. C.: Pain and contracture in poliomyelitis. *Arch. Neurol. Psychiatr., 63*:954, 1950.

Hannington-Kiff, J. G.: Pain Relief. Philadelphia, J. B. Lippincott Company, 1974.

Hardy, J. D.: The nature of pain. *J. Chronic Dis., 4*:22, 1956.

Hardy, J. D., Wolff, H. C., and Goodell, H.: Pain Sensations and Reactions. Baltimore, The Williams & Wilkins Company, 1952.

Johnson, R. H., and Spaulding, J. M.: Disorders of the autonomic nervous system. Chap 16: Pain. *Contemp. Neurol. Ser., 11*:279, 1974.

Li, C. L.: Neurological basis of pain and its possible relationship to acupuncture-analgesia. *Amer. J. Chin. Med., 1*:61, 1973.

Lim, R. K. S.: Pain. *Ann. Rev. Physiol., 32*:269, 1970.

McMasters, R. E.: A clinical approach to pain. *South. Med. J., 67*:173, 1974.

Merskey, H.: Assessment of pain. *Physiotherapy, 60*:96, 1974.

Nashold, B. S., Jr.: Central pain: its origins and treatment. *Clin. Neurosurg., 21*:311, 1974.

Newman, P. O.: Visceral Afferent Functions of the Nervous System. Monograph of the Physiological Society, No. 25. Baltimore, The Williams & Wilkins Company, 1974.

Payne, J. P., and Burt, R. A. P. (eds.): Pain. New York, Churchill Livingstone, Div. of Longman, Inc., 1972.

Perl, E. R., and Boivie, J. G.: Neural substrates of somatic sensation. *In* Guyton, A. C. (ed.): MTP International Review of Science: Physiology. Vol. 3. Baltimore, University Park Press, 1974, p. 303.

Shepherd, J. A.: Concise Surgery of the Acute Abdomen. New York, Churchill Livingstone, Div. of Longman, Inc., 1974.

Swerdlow, M.: Relief of Intractable Pain. New York, American Elsevier Publishing Co., Inc., 1974.

Way, E. L.: New Concepts in Pain and Its Clinical Management. Philadelphia, F. A. Davis Company, 1967.

Wilkins, R. H.: Neurosurgical relief of pain: recent developments. *Tex. Med., 70*:53, 1974.

Wilson, M. E.: The neurological mechanisms of pain. A review. *Anaesthesia, 29*:407, 1974.

Zotterman, Y.: Thermal sensations. *In* Magoun, H. W. (ed.): Handbook of Physiology. Sec. 1, Vol. 1. Baltimore, The Williams & Wilkins Co., 1959, p. 431.

*See also bibliographies at the ends of Chapters 47, 48, and 49.*

# 51

# Motor Functions of the Spinal Cord and the Cord Reflexes

In the discussion of the nervous system thus far, we have considered principally the input of sensory information. In the following chapters we will discuss the origin and output of motor signals, the signals that cause muscle contraction and other motor effects throughout the body. Sensory information is integrated at all levels of the nervous system and causes appropriate motor responses, beginning in the spinal cord with relatively simple reflexes, extending into the brain stem with still more complicated responses, and, finally, extending to the cerebrum where the most complicated responses are controlled. The present chapter discusses the control of motor function especially in response to spinal cord reflexes.

**Experimental Preparations for Studying Cord Reflexes.** Normally the functions of the spinal cord are strongly controlled by signals from the brain. Therefore, to study the isolated cord reflexes it is necessary to separate the cord from the higher centers. This is usually done in one of two different types of preparations: (1) the *spinal animal* or (2) the *decerebrate animal*.

The spinal animal is prepared by cutting the spinal cord at any level above the region in which the cord reflexes are to be studied. For a variable period of time, depending upon the phylogenetic level of the animal, as will be discussed later in the chapter, the cord reflexes are deeply depressed immediately after removal of the signals from the brain. However, over a period of hours or days in animals and over a period of weeks or months in the human being, the cord reflexes become progressively more active and can then be studied independently of control by the upper levels of the nervous system. Sometimes the cord reflexes become even more excitable under these conditions than normally.

In the decerebrate preparation, the brain stem is usually transected between the superior and inferior colliculi. Transecting the brain at this level removes the voluntary control centers of the forebrain, and it also removes inhibitory influences from the basal ganglia that normally suppress the activities of the lower brain stem and spinal cord; this will be discussed in detail in the following chapter. Removal of this inhibition causes an immediate increase in the activity of many of the cord reflexes, especially the extensor reflexes that help the animal hold itself up against gravity. This preparation has an advantage over the spinal animal in that many of the cord reflexes are exacerbated even in the acute preparation and, therefore, can be studied without the long and tedious delay required to prepare the spinal animal.

## ORGANIZATION OF THE SPINAL CORD FOR MOTOR FUNCTIONS

The cord gray matter is the integrative area for the cord reflexes and other motor functions. Figure 51–1 shows the typical organization of the cord gray matter in a single cord segment. Sensory signals enter the cord through the sensory roots. After entering the cord, every sensory signal travels to two separate destinations. First, either in the same segment of the cord or in nearby segments, the sensory nerve or its collaterals terminate in the gray matter of the cord and elicit local segmental responses—local excitatory effects, facilitatory effects, reflexes, and so forth. Second, the signals travel to higher levels of the nervous system—to higher levels in the cord itself, to the brain stem, or even to the cerebral cortex. It is these sensory

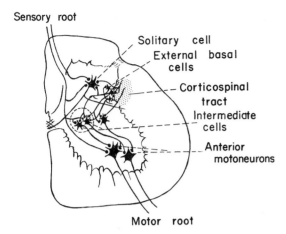

**Figure 51–1.** Connections of the sensory fibers and corticospinal fibers with the interneurons ("intermediate cells") and anterior motoneurons of the spinal cord. (Modified from Lloyd: *J. Neurophysiol.*, 4:525, 1941.)

signals that cause the conscious sensory experiences described in the past few chapters.

Each segment of the spinal cord has several hundred thousand neurons in its gray matter. Aside from the sensory relay neurons already discussed, these neurons are divided into two separate types, the *anterior motoneurons* and the *interneurons* (also called *internuncial cells* or *intermediate cells*).

**The Anterior Motoneurons.** Located in each segment of the anterior horns of the cord gray matter are several thousand neurons, 50 to 100 per cent larger than most of the others and called anterior motoneurons. These give rise to the nerve fibers that leave the cord via the anterior roots and then proceed to the muscles to innervate the skeletal muscle fibers. They can be divided into two major types, the *alpha motoneurons* and the *gamma motoneurons*.

*The Alpha Motoneurons.* The alpha motoneurons give rise to large type A alpha nerve fibers ranging from 9 to 16 microns in diameter and passing through the spinal nerves to innervate the skeletal muscle fibers. Stimulation of a single nerve fiber excites from 3 to 2000 skeletal muscle fibers (averaging about 180 fibers), which are collectively called the *motor unit*. Thus, it is the alpha motoneurons that control the contractile functions of the skeletal muscle fibers. Transmission of nerve impulses into skeletal muscles and their stimulation of the muscle fibers was discussed in Chapters 10 through 12.

*The Gamma Motoneurons.* In addition to the alpha motoneurons that excite contraction of the skeletal muscle fibers, about one-half as many much smaller motoneurons, located along with the alpha motoneurons also in the anterior horns, transmit impulses through type A gamma fibers, averaging 5 microns in diameter, to special skeletal muscle fibers called *intrafusal fibers*. These are part of the *muscle spindle*, which is discussed at length later in the chapter.

**The Interneurons.** The interneurons are present in all areas of the cord gray matter—in the dorsal horns, spread diffusely in the anterior horns, and also in the intermediate areas between these two. These cells are numerous—approximately 30 times as numerous as the anterior motoneurons. They are small and highly excitable, often exhibiting spontaneous activity and capable of firing as rapidly as 1500 times per second. They have many interconnections one with the other, and many of them directly innervate the anterior motoneurons as illustrated in Figure 51–1. The interconnections among the interneurons and anterior motoneurons are responsible for many of the integrative functions of the spinal cord that are discussed during the remainder of this chapter.

Essentially all the different types of neuronal circuits described in Chapter 47 are found in the interneuron pool of cells of the spinal cord, including the *diverging, converging,* and *repetitive-discharge* circuits. In the following sections of this chapter we will see the many applications of these different circuits to the performance of specific reflex acts by the spinal cord.

Only a few incoming sensory signals to the spinal cord or signals from the brain terminate directly on the anterior motoneurons. Instead, most of them are transmitted first through interneurons where they are appropriately processed before stimulating the anterior motoneurons. Thus, in Figure 51–1, it is shown that the corticospinal tract terminates almost entirely on interneurons, and it is through these that the cerebrum transmits most of its signals for control of muscular function.

**The Renshaw Cell Inhibitory System.** Located also in the ventral horns of the spinal cord in close association with the motoneurons are a large number of small interneurons called *Renshaw cells*. Almost immediately after the motor axon leaves a motoneuron, collateral branches pass to the Renshaw cells. These in turn are inhibitory cells that transmit inhibitory signals to the nearby motoneurons. Thus, stimulation of each motoneuron tends to inhibit the surrounding motoneurons. This effect is important for two major reasons:

1. It shows that the motor system utilizes the principle of lateral inhibition to focus, or sharpen, its

signals—that is, to allow unabated transmission of the primary signal while suppressing the tendency for signals to spread to adjacent neurons.

2. The peripheral axon of the anterior motoneuron is known to secrete acetylcholine at its nerve endings. Therefore, the ability of collateral nerve endings from the same axon to stimulate the Renshaw cells proves that acetylcholine is one of the central nervous system excitatory transmitters, acting in the same way that it acts as an excitatory transmitter at the neuromuscular junction.

### Sensory Input to the Motoneurons and Interneurons

Most of the sensory fibrils entering each segment of the spinal cord terminate on interneurons, but a very small number of large sensory fibers from the muscle spindles terminate on the anterior motoneurons. Thus, there are two pathways in the spinal cord that cord reflexes can take: either directly to the anterior motoneuron itself, utilizing a monosynaptic pathway; or through one or more interneurons first, before passing to the anterior motoneuron. The monosynaptic pathway provides an extremely rapid reflex feedback system and is the basis of the very important muscle *stretch reflex* that will be discussed later in the chapter. All other cord reflexes utilize the interneuron pathway, a pathway that can modify the signals tremendously and that can cause complex reflex patterns. For instance, the very important protective reflex called the *withdrawal reflex* utilizes this pathway.

**Multisegmental Connections in the Spinal Cord—the Propriospinal Fibers.** About half of all the nerve fibers ascending and descending in the spinal cord are *propriospinal fibers*. These are fibers that run from one segment of the cord to another. In addition, the terminal fibrils of sensory fibers as they enter the cord branch both up and down the spinal cord, some of the branches transmitting signals only a segment or two in each direction, while others transmit signals many segments. These ascending and descending fibers of the cord provide pathways for the multisegmental reflexes that will be described later in this chapter, including many reflexes that coordinate movements in both the forelimbs and hindlimbs simultaneously.

# ROLE OF THE MUSCLE SPINDLE IN MOTOR CONTROL

Muscles and tendons have an abundance of two special types of receptors: (1) *muscle spin-*dles* that detect (a) change in length of muscle fibers and (b) rate of this change in length, and (2) *Golgi tendon organs* that detect the tension applied to the muscle tendon during muscle contraction or muscle stretch.

The signals from these two receptors operate entirely at a subconscious level, causing no sensory perception at all. But they do transmit tremendous amounts of information into the spinal cord and also to the cerebellum, thereby helping these two portions of the nervous system to perform their functions for controlling muscle contraction.

## RECEPTOR FUNCTION OF THE MUSCLE SPINDLE

**Structure and Innervation of the Muscle Spindle.** The physiologic organization of the muscle spindle is illustrated in Figure 51–2. Each spindle is built around three to ten small *intrafusal muscle fibers* that are pointed at their ends and are attached to the sheaths of the surrounding *extrafusal* skeletal muscle fibers. The intrafusal fiber is a very small skeletal muscle fiber. However, the central region of each of these fibers has either no or few actin and myosin filaments. Therefore, this central portion does not contract when the ends do. The ends in turn are excited by small nerve fibers called *gamma efferent* motor nerve fibers. These average five microns in diameter and are much smaller than the sensory nerve fibers that also innervate the muscle spindle.

When the ends of the intrafusal fibers contract, this action stretches the central portion. Likewise, stretch of the entire muscle, which pulls on the ends of the intrafusal fibers where

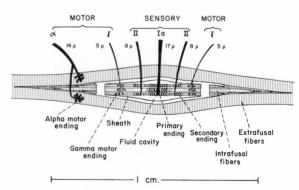

**Figure 51–2.** The muscle spindle, showing its relationship to the large extrafusal skeletal muscle fibers. Note also both the motor and the sensory innervation of the muscle spindle.

they attach to the muscle sheaths, also stretches the central portion of each intrafusal fiber. This central portion of the muscle spindle is the sensory receptor area of the muscle spindle.

The receptor portion of the muscle spindle is innervated by two different types of sensory nerves: one type Ia fiber and usually two type II fibers. The type Ia fiber averages 17 microns in diameter and transmits sensory signals from the muscle spindle to the spinal cord at a velocity approaching 100 meters per second, as rapidly as any sensory nerve fibers in the entire body. The type II nerve fibers average 8 microns in diameter and transmit signals to the central nervous system at a velocity of approximately 30 to 40 meters per second.

**The Primary Ending.** The type Ia fiber innervates the central portion of the spindle receptor. The tip of this fiber spirals around the intrafusal fibers, forming the so-called *primary ending,* also called the *annulospiral ending.*

**The Secondary Endings.** The two type II nerve fibers innervate the receptor region of the intrafusal fibers on either side of the primary ending. These fibers, like the Ia fiber, also spiral around the intrafusal fibers, and when the central portion of the intrafusal fibers is stretched, these nerve fibers are stimulated. These sensory endings are called *secondary endings.*

**Division of the Intrafusal Fibers into Nuclear Bag and Nuclear Chain Fibers—Dynamic and Static Responses of the Muscle Spindle.** There are also two different types of intrafusal fibers: (1) *nuclear bag fibers* (one to three), in which a large number of nuclei are congregated into a bag in the central portion of the receptor area, and (2) *nuclear chain fibers* (three to seven), which are about half as large in diameter and half as long as the nuclear bag fibers and have nuclei spread in a chain throughout the receptor area. The primary ending innervates both the nuclear bag intrafusal fibers *and* the nuclear chain fibers. On the other hand, the secondary endings are located almost entirely on the nuclear chain fibers. These relationships are illustrated in Figure 51–3.

**Static Response of the Secondary Receptors.** When the receptor portion of the muscle spindle is stretched, the number of impulses transmitted from the secondary endings increases almost directly in proportion to the degree of stretch, and the endings continue to transmit these impulses for many minutes. This effect is called the *static response* of the spindle receptor, meaning simply that the receptor transmits its signal for a prolonged period of time.

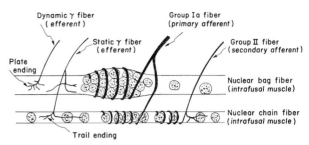

**Figure 51–3.** Details of nerve connections to the nuclear bag and nuclear chain muscle spindle fibers. (Modified from Stein: *Physiol. Rev.,* 54:225, 1974, and Boyd: *Philos. Trans. R. Soc. Lond.* [*Biol. Sci.*], 245:81, 1962.)

**Dynamic Response of the Primary Endings.** The primary ending also reacts with a static response, but it also exhibits a very strong *dynamic response,* which means that it responds extremely actively to a *change* in length. When the length of a spindle receptor area increases only a fraction of a micron, the primary receptor transmits tremendous numbers of impulses into the Ia fiber, but only *while the length is actually increasing.* As soon as the length has stopped increasing, the rate of impulse discharge returns almost back to its original level, except for the small static response that is still present in the signal.

Conversely, when the spindle receptor area shortens, this change momentarily decreases the impulse output from the primary ending; as soon as the receptor area has reached its new shortened length, the impulses reappear in the Ia fiber within a fraction of a second. Thus, the primary ending sends extremely strong signals to the central nervous system to apprise it of any change in length of the spindle receptor area.

Since both the primary and the secondary endings innervate the nuclear chain intrafusal fibers, it is assumed that it is these nuclear chain fibers that are responsible for the static response of both the primary and the secondary endings. On the other hand, only the primary endings innervate the nuclear bag fibers. Therefore, it is assumed that the nuclear bag fibers are responsible for the powerful dynamic response.

**Control of the Static and Dynamic Responses by the Gamma Efferent Nerves.** Some physiologists believe that the gamma efferent nerves to the muscle spindle can be divided into two different types: gamma-dynamic (gamma-d) and gamma-static (gamma-s). The first of these excites mainly the nuclear bag intrafusal fibers and the second excites mainly the nuclear chain intrafusal fibers. When the gamma-d fibers excite the nuclear bag fibers, the dynamic response of the muscle spindle becomes tremendously enhanced, whereas the static response remains very weak or even none at all. On the other hand, stimulation of the gamma-s fibers, which excite mainly the nuclear chain fibers, supposedly enhances the static response while having little influence on the dynamic response. We shall see in subsequent paragraphs that these two different types of responses of the muscle spindle are exceedingly important in different types of muscle control.

### Function of the Muscle Spindle in Comparing Intrafusal and Extrafusal Muscle Lengths

From the foregoing description of the muscle spindle, one can see that there are two different ways in which the spindle can be stimulated:

1. By stretching the whole muscle. This lengthens the extrafusal fibers and therefore also stretches the spindle.

2. By contracting the intrafusal muscle fibers while the extrafusal fibers remain at their normal length. Since the intrafusal fibers contract only near their two ends, this stretches the central receptor portions of the intrafusal fibers, obviously exciting the spindles.

Therefore, in effect, the muscle spindle acts as a *comparator* of the lengths of the two types of muscle fibers, the extrafusal and the intrafusal. When the length of the extrafusal fibers is greater than that of the intrafusal fibers, the spindle becomes excited. On the other hand, when the length of the extrafusal fiber is shorter than that of the intrafusal fiber, the spindle becomes inhibited.

**Continuous Discharge of the Muscle Spindles Under Normal Conditions.** Normally, particularly when there is a slight amount of gamma efferent excitation, the muscle spindles emit sensory nerve impulses all of the time. Stretching the muscle spindles increases the rate of firing, whereas shortening the spindle decreases this rate of firing. Thus, the spindles can be either excited or inhibited.

### THE MUSCLE SPINDLE REFLEX (ALSO CALLED STRETCH REFLEX OR MYOTATIC REFLEX)

From the above description of all the intricacies of the muscle spindle, one can readily see that the spindle is a very complex organ. Its function is manifested in the form of the muscle spindle reflex which, because of the static and the dynamic types of muscle spindle response, consists of a *static reflex* and a *dynamic reflex*.

**Neuronal Circuitry of the Muscle Spindle Reflex.** Figure 51–4 illustrates the basic circuit of the dynamic muscle spindle reflex, showing a type Ia nerve fiber originating in a muscle spindle and entering the dorsal root of the spinal cord. It then synapses directly with an anterior motoneuron whose motor nerve fiber transmits an appropriate reflex signal, with almost negligible delay, back to the same muscle containing the muscle spindle.

In addition to the direct synapse of the Ia muscle spindle nerve fiber with the anterior motoneuron, collaterals from this fiber, as well as the type II fibers from the secondary spindle endings, terminate on interneurons in the gray matter of the cord, and these in turn transmit more delayed signals to the anterior motoneuron.

**The Stretch Reflex.** The muscle spindle reflex is frequently called simply a *stretch reflex* because stretch of the muscle excites the muscle spindle and this in turn causes a muscle reflex contraction. This reflex has a dynamic component and a static component.

**The Dynamic Stretch Reflex.** The dynamic stretch reflex is caused by the potent dynamic signal believed to originate from the nuclear bag receptors and transmitted via the primary endings of the muscle spindles. That is, when the muscle fiber is suddenly stretched, a strong signal is transmitted to the spinal cord through the primary endings, but this signal is potent *only while the degree of stretch is increasing.* On entering the spinal cord, most of the signal goes directly to the anterior motoneurons without passing through interneurons, and it causes

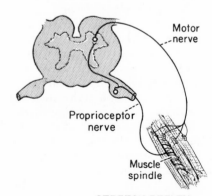

STRETCH REFLEX

**Figure 51–4.** Neuronal circuit of the stretch reflex.

reflex contraction of the same muscle from which the muscle spindle signals originated. Thus, a sudden stretch of a muscle causes reflex contraction of the same muscle, and this returns the length of the muscle back toward its original length.

**The Static Stretch Reflex.**    Though the dynamic stretch reflex is over within a fraction of a second after the muscle has been stretched to its new length, a much weaker static stretch reflex continues for a prolonged period of time thereafter. This reflex is elicited by continuous static receptor signals transmitted through both the primary and secondary endings of the muscle spindles and probably originating in the nuclear chain intrafusal fibers. The importance of the static stretch reflex is that it continues to cause muscle contraction as long as the muscle is maintained at an excessive length (for as long as several hours, but not for days). The muscle contraction in turn opposes the force that is causing the excess length.

**The Negative Stretch Reflex.**    When a muscle is suddenly shortened, exactly opposite effects occur. If the muscle is already taut, any sudden release of the load on the muscle that allows it to shorten will elicit both dynamic and static inhibition rather than reflex excitation. Thus, *this negative stretch reflex* opposes the shortening of the muscle in the same way that the positive stretch reflex opposes lengthening of the muscle. Therefore, one can begin to see that the muscle spindle reflex tends to maintain the status quo for the length of a muscle.

## FUNCTION OF THE STATIC MUSCLE SPINDLE REFLEX TO NULLIFY CHANGES IN LOAD DURING MUSCLE CONTRACTION

Let us assume that a person's biceps is contracted so that the forearm is horizontal to the earth. Then assume that a five-pound weight is put in the hand. The hand will immediately drop. However, the amount that the hand will drop is determined to a great extent by the degree of activity of the static muscle spindle reflex. If the gamma-s fibers to the muscle spindles are strongly stimulated so that the static reflex is very active, even slight lengthening of the biceps, and therefore also of the muscle spindles in the biceps, will cause a strong feedback contraction of the extrafusal skeletal muscle fibers of the biceps. This contraction in turn will limit the degree of fall of the hand, thus

automatically maintaining the forearm in a nearly horizontal position. This response is called a *load reflex.*

**Effect of Gamma Efferent Stimulation on Increasing the Effectiveness of the Load Reflex.**    If the gamma efferent nerves to the static portion of the muscle spindle (the gamma-s fibers) are not stimulated at all, then the spindle sensitivities are greatly depressed, and the effectiveness of the load reflex is almost zero. On the other hand, if these gama efferents are strongly stimulated so that the static muscle spindle reflex is an extremely potent one, then one would expect the length of the muscle to remain almost exactly constant regardless of the change in load. In many of our muscle activities a particular part of our body often is fixed in a given position, and any attempt to move that part of the body from the position is met by instantaneous reflex resistance. This probably results to a great extent from the load reflex. Furthermore, the fact that we can change the sensitivity of this load reflex by changing the intensity of gamma-s stimulation allows us to make a particular part of the body be either flail or tightly locked in place. To give an example, when a person is threading a needle he tends to fix both of his hands into precise positions, and any extra load applied to either of the two hands will hardly change its position.

**The Damping Function of the Dynamic and Static Reflexes.**    Another extremely important function of the muscle spindle reflex—indeed, probably just as important as, if not more important than, the load reflex—is the ability of the muscle spindle reflex to prevent some types of oscillation and jerkiness of the body movements. This is a damping, or smoothing, function. An example is the following:

**Use of the Damping Mechanism in Smoothing Muscle Contraction.**    Occasionally, signals from other parts of the nervous system are transmitted to a muscle in a very unsmooth form, first increasing in intensity for a few milliseconds, then decreasing in intensity, then changing to another intensity level, and so forth. When the muscle spindle apparatus is not functioning satisfactorily, the degree of muscle contraction changes jerkily during the course of such a signal. This effect is illustrated in Figure 51–5, which shows an experiment in which a sensory nerve signal entering one side of the cord is transmitted to a motor nerve from the other side of the cord to excite a muscle. In curve A the muscle spindle reflex of the excited muscle is intact. Note that the contraction is

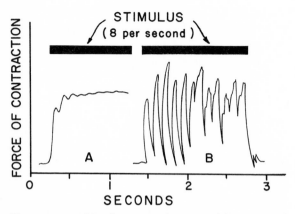

**Figure 51–5.** Muscle contraction caused by a central nervous system signal under two different conditions: (A) in a normal muscle, and (B) in a muscle whose muscle spindles had been denervated by section of the posterior roots of the cord 82 days previously. Note the smoothing effect of the muscle spindles in Part A. (Modified from Creed et al.: Reflex Activity of the Spinal Cord. Oxford University Press, 1932.)

relatively smooth even though the sensory nerve is excited at a frequency of 8 per second. Curve B, on the other hand, is the same experiment in an animal whose muscle spindle sensory nerves from the excited muscle had been sectioned three months earlier. Note the very unsmooth muscle contraction. Thus, curve A illustrates very graphically the ability of the damping mechanism of the muscle spindle to smooth muscle contractions even though the input signals to the muscle motor system may themselves be very jerky. This effect can also be called a *signal averaging* function of the muscle spindle.

*Function of the Gamma Efferent System in Controlling the Degree of Damping.* In the same way that the gamma efferent system can play a potent role in determining the effectiveness of the load reflex, so also can the gamma efferent system determine the degree of damping. For instance, there are times when a person wishes his limbs to move extremely rapidly in response to rapidly changing input signals. Under such conditions, one would wish less damping. On the other hand, there are other times when it is very important that the muscle contractions be very smooth. Under these conditions one would like a high degree of damping. Therefore, even though we are not completely sure how the nervous system can make such changes as this, psychophysiological tests do tell us that our motor system does have such capability to in-

crease or decrease the degree of damping in any given muscle response.

## ROLE OF THE MUSCLE SPINDLE IN VOLUNTARY MOTOR ACTIVITY

To emphasize the importance of the gamma efferent system, one needs to recognize that 31 per cent of all the motor nerve fibers to the muscle are gamma efferent fibers rather than large type A alpha motor fibers. Whenever signals are transmitted from the motor cortex or from any other area of the brain to the alpha motoneurons, almost invariably the gamma motoneurons are stimulated simultaneously. This causes both the extrafusal and the intrafusal muscle fibers to contract at the same time.

The purpose of contracting the muscle spindle fibers at the same time that the large skeletal muscle fibers contract is probably two-fold: First, it keeps the muscle spindle from opposing the muscle contraction. Second, it also maintains proper damping and proper load responsiveness of the muscle spindle regardless of change in muscle length. For instance, if the muscle spindle should not contract and relax along with the large muscle fibers, the receptor portion of the spindle would sometimes be flail and at other times be overstretched, in neither instance operating under optimal conditions for spindle function.

**Possible Servo Function of the Muscle Spindle Reflex.** Several physiologists have suggested that the muscle spindle reflex can also operate as a servo controller of muscle contraction in the following way: Suppose that a gamma efferent nerve fiber to a muscle spindle were stimulated. This would contract the intrafusal fibers and excite the muscle spindle receptors. The signals from these receptors would then pass to the cord and thence back to the large skeletal muscle fibers to cause them to contract also. Thus, an initial contraction of the intrafusal fibers causes a similar contraction of the surrounding large skeletal muscle fibers.

If such a servo system as this did exist, signals transmitted from the brain downward through the gamma efferent system could cause secondary contractions of the muscles in accordance with the dictates of the muscle spindles. Unfortunately, most of the data indicate that a mechanism such as this does not occur to a significant extent. For a system of this type to function properly, it would be necessary to have a static response of the muscle spindles that is very potent and also unchanging. Unfortunately, the static response of the muscle spindles is usually very weak—that is, it has a low gain. Also, it decays very slowly over a period of a few minutes to an hour so that changes in length of a muscle spindle could cause only temporary changes in length of the surrounding large skeletal muscle fibers. Therefore, despite the attractiveness of this possible type of muscle control, it is doubtful that it does actually occur in normal function.

**Probable Servo-Assist Function of the Muscle Spindle Reflex.** It is much more likely that the muscle spindle reflex acts as a servo-assist device than as a pure servo controller. But, first, let us explain what is meant by a "servo-assist mechanism."

When both the alpha and gamma motoneurons are stimulated simultaneously, if the intra- and extrafusal fibers contract equal amounts, the degree of stimulation of the muscle spindles will not change at all—either to increase or to decrease. However, in case the extrafusal muscle fibers should contract less than the intrafusal fibers (as might occur when the muscle is contracting against a great load), this mismatch will stretch the receptor portions of the spindles and, therefore, elicit a stretch reflex that will further excite the extrafusal fibers. Thus, failure of the extrafusal fibers to contract to the same degree as the intrafusal fibers contract causes an accessory neuronal signal that increases the degree of stimulation of the extrafusal fibers. This is exactly the same mechanism as that employed by power steering in an automobile. That is, the steering wheel directly turns the front wheels and will do this even when the power steering is not effective. However, if the wheels fail to follow even the slightest extra force applied to the steering wheel, a servo-assist device is activated that applies additional power to turn the wheels.

The servo-assist type of motor function has several important advantages, as follows:

1. It allows the brain to cause a muscle contraction against a load without the brain having to expend much extra nervous energy.

2. It makes the muscle contract almost the desired amount even when the load is increased or decreased between successive contractions. In other words, it makes the degree of contraction less load-insensitive.

3. It compensates for fatigue or other abnormalities of the muscle itself because any failure of the muscle to provide the proper contraction elicits an additional muscle spindle reflex stimulus to make the contraction occur.

But, unfortunately, we still do not know how important this probable function of the muscle spindle reflex actually is.

### Brain Areas for Control of the Gamma Efferent System

The gamma efferent system is excited primarily by the *bulboreticular facilitatory* region of the brain stem, and secondarily by impulses transmitted into this area from (a) the *cerebellum*, (b) the *basal ganglia*, and even (c) the *cerebral cortex*. Unfortunately, little is known about the precise mechanisms of control of the gamma efferent system. However, since the bulboreticular facilitatory area is particularly concerned with postural contractions, emphasis is given to the possible or probable important role of the gamma efferent mechanism in controlling muscle contraction for positioning the different parts of the body and for damping the movements of the different parts.

In addition to transmitting signals into the spinal cord, the muscle spindle also transmits signals up the cord to the cerebellum and thence into the bulboreticular areas, as will be discussed in greater detail in Chapter 53. However, both the signals that operate in the spinal cord and those that pass to the cerebellum are entirely subconscious, so that the conscious portion of the brain is never apprised of the immediate changes in length of the muscles. On the other hand, as has already been pointed out in Chapter 49, signals from the joint receptors constantly apprise the conscious brain of the positions of the different parts of the body, even though the muscle receptors do not.

## CLINICAL APPLICATIONS OF THE STRETCH REFLEX

**The Knee Jerk and Other Muscle Jerks.** Clinically, a method used to determine the functional integrity of the stretch reflexes is to elicit the knee jerk and other muscle jerks. The knee jerk can be elicited by simply striking the patellar tendon with a reflex hammer; this stretches the quadriceps muscle and initiates a *dynamic stretch reflex* to cause the lower leg to jerk forward. The upper part of Figure 51–6 illustrates a myogram from the quadriceps muscle recorded during a knee jerk.

Similar reflexes can be obtained from almost any muscle of the body either by striking the tendon of the muscle or by striking the belly of the muscle itself. In other words, sudden stretch of muscle spindles is all that is required to elicit a stretch reflex.

The muscle jerks are used by neurologists to assess the degree of facilitation of spinal cord centers. When large numbers of facilitatory impulses are being transmitted from the upper regions of the central nervous system into the cord, the muscle jerks are greatly exacerbated. On the other hand, if the facilitatory impulses are depressed or abrogated, the muscle jerks are considerably weakened or completely absent. These reflexes are used most frequently to determine the presence or absence of muscle spasticity following lesions in the motor areas of the brain. Ordinarily, diffuse lesions in the contralateral motor areas of the cerebral cortex cause greatly exacerbated muscle jerks.

*Clonus.* Under appropriate conditions, the muscle jerks can oscillate, a phenomenon called clonus

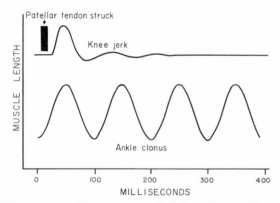

**Figure 51–6.** Myograms recorded from the quadriceps muscle during elicitation of the knee jerk and from the gastrocnemius mucle during ankle clonus.

(see lower myogram, Fig. 51–6). Oscillation can be explained particularly well in relation to ankle clonus, as follows:

If a person is standing on his tiptoes and suddenly drops his body downward to stretch one of his gastrocnemius muscles, impulses are transmitted from the muscle spindles into the spinal cord. These reflexly excite the stretched muscle, which lifts the body back up again. After a fraction of a second, the reflex contraction of the muscle dies out and the body falls again, thus stretching the spindles a second time. Again a dynamic stretch reflex lifts the body, but this too dies out after a fraction of a second, and the body falls once more to elicit still a new cycle. In this way, the stretch reflex of the gastrocnemius muscle continues to oscillate, often for long periods of time; this is clonus.

Clonus ordinarily occurs only if the stretch reflex is highly sensitized by facilitatory impulses from the cerebrum. For instance, in the decerebrate animal, in which the stretch reflexes are highly facilitated, clonus develops readily. Therefore, to determine the degree of facilitation of the spinal cord, neurologists test patients for clonus by suddenly stretching a muscle and keeping a steady stretching force applied to the muscle. If clonus occurs, the degree of facilitation is certain to be very high.

## THE TENDON REFLEX

**The Golgi Tendon Organ and Its Excitation.** Golgi tendon organs, illustrated in Figure 51–7, lie within muscle tendons immediately beyond their attachments to the muscle fibers. An average of 10 to 15 muscle fibers is usually connected in series with each Golgi tendon organ, and the organ is stimulated by the tension produced by this small bundle of muscle fibers. Thus, the major difference between the function

of the Golgi tendon apparatus and the muscle spindle is that the spindle detects relative muscle length, and the tendon organ detects muscle *tension*.

The tendon organ, like the primary receptor of the muscle spindle, responds with overexcitation to the onset of increased muscle tension, but within a small fraction of a second it settles down to a lower level of steady state firing that is almost directly proportional to the muscle tension. Thus, the Golgi tendon organs provide to the central nervous system instantaneous information of the degree of tension on each small segment of each muscle.

**Transmission of Impulses from the Tendon Organ into the Central Nervous System.** Signals from the tendon organ are transmitted through large, rapidly conducting type A alpha nerve fibers, fibers only slightly smaller than those from the primary receptors of the muscle spindle. These fibers, like those from the primary receptors, transmit signals both into local areas of the cord and through the spinocerebellar tracts into the cerebellum. The local signal is believed to excite a single inhibitory interneuron that in turn inhibits the anterior alpha motoneuron. Thus, this local circuit directly inhibits the individual muscle without affecting adjacent muscles. The signals to the cerebellum will be discussed in Chapter 53.

**Inhibitory Nature of the Tendon Reflex.** The Golgi tendon organ detects tension applied to the tendon by muscle contraction, as was discussed earlier in the chapter. Signals from the tendon organ are then transmitted into the spinal cord to cause reflex effects in the respective muscle. However, this reflex is entirely inhibitory, the exact opposite to the muscle spindle reflex. The signal from the tendon organ supposedly excites inhibitory interneurons, and these in turn inhibit the alpha motoneurons to the respective muscle.

When tension on the muscle and, therefore, on the tendon becomes extreme, the inhibitory effect from the tendon organ can be so great that it causes sudden relaxation of the entire muscle. This effect is called the *lengthening reaction;* it is probably a protective mechanism

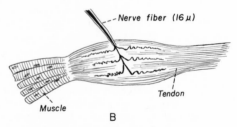

**Figure 51–7.**  Golgi tendon organ.

to prevent tearing of the muscle or avulsion of the tendon from its attachments to the bone. We know, for instance, that direct electrical stimulation of muscles in the laboratory can frequently cause such destructive effects.

However, possibly as important as this protective reaction is function of the tendon reflex as a part of the overall servo control of muscle contraction in the following manner:

**The Tendon Reflex as a Servo Control Mechanism for Muscle Tension.**  In the same way that the stretch reflex possibly operates as a feedback mechanism to control length of a muscle, the tendon reflex theoretically can operate as a servo feedback mechanism to control muscle tension. That is, if the tension on the muscle becomes too great, inhibition from the tendon organ decreases this tension back to a lower value. On the other hand, if the tension becomes too little, impulses from the tendon organ cease; and the resulting loss of inhibition allows the alpha motoneurons to become active again, thus increasing muscle tension back toward a higher level.

Very little is known at present about the function of or control of this tension servo feedback mechanism, but it is postulated to operate in the following basic manner: Signals from the brain are presumably transmitted to the cord centers to set the gain of the tendon feedback system. This can be done by changing the degree of facilitation of the neurons in the feedback loop. If the gain is high, then this system will be extremely sensitive to signals coming from the tendon organs; on the other hand, lack of excitatory signals from the brain could make the system very insensitive to the signals from the tendon organ. In this way, control signals from higher nervous centers could automatically set the level of tension at which the muscle would be maintained. If the required tension is high, then the muscle tension would be set by the servo feedback mechanism to this high level of tension. On the other hand, if the desired tension level is low, the muscle tension would automatically adjust to this low level.

This description of the tension servo system is obviously very incomplete and almost entirely theoretical at the present time, but it does depict some of the speculation on this effect. Yet, a word of caution: Despite the speculation, the measured feedback gains of the tendon-muscle feedback system have been far too little to prove that it can be useful as a tension servo controller.

**Value of a Servo Mechanism for Control of Tension.**  An obvious value of a mechanism for setting the degree of muscle tension would be to allow the different muscles to apply a desired amount of force irrespective of how far the muscles contract. An example of this would be paddling a boat, during which a person sets the amount of force that he pulls backwards on the paddle and maintains that degree of force throughout the entire movement. Were it not for some type of tension feedback mechanism, the same amount of force would not be maintained

throughout the stroke because uncontrolled muscles change their force of contraction as their lengths change. One can imagine hundreds of different patterns of muscle function that might require maintenance of constant tension rather than maintenance of constant lengths of the muscles.

# THE FLEXOR REFLEX

In the spinal or decerebrate animal, almost any type of sensory stimulus to a limb is likely to cause the flexor muscles of the limb to contract strongly, thereby withdrawing the limb from the stimulus. This is called the flexor reflex.

In its classical form the flexor reflex is elicited most frequently by stimulation of pain endings, such as by pinprick, heat, or some other painful stimulus, for which reason it is also frequently called a *nociceptive reflex*. However, even stimulation of the touch receptors can also occasionally elicit a weaker and less prolonged flexor reflex.

If some part of the body besides one of the limbs is painfully stimulated, this part, in a similar manner, will be withdrawn from the stimulus, but the reflex may not be confined entirely to flexor muscles even though it is basically the same type of reflex. Therefore, the reflex is frequently called a *withdrawal reflex*, too.

**Neuronal Mechanism of the Flexor Reflex.**  The left-hand portion of Figure 51–8 il-

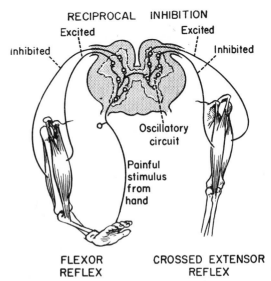

**Figure 51–8.**  The flexor reflex, the crossed extensor reflex, and reciprocal inhibition.

lustrates the neuronal pathways for the flexor reflex. In this instance, a painful stimulus is applied to the hand; as a result, the flexor muscles of the upper arm become reflexly excited, thus withdrawing the hand from the painful stimulus.

The pathways for eliciting the flexor reflex do not pass directly to the anterior motoneurons but, instead, pass first into the interneuron pool of neurons and then to the motoneurons. The shortest possible circuit is a three- or four-neuron arc; however, most of the signals of the reflex traverse many more neurons than this and involve the following basic types of circuits: (1) diverging circuits to spread the reflex to the necessary muscles for withdrawal, (2) circuits to inhibit the antagonist muscles, called *reciprocal inhibition circuits,* and (3) circuits to cause a prolonged repetitive after-discharge even after the stimulus is over.

Figure 51–9 illustrates a typical myogram from a flexor muscle during a flexor reflex. Within a few milliseconds after a pain nerve is stimulated, the flexor response appears. Then, in the next few seconds the reflex begins to *fatigue,* which is characteristic of essentially all of the more complex integrative reflexes of the spinal cord. Then, soon after the stimulus is over, the contraction of the muscle begins to return toward the base line, but, because of *after-discharge,* will not return all the way for many milliseconds. The duration of the after-discharge depends on the intensity of the sensory stimulus that had elicited the reflex; a weak stimulus causes almost no after-discharge, in contrast to an after-discharge lasting for several seconds following a strong stimulus. Furthermore, a flexor reflex initiated by nonpainful stimuli and transmitted through the large sensory fibers causes essentially no after-discharge, whereas nociceptive impulses transmitted through the small type A fibers and type C fibers cause prolonged after-discharge.

The after-discharge that occurs in the flexor reflex almost certainly results from all three types of repetitive-discharge circuits that were discussed in Chapter 47. Electrophysiological studies indicate that the immediate after-discharge, lasting for about 6 to 8 milliseconds, results from the interneuron repetitive firing mechanism and from the parallel type of circuit, with impulses being transmitted from one interneuron to another to another and all these in turn transmitting their signals successively to the anterior motoneurons. However, the prolonged after-discharge that occurs following strong pain stimuli almost certainly involves reverberating circuits in the interneurons, these transmitting impulses to the anterior motoneurons sometimes for several seconds after the incoming sensory signal is completely over.

Thus, the flexor reflex is appropriately organized to withdraw a pained or otherwise irritated part of the body away from the stimulus. Furthermore, because of the after-discharge it will hold the irritated part away from the stimulus for as long as 1 to 3 seconds even after the irritation is over. During this time, other reflexes and actions of the central nervous system can move the entire body away from the painful stimulus.

**The Pattern of Withdrawal.**   The pattern of withdrawal that results when the flexor (or withdrawal) reflex is elicited depends on the sensory nerve that is stimulated. Thus, a painful stimulus on the inside of the arm not only elicits a flexor reflex in the arm but also contracts the abductor muscles to pull the arm outward. In other words, the integrative centers of the cord cause those muscles to contract that can most effectively remove the pained part of the body from the object that causes pain. This same principle applies for any part of the body but especially to the limbs, because they have highly developed flexor reflexes.

## THE CROSSED EXTENSOR REFLEX

Approximately 0.2 to 0.5 second after a stimulus elicits a flexor reflex in one limb, the opposite limb begins to extend. This is called the *crossed extensor reflex.* Extension of the opposite limb obviously can push the entire body away from the object causing the painful stimulus.

**Neuronal Mechanism of the Crossed Extensor Reflex.**   The right-hand portion of Figure 51–8 illustrates the neuronal circuit responsible

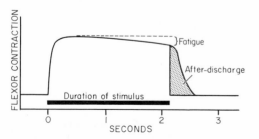

**Figure 51–9.**   Myogram of a flexor reflex, showing rapid onset of the reflex, an interval of fatigue, and finally after-discharge after the stimulus is over.

for the crossed extensor reflex, showing that signals from the sensory nerves cross to the opposite side of the cord to cause exactly opposite reactions to those that cause the flexor reflex. Because the crossed extensor reflex usually does not begin until 200 to 500 milliseconds following the initial pain stimulus, it is certain that many internuncial neurons are in the circuit between the incoming sensory neuron and the motoneurons of the opposite side of the cord responsible for the crossed extension. Furthermore, after the painful stimulus is removed, the crossed extensor reflex continues for an even longer period of after-discharge than that for the flexor reflex. Therefore, again, it is almost certain that this prolonged after-discharge results from reverberatory circuits among the internuncial cells.

Figure 51–10 illustrates a myogram recorded from a muscle involved in a crossed extensor reflex. This shows the relatively long latency before the reflex begins and also the very long after-discharge following the end of the stimulus. The prolonged after-discharge obviously would be of benefit in holding the entire body away from a painful object until other neurogenic reactions should cause the body to move away.

## RECIPROCAL INNERVATION

In the foregoing paragraphs we have pointed out several times that excitation of one group of muscles is often associated with inhibition of another group. For instance, when a stretch reflex excites one muscle, it simultaneously inhibits the antagonist muscles. This is the phenomenon of *reciprocal inhibition,* and the neuronal mechanism that causes this reciprocal relationship is called *reciprocal innervation.* Likewise, reciprocal relationships exist between the two sides of the cord as exemplified

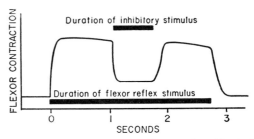

**Figure 51–11.** Myogram of a flexor reflex, illustrating crossed inhibition caused by a stronger flexor reflex in the opposite limb.

by the flexor and extensor reflexes as described above.

We will see below that the principle of reciprocal innervation is also important in most of the cord reflexes that subserve locomotion, for it helps to cause forward movement of one limb while causing backward movement of the opposite limb, and it also causes alternate movements between the forelimbs and the hindlimbs.

Figure 51–11 illustrates a typical example of reciprocal inhibition. In this instance, a moderate but prolonged flexor reflex is elicited from one limb of the body, and while this reflex is still being elicited a still stronger flexor reflex is elicited from the opposite limb, causing reciprocal inhibition of the first limb. Then removal of the strong flexor reflex allows the original reflex to reassume its previous intensity.

### FATIGUE (DECREMENT) OF REFLEXES; REBOUND

Figure 51–9 illustrated that the flexor reflex begins to *fatigue* within a few seconds after its initiation. This is a common effect in most of the cord reflexes as well as many other reflexes of the central nervous system, and it presumably results from progressive fatigue of synaptic transmission in the reflex circuits, a subject that was discussed more fully in Chapter 46.

Another effect closely allied to fatigue is *rebound.* This means that, immediately after a reflex is over, a second reflex of the same type is much more difficult to elicit for a given time thereafter. However, because of reciprocal innervation, reflexes of the antagonist muscles become even more easily elicited during that same period of time. For instance, if a flexor reflex occurs in a left limb, a second flexor reflex is more difficult to establish for a few seconds thereafter, but a crossed extensor reflex in this same limb will be greatly exacerbated. Rebound is probably one of the important

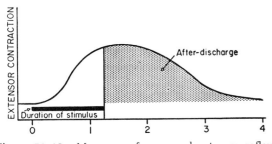

**Figure 51–10.** Myogram of a crossed extensor reflex, showing slow onset but prolonged after-discharge.

mechanisms by which the rhythmic to-and-fro movements required in locomotion are effected, which are described in more detail later in the chapter.

# THE REFLEXES OF POSTURE AND LOCOMOTION

## THE POSTURAL AND LOCOMOTIVE REFLEXES OF THE CORD

**The Positive Supportive Reaction.** Pressure on the footpad of a decerebrate animal causes the limb to extend against the pressure that is being applied to the foot. Indeed, this reflex is so strong an animal whose spinal cord has been transected for several months can often be placed on its feet, and the pressure on the footpads will reflexly stiffen the limbs sufficiently to support the weight of the body— the animal will stand in a rigid position. This reflex is called the *positive supportive reaction.*

The positive supportive reaction involves a complex circuit in the interneurons similar to those responsible for the flexor and the crossed-extensor reflexes. Furthermore, the locus of the pressure on the pad of the foot determines the position to which the limb is extended.

*The Magnet Reaction.* If pressure is applied on one side of the foot, the foot moves in that direction, or, if the pressure is applied forward, the foot moves forward, or, if it is applied backward, the foot moves backward. In this way, the bottom of the foot remains continuously extended toward the pressure that is being applied. This is called the magnet reaction, and it obviously could help in the maintenance of equilibrium, for, if an animal should tend to fall to one side, that side of the footpad would be stimulated, and the limb would automatically extend toward the falling side to push the body of the animal in the opposite direction.

**The Cord "Righting" Reflexes.** When a spinal cat or even a well-recovered young spinal dog is laid on its side, it will make incoordinate movements that indicate that it is trying to raise itself to the standing position. This is called a *cord righting reflex,* and it illustrates that relatively complicated reflexes associated with posture are at least partially integrated in the spinal cord. Indeed, a puppy with a well-healed transected thoracic cord caudal to the cord level for the forelegs can completely right itself from the lying position and can even walk on its hindlimbs. And, in the case of the opossum with a similar transection of the thoracic cord, the walking movements of the hindlimbs are hardly different from those in the normal opossum— except that the hindlimb movements are not synchronized with those of the forelimbs as is normally the case.

**The Rhythmic Stepping Reflex of a Single Limb.** Rhythmic stepping movements are frequently observed in the limbs of spinal animals. Indeed, even when the lumbar portion of the spinal cord is separated from the remainder of the cord and a longitudinal section is made down the center of the cord to block neuronal connections between the two limbs, each hind limb can still perform stepping functions. Forward flexion of the limb is followed a second or so later by backward extension. Then flexion occurs again, and the cycle is repeated over and over.

This oscillation back and forth between the flexor and extensor muscles seems to result mainly from reciprocal inhibition and rebound. That is, the forward flexion of the limb causes reciprocal inhibition of extensor muscles, but shortly thereafter the flexion begins to die out; as it does so, *rebound* inhibition of the flexors and reciprocal excitation of the extensors cause the leg to move downward and backward. After extension has continued for a time, it, too, begins to die and is followed by reciprocal excitation of the flexor muscles.

**Reciprocal Stepping of Opposite Limbs.** If the lumbar spinal cord is not sectioned down its center as noted above, every time stepping occurs in the forward direction in one limb, the opposite limb ordinarily steps backward. This effect results from reciprocal innervation between the two limbs.

**Diagonal Stepping of All Four Limbs—The "Mark Time" Reflex.** If a well healed spinal animal is held up from the table and its legs are allowed to fall downward as illustrated in Figure 51–12, the stretch on the limbs occasionally elicits stepping reflexes that involve all four limbs. In general, stepping occurs diagonally between the fore- and hindlimbs. That is, the right hindlimb and the left forelimb move backward together while the right forelimb and left hindlimb move forward. This diagonal response is another manifestation of reciprocal innervation, this time occurring the entire distance up and down the cord between the fore- and hindlimbs. Such a walking pattern is often called a *mark time reflex.*

**The Galloping Reflex.** Another type of reflex that occasionally develops in the spinal animal is the galloping reflex, in which both forelimbs move backward in unison while both

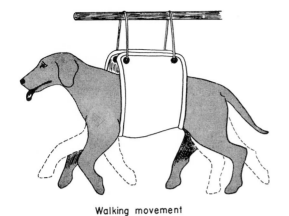

Walking movement

**Figure 51–12.** Diagonal stepping movements exhibited by a spinal animal.

hindlimbs move forward. If stretch or pressure stimuli are applied almost exactly equally to opposite limbs at the same time, a galloping reflex will likely result, whereas unequal stimulation of one side versus the other elicits the diagonal walking reflex. This is in keeping with the normal patterns of walking and of galloping, for, in walking, only one limb at a time is stimulated, and this would predispose to continued walking. Conversely, when the animal strikes the ground during galloping, the limbs on both sides are stimulated approximately equally; this obviously would predispose to further galloping and, therefore, would continue this pattern of motion in contradistinction to the walking pattern.

## THE SCRATCH REFLEX

An especially important cord reflex in lower animals is the scratch reflex, which is initiated by the *itch and tickle sensation*. It actually involves two different functions: (1) a *position sense* that allows the paw to find the exact point of irritation on the surface of the body and (2) a *to-and-fro scratching movement*.

Obviously, the scratch reflex, like the stepping movements of locomotion, involves reciprocal innervation circuits that cause oscillation. One of the important discoveries in relation to the to-and-fro movement of the scratch reflex is that it can still occur even when all the sensory roots from the oscillating limb are sectioned. In other words, feedback from the limb itself is not necessary to maintain the neuronal oscillation, which almost certainly means that the oscillation can occur intrinsically as a result of oscillating circuits within the spinal interneurons themselves.

The *position sense* of the scratch reflex is also a highly developed function, for even though a flea

might be crawling as far forward as the shoulder of a spinal animal, the hind paw can find its position. Furthermore, this can be accomplished even though 19 different muscles in the limb must be contracted simultaneously in a precise pattern to bring the paw to the exact position of the crawling flea. To make the reflex even more complicated, when the flea crosses the midline, the paw stops scratching, but the opposite paw suddenly begins the to-and-fro motion and finds the flea immediately.

## THE SPINAL CORD REFLEXES THAT CAUSE MUSCLE SPASM

In human beings, local muscle spasm is often observed. The mechanism of this has not been elucidated to complete satisfaction even in experimental animals, but it is known that pain stimuli can cause reflex spasm of local muscles, which presumably is the cause of much if not most of the muscle spasm observed at localized regions of the human body.

**Muscle Spasm Resulting from a Broken Bone.** One type of clinically important spasm occurs in muscles surrounding a broken bone. This seems to result from the pain impulses initiated from the broken edges of the bone, which cause the muscles surrounding the area to contract powerfully and tonically. Relief of the pain by injection of a local anesthetic relieves the spasm; a general anesthetic also relieves the spasm. One of these procedures is often necessary before the spasm can be overcome sufficiently for the two ends of the bone to be set back into appropriate positions.

**Abdominal Spasm in Peritonitis.** Another type of local spasm caused by a cord reflex is the abdominal spasm resulting from irritation of the parietal peritoneum by peritonitis. Here, again, relief of the pain caused by the peritonitis allows the spastic muscles to relax. Almost the same type of spasm often occurs during surgical operations; pain impulses from the parietal peritoneum cause the abdominal muscles to contract extensively and sometimes actually to extrude the gut through the surgical wound. For this reason deep surgical anesthesia is usually required for intraabdominal operations.

**Muscle Cramps.** Still another type of local spasm is the typical muscle cramp. In the past this has been ascribed to local contraction of the muscle in response to abnormal metabolic end-products in the muscle itself. However, electromyographic studies indicate that the cause of at least some muscle cramps is the following:

Any local irritating factor or metabolic abnormality of a muscle—such as severe cold, lack of blood flow to the muscle, or overexercise of the muscle—can elicit pain or other types of sensory impulses that are transmitted from the muscle to the spinal cord, thus causing reflex muscle contraction. The contraction in turn stimulates the same sensory receptors still more, which causes the spinal cord to increase the intensity of contraction still further. Thus, a positive feedback mechanism occurs so that a small amount of initial irritation causes more and more contraction until a full-blown muscle cramp ensues. Reciprocal inhibition of the muscle can sometimes completely relieve the cramp. That is, if a person purposefully contracts the muscle on the opposite side of the joint from the cramped muscle while at the same time using another hand or leg to prevent movement of the joint, the reciprocal inhibition that occurs in the cramped muscle can at times relieve the cramp immediately.

## THE AUTONOMIC REFLEXES IN THE SPINAL CORD

Many different types of segmental autonomic reflexes occur in the spinal cord, most of which are discussed in other chapters. Briefly, these include: (1) changes in vascular tone resulting from heat and cold (Chap. 72), (2) sweating, which results from localized heat at the surface of the body (Chap. 72), (3) intestino-intestinal reflexes that control some motor functions of the gut (Chap. 63), (4) peritoneo-intestinal reflexes that inhibit gastric motility in response to peritoneal irritation (Chap. 66), and (5) evacuation reflexes for emptying the bladder (Chap. 38) and the colon (Chap. 63). In addition, all the segmental reflexes can at times be elicited simultaneously in the form of the so-called mass reflex as follows:

**The Mass Reflex.** In a spinal animal or human being, the spinal cord sometimes suddenly becomes excessively active, causing massive discharge of large portions of the cord. The usual stimulus that causes this is a strong nociceptive stimulus to the skin or excessive filling of a viscus, such as overdistention of the bladder or of the gut. Regardless of the type of stimulus, the resulting reflex, called the mass reflex, involves large portions or even all of the cord, and its pattern of reaction is the same. The effects are: (1) a major portion of the body goes into strong flexor spasm, (2) the colon and bladder are likely to evacuate, (3) the arterial pressure often rises to maximal values—sometimes to a mean pressure well over 200 mm. Hg, and (4) large areas of the body break out into profuse sweating. The mass reflex might be likened to the epileptic attacks that involve the central nervous system in which large portions of the brain become massively activated.

The precise neuronal mechanism of the mass reflex is unknown. However, since its duration is prolonged, it presumably results from activation of great masses of reverberating circuits that excite large areas of the cord at once.

## SPINAL CORD TRANSECTION AND SPINAL SHOCK

When the spinal cord is suddenly transected, essentially all cord functions, including the cord re-

flexes, immediately become depressed to the point of oblivion, a reaction called *spinal shock*. The reason for this is that normal activity of the cord neurons depends to a great extent on continual tonic discharges from higher centers, particularly discharges transmitted through the vestibulospinal tract, the excitatory portions of the reticulospinal tracts, and the corticospinal tracts.

After a few days to a few months of spinal shock, the spinal neurons gradually regain their excitability. This seems to be a natural characteristic of neurons everywhere in the nervous system—that is, after they lose their source of facilitatory impulses, they increase their own natural degree of excitability to make up for the loss. In most nonprimates, the excitability of the cord centers returns essentially to normal within a few hours to a few weeks, but in man the return is often delayed for several months and occasionally is never complete; or, on the other hand, recovery is sometimes excessive, with resultant hyperexcitability of all or most cord functions.

Some of the spinal functions specifically affected during or following spinal shock are: (1) The arterial blood pressure falls immediately—sometimes to as low as 40 mm. Hg—thus illustrating that sympathetic activity becomes blocked almost to extinction. However, the pressure ordinarily returns to normal within a few days. (2) All skeletal muscle reflexes integrated in the spinal cord are completely blocked during the initial stages of shock. In lower animals, a few hours to three weeks are required for these reflexes to return to normal, and in man several months are often required. Sometimes, both in animals and man, some reflexes eventually become hyperexcitable, particularly if a few facilitatory pathways remain intact between the brain and the cord while the remainder of the spinal cord is transected. The first reflexes to return are the stretch reflex, followed in order by the progressively more complex reflexes, the flexor reflexes, the postural antigravity reflexes, and remnants of stepping reflexes. (3) The sacral reflexes for control of bladder and colon evacuation are completely suppressed in man for the first few weeks following cord transection, but they eventually return. These effects are discussed in Chapters 38 and 66.

## REFERENCES

Adams, E. B., Laurence, D. R., and Smith, J. W. G.: Tetanus. Philadelphia, J. B. Lippincott Company, 1969.
Banker, B. Q. (ed.): Research in Muscle Development and the Muscle Spindle. New York, American Elsevier Publishing Co., Inc., 1972.
Creed, R. S., Denny-Brown, D., Eccles, J. C., Liddell, E. G. T., and Sherrington, C. S.: Reflex Activity of the Spinal Cord. New York, Oxford University Press, Inc., 1932.
Easton, T. A.: On the normal use of reflexes. *Amer. Sci., 60*:591, 1972.

Gawronski, R. (ed.): Bionics: The Nervous System as a Control System. New York, American Elsevier Publishing Co., Inc., 1972.

Granit, R., Pompeiano, O., and Waltman, B.: First supraspinal control of mammalian muscle spindles: extra- and intrafusal co-activation. *J. Physiol. (Lond.)*, 147:385, 1959.

Granit, R.: Muscular tone, *J. Sport Med.*, 2:46, 1962.

Guttman, L.: Spinal Cord Injuries. Philadelphia, J. B. Lippincott Company, 1973.

Houk, J., and Henneman, E.: Responses of Golgi tendon organs to active contractions of the soleus muscle of the cat. *J. Neurophysiol.*, 30:466, 1967.

Hunt, C. C., and Perl, E. R.: Spinal reflex mechanisms concerned with skeletal muscle. *Physiol. Rev.*, 40:538, 1960.

Liddell, E. G. T.: The Discovery of Reflexes. New York, Oxford University Press, Inc., 1960.

Lippold, O.: Physiological tremor. *Sci. Amer.*, 224:65, 1971.

Lloyd, D. P.: Action in primary afferent fibers in the spinal cord. *Inter. J. Neurosci.*, 1:1, 1970.

Matthews, P. B. C.: Mammalian Muscle Receptors and Their Central Actions. Baltimore, The Williams & Wilkins Company, 1972.

Merton, P. A.: How we control the contraction of our muscles. *Sci. Amer.*, 226:30, 1972.

Porter, R.: The neurophysiology of movement performance. *In* Guyton, A. C. (ed.): MTP International Review of Science: Physiology. Vol. 3. Baltimore, University Park Press, 1974, p. 151.

Roberts, T. D. M.: Neurophysiology of Postural Mechanisms. New York, Plenum Publishing Corporation, 1967.

Sherrington, C. S.: The Integrative Action of the Nervous System. New Haven, Yale University Press, 1911.

Stein, R. B.: Peripheral control of movement. *Physiol. Rev.*, 54:215, 1974.

Wilson, D. M.: Genetic and sensory mechanisms for locomotion and orientation in animals. *Amer. Sci.*, 60:358, 1972.

*See also bibliographies at the ends of Chapters 48 and 49.*

# 52

# Motor Functions of the Brain Stem and Basal Ganglia—Reticular Formation, Vestibular Apparatus, Equilibrium, and Brain Stem Reflexes

The brain stem is a complex extension of the spinal cord. Collected in it are numerous neuronal circuits to control respiration, cardiovascular function, gastrointestinal function, eye movement, equilibrium, support of the body against gravity, and many special stereotyped movements of the body. Some of these functions—such as control of respiration and cardiovascular functions—are described in special sections of this text. The present chapter deals primarily with the control of whole body movement and equilibrium.

## THE RETICULAR FORMATION, AND SUPPORT OF THE BODY AGAINST GRAVITY

Throughout the entire extent of the brain stem—in the medulla, pons, mesencephalon, and even in portions of the diencephalon—are areas of diffuse neurons collectively known as the *reticular formation*. Figure 52–1 illustrates the extent of the reticular formation, showing it to begin at the upper end of the spinal cord and to extend (a) upward through the central portions of the thalamus, (b) into the hypothalamus, and (c) into other areas adjacent to the thalamus. The lower end of the reticular formation is continuous with the interneurons

of the spinal cord, and, indeed, the reticular formation of the brain stem functions in a manner quite analogous to many of the functions of the interneurons in the intermediolateral portions of the cord gray matter.

Interspersed throughout the reticular formation are both motor and sensory neurons; these vary in size from very small to very large. The small neurons, which constitute the greater number, have short axons that make multiple connections within the reticular formation itself. The large neurons are mainly motor in function, and their axons usually bifurcate almost immediately, with one division extending downward to the spinal cord and the other extending upward to the thalamus or other basal regions of the diencephalon or cerebrum. In general, the medial portions of the reticular formation tend to be more motor in function, whereas the more lateral portions tend to be either sensory or associative in function.

The sensory input to the reticular formation is from multiple sources, including: (1) the spinoreticular tracts and collaterals from the spinothalamic tracts, (2) the vestibular tracts, (3) the cerebellum, (4) the basal ganglia, (5) the cerebral cortex, especially the motor regions, and (6) the hypothalamus and other nearby associated areas.

Though most of the neurons in the reticular

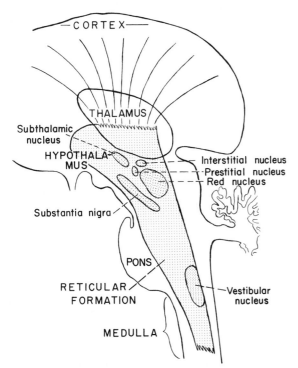

**Figure 52–1.** The reticular formation and associated nuclei.

formation are evenly dispersed, some of them are collected into *specific nuclei,* which are labeled in Figure 52–1. In general, these specific nuclei are not considered to be part of the reticular formation per se even though they do operate in association with it. In most instances they are the loci of "preprogrammed" control of stereotyped movements. As an example, the vestibular nuclei provide preprogrammed attitudinal contractions of the muscles for maintenance of equilibrium, as will be discussed later in the chapter.

## EXCITATORY FUNCTION OF THE RETICULAR FORMATION—THE BULBORETICULAR FACILITATORY AREA

By far the majority of the reticular formation is excitatory, including especially the reticular formation in the uppermost and lateral parts of the medulla and all of that in the pons, mesencephalon, and diencephalon. These areas are known collectively as the *bulboreticular facilitatory area.* Diffuse stimulation in this facilitatory area causes a general increase in muscle tone either throughout the body or in localized areas.

The bulboreticular facilitatory area and also the vestibular nuclei, which are really also part of the facilitatory system, are intrinsically excitable. That is, if they are not inhibited by signals from other parts of the nervous system, they have a natural tendency to transmit continuous nerve impulses both downward to the motor areas of the cord and upward to the cerebrum. But, in the normal animal, inhibitory signals are continually available from the basal ganglia, the cerebellum, and the cerebral cortex to keep the facilitatory system from becoming overactive. As will be noted later in the chapter in the discussion of the decerebrate animal, removal of the higher portions of the nervous system removes this inhibition and allows the reticular facilitatory area to become tonically active, which causes rigidity of the extensor, anti-gravity skeletal muscles throughout the body.

On the other hand, when the brain stem is sectioned at a level slightly lower than the vestibular nuclei, most of the motor functions of the reticular facilitatory area are then blocked, which causes almost all the musculature throughout the body to become totally flaccid.

## INHIBITORY FUNCTION OF THE LOWER RETICULAR FORMATION—THE BULBORETICULAR INHIBITORY AREA

A small portion of the reticular formation located in the ventral medial region of the lower three-fourths of the medulla has potent inhibitory functions and is known as the *bulboreticular inhibitory area.* Diffuse stimulation in this area causes decreased tone in most musculature of the body. The bulboreticular inhibitory area is not itself intrinsically excitable. However, signals arrive in this area almost all the time from the basal ganglia, the cerebral cortex, and the cerebellum to keep it excited.

The inhibitory effect of the bulboreticular inhibitory area is continually pitted against the excitatory effect of the bulboreticular facilitatory area. Consequently, when the two areas are both functioning normally, the motor functions of the spinal cord are neither excited nor inhibited. But if the excitation of the inhibitory area from the basal ganglia, cerebral cortex, and cerebellum is removed or even partially removed, then the inhibitory area becomes less active and facilitation predominates. Indeed, this is one of the usual causes of spasticity of the muscles throughout the body after a major part of the brain has been destroyed by a cerebral infarct.

## SUPPORT OF THE BODY AGAINST GRAVITY

When a person or an animal is in a standing position, continuous impulses are transmitted from the reticular formation and from closely allied nuclei, particularly from the vestibular nuclei, into the spinal cord and thence to the extensor muscles to stiffen the limbs. This allows the limbs to support the body against gravity. These impulses are transmitted mainly by way of the reticulospinal and vestibulospinal tracts.

The normal excitatory nature of the vestibular nuclei and of the bulboreticular facilitatory area provides much of the intrinsic excitation required to maintain tone in the extensor muscles. However, the degree of activity in the individual extensor muscles is determined by the equilibrium mechanisms. Thus, if an animal begins to fall to one side, the extensor muscles on that side stiffen while those on the opposite side relax. And analogous effects occur when it tends to fall forward or backward.

In essence, then, the reticular facilitatory system provides the nervous energy to support the body against gravity. But other factors, particularly the vestibular apparatuses, control the relative degree of extensor contraction in the different parts of the body, which provides the function of equilibrium.

# VESTIBULAR SENSATIONS AND THE MAINTENANCE OF EQUILIBRIUM

## THE VESTIBULAR APPARATUS

The vestibular apparatus is the sensory organ that detects sensations concerned with equilibrium. It is composed of a *bony labyrinth* containing the *membranous labyrinth,* the functional part of the apparatus. The top of Figure 52–2 illustrates the membranous labyrinth; it is composed mainly of the *cochlear duct,* the three *semicircular canals,* and the two large chambers known as the *utricle* and the *saccule.* The cochlear duct is concerned with hearing and has nothing to do with equilibrium. However, the *utricle,* the *saccule,* and the *semicircular canals* are especially important for maintaining equilibrium.

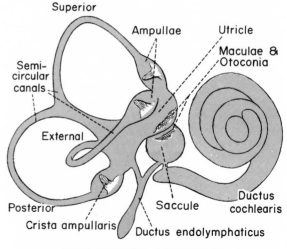

MEMBRANOUS LABYRINTH

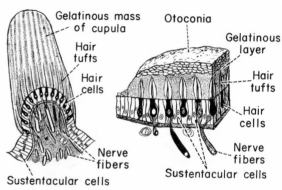

CRISTA AMPULLARIS AND MACULA

**Figure 52–2.** The membranous labyrinth, and organization of the crista ampullaris and the macula. (From Goss: Gray's Anatomy of the Human Body. Lea & Febiger; modified from Kolmer by Buchanan: Functional Neuroanatomy. Lea & Febiger.)

**The Utricle and the Saccule.** Located on the wall of both the utricle and the saccule is a small area slightly over 2 mm. in diameter called a *macula.* Each of these maculae is a sensory area for detecting the orientation of the head with respect to the direction of gravitational pull or of other acceleratory forces, as will be explained in subsequent sections of this chapter. Each macula is covered by a gelatinous layer in which many small calcium carbonate crystals called *otoconia* are imbedded. Also, in the macula are thousands of *hair cells,* which project *cilia* up into the gelatinous layer. Around the bases of the hair cells are entwined sensory axons of the vestibular nerve.

Even under resting conditions, most of the

nerve fibers surrounding the hair cells transmit a continuous series of nerve impulses. Bending the cilia of a hair cell to one side causes the impulse traffic in its nerve fibers to increase markedly; bending the cilia to the opposite side decreases the impulse traffic, often turning it off completely. Therefore, as the orientation of the head in space changes and the weight of the otoconia (whose specific gravity is about three times that of the surrounding tissues) bends the cilia, appropriate signals are transmitted to the brain to control equilibrium.

In each macula the different hair cells are oriented in different directions so that some of them are stimulated when the head bends forward, some when it bends backward, others when it bends to one side, and so forth. Therefore, a different pattern of excitation occurs in the macula for each position of the head; it is this "pattern" that apprises the brain of the head's orientation.

**The Semicircular Canals.** The three semicircular canals in each vestibular apparatus, known respectively as the *superior, posterior,* and *external* (or *horizontal*) *semicircular canals*, are arranged at right angles to each other so that they represent all three planes in space. When the head is bent forward approximately 30 degrees, the two external semicircular canals are located approximately horizontal with respect to the surface of the earth. The superior canals are then located in vertical planes that project *forward and 45 degrees outward*, and the posterior canals are also in vertical planes but project *backward and 45 degrees outward*. Thus, the superior canal on each side of the head is in a plane parallel to that of the posterior canal on the opposite side of the head, whereas the two external canals on the two sides are located in approximately the same plane.

In the *ampullae* of the semicircular canals, as illustrated in Figure 52–2, are small crests, each called a *crista ampullaris,* and on top of the crista is a gelatinous mass similar to that in the utricle and known as the *cupula.* Into the cupula are projected hairs from hair cells located along the ampullary crest, and these hair cells in turn are connected to sensory nerve fibers that pass into the *vestibular nerve.* Bending the cupula to one side, caused by flow of fluid in the canal, stimulates the hair cells, while bending in the opposite direction inhibits them. Thus, appropriate signals are sent through the vestibular nerve to apprise the central nervous system of fluid movement in the respective canal.

Recent experiments indicate that the upper end of the cupula is either attached or semi-attached to the opposite side of the ampulla and therefore might not bend as a whole unit as illustrated in Figure 52–2. Instead, the fluid accumulating on one side of the cupula simply bulges the structure in the opposite direction as if it were a diaphragm. Because of this, the hair cells in the center of the cupula will be stimulated much more powerfully at first, whereas the more lateral hair cells will be stimulated progressively more powerfully as the cupula bulges still more. This effect may account for the tremendous range of acceleration that can be detected accurately by the semicircular canal system.

**Directional Sensitivity of the Hair Cells— The Kinocilium.** Each hair cell has a large number of very small cilia plus one very large cilium called the *kinocilium.* This kinocilium is located to one side of the hair cell, always on the same side of the cell with respect to its orientation on the ampullary crest. This presumably is the cause of the directional sensitivity of the hair cells: namely, stimulation when the hairs are bent in one direction and inhibition when bent in the opposite direction.

**Neuronal Connections of the Vestibular Apparatus with the Central Nervous System.** Figure 52–3 illustrates the central connections of the vestibular nerve. Most of the vestibular nerve fibers end in the vestibular nuclei, which are located approximately at the junction of the medulla and the pons, but some fibers pass without synapsing to the fastigial nuclei, uvula, and flocculonodular lobes of the cerebellum. The fibers that end in the vestibular nuclei synapse with second order neurons that in turn send fibers into the flocculonodular lobe of the cerebellum, to the cortex of other portions of the cerebellum, into the vestibulospinal tract, into the medial

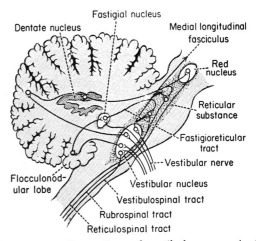

**Figure 52–3.** Connections of vestibular nerves in the central nervous system.

longitudinal fasciculus, and to other areas of the brain stem, particularly the reticular nuclei.

Note especially the very close association between the vestibular apparatus, the vestibular nuclei, and the cerebellum. The primary pathway for the reflexes of equilibrium begins in the vestibular nerves and passes next to both the vestibular nuclei and the cerebellum. Then, after much two-way traffic of impulses between these two, signals are sent into the reticular nuclei of the brain stem as well as down the spinal cord via vestibulospinal and reticulospinal tracts. In turn, the signals to the cord control the interplay between facilitation and inhibition of the extensor muscles, thus automatically controlling equilibrium.

The *flocculonodular lobes* of the cerebellum seem to be especially concerned with equilibrium functions of the semicircular canals because destruction of these lobes gives almost exactly the same clinical symptoms as destruction of the semicircular canals themselves. That is, severe injury to either of these structures causes loss of equilibrium during *rapid changes in direction of motion* but does not disturb equilibrium under static conditions, as will be discussed in subsequent sections. It is also believed that the uvula of the cerebellum plays an important role in static equilibrium.

Signals transmitted upward in the brain stem from both the vestibular nuclei and the cerebellum via the *medial longitudinal fasciculus* cause corrective movements of the eyes every time the head rotates so that the eyes can remain fixed on a specific visual object. Signals also pass upward (either through this same tract or through reticular tracts) to the cerebral cortex, probably terminating in a primary cortical center for equilibrium located near the auditory area of the superior temporal gyrus. These signals apprise the psyche of the equilibrium status of the body.

## FUNCTION OF THE UTRICLE AND THE SACCULE IN THE MAINTENANCE OF STATIC EQUILIBRIUM

It is especially important that the different hair cells be oriented in all different directions in the maculae of the utricles and saccules so that at different positions of the head, different hair cells become stimulated. The "patterns" of stimulation of the different hair cells apprise the nervous system of the position of the head with respect to the pull of gravity. In turn, the vestibular, cerebellar, and reticular motor systems reflexly excite the appropriate muscles to maintain proper equilibrium.

The maculae in the utricle and saccule are especially attuned to detecting the precise position of the head with respect to the pull of gravity when the head is in the near-vertical position. That is, a person can determine as little as ½ degree mal-equilibrium when the head leans from the precise upright position. On the other hand, as the head is leaned further and further from the upright, the determination of head orientation by the vestibular sense becomes poorer and poorer. Obviously, extreme sensitivity in the upright position is of major importance for maintenance of precise vertical static equilibrium, which is the most essential function of the vestibular apparatus.

**Detection of Linear Acceleration by the Utricle.** When the body is suddenly thrust forward, the otoconia, which have greater inertia than the surrounding fluids, fall backward on the hair cilia, and information of mal-equilibrium is sent into the nervous centers causing the individual to feel as if he were falling backward. This automatically causes him to lean his body forward until the anterior shift of the otoconia exactly equals the tendency for the otoconia to fall backward because of the linear acceleration. At this point, the nervous system detects a state of proper equilibrium and therefore shifts the body no farther forward. As long as the degree of linear acceleration remains constant and the body is maintained in this forward leaning position, the person falls neither forward nor backward. Thus, the otoconia operate to maintain equilibrium during linear acceleration in exactly the same manner as they operate in static equilibrium.

The otoconia *do not* operate for the detection of linear *motion*. When a runner first begins to run, he must lean far forward to keep from falling over backward because of acceleration, but once he has achieved running speed, he would not have to lean forward at all if he were running in a vacuum. When running in air he leans forward to maintain equilibrium only because of the air resistance against his body, and in this instance it is not the otoconia that make him lean but the pressure of the air acting on pressure end-organs in the skin, which initiate the appropriate equilibrium adjustments to prevent falling.

## THE SEMICIRCULAR CANALS AND THEIR DETECTION OF ANGULAR ACCELERATION AND ANGULAR VELOCITY

When the head suddenly *begins* to rotate in any direction, the endolymph in the membranous semicircular canals, because of its inertia, tends to remain stationary while the semicircular canals themselves turn. This causes relative fluid flow in the canals in a direction opposite to the rotation of the head.

Figure 52–4 illustrates an *ampulla* of one of the semicircular canals, showing the *cupula* and its embedded hairs bending in the direction of fluid movement. And Figure 52–5 illustrates the discharge signal from a single hair cell in the crista ampullaris when an animal is rotated for 40 seconds, showing that (1) even when the cupula is in its resting position the hair cell emits a tonic discharge of approximately 15 impulses per second; (2) when the animal is rotated, the hairs bend to one side and the rate of discharge increases greatly; and (3) with continued rotation, the excess discharge of the hair cell gradually subsides back to the resting level in about 20 seconds.

The reason for this adaptation of the receptor is that within a second or more of rotation, friction in the semicircular canal causes the endolymph to rotate as rapidly as the semicircular canal itself; then in an additional 15 to 20 seconds the cupula slowly returns to its resting position in the middle of the ampulla.

When the rotation suddenly stops, exactly the opposite effects take place: the endolymph continues to rotate while the semicircular canal stops. This time the cupula is bent in the opposite direction, causing the hair cell to stop discharging entirely. After another few seconds, the endolymph stops moving, and the cupula returns gradually to its resting position in about 20 seconds, thus allowing the discharge of the hair cell to return to its normal tonic level as shown to the right in Figure 52–5.

Thus, the semicircular canal transmits a signal when the head *begins* to rotate and a negative signal when it *stops* rotating. Furthermore,

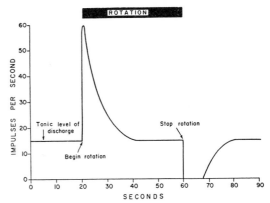

**Figure 52–5.** Response of a hair cell when a semicircular canal is stimulated first by rotation and then by stopping rotation.

at least some hair cells will always respond to rotation in any plane—horizontal, sagittal, or coronal—for fluid movement always occurs in at least one semicircular canal.

**Detection of Angular Acceleration and Angular Velocity.** Angular velocity means the rate of rotation; that is, it is the number of revolutions about an axis in a given period of time. Angular acceleration means the rate at which the angular velocity is changing.

The semicircular canals always give a person a feeling of increasing rotation when he begins to rotate or a feeling of decreasing rotation when he stops rotating—that is, they give a signal for angular acceleration. Therefore, it is commonly stated that the semicircular canals are primarily a detector of angular acceleration. However, a person also continues to feel that he is rotating for 15 to 30 seconds after he has begun to rotate. That is, he feels angular velocity during this period of time. Therefore, the semicircular canals actually detect a combination of angular acceleration and angular velocity, although the angular velocity effect may last only for 15 to 30 seconds.

To explain these effects, let us refer again to Figure 52–5. This shows that the hair cell of the semicircular canal detects neither pure angular velocity nor pure angular acceleration but, instead, detects a combination of these two. Angular acceleration occurs at the onset of the rotation, and negative angular acceleration occurs when the rotation stops. Therefore, if the signal depicted only angular acceleration, there would be a positive spike at the onset and a negative spike at the offset, but the signal would be at baseline level at all other times. On the other hand, if the hair cell detected only angular velocity, there would be a positive signal of the same intensity during the entire period of rotation. Note that the actual record of Figure 52–5 is a combination of these two types of records.

**Rate of Angular Acceleration Required to Stimulate the Semicircular Canals.** The angular

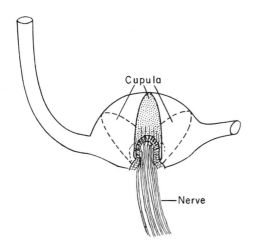

**Figure 52–4.** Movement of the cupula and its embedded hairs during rotation first in one direction and then in the opposite direction.

acceleration required to stimulate the semicircular canals in the human being averages about 1 degree per second per second. In other words, when one begins to rotate, his velocity of rotation must be as much as 1 degree per second by the end of the first second, 2 degrees per second by the end of the second second, 3 degrees per second by the end of the third second, and so forth, in order for him barely to detect that his rate of rotation is increasing.

**"Predictive" Function of the Semicircular Canals in the Maintenance of Equilibrium.** Since the semicircular canals do not detect that the body is off balance in the forward direction, in the side direction, or in the backward direction, one might at first ask: What is the function of the semicircular canals in the maintenance of equilibrium? All they detect is that the person's head is beginning to rotate or is stopping rotation in one direction or another. Therefore, the function of the semicircular canals is not likely to be to maintain static equilibrium or to maintain equilibrium during linear acceleration or when the person is exposed to steady centrifugal forces. Yet loss of function of the semicircular canals causes a person to have poor equilibrium, especially when he attempts to perform *rapid* and *intricate* body movements.

We can explain the function of the semicircular canals best by the following illustration. If a person is running forward rapidly, and then suddenly begins to turn to one side, he falls off balance a second or so later unless appropriate corrections are made *ahead of time*. But, unfortunately, the utricle cannot detect that he is off balance until *after* this has occurred. On the other hand, the semicircular canals will have already detected that the person is beginning to turn, and this information can easily apprise the central nervous system of the fact that the person *will* fall off balance within the next second or so unless some correction is made. In other words, the semicircular canal mechanism *predicts ahead of time* that mal-equilibrium is going to occur even before it occurs and thereby causes the equilibrium centers to make appropriate preventive adjustments. In this way, the person need not fall off balance before he begins to correct the situation.

Removal of the flocculonodular lobes of the cerebellum prevents normal function of the semicircular canals but does not prevent normal function of the macular receptors. It is especially interesting in this connection that the cerebellum serves as a "predictive" organ for most of the other rapid movements of the body as well as those having to do with equilibrium.

These other functions of the cerebellum are discussed in the following chapter.

## VESTIBULAR REFLEXES

**The Vestibular Phasic Postural Reflexes.** Sudden changes in the orientation of an animal in space elicits reflexes that help to maintain equilibrium and posture. For instance, if an animal is suddenly pushed to the right, even before he can fall more than a few degrees his right legs extend instantaneously. In other words, this mechanism *anticipates* that the animal will be off balance in a few seconds and makes appropriate adjustments to prevent this.

Another type of vestibular phasic postural reflex occurs when the animal's head suddenly falls downward. When this occurs, the forepaws extend forward, the extensor muscles tighten, and the muscles in the back of the neck stiffen to prevent the animal's head from striking the ground. This reflex is probably also of importance in locomotion, for, in the case of the galloping horse, the downward thrust of the head could automatically provide reflex thrust of the forelimbs to move the animal forward for the next gallop.

**The Vestibular "Righting" Reflex and Equilibrium.** In the previous chapter it was pointed out that some of the crude motions required to move an animal from a lying to a standing position can be elicited by circuits in the spinal cord. However, these *cord* "righting" reflexes cannot perform the full function of righting an animal to the upright position without signals from the brain stem. The static equilibrium organs, the maculae of the utricle and saccule, transmit impulses into the brain stem to apprise the nervous system of the status of the animal with respect to the pull of gravity, and, if the animal is in a lying position, these can elicit appropriate reflexes from the vestibular and reticular nuclei to cause the animal to climb to his feet. This is called the vestibular righting reflex.

This vestibular righting reflex requires neuronal centers in all parts of the brain stem up to as high as and including the mesencephalon.

## VESTIBULAR MECHANISM FOR STABILIZING THE EYES AND FOR NYSTAGMUS

When a person changes his direction of movement rapidly, or even leans his head sideways, forward, or backward, it would be impossible for him to maintain a stable image on the retinae of his eyes unless he had some automatic control mechanism to stabilize the direction of gaze of the eyes. In addition, the eyes would be of little use in detecting an image unless they remained "fixed" on each object long enough to gain a clear image. Fortunately, each time the head is angulated, signals from the semicircular canals cause the eyes to angulate in an equal and opposite direction to the angulation of the head. This results from reflexes transmitted through the *vestibular nuclei,* the *cerebellum,* and the *medial longitudinal fasciculus* to the *ocular nuclei.*

A special example of this stabilization of the eyes occurs when a person begins to rotate. At first the eyes remain glued to the object. But after the head has rotated far to one side, they jump suddenly in the direction of rotation of the head and "fix" on a new object; then they deviate slowly backward again as the rotation proceeds. This sudden jumping motion forward and then slow backward motion is known as *nystagmus.* The jumping motion is called the *fast component* of the nystagmus, and the slow movement is called the *slow component.*

Nystagmus always occurs automatically when the semicircular canals are stimulated. For instance, if a person's head begins to rotate to the right, backward movement of fluid in the left horizontal canal and forward movement in the right horizontal canal cause the eyes to move slowly to the left; thus, the slow component of nystagmus is initiated by the vestibular apparatuses. But, when the eyes have moved as far to the left as they reasonably can, centers located in the brain stem in close approximation with the nuclei of the abducens nerves cause the eyes to jump suddenly to the right; then the vestibular apparatuses take over once more to move the eyes again slowly to the left.

### CLINICAL TESTS FOR INTEGRITY OF VESTIBULAR FUNCTION

**Balancing Test.** One of the simplest clinical tests for integrity of the equilibrium mechanism is to have the individual stand perfectly still with his eyes closed. If he no longer has a functioning static equilibrium system of the utricles, he will waiver to one side or the other and possibly even fall. However, as noted below, some of the proprioceptive mechanisms of equilibrium are occasionally sufficiently well developed to maintain balance even with the eyes closed.

**Barany Test.** A second test that is frequently performed determines the integrity of the semicircular canals. In this instance the individual is placed in a "Barany chair" and rotated rapidly while he places his head respectively in various planes—first forward, then angulated to one side or the other. By such positioning, each pair of semicircular canals is successively placed in the horizontal plane of rotation. When the rotation of the chair is stopped suddenly, the endolymph, because of its momentum, continues to rotate in the pair of semicircular canals that has been placed in the horizontal plane, this flow of endolymph causing the cupula to bend in the direction of rotation. As a result, nystagmus occurs, with the slow component in the direction of rotation and the fast component in the opposite direction. Also, as long as the nystagmus lasts (about 15 to 20 seconds) the individual has the sensation that he is rotating in

the direction *opposite* to that in which he was actually rotated in the chair. Obviously, this test checks the semicircular canals on both sides of the head at the same time.

**Ice Water Test.** A clinical test for testing one vestibular apparatus separately from the other depends on placing ice water in one ear. The external semicircular canal lies adjacent to the ear, and cooling the ear can transfer a sufficient amount of heat from this canal to cool the endolymph. This increases the density of the endolymph, thereby causing it to sink downward and resulting in slight movement of fluid around the semicircular canal. This stimulates the canal, giving the individual a sensation of rotating and also initiating nystagmus. From these two findings one can determine whether the respective semicircular canals are functioning properly. When the semicircular canals are normal, the utricles and saccules are usually normal also, for disease usually destroys the function of all of these components at the same time.

### OTHER FACTORS CONCERNED WITH EQUILIBRIUM

**The Neck Proprioceptors.** The vestibular apparatus detects the orientation and movements *only of the head.* Therefore, it is essential that the nervous centers also receive appropriate information depicting the orientation of the head with respect to the body as well as the orientation of the different parts of the body with respect to each other. This information is transmitted from the proprioceptors of the neck and body both directly into the reticular nuclei of the brain stem and indirectly by way of the cerebellum and thence into the reticular nuclei.

By far the most important proprioceptive information needed for the maintenance of equilibrium is that derived from the *joint receptors of the neck,* for this apprises the nervous system of the orientation of the head with respect to the body. When the head is bent in one direction or the other, impulses from the neck proprioceptors keep the vestibular apparatuses from giving the person a sense of mal-equilibrium. They do this by transmitting signals that exactly oppose the signals transmitted from the vestibular apparatuses. However, *when the entire body* is changed to a new position with respect to gravity, the impulses from the vestibular apparatuses *are not opposed* by the neck proprioceptors; therefore, the person in this instance does perceive a change in equilibrium status.

***The Neck Reflexes.*** In an animal *whose vestibular apparatuses have been destroyed,* bending the neck causes immediate muscular reflexes called *neck reflexes* occurring especially in the forelimbs. For instance, bending the head forward causes both forelimbs to relax. However, when the vestibular apparatuses are *intact,* this effect does *not* occur because the vestibular reflexes function almost exactly oppositely to the neck reflexes. Thus, if the head is flexed downward, the vestibular reflex tends to extend the forelimbs, while the neck reflexes tend to relax them. Since the equilibrium of the entire body and not of the head alone must be maintained, it is easy to understand that the vestibular and neck reflexes must function oppositely. Otherwise, each time the neck should bend, the animal would immediately fall off balance.

**Proprioceptive and Exteroceptive Information from Other Parts of the Body.** Proprioceptive information from other parts of the body besides the neck is also necessary for maintenance of equilibrium because appropriate equilibrium adjustments must be made whenever the body is angulated in the chest or abdomen region or elsewhere. Presumably, all this information is algebraically added in the cerebellum and reticular substance of the brain stem, thus causing appropriate adjustments in the postural muscles.

Also important in the maintenance of equilibrium are several types of exteroceptive sensations. For instance, pressure sensations from the footpads can tell one (a) whether his weight is distributed equally between his two feet and (b) whether his weight is more forward or backward on his feet.

Another instance in which exteroceptive information is necessary for maintenance of equilibrium occurs when a person is running. The air pressure against the front of his body signals that a force is opposing the body in a direction different from that caused by gravitational pull; as a result, the person leans forward to oppose this.

**Importance of Visual Information in the Maintenance of Equilibrium.** After complete destruction of the vestibular apparatuses, and even after loss of most proprioceptive information from the body, a person can still use his visual mechanisms effectively for maintaining equilibrium. Visual images help the person maintain equilibrium simply by visual detection of the upright stance. Also, slight linear or angular movement of the body instantaneously shifts the visual images on the retina, and this information is relayed to the equilibrium centers. In this respect the optic information is similar to that from the semicircular canals and can help the equilibrium centers predict that the person will fall off balance before this actually occurs if appropriate adjustments are not immediately made. Many persons with complete destruction of the vestibular apparatus have almost normal equilibrium as long as their eyes are open and as long as they perform all motions slowly. But, when moving rapidly or when the eyes are closed, equilibrium is immediately lost.

**Conscious Perception of Equilibrium.** A cortical center for conscious perception of the state of equilibrium is believed to lie in the upper portion of the temporal lobe in close association with the primary cortical area for hearing. The sensations from the vestibular apparatuses, from the neck proprioceptors, and from most of the other proprioceptors are probably first integrated in the equilibrium centers of the brain stem before being transmitted to the cerebral cortex. Various pathologic processes in the vestibular apparatuses or in the vestibular

neuronal circuits often affect the equilibrium sensations. Thus, loss of one or both flocculonodular lobes or one of both fastigial nuclei of the cerebellum gives the person a sensation of constant mal-equilibrium, or at least this feeling lingers for the first few weeks or months after the loss. This probably results because impulses from these cerebellar areas are normally integrated into the subconscious sense of equilibrium by the reticular nuclei even before equilibrium information is sent to the cerebral cortex.

# FUNCTIONS OF THE RETICULAR FORMATION AND SPECIFIC BRAIN STEM NUCLEI IN CONTROLLING SUBCONSCIOUS, STEREOTYPED MOVEMENTS

Rarely, a child called an anencephalic monster is born without brain structures above the mesencephalic region, and some of these children have been kept alive for many months. Such a child is able to perform essentially all the functions of feeding, such as suckling, extrusion of unpleasant food from the mouth, and moving his hands to his mouth to suck his fingers. In addition, he can yawn and stretch. He can cry and follow objects with his eyes and by movements of his head. Also, placing pressure on the upper anterior parts of his legs will cause him to pull to the sitting position.

Therefore, it is obvious that many of the stereotyped motor functions of the human being are integrated in the brain stem. Unfortunately, the loci of most of these different motor control systems have not been found except for the following:

**Stereotyped Body Movements.** Most movements of the trunk and head can be classified into several simple movements, such as forward flexion, extension, rotation, and turning movements of the entire body. These types of movements are controlled by special nuclei located mainly in the mesencephalic and lower diencephalic region. For instance, *rotational movements* of the head and eyes are controlled by the *interstitial nucleus*, which is illustrated in Figure 52–1. This nucleus lies in close approximation to the *medial longitudinal fasciculus*, through which it transmits a major portion of its control impulses. The *raising movements* of the head and body are controlled by the *prestitial nucleus*, which is located approximately at the juncture of the diencephalon and mesencephalon. On the other hand, the *flexing movements* of the head and body are controlled by the *nucleus precommissuralis* located at the level of the posterior commissure. Finally, the *turning movements* of the entire body, which are much more complicated, involve both the pontile and mesencephalic reticular formation. However, for full expression of the turning movements, the caudate nucleus and the cingulate gyrus of the cerebral cortex are also required. The turning movements can cause an animal or person to continue circling around and around in one direction or the other.

**Function of the Red Nucleus.** The red nucleus is composed of two distinct parts, the *magnocellular portion* and a *small cellular portion.*

As illustrated in Figure 52–6, the magnocellular nuclei receive impulses both from the *caudate nucleus* and the *putamen* as well as from the *prestitial nucleus* and the *nucleus commissuralis.* In turn it transmits signals by way of the *rubrospinal tract* into the spinal cord and by way of collaterals from the rubrospinal tract into the *reticular nuclei* of the brain stem. Stimulation of this portion of the red nucleus causes the head and upper trunk to extend backward. Thus, this portion of the red nucleus enters into a particular type of gross body movement concerned with forward and backward deviation of the body axis.

The small cells of the red nucleus are excited principally by impulses from the *cerebellum* through the *cerebellorubral tracts.* In turn, they send impulses mainly into the reticular formation of the brain stem. These tracts are part of the overall cerebellar mechanism for control of motor function, which will be discussed in the following chapter.

**Function of the Subthalamic Areas—Forward Progression.** Much less is known about function of higher brain stem centers in posture and locomotion than of the lower centers, principally because of the complexity of the neuronal connections. However, it is known that stimulation of centers in or around the subthalamic nuclei can cause rhythmic motions, including crude forward walking reflexes. This does not mean that the individual muscles of walking are necessarily controlled from this region but simply that excitation of this region excites the appropriate patterns of action in the brain stem and cord to cause the walking movements.

A cat with its brain transected beneath the thalamus but above the subthalamus can walk in an almost completely normal fashion—so much so that the observer cannot tell the difference. However, when the animal comes to an obstruction it simply butts its head against the obstruction and tries to keep on walking. Thus, it lacks *purposefulness of locomotion.*

The function of the subthalamic region in walking is frequently described as that of controlling *forward progression.* In the foregoing discussions we have already noted that a decerebrate animal can stand perfectly well and that it can make attempts to right itself from the lying position. Unfortunately, though, the decerebrate animal, with the brain stem sectioned below the subthalamus cannot force itself to move forward in a normal walking pattern. This seems to be the function of areas either in or somewhere close to the subthalamic nuclei.

# MOTOR FUNCTIONS OF THE BASAL GANGLIA

**Physiologic Anatomy of the Basal Ganglia.** The anatomy of the basal ganglia is so complex and is so poorly known in its details that it would be pointless to attempt a complete description at this time. However, Figure 52–6 illustrates the principal structures of the basal ganglia and their neural connections with other parts of the nervous system. Anatomically, the basal ganglia are the *caudate nucleus, putamen, globus pallidus, amygdaloid nucleus,* and *claustrum.* The amygdaloid nucleus, which will be discussed in Chapter 57, and the claustrum are not concerned directly with motor functions of the central nervous system. On the other hand, the *thalamus, subthalamus, substantia nigra,* and *red nucleus* all operate in close association with the caudate nucleus, putamen, and globus pallidus and are considered to be part of the basal ganglia system for motor control.

Some important features of the different pathways illustrated in Figure 52–6 are the following:

1. Numerous nerve pathways pass from the motor portions of the cerebral cortex, particularly the so-called premotor areas, to the caudate nucleus and putamen, which together are called the *striate body.* In turn, the caudate nucleus and putamen send numerous fibers to the globus pallidus, the globus pallidus to the ventroanterior nucleus of the thalamus, and this nucleus back to the motor areas of the cerebral cortex. Thus, circular pathways are established from the motor cortical regions to the basal ganglia, the thalamus, and back to the same motor regions from which the pathways begin. These circuits obviously could operate as a feedback system of the servo control type, as will be discussed later.

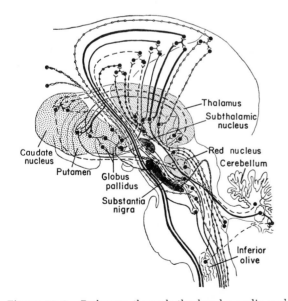

**Figure 52–6.** Pathways through the basal ganglia and related structures of brain stem, the thalamus, and cerebral cortex. (From Jung and Hassler: Handbook of Physiology, Sec. I, Vol. II, The Williams & Wilkins Co., 1960.)

2. In the following chapter it will be noted that signals also pass from the motor regions of the cerebral cortex to the pons and then into the cerebellum. In turn, signals from the cerebellum are transmitted back to the motor cortex by way of the ventrolateral nucleus of the thalamus, a nucleus closely associated with the ventroanterior thalamic nucleus through which the basal ganglial signals are transmitted. This circuit could allow integration between the basal ganglial feedback signals and the feedback signals from the cerebellum.

3. The basal ganglia have numerous short neuronal connections among themselves. Also, the lower basal ganglia, such as the globus pallidus, send tremendous numbers of nerve fibers into the lower brain stem, projecting especially onto the reticular nuclei, the red nucleus, and the inferior olive. It is presumably through these pathways that many of the so-called extrapyramidal signals for motor control are transmitted.

4. Large numbers of nerve fibers pass directly from the motor cortex to the reticular nuclei and other nuclei of the brain stem. These by-pass the basal ganglia but converge on the same brain stem nuclei that are innervated by the basal ganglia.

## FUNCTIONS OF THE DIFFERENT BASAL GANGLIA

Before attempting to discuss the functions of the basal ganglia in man, we should speak briefly of the better known functions of these ganglia in lower animals. In birds, for instance, the cerebral cortex is poorly developed while the basal ganglia are highly developed. These ganglia perform essentially all the motor functions, even controlling the voluntary movements in much the same manner that the motor cortex of the human being controls voluntary movements. Furthermore, in the cat, and to a lesser extent in the dog, decortication removes only the discrete types of motor functions and does not interfere with the animal's ability to walk, eat, fight, develop rage, have periodic sleep and wakefulness, and even participate naturally in sexual activities. However, if a major portion of the basal ganglia is destroyed, only gross stereotyped movements remain, which were discussed above in relation to the mesencephalic animal.

Finally, in the human being, decortication of very young individuals destroys the discrete movements of the body, particularly of the hands and distal portions of the lower limbs, but does not destroy the person's ability to walk crudely, to control his equilibrium, or to perform many other subconscious types of movements. However, simultaneous destruction of a major portion of the caudate nuclei almost totally paralyzes the opposite side of the body except for a few stereotyped reflexes integrated in the cord or brain stem.

With this brief background of the overall function of the basal ganglia, we can attempt to dissect the functions of the individual portions of the basal ganglia system, realizing that the system actually operates as a total unit and that individual functions cannot be ascribed completely to the different parts of the basal ganglia.

**Inhibition of Motor Tone by the Basal Ganglia.** Though it is wrong to ascribe a single function to all the basal ganglia, nevertheless, one of the general effects of diffuse basal ganglia excitation is to inhibit muscle tone throughout the body. This effect results from inhibitory signals transmitted from the basal ganglia to the bulboreticular facilitatory area and excitatory signals to the bulboreticular inhibitory area. Therefore, whenever widespread destruction of the basal ganglia occurs, the facilitatory area becomes overactive and the inhibitory area underactive, and this causes muscle rigidity throughout the body. For instance, when the brain stem is transected at the mesencephalic level, which removes the inhibitory effects of the basal ganglia, the phenomenon of decerebrate rigidity occurs.

Yet, despite this general inhibitory effect of the basal ganglia, stimulation of specific areas within the basal ganglia can at times elicit positive muscle contractions and at times even complex patterns of movements.

**Function of the Caudate Nucleus and Putamen—The Striate Body.** The caudate nucleus and putamen, because of their gross appearance on sections of the brain, are together called the *striate body*. They seem to function together to initiate and regulate gross intentional movements of the body. To perform this function they transmit impulses through two different pathways: (1) into the *globus pallidus,* thence by way of the *thalamus* to the *cerebral cortex,* and finally downward into the spinal cord through the *corticospinal* and *extracorticospinal pathways;* (2) downward through the *globus pallidus* and the *substantia nigra* by way of short axons into the *reticular formation* and finally into the spinal cord mainly through the *reticulospinal tracts.*

In summary, the striate body helps to control gross intentional movements that we normally perform unconsciously. However, this control also involves the motor cortex, with which the striate body is very closely connected.

**Function of the Globus Pallidus.** It has been suggested that the principal function of the globus pallidus is to provide background muscle tone for intended movements, whether these be initiated by impulses from the cerebral cortex or from the striate body. That is, if a person wishes to perform an exact function with one hand, he positions his body and limbs appropriately and also tenses the muscles of the upper arm. These associated tonic contractions are supposedly initiated by a circuit that strongly involves the globus pallidus. Destruction of the globus pallidus removes these associated movements and, therefore, makes it difficult or impossible for the distal portions of the limbs to perform their more discrete activities.

The globus pallidus is believed to function through two pathways: first, through feedback circuits to the thalamus, to the cerebral cortex, and thence by way of corticospinal and extracorticospinal tracts to the spinal cord; and, second, by way of short axons to the reticular formation of the brain stem and thence mainly by way of the reticulospinal tracts into the spinal cord.

Electrical stimulation of the globus pallidus while an animal is performing a gross body movement will often stop the movement in a static position, the animal holding that position for many seconds while the stimulation continues. This fits with the concept that the globus pallidus is involved in some type of servo feedback motor control system that is capable of locking the different parts of the body into specific positions. Obviously, such a circuit could be extremely important in providing the background body movements and upper limb movements when a person performs delicate tasks with his hands.

## CLINICAL SYNDROMES RESULTING FROM DAMAGE TO THE BASAL GANGLIA

Much of what we know about the function of the basal ganglia comes from study of patients with basal ganglia lesions whose brains have undergone pathologic studies after death. Among the different clinical syndromes are:

**Chorea.** Chorea is a disease in which random uncontrolled contractions of different muscle groups of the body occur continuously. Normal progression of movements cannot occur; instead, the person may perform a normal pattern of movements for a few seconds and then suddenly begin another pattern of movements; then still another pattern begins after a few seconds. Because of this peculiar progression of movements, one type of chorea is frequently called *St. Vitus' dance.*

Pathologically, chorea results from *diffuse and widespread damage of the striate body,* but the manner in which this causes the choreiform movements is not known.

**Athetosis.** In this disease, slow, writhing movements of the peripheral parts of one or more limbs—usually upper—occur continually. The movements are likely to be wormlike, first with overextension of the hands and fingers, then flexion, then rotary twisting to the side—all these continuing in a slow, rhythmic, writhing pattern. The contracting muscles exhibit a high degree of spasm, and the movements are enhanced by emotions or by excessive signals from the sensory organs. Furthermore, voluntary movements in the affected area are greatly impaired or sometimes even impossible.

The damage in athetosis is usually found in the *outer portion of the globus pallidus* or in this area and the striate body. Athetosis is usually attributed to the interruption of feedback circuits among the basal ganglia, thalamus, and cerebral cortex. The normal feedback circuits presumably allow a constant and rapid interplay between antagonistic muscle groups so that finely controlled movements can take place. However, if the feedback circuits are blocked, it is supposed that the detouring impulses may take devious routes through the basal ganglia, thalamus, and motor cortex, causing a succession of abnormal movements.

**Hemiballismus.** Hemiballismus is an uncontrollable succession of violent movements of large areas of the body. These may occur once every few seconds or sometimes only once in many minutes. For instance, an entire leg might suddenly jerk uncontrollably to full flexion, or the entire trunk might go through an extreme, sudden torsion movement, or an arm might be pulled upward suddenly with great force. Hemiballismus of the legs or trunk causes the person to fall to the ground if he is walking, and even when the person is in bed, he tosses violently when affected by these powerful and strong intermittent movements. Furthermore, attempts to perform voluntary movements frequently invoke ballistic movements in place of the normal movements.

Hemiballismus of one side of the body results from a *large lesion in the opposite subthalamus.* The smooth, rhythmic movements of the limbs or other parts of the body normally integrated in this area can no longer occur, but excitatory impulses attempting to evoke such movements elicit instead uncontrollable ballistic movements.

**Parkinson's Disease.** Parkinson's disease, which is also known as *paralysis agitans,* results almost invariably from *widespread destruction of the substantia nigra,* often associated with lesions of the *globus pallidus* and other related areas. It is characterized by (1) *rigidity* of the musculature in either widespread areas of the body or in isolated areas, (2) *tremor at rest* of the involved areas in most but not all instances, and (3) *loss of involuntary and associated* *movements, called "akinesia."* Also, there may be enhancement of salivation, sweating, and seborrheic secretion.

The rigidity in this disease is quite different from that which occurs in decerebrate rigidity, for decerebrate rigidity results from hyperactivity of the muscle spindle system, and even slight movement of a muscle is met with rather extreme reflex resistance resulting from feedback through the stretch reflex mechanism. Parkinsonian rigidity, on the other hand, is of a "plastic" type. That is, sudden movement is not met by intense resistance from the stretch reflexes as in the decerebrate type of rigidity, but instead both protagonist and antagonist muscles remain tightly contracted throughout the movement. Furthermore, the rigidity of Parkinson's disease seems to result from excess impulses transmitted in the corticospinal system, thus activating the alpha motor fibers to the muscles, whereas decerebrate rigidity results from excess activation of the gamma efferents to the muscle spindles.

Tremor usually, though not always, occurs in Parkinson's disease. Its frequency normally is 4 to 6 cycles per second. When the person performs voluntary movements, weak tremors become temporarily blocked, possibly because cortical signals to the spinal cord excite the gamma efferent fibers to the muscle spindles at the same time that they excite the alpha fibers; the gamma stimulation sensitizes the muscle spindles which theoretically could then damp out the tremor.

The mechanism of the tremor in Parkinson's disease is not known, but at least two different theories must be considered: (1) Loss of inhibitory influences from the substantia nigra or globus pallidus allows the control nuclei of antagonistic muscles in the *reticular substance of the brain stem* to oscillate. (2) Lack of inhibitory impulses in the lower basal ganglia enhances feedback in the globus pallidus-thalamus-cortical circuit, causing it to oscillate.

Another abnormality of Parkinson's disease is a high degree of "motor stiffness" and loss of automatisms (akinesia). When a person begins to perform a discrete voluntary movement with his hands, the automatic "associated" adjustments of the trunk of his body and the upper arm segments do not occur. Instead, he must voluntarily adjust these segments before he can use his hands. Furthermore, a tremendous amount of nervous effort must be made by the voluntary motor control system to overcome the motor stiffness of his musculature. Thus, the person with Parkinson's disease has a masklike face, showing almost no automatic, emotional facial expressions; he is usually bent forward because of his muscle rigidity, and all his movements of necessity are highly deliberate rather than the many casual subconscious movements that are normally a part of our everyday life. This loss of automatisms is believed to result from diminished transmission of signals into the globus pallidus and other basal ganglia from the substantia nigra. Most of the pathways from the substantia nigra to these areas are inhibitory, and

when they are absent the globus pallidus and other basal ganglia become excessively active, leading to rigidity of the muscles that are usually responsible for the automatisms. Thus, as the automatisms are lost they are replaced by rigidity.

**Treatment with L-Dopa and Anticholinergic Drugs.** Administration of L-dopa to patients with Parkinson's disease ameliorates most of the symptoms, especially the rigidity and the akinesia, in about two-thirds of the patients. The reason for this seems to be the following: The neurons that originate in the substantia nigra and then project to the globus pallidus and striate body are inhibitory neurons whose terminals probably secrete *dopamine,* a derivative of L-dopa. When the substantia nigra is destroyed and the person develops Parkinson's disease, the administered L-dopa is believed to substitute for the dopamine no longer secreted by the destroyed neurons. This causes more or less normal inhibition of the basal ganglia and relieves much or most of the akinesia and rigidity.

The same basal ganglia that are inhibited by dopamine seem to be excited by other nerve terminals derived elsewhere; these terminals probably secrete acetylcholine. Therefore, as would be expected, administration of anticholinergic drugs such as scopolamine can also benefit the parkinsonian patient.

**Coagulation of the Ventrolateral Nucleus of the Thalamus for Treatment of Parkinson's Disease.** In recent years neurosurgeons have treated Parkinson's disease patients, with varying success, by destroying portions of the basal ganglia or of the thalamus. The most prevalent treatment at present is widespread destruction of the ventrolateral nucleus of the thalamus usually by electrocoagulation. Most fiber pathways from the basal ganglia and cerebellum to the cerebral cortex pass through this nucleus so that its destruction blocks many or most of the feedback functions of the basal ganglia and cerebellum. It is presumed that blockage of some of these feedbacks limits the functions of at least certain basal ganglia and thereby removes the factors that cause the rigidity and tremor of Parkinson's disease.

# FUNCTIONS OF THE CEREBRAL CORTEX IN POSTURE AND LOCOMOTION— PURPOSEFULNESS OF LOCOMOTION

The principal function of the cerebral cortex in locomotion is to add *purposefulness* to the locomotion, as illustrated above by the subthalamic cat that would not go around an obstruction but would simply butt its head against the obstruction and continue its attempts to move forward. Such an effect obviously would not occur in an animal with a cerebral cortex. However, in addition to the purposefulness added by cortical function, several postural and righting reflexes are also integrated in the cortex. These include the *visual righting reflex,* certain aspects of *exteroceptive righting reflexes,* the *placing reaction,* and the *hopping reaction.*

**The Visual and Exteroceptive Righting Reflexes.** The visual righting reflexes depend on function of the visual cortex. That is, an animal can tell by images from the periphery whether his body is in the appropriate position with respect to its surroundings. From this information, he can raise himself from the lying position to the standing position even in the absence of the vestibular righting reflexes.

Similarly, the animal's sensory and motor cortex causes righting in response to exteroceptive information that is transmitted into the cerebrum. It has been pointed out that sensations from the skin can cause crude righting responses even in the spinal animal and even much better responses in the decerebrate animal. However, for full expression of the exteroceptive righting reflexes, the cerebral cortex is also required.

**The Placing Reaction.** A blindfolded animal held in the air and then brought in contact with a surface so that almost any part of his body makes contact with the surface immediately brings his paws toward the point of contact, thus placing his feet in an advantageous position to gain support. This is called the placing reaction. Though a weak expression of the placing reaction can occur even in animals without the cerebral cortex, here again the complete reaction can occur only when the cortex is available to provide stereognostic orientation of the different parts of the body in relation to each other, thus allowing the animal to "know" where his feet need to be moved in order to make appropriate contact.

*The Hopping Reaction.* The hopping reaction is another reaction that requires the cerebral cortex for full attainment. This reaction probably results mainly from direct transmission of proprioceptive and vestibular information to the motor cortex to cause the hopping.

# SUMMARY OF THE DIFFERENT FUNCTIONS OF THE CENTRAL NERVOUS SYSTEM IN POSTURE AND LOCOMOTION

From the discussions in the previous chapter, we could see that almost all of the discrete "patterns" of muscle movement required for posture and locomotion can be elicited by the spinal cord alone. However, coordination of these patterns to provide equilibrium, progression, and purposefulness of movement requires neuronal function at progressively higher levels of the central nervous system. Centers in the brain stem provide most of the nervous energy required to maintain postural tone and therefore to support the body against gravity. But, in ad-

dition, the brain stem centers provide especially the equilibrium reflexes. Then moving still higher we find a control center for "progression" in the subthalamic region, a center that makes the animal move forward in a normal rhythmic pattern of walking and also in a straight line. Even so, this animal still does not have purposefulness in his motion. For this the basal ganglia, the thalamus, and the cerebral cortex are required.

With this background, we can now describe the characteristics of animals with transections through their nervous systems at the different levels shown in Figure 52–7.

**The Spinal Animal.** In the spinal animal, the flexor reflex, the crossed extensor reflex, the stretch reflex, and the reflexes that control the basic rhythm and pattern of locomotion are intact after the stage of spinal shock is over. Furthermore, under the most favorable conditions the animal can even stand for a few seconds because of the positive supportive reaction.

**Transection of the Medulla Below the Vestibular Nuclei.** If the medulla is cut below the vestibular nuclei, the lower portion of the reticular substance of the brain stem will be intact; this, if anything, transmits *inhibitory* impulses into the spinal cord. Therefore, few cord reflexes can be elicited, and the animal certainly cannot stand.

**Transection of the Brain Stem Above the Vestibular Nuclei.** If the brain stem is transected above the vestibular nuclei, the vestibular nuclei and the upper reticular substance of the brain stem transmit *facilitatory* impulses into the spinal cord and actually provide far more facilitation than normal. If

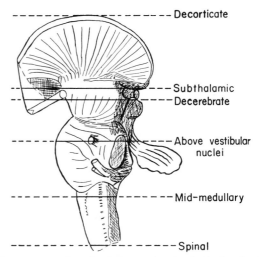

**Figure 52–7.** Levels of transection in the brain, the descriptions of which are given in the text.

the transection is well above the vestibular nuclei, sufficient postural tone can be attained for the animal to stand for minutes at a time.

**The Classical Decerebrate Animal—Decerebrate Rigidity.** The classical decerebrate animal has a transection between the superior and the inferior colliculi. In this animal the inhibitory signals from the basal ganglia and cerebral cortex to the reticular facilitatory area are removed, which allows the bulboreticular facilitatory area to become intrinsically very active, as explained earlier in the chapter. Impulses are transmitted downward to cause muscle rigidity called *decerebrate rigidity*. Especially the extensor muscles, but to some extent the flexor muscles as well, are maintained in a state of strong tonic contraction. The animal can occasionally stand, supporting its body against gravity. It also can make crude equilibrium adjustments; but for full expression of equilibrium, it needs a completely intact mesencephalon.

A major part of the increased postural tone in decerebrate rigidity results from stimulation of the *gamma activating system of the muscle spindles* and not from direct activation of the alpha motor neurons.

**The Subthalamic Animal.** With the transection above the superior colliculi, preserving the subthalamus as well as adjacent regions of the brain stem, the animal has less rigidity and is capable of walking with normal progression.

**The Decorticate Animal.** In a decorticate animal the cortex has been removed but the thalamus and basal ganglia are still intact. The capabilities of this type of animal depend on its degree of cerebral development. The cat, for instance, is capable of almost any type of motion, even to the extent of going around obstructions in his way, but, on the other hand, he still lacks much of the purposefulness of locomotion and frequently sits very still for hours at a time. The human being loses essentially all purposeful or voluntary motor functions on decortication.

# FUNCTION OF THE CORD AND BRAIN STEM REFLEXES IN MAN

Though the foregoing discussions have concerned principally reflexes in animals rather than in man, the same principles of cord and brain stem function apply almost as well to man. Yet there are quantitative differences that must always be remembered.

First, man walks in an upright position and his limbs have become straightened to the point that almost no muscular strength is required to maintain the weight of the body against gravity. For instance, the direct line between the center of mass of the body and the direction of gravitational pull runs slightly behind the axes of the hip joints so that gravity tends to extend the hips and so that the ligaments of the hip joints, rather than the muscles, support the body against gravity. Likewise, the line of gravitational

pull runs slightly in front of the mid-position of the knee joints and well in front of the ankle joints. Here again, it is mainly tension in the ligaments and uncontracted muscles that maintains the weight of the body against gravity rather than postural tone.

Therefore, in man, contraction of postural muscles occurs principally to maintain equilibrium rather than to support the body against gravity. Yet, if man assumes an animal-like position, postural tone serves essentially the same functions as in lower animals.

In man with a spinal cord transection, most of the same reflexes as those that occur in lower animals can be elicited either wholly or partially. The only reflexes that often cannot be elicited in spinal man are the rhythmic reflexes; yet the reciprocal innervation that is required for rhythmic reflexes can be demonstrated. For instance, movement of an arm in one direction often causes simultaneous and directionally similar movement of the opposite leg, thus illustrating that appropriate reciprocal innervation is still present in the cord of man to cause at least crude types of diagonal movements necessary for the mark time reflex.

Also, stimulation in the subthalamic region in man causes rhythmic walking movements similar to those that occur in animals, illustrating that this same region of the brain subserves similar functions in man as in lower animals. However, a man who has lost large portions of the motor areas of his cortex cannot use these lower regions of the brain stem effectively enough actually to provide locomotion. This is a manifestation of the process called *encephalization* that has occurred in higher orders of animals, which suppresses many functions of the lower centers while the same functions are taken over, usually in a more complicated manner, by the higher centers. Thus, in man, the motor areas of the cortex are needed to provide progression in the act of locomotion and also to provide purposefulness of locomotion.

Nevertheless, in our subsequent discussions of the motor functions of the brain in man, we must always keep in mind the basic activities of the lower centers. The cord patterns of response are integrated into the overall control of muscular activity even when this control is initiated in the cerebral cortex. It is still the anterior motoneurons that have the final decisive power in the contraction of given muscles, and it is still the patterns of organization of the interneurons in the gray matter of the cord that provide many of the patterns of muscular contraction. Impulses from above seem to control the sequence in which these patterns of contraction occur, interspersed with other patterns generated in the brain itself.

# REFERENCES

Barbeau, A., Chase, T. N., and Paulson, G. W. (eds.): Huntington's Chorea, 1872–1972. New York, Raven Press, 1973.

Brimblecombe, R. W.: Tremors and Tremorogenic Agents. Baltimore, The Williams & Wilkins Company, 1972.

Brodal, A. (ed.): Basic Aspects of Central Vestibular Mechanisms. New York, American Elsevier Publishing Co., Inc., 1971.

Calne, D. B. (ed.): Progress in the Treatment of Parkinsonism. New York, Raven Press, 1973.

Calne, D. B., Chase, T. N., and Barbeau, A. (eds.): Dopaminergic Mechanisms. (Advances in Neurology. Vol. 9.) New York, Raven Press, 1975.

Cooper, I. S.: Involuntary Movement Disorders. Hagerstown, Md., Harper & Row, Publishers, 1969.

Denny-Brown, D.: The Basal Ganglia: Their Relation to Disorders of Movement. New York, Oxford University Press, Inc., 1962.

Fischer, J. H.: The Labyrinth; Physiology and Functional Tests. New York, Grune & Stratton, Inc., 1956.

French, J. D.: The reticular formation. In Magoun, H. W. (ed.): Handbook of Physiology. Sec. 1, Vol. 2. Baltimore, The Williams & Wilkins Company, 1960, p. 1281.

Goldberg, J. M., and Fernandez, C.: Vestibular mechanisms. Ann. Rev. Physiol., 37:129, 1975.

Hornykiewicz, O.: Pharmacology and pathophysiology of dopaminergic neurons. Adv. Cytopharmacol., 1:369, 1971.

Kawamura, Y.: Neurogenesis of mastication. Front. Oral Physiol., 1:77, 1974.

Kidokoro, Y., Shuto, S., Kubota, K., and Sumino, R.: Reflex organization of cat masticatory muscles. J. Neurophysiol., 31:695, 1968.

Martin, J. P.: Basal Ganglia and Posture. Philadelphia, J. B. Lippincott Company, 1967.

Partridge, L. D., and Kim, J. H.: Dynamic characteristics of response in a vestibulomotor reflex. J. Neurophysiol., 32:485, 1969.

Precht, W.: Vestibular system. In Guyton, A. C. (ed.): MTP International Review of Science: Physiology. Vol. 3. Baltimore, University Park Press, 1974, p. 81.

Riklan, M.: L-Dopa and Parkinsonism. Springfield, Ill., Charles C Thomas, Publisher, 1973.

Sessle, B. J., and Kenny, D. J.: Control of tongue and facial motility: neural mechanisms that may contribute to movements such as swallowing and sucking. Symp. Oral Sens. Percept., 4:222, 1973.

Sherrington, C. S.: Decerebrate ridigity and reflex coordination of movements. J. Physiol. (Lond.), 22:319, 1898.

Suzuki, J. I.: Vestibular and spinal control of eye movements. Bibl. Ophthalmol., 82:109, 1972.

Thexton, A. J.: Some aspects of neurophysiology of dental interest. II. Oral reflexes and neural oscillators. J. Dent., 2:131, 1974.

Thoden, U., Magherini, P. C., and Pompeiano, O.: Cholinergic activaton of vestibular neurons leading to rapid eye movements in the mesencephalic cat. Bibl. Ophthalmol., 82:99, 1972.

Valdman, A. V. (ed.): Pharmacology and Physiology of the Reticular Formation. Progress in Brain Research. Vol. 20. New York, American Elsevier Publishing Co., Inc., 1967.

Wolfson, R. J.: The Vestibular System and Its Diseases. Philadelphia, University of Pennsylvania Press, 1966.

Young, L. R.: Cross coupling between effects of linear and angular acceleration on vestibular nystagmus. Bibl. Ophthalmol., 82:116, 1972.

# 53

# Cortical and Cerebellar Control of Motor Functions

In preceding chapters we have been concerned with many of the subconscious motor activities integrated in the spinal cord and brain stem, especially those responsible for locomotion. In the present chapter we will discuss the control of motor function by the cerebral cortex and cerebellum, much of which is "voluntary" control in contradistinction to the subconscious control effected by the lower centers. Yet we will also see that at least some motor functions of the cerebral cortex and cerebellum are not entirely "voluntary."

## PHYSIOLOGIC ANATOMY OF THE MOTOR AREAS OF THE CORTEX AND THEIR PATHWAYS TO THE CORD

Electrical stimulation anywhere within a large area of the cerebral cortex can at times cause muscle contraction. This area, illustrated by the shading in Figure 53–1, is called the *sensorimotor cortex*. This same area receives sensory signals from the somatic areas of the body.

However, the sensorimotor cortex can be divided into two separate areas: the *motor cortex,* which lies anterior to the central sulcus and the *somatic sensory cortex,* or *somesthetic cortex,* which lies posterior to the sulcus. The reason for this separation is a simple one: Electrical stimulation in the anterior region is far more likely to cause muscle contraction than stimulation in the posterior area. Conversely, stimulation in the posterior area is far more likely to cause a sensory experience. Yet, there is a very strong overlap between the two areas. This is exceedingly important, for as we shall see later in the chapter the motor

activities of the cortex are constantly controlled or modified by signals from the somatic sensory system. The area in Figure 53–1 immediately in front of the central sulcus, designated in the figure by the darkest shading, contains large numbers of *giant Betz cells,* or *pyramidal cells,* for which reason it is also called the *area pyramidalis.* This area causes motor movements following the least amount of electrical excitation and is therefore also called the *primary motor cortex.*

**The Pyramidal Tract (Corticospinal Tract).** One of the major pathways by which motor signals are transmitted from the motor areas of the cortex to the anterior motoneurons of the spinal cord is through the *pyramidal tract,* or *corticospinal tract,* which is illustrated in Figure 53–2. This tract origi-

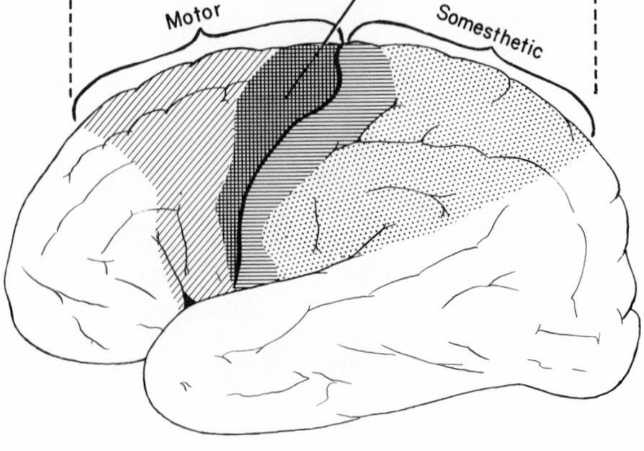

**Figure 53–1.** Relationship of the motor cortex to the somesthetic cortex.

710

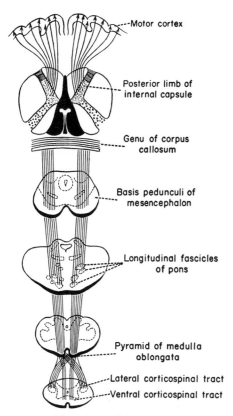

**Figure 53–2.** The pyramidal tract. (From Ranson and Clark: Anatomy of the Nervous System, 1959.)

- Motor cortex
- Posterior limb of internal capsule
- Genu of corpus callosum
- Basis pedunculi of mesencephalon
- Longitudinal fascicles of pons
- Pyramid of medulla oblongata
- Lateral corticospinal tract
- Ventral corticospinal tract

nates in all the shaded areas in Figure 53–1, including both the motor and the somesthetic areas, about three-quarters from the motor area and about one-quarter from the somesthetic regions posterior to the central sulcus. The function of the fibers from the somesthetic cortex is probably not motor, but instead is to cause feedback control of sensory input to the nervous system, as was discussed in Chapter 49.

The most impressive fibers in the pyramidal tract are the large myelinated fibers that originate in the giant Betz cells of the motor area. These account for approximately 17,000 large fibers (mean diameter of about 16 microns) in the pyramidal tract from *each* side of the cortex. However, since each corticospinal tract contains more than a million fibers, only 2 per cent of the total number of corticospinal fibers are of this large type. The other 98 per cent are mainly fibers smaller than 4 microns diameter, 60 per cent of which are myelinated and 40 per cent unmyelinated.

The pyramidal tract passes downward through the *brain stem;* then it decussates mainly to the opposite side to form the *pyramids of the medulla.* By far the majority of the pyramidal fibers then descend in the *lateral corticospinal tracts* of the cord and terminate principally on interneurons at the bases of the dorsal horns of the cord gray matter. A few fibers, however,

do not cross to the opposite side but pass on down the cord ipsilaterally. Another small percentage passes ipsilaterally down the cord in the *ventral corticospinal tracts* but then mainly decussates to the opposite side farther down the cord.

*Collaterals from the Pyramidal Tract in the Brain.* Even before the pyramidal tract leaves the brain, many collaterals are given off as follows:

1. The axons from the giant Betz cells send short collaterals back to the cortex itself. It is believed that these collaterals mainly inhibit adjacent regions of the cortex when the Betz cells discharge, thereby "sharpening" the boundaries of the excitatory signal.

2. A large number of collateral fibers pass into the *striate body,* which is composed of the *caudate nucleus* and *putamen.* From here additional pathways extend down the brain stem and then into the spinal cord through the *extrapyramidal tracts,* which are discussed later in the chapter.

3. A moderate number of collaterals pass from the pyramidal tract into the *red nuclei*. From these, additional pathways pass down the cord through the *rubrospinal tract.*

4. A moderate number of collaterals also deviate from the pyramidal tract into the *reticular substance* of the brain stem; from here signals go to the cord via *reticulospinal tracts* and others go to the cerebellum via *reticulocerebellar tracts.*

5. A tremendous number of collaterals synapse in the pontile nuclei, which give rise to the *pontocerebellar fibers.* Thus, each time signals are transmitted from the motor cortex through the pyramidal tract, simultaneous signals are transmitted into the cerebellar hemispheres.

6. Many collaterals also terminate in the *inferior olivary nuclei,* and from here secondary *olivocerebellar fibers* transmit signals to most areas of the cerebellum.

Thus, the basal ganglia, the brain stem, and the cerebellum all receive strong signals from the pyramidal tract every time a signal is also transmitted down the spinal cord to cause a motor activity.

**Importance of the Area Pyramidalis.** The giant pyramidal cells of the *area pyramidalis* are very excitable. A much less degree of electrical stimulation is required in this area to cause motor movement than in any other area of the brain. Therefore, this area is called the *primary motor area* of the cortex. A single weak shock applied to a focal point in this area is sometimes strong enough to cause a definite motor movement somewhere in the body—for instance, a flick of a finger, deviation of the lip, or some other discrete movement; and a short train of stimuli almost always elicits a movement in some discrete part of the body. In almost no other region of the cortex can one be certain of achieving a motor response with such weak stimuli.

**The Extrapyramidal Tracts.** The extrapyramidal tracts are, collectively, all the tracts besides the pyramidal tract itself that transmit motor signals from the cortex to the spinal cord. In the above discussion of the collaterals from the pyramidal tract, we have noted some of these extrapyramidal pathways that pass through the striate nuclei, red nuclei, and reticular nuclei of the brain stem. In addition to these collateral fibers from the pyramidal tract, still many other fibers course directly from the cortex to these intermediate nuclei. For instance, large numbers of neurons project directly from the motor cortex into the *caudate nucleus* and then through the *putamen, globus pallidus, subthalamic nucleus, red nucleus, substantia nigra,* and *reticular nuclei of the brain stem* before passing into the spinal cord. The multiplicity of connections within these intermediate nuclei of the basal ganglia and reticular substance was presented in the preceding chapter.

The final pathways for transmission of extrapyramidal signals into the cord are the *reticulospinal tracts,* which lie in both the ventral and lateral columns of the cord; and, to a less extent, the *rubrospinal, tectospinal,* and *vestibulospinal tracts.*

## THE FOUR MOTOR AREAS OF THE SENSORIMOTOR CORTEX

Four separate motor areas, which when stimulated are especially likely to cause contraction of specific muscles, have been found in the sensorimotor cortex. The precise locations of these in the human being have been mapped out only crudely because of the obvious impossibility of performing the necessary experiments to make accurate measurements. However, Figure 53–3 illustrates the somatotopical maps of the four separate motor areas in the monkey cortex, as described below:

1. The *primary motor area* (labeled Ms I in the figure), which is located immediately anterior to the central sulcus, with the feet areas extending down into the central sulcus, and the trunk extending toward the front of the brain. The tail and leg areas fold over the longitudinal fissure, and the head and tongue areas are located laterally in the posterior part of the frontal lobe.

2. *The supplementary motor area* (Ms II in the figure), which is located almost entirely on the medial wall of the hemisphere in the longitudinal fissure. This is illustrated in the upper portion of the figure, showing the medial hemisphere folded outward.

3. *The somatic sensorimotor area I* (Sm I), which is located immediately posterior to the central sulcus in the anterior gyrus of the parietal lobe. This area is coextensive with somatic sensory area I; the somatotopical map of the body representation is a mirror image of the map in the primary motor area immediately anterior to the central sulcus.

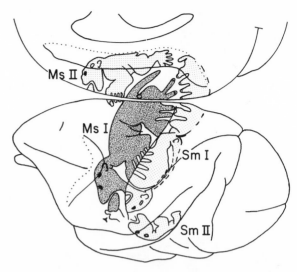

**Figure 53–3.** The different motor areas of the monkey cortex, illustrating especially the primary motor area (Ms I) located in the precentral gyrus. (The upper portion of the figure illustrates the medial surface of the cerebral hemisphere.) (From Woolsey, in Harlow and Woolsey (eds.): Biological and Biochemical Bases of Behavior. University of Wisconsin Press, 1958.)

4. *The somatic sensorimotor area II* (Sm II), which is located in the lateralmost portion of the parietal lobe immediately above the Sylvian fissure. This area is coextensive with somatic sensory area II.

**Importance of the Different Motor Representations.** Despite this multiplicity of motor representations in the cortex, very little is known about any of them except the primary motor area. Stimulation of the supplementary motor area (Ms II) seems to cause avoidance movements—that is, shrinking away from danger. Stimulation of somatic sensorimotor area II causes gross postural motor movements. Sensorimotor area I operates in association with the primary motor area, providing sensory signals that are the basis for many of the primary motor reactions.

The primary motor area is an extremely important one because even a weak stimulus usually elicits a muscle contraction somewhere in the body. Therefore, the remainder of this discussion will be concerned almost entirely with the primary motor area.

## THE PRIMARY MOTOR CORTEX OF THE HUMAN BEING

Figure 53–4 gives an approximate map of the human brain showing the points in the primary motor cortex that cause muscle contractions in

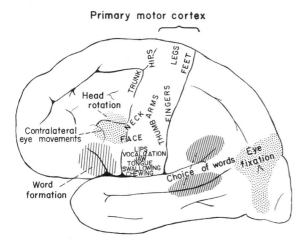

**Figure 53–4.** Representation of the different muscles of the body in the motor cortex and location of other cortical areas responsible for certain types of motor movements.

different parts of the body when electrically stimulated. These have been determined by electrical stimulation of the human brain in patients having brain operations under local anesthesia. Indeed, such stimulation is commonly used by the neurosurgeon to determine the location of the motor cortex, thus allowing him to avoid this area, if possible, during the operation.

Note in the figure that stimulation of the most lateral portions of the motor cortex causes muscular contractions related to swallowing, chewing, and facial movements, whereas stimulation of the midline portion of the motor cortex where it bends over into the longitudinal fissure, causes contraction of the legs, feet, or toes. The spatial organization is similar to that of the somatic sensory cortex I, which was shown in Figure 49–4 of Chapter 49.

**Afferent Fibers that Excite the Primary Motor Cortex.** The primary motor cortex is stimulated to action by signals from many different sources, including:

1. Subcortical fibers from adjacent regions of the cortex, especially from the somatic sensory areas and from the frontal areas—also subcortical fibers from the visual and auditory cortices.

2. Subcortical fibers that pass through the corpus callosum from the opposite cerebral hemisphere. These fibers connect corresponding areas of the motor cortices in the two sides of the brain.

3. Somatic sensory fibers derived directly from the ventrobasal complex of the thalamus. That is, every time a sensory signal is transmitted into somatic sensory area I, collateral signals are also transmitted at less intensity into the primary motor area as well.

4. Tracts from the lateral nuclei of the thalamus, which in turn receive tracts from both the basal ganglia and the cerebellum. These tracts provide signals that are necessary for coordination between the functions of the motor cortex, the basal ganglia, and the cerebellum.

5. Fibers from the nonspecific nuclei of the thalamus. These fibers control the general level of excitability of the motor cortex in the same manner that they also control the general level of excitability of most other regions of the cerebral cortex.

**Vertical Columnar Arrangement of the Cells in the Motor Cortex.** In Chapters 49 and 60 it is pointed out that the cells in the somatic sensory cortex and the visual cortex—and perhaps in all other parts of the brain as well—are organized in vertical columns of cells. In a like manner, the cells of the motor cortex are also organized in vertical columns—columns about 1 mm. in diameter and having several thousand cells in each column.

The columns of cells, in turn, are themselves arranged in six distinct layers, as is the arrangement throughout the cerebral cortex. The functions of these layers from outside inward are:

Layers I through IV are sensory; that is, they receive afferent fibers from other sources. The fibers from the nonspecific nuclei of the thalamus (from the "reticular activating system" that will be discussed in the following chapter) terminate in layers I and II. These fibers provide general enhancement of the degree of excitability of the motor cortex, as occurs also in other cortical regions of the brain when the reticular activating system is stimulated.

Layer IV is the main terminus and layer III the secondary terminus of the sensory afferents from the specific nuclei in the ventrobasal complex of the thalamus, the afferents from the sensory areas of the cortex, the afferents from the lateral nuclei of the thalamus that transmit signals from the cerebellum and the basal ganglia, and the corpus callosal afferents from the opposite cerebral hemisphere.

Layers V and VI are the origins of the efferent fibers or "output fibers," from the motor cortex. The large Betz cells that give rise to the long pyramidal tract fibers lie in layer V. Efferent fibers also pass from both layers V and VI to other areas of the brain.

*Function of the Columns of Cells.* Each column of cells seems to perform a specific motor function, such as stimulating a particular muscle or perhaps stimulating several synergistic muscles. The cells in each column seem to operate both as an integrative and an amplifying system. The integrative function of the column is its ability to combine signals from many different sources and to determine the appropriate output response. Specific cells within a column seem to be responsive to different types of input signals. For instance, some cells respond to somatic sensory signals related to joint movement, others to tactile stimuli, and still others to feedback signals from the cerebellum and basal ganglia. Also, the general level of activation of the motor cortex is provided by signals from the nonspecific nuclei of the thalamus.

The amplifying function of the column of cells allows weak input signals to cause stimulation of large numbers of output neurons. Direct electrical stimulation of a single output neuron in the motor cortex by means of a micropipette electrode will almost never excite a muscle contraction. Instead, at least several output neurons (preferably a large number) need to be excited simultaneously. Therefore, divergence of the signal within the column provides this needed motive power.

**Degree of Representation of Different Muscle Groups in the Primary Motor Cortex.** The different muscle groups of the body are not represented equally in the motor cortex. In general, the degree of representation is proportional to the discreteness of movement required of the respective part of the body. Thus, the thumb and fingers have large representations, as is true also of the lips, tongue, and vocal cords. The relative degrees of representation of the different parts of the body are illustrated in Figure 53–5, a figure constructed by Penfield and Rasmussen on the basis of stimulatory charts made of the human motor cortex during hundreds of brain operations.

When barely threshold stimuli are used, only small segments of the peripheral musculature ordinarily contract at a time. In the "finger" and "thumb" regions, which have tremendous representation in the cerebral cortex, threshold stimuli can sometimes cause single muscles or, at times, even single fasciculi of muscles to contract, thus illustrating that a high degree of control is exercised by this portion of the motor cortex over discrete muscular movement.

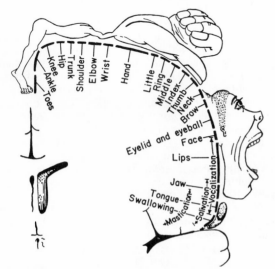

**Figure 53–5.** Degree of representation of the different muscles of the body in the motor cortex. (From Penfield and Rasmussen: The Cerebral Cortex of Man: A Clinical Study of Localization of Function. The Macmillan Co., 1968.)

On the other hand, threshold stimuli in the trunk region of the body might cause as many as 30 to 50 small trunk muscles to contract simultaneously, thus illustrating that the motor cortex does not control discrete trunk muscles but instead controls *groups* of muscles. Similarly, threshold stimuli in the lip, tongue, and vocal cord regions of the motor cortex cause contraction of minute muscular areas, whereas stimulation in the leg region ordinarily excites several synergistic muscles at a time, causing some gross movement of the leg. Also, it should be recognized that the finger muscles have far more representation in the motor cortex than do the muscles of the upper arm.

## COMPLEX MOVEMENTS ELICITED BY STIMULATING THE CORTEX ANTERIOR TO THE MOTOR CORTEX—THE CONCEPT OF A "PREMOTOR CORTEX"

Electrical stimulation of the cerebral cortex for distances 1 to 3 centimeters in front of the primary motor cortex will often elicit complex contractions of groups of muscles. Occasionally, vocalization occurs, or rhythmic movements such as alternate thrusting of a leg forward and backward, coordinate moving of the eyes, chewing, swallowing, or contortion of parts of the body into different postural positions.

Some neurophysiologists have called this area the *premotor cortex* and have ascribed special capabilities to it to control coordinated movements involving many muscles simultaneously. One might also call the premotor cortex a *motor association area*. In fact, it is peculiarly organized to perform such a function for the following reasons: (1) It has long subcortical neuronal connections with the sensory association areas of the parietal lobe; (2) It has direct subcortical connections with the primary motor cortex; (3) It connects with areas in the thalamus contiguous with the thalamic areas that connect with the primary motor cortex; (4) The premotor area has direct connections with the basal ganglia.

Still another reason for the belief that there is a motor association area (a "premotor cortex") is that damage to this area causes loss of certain coordinate skills, as follows:

**Broca's Area and Speech.** Referring again to Figure 53–4, note that immediately anterior to the primary motor cortex and immediately above the Sylvian fissure is an area labeled "word formation." This region is called *Broca's area*. Damage to it does not prevent a person from vocalizing, but it does make it impossible for the person to speak whole words other than simple utterances such as "no" or "yes." A closely associated cortical area also causes appropriate respiratory function so that the vocal cords can be activated simultaneously with the movements of the mouth and tongue during speech. Thus, the activities that are related to Broca's area are highly complex.

**The Voluntary Eye Movement Field.** Immediately above Broca's area is a locus for controlling eye movements. Damage to this area prevents a person from voluntarily moving his eyes toward different objects. Instead, the eyes tend to lock on specific objects, an effect controlled by signals from the occipital region, as explained in Chapter 60. This frontal area also controls eyelid movements such as blinking.

**Head Rotation Area.** Still slightly higher in the "premotor region," electrical stimulation will elicit head rotation. This area is closely associated with the eye movement field and is presumably related to directing the head toward different objects.

**Area for Hand Skills.** In the frontal area immediately anterior to the primary motor cortex for the hands and fingers is a region neurosurgeons have observed to be an area for hand skills. That is, when tumors or other lesions cause destruction in this area, the hand movements become incoordinate and nonpurposeful, a condition called *motor apraxia*.

**Summary of the Premotor Concept.** It is clear that at least some areas anterior to the primary motor cortex can cause complex coordinate movements, such as for speech, for eye movements, for head movements, and perhaps even for hand skills. However, it should be remembered that all of these areas are closely connected with corresponding areas in the primary motor cortex, the thalamus, and the basal ganglia. Therefore, the complex coordinate movements almost certainly result from a cooperative effort of all these substructures.

## EFFECTS OF LESIONS IN THE MOTOR AND PREMOTOR CORTEX

The motor cortex is frequently damaged— especially by the common abnormality called a "stroke," which is caused by loss of blood supply to the cortex. Also, experiments have been performed in animals to remove selectively different parts of the motor cortex.

**Ablation of the Primary Motor Cortex.** Removal of a very small portion of the primary motor cortex in a monkey causes paralysis of the represented muscles. If the sublying cau-

date nucleus is not damaged, gross postural and limb "fixation" movements can still be performed, but the animal loses voluntary control of discrete movements of the distal segments of the limbs—of the hands and fingers, for instance. This does not mean that the muscles themselves cannot contract, but that the animal's ability to control the fine movements is gone.

From these results one can conclude that the primary motor cortex is concerned mainly with voluntary initiation of finely controlled movements. On the other hand, the deeper motor areas, particularly the basal ganglia, seem to be responsible mainly for the involuntary and postural body movements.

**Muscle Spasticity Caused by Ablation of Large Areas of the Motor Cortex—Extrapyramidal Lesions as the Basis of Spasticity.** It should be recalled that the motor cortex gives rise to tracts that descend to the spinal cord through both the pyramidal tract and the extrapyramidal tract. These two tracts have opposing effects on the tone of the body muscles. The pyramidal tract causes continuous facilitation and, therefore, a tendency to increase muscle tone throughout the body. On the other hand, the extrapyramidal system transmits inhibitory signals through the basal ganglia and the bulboreticular system of the brain stem, with resultant inhibition of muscle action. When the motor cortex is destroyed, the balance between these two opposing effects may be altered. If the lesion is located discretely in the primary motor cortex where the large Betz cells lie, very little change in muscle tone results because both pyramidal and extrapyramidal elements are affected about equally. On the other hand, the usual lesion is very large and involves large portions of the sensorimotor cortex, both anterior and posterior to the primary motor area, and both of these regions normally transmit inhibitory signals through the extrapyramidal tracts. Therefore, loss of extrapyramidal inhibition is the dominant feature, thus leading to muscle spasm.

If the lesion involves the basal ganglia as well as the motor cortex itself, the spasm is even more intense because the basal ganglia normally provide additional very strong inhibition of the bulboreticular system, and loss of this inhibition further exacerbates the reticular excitation of the muscles. In patients with strokes, the lesion almost invariably affects both the motor cortex itself and the sub-lying basal ganglia, so that very intense spasm normally occurs in the muscles of the opposite side of the body.

The muscle spasm that results from damage to the extrapyramidal system is different from the spasticity that occurs in decerebrate rigidity. In decerebrate rigidity the spasm is of the antigravity muscles and is called *extensor spasm*. On the other hand, the spasm resulting from lesions of the extrapyramidal system usually involves the flexors as well, causing intense stiffening of the limbs and other parts of the body.

**The Babinski Sign.** Destruction of the foot region of the area pyramidalis or transection of the foot portion of the pyramidal tract causes a peculiar response of the foot called the *Babinski sign*. This response is demonstrated when a firm tactile stimulus is applied to the sole of the foot: The great toe extends upward and the other toes fan outward. This is in contradistinction to the normal effect in which all the toes bend downward. Also, the Babinski sign does not occur when the damage is only in the extrapyramidal system. Therefore, the sign is often used clinically to detect damage specifically in the pyramidal portion of the motor control system. Despite the clinical significance of the Babinski sign, however, its physiological mechanism is not known.

# STIMULATION OF THE SPINAL MOTONEURONS BY MOTOR SIGNALS FROM THE BRAIN

Figure 53–6 shows several different motor tracts entering a segment of the spinal cord from the brain. The corticospinal tract (pyramidal tract) terminates mainly on small interneurons in the base of the dorsal horns, although as many as 15 per cent may also terminate directly on the anterior motoneurons themselves. From the primary interneurons, most of the motor signals are transmitted through still other interneurons before finally exciting the anterior motoneurons.

Figure 53–6 shows several other descending tracts from the brain, collectively called the extrapyramidal tracts, which also carry signals to the anterior motoneurons: (1) the rubrospinal tract, (2) the reticulospinal tract, (3) the tectospinal tract, and (4) the vestibulospinal tract. In addition, sensory signals arriving through the dorsal sensory roots, as well as signals transmitted from segment to segment of the spinal cord via the propriospinal tracts, also stimulate the anterior motoneurons. In general, the extrapyramidal tracts terminate farther anteriorly in the gray matter of the spinal cord than does the corticospinal tract; a larger share of these fibers terminate directly on the anterior motoneurons. Also, some terminate on *inhibitory neurons*, which in turn inhibit the anterior motoneurons, an effect opposite to the excitatory effect of the corticospinal tract.

The corticospinal tract seems to cause very specific muscle contractions; neurophysiological recordings have demonstrated that the intensity of muscle contraction—everything else being equal—is directly proportional to the nervous energy transmitted through the corticospinal tract. On the other hand, the extrapyramidal tracts terminate in the cord gray matter near the anterior motoneurons and provide less specific muscle contractions. Instead, they provide such effects as general facilitation, general inhibition, or gross postural signals, all of which provide the background against which the corticospinal system operates. Finally, the sensory and propriospinal signals also add their input to the total melee of spinal integration.

**Patterns of Movement Elicited by Spinal Cord Centers.** From Chapter 51, recall that the spinal cord can provide specific reflex patterns of movement in response to sensory nerve stimulation. Many of these patterns are also important when the anterior motoneurons are excited by signals from the brain. For instance, the stretch reflex is functional at all times, helping to damp the motor movements initiated from the brain and probably providing at least part of the motive power required to cause the muscle contractions employing the servo-assist mechanism that was described in Chapter 51.

Also, when a brain signal excites an agonist muscle, it is not necessary to transmit an inverse signal to the antagonist at the same time;

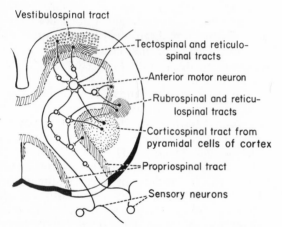

Vestibulospinal tract

Tectospinal and reticulospinal tracts

Anterior motor neuron

Rubrospinal and reticulospinal tracts

Corticospinal tract from pyramidal cells of cortex

Propriospinal tract

Sensory neurons

**Figure 53–6.** Convergence of all the different motor pathways on the anterior motoneurons.

this transmission will be achieved by the reciprocal innervation circuit that is always present in the cord for coordinating the functions of antagonistic pairs of muscles.

Finally, parts of the other reflex mechanisms such as withdrawal, stepping and walking, scratching, postural mechanisms, and so forth, are at times activated by signals from the brain. Thus, very simple signals from the brain can lead, at least theoretically, to many of our normal motor activities, particularly for such functions as walking and the attainment of different postural attitudes of the body.

On the other hand, at times it is important to suppress the cord mechanisms to prevent their interference with the performance of patterns of motor activity generated within the brain itself. It has been suggested that this is the specific function of some of the inhibitory signals transmitted through some reticulospinal fibers.

# THE CEREBELLUM AND ITS MOTOR FUNCTIONS

The cerebellum has long been called a *silent area* of the brain principally because electrical excitation of this structure does not cause any sensation and rarely any motor movement. However, as we shall see, removal of the cerebellum does cause the motor movements to become highly abnormal—so much so that some very prominent neurophysiologists consider the cerebellum to be as important as all other motor control elements of the nervous system. The cerebellum is especially vital to the control of very rapid muscular activities such as running, typing, playing the piano, and even talking. Loss of this area of the brain can play havoc with each of these activities even though its loss causes paralysis of no muscles.

But how is it that the cerebellum can be so important when it has no direct control over muscle contraction? The answer to this is that it *monitors and makes corrective adjustments in the motor activities elicited by other parts of the brain*. It receives continuously updated information from the peripheral parts of the body to determine the instantaneous status of each part of the body—its position, its rate of movement, forces acting on it, and so forth. The cerebellum *compares* the actual physical status of each part of the body as depicted by the sensory information with the status that is intended by the motor system. If the two do not compare favorably, then appropriate corrective signals are transmitted instantaneously back into the motor

system to increase or decrease the levels of activation of the specific muscles.

Since the cerebellum must make major motor corrections extremely rapidly *during the course of motor movements*, a very extensive and rapidly acting cerebellar input system is required both from the peripheral parts of the body and from the cerebral motor areas. Also, an extensive output system feeding equally as rapidly into the motor system is necessary to provide the necessary corrections of the motor signals.

## THE INPUT SYSTEM TO THE CEREBELLUM

The human cerebellum is generally divided into two major lobes, as illustrated in Figure 53–7, the *anterior lobe,* which is small, and the *posterior lobe,* which is much larger. The posterior cerebellum is greatly enlarged bilaterally to form the *cerebellar hemispheres,* which are also known as the *neocerebellum* (new cerebellum) because they represent a phylogenetically new portion of the cerebellum. The older portions of the cerebellum, including the small midline section of the posterior cerebellum and the entire anterior cerebellum, are called the *paleocerebellum* (old cerebellum).

**The Afferent Pathways.** The basic afferent pathways to the cerebellum are also illustrated in Figure 53–7. An extensive and important afferent pathway is the *corticocerebellar pathway,* which originates mainly in the *motor cortex* (but to a lesser extent, in the sensory cortex as well) and then passes by way of the *pontile nuclei* and *pontocerebellar tracts* directly to the cortex of the cerebellum. In addition, important afferent tracts originate in the

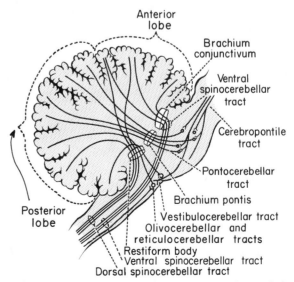

**Figure 53–7.** The principal afferent tracts to the cerebellum.

brain stem; they include: (a) an extensive *olivo-cerebellar* tract, which passes from the *inferior olive* to all parts of the cerebellum; this tract is excited by fibers from the *motor cortex,* the *basal ganglia,* widespread areas of the *reticular formation,* and the spinal cord through the *spino-olivary tract;* (b) *vestibulocerebellar fibers,* some of which originate in the vestibular apparatus itself; most of these terminate in the *flocculonodular lobe* and *fastigial nucleus* of the cerebellum; (c) *reticulocerebellar fibers,* which originate in different portions of the reticular formation and terminate mainly in the midline structures of the paleocerebellum.

The cerebellum also receives important sensory signals directly from the peripheral parts of the body, which reach the cerebellum by way of the *ventral* and *dorsal spinocerebellar* tracts illustrated in Figures 53–7 and 53–8. The signals transmitted in these tracts originate in the muscle spindles, the Golgi tendon organs, and the large tactile receptors of the skin and joints, and they apprise the cerebellum of the momentary status of muscle contraction, degree of tension on the muscle tendons, positions of the parts of the body, and forces acting on the surfaces of the body. All of this information keeps the cerebellum constantly apprised of the instantaneous physical status of the body.

Note in Figure 53–8 that the nerve fibers from the sensory receptors terminate on Clarke's cells at the bases of the dorsal horns and that most of these fibers pass ipsilaterally up to the cerebellum. These pathways can transmit impulses at velocities over 100 meters per second, which is the most rapid conduction of any pathway in the entire central nervous system. This extremely rapid conduction is also impor-

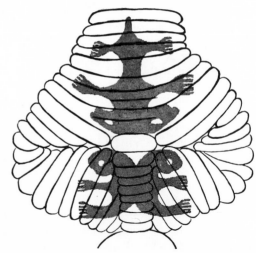

**Figure 53–9.** The sensory projection areas, called "homunculi," on the cortex of the cerebellum. (From Snider: *Sci. Amer.,* 199:4, 1958. Copyright © 1958 by Scientific American, Inc. All rights reserved.)

tant for the instantaneous apprisal of the cerebellum of changes that take place in the status of the body.

In addition to the signals in the spinocerebellar tracts, other signals are transmitted through the dorsal columns to the medulla and then relayed from there to the cerebellum. Likewise, signals are transmitted through the *spinoreticular pathway* to the reticular substance of the brain stem and through the *spino-olivary pathway* to the inferior olivary nucleus and then relayed to the cerebellum. Thus, the cerebellum collects continual information about all parts of the body even though it is operating at a subconscious level.

**Spatial Localization of Sensory Input to the Cerebellum.** Afferent fibers entering the cerebellum terminate in distinct spatially oriented areas of the cerebellar cortex. Figure 53–9 illustrates the loci of two different "homunculi," as recorded in the monkey, which show the terminations of sensory fibers from all parts of the body. Figure 53–10 illustrates the human cerebellum, showing the approximate sensory terminations based on anatomical data and extrapolation from animals. Note especially that there are two separate representations of the body in the cerebellum. One of these lies mainly in the anterior cerebellum but extends also into the upper part of the posterior cerebellum, and the other is split in half and lies entirely in the posterior cerebellum.

It is also important to note that the signals from the motor cortex project to corresponding points in (a) the peripheral muscles and joints of the body, and (b) the sensory representations for the same muscles and joints in the cerebellum. Therefore, each respective point in the cerebellum in reality represents a specific muscle or a specific joint; simultaneously it receives information directly from the motor cortex concern-

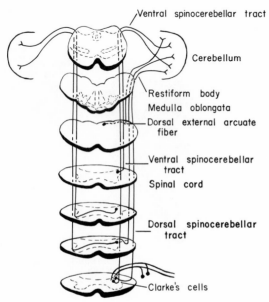

Ventral spinocerebellar tract

Cerebellum

Restiform body

Medulla oblongata

Dorsal external arcuate fiber

Ventral spinocerebellar tract

Spinal cord

Dorsal spinocerebellar tract

Clarke's cells

**Figure 53–8.** The spinocerebellar tracts.

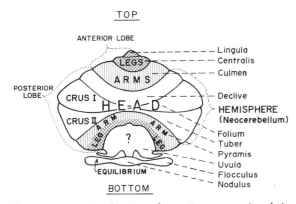

**Figure 53–10.** Localization of somatic sensory signals in the cerebellum of the human being.

ing the motor signals to activate the muscle or joint and from the muscle or joint concerning the effect actually produced. Then it projects signals back into the motor pathway to the same respective topographical part of the system.

*The Equilibrium Area of the Cerebellum.* Note in Figure 53–10 at the lowest part of the cerebellum (or most posterior part) an area labeled "equilibrium." This is the flocculonodular lobe of the cerebellum, a very old part that connects through bidirectional pathways with the vestibular nuclei. As was discussed in the previous chapter, this area is concerned almost entirely with equilibrium, and its destruction causes serious loss of equilibrium particularly when rapid movements are being performed.

### OUTPUT SIGNALS FROM THE CEREBELLUM

**The Deep Cerebellar Nuclei and the Efferent Pathways.** Located deep in the cerebellar mass are four *deep cerebellar nuclei*–the *dentate, globose, emboliform,* and *fastigial nuclei.* These nuclei receive signals from two different sources: (1) the cerebellar cortex, and (2) all the sensory afferent tracts to the cerebellum. Each time an input signal arrives in the cerebellum, it goes both to the cerebellar cortex and also directly to the deep nuclei through collateral fibers; then, a short time later, the cerebellar cortex relays its output signals also to the the deep nuclei. Thus, all the input signals that enter the cerebellum eventually end in the deep nuclei. We shall discuss this circuit in greater detail below.

All of the efferent tracts from the cerebellum arise in the deep nuclei—none from the cerebellar cortex, which transmits its output signals only through the deep nuclei. Efferent signals are transmitted to many portions of the motor system including (1) the motor cortex, (2) the basal ganglia, (3) the red nucleus, (4) the reticular formation of the brain stem, and (5) the

vestibular nuclei. The important tracts, illustrated in Figure 53–11, are the following:

*1. To the Motor Cortex.* Fibers originate in the dentate nuclei and pass to the ventrolateral nucleus of the thalamus where they synapse with second order neurons that pass directly to the motor cortex.

*2. To the Basal Ganglia.* Signals pass from the dentate nucleus to the centromedian nucleus of the thalamus from which they are relayed to the striate body of the basal ganglia.

*3. To the Red Nucleus.* Fibers originating also in the dentate nucleus terminate in the *red nucleus.* This in turn transmits signals to the spinal cord through the rubrospinal tract, and it also sends short axons into the reticular formation.

*4. To the Reticular Formation.* Fibers enter the reticular formation, probably from all four of the deep cerebellar nuclei—the dentate, globose, and emboliform nuclei through the brachium conjunctivum into the upper part of the reticular formation, and from the fastigial nuclei into the lower reticular formation via the restiform body.

*5. To the Vestibular Nuclei.* As has already been pointed out in the previous chapter, the *fastigial nuclei* have bidirectional connections with the vestibular nuclei. The fastigial nuclei subserve primarily the flocculonodular lobe of the cerebellum and other closely allied regions of the older cerebellum.

### THE NEURONAL CIRCUIT OF THE CEREBELLUM

The structure of the cerebellar cortex is entirely different from the cerebral cortex. Furthermore, each part of the cerebellar cortex has a neuronal organization almost precisely the same as that in all other parts.

The human cerebellum is actually a large folded sheet, approximately 17 cm. wide by 120

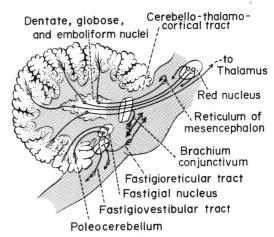

**Figure 53–11.** Principal efferent tracts from the cerebellum.

cm. long, the folds lying crosswise, as illustrated in Figures 53–9 and 53–10. Each fold is called a *folium*.

**The Functional Unit of the Cerebellar Cortex—the Purkinje Cells.** The cerebellum has approximately 30 million nearly identical functional units, one of which is illustrated in Figure 53–12. This functional unit centers around the Purkinje cell, of which there are also 30 million in the cerebellar cortex.

Note to the far right in Figure 53–12 the three major layers of the cerebellum: the *molecular layer,* the *Purkinje cell layer,* and the *granular cell layer*. In addition to these layers, the deep nuclei are located far within the center of the cerebellar mass.

**The Neuronal Circuit of the Functional Unit.** As illustrated in Figure 53–12, the output from the functional unit is from a deep nuclear cell. However, this cell is continually under the influence of both excitatory and inhibitory influences. The excitatory influences arise from direct connections with the afferent fibers that enter the cerebellum. The inhibitory influences arise entirely from the Purkinje cells in the cortex of the cerebellum.

The afferent inputs to the cerebellum are of two types, one called the *climbing fiber type* and the other called the *mossy fiber type*. There is one climbing fiber for about 10 Purkinje cells. After sending collaterals to several deep nuclear cells, the climbing fiber projects all the way to the molecular layer of the cerebellar cortex where it makes about 300 synapses with the dendrites of each Purkinje cell. This climbing fiber is distinguished by the fact that a single

impulse in it will always cause a single action potential in each Purkinje cell with which it connects. Another distinguishing feature of the climbing fibers is that they all originate in the inferior olive of the medulla.

The mossy fibers also send collaterals to excite the deep nuclear cells. Then these fibers proceed to the granular layer of the cortex where they synapse with hundreds of *granule cells*. These in turn send their very small axons, less than 1 micron in diameter, up to the outer surface of the cerebellar cortex to enter the surface-lining molecular layer. Here the axons divide into two branches that extend 1 to 2 millimeters in each direction parallel to the folia. There are literally millions of these *parallel nerve fibers* in each small segment of the cerebellar cortex (there are about 1000 granule cells for every Purkinje cell). It is into this molecular layer that the dendrites of the Purkinje cells project, and each Purkinje cell synapses with 80,000 to 200,000 of these parallel fibers. Yet, the mossy fiber input to the Purkinje cell is quite different from the climbing fiber input because stimulation of a single mossy fiber will never elicit an action potential in the Purkinje cell; instead, large numbers of mossy fibers must be stimulated simultaneously to activate the Purkinje cell. Furthermore, this activation usually takes the form of prolonged facilitation or excitation that, when it reaches threshold for stimulation, causes repetitive Purkinje cell firing rather than the single action potential occurring in response to the climbing fiber input.

Thus, the Purkinje cells are stimulated by two types of input circuits—one that causes a highly specific output in response to the incoming signal and the other that causes a less specific but tonic type of response. It should be noted that the greater proportion of the afferent input to the cerebellum is of the mossy fiber type because this represents the afferent input from all of the cerebellar afferent tracts besides those from the inferior olive.

**Balance Between Excitation and Inhibition in the Deep Cerebellar Nuclei.** Referring again to the circuit of Figure 53–12, one should note that direct stimulation of the deep nuclear cells by both the climbing and the mossy fibers excites them, whereas the signals arriving from the Purkinje cells inhibit them. Normally, there is a continual balance between these two effects so that the degree of output from the deep nuclear cell remains relatively constant. On the other hand, in the execution of rapid motor movements, the *timing* of the two effects on the

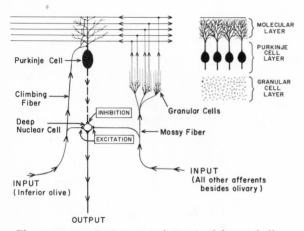

**Figure 53–12.** Basic neuronal circuit of the cerebellum. Also shown in the upper right hand corner are the three major layers of the cerebellar cortex.

deep nuclei is such that the excitation appears before the inhibition. Also, the relative balance between excitation and inhibition probably changes. In this way, very rapid excitatory and inhibitory transient signals can be fed back into the motor pathways to correct motor movements that are going awry. The inhibitory portions of these signals resemble delay-line negative feedback signals of the type that are very effective in providing damping. That is, when the motor system is excited, a negative feedback signal occurs after a short delay to stop the muscle movement from overshooting its mark, the usual cause of oscillation.

**Other Inhibitory Cells in the Cerebellar Cortex.** In addition to the granule cells and Purkinje cells, three other types of neurons are also located in the cerebellar cortex: *basket cells, stellate cells,* and *Golgi cells.* All of these are inhibitory cells with very short axons. After being excited, they inhibit either the Purkinje cells in the case of the stellate and basket cells, or the granule cells in the case of the Golgi cells. There is a strong possibility that this inhibition provides a sharpening of the signal. That is, these cells possibly or probably prevent crosstalk between individual functional units of the cerebellum.

**Special Features of the Cerebellar Neuronal Circuit.** A special feature of the cerebellum is that there are no reverberatory pathways in the cerebellar neuronal circuits, so that the input-output signals of the cerebellum are very rapid transients that never persist for long periods of time.

Another special feature is that most of the cells of the cerebellum are constantly active. This is especially true of the deep nuclear cells; they continually send output signals to the other areas of the motor system. The importance of this is that decrease of the nuclear cell firing rate can provide an inhibitory output signal from the cerebellum, while an increase in firing rate can provide an excitatory output signal.

## FUNCTION OF THE CEREBELLUM IN VOLUNTARY MOVEMENTS

The cerebellum functions only in association with motor activities initiated elsewhere in the central nervous system. These activities may be initiated originally in the spinal cord, in the reticular formation, in the basal ganglia, or in the major areas of the cerebral cortex. We will discuss, first, the operation of the cerebellum in association with the motor cortex for the control of voluntary movements.

Figure 53–13 illustrates the basic cerebellar pathways involved in cerebellar control of voluntary movements. When motor impulses are transmitted from the cerebral cortex downward through the pyramidal and extrapyramidal tracts to excite the muscles, collateral impulses are transmitted simultaneously into the cerebellum through the pontocerebellar and olivocerebellar tracts. Therefore, for every motor movement that is performed, not only do the muscles receive activating signals, but the cerebellum receives similar signals at the same time.

When the muscles respond to the motor signals, the muscle spindles, Golgi tendon apparatuses, joint receptors, and other peripheral receptors transmit signals upward mainly through the spinocerebellar and spino-olivary pathways to terminate in the cerebellum.

After the signals from the periphery and those from the motor cortex are integrated, efferent impulses are transmitted from the cerebellar cortex to the dentate nuclei and thence upward through the ventrolateral nuclei of the thalamus back to the motor cortex where the stimulus first originated.

**"Error Control" by the Cerebellum.** One will readily recognize that the circuit described above represents a complicated feedback circuit beginning in the motor cortex and also returning to the motor cortex. Furthermore,

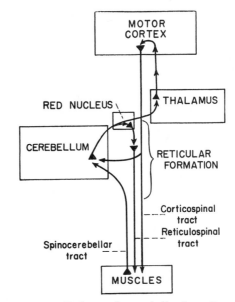

**Figure 53–13.** Pathways for cerebellar "error" control of voluntary movements.

experiments have shown that the cerebellum operates in much the same manner as a servomechanism such as that used in (a) industrial control systems, (b) the control system of an anti-aircraft gun, or (c) the control system of the automatic pilot. That is, the cerebellum seems to compare the "intentions" of the cortex with the "performance" by the parts of the body, and, if the parts are not moving according to the intentions of the cortex, the "error" between these two is calculated by the cerebellum so that appropriate and immediate corrections can be made. Thus, if the cortex has transmitted a signal intending the limb to move to a particular point, but the limb begins to move too fast and obviously is going to overshoot the point of intention, the cerebellum can initiate "braking impulses" to slow down the movement of the limb and stop it at the point of intention.

Ordinarily, the motor cortex transmits far more impulses than are needed to perform each intended movement, and the cerebellum therefore must act to inhibit the motor cortex at the appropriate time after the muscle has begun to move. The cerebellum automatically assesses the rate of movement and calculates the length of time that will be required to reach the point of intention. Then appropriate inhibitory impulses are transmitted to the motor cortex to inhibit the agonist muscle and to excite the antagonist muscle. In this way, appropriate "brakes" are applied to stop the movement at the precise point of intention.

Experiments have demonstrated two important characteristics of the cerebellar feedback system:

1. When a person performs a particular act, such as moving a heavy weight rapidly to a new position, at first he is unable to judge the degree of inertia that will be involved in the movement. Therefore, he almost always overshoots the point of intention on his first trial. However, after several trials, he can stop the movement at the precise point. This illustrates that learned knowledge of the inertia of the system is an important feature of the cerebellar feedback mechanism, though it is likely that this learning occurs in the cerebral cortex rather than in the cerebellum.

2. When a rapid movement is made toward a point of intention, the agonist muscle contracts strongly throughout the early course of movement. Then, suddenly, shortly before the point of intention is reached, the agonist muscle becomes completely inhibited while the antagonist muscle becomes strongly excited. Furthermore, the point at which this reversal of excitation occurs depends on the rate of movement and on the previously learned knowledge of the inertia of the system. The faster the movement and the greater the inertia, the earlier the reversal point appears in the course of movement.

Since all these events transpire much too rapidly for the motor cortex to reverse the excitation "voluntarily," it is evident that the excitation of the antagonist muscle toward the end of a movement is an entirely automatic and subconscious function and is not a "willed" contraction of the same nature as the original contraction of the agonist muscle. We shall see below that in patients with serious cerebellar damage, excitation of the antagonist muscles does not occur at the appropriate time but instead always too late. Therefore, it is almost certain that one of the major functions of the cerebellum is automatic excitation of antagonist muscles at the end of a movement while at the same time inhibiting agonist muscles that have started the movement.

**The "Damping" Function of the Cerebellum.** One of the by-products of the cerebellar feedback mechanism is its ability to "damp" muscular movements. To explain the meaning of "damping" we must first point out that essentially all movements of the body are "pendular." For instance, when an arm is moved, momentum develops, and the momentum must be overcome before the movement can be stopped. And, because of the momentum, all pendular movements have a tendency to overshoot. If overshooting does occur in a person whose cerebellum has been destroyed, the conscious centers of the cerebrum eventually recognize this and initiate a movement in the opposite direction to bring the arm to its intended position. But again the arm, by virtue of its momentum, overshoots, and appropriate corrective signals must again be instituted. Thus, the arm oscillates back and forth past its intended point for several cycles before it finally fixes on its mark. This effect is called an *action tremor,* or *intention tremor.*

However, if the cerebellum is intact, appropriate subconscious signals stop the movement precisely at the intended point, thereby preventing the overshoot and also the tremor. This is the basic characteristic of a damping system. All servocontrol systems regulating pendular elements that have inertia must have damping circuits built into the servomechanisms. In the motor control system of our central nervous system, the cerebellum seems to provide much of this damping function.

*Relationship of Cerebellar Damping to Damping Caused by the Stretch Reflex.* The stretch reflex, which is strictly a cord reflex, also provides much of the damping of body movements, as pointed out in Chapter 51. The cerebellar damping mechanism seems to be superimposed onto the basic cord damping mechanism. The muscle spindles excite both these functions. Rapid stretch of a muscle transmits spindle signals both directly to the anterior motoneuron to cause the damping action of the stretch reflex and simultaneously to the cerebellum to cause additional damping. The cerebellum probably provides its damping action by transmitting reflex signals into the reticular formation and vestibular nuclei and thence back down the cord to stimulate or inhibit both the alpha and the gamma motoneurons.

The cerebellar damping mechanism at times supports the stretch reflex damping system and at other times inhibits it, thus illustrating that the cerebellar system has considerable latitude of control that makes it adaptable to more complex motor activities than is true of the stretch reflex.

**Function of the Cerebellum in Prediction.** Another important by-product of the cerebellar feedback mechanism is its ability to help the central nervous system predict the future positions of moving parts of the body. Without the cerebellum a person "loses" his limbs *when they move rapidly*, indicating that feedback information from the periphery probably must be analyzed by the cerebellum if the brain is to keep up with the motor movements. Thus, the cerebellum detects from the incoming proprioceptive signals the rapidity with which the limb is moving and then seems to predict from this the projected time course of movement. This allows the cerebellum, operating through the cerebellar output circuits to inhibit the agonist muscles and to excite the antagonist muscles when the movement approaches the point of intention.

Without the cerebellum this predictive function is so deficient that moving parts of the body move much farther than the point of intention. This failure to control the distance that the parts of the body move is called *dysmetria,* which means simply poor control of the distance of movement. As would be expected, dysmetria is much more pronounced in rapid movements than in slow movements.

*Failure of Smooth "Progression of Movements."* One of the most important features of normal motor function is one's ability to progress from one movement to the next in an orderly succession. When the cerebellum becomes dysfunctional and the person loses his subconscious ability to predict ahead of time how far the different parts of his body will move in a given time, he becomes unable also to control the beginning of the next movement. As a result, the succeeding movement may begin much too early or much too late. Therefore, movements such as those required for writing, running, or even talking all become completely incoordinate, lacking completely in the ability to progress in an orderly sequence from one movement to the next.

**Extramotor Predictive Functions of the Cerebellum.** The cerebellum also plays a role in predicting other events besides simply movements of the body. For instance, the rates of progression of both auditory and visual phenomena can be predicted. As an example, a person can predict from the changing visual scene how rapidly he is approaching an object. A striking experiment that demonstrates the importance of the cerebellum in this ability is the effect of removing the "head" portion of the cerebellum in monkeys. Such a monkey occasionally charges the wall of a corridor and literally bashes its brains out because it is unable to predict when it will reach the wall.

Unfortunately, we are only now beginning to learn about these extramotor predictive functions of the cerebellum. It is quite possible that the cerebellum provides a "time base" against which signals from other parts of the central nervous system can be compared. It is often stated that the cerebellum is especially important in interpreting spatiotemporal relationships in sensory information.

## FUNCTION OF THE CEREBELLUM IN INVOLUNTARY MOVEMENTS

The cerebellum functions in involuntary, subconscious, or postural movements in almost exactly the same manner as it does in voluntary movements except that slightly different pathways are involved, as shown in Figure 53–14. The extrapyramidal signals that originate in the motor cortex, basal ganglia, or reticular formation to cause involuntary movements pass mainly into the inferior olive and thence into the cerebellum, primarily to the paleocerebellum. Then as the muscle movements occur, proprioceptive information from the muscles, joints, and other peripheral parts of the body passes to the cerebellum, thus providing the same type of "error" control for involuntary movements as is provided for voluntary movements, as discussed above. Signals then feed from the cerebellum back to the *motor cortex, basal ganglia,* and *reticular formation.* They feed back into the basal ganglia mainly by way of the *emboliform*

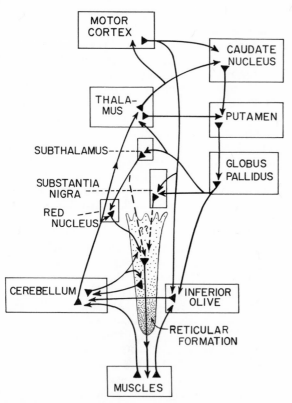

**Figure 53–14.** Pathways for cerebellar "error" control of involuntary movements.

*nuclei,* then upward through the *centromedian nuclei* of the thalamus, and thence into the *putamen.*

One of the principal pathways through which corrective signals are transmitted to the muscles is *reticulospinal fibers* that control the *gamma efferent system* to the muscle spindles. Indeed, the cerebellum is one of the most powerful of all the activators of this system, which indicates a special importance of the gamma efferents in cerebellar functions.

**Function of the Cerebellum for Control of Spinal Reflex Movements.** Even movements elicited by spinal reflexes are modified to some extent by the cerebellum. For instance, in the absence of the cerebellum, sudden movement of a limb by some extraneous force is met with almost no resistance, but in the presence of the cerebellum the stretch reflexes oppose the movement almost instantaneously. Thus, the cerebellum, operating through a cerebellar-reticular feedback system, greatly enhances the power of the spinal stretch reflex.

On the other hand, many movements integrated in the spinal cord, such as scratching and cleaning of the face with the paws, can be performed completely coordinately in the absence of a cerebellum.

## FUNCTION OF THE CEREBELLUM IN EQUILIBRIUM

In the discussion of the vestibular apparatus and of the mechanism of equilibrium in Chapter 52, it was

pointed out that the flocculonodular lobes of the cerebellum are required for proper integration of the equilibratory impulses. Removal of the *nodulus* in particular causes almost complete loss of equilibrium, and removal of the *flocculus* causes temporary loss. The symptoms of mal-equilibrium are essentially the same as those that result from destruction of the semicircular canals. This indicates that these areas of the cerebellum are particularly important for integration of *changes in direction of motion* as detected by the semicircular canals, but not so important for integrating static signals of equilibrium from the maculae in the utricles and saccules. This is in keeping with the other functions of the cerebellum; that is, the semicircular canals allow the central nervous system to *predict* ahead of time that rotational movements of the body are going to cause mal-equilibrium, and this predictive function causes appropriate muscles to contract to prevent the person from losing equilibrium even before this state develops. This is very much the same as the predictive function of the cerebellum for coordination of rapid voluntary movements.

## CLINICAL ABNORMALITIES OF THE CEREBELLUM

Destruction of small portions of the cerebellar cortex causes no detectable abnormality in motor function. In fact, several months after as much as half the cerebellar cortex has been removed, the motor functions appear to be almost entirely normal—but only as long as the person performs all movements very slowly. Thus, the remaining areas of the cerebellum compensate tremendously for loss of part of the cerebellum.

To cause serious and continuous dysfunction of the cerebellum, the cerebellar lesion must usually involve the deep cerebellar nuclei—the *dentate, emboliform, globose,* and *fastigial nuclei*–as well as the cerebellar cortex.

**Dysmetria and Ataxia.** Two of the most important symptoms of cerebellar disease are dysmetria and ataxia. It was pointed out above that in the absence of the cerebellum a person cannot predict ahead of time how far his movements will go. Therefore, the movements ordinarily overshoot their intended mark. This effect is called *dysmetria,* and it results in incoordinate movements which are called *ataxia.*

Dysmetria and ataxia can also result from lesions in the spinocerebellar tracts, for the feedback information from the moving parts of the body is essential for accurate control of the muscular movements.

*Past Pointing.* Past pointing means that in the absence of the cerebellum a person ordinarily moves his hand or some other moving part of his body considerably beyond the point of intention. This probably results from the following effect: The motor cortex normally transmits more impulses to the muscles to perform a given motor function than are actually needed. The cerebellum automatically corrects this by inhibiting the movement after it has begun. However, if the cerebellum is not available to cause this inhibition, the movement ordinarily goes far beyond the intended point. Therefore, past pointing is actually a manifestation of dysmetria.

**Failure of Progression.** *Dysdiadochokinesia.* When the motor control system fails to predict ahead of time where the different parts of the body will be at a given time, it temporarily "loses" the parts during rapid motor movements. As a result, the succeeding movement may begin much too early or much too late so that no orderly "progression of movement" can occur. One can demonstrate this readily in a patient with cerebellar damage by having him turn one of this hands upward and downward at a rapid rate. He rapidly "loses" his hand and does not know its position during any portion of the movement. As a result, a series of jumbled movements occurs instead of the normal coordinate upward and downward motions. This is called dysdiadochokinesia.

*Dysarthria.* Another instance in which failure of progression occurs is in talking, for the formation of words depends on rapid and orderly succession of individual muscular movements in the larynx, mouth, and respiratory system. Lack of coordination between these and inability to predict either the intensity of the sound or the duration of each successive sound cause jumbled vocalization, with some syllables loud, some weak, some held long, some held for a short interval, and resultant speech that is almost completely unintelligible. This is called dysarthria.

**Intention Tremor.** When a person who has lost his cerebellum performs a voluntary act, his muscular movements are jerky; this reaction is called an *intention tremor* or an *action tremor,* and it results from failure of the cerebellar system to damp the motor movements. Tremor is particularly evident when the dentate nuclei or the brachium conjunctivum is destroyed, but it is not present when the spinocerebellar tracts from the periphery to the cerebellum are destroyed. This indicates that the feedback pathway from the cerebellum to the motor cortex is a principal pathway for damping of muscular movements.

*Cerebellar Nystagmus.* Cerebellar nystagmus is a tremor of the eyeballs that occurs usually when a person attempts to fixate his eyes on a scene to the side of this head. This off-center type of fixation results in rapid tremorous movements of the eyes rather than a steady fixation, and it is probably another manifestation of the failure of damping by the cerebellum. It also occurs especially when the flocculonodular lobes are damaged; and in this instance it is associated with loss of equilibrium, presumably because of dysfunction of the pathways through the cerebellum from the semicircular canals. Nystagmus resulting from damage to the semicircular canals was discussed in Chapter 52.

**Rebound.** If a person with cerebellar disease is asked to contract his arm tightly while the physician holds it back at first and then lets go, the arm will fly back until it strikes the face instead of being automatically stopped. This is called rebound, and it results from loss of the cerebellar component of the stretch reflex. That is, the normal cerebellum ordinarily instantaneously and powerfully sensitizes the spinal cord reflex mechanism whenever a portion of the body begins to move unexpectedly in an unwilled direction. But, without the cerebellum, activation of the antagonist muscles fails to occur, thus allowing over-movement of the limb.

**Hypotonia.** Loss of the deep cerebellar nuclei, particularly the dentate nuclei, causes moderate decrease in tone of the peripheral musculature on the side of the lesion, though after several months the motor cortex usually compensates for this by an increase in its intrinsic activity. The hypotonia results from loss of facilitation of the motor cortex by the tonic discharge of the dentate nuclei.

# SENSORY FEEDBACK CONTROL OF MOTOR FUNCTIONS

Everyone who has ever studied the relationship of the somatic sensory areas to the motor areas of the cortex has been impressed by the close functional interdependence of the two areas. Anatomically, the characteristic pyramidal cells of the motor cortex extend backward into the anterior lip of the postcentral gyrus where they intermingle with the large numbers of granule cells of the somatic sensory cortex. Likewise, the granule cells of the sensory cortex extend anteriorly into the precentral gyrus, the *area pyramidalis*. Thus the two areas fade into each other with many somatic sensory fibers actually terminating directly in the motor cortex and some motor signals originating in the sensory cortex.

Furthermore, when a portion of the somatic sensory cortex in the postcentral gyrus is removed, the muscles controlled by the motor cortex immediately anterior to the removed area often lose much of their coordination. This observation illustrates that the somatic sensory area plays a major role in the control of motor functions. The overall mechanism by which sensory feedback plays this role has been postulated to be the following:

## THE SENSORY "ENGRAM" FOR MOTOR MOVEMENTS

A person performs a motor movement mainly to achieve a purpose. It is primarily in the sensory and sensory association areas that he experiences effects of motor movements and records "memories" of the different patterns of motor movements. These are called sensory engrams of the motor movements. When he wishes to achieve some purposeful act, he presumably calls forth one of these engrams and then sets the motor system of the brain into action to reproduce the sensory pattern that is laid down in the engram.

**The Proprioceptor Feedback Servomechanism for Reproducing the Sensory Engram.** From the earlier discussion of cerebellar function, it is clear how proprioceptor signals from the periphery can affect motor activity. However, in addition to the feedback pathways through the cerebellum, more slowly acting feedback pathways also pass from proprioceptors to the sensory areas of the cerebral cortex and thence back to the motor cortex. Each of these feedback pathways is capable of modifying the motor response. For instance, if a person learns to cut with scissors, the movements involved in this process cause a particular sequential pattern of proprioceptive impulses to pass to the somatic sensory area. Once this pattern has been "learned" by the sensory cortex, the memory engram of the pattern can be used to activate the motor system to perform the same sequential pattern whenever it is required.

To do this, the proprioceptor signals from the fingers, hands, and arms are compared with the engram, and if the two do not match each other, the difference, called the "error," supposedly initiates additional motor signals that automatically activate appropriate muscles to bring the fingers, hands, and arms into the necessary sequential attitudes for performance of the task. Each successive portion of the engram presumably is projected according to a time sequence, and the motor control system automatically follows from one point to the next so that the fingers go through the precise motions necessary to duplicate exactly the sensory engram of the motor activity.

Thus, one can see that the motor system in this case actually acts as a servomechanism, for it is not the motor cortex itself that controls the pattern of activity to be accomplished. Instead, the pattern is located in the sensory part of the brain, and the motor system merely "follows" the pattern, which is the definition of a servomechanism. If ever the motor system fails to follow the pattern, sensory signals are fed back to the cerebral cortex to apprise the sensorium of this failure, and appropriate corrective signals are transmitted to the muscles.

Other sensory signals besides somesthetic signals are also involved in motor control, particularly visual signals. However, these other sensory systems are often slower to recognize error than is the somatic proprioceptor system. Therefore, when the sensory engram depends on visual feedback for control purposes, the motor movements are usually considerably slowed in comparison with those that depend on somatic feedback.

An extremely interesting experiment that demonstrates the importance of the sensory engram for control of motor movements is one in which a monkey has been trained to perform some complex activity and then various portions of his cortex are removed. Removal of small portions of the motor cortex that control the muscles normally used for the activity does not prevent the monkey from performing the activity. Instead he automatically uses other muscles in place of the paralyzed ones to perform the same activity. On the other hand, if the corresponding somatic sensory cortex is removed but the motor cortex is left intact, the monkey loses all ability to perform the activity. Thus, this experiment demonstrates that the motor system acts automatically as a servomechanism to use whatever muscles are available to follow the pattern of the sensory engram, and if some muscles are missing, other muscles are substituted automatically. The experiment also demonstrates forcefully that the somatic sensory cortex is essential to at least some types of "learned" motor performance.

### ESTABLISHMENT OF "SKILLED" MOTOR PATTERNS

Many motor activities are performed so rapidly that there is insufficient time for sensory feedback signals to control these activities. For instance, the movements of the fingers during typing occur much too rapidly for somatic sensory signals to be transmitted either to the somatic sensory cortex or even directly to the motor cortex and for these then to control each discrete movement. It is believed that the patterns for control of these rapid coordinate muscular movements are established in the motor system itself, probably involving complex circuitry in the primary motor cortex, the so-called premotor area of the cortex, the basal ganglia, and even the cerebellum. Indeed, lesions in any of these areas can destroy one's ability to perform rapid coordinated muscular contractions, such as those required during the act of typing, talking, or writing by hand.

**Role of Sensory Feedback During Establishment of the Rapid Motor Patterns.** Even a highly skilled motor activity can be performed the very first time if it is performed extremely slowly—slowly enough for sensory feedback to guide the movements through each step. However, to be really useful, many skilled motor activities must be performed rapidly. This probably is achieved by successive performance of the same skilled activity until finally an engram of the skilled activity is laid down in the motor system as well as in the sensory system. This motor engram causes a precise set of muscles to go through a specific sequence of movements required to perform the skilled activity. Therefore, such an engram is called a *pattern of skilled motor function,* and the motor areas are primarily concerned with this.

After a person has performed a skilled activity many times, the motor pattern of this activity can thereafter cause the hand or arm or other part of the body to go through the same pattern of activity again and again, now entirely *without* sensory feedback control. However, even though sensory feedback control is no longer present, the sensory system still determines whether or not the act has been performed correctly. This determination is made in retrospect rather than while the act is being performed. If the pattern has not been performed correctly, information from the sensory system supposedly can help to correct the pattern the next time it is performed.

Thus, eventually, hundreds of patterns of different coordinate movements are laid down in the motor system, and these can be called upon one at a time in different sequential orders to perform literally thousands of complex motor activities.

An interesting experiment that demonstrates the applicability of these theoretical methods of muscular control is one in which the eyes are made to "follow" an object that moves around and around in a circle. At first, the eyes can follow the object only when it moves around the circle slowly, and even then the movements of the eyes are extremely jerky. Thus, sensory feedback is being utilized to control the eye movements for following the object. However, after a few seconds, the eyes begin to follow the moving object rather faithfully, and the rapidity of movement around the circle can be increased to many times per second, and still, the eyes continue to follow the object. Sensory feedback control of each stage of the eye movements at these rapid rates would be completely impossible. Therefore, by this time, the eyes have developed a pattern of movement that is not dependent upon step-by-step sensory feedback. Nevertheless, if the eyes should fail to follow the object around the circle, the sensory system would immediately become aware of this and presumably could make corrections in the pattern of movement.

## EFFECT OF DIFFERENT MUSCULAR LOADS ON THE PERFORMANCE OF SKILLED MOTOR ACTIVITIES

All of us know that a person can move his hand around and around in a circle whether the hand be empty or loaded with a 10-pound weight. That is, the load makes little difference in the faithfulness of performance of the required activity. The reason for this is that the motor activity is probably controlled by a servo feedback mechanism that monitors the performance of the motor act constantly, and, if the performance is not according to the sensory engram of the required movement, appropriate "error" signals are automatically transmitted into the motor system to achieve the desired result. One can see, therefore, the practical significance of the servomechanism type of control of muscular activity.

Unfortunately, we still do not know all the neuronal circuitry that allows the pattern of movement to be dissociated from the muscle load. If the motor activity is performed very slowly, obviously feedback through the sensory cortex can provide adjustment of muscle forces so that the effects of load can be compensated. After a pattern of motor function has been established, however, precise patterns can be executed entirely subconsciously, despite widely varying loads. Even after loss of the cerebellum this can still be achieved, though not very well for rapid movements. Therefore, by the process of elimination, we can perhaps point to the basal ganglia as likely major components of the servomechanism for dissociating load from muscle performance. Another reason for implicating the basal ganglia is that their circuitry is ideally organized for this purpose, having tremendous numbers of interlacing feedback pathways and also receiving strong feedback sensory signals from all parts of the body. This, however, is entirely hypothetical, though it is an approach to one of the most important questions in the field of motor neurophysiology.

## INITIATION OF VOLUNTARY MOTOR ACTIVITY

Because of the spectacular properties of the primary motor cortex and of the instantaneous muscle contractions that can be achieved by stimulating this area, it has become customary to think that the initial brain signals that elicit voluntary muscle contractions begin in the primary motor cortex. However, this almost certainly is far from the truth. Indeed, experiments have shown that the cerebellum and the basal ganglia are activated at almost exactly the same time that the motor cortex is activated, and all of these are activated even *before* the voluntary movement occurs. Furthermore, there is no known mechanism by which the motor cortex can conceive the entire sequential pattern that is to be achieved by the motor movements.

Therefore, we are left with an unanswered question: What is the locus of the initiation of voluntary motor activity? In the following chapter we will learn that the reticular formation of the brain stem and much of the thalamus play essential roles in activating all other parts of the brain. Therefore, it is very likely that these areas provide the initial signals that lead to subsequent activity in the motor cortex, the basal ganglia, and the cerebellum at the onset of voluntary movement. Here again, we come to a circular question: What is it that initiates the activity in the brain stem and the thalamus? We do know part of the answer to this: The activity is initiated by continual sensory input into these areas, including sensory input from the peripheral receptors of the body and from the cortical sensory storage, or memory, areas. Much of motor activity is almost reflex in nature, occurring instantly after an incoming sensory signal from the periphery. But so-called voluntary activity occurs minutes, hours, or even days after the initiating sensory input—after analysis, storage of memories, recall of the memories, and finally initiation of a motor response. We shall see in Chapter 55 that the control of motor activity is strongly influenced by these prolonged procedures of cerebration. Furthermore, damage to the essential areas of the cerebral cortex for analysis of sensory information leads to serious deficits and abnormalities of voluntary muscle control.

Therefore, for the present, let us conclude that the immediate energy for eliciting voluntary motor activity probably comes from the basal regions of the brain. These, in turn, are under the control of the different sensory inputs, the memory storage areas, and the associated areas of the brain that are devoted to the processes of analysis, a mechanism frequently called *cerebration*.

# REFERENCES

Allen, G. I., and Tsukahara, N.: Cerebrocerebellar communication systems. *Physiol. Rev.*, 54:957, 1974.

Asanuma, H.: Cerebral cortical control of movement. *Physiologist*, 16:143, 1973.

Asanuma, H.: Recent developments in the study of the columnar arrangement of neurons within the motor cortex. *Physiol. Rev.*, 55:143, 1975.

Barr, M. L: The Human Nervous System, 2nd Ed. Hagerstown, Md., Harper & Row, Publishers, 1974.

Brooks, V. B., and Stoney, S. D., Jr.: Motor mechanisms: the role of the pyramidal system in motor control. *Ann. Rev. Physiol.*, 33:337, 1971.

Cohen, B., and Highstein, S. M.: Cerebellar control of the vestibular pathways to oculomotor neurons. *Prog. Brain Res.*, 37:411, 1972.

Cooper, I. S., Riklan, M., and Snider, R. S. (eds.): The Cerebellum, Epilepsy, and Behavior. New York, Plenum Publishing Corporation, 1974.

Eccles, J. C.: The Understanding of the Brain. New York, McGraw-Hill Book Company, 1973.

Eccles, R. M., Shealy, C. N., and Willis, W. D.: Patterns of innervation of kitten motoneurones. *J. Physiol. (Lond.)*, 165:392, 1963.

Eliasson, S. G., Prensky, A. L., and Hardin, W. B., Jr. (eds.): Neurological Pathophysiology. New York, Oxford University Press, 1974.

Evarts, E. V.: Activity of pyramidal tract neurons during postural fixation. *J. Neurophysiol.*, 32:375, 1969.

Evarts, E. V.: Brain mechanisms in movement. *Sci. Amer.*, 229:96, 1973.

Evarts, E. V., and Thach, W. T.: Motor mechanisms of the CNS: cerebrocerebellar interrelations. *Ann. Rev. Physiol.*, 31:451, 1969.

Fox, C. A., and Snider, R. S.: The Cerebellum. Progress in Brain Research. Vol. 25. New York, American Elsevier Publishing Co., Inc., 1967.

Granit, R.: The Basis of Motor Control. New York, Academic Press, Inc., 1970.

Grillner, S.: Locomotion in vertebrates: central mechanisms and reflex interaction. *Physiol. Rev.*, 55:247, 1975.

Hunt, C. C., and Kuno, M.: Properties of spinal interneurones. *J. Physiol. (Lond.)*, 147:346, 1959.

Langworthy, O. R.: The Sensory Control of Posture and Movement. Baltimore, The Williams & Wilkins Company, 1970.

Lenman, J. A. R.: Clinical Neurophysiology. Philadelphia, J. B. Lippincott Company, 1975.

Llinas, R., Bloedel, J. E., and Hillman, D. E.: Functional characterization of neuronal circuitry of frog cerebellar cortex. *J. Neurophysiol.*, 32:847, 1969.

Llinas, R.: Eighteenth Bowditch lecture. Motor aspects of cerebellar control. *Physiologist*, 17:19, 1974.

Massion, J.: The mammalian red nucleus. *Physiol. Rev.*, 47:383, 1967.

Merritt, H. H.: A Textbook of Neurology. Philadelphia, Lea & Febiger, 1973.

Mettler, F. A.: Muscular tone and movement: their cerebral control in primates. *Neurosci. Res.*, 1:175, 1968.

O'Connell, A. L., and Gardner, E. B.: Understanding the Scientific Bases of Human Movement. Baltimore, The Williams & Wilkins Company, 1972.

Oscarsson, O.: Functional organization of the spino- and cuneocerebellar tracts. *Physiol. Rev.*, 45:495, 1965.

Paillard, J.: The patterning of skilled movements. *In* Magoun, H. W. (ed.): Handbook of Physiology. Sec. 1, Vol. 3. Baltimore, The Williams & Wilkins Company, 1960, p. 1679.

Penfield, W., and Rasmussen, T.: The Cerebral Cortex of Man. New York, The Macmillan Company, 1950.

Porter, R.: Functions of the mammalian cerebral cortex in movement. *Prog. Neurobiol.*, 1:3, 1973.

Porter, R.: The neurophysiology of movement performance. *In* Guyton, A. C. (ed.): MTP International Review of Science: Physiology. Vol. 3. Baltimore, University Park Press, 1974, p. 151.

Stark, L., Ida, M., and Willis, P. A.: Dynamic characteristics of the motor coordination system in man. *Biophys. J.*, 1:279, 1961.

Woody, C. D.: Aspects of the electrophysiology of cortical processes related to the development and performance of learned motor responses. *Physiologist*, 17:49, 1974.

# | 54 |

# Activation of the Brain—the Reticular Activating System; the Diffuse Thalamocortical System; Brain Waves; Epilepsy; Wakefulness and Sleep

In the previous chapters of this section on neurophysiology we have discussed, first, the somatic sensory mechanisms of the nervous system and, second, the motor mechanisms. In the present chapter we will consider the functions of the *reticular activating system,* a system that controls the overall degree of central nervous system activity, including control of wakefulness and sleep, and control of at least part of our ability to direct attention toward specific areas of our conscious minds.

Figure 54–1A illustrates the extent of this system, showing that it begins in the lower brain stem and extends upward through the mesencephalon and thalamus to be distributed throughout the cerebral cortex. Impulses are transmitted from the ascending reticular activating system to the cortex by two different pathways. One pathway passes upward from the brain stem portion of the reticular formation to the intralaminar, midline, and reticular nuclei of the thalamus and thence through diverse pathways to essentially all parts of the cerebral cortex as well as the basal ganglia. A second and probably much less important pathway is through the subthalamic, hypothalamic, and adjacent areas.

## FUNCTION OF THE RETICULAR ACTIVATING SYSTEM IN WAKEFULNESS

Diffuse electrical stimulation in the *mesencephalic, pontile and upper medullary portions of the reticular formation*—an area also called the *bulboreticular facilitatory area* and discussed in Chapter 52 in relation to the motor functions of the nervous system—causes immediate and marked activation of the cerebral cortex and even causes a sleeping animal to awaken instantaneously. Furthermore, when this brain stem portion of the reticular formation is damaged severely, as occurs (a) when a *brain tumor* develops in this region, (b) when serious *hemorrhage* occurs, or (c) in diseases such as *encephalitis lethargica* (sleeping sickness), the person passes into coma and is completely nonsusceptible to normal awakening stimuli.

**Function of the Brain Stem Portion of the Reticular Activating System.** Electrical stimuli applied to different portions of the reticular activating system have shown that the brain stem portion functions quite differently from the thalamic portion. Electrical stimulation of the

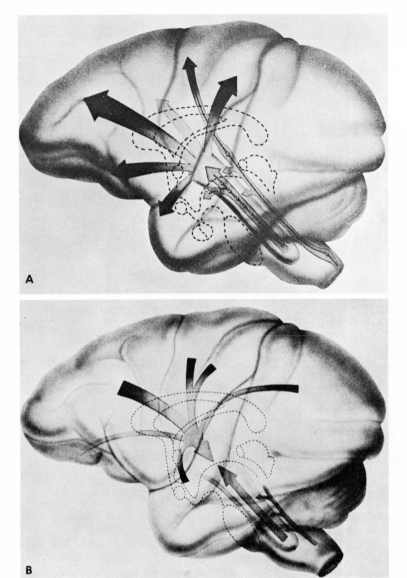

**Figure 54–1.** (A) The ascending reticular activating system schematically projected on a monkey brain. (From Lindsley: Reticular Formation of the Brain. Little, Brown and Co.) (B) Convergence of pathways from the cerebral cortex and from the spinal afferent systems on the reticular activating system. (From French, Hernandez-Peon, and Livingston: *J. Neurophysiol.*, 18:74, 1955.)

brain stem portion causes generalized activation of the entire brain, including activation of the cerebral cortex, thalamic nuclei, basal ganglia, hypothalamus, other portions of the brain stem, and even the spinal cord. Furthermore, once the brain stem portion is stimulated, the degree of activation throughout the nervous system remains high for as long as a half minute or more after the stimulation is over. Therefore, *it is believed that the brain stem portion of the reticular activating system is basically responsible for normal wakefulness of the brain.*

**Function of the Thalamic Portion of the Activating System.** Electrical stimulation in different areas of the thalamic portion of the activating system (if the stimulation is not too strong) activates specific regions of the cerebral cortex more than others. This is distinctly different from stimulation in the brain stem portion, which activates all the brain at the same time. Therefore, it is believed that the thalamic portion of the activating system has two specific functions: first, it relays most of the diffuse facilitatory signals from the brain stem to all parts of the cerebral cortex to cause generalized activation of the cerebrum; second, stimulation of selected points in the thalamic activating system causes specific activation of certain areas of the cerebral cortex in distinction to the other areas. This selective activation of specific cortical areas possibly or probably plays an important role in our ability to direct our attention to certain parts of our mental activity, which is discussed later in the chapter.

### THE AROUSAL REACTION—SENSORY ACTIVATION OF THE RETICULAR ACTIVATING SYSTEM

When an animal is asleep, the level of activity of the reticular activating system is greatly decreased; yet almost any type of sensory signal can immediately activate the system. For instance, proprioceptive signals from the joints and muscles, pain impulses from the skin, visual signals from the eyes, auditory signals from the ears, or even visceral sensations from the gut can all cause sudden activation of the reticular activating system and therefore arouse the animal. This is called the *arousal reaction.*

Some types of sensory stimuli are more potent than others in eliciting the arousal reaction, the most potent being pain and proprioceptive somatic impulses.

Anatomically, the reticular formation of the brain stem is admirably constructed to perform the arousal functions. It receives tremendous numbers of signals either directly or indirectly from the *spinoreticular tracts,* the *spinothalamic tracts,* the *spinotectal tracts,* the *auditory tracts,* the *visual tracts,* and others, so that almost any sensory stimulus in the body can activate it. The reticular formation in turn can transmit signals both upward into the brain and downward into the spinal cord. Indeed, many of the fibers originating from cells in the reticular formation divide, with one branch of the fiber passing upward and the other branch passing downward, as was explained in Chapter 52.

### STIMULATION OF THE RETICULAR ACTIVATING SYSTEM BY THE CEREBRAL CORTEX

In addition to activation of the reticular activating system by sensory impulses, the cerebral cortex can also stimulate this system. Direct fiber pathways pass into the reticular activating areas, as shown in Figure 54–1B, from almost all parts of the cerebrum but particularly from (1) the *sensorimotor cortex* of the pre- and postcentral gyri, (2) the *frontal cortex,* (3) the *cingulate gyrus,* (4) the *hippocampus* and *other structures of the rhinencephalon,* (5) the *hypothalamus,* and (6) the *basal ganglia.* Because of

an exceedingly large number of nerve fibers that pass from the motor regions of the cerebral cortex to the reticular formation, motor activity in particular is associated with a high degree of wakefulness, which partially explains the importance of movement to keep a person awake. However, intense activity of any other part of the cerebrum can also activate the reticular activating system and consequently can cause a high degree of wakefulness.

## THE DIFFUSE THALAMOCORTICAL SYSTEM

At this point it is important that we explain some of the interrelationships between the thalamus and the cerebral cortex. The thalamus is the main entryway for essentially all sensory nervous signals to the cortex, with the exception of signals from the olfactory system. In Chapters 49 and 50 we have already discussed the transmission of the somatic signals through the thalamus to the cerebral cortex, and we shall see in Chapters 60, 61, and 62 that all the signals from the visual, auditory, and taste systems are also relayed through the thalamus before reaching the cortex. In the following chapter we shall discuss the two-way communication between essentially all parts of the cerebral cortex and the thalamus. For all of these functions of the thalamus, there are very definite nuclei that are called the *specific nuclei of the thalamus;* the combination of these nuclei and their connecting areas in the cortex is called the *specific thalamocortical system.* It is this system of which we usually speak when we discuss the transmission of sensory information into the cerebral cortex.

In addition to the specific thalamocortical system, there is a separate system, partially separated from the specific system, called the *diffuse thalamocortical system,* or the *generalized thalamocortical system,* or the *nonspecific thalamocortical system.* This system is subserved by *diffuse thalamic nuclei* and also by *diffuse fiber projections* from the thalamus to the cortex. In general, the thalamic neurons of this system are not collected into discrete nuclei as are the neurons of the specific system. Instead, they lie mainly between the specific nuclei or on the outer surface of the thalamus; they are divided into three separate groups: (1) the *intralaminar nuclei,* which are diffuse collections of neurons that lie in the middle of the thalamic mass between the specific nuclei; (2)

the *midline nuclei,* which are small collections of neurons in the midline of the thalamus where the two halves of the thalamus come together and which seem to have little organization; and (3) the *reticular nuclei* of the thalamus, which form a thin shell of diffuse neurons that lies over the surface of the thalamus and through which all the pathways from the thalamus to the cerebral cortex must pass. All of these nuclei make multiple connections with the specific thalamic nuclei. They also project very small fibers to all parts of the cerebral cortex, with the exception of small portions of the temporal lobes that are mainly associated with the limbic system, a very old part of the brain which plays major roles in behavior, emotions, and so forth as will be discussed in Chapter 56.

The diffuse system of the thalamus is continuous with the upper end of the reticular formation in the brain stem and receives much of its input from this source. However, it is also an important terminus of one of the very important somatic sensory pathways, the *paleo-spinothalamic pathway for pain*—the pathway that transmits the intolerable type of burning, aching pain. This is the same pathway that transmits the major sensory signals into much of the reticular formation of the brain stem as well.

**Control Functions of the Diffuse Thalamocortical System.** One can well understand from the above description of the diffuse thalamic nuclei that they lie in propitious locations for affecting the levels of activity of all of the specific nuclei of the thalamus. It is also clear that they are properly located to relay control functions from the brain stem reticular formation to the cerebrum, and that they in turn can receive much helpful input information from the specific nuclei of the thalamus. Perhaps most important of all, however, is that these nuclei have a specific capability to control the level of activity of the entire cerebral cortex itself; this is mediated by signals that play on the cerebral cortex through the diffuse thalamocortical projections.

**Effect of the Diffuse Thalamic System on Cortical Activity.** Activation of the cortex by the diffuse thalamic system is entirely different from activation by the specific system. Some of the differences are the following:

1. Stimulation of a specific thalamic nucleus activates the cortex within 1 to 2 milliseconds, whereas stimulation of the diffuse system causes no activation for approximately the first 25 milliseconds. The activation level builds up over a period of several hundred milliseconds if the stimulation is continued.

2. At the end of stimulation of the thalamus, the activation of the cortex by the specific nuclei dies away within another few milliseconds, whereas the activation by the diffuse system sometimes continues as an "after-discharge" for as long as 30 seconds.

3. Signals from the specific nuclei to the cortex activate mainly layer IV of the cortex, as was explained in Chapter 49, whereas activation of the diffuse thalamic system activates mainly layers I and II of the cortex. Since this latter activation is prolonged and because layers I and II are the loci of many of the dendrites of the deeper cortical neurons, it is supposed that the stimulation by the diffuse thalamic system mainly causes partial depolarization of large numbers of dendrites near the surface of the cortex; this in turn causes a generalized increase in the degree of facilitation of the cortex. When the cortex is thus facilitated by the diffuse thalamic system, specific signals that enter the cortex from other sources are exuberantly received.

Figure 54–2 illustrates the major differences between the specific and the diffuse thalamocortical systems. It shows the response of the visual cortex to (a) stimulation of a specific thalamic nucleus, and (b) stimulation of the diffuse thalamocortical system. In the upper record, elicited by stimulation of the lateral geniculate body, the specific nucleus for vision, the response occurred within a few milliseconds after the stimulus. Also, it began with a record below the line, indicating positivity on the surface of the cortex; this is the type of record that occurs when the deeper layers of the cortex are stimulated first—in this case, layer IV of the cortex. In the lower record, the stimulus was

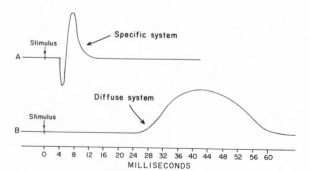

**Figure 54–2.** Comparison of the response of the visual cortex following stimulation of the lateral geniculate body (record A), which is part of the specific thalamocortical system, with the response of the visual cortex following stimulation of the visual portion of the diffuse thalamocortical system (record B). Note in record B the long latent period before appearance of the diffuse response and also the prolonged duration of the response.

applied to the visual region of the diffuse thalamic nuclei. Note the very long latent period, the long response, and the fact that the response is entirely above the line (indicating negativity) which means that the superficial layers of the cortex were stimulated by this signal.

Another feature of stimulation of the diffuse system is the phenomenon called the *recruiting response*. When this system is stimulated repetitively at a frequency of about 10 to 12 cycles per second, the first response is relatively weak, the second a little stronger, the third still stronger, and so forth, the response becoming considerably stronger for the first 4 to 10 stimuli. That is, the system "recruits" more and more cortical synapses into the pattern of response with the repetitive stimulation.

In summary, the diffuse thalamocortical system controls the overall degree of activity of the cortex. It probably can at times facilitate activity in regional areas of the cortex distinct from the remainder of the cortex. Collateral signals control the level of activity in the specific nuclei of the thalamus, the basal ganglia, the hypothalamus, and other structures of the cerebrum as well.

# ATTENTION

So long as a person is awake, he has the ability to direct his attention to specific aspects of his mental environment. Furthermore, his degree of attention can change remarkably from (a) almost no attention at all to (b) broad attention to almost everything that is going on, or to (c) intense attention to a minute facet of his momentary mental experience.

Unfortunately, the basic mechanisms by which the brain accomplishes its diverse acts of attention are not known. However, a few clues are beginning to fall into place, as follows:

**Brain Stem Control of Overall Attentiveness.** In exactly the same way that a person can change from a state of sleep to a state of wakefulness, there can be all degrees of wakefulness, from wakefulness in which a person is nonattentive to almost all his surroundings to an extremely high degree of wakefulness in which the person reacts instantaneously to almost any sensory experience. These changes in degree of *overall attentiveness* seem to be caused primarily by changes in activity of the brain stem portion of the reticular activating system. Thus, control of the general level of attentiveness is

probably exerted by the same mechanism that controls wakefulness and sleep, the control center for which is located in the mesencephalon and upper pons.

**Function of the Thalamus in Attention.** Earlier in the chapter it was pointed out that stimulation of a single specific area in the thalamic portion of the reticular activating system, when the stimulus intensity is not too strong, activates only a specific area of the cerebral cortex. Since the cerebral cortex is one of the most important areas of the brain for conscious awareness of our surroundings, one can surmise that the ability of specific thalamic areas to excite specific cortical regions might be one of the mechanisms by which a person can direct his attention to specific aspects of his mental environment, whether these be immediate sensory experiences or stored memories.

**Relation of Centrifugal Control of Sensory Information to Attention.** Nervous pathways extend centrifugally from almost all sensory areas of the brain toward the lower centers to control the intensity of sensory input to the brain. For instance, the auditory cortex can either inhibit or facilitate signals from the cochlea, the visual cortex can control the signals from the retina in the same way, and the somesthetic cortex can control the intensities of signals from the somatic areas of the body.

Thus, it is likely that activated regions of the cortex control their own sensory input under some conditions. This is another means by which the brain might direct its attention to specific phases of its mental activity.

## POSSIBLE "SCANNING" AND "PROGRAMMING" FUNCTIONS OF THE BRAIN'S ACTIVATING SYSTEM

At this point it is important to refer back to the analogy drawn in Chapter 46 between the functions of the central nervous system and of a general purpose computer. Almost all complex computers, particularly those that are capable of storing information and then recalling this information at later times, have a control center called a "programming unit" that directs the attention of different parts of the computer to specific information stored in other parts of the computer. It is quite likely that the activating system of the brain, particularly the thalamic part, operates in much the same way as the programming unit of a computer for the following reasons:

1. The ability of the activating system to direct one's attention to specific mental activities is analogous to the ability of the programming unit to call

forth information that has been stored in the memory of the computer or to call forth information that is given to the computer at its input, information that is comparable to sensory information in the human being.

2. The programming unit can "find information" when its exact locus is not known. In the computer, this is achieved by dictating certain qualities required of the information and then searching through hundreds or thousands of stored bits of information until the appropriate information is found, a function called *scanning*. We know from psychological tests that our own brain can perform this same function, though we do not know how it does so. Yet, since the thalamus perhaps plays a role in directing our attention to stored information in specific areas of the cortex, it is reasonable to postulate that the thalamus, probably operating in conjunction with other basal areas of the brain, plays a major role in this scanning operation.

3. The programming unit of a computer determines the sequence of processing of information. Our brain can also control its thoughts in orderly sequence. Here again, mainly on the basis of anatomical considerations and on the basis of the fact that the reticular activating system is the primary controller of cerebral activity, it can be surmised, until we know more about the subject, that the reticular activating system is the primary controller of our orderly sequence of thoughts.

It is obviously premature to suggest an anatomical locus for the scanning function of the brain; indeed, this function probably requires complex circuitry that involves several loci. However, good candidates for this function are the centrum medianum and the medial dorsal nucleus of the thalamus, both of which are located in the middle of each half of the thalamus. They are large nuclei surrounded on all sides by the diffuse thalamic system, and they make multiple bidirectional connections with essentially all parts of the thalamus. Obviously, they are propitiously located for playing major roles in determining attention, drive, and so forth.

### EFFECT OF BARBITURATE ANESTHESIA ON THE RETICULAR ACTIVATING SYSTEM

The barbiturates have a specific depressant effect on the brain stem portion of the reticular activating system. Therefore, barbiturates obviously can either depress brain activity or even cause sleep. Yet it is especially interesting that barbiturate anesthesia does not block transmission in most of the specific sensory systems and also does not entirely block function of the thalamic portion of the reticular activating system. It is probable that many other clinically used anesthetics also have specific depressant effects on the brain stem portion of the reticular activating system and in this way cause general anesthesia.

## BRAIN WAVES

Electrical recordings from the surface of the brain or from the outer surface of the head demonstrate continuous electrical activity in the brain. Both the intensity and patterns of this electrical activity are determined to a great extent by the overall level of excitation of the brain resulting from functions in the reticular activating system. The undulations in the recorded electrical potentials, shown in Figure 54–3, are called *brain waves,* and the entire record is called an *electroencephalogram* (EEG).

The intensities of the brain waves on the surface of the scalp range from zero to 300 microvolts, and their frequencies range from once every few seconds to 50 or more per second. The character of the waves is highly dependent on the degree of activity of the cerebral cortex, and the waves change markedly between the states of wakefulness and sleep.

Much of the time, the brain waves are irregular, and no general pattern can be discerned in the EEG. However, at other times, distinct patterns do appear. Some of these are characteristic of specific abnormalities of the brain, such as epilepsy, which is discussed later. Others occur even in normal persons and can be classified into *alpha, beta, theta,* and *delta waves,* which are all illustrated in Figure 54–3.

*Alpha waves* are rhythmic waves occurring at a frequency between 8 and 13 per second and are found in the EEG's of almost all normal persons when they are awake in a quiet, resting state of cerebration. These waves occur most intensely in the occipital region but can also be recorded at times from the parietal and frontal regions of the scalp. Their voltage usually is about 50 microvolts. During sleep the alpha waves disappear entirely, and when the awake person's attention is directed to some specific type of mental activity, the alpha waves are replaced by asynchronous, higher frequency but lower voltage waves. Figure 54–4 illustrates the effect on the alpha waves of simply opening the eyes in bright light and

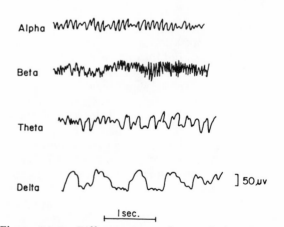

**Figure 54–3.** Different types of normal electroencephalographic waves.

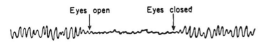

**Figure 54–4.** Replacement of the alpha rhythm by an asynchronous discharge on opening the eyes.

then closing the eyes again. Note that the visual sensations cause immediate cessation of the alpha waves and that these are replaced by low voltage, asynchronous waves.

*Beta waves* occur at frequencies of more than 14 cycles per second and as high as 25 and rarely 50 cycles per second. These are most frequently recorded from the parietal and frontal regions of the scalp, and they can be divided into two major types, *beta I* and *beta II*. The beta I waves, which are illustrated in Figure 54–3, have a frequency about twice that of the alpha waves, and these are affected by mental activity in very much the same way as the alpha waves—that is, they disappear and in their place appears an asynchronous but low voltage recording. The beta II waves, on the contrary, appear during *intense* activation of the central nervous system or during tension. Thus, one type of beta wave is inhibited by cerebral activity while the other is elicited.

*Theta waves* have frequencies between 4 and 7 cycles per second. These occur mainly in the parietal and temporal regions in children, but they also occur during emotional stress in some adults, particularly during disappointment and frustration. They can often be brought out in the EEG of a frustrated person by allowing him to enjoy some pleasant experience and then suddenly removing this element of pleasure; this causes approximately 20 seconds of theta waves. These same waves also occur in many brain disorders.

*Delta waves* include all the waves of the EEG below 3½ cycles per second and sometimes as low as 1 cycle every 2 to 3 seconds. These occur in deep sleep, in infancy, and in very serious organic brain disease. And they occur in the cortex of animals that have had subcortical transections separating the cerebral cortex from the thalamus. Therefore, delta waves can occur strictly in the cortex independently of activities in lower regions of the brain.

## ORIGIN OF THE DIFFERENT TYPES OF BRAIN WAVES

The discharge of a single neuron or single nerve fiber in the brain cannot be recorded from the surface of the head. Instead, for an electrical potential to be recorded all the way through the skull, large portions of nervous tissue must emit electrical current simultaneously. There are two ways in which this can occur. First, tremendous numbers of nerve fibers can discharge in synchrony with each other, thereby generating very strong electrical currents. Second, large numbers of neurons can partially discharge,

though not emit action potentials; furthermore, these partially discharged neurons can give prolonged periods of current flow that can undulate slowly with changing degrees of excitability of the neurons. Simultaneous electrical measurements within the brain while recording brain waves from the scalp indicate that it is the second of these that causes the usual brain waves.

To be more specific, the surface of the cerebral cortex is composed almost entirely of a mat of dendrites from neuronal cells in the lower layers of the cortex. When signals impinge on these dendrites, they become partially discharged, emitting negative potentials characteristic of excitatory postsynaptic potentials, as discussed in Chapter 46. This partially discharged state makes the neurons of the cortex highly excitable, and the negative potential is simultaneously recorded from the surface of the scalp, indicating this high degree of excitability.

One of the important sources of signals to excite the outer dendritic layer of the cerebral cortex is the ascending reticular activating system. Therefore, brain wave intensity is closely related to the degree of activity in either the brain stem or the thalamic portions of the reticular activating system.

**Origin of Delta Waves.** Transection of the fiber tracts from the thalamus to the cortex, which blocks the reticular activating system fibers, causes delta waves in the cortex. This indicates that some synchronizing mechanism can occur in the cortical neurons themselves—entirely independently of lower structures in the brain—to cause the delta waves.

Delta waves also occur in very deep "slow wave" sleep; and this suggests that the cortex is then released from the activating influences of the reticular activating system, as was explained earlier in the chapter.

**Origin of the Alpha Waves.** Alpha waves will *not* occur in the cortex without connections with the thalamus. Also, stimulation in the diffuse thalamic nuclei often sets up waves in the diffuse thalamocortical system at a frequency of between 8 and 13 per second, the natural frequency of the alpha waves. Therefore, it is assumed that the alpha waves result from spontaneous activity in the diffuse thalamocortical system, which causes both the periodicity of the alpha waves and the synchronous activation of literally millions of cortical neurons during each wave.

## EFFECT OF VARYING DEGREES OF CEREBRAL ACTIVITY ON THE BASIC RHYTHM OF THE ELECTROENCEPHALOGRAM

There is a general relationship between the degree of cerebral activity and the average frequency of the electroencephalographic rhythm, the frequency increasing progressively with higher and higher degrees of activity. This is illustrated in Figure 54–5, which shows the existence of delta waves in stupor,

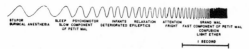

**Figure 54–5.** Effect of varying degrees of cerebral activity on the basic rhythm of the EEG. (From Gibbs and Gibbs: Atlas of Electroencephalography. Addison-Wesley, 1974.)

surgical anesthesia, and sleep; theta waves in psychomotor states and in infants; alpha waves during relaxed states; and beta waves during periods of intense mental activity. However, during periods of mental activity the waves usually become asynchronous rather than synchronous so that the voltage falls considerably, despite increased cortical activity, as illustrated in Figure 54–4.

## CLINICAL USE OF THE ELECTROENCEPHALOGRAM

One of the most important uses of the EEG is to diagnose different types of epilepsy and to find the focus in the brain causing the epilepsy. This is discussed further below. But, in addition, the EEG can be used to localize brain tumors or other space-occupying lesions of the brain and to diagnose certain types of psychopathic disturbances.

There are two means by which brain tumors can be localized. Some brain tumors are so large that they block electrical activity from a given portion of the cerebral cortex, and when this occurs the voltage of the brain waves is considerably reduced in the region of the tumor. However, more frequently a brain tumor compresses the surrounding neuronal tissue and thereby causes abnormal electrical excitation of these surrounding areas; this in turn leads to synchronous discharges of very high voltage waves in the EEG, as shown in the middle two records of Figure 54–6. Localization of the origin of these spikes on the surface of the scalp is a valuable means for locating the brain tumor.

The upper part of Figure 54–6 shows the placement of 16 different electrodes on the scalp, and the lower part of the figure shows the brain waves from four of these electrodes marked in the figure by X's. Note that in two of these, intense brain waves are recorded and, furthermore, that the two waves are essentially of reverse polarity to each other. This reverse polarity means that the origin of the spikes is somewhere in the area *between* the two respective electrodes. Thus, the excessively excitable area of the brain has been located, and this is a lead to the location of the brain tumor.

Use of brain waves in diagnosing psychopathic abnormalities is generally not very satisfactory because only a few of these cause distinct brain wave patterns. Yet by observing combinations of different types of basic rhythms, reactions of the rhythms to attention, changes of the rhythms with forced breathing to cause alkalosis, the appearance of particular characteristics in the brain waves (such as "spin-

dles" of alpha waves), and so forth, an experienced electroencephalographer can detect at least certain types of psychopathic disturbances. For instance, it was pointed out above that theta waves are frequently found in persons with brain abnormalities.

# EPILEPSY

Epilepsy is characterized by uncontrolled excessive activity of either a part of the central nervous system or all of it. A person who is predisposed to epilepsy has attacks when the basal level of excitability of his nervous system (or of the part that is susceptible to the epileptic state) rises above a certain critical threshold. But, as long as the degree of excitability is held below this threshold, no attack occurs.

Basically, the two different types of epilepsy are: *generalized epilepsy* and *focal epilepsy*. Generalized epilepsy involves essentially all parts of the brain at once, whereas focal epilepsy involves only a portion—sometimes only a minute focal spot and other times a moderate portion of the brain.

## GENERALIZED EPILEPSY

**Grand Mal.** Grand mal epilepsy is characterized by extreme neuronal discharges originating in the brain stem portion of the reticular activating system. These then spread throughout the entire central nervous system, to the cortex, to the deeper parts of the brain, and even into the spinal cord to cause generalized *tonic convulsions* of the entire body followed toward the end of the attack by alternating muscular contractions called *clonic convulsions*. The

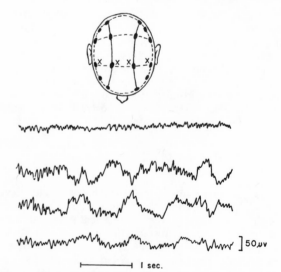

**Figure 54–6.** Localization of a brain tumor by means of the EEG, illustrating (above) the placement of electrodes and (below) the records from the four electrodes designated by X's.

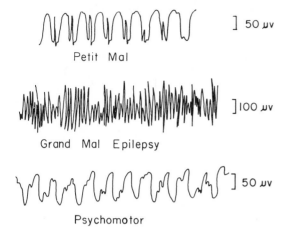

**Figure 54–7.** Electroencephalograms in different types of epilepsy.

grand mal seizure lasts from a few seconds to as long as three to four minutes and is characterized by post-seizure depression of the entire nervous system; the person remains in stupor for one to many minutes after the attack is over and then often remains severely fatigued for many hours thereafter.

The middle recording of Figure 54–7 illustrates a typical electroencephalogram from almost any region of the cortex during a grand mal attack. This illustrates that high voltage synchronous discharges occur over the entire cortex having almost the same periodicity as the normal alpha waves. Furthermore, the same type of discharge occurs on both sides of the brain at the same time, illustrating that the origin of the abnormality is in the lower centers of the brain that control the activity of the cerebral cortex and not in the cerebral cortex itself.

In experimental animals or even in human beings, grand mal attacks can be initiated by administering neuronal stimulants, such as the well-known drug Metrazol, or they can be caused by insulin hypoglycemia or by the passage of alternating electrical current directly through the brain. Electrical recordings from the thalamus and also from the reticular formation of the brain stem during the grand mal attack show typical high voltage activity in both of these areas similar to that recorded from the cerebral cortex. Furthermore, in an experimental animal, even after transecting the brain stem as low as the mesencephalon, a typical grand mal seizure can still be induced in the portion of the brain stem beneath the transection.

Presumably, therefore, a grand mal attack is caused by intrinsic overexcitability of the neurons that make up the reticular activating structures or from some abnormality of the local neuronal pathways. The synchronous discharges from this region could result from local reverberating circuits.

One might ask: What stops the grand mal attack after a given time? This is believed to result from (1)

*fatigue of the neurons* involved in precipitating the attack and (2) *active inhibition* by certain structures of the brain. The change from the tonic type of convulsion to the clonic type during the latter part of the grand mal attack has been suggested to result from partial fatigue of the neuronal system so that some of the excited neurons fade out for a moment, then return to activity after a brief rest, only to fatigue a second time and then a third time, and so forth, until the entire seizure is over. The stupor and fatigue that occur after a grand mal seizure is over is believed to result from the intense fatigue of the neurons following their intensive activity during the grand mal attack.

The *active inhibition* that helps to stop the grand mal attack is believed to result from feedback circuits through inhibitory areas of the brain. The grand mal attack undoubtedly excites such areas as the basal ganglia, which in turn emit many inhibitory impulses into the reticular formation of the brain stem. But the nature of such active inhibition is very much a matter of speculation.

**Petit Mal Epilepsy.** Petit mal epilepsy is closely allied to grand mal epilepsy in that it too almost certainly originates in the brain stem reticular activating system. It is characterized by 5 to 20 seconds of unconsciousness during which the person has several twitchlike contractions of the muscles, usually in the head region—especially blinking of the eyes; this is followed by return of consciousness and resumption of previous activities. The patient may have one such attack in many months, or in rare instances he may have a rapid series of attacks, one following the other. However, the usual course is for the petit mal attacks to appear in late childhood and then to disappear entirely by the age of 30. On occasion, a petit mal epileptic attack will initiate a grand mal attack.

The brain wave pattern in petit mal epilepsy is illustrated by the first record of Figure 54–7, which is typified by a *spike and dome pattern*. The spike portion of this recording is almost identical to the spikes that occur in grand mal epilepsy, but the dome portion is distinctly different. The spike and dome can be recorded over the entire cerebral cortex, illustrating that the seizure originates in the reticular activating system of the brain.

Since petit mal and grand mal epilepsy both originate in essentially the same locus of the brain stem, it is believed that they have the same common cause but that some influence inhibits neuronal activity during the petit mal attack so that it will not progress into the grand mal seizure.

## FOCAL EPILEPSY

Focal epilepsy can involve almost any part of the brain, either localized regions of the cerebral cortex or deeper structures of both the cerebrum and brain stem. And almost always, focal epilepsy results from some localized organic lesion of the brain, such as a scar that pulls on the neuronal tissue, a tumor that compresses an area of the brain, or a destroyed area of brain tissue. Lesions such as these can promote extremely rapid discharges in the local neurons, and when the discharge rate rises above approximately 1000 per second, synchronous waves begin to spread over the adjacent cortical regions. These waves presumably result from *localized reverberating circuits* that gradually recruit adjacent areas of the cortex into the discharge zone. The process spreads to adjacent areas at a rate as slow as a few millimeters a minute to as fast as several centimeters per second. When such a wave of excitation spreads over the motor cortex, it causes a progressive "march" of muscular contractions throughout the opposite side of the body, beginning most characteristically in the mouth region and marching progressively downward to the legs, but at other times marching in the opposite direction. This is called *jacksonian epilepsy.*

A focal epileptic attack may remain confined to a single area of the brain, but in many instances the strong signals from the convulsing cortex or other part of the brain excite the brain stem portion of the reticular activating system so greatly that a grand mal epileptic attack ensues as well.

One type of focal epilepsy is the so-called *psychomotor seizure,* which may cause (1) a short period of amnesia, (2) an attack of abnormal rage, (3) sudden anxiety, discomfort, or fear, (4) a moment of incoherent speech or mumbling of some trite phrase, or (5) a motor act to attack someone, to rub the face with the hand, or so forth. Sometimes the person cannot remember his activities during the attack, but at other times he will have been conscious of everything that he had been doing but unable to control it.

The lower tracing of Figure 54–7 illustrates a typical electroencephalogram during a psychomotor attack, showing a low frequency rectangular wave with a frequency between 2 and 4 per second and with superimposed 14 per second waves.

The electroencephalogram can often be used to localize abnormal spiking waves originating in areas of organic brain disease that might predispose to focal epileptic attacks. Once such a focal point is found, surgical excision of the focus frequently prevents future epileptic attacks.

# WAKEFULNESS AND SLEEP

**The Two Conditions Required for Wakefulness.** We have already seen in the discussions earlier in this chapter that wakefulness requires at least a certain level of activity in the reticular activating system. However, there is a second condition also necessary for wakefulness to occur: the nervous activity of the brain must be channeled in the proper directions. For instance, during a grand mal epileptic attack a person's brain is many times as active as it is during normal wakefulness, but nevertheless he is completely unconscious—certainly not a state of wakefulness. Therefore, for the state of wakefulness to occur it is not enough that the brain simply be active.

Sleep is defined as a state of unconsciousness from which a person can be aroused by appropriate sensory or other stimuli. Therefore, the unconsciousness caused by deep anesthesia, by total inactivity of the reticular activating system in diseased states (coma), and by excessive activity of the reticular activating system as occurs in grand mal epilepsy would not be considered to be sleep. However, coma and anesthesia do have many characteristics similar to those of deep sleep.

**Two Different Types of Sleep.** There are two different ways in which sleep can occur. First, it can result from decreased activity in the reticular activating system; this is called *slow wave sleep* because the brain waves are very slow. Second, sleep can result from abnormal channeling of signals in the brain even though brain activity may not be significantly depressed; this is called *paradoxical sleep* or *desynchronized sleep.*

Most of the sleep during each night is of the slow wave variety; this is the deep, restful type of sleep that the person experiences after having been kept awake for the previous 24 to 48 hours. On the other hand, short episodes of paradoxical sleep usually occur at intervals during each night, and this type of sleep seems to have purposeful functions that will be discussed later.

## SLOW WAVE SLEEP

Slow wave sleep is frequently called by different names, such as *deep restful sleep, dreamless sleep, delta wave sleep,* or *normal sleep.* However, we shall see later that paradoxical sleep is also normal and that it has some characteristics of deep sleep.

**Electroencephalographic Changes as a Person Falls Asleep.** Beginning with wakefulness and proceeding to deep slow wave sleep, the electroencephalogram changes as follows:

1. Alert wakefulness—low voltage, high frequency beta waves showing desynchrony, as illustrated by the second record in Figure 54–3

2. Quiet restfulness—predominance of alpha waves; a type of "synchronized" brain waves

3. Light sleep—slowing of the brain waves to theta or delta low-voltage variety, but interspersed with spindles of alpha waves called *sleep spindles* that last for a few seconds at a time

4. Deep slow wave sleep—high voltage delta waves occurring at a rate of 1 to 2 per second, as illustrated by the fourth record of Figure 54–3.

In paradoxical sleep the brain waves change to still a different pattern, as will be discussed below.

**Origin of the Delta Waves in Sleep.** When the fiber tracts between the thalamus and the cortex are transected, delta waves are presently generated in the isolated cortex, indicating that this type of wave probably occurs intrinsically in the cortex when it is not being driven from below. Therefore, it is assumed that this is also the origin of the high voltage delta waves during deep slow wave sleep. That is, it is assumed that the degree of activity in the reticular activating system has fallen to a level too low to maintain normal excitabiltiy of the cortex so that the cortex then becomes its own pacemaker.

**Characteristics of Deep Slow Wave Sleep.** Most of us can understand the characteristics of deep slow wave sleep by referring back to the last time that we were kept awake for more than 24 hours and then remembering the deep sleep that occurred within 30 minutes to an hour after going to sleep. This sleep is dreamless, exceedingly restful, and is associated with a decrease in both peripheral vascular tone and also most of the other vegetative functions of the body as well. There is also a 10 to 30 per cent decrease in blood pressure, respiratory rate, and basal metabolic rate.

## PARADOXICAL SLEEP ("REM" SLEEP)

Paradoxical sleep never occurs by itself, but instead is always superimposed onto slow wave sleep. In a normal night of sleep, bouts of paradoxical sleep lasting 5 to 20 minutes appear on the average of every 90 minutes, the first such period occurring 80 to 100 minutes after the person falls asleep. When the person is extremely tired, the duration of each bout of paradoxical sleep is very short, and it may even be absent. On the other hand, as the person has become rested, the duration of the paradoxical bouts greatly increases.

There are several very important characteristics of paradoxical sleep:

1. It is usually associated with active dreaming.

2. The person is even more difficult to arouse than during deep slow wave sleep.

3. The muscle tone throughout the body is exceedingly depressed, indicating strong activation of the bulboreticular inhibitory system of the lower brain stem.

4. The heart rate and respiration usually become irregular, which is characteristic of the dream state.

5. Despite the extreme inhibition of the peripheral muscles, irregular muscle movements occur. These include, in particular, rapid movements of the eyes; consequently, paradoxical sleep has often been called *REM sleep,* for "rapid eye movements."

6. The electroencephalogram shows a desynchronized pattern of low voltage beta waves similar to those that occur during wakefulness. Therefore, this type of sleep is also frequently called *desynchronized sleep,* meaning desynchronized brain waves.

In summary, paradoxical sleep is a type of sleep in which the brain is probably as active as it is during many states of wakefulness. However, the brain activity is not channeled in the proper direction for the person to be aware of his surroundings and therefore to be awake.

## BASIC THEORIES OF SLEEP

Despite the universality of the phenomenon of sleep, physiologists must admit with chagrin that they know very little about the basic mechanism of sleep. However, there are two major schools of thought about the causation of sleep: *The first of these beliefs is that sleep is a passive process,* occurring when the neuronal mechanisms that cause wakefulness become fatigued and therefore succeed to a lower level of activity. *The second theory is that active centers in the brain transmit signals into the reticular activating system to inhibit it and thereby produce sleep.* With either of these theories, it is generally assumed that the usual slow wave type of sleep results from decreased excitability of the reticular activating system, caused passively in the first instance and by active inhibition in the second instance.

**The Feedback Theory of Wakefulness and Sleep—a Passive Mechanism of Sleep.** From discussions earlier in the chapter we can recall that the reticular activating system can excite the cortex; but, also, the cortex in turn can excite the activating system. Thus, there exists a *positive* "feedback" that helps to keep the ascending reticular activating system active once it becomes excited.

There is also a second positive feedback that probably contributes to maintaining activity in the reticular activating system: The reticular activating system transmits signals not only to the cortex but also to the peripheral musculature. Any peripheral movement in turn transmits somatic sensory signals back into the reticular activating system, and many of these signals also take part in activating it. Thus, this additional positive feedback can help to maintain activity in the system once it begins.

*Onset of Wakefulness.* Keeping in mind the above two positive feedback mechanisms, we can assume that any arousal signal entering the resting reticular activating system will initiate signals that go both to the cortex and to the peripheral areas of the body. Return signals from both of these areas will further activate the reticular activating system thereby sending out more efferent signals, giving more return signals, and continuing on and on. Thus, once activity is initiated, the level of response will theoretically increase rapidly. This could explain the rapid onset of wakefulness. One of the obvious problems with such a mechanism as this is the possibility that the feedback mechanisms would become overly active and therefore produce a state similar to that of grand mal epilepsy. However, a brief consideration of the theory of positive feedback shows that once the activity reaches a critical level the neuronal discharge rates will begin to saturate so that the feedback gain will decrease to unity. Therefore, the degree of excitation of the reticular activating system will become stable. Furthermore, the level of this excitation plateau, and therefore the level of wakefulness as well, becomes a function of the excitability or fatigue of the neurons in the wakefulness circuit. Thus, this theory has a very solid foundation, and it fits in with the known characteristics of positive feedback.

*Onset of Sleep.* As neurons become fatigued, the level of positive feedback obviously will decrease in both the cortical and the peripheral circuits. Furthermore, each time one of the millions of parallel neurons in the feedback circuits falls out of activity, lack of this neuron's contribution to the feedback reduces the level of excitability of all the other neurons. Thus, there comes a point when the feedback activity cannot maintain enough excitability in the reticular activating system for the circuits to continue reverberating. Therefore, this could explain the initial slow onset of drowsiness and the final abrupt change from wakefulness to sleep.

*Cause of Fatigue of the Neurons.* If the passive mechanism for sleep is the correct one, we still need to explain the cause of fatigue of neurons after 16 hours of wakefulness and their recovery of excitability after 8 hours of sleep. We do know that this same phenomenon occurs in the spinal cord and in other areas of the nervous system that are exposed to excessive activity and then to excessive inactivity. For instance, continued facilitation of the spinal cord by fiber tracts from the brain causes decreased excitability of the spinal cord neurons. After the tracts from the brain are cut, the level of excitability of these neurons increases during the first few hours and then increases progressively less rapidly over a period of days and even months. Therefore, it is known that the excitability of neurons changes over a period of hours inversely to their degree of activity. This established fact about neuronal activity fits in well with the passive theory of sleep.

However, it is possible that the local environment of the neurons also changes. Perhaps the neurons secrete some transmitter substance into the local fluids and this in turn causes inhibition after periods of activity, or excitation after periods of inactivity. We will speak more about this shortly.

**Active Mechanisms for Causing Sleep.** Electrical stimulation under appropriate conditions in widely different areas of the brain can cause sleep. Three of the more important areas for achieving this are:

1. An area in or near the rostral portion of the solitary tracts of the brain stem located bilaterally in the pons.

2. Some areas in the diffuse nuclei of the thalamus.

3. The basal forebrain region between the hypothalamus and the supraorbital areas of the frontal lobes.

Stimulation in these areas at a frequency of 8 to 13 cycles per second (the same frequency as the alpha brain waves) is especially prone to cause sleep, whereas stimulation at other frequencies causes the reverse effect in some instances.

It has also been shown that lesions in the basal forebrain area or bilaterally in the lower pontine reticular substance can cause insomnia—sometimes to the extent that the animal cannot sleep at all, which causes lethal exhaustion within a few days.

Therefore, for obvious reasons, many research workers have come to the conclusion that sleep is caused by increased activity in one or more of these "sleep-producing" centers of the brain. The one most frequently considered to be "the" active sleep center is the bilateral center in the lower brain stem reticular formation near the solitary tracts.

The mechanism of the sleep caused by these sleep-producing centers almost certainly is active inhibition of the reticular activating system.

**Mechanism of Paradoxical Sleep.** Unfortunately, we do not know the cause of paradoxical sleep. However, the reticular activating system seems to have a periodic excitability cycle occurring once every 90 minutes: first decreased activity, then increased activity, and the system cycling throughout the 24-hour day in this manner. It is on the peaks of these cycles that paradoxical sleep occurs. Indeed, the periodicity of this mechanism has even been demonstrated during wakefulness, with periods of greater and lesser excitability throughout the day.

Thus, for the time being, we can possibly or probably consider paradoxical sleep to be caused by a period of intrinsic activity of the reticular activating system superimposed onto the natural mechanism of sleep—enough of a superimposition to cause dreams but not enough to wake the person. The mechanism seems also to activate the bulboreticular inhibitory area in the lower brain stem, which presumably is the cause of the depressed motor functions of the cord.

What is the purpose of paradoxical sleep? This we do not know, but a purpose speculated (as indeed almost all that we have to say about paradoxical sleep is speculated) would be that it is a periodic test of the intrinsic excitability of the neurons of the system. When this intrinsic excitability is great enough, one could suppose that this test of excitability might cause the person to awaken, but if the level of excitability is below a critical value, the person would pass through this phase of paradoxical sleep and fall back into another cycle of deep slow wave sleep before being tested again 90 minutes later.

**Biochemical Theories for the Production of Sleep.** Three important findings indicate that the biochemical milieu of the brain stem neurons might change during the wakefulness-sleep cycle, and furthermore, that some sleep-producing substance might be the cause of the cycle.

First, it has been demonstrated that either dialyzed blood or whole cerebrospinal fluid from animals that have been kept awake for several days contains a substance that causes sleep when injected into the ventricular system of an animal. This substance is a *small polypep-tide* with a molecular weight probably less than 500.

Second, it has been found that lesions in the midline area of the brain stem, an area known as the *nuclei of the raphe,* cause insomnia; this condition also causes destruction of a system of neurons that secrete *serotonin* into the reticular activating system. Therefore, it has been suggested that sleep is caused by secretion of serotonin from these neurons; what it is that controls the activity of these neurons is unexplained.

Third, it has been shown that destruction of small bilateral areas in the dorsal part of the upper middle pons, called the *locus ceruleus,* blocks the occurrence of paradoxical sleep. The neurons of these areas secrete norepinephrine. Therefore, it has been suggested that norepinephrine, possibly acting on the reticular activating system, is the factor that causes the paradoxical episodes during sleep. But, here again, the factors that cause the locus ceruleus to turn on and off are completely unknown.

In summary, therefore, changes in the biochemical status of the local neurons in the reticular activating system, or changes in certain specific nuclei, could increase and decrease the activity of this system and thereby could be partially responsible for inducing the states of wakefulness and sleep.

## PHYSIOLOGICAL EFFECTS OF SLEEP

Sleep causes two major types of physiological effects: first, effects on the nervous system itself and, second, effects on other structures of the body. The first of these seems to be by far the more important, for any person who has a transected spinal cord in the neck shows no physiological effects in the body beneath the level of transection that can be attributed to a sleep and wakefulness cycle; this lack of sleep and wakefulness causes neither significant harm to the bodily organs nor even any deranged function. On the other hand, lack of sleep certainly does affect the functions of the central nervous system.

Prolonged wakefulness is often associated with progressive malfunction of the mind and behavioral activities of the nervous system. We are all familiar with the increased sluggishness of thought that occurs toward the end of a prolonged wakeful period, but in addition, a person can become irritable or even psychotic following forced wakefulness for prolonged periods of time. Therefore, we can assume that sleep in some way not presently understood restores both normal sensitivities of and normal "balance" between the different parts of the central nervous system. This might be likened to the "rezeroing" of electronic analog computers after prolonged use, for all computers of this type gradually lose their "base line" of operation; it is reasonable to assume that the same effect occurs in the central nervous system, since it, too, is an analog computer (though using a digital method of information transmission). Therefore, in the absence of any definitely demonstrated functional value of sleep, we might postulate, on the basis of known psychological changes that occur with wakefulness and sleep, that sleep performs this rezeroing function for the nervous system. Yet this is but a suggestion based purely on psychological data and on analogy with computers.

Even though, as pointed out above, wakefulness and sleep have not been shown to be necessary for somatic functions of the body, the cycle of enhanced and depressed nervous excitability that follows along with the cycle of wakefulness and sleep does have moderate effects on the peripheral body. For instance, there is enhanced sympathetic activity during wakefulness and also enhanced numbers of impulses to the skeletal musculature to increase muscular tone. Conversely, during sleep sympathetic activity decreases while parasympathetic activity occasionally increases, and the muscular tone becomes almost nil. Therefore, during sleep, arterial blood pressure falls, pulse rate decreases, skin vessels dilate, activity of the gastrointestinal tract sometimes increases, muscles fall into a completely relaxed state, and overall basal metabolic rate of the body falls by about 10 to 20 per cent.

# REFERENCES

Andrew, R. J.: Arousal and the causation of behaviour. *Behaviour*, 51:135, 1974.

Ayala, G. F., Dichter, M., Gumnit, B. J., Matsumoto, H., and Spencer, W. A.: Genesis of epileptic interictal spikes. New knowledge of cortical feedback systems suggests a neurophysiological explanation of brief paroxysms. *Brain Res.*, 52:1, 1973.

Barlow, J. S. (ed.): Complementary Electrophysiological Techniques and Methods for Evaluation of the Central Nervous System. Amsterdam, Elsevier Nederland, BV, 1973.

Berger, R. J.: Bioenergetic functions of sleep and activity rhythms and their possible relevance to aging. *Fed. Proc.*, 34:97, 1975.

Brazier, M. A.: Epilepsy: Its Phenomena in Man. New York, Academic Press, 1975.

Clemente, C. D., Purpura, D. P., and Mayer, F. E. (eds.): Sleep and the Maturing Nervous System. New York, Academic Press, Inc., 1972.

Cooper, I. S., Riklan, M., and Snider, R. S. (eds.): The Cerebellum, Epilepsy, and Behavior. New York, Plenum Publishing Corporation, 1974.

Cooper, J. R., Bloom, F. E., and Roth, R. H.: The Biochemical Basis of Neuropharmacology, 2nd Ed. New York, Oxford University Press, 1974.

Costa, E., Gessa, G. L., and Sandler, M. (eds.): Serotonin, New Vistas: Biochemistry and Behavioral and Clinical Studies. New York, Raven Press, 1974.

Eliasson, S. G., Prensky, A. L., and Hardin, W. B., Jr. (eds.): Neurological Pathophysiology. New York, Oxford University Press, 1974.

Freeman, W. J.: Mass Action in the Nervous System: Examination of the Neurophysiological Basis of Adaptive Behavior Through EEG. New York, Academic Press, Inc., 1975.

Freemon, F. R.: Clinical pharmacology of sleep: a critical review of all-night electroencephalographic studies. *Behav. Neuropsychiatry*, 4:49, 1973.

Freemon, F. R.: Sleep Research. Springfield, Ill., Charles C Thomas, Publisher, 1974.

Fried, P. A.: The septum and hyper-reactivity: a review. *Brit. J. Psychol.*, 64:267, 1973.

Greenberg, B., and Pearlman, C.: Cutting the REM nerve: an approach to the adaptive role of REM sleep. *Perspect. Biol. Med.*, 17:513, 1974.

Izquierdo, I., and Nasello, A. G.: Pharmacology of the brain: the hippocampus, learning, and seizures. *Prog. Drug. Res.*, 16:211, 1972.

Jasper, H. H.: Unspecific thalamocortical relations. *In* Magoun, H. W. (ed.): Handbook of Physiology. Sec. 1, Vol. 2. Baltimore, The Williams & Wilkins Company, 1960, p. 1307.

Johnson, L. C.: Are stages of sleep related to waking behavior? *Amer. Sci.*, 61:326, 1973.

Magoun, H. W.: The Waking Brain, 2nd Ed. Springfield, Ill., Charles C Thomas, Publisher, 1962.

Mills, J. N.: Human circadian rhythms. *Physiol. Rev.*, 46:128, 1966.

Petre-Quadens, O. J., and Schlag, J. D. (eds.): Basic Sleep Mechanisms. New York, Academic Press, Inc., 1974.

Rubin, R. T., Poland, R. E., Rubin, L. E., and Gouin, P. R.: The neuroendocrinology of human sleep. *Life Sci.*, 14:1041, 1974.

Schmidt, R. F. (ed.): Fundamentals of Neurophysiology. Springer-Verlag New York, Inc., 1975.

Shellenberger, M. K., and Walaszek, E. J.: Pharmacology of the central nervous system. *Prog. Neurol. Psychiatry*, 27:67, 1972.

Stenevi, U., Björklund, A., and Moore, R. Y.: Morphological plasticity of central adrenergic neurons. *Brain Behav. Evol.*, 8:110, 1973.

Stern, W. C., and Morgane, P. J.: Theoretical view of REM sleep function: maintenance of catecholamine systems in the central nervous system. *Behav. Biol.*, 11:1, 1974.

Thompson, R. F., and Patterson, M. (eds.): Bioelectric Recording Techniques, Part A. Cellular Processes and Brain Potentials. New York, Academic Press, Inc., 1973.

Wallace, R. K., and Benson, H.: The physiology of meditation. *Sci. Amer.*, 226:84, 1972.

# 55

# The Cerebral Cortex and Intellectual Functions of the Brain

It is ironic that we know least about the mechanisms of the cerebral cortex of all parts of the brain, even though it is by far the largest portion of the nervous system. Yet, we do know the effects of destruction or of specific stimulation of various portions of the cortex, and still more is known from electrical recordings from the cortex or from the surface of the scalp. In the early part of the present chapter the facts known about cortical functions are discussed and then some basic theories of the neuronal mechanisms involved in thought processes, memory, analysis of sensory information, and so forth are presented briefly.

## PHYSIOLOGIC ANATOMY OF THE CEREBRAL CORTEX

The functional part of the cerebral cortex is comprised mainly of a thin layer of neurons 2 to 5 mm. in thickness, covering the surface of all the convolutions of the cerebrum and having a total area of about one-quarter of a square meter. The total cerebral cortex contains approximately 10 billion neurons.

Figure 55–1 illustrates the typical structure of the cerebral cortex, showing successive layers of different types of cells. Most of the cells are of three types: *granular, fusiform,* and *pyramidal,* the latter named for their characteristic pyramidal shape. To the right in Figure 55–1 is illustrated the typical organization of nerve fibers within the different layers of the cortex. Note particularly the large number of horizontal fibers extending between adjacent areas of the cortex, but note also the vertical fibers that extend to and from the cortex to lower areas of the brain stem or to distant regions of the cerebral cortex through long association bundles of fibers.

Neurohistologists have divided the cerebral cortex into almost 100 different areas which have slightly different architectural characteristics. Yet in all these different areas there still persist at least crude representations of all the six major layers of the cortex. Furthermore, to the untrained histologist, not more

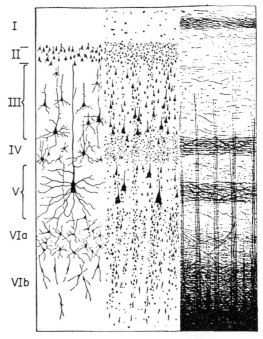

**Figure 55–1.** Structure of the cerebral cortex, illustrating: *I,* molecular layer; *II,* external granular layer; *III,* layer of pyramidal cells; *IV,* internal granular layer; *V,* large pyramidal cell layer; *VI,* layer of fusiform or polymorphic cells. (From Ranson and Clark (after Brodmann): Anatomy of the Nervous System, 1959.)

than five major architectural types of cortex can be distinguished as follows: Type 1 contains large numbers of pyramidal cells with few granular cells and, therefore, is often called the *agranular cortex*. At the other extreme, type 5 contains almost no pyramidal cells but is filled with closely packed granule cells and is called the *granular cortex*. Types 2, 3, and 4 have intermediate characteristics containing graded proportions of pyramidal and agranular cells.

The agranular cortex, which contains large numbers of pyramidal cells, is characteristic of the motor areas of the cerebral cortex; while the granular cortex, containing almost no pyramidal cells, is characteristic of the primary sensory areas. The other three intermediate types of cortex are characteristic of the association areas between the primary sensory and motor regions.

Figure 55–2 shows a map of cortical areas classified on the basis of histological studies by Brodmann. This classification provides a basis for discussion of functional areas of the brain, even though different architectural types of cortex often have similar functions.

**Anatomical Relationship of the Cerebral Cortex to the Thalamus and Other Lower Centers.** All areas of the cerebral cortex have direct afferent and efferent connections with the thalamus. Figure 55–3 shows the areas of the cerebral cortex connected with specific parts of the thalamus. These connections are in *two* directions, both from the thalamus to the cortex and then from the cortex back to essentially the same area of the thalamus. Furthermore, when the thalamic connections are cut, the functions of the corresponding cortical area become entirely or almost entirely abrogated. Therefore, the cortex operates in close association with the thalamus and can almost be considered both anatomically and functionally a large outgrowth of the thalamus; for this reason the thalamus and the cortex together are called the *thalamocortical system,* as was explained in the previous chapter. Also, all pathways from the

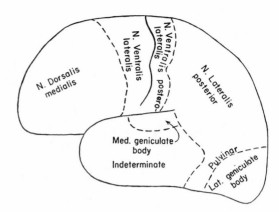

**Figure 55–3.** Areas of the cerebral cortex that connect with specific portions of the thalamus. (Modified from Elliott: Textbook of the Nervous System. J. B. Lippincott Co.)

sensory organs to the cortex pass through the thalamus, with the single exception of the sensory pathways of the olfactory tract.

# FUNCTIONS OF CERTAIN SPECIFIC CORTICAL AREAS

Studies in human beings by neurosurgeons have shown that some specific functions are localized to certain general areas of the cerebral cortex. Figure 55–4 gives a map of some of these areas as determined by Penfield and Rasmussen from electrical stimulation of the cortex or by neurological examination of patients after portions of the cortex had been removed. The

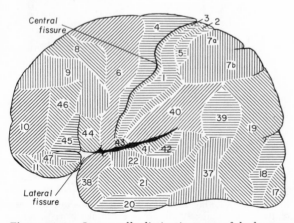

**Figure 55–2.** Structurally distinctive areas of the human cerebral cortex. (From Brodmann, modified by Buchanan: Functional Neuroanatomy. Lea & Febiger, 1966.)

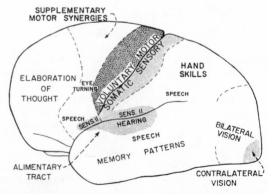

**Figure 55–4.** Functional areas of the human cerebral cortex as determined by electrical stimulation of the cortex during neurosurgical operations and by neurological examinations of patients with destroyed cortical regions. (From Penfield and Rasmussen: Cerebral Cortex of Man: A Clinical Study of Localization of Function. The Macmillan Co., 1968.)

lightly shaded areas are *primary sensory areas,* while the darkly shaded area is the *voluntary motor area* (also called *primary motor area*) from which muscular movements can be elicited with relatively weak electrical stimuli. These primary sensory and motor areas have highly specific functions, while other areas of the cortex perform more general functions that we call association or cerebration.

## SPECIFIC FUNCTIONS OF THE PRIMARY SENSORY AREAS

The primary sensory areas all have certain functions in common. For instance, somatic sensory areas, visual sensory areas, and auditory sensory areas all have spatial localizations of signals from the peripheral receptors (which is discussed in detail in Chapters 49, 60, and 61).

Electrical stimulation of the primary sensory areas in the parietal lobes in awake patients gives relatively uncomplicated sensations. For instance, in the somatic sensory area the patient states that he feels tingling in the skin, numbness, mild "electric" feeling, or, rarely, mild degrees of temperature sensations. And these sensations are localized to discrete areas of the body in accord with the spatial representation in the somatic sensory cortex, as described in Chapter 49. Therefore, it is believed that the primary somatic sensory cortex analyzes only the simple aspects of sensations and that analysis of intricate patterns of sensory experience requires adjacent parts of the parietal lobes called *sensory association areas* to operate in association with this area.

Electrical stimulation of the primary visual cortex in the occipital lobes causes the person to see flashes of light, bright lines, colors, or other simple visions. Here again, the visual images are localized to specific regions of the visual fields in accord with the portion of the primary visual cortex stimulated, as described in Chapter 60. But the visual cortex alone is not capable of complete analysis of complicated visual patterns; for this, the visual cortex must operate in association with adjacent regions of the occipital cortex, the *visual association areas.*

Electrical stimulation of the auditory cortex in the temporal lobes causes a person to hear a simple sound which may be weak or loud, have low or high frequency, or have other uncomplicated characteristics, such as a squeak or even an undulation. But never are words or any other fully intelligible sound heard. Thus, the primary auditory cortex, like the other primary sensory areas, can detect the individual elements of auditory experience but cannot analyze complicated sounds. Therefore, the primary auditory cortex alone is not sufficient to give one even the usual auditory experiences; these can be achieved, however, when the primary area operates together with the *auditory association areas* in adjacent regions of the temporal lobes.

Despite the inability of the primary sensory areas to analyze the incoming sensations fully, when these areas are destroyed, the ability of the person to utilize the respective sensations usually suffers drastically. For instance, loss of the primary visual cortex in one occipital lobe causes a person to become blind in the ipsilateral halves of both retinae, and loss of the primary visual cortices in both hemispheres causes total blindness. Likewise, loss of both primary auditory cortices causes almost total deafness. Loss of the postcentral gyri causes *depression* of somatic sensory sensations—though not total loss—presumably because of additional cortical representation of these sensations in other cortical areas: somatic sensory area II and the motor cortex, for example. (In animals far down the phylogenetic scale, loss of the visual and auditory cortices may have little effect on vision and hearing, and even an anencephalic human infant detects some visual scenes in the absence of the visual cortex, for such a child can observe a moving object and even follow it by movement of its eyes and head.)

Therefore, we can summarize the functions of the primary sensory areas of the human cerebral cortex in the following way: The lower centers of the brain relay a large part of the sensory signals to the cerebral cortex for analysis. In turn, the primary sensory areas transmit the results of their analyses back to the lower centers and to other regions of the cerebral cortex, as is discussed later in the chapter.

## THE SENSORY ASSOCIATION AREAS

Around the borders of the primary sensory areas are regions called *sensory association areas* or *secondary sensory areas.* In general, these areas extend 1 to 5 centimeters in one or more directions from the primary sensory areas; each time a primary area receives a sensory signal, secondary signals spread; after a delay of a few milliseconds into the respective association area as well. Part of this spread occurs through subcortical fiber tracts, but a major part also occurs in the thalamus, beginning in the sensory relay nuclei, passing next to corresponding *thalamic association areas,* and then traveling to the association cortex.

The general function of the sensory association areas is to provide a higher level of interpretation of the sensory experiences. The general areas for the interpretative functions for somatic, visual, and auditory experiences are illustrated in Figure 55–5.

Destruction of the sensory association areas greatly reduces the capability of the brain to analyze different characteristics of sensory experiences. For instance, damage in the temporal lobe below and behind the primary auditory area in the "dominant hemisphere" of the brain often causes a person to lose the understanding of words or of other auditory experiences even though he hears them.

Likewise, destruction of areas 18 and 19 of the occipital lobe in the dominant hemisphere or the presence of a brain tumor or other lesion in these areas, which represent the visual association areas, does not cause blindness or prevent normal activation of the primary visual cortex but does greatly reduce the person's ability to interpret what he sees. Such a person often loses his ability to recognize the meanings of words, a condition that is called *word blindness* or *alexia.*

Finally, destruction of the somatic sensory association area in the parietal cortex posterior to primary somatic area I causes the person to lose his spatial perception for location of the different parts of his body. In the case of the

hand that has been "lost," the skills of the hand are greatly reduced. Thus, this area of the cortex seems to be necessary for interpretation of somatic sensory experiences.

The functions of the association areas are described in more detail in Chapter 49 for somatic, in Chapter 60 for visual, and in Chapter 61 for auditory experiences.

## INTERPRETATIVE FUNCTION OF THE POSTERIOR TEMPORAL LOBE AND ANGULAR GYRUS REGIONS— THE GENERAL INTERPRETATIVE AREA

The somatic, visual, and auditory association areas, which can actually be called "interpretative areas," all meet one another in the posterior part of the superior temporal lobe and in the anterior part of the angular gyrus, that is, the gyrus at the posterior tip of the lateral fissure where the temporal, parietal, and occipital lobes all come together. This area of confluence of the different sensory interpretative areas plays the greatest single role of any part of the cerebral cortex in the higher levels of brain function that we call *cerebration.* Therefore, this region has frequently been called by different names suggestive of the area having almost global importance: the *general interpretative area,* the *gnostic area,* the *knowing area,* the *tertiary association area,* and so forth.

Following severe damage in the general interpretative area, a person might hear perfectly well and even recognize different words but still might be unable to arrange these words into a coherent thought. Likewise, he may be able to read words from the printed page but be unable to recognize the thought that is conveyed. In addition, he has similar difficulties in understanding the higher levels of meaning of somatic sensory experiences, even though there is no loss of sensation itself.

Electrical stimulation in the temporal lobe, particularly in its posterior part, or in the anterior angular gyrus of the conscious patient occasionally causes a highly complex thought. This is particularly true when the stimulatory electrode is passed deep enough into the brain to approach the corresponding connecting areas of the thalamus. The types of thoughts that might be experienced include memories of complicated visual scenes that one might remember from childhood, auditory hallucinations such as a specific musical piece, or even a discourse by a specific person. For this reason it is believed that complicated memory patterns

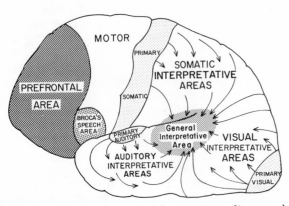

**Figure 55–5.** Organization of the somatic, auditory, and visual association areas into a general mechanism for interpretation of sensory experience. All of these feed into the *general interpretative area* located in the posterosuperior portion of the temporal lobe and the angular gyrus. Also shown in the figure are the prefrontal area and Broca's speech area.

involving more than one sensory modality are stored at least partially in the temporal lobe and the angular gyrus. This belief is in accord with the importance of the general interpretative area in interpretation of the complicated meanings of different sensory experiences.

Once again we must emphasize the global importance of this area. Loss of it in an adult usually leads to a lifetime thereafter of an almost demented existence.

**The Dominant Hemisphere.** The interpretative functions of the temporal lobe and angular gyrus are usually highly developed in only one cerebral hemisphere, the *dominant hemisphere*. At birth these regions in both hemispheres have almost the same capability of developing; we know this because removal of either hemisphere in early childhood causes the opposite to develop full dominant characteristics.

A theory that might explain hemispheric dominance in the adult is the following: The attention of the "mind" seems to be directed to one portion of the brain at a time, as was explained in the preceding chapter. Presumably, one angular gyrus region begins to be used to a greater extent than the other, and, thenceforth, because of the tendency to direct one's attention to the better developed region, the rate of learning in the cerebral hemisphere that gains the first start increases rapidly while that in the opposite side remains slight. Therefore, in the normal human being, one side becomes dominant over the other.

In more than 9 out of 10 persons the left temporal lobe and angular gyrus become dominant, and in the remaining one tenth of the population either both sides develop simultaneously to have dual dominance, or, more rarely, the right side alone becomes highly developed.

Usually associated with the dominant temporal lobe and angular gyrus is dominance of certain portions of the somesthetic cortex and motor cortex for control of voluntary motor functions. For instance, as is discussed later in the chapter, the premotor speech area, located far laterally in the intermediate frontal area, is almost always dominant also on the left side of the brain. This speech area causes the formation of words by exciting simultaneously the laryngeal muscles, the respiratory muscles, and the muscles of the mouth.

Though the interpretative areas of the temporal lobe and angular gyrus, as well as many of the motor areas, are highly developed in only a single hemisphere, they are capable of receiving sensory information from both hemispheres and are also capable of controlling motor activities in both hemispheres, utilizing mainly fiber pathways in the *corpus callosum* for communication between the two hemispheres. This unitary, cross-feeding organization prevents interference between the two sides of the brain; such interference, obviously, could create havoc with both thoughts and motor responses.

**Effect of Temporal Lobe and Angular Gyrus Destruction in the Dominant Hemisphere.** Destruction of the dominant temporal lobe and angular gyrus in an adult person normally leaves a great void in his intellect because of his inability to interpret the meanings of sensory experiences. Therefore, the neurosurgeon assiduously avoids surgery in this region. The adult can only gradually develop the interpretative functions of the nondominant temporal lobe and angular gyrus, and even then only to a slight extent. Thus, only a small amount of intellect can return. However, if this area is destroyed in a child under six years of age, the opposite side can usually develop to full extent, thus eventually returning the capabilities of the child essentially to normal.

**Role of Language in Interpretation and in Intellectual Functions.** A major share of our sensory experiences are converted into their language equivalent before being stored in the memory areas of the brain and before being processed for other intellectual purposes. For instance, when we read a book, we do not store the visual images of the words but, instead, store the information in language form. At least the information is converted to language form before its meaning is discerned.

The sensory area of the dominant hemisphere for interpretation of language is closely associated with the primary hearing area located in the auditory association areas of the temporal lobe. This locus probably results from the fact that the first introduction to language is by way of hearing. Later in life, when visual perception of language through the medium of reading develops, the visual information is then presumably channeled into the already developed language regions of the dominant temporal lobe. This probably also explains why the general interpretative area of the brain is more closely allied with the auditory association areas than with other sensory areas of the cortex.

**Interpretative Functions that Remain After Destruction of the General Interpretative Area.** One of the reasons loss of the general interpretative area is so destructive to the intellectual functions of the body is that most of our intellectual functions operate through the language medium. However, after loss of the general interpretative area, other interpretative functions of the brain still persist, such as capability to react with somatic motor activities in response to visual scenes—that is, ability to associate visual and somatic functions but loss of ability to react verbally. In addition, the temporal lobe of the opposite hemisphere still retains capabilities for associating somatic, visual, and auditory experiences of a nonlanguage nature. Thus, even though we speak of the dominant hemisphere, this dominance is primarily for the language-related intellectual functions, whereas the opposite hemisphere can actually be dominant in some other types of association.

## THE PREFRONTAL AREAS

The prefrontal areas are those portions of the frontal lobes that lie anterior to the motor regions, as shown in Figure 55–5. For years, this part of the cortex has been considered to be the locus of the higher intellect of the human being, principally because the main difference between the brain of monkeys and that of man is the great prominence of man's prefrontal areas. Yet efforts to show that the prefrontal cortex is more important in higher intellectual functions than other portions of the cortex have not been successful. Indeed, destruction of the posterior temporal lobe and angular gyrus region in the dominant hemisphere causes infinitely more harm to the intellect than does destruction of both prefrontal areas.

Psychological studies in lower animals have shown that all portions of the cortex not immediately associated with either sensory or motor functions are important in the ability of an animal to learn complicated information— such as a rat's learning to run through a maze—and that all the different areas are approximately equipotent in this regard. Therefore, the importance of the prefrontal areas to the human being is perhaps not that these areas perform some specific function different from those of other cortical areas, but primarily that they supply much additional cortical area in which cerebration can take place.

The prefrontal areas also control some of the autonomic functions of the body by transmitting impulses to the hypothalamus either directly or indirectly through the thalamus. These functions are described in the following two chapters.

**Prevention of Distractibility by the Prefrontal Areas.**   One of the outstanding characteristics of a person who has lost his prefrontal areas is the ease with which he can be *distracted* from a sequence of thoughts. Likewise, in lower animals whose prefrontal areas have been removed, the ability to concentrate on psychological tests is almost completely lost. The human being without prefrontal areas is still capable of performing many intellectual tasks, such as answering short questions and performing simple arithmetic computations (such as $9 \times 6 = 54$), thus illustrating that the basic intellectual activities of the cerebral cortex are still intact without the prefrontal areas. Yet if concerted *sequences* of cerebral functions are required of the person, he becomes completely disorganized. Therefore, the prefrontal areas seem to be important in keeping the mental functions directed toward goals.

**Possible Role of the Prefrontal Lobes for "Immediate Memory."**   Some research workers have suggested that the highly distractible nature of a person or animal after the prefrontal lobes have been destroyed results from loss of the ability to retain immediate memory. However, specific psychological tests have proved that prefrontal lobectomized human beings still have almost normal capability for memory provided that the information is presented to the person in such a way that he is not distracted from it while learning. Therefore, it seems that this loss of "immediate memory" results from inability of the brain to classify incoming information and to "code" it for storage in the memory areas. Without this ability to code the information, it is lost immediately from the mind, thus giving the effect of loss of "immediate memory" even though the memory process itself probably is not impaired.

**Elaboration of Thought by the Prefrontal Areas.**   Another function that has been ascribed to the prefrontal areas by psychologists and neurologists is *elaboration of thought*. This means simply an increase in depth and abstractness of the different thoughts. Psychological tests have shown that prefrontal lobectomized lower animals presented with successive bits of sensory information fail to store these bits in memory—probably because of their inability to code this information, the initial requirement for memory storage, as was discussed above. If the prefrontal areas are intact, many such successive bits of information can be stored in all areas of the brain and can be called forth again and again during the subsequent periods of cerebration. This ability of the prefrontal areas to cause storage—even though it be temporary—of many types of information simultaneously, and then perhaps also to cause recall of this information could well explain the many functions of the brain that we associate with higher intelligence, such as the abilities to (1) plan for the future, (2) delay action in response to incoming sensory signals so that the sensory information can be weighed until the best course of response is decided, (3) consider the consequences of motor actions even before these are performed, (4) solve complicated mathematical, legal, or philosophical problems, (5) correlate all avenues of information in diagnosing rare diseases, and (6) control one's activities in accord with moral laws.

### Effects of Destruction of the Prefrontal Areas

The person without prefrontal areas ordinarily acts precipitously in response to incoming sensory signals, such as striking an adversary too large to be beaten instead of pursuing the more judicious course of running away. Also, he is likely to lose many or most of his morals; he has little embarrassment in relation to his excretory, sexual, and social activities; and he is prone to quickly changing moods of sweetness, hate, joy, sadness, exhilaration, and rage. In short, he is a highly *distractible* person with lack of ability to pursue long and complicated thoughts.

# THOUGHTS, MEMORY, LEARNING, AND CONSCIOUSNESS

Though all of us know what a *thought* is, any attempt to describe it in abstract terms is almost impossible. The comprehension of a visual scene at a given instant would be a single thought; or overall comprehension of one's surroundings in response to other sensory information would be another thought; projection in one's mind of a mathematical equation is still a third thought. Thus, each instant of awareness could be defined as a thought. And the awareness itself can be defined as *consciousness*.

With the above definitions of thoughts and consciousness in mind, it now becomes easy to define memory and learning. *Memory* is the capability of recalling a thought at least once and usually again and again, and *learning* is the capability of the nervous system to store memories. Therefore, thoughts, consciousness, memory, and learning all go hand in hand and, in so far as the central nervous system is concerned, are almost inseparable.

Our most difficult problem in discussing con-

sciousness, thoughts, memory, and learning is that we do not know the neural mechanism of a thought. We know that destruction of large portions of the cerebral cortex does not prevent a person from having thoughts, but it does reduce his *degree* of awareness of his surroundings. On the other hand, destruction of far smaller portions of the thalamus, or especially of the mesencephalic portion of the reticular activating system can cause tremendously decreased awareness or even complete unconsciousness.

Each thought almost certainly involves simultaneous signals in portions of the cerebral cortex, thalamus, rhinencephalon, and reticular formation of the brain stem. Some crude thoughts probably depend almost entirely on lower centers; the thought of pain is probably a good example, for electrical stimulation of the human cortex rarely elicits anything more than the mildest degrees of pain, while stimulation of certain areas of the hypothalamus and mesencephalon in animals apparently causes excruciating pain. On the other hand, a type of thought pattern that requires mainly the cerebral cortex is that involving vision, because loss of the visual cortex causes complete inability to perceive visual form or color.

Therefore, we might formulate a definition of a thought in terms of neural activity as follows: A thought probably results from the momentary "pattern" of stimulation of many different parts of the nervous system at the same time, probably involving most importantly the cerebral cortex, the thalamus, the rhinencephalon, and the upper reticular formation of the brain stem. The stimulated areas of the rhinencephalon, thalamus, and reticular formation perhaps determine the crude nature of the thought, giving it such qualities as pleasure, displeasure, pain, comfort, crude modalities of sensation, localization to general parts of the body, and other gross characteristics. On the other hand, the stimulated areas of the cortex probably determine the discrete characteristics of the thought (such as specific localization of sensations in the body and of objects in the fields of vision), discrete patterns of sensation (such as the rectangular pattern of a concrete block wall or the texture of a rug), and other individual characteristics that enter into the overall awareness of a particular instant.

### MEMORY AND TYPES OF MEMORY

If we accept the above approximation of what constitutes a thought, we can see immediately that the mechanism of memory must be equally as complex as the mechanism of a thought, for, to provide memory, the nervous system must re-create the same spatial and temporal pattern of stimulation in the central nervous system at some future date. Though we cannot explain in detail what a memory is, we do know some of the basic neuronal processes that probably lead to the process of memory.

All of us know that all degrees of memory occur, some memories lasting a few seconds and others lasting hours, days, months, or years. Possibly all of these types of memory are caused by the same mechanism operating to different degrees of fulfillment. Yet, it is also possible that different mechanisms of memory do exist. Indeed, most physiologists classify memory into from two to four different types. For the purpose of the present discussion, we will use the following classification:

1. *Sensory memory*
2. *Short-term memory* or *primary memory*
3. *Long-term memory,* which itself can be divided into *secondary memory* and *tertiary memory*

The basic characteristics of these types of memory are the following:

**Sensory Memory.** Sensory memory means the ability to retain sensory signals in the sensory areas of the brain for a very short interval of time following the actual sensory experience. Usually these signals remain available for analysis for several hundred milliseconds but are replaced by new sensory signals in less than one second. Nevertheless, during the short interval of time that the instantaneous sensory information remains in the brain it can continue to be used for further processing; most important, it can be "scanned" to pick out the important points. Thus, this is the initial stage of the memory process.

**Short-Term Memory (Primary Memory).** Short-term memory (or primary memory) is the memory of a few facts, words, numbers, letters, or other bits of information for a few seconds to a minute or more at a time. This is typified by a person's memory of the digits in a telephone number for a short period of time after he has looked up the number in the telephone directory. This type of memory is usually limited to about seven bits of information, and when new bits of information are put into this *short-term store,* some of the older information is displaced. Thus, if a person looks up a second telephone number, the first is usually lost. One of the most important features of short-term memory is that the information in this memory store is instantaneously available so that the

person does not have to search through his mind for it as he does for information that has been put away in the long-term memory stores.

**Long-Term Memory.** Long-term memory is the storage in the brain of information that can be recalled at some later time—minutes, hours, days, months, or years later. This type of memory has been called *fixed memory, permanent memory,* and by several other names. Long-term memory is also usually divided into two different types, *secondary memory* and *tertiary memory,* the characteristics of which are the following:

A *secondary memory* is a long-term memory that is stored with either a weak or only a moderately strong memory trace. For this reason it is easy to forget and it is sometimes difficult to recall. Furthermore, the time required to search for the information is relatively long. This type of memory can last from several minutes to several years. When the memories are so weak that they will last for only a few minutes to a few days, they are also frequently called *recent memory.*

A *tertiary memory* is a memory that has become so well ingrained in the mind that the memory can usually last the lifetime of the person. Furthermore, the very strong memory traces of this type of memory makes the stored information available within a split second. This type of memory is typified by one's knowledge of his own name, by his ability to recall immediately the numbers from 1 to 10, the letters of the alphabet, and the words that he uses in speech, and also by the memory of his own precise physical structure and of his very familiar immediate surroundings.

## PHYSIOLOGICAL BASIS OF MEMORY

Despite the many advances in neurophysiology during the past half century, we still cannot explain what is perhaps the most important function of the brain: its capability for memory. Yet, physiological experiments are beginning to generate conceptual theories of the means by which memory could occur. Some of these are discussed in the following few sections.

**Possible Mechanisms for Short-Term Memory.** Short-term memory requires a neuronal mechanism that can hold specific information signals for a few seconds to at most a minute or more. Several such mechanisms are the following:

*Reverberating Circuit Theory of Short-Term Memory.* When a tetanizing electrical stimulus is applied directly to the surface of the cerebral cortex and then is removed after a second or more, the local area excited by this stimulus continues to emit rhythmic action potentials for short periods of time. This effect results from local reverberating circuits, the signals passing through a multistage circuit of neurons in the local area of the cortex itself or perhaps also back and forth between the cortex and the thalamus.

It is postulated that sensory signals reaching the cerebral cortex can set up similar reverberating oscillations and that these could be the basis for short-term memory. Then, as the reverberating circuit fatigues, or as new signals interfere with the reverberations, the short-term memory fades away.

One of the principal observations in support of this theory is that any factor that causes a general disturbance of brain function, such as sudden fright, a very loud noise, or any other sensory experience that attracts the person's undivided attention immediately erases the short-term memory. The memory cannot be recalled when the disturbance is over unless a portion of this memory had already been placed into the long-term memory store, as will be discussed in subsequent sections.

***Post-Tetanic Potentiation Theory of Short-Term Memory.*** In most parts of the nervous system, including even the anterior motoneurons of the spinal cord, tetanic stimulation of a neuron for a few seconds causes a subsequent increased excitability of the neuron for a few seconds to a few hours.

If during this time the neuron is stimulated again, it responds much more vigorously than normally, a phenomenon called *post-tetanic potentiation*. This is obviously a type of memory that depends on change in the excitability of the involved neurons, and it could be the basis for short-term memory. It is likely that this phenomenon results from some temporary change in the synapses of the neurons.

***DC Potential Theory of Short-Term Memory.*** Another change that often occurs in neurons following a period of excitation is a prolonged decrease in the membrane potential of the neuron lasting for from seconds to minutes. Because this changes the excitability of the neuron, it could be the basis for short-term memory. Such changes in neuronal potentials are called *DC* potentials or sometimes *electrotonic potentials*. Measurements in the cerebral cortex show that such potentials occur especially in the superficial dendritic layers of the cortex, indicating that the process of short-

term memory could result from changes in dendritic membrane potentials.

### Mechanism of Long-Term Memory, Enhancement of Synaptic Transmission Facility

Long-term memory means the ability of the nervous system to recall thoughts long after initial elicitation of the thoughts is over. We know that long-term memory does not depend on continued activity of the nervous system, because the brain can be totally inactivated by cooling, by general anesthesia, by hypoxia, by ischemia, or by any other method and yet memories that have been previously stored are still retained when the brain becomes active once again. Therefore, it is assumed that long-term memory must result from some actual alterations of the synapses, either physical or chemical.

Many different theories have been offered to explain the synaptic changes that cause long-term memory. Among the most important of these are:

**1. Anatomical Changes in the Synapses.** Cajal, more than half a century ago, discovered that the number of terminal fibrils ending on neuronal cells and dendrites in the cerebral cortex increases with age. Conversely, physiologists have shown that inactivity of regions of the cortex causes thinning of the cortex: for instance, thinning of the primary visual cortex in animals that have lost their eyesight. Also, intense activity of a particular part of the cortex can cause excessive thickening of the cortical shell in that area alone. This has been demonstrated especially in the visual cortex of animals subjected to repeated visual experiences. Finally, some neuroanatomists believe that they can observe electronmicrographic changes in presynaptic terminals that have been subjected to intense and prolonged activity.

All of these observations have led to a widely held belief that fixation of memories in the brain results from physical changes in the synapses themselves: perhaps changes in numbers of presynaptic terminals, perhaps in sizes of the terminals, perhaps in the sizes and conductivities of the dendrites, or perhaps in their chemical compositions. Also, it has been suggested that permanent changes might occur in excitability of the postsynaptic neurons. Such physical changes could cause permanent or semipermanent increase in the degree of facilitation of the synapses, thus allowing signals to pass through the synapses with progressive ease the more often the memory trace is used. This obviously would explain the tendency for memories to become more and more deeply fixed in the nervous system the more often they are recalled or the more often the person repeats the sensory experience that leads to the memory trace.

**2. Theoretical Function of RNA in Memory.** The discovery that DNA and RNA can act as codes to control reproduction, which in itself is a type of memory from one generation to another, plus the fact that these substances once formed in a cell tend to persist for the lifetime of the cell, has led to the theory that nucleic acids might be involved in memory changes of the neurons, changes that could last for the lifetime of the person. In addition, biochemical studies have shown an increase in RNA in some active neurons. Yet, a mechanism by which RNA could cause facilitation of synaptic transmission has never been found. Therefore, this theory seems to be based mainly on analogy rather than on factual evidence. The only possible supporting evidence is that several research workers have claimed that long-term memory will not occur to a significant extent when the chemical processes for formation of RNA or of protein are blocked. Since these substances are required for anatomical changes in the synapses as well as for RNA synthesis, however, these observations could support almost any type of theory for long-term memory.

**3. The Glial Theory and Other Extra-neuronal Theories.** Anatomists have demonstrated that the structures of glial cells surrounding neurons change markedly under different functional conditions. This obviously has led to the theory that glial changes might increase the facilitation of synapses and, therefore, be the basis of memory. Also, chemists have demonstrated that the quantity and chemical composition of mucopolysaccharides surrounding synapses change under different functional conditions, and this leads to still another theory for memory. However, in both of these instances there has been little more than circumstantial evidence.

**Summary.** The theory that seems most likely at present for explaining long-term memory is that some actual physical or chemical change occurs in the synaptic knobs themselves or in the postsynaptic neurons, these changes permanently facilitating the transmission of impulses at the synapses. If all the synapses are thus facilitated in a thought circuit, this circuit can be re-excited by any one of many diverse

incoming signals at later dates, thereby causing memory. The overall facilitated circuit is called a *memory engram* or a *memory trace*.

### Consolidation of Long-Term Memory

If a memory is to last in the brain so that it can be recalled days later, it must become "consolidated" in the neuronal circuits. This process requires 5 to 10 minutes for minimal consolidation and an hour or more for maximal consolidation. For instance, if a strong sensory impression is made on the brain but is then followed within a minute or so by an electrically induced brain convulsion, the sensory experience will not be remembered at all. Likewise, brain concussion, sudden application of deep general anesthesia, and other effects that temporarily block the dynamic function of the brain can prevent consolidation.

However, if the same sensory stimulus is impressed on the brain and the strong electrical shock is delayed for more than 5 to 10 minutes, at least part of the memory trace will have become established. If the shock is delayed for an hour or more, the memory will have become fully consolidated.

This process of consolidation and the time required for consolidation can probably be explained by the phenomenon of *rehearsal* of the short-term memory as follows:

**Role of Rehearsal in Transference of Short-Term Memory into Long-Term Memory.** Psychological studies have shown that rehearsal of the same information again and again accelerates and potentiates the degree of transfer of short-term memory into long-term memory, and therefore also accelerates and potentiates the process of consolidation. The brain has a natural tendency to rehearse new-found information, and especially to rehearse new-found information that catches the mind's attention. Therefore, over a period of time the important features of sensory experiences become progressively more and more fixed in the long-term memory stores. This explains why a person can remember small amounts of information studied in depth far better than he can remember large amounts of information studied only superficially. And it also explains why a person who is wide-awake will consolidate memories far better than will a person who is in a state of mental fatigue.

**Codifying of Memories During the Process of Consolidation.** One of the most important features of the process of consolidation is that memories to be placed permanently into the long-term memory storehouse are first codified into different classes of information. During this process similar information is recalled from the long-term storage bins and is used to help process the new information. The new and old are compared for similarities and for differences, and part of the storage process is to store the information about these similarities and differences rather than simply to store the information unprocessed. Thus, during the process of consolidation, the new memories are not stored randomly in the mind, but instead are stored in direct association with other memories of the same type. This is obviously necessary if one is to be able to scan the memory store at a later date to find the required information.

**Transfer of Sensory Memory into Long-Term Memory.** Some physiopsychologists believe that sensory memory can be transferred directly into long-term memory without first passing through the stage of short-term memory. This would apply to such information as visual scenes, musical tunes, tactile experiences, and so forth. Here again, the greater the number of times that the person experiences the sensory information, which is a form of rehearsal, the greater also is the degree of consolidation of the memories.

**Change of Long-Term Secondary Memory into Long-Term Tertiary Memory—Role of Rehearsal.** Rehearsal also plays an extremely important role in changing the weak trace type of long-term memory, called secondary memory, into the strong trace type, called tertiary memory. That is, each time a memory is recalled or each time the same sensory experience is repeated, a more and more indelible memory trace develops in the brain. The memory finally becomes so deeply fixed in the brain that it can be recalled within fractions of a second and it will also last for a lifetime, both of which are the characteristics of long-term tertiary memory.

*Role of Specific Parts of the Brain in the Memory Process*

**Role of the Hippocampus for Rehearsal, Codification, and Consolidation of Memories—Anterograde Amnesia.** Persons who have had both hippocampal gyri removed have essentially normal memory for information stored in the brain prior to removal of the hippocampi. However, after loss of these gyri, these persons have very little capability for transferring short-term memory into long-term

memory. That is, they do not have the ability to separate out the important information, to codify it, to rehearse it, and to consolidate it in the long-term memory store. Therefore, these persons develop serious *anterograde amnesia,* meaning simply inability to establish new memories.

Other types of lesions that frequently give anterograde amnesia include lesions of (1) the mammillary bodies, (2) the anterior nuclei of the thalamus, (3) the anterior columns of the fornix, or (4) the dorsal medial nuclei of the thalamus.

**Retrograde Amnesia—Possible Relation to Thalamic Lesions.** Excessive damage to large portions of the thalamus is incompatible with life. However, lesser thalamic damage will often lead to some degree of retrograde amnesia, which means inability to recall memories from the long-term store even though these memories are still known to be there. In the previous chapter, it was pointed out that the thalamus possibly or probably plays a major role in directing the person's attention to information in different parts of his immediate sensory sphere or of his memory storehouse. Therefore, the thalamus could easily play similar important roles in codifying, storing, and recalling memories.

# INTELLECTUAL OPERATIONS OF THE BRAIN

Thus far, we have considered the approximate nature of thoughts and possible mechanisms by which memory and learning can occur. Now we need to consider the mechanisms by which the brain performs complex intellectual operations, such as the analysis of sensory information and the establishment of abstract thoughts. About these mechanisms we know almost nothing, but experiments along these lines have established the following important facts: First, the brain can focus its attention on specific types of information (which was discussed in the preceding chapter). Second, the different qualities of each set of information signals are split away from the central signal and are transmitted to multiple areas of the brain. Third, the brain compares new information with old information in its memory loci. And, fourth, the brain determines patterns of stimulation.

**Analysis of Information by Splitting its Qualities.** When sensory information enters the nervous system, one of the first steps in its analysis is to transmit the signal to separate parts of the brain that are selectively adapted for detecting specific qualities. For instance, if the hand is placed on a hot stove, (a) pain information is sent through the spinoreticular and spinothalamic tracts into the reticular formation of the brain stem and into certain thalamic and hypothalamic nuclei, (b) tactile information is sent through the dorsal column system to the somesthetic cortex, giving the cortex a detailed description of the part of the hand that is touching the hot stove, and (c) kinesthetic information from the muscles and joints is sent to the cerebellum, the reticular formation, and to higher centers of the brain to give momentary information of the position of the hand. Thus, the different qualities of the overall information are dissected and transmitted to different parts of the brain. It is on the basis of analyses in all these parts of the brain that motor responses to the incoming sensory information are formulated.

One can see that different centers of the brain react to specific qualities of information—the reticular formation and certain regions of the thalamus and hypothalamus to pain, regions of the mesencephalon and hypothalamus to the affective nature of the sensation (that is, whether it is pleasant or unpleasant), the somatic cortex to the localization of sensation, other areas to kinesthetic activities, and others to visual, auditory, gustatory, olfactory, and vestibular information.

**Analysis of New Information by Comparison with Memories.** We all know that new sensory experiences are immediately compared with previous experiences of the same or similar types. For instance, this is the way we recognize a person whom we know. Yet, it still remains a mystery how we can make these comparisons. One theory suggests that the memory trace might be stored in one layer of the cerebral cortex and that the new incoming sensory experience might be impressed on another layer. If the new pattern of stimulation matches the memory trace, some interaction between the two gives the person a sense of recognition.

**Analysis of Patterns.** In analyzing information, the brain depends to a great extent on "patterns" of stimulation. For instance, a square is detected as a square regardless of its position or angle of rotation in the visual field. Likewise, a series of parallel bars is detected as parallel bars regardless of their orientation, or a fly is detected as a fly whether it is seen in the peripheral or central field of vision. We can ex-

tend the same logic to the somatic sensory areas, for a person can detect a cube simply by feeling it whether it be in an upright, horizontal, or angulated position. Also, it can be detected even by the feet even though the feet may have never felt a cube before.

It is believed that one of the major methods for analyzing information is to dissect it into its component patterns. For instance, repetitive auditory sensations and visual sensations of a series of lights have at least one characteristic in common: both have the pattern of repetitiveness. Yet we cannot say how patterns of sensations are detected. One theory holds that their detection results from traveling waves passing through the cortex. As these cross the brain they might cause a signal to be transmitted every time they interact with an excited point of the cortex. For instance, if a traveling wave were passing over a visual cortex that is being excited by a regular series of lights, every time this wave should come in contact with an activated point, a signal would theoretically be transmitted away from the cortex to some *analyzer portion* of the brain. The analyzer would receive a train of rhythmic impulses, each impulse denoting a separate light, and this train of impulses could then theoretically give the person the sensation of repetitiveness. Likewise, repetitive sounds entering the ears could transmit a train of impulses to the same analyzer which would also give a sensation of repetitiveness, or movement of the hand over a corrugated surface could give the analyzer the same sensation of repetitiveness.

If we use our imaginations, we can see that only a few different types of analytical patterns can go a long way toward explaining almost all types of sensory impressions.

*Processing of Information so that Patterns Can Be Determined.* Beautiful examples of the ability of the brain to process information for determination of patterns have been discovered in relation to the visual system. In Chapter 60 we point out that the retina itself processes visual information to a great extent even before it is transmitted into the brain. For instance, the retina is especially responsive to rapidly changing light intensities or to movement of images across the receptors, and the signals sent to the brain are appropriately modified to transmit these qualities. Then, when the image is transmitted through the lateral geniculate body and thence into the primary visual cortex, the visual signals are altered again in such a way that the cortex detects mainly boundaries between light and dark areas. That is, the visual scene is "differentiated" by this mechanism, bringing out strongly the contrasts of the scene but de-emphasizing the flat areas. This explains why a few lines drawn on a paper can give one the impression of a person's appearance. Certainly the lines do not represent the actual picture of a person, but they do give the visual cortex the same pattern of contrasts that his own visual system would give.

One can see, therefore, that the visual cortex utilizes transformed types of information to point up the most striking characteristics of the visual scene while neglecting unimportant information. Then, the patterns of the visual scene can be extracted from this "pre-selected" information.

**Converging of Information in Association Areas of the Brain.**   Certain areas of the brain perform specific functions, such as the primary sensory areas and the motor areas. On the other hand, some of the remainder of the brain, the *association areas,* do not necessarily perform functions only associated with a particular modality of sensation or with a specific motor function. These areas often receive information from several sensory inputs and also from other association areas of the brain. Association areas are also located in lower brain regions, such as in the *thalamus* and *caudate nuclei.* Indeed, for every association area in the cortex, there is a reciprocally connected association area of the thalamus. In addition, other areas of the thalamus, such as the centromedial and dorsomedial nuclei, interconnect widely within the thalamus itself and presumably act as association areas between the different thalamic nuclei.

It is in the association areas that information from different sources interact. Thus, the posterior temporal lobe and the angular gyrus of the cerebral cortex, where the temporal, parietal, and occipital lobes all come together, is an association area that receives auditory, visual, and tactile information. Neurophysiological experiments have demonstrated that these interacting types of information can at times inhibit each other and at other times facilitate each other. Yet, how the interactions in the association areas are analyzed beyond this point is still a mystery.

# INTELLECTUAL ASPECTS OF MOTOR CONTROL

Control of muscular movements by the cerebral cortex and other areas of the brain was discussed in Chapter 53, but we still need to consider the manner in which the intellectual operations of the brain elicit motor events. Almost all sensory and even abstract experiences of the mind are eventually expressed in some type of motor activity, such as actual muscular movements of a directed nature, tenseness of the muscles, total relaxation of the muscles, attainment of certain postures, tapping of the fingers, grimaces of the face, or speech.

Psychological tests show that the analytical portions of the brain control motor activities in the following sequence of three stages: (1) origin of the thought of the motor activity to be performed, (2) determination of the sequence of movements necessary to perform the overall task, and (3) control of the muscular movements themselves.

**Origin of the Thought of a Motor Activity in the General Interpretative Area of the Brain.**   A large number of motor movements, which we call subconscious movements, are conceived in the lower regions of the brain, but complex learned tasks can be performed only when the cerebral cortex is present. The most important cortical areas for this purpose are the posterior part of the superior temporal gyrus and the angular gyrus of the dominant hemisphere. The student will immediately recognize this area to be the *general interpretative area* of the brain. Thus, the sensory interactions that take place in this all-important sensory part of the dominant hemisphere not only play a major role in interpreting sensory experiences but also in determining the person's course of motor activities.

**Determination of Sequences of Movement.**   In our earlier discussion of cortical control of motor activities, it was pointed out that the somatic association area is necessary for performance of complex learned movements; destruction of this area both in animals and human beings causes the motor activities to lose much of their purposefulness. Therefore, it is postulated that once the general interpretative area of the brain has determined the motor activity to be performed, the actual sequences of movement are determined in the somatic association areas as discussed in Chapter 53. Signals are transmitted from here to the motor portion of the brain to control the actual motor movements, as also discussed in Chapter 53.

# FUNCTION OF THE BRAIN IN COMMUNICATION

One of the most important differences between the human being and lower animals is the facility with which human beings can communicate with one another. Furthermore, because neurological tests can easily assess the ability of a person to communicate with others, we know perhaps more about the sensory and motor systems related to communication than about any other segment of cortical function. Therefore, we will review rapidly the function of the cortex in communication, and from this one can see immediately how the principles of sensory analysis and motor control that were just discussed apply to this art.

There are two aspects to communication: first, the *sensory aspect,* involving the ears and eyes, and, second, the *motor aspect,* involving vocalization and its control.

**Sensory Aspects of Communication.** We noted earlier in the chapter that destruction of portions of the *auditory* and *visual association areas* of the cortex can result in inability to understand the spoken word or the written word. These effects are called respectively *auditory receptive aphasia* and *visual receptive aphasia* or, more commonly, *word deafness* and *word blindness* (also called *alexia).* On the other hand, some persons are perfectly capable of understanding either the spoken word or the written word but are unable to interpret their meanings when used to express a thought. This results most frequently when the *angular gyrus* or *posterior portion of the superior temporal gyrus* in the dominant hemisphere is damaged or destroyed. This is considered to be *general sensory aphasia* or *general agnosia.*

**Motor Aspects of Communication.** *Syntactical Aphasia.* The process of speech involves (a) formation in the mind of thoughts to be expressed and choice of the words to be used and (b) the actual act of vocalization. The formation of thoughts and choice of words is principally the function of the sensory areas of the brain, for we find that a person who has a destructive lesion in the same *posterior temporal lobe and angular gyrus region* that is involved in general sensory aphasia also has inability to formulate intelligible thoughts to be communicated. And at other times the thoughts can be formulated, but the person is unable to put together the appropriate words to express the thought; this results most frequently from destruction of the angular gyrus. These inabilities to formulate thoughts and word sequences are called *syntactical aphasias* or, in honor of the neurologist who first delimited the brain area responsible, *Wernicke's aphasia.*

*Motor Aphasia.* Often a person is perfectly capable of deciding what he wishes to say, and he is capable of vocalizing, but he simply cannot make his vocal system emit words instead of noises. This effect, called *motor aphasia,* almost always results from damage to *Broca's speech area,* which lies in the *premotor* facial region of the cortex—about 95 per cent of the time in the left hemisphere, as illustrated in Figure 55–5. Therefore, we assume that the *patterns* for control of the larynx, lips, mouth, respiratory system, and other accessory muscles of articulation are all controlled in this area.

*Articulation.* Finally, we have the act of articulation itself, which means the muscular movements of the mouth, tongue, larynx, and so forth, that are responsible for the actual emission of sound. The facial and laryngeal regions of the motor cortex activate these muscles, and the sensory cortex helps to control the muscle contractions by feedback servomechanisms described in Chapter 53. Destruction of these regions can cause either total or partial inability to speak distinctly.

# FUNCTION OF THE CORPUS CALLOSUM AND ANTERIOR COMMISSURE TO TRANSFER THOUGHTS, MEMORIES, AND OTHER INFORMATION TO THE OPPOSITE HEMISPHERE

Fibers in the *corpus callosum* connect the respective cortical areas of the two hemispheres with each other except for the cortices of the anterior portions of the temporal lobes; these temporal cortical areas are interconnected by fibers that pass through the *anterior commissure.* Because of the tremendous number of fibers in the corpus callosum, it was very early believed that this massive structure must have some important function to correlate activities of the two cerebral hemispheres. However, cutting the corpus callosum in experimental animals, or in occasional instances even in the human being, caused no change in the function of the brain that could be discerned by gross observation. Therefore, for a long time the function of the corpus callosum was a mystery.

However, psychological experiments have now demonstrated the function of the corpus callosum and the anterior commissure. This can be explained best by recounting one of the experiments. A monkey is first prepared by cutting the corpus callosum and splitting the optic chiasm longitudinally. Then it is taught to recognize different types of objects with its right eye while its left eye is covered. Then the right eye is covered and the monkey is tested to determine whether or not its left eye can recognize the same object. The answer to this is that the left eye *cannot* recognize the object. Yet, on repeating the same experiment with the optic chiasm still split but the corpus callosum intact, it is found invariably that recognition in one hemisphere of the brain creates recognition also in the opposite hemisphere.

Thus, one of the functions of the corpus callosum and the anterior commissure is to make information stored in the cortex of one hemisphere available to the corresponding cortical area of the opposite hemisphere. Two important examples of such cooperation between the two hemispheres are:

1. Cutting of the corpus callosum blocks transfer of information from the general interpretative area of the dominant hemisphere to the motor cortex on the opposite side of the brain. Therefore, the intellectual functions of the brain, located primarily in the dominant hemisphere, lose much of their control over the right motor cortex and therefore also of the voluntary motor functions of the left hand and arm even though the usual subconscious movements of the left hand and arm are completely normal.

2. Cutting of the corpus callosum prevents transfer of somatic and visual information from the right hemisphere into the general interpretative area of the dominant hemisphere. Therefore, somatic and visual information from the left side of the body frequently fails to reach the general interpretative area of the brain and, therefore, cannot be used for decision-making.

Yet, in one human being recently studied who had been born with congenital absence of the corpus callosum, it was found that most of these associative functions between the hemispheres could be performed. This emphasizes the importance of the lower brain centers, particularly of the reticular formation and of the thalamus, for correlation of activities between the two sides of the brain. Therefore, it should not be thought that the different parts of the cerebral cortex communicate with each other only via subcortical fiber tracts because perhaps equally as much communication occurs by way of the lower centers as well.

# REFERENCES

Atkinson, R. C., and Shiffrin, R. M.: The control of short-term memory. *Sci. Amer.*, 225:82, 1971.

Bower, T. G. R.: The object in the world of the infant. *Sci. Amer.*, 225:30, 1971.

Cole, R. A., and Scott, B.: Toward a theory of speech perception. *Psychol. Rev.*, 81:348, 1974.

Corning, W. C., Dyal, J. A., and Willows, A. O. D. (eds.): Invertebrate Learning. Vol. 3. New York, Plenum Publishing Corporation, 1974.

Darley, F. L.: Brain Mechanisms Underlying Speech and Language. New York, Grune & Stratton, Inc., 1967.

Darley, F. L., Aronson, A. E., and Brown, J. R.: Motor Speech Disorders. Philadelphia, W. B. Saunders Company, 1975.

Deutsch, J. A. (ed.): The Physiological Basis of Memory. New York, Academic Press, Inc., 1972.

Dimond, S. J.: The Double Brain. New York, Churchill Livingstone, Div. of Longman, Inc., 1972.

Flanagan, J. L.: The synthesis of speech. *Sci. Amer.*, 226:48, 1972.

Gazzaniga, M. S.: One brain—two minds? *Amer. Sci.*, 60:311, 1972.

Geschwind, N.: Language and the brain. *Sci. Amer.*, 226:76, 1972.

Geschwind, N.: The apraxias: neural mechanisms of disorders of learned movement. *Amer. Sci.*, 63:188, 1975.

Gibbs, M. E., and Mark, R. F.: Inhibition of Memory Formation. New York, Plenum Publishing Corporation, 1973.

Goldensohn, E. S., and Appel, S. H.: Scientific Approaches to Clinical Neurology. Philadelphia, Lea & Febiger, 1975.

Haber, R. N.: How we remember what we see. *Sci. Amer.*, 222:104, 1970.

Hailman, J. P.: How an instinct is learned. *Sci. Amer.*, 221:98, 1969.

Harris, A. J.: Inductive functions of the nervous system. *Ann. Rev. Physiol.*, 36:251, 1974.

Hess, W. R.: The Biology of Mind. Chicago, University of Chicago Press, 1964.

Jakobson, R.: Verbal communication. *Sci. Amer.*, 227:72, 1972.

Jakoubek, B.: Brain Function and Macromolecular Synthesis. New York, Academic Press, Inc., 1974.

Jenkins, J. J., Jimenez-Pabon, E., Shaw, R. E., and Sefer, J. W.: Schuell's Aphasia in Adults. Hagerstown, Md., Harper & Row, Publishers, 1975.

Jyont, R. J., and Goldstein, M. N.: Minor cerebral hemisphere. *Adv. Neurol.*, 7:147, 1975.

Kimura, D.: The asymmetry of the human brain. *Sci. Amer.*, 228:70, 1973.

Kinsbourne, M., and Smith, W. L. (eds.): Hemispheric Disconnection and Cerebral Function. Springfield, Ill., Charles C Thomas, Publisher, 1974.

Luria, A. R.: The functional organization of the brain. *Sci. Amer.*, 222:66, 1970.

McGuigan, F. J., and Schoonover, R. A. (eds.): The Psychophysiology of Thinking; Studies of Covert Processes. New York, Academic Press, Inc., 1973.

Maggio, E.: Psychophysiology of Learning and Memory. Springfield, Ill., Charles C. Thomas, Publisher, 1971.

Mark, R. F.: Memory and Nerve Cell Connections: Criticisms and Contributions from Developmental Neurophysiology. New York, Oxford University Press, 1974.

Mattingly, I. G.: Speech cues and sign stimuli. *Amer. Sci.*, 60:327, 1972.

Nebes, R. D.: Hemispheric specialization in commissurotomized man. *Psychol. Bull.*, 81:1, 1974.

Needham, C. W.: Neurosurgical Syndromes of the Brain. Springfield, Ill., Charles C Thomas, Publisher, 1973.

Ornstein, R. E. (ed.): The Nature of Human Consciousness: a Book of Readings. San Francisco. W. H. Freeman Company, Publishers, 1973.

Penfield, W.: The Excitable Cortex in Conscious Man. Springfield, Ill., Charles C Thomas, Publisher, 1958.

Peters, M., and Ploog, D.: Communication among primates. *Ann. Rev. Physiol.*, 35:221, 1973.

Rosenblatt, J. S.: Learning in newborn kittens. *Sci. Amer.*, 227:18, 1972.

Schultz, A.: A Theory and Consciousness. New York, Philosophical Library, 1973.

Schwartz, M.: Physiological Psychology. New York, Appleton-Century-Crofts, 1973.

Somjen, G. G.: Neurophysiology Studied in Man. (Proceedings of a Symposium held in Paris, 1971.) New York, American Elsevier Publishing Company, 1972.

Stein, D. G., and Rosen, J. J.: Learning and Memory. New York, The Macmillan Company, 1974.

Thompson, R. F.: Introduction to Physiological Psychology. New York, Harper & Row, Publishers, 1975.

Timmons, B. A., and Boudreau, J. P.: Auditory feedback as a major factor in stuttering. *J. Speech Hear. Disord.*, 37:476, 1972.

Tulving, E.: Cue-dependent forgetting. *Amer. Sci.*, 62:74, 1974.

Wassermann, G. D.: Brains and Reasoning; Brain Science as a Basis of Applied and Pure Philosophy. London, Macmillan & Co. Ltd., 1974.

White, M. J.: Hemispheric asymmetries in tachistoscopic information-processing. *Brit. J. Psychol.*, 63:497, 1972.

Wyrwicka, W.: The Mechanisms of Conditioned Behavior. Springfield, Ill., Charles C Thomas, Publisher, 1972.

# 56

# Behavioral Functions of the Brain: The Limbic System, Role of the Hypothalamus, and Control of Vegetative Functions of the Body

Behavior is a function of the entire nervous system, not of any particular portion. Even the discrete cord reflexes are an element of behavior, and the wakefulness and sleep cycle discussed in Chapter 54 is certainly one of the most important of our behavioral patterns. However, in this chapter we will deal with those special types of behavior associated with emotions, subconscious motor and sensory drives, and the intrinsic feelings of pain and pleasure. These functions of the nervous system are performed mainly by subcortical structures located in the basal regions of the brain. However, older portions of the cerebral cortex located on the medial and ventral portions of the cerebral hemispheres also play a role. This overall group of brain structures is frequently called the *limbic system*.

Portions of the limbic system, especially the hypothalamus and related structures, control many of the internal functions of the body, such as body temperature, osmolality of the body fluids, body weight, and so forth; these functions are collectively called *vegetative functions* of the body.

## THE LIMBIC SYSTEM

Figure 56–1 illustrates the anatomical structures of the limbic system, showing these to be an interconnected complex of basal brain elements. Located in the midst of all of these is the hypothalamus, which is considered by many anatomists to be a separate structure from the remainder of the limbic system, but which, from a physiological point of view, is considered to be one of the central elements of the system. Therefore, Figure 56–2 illustrates schematically this key position of the hypothalamus in the limbic system and shows that surrounding it are the other subcortical structures of the limbic system, including the *preoptic area,* the *septum,* the *paraolfactory area,* the *epithalamus,* the *anterior nuclei of the thalamus, portions of the basal ganglia,* the *hippocampus,* and the *amygdala.* Surrounding the subcortical limbic areas is the *limbic cortex* composed of a ring of cerebral cortex (a) beginning in the *orbitofrontal area* on the ventral surface of the frontal lobes, (b) extending upward in front of and over the corpus callosum onto the medial aspect of the cerebral hemisphere to the *cingulate gyrus,* and finally (c) passing posterior to the corpus callosum and downward onto the ventromedial surface of the temporal lobe to the *hippocampal gyrus, pyriform area,* and *uncus.* Thus, on the medial and ventral surfaces of each cerebral hemisphere is a ring of paleocortex that surrounds a group of deep structures intimately associated with overall behavior and with emotions.

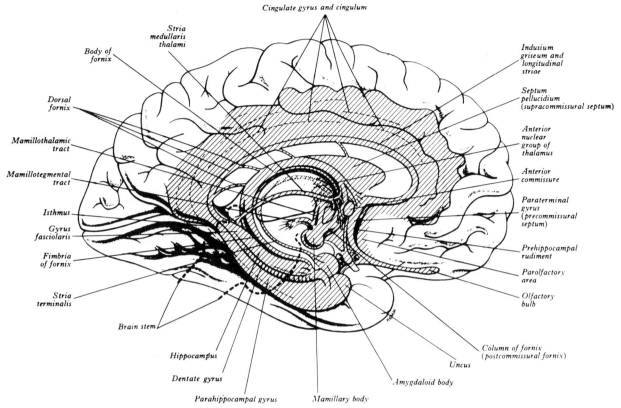

**Figure 56–1.** Anatomy of the limbic system illustrated by the shaded areas of the figure. (From Warwick and Williams: Gray's Anatomy, 35th Brit. Ed., London, Longman Group Ltd., 1973.)

In the past, many neurophysiologists have lumped the entire limbic system together as a unit structure that controls the different emotional and behavioral patterns of the nervous system. As we learn more about the various portions of this system, it is becoming evident that different areas of the system perform very specific functions.

## VEGETATIVE FUNCTIONS OF THE HYPOTHALAMUS

The hypothalamus provides the most important output pathway through which the limbic system controls many major functions of the body, especially the vegetative functions, which are the involuntary functions necessary for living. The different hypothalamic centers are so important in these controls that they are discussed in detail throughout the text in association with the specific functional systems, such as in connection with arterial pressure regulation in Chapter 21, thirst and water conservation in Chapter 36, and temperature regulation in Chapter 72. However, to illustrate the organization of the hypothalamus as a functioning unit, we will summarize some of its vegetative functions here.

Figure 56–3 summarizes most of the vegetative functions of the hypothalamus and shows, with the exception of the lateral hypothalamic areas that overlie the areas illustrated, the major nuclei or major areas which when stimulated affect respective vegetative activities. A word of caution must be issued in studying this diagram, however, for the areas that cause specific activities are not nearly so discrete as shown in the figure. The illustrated areas are merely those from which the functions are likely to be elicited. Also, it is not known whether the effects noted in the figure result from activation of specific control nuclei or whether they result merely from activation of fiber tracts leading from control nuclei located elsewhere. For instance, stimulation of the

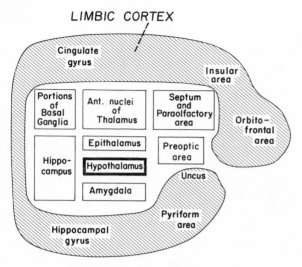

**Figure 56–2.** The limbic system.

posterior hypothalamus, through which many of the fiber pathways from other portions of the hypothalamus pass, can at times elicit many functions believed to be controlled primarily by other hypothalamic nuclei. With this caution in mind, we can give the following general description to the vegetative control functions of the hypothalamus.

**Cardiovascular Regulation.** Stimulation of widespread areas throughout the hypothalamus can cause every known type of neurogenic effect in the cardiovascular system, including increased arterial pressure, decreased arterial pressure, increased

heart rate, and decreased heart rate. In general, stimulation in the *posterior* and *lateral hypothalamus* increases the arterial pressure and heart rate, while stimulation in the *preoptic area* often has opposite effects, causing a decrease in both heart rate and arterial pressure. These effects are transmitted mainly through the cardiovascular control centers in the reticular substance of the medulla and pons.

**Regulation of Body Temperature.** Large areas in the anterior portion of the hypothalamus, especially in the *preoptic area,* are concerned with regulation of body temperature. An increase in temperature of the blood flowing through these areas increases their activity, while a decrease in temperature decreases their activity. In turn, these areas control the mechanisms for increasing or decreasing body temperature, as discussed in Chapter 72.

**Regulation of Body Water.** The hypothalamus regulates body water in two separate ways, by (1) creating the sensation of thirst, which makes an animal drink water, and (2) controlling the excretion of water into the urine. An area called the *thirst center* is located in the lateral hypothalamus. When the electrolytes inside the neurons of this small center become too concentrated, the animal develops an intense desire to drink water; he will search out the nearest source of water and drink enough to return the electrolyte concentration of the thirst center to normal.

Control of renal excretion of water is vested mainly in the *supraoptic nuclei.* When the body fluids become too concentrated, the neurons of this area become stimulated and secrete a hormone called *antidiuretic hormone.* This hormone is stored in the posterior pituitary gland and when released into the blood acts on the collecting ducts of the kidneys to cause massive reabsorption of water, thereby de-

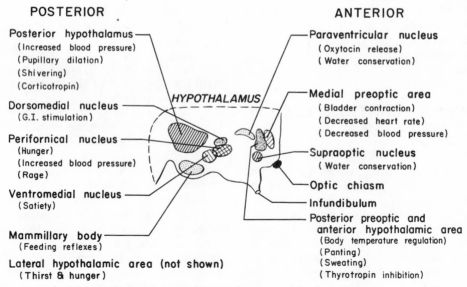

**Figure 56–3.** Autonomic control centers of the hypothalamus.

creasing the loss of water into the urine. These functions were presented in detail in Chapter 36.

**Regulation of Uterine Contractility and of Milk Ejection by the Breasts.** Stimulation of the paraventricular nuclei causes its neuronal cells to secrete the hormone *oxytocin*. This in turn causes increased contractility of the uterus and also contraction of the myoepithelial cells that surround the alveoli of the breasts, which then causes the alveoli to empty the milk through the nipples. At the end of pregnancy especially large quantities of oxytocin are secreted, and this secretion helps to promote labor contractions that expel the baby. Also, when a baby suckles the mother's breast oxytocin is released, and this release performs the necessary function of expelling the milk through the nipple so that the baby can nourish itself. These functions are discussed in Chapter 82.

**Gastrointestinal and Feeding Regulation.** Excitation of several areas of the hypothalamus causes an animal to experience extreme hunger, a voracious appetite, and an intense desire to search for food. The two areas that, when excited, are most associated with hunger are the *perifornical nucleus* and the *lateral hypothalamic area*. On the other hand, damage to these areas of the hypothalamus causes the animal to lose desire for food, sometimes causing starvation.

A center that opposes the desire for food, called the *satiety center,* is located in the *ventromedial nucleus.* When this center is stimulated, an animal that is eating food suddenly stops eating and shows complete indifference to the food. On the other hand, if this area is destroyed, the animal cannot be satiated, but instead, his hypothalamic hunger centers become overactive so that he has a voracious appetite, often resulting in tremendous obesity.

Another area of the hypothalamus that enters into the overall control of gastrointestinal activity is the *dorsomedial nucleus* which, when stimulated, increases the degree of peristalsis in the gut and also increases glandular secretion. Also, the *mammillary body* enters into many feeding reflexes, such as licking the lips and swallowing.

**Hypothalamic Control of the Anterior Pituitary Gland.** Stimulation of certain areas of the hypothalamus also causes the anterior pituitary gland to secrete its hormones. This subject will be discussed in detail in Chapter 75 in relation to the overall control of the endocrine glands, but, briefly, the basic mechanism of the control of the anterior pituitary is the following:

The adenohypophysis receives its blood supply mainly from veins that flow into the anterior pituitary sinuses from the lower part of the hypothalamus. As the blood courses through the hypothalamus before reaching the anterior pituitary, *neurosecretory substances* are secreted into the blood by various hypothalamic nuclei. They are then transported in the blood to the pituitary tissues where they act on the glandular cells to cause release of the anterior pituitary hormones.

Stimulation of the following areas of the hypothalamus has been said to cause secretion of specific hormones: (1) preoptic and anterior hypothalamic nuclei, thyrotropin; (2) posterior hypothalamus and median eminence of the infundibulum, corticotropin; (3) median eminence, follicle-stimulating hormone; and (4) anterior hypothalamus, luteinizing hormone.

**Summary.** A number of discrete areas of the hypothalamus have now been found that control specific vegetative functions. However, these areas are still poorly delimited, so much so that the above separation of different areas for different hypothalamic functions is partially artificial.

# ROLE OF THE RETICULAR FORMATION IN BEHAVIORAL FUNCTIONS OF THE BRAIN

At this point we need to recognize that many of the behavioral functions elicited from the hypothalamus and from other limbic structures are mediated through the reticular formation of the brain stem. It was pointed out in Chapters 52 and 54 that stimulation of the excitatory portion of the reticular formation can cause high degrees of somatic excitability; in Chapter 57 we will see that most of the signals for control of the autonomic nervous system either originate in or are transmitted from higher centers through nuclei located in the brain stem reticular formation.

Therefore, from a physiological point of view the reticular formation is a very important part of the limbic system even though anatomically it is considered to be a separate entity. A very important route of communication between the limbic system and the reticular formation is the *medial forebrain bundle* that extends from the septal and orbitofrontal cortical regions downward through the hypothalamus to the reticular formation. This bundle carries fibers in both directions, forming a trunk-line communication system. A second route of communication is through short pathways among the reticular formation, the thalamus, the hypothalamus, and most of the other contiguous areas of the basal brain.

# BEHAVIORAL FUNCTIONS OF THE LIMBIC SYSTEM

## PLEASURE AND PAIN; REWARD AND PUNISHMENT

In recent years, it has been learned that many hypothalamic and other limbic structures are particularly concerned with the affective nature of sensory sensations—that is, with whether the sensations are *pleasant* or *painful*. These affective qualities are also called *reward* and *punishment* or *satisfaction* and *aversion*. Electrical stimulation of certain regions pleases or satisfies the animal, whereas electrical stimulation of other regions causes extreme pain, fear, defense, escape reactions, and all the other elements of punishment. Obviously, these two oppositely responding systems greatly affect the behavior of the animal.

**Reward Centers.** Figure 56–4 illustrates a technique that has been used for localizing the specific reward and punishment areas of the brain. In this figure a lever is placed at the side of the cage and is arranged so that depressing the lever makes electrical contact with a stimulator. Electrodes are placed successively at different areas in the brain so that the animal can stimulate the area by pressing the lever. If stimulating the particular area gives the animal a sense of reward, then he will press the lever again and again, sometimes as much as 7000 times per hour. Furthermore, when offered the choice of eating some delectable food as opposed to the opportunity to stimulate the reward center, he often chooses the electrical stimulation.

By using this procedure, the major reward centers have been found to be located in the septum and hypothalamus, primarily *along the course of the medial forebrain bundle* and *in the ventromedial nuclei of the hypothalamus*. Less potent reward centers, which are probably secondary to the major ones in the septum and hypothalamus, are found in the amygdala, certain areas of the thalamus and basal ganglia, and finally extending downward into the basal tegmentum of the mesencephalon.

**Punishment Centers.** The apparatus illustrated in Figure 56–4 can also be connected so that pressing the lever turns off rather than turns on an electrical stimulus. In this case, the animal will not turn the stimulus off when the electrode is in one of the reward areas, but when it is in certain other areas he immediately learns to turn it off. Stimulation in these areas causes the animal to show all the signs of pain, displeasure, and punishment. Furthermore, prolonged stimulation for 24 hours or more causes the animal to become severely sick and actually leads to death.

By means of this technique, the principal centers for pain, punishment, and escape tendencies have been found in the *central gray area surrounding the aqueduct of Sylvius in the mesencephalon* and extending upward into the *periventricular structures of the hypothalamus and thalamus*. The *perifornical nucleus* of the hypothalamus is the most reactive of all the hypothalamic areas.

It is particularly interesting that stimulation in the pain and punishment centers can frequently inhibit the reward and pleasure centers completely, illustrating that pain can take precedence over pleasure and reward.

**Importance of Reward and Punishment in Behavior.** Almost everything that we do depends on reward and punishment. If we are doing something that is rewarding, we continue to do it; if it is punishing, we cease to do it. Therefore, the reward and punishment centers undoubtedly constitute one of the most important of all the controllers of our bodily activities, our motivations, and so forth.

**Importance of Reward and Punishment in Learning and Memory—Habituation or Reinforcement.** Animal experiments have shown that a sensory experience causing neither reward nor punishment is remembered hardly at all. Electrical recordings have shown that new and novel sensory stimuli always excite the cerebral cortex. But repetition of the stimulus over and over leads to almost complete extinction of the cortical response if the sensory experience does not elicit either a sense of reward or punishment. Thus, the animal becomes

**Figure 56–4.** Technique for localizing reward and punishment centers in the brain of a monkey.

*habituated* to the sensory stimulus, and thereafter ignores this stimulus.

If the stimulus causes either reward or punishment rather than indifference, however, the cortical response becomes progressively more and more intense with repetitive stimulation instead of fading away, and the response is said to be *reinforced*. Thus, an animal builds up strong memory traces for sensations that are either rewarding or punishing but, on the other hand, develops complete habituation to indifferent sensory stimuli. Therefore, it is evident that the reward and punishment centers of the midbrain have much to do with selecting the information that we learn.

**Effect of Tranquilizers on the Reward and Punishment Centers.** Administration of a tranquilizer, such as chlorpromazine, inhibits both the reward and punishment centers, thereby greatly decreasing the affective reactivity of the animal. Therefore, it is presumed that tranquilizers function in psychotic states by suppressing many of the important behavioral areas of the hypothalamus and its associated regions of the brain.

## THE AFFECTIVE-DEFENSIVE PATTERN— RAGE

An emotional pattern that involves the hypothalamus and has been well characterized by Hess is the affective-defensive pattern. This can be described as follows:

Stimulation of the punishment centers of the brain, especially the *perifornical nuclei of the hypothalamus,* which are also the hypothalamic regions that give the most intense sensation of punishment, causes the animal to (1) develop a defense posture, (2) extend its claws, (3) lift its tail, (4) hiss, (5) spit, (6) growl, and (7) develop pilo-erection, wide-open eyes, and dilated pupils. Furthermore, even the slightest provocation causes an immediate savage attack. This is approximately the behavior that one would expect from an animal being severely punished, and it is a pattern of behavior that is called *rage*.

Stimulation of the more rostral areas of the punishment areas—in the midline preoptic areas—causes mainly fear and anxiety, associated with a tendency for the animal to run away.

All aspects of the rage phenomenon can still be elicited after decortication of an animal; most aspects can be evoked even after the thalamus is removed but with the hypothalamus intact, except that *spontaneous* rage is more likely to occur if the thalamus is intact. Also, a still more primitive pattern of rage can occur even in animals whose brain stems are sectioned below the hypothalamus but above the mesencephalon. In these animals, extremely noxious somatic sensory stimuli will cause the animal to grimace, to develop a defense posture, and to become generally excited. Therefore, much if not most of the rage pattern is mediated through the brain stem reticular formation.

**Docility and Tameness.** Exactly the opposite emotional behavioral patterns occur when the reward centers are stimulated: docility and tameness.

## FUNCTION OF THE HYPOTHALAMUS AND ADJACENT AREAS IN SLEEP, WAKEFULNESS, ALERTNESS, AND EXCITEMENT

Stimulation of regions of the hypothalamus dorsal to the mammillary bodies greatly excites the reticular activating system and therefore causes wakefulness, alertness, and excitement. In addition, the sympathetic nervous system becomes excited in general, increasing the arterial pressure, causing pupillary dilatation, and enhancing other activities associated with sympathetic activity.

On the other hand, stimulation in the septum, in a few areas of the lateral or anterior hypothalamus, or in isolated points of the thalamic portions of the reticular activating system often inhibits the mesencephalic portion of the reticular activating system, causing somnolence and sometimes actual sleep. Thus, the hypothalamus and other limbic structures indirectly contribute much to the control of the degree of excitement and alertness.

## PSYCHOSOMATIC EFFECTS OF THE HYPOTHALAMUS AND RETICULAR ACTIVATING SYSTEM

We are all familiar with the fact that abnormal function in the central nervous system can frequently lead to serious dysfunction of the different somatic organs of the body. This is also true in experimental animals, for, as pointed out above, prolonged electrical stimulation in the punishment regions of the brain can actually lead to severe sickness of the animal, culminating in death within 24 to 48 hours. We need, therefore, to understand briefly the mechanisms by which stimulatory effects in the brain can affect the peripheral organs. Ordinarily, this occurs through three routes: (1) through the motor nerves to the skeletal muscles throughout the body, (2) through the autonomic nerves to the different internal organs of the body, and (3) through the hormones secreted by the pituitary gland in response to nervous activity in the hypothalamus, as will be explained in Chapter 75.

**Psychosomatic Disorders Transmitted Through the Skeletal Nervous System.** Abnormal psychic states can greatly alter the degree of nervous stimulation to the skeletal musculature throughout the body and thereby increase or decrease the skeletal muscular tone. During states of attention the general skeletal muscular tone as well as sympathetic tone normally increases, whereas during somnolent states skeletal muscular and sympathetic activity both decrease greatly. In neurotic and psychotic states, such as anxiety, tension, and mania, generalized overactivity of both the muscles and sympathetic system often occur throughout the body. This in turn results in intense feedback from the muscle proprioceptors to the reticular activating system, and the epinephrine circulating in the blood as a result of the sympathetic activity perhaps also excites the reticular activating system, both of which undoubtedly help to maintain an extreme degree of wakefulness and alertness that characterizes these emotional states. Unfortunately, though, the wakefulness prevents adequate sleep and also leads to progressive bodily fatigue but still inability to go to sleep.

**Transmission of Psychosomatic Effects Through the Autonomic Nervous System.** Many psychosomatic abnormalities result from hyperactivity of either the sympathetic or parasympathetic system. In general, hyperactivity of the sympathetic system occurs in widespread areas of the body at the same time rather than in focal areas, and the usual effects are (1) increased heart rate—sometimes with palpitation of the heart, (2) increased arterial pressure, (3) constipation, and (4) increased metabolic rate. On the other hand, parasympathetic signals are likely to be much more focal. For instance, stimulation of specific areas in the dorsal motor nuclei of the vagus nerves can cause more or less specifically (1) increased or decreased heart rate and palpitation of the heart, (2) esophageal spasm, (3) increased peristalsis in the upper gastrointestinal tract, or (4) increased hyperacidity of the stomach with resultant development of peptic ulcer. Stimulation of the sacral region of the parasympathetic system, on the other hand, is likely to cause extreme colonic glandular secretion and peristalsis with resulting diarrhea. One can readily see, then, that emotional patterns controlling the sympathetic and parasympathetic centers of the hypothalamus can cause wide varieties of peripheral psychosomatic effects.

**Psychosomatic Effects Transmitted Through the Anterior Pituitary Gland.** Electrical stimulation of the posterior hypothalamus increases the secretion of corticotropin by the anterior pituitary gland (as explained earlier in the chapter) and therefore indirectly increases the output of adrenocortical hormones. One of the effects of this is a gradual increase in stomach hyperacidity because of the effect of glucocorticoids on stomach secretion. Over a prolonged period of time this obviously could lead to peptic ulcer, which is a well-known effect of hypersecretion by the adrenal cortex. Likewise, activity in the anterior hypothalamus increases the pituitary secretion of thyrotropin, which in turn increases the output of thyroxine and leads to an elevated basal metabolic rate. It is well known that different types of emotional disturbances can lead to thyrotoxicosis (as will be explained in Chapter 76), this presumably resulting from overactivity in the anterior hypothalamus.

From these examples, therefore, it is evident that many types of psychosomatic diseases of the body can be caused by abnormal control of anterior pituitary secretion.

# FUNCTIONS OF OTHER PARTS OF THE LIMBIC SYSTEM

A large portion of the signals originating in other parts of the limbic system besides the hypothalamus are funneled through the hypothalamus, giving rise to various vegetative or behavioral effects that we know to result when the hypothalamus is stimulated. Other signals are funneled into the reticular activating system to control the degree of wakefulness, alertness, and attention. And still other behavioral signals are transmitted directly from the limbic structures into nonlimbic portions of the cerebral cortex to affect such diverse activities of the cortex as its analytical functions, its sensory functions, and even at times its motor functions.

Yet, unfortunately, the effects caused by many of the different limbic structures are mainly subjective, which prevents adequate experimental study in animals. Therefore, our knowledge of the functions of many of the limbic structures is very poor indeed. About the best that we can do at present is to catalogue some of the overt effects known to occur when these parts of the limbic system are stimulated.

## FUNCTIONS OF THE AMYGDALA

The amygdala is a complex of nuclei located immediately beneath the medial surface of the cerebral cortex in the pole of each temporal lobe. In lower animals, this complex is concerned primarily with association of olfactory stimuli with stimuli from other parts of the brain. Indeed, it is pointed out in Chapter 62 that one of the major divisions of the olfactory tract leads directly to a portion of the amygdala called the *corticomedial nuclei* that lie immediately beneath the cortex in the pyriform area of the temporal lobe. However, in the human being, another portion of the amygdala, the *basolateral nuclei,* has become much more highly developed than the olfactory portion and plays very important roles in many behavioral activities not generally associated with olfactory stimuli.

The amygdala receives impulses from all portions of the limbic cortex, from the orbital surfaces of the frontal lobes, from the cingulate gyrus, and from the hippocampal gyrus. In turn, it transmits signals (a) back into these same cortical areas, (b) into the hippocampus, (c) into the septum, (d) the thalamus, and (e) especially into the hypothalamus.

**Effects of Stimulating the Amygdala.** In general, stimulation in the amygdala can cause almost all the same effects as those elicited by stimulation of the hypothalamus, plus still other effects. The effects that are mediated through the hypothalamus include (1) increases or decreases in arterial pressure, (2) increases or decreases in heart rate, (3) increases or decreases in gastrointestinal motility and secretion, (4) defecation and micturition, (5) pupillary dilatation or, rarely, constriction, (6) pilo-erection, (7) secretion of the various anterior pituitary hormones, including especially the gonadotropins and corticotropin.

Aside from these effects mediated through the hypothalamus, amygdala stimulation can also cause different types of involuntary movement. These include (1) tonic movements, such as raising the head or bending the body, (2) circling movements, (3) occasionally clonic, rhythmic movements, and (4) different types of movements associated with olfaction and eating, such as licking, chewing, and swallowing. And stimulation can alter respiration or at other times stop all movements of an animal, freezing the animal in its present postural state, a phenomenon called the *arrest reaction.*

In addition, stimulation of certain amygdaloid nuclei can rarely cause a pattern of rage, escape, punishment, and pain similar to the affective-defense pattern elicited from the hypothalamus as described above. And stimulation of other nuclei can give reactions of reward and pleasure.

Finally, excitation of still other portions of the amygdala can cause sexual activities that include erection, copulatory movements, ejaculation, ovulation, uterine activity, and premature labor.

In short, stimulation of appropriate portions of the amygdaloid nuclei can give almost any pattern of behavior. It is believed that the normal function of the amygdaloid nuclei is to help control the overall pattern of behavior demanded for each occasion.

## FUNCTIONS OF THE HIPPOCAMPUS

The hippocampus is an elongated structure composed of a modified type of cerebral cortex. It folds inward to form the ventral surface of the inferior horn of the lateral ventricle. One end of the hippocampus abuts against the amygdaloid nuclei, and it also fuses along one of its borders with the hippocampal gyrus, which is the cortex of the ventromedial surface of the temporal lobe. The hippocampus has numerous connections with almost all parts of the limbic system, including especially the amygdala, hippocampal gyrus, cingulate gyrus, hypothalamus, and other areas closely related to the hypothalamus.

Unfortunately, as is true of most limbic structures, we know very little about the overall function of the hippocampus and, therefore, for the present can only describe some overt effects that occur when this area is stimulated or destroyed.

Stimulation of different regions of the hippocampus can cause *involuntary tonic or clonic movements* in different parts of the body. At times it causes *rage or other emotional reactions.* And stimulation sometimes causes different types of *sexual phenomena.* One of the most remarkable effects of stimulation in conscious man is immediate *loss of contact* with any person with whom he might be talking; indicating that the hippocampus can play a role in determining a person's attention.

A special feature of the hippocampus is that weak electrical stimuli can cause local epileptic seizures in this region. These seizures cause various psychomotor effects including olfactory, visual, auditory, tactile, and other types of hallucinations that are uncontrollable even though the person has not lost consciousness and even though he knows the hallucinations to be unreal. One of the reasons for this hyperexcitability of the hippocampus is probably that it is composed of a different type of cortex from that elsewhere in the cerebrum, having only three layers instead of the six layers found elsewhere.

**Theoretical Function of the Hippocampus.** Almost any type of sensory experience causes instantaneous activation of different parts of the hippocampus, and the hippocampus in turn distributes many outgoing signals to the hypothalamus and other parts of the limbic system through the *fornix.* Thus, one of the major sources of input signals to the hypothalamus and other limbic structures is the hippocampus. On the basis of this, it is often stated that the primary role of the hippocampus is to provide a channel through which different incoming sensory signals can excite appropriate limbic reactions.

*Function of the Hippocampus in Learning.* Earlier in the chapter (and also in the previous chapter) it was pointed out that reward and punishment centers play a major role in determining which information will be stored in memory. A person becomes rapidly habituated to indifferent stimuli but learns assiduously any sensory experience that causes either pain or pleasure. Yet what is the route by which reward and punishment centers receive their original information? It has been suggested that the hippocampus plays the role of associating the affective characteristics of different sensory signals and then in turn transmitting the correlated information into the reward or punishment areas of the hypothalamus and other limbic centers to help control the information that a person will learn. In fact, when the hippocampus has been removed bilaterally, a person develops anterograde amnesia, which means that it is difficult or almost impossible for him to store new memories. This is especially true for verbal information, perhaps because the temporal lobes are particularly concerned with this type of information.

## FUNCTION OF THE HIPPOCAMPAL, CINGULATE, AND ORBITOFRONTAL CORTEX

Probably the most poorly understood portion of the entire limbic system is the ring of cerebral cortex, called the *limbic cortex,* that surrounds the subcortical limbic structures. The limbic cortex is among the oldest of all parts of the cerebral cortex. In lower animals it plays a major role in various olfactory, gustatory, and feeding phenomena. However, in the human being, these functions of the limbic cortex are of minor importance. Instead, the limbic cortex of the human being is believed to be the cerebral association cortex for control of the lower centers that have to do primarily with behavior. For instance, the hippocampal gyrus has close interconnections with the hippocampus and thence through the fornix with almost all areas of the limbic system, especially with the hypothalamus. The cingulate gyrus likewise connects directly with the hippocampus and also with all other limbic areas. Finally, the orbitofrontal cortex interconnects profusely with the septal area, hypothalamus, preoptic area, and other associated regions.

**Effect of Stimulation.**   Stimulation of the different regions of the limbic cortex has failed to give any real idea of their functions, because the effects are just as diverse as those already recounted for stimulation of the hypothalamus, amygdala, or hippocampus, including such effects as (a) inhibition or acceleration of respiration, (b) facilitation or inhibition of spontaneous or cortical induced movements, (c) facilitation of clonic or tonic movements of the body, (d) vocalization, (e) chewing, (f) licking, (g) swallowing, (h) cardiovascular effects, such as increases or decreases in arterial pressure and increases or decreases in heart rate, (i) gastric effects, such as increases or decreases in gastrointestinal motility, (j) increases or decreases in gastrointestinal secretions, and (k) affective reactions including rage, docility, excitement, alertness, etc.

**Effect of Ablation of Different Portions of the Limbic Cortex.**   Ablation experiments also have failed to give much idea of limbic cortical function, because the effects observed by different experimenters have often been quite different from each other. Also, the effects of ablation have sometimes been so inconsequential that no changes have been observed in the animal's behavior. However, the following general effects have been observed following ablation of certain specific areas.

*Ablation of the Anterior Portion of the Cingulate Gyri.*   Bilateral removal of the anterior portion of cingulate gyri causes increased tameness of an animal to the extent that he no longer has fear of a man even though this fear might have existed preoperatively. Furthermore, previous rage reactions are suppressed, and the animal seems to lack "social consciousness."

*Ablation of the Posterior Orbital Cortex.*   Bilateral removal of the posterior portion of the fronto-orbital cortex often causes an animal to develop insomnia and an intense degree of motor restlessness, becoming unable to sit still, but instead moving continually.

*Ablation of the Anterior Temporal Region–The Klüver-Bucy Syndrome.*   When the anterior temporal cortex is removed, many important subcortical structures are also frequently damaged or removed, including especially the amygdala. Therefore, it is doubtful that what has been learned from anterior temporal ablation is of particular value in determining the function of the temporal limbic *cortex.* However, removal of the entire anterior tip of the temporal lobe in the dominant hemisphere of a monkey, and sometimes in man, causes intense changes in behavior, including (1) excessive tendency to examine objects, (2) loss of fear, (3) decreased aggressiveness, (4) tameness, (5) changes in dietary habits such that a herbivorous animal sometimes even becomes carnivorous, and (6) sometimes psychic blindness. This combination of effects is called the Klüver-Bucy syndrome.

In cats, opposite effects are sometimes observed, such as extreme savagery and rage. Yet, in other cats, removal of the anterior temporal poles causes extreme docility. Since different experimenters remove or damage different amounts of tissue in their experiments, these results show primarily that the anterior temporal poles, including the sublying amygdala, can have diverse effects on behavior, depending on the precise areas destroyed. This is to be expected because of the great diversity of effects that can be caused by stimulation of different areas of the amygdala and other closely associated regions.

**Overall Function of the Hippocampal Gyrus, Cingulate Gyrus, and Orbitofrontal Cortex.**   After the above discussion, we find ourselves immediately perplexed regarding the function of the cortical regions of the limbic system. The reason for this probably is that these regions are actually association areas that correlate information from many sources but cause no direct, overt effects that can be observed objectively. Thus, in the insular and anterior temporal cortex we find gustatory and olfactory associations in particular. In the hippocampal gyrus, there is a tendency for complex auditory associations. In the cingulate cortex, there is some reason to believe that sensorimotor associations occur. And, finally, the orbitofrontal cortex presumably aids in the analytical functions of the prefrontal lobes.

Therefore, until further information is available, it is perhaps best to state that the cortical regions of the limbic system occupy intermediate associative positions between the functions of the remainder of the cerebral cortex and the functions of the lower centers for control of behavioral patterns.

# REFERENCES

Clemente, C. D., and Chase, M. H.: Neurological substrates of aggressive behavior. *Ann. Rev. Physiol., 35*:329, 1973.

Cooper, J. B., Bloom, F. E., and Roth, R. H.: The Biochemical Basis of Neuropharmacology, 2nd Ed. New York, Oxford University Press, 1974.

Davidson, J. M., and Levine, S.: Endocrine regulation of behavior. *Ann. Rev. Physiol., 34*:375, 1972.

Defeudis, F. V.: Central Cholinergic Systems and Behavior. New York, Academic Press, Inc., 1974.

DiCara, L. (ed.): Limbic and Autonomic Nervous Systems Research. New York, Plenum Publishing Corporation, 1975.

Estes, W. K.: Reinforcement in human behavior. *Amer. Sci., 60*:723, 1972.

Frohman, L. A.: The hypothalamus and metabolic control. *Pathobiol. Ann., 1*:353, 1971.

Gazzaniga, M. S., and Blakemore, C.: Handbook of Psychobiology. New York, Academic Press, Inc, 1975.

Grenell, R. G., and Gabay, S. (eds.): Biological Foundations of Psychiatry. New York, Raven Press, 1975.

Gunderson, E. K. E., and Rahe, R. H. (eds.): Life Stress and Illness. Springfield, Ill., Charles C Thomas, Publisher, 1974.

Herd, J. A.: Physiology of strong emotions. *Physiologist, 15*:5, 1972.

Hollister, L. E.: Clinical Use of Psychotherapeutic Drugs. Springfield, Ill., Charles C Thomas, Publisher, 1974.

Hyvarinen, J.: CNS: afferent mechanism with emphasis on physiological and behavioral correlations. *Ann. Rev. Physiol., 35*:243, 1973.

Isaacson, R. L.: The Limbic System. New York, Plenum Publishing Corporation, 1974.

Kiev, A. (ed.): Somatic Manifestations of Depressive Disorders. New York, American Elsevier Publishing Co., Inc., 1975.

Lester, D.: A Physiological Basis for Personality Traits. Springfield, Ill., Charles C Thomas, Publisher, 1974.

Levi, L. (ed.): Emotions: Their Parameters and Measurement. New York, Raven Press, 1975.

Levine, S.: Stress and behavior. *Sci. Amer., 224*:26, 1971.

Luborsky, L.: New directions in research on neurotic and psychosomatic symptoms. *Amer. Sci., 58*:661, 1970.

Mandell, A. J. (ed.): Neurobiological Mechanisms of Adaptation and Behavior. (Advances in Biochemical Psychopharmacology. Vol. 13.) New York, Raven Press, 1975.

Millon, T.: Medical Behavioral Science. Philadelphia, W. B. Saunders Company, 1975.

Olds, J.: Hypothalamic substrates of reward. *Physiol. Rev., 42*:554, 1962.

Rolls, E. T.: The Brain and Reward. Oxford, Pergamon Press, 1975.

Sabelli, H. C. (ed.): Chemical Modulation of Brain Function. New York, Raven Press, 1973.

Schally, A. V., Kastin, A. J., and Arimura, A.: The hypothalamus and reproduction. *Amer. J. Obstet. Gynecol., 114*:423, 1972.

Smythies, J. R.: Brain Mechanisms and Behavior. New York, Academic Press, Inc., 1970.

Stearns, F. R.: Anger: Psychology, Physiology, Pathology. Springfield, Ill., Charles C Thomas, Publisher, 1972.

Strumwasser, F.: Neural and humoral factors in the temporal organization of behavior. *Physiologist, 16*:9, 1973.

Thomas, A., Chess, S., and Birch, H. G.: The origin of personality. *Sci. Amer., 223*:102, 1970.

Usdin, E. (ed.): Neuropsychopharmacology of Monoamines and Their Regulatory Enzymes. New York, Raven Press, 1975.

Wayner, M. J.: Specificity of behavioral regulation. *Physiol. Behav., 12*:851, 1974.

Weil, J. L.: A Neurophysiological Model of Emotional and Intentional Behavior. Springfield, Ill., Charles C Thomas, Publisher, 1974.

Widroe, H. J. (ed.): Human Behavior and Brain Function. Springfield, Ill., Charles C Thomas, Publisher, 1975.

Winter, A. (ed.): The Surgical Control of Behavior. Springfield, Ill., Charles C Thomas, Publisher, 1971.

Zimmermann, E., and George, R. (eds.): Narcotics and the Hypothalamus. (Kroc Foundation Series. Vol. 2.) New York, Raven Press, 1974.

# 57

# The Autonomic Nervous System; The Adrenal Medulla

The portion of the nervous system that controls the visceral functions of the body is called the *autonomic nervous system*. This system helps to control arterial pressure, gastrointestinal motility and secretion, urinary output, sweating, body temperature, and many other activities, some of which are controlled almost entirely by the autonomic nervous system and some only partially.

## GENERAL ORGANIZATION OF THE AUTONOMIC NERVOUS SYSTEM

The autonomic nervous system is activated mainly by centers located in the *spinal cord, brain stem,* and *hypothalamus.* Also, portions of the cerebral cortex can transmit impulses to the lower centers and in this way influence autonomic control. Often the autonomic nervous system operates by means of *visceral reflexes.* That is, sensory signals from parts of the body send impulses into the centers of the cord, brain stem, or hypothalamus, and these in turn transmit appropriate reflex responses back to the visceral organs to control their activities.

The autonomic impulses are transmitted to the body through two major subdivisions called the *sympathetic* and *parasympathetic systems,* the characteristics and functions of which follow.

### PHYSIOLOGIC ANATOMY OF THE SYMPATHETIC NERVOUS SYSTEM

Figure 57–1 illustrates the general organization of the sympathetic nervous system, showing one of the two *sympathetic chains* to the side of the spinal column and nerves extending to the different internal organs. The sympathetic nerves originate in the spinal cord between the segments T-1 and L-2. They begin in the *sympathetic motoneurons* of the *intermediolateral horns* of the spinal gray matter.

**Preganglionic and Postganglionic Sympathetic Neurons.** The sympathetic nerves are different from skeletal motor nerves in the following way: Each motor pathway to a skeletal muscle is comprised of a single fiber originating in the cord. Each sympathetic pathway is comprised of a *preganglionic neuron* and a *postganglionic neuron.* The cell body of the preganglionic neuron lies in the intermediolateral horn of the spinal cord, and its fiber passes, as illustrated in Figure 57–2, through an *anterior root* of the cord into a *spinal nerve.* From here, the preganglionic fiber leaves the spinal nerve immediately and passes through the *white ramus* to a *ganglion* of the *sympathetic chain.* Here the fiber either synapses immediately with postganglionic neurons or often passes on through the chain into one of its radiating nerves to synapse with postganglionic neurons in an outlying sympathetic ganglion. The fiber of each postganglionic neuron than travels to its destination in one of the organs.

**Sympathetic Nerve Fibers in the Skeletal Nerves.** Many of the fibers from the postganglionic neurons in the sympathetic chain pass back into the spinal nerves through *gray rami* (see Fig. 57–2) at all levels of the cord. These pathways are made up of type C fibers that extend to all parts of the body in the skeletal nerves. They control the blood vessels, sweat glands, and pilo-erector muscles of the hairs. Approximately 8 per cent of the fibers in the average skeletal nerve are sympathetic fibers, a fact that indicates their importance.

**Segmental Distribution of Sympathetic Nerves.** The sympathetic pathways originating in the different segments of the spinal cord are not necessarily distributed to the same part of the body as the spinal nerve fibers from the same segments. Instead, the

768

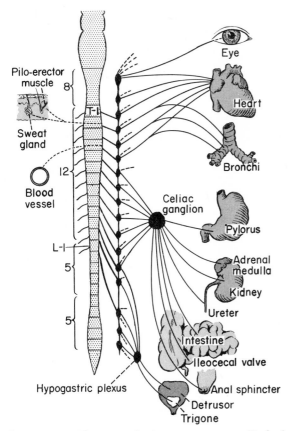

**Figure 57–1.** The sympathetic nervous system. Dashed lines represent postganglionic fibers in the gray rami leading into the spinal nerves for distribution to blood vessels, sweat glands, and pilo-erector muscles.

that secrete epinephrine and norepinephrine. These secretory cells are embryologically derived from nervous tissue and are analogous to postganglionic neurons; indeed, they even have rudimentary nerve fibers.

## PHYSIOLOGIC ANATOMY OF THE PARASYMPATHETIC NERVOUS SYSTEM

The parasympathetic nervous system is illustrated in Figure 57–3, showing that parasympathetic fibers leave the central nervous system through several of the cranial nerves, the second and third sacral spinal nerves, and occasionally the first and fourth sacral nerves. About 75 per cent of all parasympathetic nerve fibers are in the vagus nerves, passing to the entire thoracic and abdominal regions of the body. However, the vagus nerve is not entirely parasympathetic in function because it also carries some voluntary skeletal nerve fibers from the nucleus ambiguous to the laryngeal and pharyngeal muscles and some afferent nerve fibers from the baroreceptors of the arteries and from stretch receptors of the lungs to the medulla. Nevertheless, when a physiologist speaks of the parasympathetic nervous system, in the back of his mind he thinks mainly of the two vagus nerves. The vagus nerves supply parasympathetic nerves to the heart, the lungs, the esophagus, the stomach, the small intestine, the proximal half of the colon, the liver, the gallbladder, the pancreas, and the upper portions of the ureters.

Parasympathetic fibers in the *third nerve* flow to the pupillary sphincters and ciliary muscles of the eye. Fibers from the *seventh nerve* pass to the lacrimal, nasal, and submaxillary glands, and fibers from the *ninth nerve* pass to the parotid gland.

*sympathetic fibers from T-1 generally pass up the sympathetic chain into the head; from T-2 into the neck; T-3, T-4, T-5, and T-6 into the thorax; T-7, T-8, T-9, T-10, and T-11 into the abdomen; T-12, L-1, and L-2 into the legs.* This distribution is only approximate and overlaps greatly.

The distribution of sympathetic nerves to each organ is determined partly by the position in the embryo at which the organ originates. For instance, the heart receives many sympathetic nerves from the neck portion of the sympathetic chain because the heart originates in the neck of the embryo. Likewise, the abdominal organs receive their sympathetic innervation from the lower thoracic segments because the primitive gut originates in the lower thoracic area.

**Special Nature of the Sympathetic Nerve Endings in the Adrenal Medullae.** Preganglionic sympathetic nerve fibers pass, without synapsing, all the way from the intermediolateral horn cells of the spinal cord, through the sympathetic chains, through the splanchnic nerves, and finally into the adrenal medullae. There they end directly on special cells

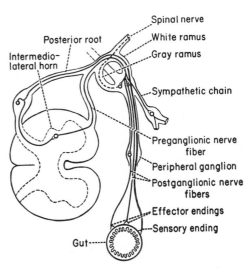

**Figure 57–2.** Nerve connections between the spinal cord, sympathetic chain, spinal nerves, and peripheral sympathetic nerves.

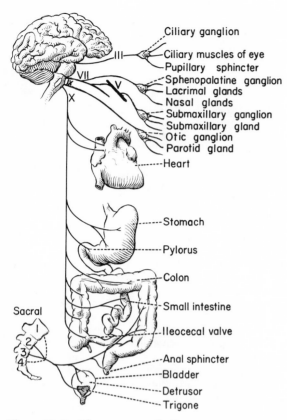

**Figure 57–3.** The parasympathetic nervous system.

The sacral parasympathetic fibers congregate in the form of the *nervi erigentes,* which leave the sacral plexus on each side of the cord and distribute their peripheral fibers to the descending colon, rectum, bladder and lower portions of the ureters. Also, this sacral group of parasympathetics supplies fibers to the external genitalia to cause various sexual reactions.

**Preganglionic and Postganglionic Parasympathetic Neurons.** The parasympathetic system, like the sympathetic, has both preganglionic and postganglionic neurons, but, except in the case of a few cranial parasympathetic nerves, the preganglionic fibers pass uninterrupted to the organ that is to be excited by parasympathetic impulses. In the wall of the organ are located the *postganglionic neurons* of the parasympathetic system. The preganglionic fibers synapse with these; and then short postganglionic fibers, 1 millimeter to several centimeters in length, leave the neurons to spread in the substance of the organ. This location of the parasympathetic postganglionic neurons in the visceral organ itself is quite different from the arrangement of the sympathetic ganglia, for the cell bodies of the sympathetic postganglionic neurons are always located in ganglia of the sympathetic chain or in various other discrete ganglia in the abdomen or thorax rather than in the excited organ itself.

# BASIC CHARACTERISTICS OF SYMPATHETIC AND PARASYMPATHETIC FUNCTION

## CHOLINERGIC AND ADRENERGIC FIBERS—SECRETION OF ACETYLCHOLINE OR NOREPINEPHRINE BY THE POSTGANGLIONIC NEURONS

It will be recalled from Chapter 12 that skeletal nerve endings secrete acetylcholine. This is also true of the *preganglionic neurons* of both the sympathetic and parasympathetic system, and it is true, too, of the *parasympathetic postganglionic neurons.* Therefore, all these fibers are said to be *cholinergic because they secrete acetylcholine at their nerve endings.*

A few of the postganglionic endings of the sympathetic nervous system also secrete acetylcholine, and these fibers, too, are cholinergic; but by far the majority of the sympathetic postganglionic endings secrete *norepinephrine.* These fibers are said to be *adrenergic,* a term derived from *noradrenalin,* which is the English name for norepinephrine. Thus, there is a basic functional difference between the respective postganglionic neurons of the parasympathetic and sympathetic systems, one secreting acetylcholine and the other principally norepinephrine. The chemical structures of these substances are the following:

$$CH_3-\underset{\underset{O}{\|}}{C}-O-CH_2-CH_2-\overset{+}{\underset{\underset{CH_3}{|}}{N}}\overset{CH_3}{\underset{CH_3}{\diagdown}}$$

Acetylcholine

$$HO-\hspace{-4pt}\bigcirc\hspace{-4pt}\overset{HO}{\phantom{|}}\hspace{-4pt}-\underset{\underset{OH}{|}}{CH}-CH_2-NH_2$$

Norepinephrine

The acetylcholine and norepinephrine secreted by the postganglionic neurons act on the different organs to cause the respective parasympathetic or sympathetic effects. Therefore, these substances are called *parasympathetic* and *sympathetic mediators,* respectively, or sometimes *cholinergic* and *adrenergic* mediators.

**Mechanism of Secretion of Acetylcholine and Norepinephrine by Autonomic Nerve Endings.** Secretion of the transmitter substances acetylcholine and norepinephrine by autonomic nerve endings occurs in the same way as secretion at other synapses, as described previously for the central nervous system in Chapter 46 and for the neuromuscular junction in Chapter 12. Briefly, the nerve endings of both the sympathetic and parasympathetic systems contain large numbers of small transmitter vesicles about 300 to 600 Ångstroms in diameter. The vesicles in the cholinergic nerve endings contain acetylcholine and in electronmicrographs have a clear appearance. The vesicles in the adrenergic nerve endings contain norepinephrine and are granular in appearance.

Some of the autonomic nerve endings are similar to but much smaller in size than those of the skeletal muscle neuromuscular junction; this is especially true of the endings on some types of smooth muscle cells, as was explained in Chapter 12. However, most of the terminal filaments merely touch the effector cells of the organs that they innervate as they pass by; in some instances they terminate in connective tissue located adjacent to the cells that are to be stimulated. Where these filaments pass over the effector cells, they usually have bulbous enlargements called *varicosities,* and it is in these varicosities that the transmitter vesicles of acetylcholine and norepinephrine are found. Also in the varicosities are large numbers of mitochondria to supply the ATP required to energize acetylcholine and norepinephrine synthesis.

When an action potential spreads over the terminal fibers, the depolarization process increases the permeability of the fiber membrane to calcium ions, thus allowing these to diffuse in moderate numbers into the nerve terminals. There they interact with the vesicles adjacent to the membrane, causing them in some way not yet understood to fuse with the nerve membrane and to empty their contents to the exterior. Thus, the transmitter substance is secreted.

**Synthesis of Acetylcholine, its Destruction After Secretion, and Duration of Action.** Acetylcholine is continually synthesized in the terminal endings of cholinergic nerve fibers. Most of this synthesis probably occurs in the axoplasm and the acetylcholine is then transported to the interior of the vesicles, though it is possible that some of it is also synthesized in the vesicles themselves. The basic chemical reaction of this synthesis is the following:

$$\text{Acetyl-CoA} + \text{Choline} \xrightarrow{\substack{choline \\ acetylase}} \text{Acetylcholine}$$

Once the acetylcholine has been secreted by the cholinergic nerve ending, most of it is split into acetate ion and choline by the enzyme *cholinesterase* that is present both in the terminal nerve ending itself and on the surface of the receptor organ. Thus, this is the same mechanism of acetylcholine destruction as occurs at the neuromuscular junctions of skeletal nerve fibers. The choline that is formed is in turn transported back into the terminal nerve ending where it is used again for synthesis of new acetylcholine. Though most of the acetylcholine is usually destroyed within a fraction of a second after its secretion, it sometimes persists for as long as several seconds, and a small amount also diffuses into the surrounding fluids. These fluids contain a different type of cholinesterase called *serum cholinesterase* that destroys the remaining acetylcholine within another few seconds. Therefore, the action of acetylcholine released by cholinergic nerve fibers usually lasts for a few seconds at most and more often for only a fraction of a second.

**Synthesis of Norepinephrine, Its Removal, and Duration of Action.** Synthesis of norepinephrine begins in the axoplasm of the terminal nerve endings of adrenergic nerve fibers but is completed inside the vesicles. The basic steps are the following:

1. Tyrosine $\xrightarrow{hydroxylation}$ DOPA

2. DOPA $\xrightarrow{decarboxylation}$ Dopamine

3. Transport of dopamine into the vesicles

4. Dopamine $\xrightarrow{hydroxylation}$ Norepinephrine

In the adrenal medulla this reaction goes still one step further to form epinephrine as follows:

5. Norepinephrine $\xrightarrow{methylation}$ Epinephrine

Following secretion of norepinephrine by the terminal nerve endings, it is removed from the secretory site in three different ways: (1) re-uptake into the adrenergic nerve endings themselves by an active transport process—accounting for removal of 50 to 80 per cent of the secreted norepinephrine; (2) diffusion away from the nerve endings into the surrounding body fluids and thence into the blood—accounting for removal of most of the remainder of the norepinephrine; and (3) destruction by enzymes to a slight extent (one of these enzymes is *monamine oxidase,* which is found in the nerve endings themselves, and another is *catechol-O-methyl transferase,* which is present diffusely in all tissues).

Ordinarily, the norepinephrine secreted directly in a tissue by adrenergic nerve endings remains active for only a few seconds, illustrating that its re-uptake and diffusion away from the tissue is rapid. However, the norepinephrine and epinephrine secreted into the blood by the adrenal medullae remain active until they diffuse into some tissue where they are destroyed by catechol-O-methyl transferase; this occurs mainly in the liver. Therefore, when secreted into the blood, both norepinephrine and epinephrine remain very active for 10 to 30 seconds, followed by decreasing activity thereafter for one to several minutes.

## RECEPTOR SUBSTANCES OF THE EFFECTOR ORGANS

The acetylcholine, norepinephrine, and epinephrine secreted by the autonomic nervous system all stimulate the effector organs by first reacting with receptor substances in the effector cells. The receptor is believed in most instances to be in the cell membrane and is probably a protein or a lipoprotein. The most likely mechanism for function of the receptor is that the transmitter substance first binds with the receptor and this causes a basic change in the molecular structure of the receptor compound. Because the receptor is an integral part of the cell membrane, this structural change often alters the permeability of the cell membrane to various ions—for instance, to allow rapid influx of sodium, chloride, or calcium ions into the cell or to allow rapid efflux of potassium ions out of the cell. These ionic changes then usually alter the membrane potential, sometimes eliciting action potentials (as occurs in smooth muscle cells) and at other times causing electrotonic effects on the cells (as occurs on glandular cells) to produce the responses. The ions themselves have direct effects on the receptor cells, such as the effect of calcium ions to promote smooth muscle contraction.

Another way that the receptor can function, besides changing the membrane permeability, is to activate an enzyme in the cell membrane—this enzyme in turn promoting chemical reactions within the cell. For instance, epinephrine increases the activity of *adenyl cyclase* in some cell membranes, and this effect causes the formation of cyclic AMP inside the cell; the cyclic AMP then initiates many intracellular activities.

**The Acetylcholine Receptors—"Muscarinic" and "Nicotinic" Receptors.** Acetylcholine activates at least two different types of receptors. These are called *muscarinic* and *nicotinic* receptors. The reason for these names is that muscarine, a poison from toadstools, also activates the muscarinic receptors but will not activate the nicotinic receptors, whereas nicotine will activate the other receptors; finally, acetylcholine activates both of them.

The muscarinic receptors are found in all the effector cells stimulated by the postganglionic neurons of the parasympathetic nervous system as well as those stimulated by the cholinergic nerve endings of the sympathetic system.

The nicotinic receptors are found in the membranes of sympathetic and parasympathetic postganglionic neurons and in the membranes of skeletal muscle fibers at the neuromuscular junction (discussed in Chapter 12).

An understanding of the two different types of receptors is especially important because specific drugs are frequently used in the practice of medicine to stimulate or to block one or the other of the two types of receptors.

**The Adrenergic Receptors—"Alpha" and "Beta" Receptors.** Research experiments using different drugs (called *sympathomimetic drugs*) to mimic the action of norepinephrine on sympathetic effector organs have shown that there are at least two—and perhaps more—different types of adrenergic receptors. At present these are generally classified as *alpha receptors* and *beta receptors*. Norepinephrine and epinephrine, both of which are secreted by the adrenal medulla, have somewhat different effects in exciting these two receptors. Norepinephrine excites mainly alpha receptors but excites the beta receptors to a slight extent as well. On the other hand, epinephrine excites both types of receptors approximately equally. Therefore, the relative effects of norepinephrine and epinephrine on different effector organs is determined by the types of receptors in the organs. Obviously, if they are all beta receptors, epinephrine will be the more effective excitant.

Table 57–1 gives the distribution of alpha and beta receptors in some of the organs and systems controlled by the sympathetics. Note that certain alpha functions are excitatory while others are inhibitory. Likewise, certain beta functions are excitatory and others are inhibitory. Therefore, alpha and beta receptors are not necessarily associated with excitation or inhibition but simply with the affinity of the hormone for the receptors in a given effector organ.

A synthetic hormone closely similar to epi-

**Table 57–1.   Relationships of Adrenergic Receptors to Function**

| Alpha Receptor | Beta Receptor |
|---|---|
| Vasoconstriction | Vasodilatation |
| Iris dilatation | Cardioacceleration |
| Intestinal relaxation | Increased myocardial strength |
| Pilomotor contraction | Myometrial relaxation |
| | Intestinal relaxation |
| | Uterus relaxation |
| | Bronchodilatation |
| | Calorigenesis |
| | Muscle glycogenolysis |
| | Lipolysis |

nephrine and norepinephrine, *isopropyl norepinephrine,* has an extremely strong action on beta receptors but almost no action on alpha receptors. Later in the chapter we will discuss various drugs that can mimic the adrenergic actions of epinephrine and norepinephrine or that will block specifically the alpha or beta receptors.

## EXCITATORY AND INHIBITORY ACTIONS OF SYMPATHETIC AND PARASYMPATHETIC STIMULATION

Table 57–2 gives the effects on different visceral functions of the body caused by stimulating the parasympathetic and sympathetic nerves. From this table it can be seen that *sympathetic stimulation causes excitatory effects in some organs but inhibitory effects in others. Likewise, parasympathetic stimulation causes excitation in some organs but inhibition in others.* Also, when sympathetic stimulation excites a particular organ, parasympathetic stimulation often inhibits it, illustrating that the two systems occasionally act reciprocally to each other. However, most organs are dominantly controlled by one or the other of the two systems, so that, except in a few instances, the two systems do not actively oppose each other.

There is no generalization one can use to explain whether sympathetic or parasympathetic stimulation will cause excitation or inhibition of a particular organ. Therefore, to understand sympathetic and parasympathetic function, one must learn the functions of these two nervous systems as listed in Table 57–2. Some of these

TABLE 57–2.  **Autonomic Effects on Various Organs of the Body**

| Organ | Effect of Sympathetic Stimulation | Effect of Parasympathetic Stimulation |
|---|---|---|
| Eye: Pupil | Dilated | Contracted |
|     Ciliary muscle | None | Excited |
| Glands: Nasal | Vasoconstriction | Stimulation of thin, copious |
|     Lacrimal | |     secretion containing many enzymes |
|     Parotid | | |
|     Submaxillary | | |
|     Gastric | | |
|     Pancreatic | | |
| Sweat glands | Copious sweating (cholinergic) | None |
| Apocrine glands | Thick, odoriferous secretion | None |
| Heart: Muscle | Increased rate | Slowed rate |
| | Increased force of contraction | Decreased force of atrial contraction |
|     Coronaries | Vasodilated | Constricted |
| Lungs: Bronchi | Dilated | Constricted |
|     Blood vessels | Mildly constricted | None |
| Gut: Lumen | Decreased peristalsis and tone | Increased peristalsis and tone |
|     Sphincter | Increased tone | Decreased tone |
| Liver | Glucose released | None |
| Gallbladder and bile ducts | Inhibited | Excited |
| Kidney | Decreased output | None |
| Ureter | Inhibited | Excited |
| Bladder: Detrusor | Inhibited | Excited |
|     Trigone | Excited | Inhibited |
| Penis | Ejaculation | Erection |
| Systemic blood vessels: | | |
|     Abdominal | Constricted | None |
|     Muscle | Constricted (adrenergic) | None |
| | Dilated (cholinergic) | |
|     Skin | Constricted (adrenergic) | Dilated |
| | Dilated (cholinergic) | |
| Blood: Coagulation | Increased | None |
|     Glucose | Increased | None |
| Basal metabolism | Increased up to 100% | None |
| Adrenal cortical secretion | Increased | None |
| Mental activity | Increased | None |
| Piloerector muscles | Excited | None |
| Skeletal muscle | Increased glycogenolysis | None |
| | Increased strength | |

functions need to be clarified in still greater detail as follows:

## EFFECTS OF SYMPATHETIC AND PARASYMPATHETIC STIMULATION ON SPECIFIC ORGANS

**The Eye.** Two functions of the eye are controlled by the autonomic nervous system: These are the pupillary opening and the focus of the lens. Sympathetic stimulation contracts the meridional *fibers of the iris* and, therefore, dilates the pupil, whereas parasympathetic stimulation contracts the *circular muscle of the iris* to constrict the pupil. The parasympathetics that control the pupil are reflexly stimulated when excess light enters the eyes; this reflex reduces the pupillary opening and decreases the amount of light that strikes the retina. On the other hand, the sympathetics become stimulated during periods of excitement and, therefore, increase the pupillary opening at these times.

Focusing of the lens is controlled entirely by the parasympathetic nervous system. The lens is normally held in a flattened state by tension of its radial ligaments. Parasympathetic excitation contracts the *ciliary muscle,* which releases this tension and allows the lens to become more convex. This causes the eye to focus on objects near at hand. The focusing mechanism is discussed in Chapters 58 and 60 in relation to function of the eyes.

**The Glands of the Body.** The *nasal, lacrimal, salivary,* and many *gastrointestinal glands* are all strongly stimulated by the parasympathetic nervous system, resulting in copious quantities of secretion. The glands of the alimentary tract most strongly stimulated by the parasympathetics are those of the upper tract, especially those of the mouth and stomach. The glands of the small and large intestines are controlled principally by local factors in the intestinal tract itself and not by the autonomic nerves.

Sympathetic stimulation has a slight direct effect on glandular cells in causing formation of a concentrated secretion. However, it also causes vasoconstriction of the blood vessels supplying the glands and in this way often reduces their rates of secretion.

The *sweat glands* secrete large quantities of sweat when the sympathetic nerves are stimulated, but no effect is caused by stimulating the parasympathetic nerves. However, the sympathetic fibers to most sweat glands are *cholinergic,* in contrast to most other sympathetic fibers, which are adrenergic. Furthermore, the sweat glands are stimulated primarily by centers in the hypothalamus that are usually considered to be parasympathetic centers. Therefore, sweating could be called a parasympathetic function.

The *apocrine glands* secrete a thick, odoriferous secretion as a result of sympathetic stimulation, but they do not react to parasympathetic stimulation. Furthermore, the apocrine glands, despite their close embryological relationship to sweat glands, are controlled by adrenergic fibers rather than by cholinergic fibers and are controlled by the sympathetic centers of the central nervous system rather than by the parasympathetic centers.

**The Gastrointestinal System.** The gastrointestinal system has its own intrinsic set of nerves known as the *intramural plexus.* However, both parasympathetic and sympathetic stimulation can affect gastrointestinal activity—parasympathetic especially. Parasympathetic stimulation, in general, increases the overall degree of activity of the gastrointestinal tract by promoting peristalsis, thus allowing rapid propulsion of contents along the tract. This propulsive effect is associated with simultaneous increases in rates of secretion by many of the gastrointestinal glands, which was described above.

Normal function of the gastrointestinal tract is not very dependent on sympathetic stimulation. However, in some diseases, strong sympathetic stimulation inhibits peristalsis and increases the tone of the sphincters. The net result is greatly slowed propulsion of food through the tract.

**The Heart.** In general, sympathetic stimulation increases the overall activity of the heart. This is accomplished by increasing both the rate and force of the heartbeat. Parasympathetic stimulation causes mainly the opposite effects, decreasing the overall activity of the heart. To express these effects in another way, sympathetic stimulation increases the effectiveness of the heart as a pump, whereas parasympathetic stimulation decreases its effectiveness. However, sympathetic stimulation unfortunately also greatly increases the metabolism of the heart while parasympathetic stimulation decreases its metabolism and allows the heart a certain degree of rest.

**Systemic Blood Vessels.** Most blood vessels, especially those of the abdominal viscera and the skin of the limbs, are constricted by sympathetic stimulation. Parasympathetic stimulation generally has almost no effects on blood vessels but does dilate vessels in certain restricted areas such as in the blush area of the face.

**Effect of Sympathetic and Parasympathetic Stimulation on Arterial Pressure.** The arterial pressure in the circulatory system is caused by propulsion of blood by the heart and by resistance to flow of this blood through the vascular system. In general, sympathetic stimulation increases both propulsion by the heart and resistance to flow, which can cause the pressure to increase greatly.

On the other hand, parasympathetic stimulation decreases the pumping effectiveness of the heart, which lowers the pressure a moderate amount, though not nearly so much as the sympathetics can increase the pressure.

**The Lungs.** In general, the structures in the lungs do not have extensive sympathetic or parasympathetic innervation. Therefore, all effects of stimulation are mild. Sympathetic stimulation can mildly dilate the bronchi and mildly constrict the blood vessels. On the contrary, parasympathetic stimulation can cause mild constriction of the bronchi and can perhaps mildly dilate the vessels.

**Effects of Sympathetic and Parasympathetic Stimulation on Other Functions of the Body.** Because of the great importance of the sympathetic and parasympathetic control systems, these are discussed many times in this text in relation to a myriad of body functions that are not considered in detail here. In general, most of the entodermal structures, such as the ducts of the liver, the gallbladder, the ureter, and the bladder, are inhibited by sympathetic stimulation but excited by parasympathetic stimulation. Sympathetic stimulation also has metabolic effects, causing release of glucose from the liver, increase in blood glucose concentration, increase in glycogenolysis in muscle, increase in muscle strength, increase in basal metabolic rate, and increase in mental activity. Finally, the sympathetics and parasympathetics are involved in regulating the male and female sexual acts, as will be explained in Chapters 80 and 81.

## FUNCTION OF THE ADRENAL MEDULLAE

Stimulation of the sympathetic nerves to the adrenal medullae causes large quantities of epinephrine and norepinephrine to be released into the circulating blood, and these two hormones in turn are carried in the blood to all tissues of the body. On the average, approximately 80 per cent of the secretion is epinephrine and 20 per cent norepinephrine, though the relative proportions of these change considerably under different physiological conditions.

The circulating norepinephrine has almost the same effects on the different organs as those caused by direct sympathetic stimulation, except that *the effects last about 10 times as long* because norepinephrine is removed from the blood slowly. For instance, norepinephrine causes constriction of essentially all the blood vessels of the body; it causes increased activity of the heart, inhibition of the gastrointestinal tract, dilation of the pupil of the eye, and so forth.

Epinephrine, also, causes almost the same effects as those caused by norepinephrine, but the effects differ in the following respects: First, epinephrine has a greater effect on cardiac activity than norepinephrine. Second, epinephrine causes only weak constriction of the blood vessels of the muscles in comparison with a much stronger constriction that results from norepinephrine. Since the muscle vessels represent a major segment of all vessels of the body, this difference is of special importance because

norepinephrine greatly increases the total peripheral resistance and thereby greatly elevates arterial pressure, whereas epinephrine raises the arterial pressure to a less extent but increases the cardiac output considerably because of its effect on the heart and veins.

A third difference between the action of epinephrine and norepinephrine relates to their effects on tissue metabolism. Epinephrine probably has several times as great a metabolic effect as norepinephrine. Indeed, the epinephrine secreted by the adrenal medullae increases the metabolic rate of the body often to as much as 100 per cent above normal, in this way increasing the activity and excitability of the whole body. It also increases the rate of other metabolic activities, such as glycogenolysis in the liver and muscle and glucose release into the blood.

In summary, stimulation of the adrenal medullae causes the release of hormones that have almost the same effects throughout the body as direct sympathetic stimulation, except that the effects are greatly prolonged. The only significant differences are caused by the epinephrine in the secretion, which increases the rate of metabolism and cardiac output to a greater extent than is caused by direct sympathetic stimulation.

**Value of the Adrenal Medullae to the Function of the Sympathetic Nervous System.** Usually, when any part of the sympathetic nervous system is stimulated, the entire system, or at least major portions of it, is stimulated at the same time. Therefore, norepinephrine and epinephrine are almost always released by the adrenal medullae at the same time that the different organs are being stimulated directly by the sympathetic nerves. Therefore, the organs are actually stimulated in two different ways simultaneously, directly by the sympathetic nerves and indirectly by the medullary hormones. The two means of stimulation support each other, and either can usually substitute for the other. For instance, destruction of the direct sympathetic pathways to the organs does not abrogate excitation of the organs because norepinephrine and epinephrine are still released into the circulating fluids and indirectly cause stimulation. Likewise, total loss of the two adrenal medullae usually has little significant effect on the operation of the sympathetic nervous system because the direct pathways can still perform almost all the necessary duties. Thus, the dual mechanism of sympathetic stimulation provides a safety factor, one mechanism substituting for the other when the second is missing.

Another important value of the adrenal medullae is the capability of epinephrine and norepinephrine to stimulate structures of the body that are not innervated by direct sympathetic fibers. For instance, the metabolic rate of every cell of the body is increased by these hormones, especially by epinephrine, even though only a small proportion of all the cells in the body are innervated by sympathetic fibers.

## RELATIONSHIP OF STIMULUS RATE TO DEGREE OF SYMPATHETIC AND PARASYMPATHETIC EFFECT

A special difference between the autonomic nervous system and the skeletal nervous system is the low frequency of stimulation required for full activation of autonomic effectors. In general, only one impulse every second or so suffices to maintain normal sympathetic or parasympathetic effect, and full activation occurs when the nerve fibers discharge 10 to 30 times per second. This compares with full activation in the skeletal nervous system at about 75 to 200 impulses per second.

## SYMPATHETIC AND PARASYMPATHETIC "TONE"

The sympathetic and parasympathetic systems are continually active, and the basal rates of activity are known, respectively, as *sympathetic tone* or *parasympathetic tone*.

The value of tone is that it allows a single nervous system to increase or to decrease the activity of a stimulated organ. For instance, sympathetic tone normally keeps almost all the blood vessels of the body constricted to approximately half their maximum diameter. By increasing the degree of sympathetic stimulation, the vessels can be constricted even more; but, on the other hand, by inhibiting the normal tone, the vessels can be dilated. If it were not for the continual sympathetic tone, the sympathetic system could cause only vasoconstriction, never vasodilatation.

Another interesting example of tone is that of the parasympathetics in the gastrointestinal tract. Surgical removal of the parasympathetic supply to the gut by cutting the vagi can cause serious and prolonged gastric and intestinal "atony," thus illustrating that in normal function the parasympathetic tone to the gut is strong. This tone can be decreased by the brain, thereby inhibiting gastrointestinal motility, or, on the other hand, it can be increased, thereby promoting increased gastrointestinal activity.

**Tone Caused by Basal Secretion of Norepinephrine and Epinephrine by the Adrenal Medullae.** The normal resting rate of secretion by the adrenal medullae is about 0.2 $\mu$gm./kg./min. of epinephrine and about 0.05 $\mu$gm./kg./min. of norepinephrine. These quantities are considerable—indeed, enough to maintain the blood pressure almost up to the normal value even if all direct sympathetic pathways to the cardiovascular system are removed. Therefore, it is obvious that much of the overall tone of the sympathetic nervous system results from basal secretion of epinephrine and norepinephrine in addition to that which results from direct sympathetic stimulation.

**Effect of Loss of Sympathetic or Parasympathetic Tone Following Denervation.** Immediately after a sympathetic or parasympathetic nerve is cut, the innervated organ loses its sympathetic or parasympathetic tone. In the case of the blood vessels, for instance, cutting the sympathetic nerves results immediately in almost maximal vasodilatation. However, over several days or weeks, the *intrinsic tone* in the smooth muscle of the vessels increases, usually restoring almost normal vasoconstriction.

Essentially the same events occur in most effector organs whenever sympathetic or parasympathetic tone is lost. That is, intrinsic compensation soon develops to return the function of the organ almost to its normal basal level. However, in the parasympathetic system the compensation sometimes requires many months. For instance, loss of parasympathetic tone to the heart increases the heart rate from 90 to 160 beats per minute in a dog, and this will still be about 120 six months later.

## DENERVATION SUPERSENSITIVITY OF SYMPATHETIC AND PARASYMPATHETIC ORGANS FOLLOWING DENERVATION

During the first week or so after a sympathetic or parasympathetic nerve is destroyed, the innervated organ becomes more and more sensitive to injected norepinephrine or acetylcholine, respectively. This effect is illustrated in Figure 57–4; the blood flow in the forearm before removal of the sympathetics was 200 ml. per minute, and a test dose of norepinephrine caused only a slight depression in flow. Then the stellate ganglion was removed, and normal sympathetic tone was lost. At first, the blood flow rose markedly

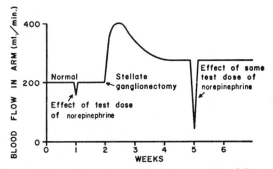

**Figure 57–4.** Effect of sympathectomy on blood flow in the arm, and the effect of a test dose of norepinephrine before and after sympathectomy, showing sensitization of the vasculature to norepinephrine.

because of the lost vascular tone, but over a period of days to weeks the blood flow returned almost to normal because of progressive increase in intrinsic tone of the vascular musculature itself, thus compensating for the loss of sympathetic tone. Another test dose of norepinephrine was then administered and the blood flow decreased much more than before, illustrating that the blood vessels had become about two to four times as responsive to norepinephrine as previously. This phenomenon is called *denervation supersensitivity*. It occurs in both sympathetic and parasympathetic organs and to a far greater extent in some organs than in others, often increasing the response as much as 10-fold.

**Mechansim of Denervation Supersensitivity.** The precise cause of denervation supersensitivity is not known, although several suggestions have been made. In the case of the sympathetic nervous system, it is supposed that destruction of the nerve endings prevents removal of norepinephrine or epinephrine by the process of re-uptake into the nerve endings. Likewise, the monamine oxidase that is found in the nerve endings is not available to help destroy the norepinephrine or epinephrine. Therefore, these two hormones can act strongly and for prolonged periods of time on the receptor organs.

In the case of acetylcholine supersensitivity, it is possible that loss of cholinesterase, particularly that which is attached to the nerve endings themselves, causes part of the supersensitivity. On the other hand, it is also possible that some functional system of the receptor cells themselves increases in activity when the cells are not continually bombarded by nerve stimuli.

# THE AUTONOMIC REFLEXES

It is mainly by means of autonomic reflexes that the autonomic nervous system regulates visceral functions. Throughout this text the functions of these reflexes are discussed in detail in relation to individual organs, but, to illustrate their importance, a few are presented here briefly.

**Cardiovascular Autonomic Reflexes.** Several reflexes in the cardiovascular system help to control the arterial blood pressure, cardiac output, and heart rate. One of these is the *baroreceptor reflex,* which was described in Chapter 22 along with other cardiovascular reflexes. Briefly, stretch receptors called *baroreceptors* are located in the walls of the major arteries, including the carotid arteries and the aorta. When these become stretched by high pressure, signals are transmitted to the brain stem, where they inhibit the sympathetic centers. This results in decreased sympathetic impulses to the heart and blood vessels, which allows the arterial pressure to fall back toward normal.

**The Gastrointestinal Autonomic Reflexes.** The uppermost part of the gastrointestinal tract and the rectum are controlled principally by autonomic reflexes. For instance, the smell of appetizing food initiates signals from the nose to the vagal, glossopharyngeal, and salivary nuclei of the brain stem. These in turn transmit signals through the parasympathetic nerves to the secretory glands of the mouth and stomach, causing secretion of digestive juices even before food enters the mouth. And when fecal matter fills the rectum at the other end of the alimentary canal, sensory impulses initiated by stretching the rectum are sent to the sacral portion of the spinal cord, and a reflex signal is retransmitted through the parasympathetics to the distal parts of the colon; these result in strong peristaltic contractions that empty the bowel.

**Other Autonomic Reflexes.** Emptying of the bladder is also controlled in the same way as emptying of the rectum; stretching the bladder sends impulses to the sacral cord, and this in turn causes contraction of the bladder as well as relaxation of the urinary sphincters, thereby promoting micturition.

Also important are the sexual reflexes which are initiated both by psychic stimuli from the brain and stimuli from the sexual organs. Impulses from these sources converge on the sacral cord and, in the male, result, first, in erection, mainly a parasympathetic function, and then in ejaculation, a sympathetic function.

Other autonomic reflexes include reflex contributions to the regulation of pancreatic secretion, gallbladder emptying, urinary excretion, sweating, blood glucose concentration, and many other visceral functions, all of which are discussed in detail at many points throughout this text.

## MASS DISCHARGE OF THE SYMPATHETIC SYSTEM VERSUS DISCRETE CHARACTERISTICS OF PARASYMPATHETIC REFLEXES

In general, large portions of the sympathetic nervous system often become stimulated simultaneously, a phenomenon called *mass discharge*. This characteristic of sympathetic action is in keeping with the usually diffuse nature of sympathetic function, such as overall regulation of arterial pressure or of metabolic rate.

However, in a few instances sympathetic activity does occur in isolated portions of the system. The most important of these are: (1) In the process of heat regulation, the sympathetics control sweating and blood flow in the skin without affecting other organs innervated by the sympathetics. (2) During muscular activity in some animals, cholinergic vasodilator fibers of the skeletal muscles are stimulated independently of all the remainder of the sympathetic system. (3) Many "local reflexes" involving the spinal cord but not the higher nervous centers affect local areas. For instance, heating a local skin area causes local vasodilatation and enhanced local sweating, while cooling causes the opposite effects.

In contrast to the sympathetic system, most reflexes of the parasympathetic system are very specific. For instance, parasympathetic cardiovascular reflexes usually act only on the heart to increase or decrease its rate of beating. Likewise, parasympathetic reflexes frequently cause secretion only in the mouth or, in other instances, secretion only by the stomach glands. Finally, the rectal emptying reflex does not affect other parts of the bowel to a major extent.

Yet there is often association between closely allied parasympathetic functions. For instance, though salivary secretion can occur independently of gastric secretion, these two often also occur together, and pancreatic secretion frequently occurs at the same time. Also, the rectal emptying reflex often initiates a bladder emptying reflex, resulting in simultaneous emptying of both the bladder and rectum. Conversely, the bladder emptying reflex can help to initiate rectal emptying.

## "ALARM" OR "STRESS" FUNCTION OF THE SYMPATHETIC NERVOUS SYSTEM

From the above discussions of the sympathetic nervous system, one can already see that mass sympathetic discharge increases in many ways the capability of the body to perform vigorous muscle activity. Let us quickly summarize these ways:

1. increased arterial pressure
2. increased blood flow to active muscles concurrent with decreased blood flow to organs that are not needed for rapid activity
3. increased rates of cellular metabolism throughout the body
4. increased blood glucose concentration
5. increased glycolysis in muscle
6. increased muscle strength
7. increased mental activity
8. increased rate of blood coagulation

The sum of these effects permits the person to perform far more strenuous physical activity than would otherwise be possible. Since it is physical *stress* that usually excites the sympathetic system, it is frequently said that the purpose of the sympathetic system is to provide extra activation of the body in states of stress: this is often called the sympathetic stress *reaction*.

The sympathetic system is also strongly activated in many emotional states. For instance, in the state of *rage,* which is elicited mainly by stimulating the hypothalamus, signals are transmitted downward through the reticular formation and spinal cord to cause massive sympathetic discharge, and all of the sympathetic events listed above ensue immediately. This is called the sympathetic *alarm reaction.* It is also frequently called the *fight or flight reaction* because an animal in this state decides almost instantly whether to stand and fight or to run. In either event, the sympathetic alarm reaction makes its subsequent activities extremely vigorous.

## MEDULLARY, PONTINE, AND MESENCEPHALIC CONTROL OF THE AUTONOMIC NERVOUS SYSTEM

Many areas in the reticular substance of the medulla, pons, and mesencephalon, as well as many special nuclei, (see Fig. 57–5) control different autonomic functions such as arterial pressure, heart rate, glandular secretion in the upper part of the gastrointestinal tract, gastrointestinal peristalsis, the degree of contraction of the urinary bladder, and many others. The con-

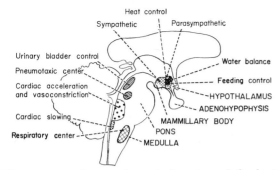

**Figure 57–5.** Autonomic control centers of the brain stem.

trol of each of these is discussed in detail at appropriate points in this text. Suffice it to point out here that the most important factors controlled in the lower brain stem are arterial pressure, heart rate, and respiration. Indeed, transection of the brain stem at the midpontile level allows normal basal control of arterial pressure to continue as before but prevents its modulation by higher nervous centers, particularly the hypothalamus. On the other hand, transection immediately below the medulla causes the arterial pressure to fall to about one-half normal. Closely associated with the cardiovascular regulatory centers in the medulla is the medullary center for regulation of respiration discussed in detail in Chapter 42. Though this is not considered to be an autonomic function, it is one of the *involuntary* functions of the body.

**Control of Lower Brain Stem Autonomic Centers by Higher Areas.** Signals from the hypothalamus and even from the cerebrum can affect the activities of almost all the lower brain stem autonomic control centers. For instance, stimulation in appropriate areas of the hypothalamus can activate the medullary cardiovascular control centers strongly enough to increase the arterial pressure to more than double normal. Likewise, higher centers can transmit signals into the lower centers to control body temperature, increase or decrease salivation and gastrointestinal activity, or to cause urinary bladder contraction. Therefore, to a great extent the autonomic centers in the lower brain stem are relay stations for control activities initiated at higher levels of the brain.

In the previous chapter it was also pointed out that many of the behavioral responses of an animal are mediated through the hypothalamus, reticular formation, and the autonomic nervous system. Indeed, the higher areas of the brain can alter the function of the whole autonomic nervous system or of portions of it strongly enough to cause severe autonomic-induced dis-

ease, such as peptic ulcer, constipation, heart palpitation, and even heart attacks.

# PHARMACOLOGY OF THE AUTONOMIC NERVOUS SYSTEM

## DRUGS THAT ACT ON ADRENERGIC EFFECTOR ORGANS—THE SYMPATHOMIMETIC DRUGS

From the foregoing discussion, it is obvious that intravenous injection of norepinephrine causes essentially the same effects throughout the body as sympathetic stimulation. Therefore, norepinephrine is called a *sympathomimetic*, or *adrenergic*, *drug*. Other sympathomimetic drugs include *epinephrine, phenylephrine, methoxamine,* and many others. These differ from each other in the degree to which they stimulate different sympathetic effector organs and in their duration of action. Norepinephrine and epinephrine have actions as short as one to three minutes, while the actions of most other commonly used sympathomimetic drugs last 30 minutes to 2 hours.

**Drugs that Cause Release of Norepinephrine from Nerve Endings.** Certain drugs have a sympathomimetic action in an indirect manner rather than by directly exciting adrenergic effector organs. These drugs include ephedrine, tyramine, and amphetamine. Their effect is to cause release of norepinephrine from its storage vesicles in the sympathetic nerve endings. The norepinephrine in turn causes the sympathetic effects.

**Drugs that Block Adrenergic Activity.** Adrenergic activity can be blocked at several different points in the stimulatory process as follows:

1. The synthesis and storage of norepinephrine in the sympathetic nerve endings can be prevented. The best known drug that causes this effect is *reserpine*.

2. Release of norepinephrine from the sympathetic endings can be blocked. This is caused by *guanethidine*.

3. The *alpha* receptors can be blocked. Two drugs that cause this effect are *phenoxybenzamine* and *phentolamine*.

4. The beta receptors can be blocked. A drug that does this is *propanolol*.

5. Sympathetic activity can be blocked by drugs that block transmission of nerve impulses through the autonomic ganglia. These are discussed in the following section, but the most important drug for blockade of both sympathetic and parasympathetic transmission through the ganglia is *hexamethonium*.

## DRUGS THAT ACT ON CHOLINERGIC EFFECTOR ORGANS

**Parasympathomimetic Drugs (Muscarinic Drugs).** Acetylcholine injected intravenously usually does not cause exactly the same effects throughout the body as parasympathetic stimulation because the acetylcholine is destroyed by cholinesterase in the blood and body fluids before it can reach all the effector organs. Yet a number of other drugs that are not so rapidly destroyed can produce typical parasympathetic effects, and these are called parasympathomimetic drugs.

Three commonly used parasympathomimetic drugs are *pilocarpine, methacholine,* and *carbamylcholine.* Although these drugs have almost the same effects on the different effector organs as does parasympathetic stimulation, there are slight variations. For instance, carbamylcholine is especially active on the bladder and gastrointestinal tract, while methacholine is especially active on the cardiovascular system and eyes.

Parasympathomimetic drugs act on the effector organs of cholinergic *sympathetic* fibers also. For instance, these drugs cause profuse sweating. Also, they cause vascular dilatation, this effect occurring even in vessels not innervated by cholinergic fibers.

**Drugs that Have a Parasympathetic Potentiating Effect.** Some drugs do not have a direct effect on parasympathetic effector organs but do potentiate the effects of the naturally secreted acetylcholine at the parasympathetic endings. These are the same drugs as those listed in Chapter 12 that potentiate the effect of acetylcholine at the neuromuscular junction—that is, *neostigmine, physostigmine,* and *diisopropyl fluorophosphate.* These inhibit cholinesterase, thus preventing rapid destruction of the acetylcholine liberated by the parasympathetic nerve endings. As a consequence, the quantity of acetylcholine acting on the effector organs progressively increases with successive stimuli, and the degree of action also increases.

**Drugs that Block Cholinergic Activity at Effector Organs.** *Atropine* and similar drugs, such as *homatropine* and *scopolamine,* block the action of acetylcholine on cholinergic effector organs. However, these drugs do not affect the nicotinic action of acetylcholine on the postganglionic neurons or on skeletal muscle.

## DRUGS THAT STIMULATE THE POSTGANGLIONIC NEURONS— "NICOTINIC DRUGS"

The preganglionic neurons of both the parasympathetic and sympathetic nervous systems secrete acetylcholine at their endings, and the acetylcholine in turn stimulates the postganglionic neurons. Therefore, injected acetylcholine can also stimulate the postganglionic neurons of both systems, thereby causing both sympathetic and parasympathetic effects in the body. *Nicotine* is a drug that can also stimulate postganglionic neurons in the same manner as acetylcholine because the neuronal membranes contain *nicotinic receptors*, but it cannot directly stimulate the autonomic effector organs, which have

muscarinic receptors as explained earlier in the chapter. Therefore, drugs that cause autonomic effects by stimulating the postganglionic neurons are frequently called nicotinic drugs. Some drugs, such as *acetylcholine* itself, *carbamylcholine* and *methacholine,* have both nicotinic and muscarinic actions, whereas pilocarpine has only muscarinic actions.

Nicotine excites both the sympathetic and parasympathetic systems at the same time, resulting in strong sympathetic vasoconstriction in the abdominal organs and limbs, but at the same time resulting in parasympathetic effects, such as increased gastrointestinal activity and, sometimes, slowing of the heart.

**Ganglionic Blocking Drugs.** Many important drugs block impulse transmission from the preganglionic neurons to the postganglionic neurons, including *tetraethyl ammonium ion, hexamethonium ion,* and *pentolinium.* These inhibit impulse transmission in both the sympathetic and parasympathetic systems simultaneously. They are often used for blocking sympathetic activity but rarely for blocking parasympathetic activity because the sympathetic blockade usually far overshadows the effects of parasympathetic blockade. The ganglionic blocking drugs have been especially important in reducing arterial pressure in patients with hypertension.

# REFERENCES

Axelsson, J.: Catecholamine functions. *Ann. Rev. Physiol.,* 33:1, 1971.

Banks, P., and Mayor, D.: Intra-axonal transport in noradrenergic neurons in the sympathetic nervous system. *Biochem. Soc. Symp.,* 36:133, 1972.

Bennett, M. R.: Autonomic Neuromuscular Transmission. Physiological Society Monograph, 1972.

Bhagat, B. D.: Recent Advances in Adrenergic Mechanisms. Springfield, Ill., Charles C Thomas, Publisher, 1971.

Carrier, O., Jr.: Pharmacology of the Peripheral Autonomic Nervous System. Chicago, Year Book Medical Publishers, 1972.

Costa, E., and Sandler, M. (eds.): Monoamine Oxidases—New Vistas. (Advances in Biochemical Psychopharmacology. Vol. 5.) New York, Raven Press, 1972.

Dahlström, A.: Aminergic transmission. Introduction and short review. *Brain Res.,* 62:441, 1973.

De Potter, W. P., Chubb, I. W., and De Schaepdryver, A. F.: Pharmacological aspects of peripheral noradrenergic transmission. *Arch. Int. Pharmacodyn. Ther.,* 196(Suppl. 196):258, 1972.

DiCara, L. V.: Learning in the autonomic nervous system. *Sci. Amer.,* 222:30, 1970.

Fonnum, F.: Recent developments in biochemical investigations of cholinergic transmission. *Brain Res.,* 62:497, 1973.

Gebber, G. L., and Snyder, D. W.: Hypothalamic control of baroreceptor reflexes. *Amer. J. Physiol.,* 218:124, 1970.

Geffen, L. B., and Livett, B. G.: Synaptic vesicles in sympathetic neurons. *Physiol. Rev.,* 51:98, 1971.

Guyton, A. C., and Gillespie, W. M., Jr.: Constant infusion of epinephrine: rate of epinephrine secretion and destruction in the body. *Amer. J. Physiol.,* 165:319, 1951.

Guyton, A. C., and Reeder, R. C.: Quantitative studies on the autonomic actions of curare. *J. Pharmacol. Exp. Ther.,* 98:188, 1950.

Häggendal, J.: Some aspects of the release of the adrenergic transmitter. *J. Neural Transm.,* Suppl. 11:135, 1974.

Hess, W. R.: The Functional Organization of the Diencephalon. New York, Grune & Stratton, Inc., 1958.

Hingerty, D., and O'Boyle, A.: Clinical Chemistry of the Adrenal Medulla. Springfield, Ill., Charles C Thomas, Publisher, 1972.

Jenkinson, D. H.: Classification and properties of peripheral adrenergic receptors. *Brit. Med. Bull., 29*:142, 1973.

Joó, F.: On the problem of the origin of nerve terminals in the sympathetic ganglia. *J. Neural Transm., Suppl. 11*:285, 1974.

Koizumi, K., and Brooks, C. M.: The integration of autonomic system reactions: a discussion of autonomic reflexes, their control and their association with somatic reactions. *Ergeb. Physiol., 67*:1, 1972.

McGeer, P. L., and McGeer, E. G.: Neurotransmitter synthetic enzymes. *Prog. Neurobiol., 2*:69, 1973.

Molinoff, P. B., Nelson, D. L., and Orcutt, J. C.: Dopamine-beta-hydroxylase and the regulation of the noradrenergic neuron. *Adv. Biochem. Psychopharmacol., 12*:95, 1974.

Nickerson, M.: Adrenergic receptors. *Circ. Res., 32(Suppl. 1)*:53, 1973.

Paton, D. M. (ed.): The Mechanism of Neuronal and Extraneuronal Transport of Catecholamines. New York, Raven Press, 1975.

Réthelyi, M.: Spinal transmission of autonomic processes. *J. Neural. Transm., Suppl. 11*:195, 1974.

Robinson, G. A., and Sutherland, E. W.: On the relation of cyclic AMP to adrenergic receptors and sympathin. *Adv. Cytopharmacol., 1*:263, 1971.

Sato, A., and Schmidt, R. F.: Somatosympathetic reflexes: afferent fibers, central pathways, and discharge characteristics. *Physiol. Rev., 53*:916, 1973.

Silver, A.: The Biology of Cholinesterases. New York, American Elsevier Publishing Co., Inc., 1974.

Smith, A. D.: Cellular control of the uptake, storage, and release of noradrenaline in sympathetic nerves. *Biochem. Soc. Symp. 36*:103, 1972.

Steer, M. L., Atlas, D., and Levitzki, A.: Inter-relations between beta-adrenergic receptors, adenylate cyclase, and calcium. *New Engl. J. Med., 292*:409, 1975.

Taxi, J.: Dynamics of transmissional ultrastructures in sympathetic neurones on the rat. *J. Neural Transm., Suppl. 11*:103, 1974.

von Euler, U. S.: Noradrenaline. Springfield, Illinois, Charles C Thomas, Publisher, 1956.

Winkler, H., Schneider, F. H., Rufener, C., Nakane, P. K., and Hörtnagl, H.: Membranes of adrenal medulla: their role in exocytosis. *Adv. Cytopharmacol., 2*:127, 1974.

# PART X

# THE SPECIAL SENSES

# | 58 |

# The Eye:
# I. Optics of Vision

## PHYSICAL PRINCIPLES OF OPTICS

Before it is possible to understand the optical systems of the eye, the student must be thoroughly familiar with the basic physical principles of optics, including the physics of refraction, and have a knowledge of focusing, depth of focus, and so forth. Therefore, in the present study of the optics of the eye, a brief review of these physical principles is first presented, and then the optics of the eye are discussed.

### *REFRACTION OF LIGHT*

**The Refractive Index of a Transparent Substance.** Light rays travel through a vacuum at a velocity of approximately 300,000 kilometers per second and practically as fast through air and other gaseous media but much slower through transparent solids and liquids. The refractive index of a transparent substance is the *ratio* of the velocity of light in air to that in the substance. Obviously, the refractive index of air itself is 1.00.

If light travels through a particular type of glass at a velocity of 200,000 kilometers per second, the refractive index of this glass is 300,000 divided by 200,000, or 1.50.

**Refraction of Light Rays at an Interface Between Two Media with Different Refractive Indices.** When light waves traveling forward in a beam, as shown in the upper part of Figure 58–1, strike an interface that is perpendicular to the beam, the waves enter the second refractive medium without deviating in their course. The only effect that occurs is decreased velocity of transmission. On the other hand, if, as illustrated in the lower part of Figure 58–1, the

light waves strike an angulated interface, the light waves bend if the refractive indices of the two media are different from each other. In this particular figure the light waves are leaving air, which has a refractive index of 1.00, and are entering a block of glass having a refractive index of 1.50. When the beam first strikes the angulated interface, the lower edge of the beam enters the glass ahead of the upper edge. The wave front in the upper portion of the beam continues to travel at a velocity of 300,000 kilometers per second while that which has entered the glass travels at a velocity of 200,000 kilometers per second. This causes the upper portion of the wave front to move ahead of the lower portion so that it is no longer vertical but is angulated to the right. Because *the direction in which light travels is always perpendicular to the wave front,* the direction of travel of the light beam now bends downward.

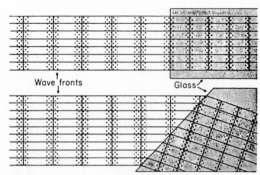

**Figure 58–1.** Wave fronts entering *(top)* a glass surface perpendicular to the light rays and *(bottom)* a glass surface angulated to the light rays. This figure illustrates that the distance between waves after they enter the glass is shortened to approximately two-thirds that in air. It also illustrates that light rays striking an angulated glass surface are refracted.

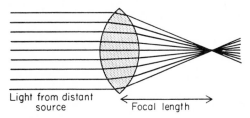

**Figure 58–2.** Bending of light rays at each surface of a convex spherical lens, showing that parallel light rays are focused to a point focus.

The bending of light rays at an angulated interface is known as *refraction*. Note particularly that the degree of refraction increases as a function of (1) the ratio of the two refractive indices of the two transparent media and (2) the degree of angulation between the interface and the wave front of the beam.

## APPLICATION OF REFRACTIVE PRINCIPLES TO LENSES

**The Convex Lens.** Figure 58–2 shows parallel light rays entering a convex lens. The light rays passing through the center of the lens strike the lens exactly perpendicular to the lens surfaces and therefore pass through the lens without being refracted at all. Toward either edge of the lens, however, the light rays strike a progressively more angulated interface. Therefore, the outer rays bend more and more toward the center.

When the rays have passed through the glass, the outer rays bend still more toward the center. Thus, parallel light rays entering an appropriately formed convex lens come to a single point focus at some distance beyond the lens.

**The Concave Lens.** Figure 58–3 shows the effect of a concave lens on parallel light rays. The rays that enter the very center of the lens strike an interface that is absolutely perpendicular to the beam and, therefore, do not refract at all. The rays at the edge of the lens enter the lens ahead of the rays toward the center. This is opposite to the effect in the convex lens, and it causes the peripheral light rays to *diverge* away from the light rays that pass through the center of the lens.

Thus, the concave lens *diverges* light rays, whereas the convex lens *converges* light rays.

**Spherical versus Cylindrical Lenses.** Figure 58–4 illustrates both a convex *spherical* lens and a convex *cylindrical* lens. Note that the convex cylindrical lens bends light rays from the two sides of the lens but not from either the top or the bottom. Therefore, parallel light rays are bent to a focal *line*. On the other hand, light rays that pass through the spherical lens are refracted at all edges of the lens toward the central ray, and all the rays come to a focal *point*.

The cylindrical lens is well illustrated by a test tube full of water. If the test tube is placed in a beam of sunlight and a piece of paper is brought progressively closer to the tube, a certain distance will be found at which the light rays come to a focal line. On the other hand, the spherical lens is illustrated by an ordinary magnifying glass. If such a lens is placed in a beam of sunlight and a piece of paper is brought progressively closer to the lens, the light rays will impinge on a common focal point at an appropriate distance.

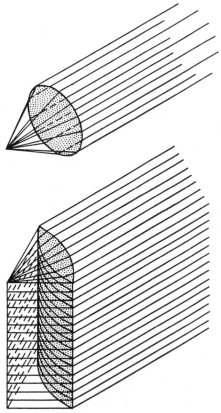

**Figure 58–4.** *Top:* Point focus of parallel light rays by a spherical convex lens. *Bottom:* Line focus of parallel light rays by a cylindrical convex lens.

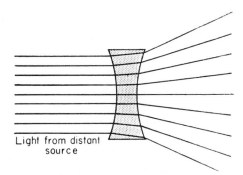

**Figure 58–3.** Bending of light rays at each surface of a concave spherical lens, illustrating that parallel light rays are diverged by a concave lens.

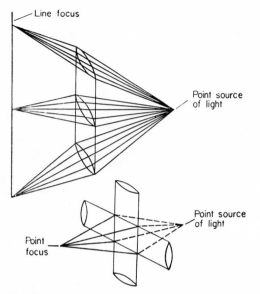

**Figure 58–5.** Two cylindrical convex lenses at right angles to each other, illustrating that one lens converges light rays in one plane and the other lens converges light rays in the plane at right angles. The two lenses combined give the same point focus as that obtained with a spherical convex lens.

*Concave* cylindrical lenses *diverge* light rays in only one plane in the same manner that convex cylindrical lenses converge light rays in one plane.

Figure 58–5 shows two convex cylindrical lenses at right angles to each other. The vertical cylindrical lens causes convergence of the light rays that pass through the two sides of the lens. The horizontal lens converges the top and bottom rays. Thus, all the light rays come to a single point focus. In other words, *two cylindrical lenses crossed at right angles to each other perform the same function as one spherical lens of the same refractive power.*

## FOCUSING BY CONVEX LENSES

The distance from a convex lens at which parallel rays converge to a common focal point is the *focal length* of the lens. The diagram at the top of Figure 58–6 illustrates this focusing of parallel light rays. In the middle diagram of Figure 58–6, the light rays that enter the convex lens are not parallel but are diverging because the origin of the light is a point source not far away from the lens itself. The rays striking the center of the lens are perpendicular to the lens surface and pass through the lens without any refraction. In the discussion above it was shown that the degree of bending of light rays is a function of (a) the ratio of refractive indices at the interface and (b) the angle of incidence of the light rays. Therefore, the edges of the convex lens bend the diverging rays approximately the same amount as they do parallel rays because the greater refraction at one interface is ap-

proximately balanced by less refraction at the other interface. However, because these rays are already diverging, even though they are bent the same extent as parallel rays, it can be seen from the diagram that they do not come to a point focus at the same distance away from the lens as do the parallel rays. In other words, when rays of light that are already diverging enter a convex lens, the distance of focus on the other side of the lens is farther from the lens than is the case when the entering rays are parallel to each other.

In the lower diagram of Figure 58–6 are shown light rays that also are diverging toward a convex lens that has far greater curvature than that of the upper two lenses of the figure. In this diagram the distance from the lens at which the light rays come to a focus is exactly the same as that from the lens in the first diagram, in which the lens was convex but the light rays entering it were parallel. This illustrates that both parallel rays and diverging rays can be focused at the same distance behind a lens provided that the lens changes its convexity. It is obvious that the nearer the point source of light is to the lens, the greater is the divergence of the light rays and the greater must be the curvature of the lens to cause focusing at the same distance.

The relationship of focal length of the lens, distance of the point source of light, and distance of focus is expressed by the following formula:

$$\frac{1}{f} = \frac{1}{a} + \frac{1}{b}$$

in which $f$ is the focal length of the lens, $a$ the distance of the point source of light from the lens, and $b$ the distance of focus from the lens.

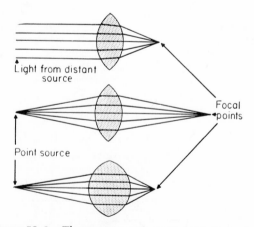

**Figure 58–6.** The upper two lenses of this figure have the same strength, but the light rays entering the top lens are parallel, while those entering the second lens are diverging; the effect of parallel versus diverging rays on the focal distance is illustrated. The bottom lens has far more refractive power than either of the other two lenses, illustrating that the stronger the lens the nearer to the lens is the point focus.

## FORMATION OF AN IMAGE BY A CONVEX LENS

The upper drawing of Figure 58–7 illustrates a convex lens with two point sources of light to the left. Because light rays from any point source pass through the center of a convex lens without being refracted in either direction, the light rays from both point sources of light are shown to pass straight through the lens center. Furthermore, the other light rays from each point source of light, passing through the upper and side edges of the lens, all come to point focus behind the lens *directly in line with each respective point source of light and the center of the lens.*

Any object in front of the lens is in reality a mosaic of point sources of light. Some of these points are very bright, some are very weak, and they vary in color. The light rays from each point source of light that enter the very center of the convex lens pass directly through this lens without any of the rays bending. Furthermore, the light rays that enter the edges of the lens finally come to focal points behind the lens in line with the rays that pass through the center. Therefore, every point source of light on the object comes to a separate point focus on the opposite side of the lens. If all portions of the object are the same distance in front of the lens, all the focal points behind the lens will fall in a common plane a certain distance behind the lens. If a white piece of paper is placed at this distance, one can see an image of the object, as is illustrated in the lower portion of Figure 58–7. However, this image is upside down with respect to the original object, and the two lateral sides of the image are reversed with respect to the original. This is the method by which the lens of a camera focuses light rays on the camera film.

## MEASUREMENT OF THE REFRACTIVE POWER OF A LENS—THE DIOPTER

The more a lens bends light rays, the greater is its "refractive power." This refractive power is measured in terms of *diopters.* The refractive power of a

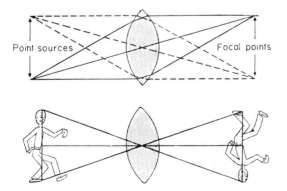

**Figure 58–7.** The top drawing illustrates two point sources of light focused at two separate points on the opposite side of the lens. The lower drawing illustrates formation of an image by a convex spherical lens.

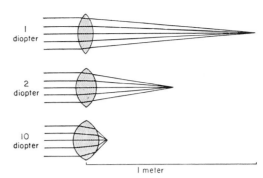

**Figure 58–8.** Effect of lens strength on the focal distance.

convex lens is equal to 1 meter divided by its focal length. Thus a spherical lens has a refractive power of +1 diopter when it is capable of converging parallel light rays to a focal point 1 meter beyond the lens, as illustrated in Figure 58–8. If the lens is capable of bending parallel light rays twice as much as a lens with a power of +1 diopter, it is said to have a strength of +2 diopters, and, obviously, the light rays come to a focal point 0.5 meter beyond the lens. A lens capable of converging parallel light rays to a focal point only 10 cm. beyond the lens has a refractive power of +10 diopters.

The refractive power of concave lenses cannot be stated in terms of the focal distance beyond the lens because the light rays diverge rather than focus to a point. Therefore, the power of a concave lens is stated in terms of its ability to diverge light rays in comparison with the ability of convex lenses to converge light rays. That is, if a concave lens diverges light rays the same amount that a 1 diopter convex lens converges them, the concave lens is said to have a dioptric strength of −1. Likewise, if the concave lens diverges the light rays as much as a +10 diopter lens converges them, it is said to have a strength of −10 diopters.

Note particularly that concave lenses can "neutralize" the refractive power of convex lenses. Thus, placing a 1 diopter concave lens immediately in front of a 1 diopter convex lens results in a lens system with zero refractive power.

The strengths of cylindrical lenses are computed in the same manner as the strengths of spherical lenses. If a cylindrical lens focuses parallel light rays to a line focus 1 meter beyond the lens, it has a strength of +1 diopter. On the other hand, if a cylindrical lens of a concave type *diverges* light rays as much as a +1 diopter cylindrical lens *converges* them, it has a strength of −1 diopter.

# THE OPTICS OF THE EYE

## THE EYE AS A CAMERA

The eye, as illustrated in Figure 58–9 is optically equivalent to the usual photographic cam-

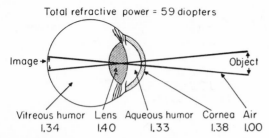

Total refractive power = 59 diopters

Image →

Object

Vitreous humor    Lens    Aqueous humor    Cornea    Air
1.34              1.40    1.33              1.38      1.00

**Figure 58–9.**   The eye as a camera. The numbers are the refractive indices.

era, for it has a lens system, a variable aperture system, and a retina that corresponds to the film. The lens system of the eye is composed of (1) the interface between air and the anterior surface of the cornea, (2) the interface between the posterior surface of the cornea and the aqueous humor, (3) the interface between the aqueous humor and the anterior surface of the lens, and (4) the interface between the posterior surface of the lens and the vitreous humor. The refractive index of air is 1; the cornea, 1.38; the aqueous humor, 1.33; the lens (on the average), 1.40; and the vitreous humor, 1.34.

*The Reduced Eye.*   If all the refractive surfaces of the eye are algebraically added together and then considered to be one single lens, the optics of the normal eye may be simplified and represented schematically as a "reduced eye." This is useful in simple calculations. In the reduced eye, a single lens is considered to exist with its central point 17 mm. in front of the retina and to have a total refractive power of approximtely 59 diopters when the lens is accommodated for distant vision.

The anterior surface of the cornea provides about 48 diopters of the eye's total dioptric strength for three reasons: (1) the refractive index of the cornea is markedly different from that of air; (2) the surface of the cornea is further away from the retina than are the surfaces of the eye lens; and (3) the curvature of the cornea is reasonably great.

The posterior surface of the cornea is concave and actually acts as a concave lens, but, because the difference in refractive index of the cornea and the aqueous humor is slight, this posterior surface of the cornea has a refractive power of only about −4 diopters, which neutralizes only a small part of the refractive power of the other refractive surfaces of the eye.

The total refractive power of the crystalline lens of the eye when it is surrounded by fluid on each side is only 15 diopters of the total refrac-

tive power of the eye's lens system. If this lens were removed from the eye and then surrounded by air, its refractive power would be about 150 diopters. Thus, it can be seen that the lens inside the eye is not nearly so powerful as it is outside the eye. The reason for this is that the fluids surrounding the lens have refractive indices not greatly different from the refractive index of the lens itself, the smallness of the differences greatly decreasing the amount of light refraction at the lens interfaces. But the importance of the crystalline lens is that its curvature and therefore also its strength can change to provide accommodation, which will be discussed later in the chapter.

**Formation of an Image on the Retina.**   In exactly the same manner that a glass lens can focus an image on a sheet of paper, the lens system of the eye can also focus an image on the retina. The image is inverted and reversed with respect to the object. However, the mind perceives objects in the upright position despite the upside down orientation on the retina because the brain is trained to consider an inverted image as the normal.

## THE MECHANISM OF ACCOMMODATION

The refractive power of the crystalline lens of the eye can be voluntarily increased from 15 diopters to approximately 29 diopters in young children; this is a total "accommodation" of 14 diopters. To do this, the shape of the lens is changed from that of a moderately convex lens to that of a very convex lens. The mechanism of this is the following:

Normally, the lens is composed of a strong elastic capsule filled with viscous proteinaceous, but transparent fibers. When the lens is in a relaxed state, with no tension on its capsule, it assumes a spherical shape, owing entirely to the elasticity of the lens capsule. However, as illustrated in Figure 58–10, approximately 70 ligaments attach radially around the lens, pulling the lens edges toward the edge of the choroid. These ligaments are constantly tensed by the elastic pull of their attachments to the choroid, and the tension on the ligaments causes the lens to remain relatively flat under normal resting conditions of the eye. At the insertions of the ligaments in the choroid is the ciliary muscle, which has two sets of smooth muscle fibers, the *meridional fibers* and the *circular fibers*. The meridional fibers extend from the corneoscleral junction to the insertions of the ligaments in the choroid approximately 2 to

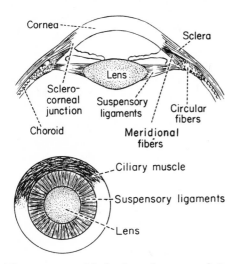

**Figure 58–10.** Mechanism of accommodation.

3 mm. behind the corneoscleral junction. When these muscle fibers contract, the insertions of the ligaments are pulled forward, thereby releasing a certain amount of tension on the crystalline lens. The circular fibers are arranged circularly all the way around the eye so that when they contract a sphincter-like action occurs, decreasing the diameter of the circle of ligament attachments and allowing the ligaments to pull less on the lens capsule.

Thus, contraction of both sets of smooth muscle fibers in the ciliary muscle relaxes the ligaments to the lens capsule, and the lens assumes a more spherical shape, like that of a balloon, because of elasticity of its capsule. When the ciliary muscle is completely relaxed, the dioptric strength of the lens is as weak as it can become. On the other hand, when the ciliary muscle contracts as strongly as possible, the dioptric strength of the lens becomes maximal.

**Autonomic Control of Accommodation.** The ciliary muscle is controlled mainly by the parasympathetic nervous system but also to a slight extent by the sympathetic system, as will be discussed in Chapter 60. Stimulation of the parasympathetic fibers to the eye contracts the ciliary muscle, which in turn relaxes the ligaments of the lens and increases its refractive power. With an increased refractive power, the eye is capable of focusing on objects that are nearer to it than is an eye with less refractive power. Consequently, as a distant object moves toward the eye, the number of parasympathetic impulses impinging on the ciliary muscle must be progressively increased for the eye to keep the object constantly in focus. Sympathetic stimulation has a weak effect in relaxing the ciliary muscle and therefore in focusing the eye for far vision.

**Presbyopia.** As a person grows older, his lens loses its elastic nature and becomes a relatively solid mass, probably because of progressive denaturation of the proteins. Therefore, the ability of the lens to assume a spherical shape progressively decreases, and the power of accommodation decreases from approximately 14 diopters shortly after birth to approximately 2 diopters at the age of 45 to 50. Thereafter, the lens of the eye may be considered to be almost totally nonaccommodating, which condition is known as "presbyopia."

Once a person has reached the state of presbyopia, each eye remains focused permanently at an almost constant distance; this distance depends on the physical characteristics of each individual's eyes. Obviously, the eyes can no longer accommodate for both near and far vision. Therefore, for an older person to see clearly both in the distance and nearby, he must wear bifocal glasses with the upper segment focused for far-seeing and the lower segment focused for near-seeing.

## THE PUPILLARY APERTURE

A major function of the iris is to increase the amount of light that enters the eye during darkness and to decrease the light that enters the eye in bright light. The reflexes for controlling this mechanism will be considered in the discussion of the neurology of the eye in Chapter 60. The amount of light that enters the eye through the pupil is proportional to the area of the pupil or to the *square of the diameter* of the pupil. The pupil of the human eye can become as small as approximately 1.5 mm. and as large as 8 mm. in diameter. Therefore, the quantity of light entering the eye may vary approximately 30 times as a result of changes in pupillary aperture size.

**Depth of Focus of the Lens System of the Eye.** Figure 58–11 illustrates two separate eyes that are exactly alike except that the diameters of the pupillary apertures are different. In the upper eye the pupillary aperture is small, and in the lower eye the aperture is large. In front of each of these two eyes are two small point sources of light, and light from each passes through the pupillary aperture and focuses on the retina. Consequently, in both eyes the retina sees two spots of light in perfect focus. It is evident from the diagrams, however, that if the

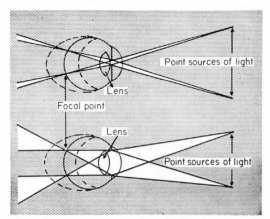

**Figure 58–11.** Effect of small and large pupillary apertures on the depth of focus.

retina is moved forward or backward to an out-of-focus position, the size of each spot will not change much in the upper eye, but in the lower eye the size of each spot will increase greatly, and it becomes a "blur circle." In other words, the upper lens system has far greater *depth of focus* than the bottom lens system. When a lens system has great depth of focus, the retina can be considerably displaced from the focal plane and still discern the various points of an image rather distinctly; whereas, when a lens system has a shallow depth of focus, moving the retina only slightly away from the focal plane causes extreme blurring of the image.

The greatest possible depth of focus occurs when the pupil is extremely small. The reason for this is that with a very small aperture the eye acts more or less as a pinhole camera. It will be recalled from the study of physics that light rays coming from an object and passing through a pinhole are in focus on any surface at all distances beyond the pinhole. In other words, the depth of focus of a pinhole camera is infinite.

## "NORMAL" ABERRATIONS OF VISION

**Spherical Aberration.** The crystalline lens of the eye is not nearly so regularly formed as lenses made by good opticians. Indeed, the light rays passing through the peripheral edges of the eye lens are not brought to really sharp focus with the other light rays, as illustrated in Figure 58–12. This effect is known as "spherical aberration," and the lens system of the human eye is quite subject to such an error. Therefore, increasing the aperture of the pupil progressively decreases the sharpness of focus. This partially explains why visual acuity is decreased at

low illumination levels which cause the pupil to dilate.

**Chromatic Aberration.** The lens of the eye is also subject to "chromatic aberration," which is illustrated in Figure 58–13. This means that the refractive power of the lens is different for the different colors, so that different colors focus at different distances behind the lens. Furthermore, the greater the aperture of the lens, the greater are the errors of chromatic aberration, for those light rays passing through the center of the lens are proportionately less affected than are the rays passing through the periphery of the lens.

**Diffractive Errors of the Eye.** Still another error in the optical system of all eyes is "diffraction" of light rays. Diffraction means bending of light rays as they pass over sharp edges; this obviously occurs as the rays pass over the edges of the pupil. Diffractive errors become especially important when the pupil becomes very small, because "interference" patterns then appear on the retina. A thorough consideration of diffraction at this point is impossible, and this phenomenon is simply noted to explain that as the pupil becomes 1.5 mm. in size the sharpness of vision becomes less than when the pupil is approximately 2.5 mm. in size despite the fact that the depth of focus is better the smaller the pupillary diameter.

**Cataracts.** Cataracts are an especially common eye abnormality that occurs in older people. A cataract is a cloudy or opaque area in the lens. In the early stage of cataract formation the proteins in the lens fibers immediately beneath the capsule become denatured. Later, these same proteins coagulate to form opaque areas in place of the normal transparent protein fibers of the lens. Finally, in still later stages, calcium is often deposited in the coagulated proteins, thus further increasing the opacity.

When a cataract has obscured light transmission so greatly that it seriously impairs vision, the condition can be corrected by surgical removal of the entire lens. When this is done, however, the eye loses a large portion of its refractive power, which must be replaced by a powerful convex lens (about +15 diopters) in front of the eye, as will be explained in the following sections.

## ERRORS OF REFRACTION

**Emmetropia.** As shown in Figure 58–14, the eye is considered to be normal or "emmetropic" if, when

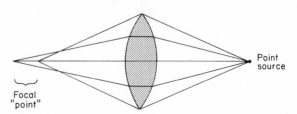

**Figure 58–12.** Spherical aberration.

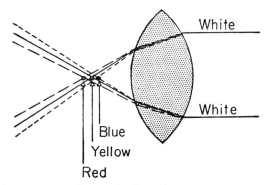

**Figure 58–13.** Chromatic aberration.

the ciliary muscle is completely relaxed, parallel light rays from distant objects are in sharp focus on the retina. This means that the emmetropic eye can, with its ciliary muscle completely relaxed, see all distant objects clearly, but to focus objects at close range it must contract its ciliary muscle and thereby provide various degrees of accommodation.

**Hypermetropia (Hyperopia).** Hypermetropia, which is also known as "far-sightedness," is due either to an eyeball that is too short or to a lens system that is too weak when the ciliary muscle is completely relaxed. In this condition, parallel light rays are not bent sufficiently by the lens system to come to a focus by the time they reach the retina. In order to overcome this abnormality, the ciliary muscle may contract to increase the strength of the lens. Therefore, the far-sighted person is capable, by using his mechanism of accommodation, of focusing distant objects on his retina. If he has used only a small amount of strength in his ciliary muscle to accommo-

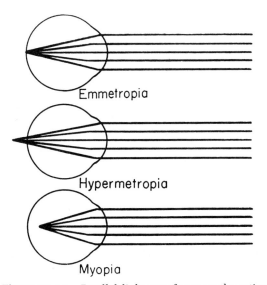

**Figure 58–14.** Parallel light rays focus on the retina in emmetropia, behind the retina in hypermetropia, and in front of the retina in myopia.

date for the distant objects, then he still has much accommodative power left, and objects closer and closer to the eye can also be focused sharply until the ciliary muscle has contracted to its limit. The distance of the object away from the eye at this point is known as the "near point" of vision.

In old age, when the lens becomes presbyopic, the far-sighted person often is not able to accommodate his lens sufficiently to focus even distant objects, much less to focus near objects.

**Myopia.** In myopia, or "near-sightedness," even when the ciliary muscle is completely relaxed, the strength of the lens is still so great that light rays coming from distant objects are focused in front of the retina. This is usually due to too long an eyeball but it can occasionally result from too much power of the lens system of the eye.

No mechanism exists by which the eye can decrease the strength of its lens beyond that which exists when the ciliary muscle is completely relaxed. Therefore, the myopic person has no mechanism by which he can ever focus distant objects sharply on his retina. However, as an object comes nearer and nearer to his eye it finally comes near enough that its image is focused on the retina. Then, when the object comes still closer to the eye, the person can use his mechanism of accommodation to keep the image focused clearly. Therefore, a myopic person has a definite limiting "far point" for acute vision as well as a "near point"; when an object comes inside the "far point," he can use his mechanism of accommodation to keep the object in focus until the object reaches the "near point."

*Correction of Myopia and Hypermetropia by Use of Lenses.* It will be recalled that light rays passing through a concave lens diverge. Therefore, if the refractive surfaces of the eye have too much refractive power, as in myopia, some of this excessive refractive power can be neutralized by placing in front of the eye a concave spherical lens, which will diverge rays. On the other hand, in a person who has hypermetropia—that is, one who has too weak a lens for the distance of the retina away from the lens—the abnormal vision can be corrected by adding refractive power with a convex lens in front of the eye. These corrections are illustrated in Figure 58–15. One usually determines the strength of the concave or convex lens needed for clear vision by "trial and error"—that is, by trying first a strong lens and then a stronger or weaker lens until the one that gives the best visual acuity is found.

**Astigmatism.** Astigmatism is a refractive error of the lens system of the eye caused usually by an oblong shape of the cornea or, rarely, by an oblong shape of the lens. A lens surface like the side of an egg lying edgewise to the incoming light would be an example of an astigmatic lens. The degree of curvature in a plane through the long axis of the egg is not nearly so great as the degree of curvature in a plane through the short axis. The same is true of an astigmatic lens. Because the curvature of the astigmatic lens along one plane is less than the curvature along

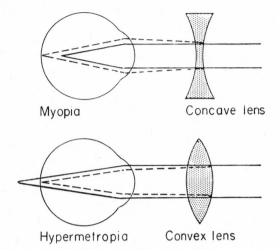

Figure 58–15. Correction of myopia with a concave lens and correction of hypermetropia with a convex lens.

the other plane, light rays striking the peripheral portions of the lens in one plane are not bent nearly so much as are rays striking the peripheral portions of the other plane.

This is illustrated in Figure 58–16, which shows what happens to rays of light emanating from a point source and passing through an oblong astigmatic lens. The light rays in the vertical plane, which is indicated by plane BD, are refracted greatly by the astigmatic lens because of the greater curvature in the vertical direction than in the horizontal direction. However, the light rays in the horizontal plane, indicated by plane AC, are bent not nearly so much as the light rays in the vertical plane. It is obvious, therefore, that the light rays passing through an astigmatic lens do not all come to a common focal point because the light rays passing through one plane of the lens focus far in front of those passing through the other plane.

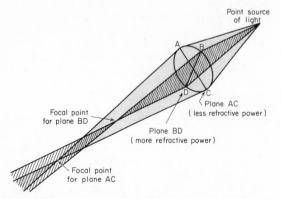

Figure 58–16. Astigmatism, illustrating that light rays focus at one focal distance in one focal plane and at another focal distance in the plane at right angles.

Placing an appropriate *spherical* lens in front of an astigmatic eye can bring the light rays that pass through *one plane* of the lens into focus on the retina, but spherical lenses can never bring *all* the light rays into complete focus at the same time. This is the reason why astigmatism is a very undesirable refractive error of the eyes. Furthermore, the accommodative powers of the eyes cannot compensate for astigmatism for the same reasons that spherical lenses placed in front of the eyes cannot correct the condition.

*Correction of Astigmatism with a Cylindrical Lens.* In correcting astigmatism with lenses it must always be remembered that two cylindrical lenses of equal strength may be crossed at right angles to give the same refractive effects as a spherical lens of the same strength. However, if one of the crossed cylindrical lenses has a different strength from that of the second lens, the light rays in one plane may not be brought to a common focal point with the light rays in the opposite plane. This is the situation that one usually finds in the astigmatic eye. In other words, one may consider an astigmatic eye as having a lens system made up of two cylindrical lenses of slightly different strengths. Another way of looking at the astigmatic lens system of the eye is that this system is a spherical lens with a superimposed cylindrical lens.

To correct the focusing of an astigmatic lens system, it is necessary to determine both the *strength* of the cylindrical lens needed to neutralize the excess cylindrical power of the eye lens and the *axis* of this abnormal cylindrical lens.

There are several methods for determining the axis of the abnormal cylindrical component of the lens system of an eye. One of these methods is based on the use of parallel black bars, as shown in Figure 58–17. Some of these parallel bars are vertical, some horizontal, and some at various angles to the vertical and horizontal axes. After placing, by trial and error, various spherical lenses in front of the astigmatic eye, a strength of lens will usually be found that will cause sharp focus of one set of these parallel bars on the retina of the astigmatic eye. If there is no accompanying myopia or hypermetropia, the spherical lens may not be necessary.

It can be shown from the physical principles of optics discussed earlier in this chapter that the axis of the *out-of-focus* cylindrical component of the optical system is parallel to the black bars that are fuzzy in appearance. Once this axis is found, the examiner tries progressively stronger and weaker positive or negative cylindrical lenses, the axes of which are placed parallel to the out-of-focus bars until the patient sees all the crossed bars with equal clarity. When this has been accomplished, the examiner directs the optician to grind a special lens having both the spherical correction plus the cylindrical correction at the appropriate axis.

**Correction of Optical Abnormalities by Use of Contact Lenses.** In recent years, either glass or plastic contact lenses have been used to fit snugly against the anterior surface of the cornea. These

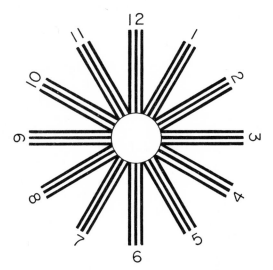

**Figure 58–17.** Chart composed of parallel black bars for determining the axis (meridian) of astigmatism.

lenses are held in place by a thin layer of tears that fill the space between the contact lens and the anterior surface.

A special feature of the contact lens is that it nullifies almost entirely the refraction that normally occurs at the anterior surface of the cornea. The reason for this is that the tears between the contact lens and the cornea have a refractive index almost equal to that of the cornea so that no longer does the anterior surface of the cornea play a significant role in the eye's optical system. Instead, the anterior surface of the contact lens now plays the major role and its posterior surface a minor role, the refraction of this lens now substituting for the cornea's usual refraction. This is especially important in persons whose eye refractive errors are caused by an abnormally shaped cornea, such as persons who have an odd-shaped, bulging cornea—a condition called *keratoconus*. Without the contact lens the bulging cornea causes such severe abnormality of vision that almost no glasses can correct the vision satisfactorily; when a contact lens is used, however, the corneal refraction is neutralized, and normal refraction by the contact lens is substituted in its place.

The contact lens has several other advantages as well, including (1) the lens turns with the eye and gives a broader field of clear vision than do usual glasses, and (2) the contact lens has little effect on the size of the object that the person sees through the lens; on the other hand, lenses placed several centimeters in front of the eye do affect the size of the image even though they do correct the focus.

## SIZE OF THE IMAGE ON THE RETINA AND VISUAL ACUITY

If the distance from an object to the eye lens is 17 meters and the distance from the center of the lens to the image is 17 millimeters, the ratio of the object size to image size is 1000 to 1. Therefore, an object 17 meters in front of the eye and 1 meter in size produces an image on the retina 1 millimeter in size.

Theoretically, a point of light from a distant point source, when focused on the retina, should be infinitely small. However, since the lens system of the eye is not perfect, such a retinal spot ordinarily has a total diameter of about 11 microns even with maximum resolution of the optical system. However, it is brightest in its very center and shades off gradually toward the edges.

The average diameter of cones *in the fovea* of the retina, the central part of the retina where vision is most highly developed, is approximately 1.5 microns, which is one-seventh the diameter of the spot of light. Nevertheless, since the spot of light has a bright center point and shaded edges, a person can distinguish two separate points if their centers lie approximately 2 microns apart on the retina, which is slightly greater than the width of a foveal cone. This discrimination between points is illustrated in Figure 58–18.

The maximum visual acuity of the human eye for discriminating between point sources of light is 26 seconds. That is, when light rays from two separate points strike the eye with an angle of at least 26 seconds between them, they can usually be recognized as two points instead of one. This means that a person with maximal acuity looking at two bright pinpoint spots of light 10 meters away can barely distinguish the spots as separate entities when they are 1 millimeter apart.

The fovea is only a fraction of a millimeter in diameter, which means that maximum visual acuity occurs in only 3 degrees of the visual field. Outside this foveal area the visual acuity

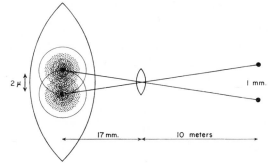

**Figure 58–18.** Maximal visual acuity for two point sources of light.

is reduced 5- to 10-fold, and it becomes progressively poorer as the periphery is approached. This is caused by the connection of many rods and cones to the same nerve fiber, as discussed in Chapter 60.

**Clinical Method for Stating Visual Acuity.** Usually the test chart for testing eyes is placed 20 feet away from the tested person, and if the person can see the letters of the size that he should be able to see at 20 feet, he is said to have 20/20 vision: that is, normal vision. If he can see only letters that he should be able to see at 200 feet, he is said to have 20/200 vision. On the other hand, if he can see at 20 feet letters that he should be able to see only at 15 feet, he is said to have 20/15 vision. In other words, the clinical method for expressing visual acuity is to use a mathematical fraction that expresses the ratio of two distances, which is also the ratio of one's visual acuity to that of the normal person.

## DETERMINATION OF DISTANCE OF AN OBJECT FROM THE EYE— DEPTH PERCEPTION

There are three major means by which the visual apparatus normally perceives distance, a phenomenon that is known as depth perception. These are (1) relative sizes of objects, (2) moving parallax, and (3) stereopsis.

**Determination of Distance by Relative Sizes.** If a person knows that a man is six feet tall and then he sees this man even with only one eye, he can determine how far away the man is simply by the size of the man's image on his retina. He does not consciously think about the size of this image, but his brain has learned to determine automatically from the image sizes the distances of objects from the eye when the dimensions of these objects are already known.

**Determination of Distance by Moving Parallax.** Another important means by which the eyes determine distance is that of moving parallax. If a person looks off into the distance with his eyes completely still, he perceives no moving parallax, but, when he moves his head to one side or the other, the images of objects close to him move rapidly across his retinae while the images of distant objects remain rather stationary. For instance, if he moves his head 1 inch and an object is only 1 inch in front of his eye, the image moves almost all the way across his retinae, whereas the image of an object 200 feet away from his eyes does not move perceptibly. Thus, by this mechanism of moving paral-

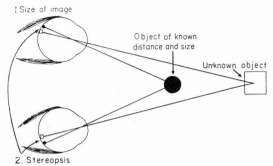

**Figure 58–19.** Perception of distance (1) by the size of the image on the retina and (2) as a result of stereopsis.

lax, one can tell the *relative distances* of different objects even though only one eye is used.

**Determination of Distance by Stereopsis.** Another method by which one perceives parallax is that of binocular vision. Because one eye is a little more than 2 inches to one side of the other eye, the images on the two retinae are different one from the other—that is, an object that is 1 inch in front of the bridge of the nose forms an image on the temporal portion of the retina of each eye, whereas a small object 20 feet in front of the nose has its image at closely corresponding points in the middle of the eye. This type of parallax is illustrated in Figure 58–19, which shows the images of a black spot and a square actually reversed on the retinae because they are at different distances in front of the eyes. This gives a type of parallax that is present all the time when both eyes are being used. It is almost entirely this binocular parallax (or stereopsis) that gives a person with two eyes far greater ability to judge relative distances *when objects are nearby* than a person who has only one eye. However, stereopsis is virtually useless for depth perception at distances beyond 200 feet.

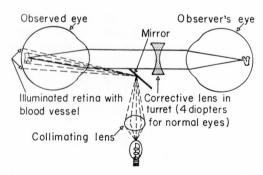

**Figure 58–20.** The optical system of the ophthalmoscope.

# OPTICAL INSTRUMENTS

## THE OPHTHALMOSCOPE

The ophthalmoscope is an instrument designed so that an observer can look through it into another person's eye and see the retina with clarity. Though the ophthalmoscope appears to be a relatively complicated instrument, its principles are simple. The basic portions of such an instrument are illustrated in Figure 58–20 and may be explained as follows:

If a bright spot of light is on the retina of an emmetropic eye, light rays from the spot diverge toward the lens system of the eye, and, after they pass through the lens system, they are parallel with each other because the retina is located in the focal plane of the lens. When these parallel rays pass into an emmetropic eye of another person, they focus to a point on the retina of the second person because his retina is in the focal plane of his lens. Therefore, any spot of light on the retina of the observed eye comes to a focal spot on the retina of the observing eye. Likewise, when the bright spot of light is moved to different portions of the observed retina, the focal spot on the retina of the observer also moves approximately an equal amount. Thus, if the retina of one person is made to emit light, the image of his retina will be focused on the retina of the observer provided that the two eyes are simply looking into each other. These principles, of course, apply only to completely emmetropic eyes.

To make an ophthalmoscope, one need only devise a means for illuminating the retina to be examined. Then, the reflected light from that retina can be seen by the observer simply by putting the two eyes close to each other. To illuminate the retina of the observed eye, an angulated mirror or a segment of a prism is placed in front of the observed eye in such a manner that light from a bulb is reflected into the observed eye. Thus, the retina is illuminated through the pupil, and the observer sees into the subject's pupil by looking over the edge of the mirror or prism, or preferably *through* an appropriately designed prism so that the light will not have to enter the pupil at an angle.

It was noted above that these principles apply only to persons with completely emmetropic eyes. If the refractive power of either eye is abnormal, it is necessary to correct this refractive power in order for the observer to see a sharp image of the observed retina. Therefore, the usual ophthalmoscope has a series of about 20 lenses mounted on a turret so that the turret can be rotated from one lens to another, and the correction for abnormal refractive power of either or both eyes can be made at the same time by selecting a single lens of appropriate strength. In normal young persons, when the two eyes come close together, a natural accommodative reflex occurs that causes approximately +2 diopters increase in the strength of the lens of each eye. To correct for this, it is necessary that the lens turret be rotated to approximately −4 diopters correction.

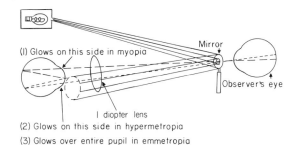

(1) Glows on this side in myopia

1 diopter lens

(2) Glows on this side in hypermetropia

(3) Glows over entire pupil in emmetropia

**Figure 58–21.** The retinoscope.

## THE RETINOSCOPE

The retinoscope, illustrated in Figure 58–21, is an instrument that can be used to determine the refractive power of an eye even though the subject cannot converse with the observer. Such a procedure is valuable for fitting glasses to an infant.

To use the retinoscope, one places a bright spot of light behind and to one side of the observed eye, and stands 1 meter away, looking through a hole in the middle of a mirror. The observer then rotates this mirror from side to side, casting a reflected beam of light into the pupil of the observed eye while the subject keeps his gaze intently on the observer's eye. If the observed eye is normal, when this beam of light travels across the pupil the entire pupil suddenly glows red. If the eye has abnormal refractive powers, the red glow appears either on the side of the pupil into which the light first shines or on the opposite side of the pupil—one or the other. *In hyperopia the first glow appears on the side of the pupil from which the light beam is being moved. In myopia the first glow appears on the opposite side of the pupil.* The cause of this difference is that the lens of the hyperopic eye is too weak and that of the myopic eye too strong to focus the light from the edge of the pupil on the retina exactly in line with the observer's eye. For a fuller understanding of this effect, however, the student is referred to texts on physiologic optics.

One can fit glasses to a patient by placing selected lenses in front of the observed eye one at a time until the glow suddenly covers the pupil over its entire extent rather than spreading from one side of the pupil to the other. However, it should be noted that in retinoscopy one tests an eye that is focused on the observer's eye at 1 meter's distance. This must be taken into consideration in prescribing glasses; 1 diopter strength must be subtracted for far vision.

# REFERENCES

Armington, J. C.: Vision. *Ann. Rev. Physiol.*, 27:163, 1965.
Bodis-Wollner, I.: On dissociated visual functions. *Mt. Sinai J. Med. N.Y.*, 41:38, 1974.
Campbell, C. J., Koester, C. J., Rittler, M. C., and Tackaberry R. B.: Physiological Optics. Hagerstown, Md., Harper & Row, Publishers, 1974.

Clayton, R. M.: Problems of differentiation in the vertebrate lens. *Curr. Top. Dev. Biol.,* 5:115, 1970.

Davson, H.: The Physiology of the Eye, 3rd Ed. New York, Academic Press, Inc., 1972.

Davson, H., and Graham, L. T., Jr.: The Eye. Vols. 1–6. New York, Academic Press, Inc. 1969–1974.

Duke-Elder, S., and Perkins, E. S. (eds.): The Transparency of the Cornea. Springfield, Ill., Charles C Thomas, Publisher, 1960.

Follmann, P.: Dynamography. *Ophthalmologica,* 169:192, 1974.

Linksz, A.: Physiology of the Eye. Vol. 1. Optics. New York, Grune & Stratton, Inc., 1950.

Paterson, C. A.: Extracellular space of the crystalline lens. *Amer. J. Physiol.,* 218:797, 1970.

Polyak, S.: The Vertebrate Visual System. Chicago, University of Chicago Press, 1957.

Rabbetts, R. B.: A comparison of astigmatism and cyclophoria in distance and near vision. *Brit. J. Physiol. Opt.,* 27:161, 1973.

Ruben, M.: Textbook of Contact Lens Practice. Baltimore, The Williams & Wilkins Company, 1973.

Sekuler, R.: Spatial vision. *Ann. Rev. Psychol.,* 25:195, 1974.

Siegel, I. M.: Optics and visual physiology. *Arch. Ophthalmol.,* 90:327, 1973.

Sloane, A. E.: Manual of Refraction. Boston, Little, Brown and Company, 1961.

Toates, F. M.: Accommodation function of the human eye. *Physiol. Rev.,* 52:828, 1972.

Tommila, V.: Stereoscopic and binocular vision. *Ophthalmologica,* 169:90, 1974.

Trevor-Roper, P.: The Eye and Its Disorders. Philadelphia, J. B. Lippincott Company, 1974.

Whitteridge, D.: Binocular vision and cortical function. *Proc. R. Soc. Med.,* 65:947, 1972.

Witkovsky, P.: Peripheral mechanisms of vision. *Ann. Rev. Physiol.,* 33:257, 1971.

# 59

# The Eye:
# II. Receptor Functions of
# the Retina

The retina is the light-sensitive portion of the eye, containing the cones, which are mainly responsible for color vision, and the rods, which are mainly responsible for vision in the dark. When the rods and cones are excited, signals are transmitted through successive neurons in the retina itself and finally into the optic nerve fibers and cerebral cortex. The purpose of the present chapter is to explain specifically the mechanisms by which the rods and cones detect both white and colored light.

## ANATOMY AND FUNCTION OF THE STRUCTURAL ELEMENTS OF THE RETINA

**The Layers of the Retina.** Figure 59–1 shows the functional components of the retina arranged in layers from the outside to the inside as follows: (1) pigment layer, (2) layer of rods and cones projecting into the pigment, (3) outer lining membrane, (4) outer nuclear layer, (5) outer plexiform layer, (6) inner nuclear layer, (7) inner plexiform layer, (8) ganglionic layer, (9) layer of optic nerve fibers, and (10) inner limiting membrane.

After light passes through the lens system of the eye and then through the vitreous humor, it enters the retina from the bottom (see Figure 59–1); that is, it passes through the ganglion cells, the plexiform layer, the nuclear layer, and the limiting membranes before it finally reaches the layer of rods and cones located all the way on the opposite side of the retina. This distance is a thickness of several hundred microns; visual acuity is obviously decreased by this passage through such nonhomogeneous tissue. However, in the central region of the retina, as will be discussed below, the initial layers are pulled aside to prevent this loss of acuity.

**The Foveal Region of the Retina and its Importance in Acute Vision.** A minute area in the center of the retina, illustrated in Figure 59–2, called the *macula* and occupying a total area of less than 1 square millimeter is especially capable of acute and detailed vision. This area is composed entirely of cones, but the cones are very much elongated and

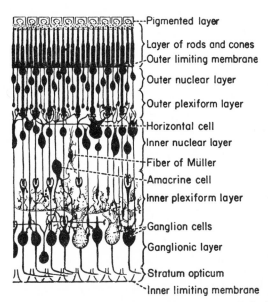

**Figure 59–1.** Plan of the retinal neurons. (From Polyak: The Retina. University of Chicago Press.)

- Pigmented layer
- Layer of rods and cones
- Outer limiting membrane
- Outer nuclear layer
- Outer plexiform layer
- Horizontal cell
- Inner nuclear layer
- Fiber of Müller
- Amacrine cell
- Inner plexiform layer
- Ganglion cells
- Ganglionic layer
- Stratum opticum
- Inner limiting membrane

**Figure 59–2.** Photomicrograph of the macula and of the fovea in its center. Note that the inner layers of the retina are pulled to the side to decrease the interference with light transmission. (From Bloom and Fawcett: A Textbook of Histology, 10th Ed., 1975; courtesy of H. Mizoguchi.)

have a diameter of only 1.5 microns in contradistinction to the very large cones located farther peripherally in the retina. The central portion of the macula, only 0.4 mm. in diameter, is called the *fovea;* in this region the blood vessels, the ganglion cells, the inner nuclear layer of cells, and the plexiform layers are all displaced to one side rather than resting directly on top of the cones. This allows light to pass unimpeded to the cones rather than through several layers of retina, which aids immensely in the acuity of visual perception by this region of the retina.

**The Rods and Cones.** Figure 59–3 is a diagrammatic representation of a photoreceptor (either a rod or a cone) though the cones may be distinguished by having a conical upper end. In general, the rods are narrower and longer than the cones, but this is not always the case. In the peripheral portions of the retina the rods are 2 to 5 microns in diameter whereas the cones are 5 to 8 microns in diameter; in the central part of the retina, in the fovea, the cones have a diameter of only 1.5 microns.

To the right in Figure 59–3 are illustrated the four major functional segments of either a rod or a cone: (1) the outer segment, (2) the inner segment, (3) the nucleus, and (4) the synaptic body. It is in the outer segment that the light-sensitive photochemical is found. In the case of the rods, this is *rhodopsin,* and in the cones it is one of several photochemicals almost exactly the same as rhodopsin except for a difference in spectral sensitivity. The cross-marks in the outer segment represent discs formed by infoldings of the cell membrane. These act as shelves to which the photosensitive pigments are attached. In this outer segment the concentration of the photosensitive pigments is approximately 40 per cent.

The inner segment contains the usual cytoplasm of the cell with the usual cytoplasmic organelles. Particularly important are the mitochondria, for we shall

see later that the mitochondria in this segment play an important role in providing most of the energy for function of the photoreceptors.

The synaptic body is the portion of the rod and cone that connects with the subsequent neuronal cells, the horizontal and bipolar cells, that represent the next stages in the vision chain.

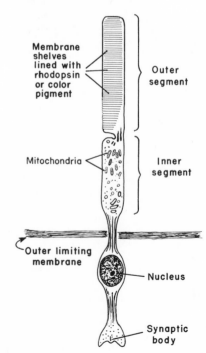

**Figure 59–3.** Schematic drawing of the functional parts of the rods and cones. (Redrawn from R. Young: General Cell Biology.)

Figure 59–4 shows in more detail the photoreceptor discs in the outer segments of the rods and cones. Note especially their attachments to the cell membrane from which they originated.

**Regeneration of Photoreceptors.** The inner segments of the rods continually synthesize new rhodopsin. This in turn migrates to the base of the outer segment, where the cell membrane is continually forming new folds that become the discs of the rod and to which the rhodopsin attaches. Several new discs are formed each hour, and these migrate outward along the rod toward the tip, carrying the new rhodopsin with them. When the discs reach the top they degenerate and are dissoluted by the pigment epithelium. Thus, the light receptor portion of the rod is continually being replaced.

On the other hand, new disc formation does not seem to occur in the cones. However, new visual pigment does migrate throughout the outer segment and local repair of the disc system does seem to occur.

These processes of disc generation and repair are probably important to the provision of continually viable photoreceptor chemical mechanisms in the photoreceptors.

**The Pigment Layer of the Retina.** The black pigment *melanin* in the pigment layer, and still more melanin in the choroid, prevents light reflection throughout the globe of the eyeball, and this is extremely important for acute vision. This pigment performs the same function in the eye as the black paint inside the bellows of a camera. Without it, light rays would be reflected in all directions within the eyeball and would cause diffuse lighting of the retina rather than the contrasts between dark and light spots required for formation of precise images.

The importance of melanin in the pigment layer and choroid is well illustrated by its absence in *albinos,* persons hereditarily lacking in melanin pigment in all parts of their bodies. When an albino enters a bright area, light that impinges on the retina is reflected in all directions by the white surface of the unpigmented choroid so that a single discrete spot of light that would normally excite only a few rods or cones is reflected everywhere and excites many of the receptors. Also, a large amount of diffuse light enters the eye through the unpigmented iris. Thus, the lack of melanin causes the visual acuity of albinos, even with the best of optical correction, to be rarely better than 20/100 to 20/200. The pigment layer

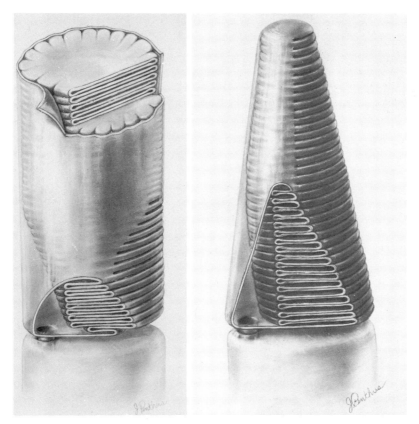

**Figure 59–4.**   Membranous structures of the outer segments of a rod (left) and a cone (right). (Courtesy of Dr. Richard Young.)

also stores large quantities of *vitamin A*. This vitamin A is exchanged back and forth through the membranes of the outer segments of the rods and cones which themselves are embedded in the pigment layer. We shall see later that vitamin A is an important precursor of the photosensitive pigments and that this interchange of vitamin A is very important for adjustment of the light sensitivity of the receptors.

**The Blood Supply of the Retina—the Retinal Arterial System and the Choroid.** The nutrient blood supply for the inner layers of the retina is derived from the central retinal artery, which enters the inside of the eye along with the optic nerve and then divides to supply the entire inner retinal surface. Thus, to a great extent, the retina has its own blood supply independent of the other structures of the eye.

However, the outer layer of the retina is adherent to the *choroid,* which is a highly vascular tissue between the retina and the sclera. The outer layers of the retina, including the outer segments of the rods and cones, depend mainly on diffusion from the choroid vessels for their nutrition, especially for their oxygen.

*Retinal Detachment.* The neural retina occasionally detaches from the pigment epithelium. In a few instances the cause of such detachment is injury to the eyeball that allows fluid or blood to collect between the retina and the pigment epithelium, but more often it is caused by contracture of fine collagenous fibrils in the vitreous humor, which pull the retina unevenly toward the interior of the globe.

Fortunately, partly because of diffusion across the detachment gap and partly because of the independent blood supply to the retina through the retinal artery, the retina can resist degeneration for a number of days and can become functional once

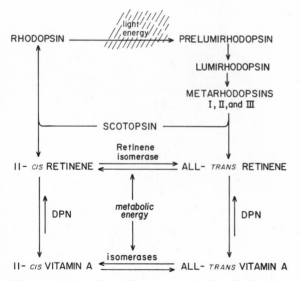

**Figure 59–5.** Photochemistry of the rhodopsin-retinene-vitamin A visual cycle.

again if surgically replaced in its normal relationship with the pigment epithelium. But, if not replaced soon, the retina finally degenerates and then is unable to function even after surgical repair.

# PHOTOCHEMISTRY OF VISION

Both the rods and cones contain chemicals that decompose on exposure to light and, in the process, excite the nerve fibers leading from the eye. The chemical in the *rods* is called *rhodopsin,* and the light-sensitive chemicals in the *cones* have compositions only slightly different from that of rhodopsin.

In the present section we will discuss principally the photochemistry of rhodopsin, but we may apply almost exactly the same principles to the photochemistry of the light-sensitive substances of the cones.

## THE RHODOPSIN-RETINENE VISUAL CYCLE, AND EXCITATION OF THE RODS

**Rhodopsin and its Decomposition by Light Energy.** The outer segment of the rod that projects into the pigment layer of the retina has a concentration of about 40 per cent of the light-sensitive pigment called *rhodopsin* or *visual purple*. This substance is a combination of the protein *scotopsin* and the carotenoid pigment *retinene*. Furthermore, the retinene is a particular type called 11-*cis* retinene (also called neo-b retinine). This *cis* form of the retinene is important because only this form can combine with scotopsin to synthesize rhodopsin.

When light energy is absorbed by rhodopsin, the rhodopsin immediately begins to decompose, as shown in Figure 59–5. The cause of this is an instantaneous change of the *cis* form of retinene into an all-*trans* form, which still has the same chemical structure as the *cis* form but has a different physical structure—a straight molecule rather than a curved molecule. Because the stereoscopic orientation of the reactive sites of the all-*trans* retinene no longer fit with that of the reactive sites on the protein scotopsin, it begins to pull away from the scotopsin. The immediate product is *prelumirhodopsin,* which is a partially split combination of the all-*trans* retinene and scotopsin. However, prelumirhodopsin is an extremely unstable compound and decays in a small fraction of a second to *lumirhodopsin*. This decays in another fraction of a second to *metarhodopsin I,* then successively to *metarhodopsin II* and *metarhodopsin III*. All of these are loose combinations of the all-*trans* retinene and scotopsin, but the metarhodopsin III, too, is unstable, and it slowly decomposes during the next few seconds into completely split products—scotopsin and all-*trans* retinene. During the process of splitting, the rods are excited, and signals are transmitted into the central nervous system.

**Reformation of Rhodopsin.** The first stage in reformation of rhodopsin, as shown in Figure 59–5, is to reconvert the all-*trans* retinene into 11-*cis* retinene. This process is catalyzed by the enzyme *retinene isomerase*. However, this process also requires active metabolism and energy transfer to the 11-*cis* retinene; once the 11-*cis* retinene is formed, it automatically recombines with the scotopsin to reform rhodopsin, an exergonic process (which means that it gives off energy). The product, rhodopsin, is a stable compound until its decomposition is again triggered by absorption of light energy.

**Excitation of the Rods When Rhodopsin Decomposes.** Although the exact way in which rhodopsin decomposition excites the rod is still speculative, it is believed to occur in the following general way: When the all-*trans* retinene begins to split away from the scotopsin, several ionized radicals are momentarily exposed, and these are believed to excite the rod in one of several ways. One suggestion has been that the instantaneous ionic field of the radicals causes changes in the membrane of the rod that are responsible for the rod excitation. The physical construction of the outer segment of the rod is compatible with this concept because the cell membrane folds inward to form tremendous numbers of shelf-like discs lying one on top of the other, each one lined on the inside of the membrane with aggregates of rhodopsin. This extensive relationship between rhodopsin and the cell membrane could explain why the rod is so exquisitely sensitive to light, being capable of detectable excitation following absorption of only one quantum of light energy.

A second theory for excitation of the rod is that the momentarily ionized all-*trans* retinene-scotopsin complex acts as an enzyme to promote some yet unspecified chemical reaction in the outer segment of the rod. This reaction in turn supposedly leads to rod excitation.

*Generation of the Receptor Potential.* Regardless of the precise mechanism by which the membrane of the outer segment becomes altered in the presence of decomposing rhodopsin, it is known that this excitation process *de-*

*creases* the conductance of this membrane for sodium ions. It has also been shown that the inner segment of the rod continually pumps sodium from inside the rod to the outside. Based on these two facts, the following theory for development of the receptor potential has been proposed:

Figure 59–6 illustrates movement of sodium ions in a complete electrical circuit through the inner and outer segments of the rod. The inner segment continually pumps sodium from inside the rod to the outside, thereby creating a negative potential on the inside of the entire cell. However, the membrane of the outer segment of the rod, in the unexcited condition, is very leaky to sodium. Therefore, sodium continually leaks back to the inside of the rod and thereby

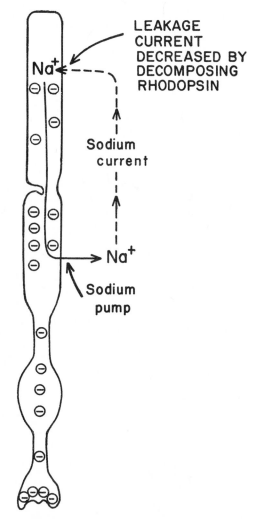

**Figure 59–6.** Theoretical basis for the generation of a hyperpolarization receptor potential caused by rhodopsin decomposition.

neutralizes much of the negativity on the inside of the entire cell. Thus, under normal conditions, when the rod is not excited, there is a reduced amount of electronegativity inside the membrane of the rod.

When the rhodopsin in the outer segment of the rod is exposed to light and begins to decompose, however, this *decreases* the leakage of sodium to the interior of the rod even though still more and more sodium continues to be pumped out. Thus, a net loss of sodium from the rod creates increased negativity inside the membrane, and the greater the amount of light energy striking the rod, the greater the electronegativity. This process is called *hyperpolarization*. Furthermore, it is exactly opposite to the effect that occurs in almost all other sensory receptors in which the degree of negativity is generally reduced during stimulation rather than increased, producing a state of depolarization rather than hyperpolarization.

***Characteristics of the Receptor Potential.*** The receptor potential has two phases: an instantaneous very low voltage phase, the electrical potential of which is directly proportional to the light energy; and a secondary phase of much greater voltage, the potential of which is proportional to the logarithm of the light energy. The first potential is believed to result from the immediate interaction of light with the rhodopsin. The second is believed to result from the sodium current mechanism described above.

This transduction of light energy into a logarithmic type of receptor potential seems to be an exceedingly important one for function of the eye, because it allows the eye to discriminate light intensities through a range many thousand times as great as would otherwise be possible. In general, regardless of the original light intensity, the eye can detect an increase in light intensity when this increase is about 0.5 to 1 per cent over the intensity already existing, which again illustrates a logarithmic type of response of the retina.

**Relationship Between Retinene and Vitamin A.** The lower part of the scheme in Figure 59–5 illustrates that each of the two types of retinene can be converted into a corresponding type of vitamin A, and in turn the two types of vitamin A can be reconverted into the two types of retinene. Thus, the two retinenes are in equilibrium with vitamin A. Most of the vitamin A of the retina is stored in the pigment layer of the retina rather than in the rods themselves, but this vitamin A is readily available to the rods.

Retinene is formed from vitamin A by dehydrogenation in accordance with the following reaction:

$$C_{19}H_{27}CH_2OH + DPN^+ \underset{\longleftarrow}{\overset{\text{Alcohol dehydrogenase}}{\longrightarrow}}$$

$$\underset{\text{Vitamin A}}{} \quad\quad C_{19}H_{27}CHO + DPNH + H^+$$
$$\underset{\text{Retinene}}{}$$

The equilibrium of this equation is strongly in favor of vitamin A. Therefore, vitamin A concentration is normally far greater than free retinene concentration in the retina.

Another important feature of the conversion of retinene into vitamin A, or the converse conversion of vitamin A into retinene, is that these processes require a much longer time to approach equilibrium than it takes for conversion of retinene and scotopsin into rhodopsin, or for conversion (under the influence of strong light energy) of rhodopsin into retinene and scotopsin. Therefore, all the reactions of the upper part of Figure 59–5 can take place relatively rapidly in comparison with the slow interconversions between retinene and vitamin A.

Yet, if the retina remains exposed to strong light for a long time, most of the stored rhodopsin will eventually be converted into retinene and then into vitamin A, thereby decreasing the concentration of all the photochemicals in the rods much more than would be true were it not for this subsequent interconversion from retinene to vitamin A.

Conversely, during total darkness, essentially all the retinene in the rods becomes converted into rhodopsin, thereby reducing the concentration of retinene almost to zero. This then allows much of the vitamin A to be converted into retinene, which also becomes rhodopsin within another few minutes. Thus, in complete darkness not only is almost all the retinene of the rods converted into rhodopsin but also much of the vitamin A stored in the pigment layer of the retina is absorbed by the rods and also converted into rhodopsin.

*Night Blindness.* Night blindness occurs in severe vitamin A deficiency. When the total quantity of vitamin A in the blood becomes greatly reduced, the quantities of vitamin A, retinene, and rhodopsin in the rods, as well as the color photosensitive chemicals in the cones, are all depressed, thus decreasing the sensitivities of the rods and cones. This condition is called night blindness because at night the amount of available light is far too little to permit adequate vision, though in daylight, sufficient light is available to excite the rods and cones despite their reduction in photochemical substances.

For night blindness to occur, a person often must remain on a vitamin A deficient diet for weeks or months, because large quantities of vitamin A are normally stored in the liver and are made available to the rest of the body in times of need. However, once night blindness does develop it can sometimes be completely cured in a half hour or more by intravenous injection of vitamin A. This results from the ready conversion of vitamin A into retinene and thence into rhodopsin.

## PHOTOCHEMISTRY OF COLOR VISION BY THE CONES

It was pointed out at the outset of this discussion that the photochemicals in the cones have almost exactly the same chemical compositions as rhodopsin in the rods. The only difference is that the protein portions, the opsins, called *photopsins* in the cones, are different from the scotopsin of the rods. The retinene portions are exactly the same in the cones as in the rods. The color-sensitive pigments of the cones, therefore, are combinations of retinene and photopsins.

In the discussion of color vision later in the chapter, it will become evident that three different types of pigments are present in different cones, thus making these cones selectively sensitive to the different colors, blue, green, and red. The absorption characteristics of the pigments in the three types of cones show peak absorbancies at light wavelengths, respectively, of 430, 535, and 575 millimicrons. These are also the wavelengths for peak light sensitivity for each type of cone, which begins to explain how the retina differentiates the colors. The approximate absorption curves for these three pigments are shown in Figure 59–7. The peak absorption for the rhodopsin of the rods, on the other hand, occurs at 505 millimicrons.

Note that the peak absorption of the so-called "red" cone is actually in the orange color band. It is called the "red" cone because it responds to red more intensely than do any of the other cones.

## AUTOMATIC REGULATION OF RETINAL SENSITIVITY—DARK AND LIGHT ADAPTATION

**Relationship of Sensitivity to Pigment Concentration.** The sensitivity of rods is approximately proportional to the antilogarithm of the rhodopsin concentration, and it is assumed that

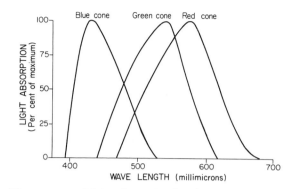

**Figure 59–7.** Light absorption by the respective pigments of the three color receptive cones of the human retina. (Drawn from curves recorded by Marks, Dobelle, and MacNichol, Jr.: *Science,* 143:1181, 1964, and by Brown and Wald: *Science,* 144:45, 1964.)

this relationship also holds true in the cones. This means that a minute decrease in rhodopsin reduces the sensitivity of rods tremendously. For instance, the sensitivity of a rod is reduced about 8.5 times when its concentration of rhodopsin is reduced from maximum by only 0.006 per cent, and the sensitivity decreases over 3000 times when the rhodopsin concentration is reduced 0.6 per cent. This antilog effect is postulated to result from the fact that the first rhodopsin to be decomposed on exposure to light lies near the surface of the membrane discs in the rod. After this portion has decomposed, far greater decomposition of the deeper layers of rhodopsin is required to excite the rods.

In summary, the sensitivity of the rods and cones can be altered up or down tremendously by only slight changes in concentrations of the photosensitive chemicals.

**Light and Dark Adaptation.** If a person has been in bright light for a long time, large proportions of the photochemicals in both the rods and cones have been reduced to retinene and opsins. Furthermore, most of the retinene of both the rods and cones has been converted into vitamin A. Because of these two effects, the concentrations of the photosensitive chemicals are considerably reduced, and the sensitivity of the eye to light is even more reduced. This is called *light adaptation.*

On the other hand, if the person remains in the darkness for a long time, essentially all the retinene and opsins in the rods and cones become converted into light-sensitive pigments. Furthermore, large amounts of vitamin A are converted into retinene, which is then changed into additional light-sensitive pigments, the final limit being determined by the amount of opsins in the rods and cones. Because of these two effects, the visual receptors gradually become so sensitive that even the minutest amount of light causes excitation. This is called *dark adaptation*.

Figure 59–8 illustrates the course of dark adaptation when a person is exposed to total darkness after having been exposed to bright light for several hours. Note that retinal sensitivity is very low when one first enters the darkness, but within 1 minute the sensitivity has increased 10-fold—that is, the retina can respond to light of one-tenth intensity. At the end of 20 minutes the sensitivity has increased about 6000-fold, and at the end of 40 minutes it has increased about 25,000-fold.

The resulting curve of Figure 59–8 is called the *dark adaptation curve*. Note, however, the inflection in the curve. The early portion of the curve is caused by adaptation of the cones, for these adapt much more rapidly than the rods because of a basic difference in the rate at which they resynthesize their photosensitive pigments. On the other hand, the cones do not achieve anywhere near the same degree of sensitivity as the rods. Therefore, despite rapid adaptation by the cones, they cease adapting after only a few minutes, while the slowly adapting rods continue to adapt for many minutes or even hours, their sensitivity increasing tremendously. However, a large share of the greater sensitivity of the rods is caused by convergence of as many as 100 rods onto a single ganglion cell in the retina; these rods summate to increase their sensitivity, as will be discussed in the following chapter.

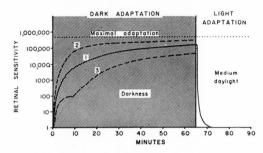

**Figure 59–9.** Dark and light adaptation. (The retina is considered to have a sensitivity of 1 when maximally light-adapted.)

Figure 59–9 illustrates dark adaptation, but in this figure it is shown that the rate of dark adaptation differs depending on the previous degree of exposure of the retina to light. Curve number 1 shows the dark adaptation after a person had been in bright light for approximately 20 minutes, curve number 2 shows the rate after he had been in bright light for only a minute or so, and curve number 3 after he had been in bright light for many hours. Note that the eye exposed to bright light for a long time adapts slowly in comparison with the eye that has been in bright light for only a few minutes. The cause of this difference is probably the slowness of the interconversions between vitamin A and retinene, particularly the necessity for vitamin A to be absorbed into the rods and cones from the pigment layer of the retina. The eye that has remained in bright light for only a minute or so has not converted significant quantities of its retinene into vitamin A. Therefore, this eye can resynthesize its photosensitive pigments in a short time. Conversely, the eye that has been exposed to prolonged bright light will have had time to convert most of its retinene into vitamin A, and since the reconversion from vitamin A to retinene is very slow, dark adaptation of the eye also is slow.

To the right in Figure 59–9 is shown a typical light adaptation curve, illustrating that light adaptation occurs much more rapidly than dark adaptation. Also, the slow component caused by retinene-vitamin A interconversion does not play as prominent a role in light adaptation as in dark adaptation. Therefore, a person exposed to extremely bright light after having been in prolonged darkness becomes adjusted to the new light conditions in only a few minutes. This is in contrast to dark adaptation, which requires 10 to 20 minutes for moderate adaptation and 10 to 18 hours for maximal adaptation.

**Other Mechanisms of Light and Dark Adaptation.** In addition to adaptation caused by changes in concentrations of rhodopsin or color photochemicals, the eye has two other mechanisms for light and dark adaptation. The first of these is the *change in pupillary size*, which was discussed in the previous chapter. This can cause a degree of adaptation by approximately 20- to 30-fold because of changes in the light allowed through the pupillary opening.

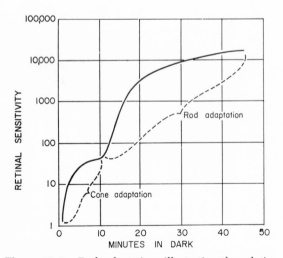

**Figure 59–8.** Dark adaptation, illustrating the relationship of cone adaptation to rod adaptation.

The other mechanism is *neural adaptation,* involving the neurons in the successive stages of the visual chain in the retina itself. That is, when the light intensity first increases, the intensity of the signals transmitted by the bipolar cells, the horizontal cells, the amacrine cells, and the ganglion cells, are all very intense. However, the intensities of these signals all decrease rapidly with time. Although the degree of this adaptation is only a few-fold rather than the many thousand-fold that occurs during adaptation of the photochemical system, this neural adaptation occurs in seconds in contrast to the many minutes required for full adaptation by the photochemicals.

**Value of Light and Dark Adaptation in Vision.** Between the limits of maximal dark adaptation and maximal light adaptation, the eye can change its sensitivity to light by as much as 500,000 to 1,000,000 times, the sensitivity automatically adjusting to changes in illumination.

Since the registration of images by the retina requires detection of both dark and light spots in the image, it is essential that the sensitivity of the retina always be adjusted so that the receptors respond to the lighter areas and not to the darker areas. An example of maladjustment of the retina occurs when a person leaves a movie theater and enters the bright sunlight, for even the dark spots in the images then seem exceedingly bright, and, as a consequence, the entire visual image is bleached, having little contrast between its different parts. Obviously, this is poor vision, and it remains poor until the retina has adapted sufficiently that the dark spots of the image no longer stimulate the receptors excessively.

Conversely, when a person enters darkness, the sensitivity of the retina is usually so slight that even the light spots in the image cannot excite the retina. But, after dark adaptation, the light spots begin to register. As an example of the extremes of light and dark adaptation, the light intensity of the sun is approximately 30,000 times that of the moon; yet the eye can function well both in bright sunlight and in bright moonlight.

**Negative After-Images.** If one looks steadily at a scene for a while, the bright portions of the image cause light adaptation of the retina while the dark portions of the image cause dark adaptation. In other words, areas of the retina that are stimulated by light become less sensitive while areas that are exposed only to darkness gain in sensitivity. If the person then moves his eyes away from the scene and looks at a bright white surface he sees exactly the same scene that he had been viewing, but the light areas of the scene now appear dark, and the dark areas appear light. This is known as the negative after-image, and

it is a natural consequence of light and dark adaptation.

The negative after-image persists as long as any degree of light and dark adaptation remains in the respective portions of the retina. Referring back to Figure 59–9, one can see from the dark adaptation curves that a negative after-image could possibly persist as long as an hour under favorable conditions.

**Photopic versus Scotopic Vision.** *Photopic vision* means vision capable of discriminating color, whereas *scotopic vision* means vision capable of discriminating only between shades of black and white. In bright light one finds that his vision is photopic, whereas below a critical light intensity his vision is scotopic. The reasons for this difference is the following: In very dim light, only rods are capable of becoming dark-adapted to a sensitivity level required for light detection. Therefore, in dim light the retina is capable only of scotopic vision. On the other hand, in very bright light, the rods become light-adapted to the point that they either become inoperative or overshadowed by the signals from the cones; in contrast, the cones find bright light especially suitable for optimal function. Some physiologists believe that the cones in bright light inhibit rod function by transmitting inhibiting signals through the horizontal cells to the synaptic bodies of the rods. At any rate, in bright light, function of the retina appears to be based almost entirely on cone detection of the light signals.

## FUSION OF FLICKERING LIGHTS BY THE RETINA

A flickering light is one whose intensity alternately increases and decreases rapidly. An instantaneous flash of light excites the visual receptors for as long as $1/10$ to $1/5$ second, and because of the *persistence* of excitation, rapidly successive flashes of light become *fused* together to give the appearance of being continuous. This effect is well known when one observes motion pictures or television. The images on the motion picture screen are flashed at a rate of 24 frames per second, while those of the television screen are flashed at a rate of 60 frames per second. As a result, the images fuse together, and continuous motion is observed.

The frequency at which flicker fusion occurs, called the *critical frequency for fusion,* varies with the light intensity. Figure 59–10 illustrates the effect of intensity of illumination on the critical fusion frequency. At a low intensity, fusion results even when the rate of flicker is as low as 2 to 6 per second. However, in bright illumina-

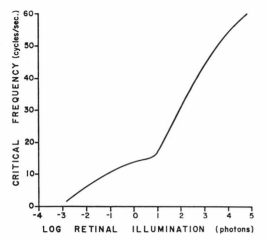

**Figure 59–10.** Relationship of intensity of illumination to the critical frequency for fusion.

tion, the critical frequency for fusion rises to as great as 60 flashes per second. This difference results at least partly from the fact that the cones, which operate mainly at high levels of illumination, can detect much more rapid alterations in illumination than can the rods, which are the important receptors in dim light.

# COLOR VISION

From the preceding sections, we know that different cones are sensitive to different colors of light. The present section is a discussion of the mechanisms by which the retina detects the different gradations of color in the visual spectrum.

## *THE TRI-COLOR THEORY OF COLOR PERCEPTION*

Many different theories have been proposed to explain the phenomenon of color vision, but they are all based on the well-known observation that the human eye can detect almost all gradations of colors when red, green, and blue monochromatic lights are appropriately mixed in different combinations.

The first important theory of color vision was that of Young, which was later expanded and given more experimental basis by Helmholtz. Therefore, the theory is known as the *Young-Helmholtz theory*. According to this theory, there are three different types of cones, each of which responds maximally to a different color.

As time has gone by, the Young-Helmholtz theory has been expanded, and more details have been worked out. It now is generally accepted as the mechanism of color vision.

**Spectral Sensitivities of the Three Types of Cones.** On the basis of psychological tests, the spectral sensitivities of the three different types of cones in human beings are essentially the same as the light absorption curves for the three types of pigment found in the respective cones. These were illustrated in Figure 59–7, and they can readily explain almost all the phenomena of color vision.

**Interpretation of Color in the Nervous System.** Referring again to Figure 59–7, one can see that a red monochromatic light with a wavelength of 610 millimicrons stimulates the red cones to a stimulus value of approximately 0.75 (75 per cent of the peak stimulation at optimum wavelength), while it stimulates the green cones to a stimulus value of approximately 0.13 and the blue cones not at all. Thus, the ratios of stimulation of the three different types of cones in this instance are 75:13:0. The nervous system interprets this set of ratios as the sensation of red. On the other hand, a monochromatic blue light with a wavelength of 450 millimicrons stimulates the red cones to a stimulus value of 0, the green cones to a value of 0.14 and the blue cones to a value of 0.86. This set of ratios—0:14:86—is interpreted by the nervous system as blue. Likewise, ratios of 100:50:0 are interpreted as orange-yellow and 50:85:15 as green.

This scheme also shows how it is possible for a person to perceive a sensation of yellow when a red light and a green light are shone into the eye at the same time, for this stimulates the red and green cones approximately equally, which gives a sensation of yellow even though no wavelength of light corresponding to yellow is present.

*Perception of White Light.* Approximately equal stimulation of all the red, green, and blue cones gives one the sensation of seeing white. Yet there is no wavelength of light corresponding to white; instead, white is a combination of all the wavelengths of the spectrum. Furthermore, the sensation of white can be achieved by stimulating the retina with a proper combination of only three chosen colors that stimulate the respective types of cones.

**Integration of Color Sensations by the Retina and the Brain.** From psychological studies, we know that the interpretation of color

**Figure 59–11.**   Two Ishihara charts: *Upper:* In this chart, the normal person reads "74," whereas the red-green color blind person reads "21." *Lower:* In this chart, the red-blind person (protanope) reads "2," while the green-blind person (deuteranope) reads "4." The normal person reads "42." (From Ishihara: Tests for Colour-Blindness. Tokyo, Kanehara and Co.)

is performed partly by the retina and partly by the brain. If a person places a monochromatic green filter in front of his left eye and a monochromatic red filter in front of his right eye the visual object appears mainly yellow. This integration of color sensations obviously could not be occurring in the retina because one retina is allowed to respond only to green light and the other only to red light. However, sensations perceived in this way are not as clearly mixed as those perceived when the two monochromatic lights are mixed in the same retina. Therefore, at least some degree of the interpretation of color occurs in the retina itself even before the light information is transmitted into the brain. Some of the known features of color signal processing in the neuronal cells of the retina will be presented in the following chapter.

## COLOR BLINDNESS

**Red-Green Color Blindness.** When a single group of color receptive cones is missing from the eye, the person is unable to distinguish some colors from others. As can be observed by studying Figure 59-7, if the red cones are missing, light of 525 to 625 millimicrons wavelength can stimulate only the green-sensitive cones, so that the *ratio* of stimulation of the different cones does not change as the color changes from green all the way through the red spectrum. Therefore, within this wavelength range, all colors appear to be the same to this "color blind" person.

On the other hand, if the green-sensitive cones are missing, the colors in the range from green to red can stimulate only the red-sensitive cones, and the person also perceives only one color within these limits. Therefore, when a person lacks either the red or green types of cones, he is said to be "red-green" color blind. However, when one or more types of cones are abnormal but still function partially, a person has "color weakness," instead of color blindness.

The person with loss of red cones is called a *protanope;* his overall visual spectrum is noticeably shortened at the long wavelength end because of lack of the red cones. The color blind person who lacks green cones is called a *deuteranope;* this person has a perfectly normal visual spectral width because the absent green cones operate in the middle of the spectrum where red or blue cones also operate.

**Blue Weakness.** Occasionally, a person has "blue weakness," which results from diminished or absent blue receptors. If we observe Figure 59-7 once again, we can see that the blue cones are sensitive to a spectral range almost entirely different from that of both the red and green cones. Therefore, if the blue receptors are completely absent, the person has a greater preponderance of green, yellow, orange, and red in his visual spectrum than he has of blue, thus producing this rarely observed type of color weakness or blindness.

**Tests for Color Blindness.** Tests for color blindness depend on the person's ability to distinguish various colors from each other and also on his ability to judge correctly the degree of contrast between colors. For instance, to determine whether or not a person is red-green color blind, he may be given many small tufts of wool whose colors encompass the entire visual spectrum. He is then asked to place those tufts that have the same colors in the same piles. If he is not color blind, he recognizes immediately that all the tufts have slightly different colors; however, if he is red-green color blind, he places the red, orange, yellow, and yellow-green colors all together as having essentially the same color.

*Stilling and Ishihara Test Charts.* A rapid method for determining color blindness is based on the use of spot-charts such as those illustrated in Figure 59-11. These charts are arranged with a confusion of spots of several different colors. In the top chart, the normal person reads "74," while the red-green color blind person reads "21." In the bottom chart, the normal person reads "42," while the red blind protanope reads "2," and the green blind deuteranope reads "4."

If one will study these charts while at the same time observing the spectral sensitivity curves of the different cones in Figure 59-7, he can readily understand how excessive emphasis can be placed on spots of certain colors by color blind persons in comparison with normal persons.

**Genetics of Color Blindness.** Color blindness is sex-linked and probably results from absence of appropriate color genes in the X chromosomes. This lack of color genes is a recessive trait; therefore, color blindness will not appear as long as another X chromosome carries the genes necessary for development of the respective color-receptive cones.

Because the male human being has only one X chromosome, all three color genes must be present in this single chromosome if he is not to be color blind. In approximately 1 out of every 50 times, the X chromosome lacks the red gene; in approximately 1 out of every 16 it lacks the green gene; and rarely it lacks the blue gene. This means, therefore, that 2 per cent of all men are red color blind (protanopes) and that 6 per cent are green color blind (deuteranopes), making a total of approximately 8 per cent who are red-green color blind. Because a female has two X chromosomes, red-green color blindness is a rare abnormality in the female, occurring in only 1 out or every 250 women.

## REFERENCES

Alpern, M.: Distal mechanisms of vertebrate color vision. *Ann. Rev. Physiol.,* 30:279, 1968.
Bitensky, M. W., Miki, N., Marcus, F. R., and Keirns, J. J.: The role of cyclic nucleotides in visual excitation. *Life Sci.,* 13:1451, 1973.

Brindley, G. S.: Physiology of the Retina and Visual Pathway, 2nd Ed. Baltimore, The Williams & Wilkins Company, 1970.

Brown, J. L.: Visual sensitivity. *Ann. Rev. Psychol., 24*:151, 1973.

Ciba Foundation: Color Vision: Physiology and Experimental Psychology. Boston, Little, Brown and Company, 1965.

Davson, H.: The Physiology of the Eye, 3rd Ed. New York, Academic Press, Inc., 1972.

Daw, N. W.: Neurophysiology of color vision. *Physiol. Rev., 53*:571, 1973.

Duke-Elder, S. (ed.): Diseases of the Retina. *In* Systems of Ophthalmology. Vol. 10. St. Louis, The C. V. Mosby Company, 1967.

Enright, J. T.: Stereopsis, visual latency, and three-dimensional moving pictures. *Amer. Sci., 58*:536, 1970.

Fine, B. S., and Yanoff, M.: Ocular Histology. Hagerstown, Md., Harper & Row, Publishers, 1972.

Gamow, R. I., and Harris, J. F.: The infrared receptors of snakes. *Sci. Amer., 228*:94, 1973.

Graymore, C. N. (ed.): Biochemistry of the Eye. New York, Academic Press, Inc., 1970.

Linksz, A.: An Essay on Color Vision and Clinical Color Vision Tests. New York, Grune & Stratton, Inc., 1964.

MacNichol, E. F., Jr.: Three-pigment color vision. *Sci. Amer., 211*:48, 1964.

Marks, W. B., Dobelle, W. H., and MacNichol, E. E., Jr.: Visual pigments of single primate cones. *Science, 143*:1181, 1964.

Maturana, H. R., Uribe, G., and Frenk, S.: A biological theory of relativistic color coding in the primate retina. *Arch. Biol. Med. Exp. (Santiago), 1*:1, 1968.

Michael, C. R.: Color vision. *New Engl. J. Med., 288*:724, 1973.

Millodot, M.: Variation of visual acuity in the central region of the retina. *Brit. J. Physiol. Opt., 27*:24, 1972.

Padgham, C. A., and Saunders, J. E.: The Perception of Light and Color. New York, Academic Press, Inc., 1975.

Ripps, H., and Weale, R. A.: The photophysiology of vertebrate color vision. *Photophysiology, 5*:127, 1970.

Stabell, B., and Stabell, U.: Chromatic rod vision. IX. A theoretical survey. *Vision Res., 13*:449, 1973.

Tomita, T., Miller, W. H., Hashimoto, Y., and Saito, T.: Electrical response of retinal cells as a sign of transport. *Exp. Eye Res., 16*:327, 1973.

Van der Tweel, L. H., and Estévez, O.: Subjective and objective evaluation of flicker. *Ophthalmologica, 169*:70, 1974.

Wald, G.: Proceedings: visual pigments and photoreceptors— review and outlook. *Exp. Eye Res., 18*:333, 1974.

Wasserman, G. S.: Invertebrate color vision and the tuned-receptor paradigm. *Science, 180*:268, 1973.

Werblin, F. S.: The control of sensitivity in the retina. *Sci. Amer., 228*:70, 1973.

Young, R. W.: Visual cells, *Sci. Amer., 223*:80, 1970.

Young, R. W.: Proceedings: Biogenesis and renewal of visual cell outer segment membranes. *Exp. Eye Res., 18*:215, 1974.

# | 60 |

# The Eye:
# III. Neurophysiology of Vision

## THE VISUAL PATHWAY

Figure 60–1 illustrates the visual pathway from the two retinae back to the *visual cortex*. After impulses leave the retinae they pass backward through the *optic nerves*. At the *optic chiasm* all the fibers from the opposite nasal halves of the retinae cross to the side where they join the fibers from the opposite temporal retinae to form the *optic tracts*. The fibers of each optic tract synapse in the *lateral geniculate body* and from here, the *geniculocalcarine fibers* pass through the *optic radiation*, or *geniculocalcarine tract*, to the *optic* or *visual cortex* in the calcarine area of the occipital lobe.

In addition, visual fibers pass to lower areas of the brain: (1) from the lateral geniculate body to the lateral thalamus, the superior colliculi, and the pretectal nuclei; (2) from the optic tracts directly to the superior colliculi; and (3) from the optic tracts directly into the pretectal nuclei of the brain stem.

## NEURAL FUNCTION OF THE RETINA

### NEURAL ORGANIZATION OF THE RETINA

The detailed anatomy of the retina was illustrated in Figure 59–1 of the preceding chapter, and Figure 60–2 illustrates the basic essentials of the retina's neural connections; to the left is the general organization of the neural elements in a peripheral retinal area and to the right the

organization of the foveal area. Note that in the peripheral region both rods and cones converge on *bipolar cells* which in turn converge on *ganglion cells*. In the fovea, where only cones exist, there is little or no convergence; instead, the cones are represented by approximately equal numbers of bipolar and ganglion cells.

Not emphasized in the figure is the fact that in both the outer and inner plexiform layers (where the neurons synapse with each other) there are many lateral connections between the different neural elements. Many of these lateral

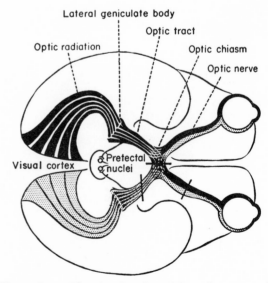

**Figure 60–1.** The visual pathways from the eyes to the visual cortex. (Modified from Polyak: The Retina. University of Chicago Press, 1941.)

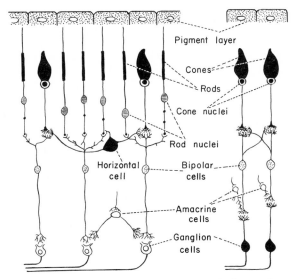

**Figure 60–2.** Neural organization of the retina: peripheral area to the left, foveal area to the right.

connections are collateral branches of the dendrites and axons of the bipolar and ganglion cells. But, in addition, two special types of cells are present in the inner nuclear layer that transmits signals laterally: (1) the *horizontal cells,* and (2) the *amacrine cells.*

Each retina contains about 125 million rods and 5.5 million cones; yet, as counted with the light microscope, only 900,000 optic nerve fibers lead from the retina to the brain. Thus, an average of 140 rods plus 6 cones converge on each optic nerve fiber. However, there are major differences between the peripheral retina and the central retina, for nearer the fovea, fewer and fewer rods and cones converge on each optic fiber, and the rods and cones both become slenderer. These two effects progressively increase the acuity of vision toward the central retina. In the very central portion, in the fovea, there are no rods at all. Also, the number of optic nerve fibers leading from this part of the retina is almost equal to the number of cones in the fovea, as shown to the right in Figure 60–2. This mainly explains the high degree of visual acuity in the central portion of the retina in comparison with the very poor acuity in the peripheral portions.

Another difference between the peripheral and central portions of the retina is that there is considerably greater sensitivity of the peripheral retina to weak light. This results partly from the fact that as many as 600 rods converge on the same optic nerve fiber in the most

peripheral portions of the retina; the signals from these rods summate to give intense stimulation of the peripheral ganglion cells. The rods are also more sensitive to weak light than are the cones.

## STIMULATION OF THE RODS AND CONES—THE RECEPTOR POTENTIAL

Stimulation of the rods and cones was discussed at length in the preceding chapter. Briefly, when light strikes either of these receptors, it decomposes a photochemical that in turn acts on the membrane of the *outer segment* of the receptor (the part that protrudes into the pigment layer) to cause a sustained *receptor potential* that lasts as long as the light continues. However, as pointed out in the previous chapter, this receptor potential is different from all other receptor potentials that have been recorded, for it is a hyperpolarization signal rather than a depolarization signal, and its intensity is proportional to the logarithm of light energy in contrast to the more linear response of many other receptors. This logarithmic response seems to be very important to vision, because it allows the eyes to detect contrasts in the image even when light intensities vary many thousand-fold, an effect that would not be possible if there were linear transduction of the signal.

Receptor potentials are transmitted unchanged through the bodies of the rods and cones. Neither the rods nor the cones generate action potentials. Instead, the receptor potentials themselves, acting at the synaptic bodies, induce signals directly in the successive neurons by *electrotonic* current flow across the synaptic gaps.

## STIMULATION OF THE BIPOLAR AND HORIZONTAL CELLS

The synaptic bodies of the rods and cones make intimate contact with the dendrites of both the bipolar and horizontal cells. The hyperpolarization that occurs in the rods and cones during stimulation (the receptor potential) is transferred directly to both the bipolar and the horizontal cells. Also, the excitation signal summates in both these cells; that is, the greater the number of rods and cones exciting a single bipolar or horizontal cell, the more intense will be the signal in each of these cells. Furthermore, in both the bipolar and the horizontal cells the transmitted signals are the hyperpolarization potentials themselves—not action potentials. Indeed, action potentials have never been recorded from horizontal cells and only under rare conditions have they been recorded in bipolar cells. Therefore, it is doubtful that any part of the normal physiological visual signal is transmitted in the form of action potentials by either of these two types of cells.

**Function of the Bipolar Cells.** The bipolar cells are the main transmitting link for the visual signal from the rods and cones to the ganglion cells. In the foveal region of the eye, the region of greatest visual acuity, there is approximately one bipolar cell for each cone, and also approximately one ganglion cell. In the peripheral region several cones converge on a single ganglion cell, and as many as several hundred rods converge. Thus, the peripheral retina is properly organized for great retinal sensitivity to very low light intensities, whereas the central retina (the foveal region) is organized for greatest possible visual acuity and faithfulness of transmission of the signals into the brain.

The bipolar cells are entirely excitatory cells. That is, they can only excite the ganglion cells.

**Function of the Horizontal Cells.** The horizontal cells lie in the inner nuclear layer and spread widely in the retina, transmitting signals laterally as far as several hundred microns. They are excited directly by the synaptic bodies of the rods and cones and in turn transmit signals mainly to bipolar cells located in areas lateral to the excited rods and cones.

The horizontal cells are *inhibitory* cells, in contrast to the bipolar cells, which are of an excitatory nature. Therefore, the signals transmitted by the horizontal cells from the excited rods and cones cause inhibition of the bipolar cells lateral to the excited point in the retina. That is, excited rods and cones transmit excitatory signals in a direct line through the bipolar cells in the area of excitation but transmit inhibitory signals through the surrounding bipolar cells.

The horizontal cells are of major importance for enhancing and helping to detect contrasts in the visual scene. They are probably also of importance in helping to differentiate colors. Both of these functions will be discussed in subsequent sections of this chapter.

**Stimulation and Function of the Amacrine Cells.** The amacrine cells, also located in the inner nuclear layer, are excited mainly by the bipolar cells but possibly also directly by the synaptic bodies of the rods and cones. These cells in turn *inhibit* the ganglion cells. However, their response is a *transient* one rather than the steady, continuous response of the bipolar and horizontal cells. That is, when the photoreceptors are first stimulated, the inhibitory signal transmitted by the amacrine cells is very intense; this signal dies away to almost nothing in a fraction of a second.

Two other differences between the amacrine cells and the bipolar and horizontal cells are that (1) the signal of the amacrine cells is a depolarization signal rather than a hyperpolarization signal, and (2) the amacrine cells transmit their signals to the ganglion cells in two different ways—either as a simple electrotonic depolarization or as occasional action potentials that occur at the onset of the initial depolarization.

## STIMULATION OF THE GANGLION CELLS

**Spontaneous, Continuous Discharge of the Ganglion Cells.** The ganglion cells transmit their signals through the optic nerve fibers to the brain in the form of action potentials. These cells, even when unstimulated, transmit continuous nerve impulses at an average rate of about 5 per second. The visual signal is superimposed onto this basic level of ganglion cell stimulation. It can be either an excitatory signal, with the number of impulses increasing to greater than 5 per second, or it can be an inhibitory signal, with the number of nerve impulses decreasing to below 5 per second—often all the way to zero.

**Summation of the Ganglion Cells by Signals from the Bipolar Cells, the Horizontal Cells, and the Amacrine Cells.** The bipolar cells transmit the main direct *excitatory* information from the rods and the cones to the ganglion cells: the horizontal cells transmit *inhibitory* information from laterally displaced rods and cones; the amacrine cells transmit direct but short-lived transient signals that signal a *change* in the level of illumination of the retina. Thus, each of these three types of cells performs a separate function in stimulating the ganglion cells.

## DIFFERENT TYPES OF SIGNALS TRANSMITTED BY THE GANGLION CELLS THROUGH THE OPTIC NERVE

**Transmission of Luminosity Signals.** The ganglion cells are of several different types, and they also have different patterns of stimulation by the bipolar, horizontal, and amacrine cells. Some of the ganglion cells respond to the intensity (*luminosity*) of the light falling on the photoreceptors. The rate of impulses from these cells remains at a level greater than the natural rate of firing as long as the luminosity is high. It is the signals from these cells that apprise the brain of the overall level of light intensity of the observed scene. The light and dark flat areas of the scene are also transmitted in this way, but far more information concerning the form of the scene is generally transmitted in terms of contrast borders, as will be discussed below.

**Transmission of Signals Depicting Contrasts in the Visual Scene—the Process of Lateral Inhibition.** A large share of the ganglion cells barely respond to the actual level of illumination of the scene; instead they respond only to contrast borders in the scene. Since it seems that this is the major means by which the form of the scene is transmitted to the brain, let us explain how this process occurs.

When flat light is applied to the entire retina—that is, when all the photoreceptors are stimulated equally by the incident light— the contrast type of ganglion cell is neither stimulated nor inhibited. The reason for this is that the signals transmitted *directly* from the photoreceptors through the bipolar cells are excitatory, whereas the signals transmitted from *laterally displaced* photoreceptors through the horizontal cells are inhibitory. These two effects neutralize each other. Now, let us examine what happens when a contrast border occurs in the visual scene. Figure 60–3 shows a spot of light shining on the retina and exciting

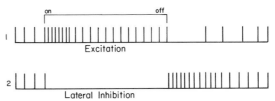

**Figure 60–4.** Responses of ganglion cells to light in (1) an area excited by a spot of light and (2) an area immediately adjacent to the excited spot; the ganglion cells in this area are inhibited by the mechanism of lateral inhibition. (Modified from Granit: Receptors and Sensory Perception: A Discussion of Aims, Means, and Results of Electrophysiological Research into the Process of Reception. Yale University Press, 1955.)

the photoreceptors. The signal is transmitted directly through the bipolar cells and thence to the ganglion cells in the same area to cause excitation. At the same time these same excited photoreceptors transmit *lateral inhibitory* signals through the horizontal cells to the surrounding bipolar and ganglion cells. Thus, when there is a contrast border, with stimulated photoreceptors on one side of the border and unstimulated photoreceptors on the opposite side of the border, the mutual cancellation of the excitatory signals through the bipolar cells and the inhibitory signals through the horizontal cells no longer occurs. Consequently, the ganglion cells on either side of this border become either excited or inhibited—excited on the light side of the border and inhibited on the dark side.

A linear contrast border acts in almost identically the same way as does a spot of light; all of the ganglion cells on the light side of the border are stimulated and all the dark side are inhibited. On the other hand, all of the contrast type of ganglion cells that lie in the flatly illuminated portion of the scene still remain unstimulated.

This process of contrasting an excitatory signal with an inhibitory signal is exactly the same as the process of *lateral inhibition* that occurs in most other types of sensory signal transmission. The process is a mechanism used by the nervous system for contrast enhancement.

When impulses are recorded from ganglion cells, it is possible to observe the effects of contrast borders on the signals transmitted in the optic nerve fibers. Figure 60–4 illustrates this effect. The upper recording is from a ganglion cell located in the area excited by a spot of light, whereas the lower recording is from a ganglion cell located a few hundred microns to the side of the spot. Note that when the spot of light is turned on the ganglion cell in the center of the

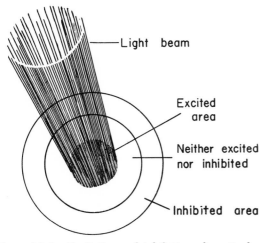

**Figure 60–3.** Excitation and inhibition of a retinal area caused by a small beam of light.

light becomes excited while the ganglion cell located laterally becomes inhibited.

**Detection of Instantaneous Changes in Light Intensity—the On-Off Response.** Many of the ganglion cells are especially excited by *change* in light intensity; this effect most often occurs in the same ganglion cells that transmit contrast border signals. For instance, the upper tracing of Figure 60–4 shows that when the light was first turned on the ganglion cell became strongly excited for a fraction of a second, and then the level of excitation diminished. The bottom tracing shows that a second ganglion cell located in the dark area lateral to the spot of light was deeply depressed at the same time. Then when the light was turned off, exactly the opposite effects occurred. Thus, the responses of these two types of cells are called the "on-off" and the "off-on" responses.

This ability of the retina to detect and transmit signals related to *change* in light intensity is caused by a rapid phase of "adaptation" of some of the neurons in the visual chain. Since this effect is extremely marked in the amacrine cells, it is believed that the amacrine cells are peculiarly adapted to detecting light intensity change.

This capability to detect change in light intensity is especially well developed in the peripheral retina. For instance, a minute gnat flying across the peripheral field of vision is instantaneously detected. On the other hand, the same gnat sitting quietly in the peripheral field of vision remains entirely below the threshold of visual detection.

**Transmission of Color Signals by the Ganglion Cells.** A single ganglion cell may be stimulated by a number of cones or by only a very few. When all three types of cones—the red, blue, and green types—all stimulate the same ganglion cell, the signal transmitted through the ganglion cell is the same for any color of the spectrum. Therefore, this signal plays no role in the detection of the different colors. Instead, it is called a "luminosity" signal.

On the other hand, many of the ganglion cells are excited by only one color type of cone but are inhibited by a second type. For instance, this frequently occurs for the red and green cones, red causing excitation and green causing inhibition—or, vice versa, that is, green causing excitation and red inhibition. The same type of reciprocal effect also occurs between some blue cones and either red or green cones.

The mechanism of this opposing effect of colors is the following: One color type of cones excites the ganglion cell by the direct excitatory route through a bipolar cell, while the other color type excites the ganglion cell by the indirect inhibitory route through a horizontal cell.

The importance of these color-contrast mechanisms is that they represent a mechanism by which the retina itself differentiates colors. Thus each color-contrast type of ganglion cell is excited by one color but inhibited by the opposite color. Therefore, the process of color analysis begins in the retina and is not entirely a function of the brain.

## REGULATION OF RETINAL FUNCTION BY CENTRIFUGAL FIBERS FROM THE CENTRAL NERVOUS SYSTEM

The sensitivity of the retina is at least partially controlled by signals from the central nervous system. Centrifugal fibers pass in the retrograde direction in the optic nerve from the brain to the retina and synapse directly with the ganglion cells. Stimulation of specific areas in the brain can enhance in some instances and diminish in other instances the sensitivities of specific areas of the retina. It is supposed that this represents a mechanism by which the central nervous system can direct one's attention to specific portions of the visual field.

# FUNCTION OF THE LATERAL GENICULATE BODY

**Anatomical Organization of the Lateral Geniculate Nuclei.** Each lateral geniculate body is composed of six nuclear layers. Layers 2, 3, and 5 (from the surface inward) receive signals from the temporal portion of the ipsilateral retina, while layers 1, 4, and 6 receive signals from the nasal retina of the opposite eye.

All layers of the lateral geniculate body relay visual information to the *visual cortex* through the *geniculocalcarine tract*.

The pairing of layers from the two eyes probably plays a major role in *fusion of vision*, because corresponding retinal fields in the two eyes connect with respective neurons that are approximately superimposed over each other in the successive layers. Also, with a little imagination, one can postulate that interaction between the successive layers could be part of the mechanism by which stereoscopic visual depth perception occurs, because this depends on comparing the visual images of the two eyes and determining their slight differences, as was discussed in Chapter 58.

### Characteristics of the Visual Signals in the Lateral Geniculate Body

The signals recorded in the relay neurons of the lateral geniculate body are similar to those recorded in the ganglion cells of the retina. A few of the neurons transmit luminosity signals, whereas the majority transmit signals depicting only contrast borders in the visual image; also, many of the neurons are particularly responsive to movement of objects across the visual scene. However, the signals of the geniculate neurons are different from those in the ganglion cells in that a greater number of the complex interactions are found. That is, a much higher percentage of the neurons respond to contrast in the visual scene or to movement. These more complex reactions presumably result from convergence of excitatory and inhibitory signals from two or more ganglion cells on the relay neurons of the lateral geniculate body.

**Color Signals in the Lateral Geniculate Body.** Signals related to black and white vision are mainly found in layers 1 and 2 of the lateral geniculate body, while color signals occur mainly in layers 3 through 6. A few of the neurons in these last four layers respond to all colors and therefore transmit "white-light" information. However, about three-quarters of the neurons respond to "opponent colors;" that is, a given cell may be excited by red but be inhibited by green. Another cell will operate exactly oppositely. These are called the "red-green" cells. Still other lateral geniculate body cells respond to the opponent colors blue and yellow, with one of these exciting the cell and the other inhibiting it. The mechanism of this effect is the same as in the retina; that is, the effect is caused by convergence of signals from red and green cones onto the same neuron, one of these signals causing inhibition while the other causes excitation.

By various interactions of this type, the color information in the visual scene is gradually dissected, and the information is then transmitted to other sections of the brain as signals that express the *ratios* of different colors in the different areas of the visual scene.

# FUNCTION OF THE PRIMARY VISUAL CORTEX

The ability of the visual system to detect spatial organization of the visual scene—that is, to detect the forms of objects, brightness of the individual parts of the objects, shading, and so forth—is dependent on function of the *primary visual cortex*, the anatomy of which is illustrated in Figure 60–5. This area lies mainly in the calcarine fissure located bilaterally on the medial aspect of each occipital cortex. Specific points of the retina connect with specific points of the visual cortex, the right halves of the two respective retinae connecting with the right visual cortex and the left halves connecting with the left visual cortex. The macula is represented at the occipital pole of the visual cortex and the peripheral regions of the retina are represented in concentric circles farther and farther forward from the occipital pole. The upper portion of the retina is represented superiorly in the visual cortex and the lower portion inferiorly. Note the large area of cortex receiving signals from the macular region of the retina. It is in this region that the fovea is represented, which gives the highest degree of visual acuity.

## DETECTION OF LINES AND BORDERS BY THE PRIMARY VISUAL CORTEX

If a person looks at a blank wall, only a few neurons of the primary visual cortex will be stimulated, whether the illumination of the wall be bright or weak. Therefore, the question must be asked: What does the visual cortex do? To answer this question, let us now place on the wall a large black cross such as that illustrated to the left in Figure 60–6. To the right is illustrated the spatial pattern of the greater majority of the excited neurons that one finds in the visual cortex. *Note that the areas of excitation occur along the sharp borders of the visual pattern*. Thus, by the time the visual signal is recorded in the primary visual cortex, it is concerned mainly with the *contrasts* in the visual

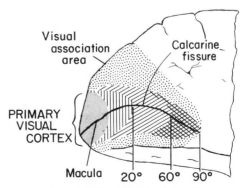

**Figure 60–5.** The visual cortex.

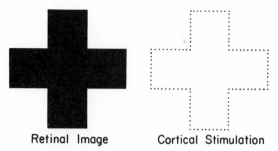

Retinal Image    Cortical Stimulation

**Figure 60–6.** Pattern of excitation occurring in the visual cortex in response to a retinal image of a black cross.

scene rather than with the flat areas. At each point in the visual scene where there is a change from dark to light or light to dark, the corresponding area of the primary visual cortex becomes stimulated. The intensity of stimulation is determined by the *gradient of contrast*. That is, the greater the sharpness in the contrast border and the greater the difference in intensities between the light and dark areas, the greater is the degree of stimulation.

Thus, the *pattern of contrasts* in the visual scene is impressed upon the neurons of the visual cortex, and this pattern has a spatial orientation roughly the same as that of the retinal image.

**Detection of Orientation of Lines and Borders.** Not only does the visual cortex detect the *existence* of lines and borders in the different areas of the retinal image, but it also detects the *orientation* of each line or border—that is, whether it is a vertical or horizontal border, or lies at some degree of inclination. The mechanism of this effect is the following:

In each small area of the visual cortex there are multiple neurons arranged in columns, each column having a diameter of about 0.5 to 1 mm. and extending downward from the surface of the cortex through its six layers. The signals arriving in the cortex from the lateral geniculate body terminate in layer 4, and from here secondary signals spread upward or downward in the column and eventually also to adjacent columns. Some columns of neurons respond to vertical lines or borders, others to lines or borders slightly angulated, still others to lines or borders even more angulated, and so forth. This process of selective stimulation of the separate columns is presumably caused by convergence of signals from two or more nerve fibers arriving from the lateral geniculate body. The effect is similar to that which occurs in the retina, one of the fibers causing inhibition and the other exci-

tation. If the orientation of the border between the dark area and the light area in the visual scene corresponds with the orientation of the neural connections of inhibitory and excitatory input fibers to a given column of the visual cortex, the column would be excited. By having many different cortical columns with slightly different synaptic connections, such interactions could lead to selective stimulation of one column for one orientation of a border, another column for another orientation, and so forth. This represents another stage in the mechanism for deciphering information in the visual scene.

**Detection of Length of Lines.** Still a higher order for deciphering information from the visual scene is the ability of yet other neuronal columns of cells to be stimulated by lines or borders of specific lengths. An interesting feature of this type of detection is that the lines can be displaced laterally from each other to a slight extent in the visual field and the detection process will continue to occur. This type of detection presumably results from convergence of signals within the visual cortex from the first-stimulated columns of neurons.

To perform all of these functions for deciphering visual information, the primary visual cortex has approximately 200 million neurons—200 times as great as the number of optic nerve fibers from each retina.

**Analysis of Color by the Visual Cortex.** One finds in the primary visual cortex specific cells that are stimulated by color intensity or by contrasts of the opponent colors, the red-green opponent colors and the blue-yellow opponent colors. These effects are almost identical to those found in the lateral geniculate body. However, the proportion of cells excited by the opponent color contrasts is vastly reduced from the proportion found in the lateral geniculate body. Since neuronal excitation by color contrasts is a means of deciphering color, it is believed that the primary visual cortex is concerned with an even higher order of detection of color than simply the deciphering of color itself, a process that seems to have been mainly completed by the time the signals have passed through the lateral geniculate body.

## PERCEPTION OF LUMINOSITY

Although most of the neurons in the visual cortex are mainly responsive to contrasts caused by lines, borders, moving objects, or opponent colors in the visual scene, a few are directly responsive to the levels of luminosity in the different areas of the visual scene. Presumably, it is these cells that detect flat areas in the scene and also the overall level of luminosity.

### EFFECT OF REMOVING THE PRIMARY VISUAL CORTEX

Removal of the primary visual cortex in human beings always causes complete blindness for detection of visual patterns. However, a few such persons have displayed very slight capability for detection of crude levels of light intensity. In lower level monkeys this ability to detect light intensity is strongly preserved, with even the capability to determine very crude signals of form, especially visual images related to movement.

### TRANSMISSION OF VISUAL INFORMATION INTO OTHER REGIONS OF THE CEREBRAL CORTEX

The primary visual cortex is located in the *calcarine fissure area*, mainly on the medial aspect of each hemisphere but also spreading slightly over the occipital pole. It is known as *Brodmann area 17* of the cortex, and it is also called the *striate area* because of its striated appearance to the naked eye.

Signals from the striate area project laterally in the occipital cortex into area 18 and then into area 19, as illustrated in Figure 60–7. These areas are frequently called the *prestriate areas*, and they are also known as *visual association areas*. In reality, they are simply the loci for additional processing of visual information.

The visual projection images in areas 18 and 19 are organized into columns of cells in the same manner as that described earlier for the primary visual cortex. There is a topographic representation of the visual field as also found in the primary visual cortex. However, the neuronal cells respond to more complex patterns than do those in the primary visual cortex. For instance, some cells are stimulated by simple geometric patterns, such as curving borders, angles, and so forth. It is presumably these progressively more complex interpretations that eventually decode the visual information, giving the person his overall impression of the visual scene that he is observing.

**Projection of Visual Information into the Temporal Cortex.** From areas 18 and 19 visual signals next proceed to the posterior portion of the inferior temporal gyrus and from here to the posterior portion of the medial temporal gyrus, areas 20 and 21 of the cortex. In these areas simple visual patterns such as lines, borders, angles, and so forth fail to cause excitation of specific neurons. The degree of integration of visual information at this level, such as interpretation of letters or words, is presumably much higher.

**Effects of Destruction of the Visual Association Areas.** Human beings who have destructive lesions of any of the visual association areas, areas 18 and 19 in the occipital cortex, or areas 20 and 21 in the temporal cortex, have difficulty with certain types of visual perception and visual learning. For instance, destruction in areas 18 and 19 generally makes it difficult to perceive form, such as shapes of objects, their sizes, and their meanings. It is believed that this type of lesion can cause the abnormality known as *dyslexia* or *word blindness*, which means that the person has difficulty understanding the meanings of words that he sees.

Destruction of the temporal projections of the visual system makes it especially difficult for animals, and presumably for human beings as well, to learn tasks that are based upon visual perceptions. For instance, the person might be able to see his plate of food perfectly well but be unable to utilize this visual information to direct his fork toward his food. Yet, when he feels the plate with his other hand, he can use stereotaxic information from the somesthetic cortex to direct the fork accurately.

**Effects of Electrical Stimulation of the Visual Areas.** Electrical stimulation in the primary visual cortex or in the parietal association areas, areas 18 and 19, causes a person to have simple optic auras—that is, flashes of light, simple colors, or simple forms such as lines, stars, or so forth—but he does not see complicated forms.

Stimulation of the visual association areas of the temporal cortex, areas 20 and 21, occasionally elicits complicated visual perceptions, sometimes causing the person to see a scene that he had known many years before. This is in keeping with the interpretative function of this area for visual function.

## THE FIELDS OF VISION; PERIMETRY

The *field of vision* is the area seen by an eye at a given instant. The area seen to the nasal side is called the *nasal field of vision*, and the area seen to the lateral side is called the *temporal field of vision*.

To diagnose blindness in specific portions of the

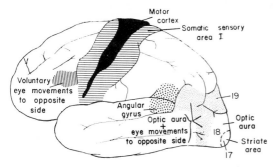

**Figure 60–7.** The visual association fields and the cortical areas for control of eye movements.

retinae, one charts the field of vision for each eye by a process known as *perimetry*. This is done by having the subject look with one eye toward a central spot directly in front of the eye. Then a small dot of light or a small object is moved back and forth in all areas of the field of vision, both laterally and nasally and upward and downward, and the person indicates when he can see the spot of light or object and when he cannot see it. At the same time, a chart (Fig. 60–8) is made for the eye, showing the areas in which the subject can see the spot and in which he cannot see it. Thus, the field of vision is plotted.

In all perimetry charts, a blind spot caused by lack of rods and cones in the retina over the optic disc is found approximately 15 degrees lateral to the central point of vision, as illustrated in the figure.

**Abnormalities in the Fields of Vision.** Occasionally blind spots are found in other portions of the field of vision besides the optic disc area. Such blind spots are called *scotomata;* they frequently result from allergic reactions in the retina or from toxic conditions, such as lead poisoning or even excessive use of tobacco.

Still another condition that can be diagnosed by perimetry is *retinitis pigmentosa*. In this disease, portions of the retina degenerate and excessive melanin pigment deposits in the degenerated areas. Retinitis pigmentosa generally causes blindness in the peripheral field of vision first and then gradually encroaches on the central areas.

*Effect of Lesions in the Optic Pathway on the Fields of Vision.* One of the most important uses of perimetry is for localization of lesions in the visual nervous system. Lesions in the optic nerve, in the optic chiasm, in the optic tract, and in the geniculocalcarine tract all cause blind areas in the visual fields, and the ''patterns'' of these blind areas indicate the location of the lesion in the optic pathway.

Destruction of an entire *optic nerve* obviously causes blindness of the respective eye. Destruction of the *optic chiasm,* as shown by the longitudinal line

across the chiasm in Figure 60–1, prevents the passage of impulses from the nasal halves of the two retinae to the opposite optic tracts. Therefore, the nasal halves are both blinded, which means that the person is blind in both temporal fields of vision; this condition is called *bitemporal hemianopsia.* Such lesions frequently result from tumors of the adenohypophysis pressing upward on the optic chiasm.

Interruption of an *optic tract,* which is also shown by a line in Figure 60–1, denervates the corresponding half of each retina on the same side as the lesion, and, as a result, neither eye can see objects to the opposite side. This condition is known as *homonymous hemianopsia.* Destruction of the *optic radiation* or the *visual cortex* of one side also causes homonymous hemianopsia. A common condition that destroys the visual cortex is thrombosis of the posterior cerebral artery, which infarcts the occipital cortex except for the foveal area, thus sparing central vision.

One can differentiate a lesion in the optic tract from a lesion in the geniculocalcarine tract or visual cortex by determining whether impulses can still be transmitted into the pretectal nuclei to initiate a pupillary light reflex. To do this, light is shown onto the blinded half of one of the retinae, and, if a pupillary light reflex can still occur, it is known that the lesion is at or beyond the lateral geniculate body, that is, in the geniculocalcarine tract or visual cortex. However, if the light reflex is also lost, then the lesion is in the optic tract itself.

# EYE MOVEMENTS AND THEIR CONTROL

To make use of the abilities of the eye, almost equally as important as the system for interpretation of the visual signals from the eyes is the cerebral control system for directing the eyes toward the object to be viewed.

**Muscular Control of Eye Movements.** The eye movements are controlled by three separate pairs of muscles shown in Figure 60–9: (1) the medial and lateral recti, (2) the superior and inferior recti, and (3) the superior and inferior obliques. The medial and lateral recti contract reciprocally to move the eyes from side to side. The superior and inferior recti contract reciprocally to move the eyes upward or downward. And the oblique muscles function mainly to rotate the eyeballs to keep the visual fields in the upright position.

**Neural Pathways for Control of Eye Movements.** Figure 60–9 also illustrates the nuclei of the third, fourth, and sixth cranial nerves and their innervation of the ocular muscles. Shown, too, are the interconnections among these three nuclei through the *medial longitudinal fas-*

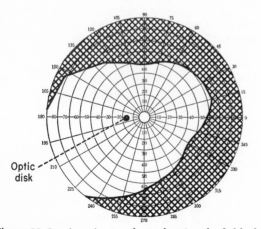

**Figure 60–8.** A perimetry chart, showing the field of vision for the left eye.

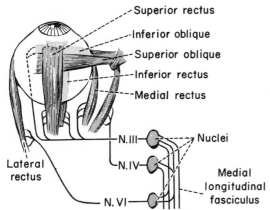

**Figure 60–9.**  The extraocular muscles of the eye and their innervation.

*ciculus.* Either by way of this fasciculus or by way of other closely associated pathways, each of the three sets of muscles to each eye is *reciprocally* innervated so that one muscle of the pair relaxes while the other contracts.

Figure 60–10 illustrates cortical control of the oculomotor apparatus, showing spread of signals from the occipital visual areas through occipitotectal and occipitocollicular tracts into the pretectal and superior colliculus areas of the brain stem. In addition, a frontotectal tract passes from the frontal cortex into the pretectal area. From both the pretectal and the superior colliculus areas, the oculomotor control signals then pass to the nuclei of the oculomotor nerves. Finally, strong signals are also transmitted into the oculomotor system from the vestibular nuclei by way of the medial longitudinal fasciculus.

## CONJUGATE MOVEMENT OF THE EYES

Simultaneous movement of both eyes in the same direction is called *conjugate movement* of the eyes. The signals to the two eyes to cause precise synchronized movement of both eyes at the same time probably originate in the reticular nuclei of the mesencephalon and pons in close association with the oculomotor nuclei. However, the conjugate

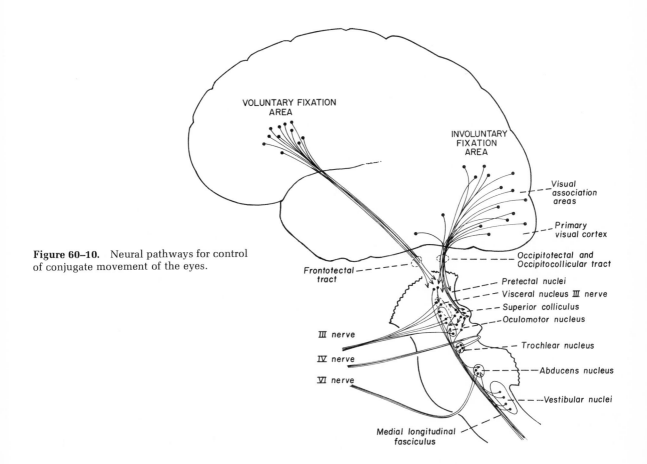

**Figure 60–10.**  Neural pathways for control of conjugate movement of the eyes.

movements are under strong control of the cerebral cortex, which transmits signals into the reticular nuclei by way of the pretectal nuclei and the superior colliculi. Signals from the vestibular nuclei also elicit conjugate movement of the eyes, probably by sending signals through the medial longitudinal fasciculi.

## FIXATION MOVEMENTS OF THE EYES

Perhaps the most important conjugate movements of the eyes are those that cause the eyes to "fix" on a discrete portion of the field of vision.

Fixation movements are controlled by two entirely different neuronal mechanisms. The first of these allows the person to move his eyes voluntarily to find the object upon which he wishes to fix his vision; this is called the *voluntary fixation mechanism*. The second is an involuntary mechanism that holds the eyes firmly on the object once it has been found; this is called the *involuntary fixation mechanism*.

The voluntary fixation movements are controlled by a small cortical field located bilaterally in the premotor cortical regions of the frontal lobes, as illustrated in Figure 60–10. Bilateral dysfunction or destruction of these areas makes it difficult or almost impossible for the person to "unlock" his eyes from one point of fixation and then move them to another point. It is usually necessary for him to blink his eyes or put his hand over his eyes for a short time, which then allows him to move the eyes.

On the other hand, the fixation mechanism that causes the eyes to "lock" on the object of attention once it is found is controlled by the *eye fields of the occipital cortex*—mainly area 19—which are also illustrated in Figure 60–10. When these areas are destroyed bilaterally, the person has difficulty or becomes completely unable to keep his eyes directed toward a given fixation point.

To summarize, the posterior eye fields automatically "lock" the eyes on a given spot of the visual field and thereby prevent movement of the image across the retina. To unlock this visual fixation, voluntary impulses must be transmitted from the "voluntary" eye fields located in the frontal areas.

**Mechanism of Fixation.** Visual fixation results from a negative feedback mechanism that prevents the object of attention from leaving the foveal portion of the retina. The eyes even normally have three types of continuous but almost imperceptible movements: (1) a *continuous tremor* at a rate of 30 to 80 cycles per second caused by successive contractions of the motor units in the ocular muscles, (2) a *slow drift* of the eyeballs in one direction or another, and (3) sudden *flicking movements* which are controlled by the involuntary fixation mechanism. When a spot of light has become fixed on the foveal region of the retina, the tremorous movements cause the spot to move back and forth at a rapid rate across the cones, and the drifting movements cause it to drift slowly across the cones. However, each time the spot of light drifts as far as the edge of the fovea, a sudden

flicking movement moves the spot away from this edge back toward the center, which is an automatic response to move the image back toward the central portion of the fovea. These drifting and flicking motions are illustrated in Figure 60–11, which shows by the dashed lines the slow drifting across the retina and by the solid lines the flicks that keep the image from leaving the foveal region.

**Fixation of the Eyes on Important Visual Highlights—Role of the Superior Colliculi.** The eyes have an automatic capability for instantaneously *fixing* on an important highlight in the visual field. However, this capability is mostly lost when the superior colliculi are destroyed even though most other conjugate movements of the eyes occur normally in the absence of the superior colliculi. The signals for fixation on the visual scene originate in the visual areas of the occipital cortex, primarily in the visual association areas, then pass to the superior colliculi, from there to reticular areas around the oculomotor nuclei, and thence into the motor nuclei themselves.

The superior colliculi are also important in causing sudden turning of the eyes to the side when a flash of light or some other sudden visual disturbance occurs on that side. In addition, signals are transmitted through the superior colliculi into the medial longitudinal fasciculus and other areas of the brain stem to cause turning of the whole head and perhaps even of the whole body toward the direction of the light. Other types of disturbances besides visual disturbances, such as strong sounds or even stroking the side of the body, will cause similar turning of the eyes, head, and body if the superior colliculi are intact. This effect, however, is absent or severely disturbed when the superior colliculi are destroyed. Therefore, it is frequently said that the superior

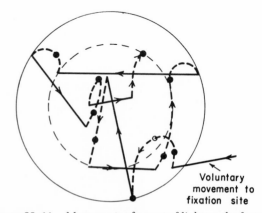

Voluntary movement to fixation site

**Figure 60–11.** Movements of a spot of light on the fovea, showing sudden "flicking" movements to move the spot back toward the center of the fovea whenever it drifts to the foveal edge. (The dashed lines represent slow drifting movements, and the solid lines represent sudden flicking movements.) (Modified from Whitteridge: Handbook of Physiology, Vol. 2, Sec. 1. Baltimore, The Williams & Wilkins Co., 1960, p. 1089.)

colliculi play an important role in orienting the eyes, the head, and the body with respect to external signals—visual signals, auditory signals, somatic signals, and perhaps even others.

**Saccadic Conjugate Movement of the Eyes.** When the visual scene is moving continually before the eyes, such as when a person is riding in a car or when he is turning around, the eyes fix on one highlight after another in the visual field, jumping from one to the next at a rate of two to three jumps per second. These jumps are called *saccades*. The saccades occur so rapidly that not more than 10 per cent of the total time is spent in moving the eyes, 90 per cent of the time being allocated to the fixation sites. Also, the brain suppresses the visual image during the saccades so that one is completely unconscious of the movements from point to point.

*Saccadic Movements During Reading.* During the process of reading, a person usually makes several saccadic movements of the eyes for each line. In this case the visual scene is not moving past the eyes, but the eyes are trained to scan across the visual scene to extract the important information. Similar saccades occur when a person observes a painting, except that the saccades occur in one direction after another from one highlight of the painting to another, then another, and so forth.

**Fixation on Moving Objects—"Pursuit Movements."** The eyes can also remain fixed on a moving object, which is called *pursuit movement*. A highly developed cortical mechanism automatically detects the course of movement of an object and then gradually develops a similar course of movement of the eyes. For instance, if an object is moving up and down in a wavelike form at a rate of several times per second, the eyes at first may be completely unable to fixate on it. However, after a second or so the eyes begin to jump coarsely in approximately the same pattern of movement as that of the object. Then after a few more seconds, the eyes develop progressively smoother and smoother movements and finally follow the course of movement almost exactly. This represents a high degree of automatic, subconscious computational ability by the cerebral cortex.

*Opticokinetic Nystagmus.* A particularly important type of pursuit movement is that called *nystagmus,* which allows the eyes to fixate on successive points in a continuously moving scene. For instance, if a person is looking out the window of a train, his eyes fix on successive points in the visual scene. To do this, the eyes fixate on some object and move slowly backward as the object also moves backward. When the eyes have moved far to the side, they automatically jump forward to fix on a new object, which is followed again by slow movement in the backward direction. This type of movement in one direction is called *opticokinetic nystagmus.* The slow movement in one direction is called the *slow component* of the nystagmus, while the rapid movement in the other direction is called the *fast component.*

Optikokinetic nystagmus obviously is a type of pursuit movement. That is, after an initial second or

so of orientation, the visual system automatically calculates the course and rate of movement of the visual scene, then follows this exactly until the eyes reach a lateral limit, at which time they jump to a new point in the scene.

**Vestibular Control of Eye Movements.** Another type of eye movement is elicited by stimulation of the vestibular apparatus. The vestibular nuclei are connected directly with the brain stem nuclei that control ocular movements, and, any time the head is accelerated in a vertical, longitudinal, lateral or angular direction, an immediate compensatory motion of the eyes occurs in the opposite direction. This allows the eyes to remain fixed on an object of attention despite rapid movements of the body or head.

Vestibular control of the eyes is especially valuable when a person is subjected to jerky motions of his body. For instance, when a person with bilateral destruction of his vestibular apparatuses rides over rough roads, he has extreme difficulty fixing his eyes on the road or on any horizontal scene. The optikokinetic type of pursuit movement is not capable of keeping the eyes fixed under such conditions because it has a latent period of about one-fifth second before the direction of movement can be detected and followed by the eyes. On the other hand, the vestibular type of eye movement has a short latent period, measured in hundredths of a second rather than one-fifth second.

Also when a person begins to rotate, his vestibular apparatuses cause a vestibular type of nystagmus. That is, the eyes lag behind the rotating head, then jump forward, then lag behind again, jump again, and so forth. This type of nystagmus is discussed in Chapter 52 in relation to vestibular function.

**Pathologic Types of Nystagmus.** Occasionally, abnormalities occur in the control system for eye movements that cause continuous nystagmus despite the fact that neither the visual scene nor the body is moving. This is likely to occur when one of the vestibular apparatuses is damaged or when severe damage is sustained in the deep nuclei of the cerebellum. This is discussed further in Chapter 53.

Another pathologic type of eye movement that is sometimes called nystagmus occurs when the foveal regions of the two eyes have been destroyed or when the vision in these areas is greatly weakened. In such a condition the eyes attempt to fix the object of attention on the foveae but always overshoot the mark because of foveal insensitivity. Therefore, they oscillate back and forth but never achieve foveal fixation. Even though this condition is known clinically as a type of nystagmus, physiologically it is completely different from the nystagmus that keeps the eyes fixed on a moving scene.

## FUSION OF THE VISUAL IMAGES

To make the visual perceptions more meaningful and also to aid in depth perception by the mechanism of stereopsis, which was discussed in Chapter 58, the visual images in the two eyes normally *fuse* with each other on "corresponding points" of the two retinae. Furthermore, three different types of fusion are required: lateral fusion, vertical fusion, and torsional fusion (same rotation of the two eyes about their optical axes.)

Fusion of the images of the two eyes probably results from two mechanisms in the visual system. First, crude fusion is inherent in the newborn baby because of inherited conjugate movements of the eyes. Second, as the visual system becomes more and more developed, the ability of the two eyes to fixate on the same object of attention causes even greater accuracy of fusion. Then, as the visual system develops its pursuit abilities, the pathways for controlling the conjugate movements of the two eyes normally develop equally. And the more the two eyes are moved in unison with each other the more "set" become the patterns of movement for the eyes, and, consequently, the more exact becomes the degree of fusion that can be maintained regardless of the position or rapidity of movement of the eyes.

Both the lateral geniculate body and the visual cortex play very important roles in this process of fusion. It was pointed out earlier in the chapter that corresponding points of the two retinae transmit visual signals, respectively, to successive nuclear layers of the lateral geniculate body. Interactions occur between the layers of the lateral geniculate body where the signals from the retinal images of the two eyes overlap each other; these cause *interference patterns of stimulation* in specific cells of the visual cortex. That is, when the two corresponding points of the retinae are not precisely in fusion, specific cells in the visual cortex become excited; this excitation presumably provides the signal that is transmitted to the oculomotor apparatus to cause convergence or divergence of the eyes so that fusion can be reestablished. Once the corresponding points of the retinae are precisely *in register* with each other, the excitation of the specific cells in the visual cortex disappears.

**The Neural Mechanism for Stereopsis.** The visual images that appear on the retina during the process of stereopsis were discussed in Chapter 58. It was pointed out that because the two eyes are a little more than 2 inches apart the images on the two retinae are not exactly the same. The closer the object is to the eye the greater is the disparity between the two images. Consequently, it is impossible for all corresponding points in the visual image to be in complete register at the same time. Furthermore, the nearer the object is to the eye the less is the degree of register. Here again, specific cells in the primary visual cortex become excited in the areas of the visual field where highlights are out of register. Presumably, this excitation is the source of the signals for detection of the distance of the object in front of the eyes; this mechanism is called *stereopsis*.

**Strabismus.** Strabismus, which is also called *squint* or *cross-eyedness,* means lack of fusion of the eyes in one or more of the coordinates described above. Three basic types of strabismus are illustrated in Figure 60–12: *horizontal strabismus, vertical strabismus,* and *torsional strabismus.* However, combinations of two or even of all three of the different types of strabismus often occur.

Strabismus is believed to be caused by an abnormal "set" of the fusion mechanism of the visual system. That is, in the early efforts of the child to fixate the two eyes on the same object, one of the eyes fixates satisfactorily while the other fails to fixate, or they both fixate satisfactorily but never simultaneously. Soon, the patterns of conjugate movements of the eyes become abnormally "set" so that the eyes never fuse.

Frequently, some abnormality of the eyes contributes to the failure of the two eyes to fixate on the same point. For instance, if at birth one eye has poor vision in comparison with the other, the good eye tends to fixate on the object of attention while the poor eye might never do so. Also, in hypermetropic infants, intense impulses must be transmitted to the ciliary muscles to focus the eyes, and some of these impulses overflow into the oculomotor nuclei to cause simultaneous convergence of the eyes, as will be discussed below. As a result, the child's fusion mechanism becomes "set" for continual inward deviation of the eyes.

***Suppression of Visual Image from a Repressed Eye.*** In most patients with strabismus the eyes alternate in fixing on the object of attention. However, in some patients, one eye alone is used all the time while the other eye becomes repressed and is never used for vision. The vision in the repressed eye develops only slightly, usually remaining 20/400 or less. If the dominant eye then becomes blinded, vision in the repressed eye can develop to only a slight extent in the adult but far more rapidly in young children. This illustrates that visual acuity is highly dependent on the proper development of the central synaptic connections from the eyes.

# AUTONOMIC CONTROL OF ACCOMMODATION AND PUPILLARY APERTURE

**The Autonomic Nerves to the Eyes.** The eye is innervated by both parasympathetic and sympathetic fibers, as illustrated in Figure 60–13. The parasym-

Horizontal          Torsional          Vertical
strabismus          strabismus         strabismus

**Figure 60–12.** The three basic types of strabismus.

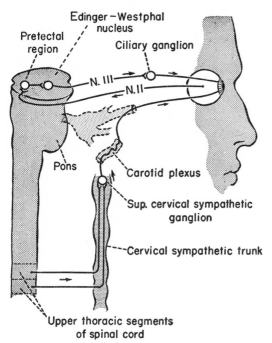

**Figure 60-13.** Autonomic innervation of the eye, showing also the reflex arc of the light reflex. (Modified from Ranson and Clark: Anatomy of the Nervous System, 1959.)

pathetic fibers arise in the *Edinger-Westphal nucleus* (the visceral nucleus of the third nerve) and then pass in the *third nerve* to the *ciliary ganglion,* which lies about 1 cm. behind the eye. Here the fibers synapse with postganglionic parasympathetic neurons that pass through the *ciliary nerves* into the eyeball. These nerves excite the ciliary muscle and the sphincter of the iris.

The sympathetic innervation of the eye originates in the *intermediolateral horn cells* of the first thoracic segment of the spinal cord. From here, sympathetic fibers enter the sympathetic chain and pass upward to the *superior cervical ganglion* where they synapse with postganglionic neurons. Fibers from these spread along the carotid artery and along successively smaller arteries until they reach the eyeball. There the sympathetic fibers innervate the radial fibers of the iris as well as several extraocular structures around the eye, which are discussed shortly in relation to Horner's syndrome. Also, they supply very weak innervation to the ciliary muscle.

## CONTROL OF ACCOMMODATION

The accommodation mechanism—that is, the mechanism which focuses the lens system of the eye—is essential to a high degree of visual acuity. Accommodation results from contrac-

tion or relaxation of the ciliary muscle, contraction causing increased strength of the lens system, as explained in Chapter 58, and relaxation causing decreased strength. The question that must be answered now is: How does one adjust his accommodation to keep his eyes in focus all the time?

Accommodation of the lens is regulated by a negative feedback mechanism that automatically adjusts the focal power of the lens for the highest degree of visual acuity. When the eyes have been fixed on some far object and then suddenly fix on a near object, the lens accommodates for maximum acuity of vision usually within one second; the precise control mechanism that causes this rapid and accurate focusing of the eye is still unclear. Some of the known features of the mechanism are the following:

First, when the eyes suddenly change the distance of their fixation point, the lens changes its strength almost invariably in the proper direction to achieve a new state of focus. In other words, the lens usually *does not hunt* back and forth on the two sides of focus in an attempt to find the focus.

Second, different types of cues that can help the lens change its strength in the proper direction include the following: (1) *Chromatic aberration* appears to be important. That is, the red light rays focus slightly posteriorly to the blue light rays. The eyes appear to be able to detect which of these two types of rays is in better focus, and this cue relays the information to the accommodating mechanism as to whether to make the lens stronger or weaker. (2) When the eyes fixate on a near object they also converge toward each other. The neural mechanisms for *convergence causes a simultaneous signal to strengthen the lens of the eye.* (3) *Since the fovea is a depressed area, the clarity of focus in the depth of the fovea versus the clarity of focus on the edges will be different.* It has been suggested that this also gives clues as to which way the strength of the lens needs to be changed. (4) It has been found that *the degree of accommodation of the lens oscillates slightly* all of the time at a frequency of approximately one-half to two times per second. It has been suggested that the visual image becomes clearer when the oscillation of the lens strength is in the appropriate direction and poorer when the lens strength is in the wrong direction. This could give a rapid cue as to which way the strength of the lens needs to change to provide appropriate focus.

It is presumed that the cortical areas that control accommodation closely parallel those that control fixation movements of the eyes, with integration of visual signals in areas 18 and 19 and transmission of motor signals to the ciliary muscle through the pretectal area and Edinger-Westphal nucleus.

## CONTROL OF THE PUPILLARY APERTURE

Stimulation of the parasympathetic nerves excites the pupillary sphincter, thereby decreasing the pupillary aperture; this is called *miosis*. On the other hand, stimulation of the sympathetic nerves excites the radial fibers of the iris and causes pupillary dilatation, which is called *mydriasis*.

**The Pupillary Light Reflex.** When light is shone into the eyes the pupils constrict, a reaction that is called the pupillary light reflex. The neuronal pathway for this reflex is illustrated in Figure 60–13. When light impinges on the retina, the resulting impulses pass through the optic nerves and optic tracts to the pretectal nuclei. From here, impulses pass to the *Edinger-Westphal nucleus* and finally back through the *parasympathetic nerves* to constrict the sphincter of the iris. In darkness, the Edinger-Westphal nucleus becomes inhibited, which results in dilatation of the pupil.

The function of the light reflex is to help the eye adapt extremely rapidly to changing light conditions, the importance of which was explained in relation to retinal adaptation in the previous chapter. The limits of pupillary diameter are about 1.5 mm. on the small side and 8 mm. on the large side. Therefore, the range of light adaptation that can be effected by the pupillary reflex is about 30 to 1.

**Pupillary Reflexes in Syphilis.** Central nervous system syphilis almost always eventually blocks the light reflex pathway between the retina and the Edinger-Westphal nucleus. It is believed that this block most likely occurs in the pretectal region of the brain stem, though it could result from destruction of small afferent fibers in the optic nerves. One of the distinguishing features of the loss of the light reflex in syphilis is that pupillary constriction, which also occurs with lens accommodation, is not lost at the same time because the parasympathetic pathways to the eyes are not involved in syphilis.

When the impulses to the Edinger-Westphal nuclei are blocked by syphilis, these nuclei are also released from inhibitory impulses arriving from some outside source. As a result, the nuclei become tonically active, causing the pupils thereafter to remain con-stricted in addition to their failure to respond to light. Such a pupil that fails to respond to light and also is very small (an *Argyll Robertson pupil*) is an important diagnostic sign of central nervous system syphilis. However, a few other conditions, including central nervous system damage from alcoholism, encephalitis, and so forth, can occasionally also cause an Argyll Robertson pupil.

**Horner's Syndrome.** The sympathetic nerves to the eye are occasionally interrupted, and this interruption frequently occurs in the cervical chain. This results in Horner's syndrome, which consists of the following effects: First, because of interruption of fibers to the pupillary dilator muscle, the pupil remains persistently constricted to a smaller diameter than that of the pupil of the opposite eye. Second, the superior eyelid droops because this eyelid is normally maintained in an open position during the waking hours partly by contraction of a smooth muscle, the *superior palpebral muscle*, which is innervated by the sympathetics. Therefore, destruction of the sympathetics makes it impossible to open the superior eyelid nearly as widely as normally. Third, the blood vessels on the corresponding side of the face and head become persistently dilated. And, fourth, sweating cannot occur on the side of the face and head affected by Horner's syndrome.

# REFERENCES

Armington, J. C.: The Electroretinogram. New York, Academic Press, Inc., 1974.
Ashworth, B.: Clinical Neuro-Ophthalmology. Philadelphia, J. B. Lippincott Company, 1973.
Bach-y-Rita, P., and Collins, C. C. (eds.): The Control of Eye Movements. New York, Academic Press, Inc., 1971.
Becker, W.: The control of eye movements in the saccadic system. *Bibl. Ophthalmol.,* 82:233, 1972.
Bouman, M. A., and Koenderink, J. J.: Psychophysical basis of coincidence mechanisms in the human visual system. *Ergeb. Physiol.,* 65:126, 1972.
Brindley, G. S.: Central pathways of vision. *Ann. Rev. Physiol.,* 32:259, 1970.
Cooper, I. S., Riklan, M., and Rakic, P. T.: The Pulvinar–LP Complex. Springfield, Ill., Charles C Thomas, Publisher, 1974.
Crone, R. A.: Diplopia. New York, American Elsevier Publishing Co., Inc., 1974.
Gerrits, H. J., and Vendrik, A. J.: Eye movements necessary for continuous perception during stabilization of retinal images. *Bibl. Ophthalmol.,* 82:339, 1972.
Glezer, V. D., Leushina, L. I., Nevskaya, A. A., and Prazdnikova, N. V.: Studies on visual pattern recognition in man and animals. *Vision Res.,* 14:555, 1974.
Gombrich, E. H.: The visual image. *Sci. Amer.,* 227:82, 1972.
Gordon, B.: The superior colliculus of the brain. *Sci. Amer.,* 227:72, 1972.
Gordon, B.: Superior colliculus: structure, physiology, and possible functions. *In* Guyton, A. C. (ed.): MTP International Review of Science: Physiology. Vol. 3. Baltimore, University Park Press, 1974, p. 185.
Guyton, J. S., and Kirkman, N.: Ocular movement: I. Mechanics, pathogenesis, and surgical treatment of alternating hypertropia (dissociated vertical divergence, double hypertropia) and some related phenomena. *Amer. J. Ophthalmol.,* 41:438, 1956.
Hagen, M. A.: Picture perception: toward a theoretical model. *Psychol. Bull.,* 81:471, 1974.

Hassler, R.: Supranuclear structures regulating binocular eye and head movements. *Bibl. Ophthalmol., 82*:207, 1972.

Hubel, D. H.: Eleventh Bowditch Lecture. *Physiologist, 10*:17, 1967.

Hubel, D. H., and Wiesel, T. N.: Receptive fields of cells in striate cortex of very young, visually inexperienced kittens. *J. Neurophysiol., 26*:994, 1963.

Hubel, D. H., and Wiesel, T. N.: Cortical and callosal connections concerned with the vertical meridian of visual fields in the cat. *J. Neurophysiol., 30*:1561, 1967.

Janisse, M. P. (ed.): Pupillary Dynamics and Behavior. New York, Plenum Publishing Corporation, 1974.

Jung, R.: How do we see with moving eyes? *Bibl. Ophthalmol., 82*:377, 1972.

Kommerell, G., and Täumer, R.: Investigations of the eye tracking system through stabilized retinal images. *Bibl. Ophthalmol., 82*:288, 1972.

Kuffler, S. W.: The single-cell approach in the visual system and the study of receptive fields. *Invest. Ophthalmol., 12*:794, 1973.

Lindenberg, R., Walsh, F. B., and Sacks, J. G.: Neuropathology of Vision. Philadelphia, Lea & Febiger, 1973.

McIlwain, J. T.: Central vision: visual cortex and superior colliculus. *Ann. Rev. Physiol., 34*:291, 1972.

Mansourian, P. G.: System analysis of the vestibulo-ocular control mechanism. *Adv. Ophthalmol., 28*:175, 1974.

Matin, E.: Saccadic suppression: a review and a analysis. *Psychol. Bull., 81*:899, 1974.

Matin, L., and Matin, E.: Visual perception of direction and voluntary saccadic eye movements. *Bibl. Ophthalmol., 82*:358, 1972.

Michael, C. R.: Retinal processing of visual images. *Sci. Amer., 220*:104, 1969.

Noton, D., and Stark, L.: Eye Movements and visual perception. *Sci. Amer., 224*:34, 1971.

Oster, G.: Phosphenes. *Sci. Amer., 222*:82, 1970.

Pettigrew, J. D.: The neurophysiology on binocular vision. *Sci. Amer., 227*:84, 1972.

Polyak, S.: The Retina. Chicago, University of Chicago Press, 1941.

Rock, I.: The perception of disoriented figures. *Sci. Amer., 230*:78, 1974.

Straschill, M., and Rieger, P.: Optomotor integration in the colliculus superior of the cat. *Bibl. Ophthalmol., 82*:130, 1972.

Szentágothai, J.: Lateral geniculate body structure and eye movement. *Bibl. Ophthalmol., 82*:178, 1972.

Toates, F. M.: Accommodation function of the human eye. *Physiol. Rev., 52*:828, 1972.

Tomita, T., Miller, W. H., Hashimoto, Y., and Saito, T.: Electrical response of retinal cells as a sign of transport. *Exp. Eye Res., 16*:327, 1973.

Westheimer, G., and Blair, S. M.: Concerning the supranuclear organization of eye movements. *Bibl. Ophthalmol., 82*:28, 1972.

Whitteridge, D.: Binocular vision and cortical function. *Proc. Roy. Soc. Med., 65*:947, 1972.

Wist, E. R.: Eye movements and space perception. *Bibl. Ophthalmol., 82*:348, 1972.

Wurtz, R. H., and Goldberg, M. E.: The role of the superior colliculus in visually-evoked eye movements. *Bibl. Ophthalmol., 82*:149, 1972.

# 61

# The Sense of Hearing

Hearing, like many somatic senses, is a mechanoreceptive sense, for the ear responds to mechanical vibration of the sound waves in the air. The purpose of the present chapter is to describe and explain the mechanism by which the ear receives sound waves, discriminates their frequencies, and finally transmits auditory information into the central nervous system.

## THE TYMPANIC MEMBRANE AND THE OSSICULAR SYSTEM

### TRANSMISSION OF SOUND FROM THE TYMPANIC MEMBRANE TO THE COCHLEA

Figure 61–1 illustrates the *tympanic membrane* (commonly called the *eardrum*) and the *ossicular system,* which transmits sound through the middle ear. The tympanic membrane is cone shaped, with its concavity facing downward toward the auditory canal. Attached to the very center of the tympanic membrane is the *handle* of the *malleus*. At its other end the malleus is tightly bound to the *incus* by ligaments so that whenever the malleus moves the incus generally moves in unison with it. The opposite end of the incus in turn articulates with the stem of the *stapes,* and the *faceplate* of the stapes lies against the membranous labyrinth in the opening of the oval window where sound waves are transmitted into the inner ear, the *cochlea.*

The ossicles of the middle ear are suspended by ligaments in such a way that the combined malleus and incus act as a single lever having its fulcrum approximately at the border of the tympanic membrane. The large *head* of the malleus, which is on the opposite side of the fulcrum from the handle, almost exactly balances the other end of the lever so that changes in position of the body will not increase or decrease the tension on the tympanic membrane.

The articulation of the incus with the stapes causes the stapes to push forward on the cochlea fluid every time the handle of the malleus moves inward and to pull backward on the fluid every time the malleus moves outward, which promotes inward and outward motion of the faceplate at the oval window.

The handle of the malleus is constantly pulled inward by ligaments and by the tensor tympani muscle, which keeps the tympanic membrane tensed. This allows sound vibrations on *any* portion of the tympanic membrane to be transmitted to the malleus, which would not be true if the membrane were lax.

**Impedance Matching by the Ossicular System.** The amplitude of movement of the stapes faceplate with each sound vibration is only three-fourths as much as the amplitude of the

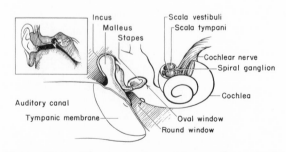

**Figure 61–1.** The tympanic membrane, the ossicular system of the middle ear, and the inner ear.

826

handle of the malleus. Therefore, the ossicular lever system does not amplify the movement distance of the stapes, as is commonly believed, but instead the system increases the *force* of movement about 1.3 times. However, the surface area of the tympanic membrane is approximately 55 sq. mm., whereas the surface area of the stapes averages 3.2 sq. mm. This 17-fold difference times the 1.3-fold ratio of the lever system, allows all the energy of a sound wave impinging on the tympanic membrane to be applied to the small faceplate of the stapes, causing approximately 22 times as much *pressure* on the fluid of the cochlea as is exerted by the sound wave against the tympanic membrane. Since fluid has far greater inertia than air, it is easily understood that increased amounts of pressure are needed to cause vibration in the fluid. Therefore, the tympanic membrane and ossicular system provide *impedance matching* between the sound waves in air and the sound vibrations in the fluid of the cochlea. Indeed, the impedance matching is probably 50 to 75 per cent of perfect for sound frequencies between 300 to 3000 cycles per second, which allows almost full utilization of the energy in the incoming sound waves.

In the absence of the ossicular system and tympanum, sound waves can travel directly through the air of the middle ear and can enter the cochlea at the oval window. However, the sensitivity for hearing is then 30 decibels less than for ossicular transmission—equivalent to a decrease from a very loud shouting voice to a barely audible voice level.

**Transmission Characteristics of the Ossicular System.** Every vibrating system that has inertia and that has an elastic component also has a natural frequency at which it can vibrate back and forth most easily. This is called its *resonant frequency*. Since the ossicular system does have inertia and since it is suspended by elastic ligaments, it has a broadly tuned natural resonating frequency between 700 and 1400 cycles per second. However, the ligaments and other structures attached to the ossicles are also viscous, which prevents excessive resonance; this is called *damping of the vibrations*. Because of the slight amount of resonance that does occur, sound waves of approximately 1200 cycles per second can be transmitted through the ossicular system with slightly greater ease than can sound waves of other frequencies.

The external auditory canal, because of its dimensions, acts as an air column resonator and has a natural resonating frequency of about 3000 cycles per second. But here again, the degree of resonance is slight and potentiates the 3000 cycle sound only a minute amount.

Combining the resonant effects of the ossicular system and of the auditory canal, transmission of sound from the air to the cochlea is excellent between the limits of 600 and 6000 cycles per second and fades both above and below these limits.

**Attenuation of Sound by Contraction of the Stapedius and Tensor Tympani Muscles.** When loud sounds are transmitted through the ossicular system into the central nervous system, a reflex occurs after a latent period of only 40 milliseconds to cause contraction of both the stapedius and tensor tympani muscles. The tensor tympani muscle pulls the handle of the malleus inward while the stapedius muscle pulls the stapes outward. These two forces oppose each other and thereby cause the entire ossicular system to develop a high degree of rigidity, thus greatly reducing the transmission of low frequency sound, frequencies below 1000 cycles per second, to the cochlea.

This *attenuation reflex* can reduce the intensity of sound transmission by as much as 30 to 40 decibels, which is about the same difference as that between a whisper and the sound emitted by a loud speaker. The function of this mechanism is probably two-fold:

1. To *protect* the cochlea from damaging vibrations caused by excessively loud sound. It is mainly low-frequency sounds (the ones that are attenuated) that are frequently loud enough to damage the basilar membrane of the cochlea. Unfortunately, because of the 40 or more millisecond latency for reaction of the reflex, the sudden loud thunderous sounds that result from explosions can still cause extensive cochlear damage.

2. To *mask* low-frequency sounds in loud environments. This usually removes a major share of the background noise and allows a person to concentrate on sounds above 1000 cycles per second frequency. It is in this upper-frequency range that most voice communication is achieved.

Another function of the tensor tympani and stapedius muscles is to decrease the person's hearing sensitivity to his own speech. This effect is activated by collateral signals transmitted to these muscles at the same time that his brain activates his voice mechanism.

### TRANSMISSION OF SOUND THROUGH THE BONE

Because the inner ear, the *cochlea*, is embedded in a bony cavity in the temporal bone, vibrations of the entire skull can cause fluid vibrations in the cochlea itself. Therefore, under

appropriate conditions, a tuning fork or an electronic vibrator placed on any bony protuber-
ance of the skull causes the person to hear the
sound if it is intense enough. Unfortunately, the
energy available even in very loud sound in
the air is not sufficient to cause hearing through
the bone except when a special electromechani-
cal sound-transmitting device is applied directly
to the bone, usually to the mastoid process.

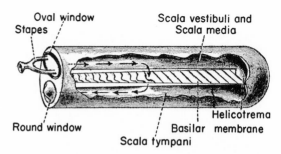

**Figure 61–3.** Movement of fluid in the cochlea following forward thrust of the stapes.

## THE COCHLEA

### FUNCTIONAL ANATOMY OF THE COCHLEA

The cochlea is a system of coiled tubes,
shown in Figure 61–1 and in cross-section in
Figure 61–2, with three different tubes coiled
side by side: the *scala vestibuli,* the *scala
media,* and the *scala tympani.* The scala vestibuli and scala media are separated from each
other by the *vestibular membrane,* and the scala
tympani and scala media are separated from
each other by the *basilar membrane.* On the
surface of the basilar membrane lies a structure,
the *organ of Corti,* which contains a series of
mechanically sensitive cells, the *hair cells.*
These are the receptive end-organs that generate nerve impulses in response to sound vibrations.

Figure 61–3 illustrates schematically the functional parts of the uncoiled cochlea for transmission of sound vibrations. First, note that the

vestibular membrane is missing from this figure.
This membrane is so thin and so easily moved
that it does not obstruct the passage of sound
vibrations from the scala vestibuli into the scala
media at all. Therefore, so far as the transmission of sound is concerned, the scala vestibuli
and scala media are considered to be a single
chamber. The importance of the vestibular
membrane is to maintain a special fluid in the
scala media that is required for normal function
of the sound receptive hair cells, as discussed
later in the chapter.

Sound vibrations enter the scala vestibuli
from the faceplate of the stapes at the oval window. The faceplate covers this window and is
connected with the window's edges by a relatively loose annular ligament so that it can move
inward and outward with the sound vibrations.
Inward movement causes the fluid to move into
the scala vestibuli and scala media, which immediately increases the pressure in the entire
cochlea and causes the round window to bulge
outward.

Note from Figure 61–3 that the distal end of
the scala vestibuli and scala tympani are continuous with each other by way of the *helicotrema.* If the stapes moves inward *very
slowly,* fluid from the scala vestibuli is pushed
through the helicotrema into the scala tympani,
and this causes the round window to bulge outward. However, if the stapes vibrates inward
and outward rapidly, the fluid simply does not
have time to pass all the way to the helicotrema,
then to the round window, and back again to the
oval window between each two successive vibrations. Instead, the fluid wave takes a
shortcut through the basilar membrane, causing
it to bulge back and forth with each sound vibration. We shall see later that each frequency of
sound causes a different "pattern" of vibration
in the basilar membrane and that this is one of

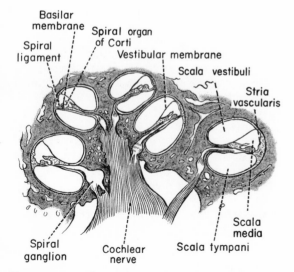

**Figure 61–2.** The cochlea. (From Goss, C. M. (ed.):
Gray's Anatomy of the Human Body. Lea & Febiger.)

the important means by which the sound frequencies are discriminated from each other.

**The Basilar Membrane and Resonance in the Cochlea.** The basilar membrane contains about 20,000 or more *basilar fibers* that project from the bony center of the cochlea, the *modiolus*, toward the outer wall. These fibers are stiff, elastic, reedlike structures that are not fixed at their distal ends except that they are embedded in the basilar membrane. Because they are stiff and free at one end, they can vibrate like reeds of a harmonica.

The lengths of the basilar fibers increase progressively from the base of the cochlea to the helicotrema, from approximately 0.04 mm. at the base to 0.5 mm. at the helicotrema, a 12-fold increase in length.

The diameters of the fibers, on the other hand, decrease from the base to the helicotrema, so that their overall stiffness decreases more than 100-fold. As a result, the stiff, short fibers near the base of the cochlea have a tendency to vibrate at a high frequency, whereas the long, limber fibers near the helicotrema have a tendency to vibrate at a low frequency.

In addition to the differences in stiffness of the basilar fibers, they are also differently "loaded" by the fluid mass of the cochlea. That is, when a fiber vibrates back and forth, all the fluid between the vibrating fiber and the oval and round windows must also move back and forth at the same time. For a fiber vibrating near the base of the cochlea, the total mass of moving fluid is slight in comparison with that for a fiber vibrating near the helicotrema. This difference, too, favors high frequency vibration near the windows and low frequency vibration near the tip of the cochlea.

Thus, high frequency resonance of the basilar membrane occurs near the base, and low frequency resonance occurs near the apex because of (1) difference in stiffness of the fibers and (2) difference in "loading."

## TRANSMISSION OF SOUND WAVES IN THE COCHLEA— THE "TRAVELING WAVE"

If the foot of the stapes moves inward instantaneously, the round window must also bulge outward instantaneously because the cochlea is bounded on all sides by bony walls. Since the fluid wave will not have time to move all the way from the oval window to the helicotrema and back to the round window, the initial effect is to cause the basilar membrane at the very

base of the cochlea to bulge in the direction of the round window. However, the elastic tension that is built up in the basilar fibers as they bend toward the round window initiates a wave that "travels" along the basilar membrane toward the helicotrema, as illustrated in Figure 61–4. Figure 61–4A shows movement of a high frequency wave down the basilar membrane, Figure 61–4B a medium frequency wave, and Figure 61–4C a very low frequency wave. Movement of the wave along the basilar membrane is comparable to the movement of a pressure wave along the arterial walls, which was discussed in Chapter 19, or it is also comparable to the wave that travels along the surface of a pond.

**Pattern of Vibration of the Basilar Membrane for Different Sound Frequencies.** Note in Figure 61–4 the different patterns of transmission for sound waves of different frequencies. Each wave is relatively weak at the outset but becomes strong when it reaches that portion of the basilar membrane that has a natural resonant frequency equal to the respective sound frequency. At this point the basilar membrane can vibrate back and forth with such great ease that the energy in the wave is completely dissipated. Consequently, the wave ceases at this point and fails to travel the remaining distance along the basilar membrane. Thus, a high frequency sound wave travels only a short distance along the basilar membrane be-

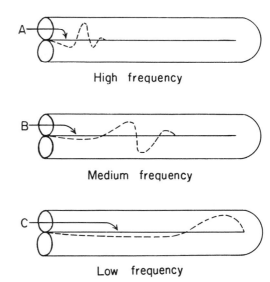

High frequency

Medium frequency

Low frequency

**Figure 61–4.** Diagrammatic representation of "traveling waves" along the basilar membrane for high, medium, and low frequency sounds.

fore it reaches its resonant point and dies out; a medium frequency sound wave travels about halfway and then dies out; and, finally, a very low frequency sound wave travels the entire distance along the membrane.

Another feature of the traveling wave is that it travels fast along the initial portion of the basilar membrane but progressively more slowly as it goes farther and farther into the cochlea. The cause of this is the high coefficient of elasticity of the basilar fibers near the stapes but a progressively decreasing coefficient farther along the membrane. This rapid initial transmission of the wave allows the high frequency sounds to travel far enough into the cochlea to spread out and separate from each other on the basilar membrane. Without this spread, all the high frequency waves would be bunched together within the first millimeter or so of the basilar membrane, and their frequencies could not be discriminated one from the other.

***Amplitude Pattern of Vibration of the Basilar Membrane.*** The dashed curves of Figure 61–5A show the position of a sound wave on the basilar membrane when the stapes *(a)* is all the way inward, *(b)* has moved back to the neutral point, *(c)* is all the way outward, and *(d)* has moved back again to the neutral point but is moving inward. The shaded area around these different waves shows the maximum extent of vibration of the basilar membrane during a

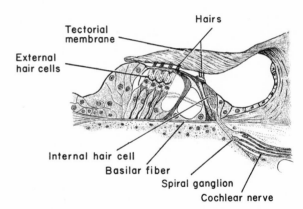

**Figure 61–6.** The organ of Corti, showing especially the hair cells and the tectorial membrane against the projecting hairs.

complete vibratory cycle. This is the amplitude pattern of vibration of the basilar membrane for this particular sound frequency.

Figure 61–5B shows the amplitude patterns of vibration for different frequencies, showing that the maximum amplitude for 8000 cycles occurs near the base of the cochlea, while that for frequencies of 50 to 100 cycles per second occurs near the helicotrema.

Note in Figure 61–5B that the basal end of the basilar membrane vibrates at least weakly for all frequencies. However, beyond the resonant area for each given frequency, the vibration of the basilar membrane cuts off sharply. The principal method by which sound frequencies are discriminated from each other is based on the "place" of maximum stimulation of the nerve fibers from the basilar membrane, as will be explained in the following section.

## FUNCTION OF THE ORGAN OF CORTI

The organ of Corti, illustrated in Figures 61–2 and 61–6, is the receptor organ that generates nerve impulses in response to vibration of the basilar membrane. Note that the organ of Corti lies on the surface of the basilar fibers and basilar membrane. The actual sensory receptors in the organ of Corti are two types of *hair cells,* a single row of *internal hair cells,* numbering about 3500 and measuring about 12 microns in diameter, and three to four rows of *external hair cells,* numbering about 20,000 and having diameters of only about 8 microns. The bases and sides of the hair cells are enmeshed by a network of cochlear nerve endings. These lead to the *spiral ganglion of Corti,* which lies in the modiolus of the cochlea. The spiral ganglion in

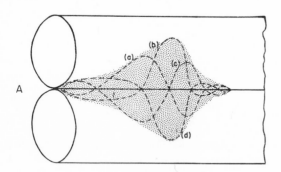

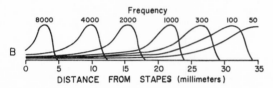

**Figure 61–5.** (A) Amplitude pattern of vibration of the basilar membrane for a medium frequency sound. (B) Amplitude patterns for sounds of all frequencies between 50 and 8000 per second, showing the points of maximum amplitude (the resonance points) on the basilar membrane for the different frequencies.

turn sends axons into the *cochlear nerve* and thence into the central nervous system at the level of the upper medulla. The relationship of the organ of Corti to the spiral ganglion and to the cochlear nerve is illustrated in Figure 61–2.

**Excitation of the Hair Cells.** Note in Figure 61–6 that minute hairs, or cilia, project upward from the hair cells and either touch or are embedded in the surface gel coating of the *tectorial membrane,* which lies above the cilia in the scala media. These hair cells are similar to the hair cells found in the macula and cristae ampullaris of the vestibular apparatus which were discussed in Chapter 52. Bending of the hairs excites the hair cells, and this in turn excites the nerve fibers enmeshing their bases.

Figure 61–7 illustrates the mechanism by which vibration of the basilar membrane excites the hair endings. This shows that the upper ends of the hair cells are fixed tightly in a structure called the *reticular lamina.* Furthermore, the reticular lamina is very rigid and is continuous with a rigid triangular structure called the *rods of Corti* that rests on the basilar fibers. Therefore, the basilar fiber, the rods of Corti, and the reticular lamina all move as a unit.

Upward movement of the basilar fiber rocks the reticular lamina upward and *inward*. Then, when the basilar membrane moves downward, the reticular lamina rocks downward and *outward*. The inward and outward motion causes the hairs to sheer back and forth in the tectorial membrane, thus exciting the cochlear nerve fibers whenever the basilar membrane vibrates.

**Mechanism by which the Hair Cells Excite the Nerve Fibers—Receptor Potentials.** Back-and-forth bending of the hairs causes alternate changes in the electrical potential across the hair cell membrane. This alternating potential is the *receptor potential* of the hair cell; and

it in turn stimulates the cochlear nerve endings that terminate on the hair cells. Most physiologists believe that the receptor potential stimulates the endings by direct electrical excitation.

When the basilar fiber bends toward the scala vestibuli (in the upward direction in Figure 61–7), the hair cell becomes depolarized, and it is this depolarization that excites an increased number of action potentials in the nerve fiber. When the basilar fiber moves in the opposite direction, the hair cell becomes hyperpolarized, and the number of action potentials decreases.

*The Endocochlear Potential.* To explain even more fully the electrical potentials generated by the hair cells, we need to explain still another electrical phenomenon called the endocochlear potential: The scala media is filled with a fluid called *endolymph* in contradistinction to the *perilymph* present in the scala vestibuli and scala tympani. The scala vestibuli and scala tympani in most young children and in some adults communicate directly with the subarachnoid space around the brain, and the perilymph is almost identical with cerebrospinal fluid. On the other hand, the endolymph that fills the scala media is an entirely different fluid probably secreted by the *stria vascularis,* a highly vascular area on the outer wall of the scala media. Endolymph contains a very high concentration of potassium and a very low concentration of sodium, which is exactly opposite to the perilymph.

An electrical potential of approximately 80 mv. exists all the time between the endolymph and the perilymph, with positivity inside the scala media and negativity outside. This is called the *endocochlear potential,* and it is believed to be generated by continual secretion of positive potassium ions into the scala media by the stria vascularis.

The importance of the endocochlear potential is that the tops of the hair cells project through the reticular lamina into the endolymph of the scala media while perilymph bathes the lower bodies of the hair cells. Furthermore, the hair cells have a negative intracellular potential $-70$ millivolts with respect to the perilymph, but $-150$ millivolts with respect to the endolymph—that is, at the upper surfaces of the hair cells where the hairs project into the endolymph. It is believed that this high electrical potential at the hair border of the cell greatly sensitizes the cell, thereby increasing its ability to respond to slight movement of the hairs.

## DETERMINATION OF PITCH—THE "PLACE" PRINCIPLE

In the minds of most persons, sound pitch and sound frequency are the same thing. However, the two are slightly different in the following way: Pitch is the conscious perception of the sound frequency and may not be the same as

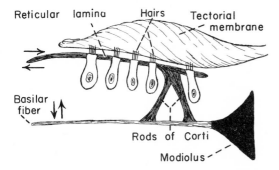

**Figure 61–7.** Stimulation of the hair cells by the to-and-fro movement of the hairs in the tectorial membrane.

the true sound frequency. The pitch is especially likely to deviate from the true sound frequency at both very high and very low frequencies; also, the pitch usually changes slightly as the intensity of the sound changes even though the frequency remains constant. Yet, from the point of view of the following discussion, we can still consider the two to be essentially the same.

From earlier discussions in this chapter it is already apparent that low pitch (or low frequency) sounds cause maximal activation of the basilar membrane near the apex of the cochlea, sounds of high pitch (or high frequency) activate the basilar membrane near the base of the cochlea, and intermediate frequencies activate the membrane at intermediate distances between these two extremes. Furthermore, there is spatial organization of the cochlear nerve fibers from the cochlea to the cochlear nuclei in the brain stem, the fibers from each respective area of the basilar membrane terminating in a corresponding area in the cochlear nuclei. We shall see later that this spatial organization continues all the way up the brain stem to the cerebral cortex. The recording of signals from the auditory tracts in the brain stem and from the auditory receptive fields in the cerebral cortex shows that specific neurons are activated by specific pitches. Therefore, the primary method used by the nervous system to detect different pitches is to determine the position along the basilar membrane that is most stimulated. This is called the *place principle* for determination of pitch.

Yet, low frequency sounds probably can be discriminated much less accurately in other ways as well. For instance, destruction of the apical half of the cochlea destroys the basilar membrane where low frequency sounds are normally detected, but low frequency sounds can still be discriminated from high frequency sounds to a slight degree. It is believed that the low frequency sounds cause synchronized low frequency volleys of impulses that can be distinguished from the high frequency signals.

## DETERMINATION OF LOUDNESS

Loudness is determined by the auditory system in at least three different ways: First, as the sound becomes louder, the amplitude of vibration of the basilar membrane and hair cells also increases so that the hair cells excite the nerve endings at more rapid rates. Second, as the amplitude of vibration increases, it causes more and more of the hair cells on the fringes of the vibrating portion of the basilar membrane to become stimulated, thus causing *spatial summation* of impulses—that is, transmission through many nerve fibers rather than through a few. Third, certain hair cells do not become stimulated until the vibration of the basilar membrane reaches a relatively high intensity, and it is believed that stimulation of these cells in some way apprises the nervous system that the sound is then very loud.

**Detection of Changes in Loudness—The Power Law.** It was pointed out in Chapter 48 that a person interprets changes in intensity of sensory stimuli approximately in proportion to a power function of the actual intensity. In the case of sound, the interpreted sensation changes approximately in proportion to the cube root of the actual sound intensity. To express this another way, the ear can discriminate changes in sound intensity from the softest whisper to the loudest possible noise of *approximately one trillion times* as much sound energy. Yet the ear interprets this much difference in sound level as approximately a 10,000-fold change. Thus, the scale of intensity is greatly "compressed" by the sound perception mechanisms of the auditory system. This obviously allows a person to interpret differences in sound intensities over an extremely wide range, a far broader range than would be possible were it not for compression of the scale.

**The Decibel Unit.** Because of the extreme changes in sound intensities that the ear can detect and discriminate, sound intensities are usually expressed in terms of the logarithm of their actual intensities. A 10-fold increase in sound energy (or a $\sqrt{10}$-fold increase in sound pressure, because energy is proportional to the square of pressure) is called 1 *bel*, and one-tenth bel is called 1 *decibel*. One decibel represents an actual increase in intensity of 1.26 times.

Another reason for using the decibel system in expressing changes in loudness is that, in the usual sound intensity range for communication, the ears can detect approximately a 1 decibel change in sound intensity.

**The "Zero" Decibel Reference Level.** The usual method for expressing the intensity of sound is to state the pressure difference between the peak of the sound compression wave and the trough of the wave. A pressure difference of *1 dyne per square centimeter* is considered to have unit intensity, and this is expressed as zero decibels when converted to the decibel scale because the logarithm of one is zero. However, it should be remembered that the decibel scale is an *energy scale*—not a pressure scale. Furthermore, energy changes are proportional to the square of the pressure changes. Therefore, if the pressure of the sound increases to 10 dynes per square centimeter, this will be equivalent to a 100-fold increase in sound energy above zero reference level. Consequently, the sound intensity will be +20 decibels. A 100-fold increase in pressure will change the sound intensity to +40 decibels.

**Threshold for Hearing Sound at Different Frequencies.** Figure 61–8 shows the energy threshold at which sounds of different frequencies can barely be heard by the ear. This figure illustrates that a 2000 cycle per second sound can be heard even when its intensity is as low as −70 decibels, which is one ten-millionth microwatt/cm.$^2$. On the other hand, a 100 cycle per second sound can be detected only if its intensity is 10,000 times as great as this—that is, at an intensity of −30 decibels.

**Frequency Range of Hearing.** The frequencies of sound that a young person can hear, before aging has occurred in the ears, is generally stated to be between 30 and 20,000 cycles per second. However, referring again to Figure 61–8, we see that the sound range depends to a great extent on intensity. If the intensity is only −60 decibels, the sound range is 500 to 5000 cycles per second, but, if the sound intensity is −20 decibels, the frequency range is about 70 to 15,000 cycles per second, and only with intense sounds can the complete range of 30 to 20,000 cycles be achieved. In old age, the frequency range falls to 50 to 8,000 cycles per second or less, as is discussed later in the chapter.

# CENTRAL AUDITORY MECHANISMS

## THE AUDITORY PATHWAY

Figure 61–9 illustrates the major auditory pathways. It shows that nerve fibers from the *spiral ganglion of the organ of Corti* enter the *cochlear nuclei* located in the upper part of the medulla. At this point, all the fibers synapse, and second order

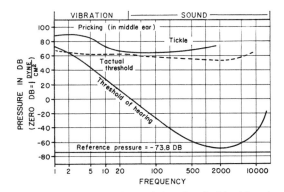

**Figure 61–8.** Relationship of the threshold of hearing and the threshold of somesthetic perception to the sound energy level at each sound frequency. (Modified from Stevens and Davis: Hearing. John Wiley & Sons.)

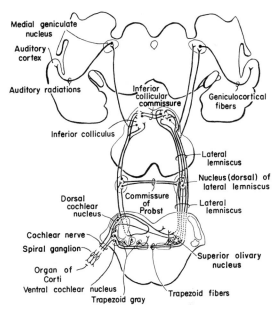

**Figure 61–9.** The auditory pathway. (Modified from Crosby, Humphrey, and Lauer: Correlative Anatomy of the Nervous System. The Macmillan Co., 1962.)

neurons pass mainly to the opposite side of the brain stem through the *trapezoid body* to the *superior olivary nucleus.* However, some of the second order fibers pass ipsilaterally to the superior olivary nucleus on the same side. Most of the fibers entering the superior olivary nucleus on either side terminate here, but some pass on through this nucleus. From the superior olivary nucleus the auditory pathway then passes upward through the *lateral lemniscus,* and many of the fibers terminate in the *nucleus of the lateral lemniscus*, but many also bypass this nucleus and pass on to the inferior colliculus where most terminate; a few pass on without terminating to higher levels. A few fibers cross from the nucleus of the lateral lemniscus through the *commissure of Probst* to the contralateral nucleus, and still other fibers cross through the *inferior collicular commissure* from one inferior colliculus to the other. From the inferior colliculus, the pathway passes through the *peduncle of the inferior colliculus* to the *medial geniculate nucleus,* where all the fibers synapse. From here, the auditory tract spreads by way of the *auditory radiation* to the *auditory cortex* located mainly in the superior temporal gyrus.

Several points of importance in relation to the auditory pathway should be noted. First, impulses from either ear are transmitted through the auditory pathways of both sides of the brain stem with only slight preponderance of transmission in the contralateral pathway. In at least three different places in the brain stem crossing-over occurs between the two pathways: (a) in the trapezoid body, (b) in the commissure of Probst, and (c) in the commissure connecting the two inferior colliculi.

Second, many collateral fibers from the auditory tracts pass directly into the reticular activating system of the brain stem. This system projects diffusely upward into the cerebral cortex and downward into the spinal cord.

Third, the pathway for transmission of sound impulses from the cochlea to the cortex consists of at least four neurons and sometimes as many as six. Neurons *may* or *may not* synapse in the superior olivary nuclei, in the nuclei of the lateral lemniscus, and in the inferior colliculi. Therefore, some of the tracts are more direct than others, which means that some impulses arrive at the cortex well ahead of others even though they might have originated at exactly the same time.

Fourth, several important pathways also exist from the auditory system into the cerebellum: (a) directly from the cochlear nuclei, (b) from the inferior colliculi, (c) from the reticular substance of the brain stem, and (d) from the cerebral auditory areas. These activate the *cerebellar vermis* instantaneously in the event of a sudden noise.

Fifth, a high degree of spatial orientation is maintained in the fiber tracts from the cochlea all the way to the cortex. In fact, there are three different spatial representations of sound frequencies in the cochlear nuclei, two representations in the inferior colliculi, one very precise representation for discrete sound frequencies in the auditory cortex, and several less precise representations in the auditory association areas.

**Firing Rates at Different Levels of the Auditory Tract.** Single nerve fibers entering the cochlear nuclei from the eighth nerve can fire at rates up to 1000 per second, the rate being determined mainly by the loudness of the sound. At low sound frequencies, the nerve impulses are usually synchronized with the sound waves but they do not necessarily occur with every wave.

In the auditory tracts of the brain stem, the firing is usually no longer synchronized with the sound frequency except at sound frequencies below 200 cycles per second. Furthermore, the firing rates are often considerably different from those in the eighth nerve. These findings demonstrate that the sound signals are not transmitted unchanged directly from the ear to the higher levels of the brain; instead, information from the sound signals begins to be dissected from the impulse traffic at levels as low as the cochlear nuclei. We will have more to say about this later, especially in relation to perception of direction from which sound comes.

Another significant feature of the auditory pathways is that low rates of impulse firing continue even in the absence of sound all the way from the cochlear nerve fibers to the auditory cortex. When the basilar membrane moves toward the scala vestibuli, the impulse traffic increases; and when the basilar membrane moves toward the scala tympani, the impulse traffic decreases. Thus, the presence of this background signal allows information to be transmitted from the basilar membrane when the membrane moves in either direction: positive information in one direction and negative information in the opposite direction. Were it not for the background signal, only the positive half of the information could be transmitted. This type of so-called "carrier wave" method for transmitting information is utilized in many parts of the brain, as has been discussed in several of the preceding chapters.

**Function of the Auditory Relay Nuclei.** Very little is known about the function of the different nuclei in the auditory pathway. However, cats and even monkeys can still detect barely threshold sounds when the cerebral cortex is removed bilaterally, which indicates that the nuclei in the brain stem and thalamus can perform many auditory functions even without the cerebral cortex. However, discrimination of pitch and tonal patterns is considerably impaired. In man, *bilateral* destruction of the cortical auditory centers probably gives a different picture: it is said to cause almost total deafness; unfortunately, this has not been studied adequately. This indicates that in man, however, the elements of the lower centers might be greatly suppressed.

One of the important features of auditory transmission through the relay nuclei is the spatial orientation of the pathways for sounds of different frequencies. For instance, in the dorsal cochlear nucleus, high frequencies are represented along the medial edge while low frequencies are represented along the lateral edge, and a similar type of spatial orientation occurs throughout the auditory pathway as it travels upward to the cortex.

## FUNCTION OF THE CEREBRAL CORTEX IN HEARING

The projection of the auditory pathway to the cerebral cortex is illustrated in Figure 61–10, which shows that the auditory cortex lies principally on the *supratemporal plane of the superior temporal gyrus* but also extends over the *lateral border of the temporal lobe,* over much of the *insular cortex,* and even into the most lateral portion of the *parietal operculum.*

Two separate areas are shown in Figure 61–10; the *primary auditory cortex* and the *au-*

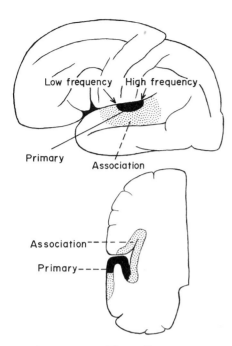

**Figure 61–10.**  The auditory cortex.

ditory association cortex. The primary auditory cortex is directly excited by projections from the medial geniculate body, while the auditory association areas are usually excited secondarily by impulses from the primary auditory cortex and by projections from thalamic association areas adjacent to the medial geniculate body.

**Locus of Sound Frequency Perception in the Primary Auditory Cortex.**  Certain parts of the primary auditory cortex are known to respond to high frequencies and other parts to low frequencies. In monkeys, the posterior part of the supratemporal plane responds to high frequencies, while the anterior part responds to low frequencies. Presumably, the same frequency localization occurs in the human cortex but this is yet unproven.

The frequency range to which each individual neuron in the auditory cortex responds is much narrower than that in the cochlear and brain stem relay nuclei. Referring back to Figure 61–5B, we note that the basilar membrane near the base of the cochlea is stimulated by all frequency sounds, and in the cochlear nuclei this same breadth of sound representation is found. Yet by the time the excitation has reached the cerebral cortex, each sound-responsive neuron responds to only a narrow range of frequencies rather than to a broad range. Therefore, some-

where along the pathway processing mechanisms in some way "sharpen" the frequency response. It is believed that this sharpening effect is caused mainly by the phenomenon of lateral inhibition, which was discussed in Chapter 47 in relation to mechanisms for transmitting information in nerves. That is, stimulation of the cochlea at one frequency causes inhibition of signals caused by sound frequencies on either side of the stimulated frequency, this effect resulting from collateral fibers angling off the primary signal pathway and exerting inhibitory influences on adjacent pathways. The same effect has also been demonstrated to be important in sharpening patterns of somesthetic images, visual images, and other types of sensations.

Some of the neurons in the auditory cortex, especially in the auditory association cortex, do not respond at all to sounds in the ear. It is believed that these neurons "associate" different sound frequencies with each other or associate sound information with information from other sensory areas of the cortex. Indeed, the parietal portion of the auditory association cortex partly overlaps somatic sensory area II, which could provide easy opportunity for association of auditory information with somatic sensory information.

**Discrimination of Sound "Patterns" by the Auditory Cortex.**  Complete bilateral removal of the auditory cortex does not prevent an animal from detecting sounds or reacting in a crude manner to the sounds. However, it does greatly reduce or sometimes even abolish his ability to discriminate different sound pitches and especially *patterns of sound*. For instance, an animal that has been trained to recognize a combination or sequence of tones, one following the other in a particular pattern, loses this ability when the auditory cortex is destroyed, and, furthermore, he cannot relearn this type of response. Therefore, the auditory cortex is important in the discrimination of *tonal patterns*.

In the human being, lesions affecting the auditory association areas but not affecting the primary auditory cortex will allow the person full capability to hear and differentiate sound tones as well as to interpret at least a few simple patterns of sound. However, he will often be completely unable to interpret the *meaning* of the sound that he hears. For instance, lesions in the posterior portion of the superior temporal gyrus often make it impossible for the person to interpret the meanings of words even though he hears them perfectly well and can often even repeat them; all the time, however, he does not know the meaning of the words. These functions of the auditory association areas and their relationship to the overall intellectual functions of the brain were discussed in detail in Chapter 55.

## DISCRIMINATION OF DIRECTION FROM WHICH SOUND EMANATES

A person determines the direction from which sound emanates by at least two different mechanisms: (1) by the time lag between the entry of sound into one ear and into the opposite ear and (2) by the difference between the intensities of the sounds in the two ears. The first mechanism functions best for frequencies below 3000 cycles per second, and the intensity mechanism operates best at higher frequencies because the head acts as a sound barrier at these frequencies. The time lag mechanism discriminates direction much more exactly than the intensity mechanism, for the time lag mechanism does not depend on extraneous factors but only on an exact interval of time between two acoustical signals. If a person is looking straight toward the sound, the sound reaches both ears at exactly the same instant, while, if the right ear is closer to the sound than the left ear, the sound signals from the right ear are perceived ahead of those from the left ear.

If a sound emanates from the right side 45 degrees from the frontal direction and if still a second sound emanates also from the right but 45 degrees from behind, the difference in time of arrival of the two sounds at the two ears will be exactly the same. For this reason, it is difficult to distinguish whether the sound is originating from the frontal or the posterior quadrant. The principal method by which a person determines this is to rotate his head quickly. If he turns his head toward the right when the sound is coming from the frontal quadrant, the time lag between the two ears becomes smaller. If the sound is originating from the posterior quadrant, the time lag becomes greater. A sudden sound that occurs so rapidly that the person does not have time to move his head often cannot be properly localized to the frontal or posterior quadrant.

**Neural Mechanisms for Detecting Sound Direction.** Destruction of the auditory cortex on both sides of the brain, in either man or lower mammals, causes loss of almost all ability to detect the direction from which sound comes. Yet, the mechanism for this detection process begins in the superior olivary nuclei,

even though it requires the neural pathways all the way from these nuclei to the cortex for interpretation of the signals. The mechanism is believed to be the following:

When the sound enters one ear slightly before it enters the other ear, it *excites* the neurons in the medial portion of the contralateral superior olivary nucleus but, at the same time, it *inhibits* the neurons in the ipsilateral superior olivary nucleus, and this inhibition lasts for a fraction of a millisecond. Therefore, for a short period of time after the sound reaches the first ear, the pathway for sound coming from the second ear is in an inhibited state. Furthermore, certain neurons of the medial superior olivary nuclei have longer time lags of inhibition than do other portions. Therefore, when the sound signal from the second ear enters the inhibited superior olivary nucleus, the signal will pass up the auditory pathway through some of the neurons but not through others. And the specific neurons through which the signal passes are determined by the time-interval of the sound between the two ears. Thus, a spatial pattern of neuronal stimulation develops, with the short lagging sounds stimulating one set of neurons maximally and the long lagging sounds stimulating another set of neurons maximally. This spatial orientation of signals is then transmitted all the way to the auditory cortex where sound direction is determined by the locus in the cortex that is stimulated maximally. It is believed that the signals for determining sound direction are transmitted through a slightly different pathway and that this pathway terminates in the cerebral cortex in a slightly different locus from the transmission pathway and the termination locus of the tonal patterns of sound.

This mechanism for detection of sound direction indicates again how the information in sensory signals is dissected out as the signals pass through different levels of neuronal activity. In this case, the "quality" of sound direction is separated from the "quality" of sound tones at the level of the superior olivary nuclei.

## CENTRIFUGAL CONDUCTION OF IMPULSES FROM THE CENTRAL NERVOUS SYSTEM

Retrograde pathways have been demonstrated at each level of the central nervous system all the way from the auditory cortex to the cochlea. The final pathway is mainly from the superior olivary nucleus to the organ of Corti.

These retrograde fibers are inhibitory. Indeed, direct stimulation of discrete points in the olivary nucleus have been shown to inhibit specific areas of the organ of Corti, reducing their sound sensitivities as much as 15 to 20 decibels. One can readily understand how this could allow a person to direct his attention to sounds of particular qualities while rejecting sounds of other qualities. This is readily demonstrated when one listens to a single instrument in a symphony orchestra.

# HEARING ABNORMALITIES

## TYPES OF DEAFNESS

Deafness is usually divided into two types; first, that caused by impairment of the cochlea or auditory nerve, which is usually classed under the heading "nerve deafness," and, second, that caused by impairment of the middle ear mechanisms for transmitting sound into the cochlea, which is usually called "conduction deafness." Obviously, if either the cochlea or the auditory nerve is completely destroyed the person is permanently deaf. However, if the cochlea and nerve are still intact but the ossicular system has been destroyed or ankylosed ("frozen" in place by fibrosis or calcification), sound waves can still be conducted into the cochlea by means of bone conduction.

**Tuning Fork Test for Differentiation Between Nerve Deafness and Conduction Deafness.** To test an ear for bone conduction with a tuning fork, one places a vibrating fork in front of the ear and the subject listens to the tone of the fork until he can no longer hear it. Then the butt of the still weakly vibrating fork is immediately placed against the mastoid process. If his bone conduction is better than his air conduction, he will again hear the sound of the tuning fork. If this occurs, his deafness may be considered to be conduction deafness. However, if on placing the butt of the fork against the mastoid process he cannot hear the sound of the tuning fork, his bone

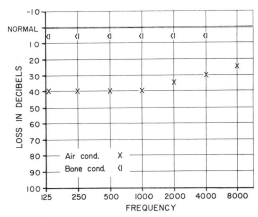

**Figure 61–12.** Audiogram of deafness resulting from middle ear sclerosis.

conduction is probably decreased as much as his air conduction, and the deafness is presumably due to damage in the cochlea or in the nervous system rather than in the ossicular system—that is, it is nerve deafness.

**The Audiometer.** To determine the nature of hearing disabilities more exactly than can be accomplished by the above method, the "audiometer" is used. This is simply an earphone connected to an electronic oscillator capable of emitting pure tones ranging from low frequencies to high frequencies. Based on previous studies of normal persons, the instrument is calibrated so that the zero intensity level of sound at each frequency is the loudness that can barely be heard by the normal person. However, a calibrated volume control can be changed to increase or decrease the loudness of each tone above or below the zero level. If the loudness of a tone must be increased to 30 decibels above normal before the subject can hear it, he is said to have a *hearing loss* of 30 decibels for that particular tone.

In performing a hearing test using an audiometer, one tests approximately 8 to 10 tones one at a time covering the auditory spectrum, and the hearing loss is determined for each of these tones. Then the so-called "audiogram" is plotted as shown in Figures 61–11 and 61–12, depicting the hearing loss for each of the tones in the auditory spectrum.

The audiometer, in addition to being equipped with an earphone for testing air conduction by the ear, is also equipped with an electronic vibrator for testing bone conduction from the mastoid process into the cochlea.

*The Audiogram in Nerve Deafness.* If a person has nerve deafness—this term including damage to the cochlea, to the auditory nerve, or to the central nervous system circuits from the ear—he has lost the ability to hear sound as tested by both the air conduction apparatus and the bone conduction apparatus. An audiogram depicting nerve deafness is illustrated in Figure 61–11. In this figure the deafness is mainly for

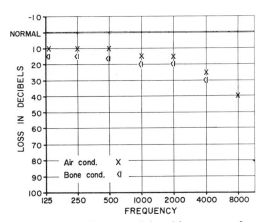

**Figure 61–11.** Audiogram of the old-age type of nerve deafness.

high frequency sound. Such deafness could be caused by damage to the base of the cochlea. This type of deafness occurs to some extent in almost all older persons.

***The Audiogram in Conduction Deafness.*** A second and frequent type of deafness is that caused by fibrosis of the middle ear following repeated infection in the middle ear or in the hereditary disease called *otosclerosis*. In this instance the sound waves cannot be transmitted easily to the oval window. Figure 61–12 illustrates an audiogram from a person with "middle ear deafness" of this type. In this case the bone conduction is essentially normal, but air conduction is greatly depressed at all frequencies, more so at the low frequencies. In this type of deafness, the faceplate of the stapes frequently becomes "ankylosed" by bony overgrowth to the edges of the oval window. In this case, the person becomes totally deaf for air conduction; but he can be made to hear again almost normally by removing the stapes and replacing it with a minute teflon or metal prosthesis that transmits the sound from the incus to the oval window.

# REFERENCES

Ainsworth, W. A.: The perception of speech signals. *Sci. Prog.*, 62:33, 1975.

Dallos, P.: The Auditory Periphery. New York, Academic Press, Inc., 1973.

Davis, H.: Biophysics and physiology of the inner ear. *Physiol. Rev.*, 37:1, 1957.

Eldredge, D. H., and Miller, J. D.: Physiology of hearing. *Ann. Rev. Physiol.*, 33:281, 1971.

Erulkar, S. D.: Comparative aspects of spatial localization of sound. *Physiol. Rev.*, 52:237, 1972.

Friedmann, I.: Pathology of the Ear. Philadelphia, J. B. Lippincott Company, 1973.

Gerber, S. E.: Introductory Hearing Science. Philadelphia, W. B. Saunders Company, 1975.

Grinnell, A. D.: Comparative physiology of hearing. *Ann. Rev. Physiol*, 31:545, 1969.

Gulick, W. L.: Hearing: Physiology and Psychophysics. New York, Oxford University Press, 1971.

Harris, J. D.: The Electrophysiology and Layout of the Auditory Nervous System. Indianapolis, The Bobbs-Merrill Co., Inc., 1974.

Harrison, J. M.: The auditory system of the medulla and localization. *Fed. Proc.*, 33:1901, 1974.

Moushegian, G., and Rupert, A. L.: Relations between the psychophysics and the neurophysiology of sound localization. *Fed. Proc.*, 33:1924, 1974.

Oster, G.: Auditory beats in the brain. *Sci. Amer.*, 229:94, 1973.

Small, A. M., Nordmark, J. O., Jeffress, L. A., Elliott, D. N., et al.: In Tobias, J. V. (ed.): Foundations of Modern Auditory Theory. Vol. 1. New York, Academic Press, Inc., 1973.

Starr, A.: Neurophysiological mechanisms of sound localization. *Fed. Proc.*, 33:1911, 1974.

Stevens, K. N., House, A. S., Tanner, W. P., Sorkin, R. D. et al.: In Tobias, J. V. (ed.): Foundations of Modern Auditory Theory. Vol. 2. New York, Academic Press, Inc., 1973.

Taylor, W. (ed.): Disorders of Auditory Function. New York, Academic Press, Inc., 1973.

Tobias, J. V. (ed.): Foundations of Modern Auditory Theory. Vols. 1 and 2. New York, Academic Press, Inc., 1970.

Warren, R. M., and Warren, R. P.: Auditory illusions and confusions. *Sci. Amer.*, 223:30, 1970.

Wever, E. G.: Electrical potentials of the cochlea. *Physiol. Rev.*, 46:102, 1966.

Wever, E. G., and Lawrence, M.: Physiological Acoustics. Princeton, Princeton University Press, 1954.

Whitfield, I. C.: Auditory Pathway. Baltimore, The Williams & Wilkins Company, 1968.

# | 62 |

# The Chemical Senses—
# Taste and Smell

## THE SENSE OF TASTE

Taste is a function of the *taste buds* in the mouth, and its importance lies in the fact it allows the person to select his food in accord with his desires and perhaps also in accord with the needs of the tissues for specific nutritive substances.

On the basis of psychologic studies, there are generally believed to be at least four *primary* sensations of taste: *sour, salty, sweet, and bitter*. Yet we know that a person can perceive literally hundreds of different tastes. These are all supposed to be combinations of the four primary sensations in the same manner that all the colors of the spectrum are combinations of three primary color sensations, as described in Chapter 59. However, there might be other less conspicuous classes or subclasses of primary sensations. Nevertheless, the following discussion is based on the usual classification of only four primary tastes.

### THE PRIMARY SENSATIONS OF TASTE

**The Sour Taste.** The sour taste is caused by acids, and the intensity of the taste sensation is approximately proportional to the logarithm of the *hydrogen ion concentration*. That is, the more acidic the acid, the stronger becomes the sensation.

**The Salty Taste.** The salty taste is elicited by ionized salts. The quality of the taste varies somewhat from one salt to another because the salts also elicit other taste sensations besides saltiness. The cations of the salts are mainly reponsible for the salty taste, but the anions also contribute at least to some extent.

**The Sweet Taste.** The sweet taste is not caused by any single class of chemicals. A list of some of the types of chemicals that cause this taste includes: sugars, glycols, alcohols, aldehydes, ketones, amides, esters, amino acids, sulfonic acids, halogenated acids, and inorganic salts of lead and beryllium. Note specifically that almost all the substances that cause a sweet taste are organic chemicals; the only inorganic substances that elicit the sweet taste at all are certain salts of lead and beryllium.

Table 62–1 shows the relative intensities of taste of certain substances that cause the sweet taste. *Sucrose*, which is common table sugar, is considered to have an index of 1. Note that one of the substances has a sweet index 5000 times as great as that of sucrose. However, this extremely sweet substance, known as *P-4000*, is unfortunately extremely toxic and therefore cannot be used as a sweetening agent. *Saccharin*, on the other hand, is also more than 600 times as sweet as common table sugar, and since it is not toxic it can be used with impunity as a sweetening agent.

**The Bitter Taste.** The bitter taste, like the sweet taste, is not caused by any single type of chemical agent, but, here again, the substances that give the bitter taste are almost entirely organic substances. Two particular classes of substances are especially likely to cause bitter taste sensations, (1) long chain organic substances and (2) alkaloids. The alkaloids include many of the drugs used in medicines such as quinine, caffeine, strychnine, and nicotine.

TABLE 62–1.    Relative Taste Indices of Different Substances

| Sour Substances | Index | Bitter Substances | Index | Sweet Substances | Index | Salty Substances | Index |
|---|---|---|---|---|---|---|---|
| Hydrochloric acid | 1 | Quinine | 1 | Sucrose | 1 | NaCl | 1 |
| Formic acid | 1.1 | Brucine | 11 | 1-propoxy-2-amino- | | NaF | 2 |
| Chloracetic acid | 0.9 | Strychnine | 3.1 | 4-nitrobenzene | 5000 | CaCl$_2$ | 1 |
| Acetyllactic acid | 0.85 | Nicotine | 1.3 | Saccharin | 675 | NaBr | 0.4 |
| Lactic acid | 0.85 | Phenylthiourea | 0.9 | Chloroform | 40 | NaI | 0.35 |
| Tartaric acid | 0.7 | Caffeine | 0.4 | Fructose | 1.7 | LiCl | 0.4 |
| Malic acid | 0.6 | Veratrine | 0.2 | Alanine | 1.3 | NH$_4$Cl | 2.5 |
| Potassium H tartrate | 0.58 | Pilocarpine | 0.16 | Glucose | 0.8 | KCl | 0.6 |
| Acetic acid | 0.55 | Atropine | 0.13 | Maltose | 0.45 | | |
| Citric acid | 0.46 | Cocaine | 0.02 | Galactose | 0.32 | | |
| Carbonic acid | 0.06 | Morphine | 0.02 | Lactose | 0.3 | | |

From Derma: *Proc. Oklahoma Acad. Sc.,* 27:9, 1947; and Pfaffman: Handbook of Physiology, Sec. I, Vol. I, p. 507, 1959. Baltimore, The Williams & Wilkins Co.

Some substances that at first taste sweet have a bitter after-taste. This is true of saccharin, which makes this substance objectionable to some people. Some substances have a sweet taste on the front of the tongue, where taste buds with special sensitivity to the sweet taste are principally located, and a bitter taste on the back of the tongue, where taste buds more sensitive to the bitter taste are located.

The bitter taste, when it occurs in high intensity, usually causes the person or animal to reject the food. This is undoubtedly an important purposive function of the bitter taste sensation because many of the deadly toxins found in poisonous plants are alkaloids, and these all cause an intensely bitter taste.

### Threshold for Taste

The threshold for stimulation of the sour taste by hydrochloric acid averages 0.0009 N; for stimulation of the salty taste by sodium chloride: 0.01 M; for the sweet taste by sucrose: 0.01 M; and for the bitter taste by quinine: 0.000008 M. Note especially how much more sensitive is the bitter taste sense to stimuli than all the others, which would be expected since this sensation provides an important protective function.

Table 62–1 gives the relative taste indices (the reciprocals of the taste thresholds) of different substances. In this table, the intensities of the four different primary sensations of taste are referred, repetively, to the intensities of taste of hydrochloric acid, quinine, sucrose, and sodium chloride, each of which is considered to have a taste index of 1.

**Taste Blindness.**   Many persons are taste blind

for certain substances, especially for different types of thiourea compounds. A substance used frequently by psychologists for demonstrating taste blindness is *phenylthiocarbamide*, for which approximately 15 to 30 per cent of all people exhibit taste blindness, the exact percentage depending on the method of testing.

## THE TASTE BUD AND ITS FUNCTION

Figure 62–1 illustrates a taste bud, which has a diameter of about $1/30$ millimeter and a length of about $1/16$ millimeter. The taste bud is composed of about 40 modified epithelial cells called *taste cells*. These cells are continually being replaced by mitotic division from the surrounding epithelial cells so that some are young cells and others mature cells that lie toward the center of the bud and soon dissolute. The life span of each taste cell is about ten days.

The outer tips of the taste cells are arranged around a minute *taste pore*, shown in Figure

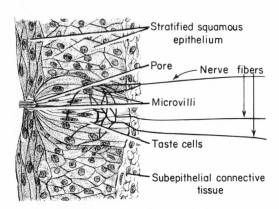

Figure 62–1.   The taste bud.

62–1. From the tip of each cell, several *microvilli*, or *taste hairs*, about 2 to 3 microns in length and 0.2 micron in width, protrude outward through the taste pore to approach the cavity of the mouth. These microvilli are believed to provide the receptor surface for taste.

Interwoven among the taste cells is a branching terminal network of several *taste nerve fibers* that are stimulated by the taste cells. These fibers invaginate deeply into folds of the taste cell membranes, so that there is extremely intimate contact between the taste cells and the nerves. Several taste buds can be innervated by the same taste fiber.

An interesting feature of the taste buds is that they completely degenerate when the taste nerve fibers are destroyed. Then, if the taste fibers regrow to the epithelial surface of the mouth, the local epithelial cells regroup themselves to form new taste buds. This illustrates the important principle of "trophic" function of nerve fibers in certain parts of the body. The cause of the tropism is unknown, but it has been postulated to be a protein trophic factor secreted by the nerve endings.

**Location of the Taste Buds.** The taste buds are found on three out of four different types of papillae of the tongue, as follows: (1) A large number of taste buds are on the walls of the troughs that surround the circumvallate papillae, which form a V line toward the posterior of the tongue. (2) Moderate numbers of taste buds are on the fungiform papillae over the front surface of the tongue. (3) Moderate numbers are on the foliate papillae located in the folds along the posterolateral surfaces of the tongue. Additional taste buds are located on the palate and a few on the tonsillar pillars and at other points around the nasopharynx. Adults have approximately 10,000 taste buds, and children a few more. Beyond the age of 45 many taste buds rapidly degenerate causing the taste sensation to become progressively less critical.

Especially important in relation to taste is the tendency for taste buds subserving particular primary sensations of taste to be localized in special areas. The sweet taste is localized *principally* on the anterior surface and tip of the tongue, the salty and sour tastes on the two lateral sides of the tongue, and the bitter taste on the circumvallate papillae on the posterior of the tongue.

**Specificity of Taste Buds for the Primary Taste Stimuli.** In the foregoing paragraphs we have discussed taste buds as if each responded to a particular type of taste stimulus and not to other taste stimuli. In a statistical sense this is true, but so far as any single taste bud is concerned, it is not true, for most taste buds respond to varying extents to at least three and usually to all four of the primary taste stimuli.

Figure 62–2 illustrates the responsiveness of four different taste buds to the different primary tastes. Figure 62–2A illustrates a bud responsive to all four types of taste stimuli, but especially to saltiness. Figure 62–2B illustrates a taste bud strongly responsive to sourness and saltiness but also responsive to a moderate degree to both bitterness and sweetness. Likewise, Figures 62–2C and 62–2D illustrate response characteristics of two other taste buds.

***Detection of Different Sensations of Taste by the Taste Buds.*** Since each taste bud responds to multiple primary taste stimuli, it is difficult to understand how a person perceives the different primary taste sensations independently of each other. However, a theory attempting to explain this is the following: Some area in the nervous system presumably is capable of detecting the *ratios* of stimulation of the different types of taste buds. For the sweet taste, a taste bud that responds strongly to sweet stimuli will transmit a stronger signal than will a taste bud that responds only weakly to the sweet taste. Thus, the ratio of the signal strength from the first bud to the signal strength from the second is high, and it is this *ratio* that elicits the sweet taste. Likewise, other ratios from other taste buds will theoretically be perceived as salty, sour, and bitter tastes.

Tastes besides the four primary tastes obviously would be detected as still other ratios of stimulation of the different taste buds, thus giv-

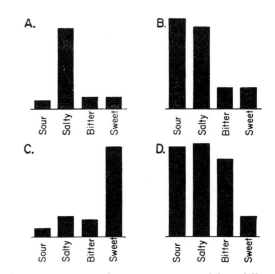

**Figure 62–2.** Specific responsiveness of four different types of taste buds, showing multiple stimulation by the different primary sensations of taste in the case of each of the taste buds.

ing all the different gradations of taste sensations that are known to occur.

**Mechanism of Stimulation of Taste Buds.** *The Receptor Potential.* The membrane of the taste cell, like that of other sensory receptor cells, normally is negatively charged on the inside with respect to the outside. Application of a taste substance to the taste hairs causes partial loss of this negative potential. The decrease in potential, within a wide range, is approximately proportional to the logarithm of concentration of the stimulating substance. This change in potential in the taste cell is the *receptor potential* for taste.

The mechanism by which the stimulating substance reacts with the taste hairs to initiate the receptor potential is unknown. It is believed by some physiologists that the substance is simply adsorbed to receptors on the surface of the taste hair and that this adsorption changes the physical characteristics of the hair membrane. This in turn makes the taste cell more permeable to ions and thus depolarizes the cell. The substance is gradually washed away from the taste hair by the saliva, thus removing the taste stimulus. Supposedly, the type of receptor substance or substances in each taste hair determine the types of taste substances that will elicit responses.

*Generation of Nerve Impulses by the Taste Bud.* The taste nerve fiber endings are encased by folds of the taste cell membranes. In some way not understood the receptor potentials of the taste cells generate impulses in the taste fibers. On first application of the taste stimulus, the rate of discharge of the nerve fibers rises to a peak in a small fraction of a second, but then it adapts within the next 2 seconds back to a much lower steady level. Thus, a strong immediate signal is transmitted by the taste nerve, and a much weaker continuous signal is transmitted as long as the taste bud is exposed to the taste stimulus.

## TRANSMISSION OF TASTE SIGNALS INTO THE CENTRAL NERVOUS SYSTEM

Figure 62–3 illustrates the neuronal pathways for transmission of taste sensations from the tongue and pharyngeal region into the central nervous system. Taste impulses from the anterior two thirds of the tongue pass first into the *fifth nerve* and then through the *chorda tympani* into the *seventh nerve*, thence into the *tractus solitarius* in the brain stem. Taste sensations

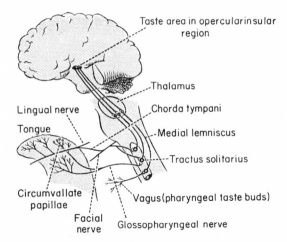

**Figure 62–3.** Transmission of taste impulses into the central nervous system.

from the circumvallate papillae on the back of the tongue and from other posterior regions of the mouth are transmitted through the *ninth nerve* also into the *tractus solitarius* but at a slightly lower level. Finally, a few taste impulses are transmitted into the *tractus solitarius* from the base of the tongue and other parts of the pharyngeal region by way of the *vagus nerve*. All taste fibers synapse in the *nuclei of the tractus solitarius* and send second order neurons to a small area of the *thalamus* located slightly medial to the thalamic terminations of the facial regions of the dorsal column-medial lemniscal system. From the thalamus, third order neurons are believed to be transmitted to the *parietal opercular-insular area* of the cerebral cortex. This lies at the very lateral margin of the postcentral gyrus in the sylvian fissure in close association with, or even superimposed on, the tongue area of somatic area I.

From this description of the taste pathways, it immediately becomes evident that they parallel closely the somatic pathways from the tongue.

**Taste Reflexes.** From the tractus solitarius a large number of impulses are transmitted directly into the *superior* and *inferior salivatory nuclei*, and these in turn transmit impulses to the submaxillary and parotid glands to help control the secretion of saliva during the ingestion of food.

**Adaptation of Taste.** Everyone is familiar with the fact that taste sensations adapt rapidly. Yet, from electrophysiological studies of taste nerve fibers, it seems that the taste buds themselves do not adapt enough by themselves to account for all or most of the taste adaptation. They have a rapid period of adaptation during the first 2 to 3 seconds after contact with the taste stimulus. The first burst of impulses from the taste bud allows one to detect extremely minute concentrations of taste substances, but normal concentrations of taste substances cause prolonged discharge of the taste fibers. Therefore, the progressive adaptation that occurs in the sensation of taste has been postulated to occur in the central nervous system itself, though the mechanism and site of this is not known. If this is true—and it is still questionable—it is a mechanism that is different from that of other sensory systems, which adapt almost entirely in the receptors.

## SPECIAL ATTRIBUTES OF THE TASTE SENSE

**Affective Nature of Taste.** *Pleasantness* and *unpleasantness* are called the "affective" attributes of a sensation. Figure 62–4 illustrates the affective effects of different types of taste at different concentrations of the stimulating substances, showing, strangely enough, that the sweet taste is likely to be unpleasant at a very low concentration but very pleasant at high concentrations. The other types of taste, on the other hand, are likely to be pleasant at low concentrations but exceedingly unpleasant at high concentrations. This is particularly true of the bitter taste.

**Importance of the Sense of Smell in Taste.** Persons with severe colds frequently state that they have lost their sense of taste. However, on testing the

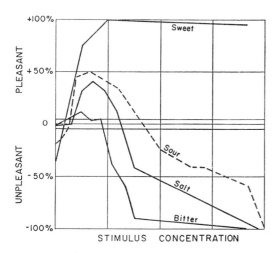

**Figure 62–4.** The affective nature of the different primary sensations of taste at progressively increasing degrees of taste stimulus. (From Engel in Woodworth and Schlosberg: Experimental Psychology. Holt, Rinehart & Winston, Inc., 1954.)

taste sensations, these are found to be completely normal. This illustrates that much of what we call taste is actually smell. Odors from the food can pass upward into the nasopharynx, often stimulating the olfactory system thousands of times as strongly as the taste system. For instance, if the olfactory system is intact, alcohol can be "tasted" in $\frac{1}{50,000}$ the concentration required when the olfactory system is not intact.

**Taste Preference and Control of the Diet.** Taste preferences mean simply that an animal will choose certain types of food in preference to others, and he automatically uses this to help control the type of diet he eats. Furthermore, to a great extent his taste preferences change in accord with the needs of the body for certain specific substances. The following experimental studies will illustrate this ability of an animal to choose food in accord with the need of his body: First, adrenalectomized animals automatically select drinking water with a high concentration of sodium chloride in preference to pure water, and this in many instances is sufficient to supply the needs of the body and prevent death as a result of salt depletion. Second, an animal injected with excessive amounts of insulin develops a depleted blood sugar, and he automatically chooses the sweetest food from among many samples. Third, parathyroidectomized animals automatically choose drinking water with high concentration of calcium chloride.

These same phenomena are also observed in many instances of everyday life. For instance, the salt licks of the desert region are known to attract animals from far and wide, and even the human being rejects any food that has an unpleasant affective sensation, which certainly in many instances protects our bodies from undesirable substances.

The phenomenon of taste preference almost certainly results from some mechanism located in the central nervous system and not from a mechanism in the taste buds themselves, because many experiments have demonstrated that taste preference can occur in animals even in the absence of changes in stimulus threshold of the taste buds for the substances involved. Another reason for believing this to be a central phenomenon is that previous experience with unpleasant or pleasant tastes plays a major role in determining one's different taste preferences. For instance, if a person becomes sick immediately after eating a particular type of food, he generally develops a negative taste preference, or taste aversion, for that particular food thereafter; the same effect can be demonstrated in animals.

## THE SENSE OF SMELL

Smell is the least well understood sense. This results partly from the location of the olfactory membrane high in the nose where it is difficult to study and partly from the fact that the sense of smell is a subjective phenomenon that cannot be studied with ease in lower animals. Still another complicating problem is the fact that the sense of smell is almost rudimentary in the human being in comparison with that of some lower animals.

## THE OLFACTORY MEMBRANE

The olfactory membrane lies in the superior part of each nostril, as illustrated in Figure 62–6. Medially it folds downward over the surface of the septum, and laterally it folds over the superior turbinate and even over a small portion of the upper surface of the middle turbinate. In each nostril the olfactory membrane has a surface area of approximately 2.4 square centimeters.

**The Olfactory Cells.** The receptor cells for the smell sensation are the *olfactory cells*, which are actually bipolar nerve cells derived originally from the central nervous system itself. There are about 100 million of these cells in the olfactory epithelium interspersed among *sustentacular cells*, as shown in Figure 62–5. The mucosal end of the olfactory cell forms a knob called the *olfactory vesicle* from which large numbers of *olfactory hairs*, or *cilia*, 0.3 micron in diameter and 50 to 150 microns in length, project into the mucus that coats the inner surface of the nasal cavity. These projecting olfactory hairs are believed to react to odors in the air and then to stimulate the olfactory cells, as is discussed below. Spaced among the olfactory cells in the olfactory membrane are many small *glands of Bowman* that secrete mucus onto the surface of the olfactory membrane.

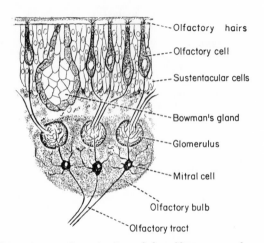

**Figure 62–5.** Organization of the olfactory membrane. (After Maximow and Bloom: A Textbook of Histology.)

Labels on figure:
- Olfactory hairs
- Olfactory cell
- Sustentacular cells
- Bowman's gland
- Glomerulus
- Mitral cell
- Olfactory bulb
- Olfactory tract

## STIMULATION OF THE OLFACTORY CELLS

**The Necessary Stimulus for Smell.** We do not know what it takes chemically to stimulate the olfactory cells. Yet we do know the physical characteristics of the substances that cause olfactory stimulation: First, the substance must be volatile so that it can be sniffed into the nostrils. Second, it must be at least slightly water soluble so that it can pass through the mucus to the olfactory cells. And, third, it must also be lipid soluble, presumably because the olfactory hairs and outer tips of the olfactory cells are composed principally of lipid materials.

Regardless of the basic mechanism by which the olfactory cells are stimulated, it is known that they become stimulated only when air blasts upward into the superior region of the nose. Therefore, smell occurs in cycles along with the inspirations, which indicates that the olfactory receptors respond in milliseconds to the volatile agents. Because smell intensity is exacerbated by blasting air through the upper reaches of the nose, a person can greatly increase his sensitivity of smell by the well known sniffing technique.

**Receptor Potentials in Olfactory Cells.** The olfactory cells are believed to react to olfactory stimuli in the same manner that most other sensory receptors react to their specific stimuli; that is, by generating a receptor potential, which in turn initiates nerve impulses in the olfactory nerve fibers. An experiment that demonstrates this property of the olfactory receptors is the following: An electrode is placed on the surface of the olfactory membrane, and its electrical potential with respect to the remainder of the body is recorded. When an odorous substance is insufflated into the nostril, the potential becomes negative and remains negative as long as the odorous air continues to pass through the nostril. This electrical recording is called the *electro-olfactogram*, and it is believed to result from summation of receptor potentials developed in the receptor olfactory cells.

Over a wide range, both the amplitude of the electro-olfactogram and the rate of olfactory nerve impulses are approximately proportional to the logarithm of the stimulus strength, which illustrates that the olfactory receptors tend to obey principles of transduction similar to those of other sensory receptors.

**Adaptation.** The olfactory receptors adapt approximately 50 per cent in the first second or so after stimulation. Thereafter, they adapt further very slowly. Yet we all know from our own experience that smell sensations adapt almost to extinction within a minute or more after one enters a strongly odorous atmosphere. Since the psychological adaptation seems to be more rapid than the adaptation of the receptors, it has been suggested that at least part of this adaptation occurs in the central nervous system, as has also been postulated for adaptation of taste sensations.

**Search for the Primary Sensations of Smell.** Most physiologists are convinced that the many smell sensations are subserved by a few rather discrete primary sensations in the same way that taste is subserved by sour, sweet, bitter, and salty sensations. Thus far, only minor success has been achieved in classifying the primary sensations of smell. Yet, on the basis of psychological tests and action potential studies from various points in the olfactory nerve pathways, it has been postulated that about seven different primary classes of olfactory stimulants preferentially excite separate olfactory cells. These classes of olfactory stimulants may be characterized as follows:

1. Camphoraceous
2. Musky
3. Floral
4. Pepperminty
5. Ethereal
6. Pungent
7. Putrid

However, it is unlikely that this list actually represents the true primary sensations of smell even though it does illustrate the results of one of the many attempts to classify them. Indeed, several clues in recent years have indicated that there may be as many as *50* or more primary sensations of smell—a marked contrast to there being only *three* primary sensations of color detected by the eyes and only *four* primary sensations of taste detected by the tongue. For instance, individual persons have been found who are specifically *odor-blind* for more than 50 different substances. Since it is presumed that odor-blindness for each substance represents a lack of the appropriate receptor for that substance, it is postulated that the sense of smell might be subserved by 50 or more primary smell sensations.

Two basic theories have been postulated to explain the abilities of different receptors to respond selectively to different types of olfactory stimulants, the *chemical theory* and the *physical theory*. The chemical theory assumes that *receptor chemicals* in the membranes of the olfactory hairs react specifically with the different types of olfactory stimulants. The type of receptor chemical determines the type of stimulant that will elicit a response in the olfactory cell. The reaction between the stimulant and the re-

ceptor substance supposedly increases the permeability of the olfactory hair membrane, and this in turn creates the receptor potential in the olfactory cell that generates impulses in the olfactory nerve fibers.

The physical theory assumes that differences in *physical receptor sites* on the olfactory hair membranes of separate olfactory cells allow specific olfactory stimulants to adsorb to the membranes of different olfactory cells. A fact that supports this theory is that many substances that have very different chemical properties but that have almost identical molecular shapes have the same odor. This indicates that a physical property of the stimulant might determine the odor.

**Affective Nature of Smell.** Smell, equally as much as taste, has the affective qualities of either pleasantness or unpleasantness. Because of this, smell is as important as, if not more important than, taste in the selection of food. Indeed, a person who has previously eaten food that has disagreed with him is often nauseated by the smell of that same type of food on a second occasion. Other types of odors that have proved to be unpleasant in the past may also provoke a disagreeable feeling; on the other hand, perfume of the right quality can wreak havoc with the masculine emotions. In addition, in some lower animals odors are the primary excitant of sexual drive.

**Threshold for Smell.** One of the principal characteristics of smell is the minute quantity of the stimulating agent in the air required to effect a smell sensation. For instance, the substance *methyl mercaptan* can be smelled when only 1/25,000,000,000 milligram is present in each milliliter of air. Because of this low threshold, this substance is mixed with natural gas to give it an odor that can be detected when it leaks from a gas pipe.

*Measurement of Smell Threshold.* One of the problems in studying smell has been difficulty in obtaining accurate measurements of the threshold stimulus required to induce smell. The simplest technique is simply to allow a person to sniff different substances in his usual manner of smelling. Indeed, some investigators feel that this is equally as satisfactory as almost any other procedure. However, to eliminate variations from person to person, more objective methods have been developed: One of these has been to place the subject's head into a box containing the volatilized agent. Appropriate precautions are taken to exclude odors from the person's own body. The person is allowed to breathe naturally, but the volatilized agent is distributed evenly in the air that is breathed.

**Gradations of Smell Intensities.** Though the threshold concentrations of substances that evoke smell are extremely slight, concentrations only 10 to 50 times above the threshold values evoke maximum intensity of smell. This is in contrast to most other sensory systems of the body, in which the ranges of detection are tremendous–for instance 500,000 to 1 in the case of the eyes and 1,000,000,000,000 to 1 in the case of the ears. This perhaps can be explained by the fact that smell is concerned more with detecting the presence or absence of odors than with quantitative detection of their intensities.

## TRANSMISSION OF SMELL SENSATIONS INTO THE CENTRAL NERVOUS SYSTEM

The function of the central nervous system in olfaction is almost as vague as the function of the peripheral receptors. However, Figures 62–5 and 62–6 illustrate the general plan for transmission of olfactory sensations into the central nervous system. Figure 62–5 shows a number of separate *olfactory cells* sending axons into the *olfactory bulb* to end on *dendrites from mitral cells* in a structure called the *glomerulus*. Approximately 25,000 axons enter each glomerulus and synapse with about 25 mitral cells that in turn send signals into the brain.

Figure 62–6 shows the major pathways for transmission of olfactory signals from the mitral cells into the brain. The fibers from the mitral cells travel through the olfactory tract and terminate either primarily or through relay neurons in two principal areas of the brain called the *medial olfactory area* and the *lateral olfactory area*, respectively. The medial olfactory

area is composed of a large group of nuclei located in the midportion of the brain superiorly and anteriorly to the hypothalamus. This group includes the *septum pellucidum*, the *gyrus subcallosus*, the *paraolfactory area*, the *olfactory trigone*, and the *medial part of the anterior perforated substance*.

The lateral olfactory area is composed of the *prepyriform area*, the *uncus*, the *lateral part of the anterior perforated substance*, and part of the *amygdaloid nuclei*.

Secondary olfactory tracts pass from the nuclei of both the medial olfactory area and the lateral olfactory area into the *hypothalamus*, *thalamus, hippocampus,* and *brain stem nuclei*. These secondary areas control the automatic responses of the body to olfactory stimuli, including automatic feeding activities and also emotional responses, such as fear, excitement, pleasure and sexual drives.

Secondary olfactory tracts also spread from the lateral olfactory area into the temporal cortex and prefrontal cortex. It is probably in this lateral olfactory area, especially in the amygdala and its associated cortical regions, that the more complex aspects of olfaction are integrated, such as association of olfactory sensations with somatic, visual, tactile, and other types of sensation. However, complete removal of the lateral olfactory area hardly affects the primitive responses to olfaction, such as licking the lips, salivation, and other feeding responses caused by the smell of food or such as the various emotions associated with smell. On the other hand, its removal does abolish the more complicated conditioned reflexes depending on olfactory stimuli. Therefore, this region is often considered to be the *primary olfactory cortex* for smell. In human beings, tumors in the region of the uncus and amygdala frequently cause the person to perceive abnormal smells.

**Centrifugal Control of the Olfactory Bulb by the Central Nervous System.** The central nervous system also transmits impulses in a backward direction to the olfactory bulb. These originate mainly from undetermined areas in the brain stem and olfactory portion of the cerebrum and terminate in the small *granule* cells in the center of the bulb, which in turn send axons to the mitral cells to inhibit these cells.

**Mechanism of Function of the Olfactory Tracts.** Electrophysiological studies of the olfactory system show that the mitral cells are continually active, and superimposed on this background are increases or decreases in impulse traffic caused by different odors. Thus, the olfactory stimuli presumably *modulate* the rate of impulses in the olfactory system and in this way transmit the olfactory information.

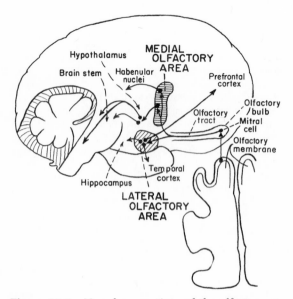

**Figure 62–6.** Neural connections of the olfactory system.

# REFERENCES

Alberts, J. R.: Producing and interpreting experimental olfactory deficits. *Physiol. Behav.*, *12*:657, 1974.

Bradley, R. M. and Mistretta, C. M.: Investigations of taste function and swallowing in fetal sheep. *Symp. Oral Sens. Percept.*, *4*:185, 1973.

Dastoli, F. R.: Taste receptor proteins. *Life Sci.*, *14*:1417, 1974.

Denton, D. A.: Salt appetite. *In* Code, C. F., and Heidel, W. (eds.): Handbook of Physiology. Sec. 6, Vol. 1. Baltimore, The Williams & Wilkins Company, 1967, p. 433.

Douek, E.: The Sense of Smell and Its Abnormalities. New York, Churchill Livingstone, Div. of Longman, Inc., 1974.

Forss, D. A.: Odor and flavor compounds from lipids. *Prog. Chem. Fats Other Lipids.*, *13*:177, 1972.

Hodgson, E. S.: Taste receptors. *Sci. Amer.*, *204*:135, 1961.

Kare, M. R., and Maller, O.: The Chemical Sense and Nutrition. Baltimore, The Johns Hopkins Press, 1967.

Lat, J.: Self-selection of dietary components. *In* Code, C. F., and Heidel, W. (eds.): Handbook of Physiology. Sec. 6, Vol. 1. Baltimore, The Williams & Wilkins Company, 1967, p. 367.

Moulton, D. G., and Beidler, L. M.: Structure and function in the peripheral olfactory system. *Physiol. Rev.*, *47*:1, 1967.

Norsiek, F. W.: The sweet tooth. *Amer. Sci.*, *60*:41, 1972.

Oakley, B., and Benjamin, R. M.: Neural mechanisms of taste. *Physiol. Rev.*, *46*:173, 1966.

Ohloff, G., and Thomas, A. F. (eds.): Gustation and Olfaction. New York, Academic Press, Inc., 1971.

Schultz, E. F., and Tapp, J. T.: Olfactory control of behavior in rodents. *Psychol. Bull.*, *79*:21, 1973.

Shepherd, G. M.: Synaptic organization of the mammalian olfactory bulb. *Physiol. Rev.*, *52*:864, 1972.

Todd, J. H.: The chemical languages of fishes. *Sci. Amer.*, *224*:98, 1971.

Wenzel, B. M., and Sieck, M. H.: Olfaction. *Ann. Rev. Physiol.*, *28*:381, 1966.

Zotterman, Y.: Olfaction and Taste. New York, The Macmillan Company, 1963.

# PART XI

# THE GASTROINTESTINAL TRACT

# ▌ 63 ▐

# Movement of Food Through the Alimentary Tract

The primary function of the alimentary tract is to provide the body with a continual supply of water, electrolytes, and nutrients, but before this can be achieved food must be moved along the alimentary tract at an appropriate rate for the digestive and absorptive functions to take place. Therefore, discussion of the alimentary system is presented in three different phases in the next three chapters: (1) movement of food through the alimentary tract, (2) secretion of the digestive juices, and (3) absorption of the digested foods, water, and the various electrolytes.

Figure 63–1 illustrates the entire alimentary tract, showing major anatomical differences between its parts. Each part is adapted for specific functions, such as: (1) simple passage of food from one point to another, as in the esophagus, (2) storage of food in the body of the stomach or fecal matter in the descending colon, (3) digestion of food in the stomach, duodenum, jejunum, and ileum, and (4) absorption of the digestive end-products in the entire small intestine and proximal half of the colon. One of the most important features of the gastrointestinal tract that is discussed in the present chapter is the myriad of autoregulatory processes in the gut that keeps the food moving at an appropriate pace—slow enough for digestion and absorption to take place but fast enough to provide the nutrients needed by the body.

## GENERAL PRINCIPLES OF INTESTINAL MOTILITY

### CHARACTERISTICS OF THE INTESTINAL WALL

Figure 63–2 illustrates a typical section of the intestinal wall, showing the following layers

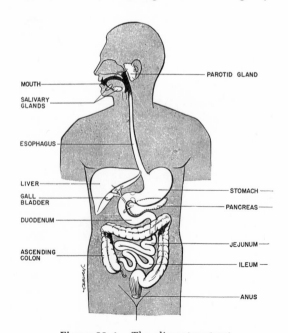

**Figure 63–1.**  The alimentary tract.

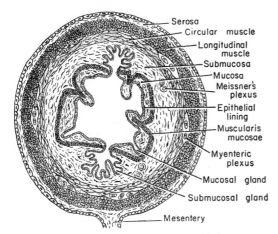

**Figure 63–2.** Typical cross-section of the gut.

from the outside inward: (1) the *serosa*, (2) a *longitudinal muscle layer*, (3) a *circular muscle layer*, (4) the *submucosa*, and (5) the *mucosa*. In addition, a sparse layer of smooth muscle fibers, the *muscularis mucosae*, lies in the deeper layers of the mucosa. The motor functions of the gut are performed by the different layers of smooth muscle.

**Characteristics of Intestinal Smooth Muscle—Electrical Activity and Contraction.** The general characteristics of smooth muscle and its function were discussed in Chapter 12. However, the specific characteristics of smooth muscle in the gut are the following:

**The Functional Syncytium.** The individual smooth muscle fibers of the gut abut against each other extremely closely. About 12 per cent of their membrane surfaces are actually fused with the membranes of adjacent muscle fibers in the form of a *nexus,* and most of the remainder of the cell membranes of adjacent fibers lie in extremely close apposition. Measurements of ionic transport through these areas of close contact demonstrate extremely low electrical resistance—so much so that intracellular electrical current can travel very easily from one smooth muscle fiber to another. Therefore, the smooth muscle of the gastrointestinal tract performs as a *functional syncytium,* which means that electrical signals originating in one smooth muscle fiber are generally propagated from fiber to fiber.

*Electrical Activity of Gastrointestinal Smooth Muscle.* The smooth muscle of the gastrointestinal tract undergoes electrical activity almost continuously. This activity is sometimes very bizarre, but it tends to have two basic types of

waves, *slow waves* and *spikes,* both of which are illustrated in the Figure 63–3. The slow waves occur at frequencies between 3 and 12 per minute, and they represent a basic continuous oscillation that occurs in the membranes of the smooth muscle. The slow waves are not all-or-nothing action potentials but, instead, can be of any graded degree of intensity.

When the muscle is stimulated by stretch, by acetylcholine, or by parasympathetic excitation, the resting membrane potential of the fiber becomes more positive, which raises the entire potential level of the slow waves from their normal mean voltage of $-50$ to $-40$ millivolts up to some less negative value by the process of *depolarization*. As the depolarization rises above $-40$ millivolts, spike potentials begin to occur on the peaks of the slow waves; the frequency of spikes increases progressively as the resting potential rises still further. However, with very strong stimulation, when the resting membrane potential rises to a value of $-15$ to $-20$ millivolts, the spikes disappear because the membrane now remains totally depolarized all of the time.

Figure 63–3 also illustrates that stimulation of the smooth muscle by either epinephrine or sympathetic excitation will *decrease* the resting membrane potential to a "hyperpolarized" value approaching $-70$ millivolts, and the electrical activity as well as the mechanical activity of the smooth muscle approaches zero.

*Excitation of Muscle Contraction.* Most gut contraction occurs in response to the spike potentials. Indeed, there is ordinarily either no contraction or only very slight contraction in

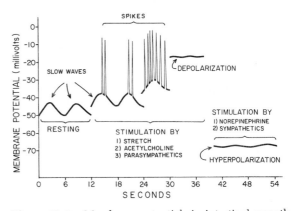

**Figure 63–3.** Membrane potentials in intestinal smooth muscle. Note the slow waves, the spike potentials, total depolarization, and hyperpolarization, all of which occur under different physiological conditions of the intestine.

response to slow waves when these have no superimposed spikes. Thus, the spikes are comparable to action potentials in skeletal muscle and are responsible for the membrane changes that excite contraction. The contraction results mainly from entry of calcium through the cell membrane to the interior of the smooth muscle where it initiates a reaction between the myosin and actin of the smooth muscle. It is possible that additional calcium might be released from the cell membrane or from mitochondria inside the cell, though this mechanism is still in doubt.

The smooth muscle of the gastrointestinal tract exhibits both *tonic contraction* and *rhythmic contraction,* both of which are characteristic of most types of smooth muscle, as discussed in Chapter 12.

Tonic contraction is continuous, lasting minute after minute or even hour after hour, sometimes increasing or decreasing in intensity but, nevertheless, continuing. It is believed to be caused by a series of spike potentials, the frequency of these determining the degree of tonic contraction. The intensity of tonic contraction in each segment of the gut determines the amount of steady pressure in the segment, and tonic contraction of the sphincters determines the amount of resistance offered at the sphincters to the movement of intestinal contents. In this way the *pyloric,* the *ileocecal,* and the *anal sphincters* all help to regulate food movement in the gut.

In different parts of the gut the rhythmic contractions of the gastrointestinal smooth muscle occur at rates as rapid as 12 times per minute or as slow as 3 times per minute. These are also the frequencies of the respective slow waves in these segments; it is these slow waves that set the frequency. These contractions are responsible for the phasic functions of the gastrointestinal tract, such as mixing of the food and peristaltic propulsion of food as discussed below.

## INNERVATION OF THE GUT— THE INTRAMURAL PLEXUS

Beginning in the esophageal wall and extending all the way to the anus is an *intramural nerve plexus.* This is composed principally of two layers of neurons and appropriate connecting fibers: the outer layer, called the *myenteric plexus* or *Auerbach's plexus,* lies between the longitudinal and circular muscular layers; and the inner layer, called the *submucosal plexus* or *Meissner's plexus,* lies in the submucosa. The myenteric plexus is mainly *motor* in function and is far more extensive than the submucosal plexus which is mainly *sensory,* receiving signals principally from the gut epithelium and from stretch receptors in the gut wall.

In general, stimulation of the myenteric plexus increases the activity of the gut, causing four principal effects: (1) increased tonic contraction, or "tone," of the gut wall, (2) increased intensity of the rhythmic contractions, (3) increased rate of rhythmic contraction, and (4) increased velocity of conduction of excitatory waves along the gut wall. On the other hand, a few myenteric plexus fibers are inhibitory rather than excitatory; these probably are adrenergic fibers, in contrast to the excitatory fibers, which are cholinergic.

The intramural plexus, including both the submucosal sensory plexus and the myenteric motor plexus, is especially responsible for many neurogenic reflexes that occur locally in the gut, such as reflexes from the mucosal epithelium to increase the activity of the gut muscle or to cause localized secretion of digestive juices by the submucosal glands. This plexus is also intimately involved in coordination of the motor movements of the gastrointestinal tract, which are discussed later in the chapter.

**Autonomic Control of the Gastrointestinal Tract.** The gastrointestinal tract receives extensive parasympathetic and sympathetic innervation that is capable of altering the overall activity of the entire gut or of specific parts of it, particularly of its upper end down to the stomach and its distal end from the mid-colon region to the anus.

*Parasympathetic Innervation.* The parasympathetic supply to the gut is divided into *cranial* and *sacral divisions,* which were discussed in Chapter 57. Except for a few parasympathetic fibers to the mouth and pharyngeal regions of the alimentary tract, the cranial parasympathetics are transmitted almost entirely in the *vagus nerves.* These fibers provide extensive innervation to the esophagus and stomach and, to much less extent, to the small intestine, gallbladder, and first half of the large intestine. The sacral parasympathetics originate in the second, third, and fourth sacral segments of the spinal cord and pass through the *nervi erigentes* to the distal half of the large intestine. The sigmoidal, rectal, and anal regions of the large intestine are considerably better supplied with parasympathetic fibers than are the other portions. These fibers function especially in the defecation reflexes, which are discussed later in the chapter.

The postganglionic neurons of the parasympathetic system are part of the myenteric plexus, so that stimulation of the parasympathetic nerves causes a general increase in activity of this plexus. This in turn excites the gut wall and facilitates most of the intrinsic excitatory nervous reflexes of the gastrointestinal tract.

*Sympathetic Innervation.* The sympathetic fibers to the gastrointestinal tract originate in the spinal cord between the segments T-8 and L-3. The preganglionic fibers, after leaving the cord, enter the sympathetic chains and pass through the chains to outlying ganglia, such as the *celiac ganglia* and various *mesenteric ganglia*. Here, the postganglionic neuron bodies are located, and postganglionic fibers spread from them along with the blood vessels to all parts of the gut. The sympathetics innervate essentially all portions of the gastrointestinal tract rather than being more extensively supplied to the most orad and most analward portions as is true of the parasympathetics.

In general, stimulation of the sympathetic nervous system inhibits activity in the gastrointestinal tract, causing effects essentially opposite to those of the parasympathetic system. However, the sympathetic system elicits excitatory effects in at least two instances: it excites (1) the ileocecal sphincter and (2) the internal anal sphincter. Thus, strong stimulation of the sympathetic system can totally block movement of food through the gastrointestinal tract, both by inhibition of the gut wall and by excitation of at least two major sphincters of the gastrointestinal tract.

*Effects of Parasympathetic and Sympathetic Denervation.* Denervating the sympathetic nervous system normally has no more than transient effects on the function of the gastrointestinal tract, which illustrates that sympathetic effects on gastrointestinal function are usually minimal.

On the other hand, denervation of the parasympathetic system almost always decreases the activity of the denervated portions of the gut for weeks, months, or years. For instance, destruction of the vagus nerves markedly reduces the tone and degree of peristaltic activity in the terminal esophagus and stomach, while destruction of the sacral parasympathetics decreases the tone of the descending colon, sigmoid, and rectum, thereby either temporarily or permanently blocking defecation.

It is evident, then, that the gastrointestinal tract is usually tonically excited by the parasympathetics, and loss of this tone reduces gastrointestinal functions. Fortunately, after several months to a year or more, the intramural nervous system of the gut usually compensates at least partially by gradually increasing its intrinsic excitability so that the excitability usually returns much of the way toward normal.

**Afferent Nerve Fibers from the Gut.** Afferent nerve fibers arise in the gut, also. Some of these have their cell bodies in the submucosal plexus and terminate mainly in the myenteric plexus. The nerve endings can be stimulated by (a) irritation of the gut mucosa, (b) excessive distention of the gut, or (c) presence of specific chemical substances in the gut. Signals transmitted through these fibers control excitation (occasionally, inhibition) of intestinal movements or intestinal secretion.

In addition to the afferent fibers that terminate in the intramural plexus, other afferent fibers whose cell bodies lie in the dorsal root ganglia of the spinal cord or in the cranial nerve ganglia transmit signals all the way to the central nervous system, traveling along with the sympathetic or parasympathetic nerves. For example, pain nerve fibers, except those from the esophagus, course along with the sympathetic nerves to the spinal cord. Also, about 80 per cent of the nerve fibers in the vagus nerves are afferent rather than efferent. These fibers transmit signals into the medulla to help initiate vagal signals that in turn control many functions of the gastrointestinal tract, as described in this and the following chapter.

# FUNCTIONAL TYPES OF MOVEMENTS IN THE GASTROINTESTINAL TRACT

Two basic types of movements occur in the gastrointestinal tract: (1) *mixing movements,* which keep the intestinal contents thoroughly mixed at all times, and (2) *propulsive movements,* which cause food to move forward along the tract at an appropriate rate for digestion and absorption.

## THE MIXING MOVEMENTS

In most parts of the alimentary tract, the mixing movements are caused by either *peristaltic contractions or local contractions of small segments of the gut wall.* These movements are modified in different parts of the gastrointestinal tract for proper performance of the respective activities of each part, as discussed separately later in the chapter.

## THE PROPULSIVE MOVEMENTS— PERISTALSIS

The basic propulsive movement of the gastrointestinal tract is peristalsis, which is illustrated in Figure 63–4. A contractile ring appears

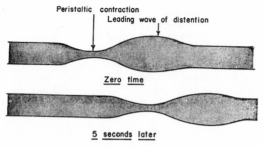

Figure 63-4.  Peristalsis.

around the gut and then moves forward; this is analogous to putting one's fingers around a thin distended tube, then constricting the fingers and moving them forward along the tube. Obviously, any material in front of the contractile ring is moved forward.

Peristalsis is an inherent property of any syncytial smooth muscle tube, and stimulation at any point causes a contractile ring to spread in both directions. Thus, peristalsis occurs in (a) the gastrointestinal tract, (b) the bile ducts, (c) other glandular ducts throughout the body, (d) the ureters, and (e) most other smooth muscle tubes of the body.

The usual stimulus for peristalsis is *distention*. That is, if a large amount of food collects at any point in the gut, the distention stimulates the gut wall 2 to 3 cm. above this point, and a contractile ring appears and initiates a peristaltic movement.

**Function of the Myenteric Plexus in Peristalsis.** Even though peristalsis is a basic characteristic of all tubular smooth muscle structures, it occurs only weakly or not at all in portions of the gastrointestinal tract that have congenital absence of the myenteric plexus. Also, it is greatly depressed or completely blocked in the entire gut when the person is treated with atropine to paralyze the myenteric plexus. Furthermore, since the myenteric plexus is principally under the control of the parasympathetic nerves, the intensity of peristalsis and its velocity of conduction can be altered by parasympathetic stimulation.

Therefore, *even though the basic phenomenon of peristalsis is not dependent on the myenteric nerve plexus, effectual* peristalsis does require an active myenteric plexus.

**Analward Peristaltic Movements.** Peristalsis, theoretically, can occur in either direction from a stimulated point, but it normally dies out rapidly in the orad direction while continuing for a considerable distance analward.

The cause of this directional transmission of peristalsis has never been ascertained, though several suggestions have been offered as follows:

*Receptive Relaxation and the "Law of the Gut."* Some physiologists believe the directional movement of peristalsis to be caused by a special organization of the myenteric plexus, which allows preferential transmission of analward signals. One reason for believing this is the effect of electrical stimulation on the gut: An electrical stimulus causes a contractile ring to appear near the point of the stimulus but at the same time sometimes causes relaxation, called "receptive relaxation," several centimeters down the gut toward the anus. It is believed that this relaxation could occur only as a result of conduction in the myenteric plexus. Obviously, a leading wave of receptive relaxation could allow food to be propelled more easily analward than in the orad direction.

This response to electrical stimulation is also called the "law of the gut" or sometimes simply the "myenteric reflex." It can be particularly well demonstrated in the esophagus and in the pyloric region of the stomach.

*The Gradient Theory for Forward Propulsion.* Another theory for forward propulsion of intestinal contents is the gradient theory. This is based on the fact that the upper part of the gastrointestinal tract usually displays a higher degree of activity than the lower part. For instance, the basic contractile rhythm in the duodenum is approximately 11 contractions per minute, while that in the ileum is 6 to 7 contractions per minute. Also, in the upper part of the gastrointestinal tract far greater quantities of secretions are formed, and these, by distending the gut, initiate far more peristaltic waves than are initiated in the lower gut. Therefore, it is postulated that a greater number of peristaltic impulses originate in the orad portions of the gut than in the more distal regions, thereby causing peristaltic waves generally to travel down rather than up the gut. This is analogous to the *pacemaker* function of the S-A node in the heart.

## INGESTION OF FOOD

The amount of food that a person ingests is determined principally by the intrinsic desire for food called *hunger,* and the type of food that he preferentially seeks is determined by his *appetite.* These mechanisms in themselves are extremely important automatic regulatory systems for maintaining an adequate nutritional supply for the body, and they will be discussed in detail in Chapter 73 in relation to nutrition of the body. The present discussion is confined to the actual mechanical aspects of food ingestion, including especially *mastication* and *swallowing*.

## MASTICATION (CHEWING)

The teeth are admirably designed for chewing, the anterior teeth (incisors) providing a strong cutting action and the posterior teeth (molars) a grinding action. All the jaw muscles working together can close the teeth with a force as great as 55 pounds on the incisors and 200 pounds on the molars. When this is applied to a small object, such as a small seed between the molars, the actual force *per square inch* may be several thousand pounds.

Most of the muscles of chewing are innervated by the motor branch of the 5th cranial nerve, and the chewing process is controlled by nuclei in the hindbrain. Stimulation of the reticular formation near the hindbrain centers for taste can cause continual rhythmic chewing movements. Also, stimulation of areas in the hypothalamus, amygdaloid nuclei, and even in the cerebral cortex near the sensory areas for taste and smell can cause chewing.

Much of the chewing process is caused by the *chewing reflex,* which may be explained as follows: The presence of a bolus of food in the mouth causes reflex inhibition of the muscles of mastication, which allows the lower jaw to drop. The sudden drop in turn initiates a stretch reflex of the jaw muscles that leads to *rebound* contraction. This automatically raises the jaw to cause closure of the teeth, but it also compresses the bolus again against the linings of the mouth, which inhibits the jaw muscles once again, allowing the jaw to drop and rebound another time, and this is repeated again and again.

Chewing of the food is important for digestion of all foods, but it is especially important for most fruits and raw vegetables, because these have undigestible cellulose membranes around their nutrient portions which must be broken before the food can be utilized. Chewing aids in the digestion of food for the following simple reason: Since the *digestive enzymes act only on the surfaces of food particles,* the rate of digestion is highly dependent on the total surface area exposed to the intestinal secretions. Also, grinding the food to a very fine particulate consistency prevents excoriation of the gastrointestinal tract and increases the ease with which food is emptied from the stomach into the small intestine and thence into all succeeding segments of the gut.

## SWALLOWING (DEGLUTITION)

Swallowing is a complicated mechanism, principally because the pharynx most of the time subserves several other functions besides swallowing and is converted for only a few seconds at a time into a tract for propulsion of food. Especially is it important that respiration not be seriously compromised during swallowing.

In general, swallowing can be divided into: (1) the *voluntary stage,* which initiates the swallowing process, (2) the *pharyngeal stage,* which is involuntary and constitutes the passage of food through the pharynx into the esophagus, and (3) the *esophageal stage,* another involuntary phase which promotes passage of food from the pharynx to the stomach.

**Voluntary Stage of Swallowing.** When the food is ready for swallowing, it is "voluntarily" squeezed or rolled posteriorly in the mouth by pressure of the tongue upward and backward against the palate, as shown in Figure 63–5. Thus, the tongue forces the bolus of food into the pharynx. From here on, the process of swallowing becomes entirely, or almost entirely, automatic and ordinarily cannot be stopped.

**Pharyngeal Stage of Swallowing.** When the bolus of food is pushed backward in the mouth, it stimulates *swallowing receptor areas* all around the opening of the pharynx, especially on the tonsillar pillars, and impulses from these pass to the brain stem to initiate a series of

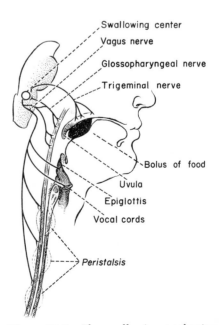

**Figure 63–5.**   The swallowing mechanism.

automatic pharyngeal muscular contractions as follows:

1. The soft palate is pulled upward to close the posterior nares, in this way preventing reflux of food into the nasal cavities.

2. The palatopharyngeal folds on either side of the pharynx are pulled medialward to approximate each other. In this way these folds form a sagittal slit through which the food must pass into the posterior pharynx. This slit performs a selective action, allowing food that has been masticated properly to pass with ease while impeding the passage of large objects. Since this stage of swallowing lasts only 1 second, any large object is usually impeded too much to pass through the pharynx into the esophagus.

3. The vocal cords of the larynx are strongly approximated, and the epiglottis swings backward over the superior opening of the larynx. Both of these effects prevent passage of food into the trachea. Especially important is the approximation of the vocal cords, but the epiglottis helps to prevent food from ever getting as far as the vocal cords. Destruction of the vocal cords or of the muscles that approximate them can cause strangulation. On the other hand, removal of the epiglottis usually does not cause serious debility in swallowing.

4. The entire larynx is pulled upward and forward by muscles attached to the hyoid bone; this movement of the larynx stretches the opening of the esophagus. At the same time, the upper 3 to 4 centimeters of the esophagus, an area called the *upper esophageal sphincter* or the *pharyngoesophageal sphincter,* relaxes, thus allowing food to move easily and freely from the posterior pharynx into the upper esophagus. This sphincter, between swallows, remains tonically and strongly contracted, thereby preventing air from going into the esophagus during respiration. The upward movement of the larynx also lifts the glottis out of the main stream of food flow so that the food usually passes on either side of the epiglottis rather than over its surface; this adds still another protection against passage of food into the trachea.

5. At the same time that the larynx is raised and the pharyngoesophageal sphincter is relaxed, the superior constrictor muscle of the pharynx contracts, giving rise to a rapid peristaltic wave passing downward over the pharyngeal muscles and into the esophagus, which also propels the food into the esophagus.

To summarize the mechanics of the pharyngeal stage of swallowing—the trachea is closed, the esophagus is opened, and a fast peristaltic wave originating in the pharynx then forces the bolus of food into the upper esophagus, the entire process occurring in 1 to 2 seconds.

***Nervous Control of the Pharyngeal Stage of Swallowing.*** The most sensitive tactile areas of the pharynx for initiation of the pharyngeal stage of swallowing lie in a ring around the pharyngeal opening, with greatest sensitivity in the tonsillar pillars. Impulses are transmitted from these areas through the sensory portions of the trigeminal and glossopharyngeal nerves into a region of the medulla oblongata closely associated with the *tractus solitarius* which receives essentially all sensory impulses from the mouth.

The successive stages of the swallowing process are then automatically controlled in orderly sequence by neuronal areas distributed throughout the reticular substance of the medulla and lower portion of the pons. The sequence of the swallowing reflex remains the same from one swallow to the next, and the timing of the entire cycle also remains constant from one swallow to the next. The areas in the medulla and lower pons that control swallowing are collectively called the *deglutition* or *swallowing center.*

The motor impulses from the swallowing center to the pharynx and upper esophagus that cause swallowing are transmitted by the 5th, 9th, 11th, and 12th cranial nerves and even a few of the superior cervical nerves.

In summary, the pharyngeal stage of swallowing is principally a reflex act. It is almost never initiated by direct stimuli to the swallowing center from higher regions of the central nervous system. Instead, it is almost always initiated by voluntary movement of food into the back of the mouth, which, in turn, elicts the swallowing reflex.

***Effect of the Pharyngeal Stage of Swallowing on Respiration.*** The entire pharyngeal stage of swallowing occurs in less than 1 to 2 seconds, thereby interrupting respiration for only a fraction of a usual respiratory cycle. The swallowing center specifically inhibits the respiratory center of the medulla during this time, halting respiration at any point in its cycle to allow swallowing to proceed. Yet, even while a person is talking, swallowing interrupts respiration for such a short time that it is hardly noticeable.

**Esophageal Stage of Swallowing.** The esophagus functions primarily to conduct food from the pharynx to the stomach, and its movements are organized specifically for this function.

Normally the esophagus exhibits two types of peristaltic movements—*primary peristalsis* and *secondary peristalsis.* Primary peristalsis is simply a continuation of the peristaltic wave that begins in the pharynx and spreads into the esophagus during the pharyngeal stage of swallowing. This wave passes all the way from the pharynx to the stomach in approximately 5 to 10 seconds. However, food swallowed by a person who is in the upright position is usually transmitted to the lower end of the esophagus even more rapidly than the peristaltic wave itself, in about 4 to 8 seconds, because of the additional effect of gravity pulling the food downward. If the primary peristaltic wave fails to move all the food that has entered the esophagus on into the stomach, secondary peristaltic waves result from distention of the esophagus by the retained food. These waves are essentially the same as the primary peristaltic waves, except that they originate in the esophagus itself rather than in the pharynx. Secondary peristaltic waves continue to be initiated until all the food has emptied into the stomach.

The peristaltic waves of the esophagus are controlled almost entirely by vagal reflexes that are part of the overall swallowing mechanism. These reflexes are transmitted through *vagal afferent fibers* from the esophagus to the medulla and then back again to the esophagus through *vagal efferent fibers.*

The musculature of the pharynx and the upper quarter of the esophagus is skeletal muscle, and, therefore, the peristaltic waves in these regions are controlled only by skeletal nerve impulses. In the lower two thirds of the esophagus, the musculature is smooth, but even this portion of the esophagus is normally under the control of the vagus nerve. However, when the vagus nerves to the esophagus are sectioned, the myenteric nerve plexus of the esophagus becomes excitable enough after several days to cause weak secondary peristaltic waves even without support from the vagal reflexes. Therefore, following paralysis of the swallowing reflex, food forced into the upper esophagus and then pulled by gravity to the lower esophagus still passes readily into the stomach.

***Receptive Relaxation of the Stomach.*** As the esophageal peristaltic wave passes toward the stomach, a wave of relaxation precedes the constriction. Furthermore, the entire stomach and, to a less extent, even the duodenum become relaxed as this wave reaches the lower end of the esophagus. Especially important, also, is relaxation of the gastroesophageal sphincter at the juncture between the esophagus and the stomach. In other words, the constrictor and the stomach are prepared ahead of time to receive food being propelled down the esophagus during the swallowing act.

## *FUNCTION OF THE GASTROESOPHAGEAL SPHINCTER*

At the lower end of the esophagus, about 5 cm. above its juncture with the stomach, the circular muscle functions as a so-called *gastroesophageal sphincter.* Anatomically this sphincter is no different from the remainder of the esophagus. However, physiologically, it remains tonically constricted, in contrast to the midportions of the esophagus which normally remain completely relaxed. However, when a peristaltic wave of swallowing passes down the esophagus, "receptive relaxation" relaxes the gastroesophageal sphincter ahead of the peristaltic wave, and allows easy propulsion of the swallowed food on into the stomach. Rarely, the gastroesophageal sphincter does not relax satisfactorily, resulting in the condition called *achalasia,* which will be discussed in detail in Chapter 66.

A principal function of the gastroesophageal sphincter is to prevent reflux of stomach contents into the upper esophagus. The stomach contents are highly acidic and contain many proteolytic enzymes. The esophageal mucosa, except in the lower eighth of the esophagus, is not capable of resisting for long the digestive action of gastric secretions in the esophagus. Fortunately, the tonic constriction of the gastroesophageal sphincter prevents significant reflux of stomach contents into the esophagus except under abnormal conditions. Indeed, increased intragastric pressure, except during vomiting, causes a vagal reflex that further constricts the sphincter to add extra insurance against reflux.

**Prevention of Reflux by Flutter-Valve Closure of the Distal End of the Esophagus.** Another factor that prevents reflux is a valvelike mechanism of that portion of the esophagus that lies immediately beneath the diaphragm. Greatly increased intra-abdominal pressure caves the esophagus inward at this point at the same time that the abdominal pressure also increases the intragastric pressure. This flutter-valve closure of the lower esophagus therefore prevents the high pressure in the stomach from forcing stomach contents into the esophagus. Otherwise, every time we should walk, cough, or breathe hard, we might expel acid into the esophagus.

## MOTOR FUNCTIONS OF THE STOMACH

The motor functions of the stomach are three-fold: (1) storage of large quantities of food until it can be accommodated in the lower portion of the gastrointestinal tract, (2) mixing of this food with gastric secretions until it forms a semifluid mixture called *chyme,* and (3) slow emptying of the food from the stomach into the small intestine at a rate suitable for proper digestion and absorption by the small intestine.

Figure 63–6 illustrates the basic anatomy of the stomach. Physiologically, the stomach can be divided into two major parts: (1) the *corpus,* or *body,* and (2) the *antrum.* The *fundus,* located at the upper end of the body of the stomach, is often considered by anatomists to be a separate entity from the body, but from a physiological point of view, the fundus is actually a functional part of the body.

### STORAGE FUNCTION OF THE STOMACH

As food enters the stomach, it forms concentric circles in the body of the stomach, the newest food lying closest to the esophageal opening and the oldest food lying nearest the wall of the stomach. Normally, the body of the stomach has relatively little tone in its muscular

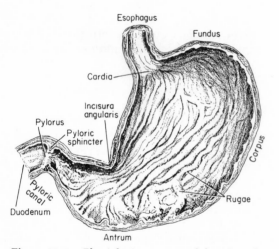

**Figure 63–6.** Physiologic anatomy of the stomach.

wall so that it can bulge progressively outward, thereby accommodating greater and greater quantities of food up to a limit of about 1 liter. The pressure in the stomach remains low until this limit is approached, for three reasons: First, the smooth muscle in the wall of the stomach exhibits a quality of *plasticity,* which means that it can increase its length greatly without significantly changing its tone. Second, the greater the diameter of the stomach, the greater also becomes the radius of curvature of the walls. At a given pressure, the stretching force on the walls increases in direct proportion to this radius of curvature (an effect called the *law of Laplace*), for which reason the pressure inside the stomach increases only slightly despite marked distention. Third, stretch of the stomach also causes a *vagal reflex* that inhibits muscle activity in the body of the stomach.

### MIXING IN THE STOMACH

The digestive juices of the stomach are secreted by the *gastric glands,* which cover almost the entire outer wall of the body of the stomach. These secretions come immediately into contact with the stored food lying against the mucosal surface of the stomach; and when the stomach is filled, weak *constrictor waves,* also called *mixing waves,* move along the stomach wall approximately once every 20 seconds. These waves begin most frequently near the midpoint of the stomach, but their origin moves further back up the stomach wall as the stomach empties. In general, the waves become more intense as they approach the antral portion of the stomach.

The mixing waves tend to move the gastric secretions and the outermost layer of food gradually toward the antral part of the stomach. On entering the antrum the waves become stronger, and the food and gastric secretions become progressively mixed to a greater and greater degree of fluidity.

In addition to the mixing caused by the waves in the body of the stomach, mixing is also caused by intense peristaltic movements in the antral portion of the stomach. These movements cause mixing in the following way: Each time a peristaltic wave passes over the antrum toward the pylorus, it digs deeply into the contents of the antrum. Yet the opening of the pylorus is small enough that only a few milliliters of antral contents are expelled into the duodenum with each peristaltic wave. Instead, most of the antral contents are squeezed backward through the peristaltic ring toward the body of the stomach. Thus, the moving peristaltic constrictive ring, combined with the reflux action, is an exceedingly important mixing mechanism of the stomach.

**Chyme.** After the food has become mixed with the stomach secretions, the resulting mixture that passes on down the gut is called chyme. The degree of fluidity of chyme depends on the relative amounts of food and stomach secretions and on the degree of digestion that has occurred. The appearance of chyme is that of a murky, milky, semifluid or paste.

**Propulsion of Food Through the Stomach.** Strong peristaltic waves occur about 20 per cent of the time in the antrum of the stomach. These waves, like the mixing waves, occur about once every 20 seconds. In fact, they are probably extensions of the mixing waves, which become potentiated as they spread from the body of the stomach into the antrum. They become intense approximately at the incisura angularis, from which they spread through the antrum. As the stomach becomes progressively more and more empty, these intense waves begin farther and farther up the body of the stomach, gradually pinching off the lowermost portions of stored food, adding this food to the chyme in the antrum.

The peristaltic waves often exert as much as 50 to 70 cm. of water pressure, which is about six times as powerful as the usual mixing waves.

**Hunger Contractions.** In addition to the mixing and peristaltic contractions of the stomach, a third type of intense contractions, called hunger contractions, often occurs when the stomach has been empty for a long time. These are usually rhythmic peristaltic contractions probably representing exacerbated mixing waves in the *body* of the stomach. However, when they become extremely strong, they often fuse together to cause a continuing tetanic contraction lasting for as long as two to three minutes.

Hunger contractions are usually most intense in young healthy persons with high degrees of gastrointestinal tonus, and they are also greatly increased by a low level of blood sugar.

When hunger contractions occur in the stomach, the person sometimes experiences a sensation of pain in the pit of his stomach called *hunger pangs*. Hunger pangs usually do not begin until 12 to 24 hours after the last ingestion of food; in starvation they reach their greatest intensity in three to four days, and then gradually weaken in succeeding days.

Hunger contractions are often associated with a feeling of hunger and therefore are perhaps an important means by which the alimentary tract intensifies the desire for food when a person is in a state of incipient starvation.

**Reflex Regulation of Stomach Contractions.** Distention of the stomach by food initiates vagal afferent signals that pass to the medulla oblongata and reflexly inhibit the tone in the storage area of the stomach but at the same time increase the rate of stomach secretion and the intensity of both the mixing and peristaltic waves. Thus, the rate of digestion and removal of the stored food is accelerated.

## EMPTYING OF THE STOMACH

**The Pyloric Pump.** Basically, stomach emptying is *promoted by peristaltic waves* in the antrum of the stomach, and it is *opposed by resistance of the pylorus* to the passage of food. The pylorus normally remains almost but not completely closed because of mild tonic contraction. A pressure of about 5 cm. water is normally exerted on the lumen of the pylorus by the pyloric sphincter. This is a very weak closing force, but it is nevertheless usually great enough to prevent flow of chyme into the duodenum except when a peristaltic wave in the antrum forces it through. Therefore, for all practical purposes *the rate of emptying of the stomach is determined principally by the degree of activity of the antral peristaltic waves.*

The antral peristaltic waves, when active, characteristically occur almost exactly three times per minute, becoming very strong near the incisura angularis and moving over the antrum, then over the pylorus, and finally into the duodenum. As the wave moves forward, the pyloric sphincter and the proximal portion of the duodenum are inhibited, which is an instance of "receptive relaxation." With each peristaltic wave, several milliliters of chyme are forced into the duodenum. This "pumping" action of the antral portion of the stomach is sometimes also called the "pyloric pump."

**Regulation of Pyloric Pump Activity.** The degree of activity of the pyloric pump is regulated both by signals from the stomach itself and also by signals from the duodenum. The gastric signals are (1) the degree of distention of the stomach by food, and (2) the presence of the hormone gastrin, which is released from the antrum of the stomach in response to stretch and to the presence of certain types of food in the stomach. Both of these signals have a positive effect on increasing the pyloric pumping force and therefore in promoting stomach emptying.

On the other hand, signals from the duodenum depress pyloric pump activity. In general, when excesses of volume of chyme or excesses of certain types of chyme have entered the duodenum, strong negative feedback signals, both nervous and hormonal, are transmitted to the stomach to depress the pyloric pump. Thus, this mechanism allows chyme to enter the duodenum only as rapidly as it can be processed by the small intestine.

*Effect of Gastric Food Volume on Rate of Emptying.* It is very easy to see how greatly enhanced food volume in the stomach could promote increased emptying from the stomach. However, this does not occur for the reasons that one would expect. It is not increased pressure in the stomach that causes the increased emptying because, in the usual normal range of volume, the increase in volume does not increase the pressure significantly, as was discussed in an earlier section. Instead, the stretch of the stomach wall elicits vagal and local myenteric reflexes in the wall of the stomach that increase the activity of the pyloric pump. In general, the rate of food emptying from the stomach is approximately proportional to the *square root* of the volume of food remaining in the stomach at any given time.

*Effect of the Hormone Gastrin on Stomach Emptying.* In the following chapter we shall see that stretch, as well as the presence of certain types of foods in the stomach—particularly meat—elicits release of a hormone called *gastrin* from the antral mucosa, and this has potent effects on causing secretion of highly acidic gastric juice by the stomach fundic glands. However, gastrin also has potent stimulatory effects on motor functions of the stomach. Most important, it enhances the activity of the pyloric pump while at the same time relaxing the pyloris itself. Thus, it is a strong influence for promoting stomach emptying. It also has a constrictor effect on the gastroesophageal sphincter at the lower end of the esophagus to prevent reflux of gastric contents into the esophagus during the enhanced gastric activity.

*The Inhibitory Effect of the Enterogastric Reflex from the Duodenum on Pyloric Activity.* Strong nervous signals are transmitted from the duodenum back to the stomach most of the time, especially when the stomach is emptying food into the duodenum. These signals probably play the most important role of all in determining the degree of activity of the pyloric pump and, therefore, also in determining the rate of emptying of the stomach. The nervous reflexes are mediated mainly by way of afferent nerve fibers in the vagus nerve to the brain stem and then back through efferent nerve fibers to the stomach, also by way of the vagi. However, some of the signals are probably transmitted directly by way of the myenteric plexus as well.

The types of factors that are continually monitored in the duodenum and that can elicit the enterogastric reflex include:

1. The degree of distention of the duodenum.
2. The presence of any degree of irritation of the duodenal mucosa.
3. The degree of acidity of the duodenal chyme.
4. The degree of osmolality of the chyme.
5. The presence of certain breakdown products in the chyme, especially breakdown products of proteins and perhaps to a lesser extent of fats.

The enterogastric reflex is especially sensitive to the presence of irritants and acids in the duodenal chyme. For instance, any time the pH of the chyme in the duodenum falls below approximately 3.5 to 4, the enterogastric reflex is immediately elicited, which inhibits the pyloric pump and reduces or even blocks further the release of acidic stomach contents into the duodenum until the duodenal chyme can be neutralized by pancreatic and other secretions.

The breakdown products of protein digestion will also elicit this reflex; by slowing the rate of stomach emptying, sufficient time is insured for adequate protein digestion in the upper portion of the small intestine.

Finally, either hypo- or hypertonic fluids (especially hypertonic) will elicit the enterogastric reflex. This effect prevents too rapid flow of nonisotonic fluids into the small intestine, thereby preventing rapid changes in electrolyte balance of the body fluids during absorption of the intestinal contents.

*Hormonal Feedback from the Duodenum in Inhibiting Gastric Emptying—Role of Fats.* When fatty foods, especially fatty acids, are present in the chyme that enters the duodenum, the activity of the pyloric pump is depressed and stomach emptying is correspondingly slowed. This plays an important role in allowing slow digestion of the fats before they proceed into the deeper recesses of the intestine.

Unfortunately, the precise mechanism by which fats cause this effect of slowing the emptying of the stomach is not completely known. Most of the effect still occurs even after the enterogastric reflex has been blocked. Therefore, it is presumed that the effect results from some hormonal feedback mechanism elicited by the presence of fats in the duodenum. In the past, this hormone has been called *enterogas-*

*trone,* but such a hormone has never yet been identified as a specific entity. On the other hand, another hormone, *secretin,* which is released from the duodenum in response to certain foods (as will be discussed in detail in the following chapter) does have an important inhibitory effect on stomach contractions. Unfortunately, fats play a relatively minor role in the extraction of secretin from the mucosa of the duodenum.

For the present, therefore, it is difficult to account for the effect of fat in the duodenum in reducing stomach emptying, even though this effect is important to the process of fat digestion and fat absorption.

*Role of Pyloric Sphincter Contraction in Stomach Emptying.* Ordinarily, the degree of contraction of the pyloric sphincter is not very great, and the contraction that does occur is usually blocked as the pyloric pump peristaltic wave approaches the pylorus. However, many of the same duodenal factors that inhibit gastric contraction can simultaneously increase the degree of contraction of the pyloric sphincter, this factor adding to the diminished stomach emptying and therefore enhancing control over the emptying process. For instance, the presence of excess acid or excess irritation in the duodenal bulb promotes a moderate degree of pyloric contraction.

*Effect of Fluidity of the Stomach Chyme on Emptying.* Obviously, the more liquid the stomach chyme the greater will be the ease of emptying. Therefore, pure fluids ingested into the stomach have rapid passage into the duodenum, while more solid foods must await mixing with the gastric secretions as well as beginning fluidization of the solids by the process of stomach digestion.

**Summary.** Emptying of the stomach is controlled to a moderate degree by stomach factors, such as the degree of filling in the stomach and the activity of stomach peristalsis. Probably the more important control of stomach emptying, however, resides in feedback signals from the duodenum, including especially the enterogastric reflex and to a less extent hormonal feedback. These two feedback signals work together to slow the rate of emptying when (a) too much fluid is already in the small intestine or (b) the chyme is excessively acid, contains too much protein or fat, is hypotonic or hypertonic, or is irritating. In this way the rate of stomach emptying is limited to that amount of chyme that the small intestine can process.

# MOVEMENTS OF THE SMALL INTESTINE

The movements of the small intestine, as elsewhere in the gastrointestinal tract, can be divided into the *mixing contractions* and the *propulsive contractions*. However, to a great extent this separation is artificial because essentially all movements of the small intestine cause at least some degree of both mixing and propulsion. Yet, the usual classification of these processes is the following:

## MIXING CONTRACTIONS (SEGMENTATION CONTRACTIONS)

When a portion of the small intestine becomes distended with chyme, this elicits localized concentric contractions spaced at intervals along the intestine. These rhythmic contractions proceed at a rate of 11 to 12 per minute in the duodenum and at progressively slower rates down to approximately 7 per minute in the terminal ileum. The longitudinal length of each one of the contractions is only about 1 cm. so that each set of contractions causes "segmentation" of the small intestine, as illustrated in Figure 63–7, dividing the intestine at times into regularly spaced segments that have the appearance of a chain of sausages. As one set of segmentation contractions relaxes a new set begins, but the contractions this time occur at new points between the previous contractions. Therefore, the segmentation contractions "chop" the chyme many times a minute, in this way promoting progressive mixing of the solid food particles with the secretions of the small intestine.

These mixing movements are dependent mainly on the myenteric plexus of the gut,

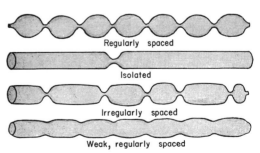

**Figure 63–7.** Segmentation movements of the small intestine.

though very weak concentric contractions can still occur even when the myenteric plexus is blocked by atropine. The intensity of the segmentation contractions is increased by parasympathetic stimulation but decreased by sympathetic stimulation.

## PROPULSIVE MOVEMENTS

Chyme is propelled through the small intestine by *peristaltic waves*. These occur in any part of the small intestine, and they move analward at a velocity of 0.5 to 5 cm. per second, much faster in the proximal intestine and much slower in the terminal intestine. However, they are normally very weak and usually die out after traveling only a few centimeters, so that movement of the chyme is much slower. As a result, the net movement of the chyme along the small intestine averages only 1 cm. per minute. This means that 3 to 10 hours are normally required for passage of chyme from the pylorus to the ileocecal valve.

Peristaltic activity of the small intestine is greatly increased after a meal. This is caused partly by the beginning entry of chyme into the duodenum but also by the so-called *gastroenteric reflex* that is initiated by distention of the stomach and conducted principally through the myenteric plexus from the stomach down along the wall of the small intestine. This reflex increases the overall degree of excitability of the small intestine, including both increased motility and secretion.

**The Peristaltic Reflex.** The usual cause of peristalsis in the small intestine is distention. Circumferential stretch of the intestine excites receptors in the gut wall, and these elicit a local myenteric reflex that begins with contraction of the longitudinal muscle over a distance of several centimeters followed by contraction of the circular muscle. Simultaneously, the contractile process spreads in an analward direction by the process of peristalsis. Movement of the peristaltic contraction down the gut is controlled by the myenteric plexus; it does not occur when this plexus has been blocked by drugs or when the plexus has degenerated.

Very intense irritation of the intestinal mucosa, such as occurs in some infectious processes, can elicit a so-called *peristaltic rush,* which is a powerful peristaltic wave that travels long distances in the small intestine in a few minutes. These waves can sweep the contents of the intestine into the colon and thereby relieve the small intestine of either irritants or excessive distention.

The function of the peristaltic waves in the small intestine is not only to cause progression

of the chyme toward the ileocecal valve but also to spread out the chyme along the intestinal mucosa. As the chyme enters the intestine from the stomach and causes initial distention of the proximal intestine, the elicited peristaltic waves begin immediately to spread the chyme along the intestine, and this process intensifies as additional chyme enters the intestine. On reaching the ileocecal valve the chyme is sometimes blocked for several hours until the person eats another meal, at which time a *gastroenteric* (also called *gastroileal*) reflex intensifies the peristalsis in the ileum and forces the remaining chyme through the ileocecal valve into the cecum.

*The Propulsive Effect of the Segmentation Movements.* The segmentation movements, though they last for only a few seconds, also travel in the analward direction and help to propel the food down the intestine. Therefore, the difference between the segmentation and the peristaltic movements is not as great as might be implied by their separation into these two classifications. The major difference is that the segmentation contractions occupy only a small longitudinal length of the small intestine at a time, whereas the peristaltic contractions occupy a considerably longer proportion of the intestinal wall.

**The Basic Electrical Rhythm of the Small Intestine and Its Control of Intestinal Contractions.** The basic rhythm of the small intestine is generated in the longitudinal muscle. Slow oscillatory electrical waves, as described earlier in the chapter, occur at the membranes of the smooth muscle, at rhythmical frequencies of 11 to 12 per minute in the duodenum and decreasing to 6 to 7 per minute in the terminal ileum. However, these electrical waves do not in themselves produce intestinal contractions. They merely set the background conditions for the contractions.

When the intestinal tract becomes overly distended or when the mucosa becomes irritated, this superimposes an additional stimulus, leading to myenteric reflexes that enhance the electrical activity of the gut. Now, during the positive phases of the slow waves, spike potentials occur. These spike potentials in turn elicit the muscle contractions. However, it is the basic rhythm of the slow wave that still determines the frequency of the contractions.

Furthermore, the slow waves travel in an analward direction. This property of the slow waves is presumably the reason, or at least part of the reason, for the analward progression of both the segmentation and the peristaltic contractions of the intestines.

The most rapid rate of rhythm of the slow

waves occurs in the intestinal muscle surrounding the point of entry of the bile duct into the duodenum. Since the slow wave travels downward along the intestine, the rate of rhythm of this point in the intestine tends also to be transmitted along the intestinal wall, thus acting as a "pacemaker" for the intestine. This is analogous to the pacemaker action of the S-A node of the heart, which controls the rate of rhythm of the entire heart. However, when this pacemaker slow wave fails to pass the entire distance down the small intestine, naturally occurring slow waves at slower frequencies in the more distal portions of the intestine then become local pacemakers.

**Movements Caused by the Muscularis Mucosae and Movements of the Villi.** The muscularis mucosae, which is stimulated by local nervous reflexes in the intramural plexus and by the sympathetic nervous system but not by the parasympathetic system, can cause short or long folds to appear in the intestinal mucosa. Also, individual fibers from this muscle extend upward into the intestinal villi and cause them to contract intermittently. The mucosal folds increase the surface area exposed to the chyme, thereby increasing the rate of absorption. The contractions of the villi—shortening, elongating, and shortening again—"milk" the villi so that lymph flows freely from the central lacteals into the lymphatic system. And both these types of contraction also agitate the fluids surrounding the villi so that progressively new areas of fluid become exposed to absorption.

Villus contractions are initiated by local nervous reflexes that occur in response to chyme in the small intestine. It is also believed that the villi are stimulated by a hormone called *villikinin*. The chyme in the intestine extracts villikinin from the mucosa, and this in turn is absorbed into the blood to excite the villi.

## CONTRACTION AND EMPTYING OF THE GALLBLADDER—CHOLECYSTOKININ

The functions of the gallbladder will be discussed specifically in relation to the overall function of the liver in Chapter 70. However, its activity is closely associated with that of the small intestine.

The liver continually secretes bile, which is then stored and concentrated in the gallbladder. The bile does not enter the small intestine until a specific stimulus causes the gallbladder to contract; this stimulus in general is initiated by the presence of fat in the small intestine as follows:

When fat and protein digestates enter the small intestine, within a few minutes these extract from the mucosa a hormone called *cholecystokinin*. This in turn passes by way of the blood to the gallbladder and causes it to contract rhythmically. However, these contractions alone also are not sufficient to cause emptying of the gallbladder, because the sphincter of Oddi, which constricts the common bile duct where it empties into the small intestine, normally remains tonically contracted despite gallbladder contraction. However, cholecystokinin also relaxes the sphincter of Oddi. In addition, every time a peristaltic wave passes over the duodenum, a wave of relaxation immediately ahead of the peristaltic contraction helps to relax the sphincter of Oddi, thus allowing small squirts of bile to flow into the duodenum. Therefore, the emptying of the gallbladder results from a combined action of cholecystokinin and intestinal peristalsis.

Regulation of gallbladder emptying by fats is a purposive effect, for bile salts help to emulsify the fats in the gut so that they can be digested by the intestinal lipases, and they also help to transport the digested fats through the intestinal epithelium.

## FUNCTION OF THE ILEOCECAL VALVE

A principal function of the ileocecal valve is to prevent backflow of fecal contents from the colon into the small intestine. As illustrated in Figure 63–8, the lips of the ileocecal valve, which protrude into the lumen of the cecum, are admirably adapted for this function. Usually the valve can resist reverse pressure of as much as 50 to 60 cm. water.

The wall of the ileum for several centimeters immediately preceding the ileocecal valve has a thickened muscular coat called the *ileocecal sphincter*. This normally remains mildly constricted and slows the emptying of ileal contents into the cecum except immediately following a meal when a gastroileal reflex (described above) intensifies the peristalsis in the ileum. Also, the hormone *gastrin,* which is liberated from the stomach mucosa in response to food in the stomach, has a direct relaxant effect on the ileocecal sphincter, thus allowing rapid emptying. Yet, even so, only about 450 ml. of chyme empties into the cecum each day. The resistance to emptying at the ileocecal valve prolongs

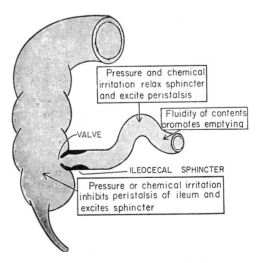

**Figure 63–8.** Emptying at the ileocecal valve.

the stay of chyme in the ileum and, therefore, facilitates absorption.

**Control of the Ileocecal Sphincter.** The degree of contraction of the ileocecal sphincter is controlled primarily by reflexes from the cecum. Whenever the cecum is distended, the degree of contraction of the ileocecal sphincter is intensified, which greatly delays emptying of additional chyme from the ileum. Also, any irritant in the cecum causes constriction of the ileocecal sphincter. For instance, when a person has an inflamed appendix, the irritation of this vestigial remnant of the cecum can cause such intense spasm of the ileocecal sphincter that it completely blocks emptying of the ileum. These reflexes from the cecum to the ileocecal sphincter are mediated by way of the myenteric plexus.

In addition, various viscero-sympathetic reflexes initiated by irritation of other portions of the gastrointestinal tract, of the kidneys, or of the peritoneum can also cause intense contraction of the sphincter and thereby delay, or even completely stop, movement of intestinal contents through the ileocecal valve.

## MOVEMENTS OF THE COLON

The functions of the colon are (1) absorption of water and electrolytes from the chyme and (2) storage of fecal matter until it can be expelled. The proximal half of the colon, illustrated in Figure 63–9, is concerned principally with absorption, and the distal half with storage; since intense movements are not required for these functions, the movements of the colon are normally sluggish. Yet in a sluggish manner, the movements still have characteristics similar to those of the small intestine and can be divided once again into mixing movements and propulsive movements.

**Mixing Movements—Haustrations.** In the same manner that segmentation movements occur in the small intestine, large circular constrictions also occur in the large intestine. At each of these constriction points, about 2.5 cm. of the circular muscle contracts, sometimes constricting the lumen of the colon to almost complete occlusion. At the same time, the longitudinal muscle of the colon, which is aggregated into three longitudinal strips called the *tineae coli,* contract. These combined contractions of the circular and longitudinal smooth muscle cause the unstimulated portion of the large intestine to bulge outward into baglike sacs called *haustrations*. The haustral contractions, once initiated, usually reach peak intensity in about 30 seconds and then disappear dur-

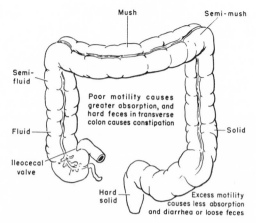

**Figure 63–9.** Absorptive and storage functions of the large intestine.

ing the next 60 seconds. They at times also move slowly analward during their period of contraction. After another few minutes, new haustral contractions occur in nearby areas but not in the same areas. Therefore, the fecal material in the large intestine is slowly "dug" into and rolled over in much the same manner that one spades the earth. In this way, all the fecal material is gradually exposed to the surface of the large intestine, and fluid is progressively absorbed until only 80 ml. of the 450 ml. daily load of chyme is lost in the feces. These contractions also help to move the fecal contents of the cecum and ascending colon into the transverse colon; and in this function, they act as very weak propulsive, as well as mixing, movements.

**Propulsive Movements—"Mass Movements."** Peristaltic waves of the type seen in the small intestine do not occur in the colon. Instead, another type of movement, called mass movements, propels the fecal contents toward the anus. These movements usually occur only a few times each day, most abundantly for about 15 minutes during the first hour or so after eating breakfast.

A mass movement is characterized by the following sequence of events: First, a constrictive point occurs at a distended or irritated point in the colon, usually in the transverse colon, and then rapidly thereafter the 20 or more cm. of colon *distal* to the constriction contracts almost as a unit, forcing the fecal material in this segment *en masse* down the colon. During this process, the haustrations disappear completely. The initiation of contraction is complete in about 30 seconds, and relaxation then occurs during the next two to three minutes. Mass movements can occur in any part of the colon, though most often they occur in the transverse or descending colon. When they have forced a mass of feces into the rectum, the desire for defecation is felt.

***Initiation of Mass Movements by the Gastrocolic and Duodenocolic Reflexes.*** The appearance of mass movements after meals is caused at least partially by so-called *gastrocolic* and *duodenocolic reflexes.* These reflexes result from distention of the stomach and duodenum. They still take place with decreased intensity when the autonomic nerves are removed; therefore, it is probable that the reflexes are basically transmitted through the myenteric plexus, though reflexes conducted through the autonomic nervous system probably reinforce this direct route of transmission.

It is likely that the hormone gastrin, which is secreted by the stomach antral mucosa in response to distention, also plays some role in this effect, because gastrin has an excitatory effect on the colon and an inhibitory effect on the ileocecal valve to allow rapid emptying of ileal contents into the cecum, this in turn eliciting increased colonic activity.

Irritation in the colon can also initiate intense mass movements. For instance, a person who has an ulcerated condition of the colon (*ulcerative colitis*) frequently has mass movements that persist almost all of the time.

Mass movements are also initiated by intense stimulation of the parasympathetic nervous system or simply by overdistention of a segment of the colon.

## DEFECATION

Most of the time the rectum is empty of feces. This results partly from the fact that a weak functional sphincter exists approximately 20 cm. from the anus at the juncture between the sigmoid and the rectum. However, when a mass movement forces feces into the rectum the process of defecation is normally initiated, including reflex contraction of the rectum and relaxation of the anal sphincters.

Continual dribble of fecal matter through the anus is prevented by tonic constriction of (1) the *internal anal sphincter,* a circular mass of smooth muscle that lies immediately inside the anus, and (2) the *external anal sphincter,* composed of striated voluntary muscle that surrounds and lies slightly distal to the internal sphincter and is controlled by the somatic nervous system.

Ordinarily, defecation results from the *defecation reflex,* which can be described as follows: When the feces enter the rectum, distention of the rectal wall initiates afferent signals that spread through the *myenteric plexus* to initiate peristaltic waves in the descending colon,

sigmoid, and rectum, forcing feces toward the anus. As the peristaltic wave approaches the anus, the internal anal sphincter is inhibited by the usual phenomenon of "receptive relaxation," and if the external anal sphincter is relaxed, defecation will occur. This overall effect is called the defecation reflex.

However, the defecation reflex itself is extremely weak, and to be effective in causing defecation it must be fortified by another reflex that involves the sacral segments of the spinal cord, as illustrated in Figure 63–10. When the afferent fibers in the rectum are stimulated, signals are transmitted into the spinal cord and thence, reflexly, back to the descending colon, sigmoid, rectum, and anus by way of parasympathetic nerve fibers in the *nervi erigentes.* These parasympathetic signals greatly intensify the peristaltic waves and convert the defecation reflex from an ineffectual weak movement into a powerful process of defecation that is sometimes effective in emptying the large bowel all the way from the splenic flexure to the anus. Also, the afferent signals entering the spinal cord initiate other effects, such as taking a deep breath, closure of the glottis, and contraction of the abdominal muscles to force downward on the fecal contents of the colon while at the same time causing the pelvic floor to pull outward and upward on the anus to evaginate the feces downward.

However, despite the defecation reflex, other effects are necessary before actual defecation occurs. First, relaxation of the internal sphincter and forward movement of feces toward the anus normally initiates an instantaneous contraction of the external sphincter which still temporarily prevents defecation. Second, except in babies and mentally inept persons, the conscious mind then takes over voluntary con-

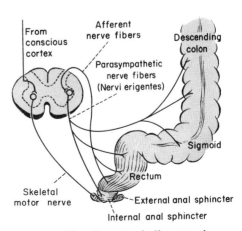

**Figure 63–10.** The afferent and efferent pathways of the parasympathetic mechanism for enhancing the defecation reflex.

trol of the external sphincter and either inhibits it to allow defecation to occur or further contracts it if the moment is not socially acceptable for defecation to occur. When the contraction is maintained, the defecation reflex dies out after a few minutes and usually will not return until an additional amount of feces enters the rectum, which may not occur until several hours thereafter.

When it becomes convenient for the person to defecate, defecation reflexes can sometimes be initiated by taking a deep breath to move the diaphragm downward and then contracting the abdominal muscles to increase the pressure in the abdomen, thus forcing fecal contents into the rectum to elicit new reflexes. Unfortunately, reflexes initiated in this way are never as effective as those that arise naturally, for which reason people who inhibit their natural reflexes too often become severely constipated.

In the newborn baby or in some persons with transected spinal cords, the defecation reflex causes automatic emptying of the lower bowel without the normal control exercised through contraction of the external anal sphincter.

## OTHER AUTONOMIC REFLEXES AFFECTING BOWEL ACTIVITY

Aside from the duodenocolic, gastrocolic, gastroileal, enterogastric, and defecation reflexes which have been discussed in this chapter, several other important nervous reflexes can affect the overall degree of bowel activity. These are the intestino-intestinal reflex, peritoneo-intestinal reflex, reno-intestinal reflex, vesico-intestinal reflex, and somato-intestinal reflex. All these reflexes are initiated by sensory signals that pass to the spinal cord and then are transmitted through the sympathetic nervous system back to the gut. And they always *inhibit* gastrointestinal activity. Thus, the *intestino-intestinal reflex* occurs when one part of the intestine becomes overdistended or its mucosa becomes excessively irritated; this blocks activity in other parts of the intestine while the local distention or irritability increases activity in the localized region and moves the intestinal contents away from the distended or irritated area.

The *peritoneo-intestinal reflex* is very much like the intestino-intestinal reflex, except that it results from irritation of the peritoneum; it causes intestinal paralysis. The *reno-intestinal* and *vesico-intestinal reflexes* inhibit intestinal activity as a result of kidney or bladder irritation. Finally, the *somato-intestinal reflex* causes intestinal inhibition when the skin over the abdomen is irritatingly stimulated. Some of these reflexes will be discussed further in Chapter 66 in relation to abnormalities of gastrointestinal function.

## REFERENCES

Barry, R. J., and Eggenton, J.: Electrical activity of the intestine. *Biomembranes*, 4B:917, 1974.

Bortoff, A.: Digestion: motility. *Ann. Rev. Physiol.*, 34:261, 1972.

Carlsson, G. E.: Bite force and chewing efficiency. *Front. Oral Physiol.*, 1:265, 1974.

Code, C. F., and Carlson, H. C.: Motor activity of the stomach. In Code, C. F., and Heidel, W. (eds.): Handbook of Physiology. Sec. 6, Vol. 4. Baltimore, The Williams & Wilkins Company, 1968, p. 1903.

Cohen, S., and Harris, L. D.: The lower esophageal sphincter. *Gastroenterology*, 63:1066, 1972.

Davenport, H.: Physiology of the Digestive Tract. 2nd Ed. Chicago, Year Book Publishers, Inc., 1966.

Drake, B.: Mastication in food science and technology. *Front. Oral Physiol.* 1:257, 1974.

Frigo, G. M., Torsoli, A., Lecchini, S., Falaschi, C. F., and Crema, A.: Recent advances in the pharmacology of peristalsis. *Arch. Inter. Pharmacodyn. Ther.*, 196(Suppl. 196):9, 1972.

Furness, J. B., and Costa, M.: The adrenergic innervation of the gastrointestinal tract. *Ergeb. Physiol.* 69:2, 1974.

Grossman, M. I.: The digestive system. *Ann. Rev. Physiol.*, 25:165, 1963.

Hunt, J. N., and Knox, M. T.: Regulation of gastric emptying. In Code, C. F., and Heidel, W. (eds.): Handbook of Physiology. Sec. 6, Vol. 4. Baltimore, The Williams & Wilkins Company, 1968, p. 1917.

Jenkins, G. N.: The Physiology of the Mouth, 3rd Ed. Philadelphia, F. A. Davis Company, 1965.

Johnson, L. R.: Gastrointestinal hormones. In Guyton, A. C. (ed.): MTP International Review of Science: Physiology. Vol 4. Baltimore, University Park Press, 1974, p. 1.

Katz, D., and Hoffman, F. (eds.): The Esophagogastric Junction. New York, American Elsevier Publishing Co., Inc., 1971.

Kawamura, Y.: Neurogenesis of mastication. *Front. Oral Physiol.*, 1:77, 1974.

Kelly, K. A.: Gastric motility after gastric operations. *Surg. Ann.*, 6:103, 1974.

Kirsner, J. B.: The stomach. In Sodeman, W. A., Jr., and Sodeman, W. A. (eds.): Pathologic Physiology: Mechanisms of Disease, 5th Ed. Philadelphia, W. B. Saunders Company, 1974, p. 709.

Makhlouf, G. M.: The neuroendocrine design of the gut. The play of chemicals in a chemical playground. *Gastroenterology*, 67:159, 1974.

Moller, E.: Action of the muscles of mastication. *Front. Oral Physiol.*, 1:121, 1974.

Shehadeh, Z., Price, W. E., and Jacobson, E. D.: Effects of vasoactive agents on intestinal blood flow and motility in the dog. *Amer. J. Physiol.*, 216:386, 1969.

Skinner, D. B.: The esophagus. In Sodeman, W. A., Jr., and Sodeman, W. A. (eds.): Pathologic Physiology: Mechanisms of Disease, 5th Ed. Philadelphia, W. B. Saunders Company, 1974, p. 697.

Sodeman, W. A., Jr., and Watson, D. W.: The large intestine. In Sodeman, W. A., Jr., and Sodeman, W. A. (eds.): Pathologic Physiology: Mechanisms of Disease, 5th Ed. Philadelphia, W. B. Saunders Company, 1974, p. 767.

Soergel, K. H.: New concepts of intestinal function. *Acta Hepatogastroenterol. (Stuttg.)*, 20:351, 1973.

Sumi, T.: Importance of pharyngeal feedback on the integration of reflex deglutition in newborn animals. *Symp. Oral Sens. Percept.*, 4:174, 1973.

Truelove, S. C.: Movements of the large intestine. *Physiol. Rev.*, 46:457, 1966.

Watson, D. W., and Sodeman, W. A., Jr.: The small intestine. In Sodeman, W. A., Jr., and Sodeman, W. A. (eds.): Pathologic Physiology: Mechanisms of Disease, 5th Ed. Philadelphia, W. B. Saunders Company, 1974, p. 734.

Weisbrodt, N. W.: Gastrointestinal motility. In Guyton, A. C. (ed.): MTP International Review of Science: Physiology. Vol. 4. Baltimore, University Park Press, 1974, p. 105.

Wood, J. D.: Neurophysiology of Auerbach's plexus and control of intestinal motility. *Physiol. Rev.*, 55:307, 1975.

Youmans, W. B.: Innervation of the gastrointestinal tract. In Code, C. F., and Heidel, W. (eds.): Handbook of Physiology. Sec. 6, Vol. 4. Baltimore, The Williams & Wilkins Company, 1968, p. 1655.

# 64

# Secretory Functions of the Alimentary Tract

Throughout the gastrointestinal tract secretory glands subserve two primary functions: First, digestive enzymes are secreted in most areas from the mouth to the distal end of the ileum. Second, mucous glands, present from the mouth to the anus, provide mucus for lubrication and protection of all parts of the alimentary tract.

Most digestive secretions are formed only in response to the presence of food in the alimentary tract, and the quantity secreted in each segment of the tract is almost exactly the amount needed for proper digestion. Furthermore, in some portions of the gastrointestinal tract even the types of enzymes and other constituents of the secretions are varied in accordance with the types of food present. The purpose of the present chapter, therefore, is to describe the different alimentary secretions, their functions, and regulation of their production.

## GENERAL PRINCIPLES OF GASTROINTESTINAL SECRETION

### ANATOMICAL TYPES OF GLANDS

Several types of glands provide the different types of secretions in the gastrointestinal tract. First, on the surface of the epithelium in most parts of the gastrointestinal tract are literally billions of *single cell mucous glands* called *goblet cells*. These function entirely by themselves without the necessity of coordination with other goblet cells, and they simply extrude their mucus directly into the lumen of the intestine.

Second, the most abundant type of gland in the gastrointestinal tract is the tubular gland, two examples of which are illustrated, in Figure 64–1A and 64–1B. Figure 64–1A shows a *crypt of Lieberkühn,* found throughout the small intestine and in modified form in the large intestine. These are simple pits lined with goblet cells to produce mucus and other epithelial cells to produce mainly serous fluids but also a small amount of enzymes. Figure 64–1B illustrates a *gastric gland* in the main body of the stomach. This is a considerably deeper tubular gland, and it is occasionally branched.

Also associated with the gastrointestinal tract are several complex glands—the *salivary glands*, the *pancreas,* and the *liver*—which provide secretions for digestion or emulsification of food. The liver has a highly specialized structure that will be discussed in Chapter 70. The salivary glands and the pancreas are compound acinous glands of the type illustrated in Figure 64–1C. These glands lie completely outside the walls of the gastrointestinal tract and, in this,

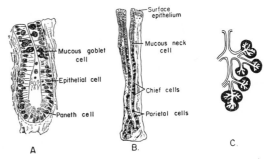

**Figure 64–1.** The anatomical types of gastrointestinal glands: (A) A simple tubular gland represented by a crypt of Lieberkühn. (B) A more elongated tubular gland represented by a gastric gland. (C) A compound acinous gland represented by the pancreas.

differ from all other gastrointestinal glands. The major secreting structures, called *acini,* are lined with secreting glandular cells; these feed into a system of ducts that finally empty through one or more portals into the intestinal tract itself.

## BASIC MECHANISMS OF STIMULATION OF THE GASTROINTESTINAL GLANDS

**Effect of Local Stimuli.** The mechanical presence of food in a particular segment of the gastrointestinal tract usually causes the glands of that region and, often of adjacent regions, to secrete moderate to large quantities of digestive juices. Part of this local effect results from direct stimulation of the surface glandular cells themselves by contact with the food. For instance, the goblet cells on the surface of the epithelium are stimulated mainly in this way. However, most local stimulation of the intestinal glands results from one of the following three methods of stimulation:

(1) Tactile stimulation or chemical irritation of the mucosa can elicit reflexes that pass through the myenteric plexus of the intestinal wall to stimulate either the mucous cells on the surface or the deeper glands of the mucosa. (2) Distention of the gut can also elicit nervous reflexes that stimulate secretion. (3) Tactile or chemical stimuli or distention can result in increased motility of the gut, as described in the preceding chapter, and the motility in turn can then increase the rate of secretion.

**Autonomic Stimulation of Secretion.** *Parasympathetic Stimulation.* Stimulation of the parasympathetic nerves to the alimentary tract almost invariably increases the rates of glandular secretion. This is especially true of the glands in the upper portion of the tract innervated by the vagus and other cranial parasympathetic nerves—including the salivary, esophageal, and gastric glands, the pancreas, and Brunner's glands in the duodenum—and also of the distal portion of the large intestine, innervated by the pelvic parasympathetic nerves. Secretion in the remainder of the small intestine and in the first two-thirds of the large intestine occurs almost entirely in response to local stimuli.

*Sympathetic Stimulation.* Stimulation of sympathetic nerves in some parts of the gastrointestinal tract causes a slight increase in secretion by the respective glands. On the other hand, sympathetic stimulation also results in constriction of the blood vessels supplying the glands. Therefore, sympathetic stimulation can have a dual effect: First, sympathetic stimulation alone can slightly increase secretion. But, second, if parasympathetic stimulation is caus-

ing copious secretion by the glands, superimposed sympathetic stimulation often reduces the secretion because of reduced blood supply.

A few intestinal glands are specifically inhibited, rather than excited, by sympathetic stimulation. Especially important is the inhibition of Brunner's glands, for this often leaves the initial portion of the duodenum unprotected against the acidic digestive juices from the stomach, thus allowing development of *peptic ulcer.*

**Regulation of Glandular Secretion by Hormones.** In the stomach and intestine several different *gastrointestinal hormones* help to regulate the volume and character of the secretions. These hormones are liberated from the gastrointestinal mucosa in response to the presence of foods in the lumen of the gut. They then are absorbed into the blood and are carried to glands where they stimulate secretion. This type of stimulation is particularly valuable in increasing the output of gastric juice and pancreatic juice when food enters the stomach or duodenum. Also, hormonal stimulation of the gallbladder wall causes it to empty its stored bile into the duodenum, as discussed in the preceding chapter. Other hormones that are still of doubtful value have also been postulated to stimulate secretion by the glands of the small intestine.

Chemically, the gastrointestinal hormones are polypeptides or polypeptide derivatives.

## BASIC MECHANISM OF SECRETION BY GLANDULAR CELLS

**Secretion of Organic Substances.** Though all the basic mechanisms by which glandular cells form different secretions and then extrude these to the exterior are not known, experimental evidence points to the following basic principles of secretion by glandular cells, as illustrated in Figure 64–2. (1) The nutrient material needed for formation of the secretion must diffuse or be actively transported from the capillary into the base of the glandular cell. (2) Many *mitochondria* located inside the cell near its base provide oxidative energy for formation of adenosine triphosphate. (3) Energy from the adenosine triphosphate (see Chap. 67), along with appropriate nutrients, is then used for synthesis of the organic substances, this synthesis occurring almost entirely on the *endoplasmic reticulum.* The *ribosomes* adherent to this reticulum are specifically responsible for formation of proteins that are to be secreted. (4) The secretory materials flow through the tubules of the endoplasmic reticulum into the

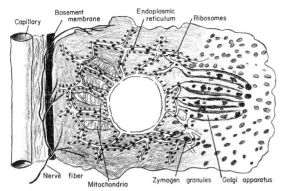

**Figure 64–2.** Basic mechanism of secretion by a glandular cell.

ity on the exterior. Parasympathetic stimulation increases this polarization voltage to values some 10 and 20 millivolts more negative than normal. This increase in polarization occurs a second or more after the nerve signal has arrived, indicating that it is caused by movement of negative ions through the membrane to the interior of the cell.

Though this mechanism for secretion is still partly theoretical, it does explain how it would be possible for nerve impulses to regulate secretion. Obviously, hormonal effects on the cell membrane could cause similar results.

vesicles of the Golgi apparatus, which lies near the secretory ends of the cells. (5) The materials then are concentrated and discharged into the cytoplasm in the form of *secretory granules*. (6) These granules are then extruded through the secretory surface into the lumen of the gland.

**Water and Electrolyte Secretion in Response to Nervous Stimulation.** A second necessity for glandular secretion is sufficient water and electrolytes to be secreted along with the organic substances. The following is a postulated method by which nervous stimulation causes water and salts to pass through the glandular cells in great profusion, which washes the organic substances through the secretory border of the cells at the same time:

(1) Nerve stimulation has a specific effect on the *basal* portion of the cell membrane to cause active transport of chloride ions to the interior. (2) The resulting increase in electronegativity inside the cell then causes positive ions also to move to the interior of the cell. (3) The excess of both of these ions inside the cell creates an osmotic force which pulls water to the interior, thereby increasing the hydrostatic pressure inside the cell and causing the cell itself to swell. (4) The pressure in the cell then results in minute ruptures of the secretory border of the cell and causes flushing of water, electrolytes, and organic materials out of the secretory end of the glandular cell and into the lumen of the gland.

In support of this theory have been the following findings: First, the nerve endings on glandular cells are principally on the bases of the cells. Second, microelectrode studies show that the electrical potential across the membrane at the base of the cell is between 30 and 40 mv., with negativity on the interior and positiv-

## LUBRICATING AND PROTECTIVE PROPERTIES OF MUCUS AND ITS IMPORTANCE IN THE GASTROINTESTINAL TRACT

Mucus is a thick secretion composed of water, electrolytes, and a mixture of several mucopolysaccharides. Mucus is slightly different in different parts of the gastrointestinal tract, but everywhere it has several important characteristics that make it both an excellent lubricant and a protectant for the wall of the gut. *First,* mucus has adherent qualities that make it adhere tightly to the food or other particles and spread as a thin film over the surfaces. *Second,* it has sufficient *body* that it coats the wall of the gut and prevents actual contact of food particles with the mucosa. *Third,* mucus has a low resistance to slippage so that the particles can slide along the epithelium with great ease. *Fourth,* mucus causes fecal particles to adhere to each other to form the fecal masses that are expelled during a bowel movement. *Fifth,* mucus is strongly resistant to digestion by the gastrointestinal enzymes. And, *sixth,* the mucopolysaccharides of mucus have amphoteric properties (as is true of all proteins) and are therefore capable of buffering small amounts of either acids or alkalies; also, mucus usually contains moderate quantities of bicarbonate ions, which specifically neutralize acids.

In summary, mucus has the ability to allow easy slippage of food along the gastrointestinal tract and also to prevent excoriative or chemical damage to the epithelium. One becomes acutely aware of the lubricating qualities of mucus when his salivary glands fail to secrete saliva, for under these circumstances it is extremely difficult to swallow solid food even when it is taken with large amounts of water.

# SECRETION OF SALIVA

**The Salivary Glands; Characteristics of Saliva.** The principal glands of salivation are the *parotid, submaxillary,* and *sublingual glands,* but in addition to these are many small *buccal glands.* The daily secretion of saliva normally ranges between 1000 and 1500 milliliters, as shown in Table 64–1.

Saliva contains two different types of secretion: (1) a *serous secretion* containing *ptyalin* (an $\alpha$-amylase), which is an enzyme for digesting starches, and (2) *mucous secretion* for lubricating purposes. The parotid glands secrete entirely the serous type, and the submaxillary glands secrete both the serous type and mucus as well. The sublingual and buccal glands secrete only mucus. Saliva has a pH between 6.0 and 7.4, a favorable range for the digestive action of ptyalin.

**Secretion of Ions in the Saliva.** Saliva contains an especially large quantity of potassium and under some conditions also a high concentration of bicarbonate ions. On the other hand, the concentrations of both sodium and chloride ions are considerably less in saliva than they are in plasma. One can understand these special concentrations of ions in the saliva from the following description of the mechanism for secretion of saliva.

Salivary secretion is a two-stage operation, the first stage involving the acini and the second the salivary ducts. The acini secrete a so-called *primary secretion* that contains the salivary enzymes in a solution of ions having compositions not greatly different from those of plasma. However, as the primary secretion flows through the ducts, two major active transport processes take place that markedly modify the ionic composition of the saliva. First, sodium ions are actively reabsorbed from the ducts, and potassium ions are actively secreted into the ducts in exchange for the sodium. Therefore, the sodium concentration of the saliva becomes reduced, as does the concentration of chloride ions, while the potassium ion concentration becomes increased. Second, bicarbonate ions are secreted into the ducts; this process is catalyzed by *carbonic anhydrase* found in the epithelial cells of the ducts. During the secretion of bicarbonate ions, still more chloride ions are passively absorbed from the ducts in exchange for the bicarbonate ions. The net result of these active transport processes is that, under resting conditions, the concentrations of sodium and chloride ions in the saliva are only about 15 mEq./liter each, approximately 1/7 to 1/10 their concentrations in plasma. On the other hand, the concentration of potassium ions is about 30 mEq./liter, about seven times as great as its concentration in plasma, and the concentration of bicarbonate ions is 50 to 90 mEq./liter, about two to four times that of plasma.

During maximal salivation, the salivary ionic concentrations change considerably because the rate of formation of primary secretion by the acini can increase as much as 20-fold. As a result of this increase the flow of this secretion through the ducts becomes so great that the ductal reconditioning of the secretion is considerably reduced. Therefore, when copious quantities of saliva are secreted, the sodium chloride concentration rises to about 1/2 to 2/3 that of plasma, while the potassium concentration falls to only four times that of plasma.

In the presence of excess aldosterone secretion, the sodium and chloride reabsorption and the potassium secretion become greatly increased so that the sodium chloride concentration in the saliva then becomes almost reduced to zero, while the potassium concentration increases still more.

Because of the high potassium ion concentration of saliva, in any abnormal state in which the saliva is lost to the exterior of the body for long periods of time, the person can develop serious depletion of potassium ions in his body, leading eventually to serious hypokalemia and paralysis that results from this condition, as was discussed in Chapter 10.

**Function of Saliva for Oral Hygiene.** Even under basal conditions, up to 1 ml./min. of saliva, almost entirely of the mucous type, is secreted all the time. This secretion plays an exceedingly important role in maintaining healthy oral tissues. The mouth is loaded with

**TABLE 64–1. Daily Secretion of Intestinal Juices**

| | Daily Volume (ml.) | pH |
|---|---|---|
| Saliva | 1200 | 6.0–7.0 |
| Gastric secretion | 2000 | 1.0–3.5 |
| Pancreatic secretion | 1200 | 8.0–8.3 |
| Bile | 700 | 7.8 |
| Succus entericus | 2000 | 7.8–8.0 |
| Brunner's gland secretion | 50(?) | 8.0–8.9 |
| Large intestinal secretion | 60 | 7.5–8.0 |
| Total | 7010 | |

pathogenic bacteria that can easily destroy tissues and can also cause dental caries. However, saliva helps to prevent the deteriorative processes in several ways. First, the flow of saliva itself helps to wash away the pathogenic bacteria. Second, the saliva also contains several factors that actually destroy bacteria. One of these is thiocyanate ions and another is an enzyme that either attacks the bacteria or aids the thiocyanate ion in entering the bacteria where it in turn becomes bactericidal.

Therefore, in the absence of salivation, the oral tissues become ulcerated and otherwise infected, and caries of the teeth becomes rampant.

**Nervous Regulation of Salivary Secretion.** Figure 64–3 illustrates the nervous pathways for regulation of salivation, showing that the submaxillary and sublingual glands are controlled principally by nerve impulses from the superior portions of the *salivatory nuclei* and the parotid gland by impulses from the inferior portions of these nuclei. The salivatory nuclei are located approximately at the juncture of the medulla and pons and are excited by both taste and tactile stimuli from the tongue and other areas of the mouth. Most taste stimuli, especially the sour taste, elicit copious secretion of saliva—often as much as 5 ml. per minute or 8 to 20 times the basal rate of secretion. Also, certain tactile stimuli, such as the presence of smooth objects in the mouth (a pebble,

for instance), cause marked salivation, whereas rough objects cause less salivation and occasionally even inhibit salivation.

Salivation can also be stimulated or inhibited by impulses arriving in the salivatory nuclei from higher centers of the central nervous system. For instance, when a person smells or eats food that he particularly likes, salivation is far greater than when he smells or eats food that he detests. The *appetite area* of the brain that partially regulates these effects is located in close proximity to the parasympathetic centers of the anterior hypothalamus, and it functions to a great extent in response to signals from the taste and smell areas of the cerebral cortex or amygdala.

Finally, salivation also occurs in response to reflexes originating in the stomach and upper intestines—particularly when very irritating foods are swallowed or when a person is nauseated because of some gastrointestinal abnormality. The swallowed saliva presumably helps to remove the irritating factor in the gastrointestinal tract by diluting or neutralizing the irritant substances.

## ESOPHAGEAL SECRETION

The esophageal secretions are entirely mucoid in character and function principally to provide lubrication for swallowing. The main body of the esophagus is lined with many simple mucous glands, but at the gastric end and, to a lesser extent, in the initial portion of the esophagus there are many compound mucous glands. The mucus secreted by the compound glands in the upper esophagus prevents mucosal excoriation by the newly entering food, while the compound glands near the esophagogastric juncture protect the esophageal wall from digestion by gastric juices that reflux into the lower esophagus. Despite this protection, a peptic ulcer at times still occurs at the gastric end of the esophagus.

## GASTRIC SECRETION

### CHARACTERISTICS OF THE GASTRIC SECRETIONS

In addition to the mucus secreting cells that line the surface of the stomach, the stomach mucosa has two entirely different types of tubular glands: the *gastric* or *fundic* glands and the *pyloric glands*. The gastric glands secrete the

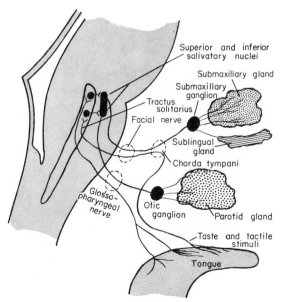

**Figure 64–3.**   Nervous regulation of salivary secretion.

Labels in figure:
Superior and inferior salivatory nuclei
Submaxillary gland
Submaxillary ganglion
Tractus solitarius
Facial nerve
Sublingual gland
Chorda tympani
Glosso-pharyngeal nerve
Otic ganglion
Parotid gland
Taste and tactile stimuli
Tongue

digestive juices, and the pyloric glands secrete almost entirely mucus for the protection of the pyloric mucosa. The gastric glands are located everywhere in the mucosa of the body and fundus of the stomach, and the pyloric glands are located in the antral portion of the stomach. In addition, a few *cardiac glands,* which are almost identical to the pyloric glands, are located about 1 cm. immediately surrounding the entry point of the esophagus.

**The Digestive Secretions from the Gastric Glands.** A typical gastric gland is shown in Figure 64–1B, which is composed of three different types of cells: the *mucous neck cells,* which secrete mucus; the *chief cells,* which secrete the digestive enzymes (*pepsin* in particular); and the *parietal cells,* which secrete hydrochloric acid and which lie mainly behind the mucous neck cells or, less often, behind the chief cells. A postulated mechanism for secretion of the enzymes was given earlier in the chapter and was depicted in Figure 64–2. However, secretion of hydrochloric acid by the parietal cells involves particular problems which require further consideration as follows:

*Basic Mechanism of Hydrochloric Acid Secretion.* The parietal cells secrete an electrolytic solution containing 160 millimols of hydrochloric acid per liter, which is almost exactly isotonic with the body fluids. The pH of this acid solution is approximately 0.8, thus illustrating its extreme acidity. At this pH the hydrogen ion concentration is about four million times that of the arterial blood. And to concentrate the hydrogen ions this tremendous amount requires over 1500 calories of energy per liter of gastric juice, as discussed in Chapter 4 in relation to membrane transport mechanisms.

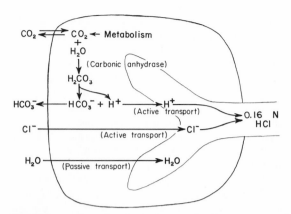

**Figure 64–5.** Postulated mechanism for the secretion of hydrochloric acid.

Figure 64–4 illustrates the basic structure of a parietal cell, showing that it contains a system of *intracellular canaliculi.* This system is a modification of the endoplasmic reticulum. Presumably the hydrochloric acid is formed at the membranes of these canaliculi and then conducted to the exterior.

Different suggestions for the precise mechanism of hydrochloric acid formation have been offered, a simple one of which is illustrated in Figure 64–5. This diagram shows that *carbon dioxide,* either formed during metabolism in the cell or entering the cell from the blood, combines with water under the influence of *carbonic anhydrase* to form *carbonic acid.* This, in turn, dissociates into *bicarbonate ion* and *hydrogen ion.* By some yet unknown active transport process, the hydrogen ion is transported through the wall of the canaliculus and into its lumen. The bicarbonate ion, in turn, diffuses backward into the blood.

Chloride ion, also, is actively transported from the blood into the canaliculi, though the basic process for this, too, is unknown. In some way that is not understood, the secretion of chloride ions is coupled with the secretion of hydrogen ions so that the two kinds of ions are secreted in equal quantities. Finally, water passes into the canaliculi by passive diffusion.

The importance of carbon dioxide in the chemical reactions for formation of hydrochloric acid is illustrated by the fact that inhibition of carbonic anhydrase by the drug *acetazolamide* almost completely blocks the formation of hydrochloride acid.

*Secretion of Pepsin.* The principal enzyme secreted by the chief cells is pepsin. This is formed inside the cells in the form of *pepsinogen,* which has no digestive activity. However,

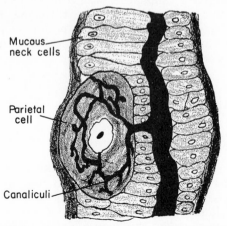

**Figure 64–4.** Anatomy of the canaliculi in a parietal cell.

once pepsinogen is secreted and comes in contact with previously formed pepsin in the presence of hydrochloric acid, it is immediately activated to form active pepsin. In this process, the pepsinogen molecule having a molecular weight of 42,500 is split to the pepsin molecule having a molecular weight of 35,000.

Pepsin is an active proteolytic enzyme in a highly acid medium (optimum pH = 2.0), but above a pH of about 5 it has little proteolytic activity and soon becomes completely inactivated. Therefore, hydrochloric acid secretion is equally as necessary as pepsin secretion for protein digestion in the stomach.

*Secretion of Other Enzymes.* Small quantities of other enzymes are also secreted in the stomach juices, including *gastric lipase* and *gastric amylase.* Gastric lipase is of little quantitative importance and is actually a *tributyrase,* for its principal activity is on tributyrin, which is butterfat; it has almost no lipolytic activity on the other fats. Gastric amylase plays a very minor role in digestion of starches.

**Secretion of Mucus in the Stomach.** The pyloric and cardiac glands are structurally similar to the gastric glands, but contain almost no chief and parietal cells. Instead, they contain almost entirely mucous cells that are identical with the mucous neck cells of the gastric glands. All these cells secrete a thin mucus, which protects the stomach wall from digestion by the gastric enzymes.

In addition, the surface of the stomach mucosa between glands has a continuous layer of mucous cells that secrete large quantities of a far more *viscid and alkaline mucus* that coats the mucosa with a mucous gel layer over 1 mm. thick, thus providing a major shell of protection for the stomach wall as well as contributing to lubrication of food transport. Even the slightest irritation of the mucosa directly stimulates the mucous cells to secrete copious quantities of this thick, viscid mucus.

## REGULATION OF GASTRIC SECRETION BY NERVOUS AND HORMONAL MECHANISMS

Gastric secretion is regulated by both nervous and hormonal mechanisms, nervous regulation being effected through the parasympathetic fibers of the vagus nerves as well as through local myenteric plexus reflexes and hormonal regulation taking place by means of the hormone *gastrin*. Thus, regulation of gastric secretion is different from the regulation of salivary secretion, which is effected entirely by nervous mechanisms.

### Vagal Stimulation of Gastric Secretion

Nervous signals to cause gastric secretion originate in the dorsal motor nuclei of the vagi and pass via the vagus nerves to the myenteric plexus of the stomach and thence to the gastric glands. In response, these glands secrete vast quantities of both pepsin and acid, but with a higher proportion of pepsin than in gastric juice elicited in other ways. Also, vagal signals to the pyloric glands, the cardiac glands, and the mucous neck cells of the gastric glands cause some increase in secretion of mucus as well.

Still another effect of vagal stimulation is to cause the antral part of the stomach mucosa to secrete the hormone gastrin. As will be explained in the following paragraphs, this hormone then acts on the gastric glands to cause additional flow of highly acid gastric juice. Thus, vagal stimulation excites stomach secretion both directly by stimulation of the gastric glands and indirectly through the gastrin mechanism.

### Stimulation of Gastric Secretion by Gastrin

When food enters the stomach, it causes the antral portion of the stomach mucosa to secrete the hormone gastrin, a heptadecapeptide. The food causes release of this hormone in two ways: (1) The actual bulk of the food distends the stomach, and this causes the hormone gastrin to be released from the antral mucosa. (2) Certain substances called secretagogues—such as food extractives, partially digested proteins, alcohol (in low concentration), caffeine, and so forth—also cause gastrin to be liberated from the antral mucosa.

Both of these stimuli—the distention and the chemical action of the secretagogues—elicit gastrin release by means of myenteric reflexes. That is, they stimulate sensory nerve fibers in the stomach epithelium which in turn synapse with the myenteric plexus. This then transmits efferent signals to special epithelial "gastrin" cells that have been identified in the gastric mucosa and that secrete the gastrin. Therefore, any factor that blocks this myenteric reflex will also block the formation of gastrin. For instance, anesthetization of the gastric mucosa to block the sensory stimuli will prevent gastrin release; administration of atropine, which blocks the action on the gastrin cells of the acetylcholine released by the myenteric plexus, will also prevent gastrin release.

Gastrin is absorbed into the blood and carried to the gastric glands where it stimulates mainly the parietal cells but to much less extent the chief cells, also. The parietal cells increase their rate of hydrochloric acid secretion as much as eight-fold, and the chief cells increase their rate of enzyme secretion two- to four-fold.

The rate of secretion in response to gastrin is somewhat less than to vagal stimulation, 200 ml. per hour in contrast to about 500 ml. per hour, indicating that the gastrin mechanism is a less potent mechanism for stimulation of stomach secretion than is vagal stimulation. However, the gastrin mechanism usually continues for several hours in contrast to a much shorter period of time for vagal stimulation. Therefore, as a whole, it is likely that the gastrin mechanism is equally as important as, if not more important than, the vagal mechanism for control of gastric secretion. Yet, when both of these work together, the total secretion is much greater than the sum of the individual secretions caused by each of the two mechanisms. In other words, *the two mechanisms multiply each other rather than simply add to each other.*

Histamine, an amino acid derivative, also stimulates gastric secretion in much the same way as gastrin. However, it is only about 80 per cent as potent as gastrin.

Below are illustrated the amino acid compositions of *gastrin* and also of *cholecystokinin* and *secretin*. These are the only three gastrointestinal hormones that have thus far been isolated with certainty. Note that all of them are polypeptides and that the last five amino acids in the gastrin and cholecystokinin molecular chains are exactly the same. It is in this terminal portion of these hormones that the principal activity resides.

**Feedback Inhibition of Gastric Acid Secretion.** When the acidity of the gastric juices increases to a pH of 2.0, the gastrin mechanism for stimulating gastric secretion becomes totally blocked. This effect probably results from two different factors. *First*, greatly enhanced acidity depresses or blocks the extraction of gastrin itself from the antral mucosa. *Second*, the acid seems to extract an inhibitory hormone from the gastric mucosa or to cause an inhibitory reflex that inhibits gastric acid secretion.

Obviously, this feedback inhibition of the gastric glands plays an important role in protecting the stomach against excessively acid secretions, which would readily cause peptic ulceration. However, in addition to this protective effect, the feedback mechanism is also important in maintaining optimal pH for function of the peptic enzymes in the digestive process.

### The Three Phases of Gastric Secretion

Gastric secretion is said to occur in three separate phases (as illustrated in Fig. 64–6): a *cephalic phase*, a *gastric phase*, and an *intestinal phase*. However, as will be apparent in the following discussion, these three phases in reality fuse together.

**The Cephalic Phase.** The cephalic phase of gastric secretion occurs even before food enters the stomach. It results from the sight, smell, thought, or taste of food; and the greater the appetite, the more intense is the stimulation. Neurogenic signals causing the cephalic phase of secretion can originate in the cerebral cortex or in the appetite centers of the amygdala or hypothalamus. They are transmitted through the dorsal motor nuclei of the vagi to the stomach. This phase of secretion accounts for

Glu- Gly- Pro- Trp- Leu- Glu- Glu- Glu- Glu- Glu- Ala- Tyr- Gly- Trp- Met- Asp- Phe- NH$_2$

$$| \quad HSO_3$$

GASTRIN

Lys- (Ala, Gly, Pro, Ser)- Arg- Val- (Ile, Met, Ser)- Lys- Asn- (Asn, Gln, His, Leu$_2$, Pro, Ser$_2$)- Arg- Ile- (Asp, Ser)- Arg- Asp- Tyr- Met- Gly- Trp- Met- Asp- Phe- NH$_2$

$$| \quad HSO_3$$

CHOLECYSTOKININ

His- Ser- Asp- Gly- Thr- Phe- Thr- Ser- Glu- Leu- Ser- Arg- Leu- Arg- Asp- Ser-
Ala- Arg- Leu- Gln- Arg- Leu- Leu- Gln- Gly- Leu- Val- NH$_2$

SECRETIN

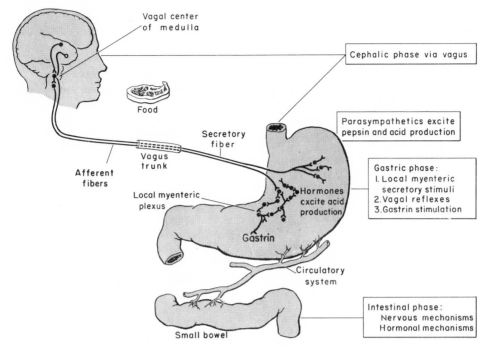

**Figure 64–6.** The phases of gastric secretion and their regulation.

perhaps one-tenth of the gastric secretion normally associated with eating a meal.

**The Gastric Phase.** Once the food enters the stomach, it excites the gastrin mechanism, which in turn causes secretion of gastric juice that continues throughout the several hours that the food remains in the stomach.

In addition, the presence of food in the stomach also causes (a) local reflexes in the myenteric plexus of the stomach and (b) vagovagal reflexes that pass all the way to the brain stem and back to the stomach; both of these reflexes cause parasympathetic stimulation of the gastric glands and add to the secretion caused by the gastrin mechanism.

The gastric phase of secretion accounts for more than two-thirds of the total gastric secretion associated with eating a meal and, therefore, accounts for most of the total daily gastric secretion of about 2000 ml.

**The Intestinal Phase.** Presence of food in the upper portion of the small intestine, particularly in the duodenum, also causes the stomach to secrete small amounts of gastric juice. This probably results from the fact that small amounts of gastrin, called *enteric gastrin*, are also released by the duodenal mucosa in response to distention or chemical stimuli of the same type as those that stimulate the stomach gastrin mechanism. However, it is possible that several other hormones released by the upper small intestinal mucosa also play minor roles in causing secretion of gastric juice. For instance, under certain conditions secretin and cholecystokinin can cause increased secretion of pepsin by the gastric glands in much the same way that gastrin acts on the gastric glands. However, under other conditions both of these hormones have significant inhibitory effects rather than excitatory effects on gastric secretion, as will be discussed in following paragraphs.

### Inhibition of Gastric Secretion by Intestinal Factors

Though chyme stimulates gastric secretion during the intestinal phase of secretion, it paradoxically often partially inhibits secretion during the gastric phase. This results from at least two different influences:

1. The presence of food in the small intestine initiates an *enterogastric reflex*, transmitted through the myenteric plexus, the sympathetic nerves, and the vagus nerves, that inhibits stomach secretion. This reflex can be initiated by distention of the small bowel, the presence of acid in the upper intestine, the presence of protein breakdown products, or irritation of the mucosa. This is part of the complex mechanism discussed in the preceding chapter for slowing down stomach emptying when the intestines are already filled.

2. The presence of acid, fat, protein breakdown products, hyper- or hypo-osmotic fluids, or any irritating factor in the small intestine causes the release of several intestinal hormones. Two of these have been indentified in pure form: *secretin* and *cholecystokinin*. Both of these are especially important for control of pancreatic secretion; cholecystokinin is particularly important in promoting emptying of the

gall bladder. However, in addition to their effects on pancreatic secretion and gall bladder emptying, they also have less potent effects on several other gastrointestinal functions. It was pointed out above that both of these can, under some conditions, cause increased gastric secretion; however, they both also can inhibit gastric secretion under the following conditions:

Secretin, although it increases the secretion of pepsin, has an inverse effect on the gastrin mechanism to decrease the secretion of acidic gastric juice. Therefore, when the gastrin mechanism is already strongly operative in causing the secretion of gastric juice, secretin exerts an important inhibitory effect. Cholecystokinin has a chemical structure that is partly similar to that of gastrin. When there is no gastrin available to stimulate gastric secretion, cholecystokinin has a weak stimulatory effect. On the other hand, when the gastrin mechanism is already strongly stimulating gastric secretion, cholecystokinin acts as a competitive inhibitor. That is, the cholecystokinin reacts with the same receptor sites of the gastric glands as does gastrin. However, cholecystokinin has only a small fraction of the stimulatory effect of gastrin. Yet, nevertheless, it prevents the gastrin action, thereby reducing the net rate of gastric juice secretion. Therefore, under usual operating conditions of the stomach, secretin and cholecystokinin have a moderate inhibitory effect on gastric secretion.

The functional purpose of the inhibition of gastric secretion by intestinal factors is probably to slow the release of chyme from the stomach when the small intestine is already filled. In fact, the enterogastric reflex as well as these inhibitory hormones reduce stomach motility at the same time that they reduce gastric secretion, as was discussed in the previous chapter.

**Secretion During the Interdigestive Period.** The stomach secretes only a few milliliters of gastric juice per hour during the "interdigestive period" when no digestion is occurring anywhere in the gut. Furthermore, the secretion that does occur is almost entirely of the so-called *nonparietal* type, meaning that it is composed mainly of mucus, containing very little pepsin and either none or very little acid. In fact, on occasion, the fluid is actually slightly alkaline. However, strong emotional stimuli frequently increase the interdigestive secretion to 50 ml. or more of highly peptic and highly acidic gastric juice per hour, in very much the same manner that the cephalic phase of gastric secretion excites secretion at the onset of a meal. This increase of secretion during the presence of emotional stimuli is believed to be one of the factors in the development of peptic ulcers, as will be discussed in Chapter 66.

# PANCREATIC SECRETION

**Characteristics of Pancreatic Juice.** Pancreatic juice contains enzymes for digesting all three major types of food: proteins, carbohydrates, and fats. It also contains large quantities of bicarbonate ions, which play an important role in neutralizing the acid chyme emptied by the stomach into the duodenum.

The proteolytic enzymes are *trypsin*, *chymotrypsin*, *carboxypolypeptidase*, *ribonuclease*, and *deoxyribonuclease*. The first three of these split whole and partially digested proteins, while the nucleases split the two types of nucleic acids: ribonucleic and deoxyribonucleic acids.

The digestive enzyme for carbohydrates is *pancreatic amylase*, which hydrolyzes starches, glycogen and most other carbohydrates except cellulose to form disaccharides.

The enzymes for fat digestion are *pancreatic lipase*, which is capable of hydrolyzing neutral fat into glycerol and fatty acids, and *cholesterol esterase*, which causes hydrolysis of cholesterol esters.

The proteolytic enzymes as synthesized in the pancreatic cells are in the inactive forms *trypsinogen*, *chymotrypsinogen*, and *procarboxypolypeptidase*, which are all enzymatically inactive. These become activated only after they are secreted into the intestinal tract. Trypsinogen is activated by an enzyme called *enterokinase*, which is secreted by the intestinal mucosa when chyme comes in contact with the mucosa. Also, trypsinogen can be activated by trypsin that has already been formed. Chymotrypsinogen is activated by trypsin to form chymotrypsin, and procarboxypolypeptidase is activated in some similar manner.

**Secretion of Trypsin Inhibitor.** It is important that the proteolytic enzymes of the pancreatic juice not become activated until they have been secreted into the intestine, for the trypsin and other enzymes would digest the pancreas itself. Fortunately, the same cells that secrete the proteolytic enzymes into the acini of the pancreas secrete simultaneously another substance called *trypsin inhibitor*. This substance is stored in the cytoplasm of the glandular cells surrounding the enzyme granules, and it prevents activation of trypsin both inside the secretory cells and in the acini and ducts of the pancreas. Since it is trypsin that activates the other pancreatic proteolytic enzymes, trypsin inhibitor also prevents the subsequent activation of all these.

However, when the pancreas becomes severely damaged or when a duct becomes blocked, large quantities of pancreatic secretion become pooled in the damaged areas of the pancreas. Under these conditions, the effect of trypsin inhibitor is sometimes overwhelmed, in which case the pancreatic secretions rapidly become activated and literally digest the entire pancreas within a few hours, giving rise to the condition called *acute pancreatitis*. This often is lethal because of accompanying shock, and even if not lethal it leads to a lifetime of pancreatic insufficiency.

**Secretion of Bicarbonate Ions.** The enzymes of the pancreatic juice are secreted entirely by the acini of the pancreatic glands. On the other hand, two other important components of pancreatic juice, water and bicarbonate ion, are secreted mainly by the epithelial cells of the small ductules leading from the acini or by special acinar cells derived from the ductules. We shall see below that the stimulatory mechanisms for (a) enzyme production and (b) production of water and bicarbonate ions are also quite different. When the pancreas is stimulated to secrete copious quantities of pancreatic juice—that is, copious quantities of the water and bicarbonate ions—the bicarbonate ion concentration can rise to as high as 145 mEq./liter, a value approximately five times that of bicarbonate ions in the plasma. Obviously, this provides a large quantity of alkaline ion in the pancreatic juice that serves to neutralize acid in the chyme emptied into the duodenum from the stomach.

## REGULATION OF PANCREATIC SECRETION

Pancreatic secretion, like gastric secretion, is regulated by both nervous and hormonal mechanisms. However, in this case, hormonal regulation is by far the more important.

**Nervous Regulation.** When the cephalic and gastric phases of stomach secretion occur, parasympathetic impulses are simultaneously transmitted along the vagus nerves to the pancreas, resulting in secretion of moderate amounts of enzymes into the pancreatic acini. However, little secretion flows through the pancreatic ducts to the intestine because little water and electrolytes are secreted along with the enzymes. Therefore, most of the enzymes are temporarily stored in the acini.

**Hormonal Regulation.** After food enters the small intestine, pancreatic secretion becomes copious, mainly in response to the hormone *secretin*. In addition, a second hormone, *cholecystokinin*, causes greatly increased secretion of enzymes.

***Stimulation of Secretion of Copious Quantities of Bicarbonate by Secretin—Neutralization of the Acidic Chyme.*** Secretin is a polypeptide containing 27 amino acids (molecular weight of about 3400) that is present in the mucosa of the upper small intestine in an inactive form *prosecretin*. When chyme enters the intestine, it causes the release and activation of secretin, which is subsequently absorbed into the blood. The one constituent of chyme that causes greatest secretin release is hydrochloric acid, though almost any type of food will cause at least some release.

Secretin causes the pancreas to secrete large quantities of fluid containing a high concentration of bicarbonate ion (up to 145 mEq./liter) but a low concentration of chloride ion. This copious flow of fluid is called *hydrelatic secretion* because the fluid is composed principally of a thin watery solution containing almost no enzymes.

The secretin mechanism is especially important for two reasons: *First*, secretin is released in especially large quantities from the mucosa of the small intestine any time the pH of the duodenal contents falls below 4.0 to 5.0. This immediately causes large quantities of pancreatic juice containing abundant amounts of sodium bicarbonate to be secreted, which results in the following reaction in the duodenum:

$$HCl + NaHCO_3 \rightarrow NaCl + H_2CO_3$$

The carbonic acid immediately dissociates into carbon dioxide and water, and the carbon dioxide is absorbed into the body fluids, thus leaving a neutral solution of sodium chloride in the duodenum. In this way, the acid contents emptied into the duodenum from the stomach become neutralized, and the peptic activity of the gastric juices is immediately blocked. Since the mucosa of the small intestine cannot withstand the intense digestive properties of gastric juice, this is a highly important protective mechanism against the development of duodenal ulcers, which will be discussed in further detail in Chapter 66.

A *second* importance of bicarbonate secretion by the pancreas is to provide an appropriate pH for action of the pancreatic enzymes. All of these function optimally in a slightly alkaline or neutral medium. The pH of the hydrelatic secretion averages 8.0.

***Cholecystokinin—Control of Enzyme Secretion by the Pancreas.*** The presence of food in the upper small intestine also causes a second hor-

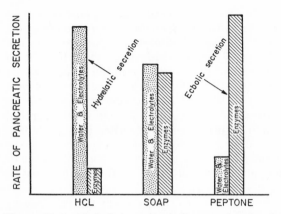

**Figure 64–7.** Hydrelatic and ecbolic secretion by the pancreas caused respectively by the presence of acids or peptone solutions in the duodenum.

mone, cholecystokinin, a polypeptide containing 33 amino acids, to be released from the mucosa. This results especially from the presence of proteoses and peptones, which are products of partial protein digestion, and of fats; however, acids will also cause its release in smaller quantities. Cholecystokinin, like secretin, passes by way of the blood to the pancreas but, instead of causing hydrelatic secretion, causes secretion of large quantities of digestive enzymes, which is similar to the effect of vagal stimulation. This type of secretion is called *ecbolic secretion* in contradistinction to hydrelatic secretion.

The differences between the effects of secretin and cholecystokinin are shown in Figure 64–7, which illustrates (a) intense hydrelatic secretion in response to acid in the duodenum, (b) a dual effect in response to soap (a fat), and (c) intense ecbolic secretion in response to peptones.

Figure 64–8 summarizes the overall regulation of pancreatic secretion. The total amount secreted each day is about 1200 ml.

# SECRETION OF BILE BY THE LIVER

The secretion of bile by the liver will be discussed in more detail in relation to the overall function of the liver in Chapter 70. Basically, bile contains no digestive enzyme and is important for digestion only because of the presence of bile salts which (1) help to emulsify fat globules so that they can be digested by the intestinal lipases and (2) help render the end-products of fat digestion soluble so that they can be absorbed through the gastrointestinal mucosa into the lymphatics.

Bile is secreted continually by the liver rather than intermittently as in the case of most other gastointestinal secretions, but the bile is stored in the gallbladder until it is needed in the gut. The gallbladder then empties the bile into the intestine in response to *cholecystokinin*, the same hormone that causes enzyme secretion by the pancreas. Therefore, bile becomes available to aid in the processes of fat digestion. This mechanism of gallbladder emptying was discussed in the preceding chapter.

The rate of bile secretion can be altered in response to four different effects: (1) vagal stimulation can sometimes more than double the secretion, (2) secretin has a moderate effect on liver secretion, effecting as much as 80 per cent increase in bile production (but without any increase in bile acid production) at the same time that it also causes the hydrelatic response of the pancreas, (3) the greater the liver blood flow (up to a point), the greater the secretion, and (4) the presence of large amounts of bile salts in the blood increases the rate of liver secretion proportionately. Most of the bile salts secreted in the bile are reabsorbed by the intestines and then resecreted by the liver over and over again, performing their actions in relation to fat digestion and absorption many times before being lost in the feces. This continual recirculation of bile salts is important for maintaining even the normal daily flow of bile. When the salts are lost to the exterior through a bile fistula rather than being reabsorbed, the rate of bile secretion becomes reduced to as little as one-third normal.

The daily volume of bile production averages 600 to 700 ml.

# SECRETIONS OF THE SMALL INTESTINE

## SECRETION OF MUCUS BY BRUNNER'S GLANDS AND BY MUCOUS CELLS OF THE INTESTINAL SURFACE

An extensive array of compound mucous glands, called *Brunner's glands*, is located in the first few centimeters of the duodenum, mainly between the pylorus and the papilla of Vater where the pancreatic juices and bile empty into the duodenum. These glands secrete

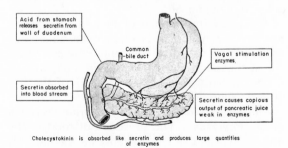

**Figure 64–8.** Regulation of pancreatic secretion.

mucus in response to: (a) direct tactile stimuli or irritating stimuli of the overlying mucosa, (b) vagal stimulation, which causes secretion concurrently with increase in stomach secretion, and (c) intestinal hormones, especially secretin. The function of the mucus secreted by Brunner's glands is to protect the duodenal wall from digestion by the gastric juice, and their rapid and intense response to irritating stimuli is especially geared to this purpose.

Brunner's glands are inhibited by sympathetic stimulation; therefore, such stimulation is likely to leave the duodenal bulb unprotected and is perhaps one of the factors that cause this area of the gastrointestinal tract to be the site of peptic ulcers in about 50 per cent of the cases.

Mucus is also secreted in large quantities by goblet cells located extensively over the surface of the intestinal mucosa. This secretion results principally from direct tactile or chemical stimulation of the mucosa by the chyme. Additional mucus is also secreted by the goblet cells in the intestinal glands, the crypts of Lieberkühn. This secretion is probably controlled mainly by local nervous reflexes.

### SECRETION OF THE INTESTINAL DIGESTIVE JUICES—THE CRYPTS OF LIEBERKÜHN

Located on the entire surface of the small intestine, with the exception of the Brunner's gland area of the duodenum, are small crypts called *crypts of Lieberkühn,* one of which is illustrated in Figure 64–1. The intestinal secretions are believed to be formed by the epithelial cells in these crypts at a rate of about 2000 ml. per day. The secretions are almost pure extracellular fluid, and they have a neutral pH in the range of 6.5 to 7.5. They are rapidly reabsorbed by the villi. This circulation of fluid from the crypts to the villi obviously supplies a watery vehicle for absorption of substances from the small intestine, which is one of the primary functions of the small intestine.

**Enzymes in the Small Intestinal Secretion.** When secretions of the small intestine are collected without cellular debris, they have almost no enzymes, the only two enzymes being *enterokinase*, which activates trypsin as explained earlier in the chapter, and a small amount of *amylase*.

However, the epithelial cells of the mucosa contain large quantities of digestive enzymes that digest food substances *while* they are being absorbed through the epithelium. These enzymes are the following: (1) several different

*peptidases* for splitting polypeptides into amino acids, (2) four enzymes for splitting disaccharides into monosaccharides—*sucrase, maltase, isomaltase,* and *lactase,* and (3) *intestinal lipase* for splitting neutral fats into glycerol and fatty acids. Most, if not all of these enzymes, are mainly in the brush border of the epithelial cells. Therefore, they presumably catalyze hydrolysis of the foods on the outside surfaces of the microvilli prior to absorption of the endproducts of digestion.

The epithelial cells deep in the crypts of Lieberkühn continually undergo mitosis, and the new cells gradually migrate along the basement membrane upward out of the crypts toward the tips of the villi where they are finally shed into the intestinal secretions. The life cycle of an intestinal epithelial cell is approximately 5 days. This rapid growth of new cells allows immediate repair of any excoriation that occurs in the mucosa.

### REGULATION OF SMALL INTESTINAL SECRETION

**Local Stimuli.** By far the most important means for regulating small intestinal secretion is various local myenteric reflexes. Especially important is distention of the small intestine, which causes copious secretion from the crypts of Lieberkühn. In addition, tactile or irritative stimuli can result in intense secretion. Therefore, for the most part, secretion in the small intestine occurs simply in response to the presence of chyme in the intestine—the greater the amount of chyme, the greater the secretion.

**Hormonal Regulation.** Crude experiments have indicated that a hormone (or hormones) called *enterocrinin* extracted from the mucosa of the small intestine by the chyme might control intestinal secretion. However, more refined experiments have thus far failed to isolate such a hormone.

## SECRETIONS OF THE LARGE INTESTINE

**Mucus Secretion.** The mucosa of the large intestine, like that of the small intestine, is lined with crypts of Lieberkühn, but the epithelial cells contain almost no enzymes. Instead, they are lined almost entirely by goblet cells. Also, on the surface epithelium of the large intestine are large numbers of goblet cells dispersed among the other epithelial cells.

Therefore, the only significant secretion in the large intestine is mucus. Its rate of secretion is regulated principally by direct, tactile stimulation of the goblet cells on the surface of the mucosa and by local myenteric reflexes to the goblet cells in the crypts of Lieberkühn. However, stimulation of the nervi erigentes, which carry the parasympathetic innervation to the distal two-thirds to one-half of the large intestine, also causes marked increase in the secretion of mucus. This occurs along with an increase in motility, which was discussed in the preceding chapter. Therefore, during extreme parasympathetic stimulation, often caused by severe emotional disturbances, so much mucus may be secreted into the large intestine that the person has a bowel movement of ropy mucus as often as every 30 minutes; the mucus contains little or no fecal material.

Mucus in the large intestine obviously protects the wall against excoriation, but, in addition, it provides the adherent medium for holding fecal matter together. Furthermore, it protects the intestinal wall from the great amount of bacterial activity that takes place inside the feces, and it, plus the alkalinity of the secretion (pH of 8.0), also provides a barrier to keep acids formed deep in the feces from attacking the intestinal wall.

**Secretion of Water and Electrolytes in Response to Irritation.** Whenever a segment of the large intestine becomes intensely irritated, such as occurs when bacterial infection becomes rampant during *enteritis*, the mucosa then secretes large quantities of water and electrolytes in addition to the normal viscid solution of mucus. This acts to dilute the irritating factors and to cause rapid movement of the feces toward the anus. The usual result is *diarrhea* with loss of large quantities of water and electrolytes but also earlier recovery from the disease than would otherwise occur.

# REFERENCES

Andersson, S.: Secretion of gastrointestinal hormones. *Ann. Rev. Physiol.*, 35:431, 1973.

Bolis, L., Keynes, R. D., and Wilbrandt, W.: Role of Membranes in Secretory Processes. New York, American Elsevier Publishing Co., Inc., 1972.

Ceccarelli, B., Clementi, F., and Meldolesi, J. (eds.): Cytopharmacology of Secretion. New York, Raven Press, 1974.

Ciba Foundation Symposium: The Exocrine Pancreas: Normal and Abnormal Function. Boston, Little, Brown and Company, 1962.

Code, C. F.: Hormonal inhibition of gastric secretion. *Proc. Inter. Union Physiol. Sci.*, 6:194, 1968.

Cooke, A. R.: The glands of Brunner. In Code, C. F., and Heidel, W. (eds.): Handbook of Physiology. Sec. 6, Vol. 2. Baltimore, The Williams & Wilkins Company, 1967, p. 1087.

Davenport, H. W.: Physiological structure of the gastric mucosa. In Code, C. F., and Heidel, W. (eds.): Handbook of Physiology. Sec. 6, Vol. 2. Baltimore, The Williams & Wilkins Company, 1967, p. 759.

Davenport, H. W.: Why the stomach does not digest itself. *Sci. Amer.*, 226:86, 1972.

Dawes, C.: Rhythms in salivary flow rate and composition. *Inter. J. Chronobiol.*, 2:253, 1974.

Goldstein, F.: Pathophysiology of the pancreas. In Sodeman, W. A., Jr., and Sodeman, W. A. (eds.): Pathologic Physiology: Mechanisms of Disease, 5th Ed. Philadelphia, W. B. Saunders Company, 1974, p. 827.

Grayson, J.: The gastrointestinal circulation. In Guyton, A. C. (ed.): MTP International Review of Science: Physiology. Vol. 4. Baltimore, University Park Press, 1974, p. 105.

Gregory, R. A.: Secretory mechanisms of the digestive tract. *Ann. Rev. Physiol.*, 27:395, 1965.

Harper, A. A.: Hormonal control of pancreatic secretion. In Code, C. F., and Heidel, W. (eds.): Handbook of Physiology. Sec. 6, Vol. 2. Baltimore, The Williams & Wilkins Company, 1967, p. 969.

Hersey, S. J.: Interactions between oxidative metabolism and acid secretion in gastric mucosa. *Biochim. Biophys. Acta*, 344:157, 1974.

Jacobson, E. D., Swan, K. G., and Grossman, M. I.: Blood flow and secretion in the stomach. *Gastroenterology*, 52:414, 1967.

Johnson, L. R.: Gastrointestinal hormones. In Guyton, A. C. (ed.): MTP International Review of Science: Physiology. Vol. 4. Baltimore, University Park Press, 1974, p. 1.

Johnson, L. R., and Grossman, M. I.: Secretin: the enterogastrone released by acid in the duodenum. *Amer. J. Physiol.*, 215:885, 1968.

Kirsner, J. B.: The stomach. In Sodeman, W. A., Jr., and Sodeman, W. A. (eds.): Pathologic Physiology: Mechanisms of Disease, 5th Ed. Philadelphia, W. B. Saunders Company, 1974, p. 709.

Konturek, S. J.: Gastric secretion. In Guyton A. C. (ed.): MTP International Review of Science: Physiology. Vol. 4. Baltimore, University Park Press, 1974, p. 227.

Lipkin, M.: Proliferation and differentiation of gastrointestinal cells. *Physiol. Rev.*, 53:891, 1973.

Mason, D. K.: Salivary Glands in Health and Disease. Philadelphia, W. B. Saunders Company, 1975.

Petersen, O. H.: Electrophysiological studies on gland cells. *Experientia*, 30:130, 1974.

Preshaw, R. M.: Pancreatic exocrine secretion. In Guyton, A. C. (ed.): MTP International Review of Science: Physiology. Vol. 4, Baltimore, University Park Press, 1974, p. 265.

Rubin, R. P.: Calcium and the Secretory Process. New York, Plenum Publishing Corporation, 1974.

Schneyer, L. H.: Salivary secretion. In Guyton, A. C. (ed.): MTP International Review of Science: Physiology. Vol. 4. Baltimore, University Park Press, 1974, p. 183.

Schneyer, L. H., Young, J. A., and Schneyer, C. A.: Salivary secretion of electrolytes. *Physiol. Rev.*, 52:720, 1972.

Sernka, T. H.: Gastrointestinal mucosal metabolism. In Guyton, A. C. (ed.): MTP International Review of Science: Physiology. Vol. 4. Baltimore, University Park Press, 1974, p. 45.

Snook, J. T.: Adaptive and nonadaptive changes in digestive enzyme capacity influencing digestive function. *Fed. Proc.*, 33:88, 1974.

Walsh, J. H.: Circulating gastrin. *Ann. Rev. Physiol.*, 37:81, 1975.

# | 65 |

# Digestion and Absorption in the Gastrointestinal Tract

The foods on which the body lives, with the exception of small quantities of substances such as vitamins and minerals, can be classified as carbohydrates, fats, and proteins. However, these generally cannot be absorbed in their natural forms through the gastrointestinal mucosa and, for this reason, are useless as nutrients without the preliminary process of digestion. Therefore, the present chapter discusses, first, the processes by which carbohydrates, fats, and proteins are digested into small enough compounds for absorption and, second, the mechanisms by which the digestive end-products, as well as water, electrolytes, and other substances, are absorbed.

## DIGESTION OF THE VARIOUS FOODS

**Hydrolysis as the Basic Process of Digestion.** Almost all the carbohydrates of the diet are large *polysaccharides* or *disaccharides,* which are combinations of *monosaccharides* bound to each other by the process of *condensation*. This means that a hydrogen ion has been removed from one of the monosaccharides, while a hydroxyl ion has been removed from the next one; the two monosaccharides then are combined with each other at these sites of removal, and the hydrogen and hydroxyl ions combine to form water. When the carbohydrates are digested back into monosaccharides, specific enzymes return the hydrogen and hydroxyl ions to the polysaccharides and thereby separate the monosaccharides from each other. This process, called hydrolysis, is the following:

$$R''-R' + H_2O \rightarrow R''OH + R'H$$

Almost the entire fat portion of the diet consists of triglycerides (neutral fats), which are combinations of three *fatty acid* molecules condensed with a single *glycerol* molecule. In the process of condensation, three molecules of water had been removed. Digestion of the triglycerides consists of the reverse process, the fat-digesting enzymes returning the three molecules of water to each molecule of neutral fat and thereby splitting the fatty acid molecules away from the glycerol. Here again, the process is one of hydrolysis.

Finally, proteins are formed from *amino acids* that are bound together by means of *peptide linkages*. In this linkage a hydroxyl ion is removed from one amino acid, while a hydrogen ion is removed from the succeeding one; thus, the amino acids also combine together by a process of condensation while losing a molecule of water. Digestion of proteins, therefore, also involves a process of hydrolysis, the proteolytic enzymes returning the water to the protein molecules to split them into their constituent amino acids.

Therefore, the chemistry of digestion is really simple, for in the case of all three major types of food, the same basic process of *hydrolysis* is involved. The only difference lies in the enzymes required to promote the reactions for each type of food.

All the digestive enzymes are proteins. Their secretion by the different gastrointestinal glands is discussed in the preceding chapter.

### DIGESTION OF CARBOHYDRATES

**The Carbohydrate Foods of the Diet.** Only three major sources of carbohydrates exist in the normal human diet. These are sucrose, which is the disaccharide known popularly as cane sugar; lactose, which is a disaccharide in milk; and starches, which are large polysaccharides present in almost all foods and particularly in the grains. Other types of carbohydrates ingested to a slight extent are glycogen, alcohol, lactic acid, pyruvic acid, pectins, dextrins, and minor quantities of other carbohydrate derivatives in meats. The diet also contains a large amount of cellulose, which is a carbohydrate. However, no enzymes capable of hydrolyzing cellulose are secreted by the human digestive tract. Consequently, cellulose cannot be considered to be a food for the human being, though it can be utilized by some lower animals.

Because starches make up the largest proportion of the carbohydrate intake of the human being, the digestive system is especially geared for their digestion. However, other polysaccharides, such as the pectins, glycogen, and the dextrins, are digested in essentially the same manner as the starches.

**Digestion of Carbohydrates in the Mouth.** When food is chewed, it is mixed with the saliva, which contains the enzyme *ptyalin* (α-amylase) secreted mainly by the parotid glands. This enzyme hydrolyzes starch into the disaccharides *maltose* and *isomaltose,* as shown in Figure 65–1; but the food remains in the mouth only a short time, and probably not more than 3 to 5 per cent of all the starches that are eaten will have become hydrolyzed into maltose and isomaltose by the time the food is swallowed. One can demonstrate the digestive action of ptyalin in the mouth by chewing a piece of bread for several minutes; after this time, the bread tastes sweet because of the maltose and isomaltose that has been liberated from the starches of the bread.

Most starches in their natural state, unfortunately, are present in the food in small globules, each of which has a thin protective cellulose covering. Therefore, most naturally occurring starches are digested only poorly by ptyalin unless the food is cooked to destroy the protective membrane.

**Digestion of Carbohydrates in the Stomach.** Even though food does not remain in the mouth long enough for ptyalin to complete the breakdown of starches into maltose, the action of ptyalin continues for as long as several hours after the food has entered the stomach; that is, until the contents of the fundus are mixed with the stomach secretions. Then the activity of the salivary amylase is blocked by the acid of the gastric secretions, for it is essentially nonactive as an enzyme once the pH of the medium falls below approximately 4.0. Nevertheless, on the average, before the food becomes completely mixed with the gastric secretions, as much as 30 to 40 per cent of the starches will have been changed into maltose and isomaltose.

***Hydrolysis of Starches and Disaccharides by the Acid of the Stomach.*** The acid of the stomach juices can, to a slight extent, hydrolyze starches and disaccharides. However, quantitatively this reaction occurs to such a slight extent that it is usually ignored as an unimportant effect.

**Digestion of Carbohydrates in the Small Intestine.** *Digestion by Pancreatic Amylase.* Pancreatic secretion, like saliva, contains a large quantity of α-amylase which is almost identical in its function with the α-amylase of saliva and is capable of splitting starches into *maltose* and *isomaltose.* Therefore, immediately after the chyme empties from the stomach into the duodenum and mixes with pancreatic juice, the starches that have not already been split are digested by amylase. In general, the starches are almost totally converted into maltose and isomaltose before they have passed beyond the jejunum.

**Hydrolysis of Disaccharides into Monosaccharides by the Intestinal Epithelial Enzymes.** The epithelial cells lining the small intestine contain the four enzymes *lactase, sucrase, maltase,* and *isomaltase,* which are capable of splitting the disaccharides lactose, sucrose, maltose, and isomaltose, respectively, into their constituent monosaccharides. These enzymes are located in the brush border of the cells lining the lumen of the intestine, and the disaccharides are digested as they come in contact with this border. The digested products, the monosaccharides, are then immediately absorbed into the portal blood. Lactose splits into a molecule of *galactose* and a molecule of *glucose.* Sucrose splits into a

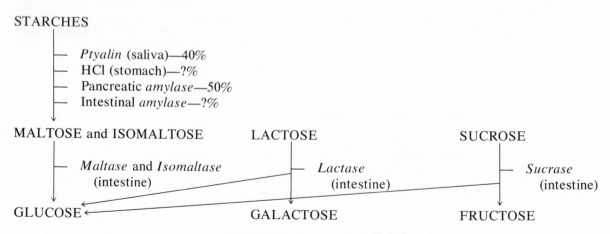

**Figure 65–1.** Digestion of carbohydrates.

molecule of *fructose* and a molecule of *glucose*. Maltose and isomaltose each split into *two molecules of glucose*. Thus, the final products of carbohydrate digestion that are absorbed into the blood are all monosaccharides.

In the ordinary diet, which contains far more starches than either sucrose or lactose, glucose represents about 80 per cent of the final products of carbohydrate digestion, and galactose and fructose each represent, on the average, about 10 per cent of the products of carbohydrate digestion.

## DIGESTION OF FATS

**The Fats of the Diet.** By far the most common fats of the diet are the neutral fats, also known as *triglycerides,* each molecule of which is composed of a glycerol nucleus and three fatty acids, as illustrated in Figure 65–2. Neutral fat is found in food of both animal origin and plant origin.

In the usual diet are also small quantities of phospholipids, cholesterol, and cholesterol esters. The phospholipids and cholesterol esters contain fatty acid, and, therefore, can be considered to be fats themselves. Cholesterol, on the other hand, is a sterol compound containing no fatty acid, but it does exhibit some of the physical and chemical characteristics of fats; it is derived from fats, and it is metabolized similarly to fats. Therefore, cholesterol is considered from a dietary point of view to be a fat.

**Digestion of Fats in the Intestine.** A small amount of short chain triglycerides of butter fat origin is digested in the stomach by gastric lipase *(tributyrase)*. However, the amount of digestion is so

slight that it is unimportant. Instead, essentially all fat digestion occurs in the small intestine as follows:

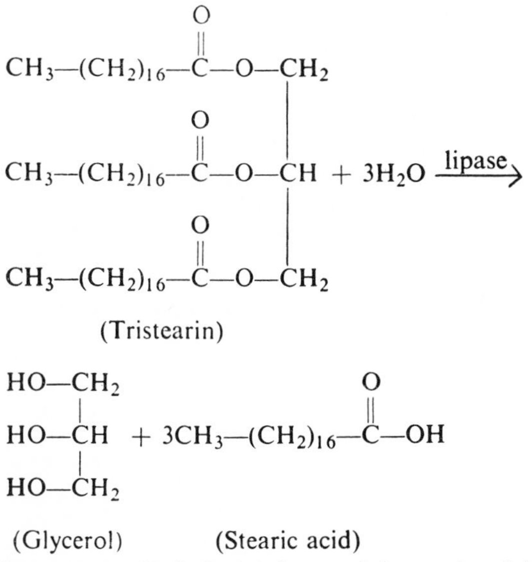

*Emulsification of Fat by Bile Acids.* The first step in fat digestion is to break the fat globules into small sizes so that the water-soluble digestive enzymes can act on the globule surfaces. This process is called emulsification of the fat, and it is achieved under the influence of *bile,* the secretion of the liver that does not contain any digestive enzymes. However, it does contain a large quantity of *bile salts* mainly in the form of ionized sodium salts which are extremely important for the emulsification of fat. The carboxyl (or polar) part of the bile salt is highly soluble in water, whereas the sterol portion of the bile salt is highly soluble in fat. Therefore, bile salts aggregate at the surfaces of the fat globules in the intestinal contents with the carboxyl portion of the bile salt projecting outward and soluble in the surrounding fluids, while the sterol portion is dissolved in the fat itself; this effect greatly decreases the interfacial tension of the fat.

When the interfacial tension of a globule of nonmiscible fluid is low, this nonmiscible fluid on agitation can be broken up into many minute particles far more easily than it can when the interfacial tension is great. Consequently, a major function of the bile salts is to make the fat globules readily fragmentable by agitation in the small bowel. This action is the same as that of many detergents that are used widely in most household cleansers for removing grease.

Each time the diameters of the fat globules are decreased by a factor of 2 as a result of agitation in the small intestine, the total surface area of the fat increases 2 times. In other words, the total surface area of the fat particles in the intestinal contents is inversely proportional to the diameters of the particles.

The lipases are water-soluble compounds and can attack the fat globules only on their surfaces. Consequently, it can be readily understood how important this detergent function of bile salts is for the digestion of fats.

**Digestion of Fats by Pancreatic Lipase and Enteric Lipase.** By far the most important enzyme for the digestion of fats is *pancreatic lipase* in the pancreatic juice. However, the epithelial cells of the small intestine also contain a small quantity of lipase known as *enteric lipase*. Both of these act alike to cause hydrolysis of fat.

*End-Products of Fat Digestion.* Most of the triglycerides of the diet are finally split into monoglycerides, free fatty acids, and glycerol, as illustrated in Figure 65–3. However, small portions are not digested at all, whereas other portions remain in a diglyceride state. Although all of these can be absorbed by the intestinal mucosa to at least some extent, the less the degree of digestion, the less also is the ease with which absorption occurs. Therefore, diseases that prevent secretion of pancreatic lipase—with resultant poor digestion of fats—also lead to absorption of only about 50 per cent of the fat in the diet.

$$CH_3-(CH_2)_{16}-\overset{\overset{\displaystyle O}{\|}}{C}-O-CH_2$$

$$CH_3-(CH_2)_{16}-\overset{\overset{\displaystyle O}{\|}}{C}-O-CH \;+\; 3H_2O \xrightarrow{\text{lipase}}$$

$$CH_3-(CH_2)_{16}-\overset{\overset{\displaystyle O}{\|}}{C}-O-CH_2$$

(Tristearin)

$$\begin{matrix} HO-CH_2 \\ | \\ HO-CH \\ | \\ HO-CH_2 \end{matrix} \;+\; 3CH_3-(CH_2)_{16}-\overset{\overset{\displaystyle O}{\|}}{C}-OH$$

(Glycerol)        (Stearic acid)

**Figure 65–2.** Hydrolysis of neutral fat catalyzed by lipase.

$$\text{Fat} \xrightarrow{\text{(Bile + Agitation)}} \text{Emulsified fat}$$

$$\text{Emulsified fat} \xrightarrow{\textit{Pancreatic lipase}} \left\{ \begin{array}{l} \text{Fatty acids} \\ \text{Glycerol} \end{array} \right\} 40\% \ (?) \\ \text{Glycerides } 60\% \ (?)$$

**Figure 65–3.**   Digestion of fats.

*Role of Bile Salts in Accelerating Fat Digestion— Formation of Micelles.*   The hydrolysis of triglycerides is a highly reversible process; therefore, accumulation of monoglycerides and free fatty acids in the vicinity of digesting fats very quickly blocks further digestion. Fortunately, the bile salts play an important role in removing the monosaccharides and the free fatty acids from the vicinity of the digesting fat globules almost as rapidly as these end-products of digestion are formed. This occurs in the following way:

Bile salts have the propensity to form *micelles,* which are small spherical globules about 25 Ångstroms in diameter and composed of 20 to 50 molecules of bile salt. These develop because each bile salt molecule is composed of a sterol nucleus that is highly fat soluble and a polar group that is highly water soluble. The sterol nuclei of the 20 to 50 bile salt molecules of the micelle aggregate together to form a small fat globule in the middle of the micelle. This aggregation causes the polar groups to project outward to cover the surface of the micelle. Since these polar groups are negatively charged, they allow the entire micelle globule to become dissolved in the water of the digestive fluids and to remain in stable solution despite the very large size of the micelle.

During the triglyceride digestion, as rapidly as the monoglycerides and free fatty acids are formed they become dissolved in the fatty portion of the micelles, which immediately removes these end-products of digestion from the vicinity of the digesting fat globules. Consequently, the digestive process can proceed unabated.

The bile salt micelles also act as a transport medium to carry the monoglycerides and the free fatty acids to the brush borders of the epithelial cells. There the monoglycerides and free fatty acids are absorbed, as will be discussed later. On delivery of these substances to the brush border, the bile salts are again released back into the chyme to be used again and again for this "ferrying" process.

**Digestion of Cholesterol Esters.**   Most of the cholesterol in the diet is in the form of cholesterol esters, which cannot be absorbed in this form, though free cholesterol is readily absorbed. A *cholesterol esterase* in the pancreatic juice hydrolyzes the esters and thus frees the cholesterol. The bile salt micelles play identically the same role in "ferrying" cholesterol as they play in "ferrying" monoglycerides and free fatty acids. Indeed, this role of the bile salt micelles is absolutely essential to the absorption of cholesterol because essentially no cholesterol

is absorbed without the presence of bile salts. On the other hand, as much as 60 per cent of the triglycerides can be digested and absorbed even in the absence of bile salts.

## DIGESTION OF PROTEINS

**The Proteins of the Diet.**   The dietary proteins are derived almost entirely from meats and vegetables. These proteins in turn are formed of long chains of amino acids bound together by *peptide linkages*. A typical linkage is the following:

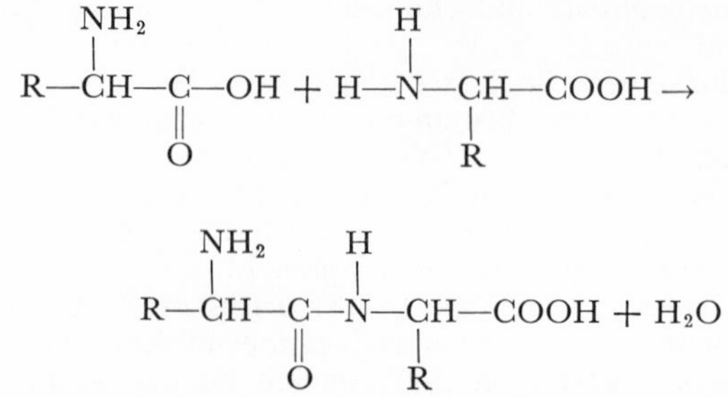

$$\underset{\substack{|| \\ O}}{R-CH-C}-OH + H-\underset{\substack{| \\ R}}{N}-CH-COOH \rightarrow$$

$$R-\underset{\substack{|| \\ O}}{C}H-C-\underset{\substack{| \\ R}}{N}-CH-COOH + H_2O$$

The characteristics of each type of protein are determined by the types of amino acids in the protein molecule and by the arrangement of these amino acids. The physical and chemical characteristics of the different proteins will be discussed in Chapter 69.

**Digestion of Proteins in the Stomach.**   Pepsin, the important peptic enzyme of the stomach, is most active at a pH of about 2 and is completely inactive at a pH above approximately 5. Consequently, for this enzyme to cause any digestive action on protein, the stomach juices must be acidic. It will be recalled from Chapter 64 that the gastric glands secrete a large quantity of hydrochloric acid. This hydrochloric acid is secreted by the parietal cells at a pH of about 0.8, but, by the time it is mixed with the stomach contents and with the secretions from the nonparietal glandular cells of the stomach, the pH ranges around 2 to 3, a highly favorable range of acidity for pepsin activity.

Pepsin is capable of digesting essentially all the different types of proteins in the diet. One of the important features of pepsin digestion is its ability to digest collagen, an albuminoid that is affected little by other digestive enzymes. Collagen is a major constituent of the intercellular connective tissue of meats, and for the digestive enzymes of the digestive tract to penetrate meats and digest the cellular proteins it is first necessary that the collagen fibers be digested. Consequently, in persons lacking peptic ac-

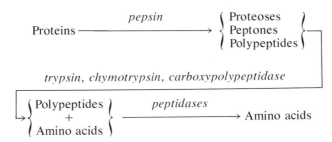

**Figure 65–4.** Digestion of proteins.

tivity in the stomach, the ingested meats are poorly penetrated by the digestive enzymes and, therefore, are poorly digested.

As illustrated in Figure 65–4, pepsin usually only begins the process of protein digestion, simply splitting the proteins into proteoses, peptones, and large polypeptides. This splitting of proteins is a process of "hydrolysis" occurring at the peptide linkages between the amino acids.

**Digestion of Proteins by Pancreatic Secretions.** When the proteins leave the stomach, they ordinarily are in the form of proteoses, peptones, large polypeptides, and about 15 per cent amino acids. Immediately upon entering the small intestine, the partial breakdown products are attacked by the pancreatic enzymes trypsin, chymotrypsin, and carboxypolypeptidase. As illustrated in Figure 65–4, these enzymes are capable of hydrolyzing all the partial breakdown products of protein to peptides, and they even hydrolyze some of them to the final stage of amino acids; most of the products, however, are dipeptides or other small polypeptides.

**Digestion of Peptides by the Epithelial Peptidases of the Small Intestine.** The epithelial cells of the small intestine contain several different enzymes for hydrolyzing the final peptide linkages of the different dipeptides and other small polypeptides as they are absorbed through the epithelium into the portal blood. The enzymes responsible for final hydrolysis of the peptides into amino acids are *amino polypeptidase* and the *dipeptidases*.

All of the proteolytic enzymes—including those of the gastric juice, the pancreatic juice, and in the brush border of the intestinal epithelial cells—are very specific for hydrolyzing individual types of peptide linkages. The linkages between certain pairs of amino acids differ in their bond energy and other physical characteristics from the linkages between other pairs. Therefore, a specific enzyme is required for each specific type of linkage. This accounts for the multiplicity of proteolytic enzymes as well as for the fact that no one single enzyme can usually digest protein all the way to its constituent amino acids.

When food has been properly masticated and is not eaten in too large a quantity at any one time, about 98 per cent of all the proteins finally become amino acids. A few molecules of protein are never digested at all, and some remain in the stages of proteoses, peptones, and varying sizes of polypeptides.

# BASIC PRINCIPLES OF GASTROINTESTINAL ABSORPTION

## *ANATOMICAL BASIS OF ABSORPTION*

The total quantity of fluid that must be absorbed each day is equal to the ingested fluid (about 1.5 liters) plus that secreted in the various gastrointestinal secretions (about 7.5 liters). This comes to a total of approximately 9 liters. About 8 to 8.5 liters of this is absorbed in the small intestine, leaving only 0.5 to 1 liter to pass through the ileocecal valve into the colon each day.

The stomach is a poor absorptive area of the gastrointestinal tract. Only a few highly lipid-soluble substances, such as alcohol and some drugs, can be absorbed in small quantities.

**The Absorptive Surface of the Intestinal Mucosa—The Villi.** Figure 65–5 illustrates

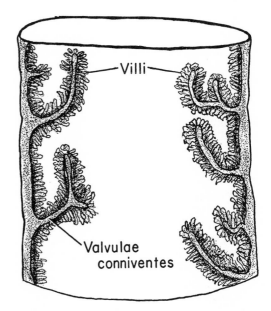

**Figure 65–5.** A longitudinal section of the small intestine, showing the valvulae conniventes covered by villi.

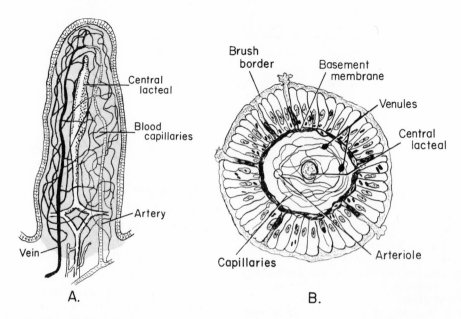

**Figure 65–6.** Functional organization of the villus: (A) Longitudinal section. (B) Cross-section showing the epithelial cells and basement membrane.

the absorptive surface of the intestinal mucosa, showing many folds called *valvulae conniventes,* which increase the surface area of the absorptive mucosa about three-fold. These folds extend circularly most of the way around the intestine and are especially well developed in the duodenum and jejunum, where they often protrude as much as 8 mm. into the lumen.

Located over the entire surface of the small intestine, from approximately the point at which the common bile duct empties into the duodenum down to the ileocecal valve, are literally millions of small *villi,* which project about 1 mm. from the surface of the mucosa, as shown on the surfaces of the valvulae conniventes in Figure 65–5 and in detail in Figure 65–6. These villi lie so close to each other in the upper small intestine that they actually touch in most areas, but their distribution is less profuse in the distal small intestine. The presence of villi on the mucosal surface enhances the absorptive area another 10-fold.

The intestinal epithelial cells are characterized by a brush border, consisting of about 600 *microvilli* $1\mu$ in length and $0.1\mu$ in diameter protruding from each cell; these are illustrated in the electron micrograph in Figure 65–7. This increases the surface area exposed to the intestinal materials another 20-fold. Thus, the combination of the valvulae conniventes, the villi, and the microvilli increases the absorptive area of the mucosa about 600-fold, making a tre-

mendous total area of about 250 square meters for the entire small intestine.

Figure 65–6A illustrates the general organization of a villus, emphasizing especially the advantageous arrangement of the vascular system for absorption of fluid and dissolved material into the portal blood, and the arrangement of the *central lacteal* for absorption into the lymphatics. Figure 65–6B shows the cross-section of a villus, and Figure 65–7 shows many small *pinocytic* vesicles, which are pinched-off portions of infolded epithelium surrounding extracellular materials that have been entrapped

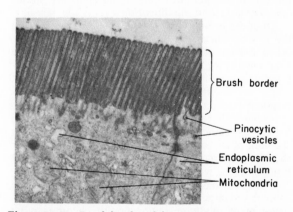

**Figure 65–7.** Brush border of the gastrointestinal epithelial cell, showing, also, pinocytic vesicles, mitochondria, and endoplasmic reticulum lying immediately beneath the brush border. (Courtesy of Dr. William Lockwood.)

inside the cells. Small amounts of substances are absorbed by this physical process of *pinocytosis,* though, as noted later in the chapter, most absorption occurs by means of single molecular transfer. Located near the brush border of the epithelial cell are many *mitochondria,* which supply the cell with oxidative energy needed for "active transport" of materials through the intestinal epithelium; this also is discussed later in the chapter.

### BASIC MECHANISMS OF ABSORPTION

Absorption through the gastrointestinal mucosa occurs by *active transport* and by *diffusion,* as is also true for other membranes. The physical principles of these processes were explained in Chapter 4.

Briefly, active transport imparts energy to the substance as it is being transported for the purpose of concentrating it on the other side of the membrane or for moving it against an electrical potential. On the other hand, the term "diffusion" means simply transport of substances through the membrane as a result of molecular movement *along,* rather than against, an electrochemical gradient.

# ABSORPTION IN THE SMALL INTESTINE

Normally, absorption from the small intestine each day consists of several hundred grams of carbohydrates, 100 or more grams of fat, 50 to 100 grams of amino acids, 50 to 100 grams of ions, and 8 or 9 liters of water. However, the absorptive *capacity* of the small intestine is far greater than this: as much as several kilograms of carbohydrates per day, 500 to 1000 grams of fat per day, 500 to 700 grams of amino acids per day, and 20 or more liters of water per day. In addition, the large intestine can absorb still more water and ions, though almost no nutrients.

### ABSORPTION OF WATER

**Isosmotic Absorption.** Water is transported through the intestinal membrane entirely by the process of *diffusion.* Furthermore, this diffusion obeys the usual laws of osmosis. Therefore, when the chyme is dilute, water is absorbed through the intestinal mucosa into the blood of the villi by osmosis.

It must be noted that water can also be transported by osmosis from the plasma into the chyme. This occurs whenever hyperosmotic solutions are discharged from the stomach into the duodenum. Usually within a few minutes, sufficient water is transferred by osmosis into the chyme to make it isosmotic with the plasma. On the other hand, if there is excess water in the chyme, the osmosis into the plasma also causes an isosmotic state within a few minutes. Thereafter, the chyme remains almost exactly isosmotic throughout its total passage through the small and large intestine.

As dissolved substances are absorbed from the lumen of the gut into the blood the absorption tends to decrease the osmotic pressure of the chyme, but water diffuses so readily through the intestinal membrane that it almost instantaneously "follows" the absorbed substances into the circulation. Therefore, as ions and nutrients are absorbed, so also is an isosmotic equivalent of water absorbed. In this way not only are the ions and nutrients almost entirely absorbed before the chyme passes through the small intestine, but so also is almost all the water absorbed.

### ABSORPTION OF THE IONS

**Active Transport of Sodium.** Twenty to 30 grams of sodium are secreted into the intestinal secretions each day. In addition, the normal person eats 4 to 5 grams of sodium each day. Combining these two, the small intestine absorbs 25 to 35 grams of sodium each day, which amounts to about one-seventh of all the sodium that is present in the body. One can well understand that whenever the intestinal secretions are lost to the exterior, as in extreme diarrhea, the sodium reserves of the body can be depleted to a lethal level within hours. Normally, this sodium is secreted and reabsorbed continually with very little lost in the feces. The sodium plays an important role in the absorption of sugars and amino acids, as we shall see in subsequent discussions.

The basic mechanism of sodium absorption from the intestine is illustrated in Figure 65–8. The principles of this mechanism, which were discussed in Chapter 4, are also essentially the same as those for absorption of sodium from the renal tubules, as discussed in Chapter 34. The motive power for the sodium absorption is provided by active transport of sodium from inside the epithelial cells through the side walls of these cells into the intercellular spaces. This is illustrated by the heavy black arrows in Figure 65–8. This active transport obeys the usual laws of active transport: it requires a carrier, it requires energy, and it is catalyzed by appropriate enzymes in or on the surface of the cell membrane.

The active transport of sodium reduces its concentration in the cell to a low value (about 50 mEq./liter), as also illustrated in Figure 65–8. Since the sodium concentration in the chyme is normally about 142 mEq./liter (that is, approximately equal to that in the plasma), sodium diffuses from the chyme through the brush border of the epithelial cell into the epithelial cell cytoplasm. This replaces the sodium that is actively transported out of the epithelial cells into the intercellular spaces.

The next step in the transport process is osmosis of water out of the epithelial cell into the intercellular spaces. This movement is caused by the osmotic gradient created by the reduced concentration of sodium inside the cell and the elevated concentration in the intercellular space. The osmotic movement of water creates a flow of fluid into the intercellular space, then through the basement membrane of the epithelium, and finally into the circulating blood of the villi. New water diffuses along with sodium through the brush border of the epithelial cell to replenish the water that flows into the intercellular spaces.

**Transport of Chloride.** In the upper part of the small intestine chloride transport is entirely by passive diffusion. The transport of sodium ions through the epithelium creates electronegativity in the chyme and electropositivity

on the basal side of the epithelial cells. Then chloride ions move along this electrical gradient to "follow" the sodium ions.

*Active Absorption of Chloride Ions and Active Secretion of Bicarbonate Ions in the Lower Ileum and in the Large Intestine.* In addition to the passive process for chloride absorption, which accounts for by far the major amount of chloride absorption in the intestines, the epithelial cells of the distal ileum and of the large intestine have the special capability of actively absorbing chloride ions. However, this occurs by means of a tightly coupled active transport mechanism in which an equivalent number of bicarbonate ions are secreted. The reason for this mechanism is probably to provide bicarbonate ions for neutralization of acidic products formed by bacteria—especially in the large intestine.

Various bacterial toxins, particularly those of cholera, colon bacilli, and staphylococci, can strongly stimulate this chloride-bicarbonate exchange mechanism. The secreted bicarbonate ion carries with it sodium ions, and the two of these together carry an isosmotic equivalent of water as well. This results in rapid flow of fluid from the distal part of the gut, thus causing diarrhea. In cholera in particular the diarrhea can be so severe that it can cause death within 24 hours.

**Absorption of Other Ions.** Calcium ions are actively absorbed, especially from the duodenum, and calcium ion absorption is exactly controlled in relation to the need of the body for calcium. One important factor controlling calcium absorption is parathyroid hormone secreted by the parathyroid glands. In the presence of this hormone, calcium can be absorbed against approximately twice as great a concentration gradient as is true in the absence of parathyroid hormone. Also, vitamin D enhances calcium absorption. These effects are discussed in Chapter 79.

Iron ions are also actively absorbed from the small intestine. The principles of iron absorption and the regulation of its absorption in proportion to the body's need for iron were discussed in Chapter 5.

Potassium, magnesium, phosphate, and probably still other ions can also be actively absorbed through the mucosa. In general, the monovalent ions are absorbed with ease and in great quantities. On the other hand, the bivalent ions are normally absorbed in only small amounts; fortunately, only small quantities of these are normally needed by the body.

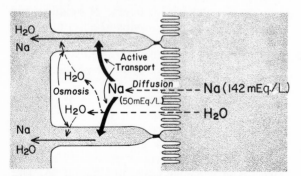

**Figure 65–8.** Absorption of sodium through the intestinal epithelium. Note also the osmotic absorption of water—that is, the water "follows" the sodium through the epithelial membrane.

## ABSORPTION OF NUTRIENTS

**Absorption of Carbohydrates.** Essentially all the carbohydrates are absorbed in the form of monosaccharides, only a small fraction of a per cent being absorbed as disaccharides and almost none as larger carbohydrate compounds. Furthermore, little carbohydrate absorption results from diffusion, for the pores of the mucosa through which diffusion occurs are essentially impermeable to water-soluble solutes with molecular weights greater than 100.

That the transport of monosaccharides is an active process is demonstrated by several important experimental observations:

1. Their transport can be blocked by metabolic inhibitors, such as iodoacetic acid, cyanides, and phlorhizin.

2. The transport is selective, specifically transporting certain monosaccharides without transporting others. The order of preference for transporting different monosaccharides and their relative rates of transport in comparison with glucose are:

| | |
|---|---|
| Galactose | 1.1 |
| Glucose | 1.0 |
| Fructose | 0.4 |
| Mannose | 0.2 |
| Xylose | 0.15 |
| Arabinose | 0.1 |

3. There is a maximum rate of transport for each type of monosaccharide. The most rapidly transported monosaccharide is galactose, with glucose running a close second. Fructose, which is also one of the three important monosaccharides for nutrition, is absorbed less than half as rapidly as either galactose or glucose; also, its mechanism of absorption is different, as will be explained below.

4. There is competition between certain sugars for the carrier system. For instance, if large amounts of galactose are being transported, the amount of glucose that can be transported simultaneously is considerably reduced.

*Mechanism of Monosaccharide Absorption.* We still do not know the precise mechanism of monosaccharide absorption, but we do know the following facts about the mechanism.

First, the chemical nature of most actively transported monosaccharide molecules (besides fructose, which has its own separate carrier) is (a) six or more carbon atoms, (b) a D-pyranose ring structure, and (c) an intact hydroxyl group at carbon 2.

Second, glucose and galactose transport becomes blocked whenever sodium transport is blocked. Therefore, it is assumed that the energy required for most if not all monosaccharide transport is actually provided by the sodium transport system. A theory that attempts to explain this is the following: It is known that the carrier for transport of glucose and some other monosaccharides, especially galactose, is present in the brush border of the epithelial cell. However, this carrier will not transport the glucose in the absence of sodium transport. Therefore, it is believed that the carrier has a receptor site for both a glucose molecule and a sodium ion, and that it will not transport the glucose to the inside of the cell if the receptor sites for both glucose and sodium are not simultaneously filled. The energy to cause movement of the carrier from the exterior of the membrane to the interior is derived from the difference in sodium concentration between the outside and inside. That is, as sodium diffuses to the inside of the cell it "drags" the carrier, and therefore the glucose along with it, thus providing the energy for transport of the glucose. For obvious reasons, this explanation is called the *sodium gradient theory* for glucose transport.

Subsequently, we will see that sodium transport is also required for transport of amino acids, suggesting a similar "carrier-drag" mechanism for amino acid transport.

*Transport of Fructose.* Transport of fructose is slightly different from that of most other actively transported monosaccharides. It is not blocked by some of the same metabolic poisons—specifically, phlorhizin—and in the process of being transported it is also converted to glucose before entering the portal blood. This conversion occurs inside the epithelial cell, the fructose first becoming phosphorylated, then converted to glucose, and finally released from the epithelial cell into the blood.

This difference between the active transport of fructose and the other monosaccharides is important for treatment of a rare genetic disorder in which the carrier for the other monosaccharides is absent. The two other major monosaccharides of diet, glucose and galactose, can in this disease be absorbed in only minute quantities, and the person normally becomes rapidly debilitated. However, by simply feeding large quantities of fructose, the necessary caloric intake is provided and the debilitation is prevented.

**Absorption of Proteins.** Almost all proteins are absorbed in the form of amino acids. However, small quantities of dipeptides are also absorbed, and extremely minute quantities of whole proteins can at times be absorbed, probably by the process of pinocytosis though not by the usual absorptive mechanisms.

The absorption of amino acids also obeys the general principles listed above for active absorption of glucose; that is, certain types of amino acids are absorbed selectively and certain ones interfere with the absorption of others, illustrating that common carrier systems exist. Finally, metabolic poisons block the absorption of amino acids in the same way that they block the absorption of glucose.

Absorption of amino acids through the intestinal mucosa can occur far more rapidly than can protein digestion in the lumen of the intestine. As a result, the normal rate of absorption is determined not by the rate at which they can be absorbed but by the rate at which they can be released from the proteins during digestion. For these reasons, essentially no free amino acids can be found in the intestine during digestion—that is, they are absorbed as rapidly as they are formed.

*Basic Mechanisms of Amino Acid Transport.* As is true for monosaccharide absorption, very little is known about the basic mechanisms of amino acid transport. However, four different carrier systems transport different amino acids—one transports *neutral amino acids,* a second transports *basic amino acids,* a third transports *acidic amino acids,* and a fourth has specificity for the two amino acids *proline* and *hydroxyproline.* Also, the transport mechanisms have far greater affinity for transporting L-stereoisomers of amino acids than D-stereoisomers. And experiments have demonstrated that *pyridoxal phosphate,* a derivative of the vitamin pyridoxine, is required for transport of many amino acids.

Amino acid transport, like glucose transport, occurs only in the presence of simultaneous sodium transport. Furthermore, the carrier systems for amino acid transport, like those for glucose transport, are in the brush border of the epithelial cell. It is believed that amino acids are transported by the same sodium gradient mechanism as that explained above for glucose transport. That is, the theory postulates that the carrier has a receptor site for both an amino acid molecule and a sodium ion. Only when both of the sites are filled will the carrier move to the interior of the cell. Because of the sodium gradient across the brush border, the sodium diffusion to the cell interior pulls the carrier and

its attached amino acids to the interior where the amino acids become trapped. Therefore, their concentrations increase within the cell, and they then diffuse through the sides or base of the cell into the portal blood.

**Absorption of Fats.** Earlier in this chapter it was pointed out that as fats are digested to form monoglycerides and free fatty acids, both of these digestive end-products become dissolved mainly in the lipid portion of the bile acid micelles. Because of the molecular dimensions of these micelles and also because of their highly charged exterior, they are soluble in the chyme. In this form the monoglycerides and the fatty acids are transported to the surfaces of the epithelial cells. On coming in contact with these surfaces, both the monoglycerides and the fatty acids immediately diffuse through the epithelial membrane, leaving the bile acid micelles still in the chyme. The micelles then diffuse back into the chyme and absorb still more monoglycerides and fatty acids, and similarly transport these also to the epithelial cells. Thus, the bile acids perform a "ferrying" function, which is highly important for fat absorption. In the presence of an abundance of bile acids, approximately 97 per cent of the fat is absorbed; in the absence of bile acids, only 50 to 60 per cent is normally absorbed.

The mechanism for absorption of the monoglycerides and fatty acids through the brush border is based on the fact that both of these substances are highly lipid soluble. Therefore, they become dissolved in the membrane and diffuse to the interior of the cell.

The undigested triglycerides and the diglycerides are both also highly soluble in the lipid membrane of the epithelial cell. However, only small quantities of these are normally absorbed because the bile acid micelles will not dissolve either triglycerides or diglycerides and therefore will not ferry them to the epithelial membrane.

After entry into the epithelial cell, many of the monoglycerides are further digested into glycerol and fatty acids by an epithelial cell lipase. Then, the free fatty acids are reconstituted by the endoplasmic reticulum into triglycerides. Almost all of the glycerol that is utilized for this purpose is synthesized *de novo* from alpha-glycerophosphate. However, minute amounts of the original glycerol from the monoglycerides do appear in the newly synthesized triglycerides.

Once formed, the triglycerides collect into globules along with absorbed cholesterol, absorbed phospholipids, and newly synthesized phospholipids. Each of these is then encased in a protein coat, utilizing proteins also synthesized by the endoplasmic reticulum. This globular mass, along with its protein coat, is extruded from the sides of the epithelial cells into the intercellular spaces, and from here it passes into the central lacteal of the villi. Such globules are called *chylomicrons*.

The protein coat of the chylomicrons makes them hydrophilic, allowing a reasonable degree of suspension stability in the extracellular fluids. Poisons or genetic disorders that prevent formation of the protein for coating the chylomicrons cause the fat to accumulate in the epithelial cell and not to be secreted into the extracellular fluid.

**Transport of the Chylomicrons in the Lymph.** From beneath the epithelial cells the chylomicrons wend their way into the central lacteal of the villi and from here are propelled along with the lymph by the lymphatic pump upward through the thoracic duct to be emptied into the great veins of the neck. Between 80 and 90 per cent of all fat absorbed from the gut is absorbed in this manner and is transported to the blood by way of the thoracic lymph in the form of chylomicrons.

**Direct Absorption of Fatty Acids into the Portal Blood.** Small quantities of short chain fatty acids, such as those from butter fat, are absorbed directly into the portal blood rather than being converted into triglycerides and absorbed into the lymphatics. The cause of this difference between short and long chain fatty acid absorption is presumably that the shorter chain fatty acids are more water-soluble, which allows direct diffusion of fatty acids from the epithelial cells into the capillary blood of the villus.

*Absorption of Bile Salts.* In the upper portion of the small intestine the bile salts are not absorbed; this failure to be absorbed requires them to remain in the chyme and to continue their function of "ferrying" free fatty acids and monoglycerides to the intestinal mucosa. Once the processes of fat digestion and fat absorption have been accomplished in the upper and midintestinal levels, however, the bile salts themselves are then absorbed from the distal ileum before the chyme empties into the large intestine. This absorption is an active process and is carrier-mediated.

After being absorbed from the distal ileum, the bile salts are again secreted in the bile by the liver and returned once more to the upper intestine. Thus, the same bile salts are resecreted several times each day and are used again and again in the process of fat absorption. Only a small portion of the bile salts (approximately 5 per cent) is lost during each cycle of this "bile salt circulation."

On occasion, the bile salts fail to be absorbed in the ileum and instead empty with the chyme into the large intestine; this occurs especially in patients whose distal ileum has been removed because of ileitis. The presence of bile salts in the large intestine frequently causes severe diarrhea, presumably because of the detergent effect of these salts acting on and irritating the large intestinal mucosa.

## ABSORPTION IN THE LARGE INTESTINE; FORMATION OF THE FECES

Approximately 500 to 1000 ml. of chyme passes through the ileocecal valve into the large intestine each day. Most of the water and electrolytes in this are absorbed in the colon, leaving only 100 to 200 ml. of fluid to be excreted in the feces.

Most of the absorption in the large intestine occurs in the proximal half of the colon, giving this portion the name *absorbing colon*, while the distal colon functions principally for storage and is therefore called the *storage colon*.

**Absorption and Secretion of Electrolytes and Water.** The mucosa of the large intestine, like that of the small intestine, has a very high capacity for active absorption of sodium, and the electrical potential created by the absorption of the sodium causes chloride absorption as well. In addition, as in the distal portion of the small intestine, the mucosa of the large intestine actively secretes bicarbonate ions while it simultaneously actively absorbs a similar amount of additional chloride ions. The bicarbonate helps to neutralize the acidic endproducts of bacterial action in the colon.

The absorption of sodium and chloride ions creates an osmotic gradient across the large intestinal mucosa, which in turn causes absorption of water.

**Bacterial Action in the Colon.** Numerous bacteria, especially colon bacilli, are present in the absorbing colon. These are capable of digesting small amounts of cellulose, in this way providing a few calories of nutrition to the body each day. In herbivorous animals this source of energy is very significant though it is of negligible importance in the human being. Other substances formed as a result of bacterial activity are vitamin K, vitamin $B_{12}$, thiamin, riboflavin, and various gases that contribute to *flatus* in the colon. Vitamin K is especially important, for the amount of this vitamin in the ingested foods is normally insufficient to maintain adequate blood coagulation.

**Composition of the Feces.** The feces normally are about three-fourths water and one-fourth solid matter composed of about 30 per cent dead bacteria, 10 to 20 per cent fat, 10 to 20 per cent inorganic matter, 2 to 3 per cent protein, and 30 per cent undigested roughage of the food and dried constituents of digestive juices, such as bile pigment and sloughed epithelial cells. The large amount of fat derives from unabsorbed fatty acids from the diet, fat formed by bacteria, and fat in the sloughed epithelial cells.

The brown color of feces is caused by *stercobilin* and *urobilin*, which are derivatives of bilirubin. The odor is caused principally by the products of bacterial action; these vary from one person to another, depending on each person's colonic bacterial flora and on the type of food he has eaten. The actual odoriferous products include *indole*, *skatole*, *mercaptans*, and *hydrogen sulfide*.

# REFERENCES

Beck, I. T.: The role of pancreatic enzymes in digestion. *Amer. J. Clin. Nutr.*, 26:311, 1973.

Borgström, B.: Fat digestion and absorption. *Biomembranes*, 4B:555, 1974.

Brindley, D. N.: The intracellular phase of fat absorption. *Biomembranes*, 4B:621, 1974.

Crane, R. K.: Absorption of sugars. In Code, C. F., and Heidel, W. (eds.): Handbook of Physiology. Sec. 6, Vol. 3. Baltimore, The Williams & Wilkins Company, 1968, p. 1323.

Crane, R. K.: Intestinal absorption of glucose. *Biomembranes*, 4A:541, 1974.

Creamer, B.: Intestinal structure in relation to absorption. *Biomembranes*, 4A:1, 1974.

Curran, P. F., and Schultz, S. G.: Transport across membranes: general principles. In Code, C. F., and Heidel, W. (eds.): Handbook of Physiology. Sec. 6, Vol. 3. Baltimore, The Williams & Wilkins Company, 1968, p. 1217.

Forth, W., and Rummel, W.: Iron absorption. *Physiol. Rev.*, 53:724, 1973.

Ganguly, J., Subbaiah, P. V., and Parthasarathy, S.: Intestinal absorption of lipids. *Biochem. Soc. Symp.*, 35:67, 1972.

Jackson, M. J.: Transport of short chain fatty acids. *Biomembranes*, 4B:673, 1974.

Johnston, J. M.: Mechanism of fat absorption. In Code, C. F., and Heidel, W. (eds.): Handbook of Physiology. Sec. 6, Vol. 3. Baltimore, The Williams & Wilkins Company, 1968, p. 1353.

Kim, Y. S., Nicholson, J. A., and Curtis, K. J.: Intestinal peptide hydrolases: peptide and amino acid absorption. *Med. Clin. N. Amer.*, 58:1397, 1974.

Kotyk, A.: Mechanisms of nonelectrolyte transport. *Biochim. Biophys. Acta*, 300:183, 1973.

Kretchmer, N.: Lactose and lactase. *Sci. Amer.*, 227:79, 1972.

Levitan, R., and Wilson, D. E.: Absorption of water soluble substances. In Guyton, A. C. (ed.): MTP International Review of Physiology. Vol. 4. Baltimore, University Park Press, 1974, p. 293.

Lipkin, M.: Proliferation and differentiation of gastrointestinal cells. *Physiol. Rev.*, 53:891, 1973.

Matthews, D. M.: Intestinal absorption of amino acids and peptides. *Proc. Nutr. Soc.*, 31:171, 1972.

Matthews, D. M.: Absorption of water-soluble vitamins. *Biomembranes*, 4B:847, 1974.

Matthews, D. M.: Absorption of amino acids and peptides from the intestine. *Clin. Endocrinol. Metabol.*, 3:3, 1974.

Ockner, R. K., and Isselbacher, K. J.: Recent concepts of intestinal fat absorption. *Rev. Physiol. Biochem. Pharmacol.*, 71:107, 1974.

Okuda, K.: Intestinal mucosa and vitamin $B_{12}$ absorption. *Digestion*, 6:173, 1972.

Olsen, W. A.: Carbohydrate absorption. *Med. Clin. N. Amer.*, 58:1387, 1974.

Rindi, G., and Ventura, U.: Thiamine intestinal transport. *Physiol. Rev.*, 52:821, 1972.

Schultz, S. G.: Principles of electrophysiology and their application to epithelial tissues. In Guyton, A. C. (ed.): MTP International Review of Science: Physiology. Vol. 4. Baltimore, University Park Press, 1974, p. 69.

Schultz, S. G., Frizzell, R. A., and Nellans, H. N.: Ion transport by mammalian small intestine. *Ann. Rev. Physiol.*, 36:51, 1974.

Silk, D. B.: Progress report. Peptide absorption in man. *Gut*, 15:494, 1974.

Simmonds, W. J.: The role of micellar solubilization in lipid absorption. *Aust. J. Exp. Biol. Med. Sci.*, 50:403, 1972.

Simmonds, W. J.: Absorption of lipids. In Guyton, A. C. (ed.): MTP International Review of Science: Physiology. Vol. 4. Baltimore, University Park Press, 1974, p. 343.

Skillman, J. J., and Silen, W.: Gastric mucosal barriers. *Surg. Ann.*, 4:213, 1972.

Smyth, D. H. (ed.): Intestinal Absorption. Vols. 4A and 4B. New York, Plenum Publishing Corporation, 1974.

Snook, J. T.: Adaptive and nonadaptive changes in digestive enzyme capacity influencing digestive function. *Fed. Proc.* 33:88, 1974.

Turnberg, L. A.: Absorption and secretion of salt and water by the small intestine. *Digestion*, 9:357, 1973.

Ugolev, A. M.: Membrane (contact) digestion. *Biomembranes*, 4A:285, 1974.

Van Campen, D.: Regulation of iron absorption. *Fed. Proc.* 33:100, 1974.

Watson, D. W., and Sodeman, W. A., Jr.: The small intestine. In Sodeman, W. A., Jr. and Sodeman, W. A. (eds.): Pathologic Physiology: Mechanisms of Disease, 5th Ed. Philadelphia, W. B. Saunders Company, 1974, p. 734.

Wiseman, G.: Absorption of protein digestion products. *Biomembranes*, 4A:363, 1974.

Wright, E. M.: The passive permeability of the small intestine. *Biomembranes*, 4A:159, 1974.

# | 66 |

# Physiology of Gastrointestinal Disorders

The logical treatment of most gastrointestinal disorders depends on a basic knowledge of gastrointestinal physiology. The purpose of this chapter, therefore, is to discuss a few representative types of malfunction that have special physiologic bases or consequences.

## DISORDERS OF SWALLOWING AND OF THE ESOPHAGUS

**Paralysis of the Swallowing Mechanism.** Damage to the 5th, 9th, or 10th nerve can cause paralysis of significant portions of the swallowing mechanism. Also, a few diseases, most commonly poliomyelitis, prevent normal swallowing by damaging the swallowing center in the brain stem. Finally, malfunction of the swallowing muscles, as occurs in *muscle dystrophy* or in failure of neuromuscular transmission in *myasthenia gravis* or *botulism*, can also prevent normal swallowing.

When the swallowing mechanism is partially or totally paralyzed, the abnormalities that can occur include: (1) complete abrogation of the swallowing act so that swallowing cannot occur at all, (2) failure of the glottis to close so that food passes into the trachea as well as into the esophagus, (3) failure of the soft palate and uvula to close the posterior nares so that food refluxes into the nose during swallowing, or (4) failure of the cricopharyngeal sphincter to remain closed during normal breathing, thus allowing large quantities of air to be sucked into the esophagus.

One of the most serious instances of paralysis of the swallowing mechanism occurs when a person is under deep anesthesia. Often he vomits large quantities of materials from the stomach into the pharynx; then instead of swallowing the materials again he simply sucks them into his trachea because the anesthetic has blocked the reflex mechanism of swallowing. As a result, such patients often choke to death on their own vomitus.

**Achalasia.** Achalasia is a condition in which the lower few centimeters of the esophagus fail to relax during the swallowing mechanism. Also, the body of the esophagus tends to contract in unison rather than peristaltically. As a result, food transmission from the esophagus into the stomach is impeded or prevented. Pathological studies have shown the physiologic basis of this condition to be either damaged or absent myenteric plexus in the lower portion of the esophagus. The musculature of the esophagus is still capable of contracting, and even exhibits incoordinate movements, but it has lost the ability to conduct a peristaltic wave, especially in its lower part, and has lost the ability to transmit a signal to cause "receptive relaxation" of the gastroesophageal sphincter as food approaches this area during the swallowing process.

## DISORDERS OF THE STOMACH

**Gastritis.** Gastritis means inflammation of the gastric mucosa. This can result from (a) action of irritant foods on the gastric mucosa, (b) excessive excoriation of the stomach mucosa by the stomach's own peptic secretions, or (c) occasionally, bacterial inflammation. One of the most frequent causes of gastritis is irritation of the mucosa by alcohol.

The inflamed mucosa in gastritis is often painful, causing a diffuse burning pain referred to the high epigastrium. Reflexes initiated in the stomach mucosa cause the salivary glands to salivate intensely, and the frequent swallowing of foamy saliva makes air accumulate in the stomach. As a result, the person usually belches profusely, a burning sensation often occurring in his throat with each belch.

*The Gastric Barrier and Its Penetration in Gastritis.* Absorption from the stomach, even by diffusion, is normally very slight. This low level of absorption is probably caused by two specific features of the gastric mucosa: (1) it is lined with highly resistant mucus cells, and (2) it has very tight junctions between the adjacent epithelial cells. Normally this barrier is so resistant to diffusion that even the highly concentrated hydrogen ions of the gastric juice, about 100,000 times as concentrated as the hydrogen ions in the plasma, barely diffuse through the epithelial membrane. However, in gastritis, this barrier becomes inflamed and its permeability is greatly increased. The hydrogen ions then do diffuse into the wall of the stomach, creating additional havoc with the stomach wall tissues, and leading to progressive stomach mucosal damage and atrophy. It also makes the mucosa susceptible to peptic digestion, thus frequently resulting in gastric ulcer.

**Gastric Atrophy.** In many persons who have chronic gastritis, the mucosa gradually becomes atrophic until little or no gastric gland activity remains. It is also believed that some persons develop autoimmunity against the gastric mucosa, this leading eventually to gastric atrophy. Loss of the stomach secretions in gastric atrophy leads to achlorhydria and, occasionally, to *pernicious anemia*.

*Achlorhydria (and Hypochlorhydria).* Achlorhydria means simply that the stomach fails to secrete hydrochloric acid, and hypochlorhydria means diminished acid secretion. Usually, when acid is not secreted, pepsin also is not secreted, and, even if it is, the lack of acid prevents it from functioning because pepsin requires an acid medium for activity. Obviously, then, essentially all digestive function in the stomach is lost when achlorhydria is present.

The method usually used to determine the degree of hypochlorhydria is to inject 0.5 milligram of histamine and then to aspirate the stomach secretions through a tube for the following hour. Each 10-minute sample of stomach secretion is titrated against sodium hydroxide to a pH of 8.5, which gives a measure of *total acid* secreted. The maximum rate of acid secretion during any 10-minute interval in the normal person rises to almost exactly 1 milliequivalent of hydrochloric acid per minute, and the total hydrochloric acid secreted during the entire hour after the injection averages about 18 milliequivalents. All degrees of hypochlorhydria—down to no acid whatsoever—occur.

The acid secretion is frequently divided into *free acid* and *combined acid*. The amount of free acid is determined by titrating the gastric secretions to a pH of 3.5, using *dimethylamino-azobenzene* as an indicator. After this titration has been performed, the same secretions are titrated to a pH of 8.5, using *phenolphthalein* as an indicator; this measures the combined acid. Gastric secretions mixed with food in the stomach usually show little or no free acid but a large amount of combined acid. On the other hand, when the stomach secretes large quantities of gastric juice while it is almost empty of food, the larger portion of the acid may then be free acid and only a small amount combined.

Though achlorhydria is associated with depressed or even no digestive capability by the stomach, the overall digestion of food in the entire gastrointestinal tract is still almost normal. This is because trypsin and other enzymes secreted by the pancreas are capable of digesting most of the protein in the diet—particularly if the food is well chewed so that no portion of the protein is protected by collagen fibers, which need peptic activity for most effective digestion.

**Pernicious Anemia in Gastric Atrophy.** Pernicious anemia, which was discussed in Chapter 5, is a common accompaniment of gastric atrophy. The normal gastric secretions contain a mucopolypeptide called *intrinsic factor*, which is secreted by the parietal cells (the HCl-producing cells) and which must be present for adequate absorption of vitamin $B_{12}$ from the ileum. The instrinsic factor combines with vitamin $B_{12}$, making it resistant to digestion and also aiding in its absorption. In the absence of intrinsic factor, an adequate amount of vitamin $B_{12}$ is not made available from the foods. As a result, maturation failure occurs in the bone marrow, resulting in pernicious anemia. This subject was discussed in more detail in Chapter 5.

Pernicious anemia also occurs frequently when most of the stomach has been removed for treatment of either stomach ulcer or gastric cancer or when the ileum, where vitamin $B_{12}$ is almost entirely absorbed, is removed.

## PEPTIC ULCER

A peptic ulcer is an excoriated area of the mucosa caused by the digestive action of gastric juice. Figure 66–1 illustrates the points in the gastrointestinal tract at which peptic ulcers frequently occur, showing that by far the most frequent site of peptic ulcers is in the first few centimeters of the duodenum. In addition, peptic ulcers frequently occur along the lesser curvature of the antral end of the stomach or, more rarely, in the lower end of the esophagus where stomach juices frequently reflux. A peptic ulcer called a *marginal ulcer* also frequently occurs wherever an abnormal opening, such as a gastrojejunostomy, is made between the stomach and some portion of the small intestine.

**Basic Cause of Peptic Ulceration.** The usual cause of peptic ulceration is too much secretion of gastric juice in relation to the degree of protection afforded by the mucous secretion and the neutralization of the gastric acid by duodenal juices. It will be recalled that all areas normally exposed to gastric juice are well supplied with mucous glands, beginning with the compound mucous glands of the lower esophagus, then including the mucous cell coating of

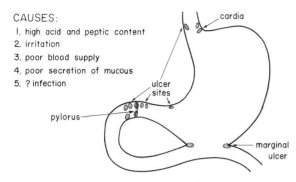

CAUSES:

1. high acid and peptic content
2. irritation
3. poor blood supply
4. poor secretion of mucous
5. ? infection

**Figure 66–1.**  Peptic ulcer.

the stomach mucosa, the mucous neck cells of the gastric glands, the deep pyloric glands that secrete almost nothing but mucus, and, finally, the coiled tubular glands of Brunner of the upper duodenum, which secrete a highly alkaline mucus.

In addition to the mucus protection of the mucosa, the duodenum is also protected by the alkalinity of the small intestinal secretions. Especially important is pancreatic secretion, which contains large quantities of sodium bicarbonate that neutralize the hydrochloric acid of the gastric juice, thus inactivating the pepsin to prevent digestion of the mucosa. Two additional mechanisms insure that this neutralization of gastric juices is complete:

1. When excess acid enters the duodenum, it reflexly inhibits gastric secretion and peristalsis, both nervously and hormonally, in the stomach, thereby decreasing the rate of gastric emptying. This allows increased time for pancreatic secretion to enter the duodenum to neutralize the acid already present. After neutralization has taken place, the reflex subsides and more stomach contents are emptied.

2. The presence of acid in the small intestine liberates secretin from the intestinal mucosa, which then passes by way of the blood to the pancreas to promote rapid secretion of pancreatic juice. Secretin also stimulates the pancreas to secrete a juice containing a high concentration of sodium bicarbonate, thus making more sodium bicarbonate available for neutralization of the acid. These mechanisms were discussed in detail in Chapters 63 and 64 in relation to gastrointestinal motility and secretion.

**Experimental Peptic Ulcer.**  Experimental peptic ulcers have been created in dogs and other animals in the following ways: (a) *Feeding ground glass* to an animal causes excoriation of the pyloric wall and allows the peptic juices to begin digesting the deeper layers of the mucosa. (b) *Transplantation of the pancreatic duct to the ileum* removes the normal neutralizing effect of pancreatic secretion in the duodenum and, therefore, allows the gastric juice to attack the mucosa of the upper duodenum. (c) *Repeated injection of histamine* causes excessive secretion of gastric juice. (d) *Continual infusion of hydrochloric acid* through a tube into the stomach causes direct irritation of the stomach mucosa and

also prevents full neutralization of the gastric juices by the pancreatic and other secretions of the small intestine. Therefore, peptic activity is enhanced, and ulcers appear in either the stomach or duodenum. (e) A portion of the *stomach is anastomosed directly with the small intestine* so that gastric juice can pass directly and rapidly into the small intestine. The mucosa of the small intestine, except at the uppermost part of the duodenum, is not sufficiently resistant to gastric juice to prevent peptic digestion. (f) *Obstruction of the blood flow* or even reduction of the blood flow to an area of the stomach or upper duodenum will cause ulcers to develop because the local area cannot produce appropriate protective secretions.

In summary, any factor that (1) *increases the rate of production of gastric juice* or (2) *blocks the normal protective mechanisms* against this juice can produce peptic ulcers. The same general principles apply to the development of peptic ulcers in the human being.

**Causes of Peptic Ulcer in the Human Being.**  Peptic ulcer occurs much more frequently in the white collar worker than in the laborer, and persons subjected to extreme anxiety for a long time seem particularly prone to peptic ulcer. For instance, the number of persons who developed peptic ulcer increased greatly during the air raids of London. Therefore, it is believed that many if not most instances of duodenal peptic ulcer in the human being result from excessive stimulation of the dorsal motor nucleus of the vagus by impulses originating in the cerebrum. The duodenal ulcers develop in the area of the duodenum where the gastric juices still have not been fully neutralized, immediately beyond the pylorus, as illustrated in Figure 66–1. Supporting this theory is the fact that duodenal ulcer patients have a high rate of gastric secretion during the interdigestive period between meals when the stomach is empty. The normal stomach secretes a total of approximately 18 milliequivalents of hydrochloric acid during the 12-hour interdigestive period through the night, while duodenal ulcer patients occasionally secrete as much as 300 milliequivalents of hydrochloric acid during this same time.

Paradoxically, gastric ulcers, in contradistinction to duodenal ulcers, often occur in patients who have normal or low secretion of hydrochloric acid. However, these patients almost invariably have an associated gastritis, indicating that ulceration in the stomach almost certainly results from reduced resistance of the stomach mucosa to digestion rather than to excess secretion of gastric juice. Stomach ulceration frequently results in patients who have ingested substances such as aspirin or alcohol that reduce the mucosal resistance. Also, reflux of duodenal contents into the stomach often leads to gastric ulceration; this response is believed to result mainly from the bile acids in the refluxed chyme because these acids have a detergent effect that reduces the mucosal resistance.

**Physiology of Treatment.**  A usual medical treatment for peptic ulcer is a diet of six or more small meals per day rather than the normal three large meals. In this way food is kept in the stomach most of the time, and the food neutralizes much if not most of the acid and dilutes the gastric juice so that its digestive action on the mucosa is minimized. Also, the composition of the diet is prescribed in accordance with the following two essential principles: (1) The diet should be bland, containing as few secretagogues as possible—that is, no alcohol, few spices, and meats boiled until the extractives have been washed away. Lack of the secretagogues reduces the secretion of gastrin by the pyloric mucosa and therefore reduces the hormonal stimulation of the gastric glands during the gastric phase of stomach secretion. This mechanism was discussed in Chapter 64. (2) The diet is formulated to contain large quantities of fat. On entering the small intestine, the fat activates feedback mechanisms, discussed in Chapters 63 and 64, that inhibit gastric secretion and also slow the rate of gastric emptying. These effects reduce the quantity of gastric juice secreted and also allow prolonged stasis of food in the stomach to buffer the gastric juice that is secreted.

Surgical treatment of peptic ulcer is effected by one or both of two procedures: (1) removal of a large portion of the stomach or (2) vagotomy. When a peptic ulcer is surgically removed, usually at least the lower three-fourths to four-fifths of the stomach is also removed and the upper stump of the stomach then anastomosed to the jejunum. If less of the stomach than this is removed, far too much gastric juice continues to be secreted, and a marginal ulcer soon develops where the stomach is anastomosed to the intestine. Since a marginal ulcer is frequently equally as debilitating as, or sometimes even more debilitating than, the original ulcer, nothing will have been accomplished. However, if only one-fifth of the stomach remains, the rate of gastric secretion is usually low enough that a marginal ulcer does not develop, and yet the stomach is still large enough to hold small meals. Furthermore, the presence of at least some functional stomach mucosa usually prevents the development of pernicious anemia.

*Vagotomy* temporarily blocks almost all secretion by the stomach and usually cures a peptic ulcer within less than a week after the operation is performed. The vagus nerves may be cut in the chest as they pass downward over the surface of the esophagus, or they may be cut as they pass through the diaphragm to spread over the surface of the stomach. Unfortunately, though, a large amount of basal stomach secretion returns three to six months after the vagotomy, and in many patients the ulcer itself also returns. Even more distressing is the *gastric atony* that usually follows vagotomy, for the motility of the stomach is often reduced to such a low level that almost no gastric emptying occurs after vagotomy. For this reason, vagotomy alone is rarely performed in the treatment of peptic ulcer. However, it is performed frequently in association with simultaneous plastic procedures to increase the size of the opening of the pylorus from the stomach into the small intestine.

## DISORDERS OF THE SMALL INTESTINE

In comparison with disorders of the stomach and colon, disorders of the small intestine are surprisingly less common when one considers the extensiveness of the small intestinal system. Perhaps the most common disorder of the small intestine and its associated structures is the development of gallstones in the gallbladder and bile ducts. (This will be discussed in Chap. 70 in relation to the function of the liver and the bile ducts.) The presence of stones in the gallbladder or bile ducts does not alter the function of the small intestine unless a stone blocks the flow of bile into the intestine. Lack of bile then prevents adequate absorption of fat from the small intestine with resultant loss of up to 40 per cent of the ingested fat in the feces.

**Abnormal Digestion of Food in the Small Intestine; Pancreatic Failure.** By far the greater portion of all digestion in the alimentary tract occurs in the upper third of the small intestine. Only rarely is digestion impaired enough to cause malnutrition. Indeed, as much as three-fourths of the small intestine has been removed without the patient's suffering serious malnutrition.

Perhaps the commonest cause of abnormal digestion is failure of the pancreas to secrete its juice into the small intestine. Lack of pancreatic secretion frequently occurs (a) in *pancreatitis*, which is discussed below, (b) when the *pancreatic duct is blocked* by a gallstone at the papilla of Vater, or (c) after the *head of the pancreas has been removed* because of malignancy. Loss of pancreatic juice means loss of trypsin, chymotrypsin, carboxypolypeptidase, pancreatic amylase, pancreatic lipase, and still a few other digestive enzymes. Without these enzymes, as much as one-half of the fat entering the small intestine may go unabsorbed and as much as one-third of the proteins and starches. As a result, large portions of the ingested food are not utilized for nutrition; and copious, fatty feces are excreted.

**Pancreatitis.** Pancreatitis means inflammation of the pancreas, and this can occur in the form of either *acute pancreatitis* or *chronic pancreatitis*. The commonest cause of acute pancreatitis is blockage of the papilla of Vater by a gallstone; this blocks the main secretory duct from the pancreas as well as the common bile duct. The pancreatic enzymes are then dammed up in the ducts and acini of the pancreas. Eventually, so much trypsinogen accumulates that it overcomes the *trypsin inhibitor* in the secretions, and a small quantity of trypsinogen becomes activated to form trypsin. Once this happens the trypsin activates still more trypsinogen as well as chymotrypsinogen and carboxypolypeptidase, resulting in a vicious cycle until all the proteolytic enzymes in the pancreatic ducts and acini become activated. These rapidly digest large portions of the pancreas itself, sometimes completely and permanently destroying the ability of the pancreas to secrete digestive enzymes.

Often in acute pancreatitis the proteolytic enzymes

eat their way all the way to the surface of the pancreas, and pancreatic juice then empties into the peritoneal cavity where additional proteolytic activity occurs, causing *chemical peritonitis*. However, as soon as large portions of the pancreas have been destroyed, the acute stage of pancreatitis subsides, leaving the person with diminished or sometimes totally absent pancreatic secretion into the gut.

Chronic pancreatitis usually occurs in much the same manner as does acute pancreatitis, beginning with blockage of the pancreatic duct but progressing slowly until pancreatic function becomes greatly diminished.

Ordinarily, the islets of Langerhans are not adversely affected by pancreatitis, so that insulin still continues to be secreted by the pancreas even though secretion into the intestine is markedly reduced.

**Malabsorption from the Small Intestine— "Sprue."** Occasionally, nutrients are not adequately absorbed from the small intestine even though the food is well digested. Several different diseases can cause decreased absorbability by the mucosa; these are often classified together under the general heading of *sprue*. Obviously, also, malabsorption can occur when large portions of the small intestine have been removed.

*Nontropical Sprue.* One type of sprue, called variously by the names *idiopathic sprue, celiac disease* (in children), or *gluten enteropathy,* results from the toxic effects of *gluten* present in certain types of grains, especially wheat and rye. It is the gliadin fraction that is responsible for this toxic effect. The gliadin causes early destruction and dissolution of newly forming epithelial cells in the crypts of Lieberkühn. Therefore, these cells fail to migrate upward onto the villi. As a result, the villi become blunted or disappear altogether, thus greatly reducing the absorptive area of the gut. Removal of wheat and rye flour from the diet, especially in children with this disease, frequently results in a miraculous cure within weeks.

*Tropical Sprue.* A different type of sprue called "tropical sprue" frequently occurs in the tropics and can often be treated with antibacterial agents. Even though no specific bacterium has ever been implicated as the cause of this tropical sprue, it is believed that this variety is often caused by inflammation of the intestinal mucosa resulting from a yet unidentified infectious agent.

*Malabsorption in Sprue.* In the early stages of sprue, the absorption of fats is more impaired than the absorption of other digestive products. The fat appearing in the stools is almost entirely in the form of soaps rather than undigested neutral fat, illustrating that the problem is one of absorption and not of digestion. In this stage of sprue, the condition is frequently called *idiopathic steatorrhea,* which means simply excess fats in the stools as a result of unknown causes.

In more severe cases of sprue the absorption of proteins, carbohydrates, calcium, vitamin K, folic acid, and vitamin $B_{12}$, as well as many other important substances, becomes greatly impaired. As a result, the person suffers (1) severe nutritional deficiency, often developing severe wasting of the tissues, (2) osteomalacia (demineralization of the bones because of calcium lack), (3) inadequate blood coagulation due to lack of vitamin K, and (4) macrocytic anemia of the pernicious anemia type, owing to diminished vitamin $B_{12}$ and folic acid absorption.

**Regional Enteritis and Appendicitis.** Regional enteritis means an inflammatory condition of the terminal portion of the ileum, and appendicitis means inflammation of the appendix. Appendicitis is usually caused by an actual infection in the appendix, while the cause of regional enteritis is not known but in some instances might also result from a low grade infectious process. Either of these two conditions causes crampy pain that is referred to the mid-abdomen. Simultaneously, they also elicit intense enterointestinal reflexes resulting in severe inhibition of gastrointestinal mobility; as a result, functional obstruction often occurs in the small bowel, causing symptoms of acute intestinal obstruction, which will be discussed later in the chapter. These are mainly distention of the upper small intestine associated with protracted cramping pains and vomiting.

# DISORDERS OF THE LARGE INTESTINE

## CONSTIPATION

Constipation means slow movement of feces through the large intestine, and it is often associated with large quantities of dry, hard feces in the descending colon which accumulate because of the long time allowed for absorption of fluid.

A frequent cause of constipation is irregular bowel habits that have developed through a lifetime of inhibition of the normal defecation reflexes. The newborn child is rarely constipated, but part of his training in the early years of life requires that he learn to control defecation, and this control is effected by inhibiting the natural defecation reflexes. Clinical experience shows that if one fails to allow defecation to occur when the defecation reflexes are excited or if one overuses laxatives to take the place of natural bowel function, the reflexes themselves become progressively less strong over a period of time and the colon becomes *atonic*. For this reason, if a person establishes regular bowel habits early in life, usually defecating in the morning after breakfast when the gastrocolic and duodenocolic reflexes cause mass movements in the large intestine, he can generally prevent the development of constipation in later life.

Constipation can also result from spasm of a small segment of the sigmoid. It should be recalled that motility, even normally, is weak in the large intestine, so that even a slight degree of spasm is often capable of causing serious constipation. This effect frequently occurs in the so-called *irritable colon syn-*

*drome* in which bowel spasm causes constipation and consequent development of small and hard feces, a condition also often associated with crampy abdominal pains. After several days, the constipation is relieved by a bowel movement that often contains a moderate to large amount of mucus. The period of constipation may then be followed by a day or so of diarrhea. Following this, the cycle begins again, with repeated bouts of alternating constipation and diarrhea.

**Megacolon.**    Occasionally, a person develops constipation which is so severe that bowel movements occur only once every several weeks. Obviously, this allows tremendous quantities of fecal matter to accumulate in the colon, causing the colon sometimes to distend to a diameter as great as 3 to 4 inches. The condition is called megacolon, or *Hirschsprung's disease.*

The most frequent cause of megacolon is congenital lack of a myenteric plexus in a segment of the sigmoid. As a consequence, neither defecation reflexes nor peristaltic motility can occur through this area of the large intestine. And the sigmoid itself becomes small and almost spastic while feces accumulate proximal to this area, causing megacolon. Thus, the diseased portion of the large intestine appears normal by x-ray, while the originally normal portion of the large intestine appears greatly enlarged.

## DIARRHEA

Diarrhea, which is the opposite of constipation, results from rapid movement of fecal matter through the large intestine. The major causes of diarrhea are: (1) infection in the gastrointestinal tract called *enteritis* and (2) excessive parasympathetic stimulation of the large intestine, which is called *psychogenic diarrhea.*

**Enteritis.**    Enteritis means infection, caused either by a virus or by bacteria, and found in the intestinal tract. In usual infectious diarrhea, the infection is most extensive in the large intestine and the distal end of the ileum. Everywhere that the infection is present, the mucosa becomes extensively irritated, and its rate of secretion becomes greatly enhanced. In addition, the motility of the intestinal wall usually increases many fold. As a result, large quantities of fluid are made available for washing the infectious agent toward the anus, and at the same time strong propulsive movements propel this fluid forward. Obviously, this is an important mechanism for ridding the intestinal tract of the debilitating infection.

Of special interest is the diarrhea caused by *cholera.* The cholera toxin directly stimulates excessive secretion of electrolytes and fluid from the crypts of Lieberkühn in the distal ileum and colon, and it specifically enhances the bicarbonate-chloride exchange mechanism, causing extreme quantities of bicarbonate ions to be secreted into the intestinal tract. The loss of fluid and electrolytes can be so debilitating within a day or so that death ensues. Therefore,

the most important basis of therapy is simply to replace the fluid and electrolytes as rapidly as they are lost. With proper therapy of this type, almost no cholera patients die, but without it, 50 per cent or more do.

**Psychogenic Diarrhea.**    Everyone is familiar with the diarrhea that accompanies periods of nervous tension, such as during examination time or when a soldier is about to go into battle. This type of diarrhea, called psychogenic or emotional diarrhea, is caused by excessive stimulation of the parasympathetic nervous system, which greatly excites both motility and secretion of mucus in the distal colon. These two effects added together can cause marked diarrhea.

The *irritable colon syndrome,* that was discussed above in relation to constipation and which often causes alternating constipation and diarrhea, is also generally believed to be a psychogenic (or emotional) disorder; it usually occurs in anxiety states.

**Ulcerative Colitis.**    Ulcerative colitis is a disease in which extensive areas of the walls of the large intestine become ulcerated. This, like psychogenic diarrhea, is frequently associated with different states of nervous tension. The motility of the ulcerated colon is often so great that mass movements occur most of the time, rather than the usual 10 to 20 minutes per day. Also, the colon's secretions are greatly enhanced. As a result, the patient has repeated diarrheal bowel movements.

Though ulcerative colitis is associated with psychogenic disorders, the precise cause of the actual ulcers is unknown. Some clinicians believe that these are caused by infectious bacteria invading the mucosa, while others believe that they result from excoriation or digestion of the mucosa by the contents of the large intestine. Regardless of the cause, if the condition has not progressed too far, removal of factors that have been causing nervous tension frequently leads to a cure. On the other hand, if the condition has progressed extremely far, the ulcers usually will not heal until an ileostomy is performed to allow the intestinal contents to drain to the exterior rather than to flow through the colon. Even then the ulcers sometimes fail to heal, and the only solution is removal of the colon itself.

## PARALYSIS OF DEFECATION IN SPINAL CORD INJURIES

From Chapter 63 it will be recalled that defecation is normally initiated by the movement of feces into the rectum, which causes a reflex to pass from the rectum to the spinal cord and then back to the descending colon, sigmoid, rectum, and anus to intensify greatly the intrinsic defecation reflex of the myenteric plexus. Frequently, this *cord* defecation reflex is blocked or altered by spinal cord injuries. For instance, destruction of the *conus medullaris* of the spinal cord destroys the sacral centers in which the cord reflex is integrated and therefore almost completely paralyzes defecation. In such instances defecation requires extensive supportive measures, such as cathartics and large enemas.

But, more frequently, the spinal cord is injured somewhere between the conus medullaris and the brain, in which case the voluntary portion of the defecation act is blocked while the basic cord reflexes for defecation are still intact. Nevertheless, loss of the voluntary aid to defecation—that is, loss of the increased abdominal pressure, the lifting of the pelvic floor, and the stretching of the anal ring by the pelvic muscles—often makes defecation a difficult process in the person with this type of cord injury. Yet, since cord defecation reflexes can still occur, a small enema given to potentiate the action of these reflexes, usually in the morning shortly after a meal, can often cause adequate defecation. In this way persons with spinal cord injuries can usually control their bowel movements each day.

# GENERAL DISORDERS OF THE GASTROINTESTINAL TRACT

## VOMITING

Vomiting is the means by which the upper gastrointestinal tract rids itself of its contents when the gut becomes excessively irritated, overdistended, or even overexcitable. The stimuli that cause vomiting can occur in any part of the gastrointestinal tract, though distention or irritation of the stomach or duodenum provides the strongest stimulus. Impulses are transmitted, as illustrated in Figure 66–2, by both vagal and sympathetic afferents to the bilateral *vomiting center* of the medulla, which lies near the tractus solitarius at approximately the level of the dorsal motor nucleus of the vagus. Appropriate motor reactions are then instituted to cause the vomiting act, and the motor impulses that cause the actual vomiting are transmitted from the vomiting center through the fifth, seventh, ninth, tenth, and twelfth cranial nerves to the upper gastrointestinal tract and through the spinal nerves to the diaphragm and abdominal muscles.

**The Vomiting Act.** Once the vomiting center has been sufficiently stimulated and the vomiting act instituted, the first effects are (1) a deep breath, (2) raising of the hyoid bone and the larynx to pull the cricoesophageal sphincter open, (3) closing of the glottis, and (4) lifting of the soft palate to close the posterior nares. Next comes a strong downward contraction of the diaphragm along with simultaneous contraction of all the abdominal muscles. This obviously squeezes the stomach between the two sets of muscles, building the intragastric pressure to a high level. Finally, the gastroesophageal sphincter relaxes, allowing expulsion of the gastric contents upward through the esophagus.

Thus, the vomiting act results from a squeezing action of the muscles of the abdomen associated with opening of the esophageal sphincters so that the gastric contents can be expelled.

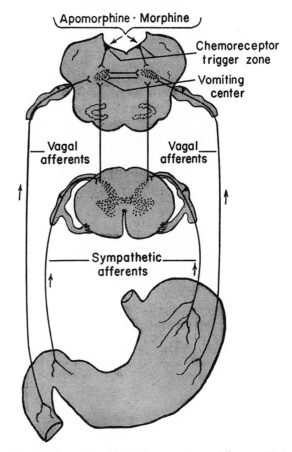

**Figure 66–2.** The afferent connections of the vomiting center.

**The Chemoreceptor Trigger Zone of the Medulla for Initiation of Vomiting by Drugs or by Motion Sickness.** Aside from the vomiting initiated by irritative stimuli in the gastrointestinal tract itself, vomiting can also be caused by impulses arising in areas of the brain outside the vomiting center. This is particularly true of a small area located bilaterally on the floor of the fourth ventricle above the area postrema and called the *chemoreceptor trigger zone*. Electrical stimulation of this area will initiate vomiting, but, more importantly, administration of certain drugs, including apomorphine, morphine, and some of the digitalis derivatives, can directly stimulate the chemoreceptor trigger zone and initiate vomiting. Destruction of this area blocks this type of vomiting but does not block vomiting resulting from irritative stimuli in the gastrointestinal tract itself.

Also, it is well known that rapidly changing motions of the body cause certain people to vomit. The mechanism for this is the following: The motion stimulates the receptors of the labyrinth, and impulses are transmitted mainly by way of the vestibular nuclei into the cerebellum. After passing through the uvula and nodule of the cerebellum, the signals are believed to be transmitted to the chemoreceptor trigger zone and thence to the vomiting center to cause vomiting.

**Cortical Excitation of Vomiting.** Various psychic stimuli, including disquieting scenes, noisome odors, and other similar psychological factors, can also cause vomiting. Stimulation of certain areas of the hypothalamus also causes vomiting. The precise neuronal connections for these effects are not known, though it is probable that the impulses pass directly to the vomiting center and do not involve the chemoreceptor trigger zone.

## NAUSEA

Everyone has experienced the sensation of nausea and knows that it is often a prodrome of vomiting. Nausea is the conscious recognition of subconscious excitation in an area of the medulla closely associated with or part of the vomiting center, and it can be caused by irritative impulses coming from the gastrointestinal tract, impulses originating in the brain associated with motion sickness, or impulses from the cortex to initiate vomiting. However, vomiting occasionally occurs without the prodromal sensation of nausea, indicating that only certain portions of the vomiting centers are associated with the sensation of nausea.

A common cause of nausea is distention or irritation of the duodenum or lower small intestine. When this occurs, the intestine contracts forcefully while the stomach relaxes, thus allowing the intestinal contents to reflux into the stomach. This is preliminary to the vomiting that often follows.

## GASTROINTESTINAL OBSTRUCTION

The gastrointestinal tract can become obstructed at almost any point along its course, as illustrated in Figure 66–3; some common causes of obstruction are: *cancer, fibrotic constriction resulting from ulceration or from peritoneal adhesions, spasm of a segment of the gut,* or *paralysis of a segment of the gut.*

The abnormal consequences of obstruction depend

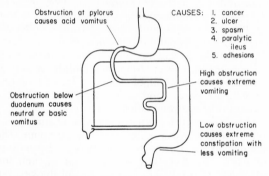

**Figure 66–3.**  Obstruction in different parts of the gastrointestinal tract.

on the point in the gastrointestinal tract that becomes obstructed. If the obstruction occurs at the pylorus, which results often from fibrotic constriction following peptic ulceration, persistent vomiting of stomach contents occurs. This obviously depresses bodily nutrition, and it also causes excessive loss of hydrogen ions from the body and can result in various degrees of alkalosis.

If the obstruction is beyond the stomach, reflux from the small intestine causes the intestinal juices to flow backward into the stomach, and these are vomited along with the stomach secretions. In this instance the person loses large amounts of water and electrolytes so that he becomes severely dehydrated, but the loss of acids and bases may be approximately equal so that little change in acid-base balance occurs. If the obstruction is near the lower end of the small intestine, then it is actually possible to vomit more basic than acidic substances; in this case acidosis may result.

Also important in obstruction of the small intestine is marked distention of the intestine proximal to the obstructed point. Large quantities of fluid and electrolytes continue to be secreted into the lumen of the small intestine and even large amounts of proteins are lost from the blood stream, partly into the intestinal lumen and partly into the gut wall, which becomes edematous as a result of the excessive distention. The plasma volume diminishes because of the protein loss, and severe circulatory shock often ensues. One might immediately ask: Why does the small intestine not absorb these fluids and electrolytes? The answer seems to be that distention of the gut greatly stimulates the secretory activity of the gut but does not correspondingly increase the rate of absorption. Normally this would act to flush the chyme farther down the small intestine and therefore relieve the distention. But, if obstruction is present, obviously this normal mechanism backfires and simply causes a vicious cycle of more and more distention.

If the obstruction is near the distal end of the large intestine, feces can accumulate in the colon for several weeks. The patient develops an intense feeling of constipation, but in the first stages of the obstruction vomiting is not severe. After the large intestine has become completely filled and it finally becomes impossible for additional chyme to move from the small intestine into the large intestine, vomiting does then begin severely. Reflux of the large intestinal contents can occasionally occur through the ileocecal valve all the way to the stomach, and the character of the vomitus becomes fecal. Some physiologists believe that this reflux is caused by reverse peristalsis, an effect not normal to the intestinal tract. Prolonged obstruction of the large intestine can finally cause rupture of the intestine itself or dehydration and circulatory shock resulting from the severe vomiting.

## GASES IN THE GASTROINTESTINAL TRACT (FLATUS)

Gases can enter the gastrointestinal tract from three different sources: (1) swallowed air, (2) gases formed as a result of bacterial action, and (3) gases that diffuse from the blood into the gastrointestinal tract.

Most gases in the stomach are nitrogen and oxygen derived from swallowed air, and a large proportion of these are expelled by belching.

Only small amounts of gas are usually present in the small intestine, and these are composed principally of air that passes from the stomach into the intestinal tract. In its transport through the small intestine, only 5 to 15 per cent of the air is absorbed, and a considerable amount of carbon dioxide actually diffuses from the blood into the air to bring it into equilibrium with the carbon dioxide of the tissue fluids.

In the large intestine, the greater proportion of the gases is derived from bacterial action; these gases include especially *carbon dioxide, methane,* and *hydrogen.* When the methane and hydrogen become suitably mixed with oxygen from swallowed air, an actual explosive mixture is occasionally formed.

Essentially all the gases in the large intestine are highly diffusible through the intestinal mucosa. Therefore, if the gases remain in the large intestine for many hours the final mixture contains approximately 75 per cent or more of nitrogen and little of the other gases. The reason for this is that nitrogen in the gut cannot easily be absorbed into the blood because of the high $P_{N_2}$ already in the blood, as explained in Chapter 43. However, if the gases are passed on through the colon rapidly, the composition of the expelled flatus may be as little as 20 per cent nitrogen, with the remaining 80 per cent composed mainly of carbon dioxide, methane, and hydrogen.

Certain foods are known to cause greater expulsion of flatus from the large intestine than others—beans, cabbage, onions, cauliflower, corn, and certain highly irritant foods such as vinegar. Some of these foods serve as a suitable medium for gas-forming bacteria, especially because of unabsorbed fermentable types of carbohydrates, but in other instances excess gas results from irritation of the large intestine, which promotes rapid expulsion of the gases before they can be absorbed.

The amount of gases entering or forming in the large intestine each day averages 7 to 10 liters, whereas the average amount expelled is usually only about 0.5 liter. The remainder is absorbed through the intestinal mucosa. Most often, a person expels large quantities of gases not because of excessive bacterial activity but because of excessive motility of the large intestine, the gases being moved on through the large intestine before they can be absorbed.

# REFERENCES

Blair, E. (ed.): Markowitz's Experimental Surgery, 6th Ed. Baltimore, The Williams & Wilkins Company, 1975.

Bockus, H. L., Berk, J. E., Haubrich, W. S., Kalser, M., Roth, J. L. A., and Vilardell, F. (eds.): Gastroenterology, 3rd Ed. Philadelphia, W. B. Saunders Company, 1975.

Brooks, F. P. (ed.): Gastrointestinal Pathophysiology. New York, Oxford University Press, 1974.

Connell, A. M.: Clinical aspects of motility. *Med. Clin. N. Amer.* 58:1201, 1974.

Csaky, T. Z. (ed.): Intestinal Absorption and Malabsorption. New York, Raven Press, 1975.

Dworken, H. J.: The Alimentary Tract. Philadelphia, W. B. Saunders Company, 1974.

Greenberger, N. J.: Effects of antibiotics and other agents on the intestinal transport of iron. *Amer. J. Clin. Nutr.,* 26:104, 1973.

Harkins, H. N., and Nyhus, L. M.: Surgery of the Stomach and Duodenum. Boston, Little, Brown and Company, 1968.

Kirsner, J. B.: The stomach. *In* Sodeman, W. A., Jr., and Sodeman, W. A. (eds.): Pathologic Physiology: Mechanisms of Disease, 5th Ed. Philadelphia, W. B. Saunders Company, 1974, p. 709.

Kirsner, J. B., and Shorter, R. G. (eds.): Inflammatory Bowel Disease. Philadelphia, Lea & Febiger, 1975.

Lindner, A. E. (ed.): Emotional Factors in Gastrointestinal Illness. New York, American Elsevier Publishing Co., Inc., 1973.

Losowsky, M. S., Walker, B. E., and Kelleher, J.: Malabsorption in Clinical Practice. New York, Churchill Livingstone, Div. of Longman, Inc., 1973.

Money, K. E.: Motion sickness. *Physiol. Rev.* 50:1, 1970.

Morson, B. C., and Dawson, I. M. P.: Gastrointestinal Pathology. Philadelphia, J. B. Lippincott Company, 1972.

Naish, J. M., and Read, A. E.: Basic Gastroenterology. Baltimore, The Williams & Wilkins Company, 1966.

Neale, G., Gompertz, D., Schönsby, H., Tabaqchali, S., and Booth, C. C.: The metabolic and nutritional consequences of bacterial overgrowth in the small intestine. *Amer. J. Clin. Nutr.,* 25:1409, 1972.

Palmer, E. D.: Functional Gastrointestinal Disease. Baltimore, The Williams & Wilkins Company, 1967.

Payne, W. S., and Olsen, A. M.: The Esophagus. Philadelphia, Lea & Febiger, 1974.

Pfeiffer, C. J. (ed.): Peptic Ulcer. Philadelphia, J. B. Lippincott Company, 1971.

Rambaud, J. C., Modigliani, R., and Bernier, J. J.: The method of intraluminal perfusion of the human small intestine. 3. Absorption studies in disease. *Digestion,* 9:343, 1973.

Rankow, R. M., and Polayes, I. M.: Diseases of the Salivary Glands. Philadelphia, W. B. Saunders Company, 1975.

Sharp, G. W.: Action of cholera toxin on fluid and electrolyte movement in the small intestine. *Ann. Rev. Med.* 24:19, 1973.

Sodeman, W. A., Jr., and Watson, D. W.: The large intestine. *In* Sodeman, W. A., Jr., and Sodeman, W. A. (eds.): Pathologic Physiology: Mechanisms of Disease, 5th Ed. Philadelphia, W. B. Saunders Company, 1974, p. 767.

Toskes, P. P., and Deren, J. J.: Vitamin $B_{12}$ absorption and malabsorption. *Gastroenterology,* 65:662, 1973.

Walker, W. A., and Isselbacher, K. J.: Uptake and transport of macromolecules by the intestine. Possible role in clinical disorders. *Gastroenterology,* 67:531, 1974.

Westergaard, H., and Dietschy, J. M.: Normal mechanisms of fat absorption and derangements induced by various gastrointestinal diseases. *Med. Clin. N. Amer.,* 58:1413, 1974.

Yoel, J.: Pathology and Surgery of the Salivary Glands. Springfield, Ill., Charles C Thomas, Publisher, 1974.

PART XII

# METABOLISM AND TEMPERATURE REGULATION

# | 67 |

# Metabolism of Carbohydrates and Formation of Adenosine Triphosphate

The next few chapters deal with metabolism in the body, which means the chemical processes that make it possible for the cells to continue living. It is not the purpose of this textbook, however, to present the chemical details of all the various cellular reactions, for this lies in the discipline of biochemistry. Instead, these chapters are devoted to: (1) a review of the principal chemical processes of the cell and (2) an analysis of their physiological implications, especially in relation to the manner in which they fit into the overall concept of homeostasis.

## RELEASE OF ENERGY FROM FOODS AND THE CONCEPT OF "FREE ENERGY"

The greater proportion of the chemical reactions in the cells is concerned with making the energy in foods available to the various physiological systems of the cell. For instance, energy is required for (a) muscular activity, (b) secretion by the glands, (c) maintenance of membrane potentials by the nerve and muscle fibers, (d) synthesis of substances in the cells, and (e) absorption of foods from the gastrointestinal tract.

**Coupled Reactions.** All the energy foods—carbohydrates, fats and proteins—can be oxidized with oxygen in the cells, and in this process large amounts of energy are released. These same foods can also be burned with pure oxygen outside the body in an actual fire, again releasing large amounts of energy. However, this time the energy is released suddenly, all in the form of heat. The energy needed by the physiological processes of the cells is not heat but instead energy (a) to cause mechanical movement in the case of muscle function, (b) to concentrate

solutes in the case of glandular secretion, and (c) to effect other functions. To provide this energy, the chemical reactions must be "coupled" with the systems responsible for these physiological functions. This coupling is accomplished by special cellular enzyme and energy transfer systems, some of which will be explained in this and subsequent chapters.

*"Free Energy."* The amount of energy liberated by complete oxidation of a food is called the free energy of the food, and this is generally represented by the symbol $\Delta F$. Free energy is usually expressed in terms of calories per mol of food substance. For instance, the amount of free energy liberated by oxidation of 1 mol of glucose (180 grams of glucose) is 686,000 calories.

## ROLE OF ADENOSINE TRIPHOSPHATE (ATP) IN METABOLISM

ATP is a labile chemical compound that is present in all cells and has the chemical structure shown in the formula on the opposite page.

From this formula it can be seen that ATP is a combination of adenine, ribose, and three phosphate radicals. The last two phosphate radicals are connected with the remainder of the molecule by so-called *high energy bonds*, which are indicated by the symbol $\sim$. The amount of free energy in each of these high energy bonds per mol of ATP is approximately 7000 calories under standard conditions and 8000 calories under the conditions of temperature and concentrations of the reactants in the body. Therefore, removal of each phosphate radical liberates 8000 calories of energy. After loss of one phosphate radical from ATP, the compound becomes *adenosine diphosphate* (ADP), and after loss of the second phosphate radical

904

the compound becomes *adenosine monophosphate* (AMP). The interconversions between ATP, ADP, and AMP are the following:

$$\text{ATP} \underset{\xrightarrow{+8000 \text{ cal.}}}{\overset{-8000 \text{ cal.}}{\rightleftharpoons}} \left\{ \begin{array}{c} \text{ADP} \\ + \\ \text{PO}_4 \end{array} \right\} \underset{\xrightarrow{+8000 \text{ cal.}}}{\overset{-8000 \text{ cal.}}{\rightleftharpoons}} \left\{ \begin{array}{c} \text{AMP} \\ + \\ 2\text{PO}_4 \end{array} \right\}$$

*ATP is present everywhere in the cytoplasm and nucleoplasm of all cells, and essentially all the physiological mechanisms that require energy for operation obtain this directly from the ATP. In turn, the food in the cells is gradually oxidized, and the released energy is used to re-form the ATP, thus always maintaining a supply of this substance; all of these energy transfers take place by means of coupled reactions.*

In summary, ATP is an intermediary compound that has the peculiar ability of entering into many coupled reactions—reactions with the food to extract energy, and reactions in relation to many physiological mechanisms to provide energy for their operation. For this reason, ATP has frequently been called the energy *currency* of the body that can be gained and spent again and again.

The principal purpose of the present chapter is to explain how the energy from carbohydrates can be used to form ATP in the cells. At least 99 per cent of all the carbohydrates utilized by the body is used for this purpose.

# TRANSPORT OF MONOSACCHARIDES THROUGH THE CELL MEMBRANE

From Chapter 65 it will be recalled that the final products of carbohydrate digestion in the alimentary tract are almost entirely glucose, fructose, and galactose, glucose representing by far the major share of these. These three monosaccharides are absorbed into the portal blood, and, after passing through the liver, are carried everywhere in the body by the circulatory system. But, before they can be used by the cells, they must be transported through the cell membrane into the cellular cytoplasm.

Monosaccharides cannot diffuse through the pores of the cell membrane, for the maximum molecular weight of particles that can do this is about 100, whereas glucose, fructose, and galactose all have molecular weights of 180. Yet these monosaccharides do pass to the interior of the cells with a reasonable degree of freedom. Therefore, it is almost certain that they are transported through the membrane by a carrier system. They presumably combine with some carrier substance that makes them soluble in the membrane and then after passing through the membrane become dissociated from this carrier. Thus far, the nature of the carrier is uncertain, though it is believed to be a protein of small molecular weight, as was discussed in Chapter 4. It is known that the carrier mechanism works in both directions, both into and out of the cell. Furthermore, the rate of transport is approximately proportional to the concentration differences of the monosaccharides on the two sides of the membrane, and the monosaccharides are *not* transported through the usual cell membrane against a concentration difference. Therefore, the transport mechanism is one of *facilitated diffusion* and not of active transport. These concepts were discussed in more detail in Chapter 4.

## FACILITATION OF GLUCOSE TRANSPORT BY INSULIN

The rate of glucose transport and also transport of some other monosaccharides is greatly increased by insulin. When large amounts of insulin are secreted by the pancreas, the rate of glucose transport into some cells increases to as much as 10 times the rate of transport when no insulin at all is secreted. And, for practical considerations, the amounts of glucose that can diffuse to the insides of most cells of the body in the absence of insulin, with the unique exception of the liver, are far too little to supply anywhere near the amount of glucose normally required for energy metabolism. Therefore, in effect, the rate of carbohydrate utilization by the cells is controlled by the rate of insulin secretion in the pancreas. The functions of insulin and its control of carbohydrate metabolism will be discussed in detail in Chapter 78.

**Failure of Disaccharides To Be Transported.** Very minute amounts of disaccharides are absorbed into the blood from the gastrointestinal tract, but none of these can be transported into the cells. Therefore, no disaccharides or larger polysaccharides are utilized for cellular metabolism. Instead, they are excreted completely in the urine.

## *PHOSPHORYLATION OF THE MONOSACCHARIDES*

Immediately upon entry into the cells, the monosaccharides combine with a phosphate radical in accordance with the following reaction:

$$\text{Glucose} \xrightarrow[\text{+ATP}]{\text{glucokinase}} \text{Glucose 6-phosphate}$$

This phosphorylation is promoted by enzymes called *hexokinases*, which are specific for each particular type of monosaccharide; thus, *glucokinase* promotes

glucose phosphorylation, *fructokinase* promotes fructose phosphorylation, and *galactokinase* promotes galactose phosphorylation.

The phosphorylation of monosaccharides is almost completely irreversible except in the liver cells, the renal tubular epithelium, and the intestinal epithelial cells in which specific phosphatases are available for reversing the reaction. Therefore, in most tissues of the body phosphorylation serves to *capture* the monosaccharide in the cell—once *in* the cell the monosaccharide will not diffuse back out except in those special cells listed above that have the necessary phosphatases.

**Conversion of Fructose and Galactose Into Glucose.** In liver cells, appropriate enzymes are available to promote interconversions between the monosaccharides, as shown in Figure 67–1; and the dynamics of the reactions are such that when the liver releases the monosaccharides back into the blood the final product of these interconversions is almost entirely glucose. In essentially all other cells glucose and fructose can be reversibly interconverted, but some of the enzymes required for conversion of galactose into the other two monosaccharides are missing. Later in the chapter it is noted that the monosaccharides must all become either *glucose 6-phosphate* or *fructose 6-phosphate* before they can be used for energy by the cells. For this reason, before galactose can be utilized by the tissues, in general it must be converted by the liver cells

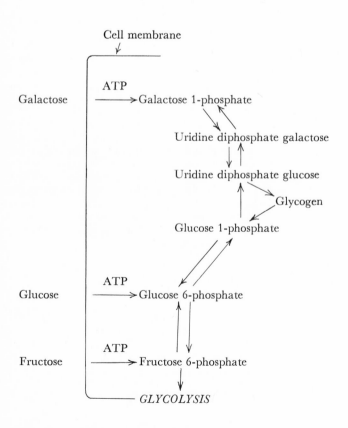

**Figure 67–1.** Interconversions of the three major monosaccharides—glucose, fructose, and galactose—in liver cells.

into glucose, which is then transported by the blood to the other cells.

In the case of fructose, most is converted into glucose as it is absorbed through the intestinal epithelial cells into the portal blood. This conversion was discussed in Chapter 65. Most of the remaining fructose is also converted by the liver into glucose.

Therefore, essentially all of the monosaccharides that circulate in the blood are the final conversion product, glucose.

# STORAGE OF GLYCOGEN IN LIVER AND MUSCLE

After absorption into the cells, glucose can be used immediately for release of energy to the cells or it can be stored in the form of *glycogen*, which is a large polymer of glucose. Also, after the other monosaccharides are converted into glucose, they too can all be polymerized into glycogen.

All cells of the body are capable of storing at least some glycogen, but certain cells can store large amounts, especially the liver cells, which can store up to 5 to 8 per cent of their weight as glycogen, and muscle cells, which can store up to 1 per cent glycogen. The glycogen molecules can be polymerized to almost any molecular weight, the average molecular weight being 5,000,000 or greater; most of the glycogen precipitates in the form of solid granules. This conversion of the monosaccharides into a high molecular weight, precipitated compound makes it possible to store large quantities of carbohydrates without significantly altering the osmotic pressure of the intracellular fluids. Obviously, high concentrations of low molecular weight, soluble monosaccharides would play havoc with the osmotic relationships between intracellular and extracellular fluids.

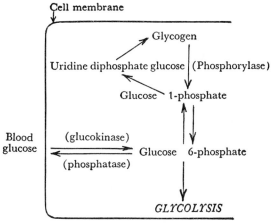

**Figure 67–2.** The chemical reactions of glycogenesis and glycogenolysis, showing also the interconversions between blood glucose and liver glycogen. (The phosphatase required for release of glucose from the cell is absent in muscle cells.)

Some physiologists believe that the glycogen stores exist in two forms, one a readily released and labile form called *free* glycogen and another, called *fixed* glycogen, that can be released only slowly.

## *GLYCOGENESIS*

Glycogenesis is the process of glycogen formation, the chemical reactions for which are illustrated in Figure 67–2. From this figure it can be seen that *glucose 6-phosphate* first becomes *glucose 1-phosphate*; then this is converted to *uridine diphosphate glucose*, which is then converted into glycogen. Several specific enzymes are required to cause these conversions, and any monosaccharide that can be converted into glucose obviously can enter into the reactions. Certain smaller compounds, including *lactic acid, glycerol, pyruvic acid,* and *some deaminated amino acids*, can also be converted into glucose or closely allied compounds and thence into glycogen.

## *GLYCOGENOLYSIS*

Glycogenolysis means the breakdown of glycogen to re-form glucose in the cells. Glycogenolysis does not occur by reversal of the same chemical reactions that serve to form glycogen; instead, each succeeding glucose molecule on each branch of the glycogen polymer is split away by a process of *phosphorylation*, catalyzed by the enzyme *phosphorylase* (several other enzymes split the glycogen molecule at the branching points).

Under resting conditions, the phosphorylase is in an inactive form so that glycogen can be stored but not reconverted into glucose. When it is necessary to re-form glucose from glycogen, therefore, the phosphorylase must first be activated. This is accomplished in the following two ways:

**Activation of Phosphorylase by Epinephrine and Glucagon.** Two hormones, epinephrine and glucagon, can specifically activate phosphorylase and thereby cause rapid glycogenolysis. The initial effect of each of these hormones is to increase the formation of cyclic adenylate in the cells, and it is this substance that activates the phosphorylase.

Epinephrine is released by the adrenal medullae when the sympathetic nervous system is stimulated. Therefore, one of the functions of the sympathetic nervous system is to increase the availability of glucose for rapid metabolism. This function of epinephrine occurs markedly both in liver cells and in muscle, thereby contributing, along with other effects of sympathetic stimulation, to preparation of the body for action, as was discussed in Chapter 57.

Glucagon is a hormone secreted by the *alpha cells* of the pancreas when the blood glucose concentration falls disastrously low. It stimulates the formation of cyclic adenylate mainly in the liver rather than elsewhere in the body. Therefore, its effect is primarily to dump glucose out of the liver into the blood, thereby elevating blood glucose concentration. The function of glucagon in blood glucose regulation is discussed in Chapter 78.

**Transport of Glucose Out of Liver Cells.** The cells of the liver, of the kidney tubules, and of the intestinal mucosa contain *phosphatases* that can split phosphate away from glucose 6-phosphate and therefore make the glucose available for retransport out of the cells into the interstitial fluids. Therefore, in these cells glucose fails to be "captured" by the usual phosphorylation mechanism of the cells but instead can diffuse rather freely in both directions through the cell membrane.

Since glucose can pass out of as well as into liver cells, when it is formed in the liver as a result of glycogenolysis, most of it immediately passes into the blood. Therefore, liver glycogenolysis causes an immediate rise in blood glucose concentration. Glycogenolysis in most other cells of the body, especially in the muscle cells, simply makes increased amounts of glucose 6-phosphate available inside the cells and increases the local rate of glucose utilization but does not release the glucose into the extracellular fluids because the glucose 6-phosphate cannot be dephosphorylated.

# RELEASE OF ENERGY FROM THE GLUCOSE MOLECULE BY THE GLYCOLYTIC PATHWAY

Since complete oxidation of 1 gram-mol of glucose releases 686,000 calories of energy, and only 8000 calories of energy are required to form 1 gram-mol of adenosine triphosphate (ATP), it would be extremely wasteful of energy if glucose should be decomposed all the way into water and carbon dioxide at once while forming only a single ATP molecule. Fortunately, cells contain an extensive series of different protein enzymes that cause the glucose molecule to split a little at a time in many successive steps, with its energy released in small packets to form one molecule of ATP at a time, forming a total of 38 mols of ATP for each mol of glucose utilized by the cells.

The purpose of the present section is to describe the basic principles by which the glucose molecule is progressively dissected and its energy released to form ATP.

## GLYCOLYSIS AND THE FORMATION OF PYRUVIC ACID

By far the most important means by which energy is released from the glucose molecule is by the process of *glycolysis* and then *oxidation of the endproducts of glycolysis*. Glycolysis means splitting of the glucose molecule to form two molecules of pyruvic acid. This occurs by 10 successive steps of chemical reactions, as illustrated in Figure 67–3. Each step is catalyzed by at least one specific protein enzyme. Note that glucose is first converted into fructose 1,6-phosphate and then split into two three-carbon atom molecules, each of which is then converted through five successive steps into pyruvic acid.

**Formation of Adenosine Triphosphate (ATP) During Glycolysis.** Despite the many chemical reactions in the glycolytic series, little energy is released, and much of the energy that is released sim-

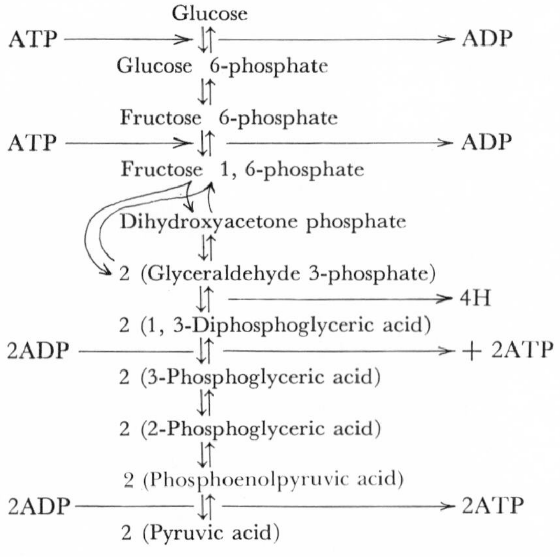

$$Glucose + 2ADP + 2PO_4^{---} \longrightarrow 2\ Pyruvic\ acid + 2ATP + 4H$$

**Figure 67–3.** The sequence of chemical reactions responsible for glycolysis.

ply becomes heat and is lost to the metabolic systems of the cells. However, between the 1,3-diphosphoglyceric acid and the 3-phosphoglyceric acid stages, and again between the phosphopyruvic acid and the pyruvic acid stages, the packets of energy released are greater than 8000 calories per mol, the amount required to form ATP, and the reactions are coupled in such a way that ATP is formed. Thus, a total of 4 mols of ATP is formed for each mol of fructose 1,6-phosphate that is split into pyruvic acid.

Yet two mols of ATP had been required to phosphorylate the original glucose to form fructose 1,6-phosphate before glycolysis could begin. Therefore, the net gain in ATP molecules by the entire glycolytic process is only two mols for each mol of glucose utilized. This amounts to 16,000 calories of energy stored in the form of ATP, but during glycolysis a total of 56,000 calories of energy is lost from the original glucose, giving an overall *efficiency* for ATP formation of 29 per cent. The remaining 71 per cent of the energy is lost in the form of heat.

**Release of Hydrogen Atoms During Glycolysis.** Note in Figure 67–3 that two hydrogen atoms are released from each molecule of glyceraldehyde during its conversion to 1,3-diphosphoglyceric acid. And, since two molecules of glyceraldehyde are formed from each glucose molecule, a total of four hydrogen atoms is released. Later in the chapter we will see that still many more hydrogen atoms are released when pyruvic acid is split into its component parts and that these hydrogen atoms plus the four released during glycolysis are oxidized to provide most of the energy used in the synthesis of ATP.

## CONVERSION OF PYRUVIC ACID TO ACETYL COENZYME A

The next stage in the degradation of glucose is conversion of its two derivative pyruvic acid molecules into two molecules of *acetyl coenzyme A* (acetyl Co-A) in accordance with the following reaction:

$$2\ CH_3\text{—}\overset{\overset{O}{\|}}{C}\text{—COOH} + 2\ Co\text{-}A\text{—SH} \longrightarrow$$
(Pyruvic acid)   (Coenzyme A)

$$2\ CH_3\text{—}\overset{\overset{O}{\|}}{C}\text{—S—Co-A} + 2CO_2 + 4H$$
(Acetyl Co-A)

From this reaction it can be seen that two carbon dioxide molecules and four hydrogen atoms are released, while the remainders of the two pyruvic acid

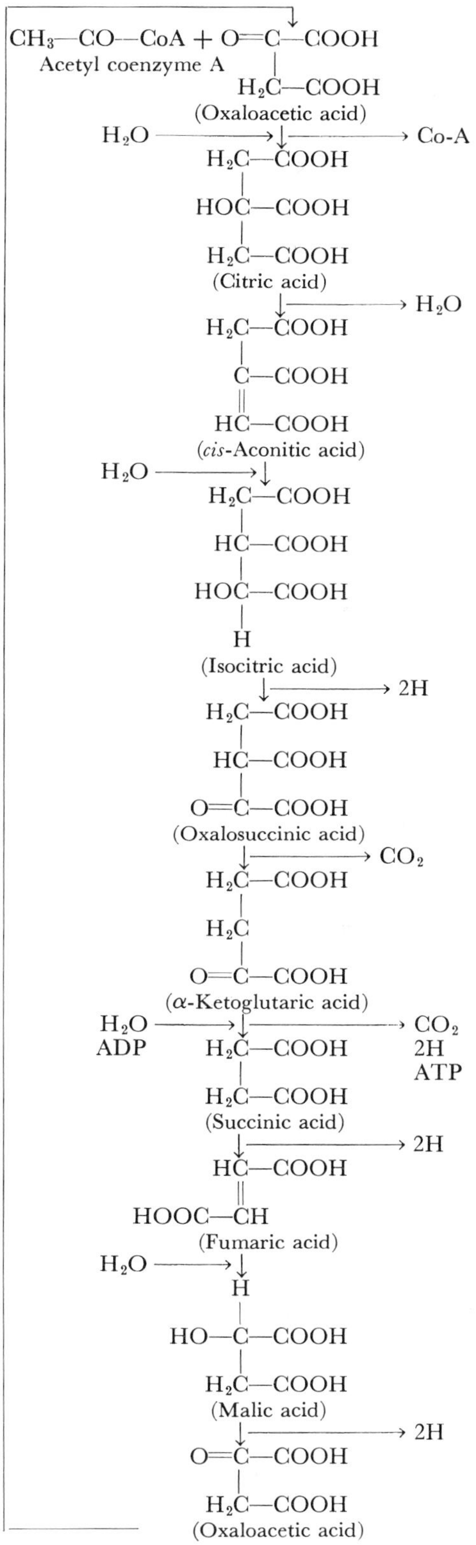

**Figure 67–4.** The chemical reactions of the citric acid cycle, showing the release of carbon dioxide and an especially large number of hydrogen atoms during the cycle.

Net reaction per molecule of glucose:
2 Acetyl Co-A + 6H₂O + 2ADP ⟶
4CO₂ + 16H + 2Co-A + 2ATP

molecules combine with coenzyme A, a derivative of the vitamin pantothenic acid, to form two molecules of acetyl Co-A. In this conversion, no ATP is formed, but six molecules of ATP are formed when the four hydrogen atoms are later oxidized, as will be discussed in a later section.

## THE CITRIC ACID CYCLE

The next stage in the degradation of the glucose molecule is called the *citric acid cycle* (also called the *tricarboxylic acid cycle* or *Krebs cycle*). This is a sequence of chemical reactions in which the acetyl portion of acetyl Co-A is degraded to carbon dioxide and hydrogen atoms. Then the hydrogen atoms are subsequently oxidized, releasing still more energy to form ATP.

The enzymes responsible for the citric acid cycle and also for the subsequent oxidation of the hydrogen to form ATP are all contained in the *mitochondria*. It is believed that the enzymes are attached to the surfaces of the intramitochondrial "shelves" and that they are arranged in a sequential order so that successive products of the reactions can be shuttled from enzyme to enzyme. This obviously would greatly enhance the speed of the chemical reactions and would promote increased rate of ATP formation. Because of the special function of the mitochondria in providing ATP to the remainder of the cell, they are frequently called the "powerhouses" of the cell.

Figure 67–4 shows the different stages of the chemical reactions in the citric acid cycle. The substances to the left are added during the chemical reactions, and the products of the chemical reactions are shown to the right. Note at the top of the column that the cycle begins with *oxaloacetic acid*, and then at the bottom of the chain of reactions *oxaloacetic acid* is formed once again. Thus, the cycle can continue over and over.

In the initial stage of the citric acid cycle, *acetyl Co-A* combines with *oxaloacetic acid* to form *citric acid*. The coenzyme A portion of the acetyl Co-A is released and can be used again and again for the formation of still more quantities of acetyl Co-A from pyruvic acid. The acetyl portion, however, becomes an integral part of the citric acid molecule. During the successive stages of the citric acid cycle, several molecules of water are added; and *carbon dioxide* and *hydrogen atoms* are released at various stages in the cycle, as shown on the right in the figure.

The net results of the entire citric acid cycle are shown at the bottom of Figure 67–4, illustrating that for each molecule of glucose originally metabolized, 2 acetyl Co-A molecules enter into the citric acid cycle along with 6 molecules of water. These then are degraded into 4 carbon dioxide molecules, 16 hydrogen atoms, and 2 molecules of coenzyme A.

**Formation of ATP in the Citric Acid Cycle.** Not a great deal of energy is released during the citric acid cycle itself; in only one of the chemical reactions—during the change from $\alpha$-ketoglutaric acid to succinic acid—is a molecule of ATP formed. Thus, for

each molecule of glucose metabolized, two acetyl Co-A molecules pass through the citric acid cycle, each forming a molecule of ATP; or, a total of two molecules of ATP is formed.

**Function of the Dehydrogenases and Nicotinamide Adenine Dinucleotide (NAD⁺) To Cause Release of Hydrogen Atoms.** As noted at several points in this discussion, hydrogen atoms are released during different chemical reactions—4 hydrogen atoms during glycolysis, 4 during the formation of acetyl Co-A from pyruvic acid, and 16 in the citric acid cycle; this makes a total of 24 hydrogen atoms. However, the hydrogen atoms are not simply turned loose in the intracellular fluid. Instead, they are released in packets of two, and in each instance the release is catalyzed by a specific protein enzyme called a *dehydrogenase*. Twenty of the twenty-four hydrogen atoms immediately combine with nicotinamide adenine dinucleotide (NAD⁺), a derivative of the vitamin niacin in accordance with the following reaction:

$$\text{Substrate} \underset{H}{\overset{H}{\diagdown}} + \text{NAD}^+ \xrightarrow{\quad \textit{dehydrogenase} \quad}$$

$$\text{NADH} + \text{H}^+ + \text{Substrate}$$

This reaction will not occur without the initial intermediation of the dehydrogenase nor without the availability of NAD⁺ to act as a hydrogen carrier. Both the free hydrogen ion and the hydrogen bound with NAD⁺ subsequently enter into the oxidative chemical reactions that form tremendous quantities of ATP, as will be discussed later.

The remaining four hydrogen atoms released during the breakdown of glucose—the four released during the citric acid cycle between the succinic and fumaric acid stages—combine with a specific dehydrogenase but are not then subsequently released to NAD⁺. Instead, they pass directly from the dehydrogenase into the oxidative processes.

**Function of Decarboxylases To Cause Release of Carbon Dioxide.** Referring again to the chemical reactions of the citric acid cycle and also to those for formation of acetyl Co-A from pyruvic acid, we find that there are three stages in which carbon dioxide is released. To cause the carbon dioxide release, other specific protein enzymes, called *decarboxylases*, split the carbon dioxide away from the substrate. The carbon dioxide in turn becomes dissolved in the fluids of the cells and is transported to the lungs where it is expired from the body (see Chap. 41).

## OXIDATION OF HYDROGEN ATOMS BY THE OXIDATIVE ENZYMES

Despite all the complexities of (a) glycolysis, (b) the citric acid cycle, (c) dehydrogenation, and (d) decarboxylation, pitifully small amounts of ATP are formed during all these processes. Instead, about 90 per cent of the final ATP is formed during subsequent oxidation of the hydrogen atoms that are released during these earlier stages of glucose degradation. Indeed, the principal function of all these earlier stages is to make the hydrogen of the glucose molecule available in a form that can be utilized for oxidation.

Oxidation of hydrogen is accomplished by a series of enzymatically catalyzed reactions that (a) change the hydrogen atoms into hydrogen ions and (b) change the dissolved oxygen of the fluids into hydroxyl ions. Then these two products combine with each other to form water. During the sequence of oxidative reactions, tremendous quantities of energy are released to form ATP. The oxidative enzymes, like the dehydrogenases and decarboxylases of the citric acid cycle, are believed to be arranged in an orderly fashion on the inner surfaces of the mitochondria, thus allowing rapid procession of the chemical reactions.

**Ionization of Hydrogen and Oxygen by the Oxidative Enzymes.** Figure 67–5 illustrates a schema of the reactions that cause ionization of both hydrogen and oxygen, with subsequent formation of water. In this schema, the two hydrogens removed

from each substrate during the degradation of glucose are shuttled from compound to compound until finally they are released into the body fluids at the cytochrome b stage in the form of two free hydrogen ions. To change a hydrogen atom into a hydrogen ion, an electron must be removed from it. This is performed by cytochrome b, which passes the electron to cytochrome c, and cytochrome c in turn passes the electron to cytochrome a, which is also known as cytochrome oxidase. At this point, two electrons are transferred from two cytochrome a molecules to one-half molecule of dissolved oxygen plus a molecule of water to form two hydroxyl ions. Thus, hydrogen ions are formed by removing electrons from hydrogen atoms, and hydroxyl ions are formed by transferring these electrons to the dissolved oxygen in the water. Once both of these ions are present in the same fluid medium, they immediately react to form water, as shown in Figure 67–5.

**Formation of ATP During Oxidation of Hydrogen—"Oxidative Phosphorylation."** The net reaction for the oxidation of each two hydrogen atoms is shown at the bottom of Figure 67–5. This illustrates that three molecules of ATP are formed

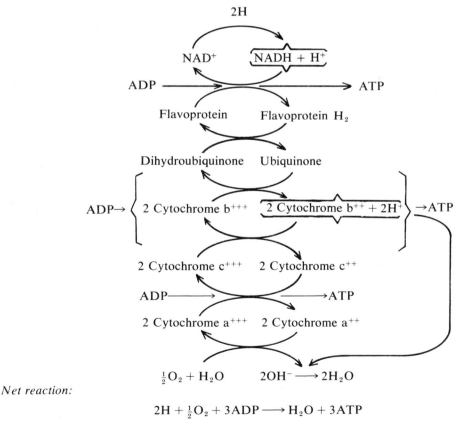

*Net reaction:*

$$2H + \tfrac{1}{2}O_2 + 3ADP \longrightarrow H_2O + 3ATP$$

**Figure 67–5.** The chemical processes of oxidative phosphorylation, showing ionization of hydrogen and oxygen and formation of water and adenosine triphosphate.

during oxidation: one between the NAD$^+$ and flavo-protein stages, one between the ubiquinone and cytochrome c stages, and the third between the cytochrome c and cytochrome a stages. This formation of ATP during the oxidation of hydrogen is called *oxidative phosphorylation*. The precise chemical means by which the oxidation of hydrogen is coupled with the conversion of adenosine diphosphate (ADP) into ATP is almost completely unknown, representing one of the most important voids in our understanding of the body's metabolic processes.

## SUMMARY OF ATP FORMATION DURING THE BREAKDOWN OF GLUCOSE

We can now determine the total number of ATP molecules formed by the energy from one molecule of glucose.

First, during glycolysis, four molecules of ATP are formed while two are expended to cause the initial phosphorylation of glucose to start the process going. This gives a net gain of *two molecules of ATP*.

Second, during each revolution of the citric acid cycle, one molecule of ATP is formed. However, because each glucose molecule splits into two pyruvic acid molecules, there are two revolutions of the cycle for each molecule of glucose metabolized, giving a net production of *two molecules of ATP*.

Third, during the entire schema of glucose breakdown, a total of 24 hydrogen atoms is released during glycolysis and during the citric acid cycle. Twenty of these atoms are oxidized by the schema of oxidative reactions shown in Figure 67–5, with the release of three ATP molecules per two atoms of hydrogen metabolized. This gives a production of *30 ATP molecules*.

Fourth, the remaining four hydrogen atoms are released by their dehydrogenase into the oxidative schema beyond the first stage of Figure 67–5, so that for these four hydrogen atoms only two ATP molecules are usually released for each two hydrogen atoms oxidized, giving a total of *four ATP molecules*.

Now, adding all the ATP molecules formed, we find a minimum of *38 ATP molecules* formed for each molecule of glucose degraded to carbon dioxide and water. Thus, 304,000 calories of energy are stored in the form of ATP, while 686,000 calories are released during the complete oxidation of each gram-mol of glucose. This represents an overall *efficiency* of energy transfer of at least 44 per cent. The remaining 56 per cent of the energy becomes heat and therefore cannot be used by the cells to perform specific functions.

## CONTROL OF GLYCOLYSIS AND OXIDATION BY ADENOSINE DIPHOSPHATE (ADP)

Continual release of energy from glucose when the energy is not needed by the cells would be an ex-tremely wasteful process. Fortunately, glycolysis and the subsequent oxidation of hydrogen atoms is continually controlled in accordance with the needs of the cells for ATP. This control is accomplished simply in the following manner:

Referring back to the various chemical reactions of the glycolytic series, the citric acid cycle, and the oxidation of hydrogen, we see that at different stages *ADP is converted into ATP. If ADP is not available at each of these stages, the reactions cannot occur, thus stopping the degradation of the glucose molecule*. Therefore, once all the ADP in the cells has been converted to ATP, the entire glycolytic and oxidative process stops. Then, when more ATP is used to perform different physiological functions in the cell, new ADP is formed, which automatically turns on glycolysis and oxidation once more. In this way, essentially a full store of ATP is automatically maintained all the time, except when the activity of the cell becomes so great that ATP is used up more rapidly than it can be formed.

## TRANSFER OF MITOCHONDRIAL ATP ENERGY TO CYTOPLASMIC ATP

Eighty per cent of the ATP in the cell is located in the mitochondria; only 20 per cent is in the cytoplasm. Furthermore, it is in the mitochondria that most of the ATP is formed. On the other hand, most of the metabolic functions of ATP are performed in the cytoplasm. Therefore, it is important for the energy of the mitochondrial ATP to be transferred through the mitochondrial membrane to the ATP in the cytoplasm. Although some physiologists believe that the ATP itself is transported through the membrane, there is much reason to believe that the energy is transferred instead by a coupled reaction at the mitochondrial membrane between the mitochondrial ATP-ADP system and the cytoplasmic ATP-ADP system. That is, whenever excess quantities of ADP appear in the cytoplasm, the ATP inside the mitochondria degrades to ADP and the energy (as well as phosphate ions released from the ATP) is passed through the mitochondrial membrane to the cytoplasmic ADP to form cytoplasmic ATP. Then very soon after, mitochondrial ATP is reconstituted by the oxidative mechanisms.

## GLUCOSE CONVERSION TO GLYCOGEN OR FAT

When glucose is not immediately required for energy, the extra glucose that continually enters the cells either is stored as glycogen or is converted into fat and stored in this form, as discussed in the following chapter. Glucose is preferentially stored as glycogen until the cells have stored as much glycogen as they can—an amount sufficient to supply the energy needs of the body for only a few hours. When the cells (primarily liver and muscle cells) approach saturation with glycogen, the additional glucose is converted into fat in the liver and in the fat cells and then is stored in the fat cells. The chemistry of this conversion is discussed in the following chapter.

## ANAEROBIC RELEASE OF ENERGY—"ANAEROBIC GLYCOLYSIS"

Occasionally, oxygen becomes either unavailable or insufficient so that cellular oxidation of glucose cannot take place. Yet, even under these conditions, a small amount of energy can still be released to the cells by glycolysis, for the chemical reactions in the glycolytic breakdown of glucose to pyruvic acid do not require oxygen. Unfortunately, this process is extremely wasteful of glucose because only 16,000 calories of energy are used to form ATP for each molecule of glucose utilized, which represents only a little over 2 per cent of the total energy in the glucose molecule. Nevertheless, this release of glycolytic energy to the cells can be a lifesaving measure for a few minutes when oxygen becomes unavailable.

**Formation of Lactic Acid During Anaerobic Glycolysis.**  The *law of mass action* states that as the end-products of a chemical reaction build up in the reacting medium, the rate of the reaction approaches zero. The two end-products of the glycolytic reactions (see Fig. 67–3) are (1) pyruvic acid and (2) hydrogen atoms, which are combined with $NAD^+$ to form NADH and $H^+$. The build-up of either or both of these would stop the glycolytic process and prevent further formation of ATP. Fortunately, when their quantities begin to be excessive these two end-products react with each other to form lactic acid in accordance with the following equation:

$$CH_3-\overset{\overset{\displaystyle O}{\|}}{C}-COOH + NADH + H^+ \xrightleftharpoons[\text{}]{\text{lactic}\atop\text{dehydrogenase}}$$
(Pyruvic acid)

$$CH_3-\overset{\overset{\displaystyle OH}{|}}{\underset{\underset{\displaystyle H}{|}}{C}}-COOH + NAD^+$$
(Lactic acid)

Thus, under anaerobic conditions, by far the major proportion of the pyruvic acid is converted into lactic acid, which diffuses readily out of the cells into the extracellular fluids and even into the intracellular fluids of other less active cells. Therefore, lactic acid represents a type of "sinkhole" into which the glycolytic end-products can disappear, thus allowing glycolysis to proceed far longer than would be possible if the pyruvic acid and hydrogen were not removed from the reacting medium. Indeed, glycolysis could proceed for only a few seconds without this conversion. Instead, it can proceed for several minutes, supplying the body with considerable quantities of ATP even in the absence of respiratory oxygen.

**Reconversion of Lactic Acid and Pyruvic Acid to Glucose in the Presence of Oxygen.**  The entire schema of glycolysis, all the way from glucose to lactic acid, is theoretically reversible, but in practice the reverse change of pyruvic acid into phosphoenolpyruvic acid is extremely slow. However, other enzymes cause the pyruvic acid to be converted first into oxaloacetic acid and then into phosphoenolpyruvic acid, after which the scheme of reconversion can proceed. Unfortunately, however, this reverse process consumes six ATP molecules in comparison with the two ATP molecules formed during glycolysis.

When a person begins to breathe oxygen again after a period of anaerobic metabolism, the extra NADH and $H^+$ as well as the extra pyruvic acid that have built up in the body fluids are rapidly oxidized, thereby greatly reducing their concentrations. As a result, the chemical reaction for formation of lactic acid immediately reverses itself, the lactic acid once again becoming pyruvic acid. Large portions of this are immediately utilized by the citric acid cycle to provide additional oxidative energy, and large quantities of ATP are formed. This excess ATP then causes as much as three-fourths of the remaining excess pyruvic acid to be converted back into glucose.

Thus, the great amount of lactic acid that forms during anaerobic glycolysis does not become lost from the body, for when oxygen is again available, the lactic acid either can be reconverted to glucose or can be used directly for energy. By far the greater proportion of this reconversion occurs in the liver, but a small amount can also occur in some other tissues.

*Use of Lactic Acid by the Heart for Energy.* Heart muscle is especially capable of converting lactic acid to pyruvic acid and then utilizing this for energy. This occurs especially in very heavy exercise, during which large amounts of lactic acid are released into the blood from the skeletal muscles.

## RELEASE OF ENERGY FROM GLUCOSE BY THE PHOSPHOGLUCONATE PATHWAY

Though essentially all the carbohydrates utilized by the muscles are degraded to pyruvic acid by glycolysis and then oxidized, the glycolytic schema is not the only means by which glucose can be degraded and then oxidized to provide energy. A second important schema for breakdown and oxidation of glucose is called the *phosphogluconate pathway,* which is responsible for as much as 30 per cent of the glucose breakdown in the liver and even more than this in fat cells. It is especially important because it can provide energy independently of all the enzymes of the tricarboxylic acid cycle and therefore is an alternate pathway for energy metabolism in case of some enzymatic abnormality of the cells.

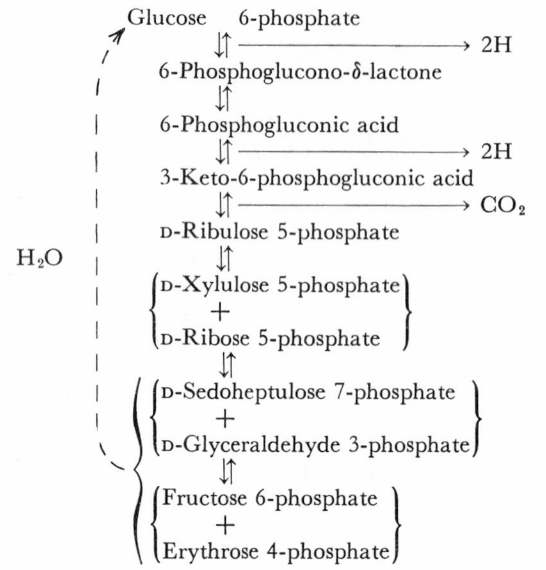

Glucose     6-phosphate
$\Updownarrow$ ————————→ 2H
6-Phosphoglucono-$\delta$-lactone
$\Updownarrow$
6-Phosphogluconic acid
$\Updownarrow$ ————————→ 2H
3-Keto-6-phosphogluconic acid
$\Updownarrow$ ————————→ $CO_2$
D-Ribulose 5-phosphate
$\Updownarrow$

$H_2O$

$\begin{Bmatrix} \text{D-Xylulose 5-phosphate} \\ + \\ \text{D-Ribose 5-phosphate} \end{Bmatrix}$
$\Updownarrow$

$\begin{Bmatrix} \text{D-Sedoheptulose 7-phosphate} \\ + \\ \text{D-Glyceraldehyde 3-phosphate} \end{Bmatrix}$
$\Updownarrow$

$\begin{Bmatrix} \text{Fructose 6-phosphate} \\ + \\ \text{Erythrose 4-phosphate} \end{Bmatrix}$

*Net reaction:*

Glucose + 12NADP$^+$ + 6$H_2O$ ⟶ 6$CO_2$ + 12H + 12NADPH

**Figure 67–6.** The phosphogluconate pathway for glucose metabolism.

**Release of Carbon Dioxide and Hydrogen by Means of the Phosphogluconate Pathway.** Figure 67–6 illustrates most of the basic chemical reactions in the phosphogluconate pathway. This shows that glucose, after several stages of conversion, releases one molecule of carbon dioxide and four atoms of hydrogen, with resultant formation of a five carbon sugar, d-ribulose. This substance in turn can change progressively into several other five-, four-, seven-, and three-carbon sugars. Finally, various combinations of these sugars can resynthesize glucose. However, *only five molecules of glucose are resynthesized for every six molecules of glucose that initially enter into the reactions.* That is, the phosphogluconate pathway is a cyclic process in which one molecule of glucose is metabolized for each revolution of the cycle. As illustrated at the bottom of Figure 67–6, the net reaction is conversion of the single molecule of glucose plus six molecules of water into six carbon dioxide molecules and 24 hydrogen atoms. Thus, by revolution of the cycle again and again, all the glucose can eventually be converted into carbon dioxide and hydrogen, and the hydrogen in turn can then be utilized for energy.

**Use of Hydrogen; The Function of Nicotinamide Adenine Dinucleotide Phosphate (NADP$^+$).** The hydrogen released during the phosphogluconate cycle does not combine with NAD$^+$ as in the glycolytic pathway, but combines with NADP$^+$, which is almost identical with NAD$^+$ except for an extra phosphate radical. This difference is extremely important because only hydrogen bound with NADP$^+$ in the form of NADPH can be used for synthesis of

fats from carbohydrates, which is discussed in the following chapter. When glycolysis becomes slowed because of cellular inactivity, the phosphogluconate pathway still remains operative (mainly in the liver) to break down any excess glucose that continues to be transported into the cells, and NADPH becomes abundant to help convert acetyl radicals into long fatty acid chains.

# FORMATION OF CARBOHYDRATES FROM PROTEINS AND FATS— "GLUCONEOGENESIS"

When the body's stores of carbohydrates decrease below normal, moderate quantities of glucose can be formed from *amino acids* and from the *glycerol* portion of fat. This process is called *gluconeogenesis.* Approximately 60 per cent of the amino acids in the body proteins can be converted into carbohydrates, while the remaining 40 per cent have chemical configurations that make this difficult. Each amino acid is converted into glucose by a slightly different chemical process. For instance, alanine can be converted directly into pyruvic acid simply by deamination; the pyruvic acid then is converted into glucose, as explained previously. Several of the more complicated amino acids can be converted into different sugars containing three-, four-, five-, or seven-carbon atoms; these can then enter the phosphogluconate pathway and eventually form glucose. Thus, by means of deamination plus several simple interconversions, many of the amino acids become glucose. Similar interconversions can change glycerol into carbohydrates.

**Regulation of Gluconeogenesis.** Diminished carbohydrates in the cells and decreased blood sugar are the basic stimuli that set off an increase in the rate of gluconeogenesis. The diminished carbohydrates can directly cause reversal of many of the glycolytic and phosphogluconate reactions, thus allowing conversion of deaminated amino acids and glycerol into carbohydrates. However, in addition, several of the hormones secreted by the endocrine glands are especially important in this regulation.

*Effect of Corticotropin and Glucocorticoids on Gluconeogenesis.* When normal quantities of carbohydrates are not available to the cells, the adenohypophysis, for reasons not completely understood at present, begins to secrete increased quantities of corticotropin, which stimulate the adrenal cortex to produce large quantities of *glucocorticoid hormones,* especially cortisol. In turn, cortisol mobilizes proteins from essentially all cells of the body, making these available in the form of amino acids in the body fluids. A high proportion of these immediately becomes deaminated in the liver and therefore provides ideal substrates for conversion into glucose. Thus, one of the most important means by which gluconeogenesis is promoted is through the release of glucocorticoids from the adrenal cortex.

*Effect of Thyroxine on Gluconeogenesis.* Thyroxine, secreted by the thyroid gland, also increases the rate of gluconeogenesis. This, too, is believed to result principally from mobilization of proteins from the cells. However, it might result to some extent also from the mobilization of fats from the fat depots, the glycerol portion of the fats being converted into glucose.

## BLOOD GLUCOSE

Except immediately after a meal, glucose is the only monosaccharide present in significant quantities in the blood and interstitial fluids. There are two reasons for this: (1) Usually 80 to 100 per cent of the monosaccharides absorbed from the gastrointestinal tract is glucose, and only rarely is more than 20 per cent of these fructose and galactose together. (2) Almost immediately both the fructose and the galactose are converted into glucose—the fructose while it is being absorbed through the intestinal epithelial cells and the galactose by the liver.

The normal blood glucose concentration in a person who has not eaten a meal within the past three to four hours is approximately 90 mg. per cent, and, even after a meal containing large amounts of carbohydrates, this rarely rises above 140 mg. per cent unless the person has diabetes mellitus, which will be discussed in Chapter 78.

The regulation of blood glucose concentration is so intimately related to insulin and glucagon that this subject will be discussed in detail in Chapter 78 in relation to the functions of insulin and glucagon.

## REFERENCES

Altszuler, N.: Actions of growth hormone on carbohydrate metabolism. *In* Greep, R. O., and Astwood, E. B. (eds.): Handbook of Physiology. Sec. 7, Vol. 4. Baltimore, The Williams & Wilkins Company, 1974, p. 233.

Ashwell, G., and Morell, A. G.: The role of surface carbohydrates in the hepatic recognition and transport of circulating glycoproteins. *Adv. Enzymol.*, 41:99, 1974.

Baltscheffsky, H., and Baltscheffsky, M.: Electron transport phosphorylation. *Ann. Rev. Biochem.*, 43:871, 1974.

Bender, A. E., and Damji, K. B.: Some effects of dietary sucrose. *World Rev. Nutr. Diet.*, 15:104, 1972.

Bondy, P. K., and Rosenberg, L. E.: Duncan's Diseases of Metabolism. Vols. 1 and 2. Philadelphia, W. B. Saunders Company, 1975.

Ciba Symposium on Control of Glycogen Metabolism. Boston, Little, Brown and Company, 1964.

Crane, F. L., and Low, H.: Quinones in energy-coupling systems. *Physiol. Rev.*, 46:662, 1966.

Der Vartanian, D. V., and LeGall, J.: A monomolecular electron transfer chain: structure and function of cytochrome $C_3$. *Biochim. Biophys. Acta.*, 346:79, 1974.

Dixon, M., and Webb, E. C.: Enzymes. New York, Academic Press, Inc., 1964.

Felig, P., and Wahren, J.: Protein turnover and amino acid metabolism in the regulation of gluconeogenesis. *Fed. Proc.*, 33:1092, 1974.

Hassinen, I. E., Ylikahri, R. H., and Kähönen, M. T.: Effect of fructose on cellular respiration in perfused rat liver. *Acta Med. Scand.*, 542 (Suppl.):105, 1972.

Henry, M. F., and Nyns, E. D.: Cyanide-insensitive respiration. An alternative mitochondrial pathway. *Subcell. Biochem.* 4:1, 1975.

Herman, R. H., Stifel, F. B., Greene, H. L., and Herman, Y. F.: Intestinal metabolism of fructose. *Acta Med. Scand.*, 542(Supple.):19, 1972.

Hess, B.: Organization of glycolysis: oscillatory and stationary control. *Symp. Soc. Exp. Biol.*, 27:105, 1973.

Krebs, H. A.: The tricarboxylic acid cycle. *Harvey Lectures*, 44:165, 1948–1949.

Lemberg, M. R.: Cytochrome oxidase. *Physiol. Rev.*, 49:48, 1969.

Levine, R., Goldstein, M. S., Huddleston, B., and Klein, S. P.: Action of insulin on "permeability" of cells to free hexoses, as studied by its effect on distribution of galactose. *Amer. J. Physiol.*, 163:79, 1950.

McGilvery, R. W.: Biochemical Concepts. Philadelphia, W. B. Saunders Company, 1975.

Mommaerts, W. F.: Energetics of muscular contraction. *Physiol. Rev.*, 49:427, 1969.

Mosbach, K.: Enzymes bound to artificial matrices. *Sci. Amer.*, 224:26, 1971.

Pyke, D. (ed.): Disorders of Carbohydrate Metabolism. Philadelphia, J. B. Lippincott Company, 1963.

Racker, E. (ed.): Energy Transducing Mechanisms. Baltimore, University Park Press, 1975.

Söling, H. D.: Interrelationship between fatty acid metabolism and hepatic gluconeogenesis. *Horm. Metab. Res.*, (Suppl. 4):56, 1974.

Toporek, M., and Maurer, P. H.: Metabolic biochemistry. *In* Sodeman, W. A., Jr., and Sodeman, W. A. (eds.): Pathologic Physiology: Mechanisms of Disease, 5th Ed. Philadelphia, W. B. Saunders Company, 1974, p. 3.

Wikström, M. K.: The different cytochrome b components in the respiratory chain of animal mitochondria and their role in electron transport and energy conservation. *Biochim. Biophys. Acta*, 301:155, 1973.

Wilson, D. F., Erecínska, M., and Dutton, P. L.: Thermodynamic relationships in mitochondrial oxidative phosphorylation. *Ann. Rev. Biophys. Bioeng.*, 3:203, 1974.

# 68

# Lipid Metabolism

A number of different chemical compounds in the food and in the body are classified as *lipids*. These include (1) *neutral fat*, known also as *triglycerides*, (2) the *phospholipids*, (3) *cholesterol*, and (4) a few others of less importance. These substances have certain similar physical and chemical properties, especially the fact that they are miscible with each other. Chemically, the basic lipid moiety of both the triglycerides and the phospholipids is *fatty acids*, which are simply long chain hydrocarbon organic acids. A typical fatty acid, palmitic acid, is the following:

$$CH_3(CH_2)_{14}COOH$$

Though cholesterol does not contain fatty acid, its sterol nucleus, as pointed out later in the chapter, is synthesized from degradation products of fatty acid molecules, thus giving it many of the physical and chemical properties of other lipid substances.

The triglycerides are used in the body mainly to provide energy for the different metabolic processes; this function they share almost equally with the carbohydrates. However, some lipids, especially cholesterol, the phospholipids, and derivatives of these, are used throughout the body to provide other intracellular functions.

**Basic Chemical Structure of Triglycerides (Neutral Fat).**   Since most of this chapter deals with utilization of triglycerides for energy, the following basic structure of the triglyceride molecule must be understood:

$$CH_3—(CH_2)_{16}—COO—CH_2$$
$$CH_3—(CH_2)_{16}—COO—CH$$
$$CH_3—(CH_2)_{16}—COO—CH_2$$

Tri-stearin

Note that three long chain fatty acid molecules are bound with one molecule of glycerol. In the human body, the three fatty acids most commonly present in neutral fat are (1) *stearic acid,* which has an 18-carbon chain and is fully saturated with hydrogen atoms, (2) *oleic acid,* which also has an 18-carbon chain but has one double bond in the middle of the chain, and (3) *palmitic acid,* which has 16-carbon atoms and is fully saturated. These or closely similar fatty acids are also the major constituents of the fats in the food.

## TRANSPORT OF LIPIDS IN THE BLOOD

### TRANSPORT FROM THE GASTROINTESTINAL TRACT— THE "CHYLOMICRONS"

It will be recalled from Chapter 65 that essentially all the fats of the diet are absorbed into the lymph (with the exception of the short chain fatty acids that can be absorbed directly into the portal blood but that normally represent only a small fraction of the fat in the diet). In the digestive tract, most triglycerides are split into glycerol and fatty acids or into monoglycerides and fatty acids. Then, on passing through the intestinal epithelial cells, these are resynthesized into new molecules of triglycerides which aggregate and enter the lymph as minute, dispersed droplets called *chylomicrons,* having a size about 0.5 micron. A small amount of protein adsorbs to the outer surfaces of the chylomicrons; this increases their suspension stability in the fluid of the lymph and prevents their adhering to the lymphatic vessel walls.

Most of the cholesterol and phospholipids absorbed from the gastrointestinal tract, as well as small amounts of phospholipids that are continually synthesized by the intestinal mucosa, also enter the chylomicrons. Thus, chylomicrons are composed principally of triglycerides, but they also contain approximately 7 per cent phospholipids, 7 per cent cholesterol, and 1 per cent protein as well. The chylomicrons are then transported up the thoracic duct and emptied into the venous blood at the juncture of the jugular and subclavian veins.

**Removal of the Chylomicrons from the Blood.** Immediately after a meal that contains large quantities of fat, the chylomicron concentration in the plasma may rise to as high as 1 to 2 per cent, and, because of the large sizes of the chylomicrons, the plasma appears turbid and sometimes yellow. However, the chylomicrons (with a half-life of only 10 to 15 minutes) are rapidly removed within an hour or so, and the plasma becomes clear once again. The fat of the chylomicrons is removed mainly in the following way:

*Hydrolysis of the Chylomicron Triglycerides by Lipoprotein Lipase; Fat Storage in the Fat Cells.* Most of the chylomicrons are removed from the circulating blood as they pass through the capillaries of adipose tissue. The membranes of the fat cells contain large quantities of an enzyme called *lipoprotein lipase*. This enzyme hydrolyzes the triglycerides into fatty acids and glycerol. The fatty acids, being highly miscible with the membranes of the cells, immediately diffuse into the fat cells. Once within these cells, the fatty acids are resynthesized into triglycerides, new glycerol being supplied by the metabolic processes of the fat cells, as will be discussed later in the chapter. The lipase also causes hydrolysis of phospholipids, this too releasing fatty acids to be stored in the fat cells in the same way. Thus, essentially all of the mass of the chylomicrons is removed from the circulating blood, and the fat absorbed from the gastrointestinal tract only a few minutes previously becomes rapidly stored in the fat cells where it waits to be used for other purposes at a later time.

*Role of Insulin in the Fat Storage Process.* Following a meal the insulin level in the blood usually increases considerably, primarily because of an increase in the blood glucose derived from the carbohydrates of the meal. (This effect will be discussed in much greater detail in Chapter 78 in relation to insulin and its functions.) The insulin has two very important effects to help promote fat storage. First, it helps to activate the lipoprotein lipase, thereby greatly accelerating the rate of hydrolysis of the chylomicron triglycerides and thus promoting their storage. It also promotes entry of glucose into the fat cells at the same time, and the fat cells convert a small amount of this into triglycerides that are stored along with the fat from the diet. Second, and even more important, it supplies a breakdown product of glucose, *α-glycerophosphate,* which provides the glycerol portion of the newly forming triglycerides; without this, no fat can be stored.

# TRANSPORT OF FATTY ACIDS IN COMBINATION WITH ALBUMIN—"FREE FATTY ACID"

When the fat that has been stored in the fat cells is to be used elsewhere in the body, usually for provid-

ing energy, it must first be transported to the other tissues. It is transported almost entirely in the form of *free fatty acid*. This is achieved by hydrolysis of the triglycerides once again into fatty acids and glycerol. Although the stimulus for initiating this hydrolysis is not completely understood, it is known that at least two factors play a role. First, decreased availability of glucose to the cell, as occurs in the period between meals, decreases the quantity of α-glycerophosphate, and this decrease removes the stimulus for new synthesis of triglycerides, thus allowing the equilibrium to shift in favor of hydrolysis. Second, a cellular lipase is activated at the same time, and this promotes rapid hydrolysis of the triglyceride.

On leaving the fat cells, the fatty acids ionize strongly in the plasma and immediately combine with albumin of the plasma proteins. The fatty acid bound with proteins in this manner is called *free fatty acid* or *nonesterified fatty acid* (or simply *FFA* or *NEFA*) to distinguish it from other fatty acids in the plasma which exist in the form of esters of glycerol, cholesterol, or other substances.

The concentration of free fatty acid in the plasma under resting conditions is about 15 mg. per 100 ml. of blood, which is a total of only 0.75 gram of fatty acids in the entire circulatory system. Yet, strangely enough, even this small amount accounts for almost all of the transport of lipids from one part of the body to another for the following reasons:

(1) Despite the minute amount of free fatty acid in the blood, its rate of "turnover" is extremely rapid, *half the plasma fatty acid being replaced by new fatty acid every two to three minutes.* One can calculate that at this rate over half of all the energy required by the body can be provided by the free fatty acid transported even without increasing the free fatty acid concentration. (2) All conditions that increase the rate of utilization of fat for cellular energy also increase the free fatty acid concentration in the blood; this sometimes increases as much as 10-fold. Especially does this occur in starvation and in diabetes when a person is not or cannot be using carbohydrates for energy.

Under normal conditions about three molecules of fatty acid combine with each molecule of albumin, but as many as 30 fatty acid molecules can combine with a single molecule of albumin when the need for fatty acid transport is extreme. This shows how variable the rate of lipid transport can be under different physiological needs.

## THE LIPOPROTEINS

In the postabsorptive state—that is, when no chylomicrons are in the blood—over 95 per cent of all the lipids in the plasma (in terms of mass, but *not* in terms of rate of transport) are in the form of lipoproteins, which are small particles much smaller than chylomicrons but similar in composition, containing mixtures of *triglycerides, phospholipids, choles-*

*terol,* and *protein.* The protein in the mixture averages about one fourth to one third of the total constituents, and the remainder is lipids. The total concentration of lipoproteins in the plasma averages about 700 mg. per 100 ml., and this can be broken down into the following average concentrations of the individual constituents:

|  | *mg./100 ml. of plasma* |
|---|---|
| Cholesterol | 180 |
| Phospholipids | 160 |
| Triglycerides | 160 |
| Lipoprotein protein | 200 |

**Types of Lipoproteins.**    Chylomicrons are usually classified as lipoproteins because they contain both lipids and protein. In addition to the chylomicrons, however, there are three other major classes of lipoprotein: (1) *very low density lipoproteins,* which contain high concentrations of triglycerides and moderate concentrations of both phospholipids and cholesterol; (2) *low density lipoproteins,* which contain relatively few triglycerides but a very high percentage of cholesterol; and (3) *high density lipoproteins,* which contain about 50 per cent protein with smaller concentrations of the lipids.

**Formation of the Lipoproteins.**    The lipoproteins are formed almost entirely in the liver, which is in keeping with the fact that most plasma phospholipids, cholesterol, and triglycerides (except those in the chylomicrons) are synthesized in the liver.

**Function of the Lipoproteins.**    The function of the lipoproteins in the plasma is poorly known, though they are known to be a means by which lipid substances can be transported throughout the body, mainly from the liver to other parts of the body. For instance, the turnover of triglycerides in the lipoproteins is as much as 1.5 grams per hour, which could account for as much as 10 to 20 per cent of the total lipids utilized by the body for energy under resting conditions. It has especially been suggested that fats synthesized from carbohydrates in the liver are transported to the adipose tissue in lipoproteins because almost no free fatty acids are present in the blood when fat synthesis is occurring, while plenty of lipoproteins are present. Even more important is the transport of cholesterol and phospholipids by the lipoproteins, because these substances are not known to be transported to any significant extent in any other form.

We shall see later in the chapter that atherosclerosis is often associated with high plasma concentrations of low density lipoprotein.

# THE FAT DEPOSITS

## *ADIPOSE TISSUE*

Large quantities of fat are frequently stored in two major tissues of the body, the adipose tissue and the liver. The adipose tissue is usually called the *fat deposits,* or simply the *fat depots.*

The major function of adipose tissue is storage of triglycerides until these are needed to provide energy elsewhere in the body. However, a subsidiary function is to provide heat insulation for the body, as will be discussed in Chapter 72.

**The Fat Cells.**    The fat cells of adipose tissue are modified fibroblasts that are capable of storing almost pure triglycerides in quantities equal to 80 to 95 per cent of their volume. The triglycerides are generally in a liquid form, and when the tissues of the skin are exposed to prolonged cold, the fatty acid chains of the triglycerides, over a period of weeks, become either shorter or more unsaturated to decrease their melting point, thereby always allowing the fat in the fat cells to remain in a liquid state. This is particularly important because only liquid fat can be hydrolyzed and then transported from the cells.

Fat cells can also synthesize fatty acids and triglycerides from carbohydrates, this function supplementing the synthesis of fat in the liver, as discussed later in the chapter.

**Exchange of Fat Between the Adipose Tissue and the Blood—Tissue Lipases.**    Large quantities of lipases are present in adipose tissue. Some of these enzymes catalyze the deposition of triglycerides from the chylomicrons and other lipoproteins. Others, when activated, cause splitting of the triglycerides of the fat cells to release free fatty acids. Because of rapid exchanges of the fatty acids, the triglycerides in the fat cells are renewed approximately once every two to three weeks, which means that the fat stored in the tissues today is not the same fat that was stored last month, thus emphasizing the dynamic state of the storage fat.

# *THE LIVER LIPIDS*

The principal functions of the liver in lipid metabolism are: (1) to degrade fatty acids into small compounds that can be used for energy, (2) to synthesize triglycerides mainly from carbohydrates and, to a lesser extent, from proteins, and (3) to synthesize other lipids from fatty acids, especially cholesterol and phospholipids.

Large quantities of triglycerides appear in the liver (a) during starvation, (b) in diabetes mellitus, or (c) in any other condition in which fat is being utilized rapidly for energy. In these conditions, the triglycerides are mobilized from the adipose tissue, transported as free fatty acids in the blood, and then redeposited as triglycerides in the liver, where the initial stages of much of the fat degradation begin. Thus, under normal physiological conditions the total amount of triglycerides in the liver is controlled to a great extent by the overall rate at which lipids are being utilized for energy.

The liver cells, in addition to containing triglycerides, contain large quantities of phospholipids and cholesterol, which are continually synthesized by the liver. Also, the liver cells are much more capable than other tissues of desaturating fatty acids so that the liver triglycerides normally are much more unsaturated than the triglycerides of the adipose tissue. This capability of the liver to desaturate fatty acids seems to be functionally important to all the tissues of the body, because many of the structural members of all cells contain reasonable quantities of desaturated fats, and their principal source seems to be the liver. This desaturation is accomplished by a dehydrogenase in the liver cells, as discussed later.

# USE OF TRIGLYCERIDES FOR ENERGY, AND FORMATION OF ADENOSINE TRIPHOSPHATE (ATP)

Approximately 40 to 45 per cent of the calories in the normal American diet are derived from fats, which is about equal to the calories derived from carbohydrates. Therefore, the use of fats by the body for energy is equally as important as the use of carbohydrates. In addition, an average of 30 to 50 per cent of the carbohydrates ingested with each meal is converted into triglycerides, then stored, and later utilized as triglycerides for energy. Therefore, as much as two thirds to three quarters of all the energy derived directly by the cells is supplied by triglycerides rather than by carbohydrates. For this reason it is equally as important, if not more important, to understand the principles of triglyceride oxidation as to understand carbohydrate oxidation.

**Hydrolysis of the Triglycerides.** The first stage in the utilization of triglycerides for energy is hydrolysis of these compounds into fatty acids and glycerol and subsequent transport of both products of hydrolysis to the active tissues where they are oxidized to give energy. Almost all cells, with the notable exception of brain tissue, can use fatty acids almost interchangeably with glucose for energy.

The glycerol, on entering the active tissue, is im-mediately changed by intracellular enzymes into glyceraldehyde, which enters the phosphogluconate pathway for glucose breakdown, and in this way is used for energy. However, before the fatty acids can be used for energy, they must be processed further in the following way:

**Entry of Fatty Acids into the Mitochondria.** The degradation and oxidation of fatty acids occur only in the mitochondria. Therefore, the first step for utilization of the fatty acids is their transport into the mitochondria. This is an enzyme-catalyzed process that employs *carnitine* as a carrier substance. Once inside the mitochondria, the fatty acid splits away from the carnitine and is then oxidized.

**Degradation of Fatty Acid to Acetyl Coenzyme-A by "Beta Oxidation."** The fatty acid molecule is degraded in the mitochondria by progressive release of 2-carbon segments in the form of acetyl coenzyme-A (acetyl Co-A). This process, which is illustrated in Figure 68–1, is called the *beta oxidation* process for degradation of fatty acids. The successive stages are the following:

1. The fatty acid molecule first combines with coenzyme A to form a *fatty acyl Co-A* molecule. This step is energized by the breakdown of an ATP molecule to AMP, a loss of two high energy phosphate bonds.

2. Fatty acyl Co-A loses two hydrogen atoms from the alpha and beta carbons, leaving a double bond at this point. The hydrogen atoms that are removed become attached to a flavoprotein (FAD) and later are oxidized, as discussed below.

3. A water molecule reacts at the site of the double bond so that a hydrogen atom from the water attaches to the alpha carbon and the remaining hydroxyl radical attaches to the beta carbon.

4. Two additional hydrogen atoms are removed, one from the beta carbon and one from the hydroxyl radical. The hydrogen atoms removed in this process combine with $NAD^+$ and are also oxidized, as discussed below.

5. The compound splits between the alpha and beta carbons, the long portion of the chain combining

(1) $RCH_2CH_2CH_2COOH + Co\text{-}A + ATP \underset{\xrightarrow{\text{thiokinase}}}{\longleftarrow} RCH_2CH_2CH_2COCo\text{-}A + AMP + Pyrophosphate$
    (Fatty acid)                                                        (Fatty acyl Co-A)

(2) $RCH_2CH_2CH_2COCo\text{-}A + FAD \xrightarrow{\text{acyl dehydrogenase}} RCH_2CH\text{=}CHCOCo\text{-}A + FADH_2$
    (Fatty acyl Co-A)

(3) $RCH_2CH\text{=}CHCOCo\text{-}A + H_2O \underset{\xrightarrow{\text{enoyl hydrase}}}{\longleftarrow} RCH_2CHOHCH_2COCo\text{-}A$

(4) $RCH_2CHOHCH_2COCo\text{-}A + NAD^+ \underset{\xrightarrow[\text{dehydrogenase}]{\text{$\beta$-hydroxyacyl}}}{\longleftarrow} RCH_2COCH_2COCo\text{-}A + NADH + H^+$

(5) $RCH_2COCH_2COCo\text{-}A + Co\text{-}A \underset{\xrightarrow{\text{thiolase}}}{\longleftarrow} RCH_2COCo\text{-}A + CH_3COCo\text{-}A$
    (Fatty acyl Co-A)(Acetyl Co-A)

**Figure 68–1.**  Beta oxidation of fatty acids to yield acetyl coenzyme A.

with a new molecule of coenzyme A, while the shorter acetyl portion remains combined with the original coenzyme A in the form of *acetyl Co-A*.

The new fatty acyl Co-A, which now has two carbon atoms less than the original fatty acyl Co-A, re-enters reaction number 2 in Figure 68–1 and proceeds through the four stages of chemical reactions until another acetyl Co-A molecule is released. This process is repeated again and again until the entire fatty acid molecule is split into acetyl Co-A. For instance, from each molecule of stearic acid, nine molecules of acetyl Co-A are formed.

**Oxidation of Acetyl Co-A.** The acetyl Co-A molecules formed by beta oxidation of fatty acids enter into the citric acid cycle as explained in the preceding chapter, combining first with oxaloacetic acid to form citric acid, which then is degraded into carbon dioxide and hydrogen atoms. The hydrogen is subsequently oxidized by the oxidative enzymes of the cells, which was also explained in the preceding chapter. The net reaction for each molecule of acetyl Co-A is the following:

$$CH_3COCo\text{-}A + Oxaloacetic\ acid + 3H_2O + ADP$$
$$\xrightarrow{Citric\ acid\ cycle}$$
$$2CO_2 + 8H + HCo\text{-}A + ATP + Oxaloacetic\ acid$$

Thus, after the initial degradation of fatty acids to acetyl Co-A, their final breakdown is precisely the same as that of the acetyl Co-A formed from pyruvic acid during the metabolism of glucose.

**ATP Formed by Oxidation of Fatty Acid.** In Figure 68–1 note also that 4 hydrogen atoms are released each time a molecule of acetyl Co-A is formed from the fatty acid chain. Therefore, for every stearic acid molecule that is split, a total of 32 hydrogen atoms are removed. In addition, for each acetyl Co-A degraded by the tricarboxylic acid cycle, 8 hydrogen atoms are removed, making an additional 72 hydrogens for each molecule of stearic acid metabolized. This added to the above 32 hydrogen atoms makes a total of 104 hydrogen atoms. Of this group 34 are removed from the degrading fatty acid by flavoproteins and 70 are removed by $NAD^+$. These two groups of hydrogen atoms are oxidized by the cells, as discussed in the preceding chapter, but they enter the oxidative system at different points, so that 1 molecule of ATP is synthesized for each of the 34 flavoprotein hydrogens and 1½ molecules of ATP are synthesized for each of the 70 DPN hydrogens. This makes 34 plus 105, or a total of 139 molecules of ATP formed by the oxidation of hydrogen derived from each molecule of stearic acid. And another 9 molecules of ATP are formed in the citric acid cycle, one for each of the 9 acetyl Co-A molecules metabolized. Thus, a total of 148 molecules of ATP are formed during the complete oxidation of one molecule of stearic acid. However, 2 high energy bonds are consumed in the initial combination of coenzyme A with the fatty acid molecule, making a net gain of 146 molecules of ATP.

## FORMATION OF ACETOACETIC ACID IN THE LIVER AND ITS TRANSPORT IN THE BLOOD

A large share of the initial degradation of fatty acids occurs in the liver. However, the liver cannot use anything like this amount of fatty acids for its own intrinsic metabolic processes. Instead, when the fatty acid chains have been split into acetyl Co-A, two molecules of acetyl Co-A condense to form one molecule of acetoacetic acid, as follows:

$$2CH_3COCo\text{-}A + H_2O \underset{other\ cells}{\overset{liver\ cells}{\rightleftharpoons}}$$
$$\underset{Acetyl\ Co\text{-}A}{}$$
$$CH_3COCH_2COOH + 2HCo\text{-}A$$
$$\underset{Acetoacetic\ acid}{}$$

Then, a large part of the acetoacetic acid is converted into *β-hydroxybutyric acid* and minute quantities to *acetone* in accord with the following reactions:

The acetoacetic acid and β-hydroxybutyric acid then freely diffuse through the liver cell membranes and are transported by the blood to the peripheral tissues. Here they again diffuse into the cells where reverse reactions occur and acetyl Co-A molecules are formed. These in turn enter the citric acid cycle and are oxidized for energy, as explained above.

Normally, the acetoacetic acid and β-hydroxybutyric acid that enter the blood are transported so rapidly to the tissues that their combined concentration in the plasma rarely rises above 3 mg. per cent. Yet despite the small quantities in the blood, large amounts are actually transported; this is analogous to the high rate of free fatty acid transport. The rapid transport of both these substances probably depends on their high degree of lipid solubility, which allows rapid diffusion through the cell membranes.

**"Ketosis" and Its Occurrence in Starvation, Diabetes, and Other Diseases.** Large quantities of acetoacetic acid, β-hydroxybutyric acid, and acetone occasionally accumulate in the blood and interstitial fluids; this condition is called *ketosis* because acetoacetic acid is a keto acid, and the three compounds are called *ketone bodies*. Ketosis occurs especially in starvation, in diabetes mellitus, or sometimes even when a person's diet is composed almost entirely of fat. In all these states, essentially no carbohydrates are metabolized—in starvation and following a high fat diet because carbohydrates are not available and in diabetes because insulin is not available to cause glucose transport into the cells.

When carbohydrates are not utilized for energy, almost all the energy of the body must come from metabolism of fats. We shall see later in the chapter that lack of availability of carbohydrates automatically increases the rate of removal of fatty acids from adipose tissues, and in addition, several hormonal responses—such as increased secretion of corticotropin by the adenohypophysis, increased secretion of glucocorticoids by the adrenal cortex, and decreased secretion of insulin by the pancreas—all further enhance the removal of fatty acids from the fat tissues. As a result, tremendous quantities of fat become available to the liver for degradation into ketone bodies. The ketone bodies in turn pour out of the liver to be carried to the cells. Yet the cells are limited in the amount of ketone bodies that can be oxidized for two reasons: First, they can be oxidized only as rapidly as adenosine diphosphate is formed in the tissues to initiate the oxidative reactions. Second, lack of carbohydrate intermediates of the glycolytic series depresses the activity of citrate lyase, one of the important rate limiting enzymes controlling the citric acid cycle, which is mainly responsible for degradation of acetoacetic acid, as explained previously. Thus, because of this limit and because of the simultaneous outpouring of tremendous quantities of acetoacetic acid and the other ketone bodies from the liver, the blood concentration of acetoacetic acid and $\beta$-hydroxybutyric acid sometimes rises to as high as 10 or more times normal, thus leading to extreme acidosis, as explained in Chapter 37.

The acetone that is formed during ketosis is a volatile substance that is blown off in small quantities in the expired air of the lungs, often giving the breath an acetone smell. This smell is frequently used as a diagnostic criterion of ketosis.

*Adaptation to a High Fat Diet.* If a person changes slowly from a carbohydrate diet to an almost completely fat diet, his body adapts to the utilization of far more acetoacetic acid than usual, and, in this instance, ketosis does not occur. For instance, the Eskimos, who sometimes live almost entirely on a fat diet, do not develop ketosis. Presumably some factor in these people enhances the rate of acetoacetic acid metabolism by the cells.

## SYNTHESIS OF TRIGLYCERIDES FROM CARBOHYDRATES

Whenever a greater quantity of carbohydrates enters the body than can be used immediately for energy or stored in the form of glycogen, the excess is rapidly converted into triglycerides and then stored in this form in the adipose tissue. Most triglyceride synthesis occurs in the liver, but smaller quantities are also synthesized in the adipose tissue. The triglycerides that are formed in the liver are then mainly transported by the lipoproteins to the adipose tissue where they too are stored until needed for energy.

**Conversion of Acetyl Co-A into Fatty Acids.** The first step in the synthesis of triglycerides is conversion of carbohydrates into acetyl Co-A. It will be recalled from the preceding chapter that this occurs during the normal degradation of glucose by the glycolytic system. It will also be remembered from earlier in this chapter that fatty acids are actually large polymers of acetic acid. Therefore, it is easy to understand how acetyl Co-A could be converted into fatty acids.

However, synthesis of by far the largest proportion of the triglycerides from acetyl Co-A is not achieved by simply reversing the oxidative degradation that was described above. Instead, its first step is conversion of acetyl Co-A into *malonyl Co-A* in accordance with step 1 of Figure 68–2. A large amount of energy is transferred from ATP to malonyl Co-A, and it is this energy that is utilized to cause the subsequent reactions required in the formation of the fatty acid molecule. Step 2 in Figure 68–2 gives the net reaction in the formation of a stearic acid molecule, showing that one acetyl Co-A molecule and eight malonyl

*Step 1:*

$$CH_3COCo\text{-}A + CO_2 + ATP$$

$\Updownarrow$ *(Acetyl Co-A carboxylase)*

$$\underset{\text{Malonyl Co-A}}{O=\overset{\overset{\displaystyle COOH}{\overset{\|}{CH_2}}}{C-CO\text{-}A}} + ADP + PO_4^{---}$$

*Step 2:*

$$1 \text{ Acetyl Co-A} + 8 \text{ Malonyl Co-A} + 16NADPH + 16H^+$$
$$\downarrow$$
$$1 \text{ Stearic Acid} + 8CO_2 + 9Co\text{-}A + 16NADP^+ + 7H_2O$$

**Figure 68–2.** Synthesis of fatty acids.

Co-A molecules combine with NADPH and hydrogen ions to form the fatty acid molecule. Carbon dioxide and Co-A are both liberated, and these are utilized again and again in the formation of malonyl Co-A.

The acetyl Co-A that is converted into fatty acid molecules is derived mainly from the *glycolytic* breakdown of glucose, and the NADPH required for fatty acid synthesis is a by-product of the *phosphogluconate pathway* of glucose degradation, which emphasizes the importance of both these pathways in fat synthesis. These two mechanisms of glucose degradation occur side-by-side in the fat and liver cells, contributing the appropriate proportions of acetyl Co-A and NADPH required for fatty acid synthesis.

**Combination of Fatty Acids with α-Glycerophosphate to Form Triglycerides.** Once the synthesized fatty acid chains have grown to contain 14 to 18 carbon atoms, they are then bound to glycerol to form triglycerides. The enzymes that cause this conversion are highly specific for fatty acids with chain lengths of 14 carbon atoms or greater, a factor that actually controls the physical quality of the triglycerides stored in the body.

As illustrated in Figure 68–3, the glycerol portion of the triglyceride is furnished by α-glycerophosphate, which is a product derived from the glycolytic schema of glucose degradation. The mechanism of this was discussed in Chapter 67.

The real importance of this mechanism for formation of triglycerides is that the whole process is controlled to a great extent by the concentration of α-glycerophosphate. When carbohydrates are available to form large quantities of α-glycerophosphate, the equilibrium shifts to promote formation and storage of triglycerides.

**Efficiency of Carbohydrate Conversion into Fat.** During triglyceride synthesis, only about 15 per cent of the original energy in the glucose is lost in the form of heat, while the remaining 85 per cent is transferred to the stored triglycerides.

**Importance of Fat Synthesis and Storage.** Fat synthesis from carbohydrates is especially important for two reasons: (1) The ability of the different cells of the body to store carbohydrates in the form of glycogen is generally slight; only a few hundred grams of glycogen are stored in the liver, the skeletal muscles, and all other tissues of the body put together. Therefore, fat synthesis provides a means by which the energy of excess ingested carbohydrates (and proteins, too) can be stored for later use. Indeed, the average person has about 200 times as much energy stored in the form of fat as stored in the form of carbohydrate. (2) Each gram of fat contains approximately 2¼ times as many calories of energy as each gram of glycogen. Therefore, for a given weight gain, a person can store far more energy in the form of fat than in the form of carbohydrate, which would be important when an animal must be highly motile to survive.

**Failure to Synthesize Fats from Carbohydrates in the Absence of Insulin.** When insulin is not available, as in diabetes mellitus, fats are poorly, if at all, synthesized by the cells. This seems to result from the following effects: First, when insulin is not available, glucose cannot enter the adipose cells satisfactorily, so that little of the acetyl Co-A and NADPH needed for fat synthesis can be derived from glucose. Second, lack of glucose in the cells greatly reduces the availabilty of α-glycerophosphate, which also makes it difficult for the tissues to form triglycerides.

## SYNTHESIS OF TRIGLYCERIDES FROM PROTEINS

Many amino acids can be converted into acetyl Co-A, as will be discussed in the following chapter. Obviously, this can be synthesized into triglycerides. Therefore, when a person has more proteins in his diet than his tissues can use as proteins, a large share of the excess is stored as fat.

## REGULATION OF ENERGY RELEASE FROM TRIGLYCERIDES

**Regulation of Energy Release by Formation of Adenosine Diphosphate (ADP) in the Tissues.** The primary factor that causes energy release from

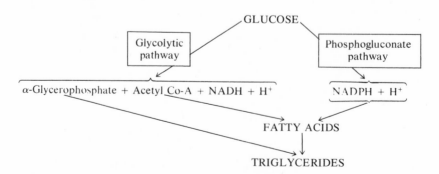

**Figure 68–3.** An overall schema for synthesis of triglycerides from glucose.

all foodstuffs is the concentration of ADP in the tissues. As explained at many points in this text, essentially all functions of the body are energized by the high energy phosphate bonds of ATP. In the process of liberating this energy the ATP becomes ADP, which in turn is a necessary substrate for the reactions responsible for energy release from essentially all foods. For instance, in the absence of ADP, acetyl Co-A cannot be oxidized, and it accumulates in the tissues. Then, in accordance with the *law of mass action,* this accumulation of acetyl Co-A brings the degradation of fatty acids to a halt. Likewise, failure to oxidize NADH builds up this end-product of fat degradation, which likewise slows fatty acid degradation.

Conversely, when the activity of the tissues accelerates so that increased quantities of ADP are formed, all the oxidative processes accelerate, and the degradation and utilization of all foodstuffs, including fats, for energy proceeds apace.

**Fat-Sparing Effect of Carbohydrates; Control of Fat Synthesis From Carbohydrates.** When adequate quantities of carbohydrates are available in the body, the utilization of triglycerides for energy is greatly depressed even though large quantities of ADP are formed. In place of fat utilization, the carbohydrates are utilized preferentially. There are several different reasons for this "fat sparing" effect of carbohydrates, but probably the most important is the following: The fats in adipose tissue cells are present in two different forms, triglycerides and small quantities of free fatty acids. These are in constant equilibrium with each other. When excess quantities of $\alpha$-glycerophosphate and NADPH are present, the equilibrium between free fatty acids and triglycerides shifts toward the triglycerides as explained earlier in the chapter; and as a result, only minute quantities of fatty acids are then available to be utilized for energy. Since $\alpha$-glycerophosphate and NADPH are both important products of glucose metabolism, the availability of large amounts of glucose automatically inhibits the use of fatty acids for energy.

Indeed, when carbohydrates are available in excess, fats are synthesized instead of being degraded. This effect is caused partially by the large quantities of acetyl Co-A formed from the carbohydrates and by the low concentration of free fatty acids in the adipose tissue, thus creating conditions appropriate for conversion of acetyl Co-A into fatty acids. However, a second effect that promotes conversion of carbohydrates to fats is the following: The first step, and the rate limiting step, in the synthesis of fatty acids is carboxylation of acetyl Co-A to form malonyl Co-A. The rate of this reaction is controlled primarily by the enzyme *acetyl Co-A carboxylase,* whose own activity is accelerated in the presence of the intermediates of the citric acid cycle. When excess carbohydrates are being utilized, these intermediates increase, thus automatically causing increased synthesis of fatty acids. Thus, an excess of carbohy-

drates in the diet not only spares the fat already in the fat stores but also increases these.

**Hormonal Regulation of Fat Metabolism.** At least seven of the hormones secreted by the endocrine glands also have marked effects on fat metabolism as follows:

As already pointed out, *insulin lack* causes depressed glucose utilization, which in turn decreases fat synthesis, promotes fat mobilization from the tissues, and increases the rate of fat utilization. In severe diabetes, a person can become extremely emaciated because of depletion of the fat stores. Conversely, an excess of insulin greatly enhances the availability of glucose to the cells, which inhibits fat utilization and enhances fat synthesis. In addition, insulin directly enhances the entry of fatty acids into fat cells, thus further increasing the fat stores while decreasing the use of fats for energy.

*Glucocorticoids* secreted by the adrenal cortex have a direct effect on the fat cells to increase the rate of fat mobilization. It has been postulated that this results from an effect of glucocorticoids to increase the cell membrane permeability of fat cells. In the absence of glucocorticoids, mobilization of fat is depressed, causing considerable depression of fat utilization. Furthermore, glucocorticoids must be available before sufficient fat can ever be mobilized to cause ketosis. Therefore, glucocorticoids are frequently said to have a *ketogenic effect.*

*Corticotropin,* especially, and *growth hormone,* to a lesser extent, from the adenohypophysis both have a fat-mobilizing effect similar to that of the adrenocortical glucocorticoids, though the precise causes of the mobilization are not known. The corticotropin also stimulates secretion of glucocorticoids which mobilize still more fat, as explained above.

*Thyroid hormone* causes rapid mobilization of fat, which is believed to result indirectly from an increased rate of energy metabolism in all cells of the body under the influence of this hormone. The resulting reduction in acetyl Co-A and other intermediates of fat metabolism in the cells would then be a stimulus to cause fat mobilization.

Finally, *epinephrine* and *norepinephrine* have direct effects on fat cells to increase their rate of fat mobilization. In times of stress, the release of epinephrine from the adrenal medullae is an important means by which fatty acids are made available for metabolism, sometimes increasing the free fatty acids in the blood as much as 10- to 15-fold.

The effects of the different hormones on metabolism are discussed further in the chapters dealing with each of them.

## OBESITY

Obesity means deposition of excess fat in the body. This subject will be discussed in detail in relation to dietary balances in Chapter 73, but briefly it is caused by ingestion of greater amounts of food than can be utilized by the body for energy. The excess food, whether fats, carbohydrates, or proteins, is then stored as fat in the adipose tissue to be used later for energy. Strains of rats have been found in which *hereditary obesity* occurs. In at least one of these strains, the obesity is caused by ineffective mobilization of fat from the adipose tissue while synthesis and storage of fat continues normally. Obviously, such a one-way process causes progressive enhancement of the fat stores, resulting in severe obesity. This problem of obesity will be discussed further in Chapter 73.

# PHOSPHOLIPIDS AND CHOLESTEROL

## *PHOSPHOLIPIDS*

The three major types of body phospholipids are the *lecithins,* the *cephalins,* and the *sphingomyelins,* typical examples of which are shown in Figure 68–4.

Phospholipids always contain one or more fatty acid molecules and one phosphoric acid radical, and they usually contain a nitrogenous base. Though the chemical structures of phospholipids are somewhat variant, their physical properties are similar, for they are all lipid soluble, are transported together in lipoproteins in the blood, and seem to be utilized similarly throughout the body for various structural purposes.

**Formation of Phospholipids.** Phospholipids are formed in essentially all cells of the body, though certain cells have a special ability to form them. Probably 90 per cent or more of the phospholipids that enter the blood are formed in the liver cells, though reasonably large quantities can also be formed by the intestinal mucosa.

The rate of phospholipid formation is governed to some extent by the usual factors that control the rate of fat metabolism, for when triglycerides are deposited in the liver, the rate of phospholipid formation increases. Also, certain specific chemical substances are needed for formation of some phospholipids. For instance, *choline,* either in the diet or synthesized in the body, is needed for the formation of lecithin because choline is the nitrogenous base of the lecithin molecule. Also, *inositol* is needed for the formation of some cephalins.

**Specific Uses of Phospholipids.** Several isolated functions of the phospholipids are the following: (1) Phospholipids possibly help to transport fatty acids through the intestinal mucosa into the lymph, possibly by making the fat particles more miscible with water. (2) Thromboplastin, which is necessary to initiate the clotting process, is composed mainly of one of the cephalins. (3) Large quantities of sphingomyelin are present in the nervous system; this substance acts as an insulator in the myelin sheath around nerve fibers. (4) Phospholipids are donors of phosphate radicals when these are needed for different chemical reactions in the tissues. (5) Perhaps the

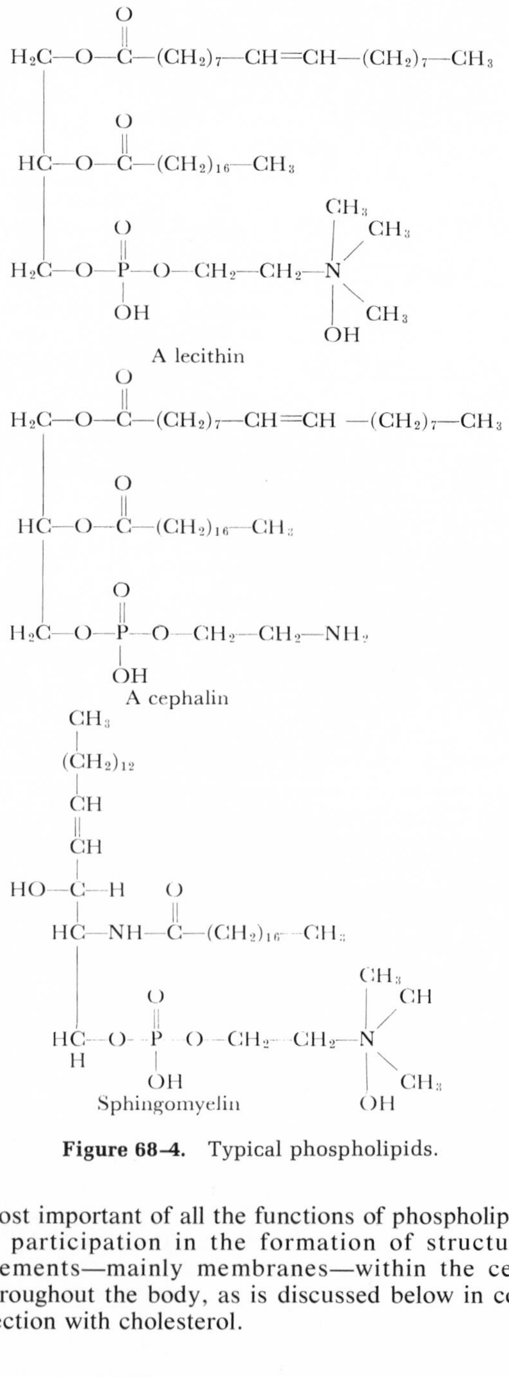

**Figure 68–4.**  Typical phospholipids.

most important of all the functions of phospholipids is participation in the formation of structural elements—mainly membranes—within the cells throughout the body, as is discussed below in connection with cholesterol.

## *CHOLESTEROL*

Cholesterol, the formula of which is illustrated, is present in the diet of all persons, and it can be absorbed slowly from the gastrointestinal tract into the intestinal lymph. It is highly fat soluble, but only slightly soluble in water, and it is capable of forming esters with fatty acids. Indeed, approximately 70 per cent of the cholesterol of the plasma is in the form of cholesterol esters.

**Formation of Cholesterol.** Besides the cholesterol absorbed each day from the gastrointestinal tract, which is called *exogenous cholesterol,* a large quantity, called *endogenous cholesterol,* is formed in the cells of the body. Essentially all the endogenous cholesterol that circulates in the lipoproteins of the plasma is formed by the liver, but all the other cells of the body form at least some cholesterol, which is consistent with the fact that many of the membranous structures of all cells are partially composed of this substance.

Cholesterol

As illustrated by the formula of cholesterol, its basic structure is a sterol nucleus. This is synthesized entirely from acetyl Co-A. In turn, the sterol nucleus can be modified by means of various side chains to form (a) cholesterol, (b) cholic acid, which is the basis of the bile acids formed in the liver, and (c) several important steroid hormones secreted by the adrenal cortex, the ovaries, and the testes (these are discussed in later chapters).

**Factors that Affect the Plasma Cholesterol Concentration—Feedback Control of Body Cholesterol.** Among the important factors that affect plasma cholesterol concentration are the following:

1. An increase in the amount of cholesterol ingested each day increases the plasma concentration slightly. However, when cholesterol is ingested, an intrinsic feedback control system for control of body cholesterol causes the liver to compensate to a great extent for this by synthesizing smaller quantities of endogenous cholesterol. As a result, plasma cholesterol concentration usually cannot be changed upward or downward more than ±15 per cent by altering the diet, though extremes of cholesterol in the diet can probably alter the level by as much as ±30 per cent.

2. A *saturated* fat diet increases blood cholesterol concentration as much as 15 to 25 per cent. This presumably results from increased fat deposition in the liver, which then provides increased quantities of acetyl Co-A in the liver cells for production of cholesterol. Therefore, to decrease the blood cholesterol concentration, it is equally as important to maintain a diet low in saturated fat as to maintain a diet low in cholesterol concentration.

3. Ingestion of fat containing highly unsaturated fatty acids usually depresses the blood cholesterol concentration a slight to moderate amount.

4. Lack of thyroid hormone increases the blood cholesterol concentration, whereas excess thyroid hormone decreases the concentration. This effect is believed to be related to the increased metabolism of all lipid substances under the influence of thyroxine.

5. The blood cholesterol also rises greatly in diabetes mellitus. This is believed to result from the general increase in lipid mobilization in this condition.

6. The female sex hormones, the *estrogens,* decrease blood cholesterol; whereas the male sex hormones, the *androgens,* increase blood cholesterol. Unfortunately, the mechanisms of these effects are unknown, but the sex effects are very important because the higher cholesterol in the male is associated with a higher incidence of heart attacks.

7. In renal retention diseases, the blood cholesterol rises greatly, along with similar increases in blood triglycerides and phosphatides. This is believed to result from inhibition of lipoprotein lipase, resulting in diminished removal of lipoproteins from the plasma, and thus causing their concentration to increase markedly.

**Specific Uses of Cholesterol.** By far the most abundant use of cholesterol in the body is to form cholic acid in the liver. As much as 80 per cent of the cholesterol is converted into cholic acid. This, in turn, is conjugated with other substances to form bile salts, which promote digestion and absorption of fats, as has already been discussed in connection with fat digestion.

A small quantity of cholesterol is used (a) by the adrenal glands to form adrenocortical hormones, (b) by the ovaries to form progesterone and estrogen, and (c) by the testes to form testosterone. However, these glands can also synthesize their own sterols and then form their hormones from these, as is discussed in the chapters on endocrinology later in the text.

A large amount of cholesterol is precipitated in the corneum of the skin. This, along with other lipids, makes the skin highly resistant to the absorption of water-soluble substances and also to the action of many chemical agents, for cholesterol and the other lipids are highly inert to such substances as acids and different solvents that might otherwise easily penetrate the body. Also, these lipid substances help to prevent water evaporation from the skin; without this protection the amount of evaporation would probably be 15 to 20 liters per day instead of the usual 300 to 400 ml.

## STRUCTURAL FUNCTIONS OF PHOSPHOLIPIDS AND CHOLESTEROL

The specific uses of phospholipids and cholesterol are probably of only minor importance in comparison with their importance for general structural purposes throughout the cells of the body.

In Chapter 2 it was pointed out that large quantities of phospholipids and cholesterol are present in the cell membrane and in the membranes of the internal organelles of all cells. It is also known that both cholesterol and the phospholipids have controlling effects on the permeability of cell membranes.

For membranes to be formed, substances that are not soluble in water must be available, and, in general, the only substances in the body that are not soluble in water (besides the inorganic substances of bone) are the lipids and some proteins. Thus, the physical integrity of cells throughout the body is based mainly on phospholipids, triglycerides, cholesterol, and certain insoluble proteins. Some phospholipids are somewhat water soluble as well as lipid soluble, which gives them the important property of helping to decrease the interfacial tension between the membranes and the surrounding fluids.

Another fact that indicates phospholipids and cholesterol to be mainly concerned with the formation of structural elements of the cells is the slow turn-over rate of these substances. For instance, radioactive phospholipids formed in the brain of mice remain in the brain several months after they are formed. Thus, these phospholipids are only slowly metabolized, and the fatty acid is not split away from them to any major extent. Consequently, the purpose of their being in the cells of the brain is presumably related to their indestructible physical properties rather than to their chemical properties—in other words, for the formation of actual physical structures within the cells of the brain.

# ATHEROSCLEROSIS

Atherosclerosis is principally a disease of the large arteries in which lipid deposits called *atheromatous plaques* appear in the subintimal layer of the arteries. These plaques contain an especially large amount of cholesterol and often are simply called cholesterol deposits. They usually are also associated with degenerative changes in the arterial wall. In a later stage of the disease, fibroblasts infiltrate the degenerative areas and cause progressive sclerosis of the arteries. In addition, calcium often precipitates with the lipids to develop *calcified plaques*. When these two processes occur, the arteries become extremely hard, and the disease is then called *arteriosclerosis,* or simply "hardening of the arteries."

Obviously, arteriosclerotic arteries lose most of their distensibility, and, because of the degenerative areas, they are easily ruptured. Also, the atheromatous plaques often protrude through the intima into the flowing blood, and the roughness of their surfaces causes blood clots to develop, with resultant thrombus or embolus formation (see Chapter 9). Almost half of all human beings die of arteriosclerosis; approximately two thirds of the deaths are caused by thrombosis of one or more coronary arteries and the remaining one third by thrombosis or hemorrhage of vessels in other organs of the body—especially the brain, kidneys, liver, gastrointestinal tract, limbs, and so forth.

Despite the extreme prevalence of atherosclerosis, little is known about its cause. Therefore, it is necessary to outline the general trends of the experimental studies rather than to present a definitive description of the mechanisms that cause atherosclerosis.

## EXPERIMENTAL PRODUCTION OF ATHEROSCLEROSIS IN ANIMALS

Severe atherosclerosis can be produced easily in rabbits by simply feeding them large quantities of cholesterol. Rabbits normally have no cholesterol in their diet, for cholesterol is not present in plants, which constitute their entire dietary intake. Therefore, the metabolic processes of the rabbit are not adapted to utilize the ingested cholesterol. Instead, the cholesterol precipitates in many areas of the body, especially in the liver and in the subintimal layer of the arteries.

In carnivorous animals, such as the dog, the induction of atherosclerosis is much more difficult, but it can be caused by excessive administration of cholesterol after the thyroid gland has been removed or inhibited by propylthiouracil. Loss of thyroid secretion greatly depresses cholesterol utilization, thus allowing much of the ingested cholesterol to deposit in the arterial walls.

As pointed out above in relation to the metabolism of cholesterol, the plasma cholesterol concentration increases in either a human being or an experimental animal that has a very high fat diet. On the other hand, substitution of unsaturated fats for saturated fats usually decreases the cholesterol level and, in some experimental animals at least, also inhibits the development of atherosclerosis.

## ATHEROSCLEROSIS IN THE HUMAN BEING

**Effect of Age, Sex, and Heredity on Atherosclerosis.** Atherosclerosis is mainly a disease of old age, but small atheromatous plaques can almost always be found in the arteries of young adults. Therefore, the full-blown disease is a culmination of a lifetime of lipid deposition rather than deposition over a few years.

Far more men die of atherosclerotic heart disease than do women. This is especially true of men younger than 50. For this reason, it is possible that the male sex hormone accelerates the development of atherosclerosis, or that the female sex hormone *protects* a person from atherosclerosis. Indeed, administration of estrogens to men who have already had coronary thromboses decreases the number of secondary coronary attacks. Furthermore, administration of estrogens to chickens with atheromatous plaques in their coronaries has in some instances actually caused the disease to regress.

Atherosclerosis and atherosclerotic heart disease

are highly hereditary in some families. In some instances, this is related to an inherited hypercholesterolemia, the excess cholesterol occurring almost entirely in the *low density lipoproteins*. It seems that either an excess of these is formed or the excess is slow to be cleared from the plasma. The presence of low density lipoproteins is associated with increased cholesterol deposition in the arterial wall. In other persons with hereditary atherosclerosis, the blood cholesterol level is completely normal. Inheritance of the tendency to atherosclerosis is sometimes caused by dominant genes, which means that once this dominant trait enters a family a high incidence of the disease occurs among the offspring.

**Other Diseases that Predispose to Atherosclerosis.** Human beings with severe *diabetes* or severe *hypothyroidism* frequently develop premature and severe atherosclerosis. In both these conditions the blood cholesterol is greatly elevated, which seems to be the cause of the atherosclerosis.

Another disease associated with atherosclerosis, in human beings, as well as in experimental animals, is *hypertension;* the incidence of atherosclerotic coronary heart disease is about twice as great in hypertensives as in normal persons. Though the cause of this is not known, it possibly results from pressure damage to the arterial walls, with subsequent deposition of cholesterol plaques.

**Relationship of Dietary Fat to Atherosclerosis in the Human Being.** A high fat diet, especially one containing cholesterol and saturated fats, greatly increases one's chances of developing atherosclerosis. Therefore, decreasing the fat can help greatly in protecting against atherosclerosis, and some experiments indicate that this can benefit even patients who have already had coronary heart attacks. Also, life insurance statistics show that the rate of mortality—mainly from coronary disease—of normal weight middle and older age persons is about half the mortality rate of overweight subjects of the same age.

## SUMMARY OF FACTORS CAUSING ATHEROSCLEROSIS

Atherosclerosis is almost certainly caused by an abnormality of lipid metabolism, but it is also exacerbated by almost any factor that injures the arterial wall. In particular, elevated blood cholesterol is often related to atherosclerosis. Finally, perhaps equally as important might be some undiscovered third factor that is inherited from generation to generation which causes increased rate of cholesterol production or deposition in the arterial walls themselves, irrespective of the blood cholesterol concentration.

## REFERENCES

Adipose tissue and metabolism (conference). *Ann. N.Y. Acad. Sci.,* 131:1, 1965.

Bierman, E. L.: Fat metabolism, atherosclerosis, and aging in man: a review. *Mech. Aging Dev.,* 2:315, 1973.

Brady, R. O.: Hereditary fat-metabolism diseases. *Sci. Amer.,* 229:88, 1973.

Casdorph, H. R. (ed.): Treatment of the Hyperlipidemic States. Springfield, Ill., Charles C Thomas, Publisher, 1971.

Folch-Pi, J.: Proteolipids. *Adv. Exp. Med. Biol.,* 32:171, 1972.

Fredrickson, D. S., and Gordon, R. S., Jr.: Transport of fatty acids. *Physiol. Rev.,* 38:585, 1958.

Fulco, A. J.: Metabolic alterations of fatty acids. *Ann. Rev. Biochem.,* 43:215, 1974.

Goodman, DeW., S.: Cholesterol ester metabolism. *Physiol. Rev.,* 45:747, 1965.

Goodman, H. M.: Regulation of lipid metabolism. *Physiologist,* 13:75, 1970.

Goodman, H. M., and Schwartz, J.: Growth hormone and lipid metabolism. *In* Greep, R. O., and Astwood, E. B. (eds.): Handbook of Physiology. Sec. 7, Vol 4. Baltimore, The Williams & Wilkins Company, 1974, p. 211.

Havel, R. J., Pernow, B., and Jones, N. L.: Uptake and release of free fatty acids and other metabolites in the legs of exercising men. *J. Appl. Physiol.,* 23:90, 1967.

Hoover, S. R.: Proceedings: Research into foods from animal sources. I. Controlling level and type of fat. *Prev. Med.* 2:346, 1973.

Kirk, J. E.: Arteriosclerosis and arterial metabolism. *In* Bittar, E. E., and Bittar, N. (eds.): The Biological Basis of Medicine. Vol. 1. New York, Academic Press, Inc. 1968, p. 493.

Myant, N. B.: Hormonal control of fatty acid metabolism. *In* Bittar, E. E., and Bittar, N. (eds.): The Biological Basis of Medicine. Vol. 2. New York, Academic Press, Inc. 1968, pp. 151, 193.

Nikkila, E. A.: Effect of drugs on plasma triglyceride metabolism. *Adv. Exp. Med. Biol.,* 26:113, 1972.

Papahadjopoulos, D.: Cholesterol and cell membrane function: a hypothesis concerning etiology of atherosclerosis. *J. Theor. Biol.,* 43:329, 1974.

Parízková, J.: Body composition and lipid metabolism. *Proc. Nutr. Soc.,* 32:181, 1973.

Robinson, D. S., and Wing, D. R.: Clearing factor lipase and its role in the regulation of triglyceride utilization. Studies on the enzyme in adipose tissue. *Adv. Exp. Med. Biol.,* 26:71, 1972.

Romsos, D. R., and Leveille, G. A.: Effect of diet on activity of enzymes involved in fatty acid and cholesterol synthesis. *Adv. Lipid Res.,* 12:97, 1974.

Spitzer, J. J.: CNS and fatty acid metabolism. *Physiologist,* 16:55, 1973.

Stary, H. C.: Proliferation of arterial cells in atherosclerosis. *Adv. Exp. Med. Biol.,* 43:59, 1974.

Steinbaugh, M., and Strong, W. B.: Primary prevention of atherosclerosis: nutritional aspects. *South. Med. J.,* 68:328, 1975.

Steinberg, D.: Hormonal control of lipolysis in adipose tissue. *Adv. Exp. Med. Biol.,* 26:77, 1972.

Thompson, S. W.: The Pathology of Parenteral Nutrition with Lipids. Springfield, Ill., Charles C Thomas, Publisher, 1974.

Tubbs, P. K.: Membranes and fatty acid metabolism. *Brit. Med. Bull.,* 24:158, 1968.

Wagner, W. D., and Clarkson, T. B. (eds.): Arterial Mesenchyme and Arteriosclerosis. New York, Plenum Publishing Corporation, 1974.

# | 69 |

# Protein Metabolism

About three quarters of the body solids are proteins. These include *structural proteins, enzymes, genes, proteins that transport oxygen, proteins of the muscle that cause contraction,* and many other types that perform specific functions both intracellularly and extracellularly throughout the body.

The basic chemical properties of proteins that explain their diverse functions are so extensive that they are a major portion of the entire discipline of biochemistry. For this reason, the present discussion is confined to the general aspects of protein metabolism.

## BASIC PROPERTIES OF PROTEINS

### THE AMINO ACIDS

The principal constituents of proteins are amino acids, 21 of which are present in the body in significant quantities. Figure 69–1 illustrates the chemical formulas of these 21 amino acids, showing that they all have two features in common: Each amino acid has an acidic group (—COOH) and a nitrogen radical that lies in close association with the acidic radical, usually represented by the amino group (—$NH_2$).

**Peptide Linkages and Peptide Chains.** In proteins, the amino acids are aggregated into long chains by means of so-called *peptide linkages,* one of which is illustrated by the following reaction:

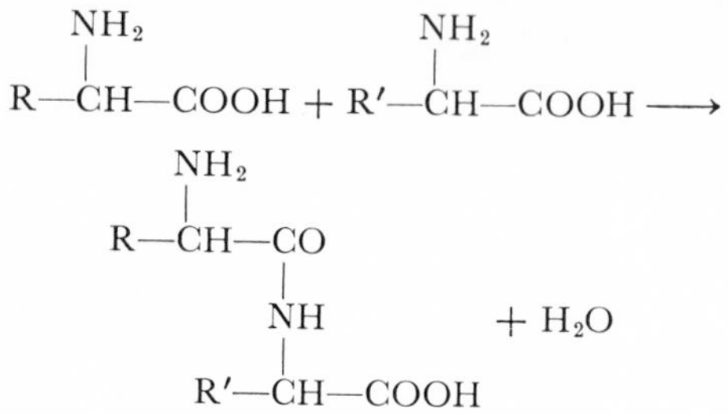

$$R—\underset{\underset{NH_2}{|}}{CH}—COOH + R'—\underset{\underset{NH_2}{|}}{CH}—COOH \longrightarrow$$

$$R—\underset{\underset{NH_2}{|}}{CH}—CO$$
$$\underset{\underset{R'—CH—COOH}{|}}{NH} \quad + H_2O$$

Note in this reaction that the amino radical of one amino acid combines with the carboxyl radical of the other amino acid. A hydrogen atom is released from the amino radical, and a hydroxyl radical is released from the carboxyl radical; and these two combine to form a molecule of water. Note that after the peptide linkage has been formed, an amino radical and a carboxyl radical are still in the new molecule, both of which are capable of combining with additional amino acids to form a *peptide chain.* Some complicated protein molecules have as many as a hundred thousand amino acids combined together principally by peptide linkages, and even the smallest protein usually has more than 20 amino acids combined together by peptide linkages.

**Other Linkages in Protein Molecules.** Some protein molecules are composed of several peptide chains rather than a single chain, and these in turn are bound with each other by other linkages, principally by *hydrogen bonds* between the CO and NH radicals of the peptides as follows:

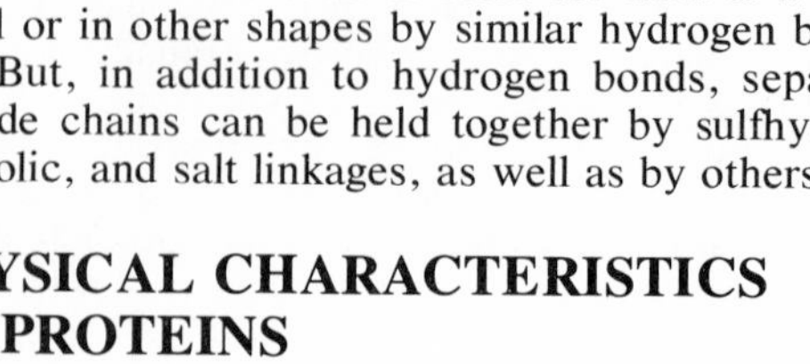

Also, many peptide chains are coiled or folded, and the successive coils or folds are held in a tight spiral or in other shapes by similar hydrogen bonding. But, in addition to hydrogen bonds, separate peptide chains can be held together by sulfhydryl, phenolic, and salt linkages, as well as by others.

## PHYSICAL CHARACTERISTICS OF PROTEINS

### GLOBULAR PROTEINS

With the exception of the fibrous proteins, which are discussed subsequently, most proteins of the

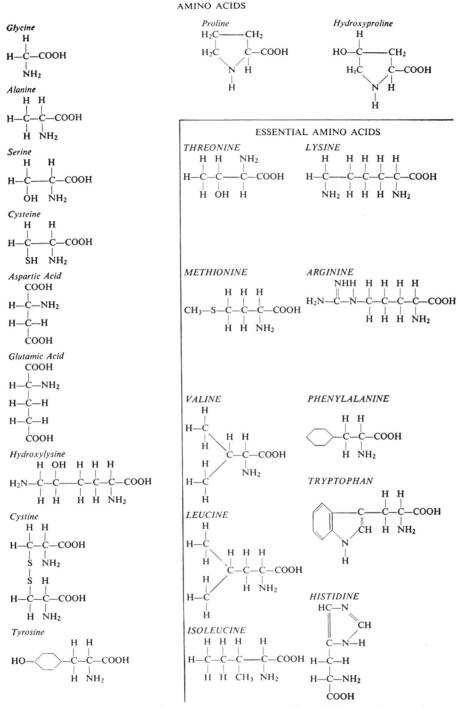

**Figure 69–1.** The amino acids, showing the 10 essential amino acids, which cannot be synthesized at all or in sufficient quantity in the body.

body assume either a globular or an elliptical shape and are called *globular proteins*. These, in general, are soluble in water or salt solutions, and they are held in a globular shape by coiling and folding of the peptide chains.

**Important Types of Globular Proteins in the Body.** There is no simple functional classification of the globular proteins for two reasons: First, these proteins perform literally thousands of different functions in the body. Second, proteins of widely varying chemical and physical characteristics many times perform very much the same function. Yet, for purposes of reference, the different globular proteins can be classified mainly on the basis of their chemical properties as follows:

The *albumins* are low molecular weight simple proteins that are readily soluble in water. They are the major proteins of the plasma but are also present in small quantities inside the cell.

*Globulins* are also simple proteins that are soluble in salt solution but poorly soluble in water. These, like albumin, are also present in large quantities in the plasma and are extensively present inside the cells. Conjugated products of globulins constitute most cellular enzymes.

*Histones* and *protamines* are highly basic proteins because of large quantities of basic amino acids in their molecules. These are the principal proteins bound with nucleic acids to form nucleoproteins.

## FIBROUS PROTEINS

Many of the highly complex proteins are fibrillar and are called fibrous proteins. In these the peptide chains are elongated, and many separate chains are held together in parallel bundles by cross-linkages. Major types of fibrous proteins are (1) *collagens,* which are the basic structural proteins of connective tissue, tendons, cartilage, and bone; (2) *elastins,* which are the elastic fibers of tendons, arteries, and connective tissue; (3) *keratins,* which are the structural proteins of hair and nails; and (4) *actins* and *myosin,* the contractile proteins of muscle.

Since fibrillar proteins are the principal structural proteins of the body, it is important to know about their physical characteristics. In general, the fibrils are extremely strong and are capable of being stretched, and then they recoil to their natural length; these are properties of typical *elastomers.* Another elastomeric characteristic of the fibrillar proteins is their tendency to *creep;* that is, if stretched for a long time, their basic length gradually becomes the stretched length, but, on the other hand, if the tension on the two ends of the fibrils is relaxed, the fibrils creep to a shorter and shorter length. For instance, a large scar that forms shortly after a severe wound gradually creeps to a smaller and smaller size if tension is not placed on the scar, but, on the other hand, such a scar can creep to a larger and larger size if it is in an area where tension is high, as occurs frequently in the skin of a person who becomes progressively more obese.

## CONJUGATED PROTEINS

In addition to the simple globular and fibrous proteins, many proteins are combined as conjugated proteins with nonprotein substances. These include the following:

*Nucleoproteins* are combinations of simple proteins and nucleic acid. The deoxyribose nucleoproteins are the principal constituents of genes, and

ribose nucleoproteins perform essential functions in the cytoplasm necessary for the synthesis of proteins, which was discussed in detail in Chapter 3.

*Mucoproteins* are a type of conjugated protein containing large quantities of complex polysaccharides. One of the most important mucoproteins, that found in mucus, is highly resistant to chemical destruction and also has exceptionally important properties that were discussed in Chapter 63 in relation to gastrointestinal physiology.

In addition to these important types of proteins are the *lipoproteins,* which contain lipid materials; *chromoproteins* (such as hemoglobin and cytochromes), which contain coloring agents; *phosphoproteins,* which contain phosphorus; and *metalloproteins,* which contain magnesium, copper, iron, zinc, or other metallic ions and which constitute many of the enzymes.

# TRANSPORT AND STORAGE OF AMINO ACIDS

## THE BLOOD AMINO ACIDS

The normal concentration of amino acids in the blood is between 35 and 65 mg. per cent. This is an average of about 2 mg. per cent for each of the 21 amino acids, though some are present in far greater concentrations than others. Since the amino acids are relatively strong acids, they exist in the blood principally in the ionized state and account for 2 to 3 milliequivalents of the negative ions in the blood. The precise distribution of the different amino acids in the blood depends to some extent on the types of proteins ingested, but the concentrations of some individual amino acids are also regulated to a certain extent by (a) selective synthesis in the different cells and (b) selective excretion by the kidneys.

**Fate of Amino Acids Absorbed from the Gastrointestinal Tract.** It will be recalled from Chapter 65 that the end-products of protein digestion in the gastrointestinal tract are almost entirely amino acids and that only rare polypeptide or protein molecules are absorbed. Immediately after a meal, the amino acid concentration in the blood rises, but the rise is usually only a few milligrams per cent for two reasons: First, protein digestion and absorption is usually extended over two to three hours, which allows only small quantities of amino acids to be absorbed at a time. Second, after entering the blood, the excess amino acids are absorbed within 5 to 10 minutes by cells throughout the entire body. Therefore, almost never do large concentrations of amino acids accumulate in the blood. Nevertheless, the turnover rate of the amino acids is so rapid that many grams of proteins can be carried from one part of the body to another in the form of amino acids each hour.

**Active Transport of Amino Acids into the Cells.** The molecules of essentially all the amino acids are much too large to diffuse through the pores of the cell membranes. Small quantities perhaps can dissolve in the matrix of the cell membrane and diffuse to the interior of the cells in this manner, but large quantities of amino acids can be transported through the membrane only by active transport utilizing carrier mechanisms. The nature of the carrier mechanisms is still poorly understood, but some of the theories are discussed in Chapter 4.

*Renal Threshold for Amino Acids.* One of the special functions of carrier transport of amino acids is to prevent loss of these in the urine. All the different amino acids can be *actively transported* through the proximal tubular epithelium, thus removing them from the glomerular filtrate and returning them to the blood. However, as is true of other active transport mechanisms in the renal tubules, there is an upper limit to the rate at which each type of amino acid can be transported. For this reason, when a particular type of amino acid rises to too high a concentration in the plasma and glomerular filtrate, the excess above that which can be actively reabsorbed is lost into the urine.

In Chapter 38 it was pointed out that appropriate enzyme systems for active reabsorption of certain amino acids are often deficient or lacking from the renal tubular epithelium. In these conditions the plasma threshold for the respective amino acids is greatly reduced. However, in the normal person, the loss of amino acids in the urine each day is insignificant, thus allowing essentially all amino acids absorbed from the gastrointestinal tract to be utilized by the cells.

## STORAGE OF AMINO ACIDS AS PROTEINS IN THE CELLS

Almost immediately after entry into the cells, amino acids are conjugated under the influence of intracellular enzymes into cellular proteins so that the concentration of amino acids inside the cells probably always remains low. Thus, so far as is known, storage of large quantities of amino acids as such probably does not occur in the cells; instead, they are mainly stored in the form of actual proteins. Yet many intracellular proteins can be rapidly decomposed again into amino acids under the influence of intracellular lysosomal digestive enzymes, and these amino acids in turn can be transported back out of the cell into the blood. The proteins that can be thus decomposed include many cellular enzymes as well as some other functioning proteins. However, the genes of the nucleus and the structural proteins such as collagen and muscle contractile proteins do not participate significantly in this reversible storage of amino acids.

Some tissues of the body participate in the storage of amino acids to a greater extent than others. Thus, the liver, which is a large organ and also has special systems for processing amino acids, stores large quantities of proteins; this is also true of the kidney and the intestinal mucosa.

**Release of Amino Acids from the Cells and Regulation of Plasma Amino Acid Concentration.** Whenever the plasma amino acid concentration falls below its normal level, amino acids are transported out of the cells to replenish the supply in the plasma. Simultaneously, intracellular proteins are degraded back into amino acids.

The plasma concentration of each type of amino acid is maintained at a reasonably constant value. Later it will be noted that the various hormones secreted by the endocrine glands are able to alter the balance between tissue proteins and circulating amino acids; growth hormone and insulin increase the formation of tissue proteins, while the adrenocortical glucocorticoid hormones increase the concentration of circulating amino acids.

**Reversible Equilibrium Between the Proteins of Different Parts of the Body.** Since cellular proteins can be synthesized rapidly from plasma amino acids and many of these in turn can be degraded and returned to the plasma almost equally as rapidly, there is constant equilibrium between the plasma amino acids and most of the proteins in the cells of the body. Therefore, it follows that there is also equilibrium between the proteins from one type of cell to the next. For instance, if any particular tissue loses proteins, it can synthesize new proteins from the amino acids of the blood; in turn, these are replenished by degradation of proteins from other cells of the body. These effects are particularly noticeable in relation to protein synthesis in cancer cells. Cancer cells are prolific users of amino acids, and, simultaneously, the proteins of the other tissues become markedly depleted.

**Upper Limit to the Storage of Proteins.** Each particular type of cell has an upper limit to the amount of proteins that it can store. After all the cells have reached their limits, the excess amino acids in the circulation are then degraded into other products and used for energy, as is discussed subsequently, or they are converted to fat and stored in this form.

## THE PLASMA PROTEINS

The three major types of protein present in the plasma are *albumin, globulin,* and *fibrinogen.* The principal function of albumin is to provide *colloid osmotic pressure,* which in turn prevents plasma loss from the capillaries, as discussed in Chapter 30. The globulins perform a number of enzymatic functions in the plasma itself, but, more important than this, they are principally responsible for both the natural and acquired immunity that a person has against invading organisms, which was discussed in Chapter 7. The fibrinogen polymerizes into long fibrin threads during blood coagulation, thereby forming blood clots that help to repair leaks in the circulatory system, which was discussed in Chapter 9.

**Formation of the Plasma Proteins.** Essentially all the albumin and fibrinogen of the plasma proteins, as well as 50 per cent or more of the globulins, are formed in the liver. The remainder of the globulins are formed in the lymphoid tissues and other cells of the reticuloendothelial system. These are mainly the gamma globulins that constitute the antibodies.

The rate of plasma protein formation by the liver can be extremely high, as great as 4 grams per hour or as much as 100 grams per day. Certain disease conditions often cause rapid loss of plasma proteins; severe burns that denude large surface areas cause loss of many liters of plasma through the denuded areas each day. The rapid production of plasma proteins by the liver is obviously valuable in preventing death in such states. Furthermore, occasionally, a person with severe renal disease loses as much as 20 to 30 grams of plasma protein in the urine each day for years. In some of these patients the plasma protein concentration may remain almost normal throughout the entire illness.

**Use of Plasma Proteins by the Tissues.**    When the tissues become depleted of proteins, the plasma proteins can act as a source for rapid replacement of the tissue proteins. Indeed whole plasma proteins can be imbibed *in toto* by the reticuloendothelial cells; then, once in the cells, these are split into amino acids that are transported back into the blood and utilized throughout the body to build cellular proteins. In this way, then, the plasma proteins function as a labile protein storage medium and represent a rapidly available source of amino acids whenever a particular tissue requires these.

**Reversible Equilibrium Between the Plasma Proteins and the Tissue Proteins.**    The rate of synthesis of plasma proteins by the liver is dependent on the concentration of amino acids in the blood, which means that the concentration of plasma proteins becomes reduced whenever an appropriate supply of amino acids is not available. On the other hand, whenever excess proteins are available in the plasma but insufficient proteins are present in the cells, the plasma proteins are used to form tissue proteins. Thus, there is a constant state of equilibrium, as illustrated in Figure 69–2, between the plasma proteins, the amino acids of the blood, and the tissue proteins. It has been estimated from radioactive tracer studies that about 400 grams of body protein are synthesized and degraded each day as part of the continual state of flux of amino acids. This illustrates once again the general principle of reversible exchange of amino acids among the different proteins of the body. Even during starvation or during severe debilitating diseases, the ratio of total tissue proteins in the body to total plasma proteins remains relatively constant at about 33 to 1.

Because of this reversible equilibrium between plasma proteins and the other proteins of the body, one of the most effective of all therapies for severe acute protein deficiency is intravenous administration of plasma protein. Within hours, the amino acids of the administered protein become distributed throughout the cells of the body to form proteins where they are needed.

## CHEMISTRY OF PROTEIN SYNTHESIS

Proteins are synthesized in all cells of the body, and the functional characteristics of each cell are dependent upon the types of protein that it can form. Basically, the genes of the cells control the protein types and thereby control the functions of the cell. This regulation of cellular function by the genes was discussed in detail in Chapter 3. Chemically, two basic processes must be accomplished for the synthesis of proteins; these are: (1) synthesis of the amino acids and (2) appropriate conjugation of the amino acids to form the respective types of whole proteins in each individual cell.

**Essential and Nonessential Amino Acids.** Eleven of the 21 amino acids normally present in animal proteins can be synthesized in the cells, while the other 10 either cannot be synthesized at all or are synthesized in quantities too small to supply the body's needs. The first group of amino acids is called *nonessential,* while the second group is called *essential amino acids*. The essential amino acids obviously must be present in the diet if protein formation is to take place in the body. Use of the word "essential" does not mean that the other 11 amino acids are not equally as essential in the formation of the proteins, but only that these others are not essential in the diet.

Synthesis of the nonessential amino acids depends on the formation first of appropriate $\alpha$-keto acids, which are the precursors of the respective amino acids. For instance, *pyruvic acid,* which is formed in large quantities during the glycolytic breakdown of glucose, is the keto acid precursor of the amino acid *alanine*. Then, by the simple process of *transamination,* an amino radical is transferred to the $\alpha$-keto acid while the keto oxygen is transferred to the donor of the amino radical. This reaction is illustrated in Figure 69–3. Note in this reaction that the amino radical is transferred to the pyruvic acid from a substance called *glutamine*. Glutamine is present in the tissues in large quantities, and it functions specifically as an amino radical storehouse. In addition, amino radicals can be transferred from *asparagine* and from some amino acids of the body fluids, particularly from glutamic acid and aspartic acid.

Transamination is promoted by enzymes called

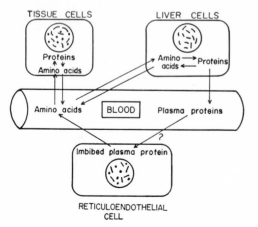

**Figure 69–2.**    Reversible equilibrium between the tissue proteins, plasma proteins, and plasma amino acids.

$$NH_2-C-CH_2-CH_2-CH-COOH + CH_3-C-COOH \quad \underrightarrow{\text{Transaminase}}$$
$$\underset{O}{\|} \qquad \underset{NH_2}{|} \qquad \qquad \underset{O}{\|}$$

(Glutamine)            (Pyruvic acid)

$$NH_2-C-CH_2-CH_2-C-COOH + CH_3-CH-COOH$$
$$\underset{O}{\|} \qquad\qquad \underset{O}{\|} \qquad\qquad\qquad \underset{NH}{|}$$

(α—Ketoglutamic acid)        (Alanine)

**Figure 69–3.** Synthesis of alanine from pyruvic acid by transamination.

*transaminases,* all of which are derivatives of pyridoxine, one of the B vitamins. Without this vitamin, the nonessential amino acids cannot be synthesized, and, therefore, protein formation cannot proceed normally.

**Formation of Proteins from Amino Acids.** Once the appropriate amino acids are present in a cell, whole proteins are synthesized rapidly. However, each peptide linkage requires from 500 to 4000 calories of energy, and this must be supplied from ATP in the cell. Protein formation proceeds through two steps: (1) "activation" of each amino acid, during which the amino acid is "energized" by energy derived from ATP, and (2) alignment of the amino acids into the peptide chains, a function that is under control of the genetic system of each individual cell. Both of these processes were discussed in detail in Chapter 3. Indeed, the formation of cellular proteins is the basis of life itself and is so important that the reader would do well to review Chapter 3.

# USE OF PROTEINS FOR ENERGY

It was pointed out earlier that there is an upper limit to the amount of protein that can accumulate in each particular type of cell. Once the cells are filled to their limits, any additional amino acids in the body fluids are degraded and used for energy or stored as fat. This degradation occurs almost entirely in the liver, and it begins with the process known as deamination.

**Deamination.** Deamination means removal of the amino groups from the amino acids. This can occur by several different means, two of which are especially important: (1) transamination, which means transfer of the amino group to some acceptor substance as explained above in relation to the synthesis of amino acids, and (2) oxidative deamination.

The greatest amount of deamination occurs by the following transamination schema:

α-Ketoglutaric acid + Amino acid

Glutamic acid + α-Keto acid

+NAD$^+$+H$_2$O

NADH+H$^+$+NH$_3$

Note from this schema that the amino group from the amino acid is transferred to ∝-ketoglutaric acid, which then becomes glutamic acid. The glutamic acid can then transfer the amino group to still other substances or can release it in the form of ammonia. In the process of losing the amino group, the glutamic acid once again becomes ∝-ketoglutaric acid, so that the cycle can be repeated again and again.

Oxidative deamination occurs to much less extent and is catalyzed by amino acid oxidases. In this process the amino acid is oxidized at the point where the amino radical attaches, which causes the amino radical to be released.

*Urea Formation by the Liver.* The ammonia released during deamination is removed from the blood almost entirely by conversion into urea, two molecules of ammonia and one molecule of carbon dioxide combining in accordance with the following net reaction:

$$2NH_3 + CO_2 \rightarrow H_2N-\underset{\underset{O}{\|}}{C}-NH_2 + H_2O$$

Essentially all urea formed in the human body is synthesized in the liver. In the absence of the liver or in serious liver disease, ammonia accumulates in the blood. This in turn is extremely toxic, especially to the brain, often leading to a state called *hepatic coma.*

The stages in the formation of urea are essentially the following:

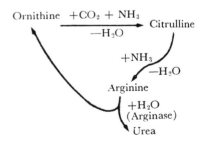

The reaction begins with the amino acid derivative *ornithine,* which combines with one molecule of carbon dioxide and one molecule of ammonia to form a second substance, *citrulline.* This in turn combines with still another molecule of ammonia to form *arginine,* which then splits into *ornithine* and *urea.* The urea diffuses from the liver cells into the body fluids and is excreted by the kidneys, while the ornithine is reused in the cycle again and again.

**Oxidation of Deaminated Amino Acids.** Once amino acids have been deaminated, the resulting keto acid products can in most instances be oxidized to release energy for metabolic purposes. This usually involves two processes: (1) the keto acid is changed into an appropriate chemical substance that can enter the tricarboxylic acid cycle and (2) this substance is then degraded by this cycle in the same manner that acetyl Co-A derived from carbohydrate and lipid metabolism is degraded.

In general, the amount of adenosine triphosphate formed for each gram of protein that is oxidized is slightly less than that formed for each gram of glucose oxidized.

**Gluconeogenesis and Ketogenesis.** Certain deaminated amino acids are similar to the breakdown products that result from glucose and fatty acid metabolism. For instance, deaminated alanine is pyruvic acid. Obviously, this can be converted into glucose or glycogen. Or it can be converted into acetyl Co-A, which can then be polymerized into fatty acids. Also, two molecules of acetyl Co-A can condense to form acetoacetic acid, which is one of the ketone bodies, as explained in the preceding chapter.

The conversion of amino acids into glucose or glycogen is called *gluconeogenesis,* and the conversion of amino acids into keto acids or fatty acids is called *ketogenesis.* Eighteen out of 21 of the deaminated amino acids have chemical structures that allow them to be converted into glucose, and 19 can be converted into fats—5 directly and the other 14 by becoming carbohydrate first and then becoming fat.

## *OBLIGATORY DEGRADATION OF PROTEINS*

When a person eats no proteins, a certain proportion of his own body proteins continues to be degraded into amino acids, then deaminated and oxidized. This involves about 30 grams of protein each day, which is called the *obligatory loss* of proteins. Therefore, to prevent a net loss of protein from the body, one must ingest at least 30 grams of protein each day, and to be on the safe side as much as 75 grams is usually recommended.

The ratios of the different amino acids in the dietary protein must be about the same as the ratios in the body tissues for the entire protein to be usable. If one particular type of essential amino acid is low in concentration, the others become unusable as well because cells form either whole proteins or none at all, as explained in Chapter 3 in relation to protein synthesis. The unusable amino acids are then deaminated and oxidized. A protein that has a ratio of amino acids different from that of the average body protein is called a *partial protein* or *incomplete protein,* and such a protein is obviously less valuable for nutrition than is the *complete protein.*

**Effect of Starvation on Protein Degradation.** Except for the excess protein in the diet or the 30 grams of obligatory protein degradation each day, the body uses almost entirely carbohydrates or fats for energy as long as these are available. However, after several weeks of starvation, when the quantity of stored fats begins to run out, the amino acids of the blood begin to be rapidly deaminated and oxidized for energy. From this point on, the proteins of the tissues degrade rapidly—as much as 100 grams daily—and the cellular functions deteriorate precipitously.

Because carbohydrate and fat utilization for energy occurs in preference to protein utilization, carbohydrates and fats are called *protein sparers.*

# HORMONAL REGULATION OF PROTEIN METABOLISM

**Growth Hormone.** Growth hormone increases the rate of synthesis of cellular proteins, causing the tissue proteins to increase. The precise mechanism by which growth hormone increases the rate of protein synthesis is not known, but growth hormone is believed to act in some direct manner either to enhance the transport of amino acids through the cell membranes or to accelerate the actual chemical processes of protein synthesis. Part of the action might also result from the effect of growth hormone on fat metabolism, for this hormone causes increased rate of fat liberation from the fat depots, making this available for energy. This in turn reduces the rate of oxidation of amino acids and consequently makes increased quantities of amino acids available to the tissues to be synthesized into proteins.

**Insulin.** Lack of insulin reduces protein synthesis almost to zero. The mechanism by which this hormone affects protein metabolism is also unknown, but insulin does slightly accelerate amino acid transport into the cells, which could be a stimulus to protein synthesis. Also, insulin increases the availability of glucose to the cells so that the use of amino acids for energy becomes correspondingly reduced. This, too, undoubtedly makes far larger quantities of amino acids available to the tissues for protein synthesis.

**Glucocorticoids.** The glucocorticoids secreted by the adrenal cortex *decrease* the quantity of protein in most tissues while increasing the amino acid concentration in the plasma. However, contrary to elsewhere in the body, these hormones *increase* both the liver proteins and the plasma proteins. It is believed that the glucocorticoids act by increasing the rate of breakdown of extrahepatic proteins, thereby making increased quantities of amino acids available in the body fluids. This in turn supposedly allows the liver to synthesize increased quantities of liver and plasma proteins.

The effects of glucocorticoids on protein metabolism are especially important in ketogenesis and gluconeogenesis, for in the absence of these hormones insufficient quantities of amino acids are usually available in the plasma to allow either significant gluconeogenesis or ketogenesis from proteins.

**Testosterone.** Testosterone, the male sex hormone, causes increased deposition of protein in the tissues throughout the body, including especially an increase in the contractile proteins of the muscles. The mechanism by which this effect comes about is unknown, but it is definitely different from the effect of growth hormone in the following way: Growth hormone causes tissues to continue growing almost indefinitely, while testosterone causes the muscles and other protein tissues to enlarge only for several months; beyond that time, despite continued administration of testosterone, further protein deposition ceases.

Estrogen, the principal female sex hormone, also causes slight deposition of protein, but its effect is relatively insignificant in comparison with that of testosterone.

**Thyroxine.** Thyroxine increases the rate of metabolism of all cells, and, as a result, indirectly affects protein metabolism. If insufficient carbohydrates and fats are available for energy, thyroxine causes rapid degradation of proteins to be used for energy. On the other hand, if adequate quantities of carbohydrates and fats are available and excesses of amino acids are also available in the extracellular fluid, thyroxine can actually increase the rate of protein synthesis. Conversely, in growing animals deficiency of thyroxine causes growth to be greatly inhibited because of lack of protein synthesis. In essence, it is believed that thyroxine has no specific effect on protein metabolism but does have an important general effect to increase the rates of both normal anabolic and normal catabolic protein reactions.

# REFERENCES

Bergen, W. G.: Protein synthesis in animal models. *J. Anim. Sci.,* 38:1079, 1974.

Borasky, R.: Ultrastructure of Protein Fibers. New York, Academic Press, Inc. 1963.

Brown, H. (ed.): Protein Nutrition. Springfield, Ill., Charles C Thomas, Publisher, 1974.

Devenyi, T., and Gergely, J.: Amino Acids, Peptides, and Proteins. New York, American Elsevier Publishing Co., Inc., 1974.

Fraser, R. D. B.: Keratins. *Sci. Amer.,* 221:86, 1969.

Goldberg, A. L., and Dice, J. F.: Intracellular protein degradation in mammalian and bacterial cells. *Ann. Rev. Biochem.,* 43:835, 1974.

Harper, A. E., Benevenga, N. J., and Wohlhueter, R. M.: Effects of ingestion of disproportionate amounts of amino acids. *Physiol. Rev.,* 50:428, 1970.

Kostyo, J. L., and Nutting, D. F.: Growth hormone and protein metabolism. *In* Greep, R. O., and Astwood, E. B. (eds.): Handbook of Physiology. Sec. 7, Vol. 4. Baltimore, The Williams & Wilkins Company, 1974, p. 187.

Kurahashi, K.: Biosynthesis of small peptides. *Ann. Rev. Biochem.,* 43:445, 1974.

Leder, P.: The elongation reactions in protein synthesis. *Adv. Protein Chem.,* 27:213, 1973.

Lipmann, F.: What do we know about protein synthesis? *Basic Life Sci.,* 1:1, 1973.

McGilvery, R. W.: Biochemical Concepts. Philadelphia, W. B. Saunders Company, 1975.

Matthews, D. M., and Payne, J. W. (eds.): Peptide Transport in Protein Nutrition. New York, American Elsevier Publishing Co., Inc., 1974.

Millward, D. J., and Garlick, P. J.: The pattern of protein turnover in the whole animal and the effect of dietary variations. *Proc. Nutr. Soc.* 31:257, 1972.

Moldave, K.: Mechanisms of mammalian protein synthesis. *Basic Life Sci.,* 1:361, 1973.

Morris, D. R., and Fillingame, R. H.: Regulation of amino acid decarboxylation. *Ann. Rev. Biochem.,* 43:303, 1974.

Munro, H., and Allison, J. B.: Mammalian Protein Metabolism. Vols. 1 and 2. New York, Academic Press, Inc., 1967.

Pestka, S.: The use of inhibitors in studies on protein synthesis. *Methods Enzymol.,* 30:261, 1974.

Ross, R. and Bornstein, P.: Elastic fibers in the body. *Sci. Amer.,* 224:44, 1971.

Rothschild, M. A., Oratz, M., Mongelli, J., and Schreiber, S. S.: Effect of albumin concentration on albumin synthesis in the perfused liver. *Amer. J. Physiol.,* 216:1127, 1969.

Schweiger, M., and Herrlich, P.: DNA-directed enzyme synthesis in vitro. *Curr. Top. Microbiol. Immunol.,* 65:59, 1974.

Siekevitz, P.: The turnover of proteins and the usage of information. *J. Theor. Biol.,* 37:321, 1972.

Truffa-Bachi, P., and Cohen, G. N.: Amino acid metabolism. *Ann. Rev. Biochem.,* 42:113, 1973.

Viets, F. G.: Fate of nitrogen under intensive animal feeding. *Fed. Proc.,* 33:1178, 1974.

Waterlow, J. C.: Protein and energy requirements of children. *Bibl. Nutr. Dieta.,* 18:6, 1973.

Weinstock, M., and Leblond, C. P.: Formation of collagen. *Fed. Proc.,* 33:1205, 1974.

Wicks, W. D.: Regulation of protein synthesis by cyclic AMP. *Adv. Cyclic Nucleotide Res.,* 4:335, 1974.

Wiessbach, H., and Brot, N.: The role of protein factors in the biosynthesis of proteins. *Cell,* 2:137, 1974.

Williams-Ashman, H. G.: Metabolic effects of testicular androgens. *In* Greep, R. O., and Astwood, E. B. (eds.): Handbook of Physiology. Sec. 7, Vol. 5. Baltimore, The Williams & Wilkins Company, 1975, p. 473.

# | 70 |

# The Liver and Biliary System

Different functions of the liver have been presented at many points in this text because the liver has so many and such varied functions that it is impossible to separate its actions from those of other organ systems. Therefore, the purpose of the present chapter is to summarize the different functions of the liver and to show how the liver operates as an individual organ.

The basic functions of the liver can be divided into: (1) its vascular functions for storage and filtration of blood, (2) its secretory function for secreting bile into the gastrointestinal tract, and (3) its metabolic functions concerned with the majority of the metabolic systems of the body.

## PHYSIOLOGIC ANATOMY
## OF THE LIVER

The basic functional unit of the liver is the liver lobule, which is a cylindrical structure several millimeters in length and 0.8 to 2 mm. in diameter. The human liver contains 50,000 to 100,000 individual lobules.

The liver lobule is constructed around a *central vein* that empties into the hepatic veins and thence into the vena cava. The lobule itself is composed principally of many *hepatic cellular plates* (two of which are illustrated in Figure 70–1) that radiate centrifugally from the central vein like spokes in a wheel. Each hepatic plate is usually two cells thick, and between the adjacent cells lie small *bile canaliculi* that empty into *terminal bile ducts* that originate in the septa between the adjacent liver lobules.

Also in the septa are small *portal venules* that receive their blood from the portal veins. From these venules blood flows into flat, branching *hepatic sinusoids* that lie between the hepatic plates, and thence into the central vein. Thus, the hepatic cells are exposed continuously to portal venous blood.

In addition to the portal venules, *hepatic arterioles* are also present in the interlobular septa. These arterioles supply arterial blood to the septal tissues,

and many of the small arterioles also empty directly into the hepatic sinusoids, most frequently emptying into these about one third of the distance away from the interlobular septa, as shown in Figure 70–1.

The venous sinusoids are lined by two types of cells: (1) typical *endothelial cells* and (2) large *Kupffer cells,* which are reticuloendothelial cells capable of phagocytizing bacteria and other foreign matter in the blood. The endothelial lining of the venous sinusoids has extremely large pores, some of which are almost 1 micron in diameter. Beneath this lining, between the endothelial cells and the hepatic cells, is a very narrow space called the *space of Disse.* Because of the large pores in the endothelium, substances in the plasma move freely into the space of Disse. Even large portions of the plasma proteins diffuse freely into this space.

In the interlobular septa are also vast numbers of *terminal lymphatics*. The spaces of Disse connect directly with the lymphatics so that excess fluid in the spaces is removed through the lymphatics.

## FUNCTION OF THE HEPATIC VASCULAR SYSTEM

The function of the hepatic vascular system was discussed in Chapter 29 in connection with the portal veins. Briefly, this can be summarized as follows:

**Blood Flow Through the Liver.** About 1000 ml. of blood flows from the portal vein through the liver sinusoids each minute, and approximately an additional 400 ml. flows into the sinusoids from the hepatic artery, the total averaging about 1400 ml. per minute.

Total hepatic blood flow per minute can be measured by a modified Fick procedure in which the dye *Bromsulphalein* or the dye *indocyanine green* is injected into the circulatory system, and its concentration is measured in the arterial blood and also in venous blood collected from the hepatic vein by means of a catheter. From these measurements, one can

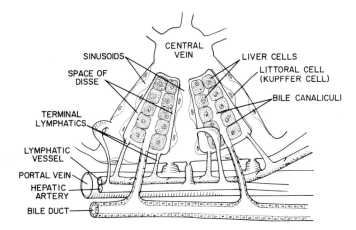

**Figure 70–1.** Basic structure of a liver lobule showing the hepatic cellular plates, the blood vessels, the bile-collecting system, and the lymph flow system comprised of the spaces of Disse and the interlobular lymphatics. (Reprinted from Guyton, Taylor, and Granger, as modified from Elias: Dynamics of the Body Fluids, 1975.)

calculate the arteriovenous difference of the dye. In addition, the rate at which the dye disappears from the blood is determined over a period of 10 minutes or more. Since the only significant route of disappearance is through the hepatic cells into the biliary tract, one can then calculate liver blood flow by the usual Fick formula as follows:

$$\text{Hepatic blood flow} = \frac{\text{Rate of blood clearance of dye}}{\text{A-V difference in dye}}$$

**Pressures and Resistance in the Hepatic Vessels.** The pressure in the hepatic vein leading from the liver into the vena cava averages almost exactly 0 mm. Hg, while the pressure in the portal vein leading into the liver averages 8 mm. Hg. This shows that the resistance to blood flow from the portal venous system to the systemic veins is normally low, especially when one considers that about 1.4 liters of blood flow by this route each minute. However, various pathologic conditions can cause the resistance to rise markedly, sometimes increasing the portal venous pressure to as high as 20 to 40 mm. Hg. The most common cause of increased hepatic vascular resistance is the disease *liver cirrhosis*, in which many of the exit pathways from the sinusoids into the central lobular venules are blocked.

**Storage of Blood in the Liver; Hepatic Congestion.** An increase in pressure in the veins draining the liver dams blood in the liver sinusoids and thereby causes the entire liver to swell markedly. The liver can store 200 to 400 ml. of blood in this way as the result of only 4 to 8 mm. Hg rise in hepatic venous pressure. For this reason, the liver is frequently said to be one of the *blood reservoirs*. Conversely, if a person hemorrhages so that large amounts of blood are lost from the circulatory system, much of the normal blood in the liver sinusoids drains into the remainder of the circulation to help replace the lost blood.

The most common cause of hepatic congestion is cardiac failure, which often increases the central venous pressure to as high as 10 to 15 mm. Hg. The continual stretching of the liver sinusoids that results and the stasis of blood caused by the hepatic congestion gradually leads to necrosis of many of the hepatic cells in the hepatic cellular plates.

**Lymph Flow from the Liver.** Because the pores in the hepatic sinusoids allow ready passage of proteins, the lymph draining from the liver usually has a protein concentration of about 6 grams per cent, which is only slightly less than the protein concentration of plasma. Also, the extreme permeability of the liver sinusoids allows large quantities of lymph to form. Indeed, between one third and one half of all the lymph formed in the body under resting conditions arises in the liver.

**Effects of High Hepatic Vascular Pressures on Fluid Transudation from the Liver Sinusoids and Portal Capillaries.** When the hepatic venous pressure rises only 3 to 5 mm. Hg above normal, excessive amounts of fluid begin to transude into the lymph and also to leak through the outer surface of the liver capsule directly into the abdominal cavity. This fluid is almost pure plasma, containing 80 to 90 per cent as much protein as normal plasma. At still higher hepatic venous pressures, 10 to 15 mm. Hg, liver lymph flow increases to as much as 20 times normal, and the "sweating" from the surface of the liver can be so great that it causes severe *ascites* (free fluid in the abdominal cavity).

Blockage of portal flow into or through the liver causes very high portal capillary pressures, resulting in edema of the gut wall and transudation of fluid through the serosa of the gut into the abdominal cavity. This, too, can cause ascites but is less likely to do so than is sweating from the liver surface, because collateral vascular channels develop rapidly from the portal veins to the systemic veins, decreasing the intestinal capillary pressure back to a safe value.

**The Hepatic Reticuloendothelial System.** The inner surfaces of all the liver sinusoids are loaded with many *Kupffer cells*, which protrude into the flowing blood as illustrated in Figure 70–1 and as shown in detail in Figure 6–4 of Chapter 6. These cells are very phagocytic, so much so that they can remove 99 per cent (or more) of bacteria in the portal venous blood before they can pass all the way through the liver sinusoids. Since the portal blood drains from the intestines, it almost always contains a reasonable number of colon bacilli. Therefore, the importance of the Kupffer cell filtration system is readily apparent. The number of Kupffer cells in the sinusoids increases markedly when increased quantities of particulate matter or other debris are present in the blood.

# SECRETION OF BILE AND FUNCTIONS OF THE BILIARY TREE

## PHYSIOLOGIC ANATOMY OF BILIARY SECRETION

All the hepatic cells continually form a small amount of secretion called *bile*. This is secreted into the minute *bile canaliculi,* which lie between the hepatic cells in the hepatic plates, and the bile then flows peripherally toward the interlobular septa where the canaliculi empty into *terminal bile ducts,* then into progressively larger ducts, finally reaching the *hepatic duct* and *common bile duct* from which the bile either empties directly into the duodenum or is diverted into the gallbladder.

**Storage of Bile in the Gallbladder.** In the discussion of bile and its function in relation to gastrointestinal secretions in Chapter 64, it was pointed out that bile secreted continually by the liver cells is normally stored in the gallbladder until needed in the duodenum. The total secretion of bile each day is some 800 to 1000 ml., and the maximum volume of the gallbladder is only 40 to 70 ml. Nevertheless, as much as 12 hours' bile secretion can be stored in the gallbladder because water, sodium, chloride, and most other small electrolytes are continually absorbed by the gallbladder mucosa, concentrating the other bile constituents, including the bile salts, cholesterol, and bilirubin. Bile is normally concentrated about five-fold, but it can be concentrated up to a maximum of 10- to 12-fold.

*Emptying of the Gallbladder.* Two basic conditions are necessary for the gallbladder to empty: (1) The sphincter of Oddi must relax to allow bile to flow from the common bile duct into the duodenum and (2) the gallbladder itself must contract to provide the force required to move the bile along the common duct. After a meal, particularly one that contains a high concentration of fat, both these effects take place in the following manner:

First, the fat (also the protein) in the food entering the small intestine causes release of a hormone called *cholecystokinin* from the intestinal mucosa, especially from the upper regions of the small intestine. The cholecystokinin in turn is absorbed into the blood and, on passing to the gallbladder, causes specific contraction of the gallbladder muscle. This provides the pressure that forces bile toward the duodenum.

Second, vagal stimulation associated with the cephalic phase of gastric secretion or with various intestino-intestinal reflexes causes an additional weak contraction of the gallbladder, which helps to force the bile from the gallbladder into the duodenum.

Third, when the gallbladder contracts, the sphincter of Oddi becomes inhibited, this effect resulting from either a neurogenic or a myogenic reflex from the gallbladder to the sphincter of Oddi. This inhibition may also, to some extent, be a direct effect of cholecystokinin on the sphincter, causing relaxation.

Fourth, the presence of food in the duodenum causes the degree of peristalsis in the duodenal wall to increase. Each time a peristaltic wave travels toward the sphincter of Oddi, this sphincter, along with the adjacent intestinal wall, momentarily relaxes because of the phenomenon of "receptive relaxation" that travels ahead of the peristaltic contraction wave, and, if the bile in the common bile duct is under sufficient pressure, a small quantity of the bile squirts into the duodenum.

In summary, the gallbladder empties its store of concentrated bile into the duodenum mainly in response to the cholecystokinin stimulus. When fat is not in the meal, the gallbladder empties poorly, but, when adequate quantities of fat are present, the gallbladder empties completely in about one hour.

Figure 70–2 summarizes the secretion of bile, its storage in the gallbladder, and its release from the bladder to the gut.

**Composition of Bile.** Table 70–1 gives the composition of bile when it is first secreted by the liver and then after it has been concentrated in the gallbladder. This table shows that the most abundant substance secreted in the bile is the *bile salts,* but also secreted or excreted in large concentrations are *bilirubin, cholesterol, lecithin,* and the usual *electrolytes* of plasma. In the concentrating process in the gallbladder, water and large portions of the electro-

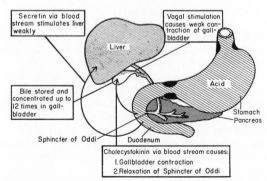

**Figure 70–2.** Mechanisms of liver secretion and gallbladder emptying.

## TABLE 70–1.    Composition of Bile

|              | Liver Bile       | Gallbladder Bile    |
|--------------|------------------|---------------------|
| Water        | 97.5  gm. %      | 92 gm. %            |
| Bile salts   | 1.1  gm. %       | 6 gm. %             |
| Bilirubin    | 0.04  gm. %      | 0.3 gm. %           |
| Cholesterol  | 0.1  gm. %       | 0.3 to 0.9 gm. %    |
| Fatty acids  | 0.12 gm. %       | 0.3 to 1.2 gm. %    |
| Lecithin     | 0.04 gm. %       | 0.3 gm. %           |
| $Na^+$       | 145 mEq./l.      | 130 mEq./l.         |
| $K^+$        | 5 mEq./l.        | 12 mEq./l.          |
| $Ca^+$       | 5 mEq./l.        | 23 mEq./l.          |
| $Cl^-$       | 100 mEq./l.      | 25 mEq./l.          |
| $HCO_3^-$    | 28 mEq./l.       | 10 mEq./l.          |

lytes are reabsorbed by the gallbladder mucosa, but essentially all the other constituents, including especially the bile salts and lipid substances such as cholesterol, are not reabsorbed and therefore become highly concentrated in the gallbladder bile.

## THE BILE SALTS AND THEIR FUNCTION

The liver cells form about 0.5 gram of *bile salts* daily. The precursor of the bile salts is *cholesterol,* which is either supplied in the diet or synthesized in the liver cells during the course of fat metabolism and then converted to *cholic acid* or *chenodeoxycholic acid* in about equal quantities. These acids then combine principally with glycine and to a lesser extent with taurine to form *glyco-* and *tauro-conjugated acids*. The salts of these acids are secreted in the bile.

The bile salts have two important actions in the intestinal tract. First, they have a detergent action on the fat particles in the food, which decreases the surface tension of the particles and allows the agitation in the intestinal tract to break the fat globules into minute sizes. This is called the *emulsifying* or *detergent function* of bile salts. Second, and even more important than the emulsifying function, bile salts help in the absorption of fatty acids, monoglycerides, cholesterol, and other lipids from the intestinal tract. They do this by forming minute complexes with the fatty acids and monoglycerides; the complexes are called *micelles,* and they are highly soluble because of the electrical charges of the bile salts. The lipids are "ferried" in this form to the mucosa where they are then absorbed; this mechanism was described in detail in Chapter 65. Without the presence of bile salts in the intestinal tract, up to 40 per cent of the lipids are lost into the stools, and the person often develops a metabolic deficit due to this nutrient loss.

Also, when fats are not absorbed adequately, the fat-soluble vitamins are not absorbed satisfactorily. Therefore, in the absence of bile salts, vitamins A, D, E, and K are poorly absorbed. Though large quantities of the first three of these vitamins are usually stored in the body, this is not true of vitamin K. Within only a few days after bile secretion ceases, the person usually develops a deficiency of vitamin K. This in turn results in deficient formation by the liver of several blood coagulation factors—prothrombin, and factors VII, IX, and X—thus resulting in serious impairment of blood coagulation.

**Enterohepatic Circulation of Bile Salts.**    Approximately 94 per cent of the bile salts are reabsorbed by the intestinal mucosa in the distal ileum. They enter the portal blood and pass to the liver. On reaching the liver the bile salts are absorbed from the venous sinusoids into the hepatic cells and then resecreted into the bile. In this way about 94 per cent of all the bile salts are recirculated into the bile, so that on the average these salts make the entire circuit some 18 times before being carried out in the feces. The small quantities of bile salts lost into the feces are replaced by new amounts formed continually by the liver cells. This recirculation of the bile salts is called the *enterohepatic circulation*.

The quantity of bile secreted by the liver each day is highly dependent on the availability of bile salts— the greater the quantity of bile salts in the enterohepatic circulation (usually a total of about 4 gm.), the greater is the rate of bile secretion. When a bile fistula forms so that bile is lost directly from the common bile duct to the exterior, the bile salts cannot be reabsorbed. Therefore, the total quantity of bile salts in the enterohepatic circulation becomes greatly depressed, and concurrently the volume of liver secretion is also depressed.

However, if a bile fistula continues to empty the bile salts to the exterior for several days to several weeks, the liver increases its production of bile salts as much as 10-fold, which increases the rate of bile secretion approximately back to normal. This also demonstrates that the daily rate of bile salt secretion is actively controlled, though the mechanism of this control is unknown.

## EXCRETION OF BILIRUBIN IN THE BILE

In addition to secreting substances synthesized by the liver itself, the liver cells also *excrete* a number of substances formed elsewhere in the body. Among the most important of these is *bilirubin,* which is one of the major end-products of hemoglobin decomposition, as was pointed out in Chapter 5.

Briefly, when the red blood cells have lived out their life span, averaging 120 days, and have become too fragile to exist longer in the circulatory system, their cell membranes rupture, and the released hemoglobin is phagocytized by reticuloendothelial cells throughout the body. Here, the hemoglobin is first split into *globin* and *heme,* and the heme ring is opened to give a straight chain of four pyrrole nuclei that is the substrate from which the bile pigments are formed. The first pigment formed is *biliverdin,* but this is rapidly reduced to *free bilirubin,* which is gradually released into the plasma. However, the free bilirubin immediately combines very strongly with the plasma albumin and is transported in this combination throughout the blood and interstitial fluids. Even when bound with the plasma protein, this bilirubin is still called "free bilirubin" to distinguish it from "conjugated bilirubin," which will be discussed below. Within hours, the free bilirubin is absorbed through the hepatic cell membrane, in this process being released from the plasma albumin but almost instantly being combined with another protein (called "Y" protein) inside the hepatic cells that traps the bilirubin inside the cells. Soon thereafter, however, the bilirubin is also removed from this protein and conjugated with other substances. About 80 per cent of it conjugates with glucuronic acid to form *bilirubin glucuronide;* an additional 10 per cent conjugates with sulfate to form *bilirubin sulfate,* and the final 10 per cent conjugates with a multitude of other substances. It is in these forms that the bilirubin is excreted by an active transport process into the bile ducts.

A small portion of the conjugated bilirubin formed by the hepatic cells returns to the plasma, either directly into the liver sinusoids or indirectly by absorption into the blood from the bile ducts or lymphatics. Regardless of the exact mechanism by which it reenters the blood, this causes a small portion of the bilirubin in the extracellular fluids always to be of the conjugated type rather than of the free type.

**Formation and Fate of Urobilinogen.** Once in the intestine, bilirubin is converted by bacterial action mainly into the substance *urobilinogen,* which is highly soluble. Some of the urobilinogen is reabsorbed through the intestinal mucosa into the blood. Most of this is re-excreted by the liver back into the gut, but about 5 per cent of it is excreted by the kidneys into the urine. After exposure to air in the urine, the urobilinogen becomes oxidized to *urobilin,* or it becomes altered and oxidized in the feces to

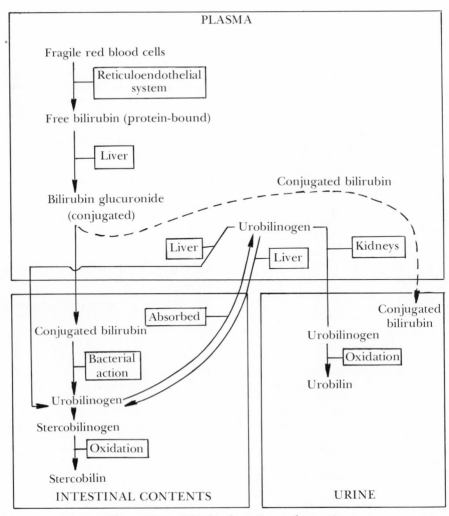

**Figure 70–3.** Bilirubin formation and excretion.

form *stercobilin*. These interrelationships of bilirubin and the other bile pigments are illustrated in Figure 70–3.

**Jaundice.** The word "jaundice" means a yellowish tint to the body tissues, including yellowness of the skin and also of the deep tissues. The usual cause of jaundice is large quantities of bilirubin in the extracellular fluids, either free bilirubin or conjugated bilirubin. The normal plasma concentration of bilirubin, including both the free and the conjugated forms, averages 0.5 mg. per 100 ml. of plasma. However, in certain abnormal conditions this can rise to as high as 40 mg. per 100 ml. The skin usually begins to appear jaundiced when the concentration rises to about three times normal—that is, above 1.5 mg. per 100 ml.

The common causes of jaundice are: (1) increased destruction of red blood cells with rapid release of bilirubin into the blood or (2) obstruction of the bile ducts or damage to the liver cells so that even the usual amounts of bilirubin cannot be excreted into the gastrointestinal tract. These two types of jaundice are called, respectively, *hemolytic jaundice* and *obstructive jaundice*. They differ from each other in the following ways:

*Hemolytic Jaundice.* In hemolytic jaundice, the excretory function of the liver is not impaired in the least, but red blood cells are hemolyzed rapidly and the hepatic cells simply cannot excrete the bilirubin as rapidly as it is formed. Therefore, the plasma concentration of *free bilirubin* rises especially high. Likewise, the rate of formation of *urobilinogen* in the intestine is greatly increased, and much of this is absorbed into the blood and later excreted by the kidneys.

*Obstructive Jaundice.* In obstructive jaundice, caused either by obstruction of the bile ducts or by damage to the liver cells, the rate of bilirubin formation is normal, but the bilirubin formed simply cannot pass from the blood into the intestines. However, the free bilirubin usually does still enter the liver cells and becomes conjugated in the usual way. This conjugated bilirubin is then returned to the blood probably by rupture of the congested bile canaliculi and direct emptying of the bile into the lymph leaving the liver. Thus, *most of the bilirubin in the plasma becomes the conjugated type* rather than the free type.

***Diagnostic Differences Between Hemolytic and Obstructive Jaundice.*** A simple test called the *van den Bergh test* can be used to differentiate between free and conjugated bilirubin in the plasma. If an immediate reaction occurs with the van den Bergh reagent, the bilirubin is of the conjugated type, and the reaction is called a "direct van den Bergh reaction." However, to demonstrate the presence of free bilirubin, one must first add alcohol to the plasma. This precipitates the protein and "frees" the free bilirubin from its protein complex so that it can then combine with the van den Bergh reagent. This result is called the "indirect van den Bergh reaction." Thus, *in hemolytic jaundice an indirect van den Bergh reaction occurs (increased free bilirubin) and in obstruc-*

*tive jaundice a direct van den Bergh reaction occurs (increased conjugated bilirubin).*

When there is total obstruction of bile flow, no bilirubin at all can reach the intestines to be converted into urobilinogen by bacteria. Therefore, urobilinogen is not reabsorbed into the blood and is not excreted by the kidneys into the urine. Consequently, in *total* obstructive jaundice, tests for urobilinogen in the urine are completely negative. Also, the stools become clay colored for lack of stercobilin and other bile pigments.

Another major difference between free and conjugated bilirubin is that kidneys can excrete conjugated bilirubin but not free bilirubin. Therefore, in severe obstructive jaundice, large quantities of conjugated bilirubin appear in the urine. This can be demonstrated simply by shaking the urine and observing the foam, which becomes colored an intense yellow.

Thus, by understanding the physiology of bilirubin excretion by the liver and by use of a few simple tests, it is possible to differentiate between obstructive and hemolytic jaundice and, often, also to determine the severity of the disease.

### SECRETION OF CHOLESTEROL; GALLSTONE FORMATION

Bile salts are formed in the hepatic cells from cholesterol, and in the process of secreting the bile salts about one-tenth as much cholesterol is also secreted into the bile. No specific function is known for the cholesterol in the bile, and it is presumed that it is simply a by-product of bile salt formation and secretion.

Cholesterol is almost insoluble in pure water, but the bile salts and lecithin in bile combine physically with the cholesterol to form ultramicroscopic *micelles* that are soluble. When the bile becomes concentrated in the gallbladder, the bile salts and lecithin become concentrated along with the cholesterol, which keeps the cholesterol in solution. Under abnormal conditions, however, the cholesterol may precipitate, resulting in the formation of *gallstones,* as shown in Figure 70–4. The different conditions that can cause cholesterol precipitation are: (1) too much absorption of water from the bile, (2) too much absorption of bile salts and lecithin from the bile, (3) too much secretion of cholesterol in the bile, and (4) inflammation of the epithelium of the gallbladder. The latter two of these require special explanation, as follows:

The amount of cholesterol in the bile is determined partly by the quantity of fat that the person eats, for the hepatic cells synthesize cholesterol as one of the products of fat metabolism in the body. For this reason, persons on a high fat diet over a period of many years are prone to the development of gallstones.

Inflammation of the gallbladder epithelium often results from low grade chronic infection; this changes the absorptive characteristics of the gallbladder mucosa, sometimes allowing excessive absorption of

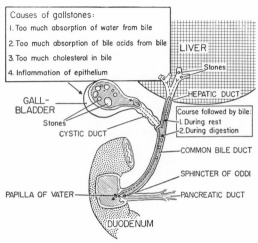

**Figure 70–4.** Formation of gallstones.

water, bile salts, or other substances that are necessary to keep the cholesterol in solution. As a result, cholesterol begins to precipitate, usually forming many small crystals of cholesterol on the surface of the inflamed mucosa. These, in turn, act as nidi for further precipitation of cholesterol, and the crystals grow larger and larger. Occasionally tremendous numbers of sand-like stones develop, but much more frequently they coalesce to form a few large gallstones, or even a single stone that fills the entire gallbladder.

**X-ray Opaque Gallstones.** Calcium sometimes precipitates in the form of calcium carbonate in the gallstones or precipitates with some of the fatty substances involved in the formation of cholesterol gallstones. In about one fourth of all cases of gallstones, the concentration of calcium in the stones is great enough to make them x-ray opaque. That is, the gallstones show up on x-ray pictures of the abdomen. In the remaining three fourths of the cases, the stones have almost no calcium and therefore have almost the same x-ray opaqueness as the tissues of the body. Consequently, the stones cannot be seen on an x-ray film. In these instances, special iodinated drugs can be administered to the patient. These drugs concentrate in the bile surrounding the stones, and an x-ray picture shows the negative outlines of the stones.

# METABOLIC FUNCTIONS OF THE LIVER

The metabolic functions of the liver are so numerous and intricate that they could not possibly be presented completely in this chapter. Therefore, for details, refer to the preceding chapters on the metabolism of carbohydrates, fats, and proteins.

Briefly, the specific roles of the liver in the different metabolic processes are described as follows:

## CARBOHYDRATE METABOLISM

In carbohydrate metabolism the liver performs the following specific functions: (1) storage of glycogen, (2) conversion of galactose to glucose, (3) gluconeogenesis, and (4) formation of many important chemical compounds from the intermediate products of carbohydrate metabolism.

The liver is especially important for maintaining a normal blood glucose concentration. For instance, storage of glycogen allows the liver to remove excess glucose from the blood, store it, and then return it to the blood when the blood glucose concentration begins to fall too low. This is called the *glucose buffer function* of the liver. As an example, immediately after a meal containing large amounts of carbohydrates, the blood glucose concentration rises about three times as much in a person with a nonfunctional liver as in a person with a normal liver.

*Gluconeogenesis* in the liver is also concerned with maintaining a normal blood glucose concentration, for gluconeogenesis occurs to a significant extent only when the glucose concentration begins to fall below normal. In such a case, large amounts of amino acids are converted into glucose, thereby helping to maintain a relatively normal blood glucose concentration.

## FAT METABOLISM

Though fat metabolism can take place in almost all cells of the body, certain aspects of fat metabolism occur much more rapidly in the liver than in the other cells. Some specific functions of the liver in fat metabolism are: (1) very high rate of beta oxidation of fatty acids and formation of acetoacetic acid, (2) formation of the lipoproteins, (3) formation of large quantities of cholesterol and phospholipids, and (4) conversion of large quantities of carbohydrates and proteins to fat.

To derive energy from neutral fats, the fat is first split into glycerol and fatty acids; then the fatty acids are split by *beta oxidation* into two-carbon acetyl radicals which form *acetyl coenzyme A* (acetyl Co-A). This in turn can then enter the tricarboxylic acid cycle and be oxidized to liberate tremendous amounts of energy. Beta oxidation can probably take place in all cells of the body, but it occurs so rapidly in the hepatic cells in comparison with the others that much of the initial oxidation of fatty acids in the body occurs in the liver. Yet the liver itself cannot utilize all the acetyl Co-A that is formed; instead, this is converted by condensation of two molecules of acetyl Co-A into *acetoacetic acid,* which is a highly soluble acid that passes from the liver cells into the extracellular fluids and then is transported throughout the body to be absorbed by the other tissues. These tissues in turn reconvert the acetoacetic acid into acetyl Co-A and then oxidize it in the usual manner. In this way, therefore, the liver is responsible for a major part of the metabolism of fats.

Except for the use of cholesterol to form bile salts, the functions of the cholesterol and phospholipids formed in the liver are still in doubt. About 80 per cent of the cholesterol is converted into bile salts, but the remainder enters the blood to be transported principally in the lipoproteins. The phospholipid lecithin likewise is transported principally in the lipoproteins. It is possible that both these substances, along with the triglyceride fractions of the lipoproteins, are absorbed by cells everywhere in the body to help form the cell membranes and intracellular structures, for it is well known that most membranous structures throughout the body contain major amounts of cholesterol, phospholipids, and triglycerides.

Most of the fat synthesis in the body from carbohydrates and proteins also occurs in the liver. After fat is synthesized in the liver it is transported in the lipoproteins to the adipose tissue to be stored.

## PROTEIN METABOLISM

Even though a large proportion of the metabolic processes for carbohydrates and fat metabolism occurs in the liver, the body could probably dispense with these functions of the liver and still survive. On the other hand, the body could not dispense with the services of the liver in protein metabolism for more than a few days without death ensuing. The most important functions of the liver in protein metabolism are: (1) deamination of amino acids, (2) formation of urea for removal of ammonia from the body fluids, (3) formation of plasma proteins, and (4) interconversions among the different amino acids and other compounds important to the metabolic processes of the body.

Deamination of the amino acids is required before these can be used for energy or before they can be converted into carbohydrates or fats. A small amount of deamination can occur in the other tissues of the body, especially in the kidneys; but the percentage of deamination occurring extrahepatically is so small that it is almost completely unimportant.

Formation of urea by the liver removes ammonia from the body fluids. Moderate amounts of ammonia are continually formed in the gut by bacteria and are then absorbed into the blood; therefore, without this function of the liver the plasma ammonia concentration rises rapidly and results in *hepatic coma* and death. Indeed, any failure of portal blood to flow through the liver—as occurs occasionally when a shunt develops between the portal vein and the vena cava—can also cause excessive ammonia in the blood, an exceedingly toxic condition.

Essentially all the plasma proteins, with the exception of part of the gamma globulins, are formed by the hepatic cells. This accounts for more than 85 per cent of all the plasma proteins. The remaining gamma globulins are the immune bodies formed mainly by the plasma cells in the lymph tissue of the body. The liver can form plasma proteins at a maximum rate of 50 to 100 grams per day. Therefore, after loss of as much as half the plasma proteins from the body, these can be replenished in approximately four to seven days. It is particularly interesting that plasma protein depletion causes rapid mitosis of the hepatic cells and actual growth of the liver to a larger size; these effects are coupled with rapid output of plasma proteins until the plasma concentration returns to normal.

Among the most important functions of the liver is its ability to synthesize certain amino acids and also to synthesize other important chemical compounds from amino acids. For instance, the so-called nonessential amino acids can be synthesized in the liver. To do this, a keto acid having the same chemical composition (except at the keto oxygen) as that of the amino acid to be formed is first synthesized. Then an amino radical is transferred through several stages of *transamination* from an available amino acid to the keto acid to take the place of the keto oxygen.

## MISCELLANEOUS METABOLIC FUNCTIONS OF THE LIVER

**Storage of Vitamins.** The liver has a particular propensity for storing vitamins and has long been known as an excellent source of certain vitamins in treating patients. The single vitamin stored to the greatest extent in the liver is vitamin A, but large quantities of vitamin D and vitamin $B_{12}$ are normally stored as well. Sufficient quantities of vitamin A can be stored to prevent vitamin A deficiency for as long as one to two years, and sufficient vitamin D and vitamin $B_{12}$ can be stored to prevent deficiency for as long as one to four months.

**Relationship of the Liver to Blood Coagulation.** The liver forms a large proportion of the blood substances utilized in the coagulation process. These are fibrinogen, prothrombin, accelerator globulin, factor VII, and several other less important coagulation factors. Vitamin K is required by the metabolic processes of the liver for the formation of prothrombin and factors VII, IX, and X. In the absence of vitamin K the concentrations of these substances fall very low and almost prevent blood coagulation.

**Storage of Iron.** Except for the iron in the hemoglobin of the blood, by far the greater proportion of the iron in the body is usually stored in the liver in the form of *ferritin*. The hepatic cells contain large amounts of a protein called *apoferritin*, which is capable of combining with either small or large quantities of iron. Therefore, when iron is available in the body fluids in extra quantities, it combines with the apoferritin to form ferritin and is stored in this form until needed by the body. When the iron in the circulating body fluids fall low, the ferritin releases the iron. Thus, the apoferritin-ferritin system of the liver acts as an *iron buffer* and also as an iron storage medium. Other functions of the liver in relation to iron metabolism are considered in Chapter 5.

# REFERENCES

Adelman, R. C., Freeman, C., and Rotenberg, S.: Impairments in hormonal control of liver enzyme activity in aging rats. *Prog. Brain Res.*, *40*:509, 1973.

Billing, B. H.: Bile pigment metabolism. *Sci. Basis Med. Ann. Rev.*, 197, 1963.

Borgström, B.: Bile salts—their physiological functions in the gastrointestinal tract. *Acta Med. Scan.*, *196*:1, 1974.

Brauer, R. W.: Liver circulation and function. *Physiol. Rev.* *43*:115, 1963.

Brown, H., and Hardwick, D. F. (eds.): Intermediary Metabolism of the Liver. Springfield, Ill., Charles C Thomas, Publisher, 1973.

Child, C. G., 3rd: Portal Hypertension as Seen Today by Seventeen Authorities. Philadelphia, W. B. Saunders Company, 1975.

Dowling, R. H.: The enterohepatic circulation. *Gastroenterology*, *62*:122, 1972.

Erlinger, S., and Dhumeaux, D.: Mechanisms and control of secretion of bile water and electrolytes. *Gastroenterology*, *66*:281, 1974.

Gall, E. A., and Mostofi, F. K. (eds.): The Liver. Baltimore, The Williams & Wilkins Company, 1973.

Goldstein, F.: Pathophysiology of gallbladder disease. *In* Sodeman, W. A., Jr., and Sodeman, W. A. (eds.): Pathologic Physiology: Mechanisms of Disease, 5th Ed. Philadelphia, W. B. Saunders Company, 1974, p. 818.

Hays, D. M.: Surgical research aspects of hepatic regeneration. *Surg. Gynecol. Obstet.*, *139*:609, 1974.

Heaton, K. W.: Bile Salts in Health and Disease. New York, Churchill Livingstone, Div. of Longman, Inc., 1972.

Iber, F. L.: Normal and pathologic physiology of the liver. *In* Sodeman, W. A., Jr., and Sodeman, W. A. (eds.): Pathologic Physiology: Mechanisms of Disease, 5th Ed. Philadelphia, W. B. Saunders Company, 1974, p. 790.

Martini, G. A., Baltzer, G., and Arndt, H.: Some aspects of circulatory disturbances in cirrhosis of the liver. *Prog. Liver Dis.*, *4*:231, 1972.

Preisig, R., Bircher, J., and Paumgartner, G.: Physiologic and pathophysiologic aspects of the hepatic hemodynamics. *Prog. Liver Dis.*, *4*:201, 1972.

Reynolds, T. B.: The role of hemodynamic measurements in portosystemic shunt surgery. *Arch. Surg.*, *108*:276, 1974.

Robinson, S. H.: The origins of bilirubin. *New Eng. J. Med.*, *279*:143, 1968.

Saba, T. M., and Di Luzio, N. R.: Reticuloendothelial blockade and recovery as function of opsonic activity. *Amer. J. Physiol.*, *216*:197, 1969.

Schein, C. J.: Acute Cholecystitis. Hagerstown, Md., Harper & Row, Publishers, 1972.

Sherlock, S.: Liver failure. *Sci. Basis Med. Ann. Rev.*, 216, 1961.

Shoemaker, W. C., and Elwyn, D. H.: Liver: functional interactions within the intact animal. *Ann. Rev. Physiol.*, *31*:227, 1969.

Viamonte, M., Jr., and Viamonte, M.: Liver circulation. *C.R.C. Crit. Rev. Clin. Radiol. Nucl. Med.*, *5*:351, 1974.

Weiner, I. M., and Lack, L.: Bile salt absorption; enterohepatic circulation. *In* Code, C. F., and Heidel, W. (eds.): Handbook of Physiology. Sec. 6, Vol. 3. Baltimore, The Williams & Wilkins Company, 1968, p. 1439.

# 71

# Energetics and Metabolic Rate

## IMPORTANCE OF ADENOSINE TRIPHOSPHATE (ATP) IN METABOLISM

In the last few chapters it has been pointed out that carbohydrates, fats, and proteins can all be used by the cells to synthesize large quantities of ATP, and that in turn the ATP can be used as an energy source for many other cellular functions. For these reasons, ATP has been called an energy ''currency'' that can be created and expended. Indeed, the cells can transfer energy from the different foodstuffs to most functional systems of the cells only through this medium of ATP. Many of the attributes of ATP were presented in Chapter 2, but others require discussion at this point.

An attribute of ATP that makes it highly valuable as a means of energy currency is the large quantity of free energy (about 8000 calories per mol under physiological conditions) vested in each of its two high energy phosphate bonds. The amount of energy in each bond, when liberated by decomposition of one molecule of ATP, is enough to cause almost any step of any chemical reaction in the body to take place if appropriate transfer of the energy is achieved. If the bonds of ATP did not have this much energy, many of the chemical reactions could not be energized by this compound. Some chemical reactions that require ATP energy use only a few hundred of the available 8000 calories, and the remainder of this energy is then lost in the form of heat. Yet even this inefficiency in the utilization of energy is better than not being able to energize the necessary chemical reactions at all.

The precise methods by which ATP is used to cause the many physical, chemical, and other types of functions in the cells lie principally in the province of biochemistry. Therefore, only the more important functions of ATP are listed here.

**Use of ATP for Synthesis of Important Cellular Components.** Probably by far the most important intracellular process that requires ATP is formation of peptide linkages between amino acids during the synthesis of proteins. The energy from two high energy bonds of ATP is used for each peptide linkage. Since an occasional protein of the body has as many as 100,000 amino acids linked together in this manner, it can readily be understood how much ATP is required for this cellular function. The different peptide linkages, depending on which types of amino acids are linked together, require from 500 to 4000 calories of energy per mol. In each instance, the amount of energy available from the two ATP high energy bonds, 16,000 calories, is always more than sufficient.

Also, it will be recalled from the preceding chapters that ATP is utilized in the synthesis of glucose from lactic acid and in the synthesis of fatty acids from acetyl Co-A. In addition, ATP is utilized in the synthesis of cholesterol, phospholipids, the hormones, and almost all other substances of the body. Even the urea excreted by the kidneys requires ATP to cause its formation. One might wonder at the advisability of expending energy to form urea which then is simply thrown away from the body. However, if he remembers the extreme toxicity of ammonia in the body fluids, he can see the value of this reaction, which keeps the ammonia concentration of the body fluids always at a low level.

**Use of ATP for Muscular Contraction.** Muscular contraction will not occur without energy from ATP. Myosin, one of the important contractile proteins of the muscle fiber, acts as an enzyme to cause breakdown of ATP into adenosine diphosphate

(ADP), thus causing the release of energy. However, the means by which this energy is coupled to the contractile process of the muscle fiber is still conjecture. Only a small amount of ATP is normally degraded in muscles when muscular contraction is not occurring, but this rate of ATP usage can rise to more than 100 times the resting level during maximal contraction. Postulated mechanisms by which ATP is utilized to cause muscle contraction were discussed in Chapter 11.

**Use of ATP for Active Transport Across Membranes.** In Chapters 4, 34, and 65, active transport of electrolytes and various nutrients across cell membranes and from the renal tubules and gastrointestinal tract was discussed. In each instance, it was noted that active transport of most electrolytes and other substances such as glucose, amino acids, and acetoacetate can occur against an electrochemical gradient, even though the natural diffusion of the substances would be in the opposite direction. Obviously, to oppose the electrochemical gradient requires energy, as was discussed in Chapter 4. This energy is provided by ATP. However, the precise means by which the energy is transferred from ATP to the active transport systems are still unknown.

*Energy for Glandular Secretion.* The same principles apply to glandular secretion as to the absorption of substances against concentration gradients, for many substances to be secreted are concentrated as they pass through the glandular cell. This concentrating process requires appropriate amounts of energy. In addition, energy is also required to synthesize the organic compounds to be secreted.

*Energy for Nerve Conduction.* The energy utilized during propagation of a nerve impulse is derived from the potential energy stored in the form of concentration differences of ions across the membranes. That is, a high concentration of potassium inside the fiber and a low concentration outside the fiber constitutes a type of energy storage. Likewise, a high concentration of sodium on the outside of the membrane and a low concentration on the inside represents another store of energy. The energy needed to pass each impulse along the fiber membrane is derived from this energy storage, with small amounts of potassium transferring out of the cell and sodium into the cell. However, active transport systems then retransport the ions back through the membrane to their former positions. Here, ATP is utilized in abundance to retransfer the sodium and potassium ions after nerve impulses have been conducted, and the rate of ATP usage increases in proportion to the number of nerve impulses transmitted.

## CREATINE PHOSPHATE AS A STORAGE DEPOT FOR ENERGY

Despite the paramount importance of ATP as a coupling agent for energy transfer, this substance is not the most abundant store of high energy phosphate bonds in the cells. On the contrary, *creatine phosphate,* which also contains high energy phosphate bonds, is several times as abundant, at least in muscle. The high energy bond of creatine phosphate contains about 8500 calories per mol under standard conditions, or 9500 calories per mol under conditions in the body (38° C. and low concentrations of the reactants). This is not greatly different from the 8000 calories per mol in each of the two high energy phosphate bonds of ATP. The formula for creatine phosphate is the following:

$$HOOC-CH_2-N-C-N \sim P-OH$$

Creatine phosphate cannot act in the same manner as ATP as a coupling agent for transfer of energy between the foods and the functional cellular systems. But it can transfer energy interchangeably with ATP. When extra amounts of ATP are available in the cell, much of its energy is utilized to synthesize creatine phosphate, thus building up this storehouse of energy. Then when the ATP begins to be used up, the energy in the creatine phosphate is transferred rapidly back to ATP and from this to the functional systems of the cells. This reversible interrelationship between ATP and creatine phosphate is illustrated by the following equation:

$$\text{Creatine phosphate} + \text{ADP} \updownarrow \text{ATP} + \text{Creatine}$$

Note particularly that the higher energy level of the high energy phosphate bond in creatine phosphate, 9500 in comparison with 8000 calories per mol, causes the reaction between creatine phosphate and ATP to proceed to an equilibrium state very much in favor of ATP. Therefore, the slightest utilization of ATP by the cells calls forth the energy from the creatine phosphate to synthesize new ATP. This effect keeps the concentration of ATP at an almost constant level for the following reason: The greater energy in the creatine phosphate bond than in the ATP bond allows such rapid transfer of energy to the ATP system that almost all the creatine phosphate must be used up in the cell before the concentration of ATP will fall significantly. For this reason we can actually call creatine phosphate an ATP "sparer" or "buffer" compound.

## ANAEROBIC VERSUS AEROBIC ENERGY

*Anaerobic energy* means energy that can be derived from the foods without the simultaneous utilization of oxygen; *aerobic energy* means energy that can be derived from the foods only by oxidative metabolism. In the discussions in the preceding three chapters it was noted that carbohydrates, fats, and proteins can all be oxidized to cause synthesis of ATP. However, carbohydrates are the only significant foods that can be utilized to provide energy without utilization of oxygen; this energy release occurs during glycolytic breakdown of glucose or glycogen to pyruvic acid. For each mol of glucose that is split into pyruvic acid, 2 mols of ATP are formed. However, when glycogen is split to pyruvic acid, each mol of glucose in the glycogen gives rise to 3 mols of ATP. The reason for this difference is that free glucose entering the cell must be phosphorylated by 1 mol of ATP before it can begin to be split, whereas this is not true of glucose derived from glycogen because it comes from the glycogen in the phosphorylated state. Thus, the best source of energy under anaerobic conditions is the stored glycogen of the cells.

**Anaerobic Energy During Hypoxia.** One of the prime examples of anaerobic energy utilization occurs in acute hypoxia. When a person stops breathing, he already has a small amount of oxygen stored in his lungs and an additional amount stored in the hemoglobin of his blood. However, these are sufficient to keep the metabolic processes functioning for only about two minutes. Continued life beyond this time requires an additional source of energy. This can be derived for another minute or so from glycolysis, the glycogen of the cells splitting into pyruvic acid and the pyruvic acid in turn becoming lactic acid, which diffuses out of the cells as described in Chapter 67.

**Anaerobic Energy Usage in Strenuous Bursts of Activity.** It is common knowledge that muscles can perform extreme feats of strength for a few seconds but are much less capable during prolonged activity. The energy used during strenuous activity is derived from: (1) ATP already present in the muscle cells, (2) stored creatine phosphate in the cells, (3) anaerobic energy released by glycolytic breakdown of glycogen to lactic acid, and (4) oxidative energy released continuously by oxidative processes in the cells. The speed of the oxidative processes cannot approach that which is required to supply all the energy demands during strenuous bursts of activity. Therefore, the other three sources of energy are called upon to their maximum extent.

The maximum amount of ATP in muscle is only about 5 millimols per liter of intracellular fluid, and this amount can maintain maximum muscle contraction for not more than a few seconds. The amount of creatine phosphate in the cells may be several times this amount, but even by utilization of all the creatine phosphate, the amount of time that maximum contraction can be maintained is still only a few seconds. Release of energy by glycolysis can occur much more rapidly than can oxidative release of energy. Consequently, most of the extra energy required during strenuous activity that lasts for more than a few seconds but less than one to two minutes is derived from anaerobic glycolysis. As a result, the glycogen content of muscles after strenuous bouts of exercise becomes greatly reduced, while the lactic acid concentration of the blood rises. Then, immediately after the exercise is over, oxidative metabolism is used to reconvert about four fifths of the lactic acid into glucose, while the remainder becomes pyruvic acid and is degraded and oxidized in the citric acid cycle. The reconversion to glucose occurs principally in the liver cells, and the glucose is then transported in the blood back to the muscles where it is stored once more in the form of glycogen.

**Oxygen Debt.** After a period of strenuous exercise, the oxidative metabolic processes continue to operate at a high level of activity for many minutes to (1) reconvert the lactic acid into glucose and (2) reconvert the decomposed ATP and creatine phosphate to their original states. The extra oxygen that must be used in the oxidative energy processes to rebuild these substances is called the *oxygen debt*.

The oxygen debt is illustrated in Figure 71–1 by the shaded area. This figure shows, first, the normal rate of energy expenditure by the body, and, second, the normal rate of oxidative metabolism. At the three-minute mark, the person exercises strongly for about four minutes. During this time the oxidative release of energy provides some energy expenditure, but a large share of energy expenditure occurs at the expense of glycolysis and at the expense of stored ATP, creatine phosphate, and oxygen bound with hemoglobin and myoglobin. At the end of exercise, the oxidative release of energy con-

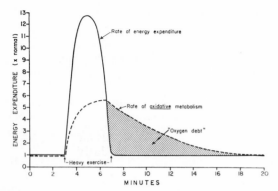

**Figure 71–1.** Oxygen debt occurring after a bout of strenuous exercise.

tinues high while the rate of energy expenditure returns to normal. The excess oxidative metabolism that must occur even after the period of exercise is over, as shown by the shaded area in the figure, represents the oxygen debt. This is usually expressed in terms of the excess oxygen that must be utilized to provide this excess metabolism.

### SUMMARY OF ENERGY UTILIZATION BY THE CELLS

With the background of the past few chapters and of the preceding discussion, we can now synthesize a composite picture of overall energy utilization by the cells as illustrated in Figure 71–2. This figure shows the anaerobic utilization of glycogen and glucose to form ATP and also the aerobic utilization of compounds derived from carbohydrates, fats, proteins, and other substances for the formation of still additional ATP. In turn, ATP is in reversible equilibrium with creatine phosphate in the cells, and, since large quantities of creatine phosphate are present in the cell, much of the stored energy of the cell is in this energy storehouse.

Energy from ATP can be utilized by the different functioning systems of the cells to provide for synthesis and growth, muscular contraction, glandular secretion, impulse conduction, active absorption, and other cellular activities. If greater amounts of energy are called forth for cellular activities than can be provided by oxidative metabolism, the creatine phosphate storehouse is first utilized, and this is followed rapidly by anaerobic breakdown of glycogen. Thus, oxidative metabolism cannot deliver energy to the cells nearly so rapidly as can the anaerobic processes, but in contrast it is quantitatively almost inexhaustible.

## CONTROL OF ENERGY RELEASE IN THE CELL

**Rate Control of Enzyme-Catalyzed Reactions.** Before it is possible to discuss the control of energy release in the cell, it is necessary to consider the basic principles of *rate control* of enzymatically catalyzed chemical reactions, which are the type of reactions that occur almost universally throughout the body.

The mechanism by which an enzyme catalyzes a chemical reaction is for the enzyme first to combine loosely with one of the substrates of the reaction. This alters the electrical forces on the substrate sufficiently that it can then react with other substances. Therefore, the rate of the overall chemical reaction is determined by both the concentration of the enzyme and the concentration of the substrate that binds with the enzyme. The basic equation expressing this concept is the following:

$$\text{Rate of reaction} = \frac{K_1 \cdot \text{Enzyme} \cdot \text{Substrate}}{K_2 + \text{Substrate}}$$

This is called the *Michaelis-Menten equation.* Figure 71–3 illustrates graphically the application of this equation.

*Role of Enzyme Concentration in the Regulation of Metabolic Reactions.* Figure 71–3 illustrates that *when the substrate is present in excess,* the rate of a chemical reaction is determined almost entirely by the concentration of the enzyme. Thus, as shown to the right in the figure, as the enzyme concentration increases from an arbitrary value of 1, 2, 4, or 8, the rate of the reaction increases proportionately. As an example of this effect, when large quantities of glucose enter the renal tubules in diabetes mellitus, the rate of reabsorption of the glucose is determined almost entirely by the concentration of the glucose transport enzymes in the proximal tubular cells because the substrate is then present in excess.

*Role of Substrate Concentration in Regulation of Metabolic Reactions.* Note also in Figure 71–3 that when the substrate concentration becomes low enough that only a small portion of the enzyme is required in the reaction, the rate of the reaction is directly proportional to the substrate concentration as well as enzyme concentration. This is the effect seen in the absorption of substances from the intestinal tract and renal tubules when their concentrations are very low. That is, the rate of absorption is then directly dependent on the concentration of the

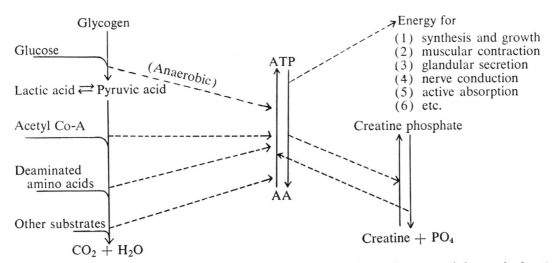

**Figure 71–2.** Overall schema of energy transfer from foods to the adenylic acid system and then to the functional elements of the cells. (Modified from Soskin and Levine: Carbohydrate Metabolism. University of Chicago Press.)

substance as well as on the concentration of the enzyme.

**Rate Limitation in a Series of Reactions.** It has become evident from the discussions in the preceding chapters that almost all chemical reactions of the body occur in series, the product of one reaction acting as a substrate of the next reaction, and so on. It is immediately obvious also that the overall rate of a complex series of chemical reactions is determined by the rate of reaction of the slowest step in the series. This is called the *rate limiting reaction* for the entire series.

**Adenosine Diphosphate (ADP) Concentration as a Rate-Controlling Factor in Energy Release.** Under *resting* conditions, the concentration of ADP in the cells is extremely slight so that the chemical reactions that depend on ADP as one of the substrates likewise are very slow. These include all the oxidative metabolic pathways as well as essentially all other pathways for release of energy in the body. Thus, *ADP is the major rate limiting factor* for almost all energy metabolism of the body.

When the cells become excessively active, regardless of the type of activity, large quantities of ATP are converted into ADP, increasing the concentration of ADP in direct proportion to the degree of activity of the cell. This automatically increases the rates of reactions of the ADP–rate-limiting steps in the metabolic release of energy. Thus, by this simple process, the amount of energy released in the cell is controlled by the degree of activity of the cell. In the absence of cellular activity, the release of energy stops, while in the presence of cellular activity the concentration of ADP increases, and all the energy-giving processes are automatically set in motion.

## THE METABOLIC RATE

The *metabolism* of the body means simply all the chemical reactions in all the cells of the body, and the *metabolic rate* is normally expressed in terms of the rate of heat liberation during the chemical reactions.

**Heat as the Common Denominator of All the Energy Released in the Body.** In discussing many of the metabolic reactions of the preceding chapters, we have noted that not all the energy in the foods is transferred to ATP; in-

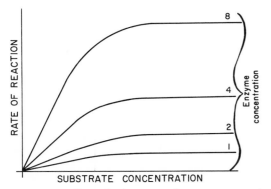

**Figure 71–3.** Effect of substrate and enzyme concentrations on the rate of enzyme-catalyzed reaction.

stead, a large portion of this energy becomes heat. On the average, about 55 per cent of the energy in the foods becomes heat during ATP formation. Then still more energy becomes heat as it is transferred from ATP to the functional systems of the cells, so that not more than about 25 per cent of all the energy from the food is finally utilized by the functional systems.

Even though 25 per cent of the energy finally reaches the functional systems of the cells, the major proportion of this also becomes heat for the following reasons: We might first consider the synthesis of protein and other growing elements of the body. When proteins are synthesized, large portions of ATP are used to form the peptide linkages, and this stores energy in these linkages. But we also noted in our discussions of proteins in Chapter 69 that there is continuous turnover of proteins, some being degraded while others are being formed. When the proteins are degraded, the energy stored in the peptide linkages is released in the form of heat into the body.

Now let us consider the energy used for muscle activity. Much of this energy simply overcomes the viscosity of the muscles themselves or of the tissues so that the limbs can move. The viscous movement in turn causes friction within the tissues, which generates heat.

We might also consider the energy expended by the heart in pumping blood. The blood distends the arterial system, the distention in itself representing a reservoir of potential energy. However, as the blood flows through the peripheral vessels, the friction of the different layers of blood flowing over each other and the friction of the blood against the walls of the vessels turns this energy into heat.

Therefore we can say that essentially all the energy expended by the body is converted into heat. The only real exception to this occurs when the muscles are used to perform some form of work outside the body. For instance, when the muscles elevate an object to a height or carry the person's body up steps, a type of potential energy is thus created by raising a mass against gravity. Also, if muscular energy is used to turn a flywheel so that kinetic energy is developed in the flywheel, this, too, would be external expenditure of energy. But, when external expenditure of energy is not taking place, it is safe to consider that all the energy released by the metabolic processes eventually becomes body heat.

**The Calorie.**   To discuss the metabolic rate and related subjects intelligently, it is necessary to use some unit for expressing the quantity of energy released from the different foods or expended by the different functional processes of the body. Most often, the *Calorie* is the unit used for this purpose. It will be recalled that 1 *calorie*, spelled with a small "c," is the quantity of heat required to raise the temperature of 1 gram of water 1° C. The calorie is much too small a unit for ease of expression in speaking of energy in the body. Consequently the large Calorie, spelled with a capital "C,"which is equivalent to 1000 calories, is the unit ordinarily used in discussing energy metabolism.

## MEASUREMENT OF THE METABOLIC RATE

**Direct Calorimetry.**   As pointed out above, it is only when the body performs some external work that energy expended within the body does not become heat. Since a person is ordinarily not performing any external work, his metabolic rate can be determined by simply measuring the total quantity of heat liberated from the body in a given time. This method is called *direct calorimetry*.

In determining the metabolic rate by direct calorimetry, one measures the quantity of heat liberated from the body in a large, specially constructed *calorimeter* as follows: The subject is placed in an air chamber so well insulated that no heat can leak through the walls of the chamber. As heat is formed by his body, it warms the air of the chamber. However, the air temperature within the chamber is maintained at a constant level by forcing the air through pipes in a cool water bath. The rate of heat gain by the water bath, a factor that can be measured with an accurate thermometer, is equal to the rate at which heat is liberated by the subject's body.

Obviously, direct calorimetry is physically difficult to perform and, therefore, is used only for research purposes.

**Indirect Calorimetry.**   Since more than 95 per cent of the energy expended in the body is derived from reaction of oxygen with the different foods, the metabolic rate can also be calculated with a high degree of accuracy from the rate of oxygen utilization. When 1 liter of oxygen is metabolized with glucose, 5.01 Calories of energy are released; when metabolized with starches, 5.06 Calories are released; with fat, 4.70 Calories; and with protein, 4.60 Calories.

From these figures it is striking how nearly equivalent are the quantities of energy liberated per liter of oxygen regardless of the type of food that is being burned. For the average diet, the *quantity of energy liberated per liter of oxygen utilized in the body averages approximately 4.825 Calories*. Using this *energy equivalent* of oxygen, one can calculate approximately the rate of heat liberation in the body from the quantity of oxygen utilized in a given period of time.

If a person should metabolize only carbohydrates during the period of the metabolic rate determination, the calculated quantity of energy liberated based on the value for the average energy equivalent of oxygen (4.825 Calories per liter), would be approximately 4 per cent too little. On the other hand, if the person were obtaining most of his energy from protein, the calculated value would be approximately 4 per cent too great, and, if he were burning almost entirely fat during the test, the error would be insignificant.

*The Metabolator.* Figure 71–4 illustrates the metabolator usually used for indirect calorimetry. This apparatus contains a floating drum, under which is an oxygen chamber connected to a mouthpiece through two rubber tubes. A valve in one of these rubber tubes allows air to pass from the oxygen chamber into the mouth, while air passing from the mouth back to the chamber is directed by means of another valve through the second tube. Before the expired air from the mouth enters the upper portion of the oxygen chamber, it flows through a lower chamber containing pellets of soda lime, which combine chemically with the carbon dioxide in the expired air. Therefore, as oxygen is used by the person's body and the carbon dioxide is absorbed by the soda lime, the floating oxygen chamber, which is precisely balanced by a weight, gradually sinks in the water, owing to the oxygen loss. This chamber is coupled to a pen that records on a moving drum the rate at which the chamber sinks in the water and thereby records the rate at which the body utilizes oxygen.

## FACTORS THAT AFFECT THE METABOLIC RATE

Factors that increase the chemical activity in the cells also increase the metabolic rate. Some of these are the following:

**Exercise.** The factor that causes by far the most dramatic effect on metabolic rate is strenuous exercise. Short bursts of maximal muscle contraction in any single muscle liberates as much as a hundred times its normal resting amount of heat for a few seconds at a time. In considering the entire body, however, maximal muscle exercise can increase the overall heat production of the body for a few seconds to about 50 times normal or sustained for several minutes to about 20 times normal in the well-trained athlete, which is an increase in metabolic rate to 2000 per cent of normal.

*Energy Requirements for Daily Activities.* When an average man of 70 kilograms lies in bed all day, he utilizes approximately 1650 Calories of energy. The process of eating increases the amount of energy utilized each day by an additional 200 or more Calories so that the same man lying in bed and also eating a reasonable diet requires a dietary intake of approximately 1850 Calories per day. If he sits in a chair all day, his total energy requirement reaches 2000 to 2250 Calories. Therefore, in round figures, it can be assumed that the daily energy requirements simply for existing (that is, performing essential functions only) is about 2000 Calories.

*Effects of Different Types of Work on Daily Energy Requirements.* Table 71–1 illustrates the rates of energy utilization while one performs different types of activities. Note that walking up stairs requires approximately 17 times as much energy as lying in bed asleep. In general over a 24-hour period a laborer can achieve a maximum rate of energy utilization as great as 6000 to 7000 Calories—in other words, as much as 3½ times the basal rate of metabolism.

**Specific Dynamic Action of Protein.** After a meal is ingested, the metabolic rate increases. This is believed to result to a very slight extent from the different chemical reactions associated with diges-

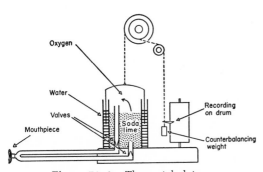

**Figure 71–4.** The metabolator.

**TABLE 71–1. Energy Expenditure per Hour During Different Types of Activity for a 70 Kilogram Man**

| Form of Activity | Calories per hour |
|---|---|
| Sleeping | 65 |
| Awake lying still | 77 |
| Sitting at rest | 100 |
| Standing relaxed | 105 |
| Dressing and undressing | 118 |
| Tailoring | 135 |
| Typewriting rapidly | 140 |
| "Light" exercise | 170 |
| Walking slowly (2.6 miles per hour) | 200 |
| Carpentry, metal working, industrial painting | 240 |
| "Active" exercise | 290 |
| "Severe" exercise | 450 |
| Sawing wood | 480 |
| Swimming | 500 |
| Running (5.3 miles per hour) | 570 |
| "Very severe" exercise | 600 |
| Walking very fast (5.3 miles per hour) | 650 |
| Walking up stairs | 1100 |

Extracted from data compiled by Professor M. S. Rose.

tion, absorption, and storage of food in the body. However, it mainly results from the action of certain of the amino acids derived from the proteins of the ingested food to stimulate directly the cellular chemical processes.

After a meal containing a large quantity of carbohydrates or fats, the metabolic rate usually increases only about 4 per cent. However, after a meal containing large quantities of protein, the metabolic rate usually begins rising within one hour, reaches a maximum about 30 per cent above normal, and lasts for as long as 3 to 12 hours. This effect of protein on the metabolic rate is called the *specific dynamic action* of protein.

**Age.**   The metabolic rate of the young child in relation to its size is almost two times that of an old person. This is illustrated in Figure 71–5, which shows the declining metabolic rates of both males and females from birth until very old age. The high metabolic rate of young children results from high rates of cellular reactions, but also partly from rapid synthesis of cellular materials and growth of the body, which require moderate quantities of energy.

**Thyroid Hormone.**   When the thyroid gland secretes maximal quantities of thyroxine, the metabolic rate sometimes rises to as much as 100 per cent above normal. On the other hand, total loss of thyroid secretion decreases the metabolic rate to as low as 50 to 60 per cent of normal. These effects can readily be explained by the basic function of thyroxine to increase the rates of activity of almost all the chemical reactions in all cells of the body. This relationship between thyroxine and metabolic rate will be discussed in much greater detail in Chapter 76 in relation to thyroid function, because one of the most useful methods for diagnosing abnormal rates of thyroid secretion is to determine the basal metabolic rate of the patient. A normal person usually has a basal metabolic rate within 10 to 15 per cent of normal, while the hyperthyroid person often has a basal metabolic rate as high as 40 to 80 per cent above normal, and a hypothyroid person can have a basal metabolic rate as low as 40 to 50 per cent below normal.

**Sympathetic Stimulation.**   Stimulation of the sympathetic nervous system with liberation of norepinephrine and epinephrine increases the metabolic rates of essentially all the tissues of the body. These hormones have a direct effect on cells to cause glycogenolysis, and this, probably along with other intracellular effects of these hormones, increases cellular activity.

Maximal stimulation of the sympathetic nervous system can increase the metabolic rate in some lower animals as much as several hundred per cent, but the magnitude of this effect in human beings is in question—probably 25 per cent or less in the adult, but as much as 100 per cent in the newborn child.

**Male Sex Hormone.**   The male sex hormone can increase the basal metabolic rate about 10 to 15 per cent, and the female sex hormone perhaps a few per cent but usually not enough to be of great significance. The difference in metabolic rates of males and females is illustrated in Figure 71–5.

**Growth Hormone.**   Growth hormone can increase the basal metabolic rate as much as 15 to 20 per cent as a result of direct stimulation of cellular metabolism.

**Fever.**   Fever, regardless of its cause, increases the metabolic rate. This is because all chemical reactions, either in the body or in the test tube, increase their rates of reaction an average of about 130 per cent for every 10° C. rise in temperature. An increase in body temperature to 110° F. increases the metabolic rate about 100 per cent.

**Climate.**   Studies of metabolic rates of persons living in the different geographic zones have shown as much as 10 to 20 per cent lower metabolic rates in tropical regions than in arctic regions. This difference is caused to a great extent by adaptation of the thyroid gland, with increased secretion in cold climates and decreased secretion in hot climates. Indeed, far more persons develop hyperthyroidism in cold regions of the earth than in tropical regions.

**Sleep.**   The metabolic rate falls approximately 10 to 15 per cent below normal during sleep. This fall is presumably due to two principal factors: (1) decreased tone of the skeletal musculature during sleep and (2) decreased activity of the sympathetic nervous system.

**Malnutrition.**   Prolonged malnutrition often decreases the metabolic rate as much as 20 to 30 per cent; this decrease is presumably caused by the paucity of necessary food substances in the cells.

In the final stages of many disease conditions, the inanition that accompanies the disease frequently causes marked premortem decrease in metabolic rate, even to the extent that the body temperature may fall a number of degrees shortly before death.

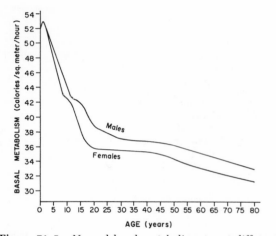

**Figure 71–5.**   Normal basal metabolic rates at different ages for each sex.

## THE BASAL METABOLIC RATE

**The Basal Metabolic Rate as a Method for Comparing Metabolic Rates Between Individuals.** It has been extremely important to establish a procedure that will measure the inherent activity of the tissues independently of exercise and other extraneous factors that would make it impossible to compare one person's metabolic rate with that of another person. To do this, the metabolic rate is usually measured under so-called *basal conditions,* and the metabolic rate then measured is called the *basal metabolic rate.*

**Basal Conditions.** The basal metabolic rate means the rate of energy utilization in the body during absolute rest but while the person is awake. The following basal conditions are necessary for measuring the basal metabolic rate:

1. The person must not have eaten any food for at least 12 hours because of the specific dynamic action of foods.

2. The basal metabolic rate is determined after a night of restful sleep, for rest reduces the activities of the sympathetic nervous system and other metabolic excitants to their minimal level.

3. No strenuous exercise is performed after the night of restful sleep, and the person must remain at complete rest in a reclining position for at least 30 minutes prior to actual determination of the metabolic rate. This is perhaps the most important of all the conditions for attaining the basal state because of the extreme effect of exercise on metabolism.

4. All psychic and physical factors that cause excitement must be eliminated, and the subject must be made as comfortable as possible. These conditions, obviously, help to reduce the degree of sympathetic activity to as little as possible.

5. The temperature of the air must be comfortable and be somewhere between the limits of 68° and 80° F. Below 68° F., the sympathetic nervous system becomes progressively more activated to help maintain body heat, and above 80° F., discomfort, sweating, and other factors increase the metabolic rate.

**Usual Technique for Determining the Basal Metabolic Rate.** The usual method for determining basal metabolic rate is first to establish the subject under basal conditions and then to measure his rate of oxygen utilization using a metabolator of the type illustrated in Figure 71–4. Then the basal metabolic rate is calculated as shown in Figure 71–6.

In the upper portion of Figure 71–6, the quantity of heat liberated in the body of the person is calculated

| | |
|---|---|
| 15 liters | — $O_2$ at standard conditions consumed in 1 hr. |
| $\times 4.825$ | — Calories liberated per liter of $O_2$ burned |
| 72.4 | — Calories liberated per hour |
| $\div 1.5$ | — Body surface area in square meters |
| 48.3 | — Calories per square meter per hour |
| $-38.5$ | — Normal value for 20-year-old man |
| 9.8 | — Excess Calories above normal |

$$\frac{9.8 \times 100}{38.5} = 25.5 \text{ per cent above normal}$$

$$BMR = +25.5$$

**Figure 71–6.** Calculation of basal metabolic rate from the rate of oxygen consumption.

from the quantity of oxygen utilized. In this figure, note that 15 liters of oxygen (corrected to standard conditions) are consumed in one hour. Multiplying this times the energy equivalent for 1 liter of oxygen (4.825 Calories), the total quantity of energy liberated in the body during the hour is 72.4 Calories.

**Expressing the Basal Metabolic Rate in Terms of Surface Area.** Obviously, if one subject is much larger than another, the total amount of energy utilized by the two subjects will be considerably different simply because of differences in body size. Experimentally, among normal persons, the average basal metabolic rate varies approximately *in proportion to the body surface area.*

Note from Figure 71–6 that the total number of Calories liberated by the person per hour is divided by his total body surface area, 1.5 square meters. This means that his basal metabolic rate is 48.3 Calories per square meter per hour.

*Method for Calculating the Total Surface Area.* The surface area of the body varies approximately in proportion to *weight$^{0.67}$*. However, measurements of

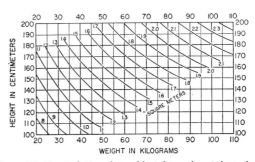

**Figure 71–7.** Relationship of height and weight to body surface area. (From DuBois: Metabolism in Health and Disease. Lea & Febiger.)

the body surface area have shown that it can be calculated more accurately by a complicated formula based on weight and height of the subject as follows:

Body surface area =
$$\text{Weight}^{0.425} \times \text{Height}^{0.725} \times 0.007184$$

Figure 71–7 presents a graph based on this formula. In the formula and in the figure, body surface area is expressed in *square meters*, weight in *kilograms*, and height in *centimeters*. The surface area of the average 70-kilogram adult is 1.73 square meters.

**Expressing the Basal Metabolic Rate in Percentage Above or Below Normal.** To compare the basal metabolic rate of any one subject with the normal basal metabolic rate, it is necessary to refer to a chart such as that in Figure 71–5, which gives the normal basal metabolic rate per square meter at each age for each sex. For example, if the person represented in the calculation of Figure 71–6 is a 20-year-old male, then we find from Figure 71–5 that his normal metabolic rate is 38.5 Calories per square meter per hour. But his actual metabolic rate, 48.3 Calories, is 9.8 Calories per square meter per hour *above* the normal mean value. It is then determined that this is 25.5 per cent above normal. Therefore, the basal metabolic rate is expressed as *plus 25.5*. Similarly, basal metabolic rates below normal are expressed as minus values.

**Constancy of the Metabolic Rate in the Same Person.** Basal metabolic rates have been measured in many subjects at repeated intervals for as long as 20 or more years. As long as a subject remains healthy, almost invariably his basal metabolic rate as expressed in percentage of normal does not vary more than 5 to 10 per cent.

**Constancy of the Basal Metabolic Rate from Person to Person.** When the basal metabolic rate is measured in a wide variety of different persons and comparisons are made within single age, weight, and sex groups, 85 per cent of normal persons have been found to have basal metabolic rates within 10 per cent of the mean. Thus, it is obvious that measurements of metabolic rates performed under basal conditions offer an excellent means for comparing the rates of metabolism from one person to another.

# REFERENCES

Chance, B.: Enzymes in action in living cells: the steady state of reduced pyridine nucleotides. *Harvey Lectures, 49*:145, 1953–1954.

Christensen, H. N., and Palmer, G. A.: Enzyme Kinetics. Philadelphia, W. B. Saunders Company, 1974.

Consolazio, C. F., Johnson, H. L., Daws, T. A., and Nelson, R. A.: Energy requirement and metabolism during exposure to extreme environments. *World Rev. Nutr. Diet, 18*:177, 1973.

Consolazio, C. F., Johnson R., and Pecora, L.: Physiological Measurements of Metabolic Functions in Man. New York, McGraw-Hill Book Company, 1963.

Energy production during exercise. *Nutr. Rev., 31*:11, 1973.

Garfinkel, D.: Computer simulation of biologically realistic metabolic networks. *Acta Biol. Med. Ger., 31*:339, 1973.

Green, D. E.: The electromechanochemical model for energy coupling in mitochondria. *Biochim. Biophys. Acta, 346*:27, 1974.

Guyton, A. C., and Farrish, C. A.: A rapidly responding continuous oxygen consumption recorder. *J. Appl. Physiol., 14*:143, 1959.

Havel, R. J.: Caloric homeostasis and disorders of fuel transport. *New Eng. J. Med., 287*:1186, 1972.

Hegsted, D. M.: Energy needs and energy utilization. *Nutr. Rev., 32*:33, 1974.

Hoch, F. L.: Metabolic effects of thyroid hormones. *In* Greep, R. O., and Astwood, E. B. (eds.): Handbook of Physiology. Sec. 7, Vol. 3. Baltimore, The Williams & Wilkins Company, 1974, p. 391.

Kagawa, Y.: Reconstitution of oxidative phosphorylation. *Biochim. Biophys. Acta, 265*:297, 1972.

Kleiber, M.: Respiratory exchange and metabolic rate. *In* Fenn, W. O., and Rahn, H. (eds.): Handbook of Physiology. Sec. 3, Vol. 2. Baltimore, The Williams & Wilkins Company, 1965, p. 927.

Mildvan, A. S.: Mechanism of enzyme action. *Ann. Rev. Biochem., 43*:357, 1974.

Myant, N. B.: Some aspects of the control of cell metabolism. *In* Bittar, E. E., and Bittar, N. (eds.): The Biological Basis of Medicine. Vol. 2. New York, Academic Press, Inc., 1968, p. 133.

Newsholme, E. A., and Start, C.: Regulation in metabolism. New York, John Wiley & Sons, 1973.

Oster, G. F., Perelson, A. S., and Katchalsky, A.: Network thermodynamics: dynamic modelling of biophysical systems. *Q. Rev. Biophys., 6*:1, 1973.

Roe, C. F.: Temperature regulation and energy metabolism in surgical patients. *Prog. Surg., 12*:96, 1973.

Shapiro, H. M.: Redox balance in the body: an approach to quantitation. *J. Surg. Res., 13*:138, 1972.

Sinclair, J. C.: Metabolic rate and body size of the newborn. *Clin. Obstet. Gynecol., 14*:840, 1971.

Toporek, M., and Maurer, P. H.: Metabolic biochemistry. *In* Sodeman, W. A., Jr., and Sodeman, W. A. (eds.): Pathologic Physiology: Mechanisms of Disease, 5th Ed. Philadelphia, W. B. Saunders Company, 1974, p. 3.

Wyndham, C. H., and Loots, H.: Responses to cold during a year in Antarctica. *J. Appl. Physiol., 27*:696, 1969.

# 72

# Body Temperature, Temperature Regulation, and Fever

The temperature of the inside of the body —the "core"—remains almost exactly constant, within ± 1° F., day in and day out except when a person develops a febrile illness. Indeed, the nude person can be exposed to temperatures as low as 55° F. or as high as 140° F. in dry air and still maintain an almost constant internal body temperature. Therefore, it is obvious that the mechanisms for control of body temperature represent a beautifully designed control system. It is the purpose of this chapter to discuss this system as it operates in health and in disease.

**"Core" Temperature versus Surface Temperature.** When speaking of the body temperature, one usually means the temperature in the interior, called the *core temperature,* and not the temperature of the skin or tissues immediately underlying the skin. The core temperature is accurately regulated, normally varying from the mean by not more than 1° F. On the other hand, the *surface temperature* rises and falls with the temperature of the surroundings. In speaking of body temperature regulation we almost always refer to the core temperature; when we refer to the ability of the skin to lose heat to the surroundings, we usually speak of the surface temperature; and when we wish to calculate the total amount of heat stored in the body, we use the *average body temperature.* The average body temperature can be approximated by the following formula:

Average temperature =
0.7 internal temperature + 0.3 surface temperature

**The Normal Body Temperature.** No single temperature level can be considered to be normal, for measurements on many normal persons have shown a *range* of normal temperatures, as illustrated in Figure 72–1, from approximately 97° F. to over 99° F. When measured by rectum, the values are approximately 1° F. greater than the oral temperatures. The average normal temperature is generally considered to be 98.6° F. (37° C.) when measured orally and approximately 1° F. or 0.6° C. higher when measured rectally.

The body temperature varies somewhat with exercise and with extremes of temperature of the surroundings, because the temperature regulatory mechanisms are not 100 per cent effective. When excessive heat is produced in the body by strenuous exercise, the rectal temperature can rise to as high as 101° to 104° F. On the other hand, when the body is exposed to extreme cold, the rectal temperature can often fall to values considerably below 98° F.

**Relationship of Body Heat to Body Temperature—Specific Heat of the Tissues.** The temperature of an object is a measure of the kinetic activity of its molecules, and this is proportional to

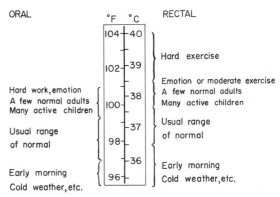

**Figure 72–1.** Estimated range of body temperature in normal persons. (From DuBois: Fever. Charles C Thomas.)

the amount of heat stored in the object. Therefore, body temperature is directly proportional to the heat in the body.

On the average, the body temperature increases 1° C. for each 0.83 Calorie stored per kilogram of body weight. In other words, the specific heat of the tissues is 0.83 Cal./kg./degree C. For a 70 kg. man, approximately 58 Calories of heat must be added to the body to raise the body temperature 1° C., or 1.8° F.

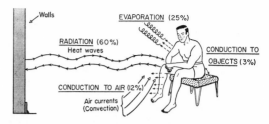

**Figure 72–3.**   Mechanisms of heat loss from the body.

# BALANCE BETWEEN HEAT PRODUCTION AND HEAT LOSS

Heat is continually being produced in the body as a by-product of metabolism, and body heat is also continually being lost to the surroundings. When the rate of heat production is exactly equal to the rate of loss, the person is said to be in heat balance. But when the two are out of equilibrium, the body heat, and the body temperature as well, will obviously be either increasing or decreasing.

Figure 72–2 illustrates this balance between heat production and heat loss. To the left are the important factors that play major roles in determining the rate of heat production and which were discussed in detail in Chapter 71; they may be listed simply as follows: (1) basal rate of metabolism of all the cells of the body; (2) increase in rate of metabolism caused by muscle activity, including that caused by shivering; (3) increase in metabolism caused by the effect of thyroxine on cells; (4) increase in metabolism caused by the effect of norepinephrine and sympathetic stimulation on cells; and (5) increase in metabolism caused by increased temperature of the body cells.

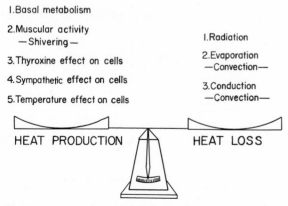

**Figure 72–2.**   Balance of heat production versus heat loss.

## HEAT LOSS

The various methods by which heat is lost from the body are depicted on the right side of the balance in Figure 72–2 and are shown pictorially in Figure 72–3. These include *radiation, conduction,* and *evaporation.* Also, the phenomenon of *convection* of the air plays a major role in the heat lost by both conduction and evaporation. However, the amount of heat lost by each of these different mechanisms varies with atmospheric conditions.

**Radiation.**   As illustrated in Figure 72–3, a nude person in a room at normal room temperature loses about 60 per cent of his total heat loss by radiation.

Loss of heat by radiation means loss in the form of infrared heat rays, a type of electromagnetic waves. Most infrared heat rays radiating from the body have wavelengths of 5 to 20 microns, 10 to 30 times the wavelengths of light rays. All mass in the universe that is not at absolute zero temperature radiates such rays. Therefore, the human body radiates heat rays in all directions. However, heat rays are also being radiated from the walls and other objects toward the body. If the temperature of the body is greater than the temperature of the surroundings, a greater quantity of heat is radiated from the body than is radiated to the body. This is the usual situation. Yet, at times, especially in the summer, the surroundings become hotter than the human body, under which circumstances more radiant heat is transmitted to the body than from the body.

Heat loss by radiation varies directly with the difference between the fourth powers of (1) the temperature of the body surface and (2) the average temperature of the surroundings. Therefore, it is impossible to state categorically exactly what percentage of the body heat will be lost by radiation unless all of the conditions surrounding the body are momentarily defined.

The surface of the human body is extremely absorbent for heat rays. A body exposed to infrared heat rays from a stove absorbs approximately 97 per cent of the rays that hit it. This rate of infrared absorption is approximately equal for human beings with either white or black skin, for at these wavelengths the different colors of the skin have no effect on absorption. On the other hand, the energy from the sun is transmitted mainly in the form of light rays rather than infrared rays. Approximately 35 per cent of these waves are reflected from the very light skin but only a small amount from the dark skin. Consequently, in sunlight, dark skin does absorb more heat than light skin.

**Conduction.** Usually, only minute quantities of heat are lost from the body by direct conduction from the surface of the body to other objects, such as a chair or a bed. When one first sits on a chair while nude, heat is conducted from the body to the chair rapidly, but within a few minutes the temperature of the chair rises almost to equal the temperature of the body, and thereafter the chair actually becomes an insulator to prevent further loss of heat. Consequently, the loss of heat by *conduction to objects* represents only a small per cent of the total loss from the body, as illustrated in Figure 72–3.

On the other hand, loss of heat by *conduction to air* does represent a sizeable proportion of the body's heat loss even under normal conditions. It will be recalled that heat is actually the kinetic energy of molecular motion, and the molecules that comprise the skin of the body are continually undergoing vibratory motion. Thus, the vibratory motion of the skin molecules can cause increased velocity of motion of the air molecules that come into direct contact with the skin. However, once the temperature of the air immediately adjacent to the skin equals the temperature of the skin, no further exchange of heat from the body to the air can occur. Therefore, conduction of heat from the body to the air is self-limited unless the heated air moves away from the skin so that new, unheated air is continually brought in contact with the skin, a phenomenon called convection.

**Convection.** Movement of air is known as convection, and the removal of heat from the body by convection air currents is commonly called "heat loss by convection." Actually, the heat must first be *conducted* to the air and then carried away by the convection currents.

A small amount of convection almost always occurs around the body because of the tendency for the air adjacent to the skin to rise as it becomes heated. Therefore, a nude person seated in a comfortable room without gross air movement still loses about 12 per cent of his heat by conduction to the air and then by convection away from the body.

*Cooling Effect of Wind.* When the body is exposed to wind, the layer of air immediately adjacent to the skin is replaced by new air much more rapidly than normally, and heat loss by convection increases accordingly. Figure 72–4 shows that the cooling effect of wind at low velocities is approximately proportional to the square root of the wind velocity. For instance, a wind of 4 miles per hour is about two times as effective for cooling as a wind of 1 mile per hour. However, when the wind velocity rises beyond a few miles per hour, additional cooling does not occur to a great extent. This is because once the wind has cooled the skin to the temperature of the air itself, the rate of heat loss cannot increase further, regardless of the wind velocity. Instead, the rate at which heat can flow from the core of the body to the skin is then the factor that determines the rapidity with which heat can be lost.

*Conduction and Convection of Heat from a Body Exposed to Water.* Water has a specific heat several thousand times as great as that of air so that each unit portion of water adjacent to the skin can absorb far greater quantities of heat than can air. Also, the conductivity of heat through water is marked in comparison with that in air. Consequently, heating a thin layer of water next to the body does not result in the formation of an "insulator zone" as occurs in air. For this reason, the rate of heat loss from the body into nonflowing water is almost as much as the rate of heat loss into rapidly flowing water.

Because of these differences in heat conduction by air and water, the rate of heat loss to water at moderate temperatures is many times as great as the rate of

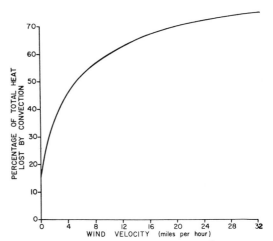

**Figure 72–4.** Effect of wind velocity on the percentage of heat loss that occurs by convection.

heat loss to air of the same temperature. However, when the water and air are extremely cold, the rate of heat loss to air becomes almost as great as to water, for both water and air are then capable of carrying away essentially all the heat that can reach the skin from the body core.

**Evaporation.** When water evaporates from the body surface, 0.58 Calorie of heat is lost for each gram of water that evaporates. Water evaporates *insensibly* from the skin and lungs at a rate of about 600 ml. per day. This causes continual heat loss at a rate of 12 to 18 Calories per hour. Unfortunately, this insensible evaporation of water directly through the skin and lungs cannot be controlled for purposes of temperature regulation because it results from continual diffusion of water molecules through the skin and respiratory surfaces regardless of body temperature. However, evaporative loss of heat can be controlled by regulating the rate of sweating, which is discussed below.

*Evaporation as a Necessary Refrigeration Mechanism at High Air Temperatures.* In the preceding discussions of radiation and conduction it was noted that as long as the body temperature is greater than that of the surroundings, heat is lost by radiation and conduction, but when the temperature of the surroundings is greater than that of the skin, instead of losing heat the body gains heat by radiation and conduction from the surroundings. Under these conditions, *the only means by which the body can rid itself of heat is by evaporation.* Therefore, any factor that prevents adequate evaporation when the surrounding temperatures are higher than body temperature permits the body temperature to rise. This occurs occasionally in human beings who are born with congenital absence of sweat glands. These persons can withstand cold temperatures as well as can normal persons, but they are likely to die of heat stroke in tropical zones, for without the evaporative refrigeration system their body temperatures must remain at values greater than those of the surroundings.

*Effect of Humid Weather on Evaporative Loss of Heat.* It is well known that hot, muggy, summer days are extremely uncomfortable, and that sweat pours from the body far more profusely on these days than normally. This is because the air is already humidified almost to its maximal extent. As a result, the rate of evaporation is greatly reduced or totally prevented so that the secreted sweat remains in the fluid state. Consequently, the body temperature approaches the temperature of the surroundings or rises above this temperature even though sweat continues to pour forth.

*Effect of Convection Air Currents on Evaporation.* As already pointed out, a thin zone of air adjacent to the skin usually remains relatively stationary and is not exchanged for new air at a rapid rate unless convection currents are present. Lack of air movement prevents effective evaporation in the same manner that it prevents effective cooling by conduction of heat to the air—that is, the local air becomes saturated with water vapor, and further evaporation cannot occur. Convection currents cause air that has become saturated with moisture to move away from the skin while unsaturated air replaces it. Indeed, convection is of even more importance for heat loss from the body by evaporation than by conduction, for the instances in which one especially needs to lose heat from the body, such as on hot days, are the same times when evaporative loss of heat from the body is far greater than conductive loss. This also explains why fans are in demand on hot days.

**Effect of Clothing on Heat Loss.** *Effect on Conductive Heat Loss.* Clothing entraps air next to the skin and in the weave of the cloth, thereby increasing the thickness of the so-called "private zone" of air adjacent to the skin and decreasing the flow of convection air currents. Consequently, the rate of heat loss from the body by conduction is greatly depressed.

A usual suit of clothes decreases the rate of heat loss from the body to about half that from a nude body, while arctic-type clothing can decrease this heat loss to as little as one sixth that of the nude state.

*Effect on Heat Loss by Radiation.* About half the heat transmitted from the skin to the clothing probably is radiated to the clothing instead of being conducted across the small intervening space, for sputtering the inside of clothing with a thin layer of gold, which reflects radiant heat, makes the insulating properties of clothing far more effective than otherwise. This is because the heat leaving the skin is reflected back from the clothing to the skin, and, consequently, little heat can then be transmitted to the clothing by radiation. As a result of this new technique, clothing for use in the arctic can be decreased in weight by about half.

*Loss of Heat Through Wet Clothing.* The effectiveness of clothing in preventing heat loss is almost completely lost when it becomes wet, for there is then no entrapped air to act as an insulator. Instead, the interstices of the clothing are filled with water which, because of its high conductivity for heat, increases the rate of heat transmission as much as 20-fold or more. One of the most important factors for protecting the body against cold in arctic regions is extreme precaution against wet clothing. Indeed, one must be careful not to overheat oneself temporarily, for sweating in one's clothes makes them much less effective thereafter as an insulator.

*Effect on Heat Loss by Evaporation.* Clothing that is pervious to moisture (cotton, but not plastic clothing) allows almost normal heat loss from the body by evaporation, for when sweating occurs the sweat itself can dampen the clothing, and evaporation then occurs from the surface of the clothing. This cools the clothing, which in turn cools the skin. Consequently, in tropical regions, light clothing that is pervious to sweat but impervious to radiant heat from the sun prevents the body from gaining radiant heat while at the same time allowing it to lose heat at a rate almost as if one were not wearing clothing.

## SWEATING AND ITS REGULATION BY THE AUTONOMIC NERVOUS SYSTEM

When the body becomes overheated, large quantities of sweat are secreted onto the surface of the skin by the eccrine sweat glands to provide rapid *evaporative cooling* of the body. Stimulation of the preoptic area in the anterior part of the hypothalamus excites sweating. The impulses from this area that cause sweating are transmitted in the autonomic pathways to the cord and thence through the sympathetic outflow to the skin everywhere in the body.

It should be recalled from the discussion of the autonomic nervous system in Chapter 57 that the eccrine sweat glands are innervated by *cholinergic* nerve fibers. However, these glands can also be stimulated by epinephrine or norepinephrine circulating in the blood even though the glands themselves, in most parts of the body, do not have adrenergic innervation. It is possible that the sweat glands of the hands and feet do have some adrenergic innervation as well as cholinergic innervation, for many emotional states that excite the adrenergic portions of the sympathetic nervous system are known also to cause local sweating of the hands and feet. Also, during muscular exercise, which normally excites adrenergic activity, localized sweating of the hands and feet occurs. In this instance, the moisture from the sweat helps the surfaces of the hands and feet to gain traction against smooth surfaces and also prevents drying of the thick, cornified layers of skin.

**Rate of Sweating.** In cold weather, the rate of sweat production is essentially zero but in very hot weather the maximum rate of sweat production is from 1.5 liters per hour in the unacclimatized person to 4 liters per hour in the person maximally acclimatized to heat. Thus, during maximal sweating, a person can lose more than 8 pounds of body weight per hour.

**Mechanism of Sweat Secretion.** The sweat glands are tubular structures consisting of two parts: (1) a deep *coiled portion* that secretes the sweat, and (2) a *duct portion* passing outward through the dermis of the skin. As is true of so many other glands, the secretory portion of the sweat gland secretes a fluid called the *precursor secretion;* then certain constituents of the fluid are reabsorbed as it flows through the duct.

The precursor secretion is an active secretory product of the epithelial cells lining the coiled portion of the sweat gland. Cholinergic sympathetic nerve fibers ending on or near the glandular cells elicit the secretion.

Since large amounts of sodium chloride are lost in the sweat, it is important especially to know how the sweat glands handle sodium and chloride during the secretory process. When the rate of sweat secretion is very low, the sodium and chloride concentrations of the sweat are very low, because these ions are reabsorbed from the precursor secretion before it reaches the surface of the body; their concentrations are sometimes as low as 5 mEq./liter each. On the other hand, when the rate of secretion becomes progressively greater, the rate of sodium chloride reabsorption does not increase commensurately, so that then the concentration of sodium in the sweat can rise almost to the level in plasma.

Other substances lost in reasonable quantities in the sweat include urea, lactic acid, and potassium ions. At low rates of sweat secretion, the concentrations of all these can be extremely high, but at high rates of secretion the concentration of urea is about two times that in the plasma, lactic acid about four times, and potassium about 1.2 times.

*Effect of Aldosterone on Sodium Loss in the Sweat.* Aldosterone decreases the sodium and chloride concentrations in sweat, thereby conserving these ions in the body fluids. Aldosterone functions in much the same way in the sweat glands that it does in the renal tubules; that is, it increases the rate of active reabsorption of sodium by the ducts. The reabsorption of sodium carries chloride ions along as well because of the electrical gradient that develops across the epithelium when sodium is reabsorbed. The importance of this aldosterone effect is to minimize loss of sodium chloride in the sweat when the blood sodium chloride concentration is already low.

**Acclimatization of the Sweating Mechanism.** A person exposed to hot weather for several weeks sweats progressively more and more profusely, sweating an average maximum of about 1.5 liters per hour at first, which rises to about double this value within 10 days and to

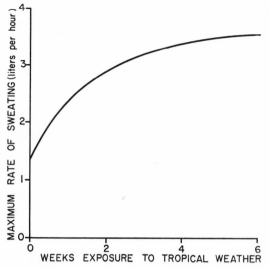

**Figure 72–5.** Acclimatization of the sweating mechanism, showing progressive increase in the maximum rate of sweating during the first few weeks of exposure to tropical weather.

about 2½ times as much within six weeks, as illustrated in Figure 72–5. This increased effectiveness of the sweating mechanism is caused by a direct increase in sweating capability of the sweat glands themselves. Associated with the increased sweating is usually decreased concentration of sodium chloride in the sweat, which allows progressively better conservation of salt. Most of this adaptation for conserving salt is probably caused by increased secretion of aldosterone, but it possibly also results in part from decreased sodium chloride concentration in the body fluids and from local changes in the sweat glands.

Extreme sweating can deplete the extracellular fluids of electrolytes, particularly of sodium and chloride. Consequently, extra sodium chloride usually must be supplied in the diet in tropical climates. A person who sweats profusely may lose as much as 15 to 20 grams of sodium chloride each day until he becomes acclimatized. On the other hand, after four to six weeks of acclimatization the loss of sodium chloride may be as little as 3 to 5 grams per day. This change occurs at least partially because of increased aldosterone secretion resulting from depletion of the salt reserves of the body.

**Long-term Acclimatization of the Sweat Apparatus.** In addition to the rapid acclimatization of the sweating mechanism just described, acclimatization can occur over many years in a different way, as follows:

A person who has lived in the tropics since childhood actually has greater numbers of active sweat glands in the body. A person is born with a considerable excess of sweat glands, but if he lives in a temperate zone, many of these become permanently inactivated during childhood. However, if he lives in the tropics they remain functional throughout life. Therefore, a person who has spent his childhood in the tropics usually possesses a much more effective sweating mechanism than does a person reared elsewhere.

**Panting as a Means of Evaporative Heat Loss.** Many lower animals do not have sweat glands, but to offset this they lose large amounts of heat by the panting mechanism. During panting, only small volumes of air pass in and out of the lungs with each breath so that mainly dead space air enters the alveoli. Because of this and because of the very rapid breathing rate, tremendous amounts of air are moved over the surfaces of the tongue, mouth, and trachea without overrespiration of the alveoli. Evaporation from these respiratory surfaces, especially of saliva on the tongue, is an important mechanism for heat control.

A special nervous center in the pons controls panting. This center modifies the normal respiratory pattern to provide the rapid and shallow breathing required for the panting mechanism.

## THE INSULATOR SYSTEM OF THE BODY

The skin, the subcutaneous tissues, and especially the fat of the subcutaneous tissues are a heat insulator for the body. The fat is especially important because it conducts heat only *one-third* as readily as other tissues. When no blood is flowing from the heated internal organs to the skin, the insulating properties of the male body are approximately equal to three-quarters the insulating properties of a usual suit of clothes. In women this insulation is still better. Obviously, the degree of insulation varies from one person to another, depending to a great extent on the quantity of adipose tissue.

Because most body heat is produced in the deeper portions of the body, the insulation beneath the skin is an effective means for maintaining normal internal temperatures, even though it allows the temperature of the skin to approach the temperature of the surroundings.

## FLOW OF BLOOD TO THE SKIN AND HEAT TRANSFER FROM THE BODY CORE

Blood vessels penetrate the subcutaneous insulator tissues and are distributed profusely in the subpapillary portions of the skin. Indeed, immediately beneath the skin is a continuous venous plexus that is supplied by inflow of blood. In the most exposed areas of the body—the hands, feet, and ears—blood is supplied through direct *arteriovenous anastomoses* from the arterioles to the veins. The rate of blood flow into this venous plexus can vary tremendously—from barely above zero to as great as 30 per cent of the total cardiac output. A high rate of blood flow causes heat to be conducted from the internal portions of the body to the skin with great efficiency, whereas reduction in the rate of blood flow decreases the efficiency of heat conduction from the internal portions of the body.

Obviously, therefore, the skin is an effective "radiator" system, and the flow of blood to the skin is the mechanism of heat transfer from the body "core" to the skin. If blood flow from the internal structures to the skin is depressed, the only means by which heat produced internally can be lost to the exterior is by heat diffusion through the insulator tissues of the skin and subcutaneous areas. This form of heat diffusion, by itself, is much too poor a means of providing the needed heat dissipation, except in cold weather.

**Control of Heat Conduction to the Skin.** Heat conduction to the skin by the blood is controlled by the degree of vasoconstriction of the arterioles and arteriovenous anastomoses that supply blood to the venous plexus of the skin, and this vasoconstriction is controlled almost entirely by the sympathetic nervous system. Ordinarily, the sympathetics remain tonically active, causing continual constriction of the arterioles supplying the skin. When the sympathetic centers of the posterior hypothalamus are stimulated, the blood vessels are constricted even more, and blood flow to the skin may almost cease; but, when these posterior centers of the hypothalamus are inhibited, decreased numbers of sympathetic impulses are transmitted to the periphery, and the blood vessels dilate.

# REGULATION OF BODY TEMPERATURE

Figure 72–6 illustrates approximately what happens to the temperature of the nude body after a few hours' exposure to dry air ranging from 30° to 170° F. Obviously, the precise dimensions of this curve vary, depending on the

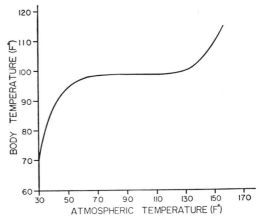

**Figure 72–6.** Effect of high and low atmospheric temperatures for several hours' duration on the internal body temperature, showing that the internal body temperature remains stable despite wide changes in atmospheric temperature.

movement of air, the amount of moisture in the air, and even the nature of the surroundings. However, in general, between approximately 60° and 130° F. in dry air, the nude body is capable of maintaining indefinitely a normal body core temperature somewhere between 98° and 100° F.

The temperature of the body is regulated almost entirely by nervous feedback mechanisms, and almost all of these operate through a *temperature regulating center* located in the *hypothalamus*. However, for these feedback mechanisms to operate, there must also exist temperature detectors to determine when the body temperature becomes either too hot or too cold. Some of these receptors are the following:

**Temperature Receptors.** Probably the most important temperature receptors for control of body temperature are many special *heat-sensitive neurons* located *in the preoptic area of the hypothalamus*. These neurons increase their impulse output as the temperature rises and decrease their output when the temperature decreases. The firing rate sometimes increases as much as 10-fold with an increase in body temperature of 10° C.

In addition to these heat sensitive neurons of the preoptic area, other receptors sensitive to temperature include: (1) a few *cold-sensitive neurons* found in different parts of the hypothalamus, the septum, and the reticular substance of the midbrain, all of which increase their rate of firing when exposed to cold (the numbers of these are few and there is doubt that they play any role in the regulation of body temperature); (2) *skin temperature receptors,* including both *warmth* and *cold receptors,* that transmit nerve impulses into the spinal cord and thence to the hypothalamic region of the brain to help control body temperature, as will be discussed later; and (3) *temperature receptors in the spinal cord, abdomen, and possibly other internal structures* of the body that also transmit signals to the central nervous system to help control body temperature.

## THERMOSTATIC DETECTION OF EXCESS BODY HEAT (EXCESS TEMPERATURE)— ROLE OF THE PREOPTIC AREA OF THE HYPOTHALAMUS

In recent years, experiments have been performed in which minute areas in the brain have been either heated or cooled by use of a socalled *thermode*. This is a device that is heated by electrical means or by passing hot water through it, or is cooled by cold water. The principal area in the brain in which heat from a thermode affects body temperature control is the preoptic area of the hypothalamus and less so in adjacent regions of the anterior hypothalamus. In these areas are located many heat-sensitive neurons whose rates of discharge increase greatly when heated; it is believed that these neurons play a decisive role in the control of body temperature.

When the preoptic area of the hypothalamus and the adjacent anterior hypothalamus are heated, changes are elicited immediately throughout the body to cause heat loss and to inhibit the rate of heat production; these effects will be explained later. However, the extent to which these changes take place is determined also by nervous signals from temperature receptors located elsewhere in the body as well, especially in the skin and spinal cord. The roles of these will be discussed in more detail in subsequent paragraphs.

## THERMOSTATIC DETECTION OF COLD—ROLE OF SKIN AND SPINAL CORD RECEPTORS

One of the ways in which the body detects cold is by reduced rates of discharge of the heat-sensitive neurons in the preoptic area. However, by the time the internal body temperature has fallen a few tenths of a degree below normal, these neurons have generally become inactive so that their signal level cannot be de-

creased any further. When the temperature falls still further, other receptors besides those in the hypothalamus seem to provide the major signals for cold. These receptors are located mainly in the spinal cord and skin, and their signals provide a major drive to make the body conserve heat as well as to produce greatly increased quantities of heat by the "shivering" process.

## FINAL INTEGRATION OF BOTH THE HEAT AND COLD THERMOSTATIC SIGNALS IN THE HYPOTHALAMUS— THE "HYPOTHALAMIC THERMOSTAT"

Even though most of the signals for cold detection arise in peripheral receptors, these signals are then transmitted to the posterior hypothalamus where they are integrated with the receptor signals from the preoptic area to give the final efferent signals for controlling heat loss and heat production. Therefore, we generally speak of the control center for temperature regulation as the *hypothalamic thermostat.*

Figure 72–7 illustrates the effectiveness of the hypothalamic thermostat in initiating temperature regulatory changes when the body temperature rises too high or falls too low. The solid

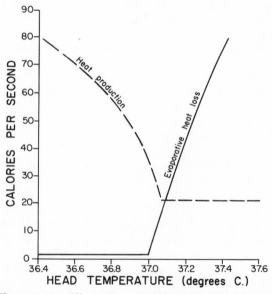

**Figure 72–7.** Effect of hypothalamic temperature on: (1) evaporative heat loss from the body and (2) heat production caused primarily by muscular activity and shivering. This figure demonstrates the extremely critical temperature level at which increased heat loss begins and increased heat production stops. (Drawn from data from Benzinger, Kitzinger, and Pratt, in Hardy (ed.): Temperature, Part 3, p. 637. Reinhold Publishing Corp.)

curve shows that almost precisely at 37° C. (98.4° F.) sweating begins and then increases rapidly as the temperature rises above this value; on the other hand, it ceases at any temperature below this same crucial level.

Likewise, the thermostat controls the rate of heat production, which is illustrated by the dashed curve. At any temperature above 37.1° C., the heat production remains almost exactly constant, but whenever the temperature falls below this level, the various mechanisms for increasing heat production become markedly activated, especially an increase in muscular activity which culminates in shivering.

## MECHANISMS OF INCREASED HEAT LOSS WHEN THE BODY BECOMES OVERHEATED

Overheating the preoptic thermostatic area increases the rate of heat loss from the body in two principal ways: (1) by stimulating the sweat glands to cause evaporative heat loss from the body and (2) by inhibiting sympathetic centers in the posterior hypothalamus; this removes the normal vasoconstrictor tone to the skin vessels, thereby allowing vasodilatation and loss of heat from the skin, and also inhibiting the shivering mechanism to prevent excessive heat production.

## MECHANISMS OF HEAT CONSERVATION AND INCREASED HEAT PRODUCTION WHEN THE BODY BECOMES COOLED

When the body core is cooled below approximately 37° C., special mechanisms are set into play to conserve the heat that is already in the body, and still other mechanisms are set into play to increase the rate of heat production, as follows:

**Heat Conservation.** *Vasoconstriction in the Skin.* One of the first effects is intense vasoconstriction of the skin vessels over the entire body. This results partly from *release of the posterior hypothalamic sympathetic areas from inhibition* by the preoptic heat signals but probably more from *the drive provided by the skin and spinal cord cold receptors.* As a result, the sympathetic areas become overactive, and intense adrenergic vasoconstriction occurs throughout the body. This vasoconstriction obviously prevents the conduction of heat from the internal portions of the body to the skin. Consequently, with maximal vasoconstriction

the only heat that can leave the body is that which can be conducted directly through the insulator layers of the skin. This effect conserves the quantity of heat in the body.

*Pilo-Erection.* A second means by which heat is conserved when the hypothalamus is cooled is pilo-erection—that is, the hairs "stand on end." Obviously, this effect in not important in the human being because of the paucity of hair, but in lower animals the upright projection of the hairs in cold weather entraps a thick layer of insulator air next to the skin so that the transfer of heat to the surroundings is greatly depressed.

*Abolition of Sweating.* Sweating is completely abolished by cooling the preoptic thermostat below about 37° C. (98.6° F.). This obviously causes evaporative cooling of the body to cease except for that resulting from insensible evaporation.

**Increased Production of Heat.** Heat production is increased in three separate ways when the temperature of the body thermostat falls below 37° C.:

*Hypothalamic Stimulation of Shivering.* Located in the dorsomedial portion of the posterior hypothalamus near the wall of the third ventricle is an area called the *primary motor center for shivering*. This area is normally inhibited by signals from the preoptic heat thermostatic area but is driven by signals from the skin and spinal cord. Therefore, in response to cold, this center becomes activated and transmits impulses through bilateral tracts down the brain stem, into the lateral columns of the spinal cord, and, finally, to the anterior motoneurons. These impulses are nonrhythmic and do not cause the actual muscle shaking. Instead, they increase the tone of the skeletal muscles throughout the body. The resulting increase in muscle metabolism increases the rate of heat production, often raising total body heat production as much as 50 per cent even before shivering occurs. Then, once the tone of the muscles rises above a certain critical level, shivering begins. This probably results from feedback oscillation of the muscle spindle stretch reflex mechanism. During maximum shivering, body heat production can rise to as high as five times normal.

*Sympathetic "Chemical" Excitation of Heat Production.* It was pointed out in Chapter 71 that either sympathetic stimulation or circulating norepinephrine and epinephrine in the blood can cause an immediate increase in the rate of cellular metabolism; this effect is called *chemical thermogenesis,* and it is believed to result at least partially from the ability of norepinephrine and epinephrine to uncouple oxidative phosphorylation, as a result of which more oxidation of foodstuffs must occur to produce the amounts of high energy phosphate compounds required for normal function of the body. Therefore, the rate of cellular metabolism increases.

The degree of chemical thermogenesis that occurs in an animal is almost directly proportional to the amount of *brown* fat that exists in the animal's tissues. This is a type of fat that contains large numbers of mitochondria in its cells, and these same cells are supplied by a strong sympathetic innervation. Upon sympathetic stimulation, the oxidative metabolism of the mitochondria is greatly stimulated, but this probably occurs in an uncoupled manner so that only small amounts of ATP are formed. Regardless of the chemical mechanism, the energy release caused by the oxidative metabolism provides a very important source of heat to the body.

The process of acclimatization greatly affects the intensity of chemical thermogenesis; some animals that have been exposed for several weeks to a very cold environment exhibit as much as a 100 to 500 per cent increase in metabolism when acutely exposed to cold, in contrast to the unacclimatized animal, which responds with an increase in metabolism of perhaps one-third as much.

In the adult man, who rarely becomes completely acclimatized to cold environments and who has almost no brown fat, it is rare that chemical thermogenesis increases the rate of heat production more than 10 to 15 per cent. However, in infants, who *do* have a small amount of brown fat in the intercapsular space, chemical thermogenesis can increase the rate of heat production as much as 100 per cent, which is probably a very important factor in maintaining normal body temperature in the newborn.

***Increased Thyroxine Output as a Cause of Increased Heat Production.*** Cooling the preoptic area of the hypothalamus also increases the production of the neurosecretory hormone *thyrotropin-releasing factor* by the hypothalamus. This hormone is carried by way of the hypothalamic portal veins to the adenohypophysis where it stimulates the secretion of *thyrotropin*. Thyrotropin, in turn, stimulates increased output of thyroxine by the thyroid gland, as will be explained in Chapter 76. The increased thyroxine increases the rate of cellular metabolism throughout the body. This in-

crease in metabolism through the thyroid mechanism does not occur immediately but requires several weeks for the thyroid gland to hypertrophy before it reaches its new level of thyroxine secretion.

Exposure of animals to extreme cold for several weeks can cause their thyroid gland to increase in size as much as 20 to 40 per cent. Unfortunately, however, the human being rarely allows himself to be exposed to the same degree of cold as that to which animals have been subjected. Therefore, we still do not know, quantitatively, how important the thyroid method of adaptation to cold is in the human being. Yet, isolated measurements have shown that military personnel residing for several months in the Arctic develop increased metabolic rates; Eskimos also have abnormally high basal metabolic rates. Also, the continuous stimulatory effect to cold on the thyroid gland can probably explain the much higher incidence of toxic thyroid goiters in persons living in colder climates than in those living in warmer climates.

### EFFECT OF THE PERIPHERAL TEMPERATURE RECEPTORS IN CHANGING THE "SET POINT" OF THE HYPOTHALMIC THERMOSTAT

The critical temperature level to which the body temperature control system attempts to control the temperature is called the "set point" of the system. Ordinarily, the set point for normal control of body temperature is 37.6° C. Above this temperature, the different heat-losing mechanisms become operative, and below this level, the heat-conserving and heat-producing mechanisms become active.

The precise temperature level of the set point is determined mainly by the hypothalamic thermostatic center. However, the signals from the peripheral temperature receptors of the body can alter the level of the set point, sometimes as much as several degrees. For instance, when the skin or the spinal cord (and perhaps some intra-abdominal temperature receptors as well) becomes greatly overheated, the set point of the thermostatic system may be reduced from a few tenths of a degree centigrade to as much as 2 to 3 degrees. For instance, sweating may now occur at any internal body temperature above 36° C. instead of above the normal temperature of 37.6° C. Conversely, increased heat production may now begin to occur only when the hypothalamic temperature falls below

36° C. rather than below the normal level of 37.6° C.

Conversely, extreme cold applied to the skin or spinal cord can increase the set point of the hypothalamic thermostat from a few tenths of a degree to as much as a degree or more above the normal level. Also, we shall see later in the chapter that fever-producing agents can sometimes raise the set point of the thermostatic system to as high as 5 to 10 degrees above normal.

## BEHAVIORAL CONTROL OF BODY TEMPERATURE

Aside from the thermostatic mechanism for body temperature control, the body has still another mechanism for body temperature control that is usually even more potent than the thermostatic system. This mechanism is behavioral control of temperature, which can be explained as follows: Whenever the internal body temperature becomes too high, signals from the preoptic area of the brain give one a psychic sensation of being overheated. Whenever the body becomes too cold, signals from the skin and perhaps from other peripheral receptors elicit the feeling of cold discomfort. Therefore, the person makes appropriate environmental adjustments to re-establish comfort. This is a much more powerful system of body temperature control than most physiologists have recognized in the past. Indeed, for man, this is the only really effective mechanism for body heat control in severely cold environs.

The obvious types of behavioral adjustments include: selecting appropriate clothing, moving the body to a different environmental setting, increasing the delivery of heat or cold from appropriate heaters or air conditioners, and so forth.

It is important to note that many other of our body's control systems utilize similar behavioral mechanisms to achieve highly refined degrees of control. For instance, even respiration is probably controlled to a great extent in this way—that is, when a person perceives that he is being subjected to air hunger, he consciously breathes more to make up the deficit. Therefore, it is not valid to think of the body's homeostatic control systems as operating only in the subconscious portions of the brain.

## LOCAL SKIN REFLEXES

When a person places his foot under a hot lamp and leaves it there for a short time, he finds *local*

*vasodilatation* and mild *local sweating*. Conversely, placing the foot in cold water causes vasoconstriction and cessation of sweating. These reactions are caused by local cord reflexes conducted from the skin receptors to the spinal cord and back to the same skin area. However, their *intensity* is controlled by the hypothalamic thermostat, so that the overall effect is approximately proportional to the hypothalamic heat control signal *times* the local signal. Such reflexes can help to prevent excessive heat exchange from locally cooled or heated portions of the body.

**Regulation of Internal Body Temperature After Cutting the Spinal Cord.** After cutting the spinal cord in the neck above the sympathetic outflow from the cord, regulation of body temperature becomes extremely poor, for the hypothalamus can then no longer control either skin blood flow or the degree of sweating anywhere in the body. On the other hand, the local temperature reflexes originating in the skin, spinal cord, and intra-abdominal receptors still exist. Unfortunately, these reflexes are not powerful. In persons with this condition, body temperature must be regulated principally by the patient's psychic response to cold and hot sensations in his head region. That is, if he feels himself becoming too hot or if he develops a headache from the heat, he knows to select cooler surroundings, and, conversely, if he has cold sensations, he selects warmer surroundings.

# ABNORMALITIES OF BODY TEMPERATURE REGULATION

## FEVER

Fever, which means a body temperature above the usual range of normal, may be caused by abnormalities in the brain itself or by toxic substances that affect the temperature regulating centers. Some causes of fever are presented in Figure 72–8. These include bacterial diseases, brain tumors, and a vicious cycle of heat production that may terminate in heat stroke.

### Resetting the Hypothalamic Thermostat in Febrile Diseases—Effect of Pyrogens

Many proteins, breakdown products of proteins, and certain other substances, such as lipopolysaccharide toxins secreted by bacteria, can cause the "set point" of the hypothalamic thermostat to rise. Substances that cause this effect are called pyrogens. It is pyrogens secreted by toxic bacteria or pyrogens released from degenerating tissues of the body that cause fever during disease conditions. When the set point of the hypothalamic thermostat becomes increased to a higher level than normal, all the

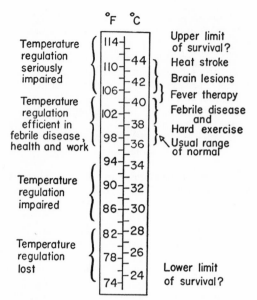

**Figure 72–8.** Body temperatures under different conditions. (From DuBois: Fever. Charles C Thomas.)

ticuloendothelial cells to form endogenous pyrogen. The endogenous pyrogen is then transmitted in the blood to the hypothalamic thermostat and increases its set point to the febrile level.

To give one an idea of the extremely powerful effect of pyrogens in resetting the hypothalamic thermostat, as little as a few nanograms of purified endogenous pyrogen injected into an animal can cause severe fever.

**Effect of Dehydration on the Hypothalamic Thermostat.** Dehydration is another factor that can cause the body temperature to rise considerably. Part of this elevation of temperature probably results from lack of available fluid for sweating, but dehydration can also cause temperature elevation even in a cold atmosphere. Consequently, dehydration almost certainly has a direct effect on the hypothalamic centers to set the hypothalamic thermostat to febrile levels.

mechanisms for raising the body temperature are brought into play, including heat conservation and increased heat production. Within a few hours after the thermostat has been set to a higher level, the body temperature also approaches this level.

**Mechanism of Action of Pyrogens in Causing Fever—Endogenous Pyrogen.** Some physiologists believe that exogenous pyrogenic substances have a direct action on the hypothalamic thermostat to increase its setting. However, there is reason to believe that most pyrogens affect the hypothalamic thermostat indirectly in the following manner:

When a body tissue degenerates or becomes diseased, polymorphonuclear leukocytes and macrophages enter the tissue; these in turn are autocatalytically destroyed as they phagocytize the degenerating tissue. As they themselves degenerate, they release a substance called *endogenous pyrogen*. This substance has a direct effect on the hypothalamic thermostat to increase its setting to febrile levels.

Likewise, when bacterial pyrogens are injected into a person they usually do not affect the hypothalamic thermostat immediately; instead, the effect is delayed for many minutes to several hours. This delay is believed to result from the fact that the pyrogens must first react with the polymorphonuclear leukocytes, monocytes, macrophages, and certain of the re-

### Characteristics of Febrile Conditions

**Chills.** When the setting of the thermostat is suddenly changed from the normal level to a higher-than-normal value as a result of tissue destruction, pyrogenic substances, or dehydration, the body temperature usually takes several hours to reach the new temperature setting. For instance, the temperature setting of the hypothalamic thermostat, as illustrated in Figure 72–9, might suddenly rise to 103° F. Because the blood temperature is less than the temperature setting of the hypothalamic thermostat, the usual autonomic responses to cause elevation of body temperature occur. During this period the person experiences chills, during which he feels extremely cold, even though his body temperature may already be above normal. Also, his skin is cold because of vasoconstriction, and he shakes all over be-

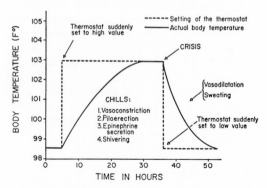

**Figure 72–9.** Effects of changing the setting of the "hypothalamic thermostat."

cause of shivering. His chills continue until his body temperature rises to the hypothalamic setting of 103° F. Then, when the temperature of the body reaches this value, he no longer experiences chills but instead feels neither cold nor hot. As long as the factor that is causing the hypothalamic thermostat to be set at a high value continues its effect, the body temperature is regulated more or less in the normal manner but at the high temperature level.

**The Crisis or "Flush."**    If the factor that is causing the high temperature is suddenly removed, the hypothalamic thermostat is suddenly set at a lower value—perhaps even back to the normal level, as illustrated in Figure 72–9. In this instance, the blood temperature is still 103° F., but the hypothalamus is attempting to regulate the body temperature at 98.6° F. This situation is analogous to excessive heating of the preoptic area, which causes intense sweating and sudden development of a hot skin because of vasodilatation everywhere. This sudden change of events in a febrile disease is known as the "crisis" or, more appropriately, the "flush." In olden days, the doctor always awaited the crisis, for once this occurred he knew immediately that the patient's temperature would soon be falling.

### Heat Stroke

The limits of extreme heat that one can stand depend almost entirely on whether the heat is dry or wet. If the air is completely dry and sufficient convection air currents are flowing to promote rapid evaporation from the body, a person can withstand several hours of air temperature at 150° F. with no apparent ill effects. On the other hand, if the air is 100 per cent humidified and evaporation cannot occur, or if the body is in water, the body temperature begins to rise whenever the surrounding temperature rises above approximately 94° F. If the person is performing heavy work, this critical temperature level may fall to 85° to 90° F.

Also, as the body temperature rises, the basal metabolism increases about 6 per cent for each degree F. rise or 10 per cent for each degree C. rise because of the intrinsic effect of heat to increase the rates of chemical reactions. Once the body temperature rises to approximately 110° F., the rate of metabolism has doubled.

Unfortunately, there is a limit to the rate at which the body can lose heat even with maximal sweating. Furthermore, when the hypothalamus becomes excessively heated, its heat regulating ability becomes greatly depressed and sweating diminishes. As a result, a vicious cycle develops: high temperature causes increased production of heat, which increases the body temperature still higher, which causes still a greater production of heat and a still higher body temperature, etc. Once the body temperature rises above 107° to 110° F., the heat regulating mechanisms often can no longer dissipate the excessive heat being produced. Therefore, the temperature

may then rise abruptly until it causes death, unless the rise is checked artificially.

**Relationship of Heat Stroke to Dehydration.**    People who work in hot and humid atmospheres can often become excessively dehydrated because of sweating and, consequently, can develop *circulatory shock*. As pointed out earlier in the chapter, dehydration probably also has a direct effect on the hypothalamus to set the thermostat at a high level, in addition to its effects to diminish sweating drastically; and these effects undoubtedly add to the vicious cycle of hyperpyrexia. This explains why it is important to maintain an adequate intake of water and sodium chloride in hot surroundings if heat stroke is to be avoided.

**Acclimatization to Heat.**    It is often extremely important to acclimatize persons to extreme heat; some examples are acclimatization of soldiers for tropical duty and acclimatization of miners for work in the two mile deep gold mines of South Africa where the temperature approaches body temperature and the humidity approaches 100 per cent. Exposure of a person to heat for several hours each day while working a reasonably heavy work load will develop increased tolerance to hot and humid conditions in about one week. Probably the most important physiologic changes that occur during this acclimatization process are an increase in plasma volume and diminished loss of salt in the sweat and urine, these effects resulting from increased secretion of aldosterone.

### Harmful Effects of High Temperature

When the body temperature rises above approximately 106° F., the parenchyma of many cells usually begins to be damaged. The pathologic findings in a person who dies of hyperpyrexia are local hemorrhages and parenchymatous degeneration of cells throughout the entire body. The brain is especially likely to suffer, because neuronal cells once destroyed can never be replaced. When the body temperature rises above 110° F., the person usually has only a few hours to live unless his temperature is brought back within normal range rapidly by sponging his body with alcohol, which evaporates and cools the body, or by bathing him in ice water.

### Effects of Chemicals and Drugs on Body Temperature

An extremely large number of foreign substances when injected into the body fluids can cause the body temperature to rise—that is, they are pyrogenic. Bacteria, pollens, dust, and vaccines are all pyrogenic because of their protein content. Ordinarily, human plasma is not pyrogenic, but it is possible for the plasma proteins and other proteins that normally do not cause pyrogenic reactions to be changed chemically by degradation or denaturization and thereafter to cause pyrogenic reactions. A few nonprotein chemicals are also known to elevate the body temperature; these include especially certain polysaccharides and some nitrated phenol compounds.

**Antipyretics.** Aspirin, antipyrine, aminopyrine, and a number of other substances known as "antipyretics" have an effect on the hypothalamic thermostat opposite to that of the pyrogens. In other words, they cause the setting of the thermostat to be lowered so that the body temperature falls, though usually not more than a degree or so. Aspirin is especially effective in lowering the hypothalamic setting when pyrogens have raised the setting, but aspirin will not lower the normal temperature. On the other hand, aminopyrine will decrease even the normal body temperature. Obviously, these drugs can be used to prevent damage to the body from excessively high body temperature.

## EXPOSURE OF THE BODY TO EXTREME COLD

A person exposed to ice water for approximately 20 to 30 minutes ordinarily dies because of heart standstill or heart fibrillation unless treated immediately. By that time, the internal body temperature will have fallen to about 77° F. Yet if he is warmed rapidly by application of external heat, his life can often be saved.

Treatment of a person whose body temperature has fallen into the seventies usually consists of application of wet heat either in the form of tub treatment or hot packs, with the water at approximately 110° F. If the temperature of the water is less than this, the rate at which heat is returned to the body is too slow for maximal benefit, and, if it is greater than this, the skin might be severely damaged because it becomes overheated while not receiving a satisfactory blood supply.

**Loss of Temperature Regulation at Low Temperatures.** As noted in Figure 72–8, once the body temperature has fallen below 85° F., the ability of the hypothalamus to regulate temperature is completely lost, and it is greatly impaired even when the body temperature falls below approximately 94° F. Part of the reason for this loss of temperature regulation is that the rate of heat production in each cell is greatly depressed by the low temperature. Also, sleepiness and even coma are likely to develop, which depress the activity of the central nervous system heat-control mechanisms and prevent shivering.

**Frostbite.** When the body is exposed to extremely low temperatures, surface areas can actually freeze; the freezing is called "frostbite." This occurs especially in the lobes of the ears and in the digits of the hands and feet. If the parts are thawed immediately, especially with water that is not above approximately 110° F., no permanent damage may result. On the other hand, prolonged freezing causes permanent circulatory impairment as well as local tissue damage. Often gangrene follows thawing, and the frost-bitten areas are lost.

***Cold-induced Vasodilatation as a Protection Against Frostbite.*** When the temperature of tissues falls almost to freezing, sudden vasodilatation occurs, often manifested by a flush of the skin. This results from local vascular reaction to the cold itself, and it overcomes the vasoconstrictor effects of the sympathetic nerve signals to the blood vessels. This mechanism, fortunately, helps to prevent frostbite by delivering warm blood to the skin. Unfortunately, it is a far less well developed mechanism in man than in most lower animals that live in the cold all the time.

**Artificial Hypothermia.** It is possible to decrease the temperature of a person by giving him a sedative to depress the hypothalamic thermostat and then packing him in ice until his temperature falls. His temperature can then be maintained below 90° F. for several days to a week or more by continual sprinkling of cool water or alcohol on the body. Such artificial cooling is often used during heart surgery so that the heart can be stopped artificially for many minutes at a time. Cooling to this extent does not cause severe physiologic results. It does slow the heart and greatly depresses body metabolism.

# REFERENCES

Benzinger, T. H.: Heat regulation: homeostasis of central temperature in man. *Physiol. Rev.*, 49:671, 1969.

Bligh, J.: Temperature Regulation in Mammals and Other Vertebrates. New York, American Elsevier Publishing Co., Inc., 1973.

Bligh, J., and Moore, R. E.,(eds.): Essays on Temperature Regulation. New York, American Elsevier Publishing Co., Inc., 1972.

Cabanac, M.: Thermoregulatory behavior. In Guyton, A. C. (ed.): MTP International Review of Science: Physiology. Vol. 7. Baltimore, University Park Press, 1974, p. 231.

Cabanac, M.: Temperature regulation. *Ann. Rev. Physiol.*, 37:415, 1975.

Chauffee, R. R. J., and Roberts, J. C.: Temperature acclimation in birds and mammals. *Ann. Rev. Physiol.*, 33:155, 1971.

Ciba Foundation Symposia: Pyrogens and Fever. New York, Churchill Livingstone, Div. of Longman, Inc., 1971.

Dill, D. B. et al. (eds.): Handbook of Physiology. Sec. 4. Baltimore, The Williams & Wilkins Company, 1965.

Dobson, R. L., and Sato, K.: The stimulation of eccrine sweating by pharmacologic agents. *Adv. Biol. Skin*, 12:447, 1972.

Folk, G. E.: Introduction to Environmental Physiology: Extremes and Mammalian Survival. Philadelphia, Lea & Febiger, 1966.

Gale, C. C.: Neuroendocrine aspects of thermoregulation. *Ann. Rev. Physiol.*, 35:391, 1973.

Hales, J. R. S.: Physiological responses to heat. In Guyton, A. C. (ed.): MTP International Review of Science: Physiology. Vol. 7. Baltimore, University Park Press, 1974, p. 107.

Hardy, J. D.: Physiology of temperature regulation. *Physiol. Rev.*, 41:521, 1961.

Hazel, J. R., and Prosser, C. L.: Molecular mechanisms of temperature compensation in poikilotherms. *Physiol. Rev.*, 54:620, 1974.

Heath, J. E.: Behavioral regulation of body temperature in poikilotherms. *Physiologist*, 13:399, 1970.

Hemingway, A.: Shivering. *Physiol. Rev.*, 43:397, 1963.

Hensel, H.: Neural processes in thermoregulation. *Physiol. Rev.*, 53:948, 1973.

Hensel, H.: Thermoreceptors. *Ann. Rev. Physiol.*, 36:233, 1974.

Jansky, L.: Non-shivering thermogenesis and its thermoregulatory significance. *Biol. Rev.*, 48:85, 1973.

Jenkinson, D. M.: Comparative physiology of sweating. *Brit. J. Dermatol.*, 88:397, 1973.

Johnson, R. H., and Spaulding, J. M.: Disorders of the autonomic nervous system. Chapter 8. Body temperature regulation and its investigation. *Contemp. Neurol. Ser., 11*:129, 1974.

Johnson, R. H., and Spaulding, J. M.: Disorders of the autonomic nervous system. Chapter 10. Sweating. *Contemp. Neurol. Ser., 11*:179, 1974.

Kerslake, D. M.: The Stress of Hot Environments. (Physiological Society Monograph, No. 29.) Cambridge, Cambridge University Press, 1972.

McRorie, R. A., and Williams, W. L.: Biochemistry of mammalian fertilization. *Ann. Rev. Biochem., 43*:777, 1974.

Meerson, F. Z.: Role of synthesis of nucleic acids and protein in adaptation to the external environment. *Physiol. Rev., 55*:79, 1975.

Mitchell, D.: Physical basis of thermoregulation. *In* Guyton, A. C. (ed.): MTP International Review of Science: Physiology. Vol. 7. Baltimore, University Park Press, 1974, p. 1.

Precht, H., Christophersen, J., Hensel, H., and Larcher, W.: Temperature and Life. New York, Springer-Verlag New York, Inc., 1973.

Riedesel, M. L.: Hibernation, Spring 1973. *Physiologist, 16*:565, 1973.

Schmidt-Nielsen, K.: Animal Physiology: Adaptation and Environment. London, Cambridge University Press, 1975.

Simon, E.: Temperature regulation: the spinal cord as a site of extrahypothalamic thermoregulatory functions. *Rev. Physiol. Biochem. Pharmacol., 71*:1, 1974.

Smith, R. E., and Horwitz, B. A.: Brown fat and thermogenesis. *Physiol. Rev., 49*:330, 1969.

Snell, E. S., and Atkins, E.: The mechanisms of fever. *In* Bittar, E. E., and Bittar, N. (eds.): The Biological Basis of Medicine. Vol. 1. New York, Academic Press, Inc., 1968, p. 397.

Swan, H.: Thermoregulation and Bioenergetics: Patterns for Vertebrate Survival. New York, American Elsevier Publishing Co., Inc., 1974.

Webster, A. J. F.: Physiological effects of cold exposure. *In* Guyton, A. C. (ed.): MTP International Review of Science: Physiology. Vol. 7. Baltimore, University Park Press, 1974, p. 33.

Webster, A. J. F.: Adaptation to cold. *In* Guyton, A. C. (ed.): MTP International Review of Science: Physiology. Vol. 7. Baltimore, University Park Press, 1974, p. 71.

Wyndham, C. H.: The physiology of exercise under heat stress. *Ann. Rev. Physiol., 35*:193, 1973.

# 73

# Dietary Balances, Regulation of Feeding; Obesity and Starvation

The intake of food must always be sufficient to supply the metabolic needs of the body and yet not enough to cause obesity. Also, since different foods contain different proportions of proteins, carbohydrates, and fats, appropriate balance must be maintained between these different types of food so that all segments of the body's metabolic systems can be supplied with the requisite materials. This chapter therefore discusses the problems of balance between the three major types of food and also the mechanisms by which the intake of food is regulated in accordance with the metabolic needs of the body.

## DIETARY BALANCES

### ENERGY AVAILABLE IN FOODS

The energy liberated from each gram of carbohydrate as it is oxidized to carbon dioxide and water is 4.1 Calories, and that liberated from fat is 9.3 Calories. The energy liberated from metabolism of the average protein of the diet as each gram is oxidized to carbon dioxide, water, and urea is 4.35 Calories. Also, these different substances vary in the average percentages that are absorbed from the gastrointestinal tract; approximately 98 per cent of the carbohydrate, 95 per cent of the fat, and 92 per cent of the protein. Therefore, in round figures the average *physiologically available energy* in each gram of the three different foodstuffs in the diet is:

|  | Calories |
|---|---|
| Carbohydrates | 4.0 |
| Fat | 9.0 |
| Protein | 4.0 |

### AVERAGE COMPOSITION OF THE DIET

The average American receives approximately 15 per cent of his energy from protein, about 40 per cent from fat, and 45 per cent from carbohydrates. In most other parts of the world the quantity of energy derived from carbohydrates far exceeds that derived from both proteins and fats. Indeed, in Mongolia the energy received from fats and proteins combined is said to be no greater than 15 to 20 per cent.

**Daily Requirement for Protein.** About 30 grams of the body proteins are degraded and used for energy daily. Therefore, all cells must continue to form new proteins to take the place of those that are being destroyed, and a supply of protein is needed in the diet for this purpose. An average man can maintain his normal stores of protein provided that his *daily intake is above 30 to 45 grams.*

*Partial Proteins.* Another factor that must be considered in analyzing the proteins of the diet is whether the dietary proteins are *complete* proteins or *partial* proteins. Complete proteins have compositions of amino acids in appropriate proportion to each other so that all the amino acids can be properly used by the human body. In general, proteins derived from animal foodstuffs are more nearly complete than are proteins derived from vegetable and grain sources. This subject is more completely discussed in Chapter 69; therefore, for the present, suffice it to say that when partial proteins are in the diet an increased minimal quantity of protein is necessary in the daily rations to maintain protein balance. A particular example of this occurs in the diet of many African natives who subsist primarily on a corn meal diet. The protein of corn is almost totally lacking in tryptophan; and this means that this diet, in effect, is almost completely protein deficient. As a result, the natives, especially the children, develop the protein deficiency syndrome called *kwashiorkor,* which consists of failure to grow, lethargy, depressed mentality, and hypoprotein edema.

**Necessity for Fat in the Diet.** The human body is capable of synthesizing some unsaturated fatty acids, especially in the liver. However, the liver cannot form others such as *linoleic, linolenic,* and *arachidonic acids,* which are essential constituents of the diet for normal operation of the body. In lower animals, scaling and exudative skin lesions appear when these unsaturated fatty acids are totally absent from the diet; they are probably also of special importance for formation of some of the structural elements of all cells throughout the body, because lack of these acids causes failure of growth. How much of the essential unsaturated fat is needed in the diet has not been determined; but because Mongolians exist on diets with fat contents as low as 10 per cent, it is presumed that only small quantities of unsaturated fats are needed. However, it was pointed out in Chapter 68 that a high *ratio* of saturated to unsaturated fats in the diet predisposes to atherogenesis, which often leads to heart attacks. Therefore, most nutritionists recommend that the fat of the diet contain a high proportion of unsaturated fatty acids.

**Necessity for Carbohydrates in the Diet.** In the discussion of carbohydrate metabolism in Chapter 67, it was pointed out that failure to metabolize carbohydrate in sufficient quantity causes rapid metabolism of fats, and ketosis is likely to develop. Also, fats and proteins alone usually cannot supply sufficient energy for all operations of the human body. Consequently, carbohydrates appear to be essential to prevent weakness. This is especially true when the body undertakes a considerable work load. Finally, as already discussed, carbohydrate is a *protein sparer;* that is, carbohydrate is burned in preference to the burning of protein, which is especially important for preserving the functional proteins in the cells.

**Composition of Different Foods.** Table 73–1 presents the compositions of selected foods, illustrating especially the high proportions of fats and proteins in meat products and the high proportions of carbohydrates in most vegetable products.

Fat is deceptive in the diet, for it often exists as 100 per cent fat undiluted by any other substances, whereas essentially all the proteins and carbohydrates of foods are mixed in watery media and often

**TABLE 73–1.   Protein, Fat, and Carbohydrate Content of Different Foods**

| Food | Protein, % | Fat, % | Carbohydrate, % | Fuel Value per 100 Grams, Calories |
|---|---|---|---|---|
| Apples | 0.3 | 0.4 | 14.9 | 64 |
| Asparagus | 2.2 | 0.2 | 3.9 | 26 |
| Bacon, fat | 6.2 | 76.0 | 0.7 | 712 |
| broiled | 25.0 | 55.0 | 1.0 | 599 |
| Beef, medium | 17.5 | 22.0 | 1.0 | 268 |
| Beets, fresh | 1.6 | 0.1 | 9.6 | 46 |
| Bread, white, milk | 9.0 | 3.6 | 49.8 | 268 |
| Butter | 0.6 | 81.0 | 0.4 | 733 |
| Cabbage | 1.4 | 0.2 | 5.3 | 29 |
| Carrots | 1.2 | 0.3 | 9.3 | 45 |
| Cashew nuts | 19.6 | 47.2 | 26.4 | 609 |
| Cheese, Cheddar, American | 23.9 | 32.3 | 1.7 | 393 |
| Chicken, total edible | 21.6 | 2.7 | 1.0 | 111 |
| Chocolate | (5.5) | 52.9 | (18.) | 570 |
| Corn (maize), entire | 10.0 | 4.3 | 73.4 | 372 |
| Haddock | 17.2 | 0.3 | 0.5 | 72 |
| Lamb, leg, intermediate | 18.0 | 17.5 | 1.0 | 230 |
| Milk, fresh whole | 3.5 | 3.9 | 4.9 | 69 |
| Molasses, medium | 0.0 | 0.0 | (60.) | 240 |
| Oatmeal, dry, uncooked | 14.2 | 7.4 | 68.2 | 396 |
| Oranges | 0.9 | 0.2 | 11.2 | 50 |
| Peanuts | 26.9 | 44.2 | 23.6 | 600 |
| Peas, fresh | 6.7 | 0.4 | 17.7 | 101 |
| Pork, ham, medium | 15.2 | 31.0 | 1.0 | 340 |
| Potatoes | 2.0 | 0.1 | 19.1 | 85 |
| Spinach | 2.3 | 0.3 | 3.2 | 25 |
| Strawberries | 0.8 | 0.6 | 8.1 | 41 |
| Tomatoes | 1.0 | 0.3 | 4.0 | 23 |
| Tuna, canned | 24.2 | 10.8 | 0.5 | 194 |
| Walnuts, English | 15.0 | 64.4 | 15.6 | 702 |

Extracted from data compiled by Chatfield and Adams, U.S. Department of Agriculture Circular No. 549, 1940.

represent less than 25 per cent of the weight of the food. Thus, the fat of one pat of butter mixed with an entire helping of potato may contain as much energy as all the potato itself. This is one of the reasons why fats are assiduously avoided in the prescription of diets for weight reduction.

## STUDY OF ENERGY BALANCES

It is important in many physiological studies to compute the intake of food and also to compute at the same time the utilization of the different types of foods by the body. Measurement of the intake of the different types of food can be accomplished by simply weighing the foods before they are eaten and analyzing these foods for their relative contents of carbohydrates, fats, and protein. Then the loss of foods in the fecal and urinary excretions is subtracted from the intake to determine the *net intake*. On the other hand, special procedures must be performed for determining *utilization* of the different types of foods for energy, as follows:

**Determination of the Rate of Protein Metabolism in the Body.** The average protein of the diet contains approximately 16 per cent nitrogen, and the remaining 84 per cent is composed of carbon, hydrogen, oxygen, and sulfur. Numerous experiments have shown that when protein is metabolized in the body, the average person excretes about 90 per cent of the protein nitrogen in the urine in the form of urea, uric acid, creatinine, and other less important nitrogen products. The remaining 10 per cent is ordinarily excreted in the feces. It is possible, therefore, to estimate relatively accurately the total quantity of protein metabolized by the body in a given time by analyzing the amount of nitrogen excreted in the urine. For instance, if 8 grams of nitrogen are excreted into the urine each day, then the total excretion, including that in the feces (an additional 10 per cent), will be about 8.8 grams. This value multiplied by 100/16 gives a calculated total protein metabolism of 55 grams per day.

*"Nitrogen Balance" in the Body.* Nitrogen balance experiments are frequently performed to determine the rate of protein increase or decrease in the body. These balance studies are performed by measuring for a week or more (a) the rate of protein intake and (b) the rate of protein utilization, the total protein utilization being estimated as was just described.

A *negative nitrogen balance* implies greater protein utilization than protein intake, causing loss of protein from the body, and a *positive nitrogen* balance implies net gain of protein in the body. Factors that cause negative balance include malnutrition, debilitating diseases, and glucocorticoid hormones. Factors that cause positive nitrogen balance include exercise, growth hormone, and testosterone.

**Relative Utilization of Fat and Carbohydrates—The Respiratory Quotient.** When one molecule of glucose is oxidized, the number of molecules of carbon dioxide liberated is exactly equal to the number of oxygen molecules necessary for the oxidative process. This fact is illustrated in Figure 73-1. Therefore the *respiratory quotient,* which is defined as the *ratio of carbon dioxide output to oxygen usage,* is 1.00. On the other hand, oxidation of triolein (the most abundant fat in the body) liberates 57 carbon dioxide molecules while 80 oxygen molecules are being utilized. Consequently, the respiratory quotient in this instance is 0.71. Finally, oxidation of alanine liberates five carbon dioxide molecules for every six oxygen molecules entering into the reaction, giving a respiratory quotient of 0.83.

The respiratory quotient for utilization of carbohydrates is always 1.00 because the quantity of oxygen in each carbohydrate molecule is always exactly sufficient to oxidize only the hydrogen within the molecule, and oxidation of each carbon atom in the molecule requires one molecule of respiratory oxygen. On the other hand, the respiratory quotient of dietary fat has been found to average 0.707, while the respiratory quotient of the average protein of the diet is 0.801.

*Use of the Respiratory Quotient for Estimating Relative Rates of Carbohydrate and Fat Metabolism.* As already pointed out, the average person receives only 15 per cent of his total energy from protein metabolism. Furthermore, the respiratory quotient of protein is approximately midway between the respiratory quotients of fat and carbohydrate (see preceding paragraph). Consequently, when the overall respiratory quotient of a person (which, in studies lasting an hour or more, is equal to the respiratory exchange ratio that was discussed in Chap. 40) is

$$C_6H_{12}O_6 + 6\ O_2 \rightarrow 6\ CO_2 + 6\ H_2O$$
Glucose

$$C_{57}H_{104}O_6 + 80\ O_2 \rightarrow 57\ CO_2 + 52\ H_2O$$
Triolein

$$2\ C_3H_7O_2N + 6\ O_2 \rightarrow (NH_2)_2CO + 5\ CO_2 + 5\ H_2O$$
Alanine

Respiratory Quotient:

$$\frac{6}{6} = 1.00$$

$$\frac{57}{80} = 0.71$$

$$\frac{5}{6} = 0.83$$

**Figure 73–1.** Utilization of oxygen and release of carbon dioxide during the oxidation of carbohydrate, fat, and protein. The respiratory quotient for each of these reactions is calculated.

measured by determining the total respiratory intake of oxygen and the total output of carbon dioxide from the lungs, one has a reasonable measure of the relative quantities of fat and carbohydrate being metabolized by the body during that particular time. For instance, if the respiratory quotient is approximately 0.71, the body is burning almost entirely fat to the exclusion of carbohydrates and proteins. If, on the other hand, the respiratory quotient is 1.00, the body is probably metabolizing almost entirely carbohydrate to the exclusion of fat and protein. Finally, a respiratory quotient of 0.85 indicates approximately equal utilization of carbohydrate and fat.

To be still more accurate in the calculation, one must consider the portion of the respiratory quotient that represents protein metabolism. To do this the rate of protein utilization in the body is first determined as discussed above; then the quantities of oxygen utilized and carbon dioxide released as a result of the protein metabolism are calculated and subtracted from the intake of oxygen and output of carbon dioxide, respectively, in order to establish the net respiratory quotient as it applies to fat and carbohydrate

*Variations in the Respiratory Quotient.* Shortly after a meal, almost all the food metabolized is carbohydrates. Consequently, the respiratory quotient approaches 1.00 at this time. Approximately 8 to 10 hours following a meal, the quantity of carbohydrate being metabolized is relatively slight, and the respiratory quotient approaches that for fat metabolism; that is, approximately 0.71.

In diabetes mellitus, little carbohydrate is utilized by the body, and consequently, most of the energy is derived from fat. Therefore, most persons with severe diabetes have respiratory quotients approaching the value for fat metabolism, 0.71.

In the human being, the respiratory quotient rarely becomes greater than 1.00, but such values have been recorded in animals in which carbohydrate is being converted rapidly into fat. When carbohydrate is being converted into fat, considerable oxygen is liberated from the carbohydrate molecule, and this oxygen becomes available for oxidation of other foodstuffs; this lessens the quantity of oxygen that must be brought into the body through the lungs. Consequently, if the animal is at the same time metabolizing only carbohydrates, the amount of carbon dioxide excreted through the lungs is greater than the intake of oxygen so that the respiratory quotient can rise to values as high as perhaps 1.10.

# REGULATION OF FOOD INTAKE

**Hunger.** The term "hunger" means a craving for food, and it is associated with a number of objective sensations. For instance, as pointed out in Chapter 63, in a person who has not had food for many hours, the stomach undergoes intense rhythmic contractions called *hunger contractions*. These cause a tight or a gnawing feeling in the pit of the stomach and sometimes actually cause pain called *hunger pangs*. In addition to the hunger pangs, the hungry person also becomes more tense and restless than usual, and often has a strange feeling throughout his entire body that might be described by the nonphysiological term "twitterness."

Some physiologists actually define hunger as the tonic contractions of the stomach. However, even after the stomach is completely removed, the psychic sensations of hunger still occur, and craving for food still makes the person search for an adequate food supply.

**Appetite.** The term "appetite" is often used in the same sense as hunger except that it usually implies desire for specific types of food instead of food in general. Therefore, appetite helps a person choose the quality of food he eats.

**Satiety.** Satiety is the opposite of hunger. It means a feeling of fulfillment in the quest for food. Satiety usually results from a filling meal, particularly when the person's nutritional storage depots, the adipose tissue and the glycogen stores, are already filled.

## NEURAL CENTERS FOR REGULATION OF FOOD INTAKE

**Hunger and Satiety Centers.** Stimulation of the *lateral hypothalamus* causes an animal to eat voraciously, while stimulation of the *ventromedial nuclei of the hypothalamus* causes complete satiety, and, even in the presence of highly appetizing food, the animal will still refuse to eat. Conversely, a destructive lesion of the ventromedial nuclei causes exactly the same effect as stimulation of the lateral hypothalamic nuclei—that is, voracious and continued eating until the animal becomes extremely obese, sometimes as large as four times normal in size. Lesions of the lateral hypothalamic nuclei cause exactly the opposite effects—complete lack of desire for food and progressive inanition of the animal. Therefore, we can label the lateral nuclei of the hypothalamus as the *hunger center* or the *feeding center,* while we can label the ventromedial nuclei of the hypothalamus as a *satiety center.*

The feeding center operates by directly exciting the emotional drive to search for food. On the other hand, it is believed that the satiety center operates primarily by inhibiting the feeding center.

**Other Neural Centers that Enter into Feeding.** If the brain is sectioned between the hypothalamus and the mesencephalon, the animal can still perform the basic mechanical features of the feeding process. He can salivate, lick his lips, chew food, and swallow. Therefore, the actual mechanics of feeding are all controlled by centers in the brain stem. The function of the hypothalamus in feeding, then, is to control the quantity of food intake and to excite the lower centers to activity.

Higher centers than the hypothalamus also play important roles in the control of feeding, particularly in the control of appetite. These centers include especially the amygdala and the cortical areas of the limbic system, all of which are closely coupled with the hypothalamus. It will be recalled from the discussion of the sense of smell that the amygdala is one of the major parts of the olfactory nervous system. Destructive lesions in the amygdala have demonstrated that some of its areas greatly increase feeding, while others inhibit feeding. In addition, stimulation of some areas of the amygdala elicits the mechanical act of feeding. However, the most important effect of destruction of the amygdala on both sides of the brain is a "psychic blindness" in the choice of foods. In other words, the animal (and presumably the human being as well) loses or at least partially loses the mechanism of appetite control of type and quality of food that he eats.

The cortical regions of the limbic system, including the infraorbital regions, the hippocampal gyrus, and the cingulate gyrus, all have areas that when stimulated can either increase or decrease feeding activities. These areas seem especially to play a role in the animal's drive to search for food when he is hungry. It is presumed that these centers are also responsible, probably operating in association with the amygdala and hypothalamus, for determining the quality of food that is eaten. For instance, a previous unpleasant experience with almost any type of food often destroys a person's appetite for that food thenceforth.

## FACTORS THAT REGULATE FOOD INTAKE

We can divide the regulation of food into (1) *nutritional regulation,* which is concerned primarily with maintenance of normal quantities of nutrient stores in the body, and (2) *alimentary regulation,* which is concerned primarily with the immediate effects of feeding on the alimentary tract and is sometimes called *peripheral regulation* or *short-term regulation.*

**Nutritional Regulation.** An animal that has been starved for a long time and is then presented with unlimited food eats a far greater quantity than does an animal that has been on a regular diet. Conversely, an animal that has been force-fed for several weeks eats little when allowed to eat according to its own desires. Thus, the feeding center in the hypothalamus is geared to the nutritional status of the body. Some of the nutritional factors that control the degree of activity of the feeding center are the following:

*Availability of Glucose to the Body Cells—The Glucostatic Theory of Hunger and of Feeding Regulation.* It has long been known that a decrease in blood glucose concentration is associated with development of hunger, which has led to the so-called glucostatic theory of hunger and of feeding regulation, as follows: When the blood glucose level falls too low, this automatically causes the animal to increase his feeding, which eventually returns the glucose concentration back toward normal. There are two other observations that also support the glucostatic theory: (1) An increase in blood glucose level increases the measured electrical activity in the satiety center in the ventromedial nuclei of the hypothalamus and simultaneously decreases the electrical activity in the feeding center of the lateral nuclei. (2) Chemical studies show that the ventromedial nuclei (the satiety center) concentrate glucose while other areas of the hypothalamus fail to concentrate glucose; therefore, it is assumed that glucose acts by increasing the degree of satiety.

*Effect of Blood Amino Acid Concentration on Feeding.* An increase in amino acid concentration in the blood also reduces feeding, and a decrease enhances feeding. In general, though, this effect is not as powerful as the glucostatic mechanism.

*Effect of Fat Metabolites on Feeding—Long-term Regulation.* The overall degree of feeding varies almost inversely with the amount of adipose tissue in the body. That is, as the quantity of adipose tissue increases, the rate of feeding decreases. Therefore, many physiologists believe that *long-term regulation* of feeding is controlled mainly by fat metabolites of undiscovered nature. This is called the "lipostatic" theory of feeding regulation. In support of this is the fact that the long-term average concentration of free fatty acids in the blood is directly proportional to the quantity of adipose tissue in the body. Therefore, it is likely that the free fatty acids or some other similar fat metabolites act in the same manner as glucose and amino acids to cause a negative feedback regulatory effect on feeding. It is also possible, if not probable, that this is by far the most important long-term regulator of feeding.

*Interrelationship Between Body Temperature and Food Intake.* When an animal is exposed to cold, he tends to overeat; and when exposed to heat, he tends to undereat. This is caused by interaction within the hypothalamus between the temperature regulating system (see Chap. 72) and the food intake regulating system. It is important because increased food intake in the cold animal (a) increases its metabolic rate and (b) provides increased fat for insulation, both of which tend to correct the cold state.

*Summary of Long-Term Regulation.* Even though our information on the different feedback factors in long-term feeding regulation is imprecise, we can make the following general statement: When the nutrient stores of the body fall below normal, the feeding center of the hypothalamus becomes highly active, and the person exhibits increased hunger; on the other hand, when the nutrient stores are abundant, the person loses his hunger and develops a state of satiety.

**Alimentary Regulation (Short-term, Non-metabolic Regulation).** The degree of hunger or satiety can be temporarily increased or decreased by habit. For instance, the normal person has the habit of eating three meals a day, and, if he misses one, he is likely to develop a state of hunger at mealtime despite completely adequate nutritional stores in his tissues. But, in addition to habit, several other short-term physiological stimuli—mainly related to the alimentary tract—can alter one's desire for food for several hours at a time as follows:

*Gastrointestinal Distention.* When the gastrointestinal tract becomes distended, especially the stomach, inhibitory signals suppress the feeding center, thereby reducing the desire for food. This effect occurs at least partially even after the vagi and the sympathetic nerves from the upper gastrointestinal tract have been severed, indicating that simple overstretching of the abdominal cavity might be the effective stimulus for this feedback response. However, nutritional signals from the liver might also be involved. Obviously, this mechanism is of particular importance in bringing one's feeding to a halt during a heavy meal.

*Metering of Food by Head Receptors.* When a person with an esophageal fistula is fed large quantities of food, even though this food is immediately lost again to the exterior, his degree of hunger is decreased after a reasonable quantity of food has passed through his mouth. This effect occurs despite the fact that the gastrointestinal tract does not become the least bit filled. Therefore, it is postulated that various "head factors" relating to feeding, such as chewing, salivation, swallowing, and tasting, "meter" the food as it passes through the mouth, and after a certain amount has passed through, the hypothalamic feeding center becomes inhibited.

**Importance of Having Both Long- and Short-term Regulatory Systems for Feeding.** The long-term regulatory system, especially the lipostatic feedback mechanism, obviously helps an animal to maintain constant stores of nutrients in his tissues, preventing these from becoming too low or too high. On the other hand, the short-term regulatory stimuli make the animal feed only when the gastrointestinal tract is receptive to food. Thus, food passes through his gastrointestinal tract fairly continuously so that his digestive, absorptive, and storage mechanisms can all work at a steady pace rather than only when the animal needs food for energy. Indeed, the digestive, absorptive, and storage mechanisms can increase their rates of activity above normal only four- to five-fold, whereas the rate of usage of stored nutrients for energy sometimes increases to 20 times normal.

It is important, then, that feeding occur rather continuously (but at a rate that the gastrointestinal tract can accommodate), regulated principally by the short-term regulatory mechanisms. However, it is also important that the intensity of the daily rhythmic feeding habits be modulated up or down by the long-term regulatory system, based principally on the level of nutrient stores in the body.

# OBESITY

**Energy Input versus Energy Output.** When greater quantities of energy (in the form of food) enter the body than are expended, the body weight increases. Therefore, obesity is obviously caused by excess energy input over energy output. For each 9.3 Calories excess energy entering the body, 1 gram of fat is stored.

Excess energy input occurs *only during the developing phase of obesity,* and once a person has become obese, all that is required of him to remain obese is that his energy input equal his energy output. For the person to reduce, the output must be *greater* than the input. Indeed, studies of obese persons have shown that their intake of food is statistically identical to or sometimes even lower than that for normal persons.

**Effect of Muscular Activity on Energy Output.** About one-third the energy used each day by the normal person goes into muscular activity, and in the laborer as much as two-thirds and occasionally three-fourths is used in this way. Since muscular activity is by far the most important means by which energy is expended in the body, it is frequently said that obesity results from *too high a ratio of food intake to daily exercise.*

### ABNORMAL FEEDING REGULATION AS A PATHOLOGIC CAUSE OF OBESITY

The preceding discussion of the mechanisms that regulate feeding emphasized that the rate of feeding is normally regulated in proportion to the nutrient stores in the body. When these stores begin to approach an optimal level in a normal person, feeding is automatically reduced to prevent overstorage. However, in many obese persons this is not true, for feeding does not slacken until body weight is far above normal. Therefore, in effect, obesity is often caused by an abnormality of the feeding regulatory mechanism. This can result from either psychogenic factors that affect the regulation or actual abnormalities of the hypothalamus itself.

**Psychogenic Obesity.** Studies of obese patients show that a large proportion of obesity results from psychogenic factors. Perhaps the most common psychogenic factor contributing to obesity is the prevalent idea that healthy eating habits require three meals a day and that each meal must be filling. Many children are forced into this habit by overly solicitous parents, and the children continue to practice it throughout life. In addition, persons are known often to gain large amounts of weight during or following stressful situations, such as the death of a parent, a severe illness, or even mental depression. It seems that eating is often a means of release from tension.

**Hypothalamic Abnormalities as a Cause of Obesity.** In the preceding discussion of feeding regulation, it was pointed out that lesions in the ventromedial nuclei of the hypothalamus cause an animal to eat voraciously, and, therefore, to become obese. Also, many persons with hypophyseal tumors that encroach on the hypothalamus develop progressive obesity, illustrating that obesity in the human being, too, can definitely result from damage to the hypothalamus. Though in the normal obese person hypothalamic damage is almost never found, it is possible that the functional organization of the feeding center is different in the obese person from that of the nonobese person. For instance, an obese person who has made himself reduce to normal weight usually develops hunger that is demonstrably far greater than that of the normal person. This indicates that the "setting" of his feeding center is at a much higher level of nutrient storage than is that of the normal person.

**Genetic Factors in Obesity.** Obesity definitely runs in families. Furthermore, identical twins usually maintain weight levels within 2 pounds of each other throughout life if they live under similar conditions, or within 5 pounds of each other if their conditions of life differ markedly. This might result partly from eating habits engendered during childhood, but it is generally believed that this close similarity between twins is genetically controlled.

The genes can direct the degree of feeding in several different ways, including (1) a genetic abnormality of the feeding center to set the level of nutrient storage high or low and (2) abnormal hereditary psychic factors that either whet the appetite or cause the person to eat as a "release" mechanism.

A genetic abnormality in the *chemistry of fat storage* is also known to cause obesity in a certain strain of rats. In these rats, fat is easily stored in the adipose tissue, but the quantity of lipoprotein lipase formed in the adipose tissue is greatly reduced, so that little of the fat can be removed. In addition, the rats develop hyperinsulinism which promotes fat storage. This combination obviously results in a one-way path, the fat continually being deposited but never released. This, too, is another possible mechanism of obesity in some human beings.

**Childhood Overnutrition as a Possible Cause of Obesity.** The number of fat cells in the adult body is determined almost entirely by the amount of fat stored in the body during early life. The rate of formation of new fat cells is especially rapid in obese infants, and this continues at a lesser rate in obese children until adolescence. Thereafter, the number of fat cells remains almost identically the same throughout the remainder of life. Thus, mainly on the basis of experiments in lower animals, it is believed that overfeeding children, especially in infancy and to a lesser extent during the older years of childhood, can lead to a lifetime of obesity. The person who has excess fat cells is thought to have a higher setting of his hypothalamic feedback autoregulatory mechanism for control of adipose tissues. In support of this belief is the fact that most extremely obese people have far more fat cells than normal people—often as much as three or more times as many. Indeed, it is rare that any single fat cell stores more than about 50 per cent more fat than normal. Therefore, a major share of obesity seems to result from excess numbers of fat cells rather than simply from enlarged fat cells.

### TREATMENT OF OBESITY

Treatment of obesity depends simply on decreasing energy input below energy expenditure. In other words, this means partial starvation. For this purpose, most diets are designed to contain large quantities of "bulk" which, in general, are made up of non-nutritive cellulose substances. This bulk distends the stomach and thereby partially appeases the hunger. In most lower animals such a procedure simply makes the animal increase his food intake still further, but the human being can often fool himself because his food intake is sometimes controlled as much by habit as by hunger. As pointed out below in connection with starvation, it is important to prevent vitamin deficiencies during the dieting period.

Various *drugs for decreasing the degree of hunger* have been used in the treatment of obesity. The most important of these is *amphetamine* (or amphetamine derivatives), which directly inhibits the feeding center in the lateral nuclei of the hypothalamus. However, there is danger in using this drug because it simultaneously overexcites the central nervous system, making the person nervous and elevating the blood pressure.

Finally, the more exercise one takes, the greater is his daily energy expenditure and the more rapidly his obesity disappears. Therefore, forced exercise is often an essential part of the treatment for obesity.

# INANITION

Inanition is the exact opposite of obesity. In addition to inanition caused by inadequate food, both psychogenic and hypothalamic abnormalities can on occasion cause greatly decreased feeding. One such condition, *anorexia nervosa*, is an abnormal psychic state in which a person loses all desire for food and even becomes nauseated by food; as a result, severe inanition occurs. Also, destructive lesions of the hypothalamus, particularly caused by vascular thrombosis, frequently cause a condition called *cachexia;* the term simply means severe inanition.

# STARVATION

**Depletion of Food Stores in the Body Tissues During Starvation.**   Even though the tissues preferentially use carbohydrate for energy over both fat and protein, the quantity of carbohydrate stores of the body is only a few hundred grams (mainly glycogen in the liver and muscles), and it can supply the energy required for body function for perhaps half a day. Therefore, except for the first few hours of starvation, the major effects are progressive depletion of tissue fat and protein. Since fat is the prime source of energy, its rate of depletion continues unabated, as illustrated in Figure 73–2, until most of the fat stores in the body are gone.

Protein undergoes three different phases of depletion: rapid depletion at first, then greatly slowed depletion, and, finally rapid depletion again shortly before death. The initial rapid depletion is caused by conversion of protein to glucose in the liver by the process of gluconeogenesis. The glucose thus formed (about two-thirds of it, that is) is used mainly to supply energy to the brain which, under normal cir-

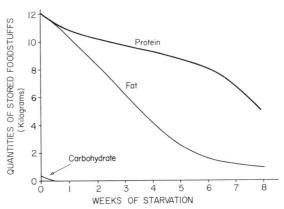

**Figure 73–2.**   Effect of starvation on the food stores of the body.

cumstances, utilizes almost no other metabolic substrate for energy besides glucose. However, after the protein stores have been partially depleted during the early phase of starvation, the remaining protein is not so easily removed from the tissues. At this time, the rate of gluconeogenesis decreases to one-third to one-fifth of its previous rate, and the rate of depletion of protein becomes greatly decreased. The lessened availability of glucose then initiates a series of events leading to *ketosis,* which was discussed in Chapter 68. Fortunately, the ketone bodies, like glucose, can cross the blood-brain barrier and can be utilized by the brain cells for energy. Therefore, approximately two-thirds of the brain's energy now is derived from these ketone bodies, principally beta-hydroxybutyrate. This sequence of events thus leads to at least partial preservation of the protein stores of the body.

However, there finally comes a time when the fat stores also are almost totally depleted, and the only remaining source of energy is proteins. At that time, protein stores once again enter a stage of rapid depletion. Since the proteins are essentially for maintenance of cellular function, death ordinarily ensues when the proteins of the body have been depleted to approximately one-half their normal level.

**Vitamin Deficiencies in Starvation.**   The stores of some of the vitamins, especially the water-soluble vitamins—the vitamin B group and vitamin C—do not last long during starvation. Consequently, after a week or more of starvation mild vitamin deficiencies usually begin to appear, and over several weeks severe vitamin deficiencies can occur. Obviously, these can add to the debility that leads to death.

# REFERENCES

Allred, J. B., and Roehrig, K. L.: Metabolic oscillations and food intake. *Fed. Proc.,* 32:1727, 1973.

Anand, B. K.: Central chemosensitive mechanisms related to feeding. In Code, C. F., and Heidel, W. (eds.): Handbook of Physiology. Sec. 6, Vol. 1. Baltimore, The Williams & Wilkins Company, 1967, p. 249.

Baile, C. A.: Putative neurotransmitters in the hypothalamus and feeding. *Fed. Proc.*, 33:1166, 1974.

Baile, C. A., and Forbes, J. M.: Control of feed intake and regulation of energy balance in ruminants. *Physiol. Rev.*, 54:160, 1974.

Björntorp, P.: Size, number, and function of adipose tissue cells in human obesity. *Horm. Metab. Res. (Suppl.),* 4:77, 1974.

Bray, G. A.: Endocrine factors in the control of food intake. *Fed. Proc.*, 33:1140, 1974.

Bray, G. A.: Nutritional factors in disease. *In* Sodeman, W. A., Jr., and Sodeman, W. A. (eds.): Pathologic Physiology: Mechanisms of Disease, 5th Ed. Philadelphia, W. B. Saunders Company, 1974, p. 839.

Bray, G. A., and Bethune, J. E. (eds.): Treatment and Management of Obesity. Hagerstown, Maryland, Harper & Row, Publishers, 1974.

Bray, G. A., and Campfield, L. A.: Metabolic factors in the control of energy stores. *Metabolism*, 24:99, 1975.

Bray, G. A., and York, D. A.: Genetically transmitted obesity in rodents. *Physiol. Rev.*, 51:598, 1971.

Buskirk, E. R.: Obesity: a brief overview with emphasis on exercise. *Fed. Proc.*, 33:1948, 1975.

Craddock, D.: Obesity and Its Management, 2nd Ed. New York, Churchill Livingstone, Div. of Longman, Inc., 1973.

Dickie, R. S.: Diet in Health and Disease. Springfield, Ill., Charles C Thomas, Publisher, 1974.

Flatt, J. P., and Blackburn, G. L.: The metabolic fuel regulatory system: implications for protein-sparing therapies during caloric deprivation and disease. *Amer. J. Clin. Nutr.*, 27:175, 1974.

Garrow, J. S.: Energy Balance and Obesity in Man. New York, American Elsevier Publishing Company, 1974.

Havel, R. J.: Caloric homeostasis and disorders of fuel transport. *New Engl. J. Med.*, 287:1186, 1972.

Hegsted, D. M.: Deprivation syndrome or protein-calorie malnutrition. *Nutr. Rev.*, 30:51, 1972.

Hegsted, D. M.: Energy needs and energy utilization. *Nutr. Rev.*, 32:33, 1974.

Heird, W. C., and Winters, R. W.: Total parenteral nutrition. The state of the art. *J. Pediatr.*, 86:2, 1975.

Hoebel, B. G.: Feeding: neural control of intake. *Ann. Rev. Physiol.*, 33:533, 1971.

Latham, M. C.: Protein-calorie malnutrition in children and its relation to physiological development and behavior. *Physiol. Rev.*, 54:541, 1974.

Lepkovsky, S.: Newer concepts in the regulation of food intake. *Amer. J. Clin. Nutr.*, 26:271, 1973.

McCance, R. A.: The composition of the body: its maintenance and regulation. *Nutr. Abstr. Rev.*, 42:1269, 1972.

Mawson, A. R.: Anorexia nervosa and the regulation of intake: a review. *Psychol. Med.*, 4:289, 1974.

Mayer, J.: General characteristics of the regulation of food intake. *In* Code, C. F., and Heidel, W. (eds.): Handbook of Physiology. Sec. 6, Vol. 1. Baltimore, The Williams & Wilkins Company, 1967, p. 3.

Nisbett, R. E.: Hunger, obesity, and the ventromedial hypothalamus. *Psychol. Rev.*, 79:433, 1972.

Novin, D., Wyrwicka, W., and Bray, G. A. (eds.): Hunger: Basic Mechanisms and Clinical Implications. New York, Raven Press, 1975.

Olson, R. E. (ed.): Protein-Calorie Malnutrition. New York, Academic Press, Inc., 1975.

Panksepp, J.: Hypothalamic regulation of energy balance and feeding behavior. *Fed. Proc.*, 33:1150, 1974.

Slavin, B. G.: The cytophysiology of mammalian adipose cells. *Inter. Rev. Cytol.*, 33:297, 1972.

Stern, J. S., and Greenwood, M. R. C.: A review of development of adipose cellularity in man and animals. *Fed. Proc.*, 33:1952, 1974.

Suskind, R. M. (ed.): Malnutrition and Immunity. New York, Raven Press, 1975.

Vague, J., Boyer, J., and Addison, G. M. (eds.): The Regulation of Adipose Tissue Mass (Proceedings of the Fourth International Meeting of Endocrinology, Marseilles, 1973). New York, American Elsevier Publishing Company, 1974.

Winick, M., Brasel, J. A., and Rosso, P.: Nutrition and cell growth. *Curr. Concepts Nutr.*, 1:49, 1972.

Wurtman, R. J. (ed.): Nutrition and the Brain. Vol. 1. New York, Raven Press, 1975.

Young, V. R., and Scrimshaw, N. S.: The physiology of starvation. *Sci. Amer.*, 225:14, 1971.

# 74

# Vitamin and Mineral Metabolism

The study of vitamin and mineral metabolism rightfully falls in the province of biochemistry. Therefore, the present discussion of this subject is greatly abbreviated. It is the purpose of this chapter only to present the major aspects of vitamin and mineral metabolism as they relate to the overall physiology of the body.

## VITAMINS

A vitamin is an organic compound needed in small quantities for operation of normal bodily metabolism and that cannot be manufactured in the cells of the body. Probably hundreds of such substances exist, most of which have not been discovered. However, a few have been studied extensively because they are present in the foods in relatively small quantities, and, as a result, dietary deficiency of one or more of them often occurs. Therefore, from a clinical point of view the agents that are generally considered to be vitamins are those organic compounds that occur in the diet in small quantities and, when lacking, can cause specific metabolic deficits.

**Daily Requirements of Vitamins.** Table 74–1 illustrates the usually recommended daily requirements of the different important vitamins for the average adult. These requirements vary considerably, depending on the nature of each person. For instance, the greater the person's size, the greater is his vitamin requirement. Second, growing persons usually require greater quantities of vitamins than do others. Third, when the person performs exercise, the vitamin requirements are increased. Fourth, during disease and fevers, the vitamin requirements are ordinarily increased. Fifth, when greater than normal quantities of carbohydrates are metabolized, the requirements of thiamine and perhaps some of the other vitamins of the B complex are increased. Sixth, during pregnancy and lactation the requirement for vitamin D by the mother is greatly increased, and the requirement for vitamin D is considerable during the period of growth in children. Finally, a number of metabolic deficits occur pathologically in which the vitamins themselves cannot be utilized properly in the body; in such conditions the requirement for one or more specific vitamins may be extreme.

**Storage of Vitamins in the Body.** Vitamins are stored to a slight extent in all the cells. However, some vitamins are stored to a major extent in the liver. For instance, the quantity of vitamin A stored in the liver may be sufficient to maintain a person without any intake of vitamin A for up to six months, and ordinarily the quantity of vitamin D stored in the liver is sufficient to maintain a person for one to two months without any additional intake of vitamin D.

The storage of vitamin K and of most water-soluble vitamins is relatively slight; this applies especially to the vitamin B compounds, for when a person's diet is deficient in vitamin B compounds, clinical symptoms of the deficiency can sometimes be recognized within a few days. Absence of vitamin C,

TABLE 74–1. Daily Requirements
of the Vitamins

| Vitamin | Daily Requirement |
|---|---|
| A | 1.7 mg. |
| Thiamine | 1.6 mg. |
| Riboflavin | 1.8 mg. |
| Niacin | 20 mg. |
| Ascorbic acid | 80 mg. |
| D (children and during pregnancy) | 11 $\mu$g. |
| E | unknown |
| K | none |
| Folic acid | 0.25 mg. |
| B$_{12}$ | 1.2 $\mu$g. |
| Inositol | unknown |
| Pyridoxine | unknown |
| Pantothenic acid | unknown |
| Biotin | unknown |
| Para-aminobenzoic acid | unknown |

another water-soluble vitamin, can cause symptoms within a few weeks and can cause death from scurvy in 20 to 30 weeks.

## VITAMIN A

Vitamin A occurs in animal tissues as *retinol,* the formula of which is illustrated. This vitamin does not occur in foods of vegetable origin, but *provitamins* for the formation of vitamin A do occur in abundance in many different vegetable foods. These are the yellow and red *carotenoid pigments,* which, since they have chemical structures similar to that of vitamin A, can be changed into vitamin A in the human body, this change occurring mainly in the intestinal mucosa and the liver cells.

Vitamin A

The basic function of vitamin A in the metabolism of the body is not known except in relation to its use in the formation of retinal pigments, which was discussed in Chapter 59. Nevertheless, some of the other physiologic results of vitamin A lack have been well documented. In addition to the need for vitamin A to form the visual pigments and therefore to prevent night blindness, it is also necessary for normal growth of most cells of the body and especially for normal growth and proliferation of the different types of epithelial cells. When vitamin A is lacking, the epithelial structures of the body tend to become stratified and keratinized. Therefore, vitamin A deficiency manifests itself by (1) scaliness of the skin and sometimes acne, (2) failure of growth of young animals, (3) failure of reproduction in many animals, associated especially with atrophy of the germinal epithelium of the testes and sometimes with interruption of the female sexual cycle, and (4) keratinization of the cornea with resultant corneal opacity and blindness.

Also, the damaged epithelial structures often become infected, for example, in the eyes, the kidneys, or the respiratory passages. Therefore, vitamin A has been called an "anti-infection" vitamin. Vitamin A deficiency also frequently causes kidney stones, probably owing to infection in the renal pelvis.

## THIAMINE (VITAMIN B₁)

Thiamine operates in the metabolic systems of the body principally as *thiamine pyrophosphate;* this compound functions as a *cocarboxylase,* operating mainly in conjunction with a protein decarboxylase for decarboxylation of pyruvic acid and other $\alpha$-keto acids, as discussed in Chapter 67.

Thiamine deficiency causes decreased utilization of pyruvic acid and some amino acids by the tissues

Thiamine chloride

but increased utilization of fats. Thus, thiamine is specifically needed for final metabolism of carbohydrates and many amino acids. Probably the decreased utilization of these nutrients is the responsible factor for the debilities associated with thiamine deficiency.

**Thiamine Deficiency and the Nervous System.** The central nervous system depends almost entirely on the metabolism of carbohydrates for its energy. In thiamine deficiency the utilization of glucose by nervous tissue may be decreased as much as 50 to 60 per cent. Therefore, it is readily understandable how thiamine deficiency could greatly impair function of the central nervous system. The neuronal cells of the central nervous system frequently show chromatolysis and swelling during thiamine deficiency, changes that are characteristic of neuronal cells with poor nutrition. Obviously, such changes as these can disrupt communication in many different portions of the central nervous system.

Also, thiamine deficiency can cause *degeneration of myelin sheaths* of nerve fibers both in the peripheral nerves and in the central nervous system. The lesions in the peripheral nerves frequently cause these nerves to become extremely irritable, resulting in "polyneuritis" characterized by pain radiating along the course of one or more peripheral nerves. Also, in severe thiamine deficiency, the peripheral nerve fibers and fiber tracts in the cord can degenerate to such an extent that *paralysis* occasionally results; and, even in the absence of paralysis, the muscles atrophy, with resultant severe weakness.

**Thiamine Deficiency and the Cardiovascular System.** Thiamine deficiency also weakens the heart muscle, so that a person with severe thiamine deficiency sometimes develops *cardiac failure.* In general, the right side of the heart becomes greatly enlarged in thiamine deficiency. Furthermore, the return of blood to the heart may be increased to as much as three times normal. This indicates that thiamine deficiency causes *peripheral vasodilation* throughout the circulatory system, possibly as a result of metabolic deficiency in the smooth muscle of the vascular system itself. Therefore, the cardiac effects of thiamine deficiency are due partly to excessive return of blood to the heart and partly to primary weakness of the cardiac muscle. *Peripheral edema* and *ascites* also occur to a major extent in some persons with thiamine deficiency because of the cardiac failure.

**Thiamine Deficiency and the Gastrointestinal Tract.** Among the gastrointestinal symptoms of thiamine deficiency are indigestion, severe constipation, anorexia, gastric atony, and hypochlorhydria. All these effects possibly result from failure of the smooth muscle and glands of the gastrointestinal tract to derive sufficient energy from carbohydrate metabolism.

The overall picture of thiamine deficiency, including polyneuritis, cardiovascular symptoms, and gastrointestinal disorders, is frequently referred to as "beriberi"—especially when the cardiovascular symptoms predominate.

## NIACIN

Niacin, also called *nicotinic acid,* functions in the body as coenzymes in the forms of nicotinamide adenine dinucleotide (NAD) and nicotinamide adenine dinucleotide phosphate (NADP), which are also known as DPN and TPN. These coenzymes are hydrogen acceptors which combine with hydrogen atoms as they are removed from food substrates by many different types of dehydrogenases. Typical operation of both of them is presented in Chapter 67. When a deficiency of niacin exists, the normal rate of dehydrogenation presumably cannot be maintained, and, therefore, oxidative delivery of energy from the foodstuffs to the functioning elements of the cells likewise cannot occur at normal rates.

Niacin

Because NAD and NADP operate in all cells of the body, it is readily understood how lack of niacin can cause multiple symptoms. Clinically, niacin deficiency causes mainly gastrointestinal symptoms, neurologic symptoms, and a characteristic dermatitis. However, it is probably much more proper to say that essentially all functions of the body are depressed or altered.

In the early stages of niacin deficiency simple physiologic changes, such as muscular weakness of all the different types of muscles and poor glandular secretion, may occur, but in severe niacin deficiency actual death of tissues ensues. Pathologic lesions appear in many parts of the central nervous system, and permanent dementia or any of many different types of psychoses may result. Also, the skin develops a cracked, pigmented scaliness in areas that are exposed to mechanical irritation or sun irradiation; thus, it seems as if the skin were unable to repair the different types of irritative damage.

Niacin deficiency causes intense irritation and inflammation of the mucous membranes of the mouth and other portions of the gastrointestinal tract, thus instituting many digestive abnormalities. It is possible that this results from generalized depression of metabolism in the gastrointestinal epithelium and failure of appropriate epithelial repair.

The clinical entity called "pellagra" and the canine disease called "black tongue" are caused mainly by niacin deficiency. Pellagra is greatly exacerbated in persons on a corn diet (such as many of the natives of Africa) because corn is very deficient in the amino acid tryptophan, which can be converted in limited quantities to niacin in the body.

## RIBOFLAVIN (VITAMIN B₂)

Riboflavin normally combines in the tissues with phosphoric acid to form two coenzymes, *flavin mononucleotide (FMN),* and *flavin adenine dinucleotide (FAD).* These in turn operate as hydrogen carriers in several of the important oxidative systems of the body. Usually, NAD, operating in association with specific dehydrogenases, accepts hydrogen removed from various food substrates and then passes the hydrogen to FMN or FAD; finally, the hydrogen is released as an ion into the surrounding fluids to become oxidized by nascent oxygen, the system for which is described in Chapter 67.

Riboflavin

Deficiency of riboflavin in lower animals causes severe *dermatitis, vomiting, diarrhea, muscular spasticity* which finally becomes muscular weakness, and then *death* preceded by coma and decline in body temperature. Thus, severe riboflavin deficiency can cause many of the same effects as lack of niacin in the diet; presumably the debilities that result in each instance are due to generally depressed oxidative processes within the cells.

In the human being riboflavin deficiency has never been known to be severe enough to cause the marked debilities noted in animal experiments, but mild riboflavin deficiency is probably common. Such deficiency causes digestive disturbances, burning sensations of the skin and eyes, cracking at the corners of the mouth, headaches, mental depression, forgetfulness, etc. Perhaps the most common characteristic lesion of riboflavin deficiency is *cheilosis,* which is inflammation and cracking at the angles of the mouth. In addition, a fine, scaly dermatitis often occurs at the angles of the nares, and keratitis of the cornea may occur with invasion of the cornea by capillaries.

Though the manifestations of riboflavin deficiency are usually relatively mild, this deficiency frequently occurs in association with deficiency of thiamine or niacin. Therefore, many deficiency syndromes, including pellagra, beriberi, sprue, and kwashiorkor, are probably due to a combined deficiency of a number of the vitamins, as well as to other aspects of malnutrition.

## VITAMIN B₁₂

Several different *cobalamin* compounds which possess the common prosthetic group illustrated below exhibit so-called "vitamin $B_{12}$" activity.

Note that this prosthetic group contains cobalt that has coordination bonds similar to those found in relation to the iron of the hemoglobin molecule. It is likely that the cobalt atom functions in much the same way that the iron atom functions.

Vitamin $B_{12}$ performs many metabolic functions, acting as a hydrogen acceptor coenzyme. For instance, it performs this function in the conversion of amino acids and similar compounds into other substances. Its most important function is probably to act as a coenzyme for reducing ribonucleotides to deoxyribonucleotides, a step that is important in the formation of genes. This could explain the two major functions of vitamin $B_{12}$, (1) promotion of growth and (2) red blood cell maturation. This latter function was described in detail in Chapter 5.

A special effect of vitamin $B_{12}$ deficiency is often demyelination of the large nerve fibers of the spinal cord, especially of the posterior columns and occasionally of the lateral columns. As a result, persons with pernicious anemia (anemia caused by failure of red cell maturation) frequently have much loss of peripheral sensation and, in severe cases, even become paralyzed.

## FOLIC ACID (PTEROYLGLUTAMIC ACID)

Several different pteroylglutamic acids, one of which is illustrated below, exhibit the "folic acid effect." Folic acid functions as a carrier of hydroxymethyl and formyl groups. Perhaps its most important use in the body is in the synthesis of purines and thymine, which are required for formation of deoxyribonucleic acid. Therefore, folic acid is required for reproduction of the cellular genes. This perhaps explains one of the most important functions of the folic acid—that is, to promote growth.

Folic acid is an even more potent growth promoter than vitamin $B_{12}$, and, like vitamin $B_{12}$, is also important for the maturation of red blood cells, as discussed in Chapter 5. However, vitamin $B_{12}$ and folic acid each perform specific and different functions in promoting growth and maturation of red blood cells.

## PYRIDOXINE (VITAMIN B₆)

Pyridoxine exists in the form of *pyridoxal phosphate* in the cells and functions as a coenzyme for many different chemical reactions relating to amino acid and protein metabolism. Its most important role is that of coenzyme in transamination for the synthesis of amino acids. As a result, pyridoxine plays many key roles in metabolism—especially in protein metabolism. Also, it is believed to act in the transport of some amino acids across cell membranes.

Pyridoxine

Dietary lack of pyridoxine in lower animals can cause dermatitis, decreased rate of growth, development of fatty liver, anemia, and evidence of mental deterioration. In the human being, pyridoxine deficiency has been known to cause convulsions, dermatitis, and gastrointestinal disturbances such as nausea and vomiting in children. However, this deficiency is rare.

## PANTOTHENIC ACID

Pantothenic acid mainly is incorporated in the body into coenzyme A, which has many metabolic roles in the cells. Two of these discussed at length in Chapters 67 and 68 are : (1) acetylation of decarboxylated pyruvic acid to form acetyl Co-A prior to its entry into the tricarboxylic acid cycle and (2) degradation of fatty acid molecules into multiple molecules of acetyl Co-A. Thus, lack of pantothenic acid can lead to depressed metabolism of both carbohydrates and fats.

Folic acid (pteroylglutamic acid)

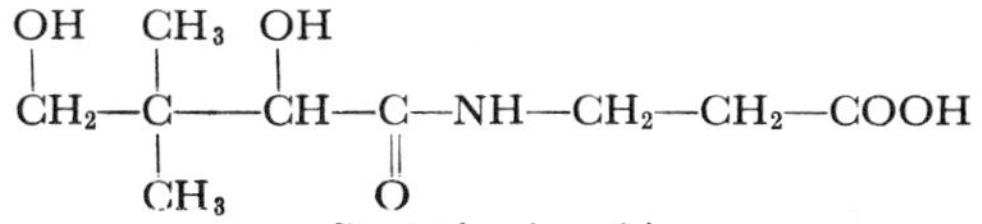

OH   CH₃  OH
|    |    |
CH₂—C———CH—C—NH—CH₂—CH₂—COOH
     |       ‖
     CH₃     O
Pantothenic acid

Deficiency of pantothenic acid in lower animals can cause retarded growth, failure of reproduction, graying of the hair, dermatitis, fatty liver, and hemorrhagic adrenal cortical necrosis. In the human being, no definite deficiency syndrome has been proved, presumably because of the wide occurrence of this vitamin in almost all foods and because small amounts of the vitamin can probably be synthesized in the body. Nevertheless, this does not mean that pantothenic acid is not of value in the metabolic systems of the body; indeed, it is perhaps as necessary as any other vitamin.

## ASCORBIC ACID (VITAMIN C)

The precise function of ascorbic acid is almost totally unknown, though many of the metabolic effects of ascorbic acid deficiency have been well catalogued. For instance, oxidation of tyrosine and phenylalanine requires an adequate supply of ascorbic acid, and ascorbic acid plays a role in formation of hydroxyproline, an integral constituent of collagen. In addition, it is almost certain from the deficiency symptoms that occur when ascorbic acid is lacking that it performs far more functions than simply these.

Though the mechanisms are not clear, ascorbic acid enhances the removal of iron from cellular ferritin, thereby increasing the concentration of iron in the body fluids. Also, in some way not clarified, ascorbic acid potentiates the effects of folic acid in at least some metabolic processes.

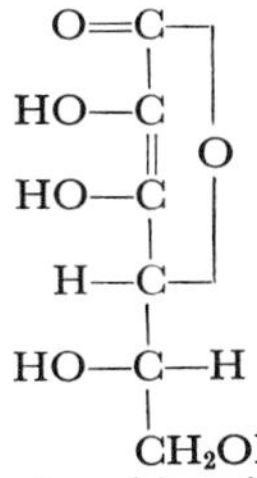

O=C
|
HO—C
‖   O
HO—C
|
H—C
|
HO—C—H
|
CH₂OH
Ascorbic acid
(vitamin C)

Physiologically, the major function of ascorbic acid appears to be maintenance of normal intercellular substances throughout the body. This includes the formation of collagen, probably because of the action of ascorbic acid in synthesis of hydroxyproline. It also enhances the intercellular cement substance between the cells, the formation of bone matrix, and the formation of tooth dentin.

Deficiency of ascorbic acid for 20 to 30 weeks, as occurred frequently during long sailing voyages in olden days, causes *scurvy,* some effects of which are the following:

One of the most important effects of scurvy is *failure of wounds to heal.* This is caused by failure of the cells to deposit collagen fibrils and intercellular cement substances. As a result, healing of a wound may require several months instead of the several days ordinarily necessary.

Lack of ascorbic acid causes *cessation of bone growth.* The cells of the growing epiphyses continue to proliferate, but no new matrix is laid down between the cells, and the bones fracture easily at the point of growth because of failure to ossify. Also, when an already ossified bone fractures in a person with ascorbic acid deficiency, the osteoblasts cannot secrete a new matrix for the deposition of new bone. Consequently, the fractured bone does not heal.

The *blood vessel walls become extremely fragile* in scurvy, presumably because of failure of the endothelial cells to be cemented together properly. Especially are the capillaries likely to rupture, and as a result many small petechial hemorrhages occur throughout the body. The hemorrhages beneath the skin cause purpuric blotches, sometimes over the entire body. To test for ascorbic acid deficiency, one can produce such petechial hemorrhages by inflating a blood pressure cuff over the upper arm; this occludes the venous return of blood, the capillary pressure rises, and red blotches occur in the skin immediately if there is a sufficiently severe ascorbic acid deficiency.

In extreme scurvy the muscle cells sometimes fragment; lesions of the gums with loosening of the teeth occur; infections of the mouth develop; vomiting of blood, bloody stools, and cerebral hemorrhage can all occur; and, finally, high fever often develops.

## VITAMIN D

Vitamin D increases calcium absorption from the gastrointestinal tract and also helps to control calcium deposition in the bone. The mechanism by which vitamin D increases calcium absorption is to promote active transport of calcium through the epithelium of the ileum. It increases the formation of a calcium-binding protein in the epithelial cells that aids in calcium absorption. The specific functions of vitamin D in relation to overall body calcium metabolism and to bone formation are presented in Chapter 79.

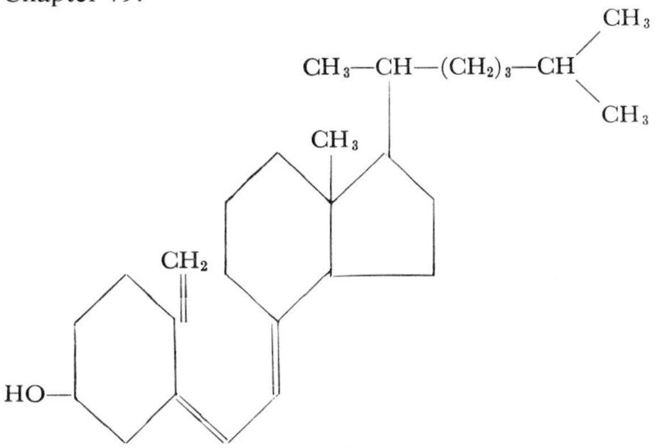

CH₃—CH—(CH₂)₃—CH(CH₃)₂

CH₃

CH₂

HO—

Vitamin D₃ (cholecalciferol)

Several vitamin D compounds exist, one of which is illustrated. This is the natural vitamin D, *cholecalciferol,* which results from ultraviolet irradiation of 7-dehydrocholesterol in the skin. A synthetic vitamin D compound, *ergocalciferol,* is formed by irradiation of ergosterol and is used to a major extent in vitamin D therapy. Both of these vitamin D's are further altered first in the liver and then in the kidneys, finally forming *1,25-dihydroxycholecalciferol,* which is the active form of the vitamin. This will be discussed in more detail in Chapter 79.

### VITAMIN E

Several related compounds, one of which is illustrated below, exhibit so-called "vitamin E activity." Only rare instances of vitamin E deficiency occur in human beings. In lower animals, lack of vitamin E can cause degeneration of the germinal epithelium in the testis and therefore can cause male sterility. Lack of vitamin E can also cause resorption of a fetus after conception in the female. Because of these effects of vitamin E deficiency, vitamin E is sometimes called the "anti-sterility vitamin."

Vitamin E deficiency in animals can also cause paralysis of the hindquarters, and pathologic changes occur in the muscles similar to those found in the disease entity "muscular dystrophy" of the human being. However, administration of vitamin E to patients with muscular dystrophy has not proved to be of any benefit.

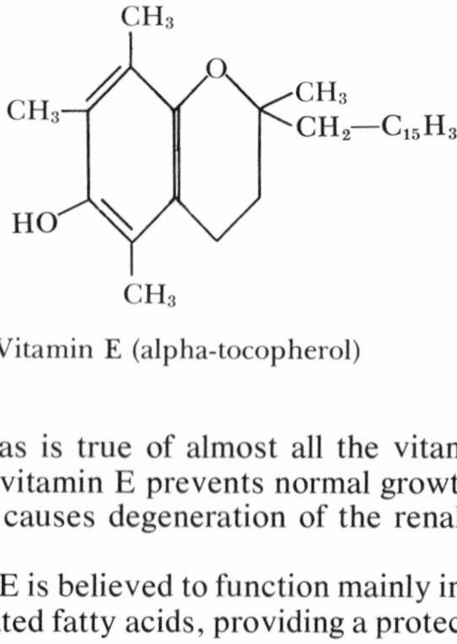

Vitamin E (alpha-tocopherol)

Finally, as is true of almost all the vitamins, deficiency of vitamin E prevents normal growth, and it sometimes causes degeneration of the renal tubular cells.

Vitamin E is believed to function mainly in relation to unsaturated fatty acids, providing a protective role to prevent oxidation of the unsaturated fats. In the absence of vitamin E, the quantity of unsaturated fats in the cells becomes diminished, causing abnormal structure and function of such cellular organelles as the mitochondria, the lysosomes, and even the cell membrane. Indeed, the muscular dystrophy-like syndrome that occurs in the vitamin E deficiency probably results from continual rupture of lysosomes and subsequent autodigestion of the muscle.

### VITAMIN K

Vitamin K is necessary for the formation by the liver of prothrombin and factor VII (proconvertin), both of which are important in blood coagulation. Therefore, when vitamin K deficiency occurs, blood clotting is retarded. The function of this vitamin and its relationships with some of the anticoagulants, such as Dicumarol, have been presented in greater detail in Chapter 9.

Several different compounds, both natural and synthetic, exhibit vitamin K activity. The chemical formula for one of the natural vitamin K compounds is illustrated below. Because vitamin K is synthesized by bacteria in the colon, a dietary source for this vitamin is not usually necessary, but, when the bacteria of the colon are destroyed by administration of large quantities of antibiotic drugs, vitamin K deficiency occurs rapidly because of the paucity of this compound in the normal diet. (See below.)

## MINERAL METABOLISM

The functions of many of the minerals, such as sodium, potassium, chloride, and so forth, have been presented at appropriate points in the text. Therefore, only specific functions of minerals not covered elsewhere are mentioned here.

The body content of the most important minerals is listed in Table 74–2, and the daily requirements of these are given in Table 74–3.

**Magnesium.** Magnesium is approximately one-eighth as plentiful in cells as potassium, and it undoubtedly performs at least some of the same intracellular functions. But magnesium is also required as a catalyst for many intracellular enzymatic reactions, particularly those relating to carbohydrate metabolism.

The extracellular magnesium concentration is slight, only 1.8 to 2.5 mEq./liter. Increased extracel-

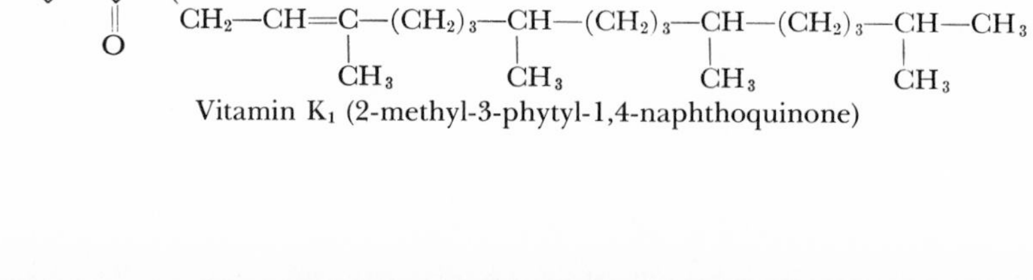

Vitamin $K_1$ (2-methyl-3-phytyl-1,4-naphthoquinone)

**TABLE 74–2.   Content in Grams of a 70 Kilogram Adult Man**

| | | | |
|---|---|---|---|
| Water | 41,400 | Mg | 21 |
| Fat | 12,600 | Cl | 85 |
| Protein | 12,600 | P | 670 |
| Carbohydrate | 300 | S | 112 |
| Na | 63 | Fe | 3 |
| K | 150 | I | 0.014 |
| Ca | 1,160 | | |

lular concentration of magnesium depresses activity in the nervous system and also depresses skeletal muscle contraction. This latter effect can be blocked by administration of calcium. Low magnesium concentration causes greatly increased irritability of the nervous system, peripheral vasodilatation, and cardiac arrhythmias.

**Calcium.**   Calcium is present in the body mainly in the form of calcium phosphate in the bone. This subject will be discussed in detail in Chapter 79, as is also the calcium content of the extracellular fluid.

Excess quantities of calcium ion in the extracellular fluids can cause the heart to stop in systole and can act as a mental depressant. On the other hand, low levels of calcium can cause spontaneous discharge of nerve fibers, resulting in tetany. This, too, will be discussed in Chapter 79.

**Phosphorus.**   Phosphate is the major anion of intracellular fluids. Phosphates have the ability to combine reversibly with a multitude of coenzyme systems and also with a multitude of other compounds that are necessary for the operation of metabolic processes. Many important reactions of phosphates have been catalogued at other points in this text, especially in relation to the function of ATP, ADP, creatine phosphate, and so forth. Suffice it to say that phosphates are perhaps the single most important mineral constituent required for cellular activity. Also, bone contains a tremendous amount of calcium phosphate, which will be discussed in Chapter 79.

**Iron.**   The function of iron in the body, especially in relation to the formation of hemoglobin, was discussed in Chapter 5. The major proportion of iron in the body is in the form of hemoglobin, though smaller quantities are present in other forms, especially in the liver and in the bone marrow. Electron carriers containing iron (especially the cytochromes) are present in all the cells of the body and are essential for most of the oxidation that occurs in the cells.

**TABLE 74–3.   Daily Requirements of Minerals**

| | | | |
|---|---|---|---|
| Na | 3.0 grams | I | 250.0 µg. |
| K | 1.0 grams | Mg | unknown |
| Cl | 3.5 grams | Co | unknown |
| Ca | 0.8 gram | Cu | unknown |
| PO₄ | 1.5 grams | Mn | unknown |
| Fe | 12.0 mg. | Zn | unknown |

Therefore, iron is absolutely essential both for transport of oxygen to the tissues and for maintenance of oxidative systems within the tissue cells, without which life would cease within a few seconds.

**Important Trace Elements in the Body.**   A few elements are present in the body in such small quantities that they are called "trace elements." Usually, the amounts of these in the foods are also minute. Yet without any one of them a specific deficiency syndrome is likely to develop.

*Iodine.*   The best known of the trace elements is iodine. This element is discussed in Chapter 76 in connection with thyroid hormone; as illustrated in Table 74–2, the entire body contains an average of only 14 milligrams. Iodine is essential for the formation of *thyroxine,* which in turn is essential for maintenance of normal metabolic rates in all the cells.

*Copper.*   Copper deficiency has been observed on rare occasions in infants receiving only a milk diet. The effect of the deficiency is a microcytic, normochromic anemia caused by failure of the intestines to absorb adequate quantities of iron. Lack of copper also decreases the quantities of certain copper-containing enzymes in the tissue cells; especially important is cytochrome oxidase, which contains copper.

*Zinc.*   Zinc is an integral part of many enzymes, one of the most important of which is *carbonic anhydrase,* present in especially high concentration in the red blood cells. This enzyme is responsible for rapid combination of carbon dioxide with water in the red blood cells of the peripheral capillary blood and for rapid release of carbon dioxide from the pulmonary capillary blood into the alveoli. Carbonic anhydrase is also present to a major extent in the gastrointestinal mucosa, in the tubules of the kidney, and in the epithelial cells of many glands of the body. Consequently, zinc in small quantities is essential for the performance of many reactions relating to carbon dioxide metabolism.

Zinc is also a component of *lactic dehydrogenase* and, therefore, is important for the interconversions between pyruvic acid and lactic acid. Finally, zinc is a component part of some *peptidases* and therefore is important for digestion of proteins in the gastrointestinal tract.

*Cobalt.*   Cobalt is an essential part of vitamin $B_{12}$, and vitamin $B_{12}$ is essential for maturation of red blood cells, as discussed earlier in this chapter.

Excess cobalt in the diet causes the opposite of anemia; i.e., polycythemia. However, the polycythemic cells formed contain relatively small concentrations of hemoglobin, so that the total quantity of hemoglobin in the circulating blood remains about normal. Thus, it appears that cobalt is concerned principally with formation of the red blood cell structure and not with formation of hemoglobin.

*Manganese.*   Lack of manganese in the diet of animals causes testicular atrophy, though the cause of this is not known. Also, manganese is necessary in the body to activate arginase, which is the principal enzyme necessary for the formation of urea. Consequently, lack of manganese in the diet might prevent the conversion of ammonium ions into urea, and excessive quantities of ammonium compounds developing in the body fluids could cause toxicity. Finally, manganese activates many of the other metabolic enzymes, including cholinesterase, muscle ATPase, and enzymes required for glycolysis.

*Fluorine.* Fluorine does not seem to be a necessary element for metabolism, but the presence of a small quantity of fluorine in the body during the period of life when the teeth are being formed subsequently protects against carious teeth. Fluorine does not make the teeth themselves stronger but, instead, has some yet unknown effect in suppressing the cariogenic process. It has been suggested that fluorine in the teeth combines with various trace metals that are necessary for activation of the bacterial enzymes, and, because the enzymes are deprived of these trace metals, they remain inactive and cause no caries.

Excessive intake of fluorine causes *fluorosis*, which is manifest in its mild state by mottled teeth and in a more severe state by enlarged bones. It has been postulated that in this condition fluorine combines with trace metals in some metabolic enzymes, including the phosphatases, so that various metabolic systems become partially inactivated. According to this theory, the mottled teeth and enlarged bones are due to abnormal enzyme systems in the odontoblasts and osteoblasts. Even though mottled teeth are highly resistant to the development of caries, the structural strength of these teeth is considerably lessened by the mottling process.

# REFERENCES

Bray, G. A.: Nutritional factors in disease. In Sodeman, W. A., Jr., and Sodeman, W. A. (eds.): Pathologic Physiology: Mechanisms of Disease, 5th Ed. Philadelphia, W. B. Saunders Company, 1974, p. 839.

Burch, R. E., Hahn, H. K., and Sullivan, J. F.: Newer aspects of the roles of zinc, manganese, and copper in human nutrition. *Clin. Chem., 21*:501, 1975.

Christensen, S.: The biological fate of riboflavin in mammals. A survey of literature and own investigations. *Acta Pharmacol. Toxicol. (Kbh) (Suppl.), 32*:3, 1973.

Davies, I. J. T.: The Clinical Significance of the Essential Biological Metals. Springfield, Ill., Charles C Thomas, Publisher, 1972.

Davis, G. K.: High-level copper feeding of swine and poultry and the ecology. *Fed. Proc., 33*:1194, 1974.

Dhar, S. K. (ed.): Metal Ions in Biological Systems. (Conference on the Role of Metal Ions in Biological Systems; Studies of Some Biochemical and Environmental Problems.) New York, Plenum Publishing Corp., 1973.

Evans, G. W.: Copper homeostasis in the mammalian system. *Physiol. Rev., 53*:535, 1973.

Fernstrom, J. D., and Wurtman, R. J.: Nutrition and the brain. *Sci. Amer., 230*:84, 1974.

Frieden, E.: The chemical elements of life. *Sci. Amer., 227*:52, 1972.

Frieden, E.: The evolution of metals as essential elements (with special reference to iron and copper). *Adv. Exp. Med. Biol., 48*:1, 1974.

Gardner, L. I., and Amacher, P. (eds.): Endocrine Aspects of Malnutrition: Marasmus, Kwashiorkor, and Psychosocial Deprivation. New York, Raven Press, 1973.

Goodhart, R. S., and Shils, M. E. (eds.): Modern Nutrition in Health and Disease, 5th Ed. Philadelphia, Lea & Febiger, 1973.

Goodman, D. S.: Vitamin A transport in human and rat plasma and its nutritional regulation. *Biochem. Soc. Symp., 35*:287, 1972.

Jennings, I. W.: Vitamins in Endocrine Metabolism. Springfield, Ill., Charles C Thomas, Publisher, 1970.

Kodicek, E.: Recommended intakes of vitamins for normal growth and development. *Bibl. Nutr. Dieta, 18*:45, 1973.

Mayer, J.: Human Nutrition. Springfield, Ill., Charles C Thomas, Publisher, 1974.

Omdahl, J. L., and DeLuca, H. F.: Regulation of vitamin D metabolism and function. *Physiol. Rev., 53*:327, 1973.

Plaut, G. W. E., Smith, C. M., and Alworth, W. L.: Biosynthesis of water-soluble vitamins. *Ann. Rev. Biochem., 43*:899, 1974.

Pories, W. J., Strain, W. H., Hsu, J. M., and Woosley, R. L. (eds.): Clinical Applications of Zinc Metabolism. Springfield, Ill., Charles C Thomas, Publisher, 1974.

Rindi, G., and Ventura, U.: Thiamine intestinal transport. *Physiol. Rev., 52*:821, 1972.

Sandstead, H. H.: Zinc nutrition in the United States. *Amer. J. Clin. Nutr., 26*:1251, 1973.

Stokstad, E. L. R., and Koch, J.: Folic acid metabolism. *Physiol. Rev., 47*:83, 1967.

Trace elements in human nutrition. Report of a WHO expert committee. *WHO Tech. Rep. Ser., 532*:1, 1973.

Underwood, E. J.: Trace Elements in Human and Animal Nutrition, 3rd Ed. New York, Academic Press, Inc., 1971.

Williams, R. J.: Physician's Handbook of Nutrition. Springfield, Ill., Charles C Thomas, Publisher, 1974.

# PART XIII

# ENDOCRINOLOGY
# AND REPRODUCTION

# 75

# Introduction to Endocrinology; and the Pituitary Hormones

The functions of the body are regulated by two major control systems: (1) the nervous system, which has been discussed, and (2) the hormonal, or endocrine, system. In general, the hormonal system is concerned principally with control of the different metabolic functions of the body, controlling the rates of chemical reactions in the cells or the transport of substances through cell membranes or other aspects of cellular metabolism, such as growth and secretion. Some hormonal effects occur in seconds, while others require several days simply to start and then continue for weeks, months, or even years.

Many interrelationships exist between the hormonal and nervous systems. For instance, at least two glands secrete their hormones only in response to appropriate nerve stimuli, the *adrenal medullae* and the *posterior pituitary gland,* and few of the adenohypophyseal hormones are secreted to a significant extent except in response to nervous activity in the hypothalamus, as is detailed later in this chapter.

## NATURE OF A HORMONE

A hormone is a chemical substance that is secreted into the body fluids by one cell or a group of cells and that exerts a physiological *control* effect on other cells of the body.

At many points in this text we have already discussed different hormones, some of which are called *local hormones* and others that are called *general hormones.* The local hormones include *acetylcholine* released at the parasympathetic and skeletal nerve endings, *secretin* released by the duodenal wall and transported in the blood to the pancreas to cause a watery pancreatic secretion, *cholecystokinin* released in the small intestine and transported to the gallbladder to cause contraction and to the pancreas to cause enzyme secretion, and many others. These hormones obviously have specific local effects, whence comes the name local hormones.

On the other hand, the general hormones are secreted by specific *endocrine glands* and are transported in the blood to cause physiologic actions at distant points in the body. A few of the general hormones affect all, or almost all, cells of the body; examples are *growth hormone* from the adenohypophysis and *thyroid hormone* from the thyroid gland. Other general hormones, however, affect specific tissues far more than other tissues; for instance, *adrenocorticotropin* from the anterior pituitary gland specifically stimulates the adrenal cortex, and the *ovarian hormones* have specific effects on the uterine endometrium. The tissues affected specifically in this way are called *target tissues.* Many examples of target organs will become apparent in the following chapters on endocrinology.

The following general hormones have proved to be of major significance and are discussed in detail in this and the following chapters:

Anterior pituitary hormones: *growth hormone, adrenocorticotropin, thyroid-stimulating hormone, follicle-stimulating hormone, luteinizing hormone, prolactin,* and *melanocyte-stimulating hormone.*

Posterior pituitary hormones: *antidiuretic hormone (vasopressin)* and *oxytocin.*

Adrenocortical hormones: especially *cortisol* and *aldosterone.*

Thyroid hormones: *thyroxine, tri-iodothyronine,* and *calcitonin.*

Pancreatic hormones: *insulin* and *glucagon.*

Ovarian hormones: *estrogens* and *progesterone.*

Testicular hormone: *testosterone.*

Parathyroid hormone: *parathormone.*

Placental hormones: *chorionic gonadotropin, estrogens, progesterone,* and *human placental lactogen.*

**Negative Feedback in the Control of Hormonal Secretion.** At many points both in this chapter and in the succeeding few chapters we will see that once a hormone accomplishes its physiological function, its rate of secretion is prevented from increasing further and at times is even decreased. This is caused by negative feedback, a phenomenon we have seen to be important in nervous control systems. In general, each gland has a basic tendency to *oversecrete* its particular hormone, but, once the normal physiological effect of the hormone has been achieved, information is transferred either directly or indirectly back to the producing gland to inhibit further secretion. On the other hand, if the gland undersecretes, the physiological effects of the hormone diminish, and the feedback decreases, thus allowing the gland to begin secreting adequate quantities of the hormone once again. In this way, the rate of secretion of each hormone is controlled in accord with the need for the hormone. The specific negative feedback mechanisms are discussed in relation to the different individual hormones.

**Chemistry of the Hormones.** Chemically, the basic types of hormones are: (1) proteins or derivatives of proteins or amino acids and (2) steroid hormones. For example, the hormones of the pancreas and anterior pituitary are proteins or large polypeptides, while the hormones of the posterior pituitary, thyroid, and adrenal medulla are derivatives of proteins or amino acids. The steroids are secreted by the glands derived from the mesenchymal zone of the embryo, including the adrenal cortex, the ovary, and the testis.

*Measurement of Hormone Concentrations*

Most hormones are present in the circulating body fluids and tissues in extremely minute quantities, some in concentrations as low as one-millionth of a milligram (one picogram) per milliliter. Therefore, except in a few instances, it has been almost impossible to measure these concentrations by usual chemical means. There are two important methods that have been employed for this purpose: (1) bioassay and (2) radioactive competitive binding methods.

**Bioassay.** Bioassay means developing an appropriate animal preparation in which one can test the action of the hormone on the animal. For instance, an appropriate bioassay for antidiuretic hormone is to measure the degree of water conservation caused by injecting plasma or a concentrated extract of the plasma from an experimental animal or human being into a test animal and to compare the animal's response with the response to a known quantity of pure antidiuretic hormone. In a similar manner, bioassay for growth hormone is based on stimulation of growth, usually of rats. Bioassays for gonadotropic hormones are based on their effects on the ovaries or other gonadotropic target tissues. By appropriate titration, a reasonable degree of accuracy can be achieved for most hormones but not for all, because appropriate animal models have not been achieved for all hormones.

**Competitive Binding Assays—Radioimmunoassay.** For measuring extremely low concentrations of hormones, a substance that specifically binds with the hormone is first found. For instance, antibodies can usually be developed that will bind specifically with a given hormone. Then a mixture is made of three different elements: (1) a fluid from the animal to be assayed, (2) the antibody, and (3) an approximate equivalent amount of purified hormone of the type to be measured but that has been tagged with a radioactive isotope. However, one specific condition must be met: There must be too little antibody for all of the hormone from the two separate sources to combine completely. Therefore, the natural hormone and the radioactive hormone *compete for the binding sites* on the antibody; the quantity of each hormone that will bind is proportional to its concentration. After binding is complete, the antibody–hormone complex is separated from the remainder of the solution, and the quantity of radioactive hormone that has bound with the antibody is measured by means of radioactive counting techniques. If a large amount of radioactive hormone has bound, then it is clear that there was only a small amount of natural hormone to compete. Conversely, if only a small amount of radioactive hormone has bound, it is clear that there was a very large amount of natural hormone to compete for the binding sites. Thus, by the use of an appropriate calibration curve, very precise measurements of the quantities of most hormones in body fluids can be achieved. As little as a fraction of a picogram (one-trillionth of a gram) of vasopressin per milliliter of assay fluid has been measured in this way.

Several other competitive binding techniques for assay of minute quantities of hormones have also been employed. One of these is to use—in place of the antibody—the specific carrier globulins of plasma that are natural binding agents for some hormones. The carrier globulin is substituted for the antibody in the assay process, and then the assay is carried out in exactly the same way as for the radioimmunoassay procedure. This technique is used mainly for assay of cortisol and thyroxin.

# MECHANISMS OF HORMONAL ACTION

The function of the different hormones is to *control* the activity levels of target tissues. To provide this control function they may alter the chemical reactions within the cells, alter the permeability of the cell membrane to specific substances, or activate some other specific cellular mechanism. The different hormones achieve these effects in many different ways. However, two important general mechanisms by which many of the hormones function are: (1) activation of the cyclic AMP system of cells, which in turn elicits the specific cellular functions, or (2) activation of the genes of the cells which causes the formation of intracellular proteins that initiate specific cellular functions. These mechanisms are described as follows:

## INTRACELLULAR HORMONAL MEDIATORS—CYCLIC AMP

Many hormones exert their effects on cells by first causing the substance *cyclic 3',5'-adenosine monophosphate* (cyclic AMP) to be formed in the cell. Once formed, the cyclic AMP causes the hormonal effects inside the cell. Thus, cyclic AMP is an intracellular hormonal mediator. It is also frequently called the *second messenger* for hormone mediation—the "first messenger" being the original stimulating hormone.

Figure 75–1 illustrates the function of cyclic AMP in more detail. It is believed that the stimulating hormone acts at the membrane of the target cell, presumably combining with a specific receptor for that particular type of hormone. The specificity of the receptor determines which hormone will affect the target cell. After binding with the receptor, the combination of hormone and receptor activates the enzyme *adenyl cyclase* in the membrane, and the portion of the adenyl cyclase that is exposed to the cytoplasm causes immediate conversion of cytoplasmic ATP into cyclic AMP. The cyclic AMP then initiates any number of cellular func-

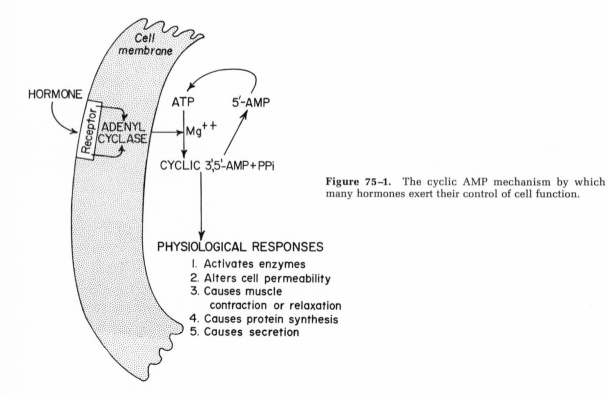

**Figure 75–1.** The cyclic AMP mechanism by which many hormones exert their control of cell function.

tions before it itself is destroyed—functions such as activating enzymes in the cell, altering the cell permeability, initiating synthesis of specific intracellular proteins, causing muscle contraction or relaxation, initiating secretion, and many other possible effects. The types of effects that will occur inside the cell are determined by the character of the cell itself. Thus, a thyroid cell stimulated by cyclic AMP forms thyroid hormones, whereas an adrenocortical cell forms adrenocortical hormones. On the other hand, cyclic AMP affects epithelial cells of the renal tubules by increasing their permeability to water.

The cyclic AMP mechanism has been shown to be an intracellular hormonal mediator for at least some of the functions of the following hormones (and perhaps many more):

1. Adrenocorticotropin
2. Thyroid-stimulating hormone
3. Luteinizing hormone
4. Follicle-stimulating hormone
5. Vasopressin
6. Parathyroid hormone
7. Glucagon
8. Catecholamines
9. Secretin
10. The hypothalamic releasing factors

It has been postulated that other types of intracellular hormonal mediators also exist. One other proposed type is the *prostaglandins,* which are a series of lipid compounds closely related to each other and widely present in cells throughout the body. These substances frequently cause intracellular inhibition, in contrast to the activation usually caused by cyclic AMP. However, such a prostaglandin system is mainly theoretical at present. Another almost certain intracellular mediator (or "second messenger") is *cyclic guanosine monophosphate,* which is a nucleotide similar to cyclic AMP and can probably catalyze at least some intracellular functions in a manner similar to that of cyclic AMP.

### ACTION OF STEROID HORMONES ON THE GENES TO CAUSE PROTEIN SYNTHESIS

A second major means by which hormones—specifically the steroid hormones secreted by the adrenal cortex, the ovaries, and the testes—act is to cause synthesis of proteins in the target cells; these proteins are probably enzymes that in turn activate other functions of the cells.

The sequence of events in steroid function is the following:

1. The steroid hormone enters the cytoplasm of the cell, where it binds with a specific *receptor protein*.
2. The combined receptor protein/hormone then diffuses into or is transported into the nucleus.
3. Somewhere along this route the receptor protein is altered to form a smaller molecular weight protein, or the steroid hormone is transferred to a second smaller protein.
4. The combination of the small protein and hormone is now the active factor that activates specific genes to form messenger RNA.
5. The messenger RNA diffuses into the cytoplasm where it promotes the translation process at the ribosomes to form new proteins.

To give an example, aldosterone, one of the hormones secreted by the adrenal cortex, enters the cytoplasm of renal tubular cells, which contain its specific receptor protein. Therefore, in these cells the above sequence of events ensues. After about 45 minutes, proteins begin to appear in the renal tubular cells that promote sodium reabsorption from the tubules and potassium secretion into the tubules. Thus, there is a characteristic delay in the final action of the steroid hormone of 45 minutes to several hours, which is in marked contrast to the almost instantaneous action of some of the peptide and peptide-derived hormones.

### OTHER MECHANISMS OF HORMONE FUNCTION

Hormones can have other direct effects on cells, though in most instances the precise mechanisms of these effects are not known. For instance, insulin increases the permeability of the cells to glucose, and growth hormone increases the transport of amino acids into cells. Also, growth hormone stimulates synthesis of proteins in several ways in addition to its effect to increase intracellular amino acids. In addition, several hormones, such as the catecholamines and acetylcholine, directly affect cell membranes by changing their permeabilities to ions and thereby exciting muscular contraction or causing other effects.

# THE PITUITARY GLAND AND ITS RELATIONSHIP TO THE HYPOTHALAMUS

The *pituitary gland* (Fig. 75–2), also called the *hypophysis,* is a small gland—less than 1 cm. in diameter and about ½ gram in weight—

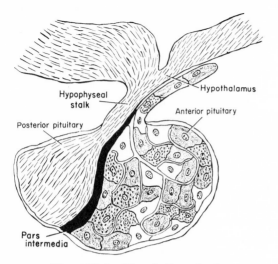

**Figure 75–2.** The pituitary gland.

that lies in the *sella turcica* at the base of the brain and is connected with the hypothalamus by the *pituitary,* or *hypophyseal, stalk.* Physiologically, the pituitary gland is divisible into two distinct portions: the *anterior pituitary,* also known as the *adenohypophysis,* and the *posterior pituitary,* also known as the *neurohypophysis.* Between these is a small, relatively avascular zone called the *pars intermedia,* which is almost absent in the human being but is much larger and much more functional in some lower animals.

Enbryologically, the two portions of the pituitary originate from different sources, the anterior pituitary from *Rathke's pouch,* which is an embryonic invagination of the pharyngeal epithelium, and the posterior pituitary from an outgrowth of the hypothalamus. The origin of the anterior pituitary from the pharyngeal epithelium explains the epithelioid nature of its cells, whereas the origin of the posterior pituitary from neural tissue explains the presence of large numbers of glial type cells in this gland.

Six very important hormones plus several less important ones are secreted by the *anterior* pituitary, and two important hormones are secreted by the *posterior* pituitary. The hormones of the anterior pituitary play major roles in the control of metabolic functions throughout the body, as shown in Figure 75–3; thus: (1) *Growth hormone* promotes growth of the animal by affecting many metabolic functions throughout the body, especially protein formation. (2) *Adrenocorticotropin* controls the secretion of some of the adrenocortical hormones, which in turn affect the metabolism of glucose, proteins,

and fats. (3) *Thyroid-stimulating hormone* controls the rate of secretion of thyroxine by the thyroid gland, and thyroxine in turn controls the rates of most chemical reactions of the entire body. (4) *Prolactin* promotes mammary gland development and milk production. And two separate gonadotropic hormones, (5) *follicle-stimulating hormone* and (6) *luteinizing hormone,* control growth of the gonads as well as their reproductive activities.

The two hormones secreted by the posterior pituitary play other roles: (1) *Antidiuretic hormone* controls the rate of water excretion into the urine and in this way helps to control the concentration of water in the body fluids. (2) *Oxytocin* (a) helps to deliver milk from the glands of the breast to the nipples during suckling; (b) and probably helps in the delivery of the baby at the end of gestation.

## CONTROL OF PITUITARY SECRETION BY THE HYPOTHALAMUS

Almost all secretion by the hypophysis is controlled by signals transmitted from the hypothalamus. Indeed, when the pituitary gland is removed from its normal position beneath the hypothalamus and transplanted to some other part of the body, its rates of secretion of the different hormones (except for prolactin) fall to low levels—in the case of some of the hormones, almost to zero.

Secretion from the posterior pituitary is controlled by nerve fibers originating in the

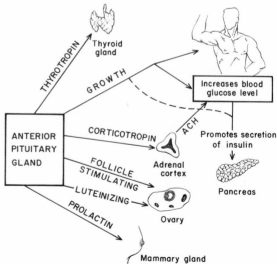

**Figure 75–3.** Metabolic functions of the anterior pituitary hormones.

hypothalamus and terminating in the posterior pituitary. In contrast, secretion by the anterior pituitary is controlled by hormones called *hypothalamic releasing* and *inhibitory factors* secreted within the hypothalamus itself and then conducted to the anterior pituitary through minute blood vessels called *hypothalamic-hypophyseal portal vessels*. In the anterior pituitary these releasing and inhibitory factors act on the glandular cells to control their secretion. This system of control will be discussed in detail later in the chapter.

The hypothalamus receives signals from almost all possible sources in the nervous system. Thus, when a person is exposed to pain, a portion of the pain signal is transmitted into the hypothalamus. Likewise, when a person experiences some powerful depressing or exciting thought, a portion of the signal is transmitted into the hypothalamus. Olfactory stimuli denoting pleasant or unpleasant smells transmit strong signal components through the amygdaloid nuclei into the hypothalamus. *Even the concentrations of nutrients, electrolytes, water, and various hormones* in the blood excite or inhibit various portions of the hypothalamus. Thus, the hypothalamus is a collecting center for information concerned with the well-being of the body, and in turn much of this information is used to control secretion by the pituitary gland.

# THE ANTERIOR PITUITARY GLAND AND ITS REGULATION BY HYPOTHALAMIC RELEASING FACTORS

## CELL TYPES OF THE ANTERIOR PITUITARY

The anterior pituitary gland is composed of several different types of cells. In general, there is one type of cell for each type of hormone that is formed in this gland; with special staining techniques these various cell types can be differentiated from one another. The only likely exception to this is that the same cell type may secrete both luteinizing hormone and follicle-stimulating hormone.

However, through the use of usual acid-base histological stains, the cell types can be separated into only three types commonly known as (1) *acidophils*, which stain strongly with acidic dyes; (2) *basophils*, which stain strongly with basic dyes; and (3) *chromophobes*, which do not stain with either. The *acidophils* produce *growth hormone* and *prolactin*;

the *basophils* produce *luteinizing hormone, follicle-stimulating hormone,* and *thyroid-stimulating hormone;* and the *chromophobes* are believed to secrete *adrenocorticotropin*.

## THE HYPOTHALAMIC-HYPOPHYSEAL PORTAL SYSTEM

The anterior pituitary is a highly vascular gland with extensive capillary sinuses among the glandular cells. Almost all of the blood that enters these sinuses passes first through a capillary bed in the tissue of the lower hypothalamus and then through small *hypothalamic-hypophyseal portal vessels* into the anterior pituitary sinuses. Thus, Figure 75–4 illustrates a small artery supplying the lowermost portion of the hypothalamus called the *median eminence*. Small vascular tufts project into the substance of the median eminence and then return to its surface, coalescing to form the hypothalamic-hypophyseal portal vessels. These in turn pass downward along the pituitary stalk to supply the anterior pituitary sinuses.

**Secretion of Hypothalamic Releasing and Inhibitory Factors into the Median Eminence.** Nerve endings have long been known to secrete acetylcholine, norepinephrine, and other transmitter substances for transmission of signals at synapses in the nervous system or for exciting skeletal muscle and autonomic structures. More recently, it has become apparent that special neurons in the hypothalamus synthesize and secrete hormones called *hypothalamic releasing* and *inhibitory factors* that control the secretion of the anterior pituitary

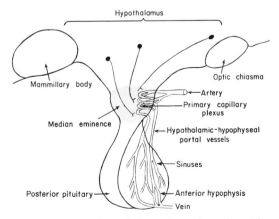

**Figure 75–4.** The hypothalamic-hypophyseal portal system.

hormones. These neurons originate in various parts of the hypothalamus and send their nerve fibers into the median eminence and the tuber cinereum, the hypothalamic tissue that extends into the pituitary stalk. The endings of these fibers are different from most endings in the central nervous system in that their function is not to transmit signals from one neuron to another but merely to secrete the hypothalamic releasing and inhibitory factors (hormones) into the tissue fluids. These hormones are immediately absorbed into the hypothalamic-hypophyseal portal capillaries and carried directly to the sinuses of the anterior pituitary gland.

**Function of the Releasing and Inhibitory Factors.** The function of the releasing and inhibitory factors is to control the secretion of the anterior pituitary hormones. For each type of anterior pituitary hormone there is a corresponding hypothalamic releasing factor; for some of the anterior pituitary hormones there is also a corresponding hypothalamic inhibitory factor. For most of the anterior pituitary hormones it is the releasing factors that are important; but, for prolactin, it is an inhibitory factor that probably exerts most control. The hypothalamic releasing and inhibitory factors that are of major importance are:

1. *Thyroid-stimulating hormone releasing factor* (TRF), which causes release of thyroid-stimulating hormone

2. *Corticotropin releasing factor* (CRF), which causes release of adrenocorticotropin

3. *Growth hormone releasing factor* (GRF), which causes release of growth hormone

4. *Luteinizing hormone releasing factor* (LRF), which causes release of luteinizing hormone

5. *Follicle-stimulating hormone releasing factor* (FRF), which causes release of follicle-stimulating hormone

6. *Prolactin inhibitory factor* (PIF), which causes inhibition of prolactin secretion

In addition to these more important hypothalamic hormones, still another excites the secretion of melanocyte-stimulating hormone, and several inhibitory factors inhibit some of the other anterior pituitary hormones. Each of these factors will be discussed in more detail at the time that the specific hormonal system controlled by them is presented in this and subsequent chapters.

**Specific Areas in the Hypothalamus that Control Secretion of Specific Hypothalamic Releasing and Inhibitory Factors.** It is believed that all or most of the hypothalamic hormones are secreted at nerve endings in the median eminence before being transported to the anterior pituitary gland. Electrical stimulation of this region excites these nerve endings and therefore causes release of essentially all of the hypothalamic factors. However, the neuronal cell bodies that give rise to these median eminence nerve endings are located in other discrete areas of the hypothalamus or in closely related areas of the basal brain. Unfortunately, the specific loci of the neurons for all the hypothalamic factors are incompletely known. However, on the basis of both electrical stimulation studies and nuclei destruction studies, the following are believed to be the major loci (also illustrated in Figure 75–5):

1. Thyroid-stimulating hormone releasing factor—paraventricular nucleus

2. Adrenocorticotropin releasing factor—anterior hypothalamus

3. Growth hormone releasing factor—lateral ventromedial nucleus

4. and 5. Luteinizing hormone releasing factor and follicle-stimulating hormone releasing factor—medial preoptic area

6. Prolactin inhibitory factor—lateral preoptic area

It must be recognized, however, that the areas just described might simply be "way stations" for relay of appropriate secretory signals. Therefore, these localizations should not be considered as undisputed fact.

# PHYSIOLOGIC FUNCTIONS OF THE ANTERIOR PITUITARY HORMONES

All of the major anterior pituitary hormones besides growth hormone exert their effects by stimulating "target glands"—the thyroid gland, the adrenal cortex, the ovaries, the testicles, and the mammary glands. The functions of each of the anterior pituitary hormones are so intimately concerned with the functions of the respective target glands that except for growth

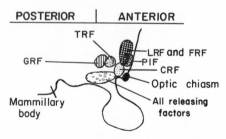

**Figure 75–5.** Loci in the hypothalamus where stimulation will cause secretion of different anterior pituitary hormones.

hormone, their functions will be discussed in subsequent chapters along with the functions of the target glands. Growth hormone, in contrast to other hormones, does not function through a target gland but instead exerts an effect on all or almost all tissues of the body.

## GROWTH HORMONE

Growth hormone (GH), also called *somatotropic hormone* (SH) or *somatotropin,* is a small protein molecule containing 188 amino acids in a single chain and having a molecular weight of 21,500. It causes growth of all tissues of the body that are capable of growing. It promotes both increased sizes of the cells and increased mitosis with development of increased numbers of cells.

Figure 75–6 illustrates weight charts of two growing rats, one of which received daily injections of growth hormone, compared with a litter-mate that did not receive growth hormone. This figure shows marked exacerbation of growth by growth hormone—both in the early days of life and even after the two rats had reached adulthood. In the early stages of development, all organs of the treated rat increased proportionately in size, but, after adulthood was reached, most of the bones ceased growing while the soft tissues continued to grow. This results from the fact that, once the epiphyses of the long bones have united with the shafts, further growth of the bones cannot occur even though most other tissues of the body can continue to grow throughout life.

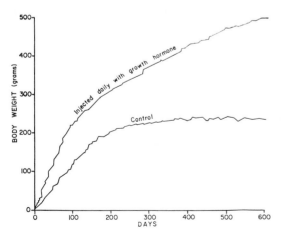

**Figure 75–6.** Comparison of weight gain of a rat injected daily with growth hormone with that of a normal rat of the same litter.

**Basic Metabolic Effects of Growth Hormone.** Growth hormone is known to have the following basic effects on the metabolic processes of the body:

1. Increased rate of protein synthesis in all cells of the body

2. Decreased rate of carbohydrate utilization throughout the body

3. Increased mobilization of fats and use of fats for energy

Thus, in effect, growth hormone enhances the body proteins, conserves carbohydrates, and uses up the fat stores. It is probable that the increased rate of growth results mainly from the increased rate of protein synthesis.

**Stimulation of Growth of Cartilage and Bone—Role of "Somatomedin."** Growth hormone does not have a *direct* effect on the growth of cartilage and bone, both of which must grow if the overall structure of the animal is to increase. However, growth hormone does indirectly stimulate their growth by causing the substance *somatomedin* to be formed in the liver and perhaps the kidney as well; this substance in turn acts directly on the cartilage and bone to promote their growth. Somatomedin is necessary for deposition of chondroitin sulfate and collagen, both of which are necessary for growth of the cartilage and bone.

Once the epiphyses of the long bones have united with the shafts, the bones can no longer increase in length, but they can continue to increase in thickness. Therefore, excess growth hormone after adolescence cannot cause further increase in height of a person but can cause disproportionate growth of the membranous bones and excessive thickening of all bones.

Somatomedin may stimulate growth of other tissues in addition to cartilage and bone, most probably causing deposition of connective tissue and thickening of the skin.

## Role of Growth Hormone in Promoting Protein Deposition

Although the most important cause of the increased protein deposition caused by growth hormone is not known, a series of different effects are known, all of which can lead to enhanced protein. These effects are:

**1. Enhancement of Amino Acid Transport Through the Cell Membranes.** Growth hormone directly enhances transport of at least some and perhaps most amino acids through the cell membranes to the interior of the cells. This increases the concentrations of the amino acids in the cells and is presumed to be at least partly responsible for the increased protein synthesis. This control of amino acid transport through the cell membrane is similar to the effect of insulin in controlling glucose transport through the membrane, as discussed in Chapters 67 and 78.

**2. Enhancement of Protein Synthesis by the Ribosomes.** Even when the amino acids are not increased in the cells, growth hormone still causes protein to be synthesized in increased amounts in the cells. This is believed to be caused partly by a direct effect on the ribosomes, making them produce greater numbers of protein molecules; the mechanism by which this effect occurs is as yet unknown.

**3. Increased Formation of RNA.** Over more prolonged periods of time, growth hormone also stimulates the transcription process in the nucleus, causing formation of increased quantities of RNA. This in turn promotes rapid protein synthesis and also promotes growth if sufficient energy, amino acids, vitamins, and other necessities for growth are available.

**4. Decreased Catabolism of Protein and Amino Acid.** In addition to the increase in protein synthesis, there is a decrease in the breakdown of protein and the utilization of protein and amino acids for energy. A possible if not probable reason for this effect is that growth hormone also mobilizes large quantities of free fatty acids from the adipose tissue, and these in turn are used to supply most of the energy for the body cells, thus acting as a potent "protein sparer." And this easy availability of fats for energy also acts as a "carbohydrate sparer," thereby decreasing the necessity to use proteins for gluconeogenesis—another factor that diminishes protein catabolism.

**Summary.** Growth hormone enhances almost all facets of amino acid uptake and protein synthesis by cells, while at the same time reducing the breakdown of proteins.

### Effect of Growth Hormone in Enhancing Fat Utilization for Energy

Growth hormone has a specific effect in causing release of fatty acids from adipose tissue and, therefore, increasing the fatty acid concentration in the body fluids. In addition, in the tissues it enhances the conversion of fatty acids to acetyl-CoA with subsequent utilization of this for energy. Therefore, under the influence of growth hormone, fat is utilized for energy in preference to both carbohydrates and proteins.

Some research workers have considered the fat mobilization and utilization effect of growth hormone to be its most important function and have also considered the protein-sparing effect to be the major factor that promotes protein deposition and growth. However, growth hormone mobilization of fat requires many minutes to hours to occur, whereas enhancement of cellular protein synthesis can begin in less than a minute under the influence of growth hormone.

*Ketogenic Effect of Growth Hormone.* Occasionally, fat mobilization under the influence of excessive amounts of growth hormone is so great that excessive quantities of acetoacetic acid are formed by the liver and are released into the body fluids, thus causing ketosis. This excessive mobilization of fat from the adipose tissue also frequently causes a fatty liver.

### Effect of Growth Hormone on Carbohydrate Metabolism

Growth hormone has three major effects on cellular metabolism of glucose. These effects are (1) decreased utilization of glucose for energy, (2) marked enhancement of glycogen deposition in the cells, and (3) diminished uptake of glucose by the cells.

**Decreased Glucose Utilization for Energy.** Unfortunately, we do not know the precise mechanism by which growth hormone decreases glucose utilization by the cells. However, the decrease probably results partially from the increased mobilization and utilization of fatty acids for energy caused by growth hormone. That is, the fatty acids form large quantities of acetyl-CoA that in turn initiate feedback effects to block the glycolytic breakdown of glucose and glycogen.

**Enhancement of Glycogen Deposition.** Since glucose and glycogen cannot be utilized for energy, the glucose that does enter the cells is rapidly polymerized into glycogen and deposited. Therefore, the cells rapidly become saturated with glycogen and can store no more.

**Diminished Uptake of Glucose by the Cells and Increased Blood Glucose Concentration.** When growth hormone is first administered to an animal, the cellular uptake of glucose is enhanced and the blood glucose concentration falls slightly. However, as the cells become saturated with glycogen and their utilization of glucose for energy decreases, further uptake of glucose then becomes greatly diminished. Without normal cellular uptake, the blood concentration of glucose increases, sometimes to as high as 50 to 100 per cent above normal.

Thus, growth hormone seems actually to enhance membrane transport of glucose, though failure of glucose utilization eventually leads to greatly diminished uptake. However, some physiologists believe that there is also a direct effect of growth hormone to reduce glucose uptake—an effect as yet unexplained.

**Necessity of Insulin and Carbohydrate for the Growth-Promoting Action of Growth Hormone.** Growth hormone fails to cause growth in a pancreatectomized animal, and it also fails to cause growth if carbohydrates are excluded from the diet. This shows that adequate insulin activity as well as adequate availability of carbohydrates is necessary for growth hormone to be effective. Part of this requirement for carbohydrates and insulin is to provide the energy needed for the metabolism of growth. But, in addition, there seem to be direct effects to promote growth. Insulin specifically increases amino acid transport into cells (which could be one such effect), but there are probably others as well.

**Diabetogenic Effect of Growth Hormone.** We have already pointed out that growth hormone leads to moderately increased blood glucose concentration. This in turn stimulates the beta cells of the islets of Langerhans to secrete extra insulin. In addition to this effect, growth hormone possibly or probably has a slight direct stimulatory effect on the beta cells as well. The combination of these two effects sometimes so greatly overstimulates insulin secretion by the beta cells that they literally "burn out." When this occurs the person develops diabetes mellitus, a disease that will be discussed in detail in Chapter 78. Therefore, growth hormone is said to have a *diabetogenic effect.*

*Diabetogenic Effects of Other Anterior Pituitary Hormones.* Growth hormone is not the only anterior pituitary hormone that increases the blood glucose concentration. At least three others can do the same: adrenocorticotropin, thyroid-stimulating hormone, and prolactin. Especially important is adrenocorticotropin, which increases the rate of cortisol secretion by the adrenal cortex. Cortisol then increases the blood glucose concentration by increasing the rate of gluconeogenesis. This effect, quantitatively, is probably equally as "diabetogenic" as the effect of growth hormone.

*Pituitary Diabetes.* From the above discussion it is readily apparent that generalized increase in secretion of all the anterior pituitary hormones causes elevated blood glucose concentration; this condition is called pituitary diabetes, and it differs from *diabetes mellitus,* which results from insulin lack, in the following ways: First, in pituitary diabetes the rate of glucose utilization by the cells is only moderately depressed, in comparison with almost no utilization in diabetes mellitus. Second, the blood glucose concentration is relatively *refractory to insulin* because adequate insulin is already available in the body to cause glucose transfer into the cell; the problem instead is failure to utilize the glucose after it enters the cell. Third, many of the side effects that result from reduced carbohydrate metabolism in diabetes mellitus are absent in pituitary diabetes.

### Regulation of Growth Hormone Secretion

For many years it was believed that growth hormone was secreted primarily during the period of growth but then disappeared from the blood at adolescence. However, this has proved to be very far from the truth, because after adolescence, secretion continues at a rate as great or almost as great as that in childhood. Furthermore, the rate of growth hormone secretion increases and decreases within minutes in relation to the person's state of nutrition or stress, such as during starvation, hypoglycemia, exercise, excitement, and trauma.

The normal concentration of growth hormone in the plasma of an adult is about 3 millimicrograms per milliliter and in the child about 5 millimicrograms per milliliter. However, these values often increase to as high as 50 millimicrograms per milliliter after depletion of the body stores of proteins or carbohydrates. Under acute conditions, hypoglycemia is a far more potent stimulator of growth hormone secretion than is a decrease in the amino acid concentration in the blood. On the other hand, in chronic conditions the degree of cellular protein depletion seems to be more correlated with the level of growth hormone secretion than is the availability of glucose. For instance, the extremely high levels of growth hormone that occur during starvation are very closely related to the amount of protein depletion. Figure 75–7 illustrates this relationship: the first column shows growth hormone levels in children with extreme protein deficiency during the malnutrition disease called *kwashiorkor;* the second column shows the levels in the same children after three days of treatment with more than adequate quantities of carbohydrates in their diets, illustrating that the carbohydrates did not lower the plasma growth hormone concentration; the

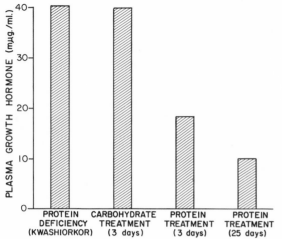

**Figure 75–7.** Effect of extreme protein deficiency on the concentration of growth hormone in the plasma in the disease kwashiorkor. The figure also shows the failure of carbohydrate treatment but the effectiveness of protein treatment in lowering growth hormone concentration. (Drawn from data in Pimstone: *Amer. J. Clin. Nutr.,* 21:482, 1968.)

third and fourth columns show the levels after treatment with protein supplement in their diet for 3 and 25 days, with concomitant decrease in the hormone. These results demonstrate that under very severe conditions of protein malnutrition, adequate calories alone are not sufficient to correct the excess production of growth hormone. Instead, the protein deficiency must also be corrected before the growth hormone concentration will return to normal.

Thus, it is almost certain that growth hormone secretion is controlled moment by moment by the nutritional and stress status of the body, and it seems that the most important factor in the control of growth hormone secretion is the level of cell protein, though changes in blood glucose concentration can also cause extremely rapid and dramatic alterations in growth hormone secretion. Consequently, it can be postulated that growth hormone operates in a feedback control system as follows: When the tissues begin to suffer from malnutrition, especially from poor protein nutrition, large quantities of growth hormone are secreted. Growth hormone, in turn, promotes the synthesis of new proteins, while at the same time conserving the protein already present in the cells. Aside from the direct effects of growth hormone on protein deposition, its effect on mobilizing fats and on increasing cellular utilization of fatty acids for energy acts as a protein sparer while it concurrently prevents oxidation of amino acids, thus making these available for protein synthesis.

## ABNORMALITIES OF GROWTH HORMONE SECRETION

**Panhypopituitarism.** This term means decreased secretion of all the anterior pituitary hormones. The decrease in secretion may be congenital (present from birth), or it may occur suddenly or slowly at any time during the life of the individual.

*Dwarfism.* Some instances of dwarfism result from deficiency of anterior pituitary secretion during childhood. In general, the features of the body develop in appropriate proportion to each other, but the rate of development is greatly decreased. A child who has reached the age of 10 may have the bodily development of a child of 4 to 5, whereas the same person on reaching the age of 20 may have the bodily development of a child of 7 to 10.

The dwarf usually does not exhibit specific thyroid deficiency or adrenocortical deficiency, for the entire body remains so small that only small quantities of thyroid-stimulating and adrenocorticotropic hormones are needed. Also, there is no mental retardation. On the other hand, the panhypopituitary dwarf does not pass through puberty and never secretes a sufficient quantity of gonadotropic hormones to develop adult sexual functions. In one-third of the dwarfs, however, the deficiency is of growth hormone alone; these individuals do mature sexually and occasionally do reproduce.

*Panhypopituitarism in the Adult.* Panhypopituitarism occurring in adulthood frequently results from one of three different abnormalities: Two tumorous conditions, craniopharyngiomas and chromophobe tumors, may compress the pituitary gland until the functioning anterior pituitary cells are totally or almost totally destroyed. The third cause is thrombosis of the pituitary blood vessels. This occurs occasionally when a mother develops circulatory shock following birth of a baby.

The effects of panhypopituitarism, in general, are (1) hypothyroidism, (2) depressed production of glucocorticoids by the adrenal glands, and (3) suppressed secretion of the gonadotropic hormones to the point that sexual functions are lost. Thus, the picture is that of a lethargic person (from lack of thyroxine) who is gaining weight because of lack of fat mobilization by growth, adrenocorticotropic, adrenocortical, and thyroid hormones and who has lost all sexual functions. Except for the abnormal sexual functions, the patient can usually be treated satisfactorily by administration of adrenocortical and thyroid hormones.

**Giantism.** Occasionally, the acidophilic cells of the anterior pituitary become excessively active, and sometimes even acidophilic tumors occur in the gland. As a result, large quantities of growth hormone are produced. All body tissues grow rapidly, including the bones, and if the epiphyses of the long bones have not become fused with the shafts before the development of the anterior pituitary acidophilia, height increases so that the person becomes a giant with heights as great as 8 to 9 feet. Thus, for giantism to occur, the acidophilia must occur prior to adolescence.

The giant ordinarily has hyperglycemia, and the beta cells of the islets of Langerhans in the pancreas eventually degenerate, partially because they become overactive owing to the hyperglycemia and partially because of a direct overstimulating effect of growth hormone on the islet cells. Consequently, about 10 per cent of the giants finally develop fullblown diabetes mellitus.

Most giants, unfortunately, eventually develop panhypopituitarism if they remain untreated, because the giantism is usually caused by a tumor of the pituitary gland that grows until the gland itself is destroyed. This general deficiency of pituitary hormones usually causes death in early adulthood. However, once giantism is diagnosed, further development can usually be blocked by gamma irradiation of the pituitary gland.

**Acromegaly.** If an acidophilic tumor occurs after adolescence—that is, after the epiphyses of the long bones have fused with the shafts—the person cannot grow taller; but his soft tissues can continue to grow, and the bones can grow in thickness. This condition is known as "acromegaly." Enlargement is especially marked in the small bones of the hands and feet and in the *membranous bones,* including the crani-um, the nose, the bosses on the forehead, the supra-orbital ridges, the lower jawbone, and portions of the vertebrae, for their growth does not cease at adolescence anyway. Consequently, the jaw protrudes forward, sometimes as much as a half inch, the forehead slants forward because of excess development of the supra-orbital ridges, the nose increases to as much as twice normal size, the foot requires a size 14 or larger shoe, and the fingers become extremely thickened so that the hand develops a size almost twice normal. In addition to these effects, changes in the vertebrae ordinarily cause a hunched back, which is known clinically as "kyphosis." Finally, many soft tissue organs, such as the tongue, liver, and especially the kidneys, become greatly enlarged. A typical acromegalic is shown in Figure 75–8.

# MELANOCYTE-STIMULATING HORMONE

Another anterior pituitary hormone normally secreted in small quantities, and perhaps in large quantities under abnormal conditions, is melanocyte-stimulating hormone (MSH). MSH stimulates the *melanocytes,* which are cells that contain the black pigment melanin and that occur in abundance between the dermis and epidermis of the skin.

In some amphibia, melanin is collected in the

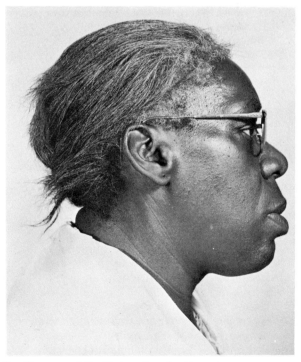

**Figure 75–8.** An acromegalic. (Courtesy of Dr. Herbert Langford.)

melanocytes in small granules called *melanosomes.* When melanocyte-stimulating hormone is not available, the melanosomes become concentrated near the nuclei of the melanocytes in such a way that the melanocytes appear light colored. Then, when MSH is secreted, the melanosomes disperse throughout the cytoplasm of the melanocytes, and the entire cell becomes almost black within a few seconds to minutes. In this way, the color of the skin can be changed from very light in the absence of the hormone to very dark in its presence.

Exposure of the human being to melanocyte-stimulating hormone over a period of 8 to 10 days causes intense darkening of the skin. It has a much greater effect in persons who have a genetically dark skin than in persons who have a genetically light skin. It is likely that this hormone acts in a different manner in the human being than in amphibia, for much of the melanin formed by human melanocytes actually leaves the melanocytes and becomes dispersed in the cells of the epidermis rather than remaining in the melanocytes.

Adrenocorticotropin (ACTH), one of the major hormones secreted by the anterior pituitary, has about one-thirtieth the melanocyte-stimulating effect of MSH, and the two hormones are often secreted in high concentrations simultaneously. This frequently occurs in Addison's disease in which the adrenal cortex is damaged so much that it cannot secrete adrenocortical hormones; lack of these hormones causes compensatory increase in ACTH secretion and, along with it, increased MSH secretion. Consequently, the skin becomes considerably darkened.

MSH occurs in two forms, an *alpha* and a *beta.* The alpha form has exactly the same chemical structure as the first portion, composed of 13 amino acids, of the 39 amino acid ACTH polypeptide chain. The beta form partially duplicates a portion of the ACTH chain, but not precisely the same portion.

# THE POSTERIOR PITUITARY GLAND AND ITS RELATION TO THE HYPOTHALAMUS

The *posterior pituitary gland,* also called the *neurohypophysis,* is composed mainly of glial-like cells called *pituicytes.* However, the pituicytes do not secrete hormones; they act simply as a supporting structure for large numbers of *terminal nerve fibers* and *terminal nerve endings* from nerve tracts that originate in the *supraoptic* and *paraventricular nuclei* of the hypothalamus, as shown in Figure 75–9. These tracts pass to the neurohypophysis through the *pituitary stalk* (hypophyseal stalk). The nerve endings are bulbous knobs that lie on the surfaces of capillaries onto which they secrete the

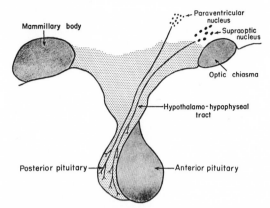

**Figure 75–9.** Hypothalamic control of the posterior pituitary.

posterior pituitary hormones: (1) *antidiuretic hormone* (ADH), also called *vasopressin,* and (2) *oxytocin.*

If the pituitary stalk is cut near the pituitary gland, leaving the entire hypothalamus intact, the posterior pituitary hormones continue, after a transient decrease for a few days, to be secreted almost normally, but they are then secreted by the cut ends of the fibers within the hypothalamus and not by the nerve endings in the posterior pituitary. The reason for this is that the hormones are initially synthesized in the cell bodies of the supraoptic and paraventricular nuclei and are then transported in combination with a "carrier" protein called *neurophysin* down to the nerve endings in the posterior pituitary gland, requiring several days to reach the gland.

*ADH is formed primarily in the supraoptic nuclei,* while *oxytocin is formed primarily in the paraventricular nuclei.* However, each of these two nuclei can synthesize approximately one-sixth as much of the second hormone as of its primary hormone.

Under resting conditions, large quantities of both ADH and oxytocin accumulate in the nerve endings of the posterior pituitary gland. Then when nerve impulses are transmitted downward along the fibers from the supraoptic and paraventricular nuclei, the hormones are immediately released from the nerve endings and are absorbed into adjacent capillaries.

## PHYSIOLOGIC FUNCTIONS OF ANTIDIURETIC HORMONE (VASOPRESSIN)

Extremely minute quantities of antidiuretic hormone (ADH)—as little as 2 millimicrograms—when injected into a person cause antidiuresis, that is, decreased excretion of water by the kidneys. This antidiuretic effect was discussed in detail in Chapter 36. Briefly, in the absence of ADH, the collecting ducts (and perhaps, to a lesser extent, parts of the distal tubules and loops of Henle) are almost totally impermeable to water, which prevents significant reabsorption of water and therefore allows extreme loss of water into the urine. On the other hand, in the presence of ADH, the permeability of these tubules to water increases greatly and allows most of the water in the tubular fluid to be reabsorbed, thereby conserving water in the body.

The precise mechanism by which ADH acts on the tubules to increase their permeability is unknown. However, the hormone first becomes fixed to the epithelium of the tubules and becomes highly concentrated in these cells. It then causes these tissues to form large quantities of cyclic AMP. This in turn is believed to open up the pores in the cell membranes and allow free diffusion of water between the tubular and peritubular fluids, though the manner in which cyclic AMP acid causes its effect on the pores is completely unknown. The water is then absorbed by osmosis, as explained in relation to the concentrating mechanism of the kidney in Chapter 36.

### Regulation of ADH Production

**Osmotic Regulation.**    When a concentrated electrolyte solution is injected into the artery supplying the hypothalamus, the supraoptic nuclei immediately transmit impulses into the posterior pituitary to release large quantities of ADH into the circulating blood. Conversely, injection of pure water into this artery causes complete cessation of impulses from the supraoptic nuclei and essentially complete cessation of ADH production. The ADH that has already been produced is destroyed by the tissues at a rate of approximately one-half every five minutes. Thus, the concentration of ADH in the body fluids can change from small amounts to large amounts, or vice versa, in only a few minutes.

It has been postulated that neurons in the supraoptic nuclei in the perinuclear region of the supraoptic area function as *osmoreceptors,* and histologists have claimed to have observed fluid chambers in some of these cells which presumably could increase and decrease in size in relation to the degree of concentration of the extracellular fluids. When the extracellular

fluids are highly dilute, osmosis of water into the cell supposedly increases the volume of the fluid chamber, while concentrated extracellular fluids supposedly reduce the volume. Regardless of whether fluid chambers do exist, concentrated body fluids do stimulate the supraoptic nuclei, while dilute body fluids inhibit them. Therefore, a feedback control system is available to control the total osmotic pressure of the body fluids, operating as follows:

When the body fluids become highly concentrated, the supraoptic nuclei become excited, impulses are transmitted to the posterior pituitary, and ADH is secreted. This passes by way of the blood to the kidneys, where it increases the permeability of the collecting tubules to water. As a result, most of the water is reabsorbed from the tubular fluid, while electrolytes continue to be lost into the urine. This effect dilutes the extracellular fluids, returning them to a reasonably normal osmotic composition.

**Role of ADH in Controlling Extracellular Fluid Sodium Ion Concentration.**    Under normal conditions, about 95 per cent of the total osmotic pressure of the extracellular fluids is determined by the sodium ion concentration of these fluids (because of the high concentration of sodium and because the anion concentration "follows" the sodium concentration). Therefore, in effect, when one says that ADH controls the osmolality of the extracellular fluids, this also means that ADH controls the sodium ion concentration of these fluids at the same time. Furthermore, recent experiments have shown that the major change that occurs in the composition of the extracellular fluid following either increased or decreased ADH secretion is a change in sodium ion concentration, with much less change in volume of water in the extracellular fluids. Therefore, it is becoming clear that ADH is a very potent controller of sodium ion concentration, a fact that was discussed in detail in Chapter 36.

**Stimulation of ADH Secretion by Low Blood Volume—Pressor Effect of ADH.**    ADH in moderate to high concentrations has a very potent effect of constricting the arterioles and therefore of increasing the arterial pressure. Also, one of the most powerful stimuli of all for increasing the secretion of ADH is severe loss of blood volume. As little as 10 per cent loss of blood will promote a moderate increase in ADH secretion, and 25 per cent or more blood loss can cause as much as 20 to 50 times normal rates of secretion.

The increased secretion is believed to result mainly from the low pressure caused in the atria of the heart by the low blood volume. The relaxation of the atrial stretch receptors supposedly elicits the increase in ADH secretion. However, the baroreceptors of the carotid, aortic, and pulmonary regions also participate in this control of ADH secretion.

It is generally stated that the normal concentration

of ADH in the body fluids, between 1 and 2 pico-grams per milliliter (trillionths of a gram per milli-liter), is too small to cause any significant pressor ef-fect. However, recent studies by Cowley in which the compensatory role of the nervous reflexes was eliminated showed that between 5 and 10 mm. Hg of the normal arterial pressure is maintained by this normal ADH. Therefore, the marked secretion of ADH following hemorrhage perhaps plays a very im-portant role in the homeostasis of arterial pressure.

Because ADH has this potent pressor effect, it is also called *vasopressin*.

**Other Factors that Affect ADH Production.** Other factors that frequently increase the output of ADH include *trauma* to the body, *pain, anxiety*, and drugs such as *morphine, nicotine, tranquilizers*, and some *anesthetics*. Each of these factors can cause retention of water in the body. This explains the fre-quent accumulation of water in many emotional states, and it also explains the diuresis that occurs when the state is over.

A substance that inhibits ADH secretion is *al-cohol*. Therefore, during an alcoholic bout, lack of ADH allows marked diuresis. Alcohol probably also dilates the afferent arterioles of the nephrons, which adds to the diuretic effect.

**Diabetes Insipidus.** Diabetes insipidus is the disease that occurs when the supraoptico-hypoph-yseal system secreting ADH fails. This will occur only when the supraoptic nuclei themselves or their nerve fibers near these nuclei are destroyed. It will not occur when the posterior pituitary gland alone is destroyed or when the pituitary stalk is cut below the median eminence, because the cut nerve fibers can continue to secrete ADH. In a person with full-blown diabetes insipidus, lack of ADH keeps his urine from ever being concentrated. The urine specific gravity remains almost constantly between 1.002 and 1.006; the urine output is usually 4 to 6 liters per day but can be as great as 12 to 15 liters per day, depending prin-cipally on how much water the person drinks. Fur-thermore, the rapid loss of fluid in the urine creates a constant thirst, which keeps the water flushing through his body.

The person with diabetes insipidus has a tendency to become dehydrated. However, this tendency is usually quite well offset by the increased thirst. Under conditions of circulatory stress or when water might not be adequately available, the fluid loss can become serious.

Diabetes insipidus can be treated easily by simply insufflating a small amount of powdered posterior pituitary gland or synthetic vasopressin into the nose several times a day. Though only a minute amount of ADH is absorbed in this way, it takes only 0.1 micro-gram of the hormone to cause a maximal antidiuretic effect.

Diabetes insipidus occurs most frequently as a re-sult of a tumor of the hypothalamus or hypophysis that destroys the portions of the hypothalamus that control ADH secretion.

**Excess Secretion of ADH—Idiopathic ADH Syn-drome.** Occasionally, excess ADH is secreted by the supraoptico-posterior pituitary system or, much more usually by certain types of tumors in the body, particularly bronchogenic carcinomas of the lungs. When this occurs, a condition called *idiopathic ADH syndrome* develops, characterized by greatly de-creased sodium ion concentration in the extracellular fluid but by only a few per cent increase in body water. The reasons for these effects were discussed in detail in Chapter 36; they especially point up the fact that ADH is much more a controller of sodium ion concentration than of body fluid volume.

### Other Actions of ADH

Large doses of ADH can also cause contraction of almost any smooth muscle tissue in the body, includ-ing contraction of most of the intestinal musculature, the bile ducts, and the uterus. However, the concen-trations required to cause these effects are far greater than that required to cause antidiuresis, and it is doubtful that these are significant physiologic effects for normal function of the body.

## OXYTOCIC HORMONE

**Effect on the Uterus.** An "oxytocic" sub-stance is one that causes contraction of the pregnant uterus. The hormone *oxytocin*, in accordance with its name, powerfully stimu-lates the pregnant uterus, especially toward the end of gestation. Therefore, many obstetricians believe that this hormone is at least partially responsible for effecting birth of the baby. This is supported by the following facts: (1) In a hypophysectomized animal, the duration of labor is considerably prolonged, thus indicating a probable effect of oxytocin during delivery. (2) The amount of oxytocin in the plasma in-creases during labor, especially during the last stage. (3) Stimulation of the cervix in a pregnant animal elicits nervous signals that pass to the hypothalamus and cause increased secretion of oxytocin.

ADH also stimulates the pregnant uterus, though less than one-hundredth as strongly as oxytocin. Likewise, oxytocin excites contrac-tion of a number of other smooth muscle struc-tures in the body besides the uterus, though usually to a much less extent than does ADH, except for the myoepithelial cells of the breasts as discussed below. The fact that the functions of the two hormones partially overlap illustrates their physiologic relation to each other; as pointed out below, their chemical structures are also similar.

**Effect of Oxytocin on Milk Ejection.** Oxytocin plays an especially important function in the process of lactation, for this hormone causes milk to be expressed from the alveoli into the ducts so that the baby can obtain it by suckling. This mechanism works as follows: The suckling stimuli on the nipple of the breast cause signals to be transmitted through the sensory nerves to the brain, the signals passing upward through the reticular areas of the brain stem and finally reaching the paraventricular nuclei in the anterior hypothalamus to cause release of oxytocin. The oxytocin then is carried by the blood to the breasts where it causes contraction of *myoepithelial cells*, which lie outside of and form a latticework that surrounds the alveoli of the mammary glands. In less than a minute after the beginning of suckling, milk begins to flow. Therefore, this mechanism is frequently called *milk letdown* or *milk ejection*.

**Possible Effect of Oxytocin in Promoting Fertilization of the Ovum.** Sexual stimulation of the female during intercourse increases the secretion of oxytocin, and the increased oxytocin has been postulated to be at least partially responsible for the uterine contractions that occur during the female orgasm. For these reasons, it has been proposed that oxytocin promotes fertilization of the ovum by causing uterine propulsion of the male semen upward through the fallopian tubes.

## CHEMICAL NATURE OF ANTIDIURETIC HORMONE (VASOPRESSIN) AND OXYTOCIN

Both oxytocin and ADH (vasopressin) are polypeptides containing eight amino acids. They have been isolated in pure form from posterior pituitary glands, and they have also been synthesized from their basic amino acid components. The amino acid compositions of these are the following:

| VASOPRESSIN | OXYTOCIN |
| --- | --- |
| Tyrosine | Tyrosine |
| Proline | Proline |
| Glutamic acid | Glutamic acid |
| Aspartic acid | Aspartic acid |
| Glycine | Glycine |
| Cystine | Cystine |
| Phenylalanine | Leucine |
| Arginine | Isoleucine |

Note that these two hormones are almost identical except that in vasopressin phenylalanine and arginine replace leucine and isoleucine of the oxytocin molecule. The similarity of the molecules explains the occasional functional similarities between these two hormones, while the slight dissimilarities of the molecules illustrate that slight chemical changes can alter the physiological properties of hormones markedly.

## REFERENCES

Acher, R.: Chemistry of the neurohypophyseal hormones: an example of molecular evolution. *In* Greep, R. O., and Astwood, E. B. (eds.): Handbook of Physiology. Sec. 7, Vol. 4, Part 1. Baltimore, The Williams & Wilkins Company, 1974, p. 119.

Barraclough, C. A.: Sex steroid regulation of reproductive neuroendocrine processes. *In* Greep, R. O., and Astwood, E. B. (eds.): Handbook of Physiology. Sec. 7, Vol. 2, Part 1. Baltimore, The Williams & Wilkins Company, 1974, p. 29.

Beach, F. A.: Behavioral endocrinology: an emerging discipline. *Amer. Sci.,* 63:178, 1975.

Berson, S. A., and Yalow, R. S.: Peptide Hormones. New York, American Elsevier Publishing Company, 1973.

Birnbaumer, L., Pohl, S. L., and Kaumann, A. J.: Receptors and acceptors: a necessary distinction in hormone binding studies. *Adv. Cyclic Nucleotide Res.,* 4:239, 1974.

Blackwell, R. E., and Guillemin, R.: Hypothalamic control of adenohypophyseal secretions. *Ann. Rev. Physiol.,* 35:357, 1973.

Cheek, D. B., and Hill, D. E.: Effect of growth hormone on cell and somatic growth. *In* Greep, R. O., and Astwood, E. B. (eds.): Handbook of Physiology. Sec. 7, Vol. 4, Part 2. Baltimore, The Williams & Wilkins Company, 1974, p. 159.

Cuatrecasas, P.: Membrane receptors. *Ann. Rev. Biochem.,* 43:169, 1974.

Daughaday; W. H., Herington, A. C., and Phillips, L. S.: The regulation of growth by endocrines. *Ann. Rev. Physiol.,* 37:211, 1975.

de Wied, D., and de Jong, W.: Drug effects and hypothalamic–anterior pituitary function. *Ann. Rev. Pharmacol.,* 14:389, 1974.

Douglas, W. W.: Mechanism of release of neurohypophyseal hormones: stimulus-secretion coupling. *In* Greep, R. O., and Astwood, E. B. (eds.): Handbook of Physiology. Sec. 7, Vol. 4, Part 1. Baltimore, The Williams & Wilkins Company, 1974, p. 191.

Everitt, A. V.: The hypothalamic-pituitary control of ageing and age-related pathology. *Exp. Gerontol.,* 8:265, 1973.

Fisher, L. (ed.): Neuroendocrine Integration: Basic and Applied Aspects. New York, Raven Press, 1975.

Flerko, B.: Hypothalamic mediation of neuroendocrine regulation of hypophyseal gonadotrophic functions. *In* Guyton, A. C. (ed.): MTP International Review of Science: Physiology. Vol. 8. Baltimore, University Park Press, 1974, p. 1.

Gold, E. R., and Balding, P.: Receptor Specific Proteins. New York, American Elsevier Publishing Company, 1974.

Greengard, P., Paoletti, R., and Robison, G. A. (eds.): Physiology and Pharmacology of Cyclic AMP. New York, Raven Press, 1972.

Gual, C., and Rosemberg, E.: Hypothalamic Hypophysiotropic Hormones. New York, American Elsevier Publishing Company, 1973.

Guillemin, R., and Burgus, R.: The hormones of the hypothalamus. *Sci. Amer.,* 227:24, 1972.

Handler, J. S., and Orloff, J.: The mechanism of action of antidiuretic hormone. *In* Orloff, F., and Berliner, R. W. (eds.): Handbook of Physiology. Sec. 8. Baltimore, The Williams & Wilkins Company, 1973, p. 791.

Hardman, J. G., Robison, G. A., and Sutherland, E. W.: Cyclic nucleotides. *Ann. Rev. Physiol.,* 33:311, 1971.

Harris, G. W., and Donovan, B. T.: The Pituitary Gland. Vols. 1–3. Berkeley, University of California Press, 1966.

Harris, G. W., Reed, M., and Fawcett, C. P.: Hypothalamic releasing factors and the control of anterior pituitary function. *Brit. Med. Bull.,* 22:266, 1967.

Hayward, J. N.: Neural control of the posterior pituitary. *Ann. Rev. Physiol.,* 37:191, 1975.

Hope, D. B., and Pickup, J. C.: Neurophysins. *In* Greep, R. O., and Astwood, E. B. (eds.): Handbook of Physiology. Sec. 7, Vol. 4, Part 1. Baltimore, The Williams & Wilkins Company, 1974, p. 173.

Knigge, K. M., and Silverman, A.: Anatomy of the endocrine hypothalamus. *In* Greep, R. O., and Astwood, E. B. (eds.): Handbook of Physiology. Sec. 7, Vol. 4, Part 1. Baltimore, The Williams & Wilkins Company, 1974, p. 1.

Konijn, T. M.: Cyclic AMP as a first messenger. *Adv. Cyclic Nucleotide Res.*, 1:17, 1972.

Laurence, K. A., and Hassouna, H.: Immunologic studies of the endocrine system in relation to reproduction. *In* Greep, R. O., and Astwood, E. B. (eds.): Handbook of Physiology. Sec. 7, Vol. 2, Part 2. Baltimore, The Williams & Wilkins Company, p. 339.

Lederis, K.: Neurosecretion and the functional structure of the neurohypophysis. *In* Greep, R. O., and Astwood, E. B. (eds.): Handbook of Physiology. Sec. 7, Vol. 4, Part 1. Baltimore, The Williams & Wilkins Company, 1974, p. 8.

Lefkowitz, R. J.: Isolated hormone receptors: physiologic and clinical implications. *New Engl. J. Med.*, 288:1061, 1973.

Levinson, G., and Shnider, S. M.: Vasopressors in obstetrics. *Clin. Anesth.*, 10:77, 1974.

LoBue, J., and Gordon, A. S. (eds.): Humoral Control of Growth and Differentiation. New York, Academic Press, Inc., 1974.

McCann, S. M., Fawcett, C. P., and Krulich, L.: Hypothalamic hypophyseal releasing and inhibiting hormones. *In* Guyton, A. C. (ed.): MTP International Review of Science: Physiology. Vol. 5. Baltimore, University Park Press, 1974, p. 31.

McCann, S. M., and Porter, J. C.: Hypothalamic pituitary stimulating and inhibiting hormones. *Physiol. Rev.*, 49:240, 1969.

McGuire, J.: Three sites for hormonal control of the pigment cell. *Adv. Biol. Skin*, 12:421, 1972.

Martin, J. B.: Neural regulation of growth hormone secretion. *New Engl. J. Med.*, 288:1384, 1973.

Muller, E. E.: Growth hormone and the regulation of metabolism. *In* Guyton, A. C. (ed.): MTP International Review of Science: Physiology. Vol. 5. Baltimore, University Park Press, 1974, p. 141.

Novales, R.R.: Actions of melanocyte-stimulating hormone. *In* Greep, R. O., and Astwood, E. B. (eds.): Handbook of Physiology. Sec. 7, Vol. 4, Part 2. Baltimore, The Williams & Wilkins Company, 1974, p. 347.

Oppenheimer, J. H., and Surks, M. I.: Quantitative aspects of hormone production, distribution, metabolism, and activity. *In* Greep, R. O., and Astwood, E. B. (eds.): Handbook of Physiology. Sec. 7, Vol. 3. Baltimore, The Williams & Wilkins Company, 1974, p. 197.

Pardee, A. B., and Palmer, L. M.: Regulation of transport systems: a means of controlling metabolic rates. *Symp. Soc. Exp. Biol.*, 27:133, 1973.

Petersen, O. H.: Cell membrane permeability change: an important step in hormone action. *Experientia*, 30:1105, 1974.

Pitot, H. C., and Yatvin, M. B.: Interrelationships of mammalian hormones and enzyme levels in vivo. *Physiol. Rev.*, 53:228, 1973.

Reichlin, S.: Regulation of somatotrophic hormone secretion. *In* Greep, R. O., and Astwood, E. B. (eds.): Handbook of Physiology. Sec. 7, Vol. 4, Part 2. Baltimore, The Williams & Wilkins Company, 1974, p. 405.

Reiter, R. J.: Comparative physiology: pineal gland. *Ann. Rev. Physiol.*, 35:305, 1973.

Reiter, R. J.: Pineal-anterior pituitary gland relationships. *In* Guyton, A. C. (ed.): MTP International Review of Science: Physiology. Vol. 5. Baltimore, University Park Press, 1974, p. 277.

Root, A. W.: Human Pituitary Growth Hormone. Springfield, Ill., Charles C Thomas, Publisher, 1972.

Saffran, M.: Chemistry of hypothalamic hypophysiotropic factors. *In* Greep. R. O., and Astwood, E. B. (eds.): Handbook of Physiology. Sec. 7, Vol. 4, Part 2. Baltimore, The Williams & Wilkins Company, 1974, p. 563.

Share, L., and Grosvenor, C. E.: The neurohypophysis. *In* Guyton, A. C. (ed.): MTP International Review of Science: Physiology. Vol. 5. Baltimore, University Park Press, 1974, p. 1.

Strada, S. J., and Robison, G. A.: Cyclic AMP as a mediator of hormonal effects. *In* Guyton, A. C. (ed.): MTP International Review of Science: Physiology. Vol. 5. Baltimore, University Park Press, 1974, p. 309.

Tager, H. S., and Steiner, D. F.: Peptide hormones. *Ann. Rev. Biochem.*, 43:509, 1974.

Tixier-Vidal, A., and Farquhar, M. G.: The Anterior Pituitary. New York, Academic Press, Inc., 1975.

Turner, M. R.: Dietary effects on the secretion and actions of growth hormone. *Proc. Nutr. Soc.*, 31:205, 1972.

Wilhelmi, A. E.: Chemistry of growth hormone. *In* Greep, R. O., and Astwood, E. B. (eds.): Handbook of Physiology. Sec. 7, Vol. 4, Part 2. Baltimore, The Williams & Wilkins Company, 1974, p. 59.

Williams, R. H.: Textbook of Endocrinology, 5th Ed. Philadelphia, W. B. Saunders Company, 1975.

# 76

# The Thyroid Hormones

The thyroid gland, which is located immediately below the larynx on either side of and anterior to the trachea, secretes *thyroxine, triiodothyronine,* and much smaller quantities of several other closely related iodinated hormones that have a profound effect on the metabolic rate of the body. It also secretes *calcitonin,* a hormone that is important for calcium metabolism and which will be considered in detail in Chapter 79. Complete lack of thyroid secretion usually causes the basal metabolic rate to fall to −30 to −40, and extreme excesses of thyroid secretion can cause the basal metabolic rate to rise as high as +60 to +100. Thyroid secretion is controlled primarily by thyroid-stimulating hormone secreted by the anterior pituitary gland.

The purpose of this chapter is to discuss the formation and secretion of the thyroid hormones, their functions in the metabolic schema of the body, and regulation of their secretion.

## FORMATION AND SECRETION OF THE THYROID HORMONES

The most abundant of the hormones secreted by the thyroid glands is *thyroxine.* However, large amounts of *triiodothyronine* are also secreted. The functions of these two hormones are qualitatively the same, but they differ in rapidity and intensity of action. Triiodothyronine is about four times as potent as thyroxine, but it is present in the blood in much smaller quantities and persists for a much shorter time than does thyroxine.

**Physiologic Anatomy of the Thyroid Gland.** The thyroid gland is composed, as shown in Figure 76–1, of large numbers of closed *follicles* (150 to 300 microns in diameter) filled with a secretory

substance called *colloid* and lined with *cuboidal epithelioid cells* that secrete into the interior of the follicles. The major constituent of colloid is the large protein *thyroglobulin,* which contains the thyroid hormones. Once the secretion has entered the follicles, it must be absorbed back through the follicular epithelium into the blood before it can function in the body. The thyroid gland, having a blood flow about five times the weight of the gland each minute, has a blood supply as rich as that of any other area of the body with the probable exception of the adrenal cortex.

## REQUIREMENTS OF IODINE FOR FORMATION OF THYROXINE

To form normal quantities of thyroxine, approximately 50 mg. of ingested iodine are required *each year,* or approximately *1 mg. per week.* To prevent iodine deficiency, common table salt is iodized with one part sodium iodide to every 100,000 parts sodium chloride.

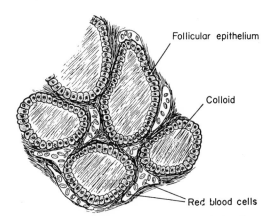

**Figure 76–1.** Microscopic appearance of the thyroid gland, showing the secretion of thyroglobulin into the follicles.

**Fate of Ingested Iodides.** Iodides ingested orally are absorbed from the gastrointestinal tract into the blood in approximately the same manner as chlorides, but iodides do not remain in the circulatory system for a prolonged time because the kidneys have a very high plasma clearance for iodide ion—about 35 ml. per minute in comparison with only 1 ml. per minute for chloride ion. Within the first three days two thirds of the ingested iodides are normally lost into the urine, and almost all of the remaining one third is selectively removed from the circulating blood by the cells of the thyroid gland and used for synthesis of thyroid hormones, which are stored in the form of *thyroglobulin* in the follicles and later secreted into the blood principally in the form of thyroxine, as is discussed below.

## THE IODIDE PUMP (IODIDE TRAPPING)

The first stage in the formation of thyroid hormones, as shown in Figure 76–2, is the transfer of iodides from the extracellular fluid into the thyroid glandular cells and thence into the follicle. The cell membranes have a specific ability to transport iodides actively to the interior of the follicle; this is called the *iodide pump, or iodide trapping.* In a normal gland, the iodide pump can concentrate the iodide to about 25 times its concentration in the blood. However, when the thyroid gland becomes maximally active, the concentration can rise to as high as 350 times that in the blood.

## THYROGLOBULIN AND CHEMISTRY OF THYROXINE AND TRIIODOTHYRONINE FORMATION

**Formation and Secretion of Thyroglobulin by the Thyroid Cells.** The thyroid cells are typical protein-secreting glandular cells. They synthesize and secrete into the follicles a large glycoprotein molecule called *thyroglobulin* with a molecular weight of 660,000.

Each molecule of thyroglobulin contains 25 tyrosine amino acids, and these are the major substrates that combine with iodine to form the thyroid hormones. These hormones form *within* the thyroglobulin molecule. That is, the tyrosine amino acid residues remain a part of the thyroglobulin molecule during the entire process of synthesis of the thyroid hormones.

In addition to secreting the thyroglobulin the glandular cells also provide the iodine, the enzymes, and other substances necessary for thyroid hormone synthesis. Only a minute portion of this synthesis occurs before the thyroglobulin is secreted into the follicles. However, all of these substances are secreted along with the thyroglobulin into the follicles where the mixture of iodine, enzymes, and thyroglobulin provides the medium in which the greater bulk of the thyroid hormones are generated during the ensuing days.

**Oxidation of the Iodides.** The first stage in the formation of the thyroid hormones is believed to be oxidative conversion of iodides either to *elemental iodine* or to *some other oxidized form* of iodine that is then capable of combining with the amino acid *tyrosine* to initiate formation of the thyroid hormones. The reason for believing this is that iodination of tyrosine can be effected readily even in vitro by elemental iodine but not by iodides. Furthermore, large amounts of the enzyme *peroxidase* as well as hydrogen peroxide are present in the thyroid glandular cells, thus providing a potent system capable of oxidizing iodides. When the peroxidase is absent from the cells, the rate of formation of thyroid hormones is greatly decreased.

**Iodination of Tyrosine and Formation of the Thyroid Hormones.** Figure 76–3 illustrates the successive stages of iodination of tyrosine and the final formation of the two important thyroid hormones, thyroxine and triiodothyronine. Tyrosine is first iodized to *monoiodotyrosine* and then to *diiodotyrosine*. Two molecules of diiodotyrosine are then *coupled,* with the loss of the amino acid alanine, to form one molecule of *thyroxine.* Or one molecule of monoiodotyrosine couples with one molecule of diiodotyrosine to form *triiodothyronine*.

**Storage of Thyroglobulin.** After synthesis of the thyroid hormones has run its course, each

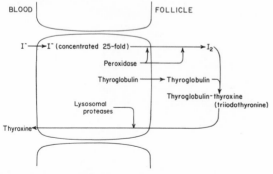

**Figure 76–2.** Mechanisms of iodine transport, thyroxine formation, and thyroxine release into the blood. (Triiodothyronine formation and release parallels that of thyroxine.)

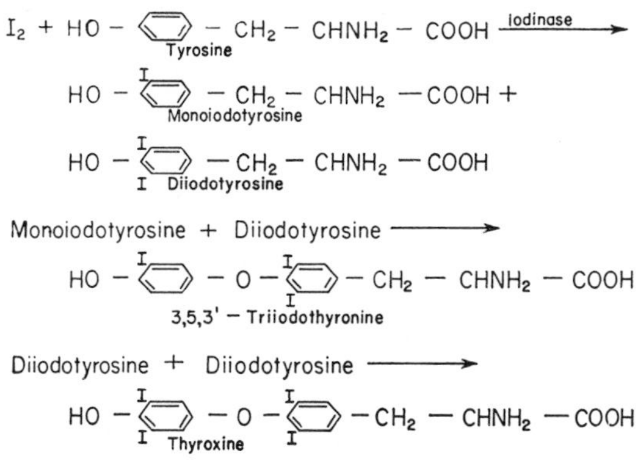

**Figure 76–3.** Chemistry of thyroxine and triiodothyronine formation.

thyroglobulin molecule contains from two to three thyroxine molecules, and there is an average of one triiodothyronine molecule for every three to four thyroglobulin molecules—about nine molecules of thyroxine for every one molecule of triiodothyronine. In this form the thyroid hormones are often stored in the follicles for several months. In fact, the total amount stored is sufficient to supply the body with its normal requirements of thyroid hormones for over three months. Therefore, even when synthesis of thyroid hormone ceases entirely, the effects of deficiency might not be observed for many months.

## RELEASE OF THYROXINE AND TRIIODOTHYRONINE FROM THYROGLOBULIN

Thyroglobulin itself is not released into the circulating blood; instead, the thyroxine and triiodothyronine are first cleaved from the thyroglobulin molecule, and then these free hormones are released. This process occurs as follows: The apical surface of the thyroid cells normally sends out pseudopodlike extensions that close around small portions of the colloid to form pinocytic vesicles. Then lysosomes immediately fuse with these vesicles to form digestive vesicles containing the digestive enzymes from the lysosomes mixed with the colloid. The *proteinases* among these enzymes digest the thyroglobulin molecules and release the thyroxine and triiodothyronine, which then diffuse through the base of the thyroid cell, through the basement membrane, and finally into the surrounding capillaries. Thus, the thyroid hormones are released into the blood.

Approximately two thirds to three fourths of

the iodinated tyrosine in the thyroglobulin never becomes thyroid hormones but instead remains monoiodotyrosine or diiodotyrosine. During the digestion of the thyroglobulin molecule to cause release of thyroxine and triiodothyronine, these iodinated tyrosines also are freed from the thyroid cells. However, they are not secreted into the blood. Instead, their iodine is cleaved from them by an *iodase enzyme* that makes this iodine available for recycling to the new thyroglobulin for formation of thyroid hormone. In the congenital absence of this iodase enzyme, persons frequently become iodine-deficient because of failure of this recycling process.

**Daily Rate of Secretion of Thyroxine and Triiodothyronine.** About 90 per cent of the thyroid hormone released from the thyroid gland is thyroxine, and about 10 per cent is triiodothyronine. However, during the ensuing few days while these hormones circulate in the blood, small portions of the thyroxine are slowly deiodinated to form additional triiodothyronine. Therefore, the quantities of the two hormones finally delivered each day to the tissues is approximately 90 micrograms of thyroxine per day and 40 micrograms of triiodothyronine per day.

Once the two hormones finally enter the peripheral tissue cells, the effect of triiodothyronine to stimulate metabolism and cause other intracellular effects is about four times as potent as that of thyroxine. On the other hand, the duration of action of thyroxine is about four times as long as the duration of action of triiodothyronine. Therefore, the integrated effect of each of the hormones over the period of its action per unit mass of hormone is probably about equal. This means, then, that roughly two thirds of the total hormone effect in the tissues is supplied by thyroxine and the remainder is supplied by triiodothyronine.

## TRANSPORT OF THYROXINE AND TRIIODOTHYRONINE TO THE TISSUES

**Binding of Thyroxine and Triiodothyronine with the Plasma Proteins.** On entering the blood, all but minute portions of the thyroxine and triiodothyronine combine immediately with several of the plasma proteins. They combine approximately as follows: two thirds with *thyroxine-binding globulin,* which is a glycoprotein; about one-quarter with *thyroxine-binding prealbumin*; and about one tenth with *albumin*. The quantity of thyroxine-binding globulin in the blood is only 1 to 1.5 milligrams per 100 milliliters of plasma, but its affinity for the thyroid hormones is so great that it still binds most of the hormones. Its affinity (and that of the other plasma proteins) is about 10 times as great for thyroxine as for triiodothyronine. Only about 0.1 per cent of the thyroxine is normally present in the plasma in the free form, whereas about 1 per cent of the triiodothyronine is present in the free form—10 times as much.

**Release of Thyroxine and Triiodothyronine to the Tissue Cells.** Because of the very high affinity of the plasma-binding proteins for the thyroid hormones, these substances—in particular, thyroxine—are released to the tissue cells only very slowly. Half of the thyroxine in the blood is released to the tissue cells approximately every seven days, whereas half of the triiodothyronine—because of its lower affinity—is released to the cells in approximately two days.

On entering the cells, both of these hormones again bind with intracellular proteins, the thyroxine once again binding more strongly than the triiodothyronine. Therefore, they are again stored, but this time in the functional cells themselves, and they are used slowly over a period of days or weeks.

**Latency and Duration of Action of the Thyroid Hormones.** After the injection of a large quantity of thyroxine into a human being, essentially no effect on the metabolic rate can be discerned for two to three days, thereby illustrating that there is a *long latent* period before thyroxine activity begins. Once activity does begin, it increases progressively and reaches a maximum in 10 to 12 days, as shown in Figure 76–4. Thereafter, it decreases with a half-life of about 15 days. Some of the activity still persists as long as 6 weeks to 2 months later.

The actions of triiodothyronine occur about four times as rapidly as those of thyroxine, with the latent period as short as 6 to 12 hours and maximum cellular activity occurring within 2 to 3 days.

A large share of the latency and prolonged period of action of these hormones is caused by their binding with proteins both in the plasma and in the tissue cells, followed by their slow release. However, we shall see in subsequent discussions that part of the latent period also results from the manner in which these hormones perform their functions in the cells themselves.

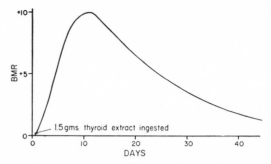

**Figure 76–4.** Prolonged effect on the basal metabolic rate of administering a large single dose of thyroid hormone.

# FUNCTIONS OF THE THYROID HORMONES IN THE TISSUES

## *GENERAL INCREASE IN METABOLIC RATE*

The principal effect of the thyroid hormones is to increase the metabolic activities of most tissues of the body (with a few notable exceptions, such as the brain, retina, spleen, testes, and lungs). The basal metabolic rate can increase to as much as 60 to 100 per cent above normal when large quantities of the hormones are secreted. The rate of utilization of foods for energy is greatly accelerated. The rate of protein synthesis is at times increased, while at the same time the rate of protein catabolism is also increased. The growth rate of young persons is greatly accelerated. The mental processes are excited, and the activity of many other endocrine glands is often increased. Yet despite the fact that we know all these many changes in metabolism under the influence of the thyroid hormones, the basic mechanism (or mechanisms) by which they act is almost completely unknown. However, some of the possible mechanisms of action of the thyroid hormones are described in the following sections.

**Effect of Thyroid Hormones to Cause Increased Protein Synthesis.** When either thyroxine or triiodothyronine is given to an animal, protein synthesis increases in almost all tissues of the body. The first stage of the increased protein synthesis begins almost immediately and results from stimulation of the translation process—that is, an increase in the formation of proteins by the ribosomes. The second stage occurs hours to days later and is caused by an almost generalized increase in RNA synthesis by the genes, the process of transcription, which leads to a generalized increase in synthesis of almost all types of proteins within the cells.

**Effect of Thyroid Hormones on the Cellular Enzyme Systems.** Within a week or so following administration of the thyroid hormones, at least 100 and probably many more intracellular enzymes are increased in quantity. This may result from the direct effect of the thyroid hormones to cause generalized increase in protein synthesis. As an example, one enzyme, $\alpha$-glycerophosphate dehydrogenase, can be increased to an activity six times its normal level. Since this enzyme is particularly important in the degradation of carbohydrates, its increase could help to explain the rapid utilization of carbohydrates under the influence of thyroxine. Also, the oxidative enzymes and the elements of the electron transport system, both of which are normally found in mitochondria, are greatly increased.

**Effect of Thyroid Hormones on Mitochondria.** When thyroxine or triiodothyronine is given to an animal, the mitochondria in most cells of the body increase in size and also in number. Furthermore, the total membrane surface of the mitochondria increases almost directly in proportion to the increased metabolic rate of the whole animal. Therefore, it is an obvious deduction that the principal function of thyroxine might be simply to increase the number and activity of mitochondria, and these in turn increase the rate of formation of ATP to energize cellular function. Unfortunately, though, the increase in number and activity of mitochondria could as well be the *result* of increased activity of the cells as be the cause of the increase.

When *extremely* high concentrations of thyroid hormones are administered, the mitochondria swell inordinately, and there is uncoupling of the oxidative phosphorylation process. However, under natural conditions, the concentration of thyroid hormones seems never to become high enough to cause this effect, even in human beings who have thyrotoxicosis.

**Effect of Thyroid Hormone to Increase Cellular Cyclic AMP.** Thyroid hormones increase cyclic AMP in some—perhaps all—cells of the body, but especially in muscle cells. Therefore, some physiologists believe that the primary action of the thyroid hormones might be simply to activate adenylcyclase, which in turn causes the formation of cyclic AMP. Then the cyclic AMP presumably acts as a *second messenger,* as was explained in the previous

chapter, to initiate all or at least many of the intracellular functions of thyroid hormones.

**Summary.** It is clear that we know many specific events that occur in the cells throughout the body under the influence of the thyroid hormones. But it is equally clear that the basic mechanisms leading to all of these effects are still almost completely unknown.

## EFFECTS OF THYROID HORMONE ON METABOLISM OF SPECIFIC DIETARY SUBSTANCES

**Effect on Protein Metabolism and on Growth.** The rates of both protein anabolism and catabolism are increased by thyroid hormone, which is the expected effect because of the increased enzymatic activities in the cells. Thus, thyroid hormone is necessary for development of structural and other proteins of the body cells and therefore is necessary for growth in the young person.

On the other hand, thyroid hormone causes rapid oxidation of carbohydrates and fats, and, when these "protein sparers" are depleted, proteins must be utilized for energy. As a result, a negative nitrogen balance then ensues. Thyroid hormone also has a specific effect on the tissues to "mobilize" protein and thereby release amino acids into the extracellular fluids. In addition to making these amino acids available for energy purposes, this effect also increases the rate of gluconeogenesis.

**Effect on Bone Growth and Calcium Metabolism.** Thyroid hormone increases the growth of bone in the same way that it increases growth of all other tissues of the body. This probably results from the effect of thyroid hormone to increase protein formation. On the other hand, thyroid hormone also causes rapid closure of the epiphyses. Therefore, a young person under the influence of thyroid hormone grows rapidly at first but then stops growing at a much younger age than his normal counterpart. Consequently, his final height may actually be less than normal.

Thyroid hormone also increases osteoclastic activity in the bones. When the concentration of the hormone is marked, the osteoclastic activity causes the bones to become porous, and greater than normal quantities of calcium and phosphate are emptied into the urine and excreted into the gastrointestinal tract. This same effect occurs when the rate of metabolism is increased as a result of fever, which indicates that the loss of calcium and phosphate from the bones following thyroid hormone administration could result simply from the increased rate of metabolism.

**Effect on Carbohydrate Metabolism.** Thyroid hormone stimulates almost all aspects of carbohydrate metabolism, including rapid uptake of glucose by the cells, enhanced glycolysis, enhanced gluconeogenesis, increased rate of absorption from the gastrointestinal tract, and even increased insulin secretion with its resultant secondary effects on car-bohydrate metabolism. All of these effects probably result from the overall increase in enzymes caused by thyroid hormone.

**Effect on Fat Metabolism.** Essentially all aspects of fat metabolism are also enhanced under the influence of thyroid hormone. However, since fats are the major source of long-term energy supplies, the fat stores of the body are depleted to a greater extent than are most of the other tissue elements; in particular, lipids are mobilized from the fat tissue, which increases the free fatty acid concentration in the plasma, and thyroid hormone also greatly accel-erates the oxidation of free fatty acids by the cells.

*Effect on Blood and Liver Fats.* Increased thyroid hormone *decreases* the quantity of cholesterol, phospholipids, and triglycerides in the blood, even though it *increases* the free fatty acids. On the other hand, decreased thyroid secretion greatly increases the concentrations of cholesterol, phospholipids, and triglycerides and almost always causes excessive deposition of fat in the liver. The large increase in circulating blood lipids in prolonged hypothyroidism is always associated with severe arteriosclerosis, which was discussed in Chapter 68.

The cause of the reduced blood cholesterol in-duced by thyroid hormone is enhancement of its excretion into the gut and its conversion to bile acids by the liver.

**Effect on Vitamin Metabolism.** Because thyroid hormone increases the quantities of many of the different enzymes and because vitamins are essential parts of some of the enzymes or coenzymes, thyroid hormone causes increased need for vitamins. There-fore, a relative vitamin deficiency can occur when excess thyroid hormone is secreted, unless at the same time increased quantities of vitamins are avail-able.

## PHYSIOLOGIC EFFECTS OF THYROID HORMONE ON DIFFERENT BODILY MECHANISMS

**Effect on Basal Metabolic Rate.** Because thyroid hormone increases metabolism in most cells of the body (with the exception of the brain, retina, spleen, testes, and lungs), excessive quantities of the hor-mone can occasionally increase the basal metabolic rate to as much as 100 per cent above normal. How-ever, in most patients with severe hyperthyroidism the basal metabolic rate ranges between 40 and 60 per cent above normal. On the other hand, when no thyroid hormone is produced, the basal metabolic rate falls almost to half normal; that is, the basal metabolic rate becomes $-30$ to $-45$, as discussed in Chapter 71. Figure 76–5 shows the approximate rela-tionship between the daily supply of thyroid hor-mones and the basal metabolic rate. Extreme amounts of the hormones are required to cause very high basal metabolic rates.

**Effect on Body Weight.** Greatly increased

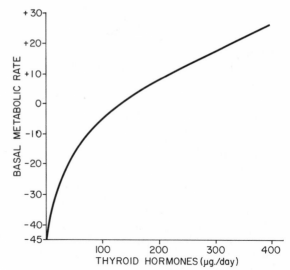

**Figure 76–5.** Relationship of thyroid hormone daily rate of secretion to the basal metabolic rate.

thyroid hormone production almost always de-creases the body weight, and greatly decreased production almost always increases the body weight; but these effects do not always occur, because thyroid hormone increases the appetite, and this may overbalance the change in the metabolic rate.

**Effect on Growth.** Because protein synthesis cannot occur normally in the absence of thyroid hormone, the growth effect of growth hormone from the pituitary gland is not significant without the con-current presence of thyroid hormone in the body fluids. There are two conditions in which this effect on growth is especially evident: First, in growing children who are hypothyroid, the rate of growth is greatly retarded. Second, in growing children who are hyperthyroid, excessive skeletal growth often occurs, causing the child to become considerably taller than otherwise. However, the epiphyses close at an early age so that the eventual height of the adult may be shortened.

**Effect on the Cardiovascular System.** Because increased metabolism induced by thyroid hormone increases the demand of the tissues for nutrient sub-stances, the following effects occur in the cardiovas-cular system in hyperthyroidism and the opposite ef-fects occur in hypothyroidism:

*Blood Flow and Cardiac Output.* Increased metabolism in the tissues causes more rapid utiliza-tion of oxygen than normally and causes greater than normal quantities of metabolic end-products to be released from the tissues. These effects cause vas-odilatation in most of the body tissues, thus increas-ing blood flow in almost all areas of the body. Espe-cially does the rate of blood flow in the skin increase because of the increased necessity for heat elimina-tion.

As a consequence of the increased blood flow to the constituent parts of the body, the cardiac output also increases, sometimes rising to 50 per cent or more above normal when excessive thyroid hormone is present.

*Heart Rate.* The heart rate increases considerably more under the influence of thyroid hormone than would be expected simply because of the increased cardiac output. Therefore, thyroid hormone probably has a direct effect on the excitability of the heart, which in turn increases the heart rate. This effect is of particular importance because the heart rate is one of the most sensitive indices that the clinician has for determining whether a patient has excessive or diminished thyroid hormone production.

*Strength of Heartbeat.* The increased enzymatic activity caused by increased thyroid hormone production apparently increases the strength of the heart when only a slight excess of thyroid hormone is secreted. This is analogous to the increase in strength of heart beat that occurs in mild fevers and during exercise. However, when thyroid hormone is increased markedly, the heart muscle strength becomes depressed because of excessive protein catabolism. Indeed, some severely thyrotoxic patients die of cardiac decompensation secondary to myocardial failure and increased cardiac load imposed by the increased output.

*Blood Volume.* Thyroid hormone causes the blood volume to increase slightly. This effect probably results at least partly from the vasodilatation which allows increased quantities of blood to collect in the circulatory system.

*Arterial Pressure.* The increased cardiac output resulting from thyroid hormone tends to increase the arterial pressure. On the other hand, dilatation of the peripheral blood vessels due to the local effects of thyroid hormone and to excessive body heat tends to decrease the pressure. Therefore, the mean arterial pressure usually is unchanged. However, because of the increased rate of run-off of blood through the peripheral vessels, the pulse pressure is increased, with the systolic pressure elevated 10 to 20 mm. Hg. and the diastolic pressure correspondingly reduced.

**Effect on Respiration.** The increased rate of metabolism caused by thyroid hormone increases the utilization of oxygen and the formation of carbon dioxide; these effects activate all the mechanisms that increase the rate and depth of respiration.

**Effect on the Gastrointestinal Tract.** In addition to increased rate of absorption of foodstuffs, which has been discussed, thyroid hormone increases both the rate of secretion of the digestive juices and the motility of the gastrointestinal tract. Often, diarrhea results. Also, associated with this increased secretion and motility is increased appetite, so that the food intake usually increases. Lack of thyroid hormone causes constipation.

**Effect on the Central Nervous System.** In general, thyroid hormone increases the rapidity of cerebration, while, on the other hand, lack of thyroid hormone decreases this function. The hyperthyroid individual is likely to develop extreme nervousness and is likely to have many psychoneurotic tendencies, such as anxiety complexes, extreme worry, or paranoias.

In general, the reaction time for various functions requiring integration by the brain is greatly reduced by administration of thyroid hormone, but, on the other hand, the rate of conduction in nerves remains normal in the presence of excess thyroid hormone. Thus, thyroid hormone seems to increase synaptic activity but does not influence peripheral nerve activity.

**Effect on the Function of the Muscles.** Slight increase in thyroid hormone usually makes the muscles react with vigor, but when the quantity of hormone becomes extreme, the muscles become weakened because of excess protein catabolism. On the other hand, lack of thyroid hormone causes the muscles to become extremely sluggish; they relax slowly after a contraction.

*Muscle Tremor.* One of the most characteristic signs of hyperthyroidism is a fine muscle tremor. This is not the coarse tremor that occurs in Parkinson's disease or in shivering, for it occurs at the rapid frequency of 10 to 15 times per second. The tremor can be observed easily by placing a sheet of paper on the extended fingers and noting the degree of vibration of the paper. The cause of this tremor is not definitely known, but it is probably due to increased activity in the areas of the cord that control muscle tone. The tremor is an excellent means for assessing the degree of thyroid hormone effect on the central nervous system.

**Effect on Sleep.** Because of the exhausting effect of thyroid hormone on the musculature and on the central nervous system, the hyperthyroid subject often has a feeling of constant tiredness; but because of the excitable effects of thyroid hormone on the synapses, it is difficult for him to sleep. On the other hand, extreme somnolence is characteristic of hypothyroidism.

# REGULATION OF THYROID HORMONE SECRETION

To maintain a normal basal metabolic rate, precisely the right amount of thyroid hormone must be secreted all the time, and, to provide this, a specific feedback mechanism operates through the hypothalamus and anterior pituitary gland to control the rate of thyroid secretion in proportion to the metabolic needs of the body. This system can be explained as follows:

**Effects of Thyroid-Stimulating Hormone (Thyrotropin) on Thyroid Secretion.** Thyroid-stimulating hormone (TSH), also known as *thyrotropin,* is an anterior pituitary hormone, a glycoprotein with a molecular weight of about 25,000, that was discussed in Chapter 75; it increases the secretion of thyroxine and triiodothyronine by the thyroid gland. Its specific effects on the thyroid gland are: (1) increased proteolysis of the thyroglobulin in the follicles, with resultant release of thyroid hormone into the circulating blood and diminishment of the follicular substance itself; (2) increased activity of the iodide pump, which increases the rate of "iodide trapping" in the glandular cells, increasing the ratio of intracellular to extracellular iodide concentration to as great as 350:1 during maximal stimulation; (3) increased iodination of tyrosine and increased coupling to form the thyroid hormones; (4) increased size and increased secretory activity of the thyroid cells; and (5) increased number of thyroid cells, plus a change from cuboidal to columnar cells and much infolding of the thyroid epithelium into the follicles. In summary, thyroid-stimulating hormone *increases all the known activities of the thyroid glandular cells*.

The most important early effect following administration of thyroid-stimulating hormone is proteolysis of the thyroglobulin, which causes release of thyroxine and triidothyronine into the blood within 30 minutes.

**Role of Cyclic AMP in the Stimulatory Effect of TSH.** In an attempt to explain the many and varied effects of thyroid-stimulating hormone on the thyroid cells, a single primary action of this hormone has been sought for years. Recent experiments have shown that the hormone almost certainly does have such a primary effect, which is to activate *adenylcyclase* in the membranes of the thyroid cells. This in turn causes formation in the cells of *cyclic AMP,* which then acts as a *second messenger* to activate essentially all systems of the thyroid cells. The result is both an immediate increase in secretion of thyroid hormones and prolonged growth of the thyroid glandular tissue itself. This method for control of thyroid cell activity is similar to the function of cyclic AMP in many other target tissues of the body.

**Hypothalamic Regulation of TSH Secretion by the Anterior Pituitary—Thyrotropin-Releasing Factor.** Electrical stimulation of several areas of the hypothalamus, but most particularly of the paraventricular area, increases the anterior pituitary secretion of TSH and correspondingly increases the activity of the thyroid gland. This control of anterior pituitary secretion is exerted by a hypothalamic hormone, *thyrotropin-releasing factor* (TRF), which is secreted by nerve endings in the me-

dian eminence of the hypothalamus and then transported from there to the anterior pituitary in the hypothalamic-hypophyseal portal blood, as was explained in Chapter 75. TRF has been obtained in pure form, and it has proved to be a very simple substance, a tripeptide amide—*pyroglutamyl-histidyl-proline-amide*. The TRF has a direct effect on the anterior pituitary gland cells to increase their output of thyroid-stimulating hormone. When the portal system from the hypothalamus to the anterior pituitary gland is completely blocked, the rate of secretion of TSH by the anterior pituitary is greatly decreased but not reduced to zero.

**Effects of Cold and Other Neurogenic Stimuli on TSH Secretion.** One of the best known stimuli for increasing the rate of TSH secretion by the anterior pituitary is exposure of an animal to cold. Exposure of rats for several weeks increases the output of thyroid hormones sometimes more than 100 per cent and can increase the basal metabolic rate as much as 50 per cent. Indeed, even human beings moving to arctic regions have been known to develop basal metabolic rates 15 to 20 per cent above normal; however, the propensity of the human being to protect himself from cold usually prevents a measurable effect.

Various emotional reactions can also affect the output of TSH and can, therefore, indirectly affect the secretion of thyroid hormone. Extreme states of excitement and anxiety—conditions that greatly stimulate the sympathetic nervous system—cause acute decrease in secretion of TSH, perhaps because these states increase the metabolic rate and the body heat.

Neither these emotional effects nor the effect of cold is observed when the hypophyseal stalk is cut, illustrating that both of these effects are mediated by way of the hypothalamus.

**Inverse Effect of Thyroid Hormone on Anterior Pituitary Secretion of TSH—Feedback Regulation of Thyroid Secretion.** Increased thyroid hormone in the body fluids decreases the secretion of TSH by the anterior pituitary. When the rate of thyroid hormone secretion rises to about 1.75 times normal, the rate of TSH secretion falls essentially to zero. Most of this depressant effect occurs even when the anterior pituitary has been completely separated from the hypothalamus, but the effect is somewhat greater if the hypothalamus and hypothalamic-hypophyseal portal system are intact. Therefore, it is probable that increased thyroid hormone inhibits anterior pituitary secretion of

TSH in two different ways: (1) by a direct effect on the anterior pituitary itself, and (2) by an indirect effect acting through the hypothalamus.

In most hormonal regulatory systems it is *not* the rate of hormone secretion itself that needs to be regulated at a constant level, but instead it is some desired *effect of the hormone* that needs to be regulated. Since thyroid hormone controls the overall metabolic activity of the body, it is tempting to believe that the factor controlled at a nearly constant rate by the TSH-thyroid hormone control system is some aspect of cell metabolism, perhaps the cellular metabolic rate itself. Thus, if the cellular metabolic rate should become too low, the TSH-thyroid system would become activated until enough thyroid hormone became available to increase the metabolic activities back to normal. Conversely, if these metabolic activities should become too great, the feedback system would become inactivated until the thyroid hormone level fell low enough to allow normal metabolic activities once again.

One of the ways by which increased metabolic rate could exert its feedback control over the TSH-thyroid system would be by causing increased heat formation in the cells of the temperature regulating center of the hypothalamus; the increased heat in this center would decrease the output of thyrotropin-releasing factor, which would decrease the output of TSH and thyroid hormone, thus returning the metabolic activities of the cells back to normal. These possible events are summarized in Figure 76–6.

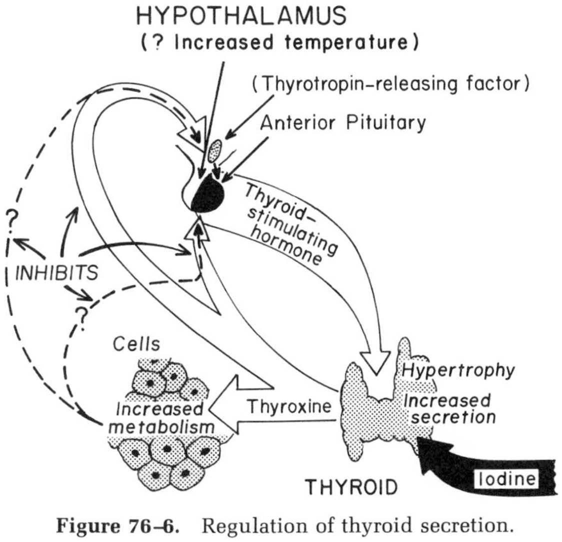

**HYPOTHALAMUS**
(? Increased temperature)

(Thyrotropin-releasing factor)

Anterior Pituitary

Thyroid-stimulating hormone

INHIBITS

Cells

Increased metabolism

Thyroxine

Hypertrophy
Increased secretion

THYROID

Iodine

**Figure 76–6.** Regulation of thyroid secretion.

## ANTITHYROID SUBSTANCES

Drugs that suppress thyroid secretion are called antithyroid substances. The three best known of these are: thiocyanate, propylthiouracil, and high concentrations of inorganic iodides. The mechanism by which each of these blocks thyroid secretion is different from the others, and they can be explained as follows:

**Decreased Iodide Trapping Caused by Thiocyanate Ions.** Administration of thiocyanates (or perchlorates and many other similar compounds) decreases the rate at which iodide is pumped into the thyroid glandular cells, and, therefore, reduces the availability of iodides to the intracellular processes for the formation of thyroxine and triiodothyronine.

Lack of formation of thyroid hormone by the gland under the influence of thiocyanates causes the thyroid gland to enlarge—that is, to become a "goiter." The mechanism of this is the following: Lack of thyroid hormone decreases the feedback inhibition of the anterior pituitary, which allows increased secretion of TSH by the pituitary. This then stimulates the glandular cells in the thyroid, making them secrete more and more thyroglobulin into the follicles, even though this does not contain significant quantities of thyroxine and triiodothyronine.

**Depression of Thyroid Hormone Formation by Propylthiouracil.** Propylthiouracil (and other similar compounds such as methimazole and carbimazole) prevents formation of thyroid hormone from iodides and tyrosine. The mechanism of this is partly to block iodination of tyrosine but principally to block the coupling of two iodinated tyrosines to form thyroxine or triiodothyronine.

Propylthiouracil does not prevent formation of thyroglobulin, but the absence of thyroxine and triiodothyronine in the thyroglobulin leads to tremendous feedback enhancement of TSH secretion by the anterior pituitary gland. Therefore, in the same manner that thiocyanates cause the thyroid gland to enlarge, so also does propylthiouracil lead to enhanced growth of the glandular tissue, thus forming a goiter.

**Decrease in Thyroid Activity Caused by Iodides.** When iodides are present in the blood *in high concentration* (100 times the normal plasma level), most activities of the thyroid gland are decreased, but often they are decreased only for a few weeks. The rate of iodide trapping is reduced, the rate of thyroid hormone formation is decreased, the secretory activity of the thyroid cells is decreased, and the rate of thyroid hormone release from the thyroglobulin is decreased. Since these are almost exactly opposite to the effects of TSH on the thyroid gland, it has been suggested that high concentrations of iodides in the blood directly inhibit the thyroid-stimulating effect of TSH.

Because iodides in high concentrations decrease all phases of thyroid activity, they decrease the size of the thyroid gland and especially decrease its blood supply, in contradistinction to the opposite effects caused by most of the other antithyroid agents. For this reason, iodides are frequently administered to patients for two or three weeks prior to surgical removal of the thyroid gland to decrease the necessary amount of surgery.

# INTERRELATIONSHIPS OF THE THYROID GLAND AND OTHER ENDOCRINE GLANDS

Increased thyroid hormone increases the rates of secretion of most other endocrine glands, but it also increases the need of the tissues for the hormones. For instance, increased thyroxine secretion increases the rate of glucose metabolism everywhere in the body and therefore causes a corresponding need for increased insulin secretion by the pancreas. Also, thyroid hormone increases many metabolic activities related to bone formation and, as a consequence, increases the need for parathyroid hormone. However, in addition to these general effects, thyroid hormone has relatively specific effects on the adrenal cortex and on the gonads.

**Effect of Thyroid Hormone on the Secretion of Adrenocortical Hormones.** Administration of thyroid hormone increases the rate of glucocorticoid secretion by the adrenal cortex in the following manner: Thyroid hormone has the potent effect of causing rapid conjugation and inactivation of glucocorticoids by the liver, which obviously decreases the quantity of circulating glucocorticoids. This in turn causes a feedback enhancement of ACTH formation by the anterior pituitary, as will be discussed in the following chapter; the ACTH in turn enhances the secretion of glucocorticoids by the adrenal cortex, thus compensating for the glucocorticoid deficiency caused by the thyroid hormone.

**Effect of Thyroid Hormone on the Gonads.** For normal sexual function to occur, thyroid secretion needs to be approximately normal—neither too great nor too little. In the male, lack of thyroid hormone is likely to cause complete loss of libido, while, on the other hand, great excesses of the hormone frequently cause impotence. In the female, lack of thyroid hormone often causes *menorrhagia* and *polymenorrhea*, which mean, respectively, excessive and frequent menstrual bleeding. In some women the lack may cause irregular periods and, occasionally, even total *amenorrhea*. A hypothyroid female, like the male, is also likely to have greatly decreased libido. Conversely, in the hyperthyroid female, *oligomenorrhea*, which means greatly reduced bleeding, is usual, and occasionally, amenorrhea results.

The action of thyroid hormone on the gonads cannot be pinpointed to a specific function but probably results from a combination of direct metabolic effects on the gonads and of excitatory and inhibitory effects operating through the anterior pituitary.

# DISEASES OF THE THYROID

## HYPERTHYROIDISM

Most effects of hyperthyroidism are obvious from the preceding discussion of the various physiologic effects of thyroid hormone. However, some specific effects should be mentioned in connection especially with the development, diagnosis, and treatment of hyperthyroidism.

**Causes of Hyperthyroidism (Toxic Goiter, Thyrotoxicosis, Graves' Disease).** In the patient with hyperthyroidism the entire thyroid gland is usually markedly hyperplastic. It is increased to two to three times normal size, with tremendous folding of the follicular cell lining into the follicles so that the number of cells is increased several times as much as the size of the gland is increased. Also, each cell increases its rate of secretion several-fold; radioactive iodine uptake studies indicate that these hyperplastic glands secrete thyroid hormone at a rate as great as 5 to 15 times normal.

These changes in the thyroid gland are similar to those caused by excessive thyroid-stimulating hormone. However, radioimmunoassay studies have shown the plasma TSH concentrations to be less than normal rather than enhanced, and often to be essentially zero. On the other hand, another substance that has an action similar to that of TSH is found in the blood of most patients. This substance, called *long-acting thyroid stimulator* (LATS), has a prolonged stimulating effect on the thyroid gland, lasting as long as 12 hours in contrast to a little over 1 hour for TSH. However, this substance probably does not come from the anterior pituitary gland, though for the present its origin is unknown. Since it is a globulin of the immunoglobulin IgG type, many investigators believe that it is an antibody formed in the immune system of the body against some antigen in the thyroid gland. Supposedly, its reaction with its antigen initiates the TSH-like effects on the thyroid gland. In turn, the high level of thyroid hormone secretion by the thyroid gland suppresses anterior pituitary formation of TSH.

Two other reasons for believing that LATS might be an immunoglobulin are the following: (1) After several months to several years of thyroid gland stimulation, the thyroid gland frequently becomes atrophic, and the person develops hypothyroidism rather than hyperthyroidism, almost as if an immunoglobulin might have destroyed the gland. (2) Many patients who have thyrotoxicosis also have other signs of autoimmune disease—in particular, inflammatory reactions in connective tissue structures and occasionally a distinct autoimmune syndrome such as myasthenia gravis.

Aside from the above variety of hyperthyroidism, the condition occasionally also results from a localized adenoma (a tumor) that develops in the thyroid tissue and secretes large quantities of thyroid hormone. This is different from the more usual type of hyperthyroidism in that it is not associated with any evidences of autoimmune disease. An interesting effect of the adenoma is that as long as it continues to secrete large quantities of thyroid hormone, function in the remainder of the thyroid gland is almost totally inhibited because the thyroid hormone from the adenoma depresses the production of TSH by the pituitary gland.

## Symptoms of Hyperthyroidism

The symptoms of toxic goiter—these are obvious from the preceding discussion of the physiology of the thyroid hormones—are intolerance to heat, increased sweating, mild to extreme weight loss (sometimes as much as 100 pounds), varying degrees of diarrhea, muscular weakness, nervousness or other psychic disorders, extreme fatigue but inability to sleep, and tremor of the hands.

**Exophthalmos and Exophthalmos-Producing Substance.** Most, but not all, persons with hyperthyroidism develop some degree of protrusion of the eyeballs, as illustrated in Figure 76–7. This condition is called *exophthalmos*. A major degree of exophthalmos occurs in about one third of the hyperthyroid patients, and the condition occasionally becomes severe enough that it causes blindness because the eyeball protrusion stretches the optic nerve. Much more often, the eyes are damaged because the eyelids do not close completely when the person blinks or when he is asleep. As a result, the dry epithelial surfaces of the eyes become irritated and often infected, resulting in ulceration of the cornea.

The cause of the protruding eyes is edematous swelling and deposition of large quantities of mucopolysaccharides in the extracellular spaces of the retro-orbital tissues; the factor or factors that initiate these changes is still in serious dispute. In some

**Figure 76–7.** Patient with exophthalmic hyperthyroidism. Note protrusion of the eyes and retraction of the superior eyelids. The basal metabolic rate was +40. (Courtesy of Dr. Leonard Posey.)

lower animals, TSH itself given in very large quantities can cause exophthalmos. Also, TSH can be split chemically to provide a separate substance called *exophthalmos-producing substance* that can cause exophthalmos without stimulating the thyroid gland. Thus, there is considerable evidence to link the exophthalmos of hyperthyroidism to thyroid-stimulating hormone, despite the fact that most patients with hyperthyroidism have considerably diminished secretion of TSH. Therefore, for the present, it has not been possible to solve the riddle of exophthalmos. It could result from pituitary secretion of an abnormal type of TSH during hyperthyroidism, or, as some investigators have suggested, the real exophthalmos-producing substance might be produced somewhere besides the pituitary.

**Diagnostic Tests for Hyperthyroidism. Protein-Bound Iodine.** Essentially all the iodine bound with proteins of the blood is thyroxine or triiodothyronine iodine, and the normal person has between 4 and 7.5 μgm. of *plasma-bound iodine* per 100 ml. of plasma. In hyperthyroidism this level usually increases in proportion to the degree of hyperthyroidism—sometimes to 15 to 20 μgm. per 100 ml. of plasma. The plasma-bound iodine test is an effective means for making a diagnosis in approximately 85 to 90 per cent of patients with hyperthyroidism, but it occasionally is in error, owing to the presence of other iodine compounds in the plasma.

*The Basal Metabolic Rate.* The basal metabolic rate is usually increased to approximately +40 to +60 in severe hyperthyroidism. A mild increase in basal metabolic rate does not necessarily mean hyperthyroidism. Indeed, an increased basal metabolic rate correctly indicates hyperthyroidism in only 60 per cent of the cases; other factors that were discussed in Chapter 71 can also cause changes in the basal metabolic rate even though thyroid hormone production is normal.

*Uptake of Radioactive Iodine.* Another method for measuring the rate of activity of the thyroid gland is to determine its rate of radioactive iodine uptake. A small test dose of radioactive iodine is given intravenously, and the rate of uptake by the thyroid gland is measured by a calibrated radioactive detector placed over the neck. In the normal person, an average of 4 per cent of the circulating radioactive iodine is assimilated by the thyroid gland per hour. In the hyperthyroid person, as much as 20 to 25 per cent is assimilated per hour.

**Physiology of Treatment in Hyperthyroidism.** The most direct treatment for hyperthyroidism is surgical removal of the thyroid gland. In general, however, it is desirable to prepare the patient for surgical removal of the gland prior to the operation. This is done by administering propylthiouracil, sometimes for as long as several months, until the basal metabolic rate of the patient has returned to normal. Then administration of high concentrations of iodides for two weeks immediately prior to operation causes the gland itself to recede in size and its blood supply to diminish. By using these preoperative procedures, the operative mortality is less than 1 in 1000 in the better hospitals, whereas prior to development of these procedures the operative mortality was as great as 1 in 25.

*Treatment of the Hyperplastic Thyroid with Radioactive Iodine.* As much as 80 to 90 per cent of an injected dose of iodide is absorbed by the hyperplastic, toxic thyroid gland within a day after injection. If this injected iodine is radioactive, it can destroy internally the secretory cells of the thyroid gland. Usually 5 millicuries of radioactive iodine is given to the patient, and then his condition is reassessed several weeks later. If he is still hyperthyroid, additional doses are repeated until he reaches a normal thyroid status. These quantities of radioiodine are about 1000 times as great as those used for diagnosis of hyperthyroidism as discussed above.

**Thyroid Storm.** Occasionally, patients with extremely severe thyrotoxicosis develop a state called *thyroid storm,* or *thyroid crisis* in which all the usual symptoms of thyrotoxicosis are excessively accentuated. The effects may be delirium, high fever, abnormal rhythm of the heart, extreme sweating, shock, vomiting, and dehydration. This condition is particularly likely to occur during the first day following surgical removal of a large hyperplastic thyroid gland, presumably because of excessive release of thyroid hormone into the circulatory system during the operative procedure. The extreme overactivity of the body's tissues can be so damaging that without treatment almost all persons entering this state die. However, by cooling the patient rapidly with ice or alcohol sponge baths and also administering large quantities of glucocorticoid hormones, it is now possible to save approximately half the patients. The glucocorticoids seem to be especially important, possibly because of their effect of reducing the breakdown of lysosomes in the damaged tissues and therefore of preventing autodigestion of the cells.

## HYPOTHYROIDISM

The effects of hypothyroidism in general are opposite to those of hyperthyroidism, but here again, a few physiological mechanisms peculiar to hypothyroidism alone are involved.

**Endemic Colloid Goiter.** The term "goiter" means a greatly enlarged thyroid gland. As pointed out in the discussion of iodine metabolism, about 50 mg. of iodine is necessary each year for the formation of adequate quantities of thyroid hormone. In certain areas of the world, notably in the Swiss Alps and in the Great Lakes region of the United States, insufficient iodine is present in the soil for the foodstuffs to contain even this minute quantity of iodine. Therefore, in days prior to iodized table salt, many persons living in these areas developed extremely large thyroid glands called *endemic goiters*.

The mechanism for development of the large endemic goiters is the following: Lack of iodine prevents production of thyroid hormone by the thyroid gland, and, as a result, no hormone is available to inhibit production of TSH by the anterior pituitary; this allows the pituitary to secrete excessively large quantities of TSH. The TSH then causes the thyroid cells to secrete tremendous amounts of thyroglobulin (colloid) into the follicles, and the gland grows larger and larger. But unfortunately, due to lack of iodine, increased thyroxine and triiodothyronine production does not occur. The follicles become tremendous in size, and the thyroid gland may increase to as large as 300 to 500 grams or more.

**Idiopathic Nontoxic Colloid Goiter.** Enlarged thyroid glands almost identical to those of endemic colloid goiter frequently develop even though the affected persons receive sufficient quantities of iodine in their diets. These goitrous glands may secrete normal quantities of thyroid hormones but more frequently the secretion of hormone is depressed, as in endemic colloid goiter.

The exact cause of idiopathic colloid goiter usually is not known, though animals fed large quantities of certain foods are known to develop enlarged thyroid glands. These foods contain so-called goitrogenic substances, the actions of which are similar to those of propylthiouracil, which was discussed previously. Especially are goitrogenic substances found in some varieties of turnips and cabbages. However, the diet only rarely contains enough quantities of goitrogenic substances to cause a large, idiopathic colloid goiter.

Another cause of this type of goiter is altered enzyme systems of the thyroid cells. That is, the cells secrete the colloid substance, but because of altered intracellular enzyme systems, the cells fail to form the thyroid hormones or are unable to release the hormones from the thyroglobulin complex.

Regardless of the cause of the goiter, the depressed thyroid hormone production induces the anterior pituitary to secrete greater than normal quantities of TSH, which then causes the gland to enlarge.

**Characteristics of Hypothyroidism.** Whether hypothyroidism is due to endemic colloid goiter, idiopathic colloid goiter, destruction of the thyroid gland by irradiation, surgical removal of the thyroid gland, or destruction of the thyroid gland by various other diseases, the physiological effects are the same. These include extreme somnolence with sleeping 14 to 16 hours a day, extreme muscular sluggishness, slowed heart rate, decreased cardiac output, decreased blood volume, increased weight, constipation, mental sluggishness, failure of many trophic functions in the body evidenced by depressed growth of hair and scaliness of the skin, development of a frog-like husky voice, and, in severe cases, development of an edematous appearance throughout the body called myxedema.

*Myxedema.* The patient with almost total lack of thyroid function develops "myxedema." Figure 76–8 shows such a patient, illustrating bagginess under the eyes and swelling of the face. In this condition, for reasons not yet explained, greatly increased quantities of mucopolysaccharides, mainly hyaluronic

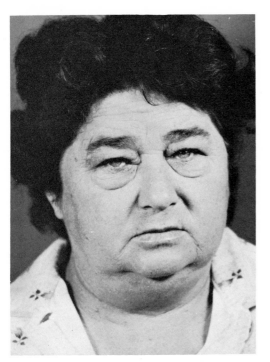

**Figure 76–8.** Patient with myxedema. (Courtesy of Dr. Herbert Langford.)

acid, collect in the interstitial spaces, and this causes the total quantity of interstitial fluid also to increase. The fluid is adsorbed to the mucopolysaccharides, thus greatly increasing the quantity of "ground substance" gel in the tissues. Because of the gel nature of the excess fluid, it is relatively immobile, and the edema is nonpitting in type.

*Arteriosclerosis in Hypothyroidism.*  As pointed out in Chapter 68, lack of thyroid hormone increases the quantity of blood lipids, most importantly cholesterol, and an increase in blood cholesterol is usually associated with atherosclerosis and arteriosclerosis. Therefore, many hypothyroid patients, particularly those with myxedema, develop severe arteriosclerosis, which results in peripheral vascular disease, deafness, and often extreme coronary sclerosis and, consequently, early demise.

**Diagnostic Tests in Hypothyroidism.**  The tests already described for diagnosis of hyperthyroidism give the opposite results in hypothyroidism. The basal metabolic rate in myxedema ranges between −30 and −40, the protein-bound iodine is less than 2 $\mu$gm./100 ml. of plasma instead of the normal 4 to 7.5, and the rate of radioactive iodine uptake by the thyroid gland (except in iodine deficiency hypothyroidism) ranges less than 1 per cent per hour rather than the normal of approximately 4 per cent per hour. Probably more important for diagnosis, however, than the various diagnostic tests are the characteristic symptoms of hypothyroidism as just discussed.

**Treatment of Hypothyroidism.**  Figure 76–4 shows the effect of thyroid hormone on the basal metabolic rate, illustrating that the hormone normally has a duration of action of more than one month. Consequently, it is easy to maintain a steady level of thyroid hormone activity in the body by daily oral ingestion of a tablet or so of desiccated thyroid gland or thyroid extract. Furthermore, proper treatment of the hypothyroid patient results in such complete normality that formerly myxedematous patients properly treated have lived into their 90's after treatment for over 50 years.

When a person with myxedema is first treated, immediate diuresis ensues, owing to removal of the myxedematous fluid from the interstitial spaces as the mucopolysaccharide in this fluid is metabolized. Obviously, treatment increases the activity in the circulatory system, and, if the person has already developed severe coronary artery disease, the increased activity of the heart may lead to anginal pain. On the other hand, lack of treatment may lead to increased severity of the coronary disease. In this instance the clinician is in a dilemma and must proceed with progressive treatment slowly if at all.

*Cretinism.*  Cretinism is the condition caused by extreme hypothyroidism during infancy and childhood, and it is characterized especially by failure of growth. Cretinism results from congenital lack of a thyroid gland *(congenital cretinism)*, from failure of the thyroid gland to produce thyroid hormone because of a genetic deficiency of the gland, or from iodine lack in the diet *(endemic cretinism)*. The severity of endemic cretinism varies greatly, depending on the amount of iodine in the diet, and whole populaces of an endemic area have been known to have cretinoid tendencies.

A newborn baby without a thyroid gland may have absolutely normal appearance and function because he has been supplied with thyroid hormone by the mother while *in utero,* but a few weeks after birth his movements become sluggish, and both his physical and mental growth are greatly retarded. Treatment of the cretin at any time usually causes normal return of physical growth, but, unless the cretin is treated within a few months after birth, his mental growth will be permanently retarded. This is probably due to the fact that physical development of the neuronal cells of the central nervous system is rapid during the first year of life so that any retardation at this point is extremely detrimental.

Skeletal growth in the cretin is characteristically more inhibited than is soft tissue growth, though both are inhibited to a certain extent. However, as a result of this disproportionate rate of growth, the soft tissues are likely to enlarge excessively, giving the cretin the appearance of an obese and stocky, short child. Indeed, occasionally the tongue becomes so large in relation to the skeletal growth that it obstructs swallowing and breathing, inducing a characteristic guttural breathing and sometimes even choking the baby.

# REFERENCES

Berberof-van Sande, J.; Schell-Frederick, E., and Dumont, J. E.: The energy metabolism of the thyroid gland. *Ann. Clin. Res.*, 4:189, 1972.

Ciba Foundation Study Group No. 18: Brain-Thyroid Relationships. Boston, Little, Brown and Company, 1964.

Danowski, T. S.: Clinical Endocrinology. Vol. 2. Thyroid. Baltimore, The Williams & Wilkins Company, 1962.

Daughaday, W. H., Herington, A. C., and Phillips, L. S.: The regulation of growth by endocrines. *Ann. Rev. Physiol.*, 37:211, 1975.

Edelman, I. S.: Thyroid thermogenesis. *New Engl. J. Med.*, 290: 1303, 1974.

Edelman, I. S., and Ismail-Beigi, F.: Thyroid thermogenesis and active sodium transport. *Recent Prog. Horm. Res.*, 30:235, 1974.

Florsheim, W. H.: Control of thyrotropin secretion. *In* Greep, R. O., and Astwood, E. B. (eds.): Handbook of Physiology. Sec. 7, Vol. 4, Part 2. Baltimore, The Williams & Wilkins Company, 1974, p. 449.

Gillie, R. B.: Endemic goiter. *Sci. Amer.*, 224:92, 1971.

Greenberg, A. H., Najjar, S., and Blizzard, R. M.: Effects of thyroid hormone on growth, differentiation, and development. *In* Greep, R. O., and Astwood, E. B. (eds.): Handbook of Physiology. Sec. 7, Vol. 3. Baltimore, The Williams & Wilkins Company, 1974, p. 377.

Greer, M. A., and Haibach, H.: Thyroid secretion. *In* Greep, R. O., and Astwood, E. B. (eds.): Handbook of Physiology. Sec. 7, Vol. 3. Baltimore, The Williams & Wilkins Company, 1974, p. 135.

McKenzie, J. M.: Humoral factors in the pathogenesis of Graves' disease. *Physiol. Rev.*, 48:252, 1968.

McKenzie, J. M.: Long-acting thyroid stimulator of Graves' disease. *In* Greep, R. O., and Astwood, E. B. (eds.): Handbook of Physiology. Sec. 7, Vol. 3. Baltimore, The Williams & Wilkins Company, 1974, p. 285.

Martin, J. B.: Regulation of the pituitary-thyroid axis. *In* Guyton, A. C. (ed.): MTP International Review of Science: Physiology. Vol. 5. Baltimore, University Park Press, 1974, p. 67.

Means, J. J., DeGroot, L. J., and Stanbury, J. B.: The Thyroid and Its Diseases, 3rd Ed. New York, McGraw-Hill Book Company, 1963.

Middlesworth, L. V.: Metabolism and excretion of thyroid hormones. *In* Greep, R. O., and Astwood, E. B. (eds.): Handbook of Physiology. Sec. 7, Vol. 3. Baltimore, The Williams & Wilkins Company, 1974, p. 215.

Nobuo, U.: Synthesis and chemistry of iodoproteins. *In* Greep, R. O., and Astwood, E. B. (eds.): Handbook of Physiology. Sec. 7, Vol. 3. Baltimore, The Williams & Wilkins Company, 1974, p. 55.

Pierce, J. G.: Chemistry of thyroid-stimulating hormone. *In* Greep, R. O., and Astwood, E. B. (eds.): Handbook of Physiology. Sec. 7, Vol. 4, Part 2. Baltimore, The Williams & Wilkins Company, 1974, p. 79.

Prange, A. J., Jr. (ed.): The Thyroid Axis, Drugs, and Behavior. New York, Raven Press, 1974.

Rawson, R. W., Sonenberg, M., and Money, W. L.: Diseases of the thyroid. *In* Duncan, G. G. (ed.): Diseases of Metabolism, 5th Ed. Philadelphia, W. B. Saunders Company, 1964, p. 1159.

Stanbury, J. B.: Cretinism and the fetal-maternal relationship. *Adv. Exp. Med. Biol.*, 30:487, 1972.

Stanbury, J. B., and Kroc, R. L. (eds.): Human Development and the Thyroid Gland. New York, Plenum Publishing Corporation, 1973.

Studer, H., Kohler, H., and Burgi, H.: Iodine deficiency. *In* Greep, R. O., and Astwood, E. B. (eds.): Handbook of Physiology. Sec. 7, Vol. 3. Baltimore, The Williams & Wilkins Company, 1974, p. 303.

Tong, W.: Actions of thyroid-stimulating hormone. *In* Greep, R. O., and Astwood, E. B. (eds.): Handbook of Physiology. Sec. 7, Vol. 3. Baltimore, The Williams & Wilkins Company, 1974, p. 255.

Werner, S. C., and Nauman, J. A.: The thyroid. *Ann. Rev. Physiol.*, 30:213, 1968.

Wolff, J.: Transport of iodine and other anions in the thyroid gland. *Physiol. Rev.*, 44:45, 1964.

# 77

# The Adrenocortical Hormones

The adrenal glands, which lie at the superior poles of the two kidneys, are each composed of two distinct parts, the *adrenal medulla* and the *adrenal cortex.* The adrenal medulla is functionally related to the sympathetic nervous system, and it secretes the hormones *epinephrine* and *norepinephrine* in response to sympathetic stimulation. In turn, these hormones cause almost the same effects as direct stimulation of the sympathetic nerves in all parts of the body. These hormones and their effects were discussed in detail in Chapter 57 in relation to the sympathetic nervous system.

The adrenal cortex secretes an entirely different group of hormones called *corticosteroids.* These hormones are all synthesized from the steroid cholesterol, and they all have similar chemical formulas. However, very slight differences in their molecular structures, which will be discussed later in the chapter, give them several very different but very important functions.

There seems to be very little direct functional relationship between the adrenal cortex and the adrenal medulla except for one possible effect: Large quantities of corticosteroids are carried into the medulla in the adrenal blood, and these hormones activate the enzyme system for converting norepinephrine into epinephrine. Since epinephrine has a potent effect on elevation of the blood glucose, this is perhaps of purposeful importance because some of the adrenal steroids also have an important effect on elevation of blood glucose, as will be discussed later in the chapter.

**Mineralocorticoids and Glucocorticoids.** The adrenocortical hormones do not all cause exactly the same effects in the body. Two major types of hormones, the *mineralocorticoids* and the *glucocorticoids,* are secreted by the adrenal cortex. In addition to these, small amounts of *androgenic hormones,* which exhibit the same effects in the body as the male sex hormone testosterone, are also secreted. These are normally unimportant, though in certain abnormalities of the adrenal cortices extreme quantities can be secreted (which is discussed later in the chapter) and can then result in masculinizing effects.

The *mineralocorticoids* have gained this name because they especially affect the electrolytes of the extracellular fluids—sodium and potassium, in particular. The *glucocorticoids* have gained this name because they exhibit an important effect in increasing blood glucose concentration. However, the glucocorticoids have additional effects on both protein and fat metabolism which may be equally as important to body function as are their effects on carbohydrate metabolism.

Over 30 different steroids have been isolated from the adrenal cortex, but only two of these are of major importance to the endocrine functions of the body—*aldosterone,* which is the principal mineralocorticoid, and *cortisol,* which is the principal glucocorticoid.

## FUNCTIONS OF THE MINERALOCORTICOIDS— ALDOSTERONE

Loss of adrenocortical secretion usually causes death within three days to two weeks unless the person receives extensive salt therapy or mineralocorticoid therapy. Without

1019

mineralocorticoids, the potassium ion concentration of the extracellular fluid rises markedly, the sodium and chloride concentrations decrease, and the total extracellular fluid volume and blood volume also become greatly reduced. The person soon develops diminished cardiac output, which proceeds to a shocklike state followed by death. This entire sequence can be prevented by the administration of aldosterone or some other mineralocorticoid. Therefore, the mineralocorticoids are said to be the "life-saving" portion of the adrenocortical hormones, while the glucocorticoids are of particular importance in helping the person resist different types of stresses, as is discussed later in the chapter.

Aldosterone exerts at least 95 per cent of the mineralocorticoid activity of the adrenocortical secretion, but cortisol, the major glucocorticoid secreted by the adrenal cortex, also provides a small amount of mineralocorticoid activity. Other adrenal steroids that occasionally have significant mineralocorticoid effects are *corticosterone*, which also exerts glucocorticoid effects, and *desoxycorticosterone*, which is secreted in extremely minute quantities but has almost the same effects as aldosterone with a potency one thirtieth that of aldosterone.

The basic actions of the mineralocorticoids are illustrated in Figure 77–1, which shows one major basic effect: *increased renal tubular reabsorption of sodium*. This basic effect, however, causes many additional secondary effects also illustrated in the figure.

## RENAL EFFECTS OF ALDOSTERONE

**Effect on Tubular Reabsorption of Sodium.** By far the most important effect of aldosterone and other mineralocorticoids is to increase the rate of tubular reabsorption of sodium. It will be recalled from Chapter 34 that sodium is reabsorbed from the renal tubules along their entire extent. Using micropuncture techniques for studying sodium reabsorption, investigators have shown that aldosterone has an especially potent effect to increase sodium reabsorption in the thick portion of the ascending limb of the loop of Henle, in the distal tubule, and in the collecting tubule. In addition, it increases sodium reabsorption in the proximal tubule but does not seem to play the all-powerful role here that it plays in the other tubules. The basic mechanisms of these reabsorption processes were presented in Chapter 34 through 36.

A high secretion rate of aldosterone can decrease the loss of sodium in the urine to as little

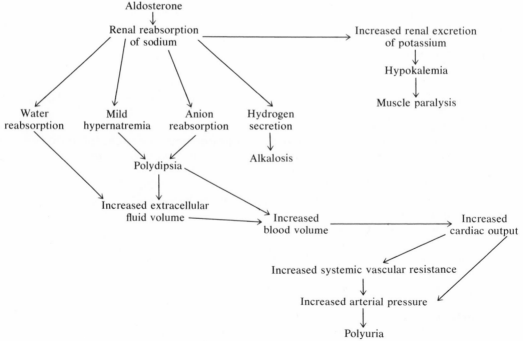

**Figure 77–1.** Direct and indirect functions of aldosterone.

as a few milligrams per day during the first few days of therapy. Beyond this time, compensatory adaptations in the body tend to return sodium excretion to normal. Conversely, total lack of aldosterone secretion can cause loss of as much as 20 grams of sodium in the urine in a day, an amount equal to one fifth of all the sodium in the body.

The increased tubular reabsorption of sodium caused by excessive aldosterone tends to increase the *quantity* of sodium in the extracellular fluids. However, significant *hypernatremia,* which means an increase in sodium *concentration,* usually does not occur because the person becomes extremely thirsty and keeps adding water to his body fluids to dilute the sodium as it is reabsorbed, thereby maintaining almost normal sodium concentration. Rarely, the concentration does rise 3 to 5 milliequivalents per liter.

**Effect on Tubular Secretion of Potassium.** At the same time that aldosterone increases sodium absorption, it also enhances potassium secretion into the distal tubules and collecting tubules of the kidneys. This is caused mainly, if not entirely, by the following mechanism: The rapid absorption of sodium from the tubules promoted by aldosterone causes extreme electronegativity in the tubules because the sodium ions carry positive charges from the tubules into the extracellular fluid. This large electrical gradient now attracts potassium from the extracellular fluid of the kidneys into the tubules, thereby promoting potassium secretion.

This effect of the kidneys to excrete large quantities of potassium under the influence of aldosterone is extremely important because it is the only potent mechanism that the body has for controlling the extracellular fluid potassium ion concentration. This effect was discussed in detail in Chapter 36.

*Hypokalemia and Muscle Paralysis; Hyperkalemia and Cardiac Toxicity.* The excessive secretion of potassium ions under the influence of aldosterone decreases the concentration of potassium in the extracellular fluids, causing the condition called *hypokalemia.* When the potassium ion concentration falls below approximately one-half normal, muscle paralysis or at least severe muscle weakness often develops. This is caused by hyperpolarization of the nerve and muscle fiber membranes (see Chapter 10), which prevents transmission of action potentials.

On the other hand, when aldosterone is deficient, the potassium ion concentration can rise far above normal. When it rises to approximately double normal, serious cardiac toxicity, including weakness of contraction and arrhythmia, becomes evident; a slightly higher concentration of potassium leads inevitably to a cardiac death.

**Alkalosis Caused by Aldosterone.** As pointed out in the discussion of acid-base balance in Chapter 37, sodium reabsorption also causes hydrogen ions to be secreted into the tubules in a manner analogous to that for secretion of potassium—that is, the hydrogen ions are pulled into the tubules by the electronegativity. Therefore, when the rate of sodium reabsorption is enhanced in response to aldosterone, the hydrogen ion concentration in the body fluids becomes reduced.

The secreted hydrogen ions originate from carbonic acid molecules, and when the hydrogen ions are secreted bicarbonate ions are left in the extracellular fluids. Thus, for each hydrogen ion secreted, one bicarbonate ion enters the body fluids, which shifts the hydrogen ion concentration in the body fluids toward the alkaline side. For these reasons, increased secretion of aldosterone promotes alkalosis, while decreased secretion promotes acidosis. The changes in body fluid pH are not excessive, however, and ordinarily can be adequately compensated by the normal acid-base regulatory mechanisms.

**Effect on Tubular Reabsorption of Anions—Especially Chloride Ions.** At the same time aldosterone causes reabsorption of sodium and creates a very negative potential in the distal and collecting tubules, it also causes greatly enhanced reabsorption of anions. The reason for this is exactly the opposite to the effects on potassium and hydrogen ions: The anions, which themselves are negatively charged, are repelled from the electronegative tubular fluid and are attracted toward the extracellular fluid.

The most important of the anions absorbed in this way is the chloride ion. Therefore, the quantity of sodium chloride in the body fluid increases under the stimulus of aldosterone.

## EFFECTS OF ALDOSTERONE ON FLUID VOLUMES AND CARDIOVASCULAR DYNAMICS

**Effect on Extracellular Fluid Volume.** From the preceding discussion, it is evident that mineralocorticoids greatly increase the quantities of sodium, chloride, and bicarbonate ions in the extracellular fluids. The net result is a marked increase in total *quantity* of electrolytes in the extracellular fluids. This in turn promotes increased water reabsorption from the tubules by creating an osmotic gradient across the tubular membrane when the electrolytes are absorbed, this in turn carrying water through the membrane in the wake of the electrolyte absorption. Also the increased electrolyte concentration causes thirst, thereby making the person drink excessive amounts of water; this effect is called *polydipsia*. The final result, therefore, is a 5 to 15 per cent increase in extracellular fluid volume, but only on rare occasions is the increase enough to cause generalized extracellular edema, which requires about a 30 per cent increase.

**Aldosterone "Escape," Polydipsia, and Polyuria.** The reason that generalized extracellular edema does not occur even when large quantities of aldosterone are administered is that the excretion of salt and water "escapes" from the aldosterone inhibition after a reasonable quantity of salt and water has accumulated in the body fluids. That is, the kidneys again excrete salt and water in the same amount as is ingested so that no additional accumulation occurs.

Though the cause of this escape has not been settled, recent research suggests that it results from the slight rise in arterial pressure that accompanies the accumulation of excess extracellular fluid volume. That is, the rise in pressure initiates increased urinary output. As explained in Chapter 22, even a few millimeters mercury rise in arterial pressure lasting several days is associated with a several-fold increase in urinary output, even though an acute rise in arterial pressure of the same amount has only a small effect.

Aldosterone also causes a person to become thirsty because the sodium ion concentration rises slightly and stimulates the thirst center. The result is excessive drinking, called *polydipsia,* and consequently also excessive urine flow, called *polyuria*.

**Effect on Blood Volume.** The plasma portion of the blood is part of the extracellular fluid, and the plasma volume increases almost proportionally with extracellular fluid volume when the increases are small (see Chapter 31). Therefore, one of the effects of aldosterone secretion is a moderate increase in blood volume.

**Effect on Cardiac Output.** Lack of aldosterone secretion has a profound effect on cardiac output, for the extracellular fluid and blood volumes decrease so greatly that venous return becomes markedly compromised. As a result, cardiac output actually falls to shock levels.

Conversely, when aldosterone secretion is excessive the resultant increase in extracellular fluid volume and blood volume increases the cardiac output. However, as discussed in Chapter 20, all tissues of the body regulate their own blood flow, thereby preventing a rise in cardiac output much above that amount required to supply the tissues adequately with nutrition. For this reason, the cardiac output rarely rises more than 10 to 20 per cent.

**Effect on Arterial Pressure and the Development of Hypertension.** When aldosterone is not secreted at all by the adrenal glands, a shocklike state of hypotension occurs because of greatly reduced body fluid volumes and cardiac output.

On the other hand, the increase in blood volume caused by excess aldosterone raises the arterial pressure. When excessive secretion of aldosterone lasts for only a few weeks, the arterial pressure rises only a few mm. Hg—rarely over 15 to 20 mm. Hg, at most. This rise is associated with "aldosterone escape" as explained above. That is, even this slight increase in pressure seems to cause enough pressure diuresis to nullify completely the salt and water retention effects of aldosterone. Therefore, the pressure effect of even large quantities of aldosterone is normally slight for the first few days to few weeks.

On the other hand, long-term secretion of large quantities of aldosterone seems to cause progressive renal damage for reasons not presently understood. Indeed, removal of an aldosterone-secreting tumor after there has been a prolonged effect on the kidneys frequently will not provide a return to normal renal excretion of fluid volume thereafter. Consequently, over long periods of time, the effect of aldosterone on arterial pressure seems to be different from the early effect—it almost invariably leads to moderate hypertension and occasionally to severe hypertension. The theoretical basis for these effects was presented in Chapters 20 through 22.

## EFFECTS OF ALDOSTERONE ON SWEAT GLANDS, SALIVARY GLANDS, AND INTESTINAL ABSORPTION

Aldosterone has almost the same effects on sweat glands and salivary glands that it has on the renal tubules. Both of these glands form a primary secretion that contains large quantities of sodium chloride, but much of the sodium chloride on passing through the excretory ducts is reabsorbed while potassium and bicarbonate ions are secreted. Aldosterone greatly increases the reabsorption of sodium chloride and the secretion of potassium. The effect on the sweat glands is important to conserve body salt in hot environments, and the effect on the salivary glands is necessary to conserve salt when excessive quantities of saliva are lost.

Aldosterone also greatly enhances sodium absorption by the intestines, which obviously prevents loss of sodium in the stools. On the other hand, in the absence of aldosterone, sodium absorption from the intestine can be very poor, leading to failure to absorb anions and water as well. The unabsorbed sodium chloride and water then lead to diarrhea, with further loss of salt from the body.

## CELLULAR MECHANISM OF ALDOSTERONE ACTION

Although for many years we have known the overall effects of mineralocorticoids on the body, the basic action of aldosterone on the tubular cells to increase transport of sodium is only now beginning to be understood. The sequence of events that leads to increased sodium reabsorption seems to be the following:

First, because of its lipid solubility in the cellular membranes aldosterone diffuses to the interior of the tubular epithelial cells.

Second, in the cytoplasm of the tubular cells aldosterone combines with a highly specific cytoplasmic *receptor protein*, a protein that has a stereomolecular configuration that will allow only aldosterone or extremely similar compounds to combine.

Third, the aldosterone-receptor complex diffuses into the nucleus where it may undergo further alterations, and then it induces specific portions of the DNA to form a type or types of messenger RNA related to the process of sodium transport.

Fourth, the messenger RNA diffuses back into the cytoplasm where it, operating in conjunction with the ribosomes, causes protein formation. The protein formed is one or more enzymes or carrier substances required for sodium transport, probably a specific ATPase that catalyzes energy transfer from cytoplasmic ATP to the sodium transport mechanism of the cell membrane.

Thus, aldosterone does not have an immediate effect on sodium transport, but must await the sequence of events that leads to the formation of the specific intracellular substance or substances required for sodium transport. Approximately 20 to 30 minutes are required before new RNA appears in the cells, and approximately 45 minutes are required before the rate of sodium transport begins to increase; the effect reaches maximum in several hours.

## REGULATION OF ALDOSTERONE SECRETION

The regulation of aldosterone secretion is so deeply intertwined with the regulation of extracellular fluid electrolyte concentrations, extracellular fluid volume, blood volume, arterial pressure, and many special aspects of renal function that it is not possible to discuss the regulation of aldosterone secretion independently of all these other factors. This subject has already been presented in great detail in Chapter 36, to which the reader is referred. However, it is important to list here as well the most important points of aldosterone secretion con-

trol. Let us note first that aldosterone is secreted by the *zona glomerulosa,* a very thin zone of cells located on the surface of the adrenal cortex immediately beneath the capsule. These cells function almost entirely independently of the deeper layers of cells in the zona reticularis and zona fasciculata, which secrete the glucocorticoids and androgens. Therefore, any stimulus that increases the secretion of aldosterone greatly enhances secretory activity in the glomerulosal cells and over a period of several days also greatly increases the number of glomerulosal cells as well as the thickness of the zona glomerulosa itself. Furthermore, the regulation of aldosterone secretion is almost entirely independent of the regulation of glucocorticoid secretion.

There are four different factors that are presently known to play essential roles in the regulation of aldosterone. In the probable order of their importance these are:

1. Potassium ion concentration of the extracellular fluid
2. Renin-angiotensin system
3. Quantity of body sodium
4. Adrenocorticotropic hormone (ACTH)

**Effect of Potassium Ion Concentration on Aldosterone Secretion.** An increase in potassium ion concentration of less than 1 mEq./liter will triple the rate of aldosterone secretion. Furthermore, this secretion will continue at an elevated level indefinitely under the stimulus of excess potassium ions.

This very potent effect of potassium ions is exceedingly important because it establishes a powerful feedback mechanism for control of extracellular fluid potassium ion concentration as follows: (1) An increase in potassium ion concentration causes increased secretion of aldosterone. (2) The aldosterone in turn has a potent effect on the kidneys, causing enhanced excretion of potassium. (3) Therefore, the potassium ion concentration returns to normal. Recent quantitative measurements of the feedback gain of this system show it to be by far the most potent of all the factors controlling aldosterone secretion, as has already been discussed in detail in Chapter 36. Serious interference with this control system can lead to a cardiac death if the potassium concentration rises too high or to a paralytic death if the potassium concentration falls too low.

This effect of potassium ions on aldosterone secretion is a direct effect of the potassium ions on the zona glomerulosa cells, though the intracellular mechanism of the effect is unknown.

**Effect of the Renin-Angiotensin System on Aldosterone Secretion.** Infusion of enough angiotensin II to increase the extracellular fluid concentration of angiotensin to about five times normal approximately triples aldosterone secretion for the first 24 hours. However, if the angiotensin infusion is continued, the rate of aldosterone secretion falls back to only 30 per cent above normal. Therefore, the percentagewise effect of angiotensin on aldosterone secretion is about 100 times less potent than the percentagewise effect of potassium ions on aldosterone secretion.

Nevertheless, elevated renin and angiotensin are found very frequently in the same clinical conditions in which aldosterone is also elevated. Therefore, the effect of angiotensin has often been stated to be the most important of all the factors controlling aldosterone secretion. Unfortunately, this conclusion was reached without consideration of all of the quantitative evidence and without comparison of the angiotensin effect on aldosterone secretion with the potassium effect. On the basis of more recent quantitive measurements, it is doubtful that the renin-angiotensin system is nearly so important. Indeed, it is now very clear that in many conditions in both human beings and animals the angiotensin concentration can proceed in one direction while the aldosterone concentration proceeds in the opposite direction. For instance, in the simple condition of dehydration, both renin and angiotensin increase markedly while the rate of aldosterone secretion occasionally falls almost to an undetectable level. Therefore, the whole question of the renin-angiotensin system in the control of aldosterone must be reassessed.

**Effect of Decreased Quantity of Total Body Sodium on Aldosterone Secretion.** When an animal or human being is placed on a sodium deficient diet, after several days the rate of aldosterone secretion increases markedly even though the sodium ion concentration of the body fluids does not fall significantly. A number of different suggestions for the cause of this effect have been the following:

1. Lack of sodium causes retention of potassium by the kidneys as explained earlier in the chapter. The elevated potassium could then cause the increased aldosterone secretion.

2. In a few experiments by McCaa and Young it has been shown that diminished sodium concentration possibly or probably causes the anterior pituitary gland to secrete some substance (not ACTH) that has an effect on the adrenal glands to increase aldosterone secretion. For the present, this substance is called the *unidentified pituitary factor*.

3. Diminished sodium leads to diminished extracellular fluid volume, blood volume, and arterial pressure. Some physiologists believe that the low arterial pressure elicits reflex effects that increase aldosterone secretion, possibly by acting through the pituitary gland or by causing release of some other neurosecretory substance from some basal region of the brain.

4. The decrease in extracellular fluid volume and the decrease in arterial pressure cause increased formation of angiotensin, and this in turn could help to elevate the rate of aldosterone secretion.

Unfortunately, the importance of each of these mechanisms is yet to be proved.

**Effect of ACTH on Aldosterone Secretion.** ACTH is the anterior pituitary hormone that controls the secretion of glucocorticoids. It also has a *permissive effect* on aldosterone secretion in the following way: All of the above regulations of aldosterone secretion will occur if there is only a minimal amount of ACTH present. However, in the total absence of ACTH, the zona glomerulosa of the adrenal cortex partially atrophies and there is almost total atrophy of the remainder of the gland. Thus, total absence of ACTH can lead to a moderate degree of aldosterone deficiency.

# FUNCTIONS OF THE GLUCOCORTICOIDS

Even though mineralocorticoids can save the life of an acutely adrenalectomized animal, the animal still is not completely normal. Instead, its metabolic systems for utilization of carbohydrates, proteins, and fats are considerably deranged. Furthermore, the animal cannot resist different types of physical or even mental stress, and minor illnesses such as respiratory tract infections can lead to death. Therefore, the glucocorticoids have functions just as important to long-continued life of the animal as do the mineralocorticoids. These are explained in the following sections.

At least 95 per cent of the glucocorticoid activity of the adrenocortical secretions results from the secretion of *cortisol*, known also as *hydrocortisone* and *compound F*. In addition to this, a small amount of glucocorticoid activity is provided by *corticosterone* and a minute amount by *cortisone*.

## EFFECT OF CORTISOL ON CARBOHYDRATE METABOLISM

**Stimulation of Gluconeogenesis.** By far the best-known metabolic effect of cortisol and other glucocorticoids on metabolism is their ability to stimulate gluconeogenesis by the liver, often increasing the rate of gluconeogenesis as much as 6- to 10-fold. This results from several different effects of cortisol:

First, cortisol increases the transport of amino acids from the extracellular fluids into the liver cells. This obviously increases the availability of amino acids for conversion into glucose.

Second, several of the enzymes required to convert amino acids into glucose are increased in the liver cells. Also, the concentration of RNA is increased in the liver cells. Therefore, it is assumed that glucocorticoids activate nuclear formation of messenger RNA's that in turn lead to the array of enzymes required for gluconeogenesis.

Third, cortisol causes mobilization of amino acids from the extrahepatic tissues, mainly from muscle. As a result, more amino acids become available in the plasma to enter into the gluconeogenesis process of the liver and thereby to promote the formation of glucose.

One of the effects of increased gluconeogenesis is a marked increase in glycogen in the liver cells.

**Decreased Glucose Utilization by the Cells.** Cortisol also causes a moderate decrease in the rate of glucose utilization by the cells. Though the cause of this decrease in unknown, most physiologists believe that somewhere between the point of entry of glucose into the cells and its final degradation cortisol directly delays the rate of glucose utilization. A suggested mechanism for this effect is based on the observation that glucocorticoids depress the oxidation of NADH. Since NADH must be oxidized to allow rapid glycolysis, this effect could account for the diminished utilization of glucose by the cells.

Also, it is known that glucocorticoids slightly depress glucose transport into the cells, which could be an additional factor that depresses cellular glucose utilization.

**Elevated Blood Glucose Concentration, and Adrenal Diabetes.** Both the increased rate of gluconeogenesis and the moderate reduction in rate of glucose utilization by the cells cause the blood glucose concentration to rise. In addition, the enzyme *glucose 6-phosphatase* is increased in the liver in response to cortisol; this enzyme catalyzes the dephosphorylation of liver glucose and thereby promotes its transport into the blood, thus further increasing the blood glucose concentration.

The increased blood glucose concentration is occasionally great enough—50 per cent or more above normal—that the condition is called *adrenal diabetes,* and it has many similarities to pituitary diabetes, which was discussed in Chapter 75. Administration of insulin lowers the blood glucose concentration only a moderate amount in adrenal diabetes, not nearly so much as it does in pancreatic diabetes. On the other hand, insulin causes greater decrease in blood glucose concentration in adrenal diabetes than in pituitary diabetes. Therefore, *pituitary diabetes is said to be weakly insulin sensitive, adrenal diabetes moderately insulin sensitive, and pancreatic diabetes strongly insulin sensitive.*

## EFFECTS OF CORTISOL ON PROTEIN METABOLISM

**Reduction in Cellular Protein.** One of the principal effects of cortisol on the metabolic systems of the body is reduction of the protein stores in essentially all body cells except those of the liver. This is caused both by decreased protein synthesis and increased catabolism of protein already in the cells. Both of these effects may possibly result from decreased amino acid transport into extrahepatic tissues, as will be discussed below, but this probably is not the only cause since cortisol also depresses the formation of RNA in many extrahepatic tissues, including especially muscle and lymphoid tissue.

*Increased Liver Protein and Plasma Proteins Caused by Cortisol.* Coincidently with the reduced proteins elsewhere in the body, the liver proteins become enhanced. Furthermore, the plasma proteins (which are produced by the liver and then released into the blood) are also increased. Therefore, these are exceptions to the protein depletion that occurs elsewhere in the body. It is believed that this effect is caused both by the ability of cortisol to enhance amino acid transport into liver cells (but not into most other cells) and by enhancement of the liver enzymes required for protein synthesis.

**Increased Blood Amino Acids, Diminished Transport of Amino Acids into Extrahepatic Cells, and Enhanced Transport into Hepatic Cells.** Recent studies in isolated tissue have demonstrated that cortisol depresses amino acid transport into muscle cells and perhaps into other extrahepatic cells. But, in contrast to this, it enhances transport into liver cells.

Obviously, the decreased transport of amino acids into extrahepatic cells decreases their intracellular amino acid concentrations and as a consequence decreases the synthesis of protein. Yet catabolism of proteins in the cells continues to release amino acids from the already existing proteins, and these diffuse out of the cells to increase the plasma amino acid concentration. Therefore, it is said that *cortisol mobilizes amino acids from the tissues.*

The increased plasma concentration of amino acids, plus the fact that cortisol enhances transport of amino acids into the hepatic cells, could also account for enhanced utilization of amino acids by the liver in the presence of cortisol— such effects as: (1) increased rate of deamination of amino acids by the liver, (2) increased protein synthesis in the liver, (3) increased formation of plasma proteins by the liver, and (4) increased conversion of amino acids to glucose—that is, enhanced gluconeogenesis.

Thus, it is possible that many of the effects of cortisol on the metabolic systems of the body can be explained very simply by changes in transport of amino acids through the cell membrane; though cortisol also increases the enzymes required for these effects, presumably by causing the formation of appropriate messenger RNA's.

## EFFECTS OF CORTISOL ON FAT METABOLISM

**Mobilization of Fatty Acids.** In much the same manner that cortisol promotes amino acid mobilization from muscle, it also promotes mobilization of fatty acids from adipose tissue. This in turn increases the concentration of free fatty acids in the plasma, which also increases their utilization for energy. Cortisol moderately enhances the oxidation of fatty acids in the cells as well, this effect perhaps resulting secondarily to the reduced availability of glycolytic products for metabolism.

The mechanism by which cortisol promotes fatty acid mobilization is not yet understood. However, in the absence of growth hormone or of ACTH, cortisol has very little fat mobilizing effect. Therefore, it is believed that cortisol simply enhances the effect of these other two hormones to activate a lipase in the fat cells, thus liberating the free fatty acids.

The increased mobilization of fats, combined with their increased oxidation in the cells, is one of the factors that helps to shift the metabolic systems of the cells in times of starvation or other stresses from utilization of glucose for energy to utilization of fatty acids instead. This cortisol mechanism, however, requires several hours to become fully developed—not nearly so rapid or nearly so powerful an effect as the similar shift elicited by a decrease in insulin. Nevertheless, it is probably an important factor for long-term conservation of body glucose and glycogen.

*Ketogenic Effects of Cortisol.* It is frequently said that cortisol has a *ketogenic effect* because ketosis usually will not develop unless cortisol is available to cause fat mobilization. However, this ketogenic effect occurs only under certain conditions, such as when insulin also is deficient.

## OTHER EFFECTS OF CORTISOL

**Function of Cortisol in Different Types of Stress.** It is amazing that almost any type of stress, whether it be physical or neurogenic, will cause an immediate and marked increase in ACTH secretion, followed within minutes by greatly increased adrenocortical secretion of cortisol. Some of the different types of stress that increase cortisol release are the following:

1. Trauma of almost any type
2. Infection
3. Intense heat or cold
4. Injection of norepinephrine and other sympathomimetic drugs
5. Surgical operations
6. Injection of necrotizing substances beneath the skin
7. Restraining an animal so that it cannot move
8. Almost any debilitating disease

Thus, a wide variety of nonspecific stimuli can cause marked increase in the rate of cortisol secretion by the adrenal cortex.

Yet, even though we know that cortisol secretion often increases greatly in stressful situations, we still are not sure why this is of significant benefit to the animal. One guess, which is probably as good as any other, is that the glucocorticoids cause rapid mobilization of amino acids and fats from their cellular stores, making these available both for energy and for synthesis of other compounds, including glucose, needed by the different tissues of the body. Indeed, it is well known that when proteins are released from most of the tissue cells, the liver cells can use the mobilized amino acids to form new proteins. It is possible that other tissues, particularly damaged tissues which are momentarily depleted of proteins, can also utilize the newly available amino acids to form new proteins that are essential to the lives of the cells. Or perhaps the amino acids are used to synthesize such essential intracellular substances as purines, pyrimidines, and creatine phosphate, which are necessary for maintenance of cellular life.

But all this is mainly supposition. It is supported only by the fact that cortisol usually does not mobilize the basic functional proteins of the cells, such as the contractile proteins of muscle fibers, until almost all other proteins have been released. This preferential effect of cortisol in mobilizing labile proteins could make amino acids available to needy cells to synthesize substances essential to life.

**Effect of Cortisol in Inflammation: Stabilization of Lysosomes.** Almost all tissues of the body respond to tissue damage, whether the damage results from simple trauma or from some cellular disease, by the process called *inflammation*. There are three stages in this process: (1) leakage of large quantities of plasmalike fluid out of the capillaries into the damaged area followed by clotting of the fluid, (2) infiltration of the area by leukocytes, (3) tissue healing, which is accomplished to a great extent by ingrowth of fibrous tissue.

The amount of cortisol normally secreted does not significantly affect inflammation and healing; but large amounts of cortisol or other glucocorticoids administered to a person block all stages of the process, even blocking the initial leakage of plasma from the capillaries into the damaged area.

Unfortunately, we still know very little about the basic mechanism of cortisol's *anti-inflammation effect*. However, it has recently been discovered that cortisol and other glucocorticoids stabilize the membranes of cellular lysosomes so that they rupture only with difficulty. It will be recalled that lysosomes contain large quantities of hydrolytic enzymes that are capable of digesting intracellular proteins. This stabilizing effect on the lysosome membrane could alone account for a major share of the anti-inflammatory effect of cortisol, because it would prevent the usual destruction of tissues that occurs in inflammation because of liberation of lysosomal enzymes.

A second factor of possible importance in the anti-inflammatory effect of cortisol is its ability to decrease the formation of bradykinin, a very powerful vasodilating substance that appears in inflamed tissues. This substance is split from an alpha globulin by the proteolytic enzyme kallikrein. It is possible that stabilization of the lysosome membrane prevents release of kallikrein into the cell from the lysosome.

Finally, cortisol decreases the permeability of the capillary membrane, and this is a significant factor in preventing the usual protein leakage into inflamed tissues. Part of this decrease in capillary permeability might result from inhibition of bradykinin formation, because bradykinin vasodilatation tends to overstretch the capillary membrane. But at least two other effects perhaps also play a role: (1) Cortisol diminishes the effects of histamine, another substance that is formed in inflamed tissue, to cause vasodilatation and increased capillary permeability. (2) Cortisol potentiates the effect of norepinephrine and epinephrine to constrict the peripheral vessels.

Regardless of the precise mechanisms by which the anti-inflammatory effect occurs, this effect of cortisol can play a major role in combating certain types of diseases, such as rheumatoid arthritis, rheumatic fever, and acute glomerulonephritis. All these are characterized by severe local inflammation, and the harmful effects to the body are caused mainly by the inflammation itself and not by other aspects of the disease. When cortisol or other glucocorticoids are administered to patients with these diseases, almost invariably the inflammation subsides within 24 to 48 hours. And even though the cortisol does not correct the basic disease condition itself but merely prevents the damaging effects of the inflammatory response, this alone can often be a lifesaving measure.

**Effect on Blood Cells and on Immunity.** Cortisol decreases the number of eosinophils and lymphocytes in the blood; this effect begins within a few minutes after the injection of cortisol and is marked within a few hours. Indeed, a finding of lymphocytopenia or eosinopenia is an important diagnostic criterion for overproduction of cortisol by the adrenal gland.

Likewise, administration of large doses of cortisol causes significant atrophy of all the lymphoid tissue throughout the body, which in turn decreases the output of antibodies from the lymphoid tissue. As a result, the level of humoral immunity for almost all foreign invaders of the body is decreased. This can occasionally lead to fulminating infection and death from diseases that otherwise would not be lethal, such as fulminating tuberculosis in a person whose disease had previously been arrested.

On the other hand, cortisol increases the production of red blood cells, the cause of which is unknown. When excess cortisol is secreted by the adrenal glands, polycythemia results, and, conversely, when the adrenal glands secrete no cortisol, anemia often results.

*Effect on Allergy.* Cortisol blocks the inflammatory response to allergic reactions in exactly the same way that it blocks other types of inflammatory responses. The basic allergic reaction between antigen and antibody is not affected by cortisol, and even some of the secondary effects of the allergic reaction, such as the release of histamine, still occur the same as ever. However, since the inflammatory response is responsible for many of the serious and sometimes lethal effects of allergic reactions, administration of cortisol can be lifesaving. For instance, cortisol effectively prevents shock or death in anaphylaxis which otherwise kills many persons, as explained in Chapter 7.

## REGULATION OF CORTISOL SECRETION— ADRENOCORTICOTROPIC HORMONE (ACTH)

**Control of Cortisol Secretion by ACTH.** Unlike aldosterone secretion by the zona glomerulosa, which is controlled mainly by potassium and angiotensin acting directly on the adrenocortical cells themselves, almost no stimuli have *direct* effects on the adrenal cells to control cortisol secretion. Instead, secretion of the glucocorticoids—both cortisol and corticosterone—is controlled almost entirely by adrenocorticotropic hormone (ACTH) secreted by the anterior pituitary gland. This hormone, also called *corticotropin* or *adrenocorticotropin,* also enhances the production of adrenal androgens by the adrenal cortex. As was pointed out earlier in the chapter, small amounts of ACTH are required for aldosterone secretion, providing a permissive role that allows the other more important factors to exert their more powerful controls.

**Control of ACTH Secretion by the Hypothalamus—Corticotropin Releasing Factor (CRF).** In the same way that the other pituitary hormones are controlled by releasing factors from the hypothalamus, so also does an important releasing factor control ACTH secretion. This is called *corticotropin releasing factor* (CRF). It is secreted into the primary capillary plexus of the hypophyseal portal system in the median eminence of the hypothalamus and then carried to the anterior pituitary gland where it induces ACTH secretion.

The anterior pituitary gland can secrete small quantities of ACTH in the absence of CRF, but most conditions that cause high ACTH secretory rates initiate this secretion by signals beginning in the hypothalamus and then transmitted by CRF to the anterior pituitary gland.

**Effect of Physiological Stress on ACTH Secretion.** It was pointed out earlier in the chapter that almost any type of physical or even mental stress can lead within minutes to greatly enhanced secretion of ACTH and the glucocorticoids, often increasing cortisol secretion as much as 20-fold. This effect is illustrated forcefully by the lowermost curve in Figure 77–2, which shows a many-fold increase in plasma corticosterone concentration in a rat within minutes after the tibia and fibula have been broken (corticosterone is the principal glucocorticoid secreted by the rat adrenal). It is believed that pain stimuli caused by the stress are first transmitted to the perifornical area of the hypo-

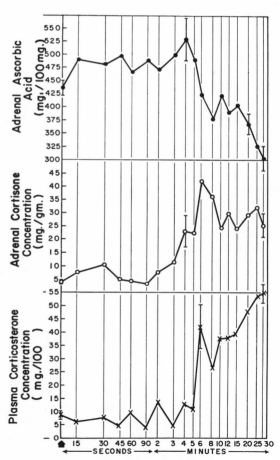

**Figure 77–2.** Rapid reaction of the adrenal cortex of a rat to stress caused by fracture of the tibia and fibula. (Courtesy of Drs. Guillemin, Dear, and Lipscomb.)

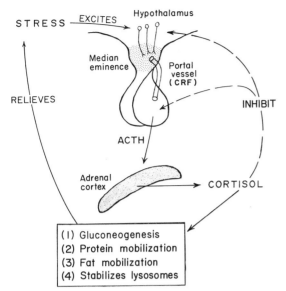

**Figure 77–3.** Mechanism for regulation of glucocorticoid secretion.

thalamus. This area in turn transmits signals into other areas of the hypothalamus and eventually to the median eminence, as shown in the schema of Figure 77–3, where CRF is secreted into the hypophyseal portal system. Within minutes the entire control sequence leads to large quantities of the glucocorticoids in the blood.

**Inhibitory Effect of Cortisol on the Hypothalamus and on the Anterior Pituitary to Cause Decreased ACTH Secretion.** Cortisol has direct negative feedback *effects* on (1) the hypothalamus to decrease the formation of CRF and (2) the anterior pituitary gland to decrease the formation of ACTH. These feedbacks help to regulate the plasma concentration of cortisol. That is, whenever the concentration becomes too great, these feedbacks automatically reduce this concentration back toward a normal control level.

**Summary of the Control System.** Figure 77–3 illustrates the overall system for control of cortisol secretion. The central key to this control is the excitation of the hypothalamus by different types of stress. These activate the entire system to cause rapid release of cortisol, and the cortisol in turn initiates a series of metabolic effects directed toward relieving the damaging nature of the stressful state. In addition, there is also direct feedback of the cortisol to the hypothalamus and anterior pituitary gland to stabilize the concentration of cortisol in

the plasma at times when the body is not experiencing stress. However, the stress stimuli are the prepotent ones; they can always break through this direct inhibitory feedback of cortisol.

**Circadian Rhythm of Glucocorticoid Secretion.** The secretory rates of CRF, ACTH, and cortisol are all high in the early morning but low in the late afternoon or evening. This effect results from a 24-hour cyclic alteration in the signals from the hypothalamus that cause cortisol secretion. When a person changes his daily sleeping habits, the cycle changes correspondingly. One of the reasons that the cycle is so important is that measurements of blood cortisol levels are meaningful only when expressed in terms of the time in the cycle at which the measurements are made.

# CHEMISTRY OF ADRENOCORTICAL SECRETION

**Locus of Hormone Formation in the Adrenal Cortex.** Aldosterone is secreted by the *zona glomerulosa* of the adrenal cortex, while the glucocorticoids and adrenal androgens are secreted by the *zona fasciculata* and *zona reticularis*. These zones are illustrated in Figure 77–4. During prolonged ACTH stimulation of the adrenal cortex, the zona fasciculata and zona reticularis both hypertrophy, whereas complete lack of ACTH causes these two zones to atrophy almost entirely while leaving the zona glomerulosa partially intact. On the contrary, conditions that increase the output of aldosterone cause hypertrophy of the zona glomerulosa while not affecting the other two zones.

**Chemistry of the Adrenocortical Hormones.** All the adrenocortical hormones are steroid compounds, and they can be formed in the adrenal cortex from acetyl coenzyme A (about 10 per cent) or from cholesterol preformed elsewhere in the body (about 90 per cent). Figure 77–5 illustrates a schema for formation of most of the important steroids found in the adrenal cortex. From this schema it can be seen that the adrenal cortex can form hormones having at least five different types of effects, two of which, the mineralocorticoids and glucocorticoids, are extremely important because they are secreted in large quantities, and three of which, the adrenal sex hormones, are normally of only minor importance because they are secreted in minute quantities. However, a change in even a single enzyme system somewhere in the scheme can cause vastly different types of hormones to be formed; occasionally, large quantities of masculinizing or, very rarely, feminizing sex hormones are secreted by adrenal tumors, as discussed later in the chapter.

Figure 77–6 illustrates the chemical formulas of four of the adrenal steroid compounds. Of these, cor-

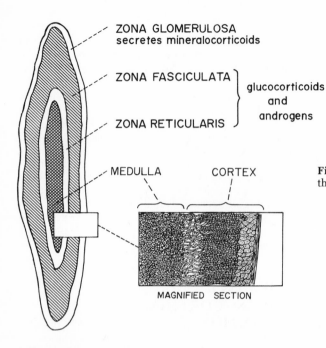

ZONA GLOMERULOSA
secretes mineralocorticoids

ZONA FASCICULATA ⎤
                  ⎬ glucocorticoids
                  ⎪     and
ZONA RETICULARIS ⎦  androgens

MEDULLA          CORTEX

MAGNIFIED SECTION

**Figure 77-4.** Secretion of adrenocortical hormones by the different zones of the adrenal cortex.

tisol and aldosterone are secreted in large quantities by the adrenal cortex. Cortisone, a glucocorticoid, and deoxycorticosterone, a mineralocorticoid, are secreted in only minor amounts, but both are used to a great extent in therapy because they can be synthesized readily.

In addition, several steroids that do not naturally occur in the adrenal cortex but that have as potent or even more potent actions than the natural hormones have been synthesized. Thus, *dexamethasone* which is a slight modification of cortisol, has thirty times as much glucocorticoid potency as cortisol but no significant increase in mineralocorticoid potency.

Therefore, it is used clinically when one wishes maximum glucocorticoid activity and yet wishes to prevent electrolyte and water retention. Likwise, *fludrocortisone* has three times as much mineralocorticoid activity as aldosterone and therefore can be used in minute quantities, only 100 micrograms per day, to treat total deficiency of mineralocorticoid secretion.

On the other hand, other substances similar in chemical structure but having almost no mineralocorticoid activity—the *spirolactones,* for instance— become firmly fixed in the renal tubular epithelial cells to the intracellular receptors to which aldo-

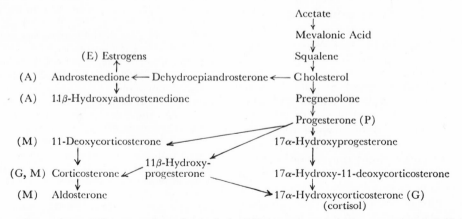

**Figure 77-5.** Schema showing the stages through which the different corticosteroids are synthesized. The physiologic characteristics of some of these corticosteroids are designated as follows: *A*, androgenic effects; *E*, estrogenic effects; *G*, glucocorticoid effects; *M*, mineralocorticoid effects; *P*, progesteronic effects.

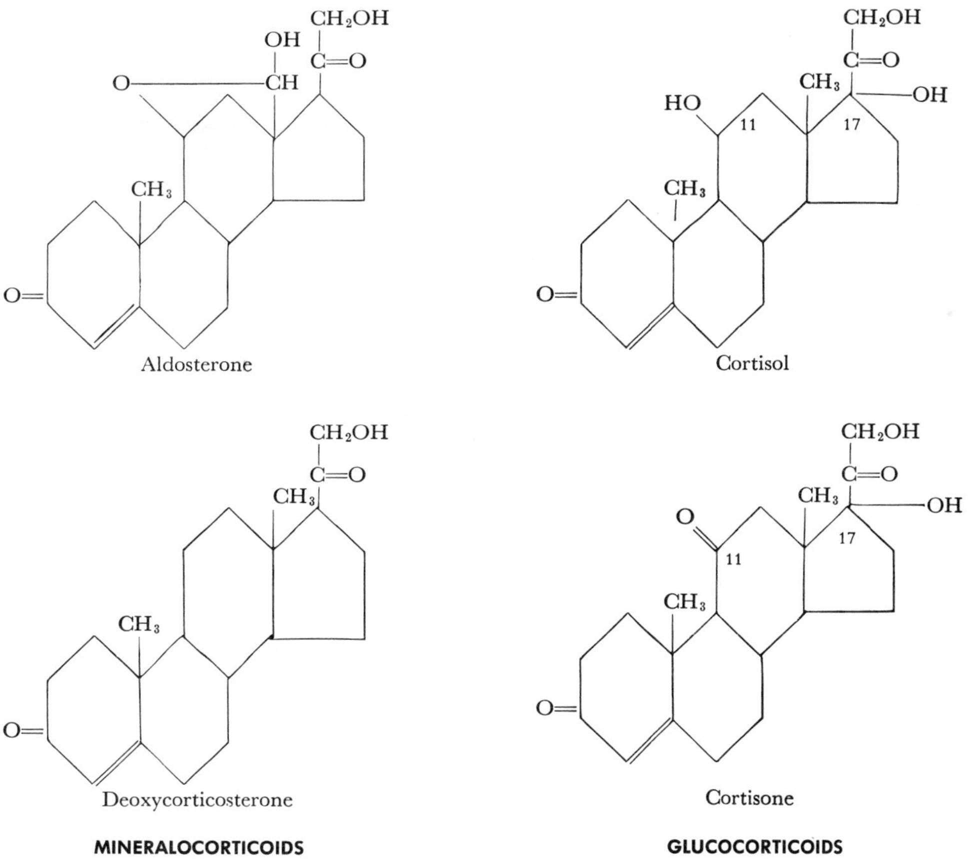

**Figure 77–6.**   Four important corticosteroids.

sterone normally binds. Therefore, these compounds block the effect of the body's own aldosterone. As a result, the kidneys then function as if no aldosterone were being secreted in the body.

**Mechanism by Which ACTH Activates Adrenocortical Cells to Produce Steroids—Function of Cyclic AMP.**   The primary effect of ACTH on the adrenocortical cells is almost certainly to activate *adenyl cyclase* in the cell membranes, and this then induces the formation of cyclic AMP in the cells, an effect that reaches maximum in three minutes. This substance, in turn, promotes the intracellular reactions for formation of adrenocortical hormones. Thus, this is another example of the function of cyclic AMP as a *second messenger* hormone.

Cyclic AMP has a specific effect to activate the phosphogluconate pathway for glucose metabolism, which was discussed in Chapter 67 and which causes the formation of large quantities of NADPH (reduced nicotinamide adenine dinucleotide phosphate). The NADPH, in turn, has an extremely high capacity for causing hydroxylation. Since the chemical reactions for formation of the adrenal steroids are mainly by a series of hydroxylations of cholesterol, it is believed

that the NADPH might be the most or at least one of the most important controllers of adrenal steroid synthesis.

**Chemistry of ACTH.**   ACTH has been isolated in pure form from the anterior pituitary. It is a large polypeptide having a chain length of 39 amino acids. A digested product of ACTH having a chain length of 24 amino acids has all the trophic effects of the total molecule.

**Chemistry of Corticotropin-Releasing Factor.**   Several small peptides that have a chemical configuration similar to that of vasopressin and oxytocin have been isolated from the hypothalamus which, upon injection into an animal, cause rapid release of corticotropin from the anterior pituitary. It is not known whether one of these is the specific corticotropin-releasing factor.

**Transport and Fate of the Adrenal Hormones.** Cortisol combines with a globulin called "transcortin" and, to a lesser extent, with albumin—about 90 per cent is normally transported in the bound form and about 10 per cent free. On the other hand, aldosterone combines only loosely with both of these so that 50 per cent is in the free form. In both the combined and free forms the hormones are transported throughout the extracellular fluid compartment. In general, the hormones become fixed in the target tissues within an hour or two for cortisol and within about 30 minutes for aldosterone.

The adrenal steroids are deactivated mainly in the liver and conjugated to form glucuronides and, to a lesser extent, sulfates, About 25 per cent of these are excreted in the bile and then in the feces and the remaining 75 per cent in the urine. The conjugated forms of these hormones are inactive.

The normal concentration of aldosterone in blood is about 10 nanograms (10 billionths of a gram) per 100 ml., and the secretory rate is about 150 $\mu$g. per day.

The concentration of cortisol in the blood averages 12 $\mu$g./100 ml., and the secretory rate averages 20 mg. per day.

# THE ADRENAL ANDROGENS

Several moderately active male sex hormones called *adrenal androgens* (two of which are illustrated in Fig.77–5) are continually secreted by the adrenal cortex. Also, progesterone and estrogens, which are female sex hormones, have been extracted from the adrenal cortex, though these are secreted in only minute quantities.

In normal physiology of the human being, even the adrenal androgens have almost insignificant effects. However, it is possible that part of the early development of the male sex organs results from childhood secretion of adrenal androgens. The adrenal androgens also exert mild effects in the female, not only before puberty but also throughout life. Some of the adrenal androgens are converted to testosterone, the major male sex hormone, in the extra-adrenal tissues, which probably accounts for much of their androgenic activity. The physiologic effects of androgens will be discussed in Chapter 80 in relation to male sexual function.

# ABNORMALITIES OF ADRENOCORTICAL SECRETION

## *HYPOADRENALISM—ADDISON'S DISEASE*

Addison's disease results from failure of the adrenal cortices to produce adrenocortical hormones, and this in turn is most frequently caused by *primary atrophy* of the adrenal cortices but also frequently by tuberculous destruction of the adrenal glands or invasion of the adrenal cortices by cancer. Basically, the disturbances in Addison's disease are:

**Mineralocorticoid Deficiency.** Lack of aldosterone secretion greatly decreases sodium reabsorption and consequently allows sodium ions, chloride ions, and water to be lost into urine in great profusion. The net result is a greatly decreased extracellu-

lar fluid volume. Furthermore, the person develops hyperkalemia and acidosis because of failure of potassium and hydrogen ions to be secreted in exchange for sodium reabsorption.

As the extracellular fluid becomes depleted, the plasma volume falls, the red blood cell concentration rises markedly, the cardiac output decreases, and the patient dies in shock, death usually occurring in the untreated patient four days to two weeks after complete cessation of mineralocorticoid secretion.

**Glucocorticoid Deficiency.** Loss of cortisol secretion makes it impossible for the person with Addison's disease to maintain normal blood glucose concentration between meals because he cannot synthesize significant quantities of glucose by gluconeogenesis. Furthermore, lack of cortisol reduces the mobilization of both proteins and fats from the tissues, thereby depressing many other metabolic functions of the body. This sluggishness of energy mobilization when cortisol is not available is one of the major detrimental effects of glucocorticoid lack. However, even when excess quantities of glucose and other nutrients are available, the person's muscles are still weak, indicating that glucocorticoids are also needed to maintain other metabolic functions of the tissues besides simply energy metabolism.

Lack of adequate glucocorticoid secretion also makes the person with Addison's disease highly susceptible to the deteriorating effects of different types of stress, and even a mild respiratory infection can sometimes cause death.

**Melanin Pigmentation.** Another characteristic of most persons with Addison's disease is melanin pigmentation in the mucous membranes and often in the skin over most of the body. This melanin is not always deposited evenly but occasionally in blotches and especially in the thin skin areas, such as the mucous membranes of the lips and the thin skin of the nipples.

The cause of the melanin deposition is believed to be the following: It is well known that the anterior pituitary gland can, on appropriate stimulation, secrete large quantities of *melanocyte-stimulating hormone* (MSH), and that this in turn increases the melanin production by the melanocytes in the skin. Therefore, it has been assumed that in persons with Addison's disease, lack of cortisol secretion exerts a feedback effect on the anterior pituitary, causing it to secrete excessive amounts of MSH, and that this in turn causes the pigmentation. In addition, lack of secretion of cortisol by the adrenal cortex also causes tremendous secretion of ACTH by the anterior pituitary, and ACTH itself has one-thirtieth as much melanocyte-stimulating effect as does MSH. Indeed, MSH has part of the same peptide structure as the ACTH molecule.

**Diagnosis of Addison's Disease.** A person with this disease has almost no urinary secretion of steroids derived from the adrenals. Also, immunoassay of the plasma aldosterone and competitive binding assay (see Chapter 75) of cortisol show either very low or unmeasurable values. In full-blown Addison's disease, less than one-tenth of the normal amount of 17-hydroxysteroids, which are derived almost entirely from cortisol secreted by the adrenals, is excreted in the urine in a 24-hour period. However, to be certain that the person has Addison's disease, ACTH is infused into the patient over a period of 8 hours or more, and the rate of secretion of 17-hydroxysteroids is again measured. If the Addison's disease is caused by total destruction of the adrenal cortices, there will be no increase in urinary 17-hydroxysteroids. If, on the other hand, the Addison's disease is caused by failure of the anterior pituitary to stimulate the adrenal cortices, successive administration of ACTH several days in a row will cause the adrenal glands to secrete normal quantities of 17-hydroxysteroids.

**Treatment of Persons with Addison's Disease.** The untreated person with Addison's disease dies within a few days because of electrolyte disturbances. Yet such a person can usually live for years if small quantities of mineralocorticoids and glucocorticoids are administered daily—about 0.1 milligram of fludrocortisone and 30 milligrams of cortisol. A high salt intake is also necessary.

**The Addisonian Crisis.** As noted earlier in the chapter, great quantities of glucocorticoids are occasionally secreted in response to different types of physical or mental stress. In the person with Addison's disease, the output of glucocorticoids does not increase during stress. Yet whenever he is specifically subjected to different types of trauma, disease, or other stresses such as surgical operations, he

is likely to develop an acute need for excessive amounts of glucocorticoids, and must be given as much as 10 or more times the normal quantities of glucocorticoids in order to prevent death.

This critical need for extra glucocorticoids and the associated severe debility in times of stress is called an Addisonian crisis.

## HYPERADRENALISM—CUSHING'S DISEASE

One type of hypersecretion by the adrenal cortex causes a complex of hormonal effects called Cushing's disease, and this usually results from general hyperplasia of both adrenal cortices. The hyperplasia in turn is most often, but not always, caused by increased secretion of ACTH by the anterior pituitary. Most abnormalities of Cushing's disease are ascribable to abnormal amounts of cortisol, but secretion of androgens is also of significance.

A special characteristic of Cushing's disease is mobilization of fat from the lower part of the body, with concomitant extra deposition of fat in the thoracic region, giving rise to a so-called "buffalo" torso. The excess secretion of steroids also leads to an edematous appearance of the face, and the androgenic potency of some of the hormones causes acne and hirsutism (excess growth of facial hair). The total appearance of the face is frequently described as a "moon face," as illustrated to the left in Figure 77–7 in a patient with Cushing's disease prior to treatment.

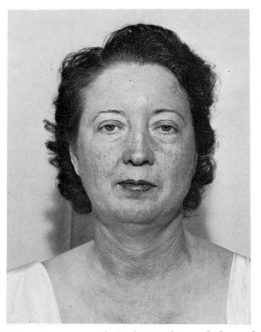 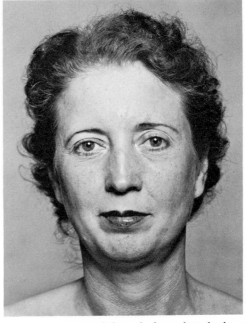

**Figure 77–7.** A person with Cushing's disease before subtotal adrenalectomy (left) and after subtotal adrenalectomy (right). (Courtesy of Dr. Leonard Posey.)

**Effects on Carbohydrate and Protein Metabolism.** The abundance of glucocorticoids secreted in Cushing's disease causes increased blood glucose concentration, sometimes to values as high as 140 to 200 mg. per cent. This effect probably results mainly from enhanced gluconeogenesis. If this "adrenal diabetes" lasts for many months, the beta cells in the islets of Langerhans occasionally "burn out" because the high blood glucose greatly overstimulates them to secrete insulin. The destruction of these cells then causes frank diabetes mellitus, which is permanent for the remainder of life.

The effects of glucocorticoids on protein catabolism are often profound in Cushing's disease, causing greatly decreased tissue proteins almost everywhere in the body with the exceptions of the liver and the plasma proteins. The loss of protein from the muscles in particular causes severe weakness. The loss of protein synthesis in the lymphoid tissues leads to a diminished immunity system, so that many of these patients die of infections. Even the collagen fibers in the subcutaneous tissue are di-

minished so that the subcutaneous tissues tear easily, resulting in development of large *purplish striae;* these are actually scars where the subcutaneous tissues have torn apart. In addition, lack of protein deposition in the bones causes *osteoporosis* with consequent weakness of the bones.

**Diagnosis and Treatment of Cushing's Disease.** Diagnosis of Cushing's disease is most frequently made on the basis of such typical findings as the buffalo torso, puffiness of the face, mild masculinizing effects, elevated blood glucose concentration that is moderately insulin resistant, several times normal plasma levels of cortisol, and secretion of 3 to 5 times the normal quantities of 17-hydroxysteroids in the urine.

Treatment in Cushing's disease consists of decreasing the secretion of ACTH, if this is possible. Hypertrophied pituitary glands that oversecrete ACTH can be surgically removed or destroyed by radiation. If ACTH secretion cannot easily be decreased, the only satisfactory treatment is usually bilateral total or partial adrenalectomy followed by administration of adrenal steroids to make up for any insufficiency that develops.

**Cushing Syndrome Caused by Adrenal Hyperplasia or Adrenal Tumor.** Some instances of Cushing syndrome result from intrinsic hyperplasia or tumor of the adrenal cortex rather than from excessive stimulation of the adrenal cortex by ACTH, and this secretion is often unresponsive to ACTH or to lack of ACTH. This condition differs from the usual Cushing syndrome principally in that there are no adrenogenic effects; instead, effects are related almost entirely to hyperglycemia and protein wasting. The therapy is usually surgical removal of the tumor or partial removal of the hyperplastic glands.

## PRIMARY ALDOSTERONISM

Occasionally a small tumor of the zona glomerulosa cells occurs and secretes large amounts of aldosterone. The effects of this are those illustrated in Figure 77–1 and discussed in detail earlier in the chapter. The most important effects are hypokalemia, slight increase in extracellular fluid volume and blood volume, very slight increase in plasma sodium concentration (usually not over a 2 to 3 per cent increase), and usually moderate hypertension. Especially interesting in primary aldosteronism are occasional periods of muscular paralysis caused by the hypokalemia. The paralysis is caused by hyperpolarization of the nerve fibers, as was explained in Chapter 10.

The failure of primary aldosteronism to cause an excessively high sodium ion concentration in the body fluids is probably caused by the following mechanism: The increase in sodium reabsorption causes simultaneous increase in water reabsorption. Also, the slight rise in extracellular fluid sodium causes polydipsia. Retention of fluid then elevates

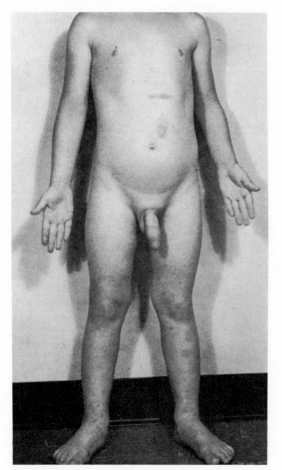

**Figure 77–8.** Adrenogenital syndrome in a 4-year-old boy. (Courtesy of Dr. Leonard Posey.)

the arterial pressure, increases the glomerular filtration rate, and leads to polyuria. This in turn washes the excess sodium chloride on through the kidneys despite the fact that the urine concentration of sodium chloride is low.

Treatment of primary aldosteronism is usually surgical removal of the tumor.

## ADRENOGENITAL SYNDROME

An occasional adrenocortical tumor secretes excessive quantities of androgens that cause intense masculinizing effects throughout the body. If this occurs in a female, she develops virile characteristics, including growth of a beard, development of a much deeper voice, occasionally development of baldness if she also has the genetic inheritance for baldness, development of a masculine distribution of hair on the body and on the pubis, growth of the clitoris to resemble a penis, and deposition of proteins in the skin and especially in the muscles to give typical masculine characteristics.

In the prepubertal male a virilizing adrenal tumor causes the same characteristics as in the female, plus rapid development of the male sexual organs and creation of male sexual desires. Typical development of the male sexual organs in a 4-year-old boy with the adrenogenital syndrome is shown in Figure 77–8.

In the adult male, the virilizing characteristics of the adrenogenital syndrome are usually completely obscured by the normal virilizing characteristics of the testosterone secreted by the testes. Therefore, it is often difficult to make a diagnosis of adrenogenital syndrome in the male adult.

In the adrenogenital syndrome, the excretion of 17-ketosteroids (which are derived from androgens) in the urine may be as much as 10 to 15 times normal.

# REFERENCES

Azarnoff, D. L.: Steroid Therapy. Philadelphia, W. B. Saunders Company, 1975.

Blair-West, J., Coghlan, J. P., Denton, D. A., and Wright, R. D.: Effect of endocrines on salivary glands. In Code, C. F., and Heidel, W. (eds.): Handbook of Physiology. Sec. 6, Vol. 2. Baltimore, The Williams & Wilkins Company, 1967, p. 633.

Bransome, E. D., Jr.: Adrenal cortex. Ann. Rev. Physiol., 30:171, 1968.

Chesley, L. C.: Renin, angiotensin, and aldosterone. Obstet. Gynecol. Ann., 3:235, 1974.

Christy, N. P.: The Human Adrenal Cortex. Hagerstown, Maryland, Harper & Row, Publishers, 1971.

Ciba Foundation Study Group No. 27: The Human Adrenal Cortex. Boston, Little, Brown and Company, 1967.

Conn, J. W.: The evolution of primary aldosteronism: 1954–1967. Harvey Lectures, 1966–1967, p. 257.

Daly, J. R., and Evans, J. I.: Daily rhythms of steroid and associated pituitary hormones in man and their relationship to sleep. Adv. Steroid Biochem. Pharmacol., 4:61, 1974.

Dempsey, M. E.: Regulation of steroid biosynthesis. Ann. Rev. Biochem., 43:967, 1974.

Denton, D. A.: Evolutionary aspects of the emergence of aldosterone secretion and salt appetite. Physiol. Rev., 45:245, 1965.

Dorfman, R. I.: Steroid Hormones. New York, American Elsevier Publishing Company, 1974.

Eisenstein, A. B.: The Adrenal Cortex. Boston, Little, Brown and Company, 1967.

Hamburger, J. I.: Nontoxic Goiter. Springfield, Ill., Charles, C Thomas, Publisher, 1973.

Hofmann, K.: Relations between chemical structure and function of adrenocorticotropin and melanocyte-stimulating hormones. In Greep, R. O., and Astwood, E. B. (eds.): Handbook of Physiology. Sec. 7, Vol. 4, Part 2. Baltimore, The Williams & Wilkins Company, 1974, p. 29.

Kastin, A. J., Viosca, S., and Schally, A. V.: Regulation of melanocyte-stimulating hormone release. In Greep, R. O., and Astwood, E. B. (eds.): Handbook of Physiology. Sec. 7, Vol. 4, Part 2. Baltimore, The Williams & Wilkins Company, 1974, p. 551.

Kornel, L.: On the effects and the mechanism of action of corticosteroids in normal and neoplastic target tissues: findings and hypotheses, with a review of information on intracellular steroid receptors in general. Acta Endocrinol. (Kbh), 74 (Suppl. 178):1, 1973.

Leung, K., and Munck, A.: Peripheral actions of glucocorticoids. Ann. Rev. Physiol., 37:245, 1975.

McCaa, R. E., Cowley, A. W., Jr., McCaa, C. S., and Guyton, A. C.: Return of plasma aldosterone concentration to control levels in nephrectomized-decapitated dogs. Inter. Cong. Endocrin., 1972.

McCaa, R. E., Young, D. B., Guyton, A. C., and McCaa, C. S.: Evidence for a role of an unidentified pituitary factor in regulatory aldosterone secretion during altered sodium balance. Suppl. 1. Circ. Res., 34:1, 1974; 35:1, 1974.

McKerns, K. W. (ed.): Functions of the Adrenal Cortex. Vols. 1 and 2. New York, Appleton-Century-Crofts, 1968.

Metabolic effects of glucocorticoids in man. Nutr. Rev., 32:301, 1974.

Munck, A., Wira, C., Young, D. A., Mosher, K. M., Hallahan, C., and Bell, P. A.: Glucocorticoid-receptor complexes and the earliest steps in the action of glucocorticoids on thymus cells. J. Steroid Biochem., 3:567, 1974.

Reid, I. A., and Ganong, W. F.: The hormonal control of sodium excretion. In Guyton, A. C. (ed.): MTP International Review of Science: Physiology. Vol. 5. Baltimore, University Park Press, 1974, p. 205.

Sharp, G. W. G., and Leaf, A.: Effects of aldosterone and its mechanism of action on sodium transport. In Handbook of Physiology. Sec. 8. Baltimore, The Williams & Wilkins Company, 1973, p. 815.

Symington, T.: Functional Pathology of the Human Adrenal Gland. New York, Churchill Livingstone, Div. of Longman, Inc., 1970.

Urquhart, J.: Physiological actions of adrenocorticotropic hormone. In Greep, R. O., and Astwood, E. B. (eds.): Handbook of Physiology. Sec. 7, Vol. 4, Part 2. Baltimore, The Williams & Wilkins Company, 1974, p. 133.

Yates, F. E.: The pituitary adrenal cortical system: stimulation and inhibition of secretion of corticotropin. In Guyton, A. C. (ed.): MTP International Review of Science: Physiology. Vol. 5. Baltimore, University Park Press, 1974, p. 109.

Yates, F. E., and Maran, J. W.: Stimulation and inhibition of adrenocorticotropin release. In Greep, R. O., and Astwood, E. B. (eds.): Handbook of Physiology. Sec. 7, Vol. 4, Part 2. Baltimore, The Williams & Wilkins Company, 1975, p. 367.

Yates, F. E., and Urquhart, J.: Control of plasma concentrations of adrenocortical hormones. Physiol. Rev., 42:359, 1962.

# 78

# Insulin, Glucagon, and Diabetes Mellitus

The pancreas, in addition to its digestive functions, secretes two important hormones, *insulin* and *glucagon*. The purpose of this chapter is to discuss the functions of these hormones in regulating glucose, lipid, and protein metabolism, as well as to discuss briefly the two diseases—*diabetes mellitus* and *hyperinsulinism*—caused, respectively, by hyposecretion of insulin and excess secretion of insulin.

**Physiologic Anatomy of the Pancreas.** The pancreas is composed of two major types of tissues, as shown in Figure 78–1: (1) the *acini*, which secrete digestive juices into the duodenum, and (2) the *islets of Langerhans*, which do not have any means for emptying their secretions externally but instead secrete insulin and glucagon directly into the blood. The digestive secretions of the pancreas were discussed in Chapter 64.

The islets of Langerhans of the human being contain two major types of cells, the *alpha* and *beta* cells, which are distinguished from one another by their morphology and staining characteristics. The beta cells secrete insulin, and the alpha cells secrete glucagon.

Beta cells are often still present in the pancreas of a person who has severe diabetes, but these cells then usually have a hyalinized appearance and contain no secretory granules; also, they do not exhibit staining reactions for insulin, in contrast to the normal beta cells, which do give staining reactions for insulin. Consequently, the hyalinized beta cells in diabetic patients are considered to be nonfunctional.

**Chemistry of Insulin.** Insulin is a small protein with a molecular weight of 5808 for human insulin. It is composed of two amino acid chains, illustrated in Figure 78–2, connected to each other by disulfide linkages. When the two amino acid chains are split apart, the functional activity of the insulin molecule is lost.

Before insulin can exert its function it must become "fixed" to the cell membranes, as explained later in the chapter. The mechanism by which insulin fixes to the tissues is to form bonds between the disulfide ring structure of the upper chain and sulfhydryl radicals of the tissues. Antidiuretic hormone fixes to the epithelial cells of the renal tubules by essentially the same type of linkage.

## EFFECT OF INSULIN ON CARBOHYDRATE METABOLISM

The earliest studies of the effect of insulin on carbohydrate metabolism showed three basic effects of the hormone: (1) enhanced rate of glucose metabolism, (2) decreased blood glu-

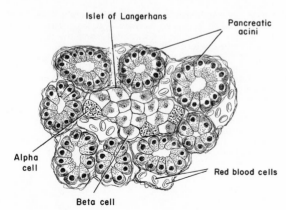

**Figure 78–1.** Physiologic anatomy of the pancreas.

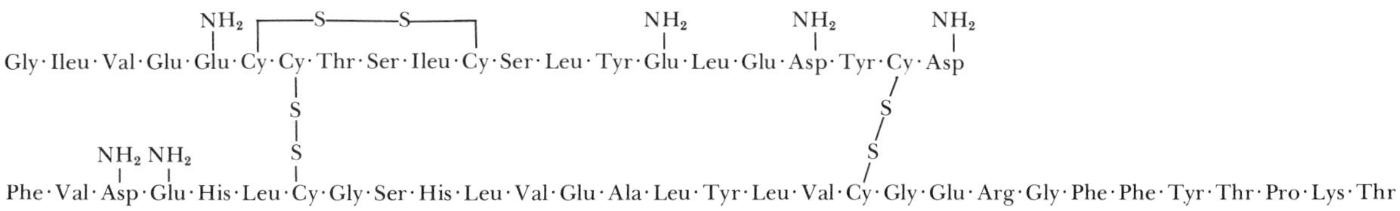

$$
\begin{array}{l}
\text{NH}_2 \quad\quad\quad \text{S————S} \quad\quad\quad \text{NH}_2 \quad\quad \text{NH}_2 \quad\quad \text{NH}_2 \\
\text{Gly·Ileu·Val·Glu·Glu·Cy·Cy·Thr·Ser·Ileu·Cy·Ser·Leu·Tyr·Glu·Leu·Glu·Asp·Tyr·Cy·Asp}
\end{array}
$$

Figure 78–2.  The human insulin molecule.

cose concentration, and (3) increased glycogen stores in the tissues. All of these effects probably result from the following basic mechanisms.

## FACILITATION OF GLUCOSE TRANSPORT THROUGH THE CELL MEMBRANE

The most important basic effect of insulin, which has been demonstrated many times, is its ability to increase the rate of glucose transport through the membranes of most cells in the body. This effect is illustrated in Figure 78–3. In the complete absence of insulin, the overall rate of glucose transport into the cells of the entire body becomes only about one-fourth the normal value. On the other hand, when great excesses of insulin are secreted and when an excess of glucose is available to be transported, the rate of glucose transport into the cells may be as great as five times normal. This means that, between the limits of no insulin at all and great excesses of insulin, the rate of glucose transport

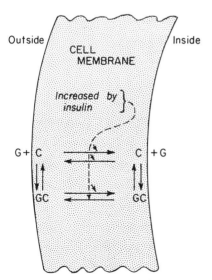

**Figure 78–3.** Effect of insulin in increasing glucose transport in either direction through the cell membrane.

for many tissues—but not for all, as will be discussed later—can be altered as much as 20-fold.

Some experiments that have been used to prove the accelerating effect of insulin on glucose transport through the cell membranes are the following:

1. Since glucose immediately becomes phosphorylated after entry into the cells and since almost no *free* glucose normally persists more than a few seconds after entering the cells, studies to prove an increase in intracellular glucose under the influence of insulin have been difficult. However, this has been accomplished in two different types of experiments: first, in experiments in which the concentration of extracellular glucose is so extreme that the glucose entering the cells cannot all be phosphorylated instantaneously and, second, in experiments in which the medium is at 4° C., a level at which the glucose-phosphorylating mechanism of the cells is almost blocked. Figure 78–4 shows increased "free" glucose inside the cells of a perfused heart under the influence of insulin, illustrating a many-fold effect of the insulin.

2. Experiments have also been conducted on the transport of other monosaccharides similar to glucose but which are not metabolized in the cells. For instance, tissues in the body cannot metabolize *l*-arabinose; therefore, its intracellular concentration can be measured without difficulty. It has been easy to demonstrate that insulin causes marked increase in *l*-arabinose transport through the cellular membrane. The same is also true of a number of other nonmetabolizable monosaccharides, including especially xylose and galactose.

3. A third group of experiments has been performed to compare the effects of insulin on glucose metabolism in (a) suspensions of whole cells and (b) cellular homogenates. In these, insulin increases the metabolism of glucose by whole cells; but in the cellular homogenates, which have no intact cell membranes, the rate of glucose metabolism is high all of the time and is not altered by the presence of insulin. This indicates that a cell membrane is necessary for full function of insulin.

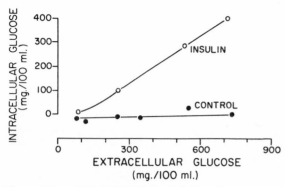

**Figure 78–4.** Effect of insulin to enhance the concentration of glucose inside muscle cells. Note that in the absence of insulin (control) the intracellular glucose concentration remained near zero despite very high extracellular concentrations of glucose. (From Park, Morgan, Kaji, and Smith, in Eisenstein (ed.): The Biochemical Aspects of Hormone Action. Little, Brown and Co.)

**Mechanism by which Insulin Accelerates Glucose Transport.** As pointed out in the discussion of the cell membrane in Chapter 4, glucose cannot pass into the cell through the cell pores but instead must enter by some transport mechanism through the membrane matrix. Figure 78–3 depicts the generally believed method by which glucose enters the cell, showing that glucose almost certainly combines with a carrier substance in the cell membrane and then is transported to the inside of the membrane where it is released to the interior of the cell. The carrier then returns to the outer surface of the membrane to transport additional quantities of glucose. This process can occur *in either direction,* as shown by the reversible arrows in the diagram.

Glucose transport through the cell membrane does not occur against a concentration gradient. That is, once the glucose concentration inside the cell rises to as high as the glucose concentration on the outside, additional glucose will not be transported to the interior. Therefore, the glucose transport process is one of *facilitated diffusion,* which means simply that glucose diffuses through the membrane by means of a facilitating carrier mechanism.

Insulin promotes glucose transport into the cell by stimulating this process of facilitated diffusion. In the complete absence of insulin, the carrier mechanism operates very poorly, whereas in the presence of insulin this mechanism is enhanced many-fold. Insulin increases the transport of glucose within seconds to minutes, indicating a direct action in the cell membrane itself.

The way in which insulin enhances facilitated diffusion is still largely unknown. All that is known is that insulin combines with a "receptor" protein in the cell membrane—a protein having a molecular weight of about 300,000. This may be the glucose carrier itself, or it may be merely the first step in a chain of events that leads to activation of the carrier system.

**Tissues in which Insulin is Effective.** Enhanced transport of glucose through the cell membrane by insulin is particularly effective in skeletal muscle and adipose tissue, and these two together make up approximately 65 per cent of the entire body by weight. In addition, insulin enhances glucose transport into the heart and into certain smooth muscle organs, such as the uterus.

On the other hand, insulin does not enhance glucose transport into the brain cells and red blood cells and does not enhance its transport through the intestinal mucosa or through the tubular epithelium of the kidney. However, these tissues together amount to less than 5 per cent of the total body mass. The effects of insulin in the remaining 30 per cent of the body are mainly unknown except for those in the liver, which will be discussed below, but there is reason to believe that most of this tissue also responds at least partially to insulin.

In summary, insulin is extremely important for glucose transport into the cells of most tissues of the body. The most important exception to this is the brain; here, glucose transport is probably more dependent on *simple diffusion* through the blood-brain barrier (see Chap. 32) than through the cell membrane.

### EFFECT OF INSULIN ON GLUCOSE UTILIZATION, GLYCOGEN STORAGE AND FAT STORAGE IN EXTRAHEPATIC TISSUES

**Glucose Utilization.** Since insulin lack decreases glucose transport into most body cells to about one-fourth normal, without insulin most tissues of the body (with the major exception of the brain) must depend on other metabolic substrates for energy. On the other hand, in the presence of excess insulin, glucose enters the cells in great abundance. The quantity of glucose that is used for energy, however, is determined by factors other than the amount of glucose that enters the cell—for instance, by the rate at which ADP becomes available to accept energy, as was explained in Chapter 67. The excess glucose (over and above that used for energy) is stored in the cells in two forms—either as glycogen or, after conversion, as fat.

**Glycogen Storage.** When there is an excess of both insulin and glucose, the glycogen stores in skeletal muscle increase markedly; also there is moderate enhancement of glycogen in the skin, glands, and at least some other tissues. This glycogen storage results from several effects: (1) the *abundance of glucose entering the cells,* (2) the inhibition of some of the glycolytic enzymes, which slows the utilization of glucose in the glycolytic pathway; this results especially from *inhibition of phosphofructokinase* by citrate ion and ATP, both of which are present in the cell in abundance when excess energy supplies are available to the cell; (3) the *activation of glycogen synthetase,* an enzyme that is required for synthesis of glycogen, and (4) the *inactivation of phosphorylase,* an enzyme that is required for breakdown of glycogen. These effects probably all result from the excess of glucose and the products of glucose metabolism in the cell. The net result is marked enhancement of glycogen storage in the tissues, especially in muscle.

*Fat Storage.* In adipose tissue, the excess glucose transported into the fat cells is largely converted into fat and stored in this form. In liver cells, after a large share of the excess glucose has been stored as glycogen and the glycogen content has reached its limit in these cells, most of the remaining excess is converted into fat. The reason for these differences among fat cells, liver cells, and muscle cells is simply that the enzyme systems are different. The conversion of glucose to fat will be discussed later in the chapter in relation to the effects of insulin on fat metabolism.

## ACTION OF INSULIN ON GLUCOSE METABOLISM IN THE LIVER

The liver plays many special roles in the metabolism of glucose, but the two most important of these are:

1. In the presence of excess insulin, excess glucose, or both, the liver takes up large quantities of glucose from the blood.

2. In the absence of insulin or when the blood glucose concentration falls very low, the liver gives glucose back to the blood.

Thus, the liver acts as an important blood glucose buffer mechanism, helping to keep the blood glucose concentration from rising too high or from falling too low. Following are some of the mechanisms by which these effects occur:

### Liver Uptake of Glucose

In the presence of excess insulin, excess glucose, or both, large quantities of glucose are taken up by the hepatic cells. However, the mechanism of this effect of insulin on hepatic cells is different from that on other cells because insulin does not enhance glucose transport through the liver cell membranes. In fact, the hepatic cell membrane is highly permeable to glucose diffusion in either direction. Instead, the enhanced glucose uptake results from the function in liver cells of a special enzyme, *glucokinase,* which is not present in other cells. This enzyme causes phosphorylation of glucose that "traps" glucose inside the liver cells. That is, once phosphorylated, the glucose cannot pass outward through the membrane.

Insulin has the special effect of increasing the concentration of glucokinase in liver cells many-fold, the increase beginning in 1 to 2 hours and becoming maximal 12 or more hours later; glucose uptake continues to increase all this while.

At the same time that there is an increase in glucose in the liver cells, there is activation of essentially all the intrahepatic enzyme systems for utilization of glucose and an increase in quantity over a period of hours. Especially important is the increase in the glycolytic enzymes, which causes the liver to use glucose to supply its energy while decreasing the use of fats and proteins for energy. In addition, large quantities of glucose are deposited as glycogen as well as being converted to fat.

**Deposition of Glycogen in the Liver.** In the presence of insulin large quantities of glycogen are deposited in the liver; this deposition results from two changes in the enzymes of the cells: (1) *increase in glycogen synthetase*, which causes synthesis of glycogen, and (2) *decrease in phosphorylase,* which causes breakdown of glycogen. Thus, more is formed and less is broken down. Once the quantity of stored glycogen reaches 5 to 6 per cent of the liver mass, the glycogen then inhibits the enzyme glycogen synthetase so that no additional glycogen will be deposited. Then the remaining excess glucose is converted almost entirely to fat.

**Formation of Fat.** Under the influence of insulin, and when an excess of glucose is available in the blood simultaneously, a major share of all the glucose that enters the body is converted by the liver into fat. This results from several effects: (1) an excess of *acetyl Co-A* is formed in the hepatic cells as an end-product of glycolysis; (2) an excess of *citrate ions* is formed by the citric acid cycle when excess acetyl Co-A enters this cycle (these ions have a direct effect on activating acetyl Co-A carboxylase, the enzyme required to initiate the first stage of lipogenesis); and (3) large quantities of $\alpha$-glycerophosphate are formed providing the glycerol nucleus that is required to combine with fatty acids during the formation of triglycerides. Once formed, the triglycerides are transported in the blood lipoproteins to the adipose tissue where they are stored as was explained in Chapter 68. In the human being the liver is the source of most of the fat that is stored in the adipose tissue, though small quantities can be synthesized in the fat cells themselves.

### Release of Glucose from the Liver in the Absence of Insulin

In the absence of insulin (or also when the blood glucose concentration falls too low) the liver releases large quantities of glucose into the blood. This glucose is derived from two sources: (1) breakdown of *stored glycogen,* and (2) formation of new glucose by the process of *gluconeogenesis.*

**Breakdown of Glycogen.**   In the absence of insulin, liver glycogen diminishes for two reasons: (1) *phosphorylase* is activated, which catalyzes the breakdown of glycogen, and (2) *glycogen synthetase,* the enzyme necessary for synthesis of glycogen is inhibited. Both of these effects probably result from increased quantities of cyclic AMP in the liver cells, an effect that occurs when insulin is deficient. The mechanism of the effect of cyclic AMP on glycogen degradation will be discussed in more detail later in the chapter in relation to glucagon.

The glycogen is degraded into glucose-6-phosphate, and this in turn is further degraded into glucose by the enzyme *glucose*-6-*phosphatase,* the activity of which is itself enhanced by the presence of glucose-6-phosphate. Finally, once the glucose is in the free form, it readily permeates the liver cell membrane and diffuses into the circulating blood.

**Enhanced Gluconeogenesis in the Absence of Insulin.**   Especially large quantities of new glucose are synthesized in the liver by the process of gluconeogenesis when insulin is lacking. Lack of insulin stimulates this process mainly by mobilizing amino acids from the protein stores of the body and glycerol from the fat stores, effects of insulin lack that will be discussed in more detail later in the chapter; these two substances are then converted into glucose by the liver. The enzymes required for gluconeogenesis are also increased, but it is not known whether this change results from the absence of insulin or is caused by the increased metabolic substrates made available to the liver. As much as 50 to 100 grams of glucose can be synthesized each day and released into the blood, an effect that occurs for weeks at a time in starvation and for months and years in patients with diabetes mellitus.

### EFFECT OF INSULIN EXCESS AND INSULIN LACK ON FAT METABOLISM

**Fat Storage in the Presence of Insulin.**   Insulin strongly enhances the transport of glucose into fat cells. The presence of excess glucose inside the fat cells has dramatic effects on promoting fat storage. Therefore, one of the most rapid and most potent effects of insulin is to promote fat storage in the adipose tissue.

The mechanism by which glucose promotes fat storage is the following: much of the glucose is immediately degraded by the process of glycolysis, and large quantities of $\alpha$-glycerophosphate, one of the degradation products, are formed inside the cell. This substance has a very strong affinity for combining with fatty acids to form triglycerides. Therefore, any fatty acids that are available, whether synthesized within the fat cell itself or split from lipoproteins by lipoprotein lipase (as explained in Chapter 68), will immediately combine with the $\alpha$-glycerophosphate and will be deposited as triglycerides. Most of the fat stored in this way in the human being is derived from the triglycerides of the lipoproteins—either the chylomicrons that have recently been absorbed from the gastrointestinal tract or the smaller lipoproteins that transport triglycerides from their site of synthesis in the liver, as explained earlier in the chapter. Smaller quantities of fatty acids are synthesized in the fat cells themselves in the same way that fatty acids are synthesized in the liver cells, as was also explained earlier.

**Release of Free Fatty Acids from Fat Cells in the Absence of Insulin.**   In the absence of insulin, fat not only is not stored in fat cells, it also immediately begins to be released in the form of free fatty acids. This results from the following mechanism: The triglycerides of fat cells are continually being split by the enzyme *lipase* in the fat cell to form fatty acid molecules and glycerol. Once split, the energetics of the cell's enzyme system are such that glycerol cannot recombine with the fatty acids to form new triglycerides; instead, the substance α-glycerophosphate is required. In the presence of insulin and glucose, this substance is always available because glucose is continually being transported into the cell and new α-glycerophosphate is being formed. Therefore, normally, the free fatty acids that are continually being formed in the fat cells are immediately converted back into triglycerides and stored again. Since in the absence of insulin no glucose is available, within minutes the concentration of α-glycerophosphate falls too low to recapture the free fatty acids. Consequently, these diffuse outward through the cell membranes into the circulating plasma where they combine with albumin and are transported throughout the body as explained in Chapter 68. The fatty acids are then used primarily for energy in almost all cells of the body, with the major exception of the brain cells, as will be discussed later in the chapter. Figure 78–5 illustrates this marked effect of insulin lack on the free fatty acid concentration in the body fluids, showing an increase in the free fatty acids occurring within minutes after the onset of insulin lack. This is associated with a much slower increase in blood glucose concentration.

**Effect of Fatty Acids on Inhibiting Uptake and Utilization of Glucose by the Cell—the Glucose–Fatty Acid Cycle.**  Lack of insulin causes severe deficiency of glucose uptake and utilization by cells because of decreased transport of glucose through the cell membrane. But in addition to this severe depression of glucose metabolism caused by a direct effect of insulin lack, the excess fatty acids in the blood, also because of insulin lack, still cause much additional indirect depression of cellular glucose utilization. This effect results from the following mechanism:

First, on entering the cells the excess fatty acids are immediately split to form acetyl Co-A. This in turn enters the citric acid cycle and provides the energy required by the cell.

Second, two of the important products of this energy system are *citrate ion* and *ATP*, both of which have a strong inhibitory effect on the enzyme *phosphofructokinase*, which is required to initiate glycolysis in the cell. Therefore, the use of glucose for energy almost entirely ceases.

Third, failure of the glycolytic process causes

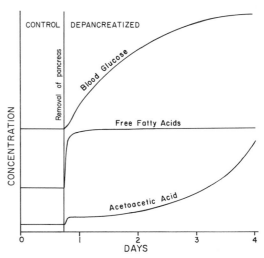

**Figure 78–5.**  Effect of removing the pancreas on the concentrations of blood glucose, plasma free fatty acids, and acetoacetic acid.

*glucose-6-phosphate* in large quantities to build up inside the cell. This in turn inhibits *hexokinase*, the enzyme that phosphorylates glucose as it first enters the cell. Consequently, the concentration of free glucose in the cell also rises considerably, which causes back-diffusion of glucose through the cell membrane, thereby greatly reducing the net uptake of glucose. Thus, glucose uptake and utilization by the cells are further depressed.

This total process is called the *glucose-fatty acid cycle*. Its net effect is approximately to double the depressive effect of insulin lack on cellular uptake and utilization of glucose.

**Effect of Insulin Lack on Blood Lipid Concentration.**  In addition to increasing the fatty acids in the circulating blood, all of the other lipid components of the plasma also greatly increase in the absence of insulin. This increase results from the transport of excess fatty acids into the liver where they are synthesized into *triglycerides, cholesterol,* and *phospholipids* and then are discharged into the blood in the form of lipoproteins. Occasionally, the blood lipids increase as much as five-fold, giving a total concentration of plasma lipids of several per cent rather than the normal 0.6 per cent. This high lipid concentration—especially the high concentration of cholesterol—leads to rapid development of atherosclerosis in persons with serious diabetes.

**Ketogenic Effect of Insulin Lack.**  Many of the fatty acids that enter the liver are also oxidized to form acetyl Co-A, and this in turn is condensed to form *acetoacetic acid*, which is released into the circulating blood. Most of this passes to the peripheral cells where it is again converted into acetyl Co-A and used for energy in the usual manner. However, in insulin lack, the acetoacetic acid concentration rises, especially when fatty acids are mobilized from the adipose tissue in far greater quantities than can be used for energy elsewhere in the body, as occurs in diabetes. Thus, Figure 78–5 illustrates a small increase in acetoacetic acid almost immediately after the onset of insulin lack followed by a much greater increase after several days of insulin lack, as often occurs in severe diabetes mellitus. As was explained in Chapter 68, some of the acetoacetic acid is also converted into *β-hydroxybutyric acid* and *acetone*. These two substances along with the acetoacetic acid are called *ketone bodies,* and their presence in large quantities in the circulating fluids is called *ketosis.* We shall see later that the acetoacetic acid and the β-hydroxybutyric acid can cause severe acidosis and coma in patients with severe diabetes, in the absence of heroic treatment, leading almost inevitably to death.

## EFFECT OF INSULIN ON PROTEIN METABOLISM AND GROWTH

**Effect of Insulin on Protein Synthesis.** The total quantity of protein in the body is increased by insulin. This results from at least three different functions of insulin that are similar to those of growth hormone:

1. Increased active transport of amino acid into the cells. (This differs from the increased glucose transport into the cells in that this transport can occur against a concentration gradient.)

2. Accelerated translation of the messenger RNA code by the ribosomes to form increased quantities of proteins.

3. Increased transcription of DNA in the cell nuclei to form increased quantities of RNA, which in turn leads to still further protein synthesis.

The basic mechanism (or mechanisms) by which insulin causes these effects is still unknown. Most physiologists believe that they are direct effects of insulin on the protein metabolic processes, but some feel that the effects could result secondarily from insulin transport of glucose into the cells, making available a multitude of glucose-derived intermediary compounds that themselves play many controlling roles in cellular metabolic processes.

**Effect of Insulin on Growth—Potentiation**

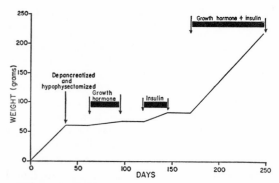

**Figure 78–6.** Effect of (a) growth hormone, (b) insulin, and (c) growth hormone plus insulin on growth in a depancreatized and hypophysectomized rat.

**of the Effects of Growth Hormone.** Insulin is nearly as essential for growth of an animal as is growth hormone. This effect is illustrated in Figure 78–6, which shows that a depancreatized, hypophysectomized rat essentially does not grow. Furthermore, neither growth hormone nor insulin alone will cause very significant growth. Yet, a combination of both of these hormones causes dramatic growth. Thus, the two hormones do not seem to affect growth by the same mechanism but instead by different mechanisms, both of which are necessary.

**Protein Depletion and Increased Plasma Amino Acids Caused by Insulin Lack.** The converse of the growth-promoting effect of insulin is the effect of insulin lack in causing extreme wasting of the body proteins, with consequent dumping of amino acids into the circulating body fluids and elevated plasma amino acid levels. These amino acids are mainly transported to the liver where they are used either for energy or for gluconeogenesis. Obviously, this degradation of large quantities of amino acids also leads to enhanced excretion of urea in the urine.

The cause of the protein degradation is the normal continuous breakdown of proteins that occurs in all tissues, but in this case the protein is not replaced by new synthesis of protein.

The protein wasting is one of the most serious of all the effects of severe diabetes mellitus. It can lead to extreme weakness as well as to many deranged functions of the organs. Also, the protein wasting and the lipolysis that occurs lead to severe weight loss.

## CONTROL OF INSULIN SECRETION

**Stimulation of Insulin Secretion by Blood Glucose.** At the normal fasting level of blood glucose of 80 to 90 mg./100 ml., the rate of insulin secretion is minimal—in the order of 10 ng./min./kg. of body weight. If the blood glucose concentration is suddenly increased to a level two to three times normal and is kept at this high level thereafter, insulin secretion increases markedly in three separate stages:

1. Insulin secretion increases up to 10-fold about 5 minutes after acute elevation of the blood glucose. However, this initial high rate of secretion is not maintained; instead, it decreases about halfway back toward normal in another 5 to 10 minutes.

2. After about 15 minutes, insulin secretion rises a second time, reaching a new plateau in 2 to 3 hours, this time usually at a rate of secretion even greater than that in the initial phase.

3. Over a period of a week or so, the rate of insulin secretion increases still more, often doubling the rate at the end of the first few hours.

These three stages of insulin secretion are believed to result from (a) an initial dumping of preformed insulin from the beta cells of the islets of Langerhans, (b) activation over the next few hours of the enzyme systems for increased production of insulin, and (c) hypertrophy of the beta cells during the next week or so.

As the concentration of blood glucose rises above 100 mg./100 ml. of blood, the rate of insulin secretion rises rapidly, reaching a peak that is some 10 to 20 times the basal level at blood glucose concentrations between 200 and 300 mg./100 ml. Thus, the increase in insulin secretion under a glucose stimulus is dramatic both in its rapidity and in the tremendous level of secretion achieved. Furthermore, the turn-off of insulin secretion is almost equally as rapid, occurring within minutes after reduction in blood glucose concentration back to the fasting level.

This response of insulin secretion to an elevated blood glucose concentration provides an extremely important feedback mechanism for regulating blood glucose concentration. That is, the rise in blood glucose increases insulin secretion, and the insulin in turn causes transport of glucose into the cells, thereby reducing the blood glucose concentration back toward the normal value. This mechanism will be discussed in more detail later in the chapter.

### Other Factors that Stimulate Insulin Secretion

**Amino Acids.** In addition to excess glucose stimulating insulin secretion most if not all of the amino acids have a similar effect. The most potent of these amino acids is *alanine*. However, this effect differs from glucose stimulation of insulin secretion in the following way: Amino acids administered in the absence of a rise in blood glucose cause only a small increase in insulin secretion. However, when administered at the same time that the blood glucose concentration is elevated, the glucose-induced secretion of insulin may be as much as doubled in the presence of the excess amino acids. Thus, the amino acids very strongly potentiate the glucose stimulus for insulin secretion.

The stimulation of insulin secretion by amino acids seems to be a purposeful response because the insulin in turn promotes transport of the amino acids into the tissue cells and also promotes intracellular formation of protein. That is, the insulin is important for proper utilization of the excess amino acids.

**Gastrointestinal Hormones.** A mixture of the three important gastrointestinal hormones—gastrin, secretin, and cholecystokinin—will cause a moderate increase in insulin secretion. These hormones are released in the gastrointestinal tract after a person eats a meal. They seem to cause an "anticipatory" increase in blood insulin in preparation for the glucose to be absorbed from the meal. These gastrointestinal hormones almost double the rate of insulin secretion following an average meal.

**Other Hormones.** Other hormones that either directly increase insulin secretion or potentiate the glucose stimulus for insulin secretion include glucagon, growth hormone, cortisol, and to a lesser extent progesterone and estrogen. The importance of the stimulatory effects of these hormones is that prolonged secretion of any one of them in large quantities can occasionally lead to exhaustion of the beta cells of the islets of Langerhans and thereby cause diabetes. Indeed, diabetes often occurs in persons maintained on high pharmacological doses of some of these hormones. It is particularly common in patients who already have a diabetic tendency, especially when glucocorticoids are administered in high concentration.

### ROLE OF INSULIN IN "SWITCHING" BETWEEN CARBOHYDRATE AND LIPID METABOLISM

From the above discussions it should already be clear that insulin promotes the utilization of carbohydrates for energy, while it depresses the utilization of fats. Conversely, lack of insulin causes fat utilization mainly to the exclusion of glucose utilization, except by brain tissue. Furthermore, the signal that controls this switching mechanism is principally the blood glucose concentration. When the glucose concentration is low, insulin secretion is suppressed and fat is utilized almost exclusively for energy; when the glucose concentration is high, insulin secretion is stimulated, and carbohydrate is utilized almost exclusively. Therefore, one of the most important functional roles of insulin in the body is to control which of these two foods from moment to moment will be utilized by the cells for energy.

At least two other hormones also play important roles in this switching mechanism: growth hormone from the anterior pituitary gland and cortisol from the adrenal cortex. Both of these are secreted in response to hypoglycemia, and both depress cellular utilization of glucose while exciting the utilization of fat.

Still another hormone that plays an important role in this overall mechanism is epinephrine. During periods of stress when the sympathetic nervous system is excited, epinephrine secreted by the adrenal medulla causes a large increase in both the blood sugar (resulting from glycogenolysis in the liver) and the blood concentration of fatty acids (caused by a direct lipolytic effect of epinephrine in the fat cells). Quantitatively, the enhancement of fatty acids is far greater than the enhancement of blood glucose. Therefore, epinephrine especially enhances the utilization of fat in such stressful states as exercise, circulatory shock, anxiety, and so forth.

# GLUCAGON AND ITS FUNCTIONS

Glucagon, a hormone secreted by the alpha cells of the islets of Langerhans, has several functions that are diametrically opposed to those of insulin. Most important of these is an increase in the blood glucose concentration, an effect which is exactly opposite to that of insulin.

Like insulin, glucagon is a small protein. It has a molecular weight of 3482 and is composed of a chain of 29 amino acids. On injection of purified glucagon into an animal, a profound *hyper*glycemic effect occurs. One microgram per kilogram of glucagon can elevate the blood glucose concentration approximately 20 mg./100 ml. of blood in about 20 minutes. For this reason, glucagon is frequently called *hyperglycemic factor*.

The two major effects of glucagon on glucose metabolism are (1) breakdown of glycogen (*glycogenolysis*) and (2) increased *gluconeogenesis*.

**Glycogenolysis and Increased Blood Glucose Concentration Caused by Gluca-**

**gon.** The most dramatic effect of glucagon is its ability to cause glycogenolysis in the liver, which in turn increases the blood glucose concentration within minutes.

It does this by the following complex cascade of events:

1. Glucagon activates *adenylcyclase* in the hepatic cell membrane,
2. Which causes the formation of *cyclic AMP,*
3. Which activates *protein kinase regulator protein,*
4. Which activates *protein kinase,*
5. Which activates *phosphorylase b kinase,*
6. Which converts *phosphorylase b* into *phosphorylase a,*
7. Which promotes the degradation of glycogen into glucose-1-phosphate,
8. Which then is dephosphorylated and the glucose released from the liver cells.

This sequence of events is exceedingly important for several reasons. First, it is one of the most thoroughly studied of all the *second messenger* functions of cyclic AMP. Second, it illustrates a cascading system in which each succeeding product is produced in greater quantity than the preceding product. Therefore, it represents a potent *amplifying* mechanism. This explains how only 1 $\mu$g./kg. of glucagon can have the extreme effect of causing hyperglycemia.

Infusion of glucagon for about four hours can cause such intensive liver glycogenolysis that all of the liver stores of glycogen become totally depleted.

**Gluconeogenesis Caused by Glucagon.** Even after all the glycogen in the liver has been exhausted under the influence of glucagon, continued infusion of this hormone causes continued hyperglycemia. This results from an effect of glucagon to increase the rate of gluconeogenesis in the liver cells. Unfortunately, the precise mechanism of this effect is unknown, but some of the processes that do seem to contribute to this are: (1) increase in the transport of amino acids into the liver; (2) increase in the conversion of amino acids to glucose precursors; (3) increase in proteolysis in extrahepatic tissues, thereby supplying additional amino acids for the gluconeogenesis process; and (4) enhancement of lipolysis in adipose tissue, thereby providing glycerol that can also be used for gluconeogenesis. Some of these effects, if not all of them, are mainly or partially promoted by the increased cyclic AMP induced in the cells by glucagon.

**Other Effects of Glucagon.** Most other effects of glucagon occur only when the concentration of this substance rises far above that normally found in the body fluids. Some of these include enhanced strength of the heart, enhanced bile secretion, enhanced secretion of calcitonin, and inhibition of gastric acid secretion. However, all of these changes are probably unimportant in the normal function of the body.

### REGULATION OF GLUCAGON SECRETION

**Effect of Blood Glucose Concentration.** Changes in blood glucose concentration have exactly the opposite effect on glucagon secretion as on insulin secretion. That is, a *decrease* in blood glucose increases glucagon secretion. When the blood glucose falls to as low as 70 mg./100 ml. of blood, the pancreas secretes very large quantities of glucagon, and this secretion rapidly mobilizes glucose from the liver. Thus, glucagon protects against hypoglycemia. (It has been claimed that patients dying of the rare disease idiopathic hypoglycemia have no alpha cells in their islets of Langerhans and therefore secrete no glucagon.)

**Effect of Exercise.** In exercise the blood glucose concentration tends to decrease, and this increases glucagon secretion. The increased glucagon in turn plays an important role in mobilizing glucose from the liver for use by the muscles.

**Effect of Starvation.** Starvation, like exercise, depletes the blood glucose. Here again, glucagon is abundantly secreted, thus helping to maintain a normal blood glucose concentration.

**Effect of Insulin Lack.** In the absence of insulin, glucose cannot enter the cells—not even the alpha cells of the islets of Langerhans. Therefore, even when the blood glucose concentration is extremely high in severe diabetic patients, glucagon secretion is still greatly elevated. This indicates that it is the glucose concentration *inside the alpha cells* and not in the extracellular fluid that is the controller of glucagon secretion.

**Effect of Amino Acids.** Amino acids enhance the secretion of glucagon, an effect exactly opposite to that of glucose. The physiological importance of this is that it is believed to prevent the hypoglycemia that would otherwise result when a meal of pure protein is ingested, because the amino acids from the protein enhance insulin secretion and thereby cause decreased blood glucose. The increased

glucagon secretion theoretically can nullify this effect.

### GLUCAGONLIKE EFFECTS OF EPINEPHRINE

Epinephrine (and to a slight extent norepinephrine as well) has an effect similar to that of glucagon in causing glycogenolysis. However, the effect of epinephrine on *liver glycogenolysis* is far less potent than that of glucagon. On the other hand, epinephrine has a much more potent effect in causing glycogenolysis in muscle cells.

Epinephrine also has a far more potent effect in causing fatty acid mobilization from adipose tissue than does glucagon, which has a similar but weaker effect.

## REGULATION OF BLOOD GLUCOSE CONCENTRATION

In the normal person the blood glucose concentration is very narrowly controlled, usually in a range between 80 and 90 mg./100 ml. of blood in the fasting person each morning before breakfast. This concentration increases to 120 to 140 mg./100 ml. during the first hour or so following a meal, but the feedback systems for control of blood glucose return the glucose concentration very rapidly back to the control level, usually within two hours after the last absorption of carbohydrates. Conversely, in starvation the gluconeogenesis function of the liver provides the glucose that is required to maintain the fasting blood glucose level.

We have already discussed several of the most important features of blood glucose regulation, including especially (1) the effect of increased insulin secretion on returning an elevated blood glucose level back toward normal, and (2) the effect of increased glucagon secretion on returning a depressed glucose concentration level back up to normal. However, there are several other important features of blood glucose regulation that still require discussion. First, however, let us explain the importance of maintaining a normal concentration of blood glucose.

**Importance of Blood Glucose Regulation.** One might ask the question: Why is it important to maintain a constant blood glucose concentration, particularly since most tissues can shift to utilization of fats and proteins for energy in the absence of glucose? The answer is that glucose is the only nutrient that can be utilized by the *brain, retina,* and *germinal epithelium* in sufficient quantities to supply them with their required energy. Therefore, it is important to maintain a blood glucose concentration at a sufficiently high level to provide this necessary nutrition.

More than half of all the glucose formed by gluconeogenesis during the interdigestive period is used for metabolism in the brain. Indeed, it is important that the pancreas not secrete any insulin during this time, for otherwise the scant supplies of glucose that are available would all go into the muscles and other peripheral tissues, leaving the brain without a nutritive source.

On the other hand, it is also important that the blood glucose concentration not rise too high for three reasons: First, glucose exerts a large amount of osmotic pressure in the extracellular fluid, and, if the glucose concentration rises to excessive values, this can cause considerable cellular dehydration. Second, an excessively high level of blood glucose concentration causes loss of glucose in the urine. And third, this causes osmotic diuresis in the kidneys, which can deplete the body of its fluids.

**Glucose-Buffer Function of the Liver.** In the preceding discussion it was pointed out that glucose can readily penetrate hepatic cells in either direction, passing from the extracellular fluid spaces into the hepatic cells to be stored as glycogen, or the glycogen can be split to glucose that then passes rapidly into the extracellular fluids. Thus, the liver acts as a large storage vault for glucose, and because of this it acts as a glucose-buffer system. When excess quantities of glucose enter the extracellular fluid, about two-thirds of this is stored almost immediately in the liver, and this prevents excessive increase in blood glucose concentration. Conversely, when the blood glucose concentration falls below normal, the stored glucose in the liver rapidly replenishes the blood glucose.

Insulin plays an important role in controlling the glucose-buffer function of the liver, for in the absence of insulin, glucokinase is greatly depressed in the liver cells, so that glucose is not readily stored in the liver in times of excess and therefore is not available for release in times of glucose need.

The liver glucose-buffer system reduces the variation in blood glucose concentration to approximately one-third what it is without the liver. In liver disease, the blood glucose concentration is poorly regulated, its concentration

rising or decreasing much more than usual upon the slightest provocation.

**Role of the Sympathetic Nervous System in Preventing Hypoglycemia.** Hypoglycemia has a direct stimulatory effect on the hypothalamus in exciting the sympathetic nervous system. This stimulation in turn causes release of epinephrine and norepinephrine from the adrenal medulla as well as release of additional norepinephrine at the sympathetic nerve terminals. The epinephrine, as well as the norepinephrine to a slight extent, has a glucagonlike effect in causing glycogenolysis in the liver. This in turn causes rapid release of glucose into the circulating blood, thus returning the blood glucose concentration back toward normal. This particular effect of epinephrine is frequently used by clinicians to treat hypoglycemic shock; that is, by injecting epinephrine they can increase the blood glucose concentration by as much as 20 to 50 milligrams per cent within a few minutes.

**Effect of Other Hormones on Blood Glucose Regulation.** In previous chapters it has already been pointed out that both growth hormone from the anterior pituitary gland and cortisol from the adrenal cortex have potent effects on increasing the blood glucose concentration. Both of these hormones diminish the utilization of glucose by the peripheral tissue cells, thereby increasing blood glucose concentration. In addition, cortisol has an especially powerful effect on promoting gluconeogenesis, which can increase the blood glucose concentration sometimes to greater than 150 mg./100 ml. of blood for long periods of time.

Both growth hormone and glucocorticoids are secreted in large amounts in prolonged hypoglycemic states such as starvation and low carbohydrate intake. Therefore, over a long period of time they help to maintain a normal blood glucose level when other factors tend to lower the blood glucose.

There are, however, two major differences between the control of blood glucose concentration by these hormones and its control by insulin and glucagon: First, the insulin-glucagon control system is much more powerful. Second, the insulin-glucagon system is activated within minutes, whereas the growth hormone-cortisol system is activated only after a period of several hours.

# DIABETES MELLITUS

**Experimental Diabetes Mellitus.** In experimental animals diabetes mellitus has been induced by many different methods. The method universally effective is *depancreatectomy,* but in order to attain significant diabetes at least 90 to 95 per cent of the pancreas usually must be removed; otherwise, the islets of Langerhans in the remaining pancreatic tissue will be able to hypertrophy sufficiently to supply enough insulin for normal metabolic needs.

A second means of producing diabetes in animals is *administration of poisons such as alloxan or streptozotocin.* These substances cause the beta cells of the islets of Langerhans to swell and finally to degenerate, but it is difficult to adjust the dosage so that most of the beta cells will be destroyed without killing the animal by other toxic effects.

A still less reliable method of inducing diabetes is *administration of anterior pituitary extract.* Especially when anterior pituitary extract is administered in conjunction with thyroxine, glucocorticoids, or even with some of the sex hormones can it result in extreme swelling and final degeneration of the beta cells of the islets of Langerhans because of "exhaustion" of the beta cells, as explained earlier in the chapter. Though this is not a satisfactory means for producing diabetes in an experimental animal, it does indicate that nonpancreatic hormonal abnormalities probably cause the development of diabetes in some patients, such as that which occurs in 10 per cent of the patients with giantism or acromegaly, in which growth hormone is overly abundant.

Most factors that are known to destroy the beta cells are especially likely to destroy these cells when an animal is on a high carbohydrate diet, for such a diet also stimulates the beta cells. Therefore, it is easy to understand why diabetes is believed to result sometimes from excessive ingestion of sweets and other carbohydrates.

**Hereditary Factors.**    Though various factors may cause the beta cells of the islets of Langerhans to "burn out," diabetes is, in general, a hereditary disease, for almost all persons who develop diabetes, especially those who develop it in early years, can trace the disease to one or more forebears. The disease is transmitted as a recessive genetic characteristic which is present in about 20 per cent of the population. This means that about 4 per cent of the population at some time during life develops at least some degree of diabetes mellitus.

Mild diabetes often occurs in older people and especially in older people who are overweight. As long as these people continue to eat excessively, the blood glucose level is elevated, but as soon as they reduce their diet their blood glucose level often reverts to normal. This diabetic effect of obesity is caused mainly by the depression of glucose metabolism in the presence of excess fatty acids in the blood.

## PATHOLOGIC PHYSIOLOGY OF DIABETES

Most of the pathology of diabetes mellitus can be attributed to one of the following three major effects

of insulin lack: (1) decreased utilization of glucose by the body cells, with a resultant increase in blood glucose concentration to as high as 300 to 1200 mg. per 100 ml.; (2) markedly increased mobilization of fats from the fat storage areas, causing abnormal fat metabolism and especially deposition of lipids in vascular walls to cause atherosclerosis; and (3) depletion of protein in the tissues of the body, caused partly by failure of glucose to be used as a "protein sparer" and partly by loss of the direct effect of insulin to promote protein anabolism.

However, in addition, some special pathologic physiological problems occur in diabetes mellitus that are not so readily apparent. These are:

**Loss of Glucose in the Urine of the Diabetic Person.**    Whenever the quantity of glucose entering the kidney tubules in the glomerular filtrate rises above approximately 225 mg. per minute, a significant proportion of the glucose begins to spill into the urine; and when the quantity rises above about 325 mg. per minute, which is the tubular maximum for glucose, as explained in Chapter 35, all the excess above this is lost into the urine. If normal quantities of glomerular filtrate are formed per minute, 225 mg. of glucose will enter the tubules each minute when the blood glucose level rises to 180 mg. per cent. Consequently, it is frequently stated that the blood "threshold" for the appearance of glucose in the urine is approximately 180 mg. per cent. When the blood glucose level rises to 300 to 500 mg. per cent—common values in persons with severe untreated diabetes—several hundred grams of glucose can be lost into the urine each day.

**Dehydrating Effect of Elevated Blood Glucose Levels in Diabetes.**    Greatly elevated blood glucose levels, as high as 1200 mg. per cent, can occur under certain conditions in diabetic persons. Yet the only significant effect of the elevated glucose is dehydration of the tissue cells, for glucose does not diffuse easily through the pores of the cell membrane, and the increased osmotic pressure in the extracellular fluids causes osmotic transfer of water out of the cells. Elevation of blood glucose from the normal level of approximately 90 mg. per cent to 400 mg. per cent increases the total osmotic pressure of the extracellular fluid 6 per cent. This alone causes slight dehydration of the tissues, and elevation to still higher values can cause considerable harm.

In addition to the direct dehydrating effect of excessive glucose, the loss of glucose in the urine causes *diuresis* because of the osmotic effect of glucose in the tubules to prevent tubular reabsorption of fluid. The overall effect is dehydration of the extracellular fluid, which then causes compensatory dehydration of the intracellular fluid for reasons discussed in Chapter 33. Thus, one of the important features of diabetes is a tendency for extracellular and intracellular dehydration to develop.

**Acidosis in Diabetes.**    The shift from carbohydrate to fat metabolism in diabetes has already been discussed. When the body depends almost entirely on fat for energy, the level of acetoacetic acid and

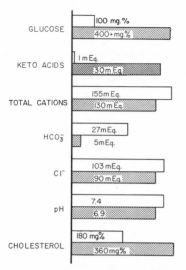

**Figure 78-7.** Changes in blood constituents in diabetic coma, showing normal values (light bars) and diabetic values (dark bars).

other keto acids in the body fluids may rise from 1 mEq./liter to as high as 30 mEq./liter. This, obviously, is likely to result in acidosis.

A second effect, which is usually even more important in causing acidosis than is the direct increase in keto acids, is a decrease in sodium concentration caused by the following effect: Keto acids have a low threshold for excretion by the kidneys; therefore, when the keto acid level rises in diabetes, as much as 100 to 200 grams of keto acids can be excreted in the urine each day. Because these are strong acids, having a pK averaging about 4.0, very little of them can be excreted in the acidic form, but instead is excreted combined with sodium derived from the extracellular fluid. As a result, the sodium concentration in the extracellular fluid usually decreases, and loss of this basic ion adds to the acidosis that is already caused by excessive keto acids in the extracellular fluid.

Obviously, all the usual reactions that occur in metabolic acidosis take place in diabetic acidosis. These include *rapid and deep breathing* called "Kussmaul respiration," which causes excessive expiration of carbon dioxide and *marked decrease in bicarbonate content of the extracellular fluids.* Likewise, *large quantities of chloride ion are excreted by the kidneys* as an additional compensatory mechanism for correction of the acidosis. These extreme effects, however, occur only in the most severe degrees of untreated diabetes. The overall changes in the electrolytes of the blood as a result of severe diabetic acidosis are illustrated in Figure 78-7.

**Relationship of Diabetic Symptoms to the Pathologic Physiology of Insulin Lack.** *Polyuria* (excessive elimination of urine), *polydipsia* (excessive drinking of water), *polyphagia* (excessive eating), *loss of weight,* and *asthenia* (lack of energy) are the earliest symptoms of diabetes. As explained above, the polyuria is due to the osmotic diuretic

effect of glucose in the kidney tubules. In turn, the polydipsia is due to dehydration resulting from polyuria. The failure of glucose utilization by the body causes loss of weight and a tendency toward polyphagia. The asthenia apparently also is caused mainly by loss of body protein.

## PHYSIOLOGY OF DIAGNOSIS

The usual methods for diagnosing diabetes are based on various chemical tests of the urine and the blood as follows:

**Urinary Sugar.** Simple office tests or more complicated quantitative laboratory tests may be used for determining the quantity of glucose lost in the urine. In general, the normal person loses undetectable amounts of glucose, whereas the diabetic loses glucose in small to large amounts, in proportion to the severity of his disease and his intake of carbohydrates. (However, a condition known as *renal glycosuria* sometimes occurs even in persons without diabetes mellitus. This condition results from a low tubular maximum for glucose, as explained in Chap. 38, so that even though the blood glucose level is perfectly normal, a large quantity of glucose may still be lost in the urine.)

**The Fasting Blood Glucose Level.** The fasting blood sugar level in the early morning, at least eight hours after any previous meal, is normally 80 to 90 mg. per cent, and 120 mg. per cent is generally considered to be the absolute upper limit of normal. A fasting blood sugar level above this value usually indicates diabetes mellitus or, less commonly, either pituitary diabetes or adrenal diabetes.

**The Glucose Tolerance Test.** As illustrated by the middle curve in Figure 78-8, when a normal fasting person ingests one gram of glucose per kilogram body weight, his blood glucose level rises from approximately 90 mg. per cent to approximately 140

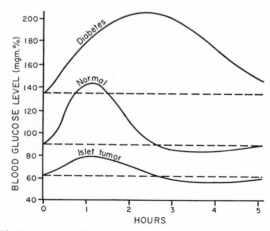

**Figure 78-8.** Glucose tolerance curve in the normal person, in a diabetic person, and in a person with an islet tumor (hyperinsulinism).

mg. per cent and falls back to below normal within two to three hours.

Though an occasional diabetic person has a normal fasting blood glucose concentration, it is usually above 120 mg. per cent, and his glucose tolerance test is almost always abnormal. On ingestion of glucose, these persons exhibit a progressive, slow rise in blood glucose level for two to three hours, as illustrated by the upper curve in Figure 78–8, and the glucose level falls back to the control value only after some five to six hours; yet it never falls below the control level. This slow fall of the curve and its failure to fall below the control level illustrates that the normal increase in insulin secretion following glucose ingestion does not occur in the diabetic, and a diagnosis of diabetes mellitus can usually be definitely established on the basis of such a curve.

**Insulin Sensitivity.**    To differentiate diabetes mellitus of pancreatic origin from high blood glucose levels resulting from excess secretion of adrenocortical or anterior pituitary hormones, an *insulin sensitivity test* can be performed. When little insulin is produced by the pancreas, a test dose of insulin causes the blood glucose level to fall markedly, indicating greatly increased "insulin sensitivity." On the other hand, when the blood glucose level is high as a result of excessive adrenocortical or anterior pituitary secretion, the glucose level responds very poorly to the test dose of insulin because the pancreas is already secreting large quantities of insulin.

**Acetone Breath.**    As pointed out in Chapter 68, small quantities of acetoacetic acid, which increase greatly in severe diabetes, can be converted to acetone, which is volatile and is vaporized into the expired air. Consequently, one frequently can make a diagnosis of diabetes mellitus simply by smelling acetone on the breath of a patient. Also, keto acids can be detected by chemical means in the urine, and their quantitation aids in determining the severity of the diabetes.

## TREATMENT OF DIABETES

The theory of treatment in diabetes mellitus is to administer enough insulin so that the patient will have normal carbohydrate metabolism. If this is done, most consequences of diabetes can be prevented.

Figure 78–9 illustrates time-activity curves of different preparations of insulin following subcutaneous injection. Regular amorphous insulin and crystalline insulin have an activity lasting six to eight hours, whereas insulin that has been precipitated slowly with zinc to form large crystalline particles (lente insulin) or insulin that has been precipitated with various protein derivatives (globin insulin and protamine zinc insulin) is relatively insoluble and is absorbed slowly. For instance, protamine zinc insulin may continue to act for as long as 36 to 48 hours. Ordinarily, the diabetic patient is given a single dose of one of the long-acting insulins each day; this in-

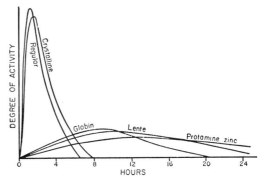

**Figure 78–9.**   Time-action curves for different types of insulin.

creases his overall carbohydrate metabolism throughout the day. Then additional quantities of regular insulin are given at those times of the day when his blood glucose level tends to rise too high, such as at meal times. Thus each patient is established on an individualized routine of treatment.

**Diet of the Diabetic.**   The insulin requirements of a diabetic are established with the patient on a standard diet containing normal, well-controlled amounts of carbohydrates, and any change in the quantity of carbohydrate intake changes the requirements for insulin. In the normal person, the pancreas has the ability to adjust the quantity of insulin produced to the intake of carbohydrate; but in the completely diabetic person, this control function is totally lost.

**Control of the Diabetic Patient in Fever and Exercise.**    Any abnormal state of metabolism in the diabetic patient alters the insulin requirement. Thus, fever, severe infection, and so forth frequently increase the requirement for insulin immensely, and failure to give the extra insulin can cause diabetic coma.

On the other hand, exercise frequently has exactly the opposite effect on insulin requirement, reducing the requirement considerably below that usually needed. The peculiar reason for this is that increased muscle activity increases the transport of glucose into the muscle cells even in the absence of insulin. Thus, *exercise actually has an insulinlike effect*. It is often remarked that a child with severe diabetes requires less insulin to control his diabetes if he lives a very active life than if he leads an overprotected life.

**Relationship of Treatment to Arteriosclerosis.**    Diabetic patients develop atherosclerosis, arteriosclerosis, and severe coronary heart disease far more easily than do normal persons. Indeed, a person who has relatively poorly controlled diabetes throughout childhood is likely to die of heart disease in his 20's.

In the early days of treating diabetes it was the tendency to reduce the carbohydrates in the diet so that the insulin requirements would be minimized. This procedure kept the blood sugar level down to normal values and prevented the loss of glucose in the urine, but it did not prevent the abnormalities of fat metabolism. Actually, it exacerbated these. Consequently, there is a tendency at present to allow the patient a normal carbohydrate diet and then to give simultaneously large quantities of insulin to metabolize the carbohydrates. This depresses the rate of fat metabolism and also depresses the high level of blood cholesterol that occurs in diabetes as a result of abnormal fat metabolism.

Because the complications of diabetes—such as atherosclerosis, greatly increased susceptibility to infection, diabetic retinopathy, cataracts, hypertension, and chronic renal disease—are more closely associated with the level of the blood lipids than with the level of blood glucose, it is the object of many clinics treating diabetes to administer sufficient glucose and insulin so that the quantity of blood lipids becomes normal.

**Treatment with Drugs that Stimulate the Release of Insulin.**   Recently, several drugs that can be taken by mouth and that cause a hypoglycemic effect have been introduced for treatment of persons with mild diabetes. The most useful drugs that have thus far been tried are *sulfonylurea compounds,* the most important of which is *tolbutamide* (Orinase).

These drugs act by stimulating insulin secretion by the islets of Langerhans. Therefore, they are effective only in patients who still have active islets which for some reason are not secreting adequate quantities of insulin. Obviously, they are of no value in the treatment of severe diabetes, for the islets have already lost all potential ability to secrete insulin.

### DIABETIC COMA

If diabetes is not controlled satisfactorily, severe dehydration and acidosis may result; and sometimes, even when the person is receiving treatment, sporadic changes in metabolic rates of the cells, such as might occur during bouts of fever, can also precipitate dehydration and acidosis.

If the pH of the body fluids falls below approximately 7.0 to 6.9, the diabetic person develops coma. Also, in addition to the acidosis, dehydration is believed to exacerbate the coma. Once the diabetic person reaches this stage, the outcome is usually fatal unless he receives immediate treatment.

**Physiological Basis of Treating Diabetic Coma.** The patient with diabetic coma is extremely refractory to insulin because acidic plasma has an *insulin antagonist,* an alpha globulin, that opposes the action of the insulin. Also, the very high free fatty acid and acetoacetic acid levels in the blood inhibit the usage of glucose, as was discussed earlier. Therefore, instead of the usual 60 to 80 units of insulin per day, which is the dosage usually necessary for control of severe diabetes, as much as 1500 to 2000 units of insulin must often be given the first day of treatment of coma. Administration of insulin alone is not likely to be sufficient to reverse the abnormal physiology and to effect a cure. In addition, it is usually neces-

sary to correct both the dehydration and acidosis immediately.

The dehydration is ordinarily corrected rapidly by administering large quantities of sodium chloride solution, and the acidosis is often corrected by administering sodium bicarbonate or sodium lactate solution; the bicarbonate is expired as carbon dioxide through the lungs, and the lactate is removed by the liver, thus providing sodium ions to neutralize the acidosis.

While correcting the dehydration of diabetic coma it is important to maintain the potassium ion of the extracellular fluids at a normal level. During the dehydration process, large quantities of potassium are removed from the tissue cells along with the removal of water. Consequently, on administering fluids and insulin to the comatose person, as these fluids and glucose are absorbed into the cells, large quantities of potassium are also reabsorbed into the cells to reestablish electrolyte balance. This may greatly depress the potassium level in the extracellular fluids and cause skeletal muscular paralysis; also, if this is carried to the extreme, excessive systolic contraction of the heart might occur. Patients have been known to die as a result of rapid loss of potassium from the extracellular fluids. Therefore, fluids administered for correction of diabetic dehydration are often fortified with a small quantity of extra potassium; yet when this is done, caution must be observed against overuse of potassium because a potassium level in the extracellular fluids greater than two times normal can cause cardiac standstill in diastole.

## HYPERINSULINISM

Though much more rare than diabetes, increased insulin production, which is known as hyperinsulinism, does occasionally occur. This usually results from an adenoma of an islet of Langerhans. About 10 to 15 per cent of these adenomas are malignant, and occasionally metastases from the islets of Langerhans spread throughout the body, causing tremendous production of insulin by both the primary and the metastatic cancers. Indeed, in order to prevent hypoglycemia, in some of these patients more than 1000 grams of glucose have had to be administered each 24 hours.

**Insulin Shock and Hypoglycemia.**   As already emphasized, the central nervous system derives essentially all its energy from glucose metabolism, and insulin is not necessary for this use of glucose. However, if insulin causes the level of blood glucose to fall to low values, the metabolism of the central nervous system becomes depressed. Consequently, in patients with hyperinsulinism or in patients who administer too much insulin to themselves, the syndrome called "insulin shock" may occur as follows:

As the blood sugar level falls into the range of 50 to 70 per cent, the central nervous system usually becomes quite excitable, for this degree of hypoglycemia seems to facilitate neuronal activity. Sometimes various forms of hallucinations result, but more often the patient simply experiences extreme nervousness, trembles all over, and breaks out in a sweat. As the blood glucose level falls to 20 to 50 mg. per cent, clonic convulsions and loss of consciousness are likely to occur. As the glucose level falls still lower, the convulsions cease, and only a state of coma remains. Indeed, at times it is difficult to distinguish between diabetic coma as a result of insulin lack and coma due to hypoglycemia caused by excess insulin. However, the acetone breath and the rapid, deep breathing of diabetic coma are not present in hypoglycemic coma.

Obviously, proper treatment for a patient who has hypoglycemic shock or coma is immediate intravenous administration of large quantities of glucose. This usually brings the patient out of shock within a minute or more. Also, administration of epinephrine can cause glycogenolysis in the liver and thereby increase the blood glucose level extremely rapidly.

If treatment is not effected immediately, permanent damage to the neuronal cells of the central nervous system occurs; this happens especially in prolonged hyperinsulinism due to pancreatic tumors. Hypoglycemic shock induced by insulin administration is frequently used for treatment of psychogenic disorders. This type of shock, like electric shock therapy, frequently benefits especially the melancholic patient.

**Depressed Glucose Tolerance Curve and Decreased Insulin Sensitivity in Hyperinsulinism.** In addition to the usual signs of hypoglycemia that appear in hyperinsulinism, a definite diagnosis of the condition can usually be made by performing a glucose tolerance test and an insulin sensitivity test. The glucose tolerance curve is often (but not always) depressed, as shown by the lower curve of Figure 78–8, illustrating that the initial glucose level is low, and the increase after ingestion of glucose is slight. When the insulin sensitivity test is performed, insulin sensitivity is found to be greatly diminished because the injected dose of insulin is so slight in relation to the quantity of insulin already present in the body that almost no effect occurs.

# REFERENCES

Cahill, G. F., Jr., Owen, O. E., and Felig, P.: Insulin and fuel homeostasis. *Physiologist*, 11:97, 1968.

Daughaday, W. H., Herington, A. C., and Phillips, L. S.: The regulation of growth by endocrines. *Ann. Rev. Physiol.*, 37:211, 1975.

Desbuquois, B., and Cuatrecasas, P.: Insulin receptors. *Ann. Rev. Med.*, 24:233, 1973.

Diabetes melitus and obesity. *Ann. N.Y. Acad. Sci.*, 148:573, 1968.

Fritz, I. B. (ed.): Insulin Action. New York, Academic Press, Inc., 1972.

Frohman, L. A.: The endocrine function of the pancreas. *Ann. Rev. Physiol.*, 31:353, 1969.

Gliemann, J., Gammeltoft, S., and Vinten, J.: The significance of insulin receptors in fat cells. *Acta Endocrinol. (Kbh), Suppl.* 77:131, 1974.

Gorden, P., Gavin, J. R., 3rd., Kahn, C. R., Archer, J. A., Lesniak, M., Hendricks, C., Neville, D. M., Jr., and Roth, J.: Application of radioreceptor assay to circulating insulin, growth hormone, and to their tissue receptors in animals and man. *Pharmacol. Rev.*, 25:179, 1973.

Greep, R. O., and Astwood, E. B. (eds.): Handbook of Physiology. Sec. 7, Vol. 1. Baltimore, The Williams & Wilkins Company, 1972.

Havel, R. J.: Lipid transport and the availability of insulin. *Horm. Metab. Res., Suppl.* 4:51, 1974.

Kryston, L. J., Shaw, R. A., and Schwager, P. (eds.): Endocrinology and Diabetes; the Thirtieth Hahnemann Symposium. New York, Grune & Stratton, Inc., 1975.

Laurent, J., Debry, G., and Floquet, J.: Hypoglycemic Tumors. New York, American Elsevier Publishing Company, 1971.

Levine, R.: Concerning the mechanisms of insulin action. *Diabetes*, 10:421, 1961.

Mahler, R. J.: The pathogenesis of pancreatic islet cell hyperplasia and insulin insensitivity in obesity. *Adv. Metab. Disord.*, 7:213, 1974.

Malaisse, W. J.: Insulin secretion and food intake. *Proc. Nutr. Soc.*, 31:213, 1972.

Marble, A., White, P., Bradley, R. F., and Krall, L. P. (eds.): Joslin's Diabetes Mellitus, 11th Ed. Philadelphia, Lea & Febiger, 1971.

Morgan, H. E., Henderson, M. J., Regan, D. M., and Park, C. R.: Regulation of glucose uptake in muscle. I. The effects of insulin and anoxia on glucose transport and phosphorylation in the isolated perfused heart of normal rats. *J. Biol. Chem.*, 236:253, 1961.

Palmer, W. K., and Tipton, C. M.: Effect of training on adipocyte glucose metabolism and insulin responsiveness. *Fed. Proc.*, 33:1964, 1974.

Park, C. R., Reinwein, D., Henderson, M. J., Cadenas, E., and Morgan, H. E.: The action of insulin on the transport of glucose through the cell membrane. *Amer. J. Med.*, 26:674, 1959.

Park, C. R., Morgan, H. E., Henderson, M. J., Regen, D. M., Cadenase, E., and Post, R. L.: The regulation of glucose uptake in muscle as studied in the perfused rat heart. *Recent Progr. Horm. Res.*, 17:493, 1961.

Pfeifer, E. F.: Obesity, islet function, and diabetes mellitus. *Horm. Metab. Res., Suppl.* 4:143, 1974.

Pilkis, S. J., and Park, C. R.: Mechanism of action of insulin. *Ann. Rev. Pharmacol.*, 14:365, 1974.

Post, R. L., Morgan, H. E., and Park, C. R.: Regulation of glucose uptake in muscle. III. The interaction of membrane transport and phosphorylation in the control of glucose uptake. *J. Biol. Chem.*, 236:269, 1961.

Role of insulin in membrane transport (symposium): *Fed. Proc.*, 24:1039, 1965.

Sims, E. A., and Danforth, E., Jr.: Role of insulin in obesity. *Isr. J. Med. Sci.*, 10:1222, 1974.

Unger, R. H.: The pancreas as a regulator of metabolism. In Guyton, A. C. (ed.): MTP International Review of Science: Physiology. Vol. 5. Baltimore, University Park Press, 1974, p. 179.

Weber, G.: Hormonal control of gluconeogenesis. In Bittar, E. E., and Bittar, N. (eds.): The Biological Basis of Medicine. Vol. 2. New York, Academic Press, Inc., 1968, p. 263.

Woods, S. C., and Porte, D., Jr.: Neural control of the endocrine pancreas. *Physiol. Rev.*, 54:596, 1974.

# 79

# Parathyroid Hormone, Calcitonin, Calcium and Phosphate Metabolism, Vitamin D, Bone and Teeth

The physiology of parathyroid hormone and of the hormone calcitonin is closely related to calcium and phosphate metabolism, the function of vitamin D, and the formation of bone and teeth. Therefore, these are discussed together in the present chapter.

## CALCIUM AND PHOSPHATE IN THE EXTRACELLULAR FLUID AND PLASMA— FUNCTION OF VITAMIN D

### ABSORPTION AND EXCRETION OF CALCIUM AND PHOSPHATE

**Intestinal Absorption of Calcium and Phosphate.** By far the major source of calcium in the diet is milk or milk products, which are also major sources of phosphate, but phosphate is also present in many other dietary foods, including especially the meats.

Calcium is poorly absorbed from the intestinal tract because of the relative insolubility of many of its compounds and also because bivalent cations are poorly absorbed through the intestinal mucosa anyway. On the other hand, phosphate is absorbed exceedingly well most of the time except when excess calcium is in the diet; the calcium tends to form almost insoluble calcium phosphate compounds that fail to be absorbed but instead pass on through the bowels to be excreted in the feces. Therefore, the rate of phosphate loss in the feces is, in general, determined by the rate of calcium loss in the feces. In other words, the major problem in the absorption of calcium and phosphate is actually a problem of calcium absorption alone, for if this is absorbed, both are absorbed.

**Excretion of Calcium in Feces and Urine; Net Rate of Absorption.** About seven-eighths of the daily intake of calcium is excreted in the feces, and the remaining one-eighth is excreted in the urine. The excretion in the feces is equal to the difference between calcium absorption and calcium entry into the intestines. An average adult ingests about 800 mg. of calcium each day, and secretes into the intestine about 600 mg. in the gastrointestinal juices, giving a total entry of 1400 mg. per day into the intestine. An average of 700 mg. of this calcium is absorbed, leaving an additional 700 mg. to be lost in the feces. Now, subtracting the 700 mg. of calcium excreted in the feces from the 800 mg. ingested, one finds that each day there is a net absorption of approximately 100 mg. of calcium. All of this, except for that portion stored in the bones, is eventually excreted in the urine.

Excretion of calcium in the urine conforms to much the same principles as those regarding the excretion of sodium. When the calcium ion concentration of the extracellular fluid is low, the rate at which calcium is excreted by the kidneys also becomes low, whereas even a minute increase in calcium ion concentration increases the calcium excretion markedly. We shall see later in the chapter that one of the factors that helps to regulate this excretion of calcium ions is parathyroid hormone.

**Intestinal and Urinary Excretion of Phosphate.** Except for the portion of phosphate that is excreted in the feces in combination with calcium, almost all the dietary phosphate is absorbed into the blood from the gut and later excreted in the urine.

Phosphate is a *threshold substance;* that is, when its concentration in the plasma is below the critical value of approximately 1 millimol/liter, no phosphate at all is lost into the urine; but, above this critical concentration, the rate of phosphate loss is directly proportional to the additional increase. Thus, the kidney regulates the phosphate concentration in the extracellular fluid by altering the rate of phosphate secretion in accordance with the plasma phosphate concentration.

As discussed later in the chapter, phosphate excretion by the kidneys is greatly increased by parathyroid hormone, thereby playing an important role in the control of plasma phosphate concentration.

## VITAMIN D AND ITS ROLE IN CALCIUM ABSORPTION

Vitamin D has a potent effect on increasing calcium absorption from the intestinal tract; it also has important effects on both bone deposition and bone reabsorption, as will be discussed later in the chapter. However, vitamin D itself is not the active substance that actually causes these effects. Instead, the vitamin D must first be converted through a succession of reactions in the liver and the kidney to the final active product, *1,25-dihydroxycholecalciferol.* Figure 79–1 illustrates the succession of steps that leads to the formation of this substance from vitamin D. This figure also shows that the conversion of vitamin D to 1,25-dihydroxycholecalciferol is very precisely controlled.

**The Vitamin D Compounds.** Several different compounds derived from sterols belong to the vitamin D family, and all these perform more or less the same functions. The most important of these, called vitamin $D_3$, is *cholecalciferol.* Most of this substance is formed in the skin as a result of irradiation of *7-dehydrocholesterol* by ultraviolet rays from the sun. Consequently, appropriate exposure to the sun prevents vitamin D deficiency.

**Conversion of Cholecalciferol to 25-Hydroxycholecalciferol in the Liver and Its Feedback Control.** The first step in the activation of cholecalciferol is to convert it to 25-hydroxycholecalciferol; this occurs in the liver. The process, however, is itself a limited one because the 25-hydroxycholecalciferol has a feedback inhibitory effect on the conversion reactions. This feedback effect is extremely important for two reasons:

First, the feedback mechanism regulates very precisely the concentration of 25-hydroxycholecalciferol in the plasma, an effect that is illustrated in Figure 79–2. Note that the intake of vitamin $D_3$ can change many-fold, and yet the concentration of 25-hydroxycholecalciferol still

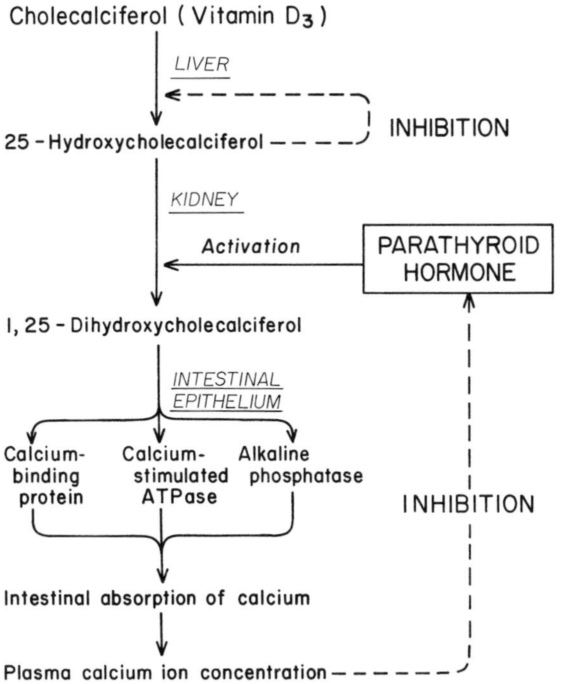

Cholecalciferol ( Vitamin D₃ )

     ↓ *LIVER*

25 - Hydroxycholecalciferol - - - - ┘ INHIBITION

     ↓ *KIDNEY*

     Activation ← PARATHYROID HORMONE

I, 25 - Dihydroxycholecalciferol

     ↓ *INTESTINAL EPITHELIUM*

Calcium-binding protein    Calcium-stimulated ATPase    Alkaline phosphatase

     ↓       INHIBITION

Intestinal absorption of calcium

Plasma calcium ion concentration - - - - - - ┘

**Figure 79–1.** Activation of vitamin $D_3$ to form *1,25-dihydroxycholecalciferol;* and the role of vitamin D in controlling the plasma calcium concentration.

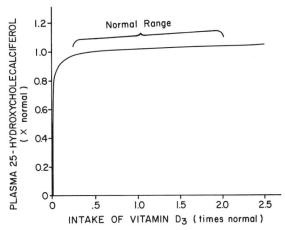

**Figure 79–2.** Effect of increasing vitamin $D_3$ intake on the plasma concentration of 25-hydroxycholecalciferol. This figure shows that tremendous changes in vitamin D intake have little effect on the final quantity of activated vitamin D that is formed.

remains within a few per cent of its normal mean value. Indeed, as much as 1000 times normal quantities of vitamin $D_3$ can be administered to a person, and the concentration of 25-hydroxycholecalciferol will still increase only three-fold. Obviously, this high degree of feedback control prevents excessive action of vitamin $D_3$ when it is present in too high a concentration.

Second, this controlled conversion of vitamin $D_3$ to 25-hydroxycholecalciferol conserves the vitamin $D_3$ for future use, because once it is converted, it persists in the body for only a short time thereafter, whereas in the vitamin D form it can be stored in the liver for as long as several months.

**Formation of 1,25-Dihydroxycholecalciferol in the Kidneys and its Control by Parathyroid Hormone.**    Figure 79–1 also illustrates the conversion in the kidneys of 25-hydroxycholecalciferol to 1,25-dihydroxycholecalciferol. This latter substance is the active form of vitamin $D_3$, and none of the previous products in the scheme of Figure 79–1 have any significant vitamin D effect. Therefore, in the absence of the kidneys vitamin D is almost totally ineffective.

Note also in Figure 79–1 that the conversion of 25-hydroxycholecalciferol to 1,25-dihydroxycholecalciferol requires parathyroid hormone. In the absence of this hormone, either none or almost none of the 1,25-dihydroxycholecalciferol is formed. Therefore, parathyroid hormone exerts a potent effect in determining the functional effects of vitamin D in the body, specifically its effects on calcium absorption in the intestines and its effects on bone.

**Hormonal Effect of 1,25-Dihydroxycholecalciferol on the Intestinal Epithelium in Promoting Calcium Absorption.**    1,25-Dihydroxycholecalciferol has several effects on the intestinal epithelium, one or all of which may play important roles in promoting intestinal absorption of calcium. Probably the most important of these effects is that this "hormone" causes formation of a *calcium-binding protein* in the cytoplasm of the intestinal epithelial cells. The rate of calcium absorption seems to be directly proportional to the quantity of this calcium-binding protein. Furthermore, this protein remains in the cells for several weeks after the 1,25-dihydroxycholecalciferol has been removed from the body, thus causing a prolonged effect on calcium absorption.

Other effects of this "hormone," 1,25-

dihydroxycholecalciferol, that might play a role in promoting calcium absorption are: (1) it causes the formation of a calcium-stimulated ATPase in the brush border of the epithelial cells; and (2) it causes the formation of an alkaline phosphatase in the epithelial cells. Unfortunately, the precise details of calcium absorption are still unknown.

**Feedback Effect of Calcium Ion Concentration on 1,25-Dihydroxycholecalciferol.** Later in the chapter we shall see that the rate of secretion of parathyroid hormone is controlled almost entirely and very potently by the plasma calcium ion concentration. When the calcium ion concentration rises, this change immediately inhibits parathyroid hormone secretion; in the absence of this secretion, 1,25-dihydroxycholecalciferol cannot be formed in the kidney. Thus, this rise in concentration provides a *negative* feedback mechanism for control of both the plasma concentration of 1,25-dihydroxycholecalciferol and also of the plasma calcium ion concentration. That is, an increase in the calcium ion concentration decreases the vitamin D effect, decreases the absorption of calcium from the intestinal tract, and thus returns the calcium ion concentration back to its normal value. We shall see later in the chapter that this is one of the very important means by which the hormonal system of the body maintains a very constant calcium ion concentration.

Figure 79–3 illustrates this feedback effect of plasma calcium concentration on the concentra-

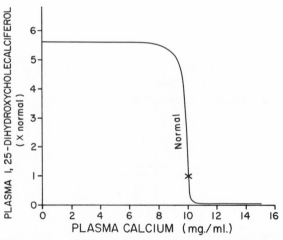

**Figure 79–3.** Effect of plasma calcium concentration on the plasma concentration of 1,25-dihydroxycholecalciferol. This figure shows that a very slight decrease in calcium concentration below normal causes marked formation of activated vitamin D, which in turn leads to greatly increased absorption of calcium from the intestine.

tion of plasma 1,25-dihydroxycholecalciferol. Note that when the calcium ion concentration is only slightly greater than 10 mg. per cent, the concentration of 1,25-dihydroxycholecalciferol falls almost to zero, obviously causing a decrease in calcium absorption from the gut to a very low level. Conversely, the concentration of this "hormone" (the 1,25-dihydroxycholecalciferol) rises markedly when the calcium ion concentration falls even slightly below 10 mg. per cent, immediately turning on the mechanism for calcium absorption from the intestines.

### THE CALCIUM IN THE PLASMA AND INTERSTITIAL FLUID

The concentration of calcium in the plasma is approximately 10 mg. per cent, normally varying between 9.2 and 10.4 mg. per cent. This is equivalent to approximately 5 mEq./liter. It is apparent from these narrow limits of normality that the calcium level in the plasma is regulated exactly—and mainly by parathyroid hormone, as discussed later in the chapter.

The calcium in the plasma is present in three different forms, as shown in Figure 79–4. (1) Approximately 50 per cent of the calcium is combined with the plasma proteins and consequently is nondiffusible through the capillary membrane. (2) Approximately 5 per cent of the calcium (0.2 mEq./liter) is diffusible through the capillary membrane but is combined with other

substances of the plasma and interstitial fluids (citrate, for instance) in such a manner that it is not ionized. (3) The remaining 45 per cent of the calcium in the plasma is both diffusible through the capillary membrane and ionized. Thus, the plasma and interstitial fluids have a normal *calcium ion concentration of approximately 2.3 mEq./liter.* This ionic calcium is important for most functions of calcium in the body, including the effect of calcium on the heart, on the nervous system, and on bone formation.

### THE INORGANIC PHOSPHATE IN THE EXTRACELLULAR FLUIDS

Inorganic phosphate in the plasma is mainly in two forms: $HPO_4^{--}$ and $H_2PO_4^{-}$. The concentration of $HPO_4^{--}$ is approximately 2.1 mEq./liter, and the concentration of $H_2PO_4^{-}$ is approximately 0.26 mEq./liter. When the total quantity of phosphate in the extracellular fluid rises, so does the quantity of each of these two types of phosphate ions. Furthermore, when the pH of the extracellular fluid becomes more acid, there is a relative increase in $H_2PO_4^{-}$ and decrease in the $HPO_4^{--}$ while the opposite occurs when the extracellular fluid becomes alkaline. These relationships were presented in the discussion of acid-base balance in Chapter 37.

Because it is difficult to determine chemically the exact quantities of $HPO_4^{--}$ and $H_2PO_4^{-}$ in the blood, ordinarily the total quantity of phosphate is expressed in terms of milligrams of *phosphorus* per 100 ml. of blood. The average total quantity of inorganic phosphorus represented by both phosphate ions is about 4 mg./100 ml., varying between normal limits of 3.5 to 4 mg./100 ml. in adults and 4 to 5 mg./100 ml. in children.

### EFFECTS OF ALTERED CALCIUM AND PHOSPHATE CONCENTRATIONS IN THE BODY FLUIDS

Changing the level of phosphate in the extracellular fluid from far below normal to as high as three to four times normal does not cause significant immediate effects on the body. There is a similar lack of effects for variations in most other anions, for even chloride ion can be substituted almost entirely by certain other anions without drastic effects.

On the other hand, elevation or depletion of calcium ion in the extracellular fluid causes extreme immediate effects. Both prolonged

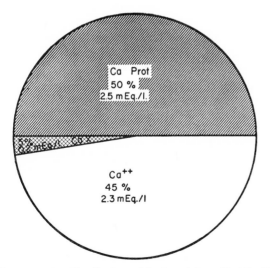

**Figure 79–4.**  Distribution of ionic calcium ($Ca^{++}$), diffusible but un-ionized calcium (*CaX*), and calcium proteinate (*Ca Prot*) in blood plasma.

hypocalcemia and hypophosphatemia greatly decrease bone mineralization, as explained later in the chapter.

**Tetany Resulting from Hypocalcemia.** When the extracellular fluid concentration of calcium ion falls below normal, the nervous system becomes progressively more and more excitable because of increased neuronal membrane permeability. This increase in excitability occurs both in the central nervous system and in the peripheral nerves, though most symptoms are manifest peripherally. The nerve fibers become so excitable that they begin to discharge spontaneously, initiating nerve impulses that pass to the peripheral skeletal muscles where they elicit tetanic contraction. Consequently, hypocalcemia causes tetany.

Figure 79–5 illustrates tetany in the hand, which usually occurs before generalized tetany develops. This is called "carpopedal spasm."

Acute hypocalcemia in the human being ordinarily causes essentially no other significant effects besides tetany because tetany kills the patient before other effects can develop. Tetany ordinarily occurs when the blood concentration of calcium reaches approximately 6 mg. per cent, which is only 40 per cent below the normal calcium concentration, and it is lethal at about 4 mg. per cent.

When the calcium in the body fluids falls to a level not quite sufficient to cause tetany, "latent tetany" results; this can be diagnosed by weakly stimulating the nerves and noting the response. For instance, tapping on the 7th nerve where it passes over the angle of the jaw causes the facial muscle to twitch. Second, placing a tourniquet on the upper arm causes ischemia of the peripheral nerves and also increases the excitability of the nerves, thus causing the muscles of the lower arm and hand to go into spasm. Finally, if the person with latent tetany hyperventilates, the resulting alkalinization of his body fluids increases the irritability of the nerves, causing overt signs of tetany.

In experimental animals, in which the level of calcium can be reduced beyond the normal lethal stage, extreme hypocalcemia can cause marked dilatation of the heart, changes in cellular enzyme activities, increased cell membrane permeability in other cells in addition to nerve cells, and impaired blood clotting.

**Hypercalcemia.** When the level of calcium in the body fluids rises above normal, the nervous system is depressed, and reflex activities of the central nervous system become sluggish. The muscles, too, become sluggish and weak, probably because of calcium effects on the muscle cell membranes. Also, increased calcium ion concentration decreases the QT interval of the heart, and it causes constipation and lack of appetite, probably because of depressed contractility of the muscular walls of the gastrointestinal tract.

The depressive effects of increased calcium level begin to appear when the blood level of calcium rises above approximately 12 mg. per cent, and they can become marked as the calcium level rises above 15 mg. per cent. When the level of calcium rises above approximately 17 mg. per cent in the body fluids, calcium phosphate is likely to precipitate throughout the body; this condition is discussed shortly in connection with parathyroid poisoning.

# BONE AND ITS RELATIONSHIPS WITH EXTRACELLULAR CALCIUM AND PHOSPHATES

Bone is composed of a tough *organix matrix* that is greatly strengthened by deposits of *calcium salts*. Average *compact bone* contains by weight approximately 30 per cent matrix and 70 per cent salts. However, *newly formed bone* may have a considerably higher percentage of matrix in relation to salts.

**The Organic Matrix of Bone.** The organic matrix of bone is approximately 95 per cent *collagen fibers,* and the remaining 5 per cent is a homogeneous medium called *ground substance*. The collagen fibers extend in all directions in the bone but primarily along the lines of tensional force. These fibers give bone its powerful tensile strength.

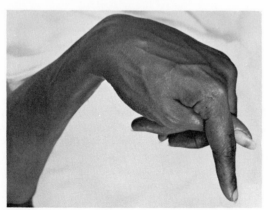

**Figure 79–5.** Hypocalcemic tetany in the hand, called "carpopedal spasm." (Courtesy Dr. Herbert Langford.)

The ground substance is composed of extracellular fluid plus *mucoprotein containing chondroitin sulfate* and *hyaluronic acid.* The precise function of these is not known, though perhaps they help to provide a medium for deposition of calcium salts.

**The Bone Salts.** The crystalline salts deposited in the organic matrix of bone are composed principally of *calcium* and *phosphate,* and the formula for the major crystalline salts, known as *hydroxyapatites,* is the following:

$$Ca^{++}{}_{10-X}(H_3O^+)_{2X} \cdot (PO_4)_6 (OH^-)_2$$

Each crystal—about 400 Å long, 10 to 30 Å thick, and 100 Å wide—is shaped like a long, flat plate. The relative ratio of calcium to phosphorus can vary markedly under different nutritional conditions, the Ca/P ratio on a weight basis varying between 1.3 and 2.0.

*Magnesium, sodium, potassium,* and *carbonate* ions are also present among the bone salts, though x-ray diffraction studies fail to show definite crystals formed by these. Therefore, they are believed to be adsorbed to the surfaces of the hydroxyapatite crystals rather than organized into distinct crystals of their own. This ability of many different types of ions to adsorb to bone crystals extends to many ions normally foreign to bone, such as *strontium, uranium, plutonium, the other transuranic elements, lead, gold, other heavy metals,* and *at least 9 out of 14 of the major radioactive products released by explosion of the hydrogen bomb.* Deposition of radioactive substances in the bone can cause prolonged irradiation of the bone tissues, and, if a sufficient amount is deposited, an osteogenic sarcoma almost invariably eventually develops.

**Tensile and Compressional Strength of Bone.** Each collagen fiber of bone is composed of repeating periodic segments every 640 Å along its length; hydroxyapatite crystals lie adjacent to each segment of the fiber bound tightly to it. This intimate bonding prevents "shear" in the bone; that is, it prevents the crystals and collagen fibers from slipping out of place, which is essential in providing strength to the bone. In addition, the segments of adjacent collagen fibers overlap each other, also causing hydroxyapatite crystals to be overlapped like bricks keyed to each other in a brick wall.

The collagen fibers of bone, like those of tendons, have great tensile strength, while the calcium salts, which are similar in physical properties to marble, have great compressional strength. These combined properties, plus the degree of bondage between the collagen fibers and the crystals, provide a bony structure that has both extreme tensile and compressional strength. Thus, bones are constructed in exactly the same way that reinforced concrete is constructed. The steel of reinforced concrete provides the tensile strength, while the cement, sand, and rock provide the compressional strength. Indeed, the compressional strength of bone is greater than that of even the best reinforced concrete, and the tensile strength approaches that of reinforced concrete.

## PRECIPITATION AND ABSORPTION OF CALCIUM AND PHOSPHATE IN BONE— EQUILIBRIUM WITH THE EXTRACELLULAR FLUIDS

**Supersaturated State of Calcium and Phosphate Ions in Extracellular Fluids with Respect to Hydroxyapatite.** The concentrations of calcium and phosphate ions in extracellular fluid are considerably greater than those required to cause precipitation of hydroxyapatite. However, because of the large number of ions required to form a single molecule of hydroxyapatite, it is very difficult for all of these ions to come together simultaneously; therefore, the precipitation process is normally extremely slow; and, without the intermediation of some active process, it requires many years to occur significantly. Furthermore, inhibitors are present in most tissues of the body, as well as in plasma, to prevent such precipitation; one such inhibitor is pyrophosphate. Therefore, hydroxyapatite crystals fail to precipitate in most normal tissues despite the state of supersaturation of the ions.

**Mechanism of Bone Calcification.** The initial stage in bone production is the secretion of collagen and ground substance by the osteoblasts. The collagen polymerizes rapidly to form collagen fibers, and the resultant tissue becomes *osteoid,* a cartilage-like material but differing from cartilage in that calcium salts precipitate in it. As the osteoid is formed, some of the osteoblasts become entrapped in the osteoid and then are called *osteocytes.* These may play an important role in the subsequent calcification process, though this is not certain.

Within a few days after the osteoid is formed, calcium salts begin to precipitate on the surfaces of the collagen fibers. The precipitates appear at periodic intervals along each collagen fiber, forming minute nidi that gradually grow over a period of days and weeks into the finished product, *hydroxyapatite crystals.*

The initial calcium salts to be deposited are not hydroxyapatite crystals but, instead, are amorphous compounds (noncrystalline) probably either $CaHPO_4$ or $Ca_3(PO_4)_2$. Then by a process of substitution and addition of atoms, these salts are reshaped into the hydroxyapatite crystals.

It is still not known what causes calcium salts to be deposited in osteoid. One theory suggests that the entrapped osteocytes in the osteoid play an important role in the following way: It is known that the osteocytes concentrate large quantities of calcium and phosphate in their mitochondria and even precipitate calcium phosphate compounds in these. Electron micrographs indicate that calcium phosphate–containing vesicles break away from the mitochondria, migrate to the walls of the cell, and then extrude the calcium phosphate into the surrounding extracellular fluid. It may be these preformed calcium phosphate salts that attach to the collagen fibers to form the initial nidi for crystallization, and it is possible that subsequent vesicles supply the necessary calcium and phosphate ions for further growth of the crystals.

However, another theory holds that at the time of formation the collagen fibers are specially constituted in advance for causing precipitation of calcium salts. One variant of this theory suggests that the osteoblasts secrete a substance into the osteoid to neutralize an inhibitor (perhaps pyrophosphate) that normally prevents hydroxyapatite crystallization. Once this neutralization has occurred, the natural affinity of the collagen fibers for calcium salts supposedly causes the precipitation. In support of this theory is the fact that purified collagen fibers prepared from other tissues of the body besides bone will also cause precipitation of hydroxyapatite crystals from plasma.

The formation of initial crystals within the collagen fibers is called *crystal seeding* or *nucleation.*

The growth of hydroxyapatite crystals in newly forming bone reaches 75 per cent completion in a few days, but it usually takes months for the bone to achieve full calcification.

*Precipitation of Calcium in Non-Osseous Tissues Under Abnormal Conditions.* Though calcium salts almost never precipitate in normal tissues besides bone, under abnormal conditions they do precipitate. For instance, they precipitate in arterial walls in the condition called arteriosclerosis and cause the arteries to become bone-like tubes. Likewise, calcium salts frequently deposit in degenerating tissues or in old blood clots. Presumably, in these instances, the inhibitor factors that normally prevent deposition of calcium salts disappear from the tissues, thereby allowing precipitation.

## EXCHANGEABLE CALCIUM

If soluble calcium salts are injected intravenously, the calcium ion concentration can be made to increase immediately to very high levels. However, within minutes to an hour or more, the calcium ion concentration returns to normal. Likewise, if large quantities of calcium ions are removed from the circulating body fluids, the calcium ion concentration again returns to normal within minutes to hours. These effects result from the fact that the body contains a type of *exchangeable* calcium that is always in equilibrium with the calcium ions in the extracellular fluids. A small portion of this exchangeable calcium is that calcium found in all tissue cells, especially in highly permeable types of cells such as those of the liver and the gastrointestinal tract. However, most of the exchangeable calcium, as shown by studies using radioactively tagged calcium, is in the bone, and it normally amounts to about 0.4 to 1.0 per cent of the total bone calcium. Most of this calcium is probably deposited in the bones by the process of adsorption or in the form of some readily mobilizable salt such as $CaHPO_4$. The radioactive calcium studies show that the exchangeable calcium occurs almost entirely at points in the bone that are currently undergoing active bone absorption.

The importance of exchangeable calcium to the body is that it provides a rapid buffering mechanism to keep the calcium ion concentration in the extracellular fluids from rising to excessive levels or falling to very low levels under transient conditions of excess or hypoavailability of calcium.

## DEPOSITION AND ABSORPTION OF BONE—REMODELING OF BONE

**Deposition of Bone by the Osteoblasts.** Bone is continually being deposited by *osteoblasts,* and it is continually being absorbed where *osteoclasts* are active. Osteoblasts are found on the outer surfaces of the bones and in the bone cavities. A small amount of osteoblastic activity occurs continually in all living bones

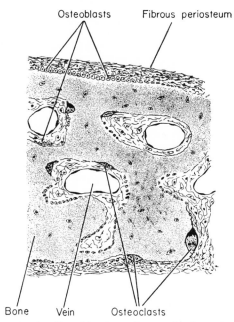

**Figure 79-6.** Osteoblastic and osteoclastic activity in the same bone.

(on about 4 per cent of all surfaces) so that at least some new bone is being formed constantly.

**Absorption of Bone—Function of the Osteoclasts.** Bone is also being continually absorbed in the presence of osteoclasts, which are normally active at any one time on about 1 per cent of the outer surfaces and cavity surfaces. Later in the chapter we will see that parathyroid hormone controls the bone absorptive activity of osteoclasts.

Histologically, bone absorption occurs immediately adjacent to the osteoclasts, as illustrated in Figure 79–6. The mechanism of this absorption is believed to be the following: The osteoclasts secrete two types of substances, (1) proteolytic enzymes, probably released from the lysosomes of the osteoclasts, and (2) several acids, including citric acid and lactic acid. The enzymes presumably digest or dissolute the organic matrix of the bone, while the acids cause solution of the bone salts.

**Formation of Osteoclasts and Osteoblasts.** New osteoclasts and osteoblasts are being formed all of the time. However, under the influence of certain stimuli, such as parathyroid hormone, the rate of formation of these cells is altered drastically, as will be discussed later in the chapter. It is especially important that these cells are formed almost always in the following orderly sequence:

1. *Mesenchymal stem cells* give rise to *osteoclasts*.
2. *Osteoclasts* give rise to *osteoblasts*.
3. *Osteoblasts* give rise to *osteocytes*.

The mesenchymal stem cells are derived from the bone marrow or from the fibrous tissue of the periosteum. When they form osteoclasts, the osteoclasts persist for a few hours to many days, depending upon stimuli such as parathyroid hormone, which will be discussed later. Likewise, other stimuli determine how long they remain in the osteoblastic stage. As the osteoblasts form new bone, they themselves gradually become entrapped in the new bone to form osteocytes.

The importance of this sequence of cellular development is the following: In adult bone, almost invariably the osteoclasts first absorb bone for a period of time and then are converted into osteoblasts, which in turn form new bone. Thus, the bone is continually being remodelled.

**Equilibrium Between Bone Deposition and Absorption.** Normally, except in growing bones, the rates of bone deposition and absorption are equal to each other so that the total mass of bone remains constant. However, osteoclasts actually eat holes in the bone in large areas; deposition of new bone soon follows. Usually, osteoclasts exist in large masses, and once a mass of osteoclasts begins to develop, it usually eats away at the bone for about three weeks, eating out a tunnel that may be as great as 1 mm. in diameter. At the end of this time the osteoclasts are converted into osteoblasts, and new bone begins to develop. Bone deposition then continues for several months, the new

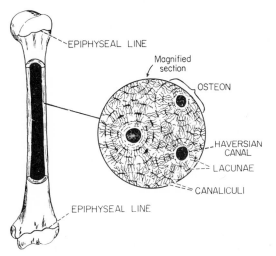

**Figure 79-7.** The structure of bone.

bone being laid down in successive layers on the inner surfaces of the cavity until the tunnel is filled. Deposition of new bone ceases when the bone begins to encroach on the blood vessels supplying the area. The canal through which these vessels run, called the *haversian canal,* therefore, is all that remains of the original cavity. Each new area of bone deposited in this way is called an *osteon,* as shown in Figure 79–7.

*Value of Continual Remodeling of Bone.* The continual deposition and absorption of bone has a number of physiologically important functions. First, bone ordinarily adjusts its strength in proportion to the degree of bone stress. Consequently, bones thicken when subjected to heavy loads. Second, even the shape of the bone can be rearranged for proper support of mechanical forces by deposition and absorption of bone in accordance with stress patterns. Third, since old bone becomes relatively weak and brittle, new organic matrix is needed as the old organic matrix degenerates. In this manner the normal toughness of bone is maintained. Indeed, the bones of children, in whom the rate of deposition and absorption is rapid, show little brittleness in comparison with the bones of old age, at which time the rates of deposition and absorption are slow.

**Control of the Rate of Bone Deposition by Bone "Stress."** Bone is deposited in proportion to the compressional load that the bone must carry. For instance, the bones of athletes become considerably heavier than those of nonathletes. Also, if a person has one leg in a cast but continues to walk on the opposite leg, the bone of the leg in the cast becomes thin and decalcified, while the opposite bone remains thick and normally calcified. Therefore, continual physical stress stimulates osteoblastic deposition of bone.

Bone stress also determines the shape of bones under certain circumstances. For instance, if a long bone of the leg breaks in its center and then heals at an angle, the compression stress on the inside of the angle causes increased deposition of bone, while increased absorption occurs on the outer side of the angle where the bone is not compressed. After many years of increased deposition on the inner side of the angulated bone and absorption on the outer side, the bone becomes almost straight. This is especially true in children because of the rapid remodeling of bone at younger ages.

The deposition of bone at points of compressional stress has been suggested to be caused by

a *piezoelectric* effect, as follows: Compression of bone causes a negative potential at the compressed site and a positive potential elsewhere in the bone. It has been shown that minute quantities of current flowing in bone cause osteoblastic activity at the negative end of the current flow, which could explain the increased bone deposition at compression sites. On the other hand, usual osteoclastic activity could account for absorption of bone at sites of tension.

**Repair of a Fracture.** A fracture of a bone in some way maximally activates all the periosteal and intraosseous osteoblasts involved in the break. Immense numbers of new osteoblasts are formed almost immediately from osteoclasts and mesenchymal stem cells. Therefore, within a short time a large bulge of osteoblastic tissue and new organic bone matrix, followed shortly by the deposition of calcium salts, develops between the two broken ends of the bone. This is called a *callus.*

Many bone surgeons utilize the phenomenon of bone stress to accelerate the rate of fracture healing. This is done by use of special mechanical fixation apparatuses for holding the ends of the broken bone together so that the patient can use his bone immediately. This obviously causes stress on the opposed ends of the broken bones, which accelerates osteoblastic activity at the break and often shortens convalescence.

**Blood Alkaline Phosphatase as an Indication of the Rate of Bone Deposition.** The osteoblasts secrete large quantities of alkaline phosphatase when they are actively depositing bone matrix. This phosphatase is believed either to increase the local concentration of inorganic phosphate or to activate the collagen fibers in such a way that they cause the deposition of calcium salts. Since some alkaline phosphatase diffuses into the blood, the blood level of alkaline phosphatase is usually a good indicator of the rate of bone formation.

The alkaline phosphatase level is below normal in only a few diseases; this includes especially hypoparathyroidism. On the other hand, it is greatly elevated (1) during growth of children, (2) following major bone fractures, and (3) in almost any bone disease that causes bone destruction and must be repaired by osteoblastic activity, such as rickets, osteomalacia, and osteitis fibrosa cystica.

## PARATHYROID HORMONE

For many years it has been known that increased activity of the parathyroid gland causes rapid absorption of calcium salts from the bones with resultant hypercalcemia in the extracellular fluid; conversely, hypofunction of the parathyroid glands causes hypocalcemia, often with resultant tetany, as described earlier in the chapter. Also, parathyroid hormone is important in phosphate metabolism as well as in calcium metabolism.

**Physiologic Anatomy of the Parathyroid Glands.** Normally there are four parathyroid glands in the human being; these are located immediately behind the thyroid gland—one behind each of the upper and each of the lower poles of the thyroid. Each parathyroid gland is approximately 6 mm. long, 3 mm. wide, and 2 mm. thick, and has a macroscopic appearance of dark brown fat; therefore, the parathyroid glands are difficult to locate during thyroid operations. For this reason, before the importance of these glands was generally recognized, total or subtotal thyroidectomy frequently resulted in total removal of the parathyroid glands. Occasionally, parathyroid glands are located in the anterior mediastinum or, rarely, in the posterior mediastinum.

Removal of half the parathyroid glands usually causes little physiologic abnormality. However, removal of three out of four normal glands usually causes transient hypoparathyroidism. But even a small quantity of remaining parathyroid tissue is usually capable of hypertrophying satisfactorily to perform the function of all the glands.

The parathyroid gland of the adult human being, illustrated in Figure 79–8, contains mainly *chief cells* and *oxyphil cells,* but oxyphil cells are absent in many animals and in young human beings. The chief cells secrete most of the parathyroid hormone. The function of the oxyphil cells is not certain; they are probably aged chief cells that still secrete some hormones.

**Chemistry of Parathyroid Hormone.** Parathyroid hormone has been isolated in a pure form. It is a small protein having a molecular weight of approximately 9500 and composed of 84 amino acids. Smaller compounds have also been isolated from the parathyroid gland that exhibit parathyroid hormone

activity, but the activity is always less than that of the larger protein molecule. Therefore, the smaller compounds are almost certainly breakdown products of the normal parathyroid hormone.

## EFFECT OF PARATHYROID HORMONE ON CALCIUM AND PHOSPHATE CONCENTRATIONS IN THE EXTRACELLULAR FLUID

Figure 79–9 illustrates the effect of injecting parathyroid hormone into a human being, showing marked elevation in calcium ion concentration in the extracellular fluid and depression of phosphate concentration. The maximum effect of parathyroid hormone on phosphate is reached within two to three hours, while the maximum effect on calcium ion concentration is reached in about eight hours and then lasts some 24 to 36 hours. The rise in calcium ion concentration is caused principally by a direct effect of parathyroid hormone to increase bone absorption. The decline in phosphate concentration, on the other hand, is caused principally by an effect of parathyroid hormone on the kidneys to increase the rate of phosphate excretion.

**Bone Absorption Caused by Parathyroid Hormone.** The effect of parathyroid hormone in causing bone absorption results from its effects on osteoclasts: (1) immediate activation of all osteoclasts that have already been formed, (2) rapid formation of new osteoclasts from mesenchymal stem cells, and (3) delay in the conversion of osteoclasts into osteoblasts. In addition, parathyroid hormone transiently depresses osteoblastic activity, though after a few days to a few weeks the great numbers of new osteoclasts formed under the influence of

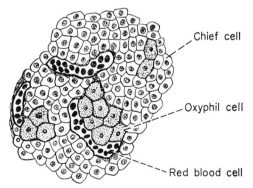

**Figure 79–8.** Histologic structure of a parathyroid gland.

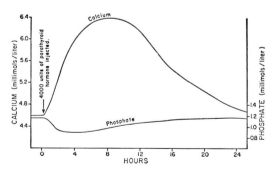

**Figure 79–9.** Effect on plasma calcium and phosphate concentrations of injecting 4000 units of parathyroid hormone into a human being.

parathyroid hormone eventually become osteoblasts. Therefore, the late effect is actually to enhance both osteoblastic and osteoclastic activity. Still, even in the late stages, there is more bone absorption than bone deposition.

Bone contains such great amounts of calcium in comparison with the total amount in all the extracellular fluids (about 1000 times as much) that even when parathyroid hormone causes a great rise in calcium concentration in the fluids, it is impossible to discern any immediate effect at all on the bones. Yet prolonged administration of parathyroid hormone finally results in evident absorption in all the bones with development of large cavities filled with very large, multinucleated osteoclasts.

Two of the experimental studies that have demonstrated the direct action of parathyroid hormone on bone are the following:

1. Implantation of small living sections of parathyroid gland adjacent to a bone causes large numbers of osteoclasts to develop in the immediate area, and this is followed by rapid local absorption of the bone.

2. Injection of parathyroid hormone into animals that have severe rickets causes reabsorption of the newly formed matrix even though no salts are present in the matrix. Therefore, parathyroid hormone acts on both the matrix and the salts at the same time, indicating once again that the primary effect of parathyroid hormone is to increase the activity of some absorptive factor related to the osteoclasts.

**Effect of Parathyroid Hormone on Phosphate and Calcium Excretion by the Kidneys.** Administration of parathyroid hormone causes immediate and rapid loss of phosphate in the urine. This effect is caused by diminished renal tubular reabsorption of phosphate ions. It was long believed that this was the most important function of parathyroid hormone. However, it is now known that this function is of secondary importance to the effect of parathyroid hormone on bone absorption.

Parathyroid hormone causes increased renal tubular reabsorption of calcium at the same time that it diminishes the rate of phosphate reabsorption. Moreover, it also increases the rate of reabsorption of magnesium ions and hydrogen ions, while it decreases the reabsorption of sodium, potassium, and amino acid ions in much the same way that it affects phosphate.

Were it not for the *long-term* effect of parathyroid hormone on the kidneys to increase calcium reabsorption, the continual loss of calcium into the urine would eventually deplete the bones of this mineral.

**Effect of Parathyroid Hormone on Intestinal Absorption of Calcium and Phosphate.** At this point we should be reminded again that parathyroid hormone greatly enhances calcium absorption from the intestines by increasing the formation of 1,25-dihydroxycholecalciferol from vitamin $D_3$, as was discussed earlier in the chapter. Also, when extra quantities of calcium are reabsorbed, this reduces the quantity of calcium phosphate eliminated in the feces, thus simultaneously increasing the absorption of phosphate ions.

**Effect of Vitamin D on Bone and its Relation to Parathyroid Activity.** Vitamin D plays important roles in both bone absorption and bone deposition. Administration of large quantities of vitamin D causes absorption of bone in much the same way that administration of parathyroid hormone does. Also, in the absence of vitamin D, the effect of parathyroid hormone to cause bone absorption is greatly reduced. Unfortunately, the mechanism of these effects is not yet known. However, the vitamin D must be converted to 1,25-dihydroxycholecalciferol by the kidneys before it has its effect on bone. The ability of this substance to promote transport of calcium ions through cell membranes, as occurs in the intestinal and renal epithelia, is probably in some way related to the ability of vitamin D to increase bone absorption.

Vitamin D also promotes bone calcification. Obviously, one of the ways in which it does this is to increase the concentrations of calcium and phosphate in the extracellular fluid. However, even in the absence of these increases, it still enhances the mineralization of bone. Here again, the mechanism of the effect is unknown, but it probably results from the ability of 1,25-dihydroxycholecalciferol to cause transport of calcium ions through cell membranes—perhaps through the osteoblastic or osteocytic cell membranes.

**Role of Cyclic AMP as a Mediator of Parathyroid Stimulation.** A large share of the effect of parathyroid hormone on its target organs is almost certainly mediated by the cyclic AMP "second messenger" mechanism. Within a few minutes after parathyroid hormone administration, the concentration of cyclic AMP increases in the osteoclasts and other target cells. This cyclic AMP, in turn, is probably responsible for such functions as osteoclastic secretion of enzymes and acids to cause bone reabsorption, formation of calcium-binding protein in intestinal and renal epithelia, and so forth.

## CONTROL OF PARATHYROID SECRETION BY CALCIUM ION CONCENTRATION

Even the slighest decrease in calcium ion concentration in the extracellular fluid causes the parathyroid glands to increase their rate of secretion and to hypertrophy. For instance, the parathyroid glands become greatly enlarged in *rickets,* in which the level of calcium is usually depressed only a few per cent; also they become greatly enlarged in pregnancy, even though the decrease in calcium ion concentration in the mother's extracellular fluid is hardly measurable; and, they are greatly enlarged during lactation because calcium is used for milk formation.

On the other hand, any condition that increases the calcium ion concentration causes decreased activity and reduced size of the parathyroid glands. Such conditions include (1) excess quantities of calcium in the diet, (2) increased vitamin D in the diet and (3) bone absorption caused by factors other than parathyroid hormone (for example, bone absorption caused by disuse of the bones).

Figure 79–10 illustrates quantitatively the relationship between plasma calcium concentration and plasma parathyroid hormone concentration. The solid line shows the acute relationship when the calcium concentration is changed over a period of a few hours. This shows that a decrease in calcium concentration from 10 mg. per cent to 9 mg. per cent approximately doubles the plasma parathyroid hormone. On the other hand, the chronic relationship that one finds when the calcium ion concentration changes over a period of many weeks is illustrated by the long-dashed line; this illustrates that about 0.1 mg. per cent decrease in plasma calcium concentration will double parathyroid hormone secretion. To state this still another way, the chronic relationship between plasma calcium and plasma parathyroid hormone shows that approximately a 1 per cent decrease in calcium can give as much as 100 per cent increase in parathyroid hormone. Obviously, this is the basis of the body's extremely potent feedback system for control of plasma calcium ion concentration.

# CALCITONIN

About ten years ago, a new hormone that has effects on blood calcium opposite to those of parathyroid hormone was discovered in several lower animals and at first was believed to be secreted by the parathyroid glands. This hormone was named *calcitonin* because it reduces the blood calcium ion concentration. Soon after its initial discovery, it was found to be secreted in the human being not by the parathyroid glands but instead by the thyroid gland, for which reason it has also been called *thyrocalcitonin*. Still more recently it was discovered that calcitonin is secreted by the *ultimobranchial glands* of fish, amphibia, reptiles, and birds. Furthermore, its concentration in these glands is extremely great. In mammals, ultimobranchial glands do not exist as such but have become incorporated into either the parathyroid or thryoid glands. The so-called *parafollicular cells*, or "C" cells, in the interstitial tissue between the follicles of the human thyroid gland are remnants of the ultimobranchial glands of lower animals, and it is these cells that secrete the calcitonin.

Calcitonin is a large polypeptide with a molecular weight of approximately 3000 and having a chain of 32 amino acids.

**Effect of Calcitonin to Decrease Plasma Calcium Concentration.** Calcitonin has the very rapid effect of decreasing blood calcium ion concentration, beginning within minutes after injection of the calcitonin. Thus, the effect of calcitonin on blood calcium ion cencentration is exactly opposite to that of parathyroid hormone, and it occurs several times as rapidly.

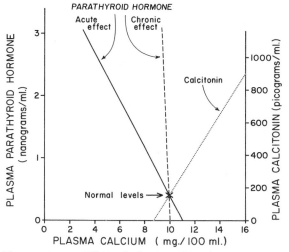

**Figure 79–10.** Effect of plasma calcium concentration on the plasma concentrations of parathyroid hormone and calcitonin. Note especially that long-term, chronic changes in calcium concentration can cause as much as a 100 per cent change in parathyroid hormone concentration for only a 1 per cent change in calcium concentration.

Calcitonin reduces plasma calcium concentration in three separate ways:

1. The immediate effect is to decrease the activity of the osteoclasts, an effect that is especially significant in growing children because of the rapid osteoclastic activity in children. A large dose of calcitonin can reduce osteoclastic activity as much as 70 per cent in 15 minutes.

2. The second effect, which can be seen within about an hour, is an increase in osteoblastic activity. That is, many of the suppressed osteoclasts are converted to osteoblasts under the influence of calcitonin. However, this effect is a transient one, lasting not more than a few days, because the supply of osteoclasts also becomes depleted with time.

3. The third and most prolonged effect of calcitonin is to prevent formation of new osteoclasts from the mesenchymal stem cells. Since osteoblasts are derived from osteoclasts, this action also has a prolonged effect of depressing osteoblastic activity. Therefore, over a long period of time the net result is simply greatly reduced osteoclastic and osteoblastic activity without any significant prolonged effect on plasma calcium ion concentration. That is, the effect on plasma calcium is mainly a transient one, lasting for a few days, at most. However, there is a prolonged effect of greatly decreasing the rate of bone remodeling.

**Importance of the Calcitonin Effect on Plasma Calcium Concentration.** Calcitonin has only a very weak effect on plasma calcium concentration in the adult human being. The reason for this is the following: In the adult, osteoclastic absorption provides only 0.8 gram of calcium to the extracellular fluid each day, and the suppression of this amount of osteoclastic activity by calcitonin has very little effect on the plasma calcium. On the other hand, the effect in children is much more marked because bone remodeling occurs rapidly in children, with osteoclastic absorption of calcium as great as 5 or more grams per day—equal to 5 to 10 times the total calcium in all the extracellular fluid. Also, in certain bone diseases in which osteoclastic activity is greatly accelerated, calcitonin then has the potent effect of reducing the calcium absorption and the blood calcium concentration.

*Effect of Plasma Calcium Concentration on the Secretion of Calcitonin*

An increase in plasma calcium concentration of about 20 per cent causes an immediate two- to three-fold increase in the rate of secretion of calcitonin, which is illustrated by the dotted line of Figure 79–10. This provides a second hormonal feedback mechanism for controlling the plasma calcium ion concentration, but one that works exactly oppositely to the parathyroid hormone system. That is, an increase in calcium concentration causes increased calcitonin secretion, and the increased calcitonin in turn reduces the plasma calcium concentration back toward normal.

However, there are two major differences between the calcitonin and the parathyroid feedback systems. First, the calcitonin mechanism operates more rapidly, reaching peak activity in less than an hour, in contrast to the several hours required for peak activity to be attained following parathyroid secretion.

The second difference is that the calcitonin mechanism acts mainly as a short-term regulator and has little long-term effect, month in and month out, on calcium ion concentration—contrary to the effect of the parathyroid hormone system. As was pointed out above, the calcitonin mechanism is a very weak one in the normal human adult, anyway. Therefore, over a prolonged period of time it is almost entirely the parathyroid system that sets the long-term level of calcium ions in the extracellular fluid.

# OVERALL CONTROL OF CALCIUM ION CONCENTRATION

It is already clear that both the parathyroid and calcitonin hormone systems provide means for control of the extracellular fluid calcium ion centration. However, the calcitonin system is a very weak one, at least in adult human beings, and the parathyroid system requires several hours to cause significant delivery of calcium ions to the extracellular fluid.

At times, however, the amount of calcium absorbed into or lost from the body fluids is as much as one gram in an hour. For instance, in cases of diarrhea, several grams of calcium can be secreted in the intestinal juices, passed into the intestinal tract, and lost into the feces each day. Conversely, after a person ingests large quantities of calcium, particularly when there is also an excess of vitamin D activity, he may absorb as much as one gram in an hour. This figure compares with a *total quantity of calcium in all the extracellular fluid of less than one gram.* The addition or subtraction of one gram from such a small amount of calcium in the extracellular fluid would obviously cause serious hyper- or hypocalcemia. However, there is a first line of defense to prevent this from occurring even before the parathyroid hormone feedback system has a chance to act. This is the following mechanism:

**Buffer Function of the Exchangeable Calcium in the Bones.**    The exchangeable calcium salts in the bones, which were discussed earlier in the chapter, are calcium phosphate compounds, probably mainly $CaHPO_4$ loosely adsorbed or bound in the bone and in reversible equilibrium with the calcium and phosphate ions in the extracellular fluid. These salts make up about 0.5 to 1 per cent of the total calcium salts of the bone. Because of the ease of deposition of these exchangeable salts and their ease of resolubility, an increase in the solubility product of extracellular fluid calcium and phosphate ion concentrations above normal causes immediate deposition of exchangeable salt. Conversely, a decrease in this solubility product causes immediate absorption of exchangeable salt. This reaction is so rapid that a single passage through a bone of blood containing a high concentration of calcium will remove almost all the excess calcium. This rapid effect results from the fact that the bone crystals are extremely small, and their total surface area exposed to the fluids of the bone is about 30 acres. Also, about 5 per cent of all the blood flows through the bones each minute—that is, slightly over 1 per cent of all the extracellular fluid each minute. Therefore, about half of any excess calcium that appears in the extracellular fluid is removed by this buffer function of the bones in approximately 70 minutes.

In addition to the buffer function of the bones, the mitochondria of many of the tissues of the body, especially of the liver and intestine, also contain a reasonable amount of exchangeable calcium that provides an additional buffer system for maintaining constancy of the extracellular fluid calcium ion concentration.

**Hormonal Control of Calcium Ion Exchange with Bone, the Second Line of Defense.**    Within 15 to 20 minutes the calcitonin mechanism for reducing calcium ion concentration can begin to function. As was pointed out above, however, this is an extremely weak mechanism in the normal human adult and, therefore, plays a significant role in controlling blood calcium concentration only when there is an already existing very rapid osteoclastic absorption of bone. For this reason, most of the hormonal control of blood calcium concentration falls on the parathyroid hormonal system.

It usually takes an hour or more before significant parathyroid hormonal effect is observed, only after many hours is the full effect reached. Thus, this mechanism for control of calcium ion concentration is a slow one to develop, but in the long run it is an extremely potent one. In fact, as was illustrated in Figure 79–10, the parathyroid glands will hypertrophy over many days, and the rate of parathyroid hormone secretion will then increase to an even greater extent. If the body's loss of calcium continues at a high rate for many months or years, the bones will become literally filled with osteoclasts. During all this time the blood calcium ion concentration will be maintained at almost exactly the normal level, until finally the bone is entirely depleted of calcium salts.

**Intestinal and Renal Control of Plasma Calcium Concentration—Role of Parathyroid Hormone.**    Though it is frequently stated that the absorption and deposition of calcium in bone is *the* long-term controller of blood calcium ion concentration, this is true only so long as the bone does not become saturated with calcium or totally depleted. However, since the bone does have these limits, it is actually a large reservoir for long-term *buffering* of calcium ion concentration over a period of months or years. It is not, however, the eventual long-term controller of plasma calcium concentration. Instead, this is achieved by the control of absorption and excretion in the intestine and kidneys.

It has already been pointed out that an increase in parathyroid hormone causes an increase in net absorption of calcium from the intestines and also causes increased reabsorption of calcium from the renal tubules. When the bone has become saturated with calcium salts and can no longer function as a depository of additional calcium ions, the slight excess of extracellular calcium ions reduces parathyroid secretion, which then decreases calcium absorption in both the intestine and the kidney tubules. Conversely, when the bone has run out of calcium salts, parathyroid secretion becomes even more increased; this increase can allow for maintenance of almost normal plasma calcium concentration by increasing calcium absorption from both the intestine and kidney tubules if the calcium is available in the intestine.

# PHYSIOLOGY OF PARATHYROID AND BONE DISEASES

## HYPOPARATHYROIDISM

When the parathyroid glands do not secrete sufficient parathyroid hormone, the osteoclasts of the bone become almost totally inactive. As a result, bone reabsorption is so depressed that the level of calcium in the body fluids decreases. Because calcium and phosphates are not being absorbed from the bone, the bone usually remains strong, and osteoblastic activity is concomitantly decreased, partly because there is less bone strain in relation to the strength of the bone and partly because very few osteoclasts are available to be converted into osteoblasts.

When the parathyroid glands are suddenly removed, the calcium level in the blood falls from the normal of 10 mg. per cent to 6 to 7 mg. per cent within two to three days. When this level is reached, the usual signs of tetany develop. Among the muscles of the body sensitive to tetanic spasm are the laryngeal muscles. Spasm of these obstructs respiration, which is the usual cause of death in tetany unless appropriate treatment is applied.

If all four parathyroid glands are removed from an animal and the animal is prevented from dying of respiratory spasm by appropriate supportive measures, the total lack of parathyroid hormone secretion normally causes the blood level of calcium to fall as low as approximately 4 to 5 mg. per cent while at the same time the level of phosphorus increases from the normal of about 4 mg. per cent to approximately 12 mg. per cent.

**Treatment of Hypoparathyroidism.** *Parathyroid Hormone (Parathormone).* Parathyroid hormone is occasionally used for treating hypoparathyroidism. However, because of the expense of this hormone, because its effect lasts only about 24 to 36 hours, and because the tendency of the body to develop immune bodies against it makes it progressively less and less active in the body, treatment of hypoparathyroidism with parathyroid hormone is rare in present-day therapy.

*Dihydrotachysterol and Vitamin D.* In addition to its ability to cause increased absorption of calcium from the gastrointestinal tract, vitamin D also causes a moderate effect similar to that of parathyroid hormone in promoting calcium and phosphate absorption from bones. Therefore, a person with hypoparathyroidism can be treated satisfactorily by administration of *large quantities* of vitamin D. One of the vitamin D compounds, dihydrotachysterol (A.T. 10), has a more marked ability to cause bone absorption than do most of the other vitamin D compounds because it can be converted directly to 1,25-dihydroxycholecalciferol by the kidneys and is not limited by the normal liver feedback mechanism that controls the conversion of vitamin $D_3$ to the active

form. Administration of calcium plus A.T. 10 three or more times a week can almost completely control the calcium level in the extracellular fluid of a hypoparathyroid person.

## HYPERPARATHYROIDISM

The cause of hyperparathyroidism ordinarily is a tumor of one of the parathyroid glands, and such tumors occur much more frequently in women than in men or children. Consequently, it is believed that pregnancy, lactation, and perhaps other causes of prolonged low calcium levels, all of which stimulate the parathyroid gland, may predispose to the development of such a tumor.

In hyperparathyroidism extreme osteoclastic activity occurs in the bones, and this elevates the calcium ion concentration in the extracellular fluid while usually (but not always) depressing slightly the concentration of phosphate ions.

**Bone Disease in Hyperparathyroidism.** Though in mild hyperparathyroidism new bone may be deposited rapidly enough to compensate for the osteoclastic reabsorption of bone, in severe hyperparathyroidism, the osteoclastic absorption soon far outstrips osteoblastic deposition, and the bone may be eaten away almost entirely. Indeed, the reason a hyperparathyroid person comes to the doctor is often a broken bone. X-ray of the bone shows extensive decalcification and occasionally large punched-out cystic areas of the bone that are filled with osteoclasts in the form of so-called giant-cell "tumors." Obviously, multiple fractures of the weakened bones result from only slight trauma, especially where cysts develop. The cystic bone disease of hyperparathyroidism is frequently called *osteitis fibrosa cystica,* or *von Recklinghausen's disease*.

As a result of increased osteoblastic activity that attempts to form new bone as rapidly as it is absorbed, the level of alkaline phosphatase in the body fluids rises markedly.

**Effects of Hypercalcemia in Hyperparathyroidism.** Hyperparathyroidism can at times cause the plasma calcium level to rise to as high as 12 to 15 mg. per cent and rarely to 15 to 20 mg. per cent. The effects of such elevated calcium levels, as detailed earlier in the chapter, are depression of the central and peripheral nervous systems, muscular weakness, constipation, abdominal pain, peptic ulcer, lack of appetite, and depressed relaxation of the heart during diastole.

*Parathyroid Poisoning and Metastatic Calcification.* When, on rare occasions, extreme quantities of parathyroid hormones are secreted, the level of calcium in the body fluids rises rapidly to very high values. Even the extracellular fluid phosphate concentration also often rises markedly instead of falling as is usually the case, probably because the kidneys cannot excrete rapidly enough all the phosphate being absorbed from the bone. Therefore, the calcium and phosphate in the body fluids become greatly supersaturated even for the deposition of calcium phosphate ($CaHPO_4$) crystals. Therefore, these crystals begin to deposit in the alveoli of the lungs, in the tubules of the kidneys, in the thyroid gland, in the acid-producing area of the stomach mucosa, and in the walls of the arteries throughout the body. This extensive *metastatic* deposition of calcium phosphate can develop within a few days.

Ordinarily, the level of calcium in the blood must rise above 17 mg. per cent before there is danger of parathyroid poisoning, but once such elevation develops along with some concurrent elevation of phosphate, death can occur in only a few days.

*Formation of Kidney Stones in Hyperparathyroidism.* Most patients with mild hyperparathyroidism show few signs of bone disease and few general abnormalities as a result of elevated calcium, but nevertheless do have an extreme tendency to form kidney stones. The reason for this is that almost all the excess calcium and phosphate mobilized from the bones in hyperparathyroidism is excreted by the kidneys, causing proportionate increase in the concentrations of these substances in the urine. As a result, crystals of calcium phosphate tend to precipitate in the kidney, forming calcium phosphate stones. Also, calcium oxalate stones develop as a result of the high level of calcium in the urine in association with normal levels of oxalate. Because the solubility of most renal stones is slight in alkaline media, the tendency for formation of renal calculi is considerably greater in alkaline urine than in acid urine. For this reason, acidotic diets and drugs are frequently used for treating renal calculi.

**Secondary Hyperparathyroidism.** Because a low level of calcium ions in the body fluids directly increases the secretion of parathyroid hormone, any factor that causes a low level of calcium initiates the condition known as secondary hyperparathyroidism. This may result from low calcium diet, pregnancy, lactation, rickets, or osteomalacia. The hyperplasia of the parathyroid glands is a corrective measure for maintaining the level of calcium in the extracellular fluids at a nearly normal value.

## RICKETS

Rickets occurs mainly in children as a result of calcium or phosphate deficiency in the extracellular fluid. Ordinarily, rickets is due to lack of vitamin D rather than to lack of calcium or phosphate in the diet. If the child is properly exposed to sunlight, the 7-dehydrocholesterol in his skin becomes activated by the ultraviolet rays and forms vitamin $D_3$, which prevents rickets by promoting calcium and phosphate absorption from the intestines as discussed earlier in the chapter.

Children who remain indoors through the winter in general do not receive adequate quantities of vitamin D without some supplementary therapy in the diet. Rickets tends to occur especially in the spring months because vitamin D formed during the preced-

ing summer can be stored for several months in the liver, and, also, calcium and phosphorus absorption from the bones must take place for several months before clinical signs of rickets become apparent.

*Calcium and Phosphate in the Blood of Patients with Rickets.* Ordinarily, the level of calcium in the blood in rickets is only slightly depressed, but the level of phosphate is greatly depressed. This is because the parathyroid glands prevent the calcium level from falling by promoting bone absorption every time the calcium level begins to fall. On the other hand, there is no good regulatory system for controlling a falling level of phosphate, and the increased parathyroid activity actually increases the excretion of phosphates in the urine.

*Effect of Rickets on the Bone.* During prolonged deficiency of calcium and phosphate in the body fluids, increased parathyroid hormone secretion protects the body against hypocalcemia by causing osteoclastic absorption of the bone; this in turn causes the bone to become progressively weaker and imposes marked physical stress on the bone, resulting in rapid osteoblastic activity. The osteoblasts lay down large quantities of organic bone matrix, osteoid, which does not become calcified because even with the "crystal seeding" effect of the matrix the product of the calcium and phosphate ions is still insufficient to cause calcification. Consequently, the newly formed, uncalcified osteoid gradually takes the place of other bone that is being reabsorbed.

Obviously, hyperplasia of the parathyroid glands is marked in rickets because of the decreased blood calcium level, and the alkaline phosphatase level in the blood is markedly increased as a result of the rapid osteoblastic activity.

*Tetany in Rickets.* In the early stages of rickets, tetany almost never occurs because the parathyroid glands continually stimulate osteoclastic absorption of bone and therefore maintain an almost normal level of calcium in the body fluids. However, when the bones become exhausted of calcium, the level of calcium may fall rapidly. As the blood level of calcium falls below 7 mg. per cent, the usual signs of tetany develop, and the child may die of tetanic respiratory spasm unless intravenous calcium is administered, which relieves the tetany immediately.

*Treatment of Rickets.* The treatment of rickets, obviously, depends on supplying adequate calcium and phosphate in the diet and also on administering vitamin D adequately. If vitamin D is not administered along with the calcium, little calcium is absorbed from the gut, and this calcium in turn carries large quantities of phosphate with it into the feces.

*Tetany Resulting from Treatment.* Occasionally, when a child is treated for rickets, tetany occurs for the following reason: Administration of vitamin D without sufficient calcium often enhances the deposition of calcium in the newly formed osteoid and further depresses the blood calcium level. This results from the direct effect of vitamin D on bone calcification, an effect that has already been discussed.

Another treatment that can frequently lead to

tetany is administration of calcium and phosphate in the diet without sufficient vitamin D. In this case the phosphate is often absorbed without absorption of calcium. The phosphate in turn causes deposition of the plasma calcium in the bone, which reduces the plasma calcium concentration.

These difficulties can be prevented by administering large quantities of calcium simultaneously with large quantities of vitamin D.

**Osteomalacia.** Osteomalacia is rickets in adults and is frequently called "adult rickets."

Normal adults rarely have dietary lack of vitamin D or calcium because large quantities of calcium are not needed for bone growth as in children. However, lack of vitamin D and calcium occasionally occurs as a result of steatorrhea (failure to absorb fat), for vitamin D is fat soluble and calcium tends to form insoluble soaps with fat; consequently, in steatorrhea vitamin D and calcium tend to pass into the feces. Under these conditions an adult occasionally has such poor calcium and phosphate absorption that "adult rickets" can occur, though this rarely proceeds to the same stages as it does in childhood.

***Osteomalacia and Rickets Caused by Renal Disease.*** "Renal rickets" is a type of osteomalacia resulting from prolonged kidney damage. The cause of this condition is believed to be failure of the damaged kidneys to form 1,25-dihydroxycholecalciferol, the active form of vitamin D. In patients whose kidneys have been completely removed or destroyed and who are being treated by hemodialysis, the problem of renal rickets is a very severe one.

Another type of renal disease that leads to rickets and osteomalacia is *congenital hypophosphatemia* resulting from congenitally reduced reabsorption of phosphates by the renal tubules. This type of rickets must be treated with phosphate compounds instead of calcium and vitamin D, and it is called *vitamin D resistant rickets.*

### *OSTEOPOROSIS*

Osteoporosis, the most common of all bone diseases, is a different disease from osteomalacia and rickets, for it results from abnormal organic matrix formation rather than abnormal bone calcification. In general, in osteoporosis the osteoblastic activity in the bone is less than normal, and consequently the rate of bone deposition is depressed. The causes of osteoporosis are (1) lack of use of the bones; (2) malnutrition to the extent that sufficient protein matrix cannot be formed; (3) lack of vitamin C, which is necessary for the secretion of intercellular substances by all cells, including the osteoblasts; (4) postmenopausal lack of estrogen secretion, for estrogens have an osteoblastic stimulating activity; (5) old age, in which many of the protein anabolic functions are poor anyway so that bone matrix cannot be deposited satisfactorily: (6) Cushing's disease, because massive quantities of glucocorticoids cause decreased deposition of protein throughout the body, cause increased catabolism of protein, and also have

the specific effect of depressing osteoblastic activity; and (7) acromegaly, possibly because of lack of sex hormones, excess of adrenocortical hormones, and often lack of insulin because of the diabetogenic effect of growth hormone. Obviously, many diseases of protein metabolism can cause osteoporosis.

## PHYSIOLOGY OF THE TEETH

The teeth cut, grind, and mix the food. To perform these functions the jaws have powerful muscles capable of providing an occlusive force between the front teeth of as much as 50 to 100 pounds and as much as 150 to 200 pounds for the jaw teeth. Also, the upper and lower teeth are provided with projections and facets which interdigitate so that each set of teeth fits with the other. This fitting is called *occlusion,* and it allows even small particles of food to be caught and ground between the tooth surfaces.

### *FUNCTION OF THE DIFFERENT PARTS OF THE TEETH*

Figure 79–11 illustrates a sagittal section of a tooth, showing its major functional parts, the *enamel, dentine, cementum,* and *pulp.* The tooth can also be divided into the *crown,* which is the portion that protrudes out of the gum into the mouth, and the *root,* which is the portion that protrudes into the bony socket of the jaw. The collar between the crown and the root where the tooth is surrounded by the gum is called the *neck.*

**Dentine.** The main body of the tooth is composed of dentine, which has a strong, bony structure. Dentine is made up principally of hydroxyapatite crystals

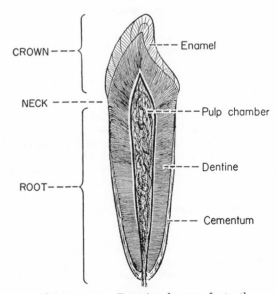

**Figure 79–11.** Functional parts of a tooth.

similar to those in the bone, but much more dense. These are embedded in a strong meshwork of collagen fibers. In other words, the principal constituents of dentine are very much the same as those of bone. The major difference is its histologic organization, for dentine does not contain any osteoblasts, osteoclasts, or spaces for blood vessels or nerves. Instead, it is deposited and nourished by a layer of cells called *odontoblasts,* which line its inner surface along the wall of the pulp cavity.

The calcium salts in dentine make it extremely resistant to compressional forces, while the collagen fibers make it tough and resistant to tensional forces that might result when the teeth are struck by solid objects.

**Enamel.**   The outer surface of the tooth is covered by a layer of enamel that is formed prior to eruption of the tooth by special epithelial cells called *ameloblasts.* Once the tooth has erupted, no more enamel is formed. Enamel is composed of small crystals of hydroxyapatite with adsorbed carbonate, magnesium, sodium, potassium, and other ions embedded in a fine meshwork of very strong and almost completely insoluble protein fibers that are similar to (but not identical with) the keratin of hair. The smallness of the crystalline structure of the salts makes the enamel extremely hard, much harder than the dentine. Also, the special protein fiber meshwork makes enamel very resistant to acids, enzymes, and other corrosive agents because this protein is one of the most insoluble and resistant proteins known.

**Cementum.**   Cementum is a bony substance secreted by cells of the *periodontal membrane,* which lines the tooth socket. Many collagen fibers pass directly from the bone of the jaw, through the periodontal membrane, and then into the cementum. These collagen fibers and the cementum hold the tooth in place. When the teeth are exposed to excessive strain, the layer of cementum becomes thicker and stronger. Also, it increases in thickness and strength with age, causing the teeth to become progressively more firmly seated in the jaws as one reaches adulthood and older.

**Pulp.**   The inside of each tooth is filled with pulp, which in turn is composed of connective tissue with an abundant supply of nerves, blood vessels, and lymphatics. The cells lining the surface of the pulp cavity are the odontoblasts, which, during the formative years of the tooth, lay down the dentine but at the same time encroach more and more on the pulp cavity, making it smaller. In later life the dentine stops growing and the pulp cavity remains essentially constant in size. However, the odontoblasts are still viable and send projections into small *dentinal tubules* that penetrate all the way through the dentine; these are of importance for providing nutrition.

## DENTITION

Each human being and most other mammals develop two sets of teeth during a lifetime. The first teeth are called the *deciduous teeth,* or *milk teeth,*

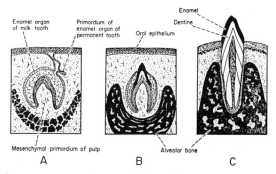

**Figure 79–12.**   (A) Primordial tooth organ. (B) The developing tooth. (C) The erupting tooth. (From Schour and Massler: Studies in tooth development: the growth pattern of human teeth. Part 1. *J. Amer. Dent. Assoc.,* 20:1778, 1940.)

and they number 20 in the human being. These erupt between the seventh month and second year of life, and they last until the sixth to the thirteenth year. After each deciduous tooth is lost, a permanent tooth replaces it, and an additional 8 to 12 molars appear posteriorly in the jaw, making the total number of permanent teeth 28 to 32, depending on whether the four *wisdom teeth* finally appear, which does not occur in everyone.

**Formation of the Teeth.**   Figure 79–12 illustrates the formation and eruption of teeth. Figure 79–12A shows invagination of the oral epithelium into the *dental lamina;* this is followed by the development of a tooth-producing organ. The epithelial cells above form ameloblasts, which form the enamel on the outside of the tooth. The epithelial cells below invaginate upward to form a pulp cavity and also to form the odontoblasts that secrete dentine. Thus, enamel is formed on the outside of the tooth, and dentine is formed on the inside, giving rise to an early tooth as illustrated in Figure 79–12B.

*Eruption of Teeth.*   During early childhood, the teeth begin to protrude upward from the jaw bone through the oral epithelium into the mouth. The cause of "eruption" is unknown, though several theories have been offered in an attempt to explain this phenomenon. One of these assumes that an increase of the material inside the pulp cavity of the tooth causes much of it to be extruded through the root canal and that this pushes the tooth upward. However, a more likely theory is that the bone underneath the tooth progressively hypertrophies and in so doing shoves the tooth forward.

*Development of the Permanent Teeth.*   During embryonic life, a tooth-forming organ also develops in the dental lamina for each permanent tooth that will be needed after the deciduous teeth are gone. These tooth-producing organs slowly form the permanent teeth throughout the first 6 to 20 years of life. When each permanent tooth becomes fully formed, it, like the deciduous tooth, pushes upward through the bone of the jaw. In so doing it erodes the root of the deciduous tooth and eventually causes it to loosen and fall out. Soon thereafter, the permanent tooth erupts to take the place of the original one.

*Metabolic Factors in Development of the Teeth.* The rate of development and the speed of eruption of teeth can be accelerated by both thyroid and growth hormones. Also, the deposition of salts in the early forming teeth is affected considerably by various factors of metabolism, such as the availability of calcium and phosphate in the diet, the amount of vitamin D present, and the rate of parathyroid hormone secretion. When all these factors are normal, the dentine and enamel will be correspondingly healthy, but, when they are deficient, the calcification of the teeth also may be defective so that the teeth will be abnormal throughout life.

## MINERAL EXCHANGE IN TEETH

The salts of teeth, like those of bone, are composed basically of hydroxyapatite with adsorbed carbonates and various cations bound together in a hard crystalline substance. Also, new salts are constantly being deposited while old salts are being reabsorbed from the teeth, as also occurs in bone. However, experiments indicate that deposition and reabsorption occur mainly in the dentine and cementum, while very little occurs in the enamel. Much of that which does occur in the enamel occurs by exchange of minerals with the saliva instead of with the fluids of the pulp cavity. The rate of absorption and deposition of minerals in the cementum is approximately equal to that in the surrounding bone of the jaw, while the rate of deposition and absorption of minerals in the dentine is only one-third that of bone. The cementum has characteristics almost identical with those of usual bone, including the presence of osteoblasts and osteoclasts, while dentine does not have these characteristics, as was explained above; this difference undoubtedly explains the different rates of mineral exchange.

The mechanism by which minerals are deposited and reabsorbed from the dentine is unknown. It is possible that the small processes of the odontoblasts that protrude into the tubules of the dentine are capable of absorbing salts and then of providing new salts to take the place of the old.

In summary, rapid mineral exchange occurs in the dentine and cementum of teeth, though the mechanism of this exchange in dentine is unknown. On the other hand, enamel exhibits extremely slow mineral exchange so that it maintains most of its original mineral complement throughout life.

## DENTAL ABNORMALITIES

The two most common dental abnormalities are *caries* and *malocclusion.* Caries means erosions of the teeth, whereas malocclusion means failure of the projections of the upper and lower teeth to interdigitate properly.

**Caries.** It is generally agreed by all research investigators of dental caries that caries result from the action of bacteria on the teeth. The first event in the development of caries is the deposit of *plaque,* a film of precipitated products of saliva and food, on the teeth. Large numbers of bacteria inhabit this plaque and are readily available to cause caries. However, these bacteria depend to a great extent on carbohydrates for their food. When carbohydrates are available, their metabolic systems are strongly activated and they also multiply. In addition, they form acids, particularly lactic acid, and proteolytic enzymes. The enzymes presumably digest the protein matrix of the enamel and dentine while the acid causes absorption of the calcium salts.

Because of the dependence of the caries bacteria on carbohydrates, it has frequently been taught that eating a diet high in carbohydrate content will lead to excessive development of caries. However, it is not the quantity of carbohydrate ingested but instead the frequency with which it is eaten that is important. If eaten in many small parcels throughout the day, such as in the form of candy, the bacteria are supplied with their preferential metabolic substrate for many hours of the day and the development of caries is extreme. If eaten in large amounts only at mealtimes, the extensiveness of the caries is greatly reduced.

Some teeth are more resistant to caries than others. Studies show that teeth formed in children who drink water containing small amounts of fluorine develop enamel that is more resistant to caries than the enamel in children who drink water not containing fluorine. Fluorine does not make the enamel harder than usual, but instead it is said to inactivate proteolytic enzymes before they digest the protein matrix of the enamel. Regardless of the precise means by which fluorine protects the teeth, it is known that small amounts of fluorine deposited in enamel make teeth about three times as resistant to caries as are teeth without fluorine.

**Malocclusion.** Malocclusion is usually caused by a hereditary abnormality that causes the teeth of one jaw to grow in an abnormal direction. In malocclusion, the teeth cannot perform their normal grinding or cutting action adequately; occasionally malocclusion results in abnormal displacement of the lower jaw in relation to the upper jaw, causing such undesirable effects as pain in the mandibular joint or deterioration of the teeth.

The orthodontist can often correct malocclusion by applying prolonged gentle pressure against the teeth with appropriate braces. The gentle pressure causes absorption of alveolar jaw bone on the compressed side of the tooth and deposition of new bone on the tensional side of the tooth. In this way the tooth gradually moves to a new position as directed by the applied pressure.

# REFERENCES

Aegerter, E., and Kirkpatrick, J. A., Jr.: Orthopedic Diseases. Philadelphia, W. B. Saunders Company, 1975.

Anderson, D. J., Hannam, A. G., and Matthews, B.: Sensory mechanisms in mammalian teeth and their supporting structures. *Physiol. Rev.,* 50:171, 1970.

Arnaud, C. D., Jr., Tenenhouse, A. M., and Rasmussen, H.: Parathyroid hormone. *Ann. Rev. Physiol., 29*:349, 1967.

Bergsma, D.: Skeletal Dysplasias. New York, American Elsevier Publishing Company, 1974.

Borle, A. B.: Calcium and phosphate metabolism. *Ann. Rev. Physiol., 36*:361, 1974.

Bourne, G. H.: The Biochemistry and Physiology of Bone, 2nd Ed., Vols. I, II, and III. New York, Academic Press, Inc., 1973.

Brickman, A. S., and Norman, A. W.: Treatment of renal osteodystrophy with calciferol (vitamin D) and related steroids. *Kidney Int., 4*:161, 1973.

Coburn, J. W., Hartenbower, D. L., and Massry, S. G.: Intestinal absorption of calcium and the effect of renal insufficiency. *Kidney Int., 4*:96, 1973.

Copp, D. H.: Endocrine regulation of calcium metabolism. *Ann. Rev. Physiol., 32*:61, 1970.

Daughaday, W. H., Herington, A. C., and Phillips, L. S.: The regulation of growth by endocrines. *Ann. Rev. Physiol., 37*:211, 1975.

David, D. S.: Calcium metabolism in renal failure. *Amer. J. Med., 58*:48, 1975.

DeLuca, H. F.: Regulation of vitamin D metabolism: a new dimension in calcium homeostasis. *Biochem. Soc. Symp., 35*:271, 1972.

DeLuca, H. F.: The kidney as an endocrine organ involved in the function of vitamin D. *Amer. J. Med., 58*:39, 1975.

Eastwood, J. B., Bordier, P. J., and de Wardener, H. E.: Some biochemical, histological, radiological, and clinical features of renal osteodystrophy. *Kidney Int., 4*:128, 1973.

Fenton, A. H.: Bone resorption and prosthodontics. *J. Prosthet. Dent., 29*:477, 1973.

Frost, H. M.: The Physiology of Cartilaginous, Fibrous, and Bony Tissue. Springfield, Ill., Charles C Thomas, Publisher, 1972.

Gray, T. K., Cooper, C. W., and Munson, P. L.: Parathyroid hormone, thyrocalcitonin, and the control of mineral metabolism. In Guyton, A. C. (ed.): MTP International review of Science: Physiology. Vol. 5. Baltimore, University Park Press, 1974, p. 239.

Gron, P.: Remineralization of enamel lesions in vivo. *Oral Sci. Rev., 3*:84, 1973.

Haddad, J. G., Jr.: Paget's disease of bone: problems and management. *Orthop. Clin. North Amer., 3*:775, 1972.

Hancox, N. M.: Biology of Bone. Cambridge, Eng., Cambridge University Press, 1972.

Harrison, H. E., and Harrison, H. C.: Calcium. *Biomembranes, 4B*:793, 1974.

Hirsch, P. F., and Munson, P. L.: Thyrocalcitonin. *Physiol. Rev., 49*:548, 1969.

Irving, J. T.: Theories of mineralization of bone. *Clin. Orthop., 97*:225, 1973.

Irving, J. T.: Calcium and Phosphorus Metabolism. New York, Academic Press, Inc., 1973.

Jenkins, G. N.: Current concepts concerning the development of dental caries. *Int. Dent. J., 22*:350, 1972.

Kreitzman, S. N.: Enzymes and dietary factors in caries. *J. Dent. Res., 53*:218, 1974.

Little, K.: Bone Behavior. New York, Academic Press, Inc., 1973.

Loomis, W. F.: Rickets. *Sci. Amer., 223*:76, 1970.

Morgan, B.: Osteomalacia, Renal Osteodystrophy, and Osteoporosis. Springfield, Ill., Charles C Thomas, Publisher, 1973.

Norman, A. W.: 1,25-Dihydroxyvitamin D$_3$: a kidney-produced steroid hormone essential to calcium homeostasis. *Amer. J. Med., 57*:21, 1974.

Omdahl, J. L., and DeLuca, H. F.: Regulation of vitamin D metabolism and function. *Physiol. Rev., 53*:327, 1973.

Osborn, J. W.: Variations in structure and development of enamel. *Oral Sci. Rev., 3*:3, 1973.

Parfitt, A. M.: Investigation of disorders of the parathyroid glands. *Clin. Endocrinol. Metabol., 3*:451, 1974.

Pindborg, J. J., and Horting-Hansen, E.: Atlas of Diseases of the Jaws. Philadelphia, W. B. Saunders Company, 1975.

Posner, A. S.: Crystal chemistry of bone mineral. *Physiol. Rev., 49*:760, 1969.

Rasmussen, H.: The Physiological and Cellular Basis of Metabolic Bone Disease. Baltimore, The Williams & Wilkins Company, 1974.

Rasmussen, H., Bordier, P., Kurokawa, K., Nagata, N., and Ogata, E.: Hormonal control of skeletal and mineral homeostasis. *Amer. J. Med., 56*:751, 1974.

Rasmussen, H., and Pechet, M. M.: Calcitonin. *Sci. Amer., 223*:42, 1970.

Searls, R. L.: Newer knowledge of chondrogenesis. *Clin. Orthop., 96*:327, 1973.

Seltzer, S., and Bender, I. B.: The Dental Pulp, 2nd Ed. Philadelphia, J. B. Lippincott Company, 1975.

Storey, E.: The nature of tooth movement. *Amer. J. Orthod., 63*:292, 1973.

Vaughan, D. J. M.: The Physiology of Bone. New York, Oxford University Press, 1970.

Weatherell, J. A., Robinson, C., and Hallsworth, A. S.: Variations in the chemical composition of human enamel. *J. Dent. Res., 53*:180, 1974.

Wheeler, R. C.: Dental Anatomy, Physiology and Occlusion, 5th Ed. Philadelphia, W. B. Saunders Company, 1975.

# 80

# Reproductive Functions of the Male and the Male Sex Hormones

The reproductive functions of the male can be divided into three major subdivisions: first, spermatogenesis, which means simply the formation of sperm; second, performance of the male sexual act; and third, regulation of male sexual functions by the various hormones. Associated with these reproductive functions are the effects of the male sex hormones on the accessory sexual organs, on cellular metabolism, on growth, and on other functions of the body.

**Physiologic Anatomy of the Male Sexual Organs.**   Figure 80–1 illustrates the various portions of the male reproductive system. Note that the testis is composed of a large number of coiled *seminiferous tubules* where the sperm are formed. (If all these tubules were placed end to end, they would be about 800 feet long.) The sperm then empty into the *epididymis,* another coiled tube approximately 20 feet long. The epididymis leads into the *vas deferens,* which enlarges into the *ampulla of the vas deferens* immediately before the vas enters the body of the *prostate gland.* A *seminal vesicle,* one located on each side of the prostate, empties into the prostatic end of the ampulla, and the contents from both the ampulla and the seminal vesicle pass into an *ejaculatory duct* leading through the body of the prostate gland to empty into the *internal urethra.* *Prostate ducts* in turn empty into the ejaculatory duct. Finally, the *urethra* is the last connecting link from the testis to the exterior. The urethra is supplied with mucus derived from a large number of small *glands of Littré* located along its entire extent and also from large bilateral *bulbo-urethral glands*

(Cowper's glands) located near the origin of the urethra.

## SPERMATOGENESIS

Spermatogenesis occurs in all the seminiferous tubules during active sexual life, beginning at an

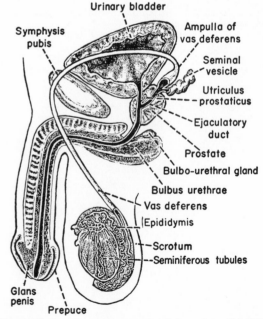

**Figure 80–1.**   The male reproductive system. (Modified from Bloom and Fawcett: Textbook of Histology, 10th Ed.)

1072

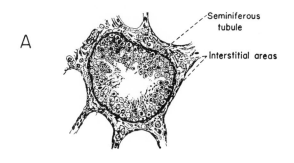

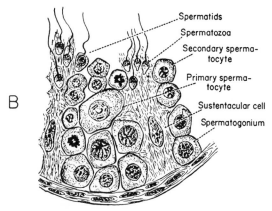

**Figure 80–2.** (A) Cross-section of a seminiferous tubule. (B) Spermatogenesis. (Modified from Arey: Developmental Anatomy, 7th Ed.)

average age of 13 as the result of stimulation by adenohypophyseal gonadotropic hormones and continuing throughout the remainder of life.

## THE STEPS OF SPERMATOGENESIS

The seminiferous tubules, one of which is illustrated in Figure 80–2A, contain a large number of small to medium-sized germinal epithelial cells called *spermatogonia,* which are located in two to three layers along the outer border of the tubular epithelium. These continually proliferate to replenish themselves, and a portion of them differentiate through definite stages of development to form sperm, as shown in Figure 80–2B.

The first stage in spermatogenesis is growth of some spermatogonia to form considerably enlarged cells called *primary spermatocytes.* Then the primary spermatocyte divides by the process of meiosis (there is no formation of new chromosomes, only separation of the chromosomal pairs) to form two *secondary spermatocytes,* each containing 23 chromosomes. Soon each of these cells divides by mitotic division to form two *spermatids,* each of which also contains only 23 chromosomes, none of them paired. Thus, one of each of the original pairs of

chromosomes is now in each spermatid. As noted in the following chapter, a similar reduction in chromosomes occurs in the ovum during its maturation. Then, when the spermatozoon combines with the ovum during the fertilization process, the original complement of 46 chromosomes is again established.

*The Sex Chromosomes.* In each spermatogonium one of the 23 pairs of chromosomes carries the genetic information that determines the sex of the eventual offspring. This pair is composed of one "X" chromosome, which is called the *female chromosome,* and one "Y" chromosome, the *male chromosome.* During meiotic division the sex-determining chromosomes divide among the secondary spermatocytes so that half of the sperm become *male sperm* containing the "Y" chromosome and the other half *female sperm* containing the "X" chromosome. The sex of the offspring is determined by which of these two types of sperm fertilizes the ovum. This will be discussed further in Chapter 82.

*Formation of Sperm.* When the spermatids are first formed, they still have the usual characteristics of epithelioid cells, but soon most of the cytoplasm disappears, and each spermatid begins to elongate into a spermatozoon, illustrated in Figure 80–3, composed of a *head, neck, body,* and *tail.* To form the head, the nuclear material is condensed into a compact mass, and the cell membrane contracts around the nucleus. It is this nuclear material that fertilizes the ovum.

At the front of the sperm head is a small structure called the *acrosome,* which is formed from the Golgi apparatus and contains hyaluronidase and proteases that play important roles in the entry of the sperm into the ovum.

The *centrioles* are aggregated in the neck of the sperm and the *mitochondria* are aggregated in the body.

Extending beyond the body is a long tail, which is an outgrowth of one of the centrioles. This has almost the same structure as a cilium, which was described in detail in Chapter 2. The tail contains two paired microtubules down the center and nine double microtubules arranged around the border. It is covered by an extension of the cell membrane, and it contains large quantities of adenosine triphosphate, which undoubtedly energize the movement of the tail. Upon release of sperm from the male genital tract into the female tract, the tail begins to wave back and forth, providing snake-like propulsion that moves the sperm forward at a maximum velocity of about 20 centimeters per hour.

**Function of the Sertoli Cells.** The Sertoli cells of the geminal epithelium, known also as the *sustentacular cells,* are illustrated in Figure 80–2B. These cells are large, extending from the base of the seminiferous epithelium all the way to the interior of the tubule. The spermatids attach themselves to the Sertoli cells, and a specific relationship exists between the spermatids and the Sertoli cells that causes the spermatids to change into spermatozoa. For this reason it is believed that the Sertoli cells provide

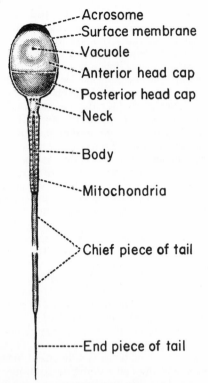

- Acrosome
- Surface membrane
- Vacuole
- Anterior head cap
- Posterior head cap
- Neck
- Body
- Mitochondria
- Chief piece of tail
- End piece of tail

**Figure 80–3.**   Structure of the human spermatozoon.

nutrient material, hormones, or enzymes that are necessary for causing appropriate changes in the spermatids.

## THE SPERM

**Maturation of Sperm in the Epididymis.** Following formation in the seminiferous tubules, the sperm pass through the *vasa recta* into the *epididymis*. Sperm removed from the seminiferous tubules are completely nonmotile, and they cannot fertilize an ovum. However, after the sperm have been in the epididymis for some 18 hours to 10 days, they develop the power of motility and also become capable of fertilizing the ovum, a process called *maturation*. It may not be any special function of the epididymis that changes the sperm from the nonmotile state into the motile and fertile state but, instead, simply an aging process. However, the epididymis secretes a copious quantity of fluid containing hormones, enzymes, and special nutrients that may be important or even essential for sperm maturation.

**Storage of Sperm.** A small quantity of sperm can be stored in the epididymis, but most sperm are stored in the vas deferens and to some extent in the ampulla of the vas deferens.

Though the sperm in these areas become motile if released to the exterior, they are relatively dormant as long as they are stored, probably for the following reasons: The sperm, as a result of their own metabolism, secrete a considerable quantity of carbon dioxide into the surrounding fluid, and the resulting acidotic state of the fluid inhibits the activity of the sperm. In addition, the fluid of the vas deferens is deficient in needed nutrients to supply the large amount of energy required for motility.

Sperm can be stored, maintaining their fertility, in the genital ducts for as long as 42 days, though it is doubtful that during normal sexual activity such prolonged storage ordinarily occurs. Indeed, with excessive sexual activity storage may be no longer than a few hours.

**Physiology of the Mature Sperm.** The usual motile and fertile sperm is capable of flagellated movement through the fluid media at a rate of approximately 1 to 4 mm. per minute. Furthermore, *normal* sperm tend to travel in a straight line rather than with a circuitous movement. The activity of sperm is greatly enhanced in neutral and slightly alkaline media, but it is greatly depressed in mildly acid media, and strong acid media can cause rapid death of sperm. The activity of sperm increases greatly with increasing temperature, but so does the rate of metabolism, causing the life of the sperm to be considerably shortened. Though sperm can live for many weeks in the genital ducts of the testes, the life of sperm in normal ejaculated semen at normal body temperature is only 24 to 72 hours.

## FUNCTION OF THE SEMINAL VESICLES

From early anatomic studies of the seminal vesicles it was erroneously believed that sperm were stored in these, whence came the name "seminal vesicles." However, these structures are only secretory glands instead of sperm storage areas.

The seminal vesicles are lined with a secretory epithelium that secretes a mucoid material containing an abundance of fructose and smaller amounts of ascorbic acid, inositol, ergothioneine, five of the amino acids, phosphorylcholine, prostaglandin, and fibrinogen. During the process of ejaculation each seminal vesicle empties its contents into the ejaculatory duct shortly after the vas deferens empties the sperm. This adds greatly to the bulk of the ejaculated semen, and the fructose and other substances in the seminal fluid are of considerable nutrient value for the ejaculated sperm until one of them fertilizes the ovum.

## FUNCTION OF THE PROSTATE GLAND

The prostate gland secretes a thin, milky, alkaline fluid containing citric acid, calcium, acid phosphate, a clotting enzyme, and a profibrinolysin. During emission, the capsule of the prostate gland contracts simultaneously with the contractions of the vas deferens and seminal vesicles so that the thin, milky fluid of the prostate gland adds to the bulk of the semen. The alkaline characteristic of the prostatic fluid may be quite important for successful fertilization of the ovum, because the fluid of the vas deferens is relatively acidic owing to the presence of metabolic end-products of the sperm and, consequently, inhibits sperm fertility. Also, the vaginal secretions of the female are acidic (pH of 3.5 to 4.0). Sperm do not become optimally motile until the pH of the surrounding fluids rises to approximately 6 to 6.5. Consequently, it is probable that prostatic fluid neutralizes the acidity of these other fluids after ejaculation and greatly enhances the motility and fertility of the sperm.

## SEMEN

Semen, which is ejaculated during the male sexual act, is composed of the fluids from the vas deferens, from the seminal vesicles, from the prostate gland, and from the mucous glands, especially the bulbo-urethral glands. The major bulk of the semen is seminal vesicle fluid (about 60 per cent), which is the last to be ejaculated and serves to wash the sperm out of the ejaculatory duct and urethra. The average pH of the combined semen is approximately 7.5, the alkaline prostatic fluid having neutralized the mild acidity of the other portions of the semen. The prostatic fluid gives the semen a milky appearance, while fluid from the seminal vesicles and from the mucous glands gives the semen a mucoid consistency. Indeed, the clotting enzyme of the prostatic fluid causes the fibrinogen of the seminal vesicle fluid to form a weak coagulum, which then dissolutes during the next 15 to 20 minutes because of lysis by fibrinolysin formed from the prostatic profibrinolysin. In the early minutes after ejaculation, the sperm remain relatively immobile, possibly because of the viscosity of the coagulum. However, after the coagulum dissolutes, the sperm simultaneously become highly motile.

Though sperm can live for many weeks in the male genital ducts, once they are ejaculated in the semen their maximal life span is only 24 to 72 hours at body temperature. At lowered temperatures, however, semen may be stored for several weeks; and when frozen at temperatures below $-100°$ C., sperm of some animals have been preserved for over a year.

**Capacitation.** In some lower animals, sperm are not capable of fertilizing the ovum immediately after being ejaculated but develop this capability during the next few hours. This phenomenon is called *capacitation*. It may be simply an aging process, or it may result from the action of those components of the semen not derived from the vas deferens or of some of the female secretions. However, it is doubtful that capacitation is required for human sperm to become fertile.

## MALE FERTILITY

The seminiferous tubular epithelium can be destroyed by a number of different diseases. For instance, bilateral orchiditis resulting from mumps usually causes sterility in a large percentage of afflicted males. Another disease that frequently localizes in the testes and can cause severe tubular damage is typhus fever. Also, many male infants are born with degenerate tubular epithelium as a result of strictures in the genital ducts or as a result of unknown causes. Finally, a cause of sterility, usually temporary, is excessive temperature of the testes, as follows:

**Effect of Temperature on Spermatogenesis.** In addition to the direct effect of increased temperature in shortening the life of sperm as noted earlier, increasing the temperature of the testes can prevent spermatogenesis by causing degeneration of all the cells of the seminiferous tubules besides the spermatogonia.

It has often been stated that the reason the testicles are located in the dangling scrotum is to maintain the temperature of these glands below the temperature of the body. On cold days scrotal reflexes cause the musculature of the scrotum to contract, pulling the testicles close to the body, whereas on warm days the musculature of the scrotum becomes almost totally relaxed so that the testicles hang far from the body. Furthermore, the scrotum is well supplied with sweat glands which presumably aid in keeping the testicles cool. Thus the scrotum apparently is designed to act as a cooling mechanism for the testicles, without which spermatogenesis is said to be deficient.

*Cryptorchidism.* Cryptorchidism means failure of a testis to descend from the abdomen into the scrotum. During the development of the male fetus, the testes are derived from the genital ridges in the abdomen. However, during the late stages of gestation, the testes descend through the inguinal canals into the scrotum. Occasionally this descent does not occur at all or occurs incompletely so that one or both testes remain in the abdomen, in the inguinal canal, or elsewhere along the route of descent.

A testicle that remains throughout life in the abdominal cavity is incapable of forming sperm. The tubular epithelium is completely degenerate, leaving only the interstitial structures of the testis. It is believed that even the few degrees higher temperature in the abdomen than in the scrotum is sufficient to cause degeneration of the tubular epithelium and consequently to cause sterility. For this reason operations to relocate the cryptorchid testes from the abdominal cavity into the scrotum prior to the beginning of adult sexual life are frequently performed on boys who have undescended testes.

Testosterone secretion by the fetal testes themselves is the stimulus that causes the testes to move into the scrotum from the abdomen. Therefore, many instances, if not most, of cryptorchidism are caused by abnormally formed testes that are unable to secrete enough testosterone.

**Effect of Sperm Count on Fertility.** The usual quantity of semen ejaculated at each coitus averages approximately 3.5 ml., and in each milliliter of semen is an average of approximately 120 million sperm, though even in "normal" persons this can vary from 35 million to 200 million. This means an average total of 400 million sperm are usually present in each ejaculate. When the number of sperm in each milliliter falls below approximately 20,000,000, the person is likely to be infertile. Thus, even though only a single sperm is necessary to fertilize the ovum, for reasons not yet completely understood, the ejaculate must contain a tremendous number of sperm for at least one to fertilize the ovum.

**Effect of Sperm Morphology and Motility on Fertility.** Occasionally a man has a completely normal number of sperm but is still infertile. When this occurs, often as many as half of the sperm are found to be abnormal, such as having two heads, abnormally shaped heads, or abnormal tails, as illustrated in Figure 80–4; at other times, the sperm appear to be completely normal but for reasons not understood are either entirely nonmotile or relatively nonmotile. Whenever most of the sperm are abnormal in morphology or are found to be nonmotile the person is likely to be infertile even though the remainder of the sperm appear to be normal.

**Function of Hyaluronidase and Proteinases Secreted by the Sperm for the Process of Fertilization.** Stored in the acrosomes of the sperm are large quantities of hyaluronidase and proteinases. Hyaluronidase is an enzyme that depolymerizes the hyaluronic acid polymers that are present in large quantities in the intercellular cement substance; proteinases can dissolute the proteins of tissues.

When the ovum is expelled from the follicle of the ovary into the abdominal cavity, it carries with it several layers of cells. Before a sperm can reach the ovum to fertilize it, these cells must be removed; it is believed that the hyaluronidase and proteinases released by the acrosomes play at least a small role (in

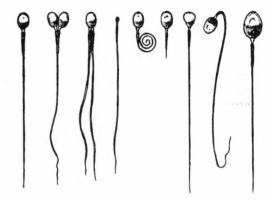

**Figure 80–4.** Abnormal sperm.

addition to a much larger role played by sodium bicarbonate in the fallopian tube secretions) in causing these cells to break away from the ovum, thus allowing the sperm to reach the surface of the ovum. When sperm are insufficient in number—below 20 million per milliliter—the man is often sterile. This sterility has been postulated to result from insufficient enzymes to help remove the cell layers from the ovum.

Another possible function of the proteinases is to allow the sperm to penetrate the mucus plug that frequently forms in the cervix of the uterus. The proteinases act as mucolytic enzymes that presumably proceed in advance of the sperm and create channels through the mucus plug. It is believed that lack of appropriate mucolytic enzyme to perform this function is also occasionally responsible for male sterility.

# THE MALE SEXUAL ACT

## NEURONAL STIMULUS FOR PERFORMANCE OF THE MALE SEXUAL ACT

The most important source of impulses for initiating the male sexual act is the glans penis, for the glans contains a highly organized sensory end-organ system that transmits into the central nervous system a special modality of sensation called *sexual sensation*. The massaging action of intercourse on the glans stimulates the sensory end-organs, and the sexual sensations in turn pass through the pudendal nerve, thence through the sacral plexus into the sacral portion of the spinal cord, and finally up the cord to undefined areas of the cerebrum. Impulses may also enter the spinal cord from areas adjacent to the penis to aid in stimulating the sexual act. For instance, stimulation of the anal epithelium, the scrotum, and perineal structures in general can all send impulses into the cord which add to the sexual sensation. Sexual sensations can even originate in internal structures, such as irritated areas of the urethra, the bladder, the prostate, the seminal vesicles, the testes, and the vas deferens. Indeed, one of the causes of "sexual drive" is probably overfilling of the sexual organs with secretions. Infection and inflammation of these sexual organs sometimes cause almost continual sexual desire, and "aphrodisiac" drugs, such as cantharides, increase the sexual desire by irritating the bladder and urethral mucosa.

**The Psychic Element of Male Sexual Stimulation.** Appropriate psychic stimuli can greatly enhance the ability of a person to perform the sexual act. Simply thinking sexual thoughts or even dreaming that the act of intercourse is being performed can cause the male sexual act to occur and to culminate in ejaculation. Indeed, *nocturnal emissions* during dreams occur in many males during some stages of sexual life, especially during the teens.

**Integration of the Male Sexual Act in the Spinal Cord.** Though psychic factors usually play an important part in the male sexual act and can actually initiate it, the cerebrum is probably not absolutely necessary for its performance, for appropriate genital stimulation can cause ejaculation in some animals and in an occasional human being after their spinal cords have been cut above the lumbar region. Therefore, the male sexual act results from inherent reflex mechanisms integrated in the sacral and lumbar spinal cord, and these mechanisms can be initiated by either psychic stimulation or actual sexual stimulation.

## STAGES OF THE MALE SEXUAL ACT

**Erection.** Erection is the first effect of male sexual stimulation, and the degree of erection is proportional to the degree of stimulation, whether this be psychic or physical.

Erection is caused by parasympathetic impulses that pass from the sacral portion of the spinal cord through the nervi erigentes to the penis. These parasympathetic impulses dilate the arteries of the penis and probably simultaneously constrict the veins, thus allowing arterial blood to flow under high pressure into the *erectile tissue* of the penis, illustrated in Figure 80–5. This erectile tissue is nothing more than large, cavernous venous sinusoids, which are normally relatively empty but which become dilated tremendously when arterial blood flows into them under pressure, since the ve-

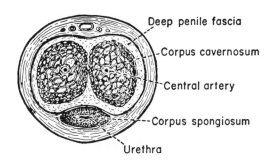

**Figure 80–5.** Erectile tissue of the penis.

nous outflow is partially occluded. Also, the erectile bodies are surrounded by strong fibrous coats; therefore, high pressure within the sinusoids causes ballooning of the erectile tissue to such an extent that the penis becomes hard and elongated.

**Lubrication.** During sexual stimulation, parasympathetic impulses, in addition to promoting erection, cause the glands of Littré and the bulbo-urethral glands to secrete mucus. Thus mucus flows through the urethra during intercourse to aid in the lubrication of coitus. However, most of the lubrication of coitus is provided by the female sexual organs rather than by the male. Without satisfactory lubrication, the male sexual act is rarely successful because unlubricated intercourse causes pain impulses which inhibit rather than excite sexual sensations.

**Emission and Ejaculation.** Emission and ejaculation are the culmination of the male sexual act. When the sexual stimulus becomes extremely intense, the reflex centers of the spinal cord begin to emit sympathetic impulses that leave the cord at L-1 and L-2 and pass to the genital organs through the hypogastric plexus to initiate emission, which is the forerunner of ejaculation.

Emission is believed to begin with contraction of the epididymis, the vas deferens, and the ampulla to cause expulsion of sperm into the internal urethra. Then, contractions in the seminal vesicles and the muscular coat of the prostate gland expel seminal fluid and prostatic fluid, forcing the sperm forward. All these fluids mix with the mucus already secreted by the bulbo-urethral glands to form the semen. The process to this point is *emission*.

The filling of the internal urethra then elicits signals that are transmitted through the pudendal nerves to the sacral regions of the cord. In turn, rhythmic nerve impulses are sent from the cord to skeletal muscles that encase the base of the erectile tissue, causing rhythmic, wavelike increases in pressure in this tissue, which "ejaculates" the semen from the urethra to the exterior. This is the process of *ejaculation*.

## TESTOSTERONE AND OTHER MALE SEX HORMONES

### *SECRETION, METABOLISM, AND CHEMISTRY OF THE MALE SEX HORMONE*

**Secretion of Testosterone by the Interstitial Cells of the Testes.**   Although several male sex hormones have been isolated from the testes, one of these, *testosterone,* is so much more abundant and potent than the others that one can consider it to be the single significant hormone responsible for the male hormonal effects caused by the testes.

Testosterone is formed by the *interstitial cells of Leydig,* which lie in the interstices between the seminiferous tubules and constitute about 20 per cent of the mass of the adult testes, as illustrated in Figure 80–6. Interstitial cells in the testes are not numerous in a child, but they *are* numerous in a newborn male infant and also in the adult male anytime after puberty; at both these times the testes secrete large quantities of testosterone. Furthermore, when tumors develop from the interstitial cells of Leydig, great quantities of testosterone are secreted. Finally, when the germinal epithelium of the testes is destroyed by x-ray treatment or by excessive heat, the interstitial cells, which are less easily destroyed, continue to produce testosterone.

**Secretion of "Androgens" Elsewhere in the body.**   The term "androgen" is used synonymously with the term male sex hormone, but it also includes male sex hormones produced elsewhere in the body besides the testes. For instance, the adrenal gland secretes at least five different androgens, though the total masculinizing activity of all these is normally so slight that they do not cause significant masculine characteristics even in women. But when an adrenal tumor of the androgen-producing cells occurs, the quantity of androgenic hormones may become great enough to cause all the usual male secondary sexual characteristics. These effects were described in connection with the adrenogenital syndrome in Chapter 77.

Rarely, embryonic rest cells in the ovary can develop into a tumor which produces androgens; one such tumor is the *arrhenoblastoma*. The normal ovary also produces minute quantities of androgens, but these are not significant.

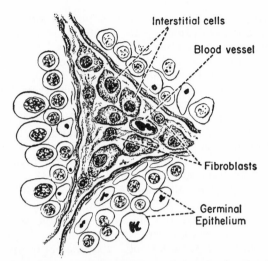

**Figure 80–6.**   Interstitial cells located in the interstices between the seminiferous tubules. (Modified from Bloom and Fawcett: Textbook of Histology, 8th Ed.)

**Chemistry of the Androgens.**   All androgens are steroid compounds, as illustrated by the formula in Figure 80–7 for *testosterone*. Both in the testes and in the adrenals, the androgens can be synthesized either from cholesterol or directly from acetyl coenzyme A.

**Metabolism of Testosterone.**   After secretion by the testes, testosterone, most of it loosely bound with plasma protein, circulates in the blood for not over 15 to 30 minutes before it either becomes fixed to the tissues or is degraded into inactive products that are subsequently excreted.

That portion of testosterone that becomes fixed to the tissues is converted within the cells to *dihydrotestosterone;* it is in this form that testosterone performs its intracellular functions. The dihydrotestosterone first binds with a receptor protein in the cytoplasm, and the resulting complex diffuses into the nucleus and binds in some way with a nuclear protein as well. In this locus, the presence of the bound dihydrotestosterone activates the DNA transcription process to

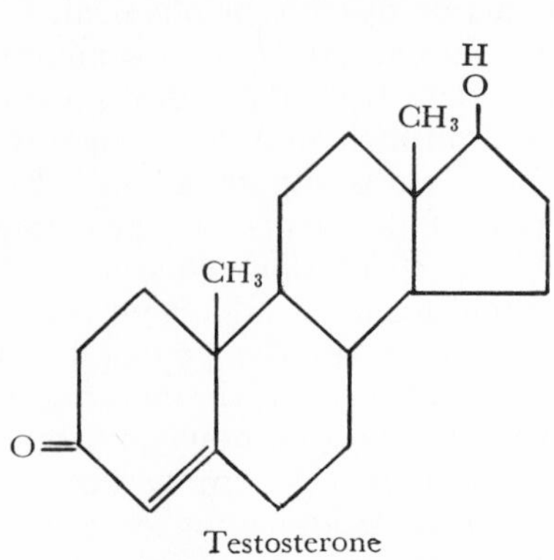

Testosterone

**Figure 80–7.**   Testosterone.

form large amounts of RNA, as will be discussed more fully later in the chapter.

*Degradation and Excretion of Testosterone.* The testosterone that does not become fixed to the tissues is rapidly converted, mainly by the liver, into *androsterone* and *dehydroepiandrosterone* and simultaneously conjugated either as glucuronides or sulfates (glucuronides, particularly). These are excreted either into the gut in the bile or into the urine.

**Production of Estrogen by the Testes.** In addition to testosterone, small amounts of estrogens are formed in the male (about one-fifth the amount in the nonpregnant female), and a reasonable quantity of these can be recovered from a man's urine. The functions of estrogens in the male are unknown.

The exact source of the estrogens in the male is also still doubtful, but the following are known: (a) The quantity of estrogens decreases when the germinal epithelium of the seminiferous tubules is destroyed. This indicates that the seminiferous tubules might synthesize estrogens in men. (b) Small amounts of estrogens are formed from testosterone during its degradation in other parts of the body. (c) The rate of estrogen secretion from the testes closely parallels the rate of testosterone secretion, for which reason it has also been suspected that the interstitial cells are the source (or a source) of the estrogens.

Thus, the problem of estrogen production in the male is unsettled except for the fact that small quantities of estrogens are produced in the testes or are formed from testosterone.

## FUNCTIONS OF TESTOSTERONE

In general, testosterone is responsible for the distinguishing characteristics of the masculine body. The testes are stimulated by chorionic gonadotropin from the placenta to produce a small quantity of testosterone during fetal development, but essentially no testosterone is produced during childhood until approximately the age of 11 to 13. Then testosterone production increases rapidly at the onset of puberty and lasts throughout most of the remainder of life as illustrated in Figure 80–8, dwindling rapidly beyond the age of 40 to become perhaps one-fifth the peak value by the age of 80.

**Functions of Testosterone During Fetal Development.** Testosterone begins to be elaborated by the male at about the second month of embryonic life. Indeed, some embryologists believe that the major functional difference between the female and the male sex chromosome is that the male chromosome causes the newly developing genital ridge to secrete testosterone while the female chromosomes causes this ridge to secrete estrogens. Injection of large quantities of male sex hormone into gravid animals causes development of male sexual organs even though the fetus is female. Also, removal of the fetal testes in a male fetus causes development of female sexual organs. Therefore, the presence or absence of testosterone in the fetus is the determining factor in the development of male or female genital organs and characteristics. That is, testosterone secreted by the genital ridges and the subsequently developing testes is responsible for the development of the male sex characteristics, including the growth of a penis and a scrotum rather than the formation of a clitoris and a vagina. Also, it causes development of the prostate gland, the seminal vesicles, and the male genital ducts, while at the same time suppressing the formation of female genital organs.

*Effect on the Descent of the Testes.* The testes usually descend into the scrotum during the last two months of gestation when the testes are secreting reasonable quantities of testosterone. If a male child is born with undescended testes, administration of testosterone causes the testes to descend in the usual manner if the inguinal canals are large enough to allow the testes to pass. Or, administration of gonadotropic hormones, which stimulate the interstitial cells of the testes to produce testosterone, also causes the testes to descend. Thus, the stimulus for descent of the testes is testosterone, indicating again that testosterone is probably an important hormone for male sexual development during fetal life.

**Effect of Testosterone on Development of Adult Primary and Secondary Sexual**

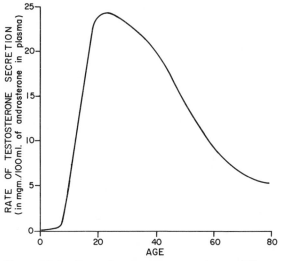

**Figure 80–8.** Rate of testosterone secretion at different ages, as judged from the concentrations of androsterone in the plasma.

**Characteristics.** Testosterone secretion after puberty causes the penis, the scrotum, and the testes all to enlarge many-fold until about the age of 20. In addition, testosterone causes the "secondary sexual characteristics" of the male to develop at the same time, beginning at puberty and ending at maturity. These secondary sexual characteristics, in addition to the sexual organs themselves, distinguish the male from the female as follows:

*Effect on the Distribution of Body Hair.* Testosterone causes growth of hair (1) over the pubis, (2) upward along the linea alba sometimes to the umbilicus and above, (3) on the face, (4) usually on the chest and (5) less often on other regions of the body, such as the back. It also causes the hair on most other portions of the body to become more prolific.

*Baldness.* Testosterone decreases the growth of hair on the top of the head; a man who does not have functional testes does not become bald. However, many virile men never become bald, for baldness is a result of two factors: first, a *genetic background* for the development of baldness and, second, superimposed on this genetic background, *large quantities of androgenic hormones.* A woman who has the appropriate genetic background and who develops a long-sustained androgenic tumor becomes bald in the same manner as a man.

*Effect on the Voice.* Testosterone secreted by the testes or injected into the body causes hypertrophy of the laryngeal mucosa and enlargement of the larynx. The effects cause at first a relatively discordant, "cracking" voice, but this gradually changes into the typical masculine bass voice.

*Effect on the Skin.* Testosterone increases the thickness of the skin over the entire body and increases the ruggedness of the subcutaneous tissues. Also, it causes increased quantities of melanin to be deposited in the skin, thereby deepening the hue of the skin.

Testosterone increases the rate of secretion by some or perhaps all the sebaceous glands. Especially important is the excessive secretion by the sebaceous glands of the face, for oversecretion of these can result in *acne.* Therefore, acne is one of the most common features of adolescence, when the male body is first becoming introduced to increased secretion of testosterone. After several years of testosterone secretion, the skin adapts itself to the testosterone in some way that allows it to overcome the acne.

*Effect on Nitrogen Retention and Muscular Development.* One of the most important male characteristics is the development of increasing musculature following puberty. This is associated with increased protein in other parts of the body as well. Many of the changes in the skin are also due to deposition of proteins in the skin, and the changes in the voice could even result from this protein anabolic function of testosterone.

Testosterone has often been considered to be a "youth hormone" because of its effect on the musculature, and it is occasionally used for treatment of persons who have poorly developed muscles.

*Effect on Bone Growth and Calcium Retention.* Following puberty or following prolonged injection of testosterone, the bones grow considerably in thickness and also deposit considerable calcium salts. Thus, testosterone increases the total quantity of bone matrix, and it also causes calcium retention. The increase in bone matrix is believed to result from the general protein anabolic function of testosterone, and the deposition of calcium salts to result from increased bone matrix available to be calcified.

Because of the ability of testosterone to increase the size and strength of bones, testosterone is often used in old age to treat osteoporosis.

When great quantities of testosterone (or any other androgen) are secreted in the still-growing child, the rate of bone growth increases markedly, causing a spurt in total body growth as well. However, the testosterone also causes the epiphyses of the long bones to unite with the shafts of the bones at an early age in life. Therefore, despite the rapidity of growth, this early uniting of the epiphyses prevents the person from growing as tall as he would have grown had testosterone not been secreted at all. Even in normal men the final adult height is slightly less than that which would have been attained had the person been castrated prior to puberty.

*Effect on Basal Metabolism.* Injection of large quantities of testosterone can increase the basal metabolic rate by as much as 15 per cent, and it is believed that even the usual quantity of testosterone secreted by the testes during active sexual life increases the rate of metabolism some 5 to 10 per cent above the value that it would be were the testes not active. This increased rate of metabolism is possibly an indirect result of the effect of testosterone on protein anabolism, the increased quantity of proteins—the enzymes especially—increasing the activities of all cells.

*Effect on the Red Blood Cells.* When normal quantities of testosterone are injected into a castrated adult, the number of red blood cells per cubic millimeter of blood increases approximately 20 per cent. Also, the average man has 500,000 to 1,000,000 more red blood cells per cubic millimeter than the average woman. However, this difference may be due partly to the increased metabolic rate following testosterone administration rather than to a direct effect of testosterone on red blood cell production.

*Effect on Electrolyte and Water Balance.* As pointed out in Chapter 77, many different steroid hormones can increase the reabsorption of sodium in the distal tubules of the kidneys. Testosterone performs this function to a slight extent but only to a minor degree in comparison with the adrenal mineralocorticoids. Nevertheless, following puberty the blood and extracellular fluid volumes of the male in relation to his weight increase to a slight extent; this effect probably results at least partly from the sodium-retaining ability of testosterone.

## BASIC MECHANISM OF ACTION OF TESTOSTERONE

Though it is not known exactly how testosterone causes all the effects just listed, it is believed that they result mainly from increased rate of protein formation in cells. This has been studied extensively in the prostate gland, one of the organs that is most affected by testosterone. In this gland, testosterone enters the cells within a few minutes after secretion, is there converted to dihydrotestosterone, and binds with a cytoplasmic "receptor protein." This combination then migrates to the nucleus where it binds with a nuclear protein and induces the DNA-RNA transcription process. Within 30 minutes the concentration of RNA begins to increase in the cells, and this is followed by progressive increase in cellular protein. After several days the quantity of DNA in the gland has also increased, and there has been a simultaneous increase in the number of prostatic cells.

Therefore, it is assumed that testosterone greatly stimulates production of proteins in general, though increasing more specifically those proteins in "target" organs or tissues responsible for the development of secondary sexual characteristics.

## CONTROL OF MALE SEXUAL FUNCTIONS BY THE GONADOTROPIC HORMONES—FSH AND LH

The anterior pituitary gland secretes two different gonadotropic hormones: (1) *follicle-stimulating hormone* (FSH); and (2) *luteinizing hormone* (LH), also called *interstitial cell-stimulating hormone* (ICSH). Both of these play major roles in the control of male sexual function.

**Regulation of Testosterone Production by LH.** Testosterone is produced by the interstitial cells of Leydig only when the testes are stimulated by LH from the pituitary gland, and the quantity of testosterone secreted varies approximately in proportion to the amount of LH available.

Injection of purified LH into a child causes fibroblasts in the interstitial areas of the testes to develop into interstitial cells of Leydig, though mature Leydig cells are not normally found in the child's testes until after the age of approximately 10. Also, simultaneous administration of a small amount of FSH along with LH greatly potentiates the effect of LH in promoting testosterone production.

*Effect of Chorionic Gonadotropin on the Fetal Testes.* During gestation the placenta secretes large quantities of chorionic gonadotropin, a hormone that has almost the same properties as LH. This hormone stimulates the formation of interstitial cells in the testes of the fetus and causes testosterone secretion. As pointed out earlier in the chapter, the secretion of testosterone during fetal life is important for promoting formation of male sexual organs.

**Regulation of Spermatogenesis by Follicle-Stimulating Hormone (FSH).** The conversion of primary spermatocytes into secondary spermatocytes in the seminiferous tubules is stimulated by FSH from the anterior pituitary gland; and in the absence of FSH, spermatogenesis will not proceed. However, FSH cannot by itself cause complete formation of spermatozoa. For spermatogenesis to proceed to completion, testosterone must be secreted simultaneously in small amounts by the interstitial cells. Thus, FSH seems to initiate the proliferative process of spermatogenesis, and testosterone diffusing from the interstitial cells into the seminiferous tubules apparently is necessary for final maturation of the spermatozoa. Because testosterone is secreted by the interstitial cells under the influence of LH, both FSH and LH must be secreted by the anterior pituitary gland if spermatogenesis is to occur.

**Regulation of LH and FSH Secretion by the Hypothalamus.** The gonadotropins, like corticotropin and thyrotropin, are secreted by the anterior pituitary gland mainly in response to nervous activity in the hypothalamus. For instance, in the female rabbit, coitus with a male rabbit elicits nervous activity in the hypothalamus that in turn stimulates the anterior pituitary to secrete FSH and LH. These hormones then cause rapid ripening of follicles in the rabbit's ovaries, followed a few hours later by ovulation.

Many other types of nervous stimuli are also known to affect gonadotropin secretion. For instance, in sheep, goats, and deer, nervous stimuli in response to changes in weather and amount of light in the day increase the quantities of gonadotropins during one season of the year, the mating season, thus allowing birth of the young during an appropriate period for survival. Also, psychic stimuli can affect fertility of the male animal, as exemplified by the fact that transporting a bull under uncomfortable conditions can often cause almost complete temporary sterility. In the human being, it is known too that various psychogenic stimuli feeding into the hypothalamus can cause marked excitatory or inhibitory effects on gonadotropin secretion, in this way sometimes greatly altering the degree of fertility.

*The Gonadotropin-Releasing Factors.* The hypothalamus controls gonadotropin secretion by way of the hypothalamic-hypophyseal portal system. Neurosecretory hormones called *gonadotropin-releasing factors* are carried in the hypophyseal portal blood to the anterior pituitary to control the secretion of LH and FSH. These factors are *luteotropin-releasing factor* (LRF) and *follicle-stimulating hormone releasing factor* (FRF). They are the same as those in the female where the interrelationships are far more complex. This will be discussed in the following chapter. These gonadotropin-releasing factors are similar to growth hormone–releasing factor, corticotropin-releasing factor, and thyrotropin-releasing factor, all of which have been noted in previous chapters to increase the secretion of respective anterior pituitary hormones.

Yet the specific areas of the hypothalamus responsible for secretion of the two gonadotropin-releasing factors are not well established. Two particular areas seem to be most important: (1) the lowermost region of the hypothalamus in the vicinity of the *median eminence* and (2) an *area in the anterior hypothalamus located in the suprachiasmal region.*

**Reciprocal Inhibition of the Hypothalamic-Anterior Pituitary Secretion of Gonadotropin Hormones by the Testicular Hormones.** Injection of testosterone into either a male or a female animal inhibits the secretion of gonadotropins. This inhibition is dependent on normal function of the hypothalamus. Therefore, it is assumed that testosterone inhibits hypothalamic stimulation of the anterior pituitary gland to produce the gonadotropins. This inhibitory effect is much more marked on the production of LH than of FSH. Therefore, one can readily see that this inhibitory effect of testosterone provides a feedback control system for maintaining testosterone secretion at a constant level; that is, excess testosterone secretion inhibits LH secretion, which in turn reduces the testosterone secretion back to normal level. Conversely, the mechanism also operates in the opposite direction to protect against too little production of testosterone.

It is known, too, that spermatogenesis in the testes in some unexplained way inhibits the secretion of FSH; conversely, failure of spermatogenesis causes marked increase in FSH. It is believed that some additional testicular hormone besides testosterone is responsible for this inhibitory feedback, and the hormone has been called *inhibin.* However, it is also possible that this inhibition results simply from estrogen secreted by the active seminiferous tubules or by the interstitial cells in response to activity in the seminiferous tubules. Obviously, this feedback from the seminiferous tubules could control the rate of spermatogenesis. For instance, it might increase the rate of spermatogenesis when the male ejaculates often.

**Puberty and Regulation of Its Onset.** During the first 10 years of life, the male child secretes almost no gonadotropins and, consequently, almost no testosterone. Then, at the age of about 10, the anterior pituitary gland begins to secrete progressively increasing quantities of gonadotropins; and this is followed by a corresponding increase in testicular function. By approximately the age of 13, the male child reaches full adult sexual capability. This period of change is called *puberty.*

Initiation of the onset of puberty has long been a mystery. In earliest history of mankind, the belief was simply that the testicles "ripened" at this time, but with the discovery of the gonadotropins the belief shifted to ripening of the anterior pituitary gland. Now it is known from experiments in which both testicular and pituitary tissues have been transplanted from infant animals into adult animals that both the testes and the anterior pituitary of the infant are capable of performing adult functions if appropriately stimulated. Therefore, it is now almost certain that *during childhood the hypothalamus, not the two glands, is at fault; that is, the hypothalamus simply does not secrete gonadotropin-releasing factors.* Also, it is known that even the minutest amount of testosterone inhibits the production of gonadotropin-releasing factors by the childhood hypothalamus. This leads to the theory that the childhood hypothalamus is so extremely sensitive to inhibition that the minutest amount of testosterone secreted by the testicles inhibits the entire system. For reasons yet unknown, the hypothalamus loses this inhibitory sensitivity at the time of puberty, which allows the secretory mechanisms to develop full activity. Thus, puberty is now believed to result from an aging process of the hypothalamic sexual control centers.

**The Male Adult Sexual Life and the Male Climacteric.** Following puberty, gonadotropic hormones are produced by the male pituitary gland for the remainder of life, and at least some spermatogenesis usually continues until death. Most men, however, begin to exhibit slowly decreasing sexual functions in the late 40's or 50's, but it is not the same rapid decrease as that which occurs in women at the menopause. This is related to decrease in testosterone secretion as depicted in Figure 80–8. The decrease in male sexual function is called the male climacteric. Occasionally, the male climacteric is associated with symptoms of hot flashes, suffocation, and psychic disorders similar to the menopausal symptoms of the female. These symptoms can be abrogated by administration of testosterone, synthetic methyl testosterone, or even estrogens that are used for treatment of menopausal symptoms in the female.

The male climacteric is associated with greatly increased secretion of FSH by the anterior pituitary gland because testosterone is no longer available to inhibit its production.

# ABNORMALITIES OF MALE SEXUAL FUNCTION

## THE PROSTATE GLAND AND ITS ABNORMALITIES

The prostate gland remains relatively small throughout childhood and begins to grow at puberty under the stimulus of testosterone. This gland reaches an almost stationary size by the age of about 20 and remains this size up to the age of approximately 40 to 50. At that time in some men it begins to degenerate along with the decreased production of testosterone by the testes. However, a benign prostatic fibroadenoma frequently develops in the prostate in older men and causes urinary obstruction. This effect is not caused by testosterone.

Cancer of the prostate gland is an extremely common cause of death, resulting in approximately 2 to 3 per cent of all male deaths.

Once cancer of the prostate gland does occur, the cancerous cells are usually stimulated to more rapid growth by testoserone and are inhibited by removal of the testes so that testosterone cannot be formed. Also, prostatic cancer can usually be inhibited by administration of estrogens. Some patients who have prostatic cancer that has already metastasized to almost all the bones of the body can be successfully treated for a few months to years by removal of the testes, by estrogen therapy, or by both; following this therapy the metastases degenerate and the bones heal. This treatment does not completely stop the cancer but does slow it down and greatly diminishes the severe bone pain. In general, the metastases return within a few months to a year or so.

## HYPOGONADISM IN THE MALE

Hypogonadism is caused by any of several abnormalities: First, the person may be born without functional testes. Second, he may have undeveloped testes, owing to failure of the anterior pituitary to secrete gonadotropic hormones. Third, cryptorchidism (undescended testes) may occur, associated with partial or total degeneration of the testes. Fourth, he may lose his testes, which is called *castration.*

When a boy loses his testes prior to puberty, a state of *eunuchism* ensues in which he continues to have infantile sexual characteristics throughout life. The height of the adult eunuch is slightly greater than that of the normal man, though the bones are quite thin, the muscles are considerably weaker than those of normal man, and, obviously, the sexual organs and secondary sexual characteristics are those of a child rather than those of an adult. The voice is child-like, there is no loss of hair on the head, and the normal masculine hair distribution on the face and elsewhere does not occur.

When a man is castrated following puberty, some male secondary sexual characteristics revert to those of a child, and others remain of masculine character. The sexual organs regress slightly in size but not to a child-like state, and the voice regresses from the bass quality only slightly. On the other hand, there is loss of masculine hair production, loss of the thick masculine bones, and loss of the musculature of the virile male.

In the castrated adult male, sexual desires are de-

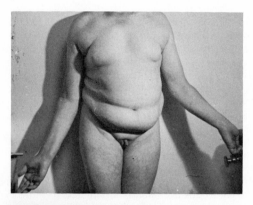

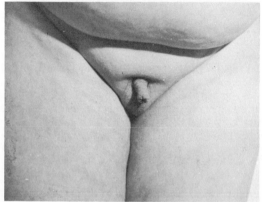

**Figure 80-9.** Adiposogenital syndrome in an adolescent male. Note the obesity and the childlike sexual organs. (Courtesy of Dr. Leonard Posey.)

creased but not totally lost, provided that sexual activities have been practiced previously. Erection can still occur as before though with less ease, but it is rare that ejaculation can take place, primarily because the semen-forming organs become degenerate, and there has been a loss of the testosterone-driven psychic desire.

**Adiposogenital Syndrome (Fröhlich's Syndrome).** Damage to certain areas of the hypothalamus greatly decreases the secretion of gonadotropic hormones by the anterior pituitary. If this occurs prior to puberty, it causes typical eunuchism. The damage often causes simultaneous overeating because of its effect on the feeding center of the hypothalamus. Consequently, the person develops severe obesity along with the eunuchism. This condition is illustrated in Figure 80–9 and is called *adiposogenital syndrome, Fröhlich's syndrome,* or hypothalamic eunuchism.

## TESTICULAR TUMORS AND HYPERGONADISM IN THE MALE

*Interstitial cell tumors* develop rarely in the testes, but when they do develop they sometimes produce as much as 100 times normal quantities of testosterone.

When such tumors develop in young children, they cause rapid growth of the musculature and bones but also early uniting of the epiphyses so that the eventual adult height actually is less than that which would have been achieved otherwise. Obviously, such interstitial cell tumors cause excessive development of the sexual organs and of the secondary sexual characteristics. In the adult male, small interstitial cell tumors are difficult to diagnose because masculine features are already present. Diagnosis can be made, however, from urine tests that show greatly increased excretion of testosterone endproducts.

Much more common than the interstitial cell tumors are tumors of the germinal epithelium. Because germinal cells are capable of differentiating into almost any type of cell, many of these tumors contain multiple tissues, such as placental tissue, hair, teeth, bone, skin, and so forth, all found together in the same tumorous mass called a *teratoma.* Often these tumors secrete no hormones, but if a significant quantity of placental tissue develops in the tumor, it may secrete large quantities of chorionic gonadotropin that has functions very similar to those of LH, which will be discussed in more detail in Chapter 82. Also, estrogenic hormones are frequently secreted by these tumors and cause the condition called *gynecomastia,* which means overgrowth of the breasts.

## REFERENCES

Armstrong, D. T.: Reproduction. *Ann. Rev. Physiol., 32*:439, 1970.

Austin, C. R., and Short, R. V. (eds.): Reproduction in Mammals Series, Books 1–5. Cambridge, Eng., Cambridge University Press, 1972.

Beck, W. W., Jr.: Artificial insemination and semen preservation. *Clin. Obstet. Gynecol., 17*:115, 1974.

Bedford, J. M.: Maturation, transport, and fate of spermatozoa in the epididymis. *In* Greep, R. O., and Astwood, E. B. (eds.): Handbook of Physiology. Sec. 7, Vol. 5. Baltimore, The Williams & Wilkins Company, 1975, p. 303.

Behrman, H. R., and Caldwell, B. V.: Role of prostaglandins in reproduction. *In* Guyton, A. C. (ed.): MTP International Review of Science: Physiology. Vol. 8. Baltimore, University Park Press, 1974, p. 63.

Blackshaw, A. W.: Testicular enzymes and spermatogenesis. *J. Reprod. Fertil. (Suppl.),* 18:55, 1973.

Brandes, D. (ed.): Male Accessory Sex Organs. New York, Academic Press, Inc., 1974.

Christensen, A. K.: Leydig cells. *In* Greep, R. O., and Astwood, E. B. (eds.): Handbook of Physiology. Sec. 7, Vol. 5. Baltimore, The Williams & Wilkins Company, 1975, p. 57.

Clermont, Y.: Kinetics of spermatogenesis in mammals: seminiferous epithelium cycle and spermatogonial renewal. *Physiol. Rev., 52*:198, 1972.

Dempsey, M. E.: Regulation of steroid biosynthesis. *Ann. Rev. Biochem., 43*:967, 1974.

Donovan, B. T.: The role of the hypothalamaus in puberty. *Prog. Brain Res., 41*:239, 1974.

Eik-Nes, K. B.: Biosynthesis and secretion of testicular steroids. *In* Greep, R. O., and Astwood, E. B. (eds.): Handbook of Physiology. Sec. 7, Vol. 5. Baltimore, The Williams & Wilkins Company, 1975, p. 95.

Fawcett, D. W.: Ultrastructure and function of the Sertoli cell. *In* Greep, R. O., and Astwood, E. B. (eds.): Handbook of Physiology. Sec. 7, Vol. 5. Baltimore, The Williams & Wilkins Company, 1975, p. 21.

Fox, C. A.: Recent studies in human coital physiology. *Clin. Endocrinol. Metabol.*, 2:527, 1973.

Glover, T. D.: Recent progress in the study of male reproductive physiology: testis stimulation, sperm formation, transport and maturation (epididymal physiology) semen analysis, storage, and artificial insemination. *In* Guyton, A. C. (ed.): MTP International Review of Science: Physiology. Baltimore, University Park Press, 1974, p. 221.

Goland, M. (ed.): Normal and Abnormal Growth of the Prostate. Springfield, Ill., Charles C Thomas, Publisher, 1975.

Gomes, W. R., and Van Demark, N. L.: The male reproductive system. *Ann. Rev. Physiol.*, 36:307, 1974.

Gondos, B.: Testicular changes associated with the initiation of spermatogenesis. *Ann. Clin. Lab. Sci.*, 5:4, 1975.

Jackson, H.: Chemical methods of male contraception. *Amer. Sci.*, 61:188, 1973.

James, V. H. T., and Martini, L. (eds.): The Endocrine Function of the Human Testis. New York, Academic Press, Inc., 1974.

Jost, A., Vigier, B., Prépin, J., and Perchellet, J. P.: Studies on sex differentiation in mammals. *Recent Prog. Horm. Res.*, 29:1, 1973.

Johnson, A. D., Gomes, W. R., and Van Demark, N. L. (eds.): The Testis. New York, Academic Press, Inc., 1971.

McRorie, R. A., and Williams, W. L.: Biochemistry of mammalian fertilization. *Ann. Rev. Biochem.*, 43:777, 1974.

Mainwaring, W. I., Mangan, F. R., Wilce, P. A., and Milroy, E. G.: Androgens. I. A review of current research on the binding and mechanism of action of androgenic steroids, notably 5-alpha-dihydrotestosterone. *Adv. Exp. Med. Biol.*, 36:197, 1973.

Mancini, R. E., and Martini, L. (eds.): Male Fertility and Sterility. New York, Academic Press, Inc., 1974.

Mann, T.: Advances in male reproductive physiology. *Fertil. Steril.*, 23:699, 1972.

Means, A. R.: Biochemical effects of follicle-stimulating hormone on the testis. *In* Greep, R. O., and Astwood, E. B. (eds.): Handbook of Physiology. Sec. 7, Vol. 5. Baltimore, The Williams & Wilkins Company, 1975, p. 203.

Phillips, D. M.: Spermiogenesis. New York, Academic Press, Inc., 1974.

Phillips, D. M.: Mammalian sperm structure. *In* Greep, R. O., and Astwood, E. B. (eds.): Handbook of Physiology. Sec. 7, Vol. 5. Baltimore, The Williams & Wilkins Company, 1975, p. 405.

Ramirez, V. D.: Endocrinology of puberty. *In* Greep, R. O., Astwood, E. B. (eds.): Handbook of Physiology. Sec. 7, Vol. 2, Part 1. Baltimore, The Williams & Wilkins Company, 1973, p. 1.

Richart, R. M., and Prager, D. J. (eds.): Human Sterilization. Springfield, Ill., Charles C Thomas, Publisher, 1972.

Root, A. W.: Endocrinology of puberty. I. Normal sexual maturation. *J. Pediatr.*, 83:1, 1973.

Sandler, M., and Gessa, G. L. (eds.): Sexual Behavior—Pharmacology and Biochemistry. New York, Raven Press, 1975.

Saxena, B. B., Beling, C. G., and Gandy, H. M.: Gonadotropins. New York, John Wiley & Sons, Inc., 1972.

Schwartz, N. B., and McCormack, C. E.: Reproduction: gonadal function and its regulation. *Ann. Rev. Physiol.*, 34:425, 1972.

Segal, S. J.: Research in fertility regulation. *New Eng. J. Med.*, 279:364, 1968.

Shahidi, N. T.: Androgens and erythropoiesis. *New Eng. J. Med.*, 289:72, 1973.

Steinberger, E.: Hormonal control of mammalian spermatogenesis. *Physiol. Rev.*, 51:1, 1971.

Steinberger, E., and Steinberger, A.: Hormonal control of testicular function in mammals. *In* Greep, R. O., and Astwood, E. B. (eds.): Handbook of Physiology. Sec. 7, Vol. 4, Part 2. Baltimore, The Williams & Wilkins Company, 1974, p. 325.

Steinberger, E., and Steinberger, A.: Spermatogenic function of the testis. *In* Greep, R. O., Astwood, E. B. (eds.): Handbook of Physiology. Sec. 7, Vol. 5. Baltimore, The Williams & Wilkins Company, 1975, p. 1.

Van Niekerk, W. A.: True Hermaphroditism. Hagerstown, Maryland, Harper & Row, Publishers, 1974.

Williams-Ashman, H. G., and Reddi, A. H.: Actions of vertebrate sex hormones. *Ann. Rev. Physiol.*, 33:31, 1971.

# | 81 |

# Sexual Functions in the Female and the Female Hormones

The sexual and reproductive functions in the female can be divided into two major phases: first, preparation of the body for conception and gestation, and second, the period of gestation itself. The present chapter is concerned with the preparation of the body for gestation, and the following chapter presents the physiology of pregnancy.

## PHYSIOLOGIC ANATOMY OF THE FEMALE SEXUAL ORGANS

Figure 81–1 illustrates the principal organs of the human female reproductive tract, including the *ovaries,* the *fallopian tubes,* the *uterus,* and the *va-*

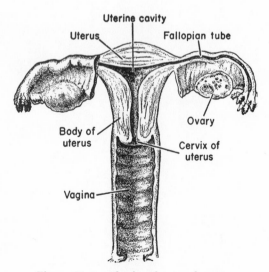

**Figure 81–1.**   The female sexual organs.

*gina.* Reproduction begins with the development of ova in the ovaries. A single ovum is expelled from an ovarian follicle into the abdominal cavity in the middle of each monthly sexual cycle. This ovum then passes through one of the fallopian tubes into the uterus, and, if it has been fertilized by a sperm, it implants in the uterus where it develops into a fetus, a placenta, and fetal membranes.

During fetal life the outer surface of the ovary is covered by a *germinal epithelium,* which embryologically is derived directly from the epithelium of the germinal ridges. As the fetus develops, *primordial ova* differentiate from the germinal epithelium and migrate into the substance of the ovarian cortex, carrying with them a layer of epithelioid granulosa cells. The ovum surrounded by a single layer of epithelioid granulosa cells is called a *primordial follicle.* At birth, about 750,000 primordial follicles are present in the two ovaries, but this number decreases rapidly so that only 400,000 remain at puberty. During all the reproductive years of the female, only about 450 of these follicles develop enough to expel their ova; the remainder degenerate (become *atretic).* At the end of reproductive capability, at the *menopause,* only a few primordial follicles remain in the ovaries, and even these degenerate soon thereafter.

## THE FEMALE HORMONAL SYSTEM

The female hormonal system, like that of the male, consists of three different hierarchies of hormones, as follows:

1. The hypothalamic releasing factors, *follicle-stimulating hormone releasing factor* (FRF) and *luteinizing hormone releasing factor* (LRF) (which may in fact be the same factor).

2. The anterior pituitary hormones, *follicle-stimulating hormone* (FSH) and *luteinizing hormone,* (LH) which are secreted in response to the releasing factor or factors from the hypothalamus.

3. The ovarian hormones, *estrogen* and *progesterone,* which are secreted by the ovaries in response to the two hormones from the anterior pituitary gland.

The various hormones are not secreted in constant, steady amounts, but instead are secreted at drastically differing rates during different parts of the female cycle. Figure 81–2 illustrates the changing concentrations of the anterior pituitary hormones, FSH and LH, and of the ovarian hormones, estradiol (estrogen) and progesterone. Although detailed measurements have not yet been made, it is reasonable to believe that the hypothalamic releasing factors also undergo cyclic variations.

Before it is possible to discuss the interplay between these different hormones, it is first necessary to describe some of their specific functions and their relationships to function of the ovaries.

# THE MONTHLY OVARIAN CYCLE AND FUNCTION OF THE GONADOTROPIC HORMONES

The normal reproductive years of the female are characterized by monthly rhythmic changes in the rates of secretion of the female hormones

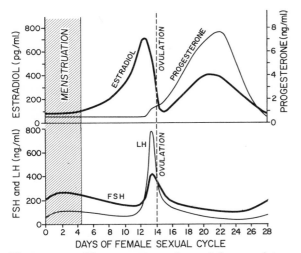

**Figure 81–2.** Plasma concentrations of the gonadotropins and ovarian hormones during the normal female sexual cycle.

and corresponding changes in the sexual organs themselves. This rhythmic pattern is called the *female sexual cycle* (or less accurately, the *menstrual cycle*). The duration of the cycle averages 28 days. It may be as short as 20 days or as long as 45 days even in completely normal women, though abnormal cycle length is occasionally associated with decreased fertility.

The two significant results of the female sexual cycles are: First, only a *single* mature ovum is normally released from the ovaries each month so that only a single fetus can begin to grow at a time. Second, the uterine endometrium is prepared for implantation of the fertilized ovum at the required time of the month.

**The Gonadotropic Hormones.** The sexual cycle is completely dependent on gonadotropic hormones secreted by the anterior pituitary gland. Ovaries that are not stimulated by gonadotropic hormones remain completely inactive, which is essentially the case throughout childhood when almost no gonadotropic hormones are secreted. However, at the age of about eight, the pituitary begins secreting progressively more and more gonadotropic hormones, which culminates in the initiation of monthly sexual cycles between the ages of 11 and 15; this culmination is called *puberty.*

(Note that the ovaries function to a slight extent during fetal life because of stimulation by gonadotropic hormones from the placenta. But immediately after birth this stimulus is lost and the ovaries become almost totally dormant until the prepubertal period.)

The anterior pituitary secretes two different hormones that are known to be essential for full function of the ovaries: (1) *follicle-stimulating hormone* (FSH), and (2) *luteinizing hormone* (LH). Both of these are small glycoproteins having molecular weights of about 30,000. The only significant effects of FSH and LH are on the ovaries in the female and the testes in the male. LH in the male is frequently called *interstitial cell-stimulating hormone* (ICSH) because it stimulates the interstitial cells.

During each month of the female sexual cycle, there is a cyclic increase and decrease of FSH and LH. These cyclic variations in turn cause the cyclic ovarian changes, which are explained in the following sections.

Both FSH and LH stimulate the ovarian cells by combining with highly specific cellular membrane receptors that in turn activate adenylcyclase. This then leads to an increase in cyclic AMP in the cells, causing growth and secretion by the specific ovarian cells. Thus, the mechanism of action of these hormones is typical of the usual cyclic AMP system for hormonal control, the details of which were explained in Chapter 75.

## FOLLICULAR GROWTH—FUNCTION OF FOLLICLE-STIMULATING HORMONE (FSH)

Figure 81–3 depicts the various stages of follicular growth in the ovaries, illustrating, first, the primordial follicle (primary follicle). Throughout childhood the primordial follicles do not grow, but at puberty, when FSH from the anterior pituitary gland begins to be secreted in large quantity, the entire ovaries and especially the follicles within them begin to grow. The first stage of follicular growth is enlargement of the ovum itself. This is followed by development of additional layers of granulosa cells around each ovum and development of several layers of theca cells around the granulosa cells. The theca cells originate from the stroma of the ovary and soon take on epithelioid characteristics; their total mass is called the *theca interna*. It is mainly these cells that are destined to secrete most of the female hormones, the estrogens and progesterone. Surrounding the theca interna, a connective tissue capsule known as the *theca externa* develops. This becomes the capsule of the developing follicle.

**The Vesicular Follicles.** At the beginning of each month of the female sexual cycle, at approximately the onset of menstruation, the concentrations of FSH and LH increase. These increases cause accelerated growth of the theca and granulosa cells in about 20 of the ovarian follicles each month. The theca cells and the granulosa cells also secrete a follicular fluid that contains a high concentration of estrogen, one of the important female sex hormones that will be discussed later. The accumulation of this fluid in the follicle causes an *antrum* to appear within the mass of theca and granulosa cells, as illustrated in Figure 81–3.

After the antrum is formed, the theca and granulosa cells continue to proliferate, the rate of secretion accelerates, and each of the growing follicles becomes a *vesicular follicle*. This accelerating growth and increasing secretion is caused by two other factors in addition to the follicle-stimulating hormone. First, a small basal amount of luteinizing hormone is also secreted by the anterior pituitary gland; this has a synergistic effect that supports the stimulatory effect of FSH. Second, the estrogen secreted into the follicle has a similar synergistic effect as well.

As the vesicular follicle enlarges, the theca and granulosa cells continue to develop at one pole of the follicle. It is in this mass that the ovum is located.

**Atresia of All Follicles but One.** After a week or more of growth—but before ovulation occurs—one of the follicles begins to outgrow all the others; the remainder begin to involute (a process called *atresia*), and these follicles are said to become *atretic*. The cause of the atresia is unknown, but it has been postulated to be the following: The one follicle that becomes more highly developed than the others secretes enough estrogen to cause feedback inhibition of secretion of the gonadotropic hormone FSH by the anterior pituitary gland. This lack of hormone does not prevent further growth of the largest follicle because the large amount of locally secreted estrogen in this follicle is already causing this follicle to continue growing. However, the lack of FSH stimulus to the less well developed follicles causes these to stop growing and, indeed, to involute.

This process of atresia obviously is important in that it allows only one of the follicles to grow large enough to ovulate. This single follicle reaches a size of approximately 1 to 1.5 centimeters at the time of ovulation.

Ovulation, in a woman who has a normal 28-day female sexual cycle, occurs 14 days after the onset of menstruation.

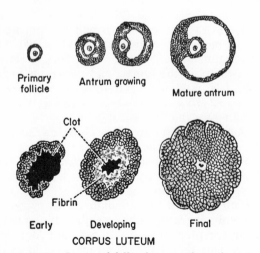

**Figure 81–3.** Stages of follicular growth in the ovary, showing also formation of the corpus luteum. (Modified from Arey: Developmental Anatomy, 7th Ed.)

The process of ovulation has never been observed in the human being, but it has been observed and studied experimentally in rats and rabbits. Shortly before ovulation, the protruding outer wall of the follicle swells rapidly, and a small area in the center of the capsule, called the *stigma,* protrudes like a nipple. In another half hour or so, fluid begins to ooze from the follicle through the stigma. About two minutes later, as the follicle becomes smaller because of loss of fluid, the stigma ruptures widely, and a more viscous fluid that has occupied the central portion of the follicle is evaginated outward into the abdomen. This viscous fluid carries with it the ovum surrounded by several layers of granulosa cells called the *corona radiata* and additional layers of loose adherent theca cells.

**Need for Luteinizing Hormone (LH) in Ovulation—Ovulatory Surge of LH.** Luteinizing hormone is necessary for final follicular growth and ovulation. Without this hormone, even though large quantities of FSH are available, the follicle will not progress to the stage of ovulation.

Approximately two days before ovulation, for reasons that are not completely known at present but which will discussed in more detail later in the chapter, the rate of secretion of LH by the anterior pituitary gland increases markedly, rising 6- to 10- fold and peaking about 18 hours before ovulation. FSH also increases about two-fold at the same time, and the two of

these hormones act synergistically to cause extremely rapid swelling of the follicle shortly before ovulation. The LH also has the specific effect on the theca and granulosa cells of changing them into *lutein cells* that in turn secrete progesterone but less estrogen. Therefore, the rate of secretion of estrogen begins to fall approximately one day prior to ovulation, while minute amounts of progesterone begin to be secreted.

It is in this environment of (a) very rapid growth of the follicle, (b) diminishing estrogen secretion after a prolonged phase of excessive estrogen secretion, and (c) beginning secretion of progesterone that ovulation occurs. Without the initial preovulatory surge of luteinizing hormone, ovulation will not take place.

**Mechanism of Ovulation.** Figure 81–4 illustrates the postulated mechanism of ovulation. It shows the initiating cause to be the large quantity of luteinizing hormone secreted by the anterior pituitary gland. The luteinizing hormone in turn causes rapid secretion of follicular steroid hormones containing a small amount of progesterone for the first time. Within a few hours two events occur, both of which are necessary for ovulation: (1) The theca externa (the capsule of the follicle) begins to form proteolytic enzymes that cause dissolution of the capsular wall and consequent weakening of the wall, resulting in further swelling of the entire follicle and degeneration at the stigma. (2) The luteinizing process of the theca and granulosa

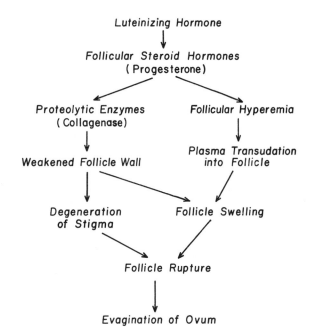

**Figure 81–4.** Postulated mechanism of ovulation. (Based primarily on the research studies of H. Lipner.)

cells is accompanied by growth of new blood vessels into the follicle wall. This in turn causes plasma transudation into the follicle, which also contributes to follicle swelling. Finally, the combined follicle swelling and simultaneous degeneration of the stigma causes follicle rupture with evagination of the ovum.

### THE CORPUS LUTEUM—THE "LUTEAL" PHASE OF THE OVARIAN CYCLE

During the first few hours after expulsion of the ovum from the follicle, the remaining theca cells of the follicle undergo rapid physical and chemical change, a process called *luteinization,* and the mass of cells becomes a *corpus luteum,* which secretes the hormones progesterone and estrogen. That is, these cells become greatly enlarged and develop lipid inclusions that give the cells a distinctive yellowish color, from which is derived the term *luteum.* A well-developed vascular supply also grows into this mass of developing *lutein cells.*

In the normal female, the corpus luteum grows to approximately 1.5 cm., reaching this stage of development approximately seven or eight days following ovulation. After this, it begins to involute and loses its secretory function, as well as its lipid characteristics, approximately 12 days following ovulation, becoming then the so-called *corpus albicans;* during the ensuing few weeks this is replaced by connective tissue.

**Luteinizing Function of Luteinizing Hormone (LH).** The change of follicular cells into lutein cells is completely dependent on the LH secreted by the anterior pituitary gland. In fact, this function gave LH its name "luteinizing."

**Secretion by the Corpus Luteum: Function of LH.** The corpus luteum is a highly secretory organ, secreting large amounts of both *progesterone* and *estrogen.* Once LH (mainly that which has been secreted during the ovulatory surge) has acted on the theca and granulosa cells to cause luteinization, the newly formed lutein cells seem to be programmed to go through a preordained sequence of (a) proliferation, (b) enlargement, and (c) secretion, then to be followed by (d) degeneration. Even in the absence of further secretion of LH by the anterior pituitary gland, this process will still occur. However, in the presence of LH the degree of growth of the corpus luteum is enhanced, its secretion is greater, and its life is extended. During this period of the ovarian cycle, the con-

centration of LH is very low; even so, however, without this LH the life of the corpus luteum is an average of only 6 to 8 days instead of the normal 12 days. We shall see in the discussion of pregnancy in the next chapter that another hormone that has almost exactly the same properties as luteinizing hormone, chorionic gonadotropin, which is secreted by the placenta, can also act on the corpus luteum to prolong its life—actually maintaining it throughout the duration of pregnancy.

**Termination of the Ovarian Cycle and Onset of the Next Cycle.** During the luteal phase of the ovarian cycle, the large amount of estrogen (and perhaps to a very slight extent, the progesterone as well) secreted by the corpus luteum causes a feedback decrease in secretion of both FSH and LH. Therefore, during this period no new follicles begin to grow in the ovary. However, when the corpus luteum degenerates completely at the end of 12 days of life (approximately also on the 26th day of the female sexual cycle), the lack of feedback suppression now allows the anterior pituitary gland to secrete several times as much FSH as well as moderately increased quantities of LH. The FSH and LH initiate growth of new follicles to begin a new ovarian cycle. At the same time, the paucity of secretion of progesterone and estrogen leads to menstruation by the uterus, as will be explained later.

### SUMMARY

Approximately each 28 days, gonadotropic hormones from the anterior pituitary gland cause new follicles to begin to grow in the ovaries, one of which finally ovulates at the 14th day of the cycle. During growth of the follicles, estrogen is secreted.

Following ovulation, the secretory cells of the follicle develop into a corpus luteum that secretes large quantities of the female hormones progesterone and estrogen. After another two weeks the corpus luteum degenerates, whereupon the ovarian hormones, estrogen and progesterone, decrease greatly and menstruation begins. A new ovarian cycle then follows.

## THE OVARIAN HORMONES— ESTROGENS AND PROGESTERONE

The two types of ovarian hormones are the *estrogens* and *progesterone.* The estrogens mainly promote proliferation and growth of specific cells in the body and are responsible for development of most secondary sexual characteristics of the female. On the other hand, progesterone is concerned almost entirely with final preparation of the uterus for pregnancy and of the breasts for lactation.

## CHEMISTRY OF THE SEX HORMONES

**The Estrogens.** In the normal, nonpregnant female, estrogens are secreted in major quantities only by the ovaries, though minute amounts are also secreted by the adrenal cortices. In pregnancy, tremendous quantities are also secreted by the placenta, indeed, up to 50 times the amount secreted by the ovaries during the normal monthly cycle.

At least six different natural estrogens have been isolated from the plasma of the human female, but only three are present in significant quantities, *β-estradiol, estrone,* and *estriol,* the formulas for which are illustrated in Figure 81–5. Both β-estradiol and estrone are present in large quantities in the venous blood from the ovaries, while estriol is an oxidative product derived from these first two, the conversion occurring mainly in the liver but also to some extent elsewhere in the body.

The estrogenic potency of β-estradiol is 12 times that of estrone and 80 times that of estriol. Considering these relative potencies, the total estrogenic effect of β-estradiol is usually many times that of the other two together. For this reason β-estradiol is considered to be the major estrogen, though the estrogenic effects of estrone are far from negligible.

*Synthesis of the Estrogens.* Note from the formulas of the estrogens in Figure 81–5 that all these are steroids. They are synthesized in the ovaries from cholesterol or acetyl coenzyme A, the acetate units of which can be conjugated to form the appropriate steroid nuclei. It is particularly interesting that progesterone as well as testosterone, the male sex hormone, are probably synthesized first, and then converted into the estrogens. Indeed, even normally minute amounts of testosterone are secreted by the ovaries.

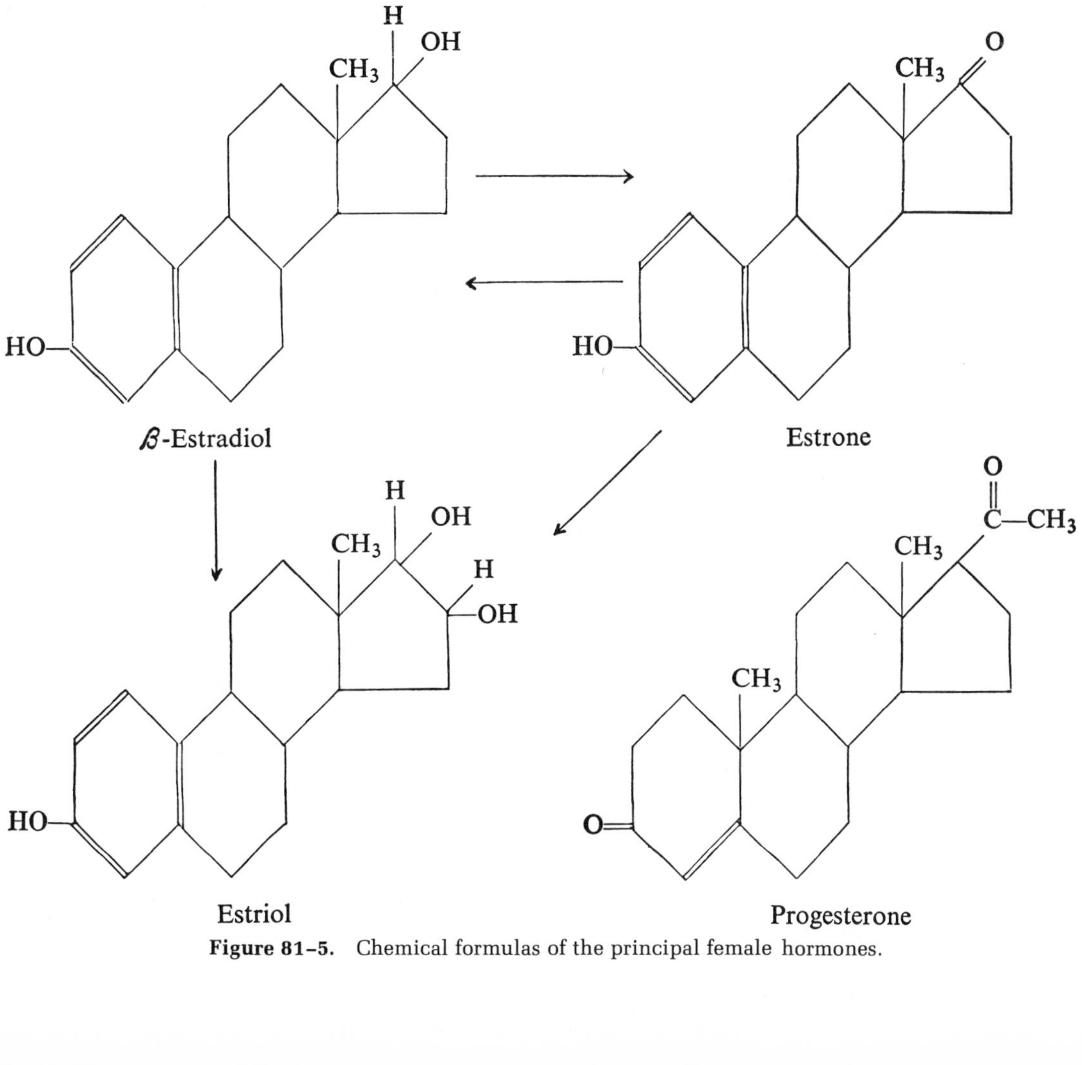

**Figure 81–5.** Chemical formulas of the principal female hormones.

*Synthetic Estrogens.* Several different synthetic sterols have estrogenic activity, one of which, *ethynyl estradiol,* has proved to be particularly valuable for therapy because it, unlike the natural estrogens, is equally as potent when administered by mouth as by injection. Still another compound which is not a sterol, *stilbestrol,* exhibits even more estrogenic activity than β-estradiol, and, furthermore, is also active when administered by mouth. Therefore, this substance, too, has had wide clinical use. Also, it has been widely used in the animal industry to effect female characteristics in male animals. It has also been shown to be carcinogenic, however, so that its use has become interdicted.

*Fate of the Estrogens; Function of the Liver in Estrogen Degradation.* Soon after they are secreted by the ovary, the estradiol and estrone that do not enter the cells to perform physiologic functions are oxidized to estriol; this oxidation occurs principally in the liver but also to a slight extent elsewhere in the body. The liver also conjugates the estrogens to form glucuronides and sulfates, and about one-fifth of these conjugated products are excreted in the bile while most of the remainder are excreted in the urine. The liver also combines estrogens loosely with a protein to form so-called *estroprotein,* and it is mainly in this form that the estrogens circulate in the extracellular fluids.

Thus, the liver plays a key role in estrogen metabolism, and, because it converts the potent estrogens, estradiol and estrone, into an almost totally impotent estrogen, estriol, and because it secretes moderate quantities into the gut, diminished liver function actually *increases* the activity of estrogens in the body, sometimes causing *hyperestrinism.*

**Progesterone.** Almost all the progesterone in the nonpregnant female is secreted by the corpus luteum during the latter half of each ovarian cycle. However, the adrenal glands form a minute quantity of progesterone or compounds that have progesteronic activity, and during pregnancy progesterone is formed in extreme quantities by the placenta, about 10 times the normal monthly amount, especially after the fourth month of gestation.

*Synthesis of Progesterone.* Progesterone is a steroid having a molecular structure not far different from those of the other steroid hormones, the estrogens, testosterone, and the corticosteroids, as illustrated in Figure 81–5. Therefore, progesterone has some functions in common with all of these.

Progesterone is probably synthesized principally from acetyl coenzyme A. However, it can also be formed from cholesterol.

At least two other "progestin" hormones are secreted by the ovaries, but the quantities of these are so slight in comparison with that of progesterone that it is usually proper to consider progesterone as the single important progestin. However, at the time of ovulation, before full development of the corpus luteum, one of these, 17 α-hydroxyprogesterone, is sometimes secreted in higher concentration than progesterone. The significance of this is not yet known.

*Fate of Progesterone.* Progesterone is secreted in far greater quantities than the estrogens by the ovaries, but its potency per unit weight is much less than that of the estrogens. Within a few minutes after secretion, almost all the progesterone is degraded to other steroids that have no progesteronic effect. Here, as is also true with the estrogens, the liver is especially important for this metabolic degradation.

The major end-product of progesterone degradation is *pregnanediol.* Approximately 10 per cent of the original progesterone is excreted in the urine in this form. One can estimate the rate of progesterone formation in the body from the rate of this excretion, but because pregnanediol exerts no progesteronic effects it can be detected in the urine only by chemical means.

## FUNCTIONS OF THE ESTROGENS —EFFECTS ON THE PRIMARY AND SECONDARY SEXUAL CHARACTERISTICS

The principal function of the estrogens is to cause cellular proliferation and growth of the tissues of the sexual organs and of other tissues related to reproduction.

**Effect on the Sexual Organs.** During childhood, estrogens are secreted only in small quantities, but following puberty the quantity of estrogens secreted under the influence of the pituitary gonadotropic hormones increases some 20-fold or more. At this time the female sexual organs change from those of a child to those of an adult. The fallopian tubes, uterus, and vagina all increase in size. Also, the external genitalia enlarge, with deposition of fat in the mons pubis and labia majora and with enlargement of the labia minora.

In addition to increase in size of the vagina, estrogens change the vaginal epithelium from a cuboidal into a stratified type, which is considerably more resistant to trauma and infection than is the prepubertal epithelium. Infections in children, such as gonorrheal vaginitis, can actually be cured by administration of estrogens simply because of the resulting increased resistance of the vaginal epithelium to infection.

During the few years following puberty, the size of the uterus increases two- to three-fold. More important, however, than increases in size are the changes that take place in the endometrium under the influence of estrogens, for estrogens cause marked proliferation of the endometrium and development of glands that will later be used to aid in nutrition of the implanting ovum. These effects are discussed later in the chapter in connection with the endometrial cycle.

**Effect on the Fallopian Tubes.** The estrogens have an effect on the mucosal lining of the fallopian tubes similar to that on the uterine endometrium: They cause the glandular tissues to proliferate; and, especially important, they cause the number of ciliated epithelial cells that line the fallopian tubes to increase. Also, the activity of the cilia is considerably enhanced, these always beating toward the uterus. This undoubtedly helps to propel the fertilized ovum toward the uterus.

**Effect on the Breasts.** The primordial breasts of both female and male are exactly alike, and under the influence of appropriate hormones, the masculine breast, at least during the first two decades of life, can develop sufficiently to produce milk in the same manner as the female breast.

Estrogens cause fat deposition in the breasts, development of the stromal tissues of the breasts, and growth of an extensive ductile system. The lobules and alveoli of the breast develop to a slight extent, but it is progesterone and prolactin that cause the determinative growth and function of these structures. In summary, the estrogens initiate growth of the breasts and the breasts' milk-producing apparatus, and they are also responsible for the characteristic external appearance of the mature female breast, but they do not complete the job of converting the breasts into milk-producing organs.

**Effect on the Skeleton.** Estrogens cause increased osteoblastic activity. Therefore, at puberty, when the female enters her reproductive years, her growth rate becomes rapid for several years. However, estrogens have another potent effect on skeletal growth—that is, they cause early uniting of the epiphyses with the shafts of the long bones. This effect is much stronger in the female than is the similar effect of testosterone in the male. As a result, growth of the female usually ceases several years earlier than growth of the male. The female eunuch who is completely devoid of estrogen production usually grows several inches taller than the normal mature female because her epiphyses do not unite early.

*Effect on the Pelvis.* Estrogens have a special effect to broaden the pelvis, changing the pelvic outlet from the narrow, funnel-like outlet of the male into a broad, ovoid outlet. The functional importance of this for the birth of a baby is obvious.

*Effect on Calcium and Phosphate Retention.* As pointed out in the preceding chapter, testosterone causes greater than normal retention of calcium and phosphate. This effect is also true of the estrogens, but to a lesser extent, because estrogens, like testosterone, promote growth of bones, entailing the deposition of increased amounts of bone matrix with subsequent retention of both calcium and phosphate.

**Effect on Protein Deposition.** Estrogens cause a slight increase in total body protein, which is evidenced by a slight positive nitrogen balance when estrogens are administered. This probably results from the growth-promoting effect of estrogen on the sexual organs and on a few other tissues of the body. The enhanced protein deposition caused by testosterone is much more general and many times as powerful as that caused by estrogens.

**Effect on Metabolism and Fat Deposition.** Estrogens increase the metabolic rate slightly but not so much as the male sex hormone testosterone. However, they cause deposition of increased quantities of fat in the subcutaneous tissues. As a result, the overall specific gravity of the female body, as judged by flotation in water, is considerably less than that of the male body which contains more protein and less fat. In addition to deposition of fat in the breasts and subcutaneous tissues, estrogens cause especially marked deposition of fat in the buttocks and thighs, causing the broadening of the hips that is characteristic of the feminine figure.

**Effect on Hair Distribution.** Estrogens do not greatly affect hair distribution. However, hair develops in the pubic region and in the axillae after puberty. It is probably androgens formed by the adrenal glands that are mainly responsible for this.

**Effect on the Skin.** Estrogens cause the skin to develop a texture that is soft and usually smooth but nevertheless thicker than that of the child or the female castrate. Also, estrogens cause the skin to become more vascular than normal; this effect is often associated with increased warmth of the skin, and it often results in greater bleeding of cut surfaces than is observed in men.

The adrenal androgens, which are secreted in increased quantities after puberty, cause increased secretion by the axillary sweat glands and also often cause acne.

**Effect on Electrolyte Balance.** The similarity of estrogenic hormones to adrenocortical hormones has been pointed out; and estrogens, like adrenocortical hormones, cause sodium and water retention by the kidney tubules. However, this effect of estrogens is slight and rarely of significance except in pregnancy, as discussed in the following chapter.

**Intracellular Functions of Estrogens.** Thus far we have discussed the gross effects of estrogens on the body. The precise functions of estrogens in causing these effects are not known other than for the following clues: Estrogens circulate in the blood for only a few minutes before they are delivered to the target cells. On entry into these cells, the estrogens combine within 10 to 15 seconds with a "receptor" protein in the cytoplasm and then, in combination with this protein, migrate to the nucleus. This immediately initiates the process of transcription, and RNA begins to be produced within a few minutes. In addition, over a period of many hours DNA also is produced, resulting eventually in division of the cell. The RNA diffuses to the cytoplasm, where it causes greatly increased protein formation and subsequently altered cellular function.

One of the principal differences between the protein anabolic effect of the estrogens and that of testosterone is that estrogen causes its effect almost exclusively in certain target organs, such as the uterus, the breasts, the skeleton, and certain fatty areas of the body; whereas testosterone has a more generalized effect throughout the body.

## FUNCTIONS OF PROGESTERONE

**Effect on the Uterus.** By far the most important function of progesterone is *to promote secretory changes in the endometrium,* thus preparing the uterus for implantation of the fertilized ovum. This function is discussed later in connection with the endometrial cycle of the uterus.

In addition to this effect on the endometrium, progesterone decreases the frequency of uterine contractions, thereby helping to prevent expulsion of the implanted ovum, an effect discussed in the following chapter.

**Effect on the Fallopian Tubes.** Progesterone also promotes secretory changes in the mucosal lining of the fallopian tubes. These secretions are important for nutrition of the fertilized, dividing ovum as it traverses the fallopian tube prior to implantation.

**Effect on the Breasts.** Progesterone promotes development of the lobules and alveoli of the breasts, causing the alveolar cells to proliferate, to enlarge, and to become secretory in

nature. However, progesterone does not cause the alveoli actually to secrete milk, for, as discussed in the following chapter, milk is secreted only after the prepared breast is further stimulated by prolactin from the pituitary.

Progesterone also causes the breasts to swell. Part of this swelling is due to the secretory development in the lobules and alveoli, but part also seems to result partly from increased fluid in the subcutaneous tissue itself.

**Effect on Electrolyte Balance.** Progesterone in very large quantity, like estrogens, testosterone, and adrenocortical hormones, can enhance sodium, chloride, and water reabsorption from the distal tubules of the kidney. Yet, strangely enough, progesterone more often causes increased sodium and water excretion. The cause of this is competition between progesterone and aldosterone, which probably occurs as follows: It is believed that these two substances combine with the same receptor proteins in the epithelial cells of the tubules. When progesterone combines with these, aldosterone cannot combine. Yet progesterone exerts many hundred times less sodium transport effect than does aldosterone. Therefore, despite the fact that under appropriate conditions progesterone can promote sodium and water retention by the renal tubules, it blocks the far more potent effect of aldosterone, thus resulting in net loss of sodium and water from the body.

**Protein Catabolic Effect.** Progesterone exerts a mild catabolic effect on the body's protein similar to that of the glucocorticoids. Though in the normal sexual cycle this effect is probably not significant, it possibly is significant during pregnancy, when proteins must be mobilized for use by the fetus.

## THE ENDOMETRIAL CYCLE AND MENSTRUATION

Associated with the cyclic production of estrogens and progesterone by the ovaries is an endometrial cycle operating through the following stages: first, proliferation of the uterine endometrium; second, secretory changes in the endometrium; and third, desquamation of the endometrium, which is known as *menstruation.* The various phases of the endometrial cycle are illustrated in Figure 81–6.

**Proliferative Phase (Estrogen Phase) of the Endometrial Cycle.** At the beginning of each menstrual cycle, most of the endometrium is desquamated by the process of menstruation.

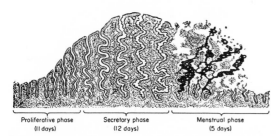

Proliferative phase  Secretory phase  Menstrual phase
(11 days)          (12 days)        (5 days)

**Figure 81–6.** Phases of endometrial growth and menstruation during each monthly female sexual cycle.

This includes complete loss of the epithelium and *stratum submucosum* and loss of most of the *stratum vasculare*. After menstruation, only a thin layer of endometrial stroma remains at the base of the original endometrium, and the only epithelial cells are those located in the remaining deep portions of the glands and crypts of the endometrium. *Under the influence of estrogens,* secreted in increasing quantities by the ovary during the first part of the ovarian cycle, the stromal cells and the epithelial cells proliferate rapidly. The endometrial surface is re-epithelialized within three to seven days after the beginning of menstruation. For the first two weeks of the sexual cycle—that is, until ovulation—the endometrium increases greatly in thickness, owing to increasing numbers of stromal cells and to progressive growth of the endometrial glands and blood vessels into the endometrium, all of which effects are promoted by the estrogens. At the time of ovulation the endometrium is approximately 2 to 3 mm. thick.

**Secretory Phase (Progestational Phase) of the Endometrial Cycle.** During the latter half of the sexual cycle, progesterone as well as estrogen is secreted in large quantity by the corpus luteum. The estrogens cause additional cellular proliferation in the endometrium during this phase of the endometrial cycle, and progesterone causes considerable swelling and secretory development of the endometrium. The glands increase in tortuosity, secretory substances accumulate in the glandular epithelial cells, and the glands secrete small quantities of endometrial fluid. Also, the cytoplasm of the stromal cells increases, lipid and glycogen deposits increase greatly in the stromal cells, and the blood supply to the endometrium further increases in proportion to the developing secretory activity, the blood vessels becoming highly tortuous. The thickness of the endometrium approximately doubles during the secretory phase so that toward the end of the monthly cycle the endometrium has a thickness of 4 to 6 mm.

The whole purpose of all these endometrial changes is to produce a highly secretory endometrium containing large amounts of stored nutrients that can provide appropriate conditions for implantation of a fertilized ovum during the latter half of the monthly cycle. From the time fertilization first takes place until the ovum implants, the fallopian and uterine secretions, called "uterine milk," provide nutrition for the early dividing ovum. Then, once the ovum implants in the endometrium, the trophoblastic cells on the surface of the blastocyst begin to digest the endometrium and to absorb the substances digested, thus making still far greater quantities of nutrients available to the early embryo.

**Menstruation.** Approximately two days before the end of the monthly cycle, the ovarian hormones, estrogens and progesterone, decrease sharply to low levels of secretion, as was illustrated in Figure 81–2, and menstruation follows.

Menstruation is caused by the sudden reduction in both progesterone and estrogens at the end of the monthly ovarian cycle. The first effect is decreased stimulation of the endometrial cells by these two hormones, followed rapidly by involution of the endometrium itself to about 65 per cent of its previous thickness. During the 24 hours preceding the onset of menstruation, the tortuous blood vessels leading to the mucosal layers of the endometrium become vasospastic, presumably because of some effect of the involution, such as release of a vasoconstrictor material, or perhaps because of a direct effect of estrogen withdrawal, since estrogens are a vasodilator of the endometrial blood vessels. The vasospasm and loss of hormonal stimulation cause beginning necrosis in the endometrium, especially of blood vessels in the stratum vasculare. As a result, blood seeps into the vascular layer of the endometrium, the hemorrhagic areas growing over a period of 24 to 36 hours. Gradually, the necrotic outer layers of the endometrium separate from the uterus at the site of the hemorrhages, until, at approximately 48 hours following the onset of menstruation, all the superficial layers of the endometrium have desquamated. The desquamated tissue and blood in the uterine vault initiate uterine contractions that expel the uterine contents.

During normal menstruation, approximately 35 ml. of blood and an additional 35 ml. of serous fluid are lost. This menstrual fluid is normally nonclotting, because a *fibrinolysin* is released along with the necrotic endometrial material. However, if excessive bleeding occurs from the uterine surface, the quantity of fibrinolysin may not be sufficient to prevent clotting. The presence of clots during menstruation ordinarily is clinical evidence of uterine pathology.

Within three to seven days after menstruation starts, the loss of blood ceases, for by this time the endometrium has become completely reepithelialized.

*Leukorrhea During Menstruation*. During menstruation tremendous numbers of leukocytes are released along with the necrotic material and blood. It is probable that some substance liberated by the endometrial necrosis causes this outflow of leukocytes. As a result of these many leukocytes and maybe still other factors, the uterus is resistant to infection during menstruation, even though the endometrial surfaces are denuded. Obviously, this is of extreme protective value.

# REGULATION OF THE FEMALE MONTHLY RHYTHM—INTERPLAY BETWEEN THE OVARIAN AND HYPOTHALAMIC-PITUITARY HORMONES

Now that we have presented the major cyclic changes that occur during the female sexual cycle, we can attempt to explain the basic rhythmic mechanism that causes these cyclic variations.

**Function of the Hypothalamus in the Regulation of Gonadotropin Secretion—the Hypothalamic Releasing Factors.** As was pointed out in Chapter 75, secretion of most of the anterior pituitary hormones is controlled by releasing factors formed in the hypothalamus and transmitted to the anterior pituitary gland by way of the hypothalamic-hypophyseal portal system. In the case of the gonadotropins, at least one releasing factor (perhaps two) are important.

The one releasing factor that is known to be important is luteinizing hormone releasing factor (LRF). This substance has been purified and has been found to be a decapeptide having the following formula:

GLU-HIS-TRP-SER-TYR-GLY-
LEU-ARG-PRO-GLY-NH$_2$

Though some research workers believe that another substance similar to this is follicle-stimulating hormone releasing factor (FRF), it has been found that the above purified releasing factor causes release not only of luteinizing hormone but also of follicle-stimulating hormone. Therefore, since there is reason to believe that this decapeptide is in reality both LRF and FRF combined in the same molecule, it is sometimes called simply LRF-FRF. It is believed further that under some conditions this substance has a more potent effect on the anterior pituitary gland to cause release of follicle-stimulating hormone and under other conditions to cause release of luteinizing hormone.

*Hypothalamic Centers for Stimulating Release of LRF-FRF.* Several different areas that profoundly influence the rate of secretion of the hypothalamic gonadotropin releasing factor LRF-FRF have been found in the hypothalamus. In monkeys—and presumably in man—the midbasal region of the hypothalamus is the area most importantly involved. In lower animals, the area around the infundibulum causes a continuous tonic secretion of gonadotropin-releasing factor, while two other areas modulate the rate of release. These areas are (1) a center in the preoptic area that causes cyclic variation in the secretory rate and (2) a center in the posterior hypothalamus that allows the psychic attitude of the animal to enhance or decrease the secretion of gonadotropin-releasing factor.

*Effect of Psychic Factors on the Female Sexual Cycle.* It is well known that the young woman on first leaving home to go to college almost as often as not experiences disruption or irregularity of the female sexual cycle. Likewise, serious stresses of almost any type can interfere with the cycle. Finally, in many lower animals no ovulation occurs at all until after copulation; the sexual excitation attendant to the sexual act initiates a sequence of events that leads first to secretion in the hypothalamus of gonadotropin-releasing factor, then to secretion of the anterior pituitary gonadotropins, and finally to ovarian secretion of hormones and ovulation. It is these effects that are believed to be mediated through the posterior hypothalamic center for modulating the output of gonadotropin-releasing factor.

*Negative Feedback Effect of Estrogen and Progesterone on Secretion of Follicle-stimulating Hormone and Luteinizing Hormone.* Estrogen in particular and progesterone very slightly inhibit the production of FSH and LH. Very large quantities of progesterone must be administered to have this effect; even then, the effect is minimal unless estrogen has been secreted or ingested prior to the progesterone.

The feedback effect of both estrogen and progesterone seem to operate mainly by the actions of these hormones on the hypothalamus, though it is possible that they have very slight direct feedback actions on the anterior pituitary gland as well. Destruction of the preoptic center that modulates secretion of gonadotropin-releasing factor from the hypothalamus will almost totally block the feedback effects of the estrogen and progesterone. Therefore, it is believed that this is the primary point of action of the feedback effect.

**Positive Feedback Effect of Estrogen Before Ovulation—the Preovulatory Luteinizing Hormone Surge.** For reasons not completely understood, the anterior pituitary gland secretes greatly increased amounts of LH on the day immediately before ovulation. This effect is illustrated in Figure 81–2, which shows a much smaller preovulatory surge of FSH as well.

Experiments have shown that infusion of estrogen into a woman for a period of two to three days during the first half of the ovarian cycle will cause rapidly accelerating growth of the follicles and also rapidly accelerating secretion of ovarian estrogens. During this period the secretion of both the follicle-stimulating hormone and luteinizing hormone by the anterior pituitary gland is at first suppressed. Then abruptly the secretion of luteinizing hormone increases about seven-fold, and the secretion of follicle stimulating hormone increases about two-fold. This abrupt increase in secretion of the gonadotropins seems to result from a positive feedback effect of the estrogen in place of the normal negative feedback that occurs during the remainder of the female sexual cycle. Its precise cause is not known, but nevertheless it is an absolutely necessary and integral part of the control mechanism. Without the normal preovulatory surge of luteinizing hormone, ovulation will never occur.

## FEEDBACK OSCILLATION OF THE HYPOTHALAMIC-PITUITARY-OVARIAN SYSTEM

Now, after discussing much of the known information about the interrelationships of the different components of the female hormonal system, we can digress from the area of proven fact into the realm of speculation and attempt to explain the feedback oscillation that controls the rhythm of the female sexual cycle. It seems to operate in approximately the following sequence of three successive events:

**1. The Postovulatory Secretion of the Ovarian Hormones and Depression of Gonadotropins.** The easiest part of the cycle to explain is the events that occur during the postovulatory phase—between ovulation and the beginning of menstruation. During this time the corpus luteum secretes very large quantities of both progesterone and estrogen. The estrogen, in particular, acts on the hypothalamus to cause strong negative feedback depression of secretion of the gonadotropins, both FSH and LH, during this period of time. These effects are illustrated in Figure 81–2.

**2. The Follicular Growth Phase.** A few days before menstruation, the corpus luteum involutes, and the secretion of both estrogen and progesterone decreases to a low ebb. This releases the hypothalamus from the feedback effect of the estrogen so that the rate of secretion of both FSH and LH increases 100 per cent or more. This initiates new follicular growth and progressive increase in the secretion of estrogen, reaching a peak of estrogen secretion at about 12½ to 13 days after the onset of menstruation. During the first 11 to 12 days of this follicular growth the rate of secretion of the gonadotropins FSH and LH decreases; then comes a sudden increase in secretion of both of these hormones, leading to the next stage of the cycle.

**3. Preovulatory Surge of LH and FSH; Ovulation.** At approximately 11½ to 12 days after the onset of menstruation, the decline in secretion of FSH and LH comes to an abrupt halt. It is believed that the high level of estrogens at this time causes a positive feedback effect, as explained earlier, which leads to a terrific surge of secretion—especially of LH and to a lesser extent of FSH. This effect may be related to the fact that the follicular secretory cells are becoming exhausted so that their rate of secretion of estrogens had already begun to fall about one day prior to the LH surge. Whatever the cause of this preovulatory LH and FSH surge, the LH leads to both ovulation and formation of the corpus luteum. Thus, the hormonal system begins a new round of the female sexual cycle.

**Anovulatory Cycles—the Sexual Cycle at Puberty.** If the preovulatory surge of luteinizing hormone is not of sufficient magnitude, ovulation will not occur, and the cycle is then said to be "anovulatory." Most of the cyclic variations of the sexual cycle continue, but they are altered in the following ways: First, lack of ovulation causes failure of development of the corpus luteum and consequently there is almost no secretion of progesterone during the entire cycle. Second, the cycle is shortened by several days, but the rhythm continues. Therefore, it is clear that progesterone has little effect on the maintenance of the cycle itself. Indeed, it is probably the estrogen secreted by the corpus luteum and not the progesterone that prolongs the cycle when ovulation occurs.

Anovulatory cycles are usual during the first few cyles following puberty and for several years prior to menopause, presumably because the LH surge is not potent enough at these times to cause ovulation.

## PUBERTY

Puberty means the onset of adult sexual life, and, as pointed out earlier in the chapter, it is caused by a gradual increase in gonadotropic hormone secretion by the pituitary, beginning approximately the eighth year of life, as illustrated in Figure 81–7.

In the female, as in the male, the infantile pituitary gland and ovaries are capable of full function if appropriately stimulated. However, the hypothalamus is extremely sensitive to the inhibitory effects of estrogens, which keeps its stimulation of the pituitary almost completely suppressed throughout childhood. Then, at puberty, for reasons not understood, the hypo-

thalamus matures; its excessive sensitivity to the negative feedback inhibition becomes greatly diminished, which allows enhanced production of gonadotropins and the onset of adult female sexual life.

Figure 81–8 illustrates (a) the increasing levels of estrogen secretion at puberty, (b) the cyclic variation during the monthly sexual cycles, (c) the further increase in estrogen secretion during the first few years of sexual life, (d) then progressive decrease in estrogen secretion toward the end of sexual life, and (e) finally almost no estrogen secretion beyond the menopause.

## THE MENOPAUSE

At an average age of approximately 45 to 50 years the sexual cycles usually become irregular, and ovulation fails to occur during many of these cycles. After a few months to a few years, the cycles cease altogether, as illustrated in Figure 81–8. This cessation of the cycles is called the *menopause*.

The cause of the menopause is "burning out" of the ovaries. In other words, throughout a woman's sexual life many of the primordial follicles grow into vesicular follicles with each sexual cycle, and eventually almost all the ova either degenerate or are ovulated. Therefore, at the age of about 45 only a few primordial follicles still remain to be stimulated by FSH and LH, and the production of estrogens by the ovary decreases as the number of primordial follicles approaches zero (also illustrated in Figure 81–8). When estrogen production falls below a critical value, the estrogens can no longer inhibit the production of FSH and LH sufficiently to cause oscillatory cycles. Consequently, as illustrated in Figure 81–7, FSH and LH (mainly FSH) are produced thereafter

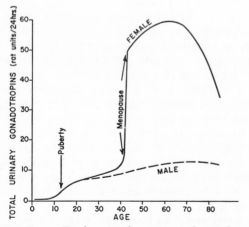

**Figure 81–7.** Total rates of secretion of gonadotropic hormones throughout the sexual lives of females and males, showing an especially abrupt increase in gonadotropic hormones at the menopause in the female.

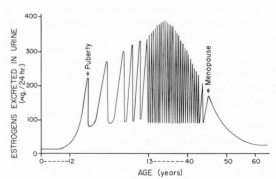

**Figure 81–8.** Estrogen secretion throughout sexual life.

in large and continous quantities. Estrogens are produced in subcritical quantities for a short time after the menopause, but over a few years, as the final remaining primordial follicles become atretic, the production of estrogens by the ovaries falls almost to zero.

**The Female Climacteric.** The term "female climacteric" means the entire time, lasting from several months to several years, during which the sexual cycles become irregular and gradually stop. In this period the woman must readjust her life from one that has been physiologically stimulated by estrogen and progesterone production to one devoid of these feminizing hormones. The secretion of estrogens decreases rapidly, and essentially no progesterone is secreted after the last ovulatory cycle. The loss of the estrogens often causes marked physiologic changes in the function of the body, including (1) "hot flashes" characterized by extreme flushing of the skin, (2) psychic sensations of dyspnea, (3) irritability, (4) fatigue, (5) anxiety, and (6) occasionally various psychotic states. These symptoms are of sufficient magnitude in approximately 15 per cent of women to warrant treatment. If psychotherapy fails, daily administration of an estrogen in small quantities will reverse the symptoms, and by gradually decreasing the dose the menopausal woman is likely to avoid severe symptoms; unfortunately, such treatment prolongs the symptoms, however.

# INTERRELATIONSHIPS OF THE OVARIES WITH OTHER GLANDS

**Relationship of the Ovaries to the Adrenal Glands.** The adrenal glands normally secrete small quantities of both estrogen and progesterone, though these quantities are usually too small to exert major effects on the body. However, rare tumors of the adrenal gland secrete specifically increased quantities of female hormones and therefore cause feminizing characteristics.

Injection of estrogens causes the adrenal cortices to hypertrophy. This effect is mediated through the pituitary, for estrogens, while inhibiting production of follicle-stimulating hormone, increase the secretion of corticotropin.

**Antagonistic Effects of Estrogens and Testosterone.** Estrogens and testosterone exert opposite effects on the sexual organs and on many secondary sexual tissues such as the breasts and the prostate gland. These antagonistic effects are in part mediated through the hypothalamic-pituitary system, because both estrogens and testosterone are capable of decreasing the production of follicle-stimulating hormone and luteinizing hormone, which in turn abrogates secretion of the normal gonadal hormones. In general, in order to antagonize the effect of estrogens on the breast, approximately 50 times as much testosterone as $\beta$-estradiol must be administered.

# ABNORMALITIES OF SECRETION BY THE OVARIES

**Hypogonadism.** Less than normal secretion by the ovaries can result from poorly formed ovaries or lack of ovaries. When ovaries are absent from birth or when they never become functional, *female eunuchism* occurs. In this condition the usual secondary sexual characteristics do not appear, and the sexual organs remain infantile. Especially characteristic of this condition is excessive growth of the long bones because the epiphyses do not unite with the shafts of these bones at as early an age as in the normal adolescent woman. Consequently, the female eunuch is essentially as tall as, or perhaps even slightly taller than, her male counterpart of similar genetic background.

When the ovaries of a fully developed woman are removed, the sexual organs regress to some extent so that the uterus becomes almost infantile in size, the vagina becomes smaller, and the vaginal epithelium becomes thin and easily damaged. The breasts atrophy and become pendulous, and the pubic hair becomes considerably thinner. These same changes occur in the woman after the menopause.

*Irregularity of Menses and Amenorrhea Due to Hypogonadism.* As pointed out in the preceding discussion of the menopause, the quantity of estrogens produced by the ovaries must rise above the critical value if they are to be able to inhibit the production of follicle-stimulating hormone sufficiently to cause an oscillatory sexual cycle. Consequently, in hypogonadism or when the gonads are secreting small quantities of estrogens as a result of other factors, such as *hypothyroidism,* the ovarian cycle likely will not occur normally. Instead, several months may elapse between menstrual periods, or menstruation may cease altogether (amenorrhea). Characteristically, prolonged ovarian cycles are frequently associated with failure of ovulation, presumably due to insufficient secretion of luteinizing hormone, which is necessary for ovulation.

**Hypersecretion by the Ovaries.** Extreme hypersecretion of ovarian hormones by the ovaries is a rare clinical entity, for excessive secretion of estrogens automatically decreases the production of gonadotropins by the pituitary, and this in turn limits the production of the ovarian hormones. Consequently, hypersecretion of feminizing hormones is recognized clinically only when a feminizing tumor develops.

A rare granulosa-theca cell tumor occasionally develops in an ovary, occurring more often after menopause than before. These tumors secrete large quantities of estrogens which exert the usual estrogenic effects, including hypertrophy of the uterine endometrium and irregular bleeding from this endometrium. In fact, bleeding is often the first indication that such a tumor exists.

**Endometriosis.** Endometriosis is the development and growth of endometrium in the peritoneal cavity, this growth usually occurring in the pelvis closely associated with the sexual organs. There are two theories for explaining the origin of intra-abdominal endometrial tissue that causes endometriosis. Some believe that this tissue results from endometrial "rests" that develop embryologically in the peritoneal cavity. Others believe that contraction of the uterus during menstruation occasionally expels viable endometrium backward through the fallopian tubes into the abdominal cavity and that this endometrial tissue then implants on the peritoneum. In support of this is the fact that much of the endometrium sloughed during menstruation is still viable and will grow easily in tissue culture.

During each ovarian cycle the endometrium in the peritoneal cavity proliferates, secretes, and desquamates in the same manner that the intrauterine endometrium does. However, when desquamation occurs within the peritoneal cavity, the tissue and the hemorrhaging blood cannot be expelled to the exterior. Consequently, the quantity of endometrial tissue in the peritoneal cavity progressively increases with each subsequent menstrual cycle.

The presence of necrotic and hemorrhage material in the abdominal cavity, and also the swelling of the endometrial tissue during each ovarian cycle, can cause considerable irritation of the peritoneum, sometimes producing severe abdominal pain. Also, fibrosis occurs in the areas of endometriosis, thereby promoting adhesions from one sexual organ to another and from the sexual organs to other intrapelvic and intra-abdominal structures. Endometriosis is one of the most prevalent causes of female infertility, owing especially to fibrotic immobilization of the sexual organs.

# THE FEMALE SEXUAL ACT

**Stimulation of the Female Sexual Act.** As is true in the male sexual act, successful performance of the female sexual act depends on both psychic stimulation and local sexual stimulation.

The psychic factors that constitute "sex drive" in women are difficult to assess. The sex hormones, and the adrenocortical hormones as well, seem to exert a direct influence on the woman to create such a sex drive, but, on the other hand, the growing female child in modern society is often taught that sex is something to be hidden and that it is immoral. As a result of this training, much of the natural sex drive is inhibited, and whether the woman will have little or no sex drive ("frigidity") or will be more highly sexed probably depends partly on a balance between natural factors and previous training.

Local sexual stimulation in women occurs in more or less the same manner as in men, for massage, irritation, or other types of stimulation of the perineal region, sexual organs, and urinary tract create sexual sensations. The *clitoris* is especially sensitive for initiating sexual sensations. As in the male, the sexual sensory signals are mediated to the sacral segments of the spinal cord through the pudendal nerve and sacral plexus. Once these signals have entered the spinal cord, they are transmitted thence to the cerebrum. Also, local reflexes that are at least partly responsible for the female orgasm are integrated in the sacral and lumbar spinal cord.

**Female Erection and Lubrication.** Located around the introitus and extending into the clitoris is erectile tissue almost identical with the erectile tissue of the penis. This erectile tissue, like that of the penis, is controlled by the parasympathetic nerves that pass through the nervi erigentes from the sacral plexus to the external genitalia. In the early phases of sexual stimulation, the parasympathetics dilate the arteries and constrict the veins of the erectile tissues, and this allows rapid accumulation of blood in the erectile tissue so that the introitus tightens around the penis; this aids the male greatly in his attainment of sufficient sexual stimulation for ejaculation to occur.

Parasympathetic impulses also pass to the bilateral Bartholin's glands located beneath the labia minora to cause secretion of mucus immediately inside the introitus. This mucus is responsible for most of the lubrication during sexual intercourse. The lubrication in turn is necessary for establishing during intercourse a satisfactory massaging sensation rather than an irritative sensation, which may be provoked by a dry vagina. A massaging sensation constitutes the optimal type of sensation for evoking the appropriate reflexes that culminate in both the male and female climaxes.

**The Female Orgasm.** When local sexual stimulation reaches maximum intensity, and especially when the local sensations are supported by appropriate psychic conditioning signals from the cerebum, reflexes are initiated that cause the female orgasm, also called the *female climax*. The female orgasm is analogous to ejaculation in the male, and it probably is important for fertilization of the ovum. Indeed, the human female is known to be somewhat more fertile when inseminated by normal sexual intercourse rather than by artificial methods, thus indicating an important function of the female orgasm. Possible effects that could result in this are:

First, during the orgasm the perineal muscles of the female contract rhythmically, which presumably results from spinal reflexes similar to those that cause ejaculation in the male. It is possible, also, that these same reflexes increase uterine and fallopian tube motility during the orgasm, thus helping to transport the sperm toward the ovum, but the information on this subject is scanty.

Second, in many lower animals, copulation causes the posterior pituitary gland to secrete oxytocin; this effect is probably mediated through the amygdaloid nuclei and then through the hypothalamus to the pituitary. The oxytocin in turn causes increased contractility of the uterus, which also is believed to cause rapid transport of the sperm. Sperm have been shown to traverse the entire length of the fallopian tube in the cow in approximately five minutes, a rate at least 10 times as fast as that which the sperm themselves could achieve. Whether or not this occurs in the human female is unknown.

In addition to the effects of the orgasm on fertilization, the intense sexual sensations that develop during the orgasm also pass into the cerebrum and in some manner lead to a sense of satisfaction characterized by relaxed peacefulness.

## FEMALE FERTILITY

**The Fertile Period of Each Sexual Cycle.** The ovum remains viable and capable of being fertilized after it is expelled from the ovary probably no longer than 24 hours. Therefore, sperm must be available soon after ovulation if fertilization is to take place. On the other hand, a few sperm can remain viable in the female reproductive tract for up to 72 hours, though most of them for not more than 24 hours. Therefore, for fertilization to take place, intercourse usually must occur some time between one day prior to ovulation up to one day after ovulation. Thus, the period of female fertility during each sexual cycle is short.

**The Rhythm Method of Contraception.** One of the often practiced methods of contraception is to avoid intercourse near the time of ovulation. The difficulty with this method of contraception is the impossibility of predicting the exact time of ovulation. Yet the interval from ovulation until the next succeeding onset of menstruation is almost always between 13 to 15 days. Therefore, if the menstrual cycle is regular, the time of ovulation averages 14 days prior to the next onset of menstruation. In other words, if the periodicity of the menstrual cycle is 28 days, ovulation usually occurs within one day of the 14th day of the cycle. If, on the other hand, the periodicity of the cycle is 40 days, ovulation usually occurs within one day of the 26th day of the cycle.

Finally, if the periodicity of the cycle is 21 days, ovulation usually occurs within one day of the 7th day of the cycle. Therefore, it is usually stated that avoidance of intercourse within 3 days on either side of the calculated day of ovulation prevents conception, and 5 days on either side of the calculated day of ovulation usually provides almost complete safety. Such a method of contraception can be used only when the periodicity of the menstrual cycle is regular, for otherwise it is impossible to determine the next onset of menstruation, and, therefore, it is impossible to predict the day of ovulation.

**Hormonal Suppression of Fertility—"The Pill."** It has long been known that administration of either estrogens or progesterone, if given in sufficient quantity, can inhibit ovulation. Though the exact mechanism of this effect is not clear, it is believed that in the presence of enough of either or both of these hormones, the hypothalamus fails to secrete the normal surge of LH-releasing factor and its stimulatory product LH that usually occurs about 13 days after the onset of the monthly sexual cycle. From this discussion earlier in the chapter, it will be recalled that this surge of LH is essential in causing ovulation.

The problem in devising methods for hormonal suppression of ovulation has been to develop appropriate combinations of estrogens and progestins that will suppress ovulation but that will not cause unwanted effects of these two hormones. For instance, too much of either of the hormones can cause abnormal menstrual bleeding patterns. However, use of a synthetic progestin in place of progesterone, especially the 19-norsteroids, along with small amounts of estrogens will usually prevent ovulation and yet, also, allow almost a normal pattern of menstruation. Therefore, almost all "pills" used for control of fertility consist of some combination of synthetic estrogens and synthetic progestins. The main reason for using synthetic estrogens and synthetic progestins is that the *natural* hormones are almost entirely destroyed by the liver immediately after they are absorbed from the gastrointestinal tract into the portal circulation. However, many of the *synthetic* hormones can resist this destructive propensity of the liver, thus allowing oral administration. Two of the most commonly used estrogens are *ethynyl estradiol* and *mestranol*. Among the most commonly used progestins are *norethindrone* and *norethynodrel*. The medication is usually begun in the early stages of the female sexual cycle and is continued beyond the time that ovulation normally would have occurred. Then the medication is stopped toward the end of the cycle, allowing menstruation to occur and a new cycle to begin.

Oral contraceptive regimens have also been devised in which very low dosage levels of estrogens and progestins are used. In these instances ovulation frequently does occur, but other effects prevent conception. These effects include (1) abnormal transport time through the fallopian tube (the usual time is almost exactly three days) so that implantation will not occur; (2) abnormal development of the endometrium so that it will not support a fertilized ovum; (3) abnormal characteristics of the cervical mucus, making this so viscous that sperm cannot enter the uterus; and possibly (4) abnormal contraction of the fallopian tubes and uterine musculature so that the ovum will be expelled rather than implanted.

**Use of Prostaglandins for Contraception.** Many compounds all similar to each other and called *prostaglandins* are found in large quantities in many different tissues of the body, including the gonads, the brain, and the kidneys. These substances have potent physiological effects, but despite tremendous amounts of research, their basic functions in the body have yet to be discovered.

When administered properly, one of the prostaglandins, $PGE_2$, can prevent implantation of the fertilized ovum on the endometrium. Therefore, $PGE_2$ has been proposed as an "after-the-fact" contraceptive; for the present, however, this is primarily speculation. This prostaglandin can cause abortion at later periods in pregnancy as well as induce labor at term.

**Abnormal Conditions Causing Female Sterility.** Approximately one out of every 6 to 10 marriages is infertile; in about 60 per cent of these, the infertility is due to female sterility.

Occasionally, no abnormality whatsoever can be discovered in the female genital organs, in which case it must be assumed that the infertility is due either to abnormal physiologic function of the genital system or to abnormal genetic development of the ova themselves.

However, probably by far the most common cause of female sterility is failure to ovulate. This can result from either hyposecretion of gonadotropic hormones, in which case the intensity of the hormonal stimuli simply is not sufficient to cause ovulation, or it can result from abnormal ovaries that will not allow ovulation. For instance, thick capsules occasionally exist on the outside of the ovaries that prevent ovulation.

Because of the high incidence of anovulation in sterile women, special methods are often utilized to determine whether or not ovulation occurs. These are all based on the effects of progesterone on the body, for the normal increase in progesterone secretion does not occur during the latter half of anovulatory cycles. In the absence of progesteronic effects, the cycle can be assumed to be anovulatory. One of these tests is simply to analyze the urine for a surge in pregnanediol, the end-product of progesterone metabolism, during the latter half of the sexual cycle, the lack of which indicates failure of ovulation. However, another common test is for the woman to chart her body temperature throughout the cycle. Secretion of progesterone during the latter half of the cycle raises the body temperature about one-half degree, the temperature rise coming abruptly at the time of ovulation. Such a temperature chart, showing the point of ovulation, is illustrated in Figure 81–9.

Lack of ovulation caused by hyposecretion of the pituitary gonadotropic hormones can be treated by

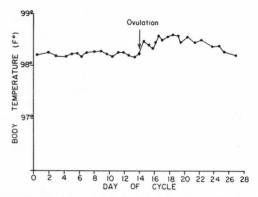

**Figure 81–9.** Elevation in body temperature shortly after ovulation.

administration of *human chorionic gonadotropin*, a hormone that will be discussed in the following chapter and that is extracted from the human placenta. This hormone, though secreted by the placenta, has almost exactly the same effects as luteinizing hormone and, therefore, is a powerful stimulator of ovulation. However, excess use of this hormone can cause ovulation from many follicles simultaneously; and this results in multiple births, an effect that has caused as many as six children to be born to mothers treated for infertility with this hormone.

One of the most common causes of female sterility is also *endometriosis,* for, as described earlier, endometriosis causes fibrosis throughout the pelvis; and this fibrosis frequently so enshrouds the ovaries that an ovum cannot be released into the abdominal cavity. Often, also, endometriosis occludes the fallopian tubes, either at the fimbriated ends or elsewhere along their extent. Another common cause of female infertility is salpingitis, that is, inflammation of the fallopian tubes; this causes fibrosis in the tubes, thereby occluding them. In past years, such inflammation was extremely common as a result of gonococcal infection, but with modern therapy this is becomming a less prevalent cause of female infertility.

Finally, still another cause of infertility that may be very important is abnormal secretion of mucus by the uterine cervix. Ordinarily, at the time of ovulation, the hormonal environment of estrogen causes secretion of a very thin mucus with special characteristics that will allow rapid mobility of sperm into the uterus and will actually guide the sperm up along mucus "threads." Abnormalities of the cervix itself, such as low-grade infection or inflammation, or normal hormonal stimulation of the cervix can lead to a viscous mucus plug that will prevent fertilization.

# REFERENCES

Anand Kumar, T. C.: Cellular and humoral pathways in the neuroendocrine regulation of reproductive function. *J. Reprod. Fertil. (Suppl.), 20:11, 1973.*

Austin, C. R.: Recent progress in the study of eggs and spermatozoa: insemination and ovulation to implantation. *In* Guyton, A. C. (ed.): MTP International Review of Science: Physiology, Vol. 8. Baltimore, University Park Press, 1974, p. 95.

Barraclough, C. A., Turgeon, J., Mann, D. R., and Cramer, O. M.: Further analysis of the CNS regulation of adenohypophysial LH release: facilitation of the ovulatory LH surge by estrogen and progesterone. *J. Reprod. Fertil. (Suppl.)*, 20:61, 1973.

Blandau, R. J.: Gamete transport in the female mammal. *In* Greep, R. O., Astwood, E. B. (eds.): Handbook of Physiology. Sec. 7, Vol. 2, Part 2. Baltimore, The Williams & Wilkins Company, 1973, p. 153.

Brenner, R. M., and West, N. B.: Hormonal regulation of the reproductive tract in female mammals. *Ann. Rev. Physiol.*, 37:273, 1975.

Briggs, M. H., and Briggs, M.: Biochemical Contraception. New York, Academic Press, Inc., 1975.

Caldwell, B. V., Auletta, F. J., and Speroff, L.: Prostaglandins in the control of ovulation, corpus luteum function, and parturition. *J. Reprod. Med.*, 10:133, 1973.

Coutinho, E. M.: Hormonal control of oviductal motility and secretory functions. *In* Guyton, A. C. (ed.): MTP International Review of Science: Physiology. Vol. 8. Baltimore, University Park Press, 1974, p. 133.

Cross, B. A.: Towards a neurophysiological basis for ovulation. *J. Reprod. Fertil. (Suppl.)*, 20:97, 1973.

Dempsey, M. E.: Regulation of steroid biosynthesis. *Ann. Rev. Biochem.* 43:967, 1974.

Dorfman, R. I.: Biosynthesis of progesterones. *In* Greep, R. O., and Astwood, E. B. (eds.): Sec. 7, Vol. 2, Part 1. Baltimore, The Williams & Wilkins Company, p. 537.

Duncan, G. W., Hilton, E. J., Kreager, P., and Lumsdaine, A. A. (eds.): Fertility Control Methods. New York, Academic Press, Inc., 1973.

Flerko, B.: Hypothalamic mediation of neuroendocrine regulation of hypophyseal gonadotropic functions. *In* Guyton, A. C. (ed.): MTP International Review of Science: Physiology. Vol. 8. Baltimore, University Park Press, 1974, p. 1.

Goldman, B. D., and Zarrow, M. X.: The physiology of progestins. *In* Greep, R. O., Astwood, E. B. (eds.): Handbook of Physiology. Sec. 7, Vol. 2, Part 1. Baltimore, The Williams & Wilkins Company, 1973, p. 547.

Greenwald, G. S.; Role of follicle-stimulating hormone and luteinizing hormone in follicular development and ovulation. *In* Greep, R. O., Astwood, E. B. (eds.): Handbook of Phsyiology. Sec. 7, Vol. 4, Part 2. Baltimore, The Williams & Wilkins Company, 1974, p. 293.

Guraya, S. S.: Morphology, histochemistry, and biochemistry of human ovarian compartments and steroid hormonal synthesis. *Physiol. Rev.*, 51:785, 1971.

Hafez, E. S. E., and Evans, T. N.: Human Reproduction. Hagerstown, Maryland, Harper & Row, Publishers, 1973.

Hasegawa, T. (ed.): Fertility and Sterility. New York, American Elsevier Publishing Company, 1973.

Jensen, E. V., DeSombre, E. R.: Effects of ovarian hormones at the subcellular level. *Curr. Top. Exp. Endocrinol.*, 1:229, 1971.

Keye, W. R., Jr., Yuen, B. H., and Jaffe, R. B.: New concepts in the physiology of the menstrual cycle. *Clin. Endocrinol. Metabol.*, 2:451, 1973.

Kistner, R. W.: The menopause. *Clin. Obstet. Gynecol.*, 16:106, 1973.

Labhsetwar, A. P.: Pituitary gonadotrophic function (FSH and LH) in various reproductive states. *Adv. Reprod. Physiol.*, 6:97, 1973.

Lawn, A. M.: The ultrastructure of the endometrium during the sexual cycle. *Adv. Reprod. Phsyiol.*, 6:61, 1973.

Linford, E.: Cervical mucus: an agent or a barrier to conception? *J. Reprod. Fertil.*, 37:239, 1974.

Lipner, H.: Mechanism of mammalian ovulation. *In*: Greep, R. O., and Astwood, E. B. (eds.): Handbook of Physiology. Sec. 7, Vol. 2, Part 1. Baltimore, The Williams & Wilkins Company, 1973, p. 409.

Lofts, B., and Lam, W. L.: Circadian regulation of gonadotrophin secretion. *J. Reprod. Fertil. (Suppl.)*, 19:19, 1973.

McCann, S. M.: Regulation of secretion of follicle-stimulating hormone and luteinizing hormone. *In* Greep, R. O., and Astwood, E. B. (eds.): Handbook of Physiology. Sec. 7, Vol. 4, Part 2. Baltimore, The Williams & Wilkins Company, 1974, p. 489.

McCann, S. M., Krulich, L., Cooper, K. J., Kalra, P. S., Kalra, S. P., Libertun, C., Negro-Vilar, A., Orias, R., Ronnekleiv, O., and Fawcett, C. P.: Hypothalamic control of gonadotrophin and prolactin secretion, implications for fertility control. *J. Reprod. Fertil. (Suppl.)*, 20:43, 1973.

Miyake, T.: Blood concentration and interplay of pituitary and gonadal hormones governing the reproduction cycle in female mammals. *In* Guyton, A. C. (ed.): MTP International Review of Science: Physiology. Vol. 8. Baltimore, University Park Press, 1974, p. 155.

Moghissi, K. S., and Hafez, E. S. E. (eds.): Biology of Mammalian Fertilization and Implantation. Springfield, Ill., Charles C Thomas, Publisher, 1972.

Moudgal, N. R. (ed.): Gonadotropins and Gonadal Function. New York, Academic Press, Inc., 1974.

Nalbandov, A. V.: Control of luteal function in mammals. *In* Greep, R. O., and Astwood, E. B. (eds.): Handbook of Physiology. Sec. 7, Vol. 2, Part 1. Baltimore, The Williams & Wilkins Company, 1973, p. 153.

Odell, W. D., and Molitch, M. E.: The pharmacology of contraceptive agents. *Ann. Rev. Pharmacol.*, 14:413, 1974.

O'Malley, B. W., and Means, A. R.: Female steroid hormones and target cell nuclei. *Science*, 183:610, 1974.

Owen, J. A., Jr.: Physiology of the menstrual cycle. *Amer. J. Clin. Nutr.*, 28:333, 1975.

Pauerstein, C. J.: The Fallopian Tube. Philadelphia, Lea & Febiger, 1974.

Richart, R. M., and Prager, D. J.: (eds.): Human Sterilization. Springfield, Ill., Charles C Thomas, Publisher, 1972.

Sandler, M., and Gessa, G. L. (eds.): Sexual Behavior—Pharmacology and Biochemistry. New York, Raven Press, 1975.

Segal, S. J., Crozier, R., Corfman, P. A., and Conliffe, P. G. (eds.): The Regulation of Mammalian Reproduction. Springfield, Ill., Charles C Thomas, Publisher, 1973.

Steinetz, B. G.: Secretion and function of ovarian estrogens. *In* Greep. R. O., and Astwood, E. B. (eds.): Handbook of Physiology. Sec. 7, Vol. 2, Part 1. Baltimore, The Williams & Wilkins Company, 1973, p. 439.

Taymor, M. L.: Induction of ovulation with gonadotropins. *Clin. Obstet. Gynecol.*, 16:201, 1973.

Wheeler, R. G., Duncan, G. W., and Speidel, J. J. (eds.): Intrauterine Devices. New York, Academic Press, Inc., 1974.

# 82

# Pregnancy and Lactation

In the preceding two chapters the sexual functions of the male and the female were described to the point of fertilization of the ovum. If the ovum becomes fertilized, a completely new sequence of events called *gestation,* or *pregnancy,* takes place, and the fertilized ovum eventually develops into a full-term fetus. The purpose of the present chapter is to discuss the early stages of ovum development after fertilization and then to discuss the physiology of pregnancy. In the following chapter some special problems of fetal and early childhood physiology are discussed.

## MATURATION OF THE OVUM AND OOGENESIS

In the fetal female, the germinal epithelium lies on the outer surfaces of the ovaries. Throughout fetal development, small ingrowths of germinal epithelium invaginate into the ovarian stroma to form *primary oocytes* and *granulosa cells* that surround the oocytes. At the time of birth, approximately 750,000 such primary oocytes are present in the two ovaries, but this declines to 200,000 to 500,000 at puberty.

Shortly before the ovum is released from the follicle, the so-called *first polar body* is expelled from the nucleus of the ovum, and this comes to lie immediately outside the cell membrane of the ovum, which itself becomes the *secondary oocyte*. In this process, each of the 23 pairs of chromosomes loses one of the partners to the polar body so that 23 *unpaired* chromosomes remain in the secondary oocyte. It is at this point that fertilization occurs. Immediately after the sperm enters the ovum, the nucleus divides again, and a *second polar* body is expelled.

## FERTILIZATION OF THE OVUM

After coitus, the first sperm are transported through the uterus to the ovarian end of the fallopian tubes within about five minutes. This is many times more rapid than the motility of the sperm themselves can account for, which indicates that propulsive movements of the uterus and fallopian tubes might be responsible for much of the sperm movement. It is known, for instance, that in some animals coitus causes the neurohypophysis to secrete oxytocin, which in turn enhances uterine contractions. Also, semen contains a *prostaglandin* that theoretically could add still further to the contractions. Even with these aids to their movement, of the ½ billion sperm deposited in the vagina only 1000 to 3000 succeed in traversing the fallopian tubes to reach the proximity of the ovum.

Sperm can remain viable in the female genital tract for 24 to 72 hours but can remain healthy and highly fertile probably for only 12 to 24 hours. Furthermore, after ovulation a mature ovum also is fertilizable probably for up to 24 hours but maximally fertilizable for only 8 to 12 hours. Therefore, if coitus occurs when these two periods of fertility overlap, fertilization can take place.

Only one sperm is required for fertilization of the ovum, the process of which is illustrated in Figure 82–1. Furthermore, almost never does more than one sperm enter the ovum for the following reason: The zona pellucida of the ovum has lattice-type structure, and once the ovum is punctured, some substance (perhaps one of the proteolytic enzymes of the sperm acrosome) seems to diffuse out of the ovum into the lattice to prevent penetration by additional sperm. Indeed, microscopic studies show that many sperm do attempt to penetrate the zona pellucida but become inactivated before traveling only part way through.

Once a sperm enters the ovum, its head swells rapidly to form a *male pronucleus,* which is also illustrated in Figure 82–1. Later, the 23

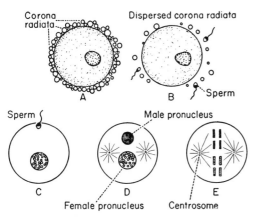

**Figure 82–1.** Fertilization of the ovum, showing (A) the mature ovum surrounded by the corona radiata, (B) dispersal of the corona radiata, (C) entry of the sperm, (D) formation of the male and female pronuclei, and (E) reorganization of a full complement of chromosomes and beginning division of the ovum. (Modified from Arey: Developmental Anatomy, 7th Ed.)

chromosomes of the male pronucleus and the 23 of the *female pronucleus* align themselves to reform a complete complement of 46 chromosomes (23 pairs) in the fertilized ovum.

**Sex Determination.** The sex of a child is determined by the type of sperm that fertilizes the ovum—that is, whether it is a male sperm or a female sperm. It will be recalled from Chapter 80 that a male sperm carries a *Y sex chromosome* and 22 *autosomal chromosomes,* while a female sperm carries the same 22 autosomal chromosomes but an *X sex chromosome.* On the other hand, the ovum always has an X sex chromosome and never a Y chromosome. After recombination of the male and female pronuclei during fertilization, the fertilized ovum then contains 44 autosomal chromosomes and either 2 X chromosomes, which causes a female child to develop, or an X and a Y chromosome, which causes a male child to develop.

## TRANSPORT AND IMPLANTATION OF THE DEVELOPING OVUM

**Entry of the Ovum into the Fallopian Tube.** When ovulation occurs, the ovum along with its attached granulosa cells, the *cumulus oophorus,* is expelled directly into the peritoneal cavity and must then enter one of the fallopian tubes. The fimbriated end of each fallopian tube falls naturally around the ovaries, and the inner surfaces of the fimbriated tentacles are lined with ciliated epithelium, the *cilia* of which continually beat toward the *abdominal ostium* of the fallopian tube. One can actually see a slow fluid current flowing toward the ostium. By this means the ovum enters one or the other fallopian tube.

It would seem likely that many ova might fail to enter the fallopian tubes. However, on the basis of conception studies it is probable that as many as 98 to 100 per cent succeed in this task. Indeed, cases are on record in which women with one ovary removed and the opposite fallopian tube removed have had as many as four children with relative ease of conception, thus illustrating that ova can even enter the opposite fallopian tube.

**Transport of the Ovum Through the Fallopian Tube.** Fertilization of the ovum normally takes place soon after the ovum enters the fallopian tube, but this process cannot occur until the cells attached to the outside of the ovum, the cumulus oophorus, are dispersed from the ovum. This dispersal probably results partly from the action of hyaluronidase and proteolytic enzymes released from the acrosomes of the sperm, but the final dissolution of these cells is caused by bicarbonate ions in the fallopian tube secretions. After fertilization has occurred, an additional three days are normally required for transport of the ovum through the tube into the cavity of the uterus. This transport is effected mainly by a feeble fluid current in the fallopian tube resulting from action of the ciliated epithelium that line the tube, the cilia always beating toward the uterus. It is possible also that weak contractions of the fallopian tube aid in the passage of the ovum.

The fallopian tubes are lined with a rugged, cryptoid surface that actually impedes the passage of the ovum despite the fluid current. Also, the *isthmus* of the fallopian tube (the last two centimeters before the uterus is entered) remains spastically contracted for the first three days following ovulation. After this time, probably under the relaxing influence of the rapidly increasing progesterone, this region loses the spasm, thus allowing entry of the ovum to the uterus. If the fertilized ovum implants in the endometrium, the endometrium does not degenerate and is not discharged by menstruation. Instead, a sequence of very rapid hormonal changes occurs, as will be explained later in the chapter; these result in still more progesterone secretion with subsequent further swelling of the endometrial cells and further collection of nutrients in the cells. The delayed transport of the ovum through the fallopian tube allows several stages of division to occur before the ovum enters the uterus. During this time, large quantities of secretions are formed by secretory cells that alternate with ciliated cells lining the fallopian tube. These secretions are for nutrition of the developing ovum. Indeed, fertilized ova

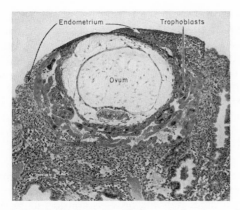

**Figure 82–2.** Implantation of the early human embryo, showing trophoblastic digestion and invasion of the endometrium. (Courtesy of Dr. Arthur Hertig.)

continue to divide in vitro as long as they are bathed in a solution of homogenized fallopian tube mucosa, but they will not divide in almost all other media.

**Implantation of the Ovum in the Uterus.** After reaching the uterus, the developing ovum usually remains in the uterine cavity an additional four to five days before it implants in the endometrium, which means that implantation ordinarily occurs on the seventh or eighth day following ovulation. During this time the ovum obtains its nutrition from the endometrial secretions, called "uterine milk." Figure 82–2 shows a very early stage of implantation, illustrating that the developing ovum is in the *blastocyst stage.*

Implantation results from the action of trophoblastic cells that develop over the surface of the blastocyst. These cells secrete proteolytic enzymes that digest and liquefy the cells of the endometrium. Simultaneously, much of the fluid and nutrients thus released is actively absorbed into the blastocyst as a result of phagocytosis by the trophoblastic cells; these absorbed substances provide the sustenance for further growth of the blastocyst. Also, at the same time, additional trophoblastic cells form cords of cells that extend into the deeper layers of the endometrium and attach to them. Thus, the blastocyst eats a hole in the endometrium and attaches to it at the same time.

Once implantation has taken place, the trophoblastic and sub-lying cells proliferate rapidly; and these, along with cells from the mother's endometrium, form the placenta and the various membranes of pregnancy.

# EARLY INTRA-UTERINE NUTRITION OF THE EMBRYO

In the previous chapter it was pointed out that the progesterone secreted during the latter half of each sexual cycle has a special effect on the endometrium to convert the endometrial stromal cells into large swollen cells that contain extra quantities of glycogen, proteins, lipids and even some necessary minerals for development of the ovum. If the fertilized ovum implants in the endometrium, the continued secretion of progesterone causes the stromal cells to swell still more and to store even more nutrients. These cells are now called *decidual cells,* and the total mass of cells is called the *decidua.*

As the trophoblastic cells invade the decidua, digesting and imbibing it, the stored nutrients in the decidua are used by the embryo for appropriate growth and development. During the first week after implantation, this is the only means by which the embryo can obtain any nutrients whatsoever, and the embryo continues to obtain a large measure of its total nutrition in this way for 8 to 12 weeks, though the placenta also begins to provide slight amounts of nutrition after approximately the sixteenth day beyond fertilization (a little over a week after implantation). Figure 82–3 depicts this trophoblastic period of nutrition, which gradually gives way to placental nutrition.

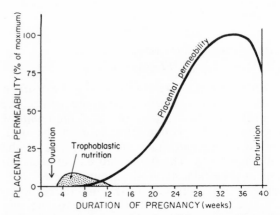

**Figure 82–3.** Nutrition of the fetus, illustrating that most of the early nutrition is due to trophoblastic digestion and absorption of nutrients from the endometrial decidua and that essentially all the later nutrition results from diffusion through the placental membrane.

# FUNCTION OF THE PLACENTA

## *DEVELOPMENTAL AND PHYSIOLOGIC ANATOMY OF THE PLACENTA*

While the trophoblastic cords from the blastocyst are attaching to the uterus, blood capillaries grow into the cords from the vascular system of the embryo, and, by the sixteenth day after fertilization, blood begins to flow. Simultaneously, blood sinuses supplied with blood from the mother develop between the surface of the uterine endometrium and the trophoblastic cords. The trophoblastic cells then gradually send out more and more projections, which become the *placental villi* in which fetal capillaries grow. Thus, the villi, carrying fetal blood, are surrounded by sinuses containing maternal blood.

The final structure of the placenta is illustrated in Figure 82–4. Note that the fetus' blood flows through two *umbilical arteries,* finally to the capillaries of the villi, and thence back through the *umbilical vein* into the fetus. On the other hand, the mother's blood flows from the *uterine arteries* into large *blood sinuses* surrounding the villi and then back into the *uterine veins* of the mother.

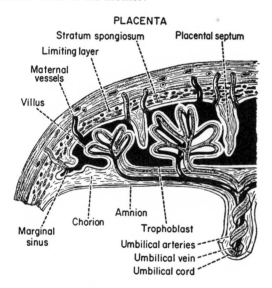

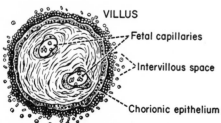

**Figure 82–4.** *Above:* Organization of the mature placenta. *Below:* Relationship of the fetal blood in the villus capillaries to the mother's blood in the intervillus spaces. (Modified from Gray and Goss: Anatomy of the Human Body. Lea & Febiger; and from Arey: Developmental Anatomy, 7th Ed.)

The lower part of Figure 82–4 illustrates the relationship between the fetal blood of the villus and the blood of the mother in the fully developed placenta. The capillaries of the villus are lined with an extremely thin endothelium and are surrounded by a layer of *mesenchymal tissue* that is covered on the outside of the villus by a layer of *syncytial trophoblastic* cells. During the first 16 weeks of pregnancy, still an additional layer of cells is present immediately beneath the syncytial trophoblastic layer. This layer is composed of distinct cuboidal cells called *cytotrophoblastic cells,* or *Langerhans' cells.*

The total villar surface area of the mature placenta is only a few square meters—many times less than the area of the pulmonary membrane. Also, remember that even at full maturity the placental membrane is still several cell layers thick, and the minimum distance between the maternal blood and the fetal blood is 3.5 microns, or almost 10 times the distance across the alveolar membranes of the lung. Nevertheless, many nutrients and other substances pass through the placental membrane by diffusion in very much the same manner as through the alveolar membranes and capillary membranes elsewhere in the body.

## *PERMEABILITY OF THE PLACENTAL MEMBRANE*

Since the major function of the placenta is to allow *diffusion* of foodstuffs from the mother's blood into the fetus' blood and diffusion of excretory products from the fetus back into the mother, it is important to know the degree of permeability of the placental membrane. *Permeability is expressed as the total quantity of a given substance that crosses the entire placental membrane in a given time for a given concentration difference.*

In the early months of development, placental permeability is relatively slight, as illustrated in Figure 82–3, for two reasons: First, the total surface area of the placental membrane is still small at that time, and, second, the villar membranes have not yet been reduced to their minimum thickness. However, as the placenta becomes older, the permeability increases progressively until the last month or so of pregnancy when it begins to decrease again. The increase in permeability is caused by both progressive enlargement of the surface area of the placental membrane and progressive thinning of the layers of the villi. On the other hand, the decrease shortly before birth results from deterioration of the placenta caused by its age and sometimes from destruction of whole segments due to infarction.

Occasionally, "breaks" occur in the placental membrane which allow fetal blood cells to pass into the mother, or more rarely, the mother's cells to pass into the fetus. Indeed, there are instances in which the fetus bleeds severely into the mother's circulation because of a ruptured placental membrane.

**Diffusion of Oxygen Through the Placental Membrane.**    Almost exactly the same principles are applicable for the diffusion of oxygen through the placental membrane as through the pulmonary membrane; these principles were discussed in Chapter 40. The dissolved oxygen in the blood of the large placental sinuses simply passes through the villar membrane into the fetal blood because of a pressure gradient of oxygen from the mother's blood to the fetus' blood. The mean $Po_2$ in the mother's blood in the placental sinuses is approximately 50 mm. Hg toward the end of pregnancy, and the mean $Po_2$ in the blood leaving the villi and returning to the fetus is about 30 mm. Hg. The mean pressure gradient for diffusion of oxygen through the placental membrane is therefore about 20 mm. Hg.

One might wonder how it is possible for a fetus to obtain sufficient oxygen when the fetal blood leaving the placenta has a $Po_2$ of only 30 mm. Hg. However, there are three different reasons why even this low $Po_2$ is capable of allowing the fetal blood to transmit almost as much oxygen to the fetal tissues as is transmitted by the mother's blood to her tissues:

First, the hemoglobin of the fetus is primarily *fetal hemoglobin,* a type of hemoglobin synthesized in the fetus prior to birth. Figure 82–5 illustrates the comparative oxygen dissociation curves of maternal hemoglobin and fetal hemoglobin, showing that the curve for fetal hemoglobin is shifted considerably to the left of that for maternal hemoglobin. This means that at a given $Po_2$, the fetal hemoglobin can carry as much as 20 to 30 per cent more oxygen than can maternal hemoglobin.

Second, the *hemoglobin concentration of the fetus is about 50 per cent greater than that of the mother;* and this is an even more important factor than the first in enhancing the amount of oxygen transported to the fetal tissues.

Third, the *Bohr effect,* which was explained in relation to the exchange of carbon dioxide and oxygen in the lung in Chapter 41, provides another factor that enhances the transport of oxygen by the fetal blood. That is, hemoglobin can carry more oxygen at a low $Pco_2$ than it can at a high $Pco_2$. The fetal blood entering the placenta carries large amounts of carbon dioxide, and this carbon dioxide diffuses from the fetal blood into the maternal blood. Loss of the carbon dioxide makes the fetal blood more alkaline while the increased carbon dioxide in the maternal blood makes this more acidic. These changes cause the combining capacity of fetal blood for oxygen to become increased while that of the maternal blood becomes decreased. This forces more oxygen from the maternal blood while enhancing the oxygen in the fetal blood. Thus, the Bohr shift operates in one direction in the maternal blood and in the other in the fetal blood, these two effects adding to make the Bohr shift twice as important here as it is for oxygen exchange in the lungs.

By these three means, the fetus is capable of receiving more than adequate oxygen through the placenta despite the fact that the fetal blood leaving the placenta has a $Po_2$ of only 30 mm. Hg.

The total diffusing capacity of the placenta for oxygen at term is about 1.2 ml. of oxygen per minute per mm. Hg oxygen gradient. This compares favorably with that of the lungs of the newborn baby.

**Diffusion of Carbon Dioxide Through the Placental Membrane.**    Carbon dioxide is continually formed in the tissues of the fetus in the same way that it is formed in maternal tissues. And the only means for excreting the carbon dioxide is through the placenta. The $Pco_2$ builds up in the fetal blood until it is about 48 mm. Hg in contrast to about 40 to 45 mm. Hg in maternal blood. Thus, a low pressure gradient for carbon dioxide develops across the placental membrane, but this is sufficient to allow adequate

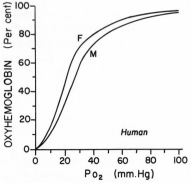

**Figure 82–5.**  Oxygen-hemoglobin dissociation curves for maternal *(M)* and fetal *(F)* bloods, illustrating the ability of the fetal blood to carry a much greater quantity of oxygen than can maternal blood for a given blood $Po_2$. (From Metcalfe, Moll, and Bartels: *Fed. Proc.,* 23:775, 1964.)

diffusion of carbon dioxide from the fetal blood into the maternal blood, because the extreme solubility of carbon dioxide in the water of the placental membrane allows carbon dioxide to diffuse through this membrane rapidly, about 20 times as rapidly as oxygen.

**Diffusion of Foodstuffs Through the Placental Membrane.** Other metabolic substrates needed by the fetus diffuse into the fetal blood in the same manner as oxygen. For instance, the glucose level in the fetal blood ordinarily is approximately 20 to 30 per cent lower than in the glucose level in the maternal blood, for glucose is being metabolized rapidly by the fetus. This in turn causes rapid diffusion of additional glucose from the maternal blood into the fetal blood.

Because of the high solubility of fatty acids in cell membranes, these also diffuse from the maternal blood into the fetal blood. Also, such substances as potassium, sodium, and chloride ions diffuse from the maternal blood into the fetal blood. As these substances are used by the fetal body, their concentrations in the fetal blood fall, and the increased concentration gradients then cause more of the same substrates to diffuse through the placental membrane.

**Active Absorption by the Placental Membrane.** As pointed out previously, early nutrition of the embryo depends on phagocytosis of fallopian tube and uterine secretions, and even on phagocytosis of the endometrial decidua. The trophoblastic cells that line the outer surface of the villi can probably also actively absorb certain nutrients from the maternal blood in the placenta at least during the first half of pregnancy and perhaps even throughout the entire pregnancy. For instance, the measured *amino acid* content of fetal blood is greater than that of maternal blood, and *calcium* and *inorganic phosphate* occur in greater concentration in fetal blood than in maternal blood, while *ascorbic acid* is as much as three times as concentrated in the fetal blood. This indicates that the placental membrane has the ability to absorb actively at least small amounts of certain substances even during the latter part of pregnancy.

**Excretion Through the Placental Membrane.** In the same manner that carbon dioxide diffuses from the fetal blood into the maternal blood, other excretory products formed in the fetus diffuse in the opposite direction into the maternal blood and then are excreted along with the excretory products of the mother. These include especially the *nonprotein nitrogens,* such as *urea, uric acid,* and *creatinine.* The level of urea in the fetal blood is only slightly greater than that in maternal blood because urea diffuses through the placental membrane with considerable ease. On the other hand, creatinine, which does not diffuse as easily, has a considerably higher concen-

tration gradient, percentagewise, between the fetal blood and maternal blood. Therefore, insofar as is known, excretion from the fetus occurs entirely as a result of diffusion gradients across the placental membrane—that is, higher concentrations of the excretory products in the fetal blood than in the maternal blood.

## STORAGE FUNCTION OF THE PLACENTA

During the first few months of pregnancy, the placenta grows tremendously in size while the fetus remains relatively diminutive. During this same time considerable quantities of metabolic substrates, including proteins, calcium, and iron, are stored in the placenta to be used in the latter months of pregnancy for growth by the fetus. Thus, in the early months of pregnancy, the placenta performs very much the same functions for the fetus that the liver performs for the adult human being, acting as a nutrient storehouse and helping to process some of the food substances that enter the fetus. For instance, in the early weeks of gestation the placenta is capable of storing glucose as glycogen and then of secreting glucose into the embryonic blood stream in much the same manner that the liver can secrete glucose into the adult blood stream. By this process, the placenta actually helps to control the fetal blood glucose concentration. Later in the growth of the fetus, these metabolic functions of the placenta become less and less important, while the fetal liver becomes progressively more important.

# HORMONAL FACTORS IN PREGNANCY

In pregnancy, the placenta forms large quantities of *chorionic gonadotropin, estrogens, progesterone,* and *human placental lactogen,* the first three of which, and perhaps the fourth as well, are essential to the continuance of pregnancy. The functions of these hormones are discussed in the following sections.

## CHORIONIC GONADOTROPIN AND ITS EFFECT IN CAUSING PERSISTENCE OF THE CORPUS LUTEUM AND IN PREVENTING MENSTRUATION

Menstruation normally occurs approximately 14 days after ovulation, at which time most of the secretory endometrium of the uterus sloughs away from the uterine wall and is expelled to the exterior. If this should happen after an ovum has implanted, the pregnancy would terminate. However, this is prevented by

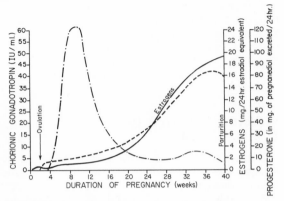

**Figure 82–6.** Rates of secretion of estrogens, progesterone, and chorionic gonadotropin at different stages of pregnancy.

the secretion of chorionic gonadotropin in the following manner:

Coincidently with the development of the trophoblastic cells from early fertilized ovum, the hormone *chorionic gonadotropin* is secreted by the syncytial trophoblastic cells into the fluids of the mother. As illustrated in Figure 82–6, the secretion of this hormone can first be measured 8 days after ovulation, just as the ovum is first implanting in the emdometrium. Then the rate of secretion rises rapidly to reach a maximum approximately seven weeks after ovulation, and decreases to a relatively low value by 16 weeks after ovulation.

**Function of Chorionic Gonadotropin.** Chorionic gonadotropin is a glycoprotein having a molecular weight of 30,000 and very much the same molecular structure and function as luteinizing hormone secreted by the pituitary. By far its most important function is to prevent the normal involution of the corpus luteum at the end of the female sexual cycle, and instead causes the corpus luteum to secrete even larger quantities of its usual hormones, progesterone and estrogens. These excess hormones cause the endometrium to continue growing and to store large amounts of nutrients rather than to be passed in the menstruum. As a result, the *decidua-like cells* that develop in the endometrium during the normal female sexual cycle become actual, nutritious *decidual cells* soon after the blastocyst implants.

Under the influence of chorionic gonadotropin, the corpus luteum grows to about two times its initial size by a month or so after pregnancy begins, and its continued secretion of estrogens and progesterone maintains the decidual nature of the uterine endometrium, which is necessary

to the early development of the placenta and other fetal tissues. If the corpus luteum is removed before approximately the eleventh week of pregnancy, spontaneous abortion usually occurs, though after this time the placenta itself secretes sufficient quantities of progesterone and estrogens to maintain pregnancy for the remainder of the gestation period.

*Effect of Chorionic Gonadotropin on the Fetal Testes.* Chorionic gonadotropin also exerts an *interstitial cell-stimulating effect* on the testes, thus resulting in the production of testosterone in male fetuses. This small secretion of testosterone during gestation is the factor that causes the fetus to grow male sex organs. Near the end of pregnancy, the testosterone secreted by the fetal testes also causes the testicles to descend into the scrotum.

Clinically, chorionic gonadotropin administered to the cryptorchid male (having undescended testes) often causes the testicles to descend into the scrotum by causing testosterone to be secreted, the testosterone in turn effecting the actual testicular descent.

## SECRETION OF ESTROGENS BY THE PLACENTA

The placenta, like the corpus luteum, secretes both estrogens and progesterone. Both histochemical and physiologic studies indicate that these two hormones are secreted by the *syncytial trophoblastic cells,* along with the secretion of chorionic gonadotropin and human placental lactogen.

Figure 82–6 shows that the daily production of placental estrogens increases markedly toward the end of pregnancy, to as much as 300 times the daily production in the middle of a normal monthly cycle.

However, the secretion of estrogen by the placenta is quite different from the secretion by the ovaries in several different ways, as follows: First, the estrogen that is secreted is about nine-tenths estriol, which, as will be recalled from Chapter 81, is formed in only small amounts in the nongravid female. Yet, because of the very low estrogenic potency of estriol, the total estrogenic activity rises only as high as 30 times normal. Second, the estrogens secreted by the placenta are not synthesized de novo from basic substrates in the placenta. Instead, the steroid compound *dehydroepiandrosterone,* which is formed in the adrenal glands of the fetus and then transported by the fetal blood to the placenta, is converted into estriol, estradiol, and estrone. (The cortices of the fetal adrenal glands are extremely large, composed almost entirely of the so-called *fetal zone,* whose primary function seems to be to secrete the dehydroepiandrosterone.)

**Function of Estrogen in Pregnancy.** In the discussions of estrogens in the preceding chapter it was pointed out that these hormones exert mainly a proliferative function on certain reproductive and associated organs. During pregnancy, the extreme quantities of estrogens cause (1) enlargement of the uterus, (2) enlargement of the breasts and growth of the breast glandular tissue, and (3) enlargement of the female external genitalia.

The estrogens also relax the various pelvic ligaments so that the sacroiliac joints become relatively limber and the symphysis pubis becomes elastic. These changes obviously make for easy passage of the fetus through the birth canal.

There is much reason to believe that estrogens also affect the development of the fetus during pregnancy, for example, by affecting the rate of cell reproduction in the early embryo.

## SECRETION OF PROGESTERONE BY THE PLACENTA

Progesterone is also a hormone essential for pregnancy. In addition to being secreted in moderate quantities by the corpus luteum at the beginning of pregnancy, it is secreted in tremendous quantities by the placenta, sometimes as much as 1 gram per day toward the end of pregnancy. Indeed, the rate of progesterone secretion increases by as much as 10-fold during the course of pregnancy, as illustrated in Figure 82–6.

The special effects of progesterone that are essential for normal progression of pregnancy are the following:

1. As pointed out earlier, progesterone causes decidual cells to develop in the uterine endometrium, and these then play an important role in the nutrition of the early embryo.

2. Progesterone has a special effect to decrease the contractility of the gravid uterus, thus preventing uterine contractions from causing spontaneous abortion.

3. Progesterone also contributes to the development of the ovum even prior to implantation, for it specifically increases the secretions of the fallopian tubes and uterus to provide appropriate nutritive matter for the developing *morula* and *blastocyst.* There are some reasons

to believe, too, that progesterone even affects cell cleavage in the early developing embryo.

4. The progesterone secreted during pregnancy also helps to prepare the breasts for lactation, which is discussed later in the chapter.

## HUMAN PLACENTAL LACTOGEN

Recently, a new hormone called *placental lactogen* has been discovered. This is a protein, having a molecular weight of about 38,000, that begins to be secreted about the fifth week of pregnancy, with progressively increasing secretion throughout the remainder of pregnancy. Human placental lactogen has two types of effects: First it has actions very similar to those of growth hormone, causing deposition of protein tissues in the same way as growth hormone. It also has the same effects as growth hormone on glucose metabolism, producing a diabetogenic glucose-tolerance curve.

Second, human placental lactogen mimics many of the functions of prolactin as well. It promotes growth of the breasts and can cause actual milk production when the breasts have been appropriately prepared by estrogen and progesterone—the same effect that prolactin from the anterior pituitary gland has, as will be discussed later in the chapter.

Therefore, it is presumed that human placental lactogen plays a role in growth of the fetus and also in preparation of the female breasts for nursing.

## OTHER HORMONAL FACTORS IN PREGNANCY

Almost all the nonsexual endocrine glands of the mother react markedly to pregnancy. This results mainly from the increased metabolic load on the mother but also to some extent from inverse effects of placental hormones on the pituitary and other glands. Some of the most notable effects are the following:

**Pituitary Secretion.** The anterior pituitary gland enlarges at least 50 per cent during pregnancy and increases its production of *corticotropin, thyrotropin,* and probably also *growth hormone.* On the other hand, production of follicle-stimulating hormone and luteinizing hormone is greatly suppressed as a result of the inhibitory effects of estrogens and progesterone from the placenta.

**Corticosteroid Secretion.** The rate of adrenocortical secretion of the *glucocorticoids* is moderately increased throughout pregnancy. It is possible that the glucocorticoids help to mobilize amino acids from the mother's tissues so that these can be used for synthesis of tissues in the fetus.

Pregnant women also usually have about a three-fold increase in secretion of *aldosterone,* reaching the peak at the end of gestation. This, along with the actions of the estrogens, causes a tendency for even the normal pregnant woman to reabsorb excess sodium from the renal tubules and therefore to retain fluid.

**Secretion by the Thyroid Gland.** The thyroid gland ordinarily enlarges about 50 per cent during pregnancy and increases its production of thyroxine a corresponding amount. The increased thyroxine production is caused primarily by increased thyrotropic hormone from the adenohypophysis.

**Secretion by the Parathyroid Glands.** The parathyroid glands also often enlarge during pregnancy; this is especially true if the mother is on a calcium deficient diet. Enlargement of these glands causes calcium absorption from the mother's bones, thereby maintaining normal calcium ion concentration in the mother's extracellular fluids as the fetus removes calcium for ossifying its own bones. This secretion of parathyroid hormone is even more intensified during lactation following the birth of the baby, because the baby requires many times more calcium than the fetus.

**Secretion of "Relaxin" by the Ovaries.** An additional substance besides the estrogens and progesterone, a hormone called relaxin, can be isolated from the corpora lutea of the ovaries and from the placenta. This hormone, when injected, causes relaxation of the ligaments of the symphysis pubis in the estrous rat and guinea pig. However, this effect is very poor in the pregnant woman. Instead, this role is probably played by both the estrogens and progesterone, which also causes relaxation of the pelvic ligaments, though many days are required for these hormones to act.

It has also been claimed that relaxin has two other effects: (1) softening of the cervix of the pregnant woman at the time of delivery and (2) inhibition of uterine motility.

Relaxin is a polypeptide having a molecular weight of about 9000.

In summary, relaxin is a substance that can be isolated from the ovaries, but its functional importance is almost totally unknown.

# RESPONSE OF THE MOTHER TO PREGNANCY

Obviously, the presence of a growing fetus in the uterus adds an extra physiologic load on the mother, and much of the response of the mother to pregnancy is due to this increased load. The hormones secreted during pregnancy either by the placenta or by the endocrine glands can also cause many reactions in the mother. Among the reactions are increased size of the various sexual organs. For instance, the uterus increases from about 30 grams to about 1100 grams, and the breasts approximately double in size. At the

same time the vagina enlarges, and the introitus opens more widely. Also, the various hormones can cause marked changes in the appearance of the mother, sometimes resulting in the development of edema, acne, and masculine or acromegalic features.

## CHANGES IN THE MATERNAL CIRCULATORY SYSTEM DURING PREGNANCY

**Blood Flow Through the Placenta.** About 625 ml. of blood flows through the maternal circulation of the placenta each minute during the latter phases of gestation. Obviously, the more rapidly this blood flows, the greater will be the concentration of oxygen and other metabolites in the fetal blood.

**Cardiac Output of the Mother.** The flow of blood through the placenta decreases the total peripheral resistance of the mother's circulatory system and, consequently, allows increased venous return of blood to the heart, which in turn tends to increase the cardiac output in the same manner that arteriovenous shunts increase the output. This factor, plus a general increase in metabolism, increases the cardiac output to 30 to 40 per cent above normal by the twenty-seventh week of pregnancy, but then, for reasons yet unexplained, the cardiac output falls to near normal during the last eight weeks of pregnancy, despite the high uterine blood flow.

**Blood Volume of the Mother.** The maternal blood volume shortly before term is approximately 30 per cent above normal. This increase occurs mainly during the latter half of pregnancy, as illustrated by the curve of Figure 82–7. The cause of the increased volume is probably mainly hormonal, for aldosterone and estrogens can both cause increased fluid retention by the kidneys, and these hormones are secreted into the maternal body in very large quantities during the latter half of pregnancy.

As a result of the increased blood volume in the mother, the hematocrit at first decreases because of dilution of the blood. However, toward the end of pregnancy the bone marrow becomes increasingly active, and the concentration of red blood cells returns almost to normal. Therefore, at the time of birth of the baby, the mother has approximately 1 to 2 liters of extra blood in her circulatory system. Only about one-fourth of this amount is normally lost during delivery of the baby, thereby allowing a considerable safety factor for the mother.

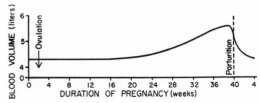

**Figure 82–7.** Effect of pregnancy on blood volume.

## WEIGHT GAIN IN THE MOTHER

During the first months of pregnancy, the mother ordinarily loses a few pounds of weight, possibly as a result of nausea, but during the entire pregnancy the average weight gain is approximately 24 pounds, most of this gain occurring during the last two trimesters. Of this increase in weight, approximately 7 pounds is fetus, and approximatly 4 pounds is amniotic fluid, placenta, and fetal membranes. This leaves 13 pounds increase in weight by the mother herself. The uterus increases approximately 2 pounds, and the breasts approximately 3 pounds, still leaving an average increase in weight of the mother herself of approximately 8 pounds. On the average, this extra 8 pounds of weight is accounted for by about 6 pounds of fluid and 2 pounds of fat.

Often during pregnancy the mother has a greatly increased desire for food, partly as a result of fetal removal of food substrates from the mother's blood and partly because of hormonal factors. Without appropriate prenatal care some mothers eat tremendous quantities of food, and the weight gain, instead of averaging 24 pounds, may be as great as 75 pounds or more. In these instances, the weight gain is due mainly to deposition of fat.

## METABOLISM IN THE MOTHER DURING PREGNANCY

As a consequence of the increased secretion of many different hormones during pregnancy, including thyroxine, adrenocortical hormones, and the sex hormones, the basal metabolic rate of the mother increases about 15 per cent during the latter half of pregnancy. As a result, the mother frequently has sensations of becoming overheated. Also, owing to the extra load that the mother is carrying, greater amounts of energy than normally must be expended for muscular activity.

**Nutrition During Pregnancy.** The supplemental food needed by the mother during pregnancy to supply the needs of the fetus and fetal membranes includes especially extra dietary quantities of the various minerals, vitamins, and proteins. The growing fetus assumes priority in regard to many of the nutritional elements in the mother's body fluids, and many portions of the fetus continue to grow even though the mother does not eat a sufficient diet. For instance, lack of adequate nutrition in the mother hardly changes the rate of growth of the fetal nervous system (except in severely depressed maternal nutrition, which can cause permanent mental damage), and the length of the fetus increases almost normally; on the other hand, lack of adequate nutrition can decrease the fetus' weight considerably, can decrease ossification of the bones, can cause anemia, hypoprothrombinemia, and decreased size of many bodily organs of the fetus.

By far the greatest growth of the fetus occurs during the last trimester of pregnancy; the weight of the child almost doubles during the last two months of pregnancy. Ordinarily, the mother does not absorb sufficient protein, calcium, phosphates, and iron from the gastrointestinal tract during the last month of pregnancy to supply the fetus. However, from the beginning of pregnancy the mother's body has been storing these substances to be used during the latter months of pregnancy. Some of this storage is in the placenta, but most of it is in the normal storage depots of the mother.

If appropriate nutritional elements are not present in the mother's diet, a number of maternal deficiencies can occur during pregnancy. Such deficiencies often occur for calcium, phosphates, iron, and the vitamins. For example, approximately 375 mg. of iron is needed by the fetus to form its blood and an additional 600 mg. is needed by the mother to form her own extra blood. The normal store of nonhemoglobin iron in the mother at the outset of pregnancy is often only 100 or so mg. and almost never over 700 mg. Therefore, without sufficient iron in the food the mother herself usually develops anemia. In general, the obstetrician supplements the diet of the mother with the needed substances. It is especially important that the mother receive vitamin D, for, even though the total quantity of calcium utilized by the fetus is small, calcium even normally is poorly absorbed by the gastrointestinal tract. Finally, shortly before birth of the baby vitamin K is often added to the diet so that the baby will have sufficient prothrombin to prevent hemorrhage, particularly brain hemorrhage, caused by the birth process.

## RESPIRATION BY THE MOTHER DURING PREGNANCY

Because of the increased basal metabolic rate of the mother, and also because of the mother's increase in size, the total amount of oxygen utilized by the mother shortly before birth of the baby is approximately 20 per cent above normal, and a commensurate amount of carbon dioxide is formed. These effects cause the minute ventilation to increase. It is also believed that the high levels of progesterone during pregnancy increase the minute ventilation still more, because progesterone increases the sensitivity of the respiratory center to carbon dioxide. The net result is an increase in minute ventilation of approximately 50 per cent and a decrease in arterial $P_{CO_2}$ to slightly below that of the normal woman. Simultaneously, the growing uterus presses upward against the abdominal contents, and these in turn press upward against the diaphragm so that the total excursion of the diaphragm is decreased. Consequently, the respiratory rate is increased to maintain adequate ventilation.

## FUNCTION OF THE MATERNAL URINARY SYSTEM DURING PREGNANCY

The rate of urine formation by the pregnant mother is usually slightly increased because of an increased load of excretory products. But, in addition, several special alterations of urinary functions are as follows:

First, reabsorption of sodium, chloride, and water by the renal tubules tends to be increased greatly as a consequence of increased production of steroid hormones by the placenta and adrenal cortex.

Second, the glomerular filtration rate often increases as much as 50 per cent during pregnancy, which tends to increase the rate of water and electrolyte loss of the urine. This factor normally almost balances the first so that the mother ordinarily has only moderate excess water and salt accumulation except when she develops *toxemia of pregnancy;* this disease is discussed later in the chapter.

Third, the ureters usually dilate during pregnancy because of two different factors: The enlarged uterus compresses the ureters as they pass over the pelvic rim, thereby increasing the intra-ureteral pressure. But, aside from this, the hormones secreted during pregnancy have a direct effect in relaxing the ureters. The two hormones that have been particularly implicated are progesterone and relaxin, though there is still much doubt about the ureteral effects of either of these. Ureteral distention and the accompanying renal pelvic distention that also results predisposes to infection in the urinary tract, frequently leading to serious renal debility, one of the common complications of pregnancy.

## THE AMNIOTIC FLUID AND ITS FORMATION

Normally, the volume of amniotic fluid is between 500 ml. and 1 liter, but, it can be only a few milliliters or as much as several liters. Studies with isotopes on the rate of formation of amniotic fluid show that on the average the water in amniotic fluid is completely replaced once every three hours, and the electrolytes sodium and potassium are replaced once every 15 hours. Yet, strangely enough, the sources of the fluid and the points of reabsorption are mainly unknown. A small portion of fluid is derived from renal excretion by the fetus. Likewise, a certain amount of absorption occurs by way of the gastrointestinal tract and lungs of the fetus. However, even after death of the fetus, the rate of turnover of the amniotic fluid is still one-half as great as it is when the fetus is normal, which indicates that much of the fluid is formed and absorbed directly through the amniotic membranes. The total volume of amniotic fluid could be regulated by the amniotic membranes themselves, for as the volume increased the pressure would rise and cause increased fluid absorption, thus returning the volume to normal.

# ABNORMAL RESPONSES OF THE MOTHER TO PREGNANCY

## HYPEREMESIS GRAVIDARUM

In the earlier months of pregnancy, the mother frequently develops hyperemesis gravidarum, a condition characterized by nausea and vomiting and commonly known as "morning sickness." Occasionally, the vomiting becomes so severe that the mother becomes greatly dehydrated, and in rare instances the condition even causes death.

The cause of the nausea and vomiting is unknown, but it occurs to its greatest extent during the same time that chorionic gonadotropin is secreted in large quantities by the placenta. Because of this coincidence, many clinicians believe that chorionic gonadotropin is in some way responsible for the nausea and vomiting; nevertheless, a causal relationship has never been proved.

On the other hand, during the first few months of pregnancy rapid trophoblastic invasion of the endometrium also takes place and, because the trophoblastic cells digest portions of the endometrium as they invade it, it is possible that degenerative products resulting from this invasion, instead of chorionic gonadotropin, are responsible for the nausea and vomiting. Indeed, degenerative processes in other parts of the body, such as following gamma ray irradiation and burns, can all cause similar nausea and vomiting.

Finally, another possible cause of the condition is the large quantity of estrogen secreted by the placenta. This theory is supported by the fact that estrogen injected daily into a person in large quantities for many weeks will often cause nausea and vomiting during the first few weeks of administration.

## TOXEMIA OF PREGNANCY

Approximately 7 per cent of all pregnant women experience rapid weight gain, edema, and often elevation of arterial pressure. This condition, known as toxemia of pregnancy, is characterized by inflammation and spasm of the arterioles in many parts of the body. It is also characterized by a slight decrease in both renal blood flow and glomerular filtration rate, which is exactly the opposite effect to that which occurs in the normal pregnant woman. Also, a fibrinoid deposit occurs in the basement membrane of the glomerular tufts, thus thickening the glomerular membrane and presumably reducing the glomerular filtration coefficient.

Various attempts have been made to prove that toxemia of pregnancy is caused by excessive secretion of placental or adrenal hormones, but proof of a hormonal basis for toxemia is yet completely lacking. Indeed, it is now believed that toxemia of pregnancy is not caused by abnormal hormonal balances. A more plausible theory is that toxemia of pregnancy results from some type of autoimmunity or allergy resulting from the presence of the fetus. Indeed, the acute symptoms disappear within a few days after birth of the baby.

Because of the diminished glomerular filtration rate in the kidneys, one of the major problems of toxemia is retention of salt and water. Therefore, it is a dictum among obstetricians that any pregnant woman who has a tendency toward toxemia must limit her salt intake. It usually is not necessary to limit the mother's water intake, for limitations of salt alone prevents excessive absorption of water by the kidney tubules for reasons discussed in Chapter 38.

**Eclampsia.** Eclampsia is a severe degree of toxemia that occurs in one out of several hundred pregnancies. Milder degrees of eclampsia are sometimes called *pre-eclampsia*.

Eclampsia is characterized by extreme vascular spasticity throughout the body, clonic convulsions followed by coma, greatly decreased kidney output, malfunction of the liver, hypertension, and a generalized toxic condition of the body. Usually, it occurs shortly before, or sometimes within a day or so, after parturition.

Even with the best treatment, some 5 per cent of eclamptic mothers still die. However, injection of vasodilator drugs plus dehydration of the patient can often reverse the vascular spasm and lower the blood pressure, bringing the mother out of the eclamptic state.

# PARTURITION

## INCREASED UTERINE IRRITABILITY NEAR TERM

Parturition means simply the process by which the baby is born. At the termination of pregnancy the uterus becomes progressively more excitable until finally it begins strong rhythmic contractions with such force that the baby is expelled. The exact cause of the increased activity of the uterus is not known, but at least two major categories of effects lead up to the culminating contractions responsible for parturition; these are, first, progressive hormonal changes that cause increased excitability of the uterine musculature, and, second, progressive mechanical changes.

**Hormonal Factors That Cause Increased Uterine Contractility.** *Ratio of Estrogens to Progesterone.* Progesterone inhibits uterine contractility during pregnancy, thereby helping to prevent expulsion of the fetus. On the other hand, estrogens have a definite tendency to increase the degree of uterine contractility. Both these hormones are secreted in progressively greater quantities throughout pregnancy, but from the seventh month onward estrogen secretion increases more than progesterone secretion, and immediately before term relatively large quantities of nonconjugated estrogens ("free estrogens") appear in the extracellular fluids. Therefore, it has been postulated that the *estrogen to progesterone ratio* increases sufficiently toward the end of pregnancy to be at least partly responsible for the increased contractility of the uterus.

*Effect of Oxytocin on the Uterus.* Oxytocin is a hormone secreted by the neurohypophysis that specifically causes uterine contraction (see Chap. 75). There are four reasons for believing that oxytocin might be particularly important in increasing the contractility of the uterus near term. (1) The uterus increases its responsiveness to a given dose of oxytocin during the latter few months of pregnancy. (2) The rate of oxytocin secretion by the neurohypophysis, as judged from thus far rather incomplete studies, seems to be considerably increased at the time of labor. (3) Though hypophysectomized animals and human beings can still deliver their young at term, labor is prolonged. (4) Recent experiments in animals indicate that irritation or stretching of the uterine cervix, as occurs during labor, can cause a neurogenic reflex to the neurohypophysis to increase the rate of oxytocin secretion.

**Mechanical Factors that Increase the Contractility of the Uterus.** *Stretch of the Uterine Musculature.* Simply stretching smooth muscle organs usually increases their contractility. Furthermore, intermittent stretch, as occurs repetitively in the uterus because of movements of the fetus, can also elicit smooth muscle contraction.

Note especially that twins are born on the average *19 days* earlier than a single child, which emphasizes the importance of mechanical stretch in eliciting uterine contractions.

*Stretch or Irritation of the Cervix.* There is much reason to believe that stretch or irritation of the uterine cervix is particularly important in eliciting uterine contractions. For instance, the obstetrician frequently induces labor by rupturing the membranes so that the head of the baby stretches the cervix more forcefully than usual or irritates it in some other way.

The mechanism by which cervical irritation excites the body of the uterus is not known. It has been supposed that stretch or irritation of neuronal cells in the cervix initiates reflexes to the body of the uterus, but the effect could also result simply from myogenic transmission of signals from the cervix to the body of the uterus.

## ONSET OF LABOR—A POSITIVE FEEDBACK THEORY FOR ITS INITIATION

During most of the months of pregnancy the uterus undergoes periodic episodes of weak and slow rhythmic contractions called *Braxton-Hicks contractions*. These become progressively stronger toward the end of pregnancy; and they eventually change rather suddenly, within hours, to become exceptionally strong contractions that start stretching the cervix and later force the baby through the birth canal, thereby causing parturition. This process is called *labor,* and the strong contractions that result in final parturition are called *labor contractions.*

Yet, strangely enough, we do not know what suddenly changes the slow and weak rhythmicity of the uterus into the strong labor contractions. Most obstetricians feel that one of the factors discussed above that increase uterine contractility suddenly becomes strong enough to promote these contractions. However, on the basis of experience during the past few years with other types of control systems, a theory has been proposed for explaining the onset of labor based on "positive feedback." This theory suggests that stretch of the cervix by the fetus' head finally becomes great enough to elicit a reflex increase in contractility of the uterine body. This pushes the baby forward, which stretches the cervix some more and initiates a new cycle. Thus, the process continues again and again until the baby is expelled. This theory is illustrated in Figure 82–8, and the data supporting it are the following:

First, labor contractions obey all the principles of positive feedback. That is, once the strength of uterine contraction becomes greater than a critical value, each contraction leads to

subsequent contractions that become stronger and stronger until maximum effect is achieved. Referring to the discussion in Chapter 1 of positive feedback in control systems, we see that this is the precise nature of all positive feedback mechanisms with a feedback gain of more than unity.

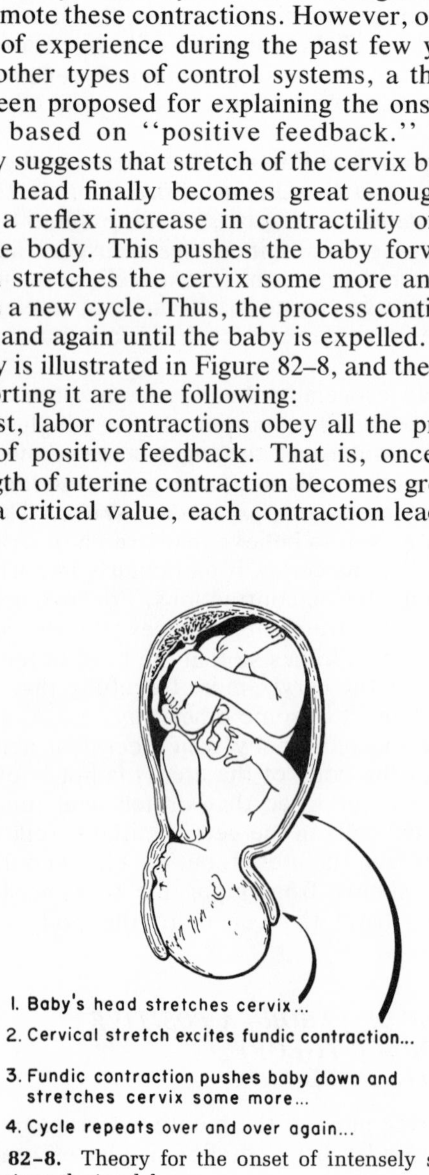

Second, the next problem is to explain how positive feedback could set off the crescendo of strength of the uterine contractions. We have two possible types of positive feedback in the uterus at term. First, the propensity for cervical irritation to cause contraction of the body of the uterus could cause positive feedback in the following way: An initial contraction of the uterine body could force the fetus' head or the fluid in the amniotic cavity against the cervix to stretch, tear, or irritate it. This stimulus of the cervix could in turn lead to feedback that would cause additional contraction of the uterine body. As a result, the next contraction would become stronger, and the process would proceed on and on with progressively stronger contractions until delivery.

A second positive feedback mechanism involves the secretion of oxytocin by the posterior pituitary gland. On the basis of studies in cows, this results from the following sequence of events: (a) uterine contraction, (b) cervical stretch or other type of cervical stimulation, (c) nerve signals from the cervix to the hypothalamic-pituitary axis, (d) increased secretion of oxytocin, and (e) increased contractility of the uterus, which leads to (f) a vicious cycle of increasing contraction.

Third, we now have to explain why the weak Braxton-Hicks contractions of midpregnancy do not lead to labor. Referring once more to the discussion of positive feedback in Chapter 1, we find that positive feedback causes a vicious cycle that goes to completion only when each feedback response is greater than the previous response. Thus, if each *increase* in strength of uterine contraction leads to a greater *increase* in strength of contraction the next time, the process would go to completion, but, if the second *increase* in contraction is not greater than the first increase, the process would fade out. Therefore, the degree of contractility of the uterus must reach a certain stage of development before the positive feedback can become a vicious cycle. Up to that time, the natural contractions can continue without developing into frank labor contractions.

1. Baby's head stretches cervix...

2. Cervical stretch excites fundic contraction...

3. Fundic contraction pushes baby down and stretches cervix some more...

4. Cycle repeats over and over again...

**Figure 82–8.** Theory for the onset of intensely strong contractions during labor.

To summarize the theory, we can assume that multiple factors increase the contractility of the uterus toward the end of pregnancy. These are additive in their effects and they cause the Braxton-Hicks contractions to become progressively stronger. Eventually, one of these becomes strong enough that the contraction itself irritates the uterus, increases its contractility because of positive feedback, and results in a second contraction stronger than the first, and a third stronger than the second, and so forth. Once these contractions become strong enough to cause this type of feedback, with each succeeding contraction greater than the one preceding, the process proceeds to completion—all simply *because positive feedback becomes a vicious cycle when the gain of the feedback is greater than unity.*

One might immediately ask about the many instances of false labor in which the contractions become stronger and stronger and then fade away. Remember that for a vicious cycle to continue, *each* new cycle of contraction must be stronger than the previous one. If, at any time after labor starts, some cycles fail to reexcite the uterus sufficiently, the positive feedback could go into a retrograde succession and the labor contractions would fade away.

## ABDOMINAL CONTRACTION DURING LABOR

Once labor contractions become strong and painful, neurogenic reflexes, mainly from the birth canal to the spinal cord and thence back to the abdominal muscles, cause intense abdominal contraction. This additional contraction of the abdominal muscles and the reflexes causing it add greatly to the uterine contraction that promotes expulsion of the fetus.

## MECHANICS OF PARTURITION

The uterine contractions during labor begin at the top of the uterine fundus and spread downward over the body of the uterus. Also, the intensity of contraction is great in the top and body of the uterus but weak in the lower segment of the uterus adjacent to the cervix. Therefore, each uterine contraction tends to force the baby downward toward the cervix.

In the early part of labor, the contractions might occur only once every 30 minutes. As labor progresses, the contractions finally appear as often as once every one to three minutes, and the intensity of contraction increases greatly with only a short period of relaxation between contractions.

The combined contractions of the uterine and abdominal musculature during delivery of the baby cause downward force on the fetus of approximately 25 pounds during each strong contraction.

It is fortunate that the contractions of labor occur intermittently because strong contractions impede or sometimes even stop blood flow through the placenta and would cause death of the fetus were the contractions continuous. Indeed, in clinical use of various uterine stimulants, such as oxytocin, overuse of the drugs can cause uterine spasm rather than rhythmic contractions and can lead to death of the fetus.

In 19 out of 20 births the head is the first part of the baby to be expelled, and in most of the remaining instances the buttocks are presented first. The head acts as a wedge to open the structures of the birth canal as the fetus is forced downward from above.

The first major obstruction to expulsion of the fetus is the uterine cervix. Toward the end of pregnancy the cervix becomes soft, which allows it to stretch when labor pains cause the body of the uterus to contract. The so-called *first stage of labor* is the period of progressive cervical dilatation, lasting until the opening is as large as the head of the fetus. This stage usually lasts 8 to 24 hours in the first pregnancy but often only a few minutes after many pregnancies.

Once the cervix has dilated fully, the fetus' head moves rapidly into the birth canal, and, with additional force from above, continues to wedge its way through the canal until delivery is effected. This is called the *second stage of labor,* and it may last from as little as a minute after many pregnancies up to half an hour or more in the first pregnancy.

## SEPARATION AND DELIVERY OF THE PLACENTA

During the succeeding 10 to 45 minutes after birth of the baby, the uterus contracts to a very small size, which causes a *shearing* effect between the walls of the uterus and the placenta, thus separating the placenta from its implantation site. Obviously, separation of the placenta opens the placental sinuses and causes bleeding. However, the amount of bleeding is limited to an average of 350 ml. by the following mechanism: The smooth muscle fibers of the uterine musculature are arranged in figures of 8 around the blood vessels as they pass through the uterine wall. Therefore, contraction of the uterus following delivery of the baby constricts the vessels that had previously supplied blood to the placenta.

## LABOR PAINS

With each uterine contraction the mother experiences considerable pain. The pain in early labor is probably caused mainly by hypoxia of the uterine muscle resulting from compression of the blood vessels to the uterus. This pain is not felt when the *hypogastric nerves,* which carry the sensory fibers leading from the uterus, have been sectioned. However, during the second stage of labor, when the fetus is being expelled through the birth canal, much more severe pain is caused by cervical stretch, perineal stretch, and stretch or tearing of structures in the vaginal canal itself. This pain is conducted by somatic nerves instead of by the hypogastric nerves.

## INVOLUTION OF THE UTERUS

During the first four to five weeks following parturition, the uterus involutes. Its weight becomes less than one-half its immediate postpartum weight within a week, and in four weeks the uterus may be as small as it had been prior to pregnancy—that is, if the mother lactates. This effect of lactation is discussed later. During early involution of the uterus the placental site on the endometrial surface autolyzes, causing a vaginal discharge known as "lochia," which is first bloody and then serous in nature, continuing in all for approximately a week and a half. After this time, the endometrial surface will have become reepithelialized and ready for normal, nongravid sex life agian.

# LACTATION

## DEVELOPMENT OF THE BREASTS

The breasts begin to develop at puberty; this development is stimulated by the estrogens of the monthly sexual cycles that stimulate growth of the stroma and ductile system plus deposition of fat to give mass to the breasts. However, much additional growth occurs during pregnancy, and the glandular tissue only then becomes completely developed for actual production of milk.

**Growth of the Ductile System—Role of the Estrogens.** All through pregnancy, the tremendous quantities of estrogens secreted by the placenta cause the ductile system of the breasts to grow and to branch. Simultaneously, the stroma of the breasts also increases in quantity, and large quantities of fat are laid down in the stroma.

Also, moderate quantities of growth hormone from the pituitary or *human placental lactogen* from the placenta are required for the estrogens to produce their effect on the breasts. These latter two hormones both cause protein deposition in the glandular cells, which is essential to the glandular growth.

**Development of the Lobule-Alveolar System—Role of Progesterone.** The synergistic action of estrogens and growth hormone can cause only a primitive lobule-alveolar system to develop in the breasts at the same time that the ducts are growing, but the additional action of progesterone causes growth of the lobules, budding of alveoli, and development of secretory characteristics in the cells of the alveoli. These changes, obviously, are analogous to the secretory effects of progesterone on the endometrium of the uterus.

**Function of Prolactin and Other Hormones in Development of the Breasts.** Prolactin is the hormone most probably concerned with causing milk secretion after birth of the baby. In lower animals prolactin has a powerful synergistic effect with estrogens and progesterone to stimulate development of the alveolar secretory system of the breast. Therefore, it is assumed that prolactin might play a similar role in the human being in aiding the final development of the breasts for lactation. It is possible that human placental lactogen from the placenta might also play such a role during pregnancy. This hormone has physiologic properties almost the same as those of prolactin, though very little is yet known about it.

In addition to prolactin, at least three of the general hormones are necessary to provide appropriate background metabolism before the breasts will develop. These are thyroid hormone, the adrenal corticosteroids, and insulin. However, these hormones seem to have only a "permissive" effect rather than a specific effect in the development of the breasts.

## INITIATION OF LACTATION— PROLACTIN

By the end of pregnancy, the mother's breasts are fully developed for nursing, but only a few milliliters of fluid are secreted each day until after the baby is born. This fluid is called *colostrum;* it contains essentially the same amounts of proteins and lactose as milk but almost no fat, and its maximum rate of production is about $1/_{100}$ the subsequent rate of milk production.

The absence of lactation during pregnancy is believed to be caused by suppressive effects of progesterone and estrogens on the milk secre-

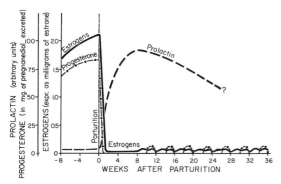

**Figure 82–9.** Changes in rates of secretion of estrogens, progesterone, and prolactin at parturition and during the succeeding weeks after parturition, showing especially the rapid increase in prolactin secretion immediately after parturition.

tory process of the breasts and also suppression of the secretion of prolactin by the pituitary.

However, immediately after the baby is born, the sudden loss of both estrogen and progesterone secretion by the placenta removes any inhibitory effects of these two hormones and presumably allows marked production of prolactin by the pituitary, as illustrated in Figure 82–9. The prolactin stimulates synthesis of large quantities of fat, lactose, and casein by the mammary glandular cells, and within two to three days the breasts begin to secrete copious quantities of milk instead of colostrum. The sudden onset of milk secretion requires, in addition to prolactin, an adequate background secretion of both growth hormone and the adrenal corticosteroids. However, only prolactin has the specific effect of causing true lactation.

(Unfortunately, prolactin has not yet been isolated in the human being. Therefore, the details of prolactin mechanisms must still be inferred from studies in lower animals. Yet, it has been shown that prolactin has an extremely powerful effect to cause lactation in monkeys, requiring little if any supportive effect from other hormones.)

**Hypothalamic Control of Prolactin Secretion.** Though secretion of most of the anterior pituitary hormones is enhanced by neurosecretory releasing factors transmitted from the hypothalamus to the anterior pituitary gland through the hypothalamic-hypophyseal portal system, the secretion of prolactin is controlled by an exactly opposite effect. That is, the hypothalamus synthesizes a prolactin inhibitory factor (PIF). Under normal conditions, large amounts of PIF are continually transmitted to the anterior pituitary gland so that the normal

rate of prolactin secretion is slight. However, during lactation the formation of PIF itself is suppressed, thereby allowing the anterior pituitary gland to secrete an uninhibited amount of prolactin. (It is interesting to note that when the pituitary stalk is cut so that neurosecretory hormones cannot be transmitted from the hypothalamus to the anterior pituitary gland, the secretion of essentially all the other hormones by the anterior pituitary gland besides prolactin is greatly depressed, while at the same time the rate of secretion of prolactin is increased several-fold.)

### THE EJECTION OR "LET-DOWN" PROCESS IN MILK SECRETION— FUNCTION OF OXYTOCIN

Milk is secreted continuously into the alveoli of the breasts, but milk does not flow easily from the alveoli into the ductile system and therefore does not continually leak from the breast nipples. Instead, the milk must be "ejected" or "let-down" from the alveoli to the ducts before the baby can obtain it. This process is caused by a combined neurogenic and hormonal reflex involving the hormone *oxytocin* as follows:

When the baby suckles the breast, sensory impulses are transmitted through somatic nerves to the spinal cord and then to the hypothalamus, there causing *oxytocin* and, to a lesser extent, *vasopressin* secretion, as described in Chapter 75. These two hormones, principally oxytocin, flow in the blood to the breasts where they cause the *myoepithelial cells* that surround the outer walls of the alveoli to contract, thereby expressing the milk from the alveoli into the ducts. Thus, within 30 seconds to a minute after a baby begins to suckle the breast, milk begins to flow. This process is called milk ejection, or milk let-down.

Suckling on one breast causes milk flow not only in that breast but also in the opposite breast. Also, it is especially interesting that the sound of the baby crying is often enough of a signal to cause milk ejection.

**Inhibition of Milk Ejection.** A particular problem in nursing the baby comes from the fact that many psychogenic factors as well as generalized sympathetic stimulation throughout the body can inhibit oxytocin secretion and consequently depress milk ejection. For this reason, the mother must have an undisturbed puerperium if she is to be successful in nursing her baby.

## CONTINUED SECRETION OF MILK FOR MANY MONTHS FOLLOWING PREGNANCY

If milk is not continually removed from the mother's breasts, the ability of the breasts to continue secreting milk is lost within one to two weeks. This is caused by cessation of prolactin secretion by the andenohypophysis. To state this another way, continued milking of the breasts causes continued secretion of prolactin and therefore continued lactation.

The stimulus that causes continued secretion of prolactin seems to be the suckling of the breasts, which, as is true in the milk ejection process, presumably results from impulses transmitted to the hypothalamus to suppress the release of *prolactin-inhibiting factor*. Decrease of this inhibitory substance allows a higher level of prolactin secretion by the anterior pituitary gland.

Milk production can continue for several years if the mother continues to be suckled, but the rate of secretion normally decreases rather markedly by seven to nine months.

## EFFECT OF LACTATION ON THE UTERUS AND ON THE SEXUAL CYCLE

The uterus involutes after parturition far more rapidly in women who lactate than in women who do not lactate. This difference probably results from greatly diminished estrogen secretion by the ovaries during the period of lactation, for estrogens are known to enlarge the uterus and presumably also to prevent rapid uterine involution. The uterus of a lactating mother usually decreases to a size even smaller than that prior to pregnancy, whereas the uterus of a nonlactating mother is likely to remain considerably larger than the pregravid size and also to remain soft for many months.

Lactation usually prevents the sexual cycle for one to five months. Presumably this is caused by preoccupation of the pituitary with production of prolactin, which reduces the rate of secretion of the other gonadotropic hormones. However, after several months of lactating, the pituitary usually begins once again to produce sufficient quantities of follicle-stimulating hormone and luteinizing hormone to re-initiate the monthly sexual cycle. The rhythmic interplay between the ovarian and pituitary hormones during the sexual month does not necessitate marked reduction in prolactin secretion.

## MILK AND THE METABOLIC DRAIN ON THE MOTHER CAUSED BY LACTATION

Table 82–1 gives the contents of human milk and cow's milk. The concentration of lactose in human

**TABLE 82–1.   Percentage Composition of Milk**

|  | Human Milk | Cow's Milk |
|---|---|---|
| Water | 88.5 | 87 |
| Fat | 3.3 | 3.5 |
| Lactose | 6.8 | 4.8 |
| Casein | 0.9 | 2.7 |
| Lactalbumin and other protein | 0.4 | 0.7 |
| Ash | 0.2 | 0.7 |

milk is approximately 50 per cent greater than that in cow's milk, but on the other hand the concentration of protein in cow's milk is ordinarily two or more times as great as that in human milk. Finally, the ash, which contains the minerals, is only one-third as much in human milk as in cow's milk.

At the height of lactation 1.5 liters of milk may be formed each day. With this degree of lactation great quantities of metabolic substrates are drained from the mother. For instance, approximately 50 grams of fat enter the milk each day, and approximately 100 grams of lactose, which must be derived from glucose, are lost from the mother each day. Also, some 2 to 3 grams of calcium phosphate may be lost each day, and, unless the mother is drinking large quantities of milk and has an adequate intake of vitamin D, the output of calcium and phosphate by the lactating mammae will be much greater than the intake of these substances. To supply the needed calcium and phosphate, the parathyroid glands enlarge greatly, and the bones become progressively decalcified. The problem of decalcification is usually not very great during pregnancy, but it can be a distinct problem during lactation.

## REFERENCES

Aref, I., and Hafez, E. S.: Utero-oviductal motility with emphasis on ova transport. *Obstet. Gynecol. Surv.,* 28:679, 1973.
Bedford, C. A., Challis, J. R., Harrison, F. A., and Heap, R. B.: The role of estrogens and progesterone in the onset of parturition in various species. *J. Reprod. Fertil. Suppl.* 16:1, 1972.
Behrman, H. R., and Caldwell, B. V.: Role of prostaglandins in reproduction. *In* Guyton, A. C. (ed.): MTP International Review of Science: Physiology. Vol. 8. Baltimore, University Park Press, 1974, p. 63.
Bindon, B. M.: Pituitary and ovarian mechanisms in implantation. *J. Reprod. Fertil. Suppl.* 18:167, 1973.
Bisset, G. W.: Milk ejection. *In* Greep, R. O., and Astwood, E. B. (eds.): Handbook of Physiology. Sec. 7, Vol. 4, Part 1. Baltimore, The Williams & Wilkins Company, 1974, p. 493.
Brackett, B. G.: Mammalian fertilization in vitro. *Fed. Proc.,* 32:2065, 1973.
Brenner, R. M., and West, N. B.: Hormonal regulation of the reproductive tract in female mammals. *Ann. Rev. Physiol.,* 37:273, 1975.
Brinster, R. L.: Nutrition and metabolism of the ovum, zygote, and blastocyst. *In* Greep, R. O., and Astwood, E. B. (eds.): Handbook of Physiology. Sec. 7, Vol. 2, Part 2. Baltimore, The Williams & Wilkins Company, 1973, p. 165.

Chard, T.: The posterior pituitary in human and animal parturition. *J. Reprod. Fertil. Suppl.* 16:121, 1972.

Coutinho, E. M., and Fuchs, F. (eds.): Physiology and Genetics of Reproduction. Basic Life Sciences Series. Vol. 4. New York, Plenum Publishing Corporation, 1974.

Cowie, A. T., and Tindal, J. S.: The Physiology of Lactation. Baltimore, The Williams & Wilkins Company, 1971.

Cross, B. A., and Dyball, R. E. J.: Central pathways for neurohypophyseal hormone release. *In* Greep, R. O., and Astwood, E. B. (eds.): Handbook of Physiology, Sec. 7, Vol. 4, Part. 1. Baltimore, The Williams & Wilkins Company, 1974, p. 269.

Finn, C. A.: Recent research on implantation in animals. *Proc. R. Soc. Med.,* 67:927, 1974.

Finn, C. A., and Martin, L.: The control of implantation. *J. Reprod. Fertil.* 39:195, 1974.

Friesen, H. G.: Placental protein and polypeptide hormones. *In* Greep, R. O., and Astwood, E. B. (eds.): Handbook of Physiology. Sec. 7. Vol. 2, Part 2. Baltimore, The Williams & Wilkins Company, 1973, p. 295.

Gould, K. G.: Application of in vitro fertilization. *Fed. Proc.,* 32:2069, 1973.

Hafez, E. S. E., and Evans, T. N.: Human Reproduction. Hagerstown, Maryland, Harper & Row, Publishers, 1973.

Hansen, J. M., and Ueland, K.: Maternal cardiovascular dynamics during pregnancy and parturition. *Clin. Anesth.,* 10:21, 1974.

Heap, R. B., Perry, J. S., and Challis, J. R. G.: The hormonal maintenance of pregnancy. *In* Greep, R. O., and Astwood, E. B. (eds.): Handbook of Physiology. Sec. 7, Vol. 2, Part 2. Baltimore, The Williams & Wilkins Company, 1973, p. 217.

Hertig, A. T., and Barton, B. R.: Fine structure of mammalian oocytes and ova. *In* Greep, R. O., and Astwood, E. B. (eds.): Handbook of Physiology. Sec. 7, Vol. 2, Part 1. Baltimore, The Williams & Wilkins Company, 1973, p. 317.

Hytten, F. E., and Leitch, I.: The Physiology of Human Pregnancy. Philadelphia, J. B. Lippincott Company, 1974.

Josimovich, J. B.: Placental protein hormones in pregnancy. *Clin. Obstet. Gynecol.,* 16:46, 1973.

Josimovich, J. B., Cobo, E., and Reynolds, M. (eds.): Lactogenic Hormones, Fetal Nutrition, and Lactation. New York, John Wiley & Sons, Inc., 1974.

Klopper, A.: The hormones of the placenta and their role in the onset of labor. *In* Guyton, A. C. (ed.): MTP International Review of Science: Physiology. Vol. 8, Baltimore, University of Park Press, 1974, p. 179.

Li, C. H.: Chemistry of ovine prolactin. *In* Greep, R. O., and Astwood, E. B. (eds.): Sec. 7, Vol. 4, Part 2. Baltimore, The Williams & Wilkins Company, 1974, p. 103.

Lincoln, D. W.: Maternal hypothalamic control of labor. *Prog. Brain Res.,* 41:289, 1974.

Linzell, J. L., and Peaker, M.: Mechanism of milk secretion. *Physiol. Rev.,* 51:564, 1971.

Longo, L. D.: Placental transfer mechanisms—an overview. *Obstet. Gynecol. Ann.,* 1:103, 1972.

McLaren, A.: Endocrinology of implantation. *J. Reprod. Fertil., Suppl.* 18:159, 1973.

Marshall, J. M.: Effects of neurohypophyseal hormones on the myometrium. *In* Greep, R. O., and Astwood, E. B. (eds.): Handbook of Physiology. Sec. 7, Vol. 4, Part 1. Baltimore, The Williams & Wilkins Company, 1974, p. 469.

Miller, R. K., and Berndt, W. O.: Mechanisms of transport across the placenta: an in vitro approach. *Life Sci.,* 16:7, 1975.

Mittwoch, U.: Genetics of Sex Differentiation. New York, Academic Press, Inc., 1973.

Moghissi, K. S., and Hafez, E. S. E. (eds.): The Placenta. Springfield, Ill., Charles C Thomas, Publisher, 1974.

Neill, J. D.: Prolactin: its secretion and control. *In* Greep, R. O., and Astwood, E. B. (eds.): Handbook of Physiology. Sec. 7, Vol. 4, Part 2. Baltimore, The Williams & Wilkins Company, 1974, p. 469.

Nicoll, C. S.: Physiological actions of prolactin. *In* Greep, R. O., and Astwood, E. B. (eds.): Handbook of Physiology. Sec. 7, Vol. 4, Part 2. Baltimore, The Williams & Wilkins Company, 1974, p. 253.

Nilsson, O.: The morphology of blastocyst implantation. *J. Reprod. Fertil.,* 39:187, 1974.

Ostergard, D. R.: Estriol in pregnancy. *Obstet. Gynecol. Surv.,* 28:215, 1973.

Page, E. W., Villee, C. A., and Villee, D. B.: Human Reproduction. Philadelphia, W. B. Saunders Company, 1972.

Parkes, A. S.: Marshall's Physiology of Reproduction, 3rd Ed., Vols. 1–3. Boston, Little, Brown and Company, 1956–1966.

Pasteels, J. L., and Robyn, C. (eds.): Human Prolactin. New York, American Elsevier Publishing Company, 1973.

Pitkin, R. M.: Calcium metabolism in pregnancy: a review. *Am. J. Obstet. Gynecol.,* 121:724, 1975.

Psychoyos, A.: Endocrine control of egg implantation. *In* Greep, R. O., and Astwood, E. B. (eds.): Handbook of Physiology. Sec. 7, Vol. 2, Part 2. Baltimore, The Williams & Wilkins Company, 1973, p. 187.

Ramsey, E. M.: Placental vasculature and circulation. *In* Greep, R. O., and Astwood, E. B. (eds.): Handbook of Physiology. Sec. 7, Vol. 2, Part 2. Baltimore, The Williams & Wilkins Company, 1973, p. 323.

Ryan, K. J.: Steroid hormones in mammalian pregnancy. *In* Greep, R. O., and Astwood, E. B. (eds.): Handbook of Physiology. Sec. 7, Vol. 2, Part 2. Baltimore, The Williams & Wilkins Company, 1973, p. 285.

Schally, A. V., Kastin, A. J., and Arimura, A.: The hypothalamus and reproduction. *Amer. J. Obstet. Gynecol.,* 114:423, 1972.

Sulman, F. G.: Hypothalamic control of lactation. *Monogr. Endocrinol.,* 3:1, 1970.

Thomson, A. M., and Hytten, F. E.: Nutrition during pregnancy. *World Rev. Nutr. Diet,* 16:23, 1973.

Tindal, J. S.: Reflex pathways controlling lactation. *Proc. R. Soc. Med.,* 65:1085, 1972.

Tindal, J. S.: Stimuli that cause the release of oxytocin. *In* Greep, R. O., and Astwood, E. B. (eds.): Handbook of Physiology. Sec. 7, Vol. 4, Part 1. Baltimore, The Williams & Wilkins Company, 1974, p. 257.

Wild, A. E.: Protein transport across the placenta. *Symp. Soc. Exp. Biol.,* 28:521, 1974.

# 83

# Special Features of Fetal and Neonatal Physiology

A complete discussion of fetal development, function of the child immediately after birth, and growth and development through the early years of life lies in the province of formal courses in obstetrics and pediatrics. However, many aspects of these are strictly physiologic problems, some of which relate to the physiologic principles that we have discussed for the adult and some of which are peculiar to the infant itself. The present chapter delineates and discusses some of the more important of these special problems.

## GROWTH AND FUNCTIONAL DEVELOPMENT OF THE FETUS

Early development of the placenta and of the fetal membranes occurs far more rapidly than development of the fetus itself. During the first two to three weeks the fetus remains almost microscopic in size, but thereafter, as illustrated in Figure 83–1, the dimensions of the fetus increase almost in proportion to age. At 12 weeks the length of the fetus is approximately 10 cm.; at 20 weeks, approximately 25 cm.; and at term (40 weeks), approximately 53 cm. (about 21 inches). Because the weight of the fetus is proportional to the cube of the length, the weight increases approximately in proportion to the cube of the age of the fetus. Note from Figure 83–1 that the weight of the fetus remains almost nothing during the first months and reaches 1 pound only at five and a half months of gestation. Then, during the last trimester of pregnancy, the fetus gains tremendously so that two months prior to birth the weight averages 3 pounds, one month prior to birth 4.5 pounds, and at birth 7 pounds, this birth weight varying from as low as 4.5 pounds to as high as 11 pounds in completely normal infants with completely normal gestational periods.

## DEVELOPMENT OF THE ORGAN SYSTEMS

Within one month after fertilization of the ovum all the different organs of the fetus have already been "blocked out," and during the next two to three months the minute details of the different organs are established. Beyond the fourth month, the organs of the fetus are grossly the same as those of the newborn child, even including most of the smaller structures of the organs. However, cellular development of these structures is usually far from complete at this time and requires the full remaining five months of pregnancy for complete development. Even at birth certain structures, particularly the nervous system, the kidneys, and the liver, still lack full development, as is discussed in more detail later in the chapter.

**The Circulatory System.** The human heart begins beating during the fourth week following fertilization, contracting at the rate of about 65 beats per minute. This increases steadily as the fetus grows and reaches a rate of approximately 140 beats per minute immediately before birth.

*Formation of Blood Cells.* Nucleated red blood cells begin to be formed in the yoke sac and mesothe-

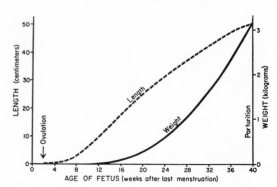

**Figure 83–1.** Growth of the fetus.

lial layers of the placenta at about the third week of fetal development. This is followed a week later by the formation of non-nucleated red blood cells by the fetal mesenchyme and by the endothelium of the fetal blood vessels. Then at approximately six weeks, the liver begins to form blood cells, and in the third month the spleen and other lymphoid tissues of the body also begin forming blood cells. Finally, from approximately the third month on, the bone marrow also forms red and white blood cells. During the midportion of fetal life, the extramarrow areas are the major sources of the fetus' blood cells, but, during the latter three months of fetal life, the bone marrow gradually takes over while these other structures lose their ability completely to form blood cells.

An especially interesting characteristic of fetal blood is that it contains an entirely different type of hemoglobin from that of adult blood, called *fetal hemoglobin*. Fetal hemoglobin combines with oxygen at a considerably lower $Po_2$ than does adult hemoglobin. This allows the fetal blood to carry as much as 30 per cent more oxygen in low $Po_2$ ranges than can adult hemoglobin. This is of special importance in the fetus because its arterial $Po_2$ is always low, as was discussed in the previous chapter.

**The Respiratory System.** Obviously, respiration cannot occur during fetal life. However, respiratory movements do take place beginning at the end of the first trimester of pregnancy. Tactile stimuli or fetal asphyxia especially cause respiratory movements.

However, during the latter three to four months of pregnancy, the respiratory movements of the fetus are mainly inhibited, for reasons unknown. This could possibly result from (1) special chemical conditions in the body fluids of the fetus, (2) the presence of fluid in the fetal lungs, or (3) other possible unexplored stimuli.

The inhibition of respiration during the latter months of fetal life prevents filling of the lungs with debris from the meconium excreted by the gastrointestinal tract into the amniotic fluid. Also, fluid is secreted into the lungs by the alveolar epithelium up until the moment of birth, thus filling the pulmonary spaces with this clean secretion.

## FUNCTION OF THE NERVOUS SYSTEM

Most of the peripheral reflexes of the fetus are well formed by the third to fourth months of pregnancy. However, some of the more important higher functions of the central nervous system are still undeveloped even at birth. Indeed, myelinization of some major tracts of the central nervous system becomes complete only after approximately a year of postnatal life.

## FUNCTION OF THE GASTROINTESTINAL TRACT

Even in midpregnancy the fetus ingests and absorbs large quantities of amniotic fluid, and during the latter two to three months, gastrointestinal function approaches that of the normal newborn infant. Small quantities of *meconium* are continually formed in the gastrointestinal tract and excreted from the bowels into the amniotic fluid. Excretion occurs particularly when the gastrointestinal tract becomes overly active as a consequence of fetal asphyxia. Meconium is composed partly of unabsorbed residue of amniotic fluid and partly of excretory products from the gastrointestinal mucosa and glands.

## FUNCTION OF THE KIDNEYS

The fetal kidneys are capable of excreting urine during at least the latter half of pregnancy, and urination occurs normally *in utero*. However, the renal control systems for regulation of extracellular fluid electrolyte balances and acid-base balance are almost nonexistent until after mid-fetal life and do not reach full development until about a month after birth.

## METABOLISM IN THE FETUS

The fetus utilizes mainly glucose for energy, and it has a high rate of storage of fat and protein, much if not most of the fat being synthesized from glucose rather than being absorbed from the mother's blood. Aside from these generalities there are some special problems of fetal metabolism in relation to calcium, phosphate, iron, and some vitamins as follows:

**Metabolism of Calcium and Phosphate.** Figure 83-2 illustrates the rate of calcium and phosphate accumulation in the fetus, showing that approximately 22.5 grams of calcium and 13.5 grams of phosphorus are accumulated in the average fetus during gestation. Approximately half of this accumulates during the last four weeks of gestation, which is also coincident with the period of rapid ossification of the fetal bones as well as with the period of rapid weight gain of the fetus.

During the earlier part of fetal life, the bones are relatively unossified, and have mainly a cartilaginous

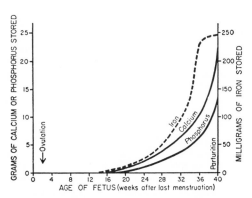

**Figure 83-2.** Calcium, phosphorus, and iron storage in the fetus at different stages of gestation.

matrix. Indeed, x-ray pictures ordinarily will not show ossification until approximately the fourth month of pregnancy.

Note especially that the total amounts of calcium and phosphate needed by the fetus during gestation represent only about one-fiftieth the quantities of these substances in the mother's bones. Therefore, this is a minimal drain from the mother. However, a great drain occurs after birth during lactation.

**Accumulation of Iron.** Figure 83–2 shows that iron accumulates in the fetus somewhat more rapidly than calcium and phosphates. Most of the iron is in the form of hemoglobin, which begins to be formed as early as the third week following fertilization of the ovum.

Small amounts of iron are concentrated in the progestational endometrium even prior to implantation of the ovum; this iron is ingested into the embryo by the trophoblastic cells for early formation of the red blood cells.

Approximately one-third of the iron in a fully developed fetus is normally stored in the liver. This iron can then be used for several months after birth by the newborn infant for formation of additional hemoglobin.

**Utilization and Storage of Vitamins.** The fetus needs vitamins equally as much as the adult and in some instances to a far greater extent. In general, the vitamins function the same in the fetus as in the adult, as discussed in Chapter 74. Special functions of several vitamins should be mentioned, however.

The B vitamins, especially vitamin $B_{12}$ and folic acid, are necessary for formation of red blood cells and also for overall growth of the fetus.

Vitamin C is necessary for appropriate formation of intercellular substances, especially the bone matrix and fibers of connective tissue.

Vitamin D probably is not necessary for fetal growth, although the mother needs it for adequate absorption of calcium from her gastrointestinal tract. If the mother has plenty of this vitamin in her body fluids, large quantities will be stored by the fetal liver to be used by the newborn child for several months after birth.

Vitamin E, though its precise function is unknown, is necessary for normal development of the early ovum. In its absence in experimental animals, spontaneous abortion usually occurs at an early age.

Vitamin K is used by the fetal liver for formation of factor VII and prothrombin. When vitamin K is insufficient in the mother, factor VII and prothrombin become deficient in the child as well as in the mother. Since most Vitamin K absorbed into the body is formed by bacterial action in the colon, the newborn child has no adequate source of vitamin K for the first week or so of life—that is, until he establishes a normal colonic bacterial flora. Therefore, prenatal storage of at least small amounts of vitamin K is helpful in preventing postnatal hemorrhage—particularly in the brain caused by birth trauma.

# ADJUSTMENTS OF THE INFANT TO EXTRA-UTERINE LIFE
*ONSET OF BREATHING*

The most obvious effect of birth on the baby is loss of the placental connection with the mother, and therefore loss of this means for metabolic support. Especially important is loss of the placental oxygen supply and placental excretion of carbon dioxide. Therefore, by far the most important immediate adjustment required of the infant is the onset of breathing.

**Cause of Breathing at Birth.** Following completely normal delivery from a mother who has not been depressed by anesthetics, the child ordinarily begins to breathe immediately and has a completely normal respiratory rhythm from the onset. The promptness with which the fetus begins to breathe indicates that breathing is initiated by sudden exposure to the exterior world, probably resulting from a slightly asphyxiated state incident to the birth process but also from sensory impulses originating in the suddenly cooled skin. However, if the infant does not breathe immediately, his body becomes progressively more hypoxic and hypercapnic, which provides additional stimulus to the respiratory center and usually causes breathing within a few seconds to a few minutes after birth.

**Delayed and Abnormal Breathing at Birth—Danger of Hypoxia.** If the mother has been depressed by an anesthetic during delivery, which at least partially also anesthetizes the child, respiration is likely to be delayed for several minutes, thus illustrating the importance of using as little obstetrical anesthesia as feasible. Also, many infants who have traumatic deliveries are slow to breathe or sometimes will not breathe at all. This can result from two possible effects: first, in a few infants, intracranial hemorrhage or brain contusion causes a concussion syndrome with greatly depressed respiratory center. Second, and probably much more important, prolonged fetal hypoxia during delivery also causes serious depression of the respiratory center. Hypoxia frequently occurs during delivery because of (a) compression of the umbilical cord; (b) premature separation of the placenta; (c) excessive contraction of the uterus, which cuts off the blood flow to the placenta; or (d) excessive anesthesia of the mother, which depresses the oxygenation even of her blood.

*Degree of Hypoxia That an Infant Can Tolerate.* In the adult, failure to breathe for only four minutes often causes death, but a newborn infant often survives as long as 15 minutes of failure to breathe after birth. Unfortunately, though, permanent brain impairment often ensues if breathing is delayed more than 8 to 10 minutes. Indeed, actual lesions develop mainly in the thalamus, the inferior colliculi, and in other brain stem nuclei, thus affecting many of the stereotype motor functions of the body.

**Expansion of the Lungs at Birth.** At birth, the walls of the alveoli are held together by the surface tension of the viscid fluid that fills them. More than 25 mm. Hg of negative pressure is required to oppose the effects of this surface tension and therefore to open the alveoli for the first time. But once the alveoli are open, further respiration can be effected with relatively weak respiratory movements. Fortunately, the first inspirations of the newborn infant are extremely powerful, usually capable of creating as much as 50 mm. Hg negative pressure in the intrapleural space.

Figure 83–3 illustrates the tremendous forces required to open the lungs at the onset of breathing. To the left is shown the pressure-volume curve (compliance curve) for the first breath after birth. Observe, first, the lowermost curve, which shows that the lungs essentially do not expand at all until the negative pressure has reached −40 cm. water (−30 mm. Hg). Then, as the negative pressure increases to −60 cm. water, only about 40 ml. air enters the lungs. Then, to deflate the lungs, considerable positive pressure is required, probably because of the viscous resistance offered by the fluid in the bronchioles.

Note that the second breath is much easier. However, breathing does not become completely normal until about 40 minutes after birth, as shown by the third compliance curve, the shape of which compares favorably with that for the normal adult, as shown in Chapter 39.

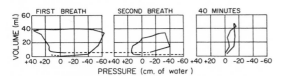

**Figure 83–3.** Pressure-volume curves of the lungs (compliance curves) of a newborn baby immediately after birth, showing (a) the extreme forces required for breathing during the first two breaths of life and (b) development of a nearly normal compliance within 40 minutes after birth. (From Smith: *Sci. Amer., 209:*32, 1963. Copyright © 1963 by Scientific American, Inc. All rights reserved.)

*Respiratory Distress Syndrome.* A small number of infants develop severe respiratory distress during the few hours to several days following birth and frequently succumb within the next day or so. The alveoli of these infants at death contain large quantities of proteinaceous fluid, almost as if pure plasma had leaked out of the capillaries into the alveoli. The fluid contains desquamated alveolar epithelial cells as well as debris from inspissated amniotic fluid. This condition is also called *hyaline membrane disease* because microscopic slides of the lung show this alveolar material to look like a hyaline membrane.

Unfortunately, the cause of the respiratory distress syndrome is not certain. It is believed, however, that it results from damage to the lungs that occurs while the fetus is still in utero. One of the most likely theories is that the lung undergoes a period of severe hypoxia with subsequent damage that leads to the respiratory distress syndrome after birth. Some of the reasons for believing this are that the lungs show pathological damage in many ways, including (1) necrotic and sloughing alveolar epithelium, (2) variable degrees of edema, (3) extensive filling of the lymphatic vessels, and (4) some extravasation of red blood cells, particularly into the lymphatics.

There is also failure to secrete adequate quantities of *surfactant,* a substance normally secreted into the alveoli and which decreases the surface tension of the alveolar fluid, therefore allowing the alveoli to open easily. Though this failure to secrete surfactant may be secondary to the lung pathology, some research workers believe it to be the basic cause of the disease and that the pathology is the secondary result. The surfactant secreting cells do not begin to secrete surfactant until the last one to three months of gestation. Therefore, many premature babies and some full-term babies are born without the capability of secreting surfactant, which therefore causes a collapse tendency of the lungs. The role of surfactant in preventing collapse of the alveoli was discussed in Chapter 39.

## CIRCULATORY READJUSTMENTS AT BIRTH

Equally as important as the onset of breathing at birth are the immediate circulatory adjustments that allow adequate blood flow through the lungs. Also, circulatory adjustments during the first few hours of life shunt more and more

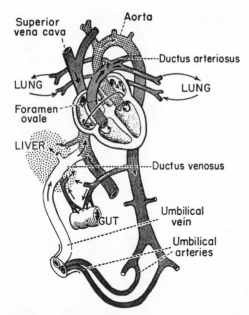

**Figure 83–4.** Organization of the fetal circulation. (Modified from Arey: Developmental Anatomy, 7th Ed.)

blood through the liver as well. To describe these readjustments we must first consider briefly the anatomic structure of the fetal circulation.

**Specific Anatomic Structure of the Fetal Circulation.** Because the lungs are mainly non-functional during fetal life and because the liver is only partially functional, it is not necessary for the fetal heart to pump much blood through either the lungs or the liver. On the other hand, the fetal heart must pump large quantities of blood through the placenta. Therefore, special anatomic arrangements cause the fetal circulatory system to operate considerably differently from that of the adult. First, as illustrated in Figure 83–4, blood returning from the placenta passes through the *ductus venosus,* mainly by-passing the liver. Then, most of the blood entering the right atrium from the inferior vena cava is directed in a straight pathway across the posterior aspect of the right atrium and thence through the *foramen ovale* directly into the left atrium. Thus, the well-oxygenated blood from the placenta enters the left side of the heart rather than the right side and is pumped by the left ventricle mainly into the vessels of the head and forelimbs.

The blood entering the right atrium from the superior vena cava is directed downward through the tricuspid valve into the right ventricle. This blood is mainly deoxygenated blood from the head region of the fetus, and it is pumped by the right ventricle into the pulmonary artery, then mainly through the *ductus arteriosus* into the descending aorta and through the two umbilical arteries into the placenta. Thus, the deoxygenated blood becomes oxygenated.

Figure 83–5 illustrates the relative proportions of the total amount of blood pumped by the heart that passes through the different vascular circuits of the fetus. This figure shows that 55 per cent of all the blood goes through the placenta, leaving only 45 per cent to pass through all the peripheral tissues of the fetus. Furthermore, during fetal life, only 12 per cent of the blood flows through the lungs; immediately after birth about three to four times this much must flow through the lungs, indicating there must be a several-fold increase at birth.

**Changes in Fetal Circulation at Birth.** The basic changes in the fetal circulation at birth were discussed in Chapter 27 in relation to congenital anomalies of the ductus arteriosus and foramen ovale that persist throughout life. Briefly, these changes are the following:

*Primary Changes in Pulmonary and Systemic Vascular Resistance at Birth.* The primary changes in circulation at birth are: First, loss of the tremendous blood flow through the placenta, which *approximately doubles the systemic vascular resistance at birth.* This obviously *increases the aortic pressure* as well as the pressures in the left ventricle and left atrium.

Second, the *pulmonary vascular resistance greatly decreases* as a result of expansion of the lungs. In the unexpanded fetal lungs, the blood vessels are compressed because of the small volume of the lungs. Immediately upon expansion these vessels are no longer compressed, and the resistance to blood flow decreases several-fold. Also, in fetal life the hypoxia and

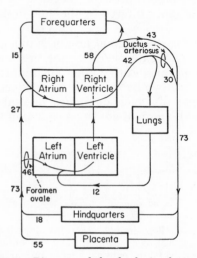

**Figure 83–5.** Diagram of the fetal circulatory system, showing relative distribution of blood flow to the different vascular areas. The numerals represent per cent of the total cardiac output flowing through the particular area. (Modified from Dawes, Mott, and Widdicombe: The fetal circulation in the lamb. *J. Physiol.*, *126:*563, 1954.)

hypercapnia of the lungs cause considerable tonic vasoconstriction of the lung blood vessels, but vasodilation takes place when aeration of the lungs eliminates the hypoxia and hypercapnia. These changes reduce the resistance to blood flow through the lungs as much as five-fold, which obviously *reduces the pulmonary arterial pressure,* the right ventricular pressure, and the right atrial pressure.

*Closure of the Foramen Ovale.* The *low right atrial pressure* and the *high left atrial pressure* that occur secondarily to the changes in pulmonary and systemic resistances at birth cause a tendency for blood to flow backward from the left atrium into the right atrium rather than in the other direction, as occurred during fetal life. Consequently, the small valve that lies over the foramen ovale on the left side of the atrial septum closes over this opening, thereby preventing further flow. In two-thirds of all persons the valve becomes adherent over the foramen ovale within a few months to a few years and forms a permanent closure. But, even if permanent closure does not occur, the left atrial pressure throughout life remains 2 to 4 mm. Hg greater than the right atrial pressure, and the back pressure keeps the valve closed.

*Closure of the Ductus Arteriosus.* Similar effects occur in relation to the ductus arteriosus, for the increased systemic resistance *elevates the aortic pressure* while the decreased pulmonary resistance *reduces the pulmonary arterial pressure.* As a consequence, within a few hours after birth, blood begins to flow backward from the aorta into the pulmonary artery rather than in the other direction as in fetal life. This backward flow can sometimes become equally as great as all the blood pumped by the right ventricle, so that as much as two times the normal amount of blood flows through the lungs. However, after only a few hours the muscular wall of the ductus arteriosus constricts markedly, and within one to eight days the constriction is sufficient to stop all blood flow. This is called *functional closure* of the ductus arteriosus. Then, sometime during the second month of life the ductus arteriosus ordinarily becomes anatomically *occluded* by growth of fibrous tissue into its lumen.

The causes of either functional closure or anatomical closure of the ductus are not completely known. However, the most likely cause is increased oxygenation of the blood flowing through the ductus. In fetal life the $P_{O_2}$ of the ductus blood is as low as 15 mm. Hg, but it increases to about 100 mm. Hg within a few hours after birth. Furthermore, many experiments have shown that the degree of contraction of the ductus is highly related to the availability of oxygen.

In one out of several thousand infants, the ductus fails to close, resulting in a *patent ductus arteriosus,* the consequences of which were discussed in Chapter 27.

*Closure of the Ductus Venosus.* In fetal life, the portal blood joins the blood from the umbilical vein and then passes through the ductus venosus directly into the vena cava, thus bypassing the liver. Immediately after birth, blood flow through the umbilical vein ceases, but most of the portal blood still flows through the ductus venosus, only a small amount passing through the channels of the liver. However, within one to three hours the muscular wall of the ductus venosus contracts strongly and closes this avenue of flow. As a consequence, the portal venous pressure rises from about 0 mm. Hg up to 6 to 10 mm. Hg, which is enough to force blood flow through the liver sinuses. Although we do not know that the ductus venosus almost never fails to close, unfortunately we know almost nothing about the cause of this closure.

## NUTRITION OF THE NEWBORN INFANT

The fetus obtains almost all of its energy from glucose obtained from the mother's blood. Immediately after birth, the amount of glucose stored in the infant's body in the form of glycogen is sufficient to supply the infant's needs for only a few hours, and unfortunately the liver of the newborn infant is still far from functionally adequate at birth, which prevents significant gluconeogenesis. Therefore, the infant's blood glucose concentration frequently falls the first day to as low as 30 to 40 mg./100 ml. of plasma, and the infant must then utilize stored fats and proteins for metabolism until he can be provided with milk from the mother two to three days later.

Special problems are also frequently associated with getting an adequate fluid supply to the newborn infant, because the infant's rate of body fluid turnover averages seven times that of an adult, and the mother's milk supply requires several days to develop. Ordinarily, the infant's weight decreases 5 to 10 per cent and sometimes as much as 20 per cent within the first two to three days of life. Most of this weight loss is loss of fluid rather than of body solids.

# SPECIAL FUNCTIONAL PROBLEMS IN THE NEONATAL INFANT

The most important characteristic of the newborn infant is instability of his various hormonal and neurogenic control systems. This results partly from the immature development of the different organs of the body and partly from the fact that the control systems simply have not become adjusted to the completely new way of life.

**The Respiratory System.** The normal rate of respiration in the newborn is about 40 breaths per minute, and his tidal air with each breath averages 16 ml. This gives a total minute respiratory volume of 640 ml. per minute, which is about two times as great in relation to the body weight as that of an adult. *The functional residual capacity of the infant is only half that of an adult in relation to his body weight.* This causes rapid changes in blood gas concentration when the normal respiration becomes altered.

**Blood Volume.** The blood volume of a newborn infant immediately after birth averages about 300 ml., but, if the infant is left attached to the placenta for a few minutes after birth or if the umbilical cord is stripped to force blood out of its vessels into the baby, an additional 75 ml. of blood enters the infant to make a total of 375 ml. Then, during the ensuing few hours, fluid is lost into the tissue spaces from this blood, which increases the hematocrit but returns the blood volume once again to normal value of about 300 ml. Some pediatricians feel that this extra blood volume in some instances causes mild pulmonary edema with some degree of respiratory distress.

**Cardiac Output.** The cardiac output of the newborn infant averages 550 ml./minute, which, like respiration and body metabolism, is about two times as much in relation to body weight as in the adult. An occasional child is born with an especially low cardiac output caused by hemorrhage through the placental membrane into the mother's blood prior to birth.

**Arterial Pressure.** The arterial pressure during the first day after birth averages about 70/50; this increases slowly during the next several months to approximately 90/60. Then there is a much slower rise during the subsequent years until the adult pressure of 120/80 is attained at adolescence.

**Blood Characteristics.** The red blood cell count in the newborn infant averages about 4 million per cubic millimeter. If blood is stripped from the cord into the infant, the red blood cell count rises an additional half to three-quarters million during the first few hours of life, giving a red blood cell count of about 4.75 million per cubic mm., as illustrated in Figure 83–6. Subsequent to this, however, few new red blood cells are formed in the infant during the first few weeks of life, presumably because the hypoxic stimulus of fetal life is no longer present to stimulate red cell production. Thus, as shown in Figure 83–6, the average red blood cell count falls to 3.25

million per cubic mm. by about 8 to 10 weeks of age. From that time on, increasing activity by the fetus provides the appropriate stimulus for returning the red blood cell count to normal within another two to three months.

Immediately after birth, the white blood cell count of the infant is about 45,000 per cubic mm., which is about five times as great as that of the normal adult.

***Neonatal Jaundice and Erythroblastosis Fetalis.*** Bilirubin formed in the fetus can cross the placenta and be excreted through the liver of the mother, but immediately after birth the only means for ridding the infant of biliribin is through the infant's own liver, which, for the first week or so of life, still functions poorly and is incapable of conjugating significant quantities of bilirubin with glucuronic acid for excretion into the bile. Consequently, the plasma bilirubin concentration rises from a normal value of less than 1 mg./100 ml. to an average of 5 mg./100 ml. during the first three days of life and then gradually falls back to normal as the liver becomes functional. This condition, called *physiologic hyperbilirubinemia,* is illustrated in Figure 83–6, and it is associated with a mild *jaundice* (yellowness) of the infant's skin and especially of the sclerae of its eyes.

However, by far the most important cause of serious neonatal jaundice is *erythroblastosis fetalis,* which was discussed in detail in Chapter 5 in relation to Rh factor incompatibility between the infant and mother. Briefly, the baby inherits an Rh positive trait from its father while the mother is Rh negative. The mother then becomes immunized against the fetus' blood, and her antibodies in turn destroy the infant's red blood cells, releasing extreme quantities of bilirubin into the plasma. This condition occurs either mildly or seriously in 1 out of every 50 to 100 newborn infants.

***Postnatal Bleeding Tendencies.*** Since most vitamin K in the body is synthesized by bacteria in the colon and since the newborn infant has not yet developed a bacterial flora in his colon, its vitamin K level falls

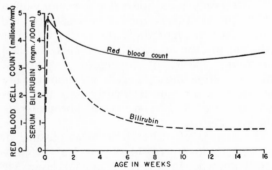

**Figure 83–6.** Changes in the red blood cell count and in the serum bilirubin concentration during the first 16 weeks of life, showing "physiologic anemia" at 6 to 12 weeks of life and "physiologic hyperbilirubinemia" during the first two weeks of life.

rapidly the first few days after birth, causing factor VII and prothrombin production by the liver to be impaired for a week or so until a bacterial flora is established. During this time the infant is likely to develop a bleeding tendency. However, if the mother is given an adequate dose of vitamin K at least four hours prior to birth of the infant, sufficient vitamin K is usually stored in the infant's liver to see it through this period of hypovitaminosis K.

**Fluid Balance, Acid-Base Balance, and Renal Function.** The rate of fluid intake and fluid excretion in the infant is seven times as great in relation to weight as in the adult, which means that even a slight alteration of fluid balance can cause rapidly developing abnormalities. Second, the rate of metabolism in the infant is two times as great in relation to body mass as in the adult, which means that two times as much acid is normally formed, which leads to a tendency toward acidosis in the infant. Third, functional development of the kidneys is not complete until the end of approximately the first month of life. For instance, the kidneys of the newborn can concentrate urine to only one and a half times the osmolality of the plasma instead of the normal three to four times as in the adult.

Therefore, considering the immaturity of the kidney, together with the marked fluid turnover in the infant and rapid formation of acid, one can readily understand that among the most important problems of infancy are acidosis, dehydration, and rare instances of overhydration.

*Changes in Total Body Water, Intracellular Fluid Volume, and Extracellular Fluid Volume During Fetal and Postnatal Life.* Figure 83–7 illustrates volume changes in the different body fluid compartments throughout fetal and postnatal life, showing especially that the early fetus is composed almost entirely of water, but the body accumulates progressively more solids throughout gestation and thereafter until adulthood. At birth the average percentage of water in the infant is 73 per cent; this compares with 58 per cent for the adult.

Another interesting aspect of the fetal body fluids is the high ratio of extracellular fluids to intracellular fluids in contrast to the opposite relationship in the adult. This change presumably results from progressive growth of the cells at the expense of the extracellular spaces. Since extracellular fluids contain mainly sodium chloride in contrast to a preponderance of potassium and magnesium phosphates in the intracellular fluids, the newborn infant has much higher total body sodium and chloride in relation to body weight and much lower total body potassium, magnesium, and phosphate than the adult.

**Liver Function.** During the first few days of life, liver function may be quite deficient, as evidenced by the following effects:

1. The liver of the newborn conjugates bilirubin with glucuronic acid poorly and therefore excretes bilirubin only slightly during the first few days of life.

2. The liver of the newborn is deficient in forming plasma proteins, so that the plasma protein concentration falls to 1 gram per cent less than that for older children. Occasionally, the protein concentration falls so low that the infant actually develops hypoproteinemic edema.

3. The gluconeogenesis function of the liver is particularly deficient. As a result, the blood glucose level of the unfed newborn infant falls to about 30 to 40 mg. per cent, and the infant must depend on its stored fats for energy until feeding can occur.

4. The liver of the newborn usually also forms too little of the factors needed for normal blood coagulation.

**Digestion, Absorption, and Metabolism of Energy Foods.** In general, the ability of the newborn infant to digest, absorb, and metabolize foods is not different from that of the older child, with the following three exceptions:

First, secretion of pancreatic amylase in the newborn infant is deficient so that the infant utilizes starches less adequately than do older children. However, the infant readily assimilates disaccharides and monosaccharides.

Second, absorption of fats from the gastrointestinal tract is somewhat less than in the older child. Consequently, milk with a high fat content, such as cow's milk, is frequently inadequately utilized.

Third, because the liver functions are imperfect during at least the first week of life, the glucose concentration in the blood is unstable and also low.

The newborn is especially capable of synthesizing proteins and thereby storing nitrogen. Indeed, with a completely adequate diet, as much as 90 per cent of the ingested amino acids are utilized for formation of body proteins. This is a much higher percentage than in adults.

**Metabolic Rate and Body Temperature.** The normal metabolic rate of the newborn in relation to body weight is about two times that of the adult, which accounts also for the two times as great cardiac output and two times as great minute respiratory volume in the infant.

However, since the body surface area is very large in relation to the body mass, heat is readily lost from

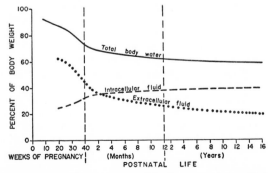

**Figure 83–7.** Changes in total body water, intracellular fluid volume, and extracellular fluid volume in the fetus and in the postnatal child. (Modified from Friis-Hansen: *Acta Paediat.,* 46:Suppl. 110, 1957.)

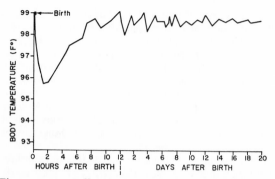

**Figure 83-8.** Fall in body temperature of the infant immediately after birth, and instability of body temperature during the first few days of life.

the body. As a result, the body temperature of the newborn infant, particularly of premature infants, falls. Figure 83-8 shows that the body temperature, of even the normal infant, falls several degrees during the first few hours after birth, but returns to normal in seven to eight hours. Still, the body temperature regulatory mechanisms remain poor during the early days of life, allowing marked deviations in temperature at first, which are also illustrated in Figure 83-8.

**Nutritional Needs during the Early Weeks of Life.** At birth, a newborn infant is usually in complete nutritional balance, provided its mother has had an adequate diet. Furthermore, function of the gastrointestinal system is usually more than adequate to digest and assimilate all the nutritional needs of the infant if these are provided in its diet. However, three specific problems do occur in the early nutrition of the infant as follows:

*Need for Calcium and Vitamin D.* The newborn infant has only just begun rapid ossification of its bones at birth so that it needs a ready supply of calcium throughout infancy. This is ordinarily supplied adequately by its usual diet of milk. Yet absorption of calcium by the gastrointestinal tract is poor in the absence of vitamin D. Therefore, the vitamin D deficient infant can develop severe rickets in only a few weeks. This is particularly true in premature babies since their gastrointestinal tracts absorb calcium even less effectively than those of normal infants.

*Necessity for Iron in the Diet.* If the mother has had adequate amounts of iron in her diet, the liver of the infant usually has stored enough iron to keep forming blood cells for four to six months after birth. But if the mother has had insufficient iron in her diet, anemia is likely to supervene in the infant after about three months of life. To prevent this possibility, early feeding of egg yolk, which contains reasonably large quantities of iron, or administration of iron in some other form is desirable by the second or third month of life.

*Vitamin C Deficiency in Infants.* Ascorbic acid (vitamin C) is not stored in significant quantities in the fetal tissues; yet it is required for proper forma-

tion of cartilage, bone, and other intercellular structures of the infant. Furthermore, milk has poor supplies of ascorbic acid, especially cow's milk, which has only one-fourth as much as mother's milk. For this reason, orange juice or other sources of ascorbic acid are usually prescribed by the third week of of life.

**Immunity.** Fortunately, the newborn inherits much immunity from its mother because many antibodies diffuse from the mother's blood through the placenta into the fetus. However, the newborn itself does not form antibodies to a significant extent. By the end of the first month, the baby's gamma globulins, which contain the antibodies, have decreased to less than one-half the original level, with corresponding decrease in immunity. Thereafter, the baby's own immunization processes begin to form antibodies, and the gamma globulin concentration returns essentially to normal by the age of 6 to 20 months.

Despite the decrease in gamma globulins soon after birth, the antibodies inherited from the mother still protect the infant for about six months against most major childhood infectious diseases, including diphtheria, measles, smallpox, and polio. Therefore, immunization against these diseases before six months is usually unnecessary. On the other hand, the inherited antibodies against whooping cough are normally insufficient to protect the newborn; therefore, for full safety the infant requires immunization against this disease within the first month or so of life.

*Allergy.* Fortunately, the newborn infant is rarely subject to allergy. Several months later, however, when it first begins to form its own antibodies, extreme allergic states can develop, often resulting in serious eczema, gastrointestinal abnormalities, or even anaphylaxis. As the child grows older and develops still higher degrees of immunity, these allergic manifestations usually disappear for the remainder of his life. This relationship of mild immunity to allergy was discussed in Chapter 7.

**Endocrine Problems.** Ordinarily the endocrine system of the infant is highly developed at birth, and the infant rarely exhibits any immediate endocrine abnormalities. However, there are special instances in which endocrinology of infancy is important.

1. If a pregnant mother bearing a female child is treated with an androgenic hormone or if she develops an androgenic tumor during pregnancy, the child will be born with a high degree of masculinization of its sexual organs, thus resulting in a type of *hermaphroditism.*

2. The sex hormones secreted by the placenta and by the mother's glands during pregnancy occasionally cause the newborn's breasts to form milk during the first few days of life. Sometimes the breasts then become inflamed or even develop infectious mastitis.

3. An infant born of a diabetic mother will have considerable hypertrophy and hyperfunction of its islets of Langerhans. As a consequence, the infant's blood glucose concentration may fall to as low as 20 mg./100 ml. or even lower shortly after birth. Fortu-

nately, the newborn infant, unlike the adult, only rarely develops insulin shock or coma from this low level of blood glucose concentration.

Because of metabolic deficits in the diabetic mother, the fetus is often stunted in growth, and growth of the newborn infant and tissue maturation are often impaired. Also, there is a high rate of intra-uterine mortality, and of those fetuses that do come to term, there is still a high mortality rate. Two-thirds of the infants who die succumb to the respiratory distress syndrome, which was described earlier in the chapter.

4. Occasionally, a child is born with hypofunctional adrenal cortices, perhaps resulting from *agenesis* of the glands or *exhaustion atrophy,* which can occur when the adrenal glands have been over-stimulated.

5. If a pregnant woman has hyperthyroidism or is treated with excess thyroid hormone, the infant is likely to be born with a temporarily hyposecreting thyroid gland. On the other hand, if prior to pregnancy a woman had had her thyroid gland removed, her pituitary may secrete great quantities of thyrotropin during gestation, and the child might be born with temporary hyperthyroidism.

# SPECIAL PROBLEMS OF PREMATURITY

All the problems just noted for neonatal life are especially exacerbated in prematurity. These can be categorized under the following two headings: (1) immaturity of certain organ systems and (2) instability of the different homeostatic control systems. Because of these effects, a premature baby rarely lives if it is born more than two and a half to three months prior to term.

## IMMATURE DEVELOPMENT OF THE PREMATURE INFANT

Almost all the organ systems of the body are immature in the premature infant, but some require particular attention if the life of the premature baby is to be saved.

**Respiration.** The respiratory system is especially likely to be underdeveloped in the premature infant. The vital capacity and the functional residual capacity of the lungs are especially small in relation to the size of the infant. Also, surfactant secretion is especially depressed. As a consequence, respiratory distress is a common cause of death. Indeed, there is a very high incidence of the respiratory distress syndrome. Also, the low functional residual capacity in the premature infant is often associated with periodic breathing of the Cheyne-Stokes type.

**Gastrointestinal Function.** Another major problem of the premature infant is to ingest and absorb adequate food. If the infant is more than two months premature, the digestive and absorptive systems are almost always inadequate. The absorption of fat is so poor that the premature infant must have a low fat diet. Furthermore, the premature infant has unusual difficulty in absorbing calcium and therefore can develop severe rickets before one recognizes the difficulty. For this reason, special attention must be paid to adequate calcium and vitamin D intake.

**Function of Other Organs.** Immaturity of other organ systems that frequently causes serious difficulties in the premature infant includes: (a) immaturity of the liver, which results in poor intermediary metabolism and often also a bleeding tendency as a result of poor formation of coagulation factors; (b) immaturity of the kidneys, which are particularly deficient in their ability to rid the body of acids, thereby predisposing to acidosis as well as to serious fluid balance abnormalities; (c) immaturity of the blood-forming mechanism of the bone marrow, which allows rapid development of anemia; and (d) depressed formation of gamma globulin by the reticuloendothelial system, which is often associated with serious infection.

## INSTABILITY OF THE CONTROL SYSTEMS IN THE PREMATURE INFANT

Immaturity of the different organ systems in the premature infant creates a high degree of instability in the homeostatic systems of the body. For instance, the acid-base balance can vary tremendously, particularly when the food intake varies from time to time. Likewise, the blood protein concentration is usually somewhat low because of immature liver development, often leading to *hypoproteinemic edema.* And inability of the infant to regulate its calcium ion concentration frequently brings on hypocalcemic tetany. Also, the blood glucose concentration can vary between the extremely wide limits of 20 mg./100 ml. to over 100 mg./100 ml., depending principally on the regularity of feeding. It is no wonder, then, with these extreme variations in the internal environment of the premature infant, that mortality is high.

**Instability of Body Temperature.** One of the particular problems of the premature infant is inability to maintain normal body temperature. Its temperature tends to approach that of its surroundings. At normal room temperature the temperature may stabilize in the low 90's or even in the 80's. Statistical studies show that a body temperature maintained below 96° F. is associated with a particularly high incidence of death, which explains the common use of the incubator in the treatment of prematurity.

## DANGER OF OXYGEN THERAPY IN THE PREMATURE INFANT

Because the premature infant frequently develops respiratory distress, oxygen therapy has often been used in treating prematurity. However, it has been discovered that use of high oxygen concentrations in

treating premature infants, especially in early prematurity, causes vascular ingrowth into the vitreous
humor of the eyes when the infant is withdrawn from
the oxygen. This is followed later by fibrosis. This
condition, known as *retrolental fibroplasia,* causes
permanent blindness. For this reason, it is particularly important to avoid treatment of premature infants with high concentrations of respiratory oxygen.
Physiologic studies indicate that the premature infant
is probably safe in up to 40 per cent oxygen, but some
child physiologists believe that complete safety can
be achieved only by normal oxygen concentration.

# GROWTH AND DEVELOPMENT
# OF THE CHILD

The major physiologic problems of the child beyond the neonatal period are related to special
metabolic needs for growth, which have been fully
covered in the sections on metabolism and endocrinology.

Figure 83–9 illustrates the changes in heights of
boys and girls from the time of birth until the age of
20 years. Note especially that these parallel each
other almost exactly until the end of the first decade
of life. Between the ages of 11 and 13 the female
estrogens cause rapid growth but early uniting of the
epiphyses at about the fourteenth to sixteenth year of
life, so that growth in height ceases. This contrasts
with the effect of testosterone in the male, which
causes growth at a slightly later age—mainly between
ages 13 and 15. The male, however, undergoes much
more prolonged growth so that his final height is considerably greater than that of the female.

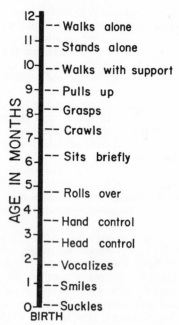

**Figure 83–10.** Behavioral development of the infant
during the first year of life.

## *BEHAVIORAL GROWTH*

Behavioral growth is principally a problem of
maturity of the nervous system. Here, it is extremely
difficult to dissociate maturity of the anatomical
structures of the nervous system from maturity
caused by training. Anatomical studies show that
certain major tracts in the central nervous system are
not completely myelinated until the end of the first
year of life. For this reason we frequently state that
the nervous system is not fully functional at birth. On
the other hand, we know that most reflexes of even
the fetus are fully developed by approximately the
third to fourth month of intra-uterine life, and that
unmyelinated nerve fibers can be equally as functional as myelinated nerve fibers. Therefore, much of
the functional immaturity of the nervous system at
birth might well be caused by lack of training rather
than actual immaturity of the anatomical structures.

Nevertheless, the brain weight of the child increases rapidly during the first year and less rapidly
during the second year, reaching almost adult proportions by the end of the second year. This is also
associated with closure of the fontanels and sutures
of the skull, which prevents much additional growth
of the cranium beyond the first two years of life.

Figure 83–10 illustrates a normal progress chart for
the infant during the first year of life. Comparison of
a baby's actual development in relation to such a
chart is frequently used for clinical assessment of a
child's mental and behavioral growth.

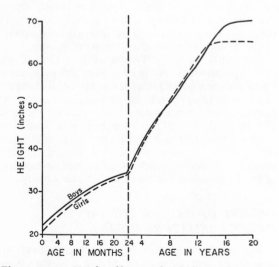

**Figure 83–9.** Height of boys and girls from infancy to 20
years of age.

# REFERENCES

Arey, L. B.: Developmental Anatomy, 7th Ed. Philadelphia, W. B. Saunders Company, 1974.

Balázs, R.: Effects of hormones and nutrition on brain development. *Adv. Exp. Med. Biol.*, 30:385, 1972.

Bartels, H.: Prenatal Respiration. New York, American Elsevier Publishing Company, 1970.

Brody, H., Harman, D., Ordy, J. M. (eds.): Aging. Vol. 1. New York, Raven Press, 1975.

Cockburn, F., and Drillien, C. M.: Neonatal Medicine. Philadelphia, J. B. Lippincott Company, 1975.

Cross, K. W. et al. (eds.): Foetal and Neonatal Physiology: Proceedings. Cambridge, Eng., Cambridge University Press, 1973.

Dallman, P. R.: Iron, vitamin E, and folate in the preterm infant. *J. Pediatr.*, 85:742, 1974.

Daughaday, W. H., Herington, A. C., and Phillips, L. S.: The regulation of growth by endocrines. *Ann. Rev. Physiol.*, 37:211, 1975.

Davis, J. A., and Dobbind, J.: Scientific Foundations of Pediatrics. Philadelphia, W. B. Saunders Company, 1975.

Dawes, G. S.: Foetal and Neonatal Physiology. Chicago, Year Book Medical Publishers, Inc., 1968.

Felig, P.: Maternal and fetal fuel homeostasis in human pregnancy. *Amer. J. Clin. Nutr.*, 26:998, 1973.

Finne, P. H., Halvorsen, S.: Regulation of erythropoiesis in the fetus and newborn. *Arch. Dis. Child.*, 47:683, 1972.

Fomon, S. J.: Infant Nutrition, 2nd Ed. Philadelphia, W. B. Saunders Company, 1975.

Gardner, L. I.: Endocrine and Genetic Diseases of Childhood and Adolescence, 2nd Ed. Philadelphia, W. B. Saunders Company, 1975.

Giroud, A.: Nutritional requirements of the embryo. *World Rev. Nutr. Diet*, 18:195, 1973.

Hafez, E. S. E. (ed.): The Mammalian Fetus. Springfield, Ill., Charles C Thomas, Publisher, 1975.

Heim, T.: Thermogenesis in the newborn infant. *Clin. Obstet. Gynecol.*, 14:790, 1971.

Hodari, A. A., and Mariona, F. (eds.): Physiological Biochemistry of the Fetus. Springfield, Ill., Charles C Thomas, Publisher, 1972.

Jelliffe, D. B.: Nutrition in early childhood. *World Rev. Nutr. Diet*, 16:1, 1973.

Kogut, M. D.: Growth and development in adolescence. *Pediatr. Clin. North Amer.*, 20:789, 1973.

Latham, M. C.: Protein-calorie malnutrition in children and its relation to phsyiological development and behavior. *Physiol. Rev.*, 54:541, 1974.

Leaf, A.: Getting old. *Sci. Amer.*, 229:44, 1973.

Lewy, J. E., and New, M. I.: Growth in children with renal failure. *Amer. J. Med.*, 58:65, 1975.

Linneweh, F. (ed.): Current aspects of perinatology and physiology of children. Berlin, Springer-Verlag, 1973.

Milunsky, A.: The Prenatal Diagnosis of Hereditary Disorders. Springfield, Ill., Charles C Thomas, Publisher, 1973.

Moghissi, K. S. (ed.): Birth Defects and Fetal Development. Springfield, Ill., Charles C Thomas, Publisher, 1974.

Moore, K. L.: The Developing Human. Philadelphia, W. B. Saunders Company, 1973.

Purves, M. J.: Onset of respiration at birth. *Arch. Dis. Child.*, 49:333, 1974.

Rudolph, A. M., and Heymann, M. A.: Fetal and neonatal circulation and respiration. *Ann. Rev. Physiol.*, 36:187, 1974.

Scarpelli, E., and Auld, P. A. M. (eds.): Pediatric Pulmonary Physiology. Philadelphia, Lea & Febiger, 1975.

Sinclair, J. C.: Thermal control in premature infants. *Ann. Rev. Med.*, 23:129, 1972.

Smith, C. A., and Nelson, N. M.: The Physiology of the Newborn Infant. Springfield, Ill., Charles C Thomas, Publisher, 1974.

Smith, C. A.: The first breath. *Sci. Amer.*, 209(4):27, 1963.

Stern, L.: The newborn infant and his thermal environment. *Curr. Probl. Pediatr.*, 1:1, 1970.

Strang, L. B.: Fetal and newborn lung. *In* Guyton, A. C. (ed.): MTP International Review of Science: Physiology. Vol. 2. Baltimore, University Park Press, 1974, p. 31.

Tanner, J. M.: Growing up. *Sci. Amer.*, 229:34, 1973.

Timiras, P. S.: Developmental Physiology and Aging. New York, The Macmillan Company, 1972.

Vaughan, V. C., McKay, R. J., and Nelson, W. E.: Textbook of Pediatrics, 10th Ed. Philadelphia, W. B. Saunders Company, 1975.

Villee, C. A., Villee, D. B., and Zuckerman, J. (eds.): Respiratory Distress Syndrome. New York, Academic Press, Inc., 1973.

Villee, D. B.: Human Endocrinology: A Developmental Approach. Philadelphia, W. B. Saunders Company, 1975.

Windle, W. F.: Brain damage by asphyxia at birth. *Sci. Amer.*, 221:76, 1969.

Windle, W. F.: Physiology of the Fetus. Springfield, Ill., Charles C Thomas, Publisher, 1971.

Winick, M.: Relation of nutrition to physical and mental development. *Bibl. Nutr. Dieta*, 18:114, 1973.

# INDEX